BIOLOGY

Exploring the Diversity of Life

Third Canadian Edition

Peter J. Russell

Paul E. Hertz

Beverly McMillan

M. Brock Fenton
Western University

Heather Addy
University of Calgary

Denis Maxwell
Western University

Tom Haffie
Western University

Bill Milsom
University of British Columbia

NELSON EDUCATION

NELSON EDUCATION

**Biology: Exploring the Diversity of Life,
Third Canadian Edition**

by Peter J. Russell, Paul E. Hertz, Beverly
McMillan, M. Brock Fenton, Heather Addy,
Denis Maxwell, Tom Haffie, and Bill Milsom

**Vice President, Editorial
Higher Education:**
Anne Williams

Senior Publisher:
Paul Fam

Marketing Manager:
Leanne Newell

Developmental Editor:
Toni Chahley

**Photo Researcher and Permissions
Coordinator:**
Kristiina Paul

**Senior Production Project
Manager:**
Natalia Denesiuk Harris

Production Service:
Integra Software Services Pvt. Ltd.

Substantive Editor:
Rosemary Tanner

Copy Editor:
Julia Cochrane

Proofreader:
Integra Software Services Pvt. Ltd.

Indexer:
Integra Software Services Pvt. Ltd.

Design Director:
Ken Phipps

Managing Designer:
Franca Amore

Interior Design:
Dianna Little

Cover Design:
Courtney Hellam and Liz
Harasymczuk

Cover Images:
(front) DAVID DOUBILET/National
Geographic Creative; (back)
kentoh/Shutterstock.com

Compositor:
Integra Software Services Pvt. Ltd.

**Library and Archives Canada
Cataloguing in Publication Data**

Russell, Peter J., author
    Biology : exploring the diversity
of life / Peter J. Russell, Paul E.
Hertz, Beverly McMillan, M. Brock
Fenton (Western University),
Heather Addy (University of
Calgary), Denis Maxwell (Western
University), Tom Haffie (Western
University), Bill Milsom (University
of British Columbia). — Third
Canadian edition.

Includes index.
Revision of: Biology : exploring
    the diversity of life / Peter
    J. Russell ... [et al.]. — 2nd
    Canadian ed. — Toronto :
    Nelson Education, [2012],
    c2013.
ISBN 978-0-17-653213-0 (bound)

    1. Biology—Textbooks. I. Hertz,
Paul E., author II. McMillan,
Beverly, author III. Fenton, M.
Brock (Melville Brockett), 1943–,
author IV. Addy, Heather D.
(Heather Dawn), 1962–, author
V. Maxwell, Denis P. (Denis Philip),
1964–, author VI. Haffie, Tom,
author VII. Milsom, Bill, 1947–,
author VIII. Title.

QH308.2.R88 2015   570
C2014   908034-4

ISBN-13: 978-0-17-653213-0
ISBN-10: 0-17-653213-7

For, and because of, our generations
of students.

About the Canadian Authors

M.B. (BROCK) FENTON received his Ph.D. from the University of Toronto in 1969. Since then, he has been a faculty member in biology at Carleton University, then at York University, and then at Western University. In addition to teaching parts of first-year biology, he has also taught vertebrate biology, animal biology, and conservation biology, as well as field courses in the biology and behaviour of bats. He has received awards for his teaching (Carleton University Faculty of Science Teaching Award; Ontario Confederation of University Faculty Associations Teaching Award; and a 3M Teaching Fellowship, Society for Teaching and Learning in Higher Education), in addition to recognition of his work on public awareness of science (Gordin Kaplan Award from the Canadian Federation of Biological Societies; Honourary Life Membership, Science North, Sudbury, Ontario; Canadian Council of University Biology Chairs Distinguished Canadian Biologist Award; The McNeil Medal for the Public Awareness of Science of the Royal Society of Canada; and the Sir Sanford Fleming Medal for public awareness of Science, the Royal Canadian Institute). He also received the C. Hart Merriam Award from the American Society of Mammalogists for excellence in scientific research. Bats and their biology, behaviour, evolution, and echolocation are the topics of his research, which has been funded by the Natural Sciences and Engineering Research Council of Canada (NSERC). In November 2014, Brock was inducted as a Fellow of the Royal Society of Canada.

HEATHER ADDY received her Ph.D. in plant–soil relationships from the University of Guelph in 1995. During this training and in a subsequent post-doctoral fellowship at the University of Alberta, she discovered a love of teaching. In 1998, she joined the Department of Biological Sciences at the University of Calgary in a faculty position that emphasizes teaching and teaching-related scholarship. In addition to teaching introductory biology classes and courses in plant and fungal biology, she is currently the coordinator of the Teaching Skills Workshop Program for the Faculty of Science. Her pedagogical focus is on collaborative learning methods such as team-based learning and the incorporation of peer mentors and class representatives in undergraduate education. She received the Faculty of Science Award for Excellence in Teaching in 2005, a Students' Union Teaching Excellence Award in 2008 and 2014, and one of the inaugural University of Calgary Teaching Awards in 2014.

DENIS MAXWELL received his Ph.D. from the University of Western Ontario in 1995. His thesis, under the supervision of Norm Huner, focused on the role of the redox state of photosynthetic electron transport in photo acclimation in green algae. Following his doctorate, he was awarded an NSERC postdoctoral fellowship. He undertook postdoctoral training at the Department of Energy Plant Research Laboratory at Michigan State University, where he studied the function of the mitochondrial alternative oxidase. After taking up a faculty position at the University of New Brunswick in 2000, he moved in 2003 to the Department of Biology at Western University. He currently serves as Associate Chair—Undergraduate Education and teaches first-year biology.

TOM HAFFIE is a graduate of the University of Guelph and the University of Saskatchewan in the area of microbial genetics. Tom has devoted his 30-year career at Western University to teaching large biology classes in lecture, laboratory, and tutorial settings. He led the development of the innovative core laboratory course in the biology program; was an early adopter of computer animation in lectures; and, most recently, has coordinated the implementation of personal response technology across campus. He holds a University Students' Council Award for Excellence in Teaching, a UWO Pleva Award for Excellence in Teaching, a Fellowship in Teaching Innovation, a Province of Ontario Award for Leadership in Faculty Teaching (LIFT), and a Canadian national 3M Fellowship for Excellence in Teaching.

Bill Milsom

BILL MILSOM (Ph.D., University of British Columbia) is a professor in the Department of Zoology at the University of British Columbia, where he has taught a variety of courses, including first-year biology, for over 30 years. His research interests include the evolutionary origins of respiratory processes and the adaptive changes in these processes that allow animals to exploit diverse environments. He examines respiratory and cardiovascular adaptations in vertebrate animals in rest, sleep, exercise, altitude, dormancy, hibernation, diving, and so on. This research contributes to our understanding of the mechanistic basis of biodiversity and the physiological costs of habitat selection. His research has been funded by NSERC, and he has received several academic awards and distinctions, including the Fry Medal of the Canadian Society of Zoologists, the August Krogh Award of the American Physiological Society, the Bidder Lecture of the Society for Experimental Biology, and the Izaak Walton Killam Award for Excellence in Mentoring. He has served as the President of the Canadian Society of Zoologists and as President of the International Congress of Comparative Physiology and Biochemistry.

About the U.S. Authors

PETER J. RUSSELL received a B.Sc. in Biology from the University of Sussex, England, in 1968 and a Ph.D. in Genetics from Cornell University in 1972. He has been a member of the Biology faculty of Reed College since 1972 and is currently a professor of biology, emeritus. Peter taught a section of the introductory biology course, a genetics course, and a research literature course on molecular virology. In 1987 he received the Burlington Northern Faculty Achievement Award from Reed College in recognition of his excellence in teaching. Since 1986, he has been the author of a successful genetics textbook; current editions are *iGenetics: A Molecular Approach, iGenetics: A Mendelian Approach*, and *Essential iGenetics*. Peter's research was in the area of molecular genetics, with a specific interest in characterizing the role of host genes in the replication of the RNA genome of a pathogenic plant virus, and the expression of the genes of the virus; yeast was used as the model host. His research has been funded by agencies including the National Institutes of Health, the National Science Foundation, the American Cancer Society, the Department of Defense, the Medical Research Foundation of Oregon, and the Murdoch Foundation. He has published his research results in a variety of journals, including *Genetics, Journal of Bacteriology, Molecular and General Genetics, Nucleic Acids Research, Plasmid*, and *Molecular and Cellular Biology*. Peter has a long history of encouraging faculty research involving undergraduates, including cofounding the biology division of the Council on Undergraduate Research in 1985. He was Principal Investigator/Program Director of a National Science Foundation Award for the Integration of Research and Education (NSF–AIRE) to Reed College, 1998 to 2002.

PAUL E. HERTZ was born and raised in New York City. He received a B.S. in Biology from Stanford University in 1972, an A.M. in Biology from Harvard University in 1973, and a Ph.D. in Biology from Harvard University in 1977. While completing field research for the doctorate, he served on the biology faculty of the University of Puerto Rico at Rio Piedras. After spending two years as an Isaac Walton Killam Postdoctoral Fellow at Dalhousie University, Paul accepted a teaching position at Barnard College, where he has taught since 1979. He was named Ann Whitney Olin Professor of Biology in 2000, and he received The Barnard Award for Excellence in Teaching in 2007. In addition to serving on numerous college committees, Paul chaired Barnard's Biology Department for eight years and served as Acting Provost and Dean of the Faculty from 2011 to 2012. He is the founding Program Director of the Hughes Science Pipeline Project at Barnard, an undergraduate curriculum and research program that has been funded continuously by the Howard Hughes Medical Institute since 1992. The Pipeline Project includes the Intercollegiate Partnership, a program for local community college students that facilitates their transfer to four-year colleges and universities. He teaches one semester of the introductory sequence for biology majors and pre-professional students, lecture and laboratory courses in vertebrate zoology and ecology, and a year-long seminar that introduces first-year students to scientific research. Paul is an animal physiological ecologist with a specific research interest in the thermal biology of lizards. He has conducted fieldwork in the West Indies since the mid-1970s, most recently focusing on the lizards of Cuba. His work has been funded by the NSF, and he has published his research in such prestigious journals as *The American Naturalist, Ecology, Nature, Oecologia*, and *Proceedings of the Royal Society*. In 2010, he and his colleagues at three other universities received funding from NSF for a project designed to detect the effects of global climate warming on the biology of Anolis lizards in Puerto Rico.

BEVERLY McMILLAN has been a science writer for more than 25 years. She holds undergraduate and graduate degrees from the University of California, Berkeley, and is coauthor of a college text in human biology, now in its tenth edition. She has also written or coauthored numerous trade books on scientific subjects and has worked extensively in educational and commercial publishing, including eight years in editorial management positions in the college divisions of Random House and McGraw-Hill.

Brief Table of Contents

Contents

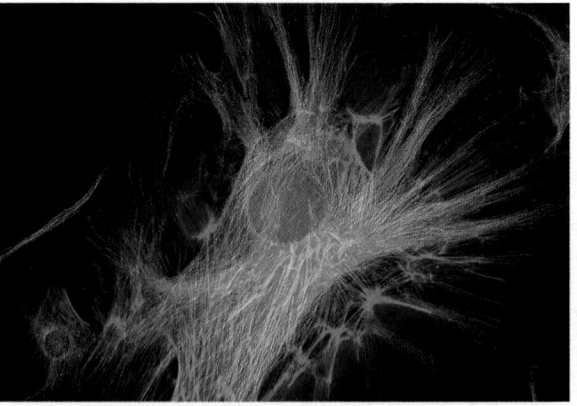

Paul Whitted/Shutterstock.com

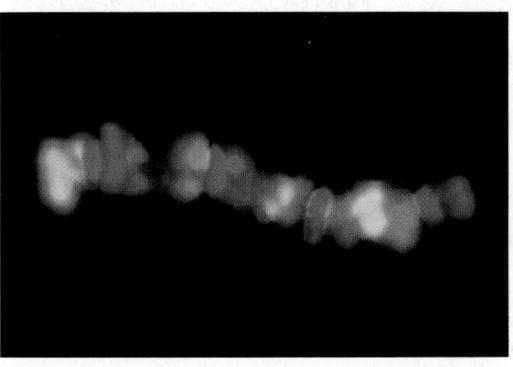

Regents of the University of California 2005/Dr. Uli Weier/Science Source

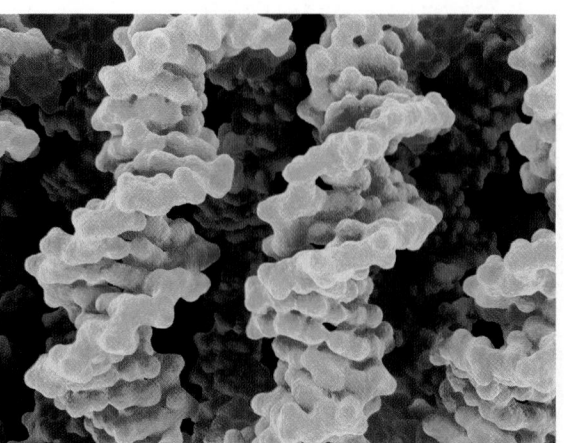

Kenneth Eward/Science Source

Ryan M. Bolton/Shutterstock.com

Daniel Hebert/Shutterstock.com

MichaelTaylor3d/Shutterstock.com

UNIT EIGHT BIOLOGY IN ACTION 780

INTRODUCTION 3: SYSTEMS AND PROCESSES 833

UNIT NINE SYSTEMS AND PROCESSES: PLANTS 835

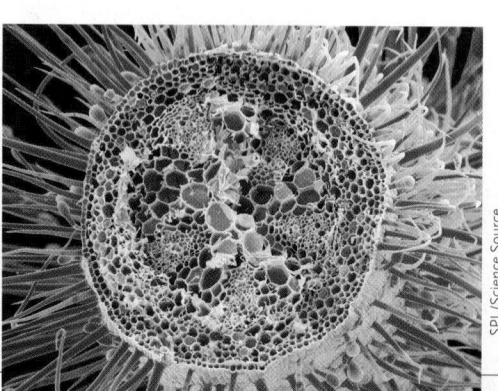

oceanbounddb/iStock/Thinkstock

SPL/Science Source

Willie Manalo/iStock/Thinkstock

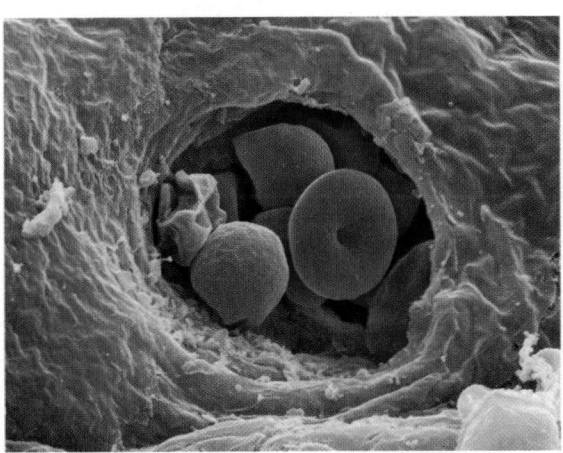

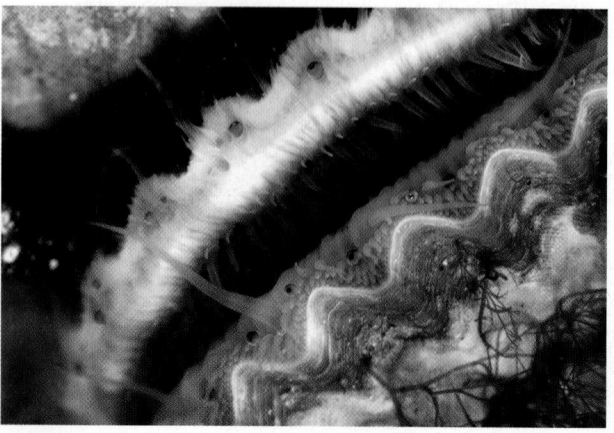

Ken Read/Science Source

Jeff Grabert/Shutterstock.com

M.B. Fenton

ostill/Shutterstock.com

CampCrazy Photography/Shutterstock.com

Preface

Welcome to an exploration of the diversity of life. The main goal of this text is to guide you on a journey of discovery about life's diversity across levels ranging from molecules to genes, cells to organs, and species to ecosystems. Along the way, we will explore many questions about the mechanisms underlying diversity as well as the consequences of diversity for our own species and for others.

At first glance, the riot of life that animates the biosphere overwhelms our minds. One way to begin to make sense of this diversity is to divide it into manageable sections on the basis of differences. Thus, in this book for instance, we highlight the divisions between prokaryotic and eukaryotic organisms, plants and animals, protostomes and deuterostomes. We also consider features found in all life forms to stress similarities as well as differences. We examine how different organisms solve the common problems of finding nutrients, energy, and mates on the third rock from our Sun. What basic evolutionary principles inform the relationships among life forms regardless of their different body plans, habitats, or life histories? Unlike many other first-year biology texts, this book has chapters integrating basic concepts such as the effects of genetic recombination, light, nutrition, and domestication across the breadth of life from microbes to mistletoe to moose. As you read this book, you will be referred frequently to other chapters for linked information that expands the ideas further.

Evolution provides a powerful conceptual lens for viewing and understanding the roots and history of diversity. We will demonstrate how knowledge of evolution helps us appreciate the changes we observe in organisms. Whether the focus is the conversion of free-living prokaryotic organisms into mitochondria and chloroplasts or the steps involved in the domestication of rice, selection for particular traits over time can explain the current condition.

We hope that Canadian students will find the subject of biology as it is presented here accessible and engaging because it is presented in familiar contexts. We have highlighted the work of Canadian scientists, used examples of Canadian species, referred to Canadian regulations and institutions, and highlighted discoveries made by Canadians.

Many biology textbooks use the first few chapters to review fundamentals of chemistry and biochemistry as well as information on the scientific method. Instead of focusing on this background information, we have used the first chapter, in particular, to immediately engage students by conveying the excitement that is modern biology. We have put important background information in the centre of the book as a distinct reference section entitled *The Chemical, Physical, and Environmental Foundations of Biology*. With their purple borders, these pages are distinct and easy to find and have become affectionately known as *The Purple Pages*. These pages enable information to be readily identifiable and accessible to students as they move through the textbook rather than being tied to a particular chapter.

In addition to presenting material about biology, this book also makes a point of highlighting particular people, important molecules, interesting contexts, and examples of life in extreme conditions. Science that appears in textbooks is the product of people who have made careful and systematic observations that led them to formulate hypotheses about these observations and, where appropriate, design and execute experiments to test these hypotheses. We illustrate this in most chapters with boxed stories about how particular people have used their ingenuity and creativity to expand our knowledge of biology. We have endeavoured to show not just the science itself but also the process behind the science.

Although biology is not simply chemistry, specific chemicals and their interactions can have dramatic effects on biological systems. From water to progesterone, amanitin, and DDT, each chapter features the activity of a relevant chemical.

To help frame the material with an engaging context, we begin each chapter with a section called "Why It Matters." In addition, several chapters include boxed accounts of organisms thriving "on the edge" at unusual temperatures, pressures, radiation dosages, salt concentrations, and so on. These brief articles explain how our understanding of "normal" can be increased through study of the "extreme."

Examining how biological systems work is another theme pervading this text and underlying the idea of diversity. We have intentionally tried to include examples that will tax your imagination, from sea slugs that steal chloroplasts for use as solar panels, to hummingbirds fuelling their hovering flight, to adaptive radiation of viruses. In each situation, we examine how biologists have explored and assessed the inner workings of organisms from gene regulation to the challenges of digesting cellulose.

Solving problems is another theme that runs throughout the book. Whether the topic is gene therapy to treat a disease in people, increasing crop production, or conserving endangered species, both the problem and the solution lie in biology. We will explore large problems facing planet Earth and the social implications that arise from them.

Science is by its nature a progressive enterprise in which answers to questions open new questions for consideration. Each chapter presents Questions for Discussion (also mentioned below) to emphasize that biologists still have a lot to learn—topics for you to tackle should you decide to pursue a career in research.

Study Breaks occur after most sections in the chapters. They contain questions written by students to identify some of the important features of the section. At the end of each chapter is a group of multiple-choice self-test questions, the answers to which can be found at the end of the book. Questions for Discussion at the end of each chapter challenge you to think more broadly about biology. You are encouraged to use these in discussions with other students and to explore potential answers by using the resources of the electronic or physical library.

To maximize the chances of producing a useful text that draws in students (and instructors), we sought the advice of colleagues who teach biology (members of the Editorial Advisory Board). We also asked students (members of the Student Advisory Boards) for their advice and comments. These groups read draft chapters, evaluated the effectiveness of important visuals in the textbook, and provided valuable feedback, but any mistakes are ours.

We hope that you are as captivated by the biological world as we are and are drawn from one chapter to another. But don't stop there—use electronic and other resources to broaden your search for understanding, and, most important, observe and enjoy the diversity of life around you.

M. Brock Fenton
Heather Addy
Denis Maxwell
Tom Haffie
Bill Milsom

London, Calgary, and Vancouver
January 2015

New to This Edition

This section highlights the changes we made to enhance the effectiveness of *Biology: Exploring the Diversity of Life*, Third Canadian Edition. Every chapter has been updated to ensure currency of information. We made organizational changes to more closely link related topics and reflect preferred teaching sequences. New features in the text have been developed to help students actively engage in their study of biology. Key chapters have been extensively revised to provide a full treatment of the subject matter that reflects new developments in these specialized fields. Enriched media offerings provide students with a broad spectrum of learning opportunities.

Organizational Changes

We divided "DNA Technologies and Genomics" into two chapters: "DNA Technologies" with a new emphasis on the role of synthetic biology and updated gene therapy, and "Genomes and Proteomes," which provides more focused coverage of genome evolution, personal genomics, and metagenomics.

In response to reviewers' comments, we moved Unit Seven: Ecology and Climate Change, and Unit Eight: Biology in Action from the end of the book to after Unit Six: Diversity of Life to enhance the flow of topics in the first half of the textbook.

The latter half of the book has been streamlined and reorganized to completely separate the discussion of the systems and processes of plants and animals. We divided "Plant and Animal Nutrition" into separate chapters dealing with plants and animals, respectively. "Plant Signals and Responses to the Environment" is a new chapter that includes more in-depth treatment of this subject matter. While students need to realize that plants and animals face the same challenges, they also need to consider that plants and animals respond to these challenges in different ways, reflecting some very fundamental differences. The material in these sections has been reorganized to fit the style in which the material is presented in most introductory biology courses. But it is our hope that as students read the chapters in Units 9 and 10, they will think about how differences in animal and plant growth, development, reproduction, and other processes relate to the fundamental difference between being motile—obtaining the carbon needed to build bodily structures from the food they eat (animals)—and being sessile—obtaining the carbon needed to build body structures from the air and sunlight they capture (plants).

Finally, we have a new Unit Eleven: Systems and Processes: Interacting with the Environment, which contains a new chapter, Chapter 52: Conservation and Evolutionary Physiology, which draws together themes from the book in a series of engaging case studies.

New Features

New Volume Introductions

Providing students with an overview of the content of each volume, our new volume introductions set the stage for the presentation of the material in the chapters that follow. These introductions enhance the flow of information across the textbook by setting the units and chapters within the broader context of the discipline as a whole.

New Chapter Dealing Exclusively with Genomes and Proteomes

In response to feedback from reviewers, and with the careful guidance of our Editorial Advisory Board, we have developed an entirely new chapter on Genomes and Proteomes. As sequencing costs continue to fall, the volume of available sequences continues to accumulate in databanks. What sequences are present in a given biological sample? What does a particular sequence mean? What can comparing sequences reveal about evolution? What can one's own sequence predict about one's future health? The new chapter addresses these questions, and more, in a Canadian context.

Extensively Revised and Rewritten Unit Five: Evolution and Classification

Given the importance and dynamism of this topic, Unit Five: Evolution and Classification has been extensively revised and rewritten. With the help and guidance of reviewers, subject matter experts, and our Editorial Advisory Board, we have rewritten this unit, reorganizing the coverage of topics, and expanding our coverage from four to five chapters. This new unit presents and carefully builds upon key concepts in evolution, providing students with a strong foundation in this essential area. Unique to this textbook, Chapter 21: Humans and Evolution forms a capstone for the unit, drawing together the key themes presented in the chapters that precede it.

Chapters in Unit 5: Evolution and Classification are as follows:

- Chapter 17: Evolution: The Development of the Theory
- Chapter 18: Microevolution: Changes within Populations
- Chapter 19: Species and Macroevolution
- Chapter 20: Understanding the History of Life on Earth
- Chapter 21: Humans and Evolution

New Chapter 52: Conservation and Evolutionary Physiology

Our ability to design conservation measures to protect species threatened by environmental change is based on an understanding of the physiological processes presented in the preceding chapters. This knowledge is required to predict the consequences of environmental change on a species' survival and of the measures that will or will not be beneficial as conservation policy. In this chapter, we discuss our understanding of how the physiological processes presented in individual chapters are integrated within organisms and then present a few case histories to illustrate how we can use this information to understand the evolution of physiological processes and to inform conservation policies. We have carefully selected only a few examples of the many that exist (such as ways to reduce mortality in salmon fisheries due to by-catch, explanations for high plant biodiversity on nutrient-poor soils, and the use of physiology to predict the behaviour of such invasive species as cheatgrass, and suggest strategies for the control of these organisms), allowing instructors to expand this treatment with other examples of their own choosing.

Improved Illustrations

A priority for the third Canadian edition was improved illustrations throughout the book. Members of our Student Advisory Board at the University of Alberta reviewed key illustrations throughout the book and provided helpful guidance on improving their clarity and effectiveness. Many figures have been revised and redrawn in response to this feedback.

Major revisions to selected chapters are listed below:

Chapter 4: Energy and Enzymes
- More accurate description of ATP hydrolysis
- More complete treatment of coupled reactions
- New section focused on the flow of energy through the biosphere
- More accurate concept of "energy spreading" in defining entropy

Chapter 7: Photosynthesis
- Added section looking at photosynthesis from a global perspective
- New "Why It Matters" focused on biofuel production
- Improved figures, including one illustrating leaf gas exchange

Chapter 8: Cell Cycles
- Refocused cell cycle control onto checkpoints
- New Concept Fix related to segregation of cytoplasmic elements
- New reference to stem cells in plants and animals as examples of asymmetric cell division (new Figure 8.19)

Chapter 14: Control of Gene Expression
- New emphasis on the role of epigenetic regulation

Chapter 15: DNA Technologies
- Updated DNA sequencing
- New emphasis on synthetic biology and updated gene therapy

Chapter 16: Genomes and Proteomes (NEW)
- Highlights human variation and personal genomics
- Discusses functional and comparative genomics, including genome evolution, metagenomics, and synthetic life
- Explains modern technologies such as next-generation sequencing and microarrays

Chapter 17: Evolution: The Development of the Theory (NEW)
- Explores the development of the theory of evolution from Aristotle's knowledge of the natural world, to Darwin, to the Modern Synthesis
- Introduces variation and selection, major components of Darwin's theory

Chapter 18: Microevolution: Changes within Populations (NEW)
- Expands the topics of variation and selection
- Introduces different types of variation, as well as a variety of selection agents such as mutation and nonrandom mating
- Discusses the role played by population biology in evolution along with the Hardy–Weinberg Principle

Chapter 19: Species and Macroevolution (NEW)
- Begins with a thorough discussion of species, species concepts, and speciation
- Explores reproductive isolating mechanisms and the variety of geographic speciation over time

Chapter 20: Understanding the History of Life on Earth (NEW)
- Outlines the history of life on Earth by exploring the geological time scale, the fossil record, and the lives of prehistoric organisms
- Ties the discussion of the history of life to phylogeny and phylogenetics as they apply to the evolution of birds

Chapter 21: Humans and Evolution (NEW)
- Draws together key themes in the previous chapters and sets them within the context of human evolution

Chapter 22: Bacteria and Archaea
- Updated "Why It Matters" captures recent research on the gut microbiome and its role in human health
- New "Life on the Edge" box deals with poikilohydric plants

Chapter 23: Viruses, Viroids, and Prions: Infectious Biological Particles
- Updated information on prions to reflect current research
- Modified colour scheme of figures to increase clarity

Chapter 24: Protists
- Updated to include the evolutionary tree of eukaryotes showing "supergoups"
- Revised classification of protist groups according to this new evolutionary tree

Chapter 25: Fungi
- Clarified placement of fungi that lack sexual stage in fungal classification

Chapter 26: Plants
- Revised section on evolutionary relationships among bryophytes to reflect current research findings

Chapter 30: Population Interactions and Community Ecology
- New "Why It Matters" dealing with morphological specializations.

Chapter 31: Ecosystems
- New table outlining the characteristics of the terrestrial ecozones of Canada

Chapter 32: Conservation of Biodiversity
- New section on white-nose syndrome and bats

Chapter 34: Organization of the Plant Body
- Revised "People behind Biology" to highlight new research with greater significance related to provision of clean drinking water

Chapter 35: Transport in Plants
- Revised and clarified material on water potential

Chapter 37: Plant Nutrition (NEW)
- Revised section detailing effects of nutrient deficiency on plant structure

Chapter 38: Plant Signals and Responses to the Environment
- Updated information on brassinosteroids and plant chemical defences

Chapter 48: Animal Nutrition
- New "People behind Biology" featuring Frederick Banting and George Best

Chapter 51: Defences against Disease
- New material on *E. coli*
- New section on parasite–host interactions and the successes and limitations of vaccines

Chapter 52: Conservation and Evolutionary Physiology (NEW)
- Integrates the different physiological processes described in the preceding chapters as they interact in plants and animals in nature
- Explores the evolution of physiological processes and the manner in which an understanding of this can inform conservation issues

THINK AND ENGAGE LIKE A SCIENTIST

 MindTap™

Engage, Adapt, and Master!

Stay organized and efficient with **MindTap**—a single destination with all the course material and study aids you need to succeed. Built-in apps leverage social media and the latest learning technology to help you succeed. Our customized learning path is designed to help you engage with biological concepts, identify gaps in your knowledge, and master the material!

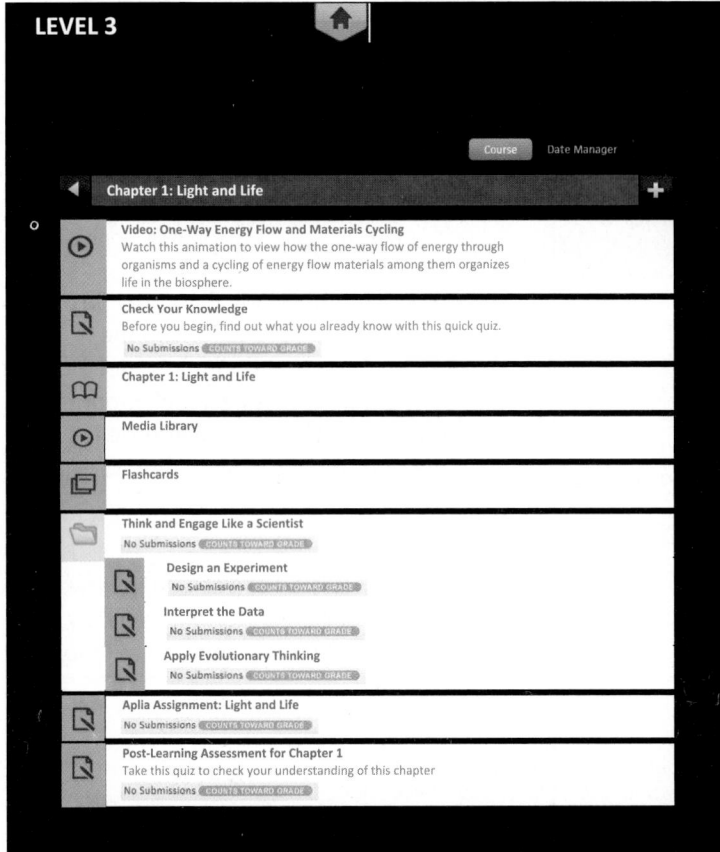

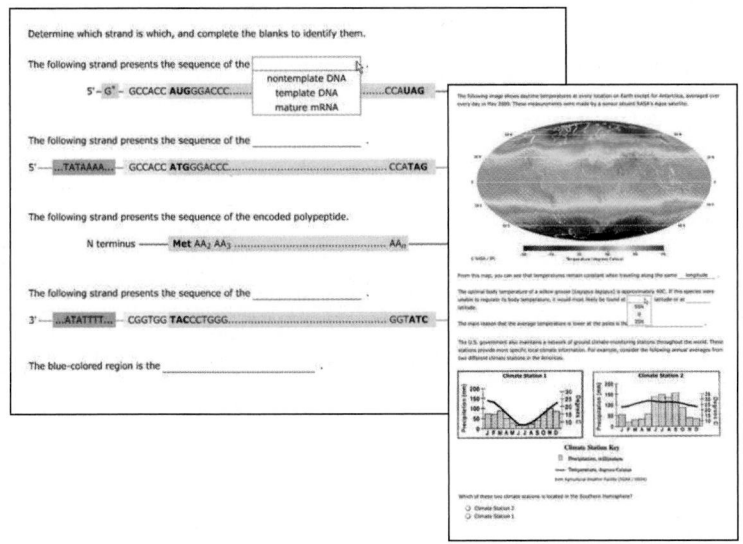

Engage!
The learning path for each chapter begins with an engaging video designed to pique your interest in the chapter contents.

Adapt!
Take the tutorial quiz to assess gaps in your knowledge. Strengthen your knowledge of concepts by reviewing the eBook, animations, and instructive videos. Reinforce your knowledge with glossary flashcards.

Master!
Think and engage like a scientist by taking short-answer quizzes:

- **Apply Evolutionary Thinking** questions ask you to interpret a relevant topic in relation to the principles of evolutionary biology.
- **Design an Experiment** challenges your understanding of the chapter and helps you deepen your understanding of the scientific method as you consider how to develop and test hypotheses about a situation that relates to a main chapter topic.
- **Interpret the Data** questions help you develop analytical and quantitative skills by asking you to interpret graphical or tabular results of experimental or observational research experiments for which the hypotheses and methods of analysis are presented.

Test your mastery of concepts with Aplia for Biology, an interactive online tool that complements the text and helps you learn and understand key concepts through focused assignments, an engaging variety of problem types, exceptional text/art integration, and immediate feedback.

And assess your knowledge of chapter concepts by taking the **Review Quiz**.

THE BIG PICTURE

Each chapter of *Biology: Exploring the Diversity of Life,* Third Canadian Edition, is carefully organized and presented in digestible chunks so you can stay focused on the most important concepts. Easy-to-use learning tools point out the topics covered in each chapter, show why they are important, and help you learn the material.

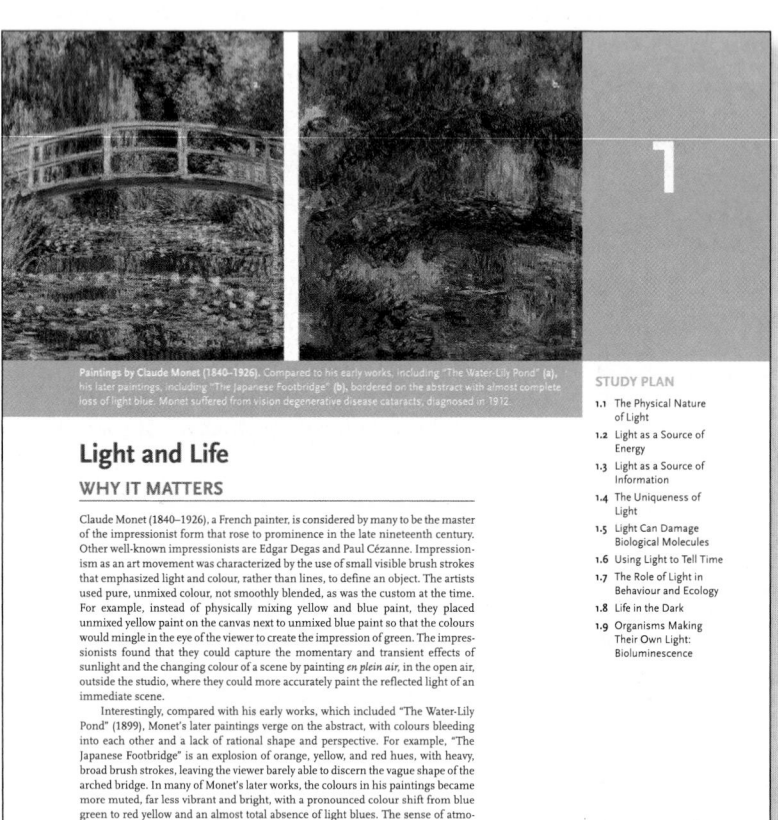

Paintings by Claude Monet (1840–1926). Compared to his early works, including "The Water-Lily Pond" (a), his later paintings, including "The Japanese Footbridge" (b), bordered on the abstract with almost complete loss of light blue. Monet suffered from vision degenerative disease cataracts, diagnosed in 1912.

Light and Life

WHY IT MATTERS

Claude Monet (1840–1926), a French painter, is considered by many to be the master of the impressionist form that rose to prominence in the late nineteenth century. Other well-known impressionists are Edgar Degas and Paul Cézanne. Impressionism as an art movement was characterized by the use of small visible brush strokes that emphasized light and colour, rather than lines, to define an object. The artists used pure, unmixed colour, not smoothly blended, as was the custom at the time. For example, instead of physically mixing yellow and blue paint, they placed unmixed yellow paint on the canvas next to unmixed blue paint so that the colours would mingle in the eye of the viewer to create the impression of green. The impressionists found that they could capture the momentary and transient effects of sunlight and the changing colour of a scene by painting *en plein air*, in the open air, outside the studio, where they could more accurately paint the reflected light of an immediate scene.

Interestingly, compared with his early works, which included "The Water-Lily Pond" (1899), Monet's later paintings verge on the abstract, with colours bleeding into each other and a lack of rational shape and perspective. For example, "The Japanese Footbridge" is an explosion of orange, yellow, and red hues, with heavy, broad brush strokes, leaving the viewer barely able to discern the vague shape of the arched bridge. In many of Monet's later works, the colours in his paintings became more muted, far less vibrant and bright, with a pronounced colour shift from blue green to red yellow and an almost total absence of light blues. The sense of atmosphere and light that he was famous for in his earlier works disappeared.

Although the change in Monet's paintings could easily be explained by an intentional change in style or perhaps an age-related change in manual dexterity, Monet himself realized that it was not his style or dexterity that had changed but, rather, his ability to see. Monet suffered from cataracts, a vision-deteriorating disease diagnosed in both eyes by a Parisian ophthalmologist in 1912 when Monet was 72. A cataract is a change in the lens of the eye, making it more opaque. The underlying cause is a progressive denaturation of one of the proteins that make up the lens. The

STUDY PLAN

1.1 The Physical Nature of Light
1.2 Light as a Source of Energy
1.3 Light as a Source of Information
1.4 The Uniqueness of Light
1.5 Light Can Damage Biological Molecules
1.6 Using Light to Tell Time
1.7 The Role of Light in Behaviour and Ecology
1.8 Life in the Dark
1.9 Organisms Making Their Own Light: Bioluminescence

NEL
3

Study Plan The Study Plan provides a list of the major sections in the chapter. Each section breaks the material into a manageable amount of information, building on knowledge and understanding as you acquire it.

Why It Matters For each chapter we provide a brief contextual overview, outlining the main points that follow.

Study Breaks The Study Breaks encourage you to pause and think about the key concepts you have just encountered before moving to the next section. The Study Break questions are written by Canadian students for their peers across the country and are intended to identify some of the important features of the section.

STUDY BREAK

1. Define light.
2. What structural feature is common to all pigments?

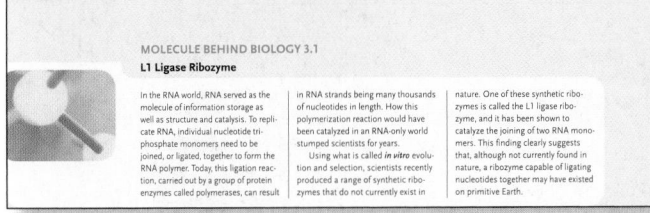

MOLECULE BEHIND BIOLOGY 3.1
L1 Ligase Ribozyme

In the RNA world, RNA served as the molecule of information storage as well as structure and catalysis. To replicate RNA, individual nucleotide triphosphate monomers need to be joined, or ligated, together to form the RNA polymer. Today, this ligation reaction, carried out by a group of protein enzymes called polymerases, can result in RNA strands being many thousands of nucleotides in length. How this polymerization reaction would have been catalyzed in an RNA-only world stumped scientists for years.

Using what is called *in vitro* evolution and selection, scientists recently produced a range of synthetic ribozymes that do not currently exist in nature. One of these synthetic ribozymes is called the L1 ligase ribozyme, and it has been shown to catalyze the joining of two RNA monomers. This finding clearly suggests that, although not currently found in nature, a ribozyme capable of ligating nucleotides together may have existed on primitive Earth.

Molecule behind Biology "Molecule behind Biology" boxes give students a sense of the exciting impact of molecular research. From water to progesterone, amanitin, and DDT, each chapter features the activity of a relevant chemical.

People behind Biology "People behind Biology" boxes in most chapters contain boxed stories about how particular people have used their ingenuity and creativity to expand our knowledge of biology. The purpose of these boxes is to recognize that advances in biology are accomplished by people.

PEOPLE BEHIND BIOLOGY 19.1
Barcode of Life Project

Today, most systematists working on living organisms include molecular characters as part of the data set when determining evolutionary relationships. Molecular data include nucleotide base sequences of DNA and RNA or the amino acid sequences of the proteins for which they code. Technological advances have automated many of the necessary laboratory techniques, and analytical software makes it easy to compare new data with information filed in data banks, for example, the Barcode of Life project.

In 2003, Paul Hebert, of the Biodiversity Institute of Ontario at the University of Guelph, proposed a new system of species identification using a sequence of approximately 600 DNA base pairs of the mitochondrial gene cytochrome oxidase I. He anticipated a database that would "provide a new master key for identifying species, one whose power will rise with increased taxon coverage and with faster, cheaper sequencing." He compared the system to the barcode system for retail products such as grocery items, calling it the Barcode of Life (**Figure 1**).

Molecular sequences have some practical advantages over organismal characters. First, they provide abundant data because every base in a nucleic acid can serve as a separate, independent character for analysis. Moreover, because many genes have been conserved by evolution, molecular sequences can be compared between distantly related organisms that share

no organismal characters. Molecular characters can also be used to study closely related species that have only minor morphological differences. Finally, many nucleic acids are not directly affected by the developmental or environmental factors that cause nongenetic morphological variations (see Chapter 18).

But there are drawbacks to using molecular characters in taxonomy. There are only four alternative character states (the four nucleotide bases) at each position in a DNA or RNA sequence (see Chapter 11). If two species have the same nucleotide base substitution at a given position in a DNA segment, their similarity may have evolved independently. As a

result, systematists often find it difficult to verify that molecular similarities were inherited from a common ancestor.

For organismal characters, biologists can establish that similarities are homologous by analyzing the characters' embryonic development or details of their function. But molecular characters have no embryonic development, and biologists still do not understand the functional significance of most molecular differences. Despite these disadvantages, molecular characters represent the genome directly, and researchers use them with great success in phylogenetic analyses and in embryology.

Figure 1

Life on the Edge "Life on the Edge" boxes provide accounts of organisms thriving "on the edge" at unusual temperatures, pressures, radiation dosages, salt concentrations, and so on. These boxes explain how our understanding of "normal" can be increased through study of the "extreme."

The Chemical, Physical, and Environmental Foundations of Biology
While many textbooks use the first few chapters to introduce and/or review, we believe that the first chapters should convey the excitement and interest of biology itself. We therefore placed important background information about biology and chemistry in the reference section entitled *The Chemical, Physical, and Environmental Foundations of Biology*, in the centre of the book. With their purple borders, these pages are distinct and easy to find and have become affectionately known as ***The Purple Pages***.

LIFE ON THE EDGE 40.3
Taking a Long Dive—How Do They Do It?

Concept Fix Icons Concept Fixes draw on the extensive research literature dealing with misconceptions commonly held by biology students. Strategically placed throughout the text, these short segments help students identify—and correct—a wide range of misunderstandings.

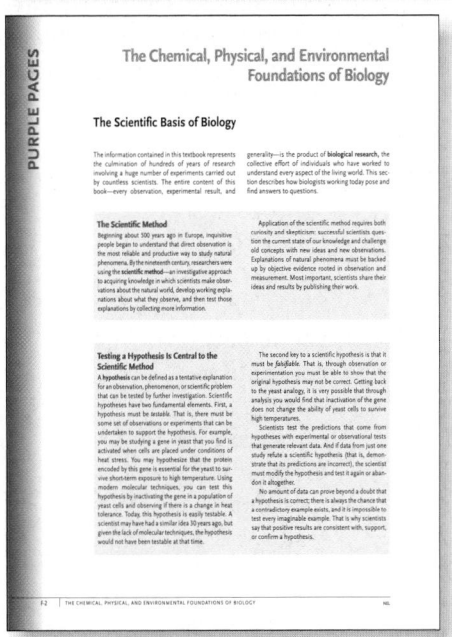

The Chemical, Physical, and Environmental Foundations of Biology

The Scientific Basis of Biology

THINK LIKE A SCIENTIST

Your study of biology focuses not only on *what* scientists now know about the living world but also on *how* they know it. Use these unique features to learn through example how scientists ask scientific questions and pose and test hypotheses.

Throughout the book, we identify recent discoveries made possible by the development of new techniques and new knowledge. In "People behind Biology" 48.1, page 1208, for example, learn how luck and good timing led to the isolation of insulin and a treatment for diabetes.

The Chemical, Physical, and Environmental Foundations of Biology Also known as *The Purple Pages*, these pages enable information to be readily identifiable and accessible to students as they move through the textbook, rather than tied to one particular chapter. *The Purple Pages* keep background information out of the main text, allowing you to focus on the bigger picture.

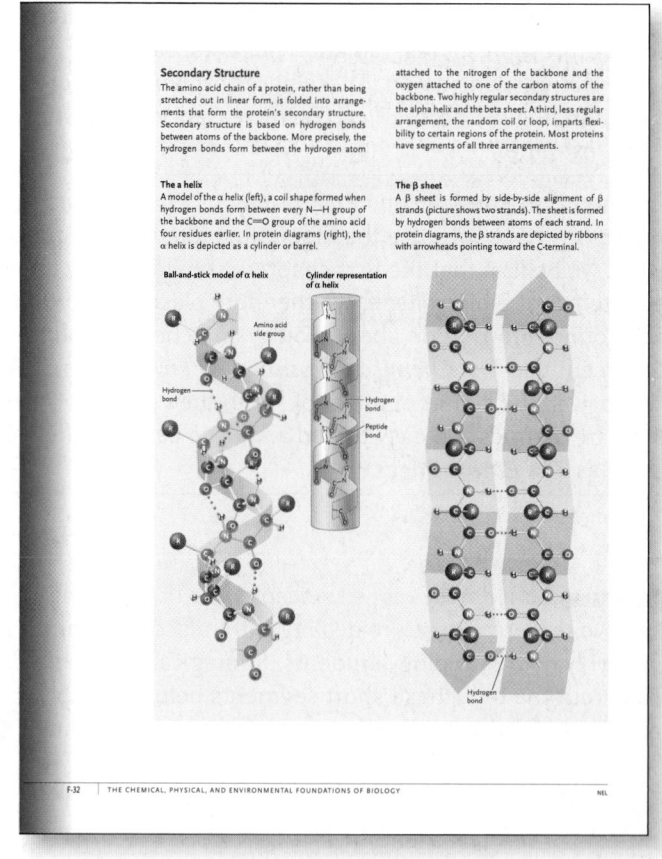

Research Figures Located throughout the book, research figures contain information about how biologists formulate and test specific hypotheses by gathering and interpreting data.

Research Method

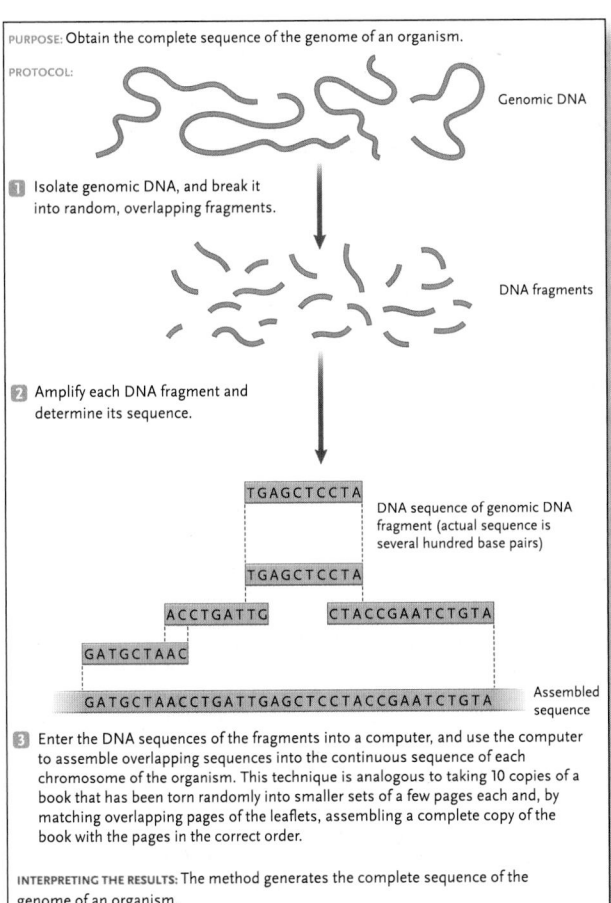

PURPOSE: Obtain the complete sequence of the genome of an organism.

PROTOCOL:

Genomic DNA

1 Isolate genomic DNA, and break it into random, overlapping fragments.

DNA fragments

2 Amplify each DNA fragment and determine its sequence.

TGAGCTCCTA

DNA sequence of genomic DNA fragment (actual sequence is several hundred base pairs)

TGAGCTCCTA
ACCTGATTG   CTACCGAATCTGTA
GATGCTAAC
GATGCTAACCTGATTGAGCTCCTACCGAATCTGTA   Assembled sequence

3 Enter the DNA sequences of the fragments into a computer, and use the computer to assemble overlapping sequences into the continuous sequence of each chromosome of the organism. This technique is analogous to taking 10 copies of a book that has been torn randomly into smaller sets of a few pages each and, by matching overlapping pages of the leaflets, assembling a complete copy of the book with the pages in the correct order.

INTERPRETING THE RESULTS: The method generates the complete sequence of the genome of an organism.

Research Method Figure 16.1 Whole-genome shotgun sequencing.

Experimental Research

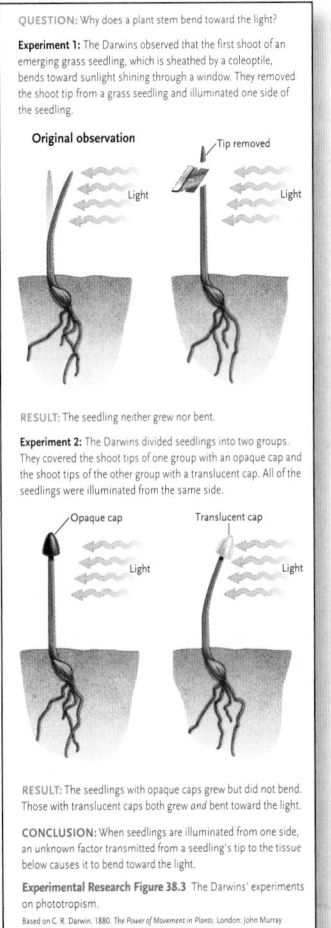

QUESTION: Why does a plant stem bend toward the light?

Experiment 1: The Darwins observed that the first shoot of an emerging grass seedling, which is sheathed by a coleoptile, bends toward sunlight shining through a window. They removed the shoot tip from a grass seedling and illuminated one side of the seedling.

Original observation   Tip removed

Light   Light

RESULT: The seedling neither grew nor bent.

Experiment 2: The Darwins divided seedlings into two groups. They covered the shoot tips of one group with an opaque cap and the shoot tips of the other group with a translucent cap. All of the seedlings were illuminated from the same side.

Opaque cap   Translucent cap

Light   Light

RESULT: The seedlings with opaque caps grew but did not bend. Those with translucent caps both grew *and* bent toward the light.

CONCLUSION: When seedlings are illuminated from one side, an unknown factor transmitted from a seedling's tip to the tissue below causes it to bend toward the light.

Experimental Research Figure 38.3 The Darwins' experiments on phototropism.

Based on C. R. Darwin. 1880. *The Power of Movement in Plants.* London: John Murray.

Observational Research

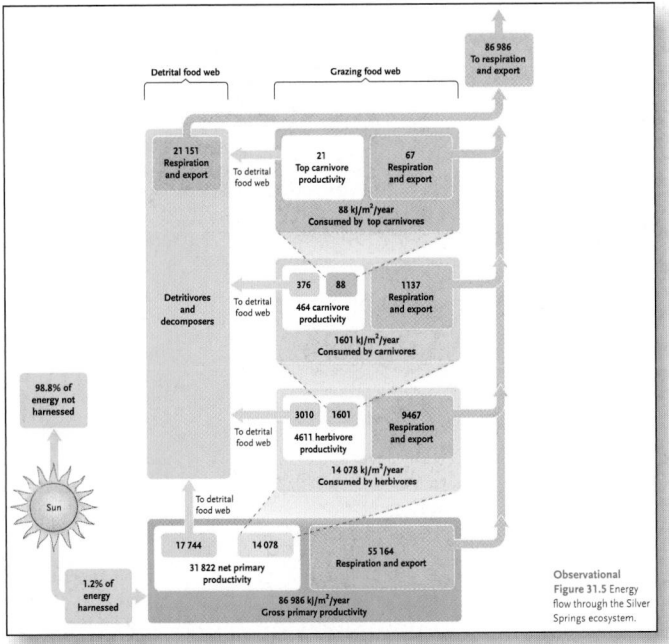

Observational Figure 31.5 Energy flow through the Silver Springs ecosystem.

Spectacular illustrations—developed with great care—help you visualize biological processes, relationships, and structures.

Illustrations of complex biological processes are annotated with numbered step-by-step explanations that lead you through all the major points. Orientation diagrams are inset on figures and help you identify the specific biological process being depicted and where the process takes place.

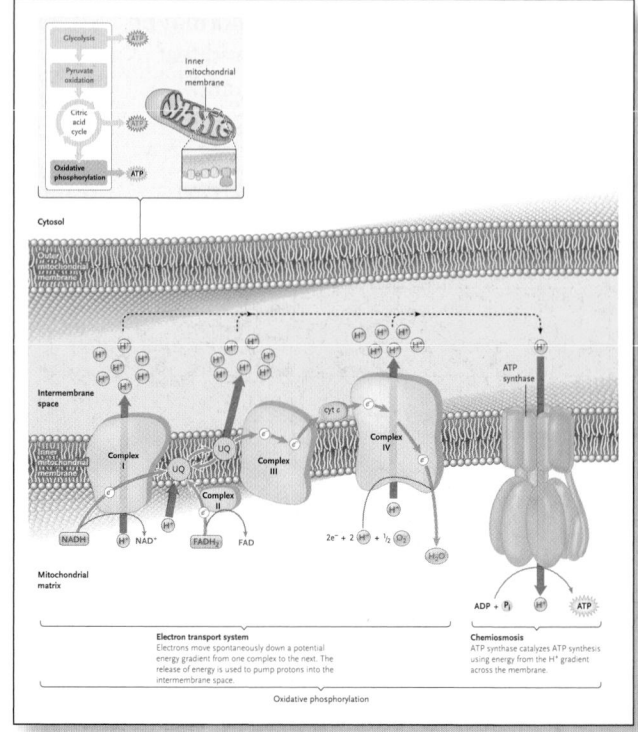

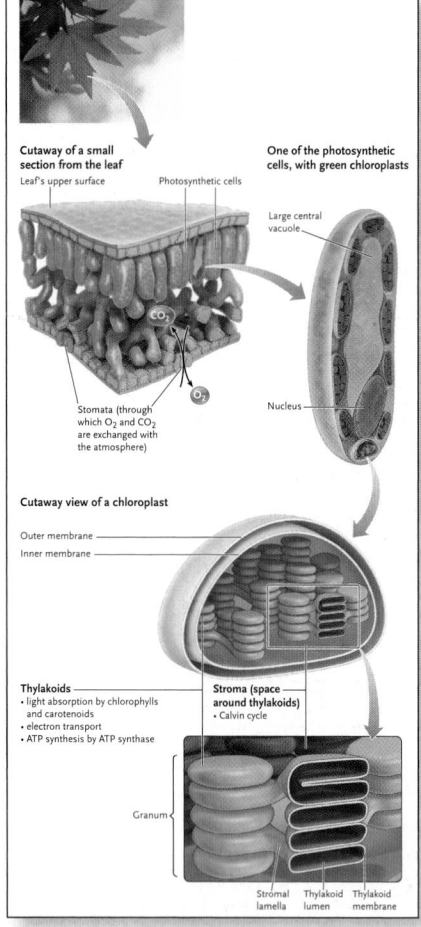

From Macro to Micro Multiple views help you visualize the levels of organization of biological structures and how systems function as a whole.

Electron micrographs are keyed to selected illustrations to help clarify biological structures.

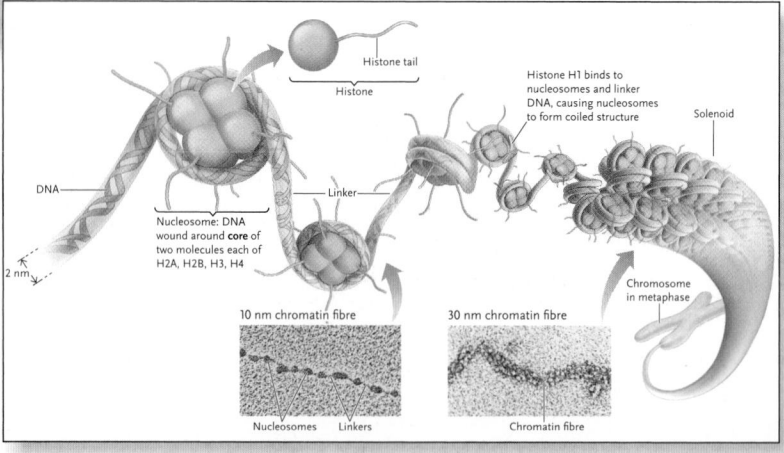

End-of-chapter material encourages you to review the content, assess your understanding, think analytically, and apply what you have learned to novel situations.

Review Key Concepts This brief review often references figures and tables in the chapter and summarizes important ideas developed in the chapter.

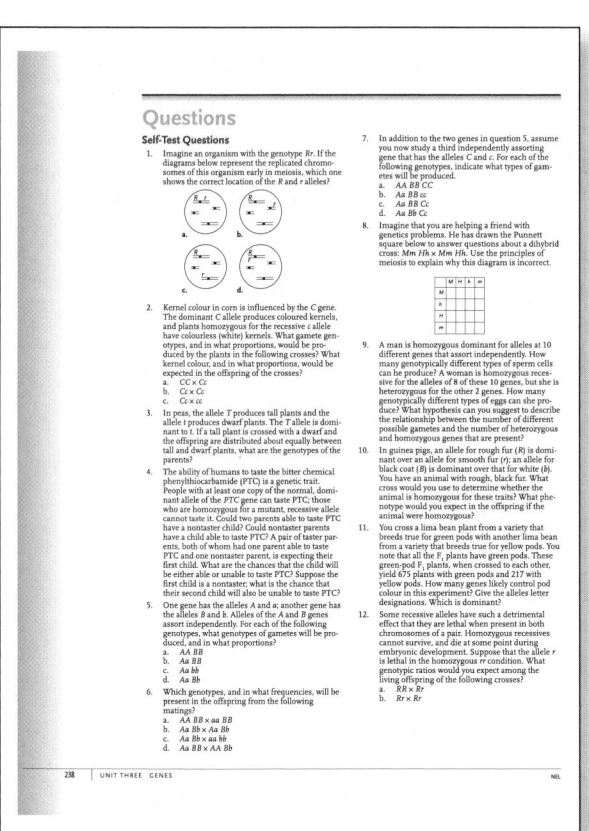

Self-Test Questions These end-of-chapter questions focus on factual content in the chapter while encouraging you to apply what you have learned. Answers to the Self-Test questions have been compiled and placed at the end of the book.

Questions for Discussion

1. The eyes of brown-eyed people are not alike but rather vary considerably in shade and pattern. What do you think causes these differences?

2. Explain how individuals of an organism that are phenotypically alike can produce different ratios of progeny phenotypes.

3. ABO blood type tests can be used to exclude paternity. Suppose a defendant who is the alleged father of a child takes a blood-type test and the results do not exclude him as the father. Do the results indicate that he is the father? What arguments could a lawyer make based on the test results to exclude the defendant from being the father? (Assume the tests were performed correctly.)

Questions for Discussion These questions enable you to participate in discussions on key questions to build your knowledge and learn from others.

Succeed in the course with these dynamic resources!

MindTap

Stay organized and efficient with MindTap—a single destination with all the course material and study aids you need to succeed. Built-in apps leverage social media and the latest learning technology.

For example:
- ReadSpeaker will read the text to you.
- Flashcards are prepopulated to provide you with a jump start for review—or you can create your own.
- You can highlight text and make notes in your MindTap Reader. Your notes will flow into Evernote, the electronic notebook app that you can access anywhere when it's time to study for the exam.

Self-quizzing allows you to assess your understanding.

Also available in the MindTap for Biology are engaging and informative videos that accompany the Purple Pages. From matter to polypeptides, Todd Nickle of Mount Royal University (pictured) will walk you through these foundational concepts, strengthening your understanding and helping you build a strong base of knowledge and understanding for biology.

Visit www.nelson.com/student to start using **MindTap**. Enter the Online Access Code from the card included with your textbook. If a code card is *not* provided, you can purchase instant access at NELSONbrain.com.

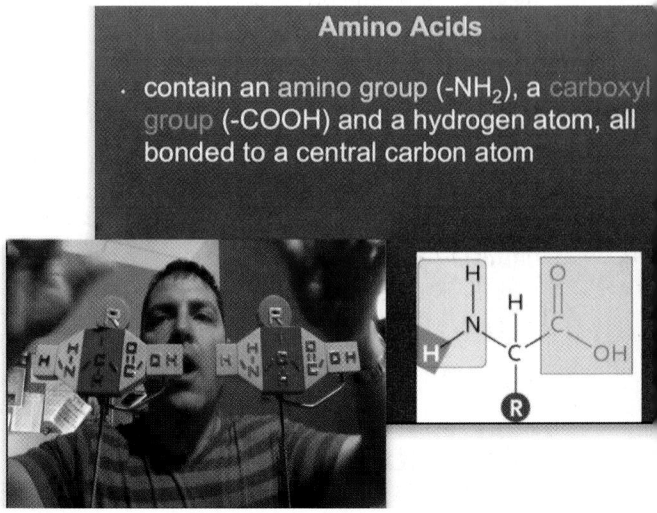

Amino Acids
- contain an amino group (-NH$_2$), a carboxyl group (-COOH) and a hydrogen atom, all bonded to a central carbon atom

Aplia for Biology

Strengthen your understanding of biology with Aplia™!

This innovative, easy-to-use, interactive technology gives you more practice, with detailed feedback to help you learn with every question!

Aplia's focused assignments and active learning opportunities (including randomized questions, exceptional text/art integration, and immediate feedback) get you involved with biology and help you think like a scientist. For more information, visit **www.aplia.com/biology**.

Interactive problems and figures help you visualize dynamic biological processes and integrate concepts, art, media, and homework practice.

Study Guide

The Study Guide (ISBN: 978-0-17-660144-7) from the second Canadian edition has been adapted for the third Canadian edition by Dora Cavallo-Medved of the University of Windsor, Roy V. Rea of the University of Northern British Columbia, and Julie Smit, of the University of Windsor. The Study Guide helps students integrate the concepts within the text and provides study strategies, interactive exercises, self-test questions, and more.

MindTap

Engage, adapt, and master!
Engage your students with the personalized teaching experience of **MindTap**. With relevant assignments that guide students to analyze, apply, and elevate thinking, **MindTap** allows instructors to measure skills and promote better outcomes with ease. Including interactive quizzing, this online tutorial and diagnostic tool identifies each student's unique needs with a pretest. The learning path then helps students focus on concepts they're having the most difficulty mastering. It refers to the accompanying MindTap Reader eBook and provides a variety of learning activities designed to appeal to diverse ways of learning. After completing the study plan, students take Aplia problem sets and then take a posttest to measure their understanding of the material. Instructors have the ability to customize the learning path, add their own content, and track and monitor student progress by using the instructor Gradebook and Progress app. The **MindTap** resources have been developed by Dora Cavallo-Medved of the University of Windor and Reehan Mirza of Nipissing University.

Aplia for Biology

Get your students engaged and motivated with Aplia for Biology.
Help your students learn key concepts via Aplia's focused assignments and active learning opportunities that include randomized, automatically graded questions, exceptional text/art integration, and immediate feedback. Aplia has a full course management system that can be used independently or in conjunction with other course management systems such as Blackboard and WebCT. **Visit www.aplia.com/biology.**

The Aplia course for *Biology: Exploring the Diversity of Life,* Third Canadian Edition, was prepared by Anna Rissanen of Memorial University, and Todd Nickle and Alexandra Farmer of Mount Royal University.

Instructor's Resource DVD: The ultimate tool for customizing lectures and presentations.

Select instructor ancillaries are provided on the *Instructor's Resource DVD* (ISBN 978-0-17-660143-0) and the Instructor Companion site.

The **Nelson Education Teaching Advantage (NETA)** program delivers research-based instructor resources that promote student engagement and higher-order thinking to enable the success of Canadian students and educators.

To ensure the high quality of these materials, all Nelson ancillaries have been professionally copy-edited.

Be sure to visit Nelson Education's **Inspired Instruction** website at **www.nelson.com/inspired/** to find out more about NETA. Don't miss the testimonials of instructors who have used NETA supplements and seen student engagement increase!

NETA Test Bank: This resource was written by Ivona Mladenovic of Simon Fraser University. It includes over 2500 multiple-choice questions written according to NETA guidelines for effective construction and development of higher-order questions. The Test Bank was copy-edited by a NETA-trained editor and reviewed by David DiBattista for adherence to NETA best practices. Also included are true/false, essay, short-answer, matching, and completion questions. Test Bank files are provided in Microsoft Word format for easy editing and in PDF for convenient printing, whatever your system.

The **NETA Test Bank** is available in a new, cloud-based platform. **Nelson Testing Powered by Cognero®** is a secure online testing system that allows you to author, edit, and manage test bank content from any place you have Internet access. No special installations or downloads are needed, and the desktop-inspired interface, with its drop-down menus and familiar, intuitive tools, allows you to create and manage tests with ease. You can create multiple test versions in an instant and import or export content into other systems. Tests can be delivered from your learning management system, your classroom, or wherever you want. Nelson Testing Powered by Cognero can be accessed through www.nelson.com/instructor.

NETA PowerPoint: Microsoft PowerPoint® lecture slides for every chapter have been created by Jane Young of the University of Northern British Columbia. There is an average of 80 slides per chapter, many featuring key figures, tables, and photographs from *Biology: Exploring the Diversity of Life,* Third Canadian Edition. The PowerPoint slides also feature "build slides"—selected illustrations with labels from the book that have been reworked to allow optimal display in PowerPoint. NETA principles of clear design and engaging content have been incorporated throughout, making it simple for instructors to customize the deck for their courses.

Image Library: This resource consists of digital copies of figures, short tables, and photographs used in the book. Instructors may use these jpegs to customize the NETA PowerPoint slides or create their own PowerPoint presentations.

NETA Instructor's Manual: This resource was written by Tamara Kelly of York University and Tanya Noel of the University of Windsor. It is organized according to the textbook chapters and addresses key educational concerns, such as typical stumbling blocks students face and how to address them. Other features include tips on teaching using cases as well as suggestions on how to present material and use technology and other resources effectively, integrating the other supplements available to both students and instructors. This manual doesn't simply reinvent what's currently in the text; it helps the instructor make the material relevant and engaging to students.

Day One: Day One—ProfInClass is a PowerPoint presentation that you can customize to orient your students to the class and their text at the beginning of the course.

TurningPoint®: Another valuable resource for instructors is **TurningPoint® classroom response software** customized for *Biology: Exploring the Diversity of Life,* Third Canadian Edition, by Jane Young at the University of Northern British Columbia. Now you can author, deliver, show, access, and grade, all in PowerPoint, with no toggling back and forth between screens! JoinIn on TurningPoint is the only classroom response software tool that gives you true PowerPoint integration. With JoinIn, you are no longer tied to your computer. You can walk about your classroom as you lecture, showing slides and collecting and displaying responses with ease. There is simply no easier or more effective way to turn your lecture hall into a personal, fully interactive experience for your students. If you can use PowerPoint, you can use JoinIn on TurningPoint! (Contact your Nelson publishing representative for details.) These contain poll slides and pre- and posttest slides for each chapter in the text.

Prospering in Biology

Using This Book

The following are things you will need to know in order to use this text and prosper in biology.

Names

What's in a name? People are very attached to names—their own names, the names of other people, the names of flowers and food and cars and so on. It is not surprising that biologists would also be concerned about names. Take, for example, our use of scientific names. Scientific names are always italicized and Latinized.

Castor canadensis Kuhl is the scientific name of the Canadian beaver. *Castor* is the genus name; *canadensis* is the specific epithet. Together they make up the name of the species, which was first described by a person called Kuhl. "Beaver" by itself is not enough because there is a European beaver, *Castor fiber,* and an extinct giant beaver, *Castoides ohioensis*. Furthermore, common names can vary from place to place (*Myotis lucifugus* is sometimes known as the "little brown bat" or the "little brown myotis").

Biologists prefer scientific names because the name (Latinized) tells you about the organism. There are strict rules about the derivation and use of scientific names. Common names are not so restricted, so they are not precise. For example, in *Myotis lucifugus, Myotis* means "mouse-eared" and *lucifugus* means "flees the light"; hence, this species is a mouse-eared bat that flees the light.

Birds can be an exception. There are accepted "standard" common names for birds. The American Robin is *Turdus migratorius*. The common names for birds are usually capitalized because of the standardization. However, the common names of mammals are not capitalized, except for geographic names or patronyms (*geographic* = named after a country, e.g., Canadian beaver; *patronym* = named after someone, e.g., Ord's kangaroo rat).

Although a few plants that have very broad distributions may have accepted standard common names (e.g., white spruce, *Picea glauca*), most plants have many common names. Furthermore, the same common name is often used for more than one species. Several species in the genus *Taraxacum* are referred to as "dandelion." It is important to use the scientific names of plants to be sure that it is clear exactly which plant we mean. The scientific names of plants also tell us something about the plant. The scientific name for the weed quack grass, *Elymus repens,* tells us that this is a type of wild rye (*Elymus*) and that this particular species spreads or creeps (*repens* = creeping). Anyone who has tried to eliminate this plant from their garden or yard knows how it creeps! Unlike for animals, plant-naming rules forbid the use of the same word for both genus and species names for a plant; thus, although *Bison bison* is an acceptable scientific name for buffalo, such a name would never be accepted for a plant.

In this book, we present the scientific names of organisms when we mention them. We follow standard abbreviations; for example, although the full name of an organism is used the first time it is mentioned (e.g., *Castor canadensis*), subsequent references to that same organism abbreviate the genus name and provide the full species name (e.g., *C. canadensis*).

In some areas of biology, the standard representation is of the genus, for example, *Chlamydomonas*. In other cases, names are so commonly used that only the abbreviation may be used (e.g., *E. coli* for *Escherichia coli*).

Units

The units of measure used by biologists are standardized (metric or SI) units, used throughout the world in science.

Definitions

The science of biology is replete with specialized terms (sometimes referred to as "jargon") used to communicate specific information. It follows that, as with scientific names, specialized terms increase the precision with which biologists communicate among themselves and with others. Be cautious about the use of terms because jargon can obscure precision. When we encounter a "slippery" term (such as *species* or *gene*), we explain why one definition for all situations is not feasible.

Time

In this book, we use CE (Common Era) to refer to the years since year 1 and BCE (Before the Common Era) to refer to years before that.

Geologists think of time over very long periods. A geological time scale (see *The Purple Pages*, pp. F-50–F-51) shows that the age of Earth could be measured in years, but it's challenging to think of billions of years expressed in days (or hours, etc.). With the advent of using the decay rates of radioisotopes to measure the age of rocks, geologists adopted 1950 as the baseline, the "Present," and the past is referred to as BP ("Before Present"). A notation of 30 000 years BP (^{14}C) indicates 30 000 years before 1950 using the ^{14}C method of dating.

Other dating systems are also used. Some archaeologists use PPNA (PrePottery Neolithic A, where A is the horizon or stratum). In deposits along the Euphrates River, 11 000 PPNA appears to be the same as 11 000 BP. In this book, we use BCE or BP as the time units, except when referring to events or species from more than 100 000 years ago. For those dates, we refer you to the geological time scale (see *The Purple Pages,* pp. F-50–F-51).

Sources

Where does the information presented in a text or in class come from? What is the difference between what you read in a textbook or an encyclopedia and the material you see in a newspaper or tabloid? When the topic relates to science, the information should be based on material that has been published in a scholarly journal. In this context, "scholarly" refers to the process of review. Scholars submit their manuscripts reporting their research findings to the editor (or editorial board) of a journal. The editor, in turn, sends the manuscript out for comment and review by recognized authorities in the field. The process is designed to ensure that what is published is as accurate and appropriate as possible. The review process sets the scholarly journal apart from the tabloid.

There are literally thousands of scholarly journals, which, together, publish millions of articles each year. Some journals are more influential than others, for example, *Science* and *Nature.* These two journals are published weekly and invariably contain new information of interest to biologists.

To collect information for this text, we have drawn on published works that have gone through the process of scholarly review. Specific references (citations) are provided, usually in the electronic resources designed to complement the book.

A citation is intended to make the information accessible. Although there are many different formats for citations, the important elements include (in some order) the name(s) of the author(s), the date of publication, the title, and the publisher. When the source is published in a scholarly journal, the journal name, its volume number, and the pages are also provided. With the citation information, you can visit a library and locate the original source. This is true for both electronic (virtual) and real libraries.

Students of biology benefit by making it a habit to look at the most recent issues of their favourite scholarly journals and use them to keep abreast of new developments.

M. Brock Fenton
Heather Addy
Denis Maxwell
Tom Haffie
Bill Milsom

London, Calgary, and Vancouver
January 2015

Acknowledgements

We thank the many people who have worked with us on the production of this text, particularly Paul Fam, Senior Publisher, whose foresight brought the idea to us and whose persistence saw the project through. Thanks go to those who reviewed the second Canadian edition text to provide us with feedback for the third Canadian edition, including

Declan Ali, University of Alberta
Eric Alcorn, Acadia University
Patricia Chow-Fraser, McMaster University
Kimberley Gilbride, Ryerson University
Roberto Quinlan, York University
Matthew Smith, Wilfrid Laurier University
Christopher Todd, University of Saskatchewan
Kenneth Wilson, University of Saskatchewan

We are also grateful to the members of the Editorial Advisory Board and the Student Advisory Boards for the third Canadian edition, who provided us with valuable feedback and alternative perspectives (special acknowledgements to these individuals are listed below). We also thank Richard Walker at the University of Calgary and Ken Davey at York University, who began this journey with us but were unable to continue. We thank Carl Lowenberger for contributing Chapter 51, Defences against Disease. We are especially grateful to Toni Chahley, Developmental Editor, who kept us moving through the chapters at an efficient pace, along with Shanthi Guruswamy and Alex Antidius, Project Managers, and Natalia Denesiuk Harris, Senior Production Project Manager. We thank Kristiina Paul, our photo researcher, for her hard work with the numerous photos in the book, and Julia Cochrane for her careful and thoughtful copy-editing. Finally, we thank Leanne Newell, Marketing Manager, for making us look good.

Brock Fenton would like to thank Allan Noon, who offered much advice about taking pictures; Laura Barclay, Jeremy McNeil, Tony Percival-Smith, C.S. (Rufus) Churcher, and David and Meg Cumming for the use of their images; and Karen Campbell for providing a critical read on Chapter 48, Putting Selection to Work.

Heather Addy would like to thank Ed Yeung for generously providing many images and assistance with revision of figures, and Cindy Graham, David Bird, Fengshan Ma, and William Huddleston for providing feedback and valuable suggestions for improving several chapters.

Tom Haffie would like to acknowledge the cheerful and insightful editorial work of Jennifer Waugh on Chapter 16.

The authors are all indebted to Johnston Miller, whose extensive background research anchored our Concept Fixes in the education literature.

We thank Rosemary Tanner, our substantive editor, for bringing her considerable skills to this edition. Her insight and expertise have brought a new level of precision and clarity to key chapters in the book. She is a consummate professional, and we are all very grateful for her patience and good humour.

It is never easy to be in the family of an academic scientist. We are especially grateful to our families for their sustained support over the course of our careers, particularly during those times when our attentions were fully captivated by bacteria, algae, fungi, parasites, snakes, geese, or bats. Saying "yes" to a textbook project means saying "no" to a variety of other pursuits. We appreciate the patience and understanding of those closest to us that enabled the temporary reallocation of considerable time from other endeavours and relationships.

Many of our colleagues have contributed to our development as teachers and scholars by acting as mentors, collaborators, and, on occasion, "worthy opponents." Like all teachers, we owe particular gratitude to our students. They have gathered with us around the discipline of biology, sharing their potent blend of enthusiasm and curiosity and leaving us energized and optimistic for the future.

Editorial and Student Advisory Boards

We were very fortunate to have the assistance of some extraordinary students and instructors of biology across Canada who provided us with feedback that helped shape this textbook into what you see before you. As such, we would like to say a very special thank-you to the following people:

Editorial Advisory Board

Peter Boag, Queen's University
Julie Clark, York University
Brett Couch, University of British Columbia
Robert Edwards, University of Calgary
Mark Fitzpatrick, University of Toronto
Jon Houseman, University of Ottawa
Andrew Laursen, Ryerson University
Todd Nickle, Mount Royal University
Cynthia Paszkowski, University of Alberta
Roy Rea, University of Northern British Columbia
Matthew Smith, Wilfrid Laurier University
Christopher Todd, University of Saskatchewan

Student Advisory Boards

University of Alberta

Paul Fam/Nelson Education Ltd.

Depicted above are members of the University of Alberta Student Advisory Board: (left to right) Shargeel Hayat, Mark Kamprath, Esperance Madera, Maxwell Douglas, Kina Tiet, Dana Miller, Jessica Lamont. Not pictured: Punit Virk.

Mount Royal University

Paul Fam/Nelson Education Ltd.

Depicted above are members of the Student Advisory Board at Mount Royal University: (left to right) Joshua Loza, Lorne Sobcyzk, Hala Al Sharbati, Nikolai Heise, Karishma Rahmat, Todd Nickle, Kathleen Malabug, Katherine Obrovac, Kristin Milloy, Michael Rudolf, Dorothy Hill, Maria Fernanda Ochoa.

ORIGINS AND ORGANIZATION

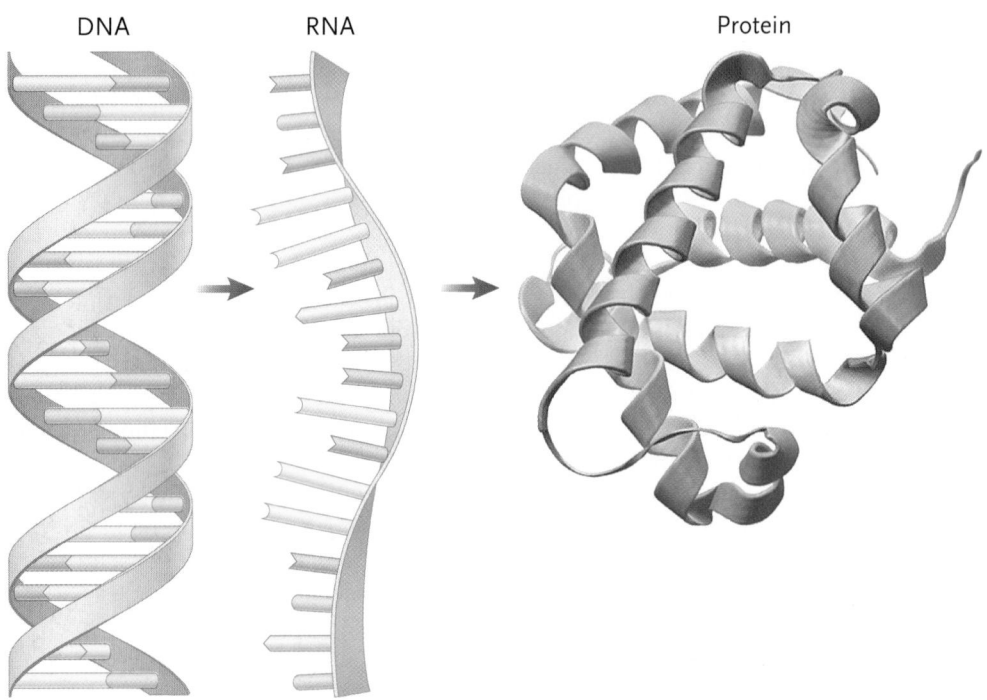

DNA RNA Protein

In all life, the molecule DNA carries information used to synthesize proteins through an RNA intermediate. Why do all life forms use this system and which one of the three molecules evolved first on the early Earth?

The opening volume of this textbook illustrates to us that, although life is truly astonishing in its diversity, it is equally astonishing in its similarity. From monkeys to mycoplasma to monocots, everything that is alive on Earth employs a variation on that remarkable innovation—the **cell.** No matter if that cell is communicating with other cells in the brain of a fruit fly, or capturing sunlight in a spruce needle, or driving the muscles of a sprinting cheetah, or thriving in the mineral-rich water of deep-sea vents— no matter what their role or activity, all cells, share a remarkably long list of common features that Volume 1 explores in detail.

All cells are surrounded by a bilayer of lipid molecules that allows for the development of an internal environment that is distinctly different from the outside world. All cells possess genes that are coded by the molecule DNA that through the process of transcription get copied into RNA. All cells contain ribosomes where some RNAs get translated into **proteins**, the fundamental structural, functional and regulatory molecule of the cell. All cells contain a specialized class of proteins called enzymes that have the astonishing ability to increase the rate of a chemical reaction by a billion times. We could go on and on.

The stark similarities present in all forms of life leads one logically to a single compelling conclusion— all living things on Earth are relatives of one another; ultimately sharing the same parent we call LUCA for Last Universal Common Ancestor. LUCA may have not been the first form of life, in fact there may have been hundreds of different forms that sprung up on the early Earth, but it was the most successful because it survived and, it alone, gave rise to the diversity of life we see today.

But hang on one minute—what is life anyway and how did it arise? Even at a time of unprecedented scientific achievements, arriving at an accurate yet succinct definition of life is frustratingly elusive. Defining life is not easy and at the same time exactly how life got started perhaps as early as 4 billion years ago remains one of the fundamental mysteries of science. Our understanding of how life got started would be helped immensely if we could find life some place other than Earth—perhaps on one of the thousands of habitable planets recently discovered orbiting distant stars. Maybe the **development** of life is incredibly rare or perhaps it is inevitable given the right conditions and enough time.

Although life occupies a unique spot in the natural world, cells are built of the same atoms and abide by the same laws of chemistry and physics as everything else. However, seemingly in defiance of the universal tendency towards increasing disorder, living organisms are able to maintain a complex and highly structured state. This requires cells to be constantly bringing in energy and matter from their surroundings, building more complex molecules then they take in and releasing less complex molecules and heat back into the environment. On Earth most of the energy used by living systems comes ultimately from sunlight—being harvested and converted into a useable chemical form through the process of photosynthesis. The conversion of carbon dioxide into sugar by photosynthesis introduces chemical energy into the biosphere. The stepwise and controlled oxidation of these molecules through cellular respiration is used to generate ATP the energy currency that is universally accepted to power cells.

Genes are stretches of DNA sequence in an organism that collectively comprise a kind of library of information about how a cell functions. Recent advances in technology have made it relatively easy to determine the entire DNA sequence of an organism, including individual humans. As a result, modern biology is awash in the As, Ts, Gs and Cs of DNA sequence revealed by thousands of sequencing projects. New insights into evolutionary history as well as gene structure and function are arising from bioinformatic analysis of such extensive data sets.

The elegant double-strandedness of DNA, whereby two long strands of nucleotides are held together by hydrogen bonds formed between complementary base pairs, affords a straightforward mechanism for replication that was recognized early on by Watson and Crick. Although conceptually simple, the mechanism for unwinding the DNA **double helix** and polymerizing new complementary bases is rather complicated and managed by a suite of interacting enzymes. Again, we see that all DNA on the planet is replicated using variations on one underlying strategy.

DNA genes provide the cell with needed RNA by transcription. One remarkable feature of all protein-coding genes is that, with minor exceptions, the information they carry is specified by a universal code. That is, a gene from one organism can be "understood" by any other organism, even if only distantly related: A gene from a spider can be expressed by a goat; A gene from a jellyfish can be expressed in a flower. The field of genetic engineering is devoted to developing the tools and applications of this technology for moving genes from one organism to another.

In a story that is about to come full circle, synthetic biologists have extensively customized naturally occurring cells and have made important advances toward their ultimate goal of creating novel life forms artificially in the lab. As students of biology in the early 21st century, you can well expect to witness a momentous event in Earth's history, the creation of one life form by another.

Paintings by Claude Monet (1840–1926). Compared to his early works, including "The Water-Lily Pond" **(a),** his later paintings, including "The Japanese Footbridge" **(b),** bordered on the abstract with almost complete loss of light blue. Monet suffered from vision degenerative disease cataracts, diagnosed in 1912.

Light and Life

WHY IT MATTERS

Claude Monet (1840–1926), a French painter, is considered by many to be the master of the impressionist form that rose to prominence in the late nineteenth century. Other well-known impressionists are Edgar Degas and Paul Cézanne. Impressionism as an art movement was characterized by the use of small visible brush strokes that emphasized light and colour, rather than lines, to define an object. The artists used pure, unmixed colour, not smoothly blended, as was the custom at the time. For example, instead of physically mixing yellow and blue paint, they placed unmixed yellow paint on the canvas next to unmixed blue paint so that the colours would mingle in the eye of the viewer to create the impression of green. The impressionists found that they could capture the momentary and transient effects of sunlight and the changing colour of a scene by painting *en plein air,* in the open air, outside the studio, where they could more accurately paint the reflected light of an immediate scene.

Interestingly, compared with his early works, which included "The Water-Lily Pond" (1899), Monet's later paintings verge on the abstract, with colours bleeding into each other and a lack of rational shape and perspective. For example, "The Japanese Footbridge" is an explosion of orange, yellow, and red hues, with heavy, broad brush strokes, leaving the viewer barely able to discern the vague shape of the arched bridge. In many of Monet's later works, the colours in his paintings became more muted, far less vibrant and bright, with a pronounced colour shift from blue green to red yellow and an almost total absence of light blues. The sense of atmosphere and light that he was famous for in his earlier works disappeared.

Although the change in Monet's paintings could easily be explained by an intentional change in style or perhaps an age-related change in manual dexterity, Monet himself realized that it was not his style or dexterity that had changed but, rather, his ability to see. Monet suffered from cataracts, a vision-deteriorating disease diagnosed in both eyes by a Parisian ophthalmologist in 1912 when Monet was 72. A cataract is a change in the lens of the eye, making it more opaque. The underlying cause is a progressive denaturation of one of the proteins that make up the lens. The increased opaqueness of the lens absorbs certain **wavelengths** of light, decreasing the transmittance of blue light. Thus, to a cataract sufferer such as Monet, the world appears more yellow.

In this, the first of the 52 chapters that make up the textbook, we introduce you to the science of biology by using light as a central connecting theme. Light is arguably the most fundamental of natural phenomena, and foundational experiments into the nature of light were a key part of the scientific revolution that took place in the sixteenth and seventeenth centuries. Beyond formally defining light and discussing its properties, in this chapter we explore the huge diversity of areas of biology that light influences, from the molecular to the ecological. This introductory tour is not intended to be complete or exhaustive but to simply set the stage for the topics that come in subsequent chapters.

1.1 The Physical Nature of Light

Light serves two important functions for life on Earth. First, it is a source of energy that directly or indirectly sustains virtually all organisms. Second, light provides organisms with information about the physical world that surrounds them. An excellent example of an organism that uses light for both energy and information is the green alga *Chlamydomonas reinhardtii* **(Figure 1.1).** *C. reinhardtii* is a single-celled photosynthetic eukaryote that is commonly found in ponds and lakes. Each cell contains a single large chloroplast that harvests light energy and uses it to make energy-rich molecules through the process of photosynthesis. In addition, each cell contains a light sensor called an eyespot that allows individual cells to gather information about the

location and intensity of a light source. With this information, cells can move toward or away from the light source, allowing them to optimize light harvesting for photosynthesis. Regardless of whether the light is used as a source of energy or as a source of information about the environment, both uses rely on the same fundamental properties of light and require the light energy to be captured by the organism.

1.1a What Is Light?

The reason there is life on Earth and, as far as we know, nowhere else in our solar system has to do with distance—specifically, the distance of 150 million kilometres separating Earth from the Sun **(Figure 1.2).** By converting hydrogen into helium at the staggering rate of some 3.4×10^{38} hydrogen nuclei per second, the Sun converts over 4 million tonnes of matter into energy every second. This energy is given off as *electromagnetic radiation,* which travels in the form of a wave at a speed of 1 079 252 848 km/h (the speed of light) and reaches Earth in just over 8 minutes. Scientists often distinguish different types of electromagnetic radiation by their wavelength, the distance between two successive peaks **(Figure 1.3).** The wavelength of electromagnetic radiation ranges from less than one picometre (10^{-12} m) for cosmic rays to more than a kilometre (10^6 m) for radio waves.

So what is light? **Light** is most commonly defined as the portion of the **electromagnetic spectrum** that humans can detect with their eyes. This is a very narrow portion of the total electromagnetic spectrum, only spanning the wavelengths from about 400 to 700 nm

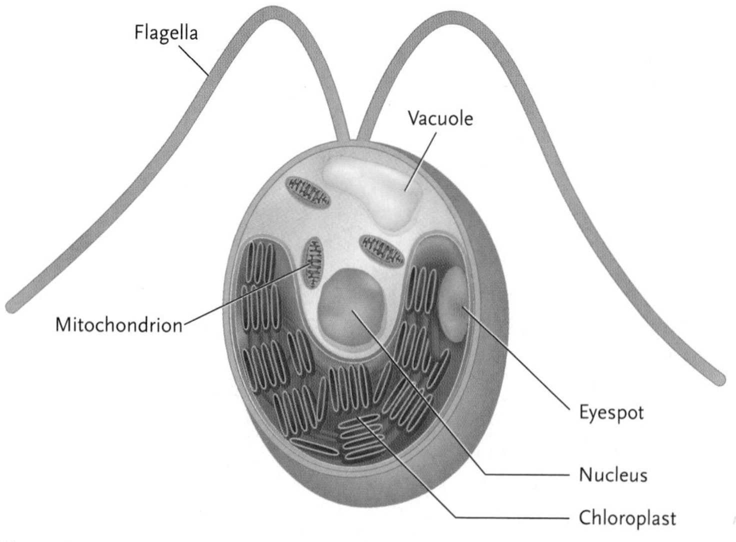

Figure 1.1
Chlamydomonas reinhardtii. Each cell contains a single chloroplast used for photosynthesis as well as an eyespot for sensing light in the environment.

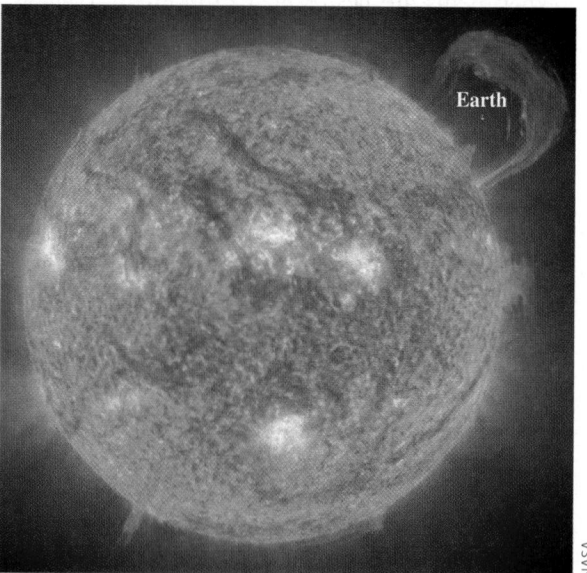

NASA

Figure 1.2
The Sun. Like most stars, the Sun generates electromagnetic radiation as a result of the nuclear fusion of hydrogen nuclei into helium. Note the superimposed image of Earth used to illustrate the relative sizes.

a. Range of the electromagnetic spectrum

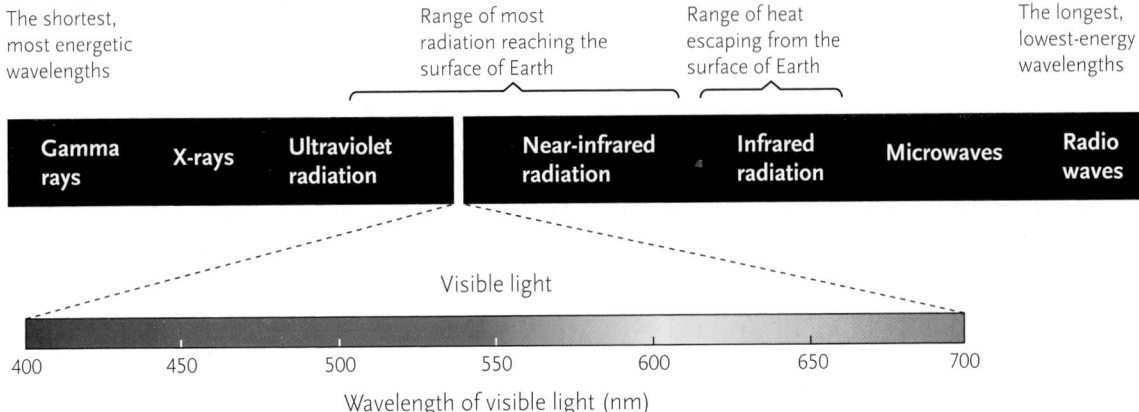

The shortest, most energetic wavelengths

Range of most radiation reaching the surface of Earth

Range of heat escaping from the surface of Earth

The longest, lowest-energy wavelengths

| Gamma rays | X-rays | Ultraviolet radiation | Near-infrared radiation | Infrared radiation | Microwaves | Radio waves |

Visible light

400   450   500   550   600   650   700

Wavelength of visible light (nm)

b. Examples of wavelengths

400 nm wavelength

700 nm wavelength

Figure 1.3
The electromagnetic spectrum. **(a)** The electromagnetic spectrum ranges from gamma rays to radio waves; visible light and the wavelengths used for photosynthesis occupy only a narrow band of the spectrum. **(b)** Examples of wavelengths, showing the difference between the longest and shortest wavelengths of visible light.

(see Figure 1.3). In physics, the definition of light often includes other regions of the electromagnetic spectrum, and thus terms such as *visible light, ultraviolet light,* and *infrared light* are commonly used.

The physical nature of light has been the focus of scientific inquiry for hundreds of years, but it is still not simple to grasp. Unlike the atoms that make up matter, light has no mass. As well, although the results of some experiments suggest that light behaves as a wave as it travels through space, the results of other experiments are best explained by light being composed of a stream of energy particles called **photons.** That light has properties of both a wave and a stream of photons is often referred to as the particle-wave duality. And so we are left with a compromise description—light is best understood as a wave of photons. The relationship between the wavelength of light and the energy of the photons it carries is an inverse one: the longer the wavelength, the lower the energy of the photons it contains. Looking at Figure 1.3, this means that shorter-wavelength blue light consists of photons of higher energy than red light, which has a longer wavelength and photons of lower energy.

1.1b Light Interacts with Matter

Although light has no mass, it is still able to interact with matter and cause change. This change is what allows the energy of light to be used by living things. When a photon of light hits an object, the photon has three possible fates: it can be reflected off the object, transmitted through the object or absorbed by the object. To be used as a source of energy or information

by an organism, absorption must take place. The absorption of light occurs when the energy of the photon is transferred to an **electron** within a molecule. This excites the electron, moving it from its ground state to a higher energy level that is referred to as an excited state **(Figure 1.4).** An important fact to remember is that a photon can be absorbed only if the energy of the photon matches the amount of energy needed to move the electron from its ground state to a specific excited state. If the energies don't match, then the photon of light is not absorbed but instead is transmitted through the molecule or reflected. It is the excited-state electron that represents the source of energy required for processes such as photosynthesis and vision.

A major class of molecules that are very efficient at absorbing photons are called **pigments (Figure 1.5, p. 6).** There is a large diversity of pigments, including

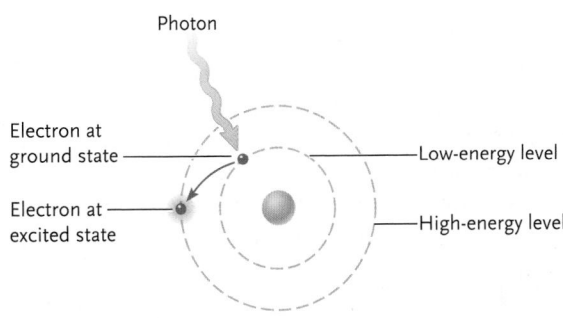

Photon

Electron at ground state ——————— Low-energy level

Electron at excited state —————— High-energy level

Figure 1.4
The absorption of a photon by a molecule results in the energy being transferred to an electron. This causes the energy to move to a higher-energy excited state.

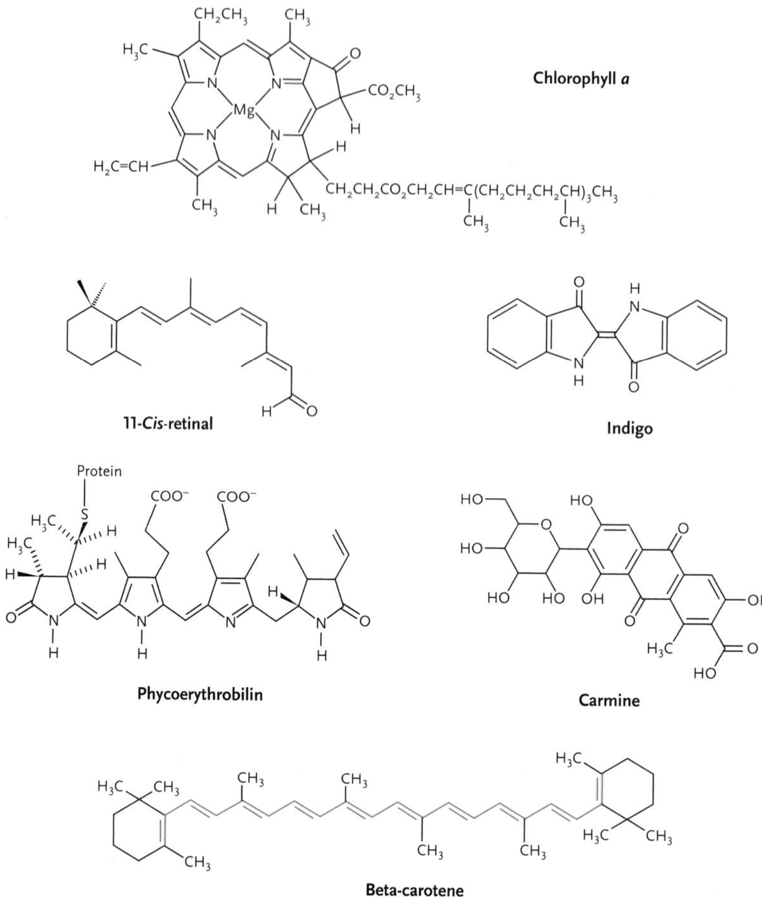

Figure 1.5

Structure of some common pigments. Chlorophyll *a*, photosynthesis. 11-*Cis*-retinal, vision. Indigo, dye. Phycoerythrobilin, red photosynthetic pigment found in red algae. Carmine, scale pigment found in some insects. Beta-carotene, an orange accessory photosynthetic pigment. A common feature of all of these pigments that is critical for light absorption is the presence of a conjugated system of double/single carbon bonds (shown in red for beta-carotene).

Chlorophyll *a*

11-*Cis*-retinal

Indigo

Phycoerythrobilin

Carmine

Beta-carotene

chlorophyll *a*, which is involved in photosynthesis; retinal, which is involved in vision; and indigo, which is used to dye jeans their distinctive blue colour.

An important question we can ask is: what is it about pigments that enable them to capture light? At first glance, the molecules shown in Figure 1.5 seem to be structurally very different from each other. However, they all have a common feature critical to light absorption: a region where carbon atoms are covalently bonded to each other with alternating single and double bonds. This bonding arrangement is called a *conjugated system,* and it results in the delocalization of electrons. None of these electrons are closely associated with a particular atom or involved in bonding and thus are available to interact with a photon of light.

Most pigments absorb light at distinctly different wavelengths. This is because the differences in chemical structure result in each pigment having distinct excited states available to its delocalized electrons. While some pigments can absorb, for example, only blue photons because they have only one high-energy excited state, others can absorb two or more different wavelengths because they have two or more excited states. Photon absorption is intimately related to the concept of colour. A pigment's colour is the result of photons of light that it *does not* absorb. Instead of being absorbed, these photons are reflected off the pigment or transmitted through the pigment to reach your eyes **(Figure 1.6).**

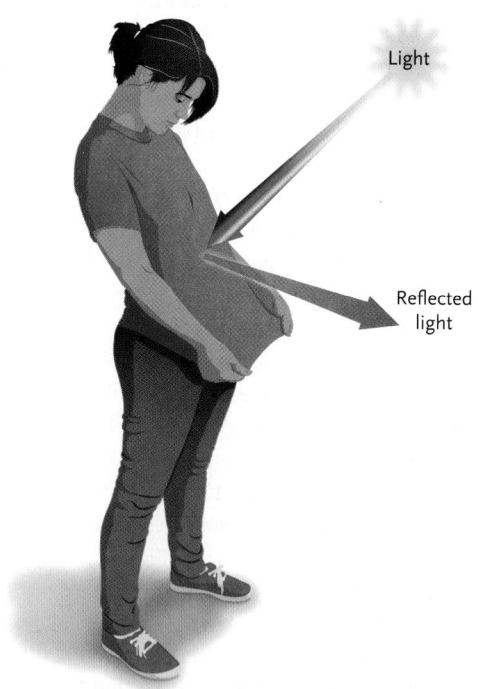

Figure 1.6

Why the t-shirt is red. Pigment molecules bound to the fabric of the shirt absorb blue, green, and yellow photons of light. Red photons are not absorbed and are instead transmitted through the shirt or reflected.

1.2 Light as a Source of Energy

As we will discuss in subsequent chapters, all living things require a constant supply of energy from their surroundings. The ultimate source of this energy is light from the Sun, which is made accessible to biological systems through the ability of plants and related organisms to convert this light energy into a chemical form. Through photosynthesis, plants absorb photons of light and use the potential energy to convert carbon dioxide into sugars (carbohydrates). Energy from the Sun enters the **biosphere** through photosynthesis.

Following light absorption, the potential energy of excited electrons within **chlorophyll** is used in photosynthetic electron **transport** to synthesize the energy-rich compounds NADPH (nicotinamide adenine dinucleotide phosphate) and **ATP (adenosine triphosphate)**. These molecules are in turn consumed in the **Calvin cycle** of photosynthesis to convert carbon dioxide into carbohydrates **(Figure 1.7)**. Although the energy of one photon is very small, the photosynthetic apparatus within the chloroplast of a single *C. reinhardtii* cell absorbs millions of photons each second. And a single cell within a typical plant leaf contains hundreds of chloroplasts!

While photosynthesis converts carbon dioxide into carbohydrates, it is the process of cellular respiration that breaks down carbohydrates and other molecules, trapping the released energy as ATP (see Figure 1.7). This in turn is used in the energy-requiring metabolic and biosynthetic processes that are fundamental to life.

Not all organisms that use light as a source of energy are classified as photosynthetic. That is, some organisms do not use the light energy to convert carbon dioxide into carbohydrates. A good example is a genus of organisms within the Archaea called *Halobacterium*. These remarkable microbes thrive in habitats that contain salt levels that are lethal to most other forms of life **(Figure 1.8, p. 8)**. Species of *Halobacterium* contain a pigment–protein complex called bacteriorhodopsin, which functions as a light-driven **proton pump**. The pigment component of bacteriorhodopsin captures photons of light that provide the energy supply needed to pump protons out of the cell. The resulting difference in H^+ concentration across the plasma membrane represents a source of potential energy that is used by the enzyme ATP synthase to generate ATP from ADP

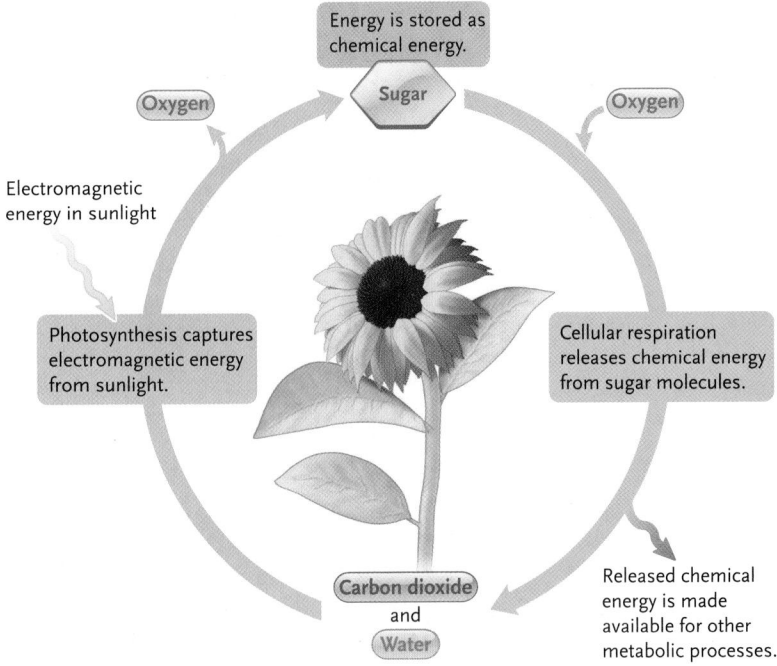

Figure 1.7

Photosynthesis converts light into a usable form of energy. Photosynthesis uses the energy in sunlight to build sugar molecules from carbon dioxide and water, releasing oxygen as a by-product. The process of cellular respiration breaks down the products of photosynthesis and releases usable energy.

(adenosine diphosphate) and inorganic phosphate (P_i) (see Figure 1.8, p. 8). We will discuss the mode of ATP generation in detail in Chapters 6 and 7. In Halobacteria the ATP synthesized through bacteriorhodopsin is used for a range of energy-requiring reactions—but not for the synthesis of carbohydrates from carbon dioxide.

1.3 Light as a Source of Information

As mentioned in "Why It Matters," the deterioration of Monet's eyesight changed the way he saw the world, thus changing the way he painted. This reminds us that many organisms use light to sense their environment—to provide them with crucial information about what is around them. The experience of trying to perform even the simplest of tasks in a dark room makes one quickly realize how important the ability to sense light has become for many forms of life. The change in Monet's eyesight during his later life also suggests that not every person, and certainly not every species, sees the world in the same way.

Figure 1.8

Halobacterium is a genus of Archaea that have a light-driven proton pump. **(a)** Electron micrograph of a colony of *Halobacterium salinarum*. **(b)** Species of *Halobacterium* thrive in hypersaline environments such as Hutt Lagoon in Australia. The pink colour of the water is due to the presence of bacteriorhodopsin within individual cells. **(c)** A model of bacteriorhodopsin shows the pigment retinal bound to a protein. **(d)** Bacteriorhodopsin functions as a light-driven proton pump, the proton gradient being used to synthesize ATP.

a. *Halobacterium salinarum*

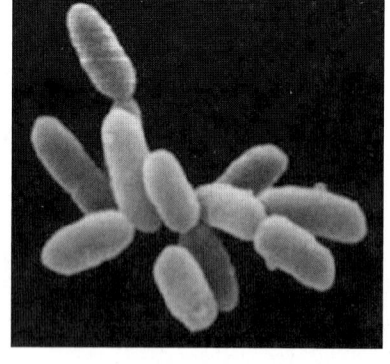

NASA

b. Hutt Lagoon, Western Australia

© J Marshall - Tribaleye Images/Alamy

c. A model of bacteriorhodopsin

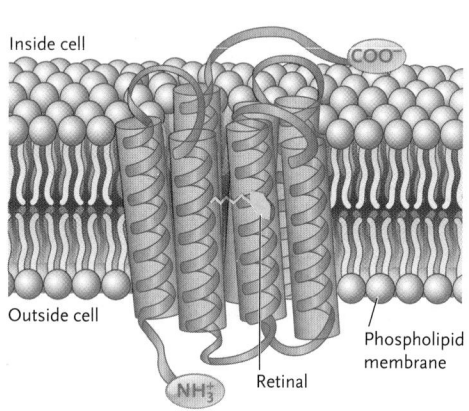

d. Bacteriorhodopsin-driven ATP formation

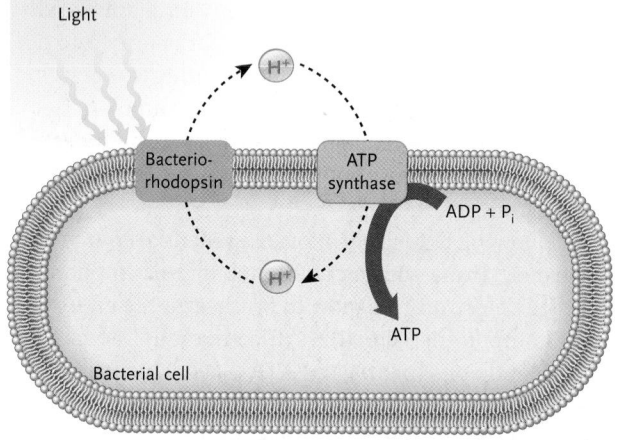

1.3a   Rhodopsin, the Universal Photoreceptor

The basic light-sensing system is termed the photoreceptor. And the most common photoreceptor in nature is rhodopsin **(Figure 1.9, p. 10)**, which is the basis of vision in all animals and insects. Each rhodopsin molecule consists of a protein called opsin that binds a single pigment molecule called retinal. Opsins are membrane proteins that span a membrane multiple times and form a complex with the retinal molecule at the centre (see Figure 1.9, p. 10).

As shown in Figure 1.9, absorption of a photon of light causes the retinal pigment molecule to change shape. This change triggers alterations to the opsin protein, which, in turn, trigger downstream events, including alterations in intracellular **ion** concentrations and electrical signals. As we will see in Chapter 45, in the case of vision, these electrical signals are sent to the visual centres of the brain. In humans, light captured by the eye involves the approximately 125 million photoreceptor cells (rods and cones) that line the retina. Each photoreceptor cell contains millions of individual rhodopsin molecules.

Bacteriorhodopsin found in Halobacterium is structurally similar to the rhodopsin in the eyes of animals. As well, the eyespot of *Chlamydomonas* contains the molecule channelrhodopsin, which is used to sense light. While rhodopsin, channelrhodopsin, and bacteriorhodopsin are structurally similar (a retinal molecule bound to a protein), they do not have a similar evolutionary history. The evolutionary path leading to rhodopsin is distinctly different from the path leading to channelrhodopsin and bacteriorhodopsin.

Rhodopsin is the most common photoreceptor found in nature, but it is not the only one. Both plants and animals have a range of other photoreceptors that absorb light of particular wavelengths. However, it remains a mystery why rhodopsin became the most common photoreceptor. Perhaps its widespread occurrence is because it developed very early during the evolution of animals. Interestingly, whereas vision and smell are different senses, proteins similar to opsins are used in olfaction, suggesting that specific aspects of opsin proteins are particularly useful for sensory perception.

1.3b   Sensing Light without Eyes

When we think about sensing light, we automatically think about our ability to see with our eyes. However, many organisms can sense the light in their surroundings even though they lack organs that we would consider to be eyes. These organisms include plants, algae, invertebrates, and even some bacteria. As an example,

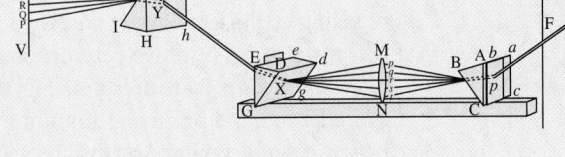
let's take a closer look at the eyespot of *C. reinhardtii*. The eyespot is a light-sensitive structure that is approxi-mately 1 μm in diameter and is found within the chlo-roplast of the cell, in a region closely associated with the cell membrane **(Figure 1.10, p. 10)**. The eyespot is com-posed of two layers of carotenoid-rich lipid globules that play a role in focusing and directing incoming light toward the photoreceptor molecule channelrhodopsin. Although the eyespot is found within the chloroplast, it does not play a role in photosynthesis. Instead, the eye-spot allows the cell to sense light direction and intensity. Using a pair of flagella, *C. reinhardtii* cells can respond to light by swimming toward or away from the light source in a process called *phototaxis*. This allows the cell to stay in the optimum light environment to maximize light capture for photosynthesis. Light absorption by the eyespot is linked to the swimming response by a **signal transduction** pathway; light absorption triggers rapid changes in the concentrations of ions, including potassium and calcium, which generate a cascade of

electrical events. These, in turn, change the beating pat-tern of the flagella used for locomotion.

In plants, a photoreceptor called phytochrome senses the light environment and is critical for *photo-morphogenesis,* the normal developmental process acti-vated when seedlings are exposed to light **(Figure 1.11, p. 10)**. Phytochrome is present in the cytosol of all plant cells, and when a seedling is exposed to wavelengths of red light, phytochrome becomes active and initiates a signal transduction pathway that reaches the nucleus. In the nucleus, these signals activate hundreds of genes, many of which code for proteins involved in photosynthesis and leaf development. Plant develop-ment is the topic of discussion in Chapter 36.

1.3c The Eye

The **eye** can be defined as the organ animals use to sense light. It is described in detail in Chapter 45. What distinguishes the eye of an **invertebrate**, for example,

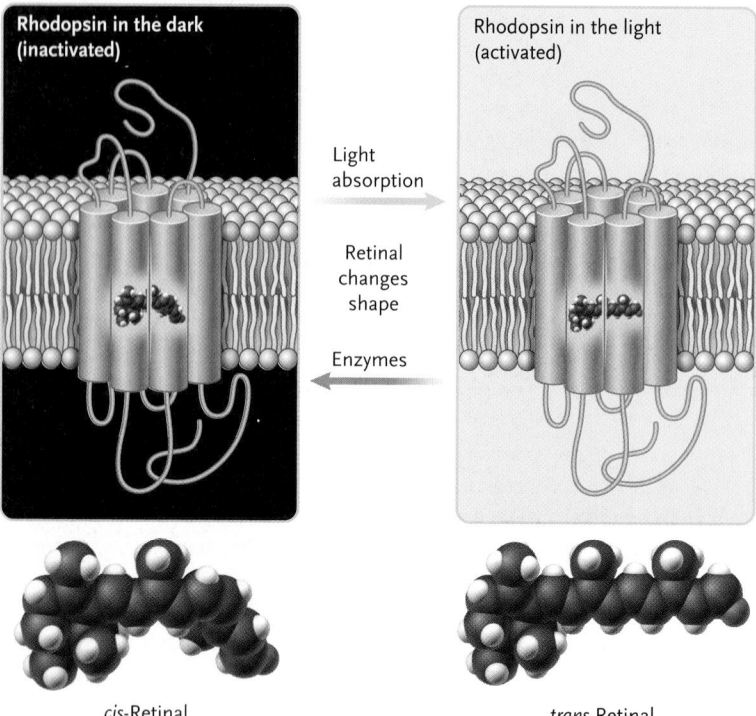

Rhodopsin in the dark
(inactivated)

Rhodopsin in the light
(activated)

Light
absorption

Retinal
changes
shape

Enzymes

cis-Retinal

trans-Retinal

Figure 1.9

Model of the photoreceptor rhodopsin. Rhodopsin consists of a protein (opsin) that binds a pigment molecule (retinal). Upon absorption of a photon of light, retinal changes shape, which triggers changes to the opsin molecule. These changes trigger signalling events, which allow the organism to "see."

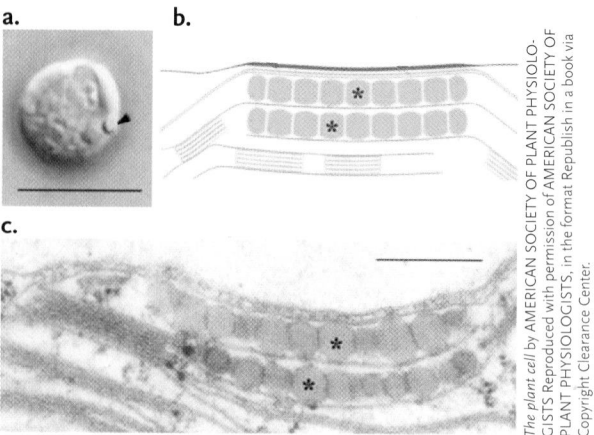

a.

b.

c.

Figure 1.10

The eyespot of *Chlamydomonas*. **(a)** Light microscope image of one *Chlamydomonas* cell. Arrowhead points to the eyespot. Bar = 10 μm. **(b)** Drawing of the eyespot apparatus with the asterisks indicating the orange pigment-rich globule layers that are found inside the chloroplast outer membrane. **(c)** Transmission electron micrograph of the same area. The eyespot contains the photoreceptor molecule channelrhodopsin (not shown). Bar = 300 nm.

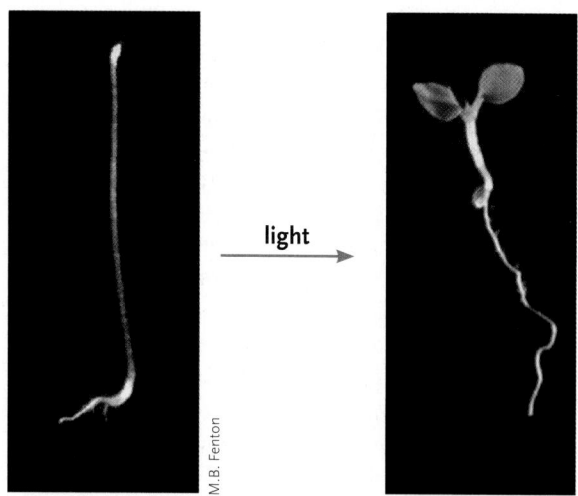

light

M.B. Fenton

M.B. Fenton

Figure 1.11

Photomorphogenesis. Shifting a seedling from darkness to light triggers a developmental program within the plant called photomorphogenesis. Light sensed by the photoreceptor phytochrome initiates the program that involves the activation of hundreds of genes.

from the eyespot of *C. reinhardtii* is vision. The process of vision requires not only an eye to focus and absorb incoming light but also a brain or at least a simple nervous system that interprets signals sent from the eye. The eye and brain are thought to have co-evolved because detailed visual processing occurs in the brain rather than in the eye. Essentially, we see not with our eyes but with our brain.

The simplest eye is the *ocellus* (plural, *ocelli*), which consists of a cup or pit lined with up to 100 photoreceptor cells. Found in all forms of true eyes, the photoreceptor cell is actually a modified nerve cell that contains thousands of individual photoreceptor molecules. A common group of organisms that contain ocelli are flatworms of the genus *Planaria* **(Figure 1.12)**. Information sent to the cerebral ganglion from individual eyes enables the worms to orient themselves so that the amount of light falling on the two ocelli remains equal and diminishes as they swim. This reaction carries them directly away from the source of the light and toward darker areas, where the risk of predation is smaller. Ocelli occur in a variety of animals, including a number of insects, arthropods, and molluscs.

In many ways, the eye of a *Planaria* (plural, *Planarians*) is not much more advanced than the eyespot of *C. reinhardtii*. In both cases, the organ is used to sense light intensity and direction to a light source, but little else. The greatest advance in eye development came with the greater sophistication that produced an actual image of the lighted environment, which allowed objects and shapes to be discerned. These image-forming eyes are found in two distinctly different types: compound eyes and single-lens eyes. *Compound eyes*, which are common in arthropods such as insects and crustaceans, are built of hundreds of individual units called ommatidia (*omma* = eye) fitted closely together **(Figure 1.13)**. Each ommatidium samples only a small part of the visual field, with incoming light being focused onto a bundle of photoreceptor cells. From these signals, the brain receives a mosaic image

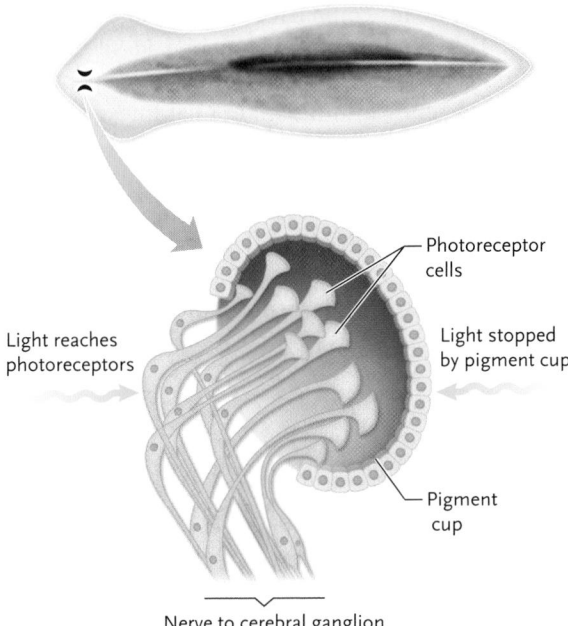

Figure 1.12
The ocellus of *Planaria*, a flatworm, and the arrangement of photoreceptor cells that allows worms to orient themselves in response to light.

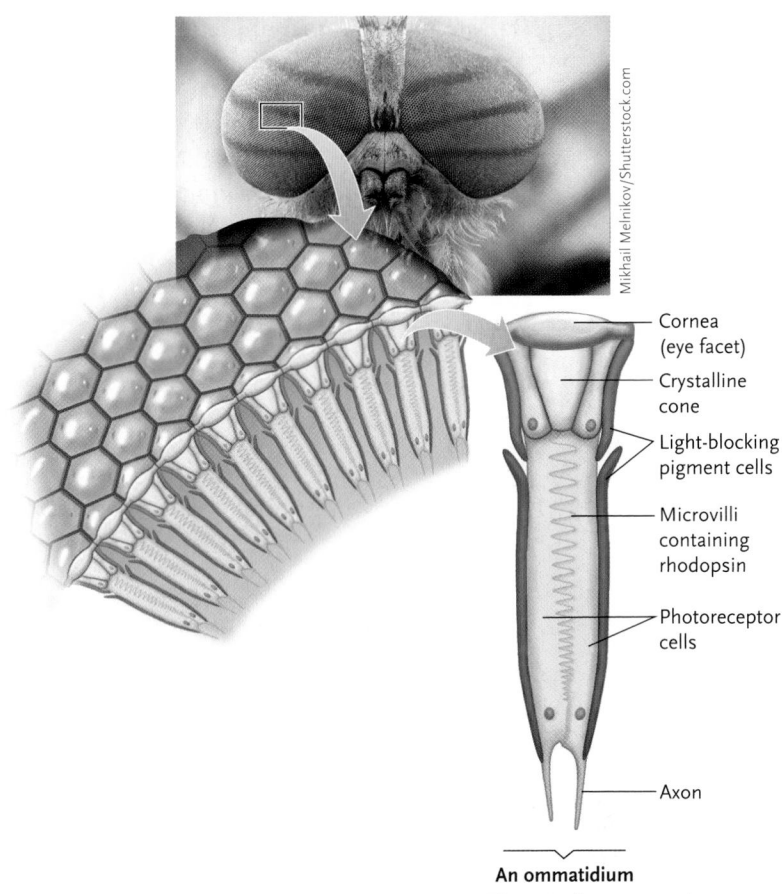

An ommatidium
The unit of a compound eye

Figure 1.13
The compound eye of a deer fly. Each ommatidium has a cornea that directs light into the crystalline cone; in turn, the cone focuses light on the photoreceptor cells. A light-blocking pigment layer at the sides of the ommatidium prevents light from scattering laterally in the compound eye.

of the world. Because even the slightest motion is detected simultaneously by many ommatidia, organisms with compound eyes are extraordinarily good at detecting movement, a lesson soon learned by fly-swatting humans.

The other major type of eye is called the *single-lens eye* **(Figure 1.14)** or camera-like eye and is found in some invertebrates and most vertebrates, including humans. Unlike compound eyes, in a single-lens eye, as light enters through the transparent cornea, a lens concentrates the light and focuses it onto a layer of photoreceptor cells at the back of the eye, the retina. The photoreceptor cells of the retina send information to the brain through the optic nerve.

1.3d Darwin and the Evolution of the Eye

When Charles Darwin presented his theory of evolution by natural selection in *On the Origin of Species by Means of Natural Selection* (1859), he recognized that

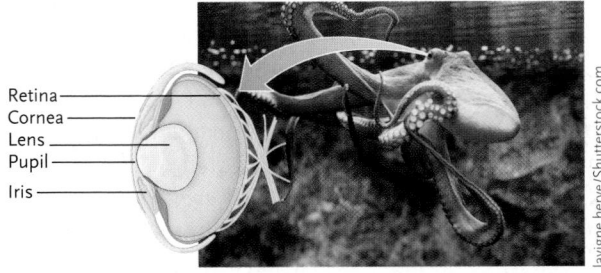

Figure 1.14
The single-lens eye of a cephalopod mollusc (an octopus).

what he called "organs of extreme perfection," such as the eye, would present a problem. He wrote:

> To suppose that the eye, with all its inimitable contrivances for adjusting the focus to different distances, for admitting different amounts of light, and for the correction of spherical and chromatic aberration, could have been formed by natural selection, seems, I freely confess, absurd in the highest possible degree. Yet reason tells me, that if numerous gradations from a perfect and complex eye to one very imperfect and simple, each grade being useful to its possessor, can be shown to exist; if further, the eye does vary ever so slightly, and the variations be inherited, which is certainly the case; and if any variation or modification in the organ be ever useful to an animal under changing conditions of life, then the difficulty of believing that a perfect and complex eye could be formed by natural selection, though insuperable by our imagination, can hardly be considered real.

Darwin proposed that the eye as it exists in humans and other animals did not appear suddenly but evolved over time from a simple, primitive eye. It now seems Darwin was very astute. Starting with a patch of light-sensitive cells on the skin, a recent study concluded that about 2000 small improvements over time would gradually yield a single-lens eye in less than 500 thousand years **(Figure 1.15).** Considering that animals with primitive eyes appeared in the **fossil** record about 500 million years ago, the single-lens eye found in humans could have evolved more than 1000 times. This kind of timing supports fossil evidence that indicates that the eye has evolved independently at least 40 times in different animal lineages before converging into a handful of fundamental designs found today.

It is somewhat surprising that something so complex as the eye could evolve 40 or more different times. However, recently it has been shown that most eyes have fundamental similarities in their

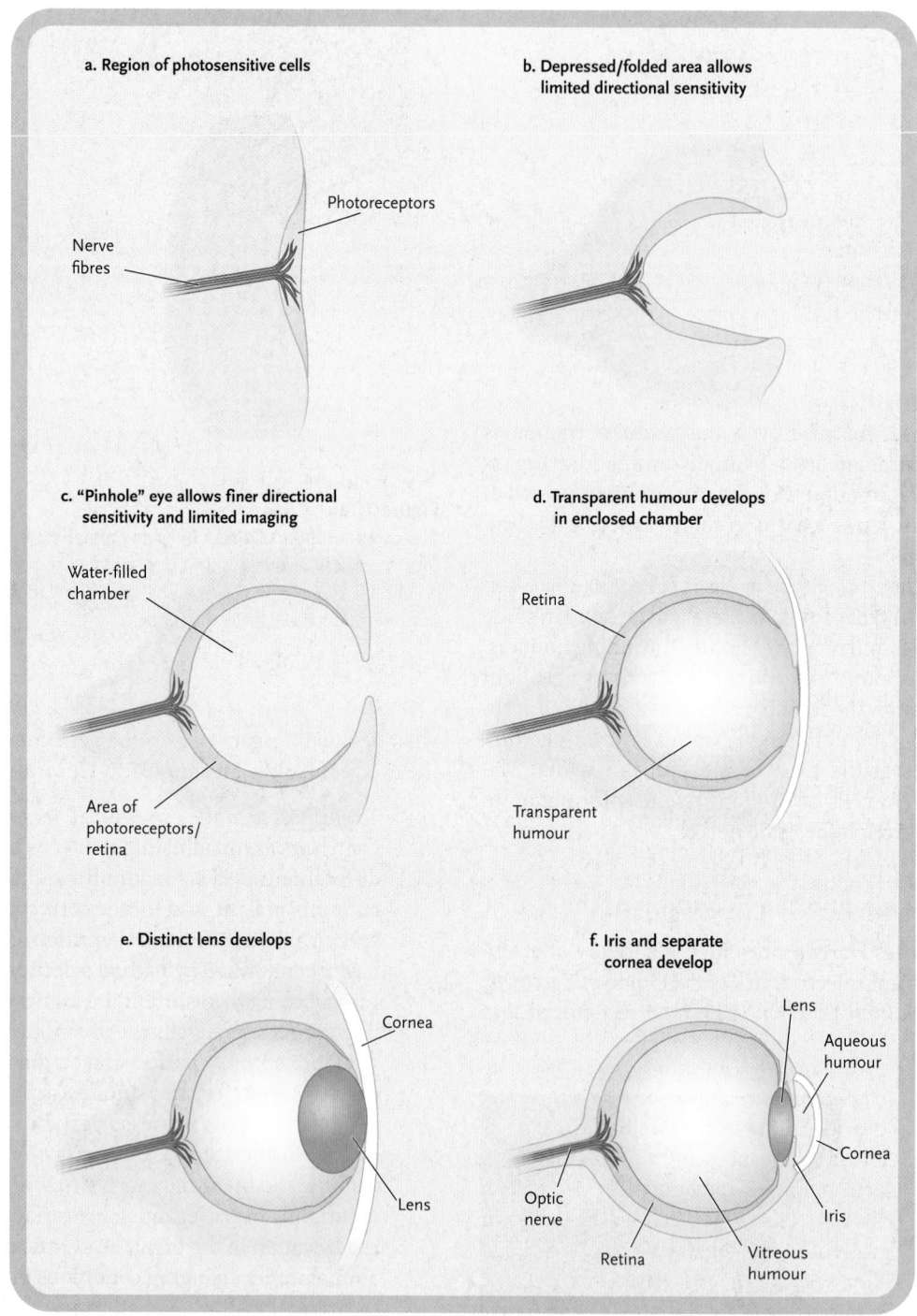

Figure 1.15

The evolution of the eye. Starting with a layer of light-sensitive cells, research suggests that a camera-like eye could evolve in less than 500 thousand years. The evolution of a more sophisticated eye can be explained by the huge advantage improved eyesight would give an organism.

underlying developmental program. For example, a diversity of organisms have recruited a similar set of highly conserved genes to orchestrate eye development. This includes a gene called *Pax6* that has been identified as a master control gene that is almost universally employed for eye formation in animals. And let's not forget that what drove eye evolution in many different animal phyla is the huge advantage eyesight, and then improved eyesight, would have to an animal. The development of heightened visual ability in a predator, as an example, would force comparable eye improvements in both prey and potentially other predators. Rapid improvements in eye development over time would therefore be critical to survival.

STUDY BREAK

1. What are the components of a photoreceptor?
2. Besides rhodopsin, what other molecules are used to sense light?
3. Differentiate between compound eyes and single-lens eyes.
4. How was Darwin able to rationalize the evolution of something so complex as the eye?

1.4 The Uniqueness of Light

Although visible light is a very small portion of the total electromagnetic spectrum, it is essential to life on Earth. In fact, it is this narrow band of energy, from a wavelength of about 400 to 700 nm, that is used for photosynthesis, vision, phototaxis, photomorphogenesis and other light-driven processes. Is it just a coincidence that all of these processes depend on such a narrow band of the electromagnetic spectrum? According to the Harvard physiologist and Nobel laureate George Wald (1906–1997), it is not a coincidence at all. Wald reasoned that visible light is used by organisms because it is the most dominant form of electromagnetic radiation reaching Earth's surface **(Figure 1.16)**. Shorter wavelengths of electromagnetic radiation are absorbed by the ozone layer high in the **atmosphere**, whereas wavelengths longer than those in the visible spectrum are absorbed by water vapour and carbon dioxide in the atmosphere.

Another reason life uses light over other wavelengths of electromagnetic radiation has to do with the energy it contains. Remember that living things are made up of molecules held together by chemical bonds (for a refresher see *The Purple Pages*). Radiation of shorter wavelengths than light contains enough energy to destroy these bonds. Alternatively, electromagnetic radiation of wavelengths longer than light are energetically relatively weak and would not supply enough

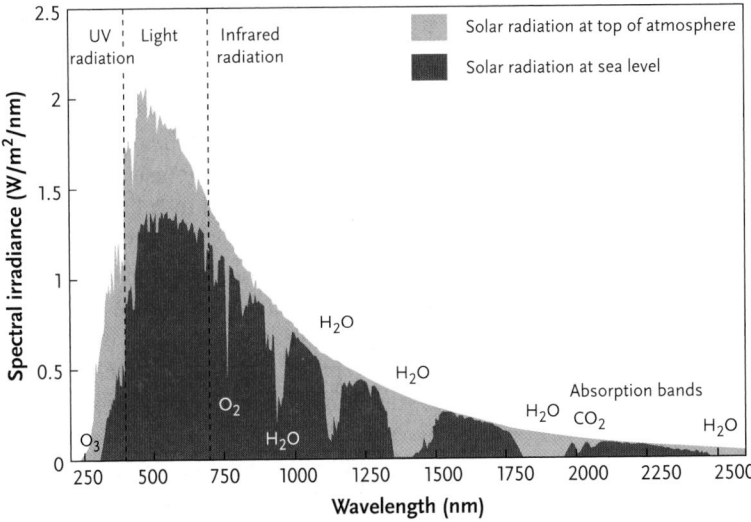

Solar radiation spectrum

Figure 1.16

Electromagnetic radiation reaching the top of Earth's atmosphere (orange) and at sea level (red). As the energy passes through the atmosphere, short-wavelength radiation (250–300 nm) gets absorbed by ozone (O_3). Other wavelengths get partially absorbed by other gases, including O_2, H_2O, and CO_2. Compared to the electromagnetic radiation that reaches the outer atmosphere, the radiation reaching Earth's surface is reduced in both short wavelengths and long wavelengths.

energy to move an electron from a ground state to a higher, excited state. Furthermore, longer wavelengths are readily absorbed by water, which is the bulk of all living things. Given these fundamental aspects of photon energy and light absorption, it would not be at all surprising to find that life on other planets within our galaxy relied on the same narrow wavelengths of the electromagnetic spectrum for a source of energy and information.

1.5 Light Can Damage Biological Molecules

Like many forms of energy, photons of light can damage biological molecules. Recall that to be used for a source of either energy or information, photons of light must be absorbed by molecules. However, absorption of excessive light energy can result in damage that in some cases may be permanent. Of particular concern is higher-energy ultraviolet radiation, which, along with visible light, reaches Earth's surface.

1.5a Damage Is an Unavoidable Consequence of Light Absorption

The photoreceptor cells that line the human retina can be damaged by exposure to bright light. The high-energy environment associated with pigment molecules and excited electrons can result in what is referred to as photo-oxidative damage. The absorption of excess light energy can result in excited electrons reacting with O_2, producing what are called reactive oxygen species. These forms of oxygen, which include the molecule hydrogen peroxide, are particularly damaging to proteins and other biological molecules, often resulting in a loss of function. Excessive damage to photoreceptor cells can lead to the death of the cell.

Unlike eyes, the photosynthetic apparatus of plants and algae is often exposed to full sunlight for hours and thus is particularly susceptible to photooxidative damage. A typical chloroplast contains hundreds of photosystems, each one trapping the energy of thousands of photons each second, converting the light into chemical energy. Compared to the photoreceptor cells of the retina, damage to photosystems can be repaired by a very efficient mechanism that involves removing damaged proteins and replacing them with newly synthesized copies. In fact, under normal light conditions, a single photosystem II **(Figure 1.17)** complex needs to be repaired about every 20 minutes. Because damage to the photosynthetic apparatus is unavoidable, a mechanism of efficient repair must have developed early during the evolution of life so that photosynthesis could be maintained even under high light conditions.

1.5b Ultraviolet Light Is Particularly Harmful

Ultraviolet light is electromagnetic radiation that has a wavelength between blue light (400 nm) and X-rays (200 nm). Because it consists of wavelengths that are shorter than visible light, the energy of the photons of ultraviolet light is greater and more damaging to biological molecules. Life on Earth is protected from the shortest-wavelength and most damaging form of ultraviolet light by the atmosphere's ozone layer. Ozone, O_3, is produced when photons of ultraviolet light interact with molecular oxygen, O_2. While short wavelengths of ultraviolet light are absorbed by ozone, longer wavelengths of ultraviolet light reach Earth's surface.

Along with shorter-wavelength X-rays and gamma rays, ultraviolet light is classified as a form of *ionizing radiation*. The photons at these wavelengths are energetic enough to remove an electron from an atom, resulting in the formation of *ions*—atoms where the total numbers of protons and electrons are not equal. Ultraviolet light can be destructive to a range of biological molecules; however, it is the structure of DNA that is particularly susceptible to damage **(Figure 1.18)**. The interaction of ultraviolet light with nucleotide bases that make up DNA can result in the formation of a dimer—when two neighbouring bases become covalently linked together. Dimers change the shape of the double-helix structure of DNA and prevent its replication, as well as hindering gene transcription. These processes are discussed in detail in Chapter 13. Nucleotide dimers are detected and repaired by a specific enzyme. Even so, the formation of nucleotide dimers can give rise to genetic mutation and has been linked to skin cancer.

For most organisms, exposure to ultraviolet light is unavoidable. Because of this, organisms use a range of behavioural, structural, and biochemical mechanisms to protect themselves from its damaging effects. For example, many animals are protected by fur or feathers covering their skin. Organisms with naked skin, such as humans and whales, are less protected and more susceptible to sunburn due to ultraviolet light exposure.

1.5c Melanin, Suntanning, and Vitamin D

To protect cells from the harmful effects of ultraviolet light, many organisms synthesize melanin, a pigment that strongly absorbs ultraviolet light. Melanin is a remarkable pigment found in all branches of the tree of life. Along with playing a key role in ultraviolet light protection in organisms as diverse as microbes and humans, it is also the major component of the ink released by cephalopods such as squid.

Melanin is very efficient at absorbing ultraviolet light and yet it dissipates over 99% of this energy harmlessly as heat. The specific wavelength of radiation that a pigment such as melanin can absorb can be determined using an instrument called a spectrophotometer. By passing light of varying wavelengths through a solution of pure pigment, the spectrophotometer detects which wavelengths of light are transmitted through the sample and thus determines which wavelengths are absorbed by the pigment. The data from

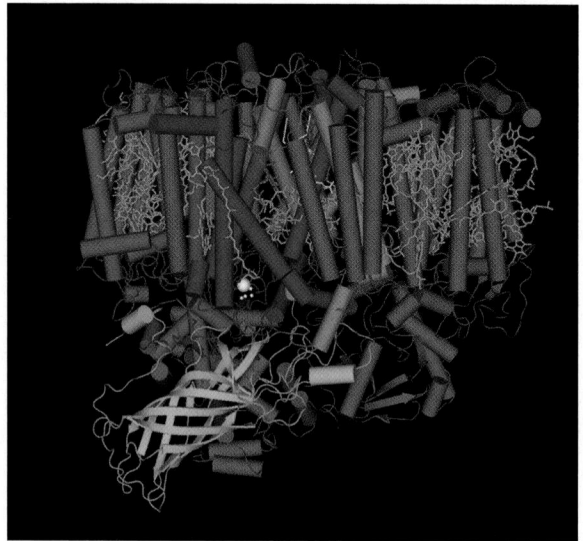

Figure 1.17
Molecular model of the structure of photosystem II. The coloured ribbons and rods represent proteins to which pigments and other cofactors are precisely bound. Light absorption results in unavoidable damage to proteins. An efficient repair system maintains photosystem function even under high light conditions.

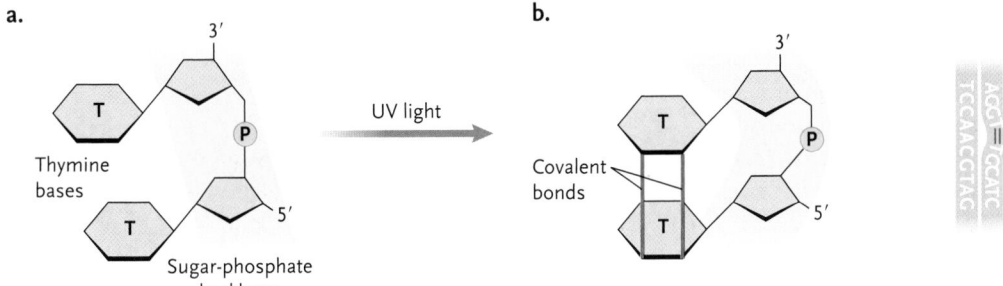

Figure 1.18

Ultraviolet light can damage DNA. Ultraviolet light absorbed by DNA can cause the formation of thymine dimers. Although cells have an efficient mechanism to repair damage to DNA, the formation of dimers can lead to mutation.

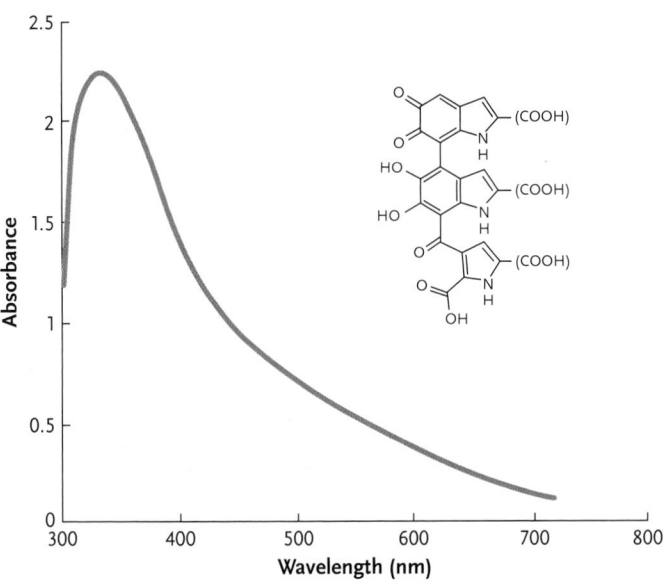

Figure 1.19

Absorption spectrum of melanin. A pure solution of melanin absorbs strongly in the ultraviolet region (300–400 nm) of the spectrum. Also shown is a portion of the chemical structure of melanin.

Figure 1.20
People differ in the amount of melanin in their skin cells.

the spectrophotometer can be used to produce an *absorption spectrum,* a plot of absorbance in relation to the wavelength of light **(Figure 1.19).**

Humans synthesize melanin in specialized skin cells called melanocytes, and melanin synthesis increases upon sun exposure, which results in the brown colour of a suntan. In general, people from countries receiving a lot of sunlight, including countries of Africa, have more melanin in their skin than people from regions receiving less direct sunlight, such as Scandinavian countries **(Figure 1.20).** Since melanin protects us from ultraviolet light, why don't all humans have high melanin levels? Although melanin filters out damaging ultraviolet wavelengths, humans require some ultraviolet radiation to synthesize vitamin D, which is critical for normal bone development. People with high melanin levels who live in regions that do not receive abundant sunlight are susceptible to vitamin D deficiency. This could occur, for example, for someone of African descent living in

Winnipeg. However, in much of the developed world, vitamin D deficiency is rare because many foods, such as milk, yogurt, and grain products, are fortified with this essential nutrient.

STUDY BREAK

1. What biological molecules are particularly susceptible to damage by ultraviolet radiation?
2. What wavelengths of electromagnetic radiation does melanin absorb?

1.6 Using Light to Tell Time

As it revolves around the Sun once a year, Earth rotates on its axis once every 24 hours. These two motions result in very predictable changes to the light and temperature at Earth's surface, giving rise to the seasons and day/night, respectively. The rhythmic and predictable nature of light and darkness during the 24 hour day has led to the evolution of many physiological and behavioural phenomena that display diurnal (*daily*) and seasonal rhythmicity.

1.6a Circadian Rhythms Are Controlled by a Biological Clock

The daily cycling of some biological phenomena is due simply to an organism responding to changes in sunlight. For example, photosynthesis and vision occur during the day and not in darkness because they both require photons of light. However, the diurnal cycling of other phenomena called **circadian** (*circa* = "around"; *diem* = "day") **rhythms** is quite different **(Figure 1.21)**. Circadian rhythms are not driven by an organism constantly detecting changes in daylight but rather are governed by an internal *biological clock* (also known as the circadian clock). Phenomena that are classified as circadian rhythms and thus are controlled by a biological clock include sleep-wake cycles, body temperature, metabolic processes, cell division, and the behaviours associated with foraging for food and **mating.**

A key attribute of all biological clocks is that while they are set by the external light environment, they can run a long time independent of external conditions—a phenomenon called *free-running*. This is analogous to winding an old-fashioned wrist watch. Once it is wound it can function for a long time without being rewound. The free-running nature of circadian rhythms was first described in 1729 by the French astronomer Jean-Jacques d'Ortous de Mairan. He found that the daily rhythmic movements of certain plant leaves continued when he placed the plants in complete darkness. In humans, the free-running nature of circadian rhythms is shown by the fact that daily fluctuations in body temperature and hormone levels, for example, will occur even if an individual is subjected to conditions of constant light or darkness.

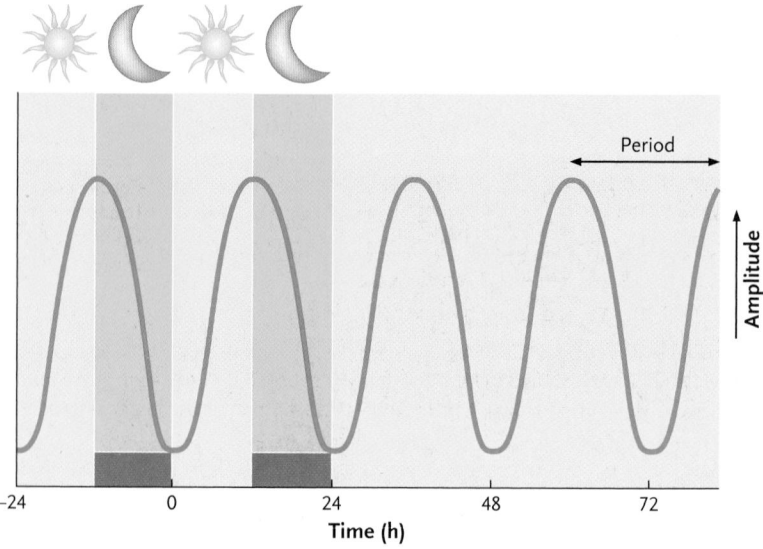

Figure 1.21

Circadian rhythms are oscillations in behaviour and physiology that have a period of approximately 24 hours. These rhythms are set by the external light environment but can run for some time (free-running) under constant conditions.

A key question we can ask at this stage is what is the physical basis of a biological clock? A requirement of anything that keeps time is the presence of something that oscillates. In the case of a traditional clock or watch it's often a crystal or a pendulum (tick, tock, tick, tock...). By comparison, a biological clock is built around a small set of so-called clock genes and clock proteins. Transcription of these genes is controlled so that the abundance of clock proteins rise and fall in a very regular pattern once every 24 hours. It is the abundance of these proteins that in turn influence various behaviours and physiological processes that show circadian rhythmicity **(Figure 1.22).**

Circadian rhythms have been found in all organisms in which they have been searched for, including species from a diverse array of phyla such as bacteria, fungi, animals, and plants. The widespread occurrence of circadian rhythms suggests that there is a selective advantage to being able to tell time. So why are circadian rhythms and the use of an underlying biological clock an advantage? The presence of biological clocks enhances an organism's ability to survive under ever-changing environments by giving them the ability to anticipate or predict when a change will occur, instead

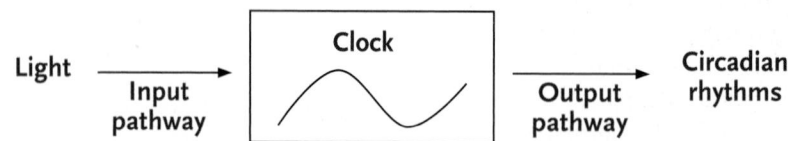

Figure 1.22

Model showing the components of circadian timekeeping. The clock is composed of a set of genes and proteins that oscillate in a very regular manner. Through an output pathway, the clock influences a wide range of behavioural and physiological phenomena. An input pathway ensures that the clock can be reset by changes to the external light environment.

of just responding after a change has occurred. This ability to predict change is seen as advantageous and increases survivability because it enables organisms to restrict their activities to specific, most beneficial, times of the day. Such activities include foraging for food, finding a mate, and avoiding predators, just to name a few. Let's work through some specific examples. In insects, emergence as adults from the pupal case is under circadian control and occurs close to dawn. This is the time of the day when the humidity in the air is highest, which is thought to prevent desiccation (drying out) of the insects, which in turn enhances their survival. In many organisms, proteins required for DNA replication are controlled by a biological clock and are synthesized at dusk. This allows for DNA replication to occur at night, which protects replicating DNA from damaging ultraviolet radiation during the day.

1.6b Biological Clocks Track the Changing Seasons

Not only are biological clocks central to diurnal behaviour and physiology, but also they have been shown to be critical to an organism's ability to keep track of the time of year. Organisms keep track of the changing seasons in part by being able to measure day length or *photoperiod*. Changes in day length and thus seasons occur because Earth is tilted on its axis as it orbits the Sun (see *The Purple Pages*). In Canada, the longest and shortest days of the year are June 21 and December 21, respectively.

Being able to determine the time of year assures that for both plants and animals certain phenomena occur under the most appropriate environmental conditions. This means that the onset of flowering occurs in the spring or summer for most angiosperm plant species, and that leaf drop followed by entrance into dormancy occurs in the autumn for trees. In animals, a huge range of phenomena are linked to being able to sense the time of year. Changes in photoperiod have been shown to provoke changes in colour of fur and feathers, and trigger migration, entry into hibernation, and changes in sexual behaviour **(Figure 1.23)**.

1.6c Jet Lag and the Need to Reset Biological Clocks

Animal cells in a range of different tissues contain clock components (genes and the proteins they encode) that regulate localized circadian-controlled processes. However, most of these so-called peripheral clocks are set by a central biological clock that is found in a very small part of the brain, the suprachiasmatic nucleus (SCN) **(Figure 1.24, p. 18)**. This central clock can receive direct light inputs through the optic nerve of the eye so that it can be reset periodically. The SCN regulates the timing of clocks in peripheral tissues in part through the release during the night of the hormone melatonin from the pineal gland.

Several conditions can interfere with normal circadian cycling. Probably the best example is jet lag, which occurs when you travel rapidly east or west across many time zones, putting your circadian cycling out of synchronization with the external light environment. As an example, let's say you take an eight-hour flight from Paris to Toronto starting at 2 p.m. **(Figure 1.25, p. 18)**. When you arrive in Toronto, your body feels like it is 10 p.m. and expects it to be dark. But because of the six-hour time zone change, when you step off the plane in Toronto it is only 4 p.m. and still daylight. The external light environment is out of synchronization with your internal biological clock. It is this confusion that results in the symptoms of jet lag, which can include lack of appetite, fatigue, insomnia, and mild depression. The clearly defined health effects of jet lag indicate that a range of behavioural and physiological processes are intimately linked to circadian timekeeping. They also show that biological clocks cannot be automatically reset to new light conditions, but instead they often take a few days to adjust. Poor synchronization between circadian clocks and the external light environment is a particular problem for shift workers (e.g., nurses, police officers, fire fighters), who usually alternate working a few weeks during the day followed by shifts at night. While there is growing evidence that this lack of synchronization is unhealthy, low-dosage melatonin given when these shift workers want to sleep seems to help.

Figure 1.23
Changes in photoperiod trigger behavioural and developmental changes. Biological clocks keep track of day length, which is critical for organisms to ensure that specific events occur only at certain times of the year. Examples of photoperiod-dependent phenomena are leaf-drop in trees and colour change in the coat of the Arctic fox (*Vulpes lagopus*).

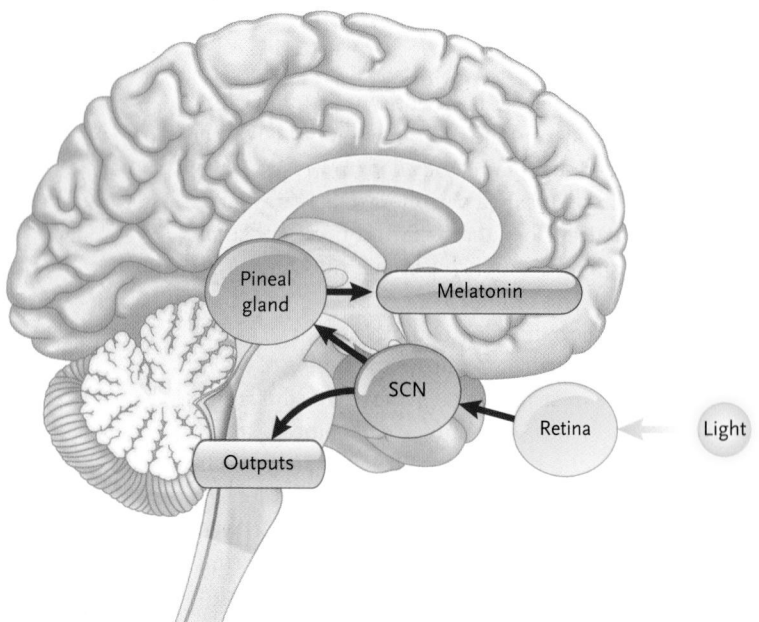

Figure 1.24

In humans central timekeeping is found in the brain. The suprachiasmatic nucleus (SCN) within the brain is the central biological clock in humans. It is set by direct light input from the eye. It controls circadian rhythms directly through an output pathway or through the synthesis of the hormone melatonin by the pineal gland.

STUDY BREAK

1. Why isn't photosynthesis considered to have a circadian rhythm?
2. What is the advantage to having a biological (circadian) clock?
3. What is the biological explanation for jet lag?

1.7 The Role of Light in Behaviour and Ecology

Nature provides a great range of light environments, ranging from the total darkness of caves and the deep ocean to the stark brightness of deserts and polar regions. Differences in the intensity and spectral composition of the light coincide with an organism's adaptations to the specific light environment of a particular habitat. For many animals, it leads to unique colorations that may serve to attract members of the same species while making them less visible to potential predators.

1.7a Using Colour as a Signal: Animals

In animals, bright coloration is thought to serve a valuable role in communication. Research suggests that what is most often communicated by the colouring is an individual's worth as a rival or as a mate. What remains unclear, and is currently being extensively studied, is what type of information is being conveyed and for which individuals the information is intended. A range of particularly colourful fish and bird species have become model systems used by ecologists investigating the role of colour in communication, while biochemists are interested in the chemistry of the actual pigments and how they are synthesized.

In the Eclectus parrot (*Eclectus roratus*), the female is more brightly coloured than the male **(Figure 1.26)**. This is an exception to the general rule that the male of a species is usually more brightly coloured and therefore more conspicuous than the female. It has been shown in a number of species, including the

Toronto
Arrive at 4 p.m.

8-hour flight

Paris
Depart at 2 p.m.

Figure 1.25

Jet lag. Flights over many time zones result in your biological clock becoming out of synchronization with the external light environment. This results in a number of unpleasant responses that are collectively referred to as *jet lag*. The effects subside as the clock within the suprachiasmatic nucleus is reset to the new light environment.

Figure 1.26

Coloured plumage of four avian species often used in studies of the role of colour in behaviour. Clockwise from top left: Eclectus parrot (*Eclectus roratus*) showing a green male and red female, European barn swallow (*Hirundo rustica rustica*), Red-winged blackbird (*Agelaius phoeniceus*), and King penguin (*Aptenodytes patagonicus*).

European barn swallow (*Hirundo rustica rustica*), that more colourful males are more likely to find a mate. An interesting finding shown for penguins has been that for both males and females, individuals with brighter yellow colouring around the eye and upper chest were found to be older and healthier and able to raise more chicks in a given year than mating pairs that were less brightly coloured (Figure 1.26).

Not only does being brightly coloured make an animal more visible, research including the study of penguins indicates that it is also a sign of being in good health. In part, this finding is based on an understanding of the pigments used for ornamentation. Many of these belong to the **carotenoid** family—the structure of beta-carotene is shown in Figure 1.5, p. 6. Unlike in plants, where carotenoids are synthesized in plant cells from precursor molecules, the carotenoids used for colouring in birds are obtained from what they eat, and then circulate in the bloodstream before being deposited in feathers. Biochemical studies have shown that carotenoids play an important role in breaking down potentially harmful reactive oxygen species. Thus, more brightly coloured individuals suggest a good diet rich in molecules that maintain good health. Besides carotenoids, different types of melanin-based pigments are also found in darker and brown colorations in birds. Finally, the dominant pigment class found in parrots, psittacofulvins, is found nowhere else in nature.

1.7b Using Colour as a Signal: Plants

Although humans marvel at the diversity of colours and patterns of flowers, botanists correctly concluded centuries ago that such displays were not designed to please humans but rather to attract pollinators. Pollination involves the movement of pollen from the anthers (male parts) of one flower to the stigma (female parts) of the same flower or other flowers to effect fertilization and production of fruit and seeds. Plant reproduction, including pollination, is discussed in more detail in Chapter 36. The goal of an insect or bird visiting a flower is not to effect pollination; it is to obtain food. This reward may be the protein-rich pollen itself, the sugar-rich nectar, or the waxes or resins found in the flower.

Plants that use animals as pollinators must attract the correct candidates to ensure efficient pollination, in part because the excess pollen or nectar can be energetically costly for the plant to produce. The dependence of a specific plant species on certain animals to act as pollinators, and the reliance of certain animals on particular flowers as a food source, has led to the co-evolution of flower–pollinator associations. Mentioned in the 1877 publication *Fertilisation of Orchids* by Charles Darwin, co-evolution refers to the fact that over evolutionary time, a change in one species triggers changes in the other. The result is that specifics of flower shape, colour, and smell make them more attractive to specific groups of potential pollinators **(Figure 1.27, p. 20)**. For example, the food reward of the flower has become an important part of the pollinator's diet, and the colour and shape of the flower coincide with the visual preferences and shape of the animal pollinator, respectively. As well, co-evolution has resulted in the breeding time of the animal often matching the flowering time of the plant.

The visual systems of pollinators differ considerably among broad groupings such as bees, bats, and birds. Thus, co-evolution has led to flower colour as a key factor that attracts specific groups of pollinators. For example, whereas hummingbirds can perceive colour across a broad range of wavelengths, bees are unable to see red. This explains why hummingbirds dominate the pollination of red-coloured flowers, whereas bees are attracted primarily to blue and yellow flowers. In

Figure 1.27
Flowering plants and their animal pollinators.

reveals patterning that is undetectable to humans. In general, it is shown that the region around the anthers and stigma is darker and thus more easily detected by the pollinating insects **(Figure 1.28)**.

1.7c Camouflage

Camouflage is a way of hiding that, in the natural world, usually involves an organism having a similar appearance to its environment. The reason for camouflage is concealment from either predators or prey. Besides simple colour, pattern and behaviour play central roles in camouflage **(Figure 1.29)**.

An excellent example of the development of camouflage is demonstrated by the peppered moth, *Biston betularia*. Before the Industrial Revolution in England, light-coloured peppered moths were far more common than the dark-coloured individuals that were prized by moth collectors. Light colour made the moths inconspicuous when resting on lichen-covered tree trunks during the day **(Figure 1.30)**. The situation changed after the Industrial Revolution, when many tree trunks became dark-coloured from deposits of soot, and air pollution killed the lichens. In this setting, light--coloured moths were easily detected by hunting birds, and dark-coloured individuals quickly became the most common form (Figure 1.30). Today, as a result of clean-air legislation and reduced air pollution, the ratio of light- to dark-coloured moths has returned to the pre-Industrial Revolution norm in some areas. The case of the peppered moth has become an often-cited example of evolution by natural selection, which is discussed further in Chapter 17.

addition, bees and some other insects can also see in the ultraviolet region of the electromagnetic spectrum and are particularly attracted to flowers that strongly reflect ultraviolet radiation. The role of ultraviolet light in flower–pollinator interactions is widely studied and has been aided by the development of photographic approaches that readily capture the ultraviolet radiation reflected off flowers. It is striking how different flowers look that were photographed using this technique than flowers photographed using visible wavelengths. The organization of distinct ultraviolet-reflecting pigments

1.7d Ecological Light Pollution

The electric light bulb is considered one of the greatest inventions because it allowed people to carry on pursuits at night that otherwise would not have been possible. However, the rapid proliferation of artificial lighting that illuminates public buildings, streets, and signs has resulted in light pollution, which has transformed the night-time environment over significant portions of Earth's surface **(Figure 1.31)**. For example, in the United States, only about 40% of people live in an area that truly gets dark at night.

Ecologists have begun to study the sometimes devastating consequences of light pollution on natural populations. The presence of artificial light disrupts orientation in nocturnal (active at night) animals otherwise accustomed to operating in the dark. For example, newly hatched sea turtles emerge from nests on sandy beaches and orient themselves and move toward the ocean because it is brighter than the silhouette of dark dunes. However, with increased beachfront lighting, hatchlings sometimes become disoriented, head inland, and die. The nocturnal lives of other animals, including many species of frogs and salamanders, have been disrupted by light pollution. As well, artificial lighting has

Figure 1.28
Two species of flowering plants (angiosperms) that are pollinated by bees. *Oenothera biennis* (top) and *Ranunculus ficaria* (bottom). Photographs capturing visible light are shown on the left, while photographs capturing only ultraviolet light are on the right.

a.

b.

c.

Figure 1.29
From a distance **(a)**, it is easy to overlook the duck (*Anas* spp.) sitting on her nest in an urban graveyard. Up close **(b)**, the pattern on her feathers breaks up her body outline, making her difficult to see, particularly when she does not move. As usual, looking for eyes can be a good way to see animals you otherwise might have overlooked, such as the Scops owl (*Otus scops*) **(c)**.

a.

b.

Figure 1.30
An example of camouflage in the peppered moth, *Biston betularia*. The moth is found in one of two forms: light-coloured and dark-coloured. During the industrial revolution, pollution darkened the bark of the trees **(a)** that were part of the moth's habitat. This resulted in increased predation of the light-coloured moth. Following antipollution measures, trees returned to being light coloured **(b)**, which resulted in an increase in the numbers of the moths that are similarly coloured.

Figure 1.31
An example of light pollution.

a negative effect on migrating birds as hundreds of thousands of migrating birds are killed each year when they collide with lighted buildings and towers. Other animals, such as bats and geckos, benefit from night lights that attract insects, effectively concentrating their prey.

1.8 Life in the Dark

Humans see very well during the day, but our visual acuity quickly falters when night approaches. With decreasing light levels, we first lose our ability to see colour, followed by our ability to distinguish shapes. This is because rod photoreceptors, which do not perceive different colours, are about 100 times as sensitive to light as cone photoreceptors (see Chapter 45 for more on this topic).

Animals that are nocturnal or live in low-light conditions often display improved visual acuity under low-light conditions compared to animals that are active during the day. A good example of a nocturnal animal is the Philippine tarsier (*Tarsius syrichta*), one of the smallest primates **(Figure 1.32).** Improved vision is often a consequence of simply having large eyes and thus being able to collect more photons, which is certainly

Figure 1.33
The blind mole rat (*Spalax* sp.) is subterranean and rarely ventures above ground. It is functionally blind.

the case for the tarsier as well as the giant squid (genus *Architeuthis*), which has eyes that measure over 30 cm in diameter! Deep-water crustaceans as well as nocturnal insects have specially designed compound eyes that enhance their light-gathering ability.

In some environments, such as caves and ocean depths, animals live in complete darkness. In fact over 90% of the ocean is at a depth where no light penetrates. Many of the animals that have become adapted to these environments cannot see even though their ancestors may have had functional eyes. A great example of this is the blind mole rat (genus *Spalax*), which spends its life in underground darkness, only rarely venturing above ground **(Figure 1.33).** Twenty-five million years of adaptation to life in the dark has resulted in the natural degeneration of the *Spalax* visual system to the point that it is effectively blind. Their eyes are not only small (less than 1 mm in diameter), but they are also covered by several layers of tissue. Behavioural and physiological studies have shown that the photoreceptors of the eye remain functional even though the image-forming part of the brain is dramatically reduced. So what purpose do these functional photoreceptors have? Since individual mole rats are exposed to brief periods of natural light, it is thought that the maintenance of functional photoreceptors allows for the proper setting of biological clocks necessary for the regulation of circadian rhythms. This is supported by the finding that while the image-forming portion of the brain is small, SCN (see Section 1.6) is well developed.

Another good example of the degeneration of the eye over time is found in the Mexican cavefish, which occurs as two morphological types: a surface-water form that has eyes and skin pigment and a cave-dwelling form that lacks eyes and pigment **(Figure 1.34).** The ancestors of the cavefish lived on the surface, and both eyes and pigment have been lost over approximately 10 000 years.

Figure 1.32
Philippine tarsier (*Tarsius syrichta*).

STUDY BREAK

1. Why do you think that the blind mole rat is still able to detect light?

a.

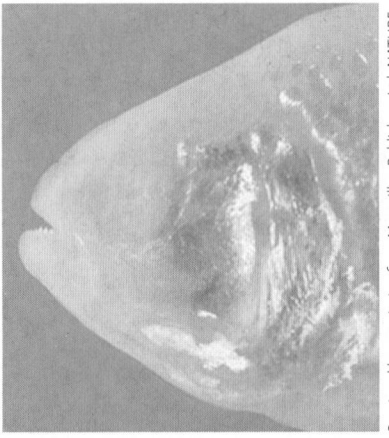

b.

Reprinted by permission from Macmillan Publishers Ltd: NATURE, Yoshiyuki Yamamoto, David W. Stock and William R. Jeffery, "Hedgehog signalling controls eye degeneration in blind cavefish," vol. 431, 844–847, copyright 2004.

Figure 1.34
An example of eye degeneration in the Mexican cave fish, *Astyanax mexicanus*. The single species exists as a surface-dwelling form **(a)** and a blind cave-dwelling form **(b)**.

1.9 Organisms Making Their Own Light: Bioluminescence

Many organisms, including certain bacteria, algae, fungi, insects, squid, and fish, are able to make their own light, a process called bioluminescence **(Figure 1.35)**. Recall from Section 1.1 that in the process of light absorption by a pigment, the energy of a photon is transferred to an electron, raising it from the ground state to an excited state. Bioluminescence is essentially the same process in reverse (see Figure 1.35). Chemical energy in the form of ATP is used to excite an electron in a **substrate** molecule from the ground state to a higher excited state, and when the electron returns to the ground state, the energy is released as a photon of light. The conversion of the chemical energy in ATP into light is very efficient. Considering that up to 95% of the energy of a light bulb is lost as heat, it is remarkable that less than 5% of the energy in ATP is lost as heat during the process of bioluminescent light production. This extraordinary efficiency is essential because high heat production would be incompatible with life.

Bioluminescent organisms generate light for a range of uses. These include attracting a mate or prey, camouflage, and communication. For example, dinoflagellates, which are unicellular algae, use bioluminescence as an alarm mechanism to scare off potential predators. In these tiny organisms, bioluminescence is triggered simply by a disturbance of the water surrounding them. When a predator such as a small fish swims close to a dinoflagellate at night, the resulting burst of light produced by all the dinoflagellates in the vicinity lights up the water around the fish. This defensive behaviour makes the fish clearly visible to its own predators.

Some marine bacteria use bioluminescence in a type of communication called *quorum sensing*. Individual bacteria often release compounds into their environment at concentrations too low to elicit a response from their neighbours. However, as a bacterial population grows, its size reaches a threshold, a quorum, whereby the concentration of compounds is high enough to elicit a physiological response in all members of the population. The response results in the activation of certain genes, including those that encode for proteins required for bioluminescence. Quorum sensing is now believed to be the basis for what are termed "milky seas" (see **Figure 1.36, p. 24**). This strange phenomenon of light on the surface of the ocean has been reported many times over the past several hundred

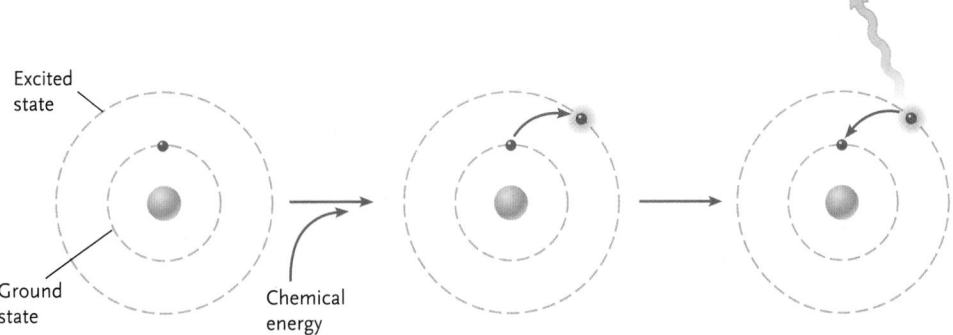

Excited state

Ground state

Chemical energy

Figure 1.35
Bioluminescence. Chemical energy is used to excite an electron in a molecule. A photon of light is released when the electron decays back down to the ground state.

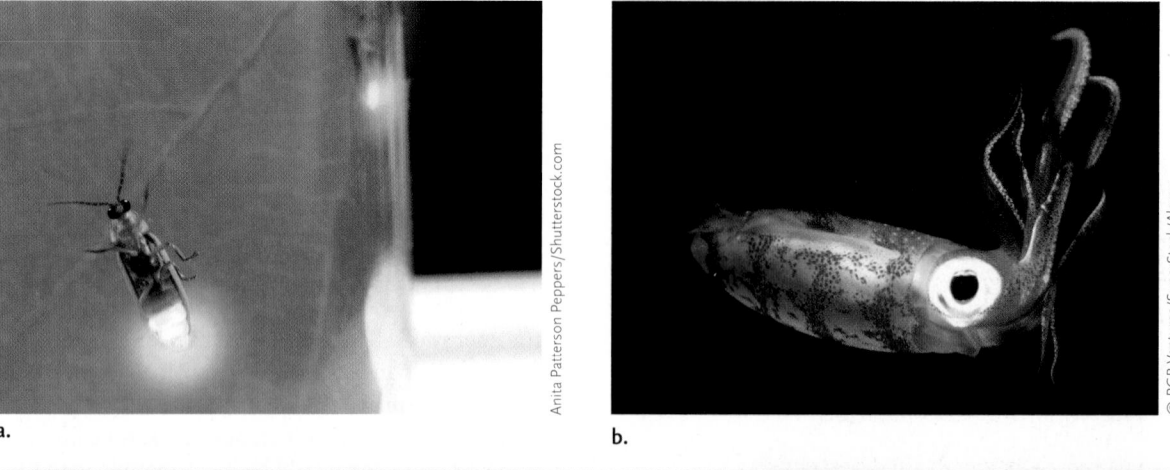

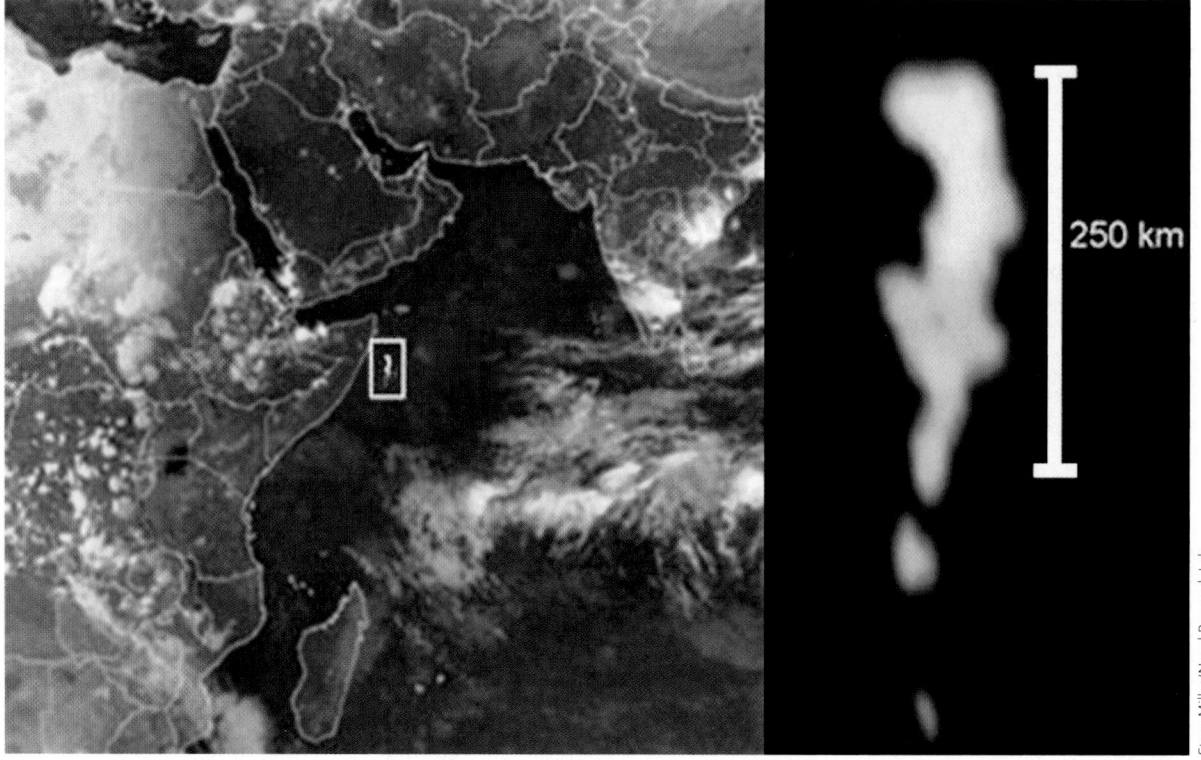

a.

b.

c.

Figure 1.36

Examples of bioluminescence. **(a)** Bioluminescent insect. **(b)** Bioluminescent squid. **(c)** Satellite image of a "milky sea," a bloom of bioluminescent bacteria off the east coast of Africa.

years by sailors and is mentioned in Jules Verne's classic book *Twenty Thousand Leagues under the Sea*.

Many bioluminescent organisms are marine and are most abundant below 800 m, a depth to which sunlight does not penetrate. Bioluminescence has not been reported in land plants or higher vertebrates. Why is bioluminescence absent in these organisms? We do not yet have the answers to this or other questions about bioluminescence, reminding us how much there is still to discover about life on Earth.

In closing, this introductory chapter discussed one phenomenon, light, and how it affects the biology of Earth. From absorption of a single photon by a pigment molecule in a single cell to affecting the composition of entire ecosystems, the influence of light spans all levels of biological organization. This chapter touched on many topics, including physics and chemistry, photosynthesis, genes and proteins, evolution and natural selection, and ecology and behaviour. As you work through the remaining chapters of this textbook, you will learn much more about these topics.

STUDY BREAK

1. Compare and contrast bioluminescence with light absorption.
2. Given that bioluminescence takes energy on the part of the organism, what are some roles that it may serve?

Review

 To access course materials such as Aplia and other companion resources, please visit www.NELSONbrain.com.

1.1 The Physical Nature of Light

- For organisms, light serves as a source of energy and as a source of information.
- Light can be defined as electromagnetic radiation that humans can detect with their eyes.
- Light can be thought of as a wave of discrete particles called photons.
- To be used, light energy must be absorbed by molecules called pigments.
- The colour of a pigment includes all wavelengths of light that are not absorbed.

1.2 Light as a Source of Energy

- The absorption of light by a pigment results in electrons becoming excited. This represents a source of potential energy.
- Photosynthesis is the dominant process on Earth that uses pigments to capture light energy and uses this energy to convert carbon dioxide into energy-rich carbohydrates.
- *Halobacteria* use bacteriorhodopsin to harvest light energy and to generate ATP.

1.3 Light as a Source of Information

- A photoreceptor (e.g., rhodopsin) consists of a pigment molecule (retinal) bound to a protein (opsin).
- The *C. reinhardtii* eyespot allows the organism to sense both light direction and intensity and respond by swimming toward or away from the light (phototaxis).
- The eye can be defined as the organ animals use to sense light. Vision requires a brain to interpret signals sent from the eye.
- The simplest eye is the ocellus found in planarians. It enables the sensing of light direction and intensity.
- Image-forming eyes include compound eyes found in arthropods and single-lens eyes found in some invertebrates and most vertebrates, including humans.
- Because the eye was thought to be an organ of "extreme perfection," Darwin initially had a difficult time explaining how it could have arisen by evolution.
- The relatively rapid evolution of the eye is explained by the huge advantage an improved eye would give an organism.

1.4 The Uniqueness of Light

- Photosynthesis, vision, and most other light-driven processes use only a narrow band of the electromagnetic spectrum. This may be because shorter wavelengths are more harmful (higher energy) and longer wavelengths tend not to reach Earth's surface.

1.5 Light Can Damage Biological Molecules

- Light is a form of energy; thus the absorption of too much light can damage biological molecules.
- The photosynthetic apparatus is constantly being damaged by light, and the damage repaired.
- Ultraviolet radiation, because of its high energy, is particularly harmful to biological molecules, particularly DNA.
- Human skin cells are protected by the pigment melanin, which absorbs ultraviolet radiation.

1.6 Using Light to Tell Time

- Many physiological and behavioural responses are geared to the daily changes in light and darkness and are called circadian rhythms.
- Circadian rhythms are found in all forms of life and evolved to enable organisms to anticipate changes in the light environment.
- Jet lag is caused when your biological clock is out of synchronization with the external light environment.

1.7 The Role of Light in Behaviour and Ecology

- Many organisms use colour to attract, warn, or hide from other organisms.
- Bright coloration is thought to convey good health.
- The widespread use of artificial lighting has been shown to disrupt numerous biological phenomena, including bird migration and the orientation of nocturnal animals.

1.8 Life in the Dark

- Unlike humans, many nocturnal animals (moths, fish, bats, frogs) see very well under dim light conditions.
- Some animals, such as the blind mole rat, are functionally blind yet are descended from ancestors that had functional eyes.

1.9 Organisms Making Their Own Light: Bioluminescence

- A range of organisms can use chemical energy to make light—this is called bioluminescence.
- Bioluminescent organisms use light to attract a mate, for camouflage, to attract prey, or to communicate.

Questions

Self-Test Questions

1. Which of the following statements about light is correct?
   a. Like sound, light is a form of electromagnetic radiation.
   b. Light of a longer wavelength contains more energy.
   c. Visible light is more energetic than radio waves.
   d. A photon of red light contains more energy than a photon of blue light.

2. For a photon of light to be used by an organism, what must occur?
   a. The photon must be absorbed.
   b. The photon must be reflected off a substance.
   c. The photon must interact with a protein in the plasma membrane.
   d. The photon must have sufficient energy to oxidize a molecule.

3. What are the components of a photoreceptor?
   a. a pigment molecule bound to a protein
   b. a protein that is involved in photosynthesis
   c. a group of many pigment molecules
   d. a molecule of chlorophyll

4. Which of the following is true for an eye, but NOT about the eyespot of *Chlamydomonas reinhardtii*?
   a. It can generate a image.
   b. It is composed of photoreceptors.
   c. It can detect changes in light intensity.
   d. It can activate a signal transduction pathway when it absorbs light.

5. Which of the following statements regarding the harmful effects of light is correct?
   a. Visible light is more harmful than ultraviolet light.
   b. Damage to the photosynthetic apparatus caused by excess light cannot be repaired.
   c. Melanin protects skins cells because it specifically absorbs ultraviolet light.
   d. Ultraviolet light specifically damages proteins.

6. Light represents only a very narrow region of the electromagnetic spectrum. However, why is it the dominant form of electromagnetic radiation used in biology?
   a. Light contains the most energy per photon.
   b. Light can excite electrons within molecules without destroying them.
   c. Light is the only form of electromagnetic radiation to reach Earth's surface.
   d. All other wavelengths of electromagnetic radiation are too destructive to biological molecules.

7. Which of the following statements is correct about a biological process that is under circadian control?
   a. A mutation to a single gene could never destroy the circadian cycling of a biological process.
   b. The amplitude of the biological process oscillates with a period of approximately 12 hours.
   c. Circadian cycling of biological processes rarely follows the actual cycling of day and night.
   d. The oscillating nature of the phenomenon continues if the organism is placed in complete darkness.

8. Which of the following statements about jet lag is correct?
   a. Someone who is blind because of non-functioning optic nerves would still experience jet lag.
   b. Jet lag occurs even when the external environment and the SCN are synchronized.
   c. Taking melatonin pills would have no effect on experiencing jet lag.
   d. Jet lag is more severe after travelling by airplane from Toronto to Hawaii than from Toronto to Lima, Peru.

9. Which of the following is illustrated by the Mexican cavefish?
   a. You don't need eyes for vision.
   b. Animals can still see in complete darkness.
   c. Eyes can still function without photoreceptors.
   d. Organs that are no longer of use can degenerate over time.

10. Which of the following is correct about bioluminescence?
    a. It requires ATP.
    b. It cannot occur in complete darkness.
    c. It is found only in bacteria and archaea.
    d. Like vision, it requires the absorption of a photon of light.

Questions for Discussion

1. Nothing ruins a coloured shirt like accidentally adding bleach when washing it. What do you think bleach does?

2. In writing this chapter, the authors found it difficult to define the "eye." Why do you think this was difficult?

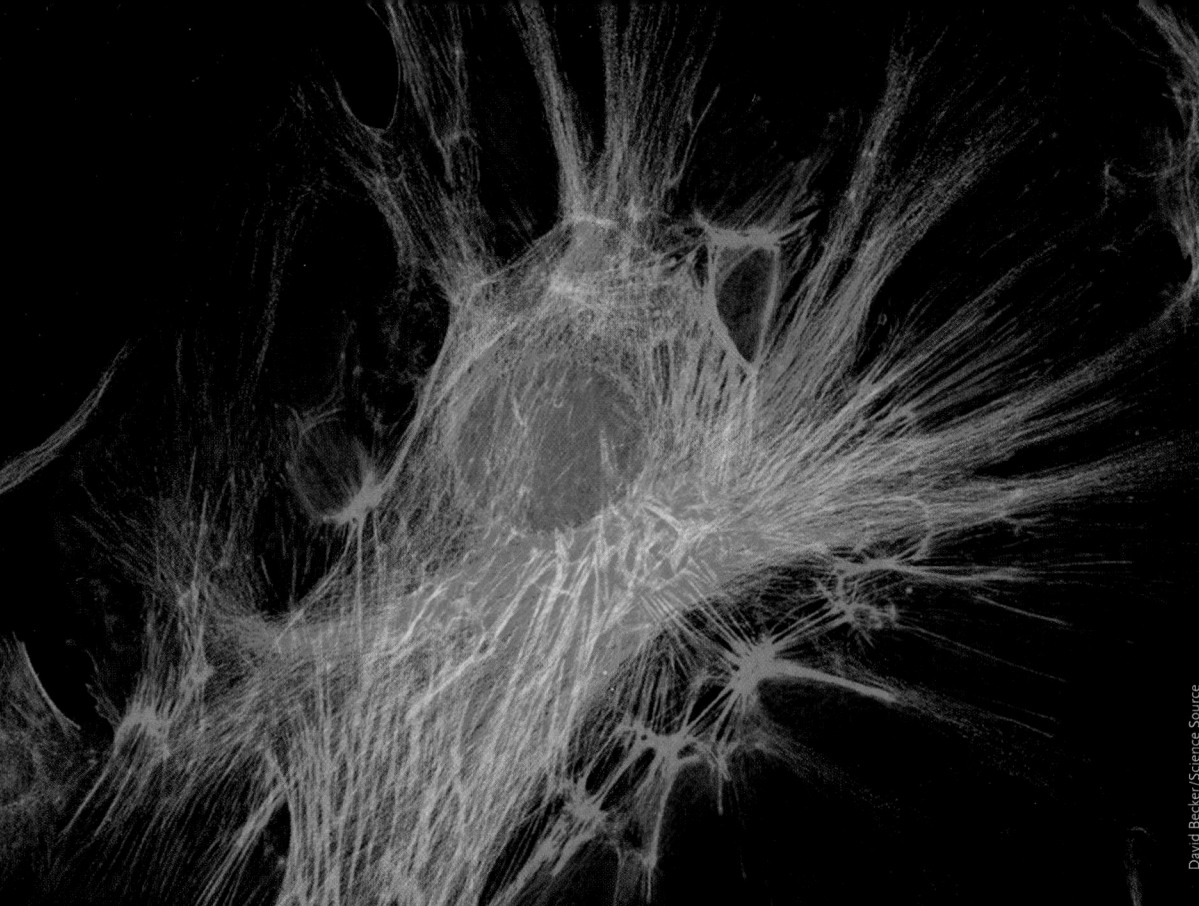

Cells fluorescently labelled to visualize their internal structure (confocal light micrograph). Cell nuclei are shown in blue and parts of the cytoskeleton in red and green.

The Cell: An Overview

WHY IT MATTERS

In the mid-1600s, Robert Hooke, Curator of Instruments for the Royal Society of England, was at the forefront of studies applying the newly invented light microscopes to biological materials. When Hooke looked at thinly sliced cork from a mature tree through a microscope, he observed tiny compartments **(Figure 2.1a, p. 28).** He gave them the Latin name *cellulae,* meaning "small rooms"—giving us the biological term *cell.* Hooke was actually looking at the walls of dead cells, which is what cork consists of.

Reports of cells also came from other sources. By the late 1600s, Anton van Leeuwenhoek (Figure 2.1b), a Dutch shopkeeper, observed "many very little animalcules, very prettily a-moving" using a single-lens microscope of his own construction. Leeuwenhoek discovered and described diverse protists, sperm cells, and even bacteria, organisms so small that they would not be seen by others for another two centuries.

In the 1820s, improvements in microscopes brought cells into sharper focus. Robert Brown, an English botanist, noticed a discrete, spherical body inside some cells; he called it a *nucleus.* In 1838, a German botanist, Matthias Schleiden, speculated that the nucleus had something to do with the development of a cell. The following year, the zoologist Theodor Schwann of Germany expanded Schleiden's idea to propose that all animals and plants consist of cells that contain a nucleus. He also proposed that even when a cell forms part of a larger organism, it has an individual life of its own. However, an important question remained: Where do cells come from? A decade later, the German physiologist Rudolf Virchow answered this question. From his studies of cell growth and reproduction, Virchow proposed that cells arise only from pre-existing cells by a process of division.

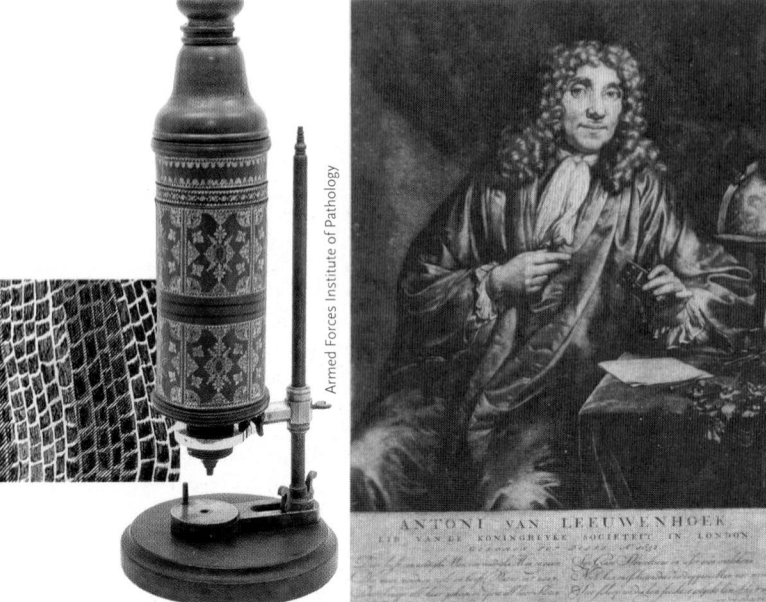

a. Hooke's microscope

National Library of Medicine

b. Leeuwenhoek and microscope

Armed Forces Institute of Pathology

Portrait of Antoni van Leeuwenhoek by Jan Verkolje (1650–1693). Public Domain

Figure 2.1

Investigations leading to the first descriptions of cells. **(a)** The cork cells drawn by Robert Hooke and the compound microscope he used to examine them. **(b)** Anton van Leeuwenhoek holding his microscope, which consisted of a single small sphere of glass fixed in a holder. He viewed objects by holding them close to one side of the glass sphere and looking at them through the other side.

Thus, by the middle of the nineteenth century, microscopic observations had yielded three profound generalizations, which together constitute what is now known as the **cell theory:**

1. All organisms are composed of one or more cells.
2. The cell is the basic structural and functional unit of all living organisms.
3. Cells arise only from the division of pre-existing cells.

These tenets were fundamental to the development of biological science.

This chapter provides an overview of our current understanding of the structure and functions of cells, emphasizing both the similarities among all cells and some of the most basic differences among cells of various organisms. The variations in cells that help make particular groups of organisms distinctive are discussed in later chapters. This chapter also introduces some of the modern microscopes that enable us to learn more about cell structure.

2.1 Basic Features of Cell Structure and Function

As the basic structural and functional units of all living organisms, cells carry out the essential processes of life. They contain highly organized systems of molecules, including the nucleic acids DNA and **RNA**, which carry hereditary information and direct the manufacture of cellular molecules. Cells use chemical molecules or light as energy sources for their activities. Cells also respond to changes in their external environment by altering their internal reactions. Further, cells duplicate and pass on their hereditary information as part of cellular reproduction. All of these activities occur in cells that, in most cases, are invisible to the naked eye.

Some types of organisms, including almost all bacteria and archaea; some protists, such as amoebas; and some fungi, such as **yeasts**, are unicellular. Each of these cells is a functionally independent organism capable of carrying out all activities necessary for its life. In more complex **multicellular organisms**, including plants and animals, the activities of life are divided among varying numbers of specialized cells. However, individual cells of multicellular organisms are potentially capable of surviving by themselves if placed in a chemical medium that can sustain them. If cells are broken open, the property of life is lost: they are unable to grow, reproduce, or respond to outside stimuli in a coordinated, potentially independent fashion. This fact confirms the second tenet of the cell theory: Life as we know it does not exist in units simpler than individual cells.

2.1a Cells Are Small and Can Only Be Seen Using a Microscope

As discussed in more detail in Chapter 3 and subsequent chapters, all forms of life are grouped into one of three domains: the Bacteria, the Archaea, and the Eukarya. Until very recently, bacteria and archaea were grouped into a single domain: the Prokaryota (**prokaryotes**); however, this domain is no longer considered to be accurate as recent research has shown that bacteria and archaea are not evolutionarily related.

As shown in **Figure 2.2,** cells representing all three domains of life assume a wide variety of forms. Individual cells range in size from tiny bacteria to an egg yolk, a single cell that can be several centimetres in diameter. Yet, all cells are organized according to the same basic plan, and all have structures that perform similar activities.

Most cells are too small to be seen by the unaided eye: Humans cannot see objects smaller than about 0.1 mm in diameter. The smallest bacteria have diameters of about 0.5 μm (a micrometre is one thousandth of a millimetre). The cells of multicellular animals range from about 5 to 30 μm in diameter. Your red blood cells are 7 to 8 μm across—a string of 2500 of these cells is needed to span the width of your thumbnail. Plant cells range from about 10 μm to a few hundred micrometres in diameter. (**Figure 2.3** explains the units of measurement used in biology to study molecules and cells.)

To see cells and the structures within them we use **microscopy**, a technique for producing visible images of objects, biological or otherwise, that are too small to be seen by the human eye **(Figure 2.4, p. 30).** The

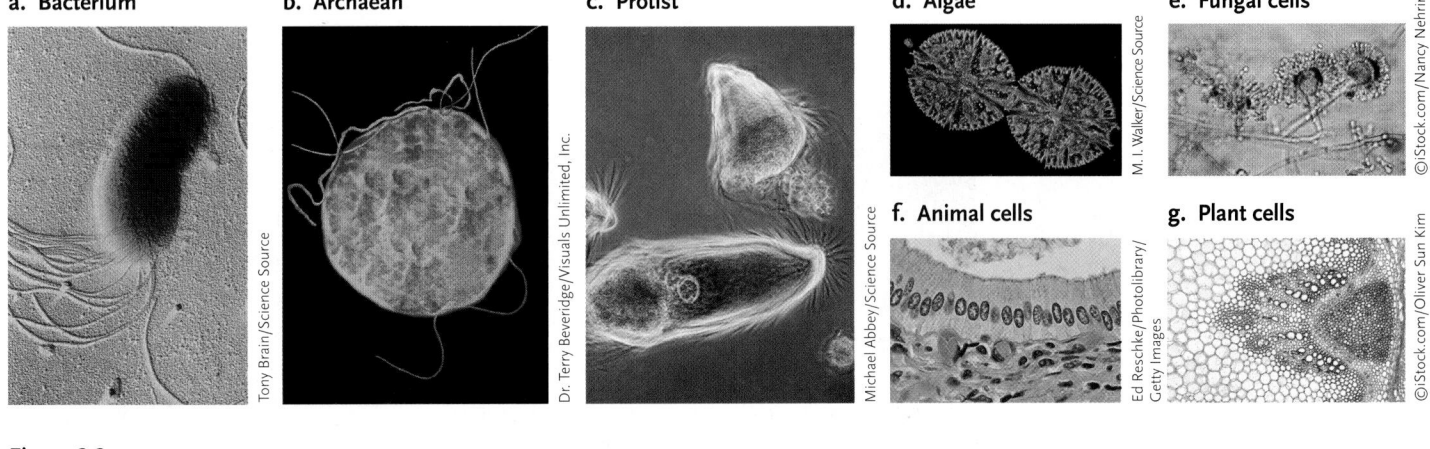

a. Bacterium **b. Archaean** **c. Protist** **d. Algae** **e. Fungal cells** **f. Animal cells** **g. Plant cells**

Figure 2.2

Examples of the various kinds of cells. **(a)** A bacterial cell with flagella, *Pseudomonas fluorescens*. **(b)** An archaean, the extremophile *Sulfolobus acidocaldarius*. **(c)** *Trichonympha*, a protist that lives in a termite's gut. **(d)** Two cells of *Micrasterias*, an algal protist. **(e)** Fungal cells of the bread mould *Aspergillus*. **(f)** Cells of a surface layer in the human kidney. **(g)** Cells in the stem of a sunflower, *Helianthus annuus*.

instrument of microscopy is the **microscope.** The two common types of microscopes are **light microscopes,** which use light to illuminate the specimen (the object being viewed), and **electron microscopes,** which use electrons to illuminate the specimen. Different types of microscopes give different magnification and resolution of the specimen. Just as for a camera or a pair of binoculars, **magnification** is the ratio of the object as viewed to its real size, usually given as something like 400:1. **Resolution** is the minimum distance by which two points in the specimen can be separated and still be seen as two points. Resolution depends primarily on the wavelength of light or electrons used to illuminate the specimen: the shorter the wavelength, the better the resolution. Hence, electron microscopes have higher resolution than light microscopes. Biologists choose the type of microscopy technique based on what they need to see in the specimen; selected examples are shown in Figure 2.4.

Why are most cells so small? The answer depends partly on the change in the surface area–to–volume ratio of an object as its size increases **(Figure 2.5, p. 31).** For example, doubling the diameter of a cell multiplies its volume by eight but multiplies its surface area by only four. The significance of this relationship is that the volume of a cell determines the amount of chemical activity that can take place within it, whereas the surface area determines the amount of substances that can be exchanged between the inside of the cell and the outside environment. Nutrients must constantly enter cells, and wastes must constantly leave; however, past a certain point, increasing the diameter of a cell gives a surface area that is insufficient to maintain an adequate nutrient–waste exchange for its entire volume.

Some cells increase their ability to exchange materials with their surroundings by flattening or by developing surface folds or extensions that increase their surface area. For example, human intestinal cells have

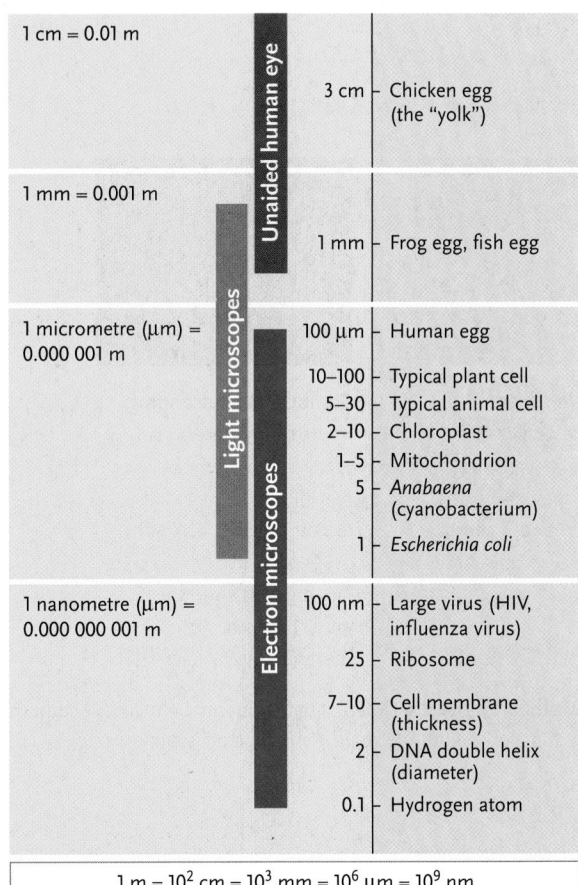

Figure 2.3

Units of measure and the ranges in which they are used in the study of molecules and cells. The vertical scale in each box is logarithmic.

closely packed, fingerlike extensions that increase their surface area, which greatly enhances their ability to absorb digested food molecules.

2.1b Cells Have a DNA-Containing Central Region That Is Surrounded by Cytoplasm

All cells are bounded by the **plasma membrane,** a **bilayer** made of lipids with embedded protein molecules **(Figure 2.6, p. 31).** The lipid bilayer is a

Light microscopy

Micrographs are of the protist *Paramecium*.

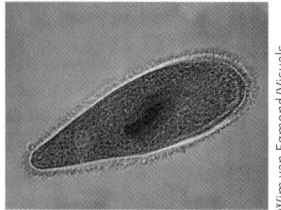

Wim van Egmond/Visuals Unlimited, Inc.

Bright field microscopy: Light passes directly through the specimen. Many cell structures have insufficient contrast to be discerned. Staining with a dye is used to enhance contrast in a specimen, as shown here, but this treatment usually fixes and kills the cells.

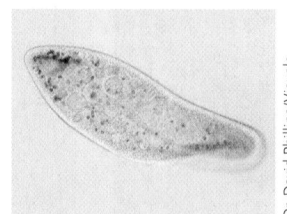

Dr. David Phillips/Visuals Unlimited, Inc.

Dark field microscopy: Light illuminates the specimen at an angle, and only light scattered by the specimen reaches the viewing lens of the microscope. This gives a bright image of the cell against a black background.

Dr. David Phillips/Visuals Unlimited, Inc.

Phase-contrast microscopy: Differences in refraction (the way light is bent) caused by variations in the density of the specimen are visualized as differences in contrast. Otherwise invisible structures are revealed with this technique, and living cells in action can be photographed or filmed.

Electron microscopy

Micrographs are of the green alga *Scenedesmus*.

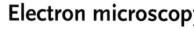

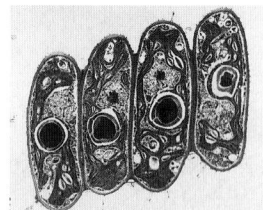

Jeremy Pickett-Heaps, University of Colorado

Transmission electron microscopy (TEM): A beam of electrons is focused on a thin section of a specimen in a vacuum. Electrons that pass through form the image; structures that scatter electrons appear dark. TEM is used primarily to examine structures within cells. Various staining and fixing methods are used to highlight structures of interest.

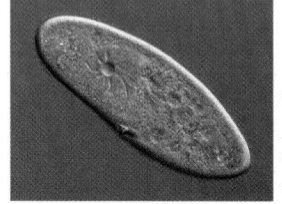

Lebendkulturen.de/ Shutterstock.com

Nomarski (differential interference contrast): Similarly to phase-contrast microscopy, special lenses enhance differences in density, giving a cell a 3D appearance.

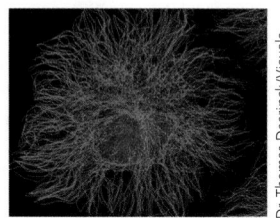

Dr. Thomas Deerinck/Visuals Unlimited, Inc.

Confocal laser scanning microscopy: Lasers scan across a fluorescently stained specimen, and a computer focuses the light to show a single plane through the cell. This provides a sharper 3D image than other light microscopy techniques.

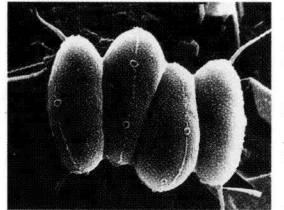
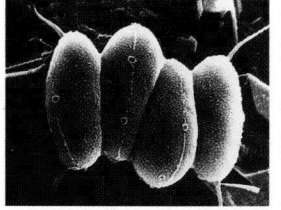

Jeremy Pickett-Heaps, University of Colorado

Scanning electron microscopy (SEM): A beam of electrons is scanned across a whole cell or organism, and the electrons excited on the specimen surface are converted to a 3D-appearing image.

Research Method Figure 2.4 Different techniques of light and electron microscopy. Each technique produces images that reveal different structures or functions of the specimen. A micrograph is a photograph of an image formed by a microscope.

hydrophobic barrier to the passage of water-soluble substances, but selected water-soluble substances can penetrate cell membranes through **transport protein** channels. The selective movement of ions and water-soluble molecules through the transport proteins maintains the specialized internal ionic and molecular environments required for cellular life. (Membrane structure and functions are discussed further in Chapter 5.)

The central region of all cells contains DNA molecules, which store hereditary information. The hereditary information is organized in the form of **genes**—segments of DNA that code for individual

proteins. The central region also contains proteins that help maintain the DNA structure and enzymes that duplicate DNA and copy its information into RNA.

All parts of the cell between the plasma membrane and the central region make up the **cytoplasm.** The cytoplasm contains the *organelles,* the *cytosol,* and the *cytoskeleton.* The **organelles** ("little organs") are small, organized structures important for cell function. The **cytosol** is an aqueous (water) solution containing ions and various **organic molecules.** The **cytoskeleton** is a protein-based framework of filamentous structures that, among other things, helps maintain proper cell shape and plays key roles in cell division and chromosome

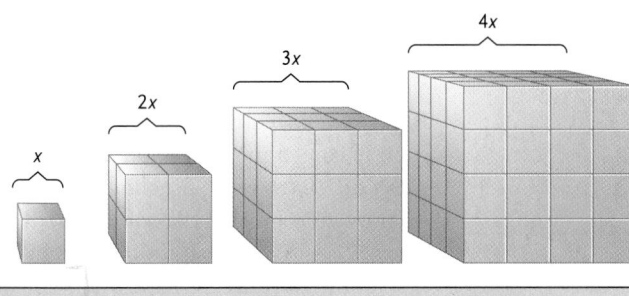

	x	2x	3x	4x
Total surface area	$6x^2$	$6(2x)^2 = 24x^2$	$6(3x)^2 = 54x^2$	$6(4x)^2 = 96x^2$
Total volume	x^3	$(2x)^3 = 8x^3$	$(3x)^3 = 27x^3$	$(4x)^3 = 64x^3$
Surface area/ volume ratio	6:1	3:1	2:1	1.5:1

Figure 2.5

Relationship between surface area and volume. The surface area of an object increases as the square of the linear dimension, whereas the volume increases as the cube of that dimension.

segregation from cell generation to cell generation. The cytoskeleton was once thought to be specific to eukaryotes, but recent research has shown that all major eukaryotic cytoskeletal proteins have functional equivalents in prokaryotes.

Many of the cell's vital activities occur in the cytoplasm, including the synthesis and assembly of most of the molecules required for growth and reproduction (except those made in the central region) and the conversion of chemical and light energy into forms that can be used by cells. The cytoplasm also conducts stimulatory signals from the outside into the cell interior and carries out chemical reactions that respond to these signals.

2.1c Cells Occur in Prokaryotic and Eukaryotic Forms, Each with Distinctive Structures and Organization

There are two fundamentally different types of cells: prokaryotic (*pro* = before; *karyon* = nucleus) and eukaryotic. As we discussed earlier in this chapter, the term *prokaryote* to describe a unique group of evolutionarily related organisms has fallen out of use by microbiologists as bacteria and archaea are seen as evolutionarily distinct. However, the term *prokaryotic cell* is still used as it refers not to a single group of organisms but rather to a particular cell architecture, that is, one lacking a nucleus. Within the prokaryotic cell that is a characteristic of both bacteria and archaea, the DNA-containing central region of the cell, the **nucleoid**, has no boundary membrane separating it from the cytoplasm. Many species of archaea and bacteria contain few if any internal membranes, but a number of other species of both groups contain extensive internal membranes.

The **eukaryotes** (*eu* = true) make up the domain Eukarya and are defined by having cells where DNA is contained within a membrane-bound compartment called the **nucleus.** The cytoplasm of eukaryotic cells typically contains extensive membrane systems that form organelles with their own distinct environments and specialized functions. As in archaea and bacteria, a plasma membrane surrounds eukaryotic cells as the outer limit of the cytoplasm.

The remainder of this chapter surveys the components of prokaryotic and eukaryotic cells in more detail.

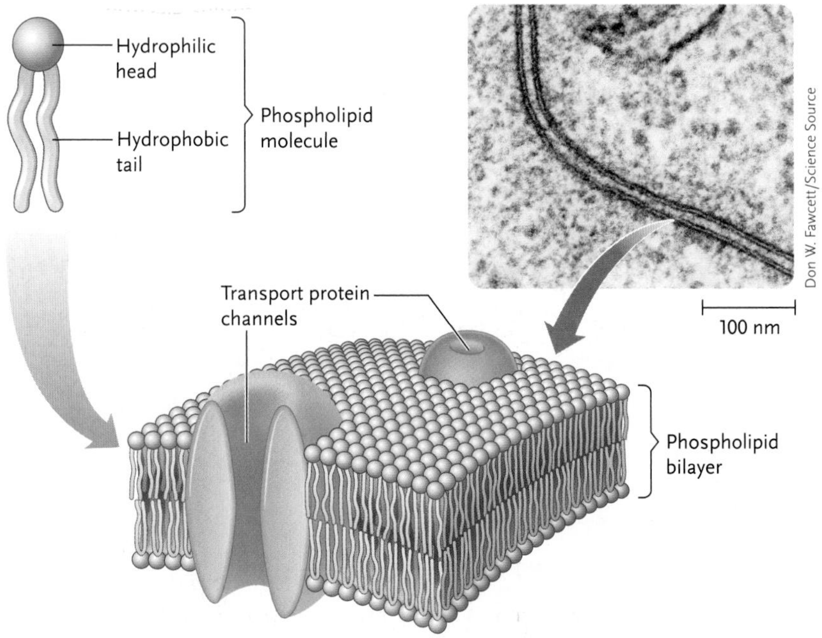

Figure 2.6

The plasma membrane consists of a phospholipid bilayer, an arrangement of phospholipids two molecules thick, which provides the framework for all biological membranes. Water-soluble substances cannot pass through the phospholipid part of the membrane. Instead, they pass through protein channels in the membrane; two proteins that transport substances across the membrane are shown. Other types of proteins are also associated with the plasma membrane. (*Inset*) Electron micrograph showing the plasma membranes of two adjacent animal cells.

Hydrophilic head

Hydrophobic tail

Phospholipid molecule

Transport protein channels

Phospholipid bilayer

100 nm

Don W. Fawcett/Science Source

2.2 Prokaryotic Cells

Most prokaryotic cells are relatively small, usually not much more than a few micrometres in length and a micrometre or less in diameter. A typical human cell has about 10 times the diameter and over 8000 times the volume of an average prokaryotic cell.

The three shapes most common among prokaryotes are spherical, rodlike, and spiral. *Escherichia coli* (*E. coli*), a normal inhabitant of the mammalian **intestine** that has been studied extensively as a model organism in genetics, molecular biology, and genomics research, is rodlike in shape. **Figure 2.7** shows an electron micrograph and a diagram of *E. coli* to illustrate the basic features of prokaryotic cell structure. More detail about prokaryotic cell structure and function, as well as about the diversity of prokaryotic organisms, is presented in Chapter 22.

The genetic material of archaea and bacteria is located in the nucleoid; in an electron microscope, that region of the cell is seen to contain a highly folded mass of DNA (see Figure 2.7). For most species, the DNA is a single, circular molecule that unfolds when released from the cell. This DNA molecule is the **prokaryotic chromosome**, the organization and regulation of which are detailed in Chapters 13 and 14.

Individual genes in the DNA molecule encode the information required to make proteins. This information is copied into a type of RNA molecule called *messenger RNA* (mRNA). Small, roughly spherical particles in the cytoplasm, the **ribosomes**, use the information in the mRNA to assemble **amino acids** into proteins. A prokaryotic ribosome consists of a large and a small subunit, each formed from a combination of **ribosomal RNA (rRNA)** and protein molecules. Each prokaryotic ribosome contains three types of rRNA molecules, which are also copied from the DNA, and more than 50 proteins.

In almost all prokaryotic cells, the plasma membrane is surrounded by a rigid external layer of material, the cell wall, which ranges in thickness from 15 to 100 nm or more (a nanometre is one-billionth of a metre). The **cell wall** provides rigidity to prokaryotic cells and, with the capsule, protects the cell from physical damage. In many prokaryotic cells, the wall is coated with an external layer of **polysaccharides** called the **glycocalyx** (a "sugar coating" from *glykys* = sweet; *calyx* = cup or vessel). When the glycocalyx is diffuse and loosely associated with the cells, it is a **slime layer;** when it is gelatinous and more firmly attached to cells, it is a **capsule.** The glycocalyx helps protect prokaryotic cells from physical damage and desiccation and may enable a cell to attach to a surface, such as other prokaryotic cells (as in forming a colony), eukaryotic cells (as in *Streptococcus pneumoniae* attaching to lung cells), or nonliving substrate (such as a rock).

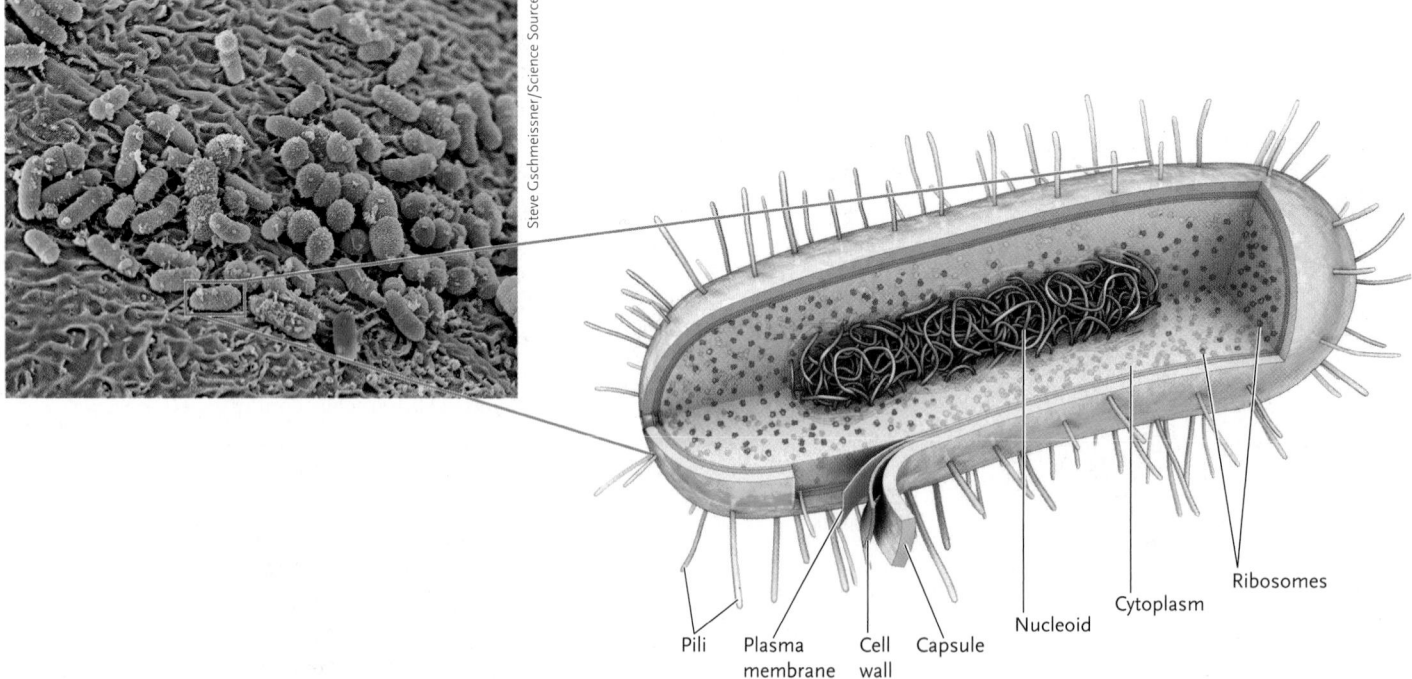

Steve Gschmeissner/Science Source

Pili   Plasma   Cell   Capsule   Nucleoid   Cytoplasm   Ribosomes
membrane   wall

Figure 2.7

Prokaryotic cell structure. An electron micrograph (left) and a diagram (right) of the bacterium *Escherichia coli*. The pili extending from the cell wall attach bacterial cells to other cells of the same species or to eukaryotic cells as a part of infection. A typical *E. coli* has four flagella.

The plasma membrane itself performs several vital functions in both bacteria and archaea. Besides transporting materials into and out of the cells, it contains most of the molecular systems that metabolize food molecules into the chemical energy of ATP. In photosynthetic bacteria, the molecules that absorb light energy and convert it to the chemical energy of ATP are also associated with the plasma membrane or with internal, saclike membranes derived from the plasma membrane.

The cells of bacteria and archaea contain few if any internal membranes; in such cells, most cellular functions occur either on the plasma membrane or in the cytoplasm. But some archaea and bacteria have more extensive internal membrane structures. For example, photosynthetic bacteria have complex layers of intracellular membranes formed by **invaginations** of the plasma membrane on which photosynthesis takes place.

As mentioned earlier, prokaryotic cells have filamentous cytoskeletal structures with functions similar to those in eukaryotes. Prokaryotic cytoskeletons play important roles in creating and maintaining the proper shape of cells; in cell division; and, for certain bacteria, in determining the polarity of the cells.

Many bacteria and archaea can move through liquids and across wet surfaces. Most commonly they do so using long, threadlike protein fibres called **flagella** (singular, *flagellum* = whip), which extend from the cell surface (see Figure 2.2a, p. 29). The **bacterial flagellum**, which is helically shaped, rotates in a socket in the plasma membrane and cell wall to push the cell through a liquid medium (see Chapter 22). In *E. coli*, for instance, rotating bundles of flagella propel the bacterium. Archaeal flagella function similarly to bacterial flagella, but the two types differ significantly in their structures and mechanisms of action. Both types of prokaryotic flagella are also fundamentally different from the much larger and more complex flagella of eukaryotic cells, which are described in Section 2.3.

Some bacteria and archaea have hairlike shafts of protein called **pili** (singular, *pilus*) extending from their cell walls. The main function of pili is attaching the cell to surfaces or other cells. A special type of pilus, the *sex pilus,* attaches one bacterium to another during mating (see Chapter 9).

STUDY BREAK

Where in a prokaryotic cell is DNA found? How is that DNA organized?

2.3 Eukaryotic Cells

The domain of the eukaryotes, Eukarya, is divided into four major groups: protists, fungi, animals, and plants. The rest of the chapter focuses on the cell components that are common to all large groups of eukaryotic organisms.

2.3a Eukaryotic Cells Have a True Nucleus and Cytoplasmic Organelles Enclosed within a Plasma Membrane

The cells of all eukaryotes have a true nucleus enclosed by membranes. The cytoplasm surrounding the nucleus contains a remarkable system of membranous organelles, each specialized to carry out one or more major functions of energy metabolism and molecular synthesis, storage, and transport. The cytosol, the cytoplasmic solution surrounding the organelles, participates in energy metabolism and molecular synthesis and performs specialized functions in support and motility.

The eukaryotic plasma membrane carries out various functions through several types of embedded proteins. Some of these proteins form channels through the plasma membrane that transport substances into and out of the cell. Other proteins in the plasma membrane act as receptors; they recognize and bind specific signal molecules in the cellular environment and trigger internal responses. In some eukaryotes, particularly animals, plasma membrane proteins recognize and adhere to molecules on the surfaces of other cells. Yet other plasma membrane proteins are important markers in the immune system, labelling cells as "self," that is, belonging to the organism. Therefore, the immune system can identify cells without those markers as being foreign, most likely *pathogens* (disease-causing organisms or viruses).

A supportive cell wall surrounds the plasma membrane of fungal, plant, and many protist cells. Because the cell wall lies outside the plasma membrane, it is an *extracellular* structure (*extra* = outside). Although animal cells do not have cell walls, they also form extracellular material with supportive and other functions.

Figure 2.8, p. 34, presents a diagram of a representative animal cell and **Figure 2.9, p. 34,** presents a diagram of a representative plant cell to show where the nucleus, cytoplasmic organelles, and other structures are located. The following sections discuss the structure and function of eukaryotic cell parts in more detail, beginning with the nucleus.

2.3b The Eukaryotic Nucleus Contains Much More DNA Than the Prokaryotic Nucleoid

The nucleus (see Figures 2.8 and 2.9, p. 34) is separated from the cytoplasm by the **nuclear envelope,** which consists of two membranes, one layered just inside the other and separated by a narrow space **(Figure 2.10, p. 35).** A network of protein filaments called *lamins* lines and reinforces the inner surface of the nuclear envelope in animal cells. Lamins are a type of

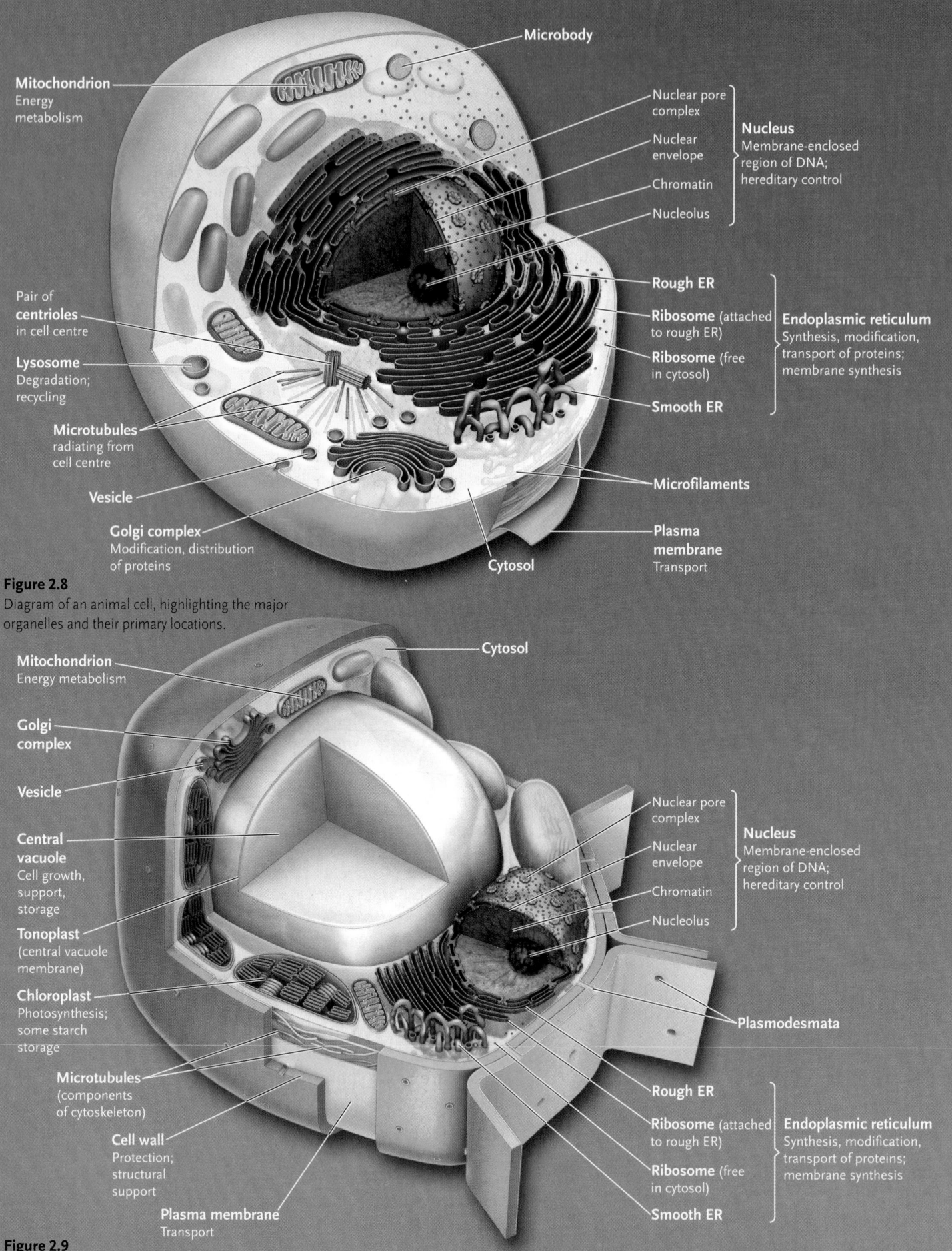

Microbody

Mitochondrion
Energy
metabolism

Nuclear pore
complex

Nuclear
envelope

Chromatin

Nucleolus

Nucleus
Membrane-enclosed
region of DNA;
hereditary control

Pair of
centrioles
in cell centre

Lysosome
Degradation;
recycling

Microtubules
radiating from
cell centre

Vesicle

Golgi complex
Modification, distribution
of proteins

Cytosol

Rough ER

Ribosome (attached
to rough ER)

Ribosome (free
in cytosol)

Smooth ER

Endoplasmic reticulum
Synthesis, modification,
transport of proteins;
membrane synthesis

Microfilaments

**Plasma
membrane**
Transport

Figure 2.8
Diagram of an animal cell, highlighting the major
organelles and their primary locations.

Mitochondrion
Energy metabolism

Cytosol

**Golgi
complex**

Vesicle

**Central
vacuole**
Cell growth,
support,
storage

Tonoplast
(central vacuole
membrane)

Chloroplast
Photosynthesis;
some starch
storage

Microtubules
(components
of cytoskeleton)

Cell wall
Protection;
structural
support

Plasma membrane
Transport

Nuclear pore
complex

Nuclear
envelope

Chromatin

Nucleolus

Nucleus
Membrane-enclosed
region of DNA;
hereditary control

Plasmodesmata

Rough ER

Ribosome (attached
to rough ER)

Ribosome (free
in cytosol)

Smooth ER

Endoplasmic reticulum
Synthesis, modification,
transport of proteins;
membrane synthesis

Figure 2.9
Diagram of a plant cell, highlighting the major organelles and their primary locations.

intermediate filament (see later in this section). Unrelated proteins line the inner surface of the nuclear envelope in protists, fungi, and plants.

Embedded in the nuclear envelope are many hundreds of nuclear pore complexes. A **nuclear pore complex** is a large, octagonally symmetrical, cylindrical structure formed of many types of proteins, called the *nucleoporins*. Probably the largest protein complex in the cell, it exchanges components between the nucleus and cytoplasm and prevents the transport of material not meant to cross the nuclear membrane. A channel through the nuclear pore complex—a **nuclear pore**—is the path for the assisted exchange of large molecules such as proteins and RNA molecules with the cytoplasm, whereas small molecules simply pass through unassisted. A protein or RNA molecule (called the *cargo*) associates with a transport protein acting as a chaperone to shuttle the cargo through the pore.

Some proteins—for instance, the enzymes for replicating and repairing DNA—must be imported into the nucleus to carry out their functions. Proteins to be imported into the nucleus are distinguished from those that function in the cytosol by the presence of a special short amino acid sequence called a nuclear localization signal. A specific protein in the cytosol recognizes and binds to the signal and moves the protein containing it to the nuclear pore complex, where it is transported through the pore into the nucleus.

The liquid or semi-liquid substance within the nucleus is called the **nucleoplasm**. Most of the space inside the nucleus is filled with **chromatin**, a combination of DNA and proteins. By contrast with most bacteria and archaea, most of the hereditary information of a eukaryote is distributed among several to many linear DNA molecules in the nucleus. Each individual DNA molecule with its associated proteins is a **eukaryotic chromosome.** The terms *chromatin* and *chromosome* are similar but have distinct meanings. *Chromatin* refers to any collection of eukaryotic DNA molecules with their associated proteins. *Chromosome* refers to one complete DNA molecule with its associated proteins.

Eukaryotic nuclei contain much more DNA than do prokaryotic nucleoids. For example, the entire complement of 46 chromosomes in the nucleus of a human cell has a total DNA length of about 2 m, compared with about 1.5 m in prokaryotic cells with the most DNA. Some eukaryotic cells contain even more DNA; for example, a single frog or salamander nucleus, although of microscopic diameter, is packed with about 10 m of DNA!

A eukaryotic nucleus also contains one or more **nucleoli** (singular, *nucleolus*), which look like irregular masses of small fibres and granules (see Figures 2.9, p. 34, and 2.10). These structures form around the genes coding for the rRNA molecules of ribosomes.

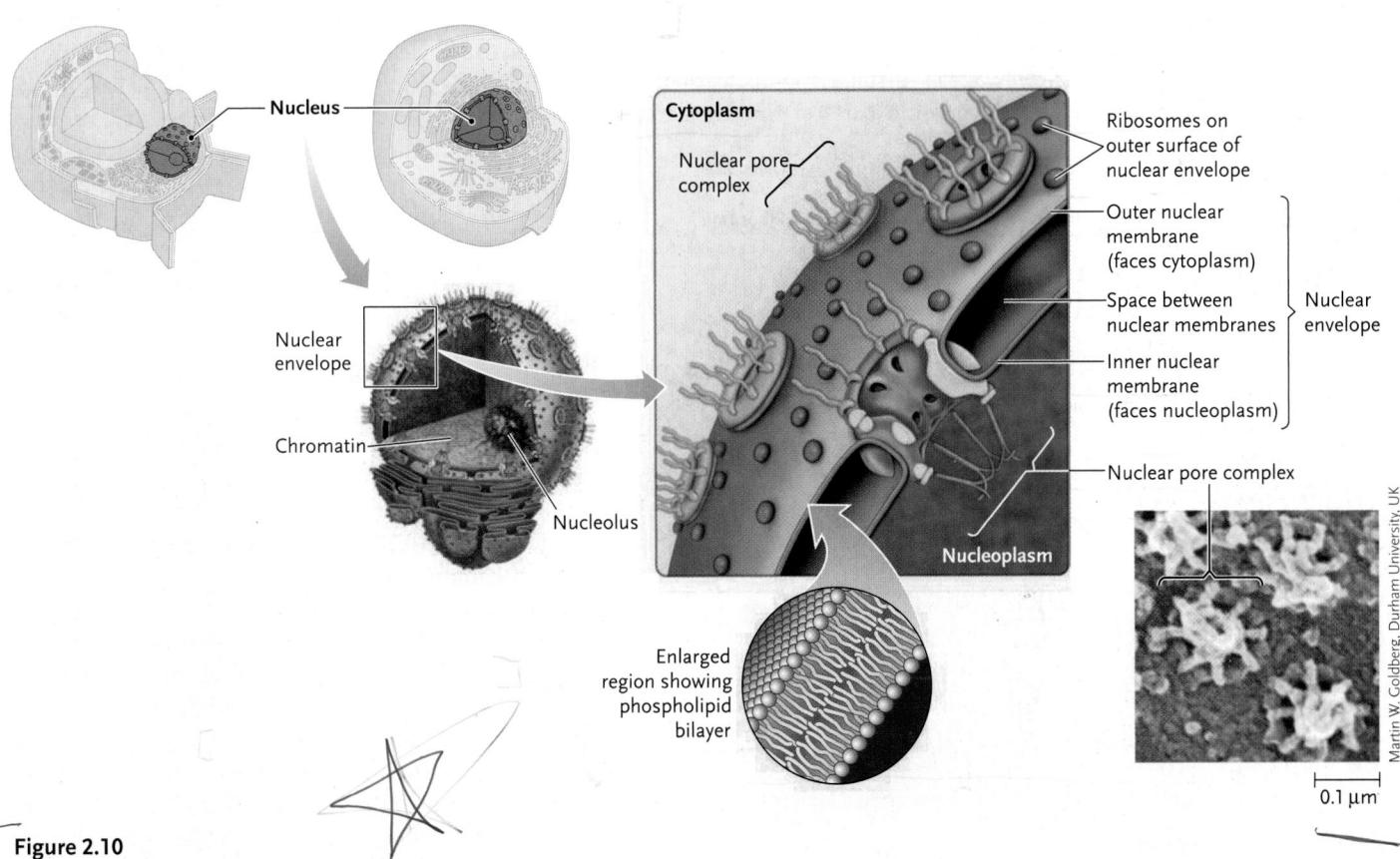

Figure 2.10

The nuclear envelope, which consists of a system of two concentric membranes with nuclear pore complexes embedded. Nuclear pore complexes are octagonally symmetrical protein structures with a channel—the nuclear pore—through the centre. They control the transport of molecules between the nucleus and the cytoplasm.

Within the nucleolus, the information in rRNA genes is copied into rRNA molecules, which combine with proteins to form ribosomal subunits. The ribosomal subunits then leave the nucleoli and exit the nucleus through the nuclear pore complexes to enter the cytoplasm, where they join on mRNAs to form complete ribosomes.

The genes for most of the proteins that the organism can make are found within the chromatin, as are the genes for specialized RNA molecules such as rRNA molecules. Expression of these genes is carefully controlled as required for the function of each cell. (The other proteins in the cell are specified by DNA in the mitochondria and chloroplasts.)

2.3c Eukaryotic Ribosomes Are Either Free in the Cytosol or Attached to Membranes

Like prokaryotic ribosomes, a eukaryotic ribosome consists of a large and a small subunit **(Figure 2.11)**. However, the structures of bacterial, archaeal, and eukaryotic ribosomes, although similar, are not identical. In general, eukaryotic ribosomes are larger than either bacterial or archaeal ribosomes; they contain 4 types of rRNA molecules and more than 80 proteins. Their function is identical to that of prokaryotic ribosomes: they use the information in mRNA to assemble amino acids into proteins.

Some eukaryotic ribosomes are freely suspended in the cytosol; others are attached to membranes. Proteins made on free ribosomes in the cytosol may remain in the cytosol; pass through the nuclear pores into the nucleus; or become parts of mitochondria,→

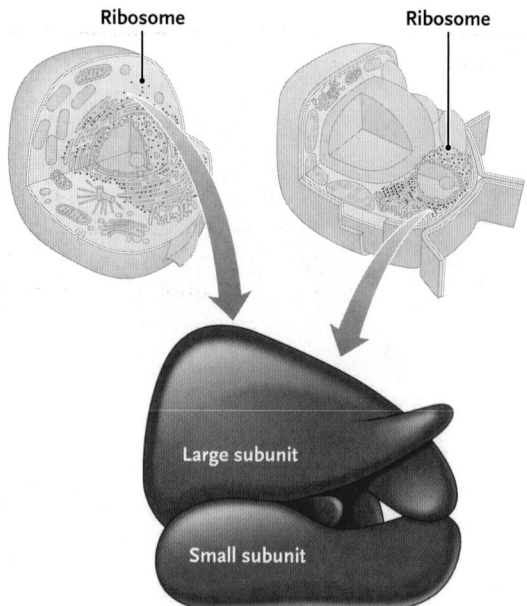

Figure 2.11
A ribosome. The diagram shows the structures of the two ribosomal subunits of mammalian ribosomes and how they come together to form the whole ribosome.

chloroplasts, the cytoskeleton, or other cytoplasmic structures. Proteins that enter the nucleus become part of chromatin, line the nuclear envelope (the lamins), or remain in solution in the nucleoplasm.

Many ribosomes are attached to membranes. Some ribosomes are attached to the nuclear envelope, but most are attached to a network of membranes in the cytosol called the *endoplasmic reticulum* (ER) (described in more detail next). The proteins made on ribosomes attached to the ER follow a special path to other organelles within the cell.

2.3d An Endomembrane System Divides the Cytoplasm into Functional and Structural Compartments

Eukaryotic cells are characterized by an **endomembrane system** (*endo* = within), a collection of interrelated internal membranous sacs that divide the cell into functional and structural compartments. The endomembrane system has a number of functions, including the synthesis and modification of proteins and their transport into membranes and organelles or to the outside of the cell, the synthesis of lipids, and the detoxification of some toxins. The membranes of the system are connected either directly in the physical sense or indirectly by **vesicles**, which are small membrane-bound compartments that transfer substances between parts of the system.

The components of the endomembrane system include the nuclear envelope, endoplasmic reticulum, Golgi complex, lysosomes, vesicles, and plasma membrane. The plasma membrane and the nuclear envelope were discussed earlier in this chapter. The functions of the other organelles are described in the following sections.

Endoplasmic Reticulum. The **endoplasmic reticulum** (ER) is an extensive interconnected network (*reticulum* = little net) of membranous channels and vesicles called **cisternae** (singular, *cisterna*). Each cisterna is formed by a single membrane that surrounds an enclosed space called the **ER lumen (Figure 2.12).** The ER occurs in two forms: rough ER and smooth ER, each with specialized structure and function.

The **rough ER** (see Figure 2.12a) gets its name from the many ribosomes that stud its outer surface. The proteins made on ribosomes attached to the ER enter the ER lumen, where they fold into their final form. Chemical modifications of these proteins, such as addition of carbohydrate groups to produce glycoproteins, occur in the lumen. The proteins are then delivered to other regions of the cell within small vesicles that pinch off from the ER, travel through the cytosol, and join with the organelle that performs the next steps in their modification and distribution. For most of the proteins made on the rough ER, the next destination is the Golgi complex,

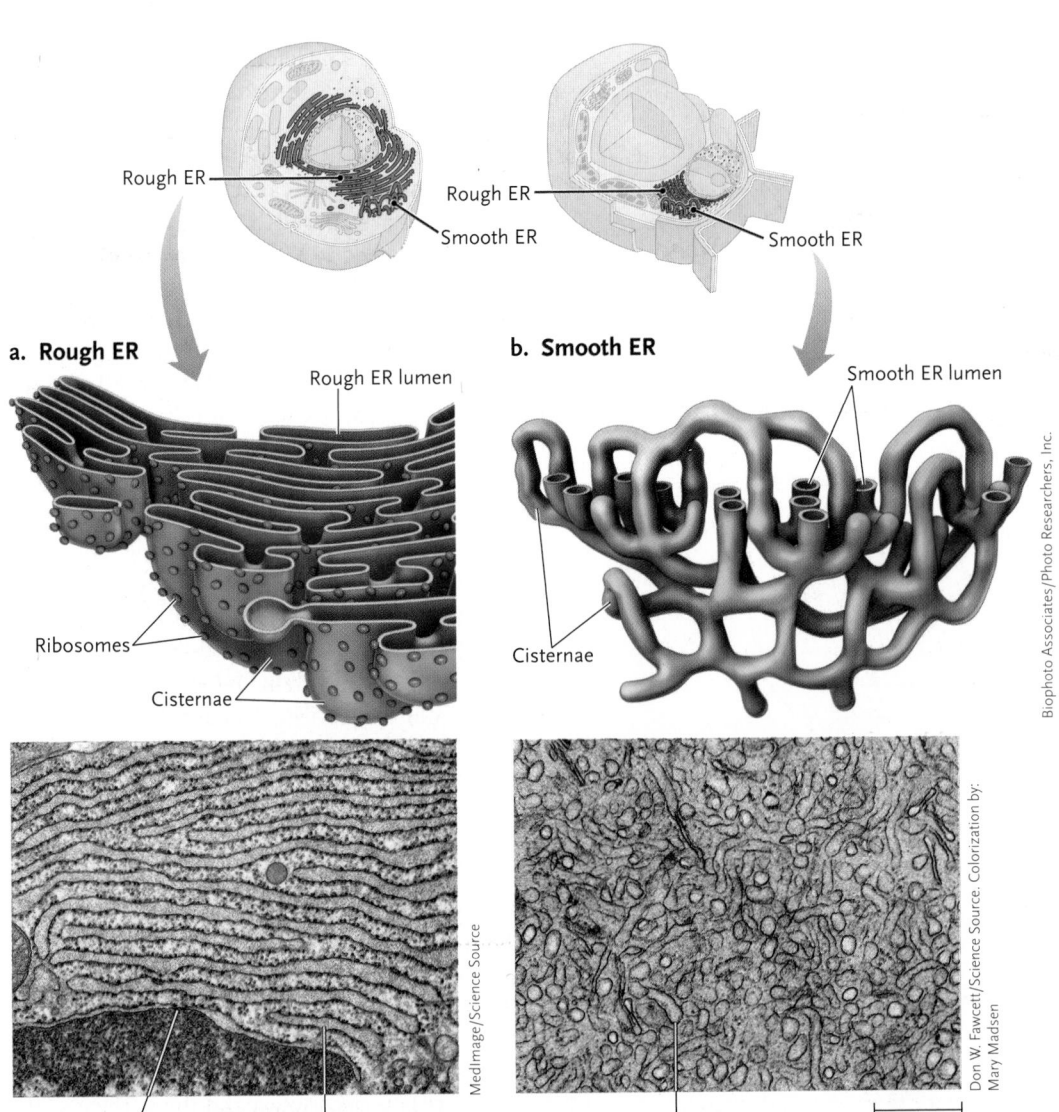

a. **Rough ER**

Rough ER lumen

Ribosomes

Cisternae

Ribosome   Vesicle budding from rough ER

Medimage/Science Source

b. **Smooth ER**

Smooth ER lumen

Cisternae

Smooth ER lumen   0.5 μm

Don W. Fawcett/Science Source. Colorization by: Mary Madsen

Biophoto Associates/Photo Researchers, Inc.

Figure 2.12
The endoplasmic reticulum.
(a) Rough ER, showing the ribosomes that stud the membrane surfaces facing the cytoplasm. Proteins synthesized on these ribosomes enter the lumen of the rough ER, where they are modified chemically and then begin their path to their final destinations in the cell. **(b)** Smooth ER membranes. Among their functions are the synthesis of lipids for cell membranes, and enzymatic conversion of certain toxic molecules to safer molecules.

which packages and sorts them for delivery to their final destinations.

The outer membrane of the nuclear envelope is closely related in structure and function to the rough ER, to which it is connected. This membrane is also a rough membrane, studded with ribosomes attached to the surface facing the cytoplasm. The proteins made on these ribosomes enter the space between the two nuclear envelope membranes. From there, the proteins can move into the ER and on to other cellular locations.

The **smooth ER** (see Figure 2.12b) is so called because its membranes have no ribosomes attached to their surfaces. The smooth ER has various functions in the cytoplasm, including synthesis of lipids that become part of cell membranes. In some cells, such as those of the liver, smooth ER membranes contain enzymes that convert drugs, poisons, and toxic by-products of cellular metabolism into substances that can be tolerated or more easily removed from the body.

The rough and smooth ER membranes are often connected, making the entire ER system a continuous network of interconnected channels in the cytoplasm. The relative proportions of rough and smooth ER reflect cellular activities in protein and lipid synthesis. Cells that are highly active in making proteins to be released outside the cell, such as pancreatic cells that make digestive enzymes, are packed with rough ER but have relatively little smooth ER. By contrast, cells that primarily synthesize lipids or break down toxic substances are packed with smooth ER but contain little rough ER.

Golgi Complex. Camillo Golgi, a late-nineteenth-century Italian neuroscientist and Nobel laureate, discovered the **Golgi complex.** The Golgi complex consists of a stack of flattened, membranous sacs (without attached ribosomes) known as cisternae **(Figure 2.13, p. 38).** In most cells, the complex looks like a stack of cupped pancakes, and like pancakes, they are separate sacs, not interconnected as the ER cisternae are. Typically there are between three and eight cisternae, but some organisms have Golgi complexes with several tens of cisternae. The number and size of Golgi complexes can vary with cell type and the metabolic activity of the cell. Some cells have a single complex, whereas cells highly active in secreting proteins from the cell can have hundreds of complexes. Golgi complexes are usually located near concentrations of rough ER membranes, between the ER and the plasma membrane.

The Golgi complex receives proteins that were made in the ER and transported to the complex in vesicles. When the vesicles contact the *cis* face of the complex (which faces the nucleus), they fuse with the Golgi membrane and release their contents directly into the cisternal (see Figure 2.13). Within the Golgi

complex, the proteins are chemically modified, for example, by removing segments of the amino acid chain, adding small functional groups, or adding lipid or carbohydrate units. The modified proteins are transported within the Golgi to the *trans* face of the complex (which faces the plasma membrane), where they are sorted into vesicles that bud off from the margins of the Golgi (see Figure 2.13). The content of a vesicle is kept separate from the cytosol by the vesicle membrane. Three quite different models have been proposed for how proteins move through the Golgi complex. The mechanism is a subject of active current research.

The Golgi complex regulates the movement of several types of proteins. Some are secreted from the cell, others become embedded in the plasma membrane as integral membrane proteins, and yet others are placed in lysosomes. The modifications of the proteins within the Golgi complex include adding "postal codes" to the proteins, tagging them for sorting to their final destinations. For instance, proteins secreted from the cell are transported to the plasma membrane in **secretory vesicles**, which release their contents to the exterior by **exocytosis (Figure 2.14a)**. In this process, a secretory vesicle fuses with the plasma membrane and spills the vesicle contents to the outside. The contents of secretory vesicles vary, including signalling molecules such as hormones and neurotransmitters (see Chapter 5), waste products or toxic substances, and enzymes (such as from cells lining the intestine). The membrane of a vesicle that fuses with the plasma membrane becomes part of the plasma membrane. In fact, this process is used to expand the surface of the cell during cell growth.

Vesicles may also form by the reverse process, called **endocytosis**, which brings molecules into the cell from the exterior (Figure 2.14b). In this process, the plasma membrane forms a pocket, which bulges inward and pinches off into the cytoplasm as an **endocytic vesicle.** Once in the cytoplasm, endocytic vesicles, which contain segments of the plasma membrane as well as proteins and other molecules, are carried to the Golgi complex or to other destinations such as lysosomes in animal cells. The substances carried to the Golgi complex are sorted and placed into vesicles for routing to other locations, which may include lysosomes. Those routed to lysosomes are digested into molecular subunits that may be recycled as building blocks for the biological molecules of the cell. Exocytosis and endocytosis are discussed in more detail in Chapters 43 and 44.

Lysosomes. **Lysosomes** (*lys* = breakdown; *some* = body) are small, membrane-bound vesicles that contain more than 30 hydrolytic enzymes for the digestion of many complex molecules, including proteins, lipids, nucleic acids, and polysaccharides **(Figure 2.15)**. The cell recycles the subunits of these molecules.

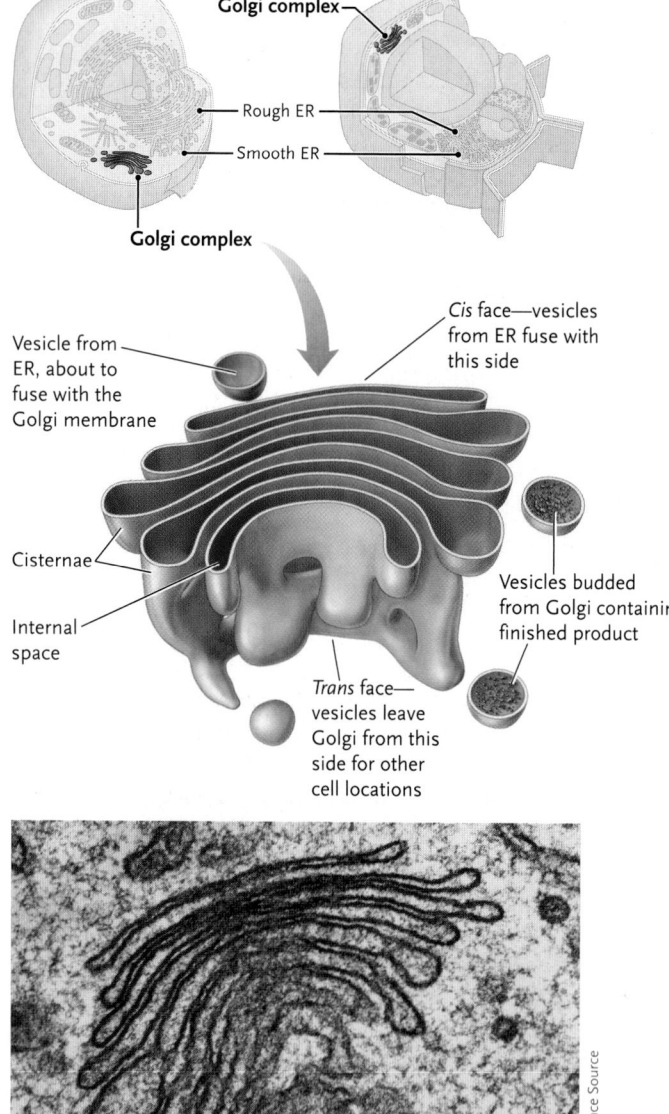

Figure 2.13
The Golgi complex.

a. Exocytosis: A secretory vesicle fuses with the plasma membrane, releasing the vesicle contents to the cell exterior. The vesicle membrane becomes part of the plasma membrane.

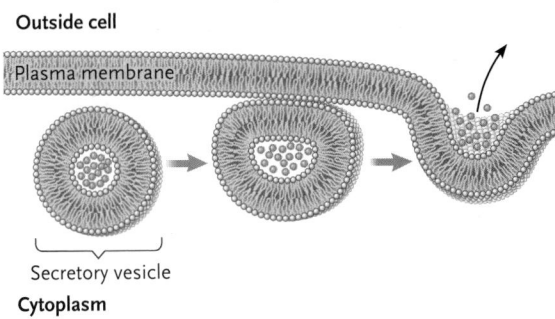

Outside cell

Plasma membrane

Secretory vesicle

Cytoplasm

b. Endocytosis: Materials from the cell exterior are enclosed in a segment of the plasma membrane that pockets inward and pinches off as an endocytic vesicle.

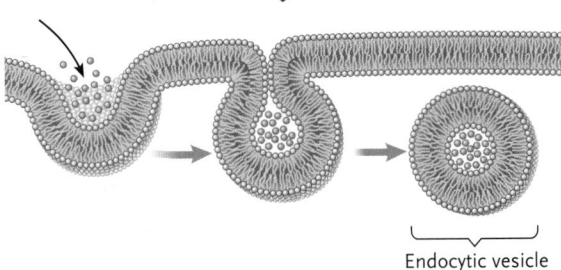

Endocytic vesicle

Figure 2.14
Exocytosis and endocytosis.

Lysosomes are found in animals but not in plants. The functions of lysosomes in plants are carried out by the central vacuole (see Section 2.4). Depending on the contents they are digesting, lysosomes assume a variety of sizes and shapes instead of a uniform struc-

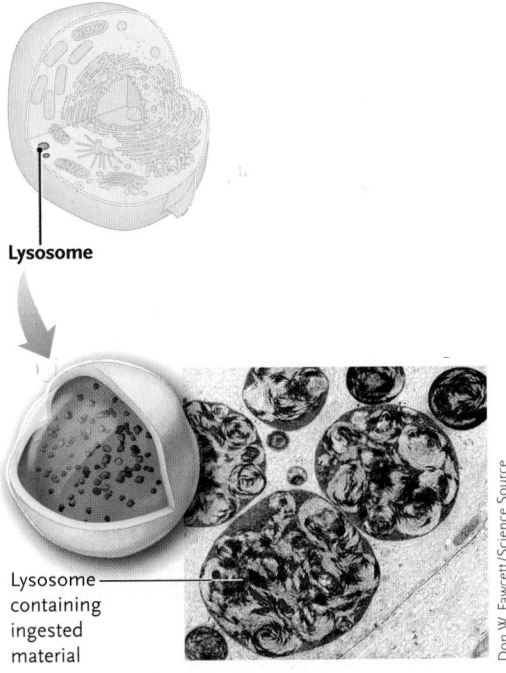

Lysosome

Lysosome containing ingested material

Don W. Fawcett/Science Source

Figure 2.15
A lysosome.

ture as is characteristic of other organelles. Most commonly, lysosomes are small (0.1–0.5 μm in diameter) oval or spherical bodies. A human cell contains about 300 lysosomes.

Lysosomes are formed by budding from the Golgi complex. Their hydrolytic enzymes are synthesized in the rough ER, modified in the lumen of the ER to identify them as being bound for a lysosome, transported to the Golgi complex in a vesicle, and then packaged in the budding lysosome.

The pH within lysosomes is acidic (pH = 5) and is significantly lower than the pH of the cytosol (pH = 7.2). The hydrolytic enzymes in the lysosomes function optimally at the acidic pH within the organelle, but they do not function well at the pH of the cytosol; this difference reduces the risk to the viability of the cell should the enzymes be released from the vesicle.

Lysosomal enzymes can digest several types of materials. They digest food molecules entering the cell by endocytosis when an endocytic vesicle fuses with a lysosome. In a process called *autophagy,* they digest organelles that are not functioning correctly. A membrane surrounds the defective organelle, forming a large vesicle that fuses with one or more lysosomes; the organelle is then degraded by the hydrolytic enzymes. They also play a role in **phagocytosis,** a process in which some types of cells engulf bacteria or other cellular debris to break them down. These cells include the white blood cells known as *phagocytes,* which play an important role in the immune system (see Chapter 5). Phagocytosis produces a large vesicle that contains the engulfed materials until lysosomes fuse with the vesicle and release the hydrolytic enzymes necessary for degrading them.

In certain human genetic diseases known as *lysosomal storage diseases,* one of the hydrolytic enzymes normally found in the lysosome is absent. As a result, the substrate of that enzyme accumulates in the lysosomes, and this accumulation eventually interferes with normal cellular activities. An example is Tay–Sachs disease, a fatal disease of the central nervous system caused by the failure to synthesize the enzyme needed for hydrolysis of **fatty acid** derivatives found in brain and nerve cells.

Summary. In summary, the endomembrane system is a major traffic network for proteins and other substances within the cell. The Golgi complex in particular is a key distribution station for membranes and proteins **(Figure 2.16, p. 40).** From the Golgi complex, lipids and proteins may move to storage or secretory vesicles, and from the secretory vesicles, they may move to the cell exterior by exocytosis. Membranes and proteins may also move between the nuclear envelope and the endomembrane system. Proteins and other materials that enter cells by endocytosis also enter the endomembrane system to travel to the Golgi complex for sorting and distribution to other locations.

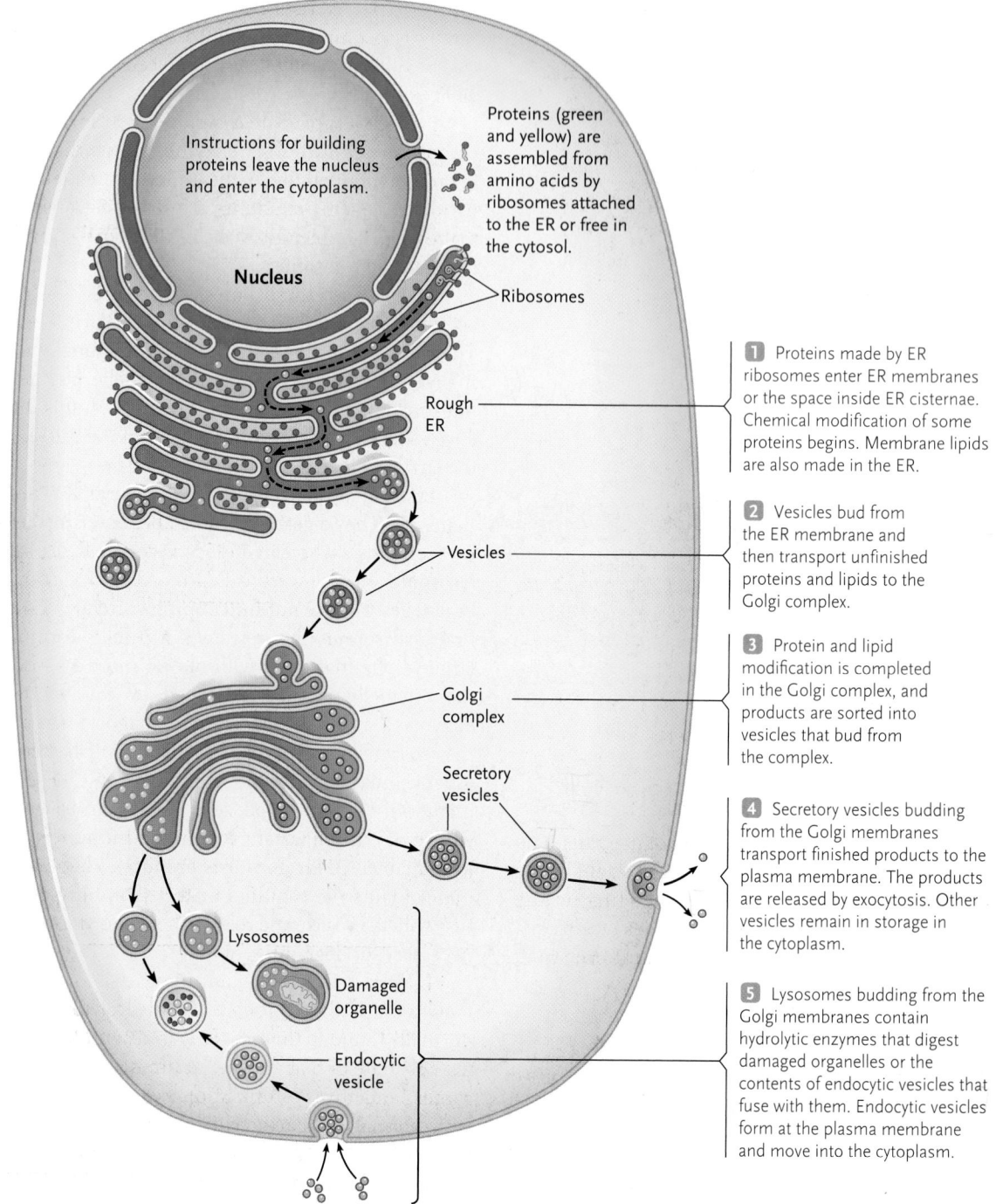

Instructions for building proteins leave the nucleus and enter the cytoplasm.

Proteins (green and yellow) are assembled from amino acids by ribosomes attached to the ER or free in the cytosol.

Nucleus

Ribosomes

Rough ER

Vesicles

Golgi complex

Secretory vesicles

Lysosomes

Damaged organelle

Endocytic vesicle

1 Proteins made by ER ribosomes enter ER membranes or the space inside ER cisternae. Chemical modification of some proteins begins. Membrane lipids are also made in the ER.

2 Vesicles bud from the ER membrane and then transport unfinished proteins and lipids to the Golgi complex.

3 Protein and lipid modification is completed in the Golgi complex, and products are sorted into vesicles that bud from the complex.

4 Secretory vesicles budding from the Golgi membranes transport finished products to the plasma membrane. The products are released by exocytosis. Other vesicles remain in storage in the cytoplasm.

5 Lysosomes budding from the Golgi membranes contain hydrolytic enzymes that digest damaged organelles or the contents of endocytic vesicles that fuse with them. Endocytic vesicles form at the plasma membrane and move into the cytoplasm.

Figure 2.16

Vesicle traffic in the cytoplasm. The ER and Golgi complex are part of the endomembrane system, which releases proteins and other substances to the cell exterior and gathers materials from outside the cell.

2.3e Mitochondria Are the Organelles in Which Cellular Respiration Occurs

Mitochondria (singular, *mitochondrion*) are the membrane-bound organelles in which cellular respiration occurs. *Cellular respiration* is the process by which energy-rich molecules such as sugars, fats, and other fuels are broken down to water and carbon dioxide by mitochondrial reactions, with the release of energy.

Much of the energy released by the breakdown is captured in ATP. In fact, mitochondria generate most of the ATP of the cell. Mitochondria require oxygen for cellular respiration—when you breathe, you are taking in oxygen primarily for your mitochondrial reactions (see Chapter 6).

Mitochondria are enclosed by two membranes **(Figure 2.17)**. The **outer mitochondrial membrane** is smooth and covers the outside of the organelle. The

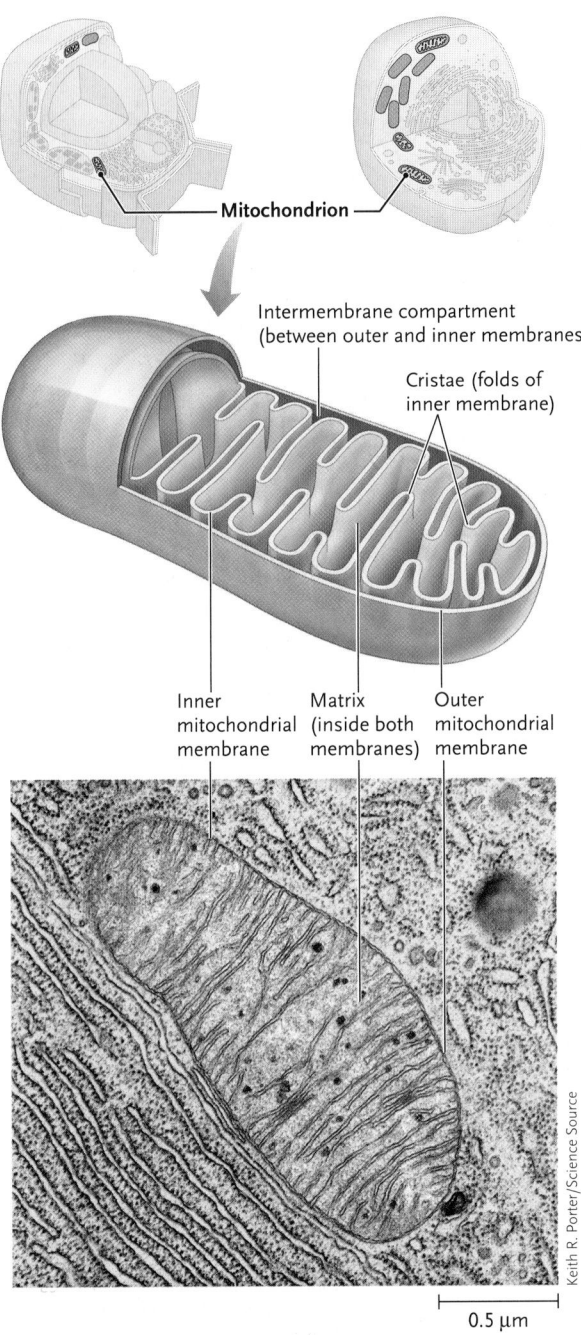

Figure 2.17

Mitochondria. The electron micrograph shows a mitochondrion from a bat pancreas, surrounded by cytoplasm containing rough ER. Cristae extend into the interior of the mitochondrion as folds from the inner mitochondrial membrane. The darkly stained granules inside the mitochondrion are probably lipid deposits.

Intermembrane compartment (between outer and inner membranes)

Cristae (folds of inner membrane)

Inner mitochondrial membrane

Matrix (inside both membranes)

Outer mitochondrial membrane

0.5 μm

Keith R. Porter/Science Source

surface area of the **inner mitochondrial membrane** is expanded by folds called **cristae** (singular, *crista*). Both membranes surround the innermost compartment of the mitochondrion, called the **mitochondrial matrix.** The ATP-generating reactions of mitochondria occur in the cristae and matrix.

The mitochondrial matrix also contains DNA and ribosomes that resemble the equivalent structures in bacteria. These and other similarities suggest that mitochondria originated from ancient bacteria that became permanent residents of the cytoplasm during the evolution of eukaryotic cells. This is discussed in more detail in Chapter 3.

2.3f The Cytoskeleton Supports and Moves Cell Structures

The characteristic shape and internal organization of each type of cell is maintained in part by its cytoskeleton, the interconnected system of protein fibres and tubes that extends throughout the cytoplasm. The cytoskeleton also reinforces the plasma membrane and functions in movement, both of structures within the cell and of the cell as a whole. It is most highly developed in animal cells, in which it fills and supports the cytoplasm from the plasma membrane to the nuclear envelope **(Figure 2.18, p. 42).** Although cytoskeletal structures are also present in plant cells, the fibres and tubes of the system are less prominent; much of cellular support in plants is provided by the cell wall and a large central vacuole (described in Section 2.4).

The cytoskeleton of animal cells contains structural elements of three major types: *microtubules, intermediate filaments,* and *microfilaments.* Plant cytoskeletons likewise contain the same three structural elements. Microtubules are the largest cytoskeletal elements, and microfilaments are the smallest. Each cytoskeletal element is assembled from proteins—microtubules from *tubulins,* intermediate filaments from a large and varied group of *intermediate filament proteins,* and microfilaments from *actins* **(Figure 2.19, p. 42).** The keratins of animal hair, nails, and claws contain a common form of intermediate filament proteins known as *cytokeratins.* For example, human hair consists of thick bundles of cytokeratin fibres extruded from hair follicle cells. The lamins that line the inner surface of the nuclear envelope in animal cells are also assembled from intermediate filament proteins.

Microtubules (Figure 2.19a, p. 42) are microscopic tubes with an outer diameter of about 25 nm and an inner diameter of about 15 nm; they function much like the tubes used by human engineers to construct supportive structures. Microtubules vary widely in length from less than 200 nm to several micrometres. The wall of the microtubule consists of 13 protein filaments arranged side by side. A filament is a linear polymer of tubulin dimers, each dimer consisting of one α-tubulin and one β-tubulin subunit bound noncovalently together. The dimers are organized head to tail in each filament, giving the microtubule a polarity, meaning that the two ends are different. One end, called the 1 (plus) end, has α-tubulin subunits at the ends of the filaments; the other end, called the 2 (minus) end, has β-tubulin subunits at the ends of the filaments. Microtubules are dynamic structures, changing their lengths as required by their functions. This is seen readily in animal cells that are changing shape. Microtubules change length by the addition or removal of tubulin dimers; this occurs

a. Microtubules

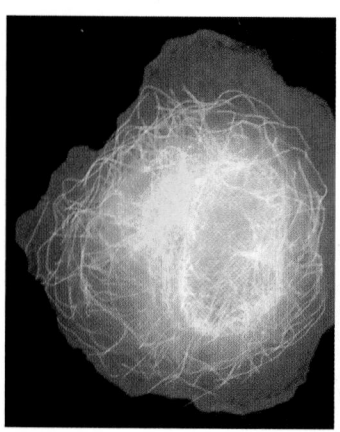

Jennifer C. Waters/Science Source

b. Intermediate filaments

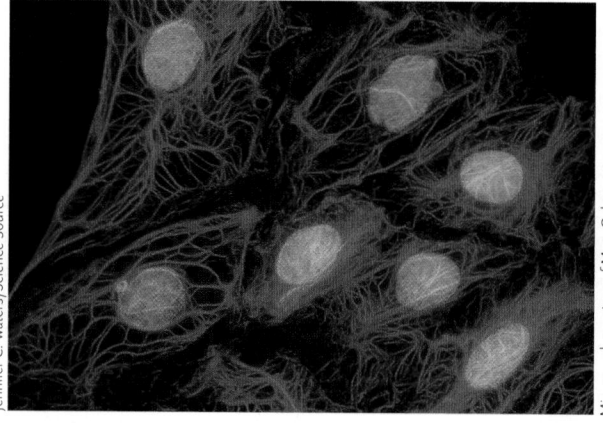

Micrograph courtesy of Mary Osborn

c. Microfilaments

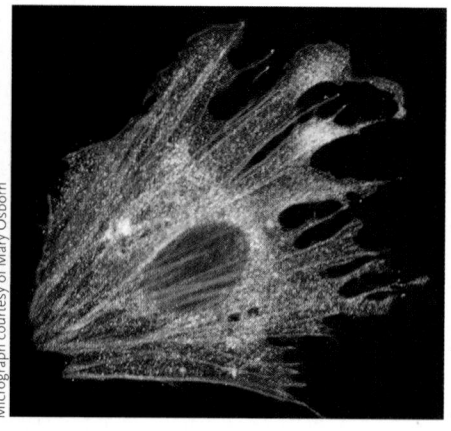

Courtesy of Dr. Vincenzo Cirulli, Diabetes and Obesity Center of Excellence, University of Washington, Department of Medicine, Institute for Stem Cells and Regenerative Medicine

Figure 2.18

Cytoskeletons of eukaryotic cells, as seen in cells stained for light microscopy. **(a)** Microtubules (yellow) and microfilaments (red) in a pancreatic cell. **(b)** Keratin intermediate filaments viewed by immunofluorescence microscopy in the rat kangaroo cell line PtK2. The nucleus is stained blue in these cells. **(c)** Microfilaments (red) in a migrating mammalian cell.

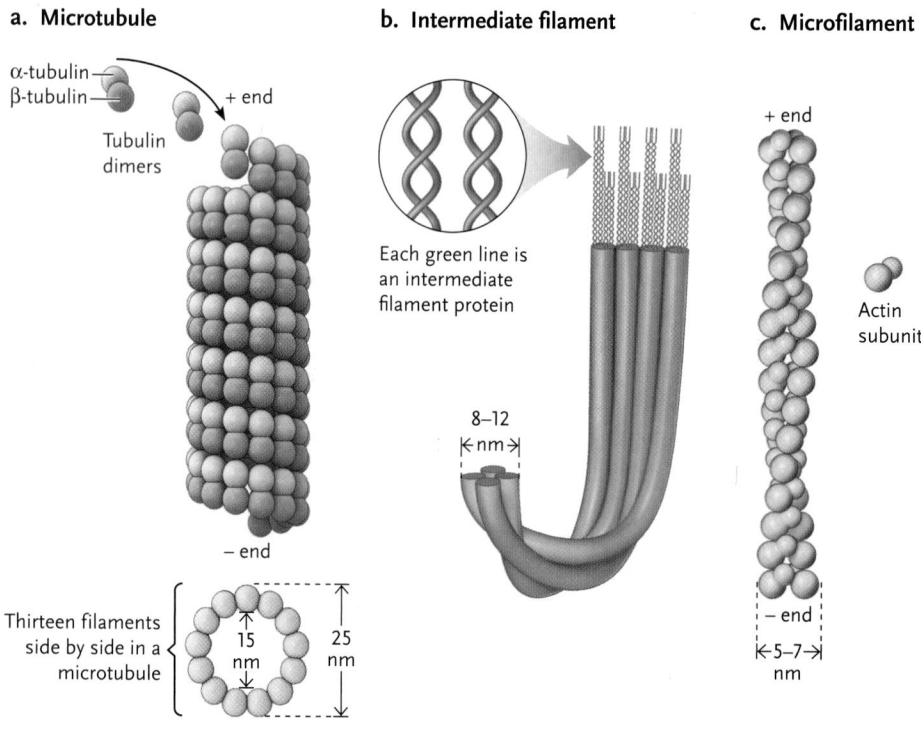

a. Microtubule

α-tubulin
β-tubulin
Tubulin dimers
+ end

Thirteen filaments side by side in a microtubule

15 nm
25 nm

− end

b. Intermediate filament

Each green line is an intermediate filament protein

8–12 nm

c. Microfilament

+ end

Actin subunit

− end

5–7 nm

Figure 2.19

The major components of the cytoskeleton. **(a)** A microtubule, assembled from dimers of α- and β-tubulin proteins. **(b)** An intermediate filament. Eight protein chains wind together to form each subunit, shown as a green cylinder. **(c)** A microfilament, assembled from two linear polymers of actin proteins wound around each other into a helical spiral.

asymmetrically, with dimers adding or detaching more rapidly at the 1 end than at the 2 end. The lengths of microtubules are tightly regulated in the cell.

Many of the cytoskeletal microtubules in animal cells are formed and radiate outward from a site near the nucleus termed the **cell centre** or **centrosome** (see Figure 2.8, p. 34). At its midpoint are two short, barrel-shaped structures also formed from microtubules called the **centrioles** (see **Figure 2.23, p. 45**). Often,

intermediate filaments also extend from the cell centre, apparently held in the same radiating pattern by linkage to microtubules. Microtubules that radiate from the cell centre anchor the ER, Golgi complex, lysosomes, secretory vesicles, and at least some mitochondria in position. The microtubules also provide tracks along which vesicles move from the cell interior to the plasma membrane and in the reverse direction. The intermediate filaments probably add support to the microtubule arrays.

Microtubules play other key roles, for instance, in separating and moving chromosomes during cell division, determining the orientation for growth of the new cell wall during plant cell division, maintaining the shape of animal cells, and moving animal cells themselves. Animal cell movements are generated by "motor" proteins that push or pull against microtubules or microfilaments, much as our muscles produce body movements by acting on bones of the skeleton. One end of a motor protein is firmly fixed to a cell structure such as a vesicle or to a microtubule or microfilament. The other end has reactive groups that "walk" along another microtubule or microfilament by making an attachment, forcefully swivelling a short distance, and then releasing **(Figure 2.20)**. ATP supplies the energy for the walking movements. The motor proteins that walk along microfilaments are called *myosins,* and the ones that walk along microtubules are called *dyneins* and *kinesins.* Some cell movements, such as the whipping motions of sperm tails, depend entirely on microtubules and their motor proteins.

a. "Walking" end of a kinesin molecule

Connects to cell structure
such as a vesicle

One "foot" of
motor protein

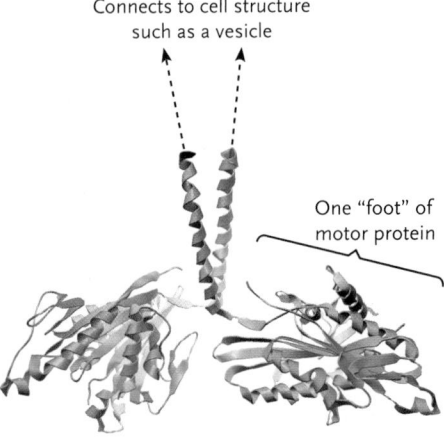

b. How a kinesin molecule "walks"

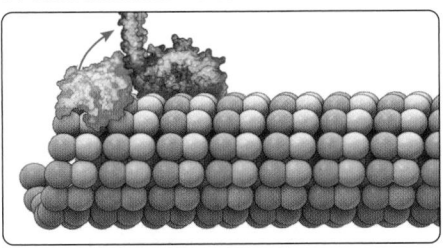

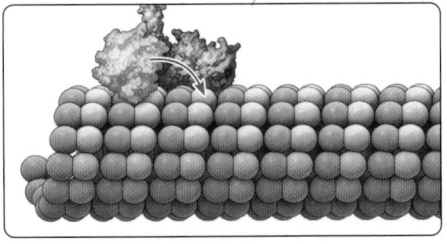

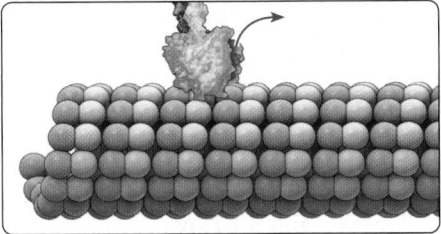

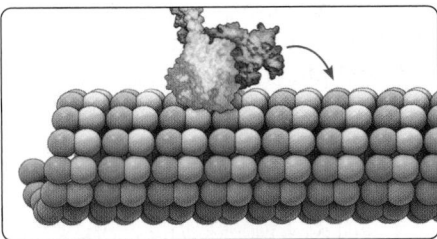

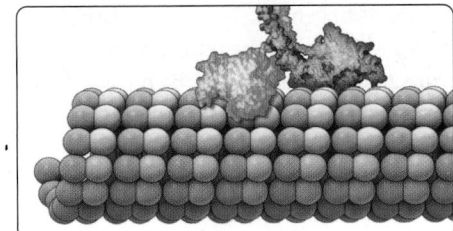

Figure 2.20

The microtubule motor protein kinesin. **(a)** Structure of the end of a kinesin molecule that "walks" along a microtubule, with α-helical segments shown as spirals and β strands as flat ribbons. **(b)** How a kinesin molecule walks along the surface of a molecule by alternately attaching and releasing its "feet."

Intermediate filaments (Figure 2.19b) are fibres with diameters of about 8 to 12 nm. ("Intermediate" signifies, in fact, that these filaments are intermediate in size between microtubules and microfilaments.) These fibres occur singly, in parallel bundles, and in interlinked networks, either alone or in combination with microtubules, microfilaments, or both. Intermediate filaments are only found in multicellular organisms. Moreover, whereas microtubules and microfilaments are the same in all tissues, intermediate filaments are tissue specific in their protein composition. Despite the molecular diversity of intermediate filaments, however, they all play similar roles in the cell, providing structural support in many cells and tissues. For example, the nucleus in epithelial cells is held within the cell by a basketlike network of intermediate filaments made of keratins.

Microfilaments (Figure 2.19c) are thin protein fibres 5 to 7 nm in diameter that consist of two polymers of actin subunits wound around each other in a long helical spiral. The actin subunits are asymmetrical in shape, and they are all oriented in the same way in the polymer chains of a microfilament. Thus, as for microtubules, microfilaments have a polarity: the two ends are designated 1 (plus) and 2 (minus). And, as for microtubules, growth and disassembly occur more rapidly at the 1 end than at the 2 end.

Microfilaments occur in almost all eukaryotic cells and are involved in many processes, including a number of structural and locomotor functions. Microfilaments are best known as one of the two components of the contractile elements in muscle fibres of vertebrates (the roles of myosin and microfilaments in muscle contraction are discussed in Chapter 46). Microfilaments are involved in the actively flowing motion of cytoplasm called **cytoplasmic streaming**, which can transport nutrients, proteins, and organelles in both animal and plant cells, and which is responsible for amoeboid movement. When animal cells divide, microfilaments are responsible for dividing the cytoplasm (see Chapter 9 for further discussion).

2.3g Flagella Propel Cells, and Cilia Move Materials over the Cell Surface

Flagella and **cilia** (singular, *cilium*) are elongated, slender, motile structures that extend from the cell surface. They are identical in structure except that cilia are usually shorter than flagella and occur on cells in greater numbers. The whiplike or oarlike movements of a flagellum propel a cell through a watery medium, and cilia move fluids over the cell surface.

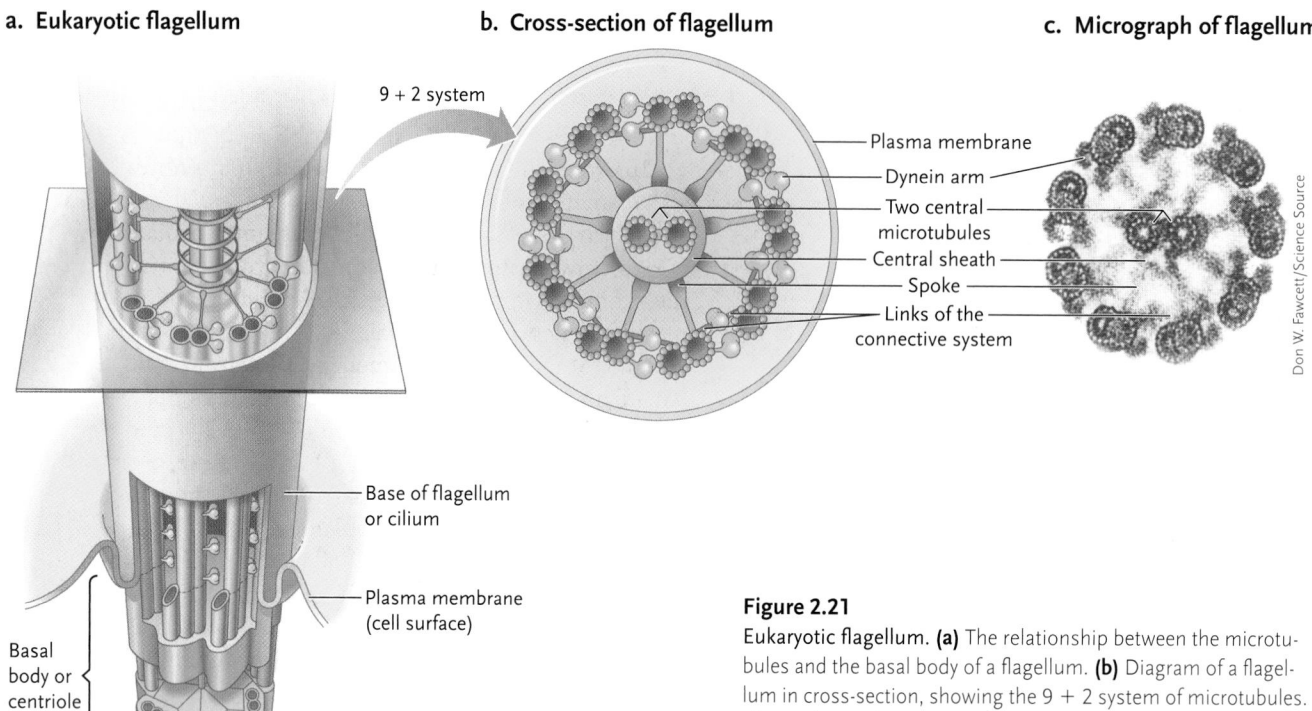

a. Eukaryotic flagellum

9 + 2 system

Base of flagellum
or cilium

Plasma membrane
(cell surface)

Basal
body or
centriole

b. Cross-section of flagellum

Plasma membrane
Dynein arm
Two central
microtubules
Central sheath
Spoke
Links of the
connective system

c. Micrograph of flagellum

Don W. Fawcett/Science Source

Figure 2.21

Eukaryotic flagellum. **(a)** The relationship between the microtubules and the basal body of a flagellum. **(b)** Diagram of a flagellum in cross-section, showing the 9 + 2 system of microtubules. The spokes and connecting links hold the system together. **(c)** Electron micrograph of a flagellum in cross-section; individual tubulin molecules are visible in the microtubule walls.

A bundle of microtubules extends from the base to the tip of a flagellum or cilium **(Figure 2.21)**. In the bundle, a circle of nine double microtubules surrounds a central pair of single microtubules, forming what is known as the 9 + 2 complex. Dynein motor proteins slide the microtubules of the 9 + 2 complex over each other to produce the movements of a flagellum or cilium **(Figure 2.22)**.

Flagella and cilia arise from the centrioles. These barrel-shaped structures contain a bundle of microtubules similar to the 9 + 2 complex, except that the central pair of microtubules is missing and the outer circle is formed from a ring of nine triple rather than double microtubules (compare Figure 2.21 and Figure 2.23, p. 45).

During the formation of a flagellum or cilium, a centriole moves to a position just under the plasma membrane. Then two of the three microtubules of each triplet grow outward from one end of the centriole to form the ring of nine double microtubules. The two central microtubules of the 9 + 2 complex also grow from the end of the centriole, but without direct

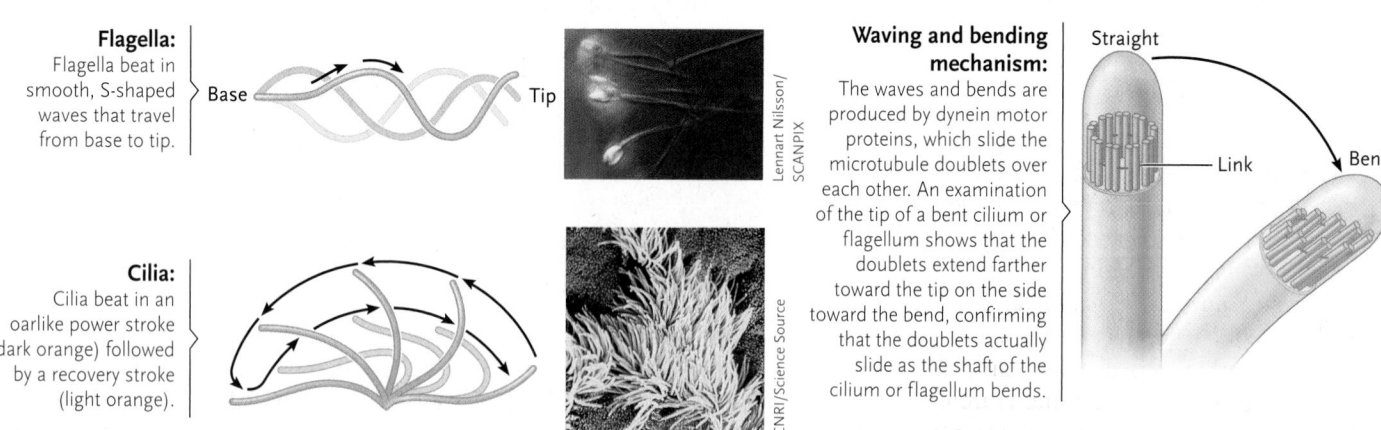

Flagella:
Flagella beat in smooth, S-shaped waves that travel from base to tip.

Base

Tip

Lennart Nilsson/ SCANPIX

Cilia:
Cilia beat in an oarlike power stroke (dark orange) followed by a recovery stroke (light orange).

CNRI/Science Source

Waving and bending mechanism:
The waves and bends are produced by dynein motor proteins, which slide the microtubule doublets over each other. An examination of the tip of a bent cilium or flagellum shows that the doublets extend farther toward the tip on the side toward the bend, confirming that the doublets actually slide as the shaft of the cilium or flagellum bends.

Straight

Link

Bent

Figure 2.22

Flagellar and ciliary beating patterns. The micrographs show a few human sperm, each with a flagellum (top), and cilia from the lining of an airway in the lungs (bottom).

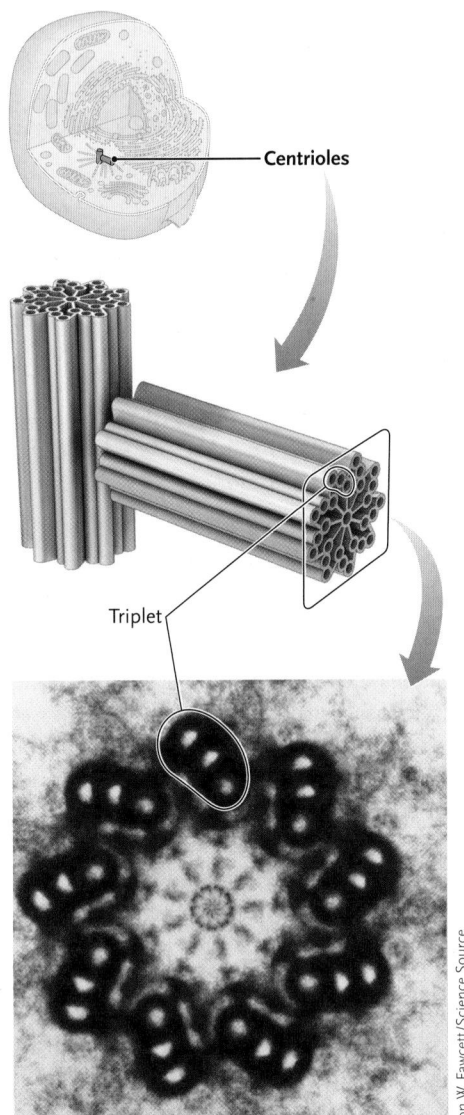

Figure 2.23
Centrioles. The two centrioles of the pair at the cell centre usually lie at right angles to each other as shown. The electron micrograph shows a centriole from a mouse cell in cross-section. A centriole gives rise to the 9 + 2 system of a flagellum and persists as the basal body at the inner end of the flagellum.

connection to any centriole microtubules. The centriole remains at the innermost end of a flagellum or cilium when its development is complete as the **basal body** of the structure (see Figure 2.21).

Cilia and flagella are found in protozoa and algae, and many types of animal cells have flagella—the tail of a sperm cell is a flagellum—as do the reproductive cells of some plants. In humans, cilia cover the surfaces of cells lining cavities or tubes in some parts of the body. For example, cilia on cells lining the ventricles (cavities) of the brain circulate fluid through the brain, and cilia in the oviducts conduct eggs from the ovaries to the uterus. Cilia covering cells that line the air passages of the lungs sweep out mucus containing bacteria, dust particles, and other contaminants.

Although the purpose of the eukaryotic flagellum is the same as that of prokaryotic flagella, the genes that encode the components of the flagellar apparatus of cells of Bacteria, Archaea, and Eukarya are different in each case. Thus, as mentioned earlier in the chapter, the three types of flagella are analogous, not homologous, structures, and they must have evolved independently.

With a few exceptions, the cell structures described so far in this chapter occur in all eukaryotic cells. The major exception is lysosomes, which appear to be restricted to animal cells. The next section describes three additional structures that are characteristic of plant cells.

STUDY BREAK

1. Where in a eukaryotic cell is DNA found? How is that DNA organized?
2. What is the nucleolus, and what is its function?
3. Explain the structure and function of the endomembrane system.
4. What are the structure and function of a mitochondrion?
5. What are the structure and function of the cytoskeleton?

2.4 Specialized Structures of Plant Cells

Chloroplasts, large and highly specialized central vacuoles, and cell walls give plant cells their distinctive characteristics, but these structures also occur in some other eukaryotes—for example, chloroplasts in algal protists and cell walls in algal protists and fungi.

2.4a Chloroplasts Are Biochemical Factories Powered by Sunlight

Chloroplasts (*chloro* = yellow-green), the sites of photosynthesis in plant cells, are members of a family of plant organelles known collectively as **plastids.** Other members of the family include amyloplasts and chromoplasts. **Amyloplasts** (*amylo* = starch) are colourless plastids that store **starch,** a product of photosynthesis. They occur in great numbers in the roots or tubers of some plants, such as the potato. **Chromoplasts** (*chromo* = colour) contain red and yellow pigments and are responsible for the colours of ripening fruits or autumn leaves. All plastids contain DNA genomes and molecular machinery for gene expression and the synthesis of proteins on ribosomes. Some of the proteins within plastids are encoded by their genomes; others are encoded by nuclear genes and are imported into the organelles.

Chloroplasts, like mitochondria, are usually lens or disc shaped and are surrounded by a smooth **outer boundary membrane** and an **inner boundary membrane**, which lies just inside the outer membrane **(Figure 2.24)**. These two boundary membranes completely enclose an inner compartment, the **stroma**. Within the stroma is a third membrane system that consists of flattened, closed sacs called **thylakoids**. In higher plants, the thylakoids are stacked, one on top of another, forming structures called **grana** (singular, *granum*).

The thylakoid membranes contain molecules that absorb light energy and convert it to chemical energy in photosynthesis. The primary molecule absorbing light is *chlorophyll,* a green pigment that is present in all chloroplasts. The chemical energy is used by enzymes in the stroma to make carbohydrates and other complex organic molecules from water, carbon dioxide, and other simple inorganic precursors. The organic molecules produced in chloroplasts, or from biochemical building blocks made in chloroplasts, are the ultimate food source for most organisms. (The physical and biochemical reactions of chloroplasts are described in Chapter 7.)

The chloroplast stroma contains DNA and ribosomes that resemble those of certain photosynthetic bacteria. Because of these similarities, chloroplasts, like mitochondria, are believed to have originated from ancient bacteria that became permanent residents of the eukaryotic cells ancestral to the plant lineage (see Chapter 3 for further discussion).

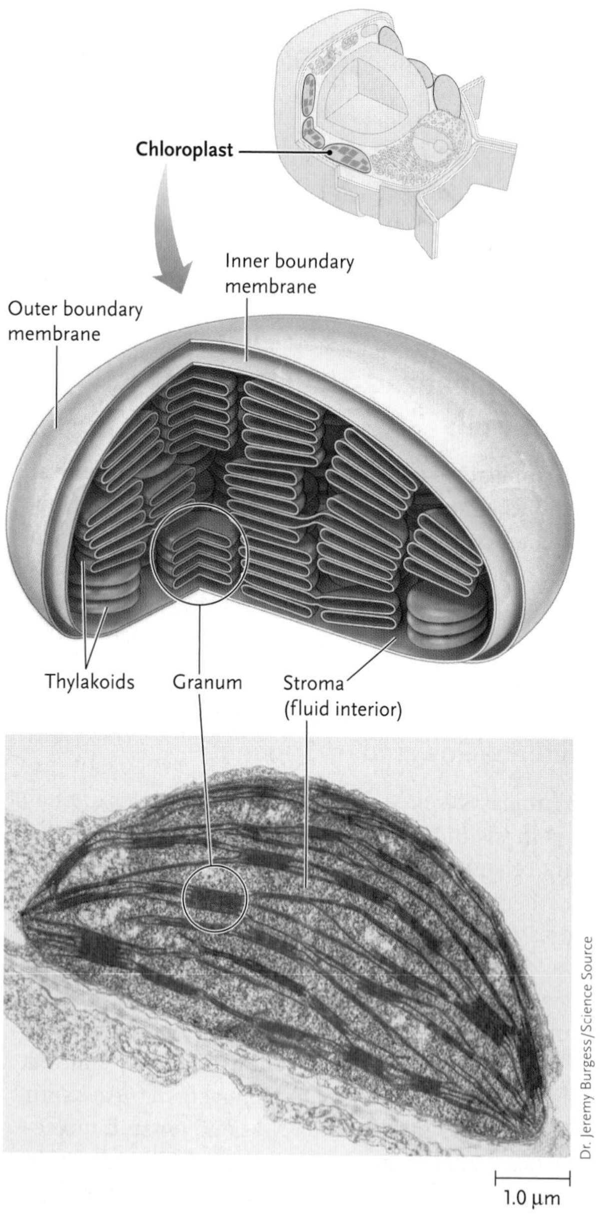

2.4b Central Vacuoles Have Diverse Roles in Storage, Structural Support, and Cell Growth

Central vacuoles (see Figure 2.9, p. 34) are large vesicles identified as distinct organelles of plant cells because they perform specialized functions unique to plants. In a mature plant cell, 90% or more of the cell's volume may be occupied by one or more large central vacuoles. The remainder of the cytoplasm and the nucleus of these cells is restricted to a narrow zone between the central vacuole and the plasma membrane. The pressure within the central vacuole supports the cells.

The membrane that surrounds the central vacuole, the **tonoplast**, contains transport proteins that move substances into and out of the central vacuole. As plant cells mature, they grow primarily by increases in the pressure and volume of the central vacuole.

Central vacuoles conduct other vital functions. They store salts, organic acids, sugars, storage proteins, pigments, and, in some cells, waste products. Pigments concentrated in the vacuoles produce the colours of many flowers. Enzymes capable of breaking down biological molecules are present in some central vacuoles, giving them some of the properties of lysosomes. Molecules that provide chemical defences against pathogenic organisms also occur in the central vacuoles of some plants.

2.4c Cell Walls Support and Protect Plant Cells

The cell walls of plants are extracellular structures because they are located outside the plasma membrane **(Figure 2.25)**. Cell walls provide support to individual cells, contain the pressure produced in the central vacuole, and protect cells against invading bacteria and fungi. Cell walls consist of cellulose fibres, which give tensile strength to the walls, embedded in a network of highly branched carbohydrates. Cell walls are perforated by minute channels,

Chloroplast

Inner boundary membrane

Outer boundary membrane

Thylakoids   Granum   Stroma (fluid interior)

1.0 μm

Dr. Jeremy Burgess/Science Source

Figure 2.24
Chloroplast structure. The electron micrograph shows a maize (corn) chloroplast.

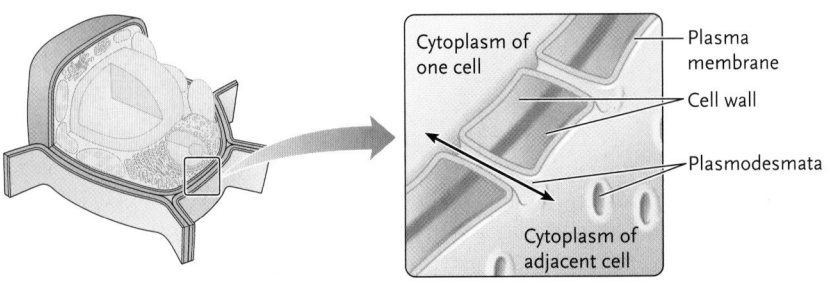
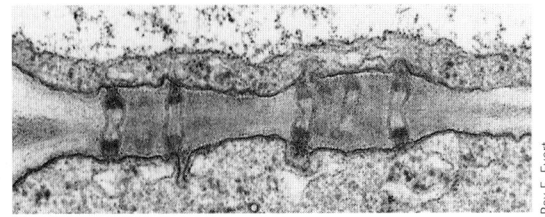

Section through five plasmodesmata that bridge the walls of two plant cells.

Figure 2.25

Cell wall structure in plants. The right diagram and electron micrograph show plasmodesmata, which form openings in the cell wall that directly connect the cytoplasm of adjacent cells.

the plasmodesmata (singular, *plasmodesma;* see Figure 2.25). A typical plant cell has between 1000 and 100 000 plasmodesmata connecting it to abutting cells. These cytosol-filled channels are lined by plasma membranes, so that connected cells essentially all have one continuous surface membrane. Most plasmodesmata also contain a narrow tubelike structure derived from the smooth ER of the connected cells. Plasmodesmata allow ions and small molecules to move directly from one cell to another through the connecting cytosol, without having to penetrate the plasma membranes or cell walls. Proteins and nucleic acids move through some plasmodesmata using energy-dependent processes.

Cell walls also surround the cells of fungi and algal protists. Carbohydrate molecules form the major framework of cell walls in most of these organisms, as they do in plants. In some, the wall fibres contain **chitin** instead of cellulose. Details of cell wall structure in the algal protists and fungi, as well as in different subgroups of the plants, are presented in later chapters devoted to these organisms. As noted earlier, animal cells do not form rigid, external, layered structures equivalent to the walls of plant cells. However, most animal cells secrete extracellular material and have other structures at the cell surface that play vital roles in the support and organization of animal body structures. The next section describes these and other surface structures of animal cells.

2.5 The Animal Cell Surface

Animal cells have specialized structures that help hold cells together, produce avenues of communication between cells, and organize body structures. Molecular systems that perform these functions are organized at three levels: individual **cell adhesion molecules** bind cells together, more complex **cell junctions** seal the spaces between cells and provide direct communication between cells, and the **extracellular matrix** (ECM) supports and protects cells and provides mechanical linkages, such as those between muscles and bone.

2.5a Cell Adhesion Molecules Organize Animal Cells into Tissues and Organs

Cell adhesion molecules are glycoproteins embedded in the plasma membrane. They help maintain body form and structure in animals ranging from sponges to the most complex invertebrates and vertebrates. Rather than acting as a generalized intercellular glue, cell adhesion molecules bind to specific molecules on other cells. Most cells in solid body tissues are held together by many different cell adhesion molecules.

Cell adhesion molecules make initial connections between cells early in embryonic development, but then attachments are broken and remade, as individual cells or tissues change position in the developing **embryo.** As an embryo develops into an adult, the connections become permanent and are reinforced by cell junctions. Cancer cells typically lose these **adhesions,** allowing them to break loose from their original locations, migrate to new locations, and form additional tumours.

Some bacteria and viruses—such as the virus that causes the common cold—target cell adhesion molecules as attachment sites during infection. Cell adhesion molecules are also partially responsible for the ability of cells to recognize one another as being part of the same individual or foreign to that individual. For example, rejection of organ transplants in mammals results from an immune response triggered by the foreign cell surface molecules.

2.5b Cell Junctions Reinforce Cell Adhesions and Provide Avenues of Communication

Three types of cell junctions are common in animal tissues **(Figure 2.26, p. 48). Anchoring junctions** form buttonlike spots, or belts, that run entirely around cells, "welding" adjacent cells together. For some anchoring junctions known as **desmosomes,** intermediate filaments anchor the junction in the underlying cytoplasm; in other anchoring junctions known as **adherens junctions,** microfilaments are the anchoring cytoskeletal component. Anchoring junctions are most common in tissues that are subject to stretching, shear,

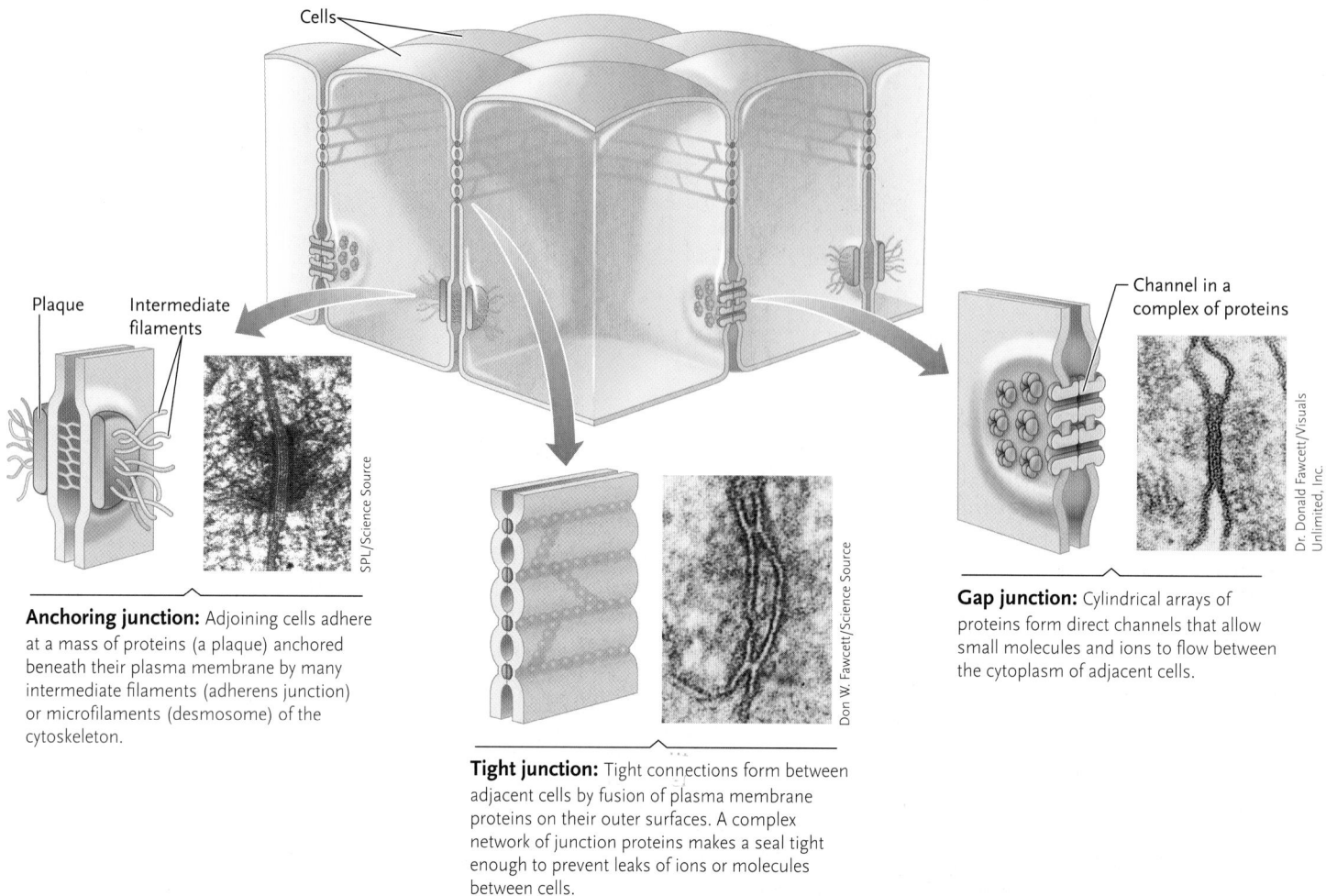

Cells

Plaque   Intermediate
          filaments

SPL/Science Source

Anchoring junction: Adjoining cells adhere at a mass of proteins (a plaque) anchored beneath their plasma membrane by many intermediate filaments (adherens junction) or microfilaments (desmosome) of the cytoskeleton.

Don W. Fawcett/Science Source

Tight junction: Tight connections form between adjacent cells by fusion of plasma membrane proteins on their outer surfaces. A complex network of junction proteins makes a seal tight enough to prevent leaks of ions or molecules between cells.

Channel in a complex of proteins

Dr. Donald Fawcett/Visuals Unlimited, Inc.

Gap junction: Cylindrical arrays of proteins form direct channels that allow small molecules and ions to flow between the cytoplasm of adjacent cells.

Figure 2.26

Anchoring junctions, tight junctions, and gap junctions, which connect cells in animal tissues. Anchoring junctions reinforce the cell-to-cell connections made by cell adhesion molecules, tight junctions seal the spaces between cells, and gap junctions create direct channels of communication between animal cells.

or other mechanical forces—for example, heart muscle, skin, and the cell layers that cover organs or line body cavities and ducts.

Tight junctions, as the name indicates, are regions of tight connections between membranes of adjacent cells (see Figure 2.26). The connection is so tight that it can keep particles as small as ions from moving between the cells in the layers.

Tight junctions seal the spaces between cells in the cell layers that cover internal organs and the outer surface of the body, or the layers that line internal cavities and ducts. For example, tight junctions between cells that line the **stomach,** intestine, and bladder keep the contents of these body cavities from leaking into surrounding tissues.

A tight junction is formed by direct fusion of proteins on the outer surfaces of the two plasma membranes of adjacent cells. Strands of the tight junction proteins form a complex network that gives the appearance of stitch work holding the cells together. Within a tight junction, the plasma membrane is not joined continuously; instead, there are regions of intercellular space. Nonetheless, the network of junction proteins

is sufficient to make the tight cell connections characteristic of these junctions.

Gap junctions open direct channels that allow ions and small molecules to pass directly from one cell to another (see Figure 2.26). Hollow protein cylinders embedded in the plasma membranes of adjacent cells line up and form a sort of pipeline that connects the cytoplasm of one cell with the cytoplasm of the next. The flow of ions and small molecules through the channels provides almost instantaneous communication between animal cells, similar to the communication that plasmodesmata provide between plant cells.

In vertebrates, gap junctions occur between cells within almost all body tissues, but not between cells of different tissues. These junctions are particularly important in heart muscle tissues and in the smooth muscle tissues that form the uterus, where their pathways of communication allow the cells of the organ to operate as a coordinated unit. Although most nerve tissues do not have gap junctions, nerve cells in dental pulp are connected by gap junctions; they are responsible for the discomfort you feel if your teeth are disturbed or damaged, or when a dentist pokes a probe into a cavity.

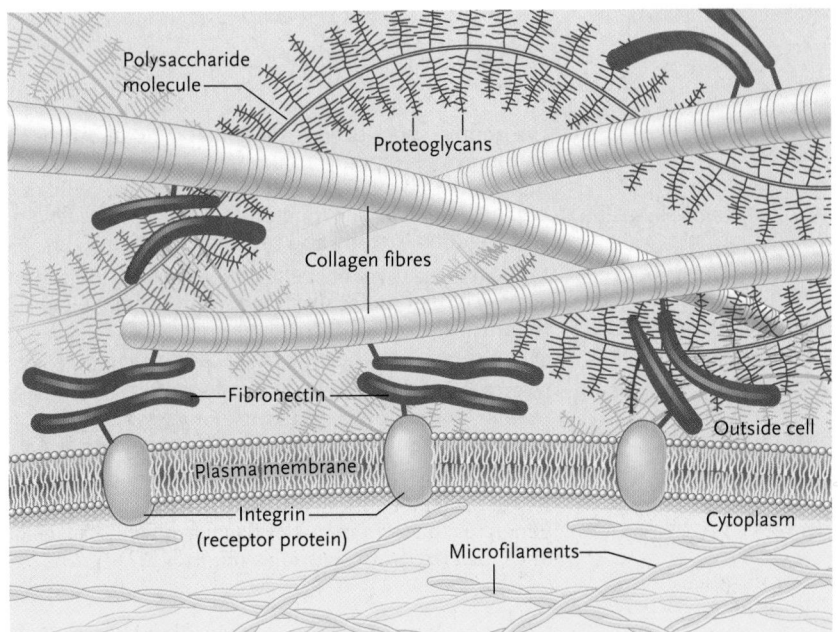

Polysaccharide molecule

Proteoglycans

Collagen fibres

Fibronectin

Outside cell

Plasma membrane

Integrin (receptor protein)

Cytoplasm

Microfilaments

Figure 2.27
Components of the ECM in an animal cell.

2.5c The Extracellular Matrix Organizes the Cell Exterior

Many types of animal cells are embedded in an ECM that consists of proteins and polysaccharides secreted by the cells themselves **(Figure 2.27)**. The primary function of the ECM is protection and support. The ECM forms the mass of skin, bones, and tendons; it also forms many highly specialized extracellular structures, such as the cornea of the eye and filtering networks in the kidney. The ECM also affects cell division, adhesion, motility, and embryonic development, and it takes part in reactions to wounds and disease.

Glycoproteins are the main component of the ECM. In most animals, the most abundant ECM glycoprotein is *collagen,* which forms fibres with great tensile strength and elasticity. In vertebrates, the collagens of tendons, cartilage, and bone are the most abundant proteins of the body, making up about half of the total body protein by weight. (Collagens and their roles in body structures are described in further detail in Chapter 39.)

The consistency of the matrix, which may range from soft and jellylike to hard and elastic, depends on a network of proteoglycans that surrounds the collagen fibres. *Proteoglycans* are glycoproteins that consist of small proteins noncovalently attached to long polysaccharide molecules. Matrix consistency depends on the number of interlinks in this network, which

determines how much water can be trapped in it. For example, cartilage, which contains a high proportion of interlinked glycoproteins, is relatively soft. Tendons, which are almost pure collagen, are tough and elastic. In bone, the glycoprotein network that surrounds collagen fibres is impregnated with mineral crystals, producing a dense and hard—but still elastic—structure that is about as strong as fibreglass or reinforced concrete.

Yet another class of glycoproteins is *fibronectins,* which aid in organizing the ECM and help cells attach to it. Fibronectins bind to **receptor proteins** called *integrins* that span the plasma membrane. On the cytoplasmic side of the plasma membrane, the integrins bind to microfilaments of the cytoskeleton. Integrins integrate changes outside and inside the cell by communicating changes in the ECM to the cytoskeleton.

Having laid the groundwork for cell structure and function in this chapter, we next take up further details of individual cell structures, beginning with the roles of cell membranes in transport in the next chapter.

STUDY BREAK

1. Distinguish between anchoring junctions, tight junctions, and gap junctions.
2. What is the structure and function of the ECM?

Review

To access course materials such as Aplia and other companion resources, please visit www.NELSONbrain.com.

2.1 Basic Features of Cell Structure and Function

- According to the cell theory, (1) all living organisms are composed of cells, (2) cells are the structural and functional units of life, and (3) cells arise only from the division of pre-existing cells.

- Cells of all kinds are divided internally into a central region containing the genetic material and the cytoplasm, which consists of the cytosol, the cytoskeleton, and organelles and is bounded by the plasma membrane.

- The plasma membrane is a lipid bilayer in which transport proteins are embedded (Figure 2.6).

- In the cytoplasm, proteins are made, most of the other molecules required for growth and reproduction are assembled, and energy absorbed from the surroundings is converted into energy usable by the cell.

2.2 Prokaryotic Cells

- Prokaryotic cells are surrounded by a plasma membrane and, in most groups, are enclosed by a cell wall. The genetic material, typically a single, circular DNA molecule, is located in the nucleoid. The cytoplasm contains masses of ribosomes (Figure 2.7).

2.3 Eukaryotic Cells

- Eukaryotic cells have a true nucleus, which is separated from the cytoplasm by the nuclear envelope perforated by nuclear pores. A plasma membrane forms the outer boundary of the cell. Other membrane systems enclose specialized compartments as organelles in the cytoplasm (Figures 2.8 and 2.9).

- The eukaryotic nucleus contains chromatin, a combination of DNA and proteins. A specialized segment of the chromatin forms the nucleolus, where ribosomal RNA molecules are made and combined with ribosomal proteins to make ribosomes. The nuclear envelope contains nuclear pore complexes with pores that allow passive or assisted transport of molecules between the nucleus and the cytoplasm (Figure 2.10).

- Eukaryotic cytoplasm contains ribosomes (Figure 2.11), an endomembrane system, mitochondria, microbodies, the cytoskeleton, and some organelles specific to certain organisms. The endomembrane system includes the nuclear envelope, the endoplasmic reticulum (ER), the Golgi complex, lysosomes, vesicles, and the plasma membrane.

- The ER occurs in two forms: rough and smooth ER. The ribosome-studded rough ER makes proteins that become part of cell membranes or are released from the cell. Smooth ER synthesizes lipids and breaks down toxic substances (Figure 2.12).

- The Golgi complex chemically modifies proteins made in the rough ER and sorts finished proteins to be secreted from the cell, embedded in the plasma membrane, or included in lysosomes (Figure 2.13).

- Lysosomes, specialized vesicles that contain hydrolytic enzymes, digest complex molecules such as food molecules that enter the cell by endocytosis, cellular organelles that are no longer functioning correctly, and engulfed bacteria and cell debris (Figure 2.15).

- Mitochondria carry out cellular respiration, the conversion of fuel molecules into the energy of ATP (Figure 2.17).

- The cytoskeleton is a supportive structure built from microtubules, intermediate filaments, and microfilaments. Motor proteins walking along microtubules and microfilaments produce most movements of animal cells (Figures 2.18–2.20).

- Motor protein–controlled sliding of microtubules generates the movements of flagella and cilia. Flagella and cilia arise from centrioles (Figures 2.21–2.23).

2.4 Specialized Structures of Plant Cells

- Plant cells contain all the eukaryotic structures found in animal cells except for lysosomes. They also contain three structures not found in animal cells: chloroplasts, a central vacuole, and a cell wall (Figure 2.9).

- Chloroplasts contain pigments and molecular systems that absorb light energy and convert it to chemical energy. The chemical energy is used inside the chloroplasts to assemble carbohydrates and other organic molecules from simple inorganic raw materials (Figure 2.24).

- The large central vacuole, which consists of a tonoplast enclosing an inner space, develops pressure that supports plant cells, accounts for much of cellular growth by enlarging as cells mature, and serves as a storage site for substances including waste materials (Figure 2.9).

- A cellulose cell wall surrounds plant cells, providing support and protection. Plant cell walls are perforated by plasmodesmata, channels that provide direct pathways of communication between the cytoplasm of adjacent cells (Figure 2.25).

2.5 The Animal Cell Surface

- Animal cells have specialized surface molecules and structures that function in cell adhesion, communication, and support.

- Cell adhesion molecules bind to specific molecules on other cells. The adhesions organize and hold together cells of the same type in body tissues.

- Cell adhesions are reinforced by various junctions. Anchoring junctions hold cells together. Tight junctions seal together the plasma membranes of adjacent cells, preventing ions and molecules from moving between the cells. Gap junctions open direct channels between the cytoplasm of adjacent cells (Figure 2.26).

- The extracellular matrix (ECM), formed from collagen proteins embedded in a matrix of branched glycoproteins, functions primarily in cell and body protection and support but also affects cell division, motility, embryonic development, and wound healing (Figure 2.27).

Questions

Self-Test Questions

1. Suppose you are examining a cell from a crime scene using an electron microscope, and you find it contains ribosomes, DNA, a plasma membrane, a cell wall, and mitochondria. What type of cell is it?
   a. a lung cell
   b. a plant cell
   c. a prokaryotic cell
   d. a cell from the surface of a human fingernail
   e. a sperm cell

2. A bacterium converts food energy into the chemical energy of ATP using proteins that are found on what part of the cell?
   a. Golgi complex
   b. flagellum
   c. ribosome
   d. cell wall
   e. plasma membrane

3. Which of the following is present in members of the domain Archaea?
   a. nuclear envelope
   b. chloroplast
   c. microtubules
   d. ribosomes
   e. plasmodesmata

4. Which statement about cell size is correct?
   a. As cell size increases, its surface area/volume ratio increases.
   b. A typical animal cell is about 1000 times the size of a bacterium.
   c. The surface area of a cell increases as the cube of the linear dimension.
   d. Increasing membrane surface area allows for a cell to maintain a larger volume.

5. How does a large nuclear-localized protein such as a transcription factor get into the nucleus?
   a. It diffuses across the lipid bilayer of the nuclear envelope.
   b. It is translated on ribosomes that are already within the nucleus.
   c. It contains a nuclear localization signal that is recognized by the nuclear pore complex.
   d. It is synthesized in the cytosol as a set of small polypeptides that diffuse into the nuceus prior to assembly.

6. Which structure is *not* used in eukaryotic protein manufacture and secretion?
   a. ribosome
   b. lysosome
   c. rough ER
   d. secretory vesicle
   e. Golgi complex

7. Suppose an electron micrograph shows that a cell has extensive amounts of smooth ER throughout. What can you deduce about the cell?
   a. The cell is synthesizing ATP.
   b. The cell is metabolically inactive.
   c. The cell is synthesizing and secreting proteins.
   d. The cell is synthesizing and metabolizing lipids.

8. Which of the following contributes to sealing the lining of the digestive track so that it can retain food?
   a. tight junctions formed by direct fusion of proteins
   b. plasmodesmata that help cells communicate their activities
   c. desmosomes forming buttonlike spots or a belt to keep cells joined
   d. gap junctions that communicate between cells of the stomach lining and its muscular wall

9. Which of the following statements about proteins is correct?
   a. Proteins are transported to the rough ER for use within the cell.
   b. Lipids and carbohydrates are added to proteins by the Golgi complex.
   c. Proteins are transported directly into the cytosol for secretion from the cell.
   d. Proteins that are to be stored by the cell are moved to the rough ER.
   e. Proteins are synthesized in vesicles.

10. Which of the following are NOT a component of the cytoskeleton?
   a. cilia
   b. actins
   c. microfilaments
   d. microtubules
   e. cytokeratins

Questions for Discussion

1. Many compound microscopes have a filter that eliminates all wavelengths except that of blue light, thereby allowing only blue light to pass through the microscope. Use the spectrum of visible light (see Figure 7.5 in Chapter 7) to explain why the filter improves the resolution of light microscopes.

2. Explain why aliens invading Earth are not likely to be giant cells the size of humans.

3. An electron micrograph of a cell shows the cytoplasm packed with rough ER membranes, a Golgi complex, and mitochondria. What activities might this cell concentrate on? Why would large numbers of mitochondria be required for these activities?

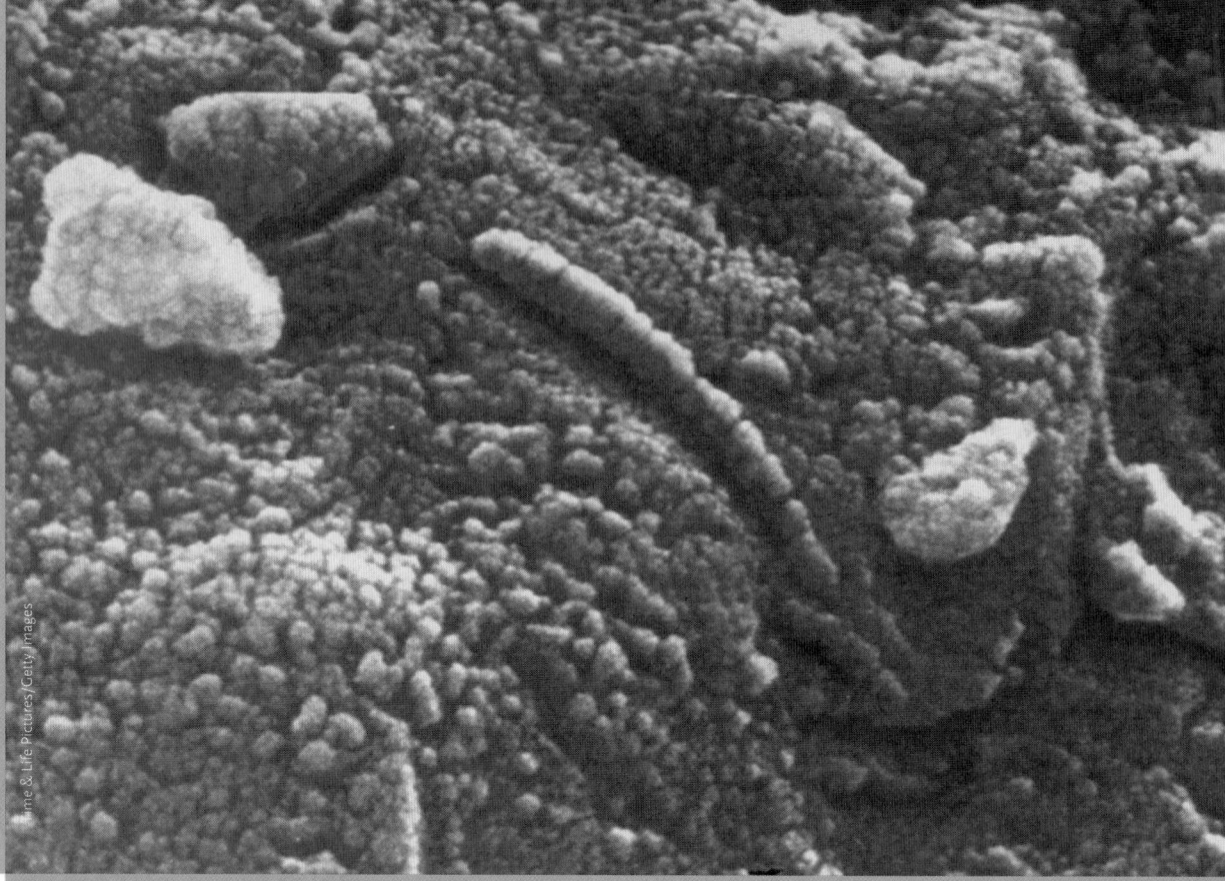

Scanning electron microscope image of a portion of the meteorite ALH84001. The elongate structure in the centre may be fossilized microorganism.

3

Defining Life and Its Origins

WHY IT MATTERS

In 1984, a group of scientists in the Antarctic discovered a 1.9 kg meteorite that they catalogued as ALH84001. Initial studies of the meteorite showed that it was about 4.5 billion years old, which is about the same age as the solar system. Its chemical composition indicated that it had originated from Mars and had impacted Earth approximately 13 thousand years ago. The meteorite garnered headlines around the world in 1996 when an article was published in the prestigious journal *Science* with evidence that ALH84001 contained distinct evidence that life had at one time existed on Mars.

Chemical analysis showed that, when on Mars, ALH84001 had at one time been fractured and subsequently infiltrated by liquid water. Using scanning electron microscopy, the coauthors of the article observed very small, elliptical, ropelike, and tubular structures in the fractured surfaces of ALH84001 that look very similar to fossilized prokaryotic cells. Furthermore, the scientists found microscopic mineral "globules," which bear strong resemblance to mineral alterations caused by primitive cells on Earth.

The analysis of meteorites for microfossils continues and remains controversial. In 2011, an article published in the *Journal of Cosmology* provided additional evidence of bacteria-like fossils within meteorites. Using sophisticated electron microscopy techniques and chemical analysis on three freshly fractured carbonaceous meteorites, data are presented in the article that show the presence of filaments that are strikingly similar in shape to cyanobacteria, a dominant form of photosynthetic bacteria on Earth. The difficulty of unequivocally assigning these structures as remnants of ancient life will continue to make the conclusions of such analyses controversial.

In this chapter, we explore two of the most basic of biological questions: What is life, and how did life evolve? After introducing the fundamental characteristics that all organisms share, we work through a discussion of the origins of life. Starting with how biologically important molecules could have been synthesized in the absence of life, we move through hypotheses regarding the first cells, to evolution of various **macromolecules** (DNA, RNA, proteins), and finally to what drove the evolution of the eukaryotic cell. The chapter closes where it begins in "Why It Matters," with a discussion on the possibility of life existing elsewhere in the galaxy.

Figure 3.1
Red-eyed tree frog on a rock.

3.1 What Is Life?

As we saw in the last chapter, all life is composed of cells—the fundamental unit of all life. But what is life? How can we define it? Picture a frog sitting on a rock, slowly shifting its head to follow the movements of insects flying nearby **(Figure 3.1).** You know instinctively that the frog is alive and the rock is not. But if you examine both at the atomic level, you will find that the difference between them is lost. The types of elements and atoms found in living things are also found in non-living forms of matter. As well, living cells obey the same fundamental laws of physics and chemistry as the **abiotic** (nonliving) world. For example, the biochemical reactions that take place within living cells, although remarkable, are only modifications of chemical reactions that occur outside cells.

3.1a Seven Characteristics Shared by All Life-Forms

Although life seems relatively easy to recognize, it is not easy to define using a single sentence or even two. Life is defined most effectively by a list of attributes that all forms of life possess. As detailed in **Figure 3.2, p. 54,** all life displays order, harnesses and utilizes energy, reproduces, responds to stimuli, exhibits homeostasis, grows and develops, and evolves.

There are a small number of biological systems that straddle the line between the **biotic** and abiotic worlds. The best example of this is a virus **(Figure 3.3, p. 54).** Viruses are very small infectious agents that you will learn more about in Chapter 23. They display many of the properties of life, including the ability to reproduce and evolve over time. However, the characteristics of life that a virus has are based on its ability to infect cells. For example, although viruses contain nucleic acids (DNA and RNA), they lack the cellular machinery and metabolism to use that genetic information to synthesize their own proteins. To make proteins, they have to infect living cells and essentially hijack their translational machinery and metabolism to reproduce. For this reason, most scientists do not consider viruses to be alive.

3.1b The Characteristics of Life Are Emergent

Each of the characteristics of life depicted in Figure 3.2 reflects a remarkable complexity resulting from a hierarchy of interactions that begins with atoms and progresses through molecules to macromolecules and cells. Depending upon the organism, this hierarchy may continue upward in complexity and include organelles, tissues, and organs. The seven properties of life shown in Figure 3.2 are called emergent because they come about, or emerge, from many simpler interactions that, in themselves, do not have the properties found at the higher levels. For example, the ability to harness and utilize energy is not a property of molecules or proteins or biological membranes in isolation; rather the ability emerges from the interactions of all three of these as part of a metabolic process. In this way, not only is the structural or functional complexity of living systems more than the sum of the parts, but it is fundamentally different.

A classic example that illustrates the concept of emergence is a type of termite nest called a cathedral **(Figure 3.4, p. 55).** These elegantly complex structures, most common in Australia, can grow to over 3 m tall and are the product of the activities of thousands of termites. Remarkably, there is no master plan that is followed or "queen" that gives instructions. Termites build up the mound cell by cell, based on local conditions, totally unaware of the overall structure that emerges.

a. Display order: All forms of life including this flower are arranged in a highly ordered manner, with the cell being the fundamental unit of life.

harmeet/StockXchng

e. Exhibit homeostasis: Organisms are able to regulate their internal environment such that conditions remain relatively constant. Sweating is one way in which the human body attempts to remove heat and thereby maintain a constant temperature.

© Tim Pannell/CORBIS

b. Harness and utilize energy: Like this hummingbird, all forms of life acquire energy from the environment and use it to maintain their highly ordered state.

Steve Byland/Shutterstock.com

f. Grow and develop: All organisms increase their size by increasing the size and/or number of cells. Many organisms also change over time.

© Karin Duthie/Alamy

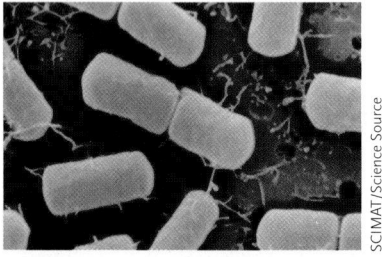

c. Reproduce: All organisms have the ability to make more of their own kind. Here, some of the bacteria have just divided into two daughter cells.

SCIMAT/Science Source

g. Evolve: Populations of living organisms change over the course of generations to become better adapted to their environment. The snowy owl illustrates this perfectly.

Stanislav Duben/Shutterstock.com

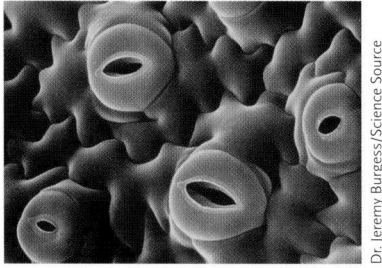

d. Respond to stimuli: Organisms can make adjustments to their structure, function, and behaviour in response to changes to the external environment. A plant can adjust the size of the pores (stomata) on the surface of its leaves to regulate gas exchange.

Dr. Jeremy Burgess/Science Source

Figure 3.2
The seven characteristics of life.

3.2 The Chemical Origins of Life

Recall from Chapter 2 that one of the tenets of the cell theory states that cells arise only from the growth and division of pre-existing cells. This tenet has probably been true for hundreds of millions of years, yet there must have been a time when this was not the case. There must have been a time when no cells existed, when there was no life. It is thought that over the course of hundreds of millions of years, cells with the characteristics of life arose out of a mixture of molecules that existed on primordial Earth. In this section we discuss the formation of the solar system and present hypotheses for how biologically important molecules could have been synthesized on early Earth in the absence of life.

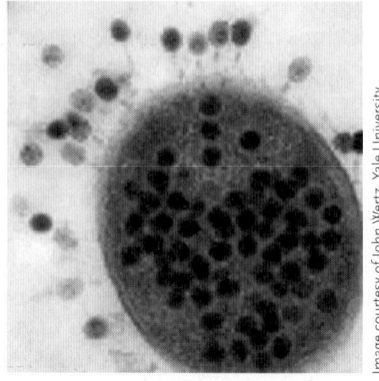

Image courtesy of John Wertz, Yale University

Figure 3.3
Bacteriophage infecting a bacterium. Notice bacteriophage on the cell surface as well as inside the bacterium. A bacteriophage is a type of virus. Viruses are generally not considered to be alive.

Figure 3.4

A termite cathedral. The sophisticated structure of a termite nest emerges from the simple work of thousands of individual termites. In a similar way, the complex properties of life emerge from much simpler molecular interactions.

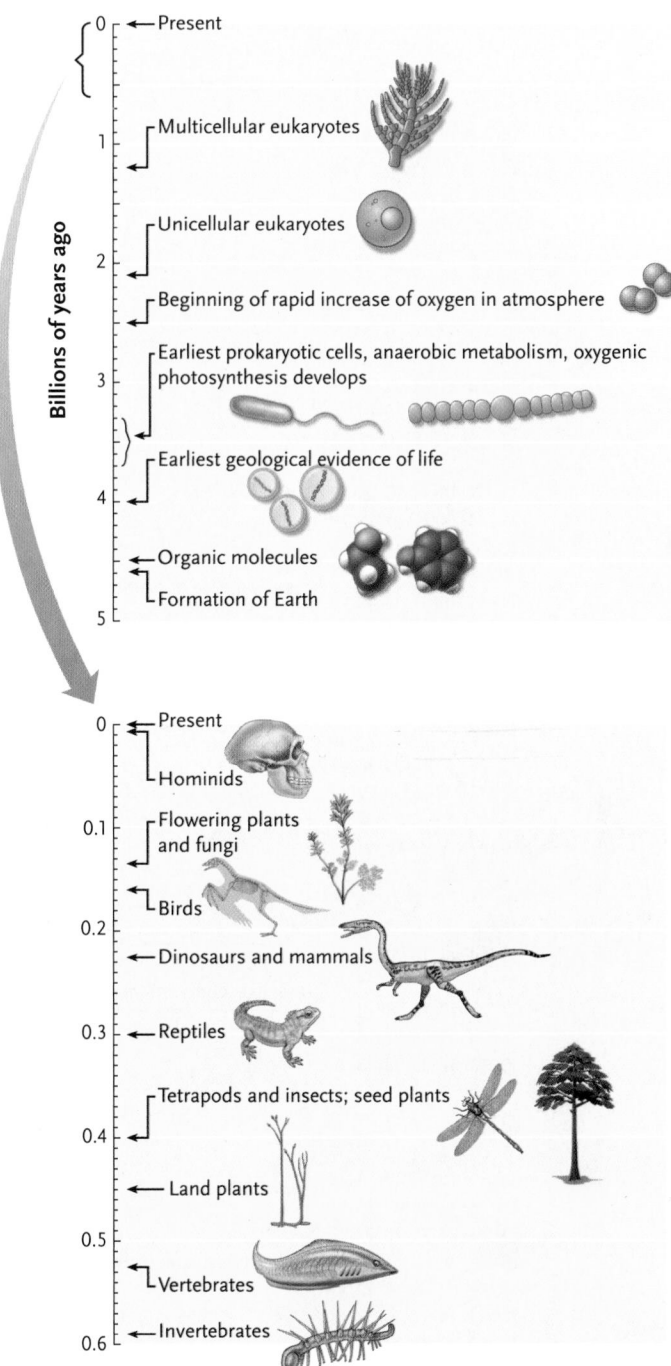

Figure 3.5

A timeline for the evolution of major forms of life. The dates presented are derived mostly from geological evidence.

3.2a Earth Is 4.6 Billion Years Old

Before we discuss the origins of life, we present in **Figure 3.5** a timeline of the evolution of major present-day organisms. Each date is still the subject of some debate within the scientific community and is based primarily on **radiometric dating** methods (see *The Purple Pages, p. F-50*). A more complete presentation of the timeline for the evolution of organismal life is found in *The Purple Pages, p. F-50*. An overview of the fossil record from which the dating and thus ordering of the major events is obtained is presented in Chapter 17. This is followed in later chapters by in-depth discussions of the evolution of the major groups of organisms, including, for example, land plants (Chapter 26), birds (Chapter 28), and mammals (Chapter 28).

Earth was formed approximately 4.6 billion years ago. To give us some sense of just how long 4.6 billion years is, as well as the relative timing of some major events in the history of life on Earth, **Figure 3.6, p. 56,** condenses the entire history of Earth into a unit of time that we are more familiar with—one year. With 4.6 billion years condensed into a single year, each day represents an interval of 12.6 million years!

Using our condensed version of the history of Earth, we set the date of the formation of Earth as January 1 at 12:00 a.m. Based on chemical evidence, life may have started as early as 4.0 billion years ago. This translates to mid-March in our one-year calendar. The first clear fossil evidence of prokaryotic cells occurs in late March or about 3.5 billion years ago. Fossil evidence of eukaryotes has been dated to about 2 billion years ago, which is not until early July using our one-year analogy. Perhaps surprisingly, animals do not make an appearance until mid-October (about 525 million years

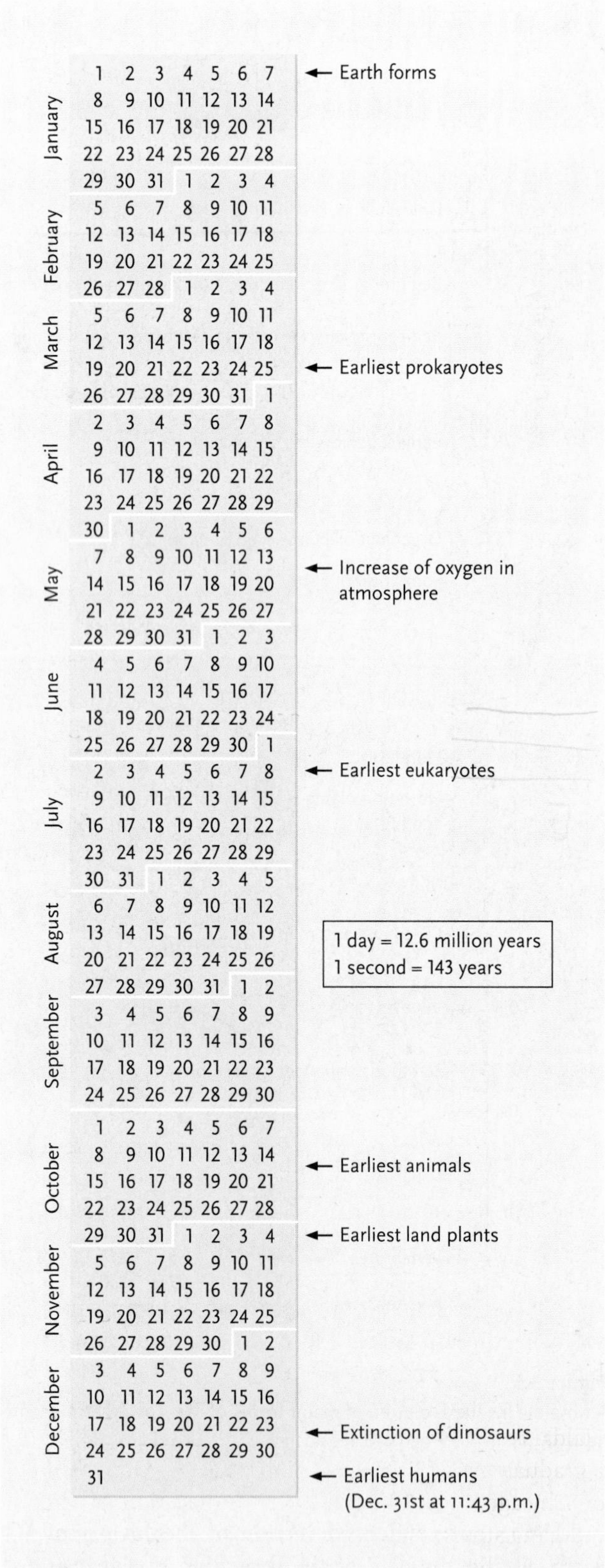

January
1 2 3 4 5 6 7   ← Earth forms
8 9 10 11 12 13 14
15 16 17 18 19 20 21
22 23 24 25 26 27 28
29 30 31 1 2 3 4

February
5 6 7 8 9 10 11
12 13 14 15 16 17 18
19 20 21 22 23 24 25
26 27 28 1 2 3 4

March
5 6 7 8 9 10 11
12 13 14 15 16 17 18
19 20 21 22 23 24 25   ← Earliest prokaryotes
26 27 28 29 30 31 1

April
2 3 4 5 6 7 8
9 10 11 12 13 14 15
16 17 18 19 20 21 22
23 24 25 26 27 28 29
30 1 2 3 4 5 6

May
7 8 9 10 11 12 13   ← Increase of oxygen in atmosphere
14 15 16 17 18 19 20
21 22 23 24 25 26 27
28 29 30 31 1 2 3

June
4 5 6 7 8 9 10
11 12 13 14 15 16 17
18 19 20 21 22 23 24
25 26 27 28 29 30 1

July
2 3 4 5 6 7 8   ← Earliest eukaryotes
9 10 11 12 13 14 15
16 17 18 19 20 21 22
23 24 25 26 27 28 29
30 31 1 2 3 4 5

August
6 7 8 9 10 11 12
13 14 15 16 17 18 19
20 21 22 23 24 25 26
27 28 29 30 31 1 2

1 day = 12.6 million years
1 second = 143 years

September
3 4 5 6 7 8 9
10 11 12 13 14 15 16
17 18 19 20 21 22 23
24 25 26 27 28 29 30

October
1 2 3 4 5 6 7
8 9 10 11 12 13 14   ← Earliest animals
15 16 17 18 19 20 21
22 23 24 25 26 27 28
29 30 31 1 2 3 4   ← Earliest land plants

November
5 6 7 8 9 10 11
12 13 14 15 16 17 18
19 20 21 22 23 24 25
26 27 28 29 30 1 2

December
3 4 5 6 7 8 9
10 11 12 13 14 15 16
17 18 19 20 21 22 23   ← Extinction of dinosaurs
24 25 26 27 28 29 30
31   ← Earliest humans (Dec. 31st at 11:43 p.m.)

Figure 3.6
The history of Earth condensed into one year.

ago) and land plants until the following month. The extinction of dinosaurs, which was completed by about 65 million years ago, does not occur until late December. What about humans? We may think humans, *Homo sapiens*, have been around a long time, but relative to other forms of life, the roughly 150 thousand years that modern humans have existed is a very short period of time—a blip on our time-scale. Using our year analogy, modern humans have existed only since December 31—more precisely, December 31 at 11:43 p.m.!

3.2b Earth Lies within the Habitable Zone around the Sun

According to the most widely accepted hypothesis, all components of the solar system were formed at the same time by the gravitational condensation of matter present in an interstellar cloud, which initially consisted mostly of hydrogen. Intense heat and pressure generated in the central region of the cloud formed the Sun, whereas the remainder of the spiralling dust and gas condensed into the planets. Astronomers agree that this series of events is probably typical for the vast majority of the stars and planetary systems in our galaxy, the Milky Way.

Once Earth was formed, its early history was marked by bombardment of rock from the still-forming solar system and extensive volcanic and seismic activity **(Figure 3.7)**. Over time, Earth radiated away some of its heat, and surface layers cooled and solidified into the rocks of the crust. Because of its size, Earth's gravitational pull was strong enough to hold an atmosphere around the planet. The atmosphere was derived partly from the original dust cloud and partly from gases released from Earth's interior as it cooled. It is estimated that it took approximately 500 million years for Earth to cool to temperatures that could nurture the development of life.

Within the Solar System, there is currently no conclusive evidence that a planet other than Earth harbours life. The primary reason for this is that, unlike the other planets, Earth is situated at a position where heat from the Sun allows for surface temperatures to be within a range that allows water to exist in a liquid state. Interestingly, recent data from NASA's Mars Reconnaissance Orbiter suggest that some liquid water may exist on Mars as well. It is the presence of liquid water that is seen as a fundamental prerequisite for the development of life (see *The Purple Pages* for more on the structure and unique properties of water). Because of the importance of water for the development of life, the region of space around a star where temperatures would allow for liquid water is termed the *habitable zone* **(Figure 3.8)**. As you would expect, the precise distance from the star that defines the habitable zone will vary depending upon the type of star and how much energy it emits.

3.2c Biologically Important Molecules Can Be Synthesized Outside Living Cells

All forms of life are composed of the major macromolecules nucleic acids, proteins, lipids, and carbohydrates (see *The Purple Pages* for an overview of these

Figure 3.7
An artist's depiction of the primordial Earth.

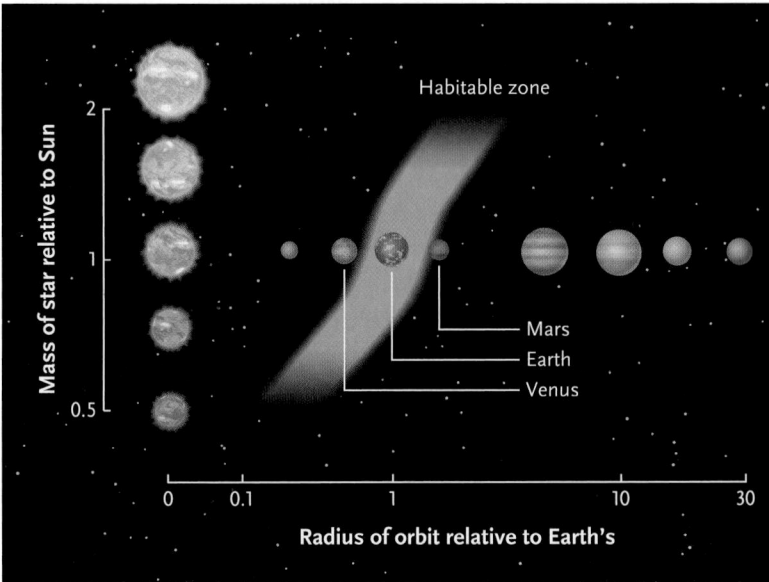

Figure 3.8
The habitable zone. The region around a star where water would exist in a liquid state and thus be conducive to the development of life is termed the *habitable zone* (shown in green). The distance from the star where the zone occurs varies depending upon the energy output of the parent star. The Sun, which is considered an average star, is shown as 1 on the scale at the left.

molecules). With the exception of lipids, these macromolecules are derived from simpler molecules such as nucleotides, amino acids, and sugars that in modern-day cells are the products of complex metabolic pathways. But how were these molecules formed in the absence of life? There are three major hypotheses.

Hypothesis 1: Reducing Atmosphere. The atmosphere of 4 billion years ago was vastly different from the one today. The primordial atmosphere probably contained an abundance of water vapour from the evaporation of water at the surface, as well as large quantities of hydrogen (H_2), carbon dioxide (CO_2), ammonia (NH_3), and methane (CH_4). There was an almost complete absence of oxygen (O_2). In the 1920s,

two scientists, Aleksander Oparin and John Haldane, independently proposed that organic molecules, essential to the formation of life, could have formed in the atmosphere of primordial Earth. A critical aspect of what is known as the Oparin–Haldane hypothesis is that the early atmosphere was a *reducing atmosphere* because of the presence of large concentrations of molecules such as hydrogen, methane, and ammonia. These molecules contain an abundance of electrons and hydrogen and would have entered into reactions with one another that would have yielded larger and more complex organic molecules.

In comparison to the proposed reducing atmosphere of primordial Earth, today's atmosphere is classified as an *oxidizing atmosphere*. The presence of high levels of oxygen prevents complex, electron-rich molecules from being formed because oxygen is a particularly strong oxidizing molecule and would itself accept the electrons from organic molecules and be reduced to water. Besides allowing for the build-up of electron-rich molecules, the lack of oxygen in the primordial atmosphere also meant that there was no ozone (O_3) layer, which only developed after oxygen levels in the atmosphere began to increase. Both Oparin and Haldane hypothesized that without the ozone layer, energetic ultraviolet light was able to reach the lower atmosphere and, along with abundant lightning, provided the energy needed to drive the formation of biologically important molecules.

Experimental evidence in support of the Oparin–Haldane hypothesis came in 1953 when Stanley Miller, a graduate student of Harold Urey's at the University of Chicago, created a laboratory simulation of the reducing atmosphere believed to have existed on early Earth. Miller placed components of a reducing atmosphere—hydrogen, methane, ammonia, and water vapour—in a closed apparatus and exposed the gases to an energy source in the form of continuously sparking electrodes **(Figure 3.9, p. 58)**. Water vapour was added to the "atmosphere" in one part of the apparatus and subsequently condensed back into water by cooling in another part. After running the experiment for one week, Miller found a large assortment of organic compounds including urea; amino acids; and lactic, formic,

Figure 3.9

The Miller–Urey experiment. Using this apparatus, Stanley Miller, a graduate student, demonstrated that organic molecules can be synthesized under conditions simulating primordial Earth.

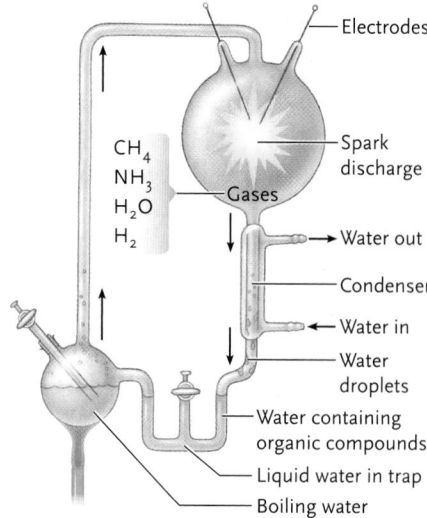

Electrodes

Spark discharge

Gases

CH_4
NH_3
H_2O
H_2

Water out

Condenser

Water in

Water droplets

Water containing organic compounds

Liquid water in trap

Boiling water

Figure 3.10

Deep-sea vent. Researchers from the Woods Hole Oceanographic Institute watch from inside the submersible Alvin as a "black smoker" chimney erupts from a seafloor vent. The regions surrounding these vents have been found to be teeming with a diversity of life.

and acetic acids after condensing the atmosphere into a liquid. In fact, as much as 15% of the carbon that was originally in the methane at the start of the experiment ended up in molecules that are common in living organisms.

Other chemicals have been tested in the Miller–Urey apparatus, including hydrogen cyanide (HCN) and formaldehyde (CH_2O), which are considered to have been among the substances formed in the primitive atmosphere. When cyanide and formaldehyde were added to the simulated primitive atmosphere in Miller's apparatus, all the building blocks of complex biological molecules were produced—amino acids; fatty acids; the **purine** and **pyrimidine** components of nucleic acids; sugars such as glyceraldehyde, ribose, glucose, and fructose; and phospholipids, which form the lipid bilayers of biological membranes.

Over the years since the Miller–Urey experiment was first conducted, considerable debate has developed in the scientific community over whether the atmosphere of primitive Earth contained enough methane and ammonia to be considered reducing. Some geologists have suggested that based on the analysis of volcanic activity, primitive Earth was probably somewhat less reactive—neither reducing nor oxidizing—with molecules including nitrogen gases (N_2), carbon monoxide (CO), and carbon dioxide (CO_2) the most dominant. Even with this composition, scientists have been able to successfully synthesize the same crucial building blocks of life in the laboratory. Regardless of the actual composition of the atmosphere on primordial Earth, the significance of the Miller–Urey experiment cannot be overstated. It was the first experiment to demonstrate the abiotic formation of molecules critical to life, such as amino acids, nucleotides, and simple sugars, and it showed that they could be produced relatively easily. At the time, this remarkable finding laid the groundwork for further research into the origins of life.

Hypothesis 2: Deep-Sea Vents. Besides the atmosphere, an **alternative hypothesis** maintains that the complex organic molecules necessary for life could have originated on the ocean floor at the site of deep-sea (hydrothermal) vents. These cracks are found around the globe near sites of volcanic or tectonic activity and release superheated nutrient-rich water at temperatures in excess of 300°C, as well as reduced molecules including methane, ammonia, and hydrogen sulfide (H_2S) **(Figure 3.10)**. Today, the areas around these vents support a remarkable diversity of life. Many of these life forms are of tremendous scientific interest because of their ability to thrive in an environment that is characterized by extreme pressure and the total absence of light.

Hypothesis 3: Extraterrestrial Origins. It is entirely possible that the key organic molecules required for life to begin came from space. Each year more than 500 meteorites impact the Earth, many of which belong to the class called carbonaceous chondrites, which are particularly rich in organic molecules. One of the most famous is the Murchison meteorite that landed in Murchison, Victoria, Australia, in 1969 **(Figure 3.11)**. Analysis of the Murchison meteorite showed that it contains an assortment of biologically important molecules including a range of amino acids such as glycine, glutamic acid, and alanine, as well as purines and pyrimidines.

3.2d Life Requires the Synthesis of Polymers

Primordial Earth contained very little oxygen, and because of this, complex organic molecules could have existed for much longer than would be possible in today's oxygen-rich world. Even if they did accumulate on early Earth, molecules such as amino acids and

Figure 3.11
The Murchison meteorite. Many meteorites have been shown to contain a range of biologically important molecules, including a number of amino acids.

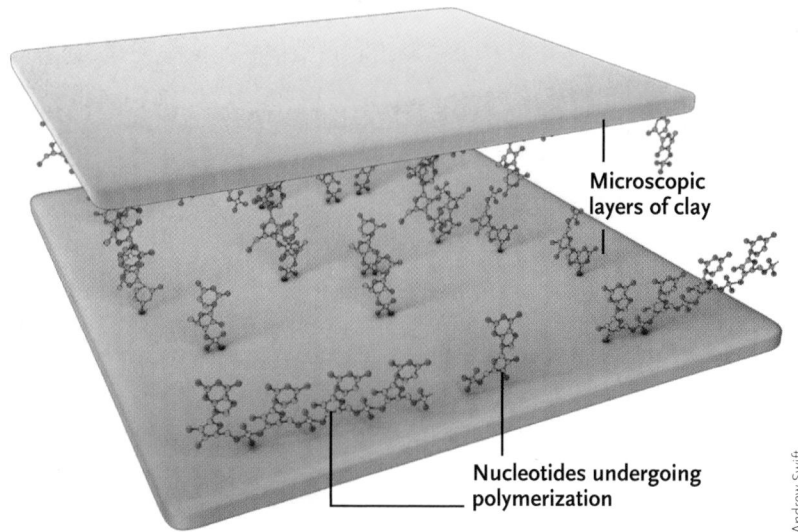

Microscopic layers of clay

Nucleotides undergoing polymerization

Figure 3.12
Clay surfaces catalyze polymerization. The charged microscopic layered structure of clay allows for the formation of relatively short polymers of proteins and nucleic acids.

nucleotides are **monomers**, which are simpler and easier to synthesize than the key chemical components of life, such as nucleic acids and proteins, which are polymers—macromolecules formed from the bonding together of individual monomers. Nucleic acids are polymers of nucleotides, proteins are polymers of amino acids, and many carbohydrates are polymers of simple sugars. Polymers are synthesized by dehydration synthesis, which is discussed in *The Purple Pages*.

Today, the synthesis of proteins and nucleic acids requires protein-based catalysts called enzymes and results in macromolecules that often consist of hundreds to many thousands of monomers linked together. So how do you make the polymers that are required for life without sophisticated enzymes? The basis of a working hypothesis to address this question must be built from the supposition that the very earliest forms of life must have been very simple—far simpler than a modern bacterium, for example. Scientists hypothesize that a polymer that consists of even 10 to 50 monomers may have been of sufficient length to impart a specific function (like a protein) or store sufficient information (like a nucleic acid) to make their formation advantageous to an organism. It is, however, doubtful that polymerization could have occurred in the aqueous environment of early Earth, as it would be very rare for monomers to interact precisely enough with one another to polymerize. It is more likely that solid surfaces, especially clays, could have provided the type of environment necessary for polymerization to occur **(Figure 3.12)**. Clays consist of very thin layers of minerals separated by layers of water only a few nanometres thick. The layered structure of clay is also

charged, allowing for molecular adhesion forces to bring monomers together in precise orientations that could more readily lead to polymer formation. Clays can also store the potential energy that may have been used for energy-requiring polymerization reactions. This *clay hypothesis* is supported by laboratory experiments that demonstrate that the formation of short nucleic acid chains and polypeptides can be synthesized on a clay surface.

STUDY BREAK

1. For understanding the origins of life, what was the significance of the Miller–Urey experiment?
2. What is the difference between a reducing atmosphere and an oxidizing atmosphere?

3.3 From Macromolecules to Life

In the previous section, we discussed how processes present on early Earth could have generated macromolecules crucial to the development of life. However, if we are to develop a comprehensive model for the origin of life, we need to explain the evolution of three key attributes of a modern cell: (1) a membrane-defined compartment—the cell, (2) a system to store genetic information and use it to guide the synthesis of specific proteins, and (3) energy-transforming pathways to bring in energy from the surroundings and harness it to sustain life. In this section, we discuss possible scenarios for the evolution of these key attributes and consider a number of hypotheses, some of which are supported by laboratory experiments.

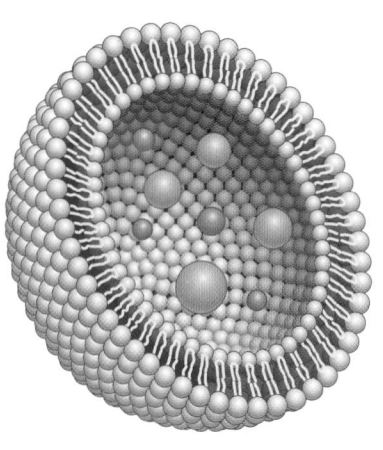

a.

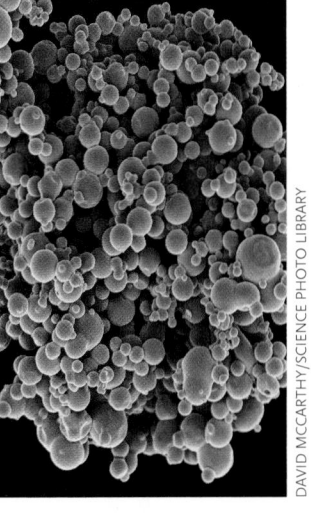

b.

DAVID MCCARTHY/SCIENCE PHOTO LIBRARY

Figure 3.13
Liposome. **(a)** An artist's rendition of a liposome, which is composed of a lipid bilayer. Liposomes can assemble spontaneously under simulated primordial conditions. **(b)** Phase micrograph of lipid vesicles assembled from phospholipids in the laboratory.

3.3a Lipid Spheres May Have Led to the Development of Cells

A critical step along the path to life is the formation of a membrane-defined compartment. Such a compartment would allow for primitive metabolic reactions to take place in an environment that is distinctly different than the external surroundings; the concentration of key molecules could be higher, and greater complexity could be maintained in a closed space. **Protobiont** is the term given to a group of abiotically produced organic molecules that are surrounded by a membrane or membranelike structure. Laboratory experiments have shown that protobionts could have formed spontaneously, that is, without any input of energy, given the conditions on primordial Earth. An early type of protobiont could have been similar to a liposome, which is a lipid vesicle in which the lipid molecules form a bilayer very similar to a cell membrane **(Figure 3.13)**. Liposomes can easily be made in the laboratory and are **selectively permeable**, allowing only some molecules to move in and out. As well, liposomes can swell and contract depending on the osmotic conditions of their environment.

Recent research from the laboratory of Jack Szostak at Harvard University has shown that the presence of clay not only catalyzes the polymerization of nucleic acids (see Section 3.2) but also accelerates the formation of lipid vesicles. As well, clay particles often become encapsulated in these vesicles, which would provide catalytically active surfaces within membrane vesicles upon which key reactions could take place. Researchers continue to experiment with producing different types of protobionts in the laboratory as a step toward understanding the origins of the first living cell. Present-day thinking is that a lipid membrane system must have evolved simultaneously with a genetic information system (see below).

3.3b RNA Can Carry Information and Catalyze Reactions

As discussed in Chapter 2, DNA is the molecule that provides every cell with the genetic instructions necessary to function. Recall as well that the information in DNA is copied into RNA, which directs protein synthesis on ribosomes. Even the simplest prokaryotic cell contains thousands of proteins, each coded by a unique DNA sequence, a gene. The flow of information from DNA to RNA to protein is common to all forms of life and is referred to as the central dogma **(Figure 3.14)**. Each step of the information flow requires the involvement of a group of proteins called enzymes, which catalyze the transcription of DNA into RNA and the translation of the RNA into protein. Enzymes are discussed in detail in Chapter 4.

A fundamental question about the flow of information from DNA to RNA to protein is: how did such a

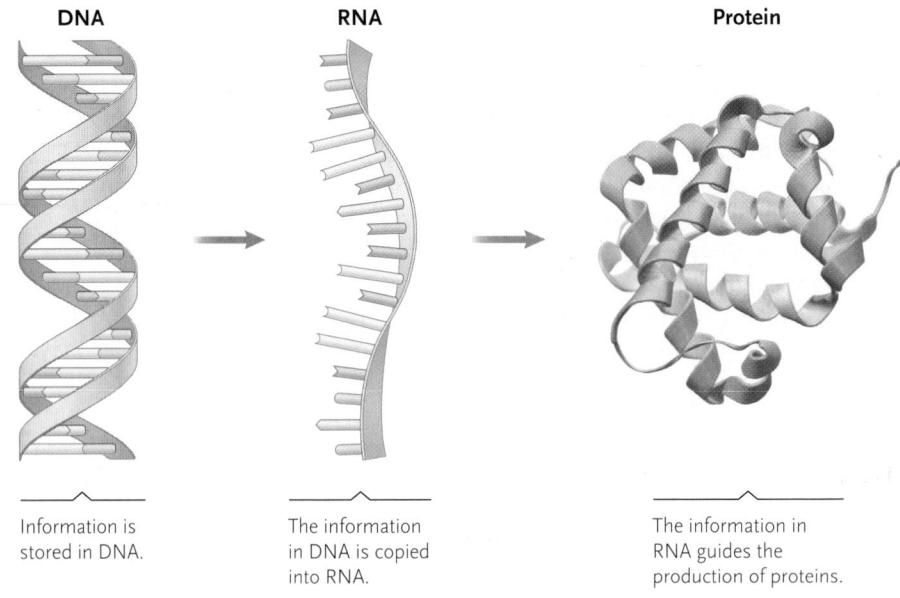

DNA

RNA

Protein

Information is stored in DNA.

The information in DNA is copied into RNA.

The information in RNA guides the production of proteins.

Figure 3.14
The central dogma. Information in DNA is used to synthesize proteins through an RNA intermediate. How did such a system evolve when the product, proteins, is required in modern-day cells to catalyze each step?

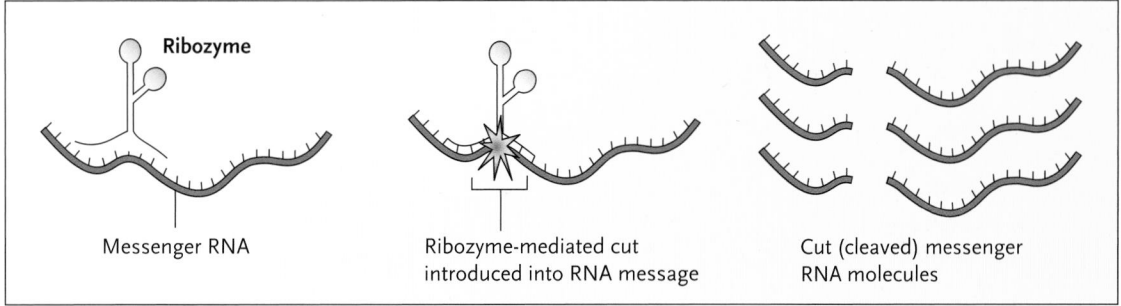

Figure 3.15

Ribozyme. An example of a ribozyme binding to an RNA molecule and catalyzing its breakage. Within a modern-day cell, such reactions may help control gene expression by altering the abundance of functional messenger RNA (mRNA) molecules.

Messenger RNA

Ribozyme-mediated cut introduced into RNA message

Cut (cleaved) messenger RNA molecules

system evolve when the products of the process, proteins, are required to catalyze each step of the process? A breakthrough in our understanding of how such a system may have evolved came in the early 1980s when Thomas Cech and Sydney Altman, working independently, discovered a group of RNA molecules that could themselves act as catalysts. This group of RNA catalysts, called **ribozymes**, can catalyze reactions on the precursor RNA molecules that lead to their own synthesis, as well as on unrelated RNA molecules **(Figure 3.15)**. Ribozymes have catalytic properties because these single-stranded molecules can fold into very specific shapes based on intramolecular hydrogen bonding or base pairing. The fact that specificity in folding imparts specificity in function is very common to proteins, especially enzymes, where precise three-dimensional shape is critical for reacting with substrate molecules. Protein folding is discussed in more detail in *The Purple Pages*.

The discovery of ribozymes revolutionized thinking about the origin of life. Instead of the contemporary system that requires all three molecules—DNA, RNA, and protein—early life may have existed in an "RNA world," where a single type of molecule, RNA, could serve as a carrier of information (due to its nucleotide sequence) and a structural/functional molecule similar to a protein (due to its ability to form unique three-dimensional shapes). Before the discovery of ribozymes, enzymes were the only known biological catalysts. For their remarkable discovery, Sydney Altman and Thomas Cech, shared the Nobel Prize in Chemistry in 1989.

3.3c RNA Is Replaced by DNA for Information Storage and Proteins for Catalysis

If life developed in an RNA world, where RNA served as both an information carrier and a catalyst, why is it that in all contemporary organisms genetic information is stored in DNA, and why do enzymes (proteins) catalyze the vast majority of biological reactions? The simple answer is that they do the respective jobs of information storage (DNA) and catalysis (protein) far better than RNA does by itself; thus, the evolution of these

molecules would have given organisms that had them a distinct advantage over others that relied solely on RNA.

A possible scenario for the development of today's system of information transfer is shown in **Figure 3.16**. The first cells may have contained only RNA, which was self-replicating and could catalyze a small number of reactions critical for survival. It is hypothesized that a small population of RNA molecules then evolved that could catalyze the formation of very short proteins before the development of ribosomes. Recall from Chapter 2 that the ribosome is the organelle in contemporary organisms required for protein synthesis. It is interesting to note that the ribosome, which plays a key role as an intermediate between RNA and protein, is composed of about two-thirds RNA and one-third protein. In fact, it is the RNA component of the ribosome, not the protein, that actually catalyzes the incorporation of amino acids onto a growing peptide chain. Thus, the ribosome can be considered a type of ribozyme.

Cells that evolved the ability to use the information present in RNA to direct the synthesis of even small proteins would be at a tremendous advantage because proteins are far more versatile than RNA molecules—for three reasons. First, the catalytic power of most enzymes is much greater than that of a ribozyme. A typical enzyme can catalyze the same reaction using

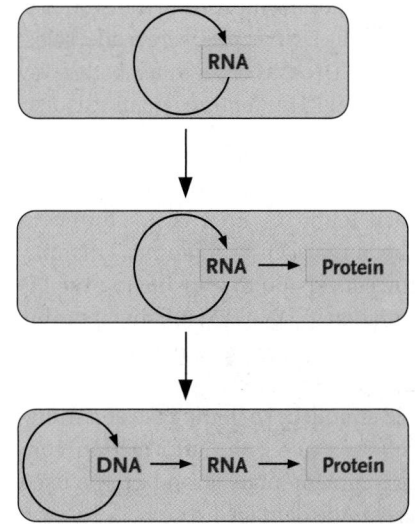

Figure 3.16

Possible scenario for the evolution of the flow of information from DNA to RNA to protein.

L1 Ligase Ribozyme

In the RNA world, RNA served as the molecule of information storage as well as structure and catalysis. To replicate RNA, individual nucleotide triphosphate monomers need to be joined, or ligated, together to form the RNA polymer. Today, this ligation reaction, carried out by a group of protein enzymes called polymerases, can result in a pool of substrate molecules many thousands of nucleotides in length. How this polymerization reaction would have been catalyzed in an RNA-only world stumped scientists for years.

Using what is called *in vitro* evolution and selection, scientists recently produced a range of synthetic ribozymes that do not currently exist in nature. One of these synthetic ribozymes is called the L1 ligase ribozyme, and it has been shown to catalyze the joining of two RNA monomers. This finding clearly suggests that, although not currently found in nature, a ribozyme capable of ligating nucleotides together may have existed on primitive Earth.

a pool of substrate molecules many thousands of times a second. By comparison, the rate of catalysis of most ribozymes is one-tenth to one-hundredth that of enzymes. Second, while the number of ribozymes is very small, a typical cell synthesizes a huge array of different proteins. Twenty different kinds of amino acids, in different arrangements, can be incorporated into a protein, whereas an RNA molecule is composed of different combinations of only four nucleotides. Third, amino acids can interact chemically with each other in bonding arrangements not possible between nucleotides. For these reasons, proteins are the dominant structural and functional molecule of a modern cell.

Continuing with the possible scenario shown in Figure 3.16, the evolution of DNA would have followed that of proteins. Compared with RNA, molecules of DNA are more structurally complex. Not only is DNA double stranded, but it also contains the sugar deoxyribose, which is more difficult to synthesize than the ribose found in molecules of RNA. A possible sequence begins with DNA nucleotides being produced by random removal of an oxygen atom from the ribose subunits of RNA nucleotides. At some point, the DNA nucleotides paired with the RNA informational molecules and were assembled into complementary copies of the RNA sequences. Some modern-day viruses carry out this RNA-to-DNA reaction using the enzyme reverse transcriptase (see Chapter 23). Once the DNA copies were made, selection may have favoured DNA, as it is a much better way to store information than RNA, for three main reasons:

- Each strand of DNA is chemically more stable, and less likely to degrade, than a strand of RNA.
- The base uracil found in RNA is not found in DNA; it has been replaced by thymine. This may be because the conversion of cytosine to uracil is a common mutation in DNA. By utilizing thymine in DNA, any uracil is easily recognized as a damaged cytosine that needs to be repaired.
- DNA is double stranded, so in the case of a mutation to one of the stands, the information contained on the complementary strand can be used to correctly repair the damaged strand.

The stability of DNA is illustrated by the fact that intact DNA can be successfully extracted from tissues that are many thousands of years old. The well-known novel and movie *Jurassic Park* are based on this demonstrated ability. By comparison, RNA needs to be quickly isolated, even from freshly isolated cells, using a strict protocol to prevent its degradation.

3.3d Simple Oxidation–Reduction Reactions Probably Preceded Metabolism

Hypotheses concerning the evolution of energy transduction and metabolism have been particularly difficult to test. Oxidation–reduction reactions were probably among the first energy-releasing reactions of the primitive cells. In our cells, we *oxidize* food molecules (e.g., sugars) and use some of the liberated energy (in the form of electrons) to *reduce* other molecules. In primitive cells, the electrons removed in an **oxidation** reaction would have been transferred in a one-step process to the substances being reduced. This, however, is not very efficient and leads to a lot of wasted energy. Over time, multi-step processes would have evolved, whereby the energy from oxidation is slowly released. A good example of the slow release of energy is cellular respiration, which is discussed in Chapter 6. The greater efficiency of stepwise energy release would have favoured development of intermediate carriers and opened the way for primitive electron transport chains.

As part of the energy-harnessing reactions, adenosine triphosphate (ATP) became established as the coupling agent that links energy-releasing reactions to those requiring energy. ATP may have first entered early cells as one of many organic molecules absorbed from the primitive environment. Initially, it was probably simply hydrolyzed into adenosine diphosphate (ADP) and inorganic phosphate, resulting in the release of energy. Later, as early cells evolved, some of the energy released during electron transfer was probably used to synthesize ATP directly from ADP and inorganic phosphate. Because of the efficiency and versatility of energy transfer by ATP, it gradually became the primary substance connecting energy-releasing and energy-requiring reactions in early cells.

STUDY BREAK

1. What are ribozymes, and what is their significance in our understanding of the origins of life?
2. In what ways is DNA better than RNA for storing genetic information?

3.4 The Earliest Forms of Life

Given hypotheses proposed for how they may have developed, what do we actually know about the earliest forms of life? In this section, we look at geological and fossil records for the earliest evidence of life. As discussed in Chapter 2, because the architecture of prokaryotic cells is the simplest known, they were most probably the first types of cells to evolve.

3.4a The Earliest Evidence of Life Is Found in Fossils

The earliest conclusive evidence of life is found in the fossilized remains of structures called stromatolites, the oldest being formed about 3.5 billion years ago. **Stromatolites** are a type of layered rock that is formed when microorganisms bind particles of sediment together, forming thin sheets **(Figure 3.17).** Stromatolites are found in habitats characterized by warm shallow water and are most common in Australia. Modern-day stromatolites are formed by the action of a specific group of photosynthetic bacteria called cyanobacteria. Because they possess a sophisticated metabolism (discussed below), cyanobacteria do not represent the earliest forms of life but rather were preceded by much simpler organisms.

Indirect (nonfossil) evidence of life existing before 3.5 billion years ago comes from research looking at the carbon composition of ancient rocks. Early photosynthetic organisms would have had the ability to take CO_2 from the atmosphere and use it to synthesize various organic molecules (sugars, amino acids, etc.). During this process, organisms would have preferentially incorporated the carbon-12 isotope (^{12}C) over other isotopes such as carbon-13 (^{13}C) (see *The Purple Pages* for a discussion of isotopes). Researchers have discovered sedimentary rocks originating from the ocean floor that contain deposits that have lower levels of the ^{13}C isotope than expected. The most likely explanation is that the deposits are actually the remains of ancient microbes. These sediments have been dated to approximately 3.9 billion years ago. If correct, this would push the origins of life to perhaps as early as 4 billion years ago—approximately 600 million years after the formation of the planet.

3.4b The First Cells Relied on Anaerobic Metabolism

The earliest forms of life were most likely **heterotrophs**, which are organisms that obtain carbon from organic molecules. Modern animals, including humans, are examples of heterotrophs; we extract energy from organic molecules such as sugars, proteins, and fats. Since the early atmosphere contained only trace amounts of oxygen, the earliest heterotrophs must have relied on anaerobic (without oxygen) forms of respiration and fermentative pathways to extract energy from organic molecules (these forms of metabolism are discussed in Chapter 6).

Compared to heterotrophs, **autotrophs** obtain carbon from the environment in an inorganic form, most often carbon dioxide. Plants and other photosynthetic organisms are the dominant autotrophs today. The earliest type of photosynthesis, which probably developed soon after heterotrophy, was *anoxygenic photosynthesis*. Today this form of autotrophy is found only in some groups of bacteria. In anoxygenic photosynthesis, compounds such as

a.

b.

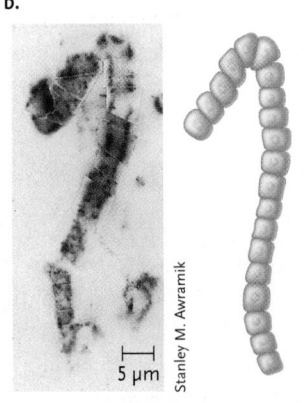

5 μm

Bill Bachmann/Science Source

Stanley M. Awramik

Figure 3.17
Early fossil evidence of life.
(a) Stromatolites exposed at low tide in Western Australia's Shark Bay. These mounds, which consist of mineral deposits made by photosynthetic cyanobacteria, are about 2000 years old; they are highly similar in structure to fossil stromatolites that formed more than 3 billion years ago. **(b)** Structures that are believed to be a strand of fossil prokaryote cells in a rock sample that is 3.5 billion years old.

hydrogen sulfide and ferrous iron (Fe^{2+}) are used as electron donors for the light reactions of photosynthesis. As we will discuss in detail in Chapter 7, the products of the light reactions, ATP and NADPH, are used to synthesize organic molecules from CO_2.

3.4c Oxygenic Photosynthesis Led to the Rise in Oxygen in the Atmosphere

Starting about 2.5 billion years ago, oxygen (O_2) levels in the atmosphere began increasing. Evidence for this comes from dating a type of sedimentary rock called banded iron **(Figure 3.18)**. Geologists believe that these distinctive striped rocks were formed in the sediments of lakes and oceans when dissolved oxygen reacted with the iron in the water, forming a red-coloured precipitate, iron oxide (rust), which ended up being incorporated into the resulting sedimentary rock formations (see Figure 3.18).

An obvious question to ask is: where did the oxygen come from? Recall from the last section that ancient stromatolites were most probably formed by the action of a group of bacteria called cyanobacteria. Unlike other autotrophs that used hydrogen sulfide or Fe^{2+} as an electron donor for photosynthesis, cyanobacteria had the remarkable ability to use an electron donor that was far more common—water **(Figure 3.19)**. The oxidation of water releases not only electrons, which can be used for photosynthetic electron transport, but also molecular oxygen. It is thought that initially the free oxygen was incorporated into mineral deposits including iron. It was only after these reservoirs became full that the oxygen started to accumulate in the atmosphere. Because it releases oxygen, photosynthesis that relies on the oxidation of water as the source of electrons is termed *oxygenic photosynthesis*.

Unlike organisms that use hydrogen sulfide or Fe^{2+}, the ability to oxidize water meant that cyanobacteria could thrive almost anywhere on the planet where there was sunlight. After all, when was the last time you crossed campus and stepped in a puddle of hydrogen sulfide? As a result of the huge ecological advantage that came with being able to oxidize water, the abundance of cyanobacteria on the planet exploded and they quickly became a dominant form of life. Astonishingly, although it evolved perhaps as early as 3.5 billion years

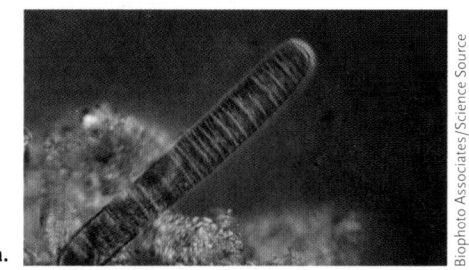

a.

b. $2H_2O + \dfrac{\text{light}}{\text{energy}} \longrightarrow 4H^+ + 4e^- + O_2$

Figure 3.19
Cyanobacteria. **(a)** Micrograph of a filamentous cyanobacterium of the genus *Nostoc*. **(b)** Ancient cyanobacteria, like modern photosynthetic organisms, were able to use water as an electron donor for photosynthesis. A consequence was the formation of oxygen (O_2), which accumulated in the atmosphere.

ago, oxygenic photosynthesis remains the dominant form of photosynthesis and is used by all plants and algae, as well as present-day cyanobacteria.

3.4d Could Life Have Come to Earth from Space?

Some scientists believe that instead of the evolution of life through processes mentioned above, life on Earth had an extraterrestrial origin. Panspermia is the name given to the hypothesis that very simple forms of life are present in outer space and that these may have "seeded" early Earth. As scientists have never found any life existing outside of Earth, there is no evidence to directly support this hypothesis. However, two points of discussion lend support to the idea of an extraterrestrial origin of life on Earth.

First, although life seems very complex, it arose relatively quickly after the formation of Earth. Earth formed 4.6 billion years ago, and we have clear fossil evidence of life dated to about 3.5 billion years ago and chemical evidence to about 3.9 billion years ago. Given that primordial Earth had to cool after being formed before life could develop, some scientists argue that this represents too short a period of time for life to develop solely by abiotic processes occurring on the cooling planet.

Second, research in the past few decades has shown that life is far more resilient than previously thought and could possibly survive for millennia in space. An *extremophile* is the general term given to an organism that is found growing in environments that are lethal to most other organisms. Most extremophiles—mainly bacteria and archaea—can thrive under conditions such as extreme temperature, high pressure, high salinity, and high radiation levels. Research on extremophiles in recent years has shown that life is far more resilient than previously thought and thus it is quite possible that organisms may be able to survive in a dormant state in interstellar space. Prolonged dormancy is a property of reproductive structures called spores, which are produced

Figure 3.18
Banded iron. The rust layers in banded iron formations provide evidence for the rise of atmospheric oxygen.

Lyle Whyte, Astrobiologist, *McGill University*

The heightened research interest into extremophiles coupled with technological advances that are driving the robotic exploration of Mars and other planets has spurred the development of the multidisciplinary science of astrobiology. Broadly defined, astrobiology is the study of the origin, evolution, distribution, and future of life in the universe. The field encompasses the search for habitable environments within and outside our solar system and the search for evidence of prebiotic chemistry and life on Mars and elsewhere. As well, astrobiology includes laboratory and field research into the origins and early evolution of life on Earth and studies of the potential for life to adapt to

challenges on Earth and in space.

Canada is at the forefront of astrobiology research, as the government has recently established the Canadian Astrobiology Training Program. The program will create the first cross-disciplinary, multi-institutional undergraduate, graduate, and postdoctoral training program in astrobiology. This initiative brings together researchers at five institutions (McGill University, McMaster University, Western University, University of Winnipeg, and University of Toronto) with expertise in fields as diverse as geology, chemistry, physics, astronomy, microbiology, and robotics.

The head of the Canadian Astrobiology Training Program is Lyle Whyte

of McGill University. Whyte is a Canada Research Chair in Environmental Microbiology. His research examines microbial biodiversity and ecology in unique Canadian high-Arctic ecosystems and is contributing significantly to the knowledge of the diversity, abundance, and critical roles played by microorganisms in polar regions. His research is providing new insights into microbial life at subzero temperatures and their role in global biogeochemical cycling. Whyte and his colleagues do considerable field research in the Canadian Arctic, which is considered by NASA and the Canadian Space Agency to be one of the best sites around the globe to mimic conditions on Mars.

by a number of bacteria and simple eukaryotes. Bacterial spores, in particular, are highly resistant to changes in the external environment and can be restored to active growth after exposure to water and moderate temperatures. Extremophiles have become a major area of research and have led, in part, to the development of the science of astrobiology (see "People behind Biology," Box 3.2).

3.4e All Present-Day Organisms Are Descended from a Common Ancestor

Based on comparing the sequence of ribosomal RNA, all present-day organisms can be categorized into one of three *domains*: Archaea, Bacteria, and Eukarya **(Figure 3.20)**. Because both archaea and bacteria share

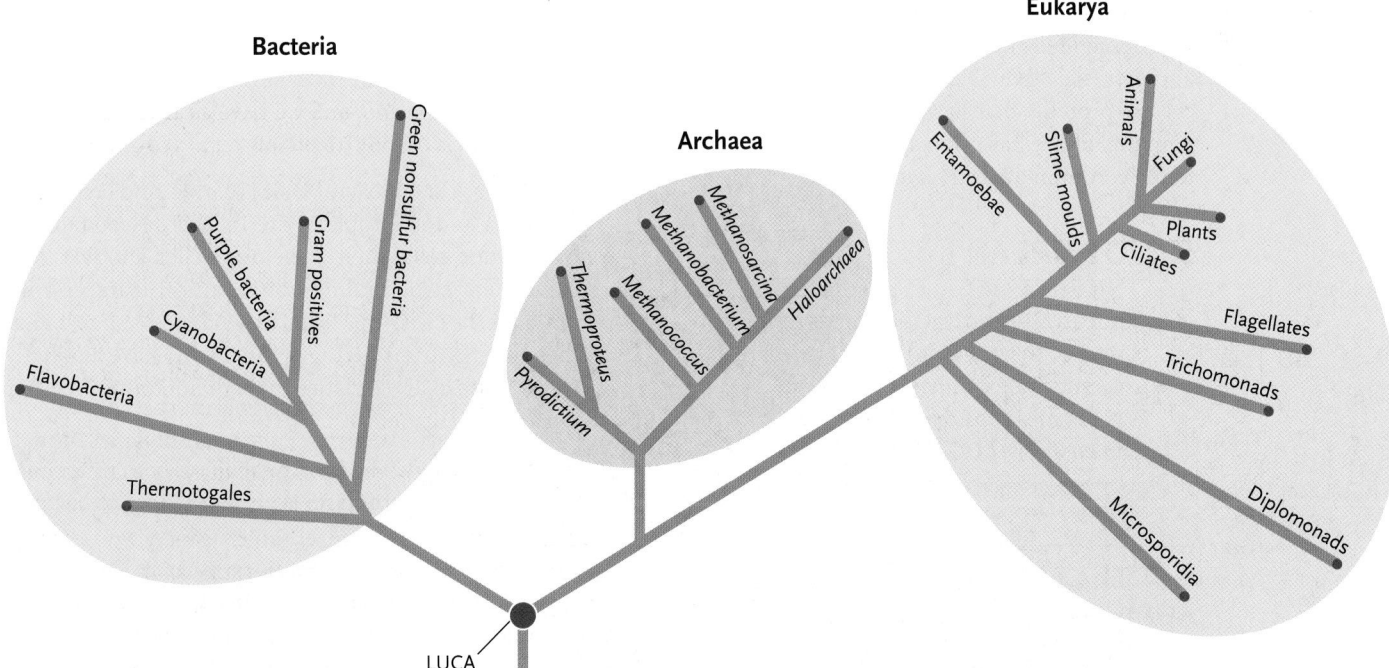

Figure 3.20

The three domains of life: Bacteria, Archaea, and Eukarya. This evolutionary tree is based on the sequencing of ribosomal RNA. All present-day organisms are thought to be descended from a common ancestor—the last universal common ancestor or LUCA.

a similar cell architecture (see Chapter 2), including the lack of a nucleus, they are often referred to as prokaryotes. But as discussed in detail in Chapter 22, archaea and bacteria are quite distinct, and evidence indicates that they are not evolutionarily closely related. In fact, molecular evidence indicates that archaea are more closely related to eukaryotes (the Eukarya) than to bacteria (see Figure 3.20, p. 65).

As will be discussed in detail in subsequent chapters, there are clear distinctions in structure and function among archaea, bacteria, and eukaryotes. This said, all life-forms currently on Earth share a remarkable set of common attributes. Perhaps the most fundamental of these are the following: (1) cells made of lipid molecules brought together forming a bilayer; (2) a genetic system based on DNA; (3) a system of information transfer—DNA to RNA to protein; (4) a system of protein assembly from a pool of amino acids by translation using messenger RNA (mRNA) and **transfer RNA** (tRNA) using ribosomes; (5) reliance on proteins as the major structural and catalytic molecule; (6) use of ATP as the molecule of chemical energy; and (7) the breakdown of glucose by the metabolic pathway of glycolysis to generate ATP.

The fact that these seven attributes are shared by all life on Earth suggests that all present-day organisms descended from a common ancestor (Figure 3.20, p. 65) that had all of these attributes. This is not to say that life evolved only once. It is quite possible that life arose many times on early Earth, each form perhaps having some of the attributes listed above. The similarities across all domains of life present today indicate, however, that only one of these primitive life-forms has descendants that survive today. We call the original life-form from which all archaea, bacteria, and eukaryotes are descended LUCA for last universal common ancestor (see Figure 3.20). Recent sequence analysis of certain proteins that have representatives in all three domains of life has given strong quantitative support to the common-ancestry hypothesis.

STUDY BREAK

1. Compared to anoxygenic photosynthesis, what is the ecological advantage to an organism of oxygenic photosynthesis?
2. What is meant by the term *last universal common ancestor (LUCA)*?

3.5 The Eukaryotic Cell and the Rise of Multicellularity

There is general agreement within the scientific community that the oldest fossils of eukaryotes are about 2.1 billion years old. Understandably, the very first eukaryotes may have appeared earlier. In fact, there is chemical evidence of eukaryotes existing as early as 2.5 billion years ago.

Present-day eukaryotic cells have two major characteristics that distinguish them from either the archaea or bacteria: (1) the separation of DNA and cytoplasm by a nuclear envelope and (2) the presence in the cytoplasm of membrane-bound compartments with specialized metabolic and synthetic functions—mitochondria, chloroplasts, the endoplasmic reticulum (ER), and the Golgi complex, among others. Hypotheses for how various eukaryotic structures and functions arose abound, some with very strong scientific evidence, others backed mostly with conjecture. In this section, we discuss how eukaryotes most probably evolved from associations of prokaryotic cells, ending with a discussion of the rise of multicellular eukaryotes.

3.5a The Theory of Endosymbiosis Suggests That Mitochondria and Chloroplasts Evolved from Ingested Prokaryotes

One feature that is found in virtually all eukaryotic cells is energy-transforming organelles: mitochondria and chloroplasts. A large amount of evidence indicates that mitochondria and chloroplasts are actually descended from free-living prokaryotic cells **(Figure 3.21)**: mitochondria are descended from aerobic (with oxygen) bacteria, while chloroplasts are descended from cyanobacteria. The established model of **endosymbiosis** states that the prokaryotic ancestors of modern mitochondria and chloroplasts were engulfed by larger prokaryotic cells, forming a mutually advantageous relationship called a **symbiosis**. Slowly, over time, the **host** cell and the endosymbionts became inseparable parts of the same organism.

3.5b Several Lines of Evidence Support the Theory of Endosymbiosis

If the theory of endosymbiosis is correct and both mitochondria and chloroplasts are indeed descendants of prokaryotic cells, then these organelles should share some clear structural and biochemical features with prokaryotic cells. Six lines of evidence suggest that these energy-transducing organelles do have distinctly prokaryotic characteristics that are not found in other eukaryotic organelles:

1. **Morphology.** The form or shape of both mitochondria and chloroplasts is similar to that of bacteria and archaea.
2. **Reproduction.** A cell cannot synthesize a mitochondrion or a chloroplast. Just like free-living prokaryotic cells, mitochondria and chloroplasts are derived only from pre-existing mitochondria and chloroplasts. Both chloroplasts and mitochondria divide by binary fission, which is how bacteria and archaea divide (see Chapter 22).

3. **Genetic information.** If the ancestors of mitochondria and chloroplasts were free-living cells, then these organelles should contain their own DNA. This is indeed the case. Both mitochondria and chloroplasts contain their own DNA, which contains protein-coding genes that are essential for organelle function. As with bacteria and archaea, the DNA molecule in mitochondria and chloroplasts is circular, while the DNA molecules in the nucleus are linear.

4. **Transcription and translation.** Both chloroplasts and mitochondria contain a complete transcription and translational machinery: genes encoded by the organelle genomes are translated into mRNA and translated on the ribosomes, messenger RNA (mRNA), and transfer RNA (tRNA) necessary to synthesize the proteins encoded by their DNA. The ribosomes of mitochondria and chloroplasts are very similar to the type found in bacteria (see point 6 on this page).

5. **Electron transport.** Similar to free-living prokaryotic cells, both mitochondria and chloroplasts have electron transport chains (ETCs) used to generate chemical energy. The ETCs of bacteria and archaea are found in the plasma membrane, and for such cells, swallowed up by endosymbiosis, this membrane is inside the membrane of the endocytic vesicle. Indeed, both mitochondria and chloroplasts have double membranes, and it is the inner membrane that contains the ETC.

6. **Sequence analysis.** Sequencing of the RNA that makes up the ribosomes of chloroplasts and mitochondria firmly establishes that they belong on the bacterial branch of the tree of life (see Figure 3.20, p. 65). Chloroplast ribosomal RNA is most similar to that of cyanobacteria, while mitochondrial ribosomal RNA is most similar to that of proteobacteria.

Whereas virtually all eukaryotic cells contain mitochondria, only plants and algae contain both mitochondria and chloroplasts. This fact indicates that endosymbiosis occurred in stages (see Figure 3.21), with the event leading to the evolution of mitochondria occurring first. Once eukaryotic cells with the ability for aerobic respiration developed, some of these became photosynthetic after taking up cyanobacteria, evolving into the plants and algae of today.

3.5c Horizontal Gene Transfer Followed Endosymbiosis

The term **genome** is defined as the complete complement of an organism's genetic material. For eukaryotes, it is common to distinguish between the DNA found in the nucleus (the nuclear genome) and the DNA that resides in either the mitochondrion (mitochondrial genome) or the chloroplast (chloroplast genome).

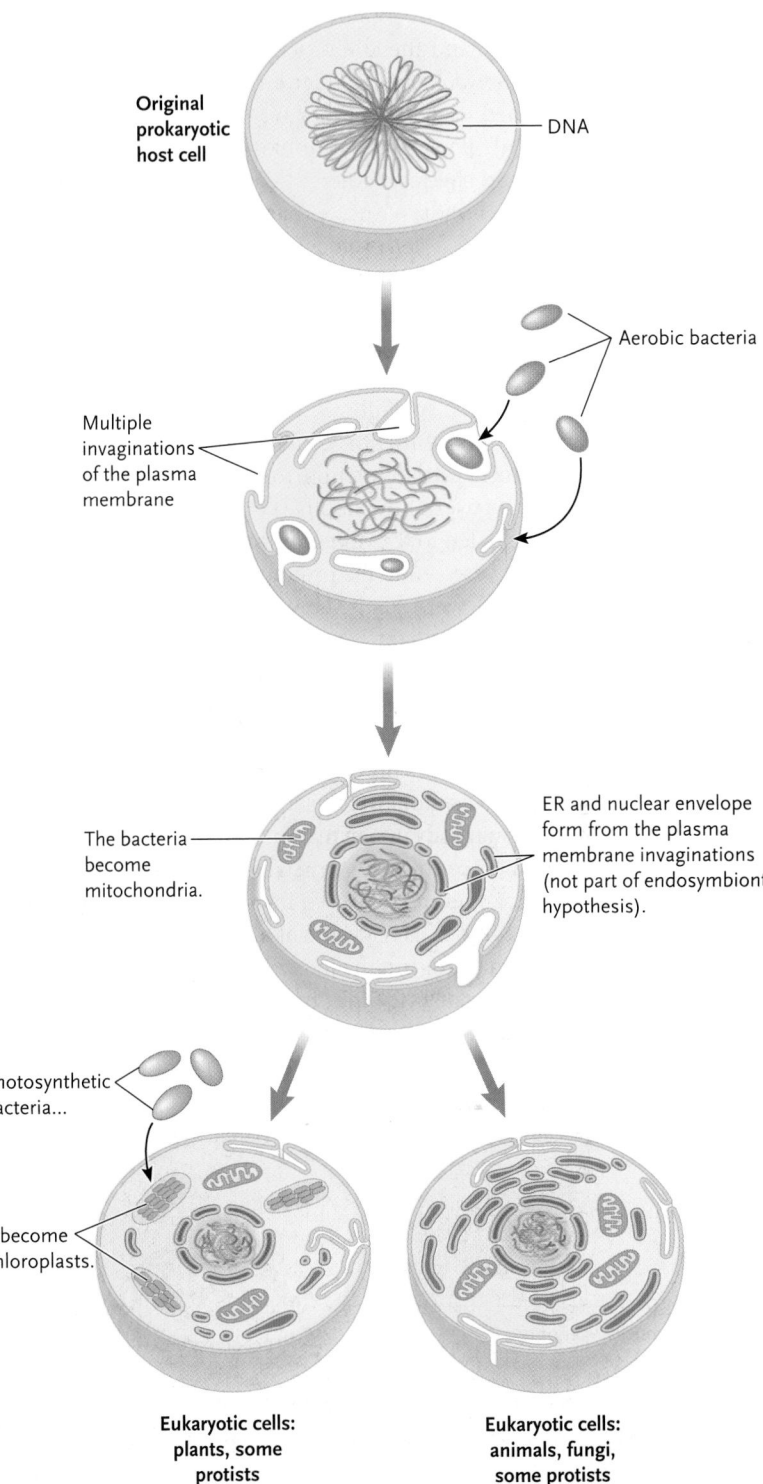

Figure 3.21
The theory of endosymbiosis. The mitochondrion is thought to have originated from an aerobic prokaryote that lived as an endosymbiont within an anaerobic prokaryote. The chloroplast is thought to have originated from a photosynthetic prokaryote that became an endosymbiont within an aerobic cell that had mitochondria.

A typical bacterium has a genome that contains about 3000 protein-coding genes. If both mitochondria and chloroplasts had once been free-living prokaryotic cells, then their genomes should have roughly the same number of genes—interestingly, they don't. The human mitochondrial genome comprises only 37 genes, and the chloroplast genome of the green alga *Chlamydomonas reinhardtii* has 99 genes. What happened to all those other genes?

Following endosymbiosis, the early eukaryotic cell would have contained at least two (nucleus and protomitochondria) and perhaps three (nucleus, protomitochondria, and protochloroplast) compartments—each with its own complete genome. These compartments and genomes would have functioned independently, each coding for proteins required for their own structure and function, just like free-living organisms. This contrasts strongly with a modern eukaryotic cell, where the function of the cell is highly integrated—mitochondrial function, for example, is strongly linked to the overall metabolism of the cell. Two major processes led to this integration. First, some of the genes that were within the protomitochondrion or protochloroplast were lost. Many of these genes would have been redundant, as the nucleus would already have genes that encode proteins with the same function. Second, many of the genes within the protomitochondria and protochloroplast were relocated to the nucleus. This process, called horizontal gene transfer (HGT), is thought to have occurred because of the evolutionary advantage that the early eukaryotic cell would gain by centralizing crucial genetic information in one place, the nucleus. It's important to realize that the outcome of HGT was not a change in gene function, only a change in the location of the gene—the nucleus as opposed to the mitochondrion or chloroplast **(Figure 3.22)**. As a side note, HGT doesn't pertain only to endosymbiotic gene transfer but to any movement of genes between organisms other than to offspring. HGT is seen as a major mechanism of diversification of genomes and thus evolution, especially in archaea and bacteria.

In a typical eukaryotic cell today, over 90% of the proteins required for mitochondrial or chloroplast function are encoded by genes that are found in the nucleus. To go along with this change of location, a large protein trafficking and sorting machinery had to evolve (Figure 3.22). Following transcription in the nucleus and translation on cytosolic ribosomes, proteins destined for the chloroplast or mitochondrion need to be correctly sorted and imported into these energy-transducing organelles, where they are trafficked to the correct location.

A major unanswered question that is being actively studied is: why do both mitochondria and chloroplasts still retain a genome—why haven't all genes moved to the nucleus? One possibility is that perhaps gene transfer is not yet complete. This seems to be the case for many chloroplast genomes. Related plant species can differ with regard to the location of a particular gene. It is not uncommon for one species to have the gene in the nucleus and a related species to have the same gene localized to the chloroplast. Another hypothesis is that since most of the genes that are retained by the mitochondrion or chloroplast code for proteins involved in electron transport, the tight regulation of these genes, which is difficult to do if the genes are in the nucleus, is essential to maintaining optimal rates of energy transformation.

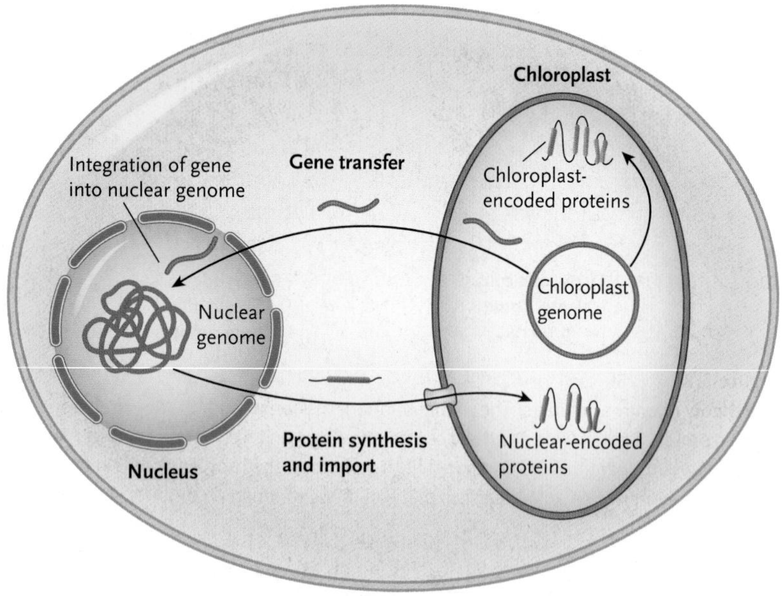

3.5d The Endomembrane System May Be Derived from the Plasma Membrane

Recall from Chapter 2 that in addition to energy-transforming organelles (mitochondria and chloroplasts), eukaryotic cells are characterized by an **endomembrane system**—a collection of internal membranes that divide the cell into structural and functional regions. These include the nuclear envelope, the ER, and the Golgi complex. As we have just seen, there is very strong evidence in support of the endosymbiotic origin of chloroplasts and mitochondria; however, the origin of the endomembrane system remains unclear. The most widely held hypothesis is that it is derived from the infolding of the plasma membrane **(Figure 3.23)**. Researchers hypothesize that in cell lines leading from prokaryotic cells to eukaryotes, pockets of the plasma membrane may have extended inward and surrounded the nuclear region. Some of these membranes fused around the DNA, forming the nuclear envelope, which defines the nucleus. The remaining membranes formed vesicles in the cytoplasm that gave rise to the ER and the Golgi complex.

Figure 3.22

Horizontal gene transfer. Over evolutionary time, some protein-coding genes that were once part of the chloroplast or mitochondrial genome have been relocated to the nuclear genome. Following transcription of these genes, translation occurs in the cytosol before protein import into the organelle (mitochondrion or chloroplast).

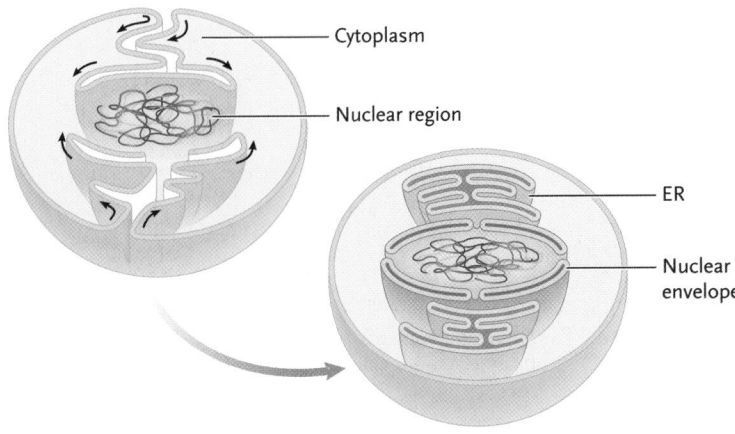

Figure 3.23
A hypothetical route for formation of the nuclear envelope and ER, through segments of the plasma membrane that were brought into the cytoplasm by endocytosis.

Labels on figure: Cytoplasm, Nuclear region, ER, Nuclear envelope

3.5e Solving an Energy Crisis May Have Led to Eukaryotes

Bacteria and archaea outnumber eukaryotes on the planet by a huge margin. Compared to eukaryotes, archaea and bacteria show remarkable biochemical flexibility, being able to use an assortment of molecules as sources of energy and carbon and thrive in harsh environments uninhabitable to eukaryotes. That said, prokaryotic cells are simple—they lack the complexity of eukaryotes, which evolved into a tremendous diversity of forms, including plants, fungi, and animals. Within each of these groups are cells with remarkable specialization in form and function. Contrast this to archaea and bacteria, which have remained remarkably simple even though they evolved as early as 4 billion years ago.

The reason that bacteria and archaea have remained very simple is that increased complexity requires increased energy, and while eukaryotic cells can generate huge amounts of it, prokaryotic cells cannot. Mitochondria, like their aerobic progenitor bacteria, undergo aerobic respiration, which generates much greater amounts of ATP from the breakdown of organic molecules than pathways of anaerobic metabolism (this is discussed further in Chapter 6). As well, while a typical aerobic bacterium relies on its plasma membrane for many functions, including nutrient and waste transport and energy production, a typical eukaryotic cell contains hundreds of mitochondria, each having a huge internal membrane surface area dedicated to generating ATP.

The ability of early eukaryotes to generate more ATP led to remarkable changes. Cells could become larger, as now there was enough energy to support a greater volume. And cells could become more complex. This complexity comes about by being able to support a larger genome that codes for a greater number of proteins. By overcoming the energy barrier, eukaryotes had the energy to support a wider variety of genes that led to what we know today to be eukaryotic-specific traits such as the **cell cycle**, sexual reproduction, phagocytosis, endomembrane trafficking, the nucleus, and multicellularity.

3.5f The Evolution of Multicellular Eukaryotes Led to Increased Specialization

One of the most profound transitions in the history of life was the evolution of multicellular eukaryotes. Clear evidence of multicellular eukaryotes, primarily small algae, appears in the fossil record starting about 1.2 billion years ago. It is easy to see how multicellularity could have developed. Perhaps a group of individual cells of a species came together to form a colony, or a single cell divided and the resulting two cells remained together. In the most simplest of multicellular organisms, all cells are structurally and functionally autonomous (independent). This gave way to a key trait of more advanced multicellular organisms: division of labour. That is, the cells were not functionally identical and thus usually not structurally similar. Some cells may specialize in harvesting energy, for example, whereas others may serve a specific role in the motility of the organism. In a multicellular system, the cells cooperate with one another for the benefit of the entire organism. Over evolutionary time, this specialization of cell function led to the development of the specialized tissues and organs that are so clearly evident in larger eukaryotes.

Like the earliest forms of life, there is little, if any, evidence in the fossil record of the earliest multicellular organisms. How they arose and developed is still an area of intense research. It is thought, however, that multicellularity arose more than once, most probably independently along the lineages leading to fungi, plants, and animals. A very useful model for the study of multicellularity is found in a group of green algae called the volvocine. All of the members of this group are evolutionarily closely related and span the full range of size and complexity, from the unicellular *Chlamydomonas*, through various colonial genera, to the multicellular *Volvox* **(Figure 3.24, p. 70)**. Unlike a true multicellular organism, a cell colony is a group of cells that are all of one type; there is no specialization in cell structure or function. *Volvox* consists of a sphere of two to three thousand small, flagellated *Chlamydomonas*-like cells that provide the individual *Volvox* with the ability to move. In addition, within the sphere lie about 16 large nonmotile cells that serve a specialized role in reproduction.

Figure 3.24

Differences in degree of multicellularity among volvocine algae.

a.
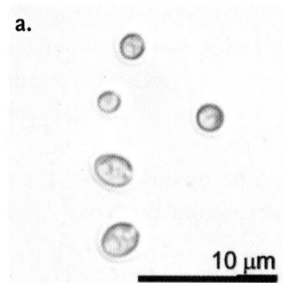
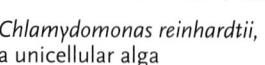
10 μm

Chlamydomonas reinhardtii, a unicellular alga

b.
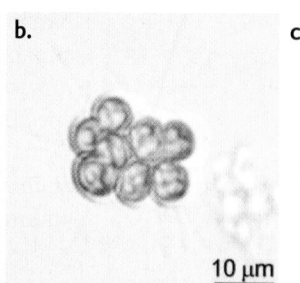
10 μm

Gonium pectorale, a group of eight undifferentiated cells

c.
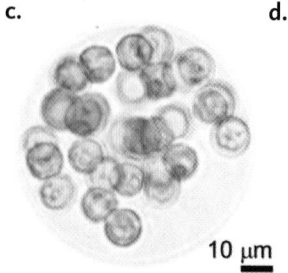

10 μm

Eudorina elegans, a spherical colony of undifferentiated cells

d.
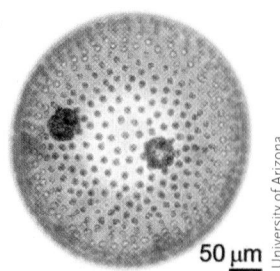
50 μm

Volvox aureu, smaller somatic cells and a few reproductive cells

Cristian A. Solari, University of Arizona

3.6 The Search for Extraterrestrial Life

Overall, the events outlined in this chapter may seem highly improbable: the formation of a habitable planet, followed by the abiotic synthesis of organic molecules, the development of the first cells, the development of DNA, RNA, proteins, and pathways of energy acquisition—all the way to the development of multicellular eukaryotes. Improbable? Perhaps. But we must keep in mind that these events took place over an almost unimaginable length of time—4.6 billion years! And as scientist and author George Wald of Harvard University put it, given so much time "the impossible becomes possible, the possible probable, and the probable virtually certain."

Most scientists maintain that the evolution of life on Earth was an inevitable outcome of the initial physical and chemical conditions on primordial Earth, brought about by its position relative to the Sun. If that is the case—that all you need is a planet in the habitable zone, then what is the probability of life existing elsewhere in our galaxy? Let's do a little arithmetic. The Milky Way galaxy contains an estimated 100 billion stars. Because the formation of planetary systems is thought to be a normal consequence of star formation, let's conservatively estimate that 50% of stars in the galaxy have planets. That would give us about 50 billion planetary systems. Given that perhaps two planets around each star would fall within the habitable zone, this would put the total number of planets able to support life within our galaxy at 100 billion! Now of course the number of those planets that actually go on to develop life would be much less and the proportion that develop intelligent life and communicating civilizations would reduce the number even further. But even still, it is distinctly possible that

a.

NASA

b.
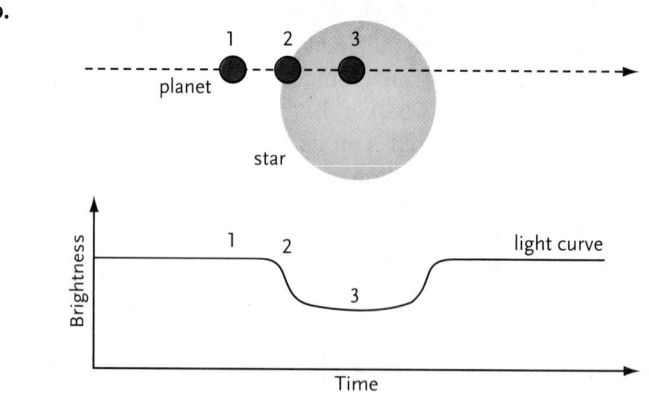
planet

star

Brightness

1   2   3

1   2

3

light curve

Time

Figure 3.25

Search for Earth-like planets. **(a)** Artist's depiction of the Kepler spacecraft, which is equipped with a photometer to detect Earth-like planets. **(b)** The transit method relies on measuring the very small but regular changes in the brightness of a star caused by an orbiting (transiting) planet.

the galaxy is teeming with advanced civilizations. Now what about finding them.

The search for extraterrestrial life is hampered by the incredible vastness of space. The closest star outside our solar system is some 40 trillion kilometres away, and using the fastest spacecraft humankind currently has, it would take a staggering 150 thousand years to get there! For this reason, the search for extraterrestrial life, which started about 50 years ago, has primarily focused on listening for the distinct signals of a communicating civilization using the science of radio astronomy. This type of detection has a huge advantage over spaceflight because radio waves travel at the speed of light. While using radio astronomy to detect signals from extraterrestrial civilizations continues, more recent initiatives are centred on the detection of Earth-like planets orbiting other stars. One method for this is employed in the NASA Kepler Mission that was launched in 2009 **(Figure 3.25, p. 70).** This project, named after the German astronomer Johannes Kepler, is a space observatory designed to continuously monitor the brightness of over 145 thousand stars. Armed with a powerful photometer, the observatory can detect the very faint but regular fluctuations in the brightness of stars. In what is termed the *transit method,* the photometer detects the slight but very regular changes in brightness of a star that are caused by a planet moving in front of and then behind the star relative to the position of the detector. Even before the Kepler Mission started, Earthbound observatories had identified over 400 extrasolar planets (planets outside our solar system). As of 2013, the Kepler mission had identified over 2740 planets, 618 of which are part of systems containing multiple planets.

STUDY BREAK

1. What is the evidence in support of endosymbiosis?
2. What are the key traits of a multicellular organism?

Review

aplia

To access course materials such as Aplia and other companion resources, please visit www.NELSONbrain.com.

3.1 What Is Life?

- All forms of life share seven characteristics: order, energy utilization, homeostasis, response to stimuli, growth and development, reproduction, and evolution.
- While a virus has some of the characteristics of life, these require it to infect living cells. Because of this it is not considered a form of life.
- The characteristics of life are referred to as emergent because they come about, or emerge, from many simpler interactions that, in themselves, do not have the properties found at the higher levels.

3.2 The Chemical Origins of Life

- Earth and the rest of the solar system were formed about 4.6 billion years ago.
- Life evolved on Earth in part because the planet is situated at a distance from the Sun so that water can exist in a liquid state. Earth is within the habitable zone of the solar system.
- The Oparin–Haldane hypothesis maintains that the organic molecules that formed the building blocks of life, such as amino acids, could have been formed given the conditions that prevailed on primitive Earth, including a reducing atmosphere that lacked oxygen.
- The Miller–Urey experiment demonstrated that abiotic synthesis of biologically important molecules is possible.

- The key macromolecules of life, such as proteins and nucleic acids, are polymers that were not formed by the Miller–Urey experiment. Instead, it is thought that polymerization reactions could have occurred on solid surfaces, such as clay.

3.3 From Macromolecules to Life

- The spontaneous formation of lipid vesicles (liposomes) may have served as the first membrane-bound compartments that developed into the first cells.
- Ribozymes are a group of RNA molecules that can catalyze specific reactions. Because they can store information and drive catalysis, it is thought that RNA was the first molecule.
- Because of their greater diversity and much higher rate of catalysis, proteins became the dominant structural and functional macromolecule of all cells.
- DNA is more stable than RNA and thus evolved as a better repository of genetic information.
- Early metabolism was probably based on simple oxidation–reduction reactions.

3.4 The Earliest Forms of Life

- Stromatolites dated to as early as 3.5 billion years ago represent the earliest fossil evidence of life. Chemical evidence suggests life may have originated 3.9 billion years ago.
- Panspermia is the hypothesis that very simple forms of life are present in space and seeded Earth soon after it cooled.
- Some early cells developed the capacity to carry out photosynthesis using water as an electron donor; the oxygen produced as a by-product accumulated, and

the oxidizing character of Earth's atmosphere increased. From this time on, organic molecules produced in the environment were quickly broken down by oxidation, and life could arise only from pre-existing life, as in today's world.

3.5 The Eukaryotic Cell and the Rise of Multicellularity

- The energy-transducing organelles—the chloroplasts and the mitochondria—are thought to have been derived from free-living prokaryotic cells.
- According to the theory of endosymbiosis, mitochondria developed from ingested aerobic bacteria; chloroplasts developed from ingested cyanobacteria.
- Following endosymbiosis, genes residing in the mitochondria and chloroplasts moved to the nucleus. This is a type of horizontal gene transfer (HGT).
- Eukaryotic cells possess an endomembrane system that probably evolved from infolding of the plasma membrane. The endomembrane system consists of the nuclear envelope, the endoplasmic reticulum (ER), and the Golgi complex.

- Eukaryotic cells are more complex than bacteria or archaea because mitochondria provide them with more energy.
- Multicellular eukaryotes probably evolved by differentiation of cells of the same species that had congregated into colonies. Multicellularity evolved several times, producing lineages of several algae and ancestors of fungi, plants, and animals.

3.6 The Search for Extraterrestrial Life

- The Milky Way galaxy contains about 100 billion stars. If 50% of those had planetary systems, and in each system 2 planets were within the habitable zone, that's potentially 100 billion planets able to sustain life.
- Interstellar distances are too great to use spacecraft to search for extraterrestrial life. Instead radio astronomy is used to listen for signals of intelligent civilization.
- Using powerful photometers, observatories in space and on the ground search for extrasolar planets by detecting the small but very regular changes in the brightness of a star caused by a transiting planet.

Questions

Self-Test Questions

1. Why are viruses not considered a form of life?
   a. They don't have a nucleus.
   b. They are not made of protein.
   c. They lack ribosomes.
   d. They cannot evolve.
   e. Thy lack nucleic acid.

2. According to the Oparin–Haldane hypothesis, what was the composition of the primordial atmosphere?
   a. water, molecular nitrogen (N_2), and carbon dioxide
   b. molecular hydrogen (H_2), water, ammonia, and methane
   c. water, molecular oxygen (O_2), and ammonia
   d. water, argon, and neon

3. Clay may have played an important role in what aspect of the development of life?
   a. formation of monomers such as amino acids
   b. formation of polymers such as short proteins or nucleic acids
   c. formation of membrane-bound compartments such as liposomes
   d. formation of multicellular organisms
   e. Both a and c are correct.

4. The Miller–Urey experiment was a huge breakthrough in our understanding of the origins of life. What was its major conclusion?
   a. Abiotic synthesis of molecules requires oxygen (O_2).
   b. Biological molecules could be formed without energy.
   c. Proteins could be synthesized without ribosomes.
   d. Abiotic synthesis of amino acids was possible.

5. Which of the following list of events is in the correct order of first appearance?
   a. O_2 in the atmosphere, anoxygenic photosynthesis, aerobic respiration, oxygenic photosynthesis
   b. oxygenic photosynthesis, anoxygenic photosynthesis, aerobic respiration, O_2 in the atmosphere
   c. anoxygenic photosynthesis, oxygenic photosynthesis, O_2 in the atmosphere, aerobic respiration
   d. aerobic respiration, O_2 in the atmosphere, oxygenic photosynthesis, anoxygenic photosynthesis

6. Which of the following statements about ribozymes is correct?
   a. They are composed of only RNA.
   b. They are able to catalyze reactions faster than enzymes.
   c. They were present only in ancient cells.
   d. Like proteins they are polymers of amino acids.

7. Why did DNA replace RNA as the means to store genetic information?
   a. The sugar ribose, present in DNA but not RNA, is less prone to breakdown.
   b. DNA contains uracil, which is more stable than the thymine present in RNA.
   c. Unlike RNA, DNA can exist in very complex three-dimensional shapes, which are very stable.
   d. The presence of complementary strands in DNA means that single base mutations can be easily repaired.

8. As part of the evolution of eukaryotic cells, infolding of the plasma membrane is thought to have led to the formation of which of the following?
   a. ribosomes
   b. microtubules
   c. mitochondria
   d. chromosomes
   e. ER

9. Which of the following statements supports the theory of endosymbiosis?
   a. Mitochondria contain proteins.
   b. Both mitochondria and chloroplasts possess their own genomes.
   c. Both mitochondria and chloroplasts are surrounded by a membrane.
   d. The nuclear envelope is derived from infolding of the plasma membrane.

10. What was an outcome of HGT?
   a. import of proteins into mitochondria
   b. increase in size of the chloroplast genome

   c. relocation of proteins once localized to the mitochondria to the nucleus
   d. decrease in the number of proteins in the chloroplast

Questions for Discussion

1. What evidence supports the idea that life originated through abiotic chemical processes?

2. Most scientists agree that life on Earth can arise only from pre-existing life, but also that life could have originated spontaneously on primordial Earth. Can you reconcile these seemingly contradictory statements?

3. What conditions would likely be necessary for a planet located elsewhere in the universe to evolve life similar to that on Earth?

4. What drove the evolution of the eukaryotic cell?

4

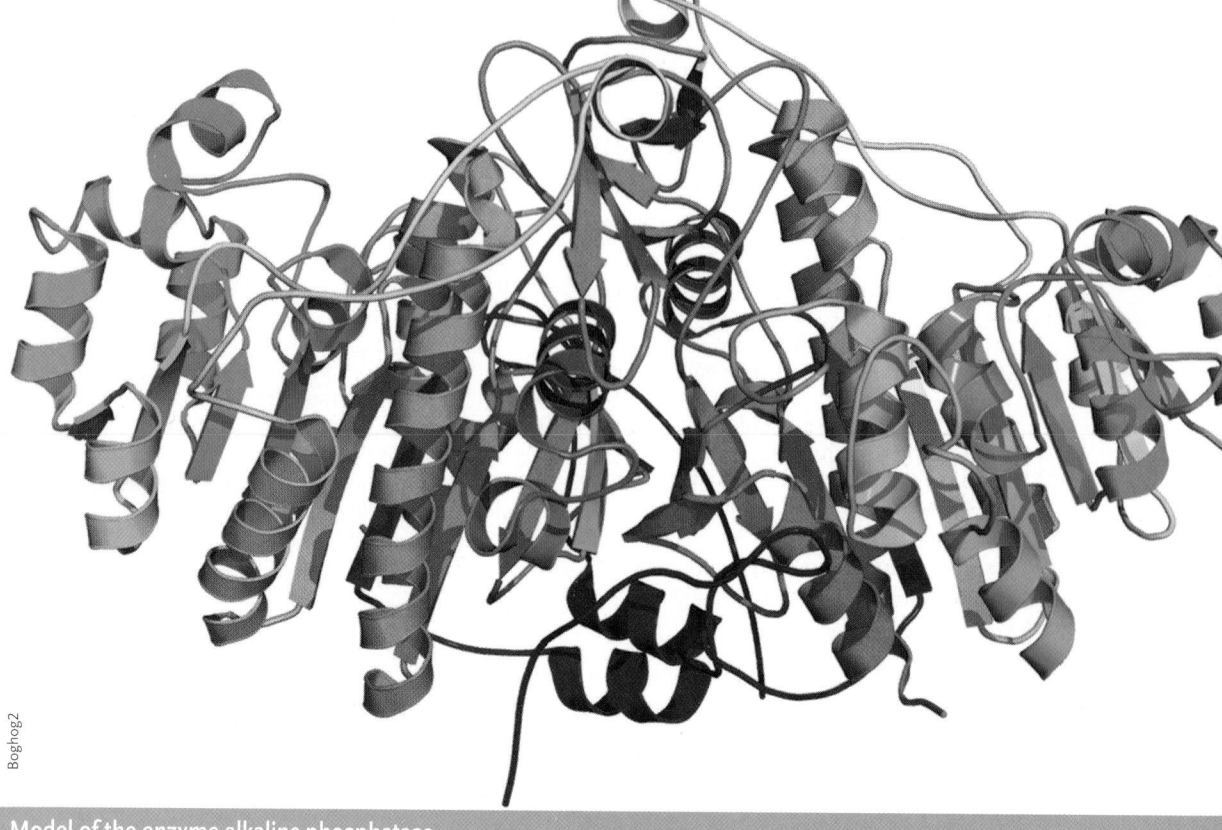

Boghog2

Energy and Enzymes

WHY IT MATTERS

Earth is a cold place—at least when it comes to chemical reactions. Life cannot survive at the high temperatures routinely used in industry for chemical synthesis. Instead, life relies on a group of catalysts called enzymes that speed up the rates of chemical reactions without the need for an increase in temperature.

Until recently, just how good enzymes are at speeding up reaction rates was not fully appreciated. Richard Wolfenden and his colleagues at the University of North Carolina experimentally measured the rates of a range of uncatalyzed and enzyme-catalyzed biochemical reactions. The prize for the greatest difference between the uncatalyzed rate and the enzyme-catalyzed rate goes to a reaction that simply removes a **phosphate group**. In the cell, a group of enzymes called phosphatases catalyze the removal of phosphate groups from a range of molecules, including proteins. The reversible addition and removal of a phosphate group from particular proteins is a central mechanism of intracellular communication in almost all cells (Chapter 5).

In the presence of the phosphatase enzyme the removal of the phosphate takes approximately 10 milliseconds. Wolfenden's research group calculated that in an aqueous environment such as within a cell, in the absence of an enzyme, the phosphate removal reaction would take over 1 trillion (10^{12}) years to occur. This exceeds the current estimate for the age of the universe! This makes the difference between the enzyme-catalyzed and uncatalyzed rates 21 orders of magnitude (10^{21}). For most reactions, the rate of the enzyme-catalyzed rate is many millions (10^{6}) of times faster than the uncatalyzed rate.

The high rates of catalysis brought about by the evolution of enzymes was of critical importance to the evolution of life on a relatively cold planet.

Enzymes are key players in the metabolic reactions that collectively accomplish the activities we associate with life, such as growth, reproduction, movement, and the ability to respond to stimuli. Central to these processes is the ability of organisms to harness and utilize energy from the surroundings, and thus this chapter starts with an overview of the principles of energy flow as governed by the laws of thermodynamics. This is followed by a focused discussion on the factors that govern chemical reactions and the central role played by **free energy**. We finish with an in-depth discussion of enzymes—the fundamental biological catalysts—which enable life to exist on this cold planet.

4.1 Energy and the Laws of Thermodynamics

Life, like all chemical and physical activities, is an energy-driven process. **Energy** is most conveniently defined as the capacity to do work. For example, it takes energy to move a car on a highway and it takes energy to climb a mountain. It also takes energy to build a protein from a group of amino acids or pump sucrose across a cell membrane.

4.1a Energy Exists in Different Forms

Energy can exist in many different forms, including chemical, electrical, and mechanical. Electromagnetic radiation, including visible, infrared, and ultraviolet light, is also a form of energy. While energy exists in many different forms, these can be transformed readily from one to another. The chemical energy present in a flashlight battery, for example, is converted into electrical energy that passes through the flashlight bulb, where it is transformed into light and heat.

All forms of energy can be grouped into one of two different types: kinetic and potential. **Kinetic energy** is the energy possessed by an object because it is in motion. Obvious examples of objects that possess kinetic energy are waves in the ocean, a falling rock, or a kicked football. A less obvious example is the kinetic energy of electricity, which is a flow of electrons. Photons of light are also a form of kinetic energy. The movement associated with kinetic energy is of use because it can perform work by making other objects move. **Potential energy** is stored energy, the energy an object has because of its position or chemical structure. A boulder at the top of a cliff has potential energy because of its position in the gravitational field of Earth. Likewise, a molecule has potential energy because of the specific arrangement of its atoms—this is also referred to as chemical energy.

So what is it about certain molecules that make them high in potential energy? The chemical energy present in a molecule has to do with the position of electrons within its atoms. In a way analogous to a boulder placed at different heights above the ground, the farther away an electron is from the nucleus of an atom the greater potential energy that electron possesses **(Figure 4.1)**. An electron, which is negatively charged, is attracted to the positively charged nucleus,

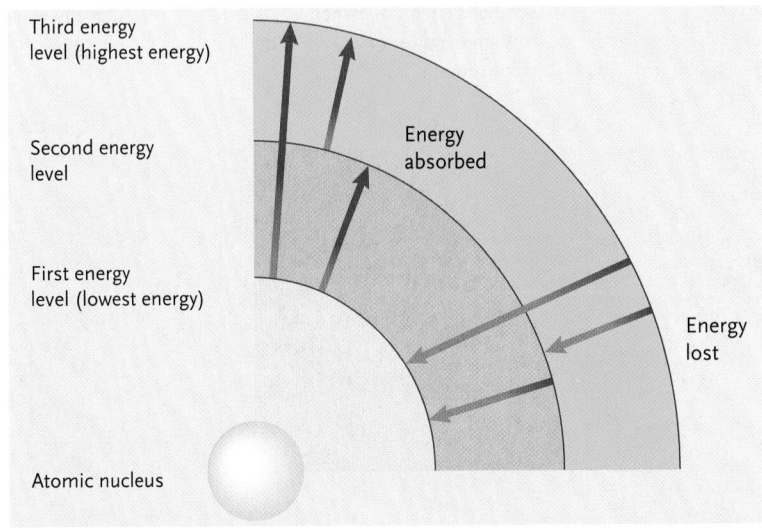

Figure 4.1

Energy levels of the electrons of an atom. Electrons can exist only in discrete energy states. When an electron gains energy it moves to a higher energy level that is farther away from the nucleus. When an electron loses energy it moves to a lower energy level closer to the nucleus.

and as it moves closer to the nucleus it loses energy. An electron moves to a higher energy level, farther from the nucleus, when it gains energy. The electronegativity of an atom strongly influences the potential energy it possesses. Atoms such as oxygen and nitrogen hold their electrons more tightly to their nucleus and thus contain less potential energy than less electronegative atoms, such as carbon and hydrogen.

4.1b The Laws of Thermodynamics Describe Energy and Its Transformation

The branch of science that concerns energy and how it changes during chemical and physical transformations is called **thermodynamics.** When discussing thermodynamics, it is important to define something called the system, which is the object being studied. A system can be anything—a single atom, one cell, or a planet. Everything outside the system is called the surroundings. The universe, in this context, is the total of the system and the surroundings. As well, it is important that we distinguish between three different types of systems: isolated, open, and closed **(Figure 4.2, p. 76).**

- An *isolated system* is one that does not exchange matter or energy with its surroundings. The only truly isolated system is the universe itself. An insulated Thermos bottle is close to being an isolated system, as very little energy or matter is exchanged with the environment.
- A *closed system* can exchange energy, but not matter, with its surroundings. A saucepan of water with a lid heating on a stove is a good example of a closed system. Earth is also considered to be a closed

Figure 4.2
Isolated, closed, and open systems in thermodynamics.

a. Isolated system: does not exchange matter or energy with its surroundings

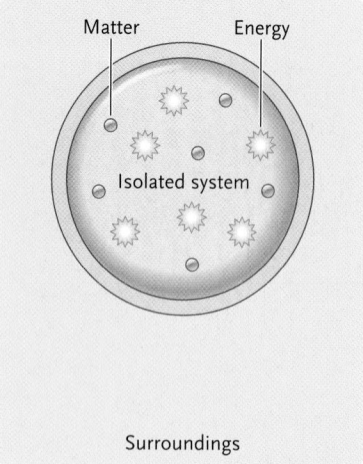

Matter   Energy

Isolated system

Surroundings

b. Closed system: exchanges energy with its surroundings

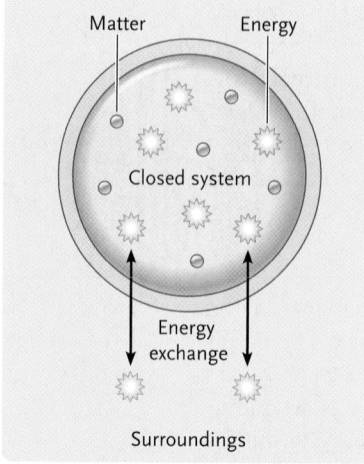

Matter   Energy

Closed system

Energy exchange

Surroundings

c. Open system: exchanges both energy and matter with its surroundings

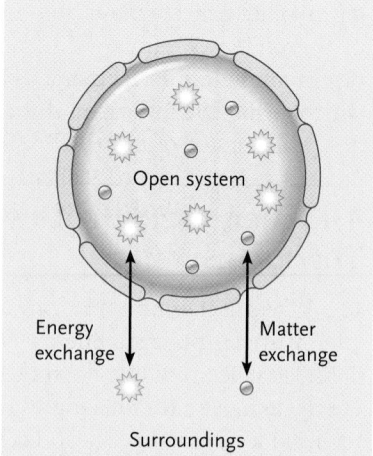

Open system

Energy exchange   Matter exchange

Surroundings

system—it takes in an enormous amount of energy generated by the Sun and releases heat, but no matter is exchanged between Earth and the rest of the universe. Each year a few meteorites hit Earth, but essentially we can consider it a closed system.

- In an *open system,* both energy and matter can move freely between the system and the surroundings. An ocean is a great example of an open system—it absorbs and releases energy, and, as a component of the hydrological cycle, water is constantly being lost and gained by the ocean through evaporation and precipitation.

4.1c The First Law of Thermodynamics: Energy Can Be Transformed but Not Created or Destroyed

Research by physicists and chemists in the nineteenth century concerning energy flow between systems and the surroundings led to the formulation of two fundamental laws of thermodynamics that apply to all systems, both living and nonliving. According to the **first law of thermodynamics,** *energy can be transformed from one form into another or transferred from one place to another, but it cannot be created or destroyed.* The first law of thermodynamics is illustrated nicely by Niagara Falls **(Figure 4.3a).** Water at the top of the falls has high potential energy because of its location in Earth's gravitational field. As the water moves over the waterfall, its potential energy is converted into kinetic

a.

© Corel

b.

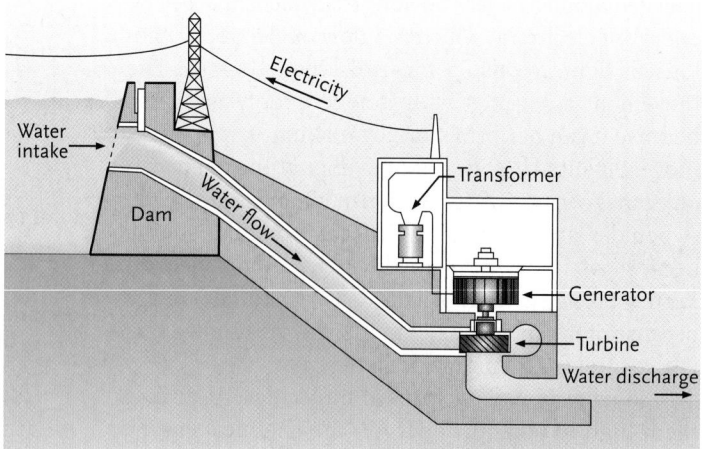

Electricity

Water intake

Transformer

Water flow

Dam

Generator

Turbine

Water discharge

Figure 4.3
Niagara Falls. **(a)** The potential energy of the water is converted into kinetic energy as it moves over the falls. **(b)** A small portion of this kinetic energy is used to turn hydroelectric turbines, converting the gravitational energy into electrical energy. In accordance with the first law of thermodynamics, energy hasn't been gained or lost but has changed form. Niagara Falls generates approximately 4.4 gigawatts of power each year.

energy. The higher the waterfall, the more kinetic energy the water will possess. When the water reaches the bottom of the waterfall, its kinetic energy is transformed into other types of energy: heat, sound, and mechanical energy (causing weathering of the rocks). For thousands of years, the kinetic energy of waterfalls has been harnessed by people to do work. At Niagara Falls today, some of the kinetic energy in the moving water is converted into electricity through the use of hydroelectric turbines (Figure 4.3b) for the use of thousands of homes and businesses.

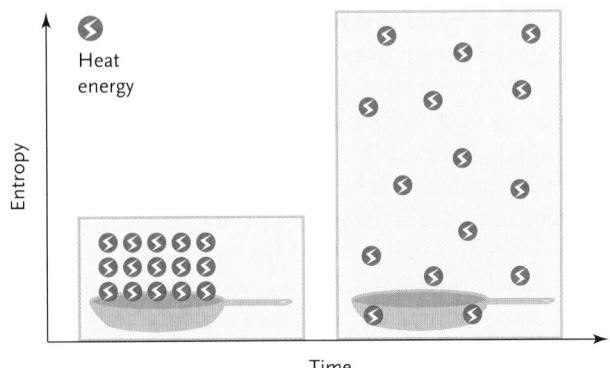

Figure 4.4
Energy tends to spread out, to become more dispersed over time. The thermodynamic measure of energy dispersal is called *entropy*.

From STARR/TAGGART/Evers/Starr, *Biology*, 12E. © 2009 Cengage Learning.

4.1d The Second Law of Thermodynamics: Energy Moves from Being Localized to Being Dispersed

Another important principle of thermodynamics is that the energy of a system tends to disperse or spread out. Many everyday situations illustrate this. For example, let's say you heat a pan on a stove and then switch the stove off **(Figure 4.4)**. At first the heat energy is concentrated very close to the pan, but slowly the heat energy disperses throughout the kitchen. This energy dispersal continues until no part of the room contains more energy than any other. This spreading out of energy is inevitable—it will happen.

In thermodynamics, the tendency of energy to become dispersed or spread out is defined as **entropy**, which is abbreviated S (think S for Spreading out). You may have been taught in high school that entropy is a measure of the *disorder* in a system. For example, a messy room has more entropy than a tidy room. But thinking of entropy as disorder is problematic for a couple of reasons. First, the idea of disorder cannot be applied scientifically in a precise way. Second, equating entropy with disorder gives you the impression that

entropy is governed by spatial regularity (order), but it isn't. Thermodynamics deals specifically with energy, not with objects.

The concept of entropy forms the basis of the **second law of thermodynamics**, *the entropy of a system and the surroundings will increase—energy will always become more spread out.* Entropy is the measure of how much energy has flowed from being localized to becoming more widely dispersed.

The tendency of energy to spread out is the underlying reason that machines can never be 100% efficient. Although energy can be transformed from one form into another, a portion of the energy will always be lost to the surroundings by the tendency of energy to spread out. For example, the engine of a car only converts a portion of the energy in gasoline to the kinetic energy that powers the wheels **(Figure 4.5a)**. Likewise, only a portion of the energy in a notebook computer battery is used to run the computer. If you touch a car engine that has just been turned off

Figure 4.5
Two examples of thermodynamic systems that display the second law of thermodynamics. **(a)** A car engine converts only about 25% of the available energy in gasoline into mechanical energy. **(b)** A runner can access only about 40% of the energy in glucose to do the work of muscle contraction. In both cases, a significant portion of the energy localized in the fuel molecules is lost to the environment as heat and gases.

or put a notebook computer on your lap for an extended period of time, it is obvious where a lot of the energy is going. It is being lost to the surroundings as *heat,* which is the energy associated with random molecular motion. The inefficiency of energy transformations also applies to living cells. For example, cellular respiration converts less than half the potential energy in glucose into a form usable for metabolism (Figure 4.5b, p. 77). We will come back to and reinforce the link between thermodynamics and life later in this chapter.

STUDY BREAK

1. What is the underlying reason that atoms differ in potential energy?
2. What is the difference between a closed and an open system? Give some examples.

4.2 Free Energy and Spontaneous Processes

The reaction between molecules of carbon dioxide and water that produces glucose and oxygen requires energy to occur, but the reverse reaction can take place all by itself. Why is that? Likewise, O_2 will readily diffuse into a region where its concentration is lower, but it will never diffuse into a region where its concentration is higher. Why not? In this section we tackle one of the basics of thermodynamics, the factors that determine if a given reaction will occur by itself or whether it needs an input of energy. A process that can occur without energy input is referred to as a *spontaneous process* (we will refine this definition later in this section). It is important to remember that in the context of thermodynamics, the term *spontaneous* does not refer to how fast a reaction will occur. As discussed in "Why It Matters," some spontaneous reactions take a millisecond to occur, others a million years. Note: We use the terms *transformation, process,* and *reaction* interchangeably to include both chemical reactions (where bonds break and form) and physical transformations such as evaporation.

4.2a Energy Content and Entropy Contribute to Making a Reaction Spontaneous

The total potential energy of a system is called its **enthalpy,** or *H.* Transformations that absorb energy from the surroundings are termed **endothermic** and result in the products having more potential energy than the starting molecules. The overall change in enthalpy ($\Delta H = H_{products} - H_{reactants}$) of an endothermic process is positive. The melting of ice is a simple example of an endothermic process. Compared to an endothermic process, a process that releases energy is called **exothermic,** as the products have less potential energy than the starting molecules (ΔH is negative). The burning of wood is a simple example of an exothermic process.

The change in enthalpy of a reaction is an important factor to evaluate to determine whether or not a reaction will occur spontaneously—however, it is not the only factor. The change in entropy is also an important consideration. Here we consider how changes in both enthalpy and entropy influence the spontaneity of a reaction.

1. *Reactions tend to be spontaneous if they are exothermic—the products have less potential energy than the reactants.* In chemical reactions the change in potential energy between products and reactants reflects a change in how tightly electrons are held by the atoms making up the molecules involved. As an example, let's look at the combustion of methane (natural gas) in air **(Figure 4.6).** The reaction is spontaneous because the reactants (methane and O_2) have greater potential energy than the products (carbon dioxide and water). The products have lower potential energy because the electrons are more tightly held by the atoms of the products than in the atoms of the reactants.

2. *Reactions tend to be spontaneous when the entropy of the products is greater than the entropy of the reactants.* Transformations tend to occur spontaneously if the energy of the products is more spread out than the energy in the starting molecules. As an example, let's look at the breakdown of glucose, a central feature of cellular respiration:

$$C_6H_{12}O_6 \text{ (s)} + 6O_2 \text{ (g)} \rightarrow 6CO_2 \text{ (g)} + 6H_2O \quad \text{(l)}$$

The reaction is spontaneous in part because of a change in enthalpy. However, its spontaneous nature is also because the entropy of the products is greater than the entropy of the reactants. But how do we know that the entropy increases simply by looking at the reaction?

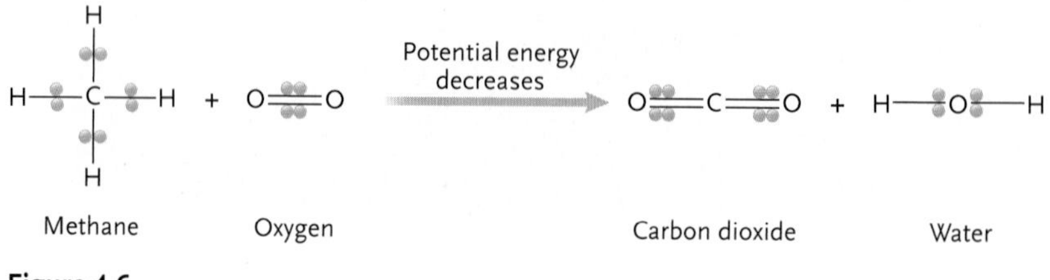

Figure 4.6

Combustion of methane is spontaneous because the products have less potential energy than the reactants. Compared to the reactants, the electronegative oxygen atoms in carbon dioxide and water hold the electrons more closely.

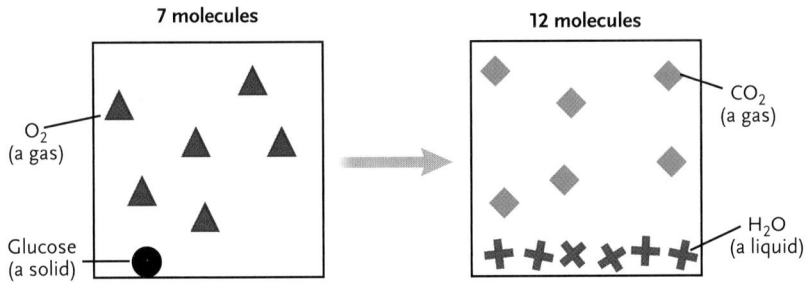

7 molecules   **12 molecules**

O_2
(a gas)

Glucose
(a solid)

CO_2
(a gas)

H_2O
(a liquid)

Figure 4.7

The breakdown of glucose results in an increase in entropy. This is because **(a)** the number of molecules increases and **(b)** a phase change has occurred (e.g., solid → liquid → gas).

Based on *Lehninger Principles of Biochemistry* (4th Ed.)
Nelson, D., and Cox, M.; W.H. Freeman and Company, New York, 2005.

As illustrated in **Figure 4.7,** whenever a chemical reaction results in an increase in the number of molecules, entropy increases. In the example of the breakdown of glucose, 7 reacting molecules are transformed into 12 molecules of product. The entropy has increased because the energy has spread out over a greater number of molecules. Entropy also increases when a solid is converted into a liquid or a liquid into a gas. Energy spreads out more readily as matter undergoes these phase changes. In the example of the breakdown of glucose, 6 molecules of a gas and one molecule of a solid are converted into six molecules of a gas and six molecules of a liquid (Figure 4.7). A phase change in the other direction—gas to liquid or liquid to solid—decreases the entropy as the energy becomes more localized.

4.2b The Change in Free Energy Indicates Whether a Process Is Spontaneous

From our previous discussion we know that both enthalpy and entropy need to be considered to determine if a reaction is spontaneous. It was the American physicist Josiah Gibbs who arrived at the mathematical relationship of how entropy and enthalpy relate to reaction spontaneity. This is referred to as Gibbs free energy (G). The change in free energy ($\Delta G = G_{products} - G_{reactants}$) is a measure of whether or not a process is spontaneous. It can be calculated for any specific transformation using the formula

$$\Delta G = \Delta H - T\Delta S$$

where ΔH is the change in the enthalpy, ΔS is the change in the entropy of the system over the course of the reaction, and T is the temperature in degrees Kelvin (K). T is part of the entropy term because energy spreading increases as temperature increases.

Recall from the introduction to Section 4.2 that we referred to a *spontaneous* process as one that will occur by itself without any external input of energy. Now that we have introduced the concept of free energy, we can be more precise with our definition: a **spontaneous reaction** is one where the free energy of the products is less than the free energy of the reactants, ΔG is negative. A spontaneous process is also referred to as an **exergonic process (Figure 4.8a)**. Similarly, a nonspontaneous process is one where the free energy of the products is greater than the free energy of the reactants, ΔG is positive. A nonspontaneous reaction is also referred to as an **endergonic process** (Figure 4.8b).

The formula $\Delta G = \Delta H - T\Delta S$ tells us that both the change in enthalpy and the change in entropy can influence the overall ΔG of a reaction. In many processes, like the breakdown of glucose, the change in enthalpy (ΔH is negative) and the change in entropy (ΔS is positive) both contribute to making the reaction exergonic. But this does not have to be the case. Let's consider a very interesting thermodynamic system: the ice cube. At room temperature, ice will spontaneously melt because the ice is absorbing energy from the surroundings. Since water has greater potential energy than ice, the melting of ice is an endothermic process (ΔH is positive)—and yet the process is exergonic (ΔG is negative). What

a. Exergonic reaction: Free energy is released, products have less free energy than reactants, and the reaction proceeds spontaneously.

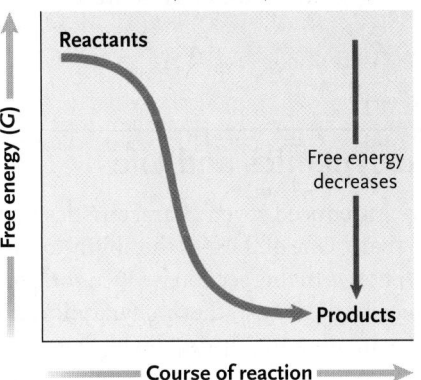

b. Endergonic reaction: Free energy is gained, products have more free energy than reactants, and the reaction is not spontaneous.

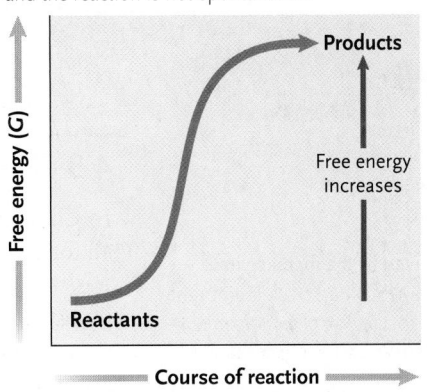

Figure 4.8

Exergonic and endergonic reactions.

explains the spontaneous melting of ice is the large increase in entropy as the solid ice changes into liquid water. The large positive $T\Delta S$ is greater than ΔH, making ΔG for the reaction negative.

Let's look at one more example of a spontaneous process: diffusion of molecules across a membrane. As illustrated in **Figure 4.9,** any molecule that can cross a membrane will move spontaneously from a compartment where it is at a higher concentration to a compartment where its concentration is lower. The spontaneous nature of diffusion is explained solely by the increase in entropy as the molecules and their associated energy spread out. Although there is no change in enthalpy (ΔH), the release of free energy during diffusion can be harnessed by the cell to do work. How this is accomplished is discussed in subsequent chapters.

4.2c Exergonic Processes Reach Equilibrium Rather Than Going to Completion

In the late nineteenth century, chemists were surprised to find that many chemical reactions never went to completion. The products were always "contaminated" with molecules of reactant. More shocking was the finding that regardless of the amount of reactants and products in the initial mixture, the system reached the same state, in which the proportion of products to reactants was a constant. As an example, consider a chemical reaction in which glucose 1-phosphate is converted into glucose 6-phosphate **(Figure 4.10)**. Starting with 0.02 M glucose 1-phosphate, the reaction will proceed spontaneously until there is 0.019 M of glucose 6-phosphate (product) and 0.001 M of glucose 1-phosphate (reactant) in the solution. In fact, regardless of the amounts of each you start with, the reaction will reach a point at which there is 95% glucose 6-phosphate and 5% glucose 1-phosphate.

The point at which there is no longer any overall change in the concentration of products and reactants is called the point of chemical equilibrium. In this state, molecules do not stop reacting; rather the rate of the forward reaction equals the rate of the backward

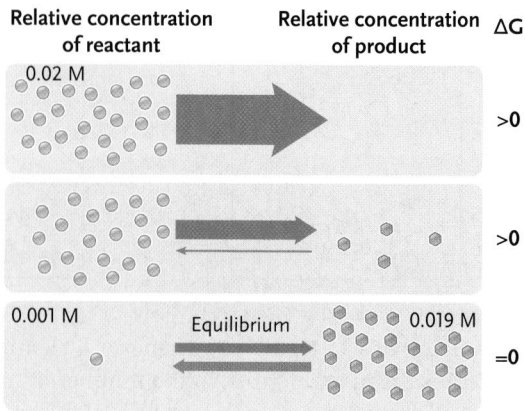

Figure 4.10
Chemical reactions run to equilibrium. No matter what quantities of glucose 1-phosphate and glucose 6-phosphate are dissolved in water, when equilibrium is attained, there will always be 95% glucose 6-phosphate and 5% glucose 1-phosphate. At equilibrium, the number of reactant molecules being converted to products equals the number of product molecules being converted back to reactants.

reaction. As a system moves toward equilibrium, the free energy of the system becomes progressively lower and reaches its lowest point when the system is at equilibrium. It is at this point that there is no tendency for spontaneous change in either the forward or the reverse direction. The system reaches a state of maximum stability, it has no capacity to do work, and $\Delta G = 0$.

The point of equilibrium is related to ΔG for the reaction, in that the more negative ΔG, the farther toward completion the reaction will move before equilibrium is established. Many reactions have a ΔG that is near zero and are thus readily **reversible** by adjusting the concentrations of products and reactants slightly.

STUDY BREAK

1. What is the difference between a reaction that is exothermic and one that is exergonic?
2. What is the thermodynamic reason that molecules spontaneously diffuse across a membrane?
3. True or false: In a reaction that has a negative ΔG, all of the reactants are converted into products.

4.3 Thermodynamics and Life

In Chapter 3 we introduced seven characteristics that all forms of life share. One of these is the ability to harness and utilize energy. In this section we focus on how living things abide by the laws of thermodynamics and yet are able to maintain a highly organized state. We expand our discussion to illustrate how energy flows

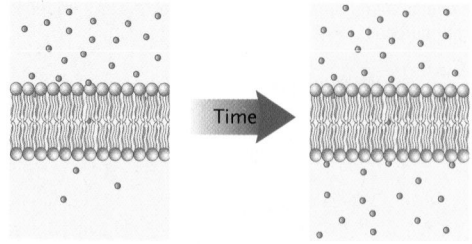

Figure 4.9
Diffusion across a membrane is driven by the increase in entropy. Molecules will move spontaneously across a membrane from a region of high concentration to a region of lower concentration because the energy associated with the molecules becomes more dispersed.

through the biosphere—the regions of Earth occupied by life.

4.3a Life Does Not Go against the Second Law of Thermodynamics

At first glance, living systems seem to go against the second law of thermodynamics and its foundational concept that the entropy of a system and its surroundings must always increase. It is easy to think this because organisms are able to maintain themselves in a highly organized state with energy being concentrated in complex molecules. Cells can synthesize molecules like proteins and nucleic acids and are filled with intricate structures such as microtubules, ribosomes, mitochondria, and chromosomes, to name a few. How is all of this possible without going against the second law?

Organisms can maintain a highly organized state because they are open systems and thus are constantly using energy and matter that they bring in from the environment to keep a low-entropy state. But according to thermodynamics isn't entropy always supposed to increase? The second law states that the entropy of a system *plus its surroundings* must increase—and this holds for living systems as well. As illustrated in **Figure 4.11**, organisms bring in energy and matter, but as a result of the thousands of chemical reactions that take place within cells, organisms also give off heat and by-products of metabolism, such as water and carbon dioxide, that spread out, increasing the entropy of the surroundings. The entropy of a system can be maintained in a low state but only because the entropy of its surroundings is constantly increasing.

But why do living systems have to keep consuming energy? Once all the proteins that are required to sustain life are synthesized, why can't the energy requiring process of protein synthesis stop? Once a human has fully developed and stopped growing, why can't we stop eating? Organisms need to constantly bring in energy and matter because at a cellular level the tendency of energy to spread out means that cellular components (proteins, organelles, etc.) are constantly becoming damaged and breaking down (Figure 4.11). Just like a car that needs to be taken to a mechanic to have new parts installed and others repaired, the breakdown of cellular systems is an inevitable consequence of the Second Law and increasing entropy. New cells need to be made and old ones maintained by the continued synthesis of proteins, carbohydrates, and a myriad of other molecules. Metabolism can never stop, and living cells never reach chemical equilibrium ($\Delta G = 0$); life requires a constant supply of energy. So, while it is easy to see why elite athletes need to eat a lot, people who don't exercise at all also need to ingest well over a thousand kilocalories every day. Although some of this food supplies us with the energy to use our muscles, much of the food energy we ingest is

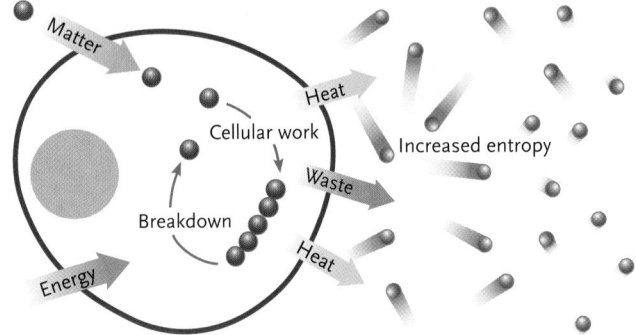

Figure 4.11
Cells are open systems. By bringing in energy and matter from their surroundings they can maintain an ordered state. The release of heat and waste gases into the environment increases the entropy of the surroundings.

used simply to maintain our low-entropy, highly organized state (Figure 4.11). We eat to maintain low entropy **(Figure 4.12)**.

4.3b The Flow of Energy through the Biosphere

Recall that Earth does not exchange matter with the rest of the universe, but it does exchange a huge amount of energy. Life exists on Earth because its position in the Solar System allowed for heat from the Sun to maintain Earth at a temperature that allowed life to evolve (see Chapter 3). But it is not the heat from the Sun that the biosphere relies on as an energy source to maintain it organized state—it's the light, which is a very concentrated form of energy that exists in packets called photons.

Figure 4.12
Why do we need to eat? The average person needs to ingest about 1500 kcal per day. A significant amount of this energy is needed to maintain the low entropy state of our cells.

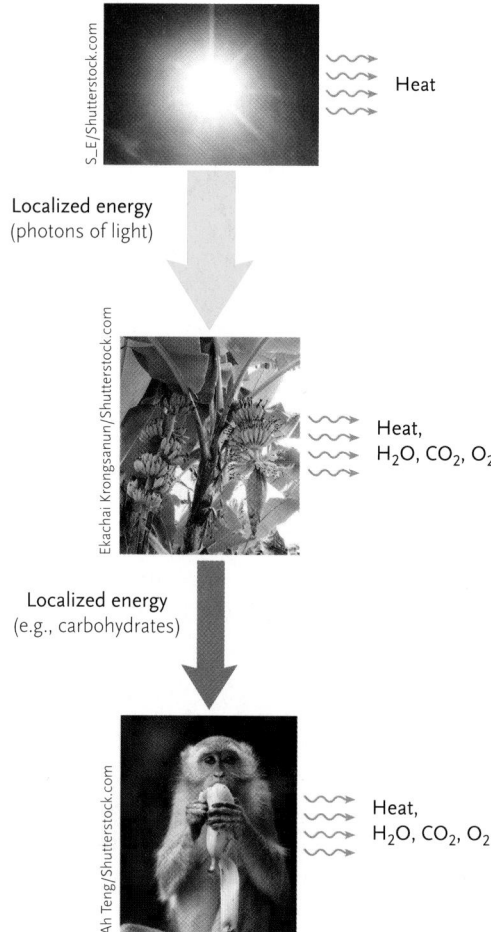

Figure 4.13

Flow of energy from the Sun through the biosphere. Living systems constantly bring in concentrated forms of energy and use them to do the work required to maintain a highly organized state. Organisms give off heat and gases (high-entropy energy).

Energy enters the biosphere when light energy from the Sun is transformed into chemical energy through the process of photosynthesis. By absorbing photons of light and using them to do work, photosynthetic organisms can transform molecules of CO_2 and water into high energy–containing molecules such as glucose. The energy in glucose in turn can drive the synthesis of a wide range of other organic molecules. As illustrated in **Figure 4.13,** the remainder of the biosphere (e.g., animals, fungi) is sustained by consuming the various forms of chemical energy produced by photosynthetic organisms.

STUDY BREAK

1. Why may someone think that life goes against the second law of thermodynamics?
2. Why can't we consume all the chemical energy produced by a plant through photosynthesis?
3. Explain in thermodynamic terms why if you stopped eating you would die.

4.4 Overview of Metabolism

The collection of all the chemical reactions present within a cell or organism is defined as **metabolism.** Many different metabolic reactions take place within a cell, resulting in the synthesis or breakdown of a huge variety of molecules. Most metabolic reactions fall into pathways and there are two fundamental types of pathways: those that require energy to build molecules and those that release energy by breaking molecules down. A lot of the metabolism of the biosphere involves the energy transformation reactions of two metabolic pathways: respiration and photosynthesis. The details of these processes, as well as others, such as protein synthesis, are the focus of later chapters. In this section, our attention is more broad as we look at the central features of how energy is transformed during metabolism.

4.4a Metabolism Consists of Catabolic and Anabolic Pathways

The individual reactions that make up metabolism are grouped into *pathways*—starting molecules undergo stepwise transformation, one reaction at a time, generating one or more final products. For example, the hormone testosterone is the end-product of a five-reaction pathway that starts with the molecule cholesterol.

A series of chemical reactions that results in the breakdown of larger, more-complex molecules into smaller, less-complex ones is called a **catabolic pathway.** Energy is released in a catabolic pathway because overall the free energy of the final product(s) of the pathway is less than the free energy of the starting molecule(s) **(Figure 4.14).** Perhaps the best example of a catabolic pathway is cellular respiration: energy-rich food molecules are converted into simpler, lower-energy molecules such as H_2O and CO_2. An **anabolic pathway,** on the other hand, is a series of reactions that results in the synthesis of larger, more-complex molecules from simpler starting molecules (Figure 4.14). Anabolic pathways, which are often called biosynthetic pathway, require energy because overall the free energy of the product(s) of the pathway is greater than the free energy of the starting molecule(s). The biosynthesis of specific carbohydrates, proteins, and nucleic acids are all products of anabolic pathways, as is photosynthesis.

A key feature of metabolism also shown in Figure 4.14 is that catabolic and anabolic pathways are linked through chemical energy. Because biosynthetic (anabolic) reactions result in the formation of new covalent bonds, they require a source of chemical energy. The energy comes from the catabolic breakdown of high-energy molecules. The specific form of chemical energy that most often links the two types of pathways is the molecule adenosine triphosphate (ATP).

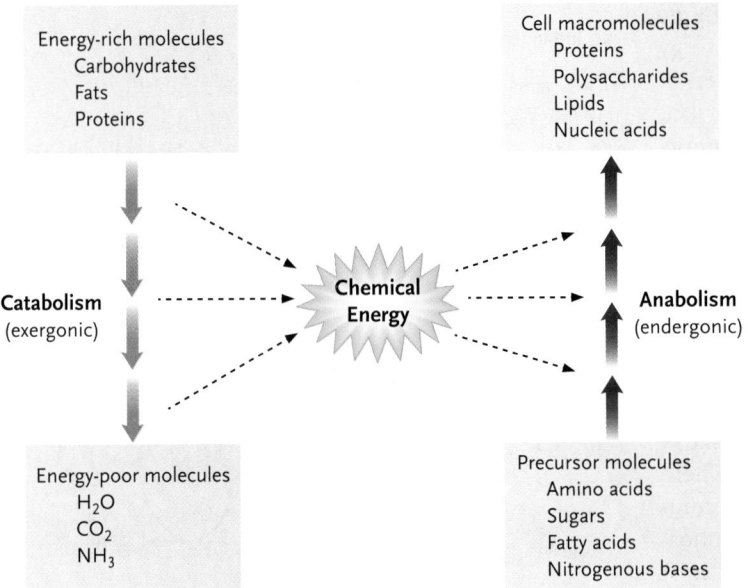

Figure 4.14

Energy relationships between the pathways of catabolism and anabolism. Energy released from the breakdown of energy-rich molecules can be harnessed by anabolic reactions, which use the energy to generate macromolecules.

From Garrett/Grisham, *Biochemistry*, 5E. © 2013 Cengage Learning.

4.4b ATP Hydrolysis Releases Free Energy

All forms of life require a readily usable form of chemical energy. Like using dollars as an accepted currency to buy goods, it would be advantageous to the cell if there were a single, widely accepted form of energy currency. Not only would this chemical currency be used for biosynthetic reactions, but ideally it could be readily transformed into mechanical energy required for muscle contraction or electrical energy required for the conduction of nerve impulses. The nucleotide ATP is that energy currency.

As shown in **Figure 4.15a,** ATP consists of a five-carbon sugar, ribose, linked to the **nitrogenous base** adenine joined to a chain of three phosphate groups. ATP is a source of free energy as a result of its reaction with water (Figure 4.15b). In this *hydrolysis* reaction, the terminal phosphate bond is broken, resulting in the formation of adenosine diphosphate and a molecule of inorganic phosphate (abbreviated P_i):

$$ATP + H_2O \rightarrow ADP + P_i$$
$$\Delta G = -7.3 \text{ kcal/mol}$$

So what is it about the chemistry of ATP that explains the negative ΔG when it is hydrolyzed? The exergonic nature of ATP hydrolysis is because of both (i) a decrease in potential energy and (ii) an increase in entropy. There is less potential energy in ADP than in ATP because the loss of the terminal phosphate has decreased the electrical repulsion among the negatively charged oxygen atoms of the phosphate groups. The hydrolysis of ATP is also spontaneous because of an increase in entropy as energy moves from being localized on one molecule (ATP) to being spread out on two molecules (ADP and P_i) (Figure 4.15b).

a. Chemical structure of ATP

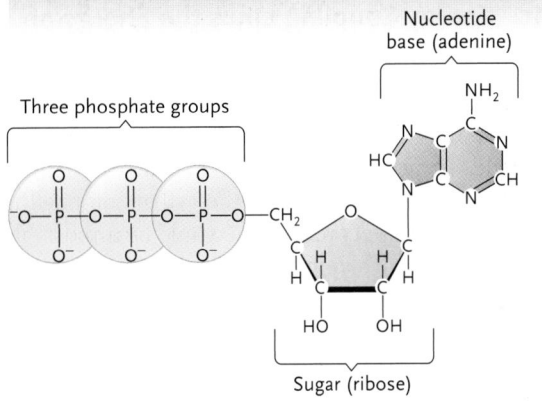

b. Hydrolysis reaction

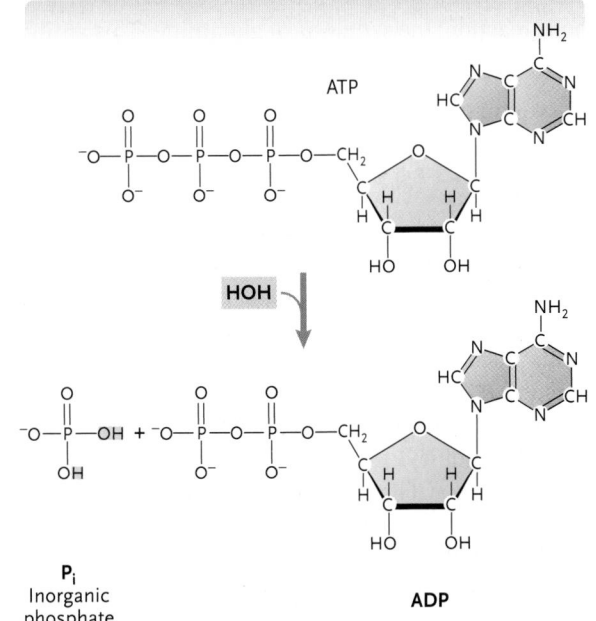

Figure 4.15
ATP, the primary molecule used to supply the energy for biosynthetic reactions. **(a)** Structure of one ATP molecule. **(b)** Reaction of ATP hydrolysis. Energy is released during the formation of ADP and P_i.

CONCEPT FIX You may have the idea that the energy associated with ATP is the result of a "high-energy phosphate bond" that releases energy when it is broken. This thinking, however, is incorrect. A foundational concept of chemistry is that energy is never released when bonds break; in fact, energy is *required*. Energy is released when ATP is hydrolyzed only because the bonds formed in the products are of lower energy than the bonds in the reactant molecules that were broken. ⬡

The fact that all forms of life use ATP as their dominant energy currency is another piece of evidence that points to all forms of life sharing a common ancestor (see Chapter 3). This is because there is nothing particularly unique to ATP. There are a number of other phosphate-containing compounds, including the other nucleotide triphosphates (GTP, CTP, TTP) (see *The Purple Pages*) that liberate comparable free energy to ATP when they are hydrolyzed. Thus, the fact that life adopted ATP doesn't reflect a unique capability of ATP but rather perhaps simply a chance event that occurred as early as 3.5 billion years ago.

4.4c Energy Coupling Links the Energy of ATP to Other Molecules

Although ATP releases free energy when it is hydrolyzed, this does not mean that it is an especially reactive molecule. In fact, the rate of ATP hydrolysis in an aqueous environment such as the cytosol of a cell is slow. If ATP were very reactive, it would be impossible for metabolism involving ATP to be tightly controlled. Its rapid hydrolysis would simply release heat, and not only can cells not use heat to do work, too much heat can cause damage and even cell death.

So how do cells harness the free energy available from ATP hydrolysis to do cellular work? To help answer this question, let's look at a very common anabolic reaction: the synthesis of glutamine. This amino acid is synthesized from glutamic acid and ammonia:

$$\text{glutamic acid} + NH_3 \rightarrow \text{glutamine} + H_2O$$
$$\Delta G = +3.4 \text{ kcal/mol}$$

The positive ΔG shows that the reaction will not occur as written—and yet molecules of glutamine are synthesized within your cells all the time. How is that possible? During metabolism, glutamine is synthesized through a process called **energy coupling**: an endergonic reaction occurs by being coupled to an exergonic reaction **(Figure 4.16)**. For the majority of energy coupling reactions, the energy is provided by the exergonic breakdown of ATP.

Looking at Figure 4.16a, it is easiest to think of energy coupling as the joining of two independent reactions, one spontaneous (exergonic) and the other nonspontaneous (endergonic). The free-energy changes of the two reactions can be added to yield the

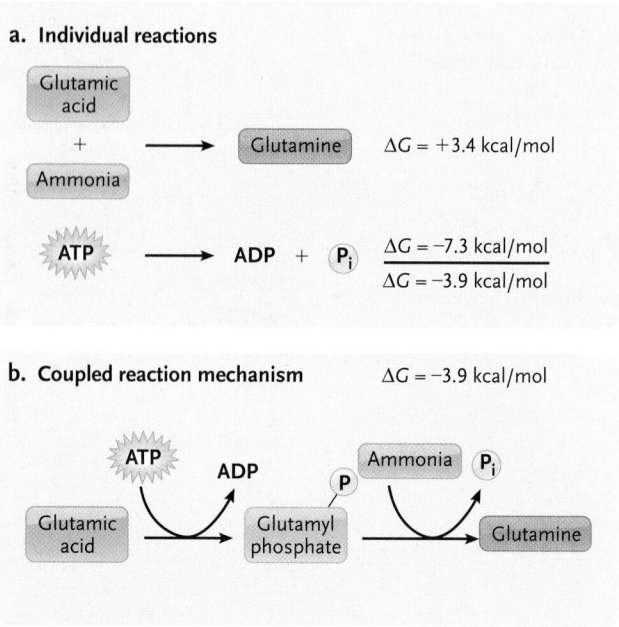

a. Individual reactions

Glutamic acid + Ammonia → Glutamine $\Delta G = +3.4$ kcal/mol

ATP → ADP + P_i $\Delta G = -7.3$ kcal/mol

$\Delta G = -3.9$ kcal/mol

b. Coupled reaction mechanism $\Delta G = -3.9$ kcal/mol

ATP → ADP

Glutamic acid → Glutamyl phosphate (P) → Glutamine + Ammonia + P_i

Figure 4.16

An example of energy coupling. The exergonic breakdown of ATP is linked to the endergonic biosynthesis of glutamine, a spontaneous process.

free-energy change of the **coupled reaction**. The sum of the two reaction free energies, ATP breakdown ($\Delta G = -7.3$ kcal/mol) and glutamine synthesis ($\Delta G = 3.4$ kcal/mol), yields -3.9 kcal/mol. This tells us that the coupled reaction will be spontaneous.

Within a cell, the actual reaction mechanism of the coupled reaction is distinctly different from the two independent reactions shown in Figure 4.16a although the overall ΔG remains simply the sum of the two reactions. Energy coupling during metabolism (Figure 4.16b) requires an enzyme that binds both a molecule of ATP and a molecule of substrate and facilitates the transfer of the terminal phosphate group from ATP to the substrate. The addition of phosphate to the substrate increases its free energy and makes it more reactive, allowing the second reaction to occur spontaneously (Figure 4.16b). An important aspect of energy coupling is that the inclusion of an enzyme facilitates the movement of potential energy from a molecule of ATP to the substrate molecule through transfer of the terminal phosphate group. The energy-wasting hydrolysis of ATP is prevented in the first reaction shown in Figure 4.16b because water cannot access the site of catalysis on the enzyme. The hydrolysis of ATP is not complete until P_i is released in the second reaction.

4.4d Cells also Couple Reactions to Regenerate ATP

We have just seen how the hydrolysis of ATP is an exergonic reaction that can be harnessed through energy coupling reactions to make biosynthetic reactions

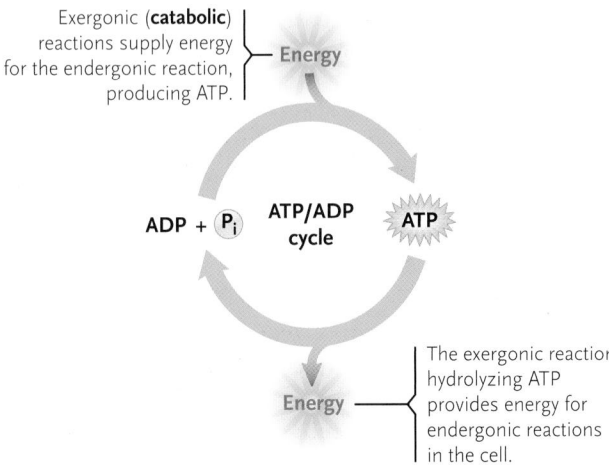

Exergonic (**catabolic**) reactions supply energy for the endergonic reaction, producing ATP.

The exergonic reaction hydrolyzing ATP provides energy for endergonic reactions in the cell.

Figure 4.17
The ATP/ADP cycle that couples reactions releasing free energy and reactions requiring free energy.

proceed spontaneously. These coupling reactions occur continuously in living cells consuming a tremendous amount of ATP. So where does the ATP for these processes come from? Some ATP is synthesized using a biosynthetic pathway that includes reactions that join ribose, adenine, and phosphate groups together. However, the vast majority of ATP is generated from recombining ADP and P_i.

If ATP breakdown is an exergonic process, then ATP synthesis from ADP and P_i is an endergonic process. The free energy required for ATP synthesis comes from the catabolism of molecules that contain an abundance of energy. For animals, these molecules are food—carbohydrates, fats, and proteins—all abundant sources of energy. In photosynthetic organisms, the capture of light energy is used to synthesize ATP from ADP and P_i.

The continuous breakdown and resynthesis of ATP is called the **ATP cycle (Figure 4.17)**. Approximately 10 million ATP molecules are broken down and resynthesized each second in a typical cell, illustrating that this cycle operates at an astonishing rate. In fact, if ATP were not regenerated from ADP and P_i, it is estimated that the average human would use an estimated 75 kg of ATP per day. It makes sense that cells should never be limited in their availability of ATP. In fact, a typical cell maintains an ATP concentration that is about 1000 times greater than that of ADP.

STUDY BREAK

1. In what ways do the end-products of catabolic pathways differ from their starting molecules?
2. In an energy coupling reaction, trace the fate of the terminal phosphate group of ATP.

4.5 The Role of Enzymes in Biological Reactions

So far in this chapter we have focused on the thermodynamics of energy transformation: exergonic and endergonic reactions and factors that determine whether a particular reaction will occur without an input of energy. We have avoided discussing anything about the rate of a reaction because, in fact, the laws of thermodynamics do not address how fast a process will occur, just whether or not it will occur. But the rate of a reaction is of fundamental importance to life, because most reactions must occur at very high rates for life to be sustained. In this section, we discuss the factors that control the rate of a chemical reaction and the central role played by enzymes in increasing reaction rates.

4.5a The Activation Energy of a Reaction Represents a Kinetic Barrier

The conversion of table sugar (sucrose) into the **monosaccharides** glucose and fructose is a spontaneous reaction, and yet a bag of sugar can sit around for decades without any detectable fructose or glucose being formed. So what is preventing this spontaneous reaction from occurring rapidly? Forget sugar, what about the planet—given the large amounts of energy trapped in the wood of trees, and the coal and oil underground, why doesn't Earth just go up in flames?

For chemical reactions to occur, established bonds need to be broken and new bonds need to be formed. For bonds to be broken, they must first be strained or otherwise made less stable, which requires a small input of energy. The initial energy investment required to start a reaction is called the **activation energy** (E_a) **(Figure 4.18a, p. 86)**. Molecules that gain the necessary activation energy occupy what is called the **transition state**, where bonds are unstable and are ready to be broken.

What provides the activation energy for chemical reactions? The molecules taking part in chemical reactions are in constant motion, and reacting molecules may periodically gain enough energy to reach the transition state. But for the molecules of sucrose on the kitchen shelf, reaching the transition state is a very rare event. Supplying larger amounts of energy would allow more molecules to gain the activation energy necessary to react. A good example of this is a propane torch **(Figure 4.19, p. 86)**. Propane is a molecule that contains an abundance of free energy. In the presence of air, it spontaneously decomposes into carbon dioxide and water. However, the reaction proceeds very slowly—the propane in a torch can sit for years and remain unchanged. This is because if left undisturbed, it is a rare event for molecules of propane to acquire the energy needed for combustion. Yet if you supply a stream of propane gas with a spark

a.

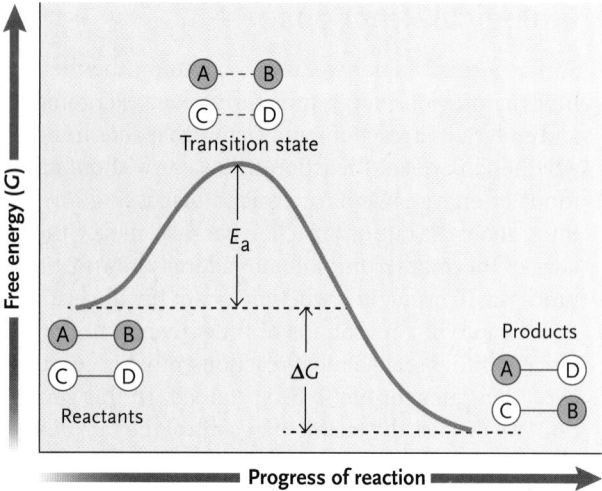

b. "Activation energy" barrier in the movement of a rock downhill

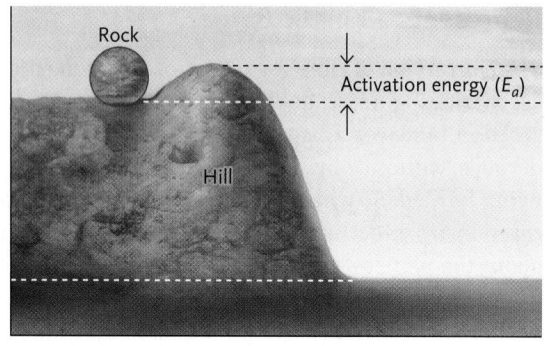

Figure 4.18
The concept of activation energy (E_a) for an exergonic reaction.

(see Figure 4.19), then you provide the molecules with the energy necessary to reach the transition state, resulting in combustion. The heat released from the initial combustion event sustains the continuous burning of the propane stream.

4.5b Enzymes Accelerate Reactions by Reducing the Activation Energy

If you walk through a typical undergraduate chemistry lab, you will find that the benches have Bunsen burners, which are used to provide the heat for a range of chemical reactions. Chemists routinely use heat to provide the energy needed for reactant molecules to get to the transition state and thus speed up the rate of a reaction. In biology, using heat to speed up a reaction is problematic for two reasons: First, high temperatures destroy the structural components of cells, particularly proteins, and can result in cell death. Second, an increase in temperature would speed up all possible chemical reactions in a cell, and thus the precise regulation of metabolic pathways would be lost. So how can you increase the rate of specific reactions without raising the temperature? You can use a **catalyst**, which is a chemical agent that speeds up the rate of a reaction without itself taking part in the reaction. The most common biological catalyst is a group of proteins called **enzymes.**

Looking back at Figure 4.18, you can think of the transition state as a kinetic barrier—it is what prevents spontaneous reactions from occurring rapidly, because so few molecules at a given time acquire the energy necessary to get to the transition state. If you could lower this energy requirement, then many more molecules would react. This is exactly what enzymes do— they increase the rate of a reaction by decreasing the activation energy **(Figure 4.20)**. Since the rate of a reaction (i.e., number of molecules of product made per second) is proportional to the number of reactant molecules that can get to the transition state, lowering the transition state results in a higher rate of reaction.

CONCEPT FIX There are two common misconceptions about the role of enzymes in biochemical reactions that we need to fix before moving on. First, although enzymes decrease the activation energy of a reaction, they do not alter the thermodynamics of a reaction. The change in

Figure 4.19
Combustion of propane. **(a)** The combustion of propane is a spontaneous reaction; however, the activation energy is a barrier that prevents its rapid breakdown. **(b)** When a spark is provided, propane obtains the energy required to attain the transition state. **(c)** The initial heat generated sustains continuous propane burning.

a.

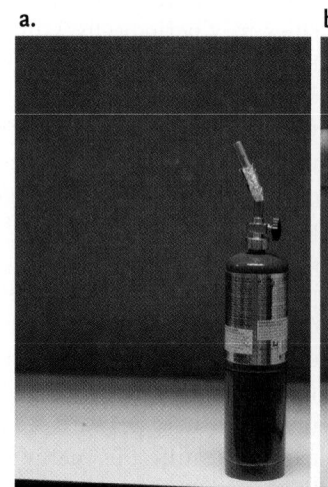

b.

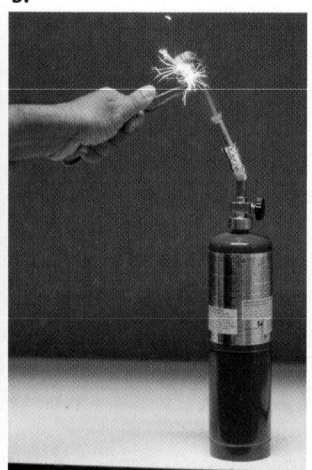

c.

Denis Maxwell

Maud Menten (1879–1960)

A fundamental topic covered in almost all introductory biochemistry courses is the Michaelis–Menten equation. First stated in 1913, the equation represents one of the fundamental concepts of biochemistry, providing a mathematical description of the kinetics of an enzyme-catalyzed reaction.

The name Menten of the equation refers to Maud Menten, who was born in 1879, in Port Lambton, Ontario. After completing secondary school, Menten attended the University of Toronto and earned a bachelor of arts degree in 1904, followed by a master's degree in physiology in 1907. In the same year, Menten was appointed a fellow at the Rockefeller Institute for Medical Research in New York City, where she studied the effect of radium bromide on cancerous tumours in rats. A year later, she returned to the University of Toronto and in 1911 became one of the first Canadian women to receive a doctor of medicine degree.

In 1912, Menten travelled to Germany to work with Leonor Michaelis, a biochemist who shared her interest in understanding enzyme kinetics. After a year of research, the two scientists coauthored a paper that put forward a description of the basis of enzyme-catalyzed chemical kinetics. The paper introduced the Michaelis–Menten equation as a tool for measuring the rates of enzyme reactions. The formula gave scientists a way to record how enzymes worked and is the standard for most enzyme-kinetic measurements. Michaelis and Menten were able to demonstrate that each enzyme, given enough substrate, has its own rate of causing that substrate to undergo chemical change. The Michaelis–Menten equation profoundly changed the study of biochemistry and earned Menten and Michaelis worldwide recognition.

When Menten returned from Berlin, she enrolled at the University of Chicago, where she obtained a Ph.D. in biochemistry in 1916. Unable to find an academic position in her native Canada, in 1918 she joined the medical school faculty at the University of Pittsburgh. While maintaining an active research program, she was also known as an avid mountain climber who went on several expeditions to the Arctic. As well, she spoke numerous languages, loved to paint, and played the clarinet. Over the years, Menten authored more than 70 publications, including discoveries related to blood sugar, hemoglobin, and kidney functions. In so-called retirement, she returned to British Columbia to do research at the British Columbia Medical Research Institute, almost until her death. A plaque commemorating the life and work of Maud Menten is located in the Medical Sciences Building, University of Toronto, Queen's Park.

free energy (ΔG) of a reaction is not altered by the presence of an enzyme. Second, enzymes do not supply energy to a reaction. Although enzymes are involved in energy coupling reactions (see Section 4.4c), the chemical energy is supplied by ATP, not by the enzyme.

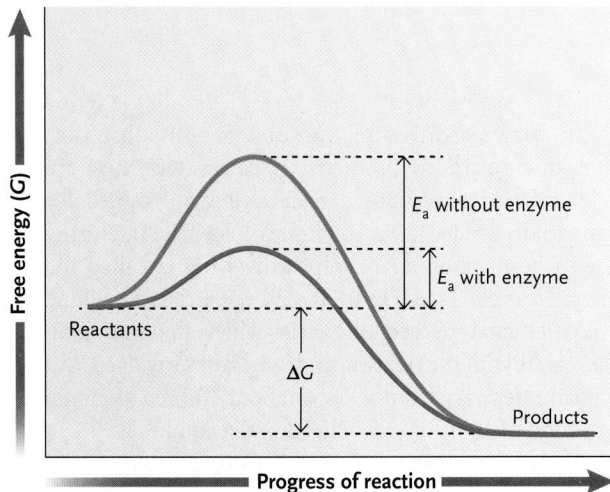

Figure 4.20

Enzymes lower the activation energy (E_a) of a reaction.

4.5c Enzymes Combine with Reactants and Are Released Unchanged

In biochemical reactions, an enzyme combines briefly with reacting molecules and is released unchanged when the reaction is complete. For example, the enzyme hexokinase **(Figure 4.21, p. 88)**, catalyzes the following reaction:

$$\text{glucose} + \text{ATP} \rightarrow \text{glucose 6-phosphate} + \text{ADP}$$

The reactant that an enzyme acts on is called the enzyme's substrate, or substrates if the enzyme binds two or more molecules. Each type of enzyme catalyzes the reaction of only a single type of molecule or a group of closely related molecules. This enzyme specificity explains why the metabolism of a typical cell involves thousands of different enzymes. Notice in Figure 4.21 that the enzyme is much larger than the substrate. As well, the substrate interacts with only a very small region of the enzyme. This region is called the **active site**—the specific site on an enzyme where catalysis takes place. The active site is usually a pocket or groove that is formed when, following protein synthesis, the newly synthesized enzyme folds into its three-dimensional shape.

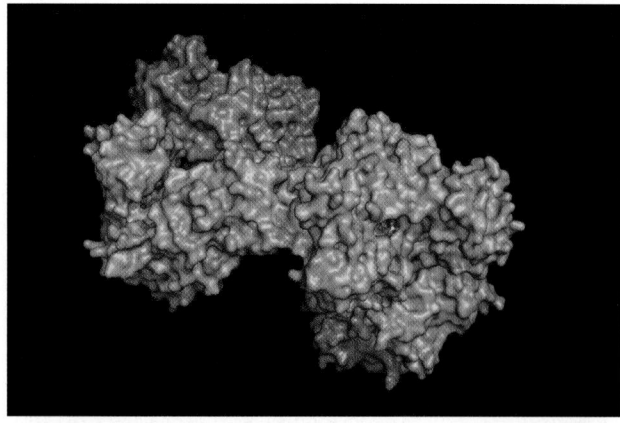

a. Like other enzymes, hexokinase has an active site where specific substrates bind and where catalysis occurs. As shown at left, the active site is a very small region of the overall enzyme.

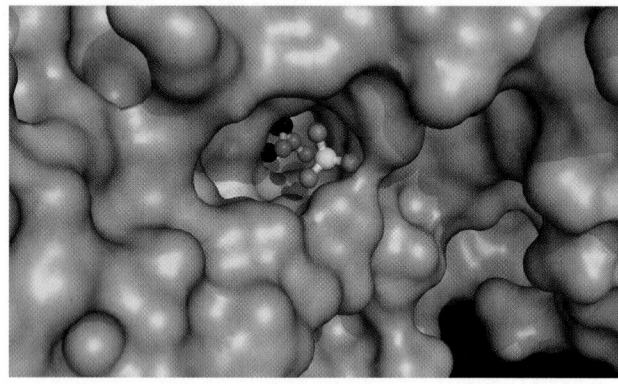

b. A close-up showing glucose and phosphate within the active site.

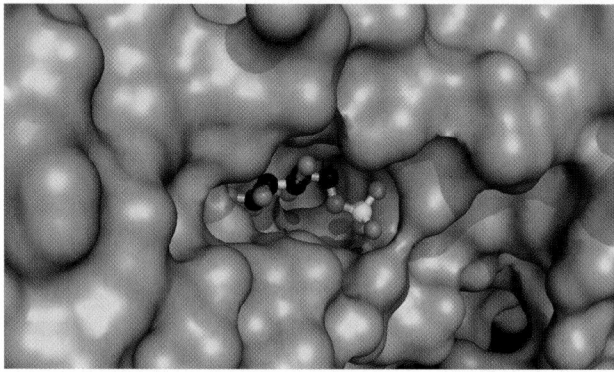

c. The glucose has bonded with the phosphate. The product of the reaction, glucose-6-phosphate, is shown leaving the active site.

Figure 4.21
Model of the enzyme hexokinase showing the catalysis of glucose with phosphate forming glucose-6-phosphate. The glucose is represented by black (carbon) and red (oxygen) spheres, while the phosphate group is shown with the phosphorus atom (yellow sphere) bonded to four oxygens (red).
From STARR/TAGGART/Evers/Starr. *Biology*, 12E. © 2009 Brooks/Cole, a part of Cengage Learning, Inc. Reproduced by permission. www.cengage.com/permissions

In the early twentieth century, biochemists proposed the *lock-and-key hypothesis* to explain the specificity of the substrate–enzyme interaction. The analogy worked well to explain how even somewhat similar substrates (keys) were unable to bind to the same enzyme (lock) to cause catalysis (unlocking of the door). However, more recently, this hypothesis has been superseded by what has become known as the *induced-fit hypothesis*. Research has shown that unlike locks, enzymes are not rigid objects but instead are flexible. Just before substrate binding, the enzyme changes its shape (**conformation**) so that the active site becomes even more precise in its ability to bind the substrate.

As shown in **Figure 4.22**, the enzyme binds to the substrate, forming an enzyme–substrate complex.

Catalysis occurs when the two are joined, with the action of the enzyme converting the substrate (or substrates) into one or more products. Because enzymes are released unchanged after a reaction, enzyme molecules can rapidly bind to other substrate molecules, catalyzing the same reaction again, repeating what is called the enzyme cycle (see Figure 4.22). The rate at which enzymes catalyze reactions varies widely depending on the specifics of the enzyme and substrates involved, but typical rates vary from a low of about 100 reactions up to a high of 10 million reactions per second.

Many enzymes require a **cofactor**, a nonprotein group that binds very precisely to the enzyme. Cofactors are often metals, such as iron, copper, zinc, or manganese. Although most cells need very small

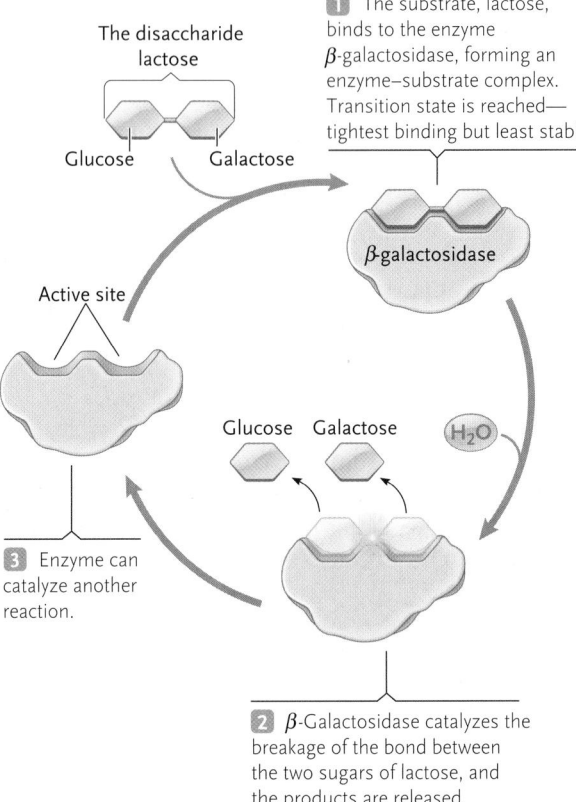

The disaccharide lactose

Glucose   Galactose

1 The substrate, lactose, binds to the enzyme β-galactosidase, forming an enzyme–substrate complex. Transition state is reached—tightest binding but least stable.

β-galactosidase

Active site

Glucose   Galactose   H₂O

3 Enzyme can catalyze another reaction.

2 β-Galactosidase catalyzes the breakage of the bond between the two sugars of lactose, and the products are released.

Figure 4.22
The catalytic cycle of an enzyme. Shown is the enzyme β-galactosidase, which cleaves the sugar lactose to produce glucose and galactose.

amounts of these metals, they are absolutely essential for the catalytic activity of the enzyme to which they bind. Some cofactors, called **coenzymes**, are organic molecules that are often derived from vitamins.

4.5d Enzymes Reduce the Activation Energy by Inducing the Transition State

We know that enzymes reduce the activation energy of a reaction, but how do they do it? An enzyme uses one of three basic mechanisms to lower the energy required to get to the transition state. These mechanisms are shown in **Figure 4.23.**

1. *Bringing the reacting molecules together.* Reacting molecules can assume the transition state only when they collide. Binding to an enzyme's active site brings the reactants together in the right orientation for catalysis to occur.

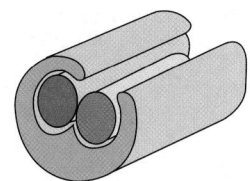

Bring reacting molecules close together

2. *Exposing the reactant molecule to altered charge environments that promote catalysis.* In some systems, the active site of the enzyme may contain ionic groups whose positive or negative charges alter the substrate in a way that favours catalysis.

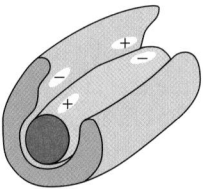

Charge interactions

3. *Changing the shape of a substrate molecule.* The active site may strain or distort substrate molecules into a conformation that mimics the transition state.

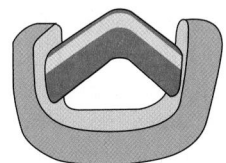

Distort or strain substrate molecules

Figure 4.23
The binding of substrate(s) to an active site results in the substrate's acquiring the transition state conformation.

Regardless of the mechanism used by a specific enzyme, the binding of the substrate to the active site results in the substrate's attaining the transition state conformation—its bonds strained and ready to be broken. Without the enzyme, substrate molecules would also be able to acquire the transition state, it's just that for most reactions would not occur very often. The addition of an enzyme simply enables many more molecules to reach the transition state. This is fundamentally why enzymes increase the rate of a reaction.

STUDY BREAK

1. Distinguish between the activation energy of a reaction and the transition state.
2. Which of the following aspects of a reaction is changed by the addition of an enzyme: free energy of products, ΔG, requirement for energy, or rate?
3. A mutation to the gene that codes for the enzyme hexokinase may result in the enzyme that is synthesized being unable to bind glucose. Why might this be the case?

4.6 Factors That Affect Enzyme Activity

Enzymes play a critical role in metabolism. Because of this, regulating how they operate is central to controlling metabolism. As you would expect, a number of factors can change the activity of a particular enzyme, including changes in the concentration of substrate and other molecules that bind to enzymes. As well, changes in environmental factors, including temperature and pH, can also effect enzyme activity.

4.6a Enzyme and Substrate Concentrations Can Change the Rate of Catalysis

Biochemists use a wide range of approaches to studying an enzyme. These span from using the tools of molecular biology and genetics to study the structure and regulation of the gene that encodes the enzyme to sophisticated computer programs for modelling the three-dimensional structure of the enzyme and its active site. However, the most fundamental and central approach has been to determine the rate of the specific reaction catalyzed by a particular enzyme and how the rate changes in response to altering certain experimental parameters. This usually requires purifying the enzyme from the remainder of the cell, incubating it in an appropriate buffered solution, and supplying the reaction mixture with substrate. With these constituents, one can then determine the rate of catalysis, which is most often done by measuring the rate at which product of the reaction is formed—so, for example, micromoles of product per second.

As shown in **Figure 4.24a,** in the presence of excess substrate, the rate of catalysis is proportional to the amount of enzyme. As enzyme concentration increases, the rate of product formation increases. In this system, where substrate concentration is high, what is limiting the rate of the reaction is the amount of enzyme in the reaction mixture. Look at what hap-

pens if we instead keep the enzyme constant at some intermediate concentration and change the substrate concentration from low to high. At very low concentrations, substrate molecules collide so infrequently with enzyme molecules that the rate at which the product is formed is slow (Figure 4.24b). As the substrate concentration increases, the reaction rate initially increases linearly as enzyme and substrate molecules collide more frequently. But as the constant number of enzyme molecules approaches the maximum rate at which they can combine with reactants and release products, increasing the substrate concentration has a smaller and smaller effect, and the rate of reaction eventually levels off. When the catalytic cycle (see Figure 4.22) is turning as fast as possible, further increases in substrate concentration have no effect on the reaction rate. At this point, the enzyme is said to be saturated with substrate.

4.6b Enzyme Activity Can Be Altered by Competitive and Noncompetitive Interactions

The rate of an enzyme-catalyzed reaction can be altered by a wide range of molecules that bind to the enzyme. A number of molecules that alter enzyme activity do so because they are structurally similar to the normal substrate of the enzyme and therefore can bind to the active site. Regulation of this type is called competitive regulation because the molecule competes with the substrate for the active site **(Figure 4.25a).** Competitive regulation is often referred to as **competitive inhibition** because the presence of the competitor decreases the rate of the normal substrate-dependent reaction.

Competitive regulators differ in how strongly they bind to the active site. Some molecules bind through covalent bonding, resulting in enzyme inhibition that is irreversible. However, many inhibitors bind to the active site weakly, through noncovalent interactions, resulting in inhibition that is readily reversible. As

Figure 4.24

Effect of increasing **(a)** enzyme concentration or **(b)** substrate concentration on the rate of an enzyme-catalyzed reaction.

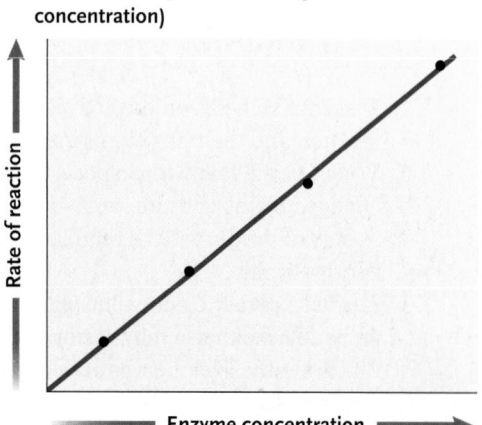

a. Rate of reaction as a function of enzyme concentration (substrate at high concentration)

Rate of reaction vs *Enzyme concentration*

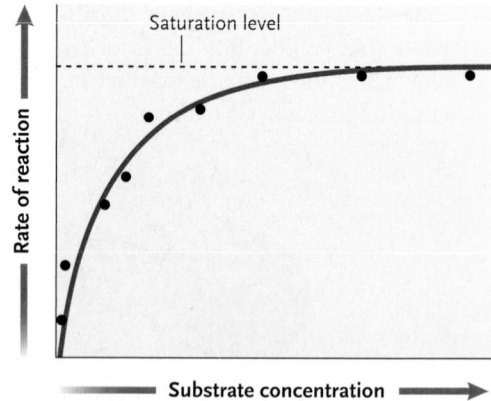

b. Rate of reaction as a function of substrate concentration (enzyme amount constant)

Saturation level

Rate of reaction vs *Substrate concentration*

Penicillin: A Competitive Inhibitor of Enzyme Action

Penicillin is an antibiotic that is used in the treatment of bacterial infections. It was first discovered by Alexander Fleming, who isolated it from the mould *Penicillium* after he accidentally found that the presence of the mould inhibited the growth of bacteria on a Petri plate. Following the development of methods for its mass production, penicillin became a true wonder drug that was effective at treating a wide range of bacterial infections that in the past often led to death.

Penicillin acts by inhibiting the synthesis of peptidoglycan, a key component of the bacterial cell wall. Peptidoglycan is a complex polymer consisting of sugars and amino acids

that forms a meshlike structure outside the plasma membrane. As such, peptidoglycan provides structural strength and protects the bacterial cell from osmotic changes that would otherwise cause the cell to burst. If a bacterium is unable to synthesize components necessary for its cell wall, it is unable to grow and divide.

A key factor that is required for the synthesis of peptidoglycan is the enzyme transpeptidase, which catalyzes the formation of a peptide bond between two amino acids, effectively linking two portions of the peptidoglycan together. Penicillin inhibits peptidoglycan synthesis because it is a competitive inhibitor of transpeptidase

activity. The structure of penicillin mimics that of the two amino acids, which are normally brought together by the active site. Penicillin binds irreversibly to the active site of transpeptidase, effectively destroying the molecule. Given the concentrations of penicillin usually administered to a patient, this leads to total inhibition of all transpeptidase activity.

Although penicillin was widely employed in the 1950s and 1960s, most infections today involve bacteria that have acquired resistance to the drug. New antibiotics are constantly being developed to try to stop the growing problem of antibiotic-resistant bacteria.

---

shown in Figure 4.25b, one trait of reversible competitive inhibition is that it can be overcome by a high substrate concentration.

Because of their ability to act on critical enzymes, many inhibitor molecules can be toxic. For example,

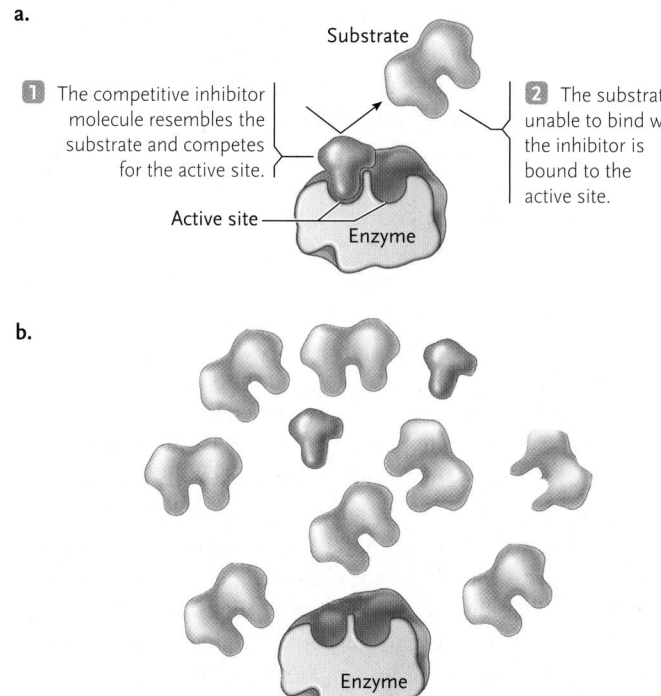

a.

1. The competitive inhibitor molecule resembles the substrate and competes for the active site.

Substrate

2. The substrate is unable to bind when the inhibitor is bound to the active site.

Active site

Enzyme

b.

Enzyme

Figure 4.25
(a) Competitive regulation of enzyme activity. **(b)** The inhibition of enzyme activity by a competitive inhibitor can be overcome by increasing the amount of substrate relative to inhibitor.

cyanide is a potent poison because it is a competitive inhibitor of cytochrome oxidase, an enzyme involved in cellular respiration (see Chapter 6). Interestingly, many drugs act by inhibiting specific enzymes. For example, a number of antibiotics, including penicillin, are effective because they target and inhibit specific bacterial enzymes (see "Molecule behind Biology").

Some regulatory molecules do not interact with the active site, but rather alter enzyme function by binding to another location on the enzyme. These regulatory molecules do not compete with substrate molecules and result in what is referred to as noncompetitive regulation **(Figure 4.26, p. 92)**. While competitive regulation results in inhibition of normal enzyme function, in noncompetitive regulation molecules that interact with the enzyme can cause an increase or decrease in enzyme function depending upon the molecule and the enzyme.

In noncompetitive regulation, enzyme activity is controlled by the reversible binding of a regulatory molecule to what is often referred to as the **allosteric site**, a location on the enzyme outside the active site. Enzymes controlled by noncompetitive regulation are often referred to as allosteric enzymes and typically have two alternative conformations controlled from the allosteric site. In one conformation, called the *high-affinity state*, the enzyme binds strongly to its substrate; in the other conformation, the *low-affinity state*,

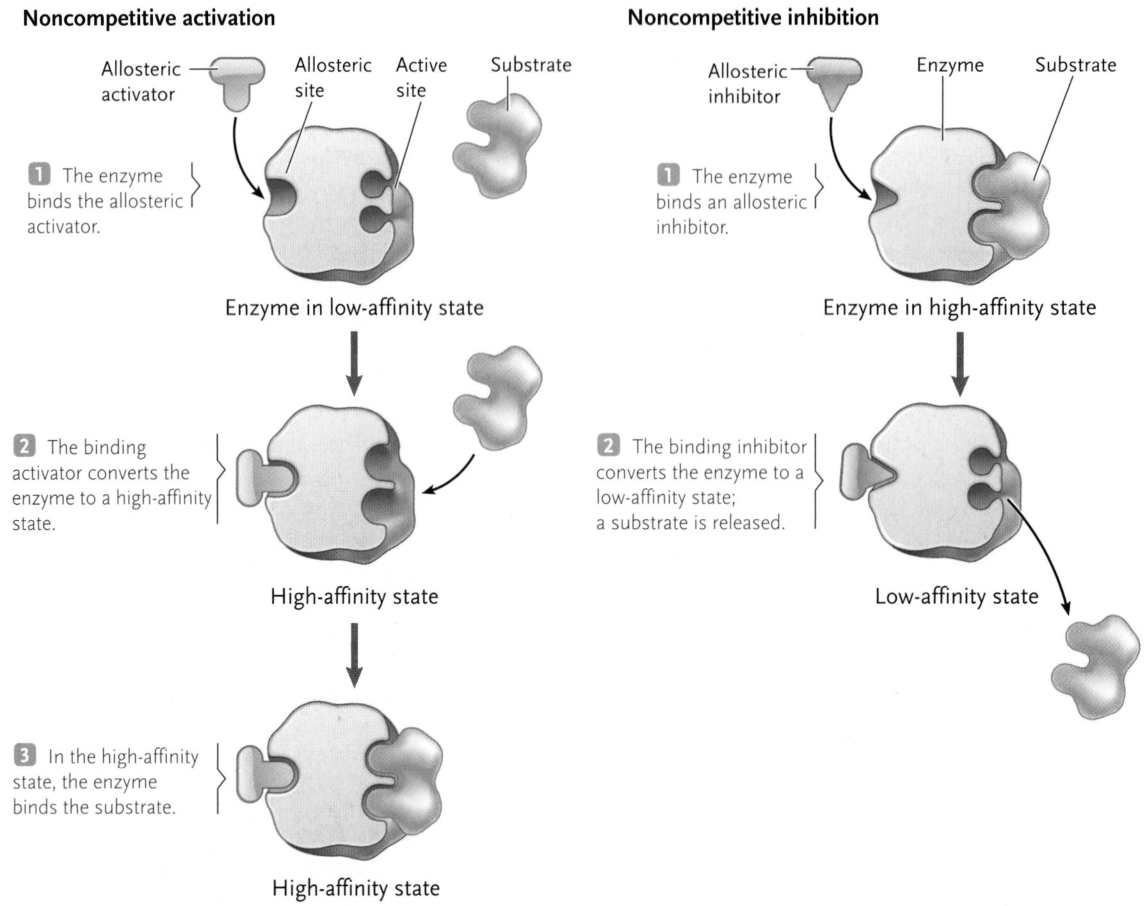

Noncompetitive activation

Allosteric activator
Allosteric site
Active site
Substrate

1 The enzyme binds the allosteric activator.

Enzyme in low-affinity state

2 The binding activator converts the enzyme to a high-affinity state.

High-affinity state

3 In the high-affinity state, the enzyme binds the substrate.

High-affinity state

Noncompetitive inhibition

Allosteric inhibitor
Enzyme
Substrate

1 The enzyme binds an allosteric inhibitor.

Enzyme in high-affinity state

2 The binding inhibitor converts the enzyme to a low-affinity state; a substrate is released.

Low-affinity state

Figure 4.26
Noncompetitive (**allosteric**) regulation.

the enzyme binds the substrate weakly or not at all. Binding with regulatory substances may induce either state: an **allosteric inhibitor** converts an enzyme from the high- to the low-affinity state, while an **allosteric activator** converts it from the low- to the high-affinity state (see Figure 4.26).

4.6c Metabolism Is Finely Controlled by Noncompetitive Regulation

For metabolism to work efficiently, the activity of enzymes needs to be adjusted upward or downward so that the amount of product synthesized by any reaction matches the needs of the cell for the product. Considering that a typical cell contains thousands of enzymes that collectively catalyze a huge range of reactions, a key question to address is how enzyme activity is controlled to regulate overall metabolism. One mechanism of regulation is to control the abundance of specific enzymes. Since enzymes are proteins, this can be facilitated by regulating gene expression (transcription) and protein synthesis (translation). But this type of regulation lacks fine control—metabolic pathways are often adjusted in seconds, yet changes in enzyme abundance can take 30 minutes or more to occur. In addition to transcriptional and translational control,

the cell is able to very rapidly regulate metabolic pathways by directly affecting enzyme activity. Two major mechanisms it uses to achieve this are allosteric control and covalent modification.

Frequently, allosteric inhibitors are a product of the metabolic pathway that they regulate. If the product accumulates in excess, its effect as an inhibitor automatically slows or stops the enzymatic reaction producing it. Accordingly, if the product becomes too scarce, the inhibition is reduced, and the product begins to accumulate again. This type of metabolic regulation, in which the product of a reaction acts to inhibit its own synthesis, is termed **feedback inhibition.** In multireaction pathways, feedback inhibition usually involves the final product inhibiting the enzyme that catalyzes one of the early reactions in the pathway. In this way, cellular resources are not wasted in producing intermediates that are not needed.

The biochemical pathway that makes the amino acid isoleucine from threonine is an excellent example of feedback inhibition. The pathway proceeds in five steps, each catalyzed by an enzyme **(Figure 4.27).** The end-product of the pathway, isoleucine, is an allosteric inhibitor of the first enzyme of the pathway, threonine deaminase. If the cell makes more isoleucine than it needs, isoleucine combines reversibly with

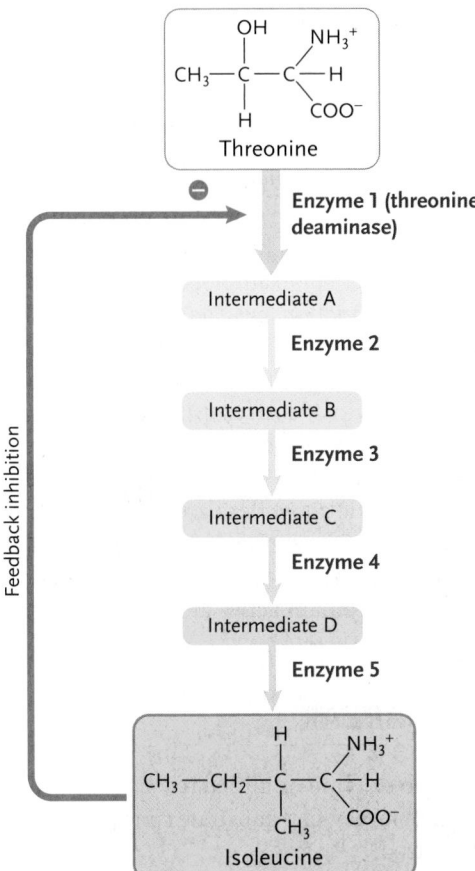

Threonine

Enzyme 1 (threonine deaminase)

Intermediate A

Enzyme 2

Intermediate B

Enzyme 3

Intermediate C

Enzyme 4

Intermediate D

Enzyme 5

Isoleucine

Feedback inhibition

Figure 4.27
Feedback inhibition in the pathway that produces isoleucine from threonine. If the product of the pathway, isoleucine, accumulates in excess, it slows or stops the pathway by acting as an allosteric inhibitor of the enzyme that catalyzes the first step in the pathway.

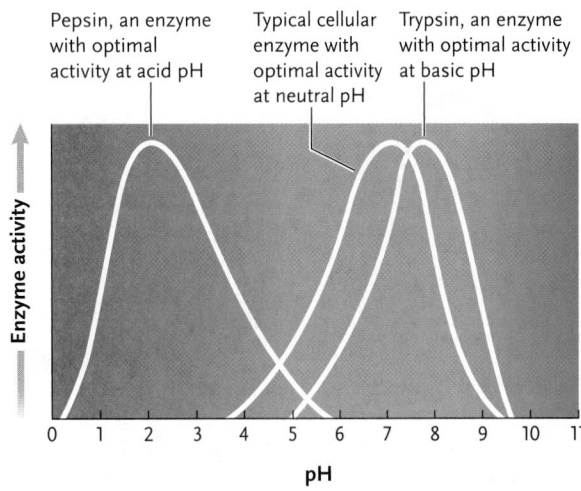

Pepsin, an enzyme with optimal activity at acid pH

Typical cellular enzyme with optimal activity at neutral pH

Trypsin, an enzyme with optimal activity at basic pH

Figure 4.28
Effects of pH on enzyme activity. An enzyme typically has an optimal pH at which it is most active; at pH values above or below the optimum, the rate of enzyme activity drops off. At extreme pH values, the rate drops to zero.

threonine deaminase at its allosteric site, converting the enzyme to the low-affinity state and inhibiting its ability to combine with threonine, the substrate for the first reaction in the pathway. If isoleucine levels drop too low, the allosteric site of threonine deaminase is vacated, the enzyme is converted to the high-affinity state, and isoleucine production increases.

4.6d Temperature and pH Are Key Factors Affecting Enzyme Activity

The activity of most enzymes is strongly altered by changes in pH and temperature. Characteristically, enzymes reach maximal activity within a narrow range of temperature or pH; at levels outside this range, enzyme activity drops off. These effects produce a typically peaked curve when enzyme activity is plotted, with the peak where temperature or pH produces maximal activity.

Effects of pH Changes. Typically, each enzyme has an optimal pH where it operates at peak efficiency in speeding the rate of its biochemical reaction **(Figure 4.28)**.

On either side of this pH optimum, the rate of the catalyzed reaction decreases because of the resulting alterations in charged groups. The effects on the structure and function of the active site become more extreme at pH values farther from the optimum, until the rate drops to zero. Most enzymes have a pH optimum near the pH of the cellular contents, about pH 7. Enzymes that are secreted from cells may have pH optima farther from neutrality. An example is **pepsin**, an enzyme secreted into the stomach. This enzyme's pH optimum is 1.5, close to the **acidity** of stomach contents. Similarly, trypsin has a pH optimum at about pH 8, allowing it to function well in the somewhat alkaline contents of the **intestine**, where it is secreted.

Effects of Temperature Changes. The effects of temperature changes on enzyme activity reflect two distinct processes. First, temperature has a general effect on chemical reactions of all kinds. As the temperature rises, the rate of chemical reactions typically increases. This effect reflects increases in the kinetic motion of all molecules, with more frequent and stronger collisions as the temperature rises. Second, temperature has a more specific effect on all proteins, including enzymes. As the temperature rises, the kinetic motions of the amino acid chains of an enzyme increase, along with the strength and frequency of collisions between enzymes and surrounding molecules. At some point, these disturbances become strong enough to denature the enzyme: the hydrogen bonds and other forces that maintain its three-dimensional structure break, making the enzyme unfold and lose its function (see *The Purple Pages* for a more detailed description of protein denaturation). The two effects of temperature act in opposition to each other to produce characteristic changes in the rate of enzymatic catalysis **(Figure 4.29, p. 94)**. In the range of

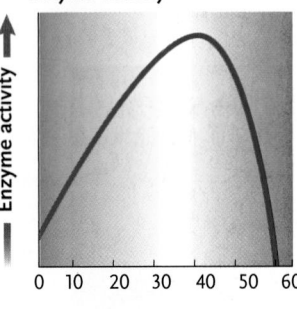

a. Effect of temperature on enzyme activity

(graph: Enzyme activity vs Temperature (°C), x-axis 0 to 60)

b.

Tony Campbell/Shutterstock.com

Figure 4.29

Effect of temperature on enzyme activity. **(a)** As the temperature rises, the rate of the catalyzed reaction increases proportionally until the temperature reaches the point at which the enzyme begins to denature. The rate drops off steeply as denaturation progresses and becomes complete. **(b)** Visible effects of environmental temperature on enzyme activity in Siamese cats. The fur on the extremities—ears, nose, paws, and tail—contains more dark brown pigment (melanin) than the rest of the body. A heat-sensitive enzyme controlling melanin production is denatured in warmer body regions, so dark pigment is not produced, but fur colour is.

0°C to about 40°C, the reaction rate doubles for every 10°C increase in temperature. Above 40°C, the increasing kinetic motion begins to denature the enzyme, reducing the rate of increase in enzyme activity. At some point, as the temperature rises, the denaturation of the enzyme causes the reaction rate to level off at a peak. Further increases cause such extensive unfolding that the reaction rate decreases rapidly to zero.

For most enzymes, the peak in activity lies between 40°C and 50°C; the drop-off becomes steep at 55°C and falls to zero at about 60°C. Thus, the rate of an enzyme-catalyzed reaction peaks at the temperature at which kinetic motion is greatest, but no significant unfolding of the enzyme has occurred. Although most enzymes have a temperature optimum between 40°C and 50°C, some have activity peaks below or above this range. For example, the enzymes of maize (corn) pollen function best near 30°C and undergo steep reductions in activity above 32°C. As a result, environmental temperatures above 32°C can seriously inhibit the growth of corn crops. Many animals living in cold regions have enzymes with much lower temperature optima than average. For example, the enzymes of arctic snow fleas are most active at 10°C. At the other extreme are the enzymes of archaeans that live in hot springs, which are so resistant to denaturation that they remain active at temperatures of 85°C or more.

STUDY BREAK

1. Why do enzyme-catalyzed reactions reach a saturation level when substrate concentration is increased?
2. Distinguish between competitive and noncompetitive inhibition.
3. Explain why the activity of an enzyme will eventually decrease to zero as the temperature rises.

Review

aplia

To access course materials such as Aplia and other companion resources, please visit www.NELSONbrain.com.

4.1 Energy and the Laws of Thermodynamics

- Energy is the capacity to do work. Kinetic energy is the energy of motion; potential energy is energy stored in an object because of its location or chemical structure. Energy may be readily converted between potential and kinetic states, but it cannot be created or destroyed.

- The potential energy of a molecule is related to the position of electrons. The farther away an electron is from the nucleus of an atom, the greater its potential energy (Figure 4.1).

- Thermodynamics is the study of energy and how it changes during chemical and physical transformations. A system that does not exchange energy or matter with its surroundings is an isolated system. A system that exchanges energy but not matter with its surroundings is a closed system. A system that

exchanges both energy and matter with its surroundings is an open system (Figure 4.2).

- The first law of thermodynamics states that energy can be transformed from one form into another or transferred from one place to another, but it cannot be created or destroyed (Figure 4.3).

- Entropy is the tendency of energy to become dispersed or spread out. The second law of thermodynamics states that the entropy of every system and the surroundings will always increase (Figure 4.4).

- Energy spreading out as evidence of the second law of thermodynamics is exhibited by both a car and a runner (Figure 4.5).

4.2 Free Energy and Spontaneous Processes

- A process that can occur without an input of energy is referred to as spontanous.

- Reactions tend to be spontaneous if they are exothermic ($\Delta H < 0$) (Figure 4.6).

- Reactions tend to be spontaneous if entropy (S) increases (Figure 4.7).

- Entropy and enthalpy are related to reaction spontaneity by Gibbs free energy (G), where $\Delta G = \Delta H - T\Delta S$.
- Processes with a negative ΔG are referred to as spontaneous; they release free energy and are known as exergonic processes. Reactions with a positive ΔG require free energy and are known as endergonic reactions (Figure 4.8).
- Diffusion is an example of a exergonic process driven by the increase in entropy as molecules spread out (Figure 4.9).
- Reactions tend to be spontaneous if they are exothermal ($\Delta H < 0$) (Figure 4.6).
- Factors that oppose the completion of spontaneous reactions, such as the relative concentrations of reactants and products, produce an equilibrium point at which reactants are converted to products and products are converted back to reactants, at equal rates (Figure 4.10).

4.3 Thermodynamics and Life

- Living systems maintain a highly organized state because they are open systems. They bring in both energy and matter from the surroundings and use them to maintain an organized state. Because they release energy and disordered molecules into the environment, the second law of thermodynamics is upheld, as the entropy of the system and its surroundings increases (Figure 4.11).
- All biological molecules break down (entropy increases). This is why a constant supply of energy (food) is required to sustain all life (Figure 4.12).
- Energy enters the biosphere when photosynthetic organisms absorb sunlight. The remainder of the biosphere is sustained by consuming the chemical energy produced by photosynthetic organisms (Figure 4.13).

4.4 Overview of Metabolism

- Metabolism is the biochemical modification and use of energy in the synthesis and breakdown of organic molecules.
- A catabolic pathway releases the potential energy of a molecule in breaking it down to a simpler molecule. An anabolic (biosynthetic) pathway uses energy to convert a simple molecule to a more complex molecule (Figure 4.14).
- The hydrolysis of ATP releases free energy that can be used as a source of energy for the cell (Figure 4.15).
- A cell can couple the exergonic reaction of ATP breakdown (not technically a hydrolysis reaction) to make an otherwise endergonic reaction proceed spontaneously. These coupling reactions require enzymes (Figure 4.16).
- The ATP used in coupling reactions is replenished by reactions that link ATP synthesis to catabolic reactions. ATP thus cycles between reactions that release free energy and reactions that require free energy (Figure 4.17).

4.5 The Role of Enzymes in Biological Reactions

- What prevents many exergonic reactions from proceeding rapidly is that they need to overcome an energy barrier (the activation energy, E_a) to get to the transition state (Figures 4.18 and 4.19).
- Enzymes are catalysts that greatly speed the rate at which spontaneous reactions occur because they lower the activation energy (Figure 4.20).
- Enzymes are usually specific: they catalyze reactions of only a single type of molecule or a group of closely related molecules (Figure 4.21).
- Catalysis occurs at the active site, which is the site where the enzyme binds to the substrate (reactant molecule). After combining briefly with the substrate, the enzyme is released unchanged when the reaction is complete (Figure 4.22).
- Three major mechanisms contribute to enzymatic catalysis by reducing the activation energy: (1) enzymes bring reacting molecules together, (2) enzymes expose reactant molecules to altered charge environments that promote catalysis, and (3) enzymes change the shape of substrate molecules (Figure 4.23).

4.6 Factors That Affect Enzyme Activity

- When substrate is abundant, the rate of a reaction is proportional to the amount of enzyme. At a fixed enzyme concentration, the rate of a reaction increases with substrate concentration until the enzyme becomes saturated with reactants. At that point, further increases in substrate concentration do not increase the rate of the reaction (Figure 4.24).
- Many cellular enzymes are regulated by nonsubstrate molecules called inhibitors. Competitive regulation (inhibition) occurs when molecules interfere with reaction rates by combining with the active site of an enzyme (Figure 4.25).
- Noncompetitive regulation (also known as allosteric regulation) occurs when molecules influence enzyme activity by binding to the enzyme at sites other than the active site (Figure 4.25). While some molecules (allosteric activators) increase enzyme activity, others (allosteric inhibitors) result in a decrease in enzyme activity.
- An example of allosteric regulation is feedback inhibition: the product of an enzyme-catalyzed pathway acts as an allosteric inhibitor of the first enzyme in the pathway (Figure 4.27).
- Typically, an enzyme has optimal activity at a certain pH and a certain temperature; at pH and temperature values above and below the optimum, the reaction rate falls off (Figures 4.28 and 4.29).

Questions

Self-Test Questions

1. Which of the following statements about energy and thermodynamics is correct?
   a. Earth is an isolated system.
   b. Living organisms are closed systems.
   c. Energy conversions can never be 100% efficient.
   d. The total amount of energy in the universe is always decreasing.

2. Which of the following statements about entropy is correct?
   a. We eat food to maintain high entropy.
   b. The entropy of any system always increases.
   c. Entropy is a measure of the total energy content of a system.
   d. The entropy of water increases as it turns from a liquid into a gas.

3. For a reaction to be exergonic, which of the following must occur?
   a. It must also be exothermic.
   b. There must be an input of energy to proceed.
   c. The products must have less enthalpy than the reactants.
   d. The products must have less free energy than the reactants.
   e. The entropy of the products must be greater than the entropy of the reactants.

4. Which of the following statements is correct?
   a. At equilibrium, ΔG is negative.
   b. Living organisms are never at equilibrium.
   c. An isolated system will never reach equilibrium.
   d. Molecules that have high free energy are very stable.
   e. Most biochemical reactions have a ΔG far from zero.

5. Instructors often mention the "hydrolysis of ATP" as the source of energy for cellular reactions. But this statement is inaccurate. Why?
   a. A molecule can never be the source of energy.
   b. ATP actually contains very little free energy.
   c. The hydrolysis of GTP is more common than ATP in cellular reactions.
   d. Water does not enter the active site of enzymes linked to ATP breakdown.

6. Propane is thermodynamically unstable; why is it kinetically stable?
   a. It is highly electronegative.
   b. Its breakdown is exergonic ($-\Delta G$).
   c. It has a high activation energy (E_a).
   d. It contains an abundance of oxygen and little hydrogen.

7. Which of the following statements about an enzyme is correct?
   a. It decreases the ΔG of an endergonic reaction.
   b. It is a protein and therefore is encoded by a gene.
   c. It can make an endergonic reaction proceed spontaneously.
   d. One enzyme molecule can only bind a single substrate molecule at any one time.

8. Compared with competitive inhibition, which of the following statements is correct only for non-competitive inhibition of an enzyme-catalyzed reaction?
   a. It changes the conformation of the enzyme.
   b. The inhibitory molecule is similar to the normal substrate.
   c. Inhibition decreases the rate at which the product is made.
   d. It results in the enzyme becoming permanently inactive.

9. Which of the following statements about allosteric enzymes is correct?
   a. An allosteric activator prevents binding at the active site.
   b. Their activity can be finely controlled by metabolites within the cell.
   c. The allosteric site of the enzyme binds additional substrate molecules.
   d. An enzyme that possesses allosteric sites does not possess an active site.

10. Which of the following explains the shape of a curve that plots enzyme activity as a function of temperature?
    a. As temperature increases, the rate of all reactions slows down.
    b. At high temperatures, the structural integrity of the enzyme breaks down.
    c. At high temperatures, the rate of catalysis stays high and constant—it saturates.
    d. At low but increasing temperatures, the rate of collisions between substrate and enzyme molecules decreases.

Questions for Discussion

1. Trees become more complex as they develop spontaneously from seeds to adults. Does this process violate the second law of thermodynamics? Why or why not?

2. Trace the flow of energy through your body. What products increase the entropy of you and your surroundings?

3. You have found a molecular substance that accelerates the rate of a particular reaction. What kind of information would you need to demonstrate that this molecular substance is an enzyme?

4. The addition or removal of phosphate groups from ATP is a fully reversible reaction. In what way does this reversibility facilitate the use of ATP as a coupling agent for cellular reactions?

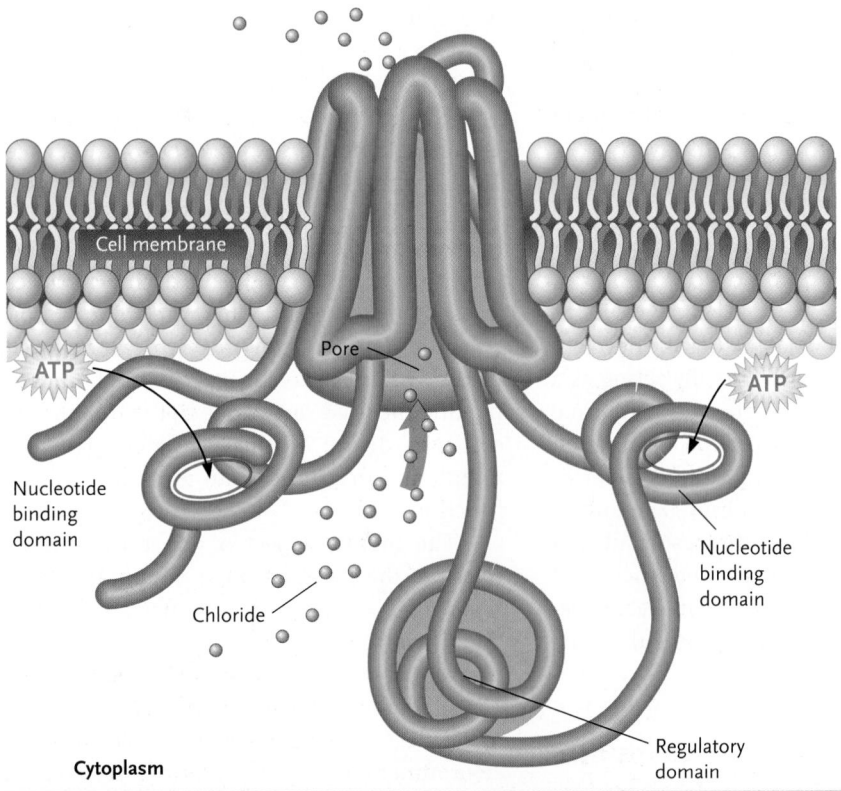

The cystic fibrosis transmembrane conductance regulator (CFTR) is a chloride pump. Mutations to the CFTR gene result in the pump being defective, causing cystic fibrosis.

Cell Membranes and Signalling

WHY IT MATTERS

Cystic fibrosis (CF) is one of the most common genetic diseases. It affects approximately 1 in 3900 children born in Canada. People with CF suffer from a progressive impairment of lung and gastrointestinal function, and although the treatment of CF patients is slowly improving, their average life span remains under 40 years. CF is caused by mutation to a gene that codes for a protein called the cystic fibrosis transmembrane conductance regulator (CFTR). In normal cells, CFTR acts as a membrane transport protein that pumps chloride ions (Cl^-) out of the cells that line the lungs and intestinal tract into the overlying mucus lining. This produces an electrical gradient across the membrane and results in the movement of positively charged sodium ions (Na^+) into the mucus lining. Because of the high ion concentration (Na^+ and Cl^-), water moves, by osmosis, out into the mucus lining, keeping it moist. Keeping the lining of the lungs and intestinal tract wet is critical to their proper functioning. In individuals with CF, the Cl^- channel CFTR does not function properly, which results in water being retained within cells, resulting in a build-up of thick, dry mucus that cannot effectively be removed by coughing. Besides obstructing airways and preventing normal breathing, the build-up of mucus in the lungs makes CF patients very susceptible to bacterial infections.

Currently, there is no cure for CF. Although treatments for CF are steadily improving, as the disease progresses in young adults, invasive procedures, including lung transplants, are often necessary. Since CF is caused by a defect to a single gene, the greatest hope is in gene therapy (see Chapter 15), which would attempt to insert normal copies of the CFTR gene into affected cells. However, many technical and ethical hurdles need to be overcome before gene therapy becomes a viable treatment option.

The structure and function of biological membranes are the focus of this chapter. We first consider the structure of membranes and then examine how membranes selectively transport substances in and out of cells and organelles. We close the chapter with a discussion of the critical role membranes play in signal transduction through the binding of molecules and the subsequent activation of intracellular signalling pathways.

5.1 An Overview of the Structure of Membranes

One of the keys to the evolution of life was the development of the cell or **plasma membrane.** By acting as a selectively permeable barrier, the plasma membrane allows for the uptake of key nutrients and elimination of waste products while maintaining a protected environment for cellular processes to occur. The subsequent development of internal membranes resulted in compartmentalization of processes and increased complexity. A good example of this is the nuclear envelope, which defines the hallmark of the eukaryotic cell—the nucleus.

5.1a A Membrane Consists of Proteins in a Fluid of Lipid Molecules

Our current view of membrane structure is based on the **fluid mosaic model (Figure 5.1).** The model proposes that membranes are not rigid with molecules

locked into place but rather consist of proteins that move around within a mixture of lipid molecules that has the consistency of olive oil.

The lipid molecules of all biological membranes exist in a double layer called a bilayer that is less than 10 nm thick. By comparison, this page is approximately 100 000 nm thick. The lipid molecules of the bilayer vibrate, flex back and forth, spin around their long axis, move sideways, and exchange places within the same bilayer half. Only rarely does a lipid molecule flip-flop between the two layers. Exchanging places within a layer occurs millions of times a second, making the lipid molecules in the membrane highly dynamic. As we will discuss later, maintaining the membrane in a fluid state is critical to membrane function.

The mosaic aspect of the fluid mosaic model refers to the fact that most membranes contain an assortment of types of proteins. This includes proteins involved in transport and attachment, signal transduction, and processes such as electron transport (Figure 5.1). Because they are larger than lipid molecules, proteins move more slowly in the fluid environment of the membrane. As well, a small number of membrane proteins anchor cytoskeleton filaments to the membrane and do not move. As also shown in Figure 5.1, a number of the lipid and protein components of some membranes have carbohydrate

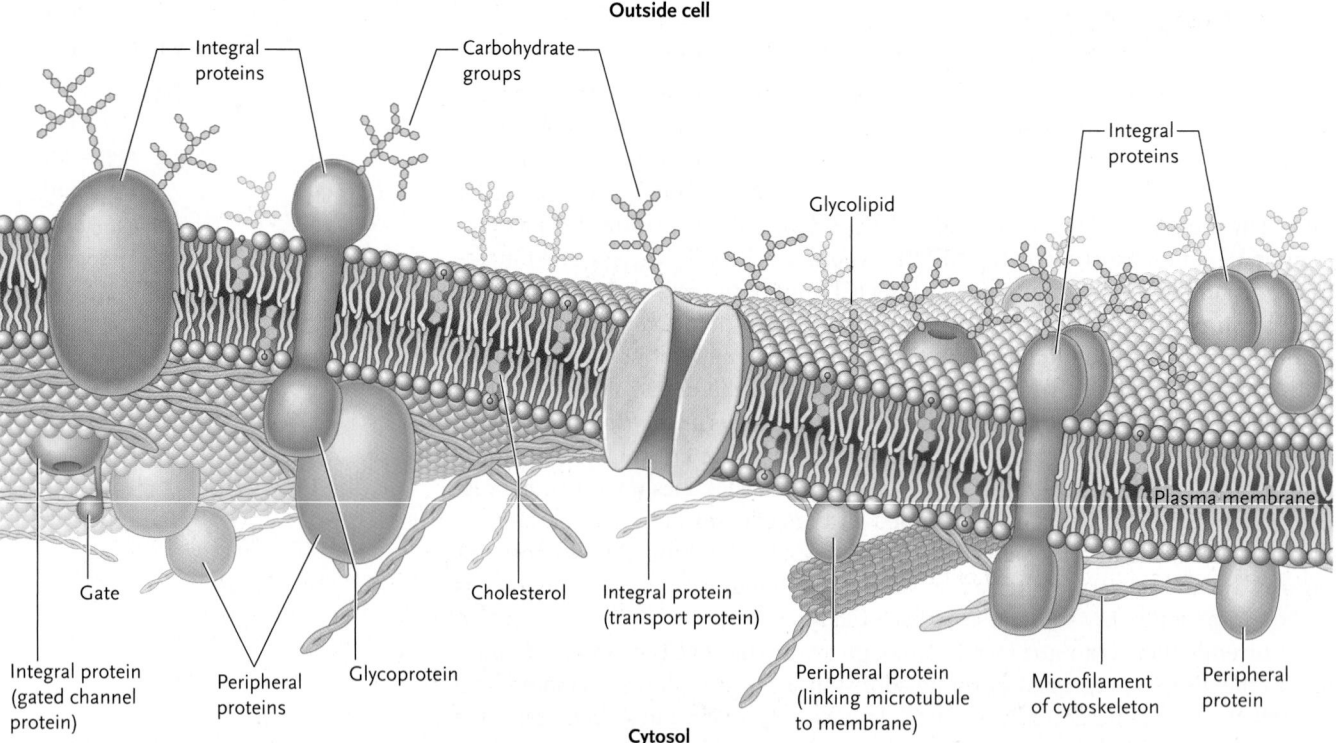

Figure 5.1

Membrane structure according to the fluid mosaic model. The model proposes that integral membrane proteins are suspended individually in a fluid lipid bilayer. Peripheral proteins are attached to integral proteins or membrane lipids mostly on the cytoplasmic side of the membrane (shown only on the inner surface in the figure). Carbohydrate groups of membrane glycoproteins and glycolipids face the cell exterior.

groups linked to them, forming glycolipids and glycoproteins.

The relative proportions of lipid and protein within a membrane vary considerably depending on the type of membrane. For example, membranes that contain protein complexes involved in electron transport, such as the inner mitochondrial membrane, contain large amounts of protein (76% protein and only 24% lipid), whereas the plasma membrane contains nearly equal amounts of protein and lipid (49% and 51%, respectively). Myelin, which is a membrane that functions to insulate nerve fibres, is composed mostly of lipids (18% protein and 82% lipid).

An important characteristic of membranes, illustrated in Figure 5.1, is that the proteins and other components of one half of the lipid bilayer are different from those that make up the other half of the bilayer. This is referred to as membrane asymmetry, and it reflects differences in the functions performed by each side of the membrane. For example, a range of glycolipids and carbohydrate groups are attached to proteins on the external side of the plasma membrane, whereas components of the cytoskeleton bind to proteins on the internal side of the plasma membrane. In addition, hormones and growth factors bind to receptor proteins that are found only on the external surface of the plasma membrane.

5.1b Experimental Evidence in Support of the Fluid Mosaic Model

The fluid mosaic model of membrane structure is supported by two major pieces of experimental evidence.

Membranes Are Fluid. In a now classic study carried out in 1970, David Frye and Michael A. Edidin grew human cells and mouse cells separately in tissue culture. They were able to tag the human or mouse membrane proteins **(Figure 5.2)** with dye molecules: the human proteins were linked to red dye molecules and the mouse proteins were linked to green. Frye and Edidin then fused the human and mouse cells. Within minutes, they found that the two distinctly coloured proteins began to mix. In less than an hour, the two colours had completely intermixed on the fused cells, indicating that the mouse and human proteins had moved around in the fused membranes.

Based on the measured rates at which molecules mix in biological membranes, the membrane bilayer appears to be about as fluid as olive oil or light machine oil.

Membrane Asymmetry. One of the key experiments revealing membrane asymmetry utilized the **freeze-fracture technique** in combination with electron microscopy **(Figure 5.3, p. 100).** In this technique, a block of cells is rapidly frozen by dipping it in liquid nitrogen (–196°C). Then the block is fractured by hitting it with a microscopically sharp knife edge. Often

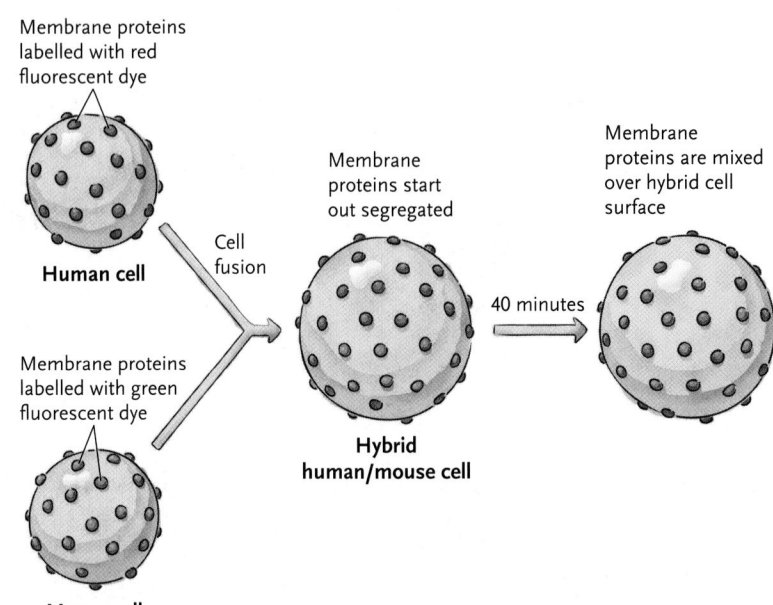

Experimental Research Figure 5.2 The Frye-Edidin experiment provided evidence that the membrane bilayer is fluid. In the experiment, membrane proteins were found to rapidly migrate over the surface of the hybrid cell.

the fracture splits bilayers into inner and outer halves, exposing the membrane interior. Using electron microscopy, the split membranes appear as smooth layers in which individual particles the size of proteins are embedded (shown in Figure 5.3c, p. 100). From these images, it is clear that the particles on either side of the membrane differ in size, number, and shape, providing evidence that the two sides are distinctly different.

STUDY BREAK

1. Describe the fluid mosaic model of membrane structure.
2. What is meant by the term *membrane asymmetry*?

5.2 The Lipid Fabric of a Membrane

The foundation or underlying fabric of all biological membranes is the lipid molecules. Collectively, the term *lipid* refers to a diverse group of water-insoluble molecules that includes **fats; phospholipids**, which are the dominant lipids in membranes; and steroids. A structural overview of these molecules is found in *The Purple Pages*. As we discuss in this section, keeping membranes in a fluid state is important to membrane function. Many organisms can adjust the types of lipids in the membranes such that membranes do not become too stiff (viscous) or too fluid (liquid).

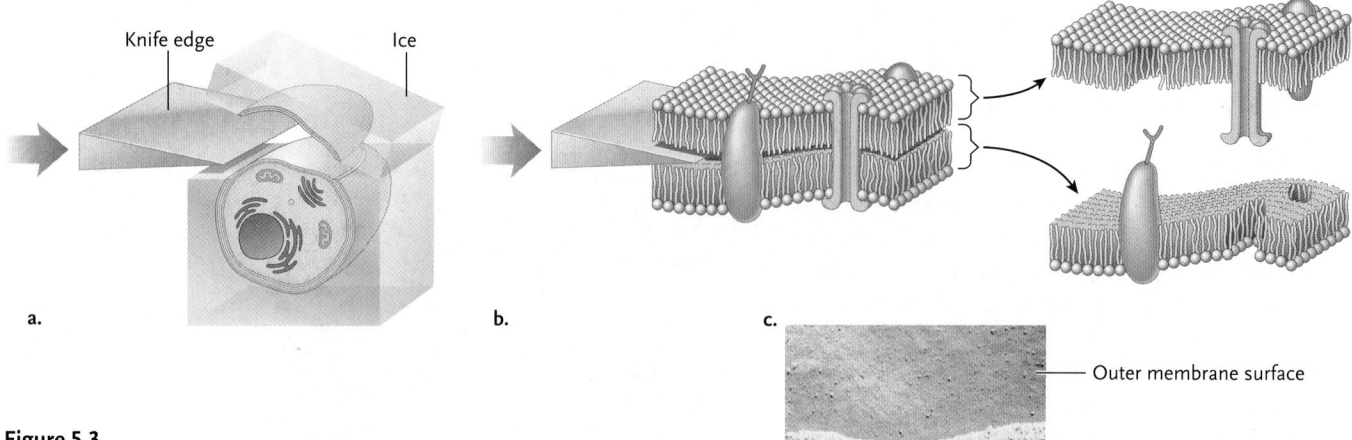

Knife edge   Ice

a.   b.   c.

Outer membrane surface

Exposed membrane interior

Don W. Fawcett/Science Source

Figure 5.3
The freeze-fracture technique allows for analysis of the membrane interior.
(a) The freeze-fracture technique. **(b)** The fracture may split the membrane bilayers into inner and outer halves. **(c)** The particles visible in the exposed membrane interior are integral membrane proteins.

5.2a Phospholipids Are the Dominant Lipids in Membranes

The lipid bilayer, which represents the foundation of biological membranes, is formed of **phospholipids**. As shown in **Figure 5.4a,** each phospholipid consists of a head group attached to two long chains of carbon and hydrogen (a **hydrocarbon**) called a fatty acid. The head group consists of glycerol linked to one of several types of **alcohols** or amino acids by a phosphate group (see Figure 5.4a). A property that all phospholipids possess, which is critical to the structure and function of membranes, is that they are **amphipathic**—the molecule contains a region that is *hydrophobic* (water fearing) and a region that is *hydrophilic* (water loving). Whereas the fatty acid chains of a lipid are nonpolar, the phosphate-containing head group is polar. Overall, polar molecules tend to be hydrophilic and nonpolar molecules hydrophobic. (For a review of molecular polarity, see *The Purple Pages*.) Laundry detergents are common amphipathic molecules—they are excellent at removing oil stains from clothing while also being soluble in water.

As illustrated in Figure 5.4a, phospholipids can differ in the degree of unsaturation of their fatty acids. Notice in Figure 5.4a that one of the fatty acids is fully saturated—all the carbons are bound to the maximum number of hydrogen atoms. The second fatty acid contains a carbon–carbon double bond (denoted by the arrow) and thus is **unsaturated**. As shown by the space-filling model, the presence of the C–C double bond imparts a kink or bend to the fatty acid tail (Figure 5.4b).

When added to water, phospholipids self-assemble into one of three structures—a **micelle**, a liposome, or

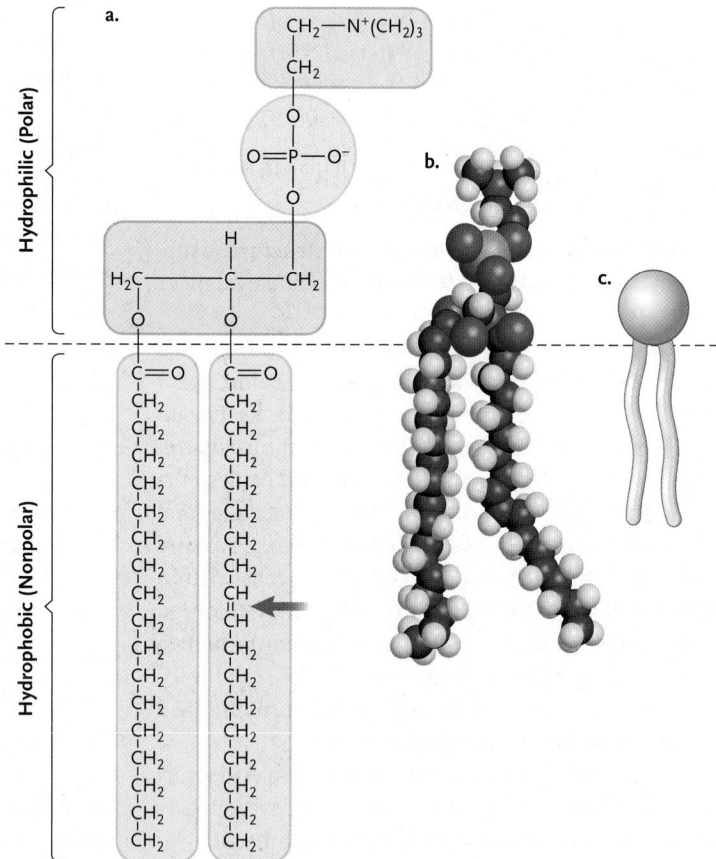

Hydrophilic (Polar)

Hydrophobic (Nonpolar)

Figure 5.4
Phospholipid structure. (a) Chemical formula of phosphatidylcholine. The polar head group consists of glycerol (shown in pink) linked to the organic molecule choline (shown in blue) by a phosphate group (shown in yellow). In addition, the glycerol is linked to two fatty acids, each 18 carbons long. The structure of phospholipids is also often represented as space-filling models **(b)** and as an icon **(c)**. As shown in the space-filling model, the presence of a carbon–carbon double bond (denoted by the arrow in (a)) imparts a bend to one of the fatty acids.

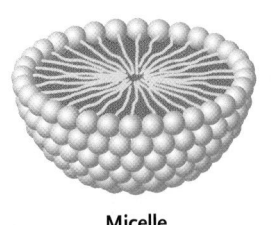

Micelle

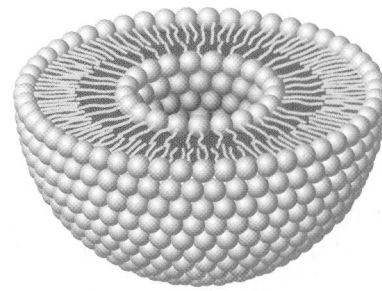

Liposome

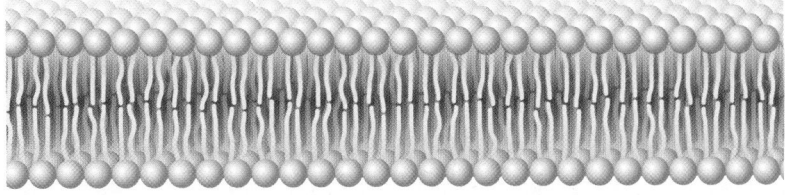

Phospholipid bilayer

Figure 5.5

In an aqueous environment, phospholipids self-assemble into micelles, liposomes, or bilayers.

a bilayer **(Figure 5.5)**. Which structure forms depends mostly on the phospholipid concentration. Phospholipids spontaneously form these structures in an aqueous environment because of the *hydrophobic effect*—the tendency of polar molecules like water to exclude hydrophobic molecules such as fatty acids. This results in the aggregation of lipid molecules in structures where the fatty acid tails interact with each other and the polar head groups associate with water. These arrangements are favoured because they represent the lowest energy state and are more likely to occur over any other arrangement.

5.2b Fatty Acid Composition and Temperature Affect Membrane Fluidity

The fluidity of the lipid bilayer is primarily influenced by two factors: the type of fatty acids that make up the lipid molecules and the temperature. Fully **saturated fatty acids** are linear, which allows lipid molecules to pack tightly together **(Figure 5.6a)**. In contrast, lipid molecules with one or more unsaturated fatty acids are prevented from packing closely together because the presence of double bonds introduces kinks in the fatty acid backbone (Figure 5.6b). As a result, the more unsaturated the fatty acids of the lipid molecules, the more fluid the membrane.

Besides fatty acid composition, the temperature can also dramatically affect membrane fluidity. As the temperature drops and the random molecular motion of lipid molecules slows down, a point is reached where fluidity is lost and the phospholipid molecules form a semisolid gel. This is exactly what happens when melted butter cools—at a certain temperature it

turns from a liquid into a solid. The temperature at which gelling occurs depends upon the fatty acid composition. The more unsaturated a group of lipid molecules, the lower the temperature at which gelling occurs. Likewise, at high temperatures, the increased molecular motion may result in membranes becoming too fluid, resulting in a loss of structure. For most membrane systems, the normal fluid state is achieved by a mixed population of saturated and unsaturated fatty acids.

5.2c Organisms Can Adjust Fatty Acid Composition

Keeping membranes in the optimal state of fluidity is absolutely essential to cell function. Exposure to low temperatures can be harmful because the resulting membrane gelling can decrease membrane permeability and inhibit the function of enzymes and receptors attached to, or localized within, the bilayer. Electron transport, for example, requires molecules to migrate rapidly within the membrane bilayer. If the membrane becomes less fluid due to low temperature, the rate of electron transport will decrease and eventually stop.

Problems also arise at high temperature. Membranes may become too fluid due to the increase in molecular motion, which can result in membrane leakage. Ions such as K^+, Na^+, and Ca^{2+} begin to freely diffuse across the membrane, resulting in an irreversible disruption of cellular ion balance that can rapidly lead to cell death.

Unlike with humans, the body temperature of most organisms closely matches that of the external environment. Examples of such organisms are plants, bacteria, protists, and insects. To live at a range of temperatures, these organisms are able to alter membrane fluidity by adjusting the relative proportion of unsaturated fatty acids.

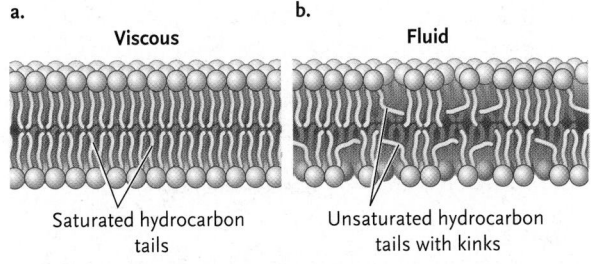

a.   **Viscous**   b.   **Fluid**

Saturated hydrocarbon tails   Unsaturated hydrocarbon tails with kinks

Figure 5.6

Lipid molecule composition affects how closely the molecules interact. Lipid molecules that contain saturated hydrocarbon tails are closely packed **(a)**, whereas unsaturated hydrocarbon tails have kinks that prevent lipid molecules from packing closely together **(b)**.

Unsaturated fatty acids are produced during fatty acid biosynthesis through the action of a group of enzymes called desaturases **(Figure 5.7a).** All fatty acids are initially synthesized as fully saturated molecules without any double bonds. Desaturases act on these saturated fatty acids by catalyzing a reaction that removes two hydrogen atoms from neighbouring carbon atoms and introducing a double bond. There are many different desaturase enzymes, each one introducing a double bond at a specific point along the fatty acid chain. Whereas some unsaturated fatty acids contain only one carbon–carbon double bond, others may contain two or more, which indicates the action of more than one desaturase.

Like many proteins, desaturase abundance is regulated at the level of gene transcription, which results in changes to desaturase transcript (mRNA) abundance.

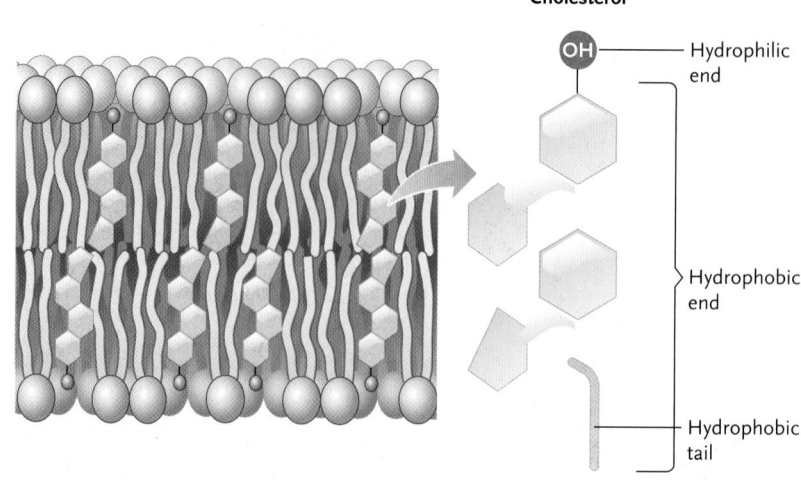

Figure 5.8

The position taken by cholesterol within a membrane. The hydrophilic–OH group at one end of the molecule extends into the hydrophilic region of the bilayer; the ring structure extends into the hydrophobic membrane interior.

Figure 5.7b shows how the abundance of a specific desaturase transcript changes with growth temperature in a cyanobacterium. As growth temperature decreases, desaturase transcript abundance goes up, which results in an increase in synthesis of the desaturase enzyme. Higher amounts of desaturases, in turn, result in an increase in the abundance of unsaturated fatty acids. By regulating desaturase abundance, many organisms can closely regulate the amount of unsaturated fatty acids that get incorporated into membranes and thereby maintain proper membrane fluidity.

Besides lipids, a group of compounds called **sterols** also influence membrane fluidity. The best example of a sterol is **cholesterol (Figure 5.8),** which is found in the membranes of animal cells but not in those of plants or prokaryotes. Sterols act as membrane buffers: at high temperatures, they help restrain the movement of lipid molecules, thus reducing the fluidity of the membrane. However, at lower temperatures, sterols disrupt fatty acids from associating by occupying space between lipid molecules, thus slowing the transition to the nonfluid gel state.

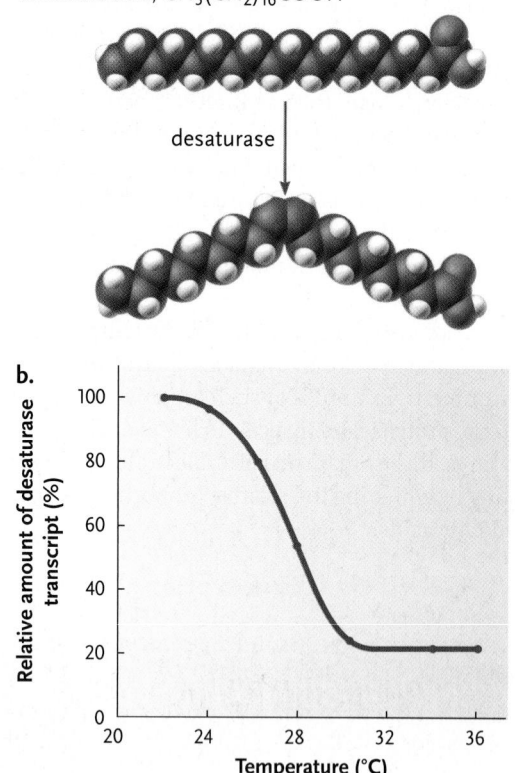

a. Stearic acid, $CH_3(CH_2)_{16}COOH$

desaturase

b.

(graph: y-axis "Relative amount of desaturase transcript (%)" from 0 to 100; x-axis "Temperature (°C)" from 20 to 36)

Figure 5.7

Organisms can regulate the degree of fatty acid unsaturation. **(a)** Desaturases are a class of enzymes that introduce double bonds into fatty acids, thereby altering the degree of unsaturation. **(b)** Graph showing relative amounts of desaturase transcript amounts (mRNA abundance) in relation to growth temperature in a cyanobacterium.

STUDY BREAK

1. Why is maintaining proper membrane fluidity important for membrane function?
2. What is the relationship between temperature and desaturase expression?

5.3 Membrane Proteins

While the lipid molecules constitute the foundation of a membrane, the unique set of proteins associated with the membrane determines its function and makes each

In the food industry, the use of fats containing saturated fatty acids is more desirable than the use of oils that contain unsaturated fatty acids. The lack of double bonds means that lipids containing saturated fatty acids are more stable and less prone to the oxidation that can decrease shelf life and affect the texture and taste of the final product. Moreover, hard fats have a higher melting temperature, which makes them useful in many applications, such as in baking and deep-frying.

Because animal-based saturated fats such as butter and lard are expensive and susceptible to spoilage, the food industry has, for many decades, used saturated fats produced through the industrial process of hydrogenation. This process removes *cis* double bonds from fatty acids by heating vegetable oil in the presence of hydrogen gas and a catalyst. In the food industry, partial hydrogenation is practised, which results in a product that is still malleable and not too hard. One of the unintended consequences of partial hydrogenation is that the *cis* double bonds that do not become hydrogenated tend to be reconfigured into the *trans* orientation. Although small amounts of *trans* fats are found naturally in the milk and meat of **ruminant** animals such as cows and sheep, through partial hydrogenation, human consumption of *trans* fats has increased tremendously over the last 70 years.

There is now clear medical evidence that the consumption of *trans* fats is unhealthy. A comprehensive review of research on *trans* fat consumption and health by the *New England Journal of Medicine* in 2006 clearly demonstrated the existence of a strong connection between *trans* fat consumption and elevated risk of coronary heart disease, a leading cause of death in North America. *Trans* fats have also been linked to increased incidence of other health problems as well. The physiological basis for the increased risk to health by increased *trans* fat consumption is not fully understood and remains a very active area of research. The increased risk may be due, in part, to the fact that a major group of enzymes called lipases, which aid in the breakdown of many types of lipids, including *cis* unsaturated fats, do not recognize the *trans* configuration. This leads to *trans* fats staying in the bloodstream longer, which may lead to increased incidence of arterial deposition, which may lead to coronary heart disease.

In response to the overwhelming medical evidence that *trans* fats are harmful, governments around the world are implementing restrictions on the amount of *trans* fats foods can contain. In Canada, the *trans* fat content of vegetable oils and soft margarines is now limited to 2% of the total fat content, whereas the *trans* fat content for all other foods is 5% of the total fat content, including ingredients sold to restaurants. Similar guidelines are in place in many European countries, as well as being implemented in the United States.

In response to these new guidelines, food manufacturers and restaurant chains have reformulated their products to be *trans* fat free. This has primarily been achieved by simply replacing hydrogenated fats with naturally saturated fats. Many nutritionists argue that these fully saturated alternatives may not offer any health benefit.

---

membrane unique. As we will discuss in this section, two major types of proteins are associated with membranes: integral and peripheral membrane proteins.

5.3a The Key Functions of Membrane Proteins

Membrane proteins can be separated into four major functional categories, as shown in **Figure 5.9, p. 104.** It should be noted that all of these functions may exist in a single membrane and that one protein or protein complex may serve more than one of these functions:

1. **Transport.** Many substances cannot freely diffuse through the membrane. Instead, a protein may provide a hydrophilic channel that allows movement of a specific molecule. Alternatively, a membrane protein may change its shape and in so doing shuttle specific molecules from one side of a membrane to the other.

2. **Enzymatic activity.** A number of enzymes are membrane proteins. The best example of this is the enzymes associated with the respiratory and photosynthetic electron transport chains (ETCs).

3. **Signal transduction.** Membranes often contain receptor proteins on their outer surface that bind to specific chemicals such as hormones. On binding, these receptors trigger changes on the inside surface of the membrane that lead to transduction of the signal through the cell.

4. **Attachment/recognition.** Proteins exposed to both the internal and external membrane surfaces act as attachment points for a range of cytoskeleton elements, as well as components involved in cell–cell recognition.

5.3b Integral Membrane Proteins Interact with the Membrane Hydrophobic Core

Proteins that are embedded in the phospholipid bilayer are called **integral membrane proteins.** Many of these traverse the entire lipid bilayer at least once and are referred to as *transmembrane proteins.* Because they

a. Transport

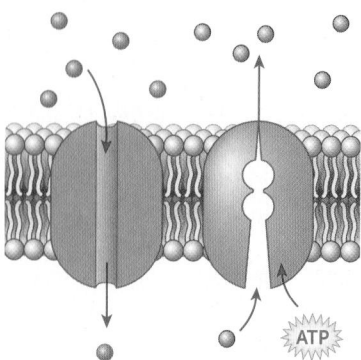

b. Enzymatic activity

Enzymes

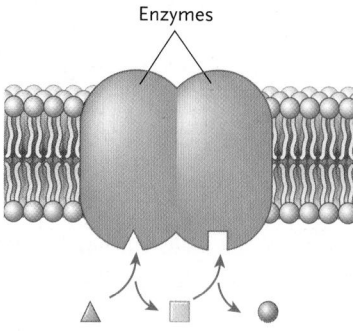

Outside cell

Channel

Alpha helix

Membrane surface

NH$_2$

Plasma membrane interior

Cytosol

COOH

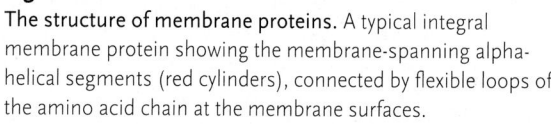

Figure 5.10

The structure of membrane proteins. A typical integral membrane protein showing the membrane-spanning alpha-helical segments (red cylinders), connected by flexible loops of the amino acid chain at the membrane surfaces.

c. Signal transduction

Signal

Receptor

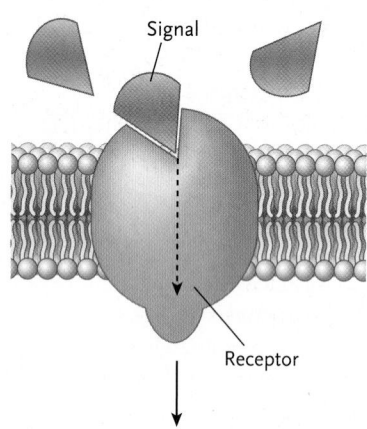

d. Attachment/recognition

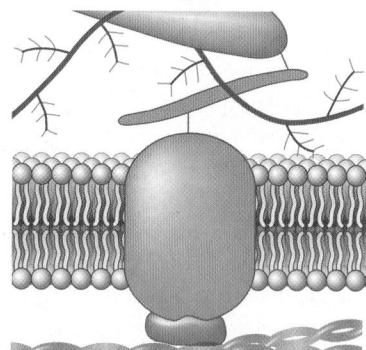

Figure 5.9

The major functions of membrane proteins.

distinct regions of predominantly nonpolar amino acids linked by regions that are dominated by polar and charged amino acids. These polar amino acids are found in the portions of the protein that are exposed to the aqueous environment on either side of the membrane (see Figure 5.11).

have to interact with both the aqueous environment on both sides of the membrane and the hydrophobic core, transmembrane proteins have distinct regions (called **domains**) that differ markedly in polarity. The domain that interacts with the lipid bilayer consists predominantly of nonpolar amino acids that collectively form a type of **secondary structure** termed an *alpha helix* **(Figure 5.10)** (see *The Purple Pages* for an overview of protein structure). By contrast, the portions of a transmembrane protein that are exposed on either side of the membrane are composed of primarily polar amino acids. (The different classes of amino acids are presented in *The Purple Pages*).

Given the amino acid sequence (**primary structure**) of a protein, it is usually quite simple to determine if it is likely a transmembrane protein. What one looks for, usually with the aid of a computer program, are stretches of primarily nonpolar amino acids. These stretches are about 17 to 20 amino acids in length, which matches the peptide length needed to span the lipid bilayer **(Figure 5.11)**. Most transmembrane proteins span the membrane more than once. So, for example, if a protein has three membrane-spanning domains, the primary sequence would show three

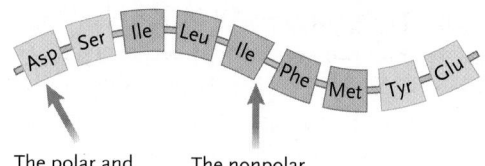

Asp — Ser — Ile — Leu — Ile — Phe — Met — Tyr — Glu

The polar and charged amino acids are hydrophilic.

The nonpolar amino acids are hydrophobic.

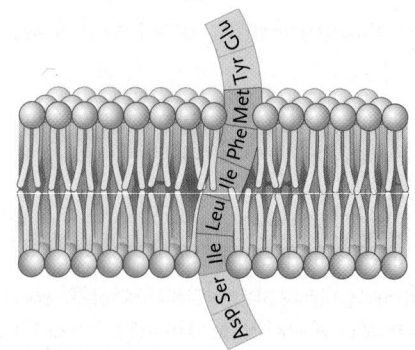

Figure 5.11

Transmembrane proteins can be identified by the presence of stretches of amino acids that are primarily nonpolar. These regions of the protein interact with the hydrophobic regions of the membrane. Usually between 17 and 20 amino acids are needed to span the membrane once. For clarity this model shows only five nonpolar amino acids spanning the membrane.

5.3c Peripheral Membrane Proteins Interact with the Membrane Hydrophilic Surface

The second major group of membrane proteins are **peripheral membrane proteins**, so called because they are positioned on the surface of a membrane and do not interact with the hydrophobic core of the membrane. Peripheral proteins are held to membrane surfaces by noncovalent bonds—hydrogen bonds and **ionic bonds**—usually by interacting with the exposed portions of integral proteins as well as directly with membrane lipid molecules. Many peripheral proteins are found on the cytoplasmic side of the plasma membrane and form part of the cytoskeleton (look back at Figure 5.1). In addition, as we will see in later chapters, key enzymes involved in both respiratory and photosynthetic electron transport are peripheral membrane proteins. Because peripheral membrane proteins do not interact with the hydrophobic core of the membrane, they are made up of a mixture of polar and nonpolar amino acids.

STUDY BREAK

1. What roles are served by membrane proteins?
2. What are the two major classes of membrane proteins?

5.4 Passive Membrane Transport

The hydrophobic nature of membranes severely restricts the free movement of many molecules into and out of cells and from one compartment to another. Molecules such as O_2 diffuse very rapidly across membranes, which is important considering the vital role O_2 plays in cellular respiration. However, a range of other molecules, including ions, charged molecules, and macromolecules, do not readily move across membranes. In this section, we consider the diffusion of molecules from one compartment to the other and the factors that influence the rate of that diffusion.

5.4a Passive Transport Is Based on Diffusion

Passive transport is defined as the movement of a substance across a membrane without the need to expend chemical energy such as ATP. What drives passive transport is **diffusion**, the net movement of a substance from a region of higher concentration to a region of lower concentration. Above absolute zero (−273°C), molecules are in constant motion, which results in molecules becoming uniformly distributed in space. Diffusion is the primary mechanism of **solute** movement within a cell.

The driving force behind diffusion is an increase in entropy (see Chapter 4). In the initial state, when molecules are more concentrated in one region or on

one side of a membrane, the energy associated with the molecules is more localized. As diffusion occurs, the entropy increases as the energy spreads out until the molecules are evenly distributed and the entropy is highest **(Figure 5.12)**. As the distribution proceeds to the state of maximum disorder, the molecules release free energy, which can accomplish work (see Section 4.2 for a discussion of entropy and free energy).

The rate of diffusion depends on the concentration difference (**concentration gradient**) between two areas or across a membrane. The larger the gradient, the faster the rate of diffusion. Similar to chemical equilibrium (see Chapter 4), when diffusing molecules reach equilibrium, there is still movement of molecules from one space to another, but no net change in concentration (see Figure 5.12).

5.4b There Are Two Types of Passive Transport: Simple and Facilitated

There are two types of passive transport: simple diffusion and facilitated diffusion. **Simple diffusion** is the movement of molecules directly across a membrane without the involvement of a transporter. The rate of simple diffusion of a molecule depends upon two factors: molecular size and lipid solubility. As shown in **Figure 5.13, p. 106**, some molecules diffuse very rapidly across the membrane, while other molecules are essentially unable to transit the membrane.

Small nonpolar molecules such as O_2 and CO_2 are readily soluble in the hydrophobic interior of a membrane and move very rapidly from one side to the other. As well, steroid hormones and many drugs that tend to be amphipathic can readily transit the lipid bilayer. Small uncharged molecules such as water and glycerol, even though they are polar, are still able to move quite rapidly across the membrane (see Figure 5.13). In contrast, the membrane is practically impermeable to charged molecules, including ions such as Cl^-, Na^+, and phosphate (PO_4^{3-}). Transport of small ions is about a billionth (10^{-9}) the rate of the transport of water. Their charge and associated hydration shell contribute to

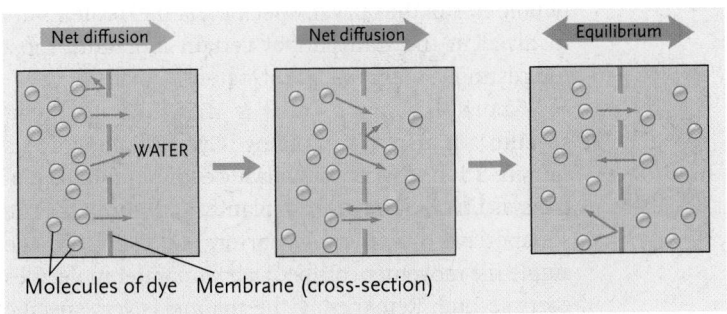

Figure 5.12

Diffusion is the movement of molecules from regions of high concentration to areas of low concentration. It is driven by the increase in entropy associated with energy becoming more dispersed.

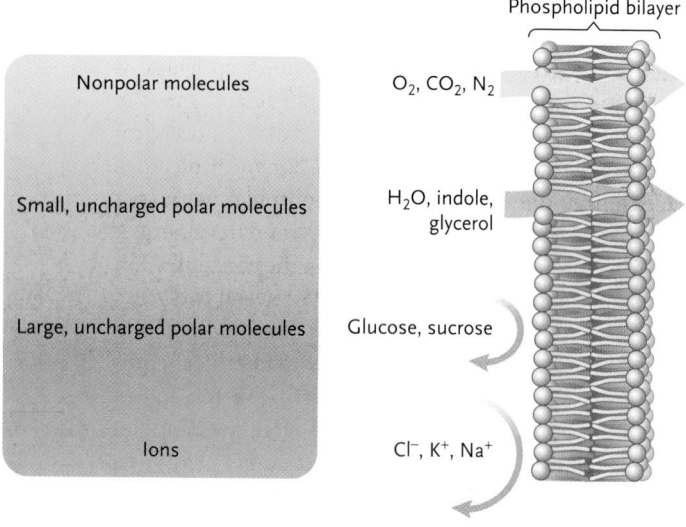

Figure 5.13

The size and charge of a molecule affect the rate of diffusion across a membrane.

ions being prevented from entering the hydrophobic core of the membrane.

The diffusion of molecules across a membrane through the aid of a transporter is called **facilitated diffusion**. The diffusion of many polar and charged molecules, such as water, amino acids, sugars, and ions, relies on specific transport complexes for their rapid movement from one compartment to another. Although facilitated diffusion involves specific transporters, just like simple diffusion, transport depends upon a concentration gradient across the membrane—when the gradient falls to zero, facilitated diffusion stops.

5.4c Two Groups of Transport Proteins Carry Out Facilitated Diffusion

Facilitated diffusion is carried out by two types of transport proteins: channel proteins and carrier proteins, both of which are transmembrane proteins **(Figure 5.14)**. **Channel proteins** form hydrophilic pathways in the membrane through which molecules can pass. The channel aids the diffusion of molecules by providing an avenue that is shielded from the hydrophobic core of the bilayer. Specific channel proteins are involved in the transport of certain ions and, most interestingly, the transport of water.

The diffusion of water is facilitated by water-specific transport proteins called aquaporins (Figure 5.14a). Aquaporins have been found in organisms as diverse as bacteria, plants, and humans. The aquaporin channel is very narrow and allows for the single-file movement of about a billion water molecules every second. Remarkably, the channel is very specific for water and does not allow for the diffusion of ions including protons. Recent three-dimensional models of aquaporin show the presence of positive charges in the centre of the channel that are thought to specifically

repel the transport of protons. For his discovery of aquaporins, Peter Agre at Johns Hopkins University received the Nobel Prize for chemistry in 2003.

Another type of channel protein that is found in all eukaryotes is the **gated channel** (Figure 5.14b). These transporters can switch between open, closed, and intermediate states and are critical to the movement of most ions, for example, sodium (Na^+), potassium (K^+), calcium (Ca^{2+}), and chlorine (Cl^-). The gates may be opened or closed by changes in voltage across the membrane, for instance, or by binding signal molecules. The opening or closing involves changes in the protein's three-dimensional shape. In animals, voltage-gated ion channels are used in nerve conduction and the control of muscle contraction (see Chapters 44 and 46). As well, CFTR, the Cl^- channel that is defective in individuals with cystic fibrosis, is a gated channel (see "Why It Matters").

The second class of transport proteins that form passageways through the lipid bilayer are **carrier proteins** (Figure 5.14). Each carrier protein binds a single specific solute, such as a sugar molecule or an amino acid, and transports it across the lipid bilayer. Because a single solute is transferred in this carrier-mediated fashion, the transfer is called *uniport transport*. In performing the transport step, the carrier protein undergoes conformational changes that progressively move the solute binding site from one side of the membrane to the other, thereby transporting the solute. This property distinguishes carrier protein function from channel protein function.

Most transport proteins display a high degree of substrate specificity, in a way similar to an enzyme. For example, transporters that carry glucose are unable to transport fructose, which is structurally similar. This specificity allows various cells and cellular compartments to tightly control what gets in and out. The kinds of transport proteins present in the plasma membrane or, for example, on the inner membrane of the mitochondrion depend ultimately on the type of cell and growth conditions.

How can you experimentally determine if a molecule is transported by facilitated diffusion and not just simple diffusion? First, with facilitated diffusion, the rate of movement across the membrane is much faster than one would predict based just on the chemical structure of the molecule being transported **(Figure 5.15, p. 108)**. Second, facilitated diffusion can be saturated in the same way as an enzyme can be saturated, by substrate. A membrane has a limited number of transporters for a particular molecule. If you measure the rate of transport at increasing concentration differences across a membrane, the rate of transport of a particular molecule (the substrate) reaches a plateau that represents a state when essentially all of the transporters are occupied all the time by substrate (they are saturated). Increasing the concentration further has no effect on the rate of transport (see Figure 5.15). By comparison, in simple diffusion, the whole membrane surface is effectively the transporter; thus, the rate of transport, although usually slower, never reaches a plateau.

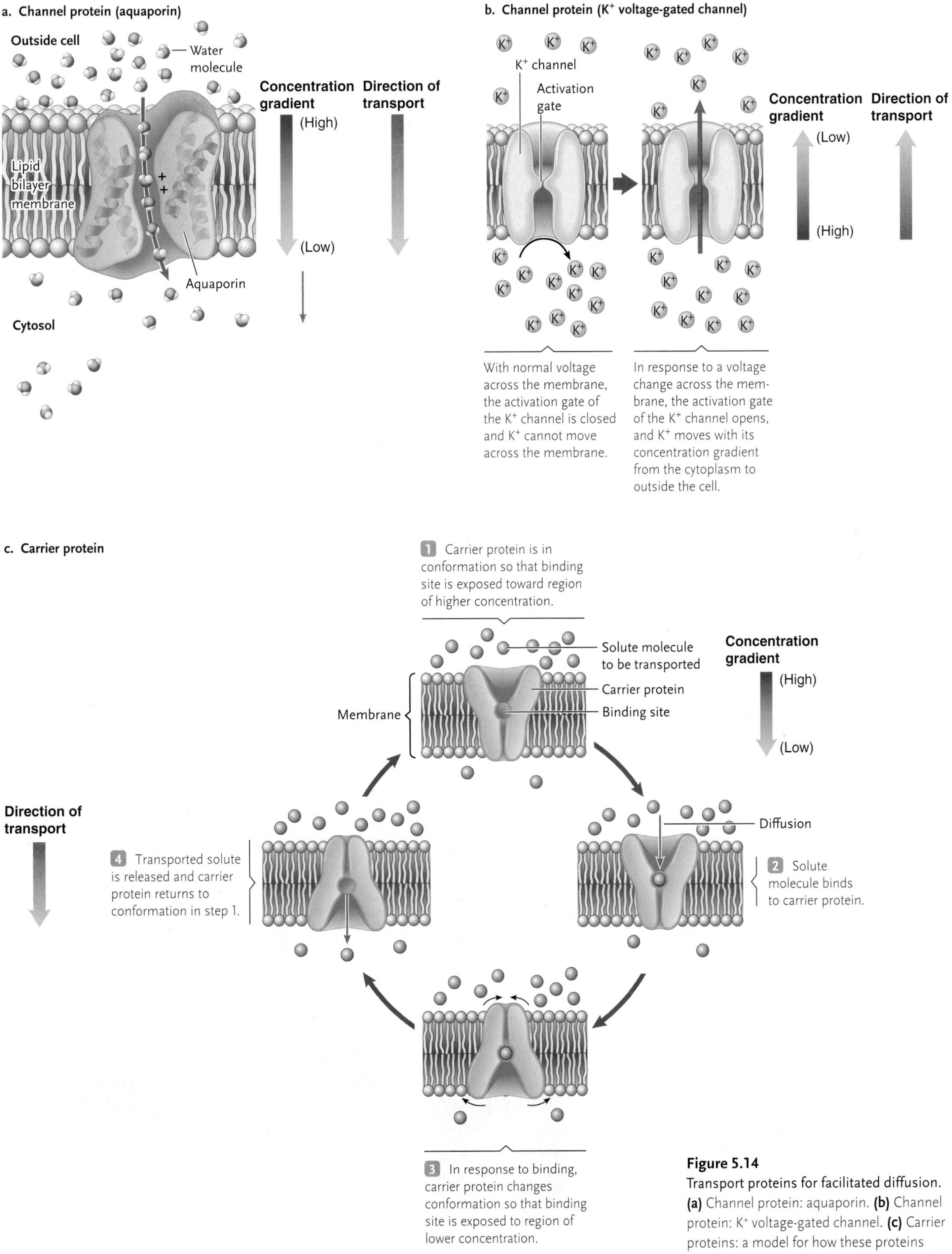

a. Channel protein (aquaporin)

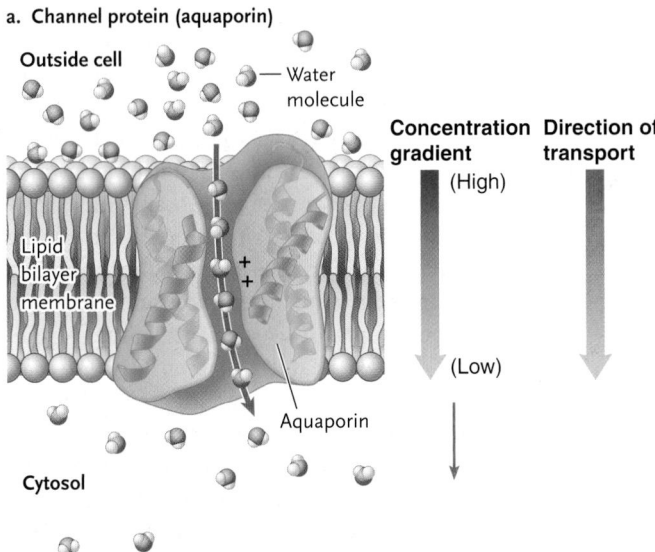

Outside cell

Water molecule

Lipid bilayer membrane

Aquaporin

Cytosol

Concentration gradient

(High)

(Low)

Direction of transport

b. Channel protein (K⁺ voltage-gated channel)

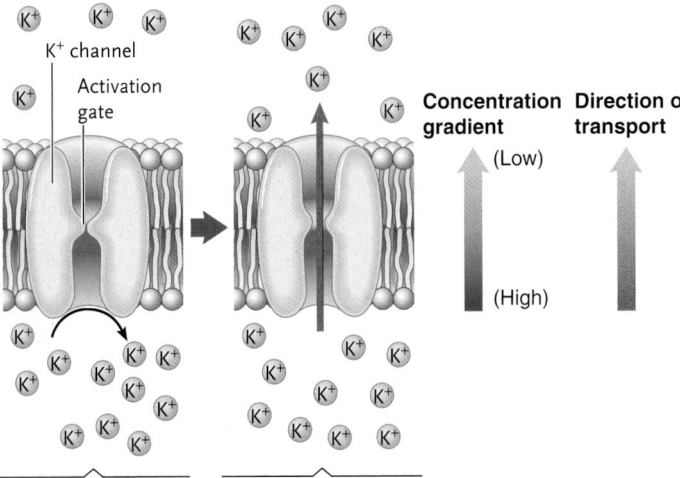

K^+ channel

Activation gate

Concentration gradient

(Low)

(High)

Direction of transport

With normal voltage across the membrane, the activation gate of the K^+ channel is closed and K^+ cannot move across the membrane.

In response to a voltage change across the membrane, the activation gate of the K^+ channel opens, and K^+ moves with its concentration gradient from the cytoplasm to outside the cell.

c. Carrier protein

1 Carrier protein is in conformation so that binding site is exposed toward region of higher concentration.

Solute molecule to be transported

Carrier protein

Binding site

Membrane

Concentration gradient

(High)

(Low)

Diffusion

2 Solute molecule binds to carrier protein.

Direction of transport

4 Transported solute is released and carrier protein returns to conformation in step 1.

3 In response to binding, carrier protein changes conformation so that binding site is exposed to region of lower concentration.

Figure 5.14
Transport proteins for facilitated diffusion.
(a) Channel protein: aquaporin. **(b)** Channel protein: K^+ voltage-gated channel. **(c)** Carrier proteins: a model for how these proteins transport solutes such as glucose.

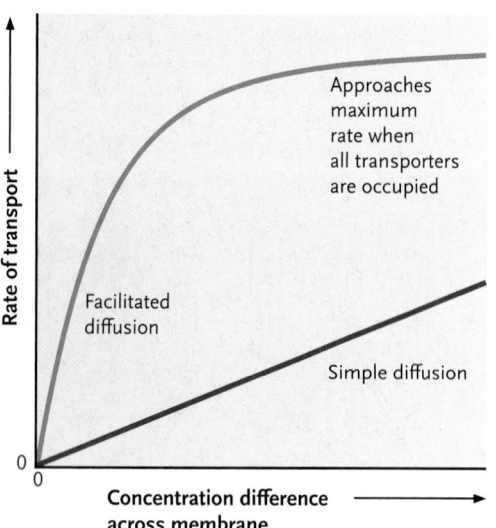

Figure 5.15
Simple diffusion and facilitated diffusion display different transport kinetics. Compared with simple diffusion, facilitated diffusion leads to higher rates of transport and displays saturation kinetics.

5.4d Osmosis Is the Passive Diffusion of Water

Like solutes, water can also move passively across membranes in a process called osmosis. The passive transport of water occurs constantly in living cells.

Inward or outward movement of water by osmosis develops forces that can cause cells to swell or shrink. Formally, **osmosis** is defined as the diffusion of water molecules across a selectively permeable membrane from a solution of lower solute concentration to a solution of higher solute concentration. For osmosis to take place, the selectively permeable membrane must allow water molecules to pass but not molecules of the solute. Osmosis occurs in cells because they contain a solution of proteins and other molecules that are retained in the cytoplasm by a membrane impermeable to them but permeable to water. Osmosis can occur by simple diffusion through the lipid bilayer or it can be facilitated by aquaporins (see Section 5.4c).

The movement of water by osmosis is dictated by solute concentration. If the solution surrounding a cell contains dissolved substances at lower concentrations than in the cell, the solution is said to be **hypotonic** to the cell (*hypo* = under or below; *tonos* = tension or tone). When a cell is in a hypotonic solution, water enters by osmosis, and the cell tends to swell **(Figure 5.16a)**. Animal cells, such as red blood cells, in a hypotonic solution may actually swell to the point of bursting. This is in contrast to plant cells, where the

Figure 5.16
Osmotic water movement. The diagrams show what happens when a cellophane bag filled with a 2 M sucrose solution is placed in **(a)** a hypotonic, **(b)** a hypertonic, or **(c)** an isotonic solution. The cellophane is permeable to water but not to sucrose molecules. The width of the arrows shows the amount of water movement. In the first beaker, the distilled water is hypotonic to the solution in the bag; net movement of water is into the bag. In the second beaker, the 10 M solution is hypertonic to the solution in the bag; net movement of water is out of the bag. In the third beaker, the solutions inside and outside the bag are isotonic; there is no net movement of water into or out of the bag. The animal cell micrographs show the corresponding effects on red blood cells placed in hypotonic, hypertonic, or isotonic solutions. (Micrographs, M. Sheetz, R. Painter, and S. Singer. *Journal of Cell Biology*, 70:493, 1976. By permission of Rockefeller University Press.)

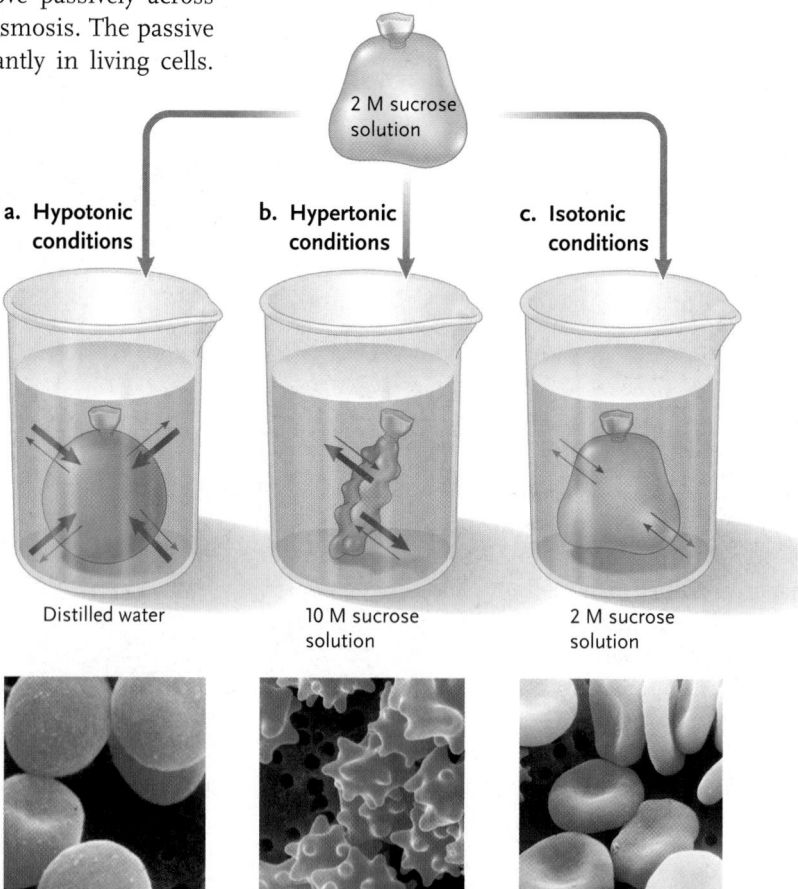

© 1976 The Rockefeller University Press. *The Journal of Cell Biology*, 1976, 70:193–203. doi: 10.1083/jcb.70.1.193.

presence of the cell wall prevents the cells from bursting in a hypotonic solution. Instead the cell pushes against the cell wall, resulting in what is called turgor pressure. This is discussed in more detail in Chapter 35.

If the solution that surrounds a cell contains solutes at higher concentrations than in the cell, then the outside solution is said to be **hypertonic** to the cell (*hyper* = over or above) (Figure 5.16b). When a cell is in a hypertonic solution, water leaves by osmosis. If the outward osmotic movement exceeds the capacity of cells to replace the lost water, both animal and plant cells will shrink (Figure 5.16b).

In animals, ions, proteins, and other molecules are concentrated in extracellular fluids, as well as inside cells so that the concentration of water inside and outside cells is usually equal or **isotonic** (*iso* = the same), as shown in Figure 5.16c. However, this comes at an energetic cost of constantly having to pump ions from one side to the other. For example, the ATP-dependent transport of Na^+ from inside to outside the cell is essential; otherwise water would move inward by osmosis and cause the cells to burst. Osmotic movement in plant cells is discussed more in depth in Chapter 35, whereas the mechanisms by which animals balance their water content are discussed in Chapter 50.

STUDY BREAK

1. How do the size and charge of a molecule influence its transport across a membrane?
2. Explain how aquaporin functions to transport water.
3. What is the difference between passive transport and active transport?

5.5 Active Membrane Transport

As shown in Figure 5.15, compared to simple diffusion, facilitated diffusion increases the rate of movement of molecules across membranes. However, this type of transport is limited to movement down a concentration gradient. Many cellular processes require molecules to be maintained in various cell compartments at very high concentrations. This is achieved by energy-dependent transport that moves molecules against a concentration gradient—from a region of lower concentration to a region of higher concentration.

5.5a Active Transport Requires Energy

The transport of molecules across a membrane against a concentration gradient requires the expenditure of energy and is referred to as **active transport.** The energy is usually in the form of ATP, and it is estimated that about 25% of a cell's ATP requirements are for the active transport of molecules. Active transport concentrates molecules such as sugars and amino acids inside cells and pushes ions in or out of cells.

The three main functions of active transport in cells and organelles are (1) uptake of essential nutrients from the fluid surrounding cells even when their concentrations are lower than in cells, (2) removal of secretory or waste materials from cells or organelles even when the concentration of those materials is higher outside the cells or organelles, and (3) maintenance of essentially constant intracellular concentrations of H^+, Na^+, K^+, and Ca^{2+}. Because ions are charged molecules, active transport of ions may contribute to voltage—an electrical potential difference—across the plasma membrane, called a membrane potential. This electrical difference across the plasma membrane is important in neurons and muscle cells and is discussed in more detail in Chapters 44 and 46, respectively.

There are two classes of active transport: primary and secondary. In **primary active transport,** the same protein that transports the molecules also hydrolyzes ATP to power the transport directly. In **secondary active transport,** the transport is indirectly driven by ATP. That is, the transport proteins use a favourable concentration gradient of ions built up by primary active transport as the energy source to drive the transport of a different molecule.

Other features of active transport (listed in **Table 5.1, p. 110**) resemble facilitated diffusion. Both processes depend on membrane transport proteins, both are specific, and the rate of both processes can plateau at high substrate concentrations.

5.5b Primary Active Transport Moves Positively Charged Ions

All primary active transport pumps move positively charged ions—H^+, Ca^{2+}, Na^+, and K^+—across membranes **(Figure 5.17, p. 110)**. The gradients of positive ions established by primary active transport pumps underlie functions that are absolutely essential for life. For example, the **proton pumps (H^+ pumps)** in plasma membranes push hydrogen ions from the cytoplasm to the cell exterior. These pumps (as in Figure 5.17) temporarily bind a phosphate group removed from ATP during the pumping cycle. Proton pumps have various functions. For example, in bacteria, archaea, and plants and fungi, proton pumps in the plasma membrane generate membrane potential. Proton pumps in lysosomes of animals and vacuoles of plants and fungi keep the pH within the organelle low, serving to activate the enzymes contained within them.

Another active transport system is the **calcium pump (Ca^{2+} pump)**, which is widely distributed among eukaryotes. It pushes Ca^{2+} from the cytoplasm to the cell exterior and from the cytosol into the vesicles of the endoplasmic reticulum (ER). As a result, Ca^{2+} concentration is typically high outside cells and inside

Table 5.1 — Characteristics of Transport Mechanisms

Characteristic	Passive Transport		Active Transport
	Simple Diffusion	Facilitated Diffusion	
Membrane component responsible for transport	Lipids	Proteins	Proteins
Binding of transported substance	No	Yes	Yes
Energy source	Concentration gradients	Concentration gradients	ATP hydrolysis or concentration gradients
Direction of transport	With gradient of transported substance	With gradient of transported substance	Against gradient of transported substance
Specificity for molecules or molecular classes	Nonspecific	Specific	Specific
Saturation at high concentrations of transported molecules	No	Yes	Yes

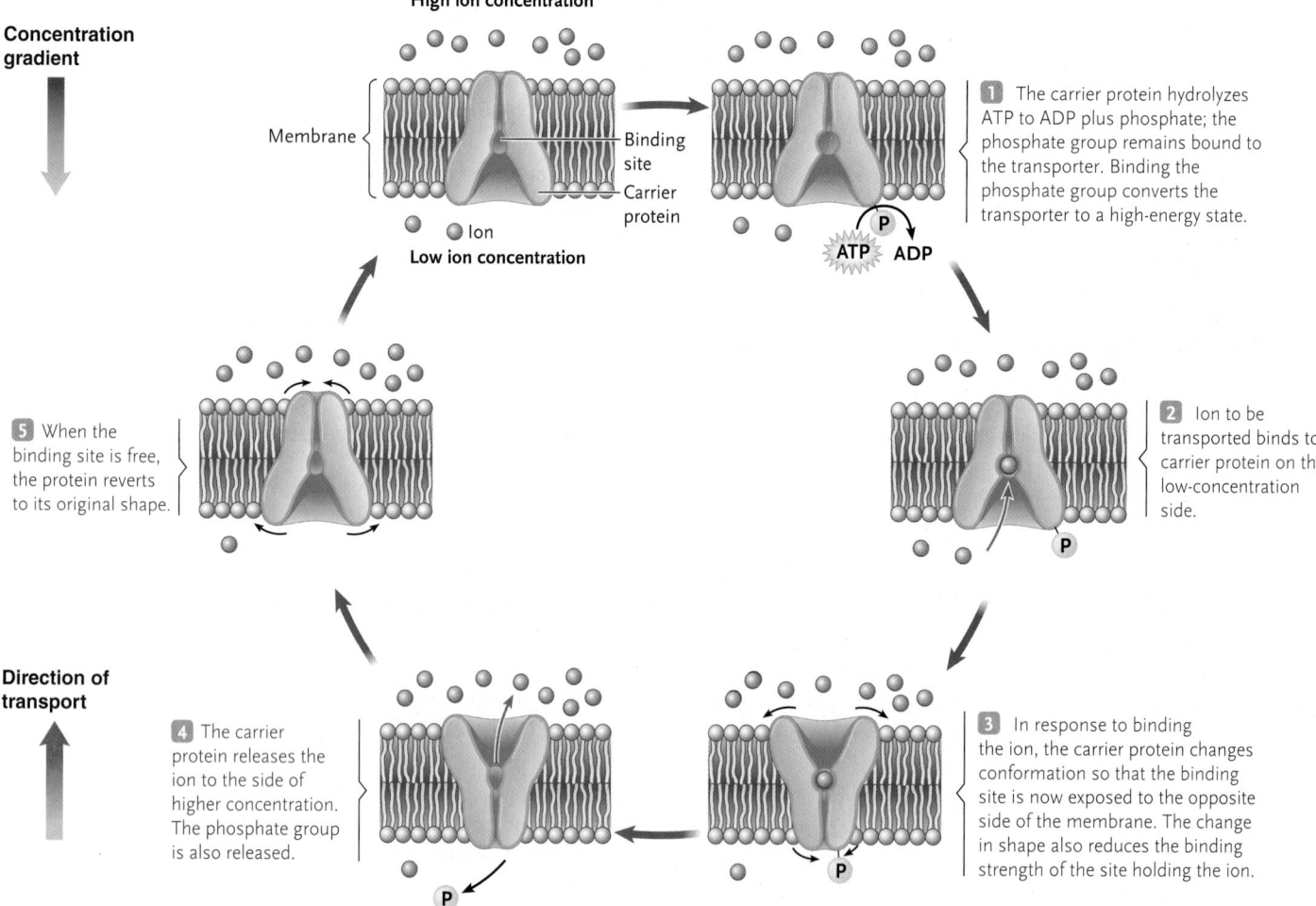

Figure 5.17
Model for how a primary active transport pump operates.

ER vesicles and low in the cytoplasmic solution. This Ca^{2+} gradient is used universally among eukaryotes as a regulatory control of cellular activities as diverse as secretion, microtubule assembly, and muscle contraction. The latter is discussed further in Chapter 46.

The **sodium–potassium pump** (or **Na⁺/K⁺ pump**), located in the plasma membrane of all animal cells, pushes three Na⁺ ions out of the cell and two K⁺ ions into the cell in the same pumping cycle (**Figure 5.18**). As a result, positive charges accumulate in excess

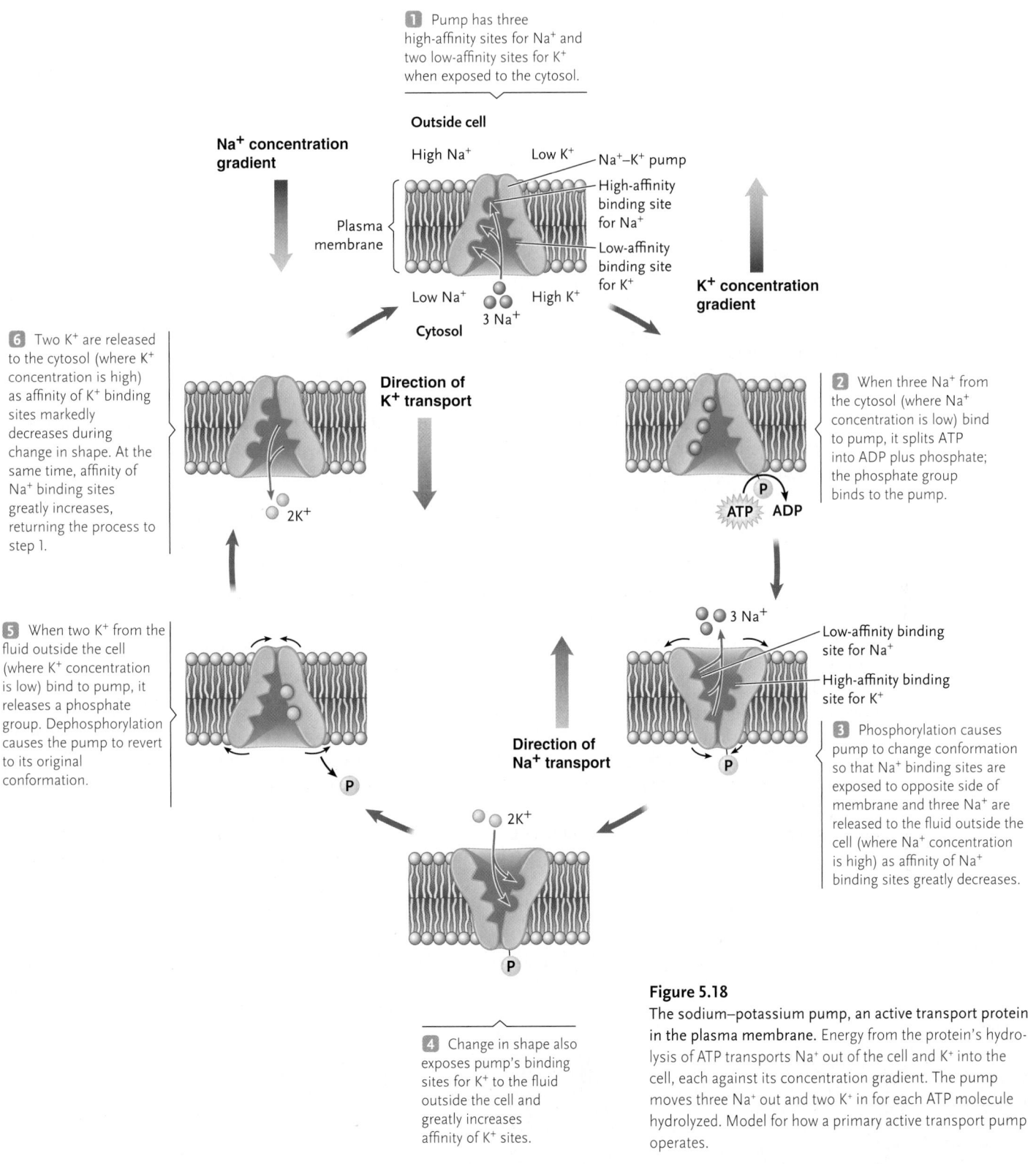

1 Pump has three high-affinity sites for Na$^+$ and two low-affinity sites for K$^+$ when exposed to the cytosol.

Na$^+$ concentration gradient

Outside cell

High Na$^+$ Low K$^+$ — Na$^+$–K$^+$ pump

High-affinity binding site for Na$^+$

Plasma membrane

Low-affinity binding site for K$^+$

Low Na$^+$ High K$^+$

3 Na$^+$

Cytosol

K$^+$ concentration gradient

6 Two K$^+$ are released to the cytosol (where K$^+$ concentration is high) as affinity of K$^+$ binding sites markedly decreases during change in shape. At the same time, affinity of Na$^+$ binding sites greatly increases, returning the process to step 1.

Direction of K$^+$ transport

2K$^+$

2 When three Na$^+$ from the cytosol (where Na$^+$ concentration is low) bind to pump, it splits ATP into ADP plus phosphate; the phosphate group binds to the pump.

P

ATP   ADP

5 When two K$^+$ from the fluid outside the cell (where K$^+$ concentration is low) bind to pump, it releases a phosphate group. Dephosphorylation causes the pump to revert to its original conformation.

3 Na$^+$

Low-affinity binding site for Na$^+$

High-affinity binding site for K$^+$

P

Direction of Na$^+$ transport

3 Phosphorylation causes pump to change conformation so that Na$^+$ binding sites are exposed to opposite side of membrane and three Na$^+$ are released to the fluid outside the cell (where Na$^+$ concentration is high) as affinity of Na$^+$ binding sites greatly decreases.

2K$^+$

P

Figure 5.18

The sodium–potassium pump, an active transport protein in the plasma membrane. Energy from the protein's hydrolysis of ATP transports Na$^+$ out of the cell and K$^+$ into the cell, each against its concentration gradient. The pump moves three Na$^+$ out and two K$^+$ in for each ATP molecule hydrolyzed. Model for how a primary active transport pump operates.

4 Change in shape also exposes pump's binding sites for K$^+$ to the fluid outside the cell and greatly increases affinity of K$^+$ sites.

outside the membrane, and the inside of the cell becomes negatively charged with respect to the outside. Voltage—an electrical potential difference—across the plasma membrane results from this difference in charge as well as from the unequal distribution of ions across the membrane created by passive transport. The voltage across a membrane, called a **membrane potential**, measures from about −50 to −200 millivolts (mV), with the minus sign indicating that the charge inside the cell is negative versus the outside. In sum, we have both a concentration difference (of the ions) and an electrical charge difference on the two sides of the membrane, constituting what is called an **electrochemical gradient.** Electrochemical gradients store energy that is used for other transport mechanisms. For instance, the electrochemical gradient across the membrane is involved with the movement of ions associated with nerve impulse transmission (described in Chapter 44). A membrane potential derived from a proton gradient across a membrane is the basis for ATP synthesis in mitochondria and chloroplasts, which will be discussed in Chapters 6 and 7, respectively.

5.5c Secondary Active Transport Moves Both Ions and Organic Molecules

As already noted, secondary active transport pumps use the concentration gradient of an ion established by a primary pump as their energy source. For example, the driving force for most secondary active transport in animal cells is the high outside/low inside Na$^+$ gradient set up by the sodium–potassium pump. In secondary active transport, the transfer of the solute across the membrane is always coupled with the transfer of the ion supplying the driving force.

Secondary active transport occurs by two mechanisms, known as *symport* and *antiport* **(Figure 5.19)**. In **symport**, the cotransported solute moves through the membrane channel in the same direction as the driving ion, a phenomenon known as **cotransport.** Sugars such as glucose and amino acids are examples of molecules actively transported into cells by symport. In **antiport**, the driving ion moves through the membrane channel in one direction, providing the energy for the active transport of another molecule in the opposite direction, a phenomenon known as **exchange diffusion.** In many cases, ions are exchanged by antiport. For example, antiport is the mechanism used in red blood cells for the coupled movement of chloride ions and bicarbonate ions through a membrane channel.

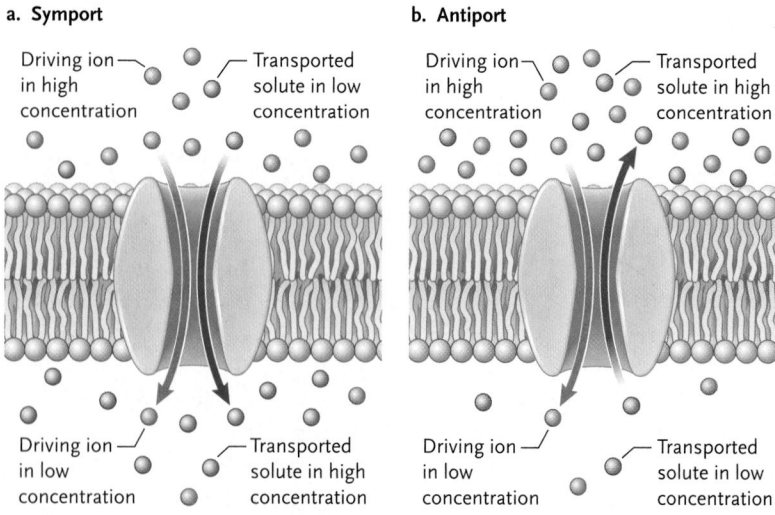

a. Symport

Driving ion in high concentration
Transported solute in low concentration

Driving ion in low concentration
Transported solute in high concentration

b. Antiport

Driving ion in high concentration
Transported solute in high concentration

Driving ion in low concentration
Transported solute in low concentration

Figure 5.19

Secondary active transport, in which a concentration gradient of an ion is used as the energy source for active transport of a solute. **(a)** In symport, the transported solute moves in the same direction as the gradient of the driving ion. **(b)** In antiport, the transported solute moves in the direction opposite to the gradient of the driving ion.

5.6 Exocytosis and Endocytosis

The largest molecules transported through cellular membranes by passive and active transport are about the size of amino acids or monosaccharides such as glucose. Eukaryotic cells import and export larger molecules by endocytosis and exocytosis. The export of materials by exocytosis primarily carries secretory proteins and some waste materials from the cytoplasm to the cell exterior. Import by endocytosis may carry proteins, larger aggregates of molecules, or even whole cells from the outside into the cytoplasm. Exocytosis and endocytosis also contribute to the back-and-forth flow of membranes between the endomembrane system and the plasma membrane. Both exocytosis and endocytosis require energy; thus, both processes stop if a cell's ability to make ATP is inhibited.

5.6a Exocytosis Releases Molecules to the Outside by Means of Secretory Vesicles

In exocytosis, secretory vesicles move through the cytoplasm and contact the plasma membrane **(Figure 5.20a)**. The vesicle membrane fuses with the plasma membrane, releasing the vesicle's contents to the cell exterior.

All eukaryotic cells secrete materials to the outside through exocytosis. For example, in animals, glandular cells secrete peptide hormones or milk proteins, and cells lining the digestive tract secrete mucus and digestive enzymes. Plant cells secrete carbohydrates by exocytosis to build a strong cell wall.

5.6b Endocytosis Brings Materials into Cells in Endocytic Vesicles

In endocytosis, proteins and other substances are trapped in pitlike depressions that bulge inward from the plasma membrane. The depression then pinches off as an endocytic vesicle. Endocytosis takes place in most eukaryotic cells by one of two distinct but related pathways. In the simpler of these mechanisms, **bulk-phase endocytosis** (sometimes called **pinocytosis**, meaning "cell drinking"), extracellular water is taken in along with any molecules that happen to be in solution in the water (Figure 5.20b). No binding by surface receptors takes place.

In the second endocytic pathway, **receptor-mediated endocytosis**, the molecules to be taken in are bound to the outer cell surface by receptor proteins

a. Exocytosis: vesicle joins plasma membrane, releases contents

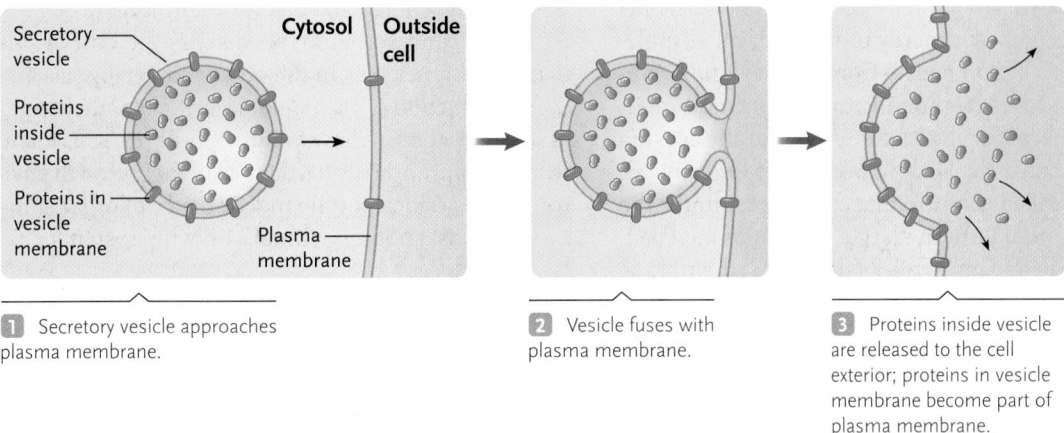

1 Secretory vesicle approaches plasma membrane.

2 Vesicle fuses with plasma membrane.

3 Proteins inside vesicle are released to the cell exterior; proteins in vesicle membrane become part of plasma membrane.

b. Bulk-phase endocytosis (pinocytosis): vesicle imports water and other substances from outside cell

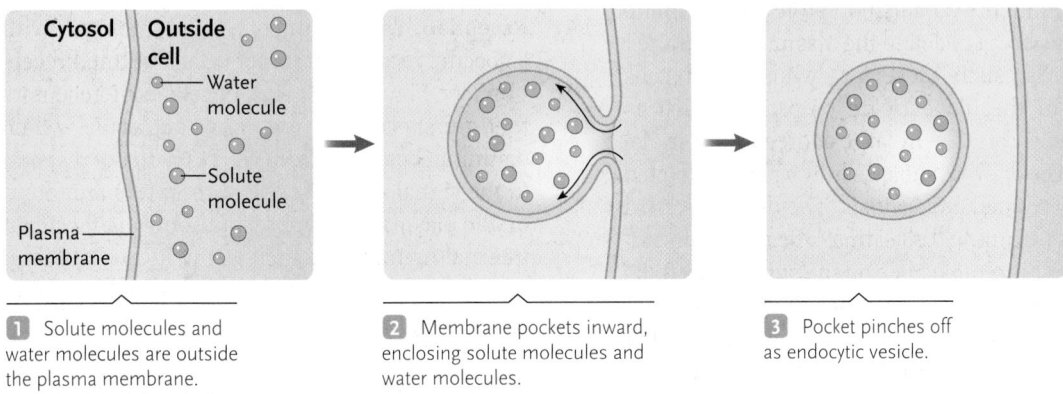

1 Solute molecules and water molecules are outside the plasma membrane.

2 Membrane pockets inward, enclosing solute molecules and water molecules.

3 Pocket pinches off as endocytic vesicle.

c. Receptor-mediated endocytosis: vesicle imports specific molecules

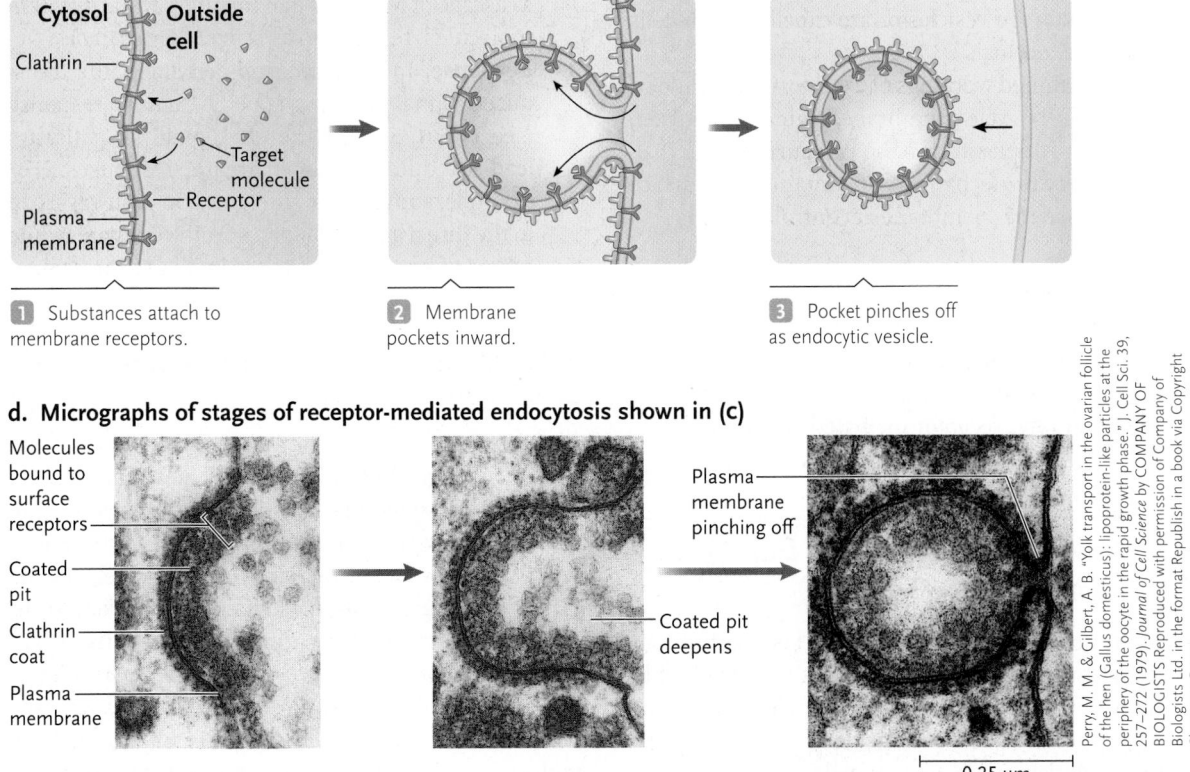

1 Substances attach to membrane receptors.

2 Membrane pockets inward.

3 Pocket pinches off as endocytic vesicle.

d. Micrographs of stages of receptor-mediated endocytosis shown in (c)

0.25 μm

Perry, M. M. & Gilbert, A. B. "Yolk transport in the ovarian follicle of the hen (Gallus domesticus): lipoprotein-like particles at the periphery of the oocte in the rapid growth phase." J. Cell Sci. 39, 257–272 (1979). *Journal of Cell Science* by COMPANY OF BIOLOGISTS Reproduced with permission of Company of Biologists Ltd. in the format Republish in a book via Copyright Clearance Center.

Figure 5.20 Exocytosis and endocytosis.

(Figure 5.20c). The receptors, which are integral proteins of the plasma membrane, recognize and bind only certain molecules—primarily proteins, or other molecules carried by proteins—from the solution surrounding the cell. After binding their target molecules, the receptors collect into a depression in the plasma membrane called a **coated pit** because of the network of proteins (called **clathrin**) that coat and reinforce the cytoplasmic side. With the target molecules attached, the pits deepen and pinch free of the plasma membrane to form endocytic vesicles. Once in the cytoplasm, an endocytic vesicle rapidly loses its clathrin coat and may fuse with a lysosome. The enzymes within the lysosome then digest the contents of the vesicle, breaking them down into smaller molecules useful to the cell. These molecular products—for example, amino acids and monosaccharides—enter the cytoplasm by crossing the vesicle membrane via transport proteins. The membrane proteins are recycled to the plasma membrane.

Some cells, such as certain white blood cells (*phagocytes*) in the bloodstream or protists such as *Amoeba proteus,* can take in large aggregates of molecules, cell parts, or even whole cells by a process related to receptor-mediated endocytosis. The process, called **phagocytosis** (meaning "cell eating"), begins when surface receptors bind molecules on the substances to be taken in **(Figure 5.21)**. Cytoplasmic lobes then extend, surround, and engulf the materials, forming a pit that pinches off and sinks into the cytoplasm as a large endocytic vesicle. The materials are then digested within the cell as in receptor-mediated endocytosis, and any remaining residues are sequestered permanently into storage vesicles or are expelled from cells as waste by exocytosis.

The combined workings of exocytosis and endocytosis constantly cycle membrane segments between the internal cytoplasm and the cell surface. The balance of the two mechanisms maintains the surface area of the plasma membrane at controlled levels.

STUDY BREAK

1. What is the mechanism of exocytosis?
2. What is the difference between bulk-phase endocytosis and receptor-mediated endocytosis?

5.7 Role of Membranes in Cell Signalling

Recall from Chapter 3 that one of the key attributes of all living things is the ability to sense and respond to changes to the environment. At the cellular level this is accomplished by the perception of signals. In multicellular organisms, signals may be derived from other cell types and tissues as well as factors external to the organism. These signals may be physical, such as changes in light and temperature, or they may be chemical, such as a hormone or growth regulator. In this section, we discuss the crucial role that membranes play in the perception of signals and the transduction of the signal to bring about changes in cell function. The ability of cells to sense and respond appropriately to changes in their growth environment is critical for the maintenance of organismal homeostasis, another hallmark of living systems.

5.7a Signal Transduction Links Signals with Downstream Cellular Responses

The steps that link the initial perception of a signal with its ultimate downstream effects is termed a signal transduction pathway or cascade. Most signal pathways involve the following three steps **(Figure 5.22)**:

1. **Reception.** The binding of a signal molecule with a specific receptor of target cells is termed reception (see Figure 5.22). Target cells have receptors that are specific for the signal molecule, which distinguishes them from cells that do not respond to the signal molecule. Most receptors are found on the plasma membrane, but some are found on internal membranes such as the ER. In addition, other receptors are soluble proteins that are found in the cytoplasm.

2. **Transduction.** The process whereby signal reception triggers other changes within the cell necessary to cause the cellular response is transduction (see Figure 5.22). Transduction typically involves a cascade of reactions that include several different molecules, referred to as a *signalling cascade*.

3. **Response.** In the third and last stage, the transduced signal causes a specific cellular response (see Figure 5.22). Different signalling pathways lead to different downstream responses. For example, some signal transduction pathways lead to the direct activation of a specific enzyme, while others often trigger changes in gene expression.

5.7b Membrane Surface Receptors

The membrane receptors that recognize and bind signal molecules are integral membrane proteins that extend through the entire membrane **(Figure 5.23a)**. Typically, the signal-binding site of the receptor is the part of the protein that extends from the outer membrane surface and is folded in a way that closely fits the signal molecule. The fit, which is similar to an enzyme–substrate interaction, is specific so that a particular receptor binds only one type of signal. When a signal molecule binds, for example, to a surface receptor associated with the plasma membrane, the molecular structure of that receptor changes so that it transmits the signal through the plasma membrane, activating the cytoplasmic end of the receptor protein.

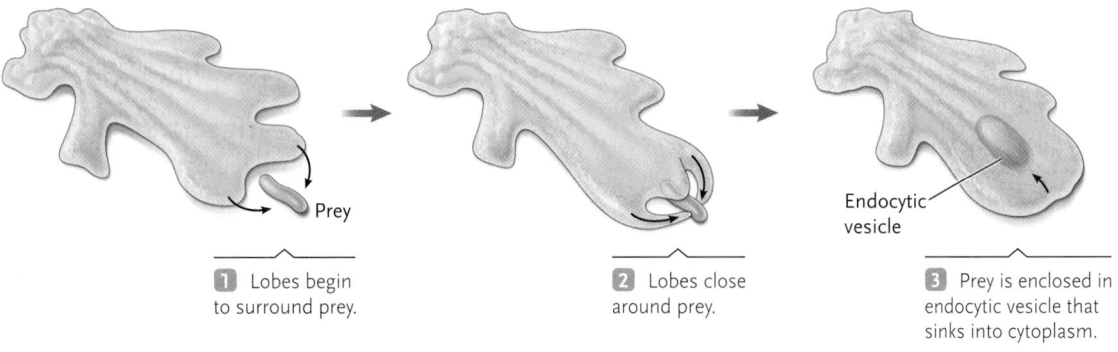

| 1 Lobes begin to surround prey. | 2 Lobes close around prey. | 3 Prey is enclosed in endocytic vesicle that sinks into cytoplasm. |

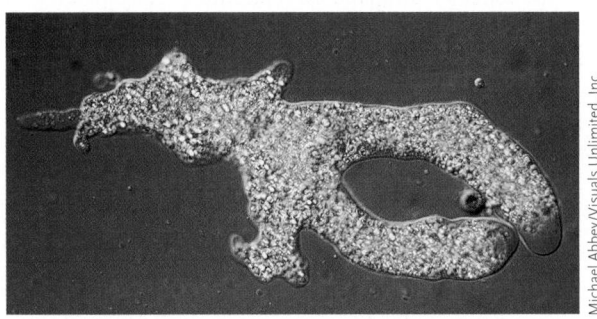

Michael Abbey/Visuals Unlimited, Inc.

Figure 5.21
Phagocytosis, in which lobes of the cytoplasm extend outward and surround a cell targeted as prey. The micrograph shows the protistan *Chaos carolinense* preparing to engulf a single-celled alga (*Pandorina*) by phagocytosis; white blood cells called phagocytes carry out a similar process in mammals.

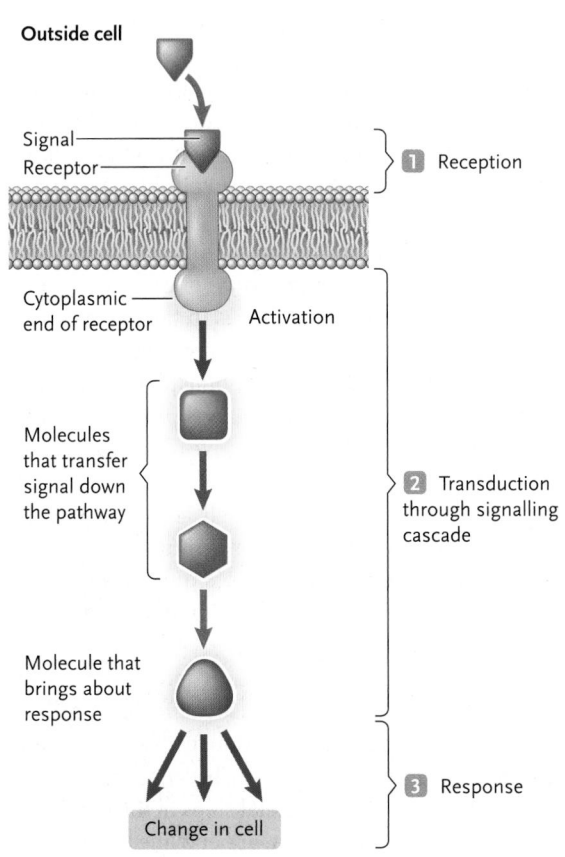

Figure 5.22
The three stages of signal transduction: reception, transduction, and response (shown for a system using a surface receptor).

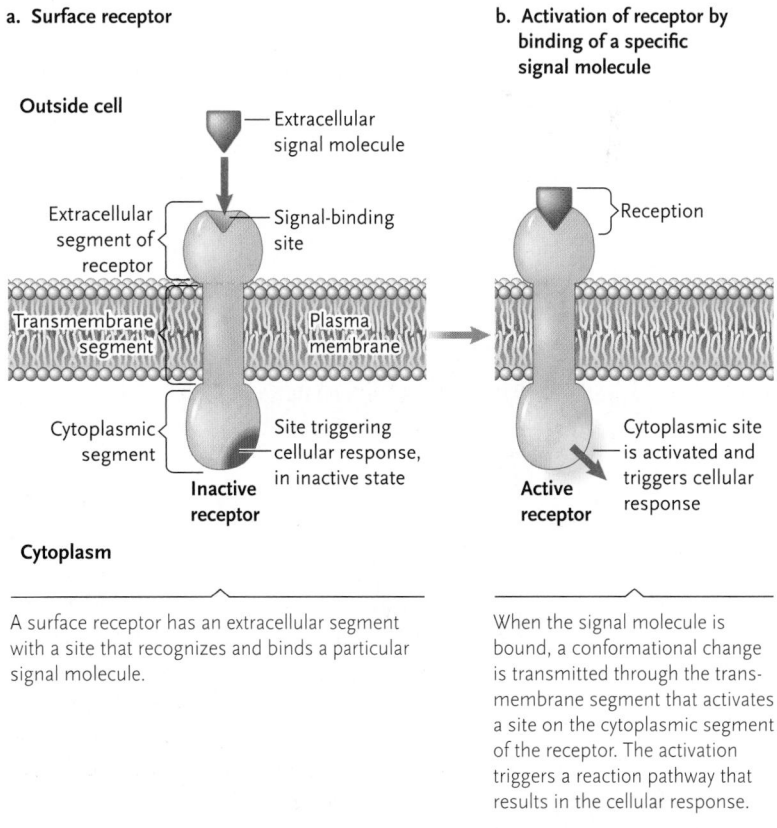

a. **Surface receptor**

b. **Activation of receptor by binding of a specific signal molecule**

A surface receptor has an extracellular segment with a site that recognizes and binds a particular signal molecule.

When the signal molecule is bound, a conformational change is transmitted through the transmembrane segment that activates a site on the cytoplasmic segment of the receptor. The activation triggers a reaction pathway that results in the cellular response.

Figure 5.23
The mechanism by which a surface receptor responds when it binds a signal molecule.

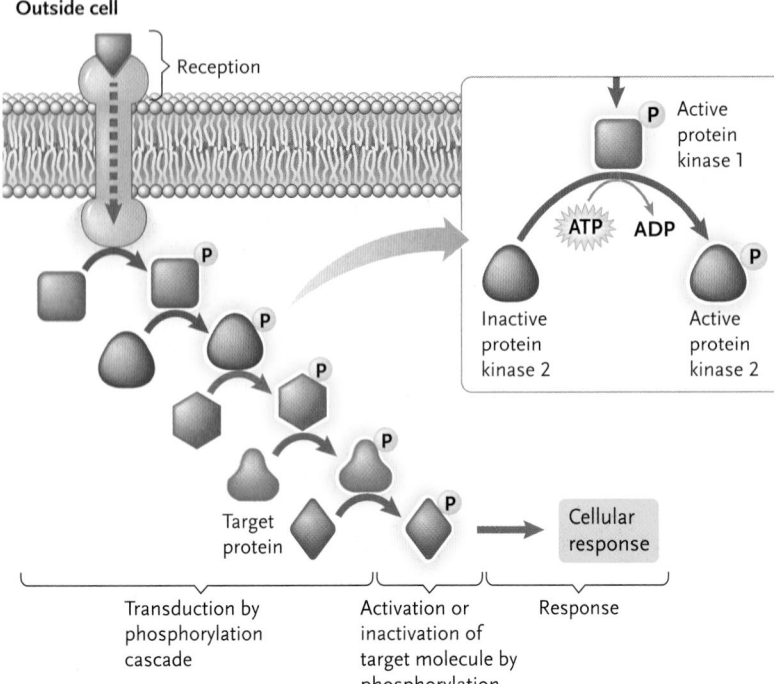

Outside cell

Reception

Active protein kinase 1

ATP   ADP

Inactive protein kinase 2   Active protein kinase 2

Target protein

Cellular response

Transduction by phosphorylation cascade | Activation or inactivation of target molecule by phosphorylation | Response

Cytoplasm

Figure 5.24
Phosphorylation, a key reaction in many signalling pathways.

The activated receptor then initiates the first step in a cascade of molecular events—the signalling cascade—that triggers the cellular response (Figure 5.23b, p. 115). The cells of most organisms typically have hundreds of membrane receptors that represent many receptor types. Receptors for a specific animal peptide hormone, for example, may number from 500 to as many as 100 000 or more per cell. Different cell types

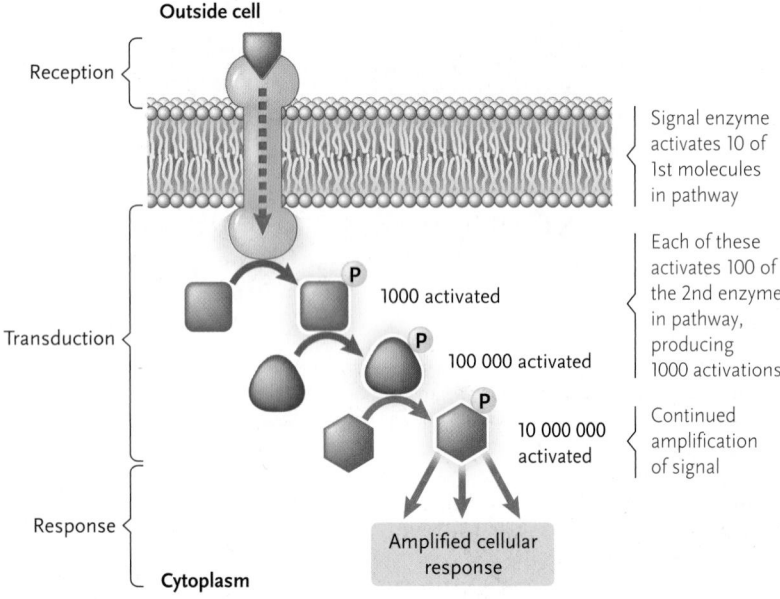

Outside cell

Reception

Transduction

Response

1000 activated

100 000 activated

10 000 000 activated

Amplified cellular response

Signal enzyme activates 10 of 1st molecules in pathway

Each of these activates 100 of the 2nd enzyme in pathway, producing 1000 activations

Continued amplification of signal

Cytoplasm

Figure 5.25
Amplification in signal transduction.

contain distinct combinations of receptors, allowing them to react individually to a diversity of signal molecules.

5.7c Signal Reception Triggers Response Pathways within the Cell

The binding of a signal molecule to a plasma membrane receptor, for example, is sufficient to trigger the activation of the signalling cascade. The signal molecule does not enter the cell. For example, experiments have shown that (1) a signal molecule produces no response if it is injected directly into the cytoplasm and (2) unrelated molecules that mimic the structure of the normal extracellular signal molecule can trigger or block a full cellular response as long as they can bind to the recognition site of the receptor. In fact, many medical conditions are treated with drugs that are signal molecule mimics.

A common characteristic of signalling mechanisms is that the signal is relayed inside the cell by **protein kinases**, enzymes that transfer a phosphate group from ATP to one or more sites on particular proteins. As shown in **Figure 5.24,** protein kinases often act in a chain, catalyzing a series of phosphorylation reactions called a *phosphorylation cascade,* to pass along a signal. The first kinase catalyzes phosphorylation of the second, which then becomes active and phosphorylates the third kinase, which then becomes active, and so on. The last protein in the cascade is the *target protein.* Phosphorylation of a target protein stimulates or inhibits its activity depending on the particular protein. This change in activity brings about the cellular response. For example, phosphorylating a target protein may alter the activity of a transcription factor that regulates the expression of a suite of genes.

The effects of protein kinases in the signal transduction pathways are balanced or reversed by another group of enzymes called **protein phosphatases,** which remove phosphate groups from target proteins. Unlike the protein kinases, which are active only when a surface receptor binds a signal molecule, most of the protein phosphatases are continuously active in cells. By continually removing phosphate groups from target proteins, the protein phosphatases quickly shut off a signal transduction pathway if its signal molecule is no longer bound at the cell surface.

Another characteristic of signal transduction pathways is **amplification**—an increase in the magnitude of each step as a signal transduction pathway proceeds **(Figure 5.25).** Amplification occurs because many of the proteins that carry out individual steps in the pathways, including the protein kinases, are enzymes. Once activated, each enzyme can activate hundreds of proteins, including other enzymes that enter the next step in the pathway. Generally, the more enzyme-catalyzed steps in a response pathway, the greater the amplification. As a

PEOPLE BEHIND BIOLOGY 5.2

Lap-Chee Tsui, *University of Hong Kong*

Identifying the gene that is defective in patients with cystic fibrosis (CF) (see "Why It Matters," p. 97) was a breakthrough in human genetics and was achieved by a research team headed by Lap-Chee Tsui (1950–) of the Department of Genetics at the Hospital for Sick Children in Toronto.

Born in Shanghai, Tsui studied biology at the Chinese University of Hong Kong and was awarded a bachelor of science degree in 1972, which was followed by a master of philosophy degree in 1974. He undertook doctoral research in the United States, completing his Ph.D. at the University of Pittsburgh in 1979. He followed this with postdoctoral training at Oak Ridge National Laboratory in Tennessee before moving in 1981 to the Department of Genetics at the Hospital for Sick Children, where he became a staff member investigating the underlying genetic cause of CF.

Although today reports of gene discovery are common, in the 1980s,

the discovery of the gene that is mutated in patients with CF was particularly noteworthy for two major reasons. First, researchers relied on DNA isolated from people with CF to identify genetic markers of the disease. Using these, researchers used the novel method of positional cloning to identify the CF gene without any knowledge of the gene itself or what it did. Second, CF is the most common single-gene disease among Caucasians; thus, much anticipation awaited this particular discovery, with many research teams worldwide trying to be the first to identify the gene.

In 1985, Tsui and his team identified the first DNA marker linked to CF, on chromosome 7. Four years later, Tsui's team, along with collaborators at the University of Michigan, finally identified the defective gene responsible for CF, defining the principal mutation (Δ*F508). This mutation is the result of a three-nucleotide deletion that results in the

loss of the amino acid phenylalanine (F) at the 508th position of the protein. As a result, the protein does not fold normally and is more quickly degraded.

The research was described in three seminal papers in the September 8, 1989, issue of *Science*. The gene was called the cystic fibrosis transmembrane conductance regulator (CFTR). *Science* named Tsui's achievement "the most refreshing scientific development of 1989," and *Maclean's* Honour Roll hailed it as one of the "discoveries of hope at the heart of human life" in the same year.

Tsui has received many honours, including fellow of the Royal Society of Canada, several honorary doctoral degrees, and the Order of Canada. Tsui is currently the vice-chancellor of the University of Hong Kong, but he remains an active researcher and is still affiliated with the Hospital for Sick Children's Program in Genetics and Genomic Biology.

result, just a few extracellular signal molecules binding to their receptors can produce a full internal response.

This chapter has introduced you to the fundamentals of membrane structure and the role membranes serve in an array of functions, from transport through cellular signalling. Membranes and the compartments they define play a fundamental role in energy metabolism, which is the central theme of the next two chapters, on respiration and photosynthesis.

Review

5.1 An Overview of the Structure of Membranes

- The fluid mosaic model proposes that the membrane consists of a fluid lipid bilayer in which proteins are embedded and float freely (Figure 5.1).
- Membranes are asymmetrical. The two halves of a membrane are not the same. The membrane proteins found on one half of the bilayer are structurally and functionally distinct from those of the other half.

5.2 The Lipid Fabric of a Membrane

- The lipid bilayer forms the structural framework of membranes and serves as a barrier preventing the passage of most water-soluble molecules.
- The structural basis of a membrane is a fluid phospholipid bilayer in which the polar regions of phospholipid molecules lie at the surfaces of the bilayer and their nonpolar tails associate in the interior (Figures 5.4 and 5.5).
- Saturated fatty acids contain the maximum number of hydrogen atoms and are linear molecules. Unsaturated fatty acids contain one or more double bonds, which cause the fatty acid to kink (Figure 5.6).

- Organisms can adjust the fatty acid composition of membrane lipids to maintain proper fluidity through the action of a group of enzymes called desaturases (Figure 5.7).

5.3 Membrane Proteins

- Proteins embedded in the phospholipid bilayer carry out most membrane functions, including transport of selected hydrophilic substances, enzymatic activity, recognition, and signal reception (Figure 5.9).
- Integral membrane proteins interact with the hydrophobic core of the membrane bilayer. Most integral membrane proteins, called transmembrane proteins, have domains that span the membrane numerous times. These domains are dominated by nonpolar amino acids (Figures 5.10 and 5.11).
- Peripheral membrane proteins associate with membrane surfaces.

5.4 Passive Membrane Transport

- Passive transport depends on diffusion, the net movement of molecules from a region of higher concentration to a region of lower concentration. Passive transport does not require cells to expend energy (Figure 5.12).
- Simple diffusion is the passive transport of substances across a membrane through the lipid molecules. Small uncharged molecules can move rapidly across membranes, whereas large or charged molecules may be strongly impeded from transiting a membrane (Figure 5.13).
- Facilitated diffusion is the diffusion of molecules across membranes by the use of specific membrane proteins—channel proteins and carrier proteins. Both channel proteins and carrier proteins are specific for certain substances. For many molecules, facilitated diffusion results in higher rates of transport than simple diffusion (Figures 5.14 and 5.15).
- Osmosis is the net diffusion of water molecules across a selectively permeable membrane in response to differences in the concentration of solute molecules.
- Water moves from hypotonic solutions (lower concentrations of solute molecules) to hypertonic solutions (higher concentrations of solute molecules). When the solutions on each side are isotonic, there is no osmotic movement of water in either direction (Figure 5.16).

5.5 Active Membrane Transport

- Active transport moves substances against their concentration gradients and requires cells to expend energy. Active transport depends on membrane proteins, is specific for certain substances, and becomes saturated at high concentrations of the transported substance.
- Active transport proteins are either primary transport pumps, which directly use ATP as their energy source, or secondary transport pumps, which use favourable concentration gradients of positively charged ions, set up by primary transport pumps, as their energy source for transport (Figure 5.17).
- Secondary active transport may occur by symport, in which the transported substance moves in the same direction as the concentration gradient used as the energy source, or by antiport, in which the transported substance moves in the direction opposite to the concentration gradient used as the energy source (Figure 5.19).

5.6 Exocytosis and Endocytosis

- Large molecules and particles are moved out of and into cells by exocytosis and endocytosis. The mechanisms allow substances to leave and enter cells without directly passing through the plasma membrane (Figure 5.20).
- In exocytosis, a vesicle carrying secreted materials contacts and fuses with the plasma membrane on its cytoplasmic side. The fusion introduces the vesicle membrane into the plasma membrane and releases the vesicle contents to the cell exterior.
- In endocytosis, materials on the cell exterior are enclosed in a segment of the plasma membrane that pockets inward and pinches off on the cytoplasmic side as an endocytic vesicle. Endocytosis occurs in two overall forms, bulk-phase endocytosis (pinocytosis) and receptor-mediated endocytosis. Most of the materials entering cells are digested into molecular subunits small enough to be transported across the vesicle membranes (Figure 5.20b, c).

5.7 Role of Membranes in Cell Signalling

- Cell communication systems based on surface receptors have three components: (1) extracellular signal molecules, (2) surface receptors that receive the signals, and (3) internal response pathways triggered when receptors bind a signal (Figure 5.22).
- Surface receptors are integral membrane proteins that extend through the plasma membrane. Binding a signal molecule induces a molecular change in the receptor that activates its cytoplasmic end (Figure 5.23).
- Many cellular response pathways operate by activating protein kinases, which add phosphate groups that stimulate or inhibit the activities of the target proteins, bringing about the cellular response (Figure 5.24). Protein phosphatases that remove phosphate groups from target proteins reverse the response. In addition, receptors are removed by endocytosis when signal transduction has run its course.
- Each step of a response pathway catalyzed by an enzyme is amplified, because each enzyme can activate hundreds or thousands of proteins that enter the next step in the pathway. Through amplification, a few signal molecules can bring about a full cellular response (Figure 5.25).

Questions

Self-Test Questions

1. Which of the following statements about the fluid mosaic model is correct?
   a. The fluid refers to the phospholipid bilayer.
   b. Plasma membrane proteins orient their hydrophilic sides toward the internal bilayer.
   c. Phospholipids often flip-flop between the inner and outer layers.
   d. The mosaic refers to proteins attached to the underlying cytoskeleton.
   e. The mosaic refers to the symmetry of the internal membrane proteins and sterols.

2. What was demonstrated by the freeze-fracture technique?
   a. The plasma membrane is fluid.
   b. Membranes remain fluid at freezing temperatures.
   c. The arrangement of membrane lipids and proteins is symmetric.
   d. The plasma membrane is a bilayer with individual proteins suspended in it.
   e. Proteins are bound to the cytoplasmic side but not embedded in the lipid bilayer.

3. Which statement is correct regarding temperature and membrane fluidity?
   a. Membrane fluidity increases with decreasing temperature.
   b. Membrane fluidity remains constant over the temperature range of 0 to 100°C.
   c. At any given temperature, membrane fluidity increases as desaturase activity increases.
   d. At any given temperature, membrane fluidity decreases as the abundance of unsaturated fatty acids increases.

4. The integral membrane protein rhodopsin, which is used in light perception, is a protein that spans the membrane seven times. Which statement about rhodopsin is correct?
   a. It contains seven hydrophobic amino acids.
   b. It is composed only of hydrophobic amino acids.
   c. It contains both hydrophobic and hydrophilic domains.
   d. It contains one long stretch of hydrophobic amino acids.
   e. It has a random assortment of both polar and nonpolar amino acids.

5. Which one of the following molecules shows the slowest rate of membrane diffusion and why?
   a. Na+, because it is small
   b. K+, because it is charged
   c. glucose, because it is large
   d. CO_2, because it contains three atoms
   e. H_2O, because water is the main component of the cytosol

6. Compared to simple diffusion, which of the following statements is correct for only facilitated diffusion?
   a. It can only transport hydrophobic molecules.
   b. It can be saturated by high substrate concentrations.
   c. It requires a source of chemical energy, such as ATP.
   d. It can transport molecules against a concentration gradient.

7. In the following diagram, assume that the setup was left unattended. Which statement is correct?

Selectively Permeable Membrane			
Inside a Cell		Extracellular Fluid	
Solvent	95%	Solvent	98%
Solute	5%	Solute	2%

   a. The cell will soon shrink.
   b. The net flow of solvent is into the cell.
   c. The cell is in a hypertonic environment.
   d. Diffusion can occur here but not osmosis.
   e. The relation of the cell to its environment is isotonic.

8. An ion moving through a membrane channel in one direction gives energy to actively transport another molecule in the opposite direction. What process does this describe?
   a. cotransport
   b. symport transport
   c. exchange diffusion
   d. facilitated diffusion
   e. primary active transport pump

9. Phagocytosis illustrates which of the following?
   a. exocytosis
   b. pinocytosis
   c. cotransport
   d. bulk-phase endocytosis
   e. receptor-mediated endocytosis

10. Many signal transduction pathways are initiated by a signal binding to a membrane receptor. Which statement about this type of signalling mechanism is correct?
    a. Signal binding may activate a protein kinase.
    b. A signal molecule injected into the cytoplasm will activate the pathway.
    c. A mutation in a single gene can disrupt an entire signalling pathway.
    d. Signal transduction pathways never include components within the nucleus.
    e. Both a and c are correct.

Questions for Discussion

1. In Chapter 4, we discussed thermodynamics. What role does thermodynamics play in membrane transport?

2. The bacterium *Vibrio cholerae* causes cholera, a disease characterized by severe diarrhea that may cause infected people to lose up to 20 L of fluid in a day. The bacterium enters the body when a person drinks contaminated water. It adheres to the intestinal lining, where it causes cells of the lining to release sodium and chloride ions. Explain how this release is related to the massive fluid loss.

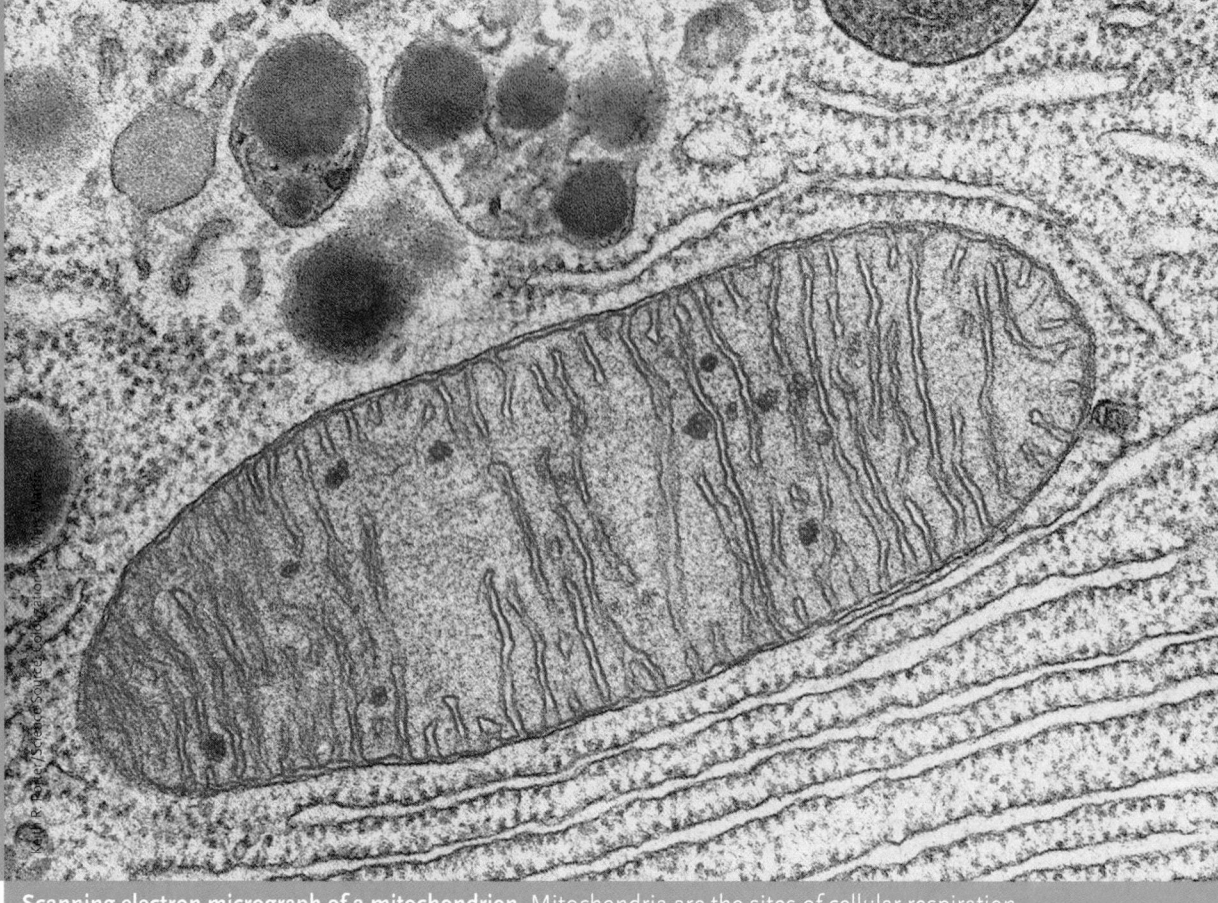

Scanning electron micrograph of a mitochondrion. Mitochondria are the sites of cellular respiration.

Cellular Respiration

WHY IT MATTERS

In the early 1960s, Swedish physician Rolf Luft mulled over some odd symptoms of a patient. The young woman was hot all the time. Even on the coldest winter days, she never stopped perspiring and her skin was always flushed. She also felt weak and was thin, despite a huge appetite.

Luft inferred that his patient's symptoms pointed to a metabolic disorder. Her cells seemed to be active, but much of their activity was being dissipated as metabolic heat. He decided to order tests to measure her metabolic rates. The patient's oxygen consumption was the highest ever recorded!

Luft also examined a tissue sample from the patient's skeletal muscles. Using a microscope, he found that her muscle cells contained many more mitochondria—the ATP-producing organelles of the cell—than normal; also, her mitochondria were abnormally shaped. Other studies showed that the mitochondria were engaged in cellular respiration—their prime function—but little ATP was being generated.

The disorder, now called *Luft syndrome*, was the first disorder to be linked directly to a defective cellular organelle. This syndrome is extremely rare and has now been shown to be due to a defect in one of the complexes of cellular respiration that links electron transport to proton pumping and subsequent ATP generation. With such a disorder, skeletal and heart muscles and the brain, the tissues with the highest energy demands, are affected the most. More than 100 mitochondrial disorders are now known. Defective mitochondria are now linked to a range of diseases and disorders including amyotrophic lateral sclerosis (ALS, also called Lou Gehrig's disease), as well as Parkinson's, Alzheimer's, and Huntington diseases.

Clearly, human health depends on mitochondria that are structurally sound and functioning properly. But of course there is nothing unique to humans here—every animal, plant, and fungus requires correctly functioning mitochondria to live. In eukaryotes, this organelle is the site of key reactions of cellular respiration—the process whereby the energy present in food molecules is extracted and converted into a form usable by the cell. In this chapter, we explore the fundamentals of cellular respiration, starting by addressing what makes a good fuel molecule.

6.1 The Chemical Basis of Cellular Respiration

We can define **cellular respiration** as the collection of metabolic reactions within cells that breaks down food molecules (e.g., carbohydrates, fats, proteins) and uses the liberated free energy to synthesize ATP. As described in detail in Chapter 4, it is ATP that is the form of chemical energy required for the thousands of biosynthetic reactions (**anabolic reactions**) that take place within a cell.

The ultimate source of the energy-rich carbon compounds found in carbohydrates, fats, and proteins is *photosynthesis,* which is the focus of the next chapter. In photosynthesis, light energy is used to extract electrons from water; the electrons then combine with hydrogen to reduce carbon dioxide into glucose, a carbohydrate. Additional biosynthetic pathways can utilize carbohydrates in the synthesis of both proteins and fats. A major by-product of photosynthesis is oxygen, a molecule needed for the most common type of cellular respiration. Thus, life and its systems are driven by a cycle of electron flow that is powered by light in photosynthesis and oxidation in cellular respiration **(Figure 6.1)**.

6.1a Food Is Fuel

Looking at **Figure 6.2, p. 122,** we can ask the following question: what is it about glucose that makes it a source of chemical energy? We could ask the same question of gasoline: what makes it good at powering a car? Both glucose and gasoline are good fuel molecules because they contain an abundance of C—H bonds. Recall from Chapter 4 that for any atom, an electron that is farther away from the nucleus contains more energy than an electron that is more closely held by the nucleus. As well, recall that as an electron moves closer to the nucleus of an atom it loses energy; as it moves away it gains energy. The electrons that form the covalent C—H bond are equidistant from both **atomic nuclei**—not strongly held by either. Because of this, the electrons can be easily removed and used to perform work. In contrast to glucose and gasoline, molecules that contain more oxygen, for example, carbon dioxide, contain less potential energy because oxygen is strongly electronegative. The more electronegative an atom, the greater the force that holds the electrons to that atom and, therefore, the greater the energy required to remove the electrons. To review the basics of electronegativity, see *The Purple Pages.* In the case of a C—H bond, neither carbon nor hydrogen is strongly electronegative. This fundamental principle of chemistry has an everyday relevance: it explains why, for example, compared to proteins and carbohydrates, fats contain more calories (energy) per unit of weight. A fat is almost entirely C—H bonds, while both proteins and carbohydrates contain varying amounts of other atoms, including oxygen. (To review the structure of these molecules, see *The Purple Pages.*)

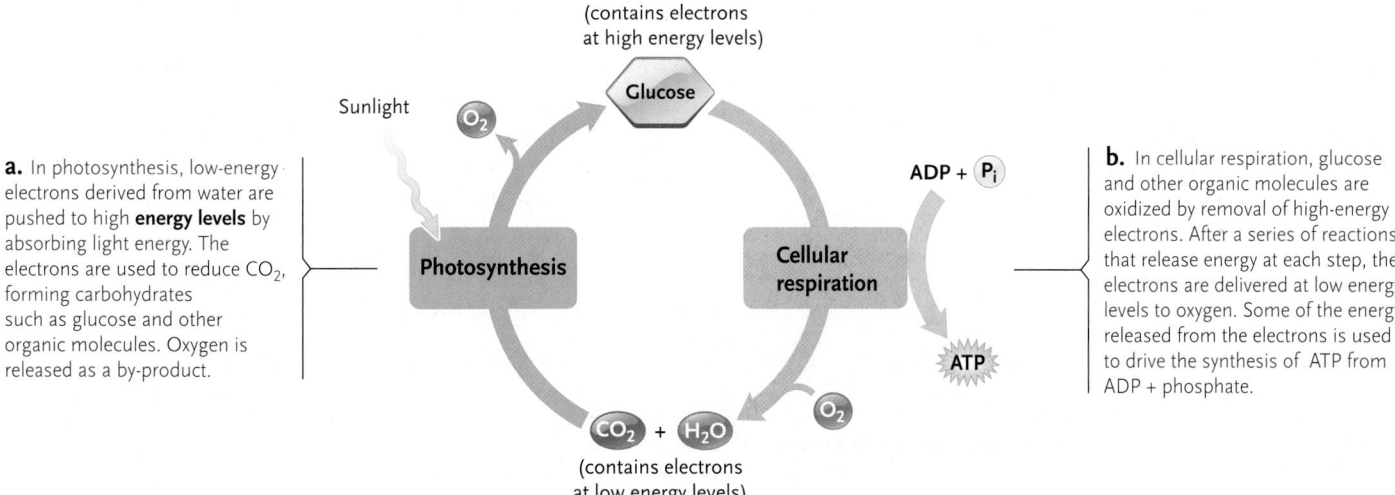

a. In photosynthesis, low-energy electrons derived from water are pushed to high **energy levels** by absorbing light energy. The electrons are used to reduce CO_2, forming carbohydrates such as glucose and other organic molecules. Oxygen is released as a by-product.

b. In cellular respiration, glucose and other organic molecules are oxidized by removal of high-energy electrons. After a series of reactions that release energy at each step, the electrons are delivered at low energy levels to oxygen. Some of the energy released from the electrons is used to drive the synthesis of ATP from ADP + phosphate.

(contains electrons at high energy levels)

Sunlight

Glucose

O_2

Photosynthesis

Cellular respiration

ADP + P_i

ATP

CO_2 + H_2O
(contains electrons at low energy levels)

O_2

Figure 6.1

Flow of energy linking photosynthesis and respiration. Photosynthesis uses light energy to convert carbon dioxide and water into energy-rich organic molecules such as glucose, which, in turn, are oxidized by cellular respiration.

a. Gasoline

H—C—C—C—C—C—C—C—C—H (octane structure with H atoms)

b. Glucose

(glucose structure)

Figure 6.2

Fuel molecules. Both gasoline (e.g., octane) and glucose are excellent fuels because electrons of C—H bonds can be easily removed.

6.1b Coupled Oxidation–Reduction Reactions Are Central to Energy Metabolism

The potential energy contained in fuel molecules is released when the molecules lose electrons, becoming **oxidized.** The electrons released from a molecule that is oxidized are gained by another molecule that becomes **reduced.** Oxidation and reduction reactions are coupled processes—one cannot happen without the other. A simple mnemonic to remember the direction of electron transport is OIL RIG—Oxidation Is Loss (of electrons), **Reduction** Is Gain (of electrons). For short, oxidation–reduction reactions are called **redox reactions.** A generalized redox reaction can be written like this:

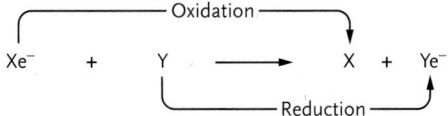

$$Xe^- \quad + \quad Y \quad \longrightarrow \quad X \quad + \quad Ye^-$$

The redox reaction describing the respiratory breakdown of glucose is as follows:

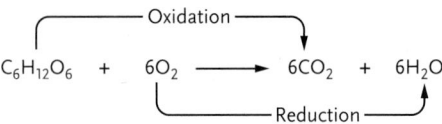

$$C_6H_{12}O_6 \quad + \quad 6O_2 \quad \longrightarrow \quad 6CO_2 \quad + \quad 6H_2O$$

Reactants   **Products**

Methane   Oxygen   Carbon dioxide   Water

becomes oxidized

$$CH_4 \quad + \quad 2\,O_2 \quad \longrightarrow \quad CO_2 \quad + \quad Energy \quad + \quad 2\,H_2O$$

becomes reduced

Figure 6.3

A redox reaction: the burning of methane in oxygen. Compare the positions of the electrons in the covalent bonds of reactants and products. In this redox reaction, methane is oxidized and oxygen is reduced.

The term *oxidation* comes from that fact that many reactions in which electrons are removed from fuel molecules involve oxygen as the atom that accepts the electrons and gets reduced. The involvement of oxygen is essential for many common oxidation reactions: a car engine requires large amounts of air (21% oxygen) to be delivered to each piston for combustion to take place; an oil fire in a pot on a stove can be rapidly extinguished by putting a lid on the pot, restricting the air supply. As we will see later in this chapter, the high affinity of O_2 for electrons (its high electronegativity) makes it ideal as the terminal electron acceptor of cellular respiration.

The concept of redox is made a little more challenging to understand by two facts. First, although many oxidation reactions involve oxygen, others, including a number involved in cellular respiration, do not. Second, the gain or loss of an electron in a redox reaction is not always complete. That is, whereas in some redox reactions electrons are transferred completely from one atom to another, in other redox reactions, what changes is the degree to which electrons are shared between two atoms. The reaction between methane and oxygen (the burning of natural gas in air) illustrates a redox reaction in which only the degree of electron sharing changes **(Figure 6.3).** The blue dots (see Figure 6.3) indicate the positions of the electrons involved in the covalent bonds of the reactants and products. Compare the reactant methane with the product carbon dioxide—in methane, the electrons are shared equally between the carbon and hydrogen atoms. In the product, carbon dioxide, electrons are closer to the oxygen than to the carbon because oxygen atoms are more electronegative. Overall, this means that the carbon atom has partially lost its shared electrons in the reaction: methane has been oxidized. Now compare the oxygen reactant with the product, water. In the oxygen molecule, the two oxygen atoms share their electrons equally. The oxygen reacts with the hydrogen from methane, producing water, in which the electrons are closer to the oxygen atom than to the hydrogen atoms. This means that each oxygen atom has partially gained electrons: oxygen has been reduced. Because of this, the reaction between methane and oxygen releases heat. The energy is released as the electrons in the C—H bonds of methane move closer to the electronegative oxygen atoms that form carbon dioxide.

6.1c Cellular Respiration Is Controlled Combustion

Like gasoline and methane, glucose can also undergo combustion and burn. The combustion of glucose releases energy as electrons are transferred to oxygen, reducing it to water, and the carbon in glucose is oxidized to carbon dioxide.

Recall from Chapter 4 that for a spontaneous reaction to proceed, the substrate molecules need to reach the transition state, which requires an energy input

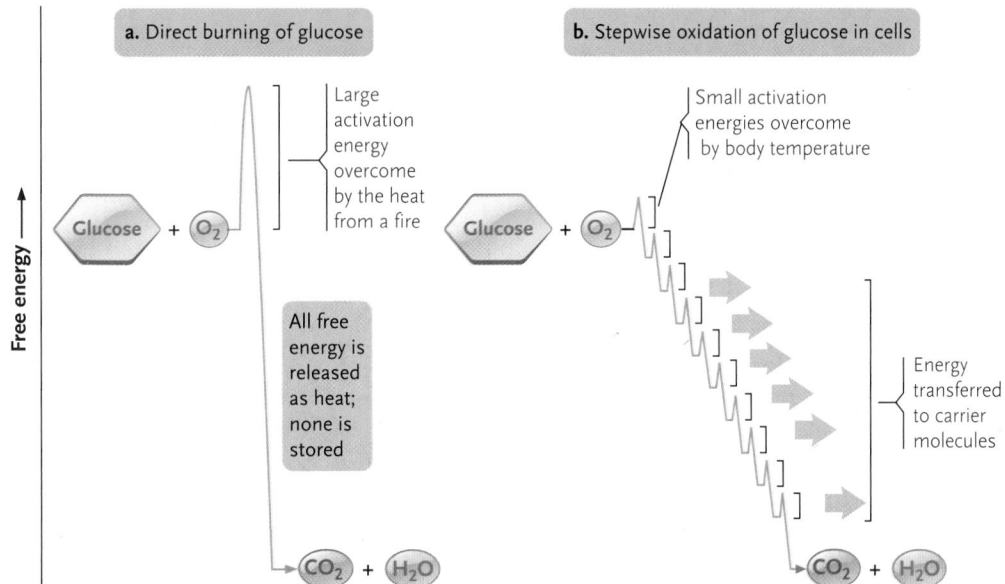

Figure 6.4
A comparison of the oxidation of glucose by combustion and cellular respiration.

a. Direct burning of glucose

Large activation energy overcome by the heat from a fire

All free energy is released as heat; none is stored

Glucose + O_2 → CO_2 + H_2O

b. Stepwise oxidation of glucose in cells

Small activation energies overcome by body temperature

Energy transferred to carrier molecules

Glucose + O_2 → CO_2 + H_2O

Free energy →

referred to as the activation energy. To get glucose to ignite, we can use a flame to provide the high activation energy **(Figure 6.4a)**. In contrast, within a cell, the oxidation of glucose occurs through a series of enzyme-catalyzed reactions (Figure 6.4b), each with a small activation energy. Thermodynamically, the two processes are identical: they are both exergonic, having the same change in free energy (ΔG) of –686 kcal/mol. The big difference is that if you simply burn glucose, the energy is released as heat and therefore not available to drive metabolic reactions. So a good way to think of the process of cellular respiration is controlled combustion—where the energy of the C—H bonds is not liberated suddenly, producing heat, but is slowly released in a stepwise fashion, with the energy being transferred to other molecules.

In cellular respiration, the oxidation of food molecules occurs in the presence of a group of enzymes called *dehydrogenases*, which facilitate the transfer of electrons from food to a molecule that acts as an energy carrier or shuttle. The most common energy carrier is the coenzyme nicotinamide adenine dinucleotide (NAD$^+$, oxidized; NADH, reduced) **(Figure 6.5)**. During respiration, the dehydrogenases remove two hydrogen atoms from a

substrate molecule and transfer the two electrons—but only one of the protons—to NAD$^+$, reducing it to NADH. The efficiency of the enzyme-catalyzed transfer of energy between food molecules and NAD$^+$ is very high. As we will see later in the chapter, the potential energy carried in NADH is used to synthesize ATP.

STUDY BREAK

1. What is it about the structure of gasoline and glucose that makes them both good fuels?
2. In the respiratory breakdown of glucose, what gets oxidized and what gets reduced?

6.2 Cellular Respiration: An Overview

At this point, let's step back and remind ourselves of the primary goal of cellular respiration: it is to transform the potential energy found in food molecules into a form that can be used for metabolic processes, ATP. We will see later in the chapter that both proteins and lipids can be oxidized by the respiratory pathway and

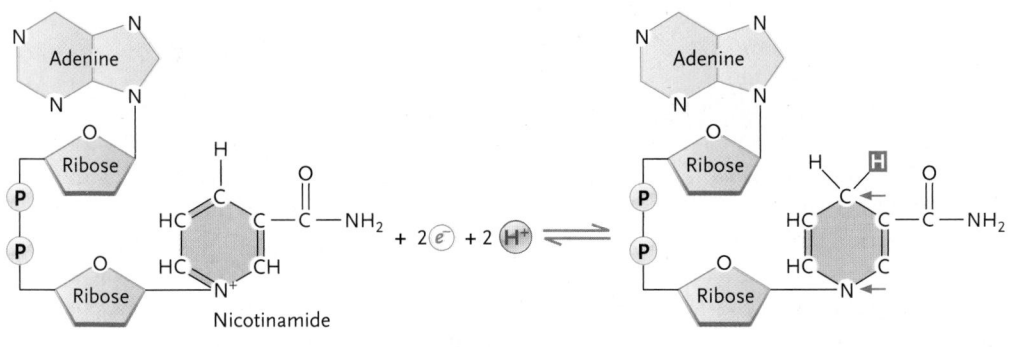

Figure 6.5
Electron carrier NAD$^+$. As the carrier is reduced to NADH, an electron is added at each of the two positions marked by a red arrow; a proton is also added at the position boxed in red. The nitrogenous base (blue) that adds and releases electrons and protons is nicotinamide, which is derived from the vitamin niacin (nicotinic acid).

Oxidized (NAD$^+$)

Reduced (NADH)

Adenine | Ribose | Nicotinamide

$+ 2 \bar{e} + 2 H^+ \rightleftharpoons$

their potential energy harnessed; however, because the oxidation of glucose utilizes the entire respiratory pathway, it is the main focus of our discussion.

6.2a Cellular Respiration Can Be Divided into Three Phases

Cellular respiration can be divided into three phases **(Figure 6.6)**:

1. *Glycolysis.* Enzymes break down a molecule of glucose into two molecules of pyruvate. Some ATP and NADH is synthesized.
2. *Pyruvate oxidation and the citric acid cycle.* Acetyl coenzyme A (acetyl-CoA), which is formed from the oxidation of pyruvate, enters a metabolic cycle, where it is completely oxidized to carbon dioxide. Some ATP and NADH is synthesized.
3. *Oxidative phosphorylation.* The NADH synthesized by both glycolysis and the citric acid cycle is oxidized, with the liberated electrons being passed along an electron transport chain (ETC) until they are transferred to oxygen, producing water. The free energy released during electron transport is used to generate a proton gradient across a membrane, which, in turn, is used to synthesize ATP.

All three stages are required to extract the maximum amount of energy that is biologically possible from a molecule of glucose; however, not all organisms undergo all three stages.

6.2b The Mitochondrion Is the Site of Cellular Respiration in Eukaryotes

In archaea and bacteria, glycolysis and the citric acid cycle occur in the cytosol, whereas oxidative phosphorylation occurs on internal membranes. By comparison, in eukaryotic cells, the citric acid cycle and oxidative phosphorylation occur in a specialized membrane-bound organelle called the mitochondrion (plural, *mitochondria*) **(Figure 6.7).** This membrane-bound organelle is often referred to as the powerhouse of the cell because, as the location of both the citric acid cycle and oxidative phosphorylation, it is the largest generator of ATP in the cell.

The mitochondrion is composed of two membranes, the outer membrane and the inner membrane, which together define two compartments (see Figure 6.7): the intermembrane space, which is found between the outer and inner membranes, and the matrix, which is the interior aqueous environment.

In the description of cellular respiration that follows, we often refer specifically to mitochondria and their various compartments, but

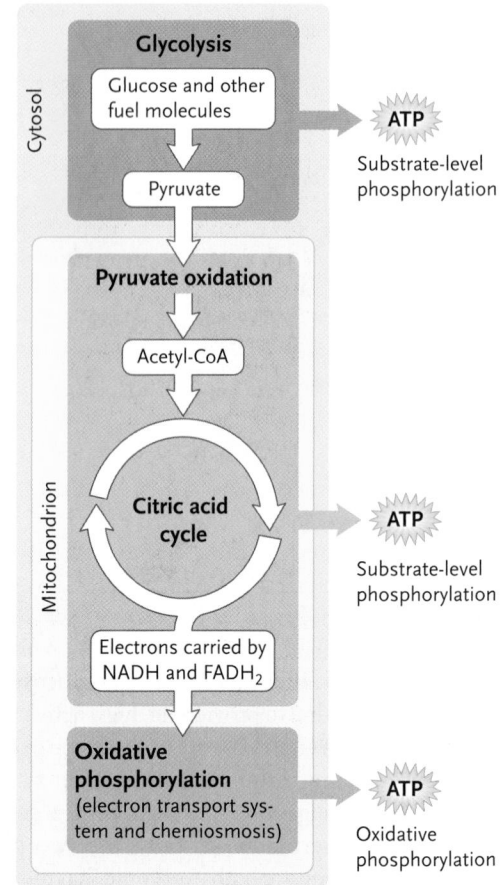

Figure 6.6
The three stages of cellular respiration: glycolysis, pyruvate oxidation and the citric acid cycle, and oxidative phosphorylation.

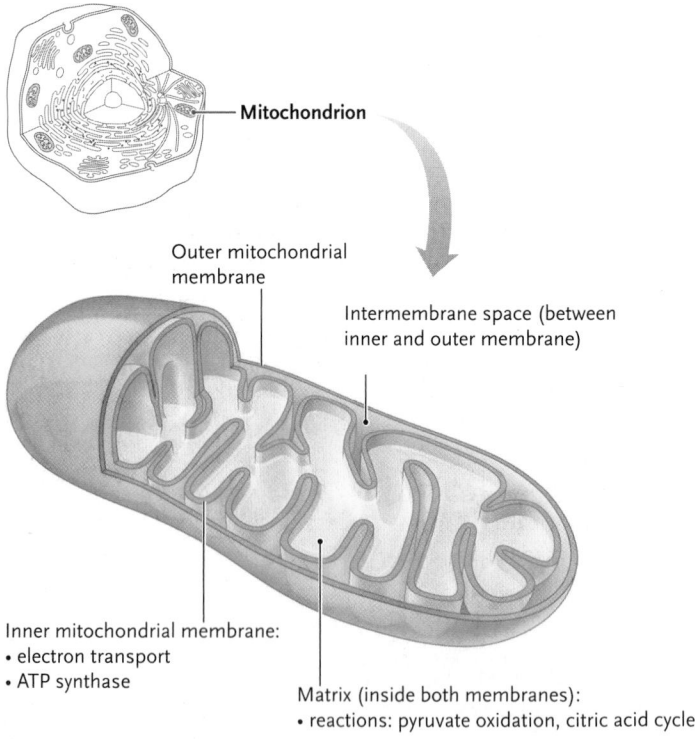

Figure 6.7
Membranes and compartments of mitochondria. Label lines that end in a dot indicate a compartment enclosed by the membranes.

it is important to remember that there is nothing uniquely eukaryotic about cellular respiration. Neither archaea nor bacteria have mitochondria, but many species of both possess the entire complement of reactions that make up cellular respiration—from glycolysis through oxidative phosphorylation.

STUDY BREAK

1. What are the three stages of cellular respiration?
2. Outline the membranes and compartments of the mitochondrion.

6.3 Glycolysis: The Splitting of Glucose

Glycolysis (*glykys* = sweet; *lysis* = breakdown) consists of 10 sequential enzyme-catalyzed reactions that lead to the oxidation of the six-carbon sugar glucose, producing two molecules of the three-carbon compound pyruvate. The potential energy released in the oxidation leads to the synthesis of both NADH and ATP.

6.3a Glycolysis Is a Universal and Ancient Metabolic Process

Glycolysis was one of the first metabolic pathways studied and is one of the best understood in terms of the enzymes involved, their mechanisms of action, and how the pathway is regulated to meet the energy needs of the cell. The first experiments investigating glycolysis took place over 100 years ago and were some of the first to show, using the extracts from yeast cells, that one could study biological reactions in an isolated, cell-free system. These experiments became the foundation of modern biochemistry.

Glycolysis is the most fundamental and probably most ancient of all metabolic pathways. This is supported by the following facts: (1) Glycolysis is universal, being found in all three domains of life—Archaea, Bacteria, and Eukarya. (2) Glycolysis does not depend upon the presence of O_2, which became abundant in Earth's atmosphere only about 2.5 billion years ago—about 1.5 billion years after scientists think life first evolved (see Chapter 3). (3) Glycolysis occurs in the cytosol of all cells using soluble enzymes and therefore does not require more sophisticated ETCs and internal membrane systems to function.

6.3b Glycolysis Includes Energy-Requiring and Energy-Releasing Steps

The key features of glycolysis are summarized in **Figure 6.8,** while **Figure 6.9, p. 126,** provides a detailed look at each reaction of the glycolytic pathway. From both figures, there are three major concepts to come away with:

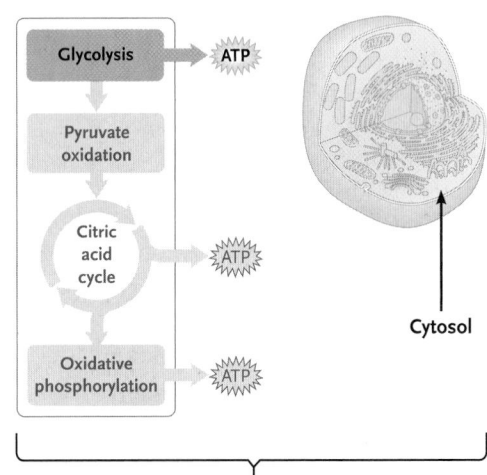

Figure 6.8
Overall reactions of glycolysis. Glycolysis, which occurs in the cytosol of all cells, splits glucose (six carbons) into pyruvate (three carbons) and yields ATP and NADH.

Cytosol

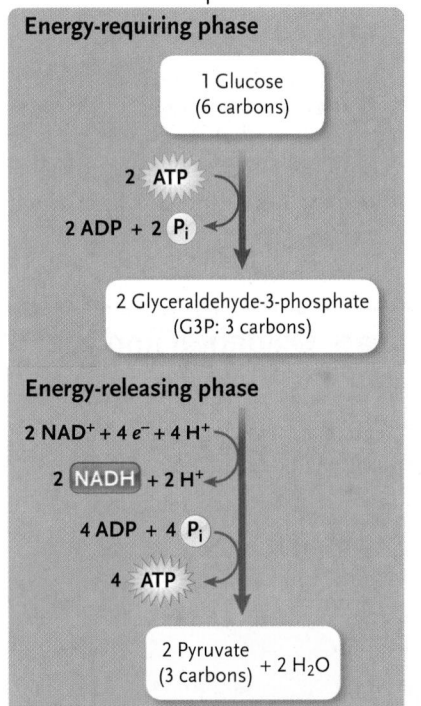

Energy-requiring phase

1 Glucose (6 carbons)

2 ATP

2 ADP + 2 P_i

2 Glyceraldehyde-3-phosphate (G3P: 3 carbons)

Energy-releasing phase

2 NAD$^+$ + 4 e^- + 4 H$^+$

2 NADH + 2 H$^+$

4 ADP + 4 P_i

4 ATP

2 Pyruvate (3 carbons) + 2 H_2O

Summary

1 glucose ⟶ 2 pyruvate + 2 H_2O

4 ATP generated – 2 ATP used ⟶ 2 ATP

2 NAD$^+$ + 4 e^- + 4 H$^+$ ⟶ 2 NADH + 2 H$^+$

1. *Energy investment followed by payoff.* Glycolysis can be looked at as consisting of two distinct phases: an initial five-step energy-requiring phase followed by a five-step energy-releasing phase. Initially, two molecules of ATP are consumed as glucose and fructose- 6-phosphate become phosphorylated. The investment of two ATP for each glucose molecule leads to an energy reward, as four ATP and two NADH molecules are produced during the energy-releasing phase.

2. *No carbon is lost.* The reactions of glycolysis convert glucose (a six-carbon molecule) into two molecules of the three-carbon compound pyruvate. Thus, no carbon is lost. However, since glucose

has been oxidized, the potential energy in two molecules of pyruvate is less than that of one molecule of glucose.

3. *ATP is generated by substrate-level phosphorylation.* During glycolysis, ATP is generated by the process of **substrate-level phosphorylation**. This mode of ATP synthesis, shown in **Figure 6.10,** involves the transfer of a phosphate group from a high-energy substrate molecule to ADP, producing ATP. Substrate-level phosphorylation, which is mediated by a specific enzyme, is also the mode of ATP synthesis used in the citric acid cycle.

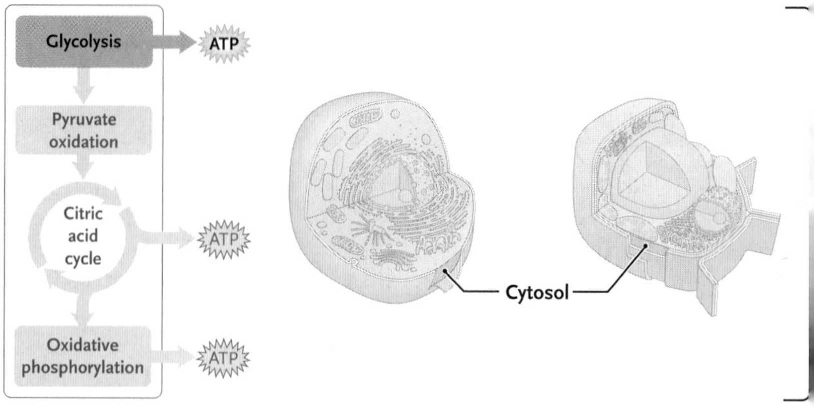

STUDY BREAK

1. What evidence suggests that glycolysis is an ancient metabolic pathway?
2. What accounts for the fact that two molecules of pyruvate have less free energy than one molecule of glucose?

6.4 Pyruvate Oxidation and the Citric Acid Cycle

The two molecules of pyruvate synthesized by glycolysis still contain usable free energy. The extraction of the remaining free energy in pyruvate and the trapping of this energy in the form of ATP and electron carriers

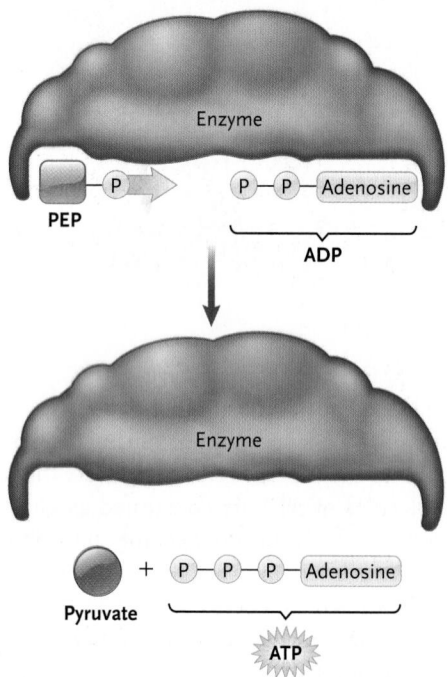

Figure 6.10
Mechanism that synthesizes ATP by substrate-level phosphorylation. A phosphate group is transferred from a high-energy donor directly to ADP, forming ATP.

Figure 6.9
Reactions of glycolysis. Because two molecules of G3P are produced in reaction 5, all the reactions from 6 to 10 are doubled (not shown). The name of the enzyme that catalyzes each reaction is in red.

such as NADH are the overarching goals of the series of reactions described in this section.

6.4a Pyruvate Oxidation Links Glycolysis and the Citric Acid Cycle

Because the reactions of the citric acid cycle are localized to the mitochondrial matrix, the pyruvate synthesized during glycolysis must pass through both the outer and inner mitochondrial membranes **(Figure 6.11, p. 128).** Large pores in the outer membrane allow pyruvate to simply diffuse through, but crossing the inner membrane requires a pyruvate-specific membrane carrier.

Once it gets into the matrix, pyruvate is converted into acetyl-CoA through a multistep process that is referred to as *pyruvate oxidation* (see Figure 6.11). The conversion of pyruvate to acetyl-CoA starts with a decarboxylation reaction whereby the carboxyl ($— COO^-$) group of pyruvate is lost as carbon dioxide. This reaction is understandable given that the carboxyl group itself contains no usable energy. The decarboxylation reaction is followed by oxidation of the remaining two-carbon molecule, producing acetate. This dehydrogenation reaction leads to the transfer of two electrons and a proton to NAD^+, forming NADH. Last, the acetyl group reacts with coenzyme A (CoA), forming the high-energy intermediate acetyl-CoA. Notice in Figure 6.11 that acetyl-CoA still contains three C—H bonds. Liberating the electrons in those bonds as a source of chemical energy is the goal of the reactions that make up the citric acid cycle.

6.4b The Citric Acid Cycle Oxidizes Acetyl Groups to Carbon Dioxide

The **citric acid cycle** consists of eight enzyme-catalyzed reactions: seven are soluble enzymes located in the mitochondrial matrix, and one enzyme is bound to the

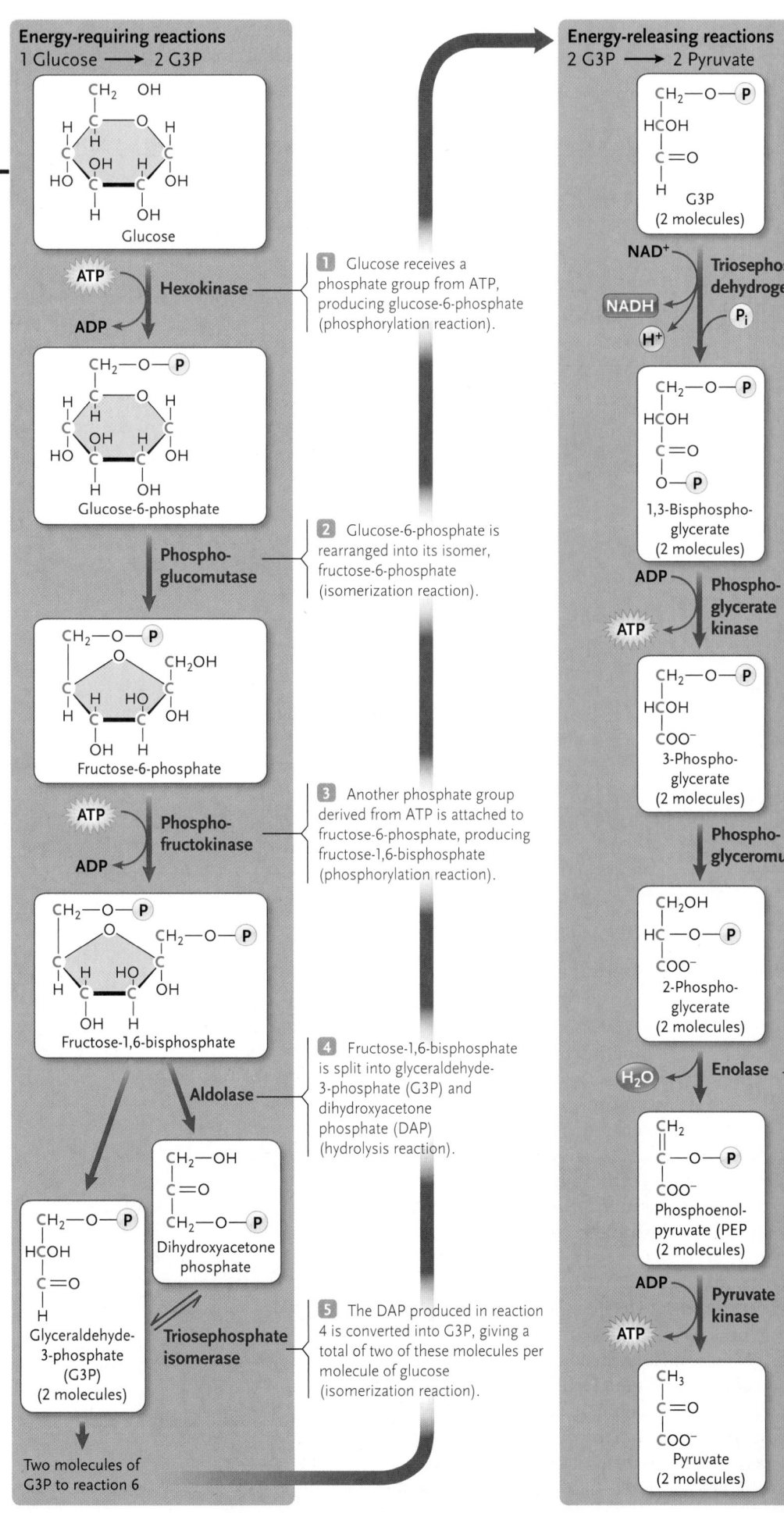

Energy-requiring reactions
1 Glucose ⟶ 2 G3P

Glucose

Hexokinase

ATP
ADP

1 Glucose receives a phosphate group from ATP, producing glucose-6-phosphate (phosphorylation reaction).

Glucose-6-phosphate

Phospho-glucomutase

2 Glucose-6-phosphate is rearranged into its isomer, fructose-6-phosphate (isomerization reaction).

Fructose-6-phosphate

Phospho-fructokinase

ATP
ADP

3 Another phosphate group derived from ATP is attached to fructose-6-phosphate, producing fructose-1,6-bisphosphate (phosphorylation reaction).

Fructose-1,6-bisphosphate

Aldolase

4 Fructose-1,6-bisphosphate is split into glyceraldehyde-3-phosphate (G3P) and dihydroxyacetone phosphate (DAP) (hydrolysis reaction).

Dihydroxyacetone phosphate

Glyceraldehyde-3-phosphate (G3P) (2 molecules)

Triosephosphate isomerase

5 The DAP produced in reaction 4 is converted into G3P, giving a total of two of these molecules per molecule of glucose (isomerization reaction).

Two molecules of G3P to reaction 6

Energy-releasing reactions
2 G3P ⟶ 2 Pyruvate

G3P (2 molecules)

NAD^+

Triosephosphate dehydrogenase

NADH
P_i
H^+

6 Two electrons and two protons are removed from G3P. Some of the energy released in this reaction is trapped by the addition of an inorganic phosphate group from the cytosol (not derived from ATP). The electrons are accepted by NAD^+, along with one of the protons. The other proton is released to the cytosol (redox reaction).

1,3-Bisphospho-glycerate (2 molecules)

ADP

Phospho-glycerate kinase

ATP

7 One of the two phosphate groups of 1,3-bisphosphoglycerate is transferred to ADP to produce ATP (substrate-level phosphorylation reaction).

3-Phospho-glycerate (2 molecules)

Phospho-glyceromutase

8 3-Phosphoglycerate is rearranged, shifting the phosphate group from the 3 carbon to the 2 carbon to produce 2-phosphoglycerate (mutase reaction—shifting of a chemical group to another within same molecule).

2-Phospho-glycerate (2 molecules)

H_2O

Enolase

9 Electrons are removed from one part of 2-phosphoglycerate and delivered to another part of the molecule. Most of the energy lost by the electrons is retained in the product, phosphoenolpyruvate (redox reaction).

Phosphoenol-pyruvate (PEP (2 molecules)

ADP

Pyruvate kinase

ATP

10 The remaining phosphate group is removed from phosphoenolpyruvate and transferred to ADP. The reaction forms ATP and the final product of glycolysis, pyruvate (substrate-level phosphorylation reaction).

Pyruvate (2 molecules)

Figure 6.11
Reactions of pyruvate oxidation. Pyruvate (three carbons) is oxidized to an acetyl group (two carbons), which is carried to the citric acid cycle by CoA. The third carbon is released as CO_2. NAD^+ accepts two electrons and one proton removed in the oxidation. The acetyl group carried from the reaction by CoA is the fuel for the citric acid cycle.

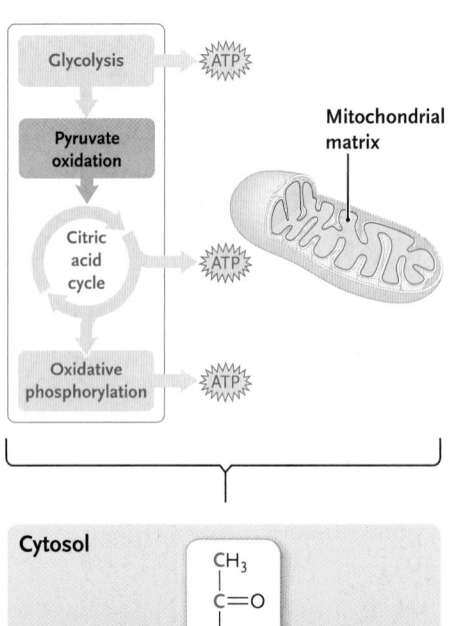

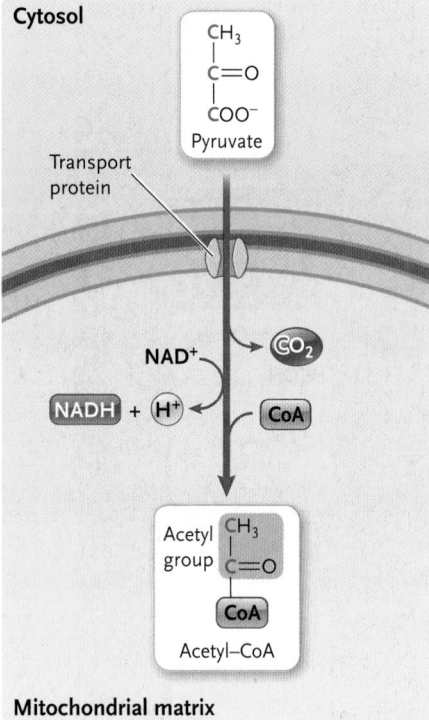

matrix side of the inner mitochondrial membrane. Combined, the reactions result in the oxidation of acetyl groups to carbon dioxide accompanied by the synthesis of ATP; NADH; and another nucleotide-based molecule, flavin adenine dinucleotide (FAD; the reduced form is $FADH_2$). A summary of the inputs and outputs of the citric acid cycle is shown in **Figure 6.12.** To put the cycle in context, the summary also includes the conversion of pyruvate to acetyl-CoA.

Looking at the stoichiometry (Figure 6.12), for one turn of the citric acid cycle, three NADH, one $FADH_2$, and a single molecule of ATP are synthesized. The energy for the synthesis of these molecules comes from the complete oxidation of one acetyl unit, resulting in the release of two molecules of carbon dioxide. The citric acid cycle is the stage of respiration where the remaining carbon atoms that were originally in glucose at the start of glycolysis are converted into carbon

dioxide. The CoA molecule that carried the acetyl group to the site of the citric acid cycle is released and participates again in pyruvate oxidation. The net reactants and products of one turn of the citric acid cycle are

$$1 \text{ acetyl-CoA} + 3 \text{ NAD}^+ + 1 \text{ FAD} + 1 \text{ ADP} + 1 \text{ P}_i + 2 \text{ H}_2\text{O} \rightarrow$$

$$2\text{CO}_2 + 3 \text{ NADH} + 1 \text{ FADH}_2 + 1 \text{ ATP} + 3 \text{ H}^+ + 1 \text{ CoA}$$

Because one molecule of glucose is converted to two molecules of pyruvate by glycolysis and each molecule of pyruvate is converted to one acetyl group, all the reactants and products in this equation should be doubled when the citric acid cycle is considered as a continuation of glycolysis and pyruvate oxidation. **Figure 6.13,** presents a detailed view of the individual reactions of the citric acid cycle.

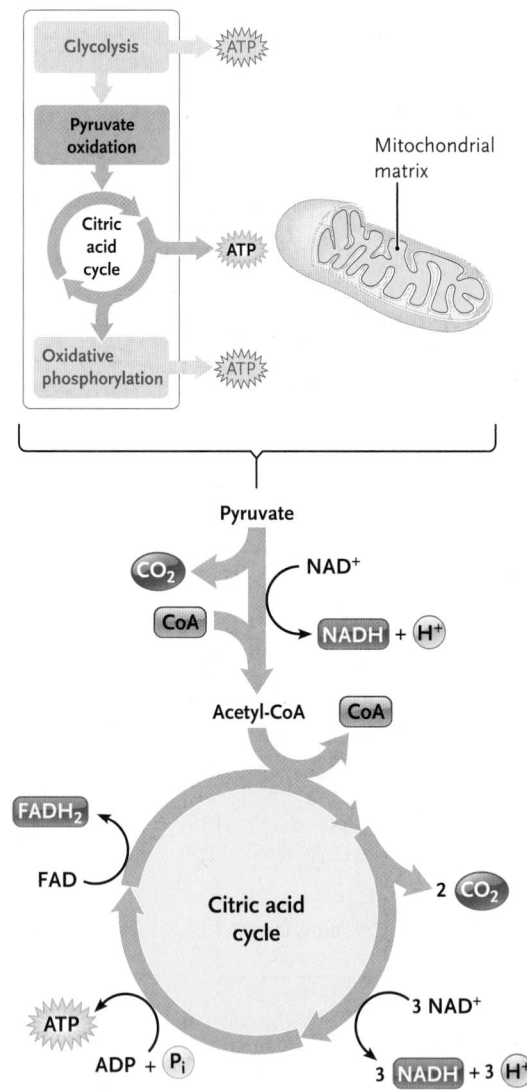

Figure 6.12
The reactions of pyruvate oxidation and the citric acid cycle. Each turn of the cycle oxidizes an acetyl group of acetyl-CoA to 2CO_2. Acetyl-CoA, NAD^+, FAD, and ADP enter the cycle; CoA, NADH, $FADH_2$, ATP, and CO_2 are released as products.

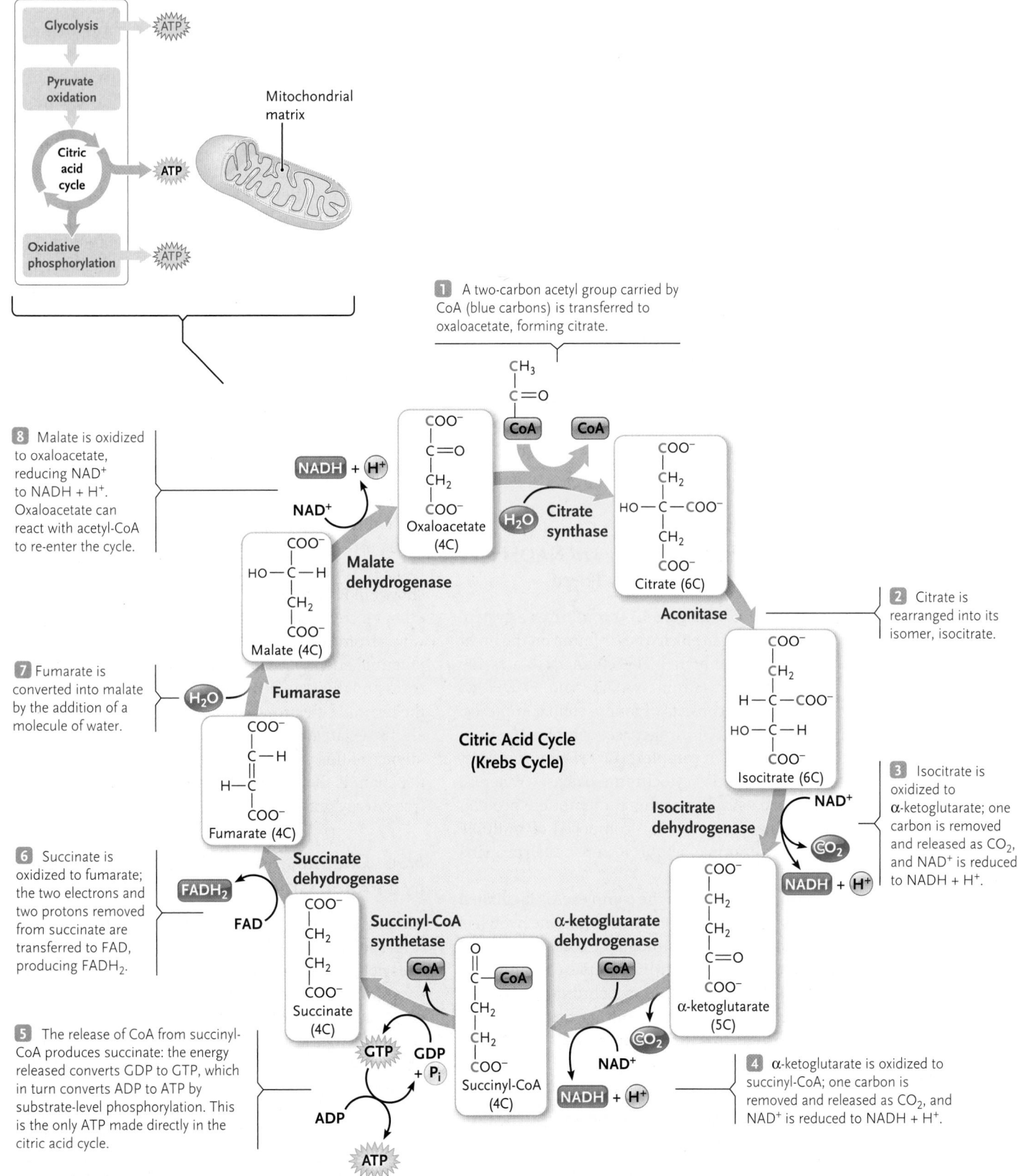

1 A two-carbon acetyl group carried by CoA (blue carbons) is transferred to oxaloacetate, forming citrate.

8 Malate is oxidized to oxaloacetate, reducing NAD⁺ to NADH + H⁺. Oxaloacetate can react with acetyl-CoA to re-enter the cycle.

2 Citrate is rearranged into its isomer, isocitrate.

7 Fumarate is converted into malate by the addition of a molecule of water.

3 Isocitrate is oxidized to α-ketoglutarate; one carbon is removed and released as CO_2, and NAD⁺ is reduced to NADH + H⁺.

6 Succinate is oxidized to fumarate; the two electrons and two protons removed from succinate are transferred to FAD, producing $FADH_2$.

4 α-ketoglutarate is oxidized to succinyl-CoA; one carbon is removed and released as CO_2, and NAD⁺ is reduced to NADH + H⁺.

5 The release of CoA from succinyl-CoA produces succinate: the energy released converts GDP to GTP, which in turn converts ADP to ATP by substrate-level phosphorylation. This is the only ATP made directly in the citric acid cycle.

Citric Acid Cycle (Krebs Cycle)

Figure 6.13

Reactions of the citric acid cycle. Acetyl-CoA, NAD⁺, FAD, and ADP enter the cycle; CoA, NADH, $FADH_2$, ATP, and CO_2 are released as products. The CoA released in reaction 1 can cycle back for another turn of pyruvate oxidation. Enzyme names are in red.

1. What are the steps involved in converting pyruvate into acetyl-CoA?
2. What purpose is served by the citric acid cycle?

6.5 Oxidative Phosphorylation: Electron Transport and Chemiosmosis

Following the citric acid cycle, all the carbon atoms originally present in glucose have been completely oxidized and released as carbon dioxide. Besides ATP formed by substrate-level phosphorylation, the potential energy originally present in glucose now exists in molecules of NADH and $FADH_2$. It is the role of the ETC coupled with the process of chemiosmosis to extract the potential energy in these molecules and synthesize additional ATP.

6.5a The Electron Transport Chain Converts the Potential Energy in NADH and $FADH_2$ into a Proton-Motive Force

The respiratory ETC **(Figure 6.14)** comprises a system of components that in eukaryotes is found on the inner mitochondrial membrane. The chain facilitates the transfer of electrons from $NADH_2$ and $FADH_2$ to oxygen. The chain consists of four protein complexes: **complex I**, NADH dehydrogenase; **complex II**, succinate dehydrogenase; **complex III**, cytochrome complex; and **complex IV**, cytochrome oxidase. Whereas complex II is a single peripheral membrane protein, the remaining complexes are composed of multiple proteins. For example, about 40 individual proteins make up complex I.

Electron flow between the complexes is facilitated by two mobile electron shuttles. Ubiquinone, which is a hydrophobic molecule found in the core of the membrane, shuttles electrons from complexes I and II to complex III. A second shuttle, cytochrome c, is located on the intermembrane space side of the membrane and transfers electrons from complex III to complex IV, cytochrome oxidase.

6.5b Electrons Move Spontaneously along the Electron Transport Chain

In an ETC it is not the proteins themselves that transfer the electrons, but rather electron transport is facilitated by nonprotein molecules called prosthetic groups. Protein subunits of each of complexes I, III, and IV bind a number of prosthetic groups very precisely to allow for electron transport **(Figure 6.15, p. 132)**. Prosthetic groups are redox-active cofactors that alternate between reduced and oxidized states as they accept electrons from upstream molecules and subsequently donate electrons to downstream molecules. A common prosthetic group is the molecule heme, which is a component of the cytochromes, including cytochrome c. Heme is a component of many biologically important compounds, including hemoglobin, where it is critical to the molecule's ability to carry oxygen. Central to its function, a heme group contains a central redox-active iron atom that alternates between Fe^{2+} and Fe^{3+}.

During electron transport (see Figure 6.15), one of the prosthetic groups of complex I, flavin mononucleotide (FMN), is reduced by electron donation from NADH on the matrix side of the inner membrane. FMN then donates the electron to the Fe/S (iron–sulfur) prosthetic group, which, in turn, donates the electron to ubiquinone. This process of reduction followed by oxidation of each carrier continues along the entire chain until, finally, the electrons are donated to oxygen (O_2), reducing it to water. The protons used in the formation of water are abundant in the aqueous environment of the cell.

A question concerning mechanism that we can ask at this stage is: what is the driving force for electron transport—why do electrons move from one complex to the next? why do electrons move down the chain? As shown in **Figure 6.16, p. 132**, the prosthetic groups and other electron carriers are organized in a very specific way—from high to low free energy. NADH has high potential energy because it contains high-energy electrons and thus can be readily oxidized. By contrast, O_2, the terminal electron acceptor of the chain, is strongly electronegative and can be easily reduced. As a consequence of this organization, electron movement along the chain is thermodynamically spontaneous, down a free energy gradient.

6.5c Chemiosmosis Powers ATP Synthesis by a Proton Gradient

Although the goal of cellular respiration is the synthesis of ATP, electron transport from NADH (or $FADH_2$) to O_2 does not actually produce any ATP. Electrons are simply passed along a chain of electron carriers until they are donated to oxygen, producing water. To understand how ATP is formed let's go back and take another look at Figure 6.14. As we have already mentioned, NADH has more potential energy than O_2, so one can ask the question: where does this energy go during electron transport? The energy that is released during electron transport is used to do work, specifically the work of transporting protons across the inner mitochondrial membrane from the matrix to the intermembrane space. As a consequence of this proton pumping across the inner membrane, which is essentially impermeable to protons, the H^+ concentration becomes much higher (and the pH lower) in the intermembrane space than in the matrix.

Proton translocation occurs at distinct sites along the ETC (see Figure 6.14). Within complexes I and IV,

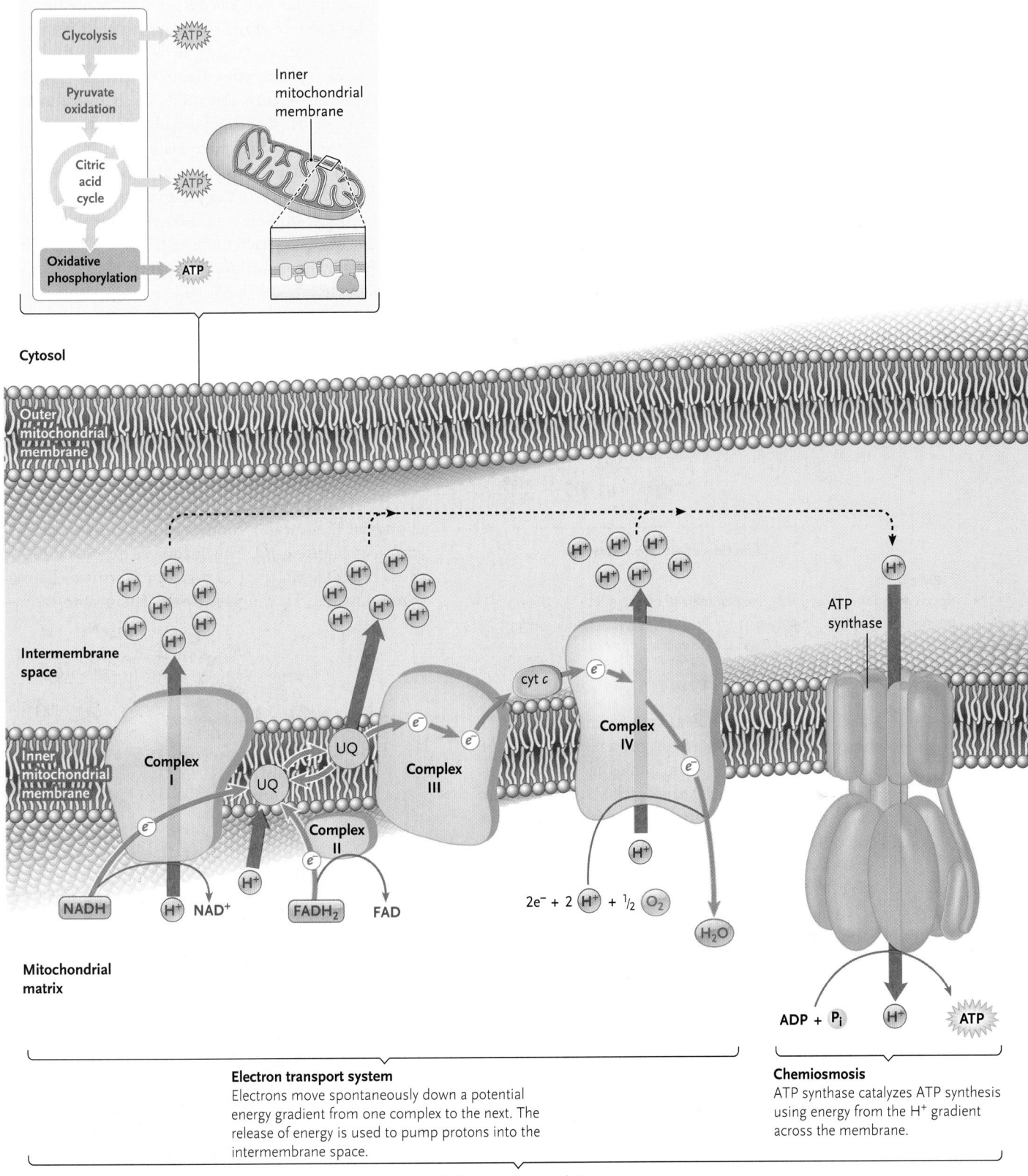

Figure 6.14

Oxidative phosphorylation: the mitochondrial ETC and ATP synthase complex. The electron transport system includes three major complexes, I, III, and IV. Two smaller electron carriers, ubiquinone (UQ) and cytochrome *c* (cyt *c*), act as shuttles between the major complexes, and succinate dehydrogenase (complex II) passes electrons to ubiquinone, bypassing complex I. Blue arrows indicate electron flow; red arrows indicate H⁺ movement. H⁺ is pumped from the matrix into the intermembrane space as electrons pass through complexes I and IV. H⁺ is also moved into the intermembrane space by the cyclic reduction/oxidation of ubiquinone. Chemiosmotic synthesis of ATP involves the ATP synthase complex that uses the energy of the proton gradient to catalyze the synthesis of ATP.

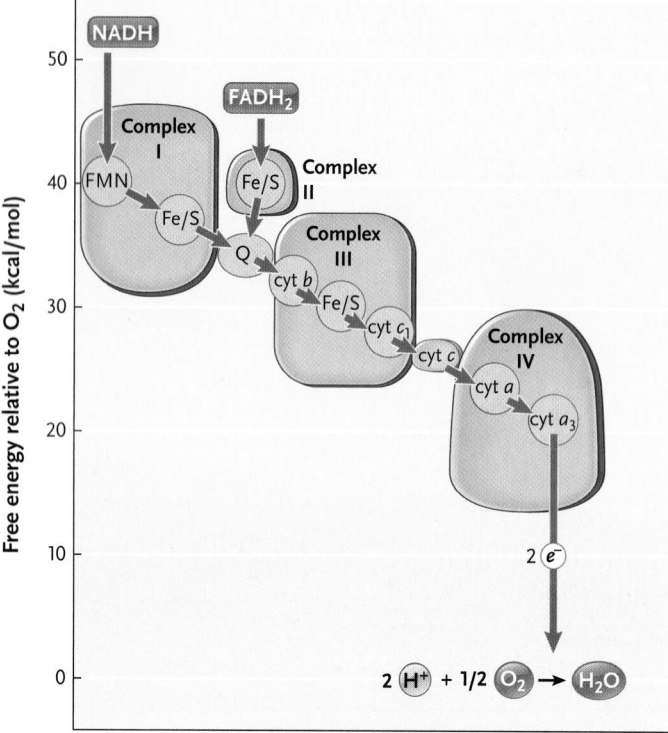

Figure 6.15

Redox components of the ETC are organized from high to low potential energy. Electron flow is spontaneous from high to low potential energy as electrons are passed from one redox molecule to the next.

"People behind Biology"). Whereas in mitochondria, the energy for chemiosmosis comes from the oxidation of energy-rich molecules such as NADH by the ETC, chemiosmosis also applies to the generation of ATP by photosynthesis energy. The utility of chemiosmosis is shown by the fact that it is not only used for ATP synthesis, the proton-motive force is also used, for example, to pump substances across membranes (see Chapter 5) and drive the rotation of flagella in bacteria.

The mode of ATP synthesis that is linked to the oxidation of energy-rich molecules by an ETC is called **oxidative phosphorylation.** Compared to the substrate-level phosphorylation that occurs during glycolysis and the citric acid cycle, oxidative phosphorylation relies on the action of a large multiprotein complex that spans the inner mitochondrial membrane called **ATP synthase** (Figure 6.16).

6.5d ATP Synthase Is a Molecular Motor

ATP synthase is a lollipop-shaped complex consisting of a basal unit, which is embedded in the inner mitochondrial membrane, connected to a headpiece by a stalk (see Figure 6.16). The headpiece extends into the mitochondrial matrix. The basal unit forms a channel through which H^+ can pass freely. The proton-motive

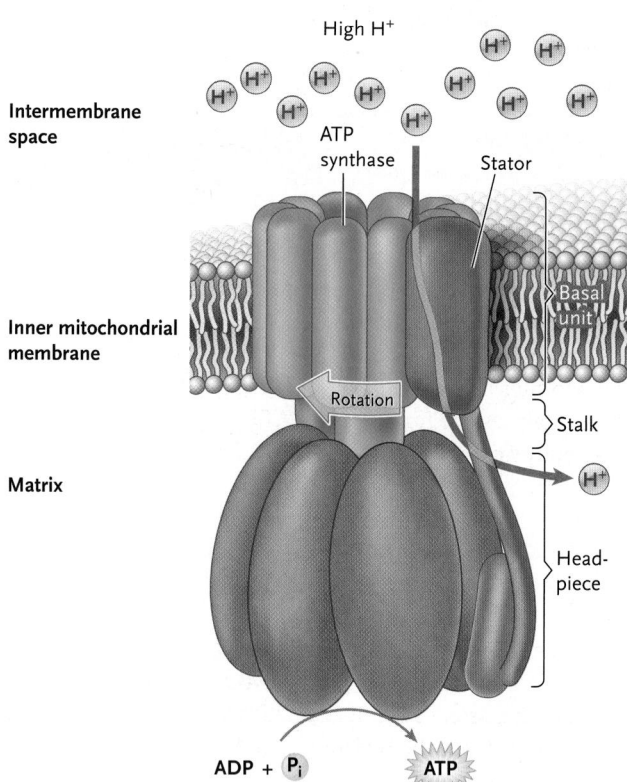

Figure 6.16

Detailed structure of ATP synthase—a molecular motor. The enzyme consists of a *basal unit*, which is embedded in the inner mitochondrial membrane, connected to a *headpiece* by a *stalk*, with the *stator* bridging the basal unit and headpiece. Protons move through a channel between the basal unit and the stator, making the stalk and headpiece spin. This results in ATP synthesis.

specific protein components use the energy released from electron transport for proton pumping. In addition, as ubiquinone molecules accept electrons from complexes I and II, they pick up protons from the matrix. After migrating through the membrane and donating electrons to complex III, ubiquinone retains a neutral charge by releasing protons into the intermembrane space.

The situation in which one side of the inner mitochondrial membrane has a higher concentration of protons than the other side represents potential energy that can be harnessed to do work. The situation is somewhat analogous to water behind a dam. The potential energy possessed by a proton gradient is derived from two factors. First, a chemical gradient exists across the membrane because the concentration of protons is not equal on both sides. Second, because protons are charged, there is an electrical difference, with the intermembrane space more positively charged than the matrix. The combination of a concentration gradient and a voltage difference across the membrane produces stored energy known as the **proton-motive force.**

Harnessing the proton-motive force to do work is referred to as **chemiosmosis.** It was first proposed as a mechanism to generate ATP by the British biochemist Peter Mitchell, who was later awarded a Nobel Prize in Chemistry for his work in this area (see

Peter Mitchell was a British biochemist who in 1978 was awarded the Nobel Prize in Chemistry for what the Royal Swedish Academy of Sciences committee stated was "his contribution to the understanding of biological energy transfer through the formulation of the chemiosmotic theory."

Mitchell completed an undergraduate degree and a Ph.D. at Cambridge University, graduating with the latter in 1951. In 1955, he was invited to set up and direct a biochemical research unit in the Department of Zoology, Edinburgh University, where he was a faculty member until 1964. From 1964 onward, he was director of the Glynn Research Institute. Glynn is a mansion that Mitchell renovated and turned into a personal research institute, located near Bodmin in Cornwall, England.

By the 1950s, it was known that both the chloroplast and the mitochondrion contained electron ETCs and made ATP, but a solid theory on how the two were linked was elusive. The dominant theories were based on substrate-level phosphorylation, which was already well understood. It was thought that ETCs passed energy to a high-energy chemical intermediate, which, in turn, passed it on to ATP through an ATP synthase that was known to exist in both the chloroplast and the mitochondrion. But the problem was that no one could find this chemical intermediate. Moreover, the substrate-level phosphorylation idea could not explain troubling findings: Why did so many different reagents act as uncouplers? Why were the enzymes of oxidative phosphorylation associated with the mitochondrial membrane? Why did coupling seem so dependent on the maintenance of membrane structure?

Mitchell proposed the chemiosmotic theory in 1961 in an elegant paper published in *Nature*. It is hard to imagine now how revolutionary the paper was at the time. It contained very little experimental evidence and was opposed by almost the entire biochemical community, which was stuck believing in the high-energy intermediate concept. The paper was based on Mitchell's realization that the movement of ions across an electrochemical membrane potential could provide the energy needed to produce ATP. The basis of chemiosmosis is that the components of the ETC are inserted into a membrane in only one way, which allows for protons to be transported in one direction during electron transport. The protons would flow back through the ATP synthase, causing synthesis of ATP. In Mitchell's model, the proton gradient across the membrane served as the high-energy intermediate—the elusive chemical intermediate could not be found because it did not exist.

force is what propels protons in the intermembrane space through the channel in the enzyme's basal unit, down their concentration gradient, and into the matrix. Evidence indicates that the binding of individual protons to sites in the headpiece causes it to rotate in a way that catalyzes the formation of ATP from ADP and P_i. The spinning of the headpiece of ATP synthase represents the smallest molecular rotary motor known in nature.

In Chapter 5, we described active transport pumps that use energy from ATP to transport ions across membranes against their concentration gradients (see Figure 5.18, p. 111). An active transport pump is, in fact, an ATP synthase that is operating in reverse. It doesn't synthesize ATP but rather uses the free energy from the hydrolysis of ATP to provide the energy necessary to pump ions (such as protons) across a membrane.

Harnessing the potential energy that is present in a proton gradient to synthesize ATP is fundamental to almost all forms of life and developed early in the evolution of life. This is shown by the fact that the ATP synthase complex found in mitochondria is structurally very similar to the ATP synthase complexes found in the thylakoid membrane of the chloroplast and the plasma membrane of many bacteria and archaea.

6.5e Electron Transport and Chemiosmosis Can Be Uncoupled

CONCEPT FIX Coming out of high school, many students think that ATP is a product of the respiratory ETC. This is a misconception that we need to fix. The generation of ATP by the ATP synthase complex is linked, or coupled, to electron transport by the proton gradient established across the inner mitochondrial membrane. But electron transport and the chemiosmotic generation of ATP are separate and distinct processes and are not always completely coupled (**Figure 6.17, p. 134**). For example, it is possible to have high rates of electron transport (and thus high rates of oxygen consumption) and yet no ATP generated by chemiosmosis. This uncoupling of the two processes occurs when mechanisms prevent the formation of a proton-motive force. ⬡

A class of chemicals called ionophores form channels across membranes through which ions, including

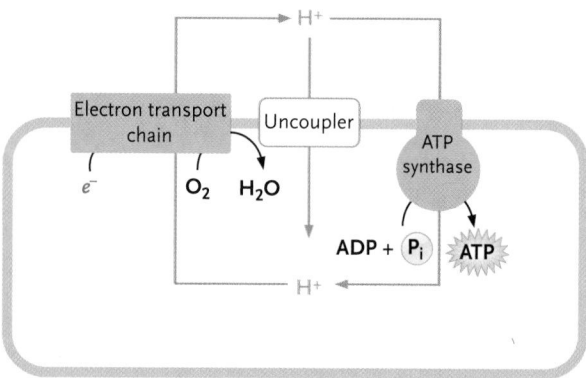

Figure 6.17

Uncoupling of electron transport and ATP synthesis. Respiratory electron transport results in the formation of a proton gradient across the membrane. Usually, this gradient is dissipated by protons flowing back to the matrix through the ATP synthase. Uncouplers, which may be specific chemicals or proteins, provide an alternative route for protons to flow back across the membrane. By circumventing the ATP synthase, no ATP is generated.

protons, can freely pass. As a consequence, in the presence of ionophores, proton pumping during electron transport is followed by the protons simply flowing back into the matrix through the ionophore channels. A proton gradient is prevented from becoming established. Often referred to as uncouplers, ionophores are very toxic because of their ability to inhibit oxidative phosphorylation. It is interesting to note that in the 1930s, low concentrations of chemical uncouplers were commonly used as diet drugs. Although people did lose weight, overdoses resulting in death were not uncommon.

When electron transport is uncoupled from the chemiosmotic synthesis of ATP, the free energy released during electron transport is not conserved by the establishment of a proton-motive force but instead is lost as heat. Many organisms take advantage of this as a means of regulating body temperature by altering the expression of a group of transmembrane proteins. These *uncoupling proteins* are localized to the inner mitochondrial membrane and, similar to chemical uncouplers, form channels through which protons can freely flow. This mechanism of regulating body temperature is especially important in animals. For example, in hibernating mammals and in newborn infants, the activity of uncoupling proteins within mitochondria of brown adipose fat is an important mechanism of heat generation.

STUDY BREAK

1. Differentiate the terms proton-motive force, chemiosmosis, and oxidative phosphorylation.
2. What does it mean that electron transport and oxidative phosphorylation are coupled processes?

6.6 The Efficiency and Regulation of Cellular Respiration

In this section, we calculate the efficiency with which cellular respiration extracts the energy from glucose. As well, we discuss how this entire multi-enzyme pathway is regulated so that it remains flexible in the face of changing cellular demands for ATP and changes in food supply.

6.6a What Are the ATP Yield and Efficiency of Cellular Respiration?

Determining the total number of ATP molecules synthesized for each molecule of glucose oxidized during cellular respiration is an important exercise that forces us to integrate all parts of the respiratory pathway. But before we look at the whole pathway, we first consider a question concerning oxidative phosphorylation: how many ATP molecules are produced by oxidative phosphorylation? Recent research suggests that for each NADH that is oxidized, and thus for each pair of electrons that travels down the ETC, 10 H^+ are pumped into the inner membrane space. (Note: Don't try to figure out how you get 10 protons pumped from 2 electrons—it is not straightforward—wait until you take an advanced biochemistry course.) We also know that somewhere between three and four H^+ are needed to flow back through the ATP synthase for the synthesis of one molecule of ATP. So that gives between 2.5 and 3.3 molecules of ATP synthesized for every NADH oxidized by the ETC. To make life easier, let's round off and say that for each NADH oxidized three ATP are synthesized. Because the oxidation of $FADH_2$ skips the proton-pumping complex I (look back at Figure 6.14, p. 131), only about two molecules of ATP are synthesized for each $FADH_2$ oxidized.

A detailed accounting of the ATP yield for each molecule of glucose oxidized is provided in **Figure 6.18**. Recall that the products of glycolysis include two molecules of ATP and two molecules of NADH. Next, the oxidation of the two molecules of pyruvate generated by glycolysis results in the synthesis of two NADH. During the citric acid cycle, the two molecules of acetyl-CoA that are oxidized result in the synthesis of two ATP, along with six NADH and two $FADH_2$. That gives us a total of ten NADH and two $FADH_2$ that can be oxidized by the ETC. Recall that about three ATP are produced by oxidative phosphorylation for each NADH oxidized by the ETC, while $FADH_2$ oxidation yields two ATP. So that gives a total of 34 ATP generated by oxidative phosphorylation as a result of the oxidation of 10 NADH and 2 $FADH_2$. So, adding up, we have 2 ATP from glycolysis, 2 ATP directly from the citric acid cycle, and 34 ATP from oxidative phosphorylation, yielding 38 molecules of ATP synthesized for each glucose oxidized!

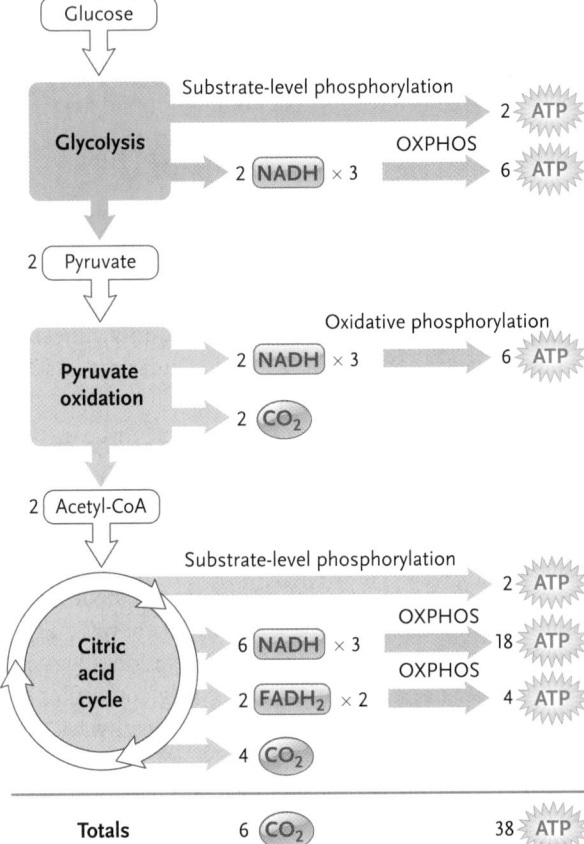

Figure 6.18

ATP yield from the oxidation of glucose. The maximum possible ATP yield from the oxidation of 1 molecule of glucose is 38. However, this yield is rarely achieved. (OXPHOS = oxidative phosphorylation)

The 38 ATP for each glucose oxidized is the maximum theoretical yield. There are three reasons, however, that this maximum is rarely achieved. First, while the maximum of 38 is true in bacteria, this is not the case in eukaryotic cells, where the theoretical maximum is only 36 ATP. This difference is due to the energy costs of transporting the NADH generated by glycolysis into the mitochondrion. The active transport system needed to move the electrons associated with NADH into the mitochondrion consumes one ATP for each molecule of NADH transported. Since two NADH are transported, the yield of ATP drops by two. The second reason the yield is less than 38 ATP is that electron transport and oxidative phosphorylation are rarely completely coupled to each other. Even under normal metabolic conditions, the inner mitochondrial membrane is somewhat leaky to protons, and thus not all the protons pumped across during electron transport pass back into the matrix through the ATP synthase. Some re-enter the matrix by slowly diffusing directly through the inner mitochondrial membrane. The third reason the theoretical maximum is not attained is that the proton-motive force generated by electron transport is used for other things besides simply generating

ATP. As a source of potential energy, the proton-motive force is used, for example, to transport the pyruvate synthesized by glycolysis into the matrix.

So how efficient is cellular respiration at extracting the energy from glucose and converting it into ATP? The phosphorylation of ADP to ATP requires about 7.3 kcal/mol. In eukaryotes, the theoretical maximum yield from that complete oxidation of a mole of glucose is 36 moles of ATP, so 36×7.3 gives a total of 263 kcal of energy. The complete oxidation of a mole of glucose releases exactly 686 kcal of energy. From these two numbers we can calculate the efficiency to be 263/686 $\times 100 = 38\%$. In other words, 38% of the energy in glucose is converted into ATP. While this value of efficiency doesn't seem very high, it is greater than the energy transformations associated with most machines engineers have developed. For example, an automobile extracts only about 25% of the energy in the fuel it burns. Recall from Chapter 4 that the second law of thermodynamics states that energy transformations can never be 100% efficient, as some of the energy is used to increase the entropy of the surroundings. We cannot forget that entropy plays a role in the energy transformations that occur during cellular respiration as well.

6.6b Fats, Proteins, and Carbohydrates Can Be Oxidized by Cellular Respiration

In addition to glucose and other six-carbon sugars, reactions leading from glycolysis through pyruvate oxidation also oxidize a range of other carbohydrates, as well as lipids and proteins, which enter the respiratory pathway at various points **(Figure 6.19, p. 136)**.

Carbohydrates such as sucrose and other disaccharides are easily broken down into monosaccharides such as glucose and fructose, which enter glycolysis at early steps. Starch is hydrolyzed by digestive enzymes into individual glucose molecules, whereas **glycogen**, a more complex carbohydrate, is broken down and converted by enzymes into glucose-6-phosphate, an early substrate molecule in glycolysis.

Among the fats, the **triglycerides** are major sources of electrons for ATP synthesis. Before entering the oxidative reactions, they are hydrolyzed into glycerol and individual fatty acids. The glycerol is converted to glyceraldehyde-3-phosphate before entering glycolysis. The fatty acids—and many other types of lipids—are split into two-carbon fragments, which enter the citric acid cycle as acetyl-CoA.

Proteins are hydrolyzed to amino acids before oxidation. The amino group ($2NH_2$) is removed, and the remainder of the molecule enters the respiratory pathway as pyruvate, acetyl units carried by coenzyme A, or intermediates of the citric acid cycle (see Figure 6.19). For example, the amino acid alanine is converted into pyruvate; leucine, into acetyl units; and phenylalanine, into fumarate, which enters the citric acid cycle.

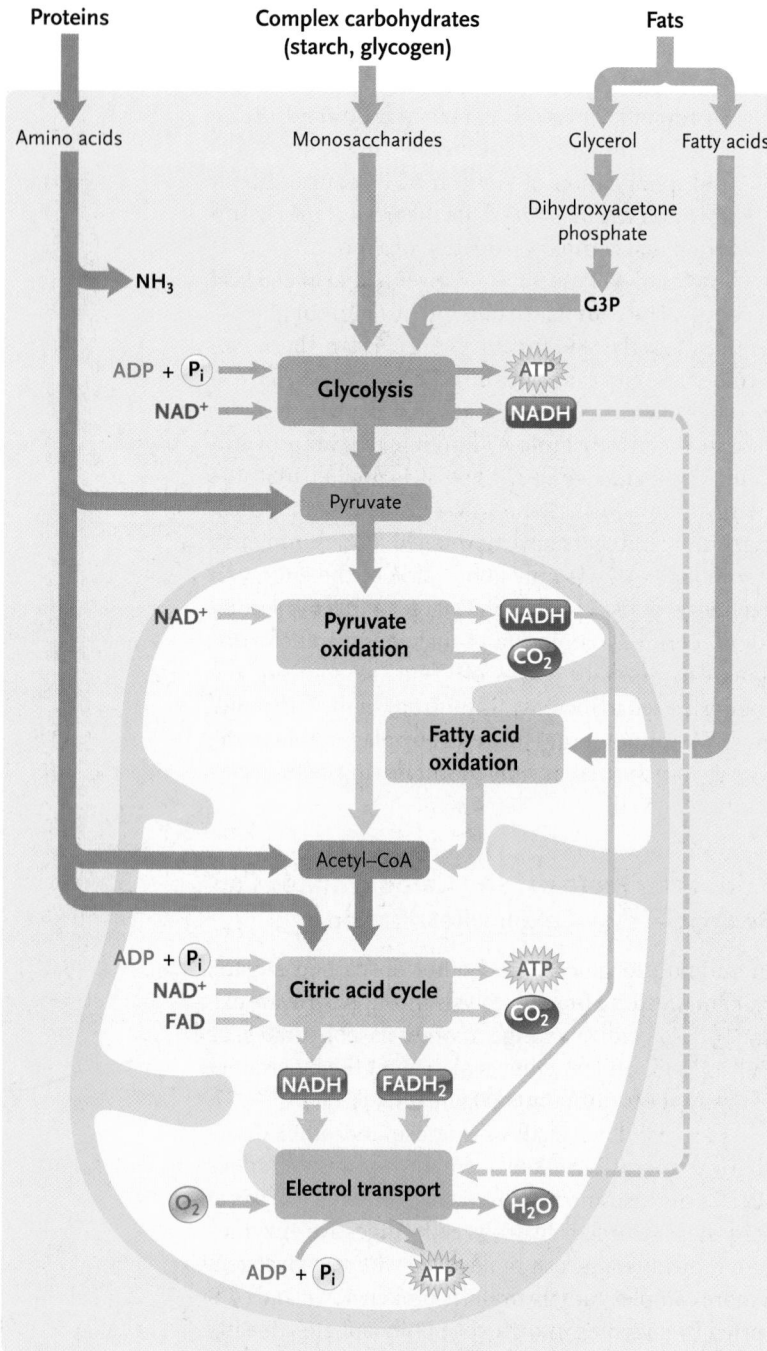

Figure 6.19

Major pathways that oxidize carbohydrates, fats, and proteins. Reactions that occur in the cytosol are shown against a tan background; reactions that occur in mitochondria are shown inside the organelle. CoA funnels the products of many oxidative pathways into the citric acid cycle.

6.6c Respiratory Intermediates Are Utilized for Anabolic Reactions

Organic molecules (carbohydrates, fats, proteins) are oxidized by cellular respiration, which is linked to the generation of ATP. Interestingly, these molecules are also the source of the carbon atoms found in a wide range of essential molecules. For example, the intermediates of glycolysis and the citric acid cycle are routinely diverted and used as the starting

substrates required to synthesize amino acids, fats, and the pyrimidine and purine bases needed for nucleic acid synthesis. As well, respiratory intermediates supply the carbon backbones for the array of hormones, growth factors, prosthetic groups, and cofactors that are essential to cell function.

The metabolic flexibility of cellular respiration allows for reactions to be adjusted rapidly. For example, whereas fatty acids can be used as a source of energy by being oxidized to acetyl-CoA, excess acetyl-CoA can be removed from the respiration and used to synthesize the fatty acids needed for a range of cellular processes.

6.6d Cellular Respiration Is Controlled by Supply and Demand

The rate at which food molecules (see Figure 6.19) are oxidized by cellular respiration is tightly controlled such that the rate of ATP generation matches the requirements of the cell for chemical energy. This illustrates the concept of supply and demand—the cell does not waste resources synthesizing molecules it already has in excess. Most metabolic pathways are regulated by supply and demand through the process of feedback inhibition: the end-products of the pathway inhibit an enzyme early in the pathway (look back at Chapter 4, Figure 4.10).

The rate of glucose oxidation by glycolysis for example, is closely regulated by several mechanisms to match the cellular demands for ATP. A key enzyme of glycolysis that is tightly regulated is phosphofructokinase, which catalyses the conversion of fructose 6-phosphate to fructose 1,6-bisphosphate **(Figure 6.20)**. Because it is an allosteric enzyme (see Chapter 4), the activity of phosphofructokinase can be adjusted by the binding of certain metabolic activators and inhibitors. Two of the key regulators of phosphofructokinase are ATP and ADP.

ATP is an allosteric inhibitor of phosphofructokinase: if excess ATP is present in the cytosol it binds to phosphofructokinase, inhibiting its activity. The resulting decrease in the concentration of fructose 1,6-bisphosphate slows or stops the subsequent reactions of glycolysis. The enzyme becomes active again when metabolic demands consume the excess ATP and the inhibition of phosphofructokinase is released. The increase in phosphofructokinase activity is not due solely to the release of ATP inhibition, however ADP, which accumulates when ATP is hydrolyzed during metabolism, is an allosteric activator of the enzyme (see Figure 6.20). Besides ATP and ADP, phosphofructokinase activity is also sensitive to the levels of citrate, which is the first product of the citric acid cycle. If the products of the citric acid cycle are in high demand, then citrate should not accumulate in the cell. Increased citrate concentrations suggest that the demand for ATP is low, which may occur, for example, under conditions of limited oxygen when the rate of oxidative phosphorylation

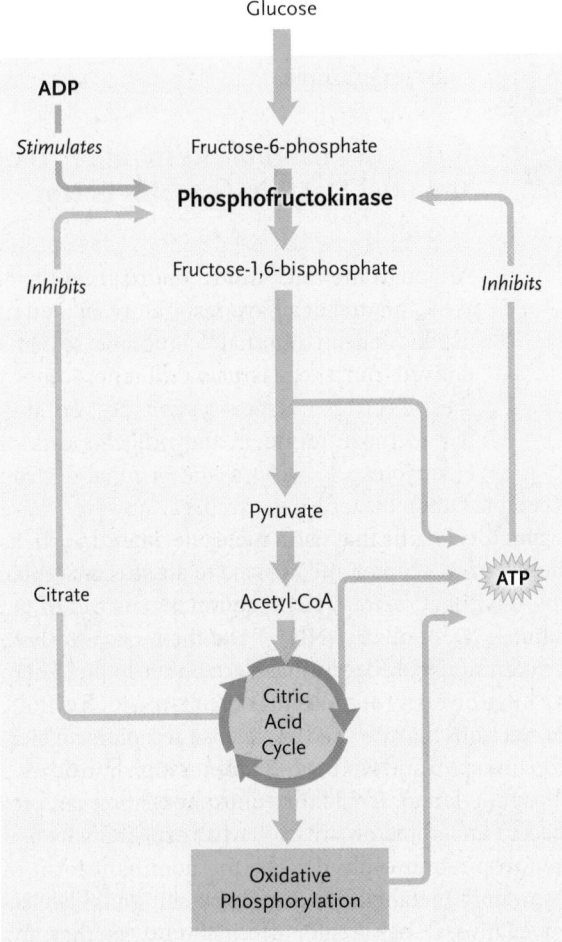

Glucose

ADP

Stimulates

Fructose-6-phosphate

Phosphofructokinase

Fructose-1,6-bisphosphate

Inhibits   *Inhibits*

Pyruvate

Citrate   Acetyl-CoA   **ATP**

Citric
Acid
Cycle

Oxidative
Phosphorylation

Figure 6.20

The control of cellular respiration. A major mechanism is allosteric control of the activity of the enzyme phospho-fructokinase, which is found early in glycolysis. High levels of ATP and the citric acid cycle intermediate citrate allosterically inhibit phosphofructokinase. Alternatively, when ATP concentrations are low, the levels of ADP increase. ADP is an allosteric activator of the enzyme.

is restricted. Alternatively, it may indicate that citrate is not required as a carbon backbone for the products of biosynthetic reactions. Overall, through various metabolic activators and inhibitors altering phosphofructokinase activity, the functional state of glycolysis and the citric acid cycle can be kept balanced.

STUDY BREAK

1. Give an accounting of the total ATP yield from the oxidation of a molecule of glucose.
2. Explain how the activity of the enzyme phospho-fructokinase is controlled.

6.7 Oxygen and Cellular Respiration

A constant supply of oxygen is required to maintain the high rates of oxidative phosphorylation necessary to supply cells with sufficient ATP. Although humans need an almost constant supply of oxygen, other organisms can survive and even thrive in the absence of oxygen.

There are two general mechanisms where cellular respiration can occur in the absence of oxygen: fermentation and anaerobic respiration. The distinction between these two processes is that fermentation does not involve the citric acid cycle or electron transport, whereas anaerobic respiration uses a molecule other than oxygen as the terminal electron acceptor of electron transport.

6.7a In Eukaryotic Cells, Low Oxygen Levels Result in Fermentation

Following glycolysis, in eukaryotic cells cellular respiration can continue along one of two distinct pathways depending on whether or not oxygen is present **(Figure 6.21, p. 138)**. When oxygen is plentiful, the pyruvate and NADH produced by glycolysis are transported into mitochondria, where they are oxidized using the citric acid cycle and the ETC. If, instead, oxygen is absent or in short supply, the pyruvate remains in the cytosol, where it is reduced, consuming the NADH generated by glycolysis by a metabolic process called **fermentation**.

Two types of fermentation exist: lactate fermentation and alcohol fermentation. In **lactate fermentation**, which is found in many bacteria and some plant and animal tissues, pyruvate is converted into the three-carbon molecule lactate **(Figure 6.22a, p. 138)**. Lactate fermentation occurs, for example, when vigorous contraction of muscle cells calls for more oxygen than the circulating blood can supply. When the oxygen content of the muscle cells returns to normal levels, the reverse of the reaction in Figure 6.22a regenerates pyruvate and NADH. Lactate is also the fermentation product of some bacteria; the sour taste of buttermilk, yogurt, and dill pickles is a sign of their activity.

Alcohol fermentation (Figure 6.22b) occurs in microorganisms such as yeasts, which are single-celled fungi. In this reaction, pyruvate is reduced to ethyl alcohol as CO_2 is released and NADH is oxidized to NAD^+. Alcoholic fermentation by yeasts has widespread commercial applications. Bakers use the yeast *Saccharomyces cerevisiae* to make bread dough rise. They mix the yeast with a small amount of sugar and blend the mixture into the dough, where oxygen levels are low. As the yeast cells convert the sugar into ethyl alcohol and CO_2, the gaseous CO_2 expands and creates bubbles that cause the dough to rise. Oven heat evaporates the alcohol and causes further expansion of the bubbles, producing a light-textured product. Alcoholic fermentation is also the mainstay of beer and wine brewing. Fruits are a natural home to wild yeasts **(Figure 6.23, p. 139)**; for example, winemakers rely on a mixture of wild and cultivated yeasts to produce wine. Alcoholic fermentation also occurs naturally in the environment; for example, overripe or rotting fruit will

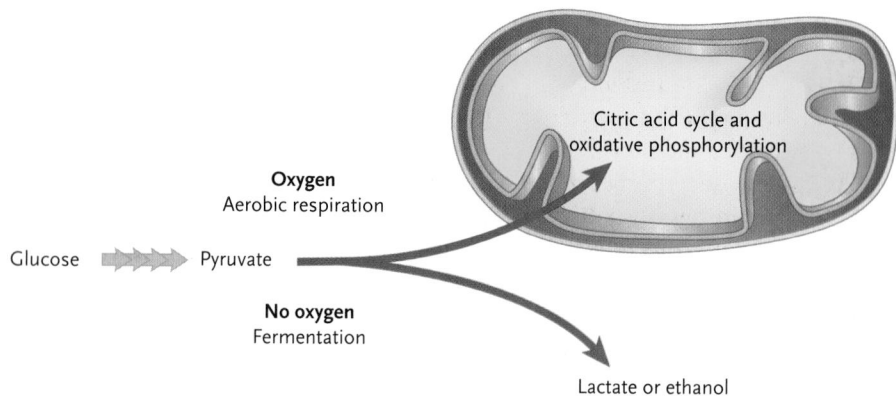

Figure 6.21

The metabolic pathway of pyruvate oxidation depends upon the presence of oxygen.

frequently start to ferment, and birds that eat the fruit may become too drunk to fly.

Overall, the reactions of fermentation play a critical role whenever organisms are exposed to conditions in which the oxygen concentration is too low to support oxidative phosphorylation. By consuming the NADH generated by glycolysis, fermentation reactions keep cytosolic NAD$^+$ levels high. This is of critical metabolic importance because NAD$^+$ is required for glycolysis (look back at Figure 6.9, step 6, p. 127). As long as there is sufficient NAD$^+$, glycolysis will continue to operate and generate ATP. Of course, the amount of ATP generated is small compared to oxidative phosphorylation,

and thus fermentation is not sufficient to support the high ATP requirement of brain cells, for example.

6.7b In Anaerobic Respiration, the Terminal Electron Acceptor Is Not Oxygen

Although they lack mitochondria, many bacteria and archaea have respiratory ETCs that are located on internal membrane systems derived from the plasma membrane. Some of these electron transport systems are very similar to those found in the mitochondria of eukaryotes and use O$_2$ as the terminal electron acceptor. Other bacteria and archaea, however, have respiratory chains that use a molecule other than O$_2$ as the electron acceptor and are said to possess anaerobic (*an* = without; *aero* = air) respiration. Instead of O$_2$, sulfate (SO$_4^{2-}$), nitrate (NO$_3^-$), and the ferric ion (Fe^{3+}) are commonly used terminal electron acceptors. There is a huge diversity of molecules that have a high affinity for electrons that are used as electron acceptors for electron transport and support ATP generation by oxidative phosphorylation. If oxidative phosphorylation can proceed in anaerobic organisms, what explains why aerobic respiration evolved to be the dominant form of respiratory metabolism? Simply by being highly electronegative, O$_2$ has greater affinity for oxygen than any

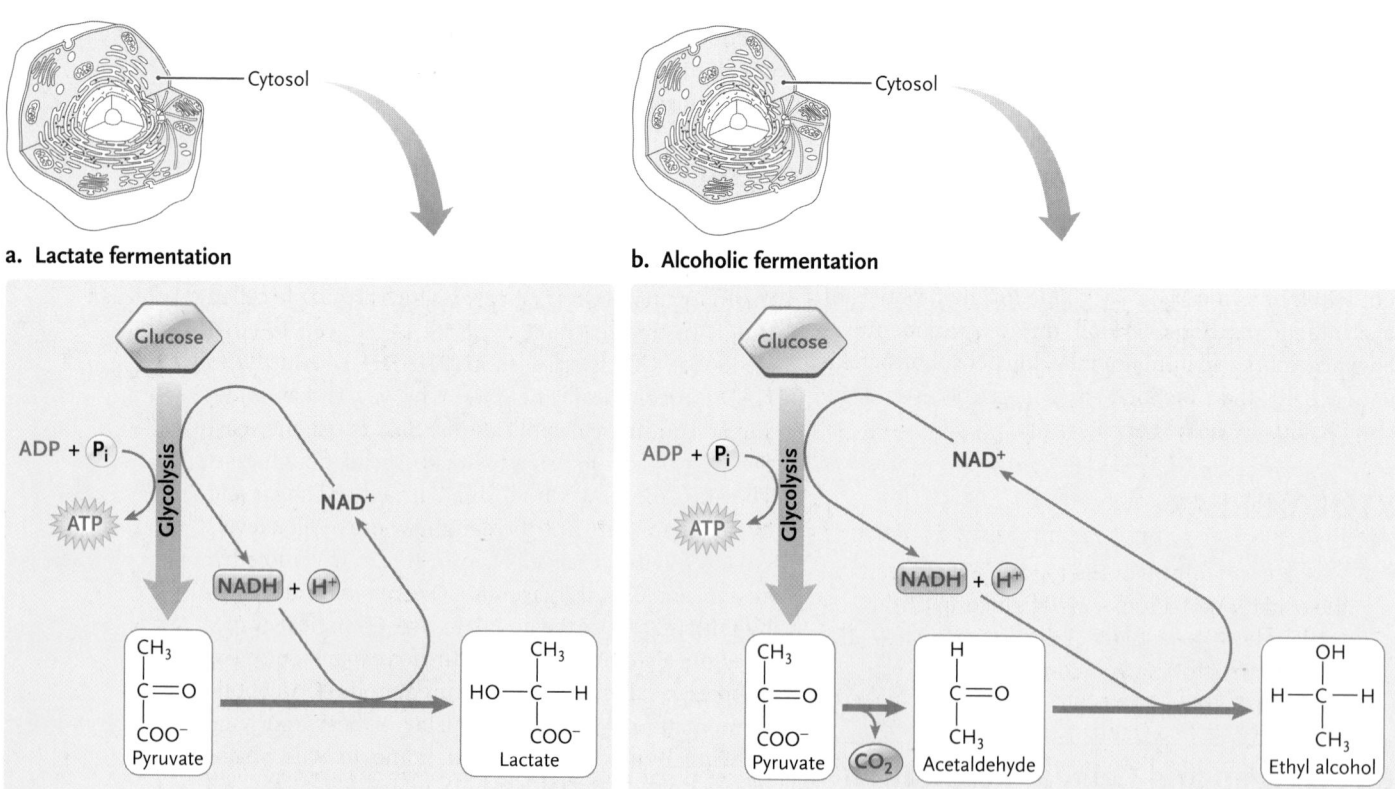

Figure 6.22

Fermentation reactions that produce **(a)** lactate and **(b)** ethyl alcohol. The fermentations, which occur in the cytosol, convert NADH to NAD$^+$, allowing the electron carrier to cycle back to glycolysis. This process keeps glycolysis running, with continued production of ATP.

Figure 6.23
Alcoholic fermentation in nature: wild yeast cells, visible as a dustlike coating on grapes.

other electron acceptor. This enables ETCs that employ O_2 to extract a greater amount of potential energy out of substrate molecules (e.g., NADH, $FADH_2$).

6.7c Organisms Differ with Respect to Their Ability to Use Oxygen

We can differentiate three lifestyles depending on the requirements of an organism for oxygen. Many archaea and bacteria and most eukaryotes are **strict aerobes**—they have an absolute requirement for oxygen for survival and are unable to live solely by fermentation. To understand why this is, look back at Figure 6.18, p. 135. In the absence of oxygen, ATP is generated solely by substrate-level phosphorylation during glycolysis: 2 ATP generated for every glucose oxidized. By comparison, in the presence of oxygen, up to 38 ATP can be generated—that's 19 times as much ATP for each glucose oxidized. As shown in Figure 6.19, the difference is explained by the huge ATP yield of oxidative phosphorylation. Humans and other animals are especially sensitive to low-oxygen environments because certain tissues, such as brain cells, have requirements for ATP that can only be met by constant and high rates of oxidative phosphorylation.

Other organisms, called **facultative anaerobes**, can switch between fermentation and aerobic respiration, depending on the oxygen supply. Facultative anaerobes include *Escherichia coli*, the bacterium that inhabits the digestive tract of humans; the *Lactobacillus* bacteria used to produce buttermilk and yogurt; and *S. cerevisiae*, the yeast used in brewing and baking. Many cell types in higher organisms, including vertebrate muscle cells, are also facultatively anaerobic. Lastly, some bacteria, some archaea, and a few fungi are classified as **strict anaerobes** because they require an oxygen-free environment to survive. Strict anaerobes gain ATP from either fermentation or anaerobic respiration. Among these organisms are the bacteria that cause botulism, tetanus, and some other serious diseases. For example, the bacterium that causes botulism thrives in the oxygen-free environment of canned foods that prevents the growth of most other microorganisms.

6.7d The Paradox of Aerobic Life Is That Oxygen Is Essential and Toxic

As we mentioned above, some microbes are strict anaerobes—they cannot live in an oxygen environment. But why can't they? Lacking the ability to use O_2 as an electron acceptor is one thing, but actually dying in the presence of O_2? The reason that strict anaerobes die in an oxygen environment is related to what is often called the *paradox of aerobic life*: although oxygen is absolutely essential to the survival of many organisms, oxygen is also potentially toxic.

It takes four electrons to completely reduce a molecule of O_2 to water **(Figure 6.24)**. Partially reduced forms of O_2 are formed when O_2 accepts fewer electrons, producing what are called *reactive oxygen species* (ROS). These molecules, which include the compounds superoxide and hydrogen peroxide (see Figure 6.24), are powerful oxidizing molecules and readily remove electrons from proteins, lipids, and DNA, resulting in oxidative damage. If ROS levels within a cell are excessive, their strong oxidizing nature can result in the destruction of many biological molecules and can be lethal. Because most cells contain an abundance of both O_2 and electron-rich molecules (e.g., proteins, lipids, nucleic acids) the formation of ROS is a consequence of aerobic life that cannot be avoided.

To survive in an oxygen-rich environment, aerobic organisms have evolved an antioxidant defence system that includes both enzymes and nonenzyme molecules that have the role of intercepting and inactivating reactive oxygen molecules as they are produced within cells. Two of the major ROS-scavenging enzymes are superoxide dismutase and catalase (see Figure 6.24). Working in concert, superoxide dismutase converts the superoxide anion to hydrogen peroxide, which in turn is reduced to water by the action of catalase. In addition to enzymes, many cells have a range of antioxidants,

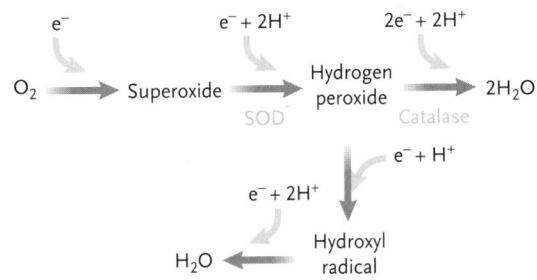

Figure 6.24
The conversion of O_2 to water is a four-electron reduction. If this occurs stepwise, it results in the formation of the intermediate ROS, which are potentially harmful. Aerobic cells contain the enzymes superoxide dismutase (SOD) and catalase, which together quickly convert superoxide and hydrogen peroxide to water.

MOLECULE BEHIND BIOLOGY 6.2
Cyanide

Cyanide is an ion that consists of a carbon atom triple-bonded to an atom of nitrogen (C — N⁻). It is a very toxic metabolic poison acting as an irreversible inhibitor of the terminal enzyme of respiratory electron transport, cytochrome oxidase. By binding to the iron atom of the heme prosthetic groups in the enzyme, cyanide prevents electron flow to O_2, essentially inhibiting electron transport and subsequent chemiosmosis. Acute cyanide poisoning can result in death within minutes of exposure.

Cyanide is produced in small amounts by a range of microorganisms and is found in small amounts in apple seeds, almonds, and the pits of fruits such as peaches. In some plants, the production of cyanide in a form bound to sugars is thought to be a deterrent to herbivory. The presence of cyanide in the potato-like root of the cassava plant is of concern because it is a staple food in a number of tropical countries. The presence of cyanide

glycosides is diminished by extensive soaking and cooking of the cassava root, but health problems associated with chronic cyanide poisoning remain quite common.

Cyanide has clear applications in a range of industries but especially in electroplating, metallurgy, and mining owing to the high solubility of gold $[Au(CN)_2]^-$ and silver $[Ag(CN)_2]^-$ cyanides in water. For these purposes, approximately 500 thousand tonnes of highly toxic sodium cyanide are produced each year. In gold mining, the addition of a solution of sodium cyanide to ore containing low amounts of gold is effective at extracting the gold by bringing the gold into solution. The resulting formation of huge amounts of cyanide-contaminated water makes this form of gold mining highly controversial, yet it remains a very effective and cheap method of extraction.

In addition to a respiratory ETC that is inhibited by cyanide, plants

contain a pathway of electron transport that is resistant to cyanide. Instead of using cytochrome oxidase, this second pathway of respiration uses a terminal oxidase called the alternative oxidase. This alternative pathway of respiration is not linked to proton pumping like the normal respiratory chain; instead, electron flow simply generates heat. Intestinally, high levels of alternative oxidase in the flowers of some plant species are used to volatilize attractants for pollinators. This includes the aptly named skunk cabbage, which tells you that the attractants for pollinators are not necessarily pleasant.

In addition to being found in all plants, the alternative oxidase has been found in algae, some fungi, and, recently, some animal phyla. The physiological role of cyanide-resistant respiration in these species is being actively investigated by a number of research groups.

including vitamin C and vitamin E, which act as reducing agents, safely and rapidly reducing reactive oxygen compounds to water. In recent years, excessive ROS formation has been implicated in a wide variety of degenerative diseases, including Parkinson's disease and Alzheimer dementia. In fact, it is thought that the progressive build-up of oxidative damage may underlie the aging process itself. This, in part, explains the huge interest in the possible protective value of a wide variety of antioxidant compounds, particularly those from certain fruits and vegetables.

So why do strict anaerobes die in the presence of oxygen? For one group, their inability to live in an oxygen environment is because they lack one or both of the enzymes superoxide dismutase and catalase, which results in a build-up of toxic ROS within their cells if they are exposed to oxygen. Interestingly, some strict anaerobes do contain these enzymes, which are highly expressed when cells are placed in an oxygen environment. The inability of this second group of anaerobes to survive in an oxygen environment seems to be linked to oxygen itself inhibiting key metabolic enzymes.

As discussed in the last section, oxidative phosphorylation generates much more ATP than the substrate-level phosphorylation that takes place during glycolysis and the citric acid cycle. From this it is clear that the evolution of the electron transport system with oxygen as the terminal electron acceptor enabled cells to extract far more energy from food molecules than other modes of metabolism. However, the evolution of the aerobic lifestyle required the development of antioxidants and enzymes such as catalase and superoxide dismutase to combat the harmful effects of oxygen, which include the inevitable formation of ROS. In addition, it required that cytochrome oxidase, the last enzyme of the mitochondrial ETC, develop a remarkable mode of catalysis. Looking back at Figure 6.14, p. 131, notice that the cytochrome oxidase complex donates electrons from the electron carrier cytochrome c to O_2. However, it does so in a way that, remarkably, leads to essentially no reactive oxygen generation. The enzyme is structurally quite complex, containing four redox centres (two hemes and two copper ions), each of which can store a single electron. When all centres are reduced, the enzyme simultaneously transfers all four electrons to O_2, producing two molecules of water. That cytochrome oxidase is the only enzyme that aerobic organisms, from bacterial to human, use as the terminal complex of electron transport indicates the chemical difficulty of carrying out the transfer of electrons to O_2 in a safe and controlled manner. Given that this single enzyme handles approximately 98% of the oxygen we metabolize, if the reaction resulted in significant amounts of partially reduced forms of oxygen (e.g., superoxide, hydrogen peroxide), aerobic life as we know it would have probably never evolved.

Review

aplia™

To access course materials such as Aplia and other companion resources, please visit www.NELSONbrain.com.

6.1 The Chemical Basis of Cellular Respiration

- Glucose, like gasoline, is a good fuel because of the presence of C—H bonds, which contain electrons that can be easily removed and used to do work (see Figure 6.2).

- Oxidation–reduction reactions, called redox reactions, partially or completely transfer electrons from donor to acceptor atoms; the donor is oxidized as it releases electrons, and the acceptor is reduced (see Figure 6.3).

- Almost all organisms obtain energy for cellular activities through cellular respiration, the process of transferring electrons from donor organic molecules to a final acceptor molecule such as oxygen; the energy that is released drives ATP synthesis.

6.2 Cellular Respiration: An Overview

- Cellular respiration occurs in three stages: (1) in glycolysis, glucose is converted to two molecules of pyruvate; (2) in pyruvate oxidation and the citric acid cycle, pyruvate is converted to an acetyl compound that is oxidized completely to CO_2; and (3) in oxidative phosphorylation, high-energy electrons produced from the first two stages pass along an electron transport chain (ETC), with the energy released being used to establish a proton gradient across the membrane. This gradient is used to synthesize ATP (see Figure 6.6).

- Both eukaryotes and prokaryotes may undergo cellular respiration. In eukaryotes, however, most of the reactions of cellular respiration occur in mitochondria (see Figure 6.7).

6.3 Glycolysis: The Splitting of Glucose

- In glycolysis, which occurs in the cytosol, glucose (six carbons) is oxidized into two molecules of pyruvate (three carbons each). Electrons removed in the oxidation are delivered to NAD^+, producing NADH. The reaction sequence produces a net gain of two ATP, two NADH, and two pyruvate molecules for each molecule of glucose oxidized (see Figures 6.8 and 6.9).

- ATP molecules produced in the energy-releasing steps of glycolysis result from substrate-level phosphorylation, an enzyme-catalyzed reaction that transfers a phosphate group from a substrate to ADP (see Figure 6.10).

6.4 Pyruvate Oxidation and the Citric Acid Cycle

- In pyruvate oxidation, which occurs inside mitochondria, one pyruvate (three carbons) is oxidized to one acetyl group (two carbons) and one CO_2. Electrons removed in the oxidation are accepted by one NAD^+ to produce one NADH. The acetyl group is transferred to coenzyme A (CoA), which carries it to the citric acid cycle (see Figure 6.11).

- In the citric acid cycle, which occurs in the matrix of the mitochondrion, acetyl groups are oxidized completely to CO_2. Electrons removed in the oxidation are accepted by NAD^+ or FAD, and substrate-level phosphorylation produces ATP. For each acetyl group oxidized by the cycle, two CO_2, one ATP, three NADH, and one $FADH_2$ are produced (see Figures 6.12 and 6.13).

6.5 Oxidative Phosphorylation: Electron Transport and Chemiosmosis

- Electrons are passed from NADH and $FADH_2$ to the ETC, which consists of four major protein complexes and two smaller shuttle carriers. As the electrons flow from one carrier to the next through the system, some of their energy is used by the complexes to pump protons across the inner mitochondrial membrane (see Figure 6.14).

- Two major protein complexes (I and IV) and the reduction/-oxidation of ubiquinone contribute to the pumping of protons from the matrix to the intermembrane space, generating a proton gradient.

- The proton gradient produced by the electron transport system is used by ATP synthase as an energy source for synthesis of ATP from ADP and P_i. The ATP synthase is embedded in the inner mitochondrial membrane together with the ETC (see Figure 6.16).

- Electron transport and the chemiosmotic synthesis of ATP are distinct and separate processes that are usually linked, or coupled, by the proton gradient. The uncoupling of the two results in high rates of electron transport (and oxygen consumption), without chemiosmotic ATP synthesis. Uncoupling results in heat generation (see Figure 6.17).

6.6 The Efficiency and Regulation of Cellular Respiration

- An estimated three ATP are synthesized as each electron pair travels from NADH to oxygen through the mitochondrial ETC; about two ATP are synthesized as each electron pair travels through the system from $FADH_2$ to oxygen.

- In glycolysis, two ATP and two NADH are synthesized; during the oxidation of pyruvate and the citric acid cycle, two ATP, eight NADH, and two $FADH_2$ are produced. That gives a total of 10 NADH and 2 $FADH_2$ that are oxidized by the ETC, leading to the synthesis of about 34 ATP. This gives a total theoretical maximum ATP yield for each glucose oxidized of 38. For a number of reasons this maximum yield is rarely reached.
- The efficiency with which the energy in glucose is conserved in the synthesis of ATP is about 30% (see Figure 6.18).
- Besides simple sugars, energy can be extracted from fats, proteins, and carbohydrates that enter the respiratory chain at different points (see Figure 6.19).
- A number of different molecules can activate and repress key steps of the respiratory pathway so that it can be controlled by supply and demand (see Figure 6.20).

6.7 Oxygen and Cellular Respiration

- Fermentation is a pathway of respiration that oxidizes fuel molecules in the absence of oxygen and does not involve oxidative phosphorylation (see Figure 6.21).

- During fermentation, pyruvate reduction in the cytosol consumes NADH. In so doing, NAD^+ is produced, which is required as a substrate for glycolysis. This allows glycolysis to continue to run, producing ATP by substrate-level phosphorylation (see Figure 6.22).
- In organisms with anaerobic respiratory pathways, the terminal electron acceptor of electron transport is a molecule other than oxygen. Anaerobic respiratory pathways are found only in archaea and bacteria.
- Many archaea, bacteria, and eukaryotes are strict aerobes. They have an absolute requirement for oxygen because they require the high-ATP yield of oxidative phosphorylation.
- Facultative anaerobes can grow in the presence of oxygen and can grow in the absence of oxygen using fermentative pathways.
- Strict anaerobes cannot grow in the presence of oxygen.
- Although oxygen is required for aerobic life, paradoxically, oxygen is toxic to cells. Reactive oxygen species (ROS) are partially reduced forms of oxygen that can damage cells. The formation of ROS by the respiratory pathway is unavoidable (see Figure 6.24).
- Cells are protected from the toxicity of oxygen by both enzymatic and nonenzymatic antioxidants that detoxify ROS.

Questions

Self-Test Questions

1. Which of the following is found in organic molecules that are good fuels?
   a. many C—H bonds
   b. many C=C double bonds
   c. an abundance of oxygen
   d. a high molecular weight

2. Which of the following general statements about cellular respiration is correct?
   a. In cellular respiration oxygen is used as an electron donor.
   b. Since bacteria lack mitochondria, they do not perform cellular respiration.
   c. Cellular respiration represents a series of reactions in which a carbon substrate is oxidized.
   d. The carbon dioxide produced during cellular respiration can be used as an energy source for metabolism.

3. Which of the following processes occurs during glycolysis?
   a. the oxidation of pyruvate
   b. the reduction of glucose
   c. oxidative phosphorylation
   d. substrate-level phosphorylation

4. Which of the following is an accurate statement about the proton-motive force?
   a. It needs to be high to synthesize ATP during the citric acid cycle.
   b. If protons were uncharged, the proton-motive force would be zero.
   c. Along with multicellularity, it was a key development in eukaryotic cells.
   d. It represents the energy associated with a proton gradient across a membrane.

5. You are reading this text while breathing in oxygen and breathing out carbon dioxide. Which two processes are the sources of the carbon dioxide?
   a. glycolysis and pyruvate oxidation
   b. glycolysis and oxidative phosphorylation
   c. pyruvate oxidation and the citric acid cycle
   d. the citric acid cycle and oxidative phosphorylation

6. Under conditions of low oxygen, what key role is played by fermentation in overall metabolism?
   a. It regenerates the NAD^+ required for glycolysis.
   b. It synthesizes additional NADH for the citric acid cycle.
   c. It allows for pyruvate to be oxidized in mitochondria.
   d. By activating oxidative phosphorylation, it allows for the synthesis of extra ATP.

7. In cellular respiration, what does the term *uncoupled* specifically refer to?
   a. when the two parts of glycolysis are running independently of each other
   b. when respiratory electron transport is operating, but chemiosmosis is inhibited
   c. when respiratory electron transport is operating, but proton pumping is inhibited
   d. when oxidative phosphorylation is occurring, but the proton-motive force remains high

8. Phosphofructokinase (PFK) is regulated by a number of metabolites. Besides the ones mentioned in the text, which one of the following would also make sense?
   a. Pyruvate could function as an activator of PFK.
   b. Glucose could function as an inhibitor of PFK.
   c. ADP could function as an activator of PFK.
   d. Acetyl-CoA could act as an activator of PFK.

9. The breakdown of fats releases fatty acids. In what form do the carbon molecules enter the respiratory pathway?
   a. as NADH
   b. a glucose
   c. as pyruvate
   d. as citrate
   e. as acetyl-CoA

10. Which of the following statements about the "paradox of aerobic life" is correct?
   a. Humans are completely protected from the toxic effects of oxygen.
   b. Hydrogen peroxide is formed when a single electron is donated to O_2.
   c. Cytochrome oxidase is a major source of reactive oxygen species.
   d. Strict anaerobes often lack the enzymes superoxide dismutase and/or catalase.

Questions for Discussion

1. Respond to this statement: Respiration occurs in animals but not in plants.

2. In your opinion, are fermentations part of cellular respiration? Why or why not?

3. Why do you think nucleic acids are not oxidized extensively as a cellular energy source?

Pascal Goetgheluck/Science Source

Bioreactors of algae being grown at the University of Arizona. Through the process of photosynthesis atmospheric CO_2 is being converted by these algae into a wide range of organic compounds. Some of these compounds are being studied for their feasibility as alternative energy sources (so-called biofuels).

7

Photosynthesis

WHY IT MATTERS

Renewable sources of energy, including solar and wind power, are meeting an ever-increasing proportion of global energy demands. Although this is good news, well over 90% of the world's energy still comes from the burning of nonrenewable reserves of coal, oil, and natural gas. The combustion of these sources of energy has been shown to be the major contributing factor to increasing atmospheric CO_2 concentrations and the acceleration of global climate change over the past 100 years.

Coal, oil, and natural gas are referred to as "fossil fuels" because they are in fact the remnants of ancient forests—formed by geological processes over millions of years. The organic carbon compounds that are burned were formed in these ancient plants through the process of photosynthesis.

In recent years, harnessing photosynthesis to generate renewable fuels has been a major aspect of the biofuel industry. Unlike coal and oil, biofuels are produced by living organisms through processes such as fermentation and photosynthesis. Biofuels include ethanol and a variety of oils that can be used, for example, to generate jet fuel.

One source of photosynthetically generated biofuels that is being intensively studied is single-celled algae grown in artificial lakes or bioreactors (small bioreactors are shown above). Culturing algae offers numerous advantages over crop plants, the traditional source of most oils. Algae grow very rapidly, and the extraction of oils is often easier than it is from plants. In addition, cultivating algae would preserve precious arable land that could be dedicated to growing food crops.

There remain a number of issues surrounding the wide implementation of algae and biofuel production. The costs associated with growing algae on a large scale are high, while the yield of oil needs to increase for it to be economically feasible. Such issues need to be addressed if biofuels are to make a serious dent in our current dependence on fossil fuels.

The focus of this chapter is photosynthesis—the process by which light energy is used to convert carbon dioxide into organic molecules. The chapter starts by laying out the basic chemistry of photosynthesis focusing on the photophysical nature of light and light absorption. This adds to the more cursory treatment of light that is found in Chapter 1. Details of the two stages of photosynthesis—the light reactions and the Calvin cycle—follow, as does a discussion of how various photosynthetic organisms have evolved mechanisms to cope with a surprising attribute of the carbon-fixing enzyme rubisco. The chapter ends with an important section on comparing photosynthesis with the topic of the previous chapter, cellular respiration.

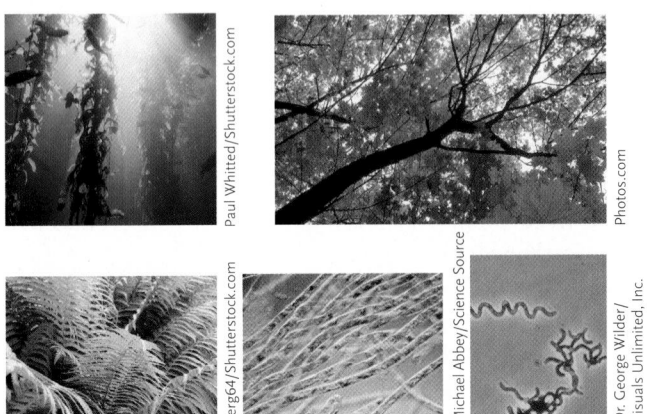

Figure 7.1
Examples of photoautotrophs.

7.1 Photosynthesis: An Overview

Photosynthesis is the use of light energy to convert carbon dioxide into organic compounds such as carbohydrates. We define an organic compound as one that contains one or more C—H bonds. Because they do not need to import already-formed organic compounds from the environment, photosynthetic organisms are called **autotrophs** (*auto* = self, *trophos* = feeding). They are more narrowly defined as **photoautotrophs** because the energy to drive the conversion of carbon dioxide into an organic form comes from light. Some organisms are autotrophic but don't use light as the energy source; instead they use compounds such as hydrogen sulfide (H_2S) and ferrous iron (Fe^{2+}). This type of metabolism is found only in some bacteria and archaeans and is termed chemoautotrophy.

Photoautotrophic organisms are known as Earth's **primary producers (Figure 7.1)**. This is because they represent the major group of organisms that generate the organic compounds that are used by other organisms—the consumers, the organisms that that live by eating plants or other animals. Eventually, the bodies of both primary producers and consumers provide the organic energy-rich molecules to support a range of **decomposers**, such as a range of bacteria and all fungi. Recall from Chapter 3 that consumers and decomposers are classified as heterotrophs because they require an already-synthesized source of organic molecules to live.

As mentioned in Chapter 3, photosynthesis is an ancient process that evolved in bacteria perhaps as early as 2.5 billion years ago. Interestingly, today photosynthesis is found in the domains Bacteria and Eukarya but is not present in the Archaea. Some archaeans, such as the Halobacteria (see Chapter 1),

do harvest light energy and convert it into chemical energy, but since this light energy is not used to convert carbon dioxide into an organic form, Halobacteria are not considered to be photosynthetic.

7.1a Photosynthesis Is an Oxidation–Reduction Process

If you studied photosynthesis in high school biology, one thing you probably memorized was the balanced chemical equation for the overall process, which is

$$6CO_2 + 12H_2O \rightarrow C_6H_{12}O_6 + 6O_2 + 6H_2O \quad (7.1)$$

As chemical equations go this seems pretty straightforward, but let's deconstruct it to see what's really going on. Taking the reaction as written, you could say that water is reacting with CO_2 (it's hydrating the carbon) to produce a six-carbon carbohydrate glucose, with O_2 and H_2O produced as by-products. As we will discuss later in the chapter, glucose is technically not the direct product of the Calvin cycle but is made soon after and so is used here for convenience. Notice from equation (7.1) that equal molar amounts of the gases CO_2 and O_2 are consumed and produced, respectively. This means that one can measure the rate of photosynthesis as either the rate at which CO_2 is consumed or the rate at which O_2 is produced or evolved.

What may not be clear to you yet is that equation 7.1 is a classic example of a oxidation–reduction or redox reaction. This type of reaction was introduced in Chapter 6, Section 6.1. To make it easy to see its redox nature let's put the equation in its simplest form by dividing through by six, giving

$$CO_2 + 2H_2O \rightarrow (CH_2O) + O_2 + H_2O \quad (7.2)$$

where the term (CH_2O) is the basic unit of a carbohydrate consisting of carbon, hydrogen, and oxygen atoms in the ratio 1C:2H:1O. A molecule of glucose is therefore six units, $(CH_2O)_6$.

Like all redox reactions, equation 7.2 is actually two half-reactions: one oxidation reaction is coupled to a reduction reaction. In equation (7.2) it is the water that is being oxidized to O_2. Removing electrons from water is not easy, and in photosynthesis it involves energy that comes from light:

$$2H_2O + \text{light energy} \rightarrow O_2 + 4H^+ + 4e^-$$

In equation (7.2), it is the CO_2 that is being reduced to carbohydrate:

$$CO_2 + 4H^+ + 4e^- \rightarrow (CH_2O) + H_2O$$

Looking back at equation (7.1), you could easily assume that the water is reacting directly with the carbon dioxide to produce the carbohydrate and cause O_2 release. As we will see in the next section, the oxidation of H_2O and the reduction of carbon dioxide are associated with distinctly different processes that are spatially separated within the chloroplast.

7.1b Photosynthesis Can Be Divided into the Light Reactions and the Calvin Cycle

The conversion of carbon dioxide into carbohydrates that defines photosynthesis requires the integration of two distinct processes **(Figure 7.2)**: the light reactions and the Calvin cycle. The light reactions involve the capture of light energy by pigment molecules and the utilization of that energy to synthesize both NADPH (nicotinamide adenine dinucleotide phosphate) and ATP. The electrons needed to reduce $NADP^+$ to NADPH come from the oxidation of H_2O, resulting in the release of O_2 (Figure 7.2). In the Calvin cycle, the electrons and protons carried by NADPH and the energy of ATP hydrolysis are used to convert

CO_2 into carbohydrate. This reduction reaction, referred to as *carbon fixation* or *CO_2 fixation*, is a reduction reaction with electrons (and protons) being added to CO_2.

It is important to realize that the carbohydrate synthesized by the Calvin cycle ends up being used for a huge range of processes. Besides sugars like glucose being a useful source of energy for cellular respiration, the carbohydrates also represent the source of the carbon skeletons used in the biosynthesis of a huge range of other molecules including lipids, amino acids, and nucleotides. In fact, one can consider all the organic molecules present in organisms as direct or indirect products of photosynthesis.

7.1c In Eukaryotes, Photosynthesis Takes Place in Chloroplasts

In photosynthetic eukaryotes, both the light reactions and the Calvin cycle take place within the chloroplast, an organelle that is comprised of three membranes that define three distinct compartments **(Figure 7.3)**. An *outer membrane* covers the entire surface of the organelle, whereas an *inner membrane* lies just inside the outer membrane. Between the outer and inner membranes is the *intermembrane space*. The aqueous environment within the inner membrane is the *stroma* of the chloroplast. Within the stroma is the third membrane system, the *thylakoid membranes*, or thylakoids, which often form flattened, closed sacs. The space enclosed by a thylakoid is called the *thylakoid lumen*.

Embedded within the thylakoid membrane are the components that carry out the light reactions of photosynthesis: proteins, pigments, electron transport carriers, and ATP synthase. The enzymes that catalyze the reactions of the Calvin cycle are found in the stroma of the chloroplast.

A number of phyla of bacteria, including the cyanobacteria, have thylakoid membranes that are formed from infoldings of the plasma membrane and carry out carbon fixation in the cytosol of the cell. In most ways, photosynthesis carried out in cyanobacteria is biochemically identical to that found in the chloroplasts of plant leaves and is part of the evidence that indicates that chloroplasts are descended from free-living cyanobacteria (see Chapter 3).

7.1d Photosynthesis from a Global Perspective

The importance of photosynthesis to life on Earth is perhaps best appreciated by getting a global perspective. A sensor onboard the OrbView-2 satellite can detect visible light at Earth's surface, including the green wavelengths that are reflected by chlorophyll, the primary photosynthetic pigment. This gives an

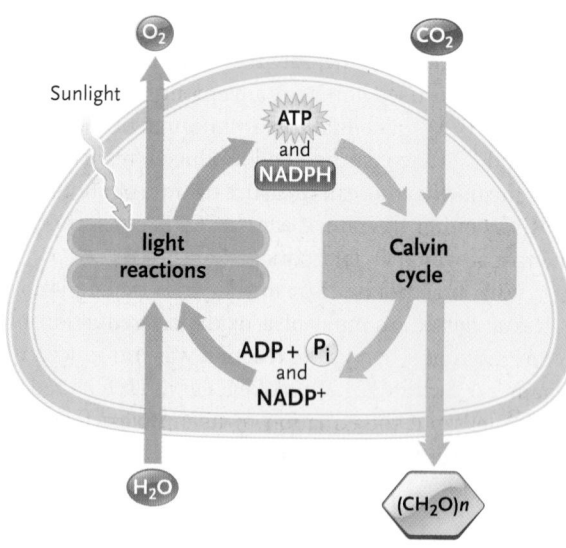

Figure 7.2

The light reactions and the Calvin cycle are the two stages of photosynthesis. The two are linked by reactants and products. Both processes occur in the chloroplasts of photoautotrophic eukaryotes (plants and algae) as well as in photosynthetic bacteria.

unprecedented global estimate of both the abundance and distribution of photosynthetic organisms **(Figure 7.4, p. 148).** Along with other information, scientists have used the satellite data to estimate that a staggering 11×10^{13} kg of CO_2 is fixed through photosynthesis each year.

While we often think about photosynthesis in terms of plants and trees, about half of the carbon is fixed by unicellular photosynthetic algae called phytoplankton that inhabit marine environments. Looking carefully at the global distribution in the oceans (Figure 7.4, p. 148), notice that phytoplankton are more abundant around the poles than they are nearer the equator. This may seem odd because the oceans near the equator receive more sunlight and are warmer. In fact, the distribution of phytoplankton is explained by the abundance of nutrients: the oceans around the equator tend to be nutrient poor, while the cold waters around the Arctic and Antarctic are nutrient rich and thus can support large phytoplankton communities.

The nutrient that limits phytoplankton growth in temperate regions of the oceans is iron. As a cofactor of many enzymes and a redox component of electron transport chains (ETCs), iron is required by most organisms in relatively high amounts. The realization that primary productivity in much of the ocean is limited by iron has spurred experiments that have seen small areas of the Pacific Ocean being fertilized with iron. These fertilized regions have seen rapid increases in phytoplankton growth. The widespread addition of iron to marine environments is seen by some as one way to combat climate change. Because climate change is associated, in part, with increased atmospheric CO_2 concentrations, increased photosynthesis in the ocean could help draw down this CO_2.

STUDY BREAK

1. Differentiate between autotroph and heterotroph; photoautotroph and chemoautotroph.
2. Define *carbon fixation*.
3. What explains the low chlorophyll concentration in the warm waters of the Pacific Ocean?

7.2 The Photosynthetic Apparatus

The photosynthetic apparatus is a series of large protein complexes found in the thylakoid membrane that are responsible for the light reactions. Central to these are two complexes built around proteins that bind pigment molecules that result in light absorption. These **photosystems** lead to the conversion of light energy into chemical energy in the initial reactions of photosynthesis.

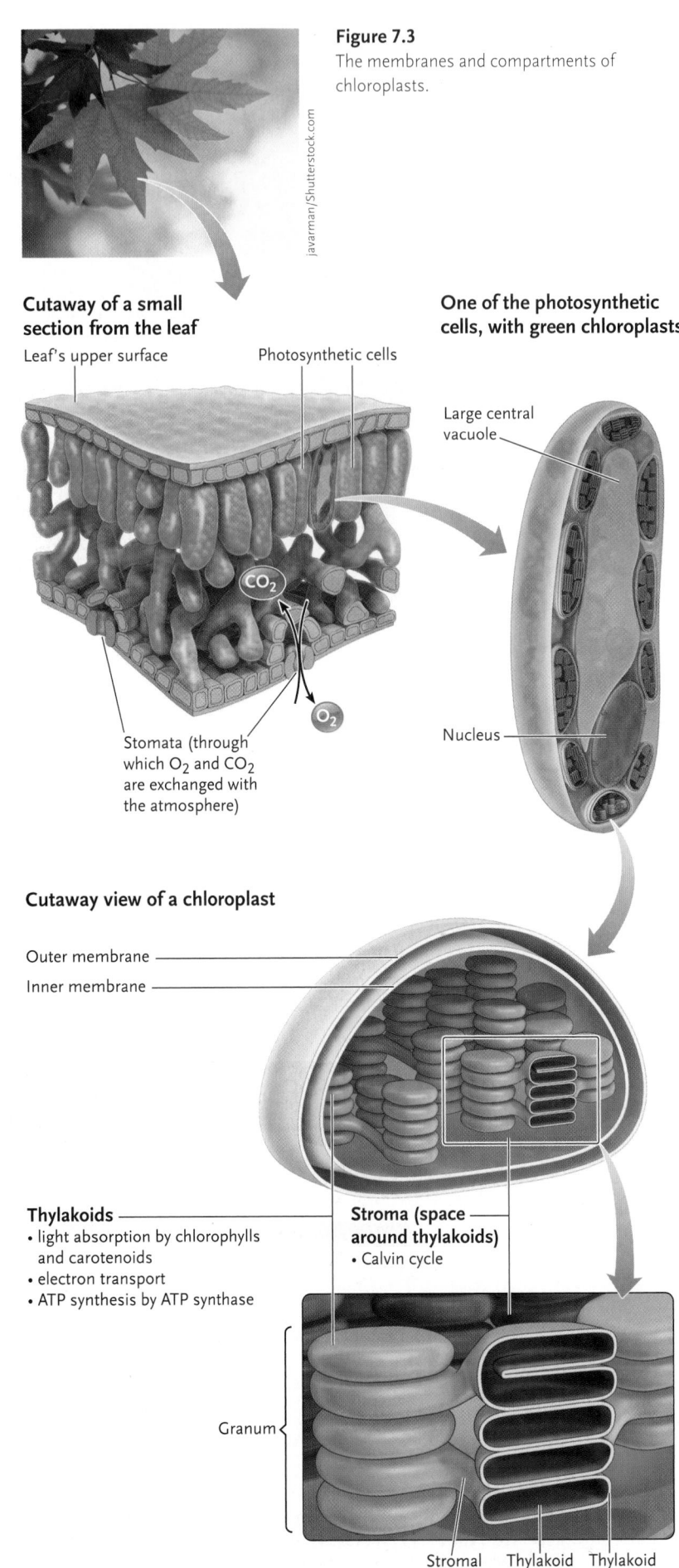

Figure 7.3
The membranes and compartments of chloroplasts.

javarman/Shutterstock.com

Cutaway of a small section from the leaf
Leaf's upper surface
Photosynthetic cells
CO_2
O_2
Stomata (through which O_2 and CO_2 are exchanged with the atmosphere)

One of the photosynthetic cells, with green chloroplasts
Large central vacuole
Nucleus

Cutaway view of a chloroplast
Outer membrane
Inner membrane

Thylakoids
• light absorption by chlorophylls and carotenoids
• electron transport
• ATP synthesis by ATP synthase

Stroma (space around thylakoids)
• Calvin cycle

Granum

Stromal lamella
Thylakoid lumen
Thylakoid membrane

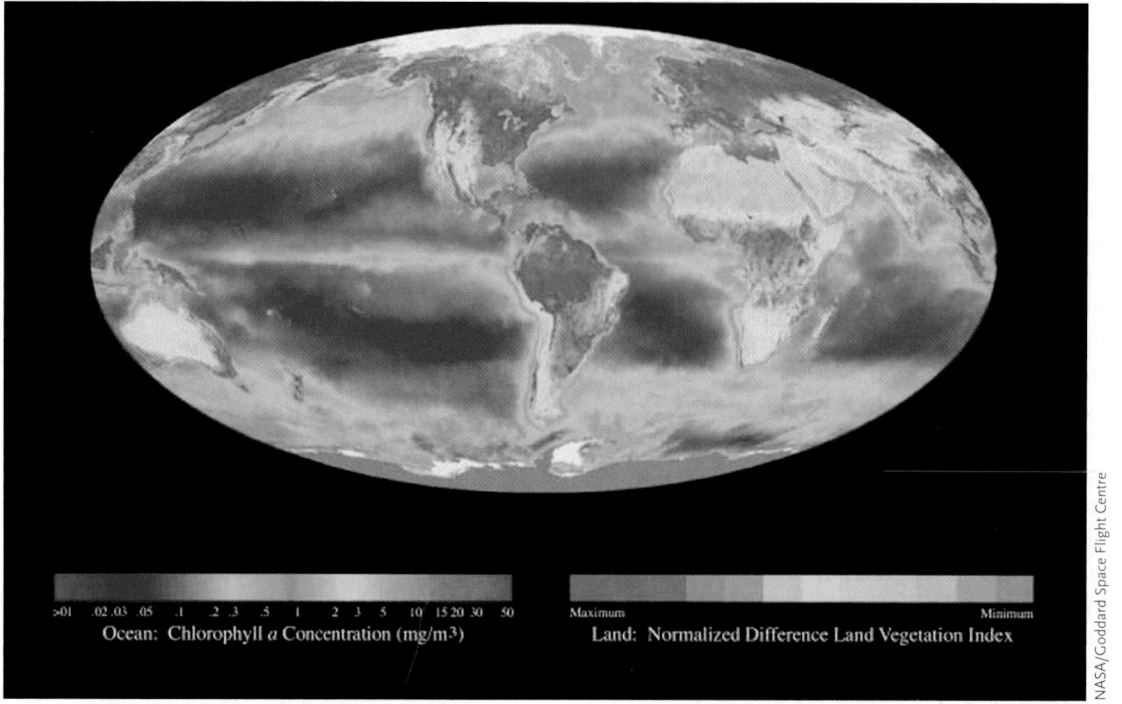

>01 .02 .03 .05 .1   .2 .3   .5   1   2   3   5   10 15 20 30 50
Ocean: Chlorophyll *a* Concentration (mg/m³)

Maximum   Minimum
Land: Normalized Difference Land Vegetation Index

Figure 7.4
Global oceanic and terrestrial photoautotroph abundance can be estimated by the satellite detection of chlorophyll at the Earth's surface. Data collected by NASA's SeaWiFS sensor onboard the OrbView-2 satellite.

7.2a Electrons in Pigment Molecules Absorb Light Energy

Photosynthesis requires the capture and utilization of light energy. As we did in Chapter 1, we can define light as that portion of the electromagnetic spectrum that humans can detect with their eyes **(Figure 7.5)**. The various forms of radiation that make up the electromagnetic spectrum differ in wavelength, ranging from very long radio waves, which have wavelengths in the

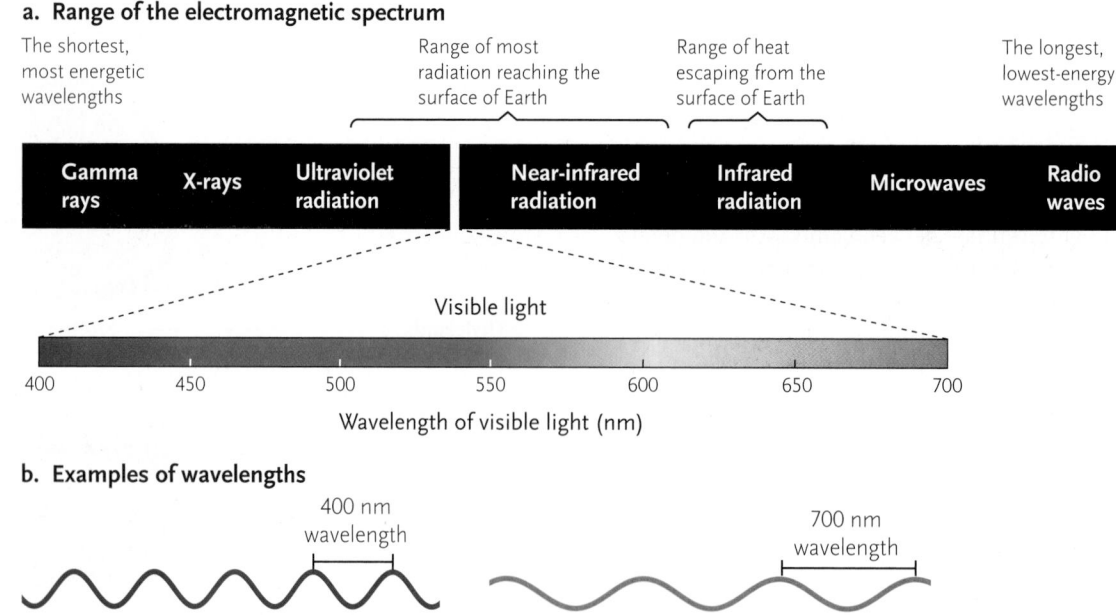

a. Range of the electromagnetic spectrum

| The shortest, most energetic wavelengths | | Range of most radiation reaching the surface of Earth | Range of heat escaping from the surface of Earth | | The longest, lowest-energy wavelengths |

| Gamma rays | X-rays | Ultraviolet radiation | Near-infrared radiation | Infrared radiation | Microwaves | Radio waves |

Visible light

400   450   500   550   600   650   700
Wavelength of visible light (nm)

b. Examples of wavelengths

400 nm wavelength

700 nm wavelength

Figure 7.5
The electromagnetic spectrum. **(a)** The electromagnetic spectrum ranges from gamma rays to radio waves; visible light, which includes the wavelengths used for photosynthesis, occupies only a narrow band of the spectrum. **(b)** Examples of wavelengths, showing the difference between the longest and shortest wavelengths of visible light.

range of 10 m to hundreds of kilometres, to gamma rays, which have wavelengths in the range of one hundredth to one millionth of a nanometre. The electromagnetic radiation that humans can detect (light or visible light) has wavelengths between about 400 nm, seen as blue light, and 700 nm, seen as red light (see Figure 7.5).

Although light can be described using the concept of a wave moving through space, the interaction of light with matter is best understood in terms of discrete packets of energy called photons (also called *quanta*). A photon of light contains a fixed amount of energy that is inversely related to its wavelength: the shorter the wavelength, the greater the amount of energy that photons of that wavelength contain. So, for example, the energy of a photon of blue light is greater than the energy found in a red photon of light.

To be used as a source of energy, photons of light must be absorbed by a molecule **(Figure 7.6a)**. Absorption occurs when the energy of a photon is transferred to an electron within a molecule, moving the electron from the ground state to an excited state. In the excited state, the electron is farther away from the nucleus and thus it contains more energy. A major class of molecules that are very efficient at absorbing visible light are pigments because their structure results in a number of excitable electrons. The structures of a diversity of pigments are presented in Chapter 1, Figure 1.5.

After a pigment molecule absorbs a photon of light, one of three possible events can occur (Figure 7.6).

1. The excited electron from the pigment molecule returns to its ground state, releasing its energy either as heat or as an emission of light of a longer wavelength—a process called *fluorescence.*
2. The energy of the excited electron (but not the electron itself) is transferred to a neighbouring pigment molecule. This transfer of energy excites an electron in the second molecule, while the electron in the first pigment molecule returns to its ground state. Very little energy is lost in this energy transfer.
3. The excited electron is transferred from the pigment molecule to a nearby electron-accepting molecule.

7.2b Chlorophylls and Carotenoids Cooperate in Light Absorption

In photosynthesis, light is absorbed by molecules of green pigments called chlorophylls and yellow-orange pigments called carotenoids. Chlorophylls are the major photosynthetic pigments in plants, green algae, and cyanobacteria. Of the chlorophylls, the most dominant types are chlorophyll *a* and *b*, which are structurally only slightly different **(Figure 7.7, p. 150).**

One can precisely determine the wavelengths of light absorbed by a pigment such as chlorophyll *a* by producing an absorption spectrum for that pigment using an instrument called a spectrophotometer and a pure sample of the pigment. An **absorption spectrum** is a plot of the absorption of light as a function of wavelength. **Figure 7.8a, p. 150,** shows that chlorophyll *a* strongly absorbs blue and red light but does not absorb green or yellow light. Why doesn't chlorophyll absorb green or yellow light? Recall from Chapter 1 that for a photon of light to be

Photon is absorbed by an electron that moves from a low-energy level to a higher energy level.

Photon

Electron at ground state — Low-energy level

Electron at excited state — High-energy level

Either   Or   Or

Pigment molecule

Electron-accepting molecule

The electron returns to its ground state by emitting a less energetic photon (fluorescence) or releasing energy as heat.

The electron returns to the ground state as its energy is transferred to an electron in a neighbouring pigment.

The high-energy electron is transferred to another molecule, an electron acceptor.

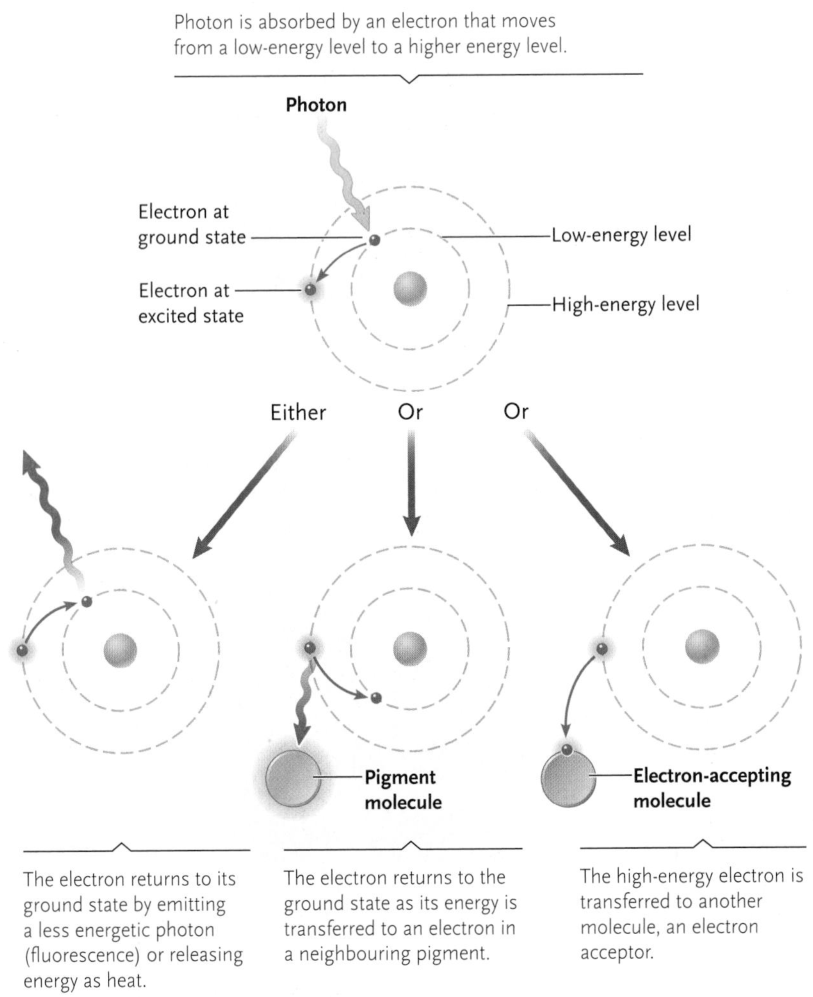

Figure 7.6
Three possible fates of an excited-state electron within a pigment molecule.

a. Chlorophyll structure

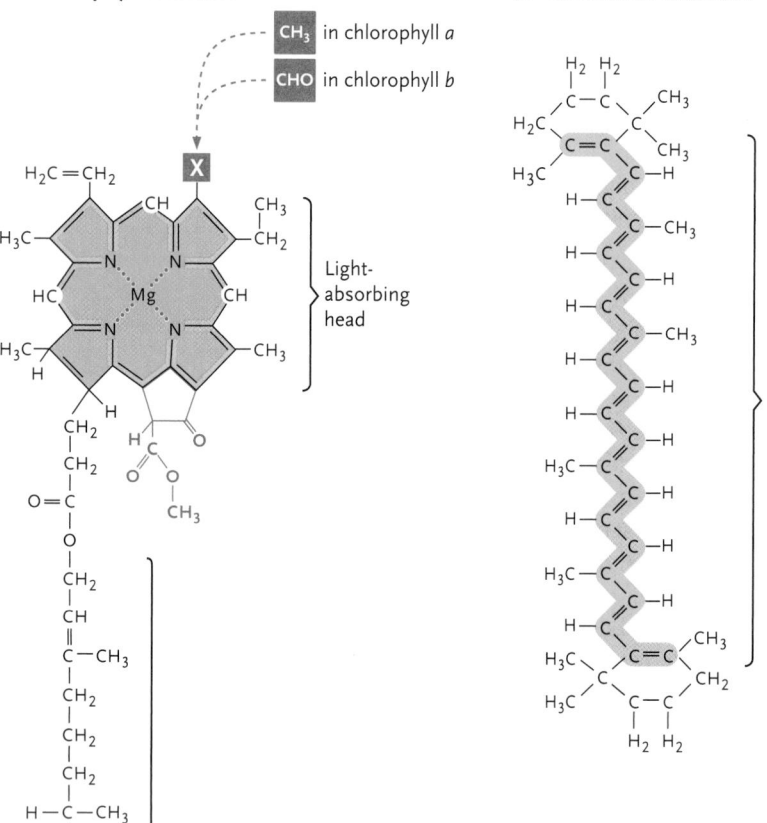

CH₃ in chlorophyll *a*

CHO in chlorophyll *b*

Light-absorbing head

Hydrophobic side chain

b. Carotenoid structure

Light-absorbing region

Figure 7.7

Pigment molecules used in photosynthesis. **(a)** Chlorophylls *a* and *b*, which differ only in the side group attached at the X. **(b)** An example of a carotenoid. In both (a) and (b), the light-absorbing electrons are distributed among the bonds shaded in orange.

effectiveness of light of particular wavelengths in driving a process (Figure 7.8b). An action spectrum for photosynthesis is usually determined by using a suspension of chloroplasts or algal cells and measuring the amount of O_2 released by photosynthesis at different wavelengths of visible light.

One of the earliest action spectra was produced in 1883 by Theodor Engelmann, who used only a light microscope and a glass prism to determine which wavelengths of light were most effective for photosynthesis **(Figure 7.9).** Engelmann placed a strand of a green alga, *Spirogyra,* on a glass microscope slide, along with water containing aerobic bacteria. He adjusted the prism so that it split a beam of light into its separate colours, which spread like a rainbow across the strand (see Figure 7.9).

a. The absorption spectra of chlorophylls *a* and *b* and carotenoids

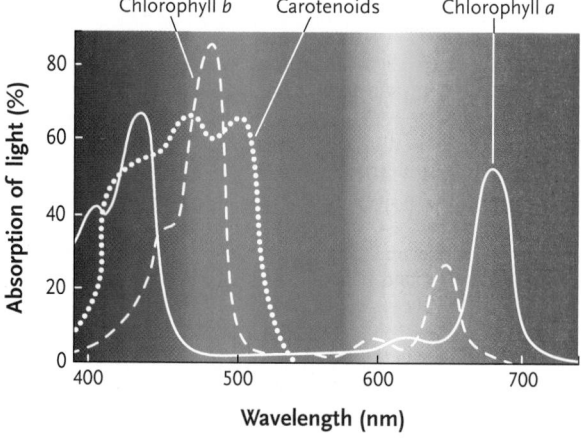

b. The action spectrum in higher plants, representing the combined effects of chlorophylls and carotenoids

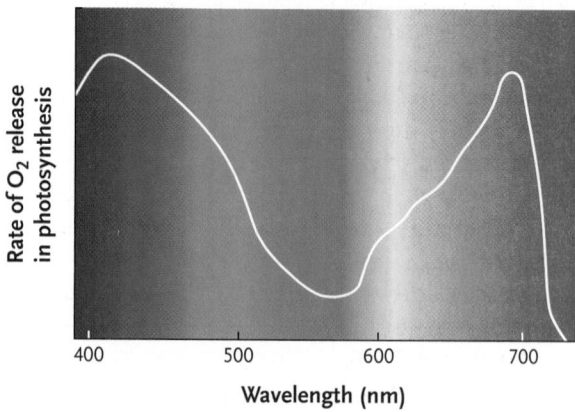

Figure 7.8

The absorption spectra of three photosynthetic pigments **(a)** and the action spectrum of photosynthesis **(b)** in plants. The absorption spectra in (a) were made from pigments that were extracted from cells and purified.

absorbed, the energy of that photon needs to match the amount of energy required to raise a pigment electron from the ground state to an excited state. If the energies do not match, then the photon is not absorbed. Chlorophyll has really only two excited states: one that matches the energy of a blue photon and one that matches that of a red photon.

The absorption spectra of the accessory pigments (chlorophyll *b* and carotenoids; see Figure 7.8a) illustrate that these pigments expand the wavelengths of light that can be effectively captured and used for photosynthesis. Figure 7.8a illustrates that the slight differences in structure between chlorophyll *a* and chlorophyll *b* are reflected in differences in their absorption spectra.

Photosynthesis depends on the absorption of light by chlorophylls and carotenoids, acting in combination. This is supported by the **action spectrum** for photosynthesis. An action spectrum is a plot of the

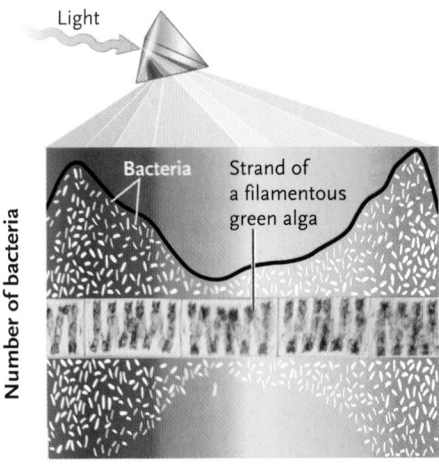

Figure 7.9
Engelmann's experiment revealed the action spectrum of light used in photosynthesis by *Spirogyra*, a green alga. The aerobic bacteria clustered along the algal strand in the regions where oxygen was released in greatest quantity—the regions in which photosynthesis proceeded at the greatest rate. Those regions corresponded to the colours (wavelengths) of light being absorbed most effectively by the alga—in this case, blue and red.

After a short time, he noticed that the bacteria had begun to cluster around the algal strand in different locations. The largest clusters were under the blue and violet light at one end of the strand and the red light at the other end. Very few bacteria were found in the green light.

7.2c Photosynthetic Pigments Are Organized into Photosystems

Photosynthetic pigments are required not only to absorb photons of light but also to transfer the energy to neighbouring molecules. To do this efficiently, pigment molecules do not float freely within the thylakoid membrane but rather are bound very precisely to specific proteins. These pigment-proteins are organized within the thylakoid membrane into complexes called photosystems **(Figure 7.10)**. Each photosystem is composed of a large **antenna complex** (also called a *light-harvesting complex*) of pigment-proteins that surrounds a central *reaction centre*. The reaction centre of a photosystem comprises a small number of proteins that bind a special chlorophyll *a* molecule and an electron-accepting molecule called the *primary electron acceptor* (Figure 7.10).

The function of a photosystem is to trap photons of light and use the energy to oxidize a reaction centre chlorophyll, with the electron being transferred to the primary electron acceptor. High rates of this oxidation–reduction reaction within the reaction centre are achieved by the large antenna complex

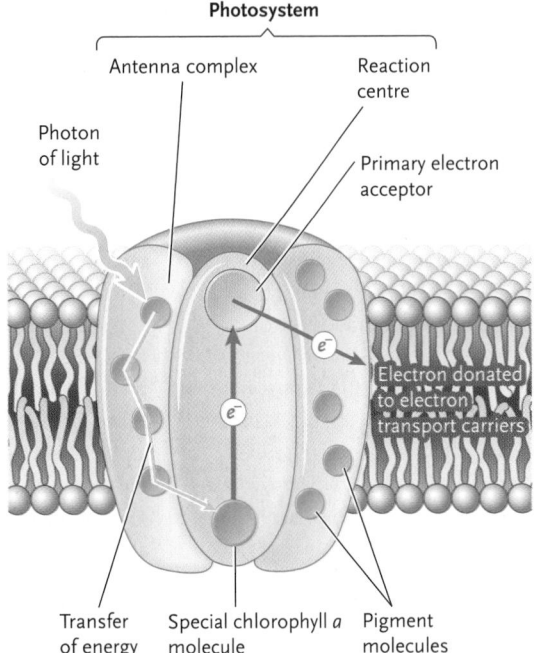

Figure 7.10
Major components of a photosystem. A group of pigment proteins form an antenna complex that surrounds a reaction centre. Light energy absorbed anywhere in the antenna complex is transferred to a special chlorophyll *a* molecule in the reaction centre. The absorbed light is converted to chemical energy when an excited electron from the chlorophyll *a* is transferred to a primary acceptor, also in the reaction centre. High-energy electrons are passed out of the photosystem to the electron transport system. The yellow arrow shows the migration of energy from one pigment to the other, while the blue arrows show movement of electrons.

of pigments absorbing light of a range of wavelengths and efficiently transferring the energy to the **reaction centre.**

There are two different types of photosystems: photosystem I and photosystem II. The specialized chlorophyll *a* in the reaction centre of **photosystem I** is called P700 (P = pigment) because its absorption maximum is at a wavelength of 700 nm. The reaction centre of **photosystem II** contains a specialized chlorophyll *a*, P680, which absorbs light maximally at 680 nm. Within a single leaf chloroplast there are thousands of photosystems (both I and II) each containing about 500 chlorophyll molecules.

STUDY BREAK

1. What contains more energy, a photon of green light or a photon of orange light?
2. What are the three possible fates of an excited-state electron?
3. How is an absorption spectrum different from an action spectrum?

The D1 Protein Keeps Photosystem II Operating

Photosystem II can be considered the most important development in the evolution of life on Earth. Unlike anything that came before it, photosystem II meant that organisms that had it could harvest the energy of the Sun and use it to extract electrons from water. These electrons were used to convert CO_2 from the atmosphere into the organic building blocks of the cell. This ability to use water meant that life could thrive almost anywhere on the planet and led to an explosion in the conversion of CO_2 into organic molecules. By splitting water, photosystem II also produced O_2, which gradually accumulated in the atmosphere and led to the development of aerobic respiration. The process of aerobic respiration extracts 18 times as much energy from sugar as the anaerobic pathway that came before it, resulting in an energy bounty that allowed the emergence of complex,

multicellular eukaryotic organisms. Because of this, photosystem II is known as the engine of life.

The splitting of water by photosystem II is the most energetically demanding reaction in all of biology. The reaction is carried out by a molecule, P680, that is found in the core of photosystem II, bound to a protein called D1. When photosystem II absorbs light, P680 is converted into the strongest known biological oxidant, $P680^+$, and this molecule is able to break apart H_2O, releasing electrons, protons, and O_2.

As a consequence of absorbing the energy of about 10 thousand photons every second and generating powerful oxidants, photosystem II is constantly being damaged, which results in its inactivation. The major site of damage is the D1 protein, which is found in the core of the complex and binds P680. Over the course of 2 billion years of

evolution, organisms that have photosystem II have been unable to prevent the damage from occurring—but they have developed a highly specialized mechanism to repair it.

It takes only 20 minutes for a newly synthesized photosystem II complex to stop working because of damage to D1. However, damaged complexes are rapidly disassembled, the damaged D1 protein is removed and degraded, a newly synthesized D1 protein is inserted, and a functional photosystem is reassembled. This repair cycle is very efficient and depends on a high rate of D1 protein synthesis. It has been estimated that in the absence of this repair system, damage to photosystem II would lower the photosynthetic productivity of the planet by 95%. Thus, life on Earth could not have evolved to present-day levels of both abundance and complexity in the absence of a D1 repair mechanism.

7.3 The Light Reactions

Photosystem I and photosystem II are the two light-trapping components involved in photosynthetic electron transport. In this section, we look in detail at how this particular electron transport chain operates and draw some analogies to respiratory electron transport.

7.3a Photosynthetic Electron Transport Synthesizes NADPH and Generates a Proton Gradient

Figure 7.11, p. 153, shows the components of photosynthetic electron transport and the ATP synthase complex within the thylakoid membrane. As in all electron transport systems, the electron carriers of the photosynthetic system consist of nonprotein cofactors that alternate between being oxidized and being reduced as electrons move through the system (see Chapter 6). The carriers, many of which are bound precisely to proteins, include the same types that act in mitochondrial electron transfer—cytochromes, quinones, and iron–sulfur centres.

Most of the electron carriers are organized into three large complexes embedded in the thylakoid membrane: photosystem II, the cytochrome complex, and photosystem I. Electron flow between photosystem II

and the cytochrome complex is facilitated by a pool of plastoquinone (PQ) molecules, which are similar in structure and function to the ubiquinone of respiratory electron transport. Electron flow from the cytochrome complex to photosystem I is linked by the mobile copper-containing protein plastocyanin.

From photosystem I, electrons are donated to an iron–sulfur protein called ferredoxin, which in turn donates electrons to the enzyme $NADP^+$ reductase found on the stromal side of the thylakoid membrane. The enzyme reduces $NADP^+$ to NADPH by using two electrons from electron transport and a proton from the surrounding aqueous environment.

7.3b Light Is Used Specifically to Oxidize Chlorophyll

Just like respiratory electron transport (Chapter 6), photosynthetic electron transport operates with electrons flowing spontaneously from molecules that are easily oxidized to molecules that are progressively more easily reduced. In the case of mitochondrial electron transport, recall that flow is from NADH, which is a source of electrons, to O_2, which has a very high affinity for electrons. In photosynthesis, electron transport occurs by the same principle; however, unlike NADH, the chlorophyll molecules in the

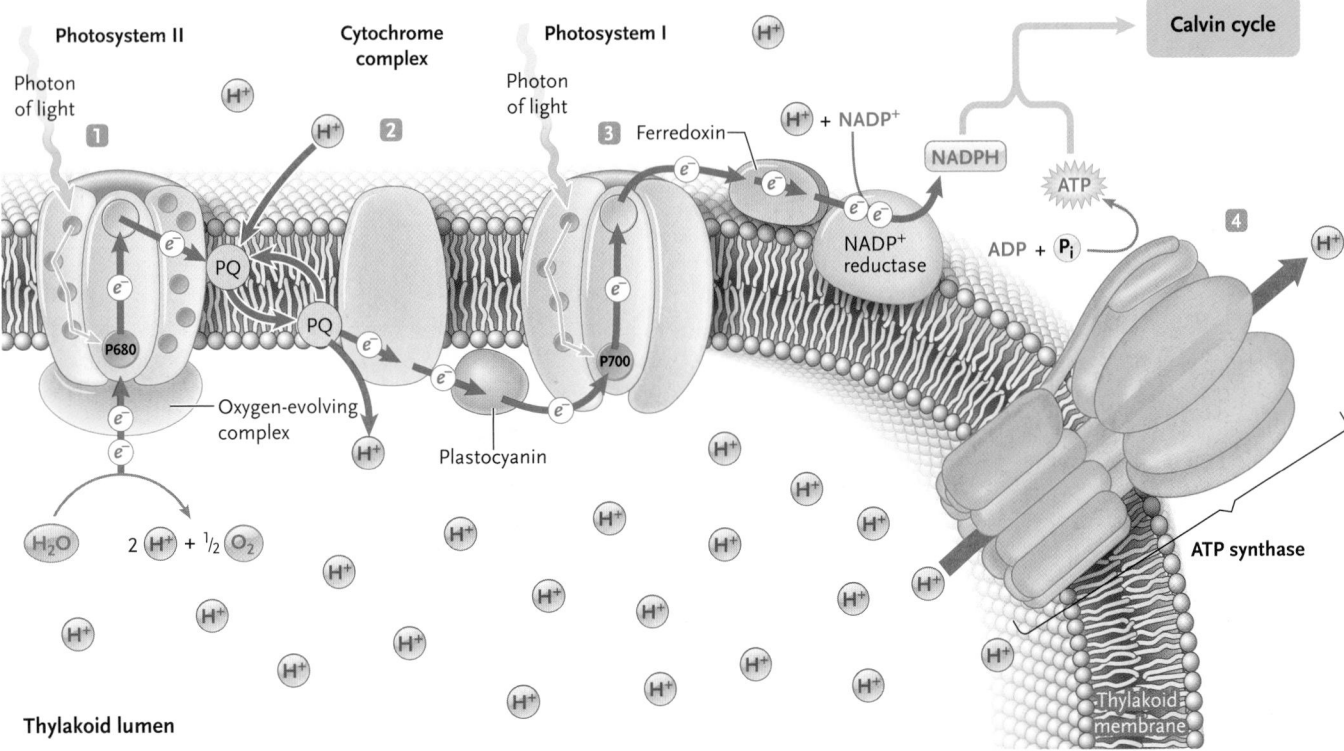

Stroma

Thylakoid lumen

Figure 7.11
A model of the eukaryotic thylakoid membrane illustrating the major protein and redox cofactors required for photosynthetic electron transport and ATP synthesis.

1 Absorption of light energy by photosystem II results in the oxidation of P680. The released electron reduces the primary acceptor molecule. P680+ is returned to the reduced state (P680) by donation of an electron from water, a process mediated by the oxygen-evolving complex.

2 From the primary acceptor, the electron is passed to the mobile carrier molecule plastoquinone (PQ). As it accepts an electron from photosystem II, PQ picks up a proton from the stroma. PQ diffuses through the membrane before binding to the cytochrome complex, at which point it donates an electron and releases a proton into the thylakoid lumen. From the cytochrome complex the electron is donated to plastocyanin.

3 Absorption of light energy by photosystem I results in the oxidation of P700. The liberated electron is used to reduce the primary acceptor before being passed to ferredoxin. This single electron is then held by the NADP+ reductase complex. P700+ is reduced back to P700 by the electron that is coming from plastocyanin. Once a second electron travels along the chain and reaches NADP+ reductase complex, NADP+ is reduced to NADPH.

4 Proton movement by the reduction–oxidation of plastoquinone (red arrows) creates a concentration gradient of H+ (a proton-motive force) across the thylakoid membrane. The gradient is dissipated as H+ diffuses back into the stroma through the ATP synthase complex, which drives the synthesis of ATP from ADP and P$_i$.

reaction centres of photosystem II and photosystem I are not easily oxidized. So what process gets a chlorophyll molecule into a state in which it readily gives up an electron? The absorption of a photon of light within photosystem II and photosystem I and the funnelling of this high energy to the reaction centre is used to excite an electron within P680 or P700 **(Figure 7.12, p. 154)**. By raising an electron in P680 to a higher excited state (denoted as P680*), the absorption of light energy produces a molecule that is easily oxidized by the ETC, and electron flow is a spontaneous process from P680* to photosystem I. A second photon of light absorbed by photosystem I results in the formation of P700*, which is easily oxidized by the primary electron acceptor of photosystem I, and in turn ferredoxin, before the electron finally being donated to NADP+

(see Figure 7.12). Oxidized P680 (P680+) is reduced by electrons donated from water (Figure 7.11), while P700+ is reduced back to P700 by electrons coming from PSII.

7.3c In the Light Reactions, ATP Is Generated by Chemiosmosis

In a way analogous to respiratory electron transport, the flow of electrons along the photosynthetic ETC is coupled to ATP synthesis by the build-up of a proton gradient. In photosynthetic electron transport, the proton gradient across the thylakoid membrane is derived from three processes (see Figure 7.11). First, protons are translocated into the lumen by the cyclic reduction and oxidation of plastoquinone as it migrates

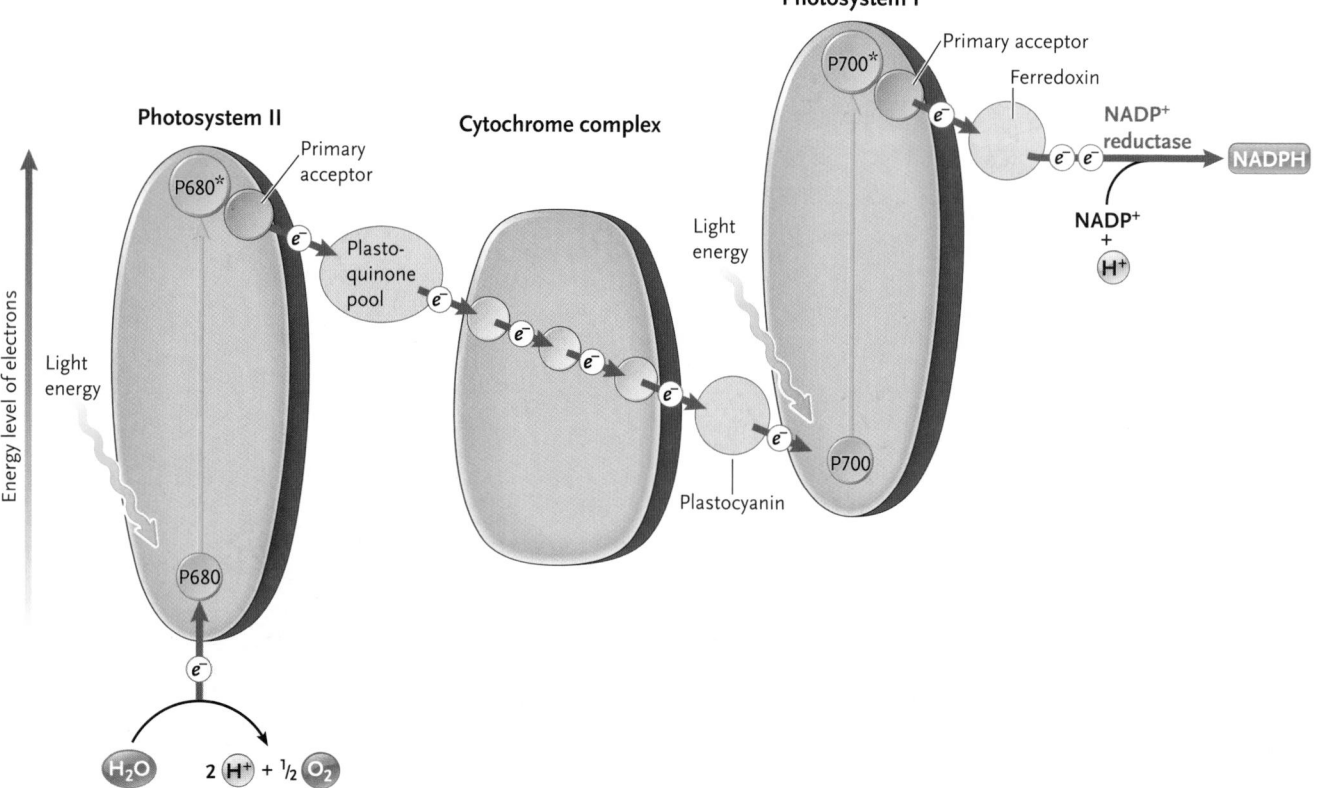

Figure 7.12

The components of the thylakoid membrane organized according to their energy level. This is also referred to as the Z scheme. When photosystem II absorbs a photon of light, an electron within the reaction centre chlorophyll of P680 gets excited to a higher energy level (P680*). This results in spontaneous electron transport to photosystem I. However, the energy level of NADP$^+$ is greater than that of P700. This energy difference is overcome by photosystem I absorbing a photon of light, producing P700*. Thus, two photons of light, one absorbed by photosystem II and another absorbed by photosystem I, are required to overcome the energy difference between H$_2$O and NADP$^+$.

from photosystem II to the cytochrome complex and back again. Second, the gradient is enhanced by the addition of two protons to the lumen from the oxidation of water, which occurs on the luminal side of photosystem II. Third, the removal of one proton from the stroma for each NADPH molecule synthesized further decreases the H$^+$ concentration in the stroma, thereby enhancing the gradient across the thylakoid membrane. The proton-motive force (see Section 6.5) established across the thylakoid membrane is used to synthesize ATP by chemiosmosis using the chloroplast ATP synthase. This multiprotein complex is structurally and functionally analogous to the ATP synthase used in oxidative phosphorylation in cellular respiration (see Figure 6.16, Chapter 6). Distinct from oxidative phosphorylation in cellular respiration, the process of using light to generate ATP is often referred to as **photophosphorylation.**

7.3d The Stoichiometry of Linear Electron Transport

We have described in detail the structure and function of the photosynthetic apparatus. Now it's time to go over the stoichiometry of the light reactions. To get a

single electron down the ETC from photosystem II (or water; it doesn't matter) to NADP$^+$ takes two photons of light, one photon absorbed by photosystem II and a second by photosystem I. Figure 7.12 shows this. But how many photons need to be absorbed by the photosynthetic apparatus to produce a single molecule of O$_2$? For all of these types of questions, we start by writing out a balanced chemical reaction, such as

$$2H_2O \rightarrow 4H^+ + 4e^- + O_2$$

The reaction shows that to produce one molecule of O$_2$ you need to oxidize two molecules of water, which results in the release of four electrons. Now to move a single electron down the ETC requires the absorption of two photons. It follows then that to get four electrons from photosystem II to NADP$^+$, the photosynthetic apparatus needs to absorb a total of eight photons of light, four by each photosystem.

7.3e Cyclic Electron Transport Generates ATP in the Absence of NADPH

The pathway of electron flow from photosystem II through photosystem I to synthesize NADPH is referred to as linear electron transport. Although this

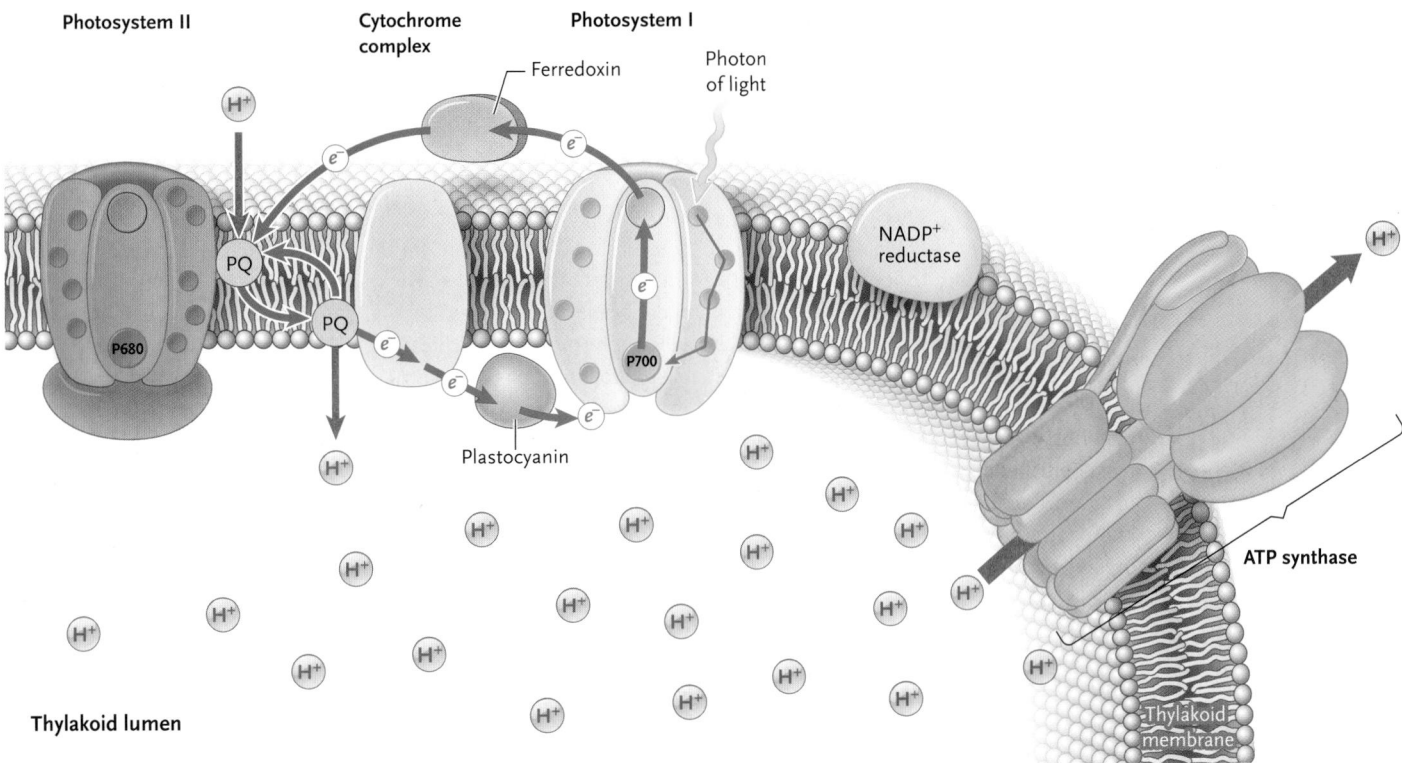

Stroma

Photosystem II   Cytochrome complex   Photosystem I

Ferredoxin

Photon of light

$NADP^+$ reductase

ATP synthase

Plastocyanin

Thylakoid membrane

Thylakoid lumen

Figure 7.13

Cyclic electron transport. Electrons move in a circular pathway from photosystem I through ferredoxin back to the plastoquinone pool, through the cytochrome complex and plastocyanin and then back to photosystem I. In cyclic electron transport, photosystem II does not operate. The pathway generates proton pumping and thus leads to ATP production but does not result in the synthesis of NADPH.

is the pathway of electron flow that occurs most often, photosystem I can function independently of photosystem II in what is called **cyclic electron transport (Figure 7.13).** In this process, electron flow from photosystem I to ferredoxin is not followed by electron donation to the $NADP^+$ reductase complex. Instead, reduced ferredoxin donates electrons back to the plastoquinone pool. In this manner, the plastoquinone pool gets continually reduced and oxidized and keeps moving protons across the thylakoid membrane without the involvement of electrons coming from photosystem II. Overall, cyclic electron transport only involves light absorption by photosystem I, with the energy being used to establish a proton-motive force and generate ATP. Unlike linear electron transport, NADPH is not formed during cyclic electron transport.

Cyclic electron transport plays an important role in overall photosynthesis. The reduction of carbon dioxide by the Calvin cycle requires more ATP than NADPH, and the additional ATP molecules are provided by cyclic electron transport. Other energy-requiring reactions in the chloroplast also depend on ATP produced by the cyclic pathway.

STUDY BREAK

1. In which compartment of the chloroplast is NADPH generated?
2. How is the proton gradient across the thylakoid membrane established during electron transport?
3. How many photons of light are required to generate one molecule of NADPH by linear electron transport?

7.4 The Calvin Cycle

Recall from Chapter 6 that carbon dioxide is a fully oxidized carbon molecule and contains no usable energy. On the other hand, carbohydrate molecules such as glucose and sucrose are an abundant source of energy because they contain many C—H bonds (see Chapter 6). In the cytosol of photosynthetic bacteria and in the stroma of the chloroplast, a series of 11 enzyme-catalyzed reactions use NADPH to reduce CO_2 into sugar. The overall process is endergonic, requiring energy supplied by ATP. These 11 enzyme-catalyzed

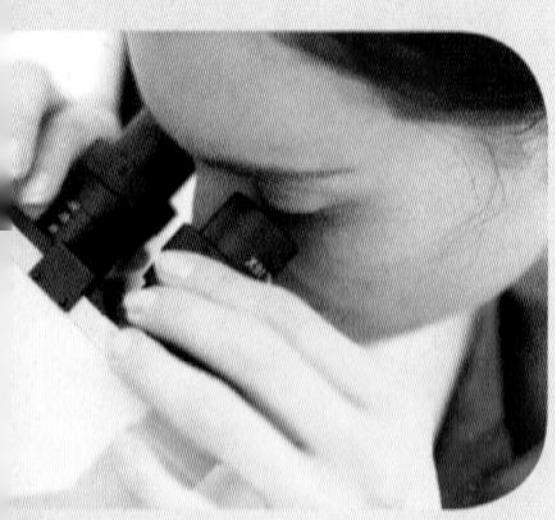

Norm Hüner, *Western University*

A number of advances to our understanding of the regulation of photosynthesis have been elucidated by the research group of Norm Hüner, who holds a Tier 1 Canada Research Chair in Environmental Stress Biology at the Western University in London, Ontario.

Hüner's research has established that the photosynthetic apparatus has a dual role: not only does it function as the primary energy transformer of the biosphere, but it also acts as a sensor of environmental change in all photoautotrophs. Using a range of organisms, including plants, green algae, and cyanobacteria, Hüner's group has discovered that the relative redox state of photosynthetic electron transport acts as a natural sensor of the balance between energy input from the Sun and the demands for that energy by the metabolic processes of the organism. The redox state of the photosynthetic apparatus can be readily assessed by measuring the *excitation pressure* on photosystem II using a fluorescence-based technique. Hüner's group has shown that changes in excitation pressure are a key trigger that initiates changes in a number of cellular processes, including gene expression, which enable photoautotrophs to readily acclimate to changes in light, temperature, and nutrient availability.

Hüner is the coauthor of an internationally acclaimed textbook entitled *Introduction to Plant Physiology*, which is currently in its fourth edition. As well, he has received more than 25 national and international awards and honours for his research, which include election as a fellow of the Academy of Science, Royal Society of Canada, and president of the Canadian Society of Plant Physiologists; an honorary degree from the University of Umea, Sweden; and an honorary professorship from Xinjiang University, China.

In recent years, Hüner has been the lead investigator in the establishment of the Biotron Experimental Climate Change Research Centre, an international research facility on the campus of Western University. The research focus of this $30 million facility is the elucidation of the mechanisms by which plants, microbes, and insects sense and adjust to climate change. One of the unique aspects of the Biotron is that it gives researchers the ability to conduct control experiments on a much larger scale than possible in a conventional laboratory. This allows for the study of how changes in temperature, light, nutrients, and carbon dioxide concentrations may affect the growth of not only individual species but also entire ecosystems.

---

reactions are collectively known as the Calvin cycle (or **light-independent reactions**), which is the most common pathway on Earth by which carbon dioxide is transformed into carbohydrates.

7.4a The Calvin Cycle Reduces Carbon Dioxide to a Carbohydrate

Like other metabolic cycles, including the citric acid cycle, the Calvin cycle generates products that are removed but it also requires that molecules be regenerated so that cycling can continue.

During each turn of the Calvin cycle, one molecule of CO_2 is converted into one reduced carbon—essentially one (CH_2O) unit of carbohydrate. To help you better understand the Calvin cycle, **Figure 7.14**, represents a summary of what occurs following three turns of the cycle. It is only after three carbon dioxide molecules get reduced that one actually generates a separate molecule—a three-carbon sugar glyceraldehyde-3-phosphate (G3P). By a reaction that is not part of the Calvin cycle, two molecules of G3P can synthesize one molecule of the six-carbon sugar glucose.

As shown in Figure 7.14, the Calvin cycle can be subdivided into three distinct phases: fixation, reduction, and regeneration. The events that take place in each of these phases during *one* turn of the cycle are as follows:

Phase 1: Fixation. This phase involves the incorporation (i.e., fixing) of a carbon atom from CO_2 (one per turn) into one molecule of the five-carbon sugar ribulose-1,5-bisphosphate (RuBP) to produce two molecules of the three-carbon compound 3-phosphoglycerate.

Phase 2: Reduction. In this phase, each molecule of 3-phosphoglycerate gets an additional phosphate added from the breakdown of ATP. This produces a total of two molecules of 1,3-bisphosphoglycerate. Each of these molecules is subsequently reduced by electrons from NADPH, producing a molecule of G3P.

Phase 3: Regeneration. For each turn of the Calvin cycle, two molecules of G3P are produced—a total of six carbon atoms. In a multistep process, five of these carbons are rearranged to regenerate the single molecule of RuBP required for the next round of carbon fixation.

Let's work through Figure 7.14 (the key is to keep track of the carbons). In three turns of the Calvin cycle, $3CO_2$ (3 carbons) are incorporated into 3 molecules of RuBP (15 carbons), which produces 6 molecules of 3-phosphoglycerate (18 carbons). Each of these molecules is phosphorylated by a phosphate donated by ATP. In total, six ATP are consumed to phosphorylate the six molecules

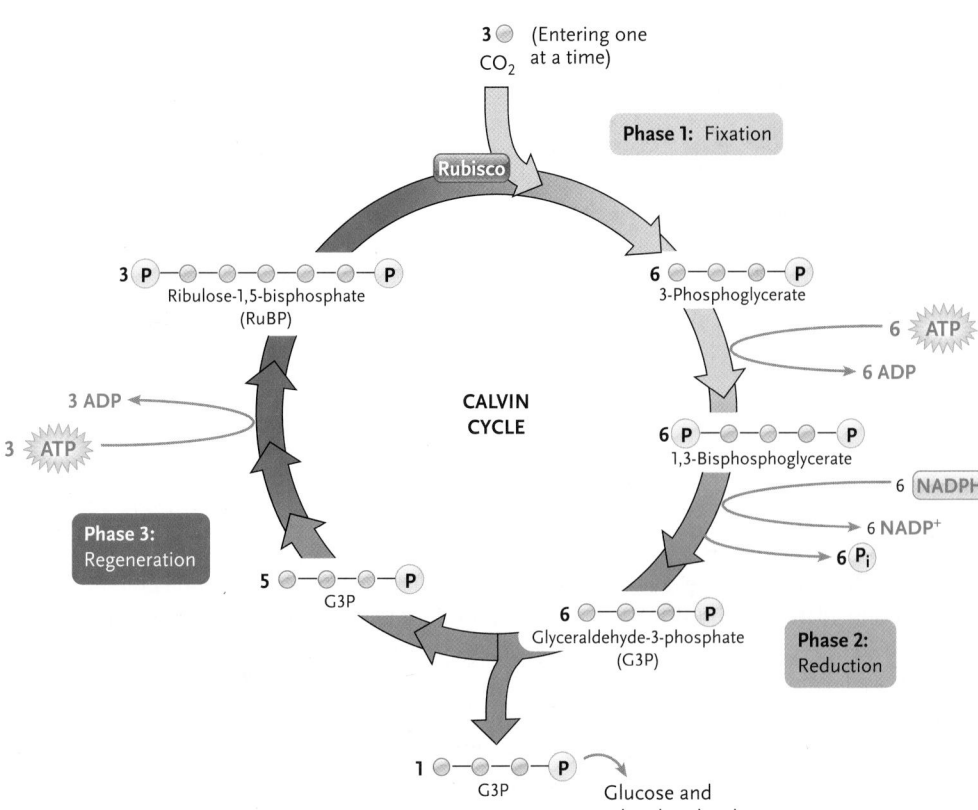

Figure 7.14

The Calvin cycle. An overview of the three phases of the Calvin cycle. The figure tracks the carbon atoms (red balls) during three turns of the cycle. For every three molecules of CO_2 that are fixed, one molecule of the three-carbon sugar G3P is synthesized.

of 3-phosphoglycerate, generating six molecules of 1,3-bisphosphoglycerate. Six molecules of NADPH are consumed in converting the 6 molecules of 1,3-bisphosphoglycerate into 6 molecules of G3P (18 carbons). Five molecules of G3P (totalling 15 carbons) are used to regenerate 3 RuBP molecules (15 carbons), which requires 3 molecules of ATP. Thus, the cycle generates one surplus molecule of G3P (three carbons) after every three turns. For the synthesis of this one extra G3P, the Calvin cycle requires a total of nine molecules of ATP and six molecules of NADPH. Both ATP and NADPH are regenerated from ADP and NADP$^+$, respectively, by the light reactions.

7.4b G3P Is the Starting Point for the Synthesis of Many Other Organic Molecules

The G3P molecule formed by three turns of the Calvin cycle is the starting point for the production of a wide variety of organic molecules. Carbohydrates, such as glucose and other monosaccharides, are made from G3P by reactions that, in effect, reverse the first half of glycolysis. Once produced, the monosaccharides may enter biochemical pathways that make disaccharides such as sucrose, polysaccharides such as starches and cellulose, and other complex carbohydrates. Other pathways that consume G3P manufacture a diverse array of other molecules, including amino acids, fatty acids, and lipids. The reactions forming these products occur both within chloroplasts and in the surrounding cytosol.

Sucrose, a disaccharide consisting of glucose linked to fructose, is the main form in which the products of photosynthesis circulate from cell to cell in plants. Organic

nutrients are stored in most plants as sucrose, starch, or a combination of the two in proportions that depend on the plant species. For example, sugar cane and sugar beets, which contain stored sucrose in high concentrations, are the main sources of the sucrose we use as table sugar.

7.4c Rubisco Is the Most Abundant Protein on Earth

Ribulose-1,5-bisphosphate (RuBP) carboxylase oxygenase, or **Rubisco,** is the enzyme of the Calvin cycle that catalyzes the fixation of CO_2 into organic form:

$$\text{Ribulose-1,5-bisphosphate (RuBP)} + CO_2 \rightarrow$$
$$\text{2 3-phosphoglycerate}$$

Rubisco is considered the most important enzyme of the biosphere because by catalyzing CO_2 fixation in all photoautotrophs, it provides the source of organic carbon for most of the world's organisms. The enzyme converts a staggering 100 billion tonnes of CO_2 into carbohydrates annually. There are so many Rubisco molecules in chloroplasts that this one enzyme accounts for about 50% of the total protein content of plant leaves. This makes Rubisco easily the planet's most abundant protein, estimated to total some 40 million tonnes worldwide. Interestingly, the high abundance of Rubisco in photosynthetic cells is explained by the fact that this very important enzyme is catalytically very slow. For most enzymes, one molecule of enzyme can react with hundreds to many thousands of molecules of substrate per second. Rubisco only processes about 3 to 10 molecules of carbon dioxide per second.

Figure 7.15

Model of Rubisco. **(a)** The functional enzyme is composed of a total of 16 subunits: 8 large subunits (LSU) (shown in white and grey) and 8 small subunits (SSU) (shown in orange and blue). The synthesis of Rubsico **(b)** requires coordinated gene expression of two genomes. Each LSU is synthesized in the stroma of the chloroplast following the transcription of a gene coded by the chloroplast chromosome. The gene that encodes the SSU is found in the nucleus, with SSU monomers being synthesized by cytosolic ribosomes before being imported into the chloroplast.

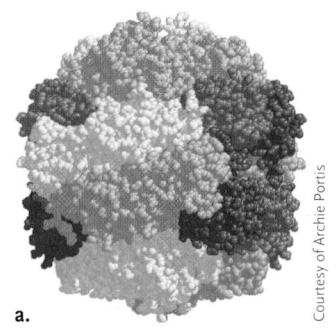

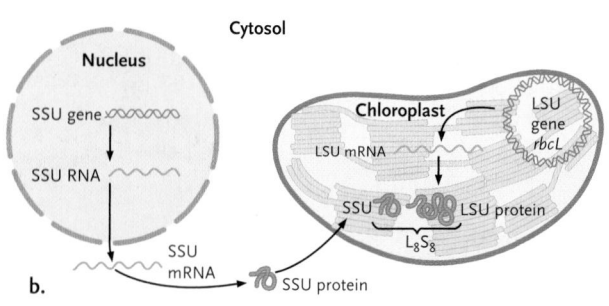

Isolation and purification of Rubisco from the chloroplast stroma has led to the elucidation of its three-dimensional structure. The molecule is cube shaped and contains eight small subunits and eight large subunits **(Figure 7.15a)**. Each of the large subunits contains an active site, which has defined binding sites for both CO_2 and RuBP. The small subunits do not have a role in catalysis but do serve an important regulatory role, although their exact function remains unknown.

The synthesis of Rubisco is quite remarkable as it requires the coordinated expression of genes of two different genomes (Figure 7.15b). While the large subunit is encoded by a gene of the chloroplast genome, the small subunit is encoded by a gene that is found in the nucleus. After the small subunit polypeptide is synthesized in the cytosol, it is imported into the chloroplast, where it associates with large subunit monomers to make the functional enzyme.

Interestingly, the vast majority of the proteins found in chloroplasts (and mitochondria) are, in fact, encoded by the nuclear genome and thus are synthesized on ribosomes in the cytosol (see Chapter 3 for further discussion on this).

STUDY BREAK

1. Glucose is not the actual product of the Calvin cycle. What is?
2. For the Calvin cycle to keep going, what compound needs to be constantly regenerated?
3. What role does the chloroplast genome play in the synthesis of Rubisco?

7.5 Photorespiration and CO_2-Concentrating Mechanisms

For being arguably the most important enzyme on the planet, Rubsico is surprisingly inefficient at fixing CO_2. The cause of this inefficiency is that the active site of

Rubisco is not specific to CO_2—a molecule of O_2 can also bind to the active site and react with RuBP. When this occurs, one of the products is a two-carbon compound that is exported from the chloroplast and actually requires the cell to consume ATP to convert it into carbon dioxide, which is simply lost. This wasteful process is called **photorespiration** because it occurs in the light and is similar to cellular respiration in that it consumes O_2 and releases CO_2. In this section, we present details on the biochemistry of the reactions that Rubsico catalyzes with O_2 and CO_2. As well, we discuss how photorespiration is exacerbated by certain environmental conditions and the key adaptations some plants and algae have evolved to minimize photorespiration.

7.5a Rubisco Is an Ancient Enzyme That Is Inhibited by Oxygen

Before we discuss the biochemistry of photorespiration, a key question we could ask is: why would natural selection have led to the evolution of an enzyme that accepts a second substrate molecule that produces a wasteful product? Rubisco and Rubisco-like proteins evolved at least 3 billion years ago as the primary enzyme in the biosphere for reducing carbon dioxide into organic form. Support for this comes, in part, from Rubisco being found in a huge diversity of organisms, including many bacteria and archaea (while they don't carry out photosynthesis, some archaea do have Rubisco). As discussed in Chapter 3, the atmosphere 3 billion years ago contained only trace amounts of O_2 and much higher levels of CO_2 than today. Under such conditions, an early form of Rubisco that could bind O_2 as well as CO_2 would not have been detrimental to an organism. Photorespiration became a problem only as the levels of oxygen in the atmosphere increased. There is evidence that over time Rubisco has slowly evolved to be more specific for CO_2, but the inhibition by O_2 remains.

O_2 can directly compete with CO_2 for the active site of Rubisco, and as such is an excellent example of a

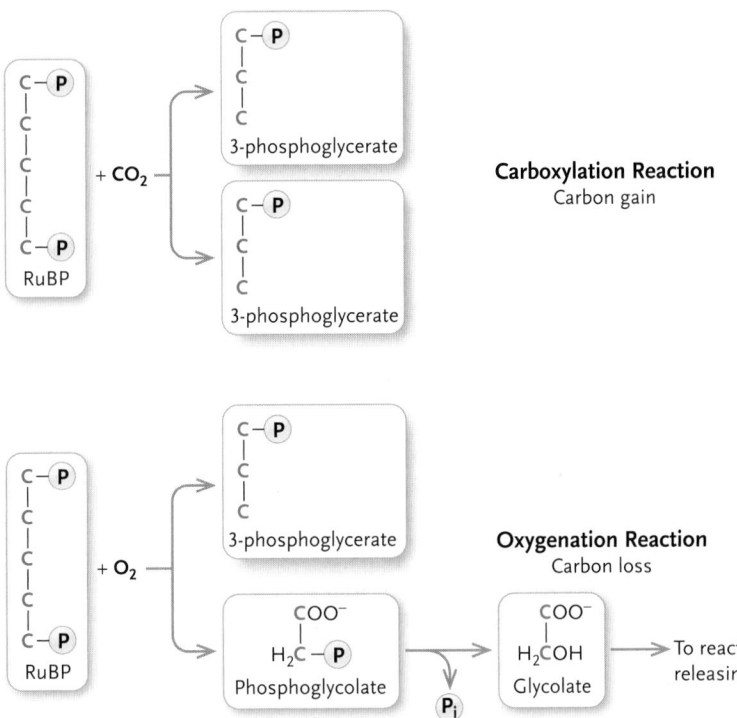

Figure 7.16
The enzyme Rubisco possesses both a carboxylase and an oxygenase activity. Compared with the usual carboxylase activity of the Calvin cycle, the oxygenase activity results in a net loss of carbon by the plant. Because oxygenase activity consumes O_2 and releases CO_2, it is also called photorespiration.

competitive inhibitor of enzyme function (see Chapter 4). When oxygen binds to the active site of Rubisco, the enzyme acts as an *oxygenase* instead of a *carboxylase*. A comparison of the products of the carboxylation reaction and the oxygenation reaction of Rubisco is shown in **Figure 7.16.** The incorporation of a CO_2 molecule into the five-carbon compound RuBP leads to a net increase in carbon by producing two molecules of the three-carbon compound 3-phosphoglycerate. By comparison, the incorporation of O_2 into RuBP in the oxygenation reaction produces a single molecule of 3-phosphoglycerate and one molecule of the two-carbon compound phosphoglycolate. There is no carbon gain—five carbons in and five carbons out. However, what makes photorespiration perhaps even more detrimental is that photoautotrophs cannot use phosphoglycolate. In the process of breaking it down to salvage the carbon, a toxic compound called glycolate is produced. The elimination of glycolate through its oxidation results in the release of carbon dioxide. Thus, whereas the carboxylation reaction leads to carbon gain, the oxygenation reaction actually results in the plant losing carbon.

If we compare the carboxylation and oxygenation reactions of Rubisco under laboratory conditions, where we can keep the concentrations of both O_2 and CO_2 equal, then the carboxylation reaction will dominate because the active site of Rubisco has a greater affinity for CO_2 than O_2. In fact, the carboxylation reaction will occur about 80 times as fast as the oxygenation reaction. However, unlike in the laboratory, the atmosphere does not contain equal amounts of the two gases—it contains approximately 21% O_2 and only about 0.04% CO_2. Because of this, under normal atmospheric concentrations and at moderate temperatures,

the oxygenation reaction can occur about once for every three times the carboxylation reaction occurs. This means that 25% of the time, the wasteful oxygenation reaction occurs, which results in net carbon loss. To counter the extent to which the oxygenation reaction occurs, many species have evolved mechanisms to try to decrease the prevalence of the oxygenation reaction. The strategies involve using mechanisms that increase the CO_2/O_2 ratio at the site of Rubisco.

7.5b Algae Pump Carbon Dioxide into Their Cells

In aquatic environments, the concentration of CO_2 dissolved in the water is usually low, well below what is needed to saturate the active site of Rubisco. Yet, interestingly, bubbling additional CO_2 into a culture of algae does not usually lead to an increase in the rate of photosynthesis, which is what you would expect. The lack of response to additional CO_2 is explained by the presence of a *carbon-concentrating mechanism* that pumps inorganic carbon into algal cells. This means that even when the concentration of CO_2 in the water is low, the amount that is actually within the cells is kept very high by this active pumping mechanism.

A model for one type of carbon-concentrating mechanism is presented in **Figure 7.17, p. 160.** In most aquatic environments, the dominant form of inorganic carbon is not CO_2 but rather the bicarbonate anion (HCO_3^-). Bicarbonate gets pumped into cells by the action of an ATP-dependent transporter on the plasma membrane. Within the cytosol, the bicarbonate is rapidly converted into CO_2 by the enzyme carbonic anhydrase. The CO_2 then diffuses into the chloroplast to the

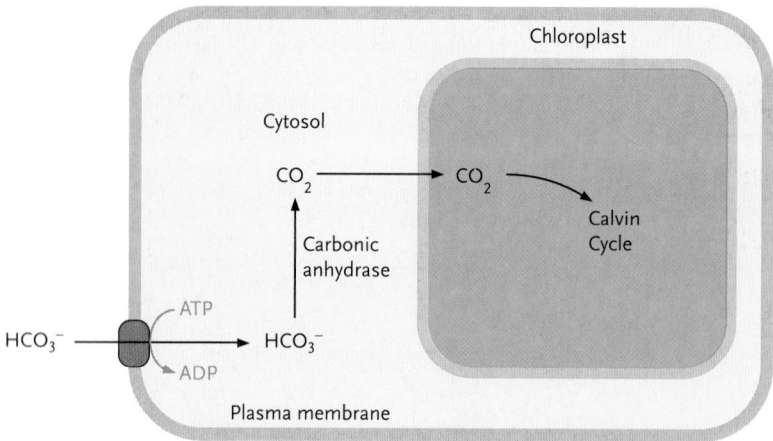

Figure 7.17

CO_2 concentration mechanism. Many aquatic photoautotrophs (e.g., algae) can increase their intracellular carbon dioxide concentrations through a mechanism that involves an ATP-dependent bicarbonate (HCO_3^-) pump on the plasma membrane. The bicarbonate is rapidly converted in the cytosol to CO_2 by the enzyme carbonic anhydrase.

site of Rubisco. This system results in a concentration of CO_2 at the site of Rubisco that is sufficiently high to essentially out-compete the O_2 that is present for the active site of Rubisco.

7.5c High Temperatures Increase Photorespiration

Like algae, land plants also face the problem of photorespiration. However, many land plants face the additional problem of trying to conserve water. Interestingly, these two problems are linked.

The surface of a plant leaf consists of a waxy cuticle that prevents evaporation of water into the air. Because this waxy cuticle also inhibits the flow of carbon dioxide, the leaf surface is covered by small pores called stomata (singular, *stoma*) that facilitate the movement of gases into and out of the leaf **(Figure 7.18).** What direction a gas moves through the stomata is governed by diffusion—movement from high to low concentration. As shown in Figure 7.18b, carbon dioxide diffuses into plant leaves. Its concentration is higher outside the leaf than it is inside because CO_2 is being consumed within the leaf by photosynthesis. Both O_2 and H_2O (water vapour) diffuse out of the leaf because their concentrations are higher in the leaf than in the outside air; O_2 is being made during photosynthesis by photosystem II, and water is moving through the plant following uptake by the roots.

Plants can regulate the size of their stomata from fully closed (to minimize water loss) to fully open (to maximize CO_2 uptake). This then illustrates the balancing act performed by plants, especially those living in dry **climates**; they need to open their stomata to let CO_2 in for photosynthesis, but to conserve water they need to keep their stomata closed. This balancing act is made more difficult in environments that are not only dry but hot as well. This is because

photorespiration becomes a bigger problem the warmer the climate. The reason for this relates to the effect of temperature on the solubility of gases in solution. As shown in **Table 7.1,** the solubility of O_2 and CO_2 (in fact all gases) *decreases* as the temperature *increases*. However, the solubility of CO_2 decreases more rapidly than that of O_2 as the temperature rises. This means that in the aqueous environment of the chloroplast stoma, the CO_2/O_2 ratio decreases as the temperature increases, and as a consequence the oxygenation reaction of Rubisco (photorespiration) becomes more common.

7.5d C₄ Plants Spatially Separate the C₄ Pathway and the Calvin Cycle

Some plant species that are adapted to hot, dry climates have evolved a mode of carbon fixation that minimizes photorespiration. In addition to the Calvin cycle, these plants have a second carbon fixation pathway called the C_4 cycle **(Figure 7.19, p. 162).** In this cycle, CO_2 initially combines with a three-carbon molecule, phosphoenolpyruvate (PEP), producing oxaloacetate, which in turn is reduced to malate. After being transported to the site of the Calvin cycle, the malate gets oxidized to pyruvate, releasing CO_2. To complete the cycle, pyruvate is converted back into PEP (for details see Figure 7.19). Because CO_2 is generated by the enzymatic conversion of malate to pyruvate, the levels of carbon dioxide at the site of the Calvin cycle are high, effectively inhibiting the oxygenation reaction of Rubisco, thereby minimizing photorespiration.

The C_4 cycle gets its name because its first product, oxaloacetate, is a four-carbon molecule rather than the three-carbon phosphoglycerate, the first product of the Calvin cycle. One often talks in terms of the C_4 pathway and the C_3 pathway when distinguishing between plants that have the C_4 cycle and those that possess only the Calvin cycle. A key distinction between C_4 and C_3 metabolism concerns the carboxylation reactions. In the C_4 cycle, the initial carboxylation reaction that incorporates CO_2 into PEP is catalyzed by the enzyme *PEP carboxylase* (Figure 7.19). Compared to Rubisco, PEP carboxylase has a rate of catalysis that is much faster, and, more importantly, O_2 cannot compete with CO_2 for its active site. It can efficiently catalyze the carboxylation of PEP regardless of the O_2 concentration near the enzyme.

C_4 metabolism is found in many tropical plants and several temperate crop species, including corn and sugar cane. In these species, the C_4 cycle occurs in mesophyll cells, which lie close to the surface of leaves and stems, where O_2 from the air is abundant (see **Figure 7.20, p. 163**). The malate intermediate of the C_4 cycle diffuses from the mesophyll cells to *bundle sheath cells,* located in deeper tissues, where O_2 concentrations are lower. In these cells, in which the Calvin cycle operates, the malate enters chloroplasts and is converted to pyruvate and CO_2.

Table 7.1	Effect of Temperature on the Solubility of O_2 and CO_2		
Temperature (°C)	$[CO_2]$ (µM in solution)	$[O_2]$ (µM in solution)	$\dfrac{[CO_2]}{[O_2]}$
5	21.93	401.2	0.0547
15	15.69	319.8	0.0491
25	11.68	264.6	0.0441
35	9.11	228.2	0.0399

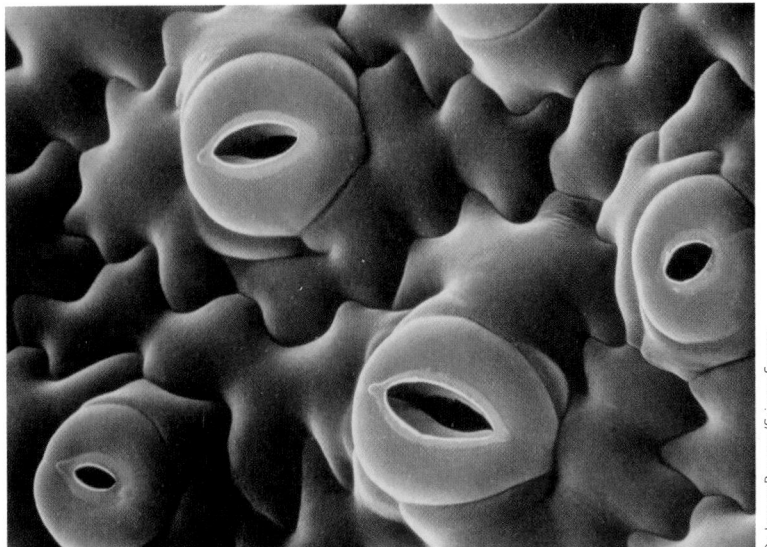

a.

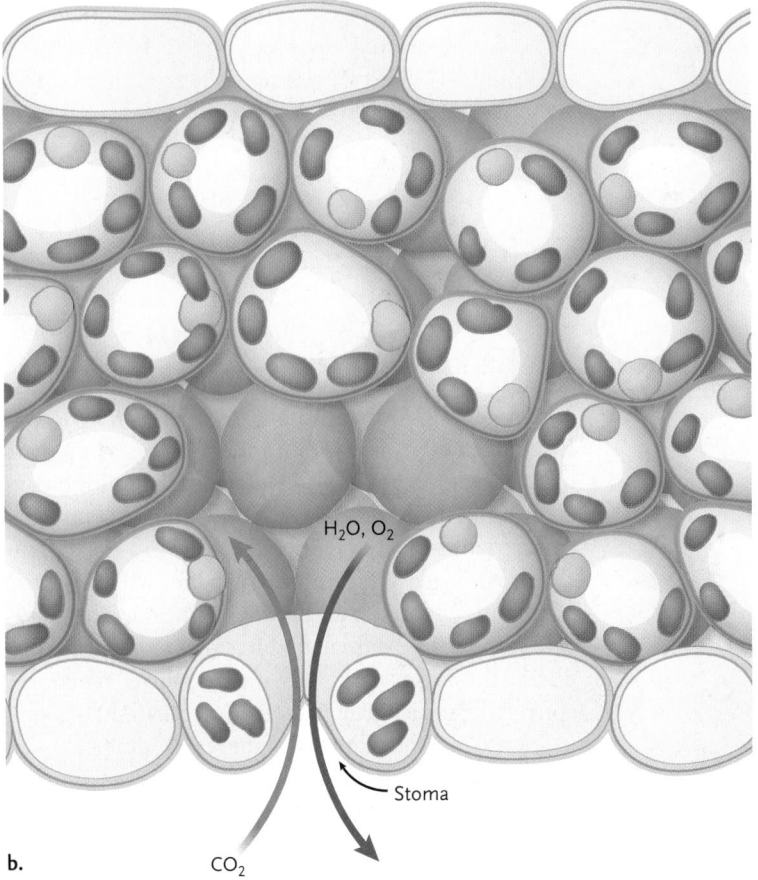

b.

Figure 7.18

Stomata. a. Micrograph of a leaf surface showing the presence of pores called stomata (sing. stoma). b. Each stoma is formed from two guard cells that control the opening and closing of the pore. This controls the movement of gases into and out of the plant and water loss.

You may ask: if C_4 metabolism is so good at preventing photorespiration, why don't all plants use it? Looking at Figure 7.19, notice that the C_4 pathway has an additional energy requirement. For each turn of the C_4 cycle, one molecule of ATP is required to regenerate PEP. In hot climates, photorespiration can decrease carbon fixation efficiency by over 50%, so the additional ATP requirement is worthwhile. As well, hot climates tend to receive a lot of sunlight, so the requirement for more ATP is easily met by absorbing more light energy and increasing the output of the light reactions.

In temperate climates (like in Canada), the lower ambient temperatures mean that photorespiration is not as big of a problem (look at Table), and the additional ATP requirement is often harder to meet given that these regions, on average, receive less sunlight. These differences in temperature and sunlight are the underlying reasons why, for example, in Florida, 70% of all native species are C_4 plants, while in Manitoba all native species are C_3 plants.

Not only do C_4 plants perform better than C_3 plants where it is hot, but they also perform better where it is dry. Because of the competing oxygenation reaction, C_3 plants need to keep their stomata open longer to fix the same number of CO_2 molecules as C_4 plants. This means that C_4 plants lose less water and are thus much better suited to arid conditions.

7.5e CAM Plants Temporally Separate the C_4 Pathway and the Calvin Cycle

Instead of running the Calvin and C_4 cycles simultaneously in different locations (spatial separation) as is the case with C4 plants, some plants, such as pineapple, run the cycles at different times (temporal separation). These plants are known as **CAM plants**, named for **crassulacean acid metabolism**, from the Crassulaceae family in which the adaptation was first observed. The plants in this group include many with thick, succulent leaves or stems, such as cactus. A comparison of C_4 and CAM metabolism is illustrated in Figure 7.20.

CAM plants typically live in regions that are hot and dry during the day and cool at night. Their fleshy leaves or stems have a low surface-to-volume ratio, and their stomata are reduced in number. Further, the stomata open only at night, when they release O_2 that accumulates from photosynthesis during the day and allow CO_2 to enter the leaves. The entering CO_2 is fixed by the C_4 pathway into malate, which accumulates throughout the night and is stored in large cell vacuoles.

Daylight initiates the second phase of the strategy. As the Sun comes up and the temperature rises, the

Figure 7.19

The C_4 cycle and its integration with the Calvin cycle. Each turn of the cycle delivers one molecule of CO_2 to the Calvin cycle. This process is energy dependent, consuming ATP.

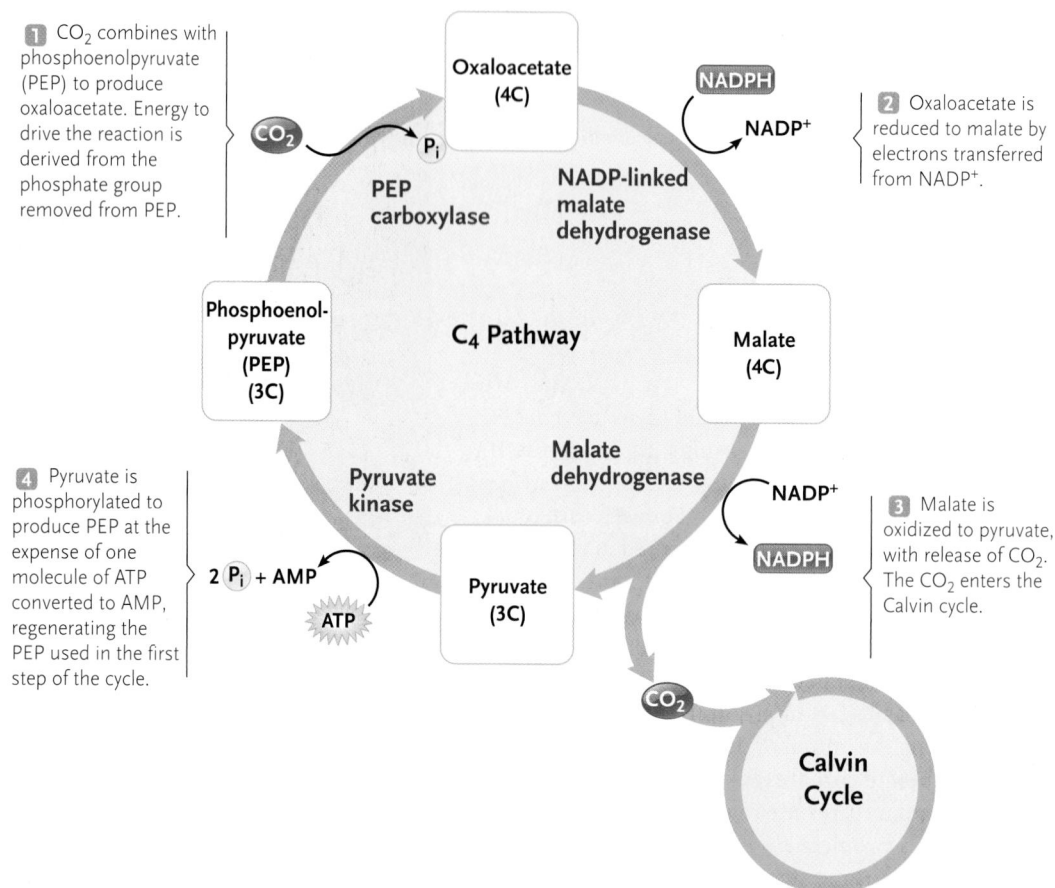

1 CO_2 combines with phosphoenolpyruvate (PEP) to produce oxaloacetate. Energy to drive the reaction is derived from the phosphate group removed from PEP.

2 Oxaloacetate is reduced to malate by electrons transferred from $NADP^+$.

4 Pyruvate is phosphorylated to produce PEP at the expense of one molecule of ATP converted to AMP, regenerating the PEP used in the first step of the cycle.

3 Malate is oxidized to pyruvate, with release of CO_2. The CO_2 enters the Calvin cycle.

stomata close, reducing water loss and cutting off the exchange of gases with the atmosphere. Malate diffuses from cell vacuoles into the cytosol, where it is oxidized to pyruvate, and CO_2 is released in high concentration. The high CO_2 concentration favours the carboxylase activity of Rubisco, allowing the Calvin cycle to proceed at maximum efficiency with little loss of organic carbon from photorespiration. The pyruvate produced by malate breakdown accumulates during the day; as night falls, it enters the C_4 reactions, converting it back to malate. During the night, oxygen is released by the plants, and more CO_2 enters.

Reduction of water loss by closure of the stomata during the hot daylight hours has the added benefit of making CAM plants highly resistant to dehydration. As a result, CAM species can tolerate extreme daytime heat and dryness.

STUDY BREAK

1. Explain how oxygenase activity suggests that Rubisco is an ancient enzyme.
2. Why is photorespiration thought to be a wasteful process?
3. What happens to the solubility of O_2 and CO_2 as the temperature is increased?

7.6 Photosynthesis and Cellular Respiration Compared

A popular misconception is that photosynthesis occurs in plants, and cellular respiration occurs only in animals. In fact, both processes occur in plants, with photosynthesis confined to tissues containing chloroplasts and cellular respiration taking place in all cells. **Figure 7.21, p. 164,** presents side-by-side schematics of photosynthesis and cellular respiration to highlight their similarities and points of connection. Note that their overall reactions are basically the reverse of each other. That is, the reactants of photosynthesis—CO_2 and H_2O—are the products of cellular respiration, and the reactants of cellular respiration—glucose and O_2—are the products of photosynthesis. Both processes have key phosphorylation reactions involving an **electron transfer system**—photophosphorylation in photosynthesis and oxidative phosphorylation in cellular respiration—followed by the chemiosmotic synthesis of ATP. Further, G3P is found in the pathways of both processes. In photosynthesis, it is a product of the Calvin cycle and is used for the synthesis of sugars and other organic fuel molecules. In cellular respiration, it is an intermediate generated in glycolysis in the conversion of glucose to pyruvate. Thus, G3P is used by anabolic pathways when it is generated

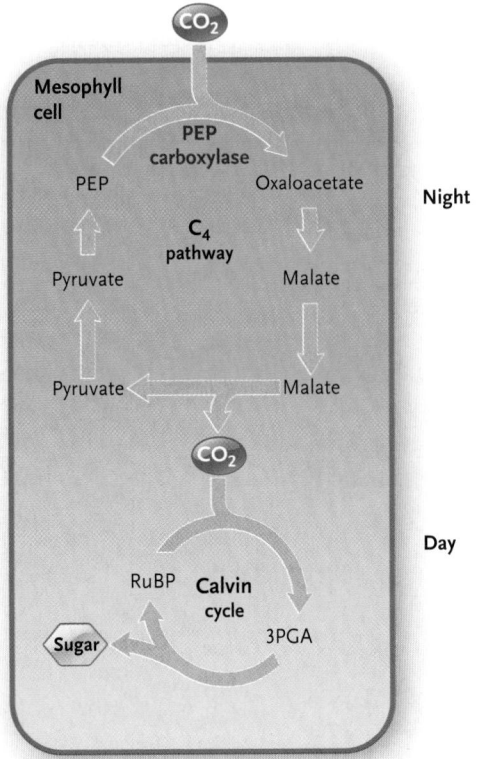

a. C$_4$—Spatial separation

CO$_2$

Mesophyll cell

PEP carboxylase

PEP → Oxaloacetate

Higher O$_2$

C$_4$ pathway

Pyruvate ← Malate

Pyruvate ← Malate

Lower O$_2$

CO$_2$

RuBP · Calvin cycle

Sugar · 3PGA

Bundle sheath cell

b. CAM—Temporal separation

CO$_2$

Mesophyll cell

PEP carboxylase

PEP → Oxaloacetate

Night

C$_4$ pathway

Pyruvate ← Malate

Pyruvate ← Malate

Day

CO$_2$

RuBP · Calvin cycle

Sugar · 3PGA

Zea mays (corn)

Photodisk/Getty Images

Opuntia basilaris (beavertail cactus)

Steve Bower/Shutterstock.com

Figure 7.20

Two alternative processes of carbon fixation to minimize photorespiration. In each case, carbon fixation produces the four-carbon oxaloacetate, which is processed to generate the CO$_2$ that feeds into the Calvin (C$_3$) cycle. **(a)** In C$_4$ plants, carbon fixation and the Calvin cycle occur in different cell types: carbon fixation by the C$_4$ pathway takes place in mesophyll cells, while the Calvin cycle takes place in bundle sheath cells. **(b)** In CAM plants, carbon fixation and the Calvin cycle occur at different times in mesophyll cells: carbon fixation by the C$_4$ pathway takes place at night, while the Calvin cycle takes place during the day.

by photosynthesis, and it is a product of a catabolic pathway in cellular respiration.

In this chapter, you have seen how photosynthesis supplies the organic molecules used as fuels by almost all the organisms of the world. It is a story of electron flow: electrons, pushed to high energy levels by the absorption of light energy, are added to CO$_2$, which is fixed into carbohydrates and other fuel molecules. The high-energy electrons are then removed from the fuel molecules by the oxidative reactions of cellular respiration, which use the released energy to power the activities of life. Among the most significant of these activities are cell growth and division, the subjects of the next chapter.

Figure 7.21
Schematic diagrams of the process of photosynthesis (left) and cellular respiration (right). Cellular respiration is shown upside down with respect to the direction of reactions to help illustrate the similarities of the process with photosynthesis.

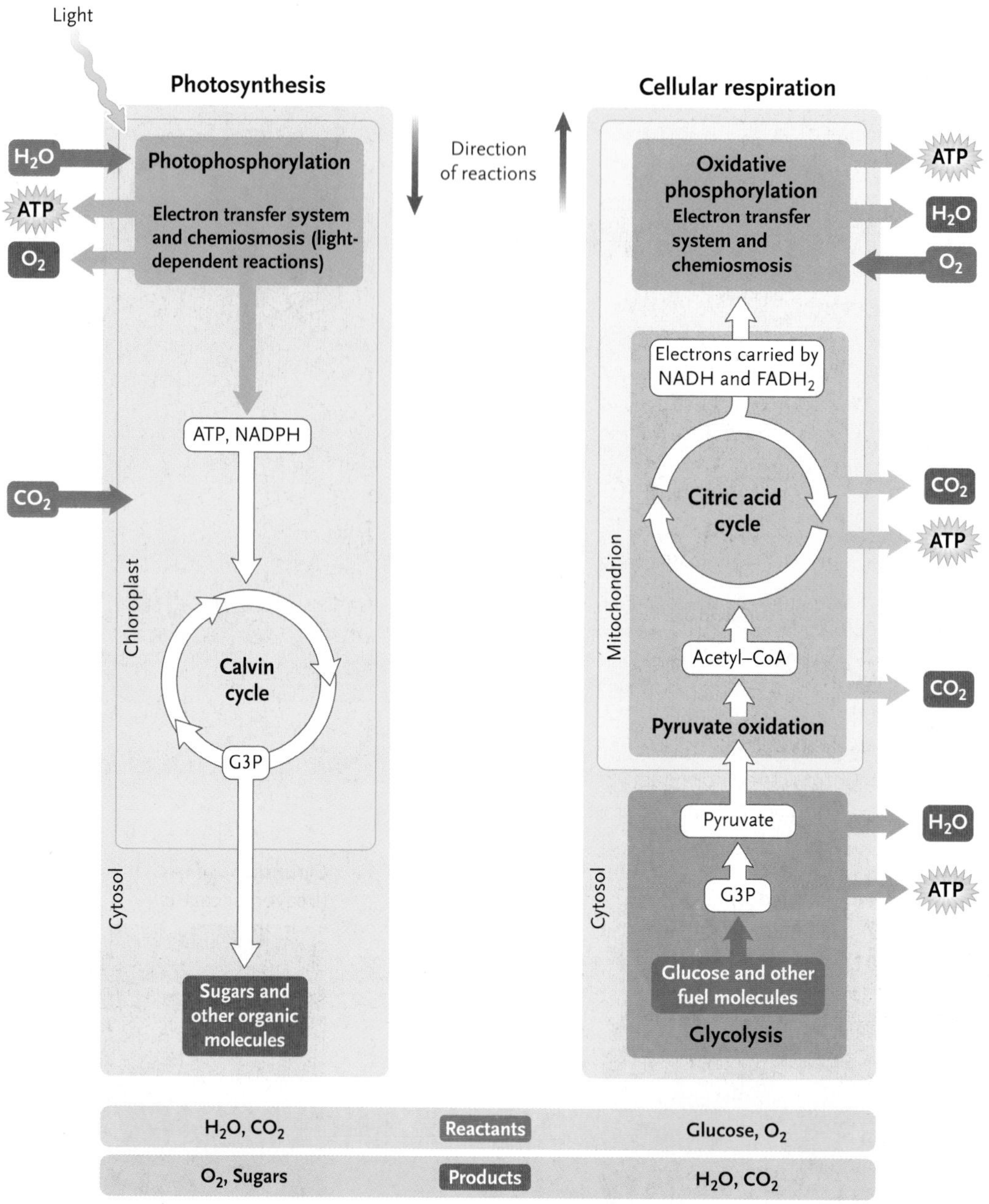

Light

Photosynthesis

H_2O → **Photophosphorylation**

ATP ←

O_2 ← **Electron transfer system and chemiosmosis (light-dependent reactions)**

Direction of reactions

↓ ↑

Cellular respiration

Oxidative phosphorylation
Electron transfer system and chemiosmosis

→ ATP

→ H_2O

O_2 →

ATP, NADPH

CO_2 →

Chloroplast

Calvin cycle

G3P

Electrons carried by NADH and $FADH_2$

Citric acid cycle

→ CO_2

→ ATP

Acetyl–CoA

→ CO_2

Pyruvate oxidation

Mitochondrion

Cytosol

Pyruvate

→ H_2O

G3P

→ ATP

Glucose and other fuel molecules

Glycolysis

Cytosol

Sugars and other organic molecules

H_2O, CO_2	Reactants	Glucose, O_2
O_2, Sugars	Products	H_2O, CO_2

Review

To access course materials such as Aplia and other companion resources, please visit www.NELSONbrain.com.

7.1 Photosynthesis: An Overview

- Photosynthesis is the use of light energy to convert carbon dioxide into an organic form.
- Photoautotrophs are the primary producers of the planet as they use the energy of sunlight to drive synthesis of organic molecules from simple inorganic molecules such as CO_2. The organic molecules are used by the photosynthetic organisms

themselves as fuels; they also form the primary energy source for animals, fungi, and other heterotrophs.

- Photosynthesis can be divided into the light reactions and the Calvin cycle. In eukaryotes, both stages take place inside chloroplasts. The light reactions, which occur on the thylakoid membrane of chloroplasts (in eukaryotes), use the energy of light to drive the synthesis of NADPH and ATP. These are consumed by the Calvin cycle, which fixes CO_2 into carbohydrates (see Figure 7.2).

7.2 The Photosynthetic Apparatus

- Visible light occurs between 400 nm (blue light) and 700 nm (red light). The energy of a photon of light is inversely related to its wavelength (see Figure 7.5).
- The absorption of light energy by pigment molecules results in electrons within the pigment being raised to a higher-energy (excited) state. There are three fates of this excited state: energy loss, transport of the energy, or oxidation of the pigment (loss of the electron) (see Figure 7.6).
- Chlorophylls and carotenoids, the photon-absorbing pigments in eukaryotes and cyanobacteria, together absorb light energy at a range of wavelengths, enabling a wide spectrum of light to be used (see Figures 7.7 and 7.8)
- Photosynthetic pigments are organized into two types of photosystems: photosystem II and photosystem I. Each photosystem consists of a reaction centre surrounded by an antenna complex. Energy trapped by pigments in the antenna is funnelled to the reaction centre, where it is used to oxidize a special reaction centre chlorophyll (denoted by P680 for photosystem II and P700 for photosystem I). The oxidation of reaction center chlorophyll is coupled to the reduction of the primary electron acceptors within each reaction centre (see Figure 7.10).

7.3 The Light Reactions

- The photosynthetic electron transport chain (ETC; the light reactions) uses the energy of light absorbed by photosystem II and photosystem I to generate reducing power in the form of NADPH. Electrons released by the oxidation of the reaction centre chlorophyll of photosystem II are passed along an ETC. To get all the way down the chain, electrons become excited again at photosystem I, and then they are delivered to $NADP^+$ as final electron acceptor. $NADP^+$ is reduced to NADPH (see Figure 7.11).
- In a way similar to respiratory electron transport, the process of photosynthetic electron transport results in the establishment of a proton gradient, in this case, across the thylakoid membrane. The proton gradient is used to generate ATP through chemiosmosis using the ATP synthase complex embedded in the thylakoid membrane (see Figure 7.11).
- Besides linear electron transport from photosystem II through photosystem I, electrons can also flow in a cycle around photosystem I, building the H^+ concentration and allowing extra ATP to be produced, but no NADPH (see Figure 7.13).

7.4 The Calvin Cycle

- In the Calvin cycle, CO_2 is reduced and converted into carbohydrate by the addition of electrons and hydrogen carried by the NADPH produced in the light-dependent reactions. ATP, also derived from the light-dependent reactions, provides energy.

The key enzyme of the light-independent reactions is Rubisco (RuBP carboxylase/oxygenase), which catalyzes the reaction that combines CO_2 with a molecule of ribulose-1,5-bisphosphate (RuBP), producing two molecules of the three-carbon compound 3-phosphoglycerate (see Figure 7.14).
- For three turns of the Calvin cycle, one molecule of the three-carbon sugar glyceraldehyde-3-phosphate (G3P) is synthesized. G3P is the starting point for synthesis of glucose (requires two G3P molecules), sucrose, starch, and other organic molecules.

7.5 Photorespiration and CO_2-Concentrating Mechanisms

- Oxygen is a competitive inhibitor of Rubisco—it can compete with CO_2 for the active site. As an oxygenase, Rubisco catalyzes the combination of RuBP with O_2 rather than CO_2, forming toxic products that cannot be used in photosynthesis. The toxic products are eliminated by reactions that release carbon in inorganic form as CO_2, greatly reducing the efficiency of photosynthesis. The entire process is called photorespiration because it uses oxygen and releases CO_2 (see Figure 7.16).
- To avoid photorespiration, a range of plants and algae have evolved mechanisms to decrease the amount of O_2 that is present at the site of Rubisco and carbon fixation.
- In aquatic photoautotrophs, an ATP-dependent process pumps bicarbonate (HCO_3^-) into the cell, which is converted to CO_2 through the action of the enzyme carbonic anhydrase. This makes the CO_2/O_2 ratio greater at the site of Rubisco (Figure 7.17).
- Some plants have evolved C_4 metabolism whereby CO_2 from the air is first fixed by a carboxylase that does not have oxygenase activity into the four-carbon compound that occurs in mesophyll cells (the site of the Calvin cycle in C_3 plants). Following transport into bundle sheath cells, decarboxylation of a related four-carbon compound releases CO_2 at the site of Rubisco. This results in the CO_2/O_2 ratio being very high (see Figures 7.19 and 7.20).
- CAM plants also first fix CO_2 into oxaloacetate and then generate CO_2 for the Calvin cycle. Here both the carbon fixation and the Calvin cycle occur in mesophyll cells, but they are separated by time; initial carbon fixation occurs at night and the Calvin cycle occurs during the day (see Figure 7.20).

7.6 Photosynthesis and Cellular Respiration Compared

- Photosynthesis occurs in the cells of plants that contain chloroplasts, whereas cellular respiration occurs in all cells. The overall reactions of the two processes are essentially the reverse of each other, with the reactants of one being the products of the other (see Figure 7.21).

Questions

Self-Test Questions

1. What is the correct definition of the term *autotroph?*
   a. an organism that uses light energy to live
   b. an organism that can synthesize carbohydrates
   c. an organism that consumes the molecules found in other organisms
   d. an organism that synthesizes organic molecules using inorganic carbon

2. Why is chlorophyll green in colour?
   a. Chlorophyll only absorbs green photons of light.
   b. Green photons of light excite electrons within chlorophyll.
   c. Chlorophyll lacks an excited state that matches the energy of green photons.
   d. Green photons of light are not of high enough energy to excite electrons in chlorophyll.

3. Which statement about light reactions is correct?
   a. They result in an increase in the pH of the thylakoid lumen.
   b. Ferredoxin shuttles electrons from the cytochrome complex to photosystem I.
   c. $P680^*$ is easier to oxidize than P680.
   d. Energy transfer from the antenna to the reaction centre of a photosystem is through electron transport.

4. What is the minimum number of photons that need to be absorbed by the photosystems to reduce three molecules of $NADP^+$ to NADPH by the photosynthetic ETC?
   a. 3
   b. 6
   c. 12
   d. 18

5. Which of the following statements correctly distinguishes between linear electron transport and cyclic electron transport?
   a. NADH is generated only during linear electron transport.
   b. Photosystem I is used only during linear electron transport.
   c. Photosystem II is required only during cyclic electron transport.
   d. A proton-motive force is generated only during cyclic electron transport.

6. The Calvin cycle is sometimes called the light-independent reactions. This is misleading since the Calvin cycle will stop operating after a plant is placed in the dark. Why will it stop?
   a. Rubisco is rapidly degraded in the dark.
   b. NAD^+ generated by the light reactions is needed to activate 3-phosphoglycerate.
   c. In the dark, oxygen builds up in the chloroplast and inhibits Rubisco activity.
   d. The Calvin cycle requires a constant supply of ATP generation by the light reactions.

7. Which of the following statements about the Calvin cycle is correct?
   a. The cycle stops if the regeneration of RuBP is prevented.
   b. After three turns, one molecule of glucose is synthesized.
   c. It takes three turns of the cycle to fix one molecule of CO_2.
   d. It takes nine molecules of ATP to fix one molecule of CO_2.

8. Why don't C_4 plants photorespire?
   a. They lack mitochondria.
   b. They express high levels of carbonic anhydrase.
   c. The CO_2/O_2 ratio in their chloroplasts is very high.
   d. The Rubisco in their chloroplasts only has affinity for CO_2.

9. Compared to C_3 plants, one often find C_4 plants in drier habitats. Why is that?
   a. They have a larger root system.
   b. They can keep their stomata closed at all times.
   c. They don't have to keep their stomata open as long.
   d. Unlike C_3 plants, the leaves of C_4 plants are covered by a waxy cuticle.

10. In what way are the light reactions of photosynthesis similar to aerobic respiration?
    a. Both processes synthesize NADPH.
    b. Both processes use substrate phosphorylation to synthesize ATP.
    c. Chemiosmosis using the proton-motive force occurs in both.
    d. Both require oxygen as the final electron acceptor of electron transport.

Questions for Discussion

1. Like other accessory pigments, the carotenoids extend the range of wavelengths absorbed in photosynthesis. They also protect plants from a potentially lethal process known as *photooxidation*. This process begins when excitation energy in chlorophylls drives the conversion of oxygen into reactive oxygen species (ROS) (see Chapter 6), substances that can damage organic compounds and kill cells. When plants that cannot produce carotenoids are grown in light, they bleach white and die. Given this observation, what molecules in the plants are likely to be destroyed by photooxidation?

2. Exposing plants to light at low temperatures increases the damage to photosystem II. Why do you think this is the case?

3. If global warming raises the temperature of our climate significantly, will C_3 plants or C_4 plants be favoured by natural selection? How will global warming change the geographical distributions of plants?

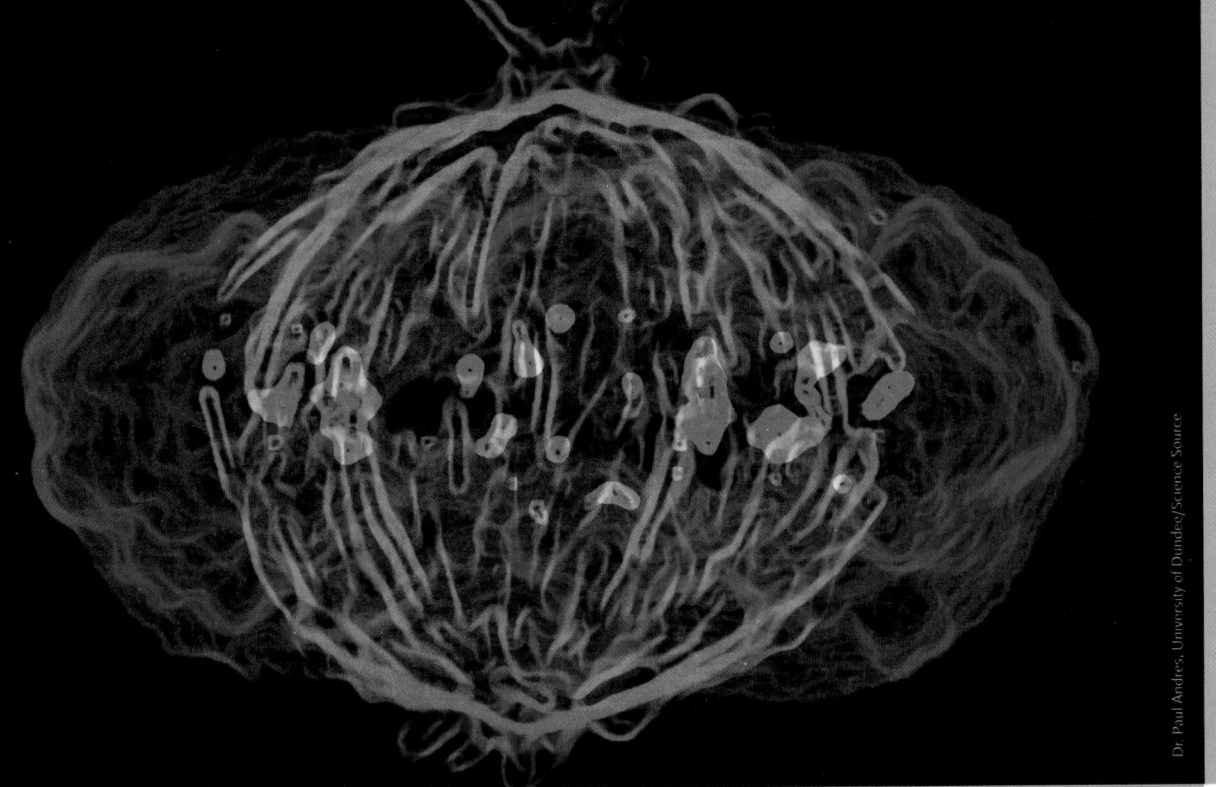

8

A cell in mitosis (fluorescence micrograph). The spindle (red) is separating copies of the cell's chromosomes (green) before cell division.

Cell Cycles

WHY IT MATTERS

As the rainy season recedes in Northern India, rice paddies and other flooded areas begin to dry. These shallow seasonal pools have provided an environment of slow-moving warm water for zebrafish (*Danio rerio*) to spawn **(Figure 8.1, p. 168).** Over the past few months, many millions of cell divisions have fuelled the growth and development of single fertilized eggs into the complex multicellular tissues and organs of these small, boldly striped fish. Most cells in the adults have now stopped dividing and are dedicated to particular functions.

Moving into the fast-running streams that feed the Ganges River, the young zebrafish encounter larger predators, such as knifefish (*Notopterus notopterus*). Imagine for a moment that a zebrafish is attacked by a knifefish; the prey narrowly escapes but not without leaving one of its fins behind in the mouth of the predator. In an amazing feat of cell cycle regulation, the entire zebrafish fin will be regenerated—skin, nerves, muscles, bones, and all—within a week!

As a model system for vertebrate development, the zebrafish has provided a popular tool for researchers to identify the stages of regeneration at the molecular level. (See *The Purple Pages* for more information about zebrafish as model organisms.) In the first step, existing skin cells migrate to close the wound and prevent bleeding. Then cells under the new skin form a temporary tissue called a blastema. Blastema cells divide up to 50 times as fast as usual, providing large numbers of daughter cells capable of maturing into new bone, nerve, muscle, and blood vessel cells in response to signal proteins produced by the skin. Once the regenerated fin has reached its normal size and shape, the new cells stop growing and dividing.

Figure 8.1
Zebrafish (*Danio rerio*).

replication, and cell division in the face of a changing environment.

Although this chapter highlights the characteristics of dividing cells, it is important for you to realize that most cells in the body of a multicellular organism are *not* destined to divide any time soon, if

The remarkable ability of eukaryotic cells to coordinate their growth and division, always yielding daughter cells in a timely fashion with a complete set of genetic material, is the focus of this chapter. In particular, we intend for you to appreciate how this coordination is achieved through the interplay of three cellular processes: (1) DNA replication, (2) a dynamically changing cytoskeleton, and (3) cell cycle "checkpoints." To provide a hint of what the ancestral cell division process may have been like, the chapter opens with a look at cell division in prokaryotic organisms and the simpler eukaryotes.

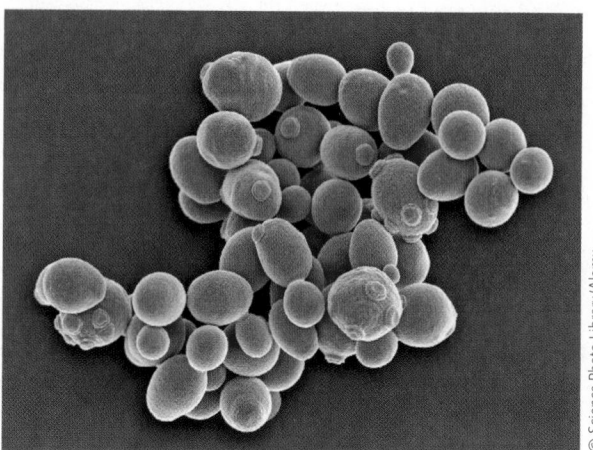

a.

b.

8.1 The Cycle of Cell Growth and Division: An Overview

While regenerating a lost body part is certainly dramatic at the scale of an individual organism, we invite you now to consider the wider, grander, view of the relevance of the cell division cycle. Scientists have known since the early nineteenth century that all life on Earth is composed of cells and their products. All cells, in all organisms that have ever lived, are descended from previous cells in an unbroken chain of cell division stretching billions of years into the past. New progeny cells are needed for expanding population size (single-celled organisms), multicellular tissue growth (new leaves), asexual reproduction, and replacement of cells lost to wear (shedding skin and gut lining) and tear (wound repair, virus infection) **(Figure 8.2)**. Although the cell division cycle is conceptually simple—grow, divide, grow, divide—repeat for billions of years—regulation of this process must be precise and complex. If cells divide too quickly, daughter cells may be too small or lacking essential cytoplasm or genetic material. If cells divide too slowly, they may grow inefficiently large or accumulate extra chromosomes. All dividing cells must meet the challenge of closely coordinating their growth, DNA

c.

Figure 8.2
Actively dividing cells provide for increased population size of yeast **(a)**, growth of skin **(b)**, and expansion of conifer needles **(c)**.

ever. In fact, some cells may even be programmed to die immediately!

8.2 The Cell Cycle in Prokaryotic Organisms

A newly formed prokaryotic cell, such as the bacterium *Escherichia coli,* must double in size, replicate its circular chromosome, and then move each of the resulting two daughter chromosomes into its own progeny cell during cell division. The entire mechanism of prokaryotic cell division, called **binary fission**, that is, splitting or dividing into two parts, can be thought of in three periods, as shown in **Figure 8.3.** Following birth, cells may grow for some time before initiating DNA synthesis (B period). Once the chromosomes are replicated and separated to opposite ends of the cell (C period), the membrane pinches together between them and two daughter cells are formed (D period).

8.2a Replication Occupies Most of the Cell Cycle in Rapidly Dividing Prokaryotic Cells

All bacteria and archaea use DNA as their hereditary information, and the vast majority of species package it all in a single, circular chromosome of double-stranded DNA (Figure 8.3, step 1). Although the chromosome is shown extended in Figure 8.3 for the purposes of illustration, it is actually compacted in a central region called the **nucleoid** throughout the cell cycle (see Figure 2.7, Chapter 2). When nutrients are abundant, prokaryotic cells have no need for a B period since they can grow quickly enough to divide their cytoplasm as soon as DNA replication is complete and chromosomes are separated. Under such optimal conditions, populations of *E. coli* cells can double every 20 minutes.

8.2b Replicated Chromosomes Are Distributed Actively to the Halves of the Prokaryotic Cell

In the 1860s, François Jacob of The Pasteur Institute in Paris, France, proposed a model for the **segregation** of bacterial chromosomes to the daughter cells in which the two chromosomes attach to the plasma membrane near the middle of the cell and separate as a new plasma membrane is added between the two sites during cell elongation. The essence of this model is that chromosome separation is passive. However, current research indicates that bacterial chromosomes rapidly separate in an active way that is linked to DNA replication events and is independent of cell elongation. The new model is summarized in the C period of Figure 8.3.

Replication of the bacterial chromosome commences at a specific region called the **origin of replication**

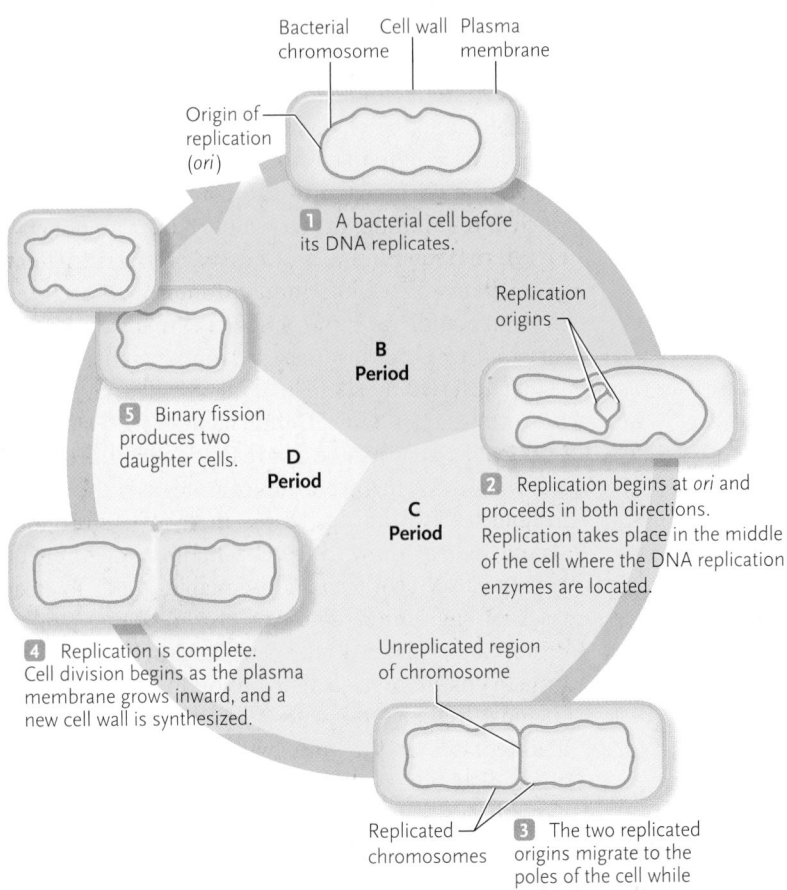

Figure 8.3

The bacterial cell cycle. During the B period, from birth to the initiation of DNA replication, the cell grows in size. The chromosome is replicated and the resulting daughter chromosomes move to opposite ends during the C period. Then the cell divides by binary fission during the D period. In very fast growing cultures, the B period may be nonexistent; cells may be born with chromosomes that are already partly replicated!

(*ori*). The *ori* is in the middle of the cell, where the enzymes for DNA replication are located. Once the *ori* has been duplicated, the two new origins migrate toward the two opposite ends (poles) of the cell as replication continues for the rest of the chromosome (see Figure 8.3, step 3). The mechanism that propels the two replicated chromosomes to their respective ends of the cell is still unknown. Next, cytoplasmic division is associated with an inward constriction of a cytokinetic ring of cytoskeletal proteins. New plasma membrane and cell wall material is assembled to divide the cell into two equal parts (see Figure 8.3, step 5).

8.2c Mitosis Has Evolved from an Early Form of Binary Fission

The prokaryotic mechanism works effectively because most prokaryotic organisms have only a single chromosome. Thus, if a daughter cell receives at least one copy of the chromosome, its genetic information is complete. By contrast, the genetic information of eukaryotes is divided among several chromosomes, with each chromosome containing a much greater length of DNA than a prokaryotic chromosome does. If a daughter cell fails to receive a copy of even one chromosome, the effects are usually lethal. It is also important to note that, during most of the cell cycle, eukaryotic chromosomes are contained within the nuclear membrane. Bacteria and archaea do not have an internal membrane around their nucleoid. Therefore, eukaryotic cellular and chromosomal architecture demands a quite different mechanism for distributing chromosomes to daughter cells. Mitosis is that different mechanism, and we will examine this process in detail later in the chapter (see **Figure 8.6, p. 172**). For now, just be aware that one of the central innovations of the evolution of mitosis is the ability to hold the two newly created molecules of double-stranded DNA (now called **chromatids**) together following DNA synthesis. This enables cells to keep track of such long replicated chromosomes and to orient them relative to the cytoskeleton at the proper time to ensure precise distribution to daughter cells. In most higher eukaryotes, the nuclear membrane disintegrates at the time when chromosomes are being distributed and then reforms around them in daughter cells.

Variations in the mitotic apparatus in modern-day organisms illuminate possible intermediates in the evolutionary pathway to mitosis from some ancestral type of binary fission. For example, in many primitive eukaryotes, such as dinoflagellates (a type of single-celled alga), the nuclear envelope remains intact during mitosis, and the chromosomes bind to the inner membrane of the nuclear membrane. When the nucleus divides, the chromosomes are segregated.

A more advanced form of the mitotic apparatus is seen in yeasts and diatoms (another type of single-celled alga). In these organisms, a **spindle** of microtubules made of polymerized tubulin protein forms and chromosomes segregate to daughter nuclei without the disassembly and reassembly of the nuclear envelope. Current evidence suggests that the type of mitosis seen in yeasts and diatoms and the type of mitosis in animals and higher plants described later in this chapter evolved separately from a common ancestral type.

STUDY BREAK

What are the three main steps in binary fission of prokaryotic organisms?

8.3 Mitosis and the Eukaryotic Cell Cycle

As long as eukaryotes require their daughter cells to be genetic copies of the **parental** cell, **mitosis** serves very well to divide the replicated DNA equally and precisely.

This is the result of three elegantly interrelated systems:

1. An elaborate master program of molecular checks and balances ensures an orderly and timely progression through the cell cycle.
2. Within the overall regulation of the cell cycle, a process of DNA synthesis replicates each DNA chromosome into two copies with almost perfect fidelity (see Section 12.3).
3. A structural and mechanical web of interwoven "cables" and "motors" of the cytoskeleton separates the replicated DNA molecules precisely into the daughter cells.

However, at a particular stage of the life cycle of sexually reproducing organisms, a cell division process called meiosis produces some cells that are genetically different from the parent cells. **Meiosis** produces daughter nuclei that are different in that they have one-half the number of chromosomes the parental nucleus had. Also, the mechanisms involved in producing the daughter nuclei produce arrangements of genes on chromosomes that are different from those in the parent cell (see Chapter 9). The cells that are the products of meiosis may function as gametes in animals (fusing with other gametes to make a zygote) and as spores in plants and many fungi (dividing by mitosis). Meiosis and its role in eukaryotic sexual reproduction are addressed in Chapter 9. We begin our discussion of mitosis with **chromosomes**, the nuclear units of genetic information divided and distributed by mitotic cell division.

8.3a Chromosomes Are the Genetic Units Divided by Mitosis

In all eukaryotes, the hereditary information of the nucleus is distributed among several linear, double-stranded DNA molecules. These DNA molecules are combined with proteins that stabilize the DNA, assist in packaging DNA during cell division, and influence the expression of individual genes. Each chromosome (*chroma* = colour, when stained with dyes used in light microscopy; *soma* = body; **Figure 8.4**) in a cell is composed of one of these DNA molecules, along with its associated proteins.

Most eukaryotes have two copies of each type of chromosome in their nuclei, and their chromosome complement is said to be **diploid**, or $2n$. For example, humans have 23 different pairs of chromosomes for a diploid number of 46 chromosomes ($2n = 46$). Other eukaryotes, mostly microorganisms, may have only one copy of each type of chromosome in their nucleus, so their chromosome complement is said to be **haploid**, or n. Baker's yeast (*Saccharomyces cerevisiae*) is an example of an organism that can grow as a diploid ($2n = 32$) and as a haploid ($n = 16$). Still others, such as many plant species, have three, four, or even more complete sets of chromosomes in each cell. The

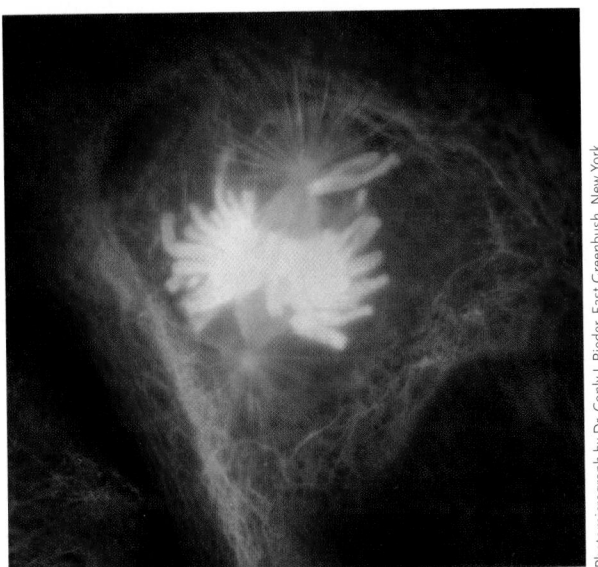

Figure 8.4
Eukaryotic chromosomes (stained blue) during mitosis.

number of chromosome sets is called the **ploidy** of a cell or species. See Chapter 19 for a look at the role of ploidy in the formation of new species.

Before a cell divides in mitosis, duplication of each chromosome produces two identical copies of each chromosome called **sister chromatids.** Duplication of a chromosome involves replicating the DNA molecule it contains, plus doubling the proteins that are bound to the DNA to stabilize it. Newly formed sister chromatids are held together tightly by sister chromatid cohesion, in which proteins called cohesins encircle the sister chromatids along their length. During mitosis, the cohesins are removed and the sister chromatids are separated, with one of each pair going to each of the two daughter nuclei. As a result of this precise division, each daughter nucleus receives exactly the same number and types of chromosomes and contains the same genetic information as the parent cell that entered the division. The equal distribution of daughter chromosomes into each of the two daughter cells that result from cell division is called **chromosome segregation.**

CONCEPT FIX The double-strandedness of DNA, the mechanism of cell division, and new vocabulary can lead to confusion about just how many chromosomes are in a particular cell at a particular time. Let's pick the fruit fly *Drosophila melanogaster* as an example organism because all of its genes are contained in only four different chromosomes. Every body cell in a fruit fly is diploid (*2n*) and therefore contains two of each of the four distinct *Drosophila* chromosomes—for a total of eight. Each of these eight chromosomes is composed of one double-stranded DNA molecule of the type shown in Figure 12.6, Chapter 12. If the cell is preparing to divide, DNA synthesis, as shown in Figure 12.7, Chapter 12, creates two identical DNA molecules from each of the eight originals. The two new DNA molecules created

from each original chromosome remain attached to one another and are now called sister chromatids. (Someone once suggested that these two new molecules should be called twin chromatids because they are identical.) Now here comes the confusing part. Since sister chromatids remain attached to each other at their centromeres following DNA synthesis, the pair of them is still referred to as just one chromosome. Before replication, one chromosome is composed of one DNA molecule; after replication, one chromosome is composed of two DNA molecules. You should see that DNA replication increases the amount of DNA in the nucleus but it does not increase the number of chromosomes. Our *Drosophila* cell has eight DNA molecules in the nucleus before DNA synthesis and eight pairs of molecules after. There are eight chromosomes before DNA synthesis and eight replicated chromosomes after. During cell division, each of the two daughter cells receives one of the two sister chromatids from each replicated chromosome. You should therefore agree to the rather counterintuitive claim that two daughter cells can each receive eight chromosomes even though there were only eight chromosomes in the original cell. ⬢

The precision of chromosome replication and segregation in the mitotic cell cycle creates a group of cells called a **clone.** Except for rare chance mutations, all cells of a clone are genetically identical. Since all the diverse cell types of a complex multicellular organism arose by mitosis from a single zygote, they should all contain the same genetic information. Forensic scientists rely on this feature of organisms when, for instance, they match the genetic profile of a small amount of tissue (e.g., cells in dog saliva recovered from a bite victim) with that of a blood sample from the suspected animal.

STUDY BREAK

1. What are the three interrelated systems that contribute to the eukaryotic cell cycle?
2. What is a chromosome composed of?

8.3b Interphase Extends from the End of One Mitosis to the Beginning of the Next Mitosis

If we set the formation of a new daughter cell as the beginning of the mitotic cell cycle, then the first and longest stage is **interphase (Figure 8.5, p. 172).** Interphase comprises three phases of the cell cycle:

1. **G$_1$ phase**, in which the cell carries out its function, and in some cases grows
2. **S phase**, in which DNA replication and chromosome duplication occur
3. **G$_2$ phase**, a brief gap in the cell cycle during which cell growth continues and the cell prepares for mitosis (the fourth phase of the cell cycle; also called M phase) and cytokinesis.

Figure 8.5

The cell cycle. The length of G_1 varies, but for a given cell type, the timing of S phase, G_2 phase, and mitosis is usually relatively uniform. Cytokinesis (red segment) usually begins while mitosis is in progress and reaches completion as mitosis ends. Cells in a state of division arrest are considered to enter a side loop (or shunt) from G_1 phase called G_0 phase.

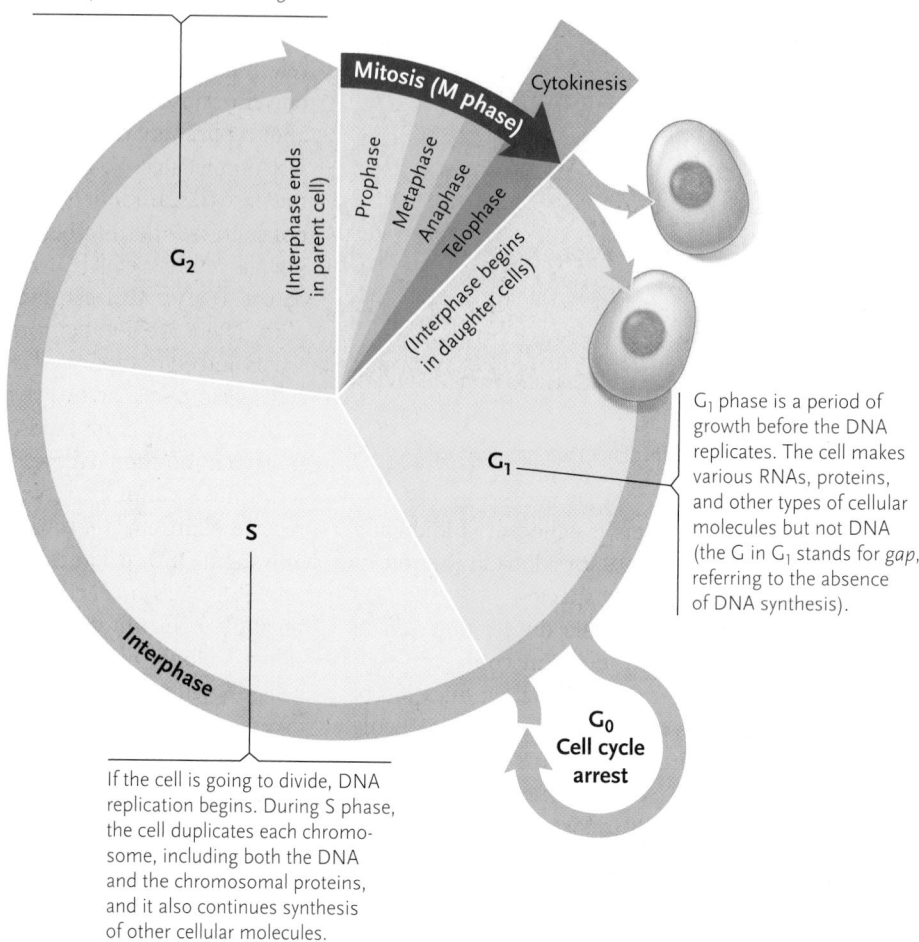

G_2 refers to the second gap in which there is no DNA synthesis. During G_2, the cell continues to synthesize RNAs and proteins, including those for mitosis, and it continues to grow. The end of G_2 marks the end of interphase; mitosis then begins.

Mitosis (M phase)

Cytokinesis

Prophase
Metaphase
Anaphase
Telophase

(Interphase ends in parent cell)

(Interphase begins in daughter cells)

G_2

G_1

G_1 phase is a period of growth before the DNA replicates. The cell makes various RNAs, proteins, and other types of cellular molecules but not DNA (the G in G_1 stands for *gap*, referring to the absence of DNA synthesis).

S

Interphase

G_0
Cell cycle arrest

If the cell is going to divide, DNA replication begins. During S phase, the cell duplicates each chromosome, including both the DNA and the chromosomal proteins, and it also continues synthesis of other cellular molecules.

Figure 8.6

Chromosomes from the muntjac deer are individually "painted" with fluorescent stain. Note that there are two cells in this picture. The metaphase cell shows two copies of each of three long, condensed chromosomes. The interphase nucleus shows the DNA of different chromosomes organized in close proximity rather than randomly distributed.

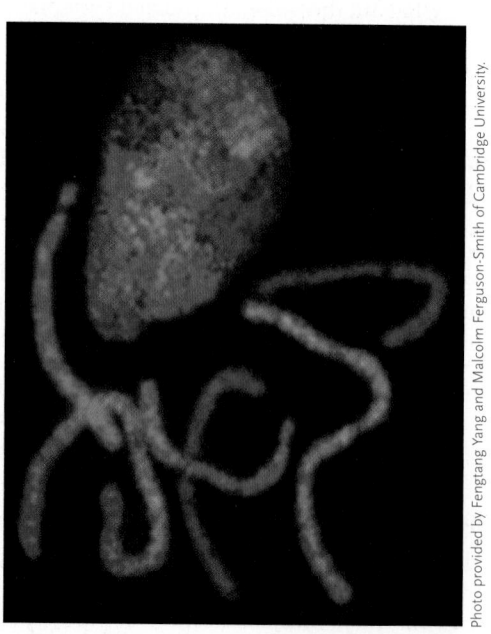

Photo provided by Fengtang Yang and Malcolm Ferguson-Smith of Cambridge University.

Usually, G_1 is the only phase of the cell cycle that varies in length. The other phases are typically uniform in length within a species. Thus, whether cells divide rapidly or slowly depends primarily on the length of G_1. Once DNA replication begins, most mammalian cells take about 10 to 12 hours to proceed through the S phase, about 4 to 6 hours to go through G_2, and about 1 hour or less to complete mitosis.

G_1 is also the stage in which many cell types stop dividing. Cells that are not destined to divide immediately enter a shunt from G_1 called the **G_0 phase**. In some cases, a cell in G_0 may start dividing again by re-entering G_1. Some cells never resume the cell cycle; for example, most cells of the human nervous system stop dividing once they are fully mature. During all the stages of interphase the chromosomes are organized, but relatively loosely packaged, within the nucleus (Figure 8.6).

Internal regulatory controls trigger each phase of the cell cycle, ensuring that the processes of one phase are

FOCUS ON RESEARCH 8.1
Research Example: Growing Cell Clones in Culture

How can investigators safely test whether a particular substance is toxic to human cells or whether it can cure or cause cancer? One widely used approach is to work with **cell cultures**—living cells grown in laboratory vessels. Many types of prokaryotic and eukaryotic cells can be grown in this way.

When cell cultures are started from single cells, they form **clones**: barring mutations, all the individuals descended from the original cell are genetically identical. Clones are ideal for experiments in genetics, biochemistry, molecular biology, and medicine because the cells lack genetic differences that could affect the experimental results.

Microorganisms such as yeasts and many bacteria are easy to grow in laboratory cultures. For example, the human intestinal bacterium *E. coli* can be grown in solutions (growth media) that contain only an organic carbon source such as glucose, a nitrogen source, and inorganic salts. The cells may be grown in liquid suspensions or on the surface of a solid growth medium such as an agar gel (agar is a polysaccharide extracted from an

alga). Many thousands of bacterial strains are used in a wide variety of experimental studies.

Many types of plant cells can also be cultured as clones in specific growth media. With the addition of plant growth hormones, complete plants can often be grown from single cultured cells. Growing plants from cultured cells is particularly valuable in genetic engineering, in which genes introduced into cultured cells can be tracked in fully developed plants. Plants that have been engineered successfully can then be grown simply by planting their seeds.

Animal cells vary in what is needed to culture them. For many types, the culture medium must contain essential amino acids—that is, the amino acids that the cells cannot make for themselves. In addition, mammalian cells require specific growth factors provided by adding blood serum, the fluid part of the blood left after red and white blood cells are removed.

Even with added serum, many types of normal mammalian cells cannot be grown in long-term cultures. Eventually, the cells stop dividing and

die. By contrast, tumour cells often form cultures that grow and divide indefinitely.

The first successful culturing of cancer cells was performed in 1951 in the laboratory of George and Margaret Gey (Johns Hopkins University, Baltimore, Maryland). Gey and Gey's cultures of normal cells died after a few weeks, but the researchers achieved success with a culture of tumour cells from a cancer patient. The cells in culture continued to grow and divide; in fact, descendants of those cells are still being cultured and used for research today. The cells were given the code name *HeLa,* from the first two letters of the patient's first and last names—Henrietta Lacks. Unfortunately, the tumour cells in Lacks's body also continued to grow, and she died within two months of her cancer diagnosis.

Other types of human cells have since been grown successfully in culture, derived from either tumour cells or normal cells that have been "immortalized" by inducing genetic changes that transformed them into tumour like cells.

---

completed successfully before the next phase can begin. Various internal mechanisms also regulate the overall number of cycles that a cell goes through. These internal controls may be subject to various external influences, such as other cells or viruses and signal molecules, including hormones, growth factors, and death signals.

CONCEPT FIX Since the emphasis of this chapter is on the behaviour of chromosomes, your attention might get focused on the events of mitosis such that you assume nothing much happens during interphase. Cells are not just "resting up" for the next round of mitosis. During interphase, an appropriate suite of genes is actively expressed to support cell growth and metabolism. It is also during interphase that DNA is replicated. ⬡

8.3c After Interphase, Mitosis Proceeds in Five Stages

If you were to watch a cell going through mitosis **(Figures 8.7, p. 174,** and **8.8, p. 176),** you would notice several dramatic changes that signal the progression

through different stages: prophase (*pro* = before), prometaphase (*meta* = between), metaphase, anaphase (*ana* = back), and telophase (*telo* = end).

Prophase. During **prophase**, the greatly extended chromosomes that were replicated during interphase begin to *condense* into compact, rodlike structures (see chromatin packing in Section 12.5). Each diploid human cell, although only about 40 to 50 µm in diameter, contains *2 m* of DNA distributed among 23 pairs of chromosomes. Condensation during prophase packs these long DNA molecules into units small enough to be divided successfully during mitosis. As they condense, the chromosomes appear as thin threads under the light microscope. The word *mitosis* (*mitos* = thread) is derived from this threadlike appearance.

While condensation is in progress, the nucleolus becomes smaller and eventually disappears in most species. The disappearance reflects a shutdown of all types of RNA synthesis, including the ribosomal RNA made in the nucleolus.

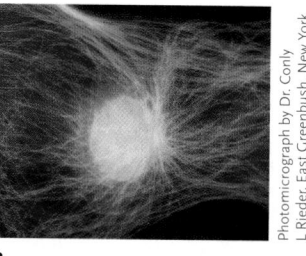

a.

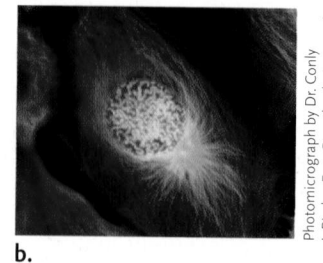

b.

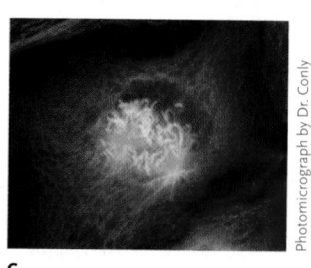

c.

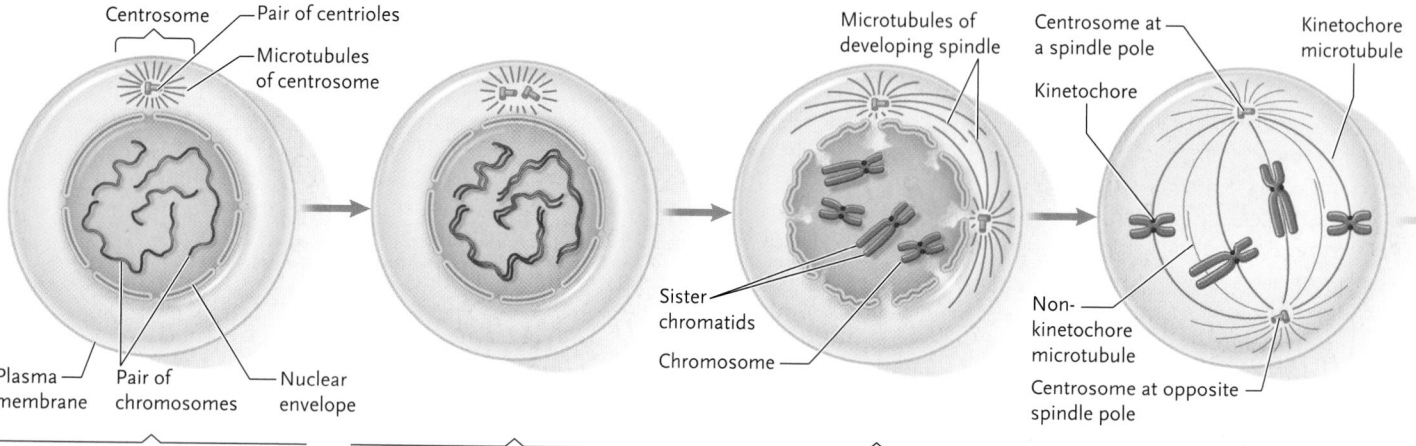

G₁ of interphase

The chromosomes are unreplicated and extend throughout the nucleus. For simplicity we show only two pairs of chromosomes. One of each pair was inherited from one parent, and the other was inherited from the other parent.

G₂ of interphase

After replication during the S phase of interphase, each chromosome is double at all points and now consists of two sister chromatids. Cohesins encircle each pair of sister chromatids along their lengths, aligning them tightly. The centrioles within the centrosome have also doubled into pairs.

Prophase

The chromosomes condense into threads that become visible under the light microscope. Each chromosome is double as a result of replication. The centrosome has divided into two parts, which are generating the spindle as they separate.

Prometaphase

The nuclear envelope has disappeared and the spindle enters the former nuclear area. Microtubules from opposite spindle poles attach to the two kinetochores of each chromosome.

Figure 8.7

The stages of mitosis. Triple-stained immunofluorescent light micrographs show mitosis in an animal cell (salamander lung). The chromosomes are blue, the spindle and cytoplasmic microtubules are yellow-green, and the intermediate filaments are red. **(a)** Interphase. Microtubules focus on the centrosome, located adjacent to the nucleus. **(b)** Prophase. Chromosomes are well condensed, the nuclear envelope is intact, and the microtubules are organized into radial arrays. **(c)** Prometaphase. The nuclear envelope has broken down to allow the chromosomes to interact with the microtubules originating from two separate centrosomes. **(d)** Metaphase. All of the replicated chromosomes are aligned on the equator of the mature mitotic spindle. **(e)** Anaphase/telophase. Chromosomes have been equally segregated and have decondensed to form two independent daughter nuclei. This cell has just begun cytokinesis. **(f)** The end result of mitosis: two genetically identical daughter cells.

In the cytoplasm, the **mitotic spindle (Figure 8.9, p. 176;** see also **Figure 8.13, p. 178)** begins to form between the two centrosomes as they start migrating toward the opposite ends of the cell to form the **spindle poles.** The spindle develops as bundles of microtubules that radiate from the spindle poles.

Prometaphase. At the end of prophase, the nuclear envelope breaks down, heralding the beginning of **prometaphase.** Bundles of spindle microtubules grow from centrosomes at the opposing spindle poles toward the centre of the cell. Some of the developing spindle enters the former nuclear area and attaches to the chromosomes.

Although replicated chromosomes are seldom visible as a double structure at this point, it is important for you to remember that each one is made up of two identical sister chromatids held together only at their **centromeres.** By this time, a complex of several proteins, a **kinetochore,** has formed on each chromatid at the centromere. Kinetochore microtubules bind to the kinetochores. These connections determine the outcome of mitosis because they attach the sister chromatids of each chromosome to microtubules leading to the opposite spindle poles (see Figure 8.9, p. 176). Microtubules that do not attach to kinetochores overlap those from the opposite spindle pole.

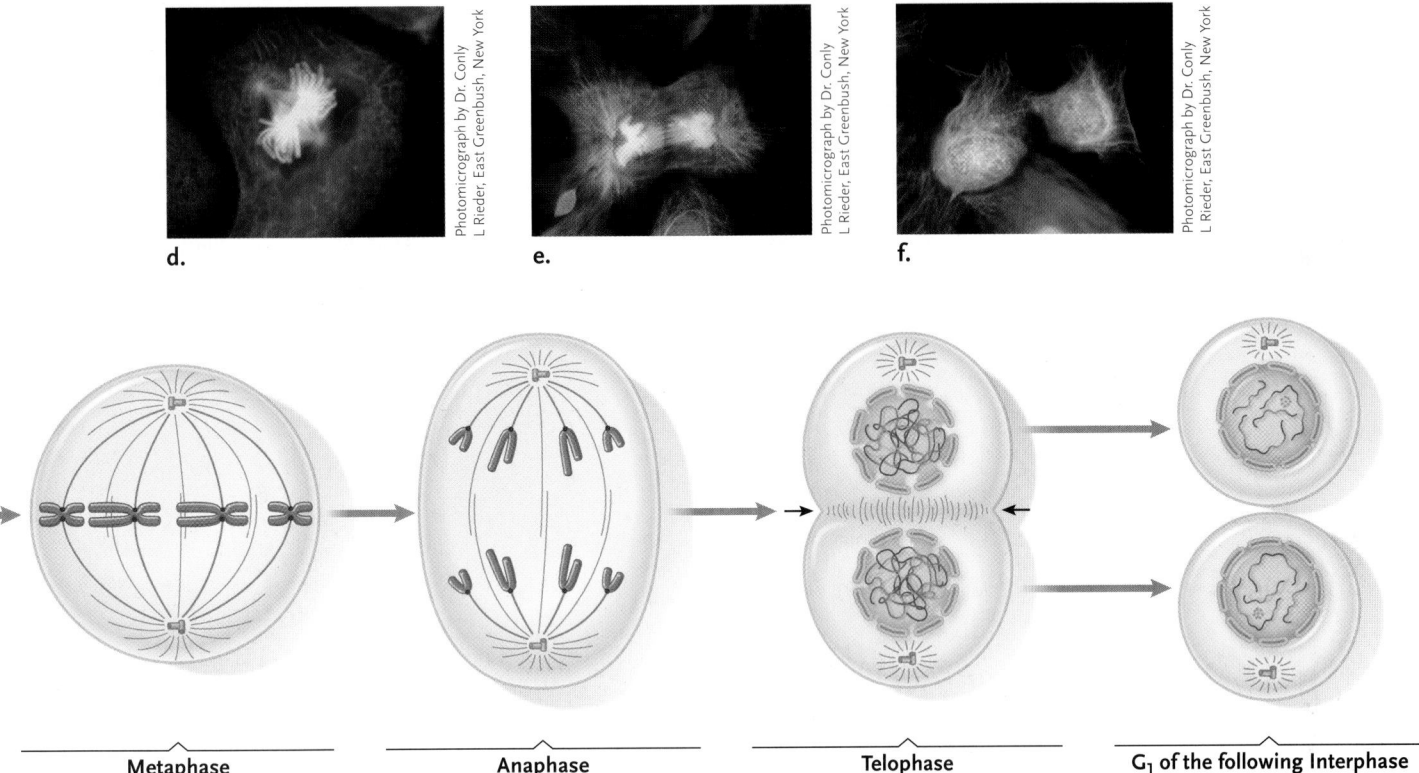

d.

Photomicrograph by Dr. Conly
L Rieder, East Greenbush, New York

e.

Photomicrograph by Dr. Conly
L Rieder, East Greenbush, New York

f.

Photomicrograph by Dr. Conly
L Rieder, East Greenbush, New York

Metaphase

The chromosomes become aligned at the spindle midpoint. Each sister chromatid pair is held in position by opposing forces: the kinetochore microtubules pulling to the poles and the cohesins binding the sister chromatids together.

Anaphase

Separase cleaves the cohesin ring holding sister chromatids together. The spindle separates the two sister chromatids of each chromosome and moves them to opposite spindle poles.

Telophase

The chromosomes unfold and return to the interphase state, and new nuclear envelopes form around the daughter nuclei. The cytoplasm is beginning to divide by furrowing at the points marked by arrows.

G₁ of the following Interphase

The two daughter cells are genetic duplicates of the parental cell that entered mitotic division.

Metaphase. During **metaphase**, the spindle reaches its final form and the spindle microtubules move the chromosomes into alignment at the spindle midpoint, also called the metaphase plate. The chromosomes complete their condensation in this stage and assume their characteristic shape as determined by the location of the centromere and the length and thickness of the chromatid arms.

Only when the chromosomes are all assembled at the spindle midpoint, with the two sister chromatids of each one attached to microtubules leading to opposite spindle poles, can metaphase give way to actual separation of chromatids.

CONCEPT FIX Although you probably think of chromosomes as X shapes, it is important to realize that few chromosomes ever actually look like this. Only chromosomes with their centromere near the middle could ever appear as an X. Only during (pro)metaphase are chromosomes condensed enough to take on any shape at all. ◼

The complete collection of metaphase chromosomes, arranged according to size and shape, forms the **karyotype** of a given species. In many cases, the karyotype is so distinctive that a species can be identified from this characteristic alone. **Figure 8.10, p. 176,** shows a human karyotype.

Anaphase. During **anaphase**, sister chromatids separate and move to opposite spindle poles. The first signs of chromosome movement can be seen at the centromeres as the kinetochores are the first sections to move toward opposite poles. The movement continues until the separated chromatids, now called daughter chromosomes, have reached the two poles. At this point, chromosome segregation has been completed.

Telophase. During **telophase**, the spindle disassembles and the chromosomes at each spindle pole decondense and return to the extended state typical of interphase. As decondensation proceeds, the nucleolus reappears, RNA transcription resumes, and a new nuclear envelope forms around the chromosomes at each pole, producing the two daughter nuclei. At this point, nuclear division is complete, and the cell has two nuclei.

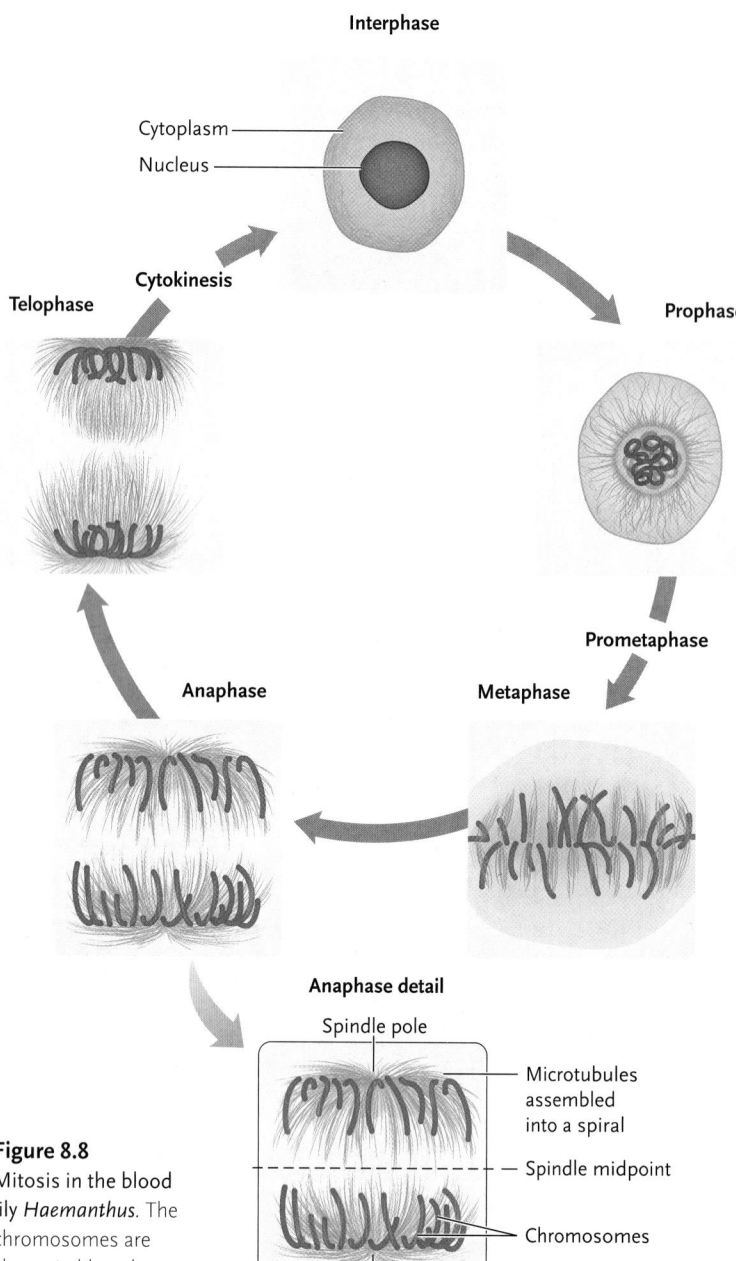

Interphase

Cytoplasm
Nucleus

Cytokinesis

Telophase

Prophase

Prometaphase

Anaphase

Metaphase

Anaphase detail

Spindle pole

Microtubules assembled into a spiral

Spindle midpoint

Chromosomes

Spindle pole

Figure 8.8
Mitosis in the blood lily *Haemanthus*. The chromosomes are shown in blue; the spindle microtubules are shown in pink.

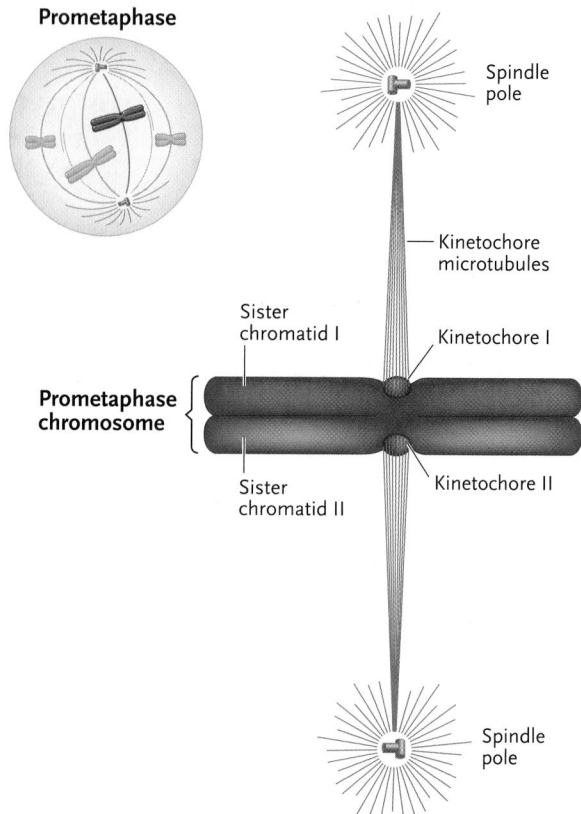

Prometaphase

Spindle pole

Kinetochore microtubules

Sister chromatid I

Kinetochore I

Prometaphase chromosome

Sister chromatid II

Kinetochore II

Spindle pole

Figure 8.9
Spindle connections made by chromosomes at mitotic metaphase in typical animal cells. The two kinetochores of the chromosome connect to opposite spindle poles, ensuring that the chromatids are separated and moved to opposite spindle poles during anaphase.

8.3d Cytokinesis Completes Cell Division by Dividing the Cytoplasm between Daughter Cells

Cytokinesis, the division of the cytoplasm, usually follows the nuclear division stage of mitosis and produces two daughter cells, each containing one of the daughter nuclei. In most cells, cytokinesis begins during telophase or even late anaphase. By the time cytokinesis is completed, the daughter nuclei have progressed to the interphase stage and entered the G_1 phase of the next cell cycle.

Cytokinesis proceeds by different pathways in the different kingdoms of eukaryotic organisms. In animals, protists, and many fungi, a groove, the **furrow,** girdles the cell and gradually deepens until it cuts the cytoplasm into two parts. In plants, a new cell wall, called the **cell plate,** forms between the daughter nuclei and grows laterally until it divides the cytoplasm. In both cases, the plane of cytoplasmic division is determined by the layer of microtubules that persist at the former spindle midpoint.

Furrowing. In furrowing, the layer of microtubules that remains at the former spindle midpoint expands laterally until it stretches entirely across the dividing

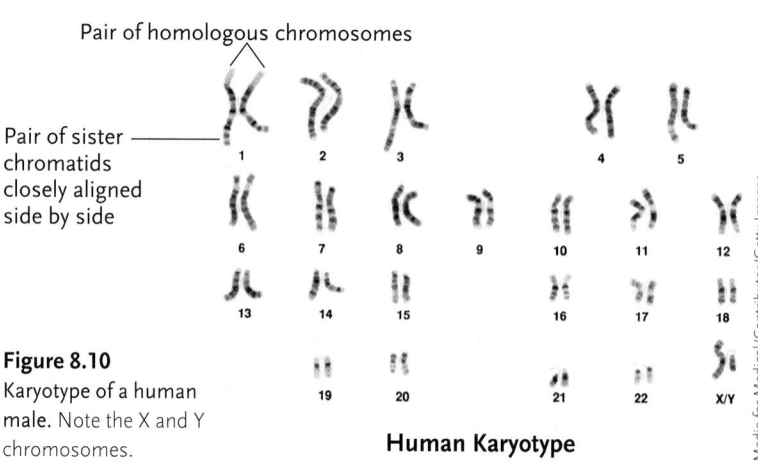

Pair of homologous chromosomes

Pair of sister chromatids closely aligned side by side

1   2   3   4   5
6   7   8   9   10   11   12
13   14   15   16   17   18
19   20   21   22   X/Y

Figure 8.10
Karyotype of a human male. Note the X and Y chromosomes.

Human Karyotype

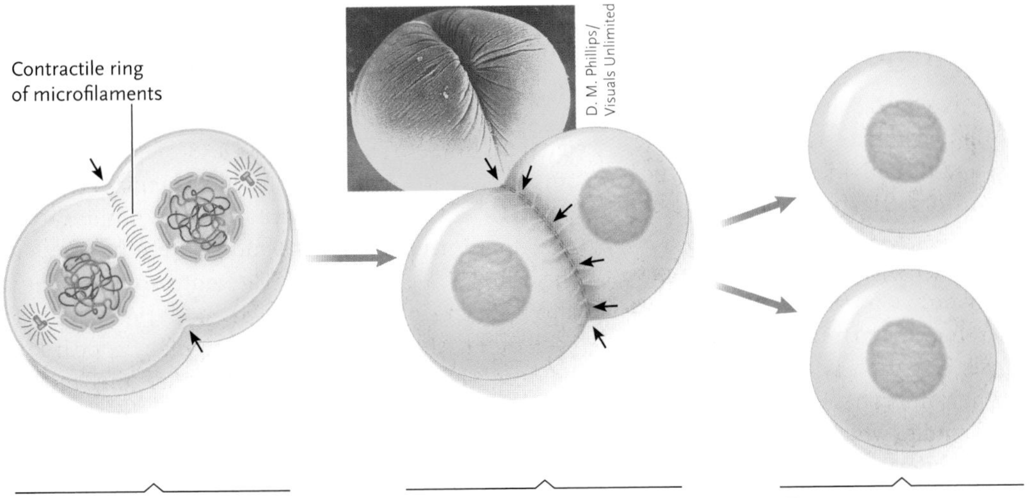

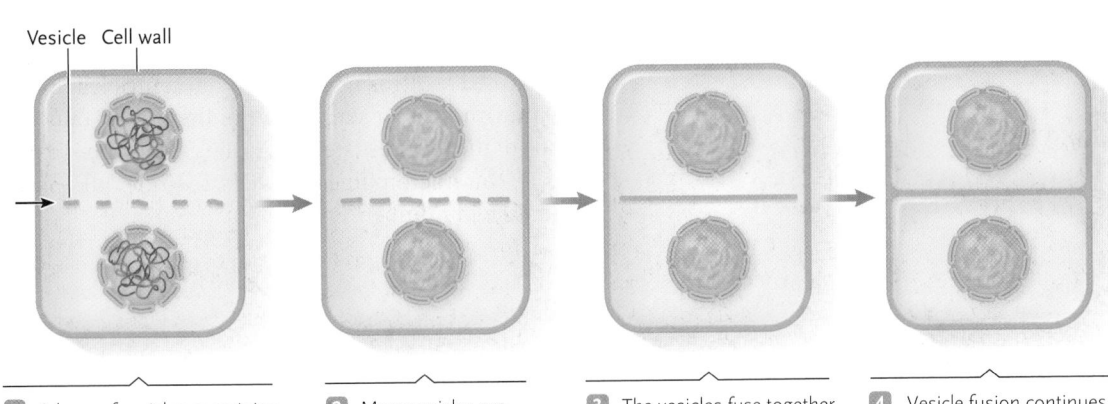

Contractile ring of microfilaments

D. M. Phillips/ Visuals Unlimited

Figure 8.11
Cytokinesis by furrowing. The micrograph shows a furrow developing in the first division of a fertilized egg cell.

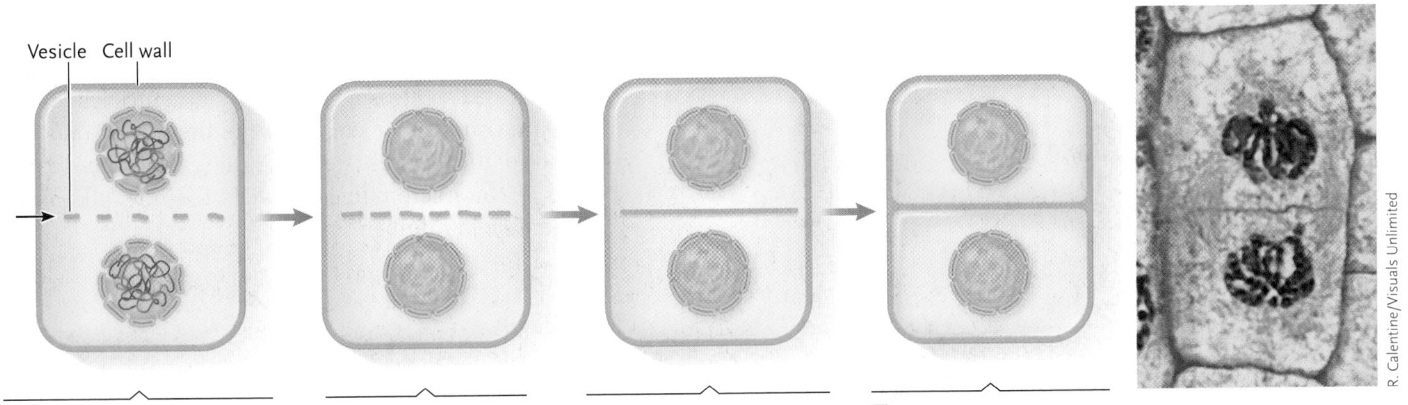

1 The furrow begins as an indentation running completely around the cell in the plane of the former spindle midpoint.

2 The furrow deepens by contraction of the micro-filaments, like a drawstring tightening around the cell.

3 Furrowing continues until the daughter nuclei are enclosed in separate cells.

Vesicle Cell wall

1 A layer of vesicles containing cell wall material collects in the plane of the former spindle midpoint (arrow).

2 More vesicles are added to the layer until it extends across the cell.

3 The vesicles fuse together, dumping their contents into a gradually expanding wall between the daughter cells.

4 Vesicle fusion continues until the daughter cells are separated by a continuous new wall, the cell plate.

R. Calentine/Visuals Unlimited

Figure 8.12
Cytokinesis by cell plate formation in plant cells.

cell **(Figure 8.11)**. As the layer develops, a band of micro-filaments forms just inside the plasma membrane, forming a belt that follows the inside boundary of the cell in the plane of the microtubule layer (microfilaments are discussed in Section 2.3f). Powered by motor proteins, the microfilaments slide together, tightening the band and constricting the cell. The constriction forms a groove—the furrow—in the plasma membrane. The furrow gradually deepens, much like the tightening of a drawstring, until the daughter cells are completely separated. The cytoplasmic division isolates the daughter nuclei in the two cells and, at the same time, distributes the organelles and other structures (which have also doubled) approximately equally.

Cell Plate Formation. In cell plate formation, the layer of microtubules that persists at the former spindle midpoint serves as an organizing site for vesicles produced by the endoplasmic reticulum (ER) and Golgi complex **(Figure 8.12)**. As the vesicles collect, the layer

expands until it spreads entirely across the dividing cell. During this expansion, the vesicles fuse together and their contents assemble into a new cell wall—the cell plate—stretching completely across the former spindle midpoint. The junction separates the cytoplasm and its organelles into two parts and isolates the daughter nuclei in separate cells. The plasma membranes that line the two surfaces of the cell plate are derived from the vesicle membranes.

STUDY BREAK

1. During which stage(s) of the cell cycle is a chromosome composed of two chromatids?
2. What are the conditions under which a chromosome could appear as an X shape under the microscope?
3. How does cytokinesis differ in plant and animal cells?

8.4 Formation and Action of the Mitotic Spindle

The mitotic spindle is central to both mitosis and cytokinesis. The spindle is made up of microtubules and their proteins, and its activities depend on their changing patterns of organization during the cell cycle.

Microtubules form a major part of the interphase cytoskeleton of eukaryotic cells. (Section 2.3f outlines the patterns of microtubule organization in the cytoskeleton.) As mitosis approaches, the microtubules disassemble from their interphase arrangement and reorganize into the spindle, which grows until it fills almost the entire cell. This reorganization follows one of two pathways in different organisms, depending on the presence or absence of a *centrosome* during interphase. However, once organized, the basic function of the spindle is the same, regardless of whether a centrosome is present.

8.4a Animals and Plants Form Spindles in Different Ways

Animal cells and many protists have a **centrosome**, a site near the nucleus from which microtubules radiate outward in all directions (**Figure 8.13,** step 1). The centrosome is the main **microtubule organizing centre (MTOC)** of the cell, anchoring the microtubule cytoskeleton during interphase and positioning many of the cytoplasmic organelles. The centrosome contains a pair of **centrioles**, usually arranged at right angles to each other. Although centrioles originally appeared to be important in the construction of the mitotic spindle, it has now been shown that they can be removed with no ill effect. The primary function of centrioles is actually to generate the microtubules needed for flagella or cilia, the whiplike extensions that provide cell motility.

When DNA replicates during the S phase of the cell cycle, the centrioles within the centrosome also duplicate, producing two pairs of centrioles (see Figure 8.13, step 2). As prophase begins in the M phase, the centrosome separates into two parts (step 3). The duplicated centrosomes, with the centrioles inside them, continue to separate until they reach opposite ends of the nucleus (step 4). As the centrosomes move apart, the microtubules between them lengthen and increase in number.

Prophase

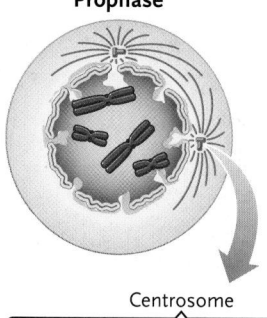

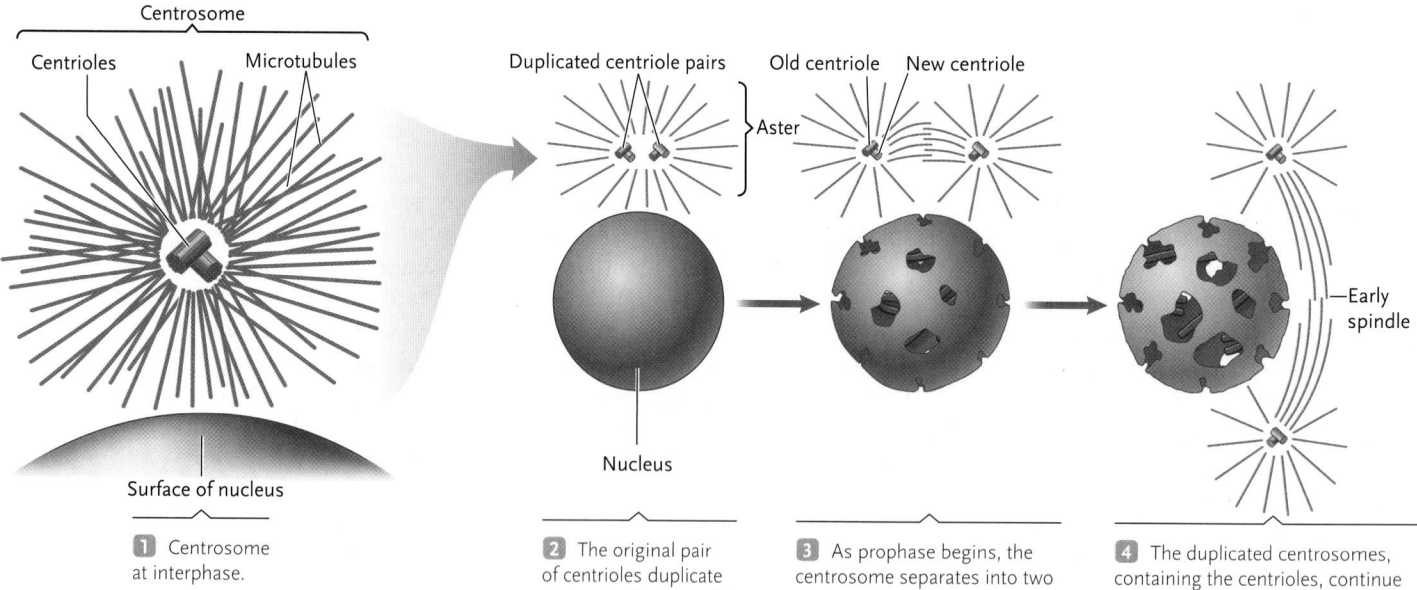

1 Centrosome at interphase.

2 The original pair of centrioles duplicate during the S phase of the cell cycle, producing two pairs of centrioles.

3 As prophase begins, the centrosome separates into two parts, each containing one "old" and one "new" centriole—one centriole of the original pair and its copy.

4 The duplicated centrosomes, containing the centrioles, continue to separate until they reach opposite sides of the nucleus. The microtubules between them lengthen and increase in number. By late prophase, the early spindle is complete, consisting of the separated centrosomes and a large mass of microtubules between them.

Figure 8.13
The centrosome and its role in spindle formation.

By late prophase, when the centrosomes are fully separated, the microtubules that extend between them form a large mass around one side of the nucleus called the early spindle. When the nuclear envelope subsequently breaks down at the end of prophase, the spindle moves into the region formerly occupied by the nucleus and continues growing until it fills the cytoplasm. The microtubules that extend from the centrosomes also grow in length and extent, producing radiating arrays that appear starlike under the light microscope. Initially named by early microscopists, **asters** are the centrosomes at the spindle tips, which form the poles of the spindle. By separating the duplicated centrioles, the spindle ensures that, when the cytoplasm divides during cytokinesis, the daughter cells each receive a pair of centrioles.

No centrosome or centrioles are present in angiosperms (flowering plants) or in most gymnosperms, such as conifers. Instead, the spindle forms from microtubules that assemble in all directions from multiple MTOCs surrounding the entire nucleus (see prophase in Figure 8.8). When the nuclear envelope breaks down at the end of prophase, the spindle moves into the former nuclear region, as in animals.

8.4b Mitotic Spindles May Move Chromosomes by a Combination of Two Mechanisms

When fully formed at metaphase, the spindle may contain from hundreds to many thousands of microtubules, depending on the species **(Figure 8.14)**. In almost all eukaryotes, these microtubules are divided into two groups. Some, called kinetochore microtubules, connect the chromosomes to the spindle poles **(Figure 8.15a)**. Others, called nonkinetochore microtubules, extend between the spindle poles without connecting to chromosomes; at the spindle midpoint, the microtubules from one pole overlap with the microtubules from the opposite pole (Figure 8.15b). The separation of the chromosomes at anaphase appears to result from a combination of separate but coordinated movements produced by the two types of microtubules.

The exact mechanism by which chromosomes move is still uncertain; at one time, it was believed that microtubules pulled the chromosomes toward the poles of dividing cells. However, subsequent data suggest that chromosomes "walk" themselves to the poles along stationary microtubules, using motor proteins in their kinetochores **(Figure 8.16, p. 180)**. The tubulin subunits of the kinetochore microtubules disassemble as the kinetochores pass along them; thus, the microtubules become shorter as the movement progresses (see Figure 8.15a). The movement is similar to pulling yourself, hand over hand, up a rope as it falls apart behind you.

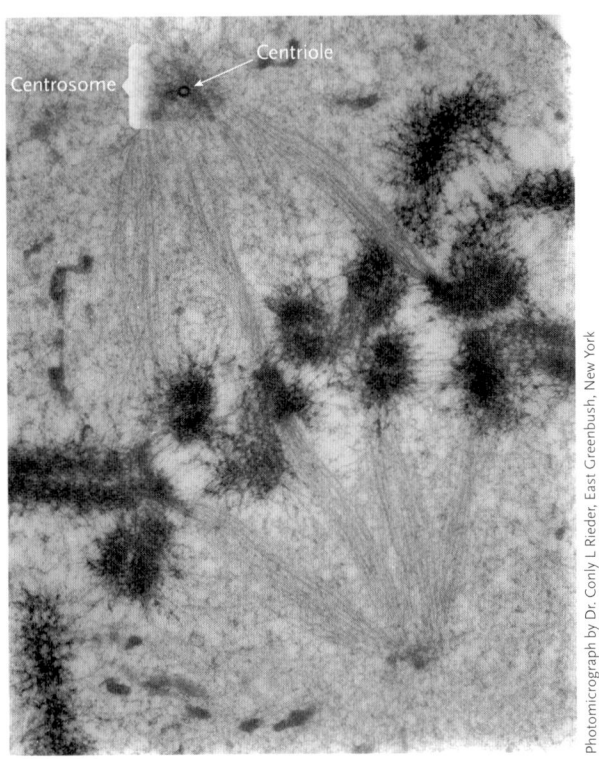

Figure 8.14
A fully developed spindle in a mammalian cell. Only microtubules connected to chromosomes have been caught in the plane of this section. One of the centrioles is visible in cross-section in the centrosome at the top of the micrograph (arrow). Original magnification × 14 000.

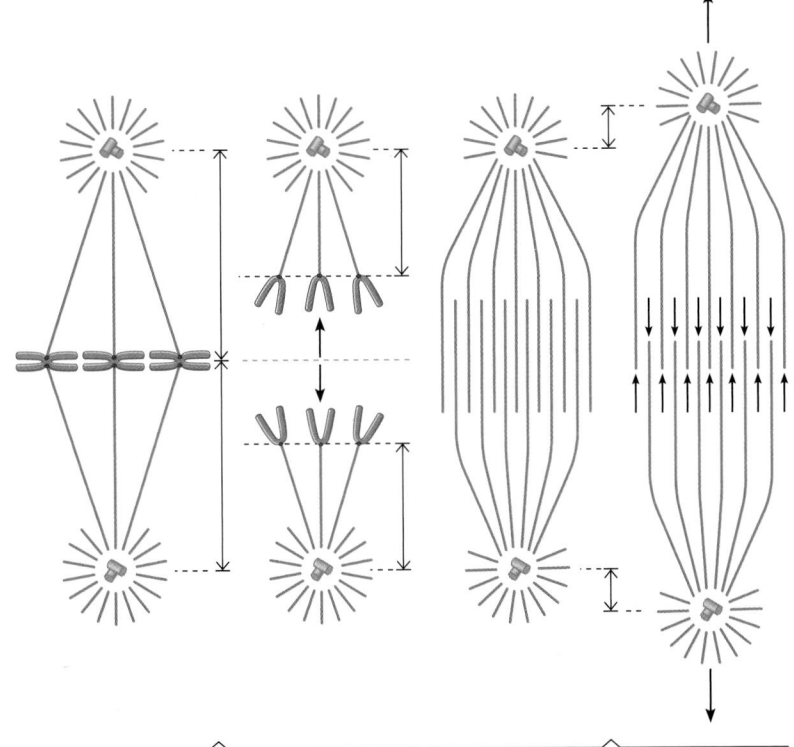

a. The kinetochore microtubules connected to the kinetochores of the chromosomes become shorter, lessening the distance from the chromosomes to the poles.

b. Sliding of the nonkinetochore microtubules in the zone of overlap at the spindle midpoint pushes poles farther apart and increases the total length of the spindle.

Figure 8.15
The two microtubule-based movements of the anaphase spindle.

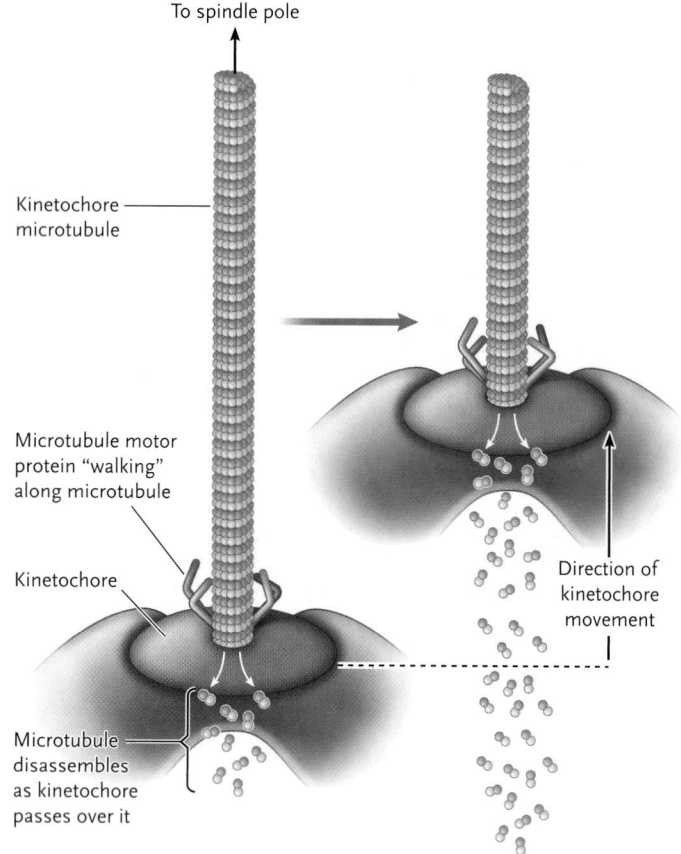

Figure 8.16
Microtubule motor proteins "walking" the kinetochore of a chromosome along a microtubule.

Chromosomes can also move toward the poles by a mechanism in which motor proteins at the spindle poles pull kinetochore microtubules poleward, disassembling those microtubules into tubulin subunits as that occurs. Both walking and pulling mechanisms are used in mitosis, although the relative contributions of the two mechanisms to chromosome movement varies among species and cell types. (The cell type in the experiment of Figure 8.16 predominantly used walking.) In nonkinetochore microtubule-based movement, the entire spindle is lengthened, elongating the cell in Metaphase and Anaphase. (see Figure 8.15b). The pushing movement is presumably produced by microtubules sliding over one another in the zone of overlap, powered by proteins acting as microtubule motors. In many species, the nonkinetochore microtubules also push the poles apart by growing in length as they slide.

STUDY BREAK

1. What is the role of the centrosome?
2. What is the role of the kinetochore?

8.5 Cell Cycle Regulation

In this section, we discuss experimental evidence for (and the operation of) regulatory mechanisms that control the mitotic cell cycle.

8.5a Cell Fusion Experiments and Studies of Yeast Mutants Identified Molecules That Control the Cell Cycle

The first insights into how the cell cycle is regulated came from experiments by Robert T. Johnson and Potu N. Rao at the University of Colorado Medical Center, Denver, published in 1970. They fused human

Evidence supporting kinetochore-based movement comes from experiments in which researchers tagged kinetochore microtubules with a microscopic beam of ultraviolet light, producing bleached sites that could be seen in the light microscope **(Figure 8.17)**. As the chromosomes moved to the spindle poles, the bleached sites stayed in the same place. This result showed that the kinetochore microtubules do not move much with respect to the poles during the anaphase movement.

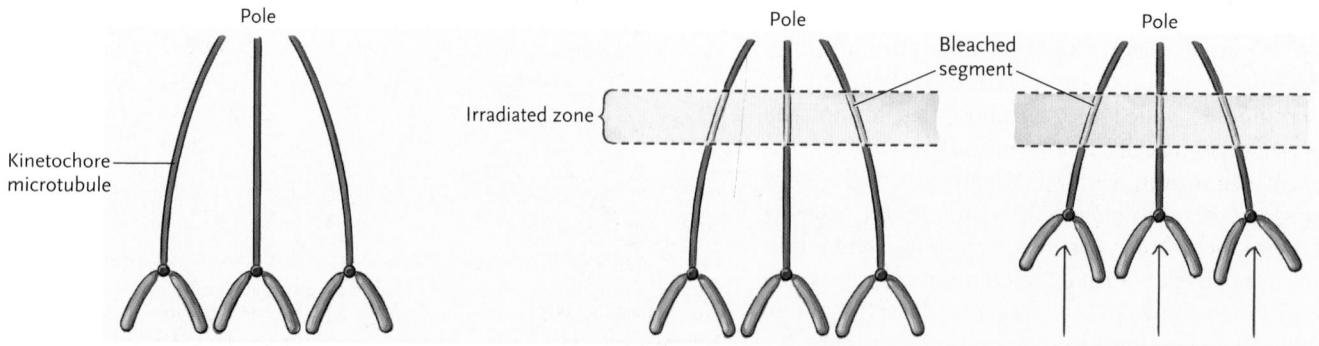

Experimental Research Figure 8.17 Experiment demonstrating that kinetochore microtubules remain stationary as chromosomes move during anaphase. A bleached region of spindle remained at the same distance from the pole as the chromosomes moved toward the pole. These observations support the hypothesis that chromosomes move by sliding over or along microtubules.

HeLa cells (a type of cancer cell that can be grown in cell culture) that were in different stages of the cell cycle and determined whether one nucleus could influence the other. Their results suggested that specific molecules in the cytoplasm cause the progression of cells from G_1 to S, and from G_2 into M.

Some key research using baker's yeast, *S. cerevisiae*, helped to identify these cell cycle control molecules and contributed to our general understanding of how the cell cycle is regulated. (*The Purple Pages* describe yeast and its role in research in more detail.) In particular, Leland Hartwell investigated yeast mutants that become stuck at some point in the cell cycle, but only when they are cultured at a high temperature. By growing the mutant cells initially at the standard temperature, and then shifting the cells to the higher temperature, Hartwell was able to use time-lapse photomicroscopy to see if and when growth and division were affected. In this way he isolated many cell division cycle, or *cdc*, mutants. By examining the mutants, he could identify the stage in the cell cycle where each mutant type was blocked by noting whether nuclei had divided, chromosomes had condensed, the mitotic spindle had formed, cytokinesis had occurred, and so on. Using this approach, Hartwell identified many genes that code for proteins involved in yeast's cell cycle and hypothesized where in the cycle these proteins operated. As might be expected, some of the proteins were involved in DNA replication, but a number of others were shown to function in cell cycle regulation. Hartwell received a Nobel Prize in 2001 for his discovery.

Paul Nurse of the Imperial Cancer Research Fund, London, United Kingdom, carried out similar research with the fission yeast *Schizosaccharomyces pombe*, a species that divides by fission rather than budding. He identified a gene called *cdc2* that encodes a protein needed for the cell to progress from G_2 to M. Nurse also made the breakthrough discovery that all eukaryotic cells studied have counterparts of the yeast *cdc2* gene, implying that this gene originated early during eukaryotic evolution and has played an essential role in cell cycle regulation in all eukaryotes since that time. The protein product of *cdc2* is a protein kinase, an enzyme that catalyzes the phosphorylation of a target protein. (Recall from Section 5.7c that phosphorylation of proteins by protein kinases can activate or inactivate proteins.) That discovery was pivotal in determining how cell cycle regulation occurs. Paul Nurse received a Nobel Prize in 2001 for his discovery.

8.5b The Cell Cycle Can Be Arrested at Specific Checkpoints

A cell has internal controls that monitor its progression through the cell cycle through the action of a particular set of control proteins called *cyclins*. Three key **checkpoints** prevent critical phases from beginning until the previous phases are completed correctly (**Figure 8.18, p. 182**):

1. The G_1/S *checkpoint* is the main point in the cell cycle at which the mechanisms governing the cell cycle determine whether the cell will proceed through the rest of the cell cycle and divide. Once it passes this checkpoint, the cell is committed to continue the cell cycle through to cell division in M. The cell cycle arrests (the cell stops proceeding through the cell cycle) at the G_1/S checkpoint if, for example, the DNA is damaged by radiation or chemicals. The G_1/S checkpoint is also the primary point at which cells "read" extracellular signals for cell growth and division. Therefore, if a hormone or growth factor required for stimulating cell growth is absent, the cells may arrest at this checkpoint. (Extracellular signals and their effects on the cell cycle are discussed in more detail later.)

2. The G_2/M *checkpoint* is at the junction between the G_2 and M phases. Passage through this checkpoint commits a cell to mitosis. Cells are arrested at the G_2/M checkpoint if DNA was not replicated fully in S, or if the DNA has been damaged by radiation or chemicals. Complete DNA replication is essential for producing genetically identical daughter cells, highlighting the importance of this checkpoint.

3. The *mitotic spindle checkpoint* is within the M phase before metaphase. This checkpoint assesses whether chromosomes are attached properly to the mitotic spindle so that they will align correctly at the metaphase plate. The checkpoint is essential for production of genetically identical daughter cells, which depends on separation of daughter chromosomes in anaphase, which, in turn, depends on the correct alignment of the chromosomes on the spindle in metaphase. Once the cell begins anaphase, it is irreversibly committed to completing M, underlining the importance of the mitotic spindle checkpoint.

The control systems that operate at the checkpoints are signals to stop; basically, they are brakes. This becomes evident when a checkpoint is inactivated by mutation or chemical treatment, allowing the cell cycle to proceed, even if DNA is damaged, DNA replication is incomplete, or the spindle did not assemble completely.

8.5c Cyclins and Cyclin-Dependent Kinases Are the Internal Controls That Directly Regulate Cell Division

The direct regulation of the cell cycle involves an internal control system consisting of proteins called **cyclins** and enzymes called **cyclin-dependent kinases (Cdks)** (see Figure 8.18, p. 182). A Cdk is a *protein kinase*, which phosphorylates and thereby regulates the

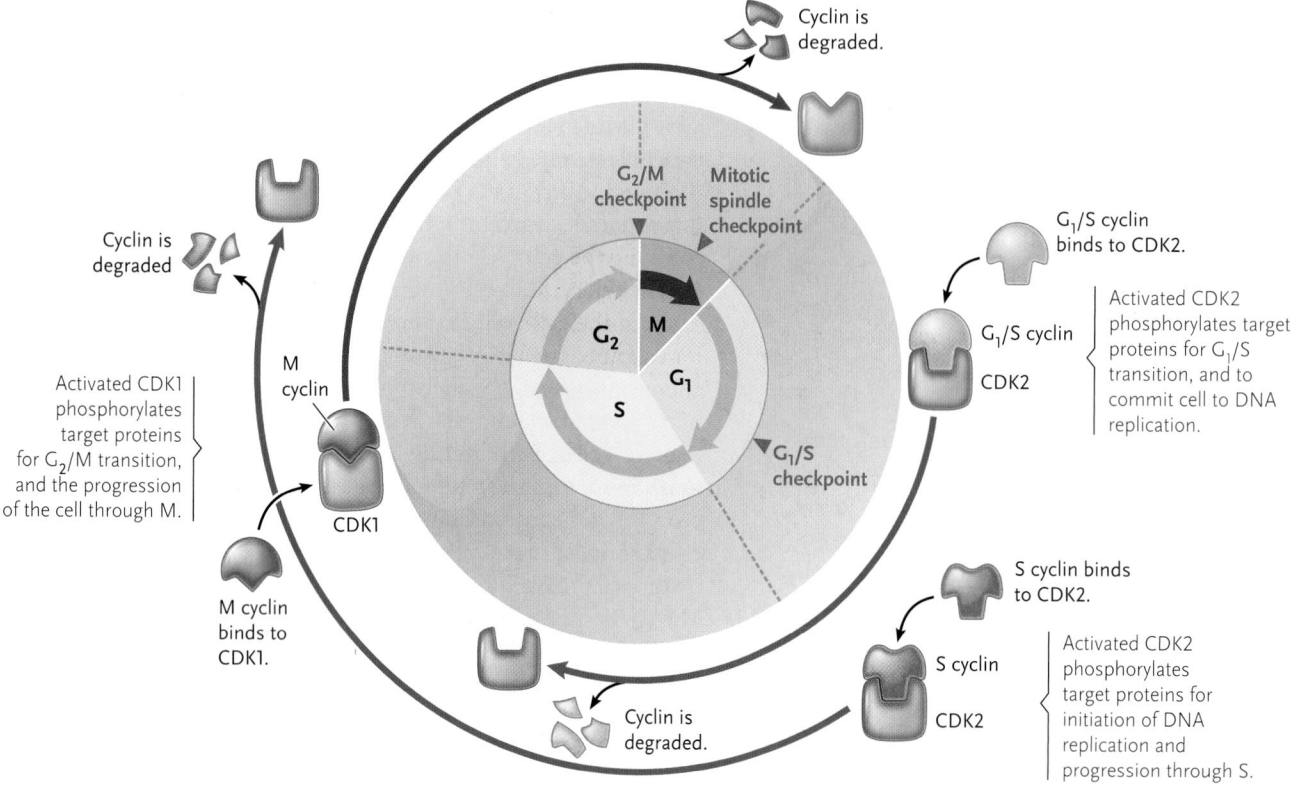

Figure 8.18

Regulation of the mitotic cell cycle by internal controls. Three key checkpoints for the G_1/S transition, for the G_2/M transition, and for the attachment of chromosomes to the mitotic spindle monitor cell cycle events to prevent crucial phases of the cell cycle from starting until previous phases are completed correctly. Complexes of cyclins and cyclin-dependent kinases (Cdks) regulate the progression of the cell through the cell cycle. The three cyclin–Cdks present in all eukaryotes are shown. The Cdks are present throughout the cell cycle, but they are active only when complexed with a cyclin (shown by the broad arrows in the figure). Each cyclin is synthesized and degraded in a regulated way so that it is present only for a particular phase of the cell cycle. During that phase, the Cdk to which it is bound phosphorylates and, thereby, regulates the activity of target proteins in the cell that are involved in initiating or regulating key events of the cell cycle.

activity of target proteins. Cdk enzymes are "cyclin dependent" because they are active *only* when bound to a cyclin molecule. Cyclins are named because their concentrations change as the cell cycle progresses. R. Timothy Hunt of the Imperial Cancer Research Fund received a Nobel Prize in 2001 for discovering cyclins. The basic control of the cell cycle by Cdks and cyclins is the same in all eukaryotes, but there are differences in the number and types of the molecules. We will focus on cell cycle regulation in vertebrates to explain how these proteins work.

The concentrations of the various Cdks remain constant throughout the cell cycle, while the concentrations of cyclins change as they are synthesized and degraded at specific stages of the cell cycle. Thus, a specific Cdk becomes active when the cell synthesizes the cyclin that binds to it and remains active until the cyclin is degraded. Each active Cdk phosphorylates particular target proteins, which play roles in initiating or regulating key events of the cell cycle. The phosphorylation regulates the activities of these proteins and keeps the cycle operating in an orderly way. These key events are DNA replication, mitosis, and cytokinesis. A succession of cyclin–Cdk complexes, each of which has specific regulatory effects, ensures that these

stages follow in sequence somewhat like a clock passing through the sequence of hours. Regulation of the activity of cyclin–Cdk complexes is integrated with the regulatory events at the cell cycle checkpoints to ensure that daughter cells do not inherit damaged DNA or abnormal numbers of chromosomes.

Three classes of cyclins, each named for the stage of the cell cycle at which they bind and activate Cdks, operate in all eukaryotes (see Figure 8.18):

1. G_1/S cyclin binds to Cdk2 near the end of G_1, forming a complex required for the cell to make the transition from G_1 to S and to commit the cell to DNA replication.
2. S cyclin binds to Cdk2 in the S phase, forming a complex required for the initiation of DNA replication and the progression of the cell through S.
3. M cyclin binds to Cdk1 in G_2, forming a complex required for the transition from G_2 and M and the progression of the cell through mitosis.

The M cyclin–Cdk1 complex is also called **M phase-promoting factor (MPF)**. In addition to initiating mitosis, the M cyclin–Cdk1 complex (MPF) also orchestrates some of its key events. When all chromosomes are correctly attached to the mitotic spindle near the end of metaphase, the M cyclin–Cdk1 complex activates another enzyme complex, the **anaphase-promoting complex (APC)**. Activated APC degrades an inhibitor of anaphase, leading to the separation of sister chromatids and the onset of daughter chromosome separation in anaphase. Later in anaphase, APC directs the degradation of the M cyclin, causing Cdk1 to lose its activity. The loss of Cdk1 activity then allows the separated chromosomes to become extended again, the nuclear envelope to reform around the two clusters of daughter chromosomes in telophase, and the cytoplasm then to divide in cytokinesis.

8.5d External Controls Coordinate the Mitotic Cell Cycle of Individual Cells with the Overall Activities of the Organism

The internal controls that regulate the cell cycle are modified by signal molecules that originate from outside the dividing cells. In animals, these signal molecules include the peptide hormones and similar proteins called growth or death factors.

Many of these external factors bind to receptors at the cell surface, which respond by triggering reactions inside the cell. These reactions often include steps that add phosphate groups to the cyclin–Cdk complexes, thereby affecting their function. The overall effect is to speed, slow, or stop the progress of cell division, depending on the particular hormone or factor and the internal pathway that is stimulated. Some growth factors are even able to break the arrest of cells shunted into the G_0 stage and return them to active division. (Hormones, growth factors, and other signal molecules are part of the cell communication system, as discussed in Chapter 5.)

Cell surface receptors in animal cells also recognize contact with other cells or with molecules of the extracellular matrix (ECM). The contact triggers internal reaction pathways that inhibit division by arresting the cell cycle, usually in the G_1 phase. The response, called **contact inhibition**, stabilizes cell growth in fully developed organs and tissues. As long as the cells of most tissues are in contact with one another or with the ECM, they are shunted into the G_0 phase and prevented from dividing. If the contacts are broken, the freed cells often enter rounds of division.

Contact inhibition is easily observed in cultured mammalian cells grown on a glass or plastic surface. In such cultures, division proceeds until all the cells are in contact with their neighbours in a continuous, unbroken, single layer. At this point, division stops. If a researcher then scrapes some of the cells from the surface, cells at the edges of the "wound" are released from inhibition and divide until they form a continuous layer and all the cells are again in contact with their neighbours.

8.5e Stem Cells Exhibit Asymmetric Cell Division

The mitotic mechanism described so far in this chapter is symmetric in that it produces two daughter cells of roughly the same size, shape, and function whose genomes are essentially identical. However, asymmetric cell division, producing daughter cells that are decidedly different in some characteristic or other, is quite common in multicellular organisms, particularly during growth and development. One of the most striking examples of asymmetric cell division is found in the populations of stem cells present in plant meristem and various animal tissues **(Figure 8.19)**.

Asymmetric cell division of stem cells provides new cells for growth and maintenance while, at the same time, maintaining the pool of stem cells. Stem cell division is asymmetric in that the two daughter cells each have a different fate. One daughter cell remains a stem cell and therefore contributes to a pool of self-renewing, relatively undifferentiated cells that are capable of multiple cell divisions. The other daughter cell, called a progenitor cell, divides a limited

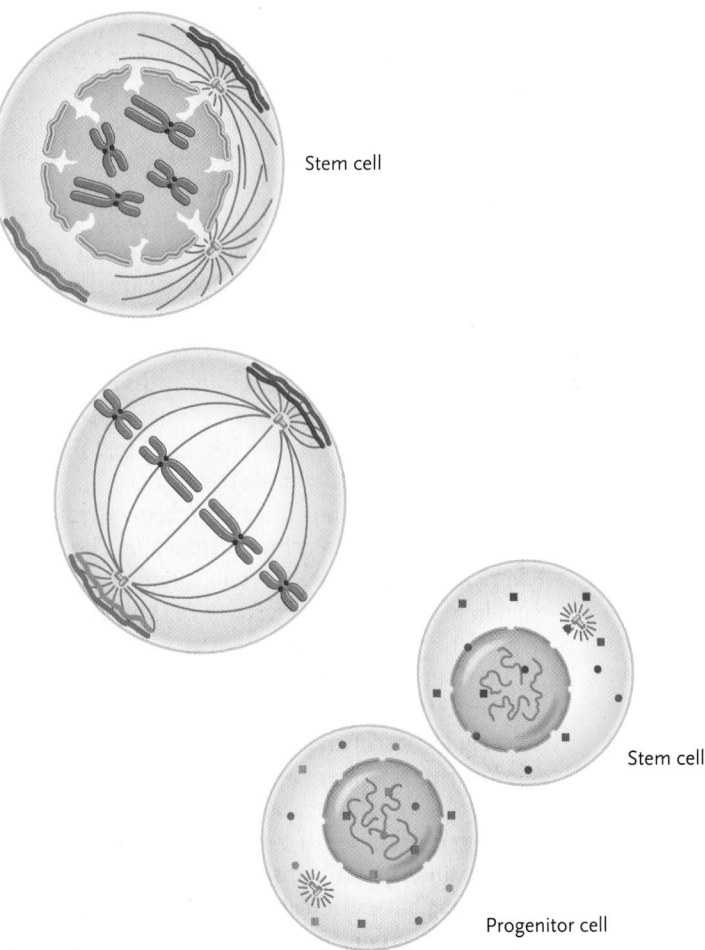

Stem cell

Stem cell

Progenitor cell

Figure 18.19

Asymmetric cell division of stem cells. The localization of different gene regulatory proteins and mRNA at the two poles during early mitosis anchors the mitotic spindle and ensures that the two daughter cells inherit a different array of regulatory elements. One cell remains a stem cell; the other becomes a differentiating progenitor cell.

number of times while undergoing subsequent differentiation into a specialized cell type needed for growth.

This asymmetry in cell fate arises from the ability of stem cells to specify the location of certain cytoplasmic components during cell division. Although mitosis divides the DNA and distributes cytoplasmic organelles relatively equally to each daughter cell, specific regulatory proteins and mRNAs (e.g., cyclins and transcription factors) are localized at only one pole or the other in the dividing cell. Therefore, the two resulting daughter cells each receive a different array of regulatory proteins that will, in turn, influence gene expression toward different cell fates.

Regulation of asymmetric division ensures the important balance between maintaining the stem cell pool and producing specialized somatic tissue. The local area where stem cells are dividing is called a niche. Signalling proteins produced by cells surrounding the niche maintain external regulation of stem cell division. Loss of this regulation is associated with cancer in brain and other tissues.

8.5f Most Cells in a Multicellular Body Cannot Divide Indefinitely

In 1961, Leonard Hayflick and Paul Moorhead reported that normal human skin cells eventually stopped dividing when grown in artificial culture. This loss of proliferative ability over time is called **cellular senescence**, and scientists have been searching for the *Hayflick factors* that are responsible for it. We consider two candidates: DNA damage and telomere shortening.

The progressive accumulation of damage to a cell's DNA sequence, or its chromosome structure, or even the genes coding for the enzyme machinery needed to repair such damage, is perhaps the most intuitive Hayflick factor. One would expect "older" cells to have diminished function if they have suffered mutations in genes controlling critical activities.

Telomeres are repetitive DNA sequences that are added to the ends of chromosomes by the enzyme telomerase. Since DNA replication machinery is unable to replicate the entire ends of linear chromosomes, telomere sequence is lost at each round of replication (see Figure 12.18, Chapter 12). Once telomeres diminish to a certain minimum length, cells stop dividing (senesce) and may die.

You might wonder why we do not just take a pill to stimulate our telomerase, rejuvenate our cells, and extend our life span. It turns out that cellular senescence is an important antitumour mechanism. Some researchers have stimulated the telomerase of cultured cells: they become "immortal" and divide out of control. Mice that have been engineered to lack telomerase, and therefore suffer faster senescence, are significantly *resistant* to cancer. It seems that by the time cells are short on telomeres, many of them are also a long way toward cancerous growth, as described below.

8.5g Cell Cycle Controls Are Lost in Cancer

Cancer occurs when cells lose the normal controls that determine when and how often they will divide. Cancer cells divide continuously and uncontrollably, producing a rapidly growing mass called a tumour **(Figure 8.20)**. Cancer cells also typically lose their adhesions to other cells and may become actively mobile. As a result, in a process called metastasis, they tend to break loose from an original tumour, spread throughout the body, and grow into new tumours in other body regions. Metastasis is promoted by changes that defeat contact inhibition and alter the cell surface molecules that link cells together or to the ECM.

Growing tumours damage surrounding normal tissues by compressing them and interfering with blood supply and nerve function. Tumours may also break through barriers such as the outer skin, internal cell layers, or the gut wall. The breakthroughs cause bleeding, open the body to infection by microorganisms, and destroy the separation of body compartments necessary for normal function. Both compression and breakthroughs can cause pain that, in advanced cases, may become extreme. As tumours increase in mass, the actively growing and dividing cancer cells may deprive normal cells of their required nutrients, leading to generally impaired body functions, muscular weakness, fatigue, and weight loss.

Cancer cells have typically accumulated mutations in a variety of genes that promote uncontrolled cell division or metastasis. Before they undergo mutation, many of these genes code for components of the cyclin–Cdk system that regulates cell division; others encode proteins that regulate gene expression, form cell surface receptors, or make up elements of the signalling pathways controlled by the receptors. When

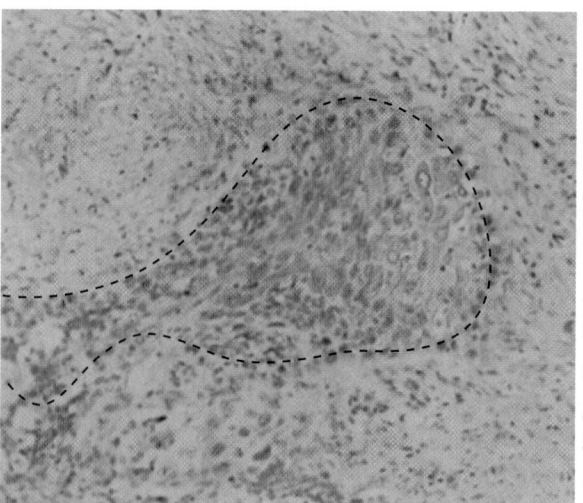

Courtesy of Professor Pierre Chambon, Institut de Génétique et de Biologie Moléculaire et Cellulaire, University of Strasbourg. Reprinted by permission from *Nature* 348:699. Copyright 1990 Macmillan Magazines, Ltd.

Figure 8.20

A mass of tumour cells (dashed line) embedded in normal tissue. As is typical, the tumour cells appear to be more densely packed because they have less cytoplasmic volume than normal cells. Original magnification × 270.

Roscovitine

Screening of a wide variety of artificially modified adenine molecules has led to the discovery of a group of compounds related to plant cytokinin hormones that selectively inhibit Cdks by competing for (and blocking) their ATP binding site. The example shown below, roscovitine, has antitumour and antiviral activity resulting from stimulation of cell death in affected cells. Note the adenine in each molecule (rectangle).

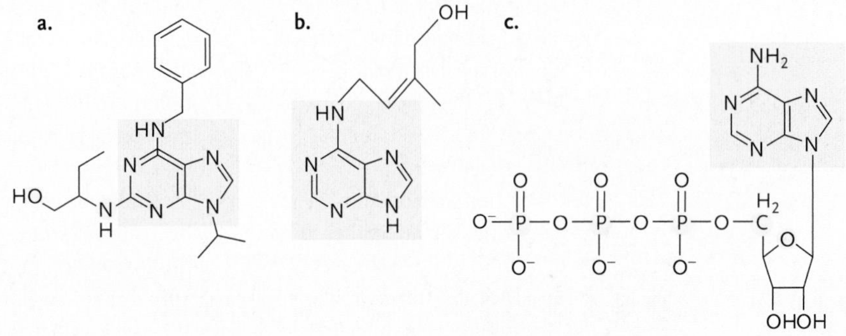

Figure 1
(a) *Roscovitine.* **(b)** *The plant cytokinin hormone zeatin.* **(c)** *ATP.*

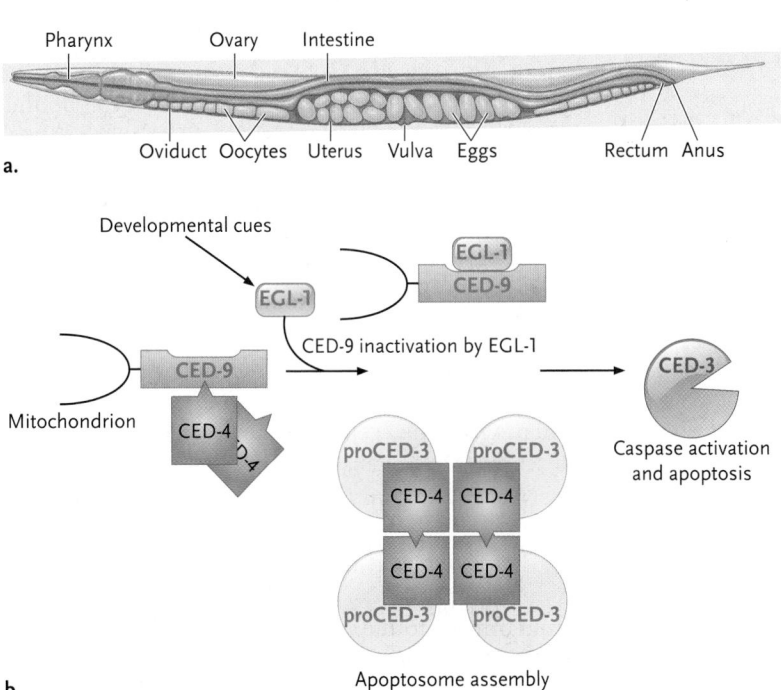

mutated, the genes, called **oncogenes**, encode altered versions of these products.

For example, a mutation in a gene that codes for a surface receptor might result in a protein that is constantly active even without binding the intended extracellular signal molecule. As a result, the internal reaction pathways triggered by the receptor, which induce cell division, are continually stimulated. Another mutation, this time in a cyclin gene, could result in increased cyclin–Cdk binding that triggers DNA replication and the rest of the cell cycle. Cancer, oncogenes, and the alterations that convert normal genes to oncogenes are discussed in further detail in Chapter 14.

8.5h Some Cells Are Programmed to Die

Normal development of multicellular organisms is a highly regulated balance between cell proliferation and cell death. Programmed cell death, called **apoptosis**, appears to be a very ancient mechanism common to all multicellular eukaryotes studied so far. Initiation of cell death can result from either internal or external signals. The nematode *Caenorhabditis elegans* is one useful model organism to study this signalling because all adult animals have exactly the same number of cells **(Figure 8.21a)**.

In addition, the fate of each of these cells, from the zygote to the adult, can be tracked with a light microscope. Detailed studies of the 1090 cells that are generated to form an adult reveal that 131 of them not only stop dividing—they stop living.

The apoptosis machinery in *C. elegans* is available in all of its cells, waiting in an inactive state for the right trigger. The main "executioner" enzyme is one of a family of normally inactive proteases called **caspases** and is coded by the *cell death abnormal* gene, *CED-3* (Figure 8.21b). If a cell is destined to die by

Figure 8.21
(a) The adult nematode "worm" *Caenorhabditis elegans* is about 1 mm long and is composed of 959 living cells. **(b)** The main cascade of programmed cell death in *C. elegans*. Cells destined to die express EGL-1 protein that, by binding to mitochondrial-bound CED-9, releases CED-4 protein. A complex of CED-4 then activates the main "executioner" caspase protease enzyme, CED-3.

PEOPLE BEHIND BIOLOGY 8.3

Dr. John Dick, *University of Toronto*

One of the best places to find actively cycling cells in a vertebrate body is in bone marrow. It is here that hematopoietic stem cells divide to produce progeny cells capable of proliferating and differentiating into the various specialized cells needed to maintain the liquid tissue called blood.

One of the best places to try to understand this developmental process, as well as the mechanisms underlying human blood disorders and cancer (leukemia), is the laboratory of John Dick at the University of Toronto. Dick is a professor of medical genetics and microbiology, a fellow of the Royal Society of Canada, and Canada Research Chair in Stem Cell Biology. Originally from rural Manitoba, he now heads a team of coworkers and international collaborators from his office in the MaRS Discovery District in downtown Toronto.

One of Dick's most powerful tools, one that gained him international recognition, is his system for modelling the production of blood by establishing human hematopoietic cells in mice. That is, he creates mice that make human blood instead of their usual mouse blood. Mice that make human blood also show human blood disorders, and Dick's team has been able to study the initiation and progression of human leukemia entirely in mice. This mouse model system provides an opportunity to better understand the genetic changes involved in leukemia and the effectiveness of emerging therapies.

Dick's sustained success arises from his ability to locate stem cells and understand their biology. His original discovery of the role of aberrant stem cells at the root of leukemia is now seen as the opening line of a story that is fundamentally altering the prevailing view of cancer and its treatment. If Dick is correct in his hypothesis that many cancers are initiated and fed by the progeny of relatively slow cycling, genetically aberrant stem cells (as is suggested by a growing body of research), then traditional therapies designed to indiscriminately kill rapidly cycling cancer cells are understandably "hit and miss." Dick has been able to identify and sort colon cancer cells into two fractions: those that can reproduce new tumours and those that cannot. This ability to specifically target cancer stem cells will be at the centre of the development of anticancer therapies.

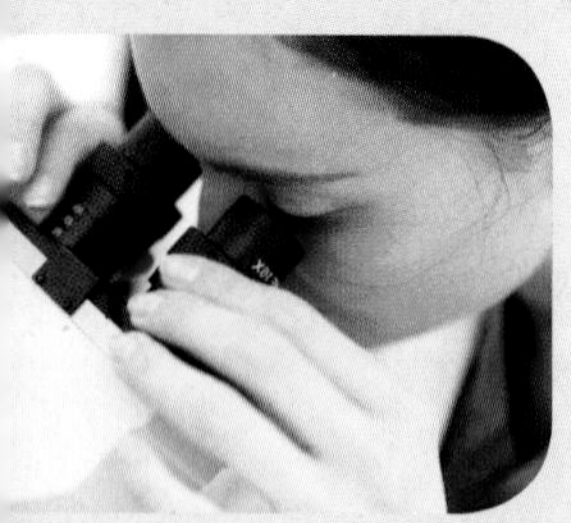

apoptosis, the cascade begins when internal developmental cues stimulate expression of a gene called *egg laying -deficient, EGL-1*. EGL-1 protein then binds to CED-9 protein, resulting in the release of bound CED-4 protein and the formation of an active apoptosome. CED-3 caspase is thus activated, and cell death ensues. The causes of death are nuclear DNA degradation and disrupted mitochondrial function. The corpses of dead cells are engulfed and eaten by neighbouring cells. The 2002 Nobel Prize in Physiology or Medicine was awarded jointly to Sydney Brenner, Robert Horvitz, and John Sulston for their discoveries concerning "genetic regulation of organ development and programmed cell death" in *C. elegans*. The Model Research Organisms section *of The Purple Pages* describes *C. elegans* and its role in research in more detail.

Removing cells that are surplus for development is one function of apoptosis, but why are other cells programmed to die? We hope you will agree that it would be beneficial for an organism to provoke apoptosis in cells suffering severe DNA damage, viral infection, or mutations leading to uncontrolled division. Sometimes perfectly normal and healthy cells die by apoptosis. For instance, the cells that make up xylem elements in the vascular tissue of woody plants actually function as "skeletons." They must die to fulfill their function as hollow, water-conducting pipes.

The overview of the cell cycle and its regulation presented in this chapter only hints at the complexity of cell growth and division. The likelihood of any given cell dividing is determined by weighing a variety of internal signals in the context of external cues from the environment. If a cell is destined to divide, then the problem of accurately replicating and partitioning its DNA requires a highly regulated, intricately interrelated series of mechanisms. Although male Australian Jack Jumper ants (*Myrmecia pilosula*) have only one chromosome to deal with, think of the problems faced by the fern *Ophioglossum pycnostichum*, which has 1260 chromosomes in each cell!

STUDY BREAK

1. Explain how the *activity* of Cdks can rise and fall with each turn of the cell cycle, whereas the *concentration* of these enzymes remains constant.
2. What observation do Hayflick factors explain?
3. What is metastasis?

Review

8.1 The Cycle of Cell Growth and Division: An Overview

- In mitotic cell division, DNA replication is followed by the equal separation—that is, segregation—of the replicated DNA molecules and their delivery to daughter cells. The process ensures that the two cell products of a division have the same genetic information as the parent cell entering division.
- Mitosis is the basis for growth and maintenance of body mass in multicelled eukaryotes and for the reproduction of many single-celled eukaryotes.
- The chromosomes of eukaryotic cells are individual, linear DNA molecules with associated proteins.
- DNA replication and the duplication of chromosomal proteins convert each chromosome into a structure composed of two exact copies known as sister chromatids.

8.2 The Cell Cycle in Prokaryotic Organisms

- Prokaryotic cells undergo a cycle of binary fission involving coordinated cytoplasmic growth, DNA replication, and cell division, producing two daughter cells from an original parent cell.
- Replication of the bacterial chromosome consumes most of the time in the cell cycle and begins at a single site called the origin through reactions catalyzed by enzymes located in the middle of the cell. Once the origin of replication (*ori*) is duplicated, the two origins actively migrate to the two ends of the cell. Division of the cytoplasm then occurs through a partition of cell wall material that grows inward until the cell is separated into two parts (see Figure 8.3).

8.3 Mitosis and the Eukaryotic Cell Cycle

- Mitosis and interphase constitute the mitotic cell cycle. Mitosis occurs in five stages. In prophase (stage 1), the chromosomes condense into short rods and the spindle forms in the cytoplasm (see Figures 8.7 and 8.8).
- In prometaphase (stage 2), the nuclear envelope breaks down, the spindle enters the former nuclear area, and the sister chromatids of each chromosome make connections to opposite spindle poles. Each chromatid has a kinetochore that attaches to spindle microtubules (see Figures 8.7 and 8.8).
- In metaphase (stage 3), the spindle is fully formed and the chromosomes, moved by the spindle microtubules, become aligned at the metaphase plate (see Figure 8.7).
- In anaphase (stage 4), the spindle separates the sister chromatids and moves them to opposite spindle poles. At this point, chromosome segregation is complete (see Figures 8.7 and 8.8).
- In telophase (stage 5), the chromosomes decondense and return to the extended state typical of interphase. A new nuclear envelope forms around the chromosomes (see Figures 8.7 and 8.8).

- Cytokinesis, the division of the cytoplasm, completes cell division by producing two daughter cells, each containing a daughter nucleus produced by mitosis (see Figures 8.7 and 8.8).
- Cytokinesis in animal cells proceeds by furrowing, in which a band of microfilaments just under the plasma membrane contracts, gradually separating the cytoplasm into two parts (see Figure 8.11).
- In plant cytokinesis, cell wall material is deposited along the plane of the former spindle midpoint; the deposition continues until a continuous new wall, the cell plate, separates the daughter cells (see Figure 8.12).

8.4 Formation and Action of the Mitotic Spindle

- In animal cells, the centrosome divides and the two parts move apart. As they do so, the microtubules of the spindle form between them. In plant cells with no centrosome, the spindle microtubules assemble around the nucleus (see Figure 8.13).
- In the spindle, kinetochore microtubules run from the poles to the kinetochores of the chromosomes, and nonkinetochore microtubules run from the poles to a zone of overlap at the spindle midpoint without connecting to the chromosomes (see Figure 8.15).
- During anaphase, the kinetochores move along the kinetochore microtubules, pulling the chromosomes to the poles. The nonkinetochore microtubules slide over each other, pushing the poles farther apart (see Figures 8.15 and 8.16).

8.5 Cell Cycle Regulation

- The cell cycle is controlled directly by complexes of cyclins and a cyclin-dependent protein kinase (Cdk). A Cdk is activated when combined with a cyclin and then adds phosphate groups to target proteins, activating them. The activated proteins trigger the cell to progress to the next cell cycle stage. Each major stage of the cell cycle begins with activation of one or more cyclin–Cdk complexes and ends with deactivation of the complexes by breakdown of the cyclins (see Figure 8.18).
- Important internal controls create checkpoints to ensure that the reactions of one stage are complete before the cycle proceeds to the next stage.
- External controls are based primarily on surface receptors that recognize and bind signals such as peptide hormones and growth factors, surface groups on other cells, or molecules of the extracellular matrix (ECM). The binding triggers internal reactions that speed, slow, or stop cell division.
- Stem cells divide asymmetrically to produce two daughter cells with different developmental fates. Although they inherit the same genetic material, they contain different arrays of gene regulatory proteins.
- Most cells in multicellular eukaryotes progressively lose the ability to divide over time by a process called cellular senescence. Factors that contribute to senescence include accumulating DNA damage and shortening telomeres.
- In cancer, control of cell division is lost and cells divide continuously and uncontrollably, forming a

rapidly growing mass of cells that interferes with body functions. Cancer cells also break loose from their original tumour (metastasize) to form additional tumours in other parts of the body.

- Certain cells may undergo a programmed cell death called apoptosis. Such a fate is appropriate for cells that are, for instance, surplus for development, damaged, infected, or functional only after death.

Questions

Self-Test Questions

1. Which of the following situations is characteristic of cell division in bacteria?
   a. Several chromatids are separated at anaphase.
   b. Binary fission produces four daughter cells.
   c. Replication begins at the *ori*, and the two new DNA molecules separate.
   d. The daughter cells receive different genetic information from the parent cell.

2. When does the mass of DNA in an elephant cell increase during the cell cycle?
   a. M phase (mitosis)
   b. G_1 phase
   c. G_2 phase
   d. S phase

3. Honeybee eggs that are not fertilized develop into fertile, haploid males called drones. Fertilized eggs can develop into diploid females, one of which might become a queen. (Fertilized eggs might also become males, but they are taken out and killed by the drones.)

   If the queen has 32 chromosomes in her body cells, how many chromatids will be present in a G_2 drone cell?
   a. 8
   b. 16
   c. 32
   d. 64

4. Which of the following is the major microtubule-organizing centre of the animal cell?
   a. the centrosome, composed of centrioles
   b. the chromatin, composed of chromatids
   c. the chromosomes, composed of centromeres
   d. the spindle, composed of actin

5. For one given oak tree cell, which of the following are more plentiful at the end of S phase than at the beginning?
   a. nuclei
   b. chromatids
   c. chromosomes
   d. Cdk2 molecules

6. Which of the following statements about mitosis is true?
   a. In prophase, the spindle separates sister chromatids and pulls them apart.
   b. Chromosomes congregate near the centre of the cell during metaphase.
   c. Both the animal cell furrow and the plant cell plate form at their former spindle poles.
   d. Cytokinesis describes the movement of chromosomes.

7. While researching an assignment on the Internet, you come across the following passage: "The cell cycle has a DNA synthesis phase (S phase) that doubles the normal full number of chromosomes from diploid ($2n$) to tetraploid ($4n$). This is followed by a G_2 cell phase that biochemically prepares the cell for the mitotic or M phase, which includes cytokinesis."

   In what way is the author of the above passage mistaken?
   a. DNA synthesis does not occur in S phase.
   b. S phase does not increase ploidy from $2n$ to $4n$.
   c. G_2 does not follow S phase.
   d. Cytokinesis is not part of mitotic cell division.

8. Which of the following statements about cell cycle regulation is true?
   a. Caspase is inactivated by cyclin binding.
   b. Cyclin binding activates Cdks to degrade target proteins.
   c. Telomere shortening promotes cell cycling.
   d. Stem cells divide more often than other somatic cells.

9. Which of the following is a characteristic of cancer cells?
   a. avoidance of metastasis
   b. avoidance of Hayflick factors
   c. contact inhibition
   d. cycle arrest at checkpoints

10. Imagine that you are in a job interview for a pharmaceutical company and are asked to suggest a good mechanism for an anticancer drug. Which of the following mechanisms would you suggest?
    a. decreased apoptosis
    b. decreased binding of cyclin to Cdk
    c. increased Cdk activity
    d. increased telomerase

Questions for Discussion

1. You have a means of measuring the amount of DNA in a single cell. You first measure the amount of DNA during G_1. At what point(s) during the remainder of the cell cycle would you expect the amount of DNA per cell to change?

2. A cell has 38 chromosomes. After mitosis and cell division, 1 daughter cell has 39 chromosomes and the other has 37. What might have caused these abnormal chromosome numbers? What effects do you suppose this might have on cell function? Why?

3. Taxol (Bristol-Myers Squibb, New York), a substance derived from the Pacific yew (*Taxus brevifolia*), is effective in the treatment of breast and ovarian cancers. It works by stabilizing microtubules, thereby preventing them from disassembling. Why would this activity slow or stop the growth of cancer cells?

4. Many chemicals in the food we eat potentially have effects on cancer cells. Chocolate, for example, contains a number of flavonoid compounds, which act as natural antioxidants. Design an experiment to determine whether any of the flavonoids in chocolate inhibit the cell cycle of breast cancer cells growing in culture.

5. The genes and proteins involved in cell cycle regulation are very different in bacteria and archaea than in eukaryotes. However, both prokaryotic and eukaryotic organisms use similar molecular regulatory reactions to coordinate DNA synthesis with cell division. What does this observation mean from an evolutionary perspective?

Mating octopuses.

Genetic Recombination

WHY IT MATTERS

A couple clearly shows mutual interest. First, he caresses her with one arm, then another—then another, and another, and another. She reciprocates. This interaction goes on for hours—a hug here, a squeeze there. At the climactic moment, the male reaches deftly under his mantle and removes a packet of sperm, which he inserts under the mantle of the female. For every one of his sperm that successfully performs its function, a fertilized egg can develop into a new octopus.

For the octopus, sex is an occasional event, preceded by a courtship ritual that involves intermingled tentacles. For another marine animal, the slipper limpet, sex is a lifelong group activity. Slipper limpets are relatives of snails. Like many other animals, a slipper limpet passes through a free-living immature stage before it becomes a sexually mature adult. When the time comes for an immature limpet to transform into an adult, it settles onto a rock or other firm surface. If the limpet settles by itself, it develops into a female. If instead it settles on top of a female, it develops into a male. If another slipper limpet settles down on that male, it, too, becomes a male. Adult slipper limpets almost always live in such piles, with the one on the bottom always being a female. All the male limpets continually contribute sperm that fertilizes eggs shed by the female. If the one female dies, the surviving male at the bottom of the pile changes into a female and reproduction continues.

The life history of these octopuses and slipper limpets illustrates a tension in biology between sameness and difference. On the one hand, the growth and repair of their multicellular tissues depend on faithful replication of DNA during mitotic cell division, as described in the previous chapter. At the level of the organism, it is important that all the individual cells in the body of a slipper limpet, for example, are genetically identical. However, on the other hand, at the level of the population, it is important that the individual limpets are genetically *different* from one another. If populations have genetic variability, they have the potential to evolve. Natural selection can act on this variability such that certain variants leave more offspring than others. Over time, the relative proportion of different variants will change. This is evolution in action.

The ultimate source of genetic diversity is mutation of the DNA sequence, often resulting from errors during DNA replication. Since mutations are relatively rare, diversity is amplified through various mechanisms that shuffle existing mutations into novel combinations. This process, of literally cutting and pasting DNA backbones into new combinations, is called **genetic recombination** and is very widespread in nature. Genetic recombination allows "jumping genes" to move, inserts some viruses into the chromosome of their hosts, underlies the spread of antibiotic resistance among bacteria and archaea, and is at the heart of meiosis in eukaryotic organisms. Genetic recombination puts the "sexual" in sexual reproduction; without genetic recombination, reproduction is asexual, and offspring are simply identical clones of their parent. We begin this chapter with a look at the basic mechanism of DNA recombination.

see Chapter 12 for a more comprehensive look at DNA structure.) Figure 9.1a shows two similar double helixes lying close together as the first step in recombination. Most of the recombination discussed in this chapter occurs between regions of DNA that are very similar, but not identical, in the sequence of bases. Such regions, which may be as short as a few base pairs or as long as an entire chromosome, are called **homologous**. Homology allows different DNA molecules to line up and recombine precisely. Once homologous regions of DNA are paired, enzymes break a covalent bond in each of the four sugar–phosphate backbones. The free ends of each backbone are then exchanged and reattached to those of the other DNA molecule, as shown in Figure 9.1b and c. The result is two recombined molecules in which the original red DNA is now covalently bound to the blue DNA, and vice versa. In this chapter, we consider all the steps shown in Figure 9.1 to make up a single recombination event. This idea is worth restating: cutting and pasting *four* DNA backbones results in *one* recombination event.

As we move through diverse examples of recombination in this chapter, from plasmids to meiotic crossing-over to transposons, the characteristics of the

9.1 Mechanism of Genetic Recombination

Biologists who study genetic recombination have developed several models to precisely explain the process in various situations. In its most general sense, genetic recombination requires the following: two DNA molecules that differ from one another, a mechanism for bringing the DNA molecules into close proximity, and a collection of enzymes to "cut," "exchange," and "paste" the DNA back together. **Figure 9.1** conveys a very simple model for recombination that, although lacking the details of more sophisticated models, highlights the basic steps involved.

The elegant double helix of DNA represented in Figure 9.1 is one of the most widely recognized biological molecules; you should be able to discern the "backbone" of the helix winding around the interior "steps" of paired bases. The sugar–phosphate backbone is held together by strong covalent bonds, whereas the bases pair with their partners through relatively weak hydrogen bonds. (If these ideas are new to you,

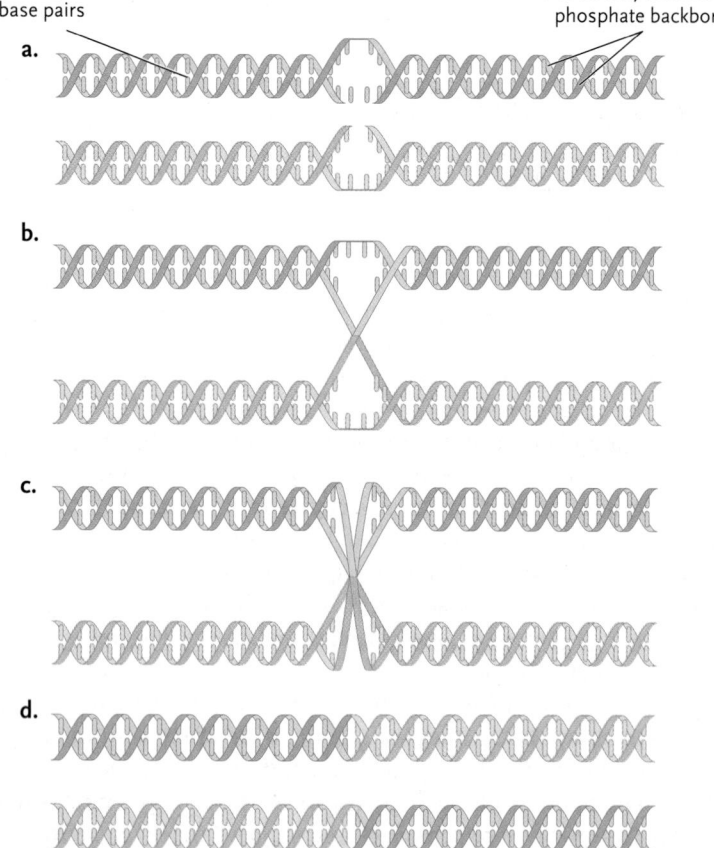

Hydrogen-bonded base pairs

Covalently bonded sugar–phosphate backbones

a.

b.

c.

d.

Figure 9.1

A simplified model of genetic recombination. **(a)** Two molecules of DNA with similar sequence are brought into close proximity. **(b)** Enzymes nick the DNA backbones, exchange the ends, and reattach them. **(c)** and **(d)** In this case, the final result is two recombined DNA molecules.

participating DNA molecules will be different, the enzymes will change, and the results of recombination will have quite different consequences for the organism in question. However, you can always return to Figure 9.1 to remind yourself of the basic underlying mechanism.

STUDY BREAK

What would happen if two circular DNA molecules were involved in a single recombination event?

9.2 Genetic Recombination in Bacteria

Genetic recombination was historically first associated with meiosis in sexually reproducing eukaryotes. Genetic and microscopic research in the early decades of the twentieth century characterized recombination and culminated in the construction of the first genetic maps of chromosomes. However, by the middle of that century, improved techniques for studying the genetics of prokaryotic organisms (and their viruses) enabled researchers to look for evidence of genetic recombination even though these organisms do not reproduce sexually by meiosis. The data showed that, for particular bacteria, there are mechanisms to bring DNA from different cells together and that this DNA recombines to create offspring that are different from either parent cell. Bacteria clearly have a type of sex in their lives. It may be surprising for you to learn that, in some types of bacterial recombination, one of the participating cells is dead. Watch for this.

Escherichia coli, the most extensively studied prokaryotic organism, is named in honour of its discoverer, a Viennese pediatrician named Theodor Escherich, who isolated it from dirty diapers during an outbreak of diarrhea in 1885. Ready availability and ease of growth in the laboratory have made *E. coli* a workhorse of bacterial genetics that has helped lay the foundations for our understanding of the role of DNA as the genetic material, as well as the molecular structure, expression, and recombination of genes. (See more information about *E. coli* as a model research organism in *The Purple Pages*.)

9.2a Genetic Recombination Occurs in *E. coli*

In 1946, two scientists at Yale University, Joshua Lederberg and Edward L. Tatum, set out to determine if genetic recombination occurs in bacteria, using *E. coli* as their experimental organism. In essence, they were testing whether bacteria have a kind of sexuality in their reproduction process. In order to understand Lederberg and Tatum's work, you first need to know how bacteria are grown in the laboratory.

Escherichia coli and many other bacteria can be grown in a **minimal medium** containing water, an organic carbon source such as glucose, and a selection of inorganic salts, including one, such as ammonium chloride, that provides nitrogen. The growth medium can be in liquid form or in the form of a gel made by adding agar to the liquid medium. (Agar is a polysaccharide material, indigestible by most bacteria, that is extracted from algae.) Since it is not practical to study a single bacterium for most experiments, researchers developed techniques for starting bacterial cultures from a single cell, generating cultures with a large number of genetically identical cells. Cultures of this type are called **clones.** To start bacterial clones, the scientist spreads a drop of a bacterial culture over a sterile agar gel in a culture dish. The culture is diluted enough to ensure that cells will be widely separated on the agar surface. Each cell divides many times to produce a clump of identical cells called a *colony.* Cells can be removed from a colony and introduced into liquid media or spread on agar and grown in essentially any quantity.

Now, for Lederberg and Tatum to detect genetic recombination, they needed some sort of detectable differences that could be shown to occur in changing combinations. The difference that proved most useful was related to nutrition. Cells require various amino acids for synthesis of proteins. Strains that are able to synthesize the necessary amino acids are called **prototrophs.** Mutant strains that are unable to synthesize amino acids are called **auxotrophs;** they can grow only if the required amino acid is provided for them in the growth medium. A strain that cannot manufacture its own arginine is represented by the genetic shorthand $argA^-$. In this shorthand, $argA$ refers to one of the genes that govern a cell's ability to synthesize arginine from simple inorganic molecules. A given strain of bacteria might carry this gene in its normal form, $argA^+$, or its mutant form, $argA^-$. These alternative forms of the gene are called alleles and might differ by as little as one base pair in their respective DNA sequences. Prokaryotic cells typically have one circular chromosome that carries one particular allele for each of their genes.

Using mutagens such as X-rays or ultraviolet light, Lederberg and Tatum isolated two different strains of *E. coli* carrying distinctive combinations of alleles for various metabolic genes. See **Figure 9.2, p. 192,** to understand how **replica plating** could isolate these auxotrophic strains. One particular strain could grow only if the vitamin biotin and the amino acid methionine were added to the culture medium. A second mutant strain did not need biotin or methionine but could grow only if the amino acids leucine and threonine were added along with the vitamin thiamine. These two multiple-mutant strains of *E. coli* were represented in genetic notation as follows:

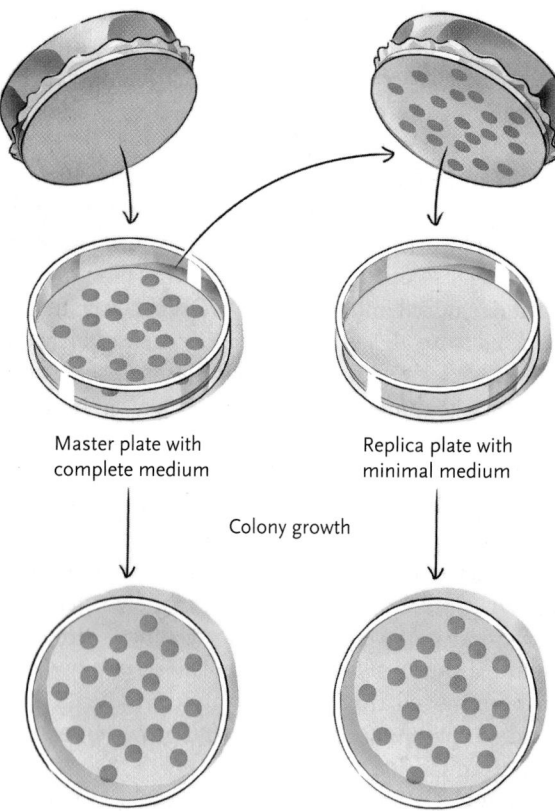

them on a minimal medium **(Figure 9.3)**. Several hundred colonies grew, even though, individually, none of the original cells carried all of the normal alleles needed for growth. You might be thinking, "They are mutants. Maybe some of the originally mutated alleles went back to normal." This possibility was easily discounted by plating large numbers of cells from each original strain onto minimal medium separately. If mutation were responsible for the initial results with mixed cultures, then colonies should have also appeared when strains were plated separately. There were none. Some form of recombination between the DNA molecules of the two parental types must have produced the necessary combination with normal alleles for each of the five genes:

$$bio^+ \ met^+ \ leu^+ \ thr^+ \ thi^+$$

9.2b Bacterial Conjugation Brings DNA of Two Cells into Close Proximity

How was DNA from two different bacterial cells able to recombine? We will see in Section 9.3 that genetic recombination in eukaryotes occurs in diploid cells by an exchange of segments between pairs of chromosomes. However, bacteria are haploid organisms; each cell typically has its own single, circular chromosome. So where do the "pairs" of chromosomes come from in bacteria? Although bacterial cells were first thought to bring their DNA together by fusing two cells together, it was later established that transfer of genetic information is unidirectional, from one donor cell to a recipient cell. Instead of fusing, bacterial cells *conjugate*. That is, cells contact each other by a long tubular structure called a *sex pilus* and then form a cytoplasmic

Master plate with complete medium

Replica plate with minimal medium

Colony growth

Research Method Figure 9.2 Replica plating transfers cells from complete media to minimal media where auxotrophs fail to grow.

Strain 1 $bio^- \ met^- \ leu^+ \ thr^+ \ thi^+$

Strain 2 $bio^+ \ met^+ \ leu^- \ thr^- \ thi^-$

Lederberg and Tatum mixed about 100 million cells of the two mutant strains together and placed

Experimental Research Figure 9.3 Experimental evidence for genetic recombination in bacteria.

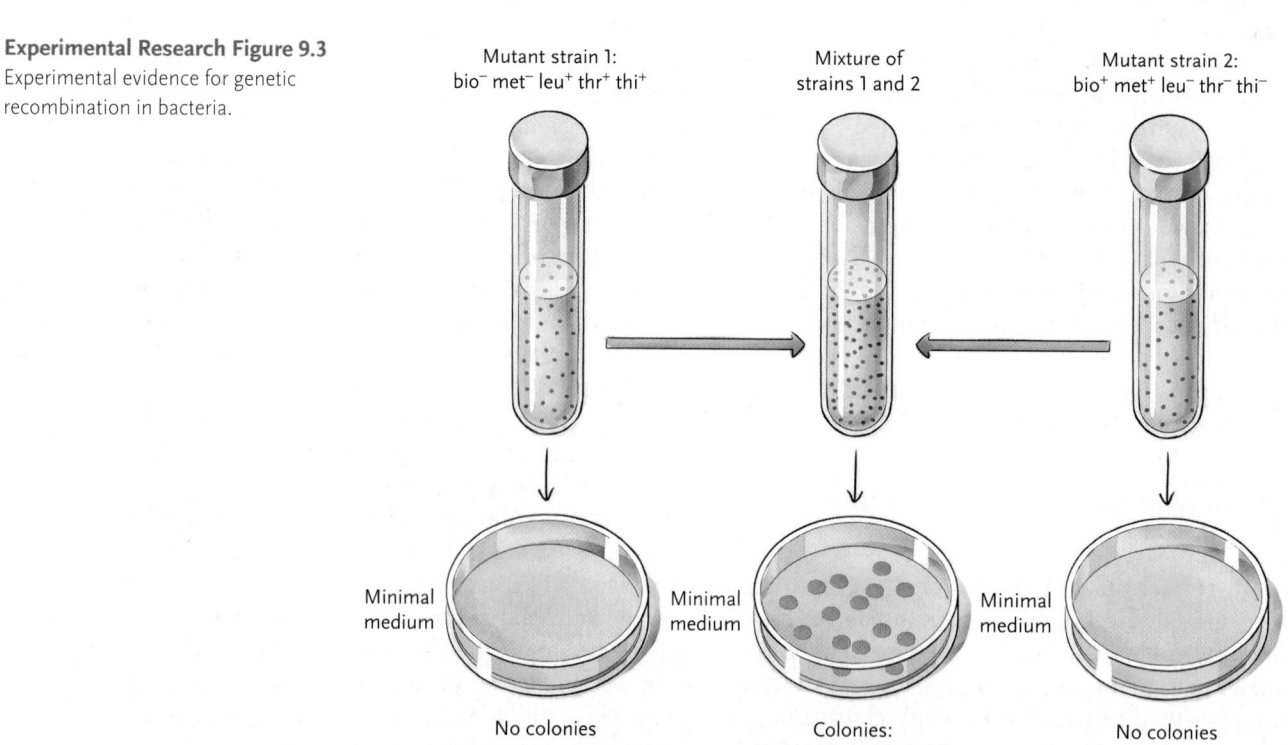

Mutant strain 1: $bio^- \ met^- \ leu^+ \ thr^+ \ thi^+$

Mixture of strains 1 and 2

Mutant strain 2: $bio^+ \ met^+ \ leu^- \ thr^- \ thi^-$

Minimal medium

Minimal medium

Minimal medium

No colonies

Colonies: $bio^+ \ met^+ \ leu^+ \ thr^+ \ thi^+$

No colonies

a. Attachment by sex pilus

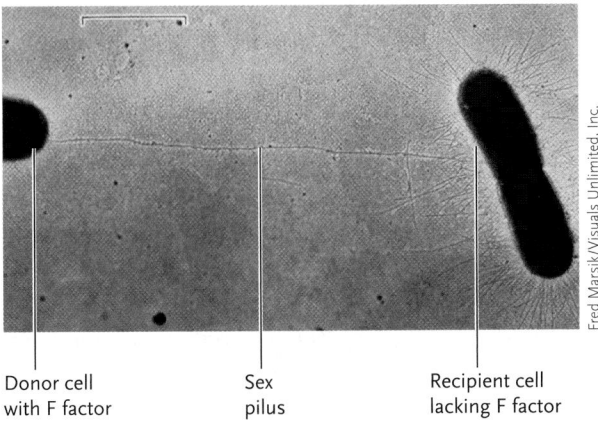

Fred Marsik/Visuals Unlimited, Inc.

Donor cell with F factor   Sex pilus   Recipient cell lacking F factor

b. Cytoplasmic bridge formed

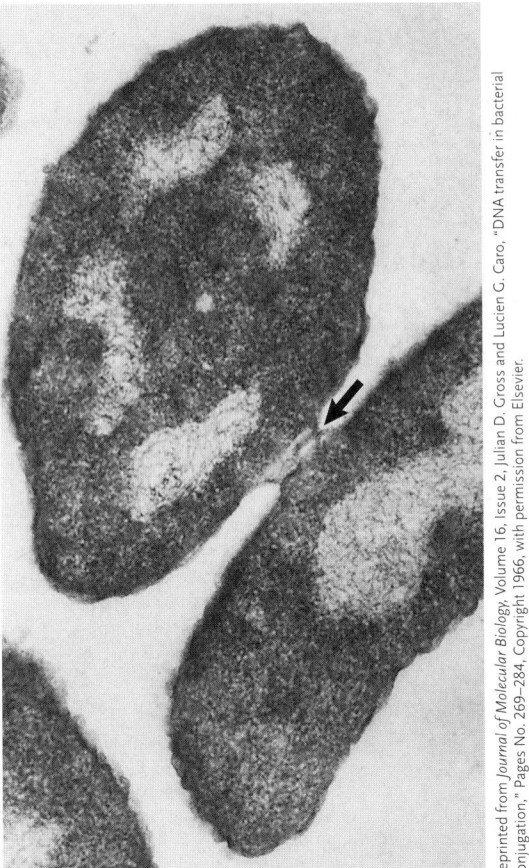

Reprinted from *Journal of Molecular Biology*, Volume 16, Issue 2, Julian D. Gross and Lucien G. Caro, "DNA transfer in bacterial conjugation," Pages No. 269–284, Copyright 1966, with permission from Elsevier.

Figure 9.4
Conjugating *E. coli* cells. **(a)** Initial attachment of two cells by the sex pilus. **(b)** A cytoplasmic bridge (arrow) has formed between the cells, through which DNA moves from one cell to the other.

bridge **(Figure 9.4a, b)**. During **conjugation**, a copy of part of the DNA of one cell moves through the cytoplasmic bridge into the other cell. Once DNA from one cell enters the other, genetic recombination can occur. Through this unidirectional transfer of a part of the chromosome, conjugation facilitates a kind of sexual reproduction in prokaryotic organisms.

The F Factor and Conjugation. Conjugation is initiated by a bacterial cell that contains a small circle of DNA in addition to the main circular chromosomal DNA **(Figure 9.5** and **Figure 9.6, p. 194)**. Such small circles are called plasmids, and this particular one is known as the *fertility* plasmid or the *F factor*. Like all plasmids, the F factor carries several genes as well as a **replication origin** that permits a copy to be passed on to each daughter cell during the usual process of bacterial cell division. This is an example of *vertical* inheritance from one generation to the next, which you should be familiar with. However, during conjugation, the F factor also can be copied and passed directly from the donor cell to the recipient cell. This is an example of *horizontal* inheritance.

Donor cells are called **F⁺ cells** because they contain the F factor. They are able to mate with recipient cells but not with other donor cells. Recipient cells lack the F factor and, hence, are called **F⁻ cells**. The F factor carries about 20 genes. Several of them encode proteins of the **sex pilus**, also called the **F pilus** (plural, *pili*) (see Figures 9.4 and 9.6a, step 1, p. 194).

a. Bacterial DNA released from cell

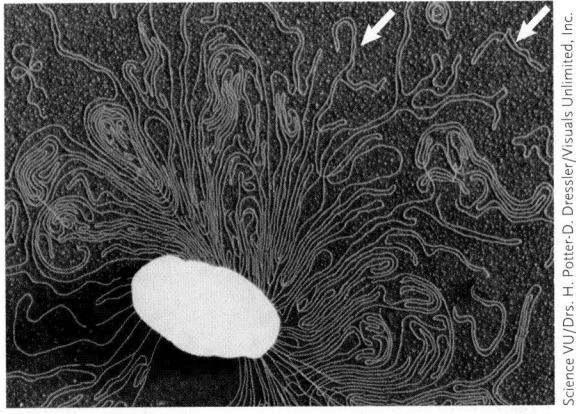

Science VU/Drs. H. Potter-D. Dressler/Visuals Unlimited, Inc.

b. Plasmid

Professor Stanley N. Cohen/Science Source

Figure 9.5
Electron micrographs of DNA released from a disrupted bacterial cell. **(a)** Plasmids (arrows) near the mass of chromosomal DNA. **(b)** A single plasmid at higher magnification (colourized).

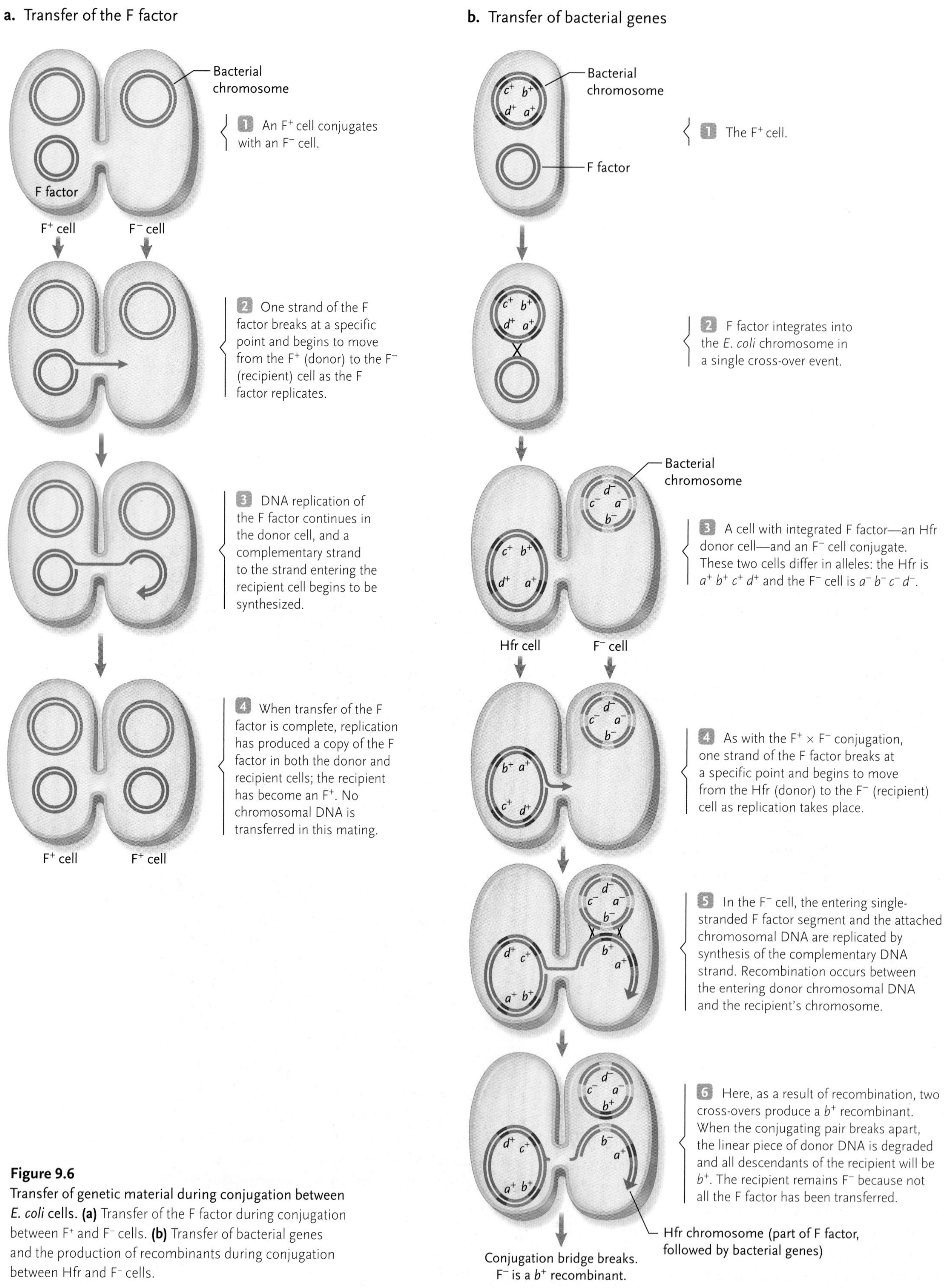

a. Transfer of the F factor

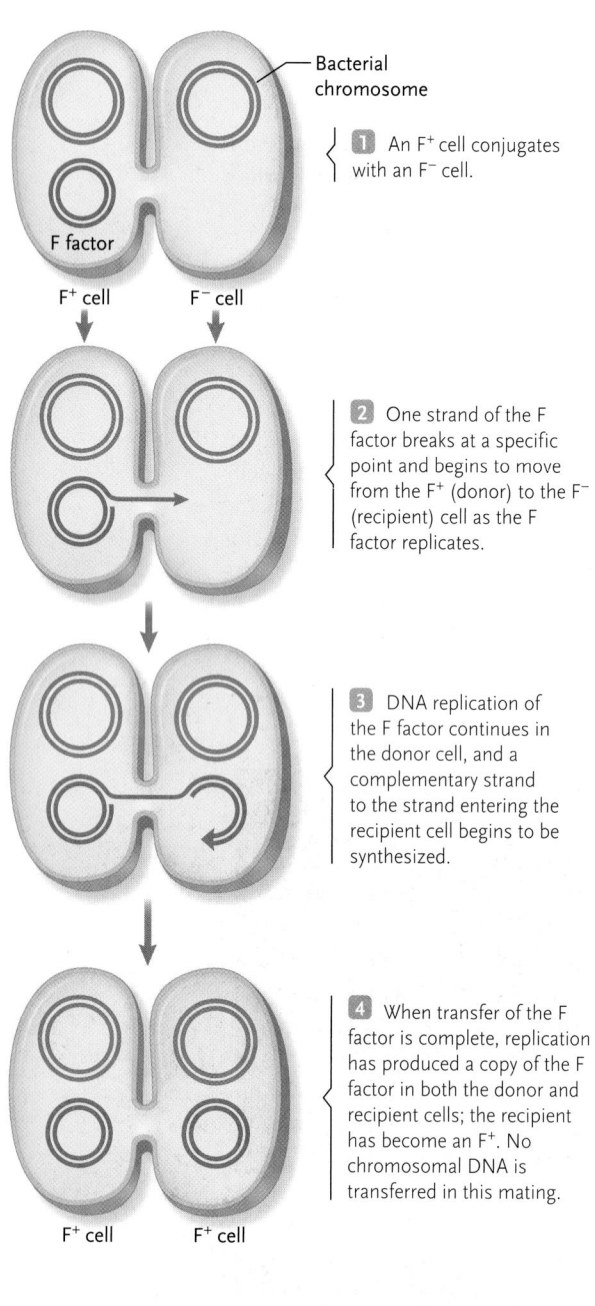

Bacterial chromosome

F factor

F⁺ cell   F⁻ cell

1 An F⁺ cell conjugates with an F⁻ cell.

2 One strand of the F factor breaks at a specific point and begins to move from the F⁺ (donor) to the F⁻ (recipient) cell as the F factor replicates.

3 DNA replication of the F factor continues in the donor cell, and a complementary strand to the strand entering the recipient cell begins to be synthesized.

4 When transfer of the F factor is complete, replication has produced a copy of the F factor in both the donor and recipient cells; the recipient has become an F⁺. No chromosomal DNA is transferred in this mating.

F⁺ cell   F⁺ cell

b. Transfer of bacterial genes

Bacterial chromosome

c^+ b^+
d^+ a^+

F factor

1 The F⁺ cell.

2 F factor integrates into the *E. coli* chromosome in a single cross-over event.

Bacterial chromosome

3 A cell with integrated F factor—an Hfr donor cell—and an F⁻ cell conjugate. These two cells differ in alleles: the Hfr is a^+ b^+ c^+ d^+ and the F⁻ cell is a^- b^- c^- d^-.

Hfr cell   F⁻ cell

4 As with the F⁺ × F⁻ conjugation, one strand of the F factor breaks at a specific point and begins to move from the Hfr (donor) to the F⁻ (recipient) cell as replication takes place.

5 In the F⁻ cell, the entering single-stranded F factor segment and the attached chromosomal DNA are replicated by synthesis of the complementary DNA strand. Recombination occurs between the entering donor chromosomal DNA and the recipient's chromosome.

6 Here, as a result of recombination, two cross-overs produce a b^+ recombinant. When the conjugating pair breaks apart, the linear piece of donor DNA is degraded and all descendants of the recipient will be b^+. The recipient remains F⁻ because not all the F factor has been transferred.

Hfr chromosome (part of F factor, followed by bacterial genes)

Conjugation bridge breaks. F⁻ is a b^+ recombinant.

Figure 9.6
Transfer of genetic material during conjugation between *E. coli* cells. **(a)** Transfer of the F factor during conjugation between F⁺ and F⁻ cells. **(b)** Transfer of bacterial genes and the production of recombinants during conjugation between Hfr and F⁻ cells.

During conjugation, the F factor replicates using a special type of DNA replication called *rolling circle*. To understand this mechanism, first picture a site, called the origin of transfer, on the F factor. Then imagine a break in just one strand of the double helix at this site. Now, imagine gently pulling the free end of the single strand of DNA away from the F factor, through the cytoplasmic bridge, and into the recipient cell. As the single strand is pulled, the remaining strand—still a circle—"rolls" like the spool on a tape dispenser. DNA synthesis fills in the complementary bases to ensure that the F factor is double stranded in both the donor and the recipient cells. When the entire F factor strand has transferred and replicated, it circularizes again (see Figure 9.6a, step 4). It is important to understand that although the recipient cell becomes F⁺, no chromosomal DNA is transferred between cells in this process. *That is, no genetic recombination occurs between the DNA of two different cells in such a mating.*

So why are we including F factor conjugation in this chapter if it does not recombine DNA of different cells? The answer lies in the Hfr cells described in the next section.

Hfr Cells and Genetic Recombination. In some F⁺ cells, the F factor comes into close proximity with the main chromosome and, lining up in a short region of homology, undergoes a recombination event. When two circular DNA molecules recombine (by the mechanism shown in Figure 9.1, p. 190), they simply fuse into one larger circle. In this way, the F factor actually becomes a part of the main bacterial chromosome (see Figure 9.6b, step 2). These special donor cells are known as **Hfr cells** (Hfr = high-frequency recombination). It is important not to be confused at this point; although a recombination event integrated the F factor into the host chromosome, this is recombination within one cell, not between the chromosomes of different cells. Hfr cells are called "high-frequency recombination" because they can promote recombination between DNA of different cells by "exporting" copies of chromosomal genes to another cell, as described below.

When the F factor is integrated into the bacterial chromosome, its genes are still available for expression. Therefore, these Hfr cells make sex pili and can conjugate with an F⁻ cell. Figure 9.6b, step 3, shows an Hfr × F⁻ mating where the two cell types differ in alleles for the genes *a*, *b*, *c*, and *d*. Note that a segment of the F factor moves through the conjugation bridge into the recipient, bringing the single-stranded chromosomal DNA behind it (see Figure 9.6b, steps 4 and 5). This is, again, rolling circle replication, in which both donor and recipient cells restore the DNA to double-strandedness. In this situation, the circle that rolls is the entire Hfr donor chromosome! Although DNA transfer often continues long enough for several genes to enter the recipient cell, the fragile

conjugation bridge soon breaks. It is rare for the entire donor chromosome to be transferred.

At this point, it is important to recall that when the F factor transfers by itself, as described in the previous section, the recipient cells often become F⁺. However, in Hfr cells, the origin of transfer is near the middle of the integrated F factor. As a result, only half of the F factor DNA is transferred at the front of the chromosomal DNA. (Think of the engine of a train.) The other half of the F factor (the dining car at the end of the train) can follow only after the rest of the entire chromosome (see Figure 9.6b, steps 4 to 6). As a result, it is very unusual for a recipient cell to obtain the entire F factor and become Hfr as well. Most likely, the recipient cell will become a **partial diploid**; it will have two copies of only those genes that came through the conjugation bridge on the donor chromosomal DNA segment.

For our example, the recipient cell in Figure 9.6b, step 5, has become, for the moment, $a^+ b^+/a^- b^-$. Although the DNA carrying + alleles for genes *a* and *b* differs slightly from that carrying − alleles, these regions are homologous and can pair for recombination. In fact, Figure 9.6 shows two recombination events, one on either side of the *b* gene, resulting in the exchange of the donor allele with that of the recipient (see Figure 9.6b, step 5). As a result, the recipient cell has become an $a^- b^+$ **recombinant.** Since enzymes in the recipient cell degrade the linear Hfr chromosome soon after recombination occurs, any incoming alleles that are not recombined onto the chromosome are lost. Following recombination, the bacterial DNA replicates and the cell divides normally, producing a clone of cells with the new combination of alleles.

In other pairs in the mating population, recombination events at different locations would lead to different recombinant recipients; perhaps the *a* gene could recombine with the homologous recipient gene, or both *a* and *b* genes could recombine to give $a^+ b^+$ recipients. The various genetic recombinants observed in the Lederberg and Tatum experiment described earlier were produced in this general way.

Mapping Genes by Conjugation. The use of conjugation for genetic mapping was discovered by two scientists, François Jacob (the same scientist who proposed the operon model for the regulation of gene expression in bacteria; see Section 14.1) and Elie L. Wollman, at the Pasteur Institute in Paris. They began their experiments by mating Hfr and F cells that differed in a number of alleles. At regular intervals after conjugation commenced, they removed some of the cells and agitated them in a blender to break apart mating pairs. They then cultured the separated cells and analyzed them for recombinants. They found that the longer they allowed cells to conjugate before separation, the greater the number of donor genes that entered the recipient and produced recombinants. By noting the order and time at which genes were transferred, Jacob and Wollman were able to map and assign the relative positions of several genes in the *E. coli* chromosome.

9.2c Transformation and Transduction Provide Additional Sources of DNA for Recombination

The discovery of conjugation and genetic recombination in *E. coli* showed that genetic recombination is not restricted to eukaryotes. Further discoveries demonstrated that DNA can transfer from one bacterial cell to another by two additional mechanisms, transformation and transduction. Like conjugation, these mechanisms transfer DNA in one direction and create partial diploids in which recombination can occur between alleles in the homologous DNA regions. Unlike conjugation, in which both donor and recipient cells are living, transformation and transduction enable recipient cells to recombine with DNA obtained from dead donors.

Transformation. In **transformation**, bacteria simply take up pieces of DNA that are released into the environment as other cells disintegrate. Fred Griffith, a medical officer in the British Ministry of Health, London, discovered this phenomenon in 1928 while trying to understand how bacteria cause pneumonia in mice. Cells of the virulent strains of *Streptococcus pneumoniae* were surrounded by a polysaccharide capsule, whereas the nonvirulent strains were not. Griffith found that a mixture of heat-killed virulent cells and living nonvirulent cells still caused pneumonia. One interpretation of this observation was that the living nonvirulent cells had been transformed to virulence by something released from the dead cells. In 1944, Oswald Avery and his colleagues at New York University found that the substance derived from the killed virulent cells, the substance capable of transforming nonvirulent bacteria to the virulent form, was DNA (discussed in Section 12.1).

Subsequently, geneticists established that in the transformation of *Streptococcus,* the linear DNA fragments taken up from disrupted virulent cells recombine with the chromosomal DNA of the non-virulent cells in much the same way as genetic recombination takes place in conjugation. In this case, the recombination introduces the normal allele for capsule formation into the DNA of the nonvirulent cells; expression of that normal allele generates a capsule around the cell and its descendants, making them virulent.

Only some species of bacteria can take up DNA from the surrounding medium by natural mechanisms, and *E. coli* is not one of them. Fortunately for molecular biologists, *E. coli* cells can be induced to take up DNA in the laboratory by a variety of artificial transformation techniques involving exposure to calcium ions and/or pulses of electric current. Artificial transformation is often used to insert recombinant DNA plasmids into *E. coli* cells as part of cloning or genetic engineering techniques. (DNA cloning and genetic engineering are discussed further in Chapter 15.)

Transduction. In **transduction**, DNA is transferred from donor to recipient cells inside the head of an infecting bacterial virus. The infection cycles of viruses that infect bacteria, called **bacteriophages** (or just phages), are described in Chapter 23. The basic details of phage infection are shown in **Figure 9.7** and **Figure 9.8, p. 198.** In general, transduction begins when new phages assemble within an infected bacterial cell; they sometimes incorporate fragments of the host cell DNA along with, or instead of, the viral DNA. After the host cell is killed, the new phages that are released may then attach to another cell and inject the bacterial DNA (and the viral DNA if it is present) into that recipient cell. The introduction of this DNA, as in conjugation and transformation, makes the recipient cell a partial diploid and allows recombination to take place. Recipients are not killed because they have received bacterial DNA rather than infective viral DNA. Lederberg and his graduate student, Norton Zinder, then at the University of Wisconsin at Madison, discovered transduction in 1952 in experiments with the bacterium *Salmonella typhimurium* and phage P22. Lederberg received a Nobel Prize in 1958 for his discovery of conjugation and transduction in bacteria.

There are two different types of transduction, generalized and specialized, arising from the different infection cycles of the phage involved. **Generalized transduction,** in which all donor genes are equally likely to be transferred, is associated with some **virulent bacteriophages,** which kill their host cells during each cycle of infection (the **lytic cycle**). Notice in Figure 9.7 that, during infection by the virulent phage, the host bacterial chromosome is degraded to provide raw material for synthesis of new phage chromosomes. However, sometimes a fragment of host chromosome avoids degradation and is packed into the head of a new phage *by mistake.* This particular phage now contains a small random sample of bacterial genes *instead of* phage genes. When the host cell bursts to release the new phage, this *transducing phage* can mechanically infect a recipient cell. However, it will deliver a linear piece of DNA from the donor cell rather than an infectious phage chromosome. The newly infected (and incredibly lucky) recipient cell will survive; incoming DNA may then pair, and recombine, with homologous regions on the recipient chromosome.

One of the most extensively studied bacteriophages is phage lambda (λ), which infects *E. coli.* Again, a mistake in the infection cycle can result in the transfer of bacterial genes from a donor to a recipient cell. However, in this case, a different type of mistake, in a different infection cycle, gives rise to a different type of transduction: **specialized transduction** (shown

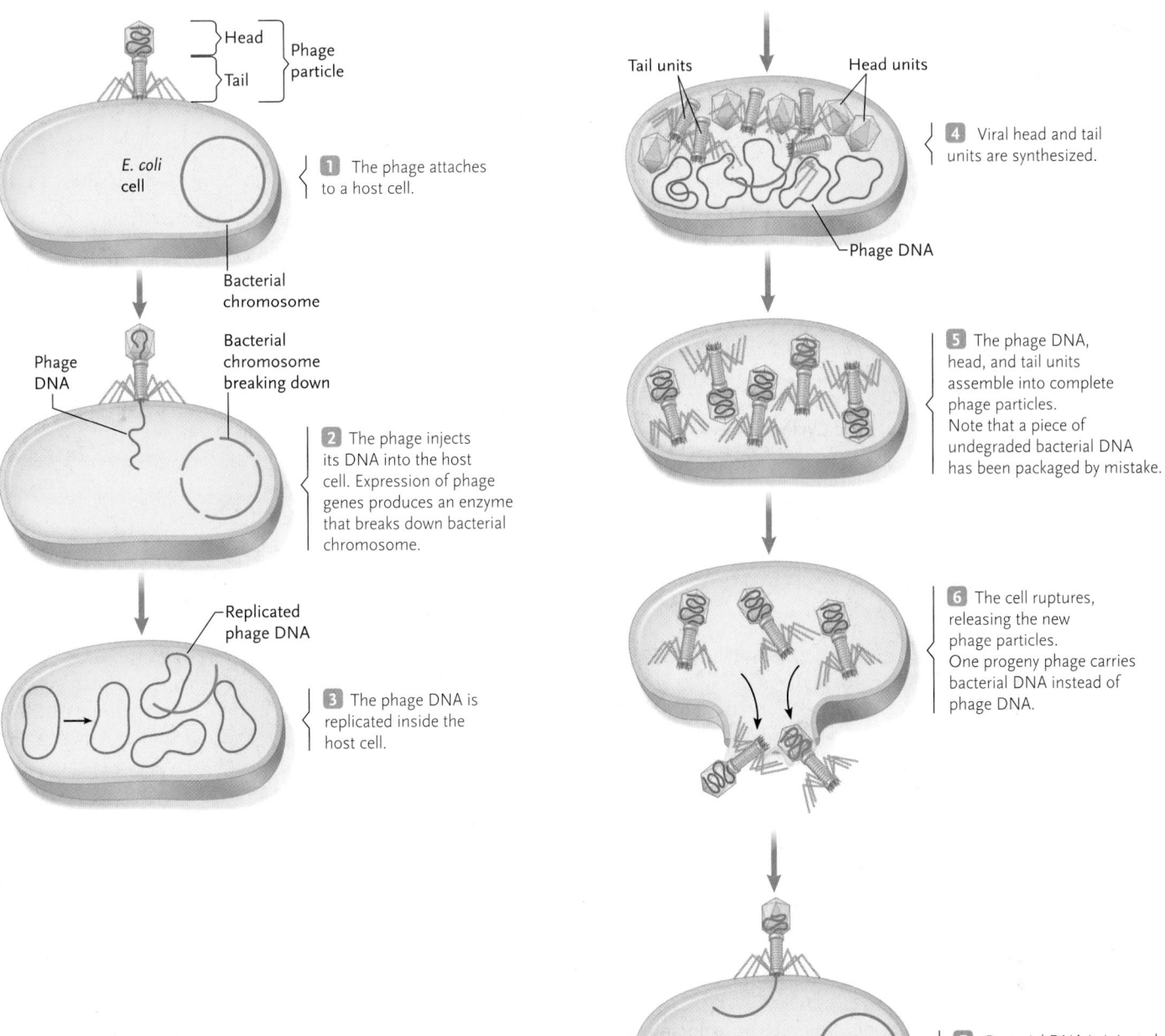

Figure 9.7

Generalized transduction. Movement of bacterial DNA from one cell to another inside the head of a lytic bacteriophage.

The figure shows the following labeled steps:

Head / **Tail** } Phage particle
E. coli cell
Bacterial chromosome

1 The phage attaches to a host cell.

Phage DNA
Bacterial chromosome breaking down

2 The phage injects its DNA into the host cell. Expression of phage genes produces an enzyme that breaks down bacterial chromosome.

Replicated phage DNA

3 The phage DNA is replicated inside the host cell.

Tail units   Head units

4 Viral head and tail units are synthesized.

Phage DNA

5 The phage DNA, head, and tail units assemble into complete phage particles. Note that a piece of undegraded bacterial DNA has been packaged by mistake.

6 The cell ruptures, releasing the new phage particles. One progeny phage carries bacterial DNA instead of phage DNA.

7 Bacterial DNA is injected into the next host, where it can recombine with similar regions on the host chromosome.

in Figure 9.8, p. 198). Lambda is a **temperate bacteriophage.** That is, when lambda first infects a new host, it determines whether this cell is likely to be a robust and long-lived host. Is it starving? Is it suffering from DNA damage? If the host cell passes this molecular health checkup, then the lambda chromosome lines up with a small region of homology on the bacterial chromosome and a phage-coded enzyme catalyzes a single recombination event. The phage is thus integrated into the host chromosomal DNA and, in this state, is called a **prophage.** (Overall, this mechanism is very similar to the integration of the F factor discussed previously.) The prophage is then replicated and passed to daughter cells along with the rest of the bacterial chromosome

as long as conditions remain favourable (the **lysogenic cycle** in Figure 9.8, p. 198).

If, however, the host cell becomes inhospitable (perhaps as a result of ultraviolet-induced DNA damage), the prophage activates several genes, releases itself from the chromosome by a recombination event, and proceeds to manufacture new phages, which are released as the cell bursts as a result of lytic growth.

In specialized transduction, the "mistake" occurs when the prophage is excised from the chromosome. Sometimes this recombination event is imprecise; bacterial DNA is removed from the host chromosome, and some prophage DNA is left behind. As a result, this bacterial DNA is packaged into new phages and carried

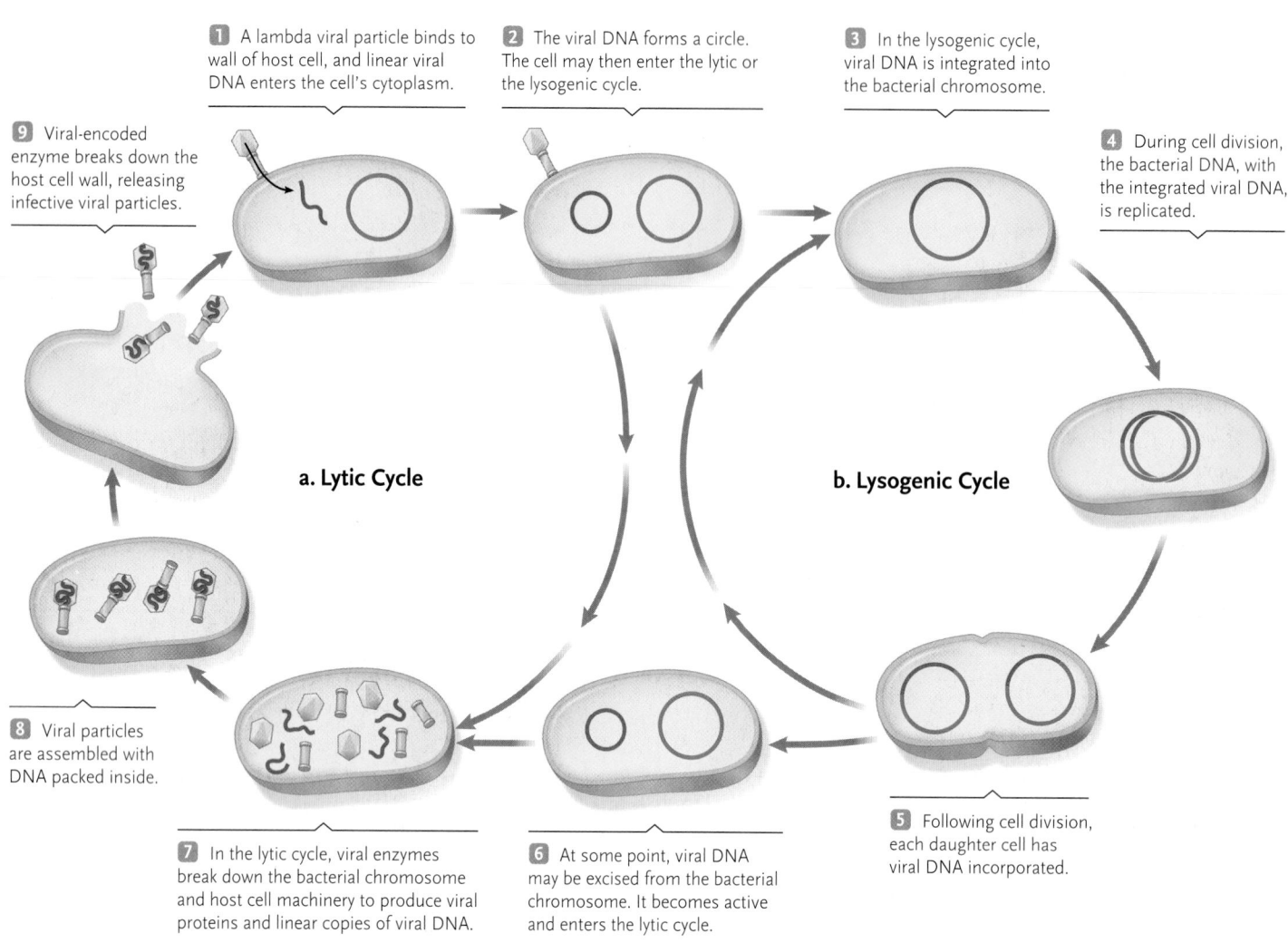

1 A lambda viral particle binds to wall of host cell, and linear viral DNA enters the cell's cytoplasm.

2 The viral DNA forms a circle. The cell may then enter the lytic or the lysogenic cycle.

3 In the lysogenic cycle, viral DNA is integrated into the bacterial chromosome.

9 Viral-encoded enzyme breaks down the host cell wall, releasing infective viral particles.

4 During cell division, the bacterial DNA, with the integrated viral DNA, is replicated.

a. Lytic Cycle

b. Lysogenic Cycle

8 Viral particles are assembled with DNA packed inside.

5 Following cell division, each daughter cell has viral DNA incorporated.

7 In the lytic cycle, viral enzymes break down the bacterial chromosome and host cell machinery to produce viral proteins and linear copies of viral DNA.

6 At some point, viral DNA may be excised from the bacterial chromosome. It becomes active and enters the lytic cycle.

Figure 9.8

The infective cycle of lambda, an example of a temperate phage, which can go through the lytic cycle **(a)** or the lysogenic cycle **(b)**.

to recipient cells. Since the transducing phage is defective, having left some of its genes behind in the host, it does not kill its new host. You should be able to see that in the case of specialized transduction only bacterial genes that are close to the integration site of the phage will likely be incorporated into the phage chromosome by the recombination mistake. Typically, only genes coding for galactose and biotin metabolism are transferred at high frequency by phage lambda.

Conjugation, transformation, and transduction are all ways in which DNA from two different bacterial cells is brought into close proximity. Homologous regions may then pair and recombine to give rise to a recipient cell that carries a different collection of alleles than it had previously. Overall, these processes create more diversity in the DNA sequence among members of a population than would arise by mutation and binary fission alone. More diversity leads to a higher likelihood that at least some individuals will be well suited to survival in a changing environment.

These basic principles also apply to single-celled and multicellular eukaryotes. The next section of this chapter introduces genetic recombination in eukaryotes as it occurs within the overall process of meiosis. Notice how DNA from two different individuals is brought close together in the same cell following fertilization. Also watch for extensive similarity of the DNA sequence (homology) that now extends the full length of large linear chromosomes. Finally, notice the genetic recombination at the centre of this process, which generates novel chromosomes with new combinations of alleles.

STUDY BREAK

1. Contrast the characteristics of F$^-$, F$^+$, and Hfr cells.
2. Explain why all genes have an equal likelihood of transfer by generalized transduction but not by specialized transduction.

9.3 Genetic Recombination in Eukaryotes: Meiosis

The octopuses and slipper limpets described at the opening of this chapter are engaged in forms of **sexual reproduction**, the production of offspring through the union of male and female **gametes**—for example, eggs and sperm cells in animals. Sexual reproduction depends on **meiosis**, a specialized process of cell division that recombines DNA sequences and produces cells with half the number of chromosomes present in the **somatic cells** (body cells) of a species. The derivation of the word *meiosis* (*meioun* = to diminish) reflects this reduction. At **fertilization**, the nuclei of an egg and a sperm cell fuse, producing a cell called the **zygote**, in which the chromosome number typical of the species is restored. Without the halving of chromosome number by the meiotic divisions, fertilization would double the number of chromosomes in each subsequent generation.

Both meiosis and fertilization also mix genetic information into new combinations; thus, none of the offspring of a mating pair are likely to be genetically identical to either their parents or their siblings. This genetic variability is the raw material for the process of evolution as described in Chapter 17.

The biological foundations of sexual reproduction are the mixing of genetic information into new combinations and the halving of the chromosome number, both of which occur through meiosis, as well as the restoration of the original chromosome number by fertilization. Intermingled tentacles in octopuses, communal sex among limpets, clouds of pollen in the wind, and the courting and mating rituals of humans are nothing more or less than variations of the means for achieving fertilization, thus bringing DNA together for recombination.

9.3a Meiosis Occurs in Different Places in Different Organismal Life Cycles

Although the life cycle of nearly all eukaryotes alternates between a stage with one basic set of chromosomes (haploid) and a stage with two basic sets of chromosomes (diploid), **Figure 9.9** shows that evolution has produced wide variety in the relative timing of mitosis, meiosis, and fertilization among different species.

🔵 CONCEPT FIX The life cycles of plants, algae, and fungi may be unfamiliar to you and can be better understood by focusing your attention on the function of the cells that are the immediate products of meiosis. You might assume that gametes are made by meiosis. This assumption is true—but only for yourself and other animals. In the life cycle of houseplants and some of the fungi living in the soil in the park, the haploid products of meiosis are spores, not gametes. These spores divide by mitosis

a. **Animal life cycles**

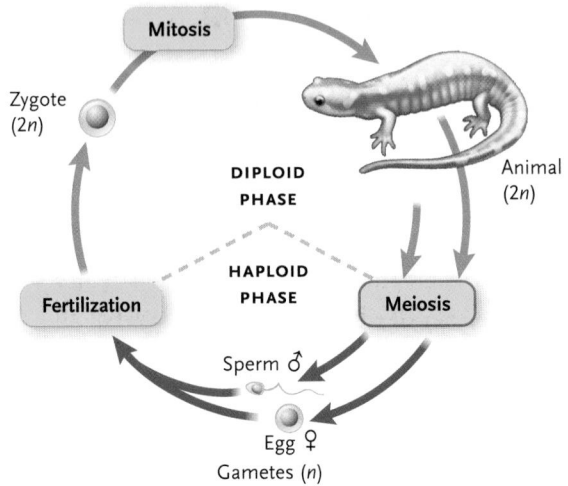

b. **All land plants and some fungi and algae (fern shown; relative length of the two phases varies widely in plants)**

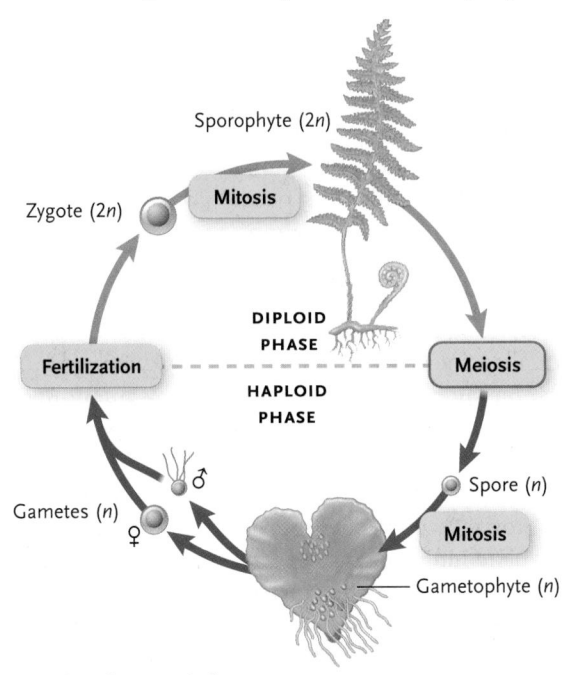

c. **Other fungi and algae**

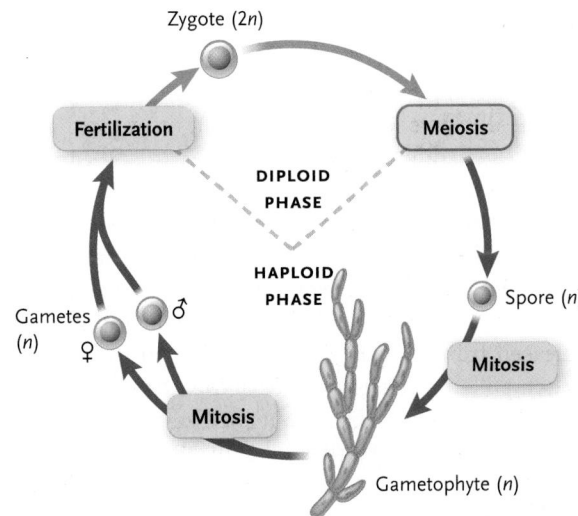

Figure 9.9

Variations in the time and place of meiosis and mitosis in the life cycle of eukaryotes. The diploid phase of the life cycles is shown in red; the haploid phase is shown in blue. *n* refers to the haploid number of chromosomes; 2*n* refers to the diploid number. **(a)** Meiosis in animal life cycles. Zygotes divide by mitosis. **(b)** Meiosis in most plants and some fungi and algae. Spores and zygotes divide by mitosis. **(c)** Meiosis in other fungi and algae. Spores divide by mitosis.

to form multicellular bodies that, in turn, make gametes by mitosis. That idea is worth repeating: many organisms make gametes by mitosis. ⬡

Animals. Animals follow the pattern in which the diploid phase dominates the life cycle (see Figure 9.9a, p. 199), the haploid phase is reduced, and meiosis is followed directly by gamete formation. (You could think of this as the "diploid life cycle" since the diploid stage is multicellular.) In male animals, each of the four nuclei produced by meiosis is enclosed in a separate cell by cytoplasmic divisions, and each of the four cells differentiates into a functional sperm cell. In female animals, only one of the four nuclei becomes functional as an egg cell nucleus.

Fertilization restores the diploid phase of the life cycle. Thus, animals are haploids only as sperm or eggs, and no mitotic divisions occur during the haploid phase of the life cycle.

Most Plants and Some Fungi. Most plants and some algae and fungi follow the life cycle pattern shown in Figure 9.9b. These organisms alternate between haploid and diploid generations in which, depending on the organism, either generation may dominate the life cycle, and mitotic divisions occur in both phases. (You could think of this as the "alternating-generations life cycle" since both the diploid and the haploid stages can be multicellular.) In these organisms, fertilization produces the diploid generation, in which the individuals are called **sporophytes** (*spora* = seed; *phyta* = plant). After the sporophytes grow to maturity by mitotic divisions, some of their cells undergo meiosis, producing haploid, genetically different, reproductive cells called **spores.** The spores are not gametes; they germinate and grow directly by mitotic divisions into a generation of haploid individuals called **gametophytes** (*gameta* = gamete). At maturity, the nuclei of some cells in gametophytes develop into egg or sperm nuclei. All the egg or sperm nuclei produced by a particular gametophyte are genetically identical because they arise through mitosis; meiosis does not occur in gametophytes. Fusion of a haploid egg and sperm nucleus produces a diploid zygote nucleus that divides by mitosis to produce the diploid sporophyte generation again.

In all plants (except bryophytes), the diploid sporophyte generation is the most visible part of the plant. The gametophyte generation is reduced to an almost microscopic stage that develops in the reproductive parts of the sporophytes—in flowering plants, in the structures of the flower. The female gametophyte remains in the flower; the male gametophyte is released from flowers as microscopic pollen grains. When pollen contacts a flower of the same species, it releases a haploid nucleus that fertilizes a haploid egg cell of a female gametophyte in the flower. The resulting cell, the zygote, reproduces by mitosis to form a sporophyte.

Sphagnum moss (commonly known as peat moss) is a good example of a plant in which the gametophyte is the most visible and familiar stage of the life cycle. In this case, the sporophyte is reduced and develops from a zygote within the body of the gametophyte. Vast peatlands of *Sphagnum* gametophytes are industrially harvested in many parts of the world for fuel and horticultural use.

Most Fungi. The life cycle of most fungi and algae follows the third life cycle pattern (see Figure 9.9c). In these organisms, the diploid phase is limited to a single cell, the zygote, produced by fertilization. Immediately after fertilization, the diploid zygote undergoes meiosis to produce the haploid phase. Mitotic divisions occur only in the haploid phase. (You could think of this as the "haploid life cycle" since the haploid stage is multicellular.)

During fertilization, two haploid gametes, usually designated simply as positive (+) and negative (−) because they are similar in structure, fuse to form a diploid nucleus. This nucleus immediately enters meiosis, producing four haploid cells. These cells develop directly or after one or more mitotic divisions into haploid spores. These spores germinate to produce haploid individuals, which grow or increase in number by mitotic divisions. Eventually, positive and negative gametes are formed in these individuals by differentiation of some of the cells produced by the mitotic divisions. Because the gametes are produced by mitosis, all the gametes of an individual are genetically identical.

CONCEPT FIX We are emphasizing that zygotes arising from fertilization contain DNA from two different parents in close proximity so that recombination may occur. However, note carefully that in the life cycles of the animals and plants you are likely familiar with, it is not this single-celled fertilized zygote that undergoes recombination. It is only after many rounds of replication by mitosis that certain cells in the resulting multicellular body are destined to divide by meiosis. That is when and where recombination occurs. ⬡

9.3b Meiosis Changes Both Chromosome Number and DNA Sequence

In order to understand the mechanism of meiosis, it is helpful to keep the big picture in mind. Chapter 8 made the point that the essence of mitotic cell division is *sameness*. That is, chromosomes are replicated and partitioned to ensure that cells produced by the process have the same number of chromosomes, with the same DNA sequence, as the cell that began the process. In this way, somatic cells are produced for most of the requirements of haploid or diploid multicellular organisms. However, the essence of meiosis is *difference*— actually two kinds of difference: halved chromosome number and recombined chromosomal DNA sequence. The products of meiosis are not intended to contribute to the body of the organisms that make them. In multicellular animals and plants, you would find that meiosis occurs only in specialized tissues that produce gametes and spores, respectively.

Both types of difference mentioned above arise from the very different behaviour of chromosomes in meiosis relative to mitosis. If you understand the significance of the chromosome pairs in diploid organisms as described below, then the differences in chromosome behaviour in meiosis and mitosis will make sense more easily.

As discussed in Section 9.1, the two representatives of each chromosome in a diploid cell constitute a *homologous pair* (*homo* = same; *logos* = information)—they have the same genes, arranged in the same order, in the DNA of the chromosomes. One chromosome of each homologous pair, the **paternal chromosome**, is derived from the male parent of the organism, and the other chromosome, the **maternal chromosome**, is derived from its female parent. Although two homologous chromosomes carry the same genes arranged in the same order, different *versions* of these genes, **alleles**, may be present on either chromosome. Recall from the bacterial conjugation material at the beginning of this chapter that different alleles of a given gene have similar, but distinct, DNA sequences. They therefore likely encode variations of the given RNA or protein gene product, which may then have a different structure, different biochemistry, or both.

For example, all the different breeds of dogs normally have 78 chromosomes in their cells, made up of 39 homologous pairs. However, each individual has a unique combination of the alleles carried by the two chromosomes of each homologous pair. The distinct set of alleles, arising from the mixing mechanisms of meiosis and fertilization in the parents, gives each individual offspring its own unique combination of inherited traits, including attributes such as size, coat colour, susceptibility to certain diseases and disorders, and aspects of behaviour and intelligence.

One of the more dramatic accomplishments of meiosis in an organism like a dog is the separation of the members of each homologous pair into different cells, thereby reducing the diploid or $2n$ number of chromosomes to the haploid or n number. Each cell produced by meiosis carries only one member of each homologous pair. An egg or sperm cell contains 39 chromosomes, one of each pair. When the egg and sperm combine in sexual reproduction to produce the zygote—the first cell of the new puppy—the diploid number of 78 chromosomes (39 pairs) is regenerated. The processes of DNA replication and mitotic cell division ensure that this diploid number is maintained in the body cells as the zygote divides and develops (see Chapter 8).

The second significant consequence of meiotic cell division is, of course, genetic recombination of the actual DNA sequence on chromosomes. Referring back to Figure 9.1, p. 190, recall that recombination involves the precise breaking of covalently bonded DNA backbones, exchanging the "ends" with those of the other homologue and reforming the bonds. As a result, each chromosome passed on to offspring is composed of a novel mixture of both maternal and paternal DNA sequence.

The following sections describe how the ability of homologues to find their respective partners, and pair intimately along their length, allows both the partitioning of homologues into separate cells and the process of recombination to occur during the first part of the two-step process of meiosis.

9.3c Meiosis Produces Four Genetically Different Daughter Cells

Cells that are destined to divide by meiosis (called **meiocytes**) move through their last turn of the cell cycle as usual, replicating DNA and making more chromosomal proteins during S phase. (See Chapter 12 for details of DNA replication.) The resulting G2 cells carry replicated chromosomes, each composed of two identical sister chromatids **(Figure 9.10, p. 202).** Following this premeiotic interphase, cells enter the first of the two meiotic divisions: meiosis I and meiosis II. During **meiosis I**, chromosomes behave dramatically differently than they do during mitosis. That is, early in meiosis I, homologous chromosomes find their partners and pair lengthwise, gene for gene, in a process called synapsis. During this intimate pairing, recombination occurs, and chromosomal segments are exchanged. As the meiocyte continues through to the end of the first division, the members of each homologous pair are moved into one or the other of the two daughter cells. These daughter cells still contain replicated chromosomes (composed of two chromatids each); however, the number of such chromosomes is only half that of the original meiocyte. That is, the cells now have the haploid number of chromosomes but each chromosome still has two chromatids.

During the second meiotic division, **meiosis II**, the sister chromatids are separated into different cells, further reducing the amount of DNA in each product of meiosis. A total of four cells, each with the haploid number of chromosomes and a novel collection of alleles, is the final result of the two meiotic divisions.

CONCEPT FIX Notice that the chromosome in the cells at the bottom of Figure 9.10 is an unreplicated, single structure. Since these cells are haploid, sometimes people come to believe that all chromosomes in haploid cells are unreplicated, single structures, while all chromosomes in diploid cells are double structures with two chromatids each. However, Figure 9.10 clearly shows that the top cell is diploid, even though its two chromosomes are unreplicated. Following meiosis I, the cells are haploid, even though their single

Figure 9.10

Production of four haploid nuclei by the two meiotic divisions. For simplicity, just one pair of homologous chromosomes is followed through the divisions.

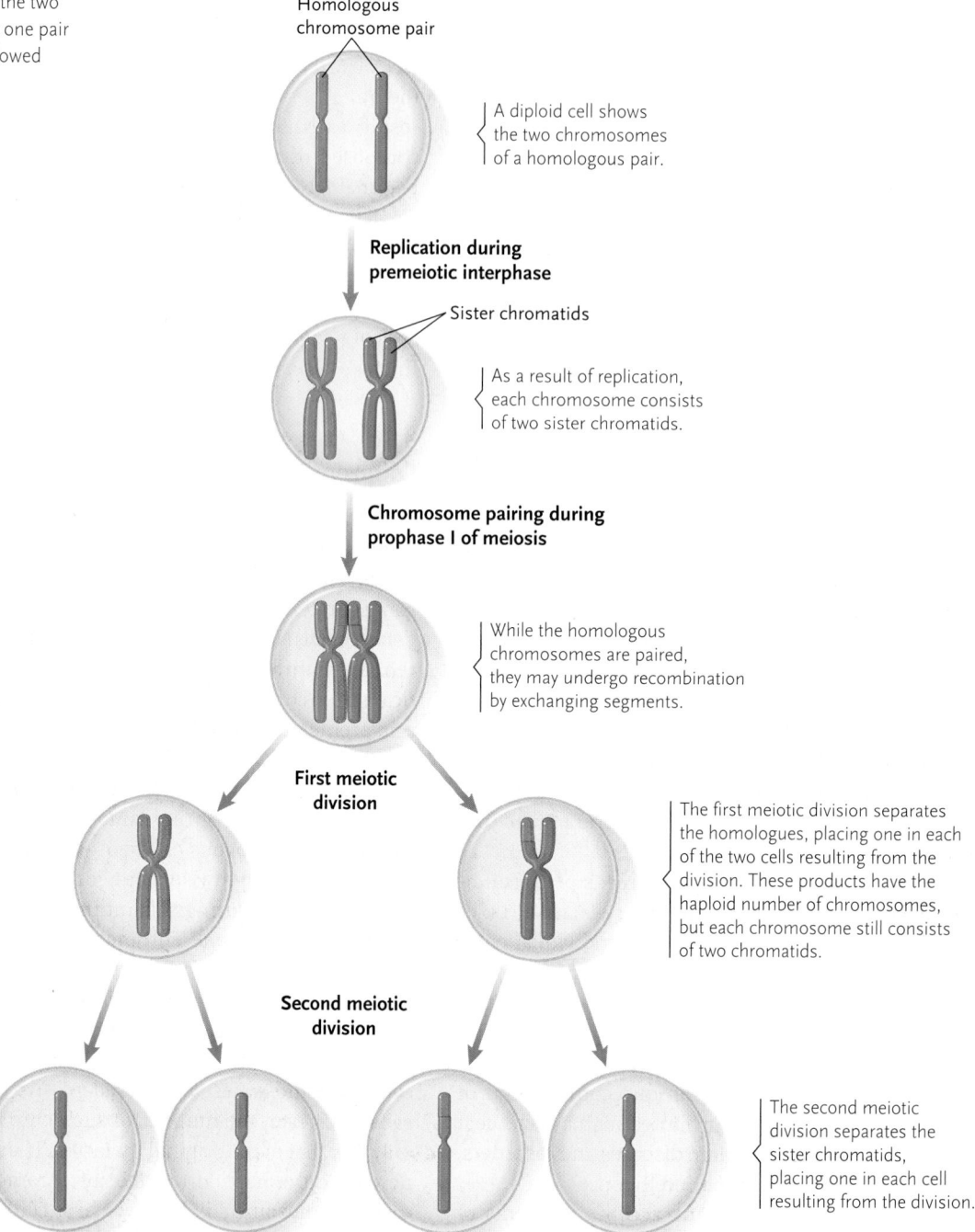

Homologous chromosome pair

A diploid cell shows the two chromosomes of a homologous pair.

Replication during premeiotic interphase

Sister chromatids

As a result of replication, each chromosome consists of two sister chromatids.

Chromosome pairing during prophase I of meiosis

While the homologous chromosomes are paired, they may undergo recombination by exchanging segments.

First meiotic division

The first meiotic division separates the homologues, placing one in each of the two cells resulting from the division. These products have the haploid number of chromosomes, but each chromosome still consists of two chromatids.

Second meiotic division

The second meiotic division separates the sister chromatids, placing one in each cell resulting from the division.

chromosome is replicated. Ploidy is determined only by the number of chromosomes; it is not influenced by whether the chromosomes are replicated or not. ⬡

For convenience, biologists separate each meiotic division into the same key stages as mitosis: prophase, prometaphase, metaphase, anaphase, and telophase. The stages are identified as belonging to the two divisions, meiosis I and meiosis II, by a I or a II, as in prophase I and prophase II. A brief interphase called **interkinesis** separates the two meiotic divisions, *but no DNA replication occurs during interkinesis.*

Prophase I. At the beginning of prophase I, the replicated chromosomes, each consisting of two sister chromatids, begin to fold and condense into threadlike structures in the nucleus (**Figure 9.11, p. 204**, step 1). The two chromosomes of each homologous pair then come together and line up side by side in a zipperlike way; this process is called **pairing** or **synapsis** (step 2). The fully paired homologues are called **tetrads**, referring to the fact that each homologous pair consists of four chromatids. *Note that chromosomes do not behave like this in mitosis.*

While they are paired, the chromatids of homologous chromosomes physically exchange segments (step 3). This physical exchange, genetic recombination, is the step that mixes the alleles of the homologous chromosomes into new combinations and contributes to the generation of variability in sexual reproduction. (This is the process, described in Chapter 11, that underlies **recombination frequency** mapping.) As prophase I finishes, a spindle forms in the cytoplasm by the same basic mechanisms described in Chapter 8.

Prometaphase I. In prometaphase I, the nuclear envelope breaks down and the spindle enters the former nuclear area (see Figure 9.11, p. 204, step 4). The two chromosomes of each pair attach to kinetochore microtubules that are anchored to opposite spindle poles. That is, both sister chromatids of one homologue attach to microtubules leading to one spindle pole, whereas both sister chromatids of the other homologue attach to microtubules leading to the opposite pole. *Notice, again, how this is different from the spindle attachments during mitosis.*

Metaphase I and Anaphase I. At metaphase I, movements of the spindle microtubules have aligned the recombined tetrads on the equatorial plane—the *metaphase plate*—between the two spindle poles (see Figure 9.11, p. 205, step 5). Then the two chromosomes of each homologous pair separate and move to opposite spindle poles during anaphase I (step 6). The movement segregates homologous pairs, delivering a haploid set of chromosomes to each pole of the spindle. However, all the chromosomes at the poles are still double structures composed of two sister chromatids joined at their centromeres.

Telophase I and Interkinesis. Telophase I is a brief, transitory stage in which there is little or no change in the chromosomes (see Figure 9.11, p. 205, step 7). New nuclear envelopes form in some species but not in others. Telophase I is followed by an interkinesis in which the single spindle of the first meiotic division disassembles and the microtubules reassemble into two new spindles for the second division. Recall that there is no DNA replication between the first and the second division.

Prophase II, Prometaphase II, and Metaphase II. During prophase of meiosis II, the chromosomes condense (see Figure 9.11, p. 204, step 8). During prometaphase II, the nuclear envelope breaks down, the spindle enters the former nuclear area, and spindle microtubules leading to opposite spindle poles attach to the two kinetochores of each chromosome. At metaphase II, movements of the chromosomes within the spindle bring them to rest at the metaphase plate (step 9).

> CONCEPT FIX Although the separation of chromatids during meiosis II is superficially similar to that in a mitotic division, it is important to remember that these two processes are quite distinct. Meiosis II is not "just like mitosis." Meiosis II occurs only in reproductive tissue, there is no immediately preceding DNA replication phase, and the resulting daughter cells are not genetically identical (see Figure 9.16, p. 210). ⬢

Anaphase II and Telophase II. Anaphase II begins as the sister chromatids of each chromosome separate from each other and move to opposite spindle poles (see **Figure 9.11, p. 205,** step 10). At the completion of anaphase II, the separated chromatids—now called chromosomes—have been segregated to the two poles. During telophase II, the chromatids decondense to the extended interphase state, the spindles disassemble, and new nuclear envelopes form around the masses of chromatin (step 11). The result is four haploid cells, each with a nucleus containing half the number of chromosomes present in the cell at the beginning of meiosis. These chromosomes all carry various new combinations of maternal and paternal alleles.

Failure in Chromosome Segregation. Rarely, chromosome segregation fails at either meiosis I or II. For example, during meiosis I, both chromosomes of a homologous pair may connect to the same spindle pole in anaphase I. In the resulting nondisjunction, as it is called, the spindle fails to separate the homologous chromosomes of the tetrad. As a result, one pole receives both chromosomes of the homologous pair, whereas the other pole has no copies of that chromosome. Meiosis II will proceed to separate the chromatids of the extra chromosome as usual, with the result that gametes will have two copies of this chromosome (instead of one). A failure at meiosis II, in which chromatids do not separate to opposite poles, also results in gametes with abnormal numbers of chromosomes. Zygotes that receive an extra chromosome from an abnormal gamete therefore have three copies of a given chromosome instead of two. In humans, most zygotes of this kind do not result in live births. One exception is Down syndrome, which can result from three copies of chromosome 21. Down syndrome involves characteristic alterations in body and facial structure, developmental delays, and significantly reduced fertility due to extra genetic information (see Chapter 11 for a more detailed discussion of Down syndrome).

Sex Chromosomes. In many eukaryotes, including most animals, one or more pairs of chromosomes, called the **sex chromosomes**, are different in male and female individuals of the same species. For example, in fruit flies, the cells of females contain a pair of sex chromosomes called the *XX pair*. Male flies contain a pair of sex chromosomes that consist of one X chromosome and a smaller chromosome called the **Y chromosome.** The two X chromosomes in females are fully homologous, whereas the male X and Y chromosomes are

Prophase I

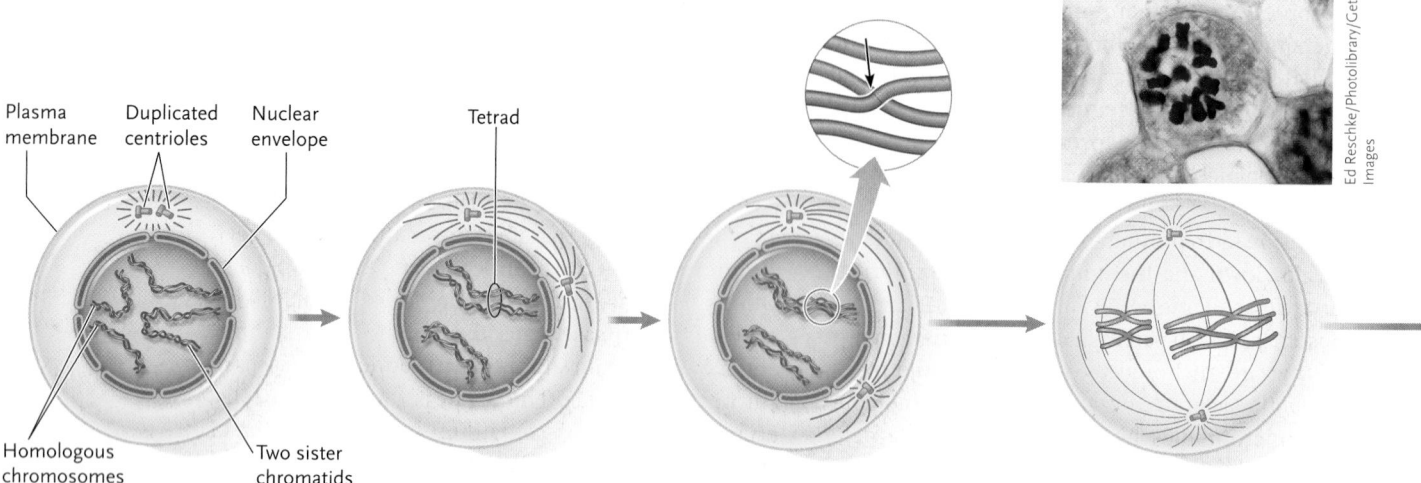

Plasma membrane

Duplicated centrioles

Nuclear envelope

Tetrad

Homologous chromosomes

Two sister chromatids

Ed Reschke/Photolibrary/Getty Images

Condensation of chromosomes

1 At the beginning of prophase I, the chromosomes begin to condense into threadlike structures. Each consists of two sister chromatids, as a result of DNA replication during premeiotic interphase. The chromosomes of two homologous pairs, one long and one short, are shown.

Synapsis

2 Homologous chromosomes come together and pair.

Recombination

3 While they are paired, the chromatids of homologous chromosomes undergo recombination by exchanging segments. The enlarged circle shows a site undergoing recombination (arrow).

Prometaphase I

4 In prometaphase I, the nuclear envelope breaks down, and the spindle moves into the former nuclear area. Kinetochore microtubules connect to the chromosomes—kinetochore microtubules from one pole attach to both sister kinetochores of one duplicated chromosome, and kinetochore microtubules from the other pole attach to both sister kinetochores of the other duplicated chromosome.

Second meiotic division

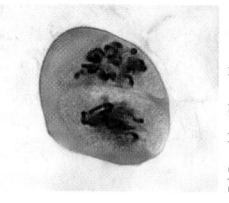

Ed Reschke/Photolibrary/Getty Images

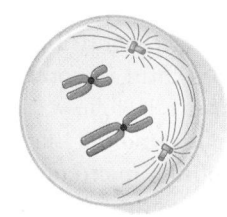

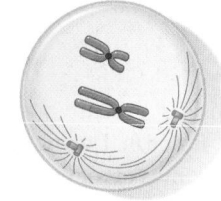

Figure 9.11

The meiotic divisions. The artwork summarizes the behaviour of chromosomes in a hypothetical animal cell having two homologous pairs of chromosomes ($2n = 4$). Photomicrographs show comparable stages in the anther cells of a lily plant.

Prophase II

8 The chromosomes condense and a spindle forms.

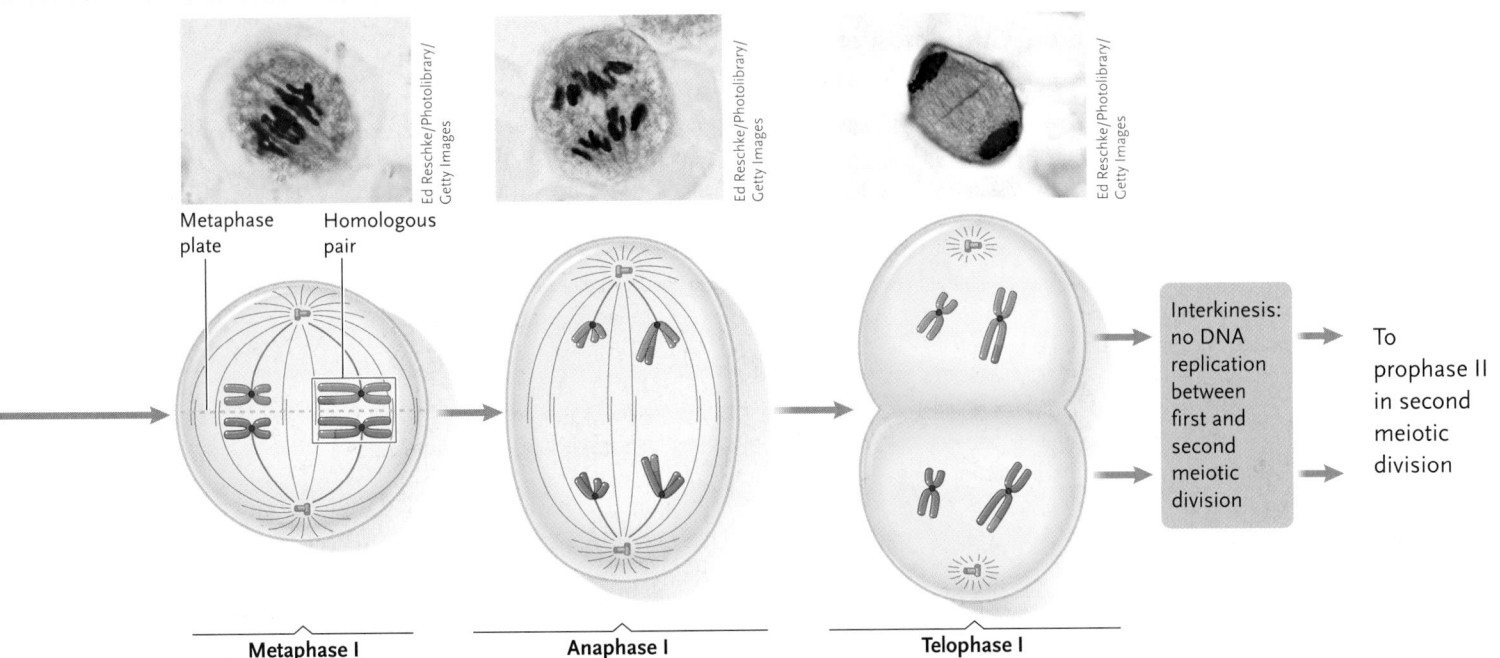

Metaphase plate   Homologous pair

Metaphase I

5 Movements of the spindle microtubules align the tetrads in the equatorial plane—metaphase plate—between the two spindle poles.

Anaphase I

6 The spindle microtubules separate the two chromosomes of each homologous pair and move them to opposite spindle poles. The poles now contain the haploid number of chromosomes. However, each chromosome at the poles still contains two chromatids.

Telophase I

7 The chromosomes undergo little or no change except for limited decondensation or unfolding in some species. The spindle of the first meiotic division disassembles, and two new spindles form for the second division.

Interkinesis: no DNA replication between first and second meiotic division

To prophase II in second meiotic division

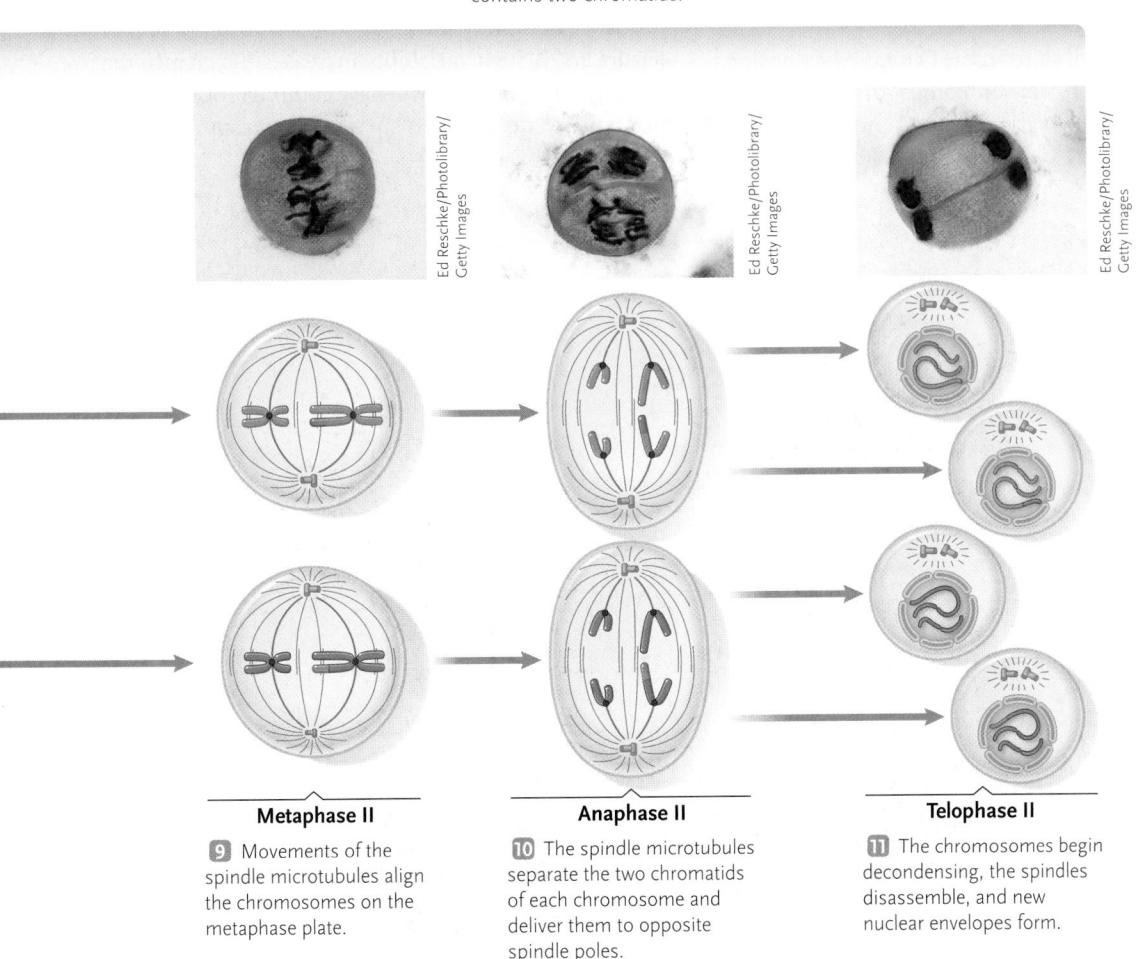

Metaphase II

9 Movements of the spindle microtubules align the chromosomes on the metaphase plate.

Anaphase II

10 The spindle microtubules separate the two chromatids of each chromosome and deliver them to opposite spindle poles.

Telophase II

11 The chromosomes begin decondensing, the spindles disassemble, and new nuclear envelopes form.

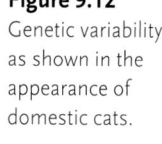

MOLECULE BEHIND BIOLOGY 9.1

Bisphenol A and the Grandmother Effect

Although this chapter documents the role of meiosis in generating genetically diverse offspring, one type of diversity that must be avoided is differences in chromosome number. Cells (or organisms) that have more, or fewer, than the normal number of chromosomes are called *aneuploid;* agents that promote this problem are known as *aneugens*. The formation of gametes by meiosis is under hormonal control in mammals, and it is not surprising to learn that synthetic chemicals influencing the action of reproductive hormones can be aneugenic. Bisphenol A, a chemical monomer used in the manufacture of polycarbonate plastics and resins, binds to estrogen receptors in mice. Exposure to relatively high concentrations has been shown to elevate the incidence of aneuploid gametes and offspring. Since meiosis is active in females before they are born, exposure of pregnant mouse mothers resulted in aneuploid gametes produced by their daughters, which, in turn, gave rise to aneuploid grandchildren.

Canada has declared BPA a toxic substance and banned its use in baby bottles.

Figure 1

Bisphenol A.

---

homologous only through a short region. The X and Y chromosomes behave as homologues (i.e., they pair where homologous, recombine, and move together to the metaphase plate) during meiosis in males. As a result of meiosis, a gamete formed by females may receive either member of the XX pair. A gamete formed by males receives either an X or a Y chromosome. (See Chapter 11 for a discussion of the inheritance of genes on sex chromosomes.)

The sequence of steps in the two meiotic divisions accomplishes the major outcomes of meiosis: the generation of genetic variability and the reduction in chromosome number. (Figure 9.16, p. 210, reviews the two meiotic divisions and compares them with the single division of mitosis.)

9.3d Several Mechanisms Contribute to Genetic Diversity

The generation of genetic variability by meiosis is a prime evolutionary advantage of sexual reproduction **(Figure 9.12)**. Such variability increases the chance that at least some offspring will have combinations of alleles that will be successful in surviving and reproducing in

Figure 9.12

Genetic variability as shown in the appearance of domestic cats.

Dr. Aurora Nedelcu, *University of New Brunswick*

Whereas the octopuses and limpets mentioned at the opening of this chapter have no choice but to undergo meiosis and follow the remaining steps of their sexual life cycle, bacteria, archaea, and many lower eukaryotes become sexual only in response to suboptimal environmental conditions, such as elevated temperature or nutrient deficiency. This observation led Aurora Nedelcu and her colleagues in The Green Lab at the University of New Brunswick to gather evidence to test the hypothesis that sex originally evolved as one of several responses available to cells dealing with stress.

A variety of external stresses all eventually cause internal oxidative stress resulting from increased concentration of damaging reactive oxygen species (ROS). Using her multicellular algal model system (*Volvox carteri*), Nedelcu has shown that stress-induced increase in ROS does indeed stimulate the expression of sex-related genes **(Figure 1).** She believes that the cells experiencing oxidative stress "turn on" their sex genes in order to benefit from the possibility that meiotic recombination will repair DNA damage caused by ROS.

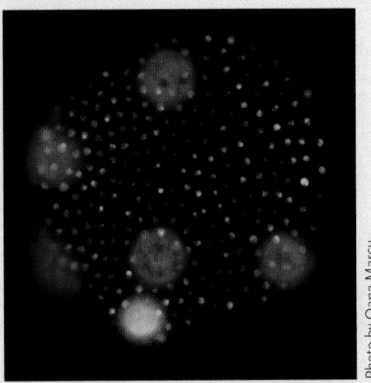

Photo by Oana Marcu

Figure 1
Volvox carteri *under heat stress. ROS indicated by green fluorescence.*

changing environments. In fact, some scientists argue that meiosis exists not to create just any variability but to generate "repaired" chromosomes to be passed on to the next generation (see "People behind Biology"). As you work through the ideas in this section, try to envision how you could pass a "perfect" copy of chromosome 6 to your children even if both copies of chromosome 6 you inherited from your parents are damaged.

The variability produced by sexual reproduction is apparent all around us, particularly in the human population. Except for identical twins, no two humans look alike, act alike, or have identical biochemical and physiological characteristics, even if they are members of the same immediate family. Other species that reproduce sexually show equivalent variability arising from meiosis.

During meiosis and fertilization, genetic variability arises from four sources: (1) genetic recombination of homologous chromosomes, (2) the differing combinations of maternal and paternal chromosomes segregated to the poles during anaphase I, (3) the differing combinations of recombinant chromatids segregated to the poles during anaphase II, and (4) the particular sets of male and female gametes that unite in fertilization. The four mechanisms, working together, produce so much total variability that no two products of meiosis produced by the same or different individuals and no two zygotes produced by union of the gametes are likely to have the same genetic makeup. Each of these sources of variability is discussed in further detail in the following sections. **Figure 9.13, p. 208,** contrasts the genetically identical daughter cells

arising from mitosis with the diverse daughter cells produced by meiosis.

Genetic Recombination. Recombination, the key genetic event of prophase I, starts when homologous chromosomes pair (**Figure 9.14, p. 209,** step 1). Recall that although homologous chromosomes have the same genes in the same order, they likely carry different versions of those genes (alleles). This means that the underlying DNA sequence is similar enough to form the basis of meiotic pairing, yet different enough to generate novel combinations after recombination. (Recall Lederberg's multiple auxotrophic *E. coli* mutants here; the idea is the same.) As the homologous chromosomes pair, they are held together tightly by a protein framework called the synaptonemal complex (**Figure 9.15, p. 209**). Supported by this framework, regions of homologous chromatids exchange segments, producing new combinations of alleles (see Figure 9.14, step 2). Recall that the exchange process is very precise and involves the breakage and rejoining of DNA molecules by enzymes (Figure 9.1, p. 190). When the exchange is complete toward the end of prophase I, the synaptonemal complex disassembles and disappears. If you now follow meiosis I and II through to the end in your mind, notice that each of the four resulting nuclei receives one of these four chromatids (see Figure 9.14, p. 209, step 3); two receive unchanged "parental" chromatids, and two receive chromatids that have new combinations of alleles due to recombination; these are called *recombinants*.

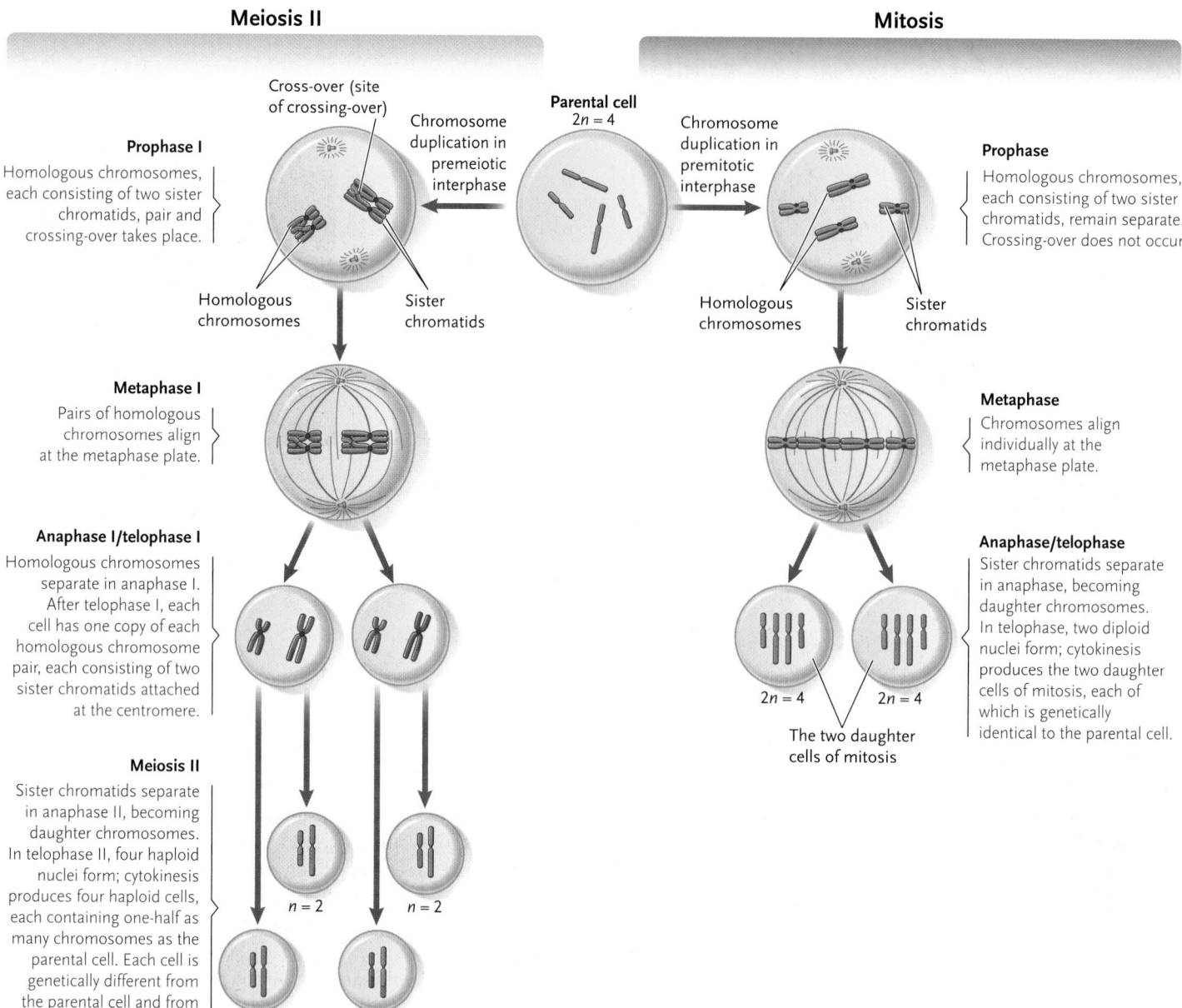

Meiosis II

Prophase I
Homologous chromosomes, each consisting of two sister chromatids, pair and crossing-over takes place.

Cross-over (site of crossing-over)

Chromosome duplication in premeiotic interphase

Homologous chromosomes

Sister chromatids

Parental cell
$2n = 4$

Mitosis

Chromosome duplication in premitotic interphase

Prophase
Homologous chromosomes, each consisting of two sister chromatids, remain separate. Crossing-over does not occur.

Homologous chromosomes

Sister chromatids

Metaphase I
Pairs of homologous chromosomes align at the metaphase plate.

Metaphase
Chromosomes align individually at the metaphase plate.

Anaphase I/telophase I
Homologous chromosomes separate in anaphase I. After telophase I, each cell has one copy of each homologous chromosome pair, each consisting of two sister chromatids attached at the centromere.

Anaphase/telophase
Sister chromatids separate in anaphase, becoming daughter chromosomes. In telophase, two diploid nuclei form; cytokinesis produces the two daughter cells of mitosis, each of which is genetically identical to the parental cell.

$2n = 4$ $2n = 4$

The two daughter cells of mitosis

Meiosis II
Sister chromatids separate in anaphase II, becoming daughter chromosomes. In telophase II, four haploid nuclei form; cytokinesis produces four haploid cells, each containing one-half as many chromosomes as the parental cell. Each cell is genetically different from the parental cell and from each other.

$n = 2$ $n = 2$

$n = 2$ $n = 2$

Figure 9.13

Comparison of key steps in meiosis and mitosis. Both diagrams use an animal cell as an example. Maternal chromosomes are shown in red; paternal chromosomes are shown in blue.

The physical effect of recombination can be seen later in prophase I, when increased condensation of the chromosomes thickens the chromosomes enough to make them visible under the light microscope (see Figure 9.11, p. 204, steps 3 and 4). Regions in which nonsister chromatids cross one another, called **cross-overs** or **chiasmata** (singular, *chiasma* = crosspiece), clearly show that two of the four chromatids have exchanged segments. Because of the shape produced, the recombination process is also called **crossing-over**.

Note that illustrations of recombination usually show chromosomes "paired" side by side, with only the closest chromatids participating in recombination (see Figure 9.14); however, chromosomes actually pair "one on top of the other" such that any two of the four chromatids can participate in a given recombination event. Recombination takes place largely at random, at almost any position along the chromosome arms. Several events likely occur at various locations along all chromatids.

CONCEPT FIX Notice in Figure 9.14 that a recombination event does not just "switch" the alleles of a given gene in a localized area. Rather, all of the DNA sequence stretching from the site of recombination to the ends of the participating chromatids is exchanged. ⬡

Random Segregation. Random segregation of chromosomes of maternal and paternal origin accounts for the second major source of genetic variability in meiosis.

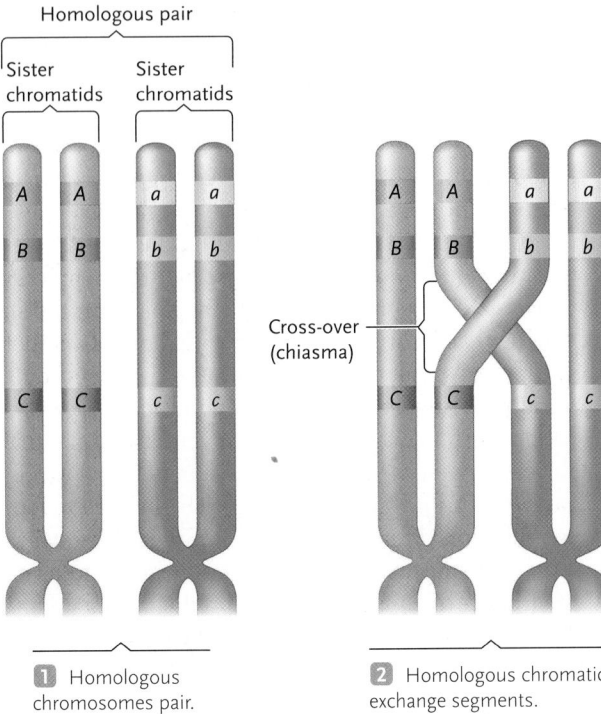

Homologous pair

Sister chromatids — Sister chromatids

A A a a
B B b b
C C c c

Cross-over (chiasma)

A A a a
B B b b
C C c c

A a
B b
C c

A a
B b
c c

1 Homologous chromosomes pair.

2 Homologous chromatids exchange segments.

3 Homologous chromosomes separate at first meiotic division.

Figure 9.14
Effects of the exchange between chromatids that accomplishes genetic recombination. Although the closest chromatids are shown crossing over, any pair of nonsister chromatids may recombine. The letters indicate two alleles (e.g., A and a) for each of three genes. In the meiocyte, the alleles are in the combination of A–B–C and a–b–c on their respective homologues. As a result of this recombination event, two of the chromatids, the recombinants, have a new combination: a–b–C and A–B–c.

Recall that the maternal and paternal members of each homologous pair are different in that they typically carry different alleles of many of the genes on that chromosome. During prometaphase I, spindle microtubules make connections to kinetochores. For each homologous pair, one chromosome makes spindle connections leading to one pole and the other chromosome connects to the opposite pole in a random choice. In making these connections, all the maternal chromosomes may connect to one pole and all the paternal chromosomes may connect to the opposite pole. Or, as is much more likely, a random combination of maternal and paternal chromosomes will be segregated to a given spindle pole **(Figure 9.16, p. 210)**.

The number of possible random combinations depends on the number of chromosome pairs in a species. For example, the 39 chromosome pairs in dogs allow 2^{39} different combinations of maternal and paternal chromosomes to be delivered to the poles, producing potentially 550 billion genetically different gametes from this source of variability alone. Note that this random partitioning of maternal and paternal chromosomes is responsible for the independent assortment of the alleles of two genes in Mendel's experiments with garden peas described in Chapter 10.

Alternative Combinations at Meiosis II. If you look carefully at the cells drawn in metaphase II in Figure 9.14, you will see that the chromosomes are still replicated, and, as a result of recombination in prophase I, each chromosome carries one recombinant chromatid and one nonrecombinant chromatid. Notice that, in this case, the chromosomes have aligned at metaphase II with both recombinant chromatids attached to the

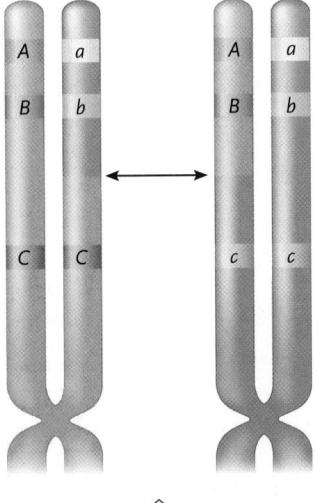

Sister chromatids of one of a homologous pair of chromosomes

Sister chromatids of the other of a homologous pair of chromosomes

Synaptonemal complex

Courtesy Diter von Wettstein

Figure 9.15
The synaptonemal complex as seen in a meiotic cell of the fungus *Neotiella*.

same spindle pole. However, since the attachment of spindles to kinetochores is random at this stage, you should be able to see that it is just as likely that these chromosomes *could* have lined up, with the smaller chromosome sending its recombinant chromatid to one pole and the larger chromosome sending its recombinant chromatid to the opposite pole. The resulting daughter cells will be genetically different, depending on how the chromosomes align in metaphase II.

Random Fertilization. The haploid products of meiosis are genetically diverse. The random combination of these cells (or their descendants) during fertilization is a matter of chance that amplifies the variability of sexual reproduction. For example, if we consider only the variability available from random separation of

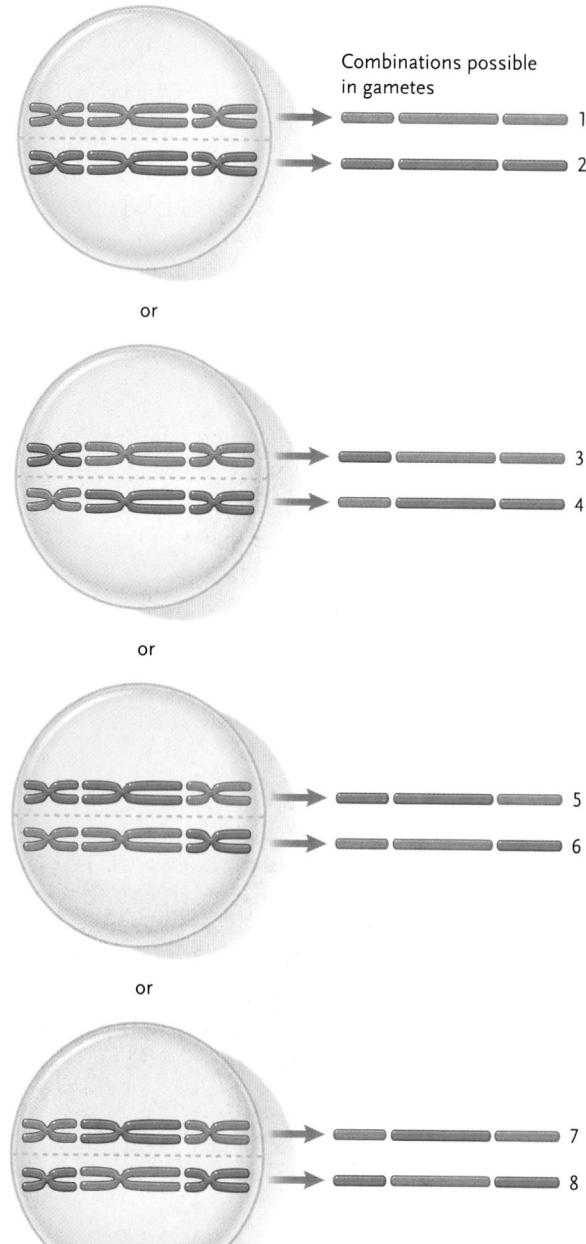

Figure 9.16

Independent assortment. Possible outcomes of the random spindle connections of three pairs of chromosomes at metaphase I of meiosis. Maternal chromosomes are red; paternal chromosomes are blue. There are four possible patterns of connections, giving eight possible combinations of maternal and paternal chromosomes in gametes (labelled 1–8).

homologous chromosomes at meiosis I along with that from random fertilization, the possibility that two children of the same human parents could receive the same combination of maternal and paternal chromosomes is 1 chance out of $(2^{23})^2$ or 1 in about 70 trillion, a number that far exceeds the number of humans who have ever lived. The further variability introduced by recombination and shuffling at meiosis II makes it practically impossible for humans and most other

sexually reproducing organisms to produce genetically identical gametes or offspring. The only exception is identical twins (or identical triplets, identical quadruplets, and so forth), which arise not from the combination of identical gametes during fertilization but from mitotic division of a single fertilized egg into separate cells that give rise to genetically identical individuals.

We have just seen that meiosis has three outcomes that are vital to sexual reproduction. This process reduces the chromosomes to the haploid number so that they can be brought together with those of another individual without doubling the usual chromosome number during fertilization. Through genetic recombination and random separation of maternal and paternal chromosomes, meiosis produces genetic variability in gametes; further variability is provided by the random combination of gametes in fertilization. These ideas form the "mechanics" that underlie the patterns of inheritance of traits in sexually reproducing organisms discovered by Mendel and described in Chapter 10.

STUDY BREAK

1. Which phase (diploid or haploid) dominates the respective life cycles of animals, plants, and fungi?
2. What are the two functions of meiosis?
3. What are the four sources of genetic variability in sexually reproducing organisms?
4. What is nondisjunction, and how does it occur?

9.4 Mobile Elements

Our examples have so far involved two participating DNA molecules that have always been at least partially homologous and that have always originated from two different individuals. However, one of the most interesting examples of genetic recombination in nature shows neither of these characteristics. All organisms appear to contain particular segments of DNA, called **mobile elements**, that can move from one place to another; they cut and paste sections of DNA using a type of recombination that does *not necessarily require homology*. Sometimes called *jumping genes,* these elements normally move from place to place *within the genome of a given cell*. The following section describes these fascinating elements in more detail.

9.4a Insertion Sequence Elements and Transposons Are the Two Major Types of Prokaryotic Mobile Elements

Mobile elements are also known by the more specific term **transposable elements (TEs)**, and their mechanism of movement, involving nonhomologous recombination, is called **transposition**. Transposition usually

a. Cut-and-paste transposition

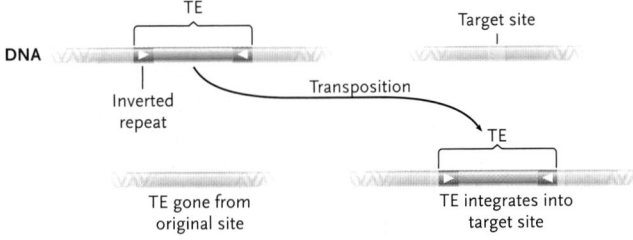

b. Copy-and-paste transposition

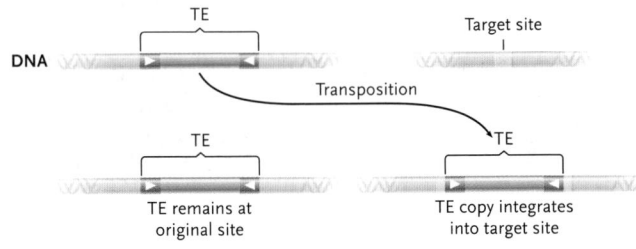

Figure 9.17

Two transposition processes for transposable elements. **(a)** Cut-and-paste transposition, in which the TE leaves one location in the DNA and moves to a new location. **(b)** Copy-and-paste transposition, in which a copy of the TE moves to a new location, leaving the original TE behind.

occurs at a low frequency in either of two ways, depending on the type of element: (1) a cut-and-paste process, in which the TE leaves its original location and transposes to a new location **(Figure 9.17a),** and (2) a copy-and-paste process, in which a copy of a TE transposes to a new location, leaving the original TE behind (Figure 9.17b). For most TEs, transposition starts with contact between the TE and the target site. This also means that TEs do not exist free of the DNA in which they are integrated; hence, the popular name *jumping genes* is actually inaccurate. TEs are never "in the air" between one location and another. TEs are important because of the genetic changes they cause. For example, they produce mutations by transposing into genes and knocking out their functions, and they increase or decrease gene expression by transposing into regulatory sequences of genes. As such, TEs are biological mutagens that increase genetic variability.

Bacterial TEs were discovered in the 1960s. They have been shown to move from site to site within the bacterial chromosome, between the bacterial chromosome and plasmids, and between plasmids. The frequency of transposition is low but constant for a given TE. Some bacterial TEs insert randomly, at any point in the DNA, whereas others recognize certain sequences as "hot spots" for insertion and insert preferentially at these locations.

The two major types of bacterial TEs are **insertion sequences (ISs)** and **transposons.** Insertion sequences are the simplest TEs. They are relatively small and contain only genes for their transposition, notably the gene for **transposase,** an enzyme that catalyzes some of the recombination reactions for inserting or removing the TE from the DNA **(Figure 9.18).** At each of the two ends of an IS is a short **inverted repeat** sequence—the same DNA sequence running in opposite directions (shown by directional arrows in the figure). The inverted repeat sequences enable the transposase enzyme to identify the ends of the TE when it catalyzes transposition. The inverted repeat sequence is an IS element on both the F factor and the bacterial chromosome that provides the homology needed for the creation of the Hfr strains described in Section 9.2.

The second type of bacterial TE, called a transposon, has an inverted repeat sequence at each end enclosing a central region with one or more genes. In a number of bacterial transposons, the inverted repeat sequences are insertion sequences, which provide the transposase for movement of the element (see Figure 9.18). Additional genes in the central region typically code for antibiotic resistance; they can originate from the main bacterial chromosome or from plasmids. These non-IS genes included in transposons are carried along as the TEs move from place to place.

Many antibiotics, such as penicillin, erythromycin, tetracycline, ampicillin, and streptomycin, which were once successful in curing bacterial infections, have lost much of their effectiveness because of resistance genes carried in transposons. Movements of the transposons, particularly to plasmids that can be transferred by conjugation within and between bacterial species, greatly increase the spread of genes, providing antibiotic resistance to infecting cells. Resistance genes have made many bacterial diseases difficult or impossible to treat with standard antibiotics.

9.4b Transposable Elements Were First Discovered in Eukaryotes

TEs were first discovered in a eukaryote, maize (corn), in the 1940s by Barbara McClintock, a geneticist working at the Cold Spring Harbor Laboratory in New York. McClintock noted that some mutations affecting kernel and leaf colour appeared and disappeared rapidly

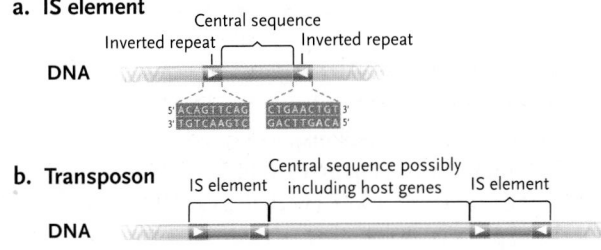

Figure 9.18

Types of bacterial TEs. **(a)** IS element. **(b)** Transposon in which the central sequence is transposed by flanking IS elements.

Figure 9.19

Barbara McClintock and corn kernels showing different colour patterns due to the movement of transposable elements. As TEs move into or out of genes, controlling pigment production in developing kernels, the ability of cells and their descendants to produce the dark pigment is destroyed or restored. The result is random patterns of pigmented and colourless (yellow) segments in individual kernels.

under certain conditions. Mapping the alleles by linkage studies produced a surprising result: the map positions changed frequently, indicating that the alleles could move from place to place in the corn chromosomes. Some of the movements were so frequent that changes in their effects could be noticed at different times in a single developing kernel (**Figure 9.19**).

When McClintock first reported her results, her findings were regarded as an isolated curiosity, possibly applying only to corn. This was because the then-prevailing opinion among geneticists was that genes are fixed in the chromosomes and do not move to other locations. Her conclusions were widely accepted only after TEs were detected and characterized in bacteria in the 1960s. By the 1970s, further examples of TEs were discovered in other eukaryotes, including yeast and mammals. McClintock was awarded a Nobel Prize in 1983 for her pioneering work, after these discoveries confirmed her early findings that TEs are probably universally distributed among both prokaryotic and eukaryotic organisms.

9.4c Eukaryotic Transposable Elements Are Classified as Transposons or Retrotransposons

Eukaryotic TEs fall into two major classes: transposons and retrotransposons. They are distinguished by the way the TE sequence moves from place to place in the DNA. Eukaryotic transposons are similar to bacterial transposons in their general structure and in the ways they transpose. However, members of the other class of eukaryotic TEs, the **retrotransposons**, transpose by a copy-and-paste mechanism that is unlike any of the other TEs we have discussed. Retrotransposons have this name because transposition occurs via an intermediate RNA copy of the TE (**Figure 9.20**). First, the retrotransposon, which is a DNA element

integrated into the chromosomal DNA, is transcribed into a complementary RNA copy. Next, an enzyme called **reverse transcriptase**, which is encoded by one of the genes of the retrotransposon, uses the RNA as a **template** to make a DNA copy of the retrotransposon.

The DNA copy is then inserted into the DNA at a new location, leaving the original in place. This insertion step involves breaking and rejoining DNA backbones, as we have seen several times in this chapter.

Once TEs are inserted into chromosomes, they become more or less permanent residents, duplicated and passed on during cell division along with the rest of the DNA. TEs inserted into the DNA of reproductive cells may be inherited, thereby becoming a permanent part of the genetic material of a species. Long-standing TEs are subject to mutation along with other sequences in the DNA. Such mutations may accumulate in a TE, gradually altering it into a nonmobile, residual sequence in the DNA. The DNA of many eukaryotes, including humans, contains a surprising amount of nonfunctional TE sequence likely created in this way.

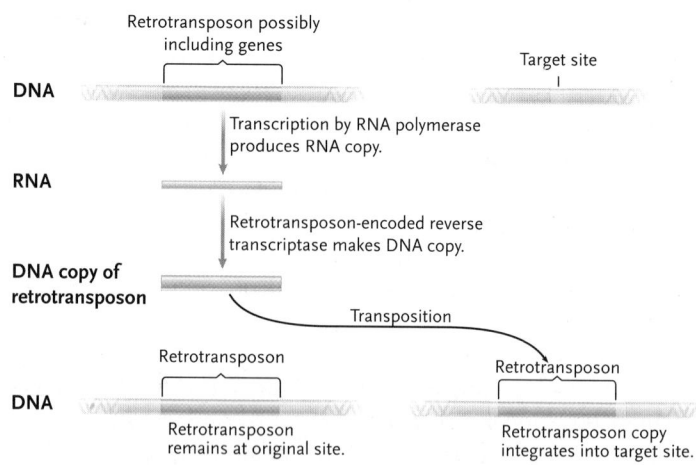

Figure 9.20

Transposition of a eukaryotic retrotransposon to a new location by means of an intermediate RNA copy.

9.4d Retrotransposons Are Similar to Retroviruses

The RNA to DNA reverse transcription associated with retrotransposon movement is strikingly similar to that employed by a class of eukaryotic viruses called **retroviruses.** When a retrovirus infects a host cell, a reverse transcriptase carried in the virus particle is released and copies the single-stranded RNA genome into a double-stranded DNA copy. The viral DNA is then inserted into the host DNA (by genetic recombination), where it is replicated and passed to progeny cells during cell division. Similar to the prophage of bacteria, the inserted viral DNA is known as a **provirus (Figure 9.21).**

Retroviruses are found in a wide range of organisms, with most so far identified in vertebrates. You, as well as other humans and mammals, contain several retroviruses in your genome as proviruses. In total, retrotransposons and retroviruses of all types occupy some 40% of the human genome!

Although many of the retroviruses do not produce infectious virus particles, they sometimes cause DNA rearrangements such as deletions and translocations. Such changes may alter the relative position of DNA sequences on the chromosome and, in turn, disturb the normal regulation of gene expression. Given your knowledge of transduction by bacterial viruses described earlier in this chapter, you will not be surprised to hear that retroviruses sometimes pick up host eukaryotic genes and move them to recipients. Such genes may become abnormally active through the effects of regulatory sequences located in the TE itself or the DNA nearby. Certain forms of cancer have been linked to this type of abnormal activation of genes that are important in regulating cell division (see Section 14.4). In one of the most dramatic examples, a cellular gene is transported to an infected cell by the avian sarcoma retrovirus. The cellular gene is overexpressed in the new environment, resulting in uncontrolled growth of infected cells, leading to tumours in infected birds.

STUDY BREAK

Among eukaryotic mobile elements, how do transposons, retrotransposons, and retroviruses differ?

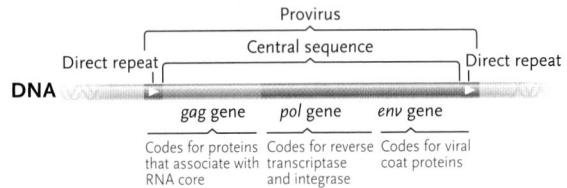

Figure 9.21

A mammalian retrovirus in the provirus form in which it is inserted into chromosomal DNA. The direct repeat at either end contains sequences capable of acting as enhancer, promoter, and termination signals for transcription. The central sequence contains genes coding for proteins, concentrated in the *gag, pol,* and *env* regions. The provirus of **human immunodeficiency virus (HIV),** the virus that causes **acquired immune deficiency syndrome (AIDS),** takes this form.

This has been a long chapter. We hope that, taken together, all of these ideas will help you understand the balance that life must strike between the stability and the plasticity of its genetic material. On the one hand, DNA must be faithfully replicated and passed to the next generation. Lack of quality control at this step would allow widespread random mutations to undermine the selection and preservation of good combinations of alleles. On the other hand, any system that made only perfectly "photocopied" DNA available for the next generation would be doomed as well; a wide variety of diverse genetic "solutions" are needed for a population to survive in constantly changing environments that are impossible to anticipate.

Genetic recombination is central to many processes that contribute changes to the sequence of DNA in all forms of life. (And we did not even discuss interesting examples of developmental genetic recombination in infecting parasites or the cells of the developing immune system, or foreign DNA taken up by rotifers.) The genetic elements discussed in this chapter, particularly plasmids and retroviruses, often act as natural genetic engineers by moving genes between species. Chapter 15 describes how human genetic engineers manipulate and clone DNA.

Review

To access course materials such as Aplia and other companion resources, please visit www.NELSONbrain.com.

9.1 Mechanism of Genetic Recombination

- Genetic recombination requires two DNA molecules that differ from one another; a mechanism for bringing the DNA molecules into close proximity; and a collection of enzymes to cut, exchange, and paste the DNA back together.
- Homology allows DNA on different molecules to line up and recombine precisely.
- Enzymatic cutting and pasting of both DNA backbones from each of the two DNA molecules is required for each recombination event (see Figure 9.1).

9.2 Genetic Recombination in Bacteria

- Study of bacterial recombination requires strains carrying different alleles.
- Lederberg and Tatum mutated bacteria to create strains that were different in their ability to manufacture certain amino acids and vitamins.
- In bacteria, the DNA of the bacterial chromosome may recombine with DNA brought into close proximity from another cell.
- Three primary mechanisms bring DNA into bacterial cells from the outside: conjugation, transformation, and transduction.
- In conjugation, two bacterial cells form a cytoplasmic bridge allowing at least some of the DNA of one cell to move into the other cell. The donated DNA can then recombine with homologous sequences of the recipient cell's DNA.
- *Escherichia coli* bacteria that are able to act as DNA donors in conjugation have an F factor, making them F⁺; recipients have no F factor and are F⁻. In Hfr strains of *E. coli,* the F factor is a part of the main chromosome. As a result, genes from the main chromosome can be transferred into F⁻ cells along with a portion of the F factor DNA (see Figure 9.6).
- In transformation, intact cells of some species absorb pieces of DNA released from cells that have disintegrated. The entering DNA fragments can recombine with the recipient cell's DNA.
- In transduction, DNA is transferred from one cell to another "by mistake" inside the head of an infecting virus (see Figure 9.7).
- Since generalized transduction transfers random fragments of the host chromosome, all host genes are transferred at equal frequency. Specialized transduction only transfers genes lying close to the point of insertion of the prophage (see Figure 9.8).

9.3 Genetic Recombination in Eukaryotes: Meiosis

- The time and place of meiosis follow one of three major pathways in the life cycles of eukaryotes, which reflect the portions of the life cycle spent in the haploid and diploid phases and whether mitotic divisions intervene between meiosis and the formation of gametes (see Figure 9.9).
- Animals have a diploid life cycle in which the diploid phase is multicellular and the haploid phase is unicellular. Meiosis is followed by gamete formation.
- Plants and some fungi have an alternating-generations life cycle in which either haploid or diploid phases are multicellular and both of which divide by mitosis. The diploid sporophytes are produced by fertilization, and the haploid gametophytes are produced by mitotic divisions of the spores formed by meiosis.
- Most fungi exhibit a haploid life cycle in which the haploid phase is multicellular and the diploid phase is limited to a single cell produced by fertilization, which then immediately undergoes meiosis.
- In animals, the products of meiosis are haploid gametes. The diploid phase of the life cycle is then restored when one gamete fuses with another at fertilization. In plants, meiosis occurs in some of the cells of the diploid sporophytes and produces a generation of haploid spores. These spores then divide by mitosis to produce multicellular gametophytes. Gametes are formed from mitotic division of specific sporophyte tissues.
- The functions of meiosis are to reduce the chromosome number (from diploid to haploid) and to generate genetic diversity in sexually reproducing organisms (see Figure 9.10).
- Meiosis occurs only in eukaryotes that reproduce sexually and only in organisms that are at least diploid—that is, organisms that have at least two representatives of each chromosome (see Figure 9.11).
- DNA replicates and the chromosomal proteins are duplicated during the premeiotic interphase, producing two copies, the sister chromatids, of each chromosome.
- During prophase I of the first meiotic division (meiosis I), the replicated chromosomes condense and come together and pair as the spindle forms in the cytoplasm.
- While they are paired, the chromatids of homologous chromosomes undergo recombination by breaking the covalent bonds of the DNA backbones, matching complementary sequences on nonsister chromatids, exchanging the ends, and restoring the bonds.
- During prometaphase I, the nuclear envelope breaks down, the spindle enters the area of the former nucleus, and kinetochore microtubules leading to opposite spindle poles attach to one kinetochore of each pair of sister chromatids of homologous chromosomes.
- At metaphase I, spindle microtubule movements have aligned the tetrads on the metaphase plate, the equatorial plane between the two spindle poles. The connections of kinetochore microtubules to opposite poles ensure that the homologous pairs separate and move to opposite spindle poles during anaphase I, reducing the chromosome number to the haploid value. Each chromosome at the poles still contains two chromatids.
- Telophase I and interkinesis are brief and transitory stages; no DNA replication occurs during interkinesis.

During these stages, the remaining single spindle of the first meiotic division disassembles and the microtubule subunits are available to reassemble into two new spindles for the second division.

- During prophase II, the chromosomes condense and the spindle reorganizes. During prometaphase II, the nuclear envelope breaks down, the spindle enters the former nuclear area, and spindle microtubules leading to opposite spindle poles attach to the two kinetochores of each chromosome. At metaphase II, the chromosomes become aligned on the metaphase plate. The connections of kinetochore microtubules to opposite spindle poles ensure that during anaphase II, the chromatids of each chromosome are separated and segregate to those opposite spindle poles.

- During telophase II, the chromosomes decondense to their extended interphase state, the spindles disassemble, and new nuclear envelopes form. The result is four haploid cells, each containing half the number of chromosomes present in a G1 nucleus of the same species. In animals, these products of meiosis function as gametes that fuse in fertilization; in plants, these products function as spores that divide by mitosis to form gametophytes.

- Meiosis II differs from mitosis in that meiosis II occurs only in reproductive tissue, is not preceded by an S phase, and results in genetically different daughter cells (see Figure 9.13).

- Nondisjunction occurs when both members of a pair of homologous chromosomes connect to spindles from the same pole. Following anaphase, one pole then receives both copies of the pair, and the other pole receives none. The overall results (following normal meiosis II) are haploid products of meiosis that have two copies of the given chromosome. After fertilization, the resulting zygote will therefore have three copies of the chromosome instead of two.

- In many eukaryotes, including most animals, one or more pairs of chromosomes, called the sex chromosomes, are different in male and female individuals of the same species.

- Recombination is the first source of the genetic variability produced by meiosis. During recombination, chromatids generate new combinations of alleles by physically exchanging segments. The exchange process involves precise breakage and joining of DNA molecules. It is catalyzed by enzymes and occurs while the homologous chromosomes are held together tightly by the synaptonemal complex. The cross-overs visible between the chromosomes at late prophase I reflect the exchange of chromatid segments that occurred during the molecular steps of genetic recombination (see Figures 9.13 and 9.14).

- The random segregation of homologous chromosomes is the second source of genetic variability produced by meiosis. The homologous pairs separate at anaphase I of meiosis, creating random combinations of maternal and paternal chromosomes travelling to each of the two spindle poles (see Figure 9.15).

- Random segregation of the chromatids of replicated chromosomes at meiosis II is a third mechanism for generating diversity.

- Random joining of male and female gametes in fertilization is the fourth source of genetic variability.

9.4 Mobile Elements

- Both prokaryotic and eukaryotic organisms contain transposable elements (TEs)—DNA sequences that can move from place to place in the DNA. The TEs may move from one location in the DNA to another or generate duplicated copies that insert at new locations while leaving the parent copy in its original location (see Figure 9.17).

- Genes of the host cell DNA may become incorporated into a TE and may be carried with it to a new location. There the genes may become abnormally active when placed near sequences that control the activity of genes within the TE or near the control elements of active host genes (see Figure 9.18).

- Eukaryotic TEs occur as transposons, which release from one location in the DNA and insert at a different site, or as retrotransposons, which move by making an RNA copy, which is then used to assemble a DNA copy that is inserted at a new location. The parent copy remains at the original location. Like retrotransposons, retroviruses make a DNA copy of their RNA genome and insert this into the host's chromosome. Retroviruses may have evolved from retrotransposons (see Figure 9.20).

- TE-instigated abnormal activation of genes regulating cell division has been linked to the development of some forms of cancer in humans and other complex animals.

Questions

Self-Test Questions

1. If recombination occurred in a bacterium undergoing transformation as shown in the figure, what would be the final genotype of the bacterial chromosome?

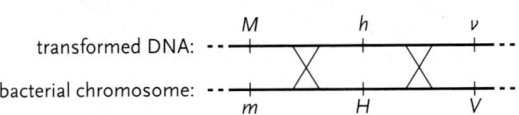

transformed DNA:
bacterial chromosome:

   a. *MHv*
   b. *MHv*
   c. *mHV*
   d. *mhV*

2. Which of the following events turns F⁺ cells into Hfr cells?
   a. replication of the F factor by rolling circle replication
   b. recombination between the F factor and the recipient chromosome
   c. transfer of the F factor to a recipient cell
   d. integration of the F factor into the host chromosome

3. Which of the following describes an aspect of bacterial conjugation?
   a. Recipient cells incorporate single-stranded DNA from donors into their chromosome.
   b. DNA from dead donor bacteria is transferred to live recipient cells.

c. Genes are transferred in a particular order from donors to recipients.

d. A virus is required for the transfer of DNA from donors to recipient cells.

4. If a virus is in the lysogenic phase of its life cycle, what is it doing?

a. bursting the host cell

b. transducing genes into a bacterial cell

c. assembling viral particles for cell rupture

d. being expressed and/or copied as a part of the host DNA

5. If the diploid number of an organism is 6, which stage of cell division does the figure below represent?

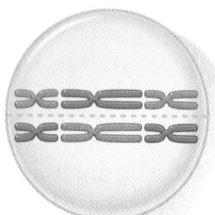

a. mitotic metaphase

b. meiotic metaphase I

c. meiotic metaphase II

d. could be either mitotic metaphase or meiotic metaphase II

6. Imagine that you are helping your younger brother with his biology homework. You notice that he has written down in his notes that "Plants are haploid and make gametes by mitosis. Animals are diploid and make gametes by meiosis." What should your response be?

a. Yes, plants make gametes by mitosis and most animals make gametes by meiosis. However, animals and plants both have a haploid and a diploid stage of their life cycle.

b. Yes, both plants and animals make gametes. The difference is that plant gametes divide by mitosis but animal gametes do not.

c. Yes, plants are simpler organisms than animals and have fewer chromosomes in their cells.

d. Yes, plants and animals use meiosis for different purposes. Animals make gametes by meiosis while plants make zygotes.

7. As a result of genetic recombination, each of the chromosomes of your family dog contains a different combination of alleles compared to those of its brothers and sisters. When did this recombination occur?

a. when your dog's parents made gametes

b. when your dog was a newly fertilized, single-celled zygote

c. when your dog grew from a zygote to a multi-cellular organism

d. when your dog reached sexual maturity

8. The number of human chromosomes in a cell in prophase I of meiosis is ___ and in telophase II is ___.

a. 92; 46

b. 46; 23

c. 23; 23

d. 23; 16

9. Consider a penguin gamete. The amount of DNA (pg) in this gamete is defined as $1C$. The number of chromosomes in this gamete is defined as $1n$.

That is, the value of C and the value of n are equal in a penguin gamete. During which other stage of penguin cell division would the value of C and the value of n also be equal?

a. during G1; both n and C equal 2

b. during G2; both n and C equal 4

c. during metaphase of meiosis II; both n and C equal 1

d. during metaphase of mitosis; both n and C equal 2

10. Which of the following is a feature of mobile elements?

a. They have no negative impact on host cells.

b. They have inverted repeat sequences at their centre.

c. They make use of recombination that does not require extensive homology.

d. They are a kind of virus.

Questions for Discussion

1. You set up an experiment like the one carried out by Lederberg and Tatum, mixing millions of *E. coli* of two strains with the following genetic constitutions.

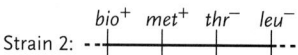

Among the bacteria obtained after mixing, you find some cells that do not require threonine, leucine, or biotin to grow but still need methionine. How might you explain this result?

2. You have a technique that allows you to measure the amount of DNA in a cell nucleus. You establish the amount of DNA in a sperm cell of an organism as your baseline. Which multiple of this amount would you expect to find in a nucleus of this organism at G2 of premeiotic interphase? At telophase I of meiosis? During interkinesis? At telophase II of meiosis?

3. Mutations are changes in DNA sequences that can create new alleles. In which cells of an individual, somatic or meiotic cells, would mutations be of greatest significance to that individual? What about to the species to which the individual belongs?

4. Sometimes pieces of chromosomes can be exchanged in a kind of rearrangement called a reciprocal translocation. Imagine the case in a diploid organism where the end of one chromosome 4 was exchanged for the end of a chromosome 12. The other chromosomes 4 and 12 remained uninvolved and normal in structure. What shape might these four chromosomes take as they tried to pair during meiosis I?

5. Experimental systems have been developed in which TEs can be induced to move under the control of a researcher. Following the induced transposition of a yeast TE, two mutants were identified with altered activities of enzyme X. One of the mutants lacked enzyme activity completely, whereas the other had five times the enzyme activity of normal cells. Both mutants were found to have the TE inserted into the gene for enzyme X. Propose hypotheses for how the two different mutant phenotypes were produced.

Rabbits, showing genetic variation in coat colour.

Ornitolog82/iStock/Thinkstock

10

Mendel, Genes, and Inheritance

WHY IT MATTERS

Parties and champagne were among the last things on Ernest Irons's mind on New Year's Eve, 1904. Irons, a medical intern, was examining a blood specimen from a new patient and was sketching what he saw through his microscope—peculiarly elongated red blood cells **(Figure 10.1, p. 218).** He and his supervisor, James Herrick, had never seen anything like them. The shape of the cells was reminiscent of a sickle, a cutting tool with a crescent-shaped blade.

The patient had complained of weakness, dizziness, shortness of breath, and pain. His father and two sisters had died from mysterious ailments that had damaged their lungs or kidneys. Did those deceased family members also have sickle-shaped red cells in their blood? Was there a connection between the abnormal cells and the ailments? How did the cells become sickled?

The medical problems that baffled Irons and Herrick killed their patient when he was only 32 years old. The patient's symptoms were characteristic of a genetic disorder now called *sickle cell disease*. This disease develops when a person has received two copies of a gene (one from each parent) that codes for an altered subunit of hemoglobin, the oxygen-transporting protein in red blood cells. When oxygen supplies are low, the altered hemoglobin forms long, fibrous, crystal-like structures that push red blood cells into the sickle shape. The altered protein differs from the normal protein by just a single amino acid.

The sickled red blood cells are too elongated and inflexible to pass through the capillaries, the smallest vessels in the circulatory system. As a result, the cells block the capillaries. The surrounding tissues become starved for oxygen and saturated

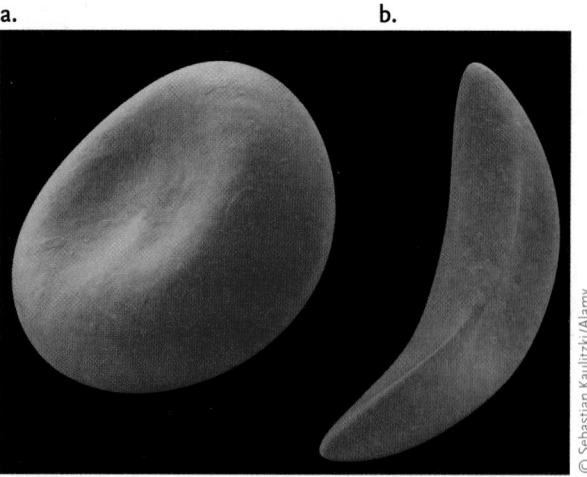

Figure 10.1
Red blood cell shape in sickle cell disease.

Figure 10.2
Gregor Mendel (1822–1884), the founder of genetics.

with metabolic wastes, causing the symptoms experienced by Irons and Herrick's patient. The problem worsens as oxygen concentration falls in tissues and more red blood cells are pushed into the sickled form. (You will learn more about sickle cell disease in this chapter and in Chapter 11.)

Researchers have studied sickle cell disease in great detail at both the molecular and the clinical level. You may find it curious, however, that our understanding of sickle cell disease—and all other heritable traits—actually began with studies of pea plants in a monastery garden.

Fifty years before Ernest Irons sketched sickled red blood cells, a scholarly monk named Gregor Mendel **(Figure 10.2)** used garden peas to study patterns of inheritance. To test his hypotheses about inheritance, Mendel bred generation after generation of pea plants and carefully observed the patterns by which parents transmit traits to their offspring. Through his experiments and observations, Mendel discovered the fundamental rules that govern inheritance. His discoveries and conclusions founded the science of genetics and still have the power to explain many of the puzzling and sometimes devastating aspects of inheritance that continue to occupy our attention.

10.1 The Beginnings of Genetics: Mendel's Garden Peas

Until about 1900, scientists and the general public believed in the **blending theory of inheritance**, which suggested that hereditary traits blend evenly in offspring through mixing of the parents' blood, much like the effect of mixing coffee and cream. Even today, many people assume that parental characteristics such as skin colour, body size, and facial features blend evenly in their offspring, with the traits of the children appearing about halfway between those of their parents. Yet if blending takes place, why don't extremes, such as very tall and very short individuals, gradually disappear over generations as repeated blending takes place? Also, why do children with blue eyes keep turning up among the offspring of brown-eyed parents?

Gregor Mendel's experiments with garden peas, performed in the 1860s, provided the first answers to these questions and many more. Mendel was an Augustinian monk who lived in a monastery in Brünn, now part of the Czech Republic. But he had an unusual education for a monk in the mid-nineteenth century. He had studied mathematics, chemistry, zoology, and botany at the University of Vienna under some of the foremost scientists of his day. He grew up on a farm and was well aware of agricultural principles and their application. He kept abreast of breeding experiments published in scientific journals. Mendel also won several awards for developing improved varieties of fruits and vegetables.

In his work with peas, Mendel studied a variety of heritable characteristics called **characters**, such as flower colour or seed shape. A variation in a character, such as purple or white flower colour, is called a **trait**. Mendel established that characters are passed to offspring in the form of discrete hereditary factors, which are now known as genes. Mendel observed that rather than blending evenly, many parental traits appear unchanged in offspring, whereas others disappear in one generation to reappear unchanged in the next. Although Mendel did not know it, the inheritance patterns he observed are the result of the segregation of chromosomes, on which the genes are located, to gametes in meiosis (see Chapter 9). Mendel's methods illustrate, perhaps as well as any experiments in the history of science, how rigorous scientific work is conducted: through observation, making hypotheses, and testing the hypotheses with experiments. Although others had studied inheritance patterns before him, Mendel's most important innovation was his quantitative approach to science, specifically his rigour and statistical analysis in an era when qualitative, purely descriptive science was the accepted practice. In this chapter, we will pay particular attention to the experimental aspect of Mendel's approach to explaining inheritance.

10.1a Mendel Chose True-Breeding Garden Peas for His Experiments

Mendel chose the garden pea (*Pisum sativum*) for his research because the plant could be grown easily in the monastery garden, without elaborate equipment. As in other flowering plants, gametes are produced in structures of the flowers **(Figure 10.3)**. The male gametes are sperm nuclei contained in the pollen, which is produced in the *anthers* of the flower. The female gametes are egg cells, produced in the *carpel* of the flowers. Normally, pea plants **self-fertilize** (also known as **self-pollinate** or, more simply, *self*): sperm nuclei in pollen produced by anthers fertilize egg cells housed in the carpel of the same flower. However, for his experiments, Mendel prevented self-fertilization by cutting off the anthers. Pollen to fertilize these flowers then had to come from a different plant. This technique is called **cross-pollination** or, more simply, a *cross*. This technique allowed Mendel to test the effects of mating pea plants of different parental types.

To begin his experiments, Mendel chose pea plants that were known to be **true-breeding** (also called *pure-breeding*); that is, when self-fertilized or, more simply, *selfed,* they passed traits without change from one generation to the next.

10.1b Mendel First Worked with Single-Character Crosses

Flower colour was among the seven characters Mendel selected for study; one true-breeding variety of peas had purple flowers, and the other true-breeding variety had white flowers (see Figure 10.3). Would these traits blend evenly if plants with purple flowers were cross-pollinated with plants with white flowers?

To answer this question, Mendel took pollen from the anthers of plants with purple flowers and placed it in the flowers of white-flowered plants. He placed the pollen on the *stigma,* the part of the carpel that receives pollen in flowers (see Figure 10.3). He also performed the reciprocal experiment by placing pollen from white-flowered plants on the stigmas of purple-flowered

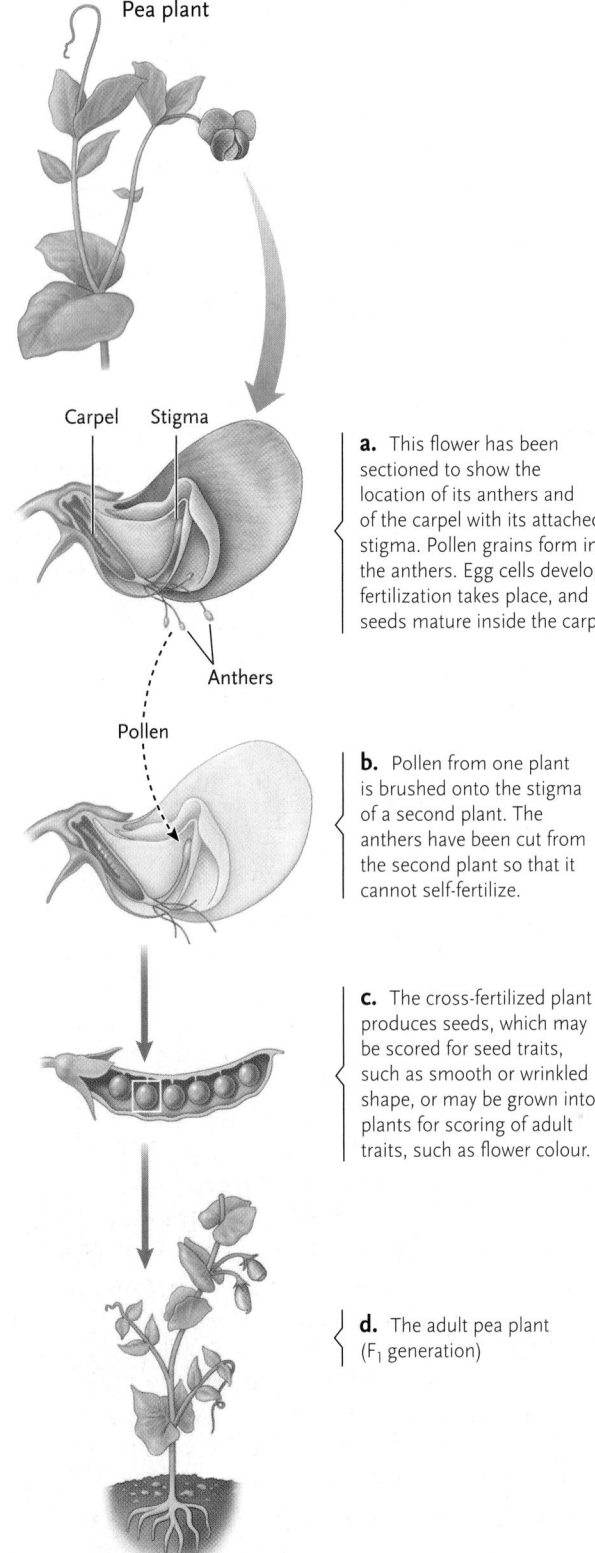

a. This flower has been sectioned to show the location of its anthers and of the carpel with its attached stigma. Pollen grains form in the anthers. Egg cells develop, fertilization takes place, and seeds mature inside the carpel.

b. Pollen from one plant is brushed onto the stigma of a second plant. The anthers have been cut from the second plant so that it cannot self-fertilize.

c. The cross-fertilized plant produces seeds, which may be scored for seed traits, such as smooth or wrinkled shape, or may be grown into plants for scoring of adult traits, such as flower colour.

d. The adult pea plant (F_1 generation)

Research Method Figure 10.3 The garden pea (*Pisum sativum*), the focus of Mendel's experiments.

plants. Seeds were the result of the crosses; each seed contains a zygote, or embryo, that will develop into a new pea plant. The plants that develop from the seeds produced by the cross—the first generation of offspring from the cross—are the **F_1 generation** (F stands for *filial; filius* = son). The plants used in the initial cross are called the parental or **P generation**. The plants that grew from the F_1 seeds all formed purple flowers, as if the trait for white flowers had disappeared. The flowers showed no evidence of blending.

Mendel then allowed the purple-flowered F_1 plants to self, producing seeds that represented the **F_2 generation**. When he planted the F_2 seeds produced by this cross, the white-flowered trait reappeared; both purple-flowered and white-flowered plants were produced. Mendel counted 705 plants with purple flowers and 224 with white flowers, in a ratio that he noted was close to 3:1, or about 75% purple-flowered plants and 25% white-flowered plants.

Mendel made similar crosses that involved six other characters, each with pairs of traits (**Figure 10.4**); for example, the character of seed colour has the traits yellow and green. In all cases, he observed a uniform F_1 generation, in which only one of the two traits was present. In the F_2 generation, the missing trait reappeared, and both traits were present among the offspring. Moreover, the trait present in the F_1 generation was present in a definite, predictable proportion among the F_2 offspring.

10.1c Mendel's Single-Character Crosses Led Him to Propose the Principle of Segregation

Using his knowledge of mathematics, Mendel developed a set of hypotheses to explain the results of his crosses. His first hypothesis was that *the adult plants carry a pair of factors that govern the inheritance of each character*. He correctly deduced that for each character, an organism inherits one factor from each parent.

In modern terminology, Mendel's factors are called *genes,* which are located on chromosomes. The different versions of a gene that produce different traits of a character are called **alleles.** Thus, there are two alleles of the gene that govern flower colour in garden peas: one allele for purple flowers and the other allele for white flowers. Organisms with two copies of each gene are now known as diploids; the two alleles of a gene in a diploid individual may be identical or different.

How can the disappearance of one of the traits, such as white flowers, in the F_1 generation and its reappearance in the F_2 generation be explained? Mendel

Figure 10.4
Mendel's crosses with seven different characters in peas, including his results and the calculated ratios of offspring.

Character	Traits crossed	F_1	F_2		Ratio
Seed shape	round × wrinkled	All round	5474 round	1850 wrinkled	2.96:1
Seed colour	yellow × green	All yellow	6022 yellow	2001 green	3.01:1
Pod shape	inflated × constricted	All inflated	882 inflated	299 constricted	2.95:1
Pod colour	green × yellow	All green	428 green	152 yellow	2.82:1
Flower colour	purple × white	All purple	705 purple	224 white	3.15:1
Flower position	axial (along stems) × terminal (at tips)	All axial	651 axial	207 terminal	3.14:1
Stem length	tall × dwarf	All tall	787 tall	277 dwarf	2.84:1

deduced that the trait that had seemed to disappear in the F₁ generation was actually present but was masked in some way by the "stronger" allele. Mendel called the masking effect **dominance.** Accordingly, Mendel's second hypothesis stated that *if an individual's pair of genes consists of different alleles, one allele is dominant over the other, **recessive,** allele.*

CONCEPT FIX What makes an allele **dominant?** In the case of flower colour in Mendel's peas, the purple allele is declared to be dominant simply because, when both alleles are present, the flowers are purple rather than white. More generally, when an organism carries two different alleles, the dominant allele is simply the one that determines the appearance of the organism. In the years since Mendel's work, the underlying mechanisms of dominance have been discovered. For example, notice the round versus wrinkled pea seed shape character shown in Figure 10.4. We now know that round seeds contain a branched form of starch called amylopectin, while wrinkled seeds do not. In the DNA of pea plants there is a gene that codes for an enzyme that produces amylopectin. At some time in the past, a mutation in this gene created an alternative, mutant, version. This mutant allele codes for an enzyme that is nonfunctional and produces no amylopectin. Therefore, plants that contain both of these alleles produce both the functional and the nonfunctional enzymes. The functional enzyme creates amylopectin, resulting in round seeds. Since the allele coding the functional enzyme determines the appearance of the seeds in such plants, it is called the dominant allele. *Notice that dominant alleles do not directly inhibit recessive alleles.* ⬡

As a third hypothesis, Mendel proposed the following: the pairs of alleles that control a character **segregate** (separate) as gametes are formed; half the gametes carry one allele, and the other half carry the other allele. This hypothesis is now known as Mendel's **principle of segregation.** During fertilization, fusion of the haploid maternal and paternal gametes produces a diploid nucleus called the zygote nucleus. The zygote nucleus receives one allele for the character from the male gamete and one allele for the same character from the female gamete, reuniting the pairs.

Mendel's three hypotheses explained the results of the crosses, as summarized in **Figure 10.5, p. 222.** Both alleles of the flower colour gene in the true-breeding parent plant with purple flowers are the same. The symbol *P* is used here to designate this allele, with the capital letter indicating that it is dominant, which gives this true-breeding parent the *PP* combination of alleles. Such an individual is called a **homozygote** (*homo* = same) and is said to be **homozygous** for the *P* allele. Therefore, when the individual produces gametes and the paired alleles separate during meiosis, all the gametes from this individual will receive a *P* allele (steps 1 and 2 in Figure 10.5).

In the original true-breeding parent with white flowers, both alleles of the flower colour gene are also the same. Here the symbol *p* is used to designate this allele, with the lowercase letter indicating that it is recessive, which gives this true-breeding plant the homozygous *pp* combination of alleles. These alleles also separate during meiosis, leading to gametes that all contain one *p* allele.

All the F₁ plants produced by crossing purple-flowered and white-flowered plants—the cross *PP* × *pp*—received the same combination of alleles: *P* from one parent and *p* from the other (step 3 in Figure 10.5, p. 222). An individual of this type, with two different alleles of a gene, is called a **heterozygote** (*hetero* = different) and is said to be **heterozygous** for the trait. Because *P* is dominant over *p*, all the *Pp* plants have purple flowers, even though they also carry the allele for white flowers. An F₁ heterozygote produced from a cross that involves a single character is called a **monohybrid** (*mono* = one; *hybrid* = an offspring of parents with different traits).

According to Mendel's hypotheses, all the *Pp* plants in the F₁ generation produce two kinds of gametes. Because the heterozygous *Pp* pair separates during meiosis I, half of the gametes receive the *P* allele and half receive the *p* allele (steps 4 and 5 of Figure 10.5, p. 222). Step 5 of Figure 10.5 shows how these gametes can combine during selfing of F₁ plants. Generally, a cross between two individuals that are each heterozygous for the same pair of alleles—*Pp* × *Pp* here—is called a **monohybrid cross.** The gametes are entered in both the rows and columns in Figure 10.5; the cells show the possible combinations. Combining two gametes that both carry the *P* allele produces a *PP* F₂ plant; combining *P* from one parent and *p* from the other produces a *Pp* plant; and combining *p* from both F₁ parents produces a *pp* F₂ plant. The homozygous *PP* and heterozygous *Pp* plants in the F₂ generation have purple flowers, the dominant trait; the homozygous *pp* offspring have white flowers, the recessive trait.

Mendel's hypotheses explain how individuals may differ genetically but still look the same. The *PP* and *Pp* plants, although genetically different, both have purple flowers. In modern terminology, **genotype** refers to the *genetic constitution of an organism,* and **phenotype** (Greek *phainein* = to show) refers to its *outward appearance.* In this case, the two different genotypes *PP* and *Pp* produce the same purple-flower phenotype.

Thus, the results of Mendel's crosses support his three hypotheses:

1. The genes that govern genetic characters are present in two copies in individuals.
2. If different alleles are present in an individual's pair of genes, one allele is dominant over the other.
3. The two alleles of a gene segregate and enter gametes singly.

**Experimental
Research Figure 10.5**
Mendel's experiment
illustrating the
principle of segregation
for flower colour in
peas.

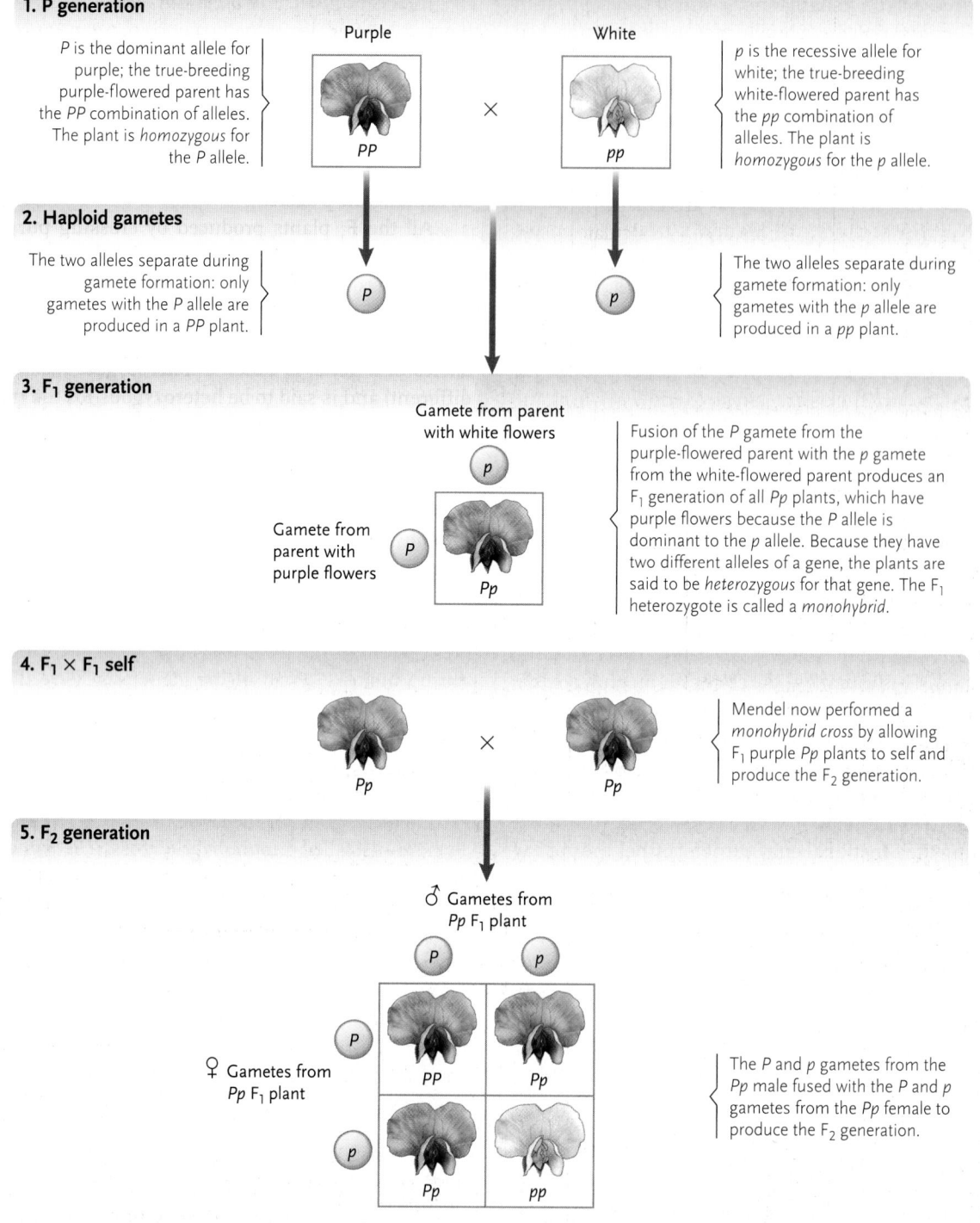

1. P generation

P is the dominant allele for purple; the true-breeding purple-flowered parent has the *PP* combination of alleles. The plant is *homozygous* for the *P* allele.

p is the recessive allele for white; the true-breeding white-flowered parent has the *pp* combination of alleles. The plant is *homozygous* for the *p* allele.

2. Haploid gametes

The two alleles separate during gamete formation: only gametes with the *P* allele are produced in a *PP* plant.

The two alleles separate during gamete formation: only gametes with the *p* allele are produced in a *pp* plant.

3. F₁ generation

Gamete from parent with white flowers

Gamete from parent with purple flowers

Fusion of the *P* gamete from the purple-flowered parent with the *p* gamete from the white-flowered parent produces an F₁ generation of all *Pp* plants, which have purple flowers because the *P* allele is dominant to the *p* allele. Because they have two different alleles of a gene, the plants are said to be *heterozygous* for that gene. The F₁ heterozygote is called a *monohybrid*.

4. F₁ × F₁ self

Mendel now performed a *monohybrid cross* by allowing F₁ purple *Pp* plants to self and produce the F₂ generation.

5. F₂ generation

♂ Gametes from *Pp* F₁ plant

♀ Gametes from *Pp* F₁ plant

The *P* and *p* gametes from the *Pp* male fused with the *P* and *p* gametes from the *Pp* female to produce the F₂ generation.

10.1d Mendel Could Predict Both Classes and Proportions of Offspring from His Hypotheses

Mendel could predict both classes and proportions of offspring from his hypotheses. To understand how Mendel's hypotheses allowed him to predict the proportions of offspring resulting from a genetic cross, let's review the mathematical rules that govern **probability**—that is, the possibility that an outcome will occur if it is a matter of chance, as in the random fertilization of an egg by a sperm cell that contains one allele or another.

In the mathematics of probability, the likelihood of an outcome is predicted on a scale of 0 to 1. An outcome that is certain to occur has a probability of 1, and an outcome that cannot possibly happen has a probability of 0. The standard game die, a cube with one of the numbers 1 through 6 on each face, is a familiar model to demonstrate working with probability. In general, we determine the probability of any given outcome (rolling a 4) by dividing that outcome by the total number of possible outcomes. For obtaining 4 in rolling a die, the probability is 1 divided by 6, or 1/6. The likelihood of rolling an even number (2 or 4 or 6) is 3/6 = 1/2. The probabilities of all the possible outcomes, when added together, must equal 1.

The Product Rule in Probability. If you roll two dice together, what is the chance of rolling double fours? Because the outcome of one die has no effect on the outcome of the other one, the two rolls are independent. When two or more events are independent, the probability that they will both occur is calculated using the **product rule**—their individual probabilities are multiplied. That is, the probability that events A and B *both* will occur equals the probability of event A *multiplied* by the probability of event B. For example, the probability of getting a 4 on the first die is 1/6; the probability of a 4 on the second die is also 1/6 **(Figure 10.6)**. Because the events are independent, the probability of getting a 4 on both dice is 1/6 × 1/6 = 1/36. Applying this principle to human families, the sex of one child has no effect on the sex of the next child; therefore, the probability of having four girls in a row is the product of their individual probabilities (very close to 1/2 for each birth): 1/2 × 1/2 × 1/2 × 1/2 = 1/16.

The Sum Rule in Probability. Another relationship, the **sum rule**, applies when several different events all give the same outcome; that is, the probability that *either* event A or event B or event C will occur equals the probability of event A *plus* the probability of event B *plus* the probability of event C. Returning to the two dice example, what is the probability of rolling a 7? Several different events all give the same total. One could make a total of 7 from a 1 on the first die and a 6 on the second, or a 5 on the first and a 2 on the second, or a 4 on the first and a 3 on the second. Each of these three combinations would be expected to occur at a frequency of 1/6 × 1/6 = 1/36. You should be able to see three more possible combinations that are just the opposite of the first three, that is, 6 on the first die and 1 on the second, and so on, for a total of six different ways to roll a 7. That is, there are six ways of obtaining the same outcome. Therefore, for the probability of rolling a 7, we sum the individual probabilities to get the final probability: 1/36 + 1/36 + 1/36 + 1/36 + 1/36 + 1/36 = 6/36 = 1/6. On average, you could expect to roll a combination of numbers totalling 7 once in every six attempts.

Probability in Mendel's Crosses. Since the randomness inherent in meiosis is comparable to the randomness inherent in rolling dice, the same rules of probability just discussed apply to genes carried on chromosomes in Mendel's crosses. For example, in the crosses that involve the purple-flowered and white-flowered traits, half of the gametes of the F_1 generation contain the P allele of the gene and half contain the p allele (see Figure 10.5, p. 222). To produce a PP zygote, two P gametes must combine. The probability of selecting a P gamete from one F_1 parent is 1/2, and the probability of selecting a P gamete from the other F_1 parent is also 1/2. Therefore, the probability of producing a PP zygote from this monohybrid cross is 1/2 × 1/2 = 1/4. That is, by the product rule, one-fourth of the offspring of the F_1 cross $Pp × Pp$ are expected to be PP, which have purple flowers **(Figure 10.7a, p. 224)**. By the same line of reasoning, one-fourth of the F_2 offspring are expected to be pp, which have white flowers (Figure 10.7b).

What about the production of Pp offspring? The cross $Pp × Pp$ can produce Pp in two different ways. A P gamete from the first parent can combine with a p gamete from the second parent (Pp), or a p gamete from the first parent can combine with a P gamete from the second parent (pP) (Figure 10.7c). Because there are two different ways to get the same outcome, we apply the sum rule to obtain the combined probability. Each of the ways to get Pp has an individual probability of 1/4; when we add these individual probabilities, we have 1/4 + 1/4 = 1/2. Therefore, half of the offspring are expected to be Pp, which have purple flowers. We could get the same result from the requirement that all of the individual probabilities must add up to 1. If the probability of PP is 1/4 and the probability of pp is 1/4, then the probability of the remaining possibility, Pp, must be 1/2, because the total of the individual probabilities must add up to 1: 1/4 + 1/4 + 1/2 = 1.

What if we want to know the probability of obtaining purple flowers in the cross $Pp × Pp$? In

a. Likelihood of rolling a double four.

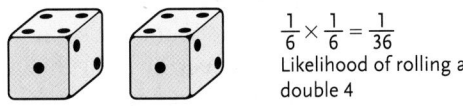

$$\frac{1}{6} \times \frac{1}{6} = \frac{1}{36}$$
Likelihood of rolling a double 4

b. Likelihood of rolling a seven in any combination.

Figure 10.6
Rules of probability. **(a)** For each die, the probability of a 4 is 1/6. Because the outcome of one die is independent of that of the other, the combined probability of rolling a 4 on both dice at the same time is calculated by multiplying the individual probabilities (product rule). **(b)** Since there are six different outcomes, each of which adds up to 7, the total likelihood of rolling a 7 is calculated by adding the individual probabilities (sum rule).

Gametes from F₁ purple *Pp* plant

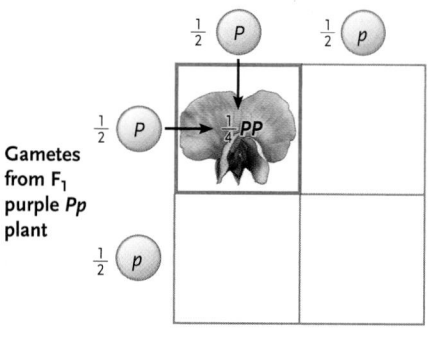

a. To produce an F₂ plant with the *PP* genotype, two *P* gametes must combine. The probability of selecting a *P* gamete from one F₁ parent is $\frac{1}{2}$, and the probability of selecting a *P* gamete from the other F₁ parent is also $\frac{1}{2}$. Using the product rule, the probability of producing purple-flowered *PP* plant from a *Pp* × *Pp* cross is $\frac{1}{2} \times \frac{1}{2} = \frac{1}{4}$.

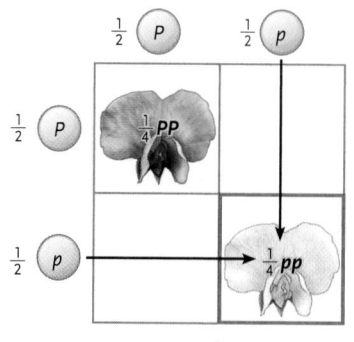

b. To produce an F₂ plant with the *pp* genotype, two *p* gametes must combine. The probability of selecting a *p* gamete from one F₁ parent is $\frac{1}{2}$, and the probability of selecting a *p* gamete from the other F₁ parent is also $\frac{1}{2}$. Using the product rule, the probability of producing a white-flowered *pp* plant from a *Pp* × *Pp* cross is $\frac{1}{2} \times \frac{1}{2} = \frac{1}{4}$.

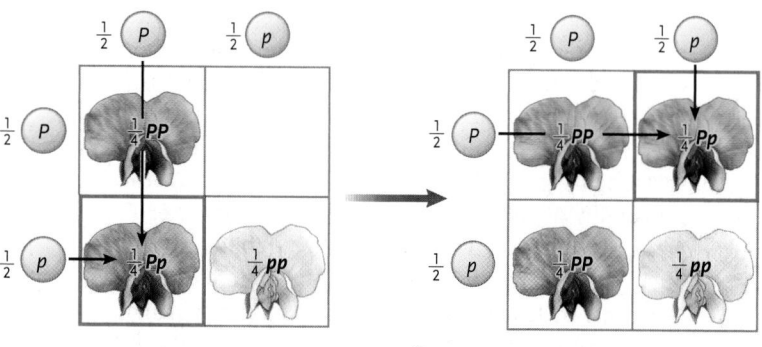

c. To produce an F₂ plant with the *Pp* genotype, a *P* gamete must combine with a *p* gamete. The cross *Pp* × *Pp* can produce *Pp* offspring in two different ways: (1) a *P* gamete from the first parent can combine with a *p* gamete from the second parent or (2) a *p* gamete from the first parent can combine with a *P* gamete from the second parent. We apply the sum rule to obtain the combined probability: each of the ways to get *Pp* has an individual probability of $\frac{1}{4}$, so the probability of *Pp*, purple-flowered offspring, is $\frac{1}{4} + \frac{1}{4} = \frac{1}{2}$.

Figure 10.7

Punnett square method for predicting offspring and their ratios in genetic crosses. The example is the F₁ × F₁ cross of purple-flowered plants from Figure 10.5, p. 222. Each cell shows the genotype and proportion of one type of F₂ plant.

this case, the rule of addition applies, because there are two ways to get purple flowers: genotypes *PP* and *Pp*. Adding the individual probabilities of these combinations, 1/4 *PP* + 1/2 *Pp*, gives a total of 3/4, indicating that three-fourths of the F₂ offspring are expected to have purple flowers. Because the total probabilities must add up to 1, the remaining one-fourth of the offspring are expected to have white flowers (1/4 *pp*). These proportions give the ratio 3:1, which is close to the ratio Mendel obtained in his cross.

In Figure 10.7 we have just stepped through the **Punnett square** method for determining the genotypes

of offspring and their expected proportions. To use the Punnett square, write the probability that meiosis will produce gametes with each type of allele from one parent at the top of the diagram and write the chance of obtaining each type of allele from the other parent on the left side. Then fill in the cells by combining the alleles from the top and from the left and multiplying their individual probabilities.

10.1e Mendel Used a Testcross to Check the Validity of His Hypotheses

Mendel realized that he could assess the validity of his hypotheses by determining whether they could be used successfully to *predict* the outcome of a cross of a different type than he had tried so far. Accordingly, he crossed an F₁ plant with purple flowers, assumed to have the heterozygous genotype *Pp*, with a true-breeding white-flowered plant, with the homozygous genotype *pp* (**Figure 10.8,** Experiment 1). There are two expected classes of offspring, *Pp* and *pp*, both with a probability of 1/2. Thus, the phenotypes of the offspring are expected to be 1 purple-flowered : 1 white-flowered. Mendel's actual results closely approach the expected 1:1 ratio. Mendel also made the same type of cross with all the other traits used in his study, including those traits affecting seed shape, seed colour, and plant height, and found the same 1:1 ratio.

A cross between an individual with the dominant phenotype and a homozygous recessive individual, such as the one described, is called a **testcross**. Geneticists use a testcross as a standard test to determine whether an individual with a dominant trait is a heterozygote or a homozygote, because these cannot be distinguished phenotypically. If the offspring of the testcross are of two types, with half displaying the dominant trait and half the recessive trait, then the individual in question must be a heterozygote (see Figure 10.8, Experiment 1). If all the offspring display the dominant trait, the individual in question must be a homozygote. For example, the cross *PP* × *pp* gives all *Pp* progeny, which show the dominant purple phenotype (see Figure 10.8, Experiment 2).

Obviously, the testcross method cannot be used for humans. However, it can be used in reverse by noting the traits present in families over several generations and working backward to deduce whether a parent must have been a homozygote or a heterozygote (see also Chapter 11).

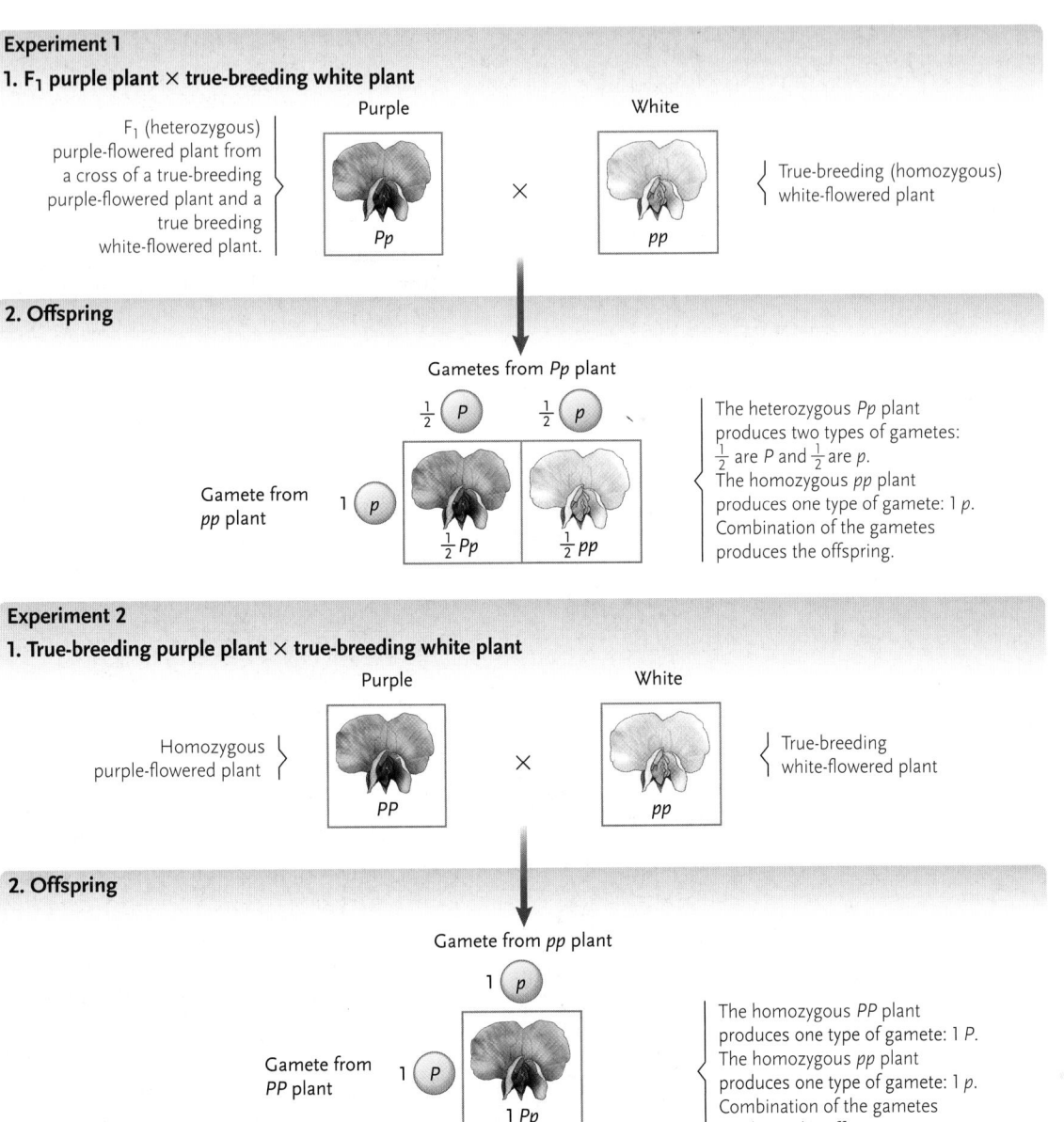

Experiment 1

1. F₁ purple plant × true-breeding white plant

F₁ (heterozygous) purple-flowered plant from a cross of a true-breeding purple-flowered plant and a true breeding white-flowered plant.

Purple

Pp

×

White

pp

True-breeding (homozygous) white-flowered plant

2. Offspring

Gametes from *Pp* plant

½ *P*   ½ *p*

Gamete from *pp* plant   1 *p*

½ *Pp*   ½ *pp*

The heterozygous *Pp* plant produces two types of gametes: ½ are *P* and ½ are *p*. The homozygous *pp* plant produces one type of gamete: 1 *p*. Combination of the gametes produces the offspring.

Experiment 2

1. True-breeding purple plant × true-breeding white plant

Homozygous purple-flowered plant

Purple

PP

×

White

pp

True-breeding white-flowered plant

2. Offspring

Gamete from *pp* plant

1 *p*

Gamete from *PP* plant   1 *P*

1 *Pp*

The homozygous *PP* plant produces one type of gamete: 1 *P*. The homozygous *pp* plant produces one type of gamete: 1 *p*. Combination of the gametes produces the offspring.

10.1f Mendel Tested the Independence of Different Genes in Crosses

Mendel next asked what happens in crosses when more than one character is involved. Would the alleles of different characters be inherited independently, or would they interact to alter their expected proportions in offspring?

To answer these questions, Mendel crossed parental stocks that had differences in two of the hereditary characters he was studying: seed shape and seed colour. His single-character crosses had shown that each was controlled by a pair of alleles. For seed shape, the *RR* or *Rr* genotype produces round seeds and the *rr* genotype produces wrinkled seeds. For seed colour, yellow is dominant. The homozygous *YY* and heterozygous *Yy* genotypes produce yellow seeds; the homozygous *yy* genotype produces green seeds.

Mendel crossed plants that bred true for the production of round and yellow seeds (*RR YY*) with plants that bred true for the production of wrinkled and green seeds (*rr yy*) **(Figure 10.9, p. 226)**. The cross, *RR YY* × *rr yy*, yielded an F₁ generation that consisted of all round yellow seeds, with the genotype *Rr Yy*. A zygote produced from a cross that involves two characters is called a **dihybrid** (*di* = two).

Mendel then planted the F₁ seeds, grew the plants to maturity, and selfed them; that is, he crossed the F₁ plants to themselves. A cross between two individuals that are heterozygous for two pairs of alleles—here *Rr Yy* × *Rr Yy*—is called a **dihybrid cross** (see Figure 10.9, p. 226). The seeds produced by these plants, representing the F₂ generation, included 315 round yellow seeds, 101 wrinkled yellow seeds, 103 round green seeds, and 32 wrinkled green seeds. Mendel noted that these numbers were close to a 9:3:3:1 ratio (3:1 for round : wrinkled, and 3:1 for yellow : green).

This 9:3:3:1 ratio was consistent with Mendel's previous findings if he added one further hypothesis: *The alleles of the genes that govern the two characters*

Experimental Research Figure 10.9
Mendel's experiment illustrating the principle of independent assortment for seed shape and seed colour in peas.

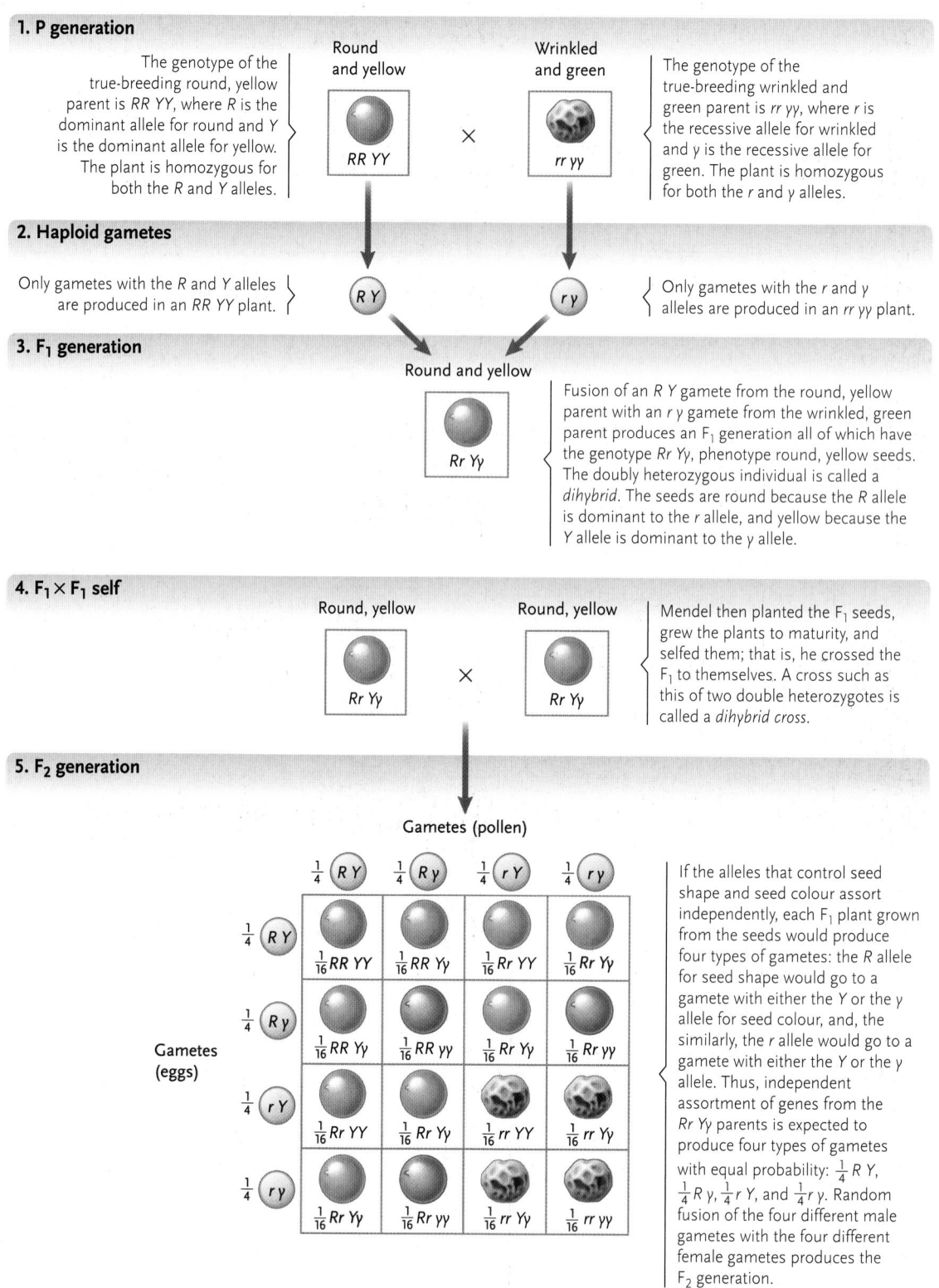

1. P generation

The genotype of the true-breeding round, yellow parent is *RR YY*, where *R* is the dominant allele for round and *Y* is the dominant allele for yellow. The plant is homozygous for both the *R* and *Y* alleles.

Round and yellow — *RR YY* × Wrinkled and green — *rr yy*

The genotype of the true-breeding wrinkled and green parent is *rr yy*, where *r* is the recessive allele for wrinkled and *y* is the recessive allele for green. The plant is homozygous for both the *r* and *y* alleles.

2. Haploid gametes

Only gametes with the *R* and *Y* alleles are produced in an *RR YY* plant.

R Y   *r y*

Only gametes with the *r* and *y* alleles are produced in an *rr yy* plant.

3. F₁ generation

Round and yellow — *Rr Yy*

Fusion of an *R Y* gamete from the round, yellow parent with an *r y* gamete from the wrinkled, green parent produces an F₁ generation all of which have the genotype *Rr Yy*, phenotype round, yellow seeds. The doubly heterozygous individual is called a *dihybrid*. The seeds are round because the *R* allele is dominant to the *r* allele, and yellow because the *Y* allele is dominant to the *y* allele.

4. F₁ × F₁ self

Round, yellow — *Rr Yy* × Round, yellow — *Rr Yy*

Mendel then planted the F₁ seeds, grew the plants to maturity, and selfed them; that is, he crossed the F₁ to themselves. A cross such as this of two double heterozygotes is called a *dihybrid cross*.

5. F₂ generation

Gametes (pollen): $\frac{1}{4}$ *R Y*, $\frac{1}{4}$ *R y*, $\frac{1}{4}$ *r Y*, $\frac{1}{4}$ *r y*

Gametes (eggs): $\frac{1}{4}$ *R Y*, $\frac{1}{4}$ *R y*, $\frac{1}{4}$ *r Y*, $\frac{1}{4}$ *r y*

	$\frac{1}{4}$ *R Y*	$\frac{1}{4}$ *R y*	$\frac{1}{4}$ *r Y*	$\frac{1}{4}$ *r y*
$\frac{1}{4}$ *R Y*	$\frac{1}{16}$ *RR YY*	$\frac{1}{16}$ *RR Yy*	$\frac{1}{16}$ *Rr YY*	$\frac{1}{16}$ *Rr Yy*
$\frac{1}{4}$ *R y*	$\frac{1}{16}$ *RR Yy*	$\frac{1}{16}$ *RR yy*	$\frac{1}{16}$ *Rr Yy*	$\frac{1}{16}$ *Rr yy*
$\frac{1}{4}$ *r Y*	$\frac{1}{16}$ *Rr YY*	$\frac{1}{16}$ *Rr Yy*	$\frac{1}{16}$ *rr YY*	$\frac{1}{16}$ *rr Yy*
$\frac{1}{4}$ *r y*	$\frac{1}{16}$ *Rr Yy*	$\frac{1}{16}$ *Rr yy*	$\frac{1}{16}$ *rr Yy*	$\frac{1}{16}$ *rr yy*

If the alleles that control seed shape and seed colour assort independently, each F₁ plant grown from the seeds would produce four types of gametes: the *R* allele for seed shape would go to a gamete with either the *Y* or the *y* allele for seed colour, and, the similarly, the *r* allele would go to a gamete with either the *Y* or the *y* allele. Thus, independent assortment of genes from the *Rr Yy* parents is expected to produce four types of gametes with equal probability: $\frac{1}{4}$ *R Y*, $\frac{1}{4}$ *R y*, $\frac{1}{4}$ *r Y*, and $\frac{1}{4}$ *r y*. Random fusion of the four different male gametes with the four different female gametes produces the F₂ generation.

segregate independently during formation of gametes. That is, the allele for seed shape that the gamete receives (*R* or *r*) has no influence on which allele for seed colour it receives (*Y* or *y*) and vice versa. The two events are completely independent. Mendel termed this assumption **independent assortment**; it is now known as Mendel's **principle of independent assortment.**

To understand the effect of independent assortment in the cross, assume that the *RR YY* parent produces only *R Y* gametes and the *rr yy* parent produces only *r y* gametes. In the F₁ generation, all possible combinations of these gametes produce only one genotype, *Rr Yy*, in the offspring. As observed, all the F₁ will be round yellow seeds.

If the alleles that control seed shape and seed colour assort independently in gamete formation, each F₁ plant grown from the seeds will produce four types of gametes. As shown in **Figure 10.10**, the random

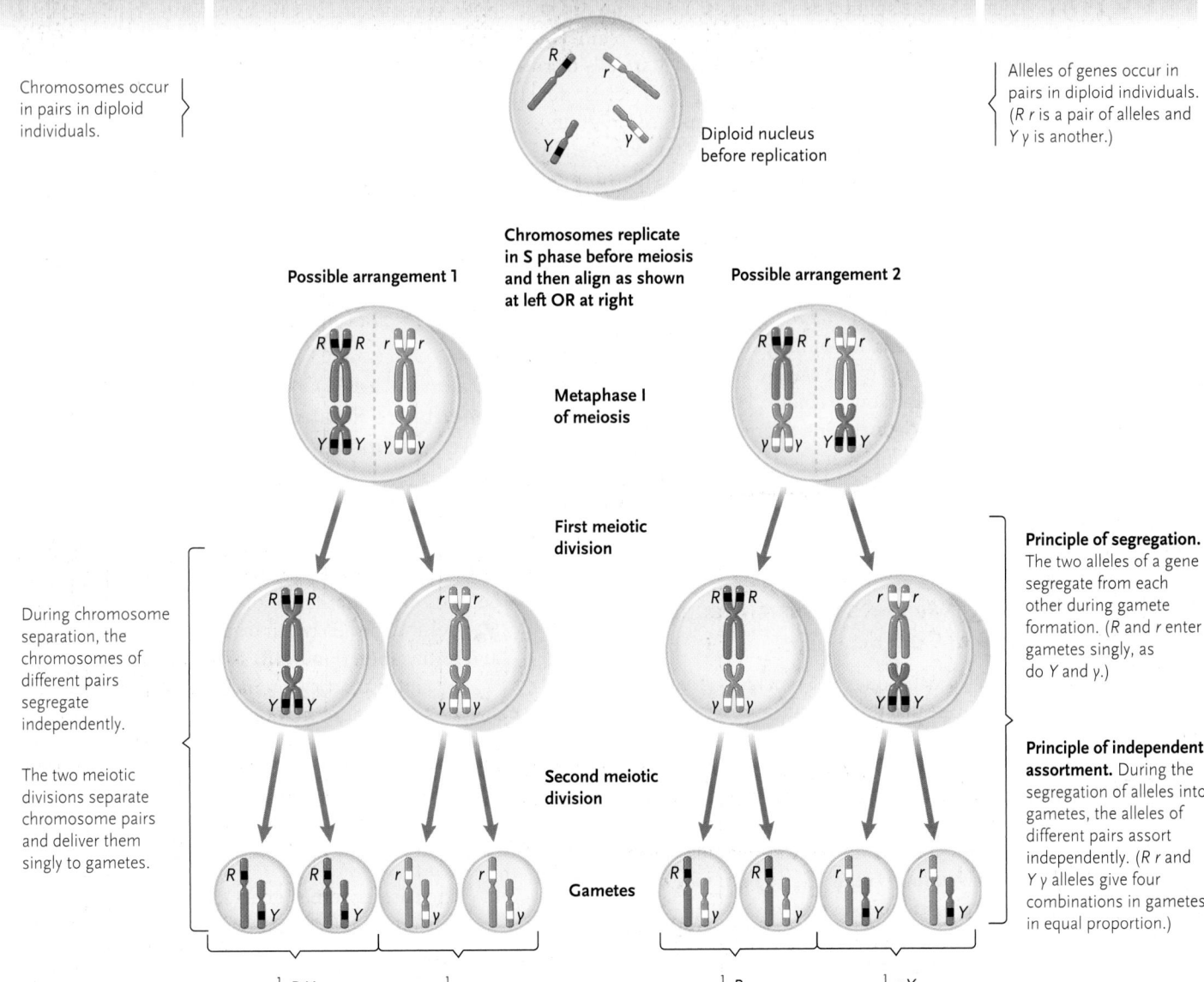

Behaviour of chromosomes in meiosis

Chromosomes occur in pairs in diploid individuals.

During chromosome separation, the chromosomes of different pairs segregate independently.

The two meiotic divisions separate chromosome pairs and deliver them singly to gametes.

Meiosis in male or female diploid parent

Diploid nucleus before replication

Chromosomes replicate in S phase before meiosis and then align as shown at left OR at right

Possible arrangement 1

Possible arrangement 2

Metaphase I of meiosis

First meiotic division

Second meiotic division

Gametes

$\frac{1}{4}RY$ $\frac{1}{4}ry$ $\frac{1}{4}Ry$ $\frac{1}{4}rY$

Behaviour of genes and alleles in meiosis and their correspondence to Mendel's principles

Alleles of genes occur in pairs in diploid individuals. (*R r* is a pair of alleles and *Y y* is another.)

Principle of segregation. The two alleles of a gene segregate from each other during gamete formation. (*R* and *r* enter gametes singly, as do *Y* and *y*.)

Principle of independent assortment. During the segregation of alleles into gametes, the alleles of different pairs assort independently. (*R r* and *Y y* alleles give four combinations in gametes in equal proportion.)

Figure 10.10

The parallels between the behaviour of chromosomes and genes and alleles in meiosis. The gametes show four different combinations of alleles produced by independent segregation of chromosome pairs.

alignment of homologous chromosome pairs in meiosis I ensures that the *R* allele for seed shape can be delivered independently to a gamete with either the *Y* or the *y* allele for seed colour, and, similarly, the *r* allele can be delivered to a gamete with either the *Y* or the *y* allele. Thus, the independent assortment of genes from the *Rr Yy* parents allows the organism to produce, overall, four types of gametes with equal probability: 1/4 *R Y*, 1/4 *R y*, 1/4 *r Y*, and 1/4 *r y*. These gametes and their probabilities are entered as the row and column headings of the Punnett square in Figure 10.9.

Filling in the cells of the diagram (see Figure 10.9) gives 16 combinations of alleles, all with an equal probability of 1 in every 16 offspring. Of these, the genotypes *RR YY*, *RR Yy*, *Rr YY*, and *Rr Yy* all have the same phenotype: round yellow seeds. These combinations occur in 9 of the 16 cells in the diagram, giving a total probability of 9/16. The genotypes *rr YY* and *rr Yy*, which produce wrinkled yellow seeds, are found in three cells, giving a probability of 3/16 for this phenotype. Similarly, the genotypes *RR yy* and *Rr yy*, which yield round green seeds, occur in three cells, giving a probability of 3/16. Finally, the genotype *rr yy*, which produces wrinkled green seeds, is found in only one cell and therefore has a probability of 1/16.

These probabilities of round yellow seeds, wrinkled yellow seeds, round green seeds, and wrinkled green seeds, in a 9:3:3:1 ratio, closely approximate the actual results of 315:101:108:32 obtained by Mendel. Thus, Mendel's first three hypotheses, with the added hypothesis of independent assortment, explain the observed results of his dihybrid cross. Mendel's testcrosses completely confirmed his hypotheses; for example, the testcross *Rr Yy* × *rr yy* produced 55 round yellow seeds, 51 round green seeds, 49 wrinkled yellow seeds, and 53 wrinkled green seeds. This distribution corresponds well to the expected 1:1:1:1 ratio in the offspring. (Try to set up a Punnett square for this cross and predict the expected classes of offspring and their frequencies.)

Mendel's first three hypotheses provided a coherent explanation of the pattern of inheritance for alternative traits of the same character, such as purple and white for flower colour. His fourth hypothesis, independent assortment, addressed the inheritance of traits for different characters, such as seed shape, seed colour, and flower colour, and showed that, instead of being inherited together, the traits of different characters were distributed independently to offspring.

10.1g Mendel's Research Founded the Field of Genetics

Mendel's techniques and conclusions were so advanced for his time that their significance was not immediately appreciated. Mendel's success was based partly on a good choice of experimental organism. He was also lucky. The characters he chose all segregate independently; that is, none of them are physically near each other on the chromosomes, a condition that would have given ratios other than 9:3:3:1, showing that they do not assort independently.

We now know that Mendel's findings demonstrated the patterns by which genes and chromosomes determine inheritance. Yet, when Mendel first reported his findings during the nineteenth century, the structure and function of chromosomes and the patterns by which they are separated and distributed to gametes were unknown; meiosis remained to be discovered. In addition, his use of mathematical analysis was a new and radical departure from the usual biological techniques of his day.

Mendel reported his results to a small group of fellow intellectuals in Brünn and presented his results in 1866 in a natural history journal published in the city. His article received little notice outside of Brünn, and those who read it were unable to appreciate the significance of his findings. His work was overlooked until the turn of the century, when three investigators—Hugo de Vries in Holland, Carl Correns in Germany, and Erich von Tschermak in Austria—independently performed a series of breeding experiments similar to Mendel's and reached the same conclusions. These investigators, in searching through previously published scientific articles, were surprised to discover Mendel's article about his experiments conducted three decades earlier. Each gave credit to Mendel's discoveries, and the quality and far-reaching implications of his work were at last realized. Mendel died in 1884, less than 20 years before the rediscovery of his experiments and conclusions; thus, he never received the recognition that he so richly deserved during his lifetime.

Mendel was unable to relate the behaviour of his "factors" (genes) to cell structures because the critical information he required was not obtained until later, through the discovery of meiosis during the 1890s. The next section describes how a genetics student familiar with meiosis was able to make the connection between Mendel's factors and chromosomes.

10.1h Sutton's Chromosome Theory of Inheritance Related Mendel's Genes to Chromosomes

By the time Mendel's results were rediscovered in the early 1900s, critical information from studies of meiosis was available. It was not long before a genetics student, Walter Sutton, recognized the similarities between the inheritance of the genes discovered by Mendel and the behaviour of chromosomes in meiosis and fertilization (Figure 10.10, p. 227).

In a historic article published in 1903, Sutton, then a graduate student at Columbia University in New York, drew all the necessary parallels between genes and chromosomes:

- Chromosomes occur in pairs in sexually reproducing, diploid organisms, as do the alleles of each gene.
- The chromosomes of each pair are separated and delivered singly to gametes, as are the alleles of a gene.
- The separation of any pair of chromosomes in meiosis and gamete formation is independent of the separation of other pairs (see Figure 10.10, p. 227), as in the independent assortment of the alleles of different genes in Mendel's dihybrid crosses.
- Finally, one member of each chromosome pair is derived in fertilization from the male parent, and the other member is derived from the female parent, in an exact parallel with the two alleles of a gene.

From this total coincidence in behaviour, Sutton correctly concluded that genes and their alleles are carried on the chromosomes, a conclusion known today as the **chromosome theory of inheritance.**

The exact parallel between the principles set forth by Mendel and the behaviour of chromosomes and genes during meiosis is shown in Figure 10.10 for an *Rr Yy* diploid. For a cross of *Rr Yy* × *Rr Yy*, when the gametes fuse randomly, the progeny will show a phenotypic ratio of 9:3:3:1. This mechanism explains the same ratio of gametes and progeny as the *Rr Yy* × *Rr Yy* cross in Figure 10.9, p. 226.

The particular site on a chromosome at which a gene is located is called the **locus** (plural, loci) of the gene. The locus is a particular DNA sequence that encodes a protein or RNA product responsible for the phenotype controlled by the gene. A locus for a gene with two alleles, *A* and *a*, on a homologous pair of chromosomes is shown in **Figure 10.11.** At the molecular level, different alleles consist of small differences in the DNA sequence of a gene, which may result in functional differences in the protein or RNA product encoded by the gene. These differences are detected as distinct phenotypes in the offspring of a cross.

All of the genetics research conducted since the early 1900s has confirmed Mendel's basic hypotheses about inheritance. This research has shown that Mendel's conclusions apply to all types of organisms, from yeast and fruit flies to humans, and has led to the rapidly growing field of human genetics. In humans, a number of easily seen traits show inheritance patterns that follow Mendelian principles **(Figure 10.12)**; for example, albinism, the lack of normal skin colour, is recessive to normal skin colour, and fingers with webs between them are recessive to normally separated fingers. Similarly, achondroplasia, the most frequent form of short-limb dwarfism, is a recessive trait that involves abnormal bone growth. Many human disorders that cannot be seen easily also show simple inheritance patterns. For instance, cystic fibrosis, in which a defect in the membrane transport of chloride ions leads to pulmonary and digestive dysfunctions and reduced life span, is a recessive trait.

The post-Mendel research has demonstrated additional patterns of inheritance (see the next section) that were not anticipated by Mendel and, in some

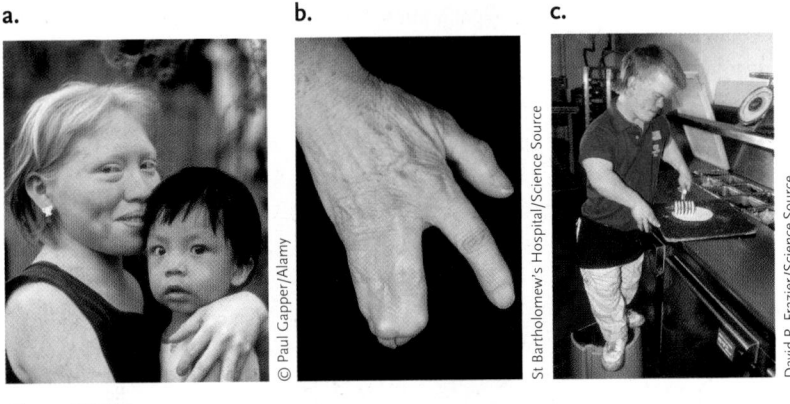

a. **b.** **c.**

© Paul Capper/Alamy
St Bartholomew's Hospital/Science Source
David R. Frazier/Science Source

Figure 10.12

Human traits showing inheritance patterns that follow Mendelian principles. **(a)** Lack of normal skin colour (albinism). **(b)** Webbed fingers. **(c)** Achondroplasia or short-limb dwarfism.

circumstances, require modifications or additions to his hypotheses.

STUDY BREAK

1. What characteristics of the garden pea made this organism a good model system for Mendel?
2. How does independent assortment explain Mendel's dihybrid cross data?
3. How is an allele related to a locus?

10.2 Later Modifications and Additions to Mendel's Hypotheses

The rediscovery of Mendel's research in the early 1900s produced an immediate burst of interest in genetics. The research that followed greatly expanded our understanding of genes and their inheritance. That research fully supported Mendel's hypotheses, but also revealed many variations on the basic principles he had outlined. The following sections discuss each of these extensions of Mendel's fundamental principles.

10.2a In Incomplete Dominance, Dominant Alleles Do Not Completely Compensate for Recessive Alleles

Incomplete dominance occurs when the effects of recessive alleles can be detected to some extent in heterozygotes. Flower colour in snapdragons shows incomplete dominance **(Figure 10.13, p. 230).** If true-breeding red-flowered and white-flowered snapdragon plants are crossed, all the F_1 offspring have pink flowers (see Figure 10.13, p. 230). The pink colour might make it appear that the pure red and white colours have blended—mixing red and white makes pink. However, when two F_1 plants are crossed, the red and white traits

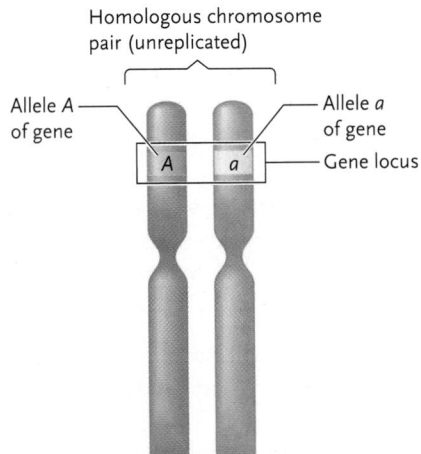

Homologous chromosome pair (unreplicated)

Allele *A* of gene — Allele *a* of gene — Gene locus

A | a

Figure 10.11

A locus, the site occupied by a gene on a pair of homologous chromosomes. Two alleles, *A* and *a*, of the gene are present at this locus in the homologous pair. These alleles have differences in the DNA sequence of the gene.

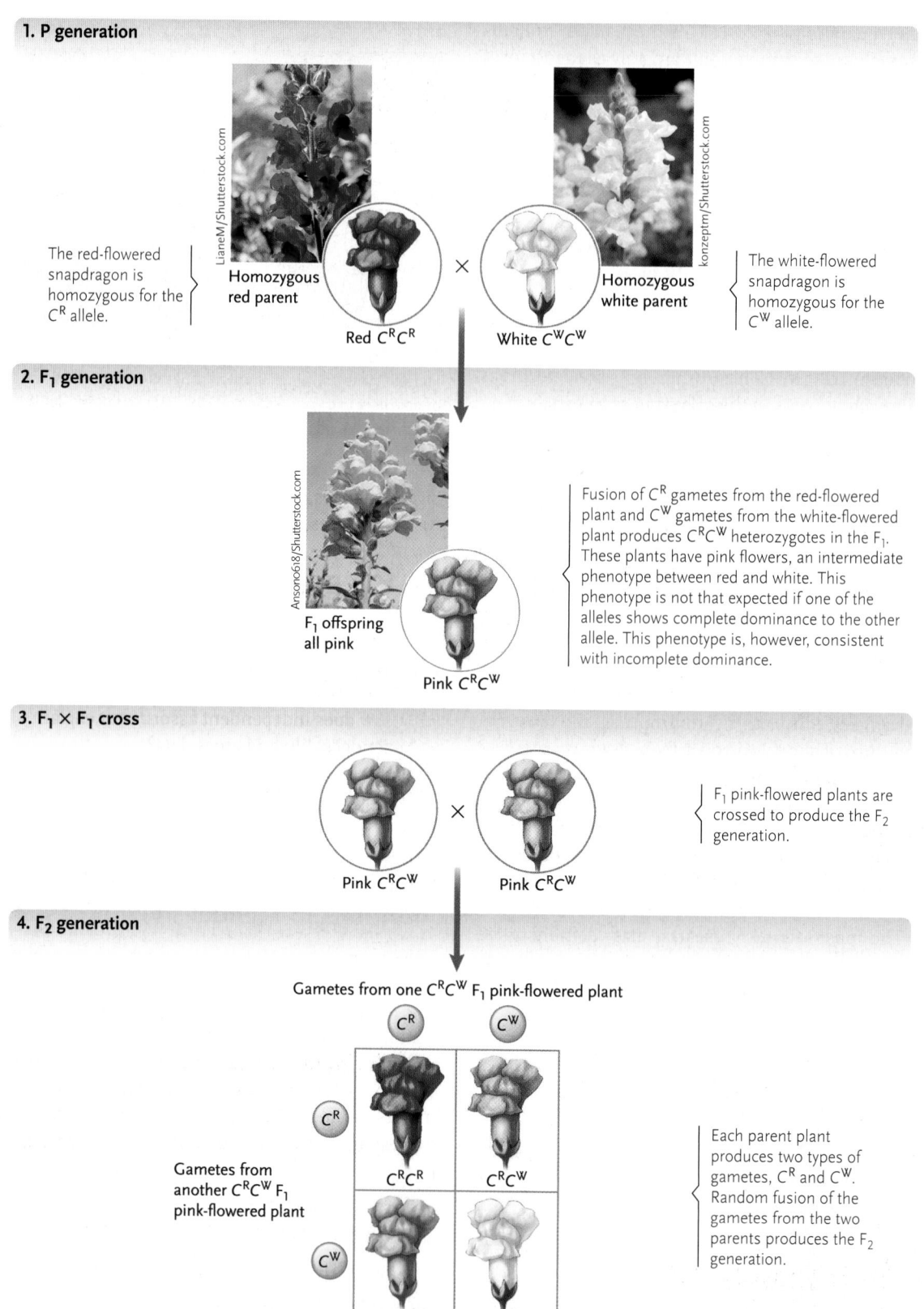

1. P generation

The red-flowered snapdragon is homozygous for the C^R allele.

Homozygous red parent

Red $C^R C^R$

×

Homozygous white parent

White $C^W C^W$

The white-flowered snapdragon is homozygous for the C^W allele.

2. F₁ generation

F₁ offspring all pink

Pink $C^R C^W$

Fusion of C^R gametes from the red-flowered plant and C^W gametes from the white-flowered plant produces $C^R C^W$ heterozygotes in the F₁. These plants have pink flowers, an intermediate phenotype between red and white. This phenotype is not that expected if one of the alleles shows complete dominance to the other allele. This phenotype is, however, consistent with incomplete dominance.

3. F₁ × F₁ cross

Pink $C^R C^W$

×

Pink $C^R C^W$

F₁ pink-flowered plants are crossed to produce the F₂ generation.

4. F₂ generation

Gametes from one $C^R C^W$ F₁ pink-flowered plant

C^R C^W

Gametes from another $C^R C^W$ F₁ pink-flowered plant

C^R

$C^R C^R$ $C^R C^W$

C^W

$C^R C^W$ $C^W C^W$

Each parent plant produces two types of gametes, C^R and C^W. Random fusion of the gametes from the two parents produces the F₂ generation.

both reappear in the **F₂** generation, which has red, pink, and white flowers in numbers approximating a 1:2:1 ratio.

This outcome can be explained by incomplete dominance between a C^R allele for red colour and a C^W allele for white colour. When one allele is not completely dominant to the other, we use a superscript to signify the character. In this case, C signifies the character for flower colour and the superscripts indicate the alleles (R for red and W for white). Therefore, the initial cross is $C^R C^R$ (red) × $C^W C^W$ (white), which produces $C^R C^W$ F₁ (pink) plants. The C^R allele encodes an enzyme that

Why Mendel's Dwarf Pea Plants Were So Short

Two independent research teams worked out the molecular basis for one of the seven characters Mendel studied—dwarfing, which is governed by stem length in garden peas. The investigators, including Diane Lester and her colleagues at the University of Tasmania in Australia and David Martin and his coworkers at Oregon State University, were interested in learning the molecular differences in the alleles of the gene that produced tall or dwarf plants. The dominant T allele (T = tall) of the gene produces plants of normal height; the recessive t allele produces dwarf plants with short stems. How can a single gene control the overall height of a plant?

Lester's team discovered that the gene encodes an enzyme that carries out a preliminary step in the synthesis of the plant hormone gibberellin, which, among other effects, causes the stems of plants to elongate. Martin's group cloned the gene and determined its complete DNA sequence. (Cloning techniques and DNA sequencing are described in Sections 15.1 and 16.2.) Comparisons of the DNA sequences from the T and t alleles revealed two versions of the enzyme that catalyzes gibberellin synthesis that differ by only a single amino acid. Lester's group found that the faulty enzyme encoded by the t allele carries out its step

(addition of a hydroxyl group to a precursor) much more slowly than the enzyme encoded by the normal T allele. As a result, plants with the t allele have only about 5% as much gibberellin in their stems as T plants. The reduced gibberellin levels limit stem elongation, producing the dwarf plants.

Thus, the methods of molecular biology allowed contemporary researchers to study a gene first discovered in the mid-nineteenth century. The findings leave little doubt that a change in a single amino acid leads to the dwarf phenotype Mendel observed in his monastery garden.

---

produces a red pigment, but two alleles ($C^R C^R$) are necessary to produce enough of the active form of the enzyme to produce fully red flowers. The enzyme is completely inactive in $C^W C^W$ plants, which produce colourless flowers that appear white because of the scattering of light by cell walls and other structures. With their single C^R allele, the $C^R C^W$ heterozygotes of the F_1 generation can produce only enough pigment to give the flowers a pink colour. When the pink $C^R C^W$ F_1 plants are crossed, the fully red and white colours reappear, together with the pink colour, in the F_2 generation, in a ratio of 1/4 $C^R C^R$ (red), 1/2 $C^R C^W$ (pink), and 1/4 $C^W C^W$ (white). This ratio is exactly the same as the ratio of genotypes produced from a cross of two heterozygotes in Mendel's experiments (e.g., see Figure 10.7, p. 224).

Some human disorders show incomplete dominance. For example, sickle cell disease (see the introduction to this chapter) is characterized by an alteration in the hemoglobin molecule that changes the shape of red blood cells when oxygen levels are low. An individual with sickle cell disease is homozygous for a recessive allele that encodes a defective form of one of the polypeptides of the hemoglobin molecule. Individuals heterozygous for that recessive allele and the normal allele have a condition known as *sickle cell trait*, which is a milder form of the disease because the individuals still produce normal polypeptides from the normal allele.

Familial hypercholesterolemia is another example of incomplete dominance. The gene involved encodes the low-density lipoprotein (LDL) receptor, a cell membrane protein responsible for removing excess cholesterol from the blood. Individuals with familial

hypercholesterolemia are homozygous for a defective LDL receptor gene, produce no LDL receptors, and have a severe form of the disease. These individuals have six times the normal level of cholesterol in the blood and therefore are very prone to atherosclerosis (hardening of the **arteries**). Many individuals with familial hypercholesterolemia have heart attacks as children. Heterozygous individuals have half the normal number of receptors, which results in a milder form of the disease. Their symptoms are twice the normal blood cholesterol level, an unusually high risk of atherosclerosis, and a high risk of heart attacks before age 35.

Many alleles that appear to be completely dominant are actually incomplete in their effects when analyzed at the biochemical or molecular level. For example, for pigments that produce fur or flower colours, biochemical studies often show that even though heterozygotes may produce enough pigment to make them look the same externally as homozygous dominants, a difference in the amount of pigment is measurable at the biochemical level. Thus, whether dominance between alleles is complete or incomplete often depends on the level at which the effects of the alleles are examined.

A similar situation occurs in humans who carry the recessive allele that causes Tay–Sachs disease. Children who are homozygous for the recessive allele do not have a functional version of an enzyme that breaks down gangliosides, a type of membrane lipid. As a result, gangliosides accumulate in the brain, leading to mental impairment and eventually to death. Heterozygotes are without symptoms of the disease, even though they have one copy of the recessive allele.

MOLECULE BEHIND BIOLOGY 10.2

Phenylthiocarbamide (PTC)

Have you ever sat down to a plate of Brussels sprouts, only to find that they taste unpleasantly bitter? This sensation arises because receptors in the membranes of taste cells on your tongue are binding to compounds such as isothiocyanate (which is toxic to your thyroid in large doses). Much of our understanding of the molecular nature of bitter taste perception has grown out of an accidental observation that a synthetic chemical, phenylthiocarbamide (PTC), tastes intensely bitter to some people and yet is tasteless to others. PTC "nontasters" make up 20 to 30% of almost all populations of humans, chimps, and gorillas studied. Although several genes influence the limits of PTC detection, one particular member of the bitter receptor gene family on human chromosome 7 is mainly responsible for PTC tasting ability. The two most common alleles of this gene, "taster" and "nontaster," show a codominant inheritance pattern in families. Since PTC is not found in nature, it is likely that the two very common alleles detect naturally occurring toxic compounds containing the bitter-tasting thiourea chemical structure shown in the diagram (N−C=S).

Figure 1

The chemical structure of phenylthiocarbamide (PTC). Note the N−C=S component.

However, at the biochemical level, reduced breakdown of gangliosides can be detected in heterozygotes, evidently due to a reduced quantity of the active enzyme.

CONCEPT FIX Since all of Mendel's traits show "simple" or "complete" dominance, you might get the idea that most traits in nature are governed by one dominant and one recessive allele. However, the above examples illustrate that only a minority of human genetic disorders show such simple dominance.

10.2b In Codominance, the Effects of Different Alleles Are Equally Detectable in Heterozygotes

Codominance occurs when alleles have approximately equal effects in individuals, making the two alleles equally detectable in heterozygotes. The inheritance of the human blood types M, MN, and N is an example of codominance. These are different blood types from the familiar blood types of the ABO blood group. The L^M and L^N alleles of the MN blood group gene that control this character encode different forms of a glycoprotein molecule located on the surface of red blood cells. If the genotype is $L^M L^M$, only the M form of the glycoprotein is present and the blood type is M; if it is $L^N L^N$, only the N form is present and the blood type is N. In heterozygotes with the $L^M L^N$ genotype, both glycoprotein types are present and can be detected, producing the blood type MN. Because each genotype has a different phenotype, the inheritance pattern for the MN blood group alleles is generally the same as for incompletely dominant alleles. That is, you would not be able to distinguish between codominance and incomplete dominance just by comparing the ratio of offspring from crosses.

The MN blood types do not affect blood transfusions and have relatively little medical importance. However, they have been invaluable in tracing human evolution and prehistoric migrations, and they are frequently used in initial tests to determine the paternity of a child. Among their primary advantages in research and paternity determination is that the genotype of all individuals, including heterozygotes, can be detected directly—and inexpensively—from their phenotype, with no requirement for further genetic tests or analysis.

10.2c In Multiple Alleles, More Than Two Alleles of a Gene Are Present in a Population

One of Mendel's major and most fundamental assumptions was that alleles occur in pairs in individuals; in the pairs, the alleles may be the same or different. After the rediscovery of Mendel's principles, it soon became apparent that although alleles do indeed occur in pairs in individuals, **multiple alleles** (more than two different alleles of a gene) may be present if all the individuals of a population are taken into account. For example, for a gene B, there could be the normal allele, B, and several alleles with alterations in the gene named, for example, $b1$, $b2$, $b3$, and so on. Some individuals in a population may have the B and $b1$ alleles of a gene; others, the $b2$ and $b3$ alleles; still others, the $b3$ and $b5$ alleles; and so on, for all possible combinations. Thus, although any one individual can have only two alleles of the gene, there are more than two alleles in the population as a whole. One of the genes that plays a part in the acceptance or rejection of organ transplants in humans has more than 200 different alleles!

The multiple alleles of a gene each contain nucleotide differences at one or more locations in their DNA sequences **(Figure 10.14)**, and these often cause detectable alterations in the structure and function of gene products encoded by the alleles. Despite the presence of multiple alleles at the population level, each diploid individual still has only two of the alleles, allowing gametes to be predicted and traced through crosses by the usual methods.

Human ABO Blood Group. The human ABO blood group provides a real example of multiple alleles, in a system that also exhibits both dominance and codominance. Karl Landsteiner, an Austrian biochemist, discovered the ABO blood group in 1901 while investigating the fact that attempts to transfer whole blood from one person to another were sometimes fatal. Landsteiner found that only certain combinations of four blood types, designated A, B, AB, and O, can be mixed safely in transfusions **(Table 10.1)**.

Landsteiner determined that, in certain combinations, red blood cells from one blood type are agglutinated or clumped by an agent in the serum of another type (the serum is the fluid in which the blood cells are suspended). The clumping was later found to depend on the action of an **antibody** in the blood serum. (Antibodies, protein molecules that interact with specific substances called antigens, are discussed in Chapter 51.)

The antigens responsible for the blood types of the ABO blood group are the carbohydrate parts of glycoproteins located on the surfaces of red blood cells (unrelated to the glycoprotein carbohydrates responsible for the blood types of the MN blood group). People with type A blood have *antigen A* on their red blood cells, and people with type B blood have *antigen B* on their red blood cells. At the same time, people with type A blood have antibodies against antigen B, and people with type B blood have antibodies against antigen A. People with type O blood have neither antigen A nor antigen B on their red blood cells, but they have antibodies against both of these antigens. People with type AB blood have neither anti-A nor anti-B antibodies, but they have both the A and B antigens, and their red blood cells are clumped by antibodies in the blood of all the other groups.

The four blood types—A, B, AB, and O—are produced by different combinations of multiple (three) alleles of a single gene I **(Figure 10.15)**. The three alleles, designated I^A, I^B, and i, produce the following blood types:

$$I^A I^A = \text{type A blood}$$
$$I^A i = \text{type A blood}$$
$$I^A I^B = \text{type AB blood}$$
$$I^B I^B = \text{type B blood}$$
$$I^B i = \text{type B blood}$$
$$ii = \text{type O blood}$$

In addition, I^A and I^B are codominant alleles that are each dominant to the i allele.

10.2d In Epistasis, Genes Interact, with the Activity of One Gene Influencing the Activity of Another Gene

The genetic characters discussed so far in this chapter, such as flower colour, seed shape, and the blood types of the ABO group, are all produced by the alleles of single genes, with each gene functioning on its own.

Table 10.1 **Blood Types of the Human ABO Blood Group**

Blood Type	Antigens	Antibodies	Blood Types Accepted in a Transfusion
A	A	Anti-B	A or O
B	B	Anti-A	B or O
AB	A and B	None	A, B, AB, or O
O	None	Anti-A, anti-B	O

B allele
5'...A T G C A G A T A C C G A T T A C A G A C C A T A G G...3'
3'...T A C G T C T A T G G C T A A T G T C T G G T A T C C...5'

b_1 allele
5'...A T G C A G A G A C C G A T T A C A G A C C A T A G G...3'
3'...T A C G T C T C T G G C T A A T G T C T G G T A T C C...5'

b_2 allele
5'...A T G C A G A T A C C G A C T A C A G A C C A T A G G...3'
3'...T A C G T C T A T G G C T G A T G T C T G G T A T C C...5'

b_3 allele
5'...A T G C A G A T A C C G A T T A C A G T C C A T A G G...3'
3'...T A C G T C T A T G G C T A A T G T C A G G T A T C C...5'

Figure 10.14

Multiple alleles. Multiple alleles consist of small differences in the DNA sequence of a gene at one or more points, which result in detectable differences in the structure of the protein encoded by the gene. The *B* allele is the normal allele, which encodes a protein with normal function. The three *b* alleles each have alterations of the normal protein-coding DNA sequence that may adversely affect the function of that protein.

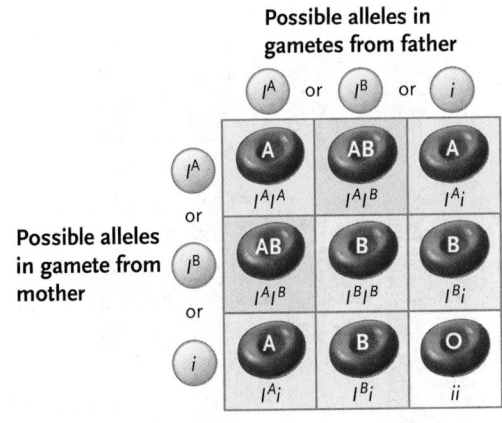

Figure 10.15

Inheritance of the blood types of the human ABO blood group. Note that although there are three possible alleles in the population, each individual parent carries only two.

This is not the case for every trait. In **epistasis** (*epi* = on or over; *stasis* = standing or stopping), genes interact, with one or more alleles of a gene at one locus inhibiting or masking the effects of one or more alleles of a gene at a different locus. The result of epistasis is that some expected phenotypes do not appear among offspring.

Labrador retrievers may have black, chocolate brown, or yellow fur **(Figure 10.16)**. The different colours result from variations in the amount and distribution in hairs of a brownish-black pigment called melanin. One gene, coding for an enzyme involved in melanin production, determines how much melanin is produced. The dominant *B* allele of this gene produces black fur colour in *BB* or *Bb* Labs; less pigment is produced in *bb* dogs, which are chocolate brown. However, another gene at a different locus determines whether the black or chocolate colour appears at all, by controlling the deposition of pigment in hairs. The dominant *E* allele of this second gene permits pigment deposition, so that the black colour in *BB* or *Bb* individuals, or the chocolate colour in *bb* individuals, actually appears in the fur. Pigment deposition is almost completely blocked in homozygous recessive *ee* individuals, so the fur lacks melanin and has a yellow colour whether the genotype for the *B* gene is *BB*, *Bb*, or *bb*. Thus, the *E* gene is said to be epistatic to the *B* gene.

Epistasis by the *E* gene eliminates some of the expected classes from crosses among Labs. Rather than two separate classes, as would be expected from a dihybrid cross without epistasis, the *BB ee*, *Bb ee*, *bB ee*, and *bb ee* genotypes produce a single yellow phenotype, giving the distribution 9/16 black, 3/16 chocolate, and 4/16 yellow. That is, the ratio is 9:3:4 instead of the expected 9:3:3:1 ratio. Many other dihybrid crosses that involve epistatic interactions produce distributions that differ from the expected 9:3:3:1 ratio.

In human biology, researchers believe that gene interactions and epistasis are common. The current thinking is that epistasis is an important factor in determining an individual's susceptibility to common human diseases. That is, the different degrees of susceptibility are the result of different gene interactions in the individuals. A specific example is insulin resistance, a disorder in which muscle, fat, and liver cells do not use insulin correctly, with the result that glucose and insulin levels become high in the blood. This disorder is believed to be determined by several genes often interacting with one another.

10.2e In Polygenic Inheritance, a Character Is Controlled by the Common Effects of Several Genes

Some characters follow a pattern of inheritance in which there is a more or less even gradation of types, forming a continuous distribution, rather than "on" or "off"

a. Black labrador

Erik Lam/Shutterstock.com

b. Chocolate brown labrador

cen/Shutterstock.com

c. Yellow labrador

cen/Shutterstock.com

d. Black × yellow labrador cross

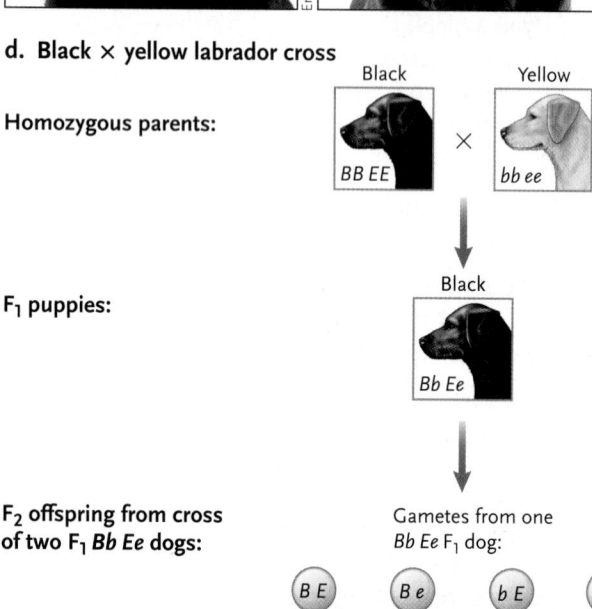

F₂ phenotypic ratio is 9 black : 3 chocolate : 4 yellow

Figure 10.16

An example of epistasis: the inheritance of coat colour in Labrador retrievers.

(discontinuous) effects such as the production of only purple or white flowers in pea plants. For example, human adults range from short to tall, in a continuous distribution of height between limits of about 1 and 2 m. Typically, a continuous distribution of this type is the result of **polygenic inheritance**, in which several to many different genes contribute to the same character. Other characters that undertake a similar continuous distribution include skin colour and body weight in humans, ear length in corn, seed colour in wheat, and colour spotting in mice. These characters are also known as **quantitative traits**.

Polygenic inheritance can be detected by defining classes of variation, such as human body height of 180 cm in one class, 181 cm in the next class, 182 cm in the next class, and so on. The number of individuals in each class is then plotted as a graph. If the plot produces a bell-shaped curve, with fewer individuals at the extremes and the greatest numbers clustered around the midpoint, it is a good indication that the trait is quantitative **(Figure 10.17)**.

The expression of a genetic phenotype can be influenced by the environment; this is particularly common with quantitative traits like body size. For example, poor nutrition during infancy and childhood is one environmental factor that can limit growth and prevent individuals from reaching the height expected from purely genetic contributions; good nutrition can have the opposite effect. Thus, the average young adult in Japan today is several inches taller than the average adult in the 1930s, when nutrition was poorer.

CONCEPT FIX At first glance, the wide variation shown in a quantitative trait might appear to support the old idea that the characteristics of parents are blended in their offspring. Commonly, people believe that the children in a family with one tall and one short parent will be of intermediate height. Although the children of such parents are indeed most likely to be of intermediate height, careful genetic analysis of hundreds of such families shows that their offspring actually range over a continuum from short to tall, forming a typical bell-shaped curve. Some children are not intermediate relative to their parents; they are either taller or shorter than both parents. Careful analysis of the inheritance of skin colour produces the same result. Although the skin colour of children is most often intermediate between that of their parents, a typical bell-shaped distribution is obtained in which some children at the extremes are lighter or darker than either parent. Thus, genetic analysis does not support the idea of blending or even mixing of parental traits in quantitative characteristics such as body size or skin colour. ⬡

a. Students at Brigham Young University, arranged according to height

Dan Fairbanks/Brigham Young University

Figure 10.17
Continuous variation in height due to polygenic inheritance.

b. Actual distribution of individuals in the photo according to height

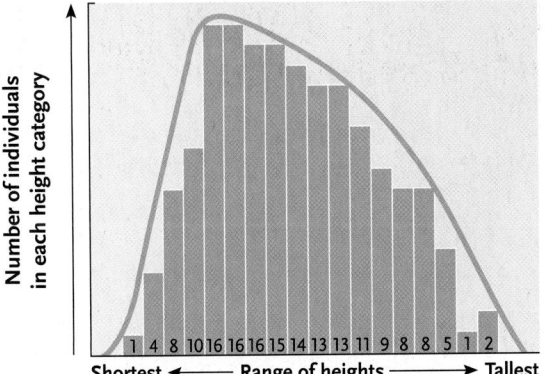

Number of individuals in each height category

1 4 8 10 16 16 16 15 14 13 13 11 9 8 8 5 1 2

Shortest ◄— Range of heights —► Tallest

c. Idealized bell-shaped curve for a population that displays continuous variation in a trait

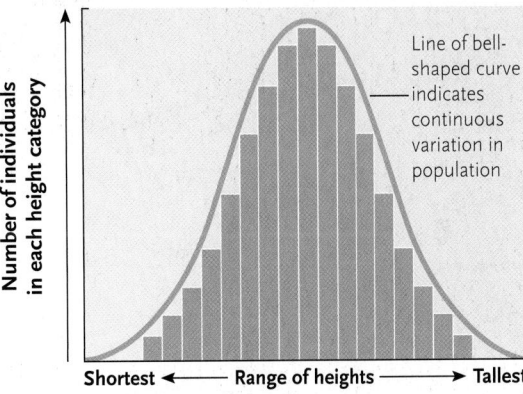

Number of individuals in each height category

Line of bell-shaped curve indicates continuous variation in population

Shortest ◄— Range of heights —► Tallest

If the sample in the photo included more individuals, the distribution would more closely approach this ideal.

10.2f In Pleiotropy, Two or More Characters Are Affected by a Single Gene

In the previous section, we saw several genes affecting the same trait. In this section we see the reverse situation: single genes affecting more than one character of an organism in a process called **pleiotropy**. For example, sickle cell disease (see earlier discussion) is caused by a recessive allele of a single gene that affects hemo~~...~~ altere~~...~~ chang~~...~~ vessel~~...~~ organ~~...~~ tions, fatigu~~...~~ pneu~~...~~ wide-~~...~~ cell di~~...~~

[Handwritten notes overlapping text:]

epistasis: one or more allele masks at the locus

polygenic inheritance: several different genes contribute to same character

quantitative traits: skin color body weight

height is quantitative because most people are average height not very short/very tall

pleiotropy: single genes affecting more than one character

sickle cell: recessive allele of single gene affects hemoglobin structure + function

Figure 10.18

Pleiotropy, as demonstrated by ~~...~~ multiple effects of the single m~~...~~ responsible for sickle cell disea~~...~~ effects are shown.)

CONCEPT FIX Although Mendel's simple, single gene experiments in peas provided a valuable scientific model for understanding inheritance, modern analyses in a wide variety of organisms are finding that many traits are quantitative and most genes have some pleiotropic effects. ⬡

The next chapter describes additional patterns of inheritance that were not anticipated by Mendel, including the effects of recombination during meiosis. These additional patterns also extend, rather than contradict, Mendel's fundamental principles.

STUDY BREAK

1. Distinguish between alleles that are incompletely dominant and those that are codominant.
2. How might you know that a trait is polygenic?

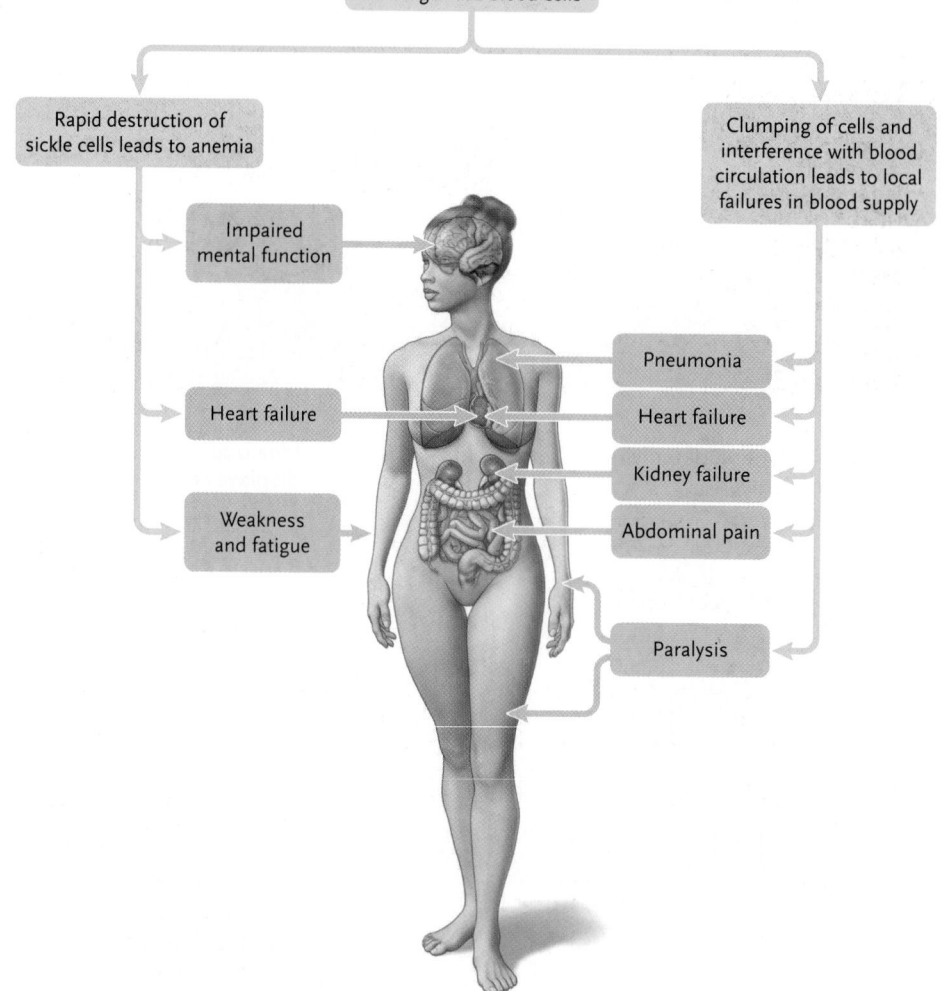

Homozygous recessive individual
↓
Abnormal hemoglobin
↓
Sickling of red blood cells

Rapid destruction of sickle cells leads to anemia

Clumping of cells and interference with blood circulation leads to local failures in blood supply

Impaired mental function

Heart failure

Weakness and fatigue

Pneumonia

Heart failure

Kidney failure

Abdominal pain

Paralysis

Review

To access course materials such as Aplia and other companion resources, please visit www.NELSONbrain.com.

10.1 The Beginnings of Genetics: Mendel's Garden Peas

- Mendel made a good choice of experimental organism in that garden peas offered simple cultivation; clearly defined, true-breeding characters (such as flower colour or seed shape); and an opportunity to make controlled pollinations.

- By analyzing his results quantitatively, Mendel showed that traits are passed from parents to offspring as hereditary factors (now called genes and alleles) in predictable ratios and combinations, disproving the notion of blended inheritance (see Figures 10.3, 10.4, and 10.5).

- Mendel realized that his results with crosses involving single characters (monohybrid crosses) could be explained if three hypotheses were true: (1) the genes that govern genetic characters occur in pairs in individuals; (2) if different alleles of a gene are present in a pair within an individual, one allele is dominant over the other; and (3) the two alleles of a gene segregate and enter gametes singly (see Figures 10.5 and 10.7).

- Mendel confirmed his hypotheses by a testcross between an F_1 heterozygote and a homozygous recessive parent. This type of testcross is still used to determine whether an individual is homozygous or heterozygous for a dominant allele (see Figure 10.8).

- To explain the results of his crosses with individuals showing differences in two characters—dihybrid crosses—Mendel added an additional hypothesis: the alleles of the genes that govern the two characters segregate independently during formation of gametes (see Figure 10.9). That is, the dihybrid cross *Aa Bb* × *Aa Bb* can be treated as two separate monohybrid crosses: *Aa* × *Aa* and *Bb* × *Bb*. The monohybrid crosses would give phenotypic ratios of 3/4 *A*__: 1/4 *aa* and 3/4 *B*__: 1/4 *bb*, respectively. The standard dihybrid ratios arise from combinations of these monohybrid ratios. That is, 9/16 *A*__ *B*__ results from 3/4 *A*__ × 3/4 *B*__, 3/16 *aa B*__ results from 1/4 *aa* × 3/4 *B*__, and so on.

- Walter Sutton was the first person to note the similarities between the inheritance of genes and the behaviour of chromosomes in meiosis and fertilization. These parallels made it obvious that genes and alleles are carried on the chromosomes, and are called the chromosome theory of inheritance (see Figure 10.10).

- A locus is the particular site where a given gene is found on the chromosomes of an organism (see Figure 10.11). An allele is just a particular version of the DNA sequence of a gene. Therefore, if an individual were heterozygous for the stem length gene of Mendel's peas, it would have a *T* allele on one homologue and a *t* allele on the other. These two alleles would each be located at exactly the same locus on their respective chromosomes.

10.2 Later Modifications and Additions to Mendel's Hypotheses

- Incomplete dominance arises when, in a heterozygote, the activity of one allele is insufficient to compensate for the inactivity of another. Codominance arises when, in a heterozygote, both alleles are equally active. In both cases, the phenotype of heterozygotes is different from that of either homozygote (see Figure 10.13).

- Many genes may have multiple alleles if all the individuals in a population are taken into account. However, any diploid individual in a population has only two alleles of these genes, which are inherited and passed on according to Mendel's principles (see Figures 10.14 and 10.15).

- In epistasis, genes interact, with alleles of one locus inhibiting or masking the effects of alleles at a different locus. The result is that some expected phenotypes do not appear among offspring (see Figure 10.16).

- A character that is subject to polygenic inheritance shows a more or less continuous variation from one extreme to another. Plotting the distribution of such characters among individuals typically produces a bell-shaped curve (see Figure 10.17).

- In pleiotropy, one gene affects more than one character of an organism (see Figure 10.18).

Questions

Self-Test Questions

1. Imagine an organism with the genotype *Rr*. If the diagrams below represent the replicated chromosomes of this organism early in meiosis, which one shows the correct location of the *R* and *r* alleles?

   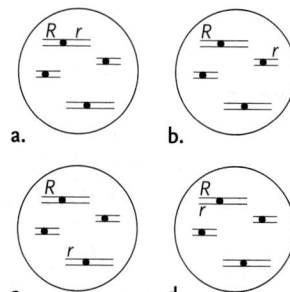

2. Kernel colour in corn is influenced by the *C* gene. The dominant *C* allele produces coloured kernels, and plants homozygous for the recessive *c* allele have colourless (white) kernels. What gamete genotypes, and in what proportions, would be produced by the plants in the following crosses? What kernel colour, and in what proportions, would be expected in the offspring of the crosses?
   a. *CC × Cc*
   b. *Cc × Cc*
   c. *Cc × cc*

3. In peas, the allele *T* produces tall plants and the allele *t* produces dwarf plants. The *T* allele is dominant to *t*. If a tall plant is crossed with a dwarf and the offspring are distributed about equally between tall and dwarf plants, what are the genotypes of the parents?

4. The ability of humans to taste the bitter chemical phenylthiocarbamide (PTC) is a genetic trait. People with at least one copy of the normal, dominant allele of the *PTC* gene can taste PTC; those who are homozygous for a mutant, recessive allele cannot taste it. Could two parents able to taste PTC have a nontaster child? Could nontaster parents have a child able to taste PTC? A pair of taster parents, both of whom had one parent able to taste PTC and one nontaster parent, is expecting their first child. What are the chances that the child will be either able or unable to taste PTC? Suppose the first child is a nontaster; what is the chance that their second child will also be unable to taste PTC?

5. One gene has the alleles *A* and *a*; another gene has the alleles *B* and *b*. Alleles of the *A* and *B* genes assort independently. For each of the following genotypes, what genotypes of gametes will be produced, and in what proportions?
   a. *AA BB*
   b. *Aa BB*
   c. *Aa bb*
   d. *Aa Bb*

6. Which genotypes, and in what frequencies, will be present in the offspring from the following matings?
   a. *AA BB × aa BB*
   b. *Aa Bb × Aa Bb*
   c. *Aa Bb × aa bb*
   d. *Aa BB × AA Bb*

7. In addition to the two genes in question 5, assume you now study a third independently assorting gene that has the alleles *C* and *c*. For each of the following genotypes, indicate what types of gametes will be produced.
   a. *AA BB CC*
   b. *Aa BB cc*
   c. *Aa BB Cc*
   d. *Aa Bb Cc*

8. Imagine that you are helping a friend with genetics problems. He has drawn the Punnett square below to answer questions about a dihybrid cross: *Mm Hh × Mm Hh*. Use the principles of meiosis to explain why this diagram is incorrect.

	M	H	h	m
M				
h				
H				
m				

9. A man is homozygous dominant for alleles at 10 different genes that assort independently. How many genotypically different types of sperm cells can he produce? A woman is homozygous recessive for the alleles of 8 of these 10 genes, but she is heterozygous for the other 2 genes. How many genotypically different types of eggs can she produce? What hypothesis can you suggest to describe the relationship between the number of different possible gametes and the number of heterozygous and homozygous genes that are present?

10. In guinea pigs, an allele for rough fur (*R*) is dominant over an allele for smooth fur (*r*); an allele for black coat (*B*) is dominant over that for white (*b*). You have an animal with rough, black fur. What cross would you use to determine whether the animal is homozygous for these traits? What phenotype would you expect in the offspring if the animal were homozygous?

11. You cross a lima bean plant from a variety that breeds true for green pods with another lima bean from a variety that breeds true for yellow pods. You note that all the F_1 plants have green pods. These green-pod F_1 plants, when crossed to each other, yield 675 plants with green pods and 217 with yellow pods. How many genes likely control pod colour in this experiment? Give the alleles letter designations. Which is dominant?

12. Some recessive alleles have such a detrimental effect that they are lethal when present in both chromosomes of a pair. Homozygous recessives cannot survive, and die at some point during embryonic development. Suppose that the allele *r* is lethal in the homozygous *rr* condition. What genotypic ratios would you expect among the living offspring of the following crosses?
   a. *RR × Rr*
   b. *Rr × Rr*

13. In garden peas, the genotypes *GG* and *Gg* produce green pods and *gg* produces yellow pods; *TT* and *Tt* plants are tall and *tt* plants are dwarfed; *RR* and *Rr* produce round seeds and *rr* produces wrinkled seeds. If a plant of a true-breeding tall variety with green pods and round seeds is crossed with a plant of a true-breeding dwarf variety with yellow pods and wrinkled seeds, what phenotypes are expected, and in what ratios, in the F$_1$ generation? What phenotypes, and in what ratios, are expected if F$_1$ individuals are crossed?

14. In chickens, a gene called *F* influences leg feathering. Feathered legs are produced by a dominant allele *F*, while featherless legs result in individuals who are homozygous for the *f* allele. A second gene, *P*, on another chromosome, influences comb shape. The dominant allele *P* produces pea combs; a recessive allele *p* of this gene causes single combs. A breeder makes the following crosses with birds 1, 2, 3, and 4; all parents have feathered legs and pea combs.

Cross	Offspring
1 × 2	all feathered, pea comb
1 × 3	3/4 feathered, 1/4 featherless, all pea comb
1 × 4	9/16 feathered, pea comb; 3/16 featherless, pea comb; 3/16 feathered, single comb; 1/16 featherless, single comb

What are the genotypes of the four birds?

15. A mixup in a hospital ward causes a mother with O and MN blood types to think that a baby given to her really belongs to someone else. Tests in the hospital show that the doubting mother is able to taste PTC (see question 4). The baby given to her has O and MN blood types and has no reaction when the bitter PTC chemical is placed on its tongue. The mother has four other children with the following blood types and tasting abilities for PTC.
   a. type A and MN blood, taster
   b. type B and N blood, nontaster
   c. type A and M blood, taster
   d. type A and N blood, taster

Without knowing the father's blood types and tasting ability, can you determine whether the child is really hers? (Assume that all her other children have the same father.)

16. In cats, the genotype *AA* produces tabby fur colour; *Aa* is also a tabby, and *aa* is black. Another independently assorting gene at a different locus is epistatic to the gene for fur colour. When present in its dominant *W* form (*WW* or *Ww*), this gene blocks the formation of fur colour and all the offspring are white; *ww* individuals develop normal fur colour. What fur colours, and in what proportions, would you expect from the cross *Aa Ww* × *Aa Ww*?

17. Having malformed hands with shortened fingers is a dominant trait controlled by a single gene; people who are homozygous for the recessive allele have normal hands and fingers. Having woolly hair is a dominant trait controlled by a different, independently assorting gene; homozygous recessive individuals have normal, nonwoolly hair. Suppose a woman with normal hands and nonwoolly hair marries a man who has malformed hands and woolly hair. Their first child has normal hands and nonwoolly hair. What are the genotypes of the mother, the father, and the child? If this couple has a second child, what is the probability that it will have normal hands and woolly hair?

Questions for Discussion

1. The eyes of brown-eyed people are not alike but rather vary considerably in shade and pattern. What do you think causes these differences?

2. Explain how individuals of an organism that are phenotypically alike can produce different ratios of progeny phenotypes.

3. ABO blood type tests can be used to exclude paternity. Suppose a defendant who is the alleged father of a child takes a blood-type test and the results do not exclude him as the father. Do the results indicate that he is the father? What arguments could a lawyer make based on the test results to exclude the defendant from being the father? (Assume the tests were performed correctly.)

11

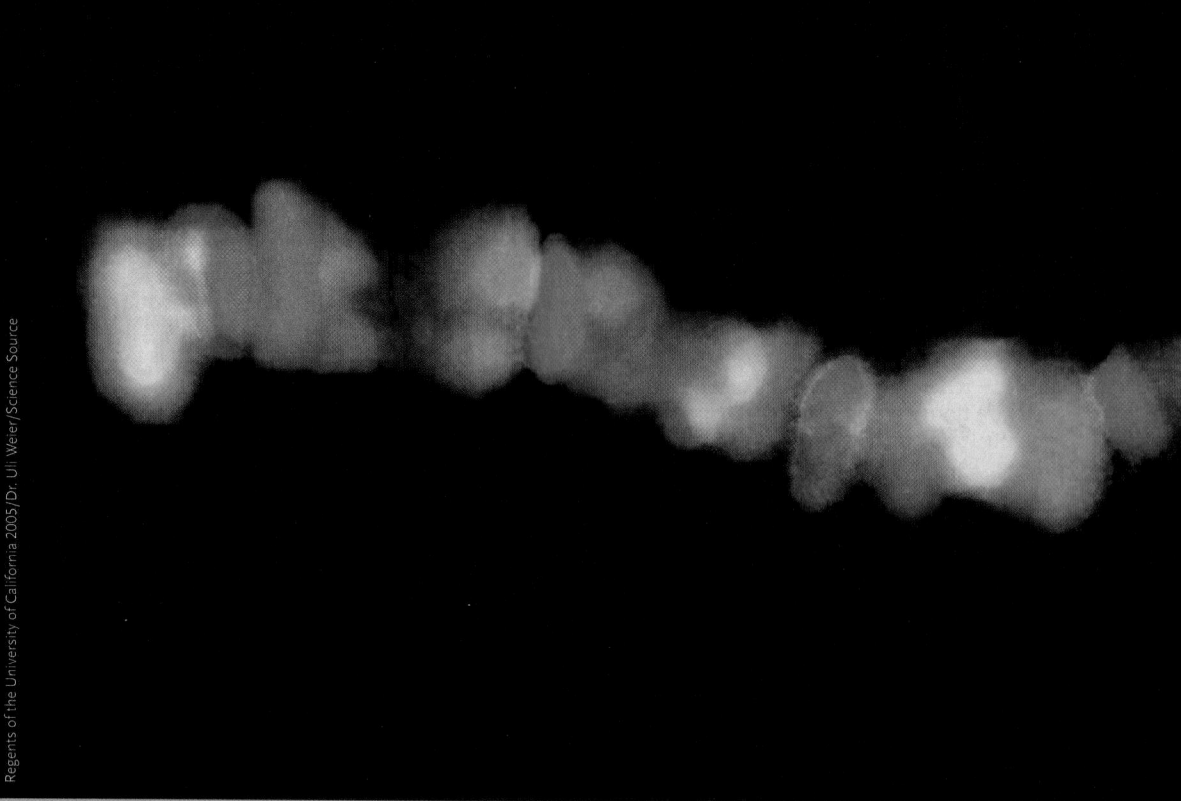

Regents of the University of California 2005/Dr. Uli Weier/Science Source

Fluorescent probes bound to specific sequences along human chromosome 10 (light micrograph). New ways of mapping chromosome structure yield insights into the inheritance of normal and abnormal traits.

Genes, Chromosomes, and Human Genetics

WHY IT MATTERS

Imagine being 10 years old and trapped in a body that each day becomes more shrivelled, frail, and old. You are just tall enough to peer over the top of the kitchen counter, and you weigh less than 16 kg. Already you are bald, and you probably have only a few more years to live. But if you are like Mickey Hayes or Fransie Geringer **(Figure 11.1),** you still have not lost your courage or your childlike curiosity about life. Like them, you still play, laugh, and celebrate birthdays.

Progeria, the premature aging that afflicts Mickey and Fransie, is caused by a genetic error that occurs once in every 8 million human births. The error is perpetuated each time cells of the embryo—then of the child—duplicate their chromosomes and divide. The outcome of that rare mistake is an acceleration of aging and a greatly reduced life expectancy.

Progeria affects both boys and girls. Usually, symptoms begin to appear before the age of 2. The rate of body growth declines to abnormally low levels. Skin becomes thinner, muscles become flaccid, and limb bones start to degenerate. Children with progeria never reach puberty, and most die in their early teens from a stroke or heart attack brought on by hardening of the arteries, a condition typical of advanced age.

The plight of Mickey and Fransie provides a telling and tragic example of the dramatic effects that gene defects can have on living organisms. The characteristics of each individual, from humans to pine trees to protozoa, depend on the combination of genes, alleles, and chromosomes inherited from its parents, as well as on environmental effects. This chapter delves into genes and the role of chromosomes in inheritance.

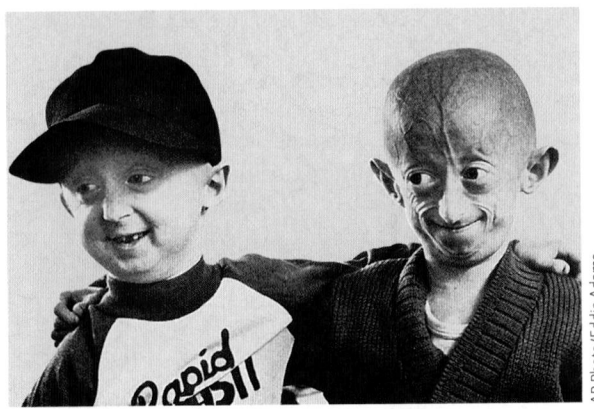

Figure 11.1

Two boys, both younger than 10, who have progeria, a genetic disorder characterized by accelerated aging and extremely reduced life expectancy.

11.1 Genetic Linkage and Recombination

In the historic experiments described in the previous chapter, Gregor Mendel carried out crosses with seven different characters in garden peas, controlled by seven different genes. He found that his observations from crosses were consistent with the hypothesis that each of the genes assorted independently of all of the others. If Mendel had extended his study to additional characters, he would soon have found exceptions to this principle. This should not be surprising because an organism has many more genes than chromosomes. Conceptually, then, chromosomes contain many genes, with each gene at a particular location, or locus. Genes located on different chromosomes assort independently during meiosis because the two chromosomes behave independently of one another during as they line up on the metaphase plate (see Chapter 9 for a review of chromosome behaviour during meiosis). Genes located on the same chromosome may be inherited together in genetic crosses—that is, they do not assort independently—because the chromosome is inherited as a single physical entity in meiosis. Genes on the same chromosome are known as **linked genes**, and the phenomenon is called **linkage**.

11.1a The Principles of Linkage and Recombination Were Determined with *Drosophila*

In the early part of the twentieth century, Thomas H. Morgan and coworkers at Columbia University used the fruit fly, *Drosophila melanogaster,* as a model organism to investigate Mendel's principles in animals. (For more information about *Drosophila* as a model research organism, see *The Purple Pages.*) Groups of genes that tended to assort together in crosses were believed to be carried on the same chromosome. It was an undergraduate student named Alfred Sturtevant, working in Morgan's lab, who developed the insight that resulted in the construction of the first genetic map showing the relative order of genes on a chromosome. This map also estimated the distance separating the genes. These brilliant and far-reaching hypotheses were typical of Morgan's group, which founded genetics research in the United States, developed *Drosophila* as a research organism, and made discoveries that were likely as significant to the development of genetics as those of Mendel.

Although it is tempting to assume that genetic maps could be made simply by looking down a microscope, finding the genes, and measuring the distance between them, the technology to do this was simply not available. Instead, Morgan's group used an indirect measure of distance. They reasoned that genes sitting relatively far apart on a chromosome would be more likely to be separated from one another during meiotic crossing-over than genes lying closer together. Figure 9.13, Chapter 9, illustrates this process of recombination occurring in the space separating two genes as they appear on chromosomes paired during meiosis. Obviously, if recombination is to be used as a measure of the distance separating genes, it must be detectable. That is why the organism used in Figure 9.13, Chapter 9, is heterozygous for all genes; the chromatids resulting from recombination are then different from the original, nonrecombinant, ones and can be identified. Following meiosis I and II, each of the four different chromatids will become a chromosome in a separate gamete (review the basic mechanisms of meiosis in Figure 9.10, Chapter 9). Which chromosome, recombinant or not, is carried by a given gamete is most clearly revealed only in offspring resulting from fertilization with a homozygous recessive gamete. That is why, in the cross originally done by Morgan in 1911, you will notice that one parent is heterozygous and the other is homozygous recessive **(Figure 11.2, p. 242)**.

To understand the following crosses, you need to learn to work with the genetic symbolism developed by Morgan instead of the *A/a* system used in Chapter 10. Although *Drosophila* notation might appear counterintuitive at first, understanding a few basic principles will help you see the logic behind it. First, note that geneticists working with fruit flies have all agreed on a "normal," or "wild-type," genotype; any change from wild type is, by definition, a mutant. Mutant alleles are named based on the altered phenotype of the organism that expresses them. The names for dominant mutant alleles are written with the first letter in uppercase, whereas those for recessive mutant alleles are written with the first letter in lowercase. For example, a dominant mutant allele transforming an antenna into a leg is called Antennapedia (*Antp*), whereas a recessive mutant allele altering eye colour is called vermilion (*v*). The notation for a wild-type allele is always made by simply adding a superscripted plus (+) sign to the

**Experimental Research
Figure 11.2** Evidence for gene linkage.

QUESTION: Do the purple-eye vestigial-wing genes of *Drosophila* assort independently?

EXPERIMENT: Morgan crossed true-breeding wild-type flies with red eyes and normal wings with purple-eyed, vestigial-winged flies. The F₁ dihybrids were all wild type in phenotype. Next he crossed the F₁ dihybrid flies with purple-eyed, vestigial-winged flies (this is a testcross) and analyzed the phenotypes of the progeny.

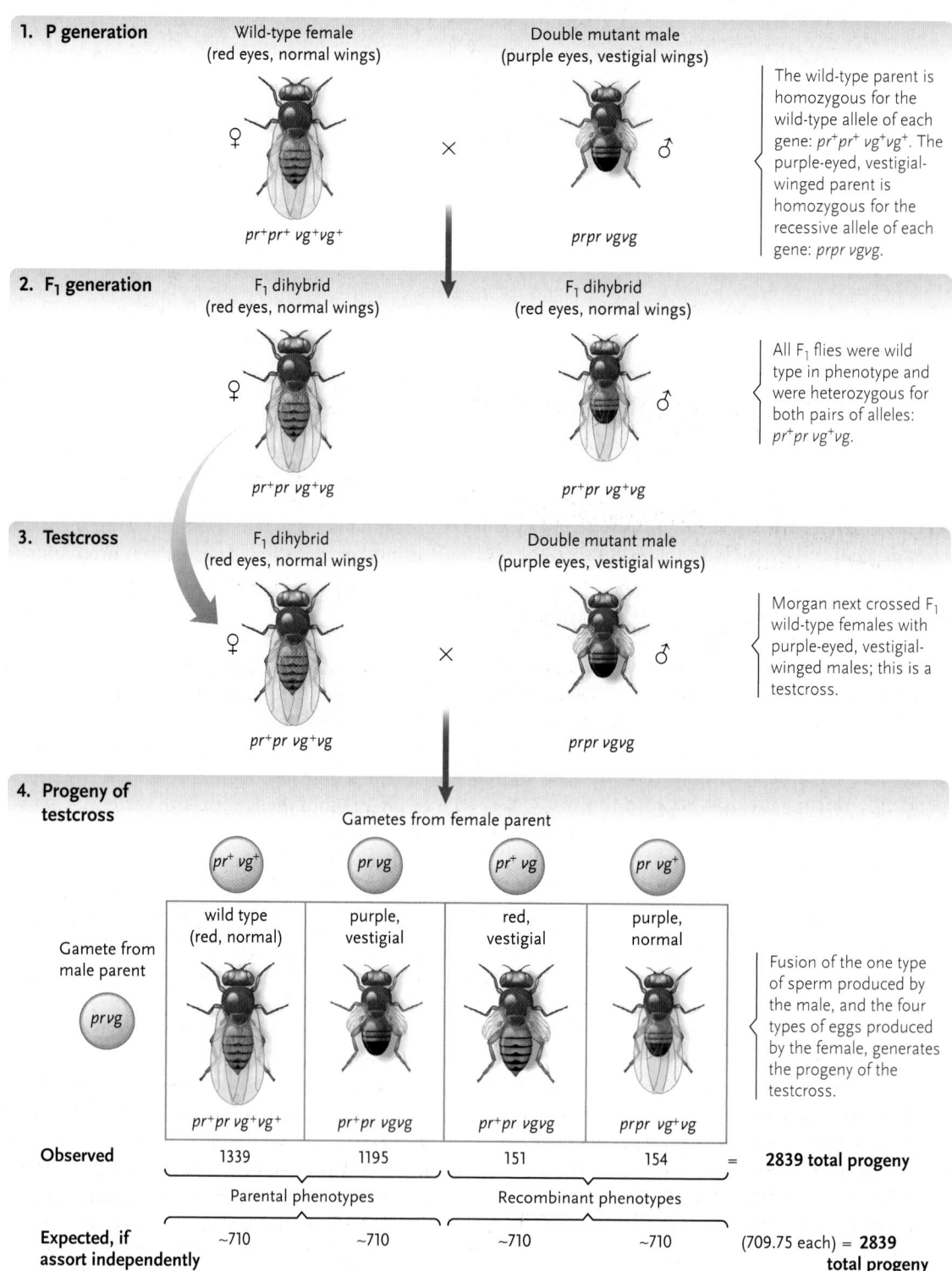

1. P generation

Wild-type female
(red eyes, normal wings)

Double mutant male
(purple eyes, vestigial wings)

$pr^+pr^+\,vg^+vg^+$ × $prpr\,vgvg$

The wild-type parent is homozygous for the wild-type allele of each gene: $pr^+pr^+\,vg^+vg^+$. The purple-eyed, vestigial-winged parent is homozygous for the recessive allele of each gene: $prpr\,vgvg$.

2. F₁ generation

F₁ dihybrid
(red eyes, normal wings)

F₁ dihybrid
(red eyes, normal wings)

$pr^+pr\,vg^+vg$ $pr^+pr\,vg^+vg$

All F₁ flies were wild type in phenotype and were heterozygous for both pairs of alleles: $pr^+pr\,vg^+vg$.

3. Testcross

F₁ dihybrid
(red eyes, normal wings)

Double mutant male
(purple eyes, vestigial wings)

$pr^+pr\,vg^+vg$ × $prpr\,vgvg$

Morgan next crossed F₁ wild-type females with purple-eyed, vestigial-winged males; this is a testcross.

4. Progeny of testcross

Gametes from female parent

$pr^+\,vg^+$ $pr\,vg$ $pr^+\,vg$ $pr\,vg^+$

Gamete from male parent

$pr\,vg$

wild type (red, normal)	purple, vestigial	red, vestigial	purple, normal
$pr^+pr\,vg^+vg^+$	$pr^+pr\,vgvg$	$pr^+pr\,vgvg$	$prpr\,vg^+vg$

Fusion of the one type of sperm produced by the male, and the four types of eggs produced by the female, generates the progeny of the testcross.

Observed	1339	1195	151	154	= **2839 total progeny**

Parental phenotypes   Recombinant phenotypes

Expected, if assort independently	~710	~710	~710	~710	(709.75 each) = **2839 total progeny**

RESULTS: 2534 of the testcross progeny flies were parental—wild-type or purple, vestigial—while 305 of the progeny were recombinant—red, vestigial or purple, normal. If the genes assorted independently, the expectation is for a 1:1:1:1 ratio for testcross progeny: approximately 1420 of both parental and recombinant progeny.

CONCLUSION: The purple-eye and vestigial-wing genes do not assort independently. The simplest alternative is that the two genes are linked on the same chromosome.

mutant allele notation. You know you understand this system if you agree that *Antp⁺* refers to a *recessive* allele giving a normal phenotype when homozygous.

Morgan began a specific breeding program using true-breeding fruit flies with normal red eyes and normal wing length (genotype *pr⁺pr⁺ vg⁺vg⁺*), along with a true-breeding fly with the recessive traits of purple eyes and vestigial (that is, short and crumpled) wings (genotype *prpr vgvg*) (Figure 11.2, step 1).

The F₁ (first-generation) offspring were all dihybrid *pr⁺pr vg⁺vg*, and because of the dominance of the wild-type alleles, they all had red eyes and normal wings (see Figure 11.2, step 2). Morgan then selected these wild-type F₁ females as the dihybrid parent and mated them to homozygous recessive males (with purple eyes and vestigial wings) as the testcross parent. If the purple and vestigial genes were carried on different chromosomes, Mendel's principle of independent assortment (see Section 10.1) would predict four classes of phenotypes in the offspring, in the approximate 1:1:1:1 ratio of red eyes, normal wings : purple, vestigial : red, vestigial : purple, normal. Given over 2800 offspring from several females, about 700 should have been in each class. However, Morgan observed two types of progeny in which the counts were much higher than 700 (red, normal and purple, vestigial) and two types with counts that were much lower (red, vestigial, and purple, normal) (see Figure 11.2, step 4).

Morgan's hypothesis to explain this non-Mendelian distribution is illustrated in **Figure 11.3, p. 244.** He suggested that the two genes are linked genetically—physically associated on the same chromosome. That is, *pr* and *vg* are linked genes. He further hypothesized that the behaviour of these linked genes is explained by *chromosome recombination* during meiosis. Furthermore, he proposed that the frequency of this recombination is a function of the distance between linked genes.

The *pr⁺pr vg⁺vg* F₁ dihybrid parents produce four types of gametes (see Figure 11.3). The two parental gametes, *pr⁺ vg⁺* and *pr vg*, are generated by simple segregation of the chromosomes during meiosis without any crossing-over (recombination) between the genes. The two recombinant gametes, *pr⁺ vg* and *pr vg⁺*, result from crossing-over between the homologous chromatids when they are paired in prophase I of meiosis (see Figures 9.10 and 9.13, Chapter 9). The offspring of the cross are produced by fusion of each of these four gametes with a *pr vg* gamete produced by the *prpr vgvg* male parent. The phenotypes of the offspring directly reflect the genotypes of the gametes produced by the dihybrid parent.

CONCEPT FIX Students of genetics sometimes assume that the wild-type and purple vestigial offspring in the above cross are called "parental" because they *look like* the parents. However, the term *parental* actually refers to genotype, not phenotype; parental offspring are the ones that *inherit chromosomes that were NOT involved in*

recombination in the dihybrid parent. Parental offspring, therefore, do not always resemble the parents of the cross. ⬢

Although Morgan could not look down a microscope and measure the distance between genes directly, he could look down a microscope and identify the phenotypes of recombinant offspring from dihybrid fruit fly testcrosses. Thus, the relative frequency of recombinant progeny became his "measure" of the distance separating genes. The example in Figure 11.3 reveals that purple eyes and vestigial wings are on the same chromosome and are separated by a recombinant offspring frequency distance of 10.7%.

11.1b Recombination Frequency Can Be Used to Map Chromosomes

The recombinant offspring frequency of 10.7% for the *pr* and *vg* genes of *Drosophila* means that 10.7% of the gametes originating from the *pr⁺pr vg⁺vg* parent contained recombined chromosomes (i.e., either *pr⁺ vg* or *pr vg⁺*). That recombinant offspring frequency is characteristic for those two genes. In other crosses that involve linked genes, Morgan found that the recombinant offspring frequency was characteristic of the two particular genes involved, and varied from less than 1% up to a maximum of 50% (see the next section).

From these observations, Alfred Sturtevant realized that the variation in recombinant offspring frequencies could be used as a means of mapping genes on chromosomes. Sturtevant himself later recalled his light bulb moment:

> I suddenly realized that the variations in the strength of linkage already attributed by Morgan to difference in the spatial separation of the gene offered the possibility of determining sequence in the linear dimensions of a chromosome. I went home and spent most of the night (to the neglect of my other homework) producing the first chromosome map.

Therefore, recombinant offspring frequencies can be used to make a **linkage map** of a chromosome showing the relative locations of genes. For example, assume that the three genes *a*, *b*, and *c* are carried together on the same chromosome. Crosses reveal a 9.6% frequency of recombinants for *a* and *b*, an 8% frequency for *a* and *c*, and a 2% frequency for *b* and *c*. These recombinant offspring frequencies allow the genes to be arranged in only one sequence on the chromosomes as follows:

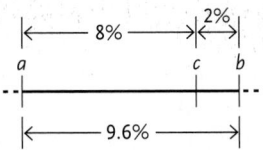

Figure 11.3

Recombinant offspring, the result of crossing-over between homologous chromosomes in the dihybrid parent. The testcross of Figure 11.2, p. 242, is redrawn here to show the two linked genes on chromosomes. The two parental homologues in the dihybrid parent (female, on the left) are coloured differently to allow us to follow them during the cross. The parental offspring inherit one or the other of the chromosomes unchanged from the dihybrid parent. The recombinant offspring inherit one or the other of the recombined chromosomes from the dihybrid parent.

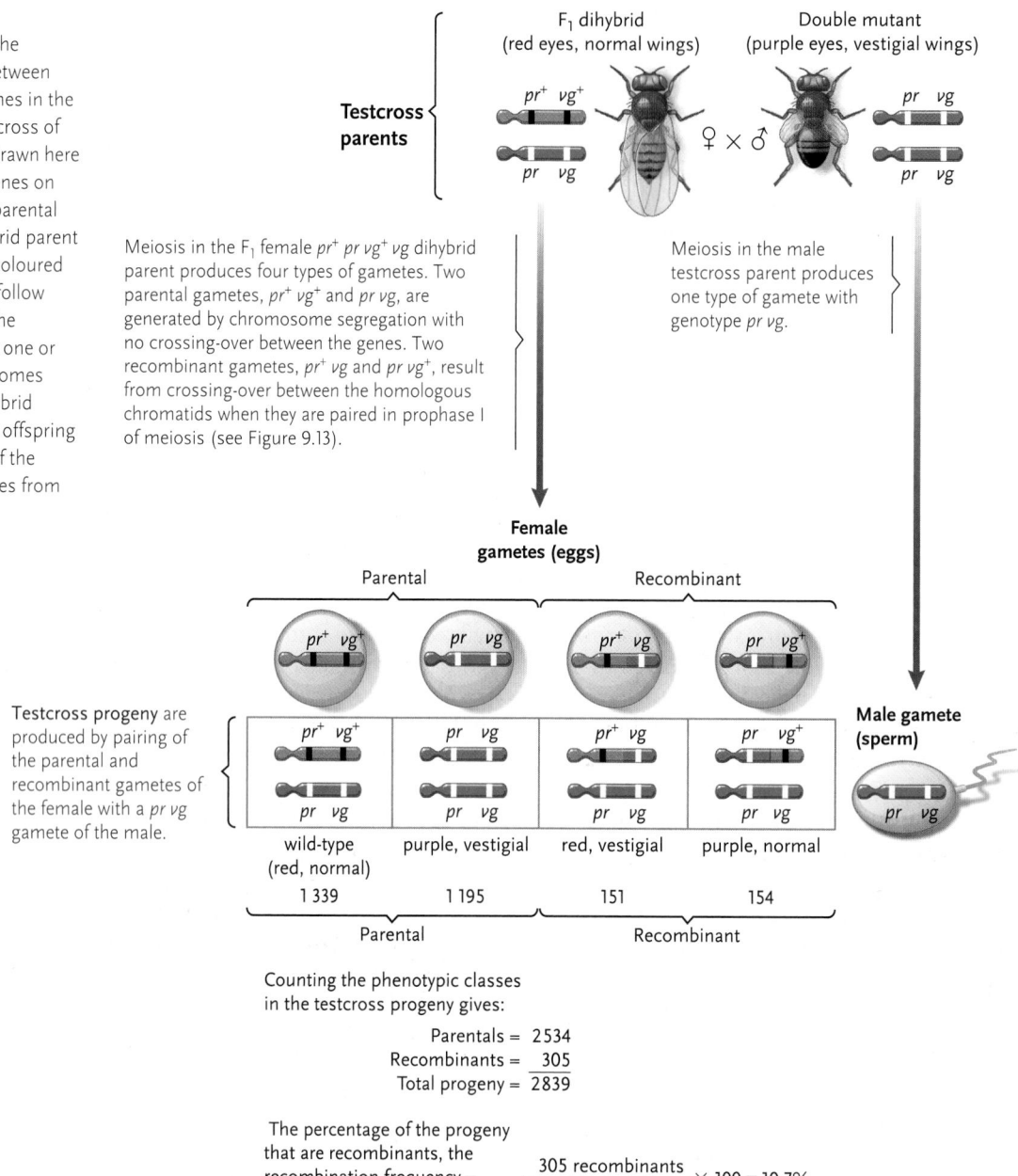

Meiosis in the F$_1$ female pr^+ pr vg^+ vg dihybrid parent produces four types of gametes. Two parental gametes, pr^+ vg^+ and pr vg, are generated by chromosome segregation with no crossing-over between the genes. Two recombinant gametes, pr^+ vg and pr vg^+, result from crossing-over between the homologous chromatids when they are paired in prophase I of meiosis (see Figure 9.13).

Meiosis in the male testcross parent produces one type of gamete with genotype pr vg.

Testcross progeny are produced by pairing of the parental and recombinant gametes of the female with a pr vg gamete of the male.

Counting the phenotypic classes in the testcross progeny gives:

Parentals = 2534
Recombinants = 305
Total progeny = 2839

The percentage of the progeny that are recombinants, the recombination frequency = $\dfrac{305 \text{ recombinants}}{2839 \text{ total progeny}} \times 100 = 10.7\%$

You will note that the *a–b* recombinant offspring frequency does not exactly equal the sum of the *a–c* and *c–b* frequencies. This is because genes farther apart on a chromosome are more likely to have more than one cross-over occur between them. Whereas a single cross-over between two genes gives recombinant chromatids, a double cross-over (two single cross-overs occurring in the same meiosis) between two genes gives the parental arrangement of alleles (and is therefore undetectable and would not be counted). You can see this simply by drawing single and double cross-overs between two genes on a piece of paper. In our example, the undetectable double cross-overs that occur between *a* and *b* have slightly decreased the overall recombinant offspring frequency between these two genes.

Using this method, Sturtevant created the first linkage map showing the arrangement of six genes on the *Drosophila* X chromosome. (A partial linkage map of a *Drosophila* chromosome is shown in **Figure 11.4.**)

Since the time of Morgan, many *Drosophila* genes and those of other eukaryotic organisms widely used for genetic research, including *Neurospora* (a fungus), yeast, maize (corn), and the mouse, have been mapped using the same approach. Recombinant offspring frequencies, together with the results of other techniques, have been used to create linkage maps of the locations of genes in the DNA of prokaryotic organisms such as *Escherichia coli*.

The unit of a linkage map, called a **map unit** (abbreviated mu), is equivalent to a recombinant offspring frequency of 1%. The map unit is also called the **centimorgan** (cM) in honour of Morgan's discoveries of linkage and recombination. Map units are not absolute physical distances such as micrometres or nanometres; rather, they are *relative*, showing the positions

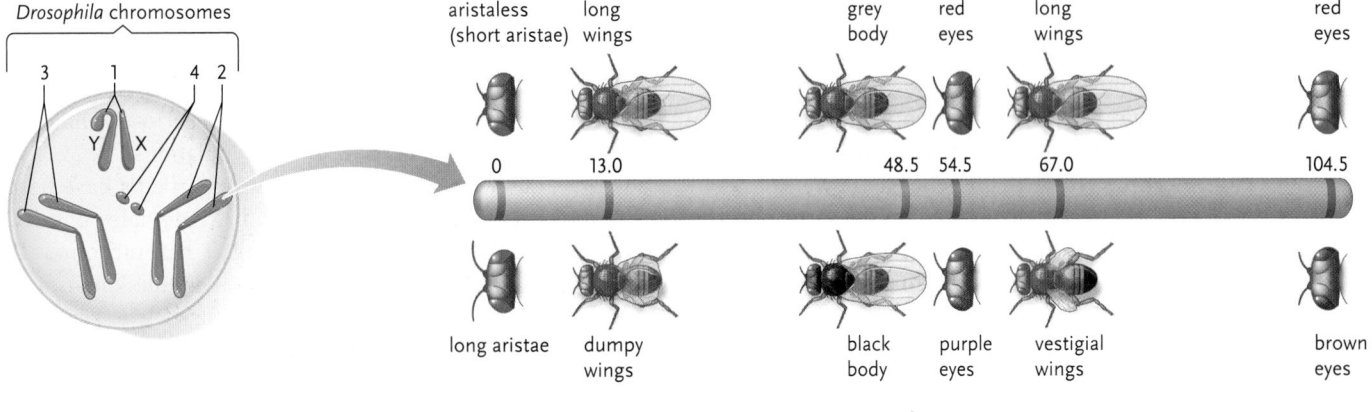

Wild-type phenotypes

aristaless (short aristae) | long wings | grey body | red eyes | long wings | red eyes

0   13.0   48.5   54.5   67.0   104.5

long aristae | dumpy wings | black body | purple eyes | vestigial wings | brown eyes

Mutant phenotypes

Figure 11.4

Relative map locations of several genes on chromosome 2 of *Drosophila*, as determined by frequencies of recombinant offspring from dihybrid testcrosses. For each gene, the diagram shows the normal or wild-type phenotype on the top and the mutant phenotype on the bottom. Mutant alleles at two different locations alter wing structure, one producing the dumpy-wing phenotype and the other the vestigial-wing phenotype; the normal allele at these locations results in normal long-wing structure. Mutant alleles at two different locations also alter eye colour.

of genes with respect to each other. One of the reasons that the units are relative and not absolute distances is that the frequency of crossing-over giving rise to recombinant offspring varies to some extent from one position to another along chromosomes.

In recent years, DNA sequencing of whole genomes has supplemented the linkage maps of a number of species. DNA sequencing shows the precise physical locations of genes right down to the number of base pairs separating them.

11.1c Widely Separated Linked Genes Assort Independently

Genes can be so widely separated on a chromosome that recombination is almost certain to occur at some point between them in every cell undergoing meiosis. When this is the case, the genes assort independently even though they are on the same chromosome. The map distance separating them will be 50 mu. (Fifty map units reflect 50% recombinant offspring. This is the same proportion of recombinant offspring observed when genes are on different chromosomes.)

To understand why this is, first recall Figure 9.13, Chapter 9, showing that a recombination event in a given cell creates 2 recombinant and 2 nonrecombinant chromatids. Next, imagine 100 meiocytes going through meiosis as usual to yield 400 gametes. If a recombination event occurred in the space separating 2 given genes in 10 of those cells, then 20 recombinant chromatids would be produced during prophase I. Twenty gametes would eventually receive recombinant chromosomes, and 20/400 = 5% of the total testcross progeny would be recombinant. We would conclude that these genes are 5 mu apart. Now assume that a recombination event occurs along the chromosome in

the space separating the two genes in *every one of the 100* cells going through meiosis. Two hundred recombinant offspring would result out of the total of 400; 50% would be recombinants; 50 mu would separate the genes.

Linkage between such widely separated genes can still be detected, however, by testing their linkage to one or more genes that lie between them. For example, the genes *a* and *c* in **Figure 11.5** are located so far apart that they assort independently and show no linkage. However, crosses that show *a* and *b* are 23 mu apart (recombinant offspring frequency of 23%), and other crosses show *b* and *c* are 34 mu apart. Therefore, *a* and *c* must

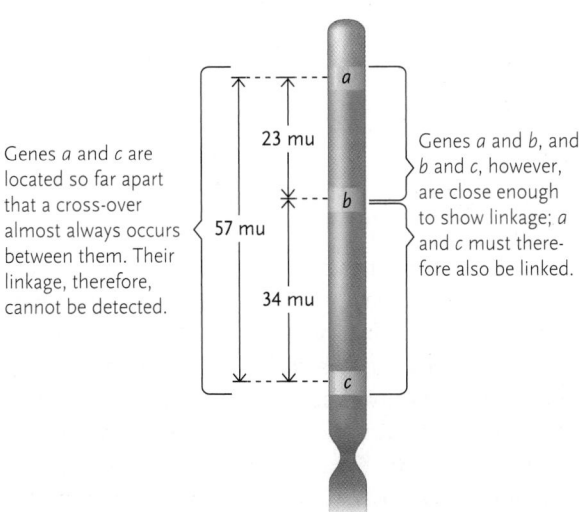

Genes *a* and *c* are located so far apart that a cross-over almost always occurs between them. Their linkage, therefore, cannot be detected.

23 mu

57 mu

34 mu

Genes *a* and *b*, and *b* and *c*, however, are close enough to show linkage; *a* and *c* must therefore also be linked.

Figure 11.5

Genes far apart on the same chromosome. Genes *a* and *c* are far apart and will not show linkage, suggesting that they are on different chromosomes. However, linkage between such genes can be established by noting their linkage to another gene or genes located between them—gene *b* here.

also be linked and carried on the same chromosome at 23 mu + 34 mu = 57 mu apart. Obviously, we could not see a recombinant offspring frequency of 57% in testcross progeny because the maximum frequency of recombinant chromatids is 50%, as described above.

We now know that some of the genes Mendel studied are actually on the same chromosome. For example, although the genes for flower colour and seed colour are actually located on the same chromosome, they are so far apart that frequent recombination between them made them assort independently in Mendel's analysis.

STUDY BREAK

1. What type of cross is typically used to discover whether two genes are linked or not?
2. How can two genes be on the same chromosome and yet assort independently (as if they were on separate chromosomes)?

11.2 Sex-Linked Genes

In many organisms, one or more pairs of chromosomes are different in males from those in females. Genes located on these chromosomes, the *sex chromosomes*, are called **sex-linked genes**; they are inherited differently in males and females.

CONCEPT FIX Note that the word "linked" in the phrase "*sex-linked gene*" means only that the gene is on a sex chromosome. The use of the word "linked" when considering two or more genes means that the genes are on the same chromosome. Linked genes might be on a sex chromosome or an autosome. ⬢

Chromosomes other than the sex chromosomes are called **autosomes**; genes on these chromosomes have the same patterns of inheritance in both sexes. In humans, chromosomes 1 to 22 are the autosomes.

11.2a Females Are XX and Males Are XY in Both Humans and Fruit Flies

In most species with sex chromosomes, females have two copies of a chromosome known as the **X chromosome,** forming a fully homologous XX pair, whereas males have only one X chromosome. Another chromosome, the Y chromosome, occurs in males but not in females. The Y chromosome has a short region of homology with the X chromosome that allows them to pair during meiosis. The XX human chromosome complement is shown in Figure 8.10, Chapter 8.

Each normal gamete produced by an XX female carries an X chromosome. Half the gametes produced by an XY male carry an X chromosome and half carry a Y. When a sperm cell carrying an X chromosome fertilizes an X-bearing egg cell, the new individual develops into an XX female. Conversely, when a sperm cell carrying a Y chromosome fertilizes an X-bearing egg cell, the combination produces an XY male. The Punnett square (see **Figure 11.6**) shows that fertilization is expected to produce females and males with an equal probability of 1/2. This expectation is closely matched in the human and *Drosophila* populations.

Other sex chromosome arrangements have been found, as in some insects with XX females and XO males (the O means there is no Y chromosome). In birds, butterflies, and some reptiles, the situation is reversed: males have a homologous pair of sex chromosomes (ZZ instead of XX), and females have the equivalent of an XY combination (ZW).

11.2b Human Sex Determination Depends on the *SRY* Gene

One gene carried on the Y chromosome, *SRY* (for sex-determining region of the Y), appears to be the master switch that directs development toward maleness at an early point in embryonic development.

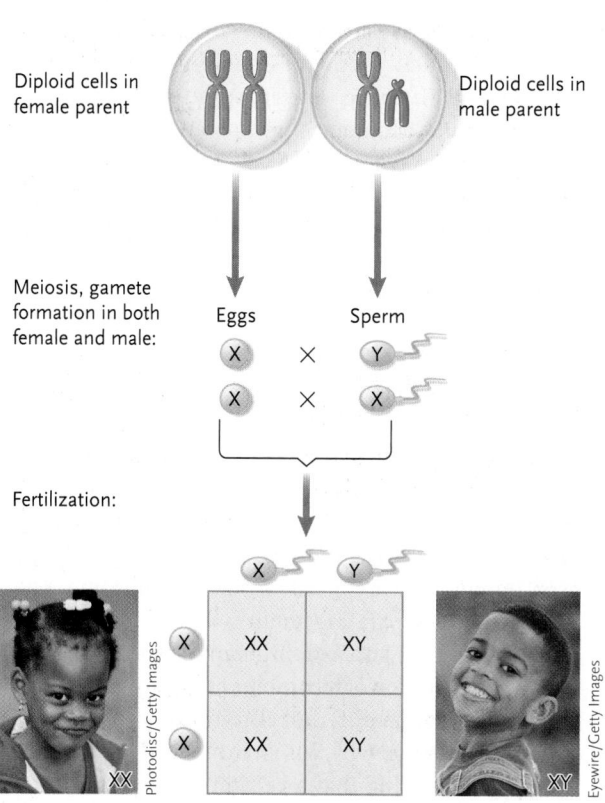

Figure 11.6

Sex chromosomes and the chromosomal basis of sex determination in humans. Females have two X chromosomes and therefore all gametes (eggs) have the X sex chromosome. Males have one X and one Y chromosome and therefore produce equal numbers of gametes containing an X chromosome versus a Y chromosome. Males transmit their Y chromosome to their sons, but not to their daughters. Males receive their X chromosome only from their mother.

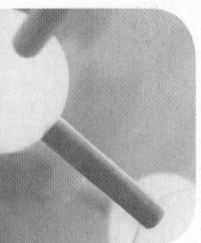

Drosopterin

The brick-red eyes of wild-type fruit flies owe their colour to a mixture of two types of pigment: bright red drosopterin and brown ommochrome. Drosopterin is the final product of a multistep biochemical pathway beginning with guanine. Mutations can alter the function of enzymes that act at different steps in this pathway to result in novel eye colours such as purple, brown, and sepia.

For the first month or so of embryonic development in humans and other mammals, the rudimentary structures that give rise to reproductive organs and tissues are the same in XX and XY embryos. After six to eight weeks, the *SRY* gene becomes active in XY embryos, producing a protein that regulates the expression of other genes, thereby stimulating part of these structures to develop as testes. As a part of stimulation by hormones secreted in the developing testes and elsewhere, tissues degenerate that would otherwise develop into female structures such as the vagina and oviducts. The remaining structures develop into the penis and scrotum. In XX embryos, which do not have a copy of the *SRY* gene, development proceeds toward female reproductive structures. The rudimentary male structures degenerate in XX embryos because the hormones released by the developing testes in XY embryos are not present. Further details of the *SRY* gene and its role in human sex determination are presented in Chapter 42.

CONCEPT FIX Although the X and Y chromosomes are called *sex chromosomes,* only a few genes they carry have any influence on sex determination or sexual function. For instance, most of the roughly 2400 known genes on the human X chromosome code for phenotypes needed by both sexes, such as colour perception, blood clotting, and DNA replication. Conversely, genes governing structures needed by only one sex or the other, such as breast development, penis structure, and facial hair, are coded on autosomes. If you are male, you have inherited the genes needed for uterine development and you will pass them along to your offspring to be used by daughters. You do not express these genes in your body. If you are female, a comparable situation is the case for genes coding for penis structure, and so on. You inherited these genes but you don't express them.

11.2c Sex-Linked Genes Were First Discovered in *Drosophila*

Since males and females have different sets of sex chromosomes, the genes carried on these chromosomes can be inherited in a distinctly non-Mendelian pattern called sex linkage. Sex linkage arises from two differences between males and females: (1) males have one X chromosome and therefore one allele for each gene on this chromosome (males are hemizygous for X-linked genes, *hemi* = half); females have two copies of the X chromosome and therefore two alleles for all genes on the X chromosome; (2) males also have one copy of the Y chromosome and one allele for each gene on this chromosome; females have no Y chromosome and therefore no Y alleles at all. Y chromosomes are present in males but not females.

Morgan discovered sex-linked genes and their pattern of sex linkage in 1910. The story of discovery started when he found a male fly in his stocks with white eyes instead of the normal red eyes **(Figure 11.7).** He crossed the white-eyed male with a true-breeding female with red eyes and observed that all the F_1 flies had red eyes **(Figure 11.8a, p. 248).** He concluded that the white-eye trait was recessive. Next, he allowed the F_1 flies to interbreed. Based on Mendel's principles, he expected that both male and female F_2 flies would show a 3:1 ratio of red-eyed flies to white-eyed flies. Morgan was surprised to find that all the F_2 females had red eyes, *but half of the F_2 males had red eyes and half had white eyes* (Figure 11.8b).

a.   **b.**

Martin Shields/Science Source

Figure 11.7

Eye colour phenotypes in *Drosophila.* **(a)** Normal, red, wild-type eye colour. **(b)** Mutant white eye colour caused by a recessive allele of a sex-linked gene carried on the X chromosome.

Experimental Research Figure 11.8
Evidence for sex-linked genes.

QUESTION: How is the white-eye gene of *Drosophila* inherited?

EXPERIMENT: Morgan crossed a white-eyed male *Drosophila* with a true-breeding female with red eyes and then crossed the F_1 flies to produce the F_2 generation. He also performed the reciprocal cross in which the phenotypes were switched in the parental flies—true-breeding white-eyed female × red-eyed male.

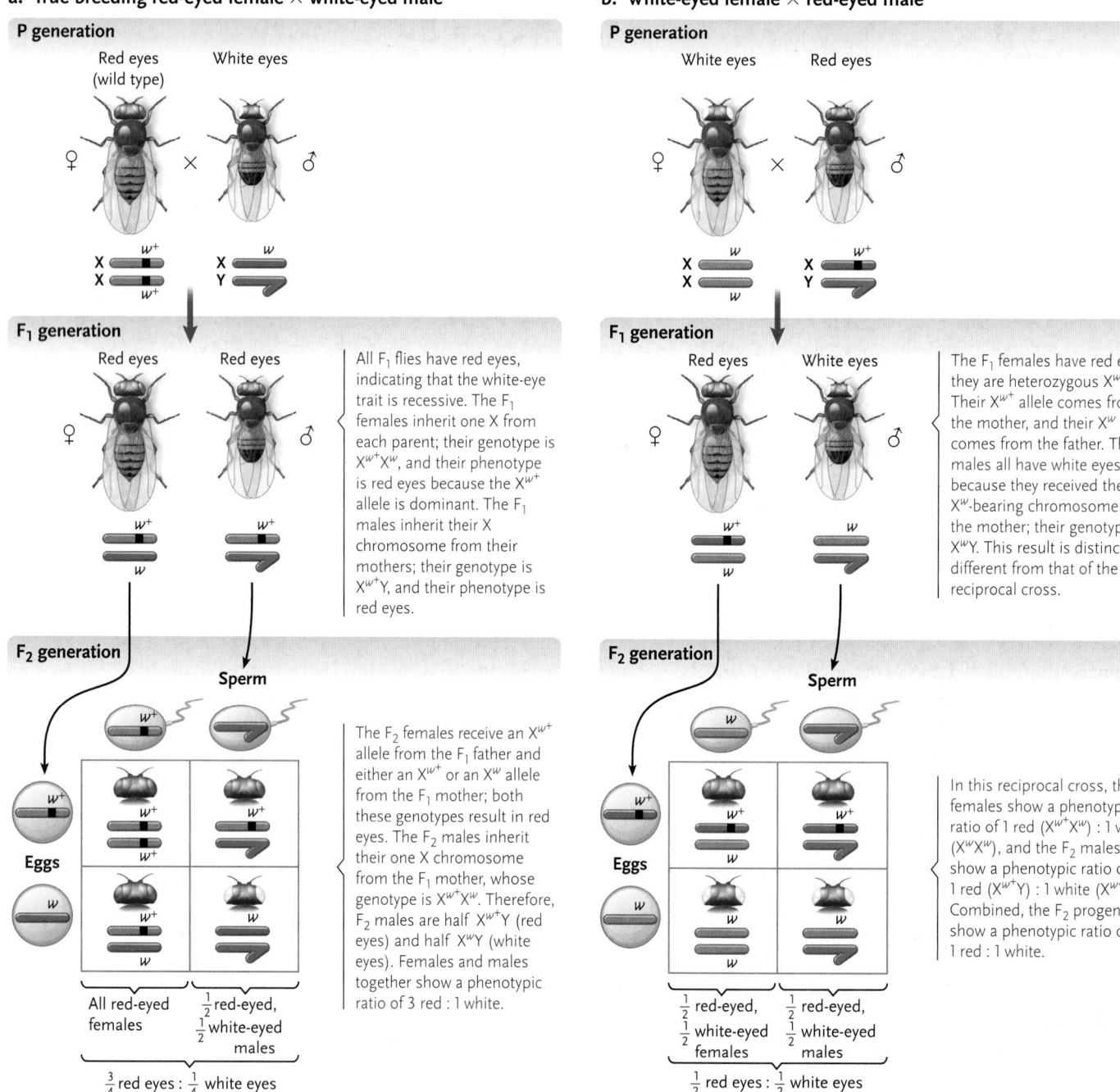

a. True-breeding red-eyed female × white-eyed male

P generation

Red eyes (wild type) × White eyes

All F_1 flies have red eyes, indicating that the white-eye trait is recessive. The F_1 females inherit one X from each parent; their genotype is $X^{w+}X^w$, and their phenotype is red eyes because the X^{w+} allele is dominant. The F_1 males inherit their X chromosome from their mothers; their genotype is $X^{w+}Y$, and their phenotype is red eyes.

F_1 generation

Red eyes   Red eyes

F_2 generation

Sperm

Eggs

All red-eyed females / $\frac{1}{2}$ red-eyed, $\frac{1}{2}$ white-eyed males

$\frac{3}{4}$ red eyes : $\frac{1}{4}$ white eyes

The F_2 females receive an X^{w+} allele from the F_1 father and either an X^{w+} or an X^w allele from the F_1 mother; both these genotypes result in red eyes. The F_2 males inherit their one X chromosome from the F_1 mother, whose genotype is $X^{w+}X^w$. Therefore, F_2 males are half $X^{w+}Y$ (red eyes) and half X^wY (white eyes). Females and males together show a phenotypic ratio of 3 red : 1 white.

b. White-eyed female × red-eyed male

P generation

White eyes × Red eyes

The F_1 females have red eyes: they are heterozygous $X^{w+}X^w$. Their X^{w+} allele comes from the mother, and their X^w allele comes from the father. The F_1 males all have white eyes because they received the X^w-bearing chromosome from the mother; their genotype is X^wY. This result is distinctly different from that of the reciprocal cross.

F_1 generation

Red eyes   White eyes

F_2 generation

Sperm

Eggs

$\frac{1}{2}$ red-eyed, $\frac{1}{2}$ white-eyed females / $\frac{1}{2}$ red-eyed, $\frac{1}{2}$ white-eyed males

$\frac{1}{2}$ red eyes : $\frac{1}{2}$ white eyes

In this reciprocal cross, the F_2 females show a phenotypic ratio of 1 red ($X^{w+}X^w$) : 1 white (X^wX^w), and the F_2 males also show a phenotypic ratio of 1 red ($X^{w+}Y$) : 1 white (X^wY). Combined, the F_2 progeny show a phenotypic ratio of 1 red : 1 white.

RESULTS: Differences were seen in both the F_1 and F_2 generations for the red ♀ × white ♂ and white ♀ × red ♂ crosses.

CONCLUSION: The segregation pattern for the white-eye trait showed that the white-eye gene is a sex-linked gene located on the X chromosome.

Morgan hypothesized that the alleles segregating in the cross were of a gene located on the X chromosome—now termed a sex-linked gene. The white-eyed male parent in the cross had the genotype X^wY—an X chromosome with a white (X^w) allele—and no other allele of that gene on the Y chromosome. The red-eyed female parent in the cross had the genotype $X^{w+}X^w$—each X chromosome carries the dominant normal allele for red eyes, X^{w+}.

We can follow the alleles in this cross (see Figure 11.8a). The F_1 flies of a cross $X^{w+}X^w \times X^wY$ are produced as follows. The X chromosome of each male comes from his mother; therefore, his genotype is $X^{w+}Y$, and his phenotype is red eyes. Each female receives one X from each parent; therefore, her genotype is $X^{w+}X^{w}$, and her phenotype is red eyes due to the dominance of the X^{w+} allele.

In the F_2 generation, each female receives an X^{w+} allele from her father (F_1) and either an X^{w+} or X^{w} allele from her mother (F_1); these genotypes result in red eyes (see Figure 11.8a). Each male receives his one X chromosome from his mother (F_1), who has the genotype $X^{w+}X^{w}$. Therefore, F_2 males are half $X^{w+}Y$ (red eyes) and half $X^{w}Y$ (white eyes).

Morgan also made a *reciprocal cross* of the one just described; that is, the phenotypes were switched between the parents. The reciprocal cross here was a white-eyed female ($X^{w}X^{w}$) with a red-eyed male ($X^{w+}Y$) (see Figure 11.8b). All F_1 males had white eyes because they received the X^{w}-bearing chromosome from their mother; thus, their genotype is $X^{w}Y$. The F_1 females have red eyes; they are all heterozygous $X^{w+}X^{w}$. *This result is clearly different from the reciprocal cross shown in Figure 11.8a.*

In the F_2 generation of this second cross, both male and female flies showed a 1:1 ratio of red eyes to white eyes (see Figure 11.8b). Again, this result differs markedly from that of the cross in Figure 11.8a.

In summary, Morgan's work showed that there is a distinctive pattern in the phenotypic ratios for reciprocal crosses in which the gene involved is on the X chromosome. A key indicator of this sex linkage is when all male offspring of a cross between a true-breeding mutant female and a wild-type male have the mutant phenotype. As we have seen, this occurs because a male receives his X chromosome from his female parent.

11.2d Sex-Linked Genes in Humans Are Inherited as They Are in *Drosophila*

For obvious reasons, experimental genetic crosses cannot be conducted with humans. However, a similar analysis can be made by interviewing and testing living members of a family and reconstructing the genotypes and phenotypes of past generations from family records. The results are summarized in a chart called a **pedigree**, which shows all parents and offspring for as many generations as possible, the sex of individuals in the different generations, and the presence or absence of the trait of interest. Females are designated by a circle and males by a square; a solid circle or square indicates the presence of the trait.

In humans, as in fruit flies, sex-linked recessive traits appear more frequently among males than females because males need to receive only one copy of the allele on the X chromosome inherited from their mothers to develop the trait. Females must receive two copies of the recessive allele, one from each parent, to express the trait. Two examples of human sex-linked traits are red–green colour-blindness, a recessive trait

in which the affected individual is unable to distinguish between the colours red and green because of a defect in light-sensing cells in the retina, and hemophilia, a recessive trait in which affected individuals have a defect in blood clotting.

CONCEPT FIX Colour-blindness does not mean that people see only black and white. The inability to see any colour at all is very rare. As shown in Figure 11.16, p. 258, colour-blindness reduces the variety of colours that can be distinguished. ⬡

People with hemophilia are "bleeders"; that is, they bleed uncontrollably if they are injured because a protein required for forming blood clots is not produced in functional form. Males are bleeders if they receive an X chromosome that carries the recessive allele. The disease also develops in females with the recessive allele on both of their X chromosomes—a rare combination. With luck and good care, affected people can reach maturity, but their lives are tightly circumscribed by the necessity to avoid injury. Even internal bleeding from slight bruises can be fatal. The disease, which affects about 1 in 7000 males, can be treated by injection of the required clotting protein.

Hemophilia has had effects reaching far beyond individuals who inherit the disease. The most famous cases occurred in the royal families of Europe descended from Queen Victoria of England **(Figure 11.9, p. 250).** The disease was not recorded in Queen Victoria's ancestors, so the recessive allele for the trait probably appeared as a spontaneous mutation in the queen or one of her parents. Queen Victoria was heterozygous for the recessive hemophilia allele; that is, she was a **carrier**, meaning that she carried the mutant allele and could pass it on to her offspring, but she did not have symptoms of the disease. A carrier is indicated in a pedigree by a male or female symbol with a central dot.

Note in Queen Victoria's pedigree in Figure 11.9 that Leopold, Duke of Albany, had hemophilia, as did his grandson, Rupert, Viscount Trematon. The trait appears in males in alternate generations (i.e., it skips a generation) because it passes with the X chromosome from mother to son. Mothers do not express the trait because they are heterozygous carriers. The sons, in turn, must pass the X chromosome with the affected allele to their daughters (and the Y chromosome to their sons), as did the Duke of Albany. The appearance of a trait in the males of alternate generations therefore suggests that the allele under study is recessive and carried on the X chromosome.

At one time, 18 of Queen Victoria's 69 descendants were affected males or female carriers. Because so many sons of European royalty were affected, the trait influenced the course of history. In Russia, Crown Prince Alexis was one of Victoria's descendants with hemophilia. His affliction drew together his parents, Czar Nicholas II and Czarina Alexandra (a granddaughter of Victoria and a carrier), and the hypnotic monk Rasputin, who manipulated the family to his advantage by

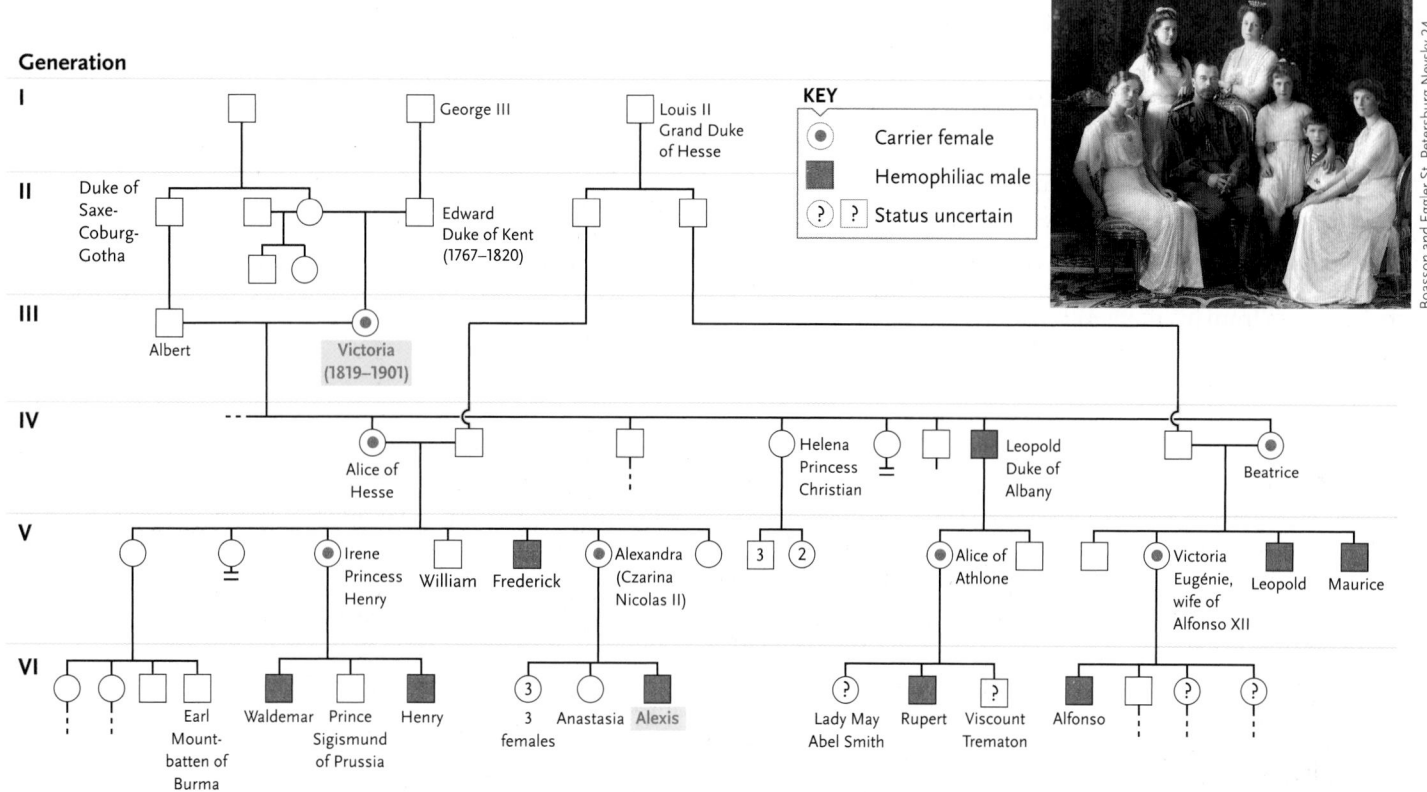

Generation

KEY
- ⊙ Carrier female
- ■ Hemophiliac male
- ⊙ ? Status uncertain

Boasson and Eggler St. Petersburg Nevsky 24

Figure 11.9

Inheritance of hemophilia in descendants of Queen Victoria of England. The photograph shows the Russian royal family in which the son, Crown Prince Alexis, had hemophilia. His mother was a carrier of the mutated gene.

convincing them that only he could control the boy's bleeding. The situation helped trigger the Russian Revolution of 1917, which ended the Russian monarchy and led to the establishment of a Communist government in the former Soviet Union, a significant event in twentieth-century history.

Hemophilia affected only sons in the royal lines but could have affected daughters if a hemophiliac son had married a carrier female. Because the disease is rare in the human population as a whole, the chance of such a mating is so low that only a few hemophiliac females have been recorded.

11.2e Inactivation of One X Chromosome Evens out Gene Effects

Although mammalian females have twice as many copies of genes carried on the X chromosome as males, it is unlikely that they require twice as much of the products of those genes. Theoretically, products from genes on the X chromosome could be equalized in males and females if (1) expression of genes on the single male X chromosome were doubled, or (2) expression of genes on both female X chromosomes were halved, or (3) one X chromosome were "turned off" in females. All of these dosage compensation

mechanisms are known in nature, but mammals use the latter; females with two X chromosomes inactivate most of the genes on one X chromosome or the other in most body cells.

As a result of the equalizing mechanism, the activity of most genes carried on the X chromosome is essentially the same in the cells of males and females. The inactivation occurs by a condensation process that folds and packs the chromatin of one of the two X chromosomes into a tightly coiled state similar to the condensed state of chromosomes during cell division. The inactive, condensed X chromosome can be seen within the nucleus in cells of females as a dense mass of chromatin called the **Barr body**.

The inactivation occurs during embryonic development. Which of the two X chromosomes becomes inactive in a particular embryonic cell line is a random event. But once one of the X chromosomes is inactivated in a cell, that same X is inactivated in all descendants of the cell. Thus, within one female, one of the X chromosomes is active in particular cells and inactive in others, and vice versa.

If the two X chromosomes carry different alleles of a gene, one allele will be active in cell lines in which one X chromosome is active, and the other allele will be active in cell lines in which the other X chromosome is active. For many sex-linked alleles, such as the recessive allele that causes hemophilia, random inactivation of either X chromosome has little overall whole-body effect in heterozygous females because the dominant allele is active in enough of the critical cells to produce a normal phenotype. However, for some genes, the inactivation of either X chromosome in heterozygotes produces recognizably different effects in distinct regions of the body.

For example, the orange and black patches of fur in calico cats result from inactivation of one of the two X chromosomes in regions of the skin of heterozygous females **(Figure 11.10)**. Males, which get only one of the two alleles, normally have either black or orange fur. Similarly, in humans, females who are heterozygous for an allele on the X chromosome that blocks development of sweat glands may have a patchy distribution of skin areas with and without the glands. Females with the patchy distribution are not seriously affected and may be unaware of the condition.

As we have seen, the discovery of genetic linkage, recombination, and sex-linked genes led to the elaboration and expansion of Mendel's principles of inheritance. Next, we examine what happens when patterns of inheritance are modified by changes in the chromosomes.

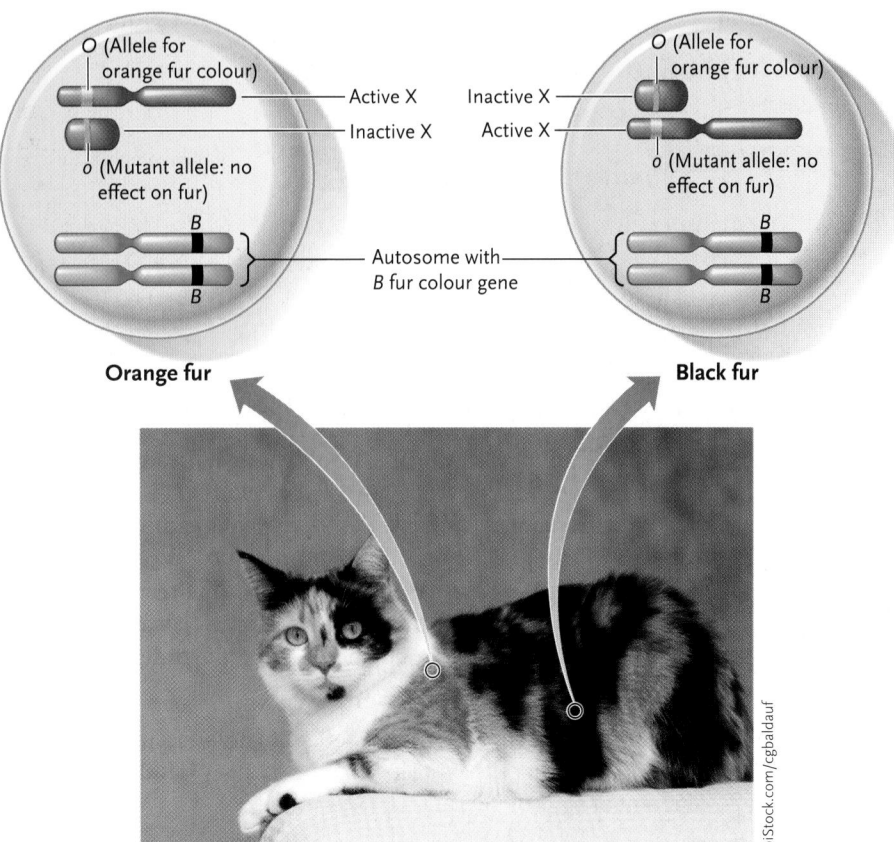

Figure 11.10

A female cat with the calico colour pattern in which patches of orange and black fur are produced by random inactivation of one of the two X chromosomes. Two genes control the black and orange colours: the *O* gene on the X chromosome is for orange fur colour, and the *B* gene on an autosome is for black fur colour. A calico cat has the genotype *Oo BB* (or *Oo Bb*). An orange patch results when the X chromosome carrying the mutant *o* allele is inactivated. In this case, the *O* gene masks the expression of the *B* gene and orange fur is produced. (This in an example of epistasis; see Section 10.2d.) A black patch results when the X chromosome carrying the *O* allele is inactivated. In this case, the mutant *o* allele cannot mask *B* gene expression and black fur results. The white patches result from interactions with a third, autosomal, gene that entirely blocks pigment deposition in the fur.

STUDY BREAK

1. What are the differences in sex chromosomes that underlie sex linkage inheritance patterns?
2. How could you determine if a given gene is sex-linked or not?

11.3 Chromosomal Alterations That Affect Inheritance

Although chromosomes are relatively stable structures, they are sometimes altered by breaks in the DNA, which can be generated by agents such as radiation or certain chemicals or by enzymes encoded in some infecting viruses. The broken chromosome fragments may be lost or they may reattach to the same or different chromosomes. The resulting changes in chromosome structure may have genetic consequences if alleles are eliminated, mixed in new combinations, duplicated, or placed in new locations by the alterations in cell lines that lead to the formation of gametes.

Genetic changes may also occur through changes in chromosome number, including addition or loss of one or more chromosomes or even entire sets of chromosomes. Both chromosomal alterations and changes in chromosome number can be a source of disease and disability, as well as a source of variability during evolution.

11.3a Deletions, Duplications, Translocations, and Inversions Are the Most Common Chromosomal Alterations

Chromosomal alterations after breakages occur in four major forms **(Figure 11.11, p. 252)**:

- A **deletion** occurs if a broken segment is lost from a chromosome.
- A **duplication** occurs if a segment is broken from one chromosome and inserted into its homologue. In the receiving homologue, the alleles in the

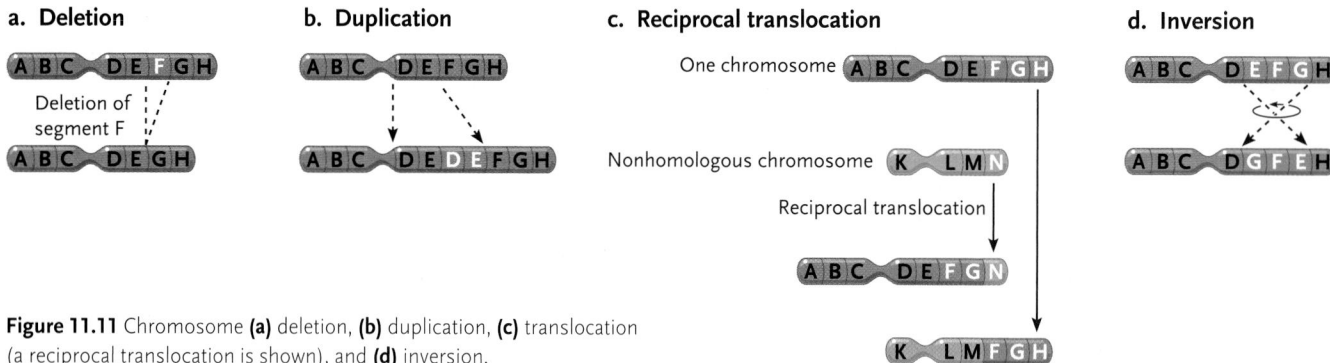

a. Deletion

A B C D E F G H

Deletion of segment F

A B C D E G H

b. Duplication

A B C D E F G H

A B C D E D E F G H

c. Reciprocal translocation

One chromosome A B C D E F G H

Nonhomologous chromosome K L M N

Reciprocal translocation

A B C D E F G N

K L M F G H

d. Inversion

A B C D E F G H

A B C D G F E H

Figure 11.11 Chromosome **(a)** deletion, **(b)** duplication, **(c)** translocation (a reciprocal translocation is shown), and **(d)** inversion.

[Handwritten notes:]
deletion: broken segment lost from chromosome
duplication: segment broken + inserted into it's homologue added to ones already there
translocation: broken segment attached to different nonhomologous chromosome
inversion: broken segment reattches to same chromose but in reverse order
reciprocal translocation: two non nomologous chromosome exchange segments which resembles genetic recombination
nondisjunction: homologous pairs don't seperate.

inserted fragment are added to the ones al[ready] there.

- A **translocation** occurs if a broken segment is att[ached] to a different, nonhomologous chromosome.
- An **inversion** occurs if a broken segment [at]taches to the same chromosome from which [it was] lost, but in reversed orientation, so that the [order] of genes is reversed.

To be inherited, chromosomal alterations must [occur] or be included in cells of the germ line leading to [devel]opment of eggs or sperm.

Deletions and Duplications. A deletion (see F[igure] 11.11a) may cause severe problems if the missing [seg]ment contains genes that are essential for n[ormal] development or cellular functions. For example, one deletion from human chromosome 5 typically leads to severe cognitive impairment and a malformed larynx. The cries of an affected infant sound more like a meow than a human cry—hence the name of the disorder, *cri-du-chat* (meaning "cat's cry").

A duplication (see Figure 11.11b) may have effects that vary from harmful to beneficial, depending on the genes and alleles contained in the duplicated region. Although most duplications are likely to be detrimental, some have been important sources of evolutionary change. That is, because there are duplicate genes, one copy can mutate into new forms without seriously affecting the basic functions of the organism. For example, mammals have genes that encode several types of hemoglobin that are not present in vertebrates such as sharks that evolved earlier; the additional hemoglobin genes of mammals are believed to have appeared through duplications, followed by mutations in the duplicates that created new and beneficial forms of hemoglobin as further evolution took place. Duplications sometimes arise during recombination in meiosis, if crossing-over occurs unequally, so that a segment is deleted from one chromosome of a homologous pair and inserted into the other.

Translocations and Inversions. In a translocation, a segment breaks from one chromosome and attaches to another, nonhomologous, chromosome. In many

[text indicating that two ... segments ... resemble ... two chromo- ... contain the ... the human ... caused by a ... human chromo- ... break does ... normal cell func- ... a gene that ... cell division ... location, it is ... can result in ... development of a] cancer in certain tissues.

In an inversion, a chromosome segment breaks and then reattaches to the same chromosome, but in reverse order (see Figure 11.11d). Inversions have essentially the same effects as translocations—genes may be broken internally by the inversion, with loss of function, or they may be transferred intact to a new location within the same chromosome, producing effects that range from beneficial to harmful.

Inversions and translocations have been important factors in the evolution of plants and some animals, including insects and primates. For example, five of the chromosome pairs of humans show evidence of translocations and inversions that are not present in one of our nearest primate relatives, gorillas. Therefore, the changes must have occurred after the gorilla and human evolutionary lineages split.

11.3b The Number of Entire Chromosomes May Also Change

At times, whole single chromosomes are lost or gained from cells entering or undergoing meiosis, resulting in a change of chromosome number. Most often, these changes occur through **nondisjunction**, the failure of homologous pairs to separate during the first meiotic division, or through misdivision, the failure of chromatids to separate during the second meiotic division (see Chapter 9 and **Figure 11.12**). As a result,

a.

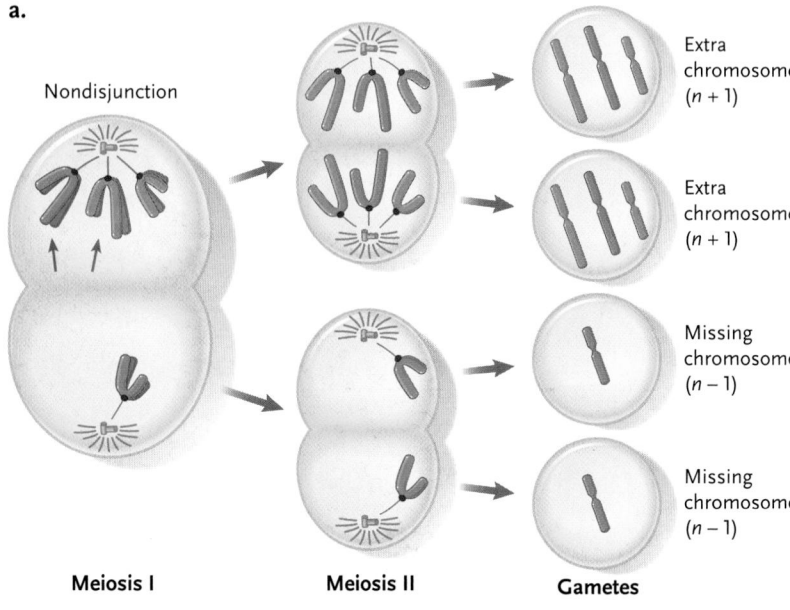

Nondisjunction during the first meiotic division causes both chromosomes of one pair to be delivered to the same pole of the spindle. The nondisjunction produces two gametes with an extra chromosome and two with a missing chromosome.

b.

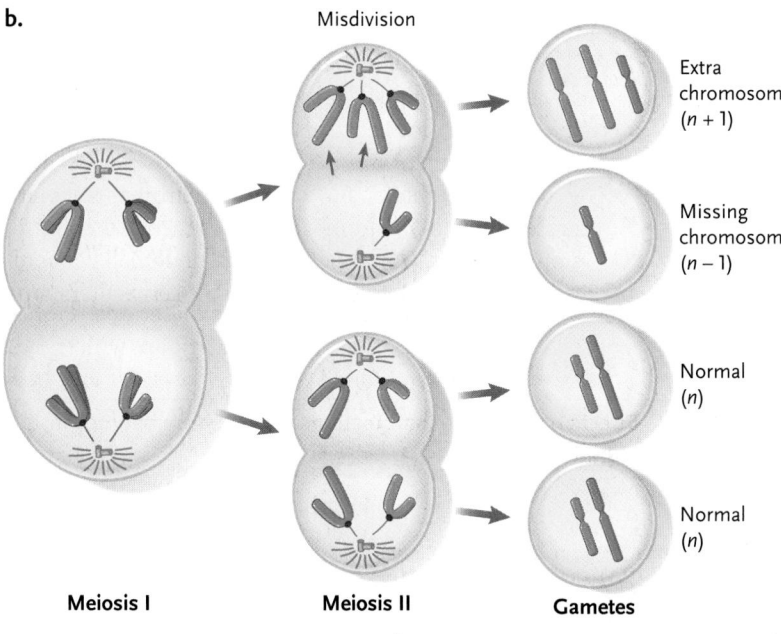

Misdivision during the second meiotic division produces two normal gametes, one gamete with an extra chromosome, and one gamete with a missing chromosome.

Figure 11.12

(a) Nondisjunction during the first meiotic division and **(b)** misdivision during the second meiotic division.

products of meiosis are produced that lack one or more chromosomes or contain extra copies of the chromosomes. *Note that failure of homologues to disjoin in meiosis I does not affect meiosis II; chromatids will most likely separate normally in meiosis II.* Fertilization by such gametes produces an individual with extra or missing chromosomes. Such individuals are called **aneuploids**,

whereas individuals with a normal set of chromosomes are called **euploids**.

Changes in chromosome number can also occur through duplication of entire sets, meaning individuals may receive one or more extra copies of the entire haploid complement of chromosomes. Such individuals are called **polyploids**. *Triploids* have three copies of each chromosome instead of two; *tetraploids* have four copies of each chromosome. Multiples higher than tetraploids also occur.

Aneuploids. The effects of addition or loss of whole chromosomes vary depending on the chromosome and the species. In animals, aneuploidy of autosomes usually produces debilitating or lethal developmental abnormalities. These abnormalities also occur in humans; addition or loss of an autosomal chromosome causes embryos to develop so abnormally that they are aborted naturally. For reasons that are not understood, aneuploidy is as much as 10 times as frequent in humans as in other mammals. Of human embryos that have been miscarried and examined, about 70% are aneuploids.

In some cases, autosomal aneuploids survive. This is the case with humans who receive an extra copy of chromosome 21—one of the smallest chromosomes **(Figure 11.13a, p. 254)**. Many of these individuals survive well into adulthood. The condition produced by the extra chromosome, called *Down syndrome* or *trisomy 21,* is characterized by short stature and some degree of cognitive impairment. About 40% of individuals with Down syndrome have heart defects, and skeletal development is slower than normal. Most do not mature sexually and remain sterile. However, with attentive care and appropriate educational opportunities, individuals with Down syndrome can successfully participate in many activities.

Most Down syndrome arises from nondisjunction or misdivision of chromosome 21 during meiosis, primarily in women (about 5% of nondisjunctions that lead to Down syndrome occur in men). The

Figure 11.13

Down syndrome.
(a) The chromosomes of a human female with Down syndrome showing three copies of chromosome 21 (circled in red). **(b)** The incidence of Down syndrome increases with the age of the mother, as determined in a study conducted in Victoria, Australia, between 1942 and 1957.

a.

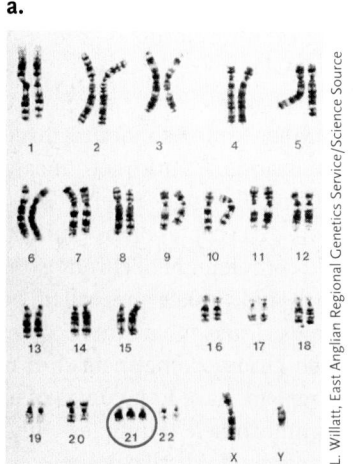

b.

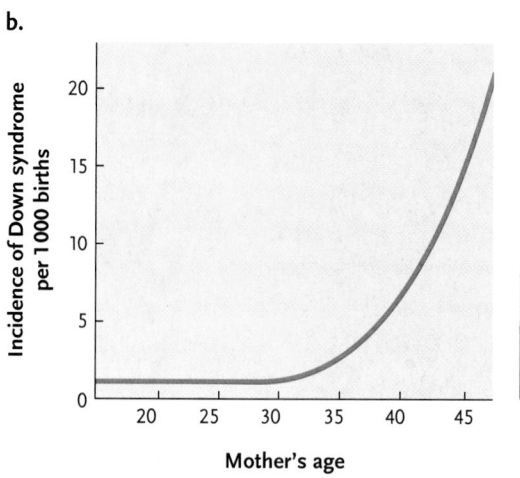

nondisjunction occurs more frequently as women age, increasing the chance that a child may be born with the syndrome (Figure 11.13b). Around the world, about 1 in every 800 children is born with Down syndrome, making it one of the most common serious human genetic defects.

Aneuploidy of sex chromosomes can also arise by nondisjunction or misdivision during meiosis (**Figure 11.14** and **Table 11.1**). Unlike autosomal aneuploidy, which usually has drastic effects on survival, altered numbers of X and Y chromosomes are often tolerated,

producing individuals who progress through embryonic development and grow to adulthood. In the case of multiple X chromosomes, the X-chromosome inactivation mechanism converts all but one of the X chromosomes to a Barr body, so the dosage of active X-chromosome genes is the same as in normal XX females and XY males. Triple X females may be taller than usual and may be at higher risk for learning disability, reduced muscle tone, and menstrual irregularities.

Because sexual development in humans is pushed toward male or female reproductive organs primarily by the presence or absence of the *SRY* gene on the Y chromosome, people with a Y chromosome are externally malelike, no matter how many X chromosomes are present. If no Y chromosome is present, X chromosomes in various numbers give rise to femalelike individuals. (Table 11.1 lists the effects of some alterations

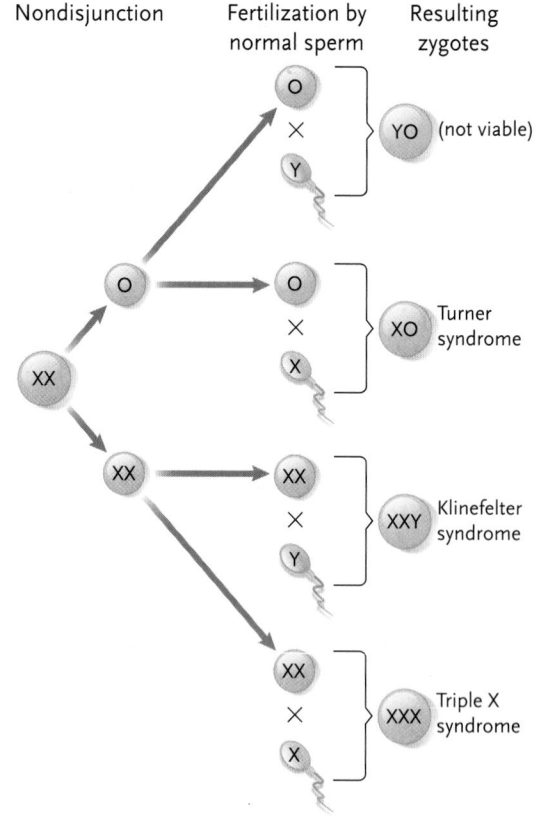

Figure 11.14

Some abnormal combinations of sex chromosomes resulting from nondisjunction of X chromosomes in females.

Table 11.1	Effects of Unusual Combinations of Sex Chromosomes in Humans		
Combination of Sex Chromosomes	**Approximate Frequency**	**Effects**	
XO	1 in 5000 births	Turner syndrome: females with underdeveloped ovaries; sterile; intelligence and external genitalia are normal; typically, individuals are short in stature with underdeveloped breasts	
XXY	1 in 2000 births	Klinefelter syndrome: male external genitalia with very small and underdeveloped testes; sterile; intelligence usually normal; sparse body hair and some development of the breasts; similar characteristics in XXXY and XXXXY individuals	
XYY	1 in 1000 births	XYY syndrome: apparently normal males but often taller than average	
XXX	1 in 1000 births	Triple-X syndrome: apparently normal female with normal or slightly delayed mental development	

Dr. Irene Ayako Uchida, *McMaster University*

In the early 1940s, the world was at war and Irene Uchida was studying English literature at the University of British Columbia and writing for a student newspaper. On returning home from a visit to Japan, she was prevented from continuing her studies. Fearing a Japanese invasion, and suspecting the Japanese people living on the Pacific coast to be a threat to national security, the Canadian government forced Uchida's family and thousands of other Japanese-Canadians to relocate to internment camps in the interior of British Columbia. During this difficult time, Uchida was head of a school for the children of internees.

She was later able to resume her education, this time at the University of Toronto, where she was encouraged to take up the field of genetics rather than her intended career in social work. Graduating with a Ph.D. in 1951, Uchida became a pioneer in the emerging field of medical cytogenetics and an international authority on the relationship between radiation and trisomies such as 18 and 21 (Down syndrome).

Uchida was made an officer of the Order of Canada in 1993, and retired in 1995 after a long and rewarding career. She died in 2013.

---

in sex chromosome number.) Similar abnormal combinations of sex chromosomes also occur in other animals, including *Drosophila,* with varying effects on viability.

Polyploids. Polyploidy often originates from failure of the spindle to function normally during mitosis in cell lines leading to germ-line cells. In these divisions, the spindle fails to separate the duplicated chromosomes, which are therefore incorporated into a single nucleus with twice the usual number of chromosomes. Eventually, meiosis takes place and produces products with two copies of each chromosome instead of one. Fusion of one such gamete with a normal haploid gamete produces a triploid zygote, and fusion of two such gametes produces a tetraploid zygote.

The effects of polyploidy vary widely between plants and animals. In plants, polyploids are often hardier and more successful in growth and reproduction than the diploid plants from which they were derived. As a result, polyploidy is common and has been an important source of variability in plant evolution. About half of all flowering plant species are polyploids, including important crop plants such as wheat and other cereals, cotton, and strawberries. One particularly widespread use of polyploids is in triploid bananas. Since triploid plants have difficulty disjoining homologues properly in meiosis, they are often sterile or, in this case, seedless.

By contrast, polyploidy is uncommon among animals because it usually has lethal effects during embryonic development. For example, in humans, all but about 1% of polyploids die before birth, and the few who are born die within a month. The lethality is probably due to disturbance of animal developmental pathways, which are typically much more complex than those of plants.

We now turn to a description of the effects of altered alleles on human health and development.

STUDY BREAK

What mechanisms are responsible for
(a) duplication of a chromosome segment,
(b) generation of a Down syndrome individual,
(c) a chromosome translocation, and
(d) polyploidy?

11.4 Human Genetics and Genetic Counselling

We have already noted a number of human genetic traits and conditions caused by mutant alleles or chromosomal alterations (see **Table 11.2, p. 256,** for a more detailed list). All of these traits are of interest as examples of patterns of inheritance that amplify and extend Mendel's basic principles. Those with harmful effects are also important because of their impact on human life and society.

11.4a In Autosomal Recessive Inheritance, Heterozygotes Are Carriers and Homozygous Recessives Are Affected by the Trait

Sickle cell disease and cystic fibrosis are examples of human traits caused by recessive alleles on autosomes. Alleles of these particular traits code for defective proteins that function poorly, if at all. Many other human genetic traits follow a similar pattern of inheritance (see Table 11.2, p. 256). These traits are passed on

Table 11.2	Examples of Human Genetic Traits
Trait	Adverse Health Effects
Autosomal Recessive Inheritance	
Albinism	Absence of pigmentation (melanin)
Attached earlobes	None
Cystic fibrosis	Excess mucus in lungs and digestive cavities
Sickle cell disease	Severe tissue and organ damage
Galactosemia	Brain, liver, and eye damage
Phenylketonuria	Severe cognitive impairment.
Tay–Sachs disease	Severe cognitive impairment, death
Autosomal Dominant Inheritance	
Free earlobes	None
Achondroplasia	Defective cartilage formation that causes dwarfism
Early balding in males	None
Campodactyly	Rigid, bent small fingers
Curly hair	None
Huntington disease	Progressive, irreversible degeneration of nervous system
Syndactyly	Webbing between fingers
Polydactyly	Extra digits
Brachydactyly	Short digits
Progeria	Premature aging
X-Linked Inheritance	
Hemophilia A	Deficient blood clotting
Red–green colour-blindness	Inability to distinguish red from green
Testicular feminizing syndrome	Absence of male organs, sterility
Changes in Chromosome Structure	
Cri-du-chat	Severe cognitive impairment, malformed larynx
Changes in Chromosome Number	
Down syndrome	Developmental delays, cognitive impairment, heart defects

according to the pattern known as **autosomal recessive inheritance,** in which individuals who are homozygous for the dominant allele are free of symptoms and are not carriers; heterozygotes are usually symptom free but are carriers. People who are homozygous for the recessive allele show the trait.

Between 10% and 15% of African Americans in the United States are carriers of sickle cell disease—that is, they have the sickle cell trait (see Section 10.2a). Although carriers make enough normal hemoglobin through the activity of the dominant allele to be essentially unaffected, the mutant, sickle cell form of the hemoglobin molecule is also present in their red blood cells. Carriers can be identified by a simple test for the mutant hemoglobin. In countries where malaria is common, including several countries in Africa, sickle cell carriers are less susceptible to contracting malaria, which helps explain the increased proportions of the recessive allele among races that originated in malarial areas.

Cystic fibrosis, one of the most common genetic disorders among people of Northern European descent, is another autosomal recessive trait **(Figure 11.15)**. About 1 in every 25 people from this line of descent is an unaffected carrier with one copy of the recessive allele, and about 1 in 2500 is homozygous for the recessive allele. The homozygous recessives have an altered membrane transport protein that results in excess Cl⁻ (chloride ions) in the extracellular fluids. Through pathways that are not completely understood, the alteration in chloride transport causes thick, sticky mucus to collect in airways of the lungs, in the ducts of glands such as the pancreas, and in the digestive tract. The accumulated mucus impairs body functions and, in the lungs, promotes pneumonia and other infections. With current management procedures, the life expectancy for a person with cystic fibrosis is about 40 years. The prevalence of cystic fibrosis alleles may have a similar explanation to those for sickle cell disease; heterozygotes may enjoy some resistance to infectious diseases such as tuberculosis or cholera.

Another autosomal recessive disease, *phenylketonuria* (PKU), appears in about 1 of every 15 000 births. Affected individuals cannot produce an enzyme that converts the amino acid phenylalanine to another amino acid, tyrosine. As a result, phenylalanine builds up in the blood and is converted into other products, including phenylpyruvate. Elevations in both phenylalanine and phenylpyruvate damage brain tissue and

© Kristina Paul

Figure 11.15
A child affected by cystic fibrosis. Daily chest thumps, back thumps, and repositioning dislodge thick mucus that collects in airways to the lungs.

Achondroplastic Dwarfing by a Single Amino Acid Change

Researchers recently found that the gene responsible for achondroplastic dwarfing is on chromosome 4. The gene codes for a receptor that binds the *fibroblast growth factor (FGF)*, a growth hormone that stimulates a wide range of mammalian cells to grow and divide. This fibroblast growth factor receptor (FGFR) gene is active in chondrocytes—cells that form cartilage and bone. Arnold Munnich and his colleagues isolated the gene that encodes the FGFR and obtained its DNA sequence. They found two versions of the gene's sequence with a single difference—one version had an adenine–thymine (A-T) base pair and the other had a guanine–cytosine (G-C) base pair at the same position in the DNA

sequence. The change substitutes arginine for glycine at one position in the amino acid sequence of the encoded protein. Arginine and glycine have very different chemical properties. The substitution occurs in a segment of the protein that extends across the membrane, connecting a hormone-binding site outside the cell with a site inside the cell that triggers the internal response.

The investigators then looked for the A-T–to–G-C substitution in the mutant form of the gene on chromosome 4 that causes achondroplastic dwarfing. The substitution was present in copies of the gene isolated from 6 families of achondroplastic dwarfs but absent in 120 people who lack the trait. This result supported the hypothesis that a

mutant allele of the FGFR on chromosome 4 is responsible for achondroplastic dwarfism.

How does the single amino acid substitution cause dwarfing? The cause is not known exactly. The change may inhibit the transmission of the signal triggered by a hormone binding to the receptor on the outer membrane. As a result, chondrocytes divide improperly and inhibit normal elongation of the limb bones. This helps explain why the achondroplasia mutation is dominant.

Identification of the gene responsible for achondroplastic dwarfing opens the future to finding a cure for the condition, possibly through gene therapy for infants or young children who carry the mutation.

can lead to cognitive impairment. However, if diagnosed early enough, an affected infant can be placed on a phenylalanine-restricted diet, which can prevent the PKU symptoms. Screening newborns for PKU is routine in the developed world and is becoming more established in the developing world as well. This is a wonderful example of how the expression of a genetic trait can be influenced by the environment.

You may have seen warnings on certain foods and drinks for phenylketonuriacs (individuals with PKU) not to use them. This is because they contain the artificial sweetener aspartame (trade name NutraSweet). Aspartame is a small molecule consisting of the amino acids aspartic acid and phenylalanine. Aspartame binds to taste receptors, signalling that the substance is sweet. Once ingested, aspartame is broken down and phenylalanine is released in amounts that might be harmful for people with PKU.

11.4b In Autosomal Dominant Inheritance, Only Homozygous Recessives Are Unaffected

Some human traits follow a pattern of **autosomal dominant inheritance** (see Table 11.2). In this case, the allele that causes the trait is dominant, and people who are either homozygous or heterozygous for the dominant allele are affected. Individuals homozygous for the recessive normal allele are unaffected.

Achondroplasia, a type of dwarfing that occurs in about 1 in 10 000 people, is caused by an autosomal dominant allele of a gene on chromosome 4. Of individuals with the dominant allele, only heterozygotes survive embryonic development; homozygous dominants are usually stillborn. When limb bones develop in heterozygous children, cartilage formation is defective, leading to disproportionately short arms and legs. The trunk and head, however, are of normal size. Affected adults are usually not much more than 122 cm tall. Achondroplastic dwarfs are of normal intelligence, are fertile, and can have children.

11.4c Males Are More Likely to Be Affected by X-Linked Recessive Traits

Red–green colour-blindness (**Figure 11.16, p. 258**) and hemophilia have already been presented as examples of human traits that demonstrate **X-linked recessive inheritance**, that is, traits due to inheritance of recessive alleles carried on the X chromosome. Another X-linked recessive human disease trait is Duchenne muscular dystrophy. In affected individuals, muscle tissue begins to degenerate late in childhood; by the onset of puberty, most individuals with this disease are unable to walk. Muscular weakness progresses, with later involvement of the heart muscle; the average life expectancy for individuals with Duchenne muscular dystrophy is 25 years.

Figure 11.16

Punnett square showing sex-linked recessive inheritance of colour-blindness in humans. Note that the mother carries the defective allele on one of her X chromosomes but is unaffected. Half of her daughters will be carriers and half of her sons will be colour-blind. Images indicate how the normal and colour-blind sons perceive a particular flower and leaves.

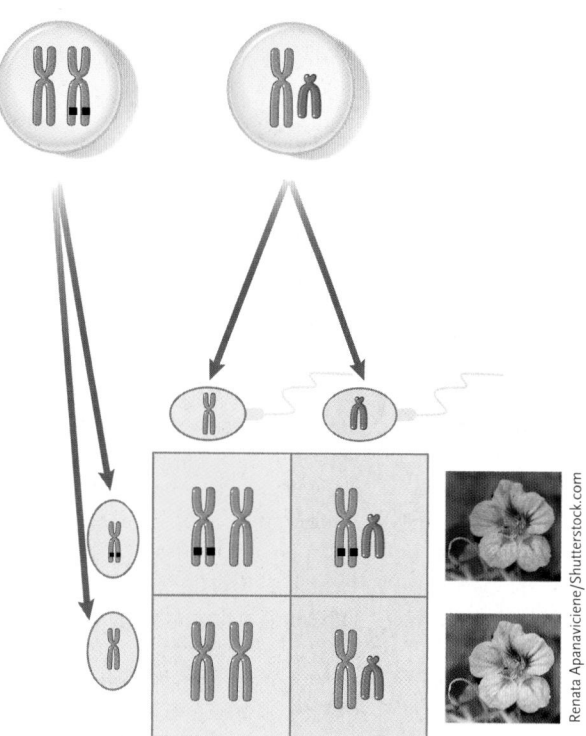

Renata Apanaviciene/Shutterstock.com

11.4d Human Genetic Disorders Can Be Predicted, and Many Can Be Treated

Each year, roughly 8 million children around the world are born with a severe disease or disability with a significant genetic component. The rate of such births in middle- and low-income countries is double that for high-income countries. Why might this be? One contributing factor has already been mentioned: in areas where malaria is endemic, the frequency of the sickle cell allele tends to be higher and the incidence of new-born sickle cell disease is higher. Nutritional deficiencies, consanguinous (blood relative) marriage practices, and higher numbers of children born to older mothers may also elevate birth defect rates in certain societies. In addition to improvements in basic financial, health, and nutritional standards, programs offering genetic counselling, prenatal diagnosis, and genetic screening can help reduce the suffering associated with genetic disorders.

Genetic counselling allows prospective parents to assess the possibility that they might have an affected child. For example, parents may seek counselling if they, a close relative, or one of their existing children has a genetic disorder. Genetic counselling begins with identification of parental genotypes through pedigrees or direct testing for an altered protein or DNA sequence. With this information in hand, counsellors can often predict the chances of having a child with the trait in question. Couples can then make an informed decision about whether to have a child.

Genetic counselling is often combined with techniques of **prenatal diagnosis**, in which cells derived from a developing embryo or its surrounding tissues or fluids are tested for the presence of mutant alleles or chromosomal alterations. In **amniocentesis**, cells are obtained from the amniotic fluid—the watery fluid surrounding the embryo in the mother's uterus (**Figure 11.17**). In **chorionic villus sampling**, cells are obtained from portions of the placenta that develop from tissues of the embryo. More than 100 genetic disorders can now be detected by these tests. If prenatal diagnosis detects a serious genetic defect, the prospective parents

Figure 11.17

Amniocentesis, a procedure used for prenatal diagnosis of genetic defects. The procedure is complicated and costly, and, therefore, it is used primarily in high-risk cases.

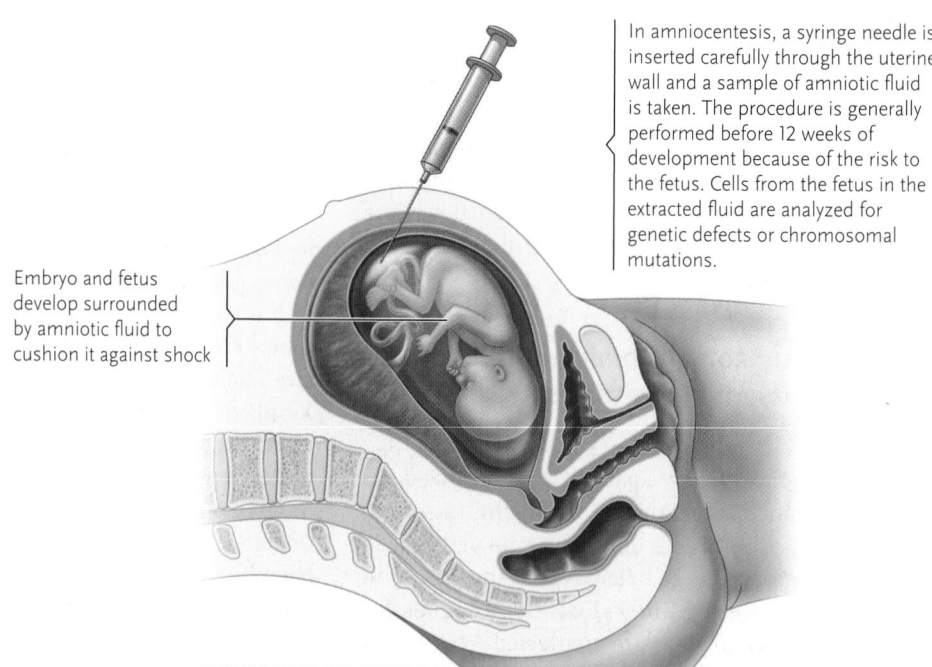

Embryo and fetus develop surrounded by amniotic fluid to cushion it against shock

In amniocentesis, a syringe needle is inserted carefully through the uterine wall and a sample of amniotic fluid is taken. The procedure is generally performed before 12 weeks of development because of the risk to the fetus. Cells from the fetus in the extracted fluid are analyzed for genetic defects or chromosomal mutations.

can reach an informed decision about whether to continue the pregnancy, including religious and moral considerations, as well as genetic and medical advice.

Once a child is born, inherited disorders are identified by **genetic screening**, in which biochemical or molecular tests for disorders are routinely applied to children and adults or to newborn infants in hospitals. The tests can detect inherited disorders early enough to start any available preventive measures before symptoms develop. As mentioned previously, worldwide newborn screening for PKU identifies affected children in time for them to avoid the debilitating symptoms of this disease. The first generation of people to survive childhood with PKU are now adults.

The characters and traits described so far in this chapter all depend on genes carried by chromosomes in the nucleus. But what of the genes located on DNA in mitochondria and chloroplasts? The following section addresses such interesting cases.

STUDY BREAK

1. What inheritance pattern would suggest that a trait is dominant and carried on an autosome?
2. How are inherited disorders detected before symptoms arise?

11.5 Nontraditional Patterns of Inheritance

We consider two examples of nontraditional patterns of inheritance in this section. In **cytoplasmic inheritance**, the pattern of inheritance follows that of genes in the cytoplasmic organelles: mitochondria or chloroplasts. In **genomic imprinting**, the expression of an allele of a particular nuclear gene is based on whether an individual organism inherits the allele from the male or female parent.

11.5a Cytoplasmic Inheritance Follows the Pattern of Inheritance of Mitochondria or Chloroplasts

Organelle DNA contains genes and alleles that, like nuclear genes, are also subject to being mutated. Mutant genes in some cases result in altered phenotypes, but the inheritance pattern of these mutant genes is fundamentally different from that of mutant genes carried on chromosomes in the nucleus. The two major differences are as follows: (1) ratios typical of Mendelian segregation are *not* found because genes are not segregating by meiosis, and (2) genes usually show uniparental inheritance from generation to generation. In *uniparental inheritance,* all progeny (both males and females) inherit the genotype of only one of the parents. For most multicellular eukaryotes, the mother's

genotype is passed on in a phenomenon called *maternal inheritance*. Maternal inheritance occurs because the amount of cytoplasm in the female gamete usually far exceeds that in the male gamete. Hence, a zygote receives most of its cytoplasm, including mitochondria and (in plants) chloroplasts, from the female ("egg" parent) and little from the male parent.

CONCEPT FIX Many people believe that they inherit half of their DNA from each of their parents. Although this idea is roughly true for nuclear DNA, recall that mitochondria also contain DNA and they are inherited exclusively from mothers. You have considerably more of your mother's DNA than your father's. ⬡

In humans, several inherited diseases have been traced to mutations in mitochondrial genes **(Table 11.3)**. Recall that the mitochondrion plays a critical role in synthesizing adenosine triphosphate (ATP), the energy source for many cellular reactions. The mutations producing the diseases in Table 11.3 are in mitochondrial genes that encode components of the ATP-generating system of the organelle. The resulting mitochondrial defects are especially destructive to the organ systems most dependent on mitochondrial reactions for energy: the central nervous system, skeletal and cardiac muscle, the liver, and the kidneys. These inherited diseases show maternal inheritance.

11.5b In Gene Imprinting, the Allele Inherited from One of the Parents Is Expressed whereas the Other Allele Is Silent

Genomic imprinting is a phenomenon in which the expression of an allele of a gene is determined by the parent that contributed it. In some cases, the paternally derived allele is expressed; in others, the maternally derived allele is expressed. The silent allele—the one that is not expressed—is called the *imprinted allele*. The imprinted allele is not inactivated by mutation. Rather, it is silenced by chemical modification (methylation) of certain bases in its sequence.

Table 11.3	Some Human Diseases Caused by Mutations in Mitochondrial Genes	
Disease	**Symptoms**
Kearns–Sayre syndrome	May include muscle weakness, mental deficiencies, abnormal heartbeat, short stature
Leber hereditary optic neuropathy	Vision loss from degeneration of the optic nerve, abnormal heartbeat
Mitochondrial myopathy and encephalomyopathy	May include seizures, strokelike episodes, hearing loss, progressive dementia, abnormal heartbeat, short stature
Myoclonic epilepsy	Vision and hearing loss, uncoordinated movement, jerking of limbs, progressive dementia, heart defects

As an example of how imprinting is involved in human disease, Prader–Willi syndrome (PWS) and Angelman syndrome (AS) in humans are both caused by genomic imprinting of a particular gene on a chromosome inherited from one parent, coincident with deletion of the same gene on the homologous chromosome inherited from the other parent. The syndromes differ with respect to the gene imprinted. Both PWS and AS occur in about 1 in 15 000 births and are characterized by serious developmental, mental, and behavioural problems. PWS individuals are compulsive overeaters (leading to obesity), have short stature, have small hands and feet, and show mild to moderate cognitive impairment. AS individuals are hyperactive, are unable to speak, have seizures, show severe cognitive impairment, and display a happy disposition with bursts of laughter.

How is genomic imprinting responsible for these two syndromes? PWS is caused when an individual has a normal maternally derived chromosome 15 and a paternally derived chromosome 15 with a deletion of a small region of several genes that includes the PWS gene. The PWS gene is imprinted, and therefore silenced, on maternally derived chromosomes. As a result, when there is no PWS gene on the paternally derived chromosome, there is no PWS gene activity, and PWS results. Similarly, AS is caused when an individual has a normal paternally derived chromosome 15 and a maternally derived chromosome 15 with a deletion of the same region; that region also includes the AS gene, the normal function of which is also required for normal development. In this case, genomic imprinting silences the AS gene on the paternally derived chromosome, and because there is no AS gene on the maternally derived chromosome, there is no AS gene activity, and AS syndrome develops.

CONCEPT FIX Although imprinted traits can show a *parent of origin effect,* imprinting is not the same as sex-linkage. Imprinted traits are not necessarily carried on sex chromosomes, and any given sex-linked allele can be inherited from either a mother or a father. ⬡

The mechanism of imprinting involves the modification of the DNA in the region that controls the expression of a gene by the binding or regulatory proteins, resulting in the addition of methyl (–CH3) groups to cytosine nucleotides. The methylation of the control region of a gene usually prevents it from being expressed. The regulation of gene expression by methylation of DNA is discussed further in Section 14.2. Genomic imprinting occurs in the gametes where the allele destined to be inactive in the new embryo after fertilization— either the father's or the mother's, depending on the gene—is methylated. That methylated (silenced) state of the gene is passed on as the cells grow and divide to produce the somatic (body) cells of the organism.

A number of cancers are associated with the failure to imprint genes. For instance, the mammalian *Igf2* (insulin growth factor 2) gene encodes a growth factor, a molecule that stimulates cells to grow and divide. *Igf2* is an imprinted gene, with the paternally derived allele "on" and the maternally derived allele "off." In some cases, the imprinting mechanism for this gene does not work, resulting in both alleles of *Igf2* being active, a phenomenon known as **loss of imprinting.** The resulting double dose of the growth factor disrupts the cell division cycle, increasing the risk of uncontrolled growth and cancer.

In this chapter, we have discussed genes and the role of chromosomes in inheritance. In the next chapter, we will turn to the molecular structure and function of the genetic material and learn about the molecular mechanism by which DNA is replicated.

STUDY BREAK

Which inheritance pattern would suggest that a trait is coded by the mitochondrial genome?

Review

To access course materials such as Aplia and other companion resources, please visit www.NELSONbrain.com.

11.1 Genetic Linkage and Recombination

- Genes consist of sequences of nucleotides in DNA and are arranged linearly in chromosomes.
- Genes carried on the same chromosome are linked in their transmission from parent to offspring. Linked genes are inherited in patterns similar to those of single genes, except for changes in the linkage due to recombination (see Figure 11.2).

- As a result of recombination, the order of a particular collection of alleles linked on any given chromosome is mixed up as a result of exchange with corresponding alleles on the other homologous chromosome. The exchanges occur while homologues pair during prophase I of meiosis.
- The likelihood of recombination between any two genes located on the same chromosome pair reflects the physical distance between them on the chromosome. The greater this distance, the greater the chance that chromatids will exchange segments at points between the genes and the greater the frequency of recombinant products of meiosis (gametes in animals, spores in plants).

- The relationship between separation and recombinant offspring frequencies is used to produce chromosome maps in which genes are assigned relative locations with respect to each other (see Figure 11.4).
- Testcrosses ($AaBb \times aabb$) can be used to detect linkage. If all progeny classes are equally frequent, then the genes are not linked.
- Genes carried on the same chromosome may not show genetic linkage (i.e., assort independently) if they are quite far apart.

11.2 Sex-Linked Genes

- Sex linkage is a pattern of inheritance produced by genes carried on sex chromosomes: chromosomes that differ between males and females. Sex-linked inheritance patterns arise because, in humans and fruit flies, females have two copies of the X chromosome and therefore two alleles for each gene. Males have only one copy of the X chromosome and therefore only one allele for each gene. Only males have an allele for genes carried on the Y chromosome.
- Sex linkage is suggested by a particular, non-Mendelian pattern of inheritance when the progeny of reciprocal crosses are different (see Figure 11.8).
- Since males have only one X chromosome, any recessive alleles that they inherit on that X chromosome will be expressed. Females must receive two copies of the recessive allele, one from each parent, to develop the trait (see Figures 11.6–11.8).
- In mammals, inactivation of one of the two X chromosomes in cells of the female makes the dosage of X-linked genes the same in males and females (see Figure 11.10).
- Parents can influence the expression of certain alleles in their offspring through DNA methylation called imprinting.

11.3 Chromosomal Alterations That Affect Inheritance

- Inheritance is influenced by processes that delete, duplicate, or invert segments within chromosomes or translocate segments between chromosomes (see Figure 11.11).
- Chromosomes also change in number by addition or removal of individual chromosomes or entire sets of chromosomes. Changes in single chromosomes usually occur through nondisjunction, in which homologous pairs fail to separate during meiosis I, or by misdivision, when sister chromatids fail to separate during meiosis II. As a result, one set of meiotic products receives an extra copy of a chromosome and the other set is deprived of the chromosome.
- Polyploids have one or more extra copies of the entire chromosome set. Polyploids usually arise when the spindle fails to function during meiosis in cell lines leading to gamete formation, producing zygotes that contain double the number of chromosomes typical for the species (see Figures 11.12–11.14).

11.4 Human Genetics and Genetic Counselling

- Three modes of inheritance are most significant in human heredity: autosomal recessive, autosomal dominant, and X-linked recessive inheritance.
- In autosomal recessive inheritance, males or females carry a recessive allele on an autosome. Heterozygotes are carriers that are usually unaffected, but homozygous individuals show symptoms of the trait. Affected children born to unaffected parents suggest autosomal recessive inheritance.
- In autosomal dominant inheritance, a dominant gene is carried on an autosome. Individuals that are homozygous or heterozygous for the trait show symptoms of the trait; homozygous recessives are normal.
- In X-linked recessive inheritance, a recessive allele for the trait is carried on the X chromosome. Male individuals with the recessive allele on their X chromosome or female individuals with the recessive allele on both X chromosomes show symptoms of the trait. Heterozygous females are carriers but usually show no symptoms of the trait.
- Genetic counselling, based on identification of parental genotypes by constructing family pedigrees and prenatal diagnosis, allows prospective parents to reach an informed decision about whether to have a child or continue a pregnancy.

11.5 Nontraditional Patterns of Inheritance

- Cytoplasmic inheritance depends on genes carried on DNA in mitochondria or chloroplasts. Cytoplasmic inheritance follows the maternal line: it parallels the inheritance of the cytoplasm in fertilization, in which most or all of the cytoplasm of the zygote originates from the egg cell. That is, all of the offspring of affected mothers would be affected; none of the offspring of affected fathers would be affected.
- Genomic imprinting is a phenomenon in which the expression of an allele of a gene is determined by the parent that contributed it. In some cases, the allele inherited from the father is expressed; in others, the allele from the mother is expressed. The silencing of the other allele is often the result of methylation of the region adjacent to the gene that is responsible for controlling the expression of that gene.

Questions

Self-Test Questions

1. In humans, red–green colour-blindness is an X-linked recessive trait. If a man with normal vision and a colour-blind woman have a son, what is the chance that the son will be colour-blind? What is the chance that a daughter will be colour-blind?

2. The following pedigree shows the pattern of inheritance of red–green colour-blindness in a family. Females are shown as circles and males as squares; the squares or circles of individuals affected by the trait are filled in black.

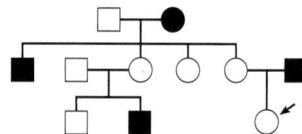

What is the chance that a son of the third-generation female indicated by the arrow will be colour-blind if the father is a normal man? If the father is colour-blind?

3. Individuals affected by a condition known as polydactyly have extra fingers or toes. The following pedigree shows the pattern of inheritance of this trait in one family:

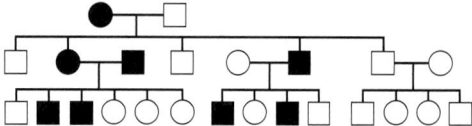

From the pedigree, can you tell if polydactyly comes from a dominant or recessive allele? Is the trait sex linked? As far as you can determine, what is the genotype of each person in the pedigree with respect to the trait?

4. A number of genes carried on the same chromosome are tested and show the following cross-over frequencies. What is their sequence in the map of the chromosome?

Genes	Cross-over Frequencies between Them
C and A	7%
B and D	3%
B and A	4%
C and D	6%
C and B	3%

5. In *Drosophila*, two genes, one for body colour and one for eye colour, are carried on the same chromosome. The wild-type grey body colour is dominant to black body colour, and wild-type red eyes are dominant to purple eyes. You make a cross between a fly with a grey body and red eyes and a fly with a black body and purple eyes. Among the offspring, about half have grey bodies and red eyes and half have black bodies and purple eyes. A small percentage have (a) black bodies and red eyes or (b) grey bodies and purple eyes. Which alleles are carried together on the chromosomes in each of the flies used in the cross? Which alleles are carried together on the chromosomes of the F_1 flies with black bodies and red eyes, and those with grey bodies and purple eyes?

6. Another gene in *Drosophila* determines wing length. The dominant wild-type allele of this gene produces long wings; a recessive allele produces vestigial (short) wings. A female that is true-breeding for red eyes and long wings is mated with a male that has purple eyes and vestigial wings.

F_1 females are then crossed with purple-eyed, vestigial-winged males. From this second cross, a total of 600 offspring are obtained with the following combinations of traits:

252 with red eyes and long wings
276 with purple eyes and vestigial wings
42 with red eyes and vestigial wings
30 with purple eyes and long wings

Are the genes linked, unlinked, or sex linked? If they are linked, how many map units separate them on the chromosome?

Drosophila with vestigial wings

7. One human gene, which is suspected to be carried on the Y chromosome, controls the length of hair on men's ears. One allele produces nonhairy ears, and another produces hairy ears. If a man with hairy ears has sons, what percentage will also have hairy ears? What percentage of his daughters will have hairy ears?

8. You conduct a cross in *Drosophila* that produces only half as many male as female offspring. What might you suspect as a cause?

Questions for Discussion

1. Can a linkage map be made for a haploid organism that reproduces sexually?

2. Crossing-over does not occur between any pair of homologous chromosomes during meiosis in male *Drosophila*. From what you have learned about meiosis and crossing-over, propose one hypothesis for why this might be the case.

3. Even though X inactivation occurs in XXY (Klinefelter syndrome) humans, they do not have the same phenotype as normal XY males. Similarly, even though X inactivation occurs in XX individuals, they do not have the same phenotype as XO (Turner syndrome) humans. Why might this be the case?

4. All mammals have evolved from a common ancestor. However, the chromosome number varies among mammals. By what mechanism might this have occurred?

5. Assume that genes *a, b, c, d, e,* and *f* are linked. Explain how you would construct a linkage map that shows the order of these six genes and the map units between them.

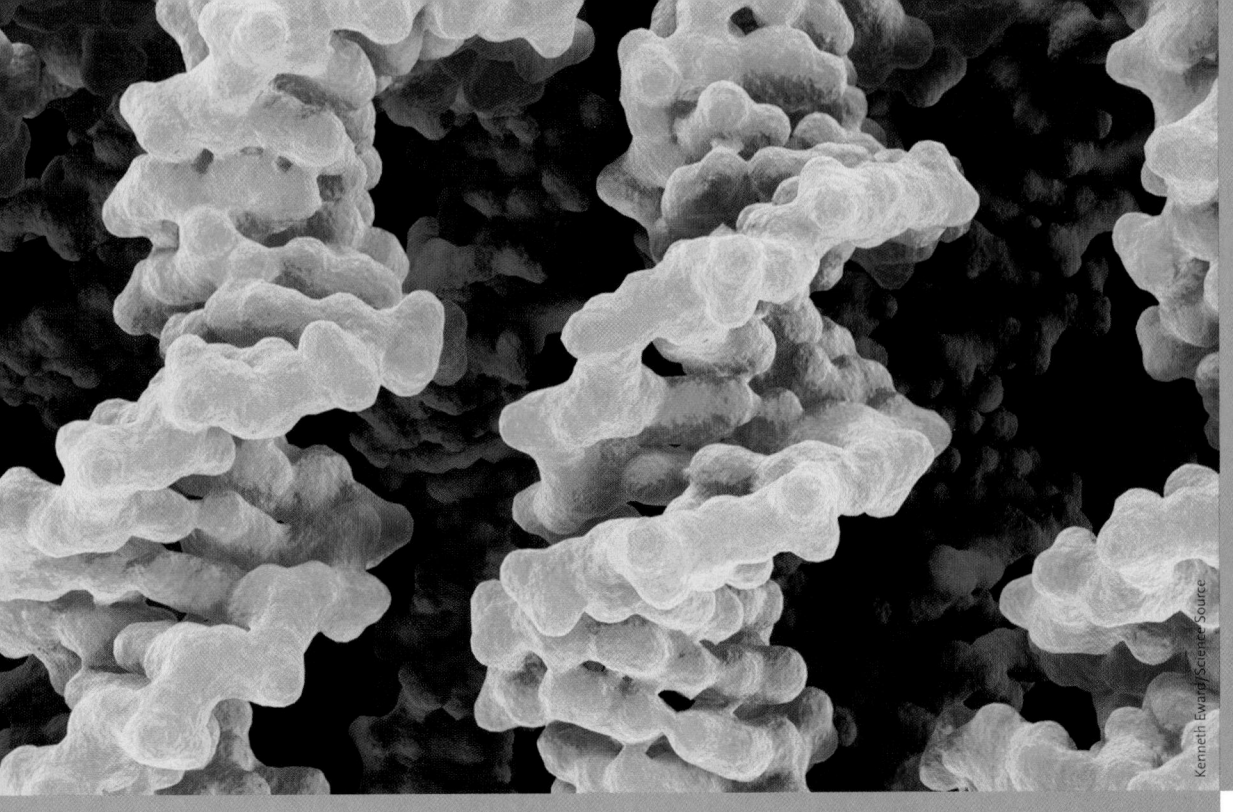

A digital model of DNA (based on data generated by X-ray crystallography).

Kenneth Eward/Science Source

12

DNA Structure, Replication, and Organization

WHY IT MATTERS

Imagine a scene 40 thousand years ago in what is now called the Drachenlock Cave in Switzerland. Flickering torchlight reflects from a collection of large bear skulls as a Neanderthal shaman arranges one, then the next, to face toward the entrance to the cave. Now fast-forward to the present to find cave bear bones and teeth once again carefully arranged by human hands, this time on the bench of a modern, ultraclean research laboratory. The scientist is completely covered by a protective gown, gloves, and a face mask. The surface of the specimens is bleached and irradiated with high-intensity ultraviolet light. A small drill bores into the interior of a molar tooth, where researchers hope to recover ancient DNA (aDNA) from *Ursus spelaeus,* a long-extinct relative of modern bears.

As much as characterization of aDNA sequences promises to enhance our understanding of the genetic history and composition of modern populations, this field is overshadowed by two significant problems: DNA damage and contamination. The double helix of DNA is subject to breakages in one or both strands in addition to inappropriate cross-linking and chemical modification of individual bases. Living cells very successfully prevent or repair most of this DNA damage, but post-mortem degradation can be extensive after thousands of years. Sustained cold temperatures preserve aDNA relatively well, facilitating successful recovery of sequences from frozen mammoths and bison in **permafrost**, penguins in ice, and the human "Ice Man" frozen in a glacier. Ancient bacterial DNA sequences have been recovered from 500 thousand-year-old sections of ice cores, and the complete genome has been sequenced from a 700 thousand-year-old horse bone recovered from permafrost.

The natural degradation of DNA over time usually means that aDNA sequences remaining in a given tissue sample are very rare and therefore prone to contamination by DNA from modern or ancient sources—hence the need for ultraclean laboratories, decontamination procedures, and authentication protocols. Suspicions of contamination have clouded some of the most dramatic reports of aDNA recovery from specimens 10 to 100 million years old.

As the future brings better techniques for the recovery and characterization of authentic aDNA sequences on Earth, we will undoubtedly turn these skills toward the search for evidence of past or present life on other planets. The Martian polar ice caps are very cold and very persistent, providing ideal conditions for preservation of DNA from any organisms that may have inhabited the Red Planet in the past.

Our current ability to find, characterize, and manipulate DNA arises ultimately from the work of a Swiss physician and physiological chemist, Johann Friedrich Miescher. In 1868, Miescher was engaged in a study of the composition of the cell nucleus. He collected pus cells from discarded bandages and extracted large quantities of an acidic substance with a high phosphorus content. He called the unusual substance *nuclein*. Nuclein is now known by its modern name, **deoxyribonucleic acid**, or **DNA**, the molecule that is the genetic material of all living organisms and, as indicated by ancient DNA studies, all extinct organisms as well.

At the time of Miescher's discovery, scientists knew nothing about the molecular basis of heredity and very little about genetics. Although Mendel had already published the results of his genetic experiments with garden peas, the significance of his findings was not widely known or appreciated. It was not known which chemical substance in cells actually carries the instructions for reproducing parental traits in offspring. Not until 1952, more than 80 years after Miescher's discovery, did scientists fully recognize that the hereditary molecule was DNA.

After DNA was established as the hereditary molecule, the focus of research changed to the three-dimensional structure of DNA. Among the scientists striving to work out the structure were James D. Watson, a young American postdoctoral student at Cambridge University in England, and the Englishman Francis H.C. Crick, then a graduate student at Cambridge University. Using chemical and physical information about DNA, in particular Rosalind Franklin's analysis of the arrangement of atoms in DNA, the two investigators assembled molecular models from pieces of cardboard and bits of wire. Eventually, they constructed a model for DNA that fit all the known data **(Figure 12.1)**. Their discovery was of momentous importance in biology. The model enabled scientists to understand key processes in cells for the first time in terms of the structure and interaction of molecules. For example, the model immediately made

Figure 12.1
James D. Watson and Francis H.C. Crick demonstrating their 1953 model for DNA structure, which revolutionized the biological sciences.

it possible to understand how genetic information is stored in the structure of DNA and how DNA replicates. Unquestionably, the discovery launched a molecular revolution within biology, making it possible for the first time to relate the genetic traits of living organisms to a universal molecular code present in the DNA of every cell. In addition, Watson and Crick's discovery opened the way for numerous advances in fields such as medicine, forensics, pharmacology, and agriculture and eventually gave rise to the current rapid growth of the biotechnology industry.

12.1 Establishing DNA as the Hereditary Molecule

In the first half of the twentieth century, many scientists believed that proteins were the most likely candidates for the hereditary molecules because they appeared to offer greater opportunities for information coding than did nucleic acids. That is, proteins contain 20 types of amino acids, whereas nucleic acids have only 4 different nitrogenous bases available for coding. Other scientists believed that nucleic acids were the hereditary molecules. In this section, we describe the experiments showing that DNA, not protein, is the genetic material.

12.1a Experiments Began When Griffith Found a Substance That Could Genetically Transform Pneumonia Bacteria

In 1928, Frederick Griffith, a British medical officer, observed an interesting phenomenon in his experiments with the bacterium *Streptococcus pneumoniae*, which causes a severe form of pneumonia in mammals. Griffith was trying to make a vaccine to prevent pneumonia infections in the epidemics that occurred after World War I. He used two strains of the bacterium in his attempts. The smooth strain, *S*, has a polysaccharide capsule surrounding each cell and forms colonies that appear smooth and glossy when grown on a culture plate. When he injected the *S* strain into mice, it was virulent (highly infective, or pathogenic), causing pneumonia and killing the mice in a day or two (**Figure 12.2**, step 1). The rough strain, *R*, does not have a polysaccharide capsule and forms colonies with a nonshiny, rough appearance. When Griffith injected the *R* strain into mice, it was avirulent (not infective, or nonpathogenic); the mice lived (step 2). Evidently, the capsule was responsible for the virulence of the *S* strain. We now know that the capsule hinders the ability of the host's immune system to detect the *Streptococcus* cells. The smooth strain could therefore live long enough to multiply and cause fatal pneumonia.

If Griffith killed the *S* bacteria by heating before injecting them into the mice, the mice remained healthy (step 3). However, quite unexpectedly, Griffith found that if he injected living *R* bacteria along with the heat-killed *S* bacteria, many of the mice died (step 4). Also, he was able to isolate living *S* bacteria with polysaccharide capsules from the infected mice. In some way, living *R* bacteria had acquired the ability to make the polysaccharide capsule from the dead *S* bacteria, and they had changed—transformed—into virulent *S* cells. The transformed bacteria were altered permanently; the smooth, infective trait was stably inherited by the descendants of the transformed bacteria. Griffith called the conversion of *R* bacteria to *S* bacteria *transformation* and the agent responsible the *transforming principle*. What was the nature of the molecule responsible for the transformation? Carbohydrates, lipids, proteins, and nucleic acids are the four main types of biological macromolecules. The structure of carbohydrates and lipids tends to be highly repetitive and therefore not very likely to carry information. However, proteins and nucleic acids are built of various combinations of different amino acids and nucleotides, respectively. This gives them a complexity of structure that makes them likely candidates for carrying the information needed for transformation.

12.1b Avery and His Coworkers Identified DNA as the Molecule That Transforms Avirulent Rough *Streptococcus* to the Virulent Smooth Form

In the 1940s, Oswald Avery, a physician and medical researcher at the hospital at the Rockefeller Institute for Medical Research in New York, and his coworkers, Colin MacLeod and Maclyn McCarty, performed an experiment designed to identify the chemical nature of the

QUESTION: What is the nature of the genetic material?

EXPERIMENT: Frederick Griffith studied the conversion of a nonvirulent (noninfective) *R* form of the bacterium *Streptococcus pneumoniae* to a virulent (infective) *S* form. The *S* form has a capsule surrounding the cell, giving colonies of it on a laboratory dish a smooth, shiny appearance. The *R* form has no capsule, so the colonies have a rough, nonshiny appearance. Griffith injected the bacteria into mice and determined how the mice were infected.

1 Mice injected with live *S* cells (control to show effect of *S* cells)

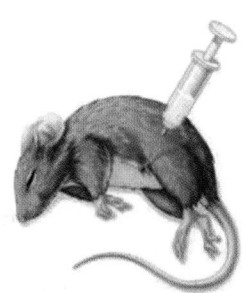

RESULT: Mice die. Live *S* cells in their blood; shows that *S* cells are virulent.

2 Mice injected with live *R* cells (control to show effect of *R* cells)

RESULT: Mice live. No live *R* cells in their blood; shows that *R* cells are nonvirulent. Evidently the capsule is responsible for virulence of the *S* strain.

3 Mice injected with heat-killed *S* cells (control to show effect of dead *S* cells)

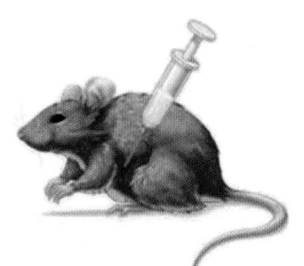

RESULT: Mice live. No live *S* cells in their blood; shows that live *S* cells are necessary to be virulent to mice.

4 Mice injected with heat-killed *S* cells plus live *R* cells

RESULT: Mice die. Live *S* cells in their blood; shows that living *R* cells can be converted to virulent *S* cells with some factor from dead *S* cells.

CONCLUSION: Griffith concluded that some molecules released when *S* cells were killed could change living nonvirulent *R* cells genetically to the virulent *S* form. He called the molecule the *transforming principle* and the process of genetic change *transformation*.

Experimental Research Figure 12.2 Griffith's experiment with infective and noninfective strains of *Streptococcus pneumoniae*.

transforming principle that can change the avirulent *rough* form of *Streptococcus* bacteria into the infective *smooth* form. Rather than working with mice, they attempted to reproduce the transformation using bacteria growing in culture tubes. They used heat to kill virulent *S* bacteria and then treated the macromolecules extracted from the cells with enzymes that break down each of the three main candidate molecules for the hereditary material—protein; DNA; and the other nucleic acid, ribonucleic acid (RNA). When they destroyed proteins or RNA, the researchers saw no effect;

the extract of *S* bacteria still transformed *R* bacteria into virulent *S* bacteria—the cells had polysaccharide capsules and produced smooth colonies on culture plates. When they destroyed DNA, however, no transformation occurred—no smooth colonies were seen on culture plates.

In 1944, Avery and his colleagues published their discovery that the transforming principle was DNA. At the time, many scientists firmly believed that the genetic material was protein. So although their findings were clearly revolutionary, Avery and his colleagues presented their conclusions in the paper cautiously, offering several interpretations of their results. Although some scientists accepted these data almost immediately, others remained unconvinced. After all, it seemed unlikely that a molecule like DNA, with only four different components (adenine, thymine, cytosine, and guanine), could hold the complex information required of the genetic material in a cell. Protein, with its 20 different amino acid components, seemed a far superior medium for coding information. Those who believed that the genetic material was protein argued that it was possible that not all protein was destroyed by Avery's enzyme treatments, and, as contaminants in their DNA transformation reaction, these remaining proteins were, in fact, responsible for the transformation. Further experiments were needed to convince all scientists that DNA is the hereditary molecule.

12.1c Hershey and Chase Found the Final Evidence Establishing DNA as the Hereditary Molecule

A final series of elegant experiments conducted in 1952 by bacteriologist Alfred D. Hershey and his laboratory assistant Martha Chase at the Cold Spring Harbor Laboratory removed any remaining doubts that DNA is the hereditary molecule. Hershey and Chase studied the infection of the bacterium *Escherichia coli* by bacteriophage T2. *Escherichia coli* is a bacterium normally found in the intestines of mammals. **Bacteriophages** (or simply **phages**; see Chapters 9 and 23) are viruses that infect bacteria. A **virus** is an infectious agent that contains either DNA or RNA surrounded by a protein coat. Viruses cannot reproduce except in a host cell. When a virus infects a cell, it can use the cell's resources to produce more virus particles.

The phage replication cycle begins when a phage attaches to the surface of a bacterium. For phages such as T2, the infected cell quickly stops producing its own molecules and instead starts making progeny phages. After about 100 to 200 phages are assembled inside the bacterial cell, a viral enzyme breaks down the cell wall, killing the cell and releasing the new phages. The whole cycle takes approximately 90 minutes.

The T2 phage that Hershey and Chase studied consists of only a core of DNA surrounded by proteins.

Therefore, one of these molecules must be the genetic material that enters the bacterial cell and directs the infective cycle within. But which one? Hershey and Chase prepared two batches of phages, one with the protein tagged with a radioactive label and the other with the DNA tagged with a radioactive label. To obtain labelled phages, they added T2 to *E. coli* growing in the presence of either the radioactive isotope of sulfur (^{35}S) or the radioactive isotope of phosphorus (^{32}P) (**Figure 12.3**, step 1). The progeny phages produced in the ^{35}S medium had labelled proteins and unlabelled DNA because sulfur is a component of proteins but not of DNA. The phages produced in the ^{32}P medium had labelled DNA and unlabelled proteins because phosphorus is a component of DNA but not of proteins.

Hershey and Chase then infected separate cultures of *E. coli* with the two types of labelled phages (step 2). After a short period to allow the genetic material to enter the bacterial cell, they mixed the bacteria in a kitchen blender. They reasoned that only the genetic material was injected into the bacterial cell, leaving the rest of the phage outside. By mixing the cells in a blender, they could shear off the phage parts that did not enter the bacteria and collect them separately for analysis.

When they infected the bacteria with phages that contained labelled protein coats, they found no **radioactivity** in the bacterial cells but could easily measure it in the material removed by the blender (step 3, top). They also found no radioactivity in the progeny phages (step 4, top). However, if the infecting phages contained radioactive DNA, they found radioactivity inside the infected bacteria but none in the phage coats removed by the blender (step 3, bottom). In addition, radioactivity *was* seen in the progeny phages (step 4, bottom). The results were unequivocal: the genetic material of the phage was DNA, not protein.

When taken together, the experiments of Griffith, Avery and his coworkers, and Hershey and Chase established that DNA, not proteins, carries genetic information. Their research also established the term *transformation*, which is still used in molecular biology. **Transformation** is the conversion of a cell's hereditary type by the uptake of DNA released by the breakdown of another cell, as in the Griffith and Avery experiments. Having identified DNA as the hereditary molecule, scientists turned next to determining its structure.

STUDY BREAK

How did Hershey and Chase exploit the reproductive cycle of a phage to gain evidence for DNA as the hereditary material?

QUESTION: Is DNA or protein the genetic material?

EXPERIMENT: Hershey and Chase performed a definitive experiment to show whether DNA or protein is the genetic material. They used phage T2 for their experiment; it consists only of DNA and protein.

1 They infected *E. coli* growing in the presence of radioactive ^{32}P or ^{35}S with phage T2. The progeny phages were labelled either in their protein with ^{35}S (top), or in their DNA with ^{32}P (bottom).

2 Separate cultures of *E. coli* were infected with the radioactively labelled phages.

3 After a short period of time to allow the genetic material to enter the bacterial cell, the bacteria were mixed in a blender. The blending sheared from the cell surface the phage coats that did not enter the bacteria. The components were analyzed for radioactivity.

4 Progeny phages were analyzed for radioactivity.

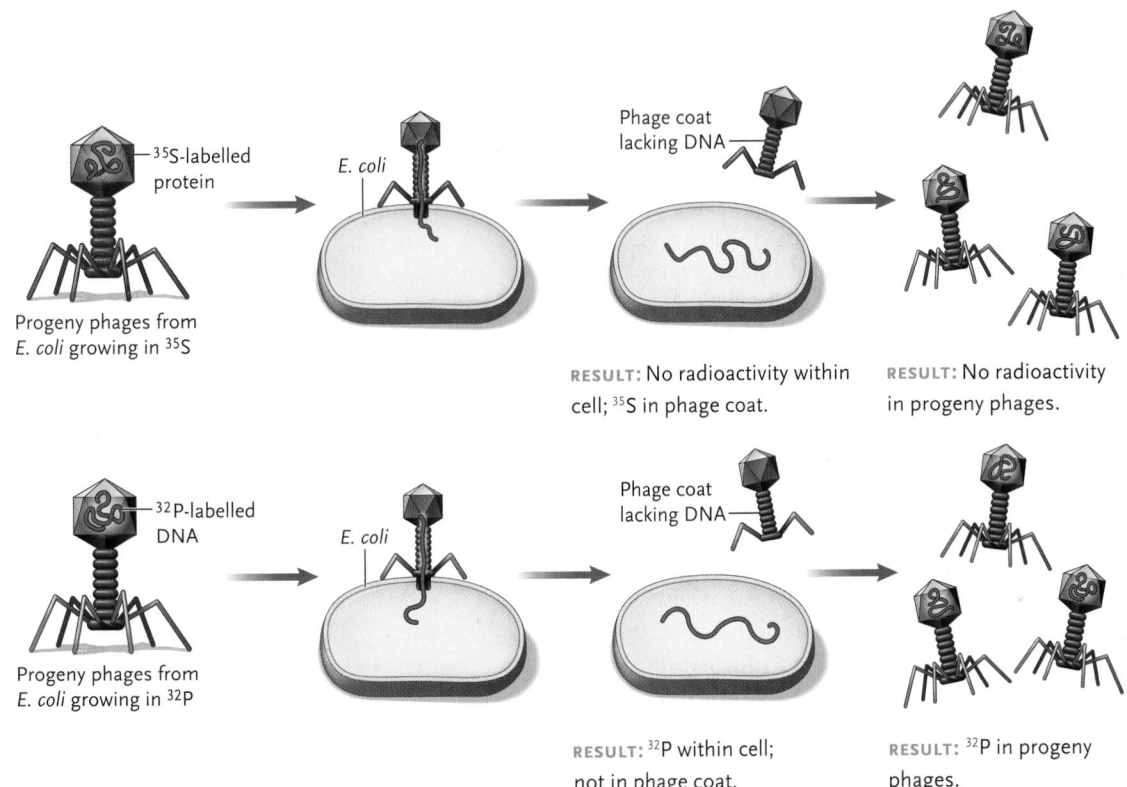

^{35}S-labelled protein

E. coli

Phage coat lacking DNA

Progeny phages from *E. coli* growing in ^{35}S

RESULT: No radioactivity within cell; ^{35}S in phage coat.

RESULT: No radioactivity in progeny phages.

^{32}P-labelled DNA

E. coli

Phage coat lacking DNA

Progeny phages from *E. coli* growing in ^{32}P

RESULT: ^{32}P within cell; not in phage coat.

RESULT: ^{32}P in progeny phages.

CONCLUSION: ^{32}P, the isotope used to label DNA, was found within phage-infected cells and in progeny phages, indicating that DNA is the genetic material. ^{35}S, the **radioisotope** used to label proteins, was found in phage coats after infection but was not found in the infected cell or in progeny phages, showing that protein is not the genetic material.

Experimental Research Figure 12.3 The Hershey and Chase experiment demonstrating that DNA is the hereditary molecule.

12.2 DNA Structure

The experiments that established DNA as the hereditary molecule were followed by a highly competitive scientific race to discover the structure of DNA. The race ended in 1953 when Watson and Crick elucidated the structure of DNA, ushering in a new era of molecular biology.

12.2a Watson and Crick Brought Together Information from Several Sources to Work Out DNA Structure

Before Watson and Crick began their research, other investigators had established that DNA contains four different nucleotides. Each nucleotide consists of the five-carbon sugar *deoxyribose* (carbon atoms on deoxyribose are numbered with primes from 1′ to 5′);

a phosphate group; and one of the four nitrogenous bases—adenine (A), guanine (G), thymine (T), and cytosine (C) **(Figure 12.4).** Two of the bases, **adenine** and **guanine**, are *purines*, nitrogenous bases built from a pair of fused rings of carbon and nitrogen atoms. The other two bases, **thymine** and **cytosine**, are *pyrimidines*, built from a single carbon ring. An organic chemist, Erwin Chargaff, measured the amounts of nitrogenous bases in DNA and discovered that they occur in definite ratios. He observed that the number of purines equals the number of pyrimidines, but, more specifically, the amount of adenine equals the amount of thymine, and the amount of guanine equals the amount of cytosine; these relationships are known as *Chargaff's rules*.

Researchers had also determined that DNA contains nucleotides joined to form a *polynucleotide chain*. In a polynucleotide chain, the deoxyribose sugars are linked by phosphate groups in an alternating sugar–phosphate–sugar–phosphate pattern, forming a **sugar–phosphate backbone** (highlighted in grey in Figure 12.4). Each phosphate group is a "bridge" between the 3′ carbon of one sugar and the 5′ carbon of the next sugar; the entire linkage, including the bridging phosphate group, is called a **phosphodiester bond**.

The polynucleotide chain of DNA has polarity, or directionality. That is, the two ends of the chain are not the same: at one end, a phosphate group is bound to the 5′ carbon of a deoxyribose sugar, whereas at the other end, a **hydroxyl group** is bonded to the 3′ carbon of a deoxyribose sugar (see Figure 12.4). Consequently, the two ends are called the **5′ end** and the **3′ end**, respectively.

These were the known facts when Watson and Crick began their collaboration in the early 1950s. However, the number of polynucleotide chains in a DNA molecule and the manner in which they fold or twist in DNA were unknown. Watson and Crick themselves did not conduct experiments to study the structure of DNA; instead, they used the research data of others for their analysis, relying heavily on data gathered by physicist Maurice H.F. Wilkins and research associate Rosalind Franklin **(Figure 12.5a)** at King's College, London. These researchers were using **X-ray diffraction** to study the structure of DNA (Figure 12.5b). In X-ray diffraction, an X-ray beam is directed at a molecule in the form of a regular solid, ideally in the form of a crystal. Within the crystal, regularly arranged atoms bend and reflect the X-rays into smaller beams that exit the crystal at definite angles determined by the arrangement of atoms in the structure of the crystal. If a photographic film is placed behind the crystal, the exiting beams produce a pattern of exposed spots. From that pattern, researchers can deduce the positions of the atoms in the crystal.

Wilkins and Franklin did not have DNA crystals to work with, but they were able to obtain X-ray diffraction

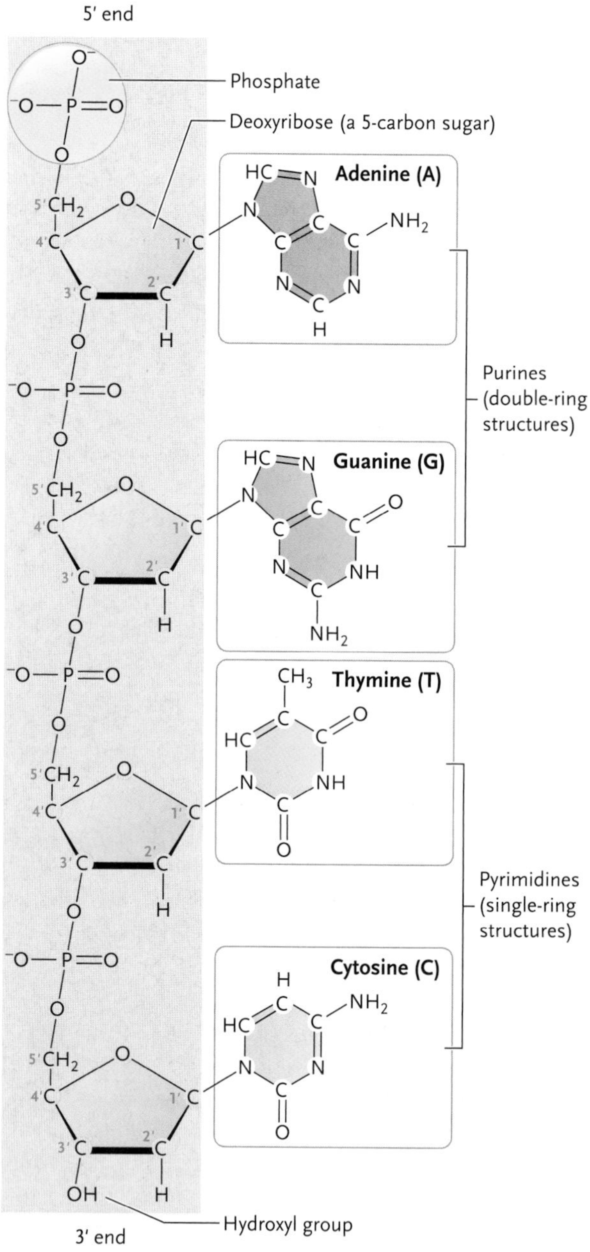

Figure 12.4
The four nucleotide subunits of DNA, linked into a polynucleotide chain. The sugar–phosphate backbone of the chain is highlighted in grey. The connection between adjacent deoxyribose sugars is a phosphodiester bond. The polynucleotide chain has polarity: at one end (5′), a phosphate group is bound to the 5′ carbon of a deoxyribose sugar, whereas at the other end (3′), a hydroxyl group is bound to the 3′ carbon of a deoxyribose sugar.

patterns from a sample of DNA molecules that had been pulled out into a fibre (see Figure 12.5). The patterns indicated that the DNA molecules within the fibre were cylindrical and about 2 nm in diameter. Separations between the spots showed that major patterns of atoms repeat at intervals of 0.34 nm and 3.4 nm within the DNA. Franklin correctly interpreted an X-shaped distribution of spots in the diffraction pattern (see dashed lines in Figure 12.5) to mean that DNA has a helical structure.

a. Rosalind Franklin

b. X-ray diffraction analysis of DNA

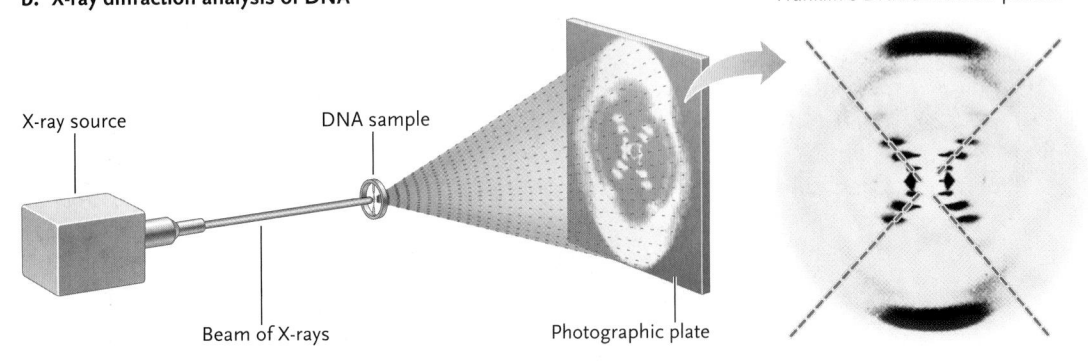

Franklin's DNA diffraction pattern

X-ray source

DNA sample

Beam of X-rays

Photographic plate

Science Source

Science Source

Figure 12.5

X-ray diffraction analysis of DNA. **(a)** Rosalind Franklin. **(b)** The X-ray diffraction method to study DNA and the diffraction pattern Rosalind Franklin obtained. The X-shaped pattern of spots (dashed lines) was correctly interpreted by Franklin to indicate that DNA has a helical structure similar to a spiral staircase.

12.2b The New Model Proposed That Two Polynucleotide Chains Wind into a DNA Double Helix

Watson and Crick constructed scale models of the four DNA nucleotides and fitted them together in different ways until they arrived at an arrangement that satisfied both Wilkins's and Franklin's X-ray data and Chargaff's chemical analysis. Watson and Crick's trials led them to a double-stranded model for DNA structure in which two polynucleotide chains twist around each other in a right-handed way, like a double-spiral staircase **(Figure 12.6, p. 270)**. They were the first to propose the famous double-helix model for DNA.

In the **double-helix model**, the two sugar–phosphate backbones are separated from each other by a regular distance. The bases extend into and fill this central space. A purine and a pyrimidine, if paired together, are exactly wide enough to fill the space between the backbone chains in the double helix. However, a purine–purine base pair is too wide to fit the space exactly, and a pyrimidine–pyrimidine pair is too narrow. From Chargaff's data, Watson and Crick proposed that the purine–pyrimidine base pairs in DNA are A-T and G-C pairs. That is, wherever an A occurs in one strand, a T must be opposite it in the other strand; wherever a G occurs in one strand, a C must be opposite it. This feature of DNA is called **complementary base-pairing**, and one strand is said to be *complementary* to the other. The base pairs, which fit together like pieces of a jigsaw puzzle, are stabilized by hydrogen bonds—two between A and T and three between G and C (see Figure 12.6; hydrogen bonds are discussed in *The Purple Pages*). The hydrogen bonds between the paired bases, repeated along the double helix, hold the two strands together in the helix.

CONCEPT FIX Although this text follows the generally common convention of referring to the DNA double helix as a *DNA molecule,* you should be aware that this terminology is, strictly speaking, inaccurate. If a molecule is defined as a collection of atoms connected by covalent bonds, then *each* of the two sugar–phosphate backbones of the double helix qualifies as a molecule. The double helix is technically composed of *two* polynucleotide molecules held together by hydrogen bonds. ◐

The base pairs lie in flat planes almost perpendicular to the long axis of the DNA helix. In this state, each base pair occupies a length of 0.34 nm along the long axis of the double helix (see Figure 12.6). This spacing accounts for the repeating 0.34 nm pattern noted in the X-ray diffraction patterns. The larger 3.4 nm repeat pattern was interpreted to mean that each full turn of the double helix takes up 3.4 nm along the length of the molecule; therefore, 10 base pairs are packed into a full turn.

Watson and Crick also realized that the two strands of a double helix fit together in a stable chemical way only if they are **antiparallel**, that is, only if they run in opposite directions (see Figure 12.6, arrows). In other words, the 3′ end of one strand is opposite the 5′ end of its complementary strand. This antiparallel arrangement is highly significant for the process of replication, which is discussed in the next section.

As hereditary material, DNA must faithfully store and transmit genetic information for the entire life cycle of an organism. Watson and Crick recognized that this information is coded into the DNA by the particular sequence of the four nucleotides. This sequence is preserved by robust covalent bonds between the molecules in a DNA double helix. Although only four different kinds of nucleotides exist, combining them in groups allows an essentially infinite number of different sequences to be "written," just as the 26 letters of the alphabet can be combined in groups to write a virtually unlimited number of words. Chapter 13 shows how taking the four nucleotides in groups of three forms enough words to spell out the structure of any conceivable protein.

Watson and Crick announced their model for DNA structure in a brief but monumental paper published in the journal *Nature* in 1953. Watson and Crick shared a Nobel Prize with Wilkins in 1962 for their discovery

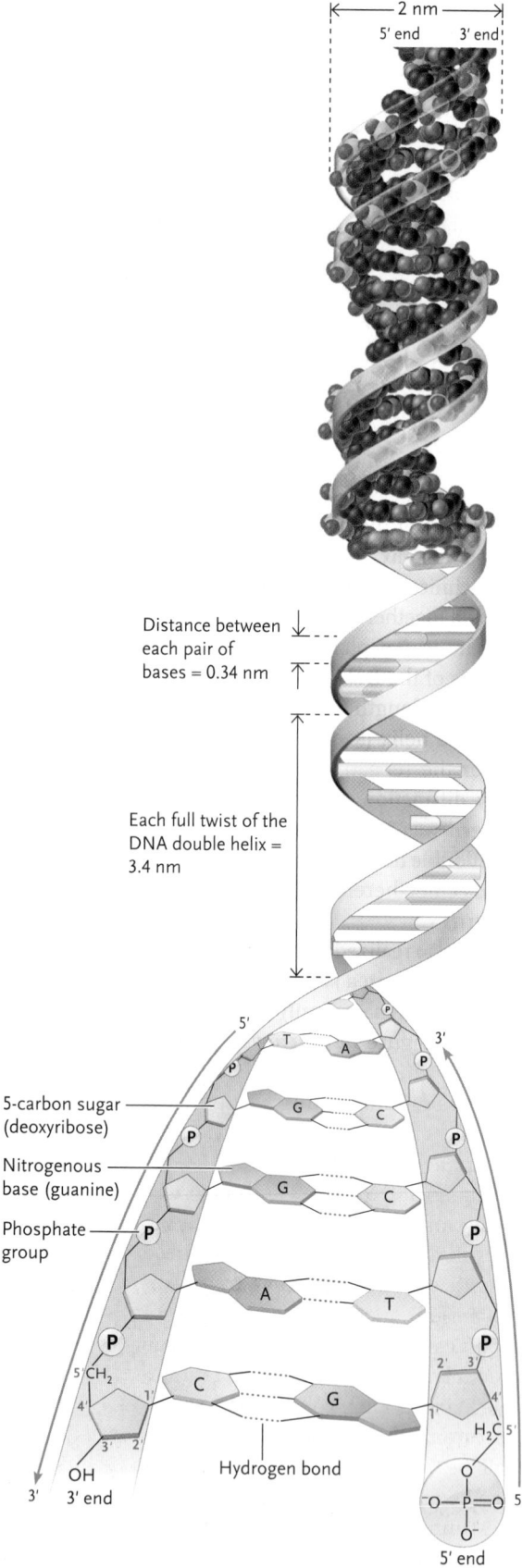

Figure 12.6

DNA double helix. Arrows and labelling of the ends show that the two polynucleotide chains of the double helix are antiparallel—that is, they have opposite polarity in that they run in opposite directions. In the space-filling model at the top, the spaces occupied by atoms are indicated by spheres. There are 10 base pairs per turn of the helix; only 8 base pairs are visible because the other 2 are obscured where the backbones pass over each other.

of the molecular structure of DNA. Rosalind Franklin might have been a candidate for a Nobel Prize had she not died of cancer at age 38 in 1958. (The Nobel Prize is normally given only to living investigators.) Unquestionably, Watson and Crick's discovery of DNA structure opened the way to molecular studies of genetics and heredity, leading to our modern understanding of gene structure and action at the molecular level.

STUDY BREAK

1. Which bases in DNA are purines? Which are pyrimidines?
2. What bonds form between complementary base pairs? Between a base and the deoxyribose sugar?
3. Which features of the DNA molecule did Watson and Crick describe?

12.3 DNA Replication

Once they had discovered the structure of DNA, Watson and Crick realized immediately that complementary base-pairing could explain how DNA replicates **(Figure 12.7)**. They imagined that, for replication, the hydrogen bonds between the two strands break, allowing them to unwind and separate. Each strand then acts as a template for the synthesis of its partner. When replication is complete, there are two double helices, each with one strand derived from the parental DNA molecule base-paired with a newly synthesized one. Most important, each of the two new double helices consists of the identical base-pair sequences as the parental DNA.

The model of replication Watson and Crick proposed is termed **semiconservative replication (Figure 12.8a, p. 272)**. Other scientists proposed two other models for replication. In the *conservative replication model,* each of the two strands of original DNA serves as a template for a new DNA double helix (Figure 12.8b). After the two complementary copies separate from their templates, they wind together into an all "new" DNA double helix. In the *dispersive replication model,* neither parental molecule remains intact; both chains of each replicated double helix contain old and new segments (Figure 12.8c, p. 272).

12.3a Meselson and Stahl Showed That DNA Replication Is Semiconservative

A definitive experiment published in 1958 by Matthew Meselson and Franklin Stahl of the California Institute of Technology demonstrated that DNA replication is semiconservative **(Figure 12.9, p. 273)**. In their experiment, Meselson and Stahl had to be able to distinguish parental DNA molecules from newly synthesized

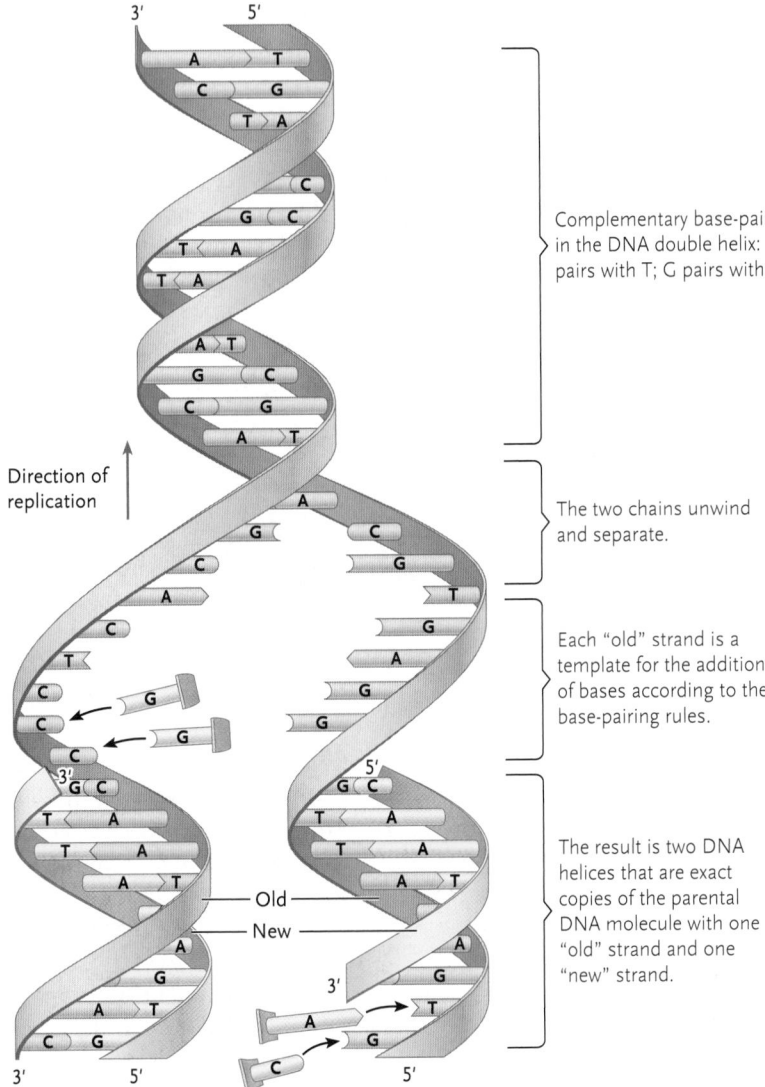

Complementary base-pairing in the DNA double helix: A pairs with T; G pairs with C.

Direction of replication

The two chains unwind and separate.

Each "old" strand is a template for the addition of bases according to the base-pairing rules.

Old

New

The result is two DNA helices that are exact copies of the parental DNA molecule with one "old" strand and one "new" strand.

Figure 12.7

Watson and Crick's model for DNA replication. The original DNA is shown in grey. A new polynucleotide chain (red) is assembled on each original chain as they unwind. The template and complementary copy chains remain together when replication is complete, producing DNA double helices that are half old and half new. The model is known as the semiconservative model for DNA replication.

DNA. To do this, they used a nonradioactive "heavy" nitrogen isotope to tag the parental DNA. The heavy isotope, ^{15}N, has one more neutron in its nucleus than the normal ^{14}N isotope. Molecules containing ^{15}N are measurably heavier (denser) than molecules of the same type containing ^{14}N.

As the first step in their experiment, Meselson and Stahl grew *E. coli* bacteria in a culture medium containing the heavy ^{15}N isotope (see Figure 12.9, step 1). The heavy isotope was incorporated into the nitrogenous bases of DNA, resulting in the entire DNA being labelled with ^{15}N. Then they transferred the bacteria to a culture medium containing the light ^{14}N isotope (step 2). All new DNA synthesized after the transfer contained the light isotope. Just before the transfer to the medium with the ^{14}N isotope, and after each round of replication following the transfer, they took a sample

of the cells and extracted the DNA (step 3).

Meselson and Stahl then mixed the DNA samples with cesium chloride (CsCl) and centrifuged the mixture at very high speed (step 4). During the centrifugation, the CsCl forms a density gradient and DNA double helices move to a position in the gradient where their density matches that of the CsCl. Therefore, DNA of different densities is separated into bands, with the densest DNA settling closer to the bottom of the tube. In Figure 12.9 "Result" shows the outcome of these experiments, and "Conclusions" shows why the results were compatible with only the semiconservative replication model.

12.3b DNA Polymerases Are the Primary Enzymes of DNA Replication

During replication, complementary polynucleotide chains are assembled from individual deoxyribonucleotides by enzymes known as **DNA polymerases.** More than one kind of DNA polymerase is required for DNA replication in all cells. *Deoxyribonucleoside triphosphates* are the substrates for the polymerization reaction catalyzed by DNA polymerases **(Figure 12.10, p. 274).** A nucleoside triphosphate is a nitrogenous base linked to a sugar, which is linked, in turn, to a chain of three phosphate groups. You have encountered a nucleoside triphosphate before, namely the ATP produced in cellular respiration (see Chapter 6). In that case, the sugar is ribose, making ATP a ribonucleoside triphosphate. The deoxyribonucleoside triphosphates used in DNA replication have the sugar *deoxyribose* rather than the sugar *ribose*. Because four different bases are found in DNA—adenine (A), guanine (G), cytosine (C), and thymine (T)—four different deoxyribonucleoside triphosphates are used for DNA replication. In keeping with the ATP naming convention, the deoxyribonucleoside triphosphates for DNA replication are given the short names dATP, dGTP, dCTP, and dTTP, where the "d" stands for "deoxyribose."

Figure 12.10, p. 274, presents a section of a DNA polynucleotide chain being replicated, showing how DNA polymerase catalyzes the assembly of a new DNA strand that is complementary to the template strand. To understand Figure 12.10, remember that the carbons

Figure 12.8

Three theoretical models for DNA replication. Experimental data support the semiconservative mechanism.

a. Semiconservative replication

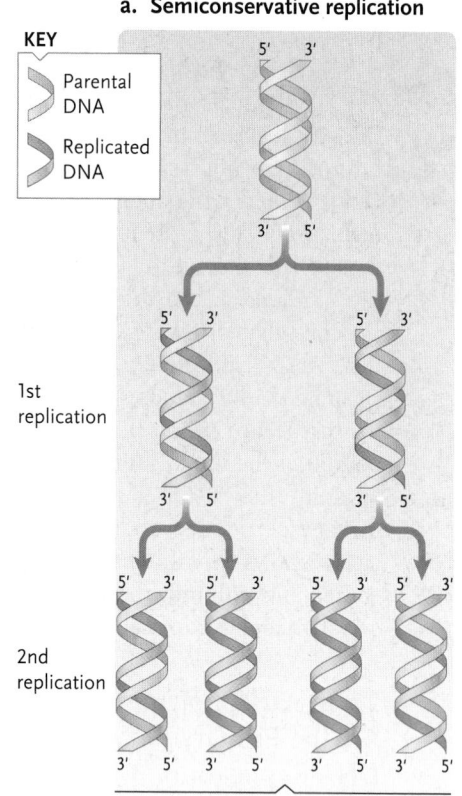

b. Conservative replication

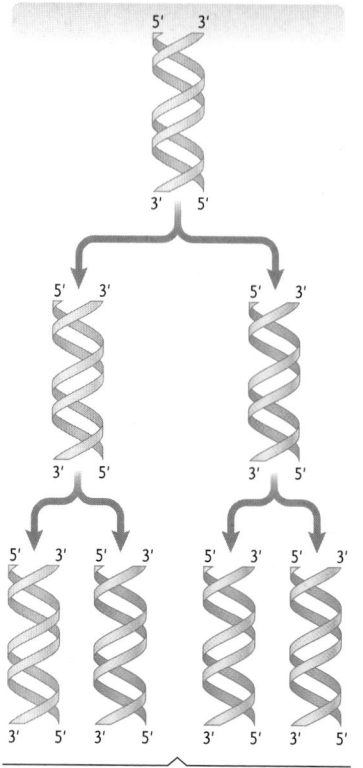

c. Dispersive replication

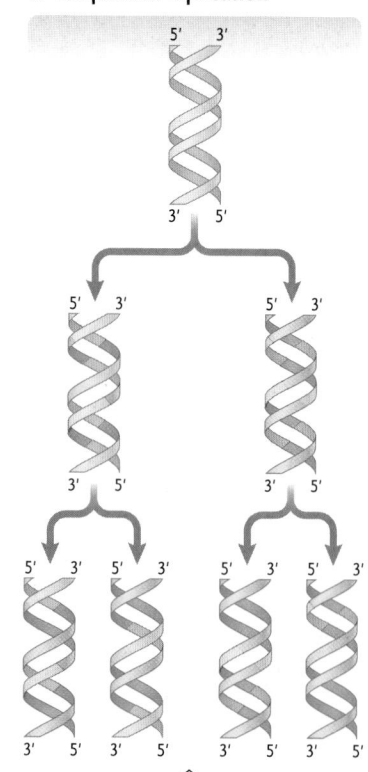

KEY

Parental DNA

Replicated DNA

1st replication

2nd replication

The two parental strands of DNA unwind, and each is a template for synthesis of a new strand. After replication has occurred, each double helix has one old strand paired with one new strand. This model was the one proposed by Watson and Crick themselves.

The parental strands of DNA unwind, and each is a template for synthesis of a new strand. After replication has occurred, the parental strands pair up again. Therefore, the two resulting double helices consist of one with two old strands, and the other with two new strands.

The original double helix splits into double-stranded segments on which new double-stranded segments form. These newly formed sections somehow assemble into two double helices, both of which are a mixture of the original double-stranded DNA interspersed with new double-stranded DNA.

in the deoxyriboses of nucleotides are numbered with primes. Each DNA strand has two distinct ends: the 5′ end has an exposed phosphate group attached to the 5′ carbon of the sugar, and the 3′ end has an exposed hydroxyl group attached to the 3′ carbon of the sugar. As you learned earlier, because of the antiparallel nature of the DNA strands within a double helix, the 5′ end of one strand is opposite the 3′ end of the other. DNA polymerase can add a nucleotide *only* to the 3′ end of an existing nucleotide chain. As a new DNA strand is assembled, a 3′ –OH group is always exposed at its "newest" end; the "oldest" end of the new chain has an exposed 5′ phosphate. DNA polymerases are therefore said to assemble nucleotide chains in the 5′ → 3′ direction.

Because of the antiparallel nature of DNA, the template strand is "read" in the 3′ → 5′ direction for this new synthesis. DNA polymerases of bacteria, archaeans, and eukaryotes all consist of several polypeptide subunits arranged to form different domains (see polypeptides in *The Purple Pages*). The polymerases share a shape that is said to resemble a partially closed human right hand in which the template DNA lies over the "palm" in a groove formed by the "fingers" and "thumb" (**Figure 12.11a, p. 275**). The palm domain is evolutionarily related among the polymerases of

bacteria, archaea, and eukaryotes, while the finger and thumb domains are different sequences in each of these three types of organisms. The template strand does not pass through the tunnel formed by the thumb and finger domains, however. Instead, the template strand and the 3′ –OH of the new strand meet at the active site for the polymerization reaction of DNA synthesis, located in the palm domain. A nucleotide is added to the new strand when an incoming dNTP enters the active site carrying a base complementary to the template strand base positioned in the active site. By moving along the template strand, one nucleotide at a time, DNA polymerase extends the new DNA strand, as we saw in Figure 12.10.

Figure 12.11b, p. 275, shows the representation of DNA polymerase used in the following DNA replication figures, and it also shows a sliding DNA clamp. The **sliding DNA clamp** is a protein that encircles the DNA and binds to the rear of the DNA polymerase in terms of the enzyme's forward movement during replication. The function of the sliding DNA clamp is to tether the DNA polymerase to the template strand. Tethering the DNA polymerase makes replication more efficient because without it, the enzyme will detach from the template after only a few dozen polymerizations.

Experimental Research Figure 12.9 The Meselson and Stahl experiment demonstrating that the semiconservative model is correct.

QUESTION: Does DNA replicate semiconservatively?

EXPERIMENT: Matthew Meselson and Franklin Stahl proved that the semiconservative model of DNA replication is correct and that the conservative and dispersive models are incorrect.

1 Bacteria grown in ^{15}N (heavy) medium. The heavy isotope is incorporated into the bases of DNA, resulting in all the DNA being heavy, that is, labelled with ^{15}N.

2 Bacteria transferred to ^{14}N (light) medium and allowed to grow and divide for several generations. All new DNA is light.

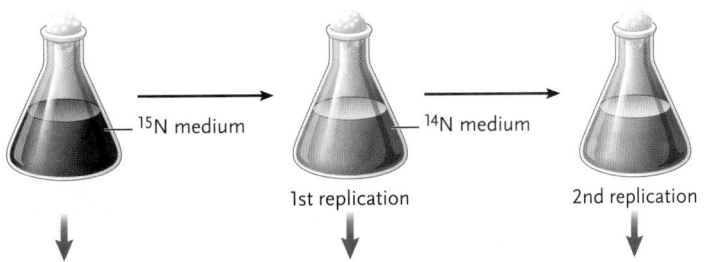

3 DNA extracted from bacteria cultured in ^{15}N medium and after each generation in ^{14}N medium.

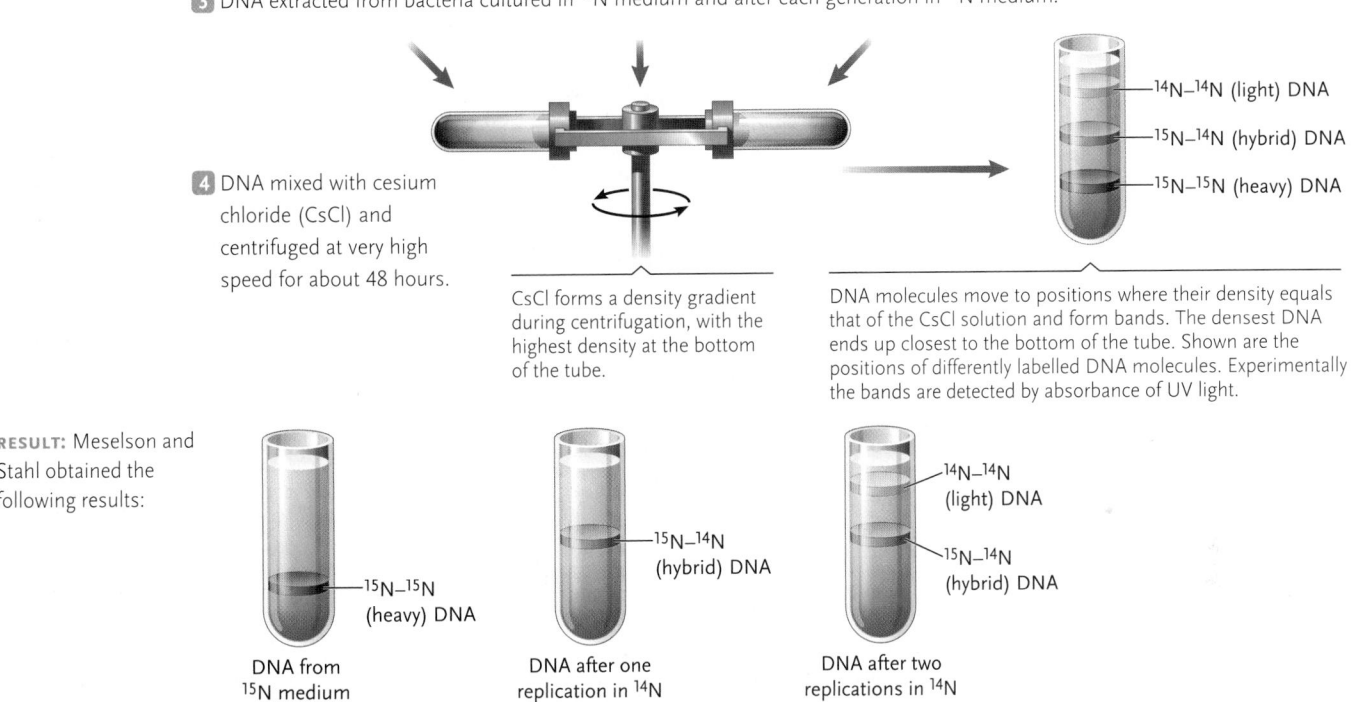

4 DNA mixed with cesium chloride (CsCl) and centrifuged at very high speed for about 48 hours.

CsCl forms a density gradient during centrifugation, with the highest density at the bottom of the tube.

DNA molecules move to positions where their density equals that of the CsCl solution and form bands. The densest DNA ends up closest to the bottom of the tube. Shown are the positions of differently labelled DNA molecules. Experimentally the bands are detected by absorbance of UV light.

$^{14}N-^{14}N$ (light) DNA
$^{15}N-^{14}N$ (hybrid) DNA
$^{15}N-^{15}N$ (heavy) DNA

RESULT: Meselson and Stahl obtained the following results:

$^{15}N-^{15}N$ (heavy) DNA

DNA from ^{15}N medium

$^{15}N-^{14}N$ (hybrid) DNA

DNA after one replication in ^{14}N

$^{14}N-^{14}N$ (light) DNA
$^{15}N-^{14}N$ (hybrid) DNA

DNA after two replications in ^{14}N

CONCLUSION: The predicted DNA banding patterns for the three DNA replication models were as follows:

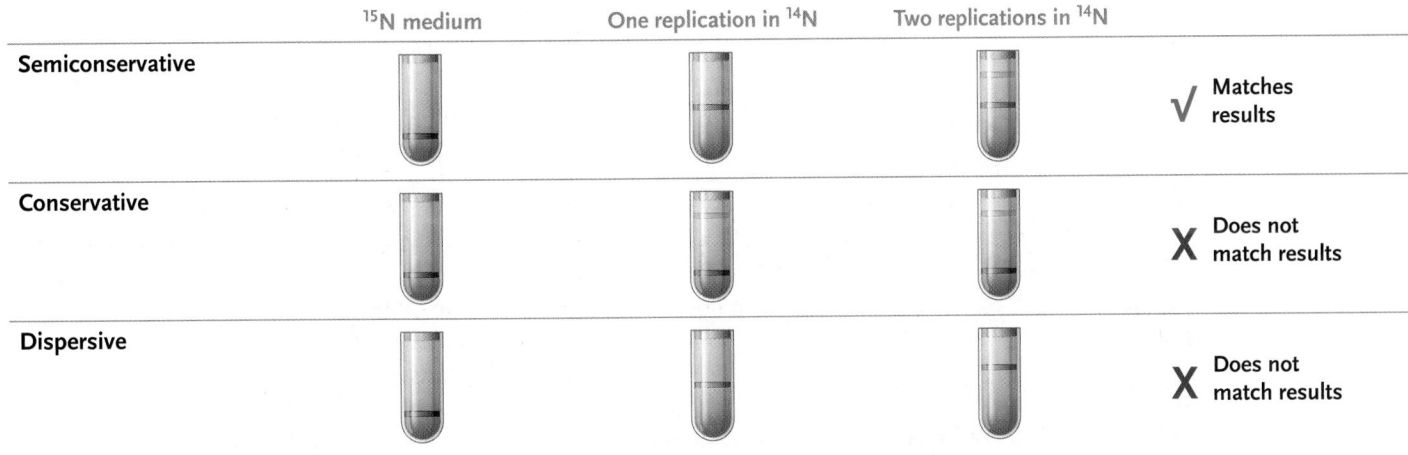

	^{15}N medium	One replication in ^{14}N	Two replications in ^{14}N	
Semiconservative				✓ Matches results
Conservative				✗ Does not match results
Dispersive				✗ Does not match results

The results support the semiconservative model.

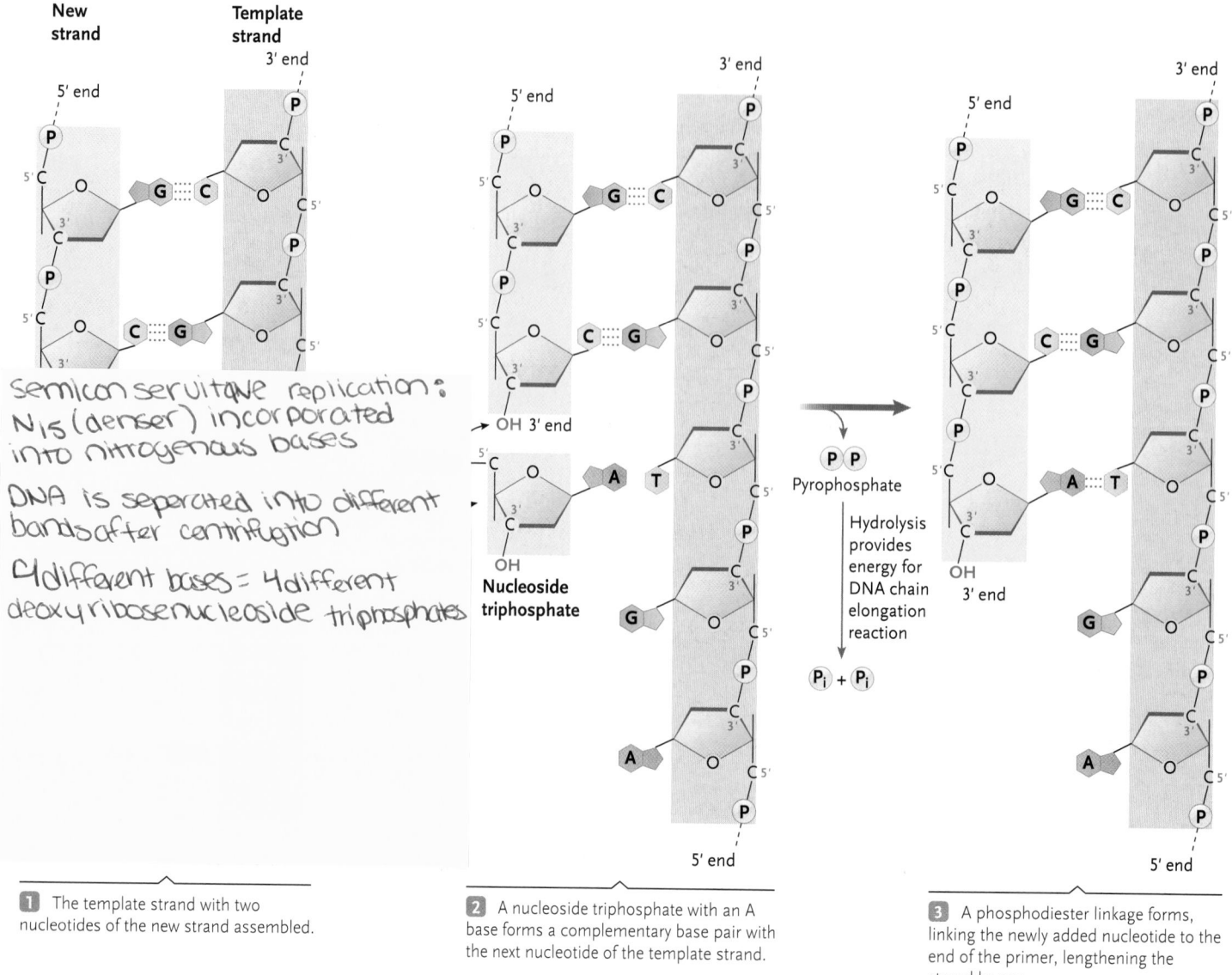

Handwritten notes (in left area of figure):

Semiconservative replication:
N15 (denser) incorporated into nitrogenous bases

DNA is separated into different bands after centrifugation

4 different bases = 4 different deoxyribose nucleoside triphosphates

New strand   **Template strand**

1 The template strand with two nucleotides of the new strand assembled.

2 A nucleoside triphosphate with an A base forms a complementary base pair with the next nucleotide of the template strand.

3 A phosphodiester linkage forms, linking the newly added nucleotide to the end of the primer, lengthening the strand by one.

Figure 12.10

Reactions assembling a complementary chain in the 5′ → 3′ direction on a template DNA strand, showing the phosphodiester linkage created when the DNA polymerase enzyme adds each nucleotide to the chain.

But, with the clamp, many tens of thousands of polymerizations occur before the enzyme detaches. Overall, the rate of DNA synthesis is much faster because of the sliding DNA clamp.

In sum, the key molecular events of DNA replication are as follows:

1. The two strands of the DNA molecule unwind for replication to occur.
2. DNA polymerase can add nucleotides only to an existing chain.
3. The overall direction of new synthesis is in the 5′ → 3′ direction, which is a direction antiparallel to that of the template strand.
4. Nucleotides enter into a newly synthesized chain according to the A-T and G-C complementary base-pairing rules.

The following sections describe how enzymes and other proteins conduct these molecular events.

focus is on the well-characterized replication system of *E. coli*. Replication in archaeans and eukaryotes is highly similar, although there are differences in the replication machinery. The replication machinery of archaeans is strikingly similar to that of eukaryotes and is clearly different from that of bacteria.

12.3c Helicases Unwind DNA for New DNA Synthesis, and Other Proteins Stabilize the DNA at the Replication Fork

In semiconservative replication, the two strands of the parental DNA molecule unwind and separate to expose the template strands for new DNA synthesis **(Figure 12.12, p. 276)**. Unwinding of the DNA for replication occurs at a small, specific sequence in the bacterial chromosome known as the **origin of replication** (*ori*). Specific proteins bind to an *ori* sequence and, in turn, promote binding of **DNA helicase**, which

a. Bacterial DNA polymerase

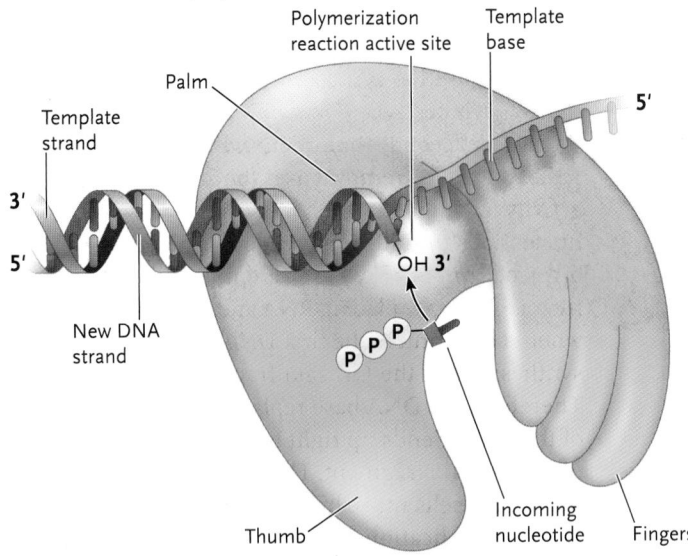

Polymerization reaction active site

Template base

Palm

Template strand

3'

5'

Polymerization reaction active site

Template base

5'

OH 3'

P P P

New DNA strand

Thumb

Incoming nucleotide

Fingers

b. How a DNA polymerase and sliding clamp are shown in the book

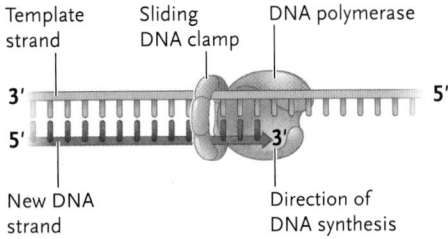

Template strand

Sliding DNA clamp

DNA polymerase

3'

5'

5'

3'

New DNA strand

Direction of DNA synthesis

Figure 12.11

DNA polymerase structure. **(a)** Stylized drawing of a bacterial DNA polymerase. The enzyme viewed from the side resembles a human right hand. The polymerization reaction site lies on the palm. When the incoming nucleotide is added, the thumb and fingers close over the site to facilitate the reaction. **(b)** How DNA polymerase is shown in subsequent figures of DNA replication. The figure also shows a sliding DNA clamp tethering the DNA polymerase to the template strand.

unwinds the DNA strands. The unwinding produces a Y-shaped structure called a **replication fork**, which consists of the two unwound template strands transitioning to double-helical DNA.

Single-stranded binding proteins (SSBs) coat the exposed single-stranded DNA segments, stabilizing the DNA and keeping the two strands from pairing back together (see Figure 12.12). The SSBs are displaced as the replication enzymes make the new polynucleotide chain on the template strands. For circular chromosomes, such as the genomes of most bacteria, unwinding the DNA will eventually cause the still-wound DNA ahead of the unwinding to become highly twisted. You can visualize this phenomenon with some string. Take two equal lengths of string and twist them around each other. Now tie the two ends of each string together. You have created a model of a circular DNA double helix. Pick anywhere in the circle and pull apart

the two pieces of string. The more you pull, the more the region where the two strings are still together becomes highly twisted. In the cell, the twisting of DNA during replication is relieved by **topoisomerase**. This enzyme cuts the DNA ahead of the replication fork, turns the DNA on one side of the break in the opposite direction of the twisting force, and rejoins the two strands (see Figure 12.12).

12.3d RNA Primers Provide the Starting Point for DNA Polymerase to Begin Synthesizing a New DNA Chain

If DNA polymerases can only add nucleotides to the 3' end of an existing strand, how can a new strand begin when there is no existing strand in place? The answer lies in a short chain a few nucleotides long called a **primer**, which is made of RNA instead of DNA **(Figure 12.13, p. 276)**. The primer is synthesized by the enzyme **primase**. Primase then leaves the template, and DNA polymerase takes over, extending the RNA primer with DNA nucleotides as it synthesizes the new DNA chain. RNA primers are removed and replaced with DNA later in replication.

12.3e One New DNA Strand Is Synthesized Continuously; the Other, Discontinuously

DNA polymerases synthesize a new DNA strand on a template strand in the 5' → 3' direction. Because the two strands of a DNA double helix are antiparallel, only one of them runs in a direction that allows DNA polymerase to make a 5' → 3' complementary copy in the direction of unwinding. That is, on this template strand—top strand in **Figure 12.14, p. 277**—new DNA is synthesized continuously in the direction of unwinding of the double helix. However, the other template strand—bottom strand in Figure 12.14—runs in the opposite direction; this means DNA polymerase has to copy it in the direction opposite to the unwinding direction. How is new DNA polymerized in the direction opposite to the unwinding? The polymerases make this strand in short lengths that are synthesized in the direction opposite to that of DNA unwinding (see Figure 12.14). The short lengths produced by this **discontinuous replication** are then covalently linked into a single continuous polynucleotide chain. The short lengths are called **Okazaki fragments**, after Reiji Okazaki, the scientist who first detected them. The new DNA strand synthesized in the direction of DNA unwinding is called the **leading strand** of DNA replication; the template for that strand is the

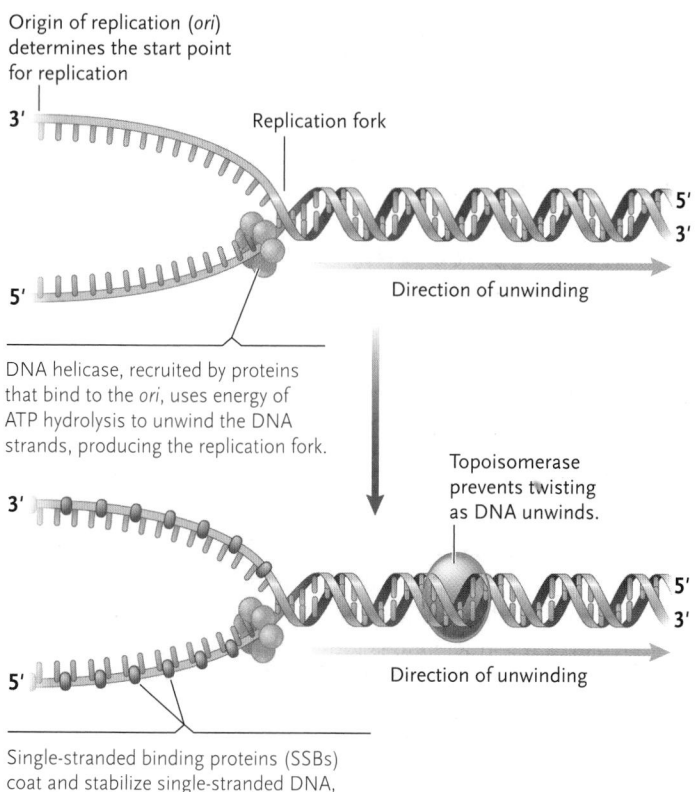

Origin of replication (*ori*) determines the start point for replication

3'

Replication fork

5'
3'

5'

Direction of unwinding

DNA helicase, recruited by proteins that bind to the *ori*, uses energy of ATP hydrolysis to unwind the DNA strands, producing the replication fork.

Topoisomerase prevents twisting as DNA unwinds.

3'

5'
3'

5'

Direction of unwinding

Single-stranded binding proteins (SSBs) coat and stabilize single-stranded DNA, preventing the two strands from re-forming double-stranded DNA.

Figure 12.12
The roles of DNA helicase, single-stranded binding proteins (SSBs), and topoisomerase in DNA replication.

Figure 12.13
Initiation of a new DNA strand by synthesis of a short RNA primer by primase, and the extension of the primer as DNA by DNA polymerase.

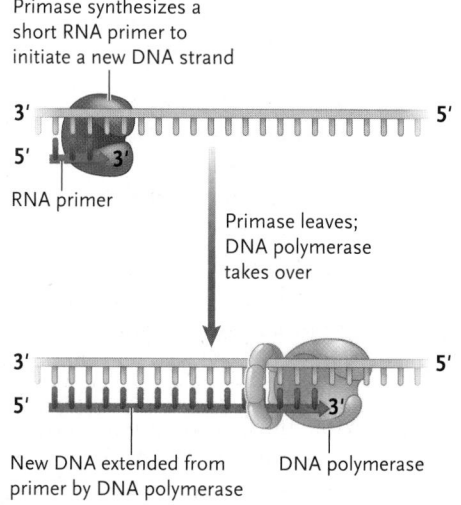

Primase synthesizes a short RNA primer to initiate a new DNA strand

3'

5'

5'
3'

RNA primer

Primase leaves; DNA polymerase takes over

3'

5'

5'
3'

New DNA extended from primer by DNA polymerase

DNA polymerase

leading strand template. The strand synthesized discontinuously in the opposite direction is called the **lagging strand**; the template strand for that strand is the **lagging strand template.**

12.3f Multiple Enzymes Coordinate Their Activities in DNA Replication

Figure 12.15, p. 278, shows how the enzymes and proteins we have introduced act in a coordinated way to replicate DNA. Primase initiates all new strands by

synthesizing an RNA primer. **DNA polymerase III**, the main polymerase, extends the primer by adding DNA nucleotides. For the lagging strand, **DNA polymerase I** removes the RNA primer at the 5′ end of the previous newly synthesized Okazaki fragment, replacing the RNA nucleotides one by one with DNA nucleotides. RNA nucleotide removal uses the 5′ → 3′ exonuclease activity of the enzyme. (An exonuclease removes nucleotides from the end of a molecule; the primer is digested from its 5′ end toward its 3′ end.) DNA polymerase I stops replacing RNA and leaves the template when it encounters the first DNA nucleotide that was synthesized in the Okazaki fragment (Figure 12.15). Therefore, the DNA base replacing the last RNA base of the primer ends up right beside the first DNA base of the Okazaki fragment. The needed covalent bond in the backbone is made by **DNA ligase** (*ligare* = to tie).

The replication process continues in the same way until the entire DNA double helix is copied. **Table 12.1, p. 279,** summarizes the activities of the major enzymes replicating DNA. Replication advances at a rate of about 500 to 1000 nucleotides per second in *E. coli* and other bacteria, and at a rate of about 50 to 100 per second in eukaryotes. The entire process is so rapid that the RNA primers and nicks left by discontinuous synthesis persist for only seconds or fractions of a second. A short distance behind the fork, the new DNA chains are fully continuous and wound into complete DNA double helices. Each helix consists of one "old" and one "new" polynucleotide. Researchers identified the enzymes that replicate DNA through experiments with a variety of bacteria and eukaryotes and with viruses that infect both types of cells. Experiments with the bacterium *E. coli* have provided the most complete information about DNA replication, particularly in the laboratory of Arthur Kornberg at Stanford University. Kornberg received a Nobel Prize in 1959 for his discovery of the mechanism for DNA synthesis.

12.3g Multiple Replication Origins Enable Rapid Replication of Large Chromosomes

Unwinding at an *ori* within a DNA molecule actually produces two replication forks: two Ys joined together at their tops to form a **replication bubble.** Typically, each of the replication forks moves away from the ori as DNA replication proceeds, with the events at each fork mirroring those in the other **(Figure 12.16, p. 280).** For small circular genomes, such as those found in many bacteria and archaeans, there is a single *ori*. Eukaryotic genomes, by contrast, are distributed among several linear chromosomes, each of which can be very long. The average human chromosome, for instance, is about 25 times as long as the *E. coli* chromosome. Nonetheless, replication of long, eukaryotic chromosomes is relatively rapid—sometimes faster than the *E. coli*

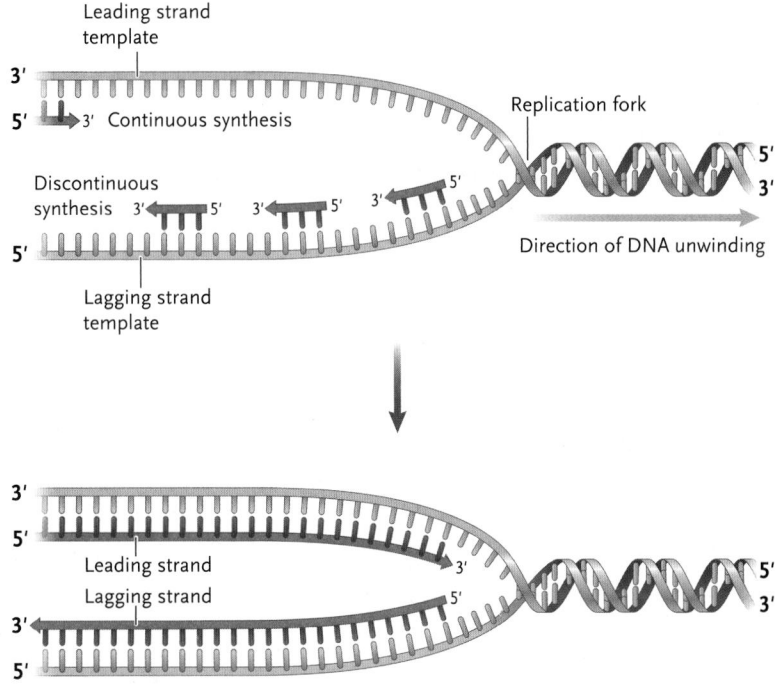

Figure 12.14

Replication of antiparallel template strands at a replication fork. Synthesis of the new DNA strand on the top template strand is continuous. Synthesis on the new DNA strand on the bottom template strand is discontinuous—short lengths of DNA are made, which are then joined into a continuous chain. The overall effect is synthesis of both strands in the direction of replication fork movement.

chromosome—because there are many, sometimes hundreds, of origins of replication along eukaryotic chromosomes. Replication initiates at each origin, forming a replication bubble at each **(Figure 12.17, p. 280)**. Movement of the two forks in opposite directions from each origin extends the replication bubbles until the forks eventually meet along the chromosomes to produce fully replicated DNA molecules. Normally, a replication origin is activated only once during the S phase of a eukaryotic cell cycle, so no portion of the DNA is replicated more than once.

CONCEPT FIX Figures 12.12 and 12.14 show that one strand of DNA is replicated continuously (leading strand), while the other is replicated discontinuously (lagging strand). This might lead you to believe that any one particular strand of DNA on a chromosome is replicated either continuously or discontinuously along its entire length. However, if

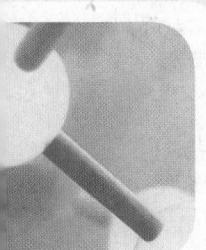

MOLECULE BEHIND BIOLOGY 12.1

Acyclic Nucleoside Phosphonates as Antiviral Drugs

Viruses are obligate parasites that exploit the cellular machinery of infected host cells for replication and gene expression. Such intimate association with host biochemistry makes it difficult for scientists to find an exclusively viral "target" for antiviral drug binding.

However, herpes viruses provide such a target when, once inside the nucleus of an infected cell, they

transcribe a gene coding for their own distinctive DNA polymerase. This novel polymerase replicates viral DNA, drawing from the cellular pool of nucleotide triphosphates.

It has been possible to selectively "poison" viral DNA replication with acyclic nucleoside phosphonates such as cidofovir (shown below) because they (1) are converted to their triphosphate form by infected cells, (2) are

then selectively incorporated into viral DNA (instead of the normal nucleotides) by viral polymerase, and (3) block further DNA synthesis.

These drugs are part of a large class of compounds called base analogues that are incorporated into DNA by "mistake." Compare the structures below with those of the standard bases shown in Figure 12.4, p. 268, and notice why these drugs are called "acyclic."

Cidofovir   Adefovir dipivoxil   Tenofovir disoproxil fumarate

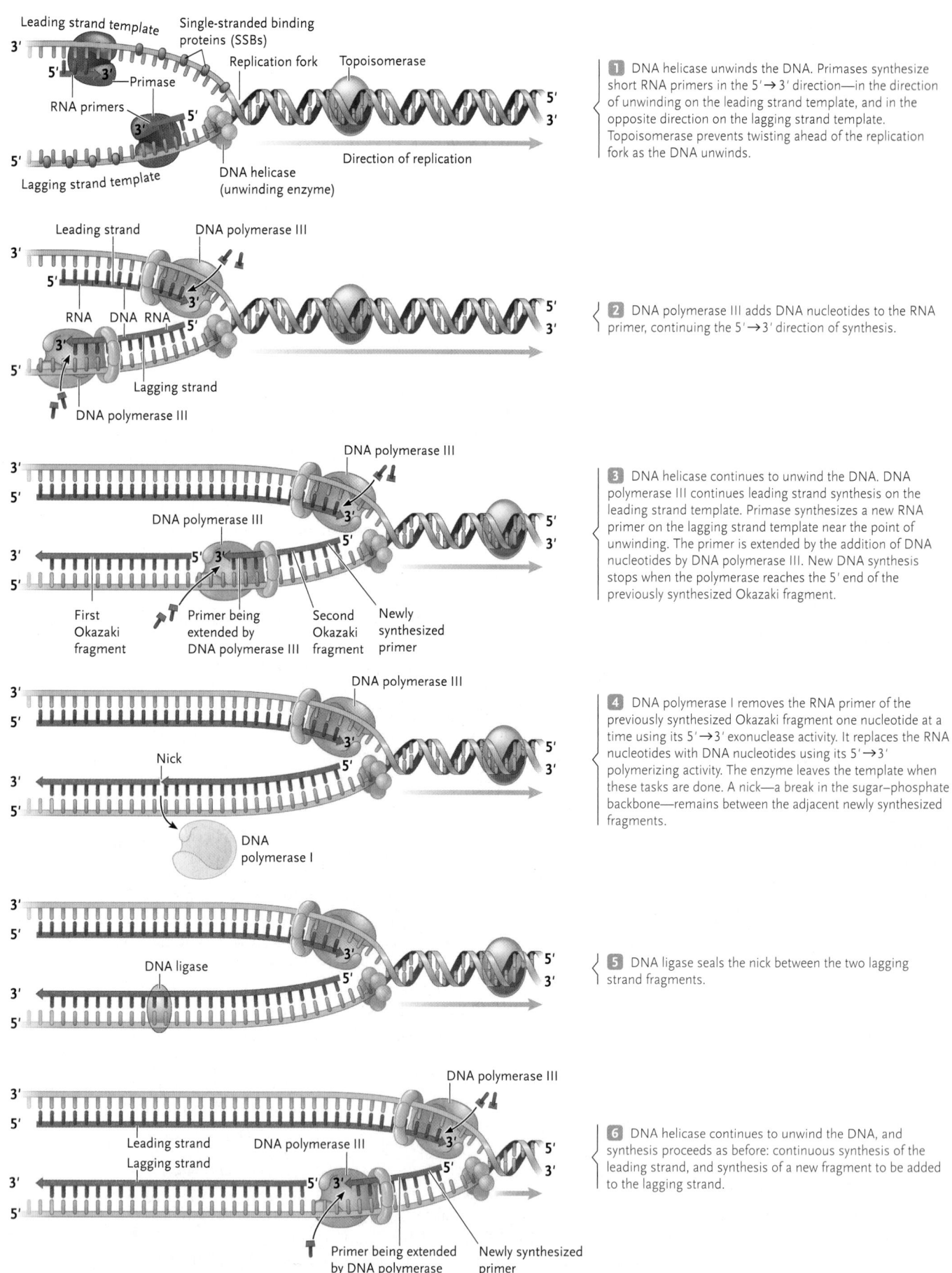

1 DNA helicase unwinds the DNA. Primases synthesize short RNA primers in the 5′→3′ direction—in the direction of unwinding on the leading strand template, and in the opposite direction on the lagging strand template. Topoisomerase prevents twisting ahead of the replication fork as the DNA unwinds.

2 DNA polymerase III adds DNA nucleotides to the RNA primer, continuing the 5′→3′ direction of synthesis.

3 DNA helicase continues to unwind the DNA. DNA polymerase III continues leading strand synthesis on the leading strand template. Primase synthesizes a new RNA primer on the lagging strand template near the point of unwinding. The primer is extended by the addition of DNA nucleotides by DNA polymerase III. New DNA synthesis stops when the polymerase reaches the 5′ end of the previously synthesized Okazaki fragment.

4 DNA polymerase I removes the RNA primer of the previously synthesized Okazaki fragment one nucleotide at a time using its 5′→3′ exonuclease activity. It replaces the RNA nucleotides with DNA nucleotides using its 5′→3′ polymerizing activity. The enzyme leaves the template when these tasks are done. A nick—a break in the sugar–phosphate backbone—remains between the adjacent newly synthesized fragments.

5 DNA ligase seals the nick between the two lagging strand fragments.

6 DNA helicase continues to unwind the DNA, and synthesis proceeds as before: continuous synthesis of the leading strand, and synthesis of a new fragment to be added to the lagging strand.

Figure 12.15

Molecular model of DNA replication. The drawings simplify the process. In reality, the enzymes assemble at the fork, replicating both strands from that position as the template strands fold and pass through the assembly.

Table 12.1	Major Proteins of DNA Replication
Protein	**Activity**
Helicase	Unwinds DNA helix
Single-stranded binding proteins	Stabilize single-stranded DNA and prevent the two strands at the replication fork from re-forming double-stranded DNA
Topoisomerase	Avoids twisting of the DNA ahead of the replication fork (in circular DNA) by cutting the DNA, turning the DNA on one side of the break in the direction opposite to that of the twisting force, and rejoining the two strands
Primase	Assembles RNA primers in the 5' → 3' direction to initiate a new DNA strand
DNA polymerase III	Main replication enzyme in *E. coli*; extends the RNA primer by adding DNA nucleotides to it 
DNA polymerase I	*Escherichia coli* enzyme that uses its 5' → 3' exonuclease activity to remove the RNA of the previously synthesized Okazaki fragment, and uses its 5' → 3' polymerization activity to replace the RNA nucleotides with DNA nucleotides 
Sliding clamp	Tethers DNA polymerase III to the DNA template, making replication more efficient
DNA ligase	Seals nick left between adjacent bases after RNA primers replaced with DNA

you look carefully at the replication bubble in Figure 12.16, you will see that a bubble consists of two replication forks travelling in opposite directions. You will see that any one particular strand of DNA on a chromosome is replicated continuously at one fork but discontinuously at the other fork. ⬡

12.3h Telomerases Solve a Special Replication Problem at the Ends of Linear DNA Molecules in Eukaryotes

The requirement for an RNA primer to initiate DNA replication (see Figures 12.13, p. 276, and 12.15, p. 278) results in the linear chromosomes of eukaryotes getting shorter at each round of replication. Think about the end of a linear DNA molecule. New DNA synthesis on the 3' → 5' template strand must be started with an RNA primer. When that primer is subsequently removed, as usual, a gap will be left in its place at the 5' end of the new DNA strand **(Figure 12.18, p. 281)**. Everywhere else on the chromosome, such gaps are filled in by DNA polymerase by elongating the 3' end of a neighbouring nucleotide. However, at the very ends of chromosomes, there is no existing nucleotide chain that can be elongated. Therefore, DNA polymerase cannot fill in the gap with the required DNA nucleotides and the resulting newly synthesized strand will be too short. (You should agree that this problem occurs on both ends of the chromosome, just on opposite strands of the double helix.) When these new, now shortened, DNA strands are used as a template for the next round of DNA replication, the resulting chromosomes will be shorter still. Indeed, when most somatic cells go through the cell cycle, their chromosomes shorten with each division. Such loss of DNA sequences can eventually have lethal consequences for the cell.

Most eukaryotic chromosomes can afford to lose some DNA sequence because a buffer of highly repetitive noncoding DNA protects genes near the ends of chromosomes. This region of noncoding DNA is called the **telomere** (*telo* = end, *mere* = segment). A telomere consists of a short DNA sequence that is repeated hundreds to thousands of times. In humans, the repeated sequence, the *telomere repeat,* is 5'-TTAGGG-3' on the template strand (the top strand in Figure 12.18, step 1, p. 281). With each replication, a fraction of the telomere repeats is lost by the mechanism described above but the genes are unaffected. The buffering fails only when the entire telomere is lost.

The length of telomeres can be maintained by the action of an unusual enzyme, called **telomerase**, which adds DNA to the ends of chromosomes. Since telomerase makes DNA, it is a type of DNA polymerase. Recall that DNA polymerases require a free 3' OH to extend, a supply of dNTPs, and a template strand. If you look closely at Figure 12.18, you might predict that telomerase elongates the 5' end of the bottom strand to fill in the gap. Although this solution appears easiest,

Figure 12.16

Synthesis of leading and lagging strands in the two replication forks of a replication bubble formed at an origin of replication.

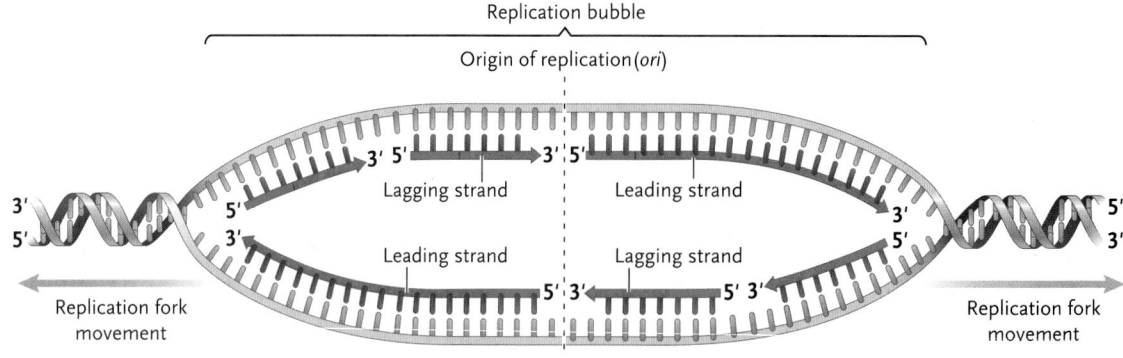

Figure 12.17

Replication from multiple origins in the linear chromosomes of eukaryotes.

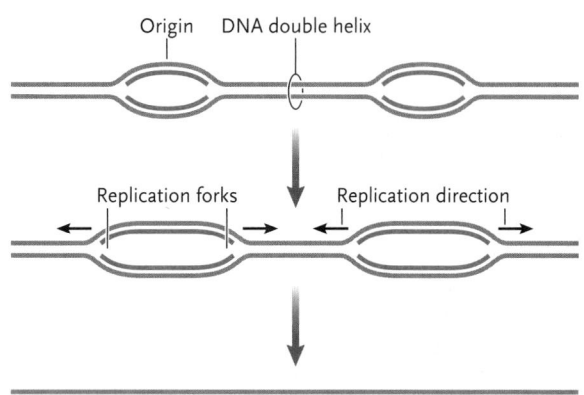

it is impossible since *there are no known polymerases capable of elongating a 5′ end.* So, instead, telomerase must elongate the available 3′ end of the top strand. But now there is a different problem—what to use for a template strand? The lack of a template on the chromosome is solved by *telomerase carrying its own template* in the form of single-stranded RNA molecule. Telomerase adds a telomere repeat to the 3′ end of the DNA using the RNA as a template (see Figure 12.18). Then it shifts toward the end of the chromosome and adds another, and another. Once several hundred repeats are added to the top strand, it is primed and used as a template as usual. When the RNA primer is removed, there will be a single-stranded region at the end of the chromosome as before.

CONCEPT FIX It is important to understand that telomerase does not directly prevent the mechanism that causes the shortening of chromosomes. Telomerase just acts against this mechanism by lengthening chromosomes. ⬡

In most multicellular organisms, telomerase is not active in somatic cells, meaning telomeres shorten when such cells divide. As a result, somatic cells are capable of only a certain number of mitotic divisions before they stop dividing and die. Telomerase is normally active only in the rapidly dividing cells of the early embryo, and in germ cells to ensure that chromosomes of gametes have telomeres restored before passing to the next generation.

Telomerase explains how cancer cells can divide indefinitely and not be limited to a certain number of divisions as a result of telomere shortening. For many cancers, as normal cells develop into cancer cells, their telomerases are reactivated, preserving chromosome length during the rapid divisions characteristic of cancer. A positive side of this discovery is that it may lead to an effective cancer treatment if a means can be found to switch off the telomerases in tumour cells. The chromosomes in the rapidly dividing cancer cells would then eventually shorten to the length at which they break down, leading to cell death and elimination of the tumour. Elizabeth Blackburn, Carol Greider, and Jack Szostak were awarded a Nobel Prize in 2009 for their discovery of how chromosomes are protected by telomeres and the enzyme telomerase.

STUDY BREAK

1. What is the importance of complementary base-pairing to DNA replication?
2. Why is a primer needed for DNA replication on both strands?
3. Two DNA polymerases are used in DNA replication. What are their roles?
4. Why are telomeres important?

12.4 Mechanisms That Correct Replication Errors

DNA polymerases make very few errors as they assemble new molecules. Most of the mistakes that do occur, called **base-pair mismatches**, are corrected, either by a proofreading mechanism carried out during replication by the DNA polymerases themselves or by a DNA repair mechanism that corrects mismatched base pairs after replication is complete.

12.4a Proofreading Depends on the Ability of DNA Polymerases to Reverse and Remove Mismatched Bases

The **proofreading mechanism**, first proposed in 1972 by Arthur Kornberg and Douglas L. Brutlag of Stanford University in California, depends on the ability of DNA

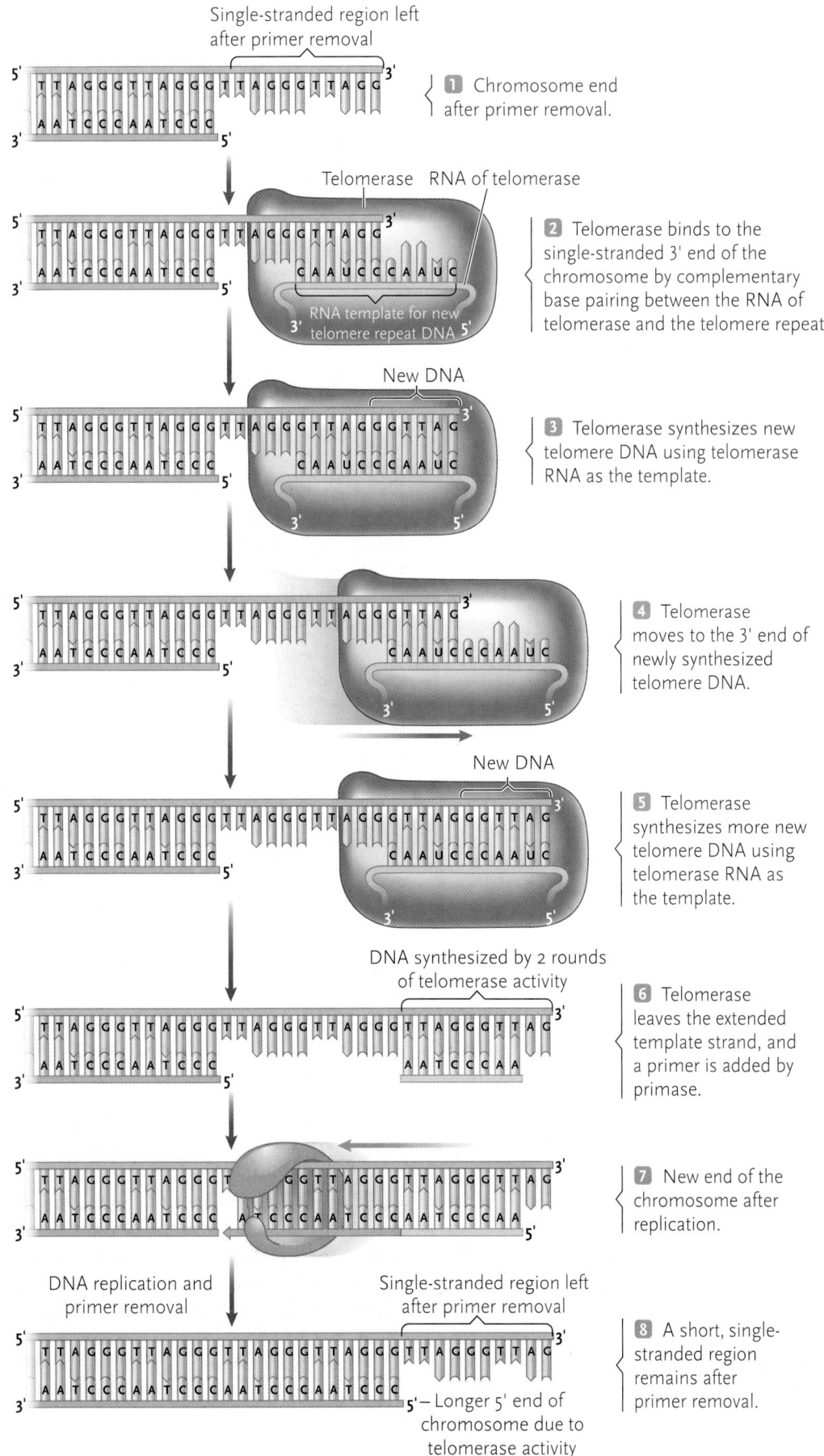

Figure 12.18

Addition of telomere repeats to the 3′ end of a eukaryotic linear chromosome by telomerase.

polymerases to back up and remove mispaired nucleotides from a DNA strand. Only when the most recently added base is correctly paired with its complementary base on the template strand can the DNA polymerases continue to add nucleotides to a growing chain. The correct pairs allow the fully stabilizing hydrogen bonds to form (**Figure 12.19,** step 1). If a newly added nucleotide is mismatched (step 2), the DNA polymerase reverses using a built-in deoxyribonuclease to remove the newly added incorrect nucleotide (step 3). The enzyme resumes working forward, now inserting the correct nucleotide (step 4).

Several experiments have confirmed that the major DNA polymerases of replication can actually proofread their work in this way. For example, when the primary DNA polymerase that replicates DNA in bacteria is intact, with its reverse activity working, its overall error rate is astonishingly low—only about 1 mis

telomerase: adds DNA to end of chromosomes

telomere: non coding DNA

telomeres shorten when cells divide once somatic cells die

cancer cells divide infinitivley their telomerases are reactivated

can benefit if we can find the switch to turn it off

proofreading depends on DNA polymerase to reverse + remove mismatches or repair mechnaisms correct base-pair mismatches

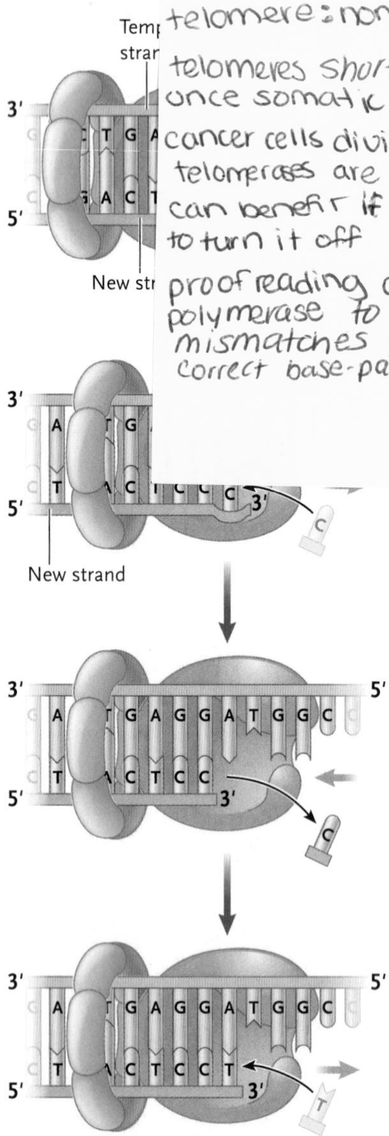

3 DNA polymerase recognizes the mismatched base pair. The enzyme reverses, using its 3′ → 5′ exonuclease to remove the mispaired nucleotide from the strand.

4 DNA polymerase resumes its polymerization activity in the forward direction, extending the new chain in the 3′ → 5′ direction.

Figure 12.19
Proofreading by a DNA polymerase.

tides assembled in the test tube. If the proofreading activity of the enzyme is experimentally inhibited, the error rate increases to about 1 mistake for every 1000 to 10 000 nucleotides assembled. Experiments with eukaryotes have yielded similar results.

12.4b DNA Repair Corrects Errors That Escape Proofreading

Any base-pair mismatches that remain after proofreading face still another round of correction by **DNA repair mechanisms.** These **mismatch repair** mechanisms increase the accuracy of DNA replication well beyond the one-in-a-million errors that persist after proofreading. As noted earlier, the "correct" A-T and G-C base pairs fit together like pieces of a jigsaw puzzle, and their dimensions separate the sugar–phosphate backbone chains by a constant distance. Mispaired bases are too large or small to maintain the correct separation, and they cannot form the hydrogen bonds characteristic of the normal base pairs. As a result, base mismatches distort the structure of the DNA helix. These distortions provide recognition sites for the enzymes catalyzing mismatch repair.

The repair enzymes move along the double helix, "scanning" the DNA for distortions in the newly synthesized nucleotide chain. If the enzymes encounter a distortion, they remove a portion of the new chain, including the mismatched nucleotides (**Figure 12.20,** step 1). The gap left by the removal (step 2) is then filled by a DNA polymerase, using the template strand as a guide (step 3). The repair is completed by a DNA ligase, which seals the nucleotide chain into a continuous DNA molecule (step 4).

The same repair mechanisms also detect and correct alterations in DNA caused by the damaging effects of chemicals and radiation, including the ultraviolet light in sunlight. Some idea of the importance of the repair mechanisms comes from the unfortunate plight of individuals with *Xeroderma pigmentosum,* a hereditary disorder in which the repair mechanism is faulty. Because of the effects of unrepaired alterations in their DNA, skin cancer can develop quickly in these individuals if they are exposed to sunlight.

The rare replication errors that remain in DNA after proofreading and DNA repair are a primary source of **mutations,** differences in DNA sequence that appear and remain in the replicated copies. When a mutation occurs in a gene, it can alter the property of the protein encoded by the gene, which, in turn, may alter how the organism functions. Hence, mutations are highly important to the evolutionary process because they are the ultimate source of the variability in offspring acted on by natural selection.

We now turn from DNA replication and error correction to the arrangements of DNA in eukaryotic and prokaryotic cells. These arrangements organize

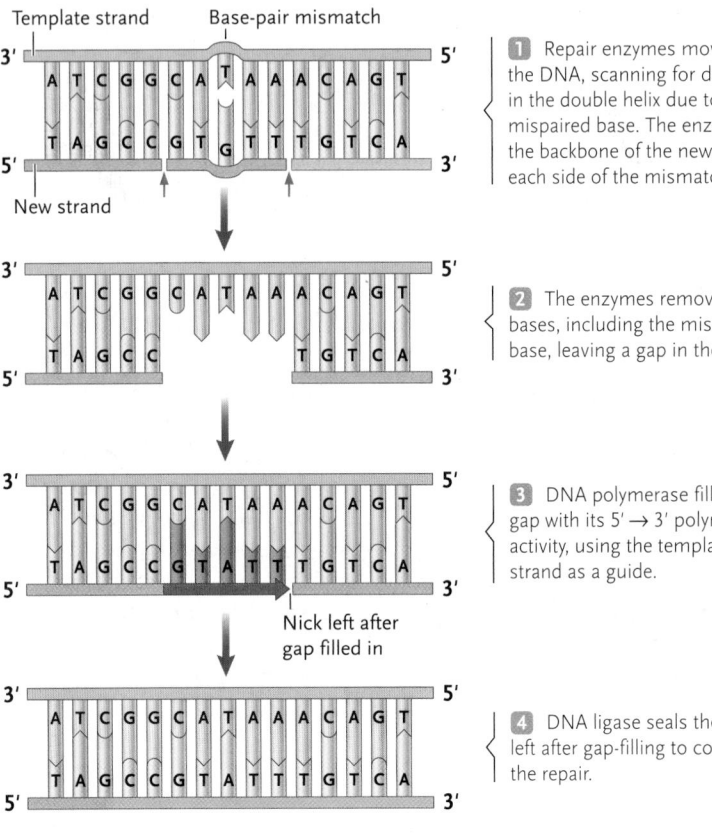

Template strand   Base-pair mismatch

1 Repair enzymes move along the DNA, scanning for distortions in the double helix due to a mispaired base. The enzymes break the backbone of the new strand on each side of the mismatch.

New strand

2 The enzymes remove several bases, including the mismatched base, leaving a gap in the DNA.

3 DNA polymerase fills in the gap with its $5' \rightarrow 3'$ polymerizing activity, using the template strand as a guide.

Nick left after gap filled in

4 DNA ligase seals the nick left after gap-filling to complete the repair.

Figure 12.20
Repair of mismatched bases in replicated DNA.

superstructures that fit the long DNA molecules into the microscopic dimensions of cells and also contribute to the regulation of DNA activity.

STUDY BREAK

Why is a proofreading mechanism important for DNA replication?

12.5 DNA Organization in Eukaryotic versus Prokaryotic Cells

Enzymatic proteins are the essential catalysts of every step in DNA replication. In addition, numerous proteins of other types organize the DNA in both eukaryotic and prokaryotic cells in addition to controlling its expression.

In eukaryotes, two major types of proteins, the histone and nonhistone proteins, are associated with DNA structure and regulation in the nucleus. These proteins are known collectively as the **chromosomal proteins** of eukaryotes. The complex of DNA and its associated proteins, termed **chromatin**, is the structural building block of a chromosome.

By comparison, the single DNA molecule of a prokaryotic cell is more simply organized and has fewer associated proteins. However, prokaryotic DNA is still

associated with two classes of proteins with functions similar to those of the eukaryotic histones and nonhistones: one class that organizes the DNA structurally and one that regulates gene activity. We begin this section with the major DNA-associated proteins of eukaryotes as they relate to packaging. The role of chromatin structure in gene regulation is addressed in Chapter 14.

12.5a Histones Pack Eukaryotic DNA at Successive Levels of Organization

The **histones** are a class of small, positively charged (basic) proteins that are complexed with DNA in the chromosomes of eukaryotes. (Most other cellular proteins are larger and are neutral or negatively charged.) The histones link to DNA by an attraction between their positive charges and the negatively charged phosphate groups of the DNA.

Five types of histones exist in most eukaryotic cells: H1, H2A, H2B, H3, and H4. The amino acid sequences of these proteins are highly similar among eukaryotes, suggesting that they perform the same functions in all eukaryotic organisms.

One function of histones is to pack DNA molecules into the narrow confines of the cell nucleus. For example, each human cell nucleus contains 2 m of DNA. Combination with the histones compacts this length so much that it fits into nuclei that are only about 10 μm in diameter.

Histones and DNA Packing. The histones pack DNA at several levels of chromatin structure. In the most fundamental structure, called a **nucleosome**, two molecules each of H2A, H2B, H3, and H4 combine to form a beadlike, eight-protein **nucleosome core particle** around which DNA winds for almost two turns **(Figure 12.21, p. 284)**. A short segment of DNA, the **linker**, extends between one nucleosome and the next. Under the electron microscope, this structure looks like beads on a string. The diameter of the beads (the nucleosomes) gives this structure its name—the **10 nm chromatin fibre** (see Figure 12.21).

Each nucleosome and linker includes about 200 base pairs of DNA. Nucleosomes compact DNA by a factor of about 7; that is, a length of DNA becomes about one-seventh the length when it is wrapped into nucleosomes.

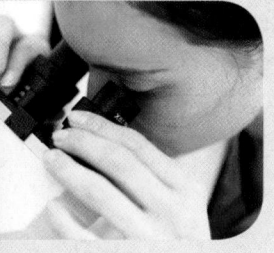

In 1944, the Austrian physicist Erwin Schrödinger published *What Is Life?* This small book speculated about the theoretical nature of the genetic material and prompted several physicists to turn their creativity to solving fundamental problems in the field of biology.

Cross-fertilization of these scientific disciplines energized research into the molecular biology of the gene and, specifically through the career of Bob Haynes, provided pioneering insights into how cells suffer and respond to DNA damage. Subsequent research has revealed that a breakdown in repair of DNA damage has important implications for cancer, aging, certain genetic diseases, and exposure to physical and chemical mutagens.

Born in 1931, Haynes earned undergraduate and Ph.D. degrees in biophysics from the University of Western Ontario before working as a postdoctoral fellow in physics at St. Bartholomew's Hospital Medical College, University of London. He joined the Biophysics Departments of the University of Chicago and the University of California at Berkeley before returning to Canada in 1968 as chair of the Biology Department at York University in Toronto. Upon his death in 1998, Haynes was a fellow of the Royal Society of Canada and an officer of the Order of Canada in recognition of his broad contributions to research in environmental mutagenesis, international leadership, and science education.

Histones and Chromatin Fibres. The fifth histone, H1, brings about the next level of chromatin packing. One H1 molecule binds both to the nucleosomes and to the linker DNA. This binding causes the nucleosomes to package into a coiled structure 30 nm in diameter, called the **30 nm chromatin fibre** or **solenoid**, with about six nucleosomes per turn (see Figure 12.21).

The arrangement of DNA in nucleosomes and solenoids compacts the DNA and probably also protects it from chemical and mechanical damage. In the test tube, DNA wound into nucleosomes and chromatin fibres is much more resistant to degradation by deoxyribonuclease (a DNA-digesting enzyme) than when it is not bound to histone proteins.

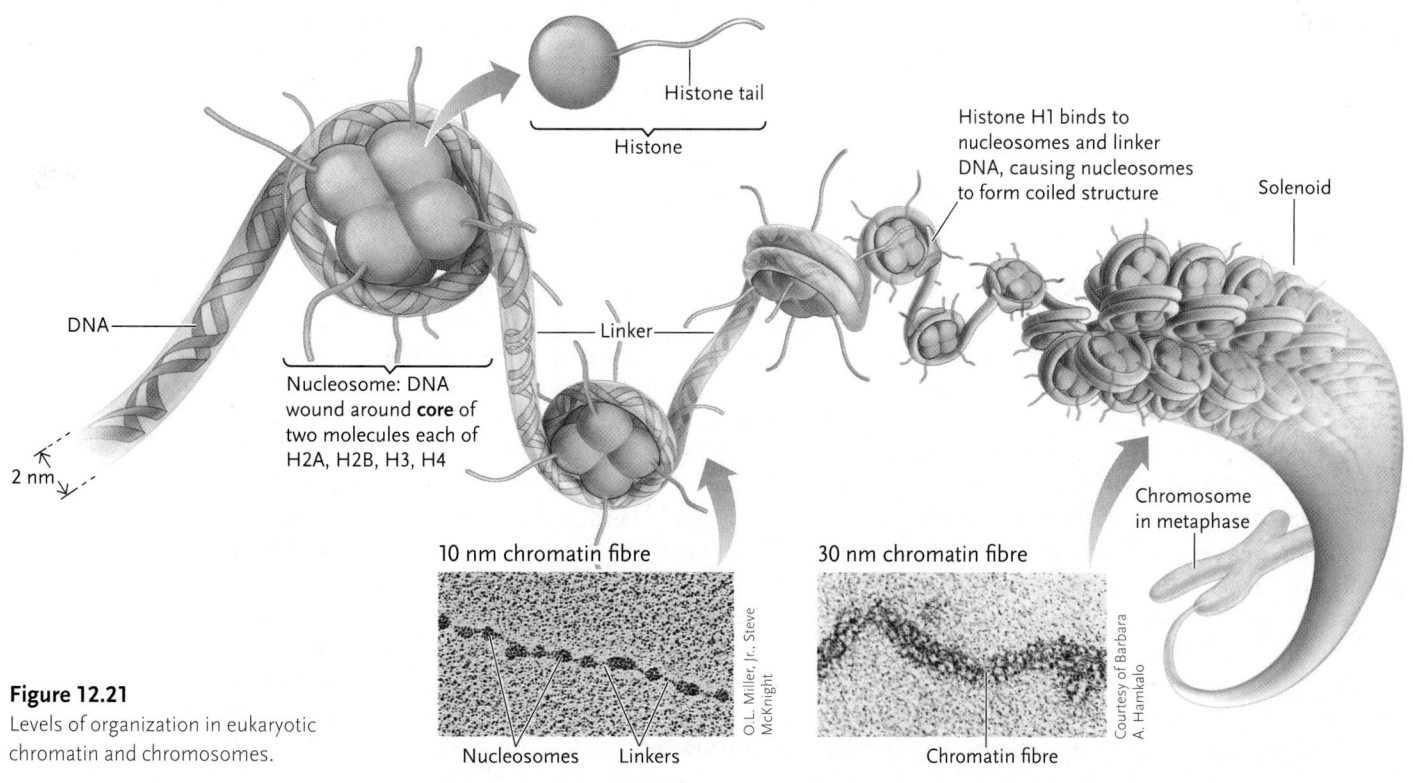

Figure 12.21

Levels of organization in eukaryotic chromatin and chromosomes.

Packing at Still Higher Levels: Euchromatin and Heterochromatin. In interphase nuclei, chromatin fibres are loosely packed in some regions and densely packed in others. The loosely packed regions are known as **euchromatin** (*eu* = true, regular, or typical), and the densely packed regions are called **heterochromatin** (*hetero* = different). Chromatin fibres also fold and pack into the thick, rodlike chromosomes that become visible during mitosis and meiosis.

Several experiments indicate that heterochromatin represents large blocks of genes that have been turned off and placed in a compact storage form. For example, recall the process of X-chromosome inactivation in mammalian females (see Section 11.2). As one of the two X chromosomes becomes inactive in cells early in development, it packs down into a block of heterochromatin called the *Barr body,* which is large enough to see under the light microscope. These findings support the idea that, in addition to organizing nuclear DNA, histones play a role in regulating gene activity.

12.5b Many Nonhistone Proteins Have Key Roles in the Regulation of Gene Expression

Nonhistone proteins are loosely defined as all the proteins associated with DNA that are not histones. Non-histones vary widely in structure; most are negatively charged or neutral, but some are positively charged. They range in size from polypeptides smaller than histones to some of the largest cellular proteins.

Many nonhistone proteins help control the expression of individual genes. (The regulation of gene expression is the subject of Chapter 14.) For example, expression of a gene requires that the enzymes and proteins for that process be able to access the gene in the chromatin. If a gene is packed into heterochromatin, it is unavailable for activation. If the gene is in the more extended euchromatin, it is more accessible. Many nonhistone proteins affect gene accessibility by modifying histones to change how they associate with DNA in chromatin, either loosening or tightening the association. Other nonhistone proteins are regulatory proteins that activate or repress the expression of a gene. Yet others are components of the enzyme–protein complexes that are needed for the expression of any gene.

12.5c DNA Is Organized More Simply in Prokaryotes than in Eukaryotes

Several features of DNA organization in prokaryotic cells differ fundamentally from eukaryotic DNA. In contrast to the linear DNA in eukaryotes, the primary DNA molecule of most prokaryotic cells is circular, with only one copy per cell. In parallel with eukaryotic terminology, the DNA molecule is called a **bacterial chromosome.** The chromosome of the best-known bacterium, *E. coli,* includes about 1360 μm of DNA, which is equivalent to 4.6 million base pairs. There are exceptions: some bacteria have two or more different chromosomes in the cell, and some bacterial chromosomes are linear.

Replication begins from a single origin in the DNA circle, forming two forks that travel around the circle in opposite directions. Eventually, the forks meet at the opposite side from the origin to complete replication **(Figure 12.22).**

Inside prokaryotic cells, the DNA circle is packed and folded into an irregularly shaped mass called the **nucleoid** (shown in Figure 2.7, Chapter 2). The DNA of the nucleoid is suspended directly in the cytoplasm with no surrounding membrane.

Figure 12.22

Replication from a single origin of replication in a circular bacterial chromosome.

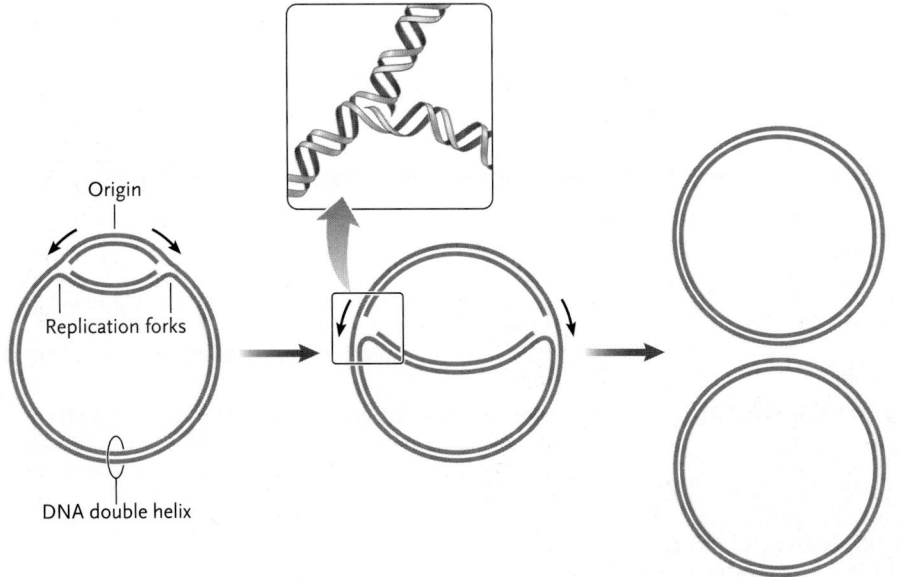

Origin

Replication forks

DNA double helix

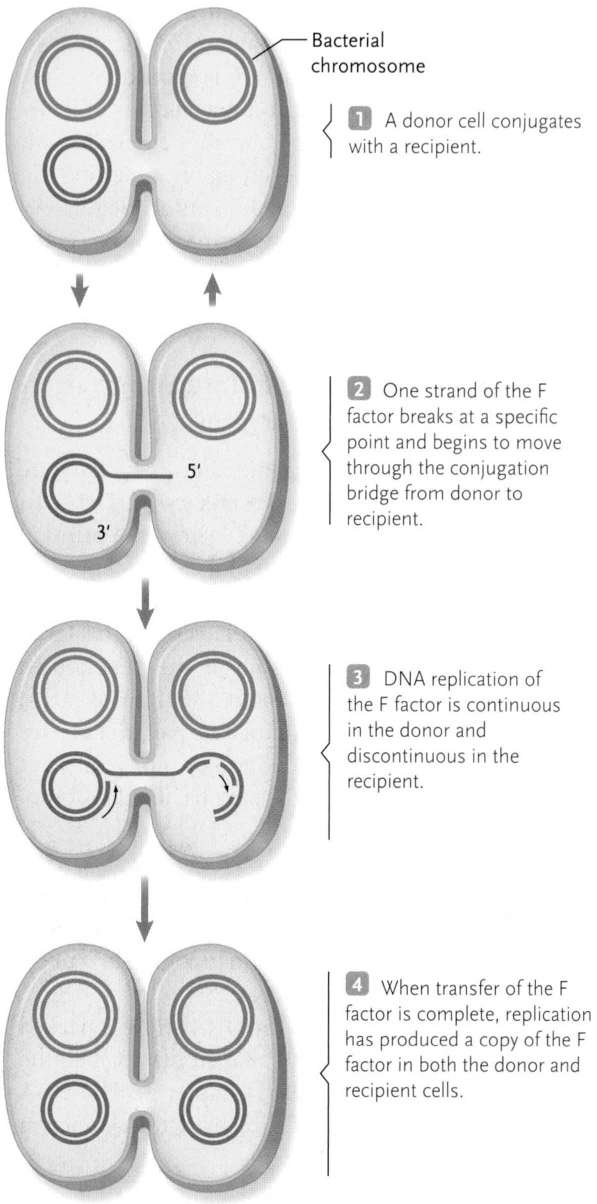

Bacterial chromosome

1 A donor cell conjugates with a recipient.

2 One strand of the F factor breaks at a specific point and begins to move through the conjugation bridge from donor to recipient.

5'

3'

3 DNA replication of the F factor is continuous in the donor and discontinuous in the recipient.

4 When transfer of the F factor is complete, replication has produced a copy of the F factor in both the donor and recipient cells.

Figure 12.23

Transfer of F factor by rolling circle replication during conjugation. One of the two strands of the plasmid is nicked and the 5' end moves from the donor into a recipient cell. The remaining circular strand "rolls" like a tape dispenser. DNA synthesis is continuous in the donor cell and discontinuous in the recipient (arrows), resulting in two complete plasmids.

Many prokaryotic cells also contain other DNA molecules, called **plasmids**, in addition to the main chromosome of the nucleoid. Most plasmids are circular, although some are linear. Plasmids have replication origins and are duplicated and distributed to daughter cells together with the bacterial chromosome during cell division. Chapter 9 describes the process of conjugation, in which plasmids are replicated while being transferred from a donor cell to a recipient cell. The DNA is replicated by a mechanism called rolling circle replication, in which one strand of the plasmid is cut and travels into a recipient cell as a linear molecule; the other strand remains circular in the donor cell **(Figure 12.23)**. DNA replication restores both strands to double-strandedness, and the linear molecule recircularizes. Although rolling circle replication follows the usual rules of DNA replication, notice that the leading and lagging strand synthesis occur in separate cells rather than at one replication fork.

Although bacterial DNA is not organized into nucleosomes, there are positively charged proteins that combine with bacterial DNA. Some of these proteins help organize the DNA into loops, thereby providing some compaction of the molecule. Bacterial DNA also combines with many types of genetic regulatory proteins that have functions similar to those of the nonhistone proteins of eukaryotes (see Chapter 14).

With this description of prokaryotic DNA organization, our survey of DNA structure and its replication and organization is complete. The next chapter revisits the same structures and discusses how they function in the expression of information encoded in DNA.

STUDY BREAK

1. What is the structure of the nucleosome?
2. What is the role of histone H1 in eukaryotic chromosome structure?

Review

To access course materials such as Aplia and other companion resources, please visit www.NELSONbrain.com.

12.1 Establishing DNA as the Hereditary Molecule

- Griffith found that a substance derived from killed virulent *Streptococcus pneumoniae* bacteria could transform nonvirulent living *S. pneumoniae* bacteria to the virulent type (Figure 12.2).

- Avery and his coworkers showed that DNA, and not protein or RNA, was the molecule responsible for transforming *S. pneumoniae* bacteria into the virulent form.

- Hershey and Chase showed that the DNA of a phage, not the protein, enters bacterial cells to direct the life cycle of the virus. Taken together, the experiments of Griffith, Avery and his coworkers, and Hershey and Chase established that DNA is the hereditary molecule (Figure 12.3).

12.2 DNA Structure

- Watson and Crick discovered that a DNA molecule consists of two polynucleotide chains twisted around each other into a right-handed double helix. Each nucleotide of the chains consists of deoxyribose; a phosphate group; and one of adenine, thymine, guanine, and cytosine. The deoxyribose sugars are linked by phosphate groups to form an alternating sugar–phosphate backbone. The two strands are held together by hydrogen bonded adenine–thymine (A-T) and guanine–cytosine (G-C) base pairs. Each full turn of the double helix involves 10 base pairs (Figures 12.4 and 12.6).
- The two strands of the DNA double helix are antiparallel.

12.3 DNA Replication

- DNA is duplicated by semiconservative replication, in which the two strands of a parental DNA molecule unwind and each serves as a template for the synthesis of a complementary copy (Figures 12.7–12.9).
- Several enzymes catalyze DNA replication. Helicase unwinds the DNA; primase synthesizes an RNA primer used as a starting point for nucleotide assembly by DNA polymerases. DNA polymerases assemble nucleotides into a polymer, one at a time, in a sequence complementary to the sequence of bases in the template. After a DNA polymerase removes the primers and fills in the resulting gaps, DNA ligase closes the remaining single-strand nicks (Figures 12.10–12.12 and 12.15).
- As the DNA helix unwinds, only one template runs in a direction that allows the new DNA strand to be made continuously in the direction of unwinding. The other template strand is copied in short lengths that run in the direction opposite to unwinding. The short lengths produced by this discontinuous replication are then linked into a continuous strand (Figures 12.14 and 12.15).
- DNA synthesis begins at sites that act as replication origins and proceeds from the origins as two replication forks moving in opposite directions (Figures 12.16 and 12.17).
- The ends of eukaryotic chromosomes consist of telomeres: short sequences repeated hundreds to thousands of times. These repeats provide a buffer against chromosome shortening during replication. Although most somatic cells show this chromosome shortening, some cell types do not because they have a telomerase enzyme that adds telomere repeats to the chromosome ends (Figure 12.18).

12.4 Mechanisms That Correct Replication Errors

- In proofreading, the DNA polymerase reverses and removes the most recently added base if it is mispaired as a result of a replication error. The enzyme then resumes DNA synthesis in the forward direction (Figure 12.19).
- In DNA mismatch repair, enzymes recognize distorted regions caused by mispaired base pairs and remove a section of DNA that includes the mispaired base from the newly synthesized nucleotide chain. A DNA polymerase then resynthesizes the section correctly, using the original template chain as a guide (Figure 12.20).

12.5 DNA Organization in Eukaryotic versus Prokaryotic Cells

- Eukaryotic chromosomes consist of DNA complexed with histone and nonhistone proteins.
- In eukaryotic chromosomes, DNA is wrapped around a core consisting of two molecules each of histones H2A, H2B, H3, and H4 to produce a nucleosome. Linker DNA connects adjacent nucleosomes. The chromosome structure in this form is the 10 nm chromatin fibre. The binding of histone H1 causes the nucleosomes to package into a coiled structure called the 30 nm chromatin fibre (Figure 12.21).
- Chromatin is distributed between euchromatin, a loosely packed region in which genes are active in RNA transcription, and heterochromatin, densely packed masses in which genes, if present, are inactive. Chromatin also folds and packs to form thick, rodlike chromosomes during nuclear division.
- Nonhistone proteins help control the expression of individual genes.
- The bacterial chromosome is a closed, circular double helix of DNA; it is packed into the nucleoid region of the cell. Replication begins from a single origin and proceeds in both directions. Many bacteria also contain plasmids, which replicate independently of the host chromosome (Figure 12.22).
- Bacterial DNA is organized into loops through interaction with proteins. Other proteins similar to eukaryotic nonhistones regulate gene activity in prokaryotic organisms.

Questions

Self-Test Questions

1. Working on the Amazon River, a biologist isolated DNA from two unknown organisms, P and Q. He discovered that the adenine content of P was 15% and the cytosine content of Q was 42%. Which of the following conclusions can be drawn?
   a. The amount of adenine in Q is 42%.
   b. The amount of guanine in P is 15%.
   c. The amount of guanine and cytosine combined in P is 70%.
   d. The amount of thymine in Q is 21%.

2. The Hershey and Chase experiment involved infecting bacterial cells with radioactively labelled viruses. What did this experiment show?
   a. ^{35}S-labelled DNA ended up inside the virus progeny.
   b. ^{32}P-labelled DNA entered bacterial cells.
   c. ^{35}S-labelled protein was incorporated into bacterial cells.
   d. ^{32}P-labelled protein was incorporated into virus coats.

3. Which of the following would appear on a list of pyrimidines?
   a. cytosine and thymine
   b. cytosine and guanine
   c. adenine and thymine
   d. adenine and guanine

4. Which of the following statements about DNA replication is true?
   a. DNA polymerase III extends an RNA primer.
   b. Some DNA polymerases can add new bases to the 5′ end of a growing strand.
   c. Each eukaryotic chromosome has a single origin of replication.
   d. Okazaki fragments are made only of RNA.

5. Which of the following statements about DNA structure is true?
   a. Each DNA strand has a 3′ OH on one end and a 5′ OH on the other end.
   b. Each strand of the double helix runs parallel to the other.
   c. The binding of adenine to thymine is through three hydrogen bonds.
   d. Bonds between components of the backbone (i.e., sugar–phosphate) are stronger than those between one strand and the other.

6. In the Meselson and Stahl experiment, the DNA in the parental generation was all $^{15}N^{15}N$, and after one round of replication, the DNA was all $^{15}N^{14}N$. What ratio of DNA would be seen after three rounds of replication?
   a. one $^{15}N^{14}N$: one $^{14}N^{14}N$
   b. one $^{15}N^{14}N$: two $^{14}N^{14}N$
   c. one $^{15}N^{14}N$: three $^{14}N^{14}N$
   d. one $^{15}N^{14}N$: four $^{14}N^{14}N$

7. Since DNA is synthesized in the 5′ → 3′ direction, which of the following must be true?
   a. The template must be read in the 5′ → 3′ direction.
   b. Polymerase must add successive nucleotides to the 3′ –OH end of the newly forming chain.
   c. Ligase must unwind the two DNA strands in opposite directions.
   d. Primase must add RNA nucleotides to the growing 5′ end.

8. Which of the following is a characteristic of telomerase?
   a. It is active in cancer cells.
   b. It is more active in adult than in embryonic cells.
   c. It has telomeres made of RNA.
   d. It shortens the ends of chromosomes.

9. What does the process of mismatch repair accomplish?
   a. It seals Okazaki fragments with ligase into a continual DNA strand.
   b. It removes RNA primers and replaces them with the correct DNA.
   c. It restores DNA sequence that is lost during the replication of the ends of chromosomes.
   d. It replaces incorrect bases that escape proofreading by DNA polymerase.

10. The DNA of prokaryotic organisms differs from that of eukaryotes. In what way?
    a. Prokaryotic DNA is surrounded by densely packed histones; eukaryotic DNA is not.
    b. Prokaryotic DNA has many sites for the initiation of DNA replication; eukaryotic DNA does not.
    c. Prokaryotic DNA is typically single stranded; eukaryotic DNA is typically double stranded.
    d. Prokaryotic DNA is rarely packaged in linear chromosomes; eukaryotic DNA is commonly packaged in linear chromosomes.

Questions for Discussion

1. Chargaff's data suggested that adenine pairs with thymine and guanine pairs with cytosine. What other data available to Watson and Crick suggested that adenine–guanine and cytosine–thymine pairs normally do not form?

2. Exposing cells to radioactive thymidine can label eukaryotic chromosomes during the S phase of interphase. If cells are exposed to radioactive thymidine during the S phase, would you expect both or only one of the sister chromatids of a duplicated chromosome to be labelled at metaphase of the following mitosis (see Section 8.3)?

3. If the cells in question 2 finish division and then enter another round of DNA replication in a medium that has been washed free of radioactive label, would you expect both or only one of the sister chromatids of a duplicated chromosome to be labelled at metaphase of the following mitosis?

4. During replication, an error uncorrected by proofreading or mismatch repair produces a DNA molecule with a base mismatch at the indicated position:

   AATTCCGACTCCTATGG

   TTAAGGTTGAGGATACC
   ↑

   This DNA molecule is received by one of the two daughter cells produced by mitosis. In the next round of replication and division, the mutation appears in only one of the two daughter cells. Develop a hypothesis to explain this observation.

5. Strains of bacteria that are resistant to an antibiotic sometimes appear spontaneously among other bacteria of the same type that are killed by the antibiotic. In view of the information in this chapter about DNA replication, what might account for the appearance of this resistance?

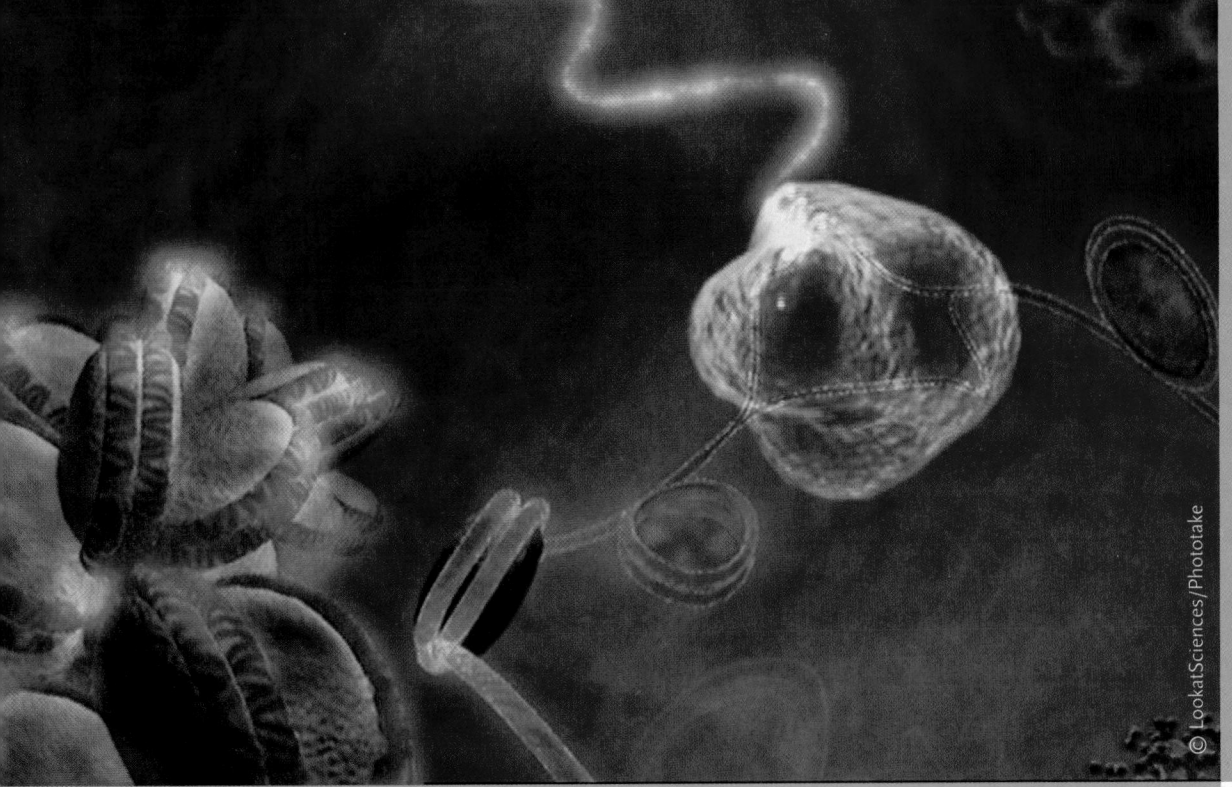

Transcription of a eukaryotic gene to produce messenger RNA (mRNA), a type of RNA that acts as a template for protein synthesis. The DNA of the gene unwinds from the nucleosome (left side) and is copied by an RNA polymerase (centre) into mRNA (exiting the top).

© LookatSciences/Phototake

Gene Structure and Expression

WHY IT MATTERS

The marine mussel *Mytilus* **(Figure 13.1, p. 290)** lives in one of the most demanding environments on Earth—it clings permanently to rocks pounded by surf day in and day out, constantly in danger of being dashed to pieces or torn loose by foraging predators. The mussel is remarkably resistant to disturbance. If you try to pry one loose, you will find how difficult it is to tear the tough, elastic fibres that hold it fast. They are even hard to cut with a knife.

The fibres holding mussels to rocks are proteins secreted by the muscular foot of the animal. The proteins include keratin (an intermediate filament protein) and another resinous protein. Along with other proteins, they form a tough, adhesive material called byssus.

Byssus is one of the world's premier underwater adhesives. It fascinates biochemists, adhesive manufacturers, dentists, and surgeons looking for better ways to hold repaired body parts together. Genetic engineers are inserting segments of mussel DNA into yeast cells, which reproduce in large numbers and serve as "factories," translating the mussel genes into proteins. Among the proteins produced are those of byssus, allowing investigators to figure out how to use or imitate the mussel glue for human needs. This exciting work, like the mussel's own byssus building, starts with one of life's universal truths: *Every protein is assembled on ribosomes according to instructions dictated by genes coded in DNA.*

In this chapter, we trace the basic process that produces proteins in all organisms, beginning with the instructions encoded in DNA and leading through RNA to the sequence of amino acids in a protein. Many enzymes and other proteins are players as well as products in this story, as are several kinds of RNA and the cell's protein-making machines, the ribosomes. As your understanding of the fundamental elements of all protein production grows, be sure to notice the differences in the kinds of information coded in DNA, differences in the mechanisms in prokaryotic

Figure 13.1

The marine mussel *Mytilus* and its natural habitat.

cells versus eukaryotes, and differences in the structure of genes that code for protein versus those that code for RNA.

13.1 The Connection between DNA, RNA, and Protein

Although the relationship between proteins and nucleic acids was once uncertain, it is now common knowledge that proteins are encoded by genes made of DNA. In this section, you will learn how that connection was discovered. We also present an overview of the molecular steps needed to go from gene to protein: transcription and translation.

13.1a Genes Specify Either Protein or RNA Products

How do we know that genes encode—that is, specify the amino acid sequence of—proteins? Two key pieces of research involving defects in metabolism illustrated this connection unequivocally. The first began in 1896 with Archibald Garrod, an English physician. He studied *alkaptonuria,* a human disease that does little harm but is detected easily by the fact that a patient's urine turns black when exposed to oxygen. Garrod and William Bateson, a geneticist, studied families of patients with the disease and concluded that it is an inherited trait. Garrod also found that people with alkaptonuria excrete a particular compound, homogentisic acid, in their urine. Garrod concluded that normal people are able to metabolize the homogentisic acid, whereas people with alkaptonuria cannot. By 1908, Garrod had concluded that the disease was an inborn error of metabolism. Garrod's work was the first to show a specific relationship between genes and metabolism.

In the second piece of research, George Beadle and Edward Tatum, working in the 1940s with the orange bread mould *Neurospora crassa,* collected data showing a direct relationship between genes and enzymes. Beadle and Tatum chose *Neurospora* for their

work because it is a haploid fungus with simple nutritional needs. That is, wild-type *Neurospora*—the form of the mould found in nature—grows readily on a minimal medium (MM) consisting of a number of inorganic salts, sucrose, and a vitamin. The researchers reasoned that the fungus uses only simple chemicals in the medium to synthesize all of the more complex molecules needed for growth and reproduction, including amino acids for proteins and nucleotides for DNA and RNA.

Beadle and Tatum exposed spores of wild-type *Neurospora* to X-rays that caused mutations. They found that some of the treated spores would not germinate and grow unless MM was supplemented with additional nutrients, such as amino acids or vitamins. Mutant strains that are unable to grow on MM are called auxotrophs (*auxo* = increased; *troph* = eater), or nutritional mutants. Beadle and Tatum hypothesized that each auxotrophic strain had a defect in a gene coding for an enzyme needed to synthesize a nutrient that now had to be added to the MM. The wild-type strain could make the nutrient for itself from raw materials in the MM, but the mutant strain could grow only if the researchers supplied the nutrient. By testing to see if each mutant strain would grow on MM supplemented with a given nutrient, Beadle and Tatum discovered which specific nutrient each mutant needed to grow and, therefore, which gene defect it had. For example, a mutant that required the addition of the amino acid arginine to grow had a defect in a gene for an enzyme involved in the synthesis of arginine. Such arginine auxotrophs are known as *arg* mutants. The assembly of arginine from raw materials is a multi-step "assembly-line" process with a different enzyme catalyzing each step. Therefore, different *arg* mutants might have defects in different enzymes and therefore have blocks at different steps in the assembly line. (This is conceptually similar to Lederberg's work with auxotrophic bacteria, described in Chapter 9.)

Beadle and Tatum determined where in the arginine synthesis pathway each of four mutants (*argE, argF, argG,* and *argH*) was blocked. They tested whether each mutant could grow on MM or on MM supplemented with one of ornithine, citrulline, argininosuccinate (three compounds known to be involved in the synthesis of arginine), and arginine itself **(Figure 13.2)**. While none of the four mutants could grow on MM because it was lacking arginine, they all grew well on MM + arginine. Each of the *arg* mutants showed a different pattern of growth on the supplemented MM (see Figure 13.2). Beadle and

Tatum deduced that the biosynthesis of arginine occurred in a number of steps, with each step controlled by a gene that encoded the enzyme for the step (see Figure 13.2). For example, the *argH* mutant grows on MM + arginine but not on MM + any of the other three compounds; this means that the mutant is blocked at the last step in the pathway, which produces arginine. Similarly, the *argG* mutant grows

QUESTION: Do genes specify enzymes?

EXPERIMENT: Test *arg* mutants of the orange bread mould *Neurospora crassa* for growth on MM (minimal medium), MM + ornithine, MM + citrulline, MM + argininosuccinate, and MM + arginine. *Arg* mutants are unable to synthesize the amino acid arginine, which is essential for growth.

Strain		Growth on MM +				
		Nothing	Ornithine	Citrulline	Argininosuccinate	Arginine
Wild type (control)	Grows on MM, and on all other supplemented media. — Growth					
argE **mutant**	Does not grow on MM; grows on all other supplemented media. — No growth					
argF **mutant**	Does not grow on MM; grows if citrulline, argininosuccinate, or arginine is in the medium, but not if ornithine is present.					
argG **mutant**	Does not grow on MM; grows if argininosuccinate or arginine is in the medium, but not if ornithine or citrulline is present.					
argH **mutant**	Does not grow on MM; grows if arginine is in the medium, but not if ornithine, citrulline, or argininosuccinate is present.					

CONCLUSION: Arginine is synthesized in a biochemical pathway. Each step of the pathway is catalyzed by an enzyme, and each enzyme is encoded by a gene.

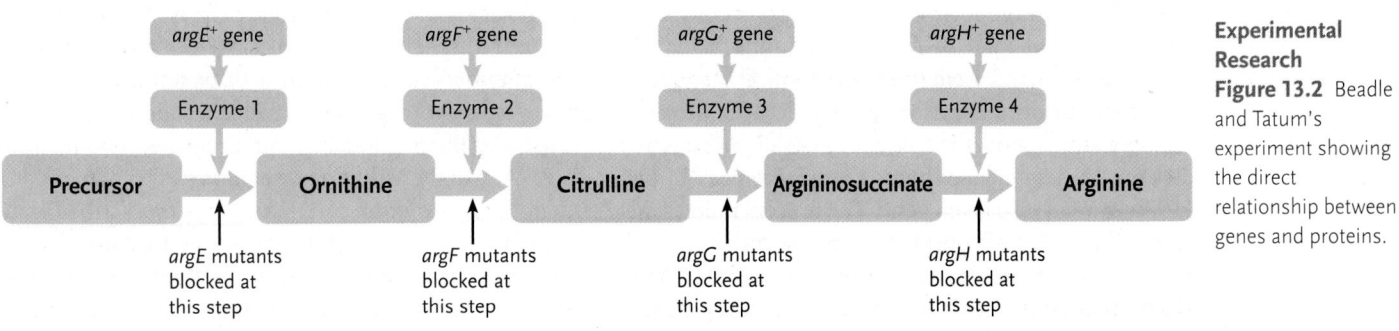

Experimental Research Figure 13.2 Beadle and Tatum's experiment showing the direct relationship between genes and proteins.

on MM + arginine or argininosuccinate but not on MM + any of the other supplements; this means that *argG* is blocked in the pathway before argininosuccinate is made (see Figure 13.2). With similar analysis, the researchers deduced the whole pathway from precursor to arginine and showed which gene encoded the enzyme that carried out each step. In sum, Beadle and Tatum had shown the direct relationship between genes and enzymes, which they put forward as the **one gene–one enzyme hypothesis**. Their experiment was a keystone in the development of molecular biology. As a result of their work, they were awarded a Nobel Prize in 1958.

It is important to understand that protein structure and function are now known to be more complex than suggested by the work of Beadle and Tatum. Many proteins consist of more than one subunit. Each of these subunits is a separate molecule, called a polypeptide, that is coded by a separate gene. Polypeptides can assemble to create a functional cluster of molecules called a protein. For instance, the protein hemoglobin is made up of four polypeptides, two each of an α-subunit and a β-subunit; this composition gives the protein its functional property of transporting oxygen rather than catalyzing a chemical reaction. Two different genes are needed to encode the hemoglobin protein: one for the α-polypeptide and one for the β-polypeptide. Beadle and Tatum's hypothesis was therefore later restated as the **one gene–one polypeptide hypothesis**. It is important to keep the distinction between protein, the functional collection of polypeptides, and polypeptide, the molecule encoded by a gene, clear in your mind as we discuss transcription and translation in the rest of this chapter.

13.1b The Pathway from Gene to Polypeptide Involves Transcription and Translation

The pathway from gene to polypeptide has two major steps, transcription and translation. **Transcription** is the mechanism by which the information encoded in DNA is made into a complementary RNA copy. It is called transcription because the information in one nucleic acid type is transferred to another nucleic acid type. **Translation** is the use of the information encoded in the RNA to assemble amino acids into a polypeptide. It is called translation because the information in a nucleic acid, in the form of nucleotides, is converted into a different kind of molecule—amino acids. In 1956, Francis Crick gave the name Central Dogma to the flow of information from DNA to RNA to protein.

In transcription, the enzyme **RNA polymerase** creates an RNA sequence that is complementary to the DNA sequence of a given gene. The process follows the same basic rules of complementary base-pairing and nucleic acid chemistry that we first encountered in DNA replication (see Chapter 12). For each of the several thousand genes that will be appropriate to express in a given cell, one DNA strand or the other is the **template strand** and is read by the RNA polymerase. The RNA transcribed from a gene encoding a polypeptide is called **messenger RNA (mRNA)**.

In translation, an mRNA associates with a **ribosome,** a particle on which amino acids are linked into polypeptide chains. As the ribosome moves along the mRNA, the amino acids specified by the mRNA are joined one by one to form the polypeptide encoded by the gene.

The processes of transcription and translation are similar in prokaryotic and eukaryotic cells **(Figure 13.3).** One key difference is that whereas prokaryotic cells can transcribe and translate a given gene simultaneously, eukaryotic cells transcribe and process mRNA in the nucleus before exporting it to the cytoplasm for translation on ribosomes.

13.1c The Genetic Code Is Written in Three-Letter Words Using a Four-Letter Alphabet

Conceptually, the transcription of DNA into RNA is straightforward. The DNA "alphabet" consists of the four letters A, T, G, and C, representing the four DNA nucleotide bases, adenine, thymine, guanine, and cytosine, and the RNA alphabet consists of the four letters A, U, G, and C, representing the four RNA bases, adenine, uracil, guanine, and cytosine. In other words, both nucleic acids share three of the four bases but differ in the other one: T in DNA is equivalent to U in RNA. But whereas there are 4 RNA bases, there are 20 amino acids. How is nucleotide information in an mRNA translated into the amino acid sequence of a polypeptide?

Breaking the Genetic Code. The nucleotide information that specifies the amino acid sequence of a polypeptide is called the **genetic code**. Scientists hypothesized that the 4 bases in an mRNA (A, U, G, C) would have to be used in combinations of at least 3 to provide the capacity to code for 20 amino acids. One- and 2-letter words were eliminated because if the code used 1-letter words, only 4 different amino acids could be specified (that is, 4^1); if 2-letter words were used, only 16 different amino acids could be specified (that is, 4^2). But if the code used 3-letter words, 64 different amino acids could be specified (that is, 4^3), more than enough to specify 20 amino acids. We know now that the genetic code is indeed a three-letter code; each three-letter word (triplet) is called a **codon**. **Figure 13.4** illustrates the relationship among a gene, codons in an mRNA, and the amino acid sequence of a polypeptide. Genetic information in DNA is first transcribed into complementary three-letter RNA codons (the RNA complement to adenine [A] in the template strand is uracil [U] instead of thymine [T]).

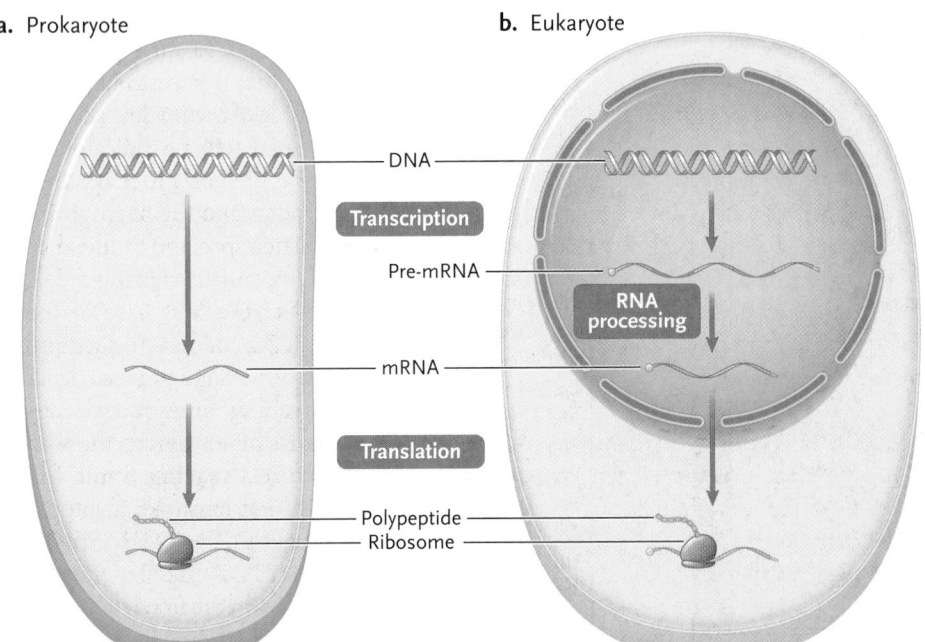

a. Prokaryote

b. Eukaryote

DNA

Transcription

Pre-mRNA

RNA processing

mRNA

Translation

Polypeptide
Ribosome

Figure 13.3

Transcription and translation in **(a)** prokaryotic and **(b)** eukaryotic cells. In prokaryotic cells, RNA polymerase synthesizes an mRNA molecule that is immediately available for translation on ribosomes. In eukaryotes, RNA polymerase synthesizes a precursor–mRNA (pre-mRNA molecule) containing extra segments that are removed by RNA processing to produce a translatable mRNA. That mRNA exits the nucleus through a nuclear pore and is translated on ribosomes in the cytoplasm. Note that only a small segment of DNA is shown. In prokaryotic cells the chromosome is circular.

CONCEPT FIX The template strand for a given gene is always read 3′ to 5′. For gene *a* in Figure 13.4, the bottom strand is the template and is therefore read left to right. However, the template for gene *b* might be the top strand; RNA polymerase would then have to be read right to left. ⬡

How do the codons correspond to the amino acids? Marshall Nirenberg and Philip Leder of the National Institutes of Health (NIH) in the United States established the identity of most of the codons in 1964. These researchers found that short, artificial mRNAs of codon length—three nucleotides— could bind to ribosomes in a test tube and cause a single transfer RNA (tRNA), with its linked amino acid, to bind to the ribosome. (As we will discuss in Section 13.4, tRNAs are a special class of RNA molecules that bring amino acids to the ribosome for assembly into the polypeptide chain.) Nirenberg and Leder then made 64 of the short mRNAs, each consisting of a different, single codon. They added the mRNAs, one at a time, to a mixture in a test tube containing ribosomes and all the different tRNAs, each linked to its own amino acid. The idea was that, from the mixture of tRNAs, each single-codon mRNA would link to the tRNA carrying the amino acid corresponding to the codon. The experiment worked for 50 of the 64 codons, allowing those codons to be assigned to amino acids definitively.

Another approach, carried out in 1966 by H. Ghobind Khorana and his coworkers, used long, artificial mRNA molecules containing only one nucleotide repeated continuously or different nucleotides in repeating patterns. Each artificial mRNA was added to ribosomes in a test tube, and the

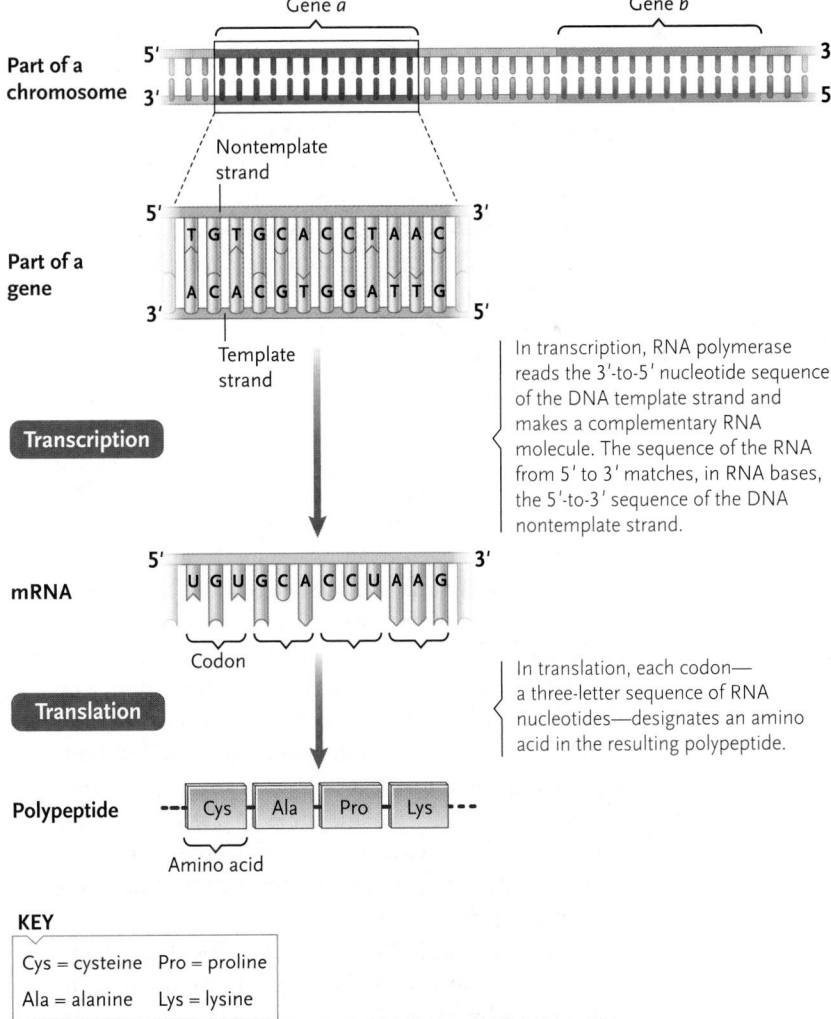

Gene *a*   Gene *b*

Part of a chromosome

Nontemplate strand

Part of a gene

TGTGCACCTAAC
ACACGTGGATTG

Template strand

In transcription, RNA polymerase reads the 3′-to-5′ nucleotide sequence of the DNA template strand and makes a complementary RNA molecule. The sequence of the RNA from 5′ to 3′ matches, in RNA bases, the 5′-to-3′ sequence of the DNA nontemplate strand.

Transcription

mRNA

UGUGCACCUAAG

Codon

In translation, each codon— a three-letter sequence of RNA nucleotides—designates an amino acid in the resulting polypeptide.

Translation

Polypeptide

Cys Ala Pro Lys

Amino acid

KEY

Cys = cysteine   Pro = proline
Ala = alanine   Lys = lysine

Figure 13.4

Relationship among a gene, codons in an mRNA, and the amino acid sequence of a polypeptide.

Second base of codon

First base of codon	U	C	A	G	Third base of codon
U	UUU Phe	UCU Ser	UAU Tyr	UGU Cys	U
	UUC Phe	UCC Ser	UAC Tyr	UGC Cys	C
	UUA Leu	UCA Ser	UAA	UGA	A
	UUG Leu	UCG Ser	UAG	UGG Trp	G
C	CUU Leu	CCU Pro	CAU His	CGU Arg	U
	CUC Leu	CCC Pro	CAC His	CGC Arg	C
	CUA Leu	CCA Pro	CAA Gln	CGA Arg	A
	CUG Leu	CCG Pro	CAG Gln	CGG Arg	G
A	AUU Ile	ACU Thr	AAU Asn	AGU Ser	U
	AUC Ile	ACC Thr	AAC Asn	AGC Ser	C
	AUA Ile	ACA Thr	AAA Lys	AGA Arg	A
	AUG Met	ACG Thr	AAG Lys	AGG Arg	G
G	GUU Val	GCU Ala	GAU Asp	GGU Gly	U
	GUC Val	GCC Ala	GAC Asp	GGC Gly	C
	GUA Val	GCA Ala	GAA Glu	GGA Gly	A
	GUG Val	GCG Ala	GAG Glu	GGG Gly	G

KEY

Ala = alanine
Arg = arginine
Asn = asparagine
Asp = aspartic acid
Cys = cysteine
Gln = glutamine
Glu = glutamic acid
Gly = glycine
His = histidine
Ile = isoleucine
Leu = leucine
Lys = lysine
Met = methionine
Phe = phenylalanine
Pro = proline
Ser = serine
Thr = threonine
Trp = tryptophan
Tyr = tyrosine
Val = valine

Figure 13.5
The genetic code, written as the codons appear in mRNA being read 5′ to 3′. The AUG initiator codon, which codes for methionine, is shown in green; the three terminator codons are boxed in red.

sequence of amino acids in the polypeptide chain made by the ribosomes was analyzed. For example, an artificial mRNA containing only uracil nucleotides in the sequence UUUUUU... resulted in a polypeptide containing only the amino acid phenylalanine; they deduced that UUU must be the codon for phenylalanine. Khorana's approach, combined with the results of Nirenberg and Leder's experiments, identified the coding assignments of all the codons. Nirenberg and Khorana received a Nobel Prize in 1968 for solving the nucleic acid code.

Features of the Genetic Code. By convention, scientists write the codons in the 5′ → 3′ direction as they appear in mRNAs, substituting U for the T of DNA **(Figure 13.5)**. Of the 64 codons, 61 specify amino acids. One of these codons, AUG, specifies the amino acid methionine. It is the first codon translated in any mRNA in both prokaryotic cells and eukaryotes. In that position, AUG is called a **start** or **initiator codon**. The three codons that do not specify amino acids—UAA, UAG, and UGA—are **stop codons** (also called **nonsense** or **termination codons**) that act as "periods" indicating the end of a polypeptide-encoding sentence. When a ribosome reaches one of the stop codons, polypeptide synthesis stops and the new polypeptide chain is released from the ribosome.

Only two amino acids, methionine and tryptophan, are specified by a single codon. All the rest are represented by at least two; some by as many as six. In other words, there are many synonyms in the nucleic acid code, a feature known as **degeneracy** (or redundancy). For example, UGU and UGC both specify cysteine, and CCU, CCC, CCA, and CCG all specify proline.

Another feature of the genetic code is that it is **commaless**; that is, the words of the nucleic acid code are sequential, with no indicators such as commas or spaces to mark the end of one codon and the beginning of the next. Therefore, the code can be read correctly only by starting at the right place—at the first base of the first three-letter codon at the beginning of a coded message (the start codon)—and reading three nucleotides at a time. In other words, there is only one correct **reading frame** for each mRNA. For example, if you read the message SADMOMHASMOPCUTOFFBOYTOT three letters at a time, starting with the first letter of the first "codon," you would find that a mother reluctantly had her small child's hair cut. However, if you start incorrectly at the second letter of the first codon, you read the gibberish message ADM OMH ASM OPC UTO FFB OYT OT.

The code is also **universal.** With a few exceptions, the same codons specify the same amino acids in all living organisms, and even in viruses. The universality of the nucleic acid code indicates that it was established in its present form very early in the evolution of life and has remained virtually unchanged through billions of years of evolutionary history. Consistency in the genetic code makes genetic engineering possible. In Chapter 15, you will see how genes from one organism can be transferred to, and interpreted, by another.

STUDY BREAK

1. On the basis of their work with auxotrophic mutants of the fungus *Neurospora crassa,* Beadle and Tatum proposed the one gene–one enzyme hypothesis. Why was this hypothesis updated subsequently to the one gene–one polypeptide hypothesis?
2. Why is the sequence of bases in the mRNA different from that in the DNA of a given gene?

13.2 Transcription: DNA-Directed RNA Synthesis

Transcription is the process by which information coded in sequential DNA bases is transferred to a complementary RNA strand. Although certain aspects of this mechanism **(Figure 13.6)** are similar to those of

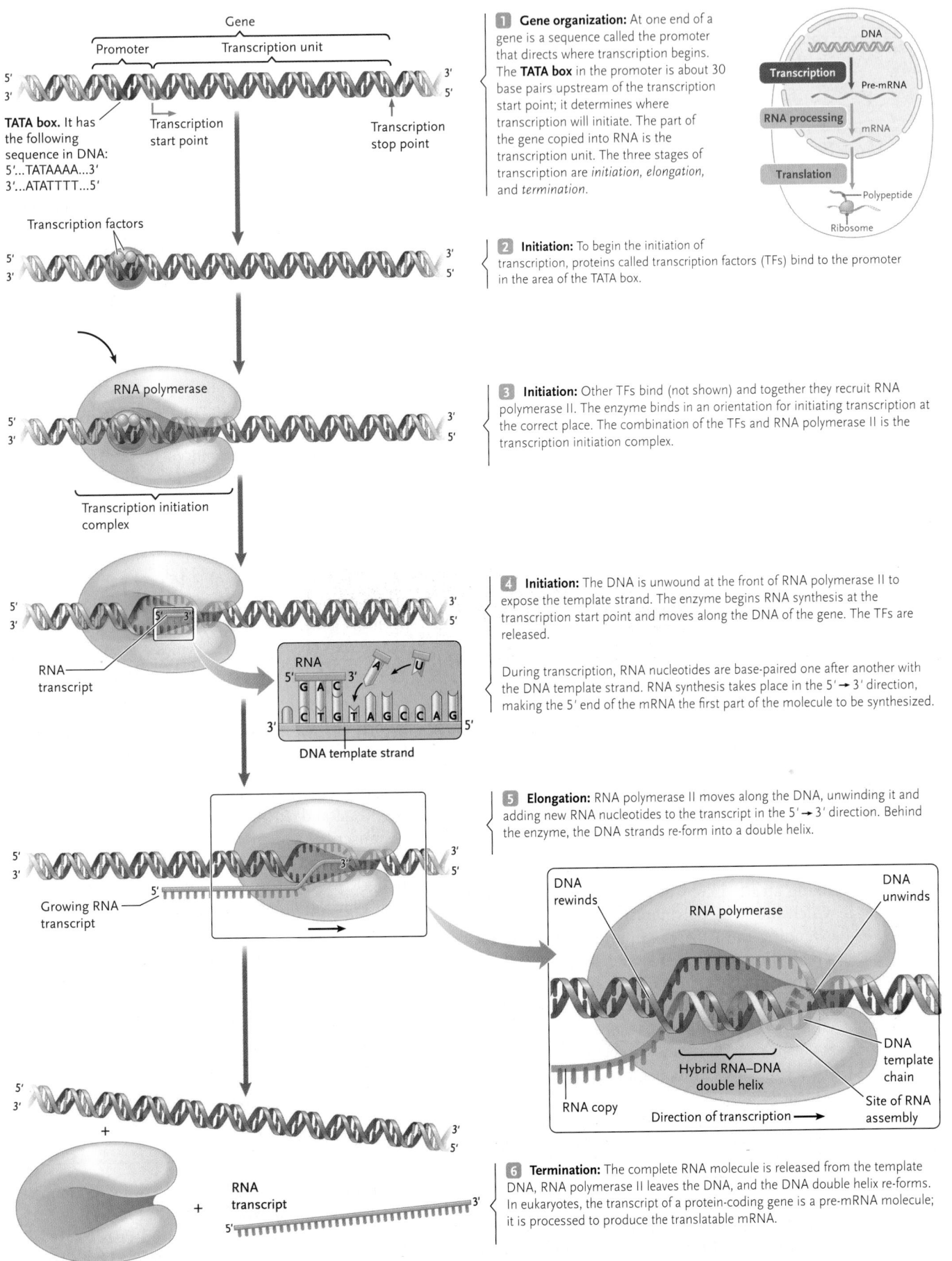

Gene organization: At one end of a gene is a sequence called the promoter that directs where transcription begins. The **TATA box** in the promoter is about 30 base pairs upstream of the transcription start point; it determines where transcription will initiate. The part of the gene copied into RNA is the transcription unit. The three stages of transcription are *initiation*, *elongation*, and *termination*.

TATA box. It has the following sequence in DNA:
5'...TATAAAA...3'
3'...ATATTTT...5'

Initiation: To begin the initiation of transcription, proteins called transcription factors (TFs) bind to the promoter in the area of the TATA box.

Initiation: Other TFs bind (not shown) and together they recruit RNA polymerase II. The enzyme binds in an orientation for initiating transcription at the correct place. The combination of the TFs and RNA polymerase II is the transcription initiation complex.

Initiation: The DNA is unwound at the front of RNA polymerase II to expose the template strand. The enzyme begins RNA synthesis at the transcription start point and moves along the DNA of the gene. The TFs are released.

During transcription, RNA nucleotides are base-paired one after another with the DNA template strand. RNA synthesis takes place in the 5' → 3' direction, making the 5' end of the mRNA the first part of the molecule to be synthesized.

Elongation: RNA polymerase II moves along the DNA, unwinding it and adding new RNA nucleotides to the transcript in the 5' → 3' direction. Behind the enzyme, the DNA strands re-form into a double helix.

Termination: The complete RNA molecule is released from the template DNA, RNA polymerase II leaves the DNA, and the DNA double helix re-forms. In eukaryotes, the transcript of a protein-coding gene is a pre-mRNA molecule; it is processed to produce the translatable mRNA.

Figure 13.6

Transcription of a eukaryotic protein-coding gene. Transcription has three stages: initiation, elongation, and termination. RNA polymerase moves along the gene, separating the two DNA strands to allow RNA synthesis in the 5' → 3' direction using the 3' → 5' DNA strand as template.

DNA replication (see Figure 12.15), it is important for you to understand how these processes are different. In transcription,

- for a given gene, *only one of the two DNA nucleotide strands acts as a template* for synthesis of a complementary copy, instead of both, as in replication.
- only a relatively small part of a DNA molecule—the sequence encoding a single gene—serves as a template, rather than all of both strands, as in DNA replication.
- RNA polymerases catalyze the assembly of nucleotides into an RNA strand, rather than the DNA polymerases that catalyze replication.
- the RNA molecules resulting from transcription are single polynucleotide chains, not double ones, as in DNA replication.
- wherever adenine appears in the DNA template chain, a uracil is matched to it in the RNA transcript instead of thymine, as in DNA replication.

Although the mechanism of transcription is similar in prokaryotic cells and eukaryotes, watch for the important differences pointed out in this section.

13.2a Transcription Proceeds in Three Steps

Figure 13.6 illustrates the general structure of a eukaryotic protein-coding gene and shows how it is transcribed. The gene consists of two main parts, a **promoter**, which is a control sequence for transcription, and a **transcription unit**, the section of the gene that is copied into an RNA molecule. Transcription takes place in three steps: (1) initiation, in which the molecular machinery that carries out transcription assembles at the promoter and begins synthesizing an RNA copy of the gene; (2) elongation, in which the RNA polymerase moves along the gene extending the RNA chain; and (3) termination, in which transcription ends and the RNA molecule—the transcript—and the RNA polymerase are released from the DNA template. Roger Kornberg of Stanford University in California received a Nobel Prize in 2006 for describing the molecular structure of the eukaryotic transcription apparatus and how it acts in transcription.

Similarities and differences in transcription of eukaryotic and bacterial protein-coding genes are as follows:

- Gene organization is the same, although the specific sequences in the promoter where the transcription apparatus assembles differ.
- In eukaryotes, RNA polymerase II, the enzyme that transcribes protein-coding genes, cannot bind directly to DNA; it is recruited to the promoter once proteins called **transcription factors** have bound. In bacteria, RNA polymerase binds directly to DNA; it is directed to the promoter by a protein factor that is then released once transcription begins.
- Elongation is essentially identical in the two types of organisms.
- In prokaryotic cells, there are two types of specific DNA sequences, called **terminators**, that signal the end of transcription of the gene. Both types of terminator sequences act *after they are transcribed*. In the first case, the terminator sequence on the mRNA uses complementary base-pairing with itself to form a "hairpin." In the second case, a protein binds to a particular terminator sequence on the mRNA. Both of these mechanisms trigger the termination of transcription and the release of the RNA and RNA polymerase from the template. In eukaryotes, there are no equivalent "transcription terminator" sequences. Instead, the 3′ end of the mRNA is specified by a different process, which is discussed in a later section.

Once an RNA polymerase molecule has started transcription and progressed past the beginning of a gene, another molecule of RNA polymerase may start transcribing as soon as there is room at the promoter. In most genes this process continues until there are many RNA polymerase molecules spaced closely along a gene, each making an RNA transcript.

13.2b Transcription of Non–Protein Coding Genes Occurs in a Similar Way

Non–protein coding genes include, for example, those for tRNAs and rRNAs. In eukaryotes, RNA polymerase II transcribes protein-coding genes, RNA polymerase III transcribes tRNA genes and the gene for one of the four rRNAs, and RNA polymerase I transcribes the genes for the three other rRNAs. The promoters for these non–protein coding genes are different from those of protein-coding genes, being specialized for the assembly of the transcription machinery that involves the correct RNA polymerase type. In bacteria a single type of RNA polymerase transcribes all types of genes. The promoters for bacterial non–protein coding genes are essentially the same as those of protein-coding genes.

STUDY BREAK

1. If the DNA template strand has the sequence 3′-CAAATTGGCTTATTACCGGATG-5′, what is the sequence of an RNA transcribed from it?
2. What is the role of the promoter in transcription?

13.3 Processing of mRNAs in Eukaryotes

Although mRNAs obviously contain regions that code for protein, they also contain noncoding regions that, although not specifying an amino acid, nevertheless play key roles in the process of protein synthesis. For instance, in prokaryotic mRNAs the coding region is flanked by untranslated ends, the 5′ untranslated region (5′ UTR) and a 3′ untranslated region (3′ UTR). These same elements are present in eukaryotic mRNAs along with additional types of noncoding elements. The following section looks at the structure and function of genes, with particular focus on the synthesis of mRNA in eukaryotes.

13.3a Eukaryotic Protein-Coding Genes Are Transcribed into Precursor mRNAs That Are Modified in the Nucleus

A eukaryotic protein-coding gene is typically transcribed into a **precursor mRNA (pre-mRNA)** that must be processed in the nucleus to produce translatable mRNA (see Figure 13.3, p. 293, **Figure 13.7,** and **Figure 13.8, p. 298**). The mature mRNA exits the nucleus and is translated by ribosomes in the cytoplasm.

Modifications of Pre-mRNA and mRNA Ends. At the 5′ end of the pre-mRNA is the 5′ guanine cap, consisting of a guanine-containing nucleotide that is reversed so that its 3′ –OH group faces the beginning rather than the end of the molecule. A capping enzyme adds this **5′ cap** to the pre-mRNA (without the need for complementary base-pairing) soon after RNA polymerase II begins transcription. The cap, which is connected to the rest of the chain by three phosphate groups, remains when pre-mRNA is processed to mRNA. The cap protects the mRNA from degradation and is the site where ribosomes attach at the start of translation.

Transcription of a eukaryotic protein-coding gene is terminated differently from that of a prokaryotic gene (Figure 13.7). The eukaryotic gene has no terminator sequence in the DNA that, after transcription into RNA, signals RNA polymerase to stop transcribing. Instead, near the 3′ end of the gene is a DNA sequence that is transcribed into the pre-mRNA. Proteins bind to this *polyadenylation signal* in the RNA and cleave it just downstream. This signals the RNA polymerase to stop transcription. Then the enzyme poly(A) polymerase adds a chain of 50 to 250 adenine nucleotides, one nucleotide at a time, to the newly created 3′ end of the pre-mRNA.

CONCEPT FIX No complementary base-pairing with a template is needed for this particular type of RNA synthesis. There is no "poly(T)" sequence in the DNA corresponding to the poly(A) sequence in the pre-mRNA.

Figure 13.7

Relationship between a eukaryotic protein-coding gene, the pre-mRNA transcribed from it, and the mRNA processed from the pre-mRNA.

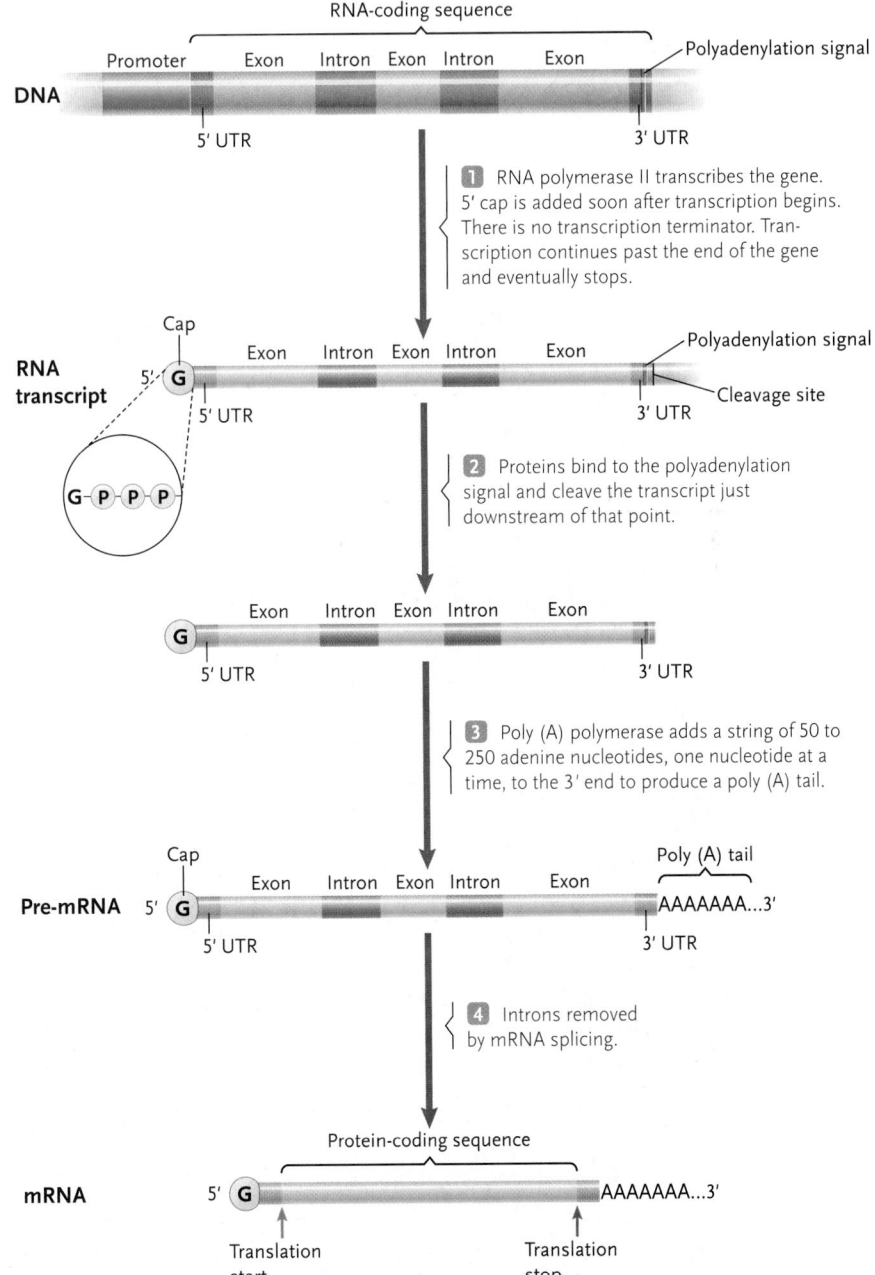

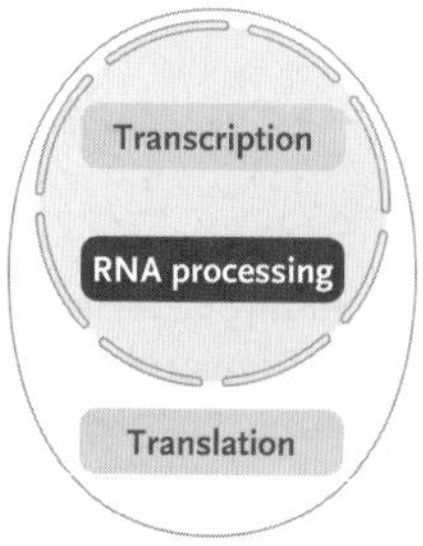

The string of adenine nucleotides, called the **poly (A) tail**, enables the mRNA produced from the pre-mRNA to be translated efficiently and protects it from attack by RNA-digesting enzymes in the cytoplasm.

Sequences Interrupting the Protein-Coding Sequence. The transcription unit of a protein-coding gene—the RNA-coding sequence—also contains non–protein coding sequences called **introns** that interrupt the protein-coding sequence (shown in Figure 13.8). The introns are transcribed into pre-mRNAs but are removed from pre-mRNAs during processing in the nucleus. The amino acid–coding sequences that are retained in finished mRNAs are called **exons**. The mechanisms by which introns originated in genes remain a mystery.

Introns were discovered by several methods, including direct comparisons between the nucleotide sequences of mature mRNAs and either pre-mRNAs or the genes encoding them. Although many eukaryotic genes do not contain introns, most have at least one; some contain more than 60. In humans, introns are, on average, 6 times the length of exons. The original discoverers of introns, Richard Roberts and Phillip Sharp, received a Nobel Prize in 1993 for their findings.

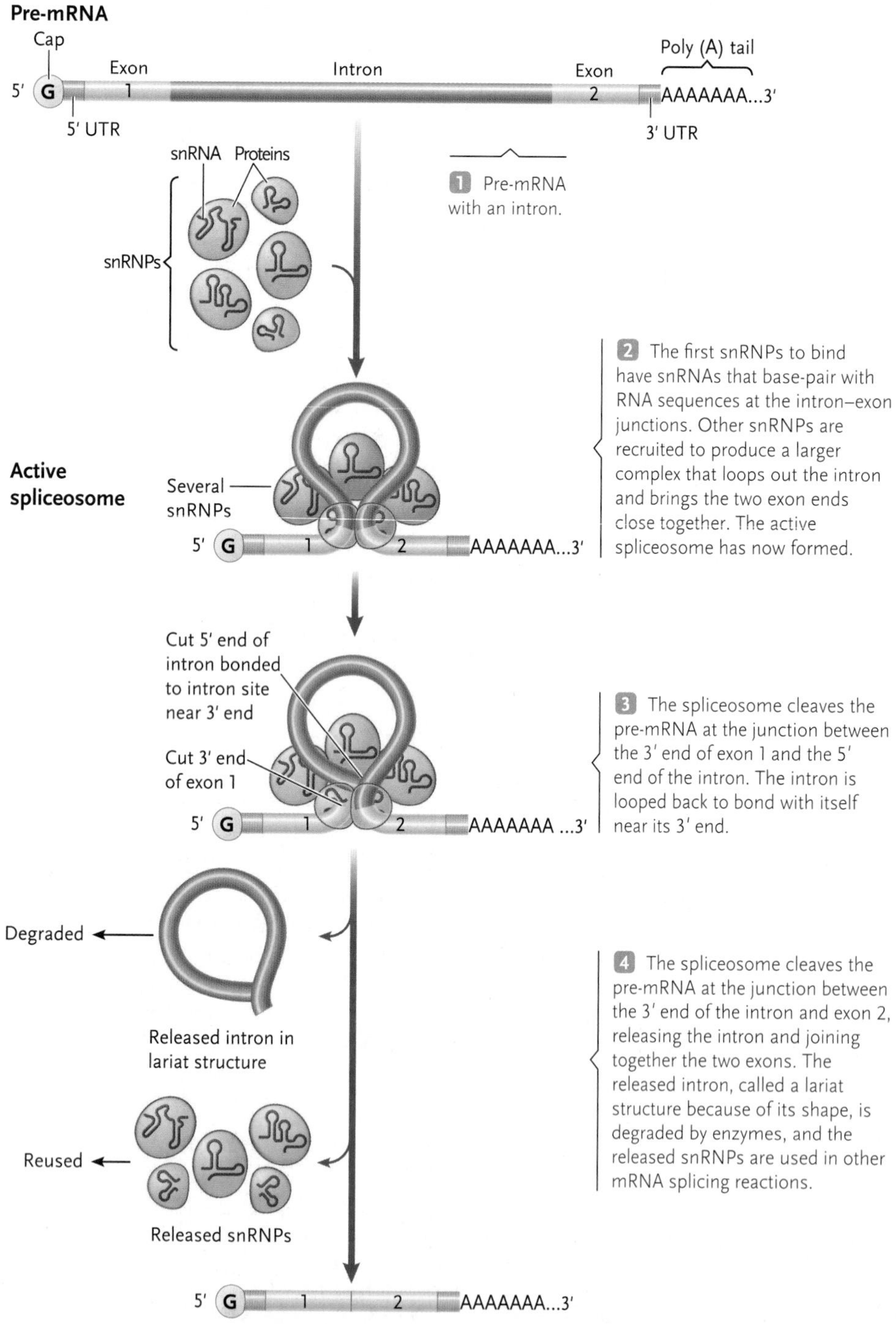

13.3b Introns Are Removed during Pre-mRNA Processing to Produce Translatable mRNA

A process called **mRNA splicing**, which occurs in the nucleus, removes introns from pre-mRNAs and joins exons together. As an illustration of one type of mRNA splicing, Figure 13.8 shows the processing of a pre-mRNA with a single intron to produce a mature mRNA. mRNA splicing takes place in a **spliceosome**, a complex formed between the pre-mRNA and a handful of **small ribonucleoprotein particles**. A ribonucleoprotein particle is a complex of RNA and proteins. The small ribonucleoprotein particles involved in mRNA splicing are located in the nucleus; each consists of a relatively short *small nuclear RNA* (snRNA) bound to a number of proteins. The particles are therefore known as snRNPs, pronounced "snurps." The snRNPs bind in a particular order to an intron in the pre-mRNA and form the active spliceosome. The spliceosome cleaves the pre-mRNA to release the intron, and joins the flanking exons.

Complementary base-pairing between regions of snRNA and mRNA ensures that the cutting and splicing are so exact that not a single base of an intron is retained in the finished mRNA, nor is a single base removed from the exons. Without this precision, removing introns would change the reading frame of the coding portion of the mRNA, producing the wrong codons from the point of a mistake onward.

13.3c Introns Contribute to Protein Variability

Introns seem wasteful in terms of the energy and raw materials required to replicate and transcribe them and the elaborate cellular machinery

Figure 13.8

mRNA splicing—the removal from pre-mRNA of introns and joining of exons in the spliceosome.

1 Pre-mRNA with an intron.

2 The first snRNPs to bind have snRNAs that base-pair with RNA sequences at the intron–exon junctions. Other snRNPs are recruited to produce a larger complex that loops out the intron and brings the two exon ends close together. The active spliceosome has now formed.

3 The spliceosome cleaves the pre-mRNA at the junction between the 3' end of exon 1 and the 5' end of the intron. The intron is looped back to bond with itself near its 3' end.

4 The spliceosome cleaves the pre-mRNA at the junction between the 3' end of the intron and exon 2, releasing the intron and joining together the two exons. The released intron, called a lariat structure because of its shape, is degraded by enzymes, and the released snRNPs are used in other mRNA splicing reactions.

junctions often fall at points dividing major functional regions in encoded proteins. The functional divisions may have allowed new proteins to evolve by exon shuffling, a process by which existing protein regions or domains, already selected for due to their useful functions, are mixed into novel combinations to create new proteins. Evolution by this mechanism would produce new proteins with novel functions much more quickly than by changes in individual nucleotides at random points (see Section 16.4b).

STUDY BREAK

1. What are the similarities and differences between pre-mRNAs and mRNAs?
2. What is the role of base-pairing in mRNA splicing?
3. How is it possible for an organism to produce more proteins than it has genes for?

13.4 Translation: mRNA-Directed Polypeptide Synthesis

Translation is the assembly of amino acids into polypeptides on ribosomes. In prokaryotic organisms, translation takes place throughout the cell, whereas in eukaryotes it occurs in the cytoplasm. (However, a few specialized genes are transcribed and translated in mitochondria and chloroplasts.)

Figure 13.10 summarizes the translation process. In prokaryotic cells, the mRNA produced by transcription is not confined within a nucleus and is therefore available immediately for translation. However, for eukaryotes the mRNA produced by splicing of the pre-mRNA first exits the nucleus and is then translated in the cytoplasm. In translation, the mRNA associates with a ribosome and another type of RNA, transfer RNA (tRNA), which brings amino acids to the complex to be joined, one by one, into the polypeptide chain. The sequence of amino acids in the polypeptide chain is determined by the sequence of codons in the mRNA. The mRNA is read from the 5′ end to the 3′ end; the polypeptide is assembled from the **N-terminal end** to the **C-terminal end**.

In this section, we will start by discussing the key players in the process, the tRNAs and ribosomes, and then walk through the translation process from a start codon to a stop codon.

13.4a tRNAs Are Small RNAs of a Highly Distinctive Structure That Bring Amino Acids to the Ribosome

Transfer RNAs (tRNAs) bring amino acids to the ribosome for addition to the polypeptide chain.

tRNA Structure. tRNAs are small RNAs, about 75 to 90 nucleotides long (mRNAs are typically hundreds of nucleotides long), with a highly distinctive structure that accomplishes their role in translation **(Figure 13.11)**. All tRNAs can base-pair with themselves to wind into four double-helical segments, forming a cloverleaf pattern in two dimensions. At the tip of one of the double-helical segments is the **anticodon**, the three-nucleotide segment that base-pairs with a codon in mRNAs. At the other end of the cloverleaf, opposite the anticodon, is a free 3′ end of the molecule that links to the amino acid corresponding to the anticodon. For example, a tRNA that

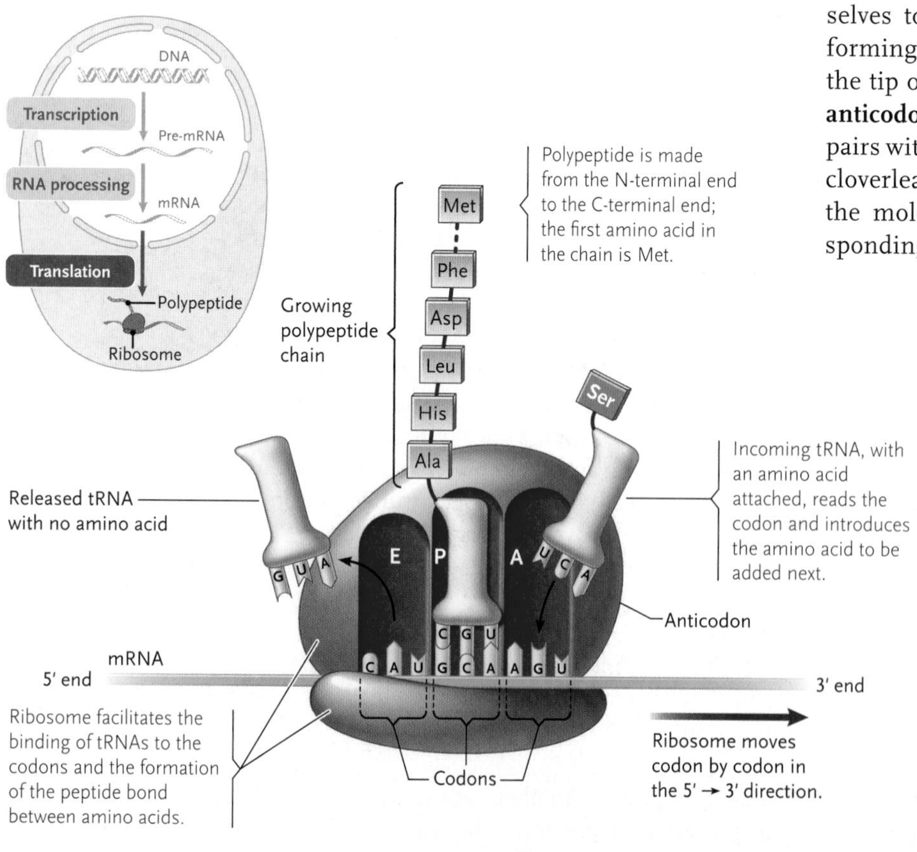

Polypeptide is made from the N-terminal end to the C-terminal end; the first amino acid in the chain is Met.

Growing polypeptide chain

Released tRNA with no amino acid

Incoming tRNA, with an amino acid attached, reads the codon and introduces the amino acid to be added next.

Anticodon

mRNA

5′ end

3′ end

Ribosome facilitates the binding of tRNAs to the codons and the formation of the peptide bond between amino acids.

Ribosome moves codon by codon in the 5′ → 3′ direction.

Codons

Figure 13.10

An overview of translation, in which ribosomes assemble amino acids into a polypeptide chain. The figure shows a ribosome in the process of translation. A tRNA molecule with an amino acid bound to it is entering the ribosome on the right. The anticodon on the tRNA will pair with the codon in the mRNA. Its amino acid will then be added to the growing polypeptide that is currently attached to the tRNA in the middle of the ribosome. As it assembles a polypeptide chain, the ribosome moves from one codon to the next along the mRNA in the 5′ → 3′ direction.

PEOPLE BEHIND BIOLOGY 13.1

Dr. Steve Zimmerly, *University of Calgary*

Steve Zimmerly and his colleagues think they know where introns came from.

Biology students (and researchers) often wonder about the origins of introns, and to investigate this question, Zimmerly collected and analyzed a large number of examples of a type of intron called "Group II" from plant organelles and bacteria. Group II introns have two fascinating abilities. First, they can splice themselves out of RNA without the need for snRNPs. Second, they are mobile; these elements can copy themselves and insert at a new location **(Figure 1)**.

The Zimmerly lab proposed a model for intron evolution that suggests the introns in nuclear genes of higher eukaryotes evolved from mobile Group II introns originating in prokaryotic cells. These Group II introns may have spread to eukaryotes at the time when their bacterial hosts

were engulfed by eukaryotic cells to become endosymbiotic mitochondria and chloroplasts. Over evolutionary time, the nuclear introns lost their mobility and became dependent on spliceosomes for accurate splicing.

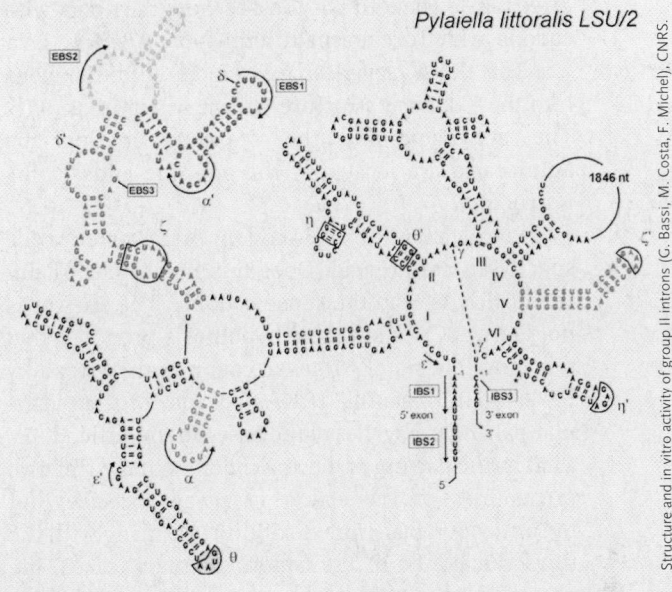

Pylaiella littoralis LSU/2

Structure and in vitro activity of group II introns (G. Bassi, M. Costa, F. Michel), CNRS.

Figure 1

Sequence of a Group II intron showing extensive complementary base-pairing with itself.

---

a. A tRNA molecule in two dimensions (yeast alanine tRNA)

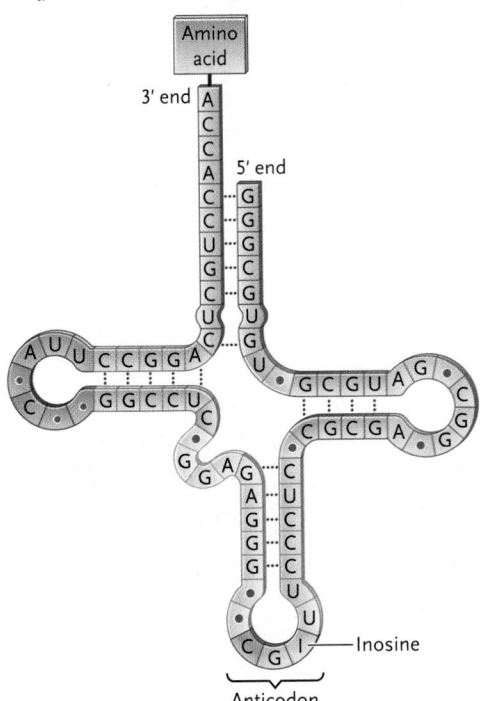

b. A tRNA molecule in three dimensions

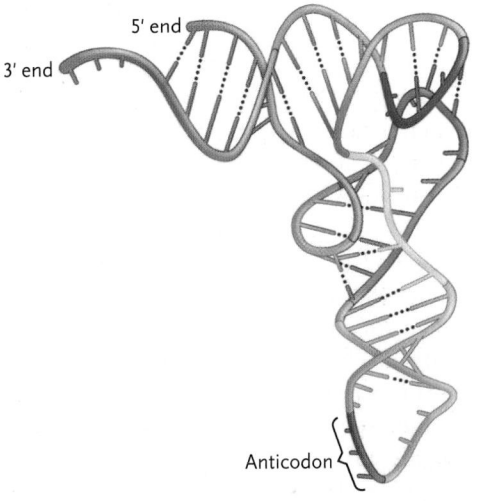

c. How an aminoacyl–tRNA complex is shown in this book

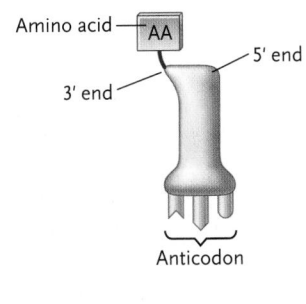

Figure 13.11

tRNA structure. The red dots show sites where bases are chemically modified into other forms; chemical modification of certain bases is typical of tRNAs. This tRNA has the purine inosine (I) in the anticodon, which has relatively loose base-pairing ability, allowing this single tRNA to pair with each of three alanine codons: 5′-GCU-3′, 5′-GCC-3′, and 5′-GCA-3′. This tRNA also has the unusual base pair G–U. Unusual base pairs, allowed by the greater flexibility of short RNA chains, are common in tRNAs.

is linked to serine (Ser) pairs with the codon 5'-AGU-3' in mRNA (see Figure 13.11). The anticodon of the tRNA that pairs with this codon is 3'-UCA-5'.

CONCEPT FIX The anticodon and codon pair in an antiparallel manner, as do the two strands in the DNA helix. We will therefore write anticodons in the 3' → 5' direction to make it easy to see how they pair with codons, which are normally written 5' → 3'.

The tRNA cloverleaf folds in three dimensions into the L-shaped structure shown in Figure 13.11b. The anticodon and the segment binding the amino acid are located at the opposite ends of the L structure.

Recall that 61 of the 64 codons of the genetic code specify an amino acid. Does this mean that 61 different tRNAs read the **sense codons?** The answer is no. Francis Crick's **wobble hypothesis** proposed that the complete set of 61 sense codons can be read by fewer than 61 distinct tRNAs because of the particular pairing properties of the bases in the anticodons. That is, the pairing of the anticodon with the first two nucleotides of the codon is always precise, but the anticodon has more flexibility in pairing with the third nucleotide of the codon. In many cases, the same tRNA's anticodon can read codons that have either U or C in the third position; for example, a tRNA carrying phenylalanine can read both codons 5'-UUU-3' and 5'-UUC-3'. Similarly, the same tRNA's anticodon can read two codons that have A or G in the third position; for example, a tRNA carrying glutamine can pair with both 5'-CAA-3' and 5'-CAG-3' codons. The special purine called inosine in the alanine tRNA shown in Figure 13.11a allows even more extensive wobble by allowing the tRNA to pair with codons that have one of U, C, and A in the third position.

Addition of Amino Acids to Their Corresponding tRNAs. The correct amino acid must be present on a tRNA if translation is to be accurate. The process of adding an amino acid to a tRNA is called **aminoacylation** (literally, the addition of an amino acid) or **charging** (because the process adds free energy as the amino acid–tRNA combinations are formed).

The finished product of charging, a tRNA linked to its "correct" amino acid, is called an **aminoacyl–tRNA**. A collection of different enzymes called **aminoacyl–tRNA synthetases** catalyzes aminoacylation, as shown in **Figure 13.12.** This energy in the aminoacyl–tRNA eventually drives the formation of the **peptide bond** linking amino acids during translation.

With the tRNAs attached to their corresponding amino acids, our attention moves to the ribosome, where the amino acids are removed from tRNAs and linked into polypeptide chains.

13.4b Ribosomes Are rRNA–Protein Complexes That Work as Automated Protein Assembly Machines

Ribosomes are ribonucleoprotein particles that carry out protein synthesis by translating mRNA into chains of amino acids. Like some automated machines, such as those forming complicated metal parts by a series of machining steps, ribosomes use an information tape—an mRNA molecule—as the directions required to accomplish a task. For ribosomes, the task is to join amino acids into ordered sequences to make a polypeptide chain.

In prokaryotic cells, ribosomes carry out their assembly functions throughout the cell. In eukaryotes, ribosomes function only in the cytoplasm, either suspended freely in the cytoplasmic solution or attached to the membranes of the endoplasmic reticulum (ER), the system of membrane-bound tubular or flattened sacs in the cytoplasm. A finished ribosome is made up of two parts of dissimilar size, called the *large* and *small ribosomal subunits* **(Figure 13.13).** Each subunit is made up of a combination of ribosomal RNA (rRNA) and ribosomal proteins.

CONCEPT FIX The endosymbiotic origin of chloroplasts and mitochondria in eukaryotic cells is reflected by the fact that these organelles still code for their own "prokaryotic" ribosomes that are distinct from those in the cytoplasm.

Prokaryotic and eukaryotic ribosomes are similar in structure and function. However, the differences in their molecular structure, particularly in the ribosomal proteins, give them distinct properties. For example, the antibiotics streptomycin and erythromycin are effective antibacterial agents because they inhibit bacterial, but not eukaryotic, ribosomes.

To fulfill its role in translation, the ribosome has special binding sites active in bringing together mRNA with aminoacyl–tRNAs (see Figure 13.13 and refer also to Figure 13.10, p. 300). One such site is where the mRNA threads a bent path through the ribosome. The **A site** (aminoacyl site) is where the incoming aminoacyl–tRNA (carrying the next amino acid to be added to the polypeptide chain) binds to the mRNA. The **P site** (peptidyl site) is where the tRNA carrying the growing polypeptide chain is bound. The **E site** (exit site) is where an exiting tRNA binds as it leaves the ribosome.

13.4c Translation Initiation Brings the Ribosomal Subunits, an mRNA, and the First Aminoacyl–tRNA Together

There are three major stages of translation: *initiation, elongation,* and *termination.* During initiation, the translation components assemble on the start codon of the mRNA. In elongation, the assembled complex reads the string of codons in the mRNA one at a time

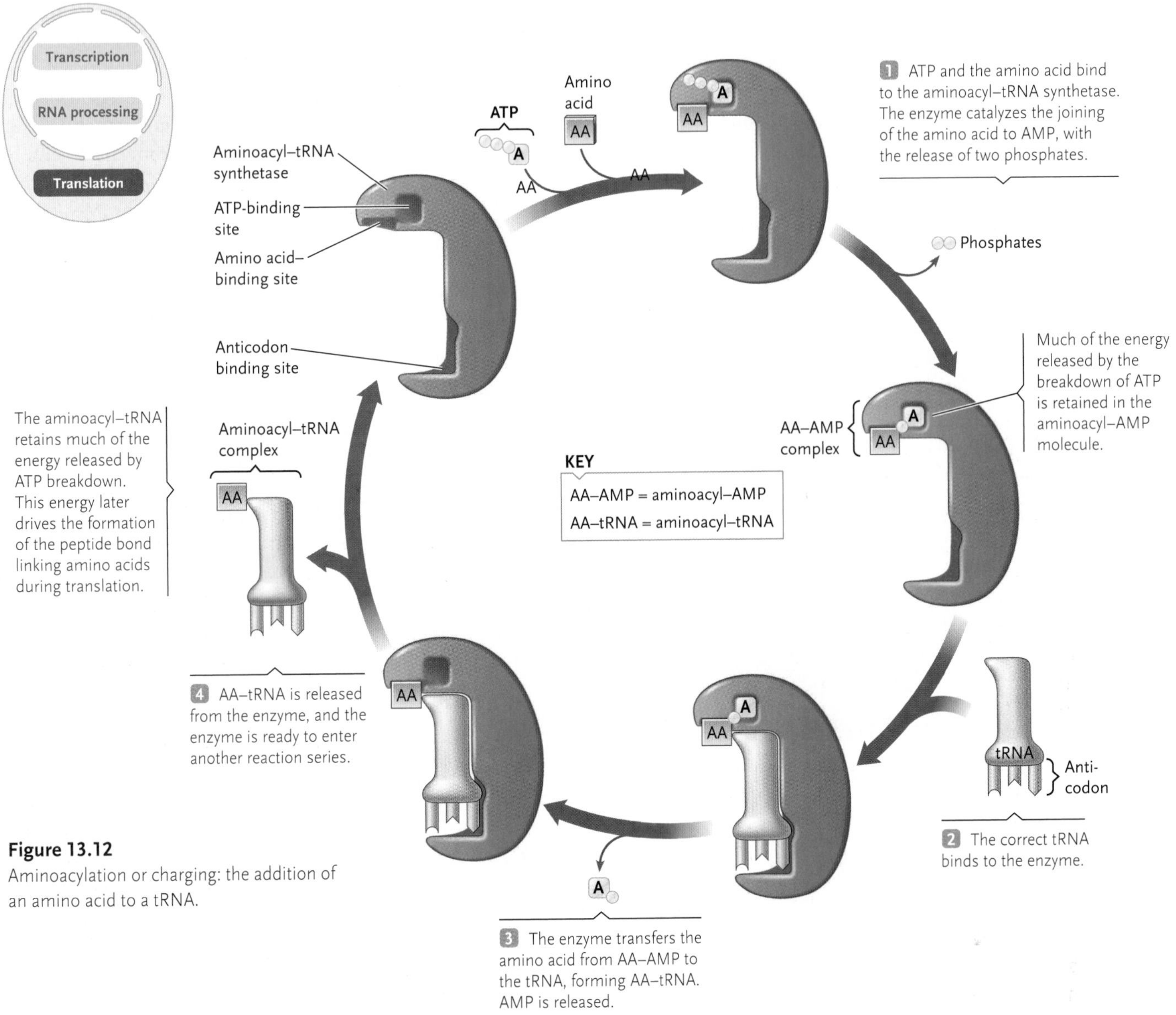

RNA processing

Translation

Aminoacyl–tRNA synthetase

ATP-binding site

Amino acid–binding site

Anticodon binding site

ATP

Amino acid

AA

AA

AA

1 ATP and the amino acid bind to the aminoacyl–tRNA synthetase. The enzyme catalyzes the joining of the amino acid to AMP, with the release of two phosphates.

Phosphates

Much of the energy released by the breakdown of ATP is retained in the aminoacyl–AMP molecule.

AA–AMP complex

The aminoacyl–tRNA retains much of the energy released by ATP breakdown. This energy later drives the formation of the peptide bond linking amino acids during translation.

Aminoacyl–tRNA complex

KEY

AA–AMP = aminoacyl–AMP

AA–tRNA = aminoacyl–tRNA

tRNA

Anti-codon

2 The correct tRNA binds to the enzyme.

4 AA–tRNA is released from the enzyme, and the enzyme is ready to enter another reaction series.

3 The enzyme transfers the amino acid from AA–AMP to the tRNA, forming AA–tRNA. AMP is released.

Figure 13.12
Aminoacylation or charging: the addition of an amino acid to a tRNA.

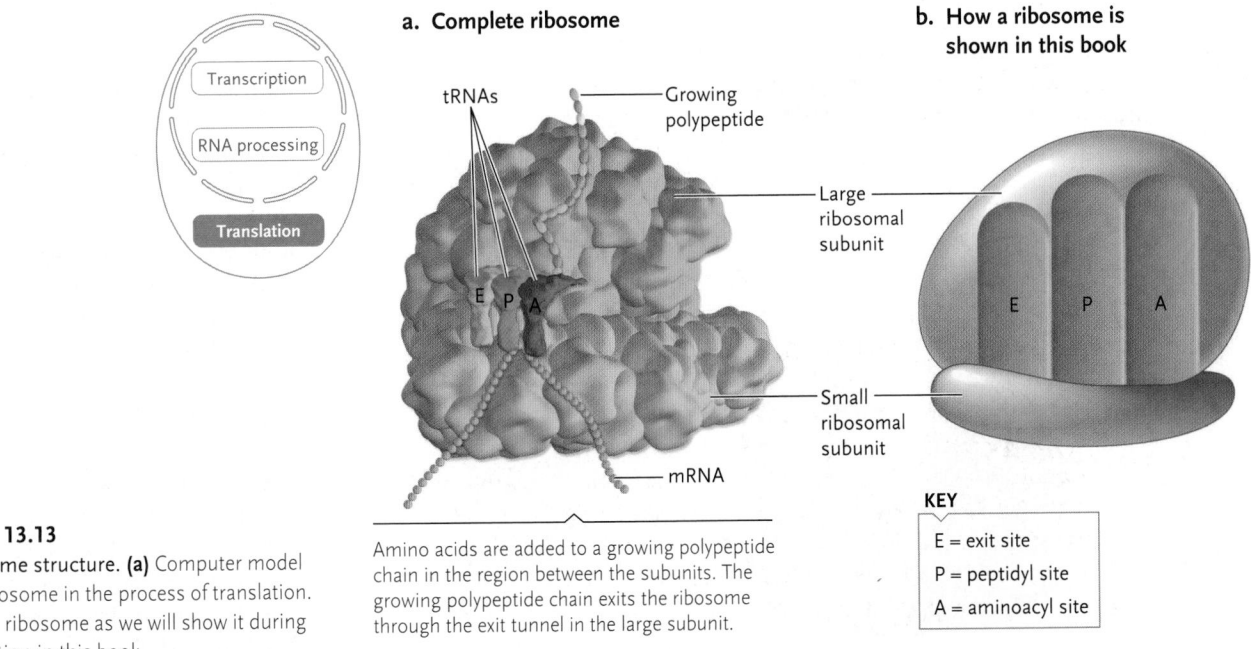

a. Complete ribosome

b. How a ribosome is shown in this book

Transcription

RNA processing

Translation

tRNAs

Growing polypeptide

E P A

mRNA

Large ribosomal subunit

Small ribosomal subunit

E P A

KEY

E = exit site

P = peptidyl site

A = aminoacyl site

Figure 13.13
Ribosome structure. **(a)** Computer model of a ribosome in the process of translation. **(b)** The ribosome as we will show it during translation in this book.

Amino acids are added to a growing polypeptide chain in the region between the subunits. The growing polypeptide chain exits the ribosome through the exit tunnel in the large subunit.

while joining the specified amino acids into the polypeptide. Termination completes the translation process when the complex disassembles after the last amino acid of the polypeptide specified by the mRNA has been added to the polypeptide.

Figure 13.14 illustrates the steps of translation initiation in eukaryotes. In bacteria, translation initiation is similar in using a special initiator Met–tRNA and GTP, but the way in which the ribosome assembles at the start codon is different than in eukaryotes. Rather than scanning from the 5′ end of the mRNA, the small ribosomal subunit, the initiator Met–tRNA and GTP bind directly to the region of the mRNA with the AUG start codon. This initiation complex is then guided by the **ribosome binding site**—a short, specific RNA sequence—just upstream of the start codon on the mRNA that base-pairs with a complementary sequence of rRNA in the small ribosomal subunit. The large ribosomal subunit then binds to the small subunit to complete the ribosome. GTP hydrolysis then begins translation.

After the initiator tRNA pairs with the AUG initiator codon, the subsequent stages of translation simply read the codons one at a time on the mRNA. The initiator tRNA–AUG pairing thus establishes the correct **reading frame**—the series of codons for the polypeptide encoded by the mRNA. This is often referred to as the open reading frame.

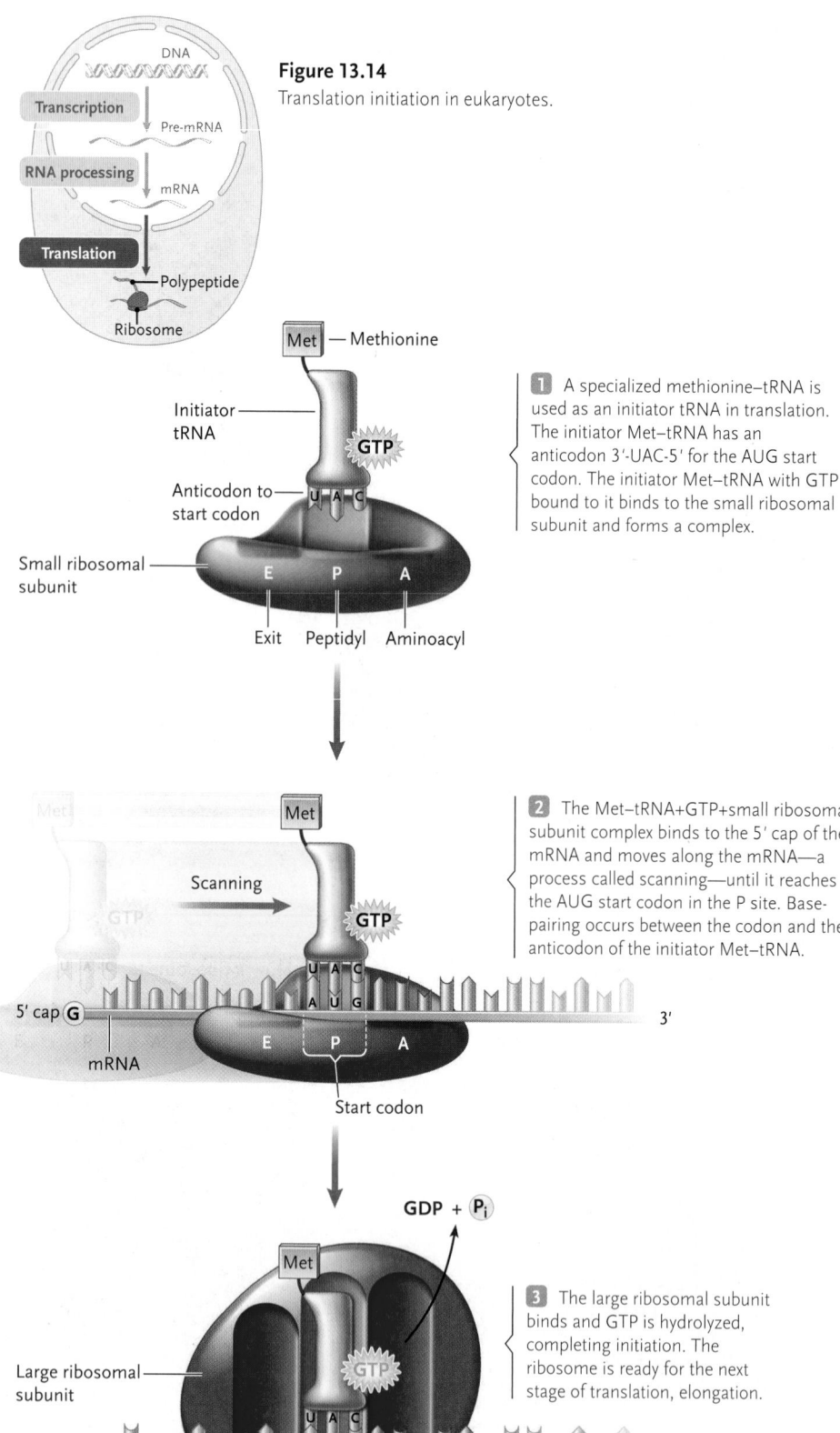

Figure 13.14
Translation initiation in eukaryotes.

1 A specialized methionine–tRNA is used as an initiator tRNA in translation. The initiator Met–tRNA has an anticodon 3′-UAC-5′ for the AUG start codon. The initiator Met–tRNA with GTP bound to it binds to the small ribosomal subunit and forms a complex.

2 The Met–tRNA+GTP+small ribosomal subunit complex binds to the 5′ cap of the mRNA and moves along the mRNA—a process called scanning—until it reaches the AUG start codon in the P site. Base-pairing occurs between the codon and the anticodon of the initiator Met–tRNA.

3 The large ribosomal subunit binds and GTP is hydrolyzed, completing initiation. The ribosome is ready for the next stage of translation, elongation.

13.4d Polypeptide Chains Grow during the Elongation Stage of Translation

The central reactions of translation take place in the elongation stage, which adds amino acids one at a time to a growing polypeptide chain. The individual steps of elongation depend on the binding properties of the P, A, and E sites of the ribosome. The P site, with one exception, can bind only to a **peptidyl–tRNA**, that is, a tRNA linked to a growing polypeptide chain containing two or more amino acids. The exception is the initiator tRNA, which is recognized by the P site as a peptidyl–tRNA even though it carries only a single amino acid, methionine. The A site can bind only to an aminoacyl–tRNA. The tRNA that was previously in the P site is shifted to the E site and then leaves the ribosome.

Figure 13.15 shows the elongation cycle of translation. The cycle begins at the point when an initiator tRNA with its attached methionine is bound to the P site, and the A site is empty (top of figure). The first step in each round of the cycle is the binding of the appropriate aminoacyl–tRNA to the codon in the A site of the ribosome (step 1). This binding is facilitated by a protein **elongation factor (EF)** that is bound to the aminoacyl–tRNA and that is released once the tRNA binds to the codon. Another EF is used when the ribosome translocates along the mRNA to the next codon (step 3). Each EF is released after its job is completed. GTP hydrolysis is used to power the ribosome along the mRNA. In elongation, a peptide bond is formed between the C-terminal end of the growing polypeptide on the P site tRNA and the amino acid on the A site tRNA (step 2). **Peptidyl transferase** catalyzes this reaction. As we noted in an earlier example of the splicing reaction, researchers have demonstrated that the enzyme activity in the large ribosomal subunit is not a protein but a ribozyme (catalytic RNA) within the large subunit.

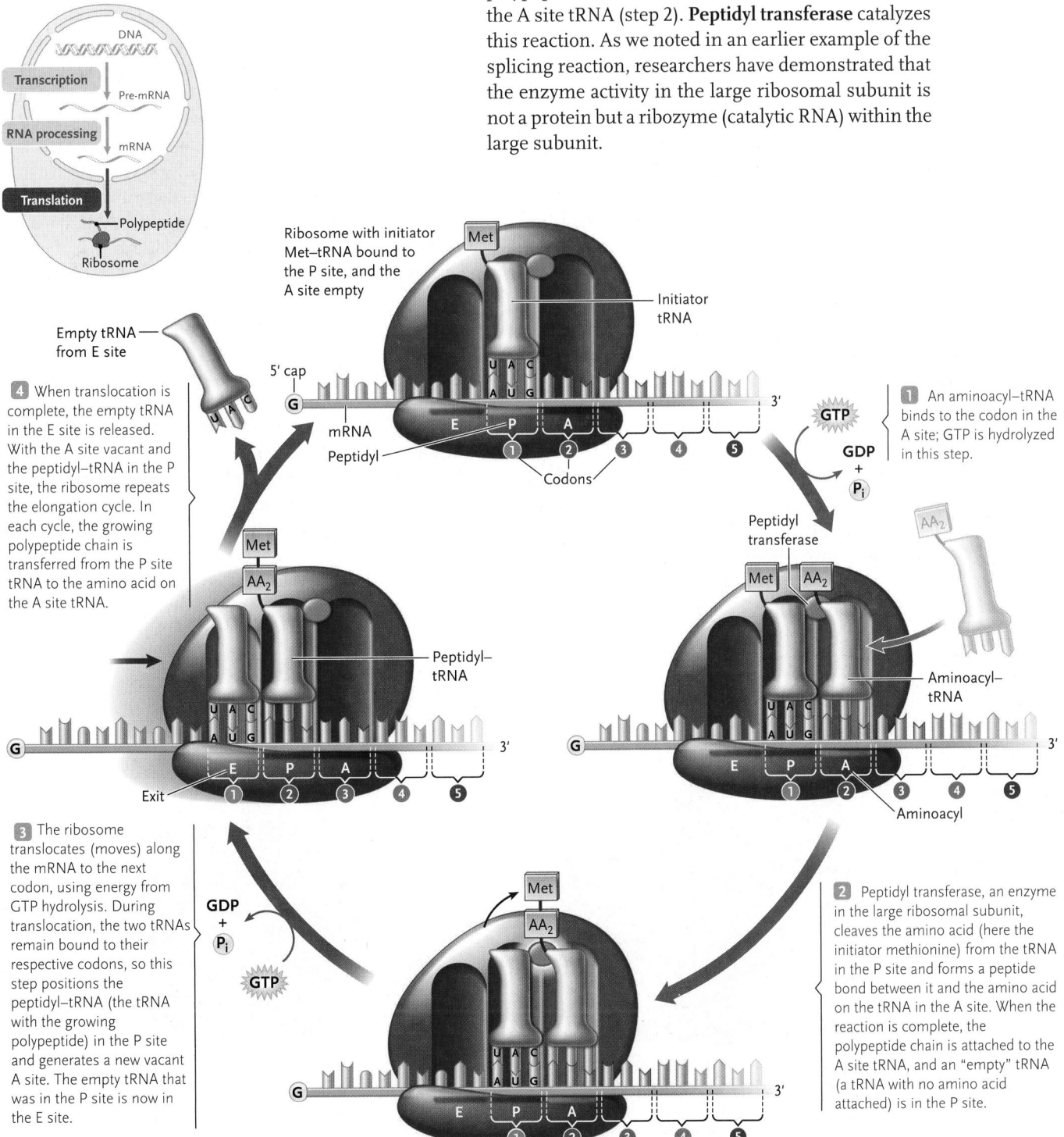

Figure 13.15

Translation elongation. A protein elongation factor (EF) complexes with the aminoacyl–tRNA to bring it to the ribosome, and another EF is needed for ribosome translocation. For simplicity, the EFs are not shown in the figure.

The elongation cycle is quite similar in prokaryotic cells and eukaryotes, turning at the rate of about 1 to 3 times per second in eukaryotes versus 15 to 20 times per second in bacteria.

13.4e Termination Releases a Completed Polypeptide from the Ribosome

Translation termination is similar in prokaryotic and eukaryotic cells; it takes place when one of the stop codons on the mRNA, UAG, UAA, or UGA arrives in the A site of a ribosome **(Figure 13.16)**. A protein **release factor (RF**; also called a **termination factor)** binds in the A site and causes the ribosome to disassemble into its subunits.

CONCEPT FIX Since the termination factor is a protein, and not a tRNA, it cannot base-pair with the stop codon. It has a similar shape as tRNA and simply wins the competition to occupy the A site of the ribosome since there are no competing tRNAs that recognize termination codons. ●

13.4f Multiple Ribosomes Simultaneously Translate a Single mRNA

Once the first ribosome has begun translation, another one can assemble with an initiator tRNA as soon as there is room at the 5′ UTR of the mRNA. Ribosomes continue to attach as translation continues and become spaced along the mRNA like beads on a string. The entire structure of an mRNA molecule and the multiple ribosomes attached to it is known as a **polysome** (a contraction of *polyribosome;* **Figure 13.17, p. 308)**. The multiple ribosomes greatly increase the overall rate of polypeptide synthesis from a single mRNA. The total number of ribosomes in a polysome depends on the length of the coding region of its mRNA molecule, ranging from a minimum of one or two ribosomes on the smallest mRNAs to as many as 100 on the longest mRNAs.

In prokaryotic cells, because of the absence of a nuclear envelope, transcription and translation are typically coupled. As soon as the 5′ end of a new mRNA emerges from the RNA polymerase, ribosomal subunits attach and initiate translation. By the time the mRNA is completely transcribed, it is covered with ribosomes from end to end, each assembling a copy of the encoded polypeptide. Meanwhile, several other RNA polymerases have likely begun transcribing the same gene, each one trailing a collection of translating ribosomes **(Figure 13.18, p. 308)**. Such a system allows prokaryotic cells to regulate the production very quickly in response to changing environmental conditions.

13.4g Newly Synthesized Polypeptides Are Processed and Folded into Finished Form

Most eukaryotic proteins are in an inactive, unfinished form when ribosomes release them. Processing reactions that convert the new proteins into the finished form include the removal of amino acids from the ends or interior of the polypeptide chain and the addition of larger organic groups, including carbohydrate or lipid structures.

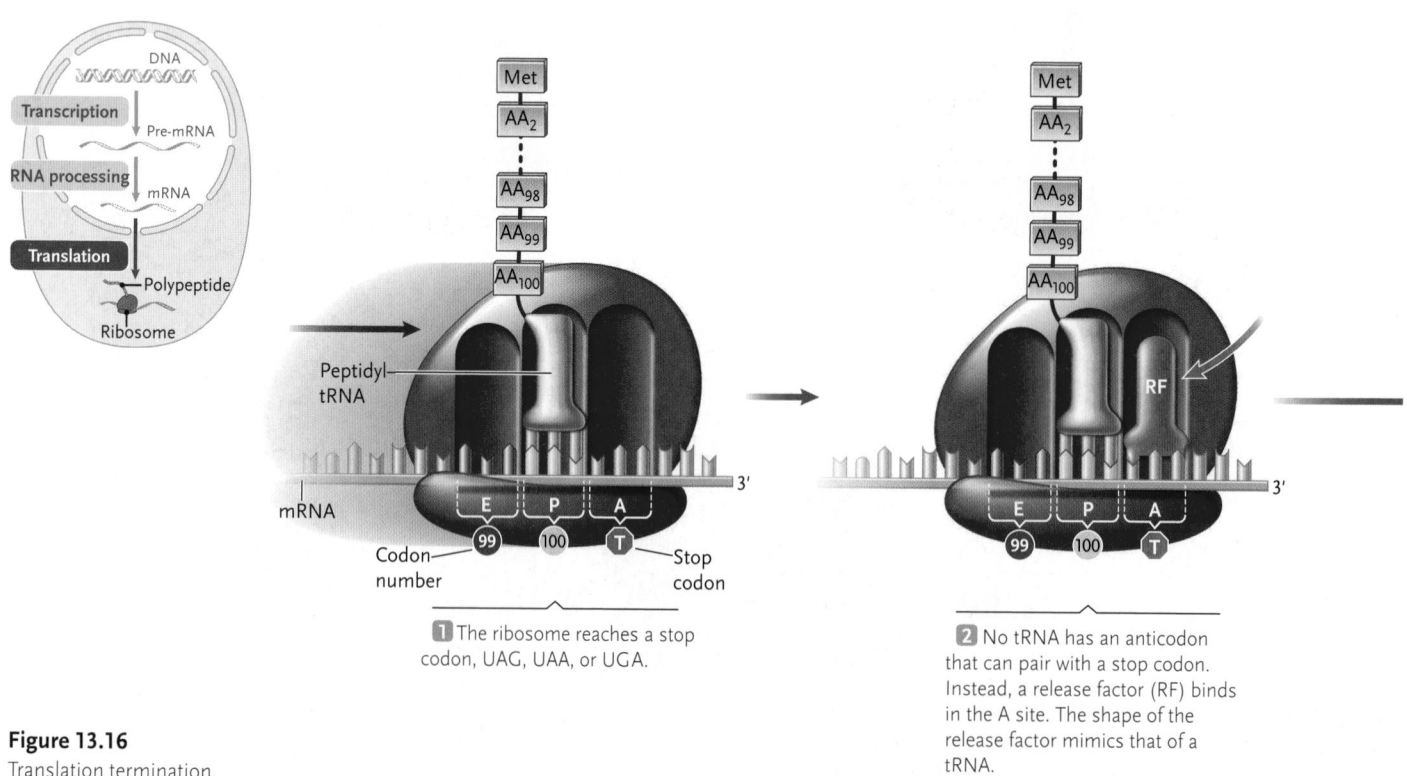

1 The ribosome reaches a stop codon, UAG, UAA, or UGA.

2 No tRNA has an anticodon that can pair with a stop codon. Instead, a release factor (RF) binds in the A site. The shape of the release factor mimics that of a tRNA.

Figure 13.16

Translation termination.

Amanitin

Alpha-amanitin is one of several potent toxins found in various species of the mushroom *Amanita* (**Figure 1**). Although composed of many amino-acid backbones linked in a ring, this interesting molecule is not a protein. It is the product of a metabolic pathway; it is not produced by translation. In the laboratory, amanitin is a useful inhibitor of eukaryotic RNA polymerase. However, on the dinner table, amanitin is a powerful poison. People suffering from amanitin poisoning show extensive, and usually fatal, liver and kidney damage.

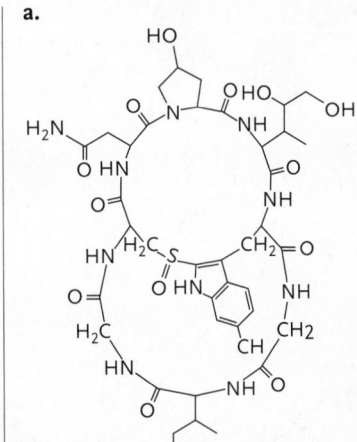

a.

b.

Phototribe/Shutterstock.com

Figure 1
(a) The very striking double circular structure of amanitin. *(b)* Amanita phalloides.

Proteins fold into their final three-dimensional shapes as the processing reactions take place. For many proteins, helper proteins called chaperones or chaperonins assist the folding process by combining with the folding protein, promoting correct three-dimensional structures and inhibiting incorrect ones.

In some cases, the same initial polypeptide may be processed by alternative pathways that produce different mature polypeptides, usually by removing different, long stretches of amino acids from the interior of the polypeptide chain. Alternative processing is another mechanism, distinct from alternative splicing of mRNA, that increases the number of proteins encoded by a single gene.

Other proteins are processed into an initial, inactive form that is later activated at a particular time or location

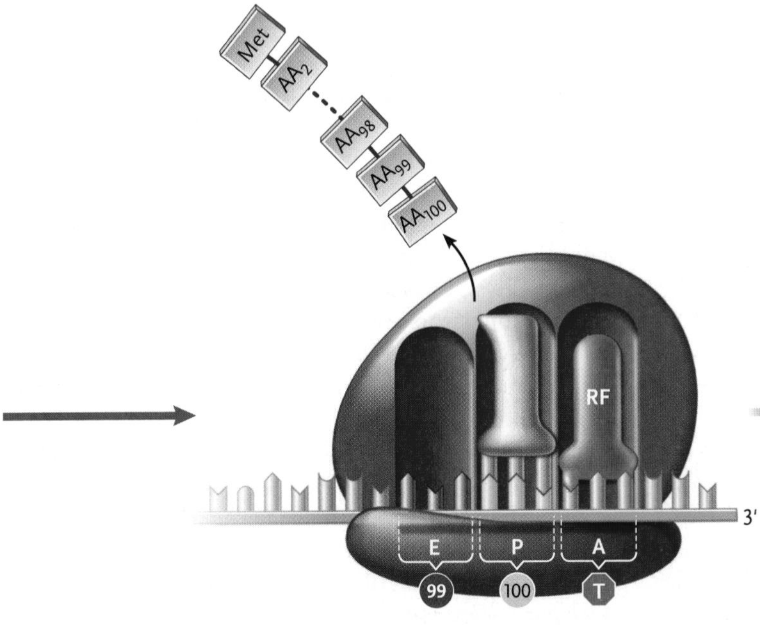

3 The RF stimulates peptidyl transferase to cleave the polypeptide from the P site tRNA. Because there is no aminoacyl–tRNA in the A for the polypeptide to be transferred to, the polypeptide is released.

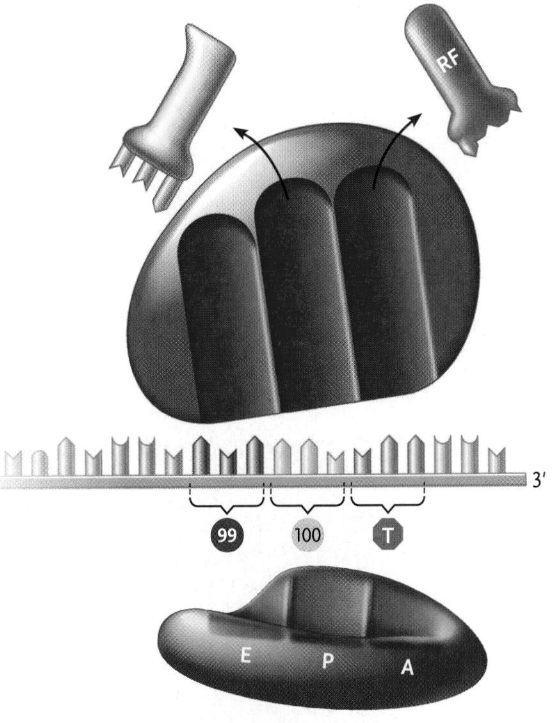

4 The empty tRNA and release factor are released, and the ribosomal subunits separate and leave the mRNA.

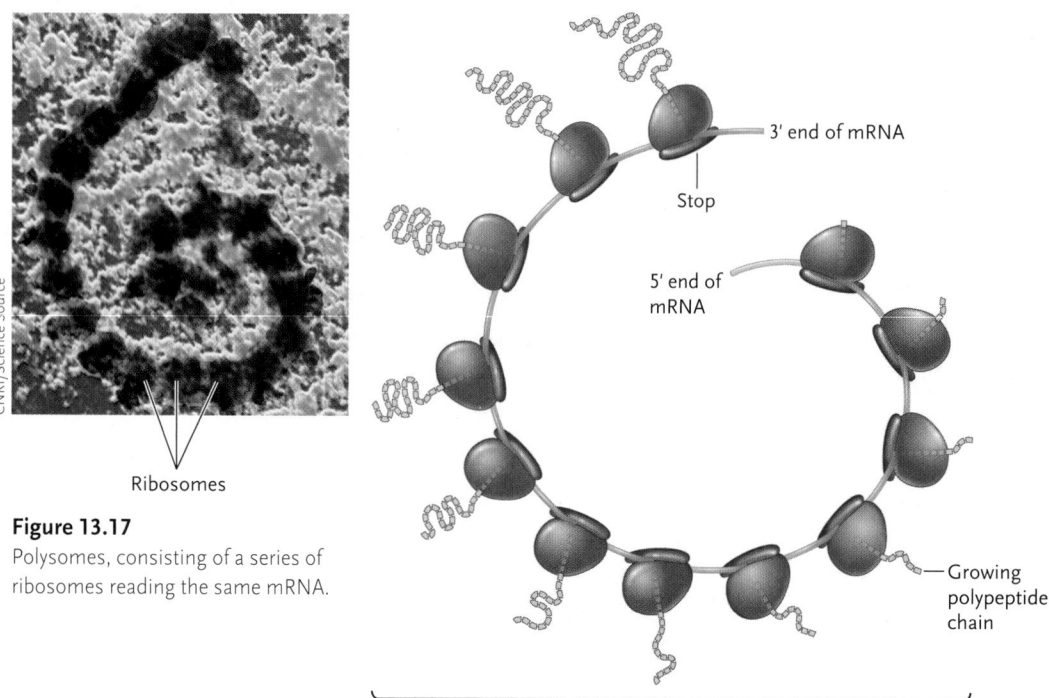

Figure 13.17
Polysomes, consisting of a series of ribosomes reading the same mRNA.

3' end of mRNA

Stop

5' end of mRNA

Growing polypeptide chain

Polysome

by removal of a covering segment of the amino acid chain. The digestive enzyme pepsin, for example, is made by cells lining the stomach in an inactive form called **pepsinogen**. When the cells secrete pepsinogen into the stomach, the high acidity of that organ triggers removal of a segment of amino acids, thus converting the enzyme into an active form that rapidly degrades food proteins in food particles. The initial production of the protein as inactive pepsinogen protects the cells that make it from having their own proteins degraded by the enzyme.

13.4h Finished Proteins Are Sorted to the Cellular Locations Where They Function

Eukaryotic cells are structurally compartmentalized, with various organelles performing specialized functions. Therefore, every protein that is made must be delivered to its appropriate compartment. Without a

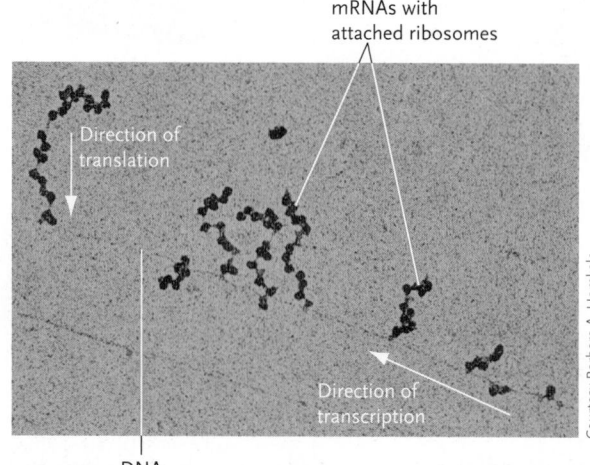

Figure 13.18
Simultaneous transcription and translation in progress in an electron microscope preparation extracted from *E. coli*, × 5 700 000.

mRNAs with attached ribosomes

Direction of translation

Direction of transcription

DNA

sorting and delivery system, cells would wind up as a jumble of proteins floating about in the cytoplasm, with none of the spatial organization that makes cellular life possible.

Although translation of all proteins begins on free ribosomes in the cytosol, there are three types of final destination compartments where the final products may be needed: (1) the cytosol; (2) the endomembrane system, which includes the Golgi complex, lysosomes, secretory vesicles, the nuclear envelope, and the plasma membrane; and (3) other membrane-bound organelles distinct from the endomembrane system, including the nucleus, mitochondria, chloroplasts, and microbodies (for example, **peroxisomes**).

Protein Sorting to the Cytoplasm. Proteins that function in the cytosol are simply released from ribosomes once translation is completed. Examples of proteins that function in the cytoplasm are cytoskeleton proteins (for example, tubulin and keratin) and the enzymes that carry out glycolysis (see Section 6.3).

Protein Sorting to the Endomembrane System. The endomembrane system is a major traffic network for proteins. Polypeptides that sort to the endomembrane system begin their synthesis on free ribosomes in the cytosol and produce a short segment of amino acids called a **signal sequence** (also called a **signal peptide**) near their N-terminal ends. As **Figure 13.19** shows, the signal sequence is recognized by a signal recognition particle that initiates a series of steps that ultimately result in the polypeptide entering the lumen (interior) of the rough ER. This mechanism is called **cotranslational import** because import of the polypeptide into the ER occurs simultaneously with translation of the mRNA encoding the polypeptide.

CONCEPT FIX Ribosomes engaged in cotranslational import stud the surface of the ER and give rise to the term *rough*. Note that ribosomes do not sit on the rough ER waiting for mRNA to translate. Rather, they only associate with the ER *after* they have begun translation as free ribosomes in the cytosol.

The signal sequence was discovered in 1975 by Günter Blobel, B. Dobberstein, and colleagues at Rockefeller University in New York when they observed that proteins sorted through the endomembrane system initially contain extra amino acids at their N-terminal ends. Blobel received a Nobel Prize

Courtesy Barbara A. Hamkalo

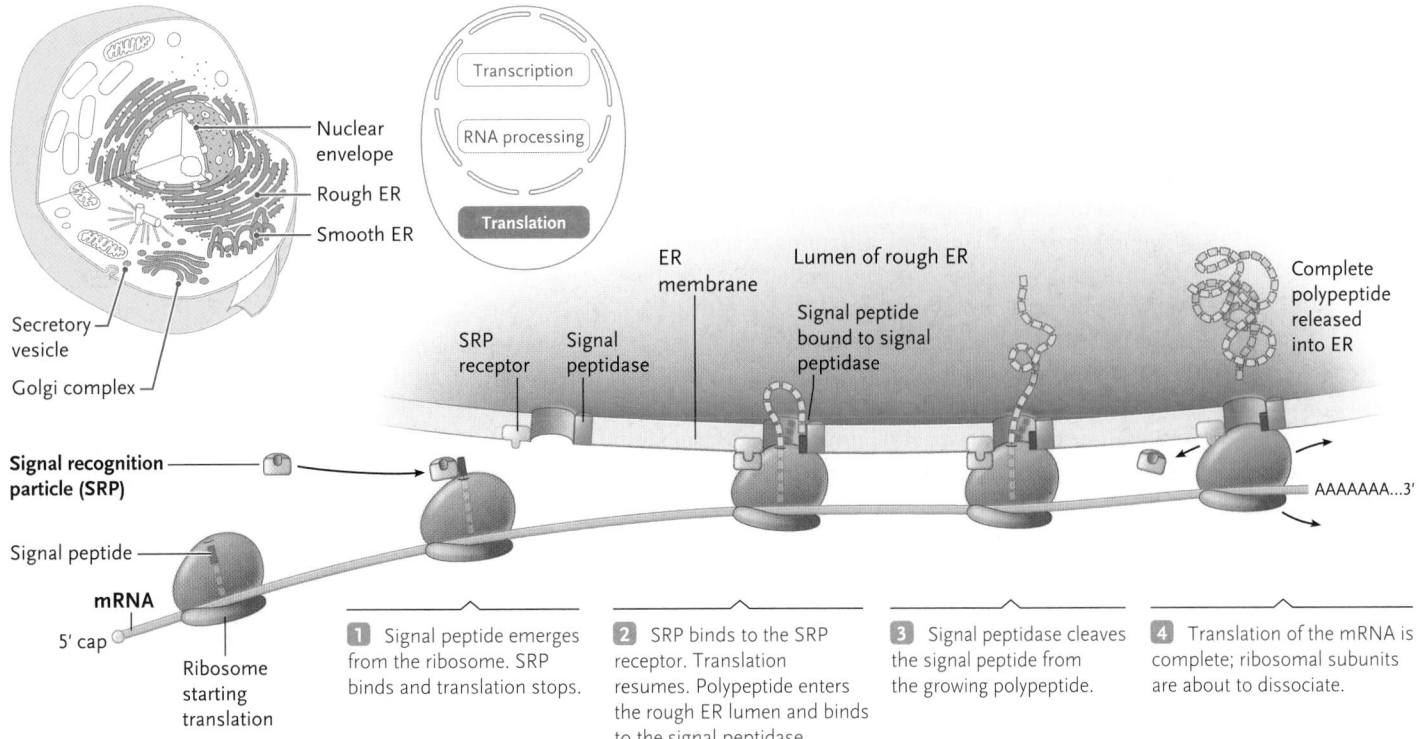

Figure 13.19

The signal mechanism directing proteins to the ER. The figure shows several ribosomes at different stages of translation of the mRNA.

in 1999 for his work with the mechanism of sorting proteins in cells.

Once inside the lumen of the rough ER, proteins fold into their final form. They also have, or obtain, a type of tag—a postal code if you will—that targets each protein for sorting to its final destination. Depending on the protein and its destination, the tag may be an amino acid sequence already coded in the protein, or a functional group or short sugar chain added to the protein in the lumen. Some proteins remain in the ER, whereas others are transported to the Golgi complex, where they may be modified further. From the Golgi complex, proteins are packaged into vesicles, which may deliver them to lysosomes, secrete them from the cell (digestive enzymes, for example), or deposit them in the plasma membrane (cell surface receptors, for instance).

Protein Sorting to the Nucleus, Mitochondria, Chloroplasts, and Microbodies. Proteins are sorted to the nucleus, mitochondria, chloroplasts, and microbodies after they have been made on free ribosomes in the cytosol. This mechanism of sorting is called **post-translational import.** Proteins destined for the mitochondria, chloroplasts, and microbodies have short amino acid sequences called **transit sequences** at their N-terminal ends that target them to the appropriate organelle. The protein is taken up into the correct organelle by interactions between its transit sequences and organelle-specific transport complexes in the membrane of the appropriate organelle. A transit peptidase enzyme within the organelle then removes the transit sequence.

Proteins sorted to the nucleus, such as the enzymes for DNA replication and RNA transcription, have short amino acid sequences called **nuclear localization signals.** A cytosolic transport protein binds to the signal and moves the nuclear protein to the nuclear pore complex, where it is then transported into the nucleus. The localization signal is never removed from nuclear proteins because they need to reenter the nucleus each time the nuclear envelope breaks down and reforms during the cell division cycle.

Although prokaryotic cells are structurally simpler than eukaryotes, the same basic system of molecular sorting signals distributes proteins throughout prokaryotic cells. In prokaryotic organisms, signals similar to the ER-directing signals of eukaryotes direct newly synthesized bacterial proteins to the plasma membrane (bacteria do not have ER membranes); further information built into the proteins keeps them in the plasma membrane or allows them to enter the cell wall or to be secreted outside the cell. Proteins without sorting signals remain in the cytoplasm. The similarity of mechanisms across all cells suggests that protein sorting is a very ancient evolutionary innovation.

13.4i Mutations Can Affect Protein Structure and Function

To this point in the chapter, we have been building an understanding of how the sequence of DNA bases in genes is directly related to the structure and function of the polypeptides that they encode. We will close the chapter with consideration of how various types of small changes in the DNA sequence might affect protein structure. (Contrast these small changes with the rather large-scale changes associated with the movement of mobile genetic elements in Chapter 9 and the chromosomal rearrangements in Chapter 11.)

Mutations are changes in the sequence of bases in the genetic material. How will mutations affect protein structure and function? Your understanding of this chapter should lead you to respond, "It depends." For instance, let's consider several different mutations in the protein-coding region of a gene, as shown in **Figure 13.20**. The normal (unmutated) DNA and amino acid sequences are shown in Figure 13.20a. **Base-pair substitution mutations** involve a change of one particular base to another in the genetic material. This will cause a change in a base in a codon in mRNA.

If a mutation alters the codon to specify a different amino acid, then the resulting protein will have a different amino acid sequence. We call this a **missense mutation** because although an amino acid is placed in the polypeptide, it is the wrong one (see Figure 13.20b). Whether the polypeptide's function is altered significantly or not depends on which amino acid is changed and what it is changed to. A missense mutation in the gene for one of the two hemoglobin polypeptides **(Figure 13.21)** results in the genetic disorder sickle cell disease, described in Chapter 10.

A second type of base-pair substitution mutation is a **nonsense mutation** (see Figure 13.20c). In this case, the mutation changes a sense (amino acid–coding) codon to a nonsense (termination) codon in the mRNA. Translation of an mRNA containing a nonsense mutation results in a premature "stop" and a shorter-than-normal polypeptide. This polypeptide will likely be partially functional at best.

Because of the degeneracy of the genetic code, some base-pair substitution mutations do not alter the amino acid specified by the gene because the changed codon specifies the same amino acid as in the normal polypeptide. Such mutations are known as **silent mutations** (see Figure 13.20d).

If a single base pair is deleted or inserted in the coding region of a gene, the reading frame of the resulting mRNA is altered. That is, after that point, the ribosome reads codons that are not the same as for the normal mRNA, typically producing a completely different amino acid sequence in the polypeptide from then on. This type of mutation is called a **frameshift mutation** (see Figure 13.20e; insertion mutation shown); the resulting polypeptide is usually nonfunctional because of the significantly altered amino acid sequence. The protein may be longer or shorter than usual, depending on where the stop codons occur in the shifted reading frame.

Figure 13.20

Effects of base-pair mutations in protein-coding genes on the amino acid sequence of the encoded polypeptide.

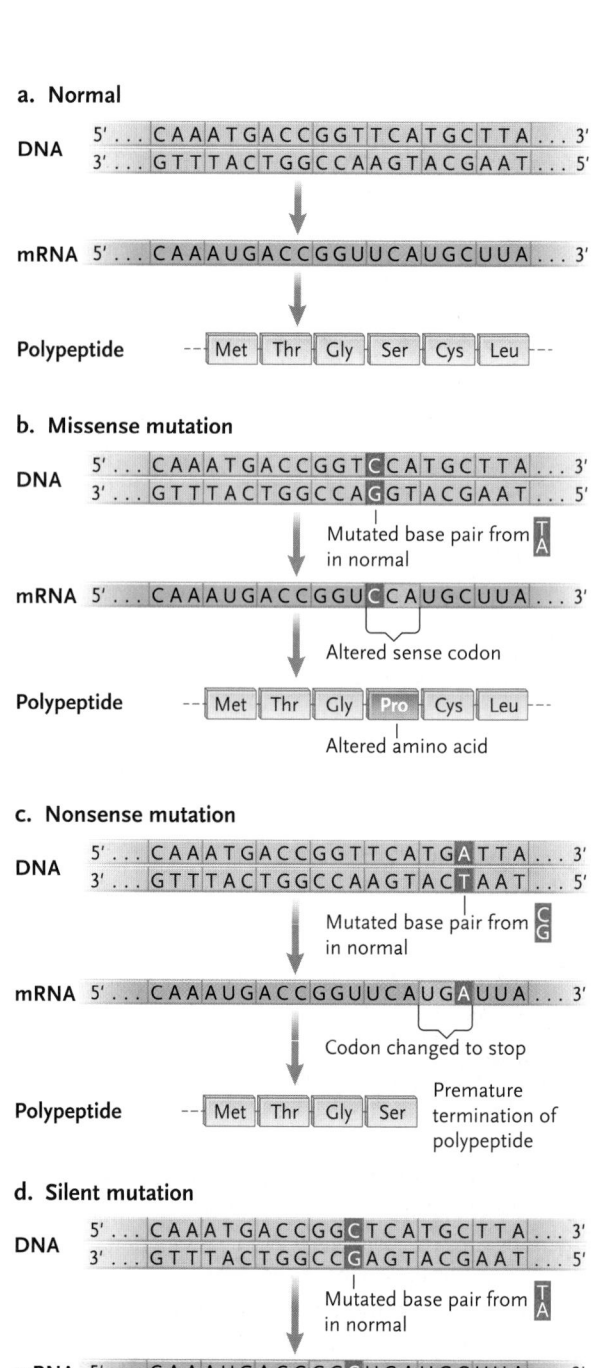
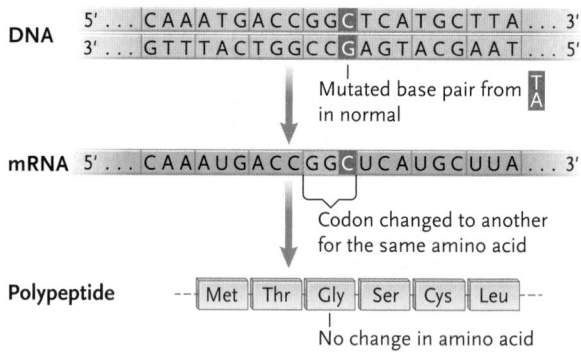
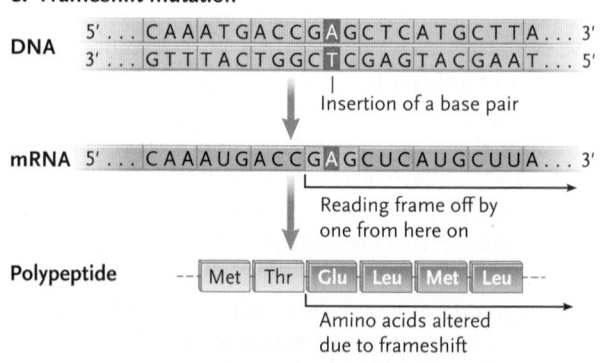

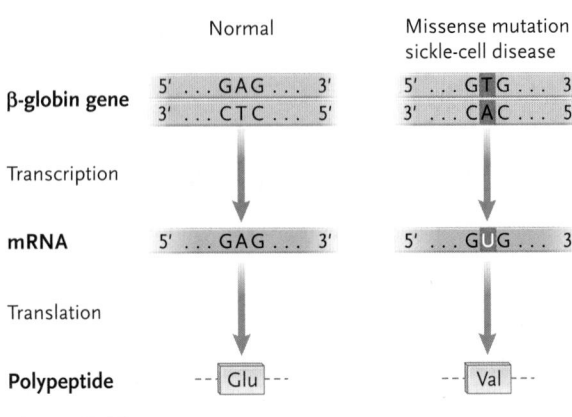

Normal | Missense mutation in sickle-cell disease

β-globin gene 5′ ... GAG ... 3′ / 3′ ... CTC ... 5′   5′ ... GTG ... 3′ / 3′ ... CAC ... 5′

Transcription

mRNA 5′ ... GAG ... 3′   5′ ... GUG ... 3′

Translation

Polypeptide -- Glu ---   -- Val ---

Figure 13.21

Missense mutation in a gene for one of the two polypeptides of hemoglobin that is the cause of sickle cell disease.

Both transcription and translation are steps in the process of gene expression, the realization of the gene's coded information in the makeup and activities of a cell. However, we will see in the next chapter that the flow of information is not one way; organisms and cells also exert control over how their genes are expressed.

STUDY BREAK

1. How does translation initiation occur in eukaryotes versus prokaryotic cells?
2. Distinguish among the E, P, and A sites of the ribosome.
3. How are proteins directed to different parts of a eukaryotic cell?

Review

To access course materials such as Aplia and other companion resources, please visit www.NELSONbrain.com.

13.1 The Connection between DNA, RNA, and Protein

- In their genetic experiments with *Neurospora crassa*, Beadle and Tatum found a direct correspondence between gene mutations and alterations of enzymes. Their one gene–one enzyme hypothesis is now restated as the one gene–one polypeptide hypothesis (Figure 13.2).

- The pathway from genes to proteins involves transcription and then translation. In transcription, a sequence of nucleotides in DNA is copied into a complementary sequence in an RNA molecule. In translation, the sequence of nucleotides in an mRNA molecule specifies an amino acid sequence in a polypeptide (Figure 13.4).

- The genetic code is a triplet code. AUG at the beginning of a coded message establishes a reading frame for reading the codons three nucleotides at a time. The code is redundant: most of the amino acids are specified by more than one codon (Figure 13.5).

- The genetic code is essentially universal.

- Aside from genes that code for protein through translation of mRNA, other genes code directly for RNA products (such as tRNA, rRNA, and snRNA) that are not translated.

13.2 Transcription: DNA-Directed RNA Synthesis

- Transcription is the process by which information coded in DNA is transferred to a complementary RNA copy (Figure 13.6).

- Transcription begins when an RNA polymerase binds to a promoter sequence in the DNA and starts synthesizing an RNA molecule. The enzyme then adds RNA nucleotides in sequence according to the DNA template. At the end of the transcribed sequence, the enzyme and the completed RNA transcript release from the DNA template. The mechanism of termination is different in eukaryotes and prokaryotic cells.

- In addition to sequences coding for amino acids, the DNA of protein-coding genes also contains several types of sequences that regulate transcription and translation.

13.3 Processing of mRNAs in Eukaryotes

- A gene encoding an mRNA molecule includes the promoter, which is recognized by the regulatory proteins and transcription factors that promote DNA unwinding and the initiation of transcription by an RNA polymerase. Transcription in eukaryotes produces a pre-mRNA molecule that consists of a 5′ cap, the 5′ untranslated region, interspersed exons (amino acid–coding segments) and introns, the 3′ untranslated regions, and the 3′ poly (A) tail. All are copied from DNA except the 5′ cap and poly (A) tail, which are added during transcription (Figure 13.7).

- Introns in pre-mRNAs are removed to produce functional mRNAs by splicing. snRNPs bind to the introns, loop them out of the pre-mRNA, clip the intron at each exon boundary, and join the adjacent exons together (Figure 13.8).

- Many pre-mRNAs are subjected to alternative splicing, a process that joins exons in different combinations to produce different mRNAs encoded by the same gene. Translation of each mRNA produced in this way generates a protein with a different function (Figure 13.9).

13.4 Translation: mRNA-Directed Polypeptide Synthesis

- Translation is the assembly of amino acids into polypeptides. Translation occurs on ribosomes. The P, A, and E sites of the ribosome are used for the stepwise addition of amino acids to the polypeptide as directed by the mRNA (Figures 13.10 and 13.13).

- Amino acids are brought to the ribosome attached to specific tRNAs. Amino acids are linked to their corresponding tRNAs by aminoacyl-tRNA synthetases. By matching amino acids with tRNAs, the reactions also provide the ultimate basis for the accuracy of translation (Figures 13.14 and 13.15).

- Translation proceeds through the stages of initiation, elongation, and termination. In initiation, a ribosome assembles with an mRNA molecule and an initiator methionine-tRNA. In elongation, amino acids linked to tRNAs are added one at a time to the growing polypeptide chain. In termination, the new polypeptide is released from the ribosome and the ribosomal subunits separate from the mRNA (Figure 13.16).

- After they are synthesized on ribosomes, polypeptides are converted into finished form by processing reactions, which include removal of one or more amino acids from the protein chains, addition of organic groups, and folding guided by chaperones.

- Proteins are distributed in cells by means of signals spelled out by amino acid sequences at the N-terminal end of the newly translated polypeptide (Figure 13.19).

- Mutations in the DNA template alter the mRNA and can lead to changes in the amino acid sequence of the encoded polypeptide. A missense mutation changes one codon to one that specifies a different amino acid, a nonsense mutation changes a codon to a stop codon, and a silent mutation changes one codon to another codon that specifies the same amino acid. A base-pair insertion or deletion is a frameshift mutation that alters the reading frame beyond the point of the mutation, leading to a different amino acid sequence from then on in the polypeptide (Figures 13.20 and 13.21).

Questions

Self-Test Questions

1. Which statement about the following pathway is true?

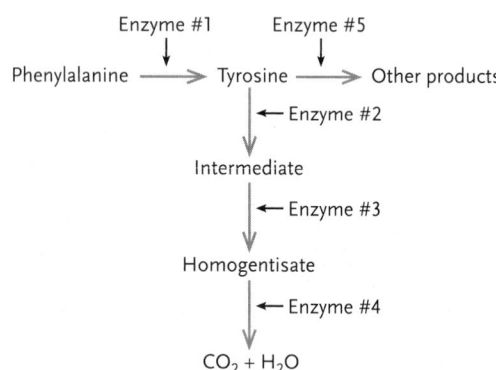

a. A mutation for enzyme #1 causes tyrosine to build up.
b. A mutation for enzyme #2 prevents tyrosine from being synthesized.
c. A mutation at enzyme #3 prevents homogentisate from being synthesized.
d. A mutation for enzyme #4 could hide a mutation in enzyme #1.

2. Which of the following statements describes *eukaryotic* mRNA?
a. It uses snRNPs to cut out introns and seal together translatable exons.
b. It is translated by ribosomes as it is being transcribed by RNA polymerase.
c. It has a guanine cap on its 3′ end and a poly (A) tail on its 5′ end.
d. It is a polymer of adenine, thymine, guanine, and cytosine bases.

3. A segment strand of DNA has a base sequence of 5′-GCATTAGAC-3′. What would be the sequence of an RNA molecule complementary to that sequence?
a. 5′-GUCTAATGC-3′
b. 5′-GCAUUAGAC-3′
c. 5′-CGTAATCTG-3′
d. 5′-GUCUAAUGC-3′

4. Which of the following statements about the initiation phase of translation in prokaryotic cells is true?
a. GTP is synthesized.
b. A region of the 5′ UTR of mRNA binds to rRNA.
c. 5′-UAC-3′ on the Met tRNA binds 3′-AUG-5′ on mRNA.
d. tRNA attaches first to the small ribosomal subunit.

5. Which of the following types of bonding involves complementary base-pairing?
a. tRNA to amino acid
b. signal peptide to signal recognition particle
c. release factor to stop codon
d. DNA to RNA during transcription of rRNA gene

6. Translation is in progress, with methionine bound to a tRNA in the P site, and a phenylalanine bound to a tRNA in the A site. What is the order of the next steps in the elongation cycle?
a. the ribosome translocates → a new aminoacyl-tRNA enters the A site → peptidyl transferase catalyzes a peptide bond between the two amino acids → empty tRNA is released from the ribosome
b. peptidyl transferase catalyzes a peptide bond between the two amino acids → a new aminoacyl-tRNA enters the A site → empty tRNA is released from the ribosome → the ribosome translocates

c. peptidyl transferase catalyzes a peptide bond between the two amino acids → the ribosome translocates → empty tRNA is released from the ribosome → a new aminoacyl-tRNA enters the A site

d. the ribosome translocates → peptidyl transferase catalyzes a peptide bond between the two amino acids → empty tRNA is released from the ribosome → a new aminoacyl-tRNA enters the A site

7. Which of the following statements about translation is true?
   a. ATP is the preferred energy source during various stages of translation.
   b. Peptide bond formation between amino acids is catalyzed by a ribozyme.
   c. When the mRNA codon UGG reaches the ribosome, there is no tRNA to bind to it.
   d. Forty-two amino acids of a protein are encoded by 84 nucleotides of the mRNA.

8. Which of the following items binds to the SRP receptor and to the signal sequence to guide a newly synthesized protein to be secreted to its proper channel?
   a. a ribosome
   b. a signal peptidase
   c. a signal recognition particle
   d. a rough ER

9. A part of an mRNA molecule with the sequence 5'-UGC GCA-3' is being translated by a ribosome. The following activated tRNA molecules are available. Which two of them can correctly bind the mRNA, resulting in a dipeptide?

tRNA Anticodon	Amino Acid
3'-GGC-5'	Proline
3'-CGU-5'	Alanine
3'-UGC-5'	Threonine
3'-CCG-5'	Glycine
3'-ACG-5'	Cysteine
3'-CGG-5'	Alanine

   a. cysteine–alanine
   b. proline–cysteine
   c. glycine–proline
   d. threonine–glycine

10. If a single base insertion mutation occurred within the first exon of a eukaryotic gene, what would be the likely result?
    a. improper splicing by spliceosome
    b. a longer mature mRNA
    c. a failure of the initiation of translation
    d. a silent mutation

Questions for Discussion

1. Would you expect rRNA genes to have start codons? Why or why not?

2. A mutation appears that alters an anticodon in a tRNA from AAU to AUU. What effect will this change have on protein synthesis in cells carrying this mutation?

3. The normal form of a gene is shown below, starting with the start codon (3' and 5' UTR are not visible):

   5'-ATGCCCGCCTTTGCTACTTGGTAG-3'

   3'-TACGGGCGGAAACGATGAACCATC-5'

   When this gene is transcribed, the result is the following mRNA molecule:

   5'-AUGCCCGCCUUUGCUACUUGGUAG-3'

   In a mutated form of the gene, two extra base pairs (underlined) are inserted:

   5'-ATGCCCGCCTAATTGCTACTTGGTAG-3'

   3'-TACGGGCGGATTAACGATGAACCATC-5'

   What effect will this mutation have on the structure of the protein encoded in the gene?

4. A geneticist is attempting to isolate mutations in the genes for four enzymes acting in a metabolic pathway in the bacterium *Escherichia coli*. The end product E of the pathway is absolutely essential for life:

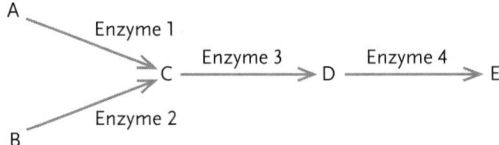

   The geneticist has been able to isolate mutations in the genes for enzymes 1 and 2, but not for enzymes 3 and 4. Develop a hypothesis to explain why.

5. How could you show experimentally that the genetic code is universal, namely, that it is the same in bacteria as it is in eukaryotes such as fungi, plants, and animals?

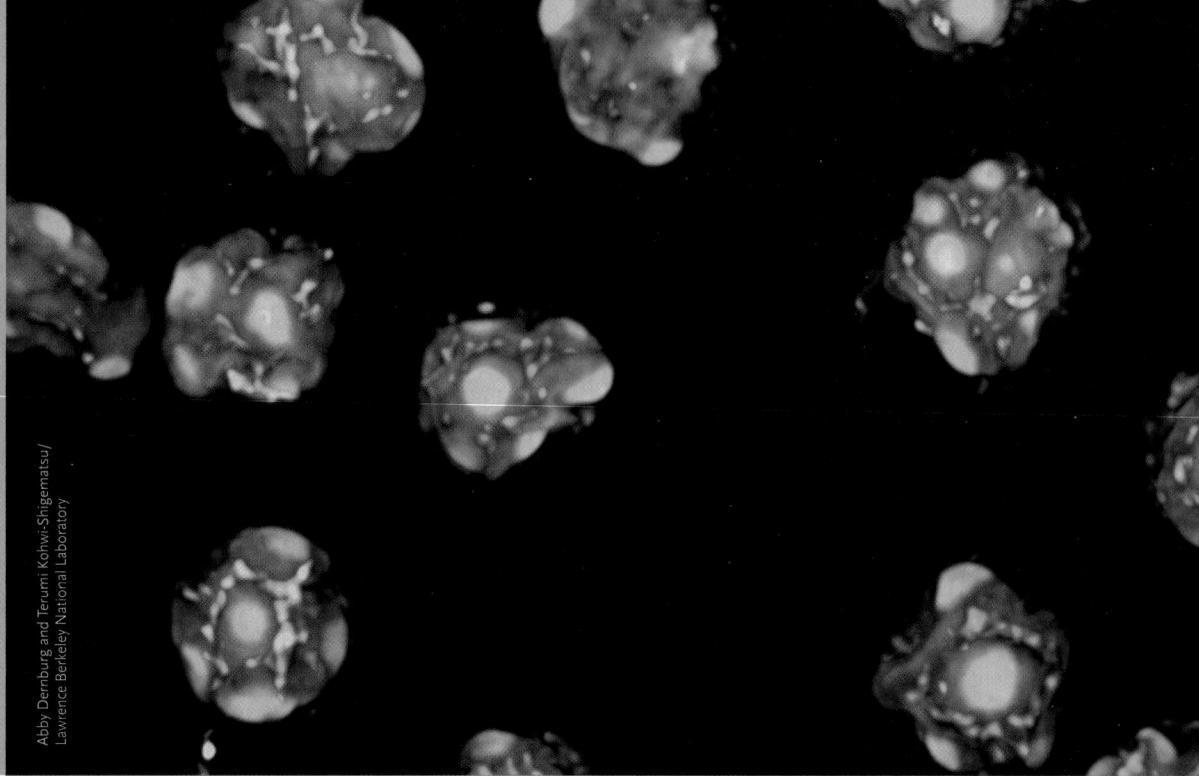

14

Abby Dernburg and Terumi Kohwi-Shigematsu/ Lawrence Berkeley National Laboratory

Chromatin remodelling proteins (gold) binding to chromatin (blue). Chromatin remodelling, a change in chromosome structure in the region of a gene, is a key step in the activation of genes in eukaryotes.

Control of Gene Expression

WHY IT MATTERS

A human egg cell is almost completely inactive metabolically when it is released from the ovary. It remains quiescent as it travels down a fallopian tube leading from the ovary to the uterus, carried along by movements of cilia lining the walls of the tube **(Figure 14.1)**. It is here, in the fallopian tube, that the egg meets sperm cells and embryonic development begins. Within seconds after the cells unite, the fertilized egg breaks its quiescent state and begins a series of divisions that continues as the egg moves through the fallopian tube and enters the uterus. Subsequent divisions produce specialized cells that *differentiate* into the distinct types tailored for the myriad specific functions in the body, from muscle cells to cells of the lens of the eye.

At first glance, you might think it most efficient for the cells in each differentiated tissue to retain only those genes needed to carry out its specific function; that is, liver cells might be expected to have a different collection of genes than bone cells. However, biochemical and cytogenetic analyses do not support this model and have, in fact, demonstrated that all nucleated cells of a developing embryo retain essentially the same set of genes that was created in the original single-celled zygote at fertilization. Structural and functional differences in cell types result from the presence or absence of the *products resulting from expression of genes* rather than the presence or absence of the genes themselves. As you saw in the previous chapter, all gene expression initially results in RNA products made by transcription. One type of RNA product, mRNA, further directs the synthesis of protein products by translation. But what determines when the product is produced, where, and how much? For example, the products of some genes, known as

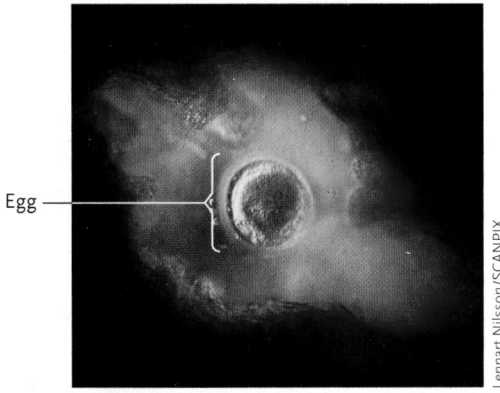

Egg

Figure 14.1

A human egg at the time of its release from the ovary. The outer layer, appearing light blue in colour, is a coat of polysaccharides and glycoproteins that surrounds the egg. Within the egg, genes and regulatory proteins are poised to enter the pathways, initiating embryonic development.

Lennart Nilsson/SCANPIX

housekeeping genes, are expressed in nearly all cells, whereas the products of other genes may be found only in certain cell types at specific times under particular environmental conditions. To illustrate this point, consider that all cells contain genes coding for the rRNA molecules needed for ribosome function, as well as genes coding for various hemoglobin polypeptides. While rRNA gene products are abundant in all cells, particular hemoglobins are found only in those cells that give rise to red blood cells in the fetus, newborn, or adult.

The material in the previous chapter on transcription and translation hinted at possible regulatory mechanisms of gene expression. Usually, when we say that a gene is "turned on," we mean that it is more likely to be transcribed actively. Beyond transcription, the expression of gene products is subject to further controls affecting the processing of ribonucleic acid (RNA), possible translation into protein, and the activity and "life span" of the product itself.

You saw in the previous chapter that transcription and translation are coincident in prokaryotic cells. This enables a rapid response to environmental conditions through regulation of transcription initiation. Eukaryotes, particularly multicellular organisms, exhibit a variety of mechanisms not used by prokaryotic organisms. In this chapter, we examine the mechanisms of transcriptional regulation and its fine-tuning by additional controls at the posttranscriptional, translational, and posttranslational levels. Our discussion begins with bacterial systems, where researchers first discovered a mechanism for transcriptional regulation, and then moves to eukaryotic systems, where the regulation of gene activity is more complex. The chapter closes with a look at the loss of regulatory controls in cancer cells. The ways in which genes regulate development are discussed in Chapters 36 and 42.

14.1 Regulation of Gene Expression in Prokaryotic Cells

Transcription and translation are closely regulated in prokaryotic cells in ways that reflect prokaryotic life histories. Prokaryotic organisms tend to be single celled and relatively simple, with generation times measured in minutes. Rather than the complex patterns of long-term cell differentiation and development typical of multicellular eukaryotes, prokaryotic cells typically undergo rapid and reversible alterations in biochemical pathways that allow them to adapt quickly to changes in their environment.

The bacterium *Escherichia coli,* for example, can find itself in the intestinal tract of a cow one minute and then in a treated municipal water supply soon after. Sugars such as lactose might be more available in the aquatic environment, and genes coding for enzymes needed to metabolize this energy source must be turned on. Other nutrients, such as the amino acid tryptophan, may be abundant in the intestinal tract. Therefore, genes coding for enzymes needed to manufacture the amino acid from scratch must be turned off. A versatile and responsive control system allows the bacterium to make the most efficient use of the particular array of nutrients and energy sources available at any given time.

14.1a The Operon Is a Unit of Transcription

For a typical metabolic process, several genes are involved, and they must be regulated in a coordinated fashion. For example, three genes encode proteins for the metabolism of lactose by *E. coli.* In the absence of lactose, the three genes are transcribed very little, whereas in the presence of lactose, the genes are transcribed quite actively. That is, the on/off control of these genes is at the level of transcription.

In 1961, François Jacob and Jacques Monod of the Pasteur Institute in Paris proposed the *operon model* for the control of the expression of genes for lactose metabolism in *E. coli.* Subsequently, data have shown the operon model to be widely applicable to the regulation of gene expression in bacteria and their viruses. Jacob and Monod received the Nobel Prize in 1965 for their explanation of bacterial operons and their regulation by repressors.

An **operon** is a cluster of prokaryotic genes and the DNA sequences involved in their regulation. The promoter, as we saw in the previous chapter, is a region where the RNA polymerase begins transcription. Another regulatory DNA sequence in the operon is the **operator**, a short segment that is a binding sequence for a **regulatory protein.** A gene that is separate from the operon encodes the regulatory protein. Some operons are controlled by a regulatory protein termed a **repressor**, which, when bound to the DNA, reduces the likelihood that genes will be transcribed. Other operons are controlled by a regulatory protein termed an **activator**, which, when bound to the DNA, increases the likelihood that genes will be transcribed. Many operons are controlled by more than one regulatory mechanism, and a number of the repressors or activators control more than one operon. The result is a complex network of superimposed controls that provides regulation of transcription, allowing almost instantaneous global responses to changing environmental conditions.

Each operon, which can contain several to many genes, is transcribed as a unit from the promoter into a single messenger RNA (mRNA), and, as a result, the mRNA contains codes for several proteins. The cluster of genes transcribed into a single mRNA is called a **transcription unit.** A ribosome translates the entire mRNA from one end to the other, sequentially making each protein encoded in the mRNA. Typically, the proteins encoded by genes in the same operon catalyze steps in the same process, such as enzymes acting in sequence in a biochemical pathway.

14.1b The *lac* Operon for Lactose Metabolism Is Transcribed When an Inducer Inactivates a Repressor

Jacob and Monod researched the genetic control of lactose metabolism in *E. coli* through a series of brilliantly creative genetic and biochemical approaches. Their studies showed that metabolism of lactose as an energy source involves three genes: *lacZ, lacY,* and *lacA* **(Figure 14.2)**. These three genes are adjacent to one another on the chromosome in the order *Z-Y-A*. The genes are transcribed as a unit into a single mRNA starting with the *lacZ* gene; the promoter for the transcription unit is upstream of *lacZ*.

The *lacZ* gene encodes the enzyme β-galactosidase, which catalyzes the conversion of the disaccharide sugar, lactose, into the monosaccharide sugars, glucose and galactose. These sugars are then further metabolized by other enzymes, producing energy for the cell by glycolysis and Kreb's cycle. The *lacY* gene encodes a permease enzyme that transports lactose actively into the cell, and the *lacA* gene encodes a transacetylase enzyme, the function of which is more relevant to metabolism of compounds other than lactose.

Jacob and Monod called the cluster of genes and adjacent sequences that control their expression the *lac operon* (see Figure 14.2). They coined the name *operon* from the key DNA sequence they discovered for regulating transcription of the operon—the operator. The operator was named because it controls the operation of the genes adjacent to it. For the *lac* operon, the operator is a short DNA sequence between the promoter and the *lacZ* gene.

These two investigators showed that the *lac* operon was controlled by a regulatory protein that they termed the *Lac repressor.* The Lac repressor is encoded by the regulatory gene *lacI*, which is nearby but separate from the *lac* operon (see Figure 14.2), and is synthesized in active form. When lactose is absent from the medium, the Lac repressor binds to the operator, thereby blocking the RNA polymerase from binding to the promoter **(Figure 14.3a).** Repressor binding is a kind of equilibrium; while it is bound to the operator most of the time, it occasionally comes off. In moments when the repressor is not bound, polymerase can successfully transcribe. As a result, *there is always a low concentration of lac operon gene products in the cell.*

When lactose is added to the medium, the *lac* operon is turned on and all three enzymes are synthesized rapidly (Figure 14.3b). How does this occur?

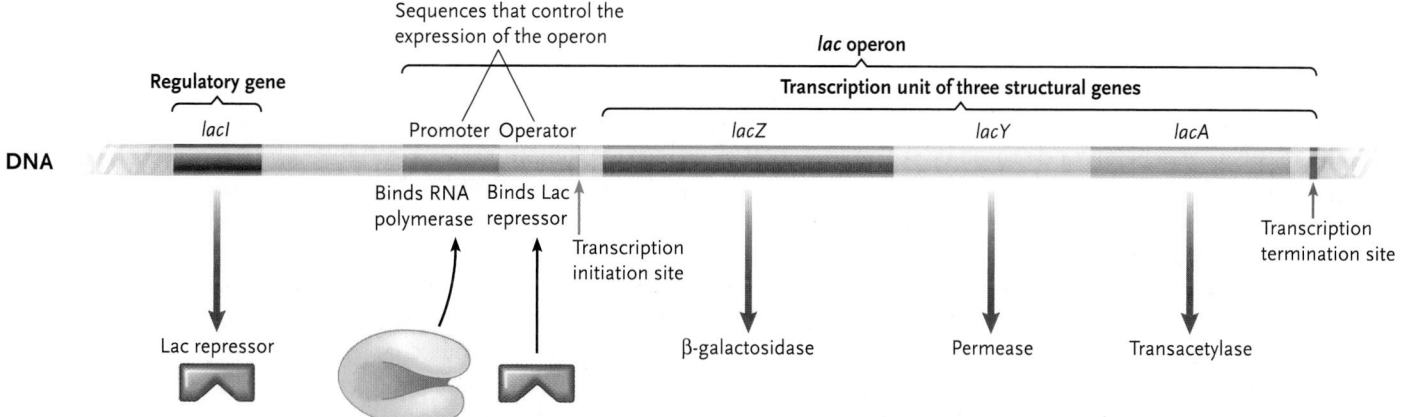

Figure 14.2

The *E. coli lac* operon. The *lacZ, lacY,* and *lacA* genes encode the enzymes taking part in lactose metabolism. The separate regulatory gene, *lacI*, encodes the Lac repressor, which plays a pivotal role in the control of the operon. The promoter binds RNA polymerase, and the operator binds the activated Lac repressor. The transcription unit, which extends from the transcription initiation site to the transcription termination site, contains the genes.

a. Lactose absent from medium: structural genes expressed at very low levels

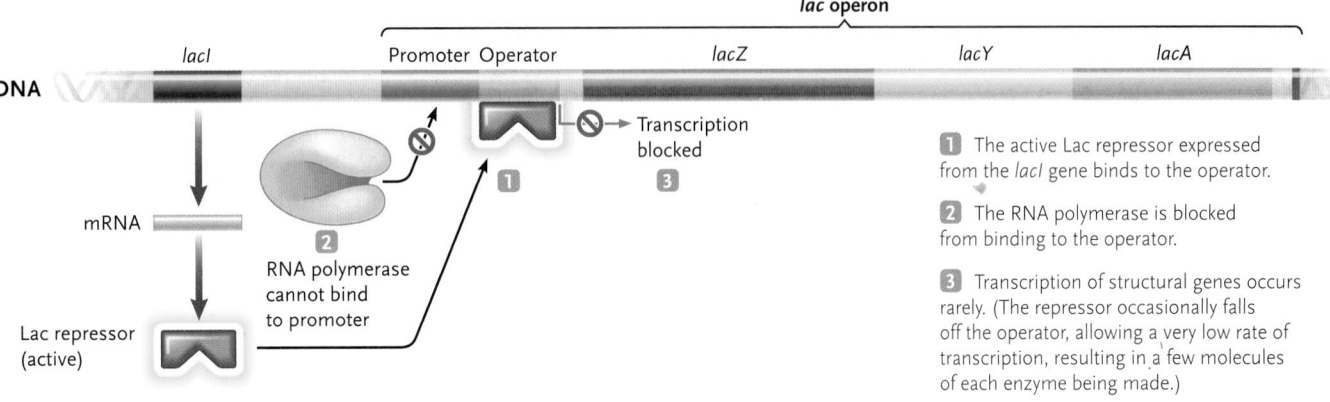

1. The active Lac repressor expressed from the *lacI* gene binds to the operator.

2. The RNA polymerase is blocked from binding to the operator.

3. Transcription of structural genes occurs rarely. (The repressor occasionally falls off the operator, allowing a very low rate of transcription, resulting in a few molecules of each enzyme being made.)

b. Lactose present in medium: structural genes expressed at high levels

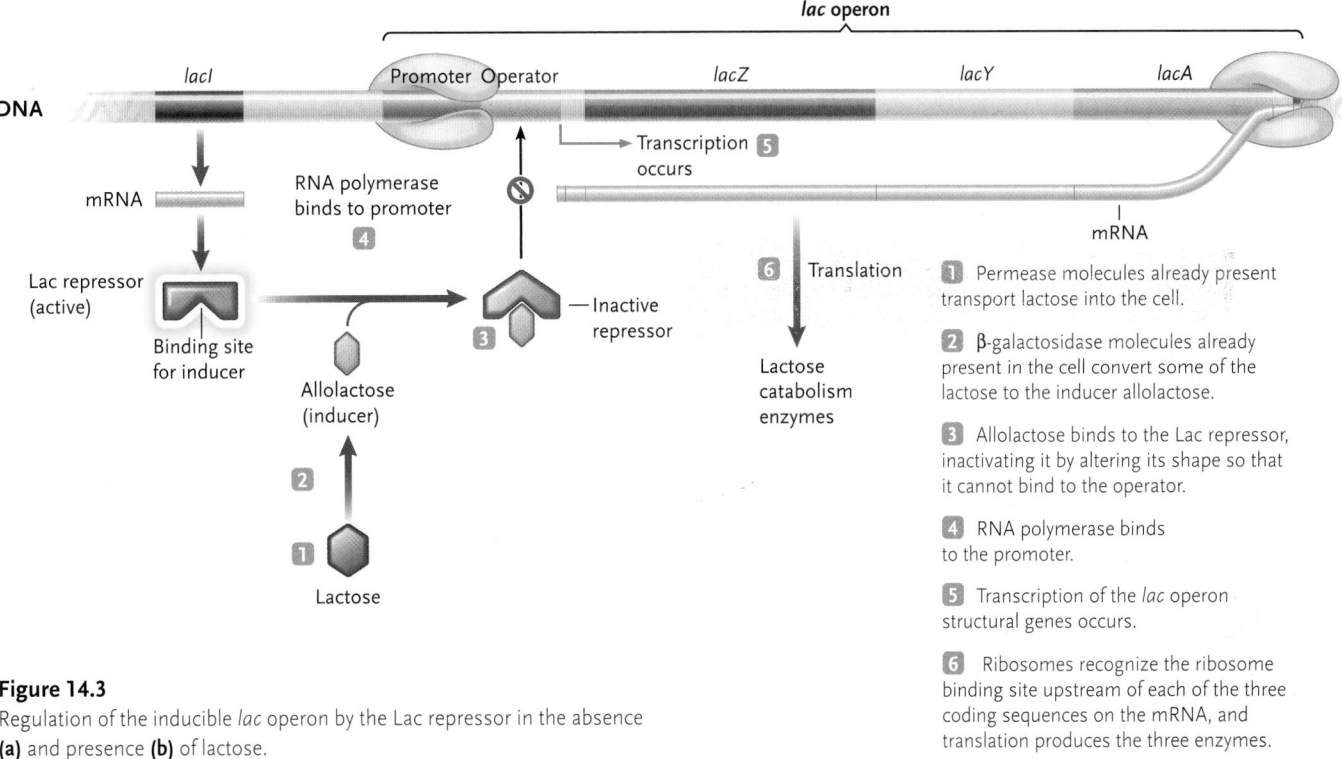

1. Permease molecules already present transport lactose into the cell.

2. β-galactosidase molecules already present in the cell convert some of the lactose to the inducer allolactose.

3. Allolactose binds to the Lac repressor, inactivating it by altering its shape so that it cannot bind to the operator.

4. RNA polymerase binds to the promoter.

5. Transcription of the *lac* operon structural genes occurs.

6. Ribosomes recognize the ribosome binding site upstream of each of the three coding sequences on the mRNA, and translation produces the three enzymes.

Figure 14.3

Regulation of the inducible *lac* operon by the Lac repressor in the absence (a) and presence (b) of lactose.

Lactose enters the cell and the low levels of β-galactosidase molecules already present convert some of it to *allolactose,* an isomer of lactose. Allolactose is an **inducer** for the *lac* operon. It binds to the Lac repressor, altering its shape so that the repressor can no longer bind to the operator DNA. With the repressor out of the way, RNA polymerase is then able to bind freely to the promoter and transcribe the three genes at a dramatically elevated rate. Because an inducer molecule increases its expression, the *lac* operon is called an **inducible operon.**

As the lactose is used up, the regulatory system switches the *lac* operon off. That is, the absence of lactose means that there are no allolactose inducer molecules to inactivate the repressor; the repressor binds to the operator, reducing transcription of the operon. These controls are aided by the fact that bacterial mRNAs are very short lived, about three minutes on average. This quick turnover permits the cytoplasm to be cleared quickly of the mRNAs transcribed from an operon. The enzymes themselves also have short lifetimes and are quickly degraded.

14.1c Transcription of the *lac* Operon Is Also Controlled by a Positive Regulatory System

Several years after Jacob and Monod proposed their negatively regulated operon model for the lactose metabolism genes, researchers found a *positive gene regulation* system that makes expression of the *lac* operon responsive to the availability of glucose. Glucose can be used directly in the glycolysis pathway to produce energy for the cell (see Chapter 6). However, lactose must first be converted into glucose by biochemical reactions that require energy. The net yield of energy from other sugars is therefore less than that for glucose, and cells will grow best if they ensure the preferential metabolism of glucose whenever it is available.

Figure 14.4 shows that the *lac* operon is sensitive to the availability of glucose through the binding of an activator protein called CAP (catabolite activator protein). The CAP binding site is on the DNA, just upstream of the *lac* promoter. When bound at this site, CAP bends the DNA in ways that make the promoter more accessible to RNA polymerase and transcription increases. To understand how CAP binding is related to the availability of glucose, you need to know that (1) CAP is synthesized in an inactive form that can only bind to DNA *after* it is activated by binding with **cyclic AMP (cAMP;** a nucleotide that plays a role in regulating cellular processes in both

a. Lactose present and glucose low or absent: structural genes expressed at very high levels

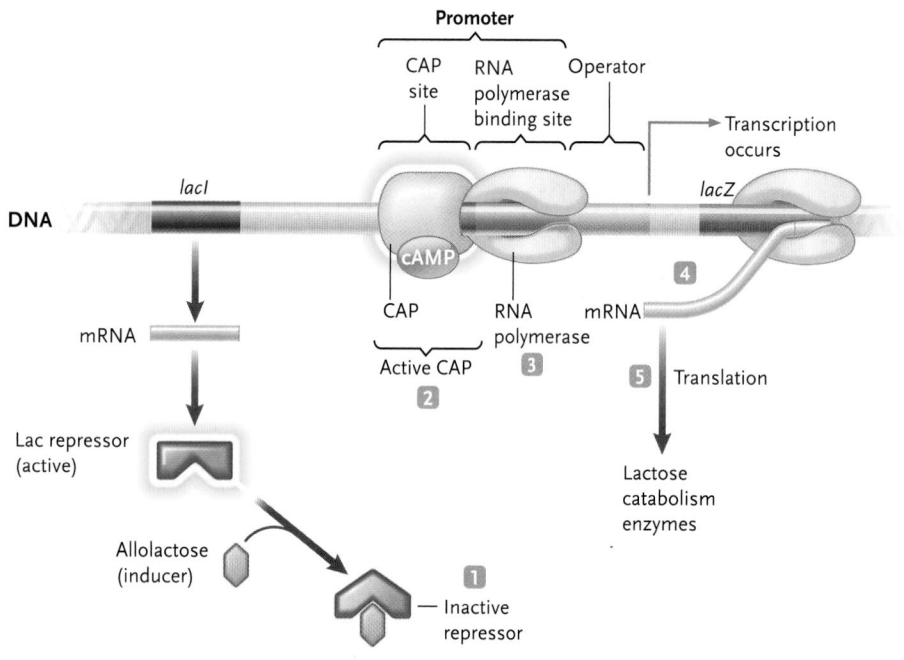

1 Lactose is converted to the inducer, allolactose, which inactivates the Lac repressor.

2 Active adenylyl cyclase synthesizes cAMP to high levels. cAMP binds to the activator CAP, activating it. Activated CAP binds to the CAP site in the promoter.

3 RNA polymerase binds efficiently to the promoter.

4 Genes of the operon are transcribed to high levels.

5 Translation produces high amounts of enzymes.

b. Lactose present and glucose present: structural genes expressed at low levels

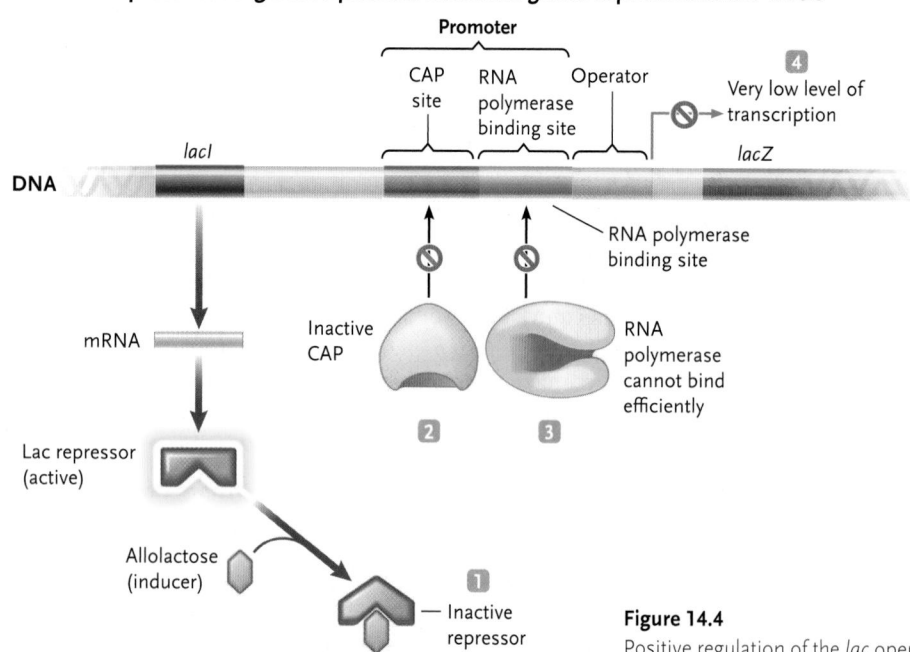

1 Lactose is converted to the inducer, allolactose, which inactivates the Lac repressor.

2 Catabolism of incoming glucose leads to inactivation of adenylyl cyclase, which causes the amount of cAMP in the cell to drop to a level too low to activate CAP. Inactive CAP cannot bind to the CAP site.

3 RNA polymerase is unable to bind to the promoter efficiently.

4 Transcription occurs at a low level: because the Lac repressor is not present to block RNA polymerase from binding to the promoter, the level of transcription is higher than when lactose is absent, but far lower than when lactose is present and glucose is absent.

Figure 14.4
Positive regulation of the *lac* operon through binding of the CAP activator protein.

prokaryotic and eukaryotic cells), and (2) cAMP levels are inversely related to the uptake of glucose from the growth medium; when glucose is abundant, cAMP levels tend to be low (meaning CAP is mostly inactive). When glucose is absent from the environment, cAMP concentration tends to be high inside the cell, leading to an increased level of activated CAP.

Taken together, the negative control by the Lac repressor and the positive control by CAP/cAMP ensure that cells express the *lac* operon most strongly only when lactose is present and glucose is not. Let's walk through one illustrative example to emphasize the interrelationships among the various players. Imagine cells growing on glucose only. In the presence of glucose, very little cAMP is available to bind to CAP. Therefore, CAP/cAMP binding will be rare and there will be very little stimulation of expression. In the absence of lactose, the Lac repressor will be bound to the operator site most of the time and very little synthesis of the *lac* genes will occur. For these two reasons, expression of the *lac* operon will be at its lowest level. If we then add lactose to the environment, it will be metabolized to the inducer, allolactose, which will bind to and inactivate the Lac repressor. RNA polymerase will then bind to the promoter and transcribe the *lac* operon genes at a low level. Expression will increase further as glucose is metabolized from the surrounding medium, allowing cAMP levels to rise, activated CAP to bind, and the *lac* promoter to become even more available to RNA polymerase.

CONCEPT FIX Inducing *lac* operon expression through negative control and repressing expression through positive control may sound confusing. How can negative control make expression increase? The answer to this apparent paradox lies in focusing your attention on the DNA-binding proteins: the Lac repressor and CAP. In general, if the binding of a protein to DNA results in decreased gene expression, that is negative control. If the binding of a protein results in increased gene expression, that is positive control. Therefore, the binding of the Lac repressor is a clear example of negative control. When this repression is *released*, the *lac* operon is induced and expression increases. *Whether gene expression is under negative or positive control depends on the impact of the respective DNA-binding proteins, not on the impact of the available substrates such as glucose or lactose.* ⬢

The same positive gene regulation system using CAP and cAMP regulates a large number of other operons that control the metabolism of many sugars. In each case, the system functions so that glucose, if it is present in the growth medium, is metabolized first. This type of regulatory system, in which several operons are under the control of a common regulator, is called a regulon.

14.1d Transcription of the *trp* Operon Genes for Tryptophan Biosynthesis Is Repressed When Tryptophan Activates a Repressor

Tryptophan is an essential amino acid used in the synthesis of proteins. If tryptophan is absent from the medium, *E. coli* must manufacture it. If tryptophan is present in the medium, then the cell will use that source rather than make its own.

The genes involved in tryptophan biosynthesis are coordinately controlled in an operon called the *trp* operon **(Figure 14.5, p. 320).** The five genes in this operon, *trpA* to *trpE*, encode the enzymes for the steps in the tryptophan biosynthesis pathway. Upstream of the *trpE* gene are the operon's promoter and operator sequences. Expression of the *trp* operon is controlled by the Trp repressor, a regulatory protein encoded by the *trpR* gene, which is located elsewhere in the genome (not nearby, as was the case for the repressor gene for the *lac* operon). In contrast to the Lac repressor, the Trp repressor is synthesized in an inactive form in which it cannot bind to the operator.

When tryptophan is absent from the medium and must be made by the cell, the *trp* operon genes are expressed (see Figure 14.5a). This is the default state; since the Trp repressor is inactive and cannot bind to the operator, RNA polymerase can bind to the promoter and transcribe the operon. The resulting mRNA is translated to produce the five tryptophan biosynthetic enzymes that catalyze the reactions for tryptophan synthesis.

If tryptophan is present, there is no need for the cell to make it, so the *trp* operon is shut off (see Figure 14.5b). This occurs because the tryptophan entering the cell binds to the Trp repressor and activates it. The active Trp repressor then binds to the operator of the *trp* operon and blocks RNA polymerase from binding to the promoter—the operon cannot be transcribed.

For the *trp* operon, then, the presence of tryptophan represses the expression of the tryptophan biosynthesis genes; hence, this operon is an example of a **repressible operon.** Here, tryptophan acts as a **corepressor,** a regulatory molecule that combines with a repressor to activate it and thus shut off the operon.

Let's compare and contrast the two operons we have discussed: (1) In the *lac* operon, the repressor is synthesized in an active form. When the inducer (allolactose) is present, it binds to the repressor and inactivates it. The operon is then transcribed. (2) In the *trp* operon, the repressor is synthesized in an inactive form. When the corepressor (tryptophan) is present, it binds to the repressor and activates it. The active repressor blocks transcription of the operon.

CONCEPT FIX Inducible and repressible operons both illustrate *negative gene regulation* because both are

a. Tryptophan absent from medium: tryptophan must be made by the cell—structural genes transcribed

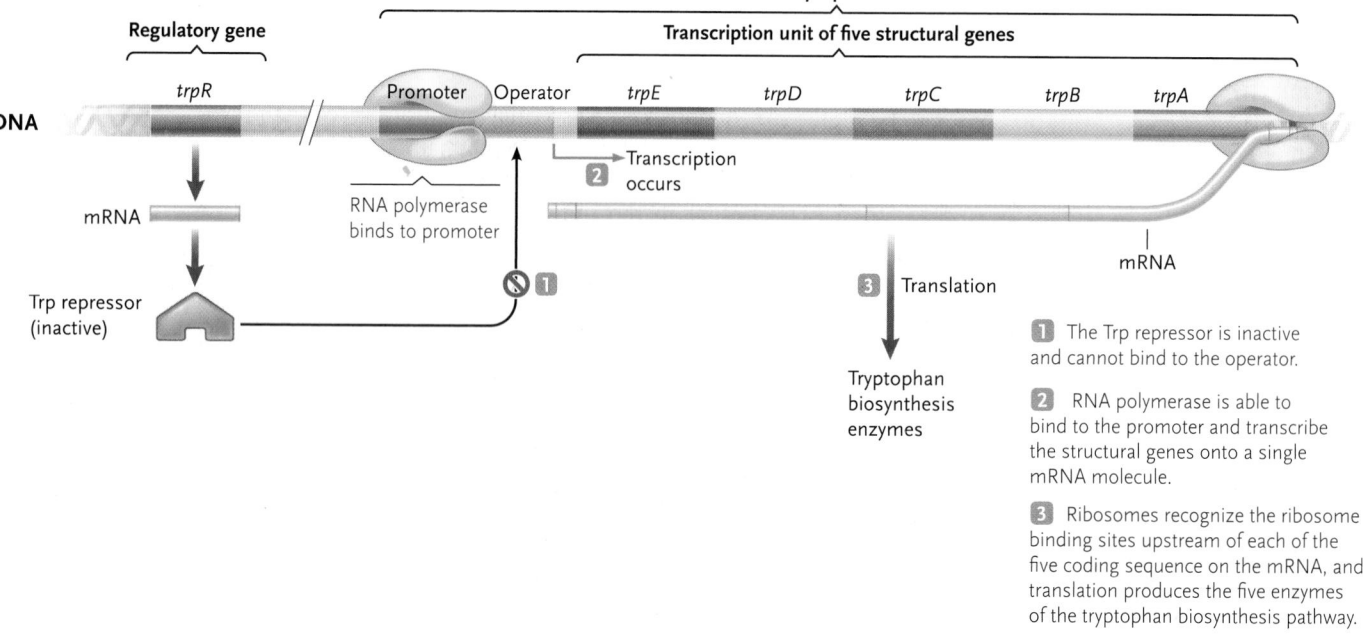

1. The Trp repressor is inactive and cannot bind to the operator.

2. RNA polymerase is able to bind to the promoter and transcribe the structural genes onto a single mRNA molecule.

3. Ribosomes recognize the ribosome binding sites upstream of each of the five coding sequence on the mRNA, and translation produces the five enzymes of the tryptophan biosynthesis pathway.

b. Tryptophan present in medium: cell uses tryptophan in medium rather than synthesizing it—structural genes not transcribed

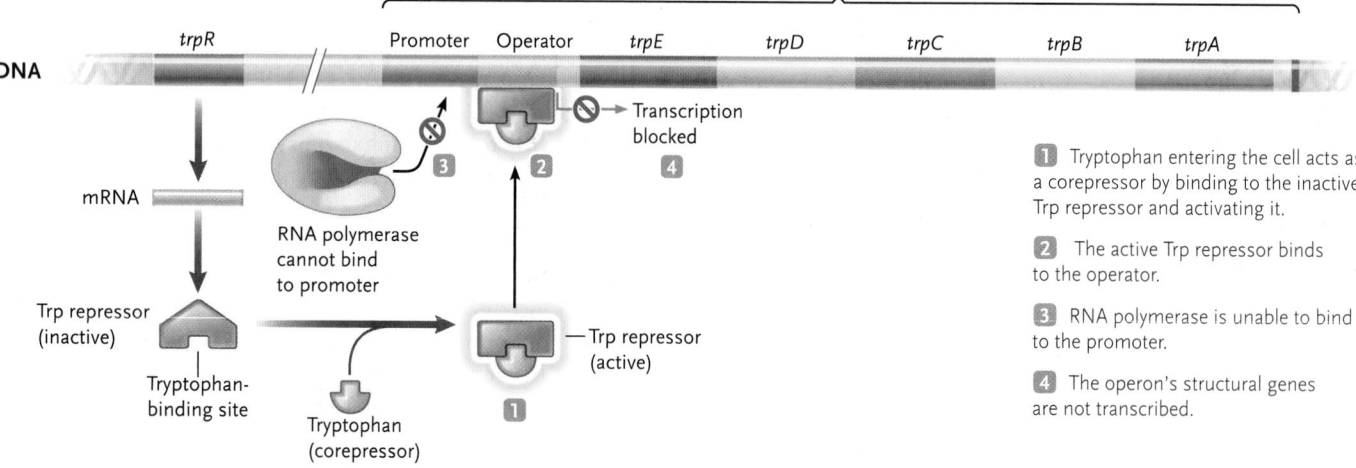

1. Tryptophan entering the cell acts as a corepressor by binding to the inactive Trp repressor and activating it.

2. The active Trp repressor binds to the operator.

3. RNA polymerase is unable to bind to the promoter.

4. The operon's structural genes are not transcribed.

Figure 14.5

Regulation of the repressible *trp* operon by the Trp repressor protein in the absence **(a)** and presence **(b)** of the amino acid tryptophan.

regulated by a repressor that turns off gene expression when it binds DNA. ⬡

In summary, regulation of gene expression in prokaryotic cells occurs primarily at the transcription level. There are also, however, some examples of regulation at the translation level. For example, some proteins can bind to the mRNAs that produce them and modulate their translation. This serves as a feedback mechanism to fine-tune the amounts of the proteins in the cell. In the remainder of the chapter, we discuss the regulation of gene expression in eukaryotes. You will see that regulation occurs at several points in the gene expression pathway and that regulatory mechanisms are more complex than those in prokaryotic cells.

STUDY BREAK

1. Suppose the *lacI* gene is mutated so that the Lac repressor is not made. How does this mutation affect the regulation of the *lac* operon?
2. Answer the equivalent question for the *trp* operon: How would a mutation that prevents the Trp repressor from being made affect the regulation of the *trp* operon?

Bacterial cells can communicate with one another through the production and detection of molecules called auto-inducers. When an autoinducer accumulates to high concentration in the local environment, it binds to membrane receptors that initiate a signal cascade, resulting in transcriptional activation of genes. This process, called quorum sensing, provides a mechanism for populations of cells to determine their density and thus coordinate gene expression as a community. For instance, although it is rather futile for an isolated single cell of *Vibrio harveyi* to express genes from its *lux* operon in order to bioluminesce,

hundreds of millions of cells, all expressing *lux* genes, collectively produce biologically significant amounts of light. Such large populations of bioluminescent bacteria are found in the light organs of squid. In a way, these populations of bacterial cells behave like multicellular organisms. Although various autoinducers are known to mediate communication among members of the same species, a novel compound, called AI-2 **(Figure 1),** has been found to facilitate communication between members of *different* species. AI-2 is unlike any other known autoinducer and is particularly interesting in that it

contains an atom of boron, an element whose function in biological systems has been quite mysterious.

Figure 1

AI-2, a universal autoinducer containing boron.

14.2 Regulation of Transcription in Eukaryotes

The molecular mechanisms in prokaryotic operon function are a simple means of coordinating synthesis of proteins with related functions. In eukaryotes, the coordinated synthesis of proteins with related functions also occurs, but without the need to organize genes under the control of a single promoter in an operon.

There are two general categories of eukaryotic gene regulation. Short-term regulation involves regulatory events in which gene sets are quickly turned on or off in response to changes in environmental or physiological conditions in the cell's or organism's environment. This type of regulation is most similar to prokaryotic gene regulation. Long-term gene regulation involves regulatory events required for an organism to develop and differentiate. Long-term gene regulation occurs in multicellular eukaryotes and not in simpler, unicellular eukaryotes. The mechanisms we discuss in this and the next section are applicable to both short-term and long-term regulation.

14.2a In Eukaryotes, Regulation of Gene Expression Occurs at Several Levels

The regulation of gene expression is more complicated in eukaryotes than in prokaryotic cells because eukaryotic cells are more complex, because the nuclear DNA is organized with histones into chromatin, and because multicellular eukaryotes produce large numbers and different types of cells. Further, the eukaryotic nuclear envelope separates the processes of transcription and translation, whereas in prokaryotic cells, translation

can start on an mRNA that is still being made. Consequently, gene expression in eukaryotes is regulated at more levels. That is, there is transcriptional regulation, posttranscriptional regulation, translational regulation, and posttranslational regulation **(Figure 14.6, p. 322).** The most important of these is transcriptional regulation.

14.2b Regulation of Transcription Initiation Involves the Effects of Proteins Binding to a Gene's Promoter and Regulatory Sites

Transcription initiation is the most common level at which the regulation of gene expression takes place.

Organization of a Eukaryotic Protein-Coding Gene. **Figure 14.7, p. 322,** shows a eukaryotic gene, emphasizing the regulatory sites involved in its expression. Immediately upstream of the transcription unit is the promoter. The promoter in the figure contains a **TATA box**, a sequence about 25 base pairs (bp) upstream of the start point for transcription that, as we will shortly see, plays an important role in transcription initiation in many promoters. The TATA box has the 7-bp consensus sequence

5'-TATAAAA-3'

3'-ATATTTT-5'

Promoters without TATA boxes have other sequence elements that play a similar role. In the following discussions, we describe transcription initiation involving a TATA box–containing promoter.

RNA polymerase II itself cannot recognize the promoter sequence. Instead, proteins called **transcription**

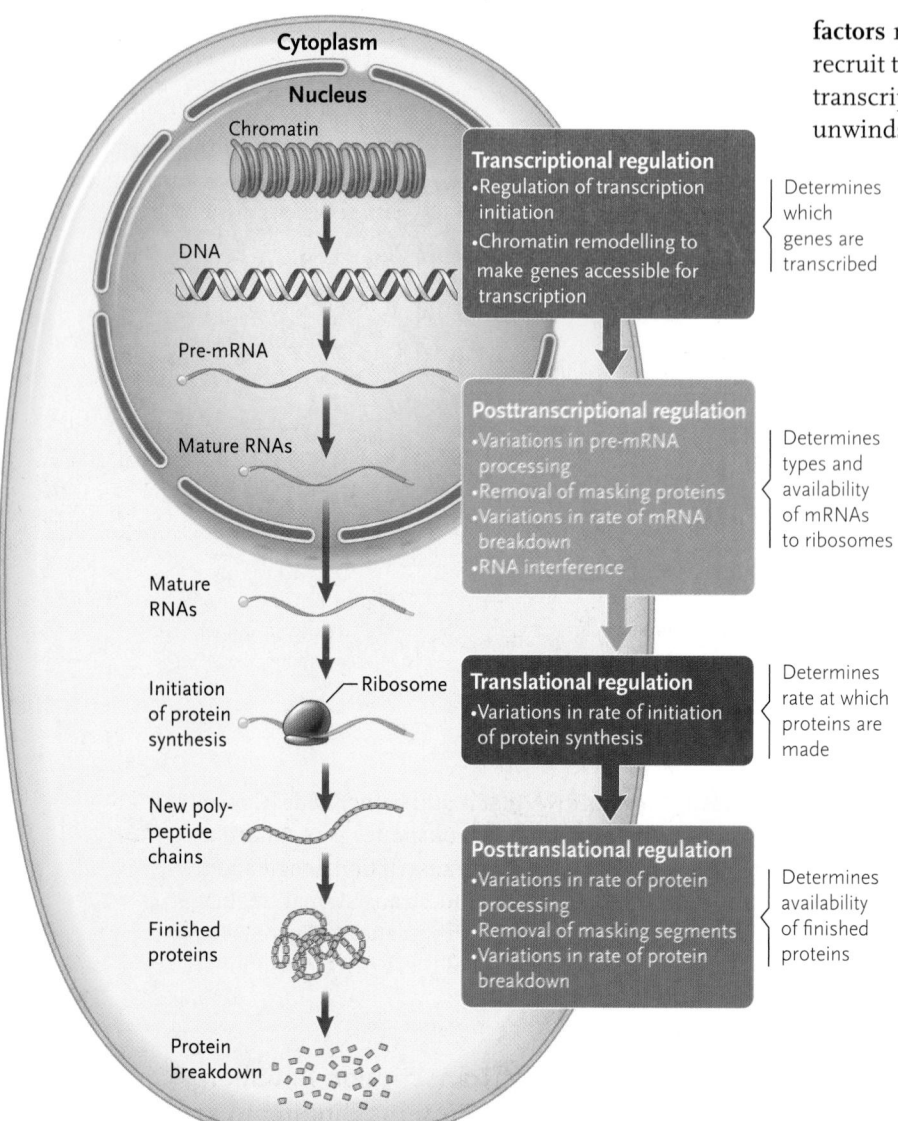

Figure 14.6

Steps in transcriptional, posttranscriptional, translational, and posttranslational regulation of gene expression in eukaryotes.

Transcriptional regulation
- Regulation of transcription initiation
- Chromatin remodelling to make genes accessible for transcription

Determines which genes are transcribed

Posttranscriptional regulation
- Variations in pre-mRNA processing
- Removal of masking proteins
- Variations in rate of mRNA breakdown
- RNA interference

Determines types and availability of mRNAs to ribosomes

Translational regulation
- Variations in rate of initiation of protein synthesis

Determines rate at which proteins are made

Posttranslational regulation
- Variations in rate of protein processing
- Removal of masking segments
- Variations in rate of protein breakdown

Determines availability of finished proteins

factors recognize and bind to the TATA box and then recruit the polymerase. Once the RNA polymerase II–transcription factor complex forms, the polymerase unwinds the DNA and transcription begins. Adjacent to the promoter, farther upstream, is the **promoter proximal region**, which contains regulatory sequences called **promoter proximal elements.** Regulatory proteins that bind to promoter proximal elements may stimulate or inhibit the rate of transcription initiation. More distant from the beginning of the gene is the **enhancer.** Regulatory proteins binding to regulatory sequences within an enhancer also stimulate or inhibit the rate of transcription initiation. Next we see more specifically how these regulatory sequences are involved in transcription initiation.

Activation of Transcription. To initiate transcription, proteins called **general transcription factors** (also called *basal transcription factors*) bind to the promoter in the area of the TATA box **(Figure 14.8).** These factors recruit the enzyme RNA polymerase II, which alone cannot bind to the promoter, and orient the enzyme to start transcription at the correct place. The combination of general transcription factors with RNA polymerase II is the **transcription initiation complex.** On its own, this complex brings about only a low rate of transcription initiation, which leads to just a few mRNA transcripts.

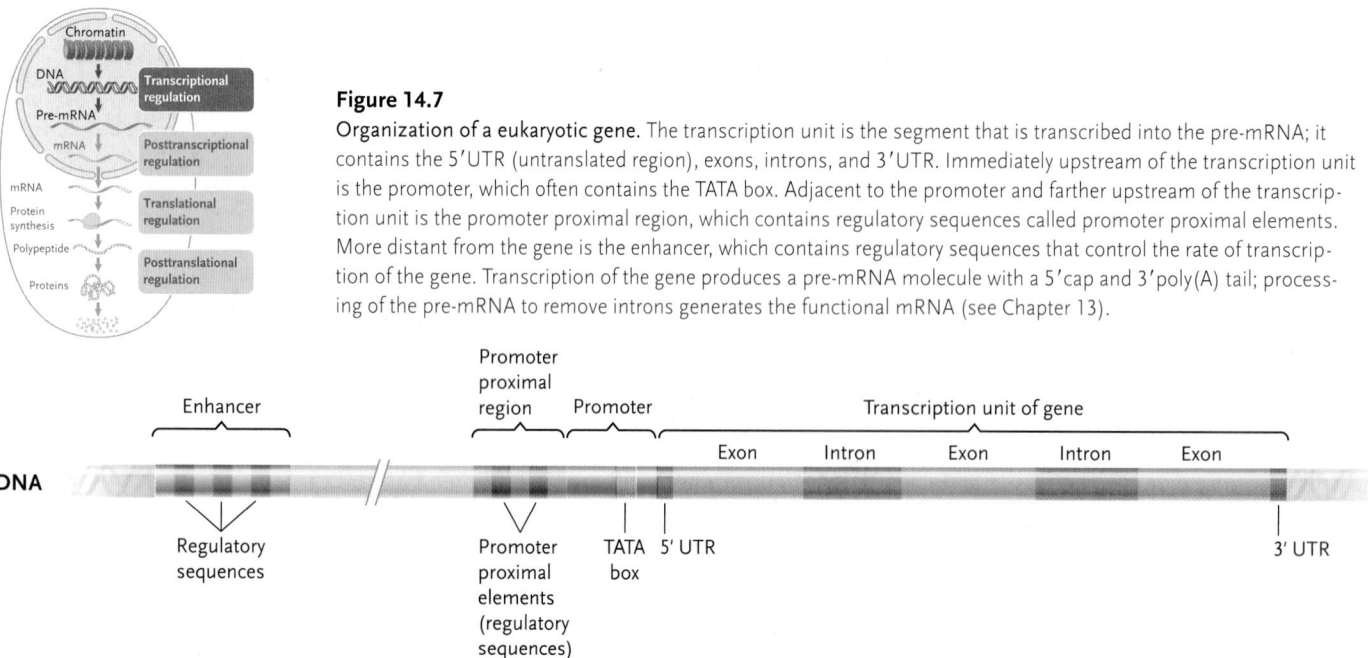

Figure 14.7

Organization of a eukaryotic gene. The transcription unit is the segment that is transcribed into the pre-mRNA; it contains the 5′UTR (untranslated region), exons, introns, and 3′UTR. Immediately upstream of the transcription unit is the promoter, which often contains the TATA box. Adjacent to the promoter and farther upstream of the transcription unit is the promoter proximal region, which contains regulatory sequences called promoter proximal elements. More distant from the gene is the enhancer, which contains regulatory sequences that control the rate of transcription of the gene. Transcription of the gene produces a pre-mRNA molecule with a 5′cap and 3′poly(A) tail; processing of the pre-mRNA to remove introns generates the functional mRNA (see Chapter 13).

Activators are regulatory proteins that play a role in a positive regulatory system that controls the expression of one or more genes. Activators that bind to the promoter proximal elements interact directly with the general transcription factors at the promoter to stimulate transcription initiation so many more transcripts are synthesized in a given time. Housekeeping genes—genes that are expressed in all cell types for basic cellular functions such as glucose metabolism—have promoter proximal elements that are recognized by activators present in all cell types. By contrast, genes expressed only in particular cell types or at particular times have promoter proximal elements that are recognized by activators found only in those cell types, or at those times when transcription of these genes needs to be activated. To turn this around, the particular set of activators present within a cell at a given time is responsible for determining which genes in that cell are expressed to a significant level.

The DNA-binding and activation functions of activators are properties of two distinct domains in the proteins. (Protein domains are introduced in *The Purple Pages*.) The three-dimensional arrangement of amino acid chains within and between domains also produces highly specialized regions called **motifs.** Several types of motifs, each with a specialized function, are found in proteins, including motifs that insert into the DNA double helix. Motifs found in the DNA-binding domains of regulatory proteins, such as activators, include the helix-turn-helix, zinc finger, and leucine zipper **(Figure 14.9).**

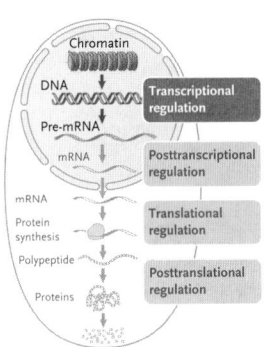

Figure 14.8
Formation of the transcription complex on the promoter of a protein-coding gene by the combination of general transcription factors with RNA polymerase. The general transcription factors are needed for RNA polymerase to bind and initiate transcription at the correct place.

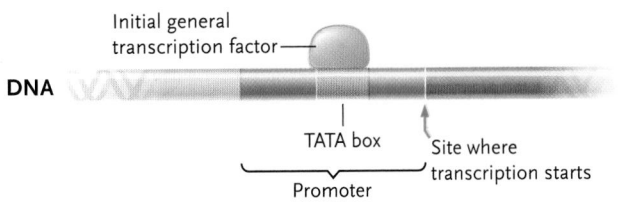

1 The first general transcription factor recognizes and binds to the TATA box of a protein-coding gene's promoter.

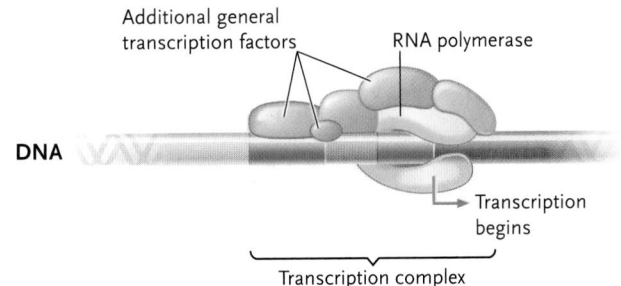

2 Additional general transcription factors and then RNA polymerase add to the complex. A general transcription factor unwinds the promoter DNA, and then transcription begins.

a. Helix-turn-helix

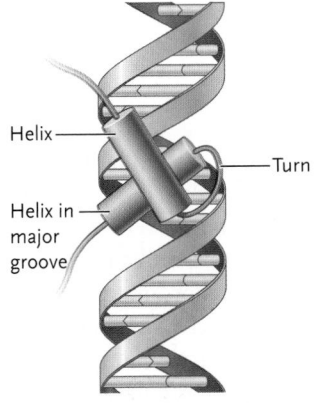

A helix-turn-helix motif is part of a protein bound to DNA. One of the α-helices binds to base pairs in the major groove of the DNA. A looped region of the protein—the turn—connects to a second α-helix, which helps hold the first helix in place.

b. Zinc finger

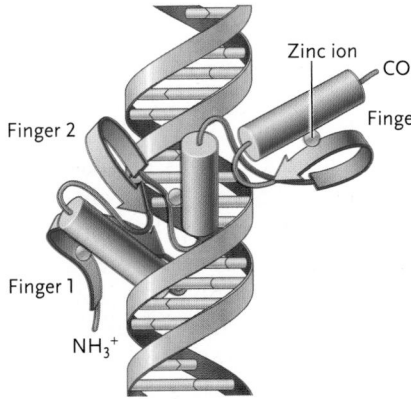

Zinc finger motifs are parts of proteins named for their resemblance to fingers projecting from a protein, and the presence of a bound zinc atom. Zinc fingers bind to specific base pairs in the grooves of DNA.

c. Leucine zipper

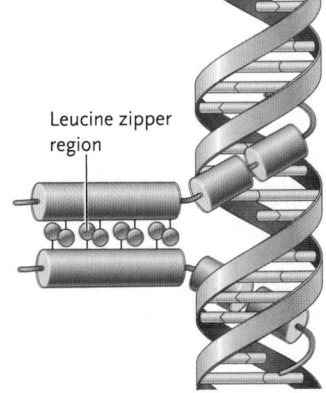

Leucine zipper proteins are dimers, with each monomer consisting of α-helical segments. Hydrophobic interactions between leucine residues within the leucine zipper motif hold the two monomers together. Other α-helices bind to DNA base pairs in the major groove.

Figure 14.9
Three DNA-binding motifs found in activators and other regulatory proteins.

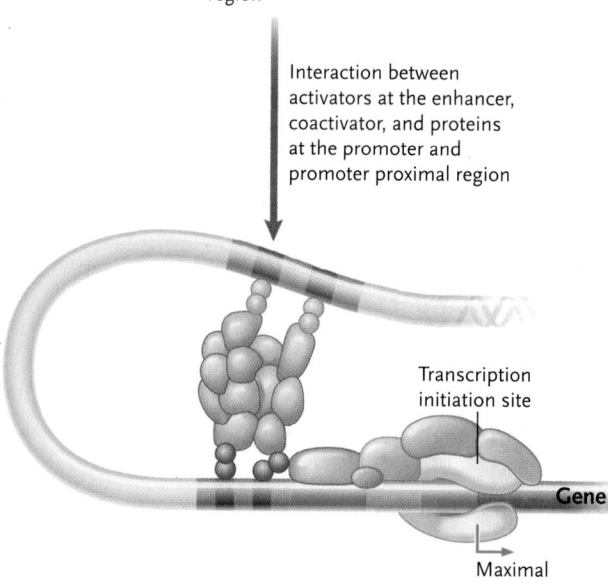

Figure 14.10

Interactions between activators at the enhancer, a coactivator, and general transcription factors at the promoter lead to maximal transcription of the gene.

Coactivator
(multiprotein complex)

Transcription
initiation site

Activator

Activator

Gene

Enhancer

Promoter
proximal
region

Promoter

Interaction between
activators at the enhancer,
coactivator, and proteins
at the promoter and
promoter proximal region

DNA
loop

Transcription
initiation site

Gene

Maximal
transcription

Activators binding at the enhancer greatly increase transcription rates **(Figure 14.10)**. The enhancers of different genes have different sets of regulatory sequences, which bind particular activators. A **coactivator** (also called a *mediator*), a large multiprotein complex, forms a bridge between the activators at the enhancer and the proteins at the promoter and promoter proximal region, causing the DNA to form a loop. The interactions between the activators at the enhancer, the coactivator, the proteins at the promoter, and the RNA polymerase greatly stimulate transcription up to its maximal rate.

Repression of Transcription. In some genes, repressors oppose the effect of activators, thereby blocking or reducing the rate of transcription. The final rate of transcription then depends on the "battle" between the activation signal and the repression signal.

Repressors in eukaryotes work in various ways. Some repressors bind to the same regulatory sequence to which activators bind (often in the enhancer), thereby preventing activators from binding to that site. Other repressors bind to their own specific site in the DNA near where the activator binds and interact with the activator so that it cannot interact with the coactivator. Yet other repressors bind to specific sites in the DNA and recruit corepressors, multiprotein complexes analogous to coactivators except that they are negative regulators, inhibiting transcription initiation.

Combinatorial Gene Regulation. Let's review the key elements of transcription regulation for a protein-coding gene. General transcription factors bind to certain promoter sequences, such as the TATA box, and recruit RNA polymerase II; this results in a basal level of transcription. Specific activators bind to promoter proximal elements and stimulate the rate of transcription initiation. Activators also bind to the enhancer to greatly stimulate transcription of the gene.

How are these events coordinated in regulating gene expression? Any given gene has a specific number and types of promoter proximal elements. In some genes, there may be only one regulatory element, but genes under complex regulatory control have many regulatory elements. Similarly, the number and types of regulatory sequences in the enhancer are specific for each gene.

Both promoter proximal regions and enhancers are important in regulating the transcription of a gene. Each regulatory sequence in these two regions binds a

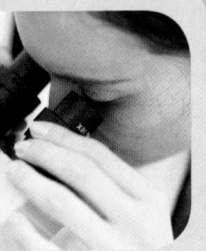
14.2c Methylation of DNA Can Control Gene Transcription

Although binding proteins to DNA is a common mechanism for regulating transcription, similar effects can also be achieved by modifying the DNA directly. In vertebrates, plants, fungi, and bacteria, **DNA methylation** enzymes add a methyl group (–CH$_3$) directly onto bases in the DNA. Methylated bases in promoter regions can prevent the binding of transcription factors, turning the gene off.

Silencing by methylation is common among vertebrates, but it is not universal among eukaryotes; very little methylation is found in the model organisms *Drosophila* and *Caenorhabditis elegans*.

14.2d Chromatin Structure Plays an Important Role in Whether a Gene Is Active or Inactive

Eukaryotic DNA is organized into chromatin by combination with histone proteins (discussed in Section 12.5). Recall that DNA is wrapped around a core of two molecules each of histones H2A, H2B, H3, and H4, forming the nucleosome (see Figure 12.21). Higher levels of chromatin organization occur when histone H1 links adjacent nucleosomes.

A eukaryotic promoter can exist in two states. In the inactive state, which is the normal state in eukaryotic cells, the nucleosomes in normal chromatin prevent general transcription factors and RNA polymerase II from binding so transcription does not occur. However, regulatory transcription factors can bind to the DNA and lead to a change in chromatin to make it active so transcription can occur. In the active state, general transcription factors and RNA polymerase II bind to the promoter and, controlled by the molecular events already discussed, transcription regulation can occur. A key regulatory event for regulating transcription initiation, then, is controlling the transition between the inactive and active states of chromatin in the region of a promoter.

Acetylation of histone tails (see Figure 12.21) is one mechanism that plays an important role in

determining whether chromatin is inactive or active. In inactive chromatin, the histone tails are not acetylated and, in this form, the tails form a tight association with the DNA wrapped around the histone octamer of a nucleosome **(Figure 14.13)**. When a regulatory transcription factor binds to a regulatory sequence associated with a gene, it can recruit protein complexes that include *histone acetyltransferase*, an enzyme that acetylates (adds acetyl groups; CH$_3$COO) to specific amino acids of the histone tails. Acetylation changes the charge of the histone tails and results in a loosening of the association of the histones with the DNA (see Figure 14.13). Usually acetylation of histones is not enough to make the chromatin completely active. Typically large multiprotein complexes bind to displace the acetylated nucleosomes in the promoter region from the DNA, or move them along the DNA away from the promoter. This type of change in chromatin structure is called **chromatin remodelling.** Then, general transcription factors and RNA polymerase II are free to bind and initiate transcription.

Inactivation of an active gene involves essentially the opposite of this process. With respect to the histones, the enzyme *histone deacetylase* catalyzes the removal of acetyl groups from the histone tails, restoring the inactive state of the chromatin in that region (see Figure 14.13).

The tails of histones can also be modified at specific positions by the enzyme-catalyzed covalent addition of methyl groups or phosphate groups. These chemical modifications can also affect chromatin structure and gene expression. Histone methylation, for instance, is associated with gene inactivation. Like acetylation, methylation and phosphorylation of histone tails are reversible. Overall, the conclusion is that the patterns of modification of histone tails are important in determining chromatin structure and gene activity. This has led to the concept of the **histone code**, which is a regulatory mechanism for altering chromatin structure and, therefore, gene activity, based on signals in histone tails represented by chemical modification patterns.

Once mRNAs are transcribed from active genes, further regulation occurs at each of the major steps in

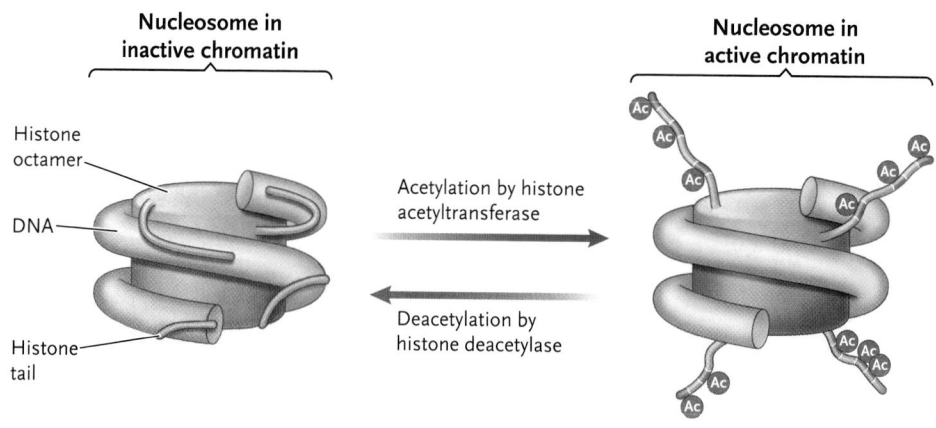

Figure 14.13

Conversion of inactive chromatin to active chromatin by acetylation of histone tails, and the reverse by deacetylation of histone tails.

Nucleosome in inactive chromatin

Histone octamer

DNA

Histone tail

Acetylation by histone acetyltransferase

Deacetylation by histone deacetylase

Nucleosome in active chromatin

the pathway from genes to proteins: during pre-mRNA processing and the movement of finished mRNAs to the cytoplasm (posttranscriptional regulation), during protein synthesis (translational regulation), and after translation is complete (posttranslational regulation). The next section takes up the regulatory mechanisms operating at each of these steps.

STUDY BREAK

1. What is the role of histones in gene expression? How does acetylation of the histones affect gene expression?
2. What are the roles of general transcription factors, activators, and coactivators in transcription of a protein-coding gene?

14.3 Posttranscriptional, Translational, and Posttranslational Regulation

The previous sections describe several mechanisms that determine which mRNAs are produced under various conditions. The following sections illustrate that once a given mRNA is made, there are several opportunities to fine-tune expression through posttranscriptional, translational, and posttranslational controls (refer again to Figure 14.6, p. 322).

14.3a Posttranscriptional Regulation Controls mRNA Availability

Posttranscriptional regulation directs translation by controlling the availability of mRNAs to ribosomes. The controls work by several mechanisms, including changes in pre-mRNA processing and the rate at which mRNAs are degraded.

Variations in Pre-mRNA Processing. In Chapter 13, we noted that mRNAs are transcribed initially as pre-mRNA molecules. These pre-mRNAs are

variously processed to produce the finished mRNAs, which then enter protein synthesis. Variations in pre-mRNA processing can regulate *which* proteins are made in cells. As described in Section 13.3, pre-mRNAs can be processed by *alternative splicing*. Alternative splicing produces different mRNAs from the same pre-mRNA by removing different combinations of exons (the amino acid–coding segments) along with the introns (the noncoding spacers). The resulting mRNAs are translated to produce a family of related proteins with various combinations of amino acid sequences derived from the exons. Alternative splicing itself is under regulatory control. Regulatory proteins specific to the type of cell control which exons are removed from pre-mRNA molecules by binding to regulatory sequences within those molecules. The outcome of alternative splicing is that appropriate proteins within a family are synthesized in cell types or tissues in which they function optimally. Perhaps three-quarters of human genes are alternatively spliced at the pre-mRNA level.

Posttranscriptional Control by Masking Proteins. Some posttranscriptional controls operate by means of *masking* proteins that bind to mRNAs and make them unavailable for protein synthesis. These controls are important in many animal eggs, keeping mRNAs in an inactive form until the egg has been fertilized and embryonic development is under way. When an mRNA is to become active, other factors—other proteins, made as part of the developmental pathway—remove the masking proteins and allow the mRNA to enter protein synthesis.

Variations in the Rate of mRNA Breakdown. The rate at which eukaryotic mRNAs break down can also be controlled posttranscriptionally. The mechanism involves a regulatory molecule, such as a steroid hormone, directly or indirectly affecting the mRNA breakdown steps, either slowing or increasing the rate of those steps. For example, in the mammary gland of the rat, the mRNA for casein (a milk protein) has a half-life of about 5 hours (meaning that it takes 5 hours for half of the mRNA present at a given time to break down). The half-life of casein mRNA changes to about 92 hours if the peptide hormone prolactin is present. Prolactin is synthesized in the brain and in other tissues, including the breast. The most important effect of prolactin is to stimulate the **mammary glands** to produce milk (that is, it stimulates lactation). During milk production, a large amount of casein must be synthesized, and this is accomplished in part by radically decreasing the rate of breakdown of the casein mRNA.

Nucleotide sequences in the 5′ UTR (untranslated region; see Section 13.3) also appear to be important in determining mRNA half-life. If the 5′ UTR is transferred experimentally from one mRNA to another, the half-life of the receiving mRNA becomes the same as that of the donor mRNA. The controlling sequences in the 5′ UTR of an mRNA might be recognized by proteins that regulate its stability.

Regulation of Gene Expression by Small RNAs.
Until relatively recently, the commonly accepted view was that regulation of gene expression in prokaryotic and eukaryotic cells involved only protein-based mechanisms. However, in 1998, Andrew Fire of the Stanford University School of Medicine and Craig Mello of the University of Massachusetts Medical School showed that RNA silenced the expression of a particular gene in the nematode worm, *C. elegans*. They called the phenomenon **RNA interference (RNAi)**. Their discovery revolutionized the way scientists thought about and studied gene regulation in eukaryotes. They now understand that posttranscriptional regulation may be carried out, not only by regulatory proteins, but also by noncoding single-stranded RNAs that can bind to mRNAs and affect their translation. We now know that RNAi is widespread among eukaryotes. Fire and Mello received a Nobel Prize in 2006 for their discovery of RNA interference.

Two major groups of small regulatory RNAs are involved in RNAi: **microRNAs (miRNAs)** and **short interfering RNAs (siRNAs)**. The transcription of an miRNA gene and the processing of the transcript to produce the functional miRNA molecule are shown in **Figure 14.14**. The miRNA, in a protein complex called the **miRNA-induced silencing complex (miRISC)**, binds to sequences in the 3′ UTRs of target mRNAs. If the miRNA and mRNA pair imperfectly, the double-stranded segment formed between the miRNA and the mRNA blocks ribosomes from translating the mRNA (shown in Figure 14.14). In this case, the target mRNA is not destroyed, but its expression is silenced. If the miRNA and mRNA pair perfectly, an enzyme in the protein complex cleaves the target mRNA where the miRNA is bound to it, destroying the mRNA and silencing its expression. RNAi by imperfect pairing and translation inhibition is the most common

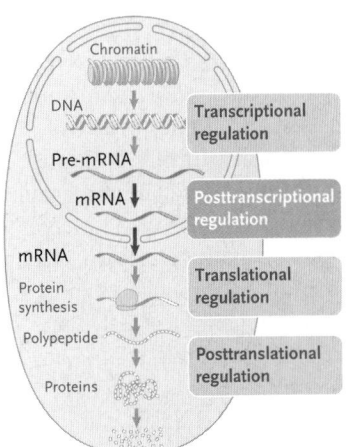

Figure 14.14
RNA interference—regulation of gene expression by microRNAs (miRNAs).

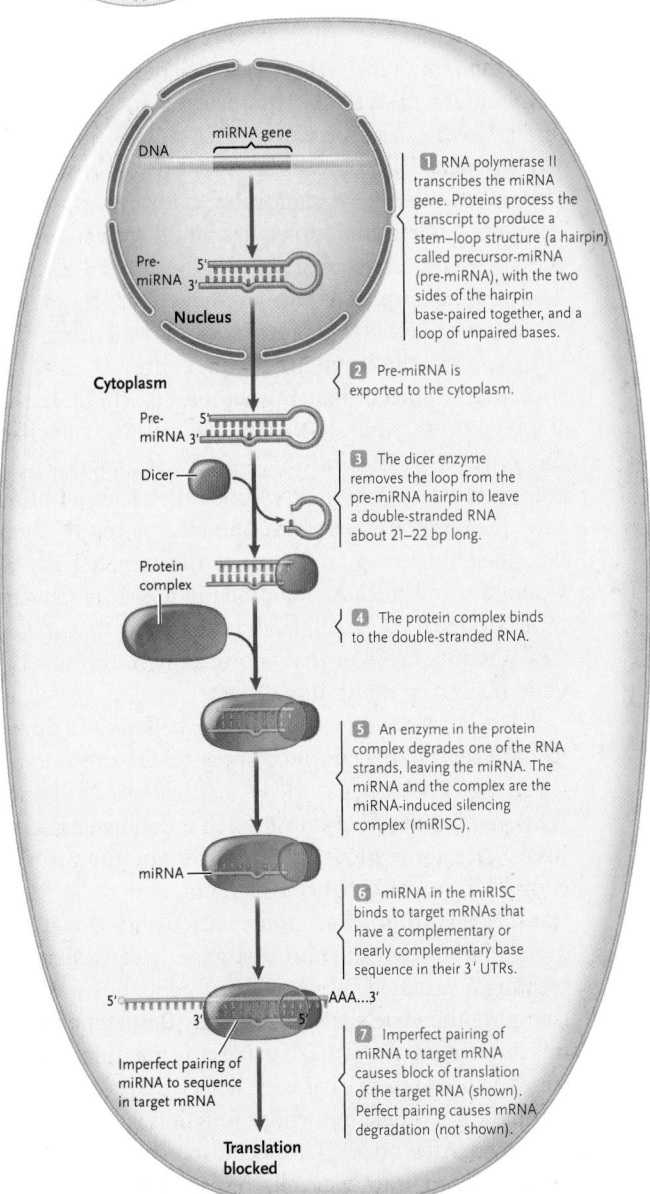

1. RNA polymerase II transcribes the miRNA gene. Proteins process the transcript to produce a stem–loop structure (a hairpin) called precursor-miRNA (pre-miRNA), with the two sides of the hairpin base-paired together, and a loop of unpaired bases.

2. Pre-miRNA is exported to the cytoplasm.

3. The dicer enzyme removes the loop from the pre-miRNA hairpin to leave a double-stranded RNA about 21–22 bp long.

4. The protein complex binds to the double-stranded RNA.

5. An enzyme in the protein complex degrades one of the RNA strands, leaving the miRNA. The miRNA and the complex are the miRNA-induced silencing complex (miRISC).

6. miRNA in the miRISC binds to target mRNAs that have a complementary or nearly complementary base sequence in their 3′ UTRs.

7. Imperfect pairing of miRNA to target mRNA causes block of translation of the target RNA (shown). Perfect pairing causes mRNA degradation (not shown).

mechanism in animals. RNAi by perfect pairing and RNA degradation is the most common mechanism in plants.

miRNA genes have been found in all multicellular eukaryotes that have been examined, and also in some unicellular ones. miRNAs play central roles in controlling gene expression in a variety of cellular, physiological, and developmental processes in animals and plants. In animals, for example, miRNAs help regulate specific developmental timing events, gene expression in neurons, brain development, cancer progression, and stem cell division.

The other major type of small regulatory RNAs is the siRNA. Whereas miRNA is produced from RNA that is encoded in the cell's genome, siRNA is produced from double-stranded RNA that is *not* encoded by nuclear genes. For example, the replication cycle of many viruses with RNA genomes involves a double-stranded RNA stage. Cells attacked by such a virus can defend themselves using siRNA that they produce from the virus's own RNA. The viral double-stranded RNA enters the cell's RNAi process in a way very similar to that described for miRNAs; double-stranded RNA is cut by Dicer (see Figure 14.14, p. 329) into short double-stranded RNA molecules, and then a protein complex binds to the molecules and degrades one of the RNA strands to produce single-stranded siRNA. The protein complex is similar to one that acts on the double-stranded RNA precursors of miRNAs. The siRNA with the protein complex in this case is the **siRNA-induced silencing complex (siRISC)**. In the RNAi process, the siRNA in the siRISC acts like the miRNA in the miRISC—single-stranded RNAs complementary to the siRNA are targeted and, in this case, the target RNA is cleaved and the pieces are then degraded. In our viral example, the targeted RNAs would be viral mRNAs for proteins the virus uses to replicate itself, or a single-stranded RNA that is the viral genome itself, or that is produced from the viral genome during replication.

The expression of any gene can be knocked down to low levels or knocked out completely in experiments involving RNAi with siRNA. To silence a gene, researchers introduce into the cell a double-stranded RNA that can be processed by Dicer and the protein complex into an siRNA complementary to the mRNA transcribed from that gene. Knocking down or knocking out the function of a gene is equivalent to creating a mutated version of that gene, but without changing the gene's DNA sequence. Researchers use this experimental approach to identify the functions of genes whose presence has been detected by sequencing complete genomes, but whose function is completely unknown. After an siRNA specific to a gene of interest is introduced into the cell, researchers look for a change in phenotype, such as properties relating to growth or metabolism. If such a change is seen, the researchers now have some insight into the gene's function, and they can investigate the gene with more focus. RNAi using siRNAs may also have some applications in medicine, perhaps to regulate the expression of genes associated with particular human diseases.

14.3b Translational Regulation Controls the Rate of Protein Synthesis

At the next regulatory level, translational regulation controls the rate at which mRNAs are used in protein synthesis. Translational regulation occurs in essentially all cell types and species. For example, translational regulation is involved in cell cycle control in all eukaryotes and in many processes during development in multicellular eukaryotes, such as red blood cell differentiation in animals. Significantly, many viruses exploit translational regulation to control their infection of cells and to shut off the host cell's own genes.

Let's consider the general role of translational regulation in animal development. During early development of most animals, little transcription occurs. The changes in protein synthesis patterns seen in developing cell types and tissues instead derive from the activation, repression, or degradation of maternal mRNAs, the mRNAs that were present in the mother's egg before fertilization. One important mechanism for translational regulation involves adjusting the length of the poly(A) tail of the mRNA. (Recall from Section 13.3 that the poly(A) tail—a string of adenine-containing nucleotides—is added to the 3' end of the pre-mRNA and is retained on the mRNA produced from the pre-mRNA after introns are removed.) That is, enzymes can change the length of the poly(A) tail on an mRNA in the cytoplasm in either direction: by shortening it or lengthening it. Increases in poly(A) tail length result in increased translation; decreases in length result in decreased translation. For example, during embryogenesis (the formation of the embryo) of the fruit fly, *Drosophila,* key proteins are synthesized when the poly(A) tails on the mRNAs for those proteins are lengthened in a regulated way. Evidence for this came from experiments in which poly(A) tail lengthening was blocked; the result was that embryogenesis was inhibited. But although researchers know that the length of poly(A) tails is regulated in the cytoplasm, how this process occurs is not completely understood.

14.3c Posttranslational Regulation Controls the Availability of Functional Proteins

Posttranslational regulation controls the availability of functional proteins in three primary ways: chemical modification, processing, and degradation. Chemical modification involves the addition or removal of chemical groups, which reversibly alters the activity of the protein. For example, you saw in Section 5.7 how the addition of phosphate groups to proteins involved in signal transduction pathways either stimulates or inhibits the activity of those proteins. Further, in Section 8.5, you learned how the addition of phosphate groups to target proteins plays a crucial role in regulating how a cell progresses through the cell division cycle. And in Section 14.2, you saw how acetylation of histones altered the properties of the nucleosome, loosening its association with DNA in chromatin.

In processing, proteins are synthesized as inactive precursors, which are converted to an active form under regulatory control. For example, you saw in Section 13.4 that the digestive enzyme pepsin is synthesized as pepsinogen, an inactive precursor that activates by removal of a segment of amino acids. Similarly, the glucose-regulating hormone insulin is synthesized as a precursor called proinsulin; processing of the precursor removes a central segment but leaves the insulin molecule, which consists of two polypeptide chains linked by disulfide bridges.

The rate of degradation of proteins is also under regulatory control. Some proteins in eukaryotic cells last for the lifetime of the individual, whereas others persist only for minutes. Proteins with relatively short cellular lives include many of the proteins regulating transcription. Typically, these short-lived proteins are marked for breakdown by enzymes that attach a "doom tag" consisting of a small protein called *ubiquitin* (**Figure 14.15,** step 1). The protein is given this name because it is indeed ubiquitous—present in almost the same form in essentially all eukaryotes. The ubiquitin tag labels the doomed proteins so that they are recognized and attacked by a *proteasome,* a large cytoplasmic complex of a number of different proteins (step 2). The proteasome unfolds the protein, and protein-digesting enzymes within the core digest the protein into small peptides. The peptides are released from the proteasome, and cytosolic enzymes further digest the peptides into individual amino acids, which are recycled for use in protein synthesis or oxidized as an energy source (step 3). The ubiquitin protein and proteasome are also recycled. Aaron Ciechanover and Avram Hershko, both of the Israel Institute of Technology in Haifa, Israel, and Irwin Rose of the University of California, Irvine, received a Nobel Prize in 2004 for the discovery of ubiquitin-mediated protein degradation.

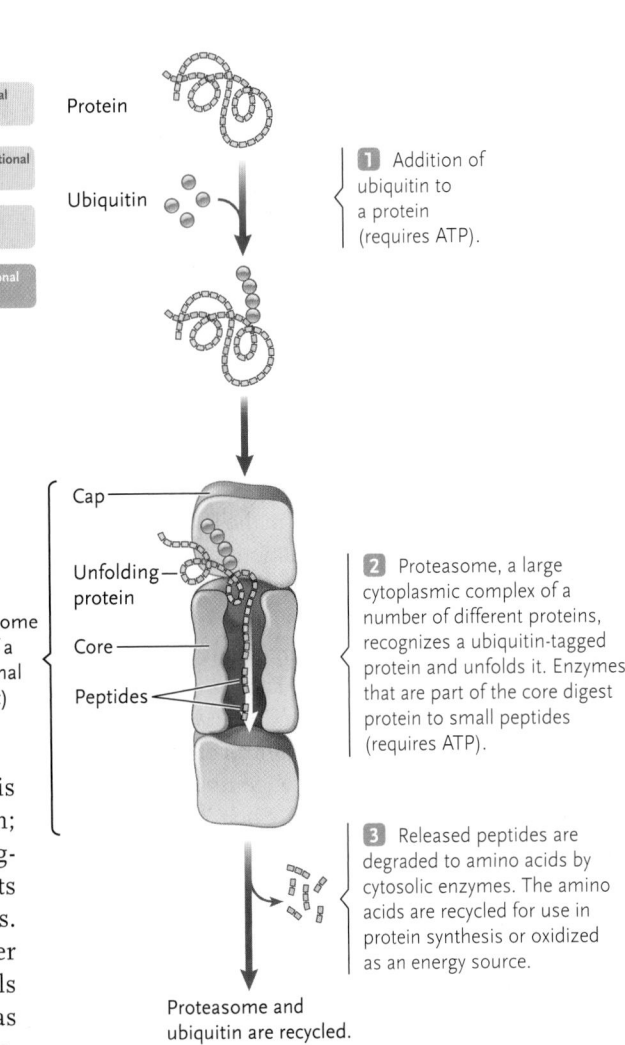

1 Addition of ubiquitin to a protein (requires ATP).

2 Proteasome, a large cytoplasmic complex of a number of different proteins, recognizes a ubiquitin-tagged protein and unfolds it. Enzymes that are part of the core digest protein to small peptides (requires ATP).

3 Released peptides are degraded to amino acids by cytosolic enzymes. The amino acids are recycled for use in protein synthesis or oxidized as an energy source.

Proteasome and ubiquitin are recycled.

Figure 14.15
Protein degradation by ubiquitin addition and enzymatic digestion within a proteasome.

We now describe cancer, a collection of diseases in which the control of gene expression goes awry.

14.3d Epigenetic Regulation Persists through Cell Division

The previous sections of this chapter outline several mechanisms for regulating changes in gene expression that are readily reversible and usually transient. *Lac* operon expression increases when lactose is present but decreases when lactose is absent. Certain genes are turned on when testosterone is present but then shut off again when hormone levels fall. However, there are many situations in which an established pattern of gene expression persists into the next cell or even organismal generation. For example, genes encoding the blood protein hemoglobin are present but inactive in most lines of vertebrate body cells. In the cell lines giving rise to red blood cells, hemoglobin genes are activated. In certain pathogenic strains of *E. coli*, genes for cell surface structures associated with virulence are turned off in most cells. Cells that happen to be involved in an active infection maintain expression of virulence genes in subsequent generations. In the case of genomic imprinting in mice (Section 11.5), an allele of a given gene is silenced during gametogenesis. This allele remains turned off in the fertilized zygote and resulting offspring. In all of these diverse examples, the appropriate pattern of gene expression is inherited from parent cells rather than having to be independently determined anew by each daughter cell.

This type of gene regulation is called **epigenetic** in that it persists through cell or organismal generations but does not result from changes in the DNA sequence. Although the notion of epigenetic regulation is not new, a wide variety of potential underlying mechanisms are still being actively investigated and debated. Two of the most well-documented epigenetic mechanisms are feedback loops and chromatin packaging.

Feedback Loops. One way that control of gene expression can be maintained over cell generations is by a self-sustaining regulatory loop. Such a loop arises when the product of a particular gene associates with its own promoter and stimulates its own transcription. A classic example of this comes from studies of phage lambda (λ), which can remain dormant in the chromosome of infected *E. coli* cells for several generations (Figure 23.5). Early in the infection cycle of this phage, the lambda repressor protein (cI) shuts down expression of other phage genes but stimulates expression of its own gene. As a result, high levels of repressor protein are maintained, even as the host cell divides, replicating the phage DNA along with the rest of its chromosome.

Comparable feedback loops are important in the differentiation of multicellular tissue types and figure prominently in the dramatic dedifferentiation of mouse adult fibroblast cells into stem cells. The creation of these induced stem cells, resulting from the introduction of only four transcription factors, won a share of the 2012 Nobel Prize in Medicine and Physiology for Shinya Yamanaka from Tokyo University, Japan.

Chromatin Packaging. Another strategy to achieve longer-term control of gene expression is to regulate the packaging of DNA and its associated proteins, known as chromatin. Section 12.5 describes the role of nucleosomes in chromatin structure. Figure 14.13 shows that this packaging is affected by chemical modifications of the histone proteins that make up nucleosomes. Densely compacted regions are called heterochromatin, and (most) genes in these regions are silenced. The specific genomic regions to be compacted into heterochromatin in a given cell are

identified by DNA-binding proteins or small RNAs that, in turn, recruit additional protein complexes responsible for modifying histones and/or DNA. Histones may be modified by acetylation, methylation, or phosphorylation. DNA modification is typically methylation of cytosine in vertebrates or adenine in bacteria (Section 14.2c).

One dramatic example of regulation that is maintained through many cell divisions is the inactivation of specific X chromosomes in mammalian females described in Section 11.2e. Although the detailed mechanism is still being discovered, it is now certain that silencing of this entire chromosome involves tightly compacting chromatin in concert with DNA methylation, histone modification, and the association of a noncoding RNA called *Xist*. Mechanisms to explain how epigenetic modifications are maintained through mitosis, binary fission, and meiosis are the subject of active research.

STUDY BREAK

1. How does miRNA silence gene expression?
2. If the poly(A) tail on an mRNA were removed, what would likely be the effect on the translation of that mRNA?

14.4 The Loss of Regulatory Controls in Cancer

Chapter 8 showed that the cell division cycle in all eukaryotes is carefully regulated by genes (see Section 8.5 and Figure 8.18). For normal cells, it is the balance between internal or external factors that stimulate cell division and corresponding factors that inhibit cell division that governs whether the cell remains in a nondividing state or whether it grows and divides.

Occasionally, differentiated cells of complex multicellular organisms deviate from their normal genetic program and begin to grow and divide inappropriately, giving rise to tissue masses called *tumours* (see Figure 8.19). Such cells have lost their normal regulatory controls and have reverted toward an embryonic developmental state in a process called *dedifferentiation* (**Figure 14.16**). If the altered cells stay together in a single mass, the tumour is *benign*. Benign tumours are usually not life-threatening, and their surgical removal generally results in a complete cure.

However, if the cells of a tumour invade and disrupt surrounding tissues, the tumour is *malignant* and is called a cancer. Sometimes, cells from malignant tumours break off and move through the blood system or lymphatic system, forming new tumours at other

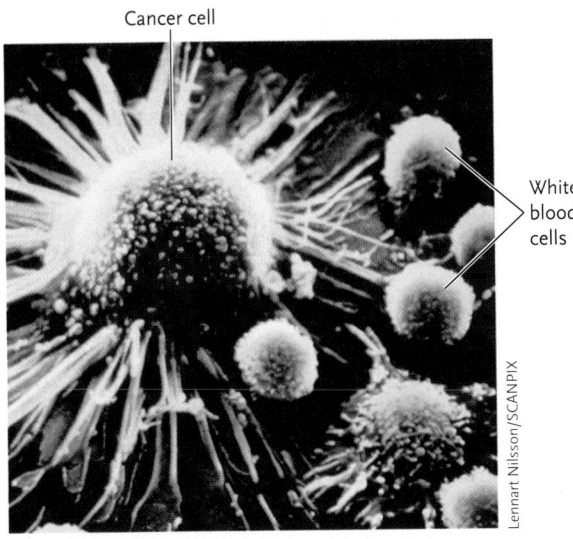

Cancer cell

White blood cells

Lennart Nilsson/SCANPIX

Figure 14.16

A scanning electron micrograph of a cancer cell surrounded by several white blood cells.

All of the characteristics of cancer cells that have been mentioned—dedifferentiation, uncontrolled division, and metastasis—reflect changes in gene expression.

14.4b Three Main Classes of Genes Are Implicated in Cancer

Three major classes of genes are altered frequently in cancers: *proto-oncogenes, tumour suppressor genes,* and miRNA genes.

Proto-oncogenes. **Proto-oncogenes** (*onkos* = bulk or mass) are genes in normal cells that encode various kinds of proteins that stimulate cell division. Examples are growth factors, receptors on target cells that are activated by growth factors (see Chapter 5), components of cellular signal transduction pathways triggered by cell division stimulatory signals (see Chapter 5), and transcription factors that regulate the expression of the structural genes for progression through the cell cycle. In cancer cells, the proto-oncogenes are deregulated and become **oncogenes**, genes that stimulate the cell to progress to the cancerous state of the unregulated cell cycle. Only one of the two proto-oncogene alleles in a cell needs to be altered for the cellular changes to occur. Alterations that can convert a proto-oncogene into an oncogene include the following:

- A mutation in a gene's promoter or other control sequences results in the gene becoming abnormally active.
- A mutation in the coding segment of the gene may produce an altered form of the encoded protein that is abnormally active.
- Translocation, a process in which a segment of a chromosome breaks off and attaches to a different chromosome (discussed in Section 11.3), may move the gene to a new location under the control of an inappropriately powerful promoter or enhancer sequence.
- Infecting viruses may introduce genes whose expression disrupts cell cycle control or alters regulatory proteins to turn genes on in the host.

Tumour Suppressor Genes. **Tumour suppressor genes** are genes in normal cells encoding proteins that inhibit cell division. The best known tumour suppressor gene is *TP53*, so called because its encoded protein, p53, has a molecular weight of 53 000 **daltons**. Among other activities, normal p53 stops cell division by combining with and inhibiting cyclin-dependent protein kinases that trigger the cell's transition from the G_1 phase to the S phase of the cell cycle (discussed in Section 8.5). This activity is particularly important if the cell has sustained DNA damage. If such a cell undergoes DNA replication and divides, the damage

locations in the body. The spreading of a malignant tumour is called *metastasis* (meaning "change of state"). Malignant tumours can result in debilitation and death in various ways, including damage to critical organs, metabolic imbalances, hemorrhage, and secondary malignancies. In some cases, malignant tumours can be eliminated from the body by surgery or be destroyed by chemicals (*chemotherapy*) or radiation.

14.4a Cancers Are Genetic Diseases

Experimental evidence of various kinds shows that cancers are genetic diseases:

1. Particular cancers can have a high incidence in some human families. Cancers that run in families are known as **familial (hereditary) cancers**. Cancers that do not appear to be inherited are known as **sporadic (nonhereditary) cancers**. Familial cancers are less frequent than sporadic cancers.
2. Descendants of cancer cells are all cancer cells. In fact, it is the cloned descendants of certain cancer cells that form a tumour.
3. The incidence of cancers increases upon exposure to mutagens, agents that cause mutations in DNA. Particular chemicals and certain kinds of radiation are effective mutagens.
4. Particular chromosomal mutations are associated with specific forms of cancer (see Section 11.3 and Figure 11.12). In these cases, chromosomal breakage affects the expression of genes associated with the regulation of cell division.
5. Some viruses can induce cancer. Some viruses carry "cancer genes" with them, while others contain viral genes that disrupt normal cell cycle control of host cells.

may result in mutation in progeny cells. Mutations can deregulate gene expression and cause a cell to progress toward cancer. However, such cancers may be avoided if p53's action to block the cell from entering S phase gives the cell time to repair the damage or, if the damage cannot be repaired, to trigger the cell to undergo programmed cell death (apoptosis: see Chapter 8). If the *TP53* gene is mutated so that the p53 protein is not produced or is produced in an inactive form, the cyclin-dependent protein kinases are continually active in triggering cell division. As a result, many mutations can result in progeny cells. Inactive *TP53* genes are found in at least 50 percent of all cancers. In general, mutations of tumour suppressor genes contribute to the onset of cancer because the mutations result in a decrease in the inhibitory action of the cell cycle controlling proteins they encode.

Both alleles of a tumour suppressor gene must be inactivated for inhibitory activity to be lost in cancer cells. **Figure 14.17** illustrates inactivation of the tumour-suppressor gene *BRCA1* (*breast cancer 1*) in sporadic and familial forms of breast cancer. *BRCA1* is involved in repair of DNA damage. Inactivating both alleles of *BRCA1* is not by itself sufficient for the development of breast cancer but is one of the gene changes typically involved. Since sporadic breast cancer requires the mutational inactivation of two normal alleles of *BRCA1*, this form of the disease typically occurs later in life than the familial form. For familial breast cancer and other familial cancers, we use the term *predisposition*

for the cancer. This term relates to the inactivation mechanism just described. That is, an individual is predisposed to develop a particular cancer if they inherit one mutant allele of an associated tumour suppressor disease, because then a mutation inactivating the other allele is all that is needed to lose the growth inhibitory properties of the tumour suppressor gene's product.

miRNA Genes. Earlier in this chapter, we discussed the role of miRNAs in regulating expression of target mRNAs. In human cancers, many miRNA genes show altered, cancer-specific expression patterns. Studying these miRNA genes has given scientists insight into the normal activities of their encoded miRNAs in cell cycle control. Some miRNAs regulate the expression of mRNAs that are the transcripts of tumour suppressor genes. If these miRNAs are overexpressed because of alterations of the genes encoding them, expression of the target mRNAs can be completely blocked, thereby removing or decreasing inhibitory signals for cell proliferation. Other miRNAs regulate the translation of mRNAs that are transcripts of particular proto-oncogenes. If these miRNA genes are inactivated, or expression of these genes is markedly reduced, expression of the proto-oncogenes is higher than normal and cell proliferation is stimulated.

CONCEPT FIX This section shows that *cancer genes* are just deregulated versions of the genes that are essential for the normally controlled growth of all cells. ⬡

Figure 14.17

Mutational inactivation of tumour suppressor gene alleles in sporadic **(a)** and familial **(b)** cancers as exemplified by the *BRCA1* gene associated with breast cancer.

a. Sporadic breast cancer. Two independent mutations of the *BRCA1* tumour suppressor.

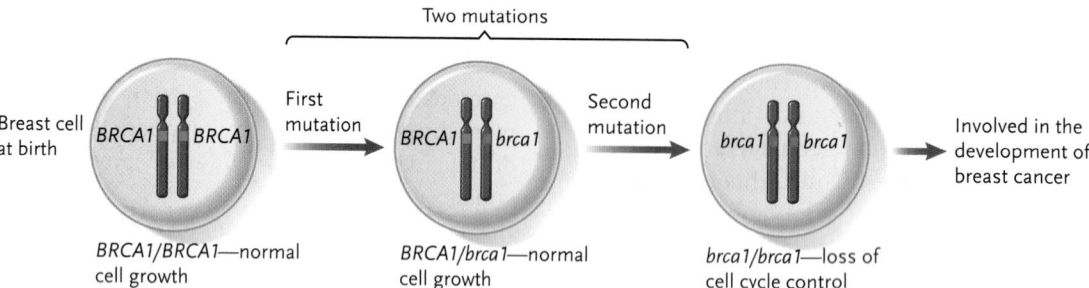

b. Familial breast cancer. An individual has a predisposition for breast cancer because of inheriting one mutated *brca1* allele; mutation of the other normal *BRCA1* allele then occurs.

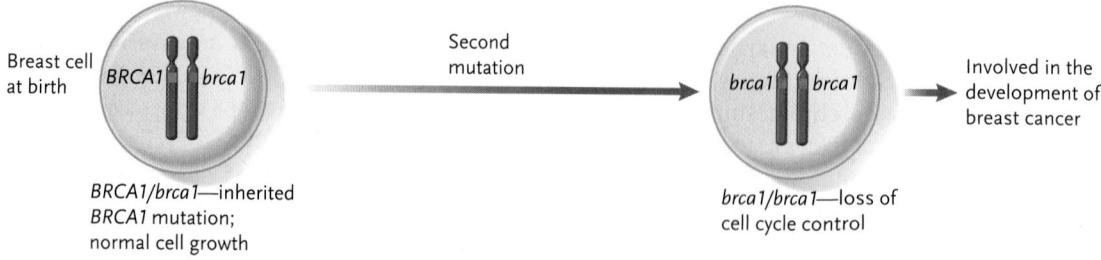

14.4c Cancer Develops Gradually by Multiple Steps

Cancer rarely develops by alteration of a single proto-oncogene to an oncogene, or inactivation of a single tumour suppressor gene. Rather, in almost all cancers, successive alterations in several to many genes gradually accumulate to transform normal cells to cancer cells. This gradual mechanism is called the *multistep progression of cancer*. One example of the steps that can occur, in this case for a form of colorectal cancer, is shown in **Figure 14.18.**

The ravages of cancer, probably more than any other example, bring home the critical extent to which humans and all other multicellular organisms depend on the mechanisms controlling gene expression to develop and live normally. In a sense, the most amazing thing about these control mechanisms is that in spite of their complexity, they operate without failures throughout most of the lives of all eukaryotes.

STUDY BREAK

1. What is the normal function of a tumour suppressor gene? How do mutations in tumour suppressor genes contribute to the onset of cancer?
2. What is the normal function of a proto-oncogene? How can mutations in proto-oncogenes contribute to the onset of cancer?
3. How can changes in expression of miRNA genes contribute to the onset of cancer?

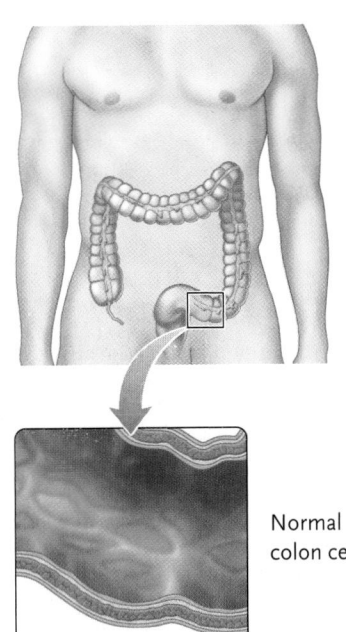

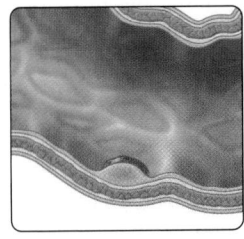

Normal colon cells

Loss of the *APC* tumour suppressor gene activity, and other DNA changes

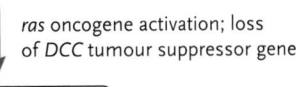

Small adenoma (benign growth)

ras oncogene activation; loss of *DCC* tumour suppressor gene

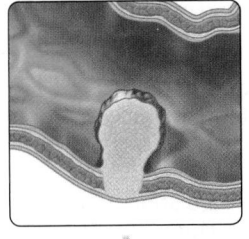

Large adenoma (benign growth)

Loss of *TP53* tumour suppressor gene activity and other mutations

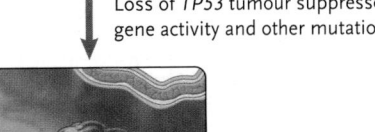

Carcinoma (malignant tumour with metastasis)

Figure 14.18
A multistep model for the development of a type of colorectal cancer.

Review

To access course materials such as Aplia and other companion resources, please visit www.NELSONbrain.com.

14.1 Regulation of Gene Expression in Prokaryotic Cells

- Transcriptional control in prokaryotic cells involves short-term changes that turn specific genes on or off in response to changes in environmental conditions. The changes in gene activity are controlled by regulatory proteins that recognize operators of operons (Figure 14.2).

- Regulatory proteins may be repressors, which slow the rate of transcription of operons, or activators, which increase the rate of transcription.

- Some repressors are made in an active form, in which they bind to the operator of an operon and inhibit its transcription. Combination with an inducer blocks the activity of the repressor and allows the operon to be transcribed (Figure 14.3).

- Activators are typically made in inactive form, in which they cannot bind to their binding site next to an operon. Combining with another molecule, often a nucleotide, converts the activator into the form in which it binds with its binding site and recruits RNA polymerase, thereby stimulating transcription of the operon (Figure 14.4).

- Other repressors are made in an inactive form, in which they are unable to inhibit transcription of an operon unless they combine with a corepressor (Figure 14.5).

14.2 Regulation of Transcription in Eukaryotes

- Operons are not found in eukaryotes. Instead, genes that encode proteins with related functions are typically scattered through the genome, while being regulated in a coordinated manner.

- Two general types of gene regulation occur in eukaryotes. Short-term regulation involves relatively rapid changes in gene expression in response to changes in environmental or physiological conditions. Long-term regulation involves changes in gene expression that are associated with the development and differentiation of an organism.

- Gene expression in eukaryotes is regulated at the transcriptional level (where most regulation occurs) and at posttranscriptional, translational, and posttranslational levels (Figures 14.6–14.8).

- Regulation of transcription initiation involves proteins binding to a gene's promoter and regulatory sites. At the promoter, general transcription factors bind and recruit RNA polymerase II, giving a very low level of transcription. Activator proteins bind to promoter proximal elements and increase the rate of transcription. Other activators bind to the enhancer and, through interaction with a coactivator, which also binds to the proteins at the promoter, greatly stimulate the rate of transcription (Figures 14.9–14.11).

- The overall control of transcription of a gene depends on the particular regulatory proteins that bind to promoter proximal elements and enhancers. The regulatory proteins are cell-type specific and may be activators or repressors. This gene regulation is achieved by a relatively low number of regulatory proteins, acting in various combinations (Figure 14.11).

- The coordinate expression of genes with related functions is achieved by each of the related genes having the same regulatory sequences associated with them.

- Transcriptionally active genes have a looser chromatin structure than transcriptionally inactive genes. The change in chromatin structure that accompanies the activation of transcription of a gene involves chromatin remodelling—specific histone modifications—particularly in the region of a gene's promoter (Figure 14.13).

- Sections of chromosomes or whole chromosomes can be inactivated by DNA methylation, a phenomenon called silencing. DNA methylation is also involved in genomic imprinting, in which transcription of either the inherited maternal or the inherited paternal allele of a gene is inhibited permanently.

14.3 Posttranscriptional, Translational, and Posttranslational Regulation

- Posttranscriptional, translational, and posttranslational controls operate primarily to regulate the quantities of proteins synthesized in cells (Figure 14.6).

- Posttranscriptional controls regulate pre-mRNA processing, mRNA availability for translation, and the rate at which mRNAs are degraded. In alternative splicing, different mRNAs are derived from the same pre-mRNA. In another process, small single-stranded RNAs complexed with proteins bind to mRNAs that have complementary sequences, and either the mRNA is cleaved or translation is blocked (Figure 14.14).

- Translational regulation controls the rate at which mRNAs are used by ribosomes in protein synthesis.

- Posttranslational controls regulate the availability of functional proteins. Mechanisms of regulation include the alteration of protein activity by chemical modification, protein activation by processing of inactive precursors, and affecting the rate of degradation of a protein.

- Epigenetic regulation of gene expression may persist through cell and organismal generations. Such regulation does not depend on changes to the DNA sequence but, rather, on transcription factors that regulate their own production in a feedback loop or on persistent changes to DNA packaging.

14.4 The Loss of Regulatory Controls in Cancer

- In cancer, cells partially or completely dedifferentiate, divide rapidly and uncontrollably, and may break loose to form additional tumours in other parts of the body.

- Proto-oncogenes, tumour suppressor genes, and miRNA genes are typically altered in cancer cells. Proto-oncogenes encode proteins that stimulate cell division. Their altered forms, oncogenes, are abnormally active. Tumour suppressor genes in their normal form encode proteins that inhibit cell division. Mutated forms of these genes lose this inhibitory activity (Figure 14.17). miRNA genes control the activity of mRNA transcripts of particular tumour suppressor genes and proto-oncogenes. Alteration of activity of such an miRNA gene can lead to a lower than normal activity of tumour suppressor gene products or a higher than normal activity of proto-oncogene products depending on the target of the miRNA. In either case, cell proliferation can be stimulated.

- Most cancers develop by multistep progression involving the successive alteration of several to many genes (Figure 14.18).

Questions

Self-Test Questions

1. Some genes are under negative regulation. Which of the following is an example of negative regulation in the *lac* operon?
   a. Binding of allolactose makes the Lac repressor unable to bind DNA.
   b. When lactose levels decrease, *lacZ* expression goes down.
   c. When Lac repressor binds the operator, *lacZ* expression goes down.
   d. When glucose levels are high, *lacZ* expression goes down.

2. For the *E. coli lac* operon, which of the following events occurs when glucose is absent and lactose is added?
   a. β-galactosidase decreases in the cell.
   b. The *lacI* gene cannot make Lac repressor protein.
   c. Allolactose binds the Lac repressor protein to remove it from the operator.
   d. The genes *lacZ, lacY,* and *lacA* are turned off.

3. Imagine a mutation in the *E. coli lac* operon that results in constitutive expression (always on). Further analysis confirms that normal amounts of functional Lac repressor protein are present. Where must the mutation be?
   a. in the *lac* promoter
   b. in the operator
   c. in the *lacZ* gene
   d. in the CAP binding site

4. Which of the following statements about the *trp* operon is correct?
   a. Tryptophan is an inducer.
   b. When end-product tryptophan binds to the Trp repressor, it stops transcription of the tryptophan biosynthesis genes.
   c. Trp repressor is synthesized in active form.
   d. Low levels of tryptophan bind to the *trp* operator and block transcription of the tryptophan biosynthesis genes.

5. How does chromatin remodelling activate gene expression?
   a. It allows repressors to disengage from the promoter.
   b. It winds genes tightly around histones.
   c. It inserts nucleosomes into chromatin.
   d. It recruits a protein complex that displaces nucleosomes from the promoter.

6. Which statement about activation of transcription is correct?
   a. RNA polymerase II binds the TATA box.
   b. A coactivator called a mediator forms a bridge between the promoter and the gene to be transcribed.
   c. Transcription factors bind the promoter and RNA polymerase.
   d. Enhancer regions bind to promoter regions.

7. The delivery of mature mRNA to the cytoplasm in eukaryotes is highly controlled. At which level of regulation is this control achieved?
   a. translational regulation
   b. posttranslational regulation
   c. transcriptional regulation
   d. posttranscriptional regulation

8. Perky ears in a certain mammal are coded by a dominant allele; the recessive allele codes for droopy ears. In males of these mammals, the gene encoding ear shape is transcribed only from the chromosome received from the female parent. This is because the gene from the male parent is silenced by methylation. What will be the result of a cross of a droopy female and a homozygous perky male?
   a. Daughters' ears will be perky and sons' ears will be droopy.
   b. All offspring will have perky ears.
   c. Sons will have one perky ear and one droopy ear.
   d. There will be equal numbers of perky-eared and droopy-eared sons and daughters.

9. Which of the following statements describes miRNA accurately?
   a. miRNA is encoded by non–protein coding genes.
   b. miRNA has a precursor that is folded and then elongated by a Dicer enzyme.
   c. miRNA forms complementary base pairs with tRNA.
   d. miRNA is translated in the cytoplasm.

10. Which of the following characteristics is exhibited by typical cancer cells?
    a. They convert oncogenes into proto-oncogenes.
    b. Oncogenes are near repressor genes.
    c. They have a balance of oncogenes and tumour suppressor genes.
    d. *TP53* mutations are one of several likely changes in DNA.

Questions for Discussion

1. In a mutant strain of *E. coli*, the CAP protein is unable to combine with its target region of the *lac* operon. How would you expect the mutation to affect transcription when cells of this strain are subjected to the following conditions?
   a. Lactose and glucose are both available.
   b. Lactose is available, but glucose is not.
   c. Both lactose and glucose are unavailable.

2. Duchenne muscular dystrophy, an inherited genetic disorder, affects boys almost exclusively. Early in childhood, muscle tissue begins to break down in affected individuals, who typically die in their teens or early twenties as a result of respiratory failure. Muscle samples from women who carry the mutation reveal some regions of degenerating muscle tissue adjacent to other regions that are normal. Develop a hypothesis explaining these observations.

3. Eukaryotic transcription is generally controlled by binding of regulatory proteins to DNA sequences rather than by modification of RNA polymerases. Develop a hypothesis explaining why this is so.

Protein microarray, a key tool of proteomics, the study of the complete set of proteins that can be expressed by an organism's genome. Each coloured dot is a protein, with a specific colour for each protein being studied.

DNA Technologies

WHY IT MATTERS

Have you ever wished you could create a new organism that would solve a real-world problem? Maybe you would create a fungus that converts toxic waste into fuel. Artificial blood requires some sort of cell to carry the oxygen; could you design such a cell? What if you could program the bacteria that normally make yogurt to produce different flavours on demand? Or maybe you could invent a multicellular organism that could measure and report the concentration of specific compounds in air.

The age of do-it-yourself (DIY) biology has arrived, and the basic tools of genetic engineering developed since the 1970s are now in the hands of undergraduates like you. Teams of students are collaborating with scientists and engineers in classrooms, labs, and informal makerspaces all over the world to develop innovative entries in the annual International Genetically Engineered Machine Competition (iGEM). In an open-source spirit of creating and sharing, teams of passionate students start with over a thousand standardized interchangeable BioBrick DNA sequences, plasmid backbones, and living cells. Adding their own custom-created parts, teams assemble parts into devices that, in turn, are combined to create innovative pathways and systems—all housed in living cells. Each year, teams present their novel living machines to address issues in categories such as art, energy, environment, food, health, measurement, and manufacturing.

The competition also highlights and promotes consideration of the broad range of risks and controversies surrounding genetic engineering, as well as the scientific, social, and ethical questions related to its rapidly expanding role in society.

In this chapter we look at the topic of **genetic engineering** as the latest addition to the broad area known as **biotechnology**, which is any technique applied to biological systems or living organisms to make or modify products or processes for a specific purpose. Thus, biotechnology includes manipulations that do not involve **DNA technologies**, such as the use of naturally occurring yeast to brew beer and bake bread and the use of bacteria to make cheese. With respect to biotechnologies that do involve the direct manipulation of genes for basic and applied research, we begin our discussion with a description of methods used to obtain genes in large quantities, an essential step for their analysis or further application.

15.1 DNA Cloning

Technologies designed to manipulate DNA must, of course, have DNA to work with. Three main sources of DNA sequences are commonly used in research:

i) DNA sequences can be extracted from the genome of cells (and viruses) using specialized enzymes, as shown in Figures 15.1 and 15.3 (pages 341 and 343).

ii) DNA can be synthesized from an mRNA template by the reverse transcriptase enzyme, as shown in Figure 15.4 (page 344).

iii) DNA sequences of choice can be created by automated chemical synthesis in the laboratory using a DNA synthesizer responding to computer input. Although existing "gene machines" can create specified sequences up to a few thousand base pairs, this capacity is expanding rapidly.

Once a DNA sequence of choice has been obtained from nature or created artificially, the number of copies of the sequence can be increased dramatically, as described in the following sections.

Remember from Chapter 8 that a *clone* is a line of genetically identical cells or individuals derived from a single ancestor. By similar reasoning, DNA cloning is a method for producing many copies of a piece of DNA; the piece of DNA is referred to as a *gene of interest*, which is a gene that a researcher wants to study or manipulate. Scientists clone DNA for many reasons. For example, a researcher might be interested in how a particular human gene functions. Each human cell contains only two copies of most genes, amounting to a very small fraction of the total amount of DNA in a diploid cell. In its natural state in the genome, the gene is extremely difficult to study. However, through DNA cloning, a researcher can produce a sample large enough for scientific experimentation.

Cloned genes are used in basic research to find out about their biological functions. For example, researchers can determine the DNA sequence of a cloned gene, giving them the ultimate information about its structure. Also, by manipulating the gene and inducing mutations in it, they can gain information about its function and about how its expression is regulated. Cloned genes can be expressed in bacteria, and the proteins encoded by the cloned genes can be produced in quantity and purified. Those proteins can be used in basic research, or, in the case of genes that encode proteins of pharmaceutical or clinical importance, they can be used in applied research.

An overview of one common method for cloning a gene of interest from a genome is shown in **Figure 15.1**; the method uses bacteria (commonly *Escherichia coli*) and plasmids, the small circular DNA molecules that replicate separately from the bacterial chromosome. The researcher first extracts DNA from cells containing the gene of interest and then cuts this DNA into fragments. One of these fragments will likely carry the desired gene. Each of the fragments is inserted into a plasmid, thus producing a collection of *recombinant DNA molecules*—**recombinant DNA** is DNA from two or more different sources that are joined together. These recombinant plasmids are then introduced into bacteria; each bacterium receives a different plasmid. The bacterium continues growing and dividing, and as it does, the recombinant plasmid DNA is also replicated. The final step is to identify which bacterium contains the plasmid carrying the gene of interest and isolate it for further study.

Cloned genes and other cloned DNA sequences from genomes are used in both basic research and applied research. In **basic research**, a researcher might want to study a cloned gene to learn about its structure, including its DNA sequence and sequences that regulate its expression. A researcher might also want to study the gene's function, including how its expression is regulated, and the nature of the gene's product. For protein-coding genes, for instance, the cloned gene could be used to produce the protein product in large quantities in a microorganism host to facilitate study of that protein's structure and function. As part of this research, the cloned gene could be manipulated in the laboratory to help dissect a gene's function.

In **applied research**, the interest in cloned genes or other cloned DNA sequences is not in the structure and function of a gene or sequence; that typically is understood, at least to a significant degree, at the beginning of research projects. Rather, cloned genes or cloned DNA sequences are used, for instance, for medical, forensic, agricultural, or commercial applications. Some examples are

- Gene therapy to correct or treat genetic diseases.
- Diagnosis of genetic diseases, such as sickle cell disease.
- DNA fingerprinting in forensics.
- Production of pharmaceuticals, such as humulin, human insulin to treat diabetes, and tissue plasminogen activator to break down blood clots.
- Generation of genetically modified animals and plants, including animals that synthesize pharmaceuticals, and plants that are nutritionally enriched, insect resistant, or herbicide resistant.
- Modification of bacteria to use in cleanup of oil spills or toxic waste.

15.1a Bacterial Enzymes Called Restriction Endonucleases Form the Basis of DNA Cloning

The key to DNA cloning is the specific joining of two DNA molecules from different sources, such as a genomic DNA fragment and a bacterial plasmid (see Figure 15.1). This specific joining of DNA is made possible, in part, by bacterial enzymes called **restriction endonucleases** (also called **restriction enzymes**), which were discovered in the late 1960s. Restriction enzymes recognize short, specific DNA sequences called *restriction sites,* typically four to eight base pairs long, and cut the DNA at specific locations within those sequences. The DNA fragments produced by cutting a long DNA molecule with a restriction enzyme are known as **restriction fragments.**

The *restriction* in the name of the enzymes refers to their normal role inside bacteria, in which the enzymes defend against viral attack by breaking down (restricting) the DNA molecules of infecting viruses. Why don't such enzymes break down the cell's own DNA? The bacterium "hides" the restriction sites in its own DNA by attaching methyl groups to bases in those sites, thereby blocking the binding of its restriction enzyme.

Hundreds of different restriction enzymes have been identified, each one cutting DNA at a specific restriction site. As illustrated by the restriction site of *Eco*RI **(Figure 15.2, p. 342),** most restriction sites are symmetrical in that the sequence of nucleotides read in the $3' \rightarrow 5'$ direction on one strand is the same as the sequence read in the $3' \rightarrow 5'$ direction on the complementary strand. A given enzyme always recognizes the same short DNA sequence as its cut site and always cuts at the same place within the sequence. The restriction enzymes most used in cloning—such as *Eco*RI—cleave the sugar–phosphate backbones of DNA to produce DNA fragments with single-stranded ends (Figure 15.2, step 1). The ends are called **sticky ends** because the short, single-stranded regions can form hydrogen bonds with complementary sticky ends

on any other DNA molecules cut with the same enzyme. For example, step 2 shows the insertion of a DNA molecule with sticky ends produced by *Eco*RI between two other DNA molecules with the same

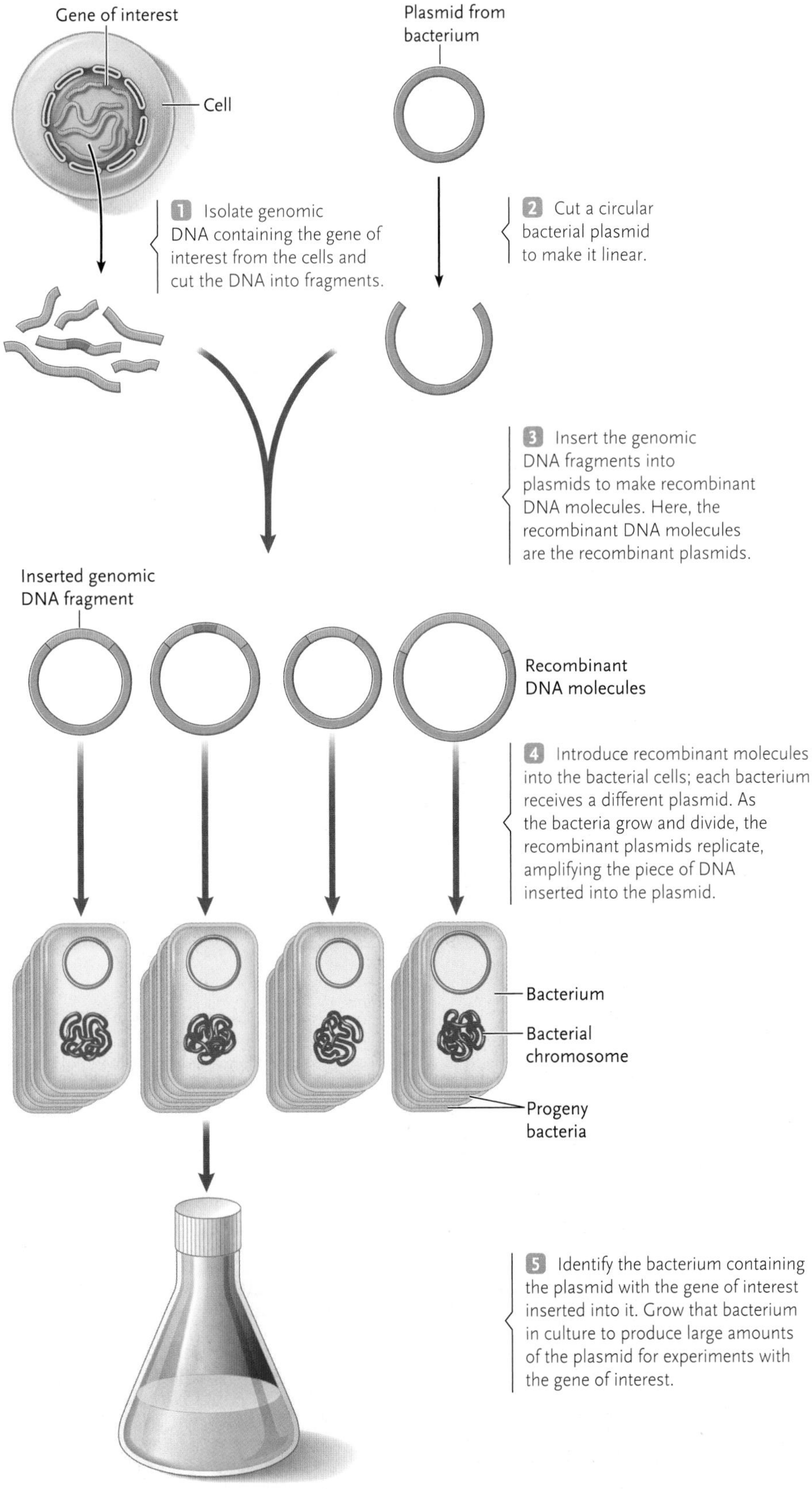

Gene of interest

Cell

Plasmid from bacterium

1 Isolate genomic DNA containing the gene of interest from the cells and cut the DNA into fragments.

2 Cut a circular bacterial plasmid to make it linear.

3 Insert the genomic DNA fragments into plasmids to make recombinant DNA molecules. Here, the recombinant DNA molecules are the recombinant plasmids.

Inserted genomic DNA fragment

Recombinant DNA molecules

4 Introduce recombinant molecules into the bacterial cells; each bacterium receives a different plasmid. As the bacteria grow and divide, the recombinant plasmids replicate, amplifying the piece of DNA inserted into the plasmid.

Bacterium

Bacterial chromosome

Progeny bacteria

5 Identify the bacterium containing the plasmid with the gene of interest inserted into it. Grow that bacterium in culture to produce large amounts of the plasmid for experiments with the gene of interest.

Figure 15.1
Overview of cloning DNA fragments in a bacterial plasmid.

sticky ends. The pairings leave nicks in the sugar–phosphate backbones of the DNA strands that are sealed by *DNA ligase,* an enzyme that has the same function in DNA replication (step 3). The result is DNA from two different sources joined together—a recombinant DNA molecule.

15.1b Bacterial Plasmids Illustrate the Use of Restriction Enzymes in Cloning

The bacterial plasmids used for cloning are examples of cloning vectors—DNA molecules into which a DNA fragment can be inserted to form a recombinant DNA molecule for cloning. Plasmid cloning vectors are usually natural plasmids that have been modified to have special features. Commonly, plasmid cloning vectors are engineered to contain two genes that are useful in the final steps of a cloning experiment for distinguishing bacteria that have recombinant plasmids from those that do not. The *amp*R gene encodes an enzyme that breaks down the antibiotic ampicillin; when the plasmid is introduced into *E. coli* and the

*amp*R gene is expressed, the bacteria become resistant to ampicillin. The *lacZ+* gene encodes β-galactosidase (part of the *lac* operon from Section 14.1), which hydrolyzes the sugar lactose, as well as a number of synthetic substrates. Restriction sites are located within the *lacZ+* gene but do not alter the gene's function. For a given cloning experiment, one of these restriction sites is chosen.

Cloning a Gene of Interest. Figure 15.3 expands on the overview of Figure 15.1 to show the steps used to clone a gene of interest using a plasmid cloning vector and restriction enzymes. In outline, the steps are as follows:

- Isolate genomic DNA and digest that DNA with a restriction enzyme (step 1).
- Digest the plasmid cloning vector with the same restriction enzyme (step 2).
- Ligate cut genomic DNA fragments and cut plasmid DNA together using DNA ligase (step 3). This produces a mixture of recombinant plasmids (plasmids with DNA fragments inserted into the cloning vector), nonrecombinant plasmids (resealed cloning vectors with no DNA fragment inserted), and joined-together pieces of genomic DNA with no cloning vector involved.
- Transform the DNA into *E. coli* (step 4). Some bacteria will take up a plasmid, whereas others will not.
- *Selection:* Spread the bacterial cells on growth medium containing ampicillin and X-gal and incubate to allow colonies to grow (step 5). Bacteria containing plasmids are selected for because of the ampicillin in the growth medium. Within each cell of a colony, the plasmids replicate until approximately 100 copies are present.
- *Screening:* The X-gal in the medium distinguishes between bacteria that have been transformed with recombinant plasmids and nonrecombinant plasmids by *blue–white screening* (see Figure 15.3, Interpreting the Results). White colonies contain recombinant plasmids, whereas blue colonies contain nonrecombinant plasmids. Among the white colonies is the one with a recombinant plasmid that contains the gene of interest. We will see a little later how we can identify that particular plasmid.

Three researchers, Paul Berg, Stanley N. Cohen, and Herbert Boyer, in 1973 pioneered the development of DNA cloning techniques using restriction enzymes and bacterial plasmids. Berg received a Nobel Prize in 1980 for his research.

15.1c DNA Libraries Contain Collections of Cloned DNA Fragments

As you have seen, the starting point for cloning a gene of interest is a large set of plasmid clones carrying fragments representing the entire DNA of an

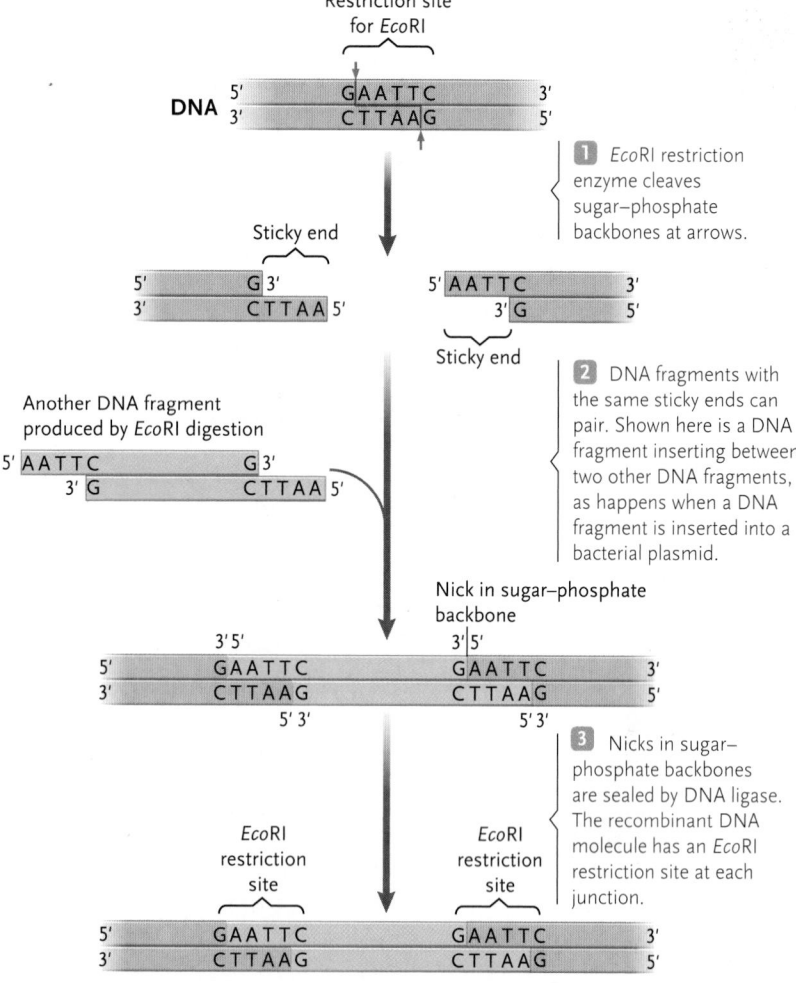

Figure 15.2

The restriction site for the restriction enzyme *Eco*RI, and the generation of a recombinant DNA molecule by complementary base-pairing of DNA fragments produced by digestion with the same restriction enzyme.

Research Method Figure 15.3 Cloning a gene of interest in a plasmid cloning vector.

PURPOSE: Cloning a gene produces many copies of a gene of interest that can be used, for example, to determine the DNA sequence of the gene, to manipulate the gene in basic research experiments, to understand its function, and to produce the protein encoded by the gene.

PROTOCOL:

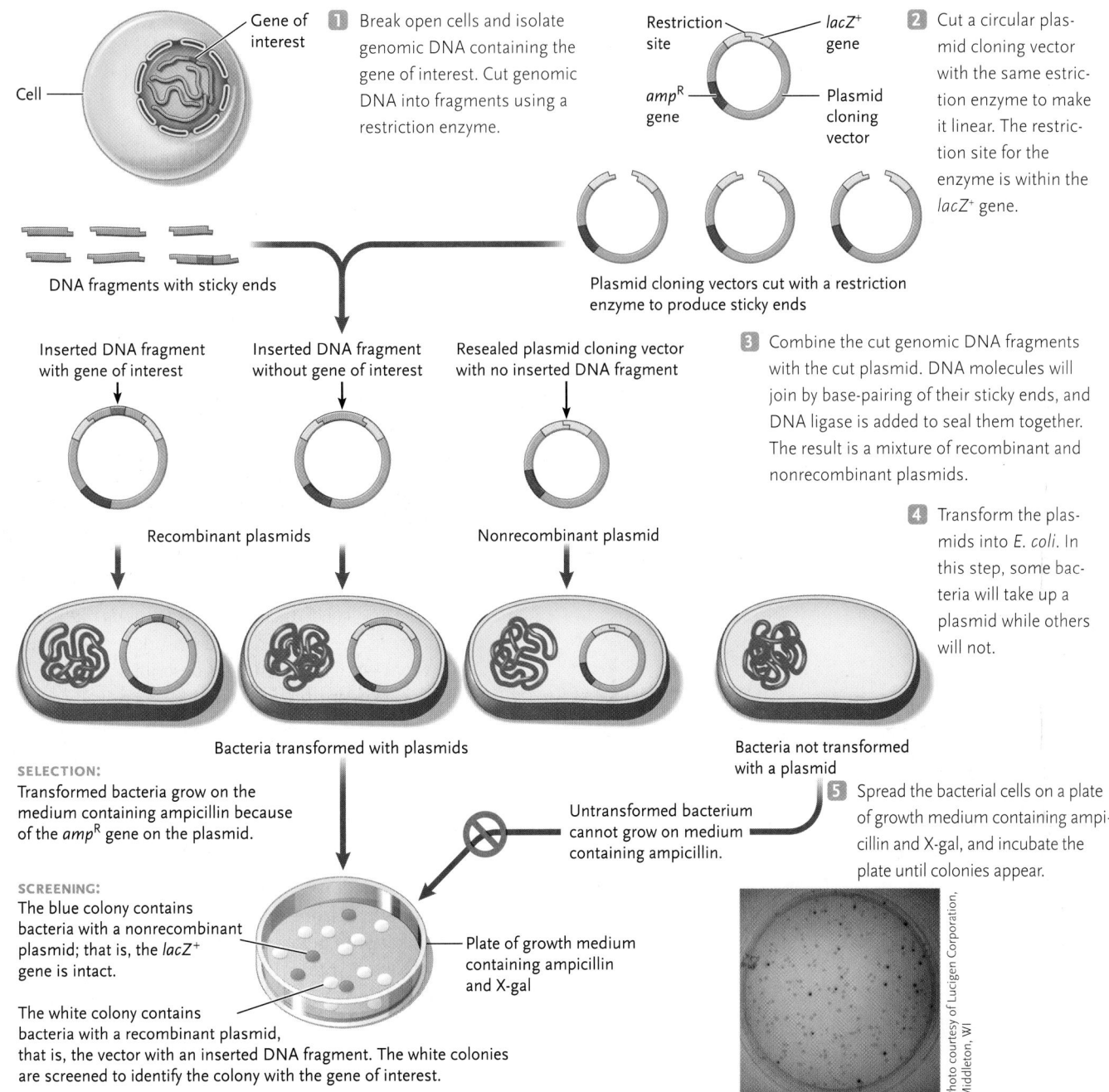

Cell — Gene of interest

1 Break open cells and isolate genomic DNA containing the gene of interest. Cut genomic DNA into fragments using a restriction enzyme.

DNA fragments with sticky ends

Restriction site — *lacZ*⁺ gene

*amp*ᴿ gene — Plasmid cloning vector

2 Cut a circular plasmid cloning vector with the same estriction enzyme to make it linear. The restriction site for the enzyme is within the *lacZ*⁺ gene.

Plasmid cloning vectors cut with a restriction enzyme to produce sticky ends

Inserted DNA fragment with gene of interest

Inserted DNA fragment without gene of interest

Resealed plasmid cloning vector with no inserted DNA fragment

3 Combine the cut genomic DNA fragments with the cut plasmid. DNA molecules will join by base-pairing of their sticky ends, and DNA ligase is added to seal them together. The result is a mixture of recombinant and nonrecombinant plasmids.

Recombinant plasmids

Nonrecombinant plasmid

4 Transform the plasmids into *E. coli*. In this step, some bacteria will take up a plasmid while others will not.

Bacteria transformed with plasmids

Bacteria not transformed with a plasmid

SELECTION:
Transformed bacteria grow on the medium containing ampicillin because of the *amp*ᴿ gene on the plasmid.

Untransformed bacterium cannot grow on medium containing ampicillin.

5 Spread the bacterial cells on a plate of growth medium containing ampicillin and X-gal, and incubate the plate until colonies appear.

SCREENING:
The blue colony contains bacteria with a nonrecombinant plasmid; that is, the *lacZ*⁺ gene is intact.

Plate of growth medium containing ampicillin and X-gal

The white colony contains bacteria with a recombinant plasmid, that is, the vector with an inserted DNA fragment. The white colonies are screened to identify the colony with the gene of interest.

Photo courtesy of Lucigen Corporation, Middleton, WI

INTERPRETING THE RESULTS: All of the colonies on the plate contain plasmids because the bacteria that form the colonies are resistant to the ampicillin present in the growth medium. Blue-white screening distinguishes bacterial colonies with nonrecombinant plasmids from those with recombinant plasmids. Blue colonies have nonrecombinant plasmids. These plasmids have intact *lacZ*⁺ genes and produce β-galactosidase, which changes X-gal to a blue product. White colonies have recombinant plasmids. These plasmids have DNA fragments inserted into the *lacZ*⁺ gene, so they do not produce β-galactosidase. As a result, they cannot convert X-gal to the blue product and the colonies are white. Among the white colonies is a colony containing the plasmid with the gene of interest. Further screening is done to identify that particular white colony (see Figure 15.4, p. 344). Once identified, the colony is cultured to produce large quantities of the recombinant plasmid for analysis or manipulation of the gene.

Figure 15.4

Synthesis of DNA from mRNA using reverse transcriptase.

PURPOSE: To produce double-stranded, complementary DNA (cDNA) copies of mRNA molecules isolated from cells.

PROTOCOL:

1 Isolate mRNAs from cells. One mRNA is shown.

2 Add primer of T DNA nucleotides (dT). The primer base-pairs to the poly (A) tail of the mRNa.

3 Reverse transcriptase uses DNA precursors to synthesize a DNA copy of the mRNA in the 5'-to-3' direction. The result is a hybrid nucleic acid molecule consisting of the mRNA base paired with a DNA strand.

4 An RNAase enzyme degrades the mRNA stand, leaving a single strand of DNA.

5 DNA polymerase uses DNA precursors to synthesize the second strand of DNA. Experimentally different methods are available for the use of primers in this reaction. The result is a double-stranded complementary DNA (cDNA) copy of the starting mRNA.

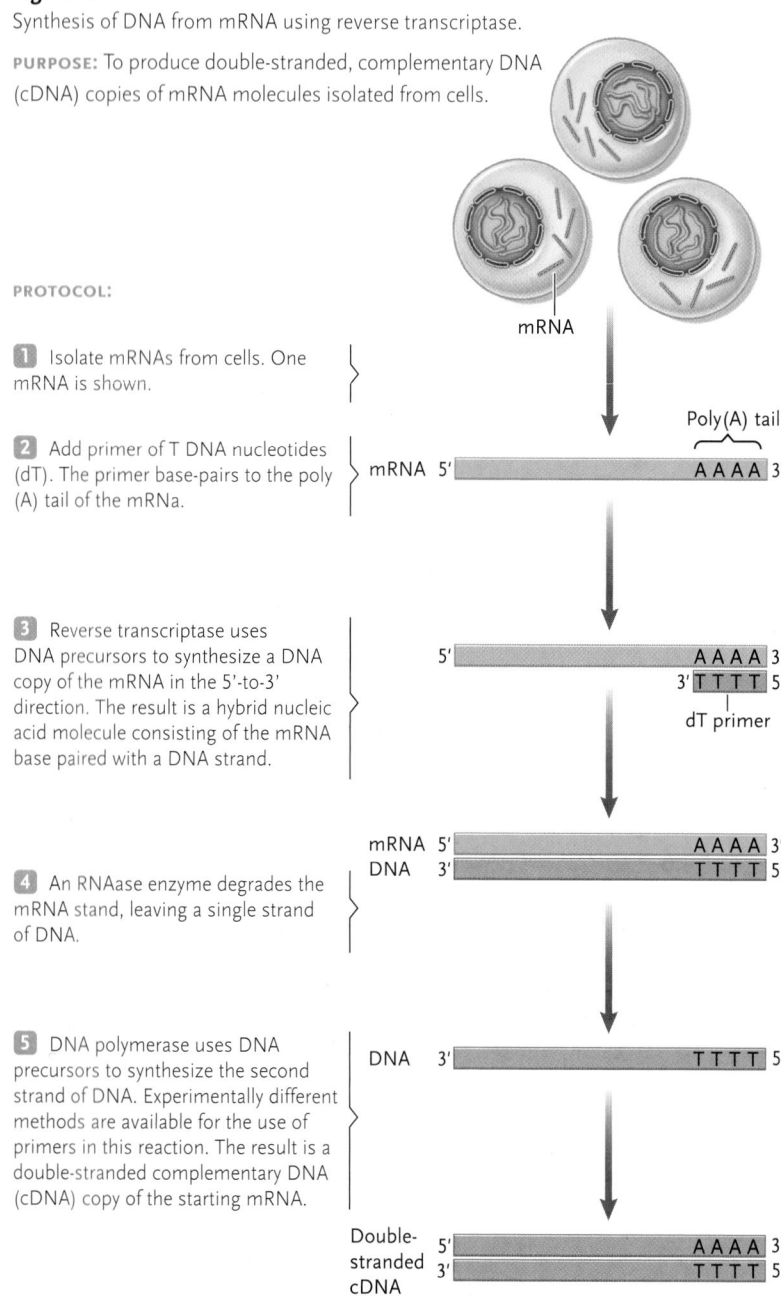

OUTCOME: The outcome is a population of double-stranded cDNA molecules that have base-pair sequences corresponding to the base sequences of the mRNA molecules isolated from the cell.

organism's genome. A collection of clones that contains a copy of every DNA sequence in a genome is called a **genomic library**. A genomic library can be made using plasmid cloning vectors or any other kind of cloning vector. The number of clones in a genomic library increases with the size of the genome. For example, a yeast genomic library of plasmid clones consists of hundreds of plasmids, whereas a human genomic library of plasmid clones consists of thousands of plasmids.

A genomic library is a resource containing the entire DNA of an organism cut into pieces. Just as for a book library, where you can search through the same

set of books on various occasions to find different passages of interest, you can search through the same genomic library on various occasions to find and isolate different genes or other DNA sequences.

Researchers also commonly use another kind of DNA library that is made starting with mRNA molecules isolated from a cell **(Figure 15.4)**. To convert single-stranded mRNA to double-stranded DNA for cloning (RNA cannot be cloned), first the researchers use the enzyme *reverse transcriptase* (made by retroviruses) to make a single-stranded DNA that is complementary to the mRNA. Then they degrade the mRNA strand with an enzyme and use DNA polymerase to make a second DNA strand that is complementary to the first. The result is **complementary DNA (cDNA)**. After adding restriction sites to each end, the researchers insert the cDNA into a cloning vector as described for the genomic library. The entire collection of cloned cDNAs made from the mRNAs isolated from a cell is a **cDNA library**.

Not all genes are active in every cell. Therefore, a cDNA library is limited in that it includes copies of only the genes that were active in the cells used as the starting point for creation of the library. This limitation can be an advantage, however, in identifying genes active in one cell type and not another. cDNA libraries are useful, therefore, for providing clues to the changes in gene activity that are responsible for cell differentiation and specialization. An ingenious method for comparing the cDNA libraries produced by different cell types—the DNA chip—is described in the next chapter.

cDNA libraries provide a critical advantage to genetic engineers who wish to insert eukaryotic genes into bacteria, particularly when the bacteria are to be used as "factories" for making the protein encoded in the gene. The genes in eukaryotic nuclear DNA typically contain many *introns*, spacer sequences that interrupt the amino acid–coding sequence of a gene (see Section 13.3). Because bacterial DNA does not contain introns, bacteria are not equipped to process eukaryotic genes correctly. However, the cDNA copy of a eukaryotic mRNA already has the introns removed, so bacteria can transcribe and translate it accurately to make eukaryotic proteins.

Screening a DNA Library for a Gene of Interest. One method used to screen a DNA library to identify a clone containing a gene of interest is based on the fact that a gene has a unique DNA sequence. In this technique, called **DNA hybridization**, a gene of interest is identified in the set of clones when it base-pairs with a short, single-stranded complementary DNA or RNA molecule called a *nucleic acid probe* **(Figure 15.5)**. The figure illustrates screening for a gene of interest in a genomic library. Similarly, the approach can be used to screen a cDNA library for the cDNA corresponding to a gene of interest. The probe is labelled with a radioactive or a nonradioactive tag, so investigators can detect it. In our

Research Method Figure 15.5 DNA hybridization to identify a DNA sequence of interest.

PURPOSE: Hybridization with a specific DNA probe allows researchers to detect a specific DNA sequence, such as a gene, within a population of DNA molecules. Here, DNA hybridization is used to screen a genomic library of plasmid clones to identify those containing a recombinant plasmid with a gene of interest.

PROTOCOL:

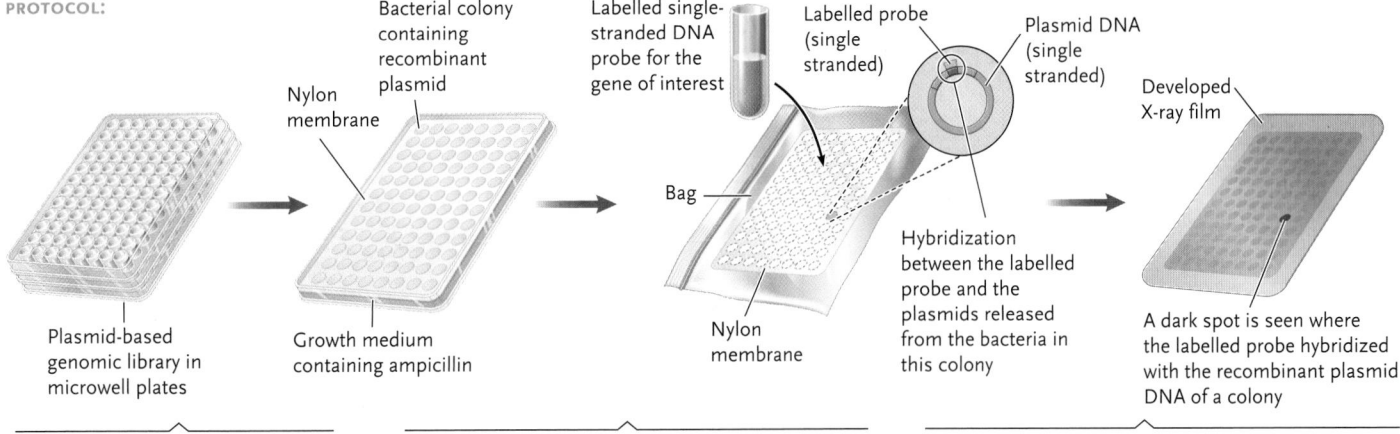

1 Replicate each genomic library microwell plate onto a plate of growth medium containing ampicillin with a nylon membrane on the surface. The pattern of colonies that grow on a membrane matches that of the clones in the microwell plate.

2 Remove the membrane and treat it to break open the cells and to denature the released DNA into single strands. The single-stranded DNA sticks to the filter in the same position as the colony from which it was derived. Place the filter in a plastic bag. Add a labelled single-stranded DNA probe (DNA or RNA) for the gene of interest and incubate. It a recombinant plasmid's inserted DNA fragment is complementary to the probe, the two will hybridize, that is form base pairs. Wash off excess labelled probe.

3 Detect the hybridization by looking for the labelled tag on the probe. If the probe was radioactively labelled, place the filter against X-ray film. The decaying radioactive compound exposes the film, giving a dark spot when the film is developed. Correlate the position of any dark spot on the film with the original microwell plate. Bacteria from that well can then be used for further study of the gene clone.

INTERPRETING THE RESULTS: DNA hybridization with a labelled probe enables a researcher to identify a sequence of interest. If the probe is for a particular gene, it allows the specific identification of a colony containing bacteria with recombinant plasmids carrying that gene. The specificity of the method depends directly on the probe used. The same collection of bacterial clones can be used again and again to search for recombinant plasmids carrying different genes or different plasmids of interest simply by changing the probe used in the experiment.

example, if we know the sequence of part of a gene of interest, we can use that information to design and synthesize a nucleic acid probe. Or, we can take advantage of DNA sequence similarities of evolutionarily related organisms. For instance, we could make a probe for the human actin gene based on the sequence of the cloned mouse actin gene and expect that the two nucleic acids would hybridize because of the evolutionary conservation of that gene. Once a colony containing plasmids with a gene of interest has been identified, that colony can be used to produce large quantities of the cloned gene.

15.1d The Polymerase Chain Reaction Amplifies DNA *In Vitro*

Producing multiple DNA copies by cloning requires a series of techniques and considerable time. A much more rapid process, **polymerase chain reaction (PCR)**, produces an extremely large number of copies of a specific DNA sequence from a DNA mixture without having to clone the sequence in a host organism. The process is called *amplification* because it increases the amount of DNA to the point where it can be analyzed or manipulated easily. Developed in 1983 by Kary B. Mullis and F. Faloona at Cetus Corporation

(Emeryville, California), PCR has become one of the most important tools in modern molecular biology, finding wide application in all areas of biology. Mullis received a Nobel Prize in 1993 for his role in the development of PCR.

How PCR is performed is shown in **Figure 15.6, p. 346.** PCR is essentially a special case of DNA replication in which a DNA polymerase replicates just a portion of a DNA molecule rather than the whole molecule. PCR takes advantage of a characteristic common to all DNA polymerases: these enzymes add nucleotides only to the 3′ end of an existing chain called the *primer* (see Section 12.3). For replication to begin, a primer must be available, base-paired to the template chain. By cycling 20 to 30 times through a series of priming and replication steps, PCR amplifies the target sequence, producing millions of copies.

Since the primers used in PCR are designed to bracket only the sequence of interest, the cycles replicate only this sequence from a mixture of essentially any DNA molecules. Thus, PCR not only finds the "needle in the haystack" among all the sequences in a mixture but also makes millions of copies of the "needle"—the DNA sequence of interest. Usually, no further purification of the amplified sequence is necessary.

Research Method Figure 15.6

The polymerase chain reaction (PCR).

PURPOSE: To amplify—produce large numbers of copies of—a target DNA sequence in the test tube without cloning.

PROTOCOL: A PCR mixture has four key elements: (1) the DNA with the target sequence to be amplified; (2) a pair of DNA primers, one complementary to one end of the target sequence and the other complementary to the other end of the target sequence; (3) the four nucleoside triphosphate precursors for DNA synthesis (dATP, dTTP, dGTP, and dCTP); and (4) DNA polymerase. Since PCR uses high temperatures that would break down normal DNA polymerases, a heat-stable DNA polymerase is used. Heat-stable polymerases are isolated from microorganisms that grow in a high-temperature area such as a thermal pool or near a deep-sea vent.

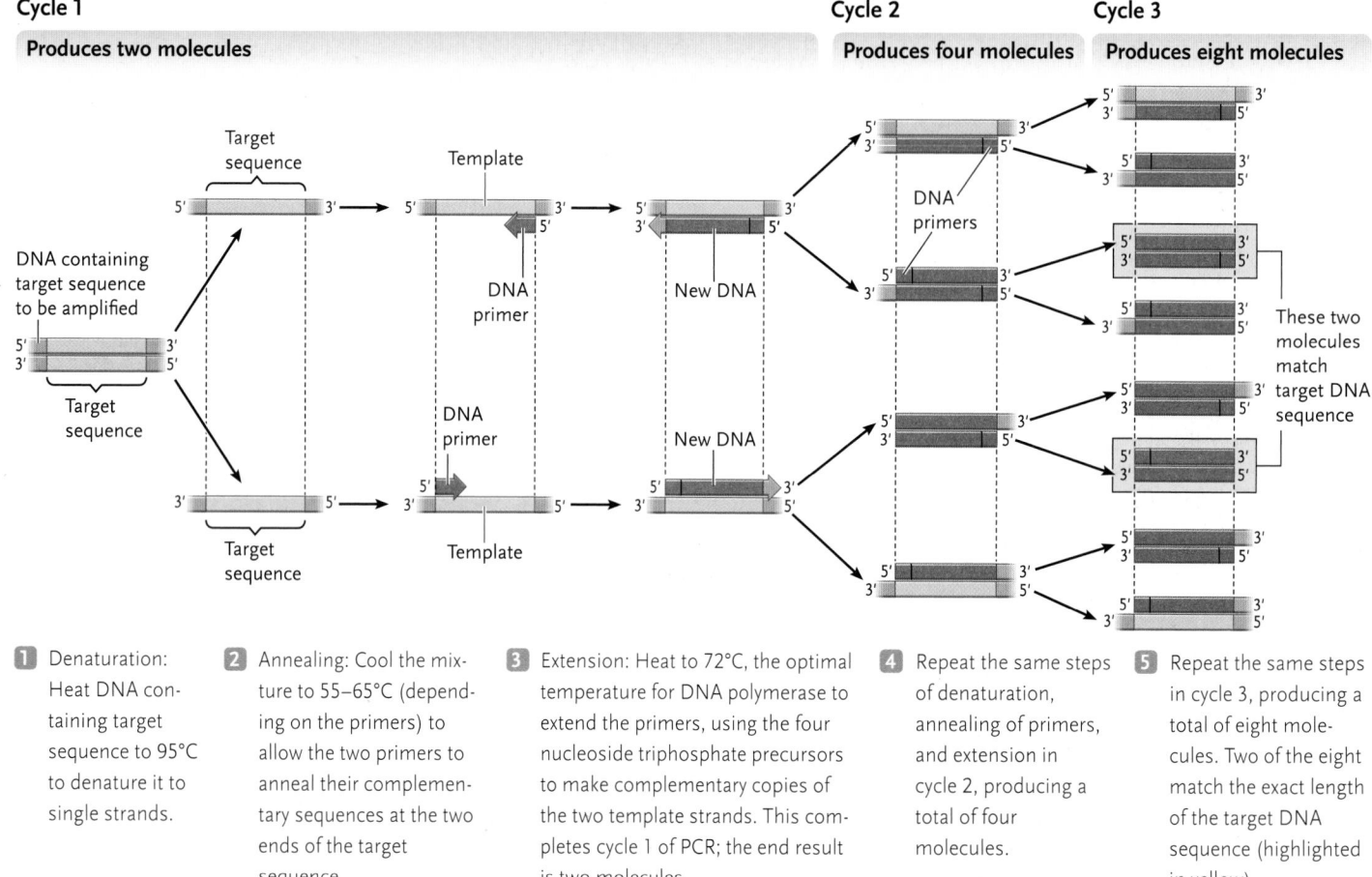

Cycle 1 — Produces two molecules

Cycle 2 — Produces four molecules

Cycle 3 — Produces eight molecules

1. Denaturation: Heat DNA containing target sequence to 95°C to denature it to single strands.

2. Annealing: Cool the mixture to 55–65°C (depending on the primers) to allow the two primers to anneal their complementary sequences at the two ends of the target sequence.

3. Extension: Heat to 72°C, the optimal temperature for DNA polymerase to extend the primers, using the four nucleoside triphosphate precursors to make complementary copies of the two template strands. This completes cycle 1 of PCR; the end result is two molecules.

4. Repeat the same steps of denaturation, annealing of primers, and extension in cycle 2, producing a total of four molecules.

5. Repeat the same steps in cycle 3, producing a total of eight molecules. Two of the eight match the exact length of the target DNA sequence (highlighted in yellow).

INTERPRETING THE RESULTS: After three cycles, PCR produces a pair of molecules matching the target sequence. Subsequent cycles amplify these molecules to the point where they outnumber all other molecules in the reaction by many orders of magnitude.

CONCEPT FIX Careful attention to Figure 15.6 can help you avoid the common pitfalls in understanding PCR. Notice that

1. the primers are made of DNA, not RNA as in natural DNA replication;
2. the left primer binds to one strand while the right primer binds to the opposite strand of the original DNA;
3. of all the DNA sequences put into the PCR reaction tube, only the *target sequence,* the sequence between the primers, is amplified; and
4. although the diagram shows DNA being synthesized left to right on the bottom strand, and right to left on the top strand, the DNA polymerase is reading the template 3′ to 5′ in both cases.

The characteristics of PCR allow extremely small DNA samples to be amplified to concentrations high enough for analysis. PCR is used, for example, to produce enough DNA for analysis from the root of a single human hair, or from a small amount of blood, semen, or saliva, such as the traces left at the scene of a crime. It is also used to extract and multiply DNA sequences from skeletal remains; ancient sources such as mammoths, Neanderthals, and Egyptian mummies; and, in rare cases, amber-entombed fossils, fossil bones, and fossil plant remains.

A successful outcome of PCR is shown by analyzing a sample of the amplified DNA using **agarose gel electrophoresis** to see if the copies are the same length as the target **(Figure 15.7).** Gel electrophoresis is a technique by which DNA, RNA, or protein molecules are separated in a gel subjected to an electric field. The type of gel and the conditions used vary with the experiment, but in each case, the gel functions as a molecular sieve to separate the macromolecules based on size, electrical charge, or other properties. To separate large DNA molecules, such as those typically produced by PCR, a gel made of agarose, a natural molecule isolated from seaweed, is used because of its large pore size.

For PCR experiments, the size of the amplified DNA is determined by comparing the position of the

Research Method Figure 15.7 Separation of DNA fragments by agarose gel electrophoresis.

PURPOSE: Gel electrophoresis separates DNA molecules, RNA molecules, or proteins according to their sizes, electrical charges, or other properties through a gel in an electric field. Different gel types and conditions are used for different molecules and types of applications. A common gel for separating large DNA fragments is made of agarose.

PROTOCOL:

1. Prepare a gel consisting of a thin slab of agarose and place it in a gel box between two electrodes. The gel has wells for placing the DNA samples to be analyzed. Add **buffer** to cover the gels.

2. Load DNA sample solutions, such as PCR products, into wells of the gel, alongside a well loaded with marker DNA fragments of known sizes. (The DNA samples, as well as the marker DNA sample, have a dye added to help see the liquid when loading the wells. The dye migrates during electrophoresis, enabling the progress of electrophoresis to be followed.)

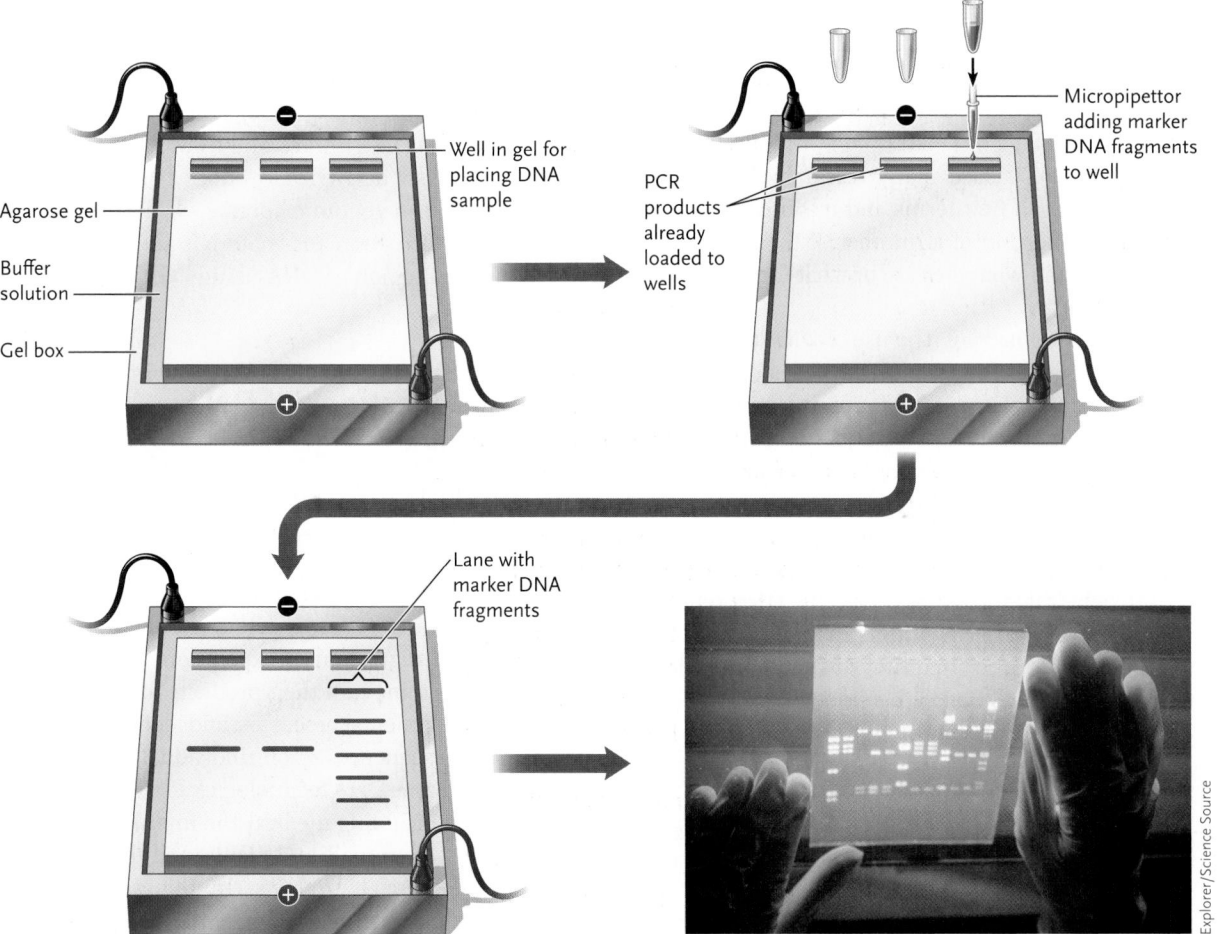

3. Apply an electric current to the gel; DNA fragments are negatively charged, so they migrate to the positive pole. Shorter DNA fragments migrate faster than longer DNA fragments. At the completion of separation, DNA fragments of the same length have formed bands in the gel. At this point, the bands are invisible.

4. Stain the gel with a dye that binds to DNA. The dye fluoresces under UV light, enabling the DNA bands to be seen and photographed. An actual gel showing separated DNA bands stained and visualized this way is shown.

INTERPRETING THE RESULTS: Agarose gel electrophoresis separates DNA fragments according to their length. The lengths of the DNA fragments being analyzed are determined by measuring their migration distances and comparing these distances to a calibration curve of the migration distances of the marker bands, which have known lengths. For PCR, agarose gel electrophoresis shows whether DNA of the correct length was amplified. For restriction enzyme digests, this technique shows whether fragments are produced as expected.

DNA band with the positions of DNA fragments of known size separated on the gel at the same time. If that size matches the predicted size for the target DNA, PCR is deemed successful. In some cases, such as DNA from ancient sources, a size prediction may not be possible; in this case, agarose gel electrophoresis analysis simply indicates whether there was DNA in the sample that could be amplified.

The advantages of PCR have made it the technique of choice for researchers, law enforcement agencies, and forensic specialists whose primary interest is in the amplification of specific DNA fragments up to a practical maximum of a few thousand base pairs. Cloning remains the technique of choice for amplification of longer fragments. The major limitation of PCR relates to the primers. To design a primer for PCR, the researcher must first have sequence information about the target DNA. By contrast, cloning can be used to amplify DNA of unknown sequence.

Review of Some of the Materials, Concepts, and Techniques Introduced in This Section. In this chapter so far we have discussed many research methods—so there are a lot of new terms and techniques to learn! Here is a collection of a number of these terms and techniques and what they are or what they do:

- *Genetic engineering.* The use of DNA technologies to alter genes for practical purposes.
- *DNA cloning.* A method for producing many copies of a piece of DNA.
- *Gene cloning.* DNA cloning that involves a gene.
- *Recombinant DNA.* DNA fragments from two or more sources that have joined together.
- *Restriction enzyme (restriction endonuclease).* An enzyme that recognizes a specific DNA sequence and cuts the DNA within that sequence. Fragments produced by cutting DNA with a restriction enzyme are *restriction fragments.*
- *Ligation.* The process of joining two or more DNA fragments together to make one DNA molecule.
- *DNA ligase.* The enzyme that seals together DNA fragments generated by restriction enzyme digestion to produce a recombinant DNA molecule.
- *Cloning vector.* DNA molecules into which a DNA fragment can be inserted to form a recombinant DNA molecule that can be replicated in a host organism for the purpose of cloning the DNA fragment.
- *Genomic DNA library.* A set of clones that collectively contains a copy of every DNA sequence in a genome.
- *cDNA (complementary DNA).* A double-stranded DNA copy of a single-stranded mRNA molecule.
- *cDNA library.* A collection of cloned cDNAs made from the mRNAs isolated from a cell.

- *DNA hybridization.* A technique to identify a gene of interest in a set of clones using a nucleic acid probe that can base-pair with the DNA sequence of the gene.
- *Polymerase chain reaction (PCR).* A DNA replication–based technique for amplifying DNA sequences, including genes, without cloning.
- *Agarose gel electrophoresis.* A technique in which an electric field passing through an agarose gel is used to separate DNA or RNA molecules on the basis of size.

STUDY BREAK

1. What features do restriction enzymes have in common? How do they differ?
2. Plasmid cloning vectors are one type of cloning vector that can be used with *E. coli* as a host organism. What features of a plasmid cloning vector make it useful for constructing and cloning recombinant DNA molecules?
3. What is a cDNA library, and from what cellular material is it derived? How does a cDNA library differ from a genomic library?
4. What information and materials are needed to amplify a region of DNA using PCR?

15.2 Applications of DNA Technologies

The ability to clone pieces of DNA—genes, especially—and to amplify specific segments of DNA by PCR revolutionized biology. These and other DNA technologies are now used for research in all areas of biology, including cloning genes to determine their structure, function, and regulation of expression; manipulating genes to determine how their products function in cellular or developmental processes; and identifying differences in DNA sequences among individuals in ecological studies. The same DNA technologies also have practical applications, including medical and forensic detection, modification of animals and plants, and the manufacture of commercial products. In this section, case studies provide examples of how the techniques are used to answer questions and solve problems.

15.2a DNA Technologies Are Used in Molecular Testing for Many Human Genetic Diseases

Many human genetic diseases are caused by defects in enzymes or other proteins that result from mutations at the DNA level. Once scientists have identified the specific mutations responsible for human genetic diseases, they can often use DNA technologies to develop molecular tests for those diseases. One example is sickle cell disease (see "Why It Matters" in Chapter 10; Section 10.2f; and Section 11.4a). People with this disease are homozygous for a DNA mutation that affects hemoglobin, the oxygen-carrying molecule of the blood. Hemoglobin consists of two copies each of the α-globin and β-globin polypeptides. The mutation, which is in the β-globin gene, alters one amino acid in the polypeptide. As a consequence, the function of hemoglobin is significantly impaired in individuals homozygous for the mutation (who have sickle cell disease) and mildly impaired in individuals heterozygous for the mutation (who have sickle cell trait).

The sickle cell mutation changes a restriction site in the DNA **(Figure 15.8)**. Three restriction sites for *Mst*II are associated with the normal β-globin gene, two within the coding sequence of the gene and one upstream of the gene. The sickle cell mutation eliminates the middle site of the three. Cutting the β-globin gene with *Mst*II produces two DNA fragments from the normal gene and one fragment from the mutated gene (see Figure 15.8). Restriction enzyme–generated DNA fragments of different lengths from the same region of the genome such as in this example are known as **restriction fragment length polymorphisms** (RFLPs, pronounced "riff-lips").

RFLPs are typically analyzed using **Southern blot analysis** (named after its inventor, researcher Edward Southern) **(Figure 15.9, p. 351)**. In this technique, genomic DNA is digested with a restriction enzyme, and the DNA fragments are separated using agarose gel electrophoresis. The fragments are then transferred—blotted—to a filter paper, and a labelled probe is used to identify a DNA sequence of interest from among the many thousands of fragments on the filter paper.

Analyzing DNA for the sickle cell mutation by *Mst*II digestion and Southern blot analysis is straightforward (see Figure 15.9). An individual with sickle cell disease will have one DNA band of 376 bp detected by the probe (lane A), a healthy individual will have two DNA bands of 175 and 201 bp (lane B), and an individual with sickle cell trait (heterozygous for normal and mutant alleles) will have three DNA bands of 376 bp (mutant allele) and 201 and 175 bp (normal allele) (lane C). The same probe detects all three RFLP fragments by binding to all or part of the sequence.

Restriction enzyme digestion and Southern blot analysis may be used to test for a number of other human genetic diseases, including phenylketonuria and Duchenne muscular dystrophy. In some cases, restriction enzyme digestion is combined with PCR for a quicker, easier analysis. The gene or region of the gene with the restriction enzyme variation is first amplified using PCR, and the amplified DNA is then cut with the diagnostic restriction enzyme. Amplification produces enough DNA so that separation by

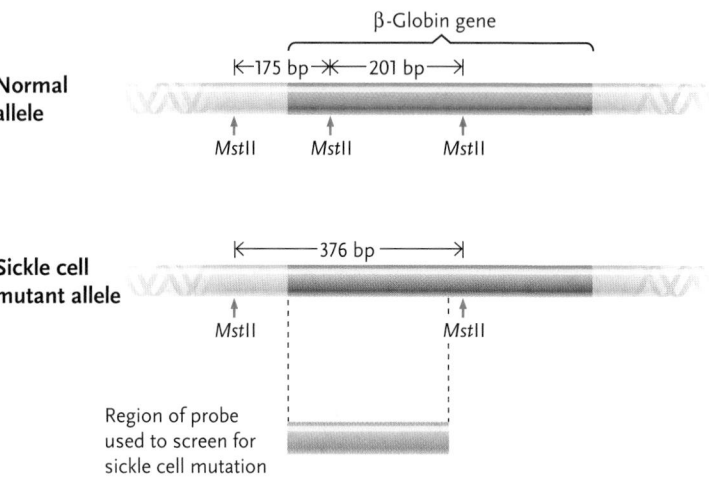

Figure 15.8
Restriction site differences between the normal and sickle cell mutant alleles of the β-globin gene. The figure shows a DNA segment that can be used as a probe to identify these alleles in subsequent analysis (see Figure 15.9, p. 351).

size on an agarose gel produces clearly visible bands, positioned according to fragment length. Researchers can then determine whether the fragment lengths match a normal or abnormal RFLP pattern. This method eliminates the need for a probe or for Southern blotting.

15.2b DNA Fingerprinting Is Used to Identify Human Individuals and Individuals of Other Species

Just as each human has a unique set of fingerprints, each also has unique combinations and variations of DNA sequences (with the exception of identical twins) known as *DNA fingerprints*. **DNA fingerprinting** is a technique used to distinguish between individuals of the same species using DNA samples. Invented by Sir Alec Jeffreys in 1985, DNA fingerprinting has become a mainstream technique for distinguishing human individuals, notably in forensics and paternity testing. Although the technique can be applied to all kinds of animals and plants, in this chapter we focus on humans.

DNA Fingerprinting Principles. In DNA fingerprinting, scientists use molecular techniques, most typically PCR, to analyze DNA variations at various loci in the genome. Several loci in noncoding regions of the genome are used for analysis. Each locus is an example of a *short tandem repeat* (STR) sequence, meaning that it has a short sequence of DNA repeated in series, with each repeat about 3 to 5 bp. Each locus has a different repeated sequence, and the number of repeats varies among individuals in a population. For example, one STR locus has the sequence AGAT repeated between 8 and 20 times. As a further source of variation, a given individual is either homozygous or heterozygous for

MOLECULE BEHIND BIOLOGY 15.1

Ethidium Bromide

Figure 15.7, p. 347, shows an agarose gel containing DNA fragments separated by electrophoresis. The fragments appear orange because a stain has bound to the DNA and is fluorescing under ultraviolet light. The stain is ethidium bromide (**Figure 1**). This relatively flat molecule slides

Figure 1
Ethidium bromide.

neatly between the bases of DNA by a process called intercalation—hence, its usefulness as a stain. However, intercalation into DNA around replication forks can increase the frequency of addition/deletion mutations in cultured cells.

---

an STR allele; perhaps you are homozygous for the 11-repeat allele or heterozygous for a 9-repeat allele and a 15-repeat allele. Likely your DNA fingerprint for this locus is different from most of the others in your class. Because each individual has an essentially unique combination of alleles (identical twins are the exception), analysis of multiple STR loci can discriminate between DNA of different individuals.

Figure 15.10, p. 352, illustrates how PCR is used to obtain a DNA fingerprint for a theoretical STR locus with three alleles of 9, 11, and 15 tandem repeats (see Figure 15.10a). Using primers that flank the STR locus, the locus is amplified from genomic DNA using PCR, and the PCR products are analyzed by gel electrophoresis (see Figure 15.10b).

CONCEPT FIX Notice in Figure 15.10b that the first lane has only one band, even though the cell that was the source of DNA was diploid and had two copies of all alleles. In this case, both alleles produce the same size fragment by PCR analysis. Therefore, fragments from both alleles migrate the same distance in the gel. Notice how this band is thicker than the others, indicating more DNA.

DNA Fingerprinting in Forensics. DNA fingerprints are routinely used to identify criminals or eliminate innocent people as suspects in legal proceedings. For example, a DNA fingerprint prepared from a hair found at the scene of a crime or from a semen sample might be compared with the DNA fingerprint of a suspect to link the suspect with the crime. Or a DNA fingerprint of blood found on a suspect's clothing or possessions might be compared with the DNA fingerprint of a victim. Typically, the evidence is presented in terms of the probability that the particular DNA sample could have come from a random individual. Hence, the media report probability values, such as one in several million, or in several billion, that a person other than the accused could have left his or her DNA at the crime scene.

Although courts initially met with legal challenges to the admissibility of DNA fingerprints, experience has shown that they are highly dependable as a line of evidence if DNA samples are collected and prepared with care and if a sufficient number of polymorphic loci are examined. There is always concern, however, about the possibility of contamination of the sample with DNA from another source during the path from crime scene to forensic lab analysis. Moreover, in some cases, criminals themselves have planted fake DNA samples at crime scenes to confuse the investigation.

There are many examples of the use of DNA fingerprinting to identify a criminal. For example, in a case in England, the DNA fingerprints of more than 4000 men were made during an investigation of the rape and murder of two teenage girls. The results led to the release of a man wrongly imprisoned for the crimes and to the confession and conviction of the actual killer. And the application of DNA fingerprinting techniques to stored forensic samples has led to the release of a number of people wrongly convicted for rape or murder.

DNA Fingerprinting in Testing Paternity and Establishing Ancestry. DNA fingerprints are also widely used as evidence of paternity because parents and their children share common alleles in their DNA fingerprints. That is, each child receives one allele of each locus from one parent and the other allele from the other parent. A comparison of DNA fingerprints for a number of loci can prove almost infallibly whether a child has been fathered or mothered by a given person. DNA fingerprints have also been used for other investigations, such as confirming that remains discovered in a remote region of Russia were actually those of Czar Nicholas II and members of his family, murdered in 1918 during the Russian revolution.

DNA fingerprinting is also widely used in studies of other organisms, including other animals, plants, and bacteria. Examples include testing for pathogenic *E. coli* in food sources such as hamburger meat, investigating cases of wildlife poaching, detecting genetically modified organisms among living organisms or in food, and comparing the DNA of ancient organisms with that of present-day descendants.

Research Method Figure 15.9 Southern blot analysis.

PURPOSE: The Southern blot technique allows researchers to identify DNA fragments of interest after separating DNA fragments on a gel. One application is to compare different samples of genomic DNA cut with a restriction enzyme to detect specific RFLPs. Here the technique is used to distinguish between individuals with sickle cell disease, individuals with sickle cell trait, and normal individuals.

PROTOCOL:

1 Isolate genomic DNA and digest with a restriction enzyme. Here, genomic DNA is isolated from three individuals: A, sickle cell disease (homozygous for the sickle cell mutant allele); B, normal (homozygous for the normal allele); and C, sickle cell trait (heterozygous for sickle cell mutant allele). Digest the DNA with *Mst*II.

2 Separate the DNA fragments by agarose gel electrophoresis. The thousands of differently sized DNA fragments produce a smear of DNA down the length of each lane in the gel, which can be seen after staining the DNA. (Gel electrophoresis and gel staining are shown in Figure 15.7, p. 347.)

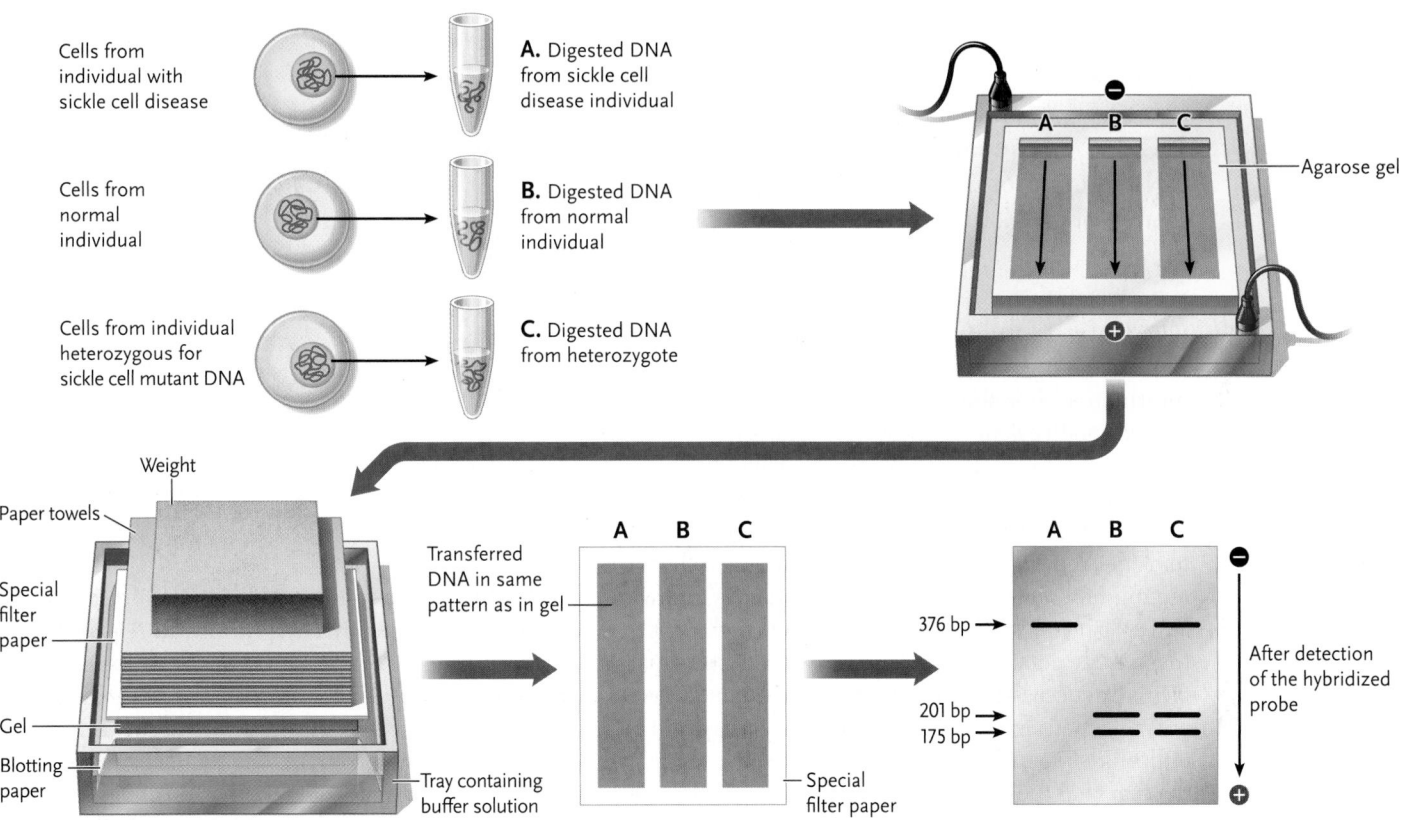

3 Hybridization with a labelled DNA probe to identify DNA fragments of interest cannot be done directly with an agarose gel. Edward Southern devised a method to transfer the DNA fragments from a gel to a special filter paper. First, treat the gel with a solution to denature the DNA into single strands. Next, place the gel on a piece of blotting paper with the ends of the paper in the buffer solution and place the special filter paper on top of the gel. Capillary action wicks the buffer solution in the tray up the blotting paper, through the gel and special filter paper, and into the weighted stack of paper towels on top of the gel. The movement of the solution transfers— blots—the single-stranded DNA fragments to the filter paper, where they stick. The pattern of DNA fragments is the same as it was in the gel.

4 To focus on a particular region of the genome, use DNA hybridization with a labelled probe. That is, incubate a labelled, single-stranded probe with the filter and, after washing off excess probe, detect hybridization of the probe with DNA fragments on the filter. For a radioactive probe, place the filter against photographic film, which, after development, will show a band or bands where the probe hybridized. In this experiment, the probe is a cloned piece of DNA from the area shown in Figure 15.8, p. 349 (the β-globin gene) that can bind to all three of the *Mst*II fragments of interest.

INTERPRETING THE RESULTS: The hybridization result indicates that the probe has identified a very specific DNA fragment or fragments in the digested genomic DNA. The RFLPs for the β-globin gene can be seen in Figure 15.8, p. 349. DNA from the sickle cell disease individual cut with *Mst*II results in a single band of 376 bp detected by the probe, while DNA from the normal individual results in two bands, of 201 bp and 175 bp. DNA from a sickle cell trait heterozygote results in three bands, of 376 bp (from the sickle cell mutant allele), and 201 bp and 175 bp (both from the normal allele). This type of analysis in general is useful for distinguishing normal and mutant alleles of genes where the mutation involved alters a restriction site.

a. Alleles at an STR locus

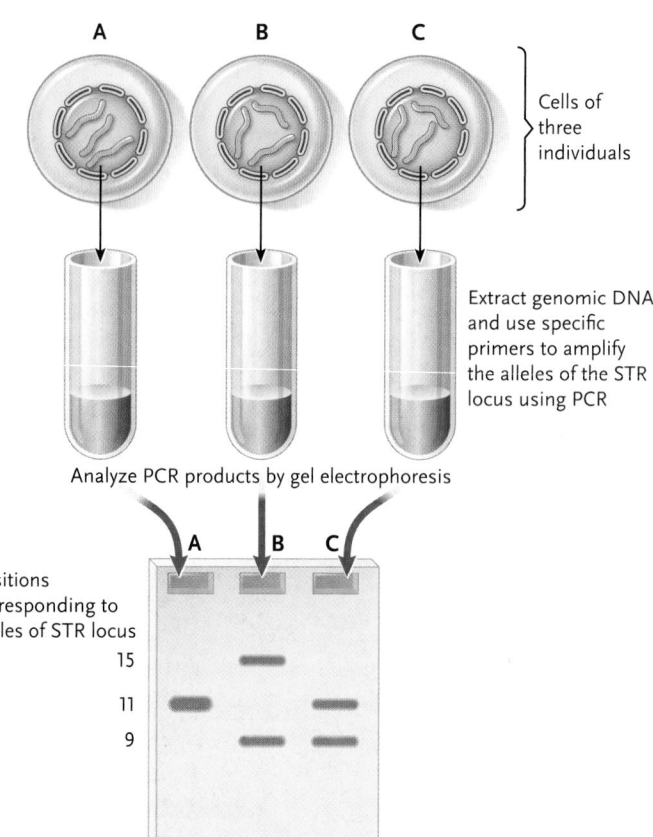

Figure 15.10
Using PCR to obtain a DNA fingerprint for an STR locus.

15.2c Genetic Engineering Uses DNA Technologies to Alter the Genes of a Cell or Organism

We have seen the many ways scientists use DNA technologies to ask and answer questions that were once completely inaccessible. Genetic engineering goes beyond gathering information; it is the use of DNA technologies to modify genes of a cell or organism. The goals of genetic engineering include using prokaryotic cells, fungi, animals, and plants as factories for the production of proteins needed in medicine and scientific research; correcting hereditary disorders; and improving animals and crop plants of agricultural importance. In many of these areas, genetic engineering has already been spectacularly successful. The successes and potential benefits of genetic engineering, however, are tempered by ethical and social concerns about its use, along with the fear that the methods may produce toxic or damaging foods or release dangerous and uncontrollable organisms to the environment.

Genetic engineering uses DNA technologies of the kind already discussed in this chapter. DNA—perhaps a modified gene—is introduced into target cells of an organism. Organisms that have undergone a gene transfer are called **transgenic**, meaning that they have been modified to contain genetic information—the *transgene*—from an external source.

The following sections discuss examples of applications of genetic engineering to bacteria, animals, and plants and assess major controversies arising from these projects.

Genetic Engineering of Bacteria to Produce Proteins. Transgenic bacteria have been made, for example, to synthesize proteins for medical applications, break down toxic wastes such as oil spills, produce industrial chemicals such as alcohols, and process minerals. *E. coli* is the organism of choice for many of these applications of DNA technologies.

Using *E. coli* to make a protein from a foreign source is conceptually straightforward **(Figure 15.11)**. First, the gene for the protein is cloned from the appropriate organism. Then the gene is inserted into an **expression vector** that, in addition to the usual features of a cloning vector, contains the regulatory sequences that allow transcription and translation of the gene. For a bacterial expression vector, this means having a promoter and a transcription terminator that are recognized by the *E. coli* transcriptional machinery and having the ribosome binding site needed for the bacterial ribosome to recognize the start codon of the transgene (see Section 13.1). The regulatory sequences flank the cluster of restriction sites of the expression vector that are used for cloning so that the inserted gene is correctly placed for transcription and translation when the recombinant plasmid is transformed into *E. coli*.

As we mentioned earlier while discussing DNA libraries, a cDNA copy of a eukaryotic gene is used when we want to use bacteria to express the protein encoded by the gene (see Figure 15.11). This is because the gene itself typically contains introns, which bacteria cannot remove when they transcribe a eukaryotic gene. However, the eukaryotic mRNA that is copied by reverse transcriptase to synthesize cDNA has had its introns removed; thus, when that cDNA is expressed in bacteria it can be transcribed and translated to make the encoded eukaryotic protein. The protein is either extracted from the bacterial cells and purified or, if the protein is secreted, purified from the culture medium.

Expression vectors are available for a number of organisms. They vary in the regulatory sequences they contain and the selectable marker they carry so that the host organism transformed with the vector carrying a gene of interest can be detected and the host can express that gene.

For example, *E. coli* bacteria have been genetically engineered to make the human hormone insulin; the commercial product is called Humulin. Insulin is required by people with some forms of diabetes. Humulin is a perfect copy of the human insulin hormone. Many other proteins, including human growth hormone to treat human growth disorders, tissue plasminogen activator to dissolve blood clots that cause heart attacks, and a vaccine against foot-and-mouth disease of cattle and other cloven-hoofed animals (a highly contagious and sometimes fatal viral disease), have been developed for commercial production in bacteria using similar methods.

A concern is that genetically engineered bacteria may be released accidentally into the environment, where possible adverse effects of the organisms are currently unknown. Scientists minimize the danger of accidental release by growing the bacteria in laboratories that follow appropriate biosafety protocols. In addition, the bacterial strains typically used are genetic mutants that cannot survive outside the growth media used in the laboratory.

Genetic Engineering of Animals. Many animals, including fruit flies, fish, mice, pigs, sheep, goats, and cows, have been altered successfully by genetic engineering. There are many purposes for these alterations, including basic research, correcting genetic disorders in humans and other mammals, and producing pharmaceutically important proteins.

Genetic Engineering Methods for Animals. Several methods are used to introduce a gene of interest into animal cells. The gene may be introduced into *germ-line cells,* which develop into sperm or eggs and thus enable the introduced gene to be passed from generation to generation. Or, the gene may be introduced into *somatic* (body) *cells,* differentiated cells that are not part of lines producing sperm or eggs, in which case the gene is not transmitted from generation to generation.

Germ-line cells of embryos are often used as targets for introducing genes, particularly in mammals **(Figure 15.12, p. 354).** The treated cells are then cultured in quantity and reintroduced into early embryos. If the technique is successful, some of the introduced cells become founders of cell lines that develop into eggs or sperm with the desired genetic information integrated into their DNA. Individuals produced by crosses using the engineered eggs and sperm then contain the

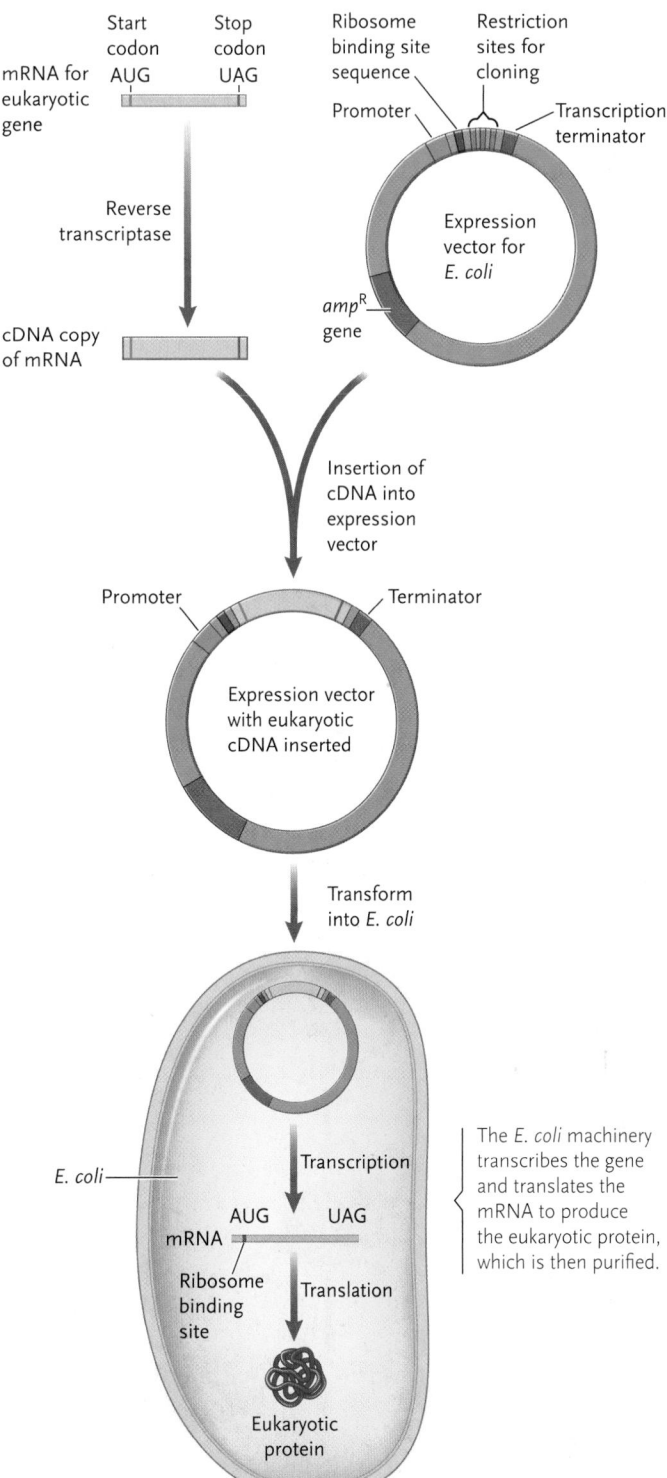

Figure 15.11

Using an expression vector to synthesize a eukaryotic protein.

introduced sequences in all of their cells. Several genes have been introduced into the germ lines of mice by this approach, resulting in permanent, heritable changes in the engineered individuals.

A related technique involves introducing desired genes into **stem cells,** which are cells capable of undergoing many divisions in an unspecialized, undifferentiated state, but which can also differentiate into

Research Method Figure 15.12 Introduction of genes into mouse embryos using embryonic germ-line cells.

PURPOSE: To make a transgenic animal that can transmit the transgene to offspring. The embryonic germ-line cells that receive the transgene develop into the reproductive cells of the animal.

PROTOCOL:

1. Introduce the desired gene into germ-line cells from an embryo by injection or electroporation.

2. Clone the cell that has the incorporated transgene to produce a pure culture of transgenic cells.

3. Inject transgenic cells into early-stage embryos (called blastocysts).

4. Implant embryos into surrogate (foster) mothers.

5. Allow embryos to grow to maturity and be born.

6. Interbreed the progeny mice.

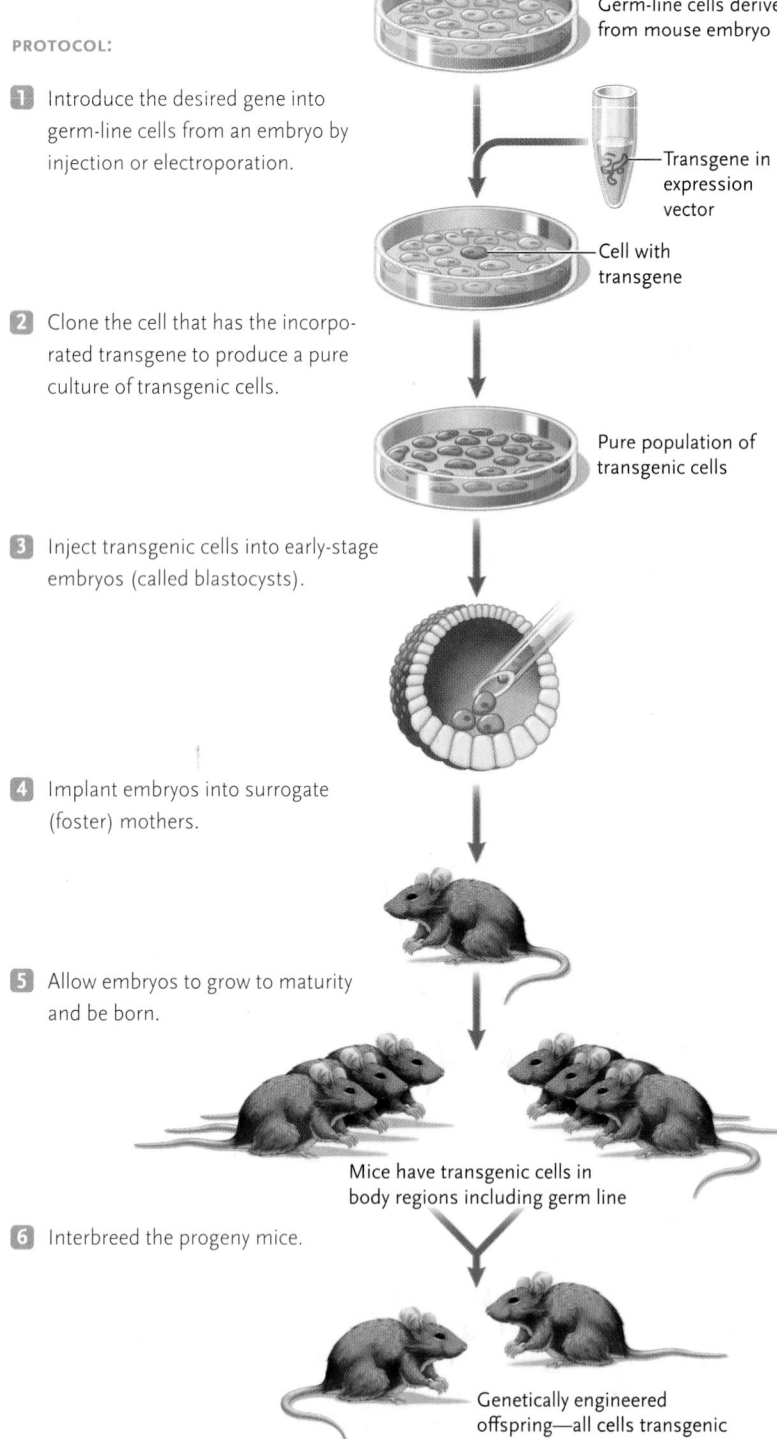

Germ-line cells derived from mouse embryo

Transgene in expression vector

Cell with transgene

Pure population of transgenic cells

Mice have transgenic cells in body regions including germ line

Genetically engineered offspring—all cells transgenic

INTERPRETING THE RESULTS: The result of the breeding is some offspring in which all cells are transgenic—a genetically engineered animal has been produced.

specialized cell types. In mammals, **embryonic stem cells** are found in a mass of cells inside an early-stage embryo (the blastocyst) (see Figure 15.12, step 3) and can differentiate into all of the tissue types of the embryo, whereas **adult stem cells** function to replace specialized cells in various tissues and organs. In mice and other nonhuman mammals, transgenes are introduced into embryonic stem cells, which are then injected into early-stage embryos as in Figure 15.12 (step 3). The stem cells then differentiate into a variety of tissues along with cells of the embryo itself, including sperm and egg cells. Males and females are then bred, leading to offspring that are either homozygotes, containing two copies of the introduced gene, or heterozygotes, containing one introduced gene and one gene that was native to the embryo receiving the engineered stem cells.

Introduction of genes into stem cells has been performed mostly in mice. One of the highly useful results is the production of a *knockout mouse,* a homozygous recessive that receives two copies of a gene altered to a nonfunctional state and thus has no functional copies. The effect of the missing gene on the knockout mouse is a clue to the normal function of the gene. In some cases, knockout mice are used to model human genetic diseases.

For introducing genes into somatic cells, somatic cells are typically removed from the body, cultured, and then transformed with DNA containing the transgene. The modified cells are then reintroduced into the body, where the transgene functions. Because germ cells and their products are not involved, the transgene remains in the individual and is not passed to offspring.

Gene Therapy: Correcting Genetic Disorders. The path to **gene therapy**—correcting genetic disorders—in humans began with experiments using mice. In 1982, Richard Palmiter at the University of Washington, Ralph Brinster of the University of Pennsylvania, and their colleagues injected a growth hormone gene from rats into fertilized mouse eggs and implanted the eggs into a surrogate mother. She gave birth to some normal-sized mouse pups that grew more quickly than normal and became about twice the size of their normal litter mates. These *giant mice* **(Figure 15.13)** attracted extensive media attention from around the world.

Palmiter and Brinster next attempted to cure a genetic disorder by gene therapy. In this experiment, they were able to correct a genetic growth hormone deficiency that produces dwarf mice. They introduced a normal copy of the growth hormone gene into fertilized eggs taken from mutant dwarf mice and implanted the eggs into a surrogate mother. The transgenic mouse pups grew to slightly larger than normal, demonstrating that the genetic defect in these mice had been corrected.

This sort of experiment, in which a gene is introduced into germ-line cells of an animal to correct a genetic disorder, is called **germ-line gene therapy.** For ethical reasons, germ-line gene therapy is not permitted with humans. Instead, humans are treated with **somatic gene therapy,** in which genes are introduced into somatic cells (as described in the previous section).

The first successful use of somatic gene therapy with a human subject who had a genetic disorder was carried out in the 1990s by W. French Anderson and his colleagues at the National Institutes of Health (NIH) in the United States. The subject was a young girl with *adenosine deaminase deficiency (ADA)*. Without the adenosine deaminase enzyme, white blood cells cannot mature (see Chapter 51); without normally functioning white blood cells, the body's immune response is so deficient that most children with ADA die of infections before reaching puberty. The researchers successfully introduced a functional ADA gene into mature white blood cells isolated from the patient. Those cells were reintroduced into the girl, and expression of the ADA gene provided a temporary cure for her ADA deficiency. The cure was not permanent because mature white blood cells, produced by differentiation of stem cells in the bone marrow, are nondividing cells with a finite lifetime. Therefore, the somatic gene therapy procedure has to be repeated every few months. Indeed, the subject of this example still receives periodic gene therapy to maintain the necessary levels of the ADA enzyme in her blood. In addition, she receives direct doses of the normal enzyme. More recent improved protocols have resulted in successful treatment of over 30 ADA patients worldwide without adverse affects or the need for ongoing therapy.

Successful somatic gene therapy has also been achieved for sickle cell disease. In December 1998, a 13-year-old boy's bone marrow cells were replaced with stem cells from the **umbilical cord** of an unrelated infant. The hope was that the stem cells would produce healthy bone marrow cells, the source of blood cells. The procedure worked, and the patient has been declared cured of the disease.

However, despite enormous efforts, human somatic gene therapy has not been the panacea people expected. Relatively little progress has been made since the first gene therapy clinical trial for ADA deficiency was described, and, in fact, there have been major setbacks. In 1999, for example, a teenage patient in a somatic gene therapy trial died as a result of a severe immune response to the viral vector being used to introduce a normal gene to correct his genetic deficiency. Furthermore, some children in gene therapy trials involving the use of retrovirus vectors to introduce genes into blood stem cells have developed a leukemia-like condition. In short, somatic gene therapy is not yet an effective treatment for human genetic disease, even though the approach has been successful in a number of cases to correct models of human genetic disorders in experimental mammals. Although no commercial human gene therapy product has been

approved for use, roughly 100 new clinical trials are approved worldwide each year, focusing on improved therapies for a long list of conditions, including cancer, congenital blindness, Parkinson's disease, malaria, multiple sclerosis, arthritis, Type I diabetes, cystic fibrosis, and muscular dystrophy, as well as various blood and immunological disorders.

Turning Domestic Animals into Protein Factories. Another successful application of genetic engineering turns animals into pharmaceutical factories for the production of proteins required to treat human diseases or other medical conditions. Most of these *pharming* projects, as they are called, engineer the animals to produce the desired proteins in milk, making the production, extraction, and purification of the proteins harmless to the animals.

One of the first successful applications of this approach was carried out with sheep engineered to produce a protein required for normal blood clotting in humans. The protein, called a *clotting factor,* is deficient in people with one form of hemophilia, who require frequent injections of the factor to avoid bleeding to death from even minor injuries. Using DNA-cloning techniques, researchers joined the gene encoding the normal form of the clotting factor to the promoter sequences of the β-lactoglobin gene, which encodes a protein secreted in milk, and introduced it into fertilized eggs. These cells were implanted into a surrogate mother, and the transgenic sheep born were allowed to mature. The β-lactoglobin promoter controlling the clotting factor gene became activated in mammary gland cells of females, resulting in the production of clotting factor. The clotting factor was then secreted into the milk. Production in the milk is harmless to the sheep and yields the protein in a form that can easily be obtained and purified.

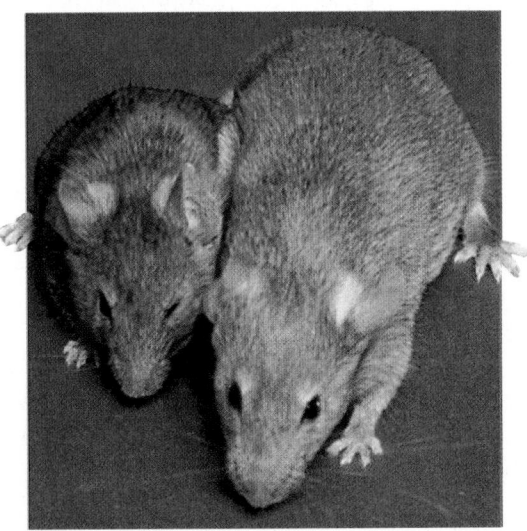

R. L. Brinster, R. E. Hammer, School of Veterinary Medicine, University of Pennsylvania

Figure 15.13
A genetically engineered giant mouse (right) produced by the introduction of a rat growth hormone gene into the animal. A mouse of normal size is on the left.

Experimental Research Figure 15.14 The first cloning of a mammal.

QUESTION: Does the nucleus of an adult mammal contain all the genetic information to specify a new organism? In other words, can mammals be cloned starting with adult cells?

EXPERIMENT: Ian Wilmut, Keith Campbell, and their colleagues fused a mammary gland cell from an adult sheep with an unfertilized egg cell from which the nucleus had been removed, and tested whether that fused cell could produce a lamb.

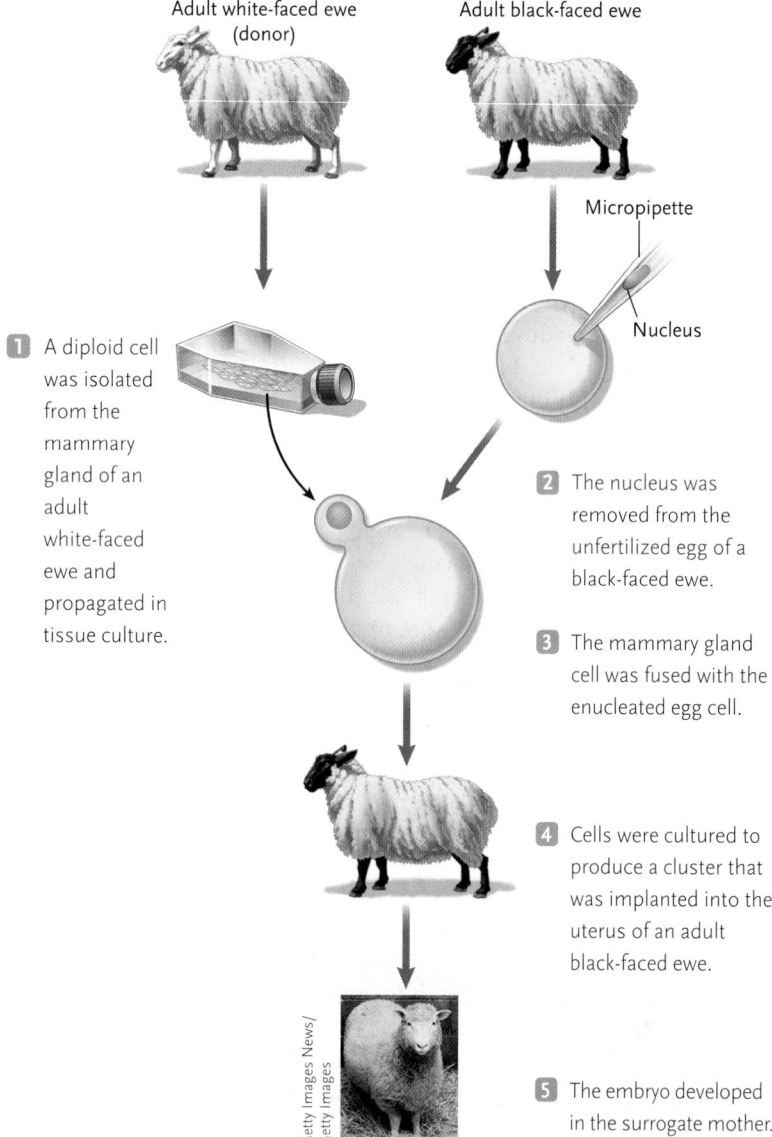

Adult white-faced ewe (donor)

Adult black-faced ewe

Micropipette

Nucleus

1 A diploid cell was isolated from the mammary gland of an adult white-faced ewe and propagated in tissue culture.

2 The nucleus was removed from the unfertilized egg of a black-faced ewe.

3 The mammary gland cell was fused with the enucleated egg cell.

4 Cells were cultured to produce a cluster that was implanted into the uterus of an adult black-faced ewe.

Getty Images News/ Getty Images

5 The embryo developed in the surrogate mother.

RESULT: Dolly was born and grew normally. She was white faced—a clone of the donor ewe. DNA fingerprinting using STR loci showed her DNA matched that of the donor ewe and neither the ewe who donated the egg nor the ewe who was the surrogate mother.

CONCLUSION: An adult nucleus of a mammal contains all the genetic material necessary to direct the development of a normal new organism, a clone of the original. Dolly was the first cloned mammal. The success rate for Wilmut and Campbell's experiment was very low—Dolly represented less than 0.4% of the fused cells they made—but its significance was huge.

Based on I. Wilmut et al. "Viable offspring derived from fetal and mammalian cells." *Nature*. 1997 Feb 27; 385(6619): 810–3.

Other similar projects are under development to produce particular proteins in transgenic mammals. These include a protein to treat cystic fibrosis, collagen to correct scars and wrinkles, human milk proteins to be added to infant formulas, and normal hemoglobin for use as an additive to blood transfusions.

Producing Animal Clones. Making transgenic mammals is expensive and inefficient. And because only one copy of the transgene typically becomes incorporated into the treated cell, not all progeny of a transgenic animal inherit that gene. Scientists reasoned that an alternative to breeding a valuable transgenic mammal to produce progeny with the transgene would be to clone the mammal. Each clone would be identical to the original, including the expression of the transgene. That this is possible was shown in 1997 when two scientists, Ian Wilmut and Keith H. S. Campbell of the Roslin Institute, Edinburgh, Scotland, announced that they had successfully cloned a sheep from a single somatic cell derived from an adult sheep **(Figure 15.14)**—the first cloned mammal.

Since the successful cloning experiment producing Dolly, many additional mammals have been cloned, including mice, goats, pigs, monkeys, rabbits, dogs, a male calf appropriately named Gene, and a domestic cat called CC (for *Copy Cat*).

Cloning farm animals has been so successful that several commercial enterprises now provide cloned copies of champion animals. One example is a clone of an American Holstein cow, Zita, who was the U.S. national champion milk producer for many years. Animal breeders estimate that there are now more than 100 cloned animals on U.S. farms, and breeders plan to produce entire herds if government approval is granted.

The cloning of domestic animals has its drawbacks. Many cloning attempts fail, leading to the death of the transplanted embryos. Cloned animals often suffer from conditions such as birth defects and poor lung development. Genes may be lost during the cloning process or may be expressed abnormally in the cloned animal. For example, molecular studies have shown that the expression of perhaps hundreds of genes in the genomes of clones is regulated abnormally.

CONCEPT FIX In studying Figures 15.12, p. 354, and 15.14, both related to manipulating animal embryos, it may be tempting to see Dolly as an advance in stem cell research. However, Dolly resulted from the transfer of a somatic cell nucleus to an egg cell that was lacking a nucleus. This cloning technique, called somatic cell nuclear transfer (SCNT), doesn't involve stem cells.

Genetic Engineering of Plants. Genetic engineering of plants has led to increased resistance to pests and disease; greater tolerance to heat, drought, and salinity; greater crop yields; faster growth; and resistance to herbicides. Another aim is to produce seeds with higher levels of amino acids. The essential amino acid lysine, for example, is present only in limited quantities in cereal grains such as wheat, rice, oats, barley, and corn; the seeds of legumes such as beans, peas, lentils, soybeans, and peanuts are deficient in the essential amino acid methionine or cysteine. Increasing the amounts of the deficient amino acids in plant seeds by genetic engineering would greatly improve the diet of domestic animals and human populations that rely on seeds as a primary food source. Efforts are also under way to increase the content of vitamins and minerals in crop plants.

Another possibility for plant genetic engineering is plant pharming to produce pharmaceutical products. Plants are ideal for this purpose because they are primary producers at the bottom rung of the **food chain** and can be grown in huge numbers with maximum conservation of the Sun's energy captured in photosynthesis.

Some plants, such as *Arabidopsis*, tobacco, potato, cabbage, and carrot, have special advantages for genetic engineering because individual cells can be removed from an adult, altered by the introduction of a desired gene, and then grown in cultures into a multicellular mass of cloned cells called a *callus*. Subsequently, roots, stems, and leaves develop in the callus, forming a young plant that can then be grown in containers or fields by the usual methods. In the plant, each cell contains the introduced gene. The gametes produced by the transgenic plants can then be used in crosses to produce offspring, some of which will have the transgene, as in the similar experiments with animals.

Methods Used to Insert Genes into Plants. Genes are inserted into plant cells by several techniques. One commonly used method takes advantage of a natural process that causes crown gall disease, which is characterized by bulbous, irregular growths—tumours, essentially—that can develop at wound sites on the trunks and limbs of deciduous trees **(Figure 15.15)**. Crown gall disease is caused by the bacterium *Rhizobium radiobacter* (formerly *Agrobacterium tumefaciens*, recently reclassified on the basis of genome analysis). This bacterium contains a large, circular plasmid called the **Ti (tumour-inducing) plasmid**. The interaction between the bacterium and the plant cell it infects stimulates the excision of a segment of the Ti plasmid called *T DNA* (for transforming DNA), which then integrates into the plant cell's genome. Genes on the T DNA are then expressed; the products stimulate the transformed cell to grow and divide and therefore to produce a tumour. The tumours provide essential nutrients for the bacterium. The Ti plasmid is used as a vector for making transgenic plants in much the same way as bacterial plasmids are used as vectors to introduce genes into bacteria **(Figure 15.16, p. 358)**.

Edward L. Barnard, Florida Department of Agriculture and Consumer Services, Bugwood.org

Figure 15.15
A crown gall tumour on the trunk of a California pepper tree. The tumour, stimulated by genes introduced from the bacterium *Rhizobium radiobacter*, is the bulbous, irregular growth extending from the trunk.

Gall

Successful Plant Genetic Engineering Projects. An early visual demonstration of the successful use of genetic engineering techniques to produce a transgenic plant is the glowing tobacco plant **(Figure 15.17, p. 358)**. The transgenic plant contained luciferase, the gene for the firefly enzyme. When the plant was soaked in the substrate for the enzyme, it became luminescent.

The most widespread application of genetic engineering of plants involves the production of transgenic crops. Thousands of such crops have been developed and field tested, and many have been approved for commercial use. If you analyze the processed plant-based foods at a national supermarket chain, you will likely find that at least two-thirds contain products made from transgenic plants.

In many cases, plants are modified to make them resistant to insect pests, viruses, or herbicides. Crops modified for insect resistance include corn, cotton, and potatoes. The most common approach to making plants resistant to insects is to introduce the gene from the bacterium *Bacillus thuringiensis* that encodes the *Bt* toxin, an organic pesticide. This toxin has been used in powder form to kill insects in agriculture for many years, and now transgenic plants making their own *Bt* toxin are resistant to specific groups of insects that feed on them. Millions of acres of crop plants planted in the United States and Canada are *Bt*-engineered varieties.

Virus infections cause enormous crop losses worldwide. Transgenic crops that are virus resistant would be highly valuable to the agricultural community. There is some promise in this area. By some unknown process, transgenic plants expressing certain viral proteins become resistant to infections by whole viruses that contain these same proteins. Two

Research Method Figure 15.16 Using the Ti plasmid of *Rhizobium radiobacter* to produce transgenic plants.

PURPOSE: To make transgenic plants. This technique is one way to introduce a transgene into a plant for genetic engineering purposes.

PROTOCOL:

1 Isolate the Ti plasmid from *Rhizobium radiobacter*. The plasmid contains a segment called T DNA (T = transforming), which induces tumours in plants.

2 Digest the Ti plasmid with a restriction enzyme that cuts within the T DNA. Mix with a gene of interest on a DNA fragment that was produced by digesting with the same enzyme. Use DNA ligase to join the two DNA molecules together to produce a recombinant plasmid.

3 Transform the recombinant Ti plasmid into a disarmed *A. Rhizobium radiobacter* that cannot induce tumours, and use the transformed bacterium to infect cells in plant fragments in a test tube. In infected cells, the T DNA with the inserted gene of interest excises from the Ti plasmid and integrates into the plant cell genome.

4 Culture the transgenic plant fragments to regenerate whole plants.

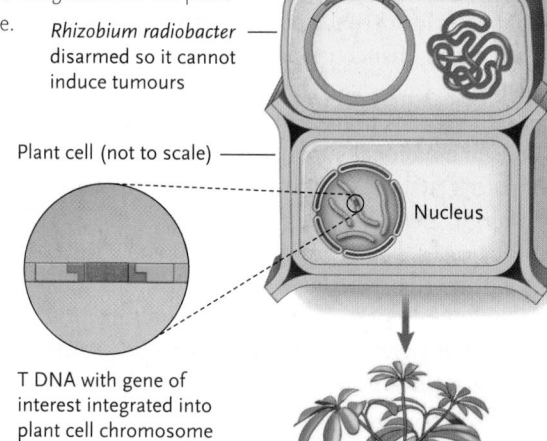

INTERPRETING THE RESULTS: The plant has been genetically engineered to contain a new gene. The transgenic plant will express a new trait based on that gene, perhaps resistance to a herbicide or production of an insect toxin, according to the goal of the experiment.

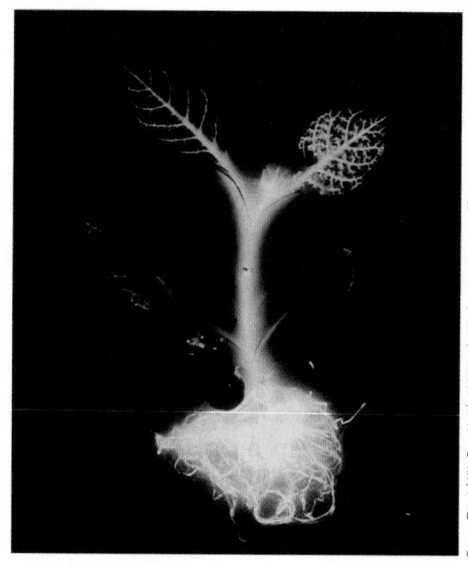

From David W. Ow, Keith V. Wood, Marlene DeLuca, Jeffrey R. de Wet, Donald R. Helinski and Stephen H. Howell, "Transient and stable expression of the firefly luciferase gene in plant cells and transgenic plants." *Science* 14 November 1986, Vol. 234 no. 4778 pp. 856–859. Reprinted with permission from AAAS.

Figure 15.17

A genetically engineered tobacco plant, made capable of luminescence by the introduction of a firefly gene coding for the enzyme luciferase.

virus-resistant genetically modified crops made so far are papaya and squash.

Several crops have also been engineered to become resistant to herbicides. For example, *glyphosate* (commonly known by its brand name, Roundup) is a highly potent herbicide that is widely used in weed control. The herbicide works by inhibiting a particular enzyme in the chloroplast. Unfortunately, it also kills crops. But transgenic crops have been made in which a bacterial form of the chloroplast enzyme has been added to the plants. The bacteria-derived enzyme is not affected by Roundup, and farmers who use these herbicide-resistant crops can spray fields of crops to kill weeds without killing the crops. Now most of the corn, soybean, canola, and cotton plants grown in North America are the genetically engineered, glyphosate-resistant ("Roundup-ready") varieties.

Crop plants are also being engineered to alter their nutritional qualities. For example, a strain of rice plants has been produced with seeds rich in β-carotene, a precursor of vitamin A, as well as iron **(Figure 15.18)**. The new rice, which is given a yellow or golden colour

Regular rice   Genetically engineered golden rice containing β-carotene

Golden Rice Humanitarian Board www.goldenrice.org

Figure 15.18

Rice genetically engineered to contain β-carotene.

by the carotene, may provide improved nutrition for the billions of people who depend on rice as a diet staple. In particular, the rice may help improve the nutrition of children younger than age 5 in southeast Asia, 70% of whom suffer from impaired vision because of vitamin A deficiency.

Plant pharming is also an active area both in university research labs and at biotechnology companies. Plant pharming involves the engineering of transgenic plants to produce medically valuable products. The approach is one described earlier: the gene for the product is cloned into a cloning vector adjacent to a promoter, in this case one active in plants, and the recombinant DNA molecule is introduced into plants. Products under development include vaccines for various bacterial and viral diseases, protease inhibitors to treat or prevent virus infections, collagen to treat scars and wrinkles, and aprotinin to reduce bleeding and clotting during heart surgery.

In contrast to animal genetic engineering, genetically altered plants have been widely developed and appear to be here to stay as mainstays of agriculture. But, as the next section discusses, both animal and plant genetic engineering have not proceeded without concerns.

15.2d DNA Technologies and Genetic Engineering Are a Subject of Public Concern

When recombinant DNA technology was developed in the early 1970s, researchers quickly recognized that in addition to the many anticipated benefits, there might be deleterious outcomes. One key concern at the time was that a bacterium carrying a recombinant DNA molecule might escape into the environment. Perhaps it could transfer that molecule to other bacteria and produce new, potentially harmful, strains. To address these concerns, the U.S. scientists who developed the technology drew up safety guidelines for recombinant DNA research in the United States. Adopted by the NIH, the guidelines listed the precautions to be used in the laboratory when constructing recombinant DNA molecules and included the design and use of host organisms that could survive only in growth media in the laboratory. Since that time, countless experiments involving recombinant DNA molecules have been done in laboratories around the world. These experiments have shown that recombinant DNA manipulations can be done safely. Over time, therefore, the recombinant DNA guidelines have become more relaxed. Nonetheless, stringent regulations still exist for certain areas of recombinant DNA research that pose significant risk, such as cloning genes from highly pathogenic bacteria or viruses, or gene therapy experiments. In essence, as the risk increases, the research facility must increase its security and must obtain more levels of approval by peer scientist groups.

Guidelines for genetic engineering also extend to research in several areas that have been the subject of public concern and debate. Although the public does not seem to be very concerned about genetically engineered microorganisms, for example, those cleaning up oil spills and hazardous chemicals, it is concerned about possible problems with **genetically modified organisms (GMOs)** used as food. A GMO is a transgenic organism; the majority of GMOs are crop plants. Issues are the safety of GMO-containing food and the possible adverse effects of the GMOs to the environment, such as by interbreeding with natural species or by harming beneficial insect species. For example, could introduced genes providing herbicide or insect resistance move from crop plants into related weed species through cross-pollination, producing "superweeds" that might be difficult or impossible to control? *Bt*-expressing corn was originally thought to have adverse effects on monarch butterflies who fed on the pollen. The most recent of a series of independent studies investigating this possibility has indicated that the risk to the butterflies is extremely low.

More broadly, different countries have reacted to GMOs in different ways. In Canada, transgenic crops are quite widely planted and harvested. Before commercialization, such GMOs are evaluated for potential risk by appropriate government regulatory agencies, including Health Canada, the Canadian Food Inspection Agency, and Environment Canada.

Political opposition to GMOs has been greater in Europe, dampening the use of transgenic crop plants in the fields and GMOs in food. In 1999, the European Union (EU) imposed a six-year moratorium on all GMOs, leading to a bitter dispute with the United States, Canada, and Argentina, the leading growers of transgenic crops. More recently, the EU has revised the GMO regulations in all member states. Basically, the EU has decided that using genetic engineering in agriculture and food production is permissible provided that the GMO or food containing it is safe for humans, animals, and the environment. All use of GMOs in the field or in food requires authorization following a careful review process.

On a global level, an international agreement, the **Cartagena Protocol on Biosafety**, "promotes biosafety by establishing practical rules and procedures for the safe transfer [between countries], handling and use of GMOs." Separate procedures have been set up for GMOs that are to be introduced into the environment and those that are to be used as food or feed or for processing. Although 167 countries have signed and implemented the Protocol, several others, mainly GMO exporters such as Canada, the United States, and Argentina, have not.

In sum, the use of DNA technologies in biotechnology has the potential for tremendous benefits to humankind. Such experimentation is not without risk, so for each experiment, researchers must assess that risk and make a judgment about whether to proceed

The discipline of genetics was originally built on the study of rare, naturally occurring mutations. Researchers routinely screened thousands (or sometimes millions) of individuals to collect a handful of useful mutations. Agents that increased the frequency of mutations were often used, but they tended to be nonspecific, and the isolation of particular mutations in specific genes remained a lottery with unfavourable odds.

Michael Smith changed all of that in the late 1970s by demonstrating that *in vitro* DNA synthesis techniques could be used to create mutated sequences. This method of site-directed mutagenesis allowed specific mutations to be introduced into any given DNA sequence. For the first time, geneticists could create the exact changes they were interested in. Smith's work was recognized with the 1993 Nobel Prize in Chemistry.

In addition to his legacy as a scientist, Smith was a generous philanthropist and a strong supporter of public education in science. He died in 2000, three years after retiring from UBC.

and, if so, how to do so safely. Furthermore, agreed-upon guidelines and protocols should ensure a level of biosafety for researchers, consumers, politicians, and governments.

15.2e Synthetic Biologists Engineer New Systems in Living Cells

Although the technologies for manipulation and transfer of genes into new hosts, as described at the beginning of this chapter, have long been known as genetic engineering, the influence of fundamental ideas and approaches of formal engineering is relatively recent. In the late 1990s, engineers, physicists, and software designers began working with molecular biologists to apply engineering design principles to biological systems. This collaboration gave rise to the new interdisciplinary field of synthetic biology. Synthetic biologists combine standardized parts (DNA sequences) to design and build modified regulatory networks to study the organization of natural systems in living cells and to create novel networks with potential benefits in a wide range of biotechnologies. (The iGEM competition described in "Why it Matters" is a good example of the approaches and potential of the field of synthetic biology.)

Using genes "mined" from organisms in the environment, or sequences created from scratch by DNA synthesizers, scientists in this field can insert new enzymes into biosynthetic pathways, resulting in cells that produce novel products such as biodiesel, gasoline, and bioplastics. This work has culminated in the creation of a yeast strain containing an engineered biochemical pathway producing artemisinin (an antimalarial drug normally produced by the wormwood plant) on an industrial scale.

The logical end point of synthetic biology is the creation of artificial life, that is, the assembly of living systems entirely from nonliving parts. Such studies would provide a powerful model for understanding the possible origin of life, as well as the dynamics of basic cellular functioning. Synthetic cells, designed entirely from off-the-shelf parts in a laboratory, could be customized to serve a staggering array of applications in biotechnology.

Since synthetic cells would require a genome, one avenue of research toward the goal of artificial life is the top-down approach, which asks, "What is the minimum collection of genes required by a living cell?" Insights into the answer to this question come from characterizing organisms that have very small genomes, as well as studies that note the effect of inactivating every gene of an organism's genome, one at a time. Although the true minimal genome has yet to be established, it is likely in the range of about 300 genes for a free-living prokaryotic cell. In 2010, Craig Venter's group published an account of a replicating strain of a bacterium, *Mycoplasma mycoides*, they had engineered to contain a 1-million-base-pair synthetic genome. The genome was synthesized in 1000-base-pair segments from digitized sequence information, and then the segments were assembled in yeast before being transferred to the *Mycoplasma*. This was the first known life form that did not obtain its genome from a parent cell, proving the concept that a cell could be "booted up" by synthesized DNA.

A complementary approach to creating synthetic life is the bottom-up interest in building a functional cell, and its various components, from nonliving parts. Early research is well under way toward the creation of various types of primitive protocells that contain simple metabolism and nucleic acid biochemistry. The rapid advance of synthetic biology has led many researchers to predict that the first truly artificial living cells will be created in the near future.

In this chapter you have learned about how individual genes can be isolated and manipulated using various DNA technologies. But a gene is just a part of a genome. Researchers also want to know about the set of genes in a complete genome, and how genes and their gene products work together in networks to control life. They also want to know more generally about the organization of the genome with respect to both genes and nongene sequences. Genomes and proteomes (the complete sets of proteins expressed by a genome) are the subjects of the next chapter.

STUDY BREAK

1. What are the principles of DNA fingerprinting?
2. What is a transgenic organism?
3. What is the difference between using germ-line cells and somatic cells for gene therapy?

Review

To access course materials such as Aplia and other companion resources, please visit www.NELSONbrain.com.

15.1 DNA Cloning

- Producing multiple copies of genes by cloning is a common first step for studying the structure and function of genes or for manipulating genes. Cloning involves cutting genomic DNA and a cloning vector with the same restriction enzyme, joining the fragments to produce recombinant plasmids, and introducing those plasmids into a living cell such as a bacterium, where replication of the plasmid takes place (see Figures 15.1–15.3).

- A clone containing a gene of interest may be identified among a population of clones by using DNA hybridization with a labelled nucleic acid probe (see Figure 15.5).

- A genomic library is a collection of clones that contains a copy of every DNA sequence in the genome. A cDNA (complementary DNA) library is the entire collection of cloned cDNAs made by reverse transcriptase from the mRNAs isolated from a cell. A cDNA library contains only sequences from the genes that are active in the cell when the mRNAs are isolated.

- PCR amplifies a specific target sequence in DNA, such as a gene, defined by a pair of primers. PCR increases DNA quantities by successive cycles of denaturing the template DNA, annealing the primers, and extending the primers in a DNA synthesis reaction catalyzed by DNA polymerase; with each cycle, the amount of DNA doubles (see Figure 15.6).

15.2 Applications of DNA Technologies

- Recombinant DNA and PCR techniques are used in DNA molecular testing for human genetic disease mutations. One approach exploits restriction site differences between normal and mutant alleles of a gene

that create restriction fragment length polymorphisms (RFLPs) detectable by DNA hybridization with a labelled nucleic acid probe (see Figures 15.8 and 15.9).

- Human DNA fingerprints are produced from a number of loci in the genome characterized by tandemly repeated sequences that vary in number in all individuals (except identical twins). To produce a fingerprint, the PCR is used to amplify the region of genomic DNA for each locus, and the lengths of the PCR products indicate the alleles an individual has for the repeated sequences at each locus. DNA fingerprints are widely used to establish paternity, ancestry, or criminal guilt (see Figure 15.10).

- Genetic engineering is the introduction of new genes or genetic information to alter the genetic makeup of humans, other animals, plants, and microorganisms such as bacteria and yeast. Genetic engineering primarily aims to correct hereditary defects; improve domestic animals and crop plants; and provide proteins for medicine, research, and other applications (see Figures 15.11, 15.12, and 15.15).

- Genetic engineering has enormous potential for research and applications in medicine, agriculture, and industry. Potential risks include unintended damage to living organisms or the environment.

- Collaborations among molecular biologists, software designers, physicists and engineers have given rise to a new discipline of synthetic biology based on formal engineering principles. Synthetic biologists can choose from a wide variety of standardized components to construct cells for use in biotechnological applications. The ultimate aim of synthetic biology is to create living systems from nonliving components. Progress is being made toward this goal through production of primitive protocells as well as determination of the minimum set of genes necessary to sustain a living cell.

Questions

Self-Test Questions

1. Restriction enzymes are used in genetic engineering to degrade DNA. How?
   a. They digest DNA, one base at a time, from the 3′ end of the DNA of interest.

   b. They remove mismatched base pairs resulting from errors in ligation.
   c. They break sugar–phosphate bonds in the DNA backbone between particular bases.
   d. They cut PCR primers away from the template DNA.

2. How are genomic libraries and cDNA libraries similar?
   a. They can both be used to express eukaryotic proteins in bacteria.
   b. They both contain the same genes: one in DNA form, one in cDNA form.
   c. They both contain all of the genes in the genome of an organism.
   d. They both depend on bacteria to reproduce the cloned DNA of interest.

3. All of the following enzymes can make nucleic acid polymers. Which one is used in PCR?
   a. DNA polymerase
   b. RNA polymerase
   c. primase
   d. reverse transcriptase

4. Which of the following statements about the separation of DNA fragments by gel electrophoresis is true?
   a. Smaller fragments travel more quickly than larger fragments.
   b. Smaller fragments travel in the opposite direction of larger fragments.
   c. Smaller fragments float to the top of the gel, while larger fragments sink to the bottom.
   d. Smaller fragments are visible under UV illumination, while larger fragments are not.

5. Recall that in gene-cloning experiments of the kind illustrated in Figure 15.3, plasmid vectors are cut open and then the DNA of interest is ligated to the resulting sticky ends. However, ligation is a random process and sometimes vectors simply recircularize without incorporating any fragments of the DNA of interest. All of these vectors in the ligation mix, those carrying the DNA of interest as well as those that are "empty," are then transformed into bacterial hosts for replication. How can colonies of bacteria transformed with empty vectors be identified relative to those colonies of bacteria transformed with "full" vectors carrying passenger DNA?
   a. Full vectors carry DNA that interrupts and inactivates the *lacZ* gene. Colonies are white.
   b. Full vectors make their hosts more antibiotic resistant. Colonies are larger.
   c. Full vectors make their hosts replicate more slowly. Colonies are smaller.
   d. Full vectors have more restriction enzymes to degrade X-Gal. Colonies are blue.

6. Which of the following statements about DNA fingerprinting is correct?
   a. It compares one particular stretch of the same DNA between two or more people.
   b. It measures different lengths of DNA produced from digestion of many repeating noncoding regions.
   c. It requires the several DNA fragments to separate on a gel with the longest lengths running the greatest distance.
   d. It can easily differentiate DNA between identical twins.

7. Dolly, a sheep, was an example of reproductive (germ-line) cloning. Which of the following examples of cell fusion was required to perform this process?
   a. the fusion of a somatic cell from one strain with an enucleated egg of another strain
   b. the fusion of an egg from one strain with the egg of a different strain
   c. the fusion of an embryonic diploid cell of one strain with an adult haploid cell (gamete) from another strain
   d. the fusion of two nucleated mammary cells from two different strains

8. Which of the following statements about somatic cell gene therapy is correct?
   a. Red blood cells can be used as a target tissue.
   b. The technique is potentially useful for all types of genetic diseases.
   c. The inserted genes are passed on to the offspring.
   d. The desired DNA can be introduced to somatic cells cultured outside the body.

9. Which of the following characteristics of the Ti plasmid makes this vector particularly useful in genetic engineering?
   a. It is circular.
   b. It carries antibiotic resistance.
   c. It can be taken up by plant cells.
   d. It has sites that are cut by restriction endonucleases.

10. What is the minimum number of genes that a free-living prokaryotic cell likely requires?
    a. about 50
    b. about 300
    c. about 10 000
    d. about 1 000 000

Questions for Discussion

1. Do you think that genetic engineering is worth the risk? Who do you think should decide whether genetic engineering experiments and projects should be carried out: scientists, judges, politicians?

2. Do you think that human germ-line cells should be modified by genetic engineering to cure birth defects? To increase intelligence or beauty?

3. Write a paragraph supporting genetic engineering and one arguing against it. Which argument carries more weight, in your opinion?

4. A forensic scientist obtained a small DNA sample from a crime scene. To examine the sample, he increased its quantity by PCR. He estimated that there were 50 000 copies of the DNA in his original sample. Derive a simple formula and calculate the number of copies he will have after 15 cycles of PCR.

5. A market puts out a bin of tomatoes that have outstanding colour, flavour, and texture. A sign posted above them identifies them as genetically engineered produce. Most shoppers pick unmodified tomatoes in an adjacent bin, even though they are pale, mealy, and nearly tasteless. Which tomatoes would you pick? Why?

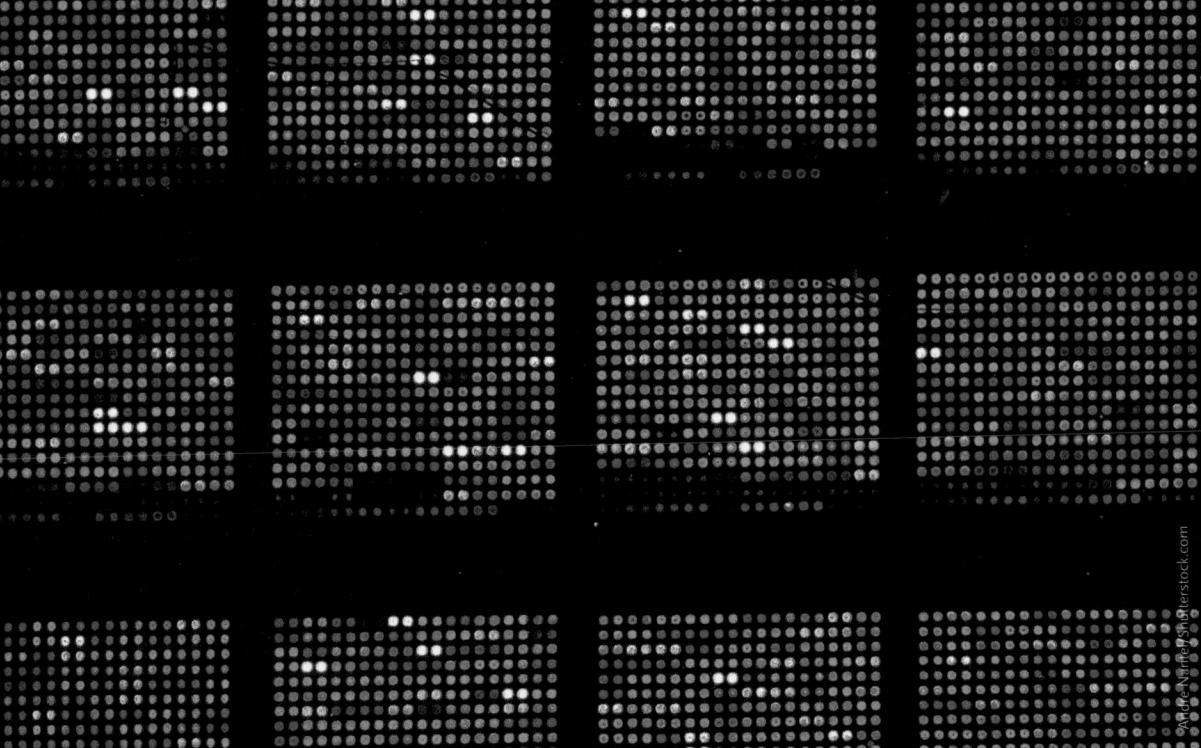

Results of DNA microarray analysis. DNA microarrays can be used, at a genomic level, to study which protein-coding genes are being expressed and the relative levels of expression of those genes.

Genomes and Proteomes

WHY IT MATTERS

You are sitting at the kitchen table with the spit kit in your hand. All you have to do is provide a small sample of saliva for the Personal Genome Project and mail it away. You want to do your part to increase knowledge of the human genome but the questions in your mind dry up your mouth. You just aren't sure this is the right thing to do.

The goal of the project sounds so important. In the years since the Human Genome Project (HGP) published the first glimpse of our 3 billion base pairs, hopes have been high that this new perspective would quickly lead to greater understanding of how genes influence our behaviour, diseases, and aging. What we got instead was greater understanding that we need to know the genome sequence of many more people. So, hundreds of volunteers have already contributed their genome sequence, along with medical and lifestyle information, to the Personal Genome Project database. Bioinformatics researchers use these large datasets to catalogue the range of human genetic variation and tease out relationships among genetics, environment, and well-being.

However, you wonder if you really want to read what is written in *your personal copies* of the human genome. What if your sequence suggests that you have an elevated risk of cancer or diabetes? Who will help you turn this knowledge into appropriate lifestyle choices? What if there is nothing you can do? You are healthy but what if you discover that you carry alleles for cystic fibrosis or Tay–Sachs disease or phenylketonuria or Mediterranean fever or G6PD deficiency? Will your fiancé(e) still be interested in you?

The project consent form explains that your genetic and personal information will be anonymous but will be publicly available online. Could someone put the pieces together to discover who you are? Could an identity thief steal your very genetic code? You know that Canada has been slow to outlaw genetic discrimination; might you be at risk of being denied health insurance or employment if your DNA sequence is public? And what about your brother and sister? They share much of your genome; should they not also share in your decision to publish family gene sequences? What if your sequence reveals a valuable variation, one that dramatically advances development of effective drugs? Will you benefit?

The Personal Genome Project is just one, definitely personal, example of research in the field of genomics, the characterization of whole genomes, including their structures (sequences), functions, and evolution. In this chapter, you will learn about the technologies underlying genomics and some of the insights that have come from genome analysis. You will also learn about proteomics, the study of the proteome, which is the complete set of proteins that can be produced by a genome. Proteomics involves characterizing the structures and functions of all expressed proteins of an organism, and the interactions among proteins in the cell. Research at the intersection of genomics and proteomics reveals highly networked interactions among genes and proteins and contributes to yet another emerging discipline, **systems biology**.

16.1 Genomics: An Overview

Genomics is the characterization of whole genomes, including their structures (sequences), functions, and evolution. Having the complete sequence of a genome makes it possible to study the complete set of genes in an organism or a virus, as well as other important sequences of their genomes. Having the complete sequence of a genome enables researchers to study the organization of genes in the genome as a whole and to determine how genes function together in networks.

Modern DNA sequencing techniques are advances on methods originally used to analyze the sequences of cloned DNA sequences and DNA sequences amplified by polymerase chain reaction (PCR; see Section 15.1).

The complete sequencing of the approximately 3-billion-base-pair human genome—the HGP—began in 1990. The task was completed in 2003 by an international consortium of researchers and by a private company, Celera Genomics (headed by J. Craig Venter). As part of the official HGP, the genomes of several important model organisms commonly used in genetic studies were sequenced for comparison: *E. coli* (representing prokaryotic cells), the yeast Saccharomyces cerevisiae (representing single-celled eukaryotes), Drosophila melanogaster and Caenorhabditis elegans (the fruit fly and a nematode worm, representing multicellular invertebrate animals), and Mus musculus (the mouse, representing nonhuman mammals). The sequences of the genomes of many organisms and viruses not part of the official HGP, including plants, have since been completed or are in progress.

Advances in many areas of scientific study have resulted from genomic approaches. For example, Genome Canada oversees diverse projects addressing topics such as Atlantic cod aquaculture, forestry breeding, agricultural crops, microorganisms active in mining and oilsands extraction, industrial production of fungal enzymes, and human health issues including autism and infectious disease.

A vast amount of DNA sequence data has been generated by genome sequencing projects. For those sequences to be useful, they need to be available centrally for access by all researchers. DNA sequences from genome sequencing projects are deposited into databases that are publicly available via the Internet. For example, GenBank® is an "annotated collection of all publicly available DNA sequences" at the National Institutes of Health (NIH). The Internet link is http://www.ncbi.nlm.nih.gov/genbank/. Currently there are over 150 billion bases in the DNA sequence records at GenBank®. Computational tools at GenBank® enable researchers and others, such as students like yourself, to perform various analyses with the sequence data.

Many other genomics databases are accessible using the Internet, with sequence data organized in different ways (perform an Internet search for "DNA sequence database"). For example, there are organism-specific sequence databases as well as databases that include summaries of particular genomics studies. The databases are available for individual researchers to use and also for collaborative efforts involving researchers all over the world. One of the main benefits of collaborative research of this kind is that researchers with different specialties can tackle a particular question or questions at a genomic or multigenomic level.

Genomics consists of three main areas of study:

1. Genome sequence determination and annotation, which means obtaining the sequences of complete genomes and analyzing them to locate putative protein-coding and noncoding RNA genes and other functionally important sequences in the genome.

2. Determining the functions of genes (a somewhat outdated term for this is **functional genomics**), which means using genome sequence data as a basis to study and understand the functions of genes and other parts of the genome. With respect to genes, this includes developing an understanding of how their expression is regulated. For protein-coding genes it also includes determining what proteins they encode, and how these proteins function in the organism's metabolic processes.

3. Studying how genomes have evolved, which means comparing genome sequence data to develop an understanding of how genes, particularly protein-coding genes, originated and genes and genomes changed over evolutionary time. Studies of genome sequences for a number of organisms represent an area of genomics known as **comparative genomics**.

Advances in each of these areas of study are accelerating as techniques are developed and improved for automating experimental procedures, and more sophisticated computer algorithms for data analysis are generated. Methods that facilitate the handling of many samples simultaneously, whether those samples are DNA molecules for sequencing or genes for analysis, are called high-throughput techniques. The next three sections of the chapter discuss each of the three areas of genomics in turn.

STUDY BREAK

What additional biological questions can be answered if we have the complete sequence of an organism's genome as compared with the sequences of individual genes?

DNA Sequencing Methods. All DNA sequencing methods have in common the following steps:

1. DNA purification;
2. DNA fragmentation;
3. amplification of fragments;
4. sequencing each fragment; and
5. assembly of fragment sequences into genome sequences.

The methods differ in how the amplification is done, the lengths of the fragments, how many fragments are sequenced simultaneously, and how the sequencing reactions themselves are done.

For decades the method devised by Frederick Sanger was by far the most common DNA sequencing

16.2 Genome Sequence Determination and Annotation

Genome sequence determination and annotation means obtaining the sequence of bases in a genome using DNA sequencing techniques and then analyzing the sequence data using computer-based approaches to identify genes and other sequences of interest, which include gene regulatory sequences, origins of replication, repetitive sequences, and transposable elements.

16.2a Genome Analysis Begins with DNA Sequencing

DNA sequencing was developed in the late 1970s by Allan M. Maxam, a graduate student, and his mentor, Walter Gilbert, of Harvard University. A few years later, Frederick Sanger, of Cambridge University, designed a method that became the one commonly used in research. Gilbert and Sanger were awarded a Nobel Prize in 1980. DNA sequencing technology has evolved since its development, and particularly rapidly in the past few years.

Whole-Genome Shotgun Sequencing. Before we discuss methods of DNA sequencing, let us consider the strategy generally used to determine the sequence of a genome, whole-genome shotgun sequencing (Figure 16.1). In this method, genomic DNA is isolated and purified, and that DNA is broken into thousands to millions of random, overlapping fragments. Each fragment is amplified to produce many copies, and then the sequence of the fragment is determined. The entire genome sequence is then assembled using computer algorithms that search for the sequence overlaps between fragments and stitch together the sequence reads to produce longer contiguous sequences.

PURPOSE: Obtain the complete sequence of the genome of an organism.

PROTOCOL:

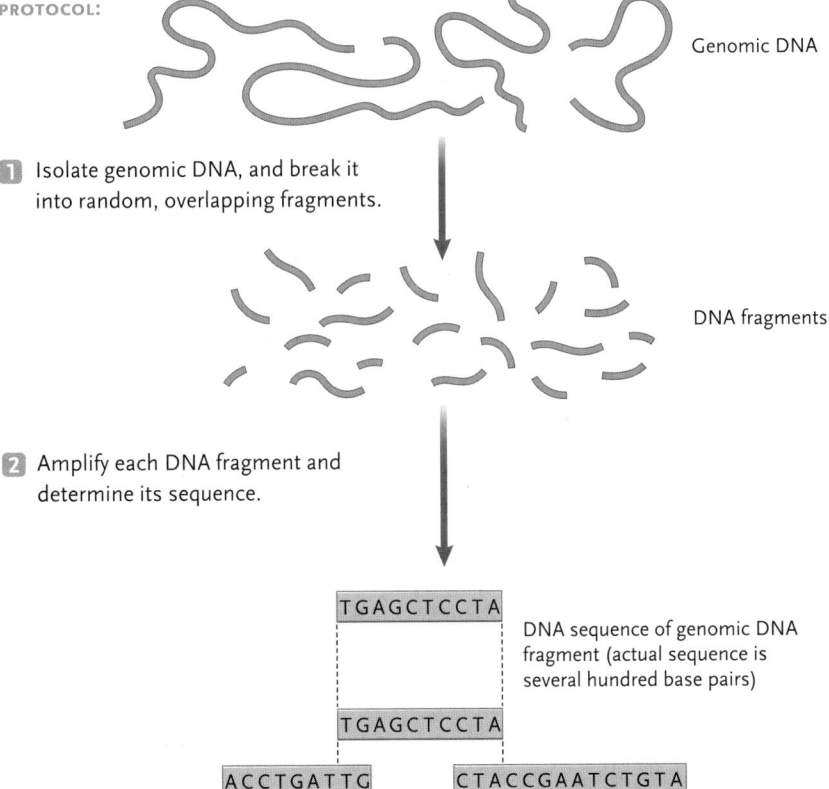

Genomic DNA

1 Isolate genomic DNA, and break it into random, overlapping fragments.

DNA fragments

2 Amplify each DNA fragment and determine its sequence.

TGAGCTCCTA

DNA sequence of genomic DNA fragment (actual sequence is several hundred base pairs)

TGAGCTCCTA

ACCTGATTG   CTACCGAATCTGTA

GATGCTAAC

GATGCTAACCTGATTGAGCTCCTACCGAATCTGTA

Assembled sequence

3 Enter the DNA sequences of the fragments into a computer, and use the computer to assemble overlapping sequences into the continuous sequence of each chromosome of the organism. This technique is analogous to taking 10 copies of a book that has been torn randomly into smaller sets of a few pages each and, by matching overlapping pages of the leaflets, assembling a complete copy of the book with the pages in the correct order.

INTERPRETING THE RESULTS: The method generates the complete sequence of the genome of an organism.

Research Method Figure 16.1 Whole-genome shotgun sequencing.

technique used. The Sanger method is a DNA synthesis–based method for DNA sequencing. It is based on the properties of nucleotides known as dideoxyribonucleotides, which have a —H on the 3′ carbon of the deoxyribose sugar instead of the —OH found in normal deoxyribonucleotides; therefore, the method, explained in **Figure 16.2,** is also called dideoxy sequencing.

PURPOSE: Obtain the sequence of a piece of DNA, such as in gene sequencing or genome sequencing. The method is shown here with a typical automated sequencing system.

PROTOCOL:

1 A dideoxy sequencing reaction contains (1) the fragment of DNA to be sequenced (denatured to single strands); (2) a DNA primer that will bind to the 3′ end of the sequence to be determined; (3) a mixture of the four deoxyribonucleotide precursors for DNA synthesis; (4) a mixture of the four dideoxyribonucleotide (dd) precursors, at about 1/100 the concentration of the deoxyribonucleotides, each labelled with a different fluorescent molecule; and (5) DNA polymerase to catalyze the DNA synthesis reaction.

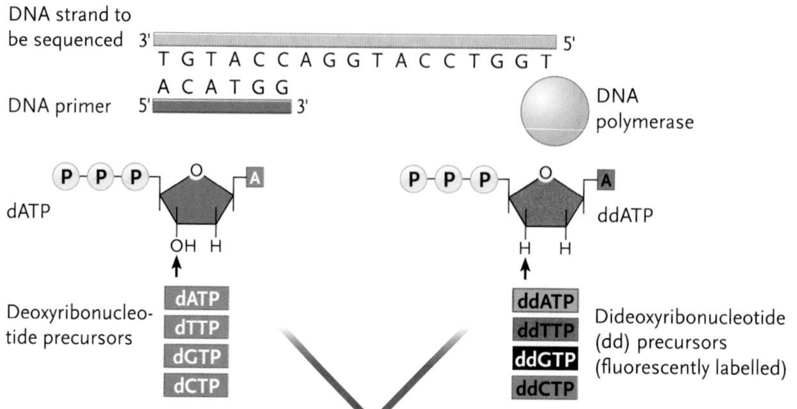

2 DNA polymerase synthesizes the new DNA strand in the 5′ → 3′ direction starting at the 3′ end of the primer. New synthesis continues until a dideoxyribonucleotide is incorporated randomly into the DNA. The dideoxyribonucleotide acts as a terminator for DNA synthesis because it has no 3′-OH group for the addition of the next base (see Section 12.3). For a large population of template DNA strands, the dideoxy sequencing reaction produces a series of new strands, with lengths from one up. At the 3′ end of each new strand is the fluorescently labelled dideoxyribonucleotide that terminated the synthesis.

3 The labelled strands are separated by electrophoresis using a polyacrylamide gel prepared in a capillary tube. The principle of separation is the same as for agarose gel electrophoresis (see Figure 15.7), but this gel can discriminate between DNA strands that differ in length by one nucleotide. As the bands of DNA fragments move near the bottom of the tube, a laser beam shining through the gel excites the fluorescent labels on each DNA fragment. The fluorescence is registered by a detector, with the wavelength of the fluorescence indicating whether ddA, ddT, ddG, or ddC is at the end of the fragment in each case.

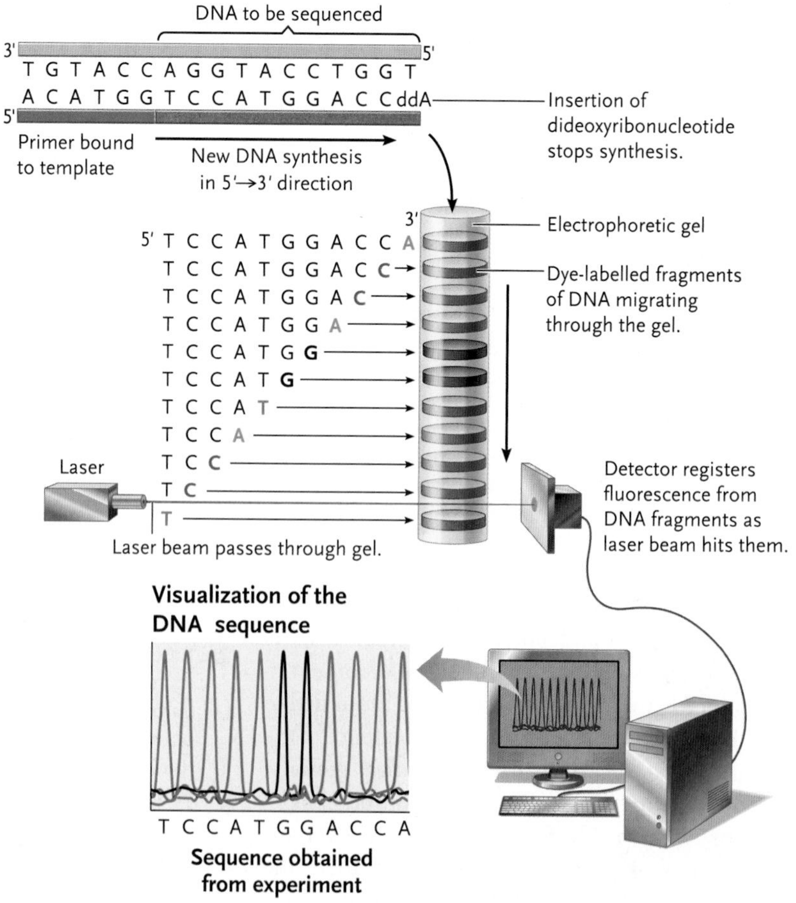

INTERPRETING THE RESULTS: The data from the laser system are sent to a computer that interprets which of the four possible fluorescent labels is at the end of each DNA strand. The results show, on a computer screen or in printouts, colours for the labels as the DNA bands passed the detector. The sequence of the newly synthesized DNA, which is complementary to the template strand, is read from left (5′) to right (3′). (The sequence shown here begins after the primer.)

Research Method Figure 16.2 Dideoxy (Sanger) method for DNA sequencing.

In recent years the dideoxy sequencing method has been replaced largely, but not completely, by faster, cheaper, and more automated techniques. In general, these newer high-throughput techniques have decreased sequencing costs by reducing the pre-paratory steps, automating more of the process, and sequencing up to a billion different DNA fragments in parallel.

Figure 16.3 outlines a next-generation DNA sequencing technique that is widely used in genome

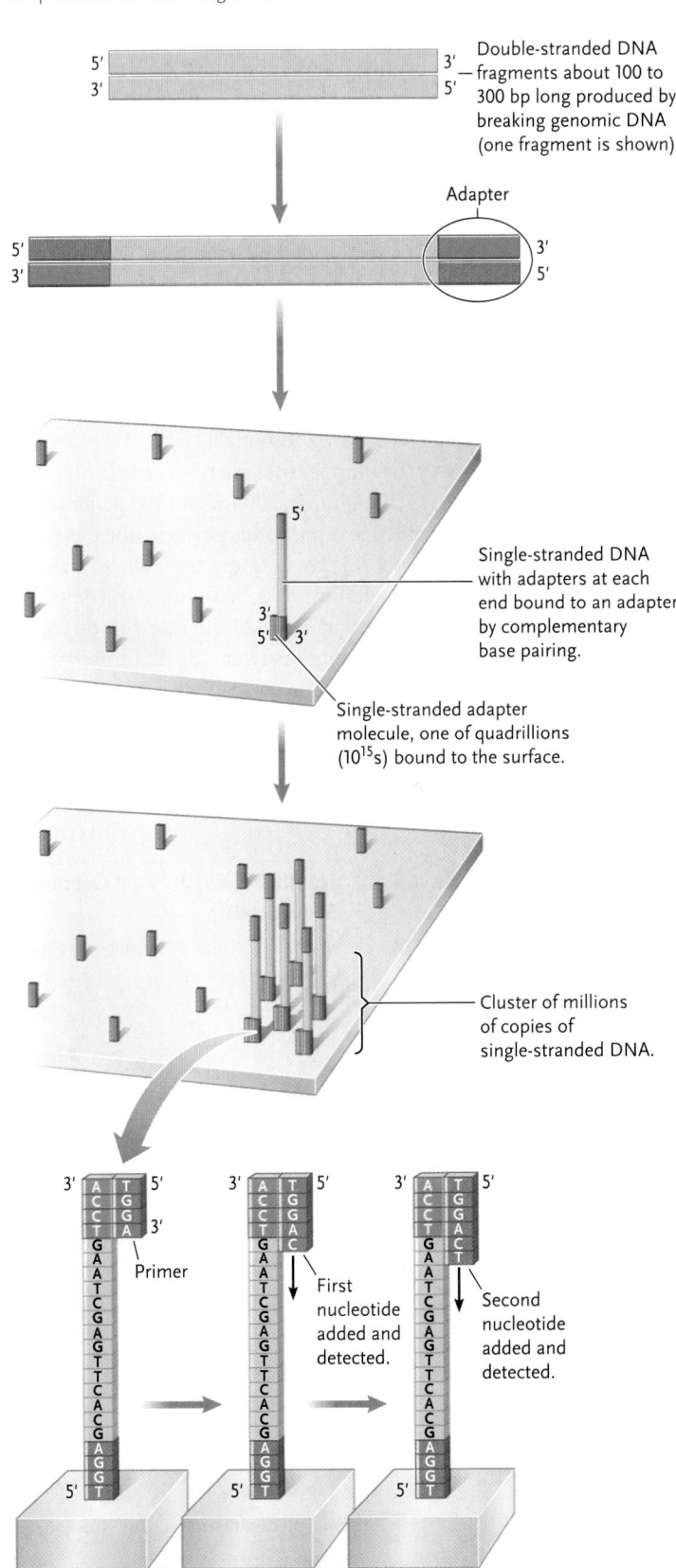

PURPOSE: Automated, massively parallel sequencing of up to a billion DNA fragments.

PROTOCOL:

1 DNA ligase attaches short double-stranded DNA adapter molecules to each end of 100- to 300-bp genomic DNA fragments.

2 The DNA fragments are denatured to single strands, which are added to a cell in an automated machine through which liquid can flow. Over the glass surface of the cell are bound about one quadrillion (1×10^{15}) single-stranded adapter molecules that have complementary sequences to those of the adapters added to the DNA in step 1. Each DNA fragment to be sequenced binds to one of the glass-bound adapter molecules by complementary base-pairing. The massive number of glass-bound adapters allows many DNA strands to be sequenced simultaneously.

3 An amplification process generates millions of copies of each of the DNA fragments that bound initially to one of the glass-bound adapters, clustered around the place on the cell where that DNA bound. Up to one billion different clusters can be produced in the sequencing cell simultaneously.

4 The DNA fragments are now ready for synthesis-based DNA sequencing. One fragment is shown for the sequencing steps.

5 DNA primers are added to the cell. A primer anneals to each DNA strand (it is the complementary strand to the adapter sequence at its end), and DNA synthesis is done in a cyclic manner one nucleotide at a time using four different fluorescently labelled DNA nucleotide precursors. Each time a labelled nucleotide is added, synthesis stops and the machine uses laser technology to measure the fluorescence so as to identify the base added. That base is the same for all strands in a cluster. By repeated cycles of addition of a nucleotide and laser detection, the sequence of the strand is obtained. Up to about 100 bases of each fragment can be sequenced.

INTERPRETING THE RESULTS: The DNA sequence obtained is complementary to the initial single-stranded DNA strand that paired with the glass-bound adapter. The DNA sequence data from all of the clusters of DNA fragments are analyzed by computer to determine overlaps between fragments, and, by the principles described in Figure 16.2 the complete sequence of a genome is assembled.

Research Method Figure 16.3 Illumina/Solexa method for DNA sequencing.

sequencing projects, the DNA synthesis–based Illumina/Solexa method. This is an example of massively parallel DNA sequencing, because up to one billion different DNA fragments can be sequenced simultaneously.

As a result of the lower cost of sequencing, and the automated, massively parallel DNA sequencing methods that are used, the genomes of many species beyond those targeted in the HGP have been determined. Although the question, "How many different genomes have been sequenced to date?" seems reasonable, it is rather hard to answer for three main reasons. One reason is that the answer depends on how you define one genome as different from another. Does "different" mean a single genome from two different species? Or can "different" mean a genome from two alternative strains, breeds, or cultivars of the same species? A second reason that this number is hard to pin down has to do with completion. Genome sequences don't need to be entirely complete in order to be informative and useful. Some genome projects stop at the "good enough" stage when the work of completing the entire sequence would be of little additional benefit. The third factor that makes the number of sequenced genomes elusive is that it keeps changing. In 2000, about 100 genome sequences were added to repositories per year, one every three days. Now, only 15 years later, the rate of new genome sequence additions is approaching 10 000 per year, more than 1 per hour! By 2020 the rate will certainly exceed 100 000 sequences per year. By that time, we will likely stop asking if our favourite genome has been sequenced and just assume that it has.

Table 16.1 presents a snapshot of completed genome sequences in one database. The values represent total numbers of completed genomes. Sometimes several genomes are included for a given species. Notice the very large number of bacterial species sequenced. This reflects both the relative ease of sequencing smaller genomes and the popularity of these organisms in research.

16.2a Genome Sequences Are Annotated to Identify Genes and Other Sequences of Importance

A raw genome sequence is simply a string of A, T, G, and C letters; it tells us practically nothing about the organism from which it derives, other than the total length of its genome. Therefore, once the complete sequence of a genome has been determined, the next step is annotation, the identification of functionally important features in the genome. These include the following:

- Protein-coding genes.
- Noncoding RNA genes. As mentioned earlier, "noncoding" means that the RNA transcript of the gene is not translated. Rather, the transcript is the final functional product of the gene. Noncoding RNA genes include genes for tRNAs, rRNAs, and snRNAs (see Chapter 13), and genes for microRNAs (miRNAs; see Chapter 14).
- Regulatory sequences associated with genes (see Chapter 15).
- Origins of replication (see Chapter 12).
- Transposable elements, viruses, and sequences related to them (see Chapter 9).
- Pseudogenes. A **pseudogene** is very similar to a functional gene at the DNA sequence level, but one or more inactivating mutations have changed the gene so that it can no longer produce a functional gene product (e.g., deletion of the promoter). Most pseudogenes are derived from protein-coding genes and are recognized by their sequence similarities to functional genes.
- Short repetitive sequences. These are sequences that are repeated a few too many times in the genome. The short tandem repeat (STR) sequences discussed in Section 15.2 are examples of short repetitive sequences.

Annotation is performed by researchers in the field of **bioinformatics,** which is the application of mathematics and computer science to extract information from biological data, including those related to genome structure, function, and evolution.

Some examples of genome annotation for protein-coding genes are illustrated in the following subsections.

Identifying Open Reading Frames by Computer Search of Genome Sequences. As outlined in Chapter 13, proteins are specified in mRNA molecules by a series of codons starting with the initiation codon AUG and

Table 16.1	Number of Organismal Genomes Sequenced*	
Organism	Number of Genomes
Prokaryotic Cells	
Archaea	330
Bacteria	17 360
Total prokaryotic cells:	17 690
Eukaryotes	
Animals	
Mammals	20
Fishes	25
Insects	77
Other animals	64
Total animals	186
Plants	69
Fungi	384
Protists	159
Total eukaryotes	798
Total all organisms	18 488

*As of May 2014, http://www.genomesonline.org.

ending with one of the three termination codons: UAG, UAA, or UGA. The span of codons from start to stop codon is called an **open reading frame (ORF)**, and ORFs that are longer than 100 codons almost always indicate the presence of a protein-coding gene. Computer algorithms are used to identify possible protein-coding genes in a genome sequence by searching for ORFs. In a DNA sequence, this means searching for ATG, separated from a stop codon (TAG, TAA, or TGA) by a multiple of three nucleotides. The search is complicated because either of the two DNA strands could be the template for a given gene. Theoretically, then, each DNA sequence has six reading frames for the three-letter genetic code, three on one strand and three on the other strand. An ORF can be in any one of these frames. This is illustrated in **Figure 16.4** for a theoretical 30-nucleotide segment of DNA. Note that each single-stranded DNA sequence generated by DNA sequencing can be used to infer the sequence of the complementary DNA strand. If an ORF is present in a particular DNA sequence, it will be in one of these frames.

We can start looking for an ORF going in the 5' → 3' direction in the top strand of Figure 16.4, starting at the leftmost A nucleotide. In this case, reading in groups of three nucleotides will lead you to the TAG at the right end, which is the stop codon. We have found an ORF in the sequence, and it is coded by the entire length of the top strand. However, if instead we start looking for an ORF in the top strand starting at either the second nucleotide (the T) or the third nucleotide (the G), we do not find a start codon, so we have not found an ORF. For the bottom strand, none of the three frames has a start codon. Computer algorithms can search easily for ORFs in all six reading frames of a DNA sequence.

CONCEPT FIX Figure 16.4 shows an ORF on the top strand, reading 5' → 3' from ATG. Although it is tempting to assume that the top strand containing the ORF is used as the template for transcription, this is not possible; RNA polymerases only read template DNA in the 3' → 5' direction. Therefore, template DNA for this ORF is the bottom strand, not the top. ⬣

Searching for protein-coding ORFs is straightforward in prokaryotic genomes because few genes have introns. Eukaryotic protein-coding genes typically have introns, so more sophisticated algorithms must be used to identify such genes. For example, the algorithms may search for particular

characteristics of protein-coding genes, such as junctions between exons and introns, sequences that are characteristic of eukaryotic promoters, and overrepresentation of certain three-base codons relative to others.

Computer identification is typically the first step in identifying protein-coding genes. However, other evidence is needed before biologists can be confident that the sequence they have annotated is a functioning protein-coding gene. Two approaches to obtaining such evidence are described in the next two subsections.

Identifying Protein-Coding Genes by Sequence Similarity Searches. One way of testing whether candidate protein-coding genes found by searching for ORFs are functioning protein-coding genes is by comparing their sequences with known, identified, and verified genes in databases. This is a sequence similarity search. Such searches can be done using an Internet browser to access the computer programs and the databases. For example, to use the BLAST (Basic Local Alignment Search Tool) program at the National Center for Biotechnology Information (http://blast.ncbi.nlm.nih.gov), a researcher pastes the putative ORF DNA sequence, or the amino acid sequence of the protein it would encode, into a browser window and sets the program to begin searching. The BLAST program searches the databases of known sequences and returns the best matches, if any. The matches are listed in order, from the closest match to the least likely match. Finding a known gene's sequence that matches the putative ORF sequence closely is good evidence that the ORF is in fact a protein-coding gene and that it encodes a protein functionally related to that of the matching gene sequence. The principle here is that genes of living organisms tend to be similar to each other because they have evolved from ancestral genes in ancestral organisms. Genes that have highly conserved sequences because they have evolved from a gene in a common ancestor are called **homologous genes.** For example, if a gene in the mouse has been characterized experimentally and its sequence is known, and that gene is evolutionarily conserved in mammals, there should be a match for that gene in the human genome sequence.

With sequence similarity searches, ORFs can be sorted into ones with known and unknown functions. For the latter, experiments are required to show whether they are real protein-coding genes and, if so, what their functions are. For example, analysis of the human genome sequence initially identified more than a thousand putative ORFs with no sequence similarities to known genes. Most of these function-unknown genes have now been shown to be pseudogenes. Such uncertainty makes it difficult to determine the exact number of protein-coding genes in a genome just from its sequence.

```
        GTC⟶
      TGT⟶
    ATG⟶
5'...ATGTCTGTTGACTGGGTTGGAAGGCAATAG...3'
3'...TACGACAATCTGACCCAACCTTCCGTTATC...5'
                      ⟵ATC
                   ⟵TAT
                ⟵TTA
```

Figure 16.4

The six reading frames of double-stranded DNA. In this particular sequence, one of them is an ORF.

Identifying Protein-Coding Genes from Sequences of Gene Transcripts. The gold standard for identifying protein-coding genes in a genome sequence is the demonstration that the sequences are transcribed in

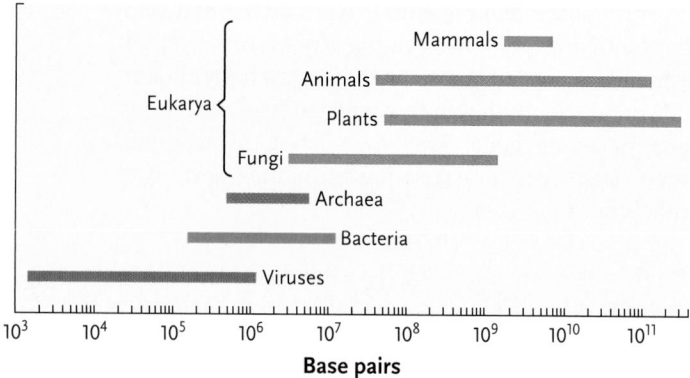

Figure 16.5
Ranges of genome sizes for viruses, bacteria, archaea, and eukaryotes. Note that the Domain Archaea has not been studied as extensively as the Domain Bacteria, so there may be representatives with substantially larger or smaller genomes than the range given.

Table 16.2	Genome Sizes and Estimated Number of Protein-Coding Genes for Selected Members of Domains Bacteria, Archaea, and Eukarya	
Domain and Organism	**Genome Size (millions of base pairs, Mb)**	**Protein-Coding Genes**
Bacteria		
Mycoplasma genitalium	0.58	475
Escherichia coli	4.6	4 146
Archaea		
Thermoplasma acidophilum	1.56	1 484
Methanosarcina acetivorans	5.75	4 540
Eukarya		
Protists		
Tetrahymena thermophila (a ciliated protist)	146	>20 000
Fungi		
Saccharomyces cerevisiae (a budding yeast)	12.1	~6 000
Neurospora crassa (orange bread mould)	40	~10 100
Plants		
Arabidopsis thaliana (thale cress)	120	~26 000
Oryza sativa (rice)	411	~56 000
Capsicum annum (hot pepper)	~3480	~34 500
Invertebrates		
Caenorhabditis elegans (a nematode worm)	100	~20 000
Drosophila melanogaster (fruit fly)	165	~13 700
Locusta migratoria (swarming locust)	6500	~17 300
Vertebrates		
Takifugu rubripes (pufferfish)	393	~27 000
Mus musculus (mouse)	2600	~22 000
Homo sapiens (human)	3200	~20 500

cells to make mRNAs. One approach to doing this makes use of the sequences of transcripts represented in cDNA libraries. Recall from Section 15.1 that a cDNA library is made starting with mRNA molecules isolated from a cell. If the mRNA molecules are isolated under different conditions and from different cell types in a multicellular organism, they will represent the activity of many of the organism's protein-coding genes. However, protein-coding genes that are rarely transcribed or that produce very few mRNA molecules are likely to be missed by this approach.

The single-stranded mRNA molecules are converted to double-stranded DNA molecules using reverse transcriptase, and these DNA molecules are cloned to produce the cDNA library. Some part of each cloned mRNA (as cDNA) is sequenced using a sequencing primer that pairs with the DNA just adjacent to the inserted cDNA fragment in the clone. Using computer algorithms, each cDNA sequence is compared with the genome sequence of the organism to map the location of the sequence and, therefore, the protein-coding gene from which the original transcript was derived.

This approach is also useful for cataloguing transcripts in humans and other eukaryotes to identify which genes are alternatively spliced. Recall from Section 14.3 that alternative splicing of pre-mRNA transcripts of protein-coding genes produces different mRNAs by using different splice sites that may remove different combinations of exons, or modify their lengths. The resulting mRNAs are translated to produce a family of related proteins having various combinations of amino acid sequences derived from the remaining exons.

16.2b Genome Landscapes Vary Markedly in Size, Gene Number, and Gene Density

With many genomes now sequenced, researchers can compare them to learn about genome sizes, the number of protein-coding genes, and the density of these genes (how widely spaced they are). A vast amount of new information is available about genome landscapes. Here we will present some generalizations and then provide a more detailed description of the *E. coli* genome and the human genome, as examples of bacterial and eukaryote genomes, respectively.

Figure 16.5 shows the ranges of genome sizes for viruses, bacteria, archaea, and different groups of eukaryotes, and **Table 16.2** gives examples of genome sizes and the number of protein-coding genes for some bacteria, archaea, and eukaryotes. We can arrive at some general conclusions about the data. Members of both the Domain Bacteria and the Domain Archaea have genomes that vary widely in size. In addition, their genes are densely packed in their genomes, with little noncoding space between them. Thus, the larger genomes of organisms in these two domains tend to reflect increased gene number.

Members of the domain Eukarya vary markedly in form and complexity, and their genomes also show great differences in size and gene density (Table 16.2). For example, the genome of yeast, *Saccharomyces cerevisiae*, is only about 0.4% the size of the human genome. However, yeast have just a little less than 30% of the number of protein-coding genes found in humans. Genes in yeast are therefore, on average, coded more closely together than in humans. Although this yeast versus human comparison might lead you to assume that there is a close relationship between organism complexity and genome size, this is not the case.

For instance, the fruit fly, *Drosophila melanogaster*, and the locust, *Schistocerca gregaria*, are both insects with similar overall physiological complexity. However, the genome of the locust, at 9300 Mb, is 52 times the size of the fruit fly genome. Even within a given genus, different species often have widely differing genome sizes. For example, there is a 50-fold variation in the genome size of the *Allium* species, which includes onions, leeks, shallots, and garlic. Among the vertebrates we again find great variation in genome size. For example, the genome of the mouse and human are about seven times as large as the pufferfish genome, *Takifugu rubripes*. And yet the pufferfish has more protein-coding genes than either the mouse or the human (Table 16.2). As we saw with yeast, the genes are coded more closely in the pufferfish genome than they are in either the mouse or the human genome.

It is important to think critically about the data presented for gene numbers in a genome. As you have learned, the determination of the protein-coding gene number involves both computer and experimental analysis. The outcomes of these analyses are therefore estimates of the number of genes present. Only when an entire genome has been studied experimentally to characterize every gene it contains can we be certain of an organism's exact gene number. For example, you have just read that the estimated number of protein-coding genes is greater than 22 thousand in the mouse genome and greater than 20 500 in the human genome. These numbers are not precise. Rather, at this point they likely reflect different extents of progress in annotating the two genomes, rather than suggesting that 2000 more protein-coding genes are necessary to create and run a mouse rather than a human.

Profile of the *E. coli* genome. *E. coli* is one of the most intensively studied model organisms, and the genome of laboratory strain K12 is one of the best annotated. In many ways, *E. coli* has a typical bacterial genome, with the vast majority of its genes on a single circular chromosome with one origin of replication **(Figure 16.6a)**, and the remainder of its genes on one or more plasmids, each of which is much smaller than the circular chromosome. With about 4.6 Mb and about 4146 protein-coding genes, the *E. coli* K12 genome is in the middle range, sizewise, of bacterial genomes (see Figure 16.5). The noncoding genes are those for rRNAs and tRNAs. There are a small number of transposable elements and repetitive sequences.

Figure 16.6b shows a close-up of a 10-kb segment of the *E. coli* genome containing a number of protein-coding genes to illustrate the following characteristics:

- The genes are close together, with little space in between. Promoters for the genes are located immediately upstream of each transcription unit (not shown in the figure).
- Some of the genes are transcribed in the left-to-right direction (using the bottom strand as the template), whereas the others are transcribed in the right-to-left direction (using the top strand as

a. Map of the circular *E. coli* K12 genome showing the genes transcribed clockwise (blue) and the genes transcribed counterclockwise (orange) and the location of the origin of replication.

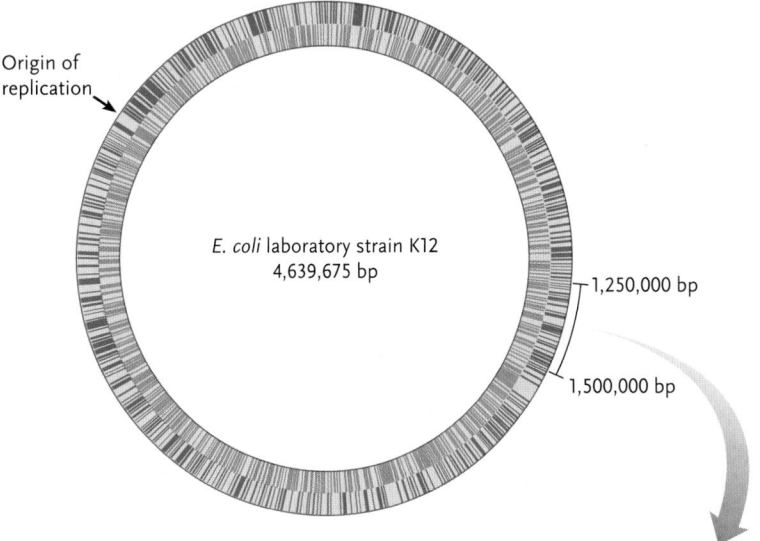

b. Detail of a 10-kb region of the *E. coli* K12 genome, from about 3:30 on the genome "clock."

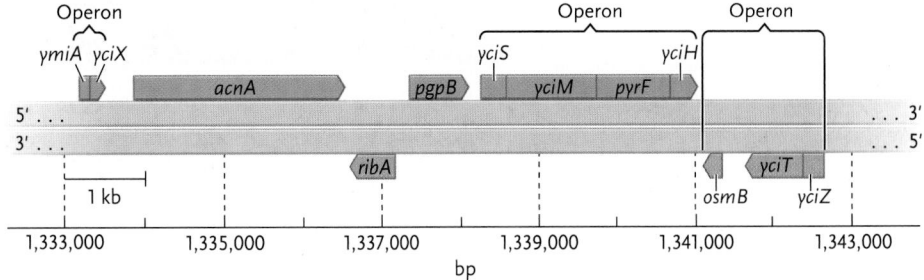

Figure 16.6

The genome of *E. coli*, laboratory strain K12.

Table 16.3 — Comparison of the *E. coli* K12 and Human Genomes

Property	*E. coli* K12 genome	*H. sapiens* genome
Chromosomes	1 circular (plus plasmids)	23 linear (pairs in diploid cells)*
Nucleotides	4.64 Mb	3 100 Mb
Protein-coding genes	4 146	~ 20 500
Noncoding RNA genes	176	~ 12 000
% coding DNA	88%	1.2%
Protein-coding genes per Mb	894	7
Introns per average gene	0	8
Average polypeptide size	330 amino acids	430 amino acids

*There are 24 different human chromosomes: 22 autosomes and the X and Y chromosomes. Each individual has 23 pairs of chromosomes.

the template). (The two template strands are transcribed in different directions because the two DNA molecules in a double helix are antiparallel; see Chapter 12.)

- Some genes are single transcription units, whereas others are organized into operons (see Chapter 14). In the genome as a whole, about one-half of protein-coding genes are organized into operons.
- The genes vary in length, reflecting the lengths of their encoded proteins.

CONCEPT FIX Look carefully at Figure 16.6, noticing that some genes are shown in blue and some in orange. Although you may have assumed that one particular strand of the double helix is used as the template

(i.e., read by RNA polymerase) for all genes on a chromosome, the figure shows this assumption to be false. One strand of the helix is used as a template for some genes (such as the blue genes in Figure 16.6), while the other strand is used as a template for other genes (the orange genes in Figure 16.6). Recall that templates have to be read 3′ → 5′, so, in Figure 16.6, the blue genes are using the bottom strand as their template. ⬢

Table 16.3 summarizes some of what has been learned about the *E. coli* K12 genome to date with respect to its physical aspects, genes, and gene products.

Other bacterial genomes may be larger or smaller than the *E. coli* K12 genome, but their genome landscapes are similar to that of *E. coli* in several ways. For example, typically there is one origin of replication, 85 to 92% of the DNA codes for proteins, there is a mixture of operons and single-gene transcription units, some genes are transcribed using one DNA strand as the template whereas others are transcribed using the other strand, and there are relatively few transposable elements or repetitive sequences.

Profile of the Human Genome. At about 3.2 billion base pairs, the human genome is about 700 times as long as the *E. coli* genome. Each human individual has 23 pairs of chromosomes. Men have 24 different chromosomes, the 22 autosomes and the X and Y chromosomes, whereas women have 23 different chromosomes, the 22 autosomes and the X chromosome. **Figure 16.7a** displays the

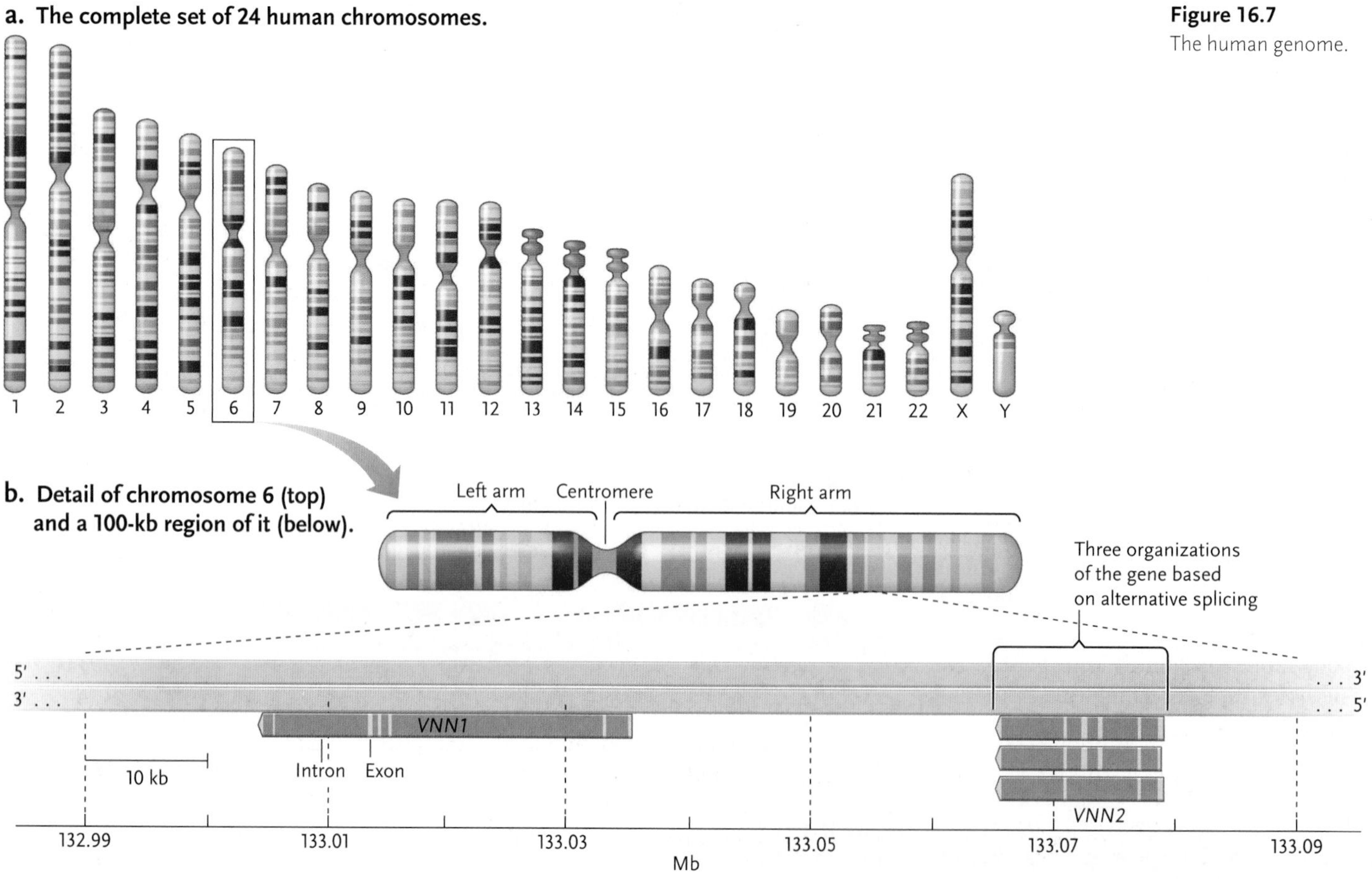

Figure 16.7

The human genome.

a. The complete set of 24 human chromosomes.

1 2 3 4 5 6 7 8 9 10 11 12 13 14 15 16 17 18 19 20 21 22 X Y

b. Detail of chromosome 6 (top) and a 100-kb region of it (below).

Left arm   Centromere   Right arm

Three organizations of the gene based on alternative splicing

5′ ... ... 3′
3′ ... ... 5′

VNN1

10 kb   Intron   Exon

VNN2

132.99   133.01   133.03   133.05   133.07   133.09

Mb

complete set of human chromosomes, depicting the banding patterns that help researchers identify regions of chromosomes. (Figure 8.10 shows the banding patterns of stained human chromosomes.) Figure 16.7b shows chromosome 6 in more detail and then a close-up of a 100-kb segment of the long arm of that chromosome to show the protein-coding genes it contains. Compare this figure with Figure 16.6b, which shows a 10-kb segment of the *E. coli* chromosome, and note the following:

- Genes are relatively far apart, with a large amount of space in between. That is, even though the human genome segment shown is 10 times as long as the *E. coli* segment shown in Figure 16.6b, it contains only 2 protein-coding genes. Each of these genes consists of transcription units that are far longer than the genes in *E. coli*, largely because they consist of about 95% introns and 5% exons. The right-hand gene in Figure 16.7b illustrates at the DNA level the alternative splicing variants for that gene; note the different exons (light pink regions) for the three gene drawings. (Alternative splicing is described in Section 14.3.)

- As in the genomes of other organisms, some genes are transcribed from one of the strands of the double helix, while other genes are transcribed from the other strand. For the particular segment shown in Figure 16.7b, both genes happen to be transcribed from the same strand.

- All of the genes are single transcription units. Eukaryotic genes are rarely organized in operons.

Table 16.3 summarizes some of what researchers have learned about the human genome to date with respect to its physical aspects, genes, and gene products. Some of the key features of the human genome are as follows:

- There are about 20 500 protein-coding genes. On average, there are 8 exons per gene, with some human genes consisting of a single exon and others having over 100 exons. Introns make up about 95% of the average transcription unit. Since about 2% of the genome consists of protein-coding sequences, introns represent about 20 to 25% of the human genome.

- On average, more than 20 regulatory sequences are associated with each protein-coding gene. These sequences are distributed over thousands to tens of thousands of base pairs upstream and downstream of each gene, as well as being scattered within introns. The regulatory sequences are widely scattered within regions that encompass 15 to 25% of the human genome.

- There are about 12 000 noncoding RNA genes, which include genes for rRNA, tRNA, small nuclear RNA, and microRNAs (see Chapter 13 and Section 14.3).

- There are thousands of pseudogenes.

- There are many inserted DNA viruses and reverse-transcribed RNA viruses, totalling nearly 10% of the entire genome.

- About 45% of the genome consists of transposable element sequences (see Section 9.4). Only a tiny fraction of these transposable elements are functionally active. The others are inactive, being the transposable element version of pseudogenes.

- The genome contains a variety of short, repeated sequences, including those at the centromeres and telomeres, as well as others scattered throughout the rest of the chromosome.

- Replication begins at hundreds to thousands of origins of replication per chromosome. However, there are no consistent, sequence-specific replication origins, as in bacteria and yeast.

CONCEPT FIX Stop for a moment to total up the previous information about the composition of your genome. Notice that over half of your DNA sequence is transposable elements and viruses that are, or once were, mobile. Although these sequences may well contain genes, they are not coding for human proteins and are often referred to as "junk DNA." Add in the thousands of pseudogenes, introns, and nonessential repetitive sequences and you may be surprised to realize that junk DNA approaches 75% of your entire genome. ⬡

Genome-Wide Association Studies Catalogue Genetic Variation. Just as we have established the basic architecture of the genome that is common to all humans (Figure 16.7), we are well into the task of discovering the various types of diversity in the genomes of different people and how this variation interacts with environmental influences to produce the overall phenotype. One type of variation in the DNA sequence of different human genomes is a change in a single nucleotide. This difference is a called a single nucleotide polymorphism (SNP, pronounced "snip"). The International HapMap Project collected, mapped, and publicly released data on well over 10 million SNPs common among people of African, Asian, and European ancestry. Thomas Hudson, at McGill University, was a Canadian member of the HapMap Consortium, contributing data for chromosomes 2 and 4. Once common DNA variants are identified, many thousands of them can be used to create DNA microarray "chips" (Section 16.3c) that enable automated screening. This is how DNA in your saliva sample in "Why It Matters" would reveal your personal profile of genetic variation.

Studies that screen a large number of people for a large number of variants are called Genome-Wide Association Studies (GWAS). Analysis of GWAS datasets can find statistical correlations between particular genetic variations and specific diseases or other traits in a population. In recent years, this type of genomic analysis has been directed toward understanding the genetic basis of complex diseases such as schizophrenia.

Although family and twins studies show a significant genetic component to schizophrenia risk, no single gene can be identified as responsible. A recent GWAS study identified 22 sites associated with elevated risk, half of them previously unknown. Variation at these sites included over 8000 SNPs as well as insertion/deletion of larger chromosomal regions, called copy number variations (CNVs).

In the hands of healthcare professionals, genomic data from patients has led to better, more personalized treatment. An example of this is that a patient's variation at two genes, VKORC1 and CYP2C9, is now recognized on product labelling as an important factor in predicting the optimal dose of the common anticoagulant drug warfarin. This field of pharmacogenomics is expanding rapidly, exploring the potential of matching medical treatments to genotypes. Another application of genomics, personal genomics, is also developing rapidly. Several entrepreneurial companies now offer affordable personal genomic screening to the public, often including additional information concerning family relationships and ancestry as well as statistical estimates for expression of hundreds of traits and medical conditions. Such personal genetic profiles should be approached with caution for the scientific and psychosocial reasons highlighted in "Why It Matters."

Up to this point, we have focused on the features of the enormous portion of the human genome encoded within the linear chromosomes of the cell's nucleus. It is also important to remember that in a eukaryote, each mitochondrion also contains a circular mitochondrial genome, or mtDNA. The human mtDNA is 16.6 kb, much smaller than even a prokaryotic genome. Its 37 genes (13 of them are protein coding) perform essential functions related to cellular respiration. In addition, photosynthetic eukaryotes, including plants and algae, have a separate circular genome up to several hundred thousand bases in each of their chloroplasts: the cpDNA. Not surprisingly, this genome contains genes involved in photosynthesis.

Other mammalian genomes are very much like the human genome. But for other eukaryotes generally, particular features can vary considerably. For example, eukaryote genomes range from having a very low percentage of protein-coding DNA, as in mammals, to almost as high a percentage as is seen for bacteria.

STUDY BREAK

1. What is the principle behind whole-genome shotgun sequencing of genomes?
2. What are the key sequences identified by genome annotation?
3. How are possible protein-coding genes identified in a genome sequence of a bacterium? Of a mammal?
4. What general differences are there in the genome landscapes of bacteria and eukaryotes?

16.3 Determining the Functions of the Genes in a Genome

Once a genome is annotated, the next step is to use the genome sequence data to understand the functions of genes and other parts of the genome. "Gene function" is considered broadly here to include regulation of gene expression, the products genes encode, and the role of these products in the function of the organism. For protein-coding genes, the gene products are proteins. We study proteins to understand their structure and function, to discover how they participate in networks of interactions with other proteins and other nonprotein molecules in the cell, and to discover how these complexes are important functionally for the organism.

We also need to determine what genes there are for noncoding RNAs in the genome, and what the functions of the RNAs are. Several of the noncoding RNA genes can be assigned functions based on evolutionary conservation principles, that is, by looking for sequence similarity with known gene sequences. An example is the rRNA genes. However, identifying and determining the functions of miRNA genes is more challenging in part because of their diversity of sequences. In this section, we focus on protein-coding genes, again because of the importance of their products in controlling the functions of cells and, therefore, of organisms. Determining the functions of protein-coding genes typically relies on computer analysis and on laboratory experiments. The following subsections present examples of these approaches.

16.3a Gene Function May Be Predicted by a Sequence Similarity Search of Sequence Databases

You learned earlier that a DNA sequence can be identified as a likely protein-coding gene by using a sequence similarity search of sequences in databases. This approach can also be used to assign the function of a gene. A high degree of similarity between the sequence of a candidate gene of unknown function and the sequence of a gene of known function likely indicates that both sequences evolved from a gene in a common ancestor and that their sequences in the present day have been conserved significantly because they code for proteins that have similar functions. As explained earlier, genes with highly conserved sequences as a result of divergence from a common ancestral gene are homologous genes.

Using sequence similarity searches to determine if a candidate gene and a known gene are homologous is by far the most common method for assigning the functions of genes. Experimental investigation of the functions of genes is considerably more expensive

Dr. Stephen Scherer, *Centre for Applied Genomics, Hospital for Sick Children, Toronto, Fellow of the Royal Society of Canada*

The idea that offspring inherit one copy of each gene from each of their parents has been a universal principle of genetics since the days when Mendel himself was explaining inheritance of traits in pea plants. Thanks to the careful genomic work of Dr. Stephen Scherer and his collaborators, we now know that this universal principle is not so universal. A breakthrough paper from Scherer's group in 2007 outlined the discovery that some genes are present in only one copy, while others show up three times in a given diploid genome. This new type of variation was called copy number variation (CNV) and has since been shown to be a surprisingly common type of genetic difference between one human and another.

The CNV discovery arose out of Scherer's determined work to understand the role of genetic variation in Autism Spectrum Disorder (ASD). Ongoing extensive international collaborative work has since identified disease-susceptibility variants on several different chromosomes, leading to the development of early diagnostic tests. Recent whole-genome sequencing studies have identified previously unknown, very rare variants in both coding and noncoding regions of the genome.

and time-consuming than DNA sequence comparisons; therefore, it is not feasible to repeat experiments in every species whose genome is sequenced. As a result, experimental data are available only for a small fraction of organisms. And because the functions of homologous protein-coding genes are so well conserved during the evolution of organisms, information about the function of a gene in one well-studied species very often applies to the homologous genes in another.

In some cases, the outcome of a sequence similarity search will indicate that the entire candidate gene's sequence is homologous to a known gene's sequence. In other cases, only part of the candidate gene sequence may match closely a sequence in a known gene. Typically this result indicates that the candidate gene encodes a protein with a domain that is related evolutionarily to a domain-encoding region of the known gene. (Protein domains are discussed in the Genome Evolution section later in this chapter.)

16.3b Gene Function May Be Determined Using Experiments That Alter the Expression of a Gene

If a researcher can determine how the phenotype of a cell or organism is affected when the expression of a gene is turned off, or reduced significantly, functional properties of the encoded protein may be inferred. In a simple example, if cells grow larger, the gene may be involved in regulating cell size.

Two main kinds of manipulations are used to turn off or reduce significantly the expression of a gene in genome-scale experiments—gene knockout and gene knockdown.

1. Gene knockout. In this approach, researchers replace a normal gene on its chromosome with a defective gene that cannot express a functional protein. Usually, the replacement lacks the ORF that encodes the gene's protein product. In effect, this is a deletion mutation that has been engineered genetically. A deletion mutation is a null mutation because there is zero expression of the gene's protein product. For a haploid organism, there is only one copy of each gene to knock out, whereas in diploid organisms both copies of each gene must be knocked out. On a genomic scale, experimental manipulations can be done to knock out each gene systematically one by one. The phenotypic consequences of zero expression of each gene can then be ascertained. Major projects have been done, or are being done, to knock out systematically the function of each gene in the genomes of several organisms, including yeast, the fruit fly, the nematode worm, and the mouse (knockout mice were introduced in Section 15.2).

2. Gene knockdown. Knocking down a gene's expression is typically done using RNA interference (RNAi). As discussed in Section 14.3, RNAi reduces the expression of a gene at the translation level. In RNAi, a small regulatory RNA (like a natural miRNA) is transcribed from an expression plasmid introduced into the cell. The sequence of that regulatory RNA forms complementary base pairs with the mRNA of a gene of interest. The base-pairing triggers the RNAi molecular mechanisms (see Figure 14.14), which knock down the expression of the gene by causing degradation of that gene's mRNA or by blocking its translation. For example, RNAi has been used to knock down gene expression of each of the approximately 20 000 genes of the nematode worm one by one. The advantage of RNAi over gene knockouts is that the decrease in function of a gene can be temporary.

Characterizing genes by studying the effects on the phenotype of knockouts or knockdowns can be very expensive and time-consuming. In a genome-wide study, thousands of knockout or knockdown strains have to be engineered genetically, and then each one has to be screened for a battery of possible phenotypic changes. The most ambitious studies of this kind have examined hundreds of phenotypes for each gene, which is only a fraction of the phenotypes that could be characterized. To make this approach really productive in the future will require further development of high-throughput methods that automate the measurement of phenotypic changes.

16.3c Transcriptomics Determines at the Genome Level When and Where Genes Are Transcribed

Some genes are transcribed in all cell types, whereas others are transcribed only when and where they are needed (see Chapter 14). Determining when and where genes are transcribed can shed light on their function. For instance, a researcher might be interested in determining at a genomic scale the gene expression patterns in different cell types, at different stages of embryonic development, at different points of the cell division cycle, or in response to mutation or changes in the environment. A medical example is identifying gene expression differences between normal cells and cells that have become cancerous. The experimental analysis itself may be qualitative—analyzing whether or not genes are expressed—or quantitative—analyzing how the level of expression of genes varies.

The complete set of transcripts in a cell is called the **transcriptome**, and the study of the transcriptome is called **transcriptomics**. Transcriptomics includes cataloguing transcripts and quantifying the changes in expression levels of each transcript during development, in different cell types, under different physiological conditions, and with other variations.

Analysis of transcriptomes is done using high-throughput hybridization or, increasingly, by sequence-based approaches. A hybridization-based approach uses **DNA microarrays**, also called **DNA chips.** The surface of a DNA microarray is divided into a microscopic grid of about 60 000 spaces. On each space of the grid, a computerized system deposits a microscopic spot containing about 10 000 000 copies of a DNA probe that is about 20 nucleotides long.

Studies of gene activity using DNA microarrays involve comparing gene expression under a defined experimental condition with expression under a reference (control) condition. As a theoretical example, **Figure 16.8** shows how a DNA microarray can be used to compare gene expression patterns in normal cells

and cancer cells in humans. mRNAs are isolated from each cell type and cDNAs are made from them, incorporating different fluorescent labels: green for one cDNA, red for the other. The two cDNAs are mixed and added to the DNA microarray, where they hybridize with whichever spots on the microarray contain complementary DNA probes. A laser excites the fluorescent labels, and the resulting green and red fluorescence is detected and quantified, enabling a researcher to see which genes are expressed in the cells. This technique is semiquantitative because it is also able to approximately quantify differences in gene expression between the two cell types (see "Interpreting the Results" in the figure).

Examples of DNA microarray analysis are screening individuals for mutations associated with genetic diseases (such as breast cancer), as well as studying changing gene expression profiles during Drosophila development, the growth of a gingivitis-causing bacterium under different conditions, flower development in plants, and the lives sea urchin larvae under different pH conditions.

A newer, sequence-based approach to analyzing transcriptomes is **RNA-seq** (whole-transcriptome sequencing). This technique uses high-throughput sequencing of cDNAs to identify and quantify RNA transcripts in a sample. For transcriptomic analysis of protein-coding genes using RNA-seq, mRNAs are isolated and converted to cDNAs (see Figure 15.4). About 30 to 400 nucleotides of each cDNA are sequenced using high-throughput techniques, and the results are aligned with the genome sequence of the organism under study. A single RNA-seq study can identify over 100 000 sequence reads, each one of which indicates the presence of a specific mRNA in the cells being studied. While a relatively new technique, RNA-seq is rapidly becoming the replacement for DNA microarrays in transcriptomics because of its decreasing cost and its greater precision for quantifying transcripts.

16.3d Proteomics Is the Characterization of All Expressed Proteins

Genome research also includes analysis of the proteins encoded by a genome, because proteins are largely responsible for cell function and, therefore, for most of the functions of an organism. The term **proteome** has been coined to refer to the complete set of proteins that can be expressed by an organism's genome. A cellular proteome is a subset of those proteins—the collection of proteins found in a particular cell type under a particular set of environmental conditions.

The study of the proteome is the field of **proteomics.** The number of possible proteins encoded by the genome is larger than the number of protein-coding genes in the

PURPOSE: DNA microarrays can be used in various experiments, including comparing the levels of gene expression in two different tissues, as illustrated here. The power of the technique is that the entire set of genes in a genome can be analyzed simultaneously.

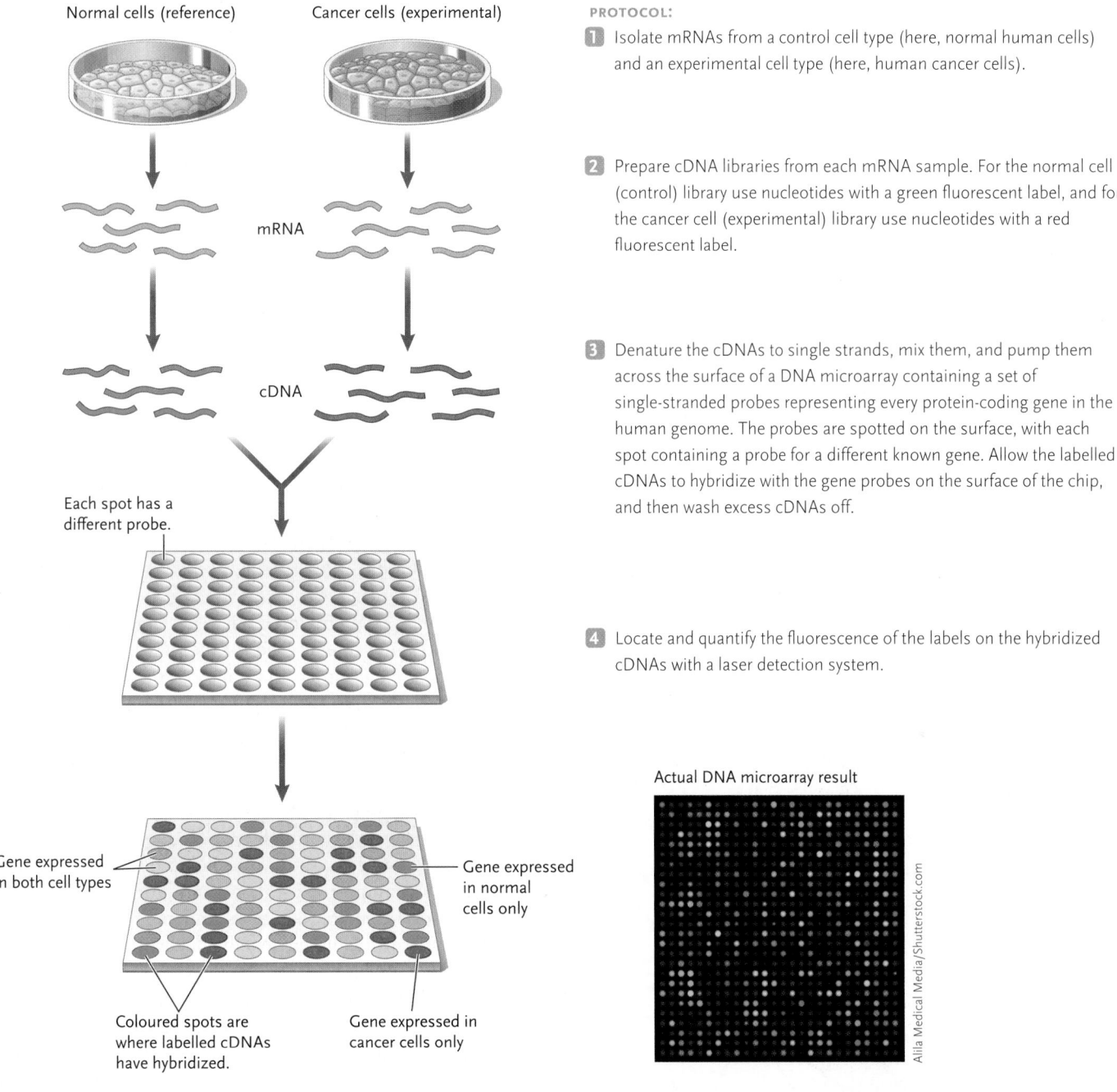

Normal cells (reference)

Cancer cells (experimental)

mRNA

cDNA

Each spot has a different probe.

Gene expressed in both cell types

Gene expressed in normal cells only

Coloured spots are where labelled cDNAs have hybridized.

Gene expressed in cancer cells only

Actual DNA microarray result

Alila Medical Media/Shutterstock.com

PROTOCOL:

1 Isolate mRNAs from a control cell type (here, normal human cells) and an experimental cell type (here, human cancer cells).

2 Prepare cDNA libraries from each mRNA sample. For the normal cell (control) library use nucleotides with a green fluorescent label, and for the cancer cell (experimental) library use nucleotides with a red fluorescent label.

3 Denature the cDNAs to single strands, mix them, and pump them across the surface of a DNA microarray containing a set of single-stranded probes representing every protein-coding gene in the human genome. The probes are spotted on the surface, with each spot containing a probe for a different known gene. Allow the labelled cDNAs to hybridize with the gene probes on the surface of the chip, and then wash excess cDNAs off.

4 Locate and quantify the fluorescence of the labels on the hybridized cDNAs with a laser detection system.

INTERPRETING THE RESULTS: The coloured spots on the microarray indicate where the labelled cDNAs have bound to the gene probes attached to the chip and, therefore, which genes were active in normal and/or cancer cells. Moreover, we can quantify the gene expression in the two cell types by the colour detected. A purely green spot indicates that the gene was active in the normal cell, but not in the cancer cell. A purely red spot indicates that the gene was active in the cancer cell, but not in the normal cell. A yellow spot indicates that the gene was equally active in the two cell types, and other colours tell us the relative levels of gene expression in the two cell types. For this particular experiment, we would be able to see how many genes have altered expression in the cancer cells, and exactly how their expression was changed.

Research Method Figure 16.8 DNA microarray analysis of gene expression levels.

genome, at least in eukaryotes. In eukaryotes, alternative splicing of gene transcripts and variation in protein processing means that expression of a gene may yield more than one protein product. The number of different proteins an organism can produce typically far exceeds the number of protein-coding genes.

Proteomics has three major goals: (1) to determine the structures and functions of all proteins, (2) to determine the location of each protein within or outside the cell, and (3) to identify physical interactions among proteins.

Determining Protein Structure. Protein structure may be determined as follows:

Clone the coding sequence of the gene into an expression vector (see Section 15.2).
↓
Transform the cloned gene into a host to express the protein.
↓
Purify the protein.
↓
Determine the structure of the protein using X-ray crystallography or nuclear magnetic resonance (NMR).

Protein structure may also be predicted nonexperimentally using computer algorithms based on the known chemistry of amino acids and how they interact.

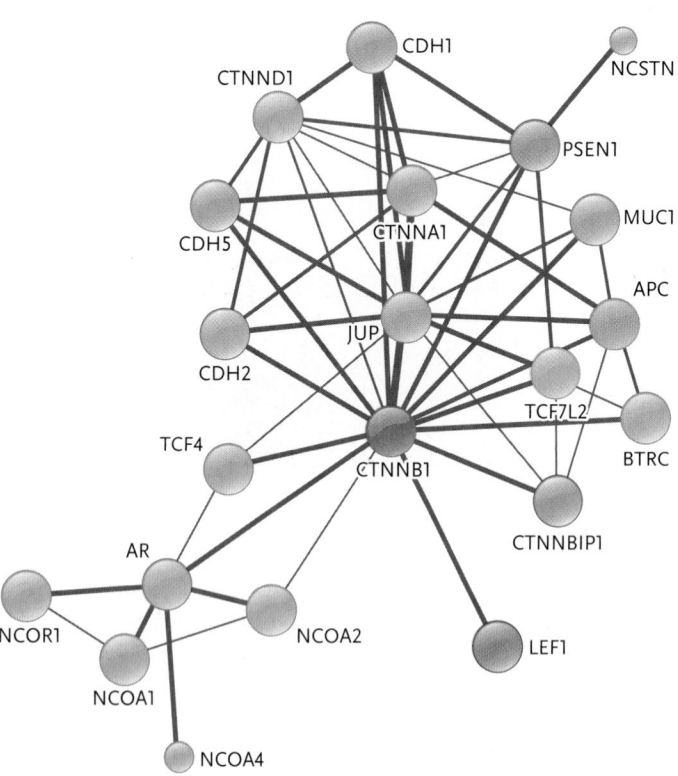

Figure 16.9

The protein-interaction network for human β-catenin (CTNNB1). Thicker lines show stronger associations between proteins.

Determining the Locations of Proteins in Cells. The location of a protein in a cell is important because it is key to its function. The cellular location of a protein can be studied by tagging the protein in some way and then visualizing the location of the tag microscopically. Different tags are used for visualization using light microscopy or electron microscopy.

Identifying Interactions among Proteins. Many proteins function by interacting with other proteins. In some cases, proteins (actually polypeptides) interact to form the quaternary structure—and therefore the functional form—of a protein (see *The Purple Pages*). Many multipolypeptide proteins exist, and you have encountered several in this book, for example, (1) hemoglobin is a four-polypeptide protein consisting of two α-globin polypeptides and two β-globin polypeptides and four associated heme groups (see "Molecule behind Biology," Box 40.1, Chapter 40); (2) RuBP carboxylase/oxygenase (rubisco), the first enzyme of the light-independent reactions of photosynthesis (see Section 9.3), consists of eight copies of a large polypeptide and eight copies of a small polypeptide; and (3) the Lac repressor protein that controls the expression of the lac operon in *E. coli* (see Section 14.1) consists of four copies of the same polypeptide.

In other interactions among proteins, the interaction is not permanent but instead serves to affect the function of one or other of the partners in the interaction. For example, in Chapter 5 you were introduced to protein kinases—enzymes that transfer a phosphate group from ATP to one or more sites on particular target proteins as part of a signal transduction pathway. The phosphorylation of the target proteins occurs as a result of the interaction between the enzymatic protein and each target protein. Once the target protein is phosphorylated, the two proteins no longer interact. Understanding the interactions among proteins is important, then, to help us understand how proteins work individually and together to determine the phenotype of a cell.

Thousands of interactions have been identified experimentally for a variety of organisms. The interaction data are assembled to produce protein-interaction networks, the analysis of which is informing us about the details and complexities of the functions of proteins in cells. **Figure 16.9** shows part of a protein-interaction network centred on the human protein β-catenin (cadherin-associated protein; the central CTNNB1 sphere in the figure). α-catenin is involved in the formation of adherens junctions (a type of cell junction; see Section 2.5) in epithelial cells where it links a-catenin (CTNNA1 in the figure) with E-cadherin (CDH1 in the figure; cadherins are discussed in Chapter 42). It also plays a key role in a signalling pathway that is important, for example, in regulating how cell fate is decided during development. Interactions with, for example, APC, LEF1, TCF4, and TCF7L2 occur as part of that signalling pathway's operation.

Just as for genomic DNA sequences, information about various properties of each protein are placed in databases, for example, Entrez (http://www.ncbi.nlm.nih.gov/Class/MLACourse/Original8Hour/Entrez/) and UniProt (http://www.uniprot.org), to create a dossier of that protein that is available to researchers worldwide.

Characterizing Protein Function. Through various approaches, researchers learn about the functions of proteins. **Figure 16.10a** shows an example of what we have learned about the functions of human protein-coding genes with respect to protein classes, and Figure 16.10b shows the functions of human protein-coding genes with respect to the biological processes involving these proteins.

STUDY BREAK

1. What are the ways by which the function of a gene identified in a genome sequence may be assigned?
2. How would you determine how a steroid hormone affects gene expression in human tissue culture cells?
3. What is the proteome, and what are the major goals of proteomics?

16.3f Systems Biology Highlights Interactions at Various Levels of Organization

The various "omics" projects have accumulated massive databases of information about the structures and functions of many thousands of individual genes, proteins, and RNAs. Fortunately, data visualization tools have enabled researchers to generate maps of interactions that reveal the structure of networks underlying the various levels of biological organization. We see that complexity is based not so much on how many genes an organism has, or how many proteins a cell can make, but rather on the networks of various interactions among these components at a given level and the interconnections among different levels.

The sum of all interactions of gene products with one another in a cell gives rise to an *interactome,* which, in turn, creates a particular array of active enzymes interacting with cofactors and substrates to create the *metabolome.* Metabolomes in all tissues and organs give rise to the overall *phenome* of an organism, which, if unhealthy, exhibits a *diseasome.* Overall, such research is leading to a more integrated and holistic understanding of living systems.

a. Protein classes

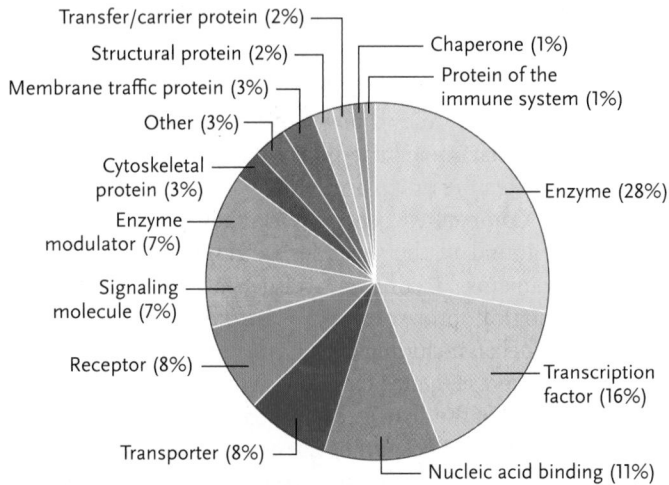

b. Biological classes

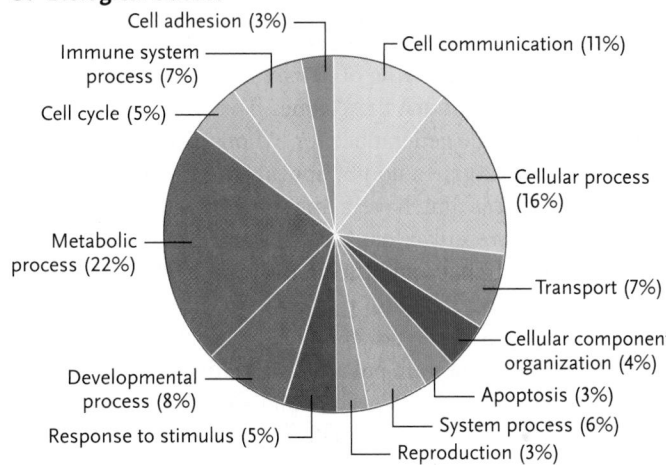

Figure 16.10

Functions of human protein-coding genes organized with respect to protein classes **(a)** and biological processes **(b)**.

16.4 Genome Evolution

DNA genomes with protein-coding genes are thought to have evolved over 3.5 billion years ago (bya), by the time of the earliest fossil microorganisms that have been discovered. These early cells probably had at most a few hundred protein-coding genes. New genes evolved as life evolved and became more complex, so that most present-day organisms have thousands or tens of thousands of protein-coding genes. In this section you will learn how genes and genomes have evolved, and how genome sequences inform us about the evolutionary history of life.

16.4a Comparative Genomics Reveals the Evolutionary History of Genes and Genomes

Understanding how genes evolved and how genomes evolved are major goals of the field of comparative genomics. Because the genes in present-day genomes evolved from ancestral genes that were in the genomes of organisms living millions to billions of years ago, we can trace the evolutionary history of genes by comparing the genomes of different groups of present-day organisms. From such comparisons, we can estimate when new genes first appeared in ancient organisms, describe how they changed over time, and gain insights into what molecular processes cause new genes to evolve in the first place.

Comparative genomics has shown that some genes are found in the genomes of almost all present-day organisms. Examples are genes involved in core biological processes like transcription and protein synthesis, including genes for some subunits of RNA polymerase, genes for many of the proteins that make up part of the structure of a ribosome, and most of the aminoacyl–tRNA synthetase enzymes that attach amino acids to tRNA molecules. The proteins coded for by these genes not only perform the same function in every organism but are also related evolutionarily. This conclusion strongly suggests that the single-celled common ancestor of all living organisms had these genes in its genome, and that these genes have been passed down through the generations for billions of years.

Most genes do not appear in the genomes of all organisms but have a more restricted distribution. There are eukaryote-specific genes, bacteria-specific genes, archaea-specific genes, animal-specific genes, plant-specific genes, primate-specific genes, human-specific genes, and so on. For example, mitosis and meiosis genes are eukaryote-specific genes, genes for flowers are plant specific, and some genes related to brain function are primate specific.

Analyzing the evolutionary history of genes provides valuable information about how life evolved on the molecular level. For example, by comparing the functions of almost 4000 evolutionarily related groups of genes in 100 genomes of bacteria, archaea, and eukaryotes, researchers have identified a period about 3 bya when many new genes evolved. By analyzing the functions of these new genes, the researchers concluded that many of the new genes evolved as adaptations to changes in the amount of oxygen in Earth's atmosphere, following the development of the oxygen-producing photosynthetic reactions (see Chapter 3).

As mentioned in Section 3.5f, the ability of otherwise solitary single cells to grow collaboratively and communally has evolved multiple times in the tree of life—giving rise to an astonishing range of multicellular life forms. Some of the genetic renovations associated with the development of multicellular lifestyles have been discovered through genomic analysis of related species of algae, comparing *Chlamydomonas* with *Volvox*, specifically. These two species last shared a unicellular common ancestor about 200 million years ago (mya). However, since that time, *Chlamydomonas* has remained unicellular, while *Volvox* has developed a relatively sophisticated multicellular colonial structure (Figure 3.24).

Aside from duplication of some cyclin genes (associated with cell cycle control) as well as some extracellular matrix (ECM) genes (associated with cell wall proteins that hold the colony together) in *Volvox*, we find relatively little difference in the number, or types, of genes in these two modern species. Therefore, it seems that the emergence of multicellularity in this case relied more heavily on repurposing and reregulating existing genes than on the acquisition of new genes.

One such repurposed gene in *Volvox* is GlsA. While unicellular *Chlamydomonas* cells always divide symmetrically, creating daughter cells with identical structure and fate, *Volvox* development depends on several rounds of asymmetric cell division resulting in few cells that are relatively large. These large cells become reproductive cells, while the thousands of smaller cells function solely as nondividing somatic cells. Mutations in GlsA result in loss of asymmetric division. A genomic DNA database search discovered a very similar gene sequence in *Chlamydomonas* called GAR1 (Figure 16.12). GlsA in *Volvox* and GAR1 in *Chlamydomonas* are orthologues, since they are derived from the same gene in their common ancestor. However, since *Chlamydomonas* only divides symmetrically, it appears that the protein product of these two genes interacts in a different network, with different effects on the plane of cell division, in the two modern species.

A family of related genes, including RegA and RlsA, code for transcription factors that provide the control of cell division necessary for cooperative growth of multicellular algae. These genes are paralogues since they arose by gene duplication in the unicellular ancestor of *Volvox*, where they likely functioned to regulate growth depending on the availability of light and nutrients. However, in modern *Volvox*, at least some of these genes are now under developmental control, regulating growth depending on reproductive versus somatic cell type.

While comparative genomics provides new insights into the evolution of known genes and genomes, it has also led to the discovery of otherwise unknown organisms and novel genes. In the mid-2000s, maverick biologist J. Craig Venter renovated a private yacht to serve as an oceanic survey laboratory, collecting seawater samples from Halifax to the Galapagos Islands. Sequencing of the total DNA extracted from such samples revealed a staggering degree of genetic diversity among the unicellular microorganisms and viruses collected from the marine environment. Hundreds of new species were discovered, along with an equally impressive number of genes otherwise unknown to science.

Venter's survey of genetic diversity in the ocean is one of the earliest examples of the now burgeoning field of **metagenomics**, in which DNA or RNA from an entire community of organisms in a particular niche is harvested collectively, sequenced, and analyzed using some of the DNA technologies described in this chapter. This approach is significant because until very recently, our understanding of the genetics of the microbial world was based almost exclusively on the very small proportion of species that can be cultivated in the laboratory. With the tools of modern metagenomics, we gain access to the genomes of a whole new world of previously inaccessible organisms and viruses.

Multicellular eukaryotes have complex relationships with rich communities of organisms inhabiting their skin, digestive, and excretory tracts. Large-scale metagenomic and metatranscriptomic analyses are significantly improving our understanding of the role of such microbial communities in human health and disease. Mining the metagenomic data harvested from microbial communities in such diverse environments as the termite gut, deep-sea hydrothermal vents, glaciers, geysers, the bovine rumen, and desert soil will certainly identify tens of thousands of novel genes that code for enzymes that may have applications in industrial biofuel production, food processing, pollution control, and drug development.

Comparative genomics has also been applied to understanding human evolution. (Human evolution is discussed in Chapter 21.) The human genome was the first mammalian genome sequenced, and researchers now have the sequences of hundreds of human genomes to study and compare to discover what makes us human and what is responsible for human variation. We also have over 40 other mammalian genomes to compare with the human genome. These include genomes of primate species that are closely related to humans, such as the common chimpanzee and the mountain gorilla, as well as less closely related mammals, such as the cow and the duck-billed platypus. Comparing the human genome with the genomes of other primates reveals which features are common to all of these primates and which are unique to humans. The human and chimpanzee genomes are strikingly similar, with 96% DNA sequence identity across the entire genome. The annotation of the chimpanzee genome is not yet complete, but it is likely that these two species share virtually all of their genes, so the genomic changes that occurred in human evolution probably involved only subtle mutations in the protein-coding sequences of genes, and mutations to regulatory sequences that determine how and when each gene is expressed (see Chapter 14). By contrast, comparisons of primate genomes with those of other mammals have identified new genes that evolved only in primates. Further studies of the functions of these genes may shed light on how primates evolved the characteristics that distinguish them from other mammals. And, interestingly, the primate-specific genes in the human genome contain the highest fraction of disease-related genes—19.4%—of any group of genes.

DNA microarrays and RNA-seq (see Section 16.3) have been used to compare which genes are transcribed in which parts of the brain in humans, chimpanzees, and rhesus monkeys. Certain groups of genes are expressed in the brain only during embryonic and early postnatal development. In both chimpanzees and rhesus monkeys, many genes involved in the formation of new synapses (communicating junctions between neurons; see Chapter 45) are transcribed only in the first year after birth, while in humans expression of these genes continues up to age five. These differences are most prominent in the prefrontal **cortex,** which is an area of the brain involved in complex decision-making (see Chapter 44). These comparative findings provide clues to how our species evolved enhanced learning abilities.

Comparative genomics also provides information about how the arrangement of genes on chromosomes has evolved. In Section 11.3 you learned about chromosomal mutations that occur when part of a chromosome is translocated to another chromosome, or inverted in place. Nondisjunction in meiosis can also cause entire chromosomes to be duplicated (see Figure 11.12). Such chromosomal mutations are uncommon, and usually they are harmful. But when a nonharmful chromosomal mutation occurs and spreads to all members of a species, the order of genes on the chromosomes of that species may then be different from the order in closely related species. Comparing the genomes of a range of related species reveals that over the course of hundreds of millions of years of biological evolution, pieces of chromosomes have changed places repeatedly by translocation and inversion, rearranging the genes on chromosomes like shuffling a deck of cards. Even so, the chromosomal arrangement of some genes is preserved after all this reshuffling, even in distantly related organisms. For example, comparisons of the human genome with the genomes of distantly related animals such as insects and sea anemones reveal blocks of homologous genes that are on the same chromosome and arranged in the same order in all of these species. This means that the order of these genes on the chromosome has been preserved from the time that all of these species evolved from a common ancestor, even though that common ancestor lived over 500 mya.

16.4b New Genes Evolve by Duplication and Exon Shuffling

The evolution of a new gene is a rare event—much less common than a mutation in an existing gene. But, nonetheless, over millions of years of biological evolution many new genes have been produced. For example, a comparison of genome sequences among four species in the Drosophila genus of fruit flies identified over 200 genes that had evolved in just the past 13 million years (a comparatively short time in evolutionary history).

Throughout the history of life on Earth, the evolution of new biological functions has almost always involved the evolution of new genes. For example, photosynthesis became possible only with the evolution of genes coding for proteins that could harness the energy in photons to synthesize ATP and electron carrier molecules. And comparative analysis of mammalian genomes has revealed genes involved in milk production and other biological functions found only in mammals. Evolutionary biologists have described a number of molecular mechanisms to explain where these new genes come from. The most common molecular mechanism to explain the origin of new genes is gene duplication, which produces multigene families after a series of duplications of genes that all derive from the same ancestral gene. New types of genes are produced by a process called **exon shuffling**, which combines parts of two or more genes.

Gene Duplication. Gene duplication is any process that produces two identical copies of a gene in an organism's genome. There are three main mechanisms by which organisms might acquire duplicate copies of a gene: (1) whole-genome duplication (WGD), (2) unequal crossing-over of homologous chromosomes during meiosis, or (3) replication of transposable elements (see Section 9.4).

a. Normal crossing-over

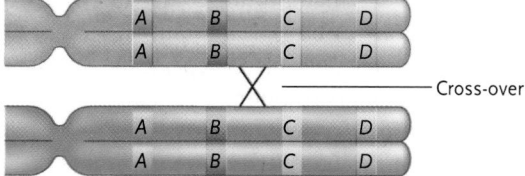

Cross-over

Crossing-over occurs between homologous chromatids during prophase I of meiosis (see Figure 11.5). Normally crossing-over occurs at the exact same point on each homolog and results in recombinant chromosomes after meiosis that have the same number of genes in each homolog.

Recombinant chromosomes (parental chromosomes not shown)

b. Unequal crossing-over

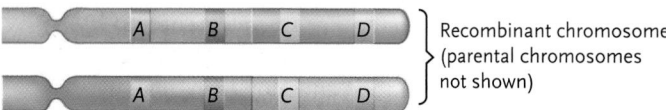

Unequal crossing-over results in recombinant chromosomes after meiosis that have a different number of genes. One (top) has duplicate genes, here B and C, whereas the other (bottom) has lost genes, here B and C.

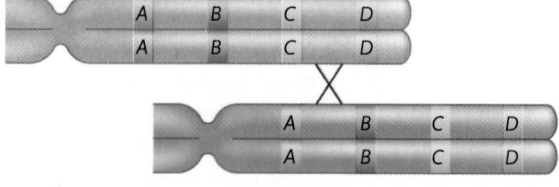

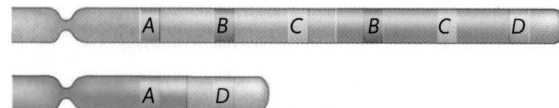

Figure 16.11
Duplication of genes by unequal crossing-over.

Widespread failure of chromosome partitioning in meiosis or mitosis can result in zygotes and daughter cells having double the normal number of chromosomes. Such cells could give rise to polyploid organisms with **whole-genome duplication (WGD)**. Although polyploidy is relatively rare, there are many thousands of modern plant species (most domestic crops) and many hundreds of animal species (many insects, amphibians, and fish) that have more than two sets of chromosomes. An ancient WGD in the yeast *Saccharomyces cerevisiae* was the first to be revealed by comparative genomics. Subsequent analyses suggest that WGDs have occurred at least twice in the evolutionary history of the vertebrates, three times in the bony fish, and up to four times in the flowering plants.

Unequal crossing-over is the rare phenomenon in meiosis in which, instead of crossing-over occurring at the exact same point on each homologue of a homologous pair of chromosomes **(Figure 16.11a)**, crossing-over occurs at different points (Figure 16.11b). The result of unequal crossing-over is that one of the recombinant chromosomes is missing one or more genes, while the other has duplicate copies of these genes. Unequal crossing-over produces tandem duplication of genes, with the duplicate copies clustered in the same region of the same chromosome.

Gene duplication may occur as a mistake when a transposable element copies itself and splices the DNA copies elsewhere in a genome. (Transposable element movement is discussed in Section 9.4.) Rarely, transposable elements copy adjacent DNA in addition to their own, producing duplicate copies of any genes in that DNA. This produces **dispersed duplication** of genes, meaning that the copies of the gene are found in different places in the genome—often on two different chromosomes.

At first, the duplicate copies of a gene have the same protein-coding sequences and encode identical proteins. The two genes are functionally redundant, meaning that one could be eliminated from the genome with no loss of biological functionality. Often, one of the redundant copies is mutated into a pseudogene, or lost by deletion. But if both genes remain functional, they will evolve slowly in different ways, as different mutations occur in each gene. Mutations in regulatory sequences may change how each duplicate gene is regulated, or mutations in protein-coding sequences may change the functional properties of the proteins produced by each gene. Over many generations, this evolutionary process can produce two homologous genes with similar but distinct functions.

For example, nitric oxide synthase enzymes catalyze a reaction that produces nitric oxide (NO), a molecule that cells use to communicate with each other. In the human genome, there are three genes for different nitric oxide synthase enzymes. One gene is expressed only in neurons, another only in the endothelial cells lining blood vessels, and the third in the liver and in a type of white blood cell called a macrophage (see

Chapter 51). Homologues of all three of these genes are found in other mammalian genomes as well as in the genomes of birds and reptiles, so the gene duplications by which these genes evolved must have happened over 200 mya. Evolution of tissue-specific expression patterns for each gene must have involved mutations to regulatory sequences that control transcription. The evolution of these genes also involved mutations to the protein-coding sequences, causing the proteins to have subtly different structures and functions. For example, the neuronal and endothelial nitric oxide synthase enzymes are regulated by Ca^{2+} ions, while the third enzyme is not. The three genes are found on three different chromosomes in the human genome, which suggests that they evolved by dispersed duplication.

Production of Multigene Families. Gene duplication is often followed by further duplication of one or both of the original duplicates. When this happens repeatedly over millions of years, a family of homologous genes called a **multigene family** evolves. The nitric oxide synthase enzymes described above are a small gene family, with three members in the human genome. Other multigene families contain tens to hundreds of genes. The members of a multigene family all evolve from one ancestral gene and therefore have similar DNA sequences and produce proteins with similar structures and functions. But because different mutations occur in each member of a multigene family, the genes gradually evolve subtly different characteristics.

Let us consider the OPT/YSL multigene family of the plant *Arabidopsis thaliana* as an example. The proteins coded for by these genes are oligopeptide transporters that shuttle short peptide molecules into the cytoplasm from organelles or the external environment. They are also important players in maintaining appropriate levels of heavy metals. This multigene family is found in other plant genomes as well as in the genomes of fungi and other eukaryotes. By comparing DNA sequences of OPT/YSL genes in the genomes of plants and other organisms, researchers have concluded that hundreds of millions of years ago and before the evolution of the plant kingdom, a gene duplication produced the ancestral OPT gene and the ancestral YSL gene. Mutations in each gene caused them to encode proteins specialized for transporting different oligopeptides. A series of more recent gene duplications occurred since the

a. Family tree showing evolutionary relationships among *OPT* and *YSL* genes

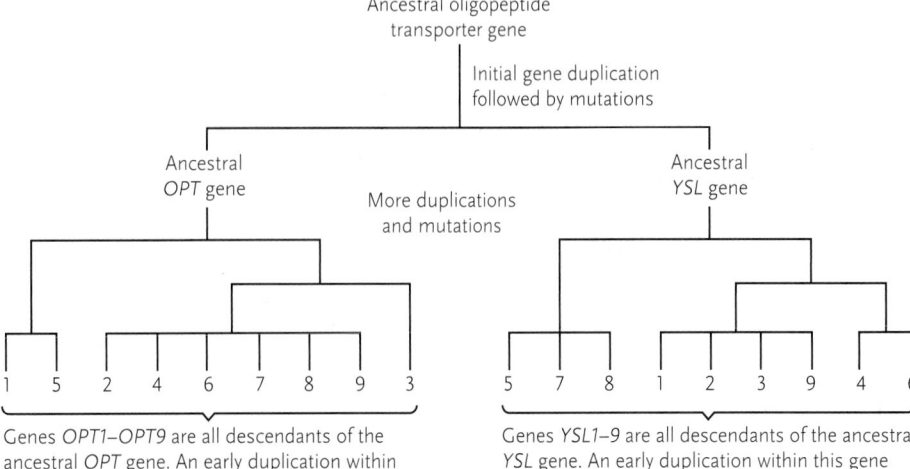

Genes *OPT1–OPT9* are all descendants of the ancestral *OPT* gene. An early duplication within this gene subfamily produced two genes, one an ancestor of *OPT1* and *OPT5* and the other an ancestor of the other seven *OPT* genes. A subsequent gene duplication in the latter group produced the ancestor of *OPT2, 4,* and *6–9,* and the ancestor of *OPT3*.

Genes *YSL1–9* are all descendants of the ancestral *YSL* gene. An early duplication within this gene subfamily produced two genes, one an ancestor of *YSL5, 7,* and *8* and the other an ancestor of the other six *YSL* genes. A subsequent gene duplication in the latter group produced the ancestor of *YSL1–3* and *9,* and the ancestor of *YSL4* and *6.*

b. Distribution of *OPT* genes on chromosomes in the *Arabidopsis thaliana* genome

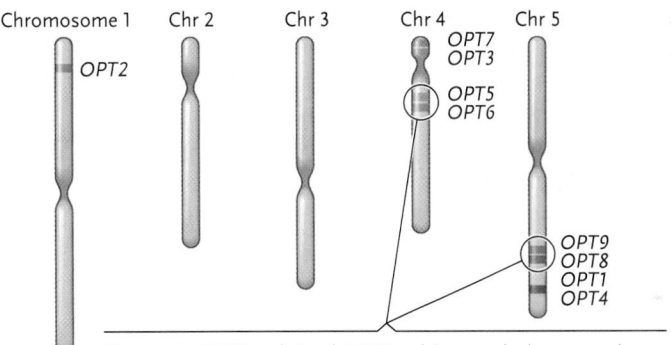

Gene pairs *OPT5* and *6* and *OPT8* and *9* are each close enough that they could have evolved by a recent tandem duplication. But only *OPT8* and *9* are near relatives in the *OPT* gene family tree (part **(a)** of figure). Therefore, we can hypothesize that *OPT8* and *9* likely resulted from a fairly recent tandem duplication, whereas *OPT6* is a dispersed duplicate of another *OPT* gene.

Figure 16.12

Evolution of the plant OPT/YSL multigene family. **(a)** Family tree showing evolutionary relationships among OPT and YSL genes. **(b)** Distribution of OPT genes on chromosomes in the Arabidopsis thaliana genome.

evolution of plants, but well before the evolution of *Arabidopsis thaliana*. Each duplicate gene in this family accumulated different mutations, producing the functionally diverse set of OPT genes and YSL genes now found in *Arabidopsis thaliana* and other plants.

Figure 16.12a illustrates the family relationships among the OPT/YSL genes using a phylogenetic tree, much as a family tree illustrates relationships in a human family (see Chapter 20 for more information on how phylogenetic trees are constructed). Figure 16.12b shows the distribution of the OPT genes on Arabidopsis thaliana chromosomes and outlines the possible evolutionary history of some of these genes.

The oligopeptide transporter gene family in *Arabidopsis thaliana* is larger than the nitric oxide synthase gene family in the human genome, but other multigene families are even larger. For example, some families of transcription factor proteins and membrane-bound receptor proteins include hundreds of genes. Some of the larger multigene families have members in all kingdoms of eukaryotes, or even in all three domains of living organisms. Such families each evolved from an ancestral gene that first appeared billions of years ago.

Exon Shuffling. The new genes produced by gene duplication evolve distinct functions, but they retain the same general function as other members of the multigene family into which they have been "born," so to speak. Our examples have illustrated that. By contrast, exon shuffling—the duplication and rearrangement of exons—is a molecular evolutionary process that combines exons of two or more existing genes to produce a gene that encodes a protein with an unprecedented function.

Remember from Section 13.3 that many protein-coding genes in eukaryotes contain introns, sequences that do not encode amino acids. The introns are present in the pre-mRNA transcripts of such genes but, by RNA processing, the introns are removed while the exons—the sequences encoding amino acids in the pre-mRNAs—are spliced together to make the mature mRNAs. In many genes, the junctions between exons fall at points within the protein-coding sequence between major functional regions in the protein. These functional regions correspond to the domains into which many proteins are divided.

Exon shuffling can occur in the following way. When a piece of DNA is cut out of a chromosome and reinserted elsewhere in the genome (through the activity of a transposable element, for example), the ends of the piece of DNA that moves may occur within the introns of a gene, causing one or more whole exons to be inserted somewhere else in a chromosome. If these exons are inserted into an intron in another gene, the amino acid sequence encoded by the exons may be added to the amino acid sequence of the encoded protein. Such a transfer of DNA can produce a new gene, coding for a protein that has one or more domains added to the other domains that it already had.

An exon shuffling event occurred very early in the evolution of animals (at least 700 mya) that produced a new gene coding for a protein that plays a key role in signalling between cells in animal tissues. Evidence for this exon shuffling event comes from comparing the genome sequence of the choanoflagellate *Monosiga brevicollis* with the sequences of a number of animal genomes, including Homo sapiens. Choanoflagellates (see Chapter 24) are single-celled or colonial protists that are thought to be related evolutionarily to animals. The evolution of multicellularity in the first animals is thought to have involved molecular mechanisms that enabled choanoflagellate-like cells to attach to and communicate with one another, so they could then specialize in performing different functions.

Figure 16.13 shows the exon shuffling event. It involves the Notch family of proteins, which are

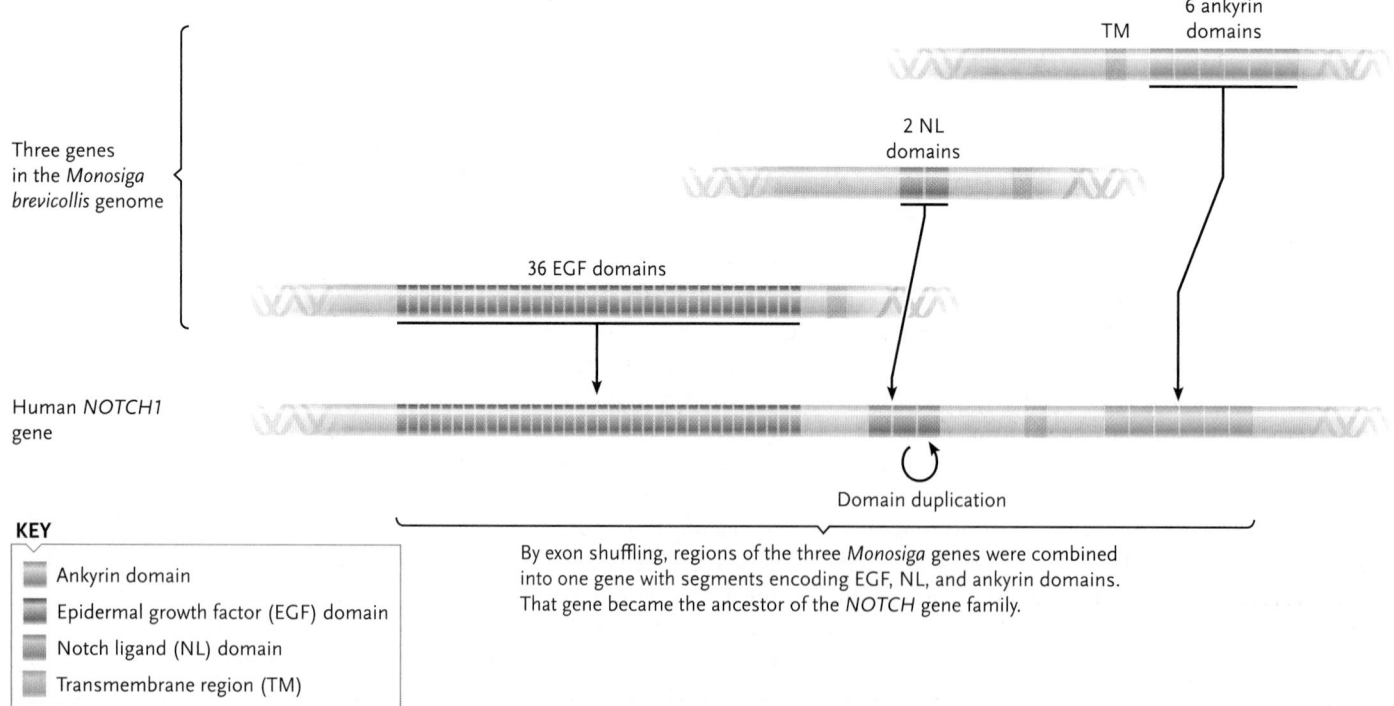

Figure 16.13

Evolution of Notch domains in animals by exon shuffling. At the top are three genes in the *Monosiga brevicollis* genome that encode three transmembrane (TM) region–containing proteins, one with epithelial growth factor (EGF) domains, a second with Notch ligand (NL) domains, and a third with ankyrin domains. At the bottom is the human NOTCH1 gene, which encodes a protein with EFG, NL, and ankyrin domains.

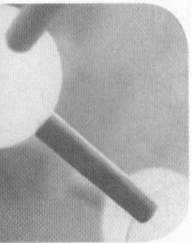

MOLECULE BEHIND BIOLOGY 16.2

Myoglobin

Mammalian myoglobin is a relatively small, extensively studied protein composed of eight alpha helices that fold to enclose an internal oxygen-binding heme group **(Figure 1).** Having a higher affinity for oxygen than hemoglobin, myoglobin acts to store oxygen for use in muscle cells. High concentrations of myoglobin in muscle tissue is an adaptation that allows marine mammals such as elephant seals and sperm whales to extend their dive time for over an hour without taking a breath. Recent studies of myoglobin chemistry and genomics have provided a glimpse into the lifestyle of the ancestors of modern whales as they ventured into the ocean depths.

Across a wide range of modern aquatic mammals, there is a strong correlation between high myoglobin concentrations and increased surface charge on the protein. Since surface charge is determined by amino acid composition, which, in turn, is determined by DNA sequence, it is therefore possible to estimate maximum myoglobin concentrations in muscle cells based on myoglobin gene sequence. Combining sequence data with body mass data allows scientists to predict maximum dive times of aquatic animals. Since comparative genomics of ancient DNA (from extinct mammoths, for example) versus modern sequences allows us to deduce the gene sequences of evolutionary ancestors, and since we have fossil records of the size of ancestral whales, we can now estimate that the maximum dive time of an ancestor of modern whales, such as the *Basilosaurus* **(Figure 2),** was the same as that of a modern dolphin (about 15 minutes).

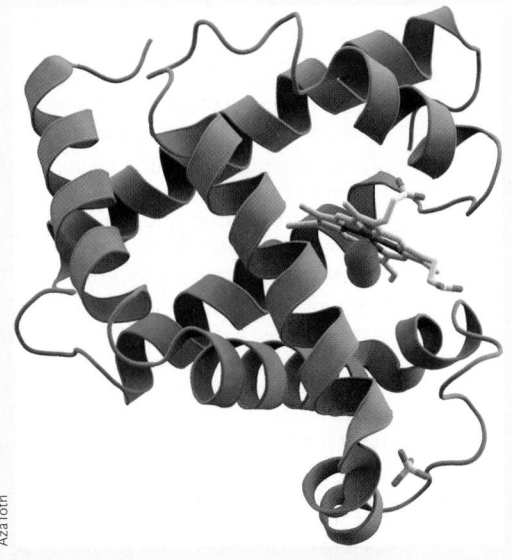

AzaToth

Figure 1
Myoglobin folded around heme group.

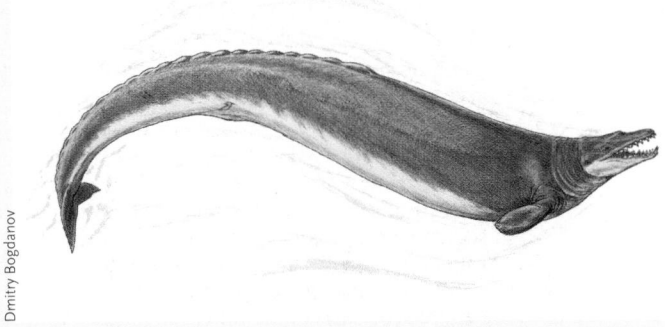

Dmitry Bogdanov

Figure 2
Basilosaurus, *an ancestor of modern whales from the Late Eocene, 40 mya.*

multidomain, membrane-spanning proteins. The human Notch1 protein, encoded by the NOTCH1 gene, contains a transmembrane (TM) region, 36 copies of an EGF domain, three copies of an NL domain, and six copies of an ankyrin domain. The TM region anchors the Notch1 protein in the plasma membrane. The three domains are key to the protein's function. The part of the protein that is outside the cell membrane includes the EGF and NL domains, which enables a Notch protein in one cell to bind to other proteins in adjacent cells in a tissue. The ankyrin domains within the cell enable it to attach to the microfilaments that make up part of the cytoskeleton (see Section 2.3).

Figure 16.13 also shows three different genes in the Monosiga brevicollis genome, one with a sequence encoding EGF domains, a second with a sequence encoding NL domains, and a third with a sequence encoding ankyrin domains. However, this organism lacks a gene homologous to the gene for the Notch protein. Researchers have hypothesized that through exon shuffling early in animal evolution, the sequences coding for the EGF, NL, and ankyrin domains in the three genes were combined in one gene, producing the ancestor of the NOTCH gene family. At some point, a duplication of the sequence coding for one NL domain occurred, since the *Monosiga brevicollis* gene has sequences coding for only two copies of that domain, while animal genes for Notch proteins have sequences coding for three copies.

The genes produced by gene duplication typically encode functional proteins because they are duplicates of an existing functional gene. Most instances of exon shuffling should theoretically produce non-functional proteins, because an existing functional protein is interrupted by one or more domains from an unrelated protein. But in a small number of cases, proteins produced by exon shuffling combine the functions of two or more proteins in a new and useful way, like the assembly of the gene for the Notch signalling protein (see Figure 16.13). Interestingly, exon shuffling is thought to account for perhaps one-third of newly evolved genes. The genes that are produced by exon shuffling have more novel functions than those produced by gene duplication. In cases like the Notch protein, they become the ancestors of

multigene families that evolve through subsequent gene duplications.

The domain structures of proteins in humans and other organisms provide evidence of how common exon shuffling has been in the evolution of proteins. The most widely used domains are found in thousands of different proteins, in dozens of different combinations with other domains.

STUDY BREAK

1. What molecular mechanisms cause tandem duplication of genes and dispersed duplication of genes?
2. Why do new genes produced by exon shuffling have more novel functions than new genes produced by gene duplication?

Review

To access course materials such as Aplia and other companion resources, please visit www.NELSONbrain.com.

16.1 Genomics: An Overview

- Genomics is the characterization of whole genomes, including their structures (sequences), functions, and evolution.

- Genome sequence data are in databases that may be accessed by researchers worldwide.

- Genomics consists of three main areas of study: genome sequence determination and annotation, the determination of complete genome sequences and identification of putative genes and other important sequences; functional genomics, the study of the functions of genes and other parts of the genome; and comparative genomics, the comparison of entire genomes or parts of them to understand evolutionary relationships and basic biological similarities and differences among species.

16.2 Genome Sequence Determination and Annotation

- The whole-genome shotgun method of sequencing a genome involves breaking up the entire genome into random, overlapping fragments, cloning each fragment, determining the sequence of the fragment in each clone, and using computer algorithms to assemble overlapping sequences into the sequence of the complete genome (Figure 16.1).

- DNA sequencing methods involve DNA purification; DNA fragmentation; amplification of fragments; sequencing each fragment; and assembly of fragment sequences into longer sequences, such as those of a genome (Figures 16.2 and 16.3).

- Once the complete sequence of a genome has been determined, it is annotated to identify key sequences, including protein-coding genes, noncoding RNA

genes, regulatory sequences associated with genes, origins of replication, transposable elements, pseudogenes, and short repetitive sequences. Annotation of a genome is the task of researchers in bioinformatics.

- Identifying protein-coding genes in a genome sequence can be done by using a computer search for open reading frames (ORFs), by searching databases for sequence similarity to genes of known function, and (the gold standard) by studying gene transcripts.

- Complete genome sequences have been obtained for many viruses, a large number of prokaryotic cells, and many eukaryotes, including humans. For organismal genomes, those of bacteria and archaea are generally smaller than those of eukaryotes, and their genes are densely packed in their genomes with little noncoding space between them. Prokaryotic genes are organized either into single transcription units or into operons. Genomes of eukaryotes vary greatly in size, but there is no correlation between genome size and type of organism. Gene density also varies, but in general it is significantly less than is seen for prokaryotic genes. Eukaryotic genes are organized into single transcription units (Figures 16.5–16.7; Tables 16.2 and 16.3).

16.3 Determining the Functions of the Genes in a Genome

- The function of a protein-coding gene may be assigned by a sequence similarity search of sequence databases, by using evidence from protein structure, or by using knockout or knockdown experiments that alter the expression of a gene.

- Transcriptomics is the study at the genome level of when and where genes are transcribed. One experimental method for studying the transcription of all or many of the genes in a genome simultaneously is the DNA microarray (DNA chip); this technique can generate qualitative information about gene transcription, such as the similarities and differences in gene expression in two cell types or in two developmental stages, as well as quantitative information about the relative levels of gene transcription (Figure 16.8).

- Proteomics is the characterization of the complete set of proteins in an organism or in a particular cell type. Protein numbers, protein structures, protein functions, protein locations, and protein interactions are all topics of proteomics (Figures 16.9 and 16.10).

16.4 Genome Evolution

- Comparative genomics traces the evolution of genomes by analyzing similarities and differences in DNA sequences in the genomes of present-day organisms.

- Comparative analysis reveals how homologous protein-coding genes have evolved in groups of organisms, how regulation of gene expression has evolved through mutations in the regulatory sequences of genes, and how chromosome structure has evolved as parts of one chromosome have broken off and been attached to other chromosomes.

- New genes evolve by whole-genome duplication (WGD) or by tandem duplication when chromosomes cross over unequally in meiosis, producing a chromosome containing two copies of the DNA coding for one or more genes (Figure 16.11). New genes evolve by dispersed duplication when transposable elements copy DNA coding for one or more genes and insert it at another location in the genome.

- Duplicate copies of genes evolve distinct functions as different mutations occur in the two copies. Mutations in protein-coding sequences produce proteins with slightly different structures, while mutations in regulatory sequences cause the genes to be expressed in different cell types and in response to different stimuli.

- Repeated cycles of duplication followed by mutation produce multigene families, which are collections of homologous genes that code for similar but functionally distinct proteins. The largest multigene families comprise hundreds of different genes (Figure 16.12).

- Exon shuffling produces functionally novel proteins by combining parts of two or more different genes. When exons that code for one or more domains of a protein are copied from one gene and inserted into the protein-coding sequence of another gene, those domains are added to the structure of the protein coded for by that gene. Adding new domains to a protein gives the protein new molecular functionalities (Figure 16.13).

Questions

Self-Test Questions

1. Why is the Solexa/Illumina DNA sequencing method faster and less expensive than the Sanger method?
   a. It sequences longer fragments of DNA.
   b. It sequences more DNA fragments at the same time.
   c. It does not require amplification of DNA fragments before sequencing.
   d. It does not require the use of computer algorithms to find places where sequence fragments overlap.

2. How do pseudogenes differ from genes?
   a. They are not transcribed.
   b. They contain longer ORFs.
   c. They do not have introns.
   d. They use a different genetic code.

3. What is the main reason that searching for ORFs is more useful for annotating bacterial protein-coding genes than it is for annotating eukaryote protein-coding genes?
   a. Eukaryote protein-coding genes contain introns.
   b. The density of protein-coding genes is much higher in eukaryote genomes.
   c. In most bacteria, all of the protein-coding genes are located on a single circular chromosome.
   d. Bacterial protein-coding genes are much longer than eukaryotic protein-coding genes.

4. Which of the following statements about genome size is true?
   a. Bacteria have genomes that vary widely in size.
   b. The human genome is the largest among eukaryotes.
   c. Organisms with large genomes tend to be more complex than organisms with small genomes.
   d. As genome size increases in a lineage, the number of genes also always increases.

5. Which of the following statements about the *E. coli* genome is true?
   a. It has a much lower gene density than the human genome.
   b. It contains longer genes than the human genome.
   c. All of the genes are transcribed from the same template strand of the DNA double helix.
   d. About half of the genes in the *E. coli* genome are grouped with other genes in operons.

6. Which of the following statements about the human genome is true?
   a. The protein-coding sequences occupy about 75% of the genome.
   b. About 45% of the genome consists of transposable element sequences.
   c. The genome sequence comprises approximately 30 million base pairs.
   d. Human cells have about 10 500 different protein-coding genes.

7. What makes up about 95% of the average human transcription unit?
   a. short repeat sequences
   b. protein-coding sequences
   c. regulatory sequences
   d. introns

8. Imagine that the DNA sequences of two protein-coding genes are similar, but only for part of the protein-coding sequence. What does this suggest?
   a. The two proteins have one or more domains in common.
   b. The two proteins were produced by duplication of an ancestral gene.
   c. The two proteins perform the same function.
   d. One of the two genes is actually a pseudogene.

9. Imagine two protein-coding genes have very similar nucleotide sequences and are located right next to each other on a chromosome. What does this suggest?
   a. One of them is a duplicate of the other, copied by a retrotransposon.
   b. One of them is a pseudogene.
   c. They were produced by unequal crossing-over.
   d. They are transcribed in the same cell types.

10. When do the proteins coded for by genes in a multigene family begin to evolve distinct functions?
   a. When gene duplication occurs.
   b. When exon shuffling occurs.
   c. When the genes are expressed by transcription and translation.
   d. When different mutations occur in each protein-coding sequence.

Questions for Discussion

1. Why are high-throughput techniques used so much in genomics research? Give examples from this chapter of different uses of high-throughput techniques.

2. Why does the Sanger DNA sequencing method work best when the concentration of dideoxyribonucleotides is much less than the concentration of deoxyribonucleotides? If you wanted to adjust the reaction mixture to produce a greater number of very long complementary sequence fragments, how would you change the relative concentration of dideoxyribonucleotides, and why?

3. Which of the methods for annotating protein-coding genes would you expect to do the best job of distinguishing functioning genes from pseudogenes, and why?

4. The genome of the yeast Saccharomyces cerevisiae is only about 0.4% the size of the human genome, yet it contains about 30% as many genes as are in the human genome. Given that, which of the features of the human genome would you expect to find many fewer of in the yeast genome?

5. How does sequencing the genomes of a greater number of animal species help in annotating and determining the functions of human protein-coding genes?

EVOLUTION, ASPECTS OF DIVERSITY, ECOLOGY, AND APPLIED BIOLOGY

Venus flytrap (*Dionaea muscipula*). A leaf specialized to trap insects shows the edges lined with bristles, black trigger hairs (trichomes) that spring the trap, and the red colour that attracts prey.

This volume consists of four units, progressing from mechanisms involved in diversification (Unit Five) to explorations of the diversity of life (Unit Six). The ecological and environmental contexts and patterns associated with diversification (Unit Seven) are considered along with two facets of applied biology (Unit Seven), specifically conservation of biodiversity and how humans have used selection to produce food. The impacts of evolution and diversification of the human evolutionary lineage establish our evolutionary roots and illustrate the fundamental impact our species has had on global diversity. On the positive side, humans have turned evolutionary processes to our advantage in domesticating other organisms to our advantage, but the negative impact of these outcomes often is alarmingly obvious.

You will see common themes in the chapters in this volume including fitness, natural selection, adaptive advantages, diversification, and evolution. The central role of the environment in evolutionary history and ongoing evolution also is apparent. The vital importance of interactions among organisms is another recurring theme.

Each chapter in this volume connects directly and indirectly to other chapters in the book, echoing the fundamental theme "exploring the diversity of life." The connections hinge on fundamental concepts and facts of life, such as those explored in Volume I. These relate to energy and metabolism, the structures of cells, and the way that cells operate. Genetics is a central pillar of biology, and advances in genetics have greatly increased our understanding of evolutionary history the relationships among organisms. Also essential are connections to the ways in which multicellular organisms operate (Volume III).

In each unit covered in this volume, it is clear how the science of biology has progressed, from observation and description, to generation of hypotheses about diversity, and to specific testable predictions arising from the hypotheses. Methodological advances and new techniques have allowed more rigorous testing of hypotheses and related predictions about how life has diversified and continues to diversify on Earth.

The solutions that plants have evolved to use animals to gain access to nitrogen illustrates the some of the principles underlying this volume. In plants, the repeated appearance of mechanisms for obtaining nitrogen by trapping and digesting animals reflects the selective advantage(s) of this way of life. The diversity of carnivorous plants and the appearance of this way of life in different evolutionary lineages of plants have always intrigued many students of biology, including Charles Darwin. In 1875, Darwin noted that there were at least 20 genera of carnivorous plants distributed in 10 families.

The diversity of animal traps in plants **(Figure 1)** is well known and there are both terrestrial and aquatic carnivorous plants. Sticky traps also appear in species in the families Droseraceae and Lentbulariaceae. Bucket traps, in which the insect (or other prey) falls into a pool of water contained in a modified leaf and drowns, have appeared independently in at least four families of plants. Active traps, where part of the plant moves to catch the animal, occur in Droseraceae and Lentbulariaceae, two families with both terrestrial and aquatic species. A recent (2012) addition to the list of carnivorous plants comes from low-nutrient, sandy soils in Brazil. Three species in the genus *Philcoxia* (family Plantaginaceae) use underground adhesive leaves to trap and digest nematode worms. In 2013, a bromeliad-like plant (*Paepalanthus bromelioides*; family Eriocaulaceae) emerged as a species on its way to becoming carnivorous. Analysis of isotopes of Nitrogen in this plant revealed 15N that was derived from termites. These plants grow on termite nests, so they had not directly captured the termites.

Isotopic analysis also revealed that some pitcher plants in the genus *Nepenthes* have specialized pitchers that allow them to collect urine and feces from mammals, specifically tree shrews and roosting bats. Isotopic analysis also revealed that many other pitcher plants acquired their nitrogen from symbionts such as rotifers, living in the pitcher water. Bacteria and rotifers consumed the animals that drowned in the pitchers and it was their excretory products that provided nitrogen to the plants.

The waterwheel plant, a relatively close relative of Venus flytraps, is widespread globally, occurring in Europe, Africa, Asia, and Australia. Venus fly-traps, however, naturally occur only in North and South Carolina in the United States. The long-term survival of both of these specialized species is a matter of concern for conservationists.

In short, studies of carnivorous plants provide rich examples of evolution, diversification, and outcomes of interactions among organisms. The many different plants with carnivorous habits provide further examples of parallel and convergent evolution that extend to details of digestion of prey and the operation of traps connect with material presented in Volume III.

a. b. c. d. e.

M.B. Fenton

Figure 1

A sampling of terrestrial insectivorous plants. Included are plants with adhesive leaves **(a, b),** and pitcher plants **(c, d and e).** Butterwort (*Pinguicula* spp. - a) has adhesive leaves that have caught insects. A leaf of a sundew (*Drosera* spp. - b) bear adhesive droplets at the ends of trichomes. The pitcher plants include c - *Darlingtonia* from California, d - *Sarracinea* from Ontario, and e - *Nepenthes* from the East Indies.

An Australian sheep blowfly (family: Calliphoridae, *Lucilia cuprina*) (6–9 mm long) (left). Maggots of Australian sheep blowflies (right).

Evolution: The Development of the Theory

WHY IT MATTERS

Australian sheep blowflies occur throughout Australia in habitats ranging from urban to rural settings, in semi-arid open lands as well as forests and woodlands. These flies have spread widely throughout the world, including to eastern Canada. They are considered pests because females deposit their eggs in open wounds on livestock such as sheep (called "fly strikes"). The eggs hatch into maggots, which eat flesh and damage the wool. A single female can produce hundreds of eggs. These flies complete their life cycle—egg, larva, pupa, and adult—in about 11 to 21 days.

Fly strikes by Australian sheep blowflies cost the wool industry over $A150 million a year ($A stands for Australian dollars). Tools for controlling the blowflies include bait traps and at least five kinds of chemical insecticides. Organophosphate insecticides are effective because they affect the nervous system and rapidly kill both adult flies and maggots.

Some Australian sheep blowflies have a mutation that makes them resistant to organophosphate insecticides. This resistance is based on a single mutation that appeared in a few individuals of this species. The DNA of resistant flies differs slightly from that of susceptible flies. When sprayed with the insecticide, resistant flies have enhanced fitness: they survive to reproduce, while the others around them die. The flies' short life cycle and high reproductive rate ensure the rapid spread of the resistant mutation in the fly population. Put another way, selection associated with the pesticide changes the population structure of sprayed flies, quickly eliminating nonresistant individuals.

This example shows how the combination of a genetic-based change (a small mutation) and strong selective pressure (lethal organophosphate insecticide) has altered the population of Australian sheep blowflies over time. This also shows how human interference can result in evolution by natural selection.

Evolution is the main unifying concept in biology, explaining how the diversity of life on Earth arose and how species change over time in response to changes in their abiotic and biotic environment. Our knowledge of how populations of organisms change over time has been enriched with data obtained using techniques of molecular biology, particularly those relating to genetics. However, we will begin with a history of the ideas and thinking that led to the work of Charles Darwin and Alfred Wallace. In 1858, these men independently published descriptions of variation within species (intraspecific variation) and proposed a process called natural selection to explain how species changed over time.

17.1 Evolution: What Is It?

Evolution means gradual change (from the Latin *evolere* = to unfold or unroll). **Biological evolution** refers to gradual change of populations or organisms over time, measuring time in generations rather than years. For example, while it appears that the sheep blowfly populations evolved quickly, it took many 11-day-long generations.

Many people associate evolution with dinosaurs and other fossils, implying that evolution happened in the past. However, organisms continue to evolve today. As in the past, new species are formed and others go extinct. In fact, ongoing evolution helps to explain the variation within and among the populations of a given species. This sometimes makes it difficult to determine when populations comprise distinct species (see Chapters 18 and 19).

Although biological evolution refers to gradual change, not all gradual change is biological evolution. Aging is a feature of life that involves gradual change. Family pictures record how you have changed since you were born. However, changes that occur over the lifetime of a single organism are not evolutionary. Gradual change is also a feature of the environment. Seasonal changes such as the fall of leaves, the growth of new sprouts, and animal reproductive behaviours are examples of gradual change. However, none of these changes is an example of evolution.

STUDY BREAK

1. What is evolution? What are selection pressures?
2. Why is the growth and development of an individual not evolution?

17.2 Pre-Darwin Knowledge of the Natural World

17.2a Early Views of Organisms

The Greek philosopher Aristotle (384–322 BCE) was a keen observer of nature and natural history. He examined the form and variety of organisms in their natural environments and believed that both inanimate objects and living organisms had fixed characteristics. Careful study of their differences and similarities enabled Aristotle to create a ladderlike classification of nature from the simplest to the most complex: minerals ranked below plants, plants below animals, animals below humans, and humans below the gods of the spiritual realm. Many of Aristotle's writings on living organisms were considered to be true for almost two millennia.

By the fourteenth century, Aristotle's classification was being merged with the biblical and other accounts of creation, at least in the Christian and Islamic worlds. At that time, Europeans thought that all of the different kinds of organisms had been specifically created by a god. The different kinds could never change or become extinct, and new kinds could never arise. Biological research was dominated by **natural theology**, which sought to name and catalogue all of God's creation. Careful study of each species could identify its position and purpose in the *Scala Naturae,* or Great Chain of Being, as Aristotle's ladder of life was called. This approach to nature and history was clear in the later work of Carolus Linnaeus (1707–1778), whose efforts were *ad majorem Dei gloriam* (for the greater glory of God) (Chapter 19).

By 1600, the English philosopher and statesman Sir Francis Bacon (1561–1626) established the importance of observation, experimentation, and finding evidence to support a proposal or theory (= inductive reasoning). Other scientists proposed theories to describe how physical events (mechanistic) worked, notably Nicolaus Copernicus (1473–1543), Galileo Galilei (1564–1642), René Descartes (1596–1650), and Sir Isaac Newton (1643–1727). The collective work of these scientists gave rise to three new disciplines—biogeography, comparative morphology, and geology—promoting a growing awareness of change.

Biogeography. As long as naturalists encountered organisms only from Europe and surrounding lands, the *Scala Naturae* was easily followed. But global explorations in the fifteenth through seventeenth centuries provided naturalists with thousands of unknown plants and animals from around the world. Although some were similar to European species, others were new and very strange.

Studies of the world distribution of plants and animals, now called **biogeography**, raised puzzling questions. Was there no limit to the number of species

African ostrich
(Struthio camelus)

Johan Swanepoel/Shutterstock.com

South American rhea
(Rhea americana)

Kenneth W. Fink/Science Source

Australian emu
(Dromaius novaehollandiae)

S.Cooper Digital/Shutterstock.com

Figure 17.1
These three species of large, flightless birds, with greatly reduced wings, appear very similar. In fact, they occupy similar habitats in geographically separated regions. Recent analysis of mitochondrial genomes reveals that all of the major lineages of large flightless birds (known as *ratites*) evolved from ancestors that had dispersed by flight.

created by God? Where did all these species fit in the *Scala Naturae*? If they had all been created in the Garden of Eden, why did some species have limited geographical distributions, whereas others were widespread? And why were some species found in Africa or Asia different from those found in Europe, whereas other species from far-flung locations were similar to each other **(Figure 17.1)**?

Comparative Morphology. When naturalists compared the morphology (anatomical structure) of organisms, they discovered interesting similarities and differences. For example, the front legs of pigs, the flippers of dolphins, and the wings of bats differ markedly in size, shape, and function **(Figure 17.2)**. But these appendages have similar locations in the mammals' bodies; all are constructed of bones, muscles, and skin; and all develop similarly in the animals' embryos. If these limbs were specially created for different means of locomotion, naturalists wondered, why didn't the

Creator use entirely different materials and structures for walking, swimming, and flying?

Natural theologians countered this argument by stating that the body plans were perfect, and there was no need to invent a new plan for every species. But a French scientist, George-Louis Leclerc (1707–1788), le Comte de Buffon, was still puzzled by the existence of body parts with no apparent function. For example, he noted that the feet of pigs and some other mammals have two toes that never touch the ground (pig digits 2 and 5 in Figure 17.2). If each species was anatomically perfect for its particular way of life, Buffon asked, why did useless structures exist?

Buffon proposed that some animals must have *changed* since their creation; he suggested that **vestigial structures**, these useless parts he observed, must have functioned in ancestral organisms. Buffon offered no explanation of how functional structures became vestigial, but he clearly recognized that some species were "conceived by Nature and produced by Time."

Geology. Georges Cuvier (1769–1832), a French zoologist, realized that the layers of fossils he found represented organisms that had lived at sequential times in the past. He suggested that abrupt changes between geologic layers marked dramatic shifts in ancient environments. Cuvier and his followers developed the theory of **catastrophism**, reasoning that each layer of fossils represented the remains of organisms that had died in a local catastrophe such as a flood. Somewhat different species then recolonized the area, and when another catastrophe struck, they became a different set of fossils in the next higher layer.

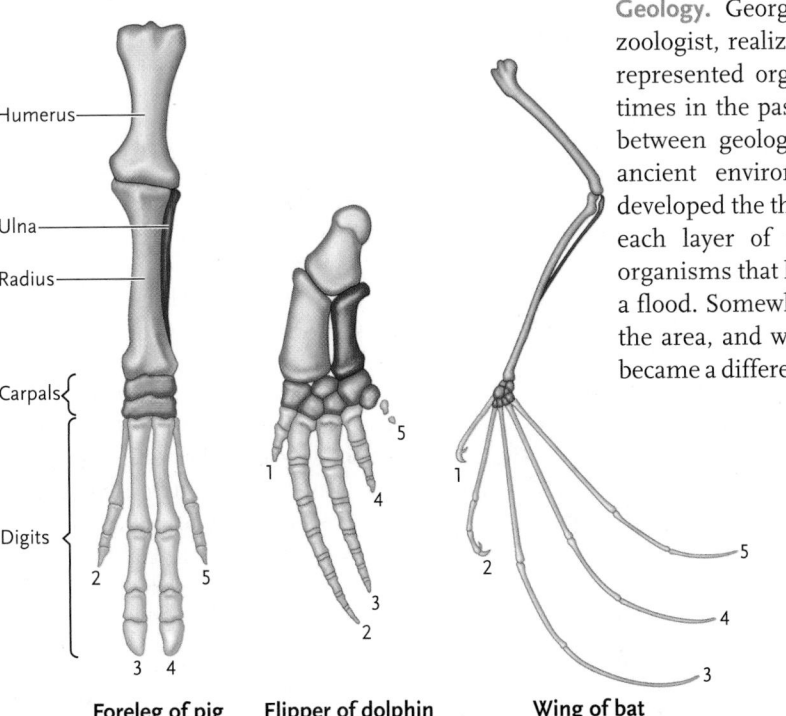

Humerus

Ulna

Radius

Carpals

Digits

Foreleg of pig   **Flipper of dolphin**   **Wing of bat**

Figure 17.2
Mammalian forelimbs and locomotion.
Pigs use their legs to walk or run, dolphins use their flippers to swim, and bats use their wings to fly. Homologous (equivalent) bones are pictured in the same colour, and digits (fingers) are numbered; pigs have lost the first digit over evolutionary time. (Limbs are not drawn to scale.)

From Fins to Fingers

The early embryonic development of the limbs of fishes and tetrapods is similar. The limbs start as buds of mesoderm (see Chapter 42), which thicken by increased cell division. As the buds elongate, cartilage is deposited at localized centres, the precursors of later limb bones. In fishes, bones develop along a central axis from base to tip **(Figure 1a)**. In tetrapods, centres of cartilage formation generate the long bones of the limb and the five digits of the foot (or hand; Figure 1b).

Biologists used molecular techniques to assess the patterns of development and determine if the digits of tetrapods were modifications of the bones radiating from the central axis in fish. In tetrapods with paired fore- and hindlimbs, groups of **homeobox** genes control their development. A comparison of *HoxD* genes in zebrafish (*Danio rerio*) and previously available data from birds and mammals revealed the details of development. Using the DNA from a rodent *HoxD* gene as a probe, researchers searched for similar genes in fragmented zebrafish DNA. After cloning and sequencing, it was clear that the *HoxD-11*, *HoxD-12*, and *HoxD-13* genes in zebrafish are arranged in the same order as they are in rodents.

Paolo Sordino, Franks van der Hoeven, and Denis Duboule tested the activity of *HoxD* genes in developing zebrafish using a nucleic acid probe that could pair with mRNA products of the genes. The probe was linked to a blue dye molecule so that the cells in which a particular *HoxD* gene was active would appear blue in the light microscope. In zebrafish, the *HoxD* genes became active in cells along the posterior side of the central axis (Figure 1c). As fin development neared completion, the activity of *HoxD* genes dropped off.

Using the same approach in tetrapods, the researchers found that the *HoxD* genes were activated in two distinct phases (Figure 1d). In phase 1, gene activity was restricted to the posterior half of the limb, as it had been in zebrafish. This period of activity corresponded to the development of long limb bones. In phase 2, the *HoxD* genes became active in a band of cells perpendicular to the central axis. Here, cartilage centres formed the bones of the digits that developed in an anterior–posterior band. These patterns differed from those in zebrafish, suggesting morphological novelty in tetrapods. These changes in *HoxD* activity must have preceded the development of tetrapod limbs.

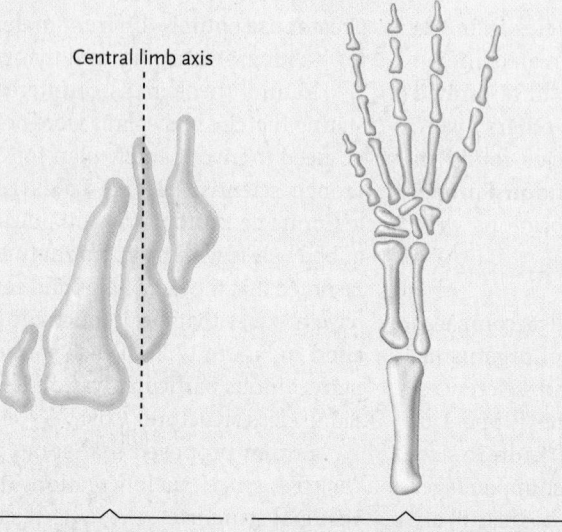

a. Fishes

Central limb axis

Bones in the fin of a fish develop from centres of cartilage formation along a central axis (dashed line).

b. Tetrapods

Bones in the limb and digits of a tetrapod also develop from centres of cartilage formation in the central axis.

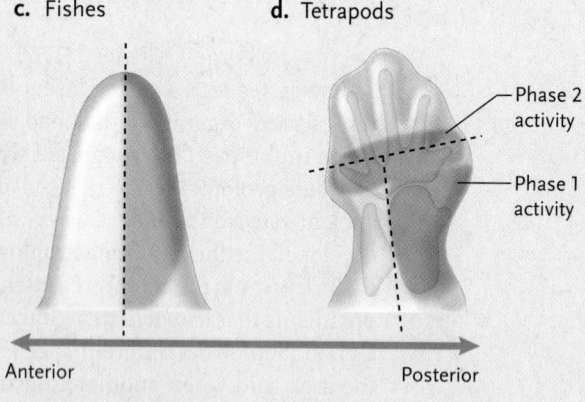

c. Fishes

d. Tetrapods

Phase 2 activity

Phase 1 activity

Anterior

Posterior

During development of the fin in fishes, *HoxD* genes become active in cells posterior to the central axis of the fin (shown in blue).

During development of the limb and digits in tetrapods, *HoxD* genes first become active in cells posterior to the central axis of the limb (blue). Later, these genes are active in a band of cells perpendicular to the central axis of the limb (green).

Figure 1

Fins versus fingers, Hox genes in action.

Today we know that huge catastrophes, such as the impact of an asteroid, can cause the extinction of tens of thousands of species. One example was the extinction event that marked the disappearance of dinosaurs and the end of the Cretaceous (66 mya [million years ago]).

17.2b Our View of Earth Changes

Bishop James Ussher (1581–1656), a theologian and scholar, calculated the age of Earth by counting the number of generations mentioned in the Bible. He came up with the year 4004 BCE as the year of Creation. Dr. John Lightfoot (1602–1675), the Vice-Chancellor of Cambridge University, continued with this research to come up with a more precise time. He concluded that Earth had been created on October 23, 4004 BCE, at 9:00 in the morning. The idea that Earth was about 6000 years old persisted for several centuries.

In 1795, the Scottish geologist James Hutton (1726–1797) argued that slow and continuous physical processes, acting over very long periods of time, produced Earth's major geologic features. The movement of water in a river slowly erodes the land and deposits sediments near the river's mouth. Given enough time, erosion creates deep canyons, and sedimentation results in thick topsoil on flood plains. Hutton's **gradualism**, the view that Earth changed *slowly* over its history, contrasted sharply with Cuvier's catastrophism.

The English geologist Charles Lyell (1797–1875) championed and extended Hutton's ideas in an influential series of books, *Principles of Geology: An Attempt to Explain the Former Changes of the Earth's Surface by Reference to Causes Now in Operation*. Lyell argued that the geologic processes that sculpted Earth's surface over long periods of time, such as volcanic eruptions, earthquakes, erosion, and the formation and movement of glaciers, are exactly the same as the processes we observe today. This concept, **uniformitarianism**, undermined any remaining notions of an unchanging Earth. Because geologic processes proceed very slowly, it must have taken millions of years, not just a few thousand, to mould the landscape into its current configuration.

STUDY BREAK

1. What did Buffon, Cuvier, and Lyell contribute to our knowledge of life?
2. How do the concepts of gradualism and uniformitarianism in geology undermine the belief that Earth is only about 6000 years old?

17.3 Biological Evolution

17.3a Lamarck

Jean Baptiste de Lamarck (1744–1829) proposed the first comprehensive theory of biological evolution based on specific mechanisms. He proposed that a metaphysical "perfecting principle" caused organisms to become better suited to their environments. In Lamarck's theory, simple organisms evolved into more complex ones, moving up the ladder of life. Microscopic organisms were replaced at the bottom by spontaneous generation (in which living organisms arise from nonliving material, such as dirt or dead organisms). Lamarck theorized that two mechanisms fostered evolutionary change. According to his principle of use and disuse, body parts grow in proportion to how much they are used, as anyone whose exercise regime includes lifting heavy weights well knows. Conversely, unused structures get weaker and shrink, like the muscles of an arm immobilized in a cast. According to his second principle—the inheritance of acquired characteristics—changes that an organism acquires during its lifetime are inherited by its offspring. Thus, Lamarck argued that long-legged wading birds, such as herons, are descended from short-legged ancestors that stretched their legs to stay dry while feeding in shallow water **(Figure 17.3)**. Their offspring inherited slightly longer legs, and after many generations, their legs became extremely long.

Today we know that Lamarck's proposed mechanisms do not cause evolutionary change. Although muscles do grow larger through continued use, structural changes acquired during an organism's lifetime are not inherited by the next generation. Despite the shortcomings of his theory, Lamarck made four important contributions to the development of an evolutionary worldview:

1. He proposed that all species change through time.
2. He recognized that changes are passed from one generation to the next.

Figure 17.3

A great blue heron (*Ardea herodias*). Like many other wading birds, herons have long, stiltlike legs. Lamarck hypothesized that as wading birds stretched their legs while feeding, successive generations of their offspring would have progressively longer legs.

3. He suggested that organisms change in response to their environments.
4. He hypothesized the existence of specific mechanisms that caused evolutionary change.

All four of these ideas became cornerstones of Darwin's later theory of evolution by natural selection. Perhaps Lamarck's most important contribution was that he fostered discussion. By the mid-nineteenth century, most educated Europeans were talking about evolutionary change, whether they believed in it or not.

17.3b Darwin

Documenting variation and describing how selection works were Charles Darwin's central contributions to our knowledge of evolution, putting it in context and explaining its significance. In 1831, Darwin set sail on HMS *Beagle* as a naturalist. The *Beagle* first sailed westward to map the coastline of South America and then on to circumnavigate the globe **(Figure 17.4)**. The *Beagle's* voyage lasted nearly five years. While on the voyage, Darwin read Charles Lyell's *Principles of Geology*. He began to see rock formations through Lyell's eyes and to apply gradualism and uniformitarianism to the living world.

Because of his seasickness, Darwin seized every chance to go ashore. He collected plants and animals in Brazilian rain forests and fossils in Patagonia. He hiked the grasslands of the pampas and climbed the Andes in Chile. He discovered fossils in Argentina that resembled organisms inhabiting the same region today. For example, despite an enormous size difference, living armadillos and fossilized glyptodonts had similar body armour, but they were unlike any other species known to science **(Figure 17.5)**. If both species had been created at the same time and both were found in South America, why didn't glyptodonts still live alongside armadillos? Darwin later wondered whether armadillos might be living descendants of the now-extinct glyptodonts.

On the Galápagos Islands **(Figure 17.6)**, Darwin found "strange and wonderful creatures," including giant tortoises, small finchlike birds, and lizards that dove into the sea to eat algae. Darwin noticed that the

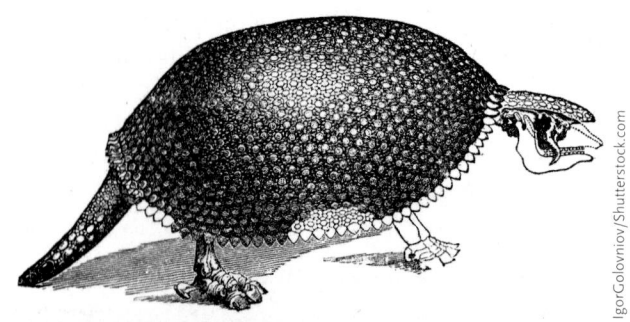

Figure 17.5

Ancestors and descendants. Darwin hypothesized that even though an extinct glyptodont (top) probably weighed 300 to 400 times as much as a living nine-banded armadillo (*Dasypus novemcinctus*), their obvious resemblance suggested that they are related.

animals and plants on different islands varied slightly in form. Indeed, experienced sailors could easily identify a tortoise's island of origin by the shape of its shell. Moreover, many species resembled those on the distant South American mainland. Why did so many different species of organisms occupy one small island cluster? Why did these species resemble others from the nearest continent? Darwin later hypothesized that the plants and animals of the Galápagos were descended from South American ancestors and that the appearance of individuals making up populations changed after being isolated on a particular island.

As he visited the different islands, Darwin collected a diverse group of finches **(Figure 17.7)**. He noticed great variability in the shapes of their bills, but he incorrectly assumed that birds on different islands belonged to the same species. Thus, he did not record the island where he captured each specimen. Luckily, the *Beagle's* captain, Robert Fitzroy, had more thoroughly documented his own collection, allowing Darwin to study the relationships and geographical distributions of a dozen species. As Darwin reviewed the data, he began to focus on two aspects of a general problem. Why were the finches on a particular island slightly different from those on nearby islands, and how did all these different species arise?

17.3c Developing the Theory of Natural Selection

Having grown up in rural England, Darwin was well aware that "like begets like"; that is, offspring resemble

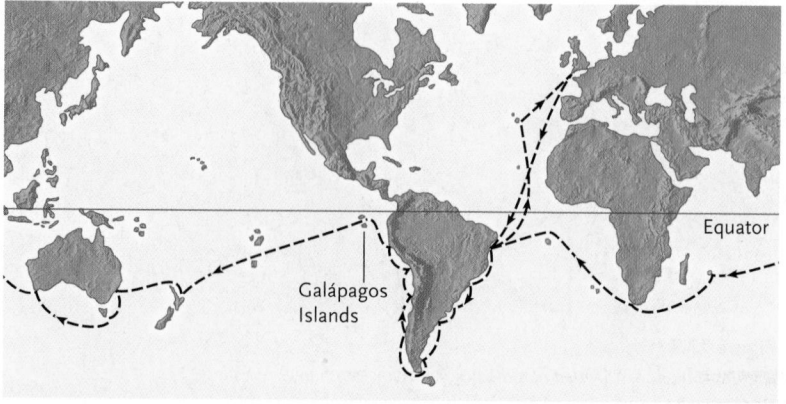

Figure 17.4

Darwin's voyage: map of the path followed by HMS *Beagle*, 1831–1836.

a. The Galápagos

Darwin

Wolf

b. Galápagos tortoise

Gabor Nedecky/Shutterstock.com

c. Marine iguana

Ryan M. Bolton/Shutterstock.com

Pinta

Marchena   Genovesa

Santiago   Equator

Bartolomé

Seymour

Rábida   Baltra

Fernandina   Pinzón

Santa Cruz

Santa Fe

Tortuga   San Cristóbal

Isabela

Española

Floreana

d. Blue-footed booby

Chris Howey/Shutterstock.com

Figure 17.6

The Galápagos Islands. Between 3 and 5 mya, volcanic eruptions created the Galápagos Islands **(a)** about 1000 km west of Ecuador. The islands were named for the giant tortoises **(b)** found there (in Spanish, *galapa* means tortoise). This tortoise (*Geochelone elephantopus*) is native to Isla Santa Cruz. **(c)** Marine iguanas (*Amblyrhynchus cristatus*) dive into the Pacific Ocean to feed on algae. **(d)** A male blue-footed booby (*Sula nebouxii*) engages in **courtship display.**

their parents. Plant and animal breeders had applied this basic truth of inheritance for thousands of years. By selectively breeding individuals with the characteristics they wanted, breeders enhanced those traits in future generations.

For example, the English people in the 1800s loved their dogs. Earlier, most dogs were working dogs, bred for hunting, rat catching, sheep herding, and guarding. In the nineteenth century, however, people wanted more decorative dogs and lap dogs. For example, dog

a. Warbler finch
(Certhidea olivacea)

© Images & Stories/Alamy

b. Common cactus-finch
(Geospiza scandens)

© Krystyna Szulecka/Alamy

c. Large ground-finch
(Geospiza magnirostris)

Stubblefield Photography/Shutterstock.com

d. Woodpecker finch
(Camarhynchus pallidus)

Tierbild Okapia/Science Source

Figure 17.7

Bill shape and food habits. The 13 finch species that inhabit the Galápagos Islands are descended from a common ancestor, a seed-eating ground finch that migrated to the islands from South America. **(a)** The warbler finch uses its slender bill to probe for insects in vegetation. **(b)** The common cactus-finch has a medium-sized bill suitable for eating cactus flowers and fruit. **(c)** The large ground-finch uses its thick, strong bill to crush cactus seeds. **(d)** The woodpecker finch hammers at tree bark with its bill; then it uses cactus spines, held in its bill, to probe for wood-boring insects such as termites.

breeders crossed Clydesdale terriers and Skye terriers with each other and with other small terrier breeds. These crosses resulted in the tiny (2–3 kg) Yorkshire terrier, a very popular breed even today.

Although the mechanism of heredity was not yet understood, selective breeding was applied countless times to produce bigger beets, plumper pigs, and prize-winning pigeons. Darwin, who bred pigeons himself, called this process **artificial selection**, since humans were selecting the characteristics they wanted in the offspring by choosing parents with those traits. Darwin could see that selection could operate in nature, but he puzzled about how it worked. He reasoned that if a person could select different characteristics when breeding organisms, then nature could do so as well. This he called **natural selection**, and he defined it as the "principle by which each slight variation [of a trait], if useful, is preserved." In other words, natural selection is the process by which characteristics that better enable organisms to adapt to specific environmental pressures will tend to increase in succeeding generations in a population. Organisms with those characteristics are better able to survive and can reproduce in greater numbers than those without the characteristics.

During the 1840s and 1850s, Darwin led a reclusive life, accumulating evidence of evolutionary change and trying to identify the mechanisms that caused it. He read Thomas Malthus's famous *Essay on the Principles of Population*. Malthus, an English clergyman and economist, observed that England's population was growing much faster than the country's agricultural capacity. This situation meant that individuals competed for food and some would inevitably starve. Darwin applied Malthus's argument to organisms in nature. He observed that many species typically produce many more offspring than are needed to replace the parent generation, yet the world is not overrun by any one species, be it sunflowers, earthworms, tortoises, or bears. Darwin calculated that if its reproduction went unchecked, a single pair of elephants (the slowest-breeding animal known) would leave roughly 19 million descendants after only 750 years. Instead, some members of every population survive and reproduce, whereas many others die without reproducing.

Darwin made several major observations (Figure 17.8):

- Individuals within populations vary in size, form, colour, behaviour, and other characteristics.
- Many of these variations are passed on from parent to offspring.
- Some of the inherited variations enable some individuals to survive and reproduce better than others. A modern example is resistance to organophosphorus insecticides, where resistant flies leave many young, and nonresistant flies leave few, if any, descendants (see "Why It Matters"). In this way, favourable hereditary traits become more common in the next generation.
- If the next generation is subjected to the same process of selection, these favourable traits will become even more common.

Because this process is analogous to artificial selection, Darwin called this mechanism natural selection, which means that individuals with certain inherited traits leave more offspring than do individuals without those traits. As an evolutionary mechanism, natural selection favours **adaptive traits**, hereditary characteristics that make organisms more likely to survive and reproduce under a given set of environmental conditions. And by favouring individuals that are well adapted to the environments in which they live, natural selection can cause populations to change through time. For example, each species of Galápagos finch (Figure 17.7) has a distinctive bill. Variations in bill size and shape make some birds better adapted for crushing seeds and others for capturing insects. Imagine an island where large seeds were the only food available; individuals

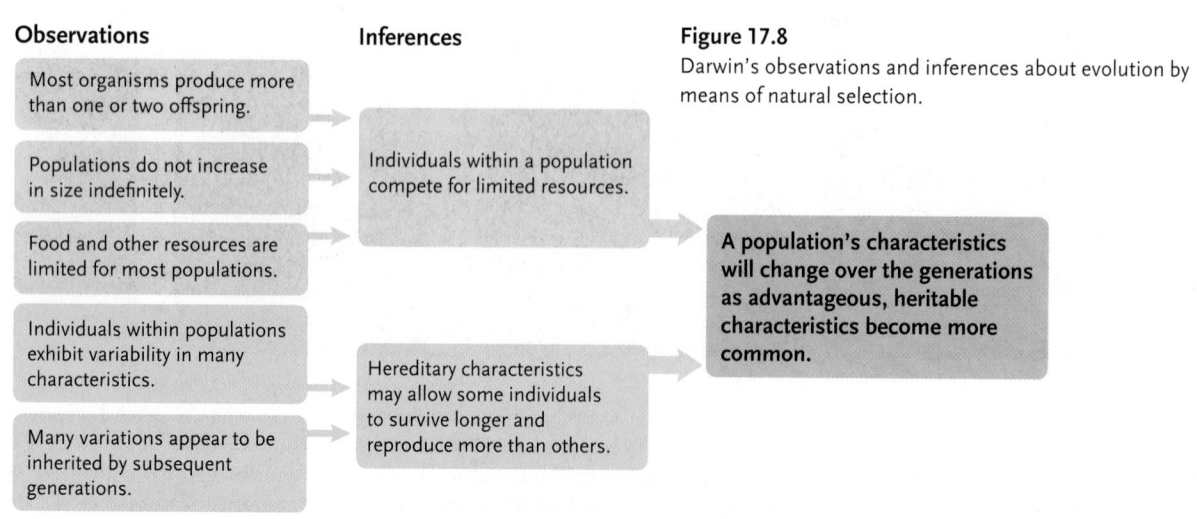

Figure 17.8
Darwin's observations and inferences about evolution by means of natural selection.

Peter and Rosemary Grant

Peter and Rosemary Grant are evolutionary biologists who work at Princeton University. They have been responsible for ground-breaking work on cactus finches (*Geospiza conirostris*), a species of Darwin's finches from Genovesa Island in the Galapagos **(Figure 1).** Of particular note are their long-term studies that have advanced our understanding of variation and how it can change over time.

The Grants analyzed a likely case of disruptive selection on the size and shape of the bills of cactus finches. During normal weather cycles, the finches eat ripe cactus fruits, seeds, and exposed insects. During drought years food is scarce and the birds also search for insects by stripping bark from the branches of bushes and trees. About 70% of cactus finches on Genovesa died during the long drought of 1977. Survivors exhibited high variability in their bills. Birds that stripped bark from branches to look for insects had particularly deep bills.

Those that opened cactus fruits to expose the fleshy interior had especially long bills. Thus, birds with extreme bill phenotypes appeared to feed efficiently on specific resources. This established disruptive selection on the size and shape of bills because the selective advantage of large bills did not apply under some environmental conditions. Intermediate bill morphologies may be favoured during nondrought years when insects and small seeds are abundant.

Geospiza conirostris

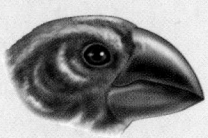

Birds with long bills open cactus fruits to feed on the fleshy pulp.

Birds with intermediate bills may be favoured during nondrought years when many types of food are available.

Birds with deep bills strip bark from trees to locate insects.

Figure 1

Disruptive selection as illustrated by cactus finches, in which these birds show variability in the sizes and shapes of their bills.

---

with a stout bill would be more likely to survive and reproduce than would birds with slender bills. These favoured individuals would pass the genes that produce stout bills to their descendants, while those with slender bills would not reproduce because of starvation. After many generations, the bills might resemble those of *Geospiza magnirostris* (see Figure 17.7c). Natural selection also changes nonmorphologic characteristics of populations; for example, insect populations that are exposed to insecticides develop resistance to these toxic chemicals over time (see "Why It Matters").

Darwin realized that natural selection could also account for striking differences between populations and, given enough time, for the production of new species. For example, suppose that small insects were the only food available to finches on a particular island. Birds with long, thin bills might be favoured by natural selection, and the population of finches might eventually possess a bill shaped like that of *Certhidea olivacea* (see Figure 17.7a). Considering that many characteristics affect survival and reproduction, natural selection would cause the populations to become more different over time, a process called **evolutionary divergence.**

17.3d Wallace

Alfred Russel Wallace was a contemporary of Darwin's who studied organisms in the Amazonian rain forest and in the East Indies. Like Darwin, Wallace was a keen observer of nature who kept careful notes and drawings of his observations. He travelled extensively in the Amazon rain forest and the Malay Archipelago (modern-day Indonesia), collecting specimens and describing the geology of the areas. Wallace's work provides one of the more interesting parts of the story of Darwin's development and publication of his theory.

On 18 June 1858, Darwin received a letter from Wallace in which he outlined his ideas about how populations of organisms change over time. Wallace's work and ideas mirrored Darwin's own research. To his credit, Darwin forwarded Wallace's manuscript to Charles Lyell, who had been encouraging Darwin to publish his theory. On 1 July 1858, papers by Darwin and Wallace were presented to the Linnaean Society of London. On 24 November 1859, Darwin's book, *On the Origin of Species by Means of Natural Selection,* was published. As evidenced by the conclusions of two different people, evolution via natural selection clearly was an idea whose time had come.

17.3e Impact of the Theory of Evolution by Natural Selection

It would be hard to overestimate the impact on western thought of Darwin's and Wallace's theory. In *The Origin*, Darwin proposed natural selection as the mechanism that drives evolutionary change. In fact, most of *The Origin* was an explanation of how natural selection acted on the variability within groups of organisms, preserving advantageous traits and eliminating disadvantageous ones.

Darwin argued that all the organisms that have ever lived arose through **descent with modification**, the evolutionary alteration and diversification of ancestral species. He envisioned this pattern of descent as a tree growing through time **(Figure 17.9)**. The base of the trunk represents the ancestor of all organisms. Branching points above it represent the evolutionary divergence of ancestors into their descendants. Each limb represents a body plan suitable for a particular way of life, smaller branches represent more narrowly defined groups of organisms, and the uppermost twigs represent living species. Biologists still apply this analogy today when studying phylogenies (see page 455, Chapter 20).

Four characteristics distinguish Darwin's theory from earlier explanations of biological diversity and adaptive traits:

1. Darwin provided purely physical, rather than spiritual, explanations for the origins of biological diversity.
2. Darwin recognized that evolutionary change occurs in groups of organisms rather than in individuals: some members of a group survive and reproduce more successfully than others.
3. Darwin described evolution as a multistage process: variations arise within groups, natural selection eliminates unsuccessful variations, and the next generation inherits successful variations.
4. Like Lamarck, Darwin understood that evolution occurs because some organisms function better than others *in a particular environment*.

Evolution was a popular topic in Victorian England, and Darwin's theory was both praised and ridiculed. Nevertheless, Darwin's painstaking logic and careful documentation convinced many readers that evolution really does take place. The major stumbling block for some readers was that Darwin had no clear idea of how a variant arose or how it was passed from one generation to the next.

Although Darwin had not speculated about the evolution of humans in *Origin of Species*, he did in another book, *The Descent of Man, and Selection in Relation to Sex*, published in 1871. Needless to say, certain influential Victorians were not amused by the suggestion that humans and apes shared a common ancestry.

STUDY BREAK

1. What observations that Darwin made on his round-the-world voyage influenced his later thoughts about evolution?
2. Describe an example of artificial selection not included in the text.
3. How did Darwin's understanding of artificial selection enable him to envision the process of natural selection?
4. What is natural selection?

17.4 Evolutionary Biology since Darwin

One of the remarkable features of Darwin's and Wallace's work is that they developed the foundational concept of natural selection without any understanding of a mechanism for how traits were inherited. This was because neither of them understood genetics, a scientific field that was also in its infancy. At about the same time as Darwin's book was published, Gregor Mendel published his work on inheritance in pea plants. However, this study was not well known in England until 1900. At that time, scientists thought that Darwin's and Mendel's theories conflicted. One problem was that Darwin had used complex characteristics, such as the structure of bird bills, to illustrate how natural selection worked. We now know that several genes often control such traits (see page 290, Chapter 13 and page 314, Chapter 14). By contrast, Mendel had studied simpler characteristics, such as the height of pea plants. A single gene often controls simple traits, which is one reason Mendel could interpret his experimental results so clearly. Biologists initially had a hard time applying Mendel's

Present

Time

Origin of life

Figure 17.9

The Tree of Life. Darwin envisioned the history of life as a tree. Branching points represent the origins of new lineages; branches that do not reach the top represent extinct groups.

straightforward experimental results to Darwin's complex examples. Today, however, it is accepted throughout the scientific community that phenotypic variation among organisms reflects genetic differences.

17.4a The Modern Synthesis

In the early twentieth century, Thomas Hunt Morgan of Columbia University determined that genes are carried on chromosomes. His experiments enabled geneticists and mathematicians to forge a critical link between Darwin's and Mendel's ideas (see pages 410 and 411, Chapter 18). The new discipline, **population genetics**, recognized the importance of genetic variation as the raw material of evolution. Population geneticists constructed mathematical models, which applied equally well to simple and complex traits, to predict how natural selection and other processes influence a population's genetics.

In the 1930s and 1940s, a unified theory of evolution, called the **modern synthesis**, integrated data from biogeography, comparative morphology, comparative embryology, genetics, paleontology, and taxonomy within an evolutionary framework. The authors of the modern synthesis focused on evolutionary change within populations. Although they considered natural selection the primary mechanism of evolution, they acknowledged the importance of other processes (such as genetic drift; see Chapter 18). Proponents of the modern synthesis also embraced Darwin's idea of gradual change and de-emphasized the significance of mutations that changed traits suddenly and dramatically.

The modern synthesis also tried to link the two levels of evolutionary change that Darwin had identified: microevolution and macroevolution. **Microevolution** describes the small-scale genetic changes that populations undergo, often in response to shifting environmental circumstances; a small evolutionary shift in the size of the bill of a Galapagos finch is an example of microevolution. **Macroevolution** describes larger-scale evolutionary changes observed in species. According to the modern synthesis, macroevolution results from the gradual accumulation of microevolutionary changes. Researchers have recently begun to unravel the genetic mechanisms that establish a relationship between these two levels of evolutionary change (see page 456, Chapter 20). Research since the discovery of the structure of DNA has led to a more thorough understanding of mutations. It also led to the science of molecular genetics.

Nowadays, biologists understand that biological evolution involves the combination of heritable changes in individuals plus selective pressures, the environmental pressures felt by organisms. Selective pressures operate on the phenotype of organisms, improving or reducing the success of individuals within a population that have a certain inherited trait. This trait can make these individuals more or less adapted to their environments and consequently more or less likely to produce viable offspring. As we have seen, these heritable traits can be expressed anywhere from the molecular level to the whole organism. Heritable changes, combined with selective pressures, can produce new types (species) of organisms from existing ancestors.

17.4b Further Evolutionary Research

Since the emergence of the modern synthesis, scientists have assembled a huge body of evidence from many biological disciplines that indicate that biological evolution is a fact of life on Earth.

Adaptation by Natural Selection. Biologists interpret the products of natural selection as evolutionary adaptations. For example, the wings of birds, which have been modified by evolutionary processes over millions of years, have an obvious function that helps these animals survive and reproduce. Throughout this book, you will encounter many examples of adaptive structures in plants and animals that have been modelled by natural selection. Sometimes, however, natural selection operates on a short time scale, as illustrated by the development of pesticide resistance in insects (see "Why It Matters").

The Fossil Record. Because evolution results from the modification of existing species, Darwin's theory proposes that all species that have ever lived are genetically related. The fossil record documents such continuity in morphological characteristics, providing clear evidence of ongoing change in **biological lineages**, evolutionary sequences of ancestral organisms and their descendants (see Chapter 20). For example, the evolution of modern birds can be traced from a dinosaur ancestor through fossils such as *Archaeopteryx lithographica* (**Figure 17.10, p. 402**). This species, discovered only two years after *The Origin* was published, resembled both dinosaurs and birds. Like the small carnivorous dinosaur *Dromaeosaurus*, *Archaeopteryx* walked on its hind legs and had teeth; long fingers with claws on its forelimbs; and a long, bony tail. Like modern birds, it had enlarged flight feathers on its forelimbs. Recently discovered fossils reveal that many bird ancestors were feathered.

Historical Biogeography. The study of the geographical distributions of plants and animals in relation to their evolutionary history is generally consistent with Darwin's theory of evolution. Species on oceanic islands often closely resemble species on the nearest mainland, suggesting that the island and mainland species have a common ancestry. Moreover, species on a continental land mass are clearly related to one another and are often distinct from those on other continents. For example, monkeys in South America have long, prehensile tails and broad noses, traits that

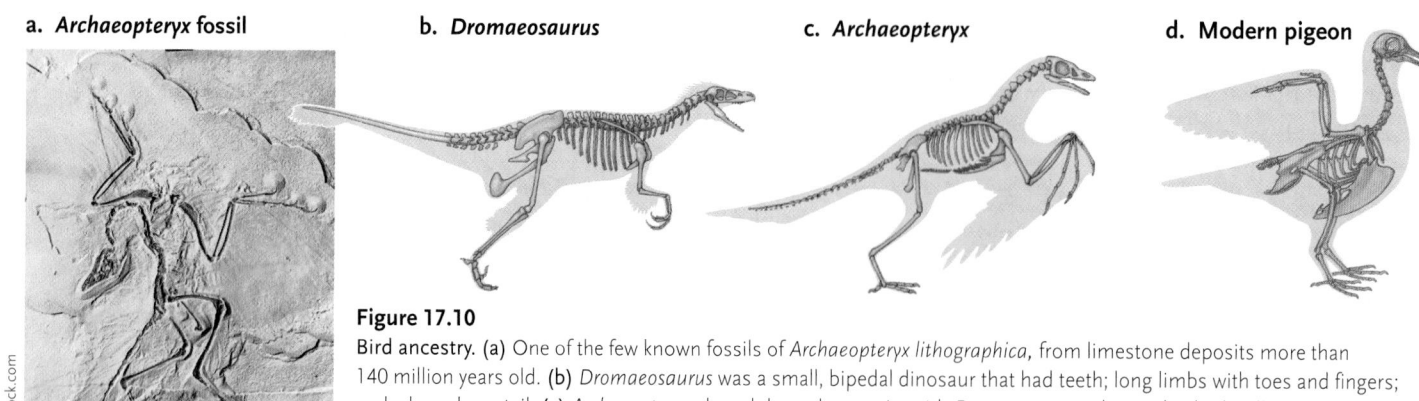

a. *Archaeopteryx* fossil   **b. *Dromaeosaurus***   **c. *Archaeopteryx***   **d. Modern pigeon**

Figure 17.10

Bird ancestry. **(a)** One of the few known fossils of *Archaeopteryx lithographica*, from limestone deposits more than 140 million years old. **(b)** *Dromaeosaurus* was a small, bipedal dinosaur that had teeth; long limbs with toes and fingers; and a long, bony tail. **(c)** *Archaeopteryx* shared these three traits with *Dromaeosaurus*, but it also had well-developed flight feathers in its forelimbs, a characteristic that it shares with modern birds. **(d)** Modern birds, such as the pigeon, have long limbs similar to those of *Dromaeosaurus* and *Archaeopteryx*, but their fingers and bony tails are greatly reduced, and a horny bill has replaced the teeth in the mouths of their ancestors.

they inherited from a shared South American ancestor. By contrast, monkeys in Africa and Asia evolved from a different common ancestor in the Old World, and their shorter tails and narrower noses distinguish them from their American cousins.

Comparative Morphology. Analyses of the structure of living and extinct organisms are based on the comparison of **homologous traits**, characteristics that are similar in two species because they inherited the genetic basis of the trait from their common ancestor. For example, the forelimbs of all four-legged vertebrates are homologous because they evolved from a common ancestor with a forelimb composed of the same component parts (see Figure 17.2, p. 393). Even though the shapes of the bones are different in pigs, dolphins, and bats, similarities in the three limbs are apparent. The differences in structural details arose over evolutionary time, allowing pigs to walk, dolphins to swim, and bats to fly. The arms of humans and the wings of birds are constructed of comparable elements, suggesting that they, too, share a common ancestor with the three species illustrated.

17.4c Molecular Techniques

Molecular techniques provide biologists with powerful tools for exploring all aspects of life—and evolutionary biology is no exception. From Darwin's time until the mid-twentieth century, biologists tried to discern the evolutionary history of animals by comparing their embryos and patterns of development **(Figure 17.11)**. The early embryos of related species are often strikingly similar, but morphological differences appear as the embryos grow and develop their adult forms (see Chapter 42). For example, the early embryos of most four-legged vertebrates (such as lizards, mammals, and birds) develop "limb buds" from which the legs or wings grow. Forelimbs and their supporting structures grow at the base of the neck, just in front of the ribcage, and hindlimbs grow right behind the ribcage (Figure 17.11a). Similarities in the limb buds and the positions of the limbs in

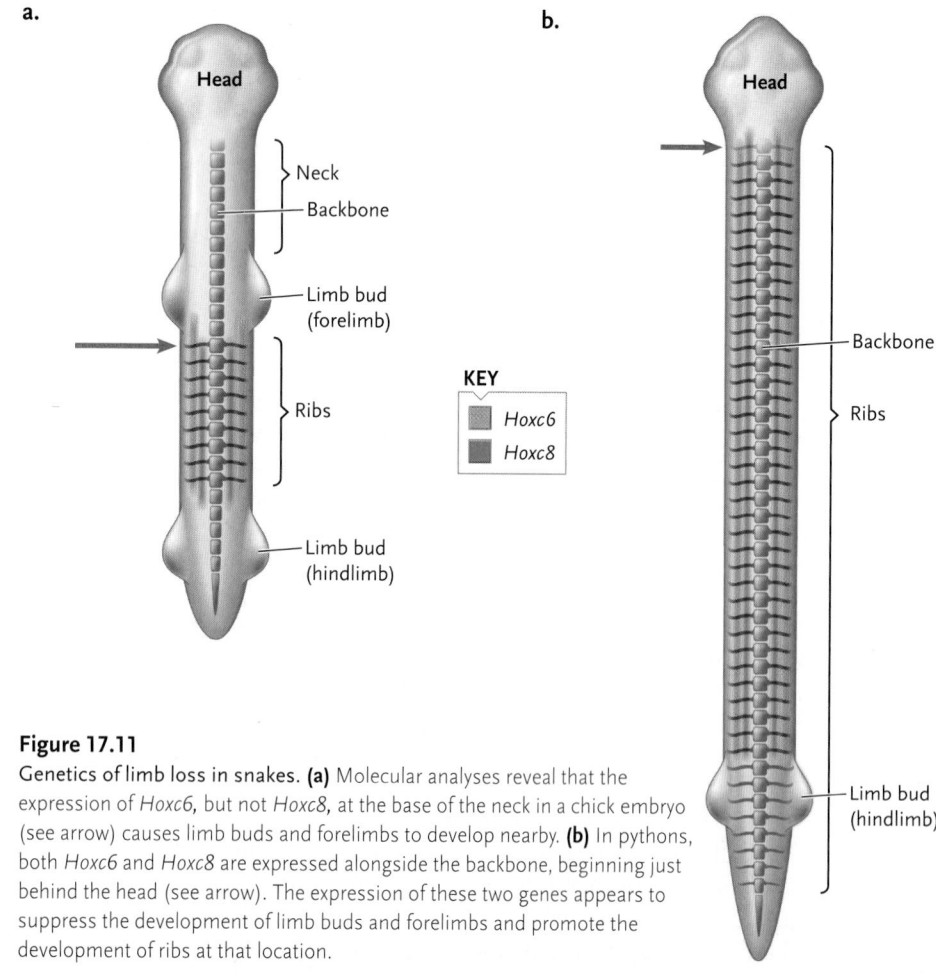

a.

Head

Neck

Backbone

Limb bud (forelimb)

Ribs

KEY

Hoxc6
Hoxc8

b.

Head

Backbone

Ribs

Limb bud (hindlimb)

Limb bud (hindlimb)

Figure 17.11

Genetics of limb loss in snakes. **(a)** Molecular analyses reveal that the expression of *Hoxc6*, but not *Hoxc8*, at the base of the neck in a chick embryo (see arrow) causes limb buds and forelimbs to develop nearby. **(b)** In pythons, both *Hoxc6* and *Hoxc8* are expressed alongside the backbone, beginning just behind the head (see arrow). The expression of these two genes appears to suppress the development of limb buds and forelimbs and promote the development of ribs at that location.

these animals provide evidence of their descent from a shared ancestor. Differences in their adult structures are caused by additional genetic instructions that have evolved over time.

However, most snakes show no traces of limbs or necks; their ribcages are positioned right behind their heads (Figure 17.11b). The fossil record shows us that snakes evolved from four-legged ancestors in stages: early snakes had small hindlimbs, and the most recently evolved snakes have no limbs at all. Only the most ancient living snakes, pythons and boas, have any traces of limbs—vestigial hindlimbs, which appear as a pair of tiny claw-like structures near the base of the tail. Observational studies of their embryos reveal that most living snakes never develop limb buds; by contrast, pythons and boas develop hind limb buds, which grow only slightly as the animal develops.

Two regulatory genes, *Hoxc6* and *Hoxc8*, determine whether forelimbs or ribs grow at a particular site along an animal's backbone. The *Hox* genes either activate or suppress other genes that direct the development of these structures. Forelimbs—but not ribs—grow just in front of the tissues where only *Hoxc6* is expressed. By contrast, ribs—but not forelimbs—grow where both *Hoxc6* and *Hoxc8* are expressed (Figure 17.11a).

Martin J. Cohn of the University of Reading and Cheryll Tickle of University College London reported that in pythons (primitive snakes), both *Hoxc6* and *Hoxc8* are expressed all along the backbone, beginning at the base of the skull; as a result, a python's ribcage develops right behind its head, and no limb buds or forelimbs develop (Figure 17.11b). Thus, snakes have no forelimbs or necks because a mutation causes the expression of *Hoxc8* to extend into a more forward region of the animal's body. All descendants of the ancestor with that original mutation now lack necks and forelimbs. Cohn and Tickle's research also suggests that the second stage in snake evolution, the reduction or complete absence of hindlimbs, is caused by other genetic variations that appeared some time after the altered expression pattern of *Hoxc8*. Thus, molecular research has identified the genetic changes that caused snakes to lose their forelimbs and necks before losing their hindlimbs.

STUDY BREAK

1. What types of data provide evidence that evolution has adapted organisms to their environments and promoted the diversification of species?
2. How have molecular techniques enhanced the study of evolutionary biology? Give an example not included in the textbook.

17.5 Variation and Selection

As Darwin observed, individuals exhibit variation in many characteristics. Natural selection acts at that individual level. The inherited characteristics of an individual may allow it to survive longer and, therefore, reproduce more than others without that characteristic (Figure 17.8). These individuals have enhanced fitness: they are better adapted to their environment. The particular traits that make these individuals more fit are considered adaptive: they help the individual survive. Not all traits are adaptive; some have no bearing on whether an individual survives or not. We will examine some examples that illustrate the relationships between variation, selection, and evolution.

17.5a Plant Poisons

Convergent evolution is the independent evolution of similar traits in unrelated species, such as the wings of insects, birds, and bats. The convergent evolution of solutions to threats to survival is a recurring theme among organisms. For example, most insects cannot ingest cardenolides (also known as cardiac glycosides), which are produced by plants such as milkweed (*Asclepias* spp.) and dogbane (*Apocyanum* spp.) to protect them from herbivorous insects. Cardenolides are toxic because they block an essential transmembrane carrier, the sodium–potassium pump, Na–K ATPase (see page 110, Chapter 5). Insects from several distinct evolutionary lineages **(Figure 17.12, p. 405)** have independently acquired the ability to ingest cardenolides, giving them access to more food resources. In monarch butterflies (*Danaus plexippus*), a simple mutation, producing a substitution of asparagine for histidine on the Na–K ATPase protein, confers protection from cardenolides. At least 17 other species of insects (representing 15 genera and 4 orders) have variations in the gene sequences that encode for the Na–K ATPase protein and also eat plants with cardenolides; these variations have evolved independently in these individual species. This situation demonstrates convergence across 300 million years of insect evolution and diversification. The convergence is one molecular solution permitting consumption of cardenolides.

Two advantages are associated with the ability to ingest cardiac glycosides. First, this ability provides herbivorous insects access to a greater variety and amount of food. Second, many insects that ingest cardiac glycosides and other toxic plant products use these molecules in their own defence. The monarch butterfly is an excellent example of protection conferred by ingested cardenolides.

17.5b Venom

Many species of animals produce venom that they use in defence (to deter or distract predators) or offence (to immobilize and digest prey). Venom that produces

The Woolly Mammoth's Closest Living Relative

Based on morphological evidence, paleobiologists suspected that woolly mammoths (*Mammuthus primigenius*) were more closely related to living Asian elephants (*Elephas maximus*) than to living African elephants (*Loxodonta africana*). When they recovered samples of mammoth DNA, researchers knew they could settle the matter.

In 2006, Hendrick Poinar of McMaster University and colleagues sequenced 13 million base pairs of mitochondrial and nuclear DNA extracted from the jawbone of a woolly mammoth (*Mammuthus primigenius*), a species that has been extinct for at least 4000 years. The mammoth, which died 27 000 years ago, had been preserved in a Siberian ice cave. When the researchers compared its DNA sequences to those from a living African elephant (*Loxodonta africana*), they discovered that more than 98% of the sequence was identical in the two species, confirming their close evolutionary relationship.

Svante Pääbo and his colleagues at the Max Planck Institute for Evolutionary Anthropology, Leipzig, Germany, and researchers at institutions in England, Germany, and the United States, used molecular techniques—specifically, PCR amplification, cloning, and sequencing of mitochondrial DNA (see Chapter 15)—to analyze the complete mitochondrial genome sequence of a woolly mammoth, and compared it to the sequences from two living elephant species.

Based on known mitochondrial DNA sequences from the African and Asian elephants, they designed and synthesized 46 pairs of PCR primers **(Figure 1).** Assuming homology with the elephant sequences, these primer pairs were predicted to amplify the entire circular mitochondrial genome of the mammoth (yellow circle) as overlapping DNA fragments. All of the primer pairs were used in a single PCR, and the resulting amplified DNA fragments were cloned and sequenced. Based on their overlaps, the sequences could be arranged in a circle (blue and red), which showed that the researchers had succeeded in amplifying the entire mitochondrial genome.

The mammoth DNA sequences were more similar to those from the Asian elephant than those from the African elephant. Thus, Asian elephants are the woolly mammoth's closest living relatives. This conclusion also makes biogeographic sense because woolly mammoths and Asian elephants occupied the same land mass (Asia), whereas African elephants live on a different continent.

Source: J. Krause et al. 2006. Multiplex amplification of the mammoth mitochondrial genome and the evolution of the Elephantidae. *Nature* 439:724–727.

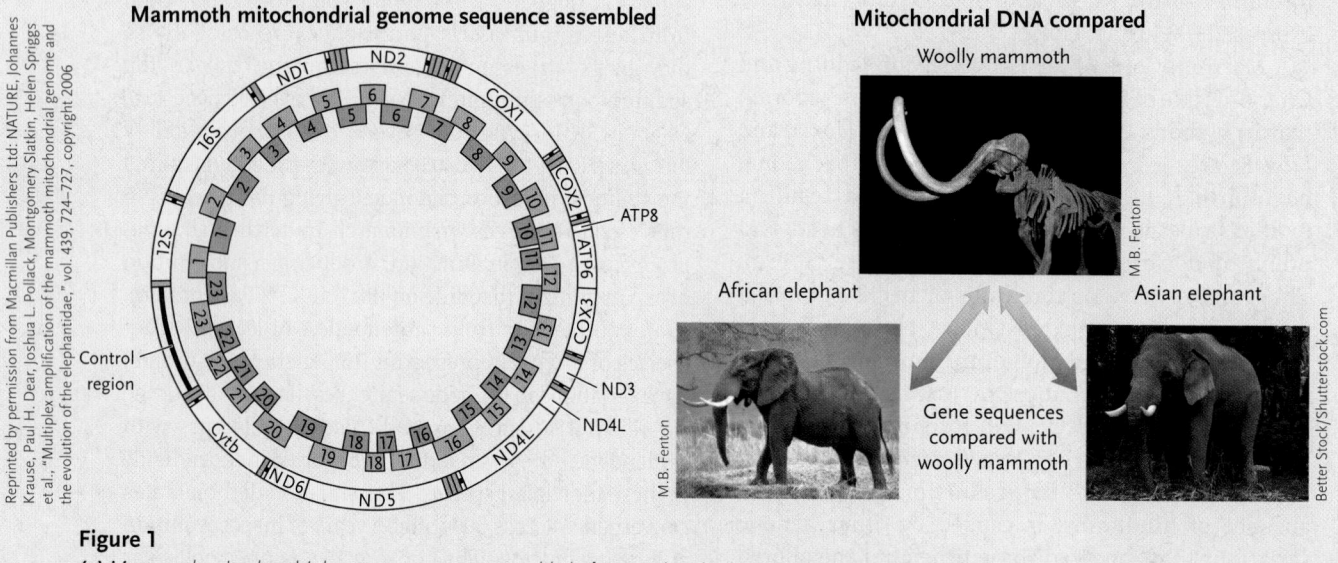

Figure 1

(a) Mammoth mitochondrial genome sequence assembled. (b) Mitochondrial DNA sequences compared.

intense, incapacitating pain is used in defence and delivered by stinging or biting. Bark scorpions (**Figure 17.13**, *Centuroides sculpturatus*) produce defensive venom that deters attacks by mammals such as house mice, rats, and humans. The scorpion's venom works by activating the Nav1.7 voltage-gated Na⁺ channels on pain receptors connected to the central nervous system. Grasshopper mice (*Onychomys torridus*) regularly catch and eat bark scorpions but show little response to their stings. In these predatory rodents, Nav1.8 acts as an analgesic by inhibiting the movement of Na⁺ ions.

In the natural world, bark scorpions have many predators, not just grasshopper mice, and these mice eat many other species of arthropods. Defensive venom therefore provides an advantage to bark scorpions in many, but not all, defensive situations. But in this example, the resistance of grasshopper mice to the venom provides them access to otherwise inaccessible

Figure 17.12
A monarch butterfly and the milkweed that its caterpillars eat. Cardenolides are found in the white latex sap. In this picture a drop of sap sits at a break in the leaf.

food, increasing their fitness and, again, changing the incidence of resistance in the population.

17.5c Sexual Selection

> With animals having separated sexes there will in most cases be a struggle between the males for possession of the females. The most vigorous individuals, or those which have most successfully struggled with their conditions of life, will generally leave most progeny. But success will often depend on having special weapons or means of defence, or on the charms of the males; and the slightest advantage will lead to victory.

So Darwin wrote in *On the Origin of Species*. In animals that are sexually dimorphic (males look different from females), heritable traits involved in competition between males or in attracting females can be subject

Figure 17.13
Bark scorpion and grasshopper mouse.

to very strong selection. **Sexual selection**, which is a type of natural selection, involves variation in reproductive success that is influenced by sexually dimorphic characters. Sexual selection can involve phenotypic variations in morphology; in behaviour; and even in **pheromones**, chemicals used to attract mates.

Variation in the morphology of males can influence the competition for females. In guppies (*Poecilia reticulata*), males with clawlike structures on their intromittent organ (penis equivalent) transfer more sperm during copulation than males lacking the "claws." In water striders (*Rheumatobates rileyi*), some males have specialized structures on their antennae that allow them to grasp and subdue resistant females. The gene family *distal-less* (DLX) controls these specialized structures, connecting morphology and function and associated advantages to underlying genetics and inheritance.

In some animals, exaggeration of morphological features that distinguish males from females are proxies for male quality, and female choice contributes to sexual selection. In stalked-eyed flies **(Figure 17.14a, p. 406)**, males with widely separated eyes are more attractive to females than males with shorter eye stalks. Male hammerheaded bats (Figure 17.14b) call to attract females to mating sites. The inflated skulls and voice boxes of males are associated with their ability to attract females. The selective pressure on these characteristics has led to their gradual enlargement or enhancement.

In other animals, sexual selection is driven by competition among males rather than female choice. Males of many species threaten or fight for dominance, such as the Siamese fighting fish (*Betta splendens*), octopus, and crabs. Using their long horns, male impala fight for access to females (Figure 17.14c). Larger, older males have larger horns and mate more often than younger, smaller males.

The act of mating, by itself, does not ensure reproduction. Furthermore, producing more offspring does not really "count" as success (fitness) until they have also reproduced. Also underlying sexual selection is the reality that females and males do not incur the same costs and benefits during reproduction (see also page 1012,

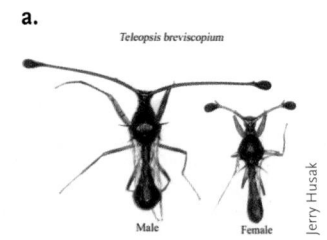

a.

Teleopsis breviscopium

Male   Female

Jerry Husak

b.

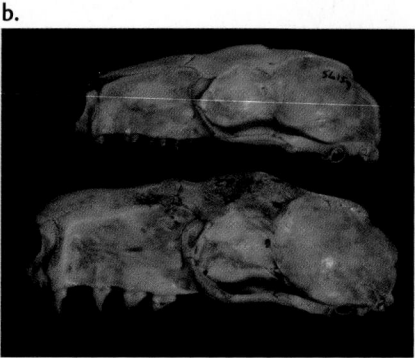

M.B. Fenton

c.

Courtesy of Naas Rautenbach

Figure 17.14

A sampling of sexual dimorphism in animals. **(a)** In stalked-eyed flies, the male's eye stalks are much longer than those of the female. **(b)** The skulls of male hammerheaded bats (*Hypsignathus monstrosus*) are much larger than female skulls (top). **(c)** Male impala (*Aepyceros melampus*) have horns but females do not.

Chapter 42). A comparison of mammals and birds illustrates the point. Birds lay eggs and both males and females may feed the young. However, female mammals alone bear the costs of pregnancy and lactation. If females mate with a strong, vigorous male, their offspring may be more adapted and better able to survive.

The tools provided by modern genetics allow biologists much more precision when it comes to measuring success in reproduction. For instance, people studying birds can now observe the behaviour of individual males and females, including copulations, and then use genetic techniques to determine which males sired which offspring. This situation shows direct connections between behaviour and reproductive output of males and of females.

STUDY BREAK

1. What is resistance? Why is it important? (See "Why It Matters.")
2. What is sexual selection? How does it relate to natural selection?
3. How do interactions between bark scorpions and grasshopper mice illustrate selection?
4. Why is heritability important in the definition of selection?

Putting Evolution in Perspective

Evolution has been the unifying theory in biology, drawing together virtually every aspect of the discipline. The scope of evolution has been introduced in this chapter, with a focus on variation, selection, selective advantages, artificial selection, and natural selection. Variation at the molecular and organismal level has been discussed in the context of selective advantages. We have also considered the contributions of Darwin and Wallace, who moved biological evolution from a vague concept to one with a clear explanation of how evolution could work. This chapter sets the stage for the balance of the unit on evolution.

Review

aplia

To access course materials such as Aplia and other companion resources, please visit www.NELSONbrain.com.

17.1 Evolution: What Is It?

- Biological evolution refers to gradual change of populations or organisms over time; measuring time in generations rather than years.
- Organisms continue to evolve today.
- Not all gradual change is biological evolution.

17.2 Pre-Darwin Knowledge of the Natural World

- Well before Darwin published his theory of evolution by natural selection, changes in scientific thought and methods paved the way for its appearance. Fields such as geology contributed through documentation of the fossil record and the discovery that Earth was very, very old.
- Remember the contributions of people such as Aristotle, Bacon, Buffon, Cuvier, Ussher, Hutton, and Lyell.

17.3 Biological Evolution

- Lamarck's ideas became cornerstones of Darwin's later theory of evolution by natural selection.
- His voyage on HMS Beagle introduced Charles Darwin to many variations in the natural world.
- Darwin understood the process of artificial selection and reasoned that if a person could select different characteristics when breeding organisms, then nature could do so as well. He defined natural selection as the "principle by which each slight variation [of a trait], if useful, is preserved."
- Remember the contributions of people such as Lamarck, Malthus, Darwin, and Wallace.

17.4 Evolutionary Biology since Darwin

- Scientists working in population genetics developed theories of evolutionary change by integrating Darwin's ideas with Mendel's research on genetics.
- A unified theory of evolution, called the modern synthesis, integrates data from biogeography, comparative morphology, comparative embryology, genetics, paleontology, and taxonomy within an evolutionary framework.
- Phenotypic variation among organisms reflects genetic differences.
- Microevolution describes the small-scale genetic changes that populations undergo, often in response to shifting environmental circumstances.

- Macroevolution describes larger-scale evolutionary changes observed in species.
- According to the modern synthesis, macroevolution results from the gradual accumulation of microevolutionary changes.
- Studies of adaptation, the fossil record, historical biogeography, and comparative morphology provide compelling evidence of evolutionary change.
- Molecular techniques have extended the achievements of the modern synthesis, allowing precise analysis of the genetic basis of evolutionary change and the genetic relatedness of living and extinct organisms.

17.5 Variation and Selection

- Darwin observed that individuals exhibit variation in many characteristics.
- Natural selection acts at that individual level.
- The inherited characteristics of an individual may allow it to survive longer and, therefore, reproduce more than others without that characteristic. These individuals have enhanced fitness: they are better adapted to their environment.
- The particular traits that make these individuals more fit are considered adaptive: they help the individual survive.
- Not all traits are adaptive; some have no bearing on whether an individual survives or not.

Questions

Self-Test Questions

1. Which of the following statements about evolutionary studies is incorrect?
   a. Biologists study the products of evolution to understand processes causing it.
   b. Biologists design molecular experiments to examine evolutionary processes operating over short time periods.
   c. Biologists study variation in homologous structures among related organisms.
   d. Biologists examine why a huge variety of species may inhabit a small island.

2. Natural selection acts on
   a. species
   b. genera
   c. individuals
   d. subspecies

3. The belief that evolution is progressive or goal oriented is called:
   a. gradualism
   b. uniformitarianism
   c. taxonomy
   d. orthogenesis
   e. the modern synthesis

4. The wings of birds, the forelegs of pigs, and the flippers of whales are examples of
   a. vestigial structures
   b. homologous structures
   c. acquired characteristics
   d. artificial selection

5. Which of the following statements is NOT compatible with Darwin's theory?
   a. Evolution has altered and diversified ancestral species.
   b. Evolution occurs in individuals rather than in groups.
   c. Natural selection eliminates unsuccessful variations.
   d. Evolution occurs because some individuals function better than others in a particular environment.

6. Which of the following does NOT contribute to the study of evolution?
   a. population genetics
   b. inheritance of acquired characteristics
   c. the fossil record
   d. comparative morphology

7. Which of the following could be an example of microevolution?
   a. a slight change in a bird population's song arising from a small genetic change in the population
   b. the evolution of many species of finch from a common ancestor
   c. the sudden disappearance of an entire genus
   d. the direct evolutionary link between living primates and humans

8. Which of the following ideas proposed by Lamarck was NOT included in Darwin's theory?
   a. Organisms change in response to their environments.
   b. Changes that an organism acquires during its lifetime are passed to its offspring.
   c. All species change with time.
   d. Changes are passed from one generation to the next.

9. Medical advances now allow many people who suffer from genetic diseases to survive and reproduce. These advances
   a. refute Darwin's theory
   b. disprove descent with modification
   c. reduce the effects of natural selection
   d. eliminate adaptive traits

10. Which of the following ideas is NOT included in Darwin's theory?
    a. All organisms that have ever existed arose through evolutionary modifications of ancestral species.
    b. The great variety of species alive today resulted from the diversification of ancestral species.
    c. Natural selection drives some evolutionary change.
    d. Natural selection preserves advantageous traits.
    e. Natural selection eliminates adaptive traits.

Questions for Discussion

1. Would Charles Darwin have had the same inspiration about natural selection and evolution had he visited other islands? Think of the situation on Hawaii, on Easter Island, or on Tristan da Cunha. Why would the island matter?

2. Explain why the characteristics we see in living organisms adapt them to the environments in which their ancestors lived rather than to the environments in which they live today. Give examples of this situation.

3. Why was there debate about the age of Earth?

Humpback whales were seriously depleted in the mid-1900s.

Microevolution: Changes within Populations

WHY IT MATTERS

As a result of more than 250 years of commercial whaling, humpback whales (*Megaptera novaeangliae*) experienced a disastrous population decline—from about 125 000 to 5000 individuals. After an international agreement limited whaling in 1966, humpbacks have rebounded strongly; about 80 000 individuals now form three distinct populations in the North Atlantic, North Pacific, and Southern oceans. Yet, conservation biologists wondered if because the population decreased to only 5000 animals, the genetic variability in the whales may have been reduced. Such a loss could have adverse effects on the population's reproductive capacity, resistance to disease, and ability to survive unfavourable environmental changes. A situation in which a population regrows from a small number is called a bottleneck (see Section 18.3c).

In the early 1990s, a large group of researchers working in Hawaii, the continental United States, Canada, Australia, South Africa, Mexico, and the Dominican Republic measured genetic variability in surviving humpback populations. They studied mitochondrial DNA (mtDNA) because it is small, easily extracted, and easily analyzed. Almost all of the variability in mtDNA comes from chance mutations that occur at a steady rate, rather than from genetic recombination (see Section 9.3). Since mtDNA is haploid, the mutation rate is more or less constant. Except for variations produced by mutations that occurred after the population bottleneck—which can be estimated from the mutation rate and subtracted from the total variation detected—the amount of variability in mtDNA should reflect the amount present in the population during the bottleneck.

The researchers obtained skin samples from 90 humpback whales distributed among the populations. They extracted mtDNA from each sample and amplified a

463-base-pair segment of mtDNA that includes most of the variable nucleotide positions. They then determined the DNA sequences of the segments. (DNA sequencing is described in Section 16.1.)

The researchers were surprised to find that the mtDNA sequence variation was relatively high in most of their samples. However, a subpopulation of the North Pacific population near Hawaii showed no variability at all in the mtDNA segment examined. One possible hypothesis for this result is that the Hawaiian subpopulation originated recently, perhaps during the twentieth century. There is indirect support for that idea: whaling records from the nineteenth century list no humpback sightings around Hawaii, and the native Hawaiian people have no legends or words describing whales of the humpback type (baleen whales). Perhaps the Hawaiian subpopulation was started by a few whales with limited genetic variability, an example of the founder effect (see Section 18.3c).

With the exception of this Hawaiian subpopulation, humpback whales appear to have retained genetic variability comparable to that seen in other animals. Humpbacks have a potential life span of about 50 years. Thus, some individuals still alive at the time of the study had been born before the most intense period of commercial hunting in the mid-twentieth century; those individuals provided a reservoir of genetic variability from the old populations. These results suggest that the hunting ban came just in time to prevent a significant loss of genetic variability in humpback whales.

The evolution of the humpback whales is an example of **microevolution**, which is a change in frequencies of alleles or heritable phenotypic variants in a population over time. A **population** of organisms includes all the individuals of a single species that live together in the same place and time. Today, when scientists study microevolution, they analyze variations—the differences between individuals—in natural populations. They determine how and why these variations are inherited. Darwin recognized the importance of heritable variation

within populations; he also realized that natural selection can change the pattern of variation in a population from one generation to the next. Scientists have since learned that microevolutionary change results from several processes, not just natural selection, and that sometimes these processes counteract each other.

In this chapter, we first examine the extensive variation that exists within natural populations. We then take a detailed look at the most important processes that alter genetic variation within populations, causing microevolutionary change. Finally, we consider how microevolution can fine-tune the functioning of populations within their environments.

18.1 Variation in Natural Populations

In some populations, individuals vary dramatically in appearance, but the members of most populations look pretty much alike **(Figure 18.1)**. However, even those that look alike, such as the *Cerion* snails in Figure 18.1b, are not identical. With a scale and ruler, you could detect differences in their weight as well as in the length and diameter of their shells. With suitable techniques, you could also document variations in their individual biochemistry, physiology, internal anatomy, and behaviour. All of these are examples of **phenotypic variation**, differences in appearance or function among individuals of a population. If a difference is heritable, it is passed from generation to generation.

18.1a Phenotypic Variation

Darwin's theory recognized the importance of heritable phenotypic variation. Today, microevolutionary studies often begin by assessing phenotypic variation within populations. Most characters exhibit **quantitative variation**: individuals differ in small, incremental ways. If you measured the height of everyone in your biology class, for example, you would see that height varies almost continuously from your shortest to your tallest classmate. Humans also exhibit quantitative variation in the length of their toes, the number of hairs on their heads, and their weight, as discussed in Section 10.2.

We usually display data on quantitative variation in a bar graph or, if the sample is large enough, as a curve **(Figure 18.2)**. The width of the curve is proportional to the variability—the amount of variation—among individuals, and the *mean* describes the average value of the character. As you will see shortly, natural selection often changes the mean value of a character or its variability within populations.

a. European garden snails
(Cepaea nemoralis)

b. Bahaman land snails
(Cerion christophei)

George Bernard/Science Source

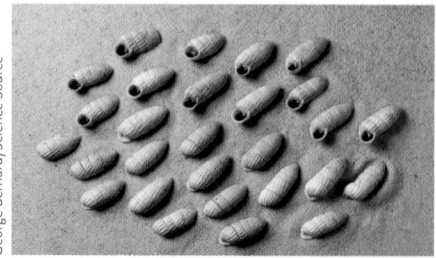
Carnegie Museum of Natural History

Figure 18.1

Phenotypic variation. **(a)** Shells of European garden snails from a population in Scotland vary considerably in appearance. **(b)** By contrast, shells of land snails from a population in the Bahamas look very similar.

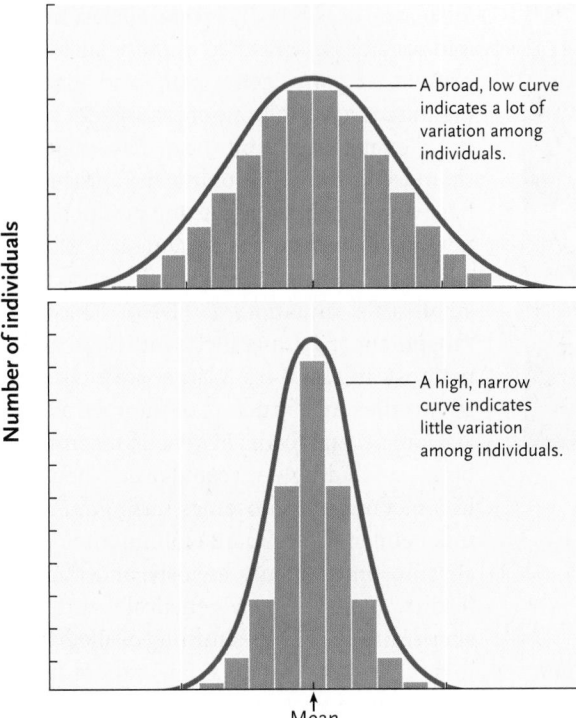

A broad, low curve indicates a lot of variation among individuals.

A high, narrow curve indicates little variation among individuals.

Mean

Measurement or value of trait

Figure 18.2

Quantitative variation. Many traits vary continuously among members of a population, and a bar graph of the data often approximates a bell-shaped curve. The mean defines the average value of the trait in the population, and the width of the curve is proportional to the variability among individuals.

Other characters, such as those Mendel studied (see Chapter 10), exhibit **qualitative variation:** they exist in two or more discrete states, and intermediate forms are often absent. Snow geese, for example, have *either* blue *or* white feathers **(Figure 18.3),** no pale blue. The existence of discrete variants of a character is called a **polymorphism** (*poly* = many, *morphos* = form); we describe such traits as *polymorphic*. The *Cepaea nemoralis* snail shells in Figure 18.1a are polymorphic

© Morales/AgeFotostock

Figure 18.3

Qualitative variation. Individual snow geese (*Chen caerulescens*) are either blue or white. Although both colours are present in many populations, geese tend to associate with others of the same colour.

in background colour, number of stripes, and colour of stripes. Biochemical polymorphisms, like the human A, B, AB, and O blood types (described in Section 10.2), are also common.

We describe phenotypic polymorphisms quantitatively by calculating the percentage or *frequency* of each trait. For example, if you counted 123 blue snow geese and 369 white ones in a population of 492 geese, the frequency of the blue phenotype would be 123/492 or 0.25, and the frequency of the white phenotype would be 369/492 or 0.75.

Phenotypic variation within populations may be caused by genetic differences between individuals, by differences in the environmental factors that individuals experience, or by an interaction between an individual's genetics and the environment. As a result, genetic and phenotypic variations may not be perfectly correlated. Organisms with different genotypes often exhibit the same phenotype. For example, recall Gregor Mendel's experiments with pea plants (see Chapter 10). Plants with homozygous recessive alleles for flower colour had white flowers, while plants with homozygous dominant alleles or with heterozygous alleles had purple flowers. Plants with purple flowers have two different genotypes, even though they exhibit the same phenotype.

Conversely, organisms with the same genotype sometimes exhibit different phenotypes. For example, the acidity of soil influences flower colour in some plants **(Figure 18.4, p. 412).** Knowing whether phenotypic variation is caused by genetic differences, environmental factors, or an interaction of the two is important because *only genetically based variation is subject to evolutionary change.* At the same time, *it is the phenotype of an individual organism, rather than its genotype, that is successful or not.* In other words, natural selection operates on the phenotype, not the genotype. And it operates on the whole phenotype, not just one gene at a time.

Knowing the causes of phenotypic variation also has important practical applications. Suppose, for example, that one field of wheat produced more grain than another. If a difference in the availability of nutrients or water caused the difference in yield, a farmer might choose to fertilize or irrigate the less productive field. But if the difference in productivity resulted from genetic differences between the plants in the two fields, a farmer might plant only the more productive genotype. Because environmental factors can influence the expression of genes, an organism's phenotype is frequently the product of an interaction between its genotype and its environment. In our hypothetical example, the farmer may maximize yield by fertilizing and irrigating the more productive genotype of wheat.

How can we determine whether phenotypic variation is caused by environmental factors or by genetic differences? We can test for an environmental cause experimentally by changing one environmental

©iStock.com/mcswin

©iStock.com/dkapp12

Figure 18.4

Environmental effects on phenotype. Soil acidity affects the expression of the gene controlling flower colour in the common garden plant *Hydrangea macrophylla*. When grown in acidic soil, it produces deep blue flowers. In neutral or alkaline soil, its flowers are bright pink.

variable and measuring the effects on genetically similar subjects. You can try this yourself by growing some cuttings from a single ivy plant in shade and other cuttings from the same plant in full sun. Although the parent plant and all its cuttings have the same genotype, the cuttings grown in sun will produce smaller leaves and shorter stems than those grown in the shade.

Breeding experiments can demonstrate the genetic basis of phenotypic variation. For example, Mendel inferred the genetic basis of qualitative traits, such as flower colour in peas, by crossing plants with different phenotypes. Although simple crosses will not reveal the genetic basis of variations in quantitative traits, these characteristics will respond to artificial selection if the variation has some genetic basis. For example, researchers observed that individual house mice (*Mus domesticus*) differ in activity levels, as measured by how much they use an exercise wheel and how fast they run. John G. Swallow and his colleagues at the University of Wisconsin, Madison, used artificial selection to produce lines of mice that exhibit increased wheel-running behaviour, demonstrating that the observed differences in these two aspects of activity level have a genetic basis.

Breeding experiments are not always practical, however, particularly for organisms with long generation times. Ethical concerns also mean these techniques are unthinkable for humans. Instead, researchers sometimes study the inheritance of particular traits by analyzing genealogical pedigrees, as discussed in Section 11.2, but this approach often provides poor results for analyses of complex traits.

18.1b Genetic Variation

An **allele** is one member of a gene pair that occupies a single location (locus) on a chromosome (see Section 10.2c). A gene can have more than one possible allele, and occasionally several alleles. In diploid organisms, only two of these alleles are present in any gene pair, and haploid organisms have only one of each type of allele.

Genetic variation, the raw material moulded by microevolutionary processes, has two potential sources: the production of new alleles and the rearrangement of existing alleles. Most new alleles probably arise from small-scale mutations in DNA. The rearrangement of *existing* alleles into new combinations can result from larger-scale changes in chromosome structure or number, as well as from several forms of genetic recombination, including crossing-over between homologous chromosomes during meiosis, independent assortment of nonhomologous chromosomes during meiosis, and random fertilizations between genetically different sperm and eggs. This shuffling of alleles into new combinations can produce an extraordinary number of novel genotypes in the next generation. By one estimate, more than 10^{600} combinations of alleles are possible in human gametes, yet there are fewer than 10^{10} humans alive today. So unless you have an identical twin, it is extremely unlikely that another person with your genotype has ever lived or ever will.

18.1c Natural Selection and Phenotypic Variation

Above, you learned that natural selection operates on an organism's phenotype. Biologists measure the effects of natural selection on phenotypic variation by recording changes in the mean and variability of characters over time (see Figure 18.2). Three modes of natural selection have been identified: directional selection, stabilizing selection, and disruptive selection **(Figure 18.5)**.

Directional Selection. Traits undergo **directional selection** when individuals near one end of the phenotypic spectrum have the highest relative fitness. Directional selection shifts a trait away from the existing mean and toward the favoured extreme (see Figure 18.5a). After selection, the trait's mean value is higher or lower than before, and variability in the trait may be reduced.

Directional selection is very common. For example, predatory fish promote directional selection for larger body size in guppies when they selectively feed on the smallest individuals in a guppy population. Most cases of artificial selection are directional, aimed at increasing or decreasing specific phenotypic traits. Humans routinely use directional selection to produce domestic animals and crops with desired characteristics, such as the small size of Chihuahuas and the intense heat of chili peppers.

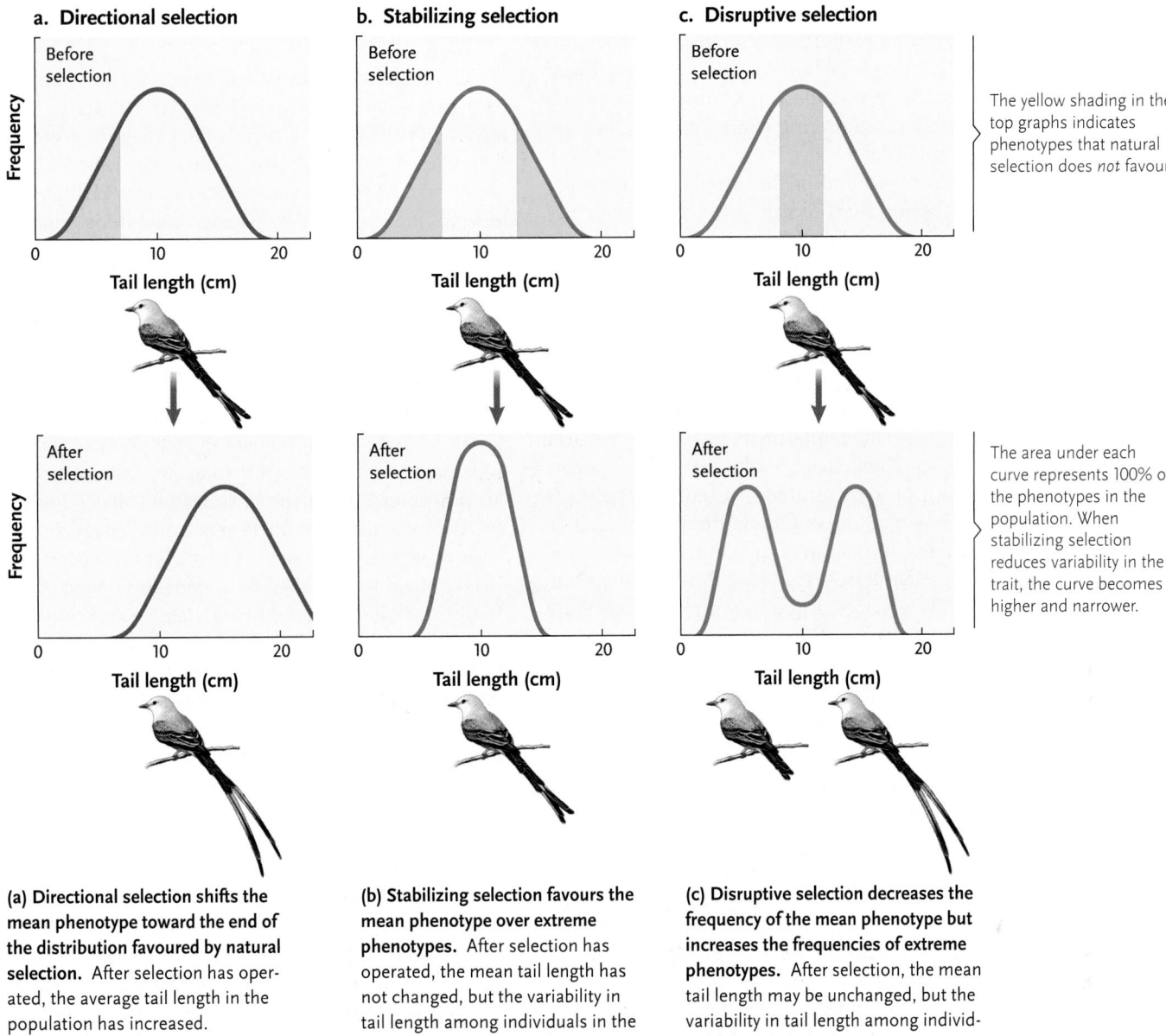

a. Directional selection

Before selection

Frequency

Tail length (cm)

0   10   20

After selection

Frequency

Tail length (cm)

0   10   20

b. Stabilizing selection

Before selection

Tail length (cm)

0   10   20

After selection

Tail length (cm)

0   10   20

c. Disruptive selection

Before selection

Tail length (cm)

0   10   20

After selection

Tail length (cm)

0   10   20

The yellow shading in the top graphs indicates phenotypes that natural selection does *not* favour.

The area under each curve represents 100% of the phenotypes in the population. When stabilizing selection reduces variability in the trait, the curve becomes higher and narrower.

(a) Directional selection shifts the mean phenotype toward the end of the distribution favoured by natural selection. After selection has operated, the average tail length in the population has increased.

(b) Stabilizing selection favours the mean phenotype over extreme phenotypes. After selection has operated, the mean tail length has not changed, but the variability in tail length among individuals in the population has decreased.

(c) Disruptive selection decreases the frequency of the mean phenotype but increases the frequencies of extreme phenotypes. After selection, the mean tail length may be unchanged, but the variability in tail length among individuals has increased.

Figure 18.5
Three modes of natural selection. A hypothetical example using tail length of birds as the quantitative trait subject to selection.

Stabilizing Selection. Traits undergo **stabilizing selection** when individuals expressing intermediate phenotypes have the highest relative fitness (Figure 18.5b). By eliminating phenotypic extremes, stabilizing selection reduces genetic and phenotypic variation and increases the frequency of intermediate phenotypes. Stabilizing selection is probably the most common mode of natural selection, affecting many familiar traits. For example, very small and very large human newborns are less likely to survive than those born at an intermediate weight.

Stabilizing selection can result from multiple selective forces acting on the same trait but in opposite directions. This pattern is seen in the gallmaking fly (*Eurosta solidaginis*), a small insect that feeds on the tall goldenrod plant (*Solidago altissima*). When a fly larva

hatches from its egg, it bores into a goldenrod stem, and the plant responds by producing a spherical growth deformity called a gall. The larva feeds on plant tissues inside the gall. Galls vary dramatically in size; genetic experiments indicate that gall size is a heritable trait of the fly, although plant genotype also has an effect.

Fly larvae inside galls are subjected to two opposing patterns of directional selection. On one hand, a tiny wasp (*Eurytoma gigantea*) parasitizes gallmaking flies by laying eggs in fly larvae inside their galls. After hatching, the young wasps feed on the fly larvae, killing them in the process. However, adult wasps are so small that they cannot easily penetrate the thick walls of a large gall; they generally lay eggs in fly larvae occupying small galls. Thus, wasps establish directional selection that favours flies that produce large galls, which are consequently less likely to be parasitized. On the other hand, several bird species open galls to feed on mature fly larvae; these predators preferentially open large galls, fostering directional selection in favour of small galls.

In about one-third of the populations surveyed, wasps and birds attacked galls with equal frequency, and flies producing galls of intermediate size had the highest survival rate. The smallest and largest galls—as well as the genetic predisposition to make very small or very large galls—were eliminated from the population.

Disruptive Selection. Traits undergo **disruptive selection** when extreme phenotypes have higher relative fitness than intermediate phenotypes (see Figure 18.5c). Thus, alleles producing extreme phenotypes become more common, promoting polymorphism. Under natural conditions, disruptive selection is much less common than directional selection and stabilizing selection. "People behind Biology" in Chapter 17 describes a good example of disruptive selection in Galápagos finches.

STUDY BREAK

1. If a population of skunks includes some individuals with stripes and others with spots, would you describe the variation as quantitative or qualitative?
2. In the experiment on house mice described above, how did researchers demonstrate that variations in activity level had a genetic basis?
3. What factors contribute to phenotypic variation in a population?
4. Which mode of natural selection increases the representation of the average phenotype in a population?

18.2 Population Genetics

To predict how certain factors may influence genetic variation, population geneticists first describe the genetic structure of a population. They then create hypotheses, formalized in mathematical models, to describe how evolutionary processes may change the genetic structure under specified conditions. Finally, researchers test the predictions of these models to evaluate the ideas about evolution that are embodied within them.

18.2a Genetic Structure of Populations

Populations are made up of individuals of the same species, each with its own genotype. In diploid organisms with pairs of homologous chromosomes, an individual's genotype includes two alleles (either two copies of the same allele or two different alleles) at each gene locus. The sum of all alleles at all gene loci in all individuals is called the population's **gene pool.**

To describe the structure of a gene pool, scientists first identify the genotypes in a representative sample and calculate **genotype frequencies**, the percentages of individuals possessing each genotype. They can then calculate **allele frequencies,** the **relative abundances** of the different alleles. For a locus with two alleles, scientists use the symbol p to identify the frequency of one allele, and q to identify the frequency of the other allele.

The calculation of genotype and allele frequencies for the two alleles at the gene locus governing flower colour in snapdragons (*Antirrhinum* spp.) is straightforward **(Table 18.1)**. This locus is easy to study because it exhibits incomplete dominance (see Section 10.2). Individuals that are homozygous for the C^R allele ($C^R C^R$) have red flowers, those homozygous for the C^W allele ($C^W C^W$) have white flowers, and heterozygotes ($C^R C^W$) have pink flowers. Genotype

Table 18.1 | **Calculation of Genotype Frequencies and Allele Frequencies for the Snapdragon Flower Colour Locus**

Because each diploid individual has two alleles at each gene locus, a sample of 1000 individuals has a total of 2000 alleles at the C locus.

Flower Colour Phenotype	Genotype	Number of Individuals	Genotype Frequencies[1]	Total Number of C^R Alleles[2]	Total Number of C^W Alleles[2]
Red	$C^R C^R$	450	450/1000 = 0.45	2 × 450 = 900	0 × 450 = 0
Pink	$C^R C^W$	500	500/1000 = 0.50	1 × 500 = 500	1 × 500 = 500
White	$C^W C^W$	50	50/1000 = 0.05	0 × 50 = 0	2 × 50 = 100
	Total	1000	0.45 + 0.50 + 0.05 = 1.0	1400	600

To calculate allele frequencies, use the total of 1400 + 600 = 2000 alleles in the sample:

$$p = \text{frequency of } C^R \text{ allele} = 1400/2000 = 0.7$$
$$q = \text{frequency of } C^W \text{ allele} = 600/2000 = 0.3$$
$$p + q = 0.7 + 0.3 = 1.0$$

[1]Genotype frequency = the number of individuals possessing a particular genotype divided by the total number of individuals in the sample.
[2]Total number of C^R or C^W alleles = the number of C^R or C^W alleles present in one individual with a particular genotype multiplied by the number of individuals with that genotype.

frequencies represent how the C^R and C^W alleles are distributed among individuals. In this example, examination of all of the plants revealed that 45% of individuals have the $C^R C^R$ genotype, 50% have the heterozygous $C^R C^W$ genotype, and the remaining 5% have the $C^W C^W$ genotype. Allele frequencies represent the commonness or rarity of each allele in the gene pool. As calculated in the table, 70% of the alleles in the population are C^R and 30% are C^W. Remember that for a gene locus with two alleles, there are three genotype frequencies but only two allele frequencies (p and q). The sum of the three genotype frequencies must equal 1; so must the sum of the two allele frequencies.

Once we have described the population, the next question might be, "is the population evolving?" or "is there evidence for evolution in the gene controlling flower colour?"

18.2b The Hardy–Weinberg Principle

When designing experiments, scientists often use control treatments to evaluate the effect of a particular factor. The control tells us what we would see if the experimental treatment had no effect. However, in studies using observational rather than **experimental data**, there is often no suitable control. In such cases, investigators develop conceptual models, called **null models**, that predict what they would see if that particular factor had no effect. Null models serve as theoretical reference points against which scientists can evaluate their observations.

Early in the twentieth century, geneticists were puzzled by the persistence of recessive traits because they assumed that natural selection replaced recessive or rare alleles with dominant or common ones. An English mathematician, G. H. Hardy, and a German physician, Wilhelm Weinberg, tackled this problem independently in 1908. Their analysis, now known as the **Hardy–Weinberg principle**, specifies the conditions under which a population of diploid organisms achieves **genetic equilibrium**, the point at which neither allele frequencies nor genotype frequencies change in succeeding generations. Their work also showed that dominant alleles need not replace recessive ones, and that the shuffling of genes in sexual reproduction does not in itself cause allele or genotype frequencies to change.

The Hardy–Weinberg principle is a mathematical model that describes how genotype frequencies are established in sexually reproducing organisms. According to this model, genetic equilibrium is possible only if *all* of the following conditions are met:

1. No mutations are occurring.
2. The population is closed to migration from other populations.
3. The population is infinite in size (i.e., there is no genetic drift; see Section 18.3c).
4. All genotypes in the population survive and reproduce equally well (selection is not acting on the trait being considered).
5. Individuals in the population mate randomly with respect to the trait being considered.

If all the conditions of the model are met, the allele frequencies of the population for an identified gene locus will never change, and the genotype frequencies will stop changing after one generation. The Hardy–Weinberg principle is thus a null model that serves as a reference point for evaluating the circumstances under which evolution *may* occur. If a population's genotype frequencies do not match the predictions of this null model, evolution may be occurring. If allele frequencies change over time, evolution is occurring. Determining which of the model's conditions are *not* met is a first step in understanding how and why the gene pool is changing (see "Using the Hardy–Weinberg Principle").

Using the Hardy–Weinberg Principle

Research Method Box

To see how the Hardy–Weinberg principle can be applied, we will analyze the snapdragon flower colour locus, using the hypothetical population of 1000 plants described in Table 18.1. This locus includes two alleles: C^R (with its frequency designated as p) and C^W (with its frequency designated as q), and three genotypes: homozygous $C^R C^R$, heterozygous $C^R C^W$, and homozygous $C^W C^W$. Table 18.1 lists the number of plants with each genotype and shows the calculation of both the genotype frequencies and the allele frequencies for the population.

Let's assume for simplicity that each individual produces only two gametes and that both gametes contribute to the production of offspring. This assumption is unrealistic, of course, but it meets the Hardy–Weinberg requirement that all individuals in the population contribute equally to the next generation. In each parent, the two alleles segregate and end up in different gametes:

450 $C^R C^R$ individuals produce $\rightarrow$ 900 C^R gametes

500 $C^R C^W$ individuals produce $\rightarrow$ 500 C^R gametes + 500 C^W gametes

50 $C^W C^W$ individuals produce $\rightarrow$ 100 C^W gametes

You can readily see that 1400 of the 2000 total gametes carry the C^R allele

(Continued)

and the other 600 carry the C^W allele. The frequency of C^R gametes is 1400/2000, or 0.7, which is equal to p; the frequency of C^W gametes is 600/2000, or 0.3, which is equal to q. Thus, the allele frequencies in the gametes are exactly the same as the allele frequencies in the parent generation. It could not be otherwise because each gamete carries one allele at each locus.

Now assume that these gametes, both sperm and eggs, encounter each other at random. In other words, individuals reproduce without regard to the genotype of a potential mate **(Figure 1)**.

We can also describe the consequences of random mating—$(p + q)$ sperm fertilizing $(p + q)$ eggs—with an equation that predicts the genotype frequencies in the offspring generation:

$$(p + q) \times (p + q) = p^2 + 2pq + q^2$$

If the population is at genetic equilibrium for this locus, p^2 is the predicted frequency of the $C^R C^R$ genotype; $2pq$, the predicted frequency of the $C^R C^W$ genotype; and q^2, the predicted frequency of the $C^W C^W$ genotype. Using the gamete frequencies determined above, we can calculate the predicted genotype frequencies in the next generation:

frequency of $C^R C^R =$
$$p^2 = (0.7 \times 0.7) = 0.49$$
frequency of $C^R C^W =$
$$2pq = 2(0.7 \times 0.3) = 0.42$$
frequency of $C^W C^W =$
$$q^2 = (0.3 \times 0.3) = 0.09$$

Notice that the predicted genotype frequencies in the offspring generation have changed from the genotype frequencies in the parent generation: the frequency of heterozygous individuals has decreased, and the frequencies of both types of homozygous individuals have increased. This result occurred because the starting population was *not in equilibrium* at this gene locus. In other words, the distribution of parent genotypes did not conform to the predicted $p^2 + 2pq + q^2$ distribution.

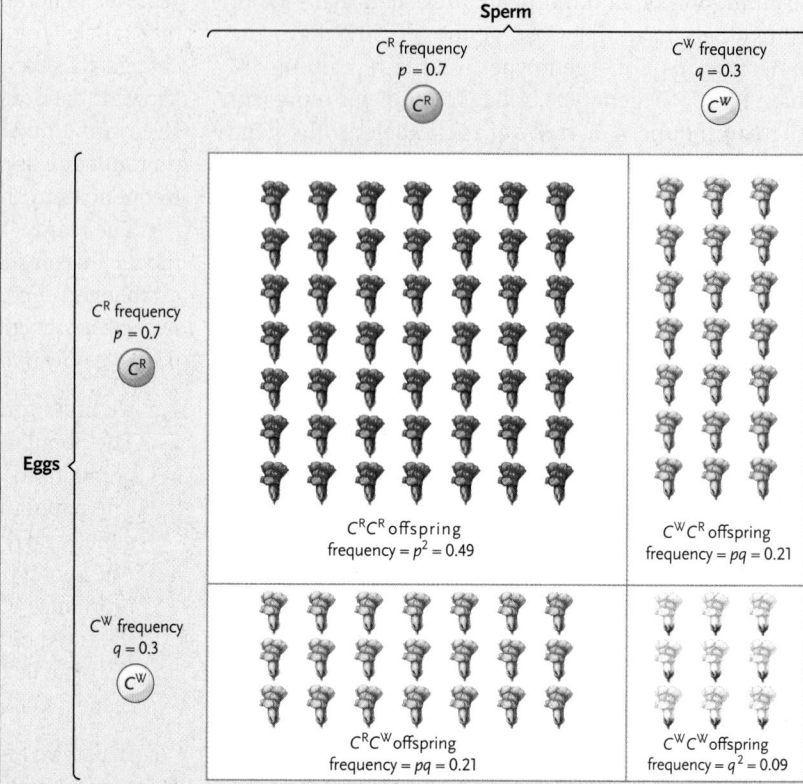

Figure 1
Mating table showing the process of random mating.

The 2000 gametes in our hypothetical population produced 1000 offspring. Using the genotype frequencies we just calculated, we can predict how many offspring will carry each genotype:

490 red ($C^R C^R$)
420 pink ($C^R C^W$)
90 white ($C^W C^W$)

In a real study, we would examine the offspring to see how well their numbers match these predictions.

What about the allele frequencies in the offspring? The Hardy–Weinberg principle predicts that they did not change. Let's calculate them and see. Using the method shown in Table 18.1 and the prime symbol (′) to indicate offspring allele frequencies, we have

$$p' = ([2 \times 490] + 420)/2000 = 1400/2000 = 0.7$$
$$q' = ([2 \times 90] + 420)/2000 = 600/2000 = 0.3$$

You can see from this calculation that the allele frequencies did not change from one generation to the next, even though the alleles were rearranged to produce different proportions of the three genotypes. Thus, the population is now at genetic equilibrium for the flower colour locus. Neither the genotype frequencies nor the allele frequencies will change in succeeding generations as long as the population meets the conditions specified in the Hardy–Weinberg model.

To verify this, you can calculate the allele frequencies of the gametes for this offspring generation and predict the genotype frequencies and allele frequencies for a third generation. You could continue calculating until you ran out of either paper or patience, but these frequencies will not change.

Researchers use calculations such as these to determine whether an actual population is near its predicted genetic equilibrium for one or more gene loci. When they discover that a population is not at equilibrium, they infer that microevolution is occurring. They then investigate the factors that might be responsible.

1. What is the difference between the genotype frequencies and the allele frequencies in a population?
2. Why is the Hardy–Weinberg principle considered a null model of evolution?
3. If the five conditions of the Hardy–Weinberg principle are all met, when will genotype frequencies stop changing?

18.3 Evolutionary Agents

In the following sections, we describe the processes that produce genetic variation and foster microevolutionary change: mutation, gene flow, genetic drift, natural selection, and nonrandom mating. These are summarized in **Table 18.2.**

18.3a Mutations

A **mutation** is a spontaneous and heritable change in DNA. Mutations are rare events: roughly one gamete in 100 thousand to one in 1 million will include a new mutation at a particular gene locus. New mutations are so infrequent, in fact, that they exert little or no immediate effect on allele frequencies in most populations. But over evolutionary time scales, their numbers are significant; mutations have been accumulating in biological lineages for billions of years. And because it creates entirely new genetic variations, a *mutation can be a major source of heritable variation.*

For most animals, only mutations in the germ line (the cell lineage that produces gametes) are heritable; mutations in other cell lineages have no direct effect on the next generation. In plants, however, mutations may occur in meristem cells, which eventually produce flowers as well as nonreproductive structures (see Chapter 34); in such cases, a mutation may be passed to the next generation and ultimately influence the gene pool.

We classify mutations based on their effect on an organism's fitness, rather than on the underlying molecular changes and the mode of inheritance of the trait (e.g., dominant, recessive). It is not immediately known whether or not a new mutation will be advantageous, deleterious, or neutral.

Deleterious mutations alter an individual's structure, function, or behaviour in harmful ways. In mammals, for example, a protein called collagen is an essential component of most extracellular structures. Several simple mutations in humans cause forms of Ehlers–Danlos syndrome, a disruption of collagen synthesis that may result in loose skin; weak joints; or sudden death from the rupture of major blood vessels, the colon, or the uterus.

Lethal mutations can cause great harm to organisms carrying them. If a lethal allele is dominant, both homozygous and heterozygous carriers will, by definition, die from its effects; if recessive, it kills only homozygous recessive individuals. Manx cats have a recessive lethal allele for taillessness: homozygous dominant cats are normal, heterozygous cats have short tails and long legs, and homozygous recessive cats do not survive embryonic development.

Neutral mutations are neither harmful nor helpful. Because of the redundancy of the genetic code, several codons with different nucleotides in the third position may specify the same amino acid in the construction of a polypeptide chain (see Section 13.1). As a result, some DNA sequence changes—especially those in the third nucleotide of the codon—do not alter the amino acid sequence of the protein under construction. Not surprisingly, mutations at the third position appear to persist longer in populations than those at the first two positions. In other instances, mutations that change the amino acid sequence in a protein or even an organism's phenotype may have no influence on its survival and reproduction. A neutral mutation might even prove to be beneficial later if the environment changes.

Sometimes a change in DNA produces an advantageous mutation, which confers some benefit on an individual that carries it. However slight the advantage, natural selection may preserve the new allele and even increase its frequency over time. Once the mutation has been passed to a new generation, other agents of microevolution determine its long-term fate.

Table 18.2	Agents of Microevolutionary Change	
Agent	Definition	Effect on Genetic Variation
Mutation	Heritable change in DNA	Introduces new genetic variation into population; does not change allele frequencies quickly
Gene flow	Change in allele frequencies as individuals join a population and reproduce	May introduce genetic variation from another population
Genetic drift	Random changes in allele frequencies caused by chance events	Reduces genetic variation, especially in small populations; can eliminate rare alleles
Natural selection	Differential survivorship or reproduction of individuals with different phenotypes	One allele replacing another or allelic variation being preserved
Nonrandom mating	Choice of mates based on their their phenotypes and genotypes	Does not directly affect allele frequencies, but usually prevents genetic equilibrium

18.3b Gene Flow

Organisms or their genetic material (in the form of pollen, spores, or fertilized eggs) sometimes move from one population to another. If the immigrants reproduce, they may introduce novel alleles into a population, shifting its allele and genotype frequencies. This phenomenon is called **gene flow** and shows that populations are not completely closed, but can be open to migration.

Gene flow is common in some animal species. For example, young male baboons typically move from one local population to another after experiencing aggressive behaviour by older males. And many marine invertebrate eggs and larvae disperse long distances, carried by ocean currents.

Dispersal agents, such as pollen-carrying wind or seed-carrying animals, are responsible for gene flow in most plant populations. For example, blue jays foster gene flow among populations of oaks by carrying acorns from nut-bearing trees to their winter caches, which may be as much as a mile away **(Figure 18.6)**. Transported acorns that go uneaten may germinate and contribute to the gene pool of a neighbouring oak population.

The movement alone of individuals from one population to another is not sufficient to foster gene flow between two populations. The immigrants must also reproduce in the population they join, thereby contributing to its gene pool. In the San Francisco Bay area, for example, Bay checkerspot butterflies (*Euphydryas editha bayensis*) rarely move from one population to another because they are poor fliers. But when adult females do change populations, it is often late in the breeding season and their offspring have virtually no chance of finding enough food to mature. Thus, many immigrant females do not foster gene flow because they do not contribute to the gene pool of the population they join.

The evolutionary importance of gene flow depends on the degree of genetic differentiation between populations and the rate of gene flow between them. If two gene pools are very different, a little gene flow may increase genetic variability within the population that receives immigrants, and it will make the two populations more similar. But if populations are already genetically similar, lots of gene flow will have little effect.

18.3c Genetic Drift

Chance events sometimes cause the allele frequencies in a population to change unpredictably. This phenomenon, known as **genetic drift**, has especially dramatic effects on small populations, clearly violating the Hardy–Weinberg assumption of an infinitely

Figure 18.6

Gene flow. Blue jays (*Cyanocitta cristata*) serve as agents of gene flow for oaks (genus *Quercus*) when the birds carry acorns from one oak population to another. An uneaten acorn may germinate and contribute to the gene pool of the population into which it was carried.

large population size. Two general circumstances, population bottlenecks and founder effects, often foster genetic drift.

Population Bottlenecks. On occasion, a stressful factor such as disease, starvation, or drought kills a large proportion of the individuals in a population, producing a **population bottleneck**, a dramatic reduction in population size. This cause of genetic drift greatly reduces genetic variation even if the population numbers later rebound **(Figure 18.7)**.

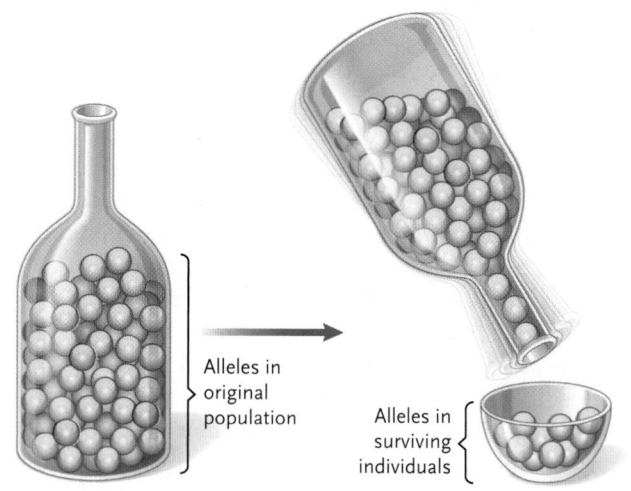

Alleles in original population

Alleles in surviving individuals

The gene pool of the original population, represented by a bottle filled with colored marbles, includes a locus with three alleles. Two of the alleles, represented by blue and green marbles, occur at high frequency; the third allele, represented by red marbles, occurs at low frequency.

If an environmental event randomly kills a large number of individuals in the population, the drastic reduction in population size is described as a population bottleneck. The process is analogous to shaking only a few of the marbles—the survivors—through the neck of the bottle. As a consequence of chance events associated with population bottlenecks, surviving individuals may not have the same allele frequencies as the original population. Rare alleles are inevitably lost.

Figure 18.7

Population bottlenecks, genetic drift, and the loss of genetic variability.

In the late nineteenth century, for example, hunters nearly wiped out northern elephant seals (*Mirounga angustirostris*) along the Pacific coast of North America. Since the 1880s, when the species received protected status, the population has increased to more than 30 000, all descended from a group of about 20 survivors. Today the population exhibits no variation in 24 proteins studied by gel electrophoresis. This low level of genetic variation, which is unique among seal species, is consistent with the hypothesis that genetic drift eliminated many alleles when the population experienced the bottleneck.

Founder Effect. When a few individuals colonize a distant locality and start a new population, they carry only a small sample of the parent population's genetic variation. By chance, some alleles may be totally missing from the new population, whereas other alleles that were rare "back home" might occur at relatively high frequencies. This change in the gene pool is called the **founder effect.**

The human medical literature provides some of the best-documented examples of the founder effect. The Old Order Amish, an essentially closed religious community in Lancaster County, Pennsylvania, have an exceptionally high incidence of Ellis–van Creveld syndrome, a genetic disorder caused by a recessive allele. In the homozygous state, the allele produces dwarfism, shortened limbs, and polydactyly (extra fingers). Genetic analysis suggests that although this syndrome affects fewer than 1% of the Amish in Lancaster County, as many as 13% may be heterozygous carriers of the allele. All of the individuals exhibiting the syndrome are descended from one couple, who helped found the community in the mid-1700s. Assuming Hardy–Weinberg, the allele frequency is 0.075 in the Amish population compared to about 0.003 in the entire population.

Small Population Implications. A simple analogy clarifies why genetic drift is more pronounced in small populations than in large ones. When individuals reproduce, male and female gametes often pair up randomly, as though the allele in any particular sperm or ovum was determined by a coin toss. Imagine that heads specifies the R allele and that tails specifies the r allele. If the two alleles are equally common (i.e., their frequencies, p and q, are both equal to 0.5), heads should be as likely an outcome as tails. But if you toss a coin 20 or 30 times to simulate random mating in a small population, you won't often see a 50:50 ratio of heads and tails. Sometimes heads will predominate and sometimes tails will—just by chance. Tossing the coin 500 times to simulate random mating in a somewhat larger population is more likely to produce a 50:50 ratio of heads and tails. And if you tossed the coin 5000 times, you would get even closer to a 50:50 ratio.

Genetic drift generally leads to the loss of alleles and reduced genetic variability; it therefore causes allele and genotype frequencies to differ from those predicted by the Hardy–Weinberg model.

Conservation Implications. Genetic drift has important implications for **conservation biology.** Because of their small population size, **endangered species** experience severe population bottlenecks, resulting in the loss of genetic variability. Moreover, the small number of individuals available for captive breeding programs may not fully represent a species' genetic diversity. Without such variation, no matter how large a population may become in the future, it will be less resistant to diseases or less able to cope with environmental change.

For example, scientists hypothesize that an environmental catastrophe produced a bottleneck in the African cheetah (*Acinonyx jubatus*) population about 10 000 years ago. Cheetahs today are remarkably uniform in genetic make-up. Their populations are highly susceptible to diseases; males also have a high proportion of sperm cell abnormalities and a reduced reproductive capacity. These observations support the hypothesis of a bottleneck resulting in a high frequency of deleterious alleles in the population. Thus, limited genetic variation, as well as small numbers, threatens populations of endangered species.

18.3d Genetic Variation in Populations

How much genetic variation actually exists within populations? In the 1960s, evolutionary biologists began to use gel electrophoresis (see Figure 15.7) to identify biochemical polymorphisms in diverse organisms. This technique separates two or more forms of a given protein if they differ significantly in shape, mass, or net electrical charge, as a result of mutation-induced changes in the underlying amino acid sequence. The identification of a protein polymorphism allowed researchers to infer genetic variation at the locus coding for that protein.

This approach revealed much more genetic variation in natural populations than anyone had imagined. For example, nearly half the loci surveyed in many populations of plants and invertebrates are polymorphic. Advances in molecular biology now allow scientists to survey genetic variation directly, and researchers have accumulated an astounding knowledge of the structure of DNA and its nucleotide sequences. In general, studies of chromosomal and mitochondrial DNA suggest that every locus exhibits some variability in its nucleotide sequence among individuals from a single population, between populations of the same species, and between related species. However, some variations detected in the protein-coding regions of DNA may not affect phenotypes because, as explained in Section 13.4a, they do not change the amino acid sequences of the proteins for which the genes code.

18.3e Natural Selection and Genetic Variability

The Hardy–Weinberg model requires all genotypes in a population to survive and reproduce equally well. But as you know, inherited traits can enable some individuals to survive better and produce more offspring than others. Natural selection is the process by which such traits become more common in subsequent generations. Thus, natural selection violates a requirement of the Hardy–Weinberg equilibrium and causes allele and genotype frequencies to differ from those predicted by the model.

Although natural selection can change allele frequencies in a population, it is the phenotype of an individual organism, rather than any particular allele, that is successful or not. When individuals survive and reproduce, their alleles, both favourable and unfavourable, are passed to the next generation. Of course, an organism with harmful or lethal dominant alleles will probably die before reproducing, and all the alleles it carries will share that unhappy fate, even those that are advantageous.

To evaluate reproductive success, evolutionary biologists consider **relative fitness**, the number of surviving offspring that an individual produces compared with the numbers left by others in the population. Thus, a particular allele will increase in frequency in the next generation if individuals carrying that allele leave *more* offspring than individuals carrying other alleles. Differences in the *relative* success of individuals are the essence of natural selection.

Natural selection tests fitness differences at nearly every stage of an organism's life cycle. One plant may be fitter than others in the population because its seeds survive colder conditions, because the arrangement of its leaves captures sunlight more efficiently, or because its flowers are more attractive to pollinators. However, natural selection exerts little or no effect on traits that appear during an individual's postreproductive life. For example, Huntington disease, a dominant-allele disorder that first strikes humans after the age of 40, is not subject to strong selection. Carriers of the disease-causing allele can reproduce before the onset of the condition, passing it to the next generation.

18.3f Nonrandom Mating

The Hardy–Weinberg model requires individuals to select mates randomly with respect to their genotypes. This requirement is, in fact, often met; humans, for example, generally marry one another in total ignorance of their genotypes for digestive enzymes or blood types.

Nevertheless, many organisms mate nonrandomly, selecting a mate with a particular phenotype. Snow geese, for example, usually select mates of their own colour (Figure 18.3), and human women are more likely to marry men who are taller than they are. If one phenotype is preferred by most potential mates, mating is not random. Hence the next generation will contain fewer heterozygous offspring—and more homozygous offspring—than the Hardy–Weinberg model predicts.

Inbreeding is a special form of nonrandom mating in which genetically related individuals mate with each other. **Self-fertilization** in plants (see Chapter 36) and a few animals (see Chapter 41) is an extreme example of inbreeding because offspring are produced from the gametes of a single parent. However, other organisms that live in small, relatively closed populations often mate with related individuals. Because relatives often carry the same alleles, inbreeding generally increases the frequency of homozygous genotypes and decreases the frequency of heterozygotes. Thus, recessive phenotypes are often expressed.

For example, the high incidence of Ellis–van Creveld syndrome among the Old Order Amish population, mentioned earlier, is partly caused by inbreeding. Although the founder effect originally established the disease-causing allele in this population, inbreeding in the small population increased the likelihood of its expression. Most human societies discourage matings between genetically close relatives, thereby reducing inbreeding and the inevitable production of recessive homozygotes that inbreeding causes.

STUDY BREAK

1. Which agents of microevolution tend to increase genetic variation within populations, and which ones tend to decrease it?
2. How does genetic drift cause the allele frequencies in a population to change?
3. In what way is sexual selection like directional selection?

18.4 Maintaining Genetic and Phenotypic Variation

Evolutionary biologists continue to discover extraordinary amounts of genetic and phenotypic variation in most natural populations. How can so much variation persist in the face of stabilizing selection and genetic drift?

18.4a Diploidy

The diploid condition reduces the effectiveness of natural selection on harmful recessive alleles. (Note that only a few recessive alleles are harmful.) Although such alleles are disadvantageous in the homozygous state, they may have little or no effect on heterozygotic individuals. Thus, recessive alleles can be protected from natural selection by the phenotypic expression of the dominant allele.

18.4b Balanced Polymorphisms

In a **balanced polymorphism**, two or more phenotypes are maintained in fairly stable proportions over many generations. Natural selection preserves balanced polymorphisms when heterozygotes have higher relative fitness, when different alleles are favoured in different environments, and when the rarity of a phenotype provides an advantage.

Heterozygote Advantage. A balanced polymorphism can be maintained by **heterozygote advantage**, when heterozygotes have higher relative fitness than either homozygote. As Darwin first discovered in his experiments on corn, the offspring of crosses between two homozygous strains of the same species often exhibit a robustness described as *hybrid vigour*. Apparently, being heterozygous at many gene loci provides some advantage, perhaps by allowing organisms to respond effectively to environmental variation.

The best-documented example of heterozygote advantage at a specific gene locus is the maintenance of the *HbS* (sickle cell) allele, which codes for a defective form of hemoglobin in humans. As you learned in Chapter 10, hemoglobin is an oxygen-transporting molecule in red blood cells. The hemoglobin produced by the *HbS* allele differs from normal hemoglobin (coded by the *HbA* allele) by just one amino acid. In *HbS/HbS* homozygotes, the faulty hemoglobin forms long, fibrous chains under low oxygen conditions, causing red blood cells to assume a sickle shape (as shown in Figure 10.1). Homozygous *HbS/HbS* individuals often die of sickle cell disease before reproducing. However, in tropical and subtropical Africa, *HbS/HbA* heterozygotes make up nearly 25% of many populations.

Why is the harmful allele maintained at such high frequency in some populations? It turns out that sickle cell disease is common in regions where malaria is prevalent **(Figure 18.8)**. Malaria is a disease transmitted by mosquitoes, in which parasites infect red blood cells. When heterozygous *HbA/HbS* individuals contract malaria, their infected red blood cells assume the same sickle shape as those of homozygous *HbS/HbS* individuals. The sickled cells lose potassium, killing the parasites, which limits their spread within the infected individual. Heterozygous individuals often survive malaria because the parasites do not multiply quickly inside them, their immune systems can effectively fight the infection, and they retain a large population of uninfected red blood cells. Homozygous *HbA/HbA* individuals are also subject to malarial infection, but because their infected cells do not sickle, the parasites multiply rapidly, causing a severe infection with a high mortality rate.

Therefore, *HbA/HbS* heterozygotes have greater resistance to malaria and are more likely to survive severe infections in areas where malaria is prevalent. Thus, natural selection preserves the *HbS* allele in these populations.

Selection in Different Environments. Genetic variability can also be maintained within a population when different alleles are favoured in different places or at different times (see "People behind Biology," p. 422). For example, the shells of European garden snails range in colour from nearly white to pink, yellow, or brown, and may be patterned by one to five coloured stripes (see Figure 18.1a). This polymorphism, which is relatively stable through time, is controlled by several

a. Distribution of *HbS* allele

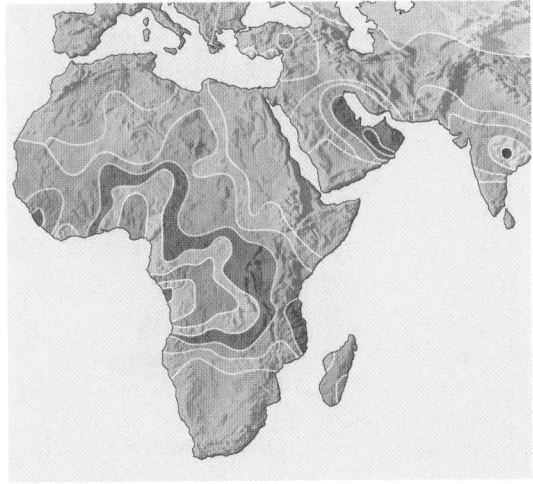

b. Distribution of malarial parasite

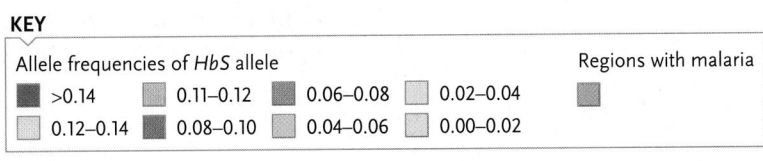

KEY

Allele frequencies of *HbS* allele				Regions with malaria
■ >0.14	0.11–0.12	■ 0.06–0.08	□ 0.02–0.04	■
0.12–0.14	■ 0.08–0.10	0.04–0.06	□ 0.00–0.02	

Figure 18.8

Heterozygote advantage. The distribution of the *HbS* allele **(a),** which causes sickle cell disease in homozygotes, roughly matches the distribution of the malarial parasite *Plasmodium falciparum* **(b)** in southern Europe, Africa, the Middle East, and India. Gene flow among human populations has carried the *HbS* allele to some malaria-free regions.

PEOPLE BEHIND BIOLOGY 18.1

Dolf Schluter

A professor of evolutionary biology and a Canada Research Chair, Dolf Schluter is in the Department of Zoology at the University of British Columbia in Vancouver. Dolf and his students use experimental approaches to study how ecological situations affect the evolution of different wild populations. Are there genetic changes associated with populations of the same species living in different settings? Schluter and six colleagues looked at the genomes of populations of threespine sticklebacks (*Gasterosteus aculeatus*) from different locations.

Phenotypic changes in populations of sticklebacks living in fresh water differ from ancestral populations in salt water (see page 412). Schluter and his colleagues asked if large genetic effect changes (quantitative trait loci) differed among populations.

Three large-effect genes, *Ectodysplasin*, *Pitx1*, and *Kitlg*, underlie the changes in these fish between fresh and salt water. Bony lateral plate armour is controlled by *Ectodysplasin*, the presence of a pelvic girdle by *Pitx1*, and pigmentation by *Kitlg*. Using fish from four freshwater populations, Schluter and colleagues measured loci associated with these quantitative traits. The four freshwater lakes fell into two categories, one with predators (prickly sculpins, *Cottus asper*), the other without. Body armour occurred at three levels across the populations. The heaviest armour occurred in sticklebacks living in marine settings. Intermediate armour occurred in fish living in fresh water with prickly sculpins. The least armour occurred in freshwater settings without prickly sculpins **(Figure 1)**.

Using selective breeding, Schluter and his colleagues explored the genetics underlying the changes they observed. The results indicated that adaptive traits associated with new environments were evident in the genomes of wild populations. There also were obvious differences among the freshwater lakes without the sculpins **(Figure 2)**.

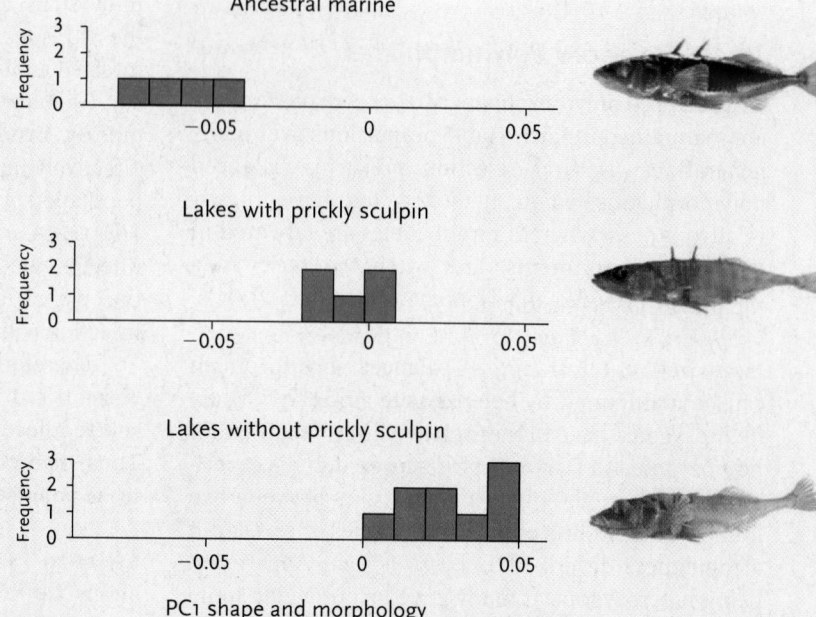

Figure 1

A comparison of the distinct phenotypic optima for threespine sticklebacks from ancestral marine populations, populations in freshwater with prickly sculpins, and those in freshwater lakes without the sculpins.

Sean M. Rogers, Patrick Tamkee, Brian Summers, Sarita Balabahadra, Melissa Marks, David M. Kingsley, and Dolph Schluter, "Genetic signature of adaptive peak shift in threespine stickleback," *Evolution*, Volume 66, Issue 8, pages 2439–2450, August 2012. John Wiley and Sons. © 2012 The Author(s). Evolution© 2012 The Society for the Study of Evolution.

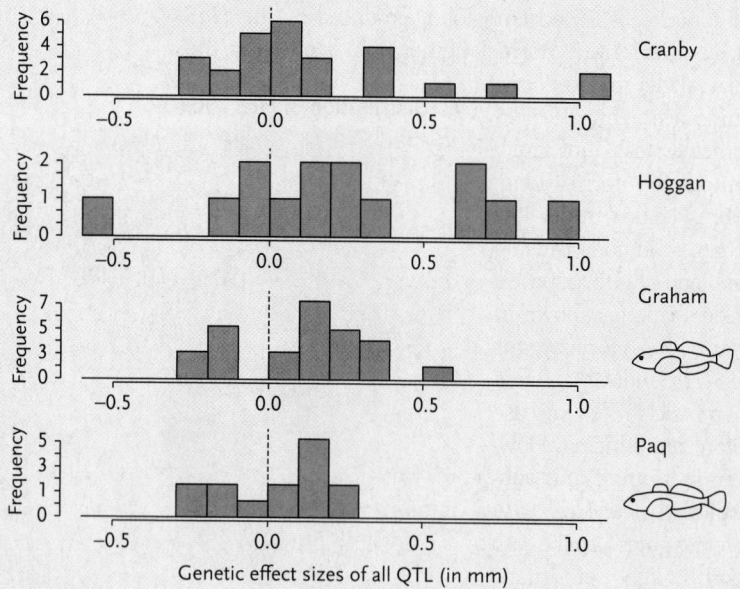

Figure 2

Variation in large-effect genetic changes in populations of threespine sticklebacks living in freshwater lakes with (Graham and Paq) and without (Cranby and Hoggan) sculpins.

Sean M. Rogers, Patrick Tamkee, Brian Summers, Sarita Balabahadra, Melissa Marks, David M. Kingsley, and Dolph Schluter, "Genetic signature of adaptive peak shift in threespine stickleback," *Evolution*, Volume 66, Issue 8, pages 2439–2450, August 2012. John Wiley and Sons. © 2012 The Author(s). Evolution© 2012 The Society for the Study of Evolution.

gene loci. The variability in colour and striping pattern can be partially explained by selection for camouflage in different habitats.

Predation by song thrushes (*Turdus ericetorum*) is a major agent of selection on the colour and pattern of these snails in England. When a thrush finds a snail, it smashes it against a rock, using it like an anvil, to break the shell. The bird eats the snail, but leaves the shell near its "anvil." Researchers collected the broken shells near an anvil and compared the phenotypes of captured snails to a random sample of the entire snail population. Their analyses indicated that thrushes are visual predators, usually capturing snails that are easy to find. Thus, well-camouflaged snails survive, and the alleles that specify their phenotypes increase in frequency.

The success of camouflage varies with habitat, however. Local subpopulations of the snail, which occupy different habitats, often differ markedly in shell colour and pattern. The predators eliminate the most conspicuous individuals in each habitat; thus, natural selection differs from place to place. In woods where the ground is covered with dead leaves, snails with unstriped pink or brown shells predominate. In hedges and fields, where the vegetation includes thin stems and grass, snails with striped yellow shells are the most common. In populations that span several habitats, selection preserves different alleles in different places, thus maintaining variability in the population as a whole.

STUDY BREAK

1. How does the diploid condition protect harmful recessive alleles from natural selection?
2. What is a balanced polymorphism?
3. Why is the allele that causes sickle cell disease very rare in human populations that are native to northern Europe?

18.5 Adaptation and Evolutionary Constraints

Although natural selection preserves alleles that confer high relative fitness on the individuals that carry them, researchers are cautious when they interpret the benefits that particular traits may provide.

18.5a The Evolution of Adaptive Traits

An **adaptive trait** is any product of natural selection that increases the relative fitness of an organism in its environment. **Adaptation** is the accumulation of adaptive traits over time, and examples range across all levels of biological organization, from the molecular to the ecological. For example, the water-retaining structures and special photosynthetic pathways of desert plants and the warning coloration of poisonous animals can be interpreted as adaptive traits. Note, however, that for adaptation to occur there must first be phenotypic variation for selection to act on.

In fact, we can concoct an adaptive explanation for almost any characteristic we observe in nature. But such explanations are just fanciful stories unless they are framed as testable hypotheses about the relative fitness of different phenotypes and genotypes. Unfortunately, evolutionary biologists cannot always conduct straightforward experiments because they sometimes study traits that do not vary much within a population or species. In such cases, they may compare variations of a trait in closely related species living in different environments. For example, one can test how the traits of desert plants are adaptive by comparing them to traits in related species from moister habitats.

When biologists try to unravel how and why a particular adaptive characteristic evolved, they must also remember that a trait they observe today may have had a different function in the past. For example, the structure of the shoulder joint in birds allows them to move their wings first upward and backward and then downward and forward during flapping flight. But analyses of the fossil record reveal that this adaptation, which is essential for flight, did not originate in birds. Some predatory nonflying dinosaurs, including the ancestors of birds, had similarly constructed shoulder joints. Researchers hypothesize that these fast-running predators may have struck at prey with a flapping motion similar to that used by modern birds. Thus, the structure of the shoulder may have evolved first as an adaptation for capturing prey, and only later proved useful for flapping flight. This hypothesis—however plausible it may be—cannot be tested by direct experimentation because the nonflying ancestors of birds have been extinct for millions of years. Instead, evolutionary biologists must use anatomical studies of birds and their ancestors, along with theoretical models about the mechanics of movement, to challenge and refine the hypothesis.

18.5b Factors Constraining Adaptive Evolution

We often marvel at how well adapted an organism is to its environment and mode of life. However, the adaptive traits of most organisms are compromises produced by competing selection pressures. Sea turtles, for example, must lay their eggs on beaches because their embryos cannot acquire oxygen under water. Although flippers allow the females to crawl to nesting sites on beaches, they are not ideally suited for terrestrial locomotion. Their structure reflects their primary function, swimming.

Moreover, no organism can be perfectly adapted to its environment because environments change over time. Natural selection preserves alleles that are successful under the prevailing environmental conditions. Thus, each generation is adapted to the environmental conditions under which its parents lived. If the environment changes from one generation to the next, adaptation will always lag behind.

Another constraint on the evolution of adaptive traits is historical. Natural selection is not an engineer that designs new organisms from scratch. Instead, it acts on new mutations and existing genetic variation. Because new mutations are fairly rare, natural selection works primarily with alleles that have been present for many generations. Thus, adaptive changes in the morphology of an organism are often based on small modifications of existing structures. The bipedal (two-footed) posture of humans, for example, evolved from the quadrupedal (four-footed) posture of our ancestors. Natural selection did not produce an entirely new skeletal design to accompany this radical behavioural shift. Instead, existing characteristics of the spinal column and the musculature of the legs and back were modified, albeit imperfectly, for an upright stance.

Evolution is the unifying theory in biology, drawing together virtually every aspect of the discipline. Microevolution focuses on variation, selection, selective advantages, and natural selection at the population level. Populations make up species, and speciation occurs when populations become diverse enough to be considered separate species. Macroevolution—evolution that occurs at or above the species level—is where we turn next.

STUDY BREAK

1. How can a biologist test whether a trait is adaptive or not?
2. Why are most organisms adapted to the environments in which their parents lived?

Review

To access course materials such as Aplia and other companion resources, please visit www.NELSONbrain.com.

18.1 Variation in Natural Populations

- Phenotypic traits exhibit either quantitative or qualitative variation within populations.
- Genetic variation, environmental factors, or an interaction between the two cause phenotypic variation within populations. Only genetically based phenotypic variation is heritable and subject to evolutionary change.
- Genetic variation arises within populations largely through mutation and genetic recombination. Artificial selection experiments and analyses of protein and DNA sequences reveal that most populations include significant genetic variation.
- Natural selection alters phenotypic variation in three ways. Directional selection increases or decreases the mean value of a trait, shifting it toward a phenotypic extreme. Stabilizing selection increases the frequency of the mean phenotype and reduces variability in the trait. Disruptive selection increases the frequencies of extreme phenotypes and decreases the frequency of intermediate phenotypes.

18.2 Population Genetics

- All the gene copies in a population make up its gene pool, which can be described in terms of allele frequencies and genotype frequencies.
- The Hardy–Weinberg principle of genetic equilibrium is a null model that describes the conditions under which microevolution, a change in allele frequencies through time, will not take place. Microevolution occurs in populations when the restrictive requirements of the model are not met.

18.3 Evolutionary Agents

- Several processes cause microevolution in populations (Table 18.2). Mutation introduces completely new genetic variation. Gene flow carries novel genetic variation into a population through the arrival and reproduction of immigrants. Genetic drift causes random changes in allele frequencies, especially in small populations. Natural selection occurs when the phenotypes of some individuals enable them to survive and reproduce more than others.
- Although nonrandom mating does not change allele frequencies, it can produce more homozygotic and fewer heterozygotic genotypes than the Hardy–Weinberg model predicts.

18.4 Maintaining Genetic and Phenotypic Variation

- Diploidy can maintain genetic variation in a population if recessive alleles are not expressed in heterozygotes and are thus hidden from natural selection.
- Polymorphisms are maintained in populations when heterozygotes have higher relative fitness than both homozygotes, when natural selection occurs in variable environments, or when the relative fitness of a phenotype varies with its frequency in the population.

18.5 Adaptation and Evolutionary Constraints

- Adaptive traits increase the relative fitness of individuals carrying them. Adaptive explanations of traits must be framed as testable hypotheses.
- Natural selection cannot result in perfectly adapted organisms because most adaptive traits represent compromises among conflicting needs, because most environments change constantly, and because natural selection can affect only existing genetic variation.

Questions

Self-Test Questions

1. Which of the following is an example of qualitative phenotypic variation?
   a. the lengths of people's toes
   b. the body sizes of pigeons
   c. human ABO blood types
   d. the birth weights of humans
   e. the number of leaves on oak trees

2. A population of mice is at Hardy–Weinberg equilibrium at a gene locus that controls fur colour. The locus has two alleles, M and m. A genetic analysis of one population reveals that 60% of its gametes carry the M allele. What percentage of mice contains both the M and m alleles?
   a. 60%
   b. 48%
   c. 40%
   d. 36%
   e. 16%

3. If the genotype frequencies in a population are 0.60 AA, 0.20 Aa, and 0.20 aa, and if the requirements of the Hardy–Weinberg principle apply, the genotype frequencies in the offspring generation will be
   a. 0.60 AA, 0.20 Aa, 0.20 aa
   b. 0.36 AA, 0.60 Aa, 0.04 aa
   c. 0.49 AA, 0.42 Aa, 0.09 aa
   d. 0.70 AA, 0.00 Aa, 0.30 aa
   e. 0.64 AA, 0.32 Aa, 0.04 aa

4. The reason spontaneous mutations do not have an immediate effect on allele frequencies in a large population is that
   a. mutations are random events, and mutations may be either beneficial or harmful
   b. mutations usually occur in males and have little effect on eggs
   c. many mutations exert their effects after an organism has stopped reproducing
   d. mutations are so rare that mutated alleles are greatly outnumbered by nonmutated alleles
   e. most mutations do not change the amino acid sequence of a protein

5. The phenomenon in which chance events cause unpredictable changes in allele frequencies is called
   a. gene flow
   b. genetic drift
   c. inbreeding
   d. balanced polymorphism
   e. stabilizing selection

6. An Eastern European immigrant carrying the allele for Tay–Sachs disease settled in a small village on the St. Lawrence River. Many generations later, the frequency of the allele in that village is statistically higher than it is in the immigrant's homeland. The high frequency of the allele in the village is probably an example of
   a. natural selection
   b. the concept of relative fitness
   c. Hardy–Weinberg genetic equilibrium
   d. phenotypic variation
   e. the founder effect

7. If a storm kills many small sparrows in a population, but only a few medium-sized and large ones, which type of selection is probably operating?
   a. directional selection
   b. stabilizing selection
   c. disruptive selection
   d. sexual selection
   e. artificial selection

8. Which of the following phenomena explains why the allele for sickle cell hemoglobin is common in some tropical and subtropical areas where the malaria parasite is prevalent?
   a. balanced polymorphism
   b. heterozygote advantage
   c. sexual dimorphism
   d. neutral selection
   e. stabilizing selection

9. The **neutral mutation hypothesis** proposes that
   a. complex structures in most organisms have not been fostered by natural selection
   b. most mutations have a strongly harmful effect
   c. some mutations are not affected by natural selection
   d. natural selection cannot counteract the action of gene flow
   e. large populations are subject to stronger natural selection than small populations

10. Phenotypic characteristics that increase the fitness of individuals are called
    a. mutations
    b. founder effects
    c. heterozygote advantages
    d. adaptive traits
    e. polymorphisms

Questions for Discussion

1. Many human diseases are caused by recessive alleles that are not expressed in heterozygotes. Some people think that eugenics—the selective breeding of humans to eliminate undesirable genetic traits—provides a way for us to rid our populations of such harmful alleles. Explain why eugenics cannot eliminate such genetic traits from human populations.

2. Using two types of beans to represent two alleles at the same gene locus, design an exercise to illustrate how population size affects genetic drift.

3. In what ways are the effects of sexual selection, disruptive selection, and nonrandom mating different? How are they similar?

4. Captive breeding programs for endangered species often have access to a limited supply of animals for a breeding stock. As a result, their offspring are at risk of being highly inbred. Why and how might zoos and conservation organizations avoid or minimize inbreeding?

19

szefei/Shutterstock.com

Birds of paradise. A male Count Raggi's bird of paradise (*Paradisaea raggiana*). There are 43 known birds of paradise species, 35 of them found only on the island of New Guinea.

Species and Macroevolution

WHY IT MATTERS

In 1927, nearly 100 years after Darwin boarded the *Beagle,* a young German naturalist named Ernst Mayr embarked on his own journey: to the highlands of New Guinea. He was searching for rare birds of paradise. These birds were known in Europe only through their ornate and colourful feathers, which were used to decorate women's hats. On his trek through the remote Arfak Mountains, Mayr identified 137 bird species (including many birds of paradise) based on differences in their size, plumage, colour, and other external characteristics.

To Mayr's surprise, the native Papuans—who were untrained in the ways of Western science, but who hunted these birds for food and feathers—had their own names for 136 of the 137 species he identified. The close match between the two lists confirmed Mayr's belief that the *species* is a fundamental level of organization in nature. Each species has a unique combination of genes underlying its distinctive appearance and habits. Thus, people who observe them closely—whether indigenous hunters or Western scientists—can often distinguish one species from another.

Mayr also discovered some remarkable patterns in the geographical distributions of the bird species in New Guinea. For example, each mountain range he explored was home to some species that lived nowhere else. Closely related species often lived on different mountaintops, separated by deep valleys of unsuitable habitat. In 1942, Mayr published the book *Systematics and the Origin of Species,* in which he described the role of geography in the evolution of new species; the book quickly became a cornerstone of the modern synthesis (see Section 17.4).

What mechanisms produce distinct species? As you discovered in Chapter 18, microevolutionary processes alter the pattern and extent of genetic and phenotypic variation within populations. When these processes differ between populations, the populations will diverge genetically, and they may eventually become so different that we recognize them as distinct species. These changes are considered macroevolution, evolution that occurs at or above the species level. Although Darwin's famous book was titled *On the Origin of Species,* he didn't dwell on the question of *how* new species arise. But the concept of **speciation**—the process of species formation—was implicit in his insight that similar species often share inherited characteristics and a common ancestry. Darwin also recognized that "descent with modification" had generated the amazing diversity of organisms on Earth.

Today, evolutionary biologists view speciation as a *process,* a series of events that occur through time. However, they usually study the *products* of speciation, species that are alive today. Because they can rarely witness the process of speciation from start to finish, scientists make inferences about it by studying organisms in various stages of species formation.

In this chapter, we consider four major topics: how biologists define and recognize species, how species maintain their genetic identity, how the geographical distributions of organisms influence speciation, and how different macroevolutionary genetic mechanisms produce new species.

19.1 What Is a Species?

A simple definition of a **species** (singular and plural, *species*) is a population of organisms capable of interbreeding and producing fertile offspring. The concept of species is based on our perception that Earth's biological diversity is packaged in discrete, recognizable units, and not as a continuum of forms grading into one another. As a group of organisms capable of interbreeding to produce fertile offspring, a species should be genetically distinct from other species. In reality, this definition does not work for organisms that reproduce asexually or those, such as some of Darwin's finches, that hybridize. The working definition of species depends upon the organisms to which it is applied. We should not be surprised that any one definition of *species* in biology is not uniformly used by all biologists. Organisms are the product of evolution, a dynamic process that does not easily accommodate rigid definitions. Our concepts of species appear to be most readily applicable to static situations but not to all species in all situations.

Although a species is a fundamental unit in biology, the diversity of living organisms makes it challenging to have one all-encompassing definition of the word *species*. In previous chapters we saw the importance of variation and selection to particular variants. Variation in reproductive patterns complicates the application of the species concept, as does **hybridization**, which is reproduction involving more than one species. Species are the products of evolution, an ongoing process, so we should not be surprised

at the diversity of biological species and of concepts about species.

19.1a Naming Species

The Swedish naturalist Carl von Linné (1707–1778), better known by his Latinized name, Carolus Linnaeus, was the first modern practitioner of **taxonomy**, the science that identifies, names, and classifies new species. A professor at the University of Uppsala, Linnaeus sent ill-prepared students around the world to gather specimens, losing perhaps a third of his followers to the rigours of their expeditions. Although he may not have been a commendable student adviser, Linnaeus developed the basic system of naming and classifying organisms that biologists have used for more than two centuries. The Linnaeus naming system holds so much information—just in an organism's name—that it's worthwhile to learn the rules and how they are applied.

19.1b Binomial Nomenclature

Linnaeus invented the system of **binomial nomenclature**, in which species are assigned a Latinized two-part name, a species name or **binomial.** The first part of the name identifies a **genus** (plural, *genera*), a group of species with similar characteristics. The second part is the **specific epithet.** The combination of the generic name and the specific epithet provides a unique name for every species. For example, *Ursus maritimus* is the binomial or species name for the polar bear and *Ursus arctos* is the brown bear. By convention, the first letter of a generic name is always capitalized, the species name is never capitalized, and the entire binomial is italicized (or underlined in handwritten work). In addition, the specific epithet is never used without the full or abbreviated generic name preceding it because the same species name is often given to species in different genera. For example, *Castor canadensis* is the North American beaver, *Papilio canadensis* is the Canadian tiger swallowtail butterfly, and *Cornus canadensis* is the Canadian bunchberry. The first use of an organism's binomial name is written in full, while subsequent mentions can use the first letter of the genus plus the specific epithet, for example, *U. maritimus*. Make sure everything is clear: *C. canadensis* refers equally to beaver and bunchberry.

Nonscientists often use different common names to identify a species. For example, *Bothrops asper,* a poisonous snake native to Central and South America, is called "barba amarilla" (in Spanish, meaning "yellow beard") in some places and "cola blanca" (meaning "white tail") in others. In fact, this species has about 50 local names. Adding to the confusion, the same common name is sometimes used for several different species. Binomials, however, allow people everywhere to discuss organisms unambiguously.

The naming of newly discovered species follows a formal process of publishing a description of the species in a scientific journal. International commissions meet periodically to settle disputes about scientific names.

Many binomials are descriptive of the organism or its habitat. *Asparagus horridus,* for example, is a spiny plant. Other species, such as the South American bird *Rhea darwinii,* are named for notable biologists. Some are named with a sense of humour; even Linnaeus called the praying mantis *Mantis religiosa. Wunderpus photogenicus* is a beautiful Indo-Malayan octopus, while *Gelae baen, Gelae belae, Gelae donut, Gelae fish,* and *Gelae rol* are a group of fungus beetles, all named by the same systematists. The submersibles used in deep-sea research—Alvin, Mir-1 and Mir-2, Nautil, and Shinkai-6500—have all had deep-sea species named after them.

19.1c The Taxonomic Hierarchy

Linnaeus described and named thousands of species on the basis of their morphological similarities and differences. Keeping track of so many species was no easy task, so he devised a **classification**, a conceptual filing system that arranges organisms into ever larger, more inclusive categories. Linnaeus' classification, called the **taxonomic hierarchy**, comprises a nested series of formal categories: **domain, kingdom, phylum, class, order, family,** genus, species, and subspecies **(Figure 19.1)**. The organisms included within any category of the taxonomic hierarchy comprise a **taxon** (plural, *taxa*). Woodpeckers, for example, are a taxon (Picidae) at the family level, and pine trees are a taxon (*Pinus*) at the genus level.

19.2 Species Concepts

The concept of species is based on our observations that Earth's biological diversity is packaged in discrete, recognizable units, and not as a continuum of forms grading into one another. As biologists have learned more about evolutionary processes—and the dazzling biodiversity these processes have produced—they have developed about 23 complementary species concepts. We discuss the most used concepts here: morphological, biological, and phylogenetic.

19.2a The Morphological Species Concept

Biologists often describe new species on the basis of visible anatomical characteristics, a process that dates back to Linnaeus' classification system. This approach is based on

Domain *Eukarya*		
Kingdom *Animalia*		
Phylum *Chordata*		
Class *Mammalia*		
Order *Monotremata*	*Rodentia*	*Docodonta*
Family *Ornithorhynchidae*	*Castoridae*	*incertae sedis*
Genus *Ornithorhynchus*	*Castor*	*Castorocauda*
Species *Ornithorhynchus anatinus*	*Castor canadensis*	*Castorocauda lutrasimilis*

M.B. Fenton

Figure 19.1

The taxonomic hierarchy. The classifications of the duck-billed platypus, the Canadian beaver, and the extinct Mesozoic beaver reflect their similarities to other species in the orders to which they belong: *Monotremata, Rodentia,* and *Docodonta,* respectively. *Incertae sedis* (Latin: of uncertain placement) indicates that we do not know in which family to place the Mesozoic beaver.

the **morphological species concept**, the idea that all individuals of a species share measurable traits that distinguish them from individuals of other species.

The morphological species concept has many practical applications. For example, paleobiologists use morphological criteria to identify the species of fossilized organisms (see Chapter 20). And because we can observe the external traits of organisms in nature, field guides to plants and animals list diagnostic (that is, distinguishing) physical characters that allow us to recognize them **(Figure 19.2)**.

Nevertheless, relying exclusively on morphology to identify species can present problems. Consider the variation in the shells of the European garden snail (*Cepaea nemoralis,* Figure 18.1a). How could anyone imagine that such a variety of shells represents just one species of snail? Conversely, morphology does not help us distinguish some closely related species that are

Yellow-throated warbler
(*Dendroica dominica*)

Myrtle warbler
(*Dendroica coronata*)

Figure 19.2

A guide book shows the diagnostic characters of yellow-throated warblers and myrtle warblers, which can be distinguished by the colour of feathers on the throat and rump.

nearly identical in appearance (like some mice). Finally, morphological species definitions tell us little about the evolutionary processes that produce new species.

19.2b The Biological Species Concept

The **biological species concept** emphasizes the dynamic nature of species. Ernst Mayr defined biological species as "groups of ... interbreeding natural populations that are reproductively isolated from [do not produce fertile offspring with] other such groups." The concept is based on reproductive criteria and is easy to apply, at least in principle: if the members of two populations interbreed and produce fertile offspring *under natural conditions,* they belong to the same species; their fertile offspring will, in turn, produce the next generation of that species. If two populations do not interbreed in nature, or fail to produce fertile offspring when they do, they belong to different species.

The biological species concept defines species in terms of population genetics and evolutionary theory. The first half of Mayr's definition notes the genetic *cohesiveness* of species: populations of the same species experience gene flow, which mixes their genetic material. Thus, we can think of a species as one large gene pool, which may be subdivided into local populations.

The second part of the biological species concept emphasizes the genetic *distinctness* of each species. Because populations of different species are reproductively isolated, they cannot exchange genetic information. In fact, the process of speciation is frequently defined as the evolution of reproductive isolation between populations.

The biological species concept also explains why individuals of a species generally look alike: members of the same gene pool share genetic traits that determine their appearance. Individuals of different species generally do not resemble one another as closely because they share fewer genetic characteristics. In practice, biologists often still use similarities or differences in morphological traits as convenient markers of genetic similarity or reproductive isolation.

However, the biological species concept does not apply to the many forms of life that reproduce asexually, including most bacteria; some protists, fungi, and plants; and a few animals. In these species, individuals don't interbreed, so it is pointless to ask whether different populations do. Similarly, we cannot use the biological species concept to study extinct organisms, because we have little or no data on their specific reproductive habits. These species must all be defined using morphological or biochemical criteria. Yet, despite its limitations, the biological species concept currently provides one of the best evolutionary definitions of a sexually reproducing species.

The diversity of species and their lifestyles partly reflects the mechanisms underlying the processes of speciation. The biological species concept defines species in terms of population genetics and evolutionary theory in a static world. The definition alludes to the genetic cohesiveness of species. Populations of the same species are said to experience gene flow that mixes their genetic material and could be the "glue" holding a species together. The second part of this concept emphasizes the genetic distinctness of each species. Because populations of different species are reproductively isolated, they cannot exchange genetic information.

The biological species concept could explain why individuals of a species generally look alike. If phenotype reflects genotype, members of the same gene pool should share genetic traits (genotype) that determine phenotype. Individuals of different species generally do not resemble one another as closely because they share fewer genetic characteristics. In practice, biologists often use similarities or differences in morphological traits as convenient markers of genetic similarity or reproductive isolation. Recently, scientists have used a short DNA sequence in a particular gene to analyze relatedness (see "People behind Biology," p. 430).

19.2c The Phylogenetic Species Concept

Recognizing the limitations of the biological species concept, biologists have developed dozens of other ways to define a species. A widely accepted alternative is the **phylogenetic species concept.** Using both morphological and genetic sequence data, scientists first construct an evolutionary tree for the organisms of interest. They then define a phylogenetic species as a cluster of populations—the tiniest twigs on this part of the tree of life—that emerge from the same small branch. Thus, a phylogenetic species comprises populations that share a recent evolutionary history. We will consider this approach for understanding the evolutionary relationships of organisms in Chapter 20.

One advantage of the phylogenetic species concept is that biologists can apply it to any group of organisms, including species that have long been extinct, as well as living organisms that reproduce asexually. Proponents of this approach also argue that the morphological and genetic distinctions between organisms on different branches of the tree of life reflect the absence of gene flow between them—one of the key requirements of the biological species definition. Nevertheless, because detailed evolutionary histories have been described for relatively few groups of organisms, biologists are not yet able to apply the phylogenetic species concept to all forms of life. Continued research on the details of evolutionary relationships will increase this concept's usefulness in the future.

PEOPLE BEHIND BIOLOGY 19.1

Barcode of Life Project

Today, most systematists working on living organisms include molecular characters as part of the data set when determining evolutionary relationships. Molecular data include nucleotide base sequences of DNA and RNA or the amino acid sequences of the proteins for which they code. Technological advances have automated many of the necessary laboratory techniques, and analytical software makes it easy to compare new data with information filed in data banks, for example, the Barcode of Life project.

In 2003, Paul Hebert, of the Biodiversity Institute of Ontario at the University of Guelph, proposed a new system of species identification using a sequence of approximately 600 DNA base pairs of the mitochondrial gene cytochrome oxidase I. He anticipated a database that would "provide a new master key for identifying species, one whose power will rise with increased taxon coverage and with faster, cheaper sequencing." He compared the system to the barcode system for retail products such as grocery items, calling it the Barcode of Life **(Figure 1).**

Molecular sequences have some practical advantages over organismal characters. First, they provide abundant data because every base in a nucleic acid can serve as a separate, independent character for analysis. Moreover, because many genes have been conserved by evolution, molecular sequences can be compared between distantly related organisms that share

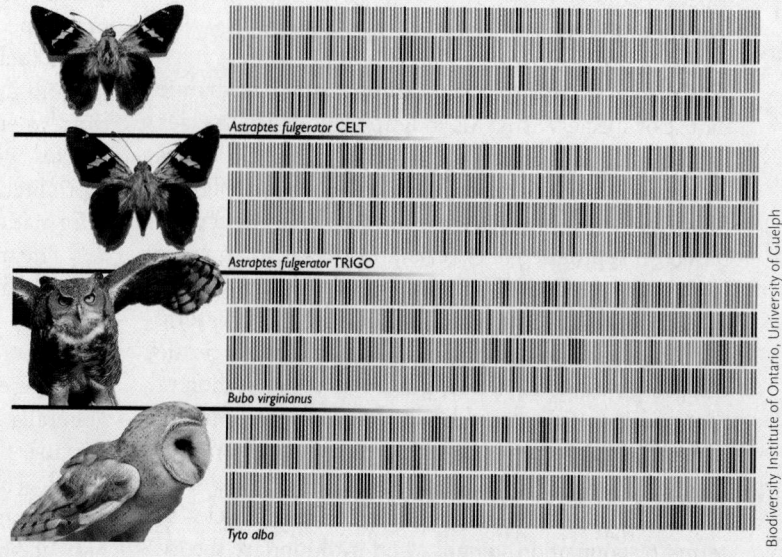

Figure 1

no organismal characters. Molecular characters can also be used to study closely related species that have only minor morphological differences. Finally, many nucleic acids are not directly affected by the developmental or environmental factors that cause nongenetic morphological variations (see Chapter 18).

But there are drawbacks to using molecular characters in taxonomy. There are only four alternative character states (the four nucleotide bases) at each position in a DNA or RNA sequence (see Chapter 11). If two species have the same nucleotide base substitution at a given position in a DNA segment, their similarity may have evolved independently. As a

result, systematists often find it difficult to verify that molecular similarities were inherited from a common ancestor.

For organismal characters, biologists can establish that similarities are homologous by analyzing the characters' embryonic development or details of their function. But molecular characters have no embryonic development, and biologists still do not understand the functional significance of most molecular differences. Despite these disadvantages, molecular characters represent the genome directly, and researchers use them with great success in phylogenetic analyses and in embryology.

STUDY BREAK

1. Why do scientists need a nomenclature system for organisms?
2. Describe the Linnean binomial system.
3. Define the morphological, biological, and phylogenetic species concepts. Compare and contrast these species concepts.

19.3 Maintaining Reproductive Isolation

Reproductive isolation is central to the biological species concept. A **reproductive isolating mechanism** is any biological characteristic that prevents the gene pools of two species from mixing. Biologists classify reproductive isolating mechanisms into two categories:

Biodiversity Institute of Ontario, University of Guelph

Figure labels: *Astraptes fulgerator* CELT, *Astraptes fulgerator* TRIGO, *Bubo virginianus*, *Tyto alba*

prezygotic isolating mechanisms, which exert their effects before the production of a zygote (fertilized egg), and **postzygotic isolating mechanisms**, which operate after zygote formation **(Table 19.1)**. These isolating mechanisms are not mutually exclusive; two or more of them may operate simultaneously. These mechanisms are considered as macroevolution.

19.3a Prezygotic Isolating Mechanisms

Biologists have identified five mechanisms that can prevent interspecific (between species) matings or fertilizations and thus prevent the production of hybrid offspring. These five prezygotic mechanisms are *ecological, temporal, behavioural, mechanical,* and *gametic isolation.*

Species living in the same geographical region may experience **ecological isolation** if they live in different habitats. For example, lions and tigers were both common in India until the mid-nineteenth century, when hunters virtually exterminated the Asian lions. However, because lions live in open grasslands and tigers in dense forests, the two species did not encounter one another and did not interbreed. Lion–tiger hybrids are sometimes born in captivity, but they do not occur under natural conditions.

Species living in the same habitat can experience **temporal isolation** if they mate at different times of day or different times of year. For example, the fruit flies *Drosophila persimilis* and *Drosophila pseudoobscura* overlap extensively in their geographical distributions, but they do not interbreed, in part because *D. persimilis* mates in the morning and *D. pseudoobscura* in the afternoon. Similarly, two species of pine in California are reproductively isolated where their geographical distributions overlap: even though both rely on the wind to carry male gametes (within pollen grains) to female gametes (ova) in other cones, *Pinus radiata* releases pollen in February and *Pinus muricata* releases pollen in April.

Many animals rely on specific signals, which may differ dramatically between species, to identify the species of a potential mate. **Behavioural isolation** results when the signals used by one species are not recognized by another. For example, female birds rely on the song, colour, and displays of males to identify members of their own species. Similarly, female fireflies identify males by their flashing patterns **(Figure 19.3)**. These behaviours (collectively called **courtship displays**) are often so complicated that signals sent by one species are like a foreign language that another species simply does not understand.

Mate choice by females and sexual selection (see Section 17.5c) generally drive the evolution of mate recognition signals. Females often spend substantial energy in reproduction, and choosing an appropriate mate—that is, a male of her own species—is critically important for the production

Table 19.1	Isolating Mechanisms	
Timing Relative to Fertilization	**Mechanism**	**Mode of Action**
Prezygotic ("premating") mechanisms	Ecological isolation	Species live in different habitats
	Temporal isolation	Species breed at different times
	Behavioural isolation	Species cannot communicate
	Mechanical isolation	Species cannot physically mate
	Gametic isolation	Species have nonmatching receptors on gametes
Postzygotic ("postmating") mechanisms	Hybrid inviability	Hybrid offspring do not complete development
	Hybrid sterility	Hybrid offspring cannot produce gametes
	Hybrid breakdown	Hybrid offspring have reduced survival or fertility

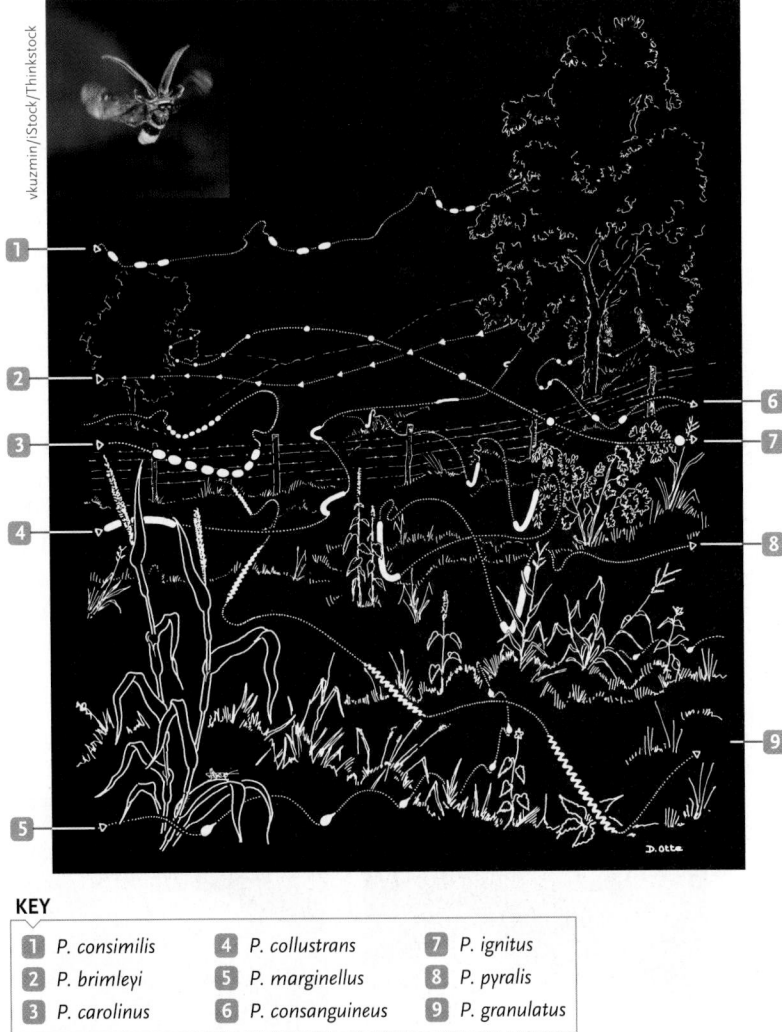

KEY

1	*P. consimilis*	**4**	*P. collustrans*	**7**	*P. ignitus*
2	*P. brimleyi*	**5**	*P. marginellus*	**8**	*P. pyralis*
3	*P. carolinus*	**6**	*P. consanguineus*	**9**	*P. granulatus*

Figure 19.3

Behavioural reproductive isolation. Male fireflies use bioluminescent signals to attract potential mates. The different flight paths and flashing patterns of males in nine North American *Photinus* species are represented here. Females respond only to the display given by males of their own species. The inset photo shows *P. pyralis.*

Illustration courtesy of James E. Lloyd. Miscellaneous Publications of the Museum of Zoology of the University of Michigan, 130:1–195, 1966.

of successful young. By contrast, a female that mates with a male from a different species is unlikely to leave any surviving offspring at all. Over time, the number of males with recognizable traits, as well as the number of females able to recognize the traits, increases in the population.

Differences in the structure of reproductive organs or other body parts—**mechanical isolation**— may prevent individuals of different species from interbreeding. In particular, many plants have anatomical features that allow only certain pollinators, usually particular bird or insect species, to collect and distribute pollen (see Chapter 36). For example, the flowers and nectar of two native California plants, the purple monkey-flower (*Mimulus lewisii*) and the scarlet monkey-flower (*Mimulus cardinalis*), attract different animal pollinators **(Figure 19.4)**. *Mimulus lewisii* is pollinated by bumblebees. The broad petals of its shallow purple flowers provide a landing platform for the bees. Bright yellow streaks on the petals serve as "nectar guides," directing bumblebees to the short nectar tube and reproductive parts, which are located among the petals. Bees enter the flowers to drink their concentrated nectar, and they pick up and deliver pollen as their legs and bodies brush against the reproductive parts of the flowers. *Mimulus cardinalis*, by contrast, is pollinated by hummingbirds. It has long red flowers with no yellow streaks, and the reproductive parts extend above the petals. The red colour attracts hummingbirds but lies outside the colour range detected by bumblebees. The nectar of *M. cardinalis* is more dilute than that of *M. lewisii* but is produced in much greater quantity, making it easier for hummingbirds to ingest. When a hummingbird visits *M. cardinalis* flowers, it pushes its long bill down the nectar tube, and its forehead touches the reproductive parts, picking up and delivering pollen. Recent research has demonstrated that where the two monkey-flower species grow side by side, animal pollinators restrict their visits to either one species or the other 98% of the time, providing nearly complete reproductive isolation.

Even when individuals of different species mate, **gametic isolation**, an incompatibility between the sperm of one species and the eggs of another, may prevent fertilization. Many marine invertebrates release gametes into the environment for external fertilization. The sperm and eggs of each species recognize one another's complementary surface proteins (see Chapter 42), but the surface proteins on the gametes of different species don't match. In animals with internal fertilization, sperm of one species may not survive within the reproductive tract of another. Interspecific matings between some *Drosophila* species, for example, induce a reaction in the female's reproductive tract that blocks "foreign" sperm from reaching eggs. Parallel physiological incompatibilities between a pollen tube and a stigma prevent interspecific fertilization in some plants.

19.3b Postzygotic Isolating Mechanisms

If prezygotic isolating mechanisms between two closely related species are incomplete or ineffective, sperm from one species may fertilize an egg of the other species. In such cases the two species will be reproductively isolated if their offspring, called hybrids, have lower fitness than those produced by intraspecific matings. Three postzygotic isolating mechanisms—*hybrid inviability*, *hybrid sterility*, and *hybrid breakdown*—can reduce the fitness of hybrid individuals.

Many genes govern the complex processes that transform a zygote into a mature organism. Hybrid individuals have two sets of developmental instructions, one from each parent species, which may not interact properly for the successful completion of embryonic development. As a result, hybrid organisms frequently die as embryos or at an early age, a phenomenon called **hybrid inviability.** For example, domestic sheep and goats can mate and fertilize one another's ova, but the hybrid embryos always die before coming to term, presumably because the developmental programs of the two parent species are incompatible.

Although some hybrids between closely related species develop into healthy and vigorous adults, they may not produce functional gametes. This **hybrid sterility** often results when the parent species differ in the number or structure of their chromosomes, which cannot pair properly during meiosis. Such hybrids have zero fitness because they leave no descendants. The most familiar example is a mule, the product of mating between a female horse ($2n = 64$) and a male donkey ($2n = 62$). Zebroids, the offspring of matings between horses and zebras, are also usually sterile **(Figure 19.5)**.

Purple monkey-flower
(Mimulus lewisii)

Scarlet monkey-flower
(Mimulus cardinalis)

TIM LAMAN/National Geographic Creative

South12th Photography/Shutterstock.com

Figure 19.4
Mechanical reproductive isolation. Because of differences in floral structure, two species of monkey-flower attract different animal pollinators. *Mimulus lewisii* attracts bumblebees, and *Mimulus cardinalis* attracts hummingbirds.

Some first-generation hybrids (F_1; see Section 10.1) are healthy and fully fertile. They can breed with other hybrids and with both parental species. However, the second generation (F_2), produced by matings between F_1 hybrids, or between F_1 hybrids and either parental species, may exhibit reduced survival or fertility, a phenomenon known as **hybrid breakdown.** For example, experimental crosses between *Drosophila* species may produce functional hybrids, but their offspring experience a high rate of chromosomal abnormalities and harmful types of genetic recombination. Thus, reproductive isolation is maintained between the species because there is little long-term mixing of their gene pools. In cases where hybrids have lower fitness, there is strong selection for mechanisms that promote assertive mating or prevent hybridization prior to mating or fertilization.

STUDY BREAK

1. What is the difference between prezygotic and postzygotic isolating mechanisms?
2. When a male duck of one species performed a courtship display to a female of another species, she interpreted his behaviour as aggressive rather than amorous. What type of reproductive isolating mechanism does this scenario illustrate?

19.4 The Geography of Speciation

Just as individuals within populations exhibit genotypic and phenotypic variation (see Figure 18.1), populations within species also differ both genetically and phenotypically. Neighbouring populations often have shared characteristics because they live in similar environments, exchange individuals, and experience comparable patterns of natural selection. Widely separated populations, by contrast, may live under different conditions and experience different patterns of selection. Because gene flow is less likely to occur between distant populations, their gene pools and phenotypes often differ.

When geographically separated populations of a species exhibit dramatic, easily recognized phenotypic variation, biologists may identify them as different subspecies **(Figure 19.6). Subspecies** are local variants of a species. Individuals from different subspecies usually interbreed where their geographical distributions meet, and their offspring often exhibit intermediate phenotypes. Various patterns of geographical variation—as well as analyses of how the variation may relate to climatic or habitat variation—have provided great insight into the speciation process. Two of the best-studied patterns are *ring species* and *clinal variation.*

Figure 19.5

Interspecific hybrids. Horses and zebroids (hybrid offspring of horses and zebras) run in a mixed herd. Zebroids are usually sterile.

Jen and Des Bartlett/Bruce Coleman/Photoshot

19.4a Ring Species

Some plant and animal species have a ring-shaped geographical distribution that surrounds uninhabitable terrain. Adjacent populations of these **ring species** can exchange genetic material directly, but gene flow between distant populations occurs only through the intermediary populations.

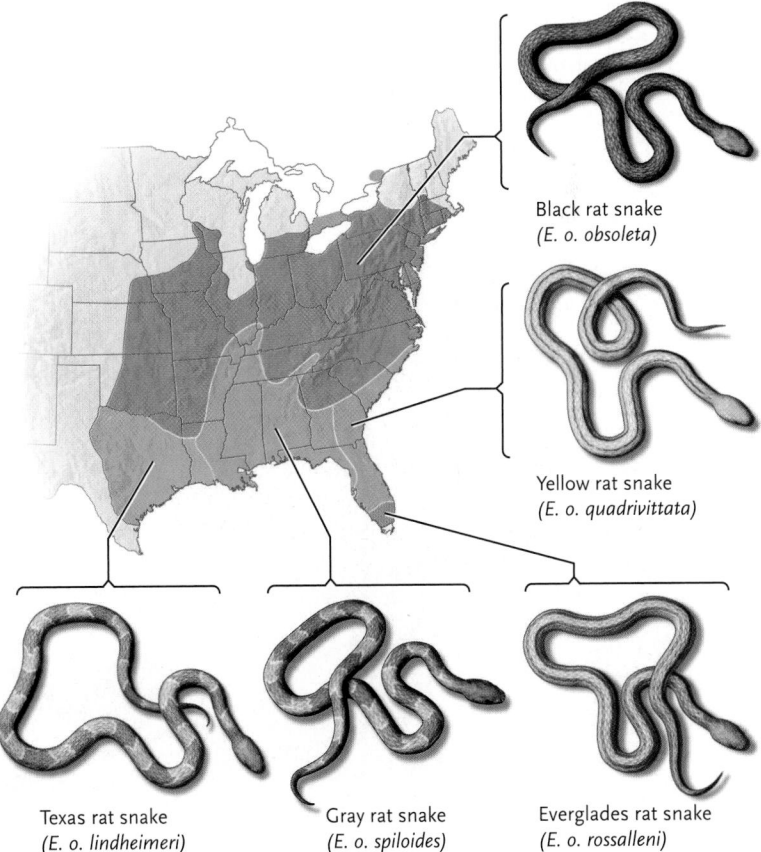

Black rat snake
(*E. o. obsoleta*)

Yellow rat snake
(*E. o. quadrivittata*)

Texas rat snake
(*E. o. lindheimeri*)

Gray rat snake
(*E. o. spiloides*)

Everglades rat snake
(*E. o. rossalleni*)

Figure 19.6

Subspecies. Five subspecies of rat snake (*Elaphe obsoleta*) in eastern North America differ in colour and in the presence or absence of stripes or blotches.

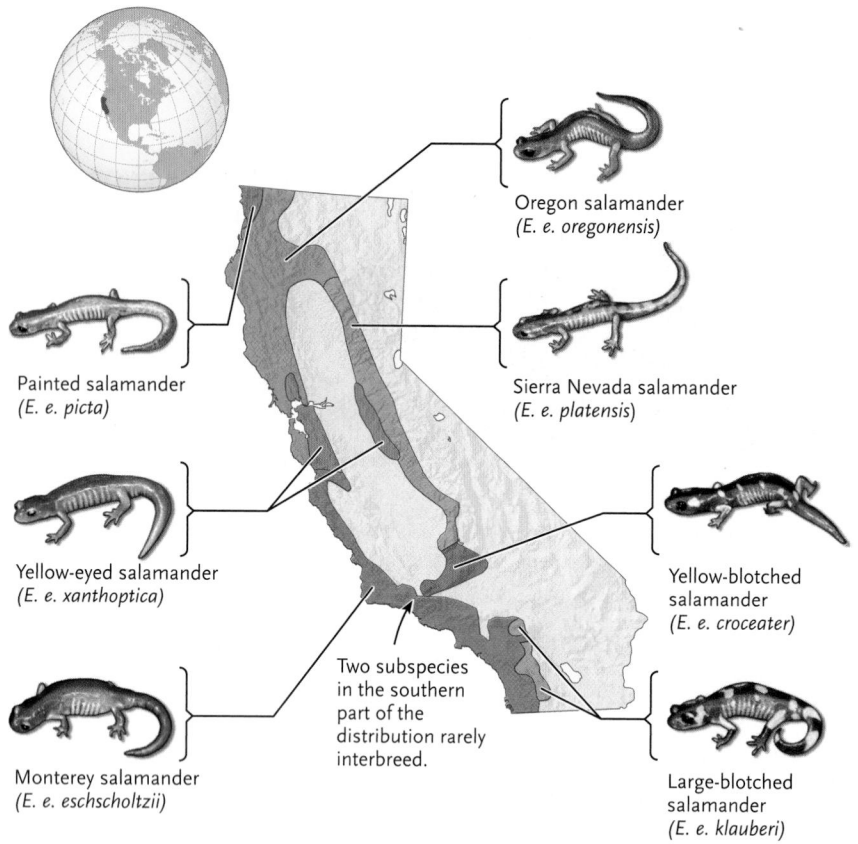

Figure 19.7

Ring species. Six of the seven subspecies of the salamander *Ensatina eschscholtzii* are distributed in a ring around California's Central Valley. Subspecies often interbreed where their geographical distributions overlap. However, the two subspecies that nearly close the ring in the south (marked with an arrow), the Monterey salamander and the yellow-blotched salamander, rarely interbreed.

Oregon salamander
(*E. e. oregonensis*)

Painted salamander
(*E. e. picta*)

Sierra Nevada salamander
(*E. e. platensis*)

Yellow-eyed salamander
(*E. e. xanthoptica*)

Yellow-blotched salamander
(*E. e. croceater*)

Two subspecies in the southern part of the distribution rarely interbreed.

Monterey salamander
(*E. e. eschscholtzii*)

Large-blotched salamander
(*E. e. klauberi*)

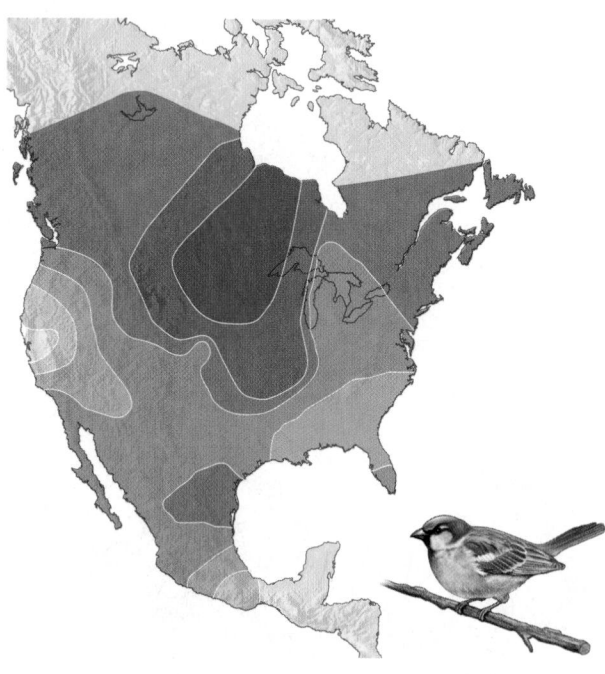

Figure 19.8

Clinal variation. House sparrows (*Passer domesticus*) exhibit clinal variation in overall body size, which was summarized from measurements of 16 skeletal features. Darker shading in the map indicates larger size.

The lungless salamander *Ensatina eschscholtzii*, an example of a ring species, is widely distributed in the coastal mountains and the Sierra Nevada of California, but it cannot survive in the hot, dry Central Valley **(Figure 19.7)**. Seven subspecies differ in biochemical traits, colour, size, and ecology. Individuals from adjacent subspecies often interbreed where their geographical distributions overlap, and intermediate phenotypes are fairly common. But at the southern end of the Central Valley, adjacent subspecies rarely interbreed. Apparently, they have differentiated to such an extent that they can no longer exchange genetic material directly.

Are the southernmost populations of this salamander subspecies or different species? A biologist who saw *only* the southern populations, which coexist without interbreeding, might define them as separate species. However, they can still exchange genetic material through the intervening populations that form the ring. Hence, biologists recognize these populations as belonging to the same species. Most likely, the southern subspecies are in an intermediate stage of species formation.

Another example is the Greenish warblers (*Phylloscopus trochiloides*) from Asia, which live surrounding the Tibetan Plateau. In these birds, courtship songs function in species recognition. If the birds do not respond to playbacks of courtship songs, they do not recognize a potential mate or competitor. Responses to playback presentations indicated that the birds recognized the songs as signals from their own species.

19.4b Clinal Variation

When a species is distributed over a large, environmentally diverse area, some traits may exhibit a **cline**, a pattern of smooth variation along a geographical gradient. For example, many birds and mammals in the northern hemisphere show clinal variation in body size **(Figure 19.8)** and the relative length of their appendages. In general, populations living in colder environments have larger bodies and shorter appendages, a pattern that is usually interpreted as a mechanism to conserve body heat (see Chapter 50).

Clinal variation usually results from gene flow between adjacent populations that are each adapting to slightly different conditions. However, if populations at opposite ends of a cline are separated by great distances, they may exchange very little genetic material through reproduction. Thus, when a cline extends over a large geographical gradient, distant populations may be genetically and morphologically distinct.

Despite the geographical variation that many species exhibit, even closely related species are genetically and morphologically different from each other. In the next section, we consider the mechanisms that maintain the genetic distinctness of closely related species by preventing their gene pools from mixing.

19.4c Allopatric Speciation

Biologists define three modes of speciation, based on the geographical relationship of populations as they become reproductively isolated: *allopatric speciation* (*allo* = different, *patria* = homeland), *parapatric speciation* (*para* = beside), and *sympatric speciation* (*sym* = together). **Allopatric speciation** may take place when a physical barrier subdivides a large population or when a small population becomes separated from a species' main geographical distribution. Allopatric speciation occurs in two stages. First, two populations become *geographically* separated, preventing gene flow between them. Then, as the populations experience distinct mutations as well as different patterns of natural selection and genetic drift, they may accumulate genetic differences that isolate them *reproductively*. Allopatric speciation is probably the most common mode of speciation in large animals.

Geographical separation sometimes occurs when a barrier divides a large population into two or more units **(Figure 19.9)**. For example, hurricanes may create new channels that divide low coastal islands and the populations inhabiting them. Uplifting mountains or landmasses as well as rivers or advancing glaciers can also produce barriers that subdivide populations. The uplift of the Isthmus of Panama, caused by movements of Earth's crust about 5 million years ago, separated a once-continuous shallow sea into the eastern tropical Pacific Ocean and the western tropical Atlantic Ocean. Populations of marine organisms were subdivided by this event. In the tropical Atlantic Ocean, populations experienced patterns of mutation, natural selection, and genetic drift that were different from those experienced by populations in the tropical Pacific Ocean. As a result, the populations diverged genetically, and pairs of closely related species now live on either side of this divide **(Figure 19.10)**.

In other cases, small populations may become isolated at the edge of a species' geographical distribution. Such peripheral populations often differ genetically from the central population because they are adapted to somewhat different environments. Once a small population is isolated, founder effects and small population size may promote genetic drift, and natural selection may favour the evolution of distinctive traits. If the isolated population experiences

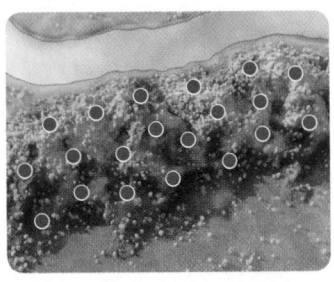 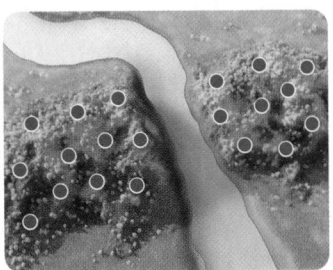 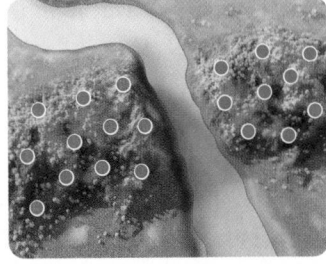 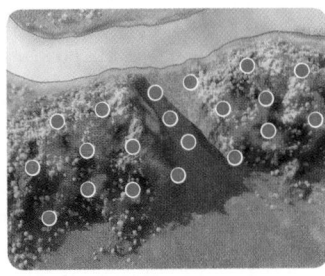

1 At first, a population is distributed over a large geographical area. A river flows along one edge of the population's geographical range.

2 A geographical change, such as a change in the river's course, separates the original population, creating a barrier to gene flow.

3 In the absence of gene flow, the separated populations evolve independently and diverge into different species.

4 When the river later changes course again, allowing individuals of the two species to come into secondary contact, they do not interbreed.

Figure 19.9
The model of allopatric speciation and secondary contact.

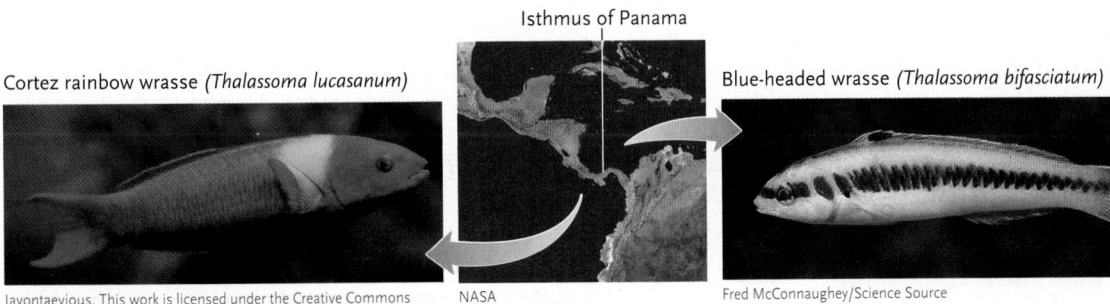

Isthmus of Panama

Cortez rainbow wrasse *(Thalassoma lucasanum)*

Blue-headed wrasse *(Thalassoma bifasciatum)*

Javontaevious. This work is licensed under the Creative Commons Attribution-ShareAlike 3.0 License, http://creativecommons.org/licenses/by-sa/3.0/

NASA

Fred McConnaughey/Science Source

Figure 19.10
Geographical separation. The uplift of the Isthmus of Panama divided an ancestral wrasse population. The Cortez rainbow wrasse now occupies the eastern Pacific Ocean, and the blue-headed wrasse now occupies the western Atlantic Ocean.

limited gene flow from the parent population, these agents of evolution will foster genetic differentiation between them. In time, the accumulated genetic differences may lead to reproductive isolation.

Populations established by colonization of oceanic islands represent extreme examples of this phenomenon. The founder effect makes the populations genetically distinct. And on oceanic archipelagos, such as the Galápagos and Hawaiian islands, individuals from one island may colonize nearby islands, founding populations that differentiate into distinct species. Each island may experience multiple invasions, and the process may be repeated many times within the archipelago, leading to the evolution of a **species cluster**, a group of closely related species recently descended from a common ancestor **(Figure 19.11).** Sometimes a species cluster can evolve relatively quickly; for example, the nearly 800 species of fruit flies now living on the Hawaiian Islands (see "Speciation in Hawaiian Fruit Flies") evolved in less than 5 million years.

Sometimes, allopatric populations reestablish contact when a geographical barrier is eliminated or breached (see Figure 19.9, step 4). This *secondary contact* provides a test of whether or not the populations have diverged into separate species. If their gene pools did not differentiate much during geographical separation, the populations will interbreed and merge. But if the populations have differentiated enough to be reproductively isolated, they have become separate species.

During the early stages of secondary contact, prezygotic reproductive isolation may be incomplete. Some members of each population may mate with individuals from the other, producing viable, fertile offspring in areas called **hybrid zones**. Although some hybrid zones have persisted for hundreds or thousands of years **(Figure 19.12, p. 438),** they are generally narrow, and ecological or geographical factors maintain the separation of the gene pools for the majority of individuals in both species.

If postzygotic isolating mechanisms cause hybrid offspring to have lower fitness than those produced within each population, natural selection will promote the evolution of prezygotic isolating mechanisms, favouring individuals that mate only with members of their own population. Recent studies of *Drosophila* suggest that this phenomenon, called **reinforcement**, enhances reproductive isolation that had begun to develop while the populations were geographically separated.

19.4d Parapatric Speciation

Sometimes a single species is distributed across a discontinuity in environmental conditions, such as a major change in soil type. Although organisms on one side of the discontinuity may interbreed freely with

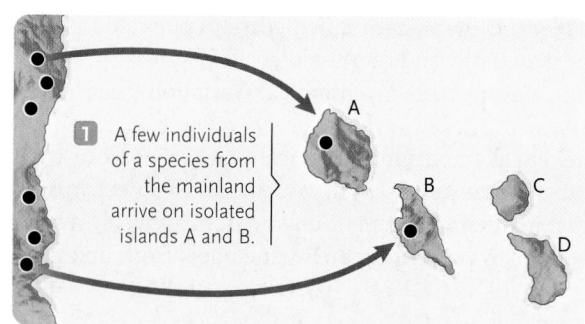

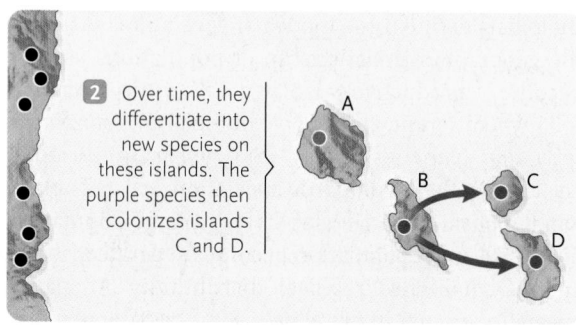

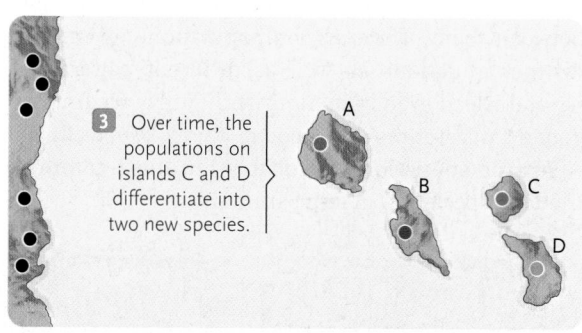

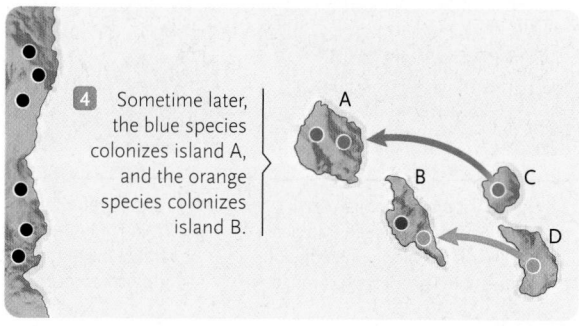

Figure 19.11

Evolution of a species cluster on an archipelago. Letters identify four islands in a hypothetical archipelago, and coloured dots represent different species. The ancestor of all the species is represented by black dots on the mainland. At the end of the process, islands A and B are each occupied by two species, and islands C and D are each occupied by one species, all of which evolved on the islands.

those on the other side, natural selection may favour different alleles on either side, limiting gene flow. In such cases, **parapatric speciation**—speciation arising between adjacent populations—may occur if hybrid offspring have low relative fitness.

Some strains of bent grass (Agrostis tenuis), a common pasture plant in the United Kingdom, have the

Speciation in Hawaiian Fruit Flies

The islands of the Hawaiian archipelago have been geographically isolated throughout their history, lying at least 3200 km from the nearest continents or other islands **(Figure 1)**. Built by undersea volcanic eruptions over millions of years, they emerged from northwest to southeast: Kauai is at least 5 million years old, and Hawaii, the "Big Island," is less than 1 million years old. Individual islands differ in maximum elevation and include diverse habitats, from sparse, dry vegetation to lush, wet forests.

Resident species must have arrived from distant mainland localities or evolved on the islands from colonizing ancestors. The islands' isolation, different ages, and geographical and ecological complexity allowed repeated interisland colonizations followed by allopatric speciation events. Thus, it is not surprising that species clusters have evolved in several groups of organisms (including flowering plants, insects, and birds).

Nearly 800 species of Hawaiian fruit flies have been discovered, most of which live on only one island. Biologists used many characters to identify the different species, including external and internal anatomy, cell structure, chromosome structure, ecology, and mating behaviour. Their data suggest that the vast majority of native Hawaiian species arose from a single ancestral species that colonized the archipelago long ago, probably from eastern Asia. The fruit flies of the Hawaiian Islands now represent more than 25% of all known fruit fly species.

Hampton Carson, of the University of Hawaii, spearheaded studies on the evolutionary relationships of Hawaiian fruit flies. He and his colleagues gathered data on hundreds of fly species—a daunting task. Most species of fruit flies are sexually dimorphic. The females of different species may be similar in appearance, but the males of even closely related species differ in virtually every aspect of their external anatomy: body size; head shape; and the structure of their eyes, antennae, mouthparts, bristles, legs, and wings. Their mating behaviour and choice of mating sites also vary dramatically.

Nevertheless, closely related species on different islands occupy comparable habitats and associate with related plant species. Carson suggested that speciation in these flies resulted from the evolution of different genetically determined *mating systems*, the behaviours and morphological characteristics that males display when seeking a mate. The mating systems serve as prezygotic isolating mechanisms.

The 100 or more species of "picture-wing" *Drosophila*, relatively large flies with patterns on their wings, illustrate the evolution of a species cluster. Carson and his colleagues used similarities and differences in the banding patterns on the giant chromosomes in the flies' salivary gland cells to trace the evolutionary origin of species on the younger islands by identifying their closest relatives on the older islands. Their analysis of 26 species on Hawaii, the youngest island, suggested that flies from the older islands colonized Hawaii at least 19 different times, and each founder population evolved into a new species there. Additional species apparently evolved when lava flows on Hawaii subdivided existing populations.

Among the picture-wing fruit flies, some interspecies matings result in hybrid sterility or hybrid breakdown. But for most species, prezygotic reproductive isolation is maintained by differences in their mating systems. For example, *Drosophila silvestris* and *D. heteroneura,* which produce healthy and fertile hybrids in the laboratory, have similar geographical distributions; however, differences in courtship behaviour and in the shape of the males' heads, a characteristic that females use to recognize males of their own species **(Figure 2),** keep these two species reproductively isolated. In nature, they hybridize only in one small geographical area.

The work of Carson and his colleagues suggests that most speciation in Hawaiian *Drosophila* has

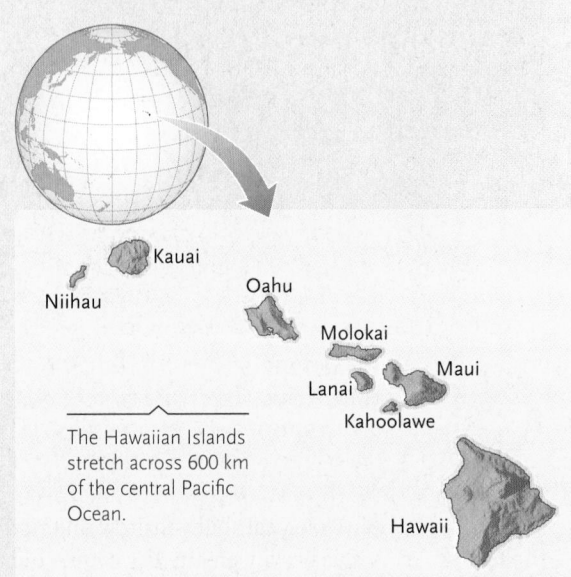

Figure 1
Geographic isolation of the Hawaiian Islands.

resulted from founder effects and genetic drift. When a fertile female—or a small group of males and females—moves to a new island, this founding population responds to novel selection pressures in its new environment. Sexual selection then exaggerates distinctive morphological and behavioural characteristics, maintaining the population's reproductive isolation from its new neighbours. The tremendous variety of Hawaiian fruit flies has undoubtedly been produced by repeated colonizations of newer islands by flies from older islands and by the back-colonization of older islands by newly evolved species. Thus, they represent what evolutionary biologists describe as an *adaptive radiation*, a cluster of closely related species that are ecologically different.

Drosophila heteroneura

Drosophila silvestris

Figure 2
Two Drosophila *species in which the males' head shapes differ.*

Bullock's oriole (*Icterus bullockii*)

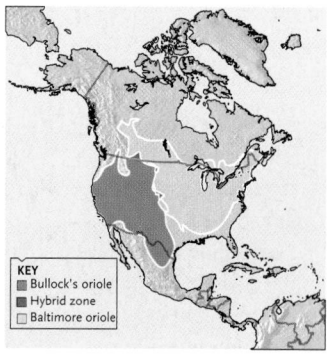

Baltimore oriole (*Icterus galbula*)

KEY
■ Bullock's oriole
■ Hybrid zone
□ Baltimore oriole

Figure 19.12

Hybrid zones. Males of the Bullock's oriole and Baltimore oriole differ in colour and courtship song. Populations of these species have maintained a hybrid zone for hundreds of years, studied by Dr. James Rising of the University of Toronto. The two oriole species now hybridize less frequently than they once did, leading some researchers to suggest that their reproductive isolation evolved recently.

physiological ability to grow on mine tailings, where the soil is heavily polluted by copper or other metals. Plants of the copper-tolerant strains grow well on polluted soils, but plants of the pasture strain do not. Conversely, copper-tolerant plants don't survive as well as pasture plants on unpolluted soils. These strains often grow within a few metres of each other where polluted and unpolluted soils form an intricate environmental mosaic. Because bent grass is wind pollinated, pollen is readily transferred from one strain to another. Laboratory tests have shown that the strains are fully interfertile. However, copper-tolerant plants flower about one week earlier than the pasture plants, which promotes prezygotic (temporal) isolation of the two strains. If the flowering times become further separated, the two strains may attain complete reproductive isolation and become separate species.

Some biologists argue that the places where parapatric populations of bent grass interbreed are really hybrid zones where previously allopatric populations have established secondary contact. Unfortunately, there is no way to determine whether the hybridizing populations were parapatric or allopatric in the past. Thus, a thorough evaluation of the parapatric speciation hypothesis must await the development of techniques that enable biologists to distinguish clearly between the products of allopatric and parapatric speciation.

19.4e Sympatric Speciation

In **sympatric speciation**, reproductive isolation evolves between distinct subgroups that arise within one population. Models of sympatric speciation do not require that the populations be either geographically or environmentally separated as their gene pools diverge. We examine below general models of sympatric speciation in animals and plants; the genetic basis of sympatric speciation is one of the topics we consider in the next section.

Insects that feed on just one or two plant species are among the animals most likely to evolve by

sympatric speciation. These insects generally carry out most important life cycle activities on or near their "host" plants. Adults mate on the host plant; females lay their eggs on it; and larvae feed on the host plant's tissues, eventually developing into adults, which initiate another round of the life cycle. Host-plant choice is genetically determined in many insect species. In others, individuals associate with the host-plant species they ate as larvae.

Theoretically, a genetic mutation could suddenly change some insects' choice of host plant. Mutant individuals would shift their life cycle activities to the new host and then interact primarily with others preferring the same new host, an example of ecological isolation. These individuals would collectively form a separate subpopulation, called a **host race.** Reproductive isolation could evolve between different host races if the individuals of each host race are more likely to mate with members of their own host race than with members of another. Some biologists criticize this model, however, because it assumes that the genes controlling two traits, the insects' host-plant choice and their mating preferences, change simultaneously. Moreover, host-plant choice is controlled by multiple gene loci in some insect species, and it is clearly influenced by prior experience in others.

The apple maggot (*Rhagoletis pomonella*) is one of the most thoroughly studied examples of possible sympatric speciation in animals **(Figure 19.13)**. This fly's natural host plant in eastern North America is the hawthorn (*Crataegus* sp.), but at least two host races have appeared in the past 150 years. The larvae of a new host race were first discovered feeding on apples in New York State in the 1860s. In the 1960s, a cherry-feeding host race appeared in Wisconsin. This is also an example of disruptive selection (see Section 18.1c).

Genetic analyses have shown that variations at just a few gene loci underlie differences in the feeding preferences of *Rhagoletis* host races; other genetic differences cause the host races to develop at different rates. Moreover, adults of the three races mate during

Figure 19.13

Sympatric speciation in animals. Male and female apple maggots (*Rhagoletis pomonella*) court on a hawthorn leaf. The female will later lay her eggs on the fruit, and the offspring will feed, mate, and lay their eggs on hawthorns as well.

different summer months. Nevertheless, individuals show no particular preference for mates of their own host race, at least under simplified laboratory conditions. Thus, although behavioural isolation has not developed between races, ecological and temporal isolation may separate adults in nature. Researchers are still not certain that the different host races are reproductively isolated under natural conditions.

In 2010, Andrew P. Michel and colleagues in the United States and Germany published a genomic analysis of the apple- and hawthorn-feeding races of *Rhagoletis*. Their results suggest that over the past 150 years, the two races have diverged at many loci in their genomes—not just at the loci that influence food choice and developmental rate—and that the divergence has largely been driven by natural selection. Ongoing genetic divergence may prevent them from interbreeding in the future.

Sympatric speciation often occurs in plants through a genetic phenomenon, **polyploidy**, in which an individual has one or more *extra* copies of the entire haploid complement of chromosomes (see Section 11.3). Polyploidy can lead to speciation because these large-scale genetic changes may prevent polyploid individuals from breeding with individuals of the parent species. Nearly half of all flowering plant species are polyploid, including many important crops and ornamental species. The genetic mechanisms that produce polyploid individuals in plant populations are well understood; we describe them in the next section as part of a larger discussion of the genetics of speciation.

STUDY BREAK

1. What are the two stages required for allopatric speciation?
2. How do the conditions leading to parapatric and sympatric speciation differ?
3. Why might insects from different host races be unlikely to mate with each other?

19.5 Genetic Mechanisms of Speciation

In this section we examine three macroevolutionary genetic mechanisms that can lead to reproductive isolation between populations. Two are related to geographical distribution: *genetic divergence* between allopatric populations and *polyploidy* in sympatric populations. The third, **chromosome alterations**, occur independently of the geographical distributions of populations.

19.5a Genetic Divergence in Allopatric Populations

In the absence of gene flow, geographically separated populations inevitably accumulate genetic differences through the actions of mutation, genetic drift, and natural selection. How much genetic divergence is necessary for speciation to occur? To understand the genetic basis of speciation in closely related species, researchers first identify the specific causes of reproductive isolation. They then use standard techniques of genetic analysis, along with new molecular approaches such as gene mapping and sequencing, to analyze the genetic mechanisms that establish reproductive isolation. These techniques now allow researchers to determine the minimum number of genes responsible for reproductive isolation in particular pairs of species.

In cases of postzygotic reproductive isolation, mutations in just a few gene loci can establish reproductive isolation. For example, if two common aquarium fishes, swordtails (*Xiphophorus helleri*) and platys (*Xiphophorus maculatus*), mate, two genes induce the development of lethal tumours in their hybrid offspring. When hybrid sterility is the primary cause of reproductive isolation between *Drosophila* spp., at least five gene loci are responsible. About 55 gene loci contribute to postzygotic reproductive isolation between the European fire-bellied toad (*Bombina bombina*) and the yellow-bellied toad (*Bombina variegata*).

In cases of prezygotic reproductive isolation, some mechanisms have a surprisingly simple genetic basis. For example, a single mutation reverses the direction of coiling in the shells of some snails (*Bradybaena* spp.). Snails with shells that coil in opposite directions cannot approach each other close enough to mate, making reproduction between them mechanically impossible.

Many traits that now function as prezygotic isolating mechanisms may originally have evolved in response to sexual selection (see Section 17.5c). In sexually dimorphic species, this evolutionary process exaggerates showy structures and courtship behaviours in males, the traits that females use to select appropriate mates. When two populations encounter one another on secondary contact, these traits may also prevent

Molecular Investigation Techniques: Monkey-Flower Speciation

Reproductive isolation is the primary criterion that biologists use to distinguish species. As noted earlier, the monkey-flower species *Mimulus lewisii* and *Mimulus cardinalis* are mechanically reproductively isolated in nature because differences in flower structure keep bumblebees or hummingbirds from carrying pollen from one species to the other (see Figure 19.4). However, the two species are easily crossed in the laboratory and produce fertile F_1 hybrids. The F_2 offspring have flowers with various forms intermediate between the two parental types, suggesting that several gene loci control the traits separating the species.

Because little was known about the genetics of the two monkey-flower species, it was not possible to use a direct genetic analysis to identify and map the flower trait genes to the chromosomes. Instead, H. D. Bradshaw and other researchers at the University of Washington used an indirect molecular approach that identified DNA sequence variations (analogous to those used in DNA fingerprinting; see Chapter 15) at various loci in the genome. Just as morphological and biochemical traits vary within a population, DNA sequences vary at particular sites in the genome. The different DNA sequence alleles can be distinguished using a polymerase chain reaction (PCR).

The researchers used a random set of 153 DNA sequence variations. They correlated the segregation of these variations with the segregation of flower traits in 93 plants of the F_2 generation. Some of the DNA sequences segregated closely with a particular flower trait. This result indicated that the particular DNA sequence variation locus was very near the gene for that flower trait on the chromosome. In other words, the flower trait locus was identified indirectly through the close linkage between the DNA variation locus and the flower trait locus.

But where are the genes? To answer this question, the researchers used the DNA sequences linked to flower trait loci as probes to find the sites on the chromosomes where they originated. For any given DNA variation locus, once its position on the chromosomes was determined, the investigators knew that the flower trait locus correlated with it must be nearby.

The investigation showed that reproductive isolation of *M. lewisii* and *M. cardinalis* results from differences in eight floral traits: the amount of (1) anthocyanin pigments and (2) carotenoid pigments in petals, (3) flower width, (4) petal width, (5) nectar volume, (6) nectar concentration, and the lengths of the stalks supporting the (7) male and (8) female reproductive parts. Although the investigators could not directly determine the number of genes controlling each trait, the characteristics of the traits, their locations at eight sites on six of the eight chromosomes, and their pattern of inheritance make it most likely that each trait is controlled by a single gene, giving a likely minimum of eight genes.

Mutations in as few as eight genes may have established reproductive isolation between these two species of monkey-flower. Thus, it appears that in some cases surprisingly little genetic change is required for the evolution of a new species.

---

interspecific mating. For example, many closely related duck species exhibit dramatic variation in the appearance of males, but not females **(Figure 19.14),** an almost certain sign of sexual selection. Yet these species hybridize readily in captivity, producing offspring that are both viable and fertile.

Reproductive isolation and speciation in ducks and other sexually dimorphic birds probably result from geographical isolation and sexual selection on just a few morphological and behavioural characteristics that influence their mating behaviour. Thus, sometimes the evolution of reproductive isolation may not require much genetic change at all. Indeed, sexual selection appears to increase the rate at which new species arise: bird lineages that are sexually dimorphic generally include more species than do related lineages in which males and females have a similar appearance.

19.5b Polyploidy

Common among plants, polyploidy may also be an important factor in the evolution of some fish, amphibian, and reptile species. Polyploid individuals can arise from chromosome duplications within a single species (autopolyploidy) or through hybridization of different species (allopolyploidy).

Mallard ducks *(Anas platyrhynchos)*

Pintail ducks *(Anas acuta)*

Figure 19.14
In closely related species, such as mallard and pintail ducks, males have much more distinctive coloration than females, a sure sign of sexual selection at work.

a. Speciation by autopolyploidy in plants

A spontaneous doubling of chromosomes during meiosis produces diploid gametes. If the plant fertilizes itself, a tetraploid zygote will be produced.

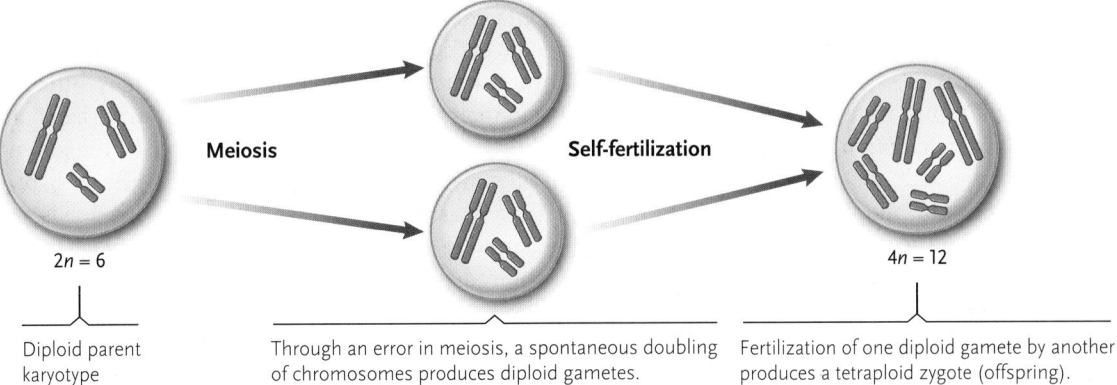

Diploid parent karyotype	Through an error in meiosis, a spontaneous doubling of chromosomes produces diploid gametes.	Fertilization of one diploid gamete by another produces a tetraploid zygote (offspring).

b. Speciation by hybridization and allopolyploidy in plants

A hybrid mating between two species followed by a doubling of chromosomes during mitosis in gametes of the hybrid can instantly create sets of homologous chromosomes. Self-fertilization can then generate polyploid individuals that are reproductively isolated from both parent species.

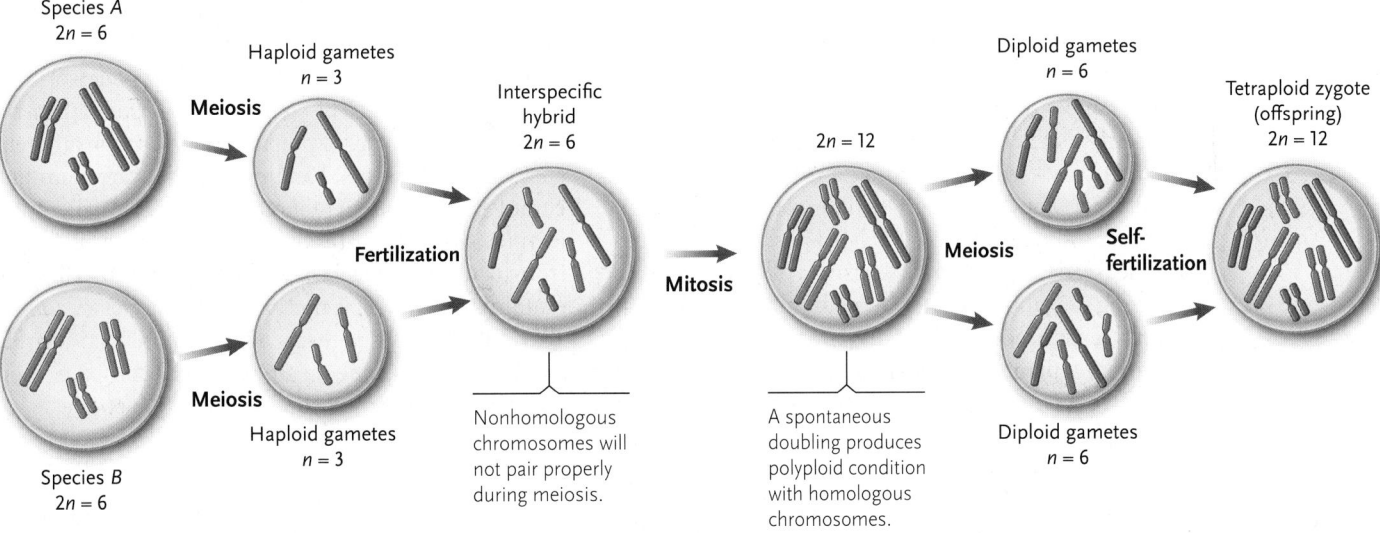

Figure 19.15

Polyploidy in plants. **(a)** Speciation by autopolyploidy in plants can occur by a spontaneous doubling of chromosomes during meiosis, producing diploid gametes. If the plant fertilizes itself, a tetraploid zygote will result. **(b)** Speciation by hybridization and allopolyploidy in plants can occur when two species mate, producing a hybrid. If chromosomes are doubled during mitosis in gametes of the hybrid, sets of homologous chromosomes are instantly created. Self-fertilization can then generate polyploid individuals that are reproductively isolated from both parent species.

In **autopolyploidy (Figure 19.15a),** a diploid ($2n$) individual may produce, for example, tetraploid ($4n$) offspring, each of which has four complete chromosome sets. Autopolyploidy often results through an error in either mitosis or meiosis, when gametes spontaneously receive the same number of chromosomes as a somatic cell. Such gametes are called **unreduced gametes** because their chromosome number has not been halved.

Diploid pollen can fertilize the diploid ovules of a self-fertilizing individual, or it may fertilize diploid ovules on another plant with unreduced gametes. The resulting tetraploid offspring can reproduce either by self-pollination or by breeding with other tetraploid individuals. However, a tetraploid plant cannot produce fertile offspring by hybridizing with its diploid parents. The fusion of a diploid gamete with a normal haploid gamete produces a triploid ($3n$) offspring, which is usually sterile because its odd number of chromosomes cannot segregate properly during meiosis. Thus, the tetraploid is reproductively isolated from the original diploid population. Many species of grasses, shrubs, and ornamental plants—including violets, chrysanthemums, and nasturtiums—are autopolyploids, having anywhere from 4 to 20 complete chromosome sets.

In **allopolyploidy** (Figure 19.15b), two closely related species hybridize and subsequently form polyploid offspring. Hybrid offspring are sterile if the two parent species have diverged enough that their chromosomes do not pair properly during meiosis. However, if the hybrid's chromosome number is doubled, the chromosome complement of the gametes is also doubled, producing homologous chromosomes that *can* pair during meiosis. The hybrid can then produce polyploid gametes and, through self-fertilization or fertilization with other doubled hybrids, establish a population of a new polyploid species. Compared with speciation by genetic divergence, speciation by allopolyploidy is extremely rapid, causing a new species to arise in one generation without geographical isolation.

Even when sterile, polyploids are often robust, growing larger than either parent species. For that reason, both autopolyploids and allopolyploids have been important to agriculture. For example, the wheat used to make flour (*Triticum aestivum*) has six sets of chromosomes **(Figure 19.16)**. Other polyploid crop plants are apples, coffee, strawberries, potatoes, oats, and tobacco.

Plant breeders often try to increase the probability of forming an allopolyploid by using chemicals that foster nondisjunction of chromosomes during mitosis. In the first such experiment, undertaken in the 1920s, scientists crossed a radish and a cabbage, hoping to develop a plant with both edible roots and leaves. Instead, the new species, *Raphanobrassica,* combined the least desirable characteristics of each parent, growing a cabbagelike root and radishlike leaves. Recent experiments have been more successful. For example, plant scientists have produced an allopolyploid grain, triticale, that has the disease resistance of its rye parent and the high productivity of its wheat parent.

In both autopolyploidy and allopolyploidy, a spontaneous doubling of chromosome number produces gametes with twice the original number of chromosomes, but the timing of doubling is different. In autopolyploidy, the doubling occurs during a meiotic cell division that produces 2n gametes in the parent. In allopolyploidy, the doubling occurs after a hybrid offspring is produced, when some of its cells are undergoing mitosis; meiosis in the polyploid hybrid then produces polyploid gametes.

19.5c Speciation from Chromosome Alterations

Other changes in chromosome structure or number may also foster speciation. Closely related species often have a substantial number of chromosome differences between them, including inversions, translocations, deletions, and duplications (described in Section 11.3). These differences, which may foster postzygotic isolation, can often be identified by comparing the *banding patterns* in stained chromosome preparations from the different species. In all species, banding patterns vary from one chromosome segment to another. When researchers find identical banding patterns in chromosome segments from two or more related species, they know that they are examining comparable portions of the species' genomes. Thus, the banding patterns allow scientists to identify specific chromosome segments and compare their positions in the chromosomes of different species.

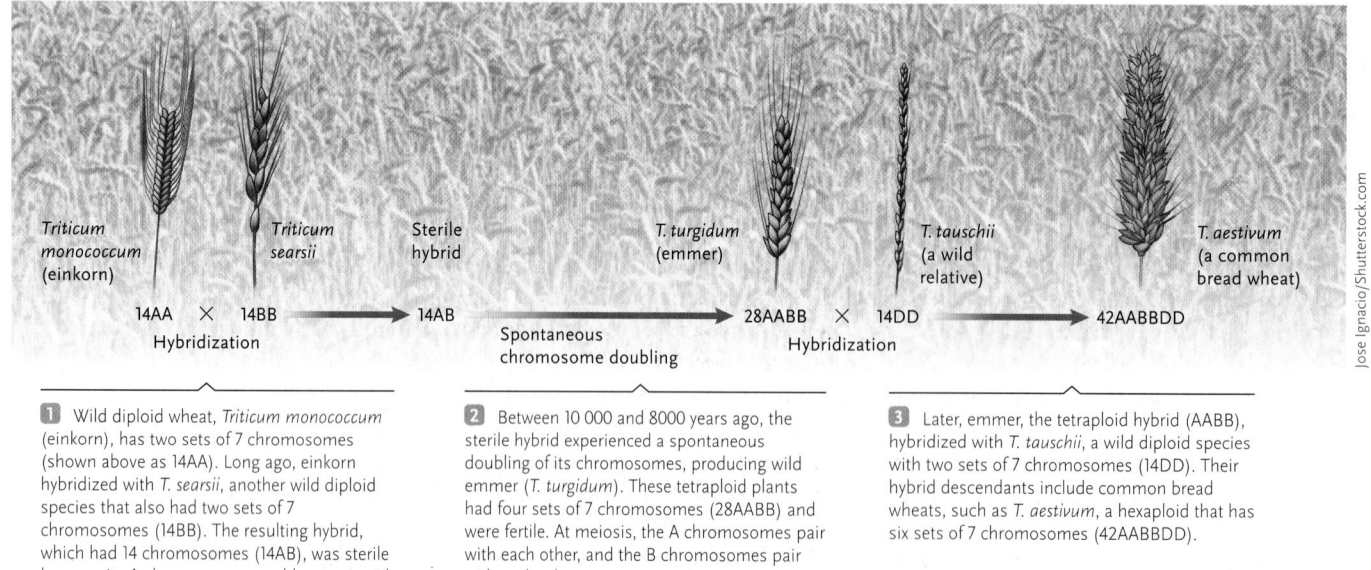

Triticum monococcum (einkorn) | Triticum searsii | Sterile hybrid | T. turgidum (emmer) | T. tauschii (a wild relative) | T. aestivum (a common bread wheat)

14AA × 14BB → 14AB | Spontaneous chromosome doubling → 28AABB × 14DD → 42AABBDD
Hybridization | Hybridization

1 Wild diploid wheat, *Triticum monococcum* (einkorn), has two sets of 7 chromosomes (shown above as 14AA). Long ago, einkorn hybridized with *T. searsii*, another wild diploid species that also had two sets of 7 chromosomes (14BB). The resulting hybrid, which had 14 chromosomes (14AB), was sterile because its A chromosomes could not pair with its B chromosomes during meiosis.

2 Between 10 000 and 8000 years ago, the sterile hybrid experienced a spontaneous doubling of its chromosomes, producing wild emmer (*T. turgidum*). These tetraploid plants had four sets of 7 chromosomes (28AABB) and were fertile. At meiosis, the A chromosomes pair with each other, and the B chromosomes pair with each other.

3 Later, emmer, the tetraploid hybrid (AABB), hybridized with *T. tauschii*, a wild diploid species with two sets of 7 chromosomes (14DD). Their hybrid descendants include common bread wheats, such as *T. aestivum*, a hexaploid that has six sets of 7 chromosomes (42AABBDD).

Jose Ignacio/Shutterstock.com

Figure 19.16

The evolution of wheat. Researchers believe the evolution of common bread wheat resulted from a hybridization between two diploid species, followed by a spontaneous doubling of the hybrid's chromosomes, and a second hybridization between the polyploid hybrid and a third diploid species.

1. How can natural selection promote reproductive isolation in allopatric populations?
2. How does polyploidy promote speciation in plants?

It is ironic that species are so fundamental to biology and evolution and yet present such a challenge when it comes to a clear articulation of an underlying concept. But this reality is hardly surprising given the diversity of living organisms and the range of situations that have led to their evolution. The study of speciation—macroevolution—is the study of evolution in action. Naming species is thus the use of fixed features of an organism to allow it to be identified. However, the application of a fixed system to a dynamic one is rarely successful. If we remember this, then topics from Hardy–Weinberg to gene flow, as well as phenotypic and genotypic variation, should all come into perspective.

Review

 aplia

To access course materials such as Aplia and other companion resources, please visit www.NELSONbrain.com.

19.1 What Is a Species?

- Most biologists define a species as a population of organisms capable of interbreeding and producing fertile offspring.
- Linnaeus invented a system of binomial nomenclature in which each species is given a unique two-part name, called a binomial.
- Species of organisms are given Latinized binomial names that are presented in italics. *Castor canadensis* is the scientific name of the Canadian beaver. *Castor* is the genus, *canadensis* is the trivial name, and the two names together, *Castor canadensis,* are the binomial or species name.
- Species are organized into a taxonomic hierarchy, comprising a nested series of formal categories: domain, kingdom, phylum, class, order, family, genus, species, and subspecies. The organisms included within any category of the taxonomic hierarchy make up a taxon (plural, *taxa*).

19.2 Species Concepts

- The morphological species concept is based on the idea that all individuals of a species share measurable traits that distinguish them from individuals of other species. This concept dates to Linnaeus' classification system.
- The biological species concept defines species as groups of interbreeding populations that are reproductively isolated from populations of other species in nature. A biological species thus represents a gene pool within which genetic material is potentially shared among populations. The biological species concept cannot be applied to organisms that reproduce only asexually, to those that are extinct, or to geographically separated populations.
- The phylogenetic species concept defines a species as a group of populations with a recently shared evolutionary history. Using both morphological and genetic sequence data, scientists first reconstruct an evolutionary tree for the organisms of interest. They then define a phylogenetic species as a cluster of populations that emerge from the same small branch.

19.3 Maintaining Reproductive Isolation

- Reproductive isolating mechanisms are biological characteristics that prevent two species from interbreeding.
- Prezygotic isolating mechanisms either prevent individuals of different species from mating or prevent fertilization between their gametes. Prezygotic isolation occurs because species live in different habitats, breed at different times, use different courtship behaviour, or differ anatomically. Prezygotic isolation can also result from genetic and physiological incompatibilities between male and female gametes.
- Postzygotic isolating mechanisms reduce the fitness of interspecific hybrids through hybrid inviability, hybrid sterility, or hybrid breakdown.

19.4 The Geography of Speciation

- Most species exhibit geographical variation of phenotypic and genetic traits. In ring species, populations are distributed in a ring around unsuitable habitat. Many species exhibit clinal variation of characteristics, which change smoothly over a geographical gradient.
- The model of allopatric speciation proposes that speciation results from divergent evolution in geographically separated populations. If allopatric populations accumulate enough genetic differences, they will be reproductively isolated upon secondary contact. Nevertheless, some species hybridize over small areas of secondary contact.
- The model of parapatric speciation suggests that reproductive isolation can evolve between parts of a population that occupy opposite sides of an environmental discontinuity.
- A model of sympatric speciation suggests that reproductive isolation may evolve between host races that rarely contact one another under natural conditions. Sympatric speciation commonly occurs in flowering plants by allopolyploidy.

19.5 Genetic Mechanisms of Speciation

- Allopatric populations inevitably accumulate genetic differences, some of which contribute to their reproductive isolation. Reproductive isolating mechanisms evolve as by-products of genetic changes that occur during divergence. Prezygotic isolating mechanisms may evolve in populations experiencing secondary contact.

- Speciation by polyploidy in flowering plants involves the duplication of an entire chromosome complement through nondisjunction of chromosomes during meiosis or mitosis. Polyploids can arise among the offspring of a single species (autopolyploidy) or, more commonly, after hybridization between closely related species (allopolyploidy).

- Chromosome alterations can promote speciation by fostering the genetic divergence of, and reproductive isolation between, populations with different numbers of chromosomes or different chromosome structure.

Questions

Self-Test Questions

1. Who is the "father" of taxonomy?
   a. Charles Darwin
   b. Charles Lyell
   c. Alfred Wallace
   d. Carolus Linnaeus
   e. Jean Baptiste de Lamarck

2. In the Linnaean hierarchy, what are the organisms classified within the same taxonomic category called?
   a. a phylum
   b. a taxon
   c. a genus
   d. a binomial
   e. an epithet

3. On what basis does the biological species concept define species?
   a. reproductive characteristics
   b. biochemical characteristics
   c. morphological characteristics
   d. behavioural characteristics
   e. all of the above

4. What is a characteristic that exhibits smooth changes in populations distributed along a geographical gradient called?
   a. ring species
   b. hybrid
   c. cline
   d. hybrid breakdown
   e. subspecies

5. If two species of holly (genus Ilex) flower during different months, how might their gene pools be kept separate?
   a. mechanical isolation
   b. ecological isolation
   c. gametic isolation
   d. temporal isolation
   e. behavioural isolation

6. Which of the following is true about prezygotic isolating mechanisms?
   a. They reduce the fitness of hybrid offspring.
   b. They generally prevent individuals of different species from producing zygotes.
   c. They are found only in animals.
   d. They are found only in plants.
   e. They are observed only in organisms that reproduce asexually.

7. In the model of allopatric speciation, which is true of the geographical separation of two populations?
   a. It is sufficient for speciation to occur.
   b. It occurs only after speciation is complete.
   c. It allows gene flow between them.
   d. It reduces the relative fitness of hybrid offspring.
   e. It inhibits gene flow between them.

8. Adjacent populations that produce hybrid offspring with low relative fitness may be undergoing which of the following?
   a. clinal isolation
   b. parapatric speciation
   c. allopatric speciation
   d. sympatric speciation
   e. geographical isolation

9. An animal breeder, attempting to cross a llama with an alpaca for finer wool, found that the hybrid offspring rarely lived more than a few weeks. What did this outcome probably result from?
   a. genetic drift
   b. prezygotic reproductive isolation
   c. postzygotic reproductive isolation
   d. sympatric speciation
   e. polyploidy

10. Which of the following could be an example of allopolyploidy?
    a. One parent has 8 chromosomes, the other has 10, and their offspring have 36.
    b. Gametes and somatic cells have the same number of chromosomes.
    c. Chromosome number increases by one in a gamete and in the offspring it produces.
    d. Chromosome number decreases by one in a gamete and in the offspring it produces.
    e. Chromosome number in the offspring is exactly half of what it is in the parents.

Questions for Discussion

1. All domestic dogs are classified as members of the species *Canis familiaris*. But it is hard to imagine how a tiny Chihuahua could breed with a gigantic Great Dane. Do you think that artificial selection for different breeds of dogs will eventually create different dog species?

2. Human populations often differ dramatically in external morphological characteristics. On what basis are all human populations classified as a single species?

3. If intermediate populations in a ring species go extinct, eliminating the possibility of gene flow between populations at the two ends of the ring, would you now identify the remaining populations as full species? Explain your answer.

4. How do human activities (such as destruction of natural habitats, diversion of rivers, and construction of buildings) influence the chances that new species of plants and animals will evolve in the future? Frame your answer in terms of the geographical and genetic factors that foster speciation.

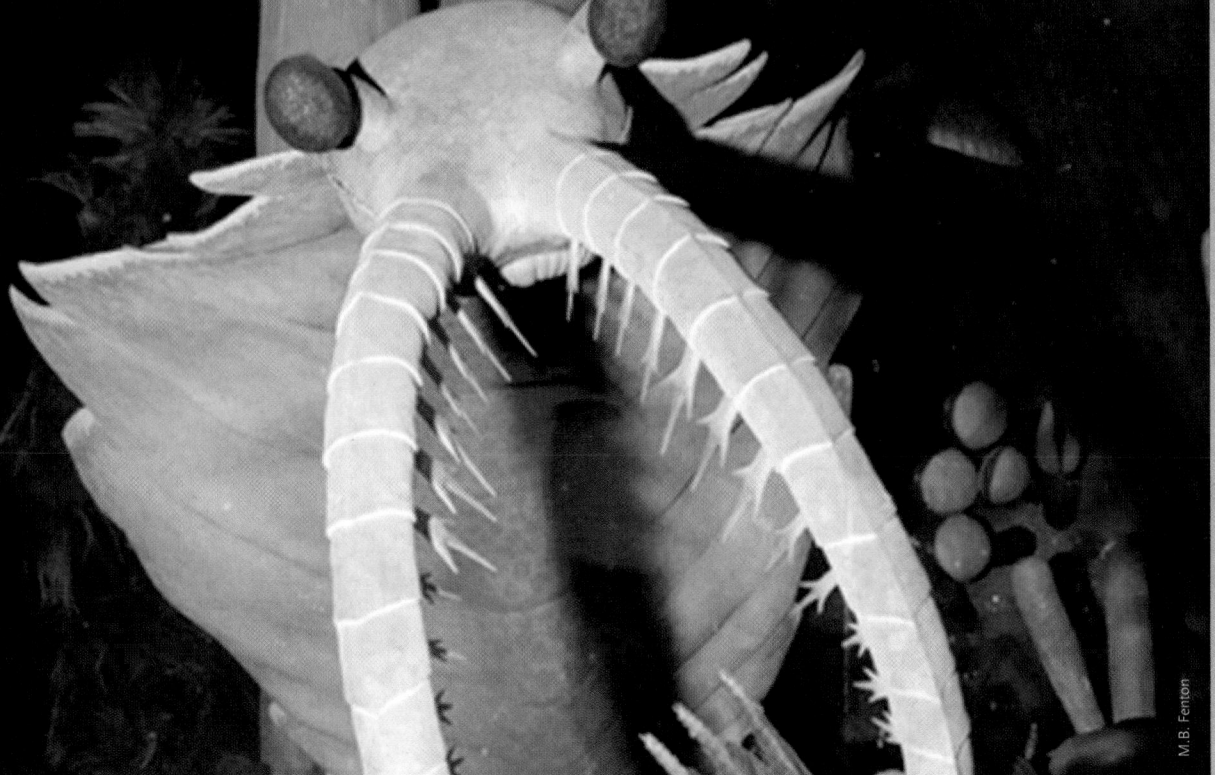

A reconstruction of a 1-m-long *Anomalocaris* that has just caught a prey gives an idea of its bizarre appearance.

M.B. Fenton

Understanding the History of Life on Earth

WHY IT MATTERS

The Burgess Shale biota, in British Columbia, is an assemblage of organisms fossilized over 500 million years ago (mya) in the Cambrian period. The organisms found there present challenges for paleontologists. Many of the specimens are incomplete and most were softbodied organisms that normally did not fossilize well. As well, many of the organisms are hard to imagine for people who grew up knowing dinosaur fossils. The Burgess Shale biota is an example of a treasure trove of fossils, or lagerstätten (German, from *Lager*, "lair or den," and *Stätte*, "place"). Such treasure troves are preserved in deposits of sedimentary rock that contain large numbers of exceptionally well-preserved fossils. The Burgess Shale is one of about 60 lagerstätten known worldwide.

Anomalocaris (meaning "abnormal shrimp," opening photograph) is one of the most spectacular animals of the Burgess Shale. Parts of *Anomalocaris* were originally described as different animals **(Figure 20.1, p. 446).** When it was discovered in 1892, the appendages of *Anomalocaris* were described as a section of the abdomen of a large shrimp. Then *Peytoia* was described from what appeared to be 32 subumbrel lobes thought to belong to a jellyfish. Another fossil, *Tuzoia*, was described as the carapace or shell of a crustacean. The last component was *Laggania*, originally thought to be part of a sea cucumber, but later recognized as another specimen with smaller frontal appendages. Over time, the discovery of more and better-preserved specimens changed our view of this animal. *Anomalocaris* and other arthropods with large frontal ("great") appendages are now classified in the order Megacheira (phylum Arthropoda). Their great appendages are modified mouthparts.

As recently as 2013, most paleontologists agreed that the parts originally described as *Anomalocaris* were the remains of two great appendages. *Peytoia* were plates around the mouth of *Anomalocaris*, while *Tuzoia* was its carapace or shell. *Laggania* was a

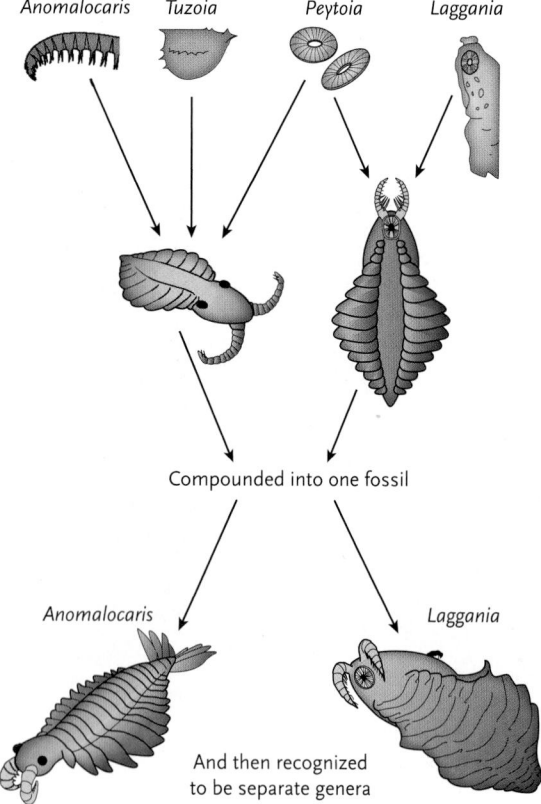

Compounded into one fossil

And then recognized
to be separate genera
within the *Anomalocaridae*

Figure 20.1

How our knowledge of *Anomalocaris* developed and emerged as more fossils were discovered. In the end, *Anomalocaris* is a composite, including parts originally described as a sea cucumber (*Laggania*), a schyphozoan medusa (jellyfish, *Peytoia*), and a bivalve arthropod (*Tuzoia*).

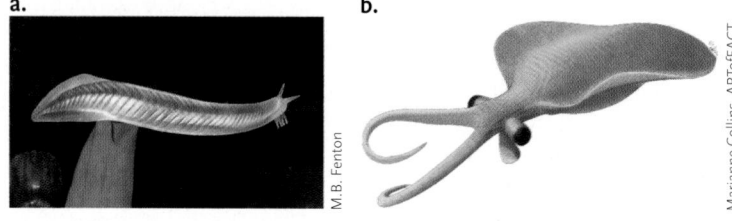

Figure 20.2

Two reconstructions of *Nectocaris*. **(a)** The earlier one presented it as a primitive chordate; **(b)** the later one as a cephalopod mollusc.

poorly preserved remains of another megacheirid, one with smaller great appendages. We presume that the two great appendages were used to seize and break the shells of prey such as trilobites. Megacheirans are known from the Cambrian through the Ordovician, but the lineage does not survive today.

Our views of other animals from the Burgess Shale have also changed considerably. Originally described as a distant relative of vertebrates **(Figure 20.2a)**, *Nectocaris* was more recently recognized as a cephalopod mollusc (Figure 20.2b). Still other animals are simply astonishing, such as *Opabinia* **(Figure 20.3)**.

The Burgess Shale and other Cambrian lagerstätten have provided the world with a glimpse of a long-extinct, highly diverse assemblage of animals. These fossils

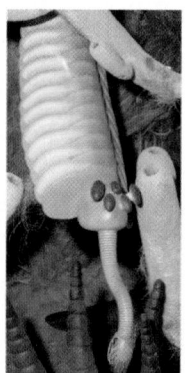

Figure 20.3

A reconstruction of the five-eyed *Opabinia* from the Burgess Shale.

have intrigued biologists and paleontologists for over 100 years. This assemblage reinforces at least three lessons:

1. We must keep an open mind when looking at life. The fossil you are looking at may be a small part of something rather than the whole thing. Some organisms, both living and fossil, are strikingly different from anything with which we are familiar.

2. Many new and astonishing organisms remain to be discovered. Some are fossils, while others are still with us.

3. How could life have diversified so much by about 600 mya? Where were all of these organisms before the Cambrian? And what happened in the millions of years since they lived?

Setting the Stage for Fossils and Phylogeny. Imagine our world without fossils. No Tyrannosaurus rex to frighten you as a child. No insects trapped in amber for hundreds of millions of years. No footprints across sandstone (Figure 21.1). Without fossils, we would understand very little about the history of life on Earth. But how do we relate these species to those that live today?

Millions of species either live or have lived on Earth. You have learned that each was formed from another species by speciation. You also know from experience that dogs and cats are more related than either is to cows or lizards. Systematists keep the binomial classification system up to date and determine the relatedness of all species. One way they do this is by drawing phylogenetic trees similar to Darwin's Tree of Life (Figure 17.9).

In this chapter, we will look at the geological history and the history of life on the planet. Then, we will use what we know about extinct animals (from the fossil record) and modern-day organisms to draw phylogenies to understand how today's species have descended from yesterday's.

20.1 The Geological Time Scale

Many fossils are found in sedimentary rock. Sediments found in any one place form distinctive strata (layers) that usually differ in colour, mineral composition, particle size, and thickness **(Figure 20.4)**. If they have not been disturbed, the strata are arranged in the order in which they formed, with the youngest layers on top. But strata have sometimes been uplifted, warped, or even inverted by geologic processes.

Geologists of the nineteenth century deduced that the fossils discovered in a particular sedimentary stratum,

a. Sedimentation

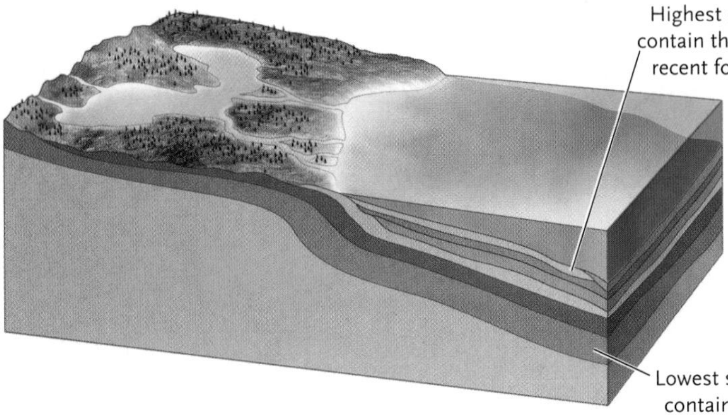

b. Geological strata in the Painted Desert, Arizona

Highest strata contain the most recent fossils.

Lowest strata contain the oldest fossils.

Nick Greaves/Alamy

Figure 20.4

Sedimentation and geological strata. **(a)** Sedimentation deposits successive layers at the bottom of a lake or sea. **(b)** Over millions of years, the upper layers compress those below them into rock. When the rocks are later exposed by uplifting or erosion, the different layers are evident as geological strata.

no matter where it was found, represent organisms that lived and died at roughly the same time in the past (see Chapter 17). Because each stratum was formed at a specific time, the sequence of fossils from lowest (oldest) to highest (newest) strata reveals their relative ages. Geologists used the sequence of strata and their distinctive fossil assemblages to establish the geologic time scale that diagrams the history of life on Earth **(Table 20.1).**

20.2 The Fossil Record

20.2a How Fossils Were Formed

We often see spectacular mounted fossil skeletons in museums **(Figure 20.5, p. 449)**. These were formed when dissolved minerals entered the spaces within the bones and then solidified. Some fossils, such as those preserved in amber (tree resin, **Figure 20.6a, p. 449**), show fine

Table 20.1			The Geological Time Scale and Major Evolutionary Events			
Eon	Era	Period	Epoch	Millions of Years Ago	Major Evolutionary Events	
Phanerozoic	Cenozoic	Quaternary	Holocene	0.01		
			Pleistocene	2.6	Origin of humans; major glaciations 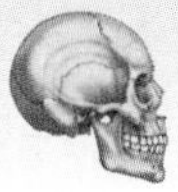	
		Neogene	Pliocene	5.3	Origin of apelike human ancestors 	
			Miocene	23.0	Angiosperms and mammals further diversify and dominate terrestrial habitats 	
		Paleogene	Oligocene	33.9	Primates diversify; origin of apes 	
			Eocene	55.8	Angiosperms and insects diversify; modern orders of mammals differentiate 	
			Paleocene	65.5	Grasslands and deciduous woodlands spread; modern birds, mammals, snakes, pollinating insects diversify; continents approach current positions 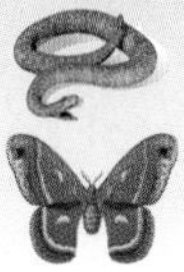	

(Continued)

Eon	Era	Period	Epoch	Millions of Years Ago	Major Evolutionary Events	
	Mesozoic	Cretaceous		145.5	Angiosperms, insects, marine invertebrates, fishes, dinosaurs diversify; asteroid impact causes mass extinction at end of period, eliminating dinosaurs and many other groups	
		Jurassic		201.6	Gymnosperms abundant in terrestrial habitats; modern fishes diversify; dinosaurs diversify and dominate terrestrial habitats; frogs, salamanders, lizards, and birds appear; continents continue to separate	
		Triassic		251.0	Predatory fishes and reptiles dominate oceans; gymnosperms dominate terrestrial habitats; radiation of dinosaurs; early mammals; Pangaea starts to break up; mass extinction at end of period	
Phanerozoic	Paleozoic	Permian		299.0	Insects and reptiles abundant and diverse in swamp forests; some reptiles colonize oceans; fishes colonize freshwater habitats; continents coalesce into Pangaea, causing glaciation and decline in sea level; mass extinction at end of period eliminates 85% of species	
		Carboniferous		359.0	Vascular plants form large swamp forests; first flying insects; amphibians diversify; first reptiles appear	
		Devonian		416.0	Terrestrial vascular plants diversify; fungi, invertebrates, amphibians colonize land; first insects and seed plants; major glaciation at end of period; mass extinction, mostly of marine life	
		Silurian		444.0	Jawless fishes diversify; first jawed fishes, arthropods, terrestrial vascular plants	
		Ordovician		488.0	Major radiations of marine invertebrates and jawless fishes; major glaciation at end of period causes mass extinction of marine life	
		Cambrian		542.0	Appearance of modern animal phyla, including earliest vertebrates (Cambrian explosion); simple marine communities	
Proterozoic				2500	High concentration of oxygen in atmosphere; origin of eukaryotic cells; evolution and diversification of "protists," fungi, softbodied animals	
Archean				3850	Evolution of prokaryotes, including anaerobic and photosynthetic bacteria; oxygen starts to accumulate in atmosphere; origin of aerobic respiration	
Hadean				4600	Formation of Earth, including crust, atmosphere, and oceans; origin of life	

Figure 20.5
Camarasaurus supremus, a herbivorous dinosaur, in the Royal Tyrrell Museum.

M.B. Fenton

details of the organisms. Some fossils, particularly plant fossils, are moulds or impressions (Figure 20.6b). These organisms may have been compressed shortly after death, so three-dimensional analysis may be difficult. Footprints may be fossilized when an animal walks across mud (Figure 20.6b and Figure 21.1). Even droppings (coprolites; Figure 20.6d) and gastroliths, stones in the stomach used to grind food, have been preserved as fossils (Figure 20.6e). Other fossilized remains, including some early humans, are frozen or mummified (Figure 20.6f). Petrified forests are trees fossilized by minerals.

Unfortunately, however, the fossil record is incomplete because few organisms fossilize completely, because some organisms are more likely

a.

b.

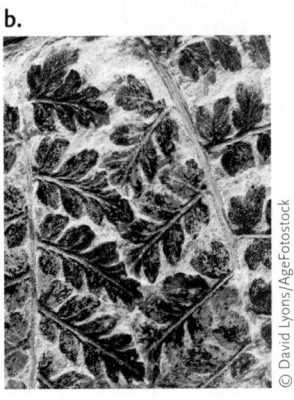

© David Lyons/AgeFotostock

c.

M.B. Fenton

d.

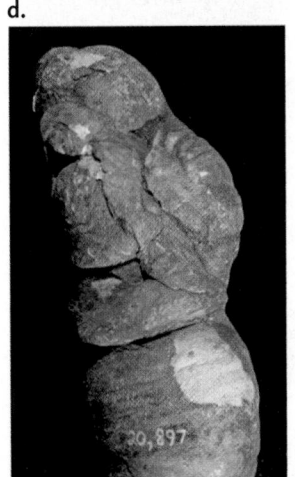

M.B. Fenton

e.

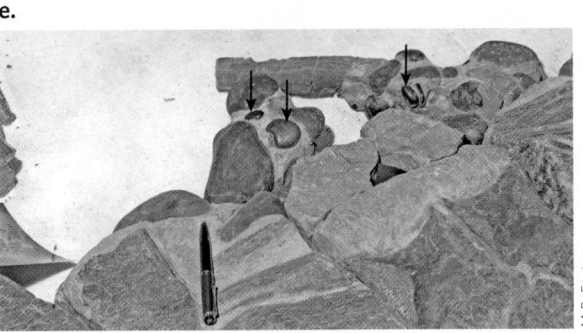

M.B. Fenton

M.B. Fenton

f.

Novosti/Science Source

Figure 20.6
A sampling of fossils. **(a)** Insects in amber; **(b)** an impression of a fern (*Sphenopteris*) from the Carboniferous period, preserved in coal; **(c)** dinosaur footprint; **(d)** coprolite; **(e)** gastroliths (the rounded, polished stones); and **(f)** a frozen baby mammoth.

X-Ray Tomography and 3-D Structure of Fossils

Tomography is the use of thin sections cut from a fossil to reconstruct the organism's appearance, for example, the structures of the wormlike mollusc Acaenoplax (**Figure 1**). X-rays allow researchers to observe fine details of specimens and otherwise invisible specimens (**Figure 2**). Applying tomographic techniques to synchrotron X-ray views allows reconstruction of the fine details of structures such as the surfaces of eggs or the nuclei (**Figure 3**).

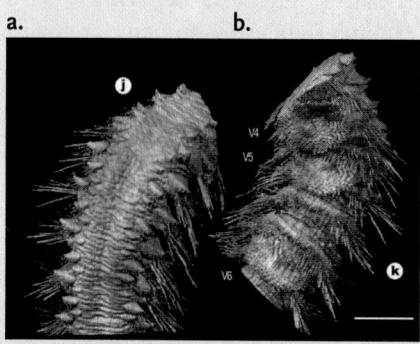

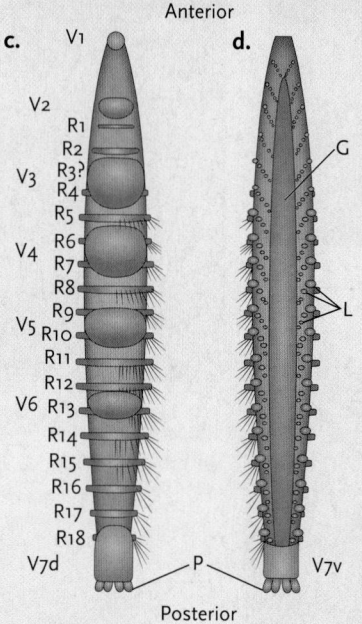

Figure 1

Dorsal (a, c) and ventral (b, d) views of reconstructions of a worm-shaped mollusc, Acaenoplax. (a) and (b) show the anterior 7 mm based on 243 slices at 30 μm intervals. (c) and (d) are overviews of the animals.

SUTTON, M. D., D. E. G. BRIGGS, et al. (2004). "Computer reconstruction and analysis of the vermiform mollusc Acaenoplax hayae from the Herefordshire Lagerstätte (Silurian, England), and implications for molluscan phylogeny," *Palaeontology* 47(2): 293–318. John Wiley and Sons. The Palaeontological Association.

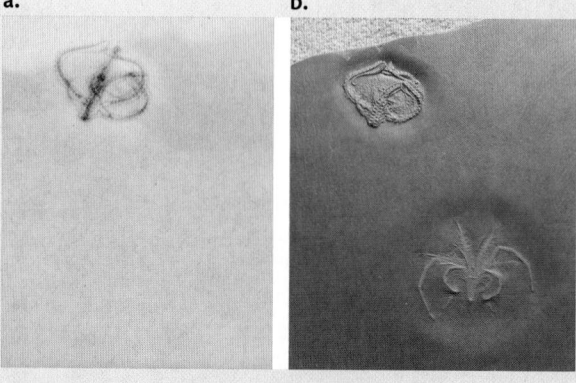

Figure 2

A Photo (a) and X-ray (b) of a 3 to 4 mm thick rock slab from the Lower Devonian. The top animal is an echinoderm (asteroid, Taeniaster beneckei), while the bottom is an arthropod (Mimetaster hexagonalis), visible only in the X-ray.

Republished with permission of British Institute of Radiology, from P. Hohenstein, "X-ray imaging for palaeontology," *British Journal of Radiology* (2004) 77, 420–425.

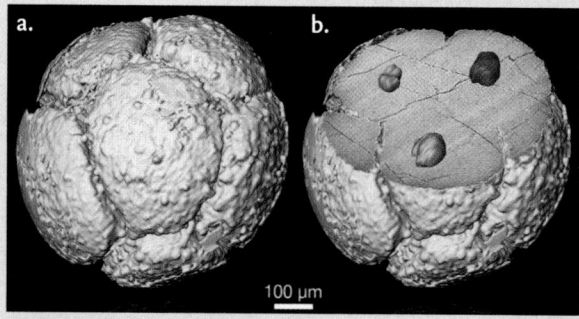

Figure 3

Synchrotron X-ray tomographic reconstructions of embryos of Tianzhushaia from the Ediacaran in China. (a) is a rendering of the surface showing six cells, while (b) shows three nuclei in a slice through the embryo.

From Huldtgren, T., J.A. Cunningham, C. Yin, M. Stampanoni, F. Marone, P.C.J. Donoghue and S. Bengtson, "Fossilized nuclei and germination structures identify Ediacaran "animal embryos" as encysting protists," *Science* 23 December 2011: Vol. 334 no. 6063 pp. 1696–1699. Reprinted with permission from AAAS.

to fossilize than others, and because natural processes destroy many fossils.

20.2b Early Fossils

Stromatolites, the first fossil evidence of life, date to about 3.5 billion years ago (bya). Oxygenic photosynthesis, by blue-green prokaryotes and plants (later), resulted in an increase in atmospheric oxygen from about 2.5 bya. The earliest unicellular eukaryotes date from just over 2 bya, and multicellular eukaryotes appeared by 1.2 bya (see Section 3.4 and Figure 3.20).

The history of life on Earth was not gradual. An "explosion" in the diversity of groups of organisms, about 600 mya, marked the beginning of the Cambrian

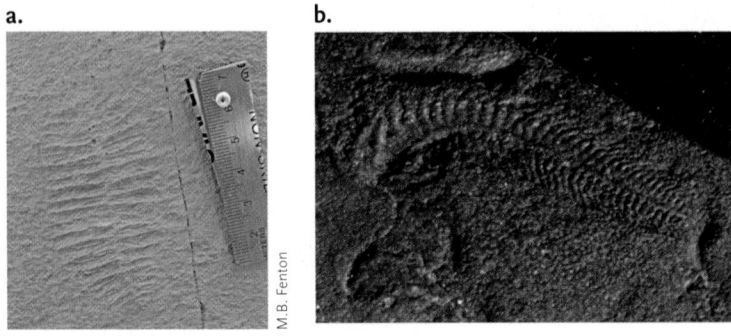

Figure 20.7
Two fossils from the Ediacaran, a rangeomorph **(a)** from Newfoundland and a *Spriggina* **(b)** from Australia.

(Table 20.1). Complex multicellular animals such as those discussed in "Why It Matters" first appeared at this time. It is often difficult to assign some of these fossil animals to any living group, for example, rangeomorphs **(Figure 20.7a)**, which may have been suspension-feeding animals, and *Spriggina* (Figure 20.7b), which appears to have been a polychaete annelid. In each case, the fossils are moulds on a rock surface. Two other explosions have occurred: land plants in the Devonian period and flowering plants in the Cretaceous.

Evolution of Compound Eyes

The eyes of *Anomalocaris* from the Burges Shales revealed previously unknown details of the visual systems of these animals. This evidence indicates an early evolutionary origin of compound eyes **(Figure 1;** see also page 445) and supports the view that *Anomalocaris* was a predator that oriented itself visually in open-water marine environments. Fossils preserved as iron oxide and as calcium phosphate both provide the same detailed picture of the surfaces of the ommatidia making up the compound eyes. Each *Anomalocaris* eye consisted of more than 16 thousand packed ommatidia, approaching the structure of modern insect eyes (Figure 1b).

Several mass extinctions have also occurred, the best known being the Cretaceous–Paleogene event, about 66 mya, in which an asteroid hit Earth near the Yucatán Peninsula, killing most of the dinosaurs along with 75% of all species. Other major mass extinction events occurred at 200, 251, 360, and 450–440 mya. In each event, at least 70% of all species went extinct. Mass extinctions have a variety of causes, such as volcanic eruptions and climate change.

After each mass extinction, many niches and other opportunities opened up for the surviving species, which underwent a period of **adaptive radiation.** Each lineage could diversify rapidly, taking advantage of the newly available niches and the reduction in competition.

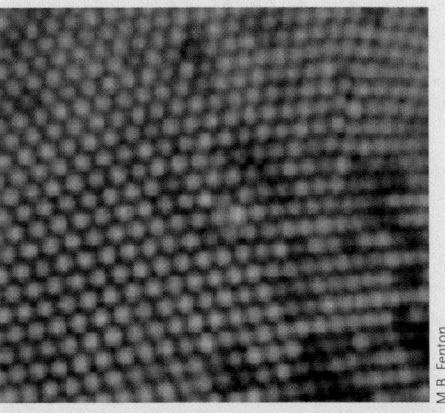

Figure 1
Ommatidia making up compound eyes in a fossil trilobite (a) and an extant deer fly (b). The ommatidia of Anomalocaris *are more like those of the fly than of a trilobite. A combination of scanning electron micrography and dispersive spectrometry was used to obtain structural details of the fossils.*

David Evans

David Evans is a young paleontologist at the Royal Ontario Museum in Toronto. Not only does he search for dinosaur fossils in places as diverse as Alberta and Mongolia, he also tries to determine the function(s) of certain structures of the extinct reptiles. With colleagues in Ohio, Dr. Evans appears to have solved a long-standing puzzle about lambeosaurines, a group of huge duck-billed dinosaurs that lived in the swampy habitats of western North America in the late Cretaceous period. Lambeosaurine heads sported bony crests with gigantic nasal passages **(Figure 1)**. Paleobiologists proposed several functions of the crests: as weapons in male combat, adornments that attracted mates, snorkels that facilitated breathing underwater, radiators that cooled the dinosaurs' bodies, structures that enhanced the sense of smell, or resonating chambers that produced honking vocalizations. Based on anatomical analyses, researchers accepted the vocalization hypothesis as the most probable: computerized acoustic models predicted that air flowing through the nasal passages would have produced low-frequency sounds (30–375 Hz).

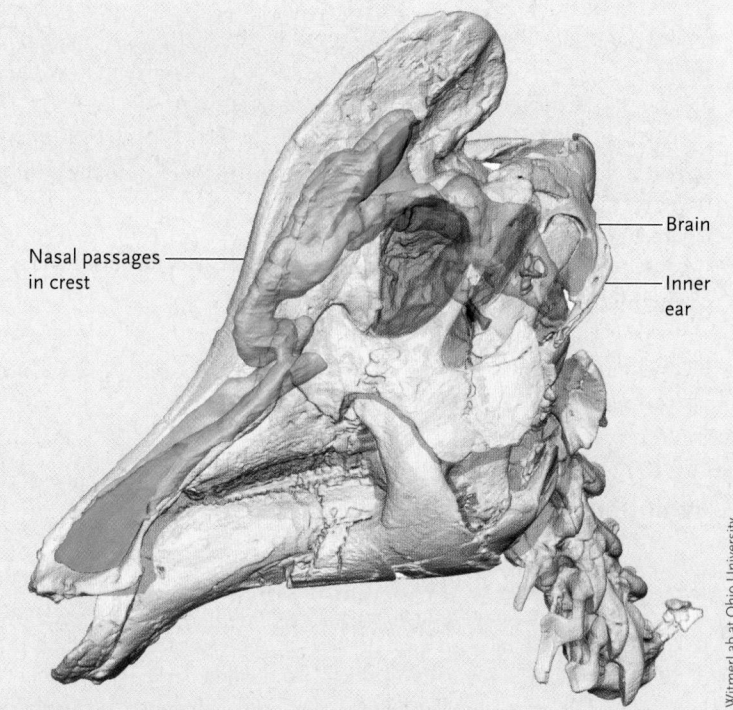

Nasal passages in crest — Brain — Inner ear

WitmerLab at Ohio University

Figure 1

Honking dinosaurs. Analysis of the sinuses and braincase of lambeosaurines (Corythosaurus spp.) revealed that the nasal passages in their crests served as resonating chambers for the production of low-frequency sounds that their inner ears could detect.

But could lambeosaurines *hear* sounds in that frequency range? In 2009, Dr. Evans and his colleagues used computed tomography and 3-D visualization software to scan and reconstruct the interior anatomy of the skulls of several lambeosaurine species. Their findings show that the inner ears of lambeosaurines were attuned to hear low-frequency sounds that matched those predicted by the earlier research. They concluded that the elaborate nasal passages in lambeosaurine crests, along with the structure of their inner ears, facilitated vocal communication.

20.2c   The Importance of Skeletons

Not surprisingly, the apparent explosion of life marking the beginning of the Cambrian coincides with the appearance of skeletons—structures that support organisms. The early fossils of bacteria and plants are recognizable due to the presence of cell walls that provided physical support and maintained the integrity of cell shape and structure. The absence of cell walls and analogous structures in invertebrate animals led to fewer fossils. However, some early animal fossils from the Ediacaran (e.g., *Spriggina*, Figure 20.7) suggest that they had supporting systems such as cuticles (see "Evolution of Compound Eyes," p. 451). Insects and other arthropods fossilized due to their exoskeleton.

When teeth and bones appear in fossil finds, it marks a sharp increase in the abundance of fossils because hard tissues lend themselves to mineral fossilization, while soft tissues are usually fossilized as moulds. Dinosaurs are particularly well represented in the fossil record because of their size and their well-developed bones (see "People behind Biology").

STUDY BREAK

1. How old are the first fossils? How old are the first fossil multicellular organisms?
2. What is a mass extinction event? Describe the Cambrian explosion.
3. Why are skeletons important to the fossil record?

20.3 Lives of Prehistoric Organisms

20.3a The Move onto Land

The movement of organisms out of the water and onto land was a momentous event in diversification. Freshwater algae existed before the first land plants appeared, during the Ordovician Period. These land plants were similar to today's liverworts and mosses. By the Silurian, lycopsids, the oldest of the vascular plants, had diverged from the main lineage that led to ferns, horsetails, and seed plants. By the middle of the Devonian, treelike lycopsids grew to at least 8 m tall, constituting the first forests and adding a vertical dimension to terrestrial ecosystems. In the Permian, some lycopsid trees were 40 m tall. These trees probably grew rapidly, perhaps achieving maturity in a few years. At maturity, lycopsids reproduced with spores and then died. Later in the Permian, tree ferns (still seen today in New Zealand and parts of Australia) replaced lycopsids as the trees of swamp forests. Today lycopsids survive as quillworts and clubmosses.

For animals to move onto land required several adaptations, such as sufficient support to allow them to maintain their integrity, and some form of protection from ultraviolet radiation. Adaptations to ensure a waterproof coating, as well as necessary changes in mechanisms of gaseous exchange, excretion, and locomotion, were also needed. Terrestrial plants provided some of the necessities for animals, such as food, oxygen, and shelter.

A variety of modern bony fishes, such as mudskippers, climbing perch, and some catfishes, regularly move onto land. The same is true of many annelids, molluscs, and arthropods. This suggests that the earliest movements of animals onto land occurred as short visits.

By the Devonian, the diversification of insects reflects the appearance of lycopsid forests. Living insects can generally be sorted into two groups. Hemimetabolous insects (e.g., dragonflies, cockroaches) have three-stage life cycles, with eggs, nymphs, and adults. Holometabolous insects have four-stage life cycles, with eggs, larvae, pupae, and adults. The latter group is the more diverse, including beetles, lepidopterans (moths and butterflies), hymenopterans (ants, bees, and wasps), and flies. In 2013, the description of previously unknown fossil insects revealed that holometabolous insects had appeared by the middle of the Carboniferous, about 300 mya, but underwent a striking diversification after the mass extinction event that marked the end of the Permian.

By the Carboniferous, terrestrial insects were diverse and included winged forms. Of the ~16 orders of insects in the Carboniferous, 11 did not survive into the Mesozoic. The wings of mid-Carboniferous dragonflies (Odonata) showed specializations that persist in living dragonflies. Coevolutionary relationships between plants and insects developed by the Late Carboniferous when some insect larvae formed galls in the internal tissue of tree fern fronds. Fossilized plant tissues show diagnostic histological and cellular details **(Figure 20.8)** typical of modern gall-forming plant–insect interactions. Amber (Figure 20.6a) had appeared in the Carboniferous, demonstrating that some trees had the biosynthetic mechanisms necessary to produce resins in defence against insects. Dissection of insects, plant parts, and other organisms in amber have revealed a great deal about their evolution.

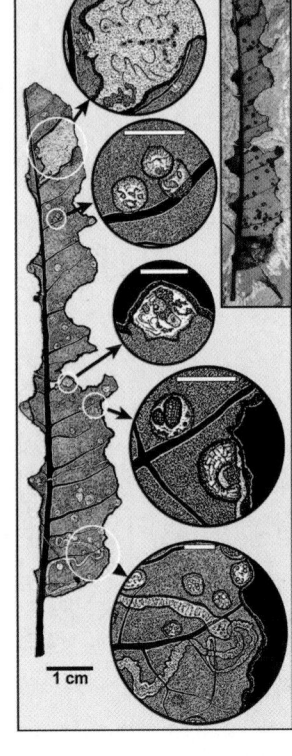

Figure 20.8
This leaf shows evidence of attack by insects, including mining, galling, and external feeding. Photographed specimen (inset upper right) is shown in detail on the left. From top to bottom: a blotch mine with insect feces (frass). Then three galls, one with an exit hole along a secondary vein, another with its margin consumed by an external feeder. Bottom, two galls, one partly consumed, along with two linear mines.

Wilf et al. 2005. "Richness of plant–insect associations in Eocene Patagonia: A legacy for South American biodiversity," *PNAS*, vol. 102 no. 25 8944–8948. Copyright 2005 National Academy of Sciences, U.S.A.

STUDY BREAK

1. Why was the movement of plants onto land fundamentally important for animals?
2. What are lycopsids and tree ferns?

20.3b The Move Back to Water

The first mammals appeared about 165 mya as small furred land animals. After the Cretaceous–Paleogenic mass extinction, the surviving mammals expanded, filling many of the now-vacant niches. Some mammals, such as otters and beavers, became semi-aquatic, bearing their young on land while living much of their lives in water.

Several lineages of mammals are marine, some of which never leave the water (Sirenia: manatees and dugongs; Cetacea: whales). Whales and toothed whales are a distinctive group of mammals whose fossil record extends back more than 50 Ma. One of the Eocene whales, *Basilosaurus cetoides,* had a pelvic girdle and leg and foot bones **(Figure 20.9, p. 454)**, features lacking in modern cetaceans. The discovery of an Eocene fossil *Indohyus,* a raccoon-sized mammal known from fossils in Pakistan, has had interesting repercussions for the classification of mammals. *Indohyus* had a groovelike structure (the involucrum; **Figure 20.10, p. 454**) at the back of its skull, a feature now known to occur in both cetaceans and even-toed ungulates or artiodactyls (pigs,

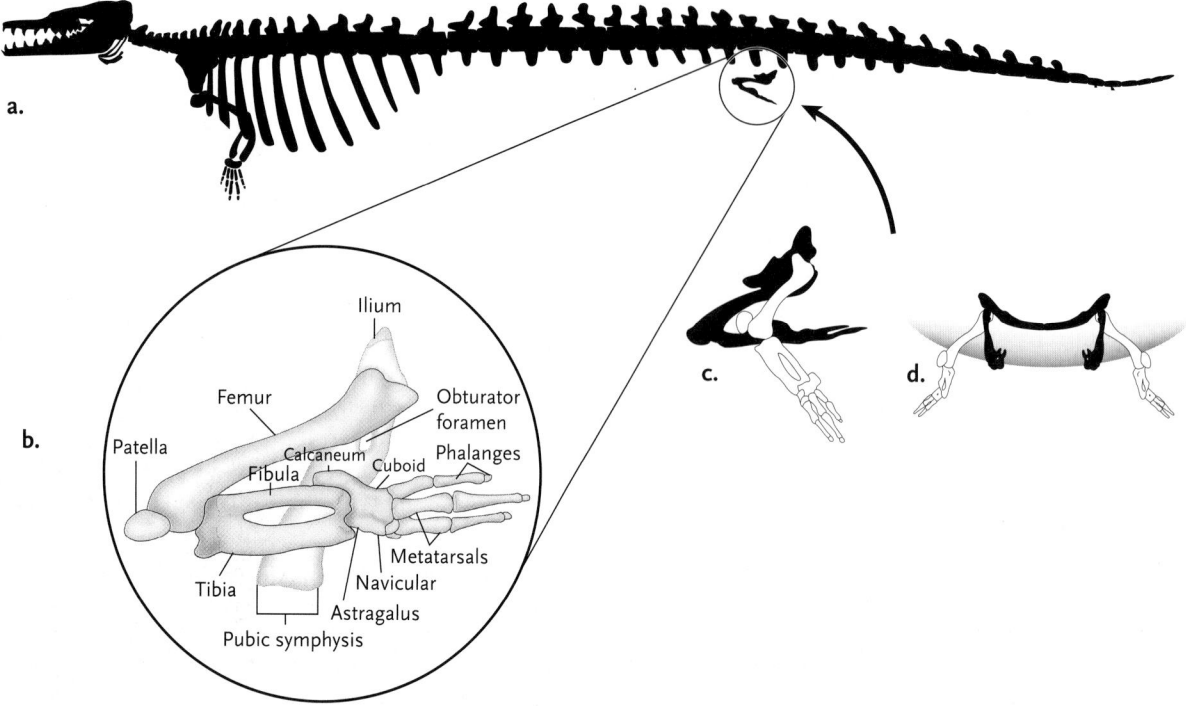

Figure 20.9

Unlike modern whales (Cetacea), the Eocene whale *Basilosaurus cetoides* from Egypt had a functional pelvic girdle and leg and foot bones. **(a)** The skeleton of the Eocene whale. **(b)** Skeletal details of the hind limb. **(c)** The pelvic girdle in resting posture; **(d)** in extended posture.

Based on Gingerich, P.D., B.H. Smith and E.L. Simons. 1990. "Hind limbs of Eocene Basilosaurus: evidence of feet in whales" *Science* 249:154–157. Gingerich, P.D., M. ul-Haq, W. von Koenigswald, W.J. Saunders, B.H. Smith and I.S. Zalmout. 2009. "New protocetid whale from the Middle Eocene of Pakistan: birth on land, precocial development, and sexual dimorphism," PLoS ONE, 4:e4366. From FENTON/DUMONT/OWEN. Integrative Animal Biology, 1E. © 2014 Nelson Education Ltd. Reproduced by permission. www.cengage.com/permissions

deer, camels, and hippopotamuses). The involucrum indicated that cetaceans and artiodactyls are more closely related to one another than either is to any other mammals. This discovery is the grounds for placing whales and artiodactyls in one order, the Cetartiodactyla.

Isotopic analysis of the teeth of *Indohyus* illustrated that over time, these animals showed a change in diet, from eating plants to eating fish. *Indohyus* apparently lived a hippopotamus-like existence, originally going ashore to graze and returning to the water, probably to avoid predators. This information provides clues to the origin of cetaceans and the fundamental changes that occurred as they became more and more aquatic. Modern whales never leave the water, unlike seals and their relatives, which haul out at least to give birth. They are divided into two main groups: toothed whales (odontocetes) that eat mainly fish and other larger marine animals, and baleen whales (mystacetes) that eat plankton (small marine organisms). The initial diversification of cetaceans involved a switch in diet from plants to fish.

The sea cows (manatees or dugongs, Sirenia) are fully aquatic herbivores apparently belonging to an African lineage of mammals, the Afrotheria, that also

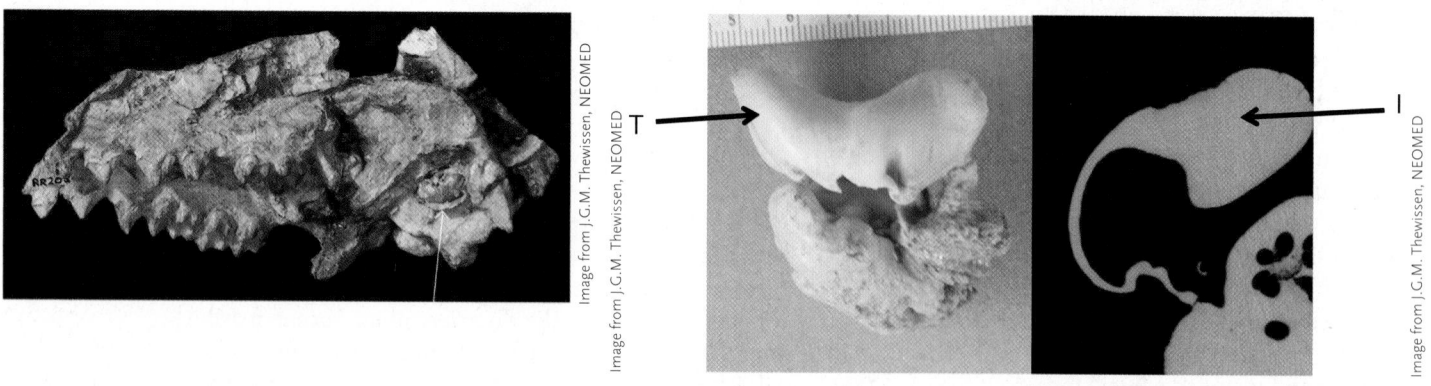

Figure 20.10

(a) The skull of an *Indohyus*, an even-toed ungulate (Artiodactyla), has a well-developed involucrum (white arrow). This feature indicates that whales and even-toed ungulates belong in a single order, Cetartiodactyla. The skull is about 15 cm long. **(b)** A lateral view of the bony part of the right ear of a modern whale (beluga, *Delphinapterus leucas*) showing the lateral wall of the tympanic plate (T). In **(c)**, a CT scan (bone in white) illustrates the thin wall of the tympanic plate and the thicker involucrum (I). Scale upper left is in millimetres. Dorsal is top of images.

includes elephants. Sirenians may be older (Palaeocene) than cetaceans and apparently originated in fresh water. The earliest complete fossil sirenians are from Jamaica. The bones indicate that the animals could move onto land but probably spent most of their lives in the water. The sirenians remained herbivores, unlike cetaceans.

A return to an aquatic existence also evolved several times in reptiles. In terms of body form, ichthyosaurs (Figure 20.11) were most similar to whales and more fishlike than the other marine reptiles. Sea turtles, mosasaurs, and plesiosaurs are other marine reptiles. The flipperlike limbs of marine mammals and reptiles (Figure 20.11) are clear evidence of anatomical convergence. The similarity extends to the flipperlike wings of penguins, marine birds that "fly" underwater.

Once again, the fossil record provides details about morphological adaptations and examples of convergence in form and function, as well as sometimes surprising information about evolutionary relationships among organisms.

STUDY BREAK

1. What is the involucrum? What does it tell us about relatedness among mammals?
2. How did the return to an aquatic existence differ between cetaceans and sirenians?

20.4 Phylogeny

Earlier, you saw Darwin's conception of branching evolution—a tree of life (Figure 17.9 and **Figure 20.12, p. 456**), as well as an evolutionary tree showing the three domains of life (Figure 3.20). Biologists call this type of diagram a phylogeny. Just like a family tree, **phylogenies** show the evolutionary history of a group of organisms. Phylogenies are presented as **phylogenetic trees**, which are formal hypotheses identifying likely relationships among groups of organisms (Figure 3.20 and **Figure 20.13, p. 457**). Like all hypotheses, they can be tested with data and are often revised as scientists gather new data. Some

a.

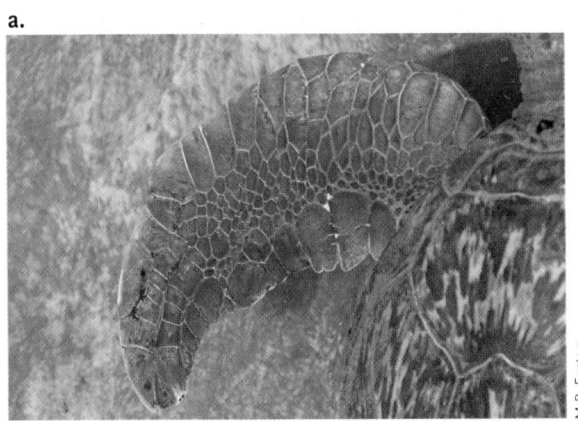

b.

c.

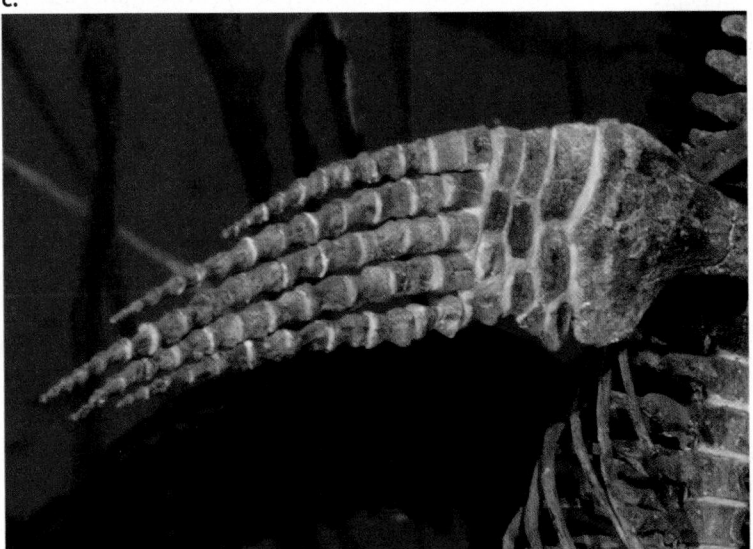

d.

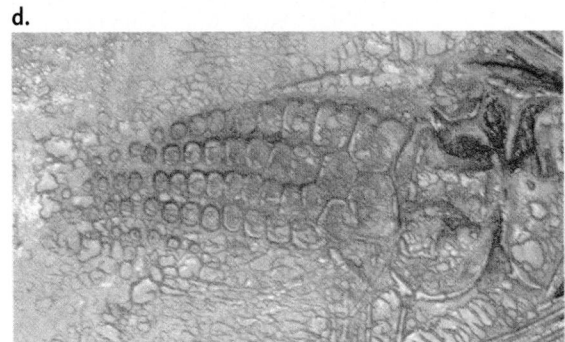

Figure 20.11
Pectoral flippers from marine reptiles. **(a)** An extant hawksbill turtle (*Eretmochelys imbricata*), **(b)** the Cretaceous turtle *Toxochelys*, **(c)** a plesiosaur, and **(d)** an ichthyosaur.

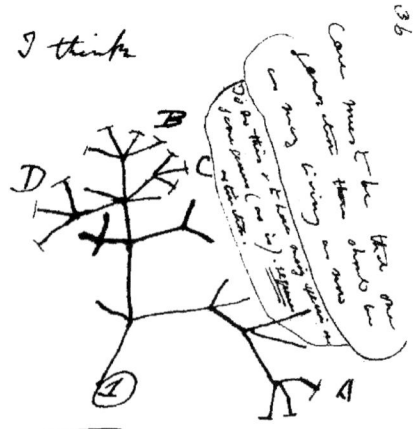

I think

Figure 20.12
Darwin and branching evolution. This entry from his notebook on the "transmutation of species" demonstrates that Charles Darwin first thought about the branching pattern of evolution in 1837, more than 20 years before he published *On the Origin of Species*.

phylogenetic trees show relatedness of large groups of organisms (phyla or classes); others show how genera or species are related. Many phylogenies include prehistoric organisms along with modern-day ones.

From before Linnaeus (see Section 19.1) until the mid-1900s, systematists classified organisms and developed phylogenies based mainly on morphological characters. Over the past 50 years, along with morphology, they considered patterns of behaviour and traits such as chromosomal anatomy, details of physiology, morphology of subcellular structures, cells, and organ systems. Modern systematists also use molecular sequences of nucleic acids and proteins as additional characters.

Accurate phylogenetic trees are essential components of the comparative method that biologists use to analyze evolutionary processes. Robust phylogenetic hypotheses allow us to distinguish similarities inherited from a common ancestor from those that evolved independently in response to similar environments.

Data collected and organized by systematists also allow biologists to select appropriate organisms for their work. Many biological experiments are first conducted with individuals of a single species (see Chapter 18), particularly a species that is a closed genetic system, where individuals do not hybridize with members of related species. If a researcher inadvertently used two species that responded differently, the mixed results would probably confuse the underlying picture.

20.4a Evaluating Systematic Characters

Systematists use guidelines to select characters for study. As we saw previously, there is more to being a beaver than being a mammal with a flattened tail. Systematists seek characters that are independent markers of underlying genetic similarity and differentiation. Ideally, systematists create phylogenetic hypotheses and classifications by analyzing the genetic changes that caused speciation and differentiation. But the fossil record is not complete, so systematists often rely on phenotypic traits as indicators of genetic similarity or divergence. Systematists study traits in which phenotypic variation reflects genetic differences, while trying to exclude differences caused by environmental conditions.

Useful systematic characters must be genetically independent, reflecting different parts of organisms' genomes. This precaution is necessary because different organismal characters can have the same genetic basis. We want to use each genetic variation only once in an analysis. For example, tropical lizards in the genus *Anolis* can climb trees because they grip the bark with small adhesive pads on the undersides of their toes. The number of pads varies from species to species and toe to toe. Researchers have used the number of pads on the fourth toe of the left hind foot as a systematic character. They do not use the number of pads on the fourth toe of the right hind foot as a *separate* character because the same genes almost certainly control the number of pads on the toes of both hind feet. The point here is not the fine-grained detail about toes, but rather the kinds of characters that can be used when assembling a picture of adaptive radiation.

Homologous Characters. The limbs of tetrapod vertebrates are homologous characters that are similar in their evolutionary history but not necessarily their function. Homologous characters are useful in preparing phylogenies because they reflect underlying genetic similarities, and the comparison of homologous characters can indicate common ancestry and genetic relatedness.

Where their functions have changed, homologous structures (inherited from a common ancestor) can differ considerably among species. The stapes, a bone in the middle ear of tetrapod vertebrates, evolved from (is homologous to) the hyomandibula, a bone that supports the jaw joint of most fishes. The structure, position, and function of the hyomandibula are different in tetrapods than they are in fishes **(Figure 20.14)**.

Homologous characters emerge from comparable embryonic structures and grow in similar ways during development. Systematists have put great stock in embryological indications of homology on the assumption that evolution has conserved the pattern of embryonic development in related organisms. Indeed, recent discoveries in **evolutionary developmental biology** have revealed that some genetic controls of developmental pathways can be very similar across a wide variety of organisms (e.g., *Pax-6* genes control the development of eyes; see Chapters 1 and 42). The same situation applies to *HoxD* genes that control the development of limbs.

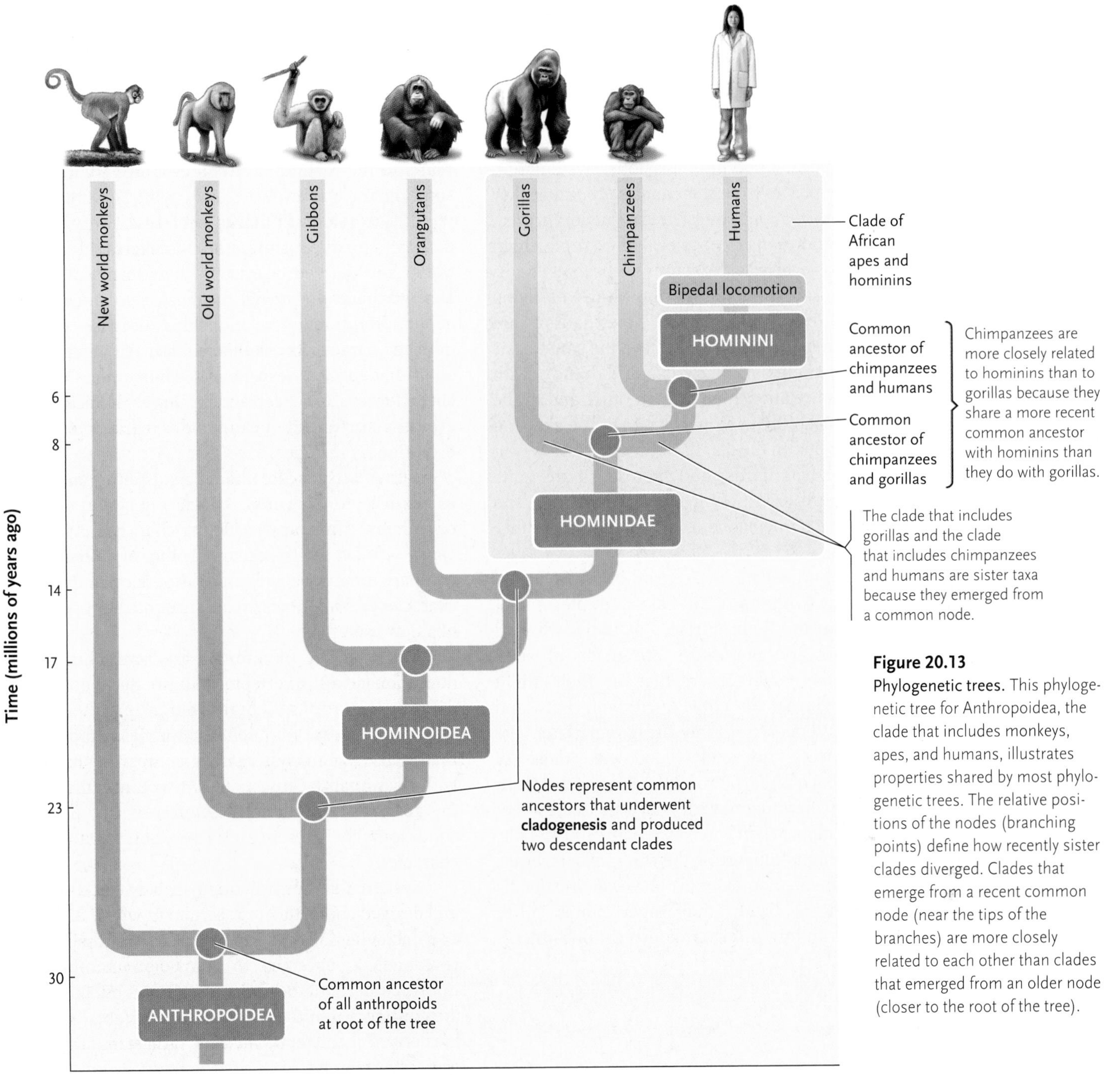

Clade of African apes and hominins

Bipedal locomotion

HOMININI

Common ancestor of chimpanzees and humans

Common ancestor of chimpanzees and gorillas

Chimpanzees are more closely related to hominins than to gorillas because they share a more recent common ancestor with hominins than they do with gorillas.

HOMINIDAE

The clade that includes gorillas and the clade that includes chimpanzees and humans are sister taxa because they emerged from a common node.

New world monkeys
Old world monkeys
Gibbons
Orangutans
Gorillas
Chimpanzees
Humans

Time (millions of years ago)

6
8
14
17
23
30

HOMINOIDEA

Nodes represent common ancestors that underwent **cladogenesis** and produced two descendant clades

ANTHROPOIDEA

Common ancestor of all anthropoids at root of the tree

Figure 20.13

Phylogenetic trees. This phylogenetic tree for Anthropoidea, the clade that includes monkeys, apes, and humans, illustrates properties shared by most phylogenetic trees. The relative positions of the nodes (branching points) define how recently sister clades diverged. Clades that emerge from a recent common node (near the tips of the branches) are more closely related to each other than clades that emerged from an older node (closer to the root of the tree).

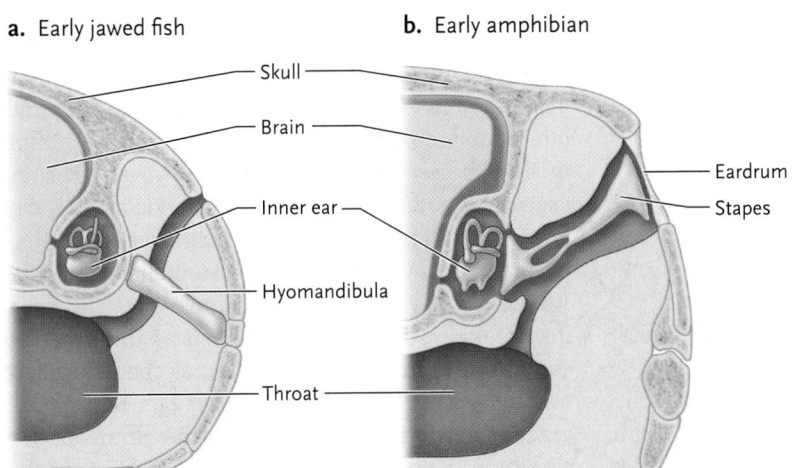

a. Early jawed fish

Skull
Brain
Inner ear
Hyomandibula
Throat

b. Early amphibian

Skull
Brain
Eardrum
Stapes

Figure 20.14
Homologous bones, different structures and functions.
(a) The hyomandibula braced the jaw joint against the skull in early jawed fishes. **(b)** The hyomandibula is homologous to the stapes, which transmits sound to the inner ear in tetrapods, exemplified here by an early amphibian. Both diagrams show a cross-section through the head just behind the jaw joint.

Analogous Characters. Analogous characters are those in different animals that serve the same function. They are homoplasious (plural noun form **homoplasies**), phenotypic similarities that evolved independently in different lineages. For example, the flattened tails of aquatic mammals such as beavers and platypuses appear to be homoplasious. Systematists exclude homoplasies from their analyses because they provide no information about shared (genetic) ancestry.

The situation can be complex. For example, flight in animals has evolved at least four times (bats, birds, insects, pterosaurs). Bones in the wings of flying vertebrates (bats, birds, and pterosaurs) are homologous **(Figure 20.15).** They have the same basic structural elements (arm, wrist, and hand), with similar spatial relationships to each other and to the bones that attach the wing to the rest of the skeleton (shoulder girdle). In the details of the bones, however, the wings of bats, birds, and pterosaurs are quite different (Figure 20.15). Wing bones of bats, birds, and pterosaurs are homologous to the forelimbs of other tetrapods.

But the large flat surfaces of bird wings are homoplasious with those of bats and pterosaurs. Feathers form the flight surfaces of birds, whereas those of bats and pterosaurs are made of skin. Therefore, one could assert that in their flight membranes, birds are convergent with bats and pterosaurs. When we extend the comparison, the wings of insects are convergent with those of vertebrates (bats, birds, pterosaurs) (Figure 20.15). In this situation, the wings of vertebrates are examples of parallel evolution. When you consider the fine details, the basic elements supporting the wings of bats, birds, and pterosaurs are homologous. However, the details of the forearm, hand, and finger bones differ substantially among these three groups of animals.

The example of wings illustrates that the distinction between parallel and convergent evolution is based on closeness of relationships, and the groups included in the comparison.

Ancestral and Derived Characters. **Mosaic evolution** refers to the reality that in all evolutionary lineages, some characters evolve slowly, while others evolve rapidly. Mosaic evolution is pervasive. Every species displays a mixture of **ancestral characters** (old forms of traits) and **derived characters** (new forms of traits). Derived characters provide the most useful information about evolutionary relationships because once a derived character is established, it usually persists in all of that species' descendants. Thus, unless derived characters are lost or replaced by newer characters over evolutionary time, they can serve as markers for entire evolutionary lineages.

Systematists score characters as either ancestral or derived only when comparing them among organisms. Thus, any particular character is derived *only in relation to* what occurs in other organisms. The comparison may be with an older version of the same character or, sometimes, its absence and the appearance of a new trait.

Most species of animals are invertebrates, by definition lacking a vertebral column. Backbones are a defining feature of vertebrates, the animal lineage that includes fishes, amphibians, reptiles, birds, and mammals. When systematists compare vertebrates with all animals lacking a vertebral column, they score the absence of a vertebral column as the ancestral condition. The presence of a backbone is a derived character.

Systematists also distinguish between ancestral and derived characters to ascertain in which direction a character has evolved. In some cases, the fossil record is detailed enough to provide unambiguous information about the direction of evolution. Biologists are confident that the presence of a vertebral column is a derived character because the earliest fossil animals were invertebrates.

Systematists use **outgroup comparison** to distinguish ancestral from derived characters. This involves comparing the group under study with more distantly related species constituting a group not otherwise included in the analysis. For example, most modern butterflies have six walking legs, but some species in two families (Nymphalidae and Papilionidae) have four walking legs and two small, nonwalking legs **(Figure 20.16).** Which is the ancestral character state? Which is derived? Outgroup comparison with other insects shows that six walking legs is the prevalent condition, representing an ancestral character. Four walking legs is a derived character. The same would apply when trying to understand the almost legless condition of female bagworms (Psychidae), another group of butterflies.

Figure 20.15

Arm, hand, and finger bones support the wings of a bat **(a)**, a bird **(b)**, and a pterosaur **(c)**. The wrist position (arrow) is more similar between bats and birds than either is to the pterosaur.

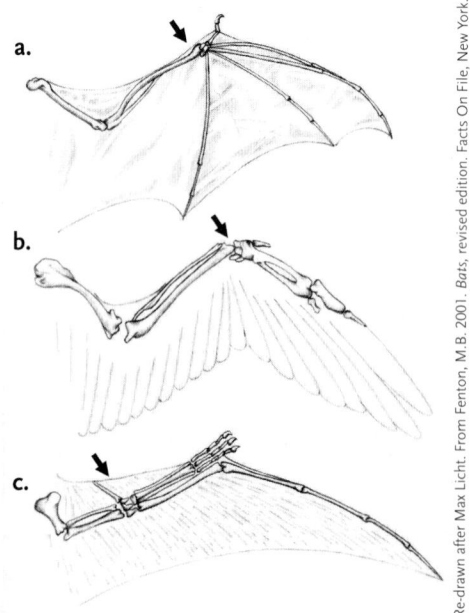

a.

b.

c.

Re-drawn after Max Licht. From Fenton, M.B. 2001. *Bats*, revised edition. Facts On File, New York.

a. Caddis fly

b. Orange palm dart butterfly

c. Monarch butterfly

Figure 20.16

In preparing a phylogeny, it is vital to include in the comparison a species that is an outgroup. In this case, **(a)** the caddis fly (order Trichoptera, family Limnephilidae) is not as closely related as the two butterflies: **(b)** orange palm dart butterfly (*Cephrenes auglades*, family Hesperiidae) and **(c)** monarch butterfly (*Danaeus plexippus*, family Nymphalidae). The comparison suggests that six walking legs (a and b) is ancestral in insects, and four walking legs (c) is a derived character state.

20.4b Phylogenetic Inference and Classification

Phylogenetic trees portray the evolutionary diversification of lineages as a hierarchy that reflects the branching pattern of evolution. Each branch represents the descendants of a single ancestral species. When converting the phylogenetic tree into a classification, systematists use the **principle of monophyly**. They try to identify **monophyletic taxa**, those derived from a single ancestral species **(Figure 20.17)**. **Polyphyletic taxa** include species from separate evolutionary lineages. If, based on the presence of wings, we placed bats, birds, pterosaurs, and insects in one taxonomic group (flying animals), it would be polyphyletic. A **paraphyletic taxon** includes an ancestor and some, but not all, of its descendants. The traditional taxon class Reptilia is paraphyletic because it includes some obvious reptiles, such as turtles, lizards, dinosaurs, and crocodiles, but not other descendants, such as mammals and birds.

Many systematists also strive to create parsimonious phylogenetic hypotheses. According to the **assumption of parsimony** (also known as Occam's Razor), the simplest explanation of an issue is usually the most accurate. Systematists assume that any particular evolutionary change is an unlikely event and presumably happened only once in any evolutionary lineage. Phylogenetic trees illustrate hypotheses that place all organisms on a single branch. Birds are portrayed as a single evolutionary branch, implying that feathered wings evolved once in their common ancestor. This hypothesis is more parsimonious than one proposing that feathered wings evolved independently in two or more vertebrate lineages. The monophyly of birds is not contradicted by the repeated evolution of flightlessness in this group. Recently discovered fossils have changed our appreciation of the evolution of birds (see Section 20.2).

20.4c Phenotypic Similarities and Differences: Traditional Evolutionary Systematics

For a century after the publication of Charles Darwin's theory of evolution by natural selection, most systematists followed Linnaeus' practice of inferring

Monophyletic taxon

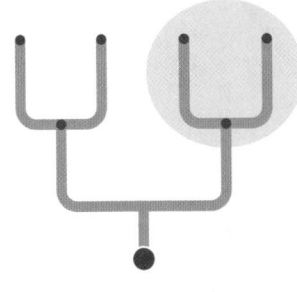

A monophyletic taxon includes an ancestral species and all of its descendants.

Polyphyletic taxon

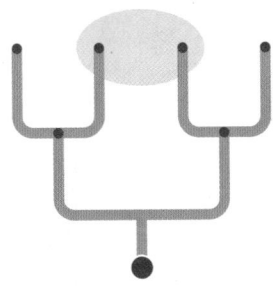

A polyphyletic taxon includes species from different evolutionary lineages.

Paraphyletic taxon

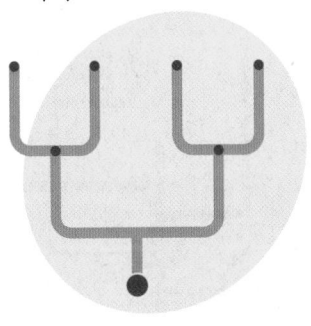

A paraphyletic taxon includes an ancestral species and only some of its descendants.

Figure 20.17

Defining taxa in a classification. Systematists can create different classifications from the same phylogenetic tree by identifying different groups of species as a single taxon (shaded).

evolutionary relationships from phenotypic similarities and differences. This represents **traditional evolutionary systematics**, which places together groups of species sharing ancestral and derived characters. Mammals are defined by their internal skeleton, vertebral column, and four limbs—ancestral characters among tetrapod vertebrates. But mammals also have derived characters such as hair, mammary glands, and a four-chambered heart (see Chapter 28). The four-chambered heart (see Section 40.1) also occurs in birds and some reptiles.

Classifications produced by traditional systematics reflect evolutionary branching and morphological divergence **(Figure 20.18a)**. Among tetrapod vertebrates, the amphibian and mammalian lineages diverged early, followed shortly by the divergence of the turtle lineage and then that of other reptiles. After this, subsequent divergences produced two groups: lepidosaurs (lizards and snakes) and archosaurs (crocodilians, dinosaurs, and birds). Although crocodilians outwardly resemble lizards, they share a more recent common ancestor with birds. Yet birds

differ from crocodilians in many morphological characters, including feathers and wings.

Even though the phylogenetic tree of tetrapod vertebrates shows six living groups, the traditional classification recognizes four classes: Amphibia, Mammalia, Reptilia, and Aves (birds). These groups (classes in classification) are given equal ranking because each represents a distinctive body plan and way of life. The class Reptilia, however, is a paraphyletic taxon because it includes some descendants of the common ancestor, namely turtles, lizards, snakes, and crocodilians (Figure 20.18a). But class Reptilia does not include other descendants of archosaurs, namely birds.

Traditional evolutionary systematists justify this definition of Reptilia because it includes morphologically similar animals with close evolutionary relationships. Crocodilians are classified with lizards, snakes, and turtles because they have a common ancestry and are covered with dry, scaly skin. Traditional systematists also argue that the key innovations thought to have initiated the adaptive radiation of

a. Traditional phylogenetic tree with classification

b. Cladogram with classification

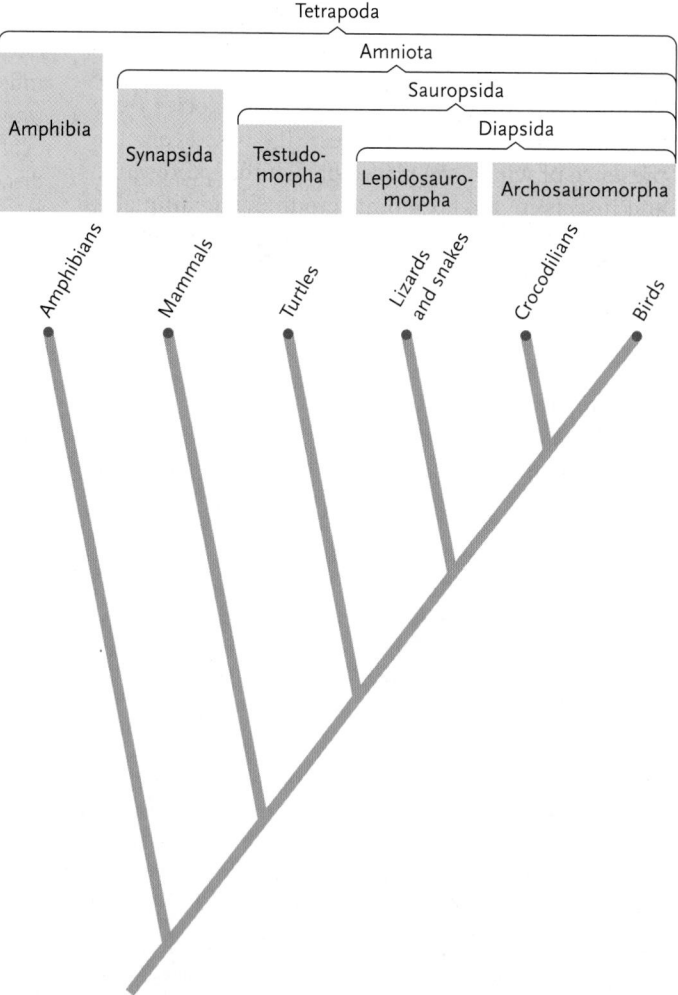

Figure 20.18

Phylogenetic trees and classifications for tetrapod vertebrates. **(a)** Traditional and **(b)** cladistic phylogenies produce different phylogenetic trees and classifications. Classifications are presented above the trees.

birds—wings, feathers, high metabolic rates, and flight—are extreme divergences from the ancestral morphology. Therefore, birds merit recognition as a separate class. The question is where to draw the line between birds and reptiles.

20.4d Cladistics: Analytical Classifications

Cladistics emerged in the 1950s and 1960s when some researchers criticized the inherent lack of clarity in classifications based on two distinct phenomena, branching evolution and morphological divergence. After all, how can we tell why two groups or organisms are classified in the same higher taxon? In some cases they have a recent common ancestor (e.g., lizards and snakes), but in other cases they do not (e.g., lizards and crocodilians).

To minimize such confusion, many systematists followed the philosophical and analytical lead of Willi Hennig, the German entomologist who wrote *Phylogenetic Systematics,* published in 1966. Hennig and his followers argued that classifications should be based solely on evolutionary relationships. **Cladistics** ignores morphological divergence, producing phylogenetic hypotheses and classifications that reflect only the branching pattern of evolution.

Cladists place species that share derived characters in one group. They argue that mammals form a monophyletic lineage, a **clade**, because they have a unique set of derived characters, including hair, mammary glands, reduction of bones in the lower jaw, and a four-chambered heart. The ancestral characters of mammals, such as an internal skeleton, a vertebral column, and four legs, do not distinguish them from other tetrapod vertebrates, so these traits are excluded from analysis.

Phylogenetic trees produced by cladists **(cladograms)** illustrate the hypothesized sequence of evolutionary branchings, with a hypothetical ancestor at each branching point (Figure 20.18b). Cladograms portray strictly monophyletic groups and are usually constructed using the assumption of parsimony. Once a researcher identifies derived, homologous characters, constructing a cladogram is straightforward (see "Constructing a Cladogram," p. 462).

Classifications produced by cladistic analysis often differ radically from those of traditional evolutionary systematics. Pairs of higher taxa are defined directly from the two-way branching pattern of the cladogram. Thus, the clade **Tetrapoda** (the traditional amphibians, reptiles, birds, and mammals) is divided into two taxa, Amphibia (tetrapods with no amnion (they lay their eggs in water); see Chapters 28 and 42) and Amniota (tetrapods with an amnion). Amniota is subdivided into two taxa on the basis of skull morphology and other characters, namely Synapsida (mammals) and Sauropsida (turtles, lizards, snakes, crocodilians, and birds). Based on cranial structure, Sauropsida is further divided into Testudomorpha (turtles) and Diapsida (lizards and snakes, crocodilians, and birds). Finally, based on anatomical details, Diapsida is subdivided into two more recently evolved taxa, Lepidosauromorpha (lizards and snakes) and Archosauromorpha (crocodilians and birds) (Figure 20.18b). The strictly cladistic classification parallels the pattern of branching evolution that produced the organisms included in the classification. These parallels are the essence and strength of the cladistic method.

Today most biologists use the cladistic approach because of its evolutionary focus, clear goals, and precise methods. Some systematists advocate abandoning the Linnaean hierarchy for classifying and naming organisms. They propose using a strictly cladistic system, called **PhyloCode**, that identifies and names clades instead of placing organisms into the familiar taxonomic groups. However, traditional evolutionary systematics has guided most laypeople's understanding of biological diversity.

20.5 Using Phylogenetics: The Evolution of Birds

The changes in the phylogeny of birds and their theropod relatives raise the obvious question, "What is a bird?" Twenty years ago, feathers were an obvious answer. The diversity of modern birds (Section 28.12, Chapter 28) makes it clear that all of them are feathered. But many dinosaurs also had feathers; indeed feathers probably evolved first as insulation for warm-blooded dinosaurs. Morphological and skeletal features associated with flight are also not the answer to "what is a bird?": not all living birds fly.

Most paleontologists accept that birds evolved from a lineage of theropod dinosaurs **(Figure 20.19, p. 463)**. The lineage leading to theropods had appeared in the early Triassic and includes well-known species such as allosaurs, tyrannosaurs, and their relatives. The following numbers correspond to numbers in Figure 20.19:

1. The long bones are hollow, and the first digit of the foot plays little role in weight support.
2. The development of a rotary wrist joint is associated with a grasping hand.
3. Bones in the shoulder girdle and the sternum (breastbone) are expanded to support chest muscles. As well, feathers appear for insulation.
4. Vaned feathers appear.
5. The trunk is shortened and the tail stiffened, resulting in better balance and maneuverability.
6. Basic perching and flight behaviour appears by the end of the Jurassic. The last three modifications are associated with powered flight.
7. A deep thorax develops.
8. A canal houses the main wing rotation muscles.
9. An elastic wishbone and a strongly keeled sternum appear.

Constructing a Cladogram

Research Method Box

Cladograms allow systematists (and others) to visualize hypothesized evolutionary relationships by grouping organisms that share derived characters. The cladogram also indicates where derived characters evolved.

Here we develop a cladogram for the nine extant groups of chordate vertebrates: lampreys (Agnatha), sharks (Chondrichthyes), bony fishes (Osteichthyes), amphibians (Amphibia), reptiles (turtles, lizards and snakes, crocodilians), birds, and mammals (see also Chapter 28). We also include lancelets (marine organisms in the subphylum Cephalochordata). Lancelets serve as the outgroup in our comparison.

We have chosen characters on which to base the cladogram **(Table 1)**, noting the presence (+) or absence (–) of 10 different characters. The characters are ancestral or derived in each group, but the outgroup (the lancelets) lacks all of these traits. We construct the cladogram from the information in the table, grouping organisms that share derived characters (right branch, **Figure 1a**), whereas the lancelets form the left branch because they lack the derived characters.

The remaining organisms except lancelets and lampreys have jaws. Now the right branch (Figure 1b) includes all living vertebrates sharing derived characters, separating them from lancelets and lampreys. The selection of different characters might give different outcomes.

a.

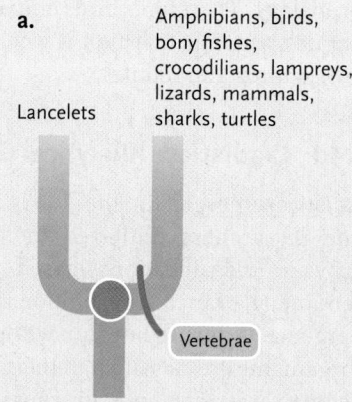

Lancelets — Amphibians, birds, bony fishes, crocodilians, lampreys, lizards, mammals, sharks, turtles

Vertebrae

b.

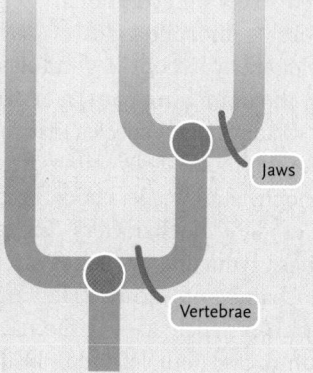

Lancelets   Lampreys — Amphibians, birds, bony fishes, crocodilians, lampreys, lizards, mammals, sharks, turtles

Jaws

Vertebrae

Figure 1
(a) A cladogram showing the separation of lancelets from living chordates. *(b) A cladogram showing the separation of lancelets and lampreys from most living chordates. Lampreys are chordates, but the cladogram suggests that they are the earliest chordates. These cladograms were prepared from the data in Table 1.*

Table 1

	Vertebrae	Jaws	Swim Bladder or Lungs	Paired Limbs	Extraembryonic Membranes	Mammary Glands	Dry, Scaly Skin	Two Openings at Back of Skull	One Opening in Front of Eye	Feathers
Lancelets	–	–	–	–	–	–	–	–	–	–
Lampreys	+	–	–	–	–	–	–	–	–	–
Sharks	+	+	–	–	–	–	–	–	–	–
Bony fishes	+	+	+	–	–	–	–	–	–	–
Amphibians	+	+	+	+	–	–	–	–	–	–
Mammals	+	+	+	+	+	–	–	–	–	–
Turtles	+	+	+	+	+	–	+	–	–	–
Lizards	+	+	+	+	+	–	+	+	–	–
Crocodilians	+	+	+	+	+	–	+	+	+	–
Birds	+	+	+	+	+	–	+	+	+	+

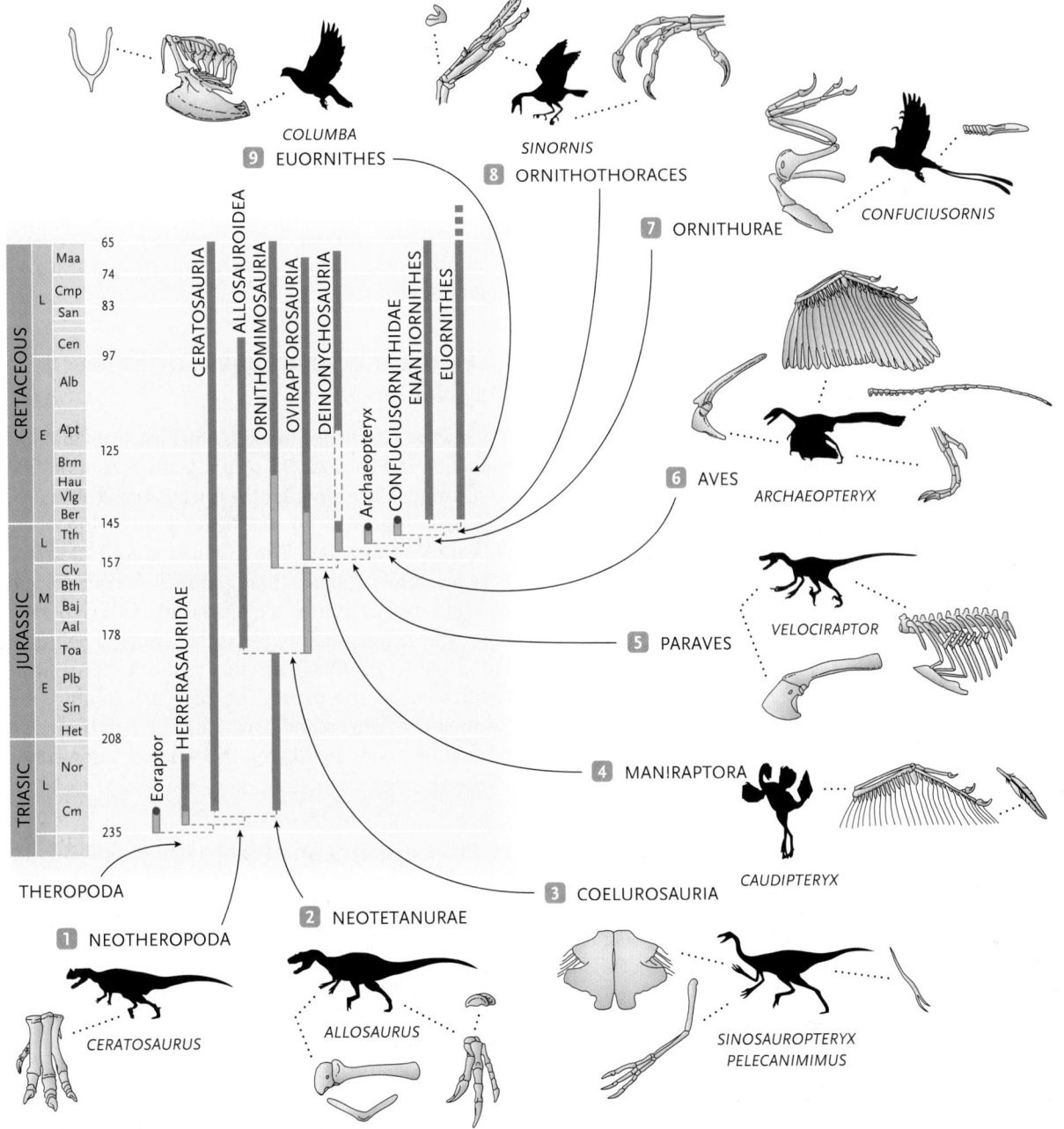

Figure 20.19

Nine morphological changes in the development of a bird from its theropod ancestors. Details in text.

From Paul C. Sereno, "The Evolution of Dinosaurs," *Science* 25 June 1999: Vol. 284 no. 5423 pp. 2137–2147. Reprinted with permission from AAAS.

In 2013, the theropod lineage Eumaniraptora was described as having three families: Dromaeosauridae, Troodontidae, and Avialae. The latter two make up the clade Paraves. Even in 2011, *Archaeopteryx* had been placed in the Avialae, arguably with other birds. Described in 1861, *Archaeopteryx lithographica* remains one of the best known fossil "birds," and there is ongoing disagreement about this species' ability to fly. Animals in the Deinonychosauria (4 and 5, Figure 20.19) would probably have been unable to fly, while *Archaeopteryx* would have had limited flight ability. In Figure 20.19, only Euornithes are "real" birds.

The arrangement of species in families changed with the discovery of fossils from a Late Jurassic lagerstätten (~150 mya) in northeast China. These deposits contain thousands of specimens preserved in lake deposits rich in volcanic ash. Fine details of soft tissues, including feathers, are preserved in these specimens. Two species are particularly important, *Xiaotingia zhengi* and *Aurornis xui*, described in 2011 and 2013, respectively. The anatomical details of these species and those of *Archaeopteryx* place all three genera in Paraves, but now *Archaeopteryx* falls within the Troodontidae; *Xiaotingia* and *Aurornis* in Avialae. If *Archaeopteryx* and other troodontids were capable of limited flight, it would appear that flight evolved once in this Paraves clade.

The story about the origin and evolution of birds is ongoing. It demonstrates the importance of fossils and modern techniques for examining them. More important, the story illustrates that phylogenies are hypotheses about evolutionary history. The hypotheses change as more data are brought to bear on the situation. In the next chapter, we will see how the same situation emerges about the history of our own species.

STUDY BREAK

1. What are the diagnostic characteristics of birds?
2. What explains the changes in the position of *Archaeopteryx* in the classification of birds?

20.6 Parallelism and Convergence Can Complicate the Scene

Understanding and documenting the diversity of life is a major challenge to biologists. The evolution of flight (above) illustrates the potential difficulty in recognizing parallel and convergent evolution. This exercise means determining evolutionary relationships among organisms and deciding if similar structures or similar-looking structures are grounds for grouping species together. In other words, if they look the same, are they closely related in an evolutionary sense?

When first discovered, ichthyosaurs **(Figure 20.20)** were thought to be fish. Later, details of their skeletons confirmed that they were not fish, but reptiles. The initial confusion is understandable because ichthyosaurs have fishlike bodies, as do mammals such dolphins and whales. Many aquatic vertebrates have fishlike bodies—think of sharks (cartilaginous fishes) and tunas (bony fishes). But these are two very different kinds of "fish." Other fish have very different bodies—think of eels, flatfish, anglerfish, and sea horses.

There is a tendency among organisms living under the same conditions to develop similar body forms. This can be called parallel or convergent evolution, depending on the evolutionary relatedness of the organisms involved. Convergent evolution is used when referring to phylogenetically more distantly related organisms; parallel evolution when referring to more closely related ones. How can you tell?

One of the most exciting and satisfying things about biology is learning to look at something, recognize it, and understand just what you are seeing. Below are two cases, examples of where even an experienced biologist might be fooled. At first, this may seem strange or even preposterous when you know what a flower looks like; surely, you will always recognize one. Read on to see that things are not always as they appear.

20.6a Case 1: What Looks Like a Flower May Not Be One

You may be astonished to realize that what you thought was a flower **(Figure 20.21a, b)** is actually a leaf modified by a fungus. The fungus, the rust *Puccinia monoica*, affects the growth of the leaves, changing their appearance and odour. The fungus-induced "flowers" have nectaries (glands that produce nectar; Figure 20.21c). Just as many biologists are fooled by the flowerlike leaves, so are insects that come to pollinate the flowers. In this way, the rust interferes with fertilization of the plant. The rust also inhibits the formation of the plant's own flowers, minimizing confusion among pollinators. So when is a flower not a flower?

20.6b Case 2: Some Plants Are Carnivorous

In some places with an abundance of water and sunlight, nitrogen for plants can be in short supply. Here we find a diversity of ways that plants trap insects to directly (or indirectly) obtain nitrogen **(Figure 20.22**; see Figure 1, "Pitcher Plant Ecosystems," Chapter 31). Like people trying to catch insects (but perhaps not for their nitrogen), plants use different methods. Insectivorous plants catch insects in sticky traps (flypaper), snap traps, and pitfall traps (pitchers). Flypaper traps have appeared in at least five evolutionary lines of plants and pitchers at least three times. For the most part, carnivorous plants do not share a close common ancestor.

20.7 Putting Fossils and Phylogeny in Perspective

The fossil record is the history book for life on Earth, and the study of extant organisms is not complete without the study of extinct ones. This chapter illustrates the richness, completeness—and incompleteness—of the fossil record. Systematists cannot complete phylogenetic trees or cladograms without considering early organisms. These phylogenetic diagrams indicate the evolutionary relatedness of organisms.

Figure 20.20

An ichthyosaur (*Stenopterygius* sp.) from Germany on display at the Royal Tyrrell Museum in Drumheller, Alberta. Although this specimen is about 2 m long, the largest ichthyosaurs were over 20 m in length.

Figure 20.21
(a) There is an obvious difference between an uninfected *Boerchera* species (left) and an infected one (right).
(b) When infected by the fungus *Puccinia monoica*, a rust, the leaves of *Boerchera* become flowerlike and **(c)** appear to produce nectar. The rust inhibits flowering so that insects visiting the "flowers" to collect nectar fertilize the rust. The insects do not visit the host plant's flowers.

a.

b.

c.

d.

e.

f.

g.

h.

Figure 20.22
Plants that catch insects in pitchers: **(a)** *Cephalotus follicularis*, **(b)** *Sarracenia purpurea*, **(c)** *Darlingtonia california*, and **(d)** *Nepenthes*; on flypaper: **(e)** *Drosera capensis*, **(f)** *Pinguicula* spp., and **(g)** *Brocchinia reducta*; or in a snap trap: **(h)** *Dioneae muscipula*.

Review

 **aplia** To access course materials such as Aplia and other companion resources, please visit www.NELSONbrain.com.

20.1 The Geological Time Scale

- Sediments found in any one place form distinctive strata (layers) that usually differ in colour, mineral composition, particle size, and thickness.

- Geologists used the sequence of strata and their distinctive fossil assemblages to establish the geologic time scale that diagrams the history of life on Earth.

20.2 The Fossil Record

- Fossils are the parts of organisms preserved in sedimentary rocks or in oxygen-poor environments.

- Fossils can include vertebrate bones, insects in amber, impression of plants and softbodied animals, footprints, coprolites, gastroliths, and frozen or mummified organisms.

- An "explosion" in the diversity of groups of organisms, about 600 mya, marked the beginning of the Cambrian. Complex multicellular animals first appeared at this time.

- Mass extinctions have occurred at least five times in the history of life, with possible causes being volcanic activity, climate change, and asteroid strikes. Following a mass extinction, the surviving organisms undergo adaptive radiation.

20.3 Lives of Prehistoric Organisms

- The first land plants, similar to today's liverworts and mosses, appeared during the Ordovician Period.

- For animals to move onto land required several adaptations, such as sufficient support to allow them to maintain their integrity, and some form of protection from ultraviolet radiation. Adaptations to mechanisms of gaseous exchange, excretion, and locomotion were also needed.

- Whales and toothed whales are a distinctive group of mammals whose fossil record extends back more than 50 mya.

20.4 Phylogeny

- Phylogenetic trees are hypotheses that portray the branching pattern of evolution. Most phylogenetic trees have an implicit or explicit time line that indicates the relative times for cladogenesis. Branch points are described as nodes, and monophyletic lineages are called clades. Two lineages that share a node are called sister clades. Clades can be rotated at nodes without changing the meaning of the tree.

- In preparing a phylogeny, systematists seek traits in which phenotypic variation reflects genetic differences rather than environmental variation. Systematists also use genetically independent traits, which reflect different parts of an organism's genome.

- Homologous characters have been inherited from a common ancestor, so phenotypic similarities between organisms reflect underlying genetic similarities. Homologous characters can differ considerably among species. Analogous (homoplasious) characters are phenotypically similar and have similar functions but evolved independently in different lineages.

- Systematists compare homologous characters to determine common ancestry and genetic relatedness. They exclude analogous structures because they provide no information about shared ancestry or genetic relatedness.

- Mosaic evolution refers to the reality that some characters evolve more slowly or more quickly than others. Ancestral characters are old forms of traits, and derived characters are new forms. Once a derived character becomes established, it occurs in all the species' descendants and is a marker for evolutionary lineages.

- An outgroup comparison can be used to identify ancestral and derived traits because it compares the group under study with more distantly related species not otherwise included in the analysis.

- A monophyletic taxon is a group of species derived from a single ancestral species. A polyphyletic taxon includes species from separate evolutionary lineages. A paraphyletic taxon includes an ancestor and some, but not all, of its descendants.

- According to the assumption of parsimony, any particular evolutionary change is a rare event, unlikely to have occurred twice in one lineage. Therefore, the fewest possible evolutionary changes should be used to account for within-lineage diversity.

- Derived character states can serve as markers of clades.

- Systematists use evidence from the fossil record as well as outgroup comparison to identify which character states are derived and which are ancestral.

- Cladistic analyses use synapomorphies (derived character states) to construct phylogenetic hypotheses.

20.5 Using Phylogenetics: The Evolution of Birds

- Most paleontologists accept that birds evolved from a lineage of theropod dinosaurs.

20.6 Parallelism and Convergence Can Complicate the Scene

- Convergent evolution refers more to distantly related organisms and parallel evolution to more closely related ones. The evolution of flight is convergent between insects (phylum Arthropoda) and vertebrates (phylum Chordata: birds, pterosaurs, and bats) and parallel within the vertebrates.

Questions

Self-Test Questions

1. The fossil record does which of the following?
   a. provides direct and indirect evidence about life in the past
   b. shows that all morphological novelties arise rapidly
   c. provides abundant data about rare species with local distributions
   d. is equally good for all organisms that ever lived
   e. provides no evidence about the physiology or behaviour of ancient organisms

2. What is the absolute age of a geological stratum determined by?
   a. the thickness of its rocks
   b. the particle size in its rocks
   c. the types of fossils found within it
   d. its position relative to other layers
   e. radiometric dating techniques

3. Fossils can be which of the following?
   a. mounted skeletons of mineralized bone
   b. spiders preserved in amber
   c. moulds or impressions of Anomalocaris
   d. coprolites or gastroliths
   e. all of the above

4. Why do adaptive radiations often follow mass extinctions?
   a. Mass extinctions limit the impact of paedomorphosis.
   b. Mass extinctions foster allometry and heterochrony.
   c. Mass extinctions decimate all forms of life on Earth.
   d. Species that form transitional fossils often survive mass extinctions.
   e. Extinctions open adaptive zones that had been previously occupied.

5. Fill in the blank. A phylogenetic tree portrays the ____ of a group of organisms.
   a. classification
   b. evolutionary history
   c. domain
   d. distribution

6. When systematists use morphological or behavioural traits to reconstruct the evolutionary history of a group of animals, what are they assuming?
   a. Phenotypic characters reflect underlying genetic similarities and differences.
   b. The animals use exactly the same traits to identify appropriate mates.
   c. The adaptive value of these traits can be explained.
   d. Variations are produced by environmental effects during development.

7. Which of the following pairs of structures are homoplasious?
   a. the wing skeleton of a bird and the wing skeleton of a bat
   b. the wing of a bird and the wing of a fly
   c. the eye of a fish and the eye of a human
   d. the wing structures of a pterosaur and those of a bird

8. Which of the following does NOT help systematists determine which version of a morphological character is ancestral and which is derived?
   a. outgroup comparison
   b. patterns of embryonic development
   c. studies of the character in more related species
   d. dating of the character by molecular clocks

9. In a cladistic analysis, a systematist groups together organisms that share which of the following?
   a. derived homologous traits
   b. derived homoplasious traits
   c. ancestral homologous traits
   d. ancestral homoplasious traits

10. How would one construct a cladogram by applying the parsimony assumption to molecular sequence data?
    a. Start by making assumptions about variations in the rates at which different DNA segments evolve.
    b. Group organisms sharing the largest number of ancestral sequences.
    c. Group organisms that share derived sequences, matching the groups to those defined by morphological characters.
    d. Group organisms sharing derived sequences, minimizing the number of hypothesized evolutionary changes.

Questions for Discussion

1. Traditional evolutionary systematists identify the Reptilia as one class of vertebrates, even though this taxon is paraphyletic. What are the advantages and disadvantages of defining paraphyletic taxa in a classification?

2. Create an imaginary phylogenetic tree for an ancestral species and its 10 descendants. Circle a monophyletic group, a polyphyletic group, and a paraphyletic group on the tree. Explain why the groups you identify match the definitions of the three types of groups.

3. Imagine that you are trying to determine the evolutionary relationships among six groups of animals that look very much alike because they have few measurable morphological characters. What data would you collect to reconstruct their phylogenetic history?

4. The geological evolution of Earth has had an obvious effect on biological evolution. You have read about how the release of oxygen by photosynthetic organisms increased atmospheric oxygen concentration. How are human activities changing the physical environment on Earth? What new selection pressures do these environmental alterations establish?

Neandertal (left) and *Homo sapiens* (right). DNA techniques, as well as fossils, are advancing our knowledge about the evolutionary history of humans.

21

Humans and Evolution

WHY IT MATTERS

From about 500 thousand to 30 thousand years ago, neandertals (*Homo neandertalensis*) roamed much of Europe and eastern and central Asia. Humans (*Homo sapiens*) arrived in the area about 40 thousand years ago. Neandertals were shorter, more heavily built, and stronger than humans, and their brains were larger as well. But the neandertals disappeared about 30 thousand years ago, and paleontologists have often wondered why humans won out. Hypotheses include interbreeding between the two species (or subspecies), larger eyes in the neandertal, and climate change. Only about 2% of the DNA of modern Europeans is neandertal, so why didn't the two species simply interbreed and blend together? Also, surprisingly, no neandertal mitochondrial DNA (mtDNA) has shown up in human mDNA. Since mDNA is passed only from mother to child, this suggests that while neandertal fathers and human mothers might have produced viable offspring, neandertal mothers and human fathers could not. Was this the only thing that separated *H. sapiens* from *H. neandertalensis*?

Recently, researchers in Seattle have developed the "Brainscan Atlas," a genetic reference about how the human brain is constructed and how it develops embryonically. Drs. Mohammed Uddin and Stephen Scherer, at Toronto's Hospital for Sick Children, used the Atlas and the Exome Variant Server (a database of all of a human's exomes) to determine which genes were responsible for autism. Uddin and Scherer determined that about 1700 genes were related only to brain development, not to any other function in the body. They then discovered that at least one-third of these 1700 genes had been implicated by other researchers in other brain and cognitive disorders.

Now, back to the neandertal conundrum: Uddin and Scherer suddenly thought that the 1700 genes could somehow be related to what makes humans uniquely human. People with autism have difficulties with communication and socialization; could these abilities also be what sets us apart from neandertals?

Dr. Ajit Varki at the Univerity of California, San Diego, thinks that despite the physical similarities between humans and neandertals, they could have been very different in **social behaviour** and in their abilities to communicate. Consequently, children with one *H. sapiens* parent and one *H. neandertalensis* parent could have been "cognitively sterile," with difficulties in communication making them likely to reproduce successfully.

This research was published in 2014, not long before this book was printed. By the time you read this, paleontologists may know a lot more about the *H. sapiens–H. neandertalensis* relationship.

DNA techniques as well as fossils are advancing our knowledge about the evolutionary history of humans. In this chapter we focus on some of the most important changes in our ancestry. We consider the implications of **bipedalism**, showing that it is much more than walking erect on two legs. We also present some of the recently discovered fossils and consider how the biological species concept applies to our own species.

21.1 The Fossil Record of Hominins

A combination of genetic and morphological analyses of living and fossil species indicates that between 5 and 10 mya in Africa, **hominoids** (superfamily Hominoidea, including apes and humans) had diverged into several lineages. One lineage, the **hominins** (family Hominidae, subfamily Homininae), includes modern humans and our bipedal ancestors (see Figure 20.13 and "The Cast of Characters: Fossil Hominins"). Where only one species of hominin (*Homo sapiens*) exists today, several species lived in the past.

The Cast of Characters: Fossil Hominins

Most of our ancestors' fossils have been found in Africa **(Figure 1)**. Brain capacity varies with overall body size, so that large individuals (typically males) have more brain capacity than smaller ones (typically females). Across the species presented below, brain capacity ranges from the size of chimpanzee brain capacity (275–500 cm³, *Orrorin tugensis*) to that of *Homo sapiens* (1000–1900 cm³). Species of *Australopithecus* have brain capacities about 400–500 cm³; *Homo habilis* about 640 cm³. *Homo erectus* brain capacity ranges from 930 to 1030 cm³; *Homo neandertalensis* from 1300 to 1600 cm³.

Orrorin tugensis: In 2000, researchers found 13 fossils of *O. tugensis* ("first man" in a local African language), a species that lived in the forests of eastern Africa about 6 mya. The thigh bones and pelvis indicate that it was bipedal.

Ardipithecus ramidus: In 1994, *Ardipithecus ramidus* was described from bone fragments (teeth and jaw fragments) collected in South Africa. These hominids stood 120 cm tall and had apelike teeth. The October 2, 2009, issue of *Science* included 11 papers about *A. ramidus* by an international team of researchers who reported data from 110 specimens. This species lived from about 4 to 6 mya, and many of its features

overturned ideas about the evolution of our own species. The structure of its pelvis and feet suggested that it was bipedal. Both males and females had small canine teeth (compared to those of other primates), which imply reduced competition between males, presumably for females in oestrus, in turn suggesting concealed ovulation (see Chapter 42). Bipedal locomotion could have enabled the hominins to exploit both land surface and trees in the search for food and shelter. Bipedal animals can carry food and be more effective provisioners. Many features of *A. ramidus* demonstrate that our ancestors showed an earlier than expected departure from a chimpanzee-like existence.

Australopithecus africanus: The first australopith to be described, *Australopithecus africanus*, was discovered by Raymond Dart in 1924. With its relatively small brain, this bipedal species was not immediately recognized as a hominin.

Australopithecus afarensis: Specimens of more than 60 individuals have been found in northern Ethiopia. The sample includes about 40% of a female's skeleton, named "Lucy" (Figure 21.5, p. 474; apparently the Beatles' song "Lucy in the Sky with Diamonds" was playing on the radio when the skeleton was first uncovered). *A. afarensis* lived

3 to 3.5 mya. This species retained several ancestral characters, including moderately large and pointed canine teeth and a relatively small brain. Males and females were 150 cm and 120 cm tall, respectively. Skeletal analyses suggest that Lucy was fully bipedal, a conclusion supported by fossilized footprints preserved in a layer of volcanic ash (Figure 21.1, p. 473). In 2010, the description of a male specimen of *Australopithecus afarensis* ("big man") provided further evidence of bipedalism, specifically details of the pelvic girdle and sacrum not preserved in Lucy. Furthermore, a well-preserved scapula (shoulder blade) provided no evidence of suspensory climbing evident in the scapulae of great apes.

Australopithecus anamensis: One of the oldest known species in the genus, *Australopithecus anamensis* lived in eastern Africa around 4 mya. Its teeth had thick enamel, which is typically a derived hominin character. A fossilized thigh bone suggests that it was bipedal.

Australopithecus sediba: In September 2011, the journal *Science* published a series of papers describing the newly discovered *Australopithecus sediba*, an "early" (2 mya) version of Lucy (*A. afarensis*). The fossils, from the Malapa site in South Africa, provided a new perspective on the evolution of humans. The ankles and feet suggest a

(Continued)

The Cast of Characters: Fossil Hominins (*Continued*)

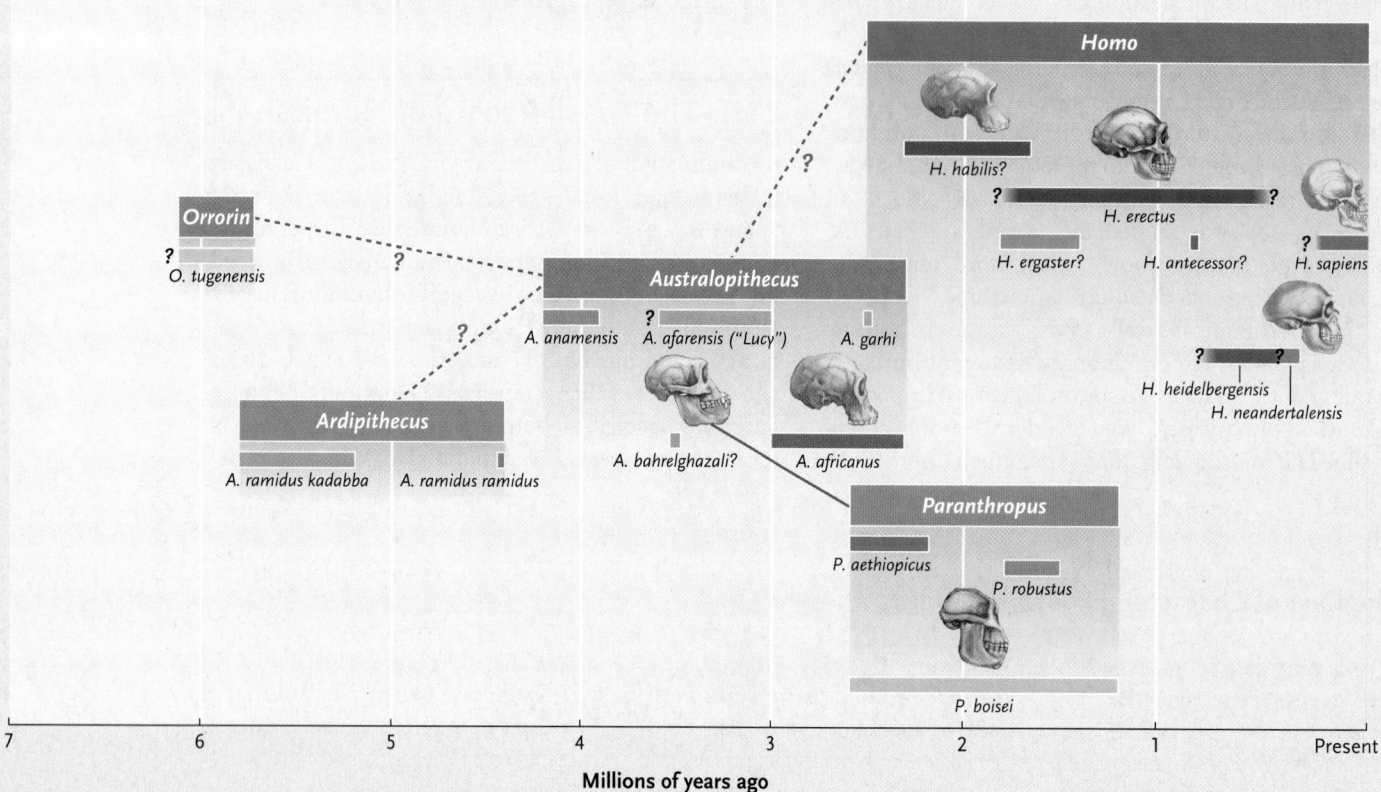

Figure 1
Hominin timeline showing several of the species described in the text. These species lived at the same place and time in eastern and southern Africa. The timeline is shown for each species, and question marks indicate uncertainty about classification and/or ages of fossils. Some skulls are reconstructions from fragments.

combination of climbing ability (arboreal life style) and bipedalism that differed from that of Lucy (see page 474). The pelvis of *A. sediba* shared features, such as the sacral and pubic areas, with those of Lucy and other australopithecines. But in the shape of its ilium, *A. sediba* resembled species of *Homo*. These features in *A. sediba* suggest that giving birth to offspring with large brains had not appeared at this stage of australopithecine evolution.

Some of the fossil *A. sediba* had nearly complete wrists and hands, indicating the capacity for strong flexion typically associated with tree climbing. But the long thumb and short fingers of *A. sediba* imply a capacity for precision gripping (see Figure 21.8, p. 476), suggesting the potential for use of tools. Construction of a virtual endocast of the skull of *A. sediba* indicates that the frontal lobes of the brain were generally like those of other australopithecines. But the features of the brain suggest its

gradual reorganization toward the appearance in species of *Homo*.

This new fossil demonstrates, once again, how such a find can influence our view of evolutionary history. This is not a "missing link" but rather a species whose features reflect the general state of evolutionary development of species in the genus *Australopithecus*. It also presages the transitions that later occur with the emergence of species in the genus *Homo*.

Homo habilis: Pliocene fossils of the earliest *Homo* are fragmentary and widely distributed in space and time. They are thought to have belonged to *Homo habilis* (meaning "handy man"). From 1.7 to 2.3 mya, *H. habilis* occupied the woodlands and savannas of eastern and southern Africa, sharing these habitats with various species of *Paranthropus*. The two genera are easy to distinguish because the brains of *H. habilis* were at least 20% larger, and their **incisors** were larger and **molars** smaller than those of *Paranthropus* spp. They ate hard-shelled nuts and

seeds, as well as soft fruits, tubers, leaves, and insects. They may also have hunted small prey or scavenged carcasses left by larger predators.

Researchers have found numerous tools dating to the time of *H. habilis* but are not sure which species made them. Many hominid species of that time probably cracked marrowbones with rocks or scraped flesh from bones with sharp stones. Paleoanthropologist Louis Leakey was the first to discover evidence of tool-*making* at eastern Africa's Olduvai Gorge, which cuts through a great sequence of sedimentary rock layers. The oldest tools at this site are crudely chipped pebbles, probably manufactured by *H. habilis*. However, humans are not the only animals to use tools (see Chapter 47).

Homo erectus: Early in the Pleistocene, about 1.8 mya, a new species of humans, *Homo erectus* ("upright man"), appeared in eastern Africa **(Figure 2).** One nearly complete skeleton suggests that *H. erectus* was taller than its ancestors and had a much larger brain, a thicker skull, and protruding brow ridges. *H. erectus* made fairly sophisticated tools, such as hand axes **(Figure 3)** used to cut food and other materials, to scrape meat from bones, and to dig for roots. *H. erectus* probably ate both plants and animals and may have hunted and scavenged animal prey. Archaeological data point to their use of fire to cook food and to keep warm. Near Lake Turkana in Kenya, fossils identified as *Homo* and dating from 1.45 to 1.55 mya were described in 2007. These suggested that *H. erectus* and *H. habilis* lived together in the same habitats for a considerable time, much as chimps and gorillas do today. Adult male *H. erectus* were much larger than adult females, suggesting a polygynous lifestyle, one male with several females (see Chapter 47).

About 1.5 mya, the pressure of growing populations apparently forced groups of *H. erectus* out of Africa. They dispersed northward from eastern Africa into both northwestern Africa and Eurasia. Some moved eastward through Asia as far as the island of Java. Judging from its geographic distribution, *H. erectus* was successful in many environments. It produced several descendant species, of which modern humans (*H. sapiens*, meaning "wise man") are the only survivors. Now-extinct descendants of *H. erectus*, archaic humans, first appeared at least 400 thousand years ago. They generally had larger brains, rounder skulls, and smaller molars than *H. erectus*.

Homo floresiensis: *H. floresiensis* was described in 2004 from Flores Island in Indonesia. Although first proposed as a distinct species, its small size was used to support the view that it was just a small individual. In 2013, analyses of various aspects of the morphology of *H. floresiensis* indicated that it was not a dwarf or microcephalic, rather a distinct species most closely related to *H. erectus*.

Homo neandertalensis: Neandertals lived in Europe and western Asia from 28 thousand to 150 thousand years ago. They are the best known of the archaic humans and sometimes have been treated as a subspecies of *Homo sapiens*. Compared with modern humans, they had a heavier build,

a. *Homo erectus*

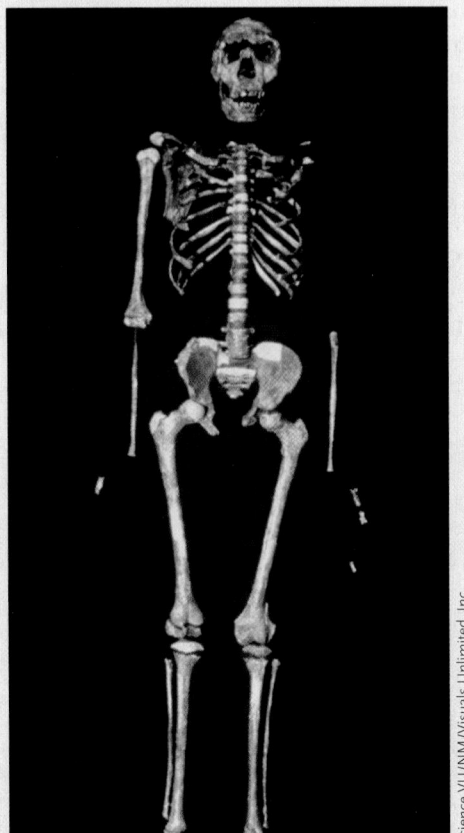

Science VU/NM/Visuals Unlimited, Inc.

Figure 2
Homo erectus, *a nearly complete skeleton from Kenya.*

b. Hand axe

©AAAC/Topham/The Image Works

Figure 3
A hand axe found at a site used by Homo erectus.

more pronounced brow ridges, and slightly larger brains. Neandertals were culturally and technologically sophisticated. They made complex tools, including wooden spears, stone axes, flint scrapers, and knives. At some sites, they built shelters of stones, branches, and animal hides, and they routinely used fire. They were successful hunters and probably ate nuts, berries, fishes, and bird eggs. Some groups buried their dead, and they may have had rudimentary speech. There is evidence that some were cannibals.

In 1997, two teams of researchers independently analyzed short segments of mtDNA extracted from the fossilized arm bone of a neandertal. Unlike nuclear DNA, which individuals inherit from both parents, only mothers pass mtDNA to offspring. mtDNA does not undergo genetic recombination (see Chapter 9) and has a high mutation rate, making it useful for phylogenetic analyses. If mutation rates in mtDNA are fairly constant, this molecule can serve as a **molecular clock.** Comparing the

neandertal sequence with mtDNA from 986 living humans revealed three times as many differences between the neandertals and modern humans as between pairs of modern humans in their sample. These data suggest that neandertals and modern humans are different species that diverged from a common ancestor 550 thousand to 690 thousand years ago.

Homo sapiens: Modern humans differ from neandertals and other archaic humans in having a slighter build, less protruding brow ridges, and a more prominent chin. The earliest fossils of modern humans found in Africa and Asia are 150 thousand years old; those from the Middle East are 100 thousand years old. Fossils from about 20 thousand years ago are known from western Europe, the most famous being those of the Cro-Magnon deposits in southern France. The widespread appearance of modern humans roughly coincided with the demise of neandertals in western Europe and the Middle East, 28 thousand to 40 thousand years ago.

Most of the hominins that lived in eastern and southern Africa from 1 to 6 mya are currently classified in the genera *Australopithecus* (*australo* = southern; *pithecus* = ape) and *Paranthropus* (*para* = beside; *anthropus* = man). With large faces, protruding jaws, and small skulls and brains, these hominins resembled apes. Between 1 and 3.7 mya, several other species of hominins occurred in eastern and southern Africa.

These adult males ranged from 40 to 50 kg in mass and from 130 to 150 cm in height; females were smaller. Most of these species had deep jaws and large molars, suggesting a diet of hard food, such as nuts, seeds, and other vegetable products. *Australopithecus africanus,* known only from southern Africa, had small jaws and teeth, suggesting a diet of softer food. The phylogenetic relationships among species in the

Chromosomal Similarities and Differences among the Great Apes

The banding patterns of humans and their closest relatives among the apes—chimpanzees, gorillas, and orangutans—reveal that whole sections of chromosomes have been rearranged over evolutionary time. For example, humans have a diploid complement of 46 chromosomes, whereas chimpanzees, gorillas, and orangutans have 48 chromosomes. The difference can be traced to the fusion (i.e., the joining together) of two ancestral chromosomes into chromosome 2 of humans; the ancestral chromosomes are separate in the other three species **(Figure 1).**

Jorge J. Yunis and Om Prakash of the University of Minnesota Medical School analyzed the banding patterns on metaphase chromosome preparations from humans, chimpanzees, gorillas, and orangutans. They identified about 1000 bands that are present in the four species. By matching the banding patterns on the chromosomes, the researchers verified that they were comparing the same segments of the genomes in the four species. They then searched for similarities and differences in the structure of the chromosomes.

Analysis of human chromosome 2 reveals that it was produced by the fusion of two smaller chromosomes that are still present in the other three species. Although the position of the centromere in human chromosome 2 matches that of the centromere in one of the chimpanzee chromosomes, in gorillas and orangutans it falls within an inverted segment of the chromosome. (Recall from Section 11.3 that an inverted chromosome segment has a reversed orientation, so the order of genes on it is reversed relative to the order in a segment that is not inverted.) Humans and chimps also differ from each other in centromeric inversions in six other chromosomes.

How might such chromosome rearrangements promote speciation? In a paper published in 2003, Arcadi Navarro of the Universitat Pompeu Fabra in Spain and Nick H. Barton of the University of Edinburgh in Scotland compared the rates of evolution in protein-coding genes that lie within rearranged chromosome segments of humans and chimpanzees to those in genes outside the rearranged segments. They discovered that proteins evolved more than twice as quickly in the rearranged chromosome segments. Navarro and Barton reasoned that because chromosome rearrangements inhibit chromosome pairing and recombination during meiosis, new genetic variations favoured by natural selection would be conserved within the rearranged segments. These variations accumulate over time, contributing to genetic divergence between populations with the rearrangement and those without it. Thus, chromosome rearrangements can be a trigger for speciation: once a chromosome rearrangement becomes established within a population, that population will diverge more rapidly from populations lacking the rearrangement. The genetic divergence eventually causes reproductive isolation.

Figure 1

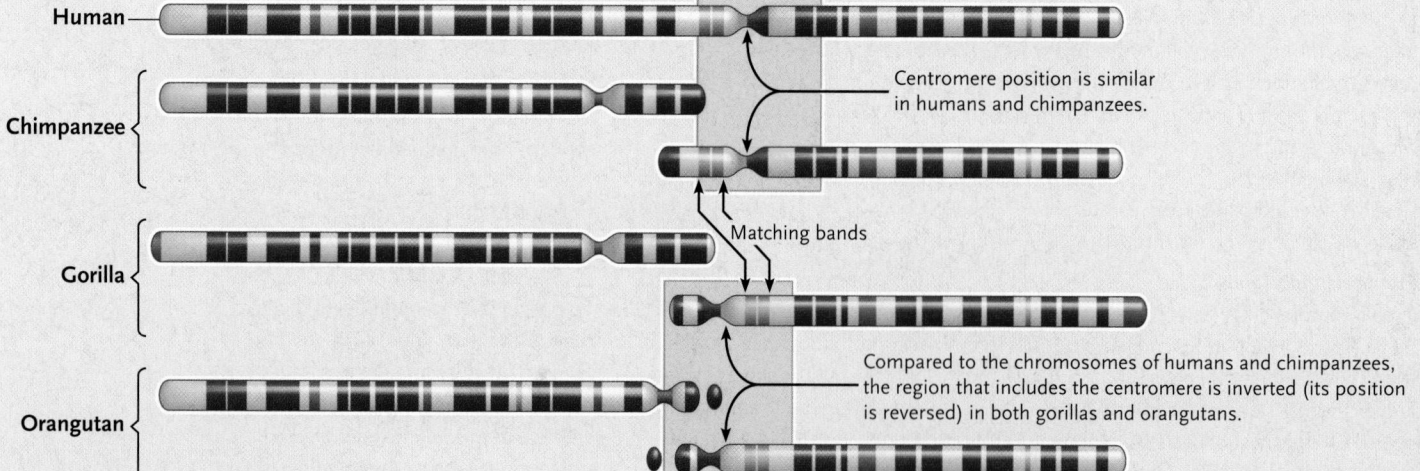

Human

Chimpanzee

Centromere position is similar in humans and chimpanzees.

Matching bands

Gorilla

Orangutan

Compared to the chromosomes of humans and chimpanzees, the region that includes the centromere is inverted (its position is reversed) in both gorillas and orangutans.

Based on J. J. Yunis and O. Prakash, "The origin of man: A chromosomal pictorial legacy." *Science* 19 March 1982: Vol. 215 no. 4539 pp. 1525–1530.

genera *Australopithecus* and *Paranthropus,* and their exact relationships to later hominids, are not yet fully understood (see Figure 20.13 and "The Cast of Characters: Fossil Hominins"). *Australopithecus* was likely ancestral to humans (various species in the genus *Homo*).

Many people believe that evolutionary biologists say that our species evolved *from* apes. But while the fossil record clearly demonstrates that the evolutionary lineage to which *H. sapiens* belongs includes chimps and gorillas (Figure 20.13), the lineage leading to humans has been distinct from the one leading to gorillas and chimps for over 6 million years. Belonging to an evolutionary lineage does not mean one species in the lineage gives rise to another. Evolutionary biologists are not proposing that humans evolved from apes, but that apes are our closest living relatives (see "Chromosomal Similarities and Differences among the Great Apes").

STUDY BREAK

1. What are three adaptations of humans that separate them from chimpanzees?
2. How old are the oldest fossils of humans? Of hominoids?
3. When did the lineage leading to chimps and orangutans diverge from the hominin lineage?

21.2 Morphology and Bipedalism

Upright posture and bipedal locomotion distinguish hominins from apes. Bipedal locomotion largely meant that the hands were not used in locomotion, allowing them to become specialized for other activities, such as carrying things and using and making tools. Sometimes paleontologists find fossilized hominin footprints, indicating bipedalism **(Figure 21.1).** But usually the fossil record of mammals such as *Homo sapiens* consists mainly of bones and teeth, the body parts most often fossilized. Bipedalism is obvious from the feet, thighs, pelvis, shoulders, and arms of these fossil and human skeletons **(Figure 21.2).** When appropriate fossils are available, it may be possible to learn when the skeletal features associated with bipedalism appeared over time.

Bipedalism in hominins involves a suite of anatomical features, not just the pattern of footfall (opening photograph) and walking and running behaviour.

21.2a Feet, Legs, and Pelvis

Paleontologists have long inferred bipedalism from the structure of the thigh bones (femora) and pelvis **(Figure 21.3, p. 474).** Both ends of the femur, at the hip and knee joints, are larger in the human than in the chimpanzee because more weight is directed through the human joints. Humans have a smaller angle at the hip end of the femur (the ball and socket joint) because

Figure 21.1
Mary Leakey discovered these fossilized footprints of an austral-opithecine made in soft, damp volcanic ash about 3.7 mya. The footprints appear to have been made by an adult and a young.

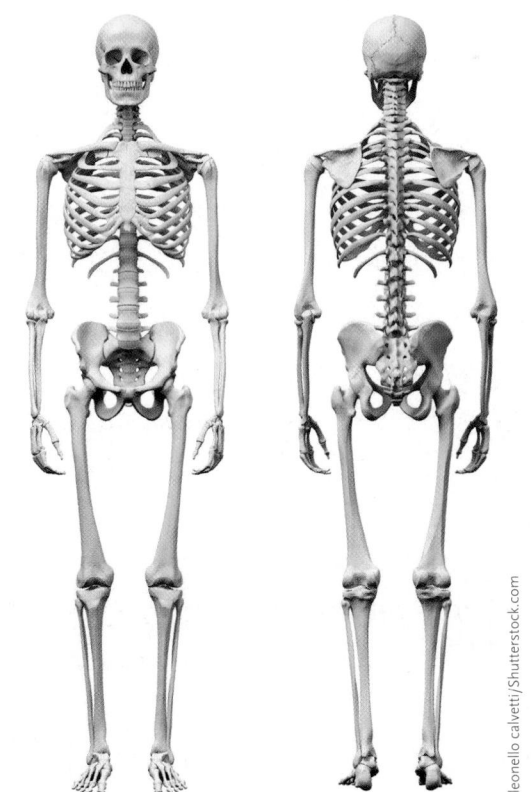

Figure 21.2
Human skeleton, front and back views.

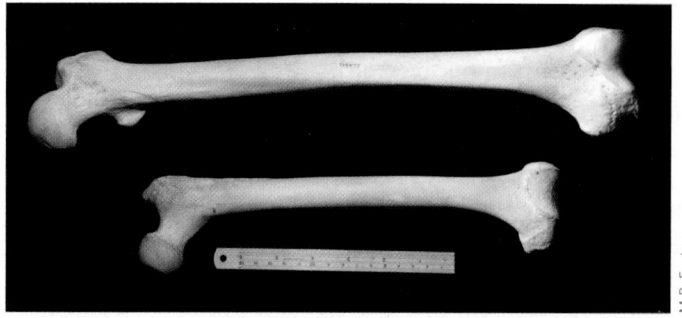

Figure 21.3

A comparison of the thigh bones (femora) of a chimpanzee and a human (top).

M.B. Fenton

of their upright stance. Also, human leg bones (both the upper leg and lower leg) are longer than in chimpanzees, while chimps have longer foreleg bones than humans do.

More recently, the metatarsals (the long foot bones) provided additional features for recognizing bipedalism in hominins. For example, the fourth metatarsals of *Australopithecus afarensis* were more like those of humans than those of either chimps or gorillas **(Figure 21.4).** Comparable features also appeared in *Ardipithecus ramidus* from 3.4 mya, indicating a longer than expected history of bipedalism.

Australopithecus sediba from the Malapa site in South Africa was an earlier (about 2 mya) version of Lucy (*Australopithecus afarensis*, **Figure 21.5**). The ankles and feet of *A. sediba* suggest a combination of climbing ability (arboreal lifestyle) and bipedalism that differed from that of Lucy. The sacral and pubic areas of the pelvis of *A. sediba* resemble those of Lucy and other australopithecines. But the shape of the ilium of *A. sediba* resembled that of species of *Homo*.

In 2007, S. K. S. Thorpe, R. L. Holder, and R. H. Crompton proposed that bipedalism arose in an arboreal setting. Specifically, they asserted that hand-assisted bipedalism allowed the ancestors of humans and great apes to move on flexible supports (branches) that would otherwise have been too small. Thorpe et al. compared human and orangutan (*Pongo abelii*) locomotion and found that orangutans walking on flexible branches increase knee and hip extension, just as humans do when running on a springy track. In a bipedal gait, humans and orangutans flex the hind limbs in a manner that differs from the gait of gorillas and chimps.

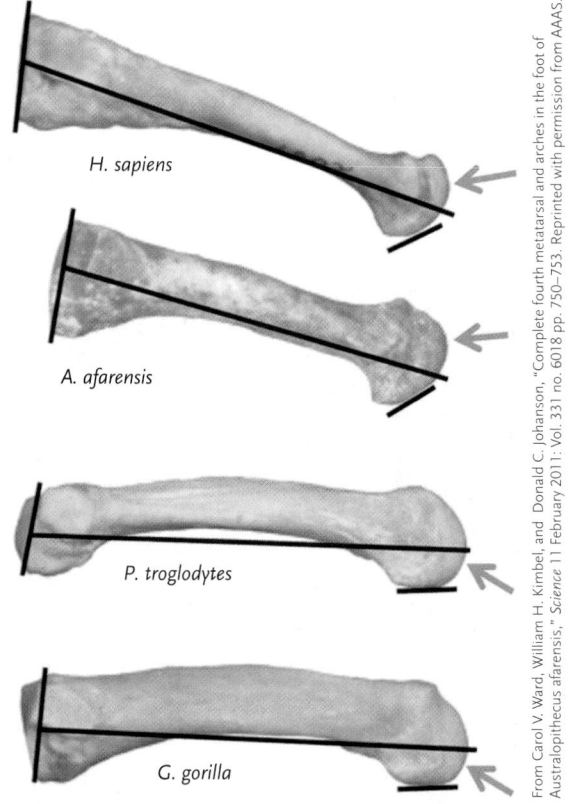

From Carol V. Ward, William H. Kimbel, and Donald C. Johanson, "Complete fourth metatarsal and arches in the foot of *Australopithecus afarensis*," *Science* 11 February 2011: Vol. 331 no. 6018 pp. 750–753. Reprinted with permission from AAAS.

Figure 21.4

The black lines indicate the angles between the proximal (on the left, the ankle end) and distal ends of the fourth metatarsal bone (arrows) of a human, *Australopithecus afarensis*, a chimpanzee (*Pan troglodytes*), and a gorilla (*Gorilla gorilla*). Note the angle in the hominins compared to the parallel lines in the ape species.

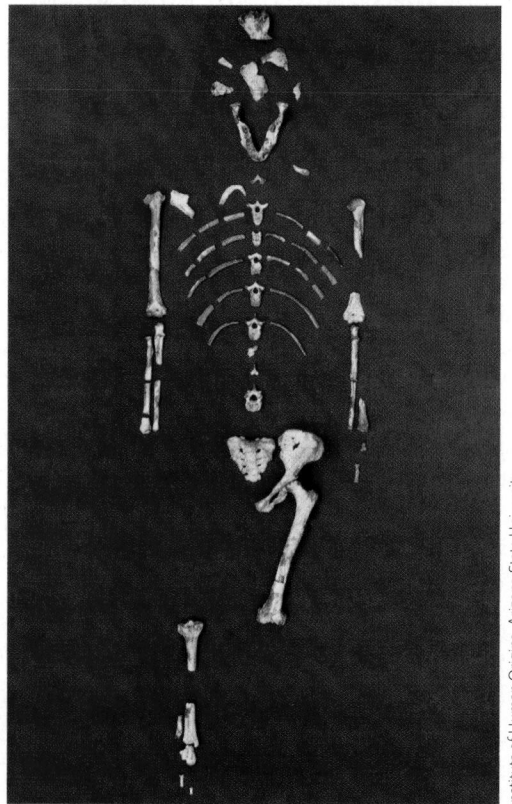

Institute of Human Origins, Arizona State University

Figure 21.5

The fossil skeleton of *Australopithecus afarensis*, popularly known as "Lucy."

21.2b Shoulders and Arms

The importance of climbing for australopiths is supported by the appearance of their shoulder blades (scapulae). The angle of the socket (glenoid fossa, **Figure 21.6**) that receives the head of the upper arm bone (humerus) faces cranially (toward the head), as it does in apes that hang from their arms. In humans, the glenoid fossa faces laterally but changes with age. A lateral-facing glenoid fossa also contributes to humans' ability to throw projectiles such as spears or stones at high speeds.

Species in the genus *Homo* have three specializations associated with throwing ability:

1. They have a "long waist" because of an increase in the number of lumbar vertebrae combined with longer individual vertebrae. The long waist allows the movement of hips and thorax to be decoupled, resulting in a large range of motion of the shoulders and the development of torque.
2. Torsion of the humerus (upper arm bone) between the orientation of its head and the axis of the elbow extends the range of motion during rotation.
3. The laterally directed glenoid fossa aligns the moment generated by flexion of a muscle, the pectoralis major, with the rotation of the torso.

This set of specializations appeared more than 2 mya in *Homo erectus*. It permits elastic storage of energy, which contributes to accurate spear throwing. Effective throwing increased the hunting potential of species in the genus *Homo*.

21.2c Hands

Species in the genus *Homo* have hands that are quite distinct from those of apes. The palms and fingers are short, while the thumbs are long, strong, and mobile **(Figure 21.7)**. This results in our ability to use two different grips **(Figure 21.8, p. 476)**, a power grip and a precision grip, allowing manipulative skills and a capacity to make precise tools. The proportions of hominin hands also deliver a performance advantage when striking with a fist. Buttressing of the elements in the hand increases the stiffness of the joint between the second metacarpal (a hand bone) and phalanges (finger

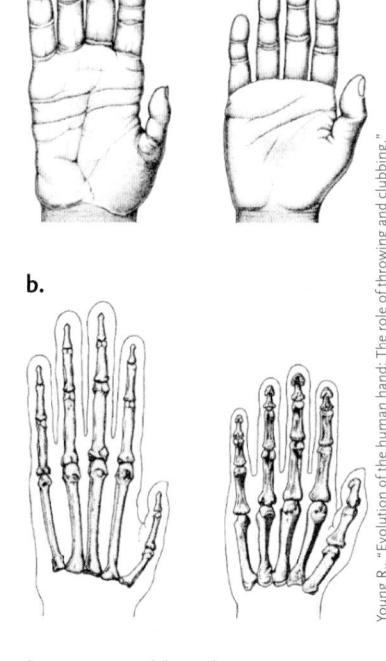

a.

b.

Figure 21.7
The chimp's hands (left) are adapted for grasping branches, while human hands (right) are adapted for precision and power grips. The relatively longer thumbs of humans also contribute to our precision grip.

bones), enabling hominins to punch with more force. The ability to present our hands palms up (supination) or palms down (pronation) also increases their versatility.

Increasing the force that can be delivered can also be achieved by the leverage associated with attaching a handle or strap (haft) to a projectile point, such as a spear or an arrow. At Kathu Pan in South Africa, hafted tools date from 500 thousand years ago. Hafted tools such as spears further enhance the impact of high-velocity throwing, while shorter stabbing blows reflect the importance of the power grip. Other tools found among hominin fossils include stone cutting tools used to take meat off bones and stone axes used for chopping trees. Through time, the tools found show improvement in the technique and the fine motor skills required to make them.

21.2d Pelvis and Birth

Female hominins suffer at least one consequence of being bipedal, namely the shift in the body's centre of mass during pregnancy. A marked posterior concavity of individual lower back (lumbar) vertebrae stabilizes the centre of mass of the upper body over the hips. Females have a derived curvature of the lumbar area and reinforcement of those vertebrae to compensate for the additional load associated with pregnancy. The anatomy of the pelvis and lower back of *Australopithecus* indicates that

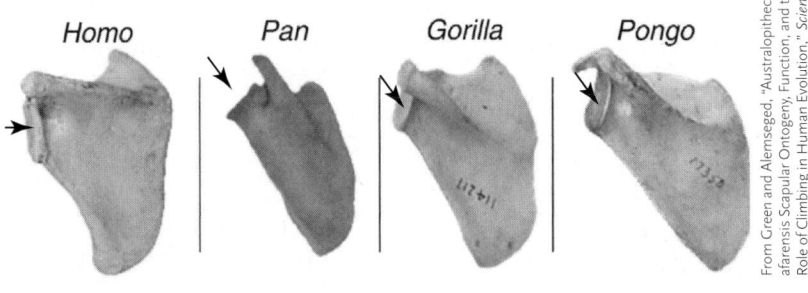

Figure 21.6
A comparison of the shoulder blades (scapulae) of a human, a chimp (*Pan*), a gorilla, and an orangutan (*Pongo*) showing the orientation of the glenoid fossa (arrows).

M.B. Fenton

M.B. Fenton

Figure 21.8

Power grip versus precision grip. Hominins grasp objects in two distinct ways. The power grip **(a)** allows us to grasp an object firmly, whereas the precision grip **(b)** allows us to manipulate objects by fine movements.

these adaptations to bipedalism preceded the evolution of species in the genus *Homo*. Compared to modern humans, the birth canals of neandertals were not as specialized. The birth process in our species is specialized and may be a relatively recent development.

STUDY BREAK

1. Name three morphological specializations associated with bipedalism.
2. What effect do these specializations have on hominins, and how do they distinguish hominins from chimps?

21.3 Human Features That Do Not Fossilize

Some features characteristic of humans are unlikely to fossilize, such as behavioural and soft tissue features associated with social organization and language. The structure of jaws and teeth as well as fossilized dung (coprolites) can be used to infer (jaws and teeth) or reveal (coprolites) what fossil animals ate. Although jaw and tooth structure can suggest the ability to eat hard food, our ancestors may have used tools to break up hard foodstuffs and fire to soften and cook them. Meanwhile, early hominins exploited a range of habitats and thrived on a diversity of food (see "Molecule behind Biology").

Humans show a great capacity for making friends—not genetically related individuals—with whom they have long-term, nonreproductive relationships that involve cooperation and mutual influence. These relationships underlie social networks, a feature of many social species. In humans, social networks may include individuals of other species, such as dogs and cats. Social networks and associated cooperative behaviour are well known from human hunter–gatherer societies, such as the Hadza of Tanzania.

One apparently unique feature of human social networks is the common use of some land areas by different groups. This leads to a pattern of movement among groups (dispersal), which enhances social learning and a cumulative culture.

Effective communication among individuals is an essential part of social networks. One aspect is an individual's ability to read and interpret the body language (see Chapter 47, page 1166) of another. Humans use both body cues and facial expressions to distinguish between intense positive and negative emotions **(Figure 21.9)**.

Language is a means of communication that involves symbolism and syntax. Although language is sometimes considered a hallmark of *Homo sapiens,* other animals also use a combination of symbolism and syntax in communication. For example, when a vervet monkey (*Chlorocebus pygerythrus*) gives the "eagle" alarm call, its fellows look skyward and move closer to the trunks of trees. When the same monkey gives a "leopard" alarm call, other monkeys in trees look down and those on the ground climb trees.

Language is not unique to *H. sapiens,* and it is not clear when language appeared in human evolution. The *FOXP2* gene is associated with speech and language in *H. sapiens,* and genomic analysis reveals that *H. sapiens* and *Homo neandertalensis* have similar *FOXP2* genes, implying that neandertals had language. A trait of many (all?) present-day human languages is a combination of simple categories (to minimize details) and a high level of informativeness to maximize communication efficiency.

STUDY BREAK

1. How does communication figure in the evolution of humans?
2. What are friends? How do they distinguish humans from other animals?

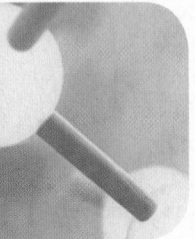

MOLECULE BEHIND BIOLOGY 21.1

Lactose Intolerance

Lactose, a sugar in milk, is broken down by the enzyme lactase. Before young mammals are weaned, they ingest milk and digest lactose. By adulthood, many mammals no longer produce lactase and therefore cannot digest lactose. In humans, a single cytosine to thymine mutation in the regulatory region of the lactase allele (LP) results in lactase persistence. Possession of the LP allele allows adults to drink milk. About 66% of adult humans are lactose intolerant (do not have the LP allele), and drinking milk results in severe intestinal distress.

Before the appearance of the LP allele, our ancestors were still able to get some nutritional value from milk. They ate fermented cheeses and yogurt, which have much lower levels of lactose than milk does. Hard cheeses such as Parmesan have very little lactose.

The LP allele appeared about 7500 years ago in what is now Hungary. It rapidly spread through the population because individuals carrying it produced about 19% more fertile offspring than those lacking the allele, a selective advantage. The LP allele illustrates how changes at the gene level can result in changes at the population level.

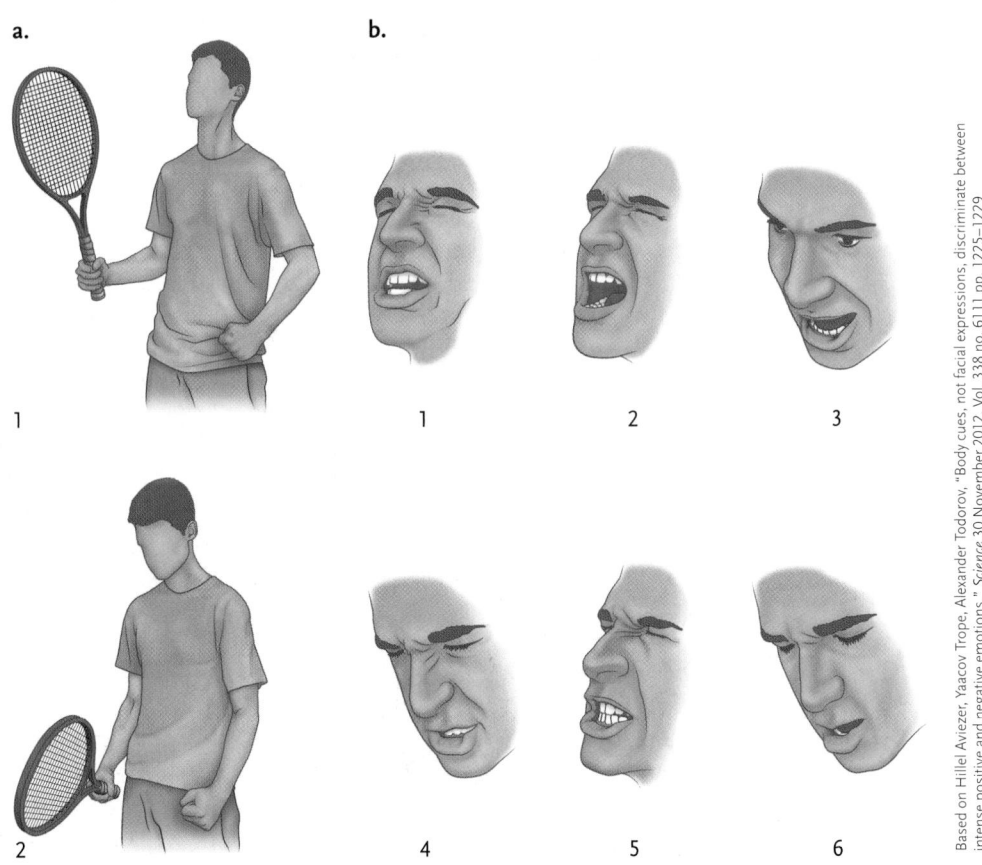

Based on Hillel Aviezer, Yaacov Trope, Alexander Todorov, "Body cues, not facial expressions, discriminate between intense positive and negative emotions," *Science* 30 November 2012, Vol. 338 no. 6111 pp. 1225–1229

Figure 21.9

(a) Body language in response to winning (1) and losing (2) a point. **(b)** Facial expressions presented as isolated views in response to winning (2, 3, 5) and losing (1, 4, 6) a point.

21.4 Dispersal of Early Humans

There are two main theories about the dispersal of our ancestors from Africa. The **African emergence hypothesis** proposes that early hominin descendants (archaic humans) left Africa and established populations in the Middle East, Asia, and Europe. Some time later, 100 thousand to 200 thousand years ago, *H. sapiens* arose in Africa and also migrated into Europe and Asia. Perhaps through competition, *H. sapiens* eventually drove archaic humans to extinction. This hypothesis suggests that all modern humans are descended from a fairly recent African ancestor.

The **multiregional hypothesis** suggests that populations of *H. erectus* and archaic humans had spread through much of Europe and Asia by 500 thousand years ago and modern humans (*H. sapiens*) evolved from descendants of these earlier dispersals. Although these geographically separated populations may have experienced some evolutionary differentiation, gene flow between them prevented reproductive isolation and maintained them as a single but variable species, *H. sapiens*.

Paleontological data do not clearly support either hypothesis, but as of 2011, genetic data **(Figure 21.10)** generally supported the African emergence hypothesis.

Further work on the Y chromosomes of thousands of men from Africa, Europe, Asia, Australia, and the Americas has confirmed that all modern humans are the descendants of a single migration out of Africa.

A rapid exodus of anatomically modern humans out of Africa may have occurred along the coast of the Indian Ocean. Archaeological material from the United Arab Emirates suggests that early emigrants may have taken advantage of lower sea levels to move along the Arabian coast around 60 thousand years ago.

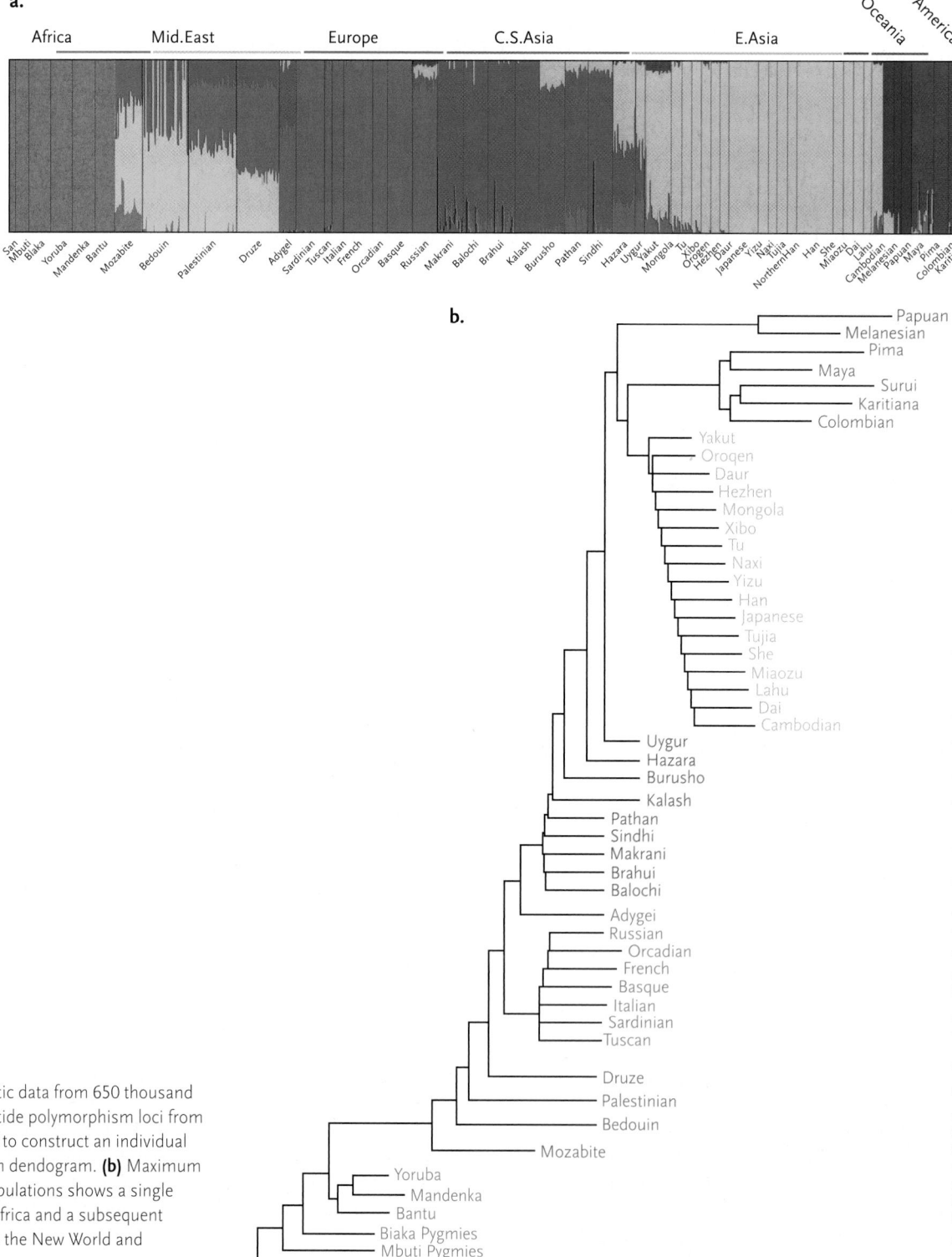

Figure 21.10

Out of Africa. **(a)** Genetic data from 650 thousand common single-nucleotide polymorphism loci from 928 humans were used to construct an individual ancestry and population dendogram. **(b)** Maximum likelihood tree of 51 populations shows a single origin in sub-Saharan Africa and a subsequent radiation across Asia to the New World and Polynesia.

From Jun Z. Li, Devin M. Absher, Hua Tang, Audrey M. Southwick, Amanda M. Casto, Sohini Ramachandran, Howard M. Cann, Gregory S. Barsh, Marcus Feldman, Luigi L. Cavalli-Sforza, and Richard M. Myers, "Worldwide Human Relationships Inferred from Genome-Wide Patterns of Variation," *Science*, February 22, 2008, vol. 319, pp. 1100–1104. Reprinted with permission from AAAS.

Earlier dispersal is clear from material found at Attirampakkam in India. These fossils indicate that Acheulian humans (probably *Homo erectus*) had occupied this site by about 1.5 mya. Acheulian cultures are typified by large cutting tools with bifaces. Other records indicate that hominins (*Homo floresiensis*) were on Flores Island (Indonesia) by 1 mya. Humans occupied sites on the highlands in New Guinea by about 49 thousand years ago. These humans exploited endemic nuts (*Pandanus* spp.) and appeared to have cleared forests to promote growth of their preferred plants. The timing of the arrival of humans in the New World is less well known. Dating of sites at caves in Oregon indicates human occupancy by about 12 thousand years ago.

21.4a The Denisovans

Genetic information in the form of DNA recovered from a finger bone of a girl who lived over 50 thousand years ago indicates that she had dark skin, brown hair, and brown eyes. The girl's fossilized bone fragments were found in Denisova Cave in Siberia. The name of the cave has been applied to the people, Denisovans, who apparently were a sister group to the neandertals. Subsequent genomic analysis revealed that the Denisovans lived in southeast Asia and interbred with the ancestors of today's Melanesians.

Vital components of our immune system (HLA class I) were acquired through the *HLA-B*73* allele inherited from Denisovans in west Asia. Genome analysis also indicates that some *HLA* haplotypes entered modern European and Oceanian human populations from both neandertals and Denisovans.

STUDY BREAK

1. What evidence supports the African emergence hypothesis?
2. By when had humans arrived in India? How do we know?

21.5 Hominins and the Species Concepts

The history of hominins clearly demonstrates the challenges inherent in recognizing species and the boundaries between them. This example is particularly illuminating because it involves paleontological, archaeological, and genomic evidence. The genomic analysis shows that the ancestors of modern humans interbred with both neandertals and Denisovans. If we apply the biological species concept (see Chapter 19), these three groups are not separate species. If we apply the phylogenetic species concept, the distinction is less clear. The morphological evidence from fossils is incomplete and does not necessarily settle the matter.

In 2013, the description of fossil hominins from Georgia (Dmanisi) changed our view of our ancestors. In 2000, these fossils from 1.7 mya were identified as *Homo georgicus*, but the 2013 presentation of data from five skulls shows the same amount of variation in morphology that we know from living populations of humans and chimpanzees. This discovery suggests a highly variable lineage and obliges us to reconsider which of the named species of *Homo* are valid.

STUDY BREAK

1. What is the impact of the fossils from Dmanisi on our view of the species diversity in the genus *Homo*?
2. How do the "species" of hominins fit the biological species concept?

With recent advances in molecular technology, we are learning a great deal more about our evolutionary history. The story continues about what sets humans apart from other animals, whether it is communication abilities, social cooperation, or even an understanding of the future. As we will see later in the book, our animal ancestry is still clear, whether the topic is population ecology (Chapter 29) or animal behaviour (Chapter 47).

Review

 aplia™

To access course materials such as Aplia and other companion resources, please visit www.NELSONbrain.com.

21.1 The Fossil Record of Hominins

- Upright posture and bipedal locomotion are key adaptations distinguishing hominins from apes.
- Evolutionary biologists are not proposing that humans evolved from apes, but that apes are our closest living relatives (see "Chromosomal Similarities and Differences among the Great Apes").

21.2 Morphology and Bipedalism

- Bipedalism may have arisen in an arboreal setting that allowed the ancestors of humans to move on flexible supports (branches). Bipedal locomotion freed the hands from locomotor functions, allowing them to become adapted for other activities, such as tool use.
- The structure and length of the femur, and the structure of the scapula (shoulder blade) can indicate whether a hominin was bipedal or not.

- Bipedal hominins were able to use their arms for carrying things as well as making and using tools.
- The hands of hominins are quite distinct from those of apes. The palms and fingers are short, while the thumbs are long, strong, and mobile. This results in our ability to use a power grip and a precision grip, allowing manipulative skills and a capacity to make precise tools.

21.3 Human Features That Do Not Fossilize

- Humans are able to develop social networks among friends and family groups.
- Effective communication among individuals is an essential part of social networks.

21.4 Dispersal of Early Humans

- The African emergence hypothesis suggests that between 0.5 and 1.5 mya, a population of *H. erectus* gave rise to several descendant species that left Africa and established populations in the Middle East, Asia, and Europe. Some time later, *H. sapiens* arose in Africa. These modern humans also migrated into Europe and Asia and eventually drove archaic humans to extinction.

- The multiregional hypothesis proposes that populations of *H. erectus* and archaic humans had spread through much of Europe and Asia by 0.5 mya. Modern humans (*H. sapiens*) then evolved from archaic humans in many regions simultaneously. Although they were geographically separated, these populations may have experienced some evolutionary differentiation; gene flow between them prevented reproductive isolation and maintained them as a single species, *H. sapiens*.

21.5 Hominins and the Species Concepts

- The history of hominins shows the challenges inherent in recognizing species and the boundaries between them. This example involves paleontological, archaeological, and genomic evidence.

Questions

Self-Test Questions

1. Which of the following is true about neandertals (*Homo neandertalensis*)?
   a. They did not occur in the same places and times as *Homo sapiens*.
   b. They did not interbreed with *Homo sapiens*.
   c. They were not behaviourally distinct from *Homo sapiens*.
   d. They were extinct by 20 thousand years ago.

2. Which describes the Hominidae?
   a. the group that includes gorillas and chimps
   b. first appear in the fossil record about 1 mya
   c. include species in the genera *Homo, Australopithecus*, and *Paranthropus*
   d. include only species in the genus *Homo*

3. In hominids, bipedalism involves specializations of which of the following?
   a. feet
   b. knees
   c. ankles
   d. pelvis and legs

4. In humans, these are specializations for throwing.
   a. feet
   b. hands
   c. shoulders
   d. waist, shoulders, and arms

5. This is true about the ability to make friends.
   a. It occurs in social organisms.
   b. It is unique to humans.
   c. It depends upon individual recognition.
   d. It occurs only in some people.

6. A variety of evidence indicates that humans evolved in which place?
   a. Africa
   b. South America
   c. Europe
   d. Australia

7. The Denisovans were which of the following?
   a. another species in the genus *Homo*
   b. larger than *Homo sapiens*
   c. originally discovered in Siberia from genetic analysis of one finger bone
   d. a variety of neandertal

8. Which of the following species of fossil hominin is the oldest?
   a. *Homo neanderetalensis*
   b. *Homo sapiens*
   c. *Ardipithecus ramidus*
   d. *Australopithecus sediba*

9. What does genetic similarity among humans, chimps, gorillas, and orangutans mean?
   a. They all belong to the same species.
   b. They have a relatively recent common ancestor compared to other primates.
   c. They are evidence of creation.
   d. They can interbreed and produce fertile offspring.

10. When did hominins move out of Africa?
    a. 2 mya
    b. 3 mya
    c. 1.5 mya
    d. 100 thousand years ago

Questions for Discussion

1. Which of the species concepts presented in Chapter 19 best fits the species of hominins currently known to us? Why?

2. What does bipedal mean? Name three groups of tetrapods that are bipedal. How do the modes of locomotion differ among these groups?

3. What is language? Is language unique to humans? What evidence would you use to support your definition of language and its appearance in humans?

4. What does lactose intolerance tell us about the evolution of humans?

5. Is tool use a characteristic of humans? Explain your answer. What advantage(s) can use of tools confer on an individual?

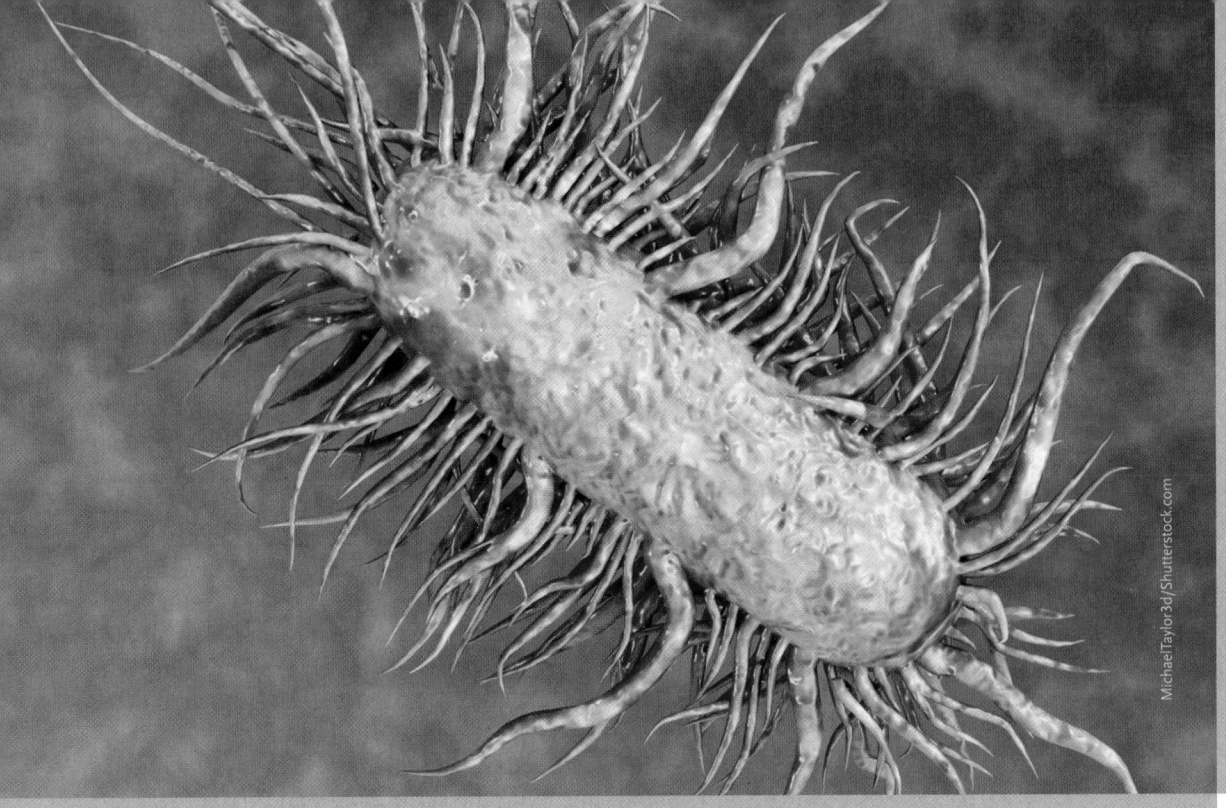

The bacterium *Escherichia coli*.

MichaelTaylor3d/Shutterstock.com

Bacteria and Archaea

WHY IT MATTERS

Who are you? What makes you "you"? Would you feel less like "you" if you knew that most of the cells in your body weren't human cells at all? The bacterial cells on and in our body outnumber our cells by ten to one. And given that, as Princeton University scientist Bonnie Bassler points out, the average person has about 30 000 human genes but more than 3 million bacterial genes, we are at most 1% human! But these bacteria aren't alien invaders—many of them may be crucial for making us unique individuals. There are about a hundred trillion bacteria of hundreds (or thousands) of different species lining your large intestine. When you were born, your gut was sterile, but immediately after birth, your intestines started to be colonized—the exact composition of these "pioneers" depends on where you were born and whether you were breastfed, among other factors. The early colonists were essential for the normal development of your gut as an infant, and throughout your life, your gut bacteria have continued to help you in many ways: they help digest your food, synthesize vitamin K for you, and produce antimicrobial factors to protect you from pathogens. The composition of your gut bacteria is more similar to that of your family than to people not related to you, but it is still unique to you; even identical twins, who share so much else, have different sets of gut bacteria. Recent research has revealed that the diversity of your gut bacteria plays a role in your odds of developing metabolic diseases and becoming obese, and may even be involved in your mental health.

CONCEPT FIX As you can tell from this introduction, not all or even most prokaryotic organisms are harmful! The idea that all bacteria cause disease is one of the major misconceptions about these organisms, but nothing could be farther from the truth: most known bacteria and members of the other group of prokaryotic organisms, archaea, play a crucial role in ecosystems, recycling nutrients and breaking down compounds that no other organisms can. Others carry out reactions important in food production, in industry (e.g., production of pharmaceutical products), and in **bioremediation** of polluted sites. ⬡

In this chapter, we first look at the structure and function of prokaryotic organisms, emphasizing the features that differentiate them from other organisms, and conclude with a look at the diversity of these fascinating organisms.

22.1 The Full Extent of the Diversity of Bacteria and Archaea Is Unknown

While reading this chapter, keep in mind that everything we know so far about bacteria and archaea is based on a tiny fraction of the total number of species. We have isolated and identified only about 6000 species, which may be as low as 1% of the total number. We know almost nothing about the prokaryotic organisms of entire habitats, such as the oceans, which make up 70% of Earth's surface. Why have we only identified so few, and why are we not even sure how many prokaryotic organisms there might be? In the past, we identified and classified bacteria and archaea based on external features (e.g., cell wall structure) and physiological differences, which meant that we had to be able to grow the organisms in culture. We have learned a great deal about the biology of some bacteria and archaea but have been unable to learn much about the majority of prokaryotic organisms, since they cannot be grown in culture (e.g., those that require extreme physicochemical conditions). Recently, molecular techniques have been developed that allow us to isolate and clone DNA from an environment and then analyze gene sequences; this means that we can now identify and characterize bacteria and archaea without having to culture them. This approach, known as **metagenomics**, now enables us to investigate the diversity of prokaryotic organisms in a wide range of environments. However, our understanding of the full extent of microbial diversity still faces other challenges, such as the fact that many environments (e.g., the deep ocean) are remote and thus very difficult and/or costly to sample.

22.1a Prokaryotic Organisms Make Up Two of the Three Domains of Life

Two of the three domains of living organisms, **Archaea** and **Bacteria**, consist of prokaryotic organisms (the third domain, **Eukarya**, includes all eukaryotes). Bacteria are the prokaryotic organisms most familiar to us, including those responsible for diseases of humans and other animals as well as those that we rely on for production of cheese, yogurt, chocolate, and other foods. Archaea are not as well known, as they were only discovered about 40 years ago. As you will see in this chapter, archaea share some cellular features with eukaryotes and some with bacteria but have still other features that are unique. Many of the archaea live under very extreme conditions that no other organisms, including bacteria, can survive.

22.2 Prokaryotic Structure and Function

We begin our survey by examining prokaryotic cellular structures and modes of reproduction, and how they obtain energy and nutrients.

In general, prokaryotic organisms are the smallest in the world **(Figure 22.1)**. Few species are more than 1 to 2 μm long (although the longest is 600 μm long, which is larger than some eukaryotes!); from 500 to 1000 of them would fit side by side across the dot on this letter "i." Despite the small size of bacteria and archaea, they dominate life on Earth: current estimates of total prokaryotic diversity are in the billions of species, and their total collective mass (their **biomass**) on Earth exceeds that of animals and may be greater than that of all plant life. Prokaryotic organisms colonize every niche on Earth that supports life, and even occur deep in the Earth's crust. They also colonize other organisms—for example, huge numbers of bacteria inhabit the surfaces and cavities of a healthy human body, including the skin, mouth and nasal passages, and large intestine. As mentioned in "Why It Matters," collectively, the bacteria in and on your body outnumber all the other cells in your body. It is not surprising that the diversity of bacteria and archaea should be so much greater than that of eukaryotes because for about 3 billion years they were the only forms of life on Earth and so had time to diversify and expand into every habitat on Earth before the first eukaryotes appeared on the scene (see Chapter 2).

22.2a Prokaryotic Cells Appear Simple in Structure Compared with Eukaryotic Cells

Three cell shapes are common among prokaryotes: spiral, spherical (or **coccoid**; *coccus* = berry), and cylindrical (known as **rods**), but some archaea even have square cells **(Figure 22.2)**.

At first glance, a typical prokaryotic cell seems much simpler than a eukaryotic cell **(Figure 22.3, p. 484)**: images taken with standard electron microscopy typically reveal little more than a cell wall and plasma membrane surrounding cytoplasm with DNA concentrated in one region and ribosomes scattered throughout. The chromosome is not contained in a membrane-bound nucleus but is packed into an area of the cell called the **nucleoid**. Prokaryotic cells have no cytoplasmic organelles equivalent to the endoplasmic reticulum (ER) or Golgi complex of eukaryotic cells (see Chapter 2). With few exceptions, the reactions carried out by organelles in eukaryotes are distributed between the plasma membrane and the cytoplasmic solution of prokaryotic cells; this means that macromolecules such as proteins are very concentrated in the cytoplasm of these cells, making the cytoplasm quite viscous. This evident simplicity of prokaryotic cells led

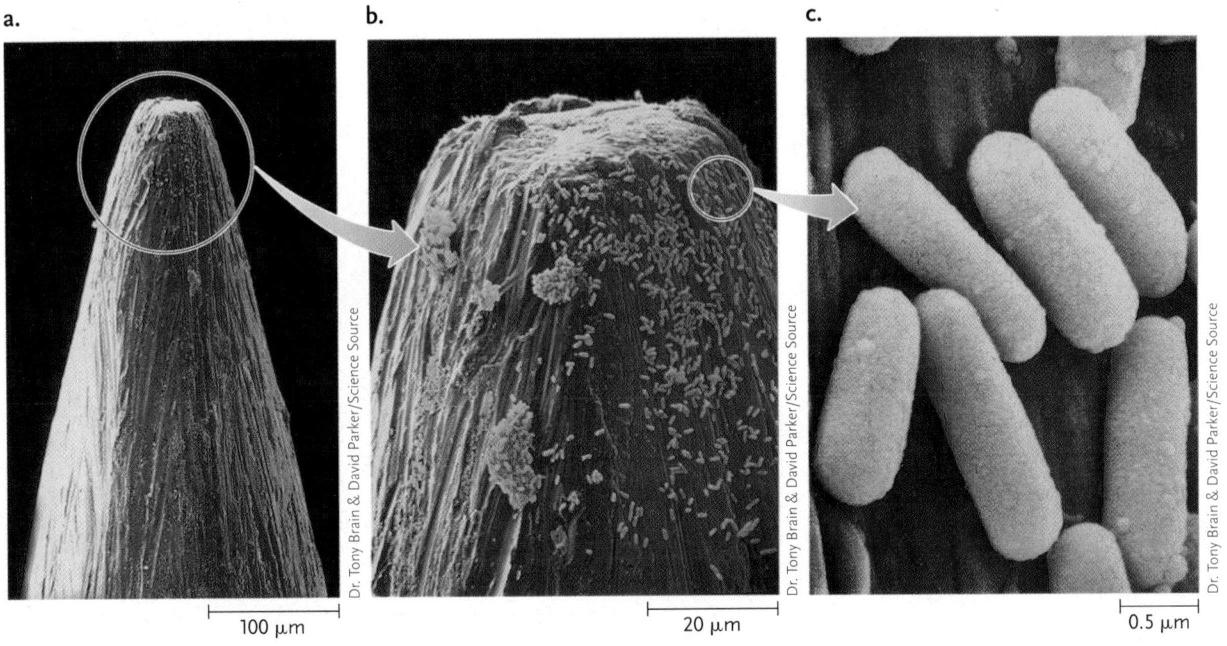

Figure 22.1

Bacillus bacteria on the point of a pin. Cells magnified **(a)** 70 times, **(b)** 350 times, and **(c)** 14 000 times.

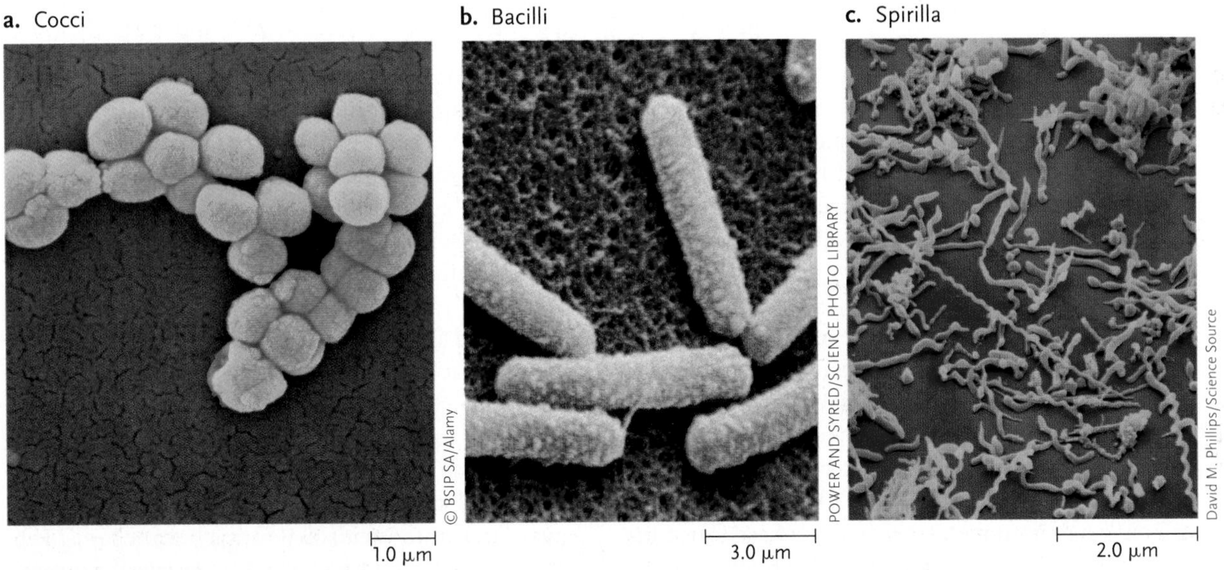

a.

Dr. Tony Brain & David Parker/Science Source

100 μm

b.

Dr. Tony Brain & David Parker/Science Source

20 μm

c.

Dr. Tony Brain & David Parker/Science Source

0.5 μm

a. Cocci

© BSIP SA/Alamy

1.0 μm

b. Bacilli

3.0 μm

c. Spirilla

POWER AND SYRED/SCIENCE PHOTO LIBRARY

David M. Phillips/Science Source

2.0 μm

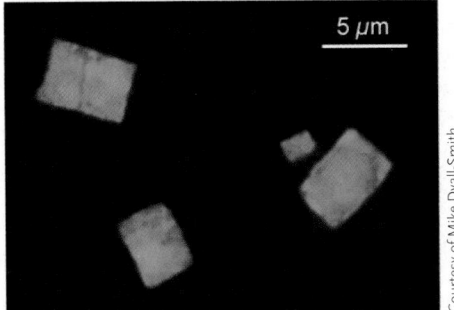

d. Square cells

5 μm

Courtesy of Mike Dyall-Smith

Figure 22.2

Common shapes of prokaryotic cells. **(a)** Scanning electron microscope (SEM) image of *Micrococcus*, a coccoid bacterium. **(b)** SEM image of *Salmonella*, a rod-shaped bacterium. **(c)** SEM image of *Spiroplasma*, a spiral bacterium. **(d)** Acridine orange–stained cells of *Haloquadratum walsbyi*, a square archaeon.

people to regard these cells as featureless and disorganized. However, the apparent simplicity of these cells is misleading. New microscopic techniques reveal that prokaryotic cells do have a cytoskeleton—not homologous to that of a eukaryote but serving some of the same functions—and have more sophisticated organization than was previously thought. In fact, recent research carried out by Laura van Niftrik of the Netherlands and her colleagues has identified a prokaryotic organelle! Certain bacteria that obtain energy by oxidizing ammonia have an internal membrane-bound compartment, where ammonia oxidation

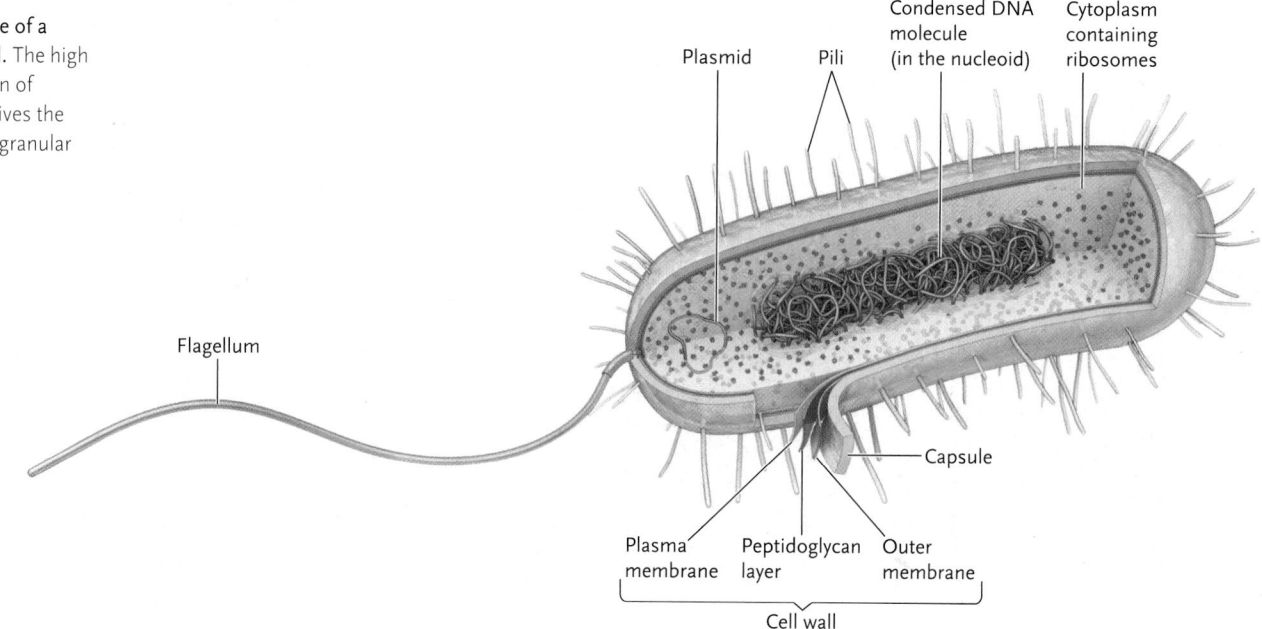

Figure 22.3

The structure of a bacterial cell. The high concentration of ribosomes gives the cytoplasm a granular appearance.

Plasmid — Pili — Condensed DNA molecule (in the nucleoid) — Cytoplasm containing ribosomes

Flagellum

Capsule

Plasma membrane — Peptidoglycan layer — Outer membrane

Cell wall

occurs. It was hypothesized that as ammonia oxidation proceeds inside this compartment, a proton-motive force could be generated across the membrane, generating ATP. Van Niftrik and her colleagues found that the membrane around this compartment does contain ATP synthase, supporting the above hypothesis. Thus, it appears that some prokaryotic cells have organelles with specialized functions.

Internal Structures. The genome of most prokaryotic cells consists of a single, circular DNA molecule, although some, such as the causative agent of Lyme disease (*Borrelia burgdorferi*), have a linear chromosome. Many prokaryotic cells also contain small circles of DNA called **plasmids (Figure 22.4)**, which generally contain genes for nonessential but beneficial functions such as antibiotic resistance. Plasmids replicate independently of the cell's chromosome and can be transferred from one cell to another, meaning that genes for antibiotic resistance are readily shared among prokaryotic cells, even among cells of different species. This *horizontal gene transfer* allows antibiotic resistance and other traits to spread very quickly in bacterial populations. Horizontal gene transfer also occurs when bacterial cells take up DNA from their environment (e.g., from other cells that have lysed) or when viruses transfer DNA from one bacterium to another (see Chapter 23). Evidence indicates that a virus transferred toxin-encoding genes from *Shigella dysenteriae* (which causes bloody diarrhea) to *E. coli*, resulting in the deadly O157:H7 strain responsible for serious illness or even death of people eating beef and other food contaminated with this bacterium.

Like eukaryotic cells, prokaryotic cells contain ribosomes. Bacterial ribosomes are smaller than eukaryotic ribosomes but carry out protein synthesis by essentially

the same mechanisms as those of eukaryotes (see Chapter 13). Archaeal ribosomes resemble those of bacteria in size but differ in structure; protein synthesis in Archaea is a combination of bacterial and eukaryotic processes, with some unique archaeal features. As a result, antibiotics that stop bacterial infections by targeting ribosome activity do not interfere with archaeal protein synthesis.

Prokaryotic Cell Walls. Most prokaryotic cells have a cell wall that lies outside their plasma membrane and protects the cell from lysing if subjected to hypotonic conditions or exposed to membrane-disrupting compounds such as detergents. The primary component of bacterial cell walls is **peptidoglycan**, a polymer of sugars and amino acids that forms linear chains. Peptide cross-linkages between the chains give the cell wall great strength and rigidity. The antibiotic penicillin prevents the formation of these cross-linkages, resulting in a weak cell wall that is easily ruptured, killing the cell **(Figure 22.5)**.

Bacteria can be divided into two broad groups, Gram-positive and Gram-negative cells, based on their reaction to the **Gram stain procedure**, traditionally used as the first step in identifying an unknown bacterium.

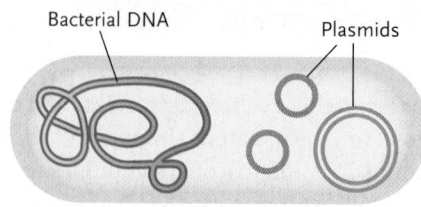

Bacterial DNA

Plasmids

Figure 22.4

Plasmids inside a prokaryotic cell.

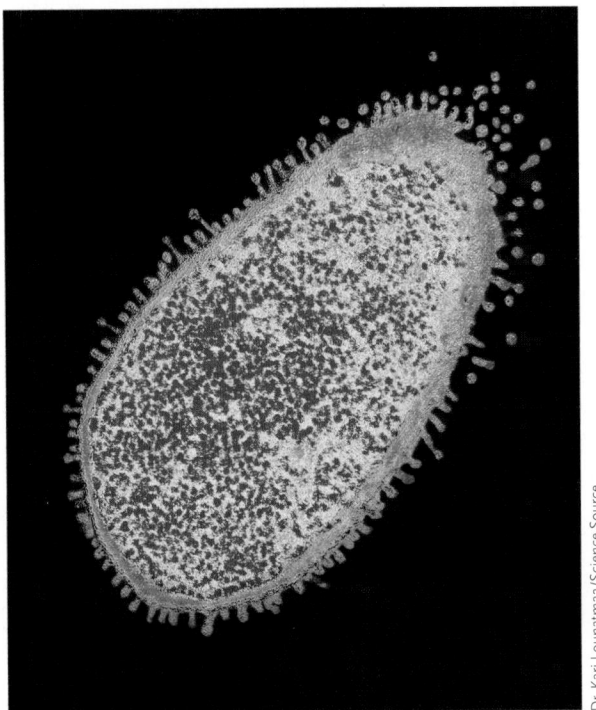

Figure 22.5

Image showing degradation of the cell wall following antibiotic treatment. The cell will eventually lyse, killing the bacterium.

Dr. Kari Lounatmaa/Science Source

Cells are first stained with crystal violet and then treated with iodine, which forms a complex with crystal violet. The cells are then rinsed with ethanol and counterstained with safranin. Some cells retain the crystal violet–iodine complex and thus appear purple when viewed under the microscope; these are termed Gram-positive cells. In other bacteria, ethanol washes the crystal violet–iodine complex out of the cells, which are colourless until counterstained with safranin; these Gram-negative cells appear pink under the microscope. The differential response to staining is related to

differences in cell wall structure: **Gram-positive** bacteria have cell walls composed almost entirely of a single, relatively thick layer of peptidoglycan (**Figure 22.6a**). This thick peptidoglycan layer retains the crystal violet–iodine complex inside the cell. **Gram-negative** cells have only a thin peptidoglycan layer in their walls, and the crystal violet–iodine complex is washed out. In contrast, the cell wall of Gram-negative bacteria has two distinct layers (Figure 22.6b), a thin peptidoglycan layer just outside the plasma membrane and an **outer membrane** external to the peptidoglycan layer. This outer membrane contains **lipopolysaccharides (LPSs)** and thus is very different from the plasma membrane. The outer membrane protects Gram-negative bacteria from potentially harmful substances in the environment; for example, it inhibits entry of penicillin. Therefore, Gram-negative cells are less sensitive to penicillin than are Gram-positive cells.

The cell walls of some archaea are assembled from a molecule related to peptidoglycan but with different molecular components and bonding structure. Others have walls assembled from proteins or polysaccharides instead of peptidoglycan. Archaea have a variable response to the Gram stain, so this procedure is not useful in identifying archaea.

The cell wall of many prokaryotic cells is surrounded by a layer of polysaccharides known as a **capsule** (**Figure 22.7, p. 486**; see also Figure 22.6). Capsules are "sticky" and play important roles in protecting cells in different environments. Cells with capsules are protected to some extent from desiccation, extreme temperatures, bacterial viruses, and harmful molecules such as antibiotics and antibodies. In many pathogenic bacteria, the presence or absence of the protective capsule differentiates infective from noninfective forms. For example, normal *Streptococcus pneumoniae* bacteria are capsulated and virulent, causing severe pneumonia in humans and other mammals. Mutant *S. pneumoniae* without capsules are nonvirulent and can easily be

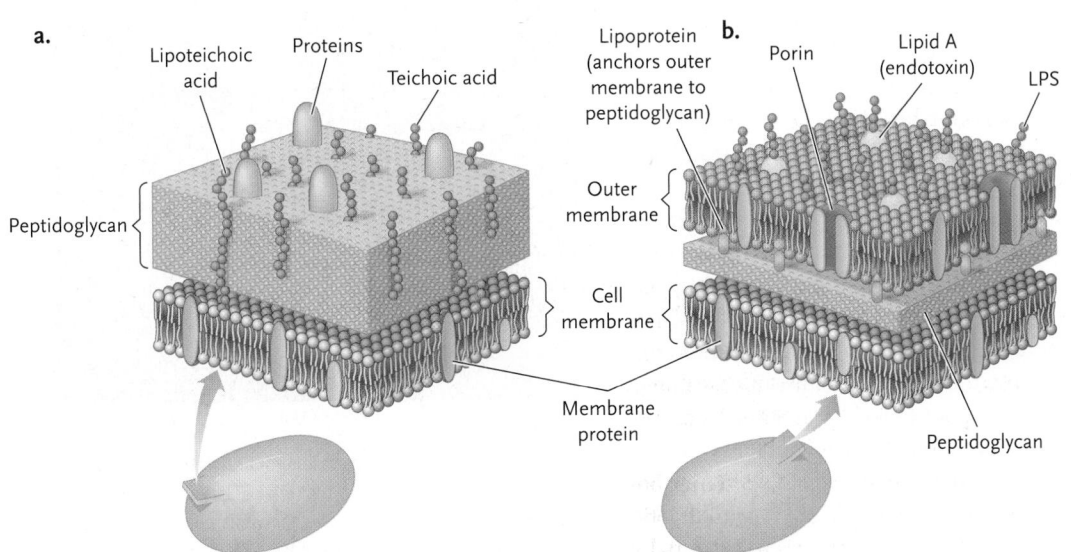

a.

Lipoteichoic acid
Proteins
Teichoic acid
Peptidoglycan

b.
Lipoprotein (anchors outer membrane to peptidoglycan)
Porin
Lipid A (endotoxin)
LPS
Outer membrane
Cell membrane
Membrane protein
Peptidoglycan

Figure 22.6

Cell wall structure in Gram-positive and Gram-negative bacteria. **(a)** The thick cell wall in Gram-positive bacteria. **(b)** The thin cell wall of Gram-negative bacteria has a thin peptidoglycan layer and outer membrane with lipopolysaccharides (LPSs). The uppermost part of LPS is the O antigen, a carbohydrate chain that elicits an antibody response in vertebrates exposed to Gram-negative bacteria such as *E. coli* O157:H7. More information on the toxic effects of lipid A, which embeds LPSs in the outer membrane, is provided on p. 489.

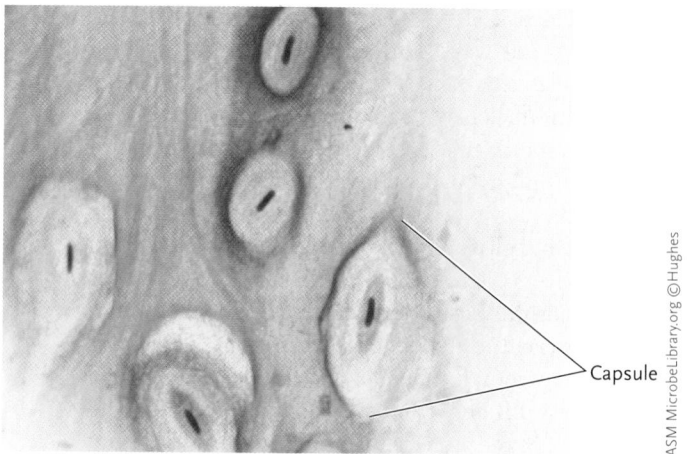

Figure 22.7

Capsules surrounding the cell wall of *Rhizobium*, a Gram-negative soil bacterium.

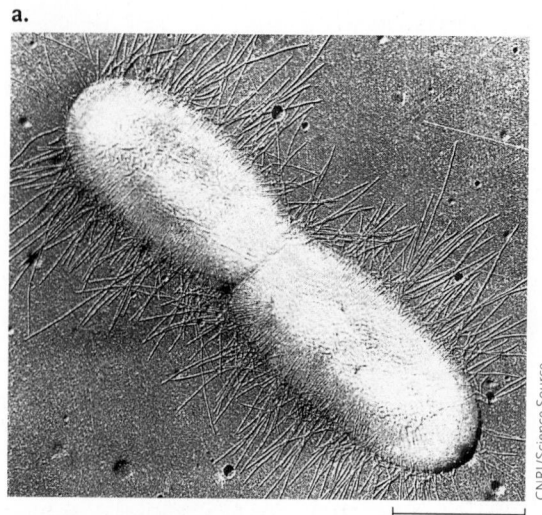

0.5 µm

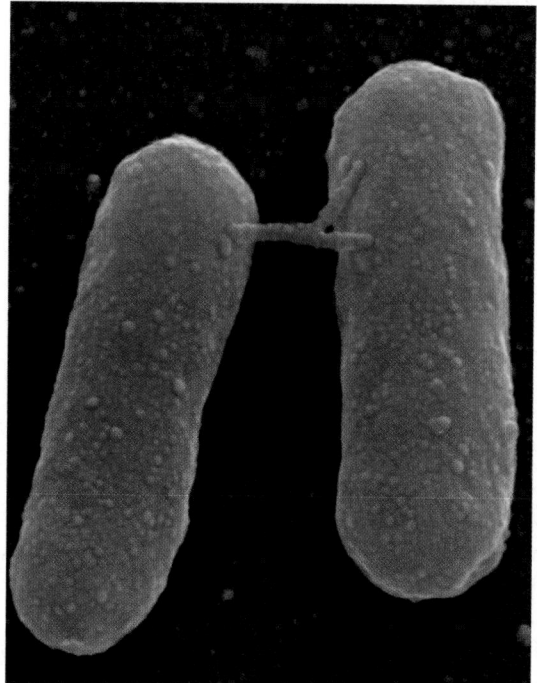

Figure 22.8

(a) Pili extending from the surface of a dividing *E. coli* bacterium. **(b)** Sex pilus connecting two bacterial cells. **(c)** Nanowires (pili that conduct electricity) on *Shewanella oneidensis*. Note that these nanowires are much longer than the cells.

eliminated by the body's immune system if they are injected into mice or other animals.

Flagella and Pili. Many prokaryotic cells can move actively through liquids and even through films of liquid on a surface, most commonly via **flagella** (singular, *flagellum* = whip) extending from the cell wall (see Figure 22.3, p. 484). As outlined in Chapter 2, prokaryotic flagella are very different from eukaryotic flagella in both structure and pattern of movement. Prokaryotic flagella are made of rigid helical proteins, some of which act as a motor, rotating the flagellum much like the propeller of a boat. Archaeal flagella are superficially similar to bacterial flagella and carry out the same function, but the two types of flagella contain different components, develop differently, and are coded for by different genes.

Some prokaryotic cells have rigid shafts of protein called **pili** (singular, *pilus* = "hair") extending from their cell walls **(Figure 22.8a)**, which enable them to adhere to or move along a surface. One type, called a *sex pilus*, not only allows bacterial cells to adhere to each other but also acts as a conduit for the transfer of plasmids from one cell to another (Figure 22.8b). Other types of pili enable bacteria to bind to animal cells. The bacterium that causes gonorrhea (*Neisseria gonorrhoeae*) uses pili to adhere to cells of the throat, eye, urogenital tract, or rectum in humans. In 2005, it was discovered that the pili of some bacteria (e.g., species of *Geobacter* and *Shewanella*) conduct electricity; these "nanowires" transfer electrons out of the cell onto minerals such as iron oxides in their environment (Figure 22.8c). Such electricity-generating bacteria hold promise for the development of microbial fuel cells as an alternative energy source.

Even though prokaryotes are simpler and less structurally diverse than eukaryotic cells, they are much more diverse metabolically, as we will now explore.

1. What features differentiate a prokaryotic cell from a eukaryotic cell? What features do both kinds of cells have?
2. How does the presence of a capsule affect the ability of the human body to mount an immune response to those bacteria?
3. How is a pilus similar to a flagellum? How is it different?
4. How does the amount of peptidoglycan in a bacterial cell wall relate to its Gram-stain reaction?

Table 22.1		Modes of Nutrition Used by Living Organisms	
Energy Source		Oxidation of Molecules*	Light
Carbon source	**CO_2**	**Chemoautotroph** Some bacteria and archaea; no eukaryotes	**Photoautotroph** Some bacteria, some protists, and most plants
	Organic molecules	**Chemoheterotroph** Some bacteria, archaea, and protists; also fungi, animals, and even some plants	**Photoheterotroph** Some bacteria

*Inorganic molecules for chemoautotrophs and organic molecules for chemoheterotrophs.

22.2b Prokaryotic Organisms Have the Greatest Metabolic Diversity of All Organisms

Organisms can be grouped into four modes of nutrition based on sources of energy and carbon (see **Table 22.1**).

In this approach to classification, we focus on carbon rather than other nutrients because carbon is the backbone of all organic molecules synthesized by an organism. Organisms such as plants that synthesize organic carbon molecules using inorganic carbon (CO_2) are **autotrophs** (*auto* = self; *troph* = nourishment). (Note that although CO_2 contains a carbon atom, oxides containing carbon are considered inorganic molecules.) All animals are **heterotrophs**, meaning that they obtain carbon from organic molecules, either from living hosts or from organic molecules in the products, wastes, or remains of dead organisms.

Organisms are also divided according to the source of the energy they use to drive biological activities. **Chemotrophs** (*chemo* = chemical) obtain energy by oxidizing inorganic or organic substances, whereas **phototrophs** obtain energy from light. Combining the carbon and energy sources allows us to group living organisms into four categories (Table 22.1).

Prokaryotic organisms (bacteria and archaea) show the greatest diversity in their modes of securing carbon and energy; they are the only representatives of two of the categories, chemoautotrophs and photoheterotrophs. **Photoheterotrophs** use light as an energy source and obtain carbon from organic molecules rather than from CO_2. **Chemoautotrophs** are commonly referred to as "lithotrophs" (*lithos* = rock, thus "rock-eaters"). As this name suggests, chemoautotrophs obtain energy by oxidizing inorganic substances such as hydrogen, iron, sulfur, ammonia, and nitrites and use CO_2 as their carbon source. Chemolithotrophs thrive in habitats such as deep-sea hydrothermal vents **(Figure 22.9),** where reduced inorganic compounds are abundant. The ability of these organisms to harness energy from these compounds makes them the foundation upon which the rest of the vent community ultimately depends, just as terrestrial organisms rely on the ability of plants and other photoautotrophs to capture light energy.

We breathe oxygen to provide the final electron acceptor for the electrons we remove from our food and pass down an electron transport chain (ETC) to make ATP via aerobic respiration (Chapter 6). Some prokaryotic organisms also use oxygen as a final electron acceptor; like us, these are aerobic organisms or **aerobes**. Aerobes may be **obligate aerobes;** that is, they cannot survive without oxygen. But some prokaryotic organisms "breathe" metals, using metals as the final electron acceptor for electrons; these organisms obtain energy via anaerobic respiration. **Anaerobic respiration** can also involve other inorganic molecules, such as nitrate or sulfate, as the final electron acceptors. Only prokaryotic organisms are capable of this type of respiration. **Obligate anaerobes** are poisoned by oxygen and survive either by fermentation, in which organic molecules are the final electron acceptors, or by anaerobic respiration. **Facultative anaerobes** use O_2 when it is present, but under anaerobic conditions, they live by fermentation or anaerobic respiration. As you learned in Chapter 6, prokaryotic organisms carry out a wider

Figure 22.9 Hydrothermal vents on the ocean floor.

Dr. Ken Macdonald/Science Source

range of fermentation reactions than do eukaryotes; many of these fermentations are economically important to humans, for example, in the production of foods such as cheese, yogurt, and chocolate.

22.2c Bacteria and Archaea Play Key Roles in Biogeochemical Cycles

The ability of prokaryotic organisms to metabolize such a wide range of substrates makes them key players in the life-sustaining recycling of elements such as carbon, oxygen, and nitrogen. The pathway by which a chemical element moves through an ecosystem is known as a **biogeochemical cycle.** As an element flows through its cycle, it is transformed from one form to another; prokaryotic organisms are crucial in many of these transformations. We will look at the nitrogen cycle as an example of the key role prokaryotic organisms play in biogeochemical cycles.

Nitrogen is a component of proteins and nucleotides and so is of vital importance for all organisms. The largest source of nitrogen on Earth is the atmosphere, which is almost 80% nitrogen. Why can't we just use this abundant atmospheric nitrogen? Most organisms cannot make use of this nitrogen because they cannot break the strong triple bond between the two nitrogen atoms. Only some bacteria and archaea can break this bond, using the enzyme nitrogenase, and convert N_2 into forms that can be used by other organisms. In this conversion process, known as **nitrogen fixation**, N_2 is reduced to ammonia (NH_3). Ammonia is quickly ionized to ammonium (NH_4^+), which prokaryotic cells then use to produce nitrogen-containing molecules such as amino acids and nucleic acids. Nitrogen fixation is the only means of replenishing the nitrogen sources used by most organisms—in other words, all organisms rely on nitrogen fixed by bacteria. Examples of nitrogen-fixing bacteria are cyanobacteria and *Rhizobium* (which is symbiotic with plants; see Chapter 38).

Other prokaryotic organisms carry out **nitrification**, the oxidation of ammonium (NH_4^+) to nitrate (NO_3^-). This oxidation process is carried out in two steps by two types of nitrifiers, ammonia oxidizers and nitrate oxidizers, present in soil and water. Ammonium oxidizers convert ammonium to nitrite (NO_2^-), whereas nitrite oxidizers convert nitrite to nitrate. Nitrate is then taken up by plants and fungi and incorporated into their organic molecules. Animals obtain nitrogen in organic form by eating other organisms or each other.

In sum, nitrification makes nitrogen available to many other organisms, including plants, animals, and bacteria that cannot metabolize ammonia. The metabolic versatility of bacteria and archaea is one factor that accounts for their abundance and persistence on the planet; another factor is their impressive reproductive capacity.

22.2d Asexual Reproduction Can Result in Rapid Population Growth

In prokaryotic organisms, asexual reproduction is the normal mode of reproduction. In this process, a parent cell divides by binary fission into two daughter cells that are exact genetic copies of the parent **(Figure 22.10)**. Reproducing by binary fission means that under favourable conditions, populations of prokaryotic organisms can have very rapid exponential growth as one cell becomes two, two become four, and so on. Some prokaryotic cells can double their population size in only 20 minutes and will even begin a second round of cell division before the first round is complete; thus, one cell, given ideal conditions, can produce millions of cells in only a few hours.

These short generation times, combined with the small genomes (roughly one-thousandth the size of the genome of an average eukaryote), mean that prokaryotic organisms have higher mutation rates than do eukaryotic organisms. This translates to roughly 1000 times as many mutations per gene, per unit time, per individual as for eukaryotes. Genetic variability in prokaryotic populations, the basis for their diversity, derives largely from mutation and to a lesser degree from horizontal gene transfer (see Chapter 9). Further, the typically much larger populations of prokaryotic organisms compared with eukaryotes contribute to the much greater genetic variability in bacteria and archaea. In short, prokaryotic organisms have an enormous capacity to adapt, which is one reason for their evolutionary success.

As we have seen, the success of bacteria is beneficial to humans in many ways but can also be detrimental to us when dealing with successful pathogenic

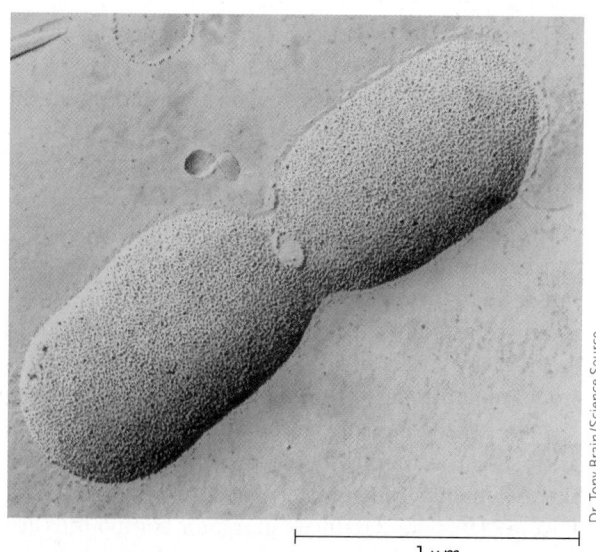

Dr. Tony Brain/Science Source

1 μm

Figure 22.10

E. coli cell dividing by binary fission. Note that a septum is forming between the two parent cells.

bacteria. In the next section, we investigate how some bacteria cause disease and how they are able to resist treatment with antibiotics.

22.2e Pathogenic Bacteria Cause Diseases by Different Mechanisms

Some bacteria produce **exotoxins**, toxic proteins that leak from or are secreted from the bacterium. For example, botulism food poisoning is caused by the exotoxin of the Gram-positive bacterium *Clostridium botulinum*, which grows in poorly preserved foods **(Figure 22.11)**. The botulism exotoxin, botulin, is one of the most poisonous substances known: just a few nanograms can cause severe illness. What makes botulin so toxic? It produces muscle paralysis that can be fatal if the muscles that control breathing are affected. Interestingly, botulin is used under the brand name Botox for the cosmetic removal of wrinkles and in the treatment of migraine headaches and some other medical conditions. Exotoxins produced by certain strains of *Streptococcus pyogenes* have "superantigen properties" (i.e., overactivation of the immune system) that cause necrotizing fasciitis ("flesh-eating disease"). In 1994, Lucien Bouchard, who was then premier of Quebec, lost a leg to this disease.

Other bacteria cause disease through **endotoxins.** Endotoxins are the lipid A portion of the LPS molecule of the outer membrane of all Gram-negative bacteria, such as *E. coli, Salmonella,* and *Shigella.* When a Gram-negative cell lyses, the LPSs of the outer membrane are released; exposure to a specific component of this layer, known as lipid A, causes endotoxic shock. When Gram-negative bacteria enter the bloodstream, endotoxin overstimulates the host's immune system, triggering inflammation and an often lethal immune response. Endotoxins have different effects depending on the bacterial species and the site of infection.

22.2f Pathogenic Bacteria Commonly Develop Resistance to Antibiotics

An **antibiotic** is a natural or synthetic substance that kills or inhibits the growth of bacteria and other microorganisms. Prokaryotic organisms and fungi produce these substances naturally as defensive molecules, and we have also developed ways to synthesize several types of antibiotics. Different types of antibiotics have different modes of action: for example, streptomycins, produced by soil bacteria, block protein synthesis in their targets, whereas penicillins, produced by fungi, target the peptide cross-linkages in peptidoglycan, as described above.

How are bacteria able to block the actions of antibiotics? There are various mechanisms by which bacteria resist antibiotics **(Figure 22.12, p. 490).** For example, some bacteria are able to pump antibiotics out of the cell using membrane-bound pumps. They can also produce molecules that bind to the antibiotic or enzymes that break down the antibiotic, rendering it ineffective against its target. Alternatively, a simple mutation can result in a change in the structure of the antibiotic's target, so that the antibiotic cannot bind to it. Finally, bacteria can develop new enzymes or pathways that are not inhibited by the antibiotic.

Bacteria can develop resistance through mutations, but they can also acquire resistance via horizontal gene transfer (e.g., plasmid transfer). Taking antibiotics routinely in mild doses, or failing to complete a prescribed dosage, contributes to the spread of resistance by selecting strains that can survive in the presence of the drug. Prescription of antibiotics for

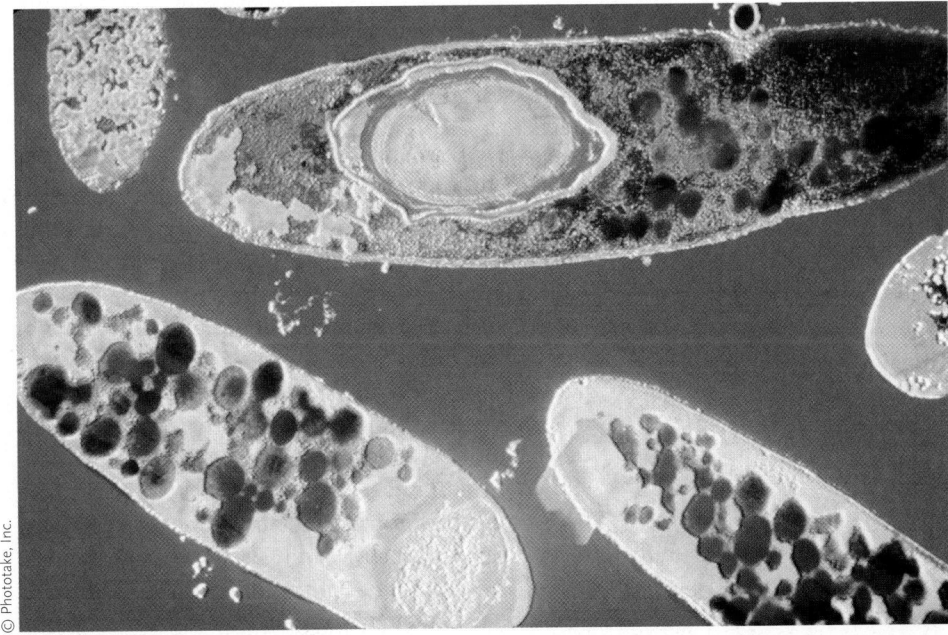

Figure 22.11
The bacterium *Clostridium butyricum*, one of the *Clostridium* species that produces the toxin botulin (colourized TEM). The large stained structure in the cells is a spore (a survival structure).

© Photorake, Inc.

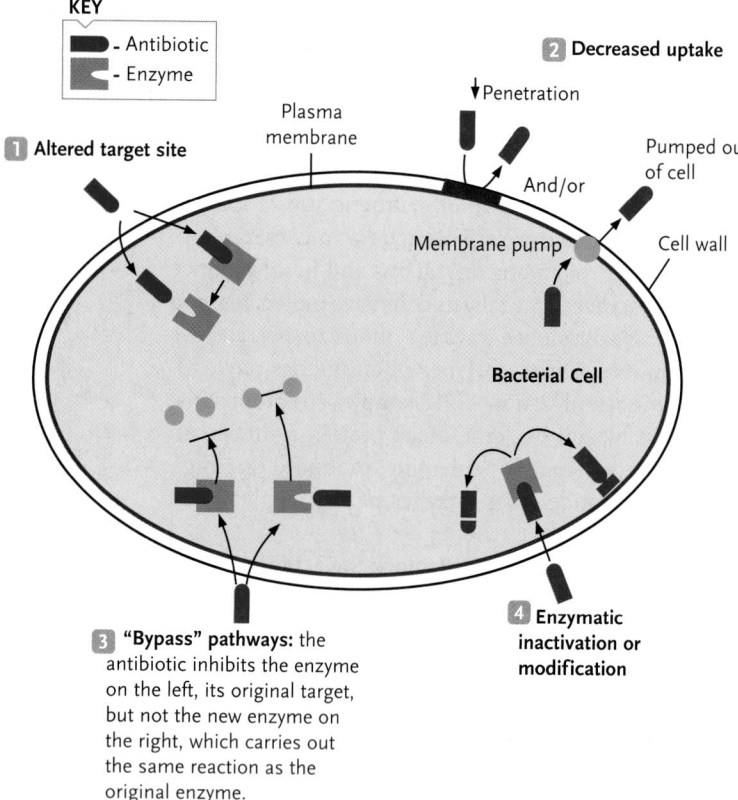

KEY

- Antibiotic
- Enzyme

1 Altered target site

Plasma membrane

2 Decreased uptake

↓ Penetration

And/or

Pumped out of cell

Membrane pump

Cell wall

Bacterial Cell

3 "Bypass" pathways: the antibiotic inhibits the enzyme on the left, its original target, but not the new enzyme on the right, which carries out the same reaction as the original enzyme.

4 Enzymatic inactivation or modification

Figure 22.12
Four major mechanisms of antibiotic resistance. See text for further explanation of each mechanism.

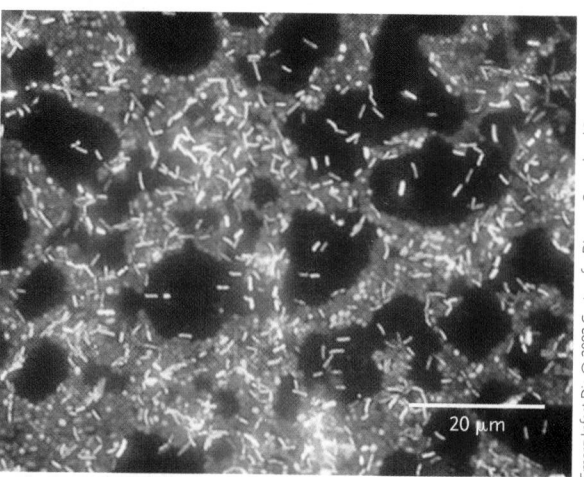

Figure 22.13
Biofilm grown on a stainless steel surface.

Emerg Infect Dis © 2002 Centers for Disease Control and Prevention (CDC)

colds and other virus-caused diseases can also promote bacterial resistance because viruses are unaffected by antibiotics, but the presence of antibiotics in your system can lead to resistance. Antibacterial agents that may promote resistance are also commonly included in such commercial products as soaps, detergents, and deodorants. Resistance is a form of evolutionary adaptation; antibiotics alter the bacterium's environment, conferring a reproductive advantage on those strains best adapted to the altered conditions.

The development of resistant strains has made tuberculosis, cholera, typhoid fever, gonorrhea, and other bacterial diseases difficult to treat with antibiotics. For example, as recently as 1988, drug-resistant strains of *Streptococcus pneumoniae*, which causes pneumonia, meningitis, and middle-ear infections, were practically unknown. Now, resistant strains of *S. pneumoniae* are common and increasingly difficult to treat.

22.2g In Nature, Prokaryotic Organisms May Live in Communities Attached to a Surface

Often, researchers grow bacteria and archaea as individuals in pure cultures. We have learned a lot about prokaryotic organisms from these pure cultures, but in nature, prokaryotic organisms rarely exist as individuals or as pure cultures. Instead, bacteria and

archaea live in communities where they interact in a variety of ways. One important type of community is known as a **biofilm**, which consists of a complex aggregation of microorganisms attached to a surface and surrounded by a film of polymers **(Figure 22.13)**. Life in a biofilm offers several benefits: organisms can adhere to hospitable surfaces, they can live on the products of other cells, conditions within the biofilm promote gene transfer between species, and the biofilm protects cells from harmful environmental conditions (see "Life on the Edge" 22.3). Biofilms form on any surface with sufficient water and nutrients. For example, you're probably familiar with how slippery rocks in a stream can be when you try to step from one to the next; the slipperiness is due to biofilms on the rocks. Dental plaque is also a biofilm; if this biofilm spreads below the gumline, it causes inflammation of the gums (gingivitis). Regular removal of plaque by brushing, flossing, and dental checkups helps prevent gingivitis.

Biofilms have practical consequences for humans, both beneficial and detrimental. On the beneficial side, for example, are the health effects each of us gains from the bacteria that live in biofilms in our gastrointestinal tracts. We also make use of biofilms in commercial applications: biofilms on solid supports are used in sewage treatment plants to process organic matter before the water is discharged, and they can be effective in bioremediating toxic organic molecules contaminating groundwater. But biofilms can also be harmful to human health. Biofilms adhere to many kinds of surgical equipment and supplies, including catheters, pacemakers, and artificial joints. Even if the bacteria colonizing these devices are not pathogenic, their presence is obviously not desirable given that these devices should be sterile. The presence of any Gram-negative bacteria is a concern, given their nature. As well, many heterotrophic bacteria will become opportunistic pathogens, given the right conditions. Biofilm infections are difficult to treat because bacteria in a biofilm

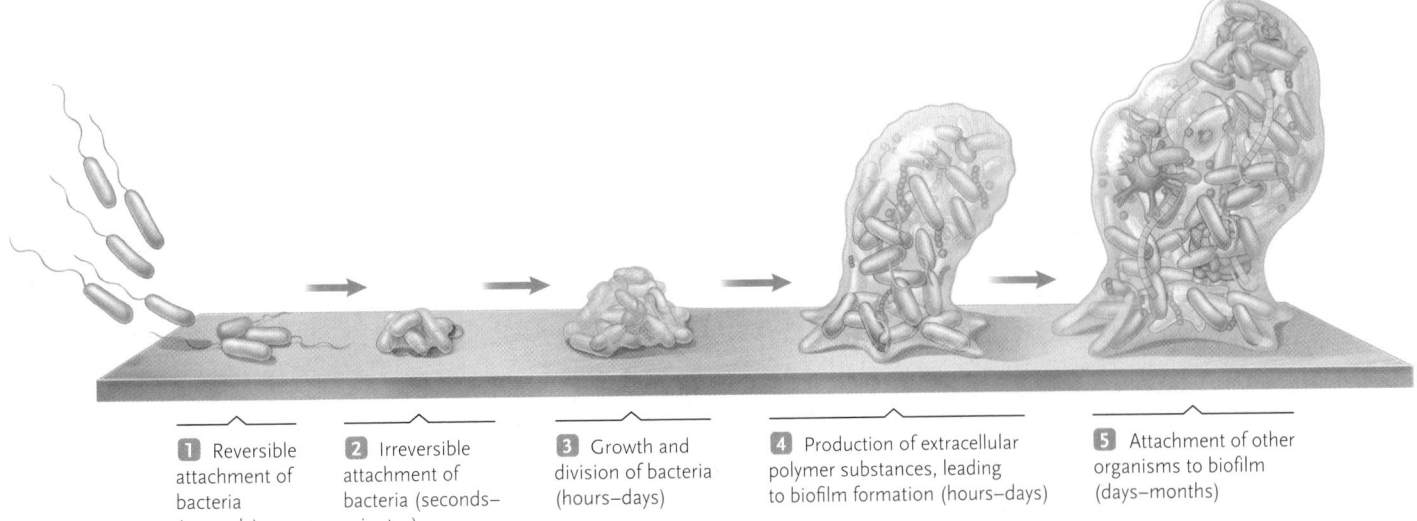

Figure 22.14

Steps in the formation of a biofilm.

1. Reversible attachment of bacteria (seconds)
2. Irreversible attachment of bacteria (seconds–minutes)
3. Growth and division of bacteria (hours–days)
4. Production of extracellular polymer substances, leading to biofilm formation (hours–days)
5. Attachment of other organisms to biofilm (days–months)

are up to 1000 times as resistant to antibiotics as are the same bacteria in liquid cultures. For example, outbreaks of the disease caused by *E. coli* O157:H7 have been caused by biofilms that are very difficult to wash off spinach, lettuce, and other produce.

How does a biofilm form? Imagine a surface, such as a rock in a stream, over which water is flowing **(Figure 22.14).** Due to the nutrients in the water, the surface rapidly becomes coated with polymeric organic molecules, such as polysaccharides or glycoproteins. Once the surface is conditioned with organic molecules, free cells attach in a reversible manner in a matter of seconds (see Figure 22.14, step 1). If the cells remain attached, the association may become irreversible (step 2), at which point the cells grow and divide on the surface (step 3). Next, the physiology of the cells changes, and they begin to secrete *extracellular polymeric substances* (EPSs), slimy, gluelike substances similar to the molecules found in bacterial capsules. EPS extends between cells in the mixture, forming a matrix that binds cells to each other and anchors the complex to the surface, thereby establishing the biofilm (step 4). The slime layer entraps a variety of materials, such as dead cells and insoluble minerals. The physiological change accompanying the formation of a biofilm results from marked changes in a prokaryotic organism's gene expression pattern—in effect, the prokaryotic cells in a biofilm become very different organisms. Over time, other organisms are attracted to and join the biofilm; depending on the environment, these may include other bacterial species, algae, fungi, or protozoa producing diverse microbial communities (step 5). As described in "Molecule behind Biology" 22.1, on p. 492, prokaryotic organisms in a biofilm communicate with each other via **quorum sensing;** in fact, this communication is part of biofilm

formation—it allows cells to start secreting EPS when a high enough cell density is reached.

Much remains to be learned about how organisms form and interact within a biofilm and how changes in gene expression during the transition are regulated.

In the next two sections, we describe the major groups of prokaryotic organisms.

STUDY BREAK

1. What is the difference between a chemoheterotroph and a photoautotroph?
2. What is the difference between an obligate anaerobe and a facultative anaerobe?
3. What is the difference between nitrogen fixation and nitrification? Why are nitrogen-fixing prokaryotic organisms important?
4. What is binary fission?
5. What is the difference between an endotoxin and an exotoxin? Explain how they differ with respect to how they cause disease.
6. Explain four mechanisms by which bacteria protect themselves from antibiotics.
7. What is a biofilm? Give an example of a biofilm that is beneficial to humans and one that is harmful. What advantages do prokaryotic cells in a biofilm gain?
8. What is quorum sensing?

22.3 The Domain Bacteria

As for other organisms, classification of bacteria and archaeans has been revolutionized by molecular techniques that allow researchers to compare nucleic acid

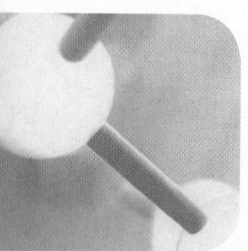

MOLECULE BEHIND BIOLOGY 22.1

N-Acyl-l-Homoserine Lactone

Most bacteria are social organisms that interact in many ways and display social behaviours, such as hunting for food in swarms, bioluminescence (see Chapter 1), biofilm formation, and virulence in pathogenic bacteria, such as the strain of *E. coli* O157:H7 responsible for "hamburger disease." These behaviours happen only when a critical population density is reached, meaning that bacteria must be able to sense the presence of other cells. How does a bacterial cell know that it is not alone? Bacteria use quorum sensing to communicate; this mechanism involves the release of signalling molecules into the environment. Accumulation of signalling molecules

enables the cell to determine the density of other cells around it and respond accordingly; the response occurs after the signalling molecule is perceived by specific receptors on the cell's membrane and triggers activation of specific genes. Different bacterial species use different signalling molecules; for example, many Gram-negative bacteria use *N*-acyl-l-homoserine lactones (a lactone is a type of cyclic ester) such as that shown in **Figure 1.** Gram-positive cells also signal each other but use small peptides rather than lactones. If we can learn to "speak" or "translate" these bacterial languages, could we interfere

with the social behaviours they control? The possibility has important implications for medical science given the role of these signals in processes such as the onset of virulence in pathogenic bacteria and communication within biofilms such as those that form on medical devices implanted in patients.

Figure 1
N-Acyl-l-homoserine lactone from *Vibrio fischeri*.

---

and protein sequences as tests of evolutionary relatedness. Ribosomal RNA (rRNA) sequences have been most widely used in the evolutionary studies of prokaryotic organisms. Researchers have identified several evolutionary branches within each prokaryotic domain **(Figure 22.15)**, but these classifications will likely change in the future when full genomic sequences can be compared. We discuss the major groups of the domain Bacteria, which is much better characterized than the domain Archaea, in this section, and those of the domain Archaea in the next section.

22.3a Molecular Studies Reveal Numerous Evolutionary Branches in the Bacteria

Bacteria as a domain is much better characterized than Archaea: sequencing studies reveal that bacteria have several distinct and separate evolutionary branches. We restrict our discussion to six particularly important groups: proteobacteria, green bacteria, cyanobacteria, Gram-positive bacteria, spirochetes, and chlamydias (see Figure 22.15).

Proteobacteria: The Purple Bacteria and Their Relatives. This highly diverse group of Gram-negative bacteria likely evolved from a purple, photosynthetic ancestor. Their purple colour comes from their photosynthetic pigment, a type of chlorophyll distinct from that of plants. Many present-day species are either photoautotrophs (the purple sulfur bacteria) or photoheterotrophs (the purple nonsulfur bacteria); both groups carry out a type of photosynthesis that does not use water as an electron donor and does not release oxygen as a by-product.

Other present-day proteobacteria are **chemoheterotrophs** that are thought to have evolved as an evolutionary branch following the loss of photosynthetic capabilities in an early proteobacterium. The evolutionary ancestors of mitochondria are considered likely to have been ancient nonphotosynthetic proteobacteria.

Among the chemoheterotrophs classified with the proteobacteria are *E. coli*; plant pathogenic bacteria; and bacteria that cause human diseases such as bubonic plague, gonorrhea, and various forms of gastroenteritis and dysentery. The proteobacteria also include both free-living and symbiotic nitrogen-fixing bacteria.

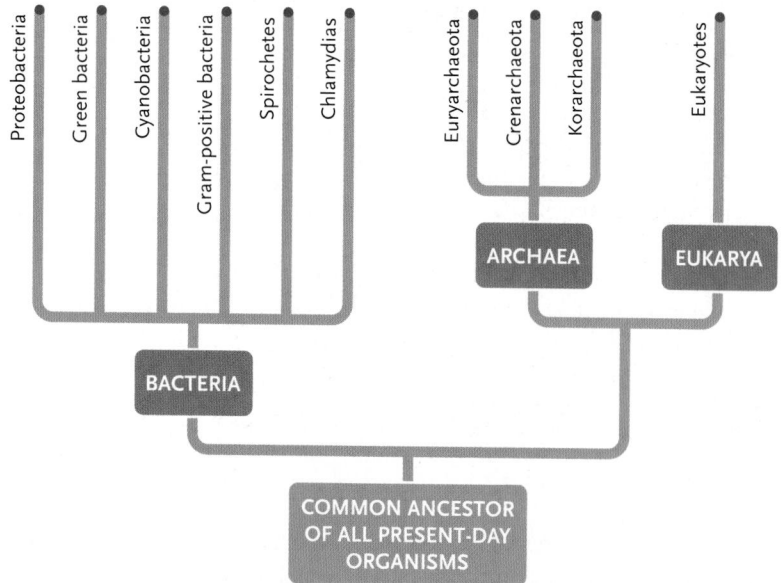

Figure 22.15
An abbreviated phylogenetic tree of Bacteria and Archaea.

Myxobacteria are an unusual group of nonphotosynthetic proteobacteria that form colonies held together by the slime they produce. Enzymes secreted by the colonies digest "prey"—other bacteria, primarily—that become stuck in the slime. When environmental conditions become unfavourable, as when soil nutrients or water are depleted, myxobacteria form a fruiting body, a differentiated multicellular stage large enough to be visible to the naked eye **(Figure 22.16)**. The fruiting body contains clusters of spores that are dispersed to form new colonies when the fruiting body bursts. Quorum sensing is involved in spore formation.

Helicobacter pylori, the cause of many gastric ulcers (see "People behind Biology" 22.2, p. 494), is also a proteobacterium.

Green Bacteria. This diverse group of photosynthetic Gram-negative bacteria is named for the chlorophyll pigments that give the cells their green colour (a different form of chlorophyll than that found in plants). Like the purple bacteria, they do not release oxygen as a by-product of photosynthesis. Also like the purple bacteria, some are photoautotrophs, whereas others are photoheterotrophs. The photoautotrophic green bacteria are fairly closely related to the Archaea and are usually found in hot springs, whereas the photoheterotrophic type is typically found in marine and high-salt environments.

Cyanobacteria. These Gram-negative photoautotrophs are blue-green in colour **(Figure 22.17)** and carry

a.

b.

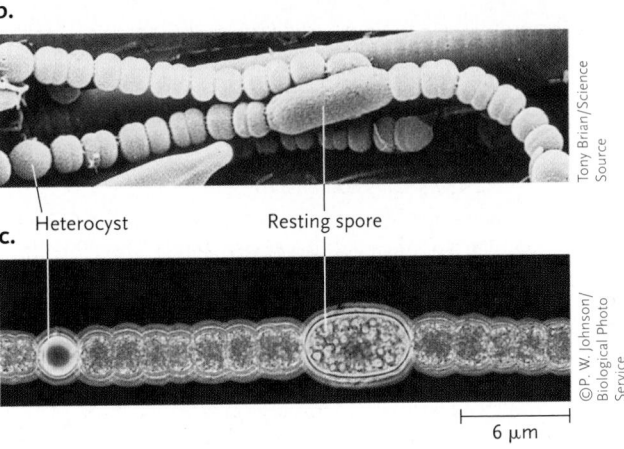

Heterocyst   Resting spore

c.

6 μm

Figure 22.17

Cyanobacteria. **(a)** A population of cyanobacteria covering the surface of a pond. **(b)** and **(c)** Chains of cyanobacterial cells. Some cells in the chains form spores. The heterocyst is a specialized cell that fixes nitrogen.

out photosynthesis by the same pathways and using the same chlorophyll as eukaryotic algae and plants. Like plants and algae, they release oxygen as a by-product of photosynthesis.

The direct ancestors of present-day cyanobacteria were the first organisms to use the water-splitting reactions of photosynthesis. As such, they were critical to the accumulation of oxygen in the atmosphere, which allowed the evolutionary development of aerobic organisms. Chloroplasts probably evolved from early cyanobacteria that were incorporated into the cytoplasm of primitive eukaryotes, which eventually gave rise to the algae and higher plants, as discussed in Chapter 26. Besides releasing oxygen, present-day cyanobacteria help fix nitrogen into organic compounds in aquatic habitats and act as symbiotic partners with fungi in lichens (see Chapter 25).

Gram-Positive Bacteria. This large group contains many species that live primarily as chemoheterotrophs. Some cause human diseases, including *Bacillus anthracis,* the causal agent of anthrax;

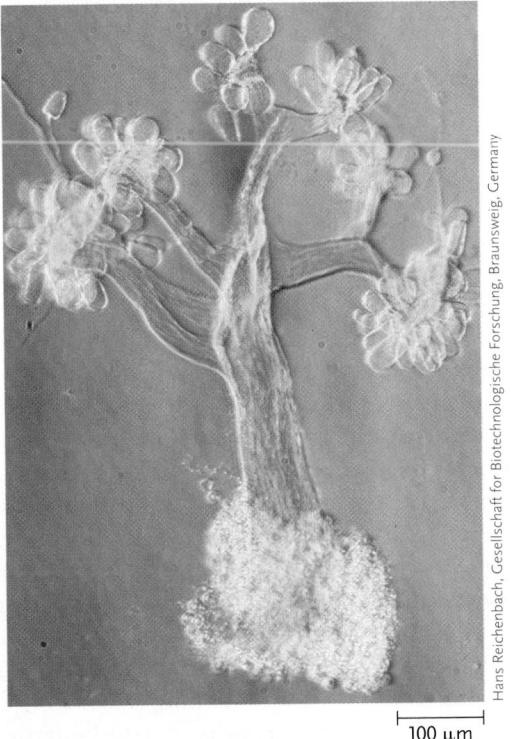

100 μm

Figure 22.16

The fruiting body of *Chondromyces* crocatus, a myxobacterium. Cells of this species collect together to form the fruiting body.

Barry Marshall, *University of Western Australia;* Robin Warren, *Royal Perth Hospital (retired)*

A few hours after you eat, you go to your doctor complaining of stomach pain, abdominal bloating, and nausea; most worryingly, you have started to vomit blood. Your doctor tells you that you have a gastric ulcer, a lesion in your stomach lining. If this visit to your doctor had occurred before the mid-1980s, your doctor would have explained that ulcers are caused by increased stomach acidity due to stress. The treatment? Drink lots of milk, take antacids, and give up alcohol and your favourite spicy foods—no more curries or chili that would aggravate your ulcer. This view of ulcers was accepted for years until two Australian physicians, Barry Marshall and Robin Warren, of the University of Western Australia, demonstrated that most ulcers are caused by a bacterial infection. Marshall and Warren observed that biopsies from patients with ulcers revealed large numbers of spiral-shaped bacterial cells in inflamed tissues. Together, the two physicians carried out a series of studies that demonstrated the link between ulcers and the presence of the bacterium; as you can see in **Table 1,** the bacterium was associated with almost all gastric and duodenal (part of the small intestine) ulcers. The bacterium was not known at the time but was later named *Helicobacter pylori*

Table 1	Association of Bacteria with Biopsy Samples	
Biopsy Appearance	Total Samples	Number (%) Associated with Bacteria
Gastric ulcer	22	18 (77%)
Duodenal ulcer	13	13 (100%)
Total	35	31 (89%)

(Figure 1). But despite Marshall and Warren having research published in respected medical journals, the medical community did not believe their findings—how could bacteria possibly survive in the very acidic conditions of the stomach? Out of frustration, and anxious to get proper treatment for his patients, Marshall drank a culture of *H. pylori!* After about a week, he developed severe abdominal pain and vomiting, and endoscopic examination of his stomach showed regions of inflammation teeming with *H. pylori*. Much to his disappointment, he did not develop ulcers, but he had made the point that *H. pylori* is pathogenic. Marshall and Warren also showed that antibiotics were effective in treating ulcers, and in 2005, they were awarded the Nobel Prize in Medicine. So how is *H. pylori* able to survive in the stomach? It is able to burrow deep into the mucus lining the stomach by means of its numerous flagella, and it produces urease, which converts urea into CO_2 and ammonia, making the region around its cells more basic.

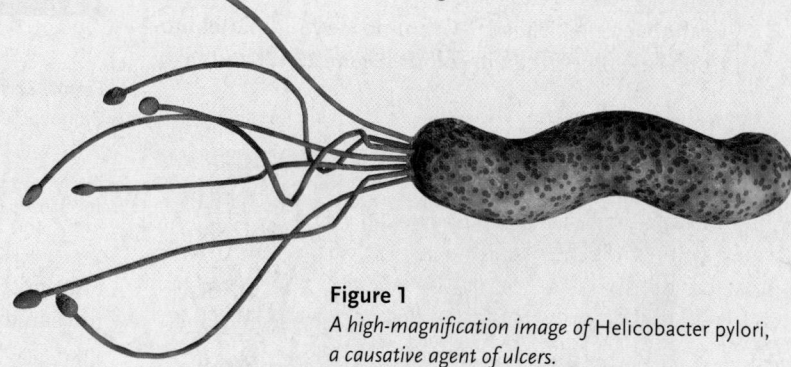

Figure 1
A high-magnification image of Helicobacter pylori, *a causative agent of ulcers.*

Figure 22.18
Streptococcus bacteria forming the long chains of cells typical of many species in this genus.

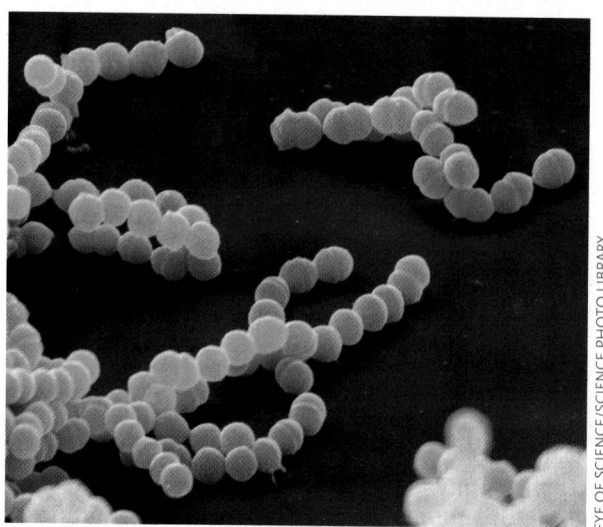

EYE OF SCIENCE/SCIENCE PHOTO LIBRARY

Staphylococcus, which causes some forms of food poisoning, toxic shock syndrome, pneumonia, and meningitis; and *Streptococcus* **(Figure 22.18),** which causes strep throat, necrotizing fasciitis, and some forms of pneumonia. However, some Gram-positive bacteria are beneficial to humans; *Lactobacillus,* for example, carries out the lactic acid fermentation used in the production of pickles, sauerkraut, and yogurt. One unusual group of bacteria, the mycoplasmas, is placed among the Gram-positive bacteria by molecular studies even though they show a Gram-negative staining reaction. This staining reaction results because they are naked cells that secondarily lost their cell walls in evolution. Some mycoplasmas, with diameters from 0.1 to 0.2 μm, are the smallest known cells.

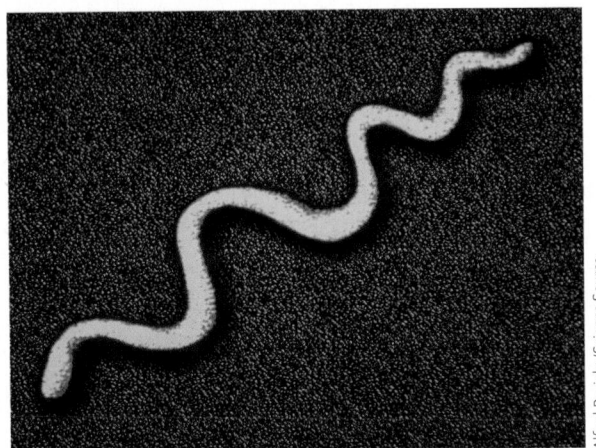

Figure 22.19
Treponema pallidum, a spirochete bacterium that causes syphilis (scanning electron microscope image).

Spirochetes. These organisms have helically spiralled flagella embedded in their cytoplasm, causing the cells to move in a twisting, corkscrew pattern **(Figure 22.19)**. Their corkscrew movements enable them to move in viscous environments such as mud and sewage, where they are common. Some spirochetes are harmless inhabitants of the human mouth; another species, *Treponema pallidum*, is the cause of syphilis. Termites have symbiotic spirochetes in their intestines that enable them to digest cellulose.

Chlamydias. These bacteria are unusual because although they are Gram-negative and have cell walls with an outer membrane, they lack peptidoglycan. All the known chlamydias are intracellular parasites that cause various diseases in animals. One bacterium of this group, *Chlamydia trachomatis*, is responsible for one of the most common sexually transmitted infections of the urinary and reproductive tracts of humans and also causes trachoma, an infection of the cornea that is the leading cause of blindness in humans.

In this section, you have seen that bacteria thrive in nearly every habitat on Earth. However, some members of the second prokaryotic domain, the Archaea, the subject of the next section, live in habitats that are too forbidding even for bacteria.

STUDY BREAK

1. What methodologies have been used to classify prokaryotic organisms?
2. What were the likely characteristics of the evolutionary ancestor of present-day proteobacteria?
3. How does photosynthesis in photosynthetic proteobacteria differ from photosynthesis in cyanobacteria?

22.4 The Domain Archaea

The first Archaea were isolated from extreme environments, such as hot springs, hydrothermal vents on the ocean floor, and salt lakes **(Figure 22.20)**. For that reason, these prokaryotes were called *extremophiles* (organisms that live in extreme environments). Subsequently, archaea have also been found living in less extreme environments.

Archaea share some cellular features with eukaryotes and some with bacteria and have other features that are unique **(Table 22.2, p. 496)**.

22.4a Unique Characteristics of Archaea

Among their unique characteristics are certain features of their plasma membranes and cell walls. The lipid molecules in archaeal plasma membranes are unlike those in the plasma membranes of the majority of bacteria: there is a different linkage between glycerol and the hydrophobic tails, and the

a.

b.

Figure 22.20
Typically extreme archaeal habitats. **(a)** Highly saline water in Great Salt Lake, Utah, coloured red purple by archaeans. **(b)** Hot, sulfur-rich water in Emerald Pool, Yellowstone National Park, coloured brightly by the oxidative activity of archaea, which convert H_2S to elemental sulfur.

Snottite Bacteria

Some of the most extreme and inhospitable environments on Earth are deep caves that have formed in sulfur-rich rocks. As water flows through these rocks, toxic H_2S gas is released at concentrations that can make the cave atmosphere toxic to humans, who cannot survive in the caves without gas masks. But extremophile bacteria and archaea thrive in these caves, including bacteria that grow in biofilms to form *snottites* **(Figure 1),** mucous stalactites that hang from the walls and ceiling of the cave. These bacteria obtain energy from H_2S and other sulfur compounds, producing sulfuric acid as a waste product that drips from the snottites. The biofilm that surrounds the bacteria protects them from the extremely acid environment (pH < 2) they have helped to create, but the acid eats away at the surrounding rock, enlarging the cave. In addition to actively contributing to cave formation, these extremophile bacteria are the foundation of the cave ecosystem. Their ability to convert inorganic chemicals such as H_2S into the organic molecules that make up their cells provides a source of organic carbon to other organisms, making them the base of the food web in such caves.

Figure 1

Snottites, Cueva de Villa Luz, Mexico.

Peter Lane Taylor, VISUALS UNLIMITED/SCIENCE PHOTO LIBRARY

Table 22.2	Characteristics of the Bacteria, Archaea, and Eukarya		
Characteristic	Bacteria	Archaea*	Eukarya
DNA arrangement	Single, circular in most, but some linear and/or multiple	Single, circular	Multiple linear molecules
Chromosomal proteins	Prokaryotic histonelike proteins	Five eukaryotic histones	Five eukaryotic histones
Genes arranged in operons	Yes	Yes	No
Nuclear envelope	No	No	Yes
Mitochondria	No	No	Yes
Chloroplasts	No	No	Yes
Peptidoglycan in cell wall	Present	Absent; some have pseudopeptidoglycan	Absent
Membrane lipids	Unbranched; linked by ester linkages	Branched; linked by ether linkage; may have polar heads at both ends	Unbranched; linked by ester linkages
RNA polymerase	Limited variations	Multiple types	Multiple types
Ribosomal proteins	Prokaryotic	Some prokaryotic, some eukaryotic	Eukaryotic
First amino acid placed in proteins	Formylmethionine	Methionine	Methionine
Aminoacyl–tRNA synthetases	Prokaryotic	Eukaryotic	Eukaryotic
Cell division proteins	Prokaryotic	Prokaryotic	Eukaryotic
Proteins of energy metabolism	Prokaryotic	Prokaryotic	Eukaryotic

*Given that very few Archaea have been identified or cultured, the information in this table is based on an extremely small data set.

tails are isoprenes rather than fatty acids (see Chapter 5). Also, some lipids have polar head groups at both ends. Why would such seemingly minor differences be significant? These unique lipids are more resistant to disruption, making the plasma membranes better suited to extreme environments. Similarly, the unique cell walls of archaea are more resistant to extremes than those of bacteria; some archaea can even survive being boiled in strong detergents!

Many archaea are chemoautotrophs, whereas others are chemoheterotrophs. Interestingly, no known member of the Archaea has been shown to be pathogenic.

22.4b Molecular Studies Reveal Three Evolutionary Branches in the Archaea

The phylogeny of Archaea is poorly developed relative to Bacteria and in quite a state of flux because a tremendous number of archaea have not been cultured, meaning that we have only metagenomic data for most of these organisms. Based on differences in rRNA sequence data, the domain Archaea is divided into three groups (see Figure 22.15, p. 492). Two major groups, the **Euryarchaeota** and the **Crenarchaeota**, contain archaea that have been cultured in the laboratory. The third group, the **Korarchaeota**, has been recognized solely on the basis of DNA taken from environmental samples.

Euryarchaeota. These organisms are found in various extreme environments. They include methanogens, extreme halophiles, and some extreme thermophiles, as described below.

Methanogens (methane generators) live in low-oxygen environments **(Figure 22.21)** and represent about one-half of all known species of Archaea. Methanogens are obligate anaerobes that live in the anoxic (oxygen-lacking) sediments of swamps, lakes, marshes, and sewage works, as well as in more moderate environments, such as the rumen of cattle and sheep, the large intestine of dogs and humans, and the hindgut of insects such as termites and cockroaches. Methanogens generate energy by converting various substrates such as carbon dioxide and hydrogen gas or acetate into methane gas, which is released into the atmosphere.

Halophiles are salt-loving organisms. Extreme halophilic Archaea live in highly saline environments such as the Dead Sea and on foods preserved by salting. They require a minimum NaCl concentration of about 1.5 M (about 9% solution) to survive and can live in a fully saturated solution (5.5 M, or 32%). Most are aerobic chemoheterotrophs, which obtain energy from sugars, alcohols, and amino acids using pathways similar to those of bacteria. Many extreme halophiles use light as a secondary energy source, supplementing the oxidations that are their primary source of energy.

Extreme thermophiles live in extremely hot environments such as hot springs and ocean floor hydrothermal vents. Their optimal temperature range for growth is 70 to 95°C, close to the boiling point of water. By comparison, no eukaryotic organism is known to live at a temperature higher than 60°C. Some extreme thermophiles are members of the Euryarchaeota, but most belong to the **Crenarchaeota**, the next group that we discuss.

Crenarchaeota. This group includes most of the extreme thermophiles, which have a higher optimal temperature range than those belonging to the Euryarchaeota. For example, the most thermophilic member of this group, *Pyrobolus*, dies below 90°C, grows optimally at 106°C, and can survive an hour of autoclaving at 121°C! *Pyrobolus* lives in ocean floor hydrothermal vents, where the pressure creates water temperatures greater than the boiling point of water on Earth's surface.

Also in this group are **psychrophiles** ("cold loving"), organisms that grow optimally in cold temperatures in the range from −10 to −20°C. These organisms are found mostly in the Antarctic and Arctic oceans, which are frozen most of the year, and in the intense cold at ocean depths.

Mesophilic members of the Crenarchaeota make up a large part of plankton found in cool, marine waters, where they are food sources for other marine organisms.

Korarchaeota. This group has been recognized solely on the basis of DNA samples obtained from marine and terrestrial hydrothermal environments. To date, no members of this group have been isolated and cultivated in the lab, and nothing is known about their physiology. Molecular data indicate that they are the oldest archaeal lineage.

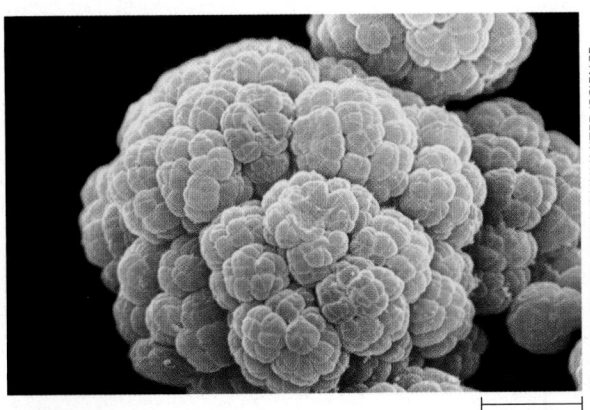

Figure 22.21

A colony of the methanogenic archaeon *Methanosarcina*, which lives in the sulfurous, waterlogged soils of marshes and swamps.

RALPH ROBINSON, VISUALS UNLIMITED/SCIENCE PHOTO LIBRARY

5 μm

Thermophilic archaea are important commercially. For example, they are very important in biotechnological applications as sources of enzymes that function under extreme physicochemical conditions (e.g., high temperature, high salinity).

In this chapter, we have focused on bacteria and archaea, whose metabolic diversity and environmental range and ecological importance belie their structural simplicity. In the next chapter, we look at still simpler entities: viruses, viroids, and prions, which are derived from living organisms and retain only some of the properties of life.

Review

To access course materials such as Aplia and other companion resources, please visit www.NELSONbrain.com.

22.1 The Full Extent of the Diversity of Bacteria and Archaea Is Unknown

- Bacteria and Archaea are the most abundant and diverse organisms on Earth; however, the vast majority of prokaryotic organisms have not been described because they cannot be cultured using standard techniques and because many environments are very difficult and/or expensive to access.
- Prokaryotic organisms make up two of the three domains of life, the Archaea and the Bacteria.

22.2 Prokaryotic Structure and Function

- Prokaryotic genomes typically consist of a single, circular DNA molecule packaged into the nucleoid (Figure 22.4). Many prokaryotic cells also contain plasmids, which replicate independently of the chromosome and can be passed to other cells.
- Gram-positive bacterial cell walls consist of a single, relatively thick peptidoglycan layer. Gram-negative bacteria have walls consisting of a relatively thin peptidoglycan sheath surrounded by an outer lipopolysaccharide (LPS) membrane (Figure 22.6).
- A polysaccharide capsule (Figures 22.3, and 22.7) surrounds many bacteria, protecting them and helping them adhere to surfaces.
- Archaea and bacteria show great diversity in their modes of obtaining energy and carbon. Two of the modes of nutrition found among eukaryotic organisms are also found in prokaryotic organisms (chemoheterotrophy and photoautotrophy), but two other modes are unique to prokaryotic organisms: chemoautotrophs obtain energy by oxidizing inorganic substrates and use carbon dioxide as their carbon source, and photoheterotrophs use light as a source of energy and obtain their carbon from organic molecules.

- Bacteria and Archaea use a range of pathways to transform energy: aerobic respiration, anaerobic respiration, and/or various forms of fermentation.
- Some bacteria and archaea are capable of nitrogen fixation, the conversion of atmospheric nitrogen to ammonia; others are responsible for nitrification, the conversion of ammonium to nitrate.
- Prokaryotic cells normally reproduce asexually by binary fission (Figure 22.10), which can result in very rapid population growth under favourable conditions.
- In nature, bacteria and archaea live in complex communities, such as biofilms (Figure 22.13).
- Pathogenic bacteria cause disease via exotoxins and endotoxins.
- Bacteria may develop resistance to antibiotics through mutation of their own genes or by acquiring resistance genes from other bacteria.

22.3 The Domain Bacteria

- Bacteria are divided into more than a dozen evolutionary branches, including Gram-negative proteobacteria, Gram-negative green bacteria, cyanobacteria, Gram-positive bacteria, spirochetes, and chlamydias.

22.4 The Domain Archaea

- A very large number of archaea have not been cultured, but we know that archaea have some features that are like those of bacteria, other features that are eukaryotic, and some that are unique (see Table 22.2).
- Archaea are classified into three groups: the Euryarchaeota (methanogens, extreme halophiles, and some extreme thermophiles); the Crenarchaeota (which includes most of the extreme thermophiles, but also some psychrophiles and mesophiles); and the Korarchaeota, known only from DNA samples.

Questions

Self-Test Questions

1. Which of the following structures is found in prokaryotic cells?
   a. cellulose cell wall
   b. ribosome
   c. mitochondria
   d. nuclear membrane

2. Which of the following statements about archaea is correct?
   a. Their cell walls contain peptidoglycan.
   b. Most are pathogens.
   c. Many are extremophiles.
   d. They have no traits in common with eukaryotic cells.

3. Which of the following statements accurately describes a plasmid?
   a. It can only replicate when the cell's chromosome replicates.
   b. It is a small circular piece of RNA outside a cell's chromosome.
   c. It is a small circular piece of DNA outside a cell's chromosome.
   d. It refers to a piece of DNA taken up from the environment by a prokaryotic cell.

4. You have isolated an unknown bacterium that produces a toxin and you are trying to determine if this is an endotoxin or an exotoxin. Which of the following features would be associated with the toxin, if it were an endotoxin?
   a. It would be secreted from the cell.
   b. It would be part of the cell wall.
   c. It would be part of the plasma membrane.
   d. It would be produced by an archaeon.

5. Place the following steps by which prokaryotic cells form a biofilm in the correct order:
   1. Cells grow and divide.
   2. The cells' physiology changes.
   3. Cells attach to a surface that is covered in organic polymers.
   4. Cells secrete extracellular polymers that "glue" the cells to the surface and to each other.
      a. 1, 2, 3, 4
      b. 2, 1,4, 3
      c. 4, 3, 1, 2
      d. 3, 1, 2, 4

6. You are growing a facultative anaerobic archaeon in culture under two conditions: one culture is in anaerobic conditions, and the other is in aerobic conditions. How would you expect the growth of the cells to compare between the two cultures?
   a. Growth would be greater in the culture in aerobic conditions.
   b. Growth would be greater in the culture in anaerobic conditions.
   c. Growth would be the same in both conditions.

7. A bacterium that oxidizes nitrite as its only energy source was found deep in a cave. How would you classify this bacterium, based on its carbon and energy source?
   a. as a chemolithotroph
   b. as a chemoheterotroph
   c. as a photoautotroph
   d. as a photoheterotroph

8. Which of the following processes converts ammonium (NH_4^+) into nitrate (NO_3^-)?
   a. nitrogen fixation
   b. ammonification
   c. nitrification
   d. denitrification

9. Which of the following groups of bacteria are all oxygen-producing photoautotrophs?
   a. spirochetes
   b. cyanobacteria
   c. proteobacteria
   d. green bacteria

10. Which of the following statements about chlamydias is correct?
    a. They lack peptidoglycan.
    b. They are Gram-positive.
    c. They are not pathogenic.
    d. They have no outer membrane in the cell wall.

Questions for Discussion

1. In the lab, you have isolated some prokaryotic cells that belong either to a Gram-positive bacterium or to an archaeon. What cellular (structural) features could you look for to determine which type of organism you have isolated? Indicate how that feature would differ between the two kinds of organism (assume that you have the necessary equipment to test for any cellular feature you want).

2. List several functions of the outer wall layer in Gram-negative bacteria.

3. You are doing research to develop new drugs and have developed a new antibiotic. This drug acts by inhibiting ribosome function in prokaryotic cells. When you test it on animal cells, however, you find that the growth of animal cells is inhibited. Explain why this drug inhibits the growth of animal cells.

4. How do bacteria resist antibiotics? Why do antibiotics lose their effectiveness so quickly?

5. In which nutritional class would you place a prokaryotic organism that uses glucose as its only energy and carbon source? What about an organism that uses elemental sulfur as an energy source and carbon dioxide as a carbon source? What is the energy source for phototrophic organisms?

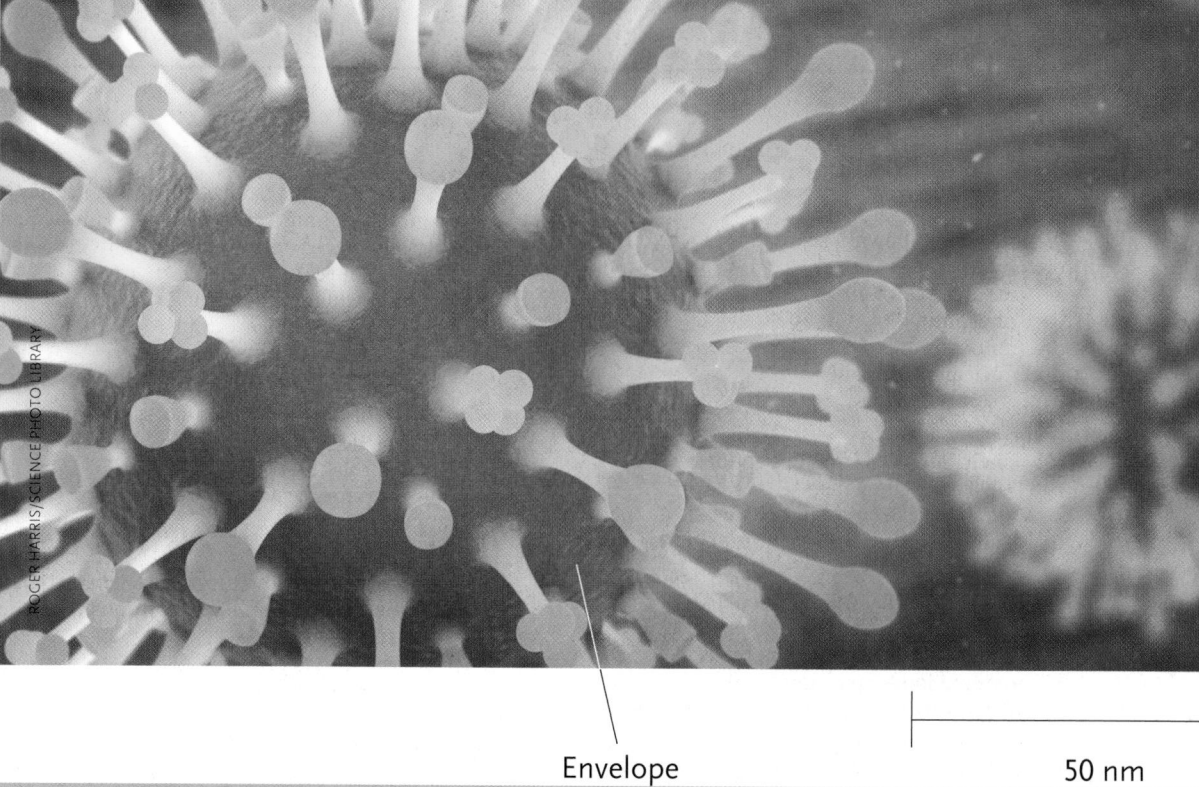

Envelope

50 nm

ROGER HARRIS/SCIENCE PHOTO LIBRARY

Influenza virus.

23

Viruses, Viroids, and Prions: Infectious Biological Particles

WHY IT MATTERS

Imagine yourself sitting in a crowded airplane bound from London, United Kingdom, to Vancouver. The person sitting beside you has a runny nose; is sneezing, coughing, and sucking on lozenges; and appears to have a fever. Recognizing that your seatmate is exhibiting many of the symptoms of influenza, a respiratory illness caused by the influenza virus shown in the micrograph above, you worry that the virus will spread to you through your seatmate's coughing and sneezing. You have just seen how influenza ("the flu") can affect people and how air travel can help it spread around the world.

At any given time, 5 to 15% of the global population of people exhibits the symptoms of influenza, and each year, about 500 000 people are killed by influenza A, one type of influenza virus. Recent research has shown that new strains of influenza A arise each year from just a few initial sources in East and Southeast Asia and then spread around the world. Colin Russell of Cambridge University in the United Kingdom and his colleagues analyzed 13 000 human influenza A viruses of the H3N2 subtype by collecting viral material from infected people on six continents between 2002 and 2007. Their analysis revealed almost continuous circulation of H3N2 in East and Southeast Asia. This regional network of overlapping epidemics appeared to be the source of influenza outbreaks elsewhere in the world. The epidemics then spread to Australia and other islands in the central and south Pacific, North America, and Europe, finally arriving in South America. As influenza viruses travel through populations around the world, they evolve, changing so much that the vaccines we developed in previous years are no longer effective, and new vaccines must be developed.

The good news is that understanding the global pattern of influenza migration will help the World Health Organization develop effective vaccines. Knowing which strains cause the initial outbreak in Asia allows scientists to formulate vaccines to target these strains, offering people in other regions some protection from the illness. Flu vaccines are commonly prepared using killed viruses (viruses that have been inactivated, for example, by chemical treatment, so that they are no longer infective) meaning that when you get your flu shot, you won't develop influenza, but your body will produce antibodies against the virus, protecting you against subsequent infection by any of those specific strains (antibodies are highly specific protein molecules produced by the immune system that recognize and bind to specific proteins of a pathogen, such as the proteins in a virus's coat).

Each winter in Canada and other countries, people line up for their annual flu shot. For many of us, this shot represents a gamble that we will be protected against the strains of the influenza virus making the rounds that winter.

For most people, getting the flu means feeling awful for a few days, but for the very young, the elderly, and people with weakened immune systems, the stakes are higher: a bout of flu can be fatal. Some flu outbreaks have been devastatingly lethal to a very large proportion of the population. The worst recorded example is the flu *pandemic* (an outbreak or epidemic that spreads around the world) of 1918. A strain of influenza virus known as the Spanish flu infected almost half of the world's population, killing about 1 in every 20 people.

Why was the Spanish flu so deadly? And why do we need to develop new flu vaccines so often? We investigate these questions later in this chapter. We also look at the beneficial roles played by viruses—not all are pathogenic—and investigate ways in which we may be able to harness the infective abilities of viruses for our own uses. For example, can we use viruses as vectors for gene therapy to fight diseases? We start with a look at the defining characteristics of viruses: how they are able to enter cells and take over the cell's machinery to make more copies of themselves. We also compare viruses with viroids and prions, other infectious particles.

23.1 What Is a Virus? Characteristics of Viruses

If you look back at the tree of life (Figure 3.20), you'll notice that viruses are not shown. That is because they lack many of the properties of life shared by all organisms (Section 3.1a) and so are not considered to be living organisms. For example, viruses cannot reproduce on their own and lack a metabolic system to provide energy for their life cycles; instead, they depend on the host cells that they infect for these functions. For this reason, viruses are infectious biological particles rather than organisms. The structure of a virus is reduced to the minimum necessary to transmit its genome from one host cell to another. A virus is simply one or more nucleic acid molecules surrounded by a protein **coat** or **capsid (Figure 23.1a, b).** Some capsids may be enclosed within a membrane or **envelope** derived from their host cell's membrane (Figure 23.1c). So a virus is not a cell—it does not have cytoplasm enclosed by a plasma membrane, as do all known living organisms.

The nucleic acid genome of a virus may be either DNA or RNA and can be composed of either a single strand or a double strand of RNA or DNA. Viral genomes range from just a few genes to over a hundred genes; all viruses have genes that encode at least their coat proteins, as well as proteins involved

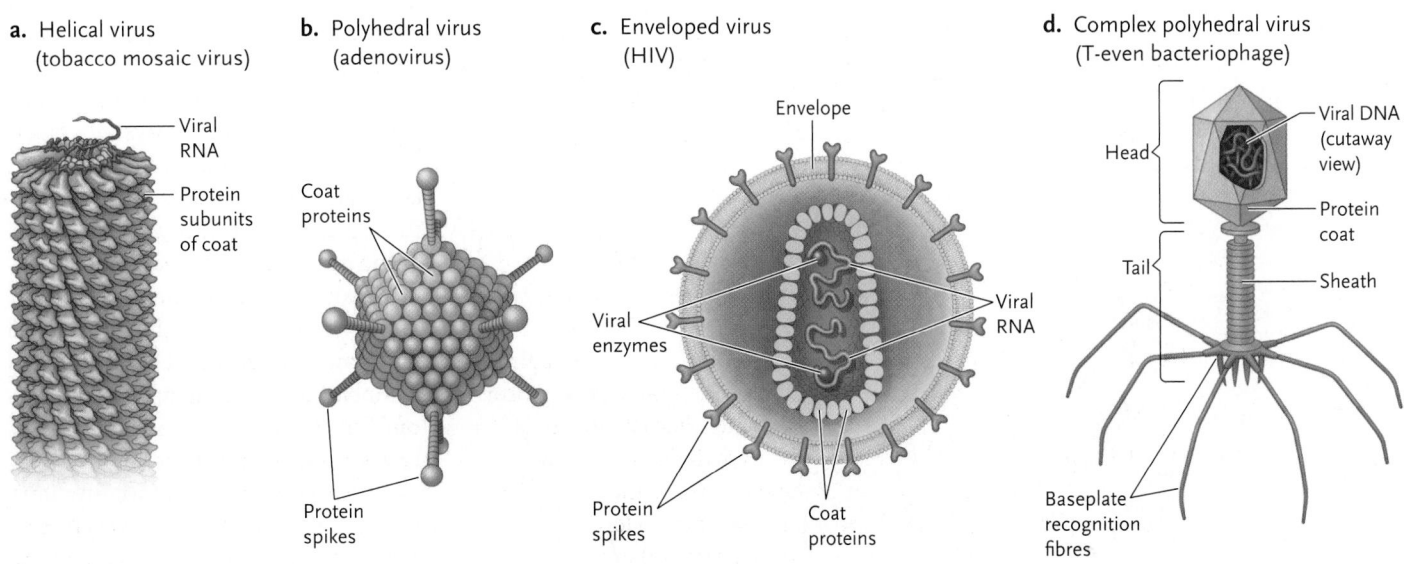

Figure 23.1
Viral structure. All viruses consist of nucleic acid surrounded by a protein coat, but they can have a very wide range of sizes and shapes.

in regulation of transcription. Genomes of **enveloped viruses** also include genes required for the synthesis of envelope proteins. Some viral genomes also include virus-specific enzymes for nucleic acid replication.

Most viruses take one of two basic structural forms, helical and polyhedral. In **helical viruses** the protein subunits assemble in a rodlike spiral around the genome (see Figure 23.1a). A number of viruses that infect plant cells are helical. In **polyhedral viruses**, the coat proteins form triangular units that fit together like the parts of a soccer ball (see Figure 23.1b). The polyhedral viruses include forms that infect animals, plants, and bacteria. In some polyhedral viruses, protein spikes that provide host cell recognition extend from the corners, where the facets fit together. Both helical and polyhedral viruses can be enveloped in a membrane derived from the host's membrane (see Figure 23.1c and **Figure 23.2**). In enveloped viruses, proteins synthesized from the viral genome in the host cell are transported to and embedded in the membrane before the virus particle buds through the host cell. These proteins allow the virus to recognize and bind to host cells.

Although they are not considered to be alive, viruses are classified into orders, families, genera, and species using several criteria, including virus size and structure, genome structure (RNA or DNA, single stranded or double stranded), and how their nucleic acid is replicated. More than 4000 species of viruses have been classified into more than 80 families. The family names end in *-viridae* and may refer either to the geographic region where the virus was first discovered or to the structure of the virus. For example, Coronaviridae, the family to which influenza virus belongs, is named for the "crown" of protein spikes on the capsid, as shown in the photomicrograph at the start of this chapter (*corona* = crown). Like some bacteria, some viruses are named for the disease they cause; these names can be one or two words, for example, herpesvirus or Ebola virus. Each type of virus is made up of many strains, differentiated by their virulence.

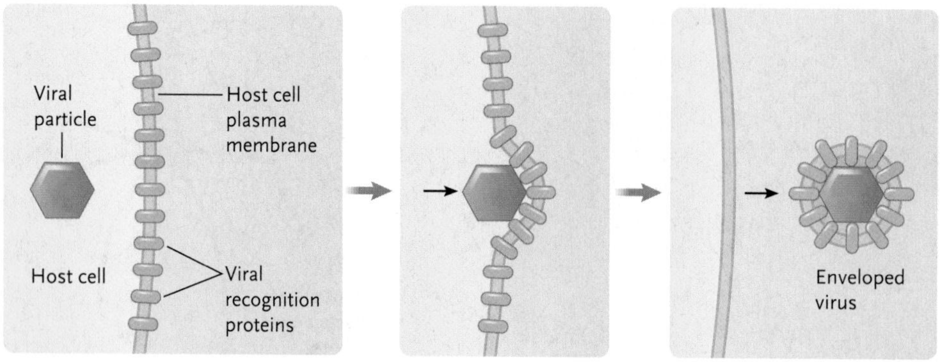

As was the case for our look at prokaryotic organisms in the previous chapter, we will just scratch the surface of viral diversity in this chapter; for example, there are millions of viruses in every millilitre of ocean water, most of which have not been identified. As we learn more about viruses, their classification will likely change.

Every living organism is likely permanently infected by one or more kinds of viruses. Usually, a virus infects only a single species or a few closely related species. A virus may even infect only one organ system or a single tissue or cell type in its host. However, some viruses are able to infect unrelated species, either naturally or after mutating.

Of the roughly 80 viral families described to date, 21 include viruses that cause human diseases. Viruses also cause diseases of wild and domestic animals. Plant viruses cause annual losses of millions of tonnes of crops, especially cereals, potatoes, sugar beets, and sugar cane. (**Table 23.1** lists some important families of viruses that infect animals.) The effects of viruses on the organisms they infect range from undetectable, to merely bothersome, to seriously debilitating or lethal. For instance, some viral infections of humans, such as those causing cold sores, chickenpox, and the common cold, are usually little more than a nuisance to healthy adults. Others cause some of the most severe and deadly human diseases, including AIDS, encephalitis, and Ebola hemorrhagic fever.

However, not all viruses are pathogens. Many viruses actually benefit their hosts; for example, infection by certain nonpathogenic viruses protects human hosts against pathogenic viruses. The "protective" viruses interfere with replication or other functions of the pathogenic viruses. Some viruses also act to defend their host cells. For example, one of the primary reasons that bacteria do not completely overrun this planet is that they are destroyed in incredibly huge numbers by viruses known as **bacteriophages**, or **phages** for short (*phagein* = to eat) (see Figure 23.1d). Viruses also provide a natural means to control some insect pests, such as spruce budworm.

Viruses are vital components of ecosystems and may be the dominant entity in some ecosystems, such as the oceans. We don't yet fully understand their roles in these ecosystems, but it is clear that they affect nutrient cycling through their effects on prokaryotic organisms. For example, in certain regions of the ocean, a few genera of cyanobacteria dominate the marine phytoplankton, making major contributions to global photosynthesis. Bacteriophages infect these cyanobacteria, causing high levels of mortality, thus influencing cyanobacterial population dynamics as well as the release of nutrients from bacterial cells. But these viruses also help keep photosynthesis

Figure 23.2
How enveloped viruses acquire their envelope.

Table 23.1   Major Animal Viruses

Viral Family	Envelope	Nucleic Acid	Diseases
Adenovirus	No	ds DNA	Respiratory infections, tumours
Flavivirus	Yes	ss RNA	Yellow fever, dengue, hepatitis C
Hepadnavirus	Yes	ds DNA	Hepatitis B
Human herpesvirus	Yes	ds DNA	
Herpes simplex I			Oral herpes, cold sores
Herpes simplex II			Genital herpes
Varicella-zoster virus			Chickenpox, shingles
Herpesvirus 4 (Epstein–Barr virus)			Infectious mononucleosis
Orthomyxovirus	Yes	ss RNA	Influenza
Papovavirus	No	ds DNA	Benign and malignant warts
Papillomavirus	No	ds DNA	Human papillomavirus (genital warts)
Paramyxovirus	Yes	ss RNA	Measles, mumps, pneumonia
Picornavirus	No	ss RNA	
Enterovirus			Polio, hemorrhagic eye disease, gastroenteritis
Rhinovirus			Common cold
Hepatitis A virus			Hepatitis A
Apthovirus			Foot-and-mouth disease in livestock
Poxvirus	Yes	ds DNA	Smallpox, cowpox
Retrovirus	Yes	ss RNA	
HTLV I, II			T-cell leukemia
HIV			AIDS
Rhabdovirus	Yes	ss RNA	Rabies, other animal diseases

ds = double-stranded; HTLV = human T lymphotropic virus; ss = single-stranded.

going in their cyanobacterial hosts, as recently discovered by Nicholas Mann and colleagues at the Univeristy of Warwick. As you read in Chapter 7, one of the proteins that make up photosystem II is very susceptible to light-induced damage and so is constantly being replaced by newly synthesized molecules. As long as the cell can make new protein quickly enough to keep up with damage, photosynthesis can continue, but if the rate of damage to photosystem II exceeds the repair rate, the rate of photosynthesis will drop. When these bacteriophages infect cyanobacteria, they shut down their host's protein synthesis. Without continued synthesis of the photosystem protein, photosynthesis should slow down following infection—but it doesn't. How is the photosynthetic rate maintained? Mann and his colleagues found that the virus's genome includes genes for this protein; expression of these viral proteins enables the repair rate to keep up with light-induced damage, allowing the cell to photosynthesize. Although the virus is doing this for "selfish" reasons (i.e., to ensure that its host has sufficient resources for the virus to complete its life cycle), the outcome of this association is that much of the carbon fixed on Earth may be facilitated by virus-controlled photosynthesis.

STUDY BREAK

1. What is a virus?
2. List three features of viruses that distinguish them from living organisms.

23.2 Viruses Infect Bacterial, Animal, and Plant Cells by Similar Pathways

Viral particles move by random molecular motions until they contact the surface of a host cell. For infection to occur, the virus or the viral genome must then enter the cell. Inside the cell, the viral genes are expressed, leading to replication of the viral genome and assembly of progeny viruses. The new viral particles or **virions**, as the extracellular form of a virus is known, are then released from the host cell, a process that often ruptures the host cell, killing it.

23.2a Bacteriophages Are Viruses That Infect Bacteria

We have learned a great deal about the infective cycles of viruses, as well as the genetics of both viruses and bacteria, from studies of the bacteriophages infecting *Escherichia coli* (*E. coli*). Some of these are **virulent bacteriophages**, which kill their host cells during each cycle of infection, whereas others are **temperate bacteriophages**. Temperate bacteriophages enter an inactive phase inside the host cell and can be passed on to several generations of daughter cells before becoming active and killing their host.

Virulent Bacteriophages. Among the virulent bacteriophages infecting *E. coli*, the **T-even bacteriophages** T2, T4, and T6 have been the most valuable in genetic studies. The coats of these phages are divided into a *head* and a *tail* (see Figure 23.1d, p. 501). A double-stranded linear molecule of DNA is packed into the head. The tail, assembled from several different proteins, has **recognition proteins** at its tip that can bind to the surface of the host cell. Once the tail is attached, it functions as a sort of syringe that injects the DNA genome into the cell **(Figure 23.3)**.

Infection begins when a T-even phage collides randomly with the surface of an *E. coli* cell and the tail attaches to the host cell wall (**Figure 23.4,** step 1). An enzyme present in the viral coat, *lysozyme,* then digests a hole in the cell wall through which the tail injects the DNA of the phage (step 2). The proteins of the viral coat remain outside. Throughout its life cycle within the bacterial cell, the phage uses host cell machinery to express its genes. One of the proteins produced early in the infection is an enzyme that breaks down the bacterial chromosome. The phage gene for a DNA polymerase that replicates the phage's DNA is also expressed early on. Eventually, 100 to 200 new viral DNA molecules are synthesized (step 3). Later in the infection, the host cell machinery transcribes the phage genes for the viral coat proteins (step 4). As the head and tail proteins assemble, the replicated viral DNA is packed into the heads (step 5).

When viral assembly is complete, the cell synthesizes a phage-encoded lysozyme that lyses the bacterial cell wall, causing the cell to rupture and releasing viral particles that can infect other *E. coli* cells (step 6). This whole series of events, from infection of a cell through to the release of progeny phages from the ruptured (or **lysed**) cell, is called the **lytic cycle.**

Some virulent phages (although not T-even phages) may package fragments of the host cell's DNA in the heads as the viral particles assemble. This transfer of bacterial genes from one bacterium to another via a virus is known as transduction. In the type of transduction described above, bacterial genes from essentially any DNA fragment can be randomly incorporated into phage particles; thus, gene transfer by this mechanism is termed generalized transduction.

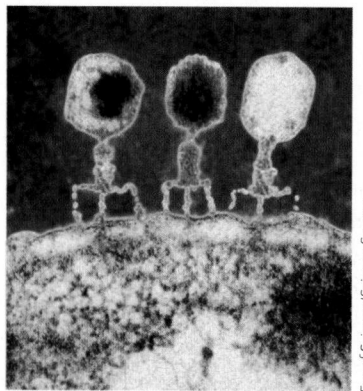

Figure 23.3
Bacteriophages injecting their DNA into *E. coli*.

Eye of Science/Science Source

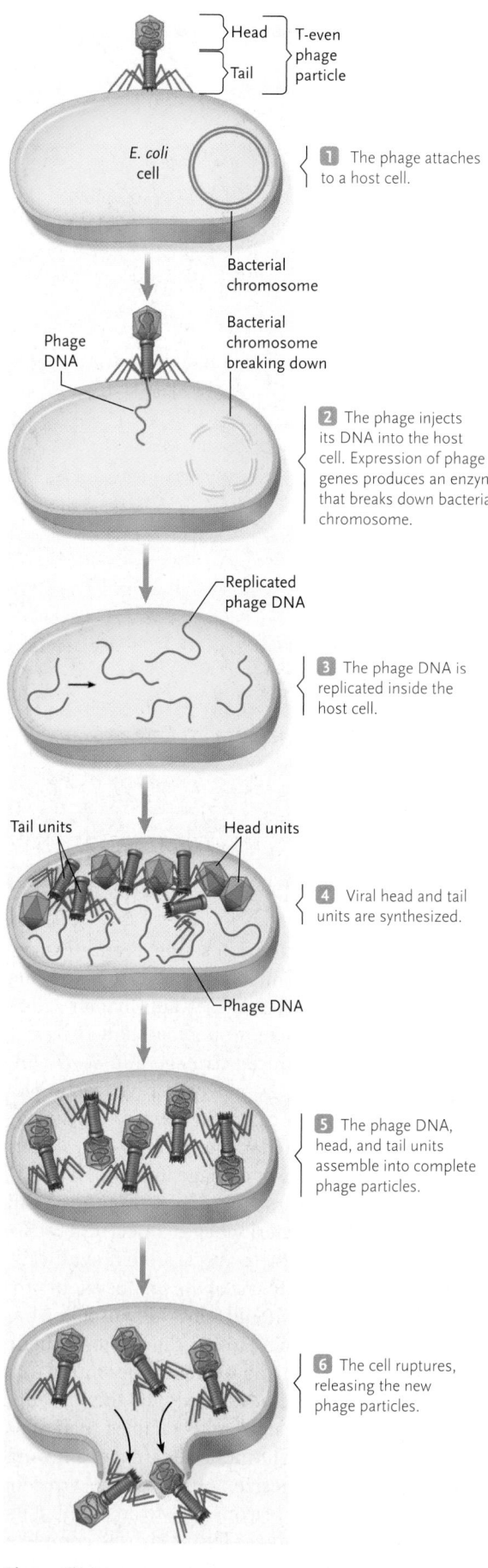

Head ⎫
Tail ⎬ T-even phage particle

E. coli cell

Bacterial chromosome

Phage DNA

Bacterial chromosome breaking down

1 The phage attaches to a host cell.

2 The phage injects its DNA into the host cell. Expression of phage genes produces an enzyme that breaks down bacterial chromosome.

Replicated phage DNA

3 The phage DNA is replicated inside the host cell.

Tail units   Head units

4 Viral head and tail units are synthesized.

Phage DNA

5 The phage DNA, head, and tail units assemble into complete phage particles.

6 The cell ruptures, releasing the new phage particles.

Figure 23.4
The infective cycle of a T-even bacteriophage, an example of a virulent phage.

A Scientist's Favourite Temperate *E. coli* Bacteriophage, Lambda. The infective cycle of the bacteriophage lambda (λ), an *E. coli* phage used extensively in research, is typical of temperate phages. Phage lambda infects *E. coli* in much the same way as the T-even phages. The phage injects its double-stranded linear DNA chromosome into the bacterium (**Figure 23.5,** step 1). Once inside, the linear chromosome forms a circle and then follows one of two paths. Sophisticated molecular switches govern which path is followed at the time of infection.

One path is the lytic cycle, which is like the lytic cycles of virulent phages. The lytic cycle (see Figure 23.5, left side) starts with steps 1 and 2 (infection) and then goes directly to steps 7 through 9 (production and release of progeny virus) and back to step 1. A second and more common path is the **lysogenic cycle** (see Figure 23.5, right side). This cycle begins when the viral chromosome integrates into the host cell's DNA by recombination (see Figure 23.5, steps 1 through 3). The DNA of a temperate phage typically inserts at one or possibly a few specific sites in the bacterial chromosome through the action of a phage-encoded enzyme that recognizes certain sequences in the host DNA. Once integrated, the lambda genes are mostly inactive, so no structural components of the phage are made. While

inserted in the host cell DNA, the virus is known as a **prophage** (*pro* = before). When the host cell DNA replicates, so does the integrated viral DNA, which is passed on to daughter cells along with the host cell DNA (see Figure 23.5, steps 4 and 5).

What triggers the integrated prophage to become active (step 6)? Certain environmental signals, such as nutrient availability and ultraviolet irradiation, stimulate this change, causing the prophage to enter the lytic cycle (see Figure 23.5, steps 6 through 9). Genes that were inactive in the prophage are now transcribed. Among the first viral proteins synthesized are enzymes that excise the lambda chromosome from the host chromosome. The result is a circular lambda chromosome that replicates itself and directs the production of linear viral DNA and coat proteins. This active stage culminates in the lysis of the host cell and the release of infective viral particles.

The excision of the prophage from its host's DNA is not always precise, resulting in the inclusion of one or more host cell genes with the viral DNA. These genes are replicated with the viral DNA and packed into the coats and may be carried to a new host cell in the next cycle of infection. Clearly, only genes that are adjacent to the integration site(s) of a temperate phage can be cut out with the viral DNA, can be included in

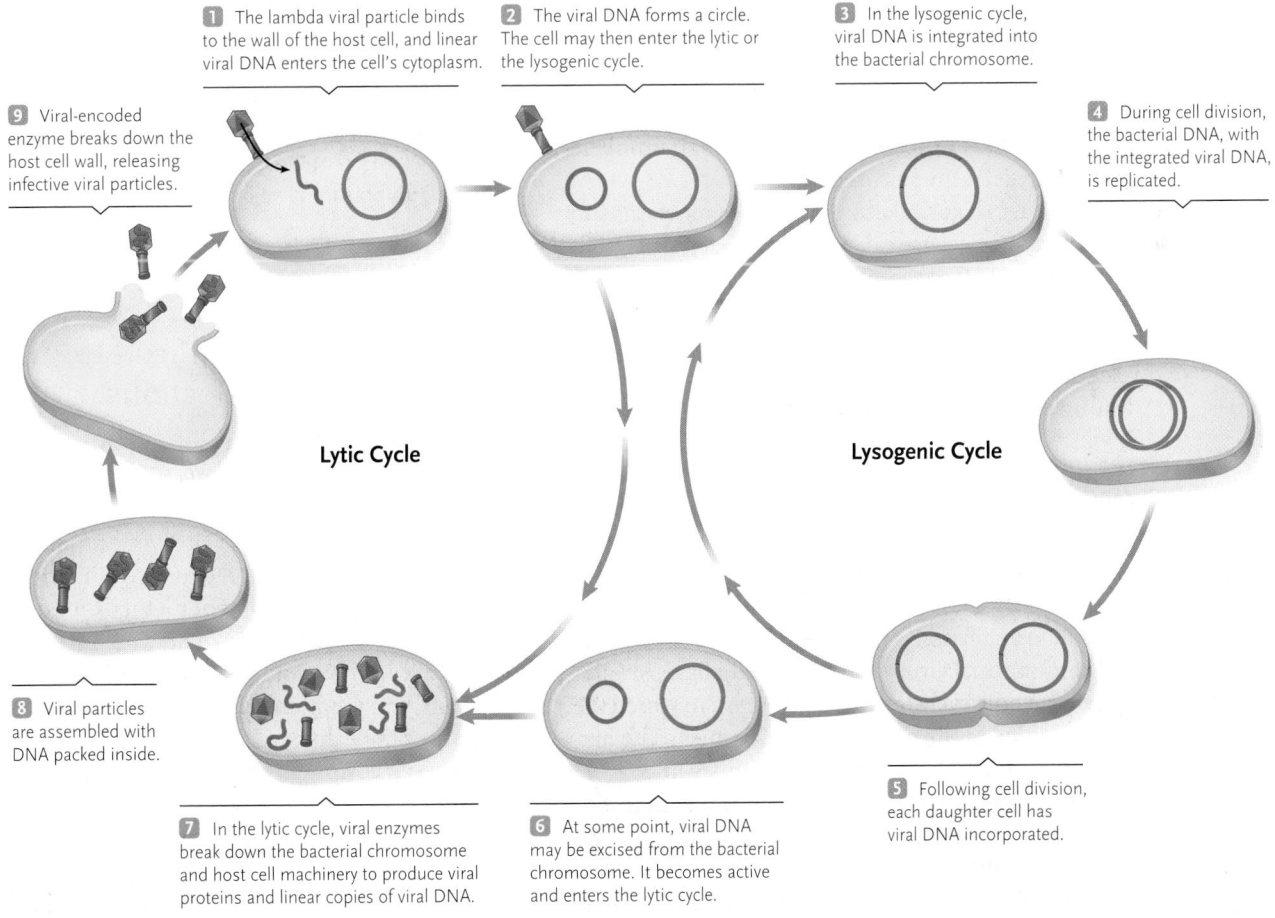

1 The lambda viral particle binds to the wall of the host cell, and linear viral DNA enters the cell's cytoplasm.

2 The viral DNA forms a circle. The cell may then enter the lytic or the lysogenic cycle.

3 In the lysogenic cycle, viral DNA is integrated into the bacterial chromosome.

4 During cell division, the bacterial DNA, with the integrated viral DNA, is replicated.

9 Viral-encoded enzyme breaks down the host cell wall, releasing infective viral particles.

Lytic Cycle

Lysogenic Cycle

8 Viral particles are assembled with DNA packed inside.

7 In the lytic cycle, viral enzymes break down the bacterial chromosome and host cell machinery to produce viral proteins and linear copies of viral DNA.

6 At some point, viral DNA may be excised from the bacterial chromosome. It becomes active and enters the lytic cycle.

5 Following cell division, each daughter cell has viral DNA incorporated.

Figure 23.5
The infective cycle of lambda, an example of a temperate phage, which can go through the lytic cycle or the lysogenic cycle.

phage particles during the lytic stage, and can undergo transduction. Accordingly, this mechanism of gene transfer is termed **specialized transduction.**

Infection of Animal Cells. Viruses infecting animal cells follow a pattern similar to that for bacterial cells, except that both the viral coat and the genome enter a host cell. Depending on the virus, removal of the coat to release the genome occurs during or after cell entry; the envelope does not enter the cell.

Viruses without an envelope, such as poliovirus, bind by their recognition proteins to the plasma membrane and are then taken into the host cell by endocytosis. The virus coat and genome of some enveloped viruses, such as herpesvirus, HIV, and the virus causing rabies, enter the host cell by fusion of their envelope with the host cell plasma membrane. Other enveloped viruses, such as influenza virus, enter host cells by endocytosis.

Once inside the host cell, the genome directs the synthesis of additional viral particles by basically the same pathways as bacterial viruses. Some animal viruses, however, replicate themselves in very complex ways; one example is HIV, the virus that causes AIDS (see "Molecule behind Biology" 23.1). Newly completed viruses that do not acquire an envelope are released by rupture of the host cell's plasma membrane, typically killing the cell. In contrast, most enveloped viruses receive their envelope as they pass through the plasma membrane, usually without breaking the membrane (see Figure 23.2, p. 502). This pattern of viral release typically does not cause immediate damage to the host cell unless very high numbers of virus particles are released.

The vast majority of animal virus infections are asymptomatic because causing disease is of no benefit to the virus. However, a number of pathogenic viruses cause diseases in a variety of ways. Some viruses (e.g., herpesvirus) cause cell death when progeny viruses are released from the cell. This can lead to massive cell death, destroying vital tissues such as nervous tissue or white or red blood cells, or causing lesions in skin and mucous membranes. Other viruses release cellular molecules when infected cells break down, which can induce fever and inflammation (e.g., influenza virus). Yet other viruses alter gene function when they insert into the host cell DNA, leading to cancer and other abnormalities.

Some animal viruses enter a **latent phase,** similar to the lysogenic cycle for bacteriophages, in which the virus remains in the cell in an inactive form. The herpesviruses that cause oral and genital ulcers in humans remain in a latent phase in the cytoplasm of some body cells for the life of the individual. At times, particularly during periods of stress, the virus becomes active in some cells, directing viral replication and causing ulcers to form as cells break down during viral release.

Plant Viruses. Plant viruses may be rodlike or polyhedral. Although most include RNA as their nucleic acid, some contain DNA. None of the known plant viruses have envelopes. They enter cells through mechanical injuries to leaves and stems; they can also be transmitted from one plant to another during pollination or via herbivorous animals such as leafhoppers, aphids, and nematodes. Plant viruses can also be transmitted from one generation to the next in seeds. Once inside a cell, plant viruses replicate via the same processes as animal viruses. However, within plants, virus particles can pass from infected to healthy cells through plasmodesmata, the openings in cell walls that interconnect the cytoplasm of plant cells, and through the vascular system.

Plant viruses are generally named and classified by the type of plant they infect and their most visible effects. *Tomato bushy stunt virus*, for example, causes dwarfing and overgrowth of leaves and stems of tomato plants, and *tobacco mosaic virus* causes a mosaic-like pattern of spots on the leaves of tobacco plants. Most species of crop plants can be infected by at least one destructive virus.

The tobacco mosaic virus was the first virus to be isolated, disassembled, and reassembled in a test tube (see Figure 23.1a, p. 501).

STUDY BREAK

1. What is the difference between a virulent phage and a temperate phage?
2. What are the two types of transduction? How do they differ from each other?
3. How do plant viruses differ from animal viruses?

23.3 It Is Typically Difficult to Treat and Prevent Viral Infections

Viral infections are typically difficult to treat because viruses are, for much of the infection, "hidden" inside host cells and use host cell machinery to replicate. Thus, there often are no obvious viral products to be targeted by drugs. Viral infections are unaffected by antibiotics and other treatment methods used for bacterial infections. As a result, many viral infections are allowed to run their course, with treatment limited to relieving the symptoms while the natural immune defences of the patient attack the virus. Some viruses, however, cause serious and sometimes deadly symptoms on infection; for these, the focus has often been on prevention through vaccine development (e.g., measles, polio). Viruses that use their own polymerases (e.g., RNA viruses such as influenza) provide more obvious targets, so researchers have spent considerable effort developing antiviral drugs to treat them. Many of

MOLECULE BEHIND BIOLOGY 23.1

Reverse Transcriptase

Acquired immune deficiency syndrome (AIDS) is a disease caused by the human immunodeficiency virus (HIV). This disease has likely already killed about 75 million people worldwide, and the epidemic continues to grow, with infection rates in some areas of Africa as high as one in three adults. Even more concerning, infection rates are increasing in south and east Asia, some of the most densely populated regions of the world. If the epidemic continues to spread at current rates, the World Health Organization has

projected AIDS as the fourth-leading cause of death by 2030 (behind heart disease, other chronic diseases, and car accidents). Although drug treatments to hold AIDS in check do exist, they are very expensive, and most people in developing countries cannot afford them. There is no cure for AIDS, so the millions of people currently infected will die prematurely.

HIV is a retrovirus that contains two copies of single-stranded RNA. It also carries several molecules of an enzyme, reverse transcriptase, in its

capsid. Replication of retroviruses is unusual: the virus's genome enters the host cell along with reverse transcriptase, which copies the viral RNA onto a complementary strand of DNA **(Figure 1).** A second strand of DNA is then synthesized, using the first strand as a template. The resulting double-stranded DNA integrates into the host cell's DNA as a provirus (comparable to the prophage described above). This DNA is transcribed by the host cell into mRNA, which is translated to produce viral

1 The glycoprotein on the surface of HIV mediates attachment to protein receptors on the host plasma membrane.

2 The viral contents enter the cell by endocytosis.

3 Reverse transcriptase catalyzes, first, the synthesis of a DNA copy of the viral RNA and, second, the synthesis of a second DNA strand complementary to the first one.

4 The double-stranded DNA is then incorporated into the host cell's DNA.

5 Transcription of the DNA results in the production of RNA. This RNA can serve as the genome for new viruses and can be translated to produce viral proteins.

6 Complete HIV particles are assembled. In macrophages, HIV buds out of the cell without rupturing the cell. In T cells, HIV exits the cell by rupturing it, effectively killing the cell.

HIV
Capsid
RNA
Reverse transcriptase enzyme
Viral RNA
Reverse transcriptase
DNA
Double-stranded DNA
Host cell's DNA
RNA
Nucleus
Ribosome
Viral exiting by budding in macrophages
Viral exit by cell lysis in T cells

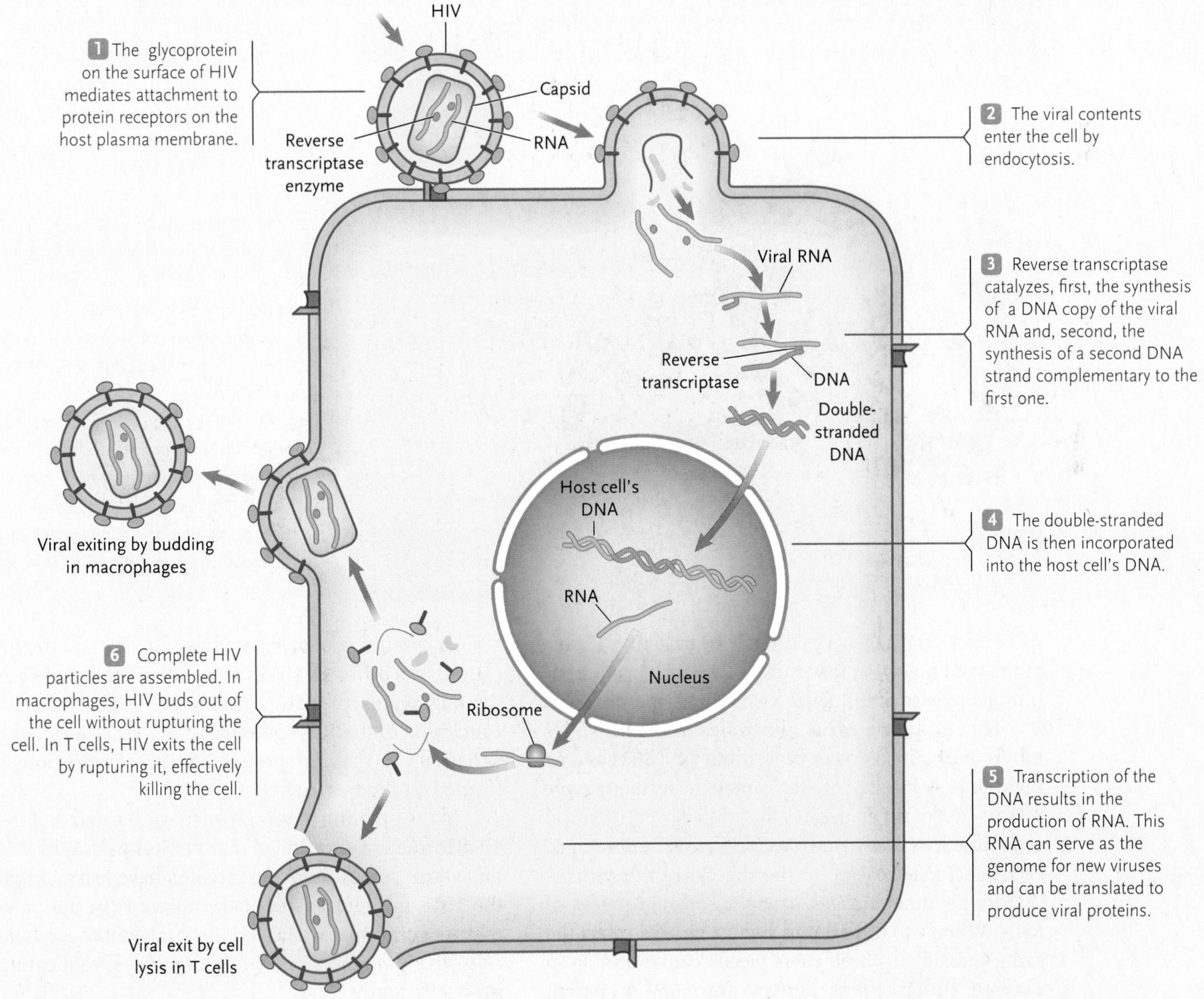

Figure 1
HIV infection cycle.

(Continued)

Reverse Transcriptase (*Continued*)

proteins, including capsid proteins and reverse transcriptase molecules. New virus particles are released from the cell to infect other cells or be passed to new hosts.

Why is HIV so lethal? It targets cells of the human immune system. Obviously, infection of these cells compromises the body's ability to fight off the virus. In addition, some of the immune system cells are not killed by the virus but instead act as a continuing source of infection.

Because reverse transcriptase **(Figure 2)** is a unique feature of HIV, it makes a good target for drug treatment (if the drugs affect only this enzyme, they will not harm the human host). Several antiretroviral drugs have been developed, although HIV has become resistant to some of these drugs. The search continues for a vaccine that would prevent HIV infection, but despite years of research, no vaccine exists yet.

Why is there no vaccine, and how does HIV become resistant so quickly to drugs? The answer to both questions is that HIV mutates quickly and extensively. In a cell's normal DNA replication process, DNA polymerase has proofreading capabilities, so the replicated DNA contains few errors. Reverse transcriptase does not have any proofreading ability, so any errors made when it catalyzes the synthesis of DNA from RNA (and there are many such errors) persist. Proteins encoded by this mutated DNA will be different from those of the original virus; for example, the proteins of the viral coat will be different and so will not be recognized by existing antibodies.

However, reverse transcriptase has also made important positive contributions to biomedical research. For example, retroviruses play an important role in gene therapy, in which new diseases are treated by the introduction of new genes into the body. Viruses are very effective vectors for introducing genes into cells. The desired genes are cloned into the viral genome, and once the virus is taken up by the cell, those genes are introduced into all cells infected by the virus. Retroviruses are particularly useful in gene therapy since the genetic material they carry is integrated into the host cell genome. Reverse transcriptase is also an important tool in molecular biology, as it can be used to synthesize complementary DNA (cDNA) from mRNA, allowing for the cloning of actively expressed genes. It can also be used in genetic engineering; for example, reverse transcriptase can be used to make cDNA out of the mRNA for insulin. This cDNA does not have introns because it is synthesized from an mRNA template and thus can be expressed in a bacterial host (which lacks the enzymes to process DNA that contains introns), allowing insulin to be produced in large quantities.

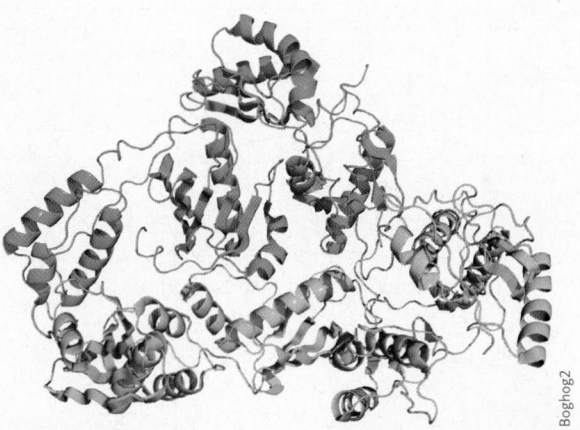

Figure 2
Reverse transcriptase.

Boghog2

these drugs fight the virus directly by targeting a stage of the viral life cycle; for example, the drug zanamivir inhibits release of influenza virus particles from cells.

The influenza virus illustrates the difficulties inherent in controlling or preventing viral diseases. As mentioned at the start of the chapter, the influenza type A virus causes flu epidemics that sweep over the world each year. Why does a new vaccine have to be developed each year? One reason for the success of this virus is that its genome consists of eight separate pieces of RNA. When two different influenza viruses infect the same individual, these RNA pieces can assemble in random combinations derived from either parent virus. The new combinations can change the protein coat of the virus, making it unrecognizable to antibodies developed against either parent virus. Being "invisible" to these antibodies means that new virus strains can infect people who have already had the flu caused by a different strain or who had flu shots effective only against the parent strains of the virus. Random mutations in the RNA genome of the virus add to the variations in the coat proteins that make previously formed antibodies ineffective.

In the opening to this chapter, we learned that the 1918 influenza virus killed many of its hosts. Why was this strain so virulent? Researchers have learned that the 1918 influenza virus had mutations in the polymerase genes that replicated the viral genome in host cells, likely making this strain capable of replicating more efficiently.

Other viruses that infect humans are also considered to have evolved from a virus that previously infected other animals. HIV is one of these; until the second half of the twentieth century, infections of this virus were apparently restricted almost entirely to chimpanzees and gorillas in Africa. Now the virus infects nearly 36 million people worldwide, with the greatest concentration of infected individuals in sub-Saharan Africa.

As illustrated by this example, our efforts to control or eliminate human diseases caused by viral pathogens are complicated when dealing with viruses that have broad host specificity and can infect other animals besides humans. Because other animals can harbour these viruses, we can never successfully eradicate the diseases they cause. For example, the influenza virus can infect birds, swine, and other animals in addition to humans.

Also, as human encroachment on wildlife habitats increases, we create the potential for the evolution of new human viruses, as strains that infect other animals mutate to infect humans. These factors, together with increasing global travel and trade, create the potential for a new human pathogenic virus to readily become a global problem, as we have experienced with HIV. A better understanding of the evolution and life cycles of viruses is crucial if we are to prevent or treat emerging viral diseases.

STUDY BREAK

What can make a viral infection more difficult to treat than a bacterial infection?

23.4 Viruses May Have Evolved from Fragments of Cellular DNA or RNA

Where did viruses come from? Several different hypotheses have been proposed to explain the origin of viruses. Some biologists have suggested that because viruses can duplicate only by infecting a host cell, they probably evolved after cells appeared. They may represent "escaped" fragments of DNA molecules that once formed part of the genetic material of living cells or an RNA copy of such a fragment. The fragments first became surrounded by a protective layer of protein with recognition functions, and then these fragments escaped from their parent cells. As viruses evolved, the information encoded in the core of the virus became reduced to a set of directions for producing more viral particles of the same kind.

More recent hypotheses suggest that viruses are very ancient, with virus like particles predating the first cells. The first viruses originated from the "primordial gene pool"—the pool of RNA that is thought to have been the first genetic material.

Regardless of when viruses originated, they do not share a common evolutionary origin. Thus, unlike cellular life, there is no common ancestor for all viruses and we cannot draw a phylogenetic tree for all viruses. However, viruses have played an important role in the evolution of cellular life because of their ability to integrate their genes into their hosts and to acquire genes from their hosts, as described above. In this way, viruses can be a source of new cellular genetic material, providing new enzymes and other proteins to a cell. Viruses may also have played a more direct role in the evolution of eukaryotic cells: some biologists have suggested that the nucleus originated from a large, double-stranded DNA virus that infected prokaryotic cells, resulting in the first eukaryotic cell.

STUDY BREAK

Why do some biologists think viruses must have originated after cells evolved, rather than predating cells?

23.5 Viroids and Prions Are Infective Agents Even Simpler in Structure than Viruses

Viroids, first discovered in 1971, are small, infectious pieces of RNA. Although the RNA is single stranded, bonding within the molecule causes it to become circular. Viroids are smaller than any virus and lack a protein coat. They also differ from viruses in that their RNA genome does not code for any proteins. Viroids are plant pathogens that can rapidly destroy entire fields of citrus, potatoes, tomatoes, coconut palms, and other crop plants. How do viroids cause such devastating diseases without synthesizing any proteins?

The manner in which viroids cause disease remains unknown. In fact, researchers believe that there is more than one mechanism. Recent research indicates that the viroid may cause disease when its RNA interacts with molecules in the cell; for example, it may disrupt normal RNA processing of the host cell: if the viroid's RNA sequence is complementary to the mRNA of the host cell, it can bind to the host's mRNA, thus preventing normal protein synthesis and causing disease.

Like viruses and virions, **prions** are small infectious particles, but they are not based on nucleic acids; instead, they are infectious protein molecules (the term "prion" is a loose acronym for *pro*teinaceous *in*fectious particle). Prions cause spongiform encephalopathies (SEs), degenerate diseases of the nervous system in mammals characterized by loss of motor control and erratic behaviour. The brains of affected animals are full of spongy holes **(Figure 23.6, p. 510)** (hence the "spongiform" designation) and deposits of proteinaceous material. Under the microscope, aggregates of misfolded proteins, called amyloid fibres, are seen in brain tissues; the accumulation of these proteins is the likely cause of the brain damage. SEs progress slowly, meaning that animals may be sick for a long time before their symptoms become obvious, but death is inevitable.

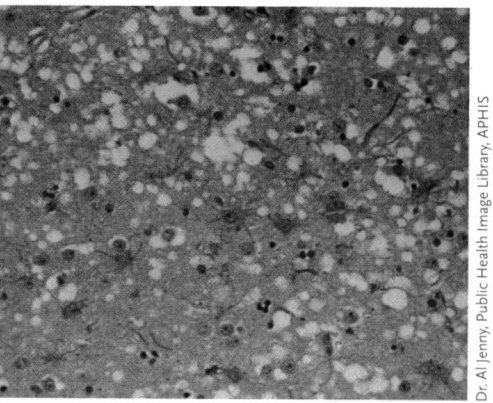

Figure 23.6

Bovine spongiform encephalopathy (BSE). The light-coloured patches in this thin section from a brain damaged by BSE are areas where tissue has been destroyed.

Dr. Al Jenny, Public Health Image Library, APHIS

One SE disease is *scrapie,* a brain disease that causes sheep to rub against fences, rocks, or trees until they scrape off most of their wool. In cattle, a similar disease is bovine spongiform encephalopathy (BSE), also known as "mad cow disease." Humans also have SE diseases, such as *kuru,* found among cannibals in New Guinea, who became infected by eating raw human brain during ritual feasts following the death of an individual. *Creutzfeldt–Jakob disease* (CJD) is a very rare SE disease that affects about one person in a million per year, globally. The symptoms of CJD include rapid mental deterioration, loss of vision and speech, and paralysis; autopsies show spongy holes and deposits in brain tissue similar to those of cattle with BSE. We don't know how CJD is transmitted naturally, but we know it can be transmitted inadvertently, for example, with corneal transplants.

SE diseases hit the headlines worldwide in the late 1980s when farmers in the United Kingdom reported a new disease, later determined to be BSE, spreading among their cattle. It is estimated that over 900 000 cows in the United Kingdom were affected, many of which entered the human food chain before they developed symptoms. Where did BSE come from? The source was determined to be meat and bone meal fed to the cows; this meal came from the carcasses of sheep and cattle. The practice of feeding animal meal to cattle had been followed for years, but a money-saving change in processing in the early 1980s (a reduction in how long rendered material was held at high temperature) allowed the infectious agent—maybe from scrapie-infected sheep—to survive in the meat and bone meal. Worse was to come when it became evident that BSE had spread to humans who had eaten contaminated beef. This new human disease, known as variant CJD, is linked to eating meat products from cattle with BSE. Between 1996, when variant CJD was first described, and 2007, there were 208 cases from 11 countries, with the vast majority of these in the United Kingdom. A 12-year study of human tissue samples removed during appendix operations in the United Kingdom suggests that about 1 in every 2000 people in the United Kingdom is a carrier for variant CJD. Will these people actually develop the disease? Evidence from studies of kuru suggests that it may take more than 50 years for prion diseases to develop, so there is some concern that a spike in variant CJD cases is still to come.

Concern about variant CJD explains why the discovery of even one cow with BSE can wreak havoc on a country's beef exports, as happened in Canada when an infected cow was found in Alberta in 2003. The United States closed its border to all beef from Canada within a day, followed shortly by border closings of 40 other countries. Loss of these markets caused serious economic hardship for Canadian ranchers and farmers.

What is the cause of BSE and other SE diseases, and how does this causative agent spread? As explained in "People behind Biology," Stanley Prusiner demonstrated that infectious proteins cause these diseases. Prions are the only known infectious agents that do not include a nucleic acid molecule, and their discovery changed some fundamental views of biology.

Our current understanding of prion infection is that prion proteins are able to survive passage through the stomach of an animal consuming them; they then enter that animal's bloodstream and proceed to the brain, where they somehow interact with normal prion proteins, causing these proteins to change shape to become abnormal and infectious. Prion proteins and the normal precursor proteins share the same amino acid sequences but differ in how they are folded. Prions are somehow able to impose their folding on normal proteins, thus "infecting" the normal proteins. As the infection spreads, neural functioning is impaired and protein fibrils accumulate, producing aggregations of fibrils that trigger apoptosis of infected cells, leading to the SE characteristic of these diseases.

What is the function of "normal" prion proteins? We don't know yet, but evidence suggests that normal prions may regulate the protein synthesis required for growth, development, and protection of brain cells. Mice lacking normal prion proteins have subtle impairments in memory and cognition. Perhaps the inability of the misfolded prion proteins to carry out their normal functions results in dementia and the other symptoms of BSE.

In this chapter, we focused on the simplest biological entities: viruses, viroids, and prions, which possess only some of the properties of life. In the next five chapters, we investigate more structurally complex organisms: the eukaryotic protists, fungi, plants, and animals.

STUDY BREAK

How do viroids and prions differ from viruses?
How do they differ from each other?

For several decades, scientists had hypothesized that a slow virus—a disease-causing virus with a long incubation period and gradual onset of pathogenicity—was responsible for scrapie and other spongiform encephalopathies. However, scientists had repeatedly examined the brains of infected animals and not found any evidence of viral infection. In 1982, Stanley Prusiner, a researcher at University of California, San Francisco, determined that the infectious agent was a protein. He pointed to the accumulation of protein fibrils in the brains of infected animals and termed this protein the prion protein (PrP). The research community mostly rejected this hypothesis because it went against all the accepted dogma of biology—genes in the form of DNA or RNA were necessary to cause disease. How could a protein make copies of itself? Prusiner located the gene for PrP and then found that prion proteins are naturally occurring membrane proteins in many types of cells, including neurons. In sheep infected with scrapie, Prusiner found "rogue" forms of the prion proteins that were abnormally folded. He proposed that these infectious prion proteins somehow interacted with "normal" prion proteins to cause misfolding of these proteins; thus, the abnormal protein structure is "infectious." The misfolded prion proteins aggregate, forming the masses of fibrils characteristic of SE diseases. In 1997, Prusiner received a Nobel Prize for his discovery of prions.

Review

aplia™

To access course materials such as Aplia and other companion resources, please visit www.NELSONbrain.com.

23.1 What Is a Virus? Characteristics of Viruses

- Viruses are nonliving infective agents. A free virus particle consists of a nucleic acid genome enclosed in a protein coat (Figure 23.1). Recognition proteins enabling the virus to attach to host cells extend from the surface of infectious viruses.

23.2 Viruses Infect Bacterial, Animal, and Plant Cells by Similar Pathways

- Viruses reproduce by entering a host cell and directing the cellular machinery to make new particles of the same kind (Figures 23.2, 23.4).

23.3 It Is Typically Difficult to Treat and Prevent Viral Infections

- Viruses are unaffected by antibiotics and most other treatment methods. As well, many viruses have great genetic variability and are located inside cells for much of the infection.

  For these reasons, viral infections are difficult to treat, which is why efforts have focused development of vaccines on preventing infection by those viruses that cause serious or fatal diseases.

23.4 Viruses May Have Evolved from Fragments of Cellular DNA or RNA

- There are several hypotheses about the origin of viruses. Viruses may have evolved after cells did and may have descended from nucleic acid fragments that "escaped" from a cell. Evidence for this hypothesis comes from the fact that viruses can duplicate only by infecting a host cell. On the other hand, a competing hypothesis suggests that viruses evolved before the first cells, with the first virus like particles originating from the pool of RNA that was the first genetic material.

- Viruses have different evolutionary origins; i.e., they do not share a common ancestor.

23.5 Viroids and Prions Are Infective Agents Even Simpler in Structure Than Viruses

- Viroids, which infect crop plants, consist of only a very small, single-stranded RNA molecule. Prions, which cause brain diseases in some animals, are infectious proteins with no associated nucleic acid. Prions are misfolded versions of normal cellular proteins, which can induce other normal proteins to misfold.

Questions

Self-Test Questions

1. Which of the following best defines a virus?
   a. a naked fragment of nucleic acid
   b. a disease-causing group of proteins
   c. an entity composed of proteins and nucleic acids
   d. an entity composed of proteins, nucleic acids, and ribosomes

2. Viruses form a capsid around their nucleic acid core. What is this capsid composed of?
   a. protein
   b. lipoprotein
   c. glycoprotein
   d. polysaccharides

3. Which of the following statements about viral envelopes is correct?
   a. They contain glycoproteins of viral origin.
   b. They are located between the virus's capsid and its nucleic acid.
   c. They are composed of a lipid bilayer derived from the viral membrane.
   d. They are composed of peptidoglycan, the same material as bacterial cell walls.

4. Which of the following characteristics distinguishes plant viruses from animal viruses?
   a. Plant viruses are easily curable.
   b. Plant viruses are covered by a membrane envelope.
   c. Plant viruses lack the ability to actively infect a host cell.
   d. Plant viruses lack the ability to replicate their RNA genome.

5. Which of the following describes what happens when a bacteriophage enters the lysogenic stage?
   a. It enters the host cell and kills it immediately.
   b. It enters the host cell, picks up host DNA, and leaves the cell unharmed.
   c. It merges with the host cell plasma membrane, forming an envelope, and then exits the cell.
   d. It injects its DNA into the host cell DNA, and the host DNA integrates viral DNA into the host genome.

6. Which of the following does reverse transcriptase synthesize?
   a. RNA from DNA
   b. DNA from RNA
   c. proteins from DNA
   d. proteins from RNA

7. Which of the following correctly describes a viroid?
   a. the smallest type of virus
   b. small infectious pieces of DNA
   c. small infectious pieces of RNA
   d. infectious pieces of RNA wrapped in a protein coat

8. Which of the following statements about temperate phages is correct?
   a. They never lyse host cells.
   b. They turn their host cell into a prophage.
   c. They integrate their DNA into the host cell chromosome.
   d. They break down the host cell's chromosome when their DNA enters the cell.

9. There are many similarities in how animal and bacterial viruses infect their host cells. Which of the following correctly states one such similarity?
   a. Both animal and bacterial viruses commonly have envelopes.
   b. For bacterial and animal viruses, only their nucleic acid enters the host cell.
   c. Both bacterial and animal viruses have a capsid divided into a head and a tail.
   d. Both animal and bacterial viruses bind to specific receptors on the host cell.

10. Which of the following statements about prions is correct?
    a. Prions can only be transmitted from animals to humans.
    b. The diseases caused by prions progress very rapidly.
    c. Prions have a different amino acid sequence from the normal protein.
    d. Prion proteins reproduce by misfolding normal proteins.

Questions for Discussion

1. From what you have read in this chapter, would you consider viruses to be alive? Why or why not?

2. Why do animal viruses have envelopes, whereas bacteriophages do not?

3. Why is it difficult to design an effective, long-lasting vaccine for the flu virus and the HIV virus?

4. What is a retrovirus, and why have these viruses been given this name?

5. From an evolutionary standpoint, why would the lysogenic state be favoured?

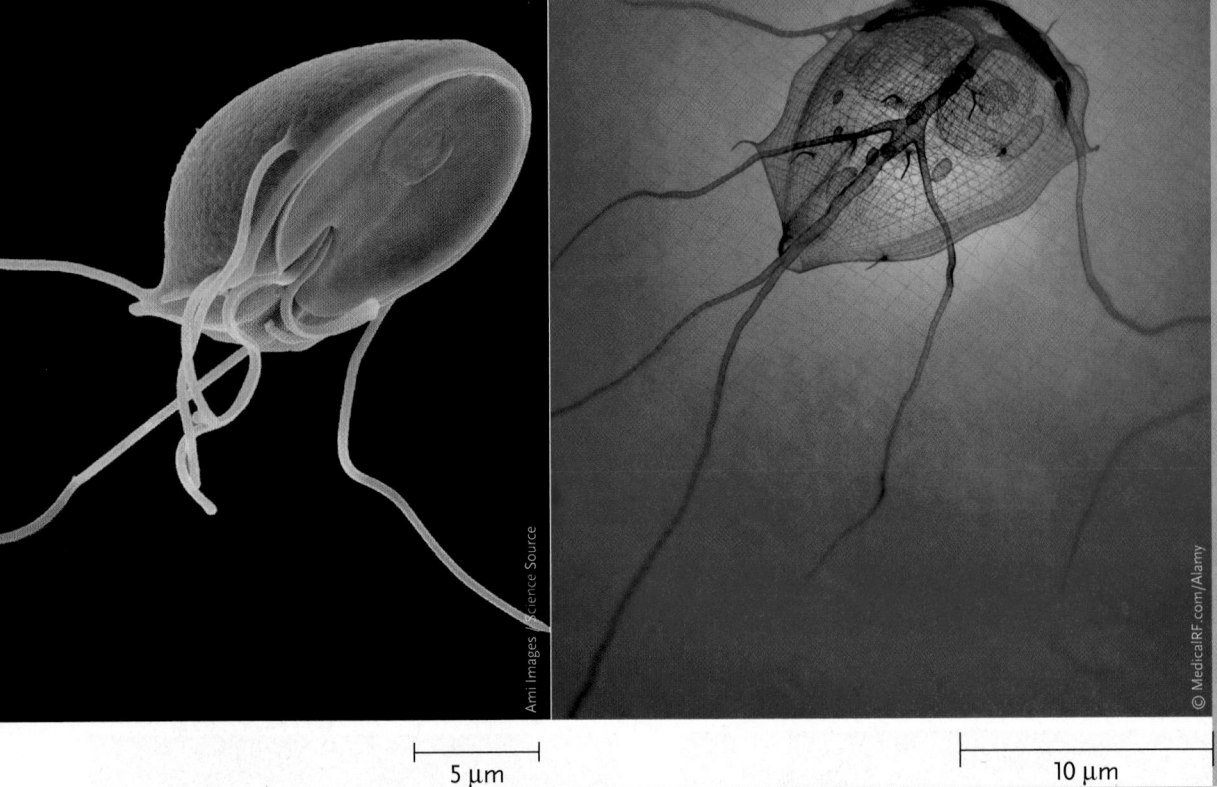

├─ 5 μm ─┤   ├─ 10 μm ─┤

Giardia lamblia. **(left)** Scanning electron microscope image. **(right)** Light microscope image.

24

Protists

WHY IT MATTERS

You are on a backpacking trip in your favourite wilderness area on a hot and sunny day. You pause to take a drink of water from your water bottle but discover it is almost empty. You are very thirsty, so you refill your bottle from a nearby stream. The water is clear and cold and looks clean, and, besides, you're out in the middle of nowhere, so it must be safe to drink, right? You continue on the hike and feel fine. But a few days after you get home, you don't feel so great: you have abdominal pain, cramps, and diarrhea. Your doctor says that you have giardiasis, or "beaver fever," caused by *Giardia lamblia,* the most common intestinal parasite in North America (it is very prevalent in water bodies formed by beaver dams). What is *Giardia,* and how does it make you sick?

Giardia is a single-celled eukaryote that can exist in two forms: a dormant cyst and a motile feeding stage. When you drank from that seemingly clean stream, you ingested some cysts. The cysts can survive for months, so it is important to boil or filter water when you are out hiking or camping. As the swallowed cysts moved from your stomach into your small intestine, the cysts released the motile feeding stage, **trophozoites** (*troph* = food; *zoon* = animal), shown in the photographs at the top of the previous page. Using their multiple flagella, the trophozoites were able to swim about in your intestinal space and attach themselves to the epithelial cells of your intestine. Infection with *Giardia* can become chronic, causing inflammation and reduction of the absorptive capacity of the gut. So why doesn't your immune system detect the presence of *Giardia* and get rid of the parasite? *Giardia* can alter the proteins on its surface that your immune system relies on to recognize an invader, and so it escapes recognition; thus, *Giardia* infections can be persistent or recur.

Giardia is a **protist** (Greek, *protistos* = the very first). Protists are a very heterogeneous collection of about 200 000 eukaryotes. Most are unicellular and microscopic, but some are large, multicellular organisms. Like their most ancient

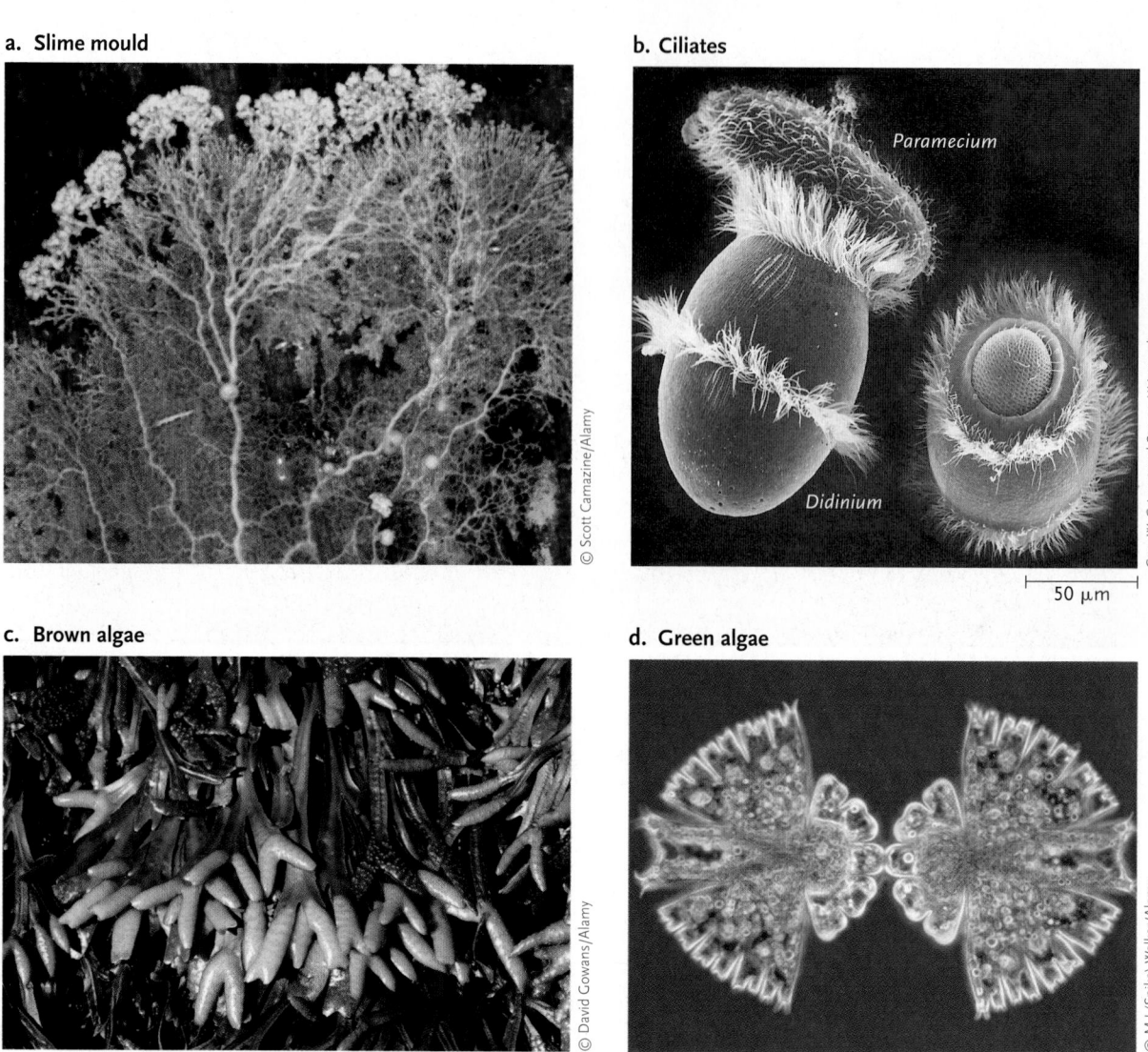

a. Slime mould

b. Ciliates

Paramecium

Didinium

50 μm

c. Brown algae

d. Green algae

25 μm

Figure 24.1

A sampling of protist diversity. **(a)** *Physarum*, a plasmodial slime mould. **(b)** *Didinium*, a ciliate, consuming another ciliate, *Paramecium*. **(c)** *Fucus gardneri* (common rockweed), a brown alga growing in rocky intertidal zones. **(d)** *Micrasterias*, a single-celled green alga, here shown dividing in two.

ancestors, almost all of these eukaryotic species are aquatic. **Figure 24.1** shows a number of protists, illustrating their great diversity.

24.1 The Vast Majority of Eukaryotes Are Protists

The diversity among protists makes it very difficult to define what a protist is. The simplest definition, and the one we will use in this book, is that a protist is any eukaryotic organism that is not an animal, a land plant, or a fungus. Earlier classifications grouped all of these "other" eukaryotes together in one kingdom, Protista. This oversimplified classification reflected our earlier understanding of eukaryote biology, which traditionally has been almost entirely based on the study of animals, land plants, and fungi—the multicellular eukaryotes. But these groups are only three branches of the very

large and diverse tree of living eukaryotes **(Figure 24.2)**. This evolutionary tree is based on molecular data, which are considered the most informative data for determining evolutionary relationships. The tree shows that eukaryotic organisms are divided into approximately five "supergroups," a taxonomic level above that of "kingdom." As you can see by looking at Figure 24.2, the vast majority of eukaryotes are not land plants, animals, or fungi but protists. The tree shown here represents our current understanding of the relationships among eukaryotic organisms, which is actively changing as researchers continue to investigate the evolutionary history of eukaryotes; the actual number of supergroups is still being debated, with some researchers dividing eukaryotes into additional supergroups to those shown in Fig 24.2. You may notice that the "root" of the tree—the last eukaryote common ancestor (LECA)—is not identified. The identity of this ancestral group is a major mystery that researchers are actively working to unravel.

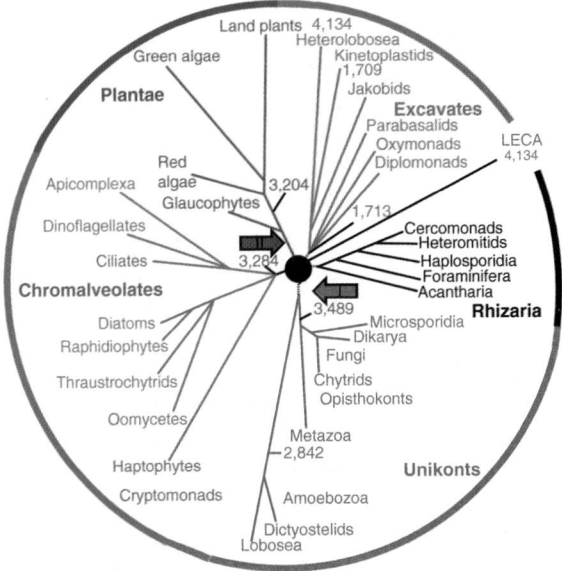

Figure 24.2

Major lineages of protists within the supergroups of eukaryotic organisms. The current evolutionary tree for eukaryotes divides these organisms among approximately five supergroups: Excavates, Unikonts (including animals and fungi), Plantae (including land plants and their algal relatives), Chromalveolates, Excavates and Rhizaria. Selected lineages of protists in each supergroup are discussed in this chapter.

The first eukaryotes likely evolved about 1.5–2 billion years ago (Ba) ago. While we don't fully understand how they evolved, we know that endosymbiosis played an important role in the process. Eukaryotes contain mitochondria (although some have very reduced versions of this organelle) and many also contain chloroplasts. As outlined in Chapter 2, mitochondria and chloroplasts are the descendants of free-living prokaryotes that, over evolutionary time, became organelles. All mitochondria are thought to have arisen from a single endosymbiotic event, but the history of chloroplasts is more complex. We will return to the evolution of chloroplasts at the end of this chapter, once you have had a chance to become familiar with the various groups of protists.

In this chapter, we will start with an overview of features of protists and then focus on key protist lineages in each of the eukaryotic supergroups. In this way, you will gain an understanding of how diverse protists are morphologically, functionally, and ecologically. As you read about the various groups of protists, think about how they differ from animals and plants, and how learning about these "other" eukaryotes changes your understanding of eukaryote biology. Protists are sometimes called the "rule-breakers" of the eukaryotic world: many of the general rules or "facts" we think we know about eukaryotic organisms are revealed as not being generally true at all once protists are considered, forcing us to rethink what is "typical" or "normal" in eukaryote biology.

STUDY BREAK

By what process did eukaryotes such as protists acquire mitochondria and chloroplasts?

24.2 Characteristics of Protists

Because protists are eukaryotes, the boundary between them and prokaryotic organisms is clear and obvious. Unlike bacteria and archaea, protists have a membrane-bound nucleus, with multiple, linear chromosomes. In addition to cytoplasmic organelles, including mitochondria and chloroplasts (in some species), protists have microtubules and microfilaments, which provide motility and cytoskeletal support. As well, they share characteristics of transcription and translation with other eukaryotes.

The phylogenetic relationship between protists and other eukaryotes is more complex (Figure 24.2). Over evolutionary time, the eukaryotic family tree branched out in many directions. All of the organisms in the eukaryotic lineages consist of protists except for three groups, the animals, land plants, and fungi, which arose from protist ancestors. Although some protists have features that resemble those of the fungi, plants, or animals, several characteristics are distinctive. In contrast to fungi, most protists are motile or have motile stages in their life cycles, and their cell walls are made of cellulose, not chitin.

How do photosynthesizing protists differ from plants? Unlike plants, many photoautotrophic protists can also live as heterotrophs, and some regularly combine both modes of nutrition. Protists do not retain developing embryos in parental tissue, as plants do, nor do they have highly differentiated structures equivalent to roots, stems, and leaves. Photosynthetic protists are sometimes referred to as *algae*; these protists are generally aquatic and often unicellular and microscopic (although many are multicellular). However, the different groups of algae are not closely related to each other (see Figure 24.2), so the term *algae* does not indicate any sort of relatedness among organisms referred to by that term.

How do protists differ from animals? Unlike protists, all animals are multicellular and have features such as an internal digestive tract and complex developmental stages. Protists also lack features that characterize many animals, including nerve cells; highly differentiated structures such as limbs and a heart; and collagen, an extracellular support protein.

STUDY BREAK

What features distinguish protists from prokaryotic organisms? What features distinguish them from fungi, plants, and animals?

24.3 Protists' Diversity Is Reflected in Their Metabolism, Reproduction, Structure, and Habitat

As you might expect from looking at Figure 24.2, protists are highly diverse in metabolism, reproduction, structure, and habitat.

Habitat. Protists live in aqueous habitats, including aquatic or moist terrestrial locations, such as oceans, freshwater lakes, ponds, streams, and moist soils, and within host organisms. In bodies of water, small photosynthetic protists collectively make up the **phytoplankton** (*phytos* = plant; *planktos* = drifting), the organisms that capture the energy of sunlight in nearly all aquatic habitats. These phototrophs provide organic substances and oxygen for heterotrophic bacteria, other protists, and the small crustaceans and animal larvae that are the primary constituents of **zooplankton** (*zoe* = life, usually meaning animal life). Although protists are not animals, biologists often include them among the zooplankton. Phytoplankton and larger multicellular protists forming seaweeds collectively account for about half of the total organic matter produced by photosynthesis.

In the moist soils of terrestrial environments, protists play important roles among the detritus feeders that recycle matter from organic back to inorganic form. In their roles in phytoplankton, in zooplankton, and as detritus feeders, protists are enormously important in world ecosystems.

Protists that live in host organisms are **parasites**, obtaining nutrients from the host. Indeed, many of the parasites that have significant effects on human health are protists, causing diseases such as malaria, sleeping sickness, and amoebic dysentery.

Structure. Whereas most protists are single cells, others live as **colonies (Figure 24.3)** in which individual cells show little or no differentiation and are potentially independent. Within colonies, individuals use cell signalling to cooperate on tasks such as feeding and movement. Some protists are large multicellular organisms; for example, the giant kelp of coastal waters can rival forest trees in size.

Many single-celled and colonial protists have complex intracellular structures, some found nowhere else among living organisms **(Figure 24.4)**. These unique structures reflect key aspects of the habitats in which protists live. For example, consider a single-celled protist living in a freshwater pond. Its cytoplasm is

Figure 24.3
Colonial protist (*Dinobryon*).

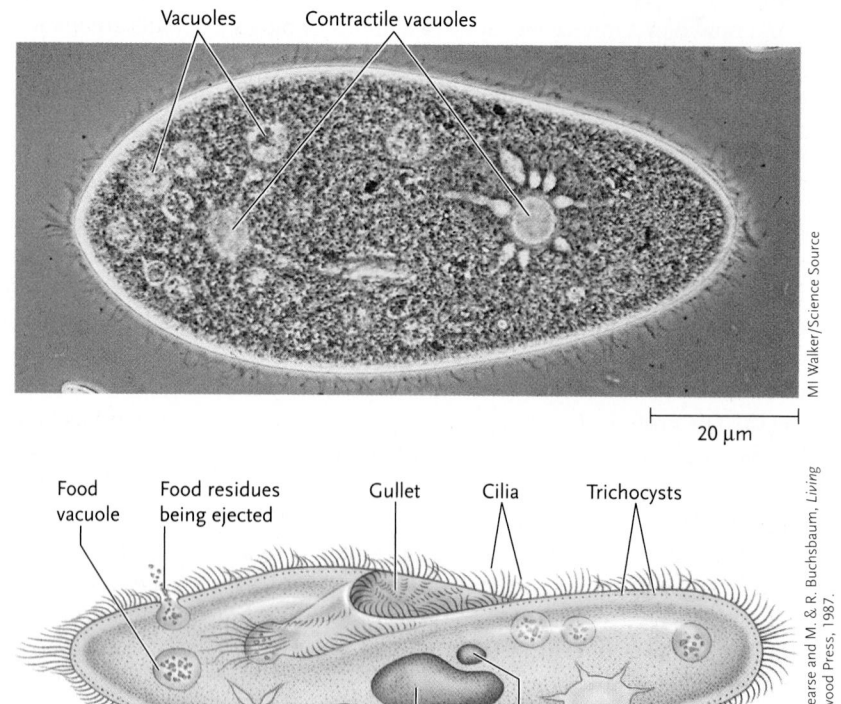

Figure 24.4
A ciliate, *Paramecium*, showing the cytoplasmic structures typical of many protists.

hypertonic to the water surrounding it, meaning that water flows into the cell by osmosis (see Section 5.5). How can the protist stop itself from bursting? A specialized cytoplasmic organelle, the **contractile vacuole**, gradually fills with fluid. When this vacuole reaches its maximum size, it moves to the plasma membrane and forcibly contracts, expelling the fluid to the outside through a pore in the membrane.

The cells of some protists are supported by an external cell wall or by an internal or external shell built up from organic or mineral matter; in some, the shell takes on highly elaborate forms. Instead of a cell wall, other protists have a **pellicle**, a layer of supportive protein fibres located inside the cell just under the plasma membrane, providing strength and flexibility **(Figure 24.5)**.

At some time during their lives, almost all protists move. Some move by amoeboid motion, in which the cell extends one or more lobes of cytoplasm called **pseudopodia** ("false feet"; see **Figure 24.6**). The rest of the cytoplasm and the nucleus then flow into the pseudopodium, completing the movement. Other protists move by the beating of flagella or cilia. In some protists, cilia are arranged in complex patterns, with an equally complex network of microtubules and other cytoskeletal fibres supporting the cilia under the plasma membrane.

Many protists can exist in more than one form, for example, as a motile form and as a nonmotile cyst that can survive unfavourable conditions. This morphological variability allows the species to live in different habitats at different stages in its life.

Metabolism. Almost all protists are aerobic organisms that live either as heterotrophs—obtaining carbon from organic molecules produced by other organisms—or as photoautotrophs, by producing organic molecules for themselves by photosynthesis (see Chapter 7). Some heterotrophic protists obtain organic molecules by engulfing part or all of other organisms (*phagocytosis*) and digesting them internally. Others absorb small organic molecules from their environment by diffusion. Some protists can live as either heterotrophs or autotrophs.

Reproduction. Reproduction may be asexual, by mitosis, or sexual, through meiotic cell division and formation of gametes. In protists that reproduce by both mitosis and meiosis, the two modes of cell division are often combined into a **life cycle** that is highly distinctive among the different protist groups. We do not yet have a complete understanding of the reproductive biology of many protists.

STUDY BREAK

Define each of the following terms in your own words, and indicate the role that each plays in the life of a protist: *pellicle, pseudopodia, contractile vacuole.*

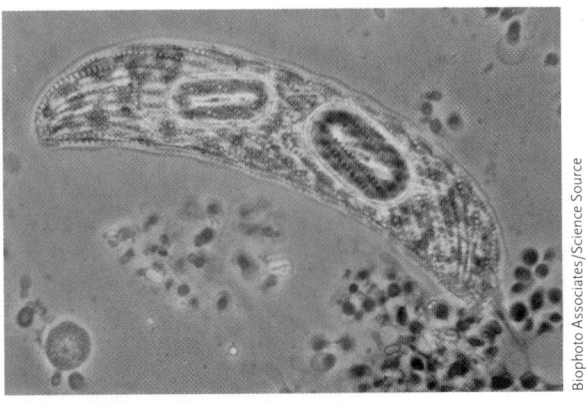

Figure 24.5
Euglena spirogyra, showing pellicle (strips of protein fibres).

Biophoto Associates/Science Source

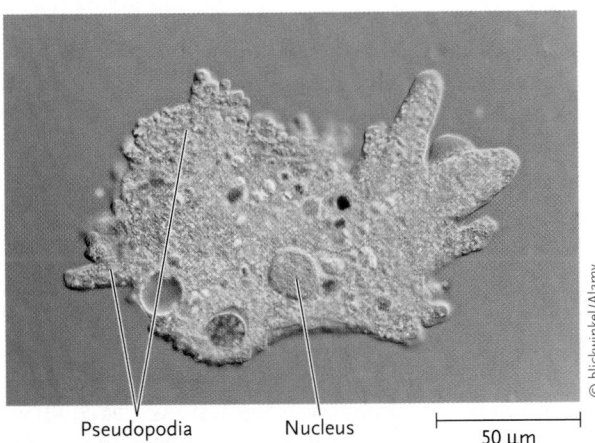

Pseudopodia   Nucleus   50 μm

Figure 24.6
Amoeba proteus of the Amoebozoa is perhaps the most familiar protist of all.

© blickwinkel/Alamy

24.4 The Eukaryotic Supergroups and Their Key Protist Lineages

In this section, we look at the biological features of the major protist lineages in each eukaryote supergroup shown in Figure 24.2, p. 515. Our focus is the ecological or economic importance of each lineage, the habitats in which you would find these organisms, and key features that differentiate the group from other protists. As you read through the information on each lineage, think about how the structural features of that group relate to its habitat and lifestyle.

24.4a Excavata Are Unicellular, Flagellated Protists, Many of Which Lack Mitochondria

This supergroup takes its name from the hollow (excavated) ventral feeding groove found in most members. Protists of this supergroup are sometimes referred to as protozoa (*proto* = first; *zoon* = animal) because, like animals, they ingest their food and move by themselves. We will consider four lineages of Excavates: Diplomonads, Parabasalids, Euglenoids, and Kinetoplastids.

Euglenoids. You have probably seen an example of one genus of euglenoids, *Euglena,* in your earlier biology classes **(Figure 24.7),** as they are often used to illustrate how some protists have plantlike features (photosynthesis) combined with features that we consider animal-like (movement). Euglenoids are important primary producers in freshwater ponds, streams, and lakes, and even some marine habitats. Most are autotrophs that carry out photosynthesis using the same photosynthetic pigments and mechanisms as plants. If light is not available, many of the photosynthetic euglenoids can also live as heterotrophs by absorbing organic molecules through the plasma membrane or by engulfing small particles. Other euglenoids lack chloroplasts and live entirely as heterotrophs.

The name *Euglena* roughly translates as "eyeball organism," a reference to the large *eyespot* that is an obvious feature of photosynthetic euglenoids (see Figure 24.7). The eyespot contains pigment granules in association with a light-sensitive structure and is part of a sensory mechanism that stimulates cells to swim toward moderately bright light or away from intensely bright light so that the organism finds optimal conditions for photosynthetic activity. In addition to an eyespot, euglenoids contain numerous organelles, including a contractile vacuole.

Rather than an external cell wall, euglenoids have a spirally grooved pellicle formed from strips of transparent, protein-rich material underneath the membrane (see Figure 24.5, p. 517). In some euglenoids, the strips are arranged in a spiral pattern, allowing the cell to change its shape in a wriggling sort of motion (known as euglenoid movement) that allows the cell to change direction. Euglenoids can also swim by whiplike movements of flagella that extend from one end of the cell. Most have two flagella: one rudimentary and short, the other long.

Kinetoplastids. Sleeping sickness is a fatal disease endemic to sub-Saharan Africa. Although the disease was almost eradicated about 40 years ago, it has been making a comeback due to wars and the subsequent refugee movement and damage to healthcare systems. Sleeping sickness is caused by various subspecies of *Trypanosoma brucei* **(Figure 24.8)** that are transmitted from one host to another by bites of the tsetse fly. Early symptoms include fever, headaches, rashes, and anemia. Untreated, the disease damages the central nervous system, leading to a sleeplike coma and eventual death. The disease has proved difficult to control because the same trypanosomes infect wild mammals, providing an inexhaustible reservoir for the parasite. Other trypanosomes, also transmitted by insects, cause Chagas disease in Central and South America and leishmaniasis in many tropical countries. Humans with Chagas disease have an enlarged liver and spleen and may experience severe brain and heart damage; leishmaniasis causes skin sores and ulcers, as well as liver and spleen damage.

Like trypanosomes, other kinetoplastids are heterotrophs that live as animal parasites. Kinetoplastid cells are characterized by a single mitochondrion that contains a large DNA-protein deposit called a *kinetoplast* (see Figure 24.8). Most kinetoplastids also have a leading and a trailing flagellum, which are used for movement. In some cases, the trailing flagellum is attached to the side of the cell, forming an undulating membrane that allows the organism to glide along or attach to surfaces.

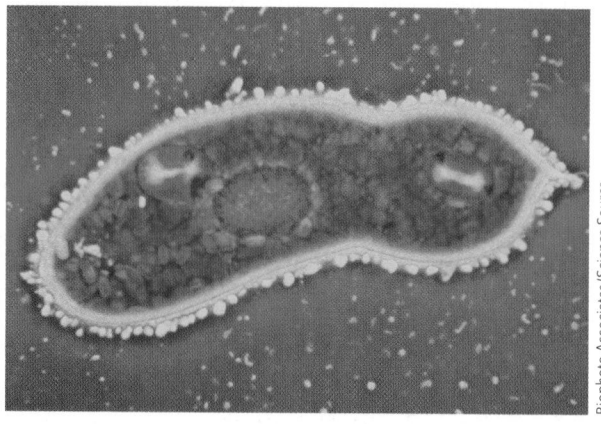

Biophoto Associates/Science Source

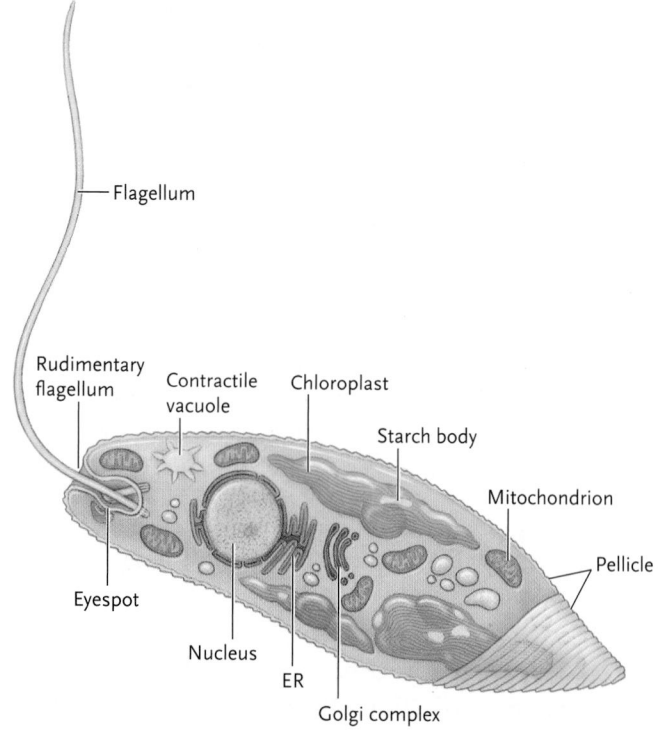

Flagellum

Rudimentary flagellum
Contractile vacuole
Chloroplast
Starch body
Mitochondrion
Pellicle
Eyespot
Nucleus
ER
Golgi complex

Figure 24.7

Body plan and a colour photo *of Euglena gracilis.*

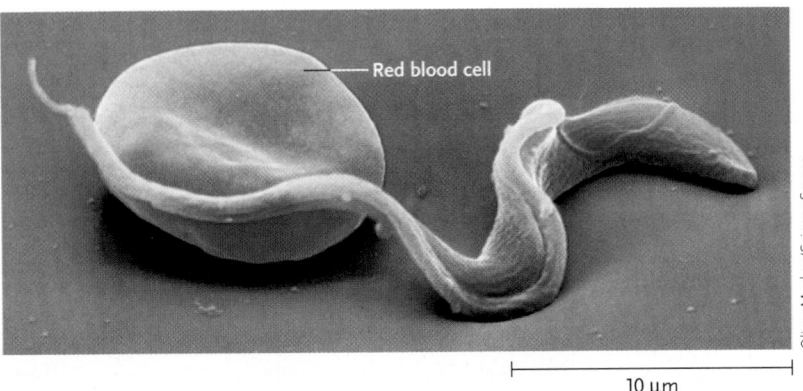

Red blood cell

10 μm

Oliver Meckes/Science Source

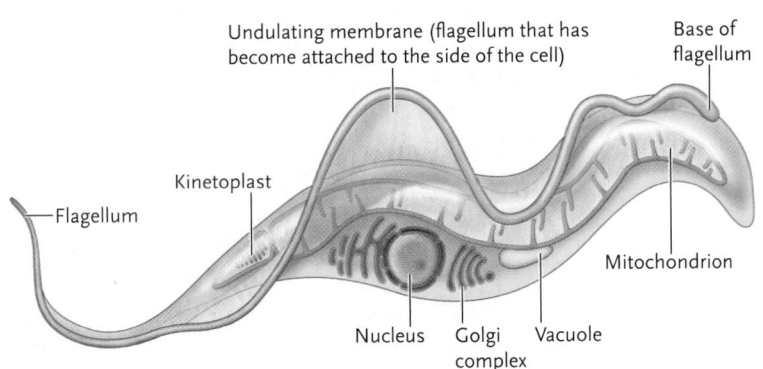

Undulating membrane (flagellum that has become attached to the side of the cell)

Base of flagellum

Kinetoplast

Flagellum

Mitochondrion

Nucleus   Golgi   Vacuole
           complex

Figure 24.8

Trypanosoma brucei, the parasitic kinetoplastid that causes African sleeping sickness.

group did have mitochondria. The nuclei of Excavata that lack mitochondria contain genes derived from mitochondria, and they also have organelles that likely evolved from mitochondria. These Excavata may have lost their mitochondria as an adaptation to the parasitic way of life, in which oxygen is in short supply.

Diplomonads. Diplomonad means *double cell,* and these organisms do look like two cells together (see the figure at the beginning of the chapter), with their two apparently identical, functional nuclei and multiple flagella arranged symmetrically around the cell's longitudinal axis. The best-known diplomonad is *Giardia lamblia,* profiled at the beginning of this chapter. Some are free living, but many live in animal intestines; some diplomonads do not cause harm to the host, whereas others, like *Giardia,* live as parasites.

Like many Excavata, Diplomonads and Parabasalids are single-celled animal parasites that lack mitochondria and move by means of flagella. Because they lack mitochondria, these organisms are limited to producing ATP via glycolysis (see Chapter 6). Originally, the lack of mitochondria in many Excavata led biologists to consider this group as the most ancient line of protists; however, it now appears that the ancestor of this

Parabasalids. The sexually transmitted disease trichomoniasis is caused by the parabasalid *Trichomonas vaginalis* (**Figure 24.9a**). The infection is usually symptomless in men, but in women, *T. vaginalis* can cause severe inflammation and irritation of the vagina and vulva. If untreated, trichomoniasis can cause infection of the uterus and fallopian tubes that can result in infertility. Luckily, drugs can easily cure the infection.

a. *Trichomonas vaginalis*

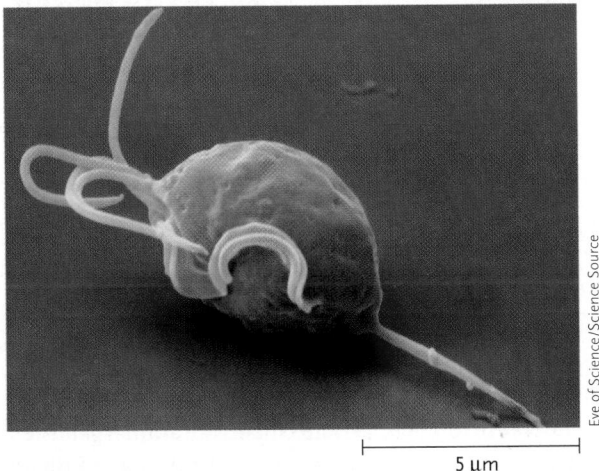

5 μm

Eye of Science/Science Source

b. *Trichonympha*

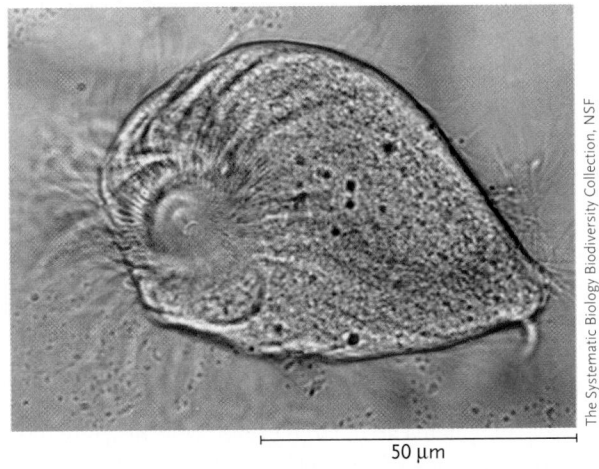

50 μm

The Systematic Biology Biodiversity Collection, NSF

Figure 24.9

Examples of parabasalids (Excavata). **(a)** A parabasalid, *Trichomonas vaginalis,* that causes a sexually transmitted disease, trichomoniasis. **(b)** *Trichonympha,* a parabasalid that lives in the guts of termites.

Parabasalids take their names from cytoplasmic structures associated with the nucleus, *parabasal bodies;* some biologists consider these structures to be the Golgi apparatus of these cells. Parabasalids are also characterized by a sort of fin called an **undulating membrane,** formed by a flagellum buried in a fold of the cytoplasm, in addition to freely beating flagella. The buried flagellum allows parabasalids to move through thick, viscous fluids, such as those lining human reproductive tracts.

Other parabasalids (e.g., *Trichonympha;* Figure 24.9b) are symbionts that live in the guts of termites and other wood-eating insects, digesting the cellulose in the wood for their hosts. As if this endosymbiotic relationship were not complex enough, biologists recently discovered that the protists themselves cannot produce the enzymes necessary to break down cellulose but instead rely on bacterial symbionts to do it.

24.4b Chromalveolates Have Complex Cytoplasmic Structures and Move via Flagella or Cilia

This group is named for the small, membrane-bound vesicles called *alveoli* (*alvus* = belly) in a layer just under the plasma membrane. The Chromalveolate supergroup includes two motile, free-living lineages as well as a motile parasitic group. We will take a closer look at some representative lineages over the next few pages.

Ciliates. This group of protists has helped us understand key aspects of eukaryotic cells, such as the existence of telomeres at the ends of eukaryotic chromosomes and the function of telomerase. These protists are examples of model organisms—organisms that are easily manipulated and easily raised in the lab and for which we have abundant data, for example, genome sequences (see *The Purple Pages*). Several protists are ideal model organisms because, even though they are single celled, the complexity of their structures and functions is comparable to that of humans and other animals. One ciliate, *Tetrahymena* **(Figure 24.10),** was the organism in which telomeres and telomerase were discovered; it was also the cell in which the first motor protein was identified, cell cycle control mechanisms were first described, and ribozymes were discovered. The involvement of ciliates with scientific research dates back several centuries—they were among the first organisms observed in the seventeenth century by the pioneering microscopist Anton van Leeuwenhoek.

The ciliates are a large group, with nearly 10 000 known species of primarily single-celled but highly complex heterotrophic organisms that swim by means of cilia (see Figures 24.4, p. 516, and 24.10). Any sample of pond water or bottom mud contains a wealth of

a. Ciliate

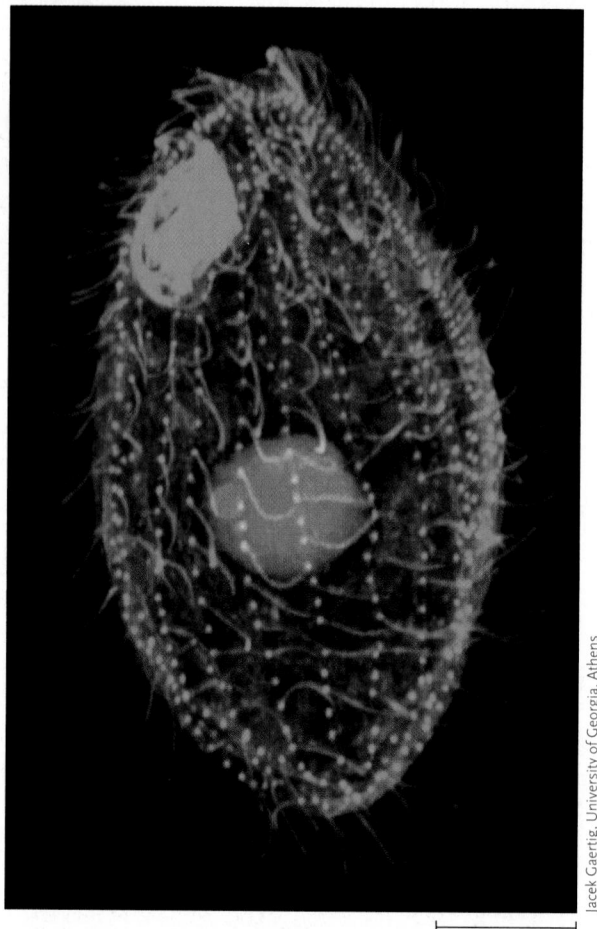

10 μm

Jacek Gaertig, University of Georgia, Athens

b. Cilia

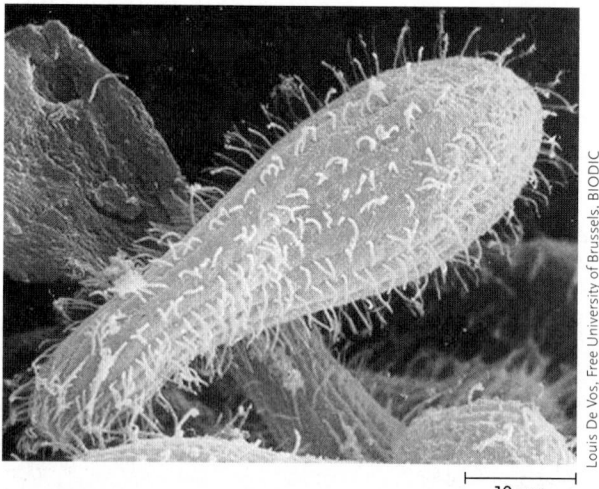

10 μm

Louis De Vos, Free University of Brussels, BIODIC

Figure 24.10

Tetrahymena, a ciliate: **(a)** stained with fluorescent dye to show cilia and microtubules; **(b)** SEM image showing cilia.

these creatures. Some ciliates live individually, whereas others are colonial. Certain ciliates are animal parasites; others live and reproduce in their hosts as mutually beneficial symbionts. A compartment of the stomach of cattle and other grazing animals contains large numbers of symbiotic ciliates that digest the cellulose in their hosts' plant diet. The host animals then digest the excess ciliates.

Ciliates have many highly developed organelles, including a mouthlike gullet lined with cilia, structures that exude toxins and other defensive materials from the cell surface, contractile vacuoles, and a complex system of food vacuoles. A pellicle reinforces the cell's shape. A complex cytoskeleton anchors the cilia just below the pellicle and coordinates the ciliary beating. The cilia can stop and reverse their beating in synchrony, allowing ciliates to stop, back up, and turn if they encounter negative stimuli.

Ciliates are the only eukaryotes that have two types of nuclei in each cell: one or more small nuclei called *micronuclei* and a single larger *macronucleus* (see Figure 24.4, p. 516). A **micronucleus** is a diploid nucleus that contains a complete complement of genes. It functions primarily in cellular reproduction, which may be asexual or sexual. The number of micronuclei present depends on the species. The **macronucleus** develops from a micronucleus but loses all genes except those required for basic functions (e.g., feeding, metabolism) of the cell and for synthesis of ribosomal RNA. The macronucleus contains numerous copies of these genes, allowing it to synthesize large quantities of proteins and rRNA.

Ciliates abound in freshwater and marine habitats, where they feed voraciously on bacteria, algae, and each other. *Paramecium* and *Tetrahymena* are typical of the group (see Figures 24.4, p. 516 and 24.10). Their rows of cilia drive them through their watery habitat, rotating the cell on its long axis while it moves forward or back and turns. The cilia also sweep water laden with prey and food particles into the gullet, where food vacuoles form. The ciliate digests food in the vacuoles and eliminates indigestible material through an anal pore. Contractile vacuoles with elaborate, raylike extensions remove excess water from the cytoplasm and expel it to the outside. When under attack or otherwise stressed, *Paramecium* discharges many dartlike protein threads from surface organelles called **trichocysts**.

Dinoflagellates. In spring and summer, the coastal waters of Canada sometimes turn reddish in colour **(Figure 24.11a)**. These **red tides** are caused by a population explosion, or *bloom,* of certain dinoflagellates that make up a large proportion of marine phytoplankton. These protists typically have a shell formed from cellulose plates (Figure 24.11b). The beating of flagella, which fit into grooves in the plates, makes dinoflagellates spin like a top (*dinos* = spinning) as they swim.

Red tides are caused by conditions such as increased nutrient runoff into coastal waters (particularly from farms and industrial areas), warm ocean surface temperatures, and calm water. Red tides occur in the waters of many other countries besides Canada and are more common in warmer waters. Some red tide

dinoflagellates produce a toxin that interferes with nerve function in animals that ingest them (see "Molecule behind Biology" 24.1).

More than 4000 dinoflagellate species are known, and most, like those that cause red tides, are single-celled organisms in marine phytoplankton. Their abundance in phytoplankton makes dinoflagellates a major primary producer of ocean ecosystems. You can sometimes see their abundance because some are **bioluminescent,** that is, they glow or release a flash of light, particularly when disturbed. Dinoflagellate luminescence can make the sea glow in the wake of a boat at night and coat nocturnal surfers and swimmers with a ghostly light **(Figure 24.12).** Why do these organisms emit light? One explanation is that this burst of light would be likely to scare off predators. The production of light depends on the enzyme *luciferase* and its substrate *luciferin* in forms similar to the system that produces light in fireflies.

Dinoflagellates live as heterotrophs or autotrophs; many can carry out both modes of nutrition. Some dinoflagellates live as symbionts in the tissues of other marine organisms, such as jellyfish, sea anemones, corals, and molluscs, and give these organisms their distinctive colours. Dinoflagellates in coral use the coral's carbon dioxide and nitrogenous waste while supplying 90% of the coral's carbon. The vast numbers of dinoflagellates living as photosynthetic symbionts in tropical coral reefs allow the

a.   b.

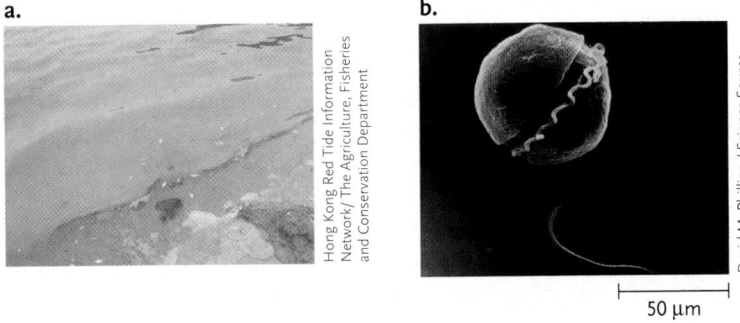

50 μm

Figure 24.11
(a) Red tide caused by dinoflagellate bloom. **(b)** *Karenia brevis*, a toxin-producing dinoflagellate.

Figure 24.12
Bioluminescent dinoflagellates (*Lingulodinium polyedrum*) lighting a breaking wave at midnight.

Saxitoxin

Some dinoflagellates that cause red tides also produce neurotoxins. Fish that feed on the dinoflagellates and birds that feed on the fish may be killed in huge numbers by the toxins. Dinoflagellate toxins do not noticeably affect clams, oysters, and other molluscs but become concentrated in their tissues. Eating the tainted molluscs can cause paralytic shellfish poisoning in humans and other animals, characterized by nausea, vomiting, shortness of breath, and a choking feeling. The main toxin responsible is saxitoxin **(Figure 1),** a neurotoxic alkaloid that is the most lethal nonprotein toxin known—a dose of just 0.2 mg is

Figure 1

Saxitoxin, one of the neurotoxins produced by dinoflagellates.

enough to kill an average-weight person. Saxitoxin acts by binding to sodium channels of nerve cells, thus preventing the normal movement of sodium ions through the channel and blocking the transmission of nerve impulses.

Saxitoxin is especially deadly for mammals because it paralyzes the diaphragm and other muscles required for breathing. There is no cure, and death can occur within minutes if the person is not treated quickly; treatment involves artificial respiration to support breathing. Saxitoxin has been experimented with as a chemical weapon but also has more constructive uses. For example, it has been used to determine the components of sodium channels in cell membranes and in studies of various nerve disorders.

Other photosynthetic protists also produce blooms, and some of these also produce toxins.

---

reefs to reach massive sizes; without dinoflagellates, many coral species would die. When stressed, corals eject their endosymbionts, a phenomenon known as coral bleaching because the absence of the pigmented dinoflagellates allows the coral's calcareous skeleton to be visible **(Figure 24.13).** What causes the coral to become stressed? Increased water temperatures appear to be the main cause, although exposure to contaminants such as oil can also cause bleaching. If the stress causing the bleaching is transient, the coral usually regains its endosymbionts, but if the stress persists, the coral will die. The severity and spatial extent of coral bleaching has been increasing over the past few decades such that it is now a global problem. In 1998, a serious bleaching event destroyed 16% of the world's reefs. Localized high ocean temperatures

in the Caribbean in 2005 resulted in more than 80% of corals bleaching, with more than 40% of these being killed.

Apicomplexans. Apicomplexans are nonmotile parasites of animals. They take their name from the *apical complex,* a group of organelles at one end of a cell, which helps the cell attach to and invade host cells. Apicomplexans absorb nutrients through their plasma membranes (rather than by engulfing food particles) and lack food vacuoles. One genus, *Plasmodium,* is responsible for malaria, one of the most widespread and debilitating human diseases. About 500 million people are infected with malaria in tropical regions, including Africa, India, Southeast Asia, the Middle East, Oceania, and Central and South America. In 2012, malaria killed an estimated 627 000 people, about half as many as were killed by AIDS that year. It is particularly deadly for children younger than six. In many countries where malaria is common, people are often infected repeatedly, with new infections occurring alongside preexisting infections.

Plasmodium is transmitted by 60 different species of mosquitoes, all members of the genus *Anopheles.* Infective cells develop inside the female mosquito, which transfers the cells to human or bird hosts **(Figure 24.14).** The infecting parasites divide repeatedly by asexual reproduction in their hosts, initially in liver cells and then in red blood cells. Their growth causes red blood cells to rupture in regular cycles every 48 or 72 hours, depending on the *Plasmodium* species. The ruptured red blood cells clog vessels and release the parasite's metabolic wastes, causing cycles of chills and fever.

Figure 24.13

Bleached elkhorn coral (*Acropora palmata*).

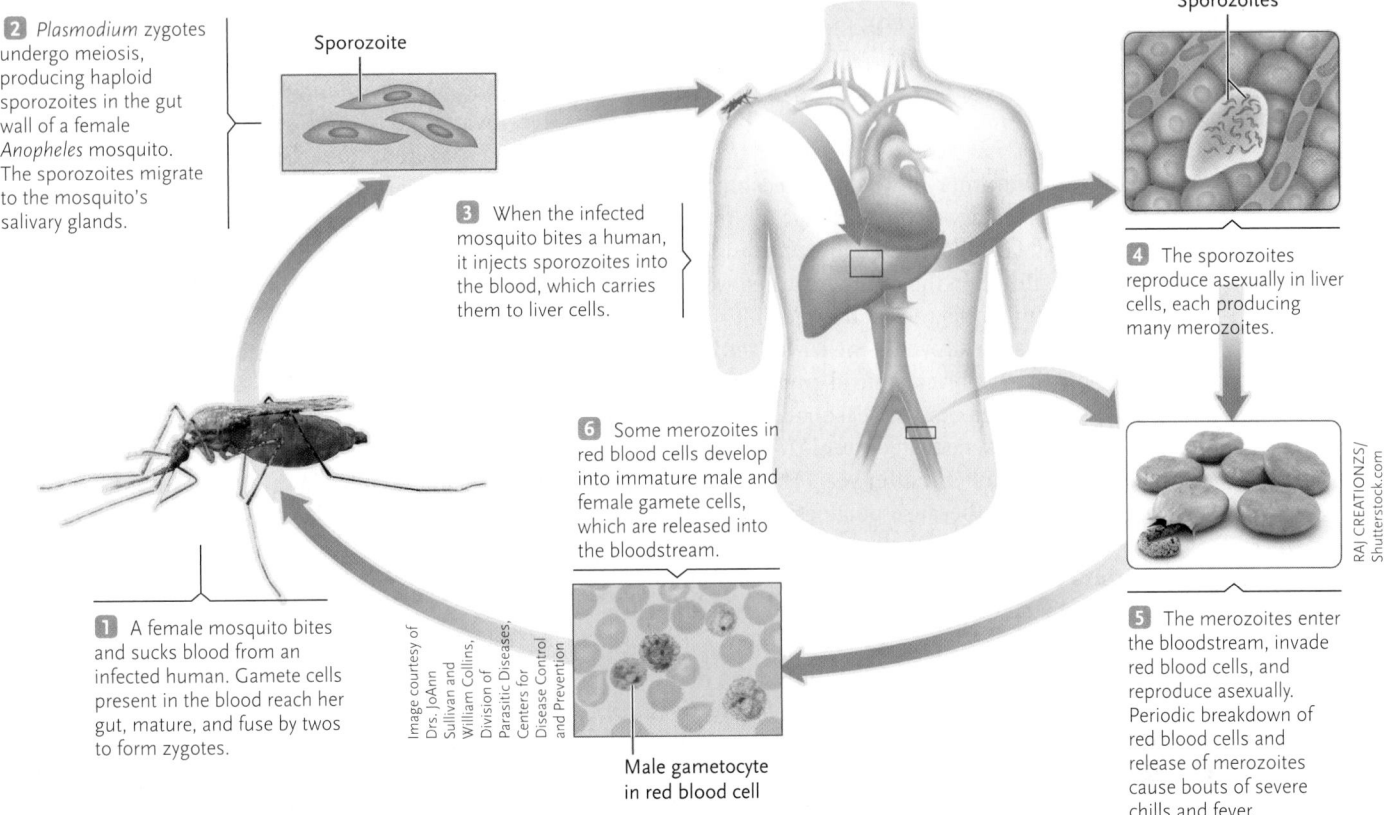

② *Plasmodium* zygotes undergo meiosis, producing haploid sporozoites in the gut wall of a female *Anopheles* mosquito. The sporozoites migrate to the mosquito's salivary glands.

Sporozoite

③ When the infected mosquito bites a human, it injects sporozoites into the blood, which carries them to liver cells.

Sporozoites

④ The sporozoites reproduce asexually in liver cells, each producing many merozoites.

RAJ CREATIONZS/ Shutterstock.com

⑥ Some merozoites in red blood cells develop into immature male and female gamete cells, which are released into the bloodstream.

⑤ The merozoites enter the bloodstream, invade red blood cells, and reproduce asexually. Periodic breakdown of red blood cells and release of merozoites cause bouts of severe chills and fever.

① A female mosquito bites and sucks blood from an infected human. Gamete cells present in the blood reach her gut, mature, and fuse by twos to form zygotes.

Image courtesy of Drs. JoAnn Sullivan and William Collins, Division of Parasitic Diseases, Centers for Disease Control and Prevention

Male gametocyte in red blood cell

Figure 24.14
Life cycle of a *Plasmodium* species that causes malaria.

The victim's immune system is ineffective because during most of the infective cycle, the parasite is inside body cells and thus "hidden" from antibodies. Furthermore, like *Giardia*, *Plasmodium* regularly changes its surface molecules, continuously producing new forms that are not recognized by antibodies developed against a previous form. In this way, the parasite keeps one step ahead of the immune system, often making malarial infections essentially permanent. For a time, malaria was controlled in many countries by insecticides such as DDT. However, the mosquitoes developed resistance to the insecticides and have returned in even greater numbers than before the spraying began.

In addition to the asexual reproduction described above for *Plasmodium*, apicomplexans also reproduce sexually, forming gametes that fuse to produce cysts. As in *Giardia*, when a host organism ingests the cysts, they divide to produce infective cells. Many apicomplexans use more than one host species for different stages of their life cycle. For example, another organism in this group, *Toxoplasma*, has the sexual phase of its life cycle in cats and the asexual phases in humans, cattle, pigs, and other animals. Feces of infected cats contain cysts; humans ingesting or inhaling the cysts develop toxoplasmosis, a disease that is usually mild in adults but can cause severe brain damage or even death to a fetus. Because of the danger of toxoplasmosis,

pregnant women should avoid emptying litter boxes or otherwise cleaning up after a cat.

The groups of Chromalveolates discussed below all share a distinctive arrangement of flagella at some stage of their life cycles. As indicated in Figure 24.15, motile cells in these organisms have two different flagella: one smooth and a second covered with bristles, giving it a "hairy" appearance (**Figure 24.15**). In many of these chromalveolates, the flagella occur only on reproductive cells such as eggs and sperm. This group of Chromalveoloates includes the oomycetes (water moulds), Diatoms, Golden algae, and Brown algae.

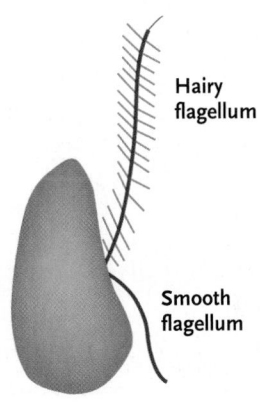

Hairy flagellum

Smooth flagellum

Figure 24.15
Stramenopile protist, with "smooth" and "hairy" flagella.

Recall that algae is a general term for photosynthetic protists, but the different groups of algae are not closely related to each other, so the term does not imply a phylogenetic grouping.

Oomycetes: Water Moulds and Downy Mildews. In Ireland, the summer of 1846 started off warm and sunny. This was a welcome change, as the previous summer had been cool and damp, causing the potato crop to fail. But then the weather turned wet and cold again and within one week at the end of July, the entire potato crop was destroyed—the leaves rotting and the tubers turning to black, putrid mush (Figure 24.16). Worse was to come: the unseasonably cool and damp growing seasons persisted until 1860, causing the potato crops to fail year after year. These crop failures were catastrophic because potatoes were virtually the only food source for most people. Altogether, about one-third of the Irish population died or emigrated (to Canada and the United States, among other countries) due to the potato famines.

In 1861, the organism that caused the blight was identified as a water mould, *Phytophthora infestans*. Originally thought to be a fungus, *P. infestans* produces infective cells that are easily dispersed by wind and water. The blight caused by this organism has recently re-emerged as a serious disease in potato-growing regions of Canada and the United States due to the migration of new strains from Mexico that are resistant to existing pesticides.

Water moulds are not fungi; they are oomycetes (Figure 24.17a), but they do share some features with fungi. Like fungi, oomycetes grow as microscopic, nonmotile filaments called **hyphae** (singular, *hypha*), forming a network called a **mycelium** (Figure 24.17b). Also like fungi, they are heterotrophs, which secrete enzymes that digest the complex molecules of surrounding organic matter or living tissue into simpler molecules that are small enough to be absorbed into their cells. Other features, however, set the Oomycota apart from the fungi; chief among them are differences in nucleotide sequence, which clearly indicate close evolutionary relationships to other heterokonts rather than to the fungi.

The water moulds live almost exclusively in freshwater lakes and streams or moist terrestrial habitats, where they are key decomposers. Dead animal or plant material immersed in water commonly becomes coated with cottony water moulds. Other water moulds, such as the mould growing on the fish shown in Figure 24.17b, parasitize living aquatic animals. The downy mildews are parasites of land plants (Figure 24.17c). Oomycetes may reproduce asexually or sexually.

Diatoms. The organisms shown in **Figure 24.18** may not look like living organisms at all but instead like artwork or jewels. These are **diatoms**, single-celled

a. Water mould

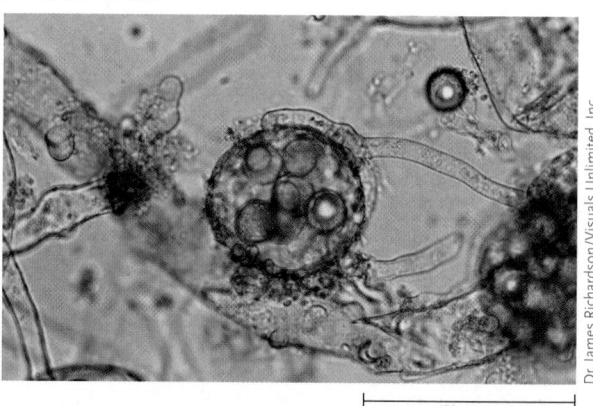

50 μm

Dr. James Richardson/Visuals Unlimited, Inc.

b. Water mould infecting fish

Bill Banaszewski/Visuals Unlimited, Inc.

c. Downy mildew

alybaba/Shutterstock.com

Figure 24.17

Oomycetes. (a) The water mould *Saprolegnia parasitica*. **(b)** *S. parasitica* growing as cottony white fibres on the tail of an aquarium fish. **(c)** Downy mildew, *Plasmopara viticola*, growing on grapes. At times, it has nearly destroyed vineyards in Europe and North America.

William M. Brown Jr., Bugwood.org

Figure 24.16

Blight caused by *Phytophthora infestans* in a potato crop.

Figure 24.18

Diatoms. Depending on the species, the shells are either radially or bilaterally symmetrical, as seen in this sample.

organisms with a glassy silica shell, which is intricately formed and beautiful in many species. The two halves of the shell fit together like the top and bottom of a Petri dish or box of chocolates (see Figure 24.18). Substances move in and out of the cell through elaborately patterned perforations in the shell. Diatom shells are common in fossil deposits. In fact, more diatoms are known as fossils than as living species—some 35 thousand extinct species have been described compared with 7000 living species. For about 180 million years, diatom shells have been accumulating into thick layers of sediment at the bottom of lakes and seas.

In fact, you probably use diatoms—or their remnants—a couple of times a day when you brush your teeth. Most toothpaste contains a mild abrasive to assist in removing plaque, a bacterial biofilm that forms on your teeth. This abrasive is commonly made from grinding the fossilized shells of diatoms into a fine powder, called *diatomaceous earth*. In addition to toothpaste, diatomaceous earth is used in filters, as an insulating material, and as a pesticide. Diatomaceous earth kills crawling insects and insect larvae by abrading their exoskeleton, causing them to dehydrate and die. Insects also die when they eat the powder, but larger animals, including humans, are unaffected by it.

Diatoms are photoautotrophs that carry out photosynthesis by pathways similar to those of plants. They are among the primary photosynthetic organisms in marine plankton and are also abundant in freshwater habitats as both phytoplankton and bottom-dwelling species. Although most diatoms are free living, some are symbionts inside other marine protists. One diatom, *Pseudonitzschia,* produces a toxic amino acid that can accumulate in shellfish. The amino acid, which acts as a nerve poison, causes amnesic shellfish poisoning when ingested by humans; the poisoning can be fatal.

Asexual reproduction in diatoms occurs by mitosis followed by a form of cytoplasmic division in which each daughter cell receives either the top or the bottom half of the parent shell. The daughter cell then secretes the missing half, which becomes the smaller, inside shell of the box. The daughter cell receiving the larger top half grows to the same size as the parent shell, but the cell receiving the smaller bottom half is limited to the size of this shell. As asexual divisions continue, the cells receiving bottom halves become progressively smaller. When a minimum size is reached, sexual reproduction is triggered. The cells produce flagellated gametes, which fuse to form a zygote. The zygote grows to normal size before secreting a completely new shell with full-sized top and bottom halves.

Although flagella are present only in gametes, many diatoms move by an unusual mechanism in which a secretion released through grooves in the shell propels them in a gliding motion.

Golden Algae. Nearly all golden algae are autotrophs and carry out photosynthesis using pathways similar to those of plants. Their colour is due to a brownish carotenoid pigment, fucoxanthin, which masks the green colour of the chlorophylls **(Figure 24.19a, p. 526).** However, most of these organisms can also live as heterotrophs if there is insufficient light for photosynthesis. They switch to feeding on dissolved organic molecules or preying on bacteria and diatoms. Golden algae are important in freshwater habitats and in *nanoplankton,* a community of marine phytoplankton composed of huge numbers of extremely small cells. During the spring and fall, blooms of golden algae are responsible for the fishy taste of many cities' drinking water.

Most golden algae are colonial forms (see Figures 24.3, p. 516, and 24.19a) in which each cell of the colony bears a pair of flagella. The golden algae have glassy shells, but in the form of plates or scales rather than in the Petri dish form of the diatoms.

Brown Algae. If you were asked where in Canada you'd find forests of giant trees, you'd likely think of the **temperate rain forests** in British Columbia. But there are also vast underwater forests in the waters off the British Columbia coast, formed not by trees but by a type of brown algae known as kelp (*Macrocystis integrifolia*), which can grow to lengths of 30 m. A related species, giant kelp (*M. pyrifera*) (Figure 24.19b–d), can grow up to 60 m long. Kelps are the largest and most complex of all protists. Their tissues are differentiated into leaflike *blades,* stalklike *stipes,* and rootlike *holdfasts* that anchor them to the bottom. Hollow, gas-filled bladders give buoyancy to the stipes and blades and help keep them upright and oriented toward the sunlit upper layers of water (Figure 24.19b). The stipes of some kelps contain tubelike vessels, similar to the vascular elements of plants, which rapidly distribute the products of photosynthesis throughout the body of the alga.

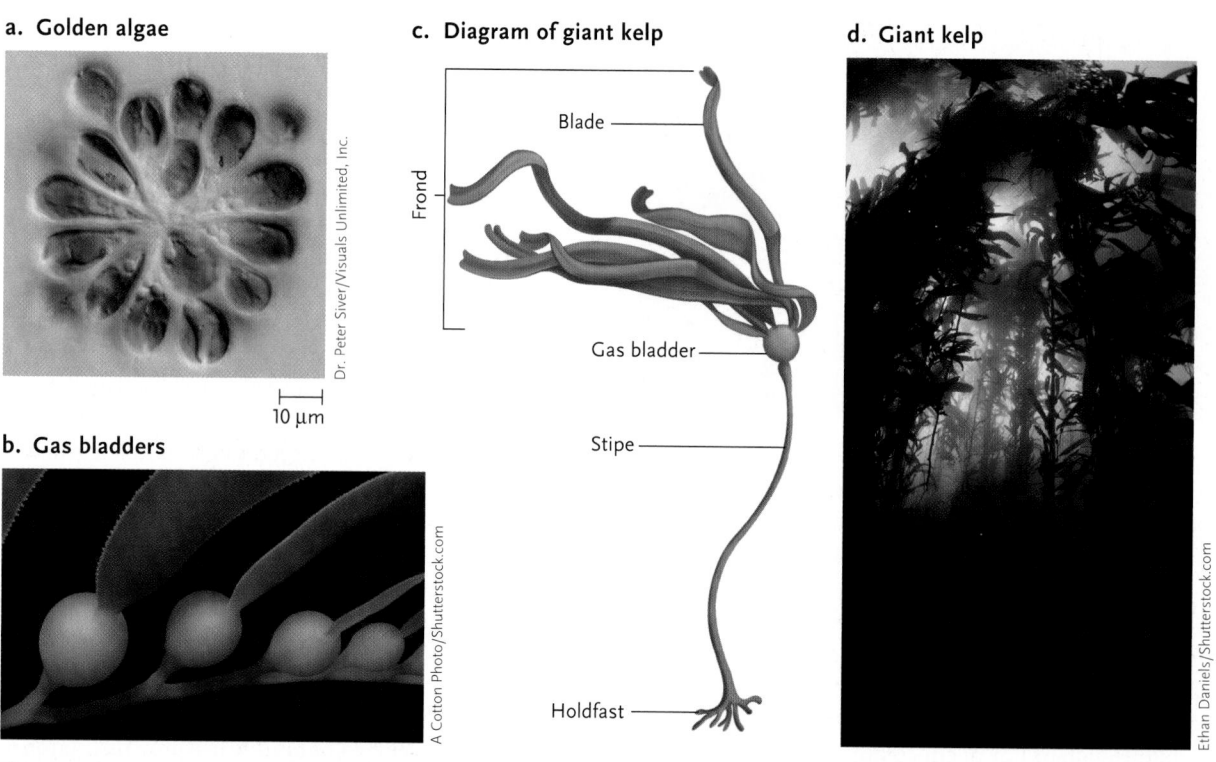

a. Golden algae

Dr. Peter Siver/Visuals Unlimited, Inc.

10 μm

b. Gas bladders

A Cotton Photo/Shutterstock.com

c. Diagram of giant kelp

Frond

Blade

Gas bladder

Stipe

Holdfast

d. Giant kelp

Ethan Daniels/Shutterstock.com

Figure 24.19

Golden and brown algae. **(a)** A microscopic swimming colony of Synura. Each cell bears two flagella, which are not visible in this light micrograph. **(b)** Gas bladders connect kelp's stipes ("stems") to its blades ("leaves"). **(c)** The fronds of giant kelp are borne on stalks known as stipes, which are anchored to the substrate by holdfasts. **(d)** A forest of *Macrocystis pyrifera* (giant kelp).

Kelps have an astonishingly fast growth rate—giant kelp can grow up to 30 cm per day!

Just as for terrestrial forests, kelp forests provide food and habitat for many marine organisms. Herds of sea otters (*Enhydra lutris*), for example, tend to live in and near kelp forests. When sea otters sleep at sea, they wrap kelp around themselves to keep from drifting away **(Figure 24.20)**. Although the forest is an important habitat for the sea otters, the otters, in turn, are critical for the survival of these forests. Sea otters are one of the few predators of sea urchins, which graze on the kelp and can cause deforestation if their populations get very large. Predation by sea otters keeps sea urchin populations in control, preventing destruction of kelp forests.

All brown algae are photoautotrophs, but not all are as large as kelps. Nearly all of the 1500 known species inhabit temperate or cool coastal marine waters. Like golden algae, brown algae contain fucoxanthin, which gives them their characteristic colour. Their cell walls contain cellulose and a mucilaginous polysaccharide, alginic acid. This alginic acid, called **algin** when extracted, is an essentially tasteless substance used to thicken such diverse products as ice cream, salad dressing, jellybeans, cosmetics, and floor polish. Brown algae are also harvested as food crops and fertilizer.

Life cycles among the brown algae are typically complex and in many species consist of alternating haploid and diploid generations **(Figure 24.21)**. The large structures that we recognize as kelps and other brown seaweeds are diploid **sporophytes**, so called because they give rise to haploid spores by meiosis. The spores, which are flagellated swimming cells, germinate and divide by mitosis to form an independent, haploid **gametophyte** generation. The gametophytes give rise to haploid gametes, the egg and sperm cells. Most brown algal gametophytes are multicellular structures only a few centimetres in diameter. Cells in the gametophyte, produced by mitosis, differentiate to form nonmotile eggs or flagellated, swimming sperm cells. The sperm cells have the two different types of

worldswildlifewonders/Shutterstock.com

Figure 24.20

A sea otter (*Enhydra lutris*) wrapped in kelp.

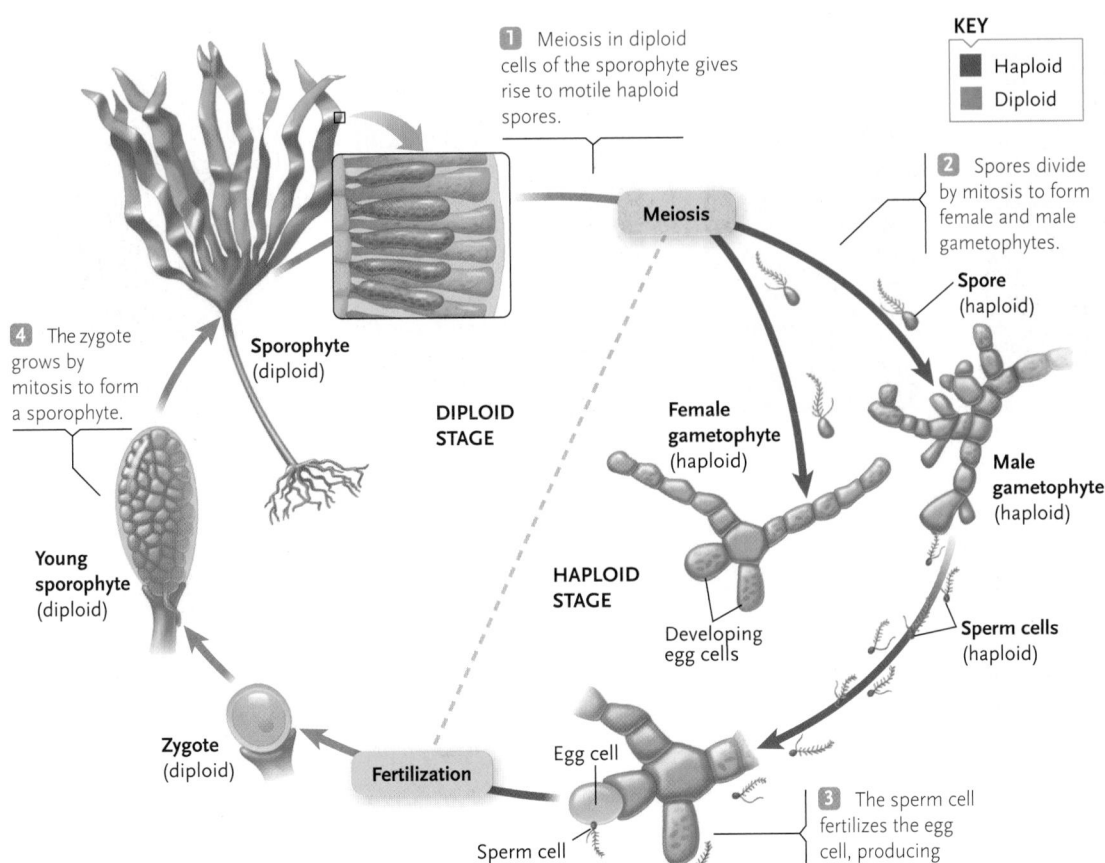

1 Meiosis in diploid cells of the sporophyte gives rise to motile haploid spores.

KEY

| | Haploid |
| | Diploid |

2 Spores divide by mitosis to form female and male gametophytes.

Meiosis

Spore (haploid)

Sporophyte (diploid)

DIPLOID STAGE

Female gametophyte (haploid)

Male gametophyte (haploid)

4 The zygote grows by mitosis to form a sporophyte.

HAPLOID STAGE

Young sporophyte (diploid)

Developing egg cells

Sperm cells (haploid)

Zygote (diploid)

Egg cell

Fertilization

Sperm cell

3 The sperm cell fertilizes the egg cell, producing a diploid zygote.

flagella characteristic of the heterokont protists. Fusion of egg and sperm produces a diploid zygote that grows by mitotic divisions into the sporophyte generation. This complex life cycle is very similar to that of land plants (see Chapter 26).

24.4c Rhizara Are Eukaryotes with Filamentous Pseudopods

Amoeba (*amoibe* = change) is a descriptive term for a single-celled protist that moves by means of pseudopodia, as described earlier in this chapter (see Figure 24.6, p. 517). Several major groups of protists contain amoebas, which are similar in form but are not all closely related. Amoebas in the Rhizaria produce stiff, filamentous pseudopodia, and many produce hard outer shells, also called *tests*. We consider here two heterotrophic groups of amoebas, the Radiolaria and the Foraminifera, and a third, photosynthesizing group, the Chlorarachniophyta.

Radiolaria. Radiolarians (*radiolus* = small sunbeam) are marine organisms characterized by a glassy internal skeleton and **axopods,** slender, raylike strands of cytoplasm supported internally by long bundles of microtubules **(Figure 24.22a, b, p. 528).** This glassy skeleton is heavy—when radiolarians die, their skeletons sink to the ocean floor—so how do radiolarians

keep afloat? The axopods provide buoyancy, as do the numerous vacuoles and lipid droplets in the cytoplasm. Axopods are also involved in feeding: prey stick to the axopods and are then engulfed, brought into the cell, and digested in food vacuoles.

Radiolarian skeletons that accumulate on the ocean floor become part of the sediment, which, over time, hardens into sedimentary rock. The presence of radiolarians in such rocks is very useful to the oil industry as indicators of oil-bearing strata.

Foraminifera: Forams. These organisms take their name from the perforations in their shells (*foramen* = little hole), through which extend long, slender strands of cytoplasm supported internally by a network of needlelike spines. Their shells consist of organic matter reinforced by calcium carbonate (Figure 24.22c–e). Most foram shells are chambered, spiral structures that, although microscopic, resemble those of molluscs.

Like radiolarians, forams live in marine environments. Some species are planktonic, but they are most abundant on sandy bottoms and attached to rocks along the coasts. Forams feed in a manner similar to that of radiolarians: they engulf prey that adhere to the strands and conduct them through the holes in the shell into the central cytoplasm, where they are digested in food vacuoles. Some forams have algal symbionts that carry out

a. Radiolarian

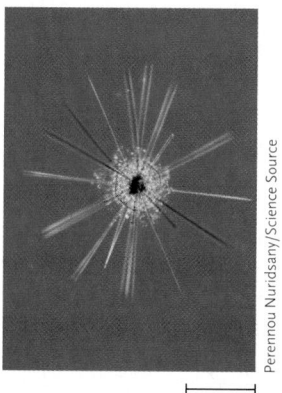

Perennou Nuridsany/Science Source

|← 10 μm →|

b. Radiolarian skeleton

Eye of Science / Science Source

c. Living foram

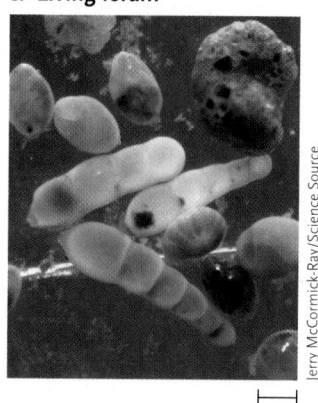

Jerry McCormick-Ray/Science Source

|← 10 μm →|

d. Foram shells

Eric V. Grave/Science Source

e. Foram body plan

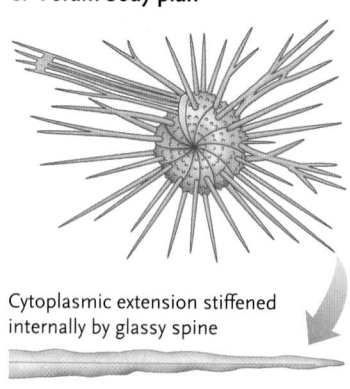

Redrawn from V. & J. Pearse and M. & R. Buchsbaum, *Living Invertebrates*, The Boxwood Press, 1987.

Cytoplasmic extension stiffened internally by glassy spine

Figure 24.22

(a) A living radiolarian. **(b)** The internal skeletons of a radiolarian. Bundles of microtubules support the cytoplasmic extensions of the radiolarian. **(c)** A living foram, showing the cytoplasmic strands extending from its shelf. **(d)** Empty foram shells. **(e)** The body plan of a foram. Needlelike, glassy spines support the cytoplasmic extensions of the forams.

photosynthesis, allowing them to live as both heterotrophs and autotrophs.

Marine sediments are typically packed with the shells of dead forams. The sediments may be hundreds of feet thick: the White Cliffs of Dover in England are composed primarily of the shells of ancient forams. Most of the world's deposits of limestone and marble contain foram shells; the great pyramids of ancient Egypt are built from blocks cut from fossil foram deposits. Because distinct species lived during different geologic periods, they are widely used to establish the age of sedimentary rocks containing their shells. As they do with radiolarian species, oil prospectors use forams as indicators of hydrocarbon deposits because layers of forams often overlie oil.

Chlorarachniophyta. Chlorarachniophytes are amoebas that contain chloroplasts and thus are photosynthetic. However, they combine this mode of nutrition with heterotrophy, engulfing food with the many filamentous pseudopodia that extend from the cell surface.

24.4d The Unikont Supergroup Includes Slime Moulds and Most Amoebas

The Unikonts include most of the amoebas other than those in Rhizaria, as well as the cellular and plasmodial slime moulds. All members of this group use pseudopods for locomotion and feeding for all or part of their life cycles.

Amoebas. Amoebas of the Unikonts are single-celled organisms that are abundant in marine and freshwater environments and in the soil. All amoebas are microscopic, although some species can grow to 5 mm in size and so are visible with the naked eye. Some amoebas are parasitic, such as the 45 species that infect the human digestive tract. One of these parasites, *Entamoeba histolytica,* causes amoebic dysentery. Cysts of this amoeba contaminate water supplies and soil in regions with inadequate sewage treatment. When ingested, a cyst breaks open to release an amoeba that feeds and divides rapidly in the digestive tract. Enzymes released by the amoebas destroy cells lining the intestine, producing the ulcerations, painful cramps, and debilitating diarrhea characteristic of the disease. Amoebic dysentery afflicts millions of people worldwide; in less developed countries, it is a leading cause of death among infants and small children.

However, most amoebas are heterotrophs that feed on bacteria, other protists, and bits of organic matter. Unlike the stiff, supported pseudopodia of Rhizaria, pseudopods of amoebas extend and retract at any point on their body surface and are unsupported by any internal cellular organization—amoebas are thus "shape-shifters." How can an amoeba capture a

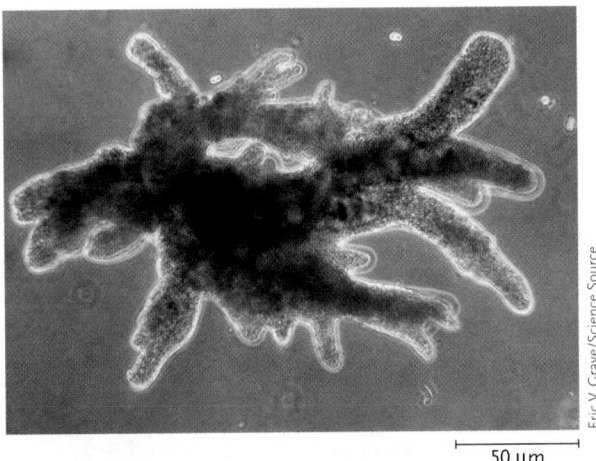

Figure 24.23
An amoeba capturing prey with pseudopods.

fast-moving organism? As an amoeba moves, its cytoplasm doesn't just move but also changes state, from a more liquid state to a more solid state and back again, allowing the amoeba to send out pseudopodia in different directions very quickly. These fast-moving pseudopods can capture even fast-swimming prey such as ciliates **(Figure 24.23)**.

Amoebas reproduce only asexually, via binary fission. In unfavourable environmental conditions, some amoebas can form a cyst, essentially by rolling up and secreting a protective membrane. They survive as cysts until favourable conditions return.

Slime Moulds. After a very wet spring in 1973, residents of Dallas, Texas, were alarmed to see large, yellow blobs that resembled scrambled eggs *crawling* on their lawns. People thought it was an alien invasion. Luckily, a local biologist was able to prevent mass panic by identifying the blobs as slime moulds, unusual heterotrophic protists. Slime moulds exist for part of their lives as individuals that move by amoeboid motion but then come together in a coordinated mass—essentially, a large amoeba—that ultimately differentiates into a stalked structure called a **fruiting body**, in which spores are formed.

There are two major evolutionary lineages of slime moulds: the **cellular slime moulds** and the plasmodial slime moulds, which differ in cellular organization. Both types of slime moulds have been of great interest to scientists because of their ability to differentiate into fruiting bodies with stalks and spore-bearing structures. This differentiation is much simpler than the complex developmental pathways of other eukaryotes, providing a unique opportunity to study cell differentiation at its most fundamental level. Slime moulds also respond to stimuli in their environment, moving away from bright light and toward food. We have learned a great deal about eukaryotic signalling pathways, cell differentiation, and cell movement from studies of slime moulds.

Slime moulds live on moist, rotting plant material such as decaying leaves and bark. The cells engulf particles of dead organic matter, along with bacteria, yeasts, and other microorganisms, and digest them internally. They can be a range of colours: brown, yellow, green, red, and even violet or blue.

These organisms exist primarily as individual cells, either separately or as a coordinated mass. Among the 70 or so species of cellular slime moulds, *Dictyostelium discoideum* is best known. Its life cycle begins when a haploid spore lands in a suitably moist environment containing decaying organic matter **(Figure 24.24, p. 530)**. The spore germinates into an amoeboid cell that grows and divides mitotically into separate haploid cells as long as the food source lasts. When the food supply dwindles, some of the cells release a **chemical signal** in pulses; in response, the amoebas move together and form a sausage-shaped mass that crawls in coordinated fashion like a slug. Some "slugs," although not much more than a millimetre in length, contain more than 100 thousand individual cells. At some point, the "slug" stops moving and differentiates into a stalked fruiting body, with some cells becoming spores, whereas others form the stalk. The cells that form the stalk die in the process, essentially sacrificing themselves so that a stalk can form. Why is formation of a stalk so crucial? Raising the spore-forming cells higher up in the air increases the likelihood that spores will be carried away by air currents and dispersed farther away from the parent. Because the cells forming the "slug" and fruiting body are all products of mitosis, this is asexual reproduction.

Cellular slime moulds also reproduce sexually: two haploid cells fuse to form a diploid zygote (also shown in Figure 24.24) that enters a dormant stage. Eventually, the zygote undergoes meiosis, producing four haploid cells that may multiply inside the spore by mitosis. When conditions are favourable, the spore wall breaks down, releasing the cells. These grow and divide into separate amoeboid cells.

Plasmodial Slime Moulds. **Plasmodial slime moulds** exist primarily as a multinucleate **plasmodium**, in which individual nuclei are suspended in a common cytoplasm surrounded by a single plasma membrane. (This is not to be confused with *Plasmodium*, the genus of apicomplexans that causes malaria.) There are about 500 known species of plasmodial slime moulds. The plasmodium **(Figure 24.25a, b, p. 530)** flows and feeds by phagocytosis like a single huge amoeba—a single cell that contains thousands to millions or even billions of diploid nuclei surrounded by a single plasma membrane. The plasmodium, which may range in size from a few centimetres to more than a metre in diameter, typically moves in thick, branching strands connected by thin sheets. The movements occur by cytoplasmic streaming, driven by actin microfilaments and myosin.

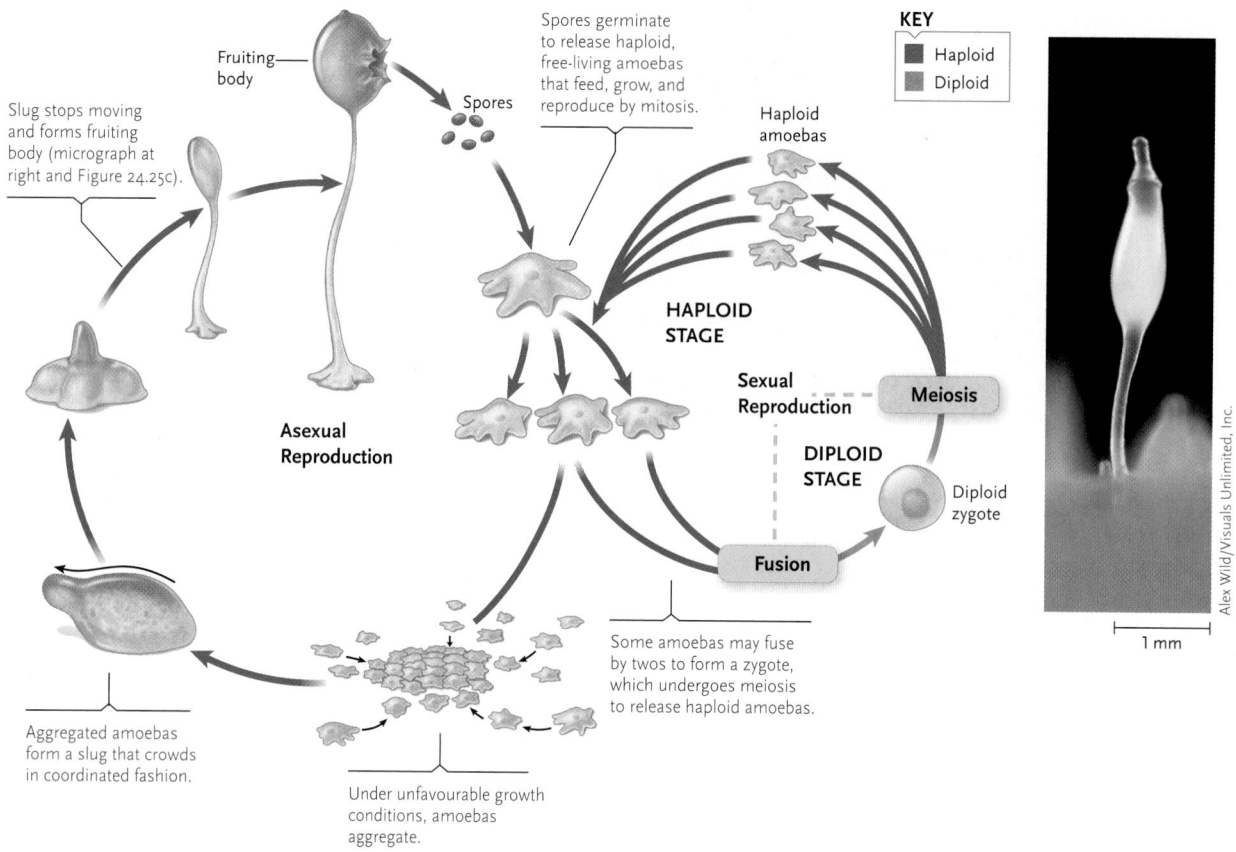

Figure 24.24
Life cycle of the cellular slime mould *Dictyostelium discoideum*. The light micrograph shows a mature fruiting body.

Fruiting body

Slug stops moving and forms fruiting body (micrograph at right and Figure 24.25c).

Spores germinate to release haploid, free-living amoebas that feed, grow, and reproduce by mitosis.

Spores

Haploid amoebas

KEY
- Haploid
- Diploid

HAPLOID STAGE

Sexual Reproduction

Meiosis

Asexual Reproduction

DIPLOID STAGE

Diploid zygote

Fusion

Some amoebas may fuse by twos to form a zygote, which undergoes meiosis to release haploid amoebas.

Aggregated amoebas form a slug that crowds in coordinated fashion.

Under unfavourable growth conditions, amoebas aggregate.

Alex Wild/Visuals Unlimited, Inc.

1 mm

a.

George Barron

b.

George Barron

c.

Greg Thorn

Figure 24.25
Slime moulds. (a and b) Plasmodia of slime moulds. (c) Fruiting bodies of slime mould.

PEOPLE BEHIND BIOLOGY 24.2

Klaus-Peter Zauner, *University of Southampton*; Soichiro Tsuda, *University of Glasgow*; and Yukio-Pegio Gunjia, *Kobe University*

Robots controlled by slime moulds? Far from being a bizarre science fiction story, slime moulds may be the future of robotics, as demonstrated in research carried out by Klaus-Peter Zauner (University of Southampton, United Kingdom) and his collaborators at Kobe University in Japan, Soichiro Tsuda and Yukio-Pegio Gunjia. Zauner and colleagues grew *Physarum polycephalum* in a six-pointed star shape on an electrical circuit and connected the circuit to a six-legged robot with a computer interface. Each point of the *Physarum* plasmodium star corresponded to one leg of the

Klaus-Peter Zauner/New Scientist

Figure 1

A slime mould (image on screen) is able to direct the movement of a robot (in the foreground).

robot **(Figure 1)**. When light was shone on certain parts of the robot, sensors mounted on the robot detected the light and illuminated the corresponding part of the plasmodium, which responded by moving away from the light, sending the robot scuttling into dark corners. One goal of Zauner's research in molecular computing is to incorporate this biological control right into the robot. Harnessing the capacity of living organisms to sense and respond to complex environments would give robots greater autonomy than is possible with control by computer programs.

These plasmodia are what the people in Dallas thought were aliens invading; after a period of heavy rain, plasmodia will sometimes crawl out of the woods to appear on lawns or the mulch of flowerbeds.

At some point, often in response to unfavourable environmental conditions, fruiting bodies form on the plasmodium. At the tips of the fruiting bodies, nuclei become enclosed in separate cells. These cells undergo meiosis, forming haploid, resistant spores that are released from the fruiting bodies and carried by water or wind. If they reach a favourable environment, the spores germinate to form gametes that fuse to form a diploid zygote. The zygote nucleus then divides repeatedly without an accompanying division of the cytoplasm, forming many diploid nuclei suspended in the common cytoplasm of a new plasmodium.

Plasmodial slime moulds are particularly useful in research because they become large enough to provide ample material for biochemical and molecular analyses. Actin and myosin extracted from *Physarum polycephalum,* for example, have been much used in studies of actin-based motility. A further advantage of plasmodial slime moulds is that the many nuclei of a plasmodium usually replicate and pass through mitosis in -synchrony, making them useful in research that tracks the changes that take place in the cell cycle. More recently, slime moulds have been used in robotics research, as outlined in "People behind Biology," 24.2.

Also included in the Unikonts are the choanoflagellates. Opisthokonta (*opistho* = posterior; *kontos* =

flagellum) are named for the single, posterior flagellum found at some stage in the life cycle of these organisms. This diverse group includes the choanoflagellates, protists thought to be the ancestors of fungi and animals.

Choanoflagellata (*choanos* = collar) are named for the collar surrounding the flagellum that the protist uses to feed and, in some species, to swim **(Figure 24.26)**. The collar resembles an upside-down lampshade and is made up of small, fingerlike projections (microvilli) of the plasma membrane. As the flagellum moves water through the collar, these projections engulf bacteria and particles of organic matter in the water.

About 150 species of choanoflagellates live in either marine or freshwater habitats. Some species are mobile, with the flagellum pushing the cells along (in the same way that animal sperm are propelled by their flagella), but most choanoflagellates are *sessile* (attached by a stalk to a surface). A number of species are colonial

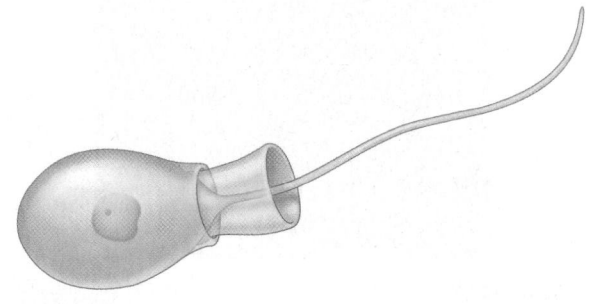

Figure 24.26

A choanoflagellate.

with a cluster of cells on a single stalk; these colonial species are of great interest to biologists studying the evolution of multicellularity in animals.

Why are choanoflagellates thought to be the ancestor of animals? Both molecular and morphological data indicate that a choanoflagellate type of protist gave rise to animals: For example, there are many morphological similarities between choanoflagellates and the collar cells (choanocytes) of sponges as well as the cells that act as excretory organisms in flatworms and rotifers (see Chapter 27). Comparisons of nucleic acid sequences done to date also support the hypothesis that choanoflagellates are the closest living relatives to animals. Molecular data also indicate that a choanoflagellate-like organism was also likely the ancestor of the fungi (see Chapter 25).

24.4e Plantae Include the Red and Green Algae and Land Plants

The Plantae supergroup consists of the red and green algae, which are protists, and the land plant. These three groups of photoautotrophs share a common evolutionary origin. Here we describe the two types of algae; we discuss land plants and how they evolved from green algae in Chapter 26.

Rhodophyta: The Red Algae. Nearly all of the 4000 known species of red algae, which are also known as the Rhodophyta (*rhodon* = rose), are small marine seaweeds **(Figure 24.27)**. Fewer than 200 species are found in freshwater lakes and streams or in soils. If you have had sushi, then you have eaten red algae: *Porphyra* is harvested for use as the *nori* wrapped around fish and rice.

Rhodophyte cell walls contain cellulose and mucilaginous pectins that give red algae a slippery texture. These pectins are widely used in industry and science. Extracted **agar** is used as a culture medium in the laboratory and as a setting agent for jellies and desserts. **Carrageenan** is used to thicken and stabilize paints, dairy products such as ice cream, and many other emulsions.

Some species secrete calcium carbonate into their cell walls; these coralline algae are important in building coral reefs—in some places, they play a bigger role in reef building than do corals.

Red algae are typically multicellular organisms, with diverse morphologies, although many have plantlike bodies composed of stalks bearing leaflike blades. Although most are free-living autotrophs, some are parasites that attach to other algae or plants.

Although most red algae are reddish in colour, some are greenish purple or black. The colour differences are produced by accessory pigments, *phycobilins*, that mask the green colour of their chlorophylls. Phycobilins absorb the shorter wavelengths of light (green and blue-green light) that penetrate to the ocean depths, allowing red algae to grow at deeper levels than any other algae. Some red algae live at depths up to 260 m if the water is clear enough to transmit light to these levels.

Red algae have complex reproductive cycles involving alternation between diploid sporophytes and haploid gametophytes. No flagellated cells occur in the red algae; instead, gametes are released into the water to be brought together by random collisions in currents.

Chlorophyta: The Green Algae. The green algae or Chlorophyta (*chloros* = green) carry out photosynthesis using the same pigments as plants, whereas other photosynthetic protists contain pigment combinations that are very different from those of land plants. This shared pigment composition is one line of evidence that one lineage of green algae was the ancestor of land plants. With at least 16 thousand species, green algae show more diversity than any other algal group. They also have very diverse morphologies, including single-celled, colonial, and multicellular species **(Figure 24.28, p. 534; see also Figure 24.1d, p. 514)**. Multicellular forms have a range of morphologies, including filamentous, tubular, and leaflike forms. Most green algae are microscopic, but some range upward to the size of small seaweeds.

Most green algae live in freshwater aquatic habitats, but some are marine, whereas others live on rocks, soil surfaces, or tree bark, or even in snow. Other organisms rely on green algae to photosynthesize for them by forming symbiotic relationships. For example, lichens are symbioses between green algae and fungi (see Chapter 25), and many animals, such as the sea slugs described in "Life on the Edge" 24.3, contain green algal chloroplasts, or entire green algae, as symbionts in their cells.

Life cycles among the green algae are as diverse as their body forms. Many can reproduce either

Filamentous red alga

© Sabena Jane Blackbird/Alamy

Figure 24.27

Red algae. *Antithamnion plumula*, showing the filamentous and branched body form most common among red algae.

LIFE ON THE EDGE 24.3

Solar-Powered Animals

We associate photosynthesis with plants and some algae, but in a few exceptional cases, animals steal chloroplasts and use them to capture solar energy through photosynthesis. In a sense, these animals are doing the same as the ancestors of the eukaryotic cells that captured photosynthetic prokaryotes and thus acquired the ability to carry out oxygenic photosynthesis (see Chapter 7). There are many examples of sequestration of chloroplasts by species of the animal phyla Platyhelminthes, Porifera, Cnidaria, Mollusca, and Urochordata.

Some sea slugs, such as the *Elysia chlorotica* shown in **Figure 1,** use specialized teeth on their radulae (*radula* = tonguelike organ) to cut into algal cells so that they can suck out chloroplasts and other cell contents. In the slug's stomach, chloroplasts are engulfed by phagocytosis and then moved to areas below the epidermis. Slugs such as E. chlorotica can live for at least five months on the energy generated by these "solar panels"—the chloroplasts they have engulfed. This story is an example of how a chemoheterotrophic organism (the slug) acquires the ability to be photoautotrophic (engage in photosynthesis). How the slugs control the functions of the chloroplasts remains unknown. Have slugs also taken over some of the chloroplast or algal genome? At least one species, *Elysia crispata*, has genes from chloroplasts in its genomic DNA. For more about these slugs, see Chapter 27.

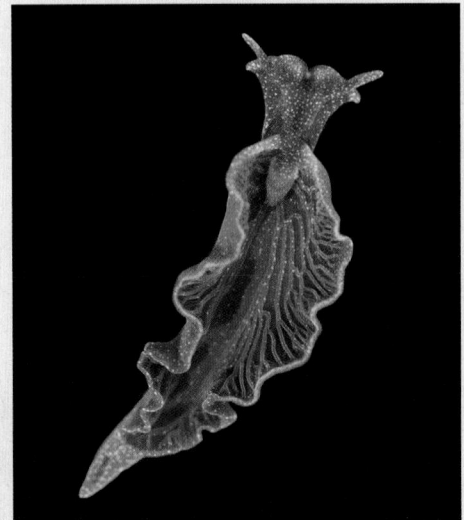

Figure 1

The solar-powered sea slug, Elysia chlorotica, *an animal that extracts chloroplasts from algae and uses them to produce food.*

Such solar-powered symbioses also occur between algae and vertebrate animals: Ryan Kerney, a biologist at Dalhousie University in Halifax, Nova Scotia, recently discovered that green algae (*Oophila amblystomatis*) colonize embryos of spotted salamanders (*Ambystoma gracilis*), becoming part of the animals' bodies. The presence of the algae allows the salamanders to capture solar energy using the chloroplasts in the algal cells. The presence of algae in these salamanders' eggs was known for over 100 years, but Kerney was the first to show that the algae actually invade the developing embryos. Through photosynthesis, the algae provide nutrients and oxygen to the embryos, and, in turn, they feed on the salamanders' waste products. This intimate symbiosis doesn't persist in mature salamanders, which have opaque skin and live mostly underground. Many questions remain to be answered: How do the algae get into the eggs? Why doesn't the salamander's immune system perceive the algae as invaders and destroy them? And are there other solar-powered vertebrates?

---

sexually or asexually, and some alternate between haploid and diploid generations. Gametes in different species may be undifferentiated flagellated cells or differentiated as a flagellated sperm cell and a non-motile egg cell. Most common is a life cycle with a multicellular haploid phase and a single-celled diploid phase **(Figure 24.29, p. 534).**

Among all the algae, the green algae are the most closely related to land plants, based on molecular, biochemical, and morphological data. Evidence of this close relationship includes not only the shared photosynthetic pigments, but also the use of starch as storage reserve and the same cell wall composition.

Which green alga might have been the ancestor of modern land plants? The evidence points to a group known as the **charophytes** as being most similar to the algal ancestors of land plants. This does not mean that modern-day charophytes are the ancestors of land plants but rather that the two groups have a common ancestor. Charophytes, including *Chara* **(Figure 24.30, p. 535),** *Spirogyra, Nitella,* and *Coleochaete,* live in freshwater ponds and lakes. Their ribosomal RNA and chloroplast DNA sequences are more closely related to plant sequences than those of any other green alga. We discuss the evolution of land plants from an algal ancestor more thoroughly in Chapter 26.

a. Single-celled green alga

c. Multicellular green alga

b. Colonial green alga

1 cm

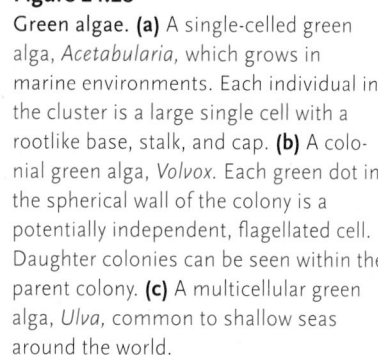

200 μm

Figure 24.28

Green algae. **(a)** A single-celled green alga, *Acetabularia,* which grows in marine environments. Each individual in the cluster is a large single cell with a rootlike base, stalk, and cap. **(b)** A colonial green alga, *Volvox.* Each green dot in the spherical wall of the colony is a potentially independent, flagellated cell. Daughter colonies can be seen within the parent colony. **(c)** A multicellular green alga, *Ulva,* common to shallow seas around the world.

Figure 24.29

The life cycle of the green alga *Ulothrix,* in which the haploid stage is multicellular and the diploid stage is a single cell, the zygote. + and − are morphologically identical mating types ("sexes") of the alga.

Reproductive structures

Dr. John Clayton, National Institute of Water and Atmospheric Research, New Zealand

Figure 24.30

The charophyte *Chara*, representative of a group of green algae that may have given rise to the plant kingdom.

STUDY BREAK

1. For each of the protist groups listed below, indicate the cell structure that characterizes the group: apicomplexans, dinoflagellates, euglenoids, radiolarians.
2. Which eukaryotic supergroups contain amoeboid forms?
3. What is the major difference between cellular slime moulds and plasmodial slime moulds?

24.5 Some Protist Lineages Arose from Primary Endosymbiosis and Others from Secondary Endosymbiosis

We have encountered chloroplasts in a number of eukaryotic organisms in this chapter: red and green algae, euglenoids, dinoflagellates, stramenopiles, chlorarachniophytes, and land plants. How did these chloroplasts evolve? Unlike the endosymbiotic event that gave rise to mitochondria, endosymbiosis involving photoautotrophs happened more than once, resulting in the formation of a wide range of photosynthetic eukaryotes.

About 1 bya, the first chloroplasts evolved from free-living photosynthetic prokaryotic organisms (cyanobacteria) ingested by eukaryote cells that had

already acquired mitochondria (see Chapter 3). In some cells, the cyanobacterium was not digested but instead formed a symbiotic relationship with the engulfing host cell—it became an endosymbiont, an independent organism living inside another organism. Over evolutionary time, the prokaryotic organism lost genes no longer required for independent existence and transferred most of its genes to the host's nuclear genome. In this process, the endosymbiont became an organelle. As explained in Chapter 3, moving genes from the endosymbiont to the nucleus would have given the host cell better control of cell functioning.

The chloroplasts of red algae, green algae, and land plants result from evolutionary divergence of the p-hotosynthetic eukaryotes formed from this primary endosymbiotic event (as shown in the top part of **Figure 24.31, p. 536**). Organisms that originated from this event have chloroplasts with two membranes, one from the plasma membrane of the engulfing eukaryote and the other from the plasma membrane of the cyanobacterium.

This **primary endosymbiosis** was followed by at least three **secondary endosymbiosis** events, each time involving different heterotrophic eukaryotes engulfing a photosynthetic eukaryote, producing new evolutionary lineages (see Figure 24.31). For example, s-econdary endosymbiosis involving red algae engulfed by a nonphotosynthetic eukaryote gave rise to the stramenopile algae and the alveolates.

Independent endosymbiotic events involving green algae and nonphotosynthetic eukaryotes produced euglenoids and chlorarachniophytes.

Organisms that formed via secondary endosymbiosis have chloroplasts surrounded by additional membranes acquired from the new host. For example, chlorarachniophytes have plastids with four membranes (see Figure 24.31). The new membranes correspond to the plasma membrane of the engulfed phototroph and the food vacuole membrane of the host.

In sum, the protists are a highly diverse and ecologically important group of organisms. Their complex evolutionary relationships, which have long been a subject of contention, are now being revised as new information is discovered, including more complete genome sequences. A deeper understanding of protists is also contributing to a better understanding of their recent descendants, the fungi, plants, and animals. We turn to these descendants in the next four chapters, beginning with the fungi.

STUDY BREAK

In primary endosymbiosis, a nonphotosynthetic eukaryotic cell engulfed a photosynthetic cyanobacterium. How many membranes surround the chloroplast that evolved?

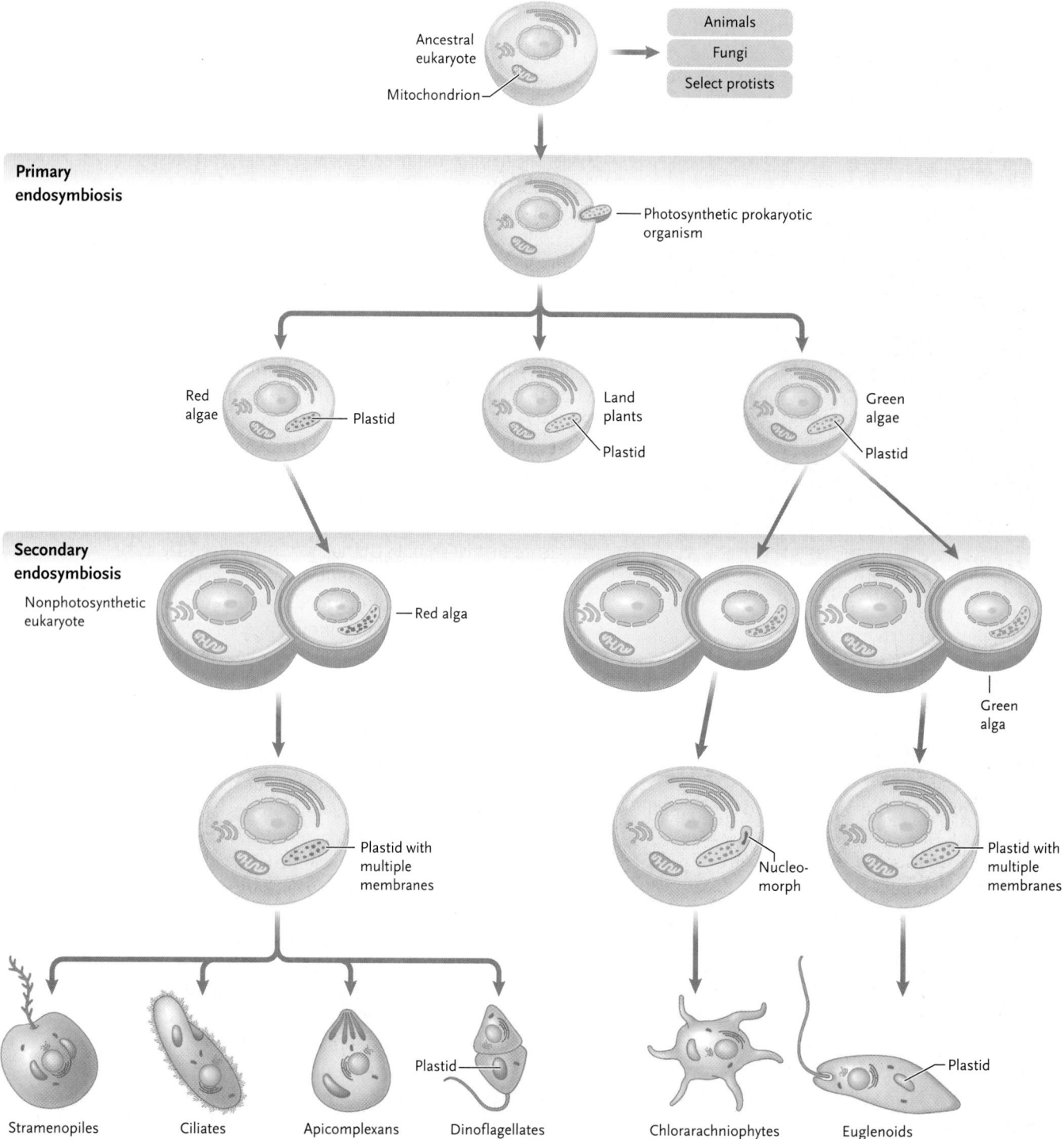

Figure 24.31
The origin and distribution of plastids among the eukaryotes by primary and secondary endosymbiosis.

Review

24.1 The Vast Majority of Eukaryotes Are Protists

- As eukaryotes, protists contain organelles including mitochondria and, sometimes, chloroplasts. Mitochondria evolved once via primary endosymbiosis, the engulfing of a free-living prokaryotic organism that became an organelle over evolutionary time. Some photosynthetic protists were formed via primary endosymbiosis involving a photosynthetic prokaryotic organism; others arose via more complex endosymbiotic events (see Section 24.5).

24.2 Characteristics of Protists

- Eukaryotes are divided into supergroups, a taxonomic level above kingdom. The vast majority of lineages within these supergroups are protists. Protists are eukaryotes that differ from fungi in having motile stages in their life cycles and cellulose cell walls. Unlike land plants, they lack roots, stems, and leaves and do not retain embryos in parental tissue. Unlike animals, protists are often unicellular; they lack collagen, nerve cells, and an internal digestive tract.

24.3 Protists' Diversity Is Reflected in Their Metabolism, Reproduction, Structure, and Habitat

- Most protists are aerobic organisms that live as autotrophs or as heterotrophs or by a combination of both nutritional modes. Some are symbionts living in or among the cells of other organisms.
- Protists live in aquatic or moist terrestrial habitats or as parasites within animals. They may be single-celled, colonial, or multicellular organisms, and they range in size from microscopic to some of Earth's largest organisms.
- Some protists are the most complex single cells known because of the wide variety of cytoplasmic structures they have; most are able to move by means of flagella, cilia, or pseudopodia.
- Reproduction may be asexual, by mitotic divisions, or sexual, involving meiosis and the union of gametes in fertilization.

24.4 The Eukaryotic Supergroups and Their Key Protist Lineages

- Excavates are almost all single-celled, heterotrophic protists that swim using flagella (Figures 24.7, 24.8). One lineage, the euglenoids, obtained plastids via secondary endosymbiosis. Other lineages, such as the diplomonads and parabasalids (Figure 24.9), lack "typical" mitochondria but often have organelles derived from mitochondria.
- Chromalveolates include ciliates (Figure 24.10), which swim using cilia and have complex cytoplasmic structures, including both micronuclei and macronuclei; apicomplexans, nonmotile parasites of animals; and dinoflagellates, which have two flagella that propel them in a "whirling" motion and are primarily marine organisms (Figure 24.11). Many alveolates are photosynthetic.
- Chromalveolates also include diatoms, photosynthetic single-celled organisms covered by a glassy silica shell (Figure 24.18); golden algae, photosynthetic, mostly colonial forms (Figure 24.3); brown algae, primarily multicellular marine forms that include large seaweeds (Figure 24.19); and the funguslike Oomycota, which often grow as masses of microscopic filaments and live as saprophytes or parasites, secreting enzymes that digest organic matter in their surroundings (Figure 24.17). Many stramenopiles (heterokonts) have flagella only on reproductive cells.
- Rhizaria are amoebas with filamentous pseudopods supported by internal cellular structures (Figure 24.22). Many produce hard outer shells. Radiolarians are primarily marine organisms that secrete a glassy internal skeleton. Foraminifera are marine, single-celled organisms that form chambered, spiral shells containing calcium. Both groups engulf prey that adhere to thin extensions of their cells. Chlorarachniophytes engulf food using their pseudopodia.
- Unikonts include most amoebas (Figure 24.23) and two types of slime moulds, cellular (which move as individual cells; Figure 24.24) and plasmodial (which move as large masses of nuclei sharing a common cytoplasm; Figure 24.25). Amoebas in this group are heterotrophs that are abundant in marine and freshwater environments and in the soil. They move by extending pseudopodia. Also included in the Unikont supergroup are opisthokonts, a broad group of eukaryotes including the choanoflagellates. Choanoflagellates have a single flagellum surrounded by a collar of fingerlike membrane projections. A choanoflagellate type of protist was likely the ancestor of animals.
- Plantae includes the red and green algae, as well as land plants. Red algae are typically multicellular, primarily photosynthetic organisms of marine environments with complex life cycles. Green algae are single-celled, colonial, multicellular species that live primarily in freshwater habitats (Figure 24.28) and carry out photosynthesis by mechanisms similar to those of plants.
- Opisthokonts are a broad group of eukaryotes that includes the choanoflagellates, which have a single flagellum surrounded by a collar of fingerlike membrane projections. A choanoflagellate type of protist was likely the ancestor of animals.

24.5 Some Protist Lineages Arose from Primary Endosymbiosis and Others from Secondary Endosymbiosis

- Several groups of protists, as well as land plants, contain chloroplasts, which arose via endosymbiosis events (Figure 24.31). In a primary endosymbiosis event, a eukaryotic cell engulfed a cyanobacterium, which became an organelle, the chloroplast. Evolutionary divergence from this ancestral phototrophic organism produced the red algae, green algae, and land plants. Other photosynthetic protists were produced by secondary endosymbiosis, in which a nonphotosynthetic eukaryote engulfed a photosynthetic eukaryote.

Questions

Self-Test Questions

1. Which group of protists move through viscous fluids using both freely beating flagella and a flagellum buried in a fold of cytoplasm, and cause a sexually transmitted disease in humans?
   a. Ciliates
   b. Parabasalids
   c. Euglenoids
   d. Diplomonads

2. Diplomonads are characterized by which of the following features?
   a. Cells with two functional nuclei and multiple flagella. *Giardia* is an example.
   b. A mouthlike gullet and a hairlike surface. *Paramecium* is an example.
   c. Nonmotility, parasitism, and sporelike infective stages. *Toxoplasma* is an example.
   d. Large protein deposits. Movement is by two flagella, which are part of an undulating membrane. *Trypanosoma* is an example.

3. Which of the following groups is the greatest contributor to protist fossil deposits?
   a. oomycetes
   b. brown algae
   c. golden algae
   d. diatoms

4. Which of the following groups has the distinguishing characteristic of gas-filled bladders and a cell wall composed of alginic acid?
   a. oomycetes
   b. brown algae
   c. golden algae
   d. diatoms

5. *Plasmodium* is transmitted to humans by the bite of a mosquito (*Anopheles*) and engages in a life cycle with infective spores, gametes, and cysts. To which group does this infective protist belong?
   a. oomycetes
   b. euglenoids
   c. dinoflagellates
   d. apicomplexans

6. The ancestor of land plants is thought to have belonged to which group of protists?
   a. red algae
   b. diatoms
   c. green algae
   d. golden algae

7. In oil exploration, the presence of shells is an indicator of oil-rich rock layers. To which group of protists would these shells belong?
   a. diatoms
   b. foraminiferans
   c. golden algae
   d. red algae

8. To which supergroup do the living representatives of the group of organisms thought to be ancestral to animals belong?
   a. Excavates
   b. Rhizaria
   c. Unikonts
   d. Chromalveolates

9. Which of the following statements about cellular slime moulds is correct?
   a. They are autotrophs.
   b. They move using cilia.
   c. They reproduce only asexually.
   d. They form a fruiting body that produces spores.

10. The latest stage for evolving the double membrane seen in modern-day algal chloroplasts is thought to involve the combining of two organisms. What are the two organisms?
    a. two ancestral photosynthetic prokaryotic organisms
    b. two ancestral nonphotosynthetic prokaryotic organisms
    c. a nonphotosynthetic eukaryote with a photosynthetic eukaryote
    d. a photosynthetic prokaryotic organism with a nonphotosynthetic eukaryote

Questions for Discussion

1. We have seen that, as a group, protists use three kinds of motility: flagella, cilia, and amoeboid movement. There are some cells in your body that also use each of these three forms of motility. Name an example of cells that use each form.

2. Photosynthetic protists (sometimes referred to as algae) are often thought of as single-celled plants. What features differentiate these protists from land plants? Would it be correct to consider the protists known as protozoa as single-celled animals? Explain why or why not.

3. Why is it harder to treat human diseases caused by protists, such as *Giardia,* than diseases caused by bacteria?

4. Many protists are able to produce cysts or other resting stages in their life cycle. What is the advantage of producing these resting structures?

5. You place a marine green algal cell and a *Paramecium* into fresh water, which is hypotonic to both cells. Water will flow into both cells. Which cell, if either, will burst? Explain your rationale.

The mushroom-forming fungus *Inocybe fastigiata*, a forest-dwelling species that commonly lives in close association with conifers and hardwood trees.

Fungi

WHY IT MATTERS

If you were asked what the first crop on Earth was and which organisms grew it, you would probably think of corn, wheat, or some other crop plant grown by humans. But you'd be wrong—the first domesticated crop was a fungus, and the first farmers were a certain group of ants over 50 million years ago (mya), whereas humans did not start farming until about 10 thousand years ago. Researchers have used molecular data combined with fossil evidence to determine when ants first domesticated their fungal crop. Today, these leaf-cutter ants (Tribe Attini) of Central and South America **(Figure 25.1a, p. 540)** still grow certain fungi in gardens. Just as humans do, the ants plant their crop, fertilize it, weed it, and then feed on it. The ants harvest leaves, flowers, and other plant parts and carry these back to their nests, where they are added to the fungal gardens in the nests (Figure 25.1b).

The ants plant small pieces of the plant material in the garden, placing bits of the fungus on each piece. They fertilize the garden with their excrement and graze on the fungal filaments. In fact, although the ants collect a wide range of plant matter, they never eat any of it directly—their sole food source is the fungus. When a queen ant leaves her birth nest to start a new nest, she carries a bit of fungus in her mouth and uses it to start a garden in the new nest. The ants' habitat contains ample supplies of other foods, so why have these ants developed this complex and rather bizarre lifestyle? What benefit do they gain by devoting their lives to looking after a fungus? The answer lies in the ability of the fungus to unlock the nutrients tied up in plant tissue. Cellulose is the most abundant organic molecule on Earth, but most organisms cannot get at the carbon it contains as they lack the enzymes needed to break apart the bonds in the molecule. Fungi are among the few organisms that can digest cellulose, so by forming a partnership with fungi, these ants gain access to a continuous source of carbon. In return, the fungus gains a secure habitat in which it doesn't have to compete with other organisms for a food source. Recent

Figure 25.1
(a) Leaf-cutter ants.
(b) Fungal garden of
leaf-cutter ants.

a.

b.

research has revealed that this ancient symbiosis is more complex than previously known (see "People behind Biology," 25.1, p. 541).

Although we often associate fungi with decay and decomposition, many fungi, such as those cultivated by leaf-cutter ants, instead live by forming symbiotic associations with other organisms. The vast majority of plants obtain soil mineral nutrients via a symbiotic relationship with soil fungi. Humans have also harnessed the metabolic activities of certain fungi to obtain substances ranging from flavourful cheeses and wine to bread and therapeutic drugs such as penicillin and the immunosuppressant cyclosporine. And, as you know from previous chapters, species such as the yeast *Saccharomyces cerevisiae* and the mould *Neurospora crassa* have long been pivotal model organisms in studies of DNA structure and function and in the development of genetic engineering methods. On the other hand, fungi collectively are the single greatest cause of plant diseases, and many species cause disease in humans and other animals. Some even produce carcinogenic toxins.

Evidence suggests that fungi were present on land at least 760 mya and possibly much earlier. Their presence on land was likely crucial for the successful colonization of land by plants, which relied on symbiotic associations with the fungi to obtain nutrients from the nutrient-poor soils of early land environments. In the course of the intervening millennia, evolution equipped fungi with a remarkable ability to break down a wide range of compounds, ranging from living and dead organisms and animal wastes to groceries, clothing, paper, and wood—even photographic film. Along with heterotrophic bacteria, they have become Earth's premier decomposers **(Figure 25.2)**. Despite their profound impact on ecosystems and other life forms, most of us have only a passing acquaintance with the fungi—perhaps limited to the mushrooms on our pizza or the invisible but annoying types that cause skin infections, such as athlete's foot. This chapter provides you with an overview of fungal biology. We begin with the features that set fungi apart from all other organisms and discuss the diversity of fungi existing today before revisiting associations between fungi and other organisms.

25.1 General Characteristics of Fungi

We begin our survey of fungi by examining the features that distinguish fungi from other forms of life, how fungi obtain nutrients, and adaptations for reproduction and growth that enable fungi to spread far and wide through the environment.

Fungi are heterotrophic eukaryotes that obtain carbon by breaking down organic molecules synthesized by other organisms. Although all fungi are heterotrophs, fungi can be divided into two broad groups based on how they obtain carbon. If a fungus obtains carbon from nonliving material, it is a **saprotroph**. Fungi that decompose dead plant and animal tissues, for example, are saprotrophs. If a fungus obtains carbon from living organisms, it is a **symbiont**. Symbiosis is the living together of two (or sometimes more)

Figure 25.2
Example of a wood decay fungus: sulfur shelf fungus (*Polyporus*).

PEOPLE BEHIND BIOLOGY 25.1

Cameron Currie, *University of Wisconsin-Madison*

Experimental Research Box

Discovery has been defined as "seeing what everyone else has seen and thinking what no one else has thought." Even though ant–fungal mutualism has been studied since 1874, in 1999 a graduate student at the University of Toronto, Cameron Currie, discovered a whole new dimension to this mutualism. For years, researchers studying this symbiosis had wondered how the ants kept their gardens free of competing fungi. The conditions created by the ants are ideal for many other fungi besides the garden fungus, and, as you know, fungal spores are everywhere— yet the ant gardens are pure monocultures of a single fungus. What prevents other fungi from invading the gardens? Biologists had thought that the ants kept other fungi out simply by weeding the gardens, removing all traces of invading fungi, and so keeping fungal competitors at bay. But Currie discovered that there is a third symbiont at work, and this organism keeps out other fungi. Currie noticed that ants had a whitish substance on their abdomens; other researchers had previously noticed this crust as well but assumed it was just part of the ant's exoskeleton **(Figure 1a).** When Currie took a closer look, he discovered that the crust was actually a bacterium of the genus *Pseudonocardia;* these bacteria are actinomycetes, which are known to produce antibiotics (e.g., streptomycin). On further exploration, Currie found that this bacterium produced an antibiotic that specifically and completely inhibited growth of a parasitic fungus, *Escovopsis*, which is the greatest threat to the gardens. If *Escovopsis* isn't stopped, it will overgrow the desirable fungus and take over the garden. This groundbreaking research, demonstrating that mutualisms do not necessarily involve just two species, was done while Currie was still a student. Since finishing his Ph.D., Currie has gone on to show that the ants' bodies have changed over evolutionary time to create and maintain a favourable environment for their bacteria. The bacteria live in specialized crypts on the ant's body that are associated with glands; secretions from these glands provide nutrients for the bacteria (Figure 1b, c). So not only did ants invent agriculture long before humans, but also they used microbes to produce antibiotics long before we thought of doing so. An interesting aspect of antibiotic use by the ants is that even though the parasitic fungus has been exposed to the antibiotic produced by the bacterium for a very long time, it has not become resistant to the antibiotic. Perhaps we can learn something from the ants about preventing antibiotic resistance, which has rendered many of the antibiotics we rely on ineffective (you read about antibiotic resistance in Chapter 22).

a.

b.

c.
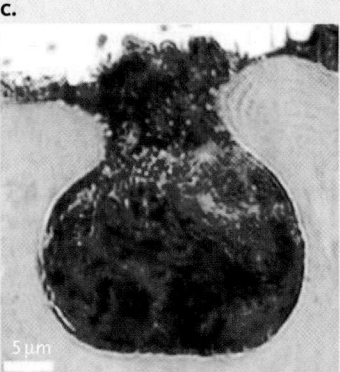

Figure 1

(a) *Leaf-cutter ant showing bacterial "crust."* **(b)** *Crypts (white spots) on the ant's body house bacteria.* **(c)** *A crypt on the body of a leaf-cutter ant.*

Photos a–c: From Cameron R. Currie, Michael Poulsen, John Mendenhall, Jacobus J. Boomsma, and Johan Billen, "Coevolved crypts and exocrine glands support mutualistic bacteria in fungus-growing ants," *Science*, Vol. 311, 6 January 2006, pp. 81–83. Reprinted with permission from AAAS.

organisms for extended periods; symbiotic relationships range along a continuum from **parasitism**, in which one organism benefits at the expense of the other, to **mutualism**, in which both organisms benefit. Although we often think of fungi as decomposers, fully half of all identified fungi live as symbionts with another organism.

Regardless of their nutrient source, fungi feed by **absorptive nutrition:** they secrete enzymes into their environment, breaking down large molecules into smaller soluble molecules that can then be absorbed into their cells. This mode of nutrition means that fungi cannot be stationary, as they would then deplete all of the food in their immediate environment.

a. Fungal hyphae

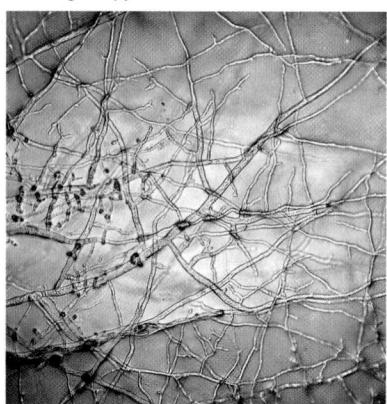

b. Mycelium and fruiting body of a mushroom-forming fungus

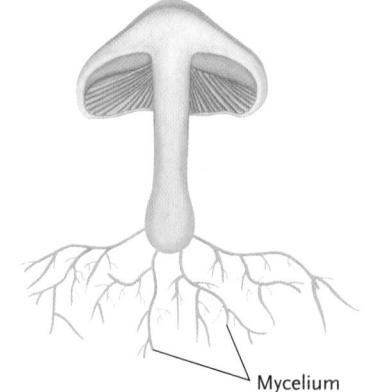

Mycelium

c. Mycelium on leaf litter

Figure 25.3

(a) Fungal hyphae forming a mycelium in culture in a Petri dish. **(b)** Sketch of the mycelium of a mushroom-forming fungus, which consists of branching septate hyphae. **(c)** Mycelium on leaf litter.

Instead, fungi have evolved the ability to proliferate quickly through their environment, digesting nutrients as they grow. How can fungi proliferate so quickly? Although some fungi grow as unicellular yeasts, which reproduce asexually by **budding** or binary fission (see Figure 25.14, p. 549), most are composed of **hyphae** ("web"; singular, *hypha*) **(Figure 25.3a)**, fine filaments that spread through whatever substrate the fungus is growing in—soil, decomposing wood, your skin—forming a network or **mycelium** (Figure 25.3b, c). Hyphae are essentially tubes of cytoplasm surrounded by cell walls made of chitin, a polysaccharide also found in the exoskeletons of insects and other arthropods.

Hyphae grow only at their tips, but because a single mycelium contains many, many tips, the entire mycelium grows outward very quickly. Together, this **apical growth** and absorptive nutrition account for much of the success of fungi. As the hyphal tips extend, they exert a mechanical force, allowing them to push through their substrate, releasing enzymes and absorbing nutrients as they go. Fungal species differ in the particular digestive enzymes they synthesize, so a substrate that is a suitable food source for one species may be unavailable to another. Although there are exceptions, fungi typically thrive only in moist environments, where they can directly absorb water, dissolved ions, simple sugars, amino acids, and other small molecules. When some of a mycelium's hyphal filaments contact a source of food, growth is channelled in the direction of the food source.

Nutrients are absorbed at the porous tips of hyphae; small atoms and molecules pass readily through these tips, and then transport mechanisms move them through the underlying plasma membrane. Some hyphae have regular cross-walls or **septa** ("fences" or "walls"; singular, *septum*) that separate a

hypha into compartments, whereas others lack septa and are effectively one large cell **(Figure 25.4)**. But even septate hyphae should be thought of as interconnected compartments rather than separate cells, as all septa have pores that allow cytoplasm and, in some fungi, even nuclei and other large organelles to flow through the mycelium. By a mechanism called *cytoplasmic streaming* (flowing of cytoplasm and organelles around a cell or, in this case, a mycelium), nutrients obtained by one part of a mycelium can be translocated to other nonabsorptive regions, such as reproductive structures.

When a fungus releases enzymes into its substrate, it faces competition from bacteria and other organisms for the nutrients that are now available. How can a fungus prevent these competitors from stealing the nutrients that it has just expended energy and resources to obtain? Many fungi produce antibacterial compounds and toxins that inhibit the growth of competing organisms. Many of these compounds are **secondary metabolites**, which are not required for day-to-day survival but are beneficial to the fungus. As we will see, many of these compounds not only are important in the life of a fungus but also benefit organisms

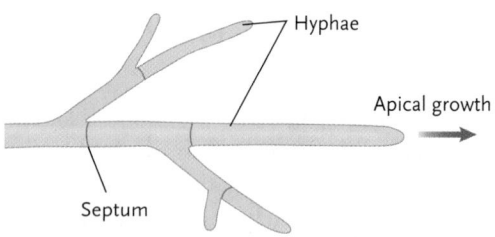

Hyphae

Apical growth

Septum

Figure 25.4

Septa. In some fungi, septa divide each hypha into separate compartments.

associated with the fungus. Many are also of commercial or medical importance to humans; for example, the antibiotic penicillin is a secondary metabolite produced by a species of *Penicillium*.

Fungi reproduce by spores, and this spore production can be amazingly prolific, with some species of fungi producing billions of spores per day **(Figure 25.5)**. These spores are microscopic, featherlight, and able to survive in the environment for extended periods after they are released. Reproducing via such spores allows fungi to be opportunists, germinating only when favourable conditions exist and quickly exploiting food sources that occur unpredictably in the environment. Releasing vast numbers of spores, as some fungi do, improves the odds that the spores will germinate and produce a new individual.

Spores can be produced asexually or sexually; some fungi produce both asexual and sexual spores at different stages of their lives. Sexual reproduction in fungi is quite complex. In all organisms, sexual reproduction involves three stages: the fusion of two haploid cells (**plasmogamy**), bringing together their two nuclei in one common cytoplasm; this cytoplasmic fusion is usually quickly followed by nuclear fusion (**karyogamy**) in most organisms; nuclear fusion is followed by meiosis to produce genetically distinct haploid cells. As we will see, fungi are unique in that these events can be separated in time for durations ranging from seconds to many years.

STUDY BREAK

1. What physical features distinguish fungi from other organisms?
2. How do fungi reduce competition for resources?
3. By what means do fungi reproduce? Why is this mode of reproduction advantageous?
4. What is the advantage of mycelial growth?

25.2 Evolution and Diversity of Fungi

25.2a Fungi Were Present on Earth by at Least 760 Million Years Ago

For many years, fungi were classified as plants because the earliest classification schemes had only two kingdoms, plants and animals. Fungi, like plants, have cell walls and do not move as animals did, so they were grouped with plants. As biologists learned more about the distinctive characteristics of fungi, however, it became clear that fungi should be treated as a separate kingdom.

CONCEPT FIX The idea that fungi are most closely related to plants has persisted, but this is a misconception! The discovery of chitin in fungal cells and recent comparisons of DNA and RNA sequences all indicate that fungi and animals are more closely related to each other than they are to other eukaryotes. The close biochemical relationship between fungi and animals may explain why fungal infections are typically so resistant to treatment and why it has proved rather difficult to develop drugs that kill fungi without damaging their human or other animal hosts. ⬢

Analysis of the sequences of several genes suggests that the lineages leading to animals and fungi likely diverged between 760 mya and 1 billion years ago (bya). What were the first fungi like? We do not know for certain: phylogenetic studies indicate that fungi first arose from a single-celled, flagellated protist similar to choanoflagellates (see Chapter 24)—the sort of organism that does not fossilize well. Although traces of what may be fossil fungi exist in rock formations nearly 1 billion years old, the oldest fossils that we can confidently assign to the modern **kingdom Fungi** appear in rock strata laid down in the late Proterozoic (900–570 mya).

Figure 25.5
Spore production by fungal fruiting bodies. Some fruiting bodies can release billions of spores per day.

© Andrew Darrington / Alamy

Table 25.1 — Summary of Fungal Phyla

Phylum	Body Type	Key Feature	
Chytridiomycota (chytrids)	One to several cells	Motile spores propelled by flagella; usually asexual	
Zygomycota (zygomycetes)	Hyphal	Sexual stage in which a resistant zygospore forms for later germination	
Glomeromycota (glomeromycetes)	Hyphal	Hyphae associated with plant roots, forming arbuscular mycorrhizas	
Ascomycota (ascomycetes)	Hyphal	Sexual spores produced in sacs called asci	
Basidiomycota (basidiomycetes)	Hyphal	Sexual spores formed on club-shaped cells called basidia of a prominent fruiting body (basidiocarp)	

25.2b Once They Appeared, Fungi Radiated into Several Major Lineages

Most likely, the first fungi were aquatic. When other kinds of organisms began to colonize land, they may well have brought fungi along with them. For example, researchers have discovered what appear to be mycorrhizas—symbiotic associations of a fungus and a plant—in fossils of some of the earliest known land plants. The final section of this chapter examines mycorrhizas more fully.

Over time, fungi diverged into the strikingly diverse lineages that we consider in the rest of this section **(Table 25.1)**. Today, there are over 60 thousand described species of fungi, with at least 1.6 million more that have not yet been described.

As the lineages diversified, different adaptations associated with reproduction arose. For example, you'll notice that the structures in which sexual spores are formed and mechanisms by which spores are dispersed became larger and more elaborate over evolutionary time. Traditionally, therefore, biologists have classified fungi primarily by the distinctive structures produced in sexual reproduction. These features are

still useful indicators of the phylogenetic standing of a fungus, but the powerful tools of molecular analysis are bringing many revisions to our understanding of the evolutionary journey of fungi.

The evolutionary origins and lineages of fungi have been obscure ever since biologists began puzzling over the characteristics of this group. With the advent of molecular techniques for research, these topics have become extremely active and exciting areas of biological research that may shed light on fundamental events in the evolution of all eukaryotes. Currently, we recognize five phyla of fungi, known formally as the Chytridiomycota, Zygomycota, Glomeromycota, Ascomycota, and Basidiomycota **(Figure 25.6)**. However, we know now that two of these phyla, the chytridiomycota and the zygomycota, are not monophyletic (i.e., they are taxa that do not contain only one ancestor and all of its descendants), so the classification scheme presented in Figure 25.6 will soon change to reflect this new information. Why do classifications of organisms change so often? Bear in mind that classification schemes such as those presented here are hypotheses that explain our best understanding of evolutionary relationships among organisms at any one time; like any other hypotheses, classification schemes are open

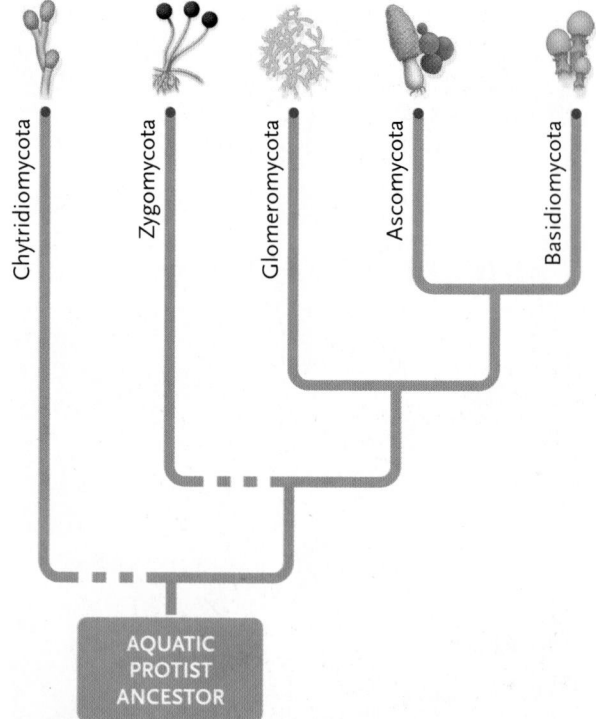

Figure 25.6

A phylogeny of fungi. This scheme represents a widely accepted view of the general relationships between major groups of fungi, but it may well be revised as new molecular findings provide more information. The dashed lines indicate that two groups, the chytrids and the zygomycetes, are probably paraphyletic—they include subgroups that are not all descended from a single ancestor.

to revision as we find out more about the organisms. Molecular data also suggest that some other eukaryotic organisms currently classified elsewhere may actually be fungi; we haven't included those organisms in this chapter but instead will focus on the groups of fungi that are best understood. Even though fungal classification will change greatly over the next few years, we summarize the major phyla recognized today as a way of illustrating the diversity of this group of organisms.

Phylum Chytridiomycota. The Chytridiomycota are likely the most ancient group of fungi, as they retain several traits characteristic of an aquatic lifestyle. For example, chytrids (as they are commonly called) are the only fungi that produce flagellated, motile spores **(Figure 25.7a);** these spores use chemotaxis (movement in response to a chemical gradient) to locate suitable substrates. Chytrids live in soil or freshwater habitats, wherever there is at least a film of water through which their motile spores can swim.

Most chytrids are saprotrophs, organisms that obtain nutrients by breaking down dead organic matter, although some are symbionts in the guts of cattle and other herbivores, where they break down cellulose to provide carbon for their hosts, and still others are parasites of animals, plants, algae, or other fungi. These tiny fungi also cause a disease, chytridiomycosis, that is one cause of the decline in amphibian species worldwide. Globally, at least 43% of all amphibian species are declining in population, and nearly 33% are threatened with extinction. Although many factors contribute to amphibian decline, including habitat loss, fragmentation, and increasing levels of environmental pollutants, chytridiomycosis has been linked to the decline of amphibian populations in Australia, New Zealand, central and South America, and parts of Europe. This disease has wiped out an estimated two-thirds of the species of harlequin frogs (*Atelopus*) in the American tropics (Figure 25.7b). The epidemic has correlated with the rising average temperature in the frogs' habitats, an increase credited to global warming. Studies show that the warmer environment provides optimal growing temperatures for the chytrid pathogen. How does infection by a chytrid kill these animals? The fungus colonizes the skin of amphibians (Figure 25.7b), which interferes with the electrolyte balance and functioning of organs because amphibians take up water and exchange gases through their skins.

Most chytrids are unicellular, although some live as chains of cells and have rhizoids (branching filamentous extensions) that anchor the fungus to its substrate and that may also absorb nutrients from the substrate (Figure 25.7a). The vegetative stage of most chytrids is haploid; asexual reproduction involves the formation of a **sporangium**, in which motile spores are formed. A few chytrids reproduce sexually, via male and female gametes that fuse to form a diploid zygote. This cell may form a mycelium that gives rise to sporangia, or it may directly give rise to either asexual or sexual spores.

Zygomycota. This group of fungi includes the moulds on fruit and bread familiar to many of us and takes its name from the structure formed in sexual reproduction, the **zygospore (Figure 25.8, p. 546).** Many zygomycetes are saprotrophs that live in soil, feeding on organic matter. Their metabolic activities release mineral nutrients in forms that plant roots can take up. Some zygomycetes are parasites of insects (and even other zygomycetes), and some wreak havoc on human food supplies, spoiling stored grains, bread, fruits, and vegetables **(Figure 25.9, p. 546).** Others, however, have become major players in commercial enterprises, where they are used in manufacturing products that range from industrial pigments to pharmaceuticals

a. Chytridiomycosis in a frog

Skin surface

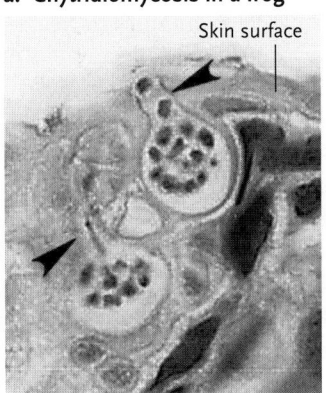

Centers for Disease Control and Prevention

b. Harlequin frog

Pedro Bernardo/Shutterstock.com

Figure 25.7

Chytrids. **(a)** Chytridiomycosis, a fungal infection, shown here in the skin of a frog. The two arrows point to flask-shaped spore-producing cells of the parasitic chytrid *Batrachochytrium dendrobatis*, which has devastated populations of harlequin frogs **(b)**.

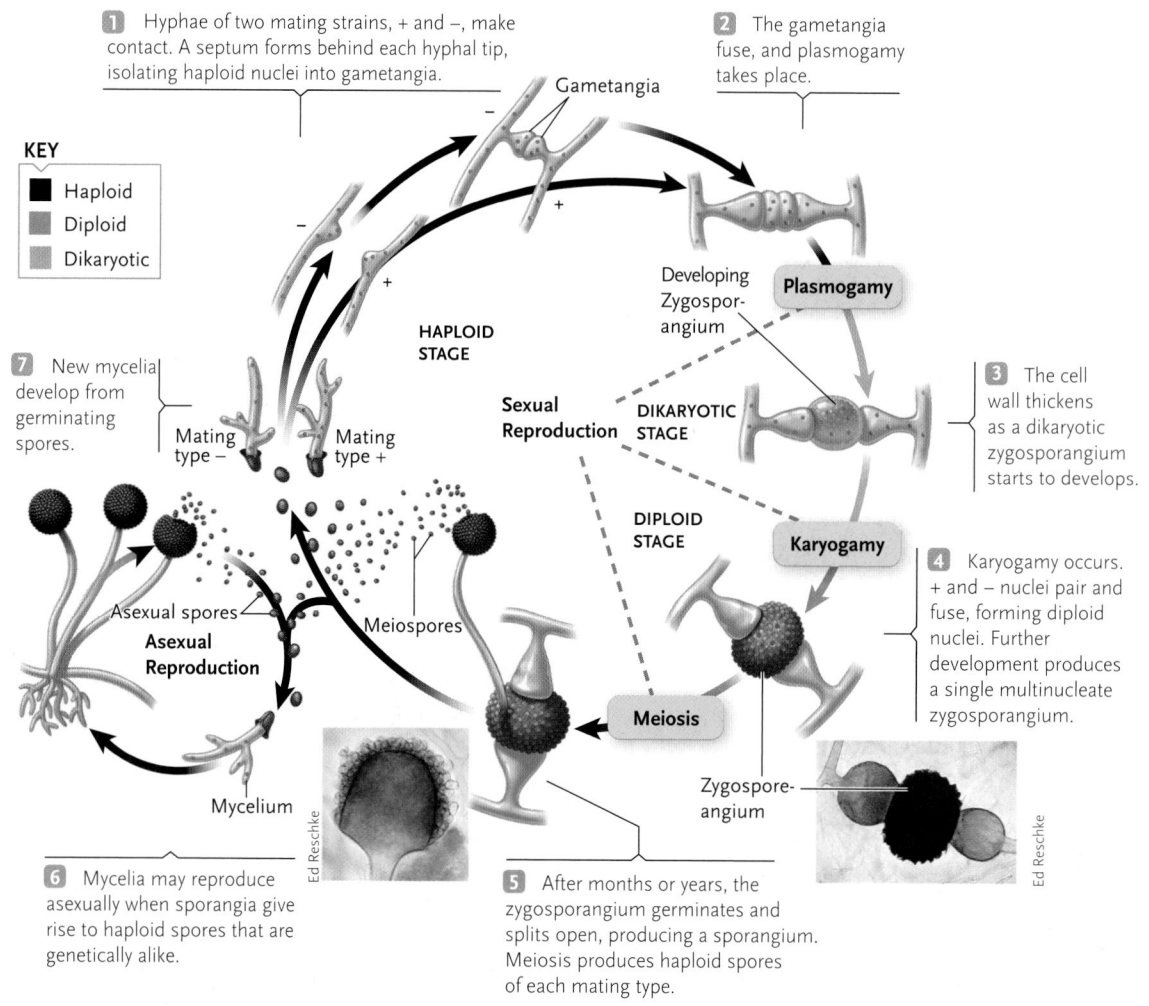

1 Hyphae of two mating strains, + and −, make contact. A septum forms behind each hyphal tip, isolating haploid nuclei into gametangia.

2 The gametangia fuse, and plasmogamy takes place.

Gametangia

KEY
- ■ Haploid
- ■ Diploid
- ■ Dikaryotic

Developing Zygosporangium

Plasmogamy

HAPLOID STAGE

7 New mycelia develop from germinating spores.

Mating type − | Mating type +

Sexual Reproduction

DIKARYOTIC STAGE

3 The cell wall thickens as a dikaryotic zygosporangium starts to develops.

DIPLOID STAGE

Asexual spores

Asexual Reproduction

Meiospores

Karyogamy

4 Karyogamy occurs. + and − nuclei pair and fuse, forming diploid nuclei. Further development produces a single multinucleate zygosporangium.

Meiosis

Mycelium

Zygospore-angium

Ed Reschke

6 Mycelia may reproduce asexually when sporangia give rise to haploid spores that are genetically alike.

5 After months or years, the zygosporangium germinates and splits open, producing a sporangium. Meiosis produces haploid spores of each mating type.

Ed Reschke

Figure 25.8
Life cycle of the bread mould *Rhizopus stolonifer,* a zygomycete. Asexual reproduction is common, but different mating types (+ and −) also reproduce sexually. In both cases, haploid spores are formed and give rise to new mycelia.

Figure 25.9
Zygomycete fungus growing on strawberries.

humbak/Shutterstock.com

such as steroids (e.g., anti-inflammatory drugs). Zygomycetes are also used in the production of fermented foods such as tempeh.

Most zygomycetes consist of a haploid mycelium that lacks regular septa, although some groups have septa, and in others, septa form to wall off reproductive structures and aging regions of the mycelium. Sexual reproduction occurs when mycelia of different **mating types** (known as + and − types, rather than male and female) produce specialized hyphae that grow toward each other and form sex organs (**gametangia**) at their tips (see Figure 25.8, steps 1 and 2). How do the

gametangia find each other? Pheromones secreted by each mycelium stimulate the development of sexual structures in the complementary strain and cause gametangia to grow toward each other. The gametangia fuse, forming a thick-walled structure, a **zygosporangium** (see Figure 25.8, step 3), which can remain dormant for months or years, allowing the zygomycete to survive unfavourable environmental conditions. Eventually, meiosis occurs in the zygosporangium, forming a meiosporangium that will produce haploid spores by meiosis. (see Figure 25.8, step 5). Note that meiosis does not always produce gametes! We often tend to characterize meiosis as the formation of gametes, probably because we are so familiar with how sexual reproduction occurs in humans and other animals. But in many organisms, such as fungi and plants, meiosis results in the formation of haploid spores.

Like other fungi, however, zygomycetes also reproduce asexually, as shown in steps 6 and 7 of Figure 25.8. When a haploid spore lands on a favourable substrate, it germinates and gives rise to a branching mycelium.

a. Sporangia of Rhizopus nigricans

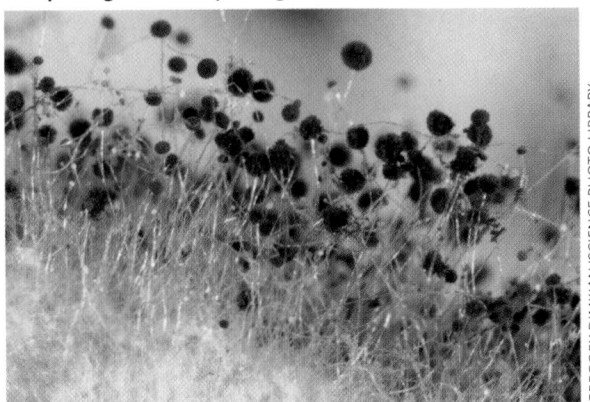

GREGORY DIMIJIAN/SCIENCE PHOTO LIBRARY

b. Sporangia (dark sacs) of *Pilobolus*

POWER AND SYRED/SCIENCE PHOTO LIBRARY

500 μm

Figure 25.10

Two of the numerous strategies for spore dispersal by zygomycetes. **(a)** The sporangia of *Rhizopus stolonifer*, shown here on a slice of bread, release powdery spores that are easily dispersed by air currents. **(b)** In *Pilobolus*, the spores are contained in a sporangium (the dark sac) at the end of a stalked structure. When incoming rays of sunlight strike a light-sensitive portion of the stalk, turgor pressure (pressure against a cell wall due to the movement of water into the cell) inside a vacuole in the swollen portion becomes so great that the entire sporangium may be ejected outward as far as 2 m—a remarkable feat given that the stalk is only 5 to 10 mm tall.

Some of the hyphae grow upward, and saclike sporangia form at the tips of these aerial hyphae. Inside the sporangia, the asexual cycle comes full circle as new haploid spores arise through mitosis and are released.

The black bread mould *Rhizopus stolonifer* may produce so many charcoal-coloured sporangia in asexual reproduction **(Figure 25.10a)** that mouldy bread looks black. The spores released are lightweight, dry, and readily wafted away by air currents. In fact, winds have dispersed *R. stolonifer* spores just about everywhere on Earth, including the Arctic. Another zygomycete, *Pilobolus* (Figure 25.10b), forcefully spews its sporangia away from the dung in which it grows. A grazing animal may eat a sporangium on a blade of grass; the spores then pass through the animal's gut unharmed and begin the life cycle again in a new dung pile.

Glomeromycota. Until recently, fungi in the phylum Glomeromycota were classified as zygomycetes based on morphological similarities such as the lack of regular septa. However, these fungi are quite dissimilar to zygomycetes in many ways—for example, sexual reproduction is unknown in this group of fungi, with spores usually forming asexually simply by walling off a section of a hypha **(Figure 25.11)**—causing many

a. Arbuscules (black) in leek root colonized by arbuscular mycorrhizal fungus

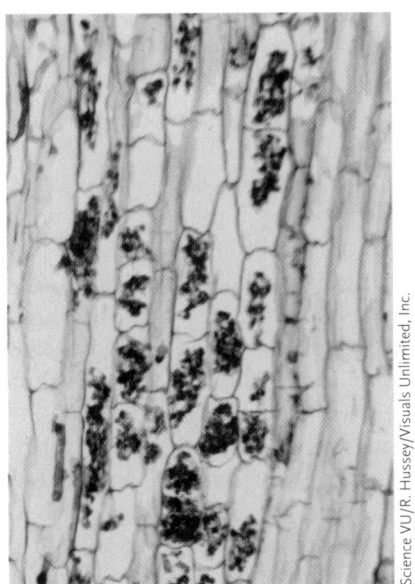

Science VU/R. Hussey/Visuals Unlimited, Inc.

b. Arbuscules inside root

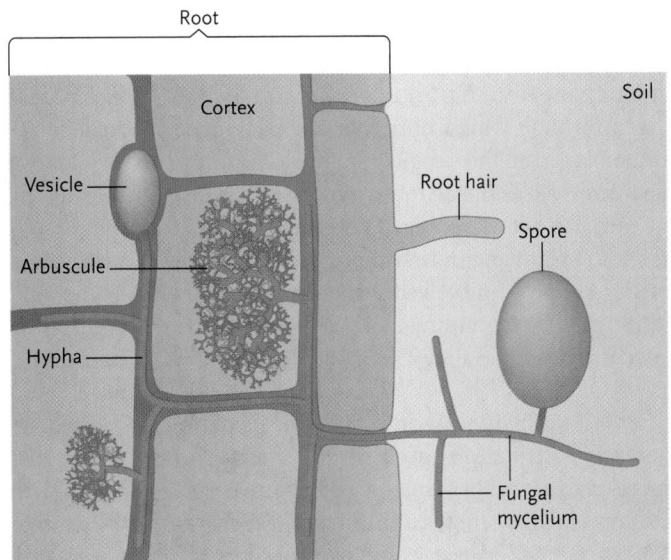

Root

Cortex

Soil

Vesicle

Arbuscule

Hypha

Root hair

Spore

Fungal mycelium

Figure 25.11

Glomeromycete fungus forming a mycorrhiza. **(a)** In this instance, the roots of leeks are growing in association with the glomeromycete fungus *Glomus versiforme* (longitudinal section). Notice the arbuscules that have formed as fungal hyphae branch after entering the root **(b)**.

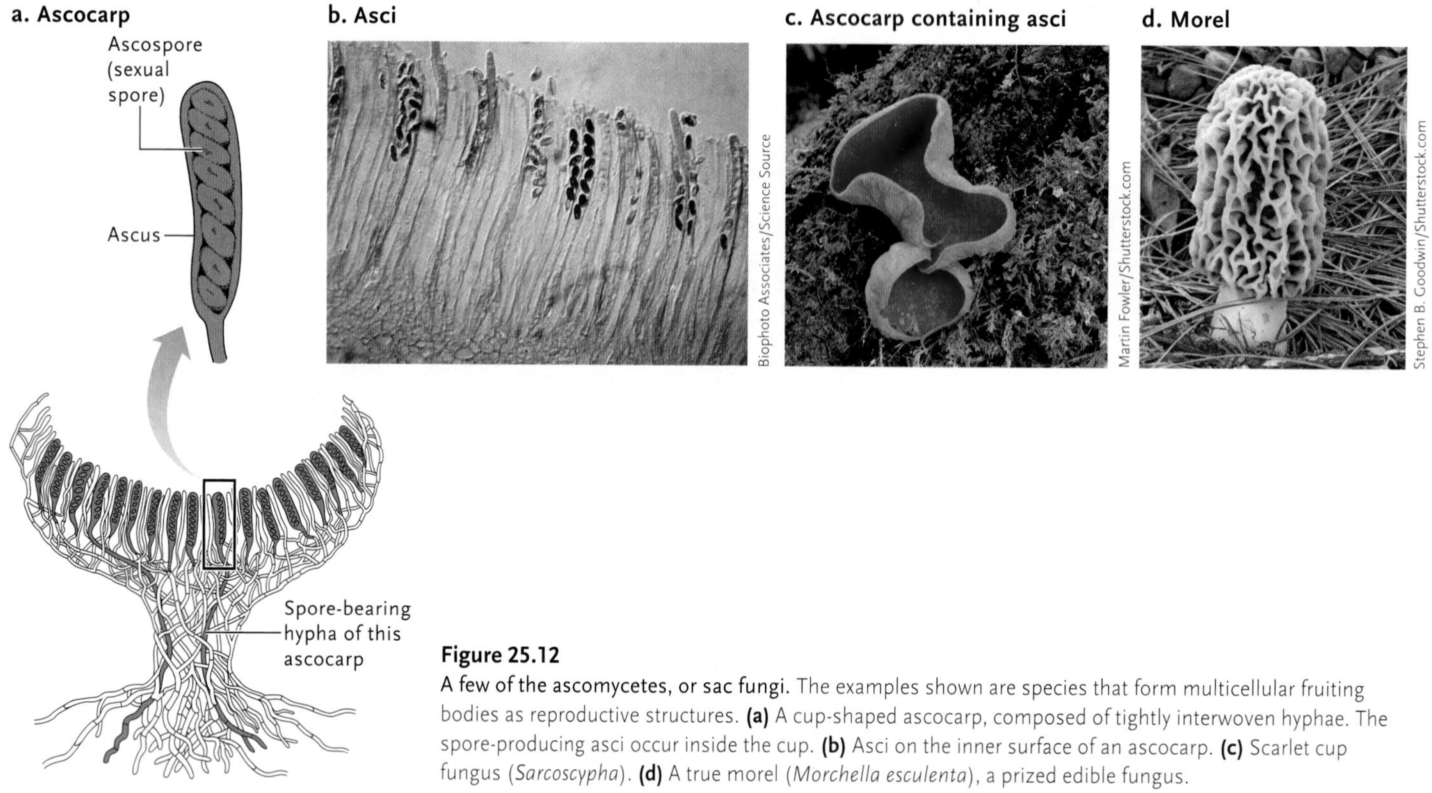

a. Ascocarp

Ascospore (sexual spore)

Ascus

Spore-bearing hypha of this ascocarp

b. Asci

Biophoto Associates/Science Source

c. Ascocarp containing asci

Martin Fowler/Shutterstock.com

d. Morel

Stephen B. Goodwin/Shutterstock.com

Figure 25.12

A few of the ascomycetes, or sac fungi. The examples shown are species that form multicellular fruiting bodies as reproductive structures. **(a)** A cup-shaped ascocarp, composed of tightly interwoven hyphae. The spore-producing asci occur inside the cup. **(b)** Asci on the inner surface of an ascocarp. **(c)** Scarlet cup fungus (*Sarcoscypha*). **(d)** A true morel (*Morchella esculenta*), a prized edible fungus.

researchers to question the inclusion of these fungi in the phylum Zygomycota. Recent evidence from molecular studies resulted in these fungi being placed in their own phylum.

The 160 known members of this phylum are all specialized to form **mycorrhizas**, or symbiotic associations with plant roots. This group of fungi has a tremendous ecological importance as they collectively make up roughly half of the fungi in soil and form mycorrhizas with many land plants, including most major crop species, such as wheat and maize. Mycelia of these fungi colonize the roots of host plants and also proliferate in the soil around the plants. Inside the roots, hyphae penetrate cell walls and branch repeatedly to form **arbuscules** ("little trees") (see Figure 25.11). The branches of each arbuscule are enfolded by the cell's plasma membrane, forming an interface with a large surface area through which nutrients are exchanged between the plant and the fungus. Some glomeromycetes also form vesicles inside roots, which store nutrients and can also act as spores. The fungus obtains sugars from the plant and in return provides the plant with a steady supply of dissolved minerals that it has obtained from the surrounding soil. We take a closer look at mycorrhizas in Section 25.3.

Ascomycota. The phylum Ascomycota takes its name from the saclike structures (**asci**; singular, *ascus*) in which spores are formed in sexual reproduction. These asci are often enclosed in a fruiting body (**ascocarp**) **(Figure 25.12a, b, c)**. However, some ascomycetes are

yeasts or filamentous fungi with a yeast stage, which reproduce asexually by budding or binary fission (see Figure 25.14). Ascomycetes are much more numerous than chytrids, zygomycetes, or glomeromycetes, with more than 30 thousand identified species.

Some ascomycetes are very useful to humans. One species, the orange bread mould *Neurospora crassa*, has been important in genetic research, including the elucidation of the one gene–one enzyme hypothesis (see Chapter 13). *Saccharomyces cerevisiae*, which produces the ethanol in alcoholic beverages and the carbon dioxide that leavens bread, is also a model organism used in genetic research. By one estimate, it has been the subject of more genetic experiments than any other eukaryotic microorganism. This multifaceted phylum also includes gourmet delicacies such as truffles (*Tuber melanosporum*) and the succulent morel *Morchella esculenta* (Figure 25.12d).

Many ascomycetes are saprotrophs, playing a key role in the breakdown of cellulose and other polymers. Ascomycetes are also common in symbiotic associations, forming mycorrhizas and lichens (see Section 25.3). A few ascomycetes prey on various agricultural insect pests—some are even carnivores that trap their prey in nooses **(Figure 25.13a)**—and thus have potential for use as biological pesticides.

However, other ascomycetes are devastating plant pathogens, including the blue-stain fungi that are associated with mountain pine beetles and contribute to the death of beetle-infested trees (Figure 25.13b). Several ascomycetes can be serious pathogens of humans.

a. A trapping ascomycete

George Barron

b. Stump of pine tree infected with blue-stain fungus

U.S. Forest Service

Figure 25.13

(a) Nematode-trapping fungus. Hyphae of this ascomycete (*Arthobotrys*) form nooselike rings. When a prey organism enters the loop, rapid changes in ion concentration draw water into the loop by osmosis. The increased turgor pressure causes the noose to tighten, trapping its prey. Enzymes produced by the fungus then break down the nematode's tissues. **(b)** Stump of a pine tree infected with blue-stain fungus; the fungus grows into the tree's water-conducting tissue, blocking the flow of water.

Yeast cells

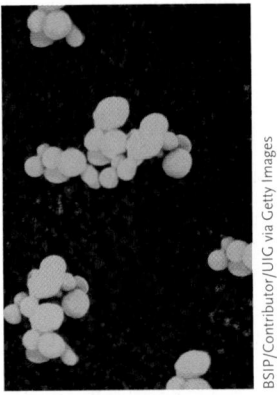

BSIP/Contributor/UIG via Getty Images

Figure 25.14

Candida albicans, the cause of yeast infections of the mouth and vagina.

The yeast *Candida albicans* **(Figure 25.14)** infects mucous membranes, especially of the vagina and mouth, causing a condition called thrush. Another yeast, *Pneumocystis jirovecii*, causes virulent pneumonia in AIDS patients and other immunocompromised people.

Claviceps purpurea, a parasite on rye and other grains, causes ergotism, a disease marked by vomiting; hallucinations; convulsions; and, in severe cases, gangrene and even death. It has even been suggested that this fungus was the cause of the Salem witch hunts of seventeenth century New England, as discussed in "Molecule behind Biology," 25.2, p. 550. Other ascomycetes cause nuisance infections, such as athlete's foot and ringworm.

Most ascomycetes grow as haploid mycelia with regular septa; large pores in the septa allow organelles, including nuclei, to move with cytoplasm through the mycelium. Sexual reproduction generally involves fusion of hyphae from mycelia of + and − mating types **(Figure 25.15, p. 551)**. The cytoplasms of the two hyphae fuse, but fusion of the nuclei is delayed, resulting in the formation of **dikaryotic hyphae** that contain two separate nuclei and thus are referred to as $n + n$ rather than n or $2n$. Sacs (asci) form at the tips of these dikaryotic hyphae; inside the asci, the two nuclei fuse, forming a diploid zygote nucleus, which then undergoes meiosis to produce four haploid nuclei. Mitosis usually follows, resulting in the formation of eight haploid spores (**ascospores**).

Unlike zygomycetes, ascomycetes do not produce asexual spores in sporangia. Instead, modified hyphae produce numerous asexual spores called **conidia** ("dust"; singular, *conidium*), such as those seen when powdery mildew attacks grasses, roses, and other common garden plants **(Figure 25.16a, p. 551)**. The mode of conidial production varies from species to species, with some ascomycetes producing chains of conidia, whereas in others, the conidia are produced on a hypha in a series of "bubbles," rather like a string of detachable beads (Figure 25.16b). Either way, conidia are formed and released much more quickly than zygomycete spores.

As you can see from Figure 25.15, these asexual reproductive structures look very different from the sexual stages and are often not formed at the same time or under the same conditions as the sexual stage of the life cycle. These differences resulted in the asexual stages of many ascomycetes being classified as separate organisms from the sexual stages of the same species. Since fungal classification traditionally relied on features produced in sexual reproduction, these asexual stages could not be placed in any of the phyla; instead, researchers grouped them together in an artificial group called the Deuteromycota (also known as Fungi Imperfecti, or the "imperfect fungi"— imperfect meaning that a sexual stage is absent). Well-known examples of fungi once classified as deuteromycetes are *Penicillium* and *Aspergillus*. Certain species of *Penicillium* (Figure 25.16b) are the source

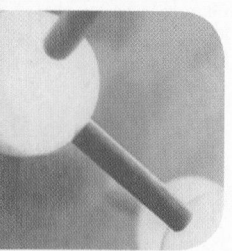

MOLECULE BEHIND BIOLOGY 25.2

Lysergic Acid

Was a fungus responsible for the Salem witch trials? In Salem, Massachusetts, in 1692–1693, several women were tried and found guilty of witchcraft. Their accusers were young women who had been experiencing bizarre symptoms: hallucinations, convulsions, a sensation of "prickling" of the skin, and even paralysis. Further evidence of witchcraft was the fact that cattle and other animals also suffered these symptoms. What was the real cause of these symptoms? Were they an example of mass hysteria? Or is there a biological explanation?

The symptoms reported by the "bewitched" girls match those of someone who has eaten flour made from wheat infected by the ascomycete fungus *Claviceps purpurea*. Ascospores of this fungus germinate when they land on the flower of a grass plant, such as wheat. The fungus grows quickly and, by the end of the growing season, forms a tough mass of hyphae known as a **sclerotium** in the seed head of the grass **(Figure 1)**. If the seed head isn't harvested, the sclerotia will fall to the ground, where they remain over winter. In the spring, the sclerotia will germinate, producing numerous fruiting bodies borne on stalks. However, if the fungus has infected a commercial grain crop, such as wheat, sclerotia are easily harvested along with the plants' seed heads and often end up being ground into flour along with the grain. In medieval times, if the weather favoured development of the fungus, up to 30% of some grain harvests were evidently not grain but sclerotia!

Figure 1
Sclerotium of Claviceps purpurea *in a grass seed head.*

Sclerotia produce many alkaloids, including lysergic acid **(Figure 2)**, which causes a range of symptoms, including hallucinations, convulsions, a sensation of ants crawling over the body, limb distortions, and dementia. These symptoms match those of the supposedly bewitched people of Salem in 1692. Further support for ergotism being the cause of the bewitching is the fact that most of the victims were adolescents, who are most suspectible to the effects of ergot alkaloids. Furthermore, the fact that cattle and other domestic animals would also have eaten infected grain and also presented the same symptoms as the "victims" suggests that ergot, not mass hysteria, was involved. Lysergic acid was purified in 1943 by a chemist (Albert Hofmann) to produce the

Figure 2
Structure of lysergic acid.

psychoactive drug LSD. Researchers hoped that this drug would be useful in psychotherapy, but its negative effects outweighed the benefits, and this line of research was dropped. However, other ergot alkaloids are used as treatment for migraine headaches.

---

of the penicillin family of antibiotics, whereas others produce the aroma and distinctive flavours of Camembert and Roquefort cheeses. Strains of *Aspergillus* grow in damp grain or peanuts. Their metabolic wastes, known as aflatoxins, can cause cancer in humans who eat the poisoned food over an extended period. With the development of molecular sequencing techniques, many fungi that were classified as deuteromycetes can now be reassigned to the appropriate phylum; most are ascomycetes, but some are basidiomycetes, which also produce conidia in asexual reproduction.

Basidiomycota. The 24 thousand or so species of fungi in the phylum Basidiomycota include the mushroom-forming species, bracket fungi, stinkhorns, smuts, rusts, and puffballs **(Figure 25.17, p. 552).** The common name for this group is club fungi, due to the club-shaped cells (**basidia**; singular, *basidium*) in which sexual spores are produced.

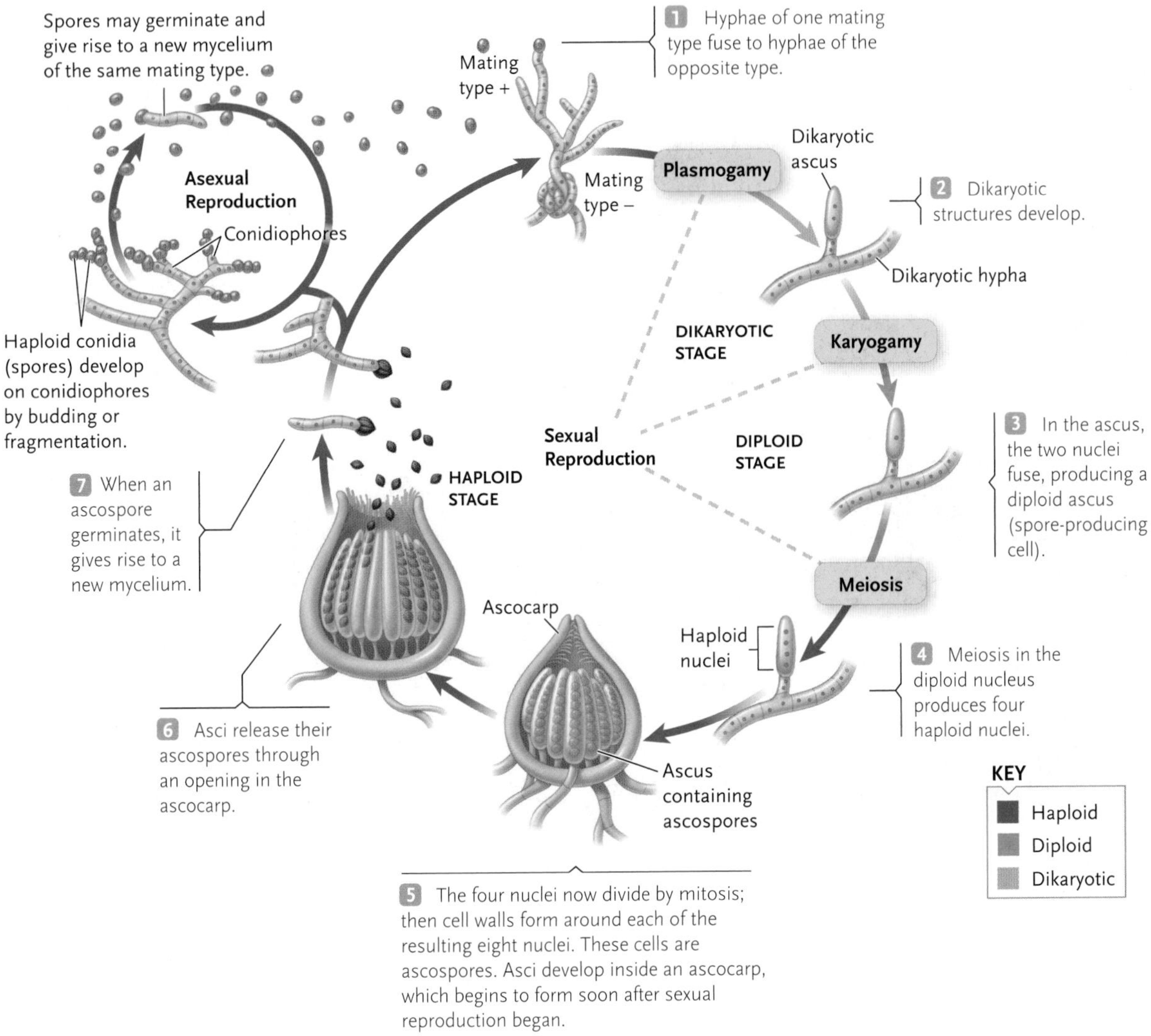

Spores may germinate and give rise to a new mycelium of the same mating type.

1 Hyphae of one mating type fuse to hyphae of the opposite type.

Mating type +

Mating type −

Plasmogamy

Dikaryotic ascus

Asexual Reproduction

Conidiophores

2 Dikaryotic structures develop.

Dikaryotic hypha

Haploid conidia (spores) develop on conidiophores by budding or fragmentation.

DIKARYOTIC STAGE

Karyogamy

3 In the ascus, the two nuclei fuse, producing a diploid ascus (spore-producing cell).

7 When an ascospore germinates, it gives rise to a new mycelium.

Sexual Reproduction

DIPLOID STAGE

HAPLOID STAGE

Meiosis

Ascocarp

Haploid nuclei

4 Meiosis in the diploid nucleus produces four haploid nuclei.

6 Asci release their ascospores through an opening in the ascocarp.

Ascus containing ascospores

KEY

■ Haploid
■ Diploid
■ Dikaryotic

5 The four nuclei now divide by mitosis; then cell walls form around each of the resulting eight nuclei. These cells are ascospores. Asci develop inside an ascocarp, which begins to form soon after sexual reproduction began.

Figure 25.15
Life cycle of the ascomycete *Neurospora crassa*.

Many basidiomycetes produce enzymes for digesting cellulose and lignin and are important decomposers of woody plant debris. Very few organisms can degrade lignin due to its very complex, irregular structure **(Figure 25.18, p. 552)**. The ability to degrade lignin also enables some basidiomycetes to break down complex organic compounds such as DDT, PCBs, and other persistent environmental pollutants that are structurally similar to lignin. Bioremediation of contaminated sites by these fungi is a very active research area.

A surprising number of basidiomycetes, including the prized edible oyster mushrooms (*Pleurotus ostreatus*), can also trap and consume small animals such as rotifers and nematodes by secreting paralyzing toxins or gluey substances that immobilize the prey, in a manner similar to that shown earlier for ascomycetes (Figure 25.13, p. 549). As is the case for insectivorous plants, such as the pitcher plants (*Saracenea purpurea*), discussed in Chapters 20 and 31, this adaptation gives the fungus access to a rich source of molecular nitrogen, an essential nutrient that is often scarce in terrestrial habitats. For example, the wood that is the substrate for many basidiomycetes is high in carbon but low in nitrogen; many wood-decay fungi have been found to be carnivorous, obtaining supplemental nitrogen from various invertebrates.

Some basidiomycetes form mycorrhizas with the roots of forest trees, as discussed later in this chapter. Recent research has shown that these mycorrhizas can be drawn into associations with achlorophyllous plants

Figure 25.16
(a) Powdery mildew on leaves. **(b)** Conidia of *Penicllium*. Note the rows of conidia (asexual spores) atop the elongate cells that produce them.

a.

NIGEL CATTLIN/SCIENCE PHOTO LIBRARY

b.

Biophoto Associates/Science Source

a. Coral fungus

b. Shelf fungus

c. White-egg bird's nest fungus

d. Fly agaric mushroom

e. Scarlet hood

Figure 25.17

Examples of basidiomycetes, or club fungi. **(a)** The light red coral fungus *Ramaria*. **(b)** The shelf fungus *Polyporus*. **(c)** The white-egg bird's nest fungus *Crucibulum laeve*. Each tiny "egg" contains spores. Raindrops splashing into the "nest" can cause "eggs" to be ejected, thereby spreading spores into the surrounding environment. **(d)** The fly agaric mushroom *Amanita muscaria*, which causes hallucinations. **(e)** The scarlet hood *Hygrophorus*.

(plants that lack chlorophyll and so cannot carry out photosynthesis), which thus obtain nutrients from the trees via shared mycorrhizal fungi. Other basidiomycetes, the rusts and smuts, are parasites that cause serious diseases in wheat, rice, and other plants. Still others produce millions of dollars worth of the common edible button mushroom (*Agaricus bisporus*) sold in grocery stores. *Amanita muscaria* (Figure 25.17d) has been used in the religious rituals of ancient societies in Central America, Russia, and India. Other species of this genus, including the death cap mushroom *Amanita phalloides*, produce deadly toxins. The *A. phalloides* toxin, called α-amanitin, halts gene transcription, and hence protein synthesis, by inhibiting the activity of RNA polymerase. Within 8 to 24 hours of ingesting as little as 5 mg of the mushroom, vomiting and diarrhea begin. Later, kidney and liver cells start to degenerate; without intensive

Figure 25.18

A portion of a lignin molecule. Unlike most other biopolymers, lignin is not composed of regularly repeating monomers but instead is a complex polymer of various phenylpropane units, joined together by a range of diverse bonds, making it very difficult to degrade.

medical care, death can follow within a few days. You can read more about the effect of amanitin on gene expression in Chapter 13.

Most basidiomycetes are mycelial, although some grow as yeasts. The mycelium of many basidiomycetes contains two different, separate nuclei as a result of fusion between two different haploid mycelia and is termed a **dikaryon** ($n + n$) **(Figure 25.19).** A dikaryotic mycelium is formed following fusion of the two haploid mycelia when both types of nuclei divide and migrate through the mycelium such that each hyphal compartment contains two dissimilar nuclei.

Basidiomycete fungi can grow for most of their lives as dikaryon mycelia—a major departure from an ascomycete's short-lived dikaryotic stage. After an extensive mycelium develops, favourable environmental conditions trigger the formation of fruiting bodies (**basidiocarps**), in which basidia develop. A basidiocarp consists of tight clusters of hyphae; the feeding mycelium is buried in the substrate. The shelf-like bracket fungi visible on trees are basidiocarps, as are the structures we call mushrooms and toadstools. Each mushroom is a short-lived reproductive body consisting of a stalk and a cap; basidia develop on "gills," the sheets of tissue on the underside of the cap. Inside each basidium, the two nuclei fuse; meiosis follows, resulting in the formation of four haploid **basidiospores** on the outside of the basidium (see Figure 25.19). Why does the fungus expend energy and resources on such elaborate spore-dispersal structures? A layer of still air

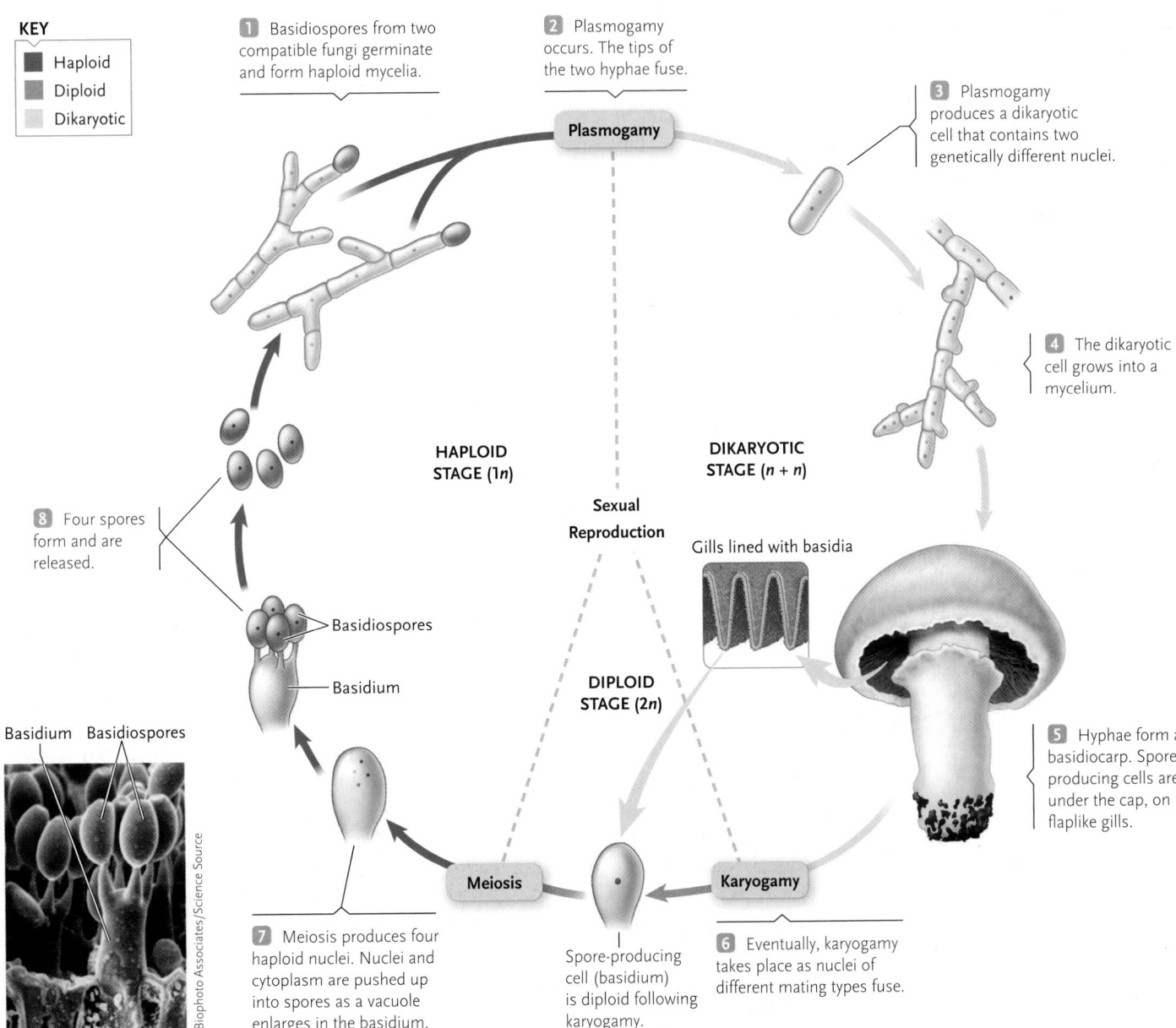

KEY
- Haploid
- Diploid
- Dikaryotic

1 Basidiospores from two compatible fungi germinate and form haploid mycelia.

2 Plasmogamy occurs. The tips of the two hyphae fuse.

Plasmogamy

3 Plasmogamy produces a dikaryotic cell that contains two genetically different nuclei.

4 The dikaryotic cell grows into a mycelium.

HAPLOID STAGE (1n)

DIKARYOTIC STAGE (n + n)

Sexual Reproduction

Gills lined with basidia

8 Four spores form and are released.

Basidiospores

Basidium

DIPLOID STAGE (2n)

5 Hyphae form a basidiocarp. Spore-producing cells are under the cap, on flaplike gills.

Basidium   Basidiospores

Biophoto Associates/Science Source

Meiosis

Karyogamy

7 Meiosis produces four haploid nuclei. Nuclei and cytoplasm are pushed up into spores as a vacuole enlarges in the basidium.

Spore-producing cell (basidium) is diploid following karyogamy.

6 Eventually, karyogamy takes place as nuclei of different mating types fuse.

Figure 25.19
Generalized life cycle of the basidiomycete *Agaricus bisporus,* a species known commonly as the button mushroom. During the dikaryotic stage, cells contain two genetically different nuclei, shown here in different colours. Inset: Micrograph showing basidia and basidiospores.

occurs just above the ground (and any other surface); by elevating the basidia above this layer, the fungus increases the likelihood that its spores will be carried away by the wind.

CONCEPT FIX People often assume, when they see mushrooms sprouting from the ground, that each mushroom is an individual. But that's not true—mushrooms and other fungal fruiting bodies are produced by mycelia growing through their substrate. A mycelium, not a mushroom or fruiting body, is the "individual." ●

The prolonged dikaryon stage in basidiomycetes allows them many more opportunities for producing sexual spores than in ascomycetes, in which the dikaryon state is short lived. Basidia can produce huge numbers of spores—some species can produce 100 million spores *per hour* during reproductive periods, day after day! Basidiomycete mycelia can live for many years and spread over large areas. The largest organism on Earth could be the mycelium of a single individual of the basidiomycete *Armillaria ostoyae*, which spreads over 8.9 km² of land in eastern Oregon. This organism weighs at least 150 tonnes and is likely at least 2400 years old, making it not only the largest but also one of the heaviest and oldest organisms on Earth.

As for ascomycetes, asexual reproduction in basidiomycetes involves formation of conidia or budding in yeast forms such as *Cryptococcus gattii,* which causes cryptococcal disease in humans. A virulent strain of *C. gattii*, first reported from Vancouver Island in 1999, has since spread to the northwestern United States and California. Normally, only people with weakened immune systems, such as transplant recipients and cancer patients, are at risk from fungal pathogens, but *C. gattii* is different, causing disease in healthy people. The disease starts when spores of the fungus, which lives in trees and soil, are inhaled. In the lungs, the spores germinate to produce yeast cells that proliferate by budding in the warm, moist lung environment; the yeast cells then spread to the central nervous system via the bloodstream. The disease is characterized by a severe cough, fever, and, if the nervous system is affected, seizures and other neurological symptoms.

STUDY BREAK

1. What evidence is there that fungi are more closely related to animals than to plants?
2. Name the five phyla of the kingdom Fungi, and describe the reproductive adaptations that distinguish them.
3. What are the two main differences between asexual spores produced by zygomycetes and asexual spores produced by ascomycetes?
4. Fungi reproduce sexually or asexually, but for many species, the life cycle includes an unusual stage not seen in other organisms. What is this genetic condition, and what is its role in the life cycle?

25.3 Fungal Lifestyles

As mentioned earlier, fungi can be categorized as saprotrophs or symbionts, depending on whether they obtain nutrients from living organisms or from dead organic matter. It is important to remember that the categories of *saprotroph* and *symbiont* were created as separate categories to classify fungi, but fungi are very versatile organisms, and many fungi are capable of acting as both symbionts and saprotrophs at different times or under different conditions. Most people are more familiar with the role of fungi as saprotrophs (decomposers) rather than as symbionts, so in this section, we take a brief look at saprotrophy and then spend more time looking at fungal symbioses.

25.3a Some Fungi Are Saprotrophs

With their adaptations for efficient extracellular digestion, fungi are masters of the decay so vital to terrestrial ecosystems (see Figure 25.2, p. 540). For instance, in a single autumn, one elm tree can shed 200 kg of withered leaves! Without the metabolic activities of saprotrophic fungi and other decomposers such as bacteria and other organisms (e.g., earthworms), natural communities would rapidly become buried in their own detritus (dead organic matter). Even worse, without decomposers to break down this detritus, the soil would become depleted of nutrients, making further plant growth impossible. As fungi (and other decomposers) digest the dead tissues of other organisms, they also make a major contribution to the recycling of the chemical elements those tissues contain. For instance, over time, the degradation of organic compounds by saprotrophic fungi helps return key nutrients such as nitrogen and phosphorus to ecosystems. But the prime example of this recycling virtuosity involves carbon. The respiring cells of fungi and other decomposers give off carbon dioxide, liberating carbon that would otherwise remain locked in the tissues of dead organisms. Each year, this activity recycles a vast amount of carbon to plants, the primary producers of nearly all ecosystems on Earth.

However, there is a downside to the impressive enzymatic abilities of saprotrophic fungi; for example, when they decompose materials that are part of our houses, they can cause major economic and health problems. Fungi growing on wood and drywall following flooding or water damage to a building **(Figure 25.20a)** not only weaken the structural integrity of the building but also can be health hazards. The airborne spores of these fungi act as allergens, and some can also cause more serious health problems— for example, some fungi can colonize and grow in sinus cavities. Another example is dry rot, which causes millions of dollars in damage to buildings in Europe, Asia, and Australia (Figure 25.20b). Dry rot is notorious not only because it causes widespread and

a.

b.

Figure 25.20

(a) Mould growth following flooding. **(b)** Mycelium of dry rot (*Serpula lacrymans*) emerging through a wall.

costly damage but also because the responsible fungus, *Serpula lacrymans,* seems to have the mysterious ability to break down dry wood completely, which should not be possible—as described above, wood decay usually happens once wood becomes wet. Does this fungus really have the amazing ability to break down dry wood? In fact, this fungus is as dependent on water for growth as any other, but it can form specialized mycelial cords, which very efficiently transport water and nutrients over long distances through concrete, bricks, and other unfavourable substrates until the fungus at last finds wood. Then the mycelial cords release water into the substrate, allowing the fungus to spread through the wood and begin the process of decay.

25.3b Some Fungi Are Symbionts

Symbiotic associations range from mutualism, in which both partners benefit, to parasitism, in which one partner benefits at the expense of the other. Many fungal parasites are pathogens, parasites that cause disease symptoms in their hosts. We have discussed several examples of fungal diseases in humans and other animals earlier in this chapter. In this section, we will focus on fungi as mutualists. Many fungi are partners in mutually beneficial interactions with animals or photosynthetic organisms; some of these associations shaped the evolution of life on Earth and still play major roles in the functioning of ecosystems today. Chapter 30 discusses the general features of symbiotic associations more fully; here we are interested in some examples of the symbioses fungi form with other organisms.

Lichens Are Associations between a Fungus and One or More Photosynthetic Organisms. A **lichen** is a compound organism formed by an association between a fungus, an ascomycete or sometimes a basidiomycete, and a green alga and/or a cyanobacterium. Lichens may grow as crusts on rocks, bark, or soil; as flattened leaflike forms; or as radially symmetrical cups, treelike structures, or hairlike strands **(Figure 25.21, p. 556)**. Lichens have vital ecological roles and important human uses. Lichens secrete acids that eat away at rock, breaking it down and converting it to soil that can support plants. Many animals, such as caribou (*Rangifer tarandus*), rely on lichens for their winter forage. Some environmental chemists monitor air pollution by monitoring lichens, most of which cannot grow in heavily polluted air because they cannot discriminate between pollutants and mineral nutrients present in the atmosphere. Just as they do for mineral nutrients, lichens efficiently absorb airborne pollutants and concentrate them in their tissues. Humans use lichens as sources of dyes and perfumes, as well as medicines. Lichen chemicals are currently being explored as a source of natural pesticides.

The fungus (called the **mycobiont**) makes up most of the body (**thallus**) of the lichen, with the photosynthetic partner (**photobiont**) usually confined to a thin layer inside the lichen thallus (see Figure 25.21a). Some lichens have a green algal photobiont inside the thallus and a cyanobacterial photobiont contained in "pockets" on or in the thallus. Because lichens are composite organisms, it may seem odd to talk of lichen "species," but biologists do give lichens binomial names, based on the mycobiont. More than 13 500 different lichen species are recognized, each a unique combination of a particular species of fungus and one or more species of photobiont. As you might expect for a compound organism made up of two (or even three) organisms, reproduction can be complicated: it is not enough for each organism to reproduce itself, because formation of a new lichen requires that both partners be dispersed and end up together. Many lichens reproduce asexually, by specialized fragments such as the **soredia** (singular, *soredium*), shown in Figure 25.21b. Each soredium consists of photobiont cells wrapped in hyphae; the soredia can be dispersed by water, wind, or passing animals.

Inside the thallus, specialized hyphae wrap around and sometimes penetrate photobiont cells, which become the fungus's sole source of carbon. Often the

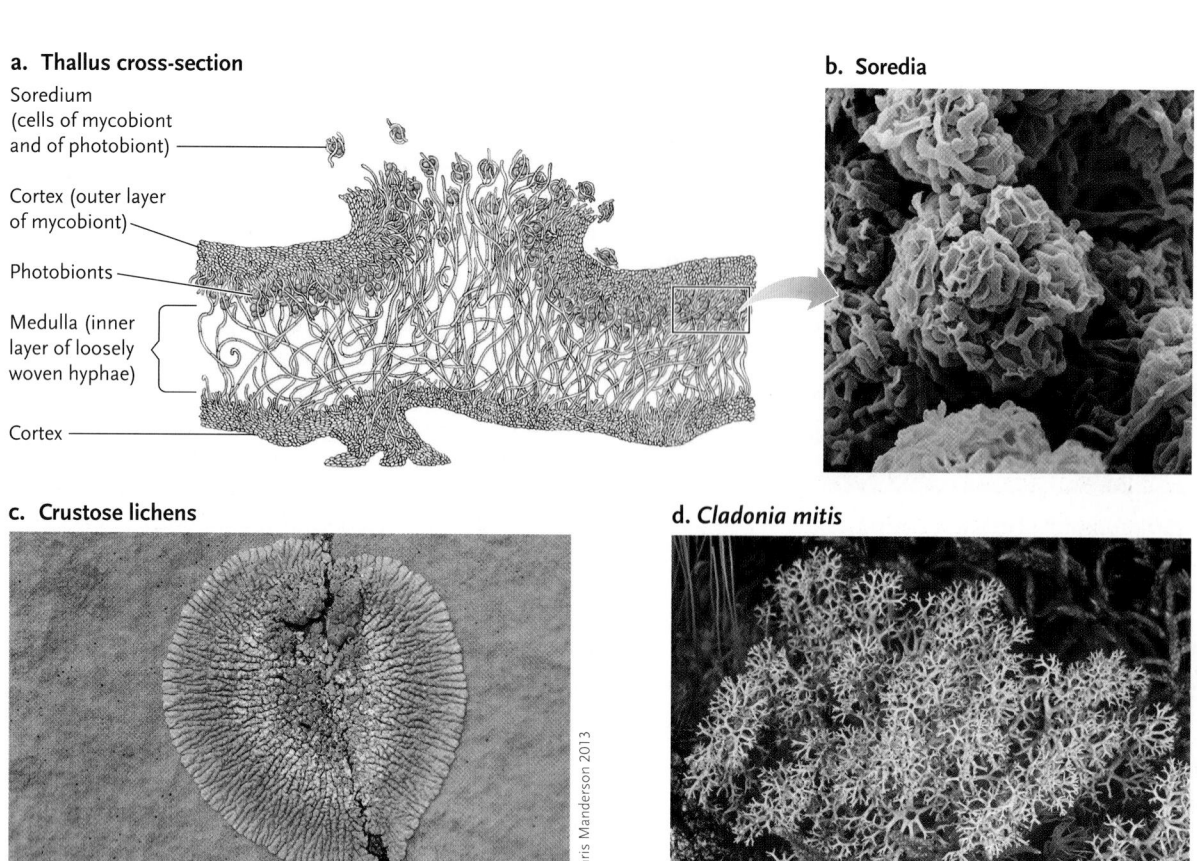

a. Thallus cross-section

Soredium
(cells of mycobiont
and of photobiont)

Cortex (outer layer
of mycobiont)

Photobionts

Medulla (inner
layer of loosely
woven hyphae)

Cortex

b. Soredia

EYE OF SCIENCE/SCIENCE PHOTO LIBRARY

c. Crustose lichens

© Chris Manderson 2013

d. *Cladonia mitis*

Stephen Sharnoff/Visuals Unlimited, Inc.

Figure 25.21

Lichens. **(a)** Diagram of a cross-section through the thallus of the foliose lichen *Lobaria verrucosa*. Soredia **(b),** which contain both hyphae and algal cells, are a type of dispersal fragment by which lichens reproduce asexually. **(c)** Crustose lichens. **(d)** *Cladonia mitis*, a branching, treelike lichen.

mycobiont absorbs up to 80% of the carbohydrates produced by the photobiont. Benefits for the photobiont are less clear cut, in part because the drain on nutrients hampers its growth and because the mycobiont often controls reproduction of the photobiont. In one view, many and possibly most lichens are parasitic symbioses, with the fungus enslaving the photobiont. On the other hand, although it is relatively rare to find a lichen photobiont species living independently in the same conditions under which the lichen survives, it may eke out an enduring existence as part of a lichen; some lichens have been dated as being more than 4000 years old! Studies have also revealed that at least some green algae clearly benefit from the relationship. Such algae are sensitive to desiccation and intense ultraviolet radiation. Sheltered by the lichen's fungal tissues, a green alga can thrive in locales where alone it would perish. Clearly, we still have much to learn about the physiological interactions between lichen partners.

Lichens often live in harsh, dry microenvironments, including on bare rock and wind-whipped tree trunks. Some lichens actually live *inside* rocks (see "Life on the Edge," Box 25.3, p. 558). Unlike plants, lichens do not control water loss from their tissues; instead, their water status reflects that of their environment,

and some lichens may dry out and re-wet several times a day. Lichens are very slow growing, even though the photobiont may have photosynthetic rates comparable to those of free-living species. What happens to all of the carbohydrates made in photosynthesis if they are not used to fuel growth? The mycobiont takes much of the carbohydrate made by the photobiont and uses it to synthesize secondary metabolites and other compounds that allow the lichen to survive the repeated wet–dry cycles and extreme temperatures common in their habitats. These compounds give lichens their vibrant colours and may also inhibit grazing on lichens by slugs and other invertebrates. The mycobiont uses other lichen chemicals to control the photobiont; some chemicals regulate photobiont reproduction, whereas others cause photobiont cells to "leak" carbohydrates to the mycobiont.

Mycorrhizas Are Symbiotic Associations between Fungi and Plant Roots. You might have learned in previous courses that plant roots are responsible for taking up soil nutrients. For most plants, however, this is not true: the roots of most plants are colonized by mycorrhizal fungi, which have mycelia that extend out into the soil far beyond the root zone of the plant and which take up most of the nutrients used by the plant **(Figure 25.22)**. Mycorrhizas, or "fungus roots," are mutualistic symbioses between certain soil-dwelling fungi and plant roots. Mycorrhizal plants greatly enhance the uptake of various nutrients, especially phosphorus and nitrogen, from soil (as discussed in Chapter 37) because the fungal mycelium has a tremendous surface area for absorbing mineral ions from a large volume of the surrounding soil.

As well, some mycorrhizal fungi can access sources of nutrients that are not available to plants; for example, certain basidiomycete fungi can penetrate directly into rocks and extract nutrients, which are then transported to their plant hosts. Other mycorrhizal associations involve the carnivorous basidiomycetes described above, which can obtain nitrogen by trapping and killing soil invertebrates and then transferring nitrogen from their prey to their host plants. By forming partnerships with these fungi, mycorrhizal plants gain access to nutrient sources that nonmycorrhizal plants do not. In exchange for soil nutrients, the plants provide the mycorrhizal fungi with sugars produced through photosynthesis. Mycorrhizas are generally mutualisms, representing a win–win situation for the partners. For plants that inhabit soils poor in mineral ions, such as in **tropical rain forests**, mycorrhizal associations are crucial for survival. Likewise, in temperate forests, species of spruce, oak, pine, and some other trees die unless mycorrhizal fungi are present **(Figure 25.23)**. There are at least seven different types of mycorrhizas, but the most common types are ectomycorrhizas and arbuscular mycorrhizas.

Arbuscular mycorrhizas are the oldest and most abundant type of mycorrhiza, formed by glomeromycete fungi and a wide range of plants, including nonseed plants and most flowering plants. In this type of mycorrhiza, fungal hyphae penetrate the cells of the root, forming arbuscules as described above (see Figure 25.11, p. 547). Fossils show that arbuscular mycorrhizas were common among ancient land plants, and some biologists have speculated that they might have been crucial for the colonization of land by plants by enhancing the transport of water and minerals to the plants.

Ectomycorrhizas evolved more recently and involve basidiomycetes and some ascomycetes. In these mycorrhizas, fungal hyphae form a sheath or mantle around a root (see Figure 25.22) and also grow between, but not inside, the root cells of their plant hosts. Ectomycorrhizal associations are very common with trees, such as the conifers of Canada's **boreal forest** and coastal rain forests. The extensive root system of a single mature pine may be studded with ectomycorrhizas involving dozens of fungal species.

a. Mycorrhizal symbiosis between Lodgepole pine and mycorrhizal fungus.

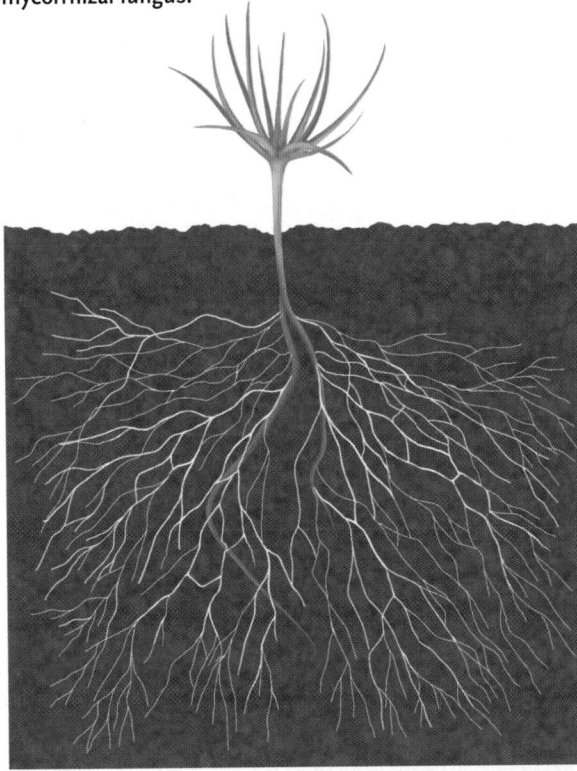

b. Mycorrhiza

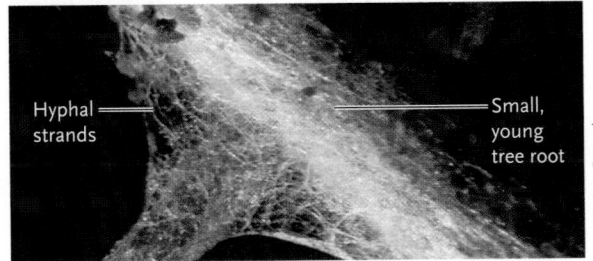

Hyphal strands

Small, young tree root

© 1999 Gary Braasch

Figure 25.22

Ectomycorrhizas. **(a)** Lodgepole pine, *Pinus contorta*, seedling grown in symbiosis with an ectomycorrhizal fungus. Notice the extent of the mycorrhizal fungal mycelium compared with the above-ground portion of the seedling, which is only about 4 cm tall. **(b)** Mycorrhiza of a hemlock tree.

Science VU/R.Roncadori/Visuals Unlimited, Inc.

Figure 25.23

Effect of mycorrhizal fungi on plant growth. The six-month-old juniper seedlings on the left were grown in sterilized low-phosphorus soil inoculated with a mycorrhizal fungus.

Cryptoendolithic Lichens

We tend to think of Antarctica as completely covered in ice, but some valleys of this continent are completely lacking in ice **(Figure 1a)**. These dry valleys may look barren, but they are home to many endoliths—organisms that live in a narrow band under the surface of porous rocks. Predominant among these endoliths are cryptoendo-lithic lichens ("crypto" = hidden; "endo" = inside; "lith" = rock) (Figure 1b). These lichens lack the stratified layers typical of most other lichens; instead, hyphae and clusters of photobiont cells grow around and between the rock crystals, and the lichen that forms is embedded inside the rock. Enough light penetrates the translucent surface layer of rock to allow photosynthesis. Studying endolithic organisms not only helps us understand the diversity of life on Earth but also may be a model for life on other planets. If some organisms can live in such extreme conditions here on Earth, could similar kinds of organisms also exist elsewhere in the universe?

a.

© University of Canterbury—Christchurch, New Zealand

b.

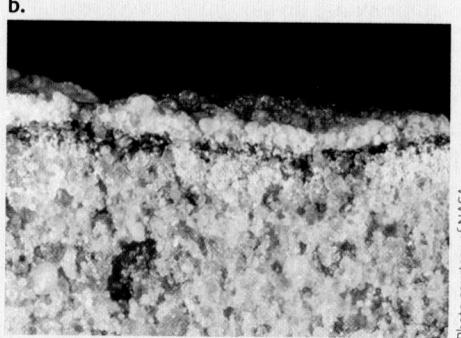

Photo courtesy of NASA

Figure 1
(a) *Antarctic dry valley.* **(b)** *Cryptoendolithic lichen.*

The musky-flavoured truffles (*Tuber melanosporum*) prized by gourmets are ascomycetes that form ectomy-corrhizal associations with oak trees (*Quercus* spp.).

For plants, the benefits of being mycorrhizal extend beyond enhanced uptake of soil nutrients. In some cases, mycorrhizal fungi enhance a plant's defences against pathogens, and nutrients can be transferred among mycorrhizal plants via shared mycorrhizal fungal hyphae. Mycorrhizal fungi may, in fact, play a major role in shaping plant communities and ecosystems.

Endophytes Are Fungi Living in the Above-Ground Tissues of Plants. Just as the roots of many plants are colonized by fungi, so too are leaves and shoots **(Figure 25.24)**. Although some of these fungi are pathogens, many others evidently peacefully coexist with their plant hosts.

Biologists have known about the presence of these leaf endophytes for some time, but recent discoveries have revealed a startling diversity of these fungi, sometimes within a single plant. Samples of plants from temperate regions have been revealed to have tens of different species of endophytes in a single plant, but tropical plants are truly impressive, with several reports of hundreds of different types of endophytes being isolated from a single plant. Most of these endophytes have not yet been identified to species as researchers have not yet observed sexual stages, so it is difficult to know how many species of endophytes are really living in these tropical plants. A bigger question is, what are these endophytes doing in these leaves? Are they mutualists, like mycorrhizal fungi? In many cases, we simply don't know enough about the interaction between the fungus and its host to answer these questions, but in some cases, the fungi do benefit their plant hosts by producing toxins that deter herbivores. Synthesis of toxins and other secondary metabolites has made these endophytes of great potential importance to humans. For example, the anticancer drug taxol (sold under the tradename Taxol) was originally isolated from the bark of the Pacific yew tree (*Taxus brevifolia*). Production of taxol from this source was limited since the tree is quite rare and makes only a small amount of taxol. However, researchers later discovered that a fungal endophyte living in the needles of the Pacific yew also makes taxol—as do other endophytes living in completely different tree species. Evidence indicates that taxol inhibits the growth of other fungi, so these endophytes may be producing it to protect themselves. Did the genes to produce taxol get transferred from the fungi to the plant? Such horizontal gene transfer is known to have occurred in the evolution of organelles such as mitochondria. The possibility that the genes necessary for biosynthesis of taxol were transferred from the endophyte to its host plant is intriguing, but, as of yet, there is no conclusive evidence to support this idea. Unlike the yew trees that

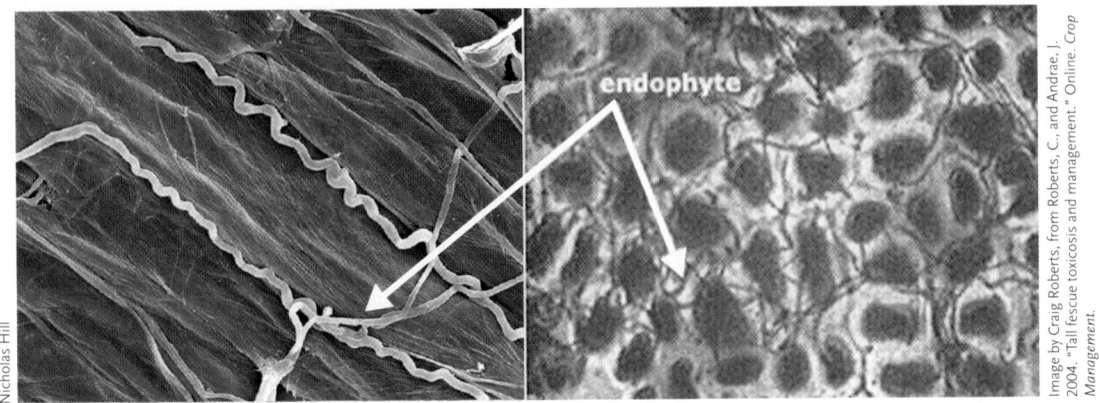

Figure 25.24
Leaf endophytes growing inside plant leaves.

were the original source of taxol, these endophytic fungi can be grown very easily in the lab, so we may be able to produce large amounts of this promising anti-cancer drug very easily. What other sources of medicines are out there, hiding inside plants? The possibility of finding new antibiotics and medicinal compounds makes saving rain forests even more urgent as not only the trees are disappearing but the endophytes inside them as well.

Even though fungi are not closely related to plants in an evolutionary sense, you can see that relationships between fungi and plants play important roles in the lives of both types of organisms. Many saprotrophic and parasitic fungi depend on plants or their products as a source of carbon. Plants rely on fungi for nutrients either directly through mycorrhizal relationships or indirectly through the role of fungi as decomposers. The very first land plants likely relied on mycorrhizal

associations to survive in the new harsh environments they faced. In the next chapter, we look at how land plants evolved and diversified.

STUDY BREAK

1. Describe the difference between a saprotroph and a symbiont.
2. What is a lichen? Explain how each partner contributes to the whole organism.
3. What benefit does a plant derive from being mycorrhizal?
4. What are the two most common types of mycorrhizas? How do they differ?
5. What is an endophyte? Why is its relationship with its plant hosts of interest to medical researchers?

Review

25.1 General Characteristics of Fungi

- Fungi can occur as single-celled yeasts or as multicellular filamentous organisms.

- A fungal mycelium consists of filamentous hyphae that grow throughout the substrate on which the fungus feeds (Figure 25.3). A cell wall of chitin surrounds the plasma membrane, and in most species, septa partition the hyphae into cell-like compartments. Pores in septa permit cytoplasm and sometimes organelles to move between hyphal cells.

- Fungi gain nutrients by extracellular digestion and absorption at hyphal tips. Saprotrophic species feed on nonliving organic matter and are key decomposers contributing to the recycling of carbon and other nutrients in ecosystems. Many fungi are symbionts, obtaining nutrients from organic matter of living hosts; these symbioses range from parasitism, in which the fungus benefits at the expense of its host, to mutualism, in which both the fungus and its host benefit.

- All fungi reproduce via spores generated either asexually or sexually (Figure 25.5). Some types also may reproduce asexually by budding or fragmentation of the parent body. Sexual reproduction usually has two stages. First, in plasmogamy, the cytoplasm of two haploid cells fuses, producing a cell that contains a haploid nucleus from each parent. In karyogamy, the nuclei fuse and form a diploid zygote; this stage is delayed in some phyla, resulting in a prolonged dikaryon ($n + n$) condition. Meiosis then generates haploid spores.

25.2 Evolution and Diversity of Fungi

- Fungi have traditionally been classified mainly on the basis of the structures formed in sexual reproduction. When a sexual phase cannot be detected, or is absent from the life cycle, the specimen is assigned to an informal grouping, the Deuteromycete fungi. Currently, five main phyla of fungi are recognized (Figure 25.6):

- Chytridiomycetes are the only fungi that produce motile, flagellated spores. Many are parasites, including the species responsible for chytridiomycosis, a disease contributing to the worldwide decline in amphibian populations (Figure 25.7).

- Zygomycetes have aseptate hyphae. Asexual reproduction involves production of spores by sporangia. Sexual reproduction occurs by way of hyphae that occur in + and − mating types; haploid nuclei in the hyphae function as gametes. Further development produces the zygosporangium, which may remain dormant for a time. When the zygosporangium breaks dormancy, it produces a stalked sporangium containing haploid spores of each mating type, which are released (Figure 25.8).

- Glomeromycetes form arbuscular mycorrhizas, the most widespread type of mycorrhiza (Figure 25.11). They reproduce asexually, by way of spores formed from hyphae.

- Ascomycetes reproduce both asexually, via chains of haploid asexual spores called conidia, and sexually, via production of haploid ascospores in saclike cells called asci. In the most complex species, asci are produced in reproductive bodies called ascocarps (Figure 25.12).

- Most Basidiomycete species reproduce only sexually. Club-shaped basidia develop on a basidiocarp (the fruiting body or mushroom) and bear sexual spores on their surface. When dispersed, these basidiospores may germinate and give rise to a haploid mycelium (see Figure 25.19).

25.3 Fungal Lifestyles

- All fungi are heterotrophs but can obtain carbon by degrading dead organic matter (as saprotrophs) or from living hosts (as symbionts). The two lifestyles are not mutually exclusive, with many fungi—such as the mycorrhizal fungi that also prey on invertebrates—combining these two modes of nutrition.

- Some basidiomycete fungi form a mutualistic symbiosis with leaf-cutter ants (see Figure 1, "People behind Biology," 25.1); the ants raise the fungi, which is the sole crop on which they feed. Recently, it was discovered that there is another partner in this ancient symbiosis, an actinomycete bacterium that lives on the ants' bodies and contributes to keeping parasitic fungi out of their fungal gardens.

- Many ascomycetes and a few basidiomycetes enter into symbioses with green algae and/or cyanobacteria to produce a compound organism known as a lichen. Fungal hyphae form the bulk of the lichen body (thallus); the hyphae entwine the algal cells that supply the lichen's carbohydrates, most of which are absorbed by the fungus (Figure 25.21).

- Fungi in the Glomeromycota, Ascomycota, and Basidiomycota form symbiotic associations known as mycorrhizas with plant roots. Hyphae of mycorrhizal fungi proliferate in the soil beyond plant roots and make mineral ions and, in some cases, organic forms of nutrients available to the plant. Some mycorrhizal associations also increase plant defences against pathogens. In turn, the fungus obtains carbohydrates and possibly other growth-enhancing substances from the plant (Figures 25.22 and 25.23).

- Endophytic fungi occur in the above-ground parts of many plants (Figure 25.24); this type of plant–fungus symbiosis is not as well understood as are mycorrhizas, but at least some endophytic fungi are known to produce toxins that deter herbivores.

Questions

Self-Test Questions

1. Which of the following traits is common to all fungi?
   a. parasitism
   b. septate hyphae
   c. reproduction via spores
   d. a prolonged dikaryotic phase

2. Which of the following is/are the chief characteristic(s) traditionally used to classify fungi into the major fungal phyla?
   a. cell wall features
   b. sexual reproductive structures
   c. adaptations for obtaining water
   d. nutritional dependence on nonliving organic matter

3. At lunch, you eat a mushroom, some truffles, a little Camembert cheese, and a bit of mouldy bread. Which group of fungi is NOT represented in this meal?
   a. Zygomycota
   b. Ascomycota
   c. Basidiomycota
   d. Glomeromycota

4. Which of the following fungal reproductive structures is diploid?
   a. ascospore
   b. zygospor-angium
   c. basidiocarp
   d. gametangium

5. Which of the following features characterizes a zygomycete?
   a. septate hyphae
   b. + and − mating strains
   c. mostly sexual reproduction
   d. a life cycle in which karyogamy does not occur

6. What is the reason that some fungi were placed in the Deuteromycetes or Fungi Imperfecti, rather than in a phylum?
   a. They form flagellated spores.
   b. They grow as single cells, rather than as hyphae.
   c. They lack a sexual reproductive stage in their life cycle.
   d. They lack an asexual reproductive stage in their life cycle.

7. Which of the following is the most accurate definition of a mushroom?
   a. a collection of saclike cells called asci
   b. the nutrient-absorbing region of an ascomycete
   c. the nutrient-absorbing region of a basidiomycete
   d. a reproductive structure formed only by basidiomycetes

8. What does it mean to classify a fungus as a saprotroph?
   a. The fungus has external digestion.
   b. The fungus forms extensive mycelia in the soil.
   c. The fungus obtains nutrients from organic matter.
   d. The fungus obtains nutrients from a living organism.

9. Which of the following best describes a lichen?
   a. an association between a green alga and a fungus
   b. an association between a basidiomycete and an ascomycete
   c. a fungus that breaks down rock to provide nutrients for an alga
   d. an organism that spends half of its life cycle as a photosymbiont and the other half as a mycobiont

10. What benefit do mycorrhizal fungi obtain from the plants with which they associate?
    a. increased nitrogen uptake
    b. a regular supply of water
    c. carbon in the form of sugars
    d. the ability to decompose organic material

Questions for Discussion

1. A mycologist wants to classify a specimen that appears to be a new species of fungus. To begin the classification process, what kinds of information on structures and/or functions must the researcher obtain to assign the fungus to one of the major fungal groups?

2. In a natural setting—a pile of horse manure in a field, for example—the sequence in which various fungi appear illustrates **ecological succession**, the replacement of one species by another in a community. The earliest fungi are the most efficient opportunists because they can form and disperse spores most rapidly. In what order would you expect representatives from each phylum of fungi to appear on the manure pile? Why?

3. As the text noted, conifers and some other types of plants cannot grow properly if their roots do not form associations with fungi. These associations provide the plant with minerals such as nitrogen and phosphate and in return fungi receive carbohydrates synthesized by the plant. In some instances, however, the plant receives proportionately more nutrients than the fungus does. Would you still classify such associations as mutualisms?

4. What evidence would you look for to determine whether the association between a plant and an endophyte was mutualistic?

5. Why is it more difficult to develop drugs against fungal infections of humans than bacterial infections?

26

Monotropa uniflora, a heterotrophic plant that lacks chlorophyll.

Plants

WHY IT MATTERS

You are out for a walk in a forest near your home; you are busy thinking about other things and so are not paying close attention to the plants that you're walking by—they are just a pleasing green background for your walk. Suddenly, a small white plant, like the one shown in the photo above, catches your eye. At least you think it's a plant. But aren't all plants green? How can there be a completely white plant?

What you have found is a plant known as ghost flower or Indian pipe (*Monotropa uniflora*), which does not produce chlorophyll and so cannot photosynthesize.

CONCEPT FIX We often assume that all plants are photoautotrophs, making their own organic carbon molecules from atmospheric CO_2 and sunlight. But some plants, such as *Monotropa*, are completely heterotrophic, living on organic carbon obtained from other plants. And other plants that do have chlorophyll supplement their carbon supply by being heterotrophic in low light levels or under other conditions that limit photosynthesis. How do heterotrophic plants get carbon? Some directly parasitize green plants, but others, like *Monotropa*, feed on neighbouring photosynthetic plants through shared root-colonizing fungi (mycorrhizal fungi; see Chapter 25). So, contrary to popular belief, not all plants are photosynthetic and green. ⬡

So if being green isn't an unifying feature of all plants, what is? What features could you look for to determine whether this *Monotropa* is a plant? What characteristics set plants apart from other organisms? And how did plants evolve? In this chapter, we investigate these questions and look at the adaptations to terrestrial life that have made plants so successful. And land plants *are* very successful: they can thrive in habitats where no animal can survive for long and some are able to grow much larger and live much longer than any animal. Together with photosynthetic bacteria and protists, plant tissues provide the nutritional foundation for nearly all ecosystems on Earth. Humans also use plants as sources of medicinal drugs, wood for building, fibres used in paper and clothing, and a wealth of other products. The partnership between humans and plants has a long evolutionary history: we first

562

domesticated cereal plants 9000 years ago, but this was not the earliest relationship between humans and plants. Our early ancestors, like modern-day primates, would have relied heavily on plants in their diet.

Despite the long history between plants and humans, there is still much about plant biology that we don't understand and many questions that remain to be answered.

We start this chapter by considering the defining characteristics of plants and then look at the evolution of plants and their adaptations to life on land; we conclude by looking at the diversity of land plants.

26.1 Defining Characteristics of Land Plants

Land plants are eukaryotes; as we learned from the *Monotropa* example, not all are capable of photosynthesizing, but almost all plants are photoautotrophs (organisms that use light as their energy source and carbon dioxide as their carbon source; see Chapter 7). Like animals, all land plants are multicellular, but if you took a piece of tissue from *Monotropa* and looked at it under the microscope, you'd see that, unlike animal cells, plant cells have walls, which are made of cellulose. All plants are sessile or stationary (not able to move around); no terrestrial animals are sessile, although some aquatic ones are. Plants also differ from animals in having an **alternation of generations** life cycle.

In most animals, the diploid stage dominates the life cycle and produces gametes—sperm or eggs—by meiosis. Gametes are the only haploid stage, and it is short-lived: fusion of gametes produces a new diploid organism (some animals, for example, social insects such as bees and wasps, have a different life cycle). In other organisms, such as many green algae, the haploid stage dominates the life cycle; the haploid alga spends much of its life producing and releasing gametes into the surrounding water. The single-celled zygote is the only diploid stage and divides by meiosis to produce spores that give rise to the haploid stage again.

In contrast, land plants have two multicellular stages in their life cycles, one diploid and one haploid **(Figure 26.1)**. The diploid generation produces spores and is called a **sporophyte** (*phyte* = plant, hence "spore-producing plant"). The haploid generation produces gametes by mitosis and is called a **gametophyte** ("gamete-producing plant"). The haploid phase of the plant life cycle begins in specialized cells of the sporophyte, where haploid spores are produced by meiosis. So in plants meiosis produces spores, not gametes. Spores are single haploid cells with fairly thick cell walls. When a spore germinates, it divides by mitosis to produce a multicellular haploid gametophyte. A gametophyte's function is to nourish and protect the forthcoming sporophyte generation. Each generation

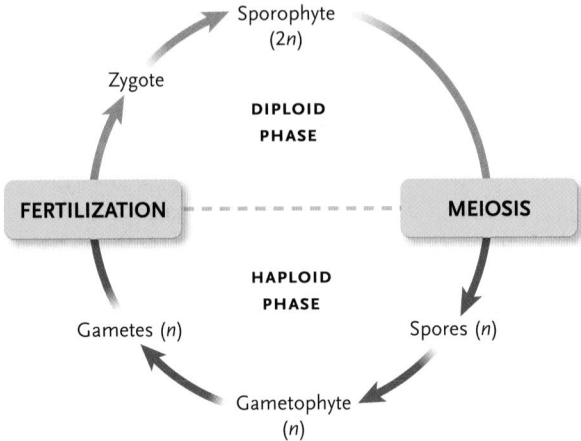

Figure 26.1

Overview of the alternation of generations, the basic pattern of the plant life cycle. The relative dominance of haploid and diploid phases is different for different plant groups.

gives rise to the other—hence the name *alternation of generations* for this life cycle.

The final defining feature of land plants is that the embryo (new sporophyte generation) is retained inside gametophyte tissue. The reasons for retention of embryos in parental tissue and for the rather complex life cycle will become clearer after we've looked at the evolution of plants and their transition onto land.

STUDY BREAK

1. What features of land plants differentiate them from other eukaryotes, for example, from fungi? From animals?
2. What is an alternation of generations life cycle? How does this differ from the life cycle of most animals?
3. What does meiosis produce in plants?
4. Differentiate between a gametophyte and a sporophyte in terms of ploidy and what is produced.

26.2 The Transition to Life on Land

Ages ago, along the shores of the ancient ocean, the only sound was the rhythmic muffled crash of waves breaking in the distance. There were no birds or other animals, no plants with leaves rustling in the breeze. In the preceding eons, cells that produce oxygen as a by-product of photosynthesis had evolved, radically changing Earth's atmosphere. Solar radiation had converted much of the oxygen into a dense ozone layer—a shield against lethal doses of ultraviolet radiation, which had kept early organisms below the water's surface. Now, they could populate the land.

Cyanobacteria were probably the first to adapt to **intertidal zones** and then to spread into shallow, coastal streams. Later, green algae and fungi made the same journey. Around 480 million years ago (mya), one group of green algae, living near the water's edge, or perhaps in a moist terrestrial environment, became the ancestors of modern plants. Several lines of evidence indicate that these algae were charophytes (a group discussed in Chapter 24): both groups have cellulose cell walls, they store energy captured during photosynthesis as starch, and their light-absorbing pigments include both chlorophyll *a* and chlorophyll *b*. Molecular data also support the relationship between the charophytes and the land plants. Like other green algae, the charophyte lineage that produced the ancestor of land plants arose in water and has aquatic descendants today (Figure 26.2). Yet because terrestrial environments pose very different challenges than aquatic environments, evolution in land plants produced a range of adaptations crucial to survival on dry land.

The algal ancestors of plants probably invaded land about 450 mya. We say "probably" because the fossil record is sketchy in pinpointing when the first truly terrestrial plants appeared, and many important stages in evolution are not represented in the fossil record. Even in more recent deposits, the most commonly found plant fossils are just microscopic bits and pieces; easily identifiable parts such as leaves, stems, roots, and reproductive parts seldom occur together. Whole fossilized plants are extremely rare. Adding to the challenge, some chemical and structural adaptations to life on land arose independently in several plant lineages. Despite these problems, botanists have been able to gain insight into several innovations and overall trends in plant evolution.

While the ancestors of land plants were making the transition to a fully terrestrial life, some remarkable adaptive changes unfolded. For example, the earliest land plants were exposed to higher levels of harmful UV radiation than their aquatic ancestors had experienced. Gradual changes in existing metabolic

pathways resulted in the ability to synthesize simple phenylpropanoids, molecules that absorb UV radiation, which enhanced the plants' ability to live on land. Where did these new metabolic pathways and associated enzyme functions come from? They did not simply appear because the plants needed them.

CONCEPT FIX The idea that evolution involves organisms "trying" to adapt or that natural selection gives organisms what they need to survive is one of the major misconceptions about evolution. Natural selection cannot sense what a species "needs," and organisms cannot try to adapt: if some individual organisms in the population have traits that allow them to survive and reproduce more in that environment than other individuals, then they will pass on these traits to more offspring, and the frequency of the traits in the population will increase. But the organism cannot "try" to get the right genes. Research shows that new enzyme functions usually follow duplication of genes, which can occur in various ways (e.g., an error during crossing-over of meiosis). Mutations in the second copy of a gene will not have negative effects on the host because the other copy retains its original function; thus over time the second copy tends to accumulate mutations. If the changes in this gene provide advantages to the host plant, then that gene is selected for. In this way, new enzyme functions and metabolic pathways evolve. ⬡

Eons of natural selection sorted out solutions to fundamental problems, among them avoiding desiccation, physically supporting the plant body in air, obtaining nutrients from soil, and reproducing sexually in environments where water would not be available for dispersal of eggs and sperm. With time, plants evolved features that not only addressed these problems but also provided access to a wide range of terrestrial environments. Those ecological opportunities opened the way for a dramatic radiation (rapid evolution and divergence; see Chapter 20) of varied plant species—and for the survival of plant-dependent organisms such as humans. Today the **kingdom Plantae** encompasses more than 300 thousand living species, organized in this textbook into 10 phyla. These modern plants range from mosses, horsetails, and ferns to conifers and flowering plants (Figure 26.3).

26.2a Early Biochemical and Structural Adaptations Enhanced Plant Survival on Land

The greatest challenge plants had to overcome to survive on land was how to survive in the dry terrestrial conditions. Unlike most modern-day plants, the earliest land plants had neither a waterproof **cuticle** (a outer waxy layer that prevents water loss from plant tissues) nor tissues with sufficient mechanical strength to allow for upright growth. These limitations restricted

Figure 26.2

Chara, a stonewort. This representative of the charophyte lineage is known commonly as a stonewort due to the calcium carbonate that accumulates on its surface.

BOB GIBBONS/SCIENCE PHOTO LIBRARY

a. Mosses growing on rocks

© Chris Manderson 1995

b. A jack pine

Michael P. Gadomski/Science Source

c. An orchid

© iStockphoto.com/Don Enright

a. Cuticle on the surface of a leaf

Cuticle   Epidermal cell

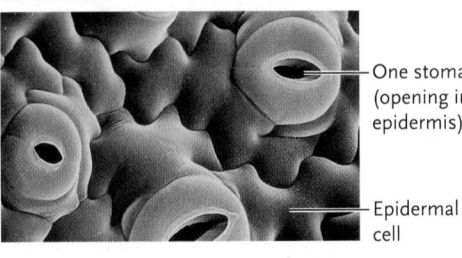

Epidermis

Jubal Harshaw/Shutterstock.com

b. Stomata

One stoma (opening in epidermis)

Epidermal cell

Jeremy Burgess/Science Source

Figure 26.3

Representatives of the kingdom Plantae. (a) Mosses growing on rocks. Mosses evolved relatively soon after plants made the transition to land. **(b)** A jack pine (*Pinus banksiana*). This species and other conifers belonging to the phylum Coniferophyta represent the gymnosperms. **(c)** An orchid, *Calypso bulbosa*, a showy example of a flowering plant.

Figure 26.4

Adaptations for limiting water loss. (a) A waxy cuticle, which covers the epidermis of land plants and helps reduce water loss. **(b)** Surface view of stomata in the epidermis (surface layer of cells) of a leaf. Stomata allow carbon dioxide to enter plant tissues and oxygen and water to leave.

these early plants to moist habitats and made it necessary for them to stay small and grow close to the ground. Like modern-day mosses, these plants were **poikilohydric** (*poikilo* = variable; *hydric* = relating to water), meaning that they have little control over their internal water content and do not restrict water loss. Instead, their water content fluctuates with moisture levels in their environment: as their habitat dries out, so do their tissues, and their metabolic activities virtually cease. When external moisture levels rise, they quickly rehydrate and become metabolically active. In other words, poikilohydric plants are drought tolerators that can survive drying out, while vascular plants, which regulate their internal water content and restrict water loss, are drought avoiders, with numerous adaptations to avoid drying out or with plant parts (e.g., underground stems) that can survive if the rest of the plant dries out. How are poikilohydric plants able to survive prolonged dehydration that would be lethal to most plants? This question is explored further in "Life on the Edge," Box 26.1.

Later-evolving plants were able to regulate water content and restrict water loss because they had cuticles covering their outer surfaces **(Figure 26.4a),** as well as **stomata** (singular, *stoma*; *stoma* = mouth), pores in the cuticle-covered surfaces (Figure 26.4b) that open and close to regulate water loss (and are the main route for carbon dioxide to enter leaves; see Chapter 35). These plants also had water-transport tissues that also provided support for upright growth, described further in Section 26.2c.

26.2b Symbiotic Associations with Fungi Were Likely Required for Evolution of Land Plants

The ancestor of land plants was not the first organism to colonize terrestrial habitats; certain bacteria, protists, and fungi had been present at least since the late Proterozoic (around 540 mya). Almost all modern-day plants form symbiotic associations, known as mycorrhizas, with certain soil fungi (see Chapter 25). In these associations, the fungus colonizes the plant's roots and grows prolifically in the soil beyond the root system, producing a very large network that takes up soil nutrients. **(Figure 26.5, p. 566).** Both partners generally benefit by a two-way exchange of nutrients: the plant provides the fungus with carbon, and the fungus increases the plant's supply of soil nutrients, which it is able to obtain much more efficiently than do the plant's own roots. Such mutually beneficial relationships may have been essential to the evolution of land plants and to their success in terrestrial habitats (see "People behind Biology," Box 26.2, p. 567), given that the first land plants lacked roots and that the soils of early Earth were nutrient poor.

26.2c Lignified Water-Conducting Cells Provided Strength and Support for Plants to Grow Upright

The earliest land plants remained small because they lacked the mechanical support necessary to grow taller.

LIFE ON THE EDGE 26.1

Poikilohydric Plants

Most land plants, including our major crops, are killed if they dry out to the point of equilibrium with the water content of the air around them; this point is all too clearly illustrated by the terrible famines that result from drought in Africa and other regions of the world. But some plants are able to survive drying out to 10% absolute water content or less for months, and even years, in some cases **(Figure 1)**.

This ability is widespread among bryophytes but much less common in vascular plants: only about 50 species of seedless vascular plants have this ability in their sporophyte stage, along with about 300 species of angiosperms. Most of these desiccation-tolerant vascular plants, known as resurrection plants, live on rock outcrops in regions of southern Africa and Australia that receive only seasonal and sporadic rainfall. As far as we know, no gymnosperms have this ability.

How does dehydration kill a plant? Cellular water maintains membrane structure as well as the shapes of macromolecules such as enzymes and other proteins. Dehydration thus results in lethal changes to both membrane structure and

With kind permission from Springer Science + Business Media: *Planta*, "Molecular cloning of abscisic acid-modulated genes which are induced during desiccation of the resurrection plant," volume 181, April 1, 1990, pp. 27–34, Dorothea Bartels, figure: An illustration of the remarkable ability for extreme vegetative desiccation tolerance in an angiosperm species.

Figure 1
Desiccation-tolerant plant shown in a dehydrated state and following re-wetting.

macromolecular shape. A cell's metabolism also relies on water; as a cell dries out, metabolism first decreases and then ceases altogether. How do poikilohydric plants survive these changes that kill all other plants? We don't understand all of the

mechanisms at play, but we do know that part of the answer is accumulation of sugars (e.g., sucrose) in cells. These sugars and certain proteins replace the water in membranes and around macromolecules, preventing lethal changes in conformation. The high sugar content also converts the cytoplasm from its normal consistency to a thick, slow-moving liquid known as glass, immobilizing the cytoplasm. The cells are able to survive in a dehydrated state with metabolism slowed to a state of dormancy or "suspended animation." The cell walls of desiccation-tolerant plants are also more flexible, able to fold as the cell dries, allowing the entire cell to contract as it dries out.

These mechanisms come at a cost to the plant, limiting their growth and reproduction. We don't yet understand how tolerance restricts growth; once we have a better understanding of this relationship, we might be able to uncouple tolerance from slow growth and develop drought-tolerant plants with a higher productivity. This very active area of research clearly has practical applications in maintaining our food supply in the face of droughts and climate change.

Figure 26.5
Mycorrhizal fungus colonizing plant root and soil around the root.

Growing low to the ground helped them stay moist but was not very effective in capturing light: since all early land plants were low growing, there would have been intense competition for light. If any plant had been able to grow taller than its neighbours, it would have had a major advantage. But how could a plant support upright growth against the force of gravity? Plants require strengthening tissue to grow upright. And growing up and away from the ground surface also requires an internal water circulation system, since diffusion is not effective over longer distances. Some of the early land plants did have specialized water-conducting cells that transported water through the plant body, but these cells did not provide mechanical strength. Later land plants were able to synthesize lignin, a polymer of phenylpropanoids (the molecules mentioned earlier that absorb UV radiation). Why were these plants able to make lignin when earlier plants did not? Changes in Earth's atmosphere and climate altered certain biochemical pathways in plants, resulting in the excess formation of lignin; see

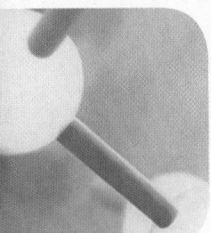

PEOPLE BEHIND BIOLOGY 26.2

Kris Pirozynski and David Malloch, *Agriculture Canada*

If almost all land plants today depend to some extent on mutually beneficial relationships with mycorrhizal fungi, would the first land plants have been any different? The hypothesis that a mutualistic relationship with soil fungi was required for the evolution of land plants was first put forward in 1975 by two researchers at the Biosystematics Research Institute of Agriculture Canada.

In their 1975 paper outlining this hypothesis, Kris Pirozynski and David Malloch pointed out that associations with fungi would have helped the earliest land plants avoid starvation: early soils would not have been as fertile as most modern-day soils, and nutrients would certainly not have been as abundant as in the aquatic environments in which the algal ancestor of land plants lived. Fungi are very adept at proliferating in their substrates and foraging for nutrients, which they take up via extracellular enzymatic digestion (see Chapter 5). The earliest plants did not have roots, so forming a partnership with fungi would have greatly enhanced their uptake of nutrients. The fungi might also have protected the roots of its plant partner from root pathogens, as do modern-day mycorrhizal fungi.

Pirozynski and Malloch's hypothesis has since received strong support from both the fossil record and molecular data.

"Molecule behind Biology," Box 26.3. This lignin was deposited in cell walls, particularly in the water-conducting cells, providing support and rigidity to those tissues and allowing the plants to grow upright. These lignified water-conducting cells make up a tissue called xylem.

Xylem is one type of **vascular tissue** (*vas* = duct or vessel). Plants with this tissue (and the other type of vascular tissue, **phloem**, which conducts sugars through the plant body) are known as **vascular plants.** It is important to note that some plants, such as some mosses, that lack vascular tissues do have tissues that conduct water and sugars through their bodies. These tissues are not the same as xylem and phloem—for example, their water-conducting cells do not have walls reinforced with lignin—and are likely not homologous with xylem and phloem, so they are not called vascular tissues. Thus, these plants are referred to as **nonvascular plants.** Chapter 35 explains how xylem and phloem perform these key internal transport functions.

Clearly, plants with lignified tissues had a clear benefit over plants lacking lignin and over time evolved to become the dominant plants in most habitats on

---

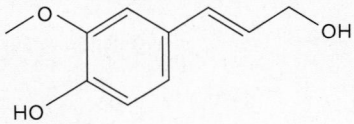

MOLECULE BEHIND BIOLOGY 26.3

Coniferyl Alcohol, a Building Block of Lignin

Lignin is a polymer of several different monomers, including coniferyl alcohol **(Figure 1).** These molecules are synthesized from the amino acid phenylalanine in a series of reactions in plant cell cytoplasm. The monomers are then transported through the cell membrane, where polymerization happens. We still do not fully understand how lignin is formed from monomers, but we do know that oxidative enzymes are involved in polymerization; thus, oxygen is required for the process. Lignin is thought to have evolved due to the high oxygen levels in the atmosphere around 430 mya, which would have favoured the polymerization reaction.

Lignin is very difficult to degrade, with only a few fungi and bacteria able to break it down (see Chapter 25). Its accumulation in plant tissues would have meant that dead vascular plants, especially if large, would have decomposed more slowly than the earlier land plants, contributing to the formation of coal, one of today's fossil fuels. The forests of the Carboniferous period were dominated by large vascular seedless plants, which were abundant in lignin. When these plants died and fell to the ground, they became buried in anaerobic sediments; even those that were not buried in such sediments would have been fairly slow to decompose due to their lignin content. Over geologic time, these buried remains became compressed and fossilized; today they form much of the world's coal reserves. This is why coal is called a "fossil fuel" and the Carboniferous period is called the Coal Age. Characterized by a moist climate over much of the planet and by the dominance of seedless vascular plants, the Carboniferous period continued for 150 million years, ending when climate patterns changed during the Paleozoic era.

Figure 1
Coniferyl alcohol, one of the monomers of lignin.

Earth. Ferns, conifers, and flowering plants—most of the plants you are familiar with—are vascular plants. Supported by lignin and with a well-developed vascular system, the body of a plant can grow very large. Extreme examples are the giant redwood trees of the northern California coast, some of which are more than 90 m tall. By contrast, non vascular plants lack lignin, although some do have simple internal transport systems, and are generally small **(Table 26.1)**.

Vascular plants also have **apical meristems**, regions of constantly dividing cells near the tips of shoots and roots that produce all tissues of the plant body. Meristem tissue is the foundation for a vascular plant's extensively branching stem and root systems and is a central topic of Chapter 34.

26.2d Root and Shoot Systems Were Adaptations for Nutrition and Support

The body of a nonvascular plant is not differentiated into true roots and stems—structures that are fundamental adaptations for absorbing nutrients from soil and for support of an erect plant body. The evolution of sturdy stems—the basis of an aerial *shoot system*—went hand in hand with the capacity to synthesize lignin. To become large, land plants also require a means of anchoring aerial parts in the soil, as well as effective strategies for obtaining soil nutrients. **Roots**—anchoring structures that also absorb water and nutrients in association with mycorrhizal fungi—were the eventual solution to these problems. The earliest fossils showing clear evidence of roots are

from vascular plants, although the exact timing of this change is uncertain. Ultimately, vascular plants developed specialized **root systems**, which generally consist of underground, cylindrical absorptive structures with a large surface area that favours the rapid uptake of soil water and dissolved mineral ions. The root system has been called "the hidden half" of a plant: "half" refers to the fact that there is as much plant biomass below ground as there is above ground. And there are other similarities between above- and below-ground parts of plants: the fine roots of a root system go through regular cycles of growth and death, just as do the leaves of most plants. "Hidden" refers to the fact that the root system is hidden from our sight below ground, meaning that we cannot study it very easily. For this reason, we know less about root systems than about the above-ground parts of plants, although recent technological advances are changing this situation.

Above ground, the simple stems of early land plants also became more specialized, evolving into **shoot systems** in vascular plants. Shoot systems have stems and leaves that arise from apical meristems and that function in the absorption of light energy from the Sun and carbon dioxide from the air. Stems grew larger and branched extensively after the evolution of lignin. The mechanical strength of lignified tissues almost certainly provided plants with several adaptive advantages. For instance, a strong internal scaffold could support upright stems bearing leaves and other photosynthetic structures and so help increase the surface area for intercepting sunlight. Also, reproductive

Table 26.1	Trends in Plant Evolution Traits Derived from Algal Ancestor: Cell Walls with Cellulose, Energy Stored in Starch, Two Forms of Chlorophyll (*a* and *b*)				
Bryophytes	Ferns and Their Relatives	Gymnosperms	Angiosperms		Functions of This Trait in Land Plants
Cuticle	———————————————————————→				Protection against water loss, pathogens
Stomata	———————————————————————→				Regulation of water loss and gas exchange (CO_2 in, O_2 out)
Nonvascular (although some have specialized water-conducting cells without lignin) →	Vascular (have xylem and phloem) ———————————————→				Internal tubes that transport water, nutrients
	Lignin ———————————————————→				Mechanical support for vertical growth
	Apical meristem ——————————→				Branching shoot system
	Roots, stems, leaves ————————→				Enhanced uptake, transport of nutrients, and enhanced photosynthesis
Haploid phase dominant ———→	Diploid phase dominant ———————→				Genetic diversity
One spore type (homospory) ———→	Homospory in most but heterospory (two spore types) in some → Heterospory ————————→				Promotion of genetic diversity
Motile sperm ————————→		Nonmotile sperm ————————→			Protection of gametes within parent body
Seedless ————————→		Seeds ————————————→			Protection of embryo

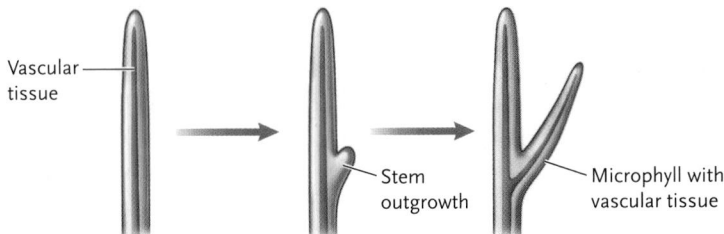

a. Development of microphylls as an offshoot of the main vertical axis

Vascular tissue

Stem outgrowth

Microphyll with vascular tissue

b. Development of megaphylls in a branching pattern

Equal branches

Unequal branching growth

Side branches fan out in same plane

Megaphyll

"Web" of photosynthetic tissue fills in space

Thick main stem with vascular tissue

Figure 26.6
Evolution of leaves. **(a)** One type of early leaflike structure may have evolved as offshoots of the plant's main vertical axis; there was only one vein (transport vessel) in each leaf. Today, the seedless vascular plants known as lycophytes (club mosses) have this type of leaf. **(b)** In other groups of seedless vascular plants, leaves arose in a series of steps that began when the main stem evolved a branching growth pattern. Small side branches then fanned out and photosynthetic tissue filled the space between them, becoming the leaf blade. With time, the small branches modified into veins.

structures borne on aerial stems might serve as platforms for more efficient launching of spores from the parent plant.

Structures we think of as "leaves" arose several times during plant evolution. In general, leaves represent modifications of stems and can be divided into two types. Microphylls are narrow leaves with only one vein or strand of vascular tissue, while megaphylls are broader leaves with multiple veins. **Figure 26.6** illustrates the basic steps of possible evolutionary pathways by which these two types of leaves evolved. In some early plants, microphylls may have evolved as flaplike extensions of the main stem. In contrast, megaphylls likely evolved from modified branches when photosynthetic tissue filled in the gaps between neighbouring branches.

Other land plant adaptations were related to the demands of reproduction in a dry environment. As described in more detail shortly, these adaptations included multicellular chambers that protect developing gametes and a multicellular embryo that is sheltered inside the tissues of a parent plant.

26.2e In the Plant Life Cycle, the Diploid Phase Became Dominant

As early plants moved into drier habitats, their life cycles were also modified considerably. The haploid gametophyte phase became physically smaller and less complex and had a shorter life span, whereas the opposite occurred with the diploid sporophyte phase. In mosses and other nonvascular plants, the

sporophyte is a little larger and longer lived than in green algae, and in vascular plants, the sporophyte is clearly larger and more complex and lives much longer than the gametophyte **(Figure 26.7)**. When you look at a pine tree, for example, you see a large,

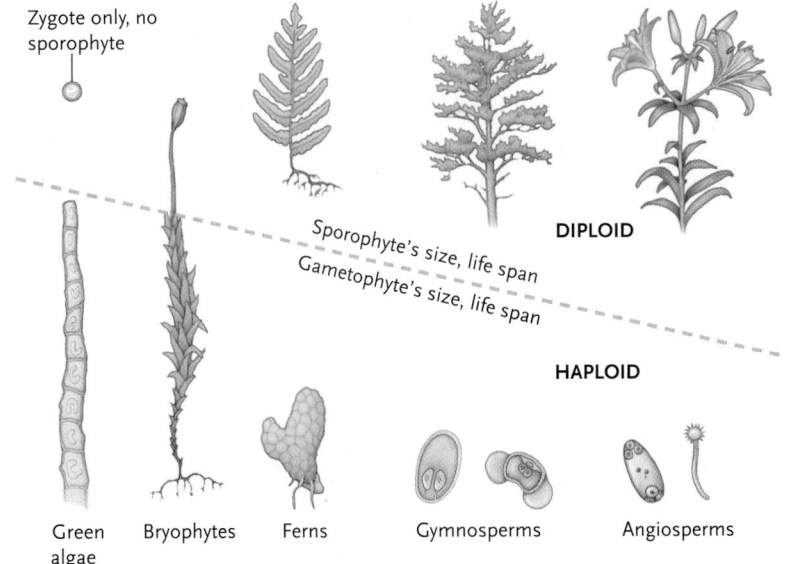

Zygote only, no sporophyte

Sporophyte's size, life span

Gametophyte's size, life span

DIPLOID

HAPLOID

Green algae

Bryophytes

Ferns

Gymnosperms

Angiosperms

Figure 26.7
Evolutionary trend from dominance of the gametophyte (haploid) generation to dominance of the sporophyte (diploid) generation, represented here by existing species ranging from a green alga (*Ulothrix*) to a flowering plant. This trend developed as early plants colonized habitats on land. In general, the sporophytes of vascular plants are larger and more complex than those of bryophytes, and their gametophytes are smaller and less complex. In this diagram, the fern represents seedless vascular plants.

long-lived sporophyte. The sporophyte generation begins after fertilization, when the zygote divides by mitosis to produce a multicellular diploid organism. Its body will eventually develop capsules called **sporangia** (*angium* = vessel or chamber, hence, "spore-producing chambers"; singular, *sporangium*), which produce spores by meiosis.

Why did the diploid phase become dominant over evolutionary time? Many botanists hypothesize that the trend toward "diploid dominance" reflects the advantage of being diploid in land environments; if there is only one copy of DNA, as in a haploid plant, and if a deleterious mutation occurs or if the DNA is damaged (e.g., by UV radiation, which is a greater problem on land than in aquatic habitats), the consequences could be fatal. In contrast, the sporophyte phase of that plant is diploid and so has a "backup" copy of the DNA that can continue to function normally even if one strand is damaged. However, it is important to remember that the land plants that do have a dominant haploid stage, such as mosses, are very successful plants in certain habitats. The lack of a dominant diploid stage has certainly not caused them to become extinct.

26.2f Some Vascular Plants Evolved Separate Male and Female Gametophytes

When a plant makes only one type of spore, it is said to be **homosporous** ("same spore") **(Figure 26.8a).**

a. *Lycopodium*

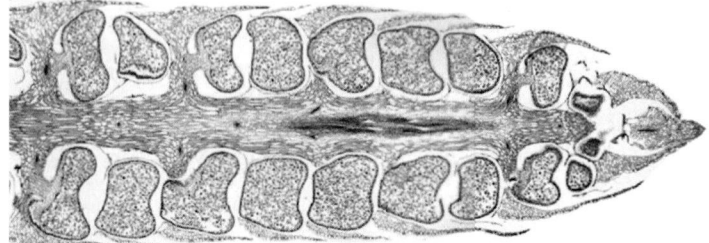

b. *Selaginella*

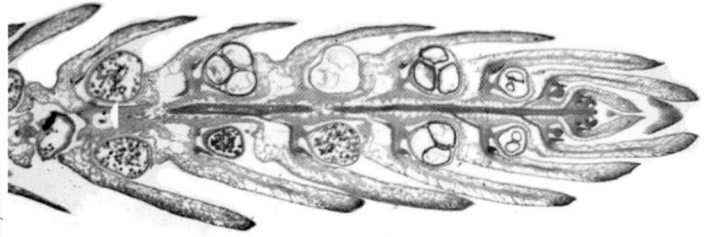

Photographer: Michael Clayton. University of Wisconsin Plant Teaching Collection, http://botit.botany.wisc.edu

Figure 26.8

Longitudinal sections through strobili of two lycophytes, **(a)** *Lycopodium* and **(b)** *Selaginella*. *Lycopodium* is a homosporous plant that produces spores of only one type, as can be seen in (a). Note that the sporangia of *Lycopodium* are all the same. The *Selaginella* strobilus shown here is from a heterosporous plant, which produces megasporangia (containing a few large megaspores) and microsporangia (containing numerous small microspores) in the same strobilus.

Usually, a gametophyte that develops from such a spore is bisexual—it can produce both sperm and eggs. However, some homosporous plants have ways to produce male and female sex organs on different gametophytes or to otherwise prevent self-fertilization, as described below in ferns. The sperm have flagella and are motile because they must swim through liquid water to encounter eggs.

Other vascular plants, including gymnosperms and angiosperms, are **heterosporous** (Figure 26.8b). They produce two types of spores—one type is smaller than the other—in two different types of sporangia. The smaller spores are **microspores**, which develop into male gametophytes, and the larger **megaspores** will develop into female gametophytes. Heterospory and the development of gametophytes inside spore walls are important steps in the evolution of the seed, as we will see further on.

As you will read in a later section, the evolution of seeds and related innovations, such as pollen grains and pollination, helped spark the rapid diversification of plants in the Devonian period, 408 to 360 mya. In fact, so many new fossils appear in Devonian rocks that paleobotanists—scientists who specialize in the study of fossil plants—have thus far been unable to determine which fossil lineages gave rise to the modern plant phyla. Clearly, however, as each major lineage came into being, its characteristic adaptations included major modifications of existing structures and functions **(Figure 26.9).** The next sections fill out this general picture, beginning with the plants that are the living representatives of the earliest land plants.

STUDY BREAK

1. What features do land plants share with their closest living relatives, the charophyte algae? What features differentiate the two groups?
2. How did mycorrhizal fungi fulfill the role we associate with roots in early land plants?
3. What is the main difference between the specialized water-conducting cells present in some nonvascular plants and those of vascular plants? How did this difference influence the evolution of vascular plants?
4. How did plant adaptations such as a root system, a shoot system, and a vascular system collectively influence the evolution of land plants?
5. Describe the difference between homospory and heterospory, and explain how heterospory paved the way for other reproductive adaptations in land plants.

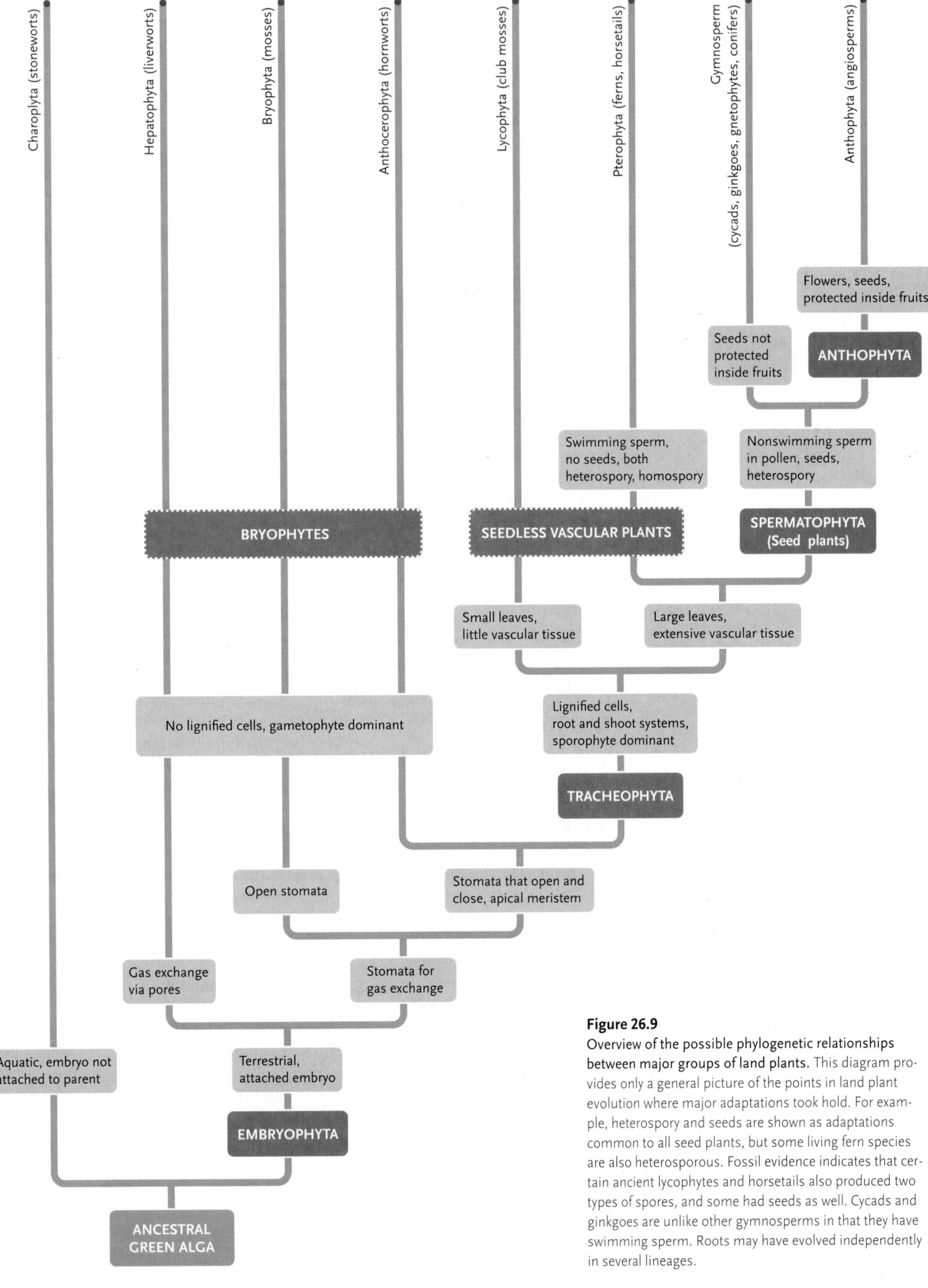

Figure 26.9

Overview of the possible phylogenetic relationships between major groups of land plants. This diagram provides only a general picture of the points in land plant evolution where major adaptations took hold. For example, heterospory and seeds are shown as adaptations common to all seed plants, but some living fern species are also heterosporous. Fossil evidence indicates that certain ancient lycophytes and horsetails also produced two types of spores, and some had seeds as well. Cycads and ginkgoes are unlike other gymnosperms in that they have swimming sperm. Roots may have evolved independently in several lineages.

Labels in figure:
- Charophyta (stoneworts)
- Hepatophyta (liverworts)
- Bryophyta (mosses)
- Anthocerophyta (hornworts)
- Lycophyta (club mosses)
- Pterophyta (ferns, horsetails)
- Gymnosperm (cycads, ginkgoes, gnetophytes, conifers)
- Anthophyta (angiosperms)

- Flowers, seeds, protected inside fruits
- Seeds not protected inside fruits
- ANTHOPHYTA
- Swimming sperm, no seeds, both heterospory, homospory
- Nonswimming sperm in pollen, seeds, heterospory
- BRYOPHYTES
- SEEDLESS VASCULAR PLANTS
- SPERMATOPHYTA (Seed plants)
- Small leaves, little vascular tissue
- Large leaves, extensive vascular tissue
- No lignified cells, gametophyte dominant
- Lignified cells, root and shoot systems, sporophyte dominant
- TRACHEOPHYTA
- Open stomata
- Stomata that open and close, apical meristem
- Gas exchange via pores
- Stomata for gas exchange
- Aquatic, embryo not attached to parent
- Terrestrial, attached embryo
- EMBRYOPHYTA
- ANCESTRAL GREEN ALGA

26.3 Bryophytes: Nonvascular Land Plants

The **bryophytes** (*bryon* = moss)—liverworts, hornworts, and mosses—are important both ecologically and economically. As colonizers of bare land, their small bodies trap particles of organic and inorganic matter, helping to build soil on bare rock and stabilizing soil surfaces with a biological crust in harsh places such as coastal dunes, inland deserts, and embankments created by road construction. In boreal forests and arctic tundras, bryophytes constitute as much as half of the biomass, and they are crucial components of the **food web** that supports animals in these ecosystems. People have long used *Sphagnum* and other absorbent "peat" mosses (which typically grow in bogs and fens) for everything from primitive diapers and filtering whiskey to increasing the water-holding capacity of garden soil. Peat moss has also found use as a fuel; each day, the Rhode generating station in Ireland, one of several that use peat in that nation, burns 2000 tonnes of peat to produce electricity.

Bryophytes have a combination of traits that allow them to bridge aquatic and land environments. Because bryophytes lack cells strengthened by lignin and are poikilohydric, it is not surprising that they are small and commonly grow on wet sites along creek banks (see Figure 26.3a, p. 565); in bogs, swamps, or the dense shade of damp forests; and on moist tree trunks or rooftops. However, some mosses live in very dry environments, such as **alpine tundra** and **arctic tundra (Figure 26.10).** Being poikilohydric enables them to live in such seemingly inhospitable habitats (see "Life on the Edge," Box 26.1, p. 566).

Bryophytes retain many of the features of their algal ancestors: they produce flagellated sperm that must swim through water to reach eggs, which is another reason they are small: the sperm must be able to swim between plants in a film of water (e.g., from rain or dew), which is only possible if the plants are relatively close to the ground. They also lack xylem and phloem (although some do have specialized conductive tissues). Bryophytes have parts that are rootlike, stemlike, and leaflike. However, the "roots" are **rhizoids** that serve only to anchor the plant to its substrate and do not take up any water or nutrients from the substrate. Bryophyte "stems" and "leaves" are not considered to be true stems and leaves like those of vascular plants because they lack vascular tissue and because they did not evolve from the same structures as vascular plant stems and leaves did. (Said another way, stems and leaves are not homologous in bryophytes and vascular plants.)

In other ways, bryophytes are clearly adapted to land. The sporophytes (but not the longer-lived gametophytes) of some species have a water-conserving cuticle and stomata. And, as is true of all plants, the

a.

b.

Figure 26.10
Bryophytes of arid habitats: **(a)** moss growing on exposed rock; **(b)** mosses and other plants in alpine tundra.

bryophyte life cycle has both multicellular gametophyte and sporophyte phases, but the sporophyte is permanently associated with the gametophyte (it never becomes independent of the gametophyte) and lives for a shorter time than the gametophyte. **Figure 26.11** shows the green, leafy gametophyte of a moss plant, with diploid sporophytes attached to it by slender stalks. Bryophyte gametophytes produce gametes inside a protective organ called a **gametangium** (plural, *gametangia*). The gametangia in which bryophyte eggs form are flask-shaped structures called **archegonia** (*archi* = first; *gonos* = seed). Flagellated sperm form in rounded gametangia called **antheridia** (*antheros* = flowerlike; singular, *antheridium*). The sperm swim through a film of water to the archegonia to fertilize eggs. Each fertilized egg gives rise to a diploid embryo sporophyte, which stays attached to the gametophyte and produces spores—and the cycle repeats.

Despite these similarities to more complex plants, bryophytes are unique in several ways. Unlike vascular plants, the gametophyte is much longer lived than the sporophyte and is photosynthetic, whereas the sporophyte remains attached to the gametophyte and depends on the gametophyte for much of its nutrition.

Bryophytes are not a monophyletic group (i.e., they did not all evolve from a common ancestor); instead, the various bryophytes evolved as separate lineages, in parallel with vascular plants.

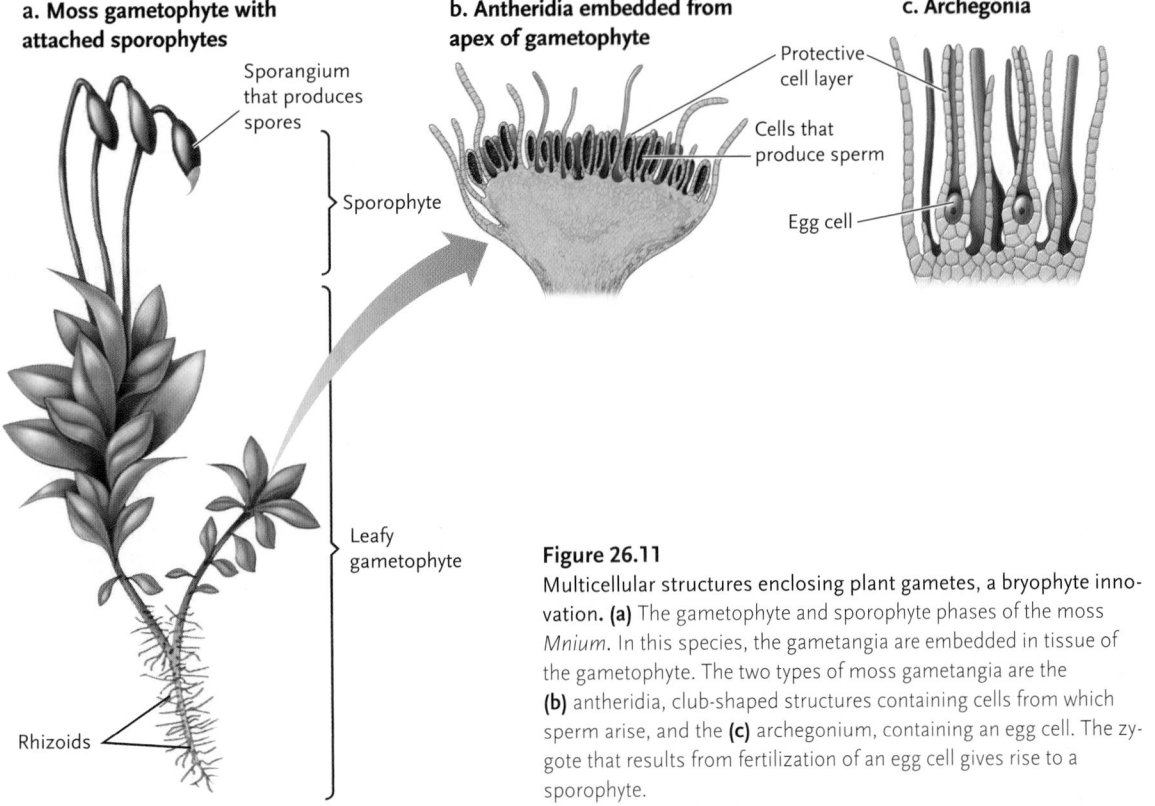

a. Moss gametophyte with attached sporophytes

Sporangium that produces spores

Sporophyte

Leafy gametophyte

Rhizoids

b. Antheridia embedded from apex of gametophyte

c. Archegonia

Protective cell layer

Cells that produce sperm

Egg cell

Figure 26.11

Multicellular structures enclosing plant gametes, a bryophyte innovation. **(a)** The gametophyte and sporophyte phases of the moss *Mnium*. In this species, the gametangia are embedded in tissue of the gametophyte. The two types of moss gametangia are the **(b)** antheridia, club-shaped structures containing cells from which sperm arise, and the **(c)** archegonium, containing an egg cell. The zygote that results from fertilization of an egg cell gives rise to a sporophyte.

26.3a Liverworts Resemble the First Land Plants

Liverworts make up the phylum **Hepatophyta**, so called because early herbalists thought that these small plants were shaped like the lobes of the human liver (*hepat* = liver; *wort* = herb). The resemblance might be a little vague to modern eyes: while some of the 6000 species of liverworts consist of a flat, branching, ribbonlike plate of tissue closely pressed against damp soil, other liverworts are leafy and superficially resemble mosses, although the arrangement of leaves is different

(Figure 26.12). This simple body, called a **thallus** (plural, *thalli*), is the gametophyte generation. Threadlike rhizoids anchor the gametophytes to their substrate. None have true stomata, the openings that regulate gas exchange in most other land plants, although some species do have pores. They lack some features present in the other two groups of bryophytes; this evidence, together with molecular data, suggests that the first land plants likely resembled modern-day liverworts.

We will look at one genus, *Marchantia* (see Figure 26.12), as an example of liverwort reproduction.

a. Thallus of *Calypogeia muelleriana*

b. Thallus of *Marchantia*

c. Male gametophyte

Male gametophyte

d. Female gametophyte

Female gametophyte

e. Asexual reproductive structures

Gemmae

Figure 26.12

Examples of liverworts. **(a)** Thallus of a leafy liverwort, *Calypogeia muelleriana*. **(b)** Thallus of the thalloid liverwort *Marchantia*, the only liverwort to produce **(c)** male and **(d)** female gametophytes on separate plants. *Marchantia* and some other liverworts also reproduce asexually by way of **(e)** gemmae, multicellular vegetative bodies that develop in tiny cups on the plant body. Gemmae can grow into new plants when splashing raindrops transport them to suitable sites.

Separate male and female gametophytes produce sexual organs (antheridia and archegonia) on tall stalks (Figure 26.12c, d). The motile sperm released from antheridia swim through surface water to reach the eggs inside archegonia. After fertilization, a small, diploid sporophyte develops inside the archegonium, matures there, and produces haploid spores by meiosis. During meiosis, sex chromosomes segregate, so some spores have the male genotype and others the female genotype. As in other liverworts, the spores develop inside jacketed sporangia that split open to release the spores. A spore that is carried by air currents to a suitable location germinates and gives rise to a haploid gametophyte, which is either male or female. *Marchantia* and some other liverworts can also reproduce asexually by way of **gemmae** (*gem* = bud; singular, *gemma*), small cell masses that form in cuplike growths on a thallus (Figure 26.12e). Gemmae can grow into

new thalli when rainwater splashes them out of the cups and onto an appropriately moist substrate.

26.3b Many Mosses Have Specialized Cells for Water and Nutrient Transport

Chances are that you have seen, touched, or sat on at least some of the approximately 10 thousand species of mosses, and the use of the name **Bryophyta** for this phylum underscores the fact that mosses are the best-known bryophytes, forming tufts or carpets of vegetation on the surface of rocks, soil, or bark.

The moss life cycle, diagrammed in **Figure 26.13**, begins when a haploid (*n*) spore lands on a wet soil surface. After the spore germinates, it elongates and branches into a filamentous web of tissue called a **protonema** ("first thread"), which can become dense enough to colour the surface of soil, rocks, or bark

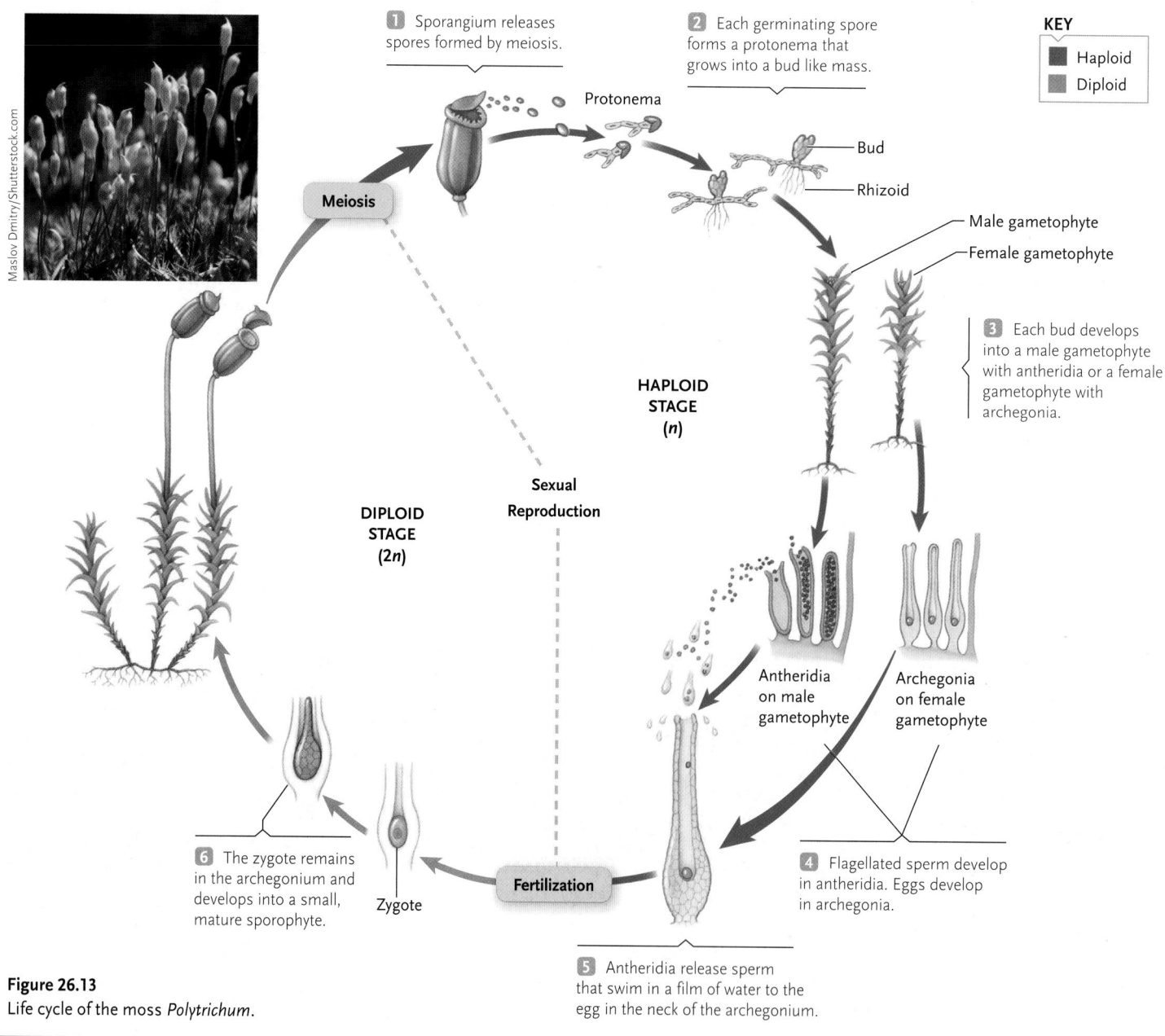

1 Sporangium releases spores formed by meiosis.

2 Each germinating spore forms a protonema that grows into a bud like mass.

KEY
- Haploid
- Diploid

Protonema

Meiosis

Bud

Rhizoid

Male gametophyte

Female gametophyte

3 Each bud develops into a male gametophyte with antheridia or a female gametophyte with archegonia.

HAPLOID STAGE (*n*)

DIPLOID STAGE (*2n*)

Sexual Reproduction

Antheridia on male gametophyte

Archegonia on female gametophyte

4 Flagellated sperm develop in antheridia. Eggs develop in archegonia.

6 The zygote remains in the archegonium and develops into a small, mature sporophyte.

Zygote

Fertilization

5 Antheridia release sperm that swim in a film of water to the egg in the neck of the archegonium.

Maslov Dmitry/Shutterstock.com

Figure 26.13
Life cycle of the moss *Polytrichum*.

visibly green. After several weeks of growth, the bud-like cell masses on a protonema develop into leafy, green gametophytes anchored by rhizoids. A single protonema can be extremely prolific, producing bud after bud, thus giving rise to a dense clone of genetically identical gametophytes. Leafy mosses may also reproduce asexually by gemmae produced at the surface of rhizoids and on above-ground parts.

Antheridia and archegonia are produced at the tips of male and female gametophytes, respectively. Propelled by flagella, sperm released from antheridia swim through a film of dew or rainwater and down a channel in the neck of the archegonium, attracted by a chemical gradient secreted by each egg. Fertilization produces the new sporophyte generation inside the archegonium, in the form of diploid zygotes that develop into small, mature sporophytes, each consisting of a sporangium on a stalk. Moss sporophytes may eventually develop chloroplasts and nourish themselves photosynthetically, but initially they depend on the gametophytes for food. Even after a moss sporophyte begins photosynthesis, it still must obtain water, carbohydrates, and some other nutrients from the gametophyte.

Certain moss gametophytes are structurally complex, with features similar to those of higher plants. For example, some species have a central strand of conducting tissue. One kind of tissue is made up of elongated, thin-walled, dead and empty cells that conduct water. In a few mosses, the water-conducting cells are surrounded by sugar-conducting tissue resembling the phloem of vascular plants. These tissues did not give rise to the xylem and phloem of vascular plants, however.

26.3c Hornworts Share a More Recent Ancestor with Vascular Plants

Roughly 100 species of hornworts make up the phylum **Anthocerophyta**. Like some liverworts, a hornwort gametophyte has a flat thallus, but the sporangium of the sporophyte phase is long and pointed, like a horn **(Figure 26.14)**, and splits into two or three ribbonlike sections when it releases spores. Sexual reproduction occurs in basically the same way as in liverworts, and hornworts also reproduce asexually by fragmentation as pieces of a thallus break off and develop into new individuals. While the gametophyte is the dominant stage of the hornwort life cycle, hornworts differ from other nonvascular plants in that their sporophytes can become free-living plants that are independent of the gametophyte! Recent genetic research into evolutionary relationships among the major groups of land plants indicates that hornworts are the group of bryophytes that have a more recent common ancestor with vascular plants.

In the next section, we turn to the vascular plants, which have lignified water-conducting tissue. Without

Figure 26.14
The hornwort *Anthoceros*. The base of each long, slender sporophyte is embedded in the flattened, leafy gametophyte.

the strength and support provided by this tissue, as well as its capacity to move water and minerals efficiently throughout the plant body, large sporophytes could not have survived on land. Unlike bryophytes, modern vascular plants are monophyletic—all groups are descended from a common ancestor.

STUDY BREAK

1. Give some examples of bryophyte features that bridge aquatic and terrestrial environments.
2. Summarize the main similarities and differences among liverworts, hornworts, and mosses.
3. How do specific aspects of a moss plant's anatomy resemble those of vascular plants?

26.4 Seedless Vascular Plants

The first vascular plants did not produce seeds and were the dominant plants on Earth for almost 200-million years, until seed plants became abundant. The fossil record shows that seedless vascular plants were well established by the late Silurian, about 428 mya, and they flourished until the end of the Carboniferous, about 250 mya. Some living seedless vascular plants have certain bryophyte-like traits, whereas others have some characteristics of seed plants. On the one hand, like bryophytes, seedless vascular plants disperse themselves by releasing spores, and they have swimming sperm that require free water to reach eggs. On the other hand, as in seed plants, the sporophyte of a seedless vascular plant becomes independent of the gametophyte at a certain point in its development and has well-developed vascular tissues (xylem and phloem). Also, the sporophyte is the larger, longer-lived stage of the life cycle and the gametophytes are very small, with some even lacking chlorophyll.

Table 26.2 summarizes these characteristics and gives an overview of seedless vascular plant features within the larger context of modern plant phyla.

In the late Paleozoic era, seedless vascular plants were Earth's dominant vegetation. Some lineages have endured to the present, but, collectively, these survivors total fewer than 14 thousand species. The taxonomic relationships between various lines are still under active investigation, and comparisons of gene sequences from the genomes in chloroplasts, nuclei, and mitochondria are revealing previously unsuspected links between some of them. In this book, we assign seedless vascular plants to two phyla, the Lycophyta (club mosses and their close relatives; the common name "club moss" for lycophytes is misleading, as they are vascular plants, not mosses) and the Pterophyta (ferns, whisk ferns, and horsetails).

26.4a Early Seedless Vascular Plants Flourished in Moist Environments

What did the first vascular plant look like? There are no living relatives of the earliest vascular plants, so we rely on fossil data to answer this question. The extinct

Table 26.2	Plant Phyla and Major Characteristics		
Phylum	Common Name	Number of Species	Common General Characteristics
Bryophytes: nonvascular plants. Gametophyte dominant, free water required for fertilization, cuticle and stomata present in some.			
Hepatophyta	Liverworts	6000	Leafy or simple flattened thallus, rhizoids; spores in capsules. Moist, humid habitats.
Bryophyta	Mosses	10 000	Simple flattened thallus, rhizoids; hornlike sporangia. Moist, humid habitats.
Anthocerophyta	Hornworts	100	Feathery or cushiony thallus; some have hydroids; spores in capsules. Moist, humid habitats; colonizes bare rock, soil, or bark.
Seedless vascular plants: sporophyte dominant, free water required for fertilization, cuticle and stomata present.			
Lycophyta	Club mosses	1000	Small simple leaves, true roots; most species have sporangia on sporophylls. Mostly wet or shady habitats.
Pterophyta	Ferns, whisk ferns, horsetails	13 000	*Ferns:* Finely divided large leaves, sporangia often in sori. Habitats from wet to arid. *Whisk ferns:* Branching stem from rhizomes; sporangia on stem scales. Tropical to subtropical habitats. *Horsetails:* Hollow photosynthetic stem, scalelike leaves, sporangia in strobili. Swamps, disturbed habitats.
Gymnosperms: vascular plants with "naked" seeds. Sporophyte dominant, fertilization by pollination, cuticle and stomata present.			
Cycadophyta	Cycads	185	Shrubby or treelike with palmlike leaves, pithy stems; male and female strobili on separate plants. Widespread distribution.
Ginkgophyta	Ginkgo	1	Woody-stemmed tree, deciduous fan-shaped leaves. Male, female structures on separate plants. Temperate areas of China.
Gnetophyta	Gnetophytes	70	Shrubs or woody vines; one has strappy leaves. Male and female strobili on separate plants. Limited to deserts, tropics.
Coniferophyta	Conifers	550	Mostly evergreen, woody trees and shrubs with needlelike or scalelike leaves; male and female cones usually on same plant.
Angiosperms: plants with flowers and seeds protected inside fruits. Sporophyte dominant, fertilization by pollination, cuticle and stomata present. Major groups: monocots, eudicots.			
Anthophyta	Flowering plants	268 500+ (including monocots and dicots, as well as magnoliids, other basal angiosperms)	Wood and herbaceous plants. Nearly all land habitats, some aquatic.
Monocots	Grasses, palms, lilies, orchids, and others	(60 000)	One cotyledon; parallel-veined leaves common; bundles of vascular tissue scattered in stem; flower parts in multiples of three.
Eudicots	Most fruit trees, roses, beans, potatoes, and others	(200 000)	Most species have two cotyledons; net-veined leaves common; central core of vascular tissue in stem; flower parts in multiples of four or five.

genus *Rhynia* was one of the earliest ancestors of modern seedless vascular plants. Based on fossil evidence, the sporophytes of the first vascular plants, such as *Rhynia* and related genera **(Figure 26.15)**, lacked leaves and roots. Above-ground photosynthetic stems produced sporangia at the tips of branches. Below ground, the plant body was supported by **rhizomes**, horizontal modified stems that can penetrate a substrate and anchor the plant. *Rhynia*'s simple stems had a central core of xylem, an arrangement seen in many existing vascular plants. Mudflats and swamps of the damp Devonian period were dominated by *Rhynia* and related plants (Figure 26.15). Although these and other now-extinct phyla came and went, ancestral forms of both modern phyla of seedless vascular plants appeared.

Carboniferous forests were swampy places dominated by members of the phylum **Lycophyta**, and fascinating fossil specimens of this group have been unearthed in North America and Europe. One example is *Lepidodendron,* which had broad, straplike leaves and sporangia near the ends of the branches **(Figure 26.16a, p. 578)**. It also had xylem and other tissues typical of all modern vascular plants. Also abundant at the time were representatives of the phylum **Pterophyta**, including ferns and giants such as *Calamites*—huge horsetails that could have a trunk diameter of 30 cm. Some early seed plants were also present, including now-extinct fern like plants, called seed ferns, that bore seeds at the tips of their leaves (Figure 26.16b).

Characterized by a moist climate over much of the planet and by the dominance of seedless vascular plants, the Carboniferous period continued for 150 million years, ending when climate patterns changed during the Paleozoic era. Most modern seedless vascular plants are confined largely to wet or humid environments because they require external water for reproduction. However, some are poikilohydric and can survive in a dehydrated state for long periods of time (see "Life on the Edge," Box 26.1, p. 566).

26.4b Modern Lycophytes Are Small and Have Simple Vascular Tissues

Lycophytes were highly diverse 350 mya, when some tree-sized forms inhabited lush swamp forests. Today, however, such giants are no more. The most familiar of the 1000 or so living species of lycophytes are club mosses (e.g., species of *Lycopodium* and *Selaginella*), which grow on forest floors, in alpine meadows, and in some prairie habitats. **(Figure 26.17, p. 578)**. For example, *Selaginella densa* (Figure 26.17b) is a dominant plant in shortgrass prairies of western North America. Club moss sporophytes have upright or horizontal stems that contain xylem and bear small green leaves and roots. Sporangia are clustered at the bases of specialized leaves, called **sporophylls** (*phyll* = leaf; thus, sporophyll = "spore-bearing leaf"). Sporophylls are clustered into a **cone** or **strobilus** (plural, *strobili*) at the tips of stems. Most lycophytes are homosporous, but some are heterosporous, producing two types of spores that will in turn produce separate male and female gametophytes.

26.4c Ferns, Whisk Ferns, Horsetails, and Their Relatives Make Up the Diverse Phylum Pterophyta

Second in size only to the flowering plants, the phylum Pterophyta (*pteron* = wing) contains a large and diverse group of vascular plants—the 13 thousand or so species of ferns, whisk ferns, and horsetails. Most ferns, including some that are popular houseplants, are native to tropical and temperate regions. Some floating species are less than 1 cm across, whereas some tropical tree ferns grow to 25 m tall. Other species are adapted to life in arctic and alpine tundras, salty mangrove swamps, and semi-arid deserts.

Features of Ferns. The familiar plant body of a fern is the sporophyte phase **(Figure 26.18, p. 579)**, which produces an above-ground clump of leaves. Young leaves are tightly coiled, and as they emerge above the soil,

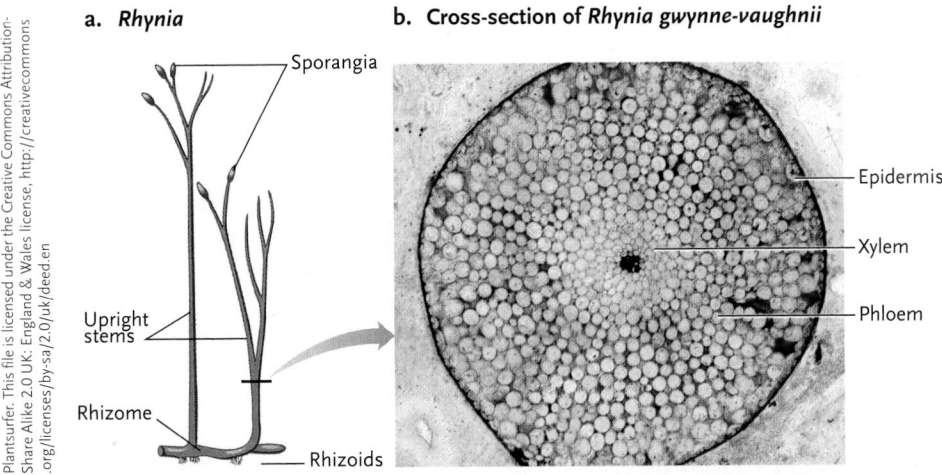

a. *Rhynia*

Sporangia

Upright stems

Rhizome

Rhizoids

b. Cross-section of *Rhynia gwynne-vaughnii*

Epidermis

Xylem

Phloem

Figure 26.15

Rhynia, an early seedless vascular plant. **(a)** Fossil-based reconstruction of the entire plant, about 30 cm tall. **(b)** Cross-section of the stem, approximately 3 mm in diameter. This fossil was embedded in chert approximately 400 mya. Still visible in it are traces of the transport tissues xylem and phloem, along with other specialized tissues.

a. The lycophyte tree (*Lepidodendron*)

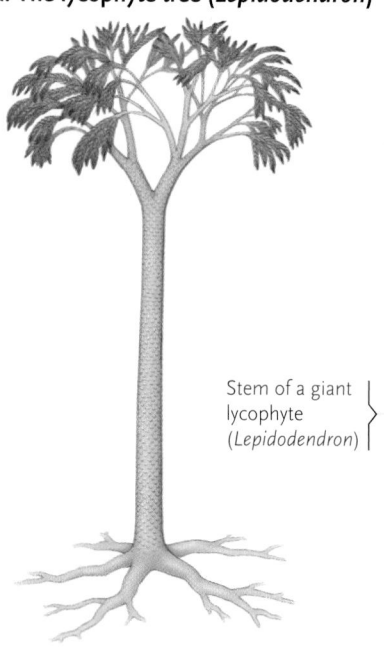

Stem of a giant
lycophyte
(*Lepidodendron*)

b. Artist's depiction of a Coal Age forest

John Weinstein/Field Museum Library/Getty Images

Seed fern (*Medullosa*); probably
related to the progymnosperms,
which may have been among the
earliest seed-bearing plants

Stem of a giant
horsetail (*Calamites*)

Figure 26.16

Reconstruction of a lycophyte tree (*Lepidodendron*) and its environment. **(a)** Fossil evidence suggests that *Lepidodendron* grew to be about 35 m tall with a trunk 1 m in diameter. **(b)** Artist's depiction of a Coal Age forest.

Figure 26.17

Lycophytes. **(a)** *Lycopodium* sporophyte, showing the conelike strobili in which spores are produced. **(b)** *Selaginella densa* sporophytes.

a. *Lycopodium* sporophyte

Strobilus

Ed Reschke

b. *Selaginella densa* sporophytes

Dave Powell, USDA Forest Service, Bugwood.org

these fiddleheads (so named because they resemble the scrolled pegheads of violins) unroll and expand. The fiddleheads of some species are edible when cooked, tasting similar to fresh asparagus, but be sure you have collected the right type of fiddlehead—some species contain a carcinogen.

Sporangia are produced on the lower surface or margins of leaves. Often several sporangia are clustered into a rust-coloured **sorus** ("heap"; plural, *sori*) (see Figure 26.17). Spores released from sporangia develop into gametophytes, which are typically small, heart-shaped plants anchored to the soil by rhizoids. Antheridia and archegonia develop on the underside of gametophytes, where moisture is trapped. Inside an

antheridium is a globular packet of haploid cells, each of which develops into a helical sperm with many flagella. When water is present, the antheridium bursts, releasing the sperm. If mature archegonia are nearby, the sperm swim toward them, drawn by a chemical attractant that diffuses from the neck of the archegonium, which is open when free water is present.

In some ferns, antheridia and archegonia are produced on a single bisexual gametophyte. In other ferns, the first spores to germinate develop into bisexual gametophytes, which produce a chemical (antheridiogen) that diffuses through the substrate and causes all later-germinating spores to develop into male gametophytes. What is the advantage of producing a

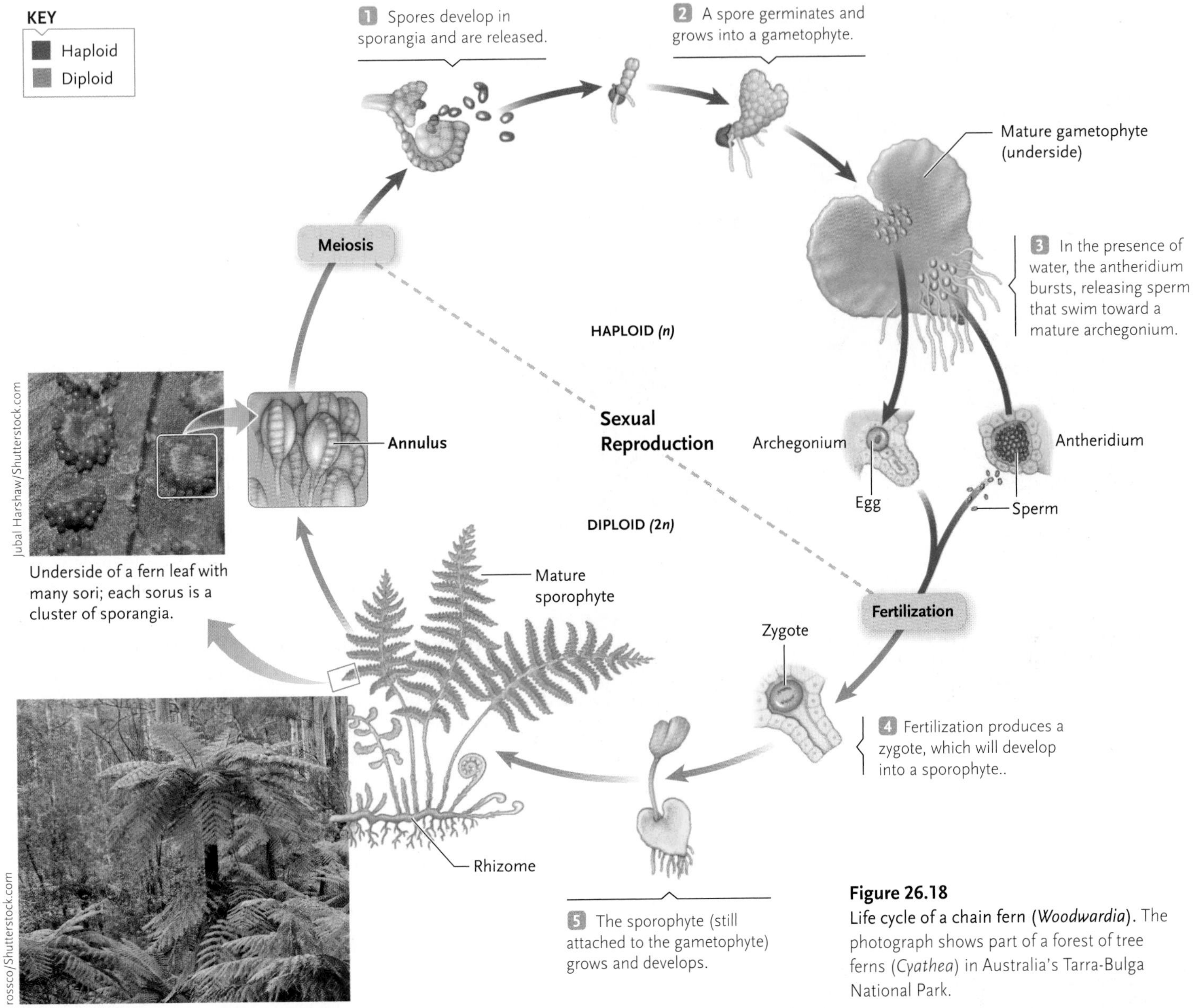

1 Spores develop in sporangia and are released.

2 A spore germinates and grows into a gametophyte.

Mature gametophyte (underside)

Meiosis

3 In the presence of water, the antheridium bursts, releasing sperm that swim toward a mature archegonium.

HAPLOID (n)

Archegonium
Antheridium

Egg
Sperm

Sexual Reproduction

Annulus

DIPLOID (2n)

Jubal Harshaw/Shutterstock.com

Underside of a fern leaf with many sori; each sorus is a cluster of sporangia.

Mature sporophyte

Fertilization

Zygote

4 Fertilization produces a zygote, which will develop into a sporophyte..

Rhizome

rossco/Shutterstock.com

5 The sporophyte (still attached to the gametophyte) grows and develops.

Figure 26.18

Life cycle of a chain fern (*Woodwardia*). The photograph shows part of a forest of tree ferns (*Cyathea*) in Australia's Tarra-Bulga National Park.

few bisexual gametophytes followed by many male gametophytes? If a bisexual gametophyte is surrounded by several male gametophytes that developed from other spores, it is more likely that eggs will be fertilized by sperm from one of the male gametophytes rather than by its own sperm, thus increasing the genetic diversity of the resulting zygote.

An embryo is retained on and nourished by the gametophyte for the first part of its life but soon develops into a young sporophyte larger than the gametophyte, with its own green leaf and root system. Once the sporophyte is nutritionally independent, the parent gametophyte degenerates and dies.

Features of Whisk Ferns. The whisk ferns and their relatives are represented by only 2 genera, with about 10 species in total; we look at just one genus, *Psilotum* **(Figure 26.19).** Whisk ferns grow in tropical and subtropical regions, often as epiphytes.

Figure 26.19
Sporophytes of a whisk fern (*Psilotum*), a seedless vascular plant. Three-lobed sporangia occur at the ends of stubby branchlets; inside the sporangia, meiosis gives rise to haploid spores.

William Ormerod/Visuals Unlimited, Inc.

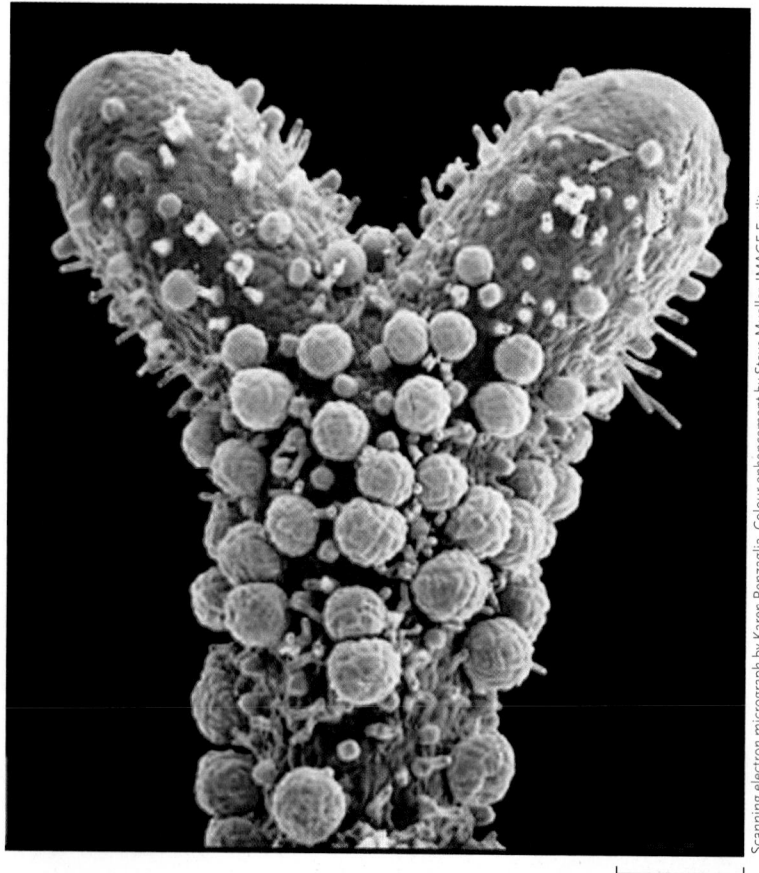

Scanning electron micrograph by Karen Renzaglia. Colour enhancement by Steve Mueller, IMAGE Facility

0.5 mm

Figure 26.20

Scanning electron micrograph image of the subterranean gametophyte of *Psilotum*. Antheridia have been coloured blue, and the smaller archegonia have been coloured pink.

The sporophytes of *Psilotum* resemble the extinct vascular plants in that they lack true roots and leaves. Instead, small, leaflike scales adorn an upright, green, branching stem, which arises from a horizontal rhizome system anchored by rhizoids. Symbiotic fungi colonize the rhizoids, increasing the plant's uptake of soil nutrients (read more about these mycorrhizal fungi in Chapter 25). The stem is photosynthetic and bears sporangia above the small scales. Gametophytes of *Psilotum* are nonphotosynthetic and live underground **(Figure 26.20)**; like the sporophyte, they obtain nutrients via symbioses with mycorrhizal fungi.

Features of Horsetails. The ancient relatives of modern-day horsetails included treelike forms taller than a two-storey building. Only 15 species in a single genus, *Equisetum,* have survived to the present **(Figure 26.21)**. Horsetails grow in moist soil along streams and in disturbed habitats, such as roadsides and beds of railway tracks. Their sporophytes typically have underground rhizomes and roots that anchor the rhizome to the soil. Small, scalelike leaves are arranged in whorls about a photosynthetic stem that is stiff and gritty because horsetails accumulate silica in their tissues. Pioneers used them to scrub out pots and pans—hence their other common name, "scouring rushes."

As in lycophytes, *Equisetum* sporangia are borne in strobili. Haploid spores germinate within a few days to produce gametophytes, which are free-living plants about the size of a small pea.

a. Sporophyte stem   **b.** Sporangia

Strobilus, an aggregation of sporangia and sporophylls at the tip of the horsetail sporophyte

c. This longitudinal section through a horsetail's strobilus shows sporangia containing spores formed by meiosis.

Figure 26.21

A species of *Equisetum*, the horsetails. **(a)** Vegetative stem. **(b)** Strobili, which bear sporangia. **(c)** Close-up of sporangium and associated structures on a strobilus.

26.4d Some Seedless Vascular Plants Are Heterosporous

Most seedless vascular plants are homosporous, but some (e.g., some lycophytes and some ferns) are heterosporous, producing microspores and megaspores in separate sporangia (see Figure 26.8, p. 570). Both types of spores are usually shed from sporangia and germinate on the ground some distance from the parent plant. In many heterosporous plants, the gametophytes produced by the spores develop inside the spore wall; this **endosporous** development provides increased protection for the gametes and, later, for the developing embryo. The microspore gives rise to a male gametophyte, which produces motile sperm. At maturity, the microspore wall will rupture, releasing the sperm, which swim to the female gametophyte; water is thus still required for fertilization in these plants. The megaspore produces a female gametophyte inside the spore wall; archegonia of this gametophyte produce eggs, as in other seedless plants.

1. Compare the lycophyte and bryophyte life cycles with respect to the sizes and longevity of gametophyte and sporophyte phases.
2. Summarize the main similarities and differences in the life cycles of lycophytes, horsetails, whisk ferns, and ferns.
3. Define *sorus* and *strobilus*. How are these two structures similar?

26.5 Gymnosperms: The First Seed Plants

Gymnosperms are the conifers and their relatives. The earliest fossils identified as gymnosperms are found in Devonian rocks. By the Carboniferous, when nonvascular plants were dominant, many lines of gymnosperms, including conifers, had also evolved. These radiated during the Permian period; the Mesozoic era that followed, 65 to 248 mya, was the age not only of the dinosaurs but of the gymnosperms as well.

The evolution of gymnosperms marked sweeping changes in plant structures related to reproduction. The evolution of gymnosperms included important reproductive adaptations—pollen and pollination, the ovule, and the seed. The fossil record has not revealed the sequence in which these changes arose, but all of them contributed to the radiation of gymnosperms into land environments.

As a prelude to our survey of modern gymnosperms, we begin by considering some of these innovations.

26.5a Major Reproductive Adaptations Occurred as Gymnosperms Evolved

The word *gymnosperm* is derived from the Greek *gymnos,* meaning naked, and *sperma,* meaning seed. As this name indicates, gymnosperms produce seeds that are exposed, not enclosed in fruit, as are the seeds of other seed plants.

Ovules: Increased Protection for Female Gametophyte and Egg. How did seeds first arise? Think about the heterosporous plants described in the previous section and picture two steps that would lead us toward the development of a seed. In the first step, spores are not shed from the plant but instead are retained inside sporangia on the sporophyte. In the second step, the number of megaspores is reduced to just one per sporangium (i.e., four megaspores are produced by meiosis, but only one survives). These two steps result in retention of a single megaspore inside a megasporangium on a plant **(Figure 26.22)**. As in all land plants, the megaspore will give rise to a female gametophyte; because this is a heterosporous plant, the gametophyte will develop inside the megaspore wall and inside the megasporangium. Physically connected to the sporophyte and surrounded

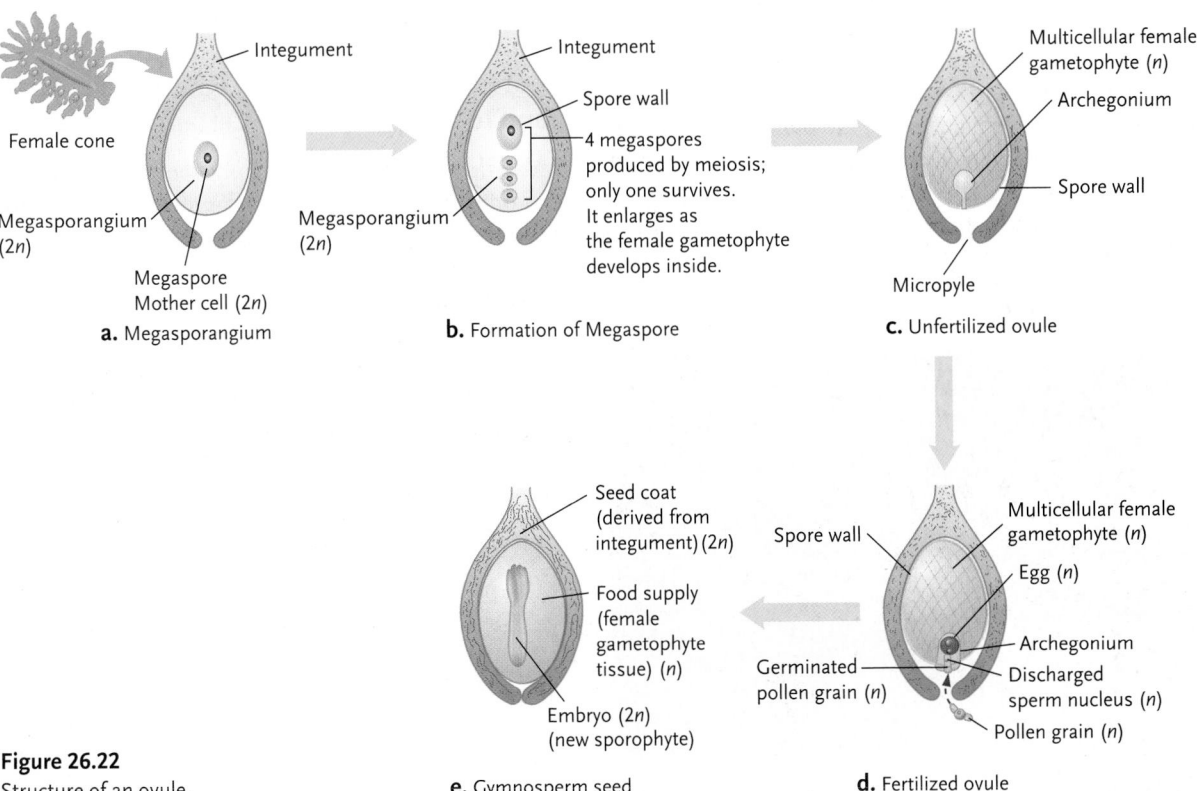

Figure 26.22
Structure of an ovule.

a. Megasporangium

Integument
Female cone
Megasporangium (2n)
Megaspore Mother cell (2n)

b. Formation of Megaspore

Integument
Spore wall
4 megaspores produced by meiosis; only one survives. It enlarges as the female gametophyte develops inside.
Megasporangium (2n)

c. Unfertilized ovule

Multicellular female gametophyte (n)
Archegonium
Spore wall
Micropyle

d. Fertilized ovule

Multicellular female gametophyte (n)
Egg (n)
Archegonium
Discharged sperm nucleus (n)
Pollen grain (n)
Germinated pollen grain (n)
Spore wall

e. Gymnosperm seed

Seed coat (derived from integument) (2n)
Food supply (female gametophyte tissue) (n)
Embryo (2n) (new sporophyte)

by protective layers, a female gametophyte no longer faces the same risks of predation or environmental assault that can threaten a free-living gametophyte.

This new structure, of an egg developing inside a gametophyte that is retained not only inside the spore wall but also inside megasporangial tissue, is an **ovule**. When fertilized, an ovule becomes a **seed**: the fertilized egg will produce an embryo surrounded by nutritive tissue, all encased in sporangial tissue that has become a seed coat.

When you look at Figure 26.22, p. 581, you can see that the megasporangium is surrounded by extra layers of sporophyte tissue, which would add additional protection for gametes and embryos, but this tissue, along with that of the megasporangium, has also created a problem: how can sperm get to the egg now that the gametophyte is enclosed inside these layers of tissue? The solution is similar to that of internal fertilization in animals: there needs to be a male structure that can penetrate sporophyte tissue and release sperm inside the female gametophyte. In the next section, we look at the male gametophyte in seed plants.

Pollen: Eliminating the Need for Water in Reproduction. As for megaspores, the microspores of gymnosperms (and other seed plants) are not dispersed. Instead, they are retained inside microsporangia and are enveloped in additional layers of sporophyte tissue. As in other heterosporous plants, each microspore produces a male gametophyte, which develops inside the microspore wall. This male gametophyte is very small relative to those of nonseed plants—it is made of only a few cells—and is called a **pollen grain**. Pollen grains are transferred to female reproductive parts via air currents or on the bodies of animal pollinators; this transfer is known as **pollination**. When the pollen grain lands on female tissue, the pollen grain germinates to produce a **pollen tube (Figure 26.23)**, a cell that grows through female gametophyte tissue by invasive growth and carries the nonmotile sperm to the egg.

Pollen and pollination were enormously important adaptations for gymnosperms because the shift to nonswimming sperm, along with a means for delivering them to female gametes, meant that reproduction no longer required liquid water. The only gymnosperms that have retained swimming sperm are the cycads and ginkgoes described below, which have relatively few living species and are restricted to just a few native habitats.

Seeds: Protecting and Nourishing Plant Embryos. As described above, a seed is the structure that forms when an ovule matures, after a pollen grain reaches it and a sperm fertilizes the egg. Seeds consist of three basic parts: (1) the embryo sporophyte; (2) the tissues surrounding the embryo containing nutrients that nourish it until it becomes established as a seedling with leaves and roots; and (3) a tough, protective outer seed coat **(Figure 26.24)**. This complex structure makes seeds ideal packages for sheltering an embryo from drought, cold, or other adverse conditions. As a result, seed plants enjoy a tremendous survival advantage over species that simply release spores to the environment. Encased in a seed, the embryo can also be transported far from its parent, as when ocean currents carry coconut seeds ("coconuts" protected in large, buoyant fruits) hundreds of kilometres across the sea. As discussed in Chapter 36, some plant embryos housed in seeds can remain dormant for months or years before environmental conditions finally prompt them to germinate and grow.

26.5b Modern Gymnosperms Include Conifers and a Few Other Groups Represented by Relatively Few Species That Tend to Be Restricted to Certain Climates

Today there are about 800 gymnosperm species. The sporophytes of nearly all are large trees or shrubs, although a few are woody vines.

Economically, gymnosperms, particularly conifers, are vital to human societies. They are sources of lumber, paper pulp, turpentine, and resins, among other products. They also have huge ecological importance. Their habitats range from **tropical forests** to deserts, but gymnosperms are most dominant in the cool-temperate

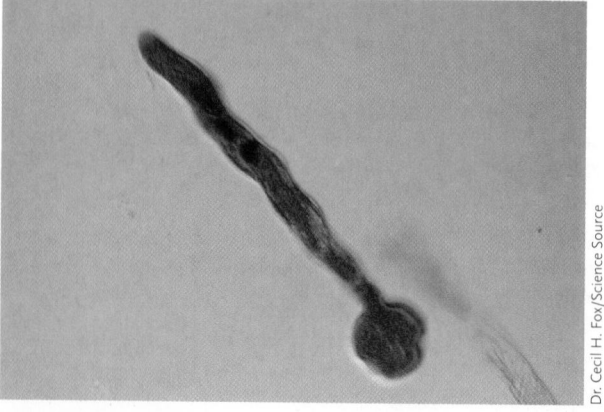

Figure 26.23
Pollen tube extending from germinating pollen grain at top right.

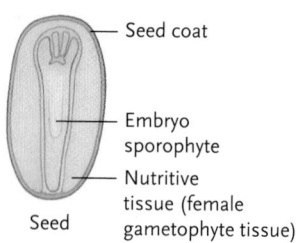

Figure 26.24
Generalized view of the seed of a pine, a gymnosperm.

zones of the Northern and Southern Hemispheres. They flourish in poor soils, where flowering plants don't compete as well. In Canada, for example, gymnosperms make up most of the boreal forests that cover about one-third of the country's landmass. Our survey of gymnosperms begins with the conifers, and then we will look at the cycads, ginkgoes, and gnetophytes—the latter two groups are remnants of lineages that have all but vanished from the modern scene.

Conifers Are the Most Common Gymnosperms. About 80% of all living gymnosperm species are members of one phylum, the **Coniferophyta**, or conifers ("cone-bearers"). Examples are pines, spruces, and firs. Coniferous trees and shrubs are longer lived, and anatomically and morphologically more complex, than any sporophyte phase we have discussed so far. Characteristically, they form woody cones, and most have needlelike leaves that are adapted to dry environments. For instance, needles have a thick cuticle, sunken stomata, and a fibrous epidermis, all traits that reduce the loss of water vapour.

Pines and many other gymnosperms produce resins, a mix of organic compounds that are by-products of metabolism. Resin accumulates and flows in long resin ducts through the wood, inhibiting the activity of wood-boring insects and certain microbes. Pine resin extracts are the raw material of turpentine and (minus the volatile terpenes) the sticky rosin used to treat violin bows. Fossil resin is known as amber and is commonly used in jewellery; amber often contains fossilized insects or even small animals.

We know a great deal about the pine life cycle **(Figure 26.25, p. 584)**, so it is a convenient model for gymnosperms. Male cones are relatively small and delicate (about 1 cm long) and are borne on the lower branches. Each cone consists of many sporophylls with two microsporangia on their undersides. Inside the microsporangia, **microspores** are produced by meiosis. Each microspore then undergoes mitosis to develop into a winged pollen grain—an immature male gametophyte. At this stage, the pollen grain consists of four cells, two that will degenerate and two that will function later in reproduction.

Young female cones develop higher in the tree, at the tips of upper branches. Ovules are produced on modified sporophylls. Inside each ovule, four megaspores are produced by meiosis, but only one survives to develop into a megagametophyte. This female gametophyte develops slowly, becoming mature only when pollination is under way; in a pine, this process takes well over a year. The mature female gametophyte is a small oval mass of cells with several archegonia at one end, each containing an egg.

Each spring, air currents release vast numbers of pollen grains from male cones—by some estimates, billions may be released from a single pine tree. The extravagant numbers ensure that at least some pollen grains will land on female cones. The process is not as random as it might seem: studies have shown that the contours of female cones create air currents that can favour the "delivery" of pollen grains near the cone scales. After pollination, the two remaining cells of the pollen grain divide, one producing sperm by mitosis, the other producing the pollen tube that grows toward the developing gametophyte. When a pollen tube reaches an egg, the stage is set for fertilization, the formation of a zygote, and early development of the plant embryo. Often fertilization occurs months to a year after pollination. Once an embryo forms, a pine seed—which, remember, includes the embryo, female gametophyte tissue, and seed coat—is eventually shed from the cone. The seed coat protects the embryo from drying out, and the female gametophyte tissue serves as its food reserve. This tissue makes up the bulk of a "pine nut."

Cycads Are Restricted to Warmer Climates. During the Mesozoic era, the **Cycadophyta** (*kykas* = palm), or cycads, flourished along with the dinosaurs. About 185 species have survived to the present, but they are confined to the tropics and subtropics.

At first glance, you might mistake a cycad for a small palm tree **(Figure 26.26, p. 585)**. Some cycads have massive cones that bear either pollen or ovules. Air currents or crawling insects transfer pollen from male plants to the developing gametophyte on female plants. Poisonous alkaloids that may help deter insect predators occur in various cycad tissues. In tropical Asia, some people consume cycad seeds and flour made from cycad trunks, but only after rinsing away the toxic compounds. Much in demand from fanciers of unusual plants, cycads in some countries are uprooted and sold in what amounts to a black-market trade, greatly diminishing their numbers in the wild.

Ginkgoes Are Limited to a Single Living Species. The phylum **Ginkgophyta** has only one living species, the ginkgo (or maidenhair) tree (*Ginkgo biloba*), which grows wild today only in warm-temperate forests of central China. Ginkgo trees are large, diffusely branching trees with characteristic fan-shaped leaves **(Figure 26.27, p. 585)** that turn a brilliant yellow in autumn. Nursery-propagated male trees are often planted in cities because they are resistant to insects, disease, and air pollutants. The female trees are equally pollution resistant, but gardeners avoid them because their seeds produce a foul odour that only a ginkgo could love. The leaves and seeds have been used in traditional Chinese medicine for centuries. The extract of the leaves is one of the most intensely investigated herbal medicines; although studies have not found any conclusive evidence for claims that the extract improves memory, there is some evidence that it does assist in blood flow and so may be effective in the treatment of circulatory disorders.

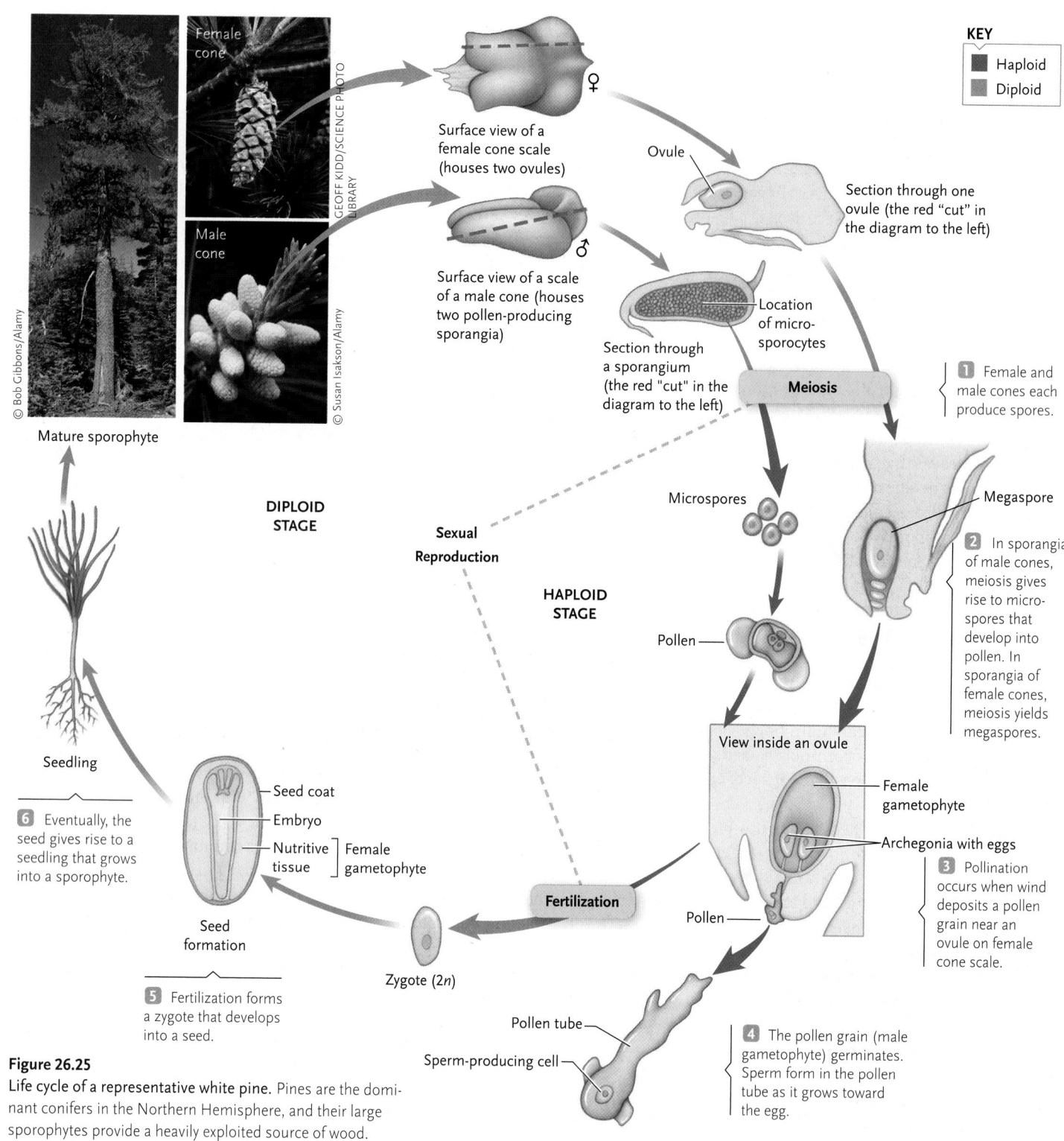

Figure 26.25

Life cycle of a representative white pine. Pines are the dominant conifers in the Northern Hemisphere, and their large sporophytes provide a heavily exploited source of wood.

Image credits: © Bob Gibbons/Alamy; GEOFF KIDD/SCIENCE PHOTO LIBRARY; © Susan Isakson/Alamy

Female cone

Male cone

Mature sporophyte

Seedling

6 Eventually, the seed gives rise to a seedling that grows into a sporophyte.

DIPLOID STAGE

Seed coat
Embryo
Nutritive tissue } Female gametophyte

Seed formation

5 Fertilization forms a zygote that develops into a seed.

Surface view of a female cone scale (houses two ovules)

Surface view of a scale of a male cone (houses two pollen-producing sporangia)

Ovule

Section through one ovule (the red "cut" in the diagram to the left)

Location of micro-sporocytes

Section through a sporangium (the red "cut" in the diagram to the left)

Sexual Reproduction

Meiosis

Microspores

Megaspore

1 Female and male cones each produce spores.

2 In sporangia of male cones, meiosis gives rise to micro-spores that develop into pollen. In sporangia of female cones, meiosis yields megaspores.

HAPLOID STAGE

Pollen

View inside an ovule

Female gametophyte

Archegonia with eggs

3 Pollination occurs when wind deposits a pollen grain near an ovule on female cone scale.

Fertilization

Pollen

Zygote (2n)

Pollen tube

Sperm-producing cell

4 The pollen grain (male gametophyte) germinates. Sperm form in the pollen tube as it grows toward the egg.

KEY
■ Haploid
■ Diploid

Gnetophytes Include Simple Seed Plants with Intriguing Features. The phylum Gnetophyta contains three genera—*Gnetum, Ephedra,* and *Welwitschia*—that together include about 70 species. Moist, tropical regions are home to about 30 species of *Gnetum,* which includes both trees and leathery-leafed vines (lianas). About 35 species of *Ephedra* grow in desert regions of the world **(Figure 26.28a–c).**

Of all the gymnosperms, *Welwitschia* is the most bizarre. This seed-producing plant grows in the hot deserts of southwest Africa. The bulk of the plant is a deep-reaching taproot. The only exposed part is a woody, disk-shaped stem that bears cone-shaped strobili and leaves. The plant never produces more than two strap-shaped leaves, which split lengthwise repeatedly as the plant grows older, producing a rather scraggly pile (Figure 26.28d).

1. What are the four major reproductive adaptations that evolved in gymnosperms?
2. What are the basic parts of a seed, and how is each one adaptive?
3. Summarize the main similarities and differences among ginkgoes, cycads, gnetophytes, and conifers.
4. Describe some features that make conifers structurally more complex than other gymnosperms.

a. *Ephedra* plant

b. *Ephedra* male cone

© William Ferguson

c. *Ephedra* female cone

Nature's Images/Science Source

Robert & Linda Mitchell Photography

d. *Welwitschia* plant with female cones

Fletcher & Baylis/Science Source

Dr. Carleton Ray/Science Source

Figure 26.26
The cycad *Zamia* showing a large, terminal female cone and fernlike leaves.

Figure 26.28
Gnetophytes. **(a)** Sporophyte of *Ephedra*, with close-ups of its **(b)** pollen-bearing cones and **(c)** seed-bearing cone, which develop on separate plants. **(d)** Sporophyte of *Welwitschia mirabilis*, with seed-bearing cones.

Figure 26.27
Ginkgo biloba. **(a)** A ginkgo tree. **(b)** A fossilized ginkgo leaf compared with a leaf from a living tree. The fossil formed at the Cretaceous–Tertiary boundary. Even though 65 million years have passed, the leaf structure has not changed much. **(c)** Pollen-bearing cones and **(d)** fleshy-coated seeds of the *Ginkgo.*

a. Ginkgo tree

b. Fossil and modern ginkgo leaves

Sinclair Stammers/Science Source

Nenov Brothers Images/Shutterstock.com

c. Male cone

A. B. Joyce/Science Source

d. Ginkgo seeds

© William Ferguson

Jiang Zhongyan/Shutterstock.com

26.6 Angiosperms: Flowering Plants

Of all plant phyla, the flowering plants, or **angiosperms**, are the most successful today. At least 260 thousand species are known (**Figure 26.29,** shows a few examples), and botanists regularly discover new ones in previously unexplored regions of the tropics. The word *angiosperm* is derived from the Greek *angeion* ("vessel") and *sperma* ("seed"). The "vessel" refers to the modified sporophyll, called a *carpel,* which surrounds and protects the ovules. Carpels are located in the centre of **flowers**, reproductive structures that are a defining feature of angiosperms. Another defining feature is the **fruit**—botanically speaking, a structure that helps protect and disperse seeds.

In addition to having flowers and fruits, angiosperms are the most ecologically diverse plants on Earth, growing on dry land and in **wetlands**, fresh water, and the seas. Angiosperms range in size from tiny duckweeds that are about 1 mm long to towering *Eucalyptus* trees more than 60 m tall.

26.6a The Fossil Record Provides Little Information about the Origin of Flowering Plants

The evolutionary origin of angiosperms has confounded plant biologists for well over a hundred years. Charles Darwin called it the "abominable mystery" because flowering plants appear suddenly in the fossil record, without a fossil sequence that links them to any other plant groups. As with gymnosperms, attempts to reconstruct the earliest flowering plant lineages have produced several conflicting classifications and family trees. Some paleobotanists hypothesize that flowering plants arose in the Jurassic period; others propose that

a. Flowering plants in a desert

b. Alpine angiosperms

c. Triticale, a grass

d. The carnivorous plant Venus flytrap

Figure 26.29

Flowering plants. Diverse photosynthetic species are adapted to nearly all environments, ranging from **(a)** deserts to **(b)** snowlines of high mountains. **(c)** Triticale, a hybrid grain derived from parental stocks of wheat (*Triticum*) and rye (*Secale*), is one example of the various grasses used by humans. **(d)** The carnivorous plant Venus flytrap (*Dionaea muscipula*) grows in nitrogen-poor soils and traps insects as an additional source of nitrogen.

they evolved in the Triassic from now-extinct gymnosperms or from seed ferns. However, progress in this area does not rely solely on fossil evidence; molecular data can be used to test hypotheses, and the combination of molecular, morphological, and fossil evidence offers great promise in solving this mystery.

The fossil record has yet to reveal obvious transitional organisms between flowering plants and either gymnosperms or seedless vascular plants. As the Mesozoic era ended and the modern Cenozoic era began, great extinctions occurred among both plant and animal kingdoms. Gymnosperms declined, and dinosaurs disappeared. Flowering plants, mammals, and social insects flourished, radiating into new environments. Today we live in what has been called "the age of flowering plants."

26.6b Angiosperms Are Subdivided into Several Groups, Including Monocots and Eudicots

Angiosperms are assigned to the phylum **Anthophyta**, a name that derives from the Greek *anthos,* meaning flower. The great majority of angiosperms are classified as either monocots or eudicots, which are differentiated on the basis of morphological features such as the number of flower parts and the pattern of vascular tissue in stems and leaves. The two groups also differ in terms of the morphology of their embryos: **monocot** embryos have a single leaf like structure called a cotyledon, whereas **eudicot** ("true dicots") embryos generally have two cotyledons (see Table 26.1, p. 568).

Botanists currently recognize several other groups of plants in addition to eudicots and monocots, but figuring out the appropriate classification for and relationships among these other groups is an ongoing challenge and an extremely active area of plant research. In this chapter, we focus only on monocots and eudicots.

There are at least 60 thousand species of monocots, including 10 thousand grasses and 20 thousand orchids. **Figure 26.30a** gives some idea of the variety of living monocots, which include grasses, palms, lilies, and orchids. The world's major crop plants (wheat, corn, rice, rye, sugar cane, and barley) are all monocots and are all domesticated grasses. Eudicots are even more diverse, with nearly 200 thousand species (Figure 26.30b). They include flowering shrubs and trees, most

a. Representative monocots

Wheat (*Triticum*)

Trillium (*Trillium*)

Western wood lily (*Lilium philadelphicum*)

b. Representative eudicots

Wild rose (*Rosa acicularis*)

Twinflower (*Linnaea borealis*)

Claret cup cactus (*Echinocereus triglochidiatus*)

Figure 26.30
Examples of monocots and eudicots. **(a)** Representative monocots: wheat (*Triticum*), trillium (*Trillium*), and Western wood lily (*Lilium philadelphicum*). **(b)** Representative eudicots: wild rose (*Rosa acicularis*), twinflower (*Linnaea borealis*), and cactus (*Echinocereus triglochidiatus*).

nonwoody (herbaceous) plants, and cacti. We will take a closer look at angiosperms in Chapter 36, which focuses on the structure and function of flowering plants.

26.6c Many Factors Contributed to the Adaptive Success of Angiosperms

Flowering plants likely originated about 140 mya. It took only about 40 million years—a short span in geologic time—for angiosperms to eclipse gymnosperms as the prevailing form of plant life on land. Several factors fuelled this adaptive success. As with other seed plants, the large, diploid sporophyte phase dominates a flowering plant's life cycle, and the sporophyte retains and nourishes the much smaller gametophytes. But flowering plants also show some evolutionary innovations not seen in gymnosperms.

More Efficient Transport of Water and Nutrients. Where gymnosperms have only one type of water-conducting cell in their xylem, angiosperms have an additional, more specialized type of cell that is larger and open ended and thus moves water more rapidly from roots to shoots (see Chapter 34). Also, modifications in angiosperm phloem tissue allow it to more efficiently transport sugars produced in photosynthesis through the plant body.

Enhanced Nutrition and Physical Protection for Embryos. Other changes in angiosperms increased the likelihood of successful reproduction and dispersal of offspring. For example, a two-step double-fertilization process in the ovules of flowering plants produces both an embryo and a unique nutritive tissue (called endosperm) that nourishes the embryonic sporophyte **(Figure 26.31)**. The ovule containing a female gametophyte is enclosed within an ovary, part of the carpel, which shelters the ovule against desiccation and against attack by herbivores or pathogens. After fertilization, an ovary develops into a fruit that not only protects seeds but also helps disperse them—for instance, when an animal eats a fruit, seeds may pass through the animal's gut none the worse for the journey and be released in a new location in the animal's feces. Above all, angiosperms have flowers, the unique reproductive organs that you will read much more about in Chapter 36.

26.6d Angiosperms Coevolved with Animal Pollinators

The evolutionary success of angiosperms is due not only to the adaptations just described but also to the efficient mechanisms of transferring pollen to female reproductive parts. Whereas a conifer depends on air currents to disperse its pollen, as do such angiosperms as grasses, many angiosperms coevolved with pollinators—insects, bats, birds, and other animals that transfer pollen from male floral structures to female reproductive parts, often while obtaining nectar. Nectar is a sugar-rich liquid secreted by flowers to attract pollinators. Pollen itself is a reward for some pollinators, such as bees, that use it as a food resource. So, while plants benefit from their animal pollinators, there is also a cost to the plant in providing a reward to the pollinator. **Coevolution** occurs when two or more species interact closely in the same ecological setting. A heritable change in one species affects selection pressure operating between them, so that the other species evolves as well. Over time, plants have coevolved with their pollinating animals.

In general, a flower's reproductive parts are positioned so that visiting pollinators will brush against them. In addition, many floral features correlate with the morphology and behaviour of specific pollinators. For example, reproductive parts may be located above nectar-filled floral tubes that are the same length as the feeding structure of a preferred pollinator. Nectar-sipping bats **(Figure 26.32a, p. 590)** and moths forage by night. They pollinate intensely sweet-smelling flowers with white or pale petals that are more visible than coloured petals in the dark. The long, thin mouthparts of moths and butterflies reach nectar in narrow floral tubes or floral spurs. The Madagascar hawkmoth uncoils a mouthpart the same length—an astonishing 22 cm—as the narrow flower of the orchid it pollinates, *Angraecum sesquipedale* (Figure 26.32b). Red and yellow flowers attract birds (Figure 26.32c), which have good daytime vision but a poor sense of smell. Hence, bird-pollinated plants do not squander metabolic resources to make fragrances. By contrast, flowers of species that are pollinated by beetles or flies may smell like rotten meat, dung, or decaying matter. This trickery by the plants is known as signal mimicry: the plant uses visual and olfactory signals to trick a pollinator into visiting it; some of these plants provide no nutritional reward for their pollinators at all. Daisies and other fragrant flowers with distinctive patterns, shapes, and red or orange components attract butterflies, which forage by day.

Bees see ultraviolet light and visit flowers with sweet odours and parts that appear to humans as yellow, blue, or purple (Figure 26.32d). Produced by pigments that absorb ultraviolet light, the colours form patterns called "nectar guides" that attract bees—which may pick up or drop off pollen during the visit. Here, as in our other examples, flowers contribute to the reproductive success of plants that bear them.

In this chapter, we have introduced some of the strategies that plants use to meet the challenges of life on Earth; they face the same challenges as animals and other terrestrial organisms (attract a mate,- reproduce, disperse offspring, and survive unfavourable conditions) but have had to find ways to do all of these without being able to move around (they are sessile).

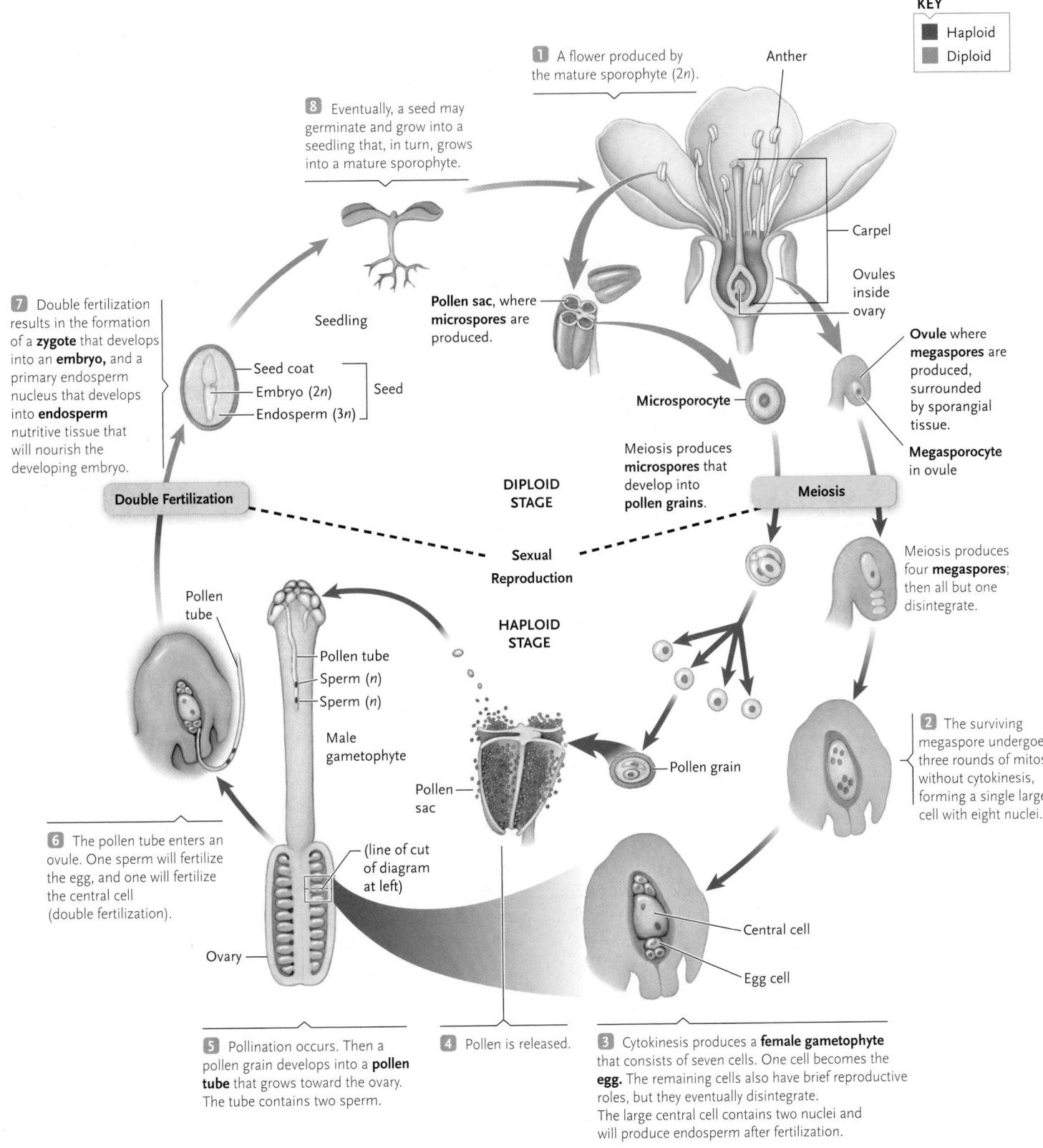

KEY
■ Haploid
■ Diploid

1 A flower produced by the mature sporophyte (2*n*).

Anther

Carpel

Ovules inside ovary

Ovule where **megaspores** are produced, surrounded by sporangial tissue.

Megasporocyte in ovule

8 Eventually, a seed may germinate and grow into a seedling that, in turn, grows into a mature sporophyte.

Seedling

Pollen sac, where **microspores** are produced.

Microsporocyte

Meiosis produces **microspores** that develop into **pollen grains**.

DIPLOID STAGE

Meiosis

Meiosis produces four **megaspores**; then all but one disintegrate.

7 Double fertilization results in the formation of a **zygote** that develops into an **embryo**, and a primary endosperm nucleus that develops into **endosperm** nutritive tissue that will nourish the developing embryo.

Seed coat
Embryo (2*n*)
Endosperm (3*n*)
Seed

Double Fertilization

Sexual Reproduction

HAPLOID STAGE

Pollen grain

2 The surviving megaspore undergoes three rounds of mitosis without cytokinesis, forming a single large cell with eight nuclei.

Pollen tube

Pollen tube
Sperm (*n*)
Sperm (*n*)

Male gametophyte

Pollen sac

Central cell

6 The pollen tube enters an ovule. One sperm will fertilize the egg, and one will fertilize the central cell (double fertilization).

(line of cut of diagram at left)

Egg cell

Ovary

5 Pollination occurs. Then a pollen grain develops into a **pollen tube** that grows toward the ovary. The tube contains two sperm.

4 Pollen is released.

3 Cytokinesis produces a **female gametophyte** that consists of seven cells. One cell becomes the **egg.** The remaining cells also have brief reproductive roles, but they eventually disintegrate. The large central cell contains two nuclei and will produce endosperm after fertilization.

Figure 26.31
Life cycle of a typical flowering plant. Double fertilization is a notable feature of the cycle. The male gametophyte delivers two sperm to an ovule. One sperm fertilizes the egg, forming the embryo, and the other fertilizes the endosperm-producing cell, which nourishes the embryo.

a. Bat pollinating a giant saguaro

b. Hawkmoth pollinating an orchid

c. Hummingbird visiting a hibiscus flower

d. Bee-attracting pattern of a marsh marigold

Visible light   UV light

Figure 26.32

Coevolution of flowering plants and animal pollinators. The colours and configurations of some flowers, and the production of nectar or odours, have coevolved with specific animal pollinators. **(a)** At night, nectar-feeding bats sip nectar from flowers of the giant saguaro cactus (*Carnegia gigantea*), transferring pollen from flower to flower in the process. **(b)** The hawkmoth (*Xanthopan morganii praedicta*) has a proboscis long enough to reach nectar at the base of the equally long floral spur of the orchid *Angraecum sesquipedale*. **(c)** A ruby-throated hummingbird (*Archilochus colubris*) sipping nectar from a hibiscus blossom (*Hibiscus*). The long, narrow bill of hummingbirds coevolved with long, narrow floral tubes. **(d)** Under ultraviolet light, the bee-attracting pattern of a gold-petalled marsh marigold becomes visible to human eyes.

Many of these topics are followed up in more detail in the chapters dealing with plant biology (Chapters 34 to 38).

The next two chapters introduce animals. As you read these chapters, look for similarities and differences in how they have addressed the challenges of life compared to plants.

STUDY BREAK

1. What are the advantages and costs to plants of using animals to disperse their pollen?
2. List at least three adaptations that have contributed to the evolutionary success of angiosperms as a group.

Review

26.1 Defining Characteristics of Land Plants

- Land plants are multicellular eukaryotes with cellulose cell walls. Most, but not all, are photoautotrophs. All have an alternation of generations life cycle (Figure 26.1), although which generation is dominant varies among groups of plants, and all retain embryos inside parental tissue.

26.2 The Transition to Life on Land

- Plants are thought to have evolved from charophyte green algae between 425 and 490 mya.

- Adaptations to terrestrial life in the earliest land plants include poikilohydry, multicellular chambers that protect developing gametes, and an embryo sheltered inside a parent plant.

- Other key evolutionary trends among land plants included symbiotic associations with fungi (Figure 26.5); the development of vascular tissues (including lignified water-conducting tissue), root systems, and shoot systems; lignified stems and leaves equipped with stomata (Figure 26.4b); increasing dominance by the diploid sporophyte generation; and a shift from homospory to heterospory.

- Gametophytes became reduced in size (Figure 26.7), male gametophytes (pollen) became specialized for dispersal without liquid water, and female gametophytes became increasingly protected inside sporophyte tissues.

26.3 Bryophytes: Nonvascular Land Plants

- Existing nonvascular land plants, or bryophytes, include the liverworts, hornworts, and mosses Figures 26.12–26.14).
- Bryophytes produce flagellated sperm that swim through free water to reach eggs. They lack xylem and phloem (although some have specialized conductive tissues), lignified tissues, roots, stems, and leaves. The gametophyte phase is dominant.

26.4 Seedless Vascular Plants

- Existing seedless vascular land plants include the lycophytes (club mosses), whisk ferns, horsetails, and ferns (Figures 26.16–26.21). Like bryophytes, they release spores and have swimming sperm. Unlike bryophytes, they have well-developed vascular tissues. The sporophyte generation is dominant and independent of the gametophyte.

- Most seedless vascular plants are homosporous, but some are heterosporous (Figure 26.8).

26.5 Gymnosperms: The First Seed Plants

- Gymnosperms (conifers and their relatives; Figures 26.22–26.28), together with angiosperms (flowering plants), are the seed-bearing vascular plants. Reproductive innovations include pollination, the ovule, and the seed. Liquid water is not required for reproduction.
- During the Mesozoic, gymnosperms were the dominant land plants. Today conifers are the primary vegetation of forests at higher latitudes and elevations and have important economic uses as sources of lumber and other products.

26.6 Angiosperms: Flowering Plants

- Angiosperms (Anthophyta) have dominated the land for more than 100 million years and are currently the most diverse plant group (Figures 26.29 and 26.30).
- The angiosperm vascular system moves water and sugars through the plant body more efficiently than that of gymnosperms. Reproductive adaptations include a protective ovary around the ovule, endosperm, flowers that attract pollinators, and fruits that protect and disperse seeds.

Questions

Self-Test Questions

1. Which of the following correctly describes an evolutionary trend that occurred as land plants evolved?
   a. becoming seedless
   b. producing only one type of spore
   c. producing nonmotile gametes
   d. haploid generation becoming dominant

2. Which of the following occurs in the life cycle of both mosses and angiosperms?
   a. The sporophyte is the dominant generation.
   b. The gametophyte is the dominant generation.
   c. Spores develop into sporophytes.
   d. The sporophyte produces spores.

3. The evolution of which of the following features freed land plants from requiring water for reproduction?
   a. lignified stems
   b. fruits and roots
   c. seeds and pollen
   d. flowers and leaves

4. Which of the following statements about archegonia is correct?
   a. They are found in all land plants.
   b. They are found in all land plants except seed plants.
   c. They are found in all land plants except angiosperms.
   d. They are found in nonvascular land plants but not in vascular plants.

5. Which of the following statements about antheridia is correct?
   a. They are found in all land plants.
   b. They are found in all land plants except seed plants.
   c. They are found in all land plants except angiosperms.
   d. They are found in nonvascular land plants but not in vascular plants.

6. Which of the options correctly pairs a plant group with its phylum?
   a. Hepatophyta: cycads
   b. Pterophyta: horsetails
   c. Bryophyta: gnetophytes
   d. Coniferophyta: angiosperms

7. A homeowner notices moss growing between bricks on his patio. Closer examination reveals tiny brown stalks with cuplike tops emerging from green leaflets. Which of the following are these brown structures?
   a. antheridia
   b. archegonia
   c. the gametophyte generation
   d. the sporophyte generation

8. Horsetails are most closely related to which of the following plant groups?
   a. club mosses and ferns
   b. mosses and whisk ferns
   c. liverworts and hornworts
   d. gnetophytes and gymnosperms

9. In which of the following groups is the evolution of true roots first seen?
   a. mosses
   b. conifers
   c. liverworts
   d. seedless vascular plants

10. Arrange the following adaptations to terrestrial life in the order in which they first appeared during the evolution of land plants:

1. seeds	a. 1, 2, 3, 4
2. vascular tissue	b. 2, 3, 4, 1
3. gametangia	c. 2, 3, 1, 4
4. flowers	d. 3, 2, 1, 4

Questions for Discussion

1. Working in the field, you discover a fossil of a previously undescribed plant species. The specimen is small and may not be complete; the parts you have do not include any floral organs. What evidence would you need to classify the fossil as a seedless vascular plant with reasonable accuracy? What evidence would you need to distinguish between a fossil lycopod and a fern?

2. Compare the size, anatomical complexity, and degree of independence of a moss gametophyte, a fern gametophyte, a Douglas fir (conifer) female gametophyte, and a dogwood (angiosperm) female gametophyte. Which one is the most protected from the external environment? Which trends in plant evolution does your work on this question bring to mind?

3. How has the relative lack of fossil early angiosperms affected our understanding of this group?

4. One of the major challenges plants faced in living on land was coping with the reduced availability of water. Contrast the water-use strategy that the earliest land plants (and mosses living today) use with that used by vascular plants. Suggest some advantages and disadvantages of each strategy.

5. Some vascular plants have motile sperm and some do not. How do male and female gametes meet in (a) plants that have swimming sperm and (b) those that do not?

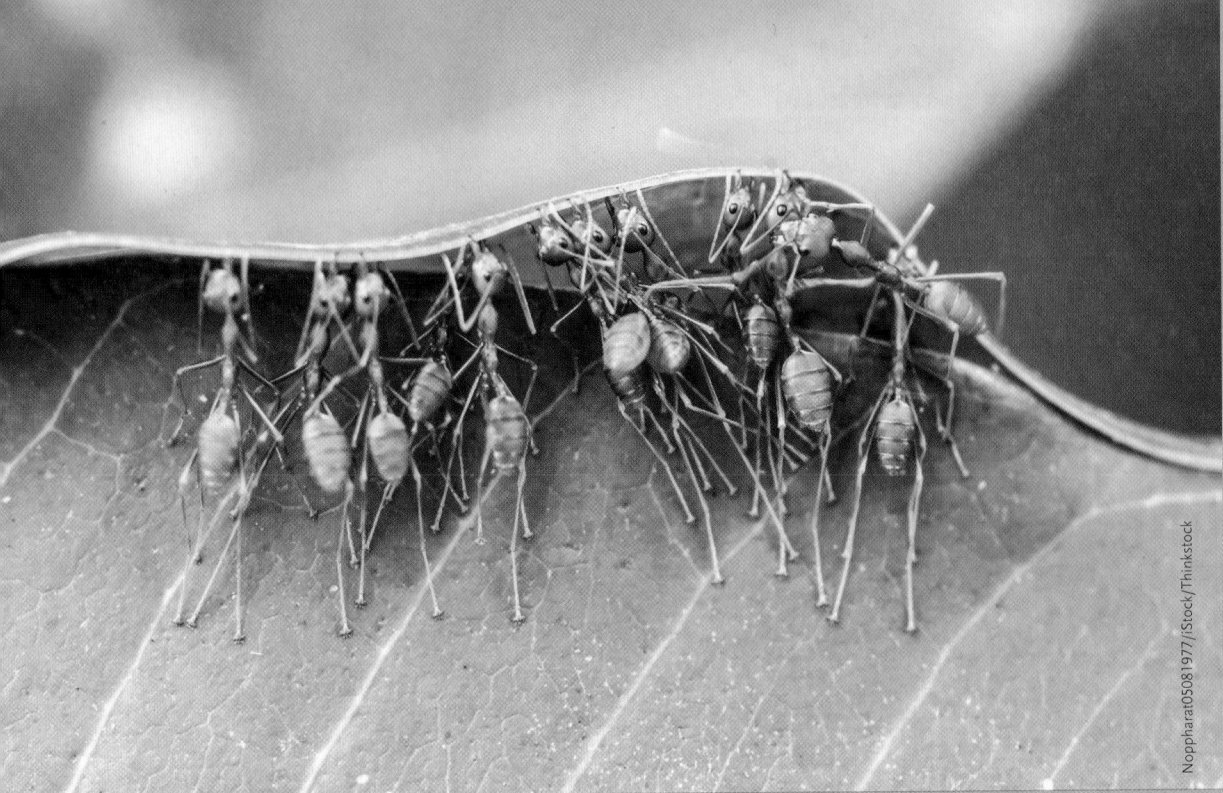

Workers of the weaver ant, *Oecophylla*, engaged in the construction of their nest, which is formed from living leaves, curled or folded to form an envelope held together by silk secreted by the larvae.

27

Diversity of Animals 1: Sponges, Radiata, Platyhelminthes, and Protostomes

WHY IT MATTERS

Beginning about 540 million years ago (mya) in the Early Cambrian, conditions were ripe for rapid development of the marine fauna, and an explosion of new forms of animals appeared, particularly in warm and shallow seas bordering continents. About 505 mya, a series of mud slides carried the animals that lived at the edge of a submarine cliff, the Cathedral Escarpment, over its edge and buried them in fine silt. That mud and its contained fossils formed shale, a sedimentary rock with layers that are easily split apart. During continent and mountain building, the shale beds, now known as the Burgess Shale, came to lie in the Canadian Rocky Mountains of eastern British Columbia in what is now Yoho National Park.

In 1909, Charles Walcott, an American paleontologist working in the Burgess Shale area, located a rich bed of very strange fossils that were exquisitely preserved, including the soft parts. Reconstructions have revealed not only familiar animals such as trilobites and sponges but also many truly bizarre animals **(Figure 27.1, p. 594)**. For example, *Opabinia* was about as long as a tube of lipstick and had five eyes on its head and a single anterior grasping organ, which was probably used to catch prey. The smaller *Hallucigenia* had seven pairs of hard spines on its back and what appear to have been seven pairs of softer ventral protuberances that probably functioned for locomotion. Some of the other organisms look like early chordates, but many do not resemble any living animals and may represent phyla never previously described. Furthermore, some of the animals from the Burgess Shale have moved between phyla as more and more details about them became available.

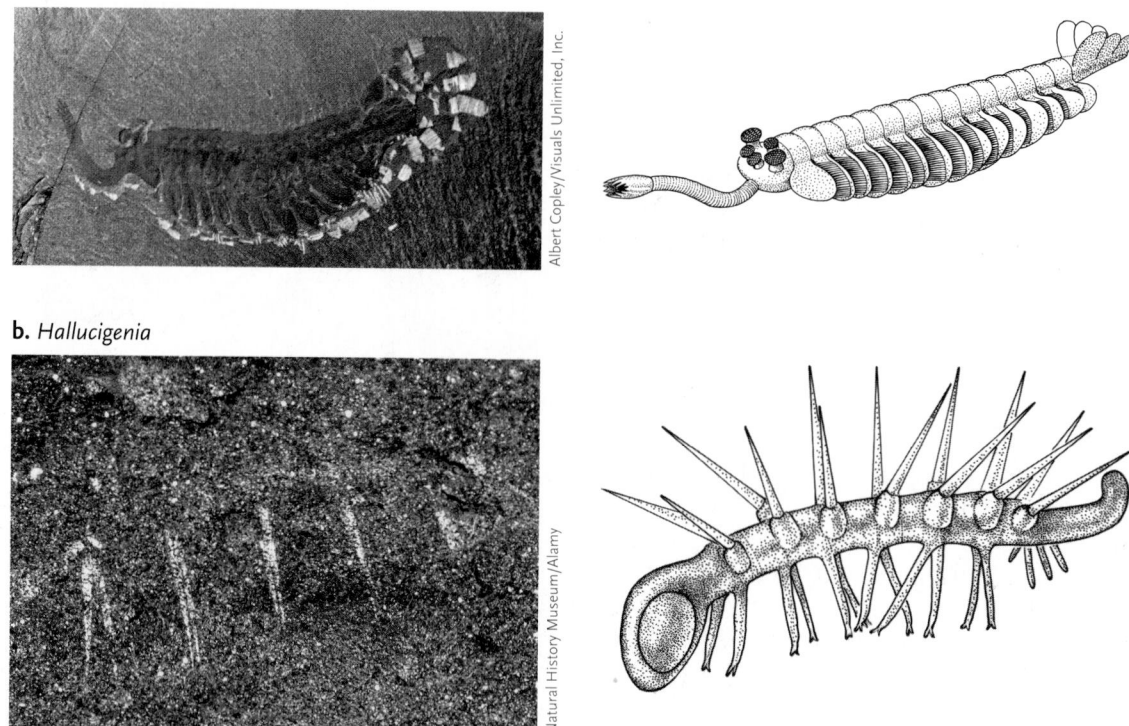

a. *Opabinia*

b. *Hallucigenia*

Albert Copley/Visuals Unlimited, Inc.

© The Natural History Museum/Alamy

Figure 27.1

Animals of the Burgess Shale. **(a)** *Opabinia* had five eyes and a grasping organ on its head. **(b)** *Hallucigenia* had seven pairs of spines and soft protuberances.

The Burgess Shale and other similar sites elsewhere provide us with a snapshot of some of the animals that inhabited the coastal waters at the time of the Cambrian Explosion. Most of the bizarre forms did not survive the extinctions that were to come. Without these extinctions, some of the forms might have survived to found lineages completely different from those living today.

In this chapter, we introduce the general characteristics of animals and a phylogenetic hypothesis about their evolutionary history and classification. We also survey the major invertebrate phyla belonging to one lineage, the Protostomia. In Chapter 19, we defined the various levels used in the Linnaean system of classifying animals. In Chapter 28, we examine the other major animal lineage, the Deuterostomia, which includes the phylum Chordata and their nearest invertebrate relatives.

In this chapter we begin our consideration of animal diversity with protostomes, metazoans in which the blastopore develops into the mouth and the anus appears as a second opening. Protostomes include molluscs, annelids, and arthropods, as well as other smaller phyla not usually familiar to biology students. While some protostomes have simple body plans with little evidence of organ systems, others have fully developed organ systems. Some protostomes have shells and an extensive fossil record, while others are soft-bodied and not well known as fossils. Protostomes include species with lifestyles across the trophic spectrum from herbivores to carnivores and **detritivores.** Some protostomes are infamous as parasites, usually because they pose important health risks to humans. The other animals we include in this chapter are sponges (poriferans), radiatia, the jellyfish (cnidarians), and flatworms (platyhelminths).

27.1 What Is an Animal?

Most biologists recognize the **kingdom Animalia** as a monophyletic group that is easily distinguished from the other kingdoms.

27.1a All Animals Share Certain Structural and Behavioural Characteristics

Animals are eukaryotic, multicellular organisms. The cell membranes of adjacent animal cells are in direct contact with one another. This is different from plants and fungi, which have cell walls around the cells. Animal cells may be organized into different morphological types, reflecting their role in the functioning of the animal as a single unit.

All animals are **heterotrophs:** they depend on other life forms for their food, either by eating them directly or by living in a parasitic association with them. They use oxygen to metabolize their food through aerobic respiration, and most store excess energy as glycogen, oil, or fat.

All animals are **motile** (able to move from place to place) at some time in their lives. Most familiar animals are motile as adults. However, in some species, such as mussels and barnacles, only the young are motile; they eventually settle down as **sessile** (unable to move from one place to another) adults. All animals are able to perceive and respond to information about the environment in which they live.

Animals reproduce either asexually or sexually; in many groups, they switch from one mode to the other. Sexually reproducing species produce haploid **gametes** (eggs and sperm) that fuse to form diploid **zygotes** (fertilized eggs). For many invertebrates, development to the adult involves one or more **larval forms**. This pattern of development, in which a species can exist during development in two or more distinct forms, is referred to as **polymorphic development**. This developmental strategy is important for the success of many of the invertebrate groups, particularly those with a parasitic lifestyle.

27.1b The Animal Lineage Probably Arose from a Colonial Choanoflagellate Ancestor

Most biologists agree that the common ancestor of all animals was probably a colonial, flagellated protist that lived at least 700 Ma ago, during the Precambrian. It may have resembled the minute, sessile choanoflagellates (see Chapter 24) that live in both freshwater and marine habitats today. In 1874, the German embryologist Ernst Haeckel proposed a colonial, flagellated ancestor, suggesting that it was a hollow, ball-shaped organism with unspecialized cells. Its cells became specialized for particular functions, and a developmental reorganization produced a double-layered, sac-within-a-sac body plan **(Figure 27.2)**. The embryology of many living animals roughly parallels this hypothetical evolutionary transformation. He included this hypothetical organism among what he called the Metazoa (*meta* = more developed; *zoon* = animal) to distinguish them from the Protozoa.

STUDY BREAK

1. What characteristics distinguish animals from plants?
2. What early steps may have led to the first metazoans?

27.2 Key Innovations in Animal Evolution

Once established, the animal lineage diversified quickly into an amazing array of body plans. Biologists have used several key morphological innovations to unravel the evolutionary relationships of the major animal groups.

27.2a Tissues and Tissue Layers Appeared Early in Animal Evolution

In most Metazoans, the process of development gives rise to two or three layers of cells that eventually form **tissues**, groups of similar differentiated cells specialized for particular functions.

In most metazoans, embryonic tissues form as either two or three concentric **germ layers** (see Chapter 42). The innermost layer, the **endoderm**, eventually develops into the lining of the gut (digestive system) and, in some animals, respiratory organs. The outermost layer, the **ectoderm**, forms the external covering and nervous system. Between the two, the **mesoderm** forms the muscles of the body wall and most other structures between the gut and the external covering. Some animals have a **diploblastic** body plan based on two embryonic layers, endoderm and ectoderm, but most are **triploblastic**, having all three germ layers.

27.2b Most Animals Exhibit Either Radial or Bilateral Symmetry

The most obvious feature of an animal's body plan is its shape **(Figure 27.3, p. 596)**. Most animals are **symmetrical**; in other words, their bodies can be divided by

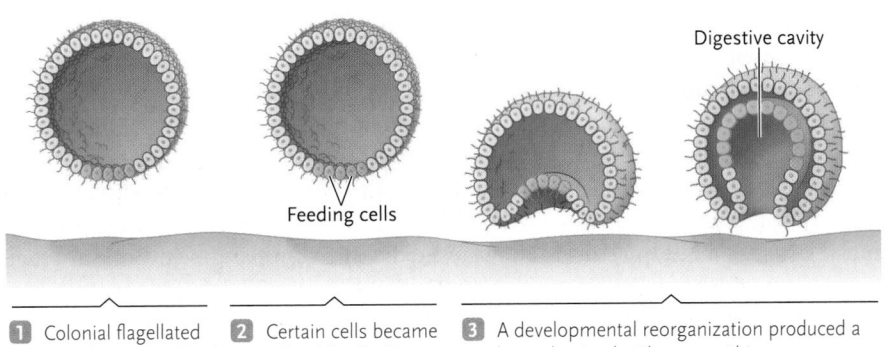

Digestive cavity

Feeding cells

1 Colonial flagellated protist with unspecialized cells. **2** Certain cells became specialized for feeding and other functions. **3** A developmental reorganization produced a two-layered animal with a sac-within-a-sac body plan.

Figure 27.2

Animal origins. Many biologists believe that animals arose from a colonial, flagellated protist in which cells became specialized for specific functions and a developmental reorganization produced two cell layers. The cell movements illustrated here are similar to those that occur during the development of many animals, as described in Chapter 42.

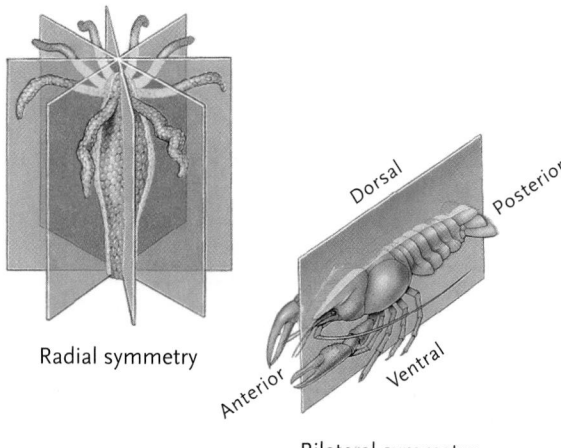

Radial symmetry

Dorsal
Posterior
Anterior
Ventral

Bilateral symmetry

Figure 27.3

Patterns of body symmetry. Most animals have either radial or bilateral symmetry.

a plane into mirror-image halves. By contrast, most sponges have irregular shapes and are therefore **asymmetrical**.

All other phyla exhibit one of two body symmetry patterns (see Figure 27.3). The Radiata includes two phyla, Cnidaria (hydras, jellyfishes, and sea anemones) and Ctenophora (comb jellies), that are radially symmetrical. Their body parts are arranged regularly around a central axis, like spokes on a wheel. Thus, any cut down the long axis of a radially symmetrical animal divides it into matching halves.

All other metazoan phyla fall within the Bilateria, animals that have **bilateral symmetry**. In other words, on either side of the body's midline they have left and right sides that are mirror images of each other. Bilaterally symmetrical animals also have front (**anterior**) and back (**posterior**) ends, as well as upper (**dorsal**) and lower (**ventral**) surfaces. As they move through the environment, the anterior end encounters food, shelter, or enemies first. In bilaterally symmetrical animals, natural selection favoured **cephalization**, the development of an anterior head where sensory organs and nerve tissue are concentrated.

27.2c Many Animals Have Body Cavities That Surround Their Internal Organs

The body plans of many bilaterally symmetrical animals include a body cavity that separates the gut from the muscles of the body wall. **Acoelomate** animals (a = without; $koiloma$ = cavity), such as flatworms (phylum Platyhelminthes), do not have such a cavity, rather a mass of cells derived largely from mesoderm that packs the region between gut and body wall **(Figure 27.4a)**.

Pseudocoelomate animals ($pseudo$ = false), including the roundworms (phylum Nematoda) and wheel animals (phylum Rotifera), have a **pseudocoelom**, a fluid-filled space between gut and muscles of the body wall that has no mesodermal lining around the endoderm

(Figure 27.4b). The muscles of the body wall, derived from mesoderm, form the outer lining of the pseudocoelom, and its inner lining is the gut, which lacks muscles. Internal organs lie within the pseudocoelom and are bathed by its fluid.

Coelomate animals have a **coelom**, a fluid-filled body cavity completely lined by mesoderm. In vertebrates, this lining takes the form of the **peritoneum**, a thin tissue derived from mesoderm (Figure 27.4c). The inner and outer layers of the peritoneum connect, forming **mesenteries**, membranes that surround the internal organs and suspend them within the coelom. In some arthropods and molluscs, the coelom has been displaced by the development of a **hemocoel**, resulting from an open circulatory system. This can be envisaged as consisting of a single large blood vessel that has expanded to fill the coelom. In these animals, the coelom persists around the gonads and, in some cases, the heart.

Biologists describe the body plan of pseudocoelomate and coelomate animals as a "tube-within-a-tube." The digestive system forms the inner tube, and the body wall forms the outer tube. The body cavity may serve a number of functions, such as the transport of nutrients and the products of metabolism, provision of an environment in which eggs and sperm can develop, a **hydrostatic skeleton** that provides a basis for locomotion (see Chapter 46), and an appropriate environment for the functioning of internal organs.

27.2d Developmental Patterns Mark a Major Divergence in Animal Ancestry

Embryological evidence suggests that bilaterally symmetrical animals are divided into two lineages, the **protostomes** and the **deuterostomes**, that differ in several developmental characteristics **(Figure 27.5, p. 598)**.

Shortly after fertilization, an egg undergoes a series of cell divisions called **cleavage**. The first two cell divisions divide a zygote as you might slice an apple, cutting it into four wedges from top to bottom. In some animals, subsequent cell divisions occur at oblique angles to the vertical axis of the embryo, ultimately producing a mass in which each cell at the top of the embryo lies in the groove between the pair of cells below it (see the left side of Figure 27.5a). This pattern is called **spiral cleavage**. It is characteristic of most protostomes, although cleavage patterns in arthropods and some other groups are highly specialized. In deuterostomes, by contrast, the third cell division is perpendicular to the vertical axis of the embryo, cutting each of the four cells near its midsection. The fourth cell division is vertical, producing a mass of cells that are stacked directly above and below one another (see the right side of Figure 27.5a). This pattern is called **radial cleavage**.

Protostomes and deuterostomes often differ in the timing of important developmental events. During

a. Acoelomate animals

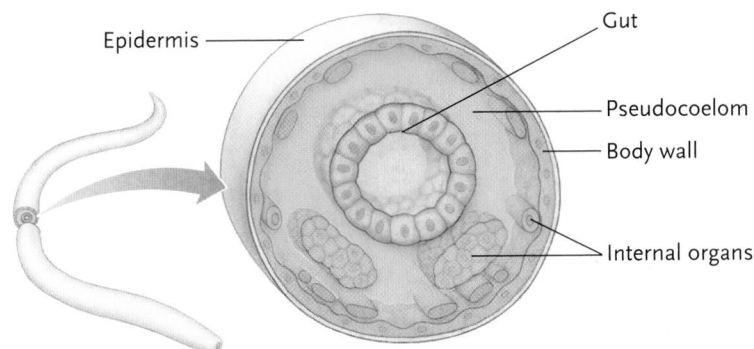

Epidermis

Gut

Internal organs

Body wall

Figure 27.4

Body plans of triploblastic animals.
(a) In acoelomate animals, no body cavity separates the gut and body wall.
(b) In pseudocoelomate animals, the pseudocoelom forms between the gut (a derivative of endoderm) and the body wall (a derivative of mesoderm).
(c) In coelomate animals, the coelom is completely lined by peritoneum (a derivative of mesoderm).

b. Pseudocoelomate animals

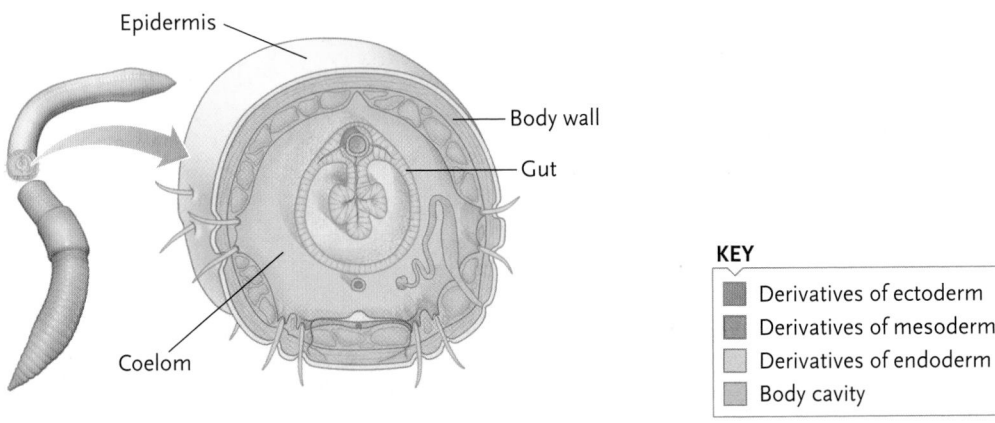

Epidermis

Gut

Pseudocoelom

Body wall

Internal organs

c. Coelomate animals

Epidermis

Body wall

Gut

Coelom

KEY

▮	Derivatives of ectoderm
▮	Derivatives of mesoderm
▮	Derivatives of endoderm
▮	Body cavity

cleavage, certain genes are activated at specific times, determining a cell's developmental path and ultimate fate. Many protostomes undergo **determinate cleavage:** each cell's developmental path is determined as the cell is produced. Thus, one cell isolated from a two- or four-cell protostome embryo cannot develop into a functional embryo or larva. By contrast, many deuterostomes have **indeterminate cleavage:** the developmental fates of cells are determined later. A cell isolated from a four-cell deuterostome embryo will develop into a functional embryo. In humans, the two cells produced by the first cleavage division sometimes separate and develop into identical twins.

As development proceeds, the **blastopore,** an opening on the surface of the embryo, eventually connects the **archenteron** (developing gut) to the outside environment (see Figure 27.5b). Later in development, a second opening at the opposite end of the embryo transforms the pouchlike gut into a digestive tube (see Figure 27.5c). The traditional view of the difference between protostomes (*proto* = first; *stoma* = mouth) and deuterostomes (*deuteros* = second) is that in protostomes the blastopore develops into the mouth and the second opening forms the anus. In deuterostomes, the blastopore develops into the anus and the second opening becomes the mouth. But an alternative view has emerged from recent genetic studies. Specifically, the genes *brachyury* and *goosecoid* are expressed as the mouth develops in the acoel *Convolutriloba longifissura,* and the mouths of other bilaterians develop the same way. This

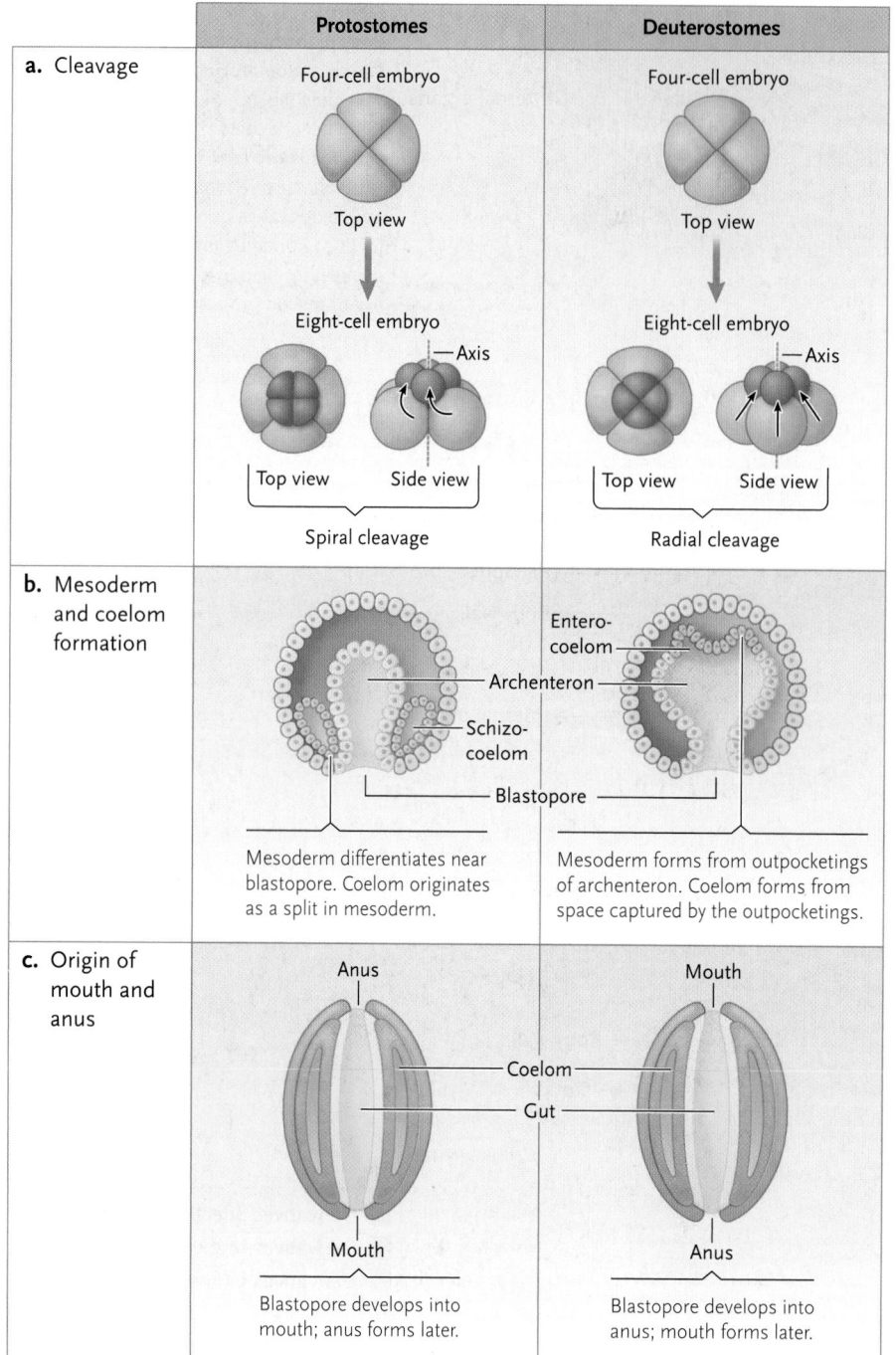

	Protostomes	Deuterostomes
a. Cleavage	Four-cell embryo — Top view → Eight-cell embryo — Axis — Top view / Side view — **Spiral cleavage**	Four-cell embryo — Top view → Eight-cell embryo — Axis — Top view / Side view — **Radial cleavage**
b. Mesoderm and coelom formation	Archenteron — Schizo-coelom — Blastopore — Mesoderm differentiates near blastopore. Coelom originates as a split in mesoderm.	Entero-coelom — Archenteron — Blastopore — Mesoderm forms from outpocketings of archenteron. Coelom forms from space captured by the outpocketings.
c. Origin of mouth and anus	Anus — Coelom — Gut — Mouth — Blastopore develops into mouth; anus forms later.	Mouth — Coelom — Gut — Anus — Blastopore develops into anus; mouth forms later.

KEY
- Derivatives of ectoderm
- Derivatives of mesoderm
- Derivatives of endoderm
- Body cavity

Figure 27.5

Protostomes and deuterostomes. The two lineages of coelomate animals differ in **(a)** cleavage patterns, **(b)** the origin of mesoderm and the coelom, and **(c)** the polarity of the digestive system.

makes the mouths in acoels and other bilaterians homologous. Furthermore, other genes, such as *caudal, orthopedia*, and *brachyury*, are expressed in a small area of the hind gut. This finding raises the possibility that the development of the anus (and a through gut) may have evolved independently in different lineages of bilaterians.

Protostomes and deuterostomes differ in the origin of mesoderm and coelom (see Figure 27.5b). In most protostomes, mesoderm originates from a few specific cells near the blastopore. As the mesoderm

grows and develops, it splits into inner and outer layers. The space between the layers forms a **schizocoelom** (*schizo* = split). In deuterostomes, mesoderm forms from outpocketings of the archenteron. The space pinched off by the outpocketings forms an **enterocoelom** (*entero* = intestine).

Several other characteristics differ in protostomes and deuterostomes. For example, the nervous system of protostomes is positioned on the ventral side of the body, and their brain surrounds the opening of the digestive tract. By contrast, the nervous system and brain of deuterostomes lie on the dorsal side of the body.

27.2e Segmentation Divides the Bodies of Some Animals into Repeating Units

Some phyla in both protostome and deuterostome lineages exhibit varying degrees of **segmentation**, the production of body parts as repeating units. During development, segmentation first arises in the mesoderm, the middle tissue layer that produces most of the body's bulk. In vertebrates, segmentation is obvious in the embryo, and in the adult there is evidence of segmentation in the vertebral column (backbone), ribs, and associated muscles, as well as the nervous system. Among invertebrates, segmentation is pronounced in annelids (earthworms and their relatives), where each segment, visible externally as a ring, has its own set of muscles, ganglion (collection of nerve cells), and excretory structures. Arthropods (insects and their relatives) are also segmented, although some segments may be specialized, bearing, for example, wings or reproductive structures.

The advantages of segmentation lie principally in movement, but to different degrees. In vertebrates, with their articulated backbone and with each segment having its own muscles, segmentation permits the S-shaped side-to-side motion—think of fish or snakes. Annelids are capable of similar motion, but many of them live in burrows or tubes. The ability to expand segments by contracting muscles of adjacent segments assists this lifestyle. The articulated stiffened cuticle of arthropods serves as a point of attachment for muscles, providing significant leverage and strength (see Chapter 46). Arthropods have taken advantage of the existence of segmental appendages to assign special functions, such as locomotion, reproduction, or gas exchange, to particular appendages.

CONCEPT FIX Many people believe that invertebrate animals are protostomes. The truth is that the deuterostomes include many species (some whole phyla) that lack backbones (they are invertebrates). No members of the phylum Echinodermata have a backbone, and not all of the Chordata, the phylum including the vertebrates, have backbones. ◉

STUDY BREAK

1. What is a tissue, and what three primary tissue layers are present in the embryos of most animals? Explain the function of each layer.
2. What kind of symmetry does an earthworm have?
3. What is the function of the coelom, and what is the importance of the fluid?

27.3 An Overview of Animal Phylogeny and Classification

For many years, biologists used the morphological innovations and embryological patterns described above, together with evidence from the fossil record, to trace the phylogenetic history of animals (see Chapter 20). That evidence led to the construction of phylogenetic trees, which were broadly accepted as reasonable hypotheses about the relatedness of various phyla. Thus, phyla with similar developmental and morphological patterns were regarded as sharing common ancestries. For example, annelids and arthropods, both schizocoelous, segmented coelomates, were seen as sharing a common ancestor, a view supported by the fossil record and by the existence of the Phylum Onychophora, which has some of the characteristics of both phyla. Increasingly, however, biologists are using molecular sequence data to reanalyze animal relationships.

27.3a Molecular Analyses Have Refined Our Understanding of Animal Phylogeny

Molecular analyses of animal relationships are often based on nucleotide sequences in small subunit ribosomal RNA and mitochondrial DNA and, more recently, the sequences of specific genes. These analyses are used to construct molecular cladograms (see Chapter 20). **Figure 27.6, p. 600,** is a phylogenetic tree developed from a number of cladograms based on molecular sequences. It represents, as do all such trees, a working hypothesis that explains the information that is now available.

The phylogenetic tree based on molecular characters includes the major lineages that biologists had defined using the morphological innovations and embryological characters described above. For example, molecular data confirm the distinctions between the Radiata and the Bilateria. They also confirm the separation of deuterostome phyla from all others within the Bilateria.

The sponges are to some degree a special case. The way that tissues form during embryology in the most primitive sponges differs from the other Metazoa and therefore sponges had, in the past, been regarded as lacking "true" tissues. Sponges were therefore placed in the Parazoa, a separate subkingdom, distinct from the Eumetazoa, the remaining metazoans. But the most recent molecular evidence includes sponges with the other Metazoa in a single monophyletic lineage. Sponges are distinct from other metazoans because they do not form distinct nervous tissue and are asymmetrical. Nonetheless, molecular data confirm that sponges and other Metazoa have a common ancestor.

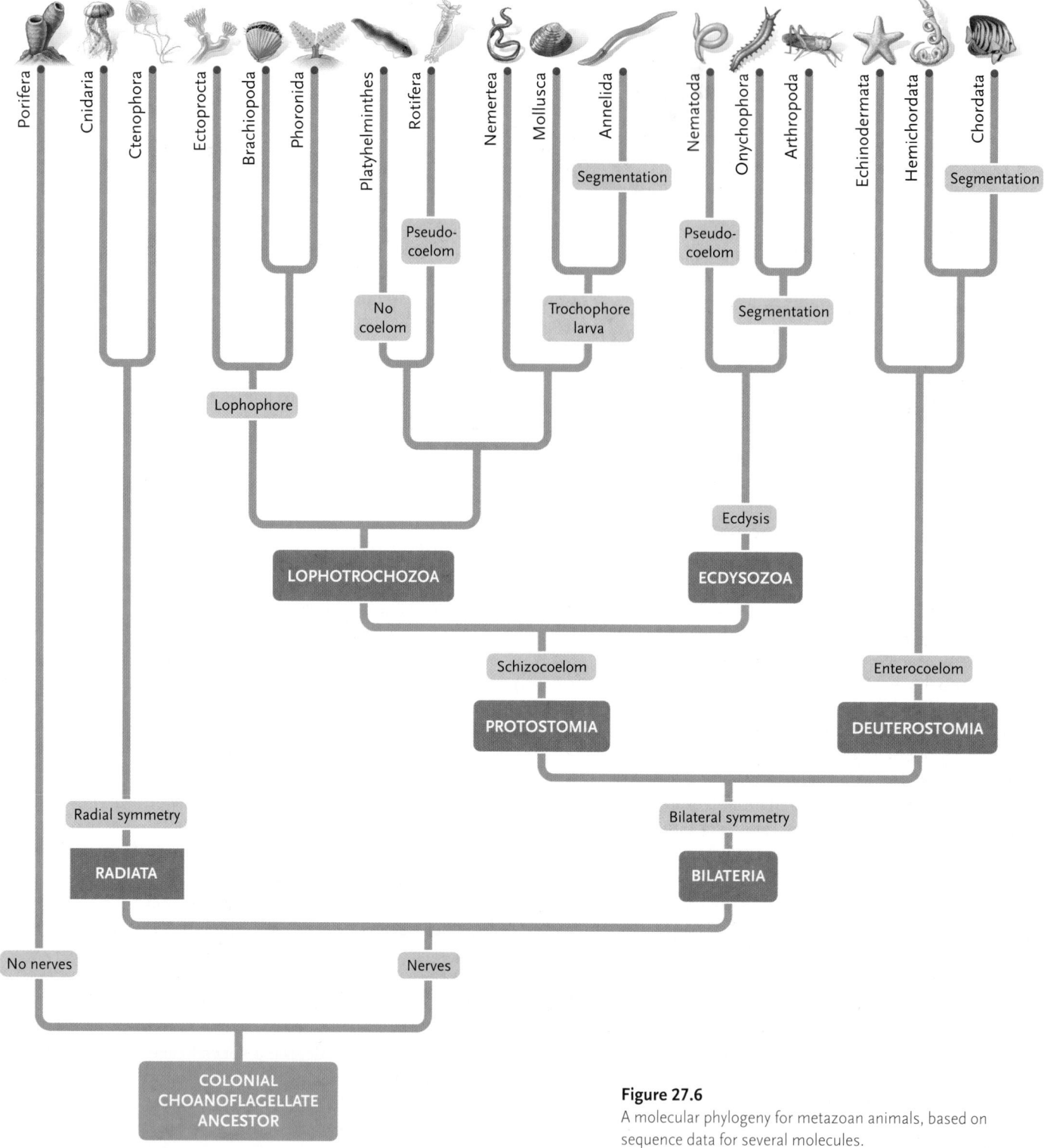

Porifera
Cnidaria
Ctenophora
Ectoprocta
Brachiopoda
Phoronida
Platyhelminthes
Rotifera
Nemertea
Mollusca
Annelida
Nematoda
Onychophora
Arthropoda
Echinodermata
Hemichordata
Chordata

Segmentation

Segmentation

Pseudo-coelom

Pseudo-coelom

No coelom

Trochophore larva

Segmentation

Lophophore

Ecdysis

LOPHOTROCHOZOA

ECDYSOZOA

Schizocoelom

Enterocoelom

PROTOSTOMIA

DEUTEROSTOMIA

Bilateral symmetry

Radial symmetry

RADIATA

BILATERIA

No nerves

Nerves

COLONIAL CHOANOFLAGELLATE ANCESTOR

Figure 27.6

A molecular phylogeny for metazoan animals, based on sequence data for several molecules.

Molecular phylogeny confirms the Protostomia and Deuterostomia as separate lineages within the Metazoa. Protostomia is, in turn, subdivided into two major lineages, Lophotrochozoa and Ecdysozoa, groupings not previously recognized. The name Lophotrochozoa (*lophos* = crest; *troch* = wheel; *zoa* = animals, plural of *zoon*) refers to both the "lophophore," a feeding structure found in three phyla (illustrated in Figure 27.15, p. 607), and the "trochophore," a type of larva found in annelids and molluscs (illustrated in Figure 27.23, p. 614). The name Ecdysozoa (*ekdero* = strip off the skin) refers to the cuticle that these species secrete and periodically replace; the shedding of the cuticle is called **ecdysis**.

27.3b Molecular Phylogeny Reveals Surprising Patterns in the Evolution of Key Morphological Innovations

Molecular phylogeny has forced biologists to reevaluate the evolution of several important morphological innovations. Traditional phylogenies based on morphology and embryology implied that the absence of a body cavity, the acoelomate condition, was ancestral and that the presence of a body cavity, the pseudocoelomate or coelomate condition, was derived. But the molecular tree provides a very different view. It suggests that the schizocoelomate condition is ancestral, having evolved in the common ancestor of the lineage. If that hypothesis is correct, then the acoelomate condition of flatworms may represent the evolutionary *loss* of the schizocoelom, *not* an ancestral condition. Similarly, the molecular tree hypothesizes that the pseudocoelom evolved independently in rotifers (Lophotrochozoa, phylum Rotifera) and in roundworms (Ecdysozoa, phylum Nematoda) as modifications of the ancestral schizocoelom.

Traditional phylogenies also suggested that the segmented body plan of several protostome phyla was inherited from a segmented common ancestor and that segmentation arose independently in the chordates by convergent evolution. The molecular tree, by contrast, suggests that segmentation evolved independently in *three* lineages: segmented worms (Lophotrochozoa, phylum Annelida), arthropods and velvet worms (Ecdysozoa, phyla Arthropoda and Onychophora), and chordates (Deuterostomia, phylum Chordata).

The hypothesis based on molecular studies and represented in Figure 27.6, is the framework we use for our consideration of the major invertebrate phyla. It is important to recognize, however, that phylogenetic trees are always provisional. In the future, new data may lead to revisions of the phylogeny.

STUDY BREAK

1. How is molecular analysis used in creating phylogenetic trees?
2. Describe the way molecular phylogeny has changed how biologists view the absence of the coelom.

27.4 Phylum Porifera

Sponges **(Figure 27.7)** are mostly marine, with a small number of species living in fresh water. They have no particular symmetry, are completely sessile as adults, and obtain their food by filtering it from the water. Sponges have been abundant since the Cambrian, and about 8000 living species are known. They range in size from 1 cm to 2 m.

Figure 27.7
Asymmetry in sponges. The shapes of sponges vary with their habitats. Those that occupy calm waters, such as this stinker vase sponge (*Ircinia campana*), may be lobed, tubular, cuplike, or vaselike.

© Jeff Rotman/Alamy

Their body plan **(Figure 27.8, p. 602)** is simple: sponges can be regarded as sacs, with a cavity, the **spongocoel**, opening to the environment via an **osculum** (osteopore). There are two layers of organized cells. The **pinacoderm** (epithelium) consists of the cells on the outside of the sponge, pinacocytes. The inner layer of cells, lining the cavity, are **choanocytes**, each with a **flagellum** surrounded by microvilli. The two layers are separated by a gelatinous matrix, the **mesohyl**, with **amoeboid** cells called **archaeocytes** that move throughout the mesohyl by typical amoeboid movement. The wall of the bag is perforated by a number of pores lined by porocytes, specialized derivatives of the pinacocytes.

Almost all sponges are **suspension (or filter) feeders.** The action of choanocytes sets up a unidirectional current by which water enters the spongocoel through the porocytes and leaves via the osculum. Flow rates can be adjusted by the porocytes, which are capable of contraction, suggesting communication among the cells in spite of the absence of nerves. Particles of food are captured by choanocytes and passed to the mesohyl, where they are ingested and digested within archaeocytes, which may also store reserves.

Some archaeocytes may become specialized to form spicules, extracellular rigid supporting structures that give shape to the sponge. Spicules are microscopic rigid structures of various shapes (depending on the species) composed of a calcareous or siliceous (silicon) material. Collagen is also found in the mesohyl, as is spongin, a collagen-like protein. Collagens and spongin are produced by archaeocytes, which are **totipotent** (like stem cells), with the capacity to differentiate into any of the cell types found in sponges, including eggs and sperm.

Most sponges are **monoecious**: individuals produce both sperm and eggs. Sperm are released into the spongocoel and then out into the environment; eggs **(oocytes)** remain in the mesohyl, where sperm from other sponges, drawn in with water, are captured by choanocytes and carried to oocytes. Early development occurs within the sponge and produces a ciliated larva, the **dispersal** stage. Sponges have various types of

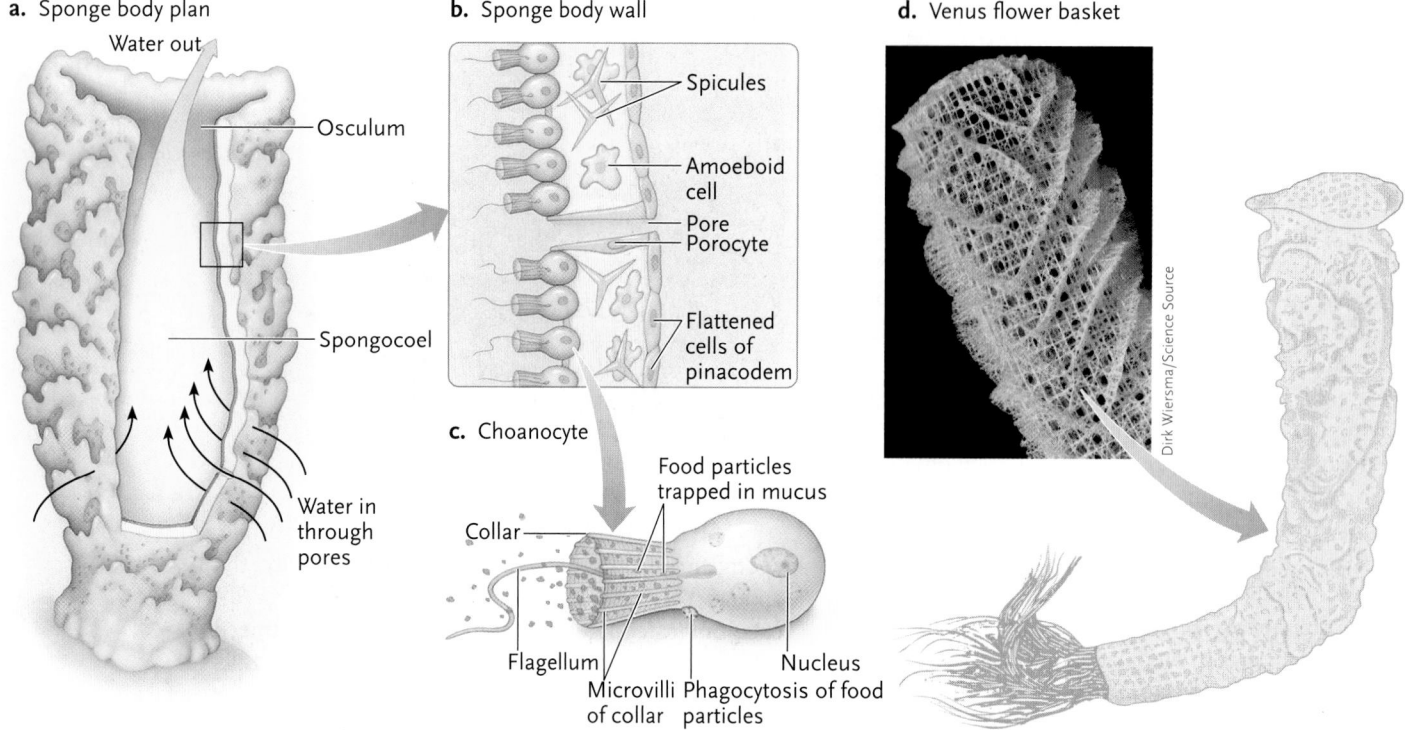

a. Sponge body plan

Water out

Osculum

Spongocoel

Water in through pores

b. Sponge body wall

Spicules

Amoeboid cell

Pore

Porocyte

Flattened cells of pinacodem

c. Choanocyte

Collar

Food particles trapped in mucus

Flagellum

Microvilli of collar

Phagocytosis of food particles

Nucleus

d. Venus flower basket

Dirk Wiersma/Science Source

Figure 27.8

The body plan of sponges. Most sponges have **(a)** simple body plans and **(b)** relatively few cell types. **(c)** Beating flagella on the choanocytes create a flow of water through incurrent pores, into the spongocoel, and out through the osculum. **(d)** Venus flower basket (Euplectella species), a marine sponge, has spicules of silica fused into a rigid framework.

larvae. While some are free swimming, others use their cilia to crawl over the substrate. There is evidence that some larvae avoid light to select a location to settle, where they undergo **metamorphosis** (a reorganization of form) into sessile adults. Some sponges also reproduce asexually; small fragments break off an adult and grow into new sponges. Many species, particularly those in fresh water, also produce **gemmules,** clusters of cells with a resistant covering that allows them to survive unfavourable conditions. Gemmules germinate into new sponges when conditions improve.

Even with a very simple basic body plan, sponges have achieved remarkable diversity. Sponges formed very large reefs during the Mesozoic, and a modern reef of sponges, originating at the end of the last ice age, has been found off the west coast of Canada. It is being studied by Verena Tunnicliffe's lab at the University of Victoria.

Many sponges serve as refuges for other species. Bacteria and cyanobacteria can be found in the mesohyl and, in some species, within archaeocytes. A curious relationship with another species occurs in the Venus flower basket, *Euplectella aspergillum* (see Figure 27.8d). Male and female shrimp (*Spongicola* species) may enter the spongocoel when small, feed on material brought in by the sponge, and grow large enough that they are unable to leave. The pair of shrimp spend their entire lives in the prison formed by the elaborate basket of spicules.

One species, *Asbestopluma hypogea,* catches small arthropods that become entangled in hook-shaped spicules on the surface. The prey are then encased in filamentous structures and digested. Choanocytes are absent in this sponge.

STUDY BREAK

1. Do sponges exhibit symmetry? If so, what type?
2. How does a sponge gather food from its environment?

27.5 Metazoans with Radial Symmetry

Unlike sponges, the remaining metazoans have some form of symmetry and well-differentiated tissues, including nerves, that develop from distinct layers in the embryo. In this section, we describe metazoans with **radial symmetry,** a body plan that permits the detection of stimuli from all directions. This is an effective adaptation for life in open water, freshwater or marine.

Two phyla of soft-bodied organisms, Cnidaria and Ctenophora, have radial symmetry and nerves. Both phyla possess a **gastrovascular cavity** with a single opening, the mouth. Gas exchange and excretion can occur by diffusion because no cell is far from a body surface.

The radiate phyla have a diploblastic body plan with only inner and outer tissue layers, the **gastrodermis** (an endoderm derivative) and the **epidermis** (an ectoderm derivative), respectively. Most species also possess a gelatinous **mesoglea** (*meso* = middle; *glea* = glue) between the two layers. The mesoglea contains widely dispersed fibrous and amoeboid cells, recalling the organization of the mesohyl in sponges.

27.5a Phylum Cnidaria

Nearly all of the 8900 species in the phylum Cnidaria (*cnid* = stinging nettle, a plant with irritating hairs) live in the sea. Their body plan is organized around a saclike gastrovascular cavity, and the mouth is ringed with tentacles, which push food into it. Cnidarians may be vase-shaped, upward-pointing **polyps** or bell-shaped, downward-pointing **medusae (Figure 27.9).** Most polyps attach to a substrate at the *aboral* (opposite the mouth) end, while medusae are unattached and float.

Cnidarians are the simplest animals that exhibit a division of labour among irreversibly specialized tissues (see Figure 27.9c) and that have nerve cells. The gastrodermis includes sensory receptor cells, gland cells, and phagocytic nutritive cells. Gland cells secrete enzymes for the **extracellular digestion** of food, which is then engulfed by nutritive cells and exposed to **intracellular digestion.** The epidermis includes sensory cells, contractile cells, and cells specialized for prey capture.

Cnidarians prey on crustaceans, fishes, and other animals. The epidermis includes unique cells, **cnidocytes**, each armed with a stinging **nematocyst (Figure 27.10, p. 604).** The nematocyst contains an encapsulated, coiled thread that is fired at prey or predators, sometimes releasing a toxin through its tip. Discharge of nematocysts may be triggered by touch, vibrations, or chemical stimuli. The toxin can paralyze small prey by disrupting nerve cell membranes. The painful stings of some jellyfishes and certain corals result from the discharge of nematocysts.

Cnidarians engage in directed movements by contracting specialized ectodermal cells with fibres that resemble those in muscles. In medusae, the mesogleal jelly serves as a deformable skeleton against which contractile cells act. Rapid contractions narrow the bell, forcing out jets of water that propel the animal. Polyps use their water-filled gastrovascular cavity as a hydrostatic skeleton. When some cells contract, fluid within the chamber is shunted about, changing the body's shape and moving it in a particular direction.

The **nerve net**, which threads through both tissue layers, is a simple nervous system that coordinates responses to stimuli (see Chapter 44). Although there is no recognizable "brain," there are control and coordination centres, particularly in a ring of nerves encircling the mouth. In spite of its structural simplicity, the nerve net permits directed swimming movements so the animal can escape predators.

Many cnidarians exist in only the polyp or the medusa form, but some have a life cycle that alternates between them **(Figure 27.11, p. 605).** In the alternating type, the polyp often produces new individuals

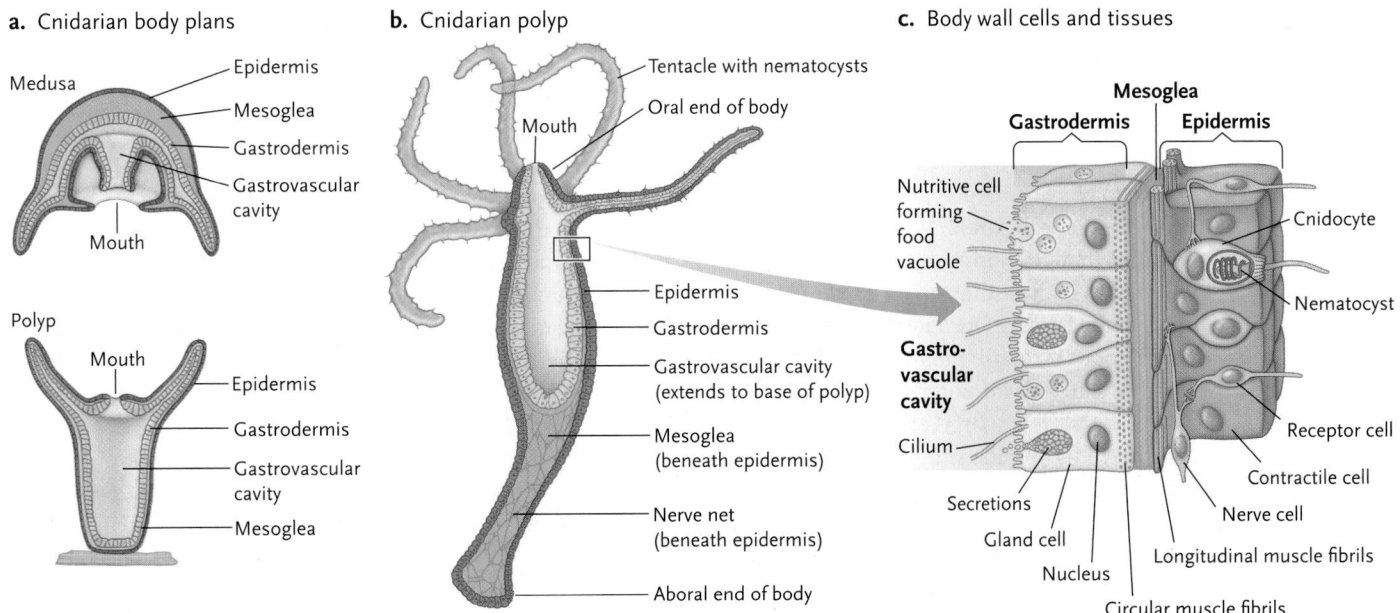

a. Cnidarian body plans

Medusa — Epidermis, Mesoglea, Gastrodermis, Gastrovascular cavity, Mouth

Polyp — Mouth, Epidermis, Gastrodermis, Gastrovascular cavity, Mesoglea

b. Cnidarian polyp

Tentacle with nematocysts
Oral end of body
Mouth
Epidermis
Gastrodermis
Gastrovascular cavity (extends to base of polyp)
Mesoglea (beneath epidermis)
Nerve net (beneath epidermis)
Aboral end of body

c. Body wall cells and tissues

Mesoglea
Gastrodermis
Epidermis
Nutritive cell forming food vacuole
Cnidocyte
Nematocyst
Gastrovascular cavity
Receptor cell
Cilium
Secretions
Contractile cell
Gland cell
Nerve cell
Nucleus
Longitudinal muscle fibrils
Circular muscle fibrils

Figure 27.9

The cnidarian body plan. (a) Cnidarians exist as either polyps or medusae. **(b)** The body of both forms is organized around a gastrovascular cavity, which extends all the way to the aboral end of the animal. **(c)** The two tissue layers in the body wall, the gastrodermis and the epidermis, include a variety of cell types.

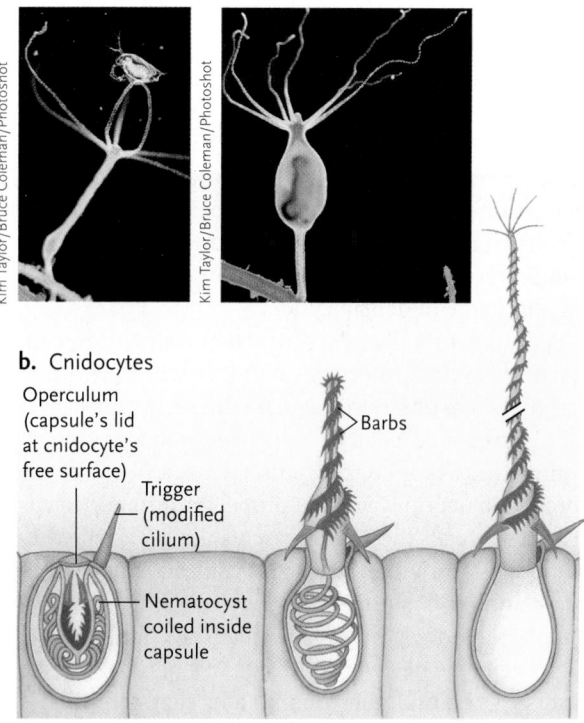

a. *Hydra* consuming a crustacean

b. Cnidocytes

Operculum (capsule's lid at cnidocyte's free surface)

Trigger (modified cilium)

Barbs

Nematocyst coiled inside capsule

Figure 27.10

Predation by cnidarians. **(a)** A polyp of a freshwater *Hydra* captures a small crustacean with its tentacles and swallows it whole. **(b)** Cnidocytes, special cells on the tentacles, encapsulate nematocysts, which are discharged at prey.

asexually from buds that break free of the parent (see Chapter 47). The medusa is often the sexual stage, producing sperm and eggs, which are released into the water. Sexual reproduction results in a ciliated, nonfeeding larval stage, the planula, that eventually settles and undergoes metamorphosis into the polyp form. The four classes of Cnidaria differ in the form that predominates in the life cycle.

Class Hydrozoa. Most of the 2700 species in the class Hydrozoa have both polyp and medusa stages in their life cycles (see Figure 27.11). The polyps form sessile colonies that develop asexually from one individual. A colony can include thousands of polyps, which may be specialized for feeding, defence, or reproduction. They share food through their connected gastrovascular cavities. A few warm-water species secrete a calcareous skeleton and form large colonies. These hydrocorals are different from the anthozoans that form **coral reefs** (see Class Anthozoa, below).

Some pelagic hydrozoans have both polyp and medusoid forms present in the same colony, which functions as an individual organism. The majestic Portuguese man-of-war jellyfish, for example, has the medusoid bell modified to form a gas-filled sail (see "Nematocycsts," p. 720). The hydroid form is represented by feeding and reproductive polyps dangling from the sail (see Chapter 30).

Unlike most Hydrozoa, freshwater species of *Hydra* (see Figure 27.10a) live as solitary polyps that attach temporarily to rocks, twigs, and leaves. Under favourable conditions, hydras reproduce by budding. Under adverse conditions, they produce eggs and sperm. Zygotes, formed by fertilization, are encapsulated in a protective coating but develop and grow when conditions improve. There is no larval stage; the eggs hatch into small *Hydra*.

Class Scyphozoa. The medusa stage predominates in the 200 species of the class Scyphozoa or jellyfish **(Figure 27.12a).** They range from 2 cm to more than 2 m in diameter. Nerve cells near the margin of the bell control their tentacles and coordinate the rhythmic activity of contractile cells, which move the animal. Specialized sensory cells are clustered at the edge of the bell: statocysts (see Chapter 45) sense gravity, and ocelli are sensitive to light. Scyphozoan medusae are either male or female, releasing gametes into the water, where fertilization takes place.

Class Cubozoa. Most of the 20 known species of box jellyfish, the Cubozoa (Figure 27.12b), exist as cube-shaped medusae only a few centimetres tall; the largest species grows to 25 cm in height. Nematocyst-rich tentacles grow in clusters from the four corners of the box-like medusa, and groups of light receptors and image-forming eyes occur on the four sides of the bell. The eyes have lenses and retinas. Unlike the scyphozoan jellyfish, cubozoans are active swimmers. They eat small fish and invertebrates, immobilizing their prey with one of the deadliest toxins produced by animals. Cubozoans live in tropical and subtropical coastal waters, where they sometimes pose a serious threat to swimmers: the nematocysts of some species can inflict considerable pain to, and may kill, humans.

Class Anthozoa. The Anthozoa includes 6000 species of corals and sea anemones **(Figure 27.13, p. 606).** Anthozoans exist only as polyps, which have a more complex structure than the Hydrozoa. A muscular pharynx leads into the gastrovascular cavity, and the body often consists of compartments partially separated by vertical membranes called septa. They reproduce by budding or fission. Most also reproduce sexually, producing eggs that develop into ciliated larvae. Corals (see Figure 27.13a) are always sessile and colonial. Their ciliated larvae settle and metamorphose into polyps that produce colonies by budding. Most species of corals build calcium carbonate skeletons that sometimes accumulate into gigantic underwater reefs. A coral reef usually contains more than one species of anthozoan. The energy needs of corals are partly fulfilled by the photosynthetic activity of symbiotic protists that live within the anthozoans. For this reason, corals are restricted to shallow water, where sunlight can penetrate.

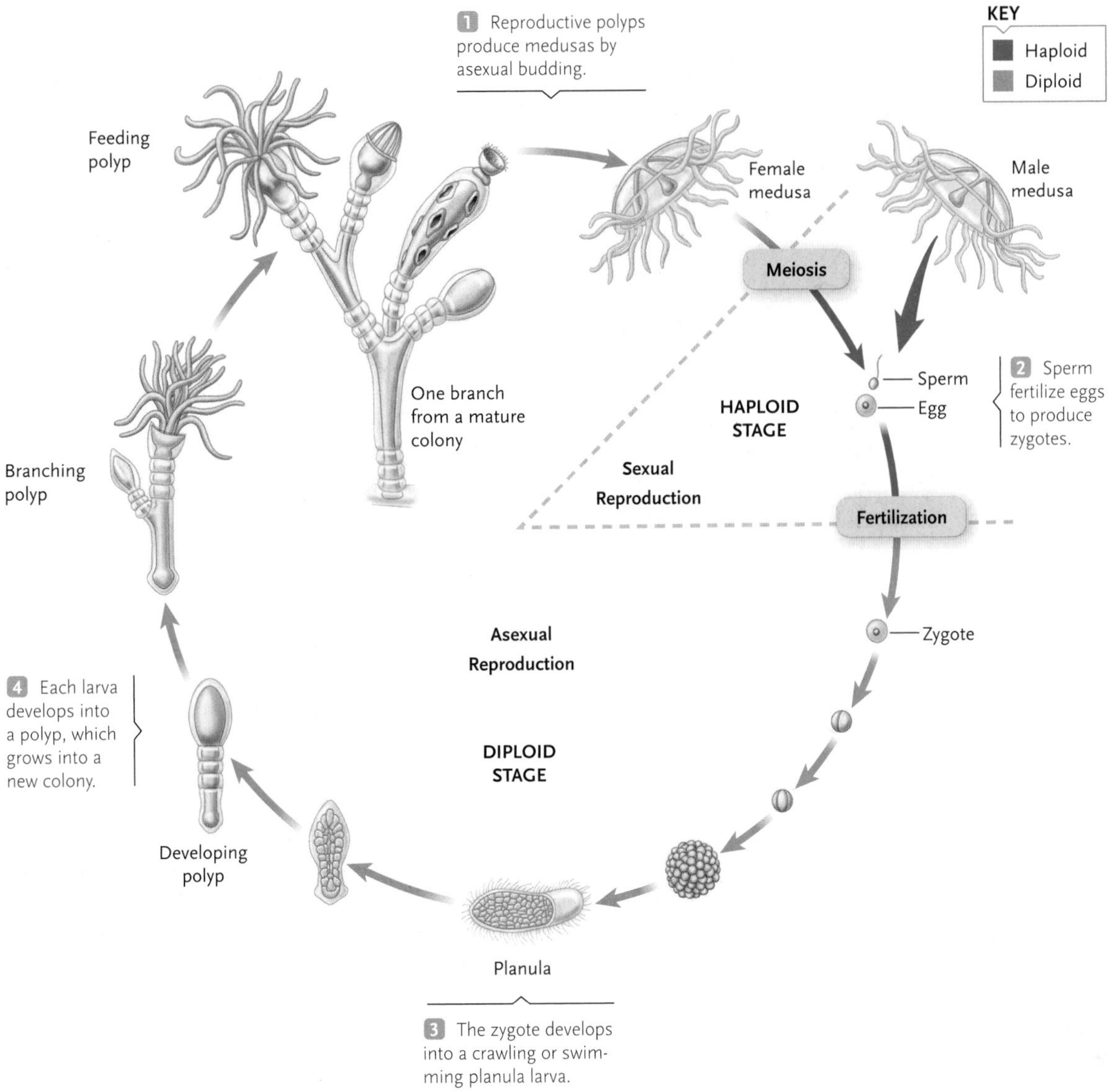

1 Reproductive polyps produce medusas by asexual budding.

KEY
■ Haploid
■ Diploid

Feeding polyp

Female medusa

Male medusa

Meiosis

One branch from a mature colony

HAPLOID STAGE

— Sperm
— Egg

2 Sperm fertilize eggs to produce zygotes.

Branching polyp

Sexual Reproduction

Fertilization

— Zygote

Asexual Reproduction

4 Each larva develops into a polyp, which grows into a new colony.

DIPLOID STAGE

Developing polyp

Planula

3 The zygote develops into a crawling or swimming planula larva.

Figure 27.11
Life cycle of *Obelia*. The life cycle of *Obelia*, a colonial hydrozoan, includes both polyp and medusa stages.

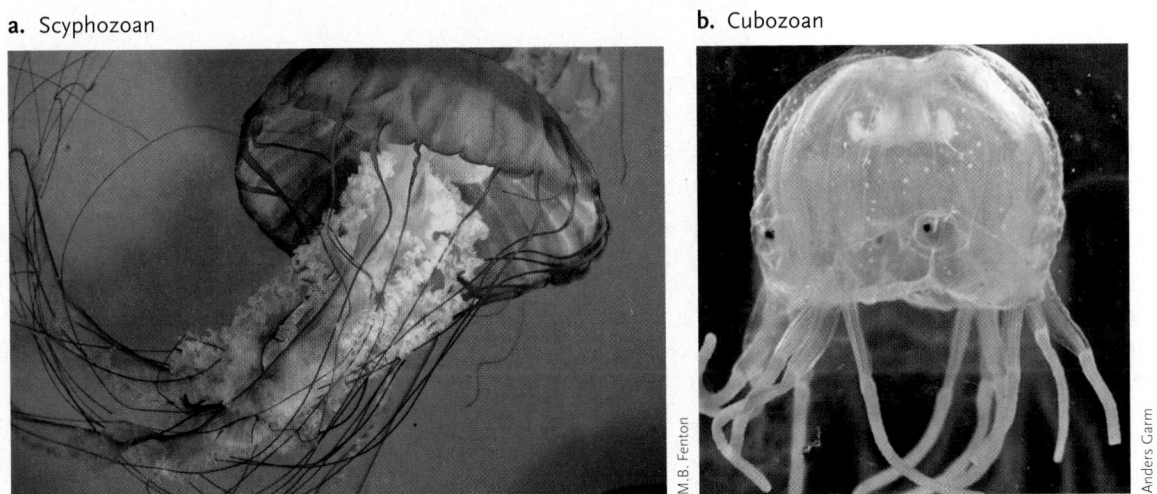

a. Scyphozoan

b. Cubozoan

M.B. Fenton

Anders Garm

Figure 27.12
Scyphozoans and cubozoans. **(a)** Most scyphozoans, like the sea nettle (*Chrysaora* species), live as floating medusae. Their tentacles trap prey, and the long oral arms transfer it to the mouth on the underside of the bell. **(b)** Cubozoans, unlike most jellyfish, are active swimmers and can change direction abruptly. They have several light-sensitive organs, but only four of them, two of which are clearly visible here, form images.

Coral

Tentacle of one polyp

Interconnected
skeletons of polyps
of a colonial coral

scubaluna/iStock/Thinkstock

Figure 27.13

Anthozoans. **(a)** Many corals are colonial, and their polyps build a hard skeleton of calcium carbonate. The skeletons accumulate to form coral reefs in shallow tropical waters. **(b)** A sea anemone detaches from its substrate to escape from a predatory sea star.

Sea anemones (see Figure 27.13b), by contrast, are soft-bodied, solitary polyps, ranging from 1 to 10 cm in diameter. They occupy shallow coastal waters. Most species are sessile, but some move by crawling slowly or by using the gastrovascular cavity as a hydrostatic skeleton.

27.5b Phylum Ctenophora

The 100 species of comb jellies in the marine phylum Ctenophora (*ctenos* = comb; *phor* = to carry) also have radial symmetry, mesoglea, and feeding tentacles. However, they differ from Cnidaria in significant ways. They lack nematocysts, they expel some waste through anal pores located at the opposite end to the mouth, and certain of their tissues appear to be of mesodermal origin. These transparent and often luminescent (light-producing) animals range in size from a few millimetres to 30 cm in diameter, with tentacles up to 1 m or more in length **(Figure 27.14)**.

Ctenophores move by beating cilia arranged on eight longitudinal plates that resemble combs. They are the largest animals to use cilia for locomotion, but they are feeble swimmers. Nerve cells coordinate the animals' movements, and a gravity-sensing statocyst helps them maintain an upright position. Most species have two tentacles with specialized cells that discharge sticky filaments to entrap small animals floating in the sea, particularly small crustaceans. The food-laden tentacles are drawn across the mouth. Others lack tentacles and take large prey by a single gulp of the mouth. Some species that attack Cnidaria incorporate the nematocysts from the prey and use them in feeding (see Chapter 30). Ctenophores are hermaphroditic, producing gametes in cells that line the gastrovascular cavity. Eggs and sperm are expelled through the mouth or from special pores, and fertilization occurs in the open water.

STUDY BREAK

1. How do cnidarians capture, consume, and digest their prey?
2. Describe the differences between a polyp and a medusa.
3. Which group of cnidarians has only a polyp stage in its life cycle?
4. What do ctenophores eat, and how do they collect their food?

27.6 Lophotrochozoan Protostomes

The remaining organisms described in this chapter are in the group Bilateria because of their bilateral symmetry. They have a greater variety of tissues, some of

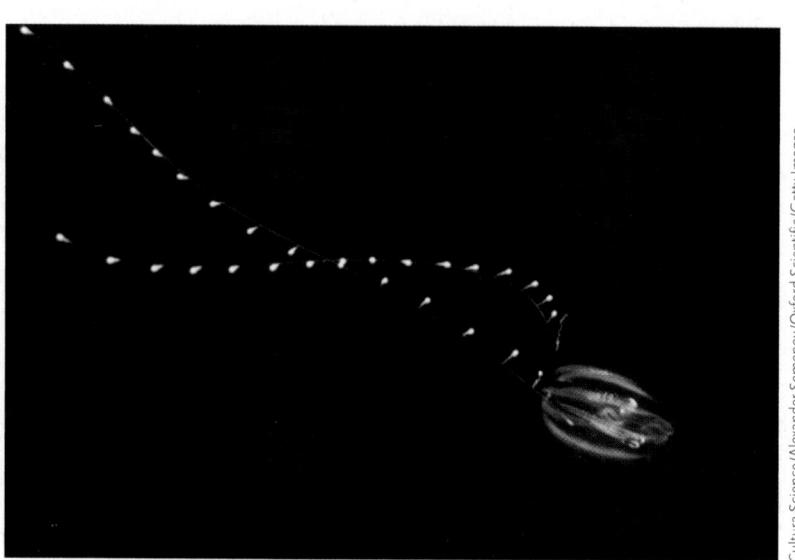

Cultura Science/Alexander Semenov/Oxford Scientific/Getty Images

Figure 27.14

Ctenophores. The comb jelly *Pleurobrachia* collects microscopic prey on its two long sticky tentacles and then wipes the food-laden tentacles across its mouth.

which are developed into organ systems. Most of the phyla have a coelom or pseudocoelom. With bilateral symmetry and sensory organs that are concentrated at the anterior end of the body, most bilaterians can make directed movements in pursuit of food or mates or to escape danger. Organ systems can operate more efficiently than simple tissues. For example, animals that have a tubular digestive system surrounded by a space (the coelom) use muscular contractions of the digestive system to move ingested food past specialized epithelial cells that break it down and absorb the breakdown products.

Molecular analyses group eight of the Bilateria phyla into the Lophotrochozoa, one of the two main protostome lineages (see Figure 27.6, p. 600).

27.6a Three Lophophorate Phyla Share a Distinctive Feeding Structure

Three small groups of mostly marine (a few also occur in fresh water) coelomate animals, the phyla Brachiopoda, Ectoprocta, and Phoronida, have a **lophophore**, a circular or U-shaped fold with one or two rows of hollow, ciliated tentacles surrounding the mouth **(Figure 27.15)**. Molecular sequence data and the lophophore suggest that these phyla have a common ancestry.

The coelomic cavity extends into the lophophore, which looks like a crown of tentacles at the anterior end of the animal. The lophophore is involved in the capture of food and serves as a site for gas exchange. Most lophophorates are sessile suspension feeders (see Chapter 48) as adults. Movement of cilia on the tentacles brings food-laden water toward the lophophore, where the tentacles capture small organisms and debris, and the cilia transport them to the mouth. The lophophorates have a complete digestive system, which is U-shaped in most species, with the anus lying outside the ring of tentacles.

Phylum Ectoprocta. The Ectoprocta (sometimes called Bryozoa or Polyzoa) are tiny colonial animals that occupy mainly marine habitats (see Figure 27.15a). They secrete a hard covering over their soft bodies. The lophophore is normally retracted into a chamber at the anterior end of the animal and extended when the animal feeds. Each colony, which may include more than a million individuals, is produced asexually by a single animal. Ectoproct colonies are permanently attached to solid substrates, where they form encrusting mats, bushy upright growths, or jellylike blobs. Sexual reproduction involves the production of eggs and sperm in the coelom. The sperm are shed through special pores. Fertilization may be internal or external, and the zygote gives rise to a ciliated larva that eventually settles and undergoes metamorphosis. Nearly 5000 living species are known, and about 50 of them live in fresh water.

Phylum Brachiopoda. The brachiopods, or lampshells, have two calcified shells that are secreted on the animal's dorsal and ventral sides (see Figure 27.15b). Most species attach to substrates with a stalk that protrudes through one of the shells. The lophophore is held within the two shells, and the animal feeds by opening its shell and drawing water over its tentacles. The animal has well-developed organs, such as a heart that propels blood through a number of interconnected sinuses and specialized excretory organs. Eggs and sperm are produced in different individuals (dioecious), and fertilization is external. The zygote gives rise to a ciliated larva.

Phylum Phoronida. The 18 or so species of phoronid worms vary in length from a few millimetres to 25 cm (see Figure 27.15c). They usually build tubes of chitin, a polymer of N-acetylglucosamine (see "Molecule behind Biology," Box 46.1, p. 1145), in soft ocean sediments or on hard substrates and feed by protruding the lophophore from the top of the tube. Phoronids reproduce both sexually and by budding. The animals are monoecious (both eggs and sperm produced by one individual). A ciliated feeding larva is produced that settles, undergoes metamorphosis, secretes a tube, and develops into an adult.

a. Ectoprocta (*Plumatella repens*)

© blickwinkel/Hecker/Alamy

b. Brachiopoda (*Terebraulina septentrionalis*)

Andrew J. Martinez/Science Source

c. Phoronida (*Phoronis*)

Andrew J. Martinez/Science Source

Figure 27.15

Lophophorate animals. Although the lophophorate animals differ markedly in appearance, they all use a lophophore to acquire food.

27.6b Phylum Platyhelminthes

The 13 thousand flatworm species in the phylum Platyhelminthes (*plat* = flat; *helminth* = worm) live in aquatic (freshwater and marine) and moist terrestrial habitats. Some are parasitic. Like cnidarians, flatworms can swim or float in water, but they are also able to crawl over surfaces. They range from less than 1 mm to more than 20 m in length, and most are just a few millimetres thick. Free-living species eat live prey or decomposing carcasses, whereas parasitic species derive their nutrition from the tissues of living hosts.

Like the radiate phyla, flatworms are acoelomate, but they have a complex structural organization that reflects their triploblastic construction **(Figure 27.16)**. In those with a gut (some parasitic forms lack this organ), endoderm lines the digestive cavity with cells specialized for the chemical breakdown and absorption of ingested food. A single opening serves as both mouth and anus. Mesoderm, the middle tissue layer, produces muscles and reproductive organs. Ectoderm produces a ciliated epidermis, the nervous system, and

the **flame cell** system, a simple excretory system (see Chapter 50). Flatworms lack circulatory or respiratory systems, but because all cells of their dorsoventrally (top-to-bottom) flattened bodies are near an interior or exterior surface, diffusion supplies them with nutrients and oxygen.

The flatworm nervous system includes two or more longitudinal ventral nerve cords interconnected by numerous smaller nerve fibres, like rungs on a ladder. An anterior **ganglion**, a concentration of nervous system tissue that serves as a primitive "brain," integrates their behaviour (see Chapter 44). Most free-living species have **ocelli** or "eye spots" that distinguish light from dark and chemoreceptor organs that sense chemical cues.

The phylum Platyhelminthes includes four classes, defined largely by their anatomical adaptations to free-living or parasitic habits. One class, Turbellaria, is free living, whereas the remaining three classes are parasitic, obtaining their nutrition from the tissues of another animal, the host.

Class Turbellaria. Most free-living flatworms (class Turbellaria) live in the sea **(Figure 27.17)**, where they may be brightly coloured. The familiar planarians and a few others live in fresh water or on land and are drab. Turbellarians swim by undulating the body wall musculature or crawl across surfaces by using muscles and cilia to glide on mucus trails produced by the ventral epidermis. Some terrestrial turbellarians are relatively large and prey on other invertebrates. For example, *Microplana termitophaga* waits at the entrance to termite colonies in Africa and entangles the prey in the slime it produces. Other species may gang up on large snails or other animals (see Figure 32.9, Chapter 32).

The gastrovascular cavity in free-living flatworms is similar to that in cnidarians. Food is ingested and wastes are eliminated through a single opening, the

Digestive system

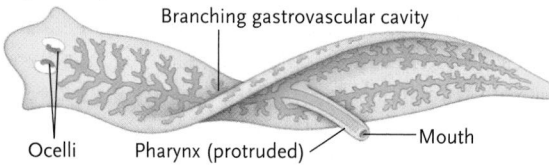

Nervous system

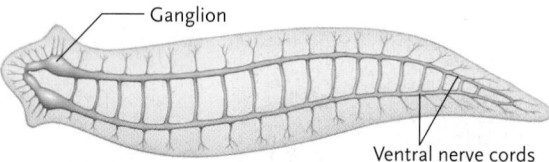

Reproductive system

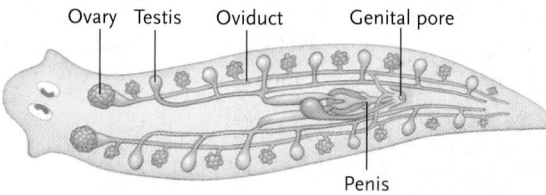

Excretory system

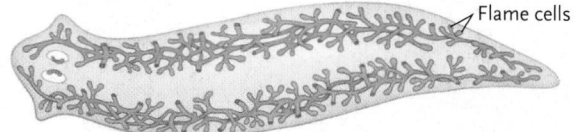

Figure 27.16

Flatworms. The phylum Platyhelminthes, exemplified by a freshwater planarian, have well-developed digestive, excretory, nervous, and reproductive systems. Because flatworms are acoelomate, their organ systems are embedded in a solid mass of tissue between the gut and the epidermis.

Figure 27.17

Turbellaria. A few turbellarians, such as *Pseudoceros dimidiatus*, are colourful marine worms.

mouth, located on the ventral surface. Most turbellarians acquire food with a muscular **pharynx** that connects the mouth to the digestive cavity (see Figure 27.16, top). Chemicals secreted into the saclike cavity digest ingested items, after which cells throughout the gastrovascular surface engulf food particles and subject them to intracellular digestion. In some species, the digestive cavity is highly branched, increasing the surface area for digestion and absorption.

Nearly all turbellarians are hermaphroditic, with complex reproductive systems (see Figure 27.16 second from bottom). When they mate, each partner functions simultaneously as a male and a female. The eggs of most species hatch directly into small worms, but ciliated larvae occur in a few marine turbellarians. Many free-living species also reproduce asexually by simply separating the anterior half of the animal from the posterior half. Both halves subsequently regenerate the missing parts.

Class Monogenea. Flukes (classes Trematoda and Monogenea) are parasites that obtain nutrients from host tissues. Monogenea flukes are **ectoparasites** that attach to the gills or skin of aquatic vertebrates. They have an anterior sucker surrounding the mouth and a more posterior sucker. The suckers may be equipped with hooks.

Reproduction occurs by internal fertilization. The eggs are released into water and hatch as ciliated larvae. The larvae attach to a new host and undergo metamorphosis.

Class Trematoda. Adult trematodes **(Figure 27.18)** are all internal parasites of vertebrates, but their development involves two or more host species in their life cycle. They are sometimes called digenean (two hosts) flukes. The host species in which sexual reproduction occurs is the primary host, and other hosts, usually invertebrates, are secondary hosts. Thus, the same individual will encounter very different environments during its life and may have two or more very different larval stages during development (see "Polymorphic

Development," p. 612). Like the monogeneans, trematodes normally have two suckers, one of which is around the mouth. The unciliated epidermis is a syncytium (the cells are interconnected without separating membranes). Trematodes can be found in many vertebrates, including humans, where they may cause some serious diseases. Infestations of trematodes may alter the behaviour of afflicted animals, making them more vulnerable to predators, often the host for the adult stage of the trematode.

Class Cestoda. Tapeworms **(Figure 27.19, p. 610)** are parasitic in the intestines of vertebrates, their primary host. They lack a mouth or digestive system and absorb nutrients from the host's intestinal contents across the syncytial epithelium. The anterior end is modified as a **scolex**, consisting of hooks and/or suckers that allow it to attach to the wall of the intestine. The remainder of the worm consists of a series of identical units, **proglottids**, each with its own reproductive system. Proglottids are generated just posterior to the scolex and become progressively more mature near the tail. The posterior, fully mature units break off or burst and are passed out with the feces. Worms may consist of only a few proglottids, but many species have 2000 to 3000, and such worms may be 10 m in length, occupying the entire length of the human small intestine.

Each proglottid contains a complete set of reproductive organs producing both sperm and eggs. Fertilization is internal and may involve a neighbouring worm, or the worm may be self-fertilizing. Each proglottid may contain as many as 50 thousand eggs. Further development varies with the species, but, typically, the egg must be eaten by an appropriate intermediate host, usually an arthropod, in which it undergoes development into a series of larval stages. The life cycle is completed when an appropriate primary host eats an infected intermediate host. For example, the adult tapeworm *Hymenolepis diminuta* lives in rat intestines. The eggs in rat feces are eaten by flour beetles, where larvae develop and form cysts. Rats become infected when they eat the beetles. Humans can also become infected by unwittingly consuming infected beetles that may live in dry breakfast cereals. The tapeworms that infest our domestic animals, notably dogs and cats, can pose a serious health threat to humans.

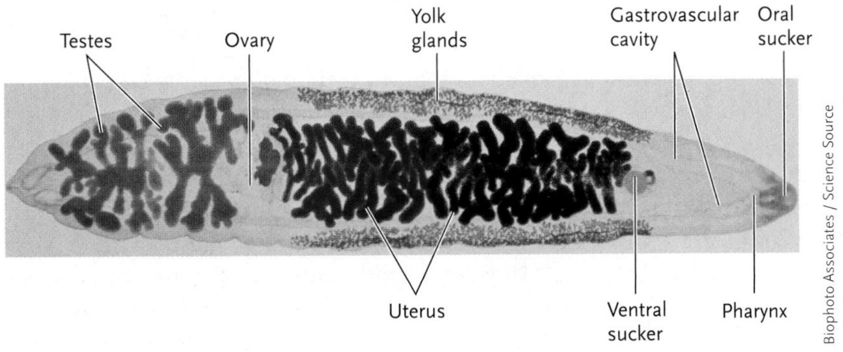

27.6c Phylum Rotifera

Most of the 1800 species in the pseudocoelomate phylum Rotifera (*rota* = wheel; *fera* = to bear) live in fresh water, and a few are marine **(Figure 27.20, p. 610)**. Most are less than 0.5 mm long, but a few range up

Figure 27.18

Trematoda. The hermaphroditic Chinese liver fluke (*Opisthorchis sinensis*) uses a well-developed reproductive system to produce thousands of eggs.

Testes   Ovary   Yolk glands   Gastrovascular cavity   Oral sucker

Uterus   Ventral sucker   Pharynx

Biophoto Associates / Science Source

Figure 27.19

Cestoda. **(a)** Tapeworms have long bodies composed of a series of proglottids, each of which produces thousands of fertilized eggs. **(b)** The anterior end is a scolex with hooks and suckers that attach to the host's intestinal wall.

a. Tapeworm

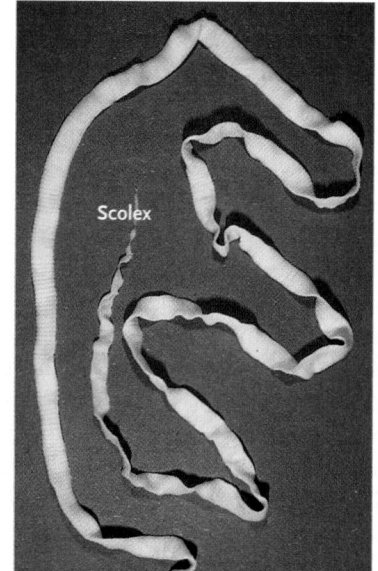

Scolex

b. Scolex

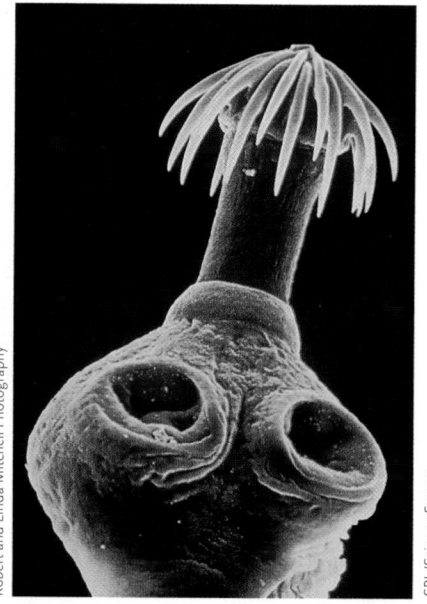

Figure 27.20

Phylum Rotifera.

(a) Despite their small size, rotifers, such as *Philodina roseola*, have complex body plans and organ systems. **(b)** This rotifer, another *Philodina* species, is laying eggs.

a. Rotifer body plan

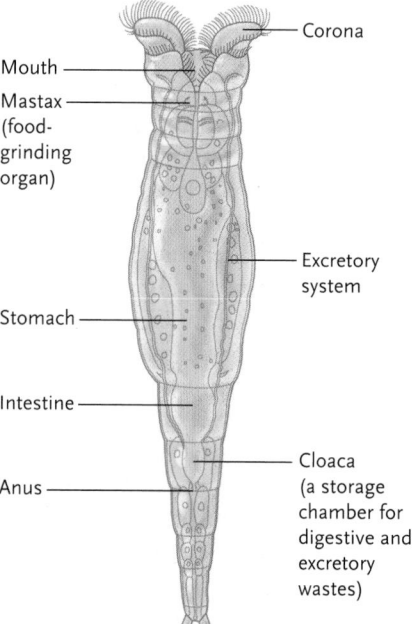

Corona
Mouth
Mastax (food-grinding organ)
Excretory system
Stomach
Intestine
Anus
Cloaca (a storage chamber for digestive and excretory wastes)

b. Rotifer laying eggs

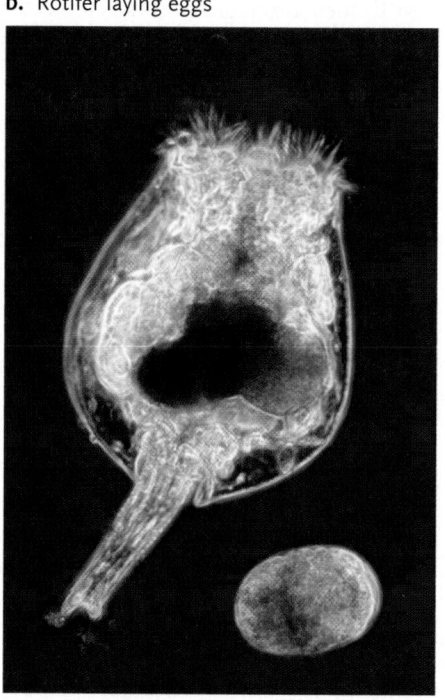

to 3 mm. They exhibit *eutely*, a mode of development in which cell division ceases early and subsequent growth is by cell enlargement. In spite of their size, they have well-developed digestive, reproductive, excretory, and nervous systems. In some habitats, rotifers make up a large part of the zooplankton (tiny animals that float in open water). Some species, however, are attached to the substrate and move only a little. Others may form colonies, and some live in pitcher plants (see Chapter 31).

Rotifers use coordinated movements of cilia, arranged in a wheel-like **corona** around the head, to propel themselves in the environment. Cilia also bring

food-laden water to their mouths. Ingested microorganisms are conveyed to the **mastax**, a toothed grinding organ, and then passed to the stomach and intestine. Rotifers have a **complete digestive system:** food enters through the mouth, and undigested waste is voided through a separate anus.

The life history patterns of some rotifers are adapted to the ever-changing environments in small bodies of water. During most months, rotifer populations of these species include only females that reproduce by **parthenogenesis** (the development of unfertilized eggs; see Chapter 42). In this particular form of parthenogenesis, females produce diploid eggs by mitosis that develop into females. When environmental conditions deteriorate, females produce eggs by meiosis. If these eggs remain unfertilized, they develop into haploid males that produce sperm. If the haploid eggs are fertilized, they produce diploid female zygotes. The fertilized eggs have durable shells and food reserves to survive drying or freezing.

27.6d Phylum Nemertea

The 650 species of ribbon worms or proboscis worms vary from less than 1 cm to 30 m in length (**Figure 27.21**). Most species are marine, but a few occupy moist terrestrial habitats. The often brightly coloured ribbon worms have no obvious coelom and use a ciliated epidermis to glide over a film of secreted mucus. Ribbon worms have a complete digestive tract with a mouth and an anus. They have a circulatory system in which fluid flows through **circulatory vessels** that carry nutrients and oxygen to tissues and remove wastes. They have a muscular, mucus-covered proboscis, a tube that can be everted (turned inside out) through a separate pore to capture prey. The proboscis is housed within a chamber, the **rhynchocoel**, which is unique to this phylum (see Figure 27.21b).

Nemerteans are aggressive predators. The proboscis may have a barb that is used to impale the prey, or the proboscis may wrap around the prey in a form of stranglehold. Many nemerteans are burrowing animals, living in tubes that protect them from predators. The life cycle includes a microscopic ciliated larva.

27.6e Phylum Mollusca

Most of the 100 thousand species of fleshy molluscs in the coelomate phylum Mollusca (*moll* = soft), including clams, snails, octopuses, and their relatives, are marine. However, many clams and snails occupy fresh water habitats, and some snails live on land. Molluscs vary in length from clams less than 1 mm across to the giant squids that can exceed 18 m in length.

The mollusc body is divided into three regions: the visceral mass, head-foot, and mantle (**Figure 27.22, p. 613**). The **visceral mass** contains the digestive, excretory, and reproductive systems and the heart. The muscular **head-foot** often provides the major means of locomotion. In the more active groups, the head area of the head-foot region is well defined and carries sensory organs and a brain. The mouth often includes a toothed **radula**, which scrapes food into small particles or drills through the shells of prey.

Many molluscs are covered by a protective shell of calcium carbonate secreted by the **mantle**, a folding of the body wall that may enclose the visceral mass. The mantle also defines a space, the **mantle cavity**, housing the **gills**, delicate respiratory structures with an enormous surface area (see Chapter 49). In most molluscs, cilia on the mantle and gills generate a steady flow of water into the mantle cavity.

Most molluscs have an **open circulatory system** in which **hemolymph**, a bloodlike fluid, leaves the circulatory vessels and bathes tissues directly. Hemolymph pools in spaces called **sinuses** and then drains into vessels that carry it back to the heart (see Figure 40.3, Chapter 40).

The sexes are usually separate, although many snails are hermaphroditic. Fertilization may be internal or external. In some snails, eggs and sperm are produced simultaneously in the same organ, an ovotestis. In others, the hermaphroditism is serial, with younger snails producing sperm and older individuals switching to egg production. Fertilization is often internal in these organisms, and in simultaneous hermaphrodites, there is a mutual exchange of sperm during copulation. Sperm may be stored for long periods before being used. In some terrestrial snails, a calcium "love dart" may be fired into one of the partners preceding a mutual exchange of sperm. Dr. Ron Chase of McGill University has shown that mucus coating of the dart makes it more likely that the shooter's sperm will be used to fertilize the eggs.

a. Ribbon worm

b. Ribbon worm anatomy

Proboscis pore

Proboscis

Rhynchocoel

Mouth

Intestine

Proboscis retractor muscle

Anus

Everted proboscis

Kjell B. Sandved/Science Source

Figure 27.21

(a) The flattened, elongated bodies of ribbon worms, such as genus *Lineus*, are often brightly coloured. **(b)** Ribbon worms have a complete digestive system and a specialized cavity, the rhynchocoel, that houses a protrusible proboscis.

Polymorphic Development

Most of the protostomes have a capacity for developmental polymorphism. During development from egg to adult, the organism may assume different morphologies. Most commonly, the immature form is referred to as a larva. In insects, for example, the caterpillar that hatches from the egg and and grows through a number of moults is a feeding stage very different from the adult butterfly, the distributive and reproductive stage. The transformation from larva to adult stage is accomplished by metamorphosis.

In other cases, particularly in sessile marine animals, the larval stage functions in distribution. Often this is an inconspicuous ciliated stage, such as the trochophore larva of molluscs and marine annelids **(see Figure 27.23, p. 614),** that drifts with ocean currents. It settles to the ocean floor in response to some signal and metamorphoses

into the form that will eventually become the adult. Some Cnidaria may have three distinct forms (see Figure 27.11, p. 605). This capacity for assuming different forms during development is particularly important for two lifestyles, parasitism and social insects.

Populations of animals that live in other organisms are faced with a particular challenge. The environment in which they live, the host, is discontinuous in space: one host is not connected to another. It is also discontinuous in time: the host eventually dies. It is thus essential for a parasite to move from one part of this discontinuous environment to another if the parasite population is to survive. Moving to another host may involve a period as a free-living form or further development in an alternative host. During its life cycle, the parasite is thus obliged to experience two or

more different environments. These different environments have favoured different developmental stages, often a series of morphologically distinct larvae. In the Chinese liver fluke (*Opisthorchis sinensis*), for example **(Figure 1),** the egg is eaten by a snail and hatches into a ciliated larva, the miracidium.

Almost immediately, metamorphosis occurs, involving extensive reorganization of the larva. This produces the sporocyst. Groups of embryonic cells, called "germ balls" **(Figure 2),** within the body cavity of the larva develop to produce additional larval stages, which are morphologically distinct from the sporocyst. In each stage, groups of embryonic germ balls are reserved to give rise to increased numbers of the next stage. The snail eventually releases enormous numbers of another larval form, the free-swimming cercaria, that

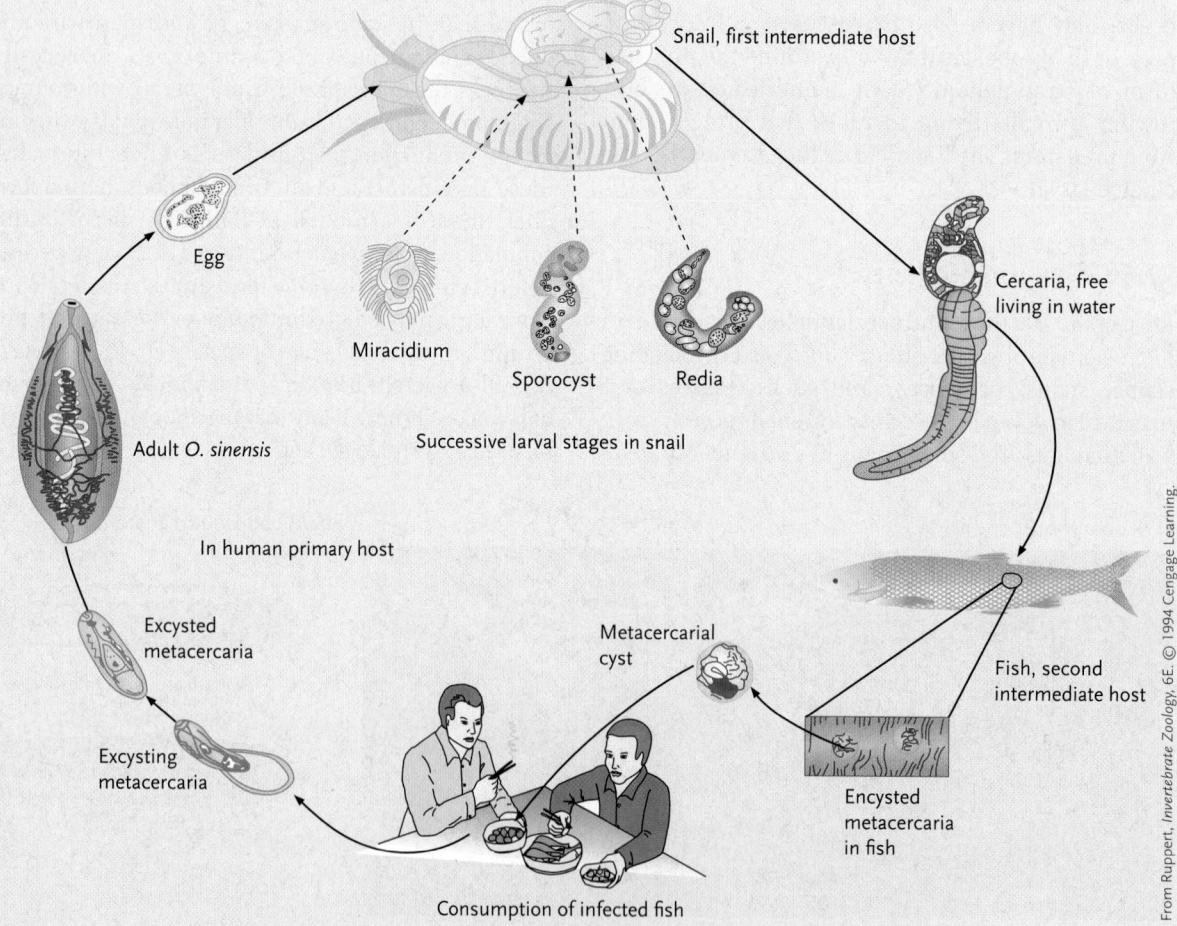

Figure 1
The life cycle of the liver fluke Opisthorchis sinensis. *Humans become infected by eating raw fish, and the adult fluke lives in the liver and bile duct. Estimates of the number of persons infected range up to 30 million.*

From Ruppert, *Invertebrate Zoology*, 6E. © 1994 Cengage Learning.

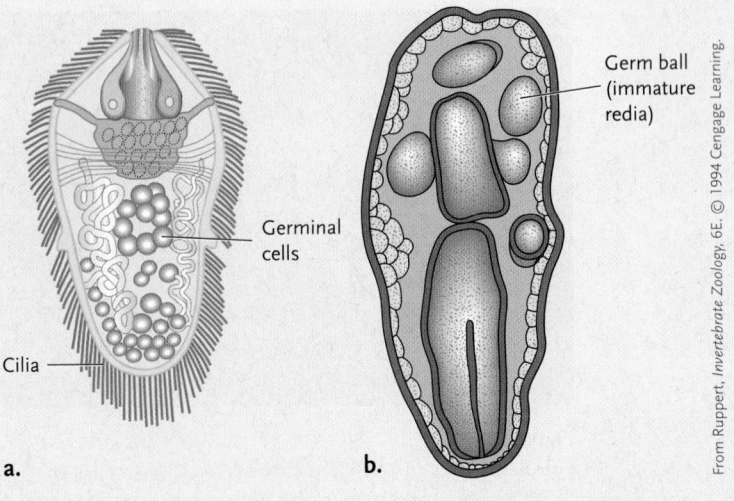

a.

Germinal cells

Cilia

b.

Germ ball (immature redia)

Germinal cells

From Ruppert, *Invertebrate Zoology*, 6E. © 1994 Cengage Learning.

Figure 2
The transformation of a miracidium **(a)** *into a sporocyst* **(b),** *involving the development of totipotent germ cells (stem cells) into germ balls that will form several copies of the next larval stage.*

enters another intermediate host, a fish, where it forms a cyst in the muscle. This cyst will develop into the adult fluke if it is consumed by a human. The existence of populations of totipotent stem cells in flatworms has made possible the developmental polymorphism on which parasitism depends. The development of the cells is directed into different pathways appropriate for each parasitic stage.

Developmental polymorphism is also a feature of social insects. Social insects live in colonies and are characterized by different castes (individuals that perform particular tasks on behalf of the entire colony), which usually differ in morphology. The difference among the castes is particularly pronounced in termites **(Figure 3).**

Entomological Society of America

Figure 3
Termite castes. The queen termite in the centre of the photograph is surrounded by workers, called pseudergates. One soldier, with an enlarged head and mandibles, is also visible. At each moult, depending on the conditions in the colony, a pseudergate may develop into another pseudergate, embark on development to a soldier, or develop into a winged supplementary reproductive caste that will leave the colony to found a new one.

Chitons

Shell Digestive system Mantle

Head Radula Foot Gill Mantle cavity

Gastropods

Sensory tentacles

Shell

Mantle

Head Radula Gill Foot Digestive system

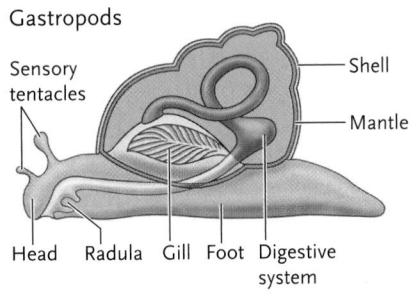

Bivalves

Digestive system Shell Mantle

Foot Gill Mantle cavity

Cephalopods

Head-foot Mantle Digestive system

Arms

Radula Mantle cavity Gill Internal shell

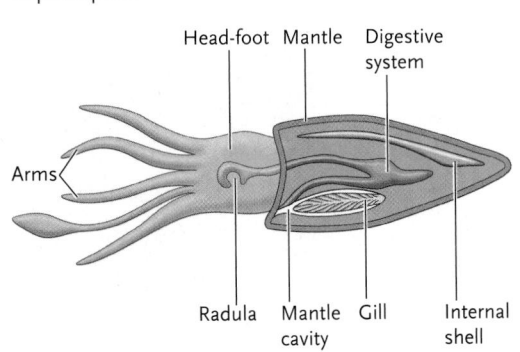

KEY

- Head-foot
- Visceral mass
- Mantle

Figure 27.22
Molluscan body plans. The bilaterally symmetrical body plans of molluscs include a muscular head-foot, a visceral mass, and a mantle.

Figure 27.23

Trochophore larva. At the conclusion of their embryological development, both molluscs and annelids typically pass through a trochophore stage. The top-shaped trochophore larva has a band of cilia just anterior to its mouth.

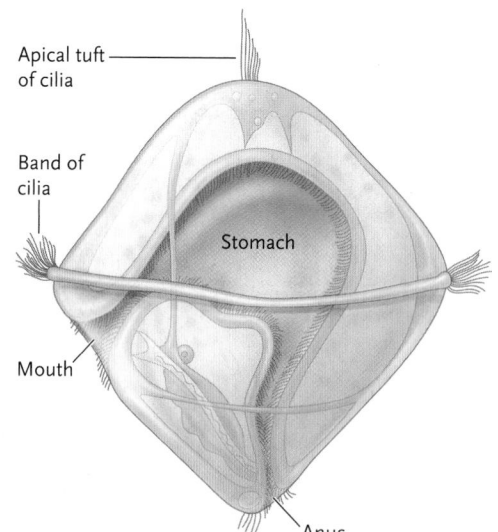

Apical tuft of cilia

Band of cilia

Stomach

Mouth

Anus

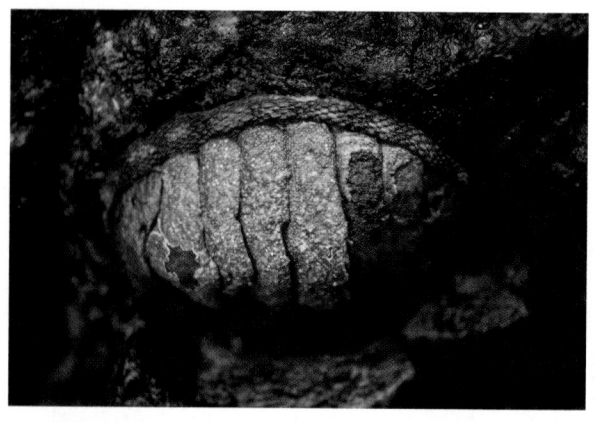

KMahelona/iStock/Thinkstock

Figure 27.24

Polyplacophora. Chitons live on rocky shores, where they use their foot and mantle to grip rocks and other hard substrates. This chiton (*Mopalia ciliata*) lives in Monterey Bay, California.

The zygotes of marine species often develop into free-swimming, ciliated **trochophore** larvae (Figure 27.23), typical of both this phylum and the phylum Annelida, which we describe in Section 27.6f. In some molluscs, the trochophore develops into a second larval stage, called a **veliger**, before metamorphosing into an adult. In some snails, the larval stage may occur only within the egg. Squids and octopuses have no larval stage, and eggs hatch into miniature replicas of the adult. Although members of the phylum have common characteristics, they have evolved an extraordinary diversity in form and lifestyle, ranging from sessile clams to the agile octopus capable of learned behaviour. The phylum includes seven classes. We examine the four most commonly encountered classes below.

Class Polyplacophora. The 600 species of chitons (Polyplacophora; *poly* = many; *plak* = plate) are sedentary molluscs that graze on algae along rocky marine coasts. The oval, bilaterally symmetrical body has a dorsal shell divided into eight plates that allow it to conform to irregularly shaped surfaces **(Figure 27.24)** and to roll into a ball when disturbed or theatened. When exposed to strong wave action, a chiton uses the muscles of its broad foot to maintain a tenacious grip, and the mantle's edge functions like a suction cup to hold fast to the substrate.

Class Gastropoda. Snails and slugs (Gastropoda; *gaster* = belly; *pod* = foot) are the largest molluscan group, numbering 40 thousand species **(Figure 27.25)**. The class exhibits a wide range of morphologies and lifestyles. Aquatic and marine species use gills to acquire oxygen, but in terrestrial species, a modified mantle cavity functions as an air-breathing lung. Some snails have the opening into the mantle cavity extended as a tubular siphon. Gastropods feed on algae, vascular plants, or animal prey. Some are scavengers, and a few are parasites.

The visceral mass of most snails is housed in a coiled or cone-shaped shell that is balanced above the rest of the body, much as you balance a backpack full of books (see Figure 27.25a, b). Most shelled species

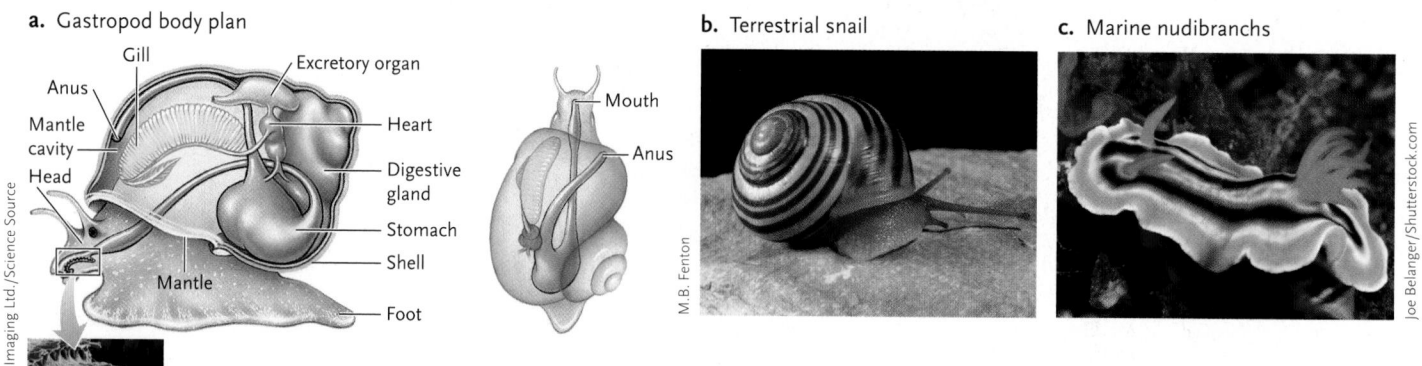

a. Gastropod body plan

Gill
Anus
Mantle cavity
Head

Excretory organ
Heart
Digestive gland
Stomach
Shell
Foot

Mantle

Clouds Hill Imaging Ltd./Science Source

Radula

b. Terrestrial snail

Mouth
Anus

M.B. Fenton

c. Marine nudibranchs

Joe Belanger/Shutterstock.com

Figure 27.25

Gastropoda. **(a)** Most gastropods have a coiled shell that houses the visceral mass. A developmental process called torsion causes the digestive and excretory systems to eliminate wastes into the mantle cavity, near the animal's head. **(b)** The terrestial snail (*Helix pomatia*) is a typical terrestrial gastropod. **(c)** Nudibranchs, like this pair of Spanish shawl nudibranchs (*Flabellina iodinea*), are shell-less marine snails.

undergo **torsion** during development. Differential growth rates and muscle contractions twist the developing visceral mass and mantle a full 180° relative to the head and foot. These events begin in the larva before the shell is established and thus are not dictated by the coiling of the shell. Indeed, a snail that has undergone torsion to the right may exist in a left-handed shell. Among the many results of this developmental manoeuvre is the relocation of the mantle cavity to the anterior, allowing the head and foot to be withdrawn into the shell. In some snails, the **operculum** is an ovoid disk of protein fortified with calcium that can be used to close the entrance to the shell. This permits the snail to survive unfavourable conditions.

Some gastropods, including terrestrial slugs and colourful nudibranchs (sea slugs), are shell-less, a condition that leaves them somewhat vulnerable to predators (see Figure 27.25c). Some nudibranchs consume cnidarians and then transfer undischarged nematocysts to projections on their dorsal surface, where these "borrowed" stinging capsules provide protection (see Chapter 30).

Because many of its neurons are large and easily accessed and identifiable, the nudibranch *Aplysia* has been widely used to explore fundamental questions in neurobiology. For example, Dr. Wayne Sossin's lab at the Montreal Neurological Institute at McGill University is examining the biochemical and molecular basis of memory and learning in *Aplysia*.

The nervous and sensory systems of gastropods are well developed. Tentacles on the head include chemical and touch receptors; the eyes detect changes in light intensity but do not form images. The importance and relative sophistication of the nervous system is well illustrated by some limpets. *Patella vulgaris* is a gastropod with a conical shell **(Figure 27.26)** that lives on rocks in the intertidal zone. During low tide, it is exposed to the air and its foot and mucus secretions combine to fasten it closely to the rock. During development, the edges of its shell grow to conform to irregularities in the rock, increasing the protection against drying. As the rising tide covers the limpet, it moves about, foraging and feeding on algae. About an hour before the falling tide would once again expose it to desiccation, it returns to its precise location so that it can seal itself against exposure to air. This involves not only precise navigation but also a sense of time attuned to the tides.

Class Bivalvia. The 8000 species of clams, scallops, oysters, and mussels (Bivalvia; *bi* = two; *valv* = folding door) are restricted to aquatic habitats. They are enclosed within a pair of shells, hinged together dorsally by an elastic ligament. Contraction of the **adductor muscles** closes the shell and stretches the ligament. When the muscles relax, the stretched ligament opens the shell **(Figure 27.27, p. 616)**. Although some bivalves are tiny, the giant clams of the South Pacific can be more than 1 m across and weigh 225 kg.

Adult mussels and oysters are sessile and permanently attached to hard substrates. However, many clams are mobile and use their muscular foot to burrow in sand or mud. Some bivalves, such as young scallops, swim by rhythmically clapping their valves together, forcing a current of water out of the mantle cavity (see Figure 27.27b). The "scallops" that we eat are their well-developed adductor muscles.

Bivalves have a reduced head and lack a radula. Part of the mantle forms two tubes called *siphons* (see Figure 27.27a). Beating of cilia on the gills and mantle carries water into the mantle cavity through the **incurrent siphon** and out through the excurrent siphon. Incurrent water carries dissolved oxygen and particulate food to the gills, where oxygen is absorbed. Mucus strands on the gills trap food, which is then transported by cilia to *palps*, where final sorting takes place; acceptable bits are carried to the mouth. The excurrent water carries away metabolic wastes and feces.

Despite their sedentary existence, bivalves have moderately well-developed nervous systems: sensory organs that detect chemicals, touch, and light and statocysts to sense their orientation. When they encounter pollutants, many bivalves stop pumping water and close their shells. When confronted by a predator, some burrow into sediments or swim away.

Class Cephalopoda. The 600 living species of octopuses, squids, and nautiluses constituting the class Cephalopoda (*cephal* = head; *pod* = foot) are active marine predators and include the fastest and most intelligent invertebrates **(Figure 27.28, p. 616)**. They vary in length from a few centimetres to 18 m. Ammonites are well-known fossil cephalopods.

M.B. Fenton

Figure 27.26
The common limpet, *Patella vulgata*.

a. Bivalve body plan

Mouth

Anterior adductor muscle

Ligament (connects to opposite shell)

Left mantle

Posterior adductor muscle

Water flows out through excurrent siphon

Water flows in through incurrent siphon

Foot Palps Left gill Right shell

b. Bivalve locomotion

© age fotostock Spain, S.L. /Alamy

c. Geoduck

Tom McHugh/Science Source

Figure 27.27

Bivalvia. **(a)** Bivalves are enclosed in a hinged two-part shell. Part of the mantle forms a pair of water-transporting siphons. **(b)** When threatened by a predator (in this case, a sea star), some scallops clap their shells together rapidly, propelling the animal away from danger. **(c)** The geoduck (*Panope generosa*) is a clam with enormous muscular siphons.

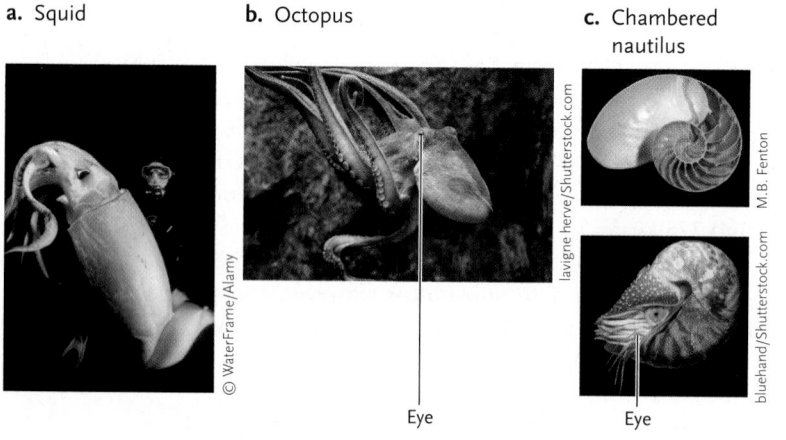

a. Squid

© WaterFrame/Alamy

b. Octopus

lavigne herve/Shutterstock.com

Eye

c. Chambered nautilus

M.B. Fenton

bluehand/Shutterstock.com

Eye

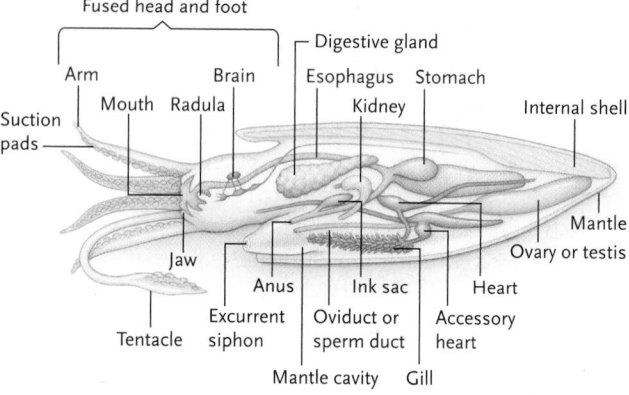

d. Internal anatomy of a squid

Fused head and foot

Arm

Mouth Radula

Suction pads

Brain

Digestive gland

Esophagus Stomach

Kidney

Internal shell

Mantle

Ovary or testis

Jaw

Anus Ink sac Heart

Tentacle

Excurrent siphon

Oviduct or sperm duct

Accessory heart

Mantle cavity Gill

Figure 27.28

Cephalopoda. **(a)** Squids, such as *Dosidicus gigas*, and **(b)** octopuses, such as *Octopus vulgaris*, are the most familiar cephalopods. **(c)** The chambered nautilus (*Nautilus macromphalus*) and its relatives retain an external shell. **(d)** Like other cephalopods, the squid body includes a fused head and foot; most organ systems are enclosed by the mantle.

The cephalopod body has a fused head and foot. The head comprises the mouth and eyes. The ancestral "foot" forms a set of arms, which are equipped with suction pads, adhesive structures, or hooks. Cephalopods use their arms to capture prey and a pair of beaklike jaws to bite or crush it. Venomous secretions often speed the captive's death. Some species use their radula to drill through the shells of other molluscs.

Cephalopods have a highly modified shell. Octopuses have no remnant of a shell at all. In squids and cuttlefishes, the shell is reduced to a stiff internal support. Only the chambered nautilus (see Figure 27.28c) and its relatives retain an external shell; spaces (chambers) in the shell regulate the animal's buoyancy. Species in the genus *Nautilus* are clearly cephalopods because the foot is modified in a way that is characteristic of that class. But they have retained an elegant, chambered shell, a body plan that is very successful, since essentially identical animals can be found among the Cambrian fossils.

Squids (see Figure 27.28a, d) move by a kind of jet propulsion. When muscles in the mantle relax, water enters the mantle cavity. When they contract, a jet of water is squeezed out through a funnel. By manipulating the position of the mantle and funnel, the animal can control the rate and direction of its locomotion. While escaping, many species simultaneously release a dark fluid ("ink") that obscures their direction of movement. Octopuses and squids are able to change colour rapidly by the migration of various pigments in special pigment cells called chromatophores. Many squids have light-emitting cells called photophores.

Cephalopods are the only molluscs to have a **closed circulatory system.** The heart and accessory hearts speed the flow of hemolymph through blood vessels and gills, enhancing the uptake of oxygen and release of carbon dioxide.

Cephalopods have larger brains than other molluscs, and their brains are more complex than any other invertebrate. Giant nerve fibres connect the brain with the muscles of the mantle, enabling quick responses to food or danger (see Chapter 44).

The image-forming eyes of cephalopods, complete with lens and retina, are similar to those of vertebrates (see Chapter 45). The same basic plan for an eye has arisen independently in the cubozoan Cnidaria, the cephalopods, and the vertebrates and represents an example of convergent evolution. Cephalopods are also highly intelligent. Octopuses, for example, learn to recognize objects with distinctive shapes or colours and can be trained to approach or avoid them.

Cephalopods have separate sexes and elaborate courtship rituals. Males store sperm within the mantle cavity and use a specialized tentacle to transfer packets of sperm into the female's mantle cavity, where fertilization occurs. The young hatch with an adult body form.

27.6f Phylum Annelida

The 15 thousand species of segmented worms in the phylum Annelida (*annelis* = ring) occupy marine, freshwater, and moist terrestrial habitats. They range from a few millimetres to as much as 3 m in length. Terrestrial annelids eat organic debris, whereas aquatic species consume algae, microscopic organisms, detritus, or other animals. They have a complete digestive system, with the mouth at the anterior end and the anus at the rear.

The annelid body is highly segmented: the body wall muscles and some organs, including respiratory surfaces; parts of the nervous, circulatory, and excretory systems; and the coelom are divided into similar repeating units (**Figure 27.29**). Body segments are separated by transverse partitions called **septa** (singular, *septum*). The digestive system and major blood vessels are not segmented and run the length of the animal.

The body wall muscles of annelids have both circular and longitudinal layers. Alternate contractions of these muscle groups allow annelids to make directed movements, using the coelom as a hydrostatic skeleton. The outer covering of annelids is a flexible cuticle that grows with the animal; it is not moulted. All annelids except leeches also have chitin-reinforced bristles, called **setae** (sometimes written *chaetae*; singular, *seta*), which protrude outward from the body wall. Setae anchor the worm against the substrate, providing traction.

Annelids have a closed circulatory system. The blood of most annelids contains hemoglobin or another oxygen-binding pigment. Oxygen diffusing across the cuticle may be picked up by capillaries in the skin to be transported to the tissues.

The excretory system is composed of paired **metanephridia** (singular, *metanephridium*) (see Figure 27.29d and Chapter 50), which usually occur in all body segments posterior to the head. The nervous system is well developed, with local control centres (ganglia) in every segment; a simple brain in the head; and sensory organs that detect chemicals, moisture, light, and touch.

Most freshwater and terrestrial annelids are hermaphroditic, and worms exchange sperm when they mate. Newly hatched worms have an adult morphology.

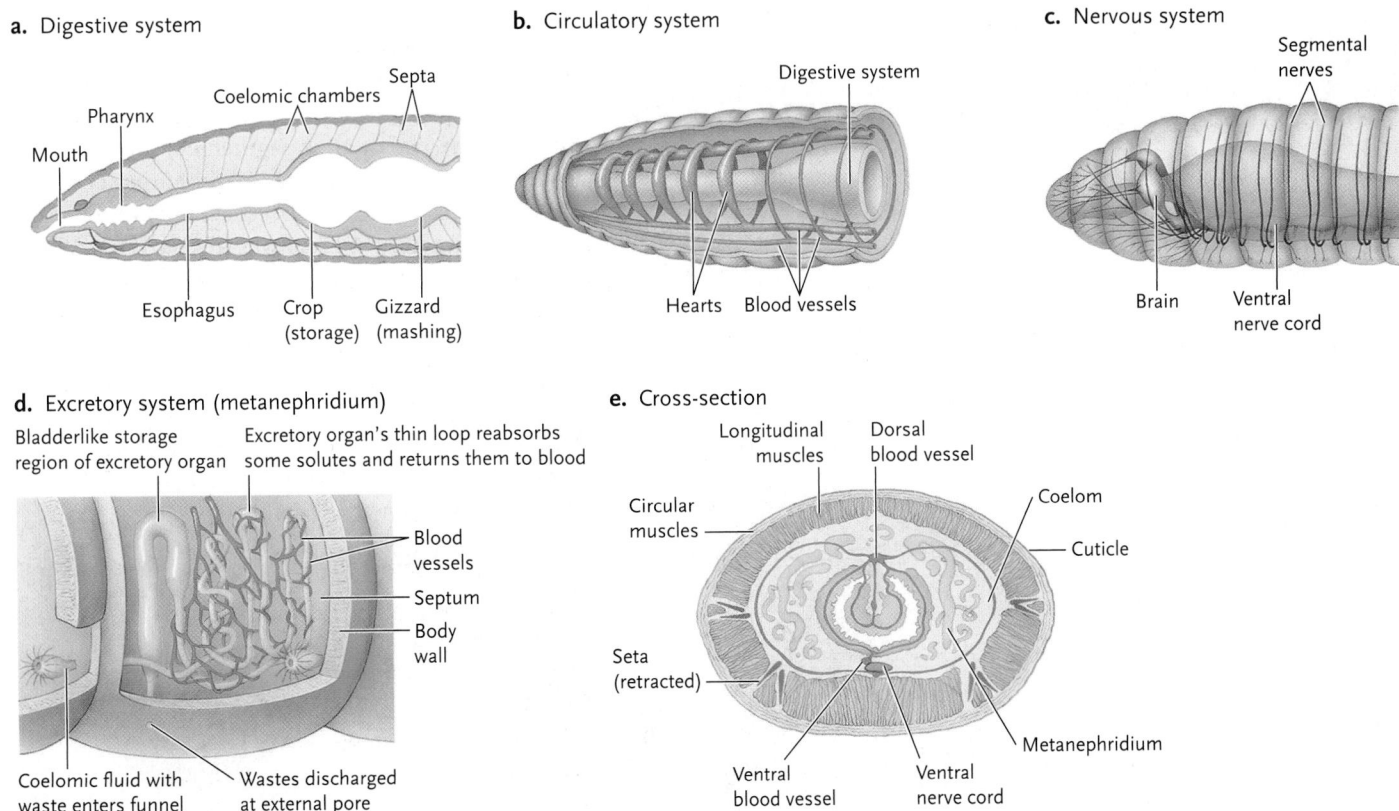

a. Digestive system

Septa
Coelomic chambers
Pharynx
Mouth
Esophagus
Crop (storage)
Gizzard (mashing)

b. Circulatory system

Digestive system
Hearts   Blood vessels

c. Nervous system

Segmental nerves
Brain   Ventral nerve cord

d. Excretory system (metanephridium)

Bladderlike storage region of excretory organ
Excretory organ's thin loop reabsorbs some solutes and returns them to blood
Blood vessels
Septum
Body wall
Coelomic fluid with waste enters funnel
Wastes discharged at external pore

e. Cross-section

Longitudinal muscles
Dorsal blood vessel
Circular muscles
Coelom
Cuticle
Seta (retracted)
Ventral blood vessel
Ventral nerve cord
Metanephridium

Figure 27.29

Segmentation in the phylum Annelida. Although the digestive system **(a)**, the longitudinal blood vessels **(b)**, and the ventral nerve cord **(c)** form continuous structures, the coelom **(a)**, blood vessels **(b)**, nerves **(c)**, and excretory organs **(d)** appear as repeating structures in most segments. The body musculature **(e)** includes both circular and longitudinal layers that allow these animals to use the coelomic chambers as a hydrostatic skeleton.

Some terrestrial annelids also reproduce asexually by fragmenting and regenerating missing parts. Marine annelids usually have separate sexes and release gametes into the sea for fertilization. The zygotes develop into trochophore larvae that add segments, gradually assuming an adult form.

Annelids are divided into three classes.

Class Polychaeta. The 10 000 species of bristle worms (Polychaeta; *poly* = many; *chaeta* = bristles) are primarily marine **(Figure 27.30)**. Many live under rocks or in tubes constructed from mucus, calcium carbonate secretions, grains of sand, and small shell fragments. Their setae project from well-developed **parapodia** (singular, *parapodium* = closely resembling a foot), fleshy lateral extensions of the body wall used for locomotion and gas exchange. Sense organs are concentrated on a well-developed head.

Many crawling or swimming polychaetes are predatory, using sharp jaws in a protrusible muscular pharynx to grab small invertebrate prey. Other species graze on algae or scavenge organic matter. A few tube dwellers draw food-laden water into the tube by beating their parapodia; most others collect food by extending feathery, ciliated, mucus-coated tentacles.

Class Oligochaeta. Most of the 3500 species of oligochaete worms (*oligo* = few) are terrestrial **(Figure 27.31)**, but they are restricted to moist habitats because they quickly dehydrate in dry air or soil. They range in length from a few millimetres to more than 3 m. Terrestrial oligochaetes, the earthworms, are nocturnal, spending their days in burrows that they excavate. They are important scavengers, assisting in mixing and aerating soil and converting plant and animal debris to nutrients useful to plants. Aquatic species live in mud or detritus at the bottom of lakes and rivers. Earthworms have complex organ systems (see Figure 27.29, p. 617), and they sense light and touch at both ends of the body. In addition, they have moisture receptors, an important adaptation in organisms that must stay wet to allow gas exchange across the skin.

Class Hirudinea. Most of the 500 species of leeches (*hirudo* = leech) live in fresh water and suck the blood of vertebrates. These blood feeders have dorsoventrally flattened, tapered bodies with a sucker at each end. Although the body wall is segmented, the coelom is reduced and not partitioned. About a quarter of the known species are not blood feeders but prey on other invertebrates. Almost all of the leeches live in fresh water, but a few marine species are known. Some leeches are terrestrial, living in the moist tropics and feeding on warm-blooded vertebrates.

a. Feather duster worm

orlandin/Shutterstock.com

b. Polychaete feeding structures

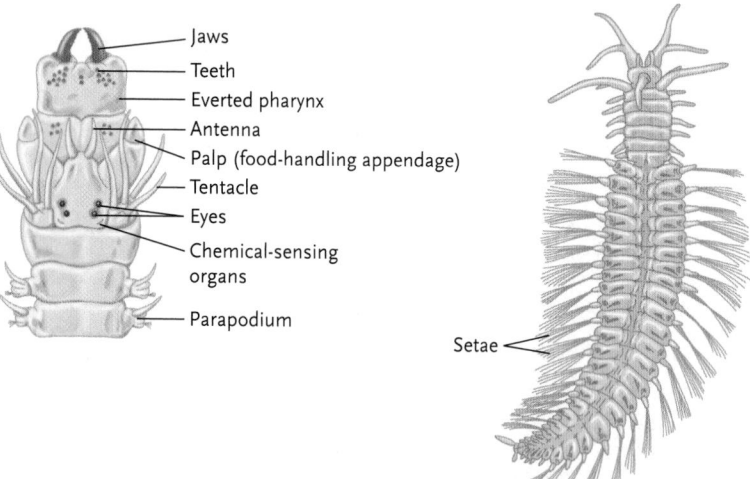

Jaws
Teeth
Everted pharynx
Antenna
Palp (food-handling appendage)
Tentacle
Eyes
Chemical-sensing organs
Parapodium

c. Polychaete setae

Setae

Figure 27.30

Polychaeta. **(a)** The tube-dwelling feather duster worm (*Sabella melanostigma*) has mucus-covered tentacles that trap small food particles. **(b)** Some polychaetes, such as *Nereis*, actively seek food; when they encounter a suitable tidbit, they evert their pharynx, exposing sharp jaws that grab the prey and pull it into the digestive system. **(c)** Many marine polychaetes (such as *Proceraea cornuta*, shown here) have numerous setae, which they use for locomotion.

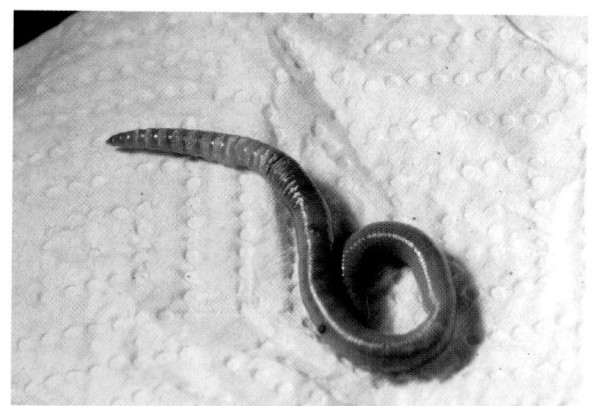

M.B. Fenton

Figure 27.31

Oligochaeta. Earthworms (genus *Lumbricus*) generally move across the ground surface at night.

Hydrothermal Vents

Hydrothermal vents on the sea bed, discovered in 1970, are equivalent to miniature undersea volcanoes. Superheated water emerges from them at temperatures up to 400°C, laden with sulfides **(Figure 1)**. The Endeavour Hot Vent Area, over 2000 m deep, lies 250 km off the coast of Vancouver Island. It has been explored by Verena Tunnicliffe at the University of Victoria and was declared a Marine Protected Area by the Canadian government in 2003. A range of invertebrates, particularly molluscs and annelids, flourish near the vents. Many of these species are new to science, and they are larger and more numerous than the fauna nearby. They experience at least brief exposure to temperatures as high as 50°C. More important, however, is the absence of sunlight, which deprives them of a source of food from plants, and the presence of sulfides. Sulfide is normally toxic, but associated with the vents are mats of bacteria that utilize sulfide for energy production and growth. The metazoans feed on these, and some of the invertebrates have symbiotic bacteria that rely on the sulfides.

Figure 1
A hot vent smoker, surrounded by organisms specific to the environment. Inset: Tube worms, originally placed in a separate Phylum Vestimentifera. Molecular analysis places them among the annelids.

Canadian Scientific Submersible Facility

Blood-feeding leeches attach to the host with the posterior sucker and use their sharp jaws on the anterior sucker to make a small, often painless, triangular incision. A sucking apparatus draws blood from the prey, and a special secretion prevents the host's blood from coagulating. Leeches have a highly branched gut that allows them to consume huge blood meals **(Figure 27.32)**. For centuries, doctors used medicinal leeches (*Hirudo medicinalis*) to "bleed" patients; today, surgeons still use them to drain excess fluid from tissues after reconstructive surgery, reducing swelling until the patient's blood vessels regenerate and resume this function.

Leech before feeding

Leech after feeding

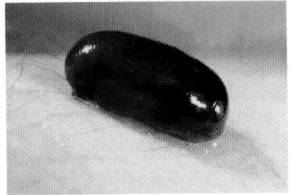

M. B. Fenton

Oxford Scientific/Getty Images

Figure 27.32
Hirudinea. Parasitic leeches consume huge blood meals, as shown by these before and after photos of a medicinal leech (*Hirudo medicinalis*). Because suitable hosts are often hard to locate, gorging allows a leech to take advantage of any host it finds.

STUDY BREAK

1. What characteristic reveals the close evolutionary relationship of ectoprocts, brachiopods, and phoronid worms?
2. Describe the three regions of the mollusc body.
3. Which organ systems exhibit segmentation in most annelid worms?

27.7 Ecdysozoan Protostomes

The three phyla in the protostome group Ecdysozoa all have an external cuticle secreted by epidermal cells. The cuticle serves as protection from harsh environmental conditions and helps parasitic species resist host defences. It also permits these animals to change the nature of the covering, which is important if the life stages live in different environments. Although many of these animals live in aquatic or moist terrestrial habitats, a tough exoskeleton allows many, particularly the insects, to thrive on dry land.

27.7a Phylum Nematoda

Members of the phylum Nematoda (*nemata* = thread) are round worms, often tapered at each end. The numerous species of free-living worms are

Figure 27.33

Phylum Nematoda. Many roundworms are animal parasites, like these *Anguillicola crassus*, shown here inside the swim bladder of an eel.

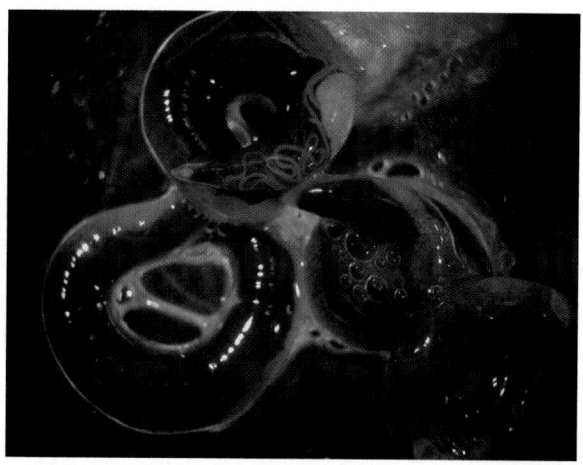

microscopic, reaching a size of at most a few millimetres. Some parasitic species, however, are larger **(Figure 27.33)**, and some are very large. The record is held by *Placentonema gigantissima*, a parasite of the placenta of the sperm whale: it may reach 9 m in length! Although superficially very similar in morphology, nematodes have achieved remarkable diversity. One species lives only in vinegar, and another is found in beer vats. In marine sediments, concentrations of a million or more worms per square metre have been reported. A hectare of farm soil may contain a billion or more nematodes. A single rotting fruit on the ground will contain tens of thousands of worms. Many species are parasitic in plants and animals, and some cause serious diseases in humans. Although fewer than 20 thousand species have been described, it is generally agreed that the number of living species is at least 100 thousand.

The nematode cuticle, often complex in structure, is composed of collagen-like proteins secreted by an epidermis that is often syncytial **(Figure 27.34)**. The cuticle is replaced four times during the life of the nematode, and in some cases, the characteristics of each of the cuticles differ. Moulting of the cuticle is not

necessary for the worms to increase in size, and growth usually occurs between moults. *Ascaris lumbricoides,* a common intestinal parasite in humans, represents an extreme case, growing in length from about 6 mm after the final moult to an adult of more than 20 cm. Nematodes, like rotifers, exhibit eutely, having few or no cell divisions in somatic cells after hatching.

This characteristic, together with a transparent cuticle, has made *Caenorhabditis elegans,* which has fewer than 1000 cells, a very useful model for studying development. It eats bacteria such as *E. coli* and can be reared easily in the lab. Because the number of cells is so small, the developmental fate of each cell in the embryo has been documented using techniques such as microinjection and laser microsurgery. The genome is small and has been completely sequenced, yielding about 17 thousand genes. *C. elegans* is widely used as a model to explore general questions in developmental biology. For example, Dr. David Baillie at Simon Fraser University in Vancouver uses *C. elegans* to ask questions such as how many of the genes are essential to normal development. He has determined which mutations are lethal and estimates that about 4000 to 5000 genes of the 17 thousand are essential for development in *C. elegans.*

Growth in nematodes occurs by an increase in cell size, and in a large nematode such as *Ascaris,* a muscle cell may be more than a centimetre in length. There are no cilia or flagella in nematodes; the spermatozoa move by amoeboid motion. A single layer of muscles forms part of the body wall. These muscles, like those of flatworms, do not receive nerves but make contact with the ventral, dorsal, or lateral nerve cords by long extensions of the muscle (see Figure 27.34).

The nervous system consists of dorsal, lateral, and particularly prominent ventral nerve cords, with a nerve ring surrounding the pharynx at the anterior. Nematodes respond to various chemicals, likely through sense organs located in pits at the anterior end. Some are sensitive to light, possibly through a general sensitivity of their nerves, but others have pigmented eye spots.

The gut of nematodes is a simple tube consisting of a single layer of epithelial cells. There are no muscles surrounding the gut, and food is propelled through the digestive system by a muscular pharynx. The pumping action of the pharynx is also responsible for maintaining pressure in the pseudocoelom. The resulting stiffness produces a hydrostatic skeleton. The cuticle has some elasticity, so that a contraction of muscles in the dorsal part of the worm results in an expansion on the ventral side. Alternate contraction and relaxation produces the dorsoventral wave that characterizes the movement of most nematodes.

The sexes of most nematodes are separate, and fertilization is internal. A few nematodes are hermaphroditic and may be self-fertilizing. Others may

Figure 27.34

Cross-section of a female *Ascaris*, a typical nematode.

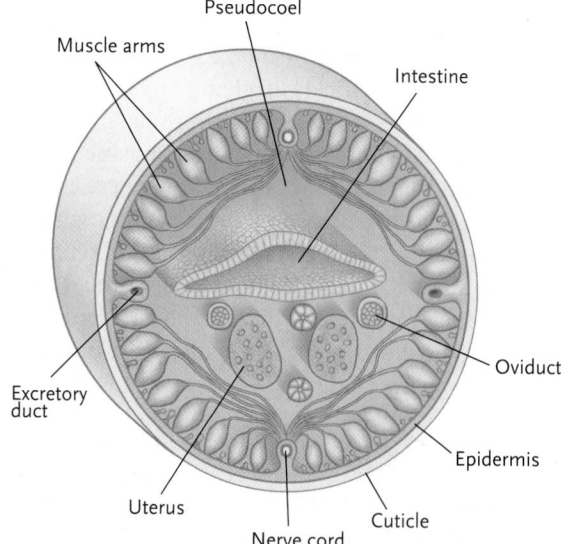

Pseudocoel
Muscle arms
Intestine
Oviduct
Epidermis
Cuticle
Nerve cord
Uterus
Excretory duct

be parthenogenetic. Fertilized females, particularly those of parasitic species, produce huge numbers of eggs. The intestinal parasite *A. lumbricoides* may produce 200 thousand eggs per day for about 10 months.

Nematodes are particularly successful as parasites and have invaded representatives of most other phyla. *A. lumbricoides* infects about a quarter to a third of the entire human population. Most wild vertebrates harbour one or more species of nematode parasites. The replaceable cuticle contributes to this success for it allows the worm to produce a different cuticle in each different environment that it encounters as it moves from a free-living form to a parasitic form or from one host to another.

27.7b Phylum Onychophora

The 65 living species of velvet worms (Onychophora; *onux* = claw; *phor* = to bear) live under stones, logs, and forest litter in moist temperate and tropical habitats in the southern hemisphere. They range in size from 15 mm to 15 cm and feed on small invertebrates and plants. Living onychophorans are all terrestrial, but fossils are known from marine environments.

Onychophorans have a flexible cuticle, superficially segmented bodies, and numerous pairs of unjointed legs **(Figure 27.35)**. Like annelids, they have pairs of excretory organs in most segments. But unlike annelids, no internal septa separate the segments; they have an open circulatory system, a specialized respiratory system similar to that of insects, and relatively large brains, jaws, and tiny claws on their feet. Many produce live young, which, in some species, are nourished within a uterus (see Chapter 39). The sexes are separate, and fertilization is internal.

Fossil onychophorans are known from the Cambrian (they are represented in the Burgess Shale), and the body plan has not changed much since then, suggesting that this highly specialized group of animals represented one of the successes in the experiments of the Cambrian speciation.

27.7c Phylum Arthropoda

If the Mesozoic was the age of the dinosaurs, we are living in the age of the arthropods (*arthros* = joint; *poda* = feet). About three-quarters of all living species of animals are arthropods, a phylum that includes insects, spiders, scorpions, crustaceans, centipedes, millipedes, and the extinct trilobites.

Arthropods have a segmented body encased in a rigid **exoskeleton.** This external covering is a complex of chitin (see "Molecule behind Biology," Box 46.1, p. 1145) and glycoproteins. In some marine and freshwater groups, such as crabs and lobsters, the cuticle is hardened with calcium carbonate. In terrestrial forms, such as insects, a surface layer of wax provides protection from dehydration. The exoskeleton is thin and flexible at the joints between body segments and at the joints of appendages. Contractions of muscles attached to the exoskeleton move individual body parts like levers, allowing highly coordinated movements and patterns of locomotion.

Although the exoskeleton has obvious advantages, it is nonexpandable and could limit the growth of the animal. Arthropods periodically develop a new cuticle beneath the old one, which they shed in the complex process of ecdysis **(Figure 27.36)**. The new cuticle is soft

Figure 27.36
Ecdysis in insects. Like all other arthropods, this cicada (*Graptopsalatsia nigrofusca*) sheds its old exoskeleton as it grows.

Figure 27.35
Phylum Onychophora. Members of the small phylum Onychophora, such as species in the genus *Dnycophor*, have segmented bodies and unjointed appendages.

and usually pleated, allowing for expansion after ecdysis. After shedding the old cuticle, arthropods swell with water or air before the new one hardens. They are especially vulnerable to predators at these times.

Primitively, each segment had a pair of lateral appendages, often specialized for locomotion, gas exchange, eating, or reproduction. As arthropods evolved, however, body segments became grouped in various ways. Each region, along with its highly modified paired appendages, is specialized, and the structure and function of the regions vary greatly among groups.

The coelom of arthropods is greatly reduced, and another cavity, the hemocoel, is filled with bloodlike hemolymph. The heart pumps the hemolymph through an open circulatory system, bathing tissues directly.

Because the hardened cuticle does not permit the easy passage of O_2 and CO_2, arthropods have specialized mechanisms for gas exchange. Marine and freshwater species, such as crabs and lobsters, rely on diffusion across gills that are specialized appendages, usually assisted by currents established by the appendages. The terrestrial groups have developed unique respiratory systems (see Chapter 49).

Many arthropods are equipped with a highly organized central nervous system, touch receptors, chemical sensors, image-forming **compound eyes**, and, in some, hearing organs. These are described in Chapters 44 and 45.

The phylogeny of this huge and diverse phylum has been a difficult and disputed subject for many years. Hexapods, and specifically insects, have been regarded by some as most closely related to myriapods, based largely on shared anatomical characters such as Malpighian tubules for excretion and one pair of antennae. Others, however, relate insects more closely to Crustacea, based on other morphological characteristics such as similarities in mouthparts and walking appendages. Molecular studies, including analysis of mitochondrial DNA and *Hox* genes, support the view that hexapods and crustacea are paraphyletic (see Chapter 20, for a definition of paraphyletic) and that myriapods and chelicerates form a separate paraphyletic grouping. This is an active area in research, and other hypotheses may be developed.

We follow the traditional definition of five *subphyla*, partly because this classification adequately reflects arthropod diversity and partly because no alternative hypothesis has been widely adopted by experts.

Subphylum Trilobita. The trilobites (*tri* = three; *lob* = lobed), now extinct, were among the most numerous animals in the shallow Paleozoic seas. Most were ovoid, dorsoventrally flattened, and heavily armoured, with two deep longitudinal grooves that divided the body into one median and two lateral lobes **(Figure 27.37)**. The head included a pair of sensory

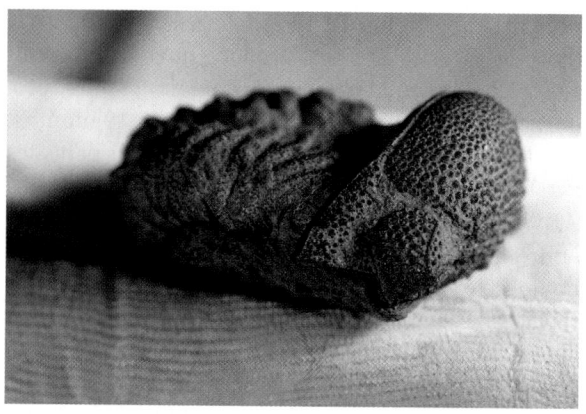

Figure 27.37
Subphylum Trilobita. Trilobites, such as the *Olenellus gilberti*, bore many pairs of relatively undifferentiated appendages.

antennae and compound eyes, and the segmented thorax and abdomen had pairs of identical appendages, each with two branches. The inner branch was used for locomotion, and the outer, consisting of a number of fine filaments, was used as a gill or in filter feeding.

The position of trilobites in the fossil record indicates that they were among the earliest arthropods. Thus, biologists are confident that their three body regions and unspecialized appendages represent ancestral traits in the phylum. Although there were numerous species, indicating a high degree of success, trilobites disappeared in the Permian mass extinction.

Subphylum Chelicerata. In spiders, ticks, mites, scorpions, and horseshoe crabs (subphylum Chelicerata; *cheol* = claw; *cera* = horn), the first pair of appendages, the **chelicerae**, are fanglike structures used for biting prey. The second pair of appendages, the **pedipalps**, serve as grasping organs, sensory organs, or walking legs. All chelicerates have two major body regions, the **cephalothorax** (a fused head–thorax) and the **abdomen**. The group originated in shallow Paleozoic seas, but most living species are terrestrial. They vary in size from less than 1 mm to 20 cm; all are predators or parasites.

The 60 thousand species of spiders, scorpions, mites, and ticks (class Arachnida) represent the vast majority of chelicerates **(Figure 27.38)**. Arachnids have four pairs of walking legs on the cephalothorax and highly modified chelicerae and pedipalps. In some spiders, males use their pedipalps to transfer packets of sperm to females. Scorpions use them (the "claws") to shred food and to grasp one another during courtship. Many predatory arachnids have excellent vision, provided by up to four pairs of simple eyes on the cephalothorax. Scorpions and some spiders also have unique pocket like respiratory organs called **book lungs** (see Chapter 49), derived from abdominal appendages.

a. Wolf spider

b. Spider anatomy

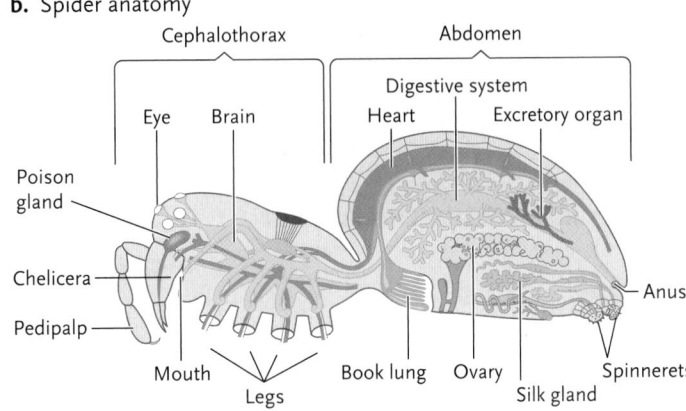

Cephalothorax | Abdomen

Digestive system
Heart | Excretory organ
Eye | Brain
Poison gland
Chelicera
Pedipalp
Mouth
Legs
Book lung | Ovary
Silk gland
Anus
Spinnerets

c. Scorpion

d. House dust mite

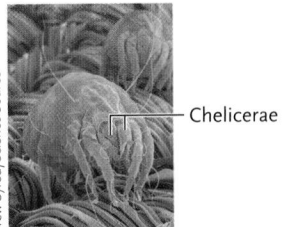

Chelicerae

Figure 27.38
Subphylum Chelicerata, Class Arachnida. **(a)** The wolf spider (*Lycosa* species) is harmless to humans. **(b)** The arachnid body plan includes a cephalothorax and an abdomen. **(c)** Scorpions have a stinger at the tip of the segmented abdomen. Many, such as *Centruroides sculpuratus*, protect their eggs and young. **(d)** House dust mites (*Dermatophagoides pteronyssinus*), shown in a scanning electron micrograph, feed on microscopic debris.

Like most other arachnids, spiders subsist on a liquid diet. They use their chelicerae to inject paralyzing poisons and digestive enzymes into prey and then suck up the partly digested tissues. Many spiders are economically important predators, helping to control insect pests. Only a few are a threat to humans. The toxin of a black widow (*Latrodectus mactans*) causes paralysis, and the toxin of the brown recluse (*Loxosceles reclusa*) destroys tissues around the site of the bite.

Although many spiders hunt actively, others capture prey on silken threads secreted by **spinnerets**, which are modified abdominal appendages. Some species weave the threads into complex, netlike webs. The silk is secreted as a liquid protein but quickly polymerizes. Spiders also use silk to make nests, to protect their egg masses, as a safety line when moving through the environment, and to wrap prey for later consumption. Spider silk is extremely tough, and the material from some spiders exceeds the tensile strength of steel. It is also highly elastic. These properties have led to proposals for its use in fabrics. A Canadian company has developed transgenic goats that produce spider silk in their milk.

Most mites are tiny, but they have a big impact. Some are serious agricultural pests that feed on plant sap. Others cause mange (patchy hair loss) or painful and itchy welts on animals. House dust mites, which feed on the dried skin cast off by humans, cause allergic reactions in many people. Ticks, which are generally larger than mites, are blood feeders that often transmit pathogens, such as those causing Rocky Mountain spotted fever and Lyme disease.

The subphylum Chelicerata also includes five species of horseshoe crabs (class Merostomata), an ancient lineage that has not changed much over its 350-Ma history **(Figure 27.39)**. Horseshoe crabs are carnivorous bottom feeders in shallow coastal waters. Beneath their characteristic shell, they have one pair of chelicerae; a pair of pedipalps; four pairs of walking legs; and a set of paperlike gills, derived from ancestral walking legs. A component of horseshoe crab blood is important to the pharmaceutical industry, where it is used to test for the presence of endotoxins resulting from bacterial contamination during manufacture.

Subphylum Crustacea. The 35 thousand species of shrimps, lobsters, crabs, and their relatives in the subphylum Crustacea (*crusta* = shell) represent a lineage that emerged more than 500 Ma ago **(Figure 27.40, p. 624)**. They are abundant in marine and freshwater habitats. A few species, such as sowbugs and pillbugs, live in moist, sheltered terrestrial environments. In many crustaceans, two and, in some cases, all three of

Figure 27.39
Marine chelicerates. Horseshoe crabs, such as *Limulus polyphemus*, are included in the Merostomata.

a. Crab

b. Lobster

c. Lobster anatomy

Eyes (one pair)
Fused segments of cephalothorax
Segmented abdomen
Antennae (two pairs)
Carapace
Maxillipeds (three pairs)
Telson
Cheliped
Swimmerets
Uropods
Four pairs of walking legs
Sperm-transfer appendage

Figure 27.40

Decapod crustaceans. **(a)** Crabs, such as this ghost crab in the genus *Ocypode*, and **(b)** lobsters (*Homarus americanus*) are typical decapod crustaceans. The abdomen of a crab is shortened and wrapped under the cephalothorax, producing a compressed body. **(c)** Lobsters bear 19 pairs of distinctive appendages; one pair of mandibles and two pairs of maxillae are not illustrated in this lateral view.

the arthropod body regions (head, **thorax**, and abdomen) may be fused. Fusion of the head and thorax into a cephalothorax is a common pattern. In some, the exoskeleton forms a **carapace**, a protective covering that extends backward from the head. Crustaceans vary in size from water fleas less than 1 mm long to lobsters that can grow to 60 cm in length and weigh as much as 20 kg.

Crustaceans generally have five characteristic pairs of appendages on the head (see Figure 27.40c). Most have two pairs of sensory antennae and three pairs of mouthparts. The latter include one pair of **mandibles**, which move laterally to bite and chew, and two pairs of **maxillae** (singular, *maxilla*), which hold and manipulate food. Numerous paired appendages posterior to the mouthparts vary among groups. Ancestrally, crustacean appendages were divided into two branches at the base, but many living species have unbranched appendages.

Most crustaceans are active animals that exhibit complex movements during locomotion and in the performance of other behaviours. These activities are coordinated by elaborate sensory and nervous systems, including chemical and touch receptors in the antennae, compound eyes, statocysts on the head, and sensory hairs embedded in the exoskeleton throughout the body. The nervous system is similar to that in annelids, but the ganglia, particularly those forming the brain, are larger and more complex. Larger species have complex, feathery gills derived from appendages tucked beneath the carapace. Metabolic wastes such as ammonia are excreted by diffusion across the gills or, in larger species, by **antennal glands**, located in the head.

The sexes are typically separate, and courtship rituals are often complex. Eggs are usually brooded on the surface of the female's body or beneath the carapace. Many have free-swimming larvae that, after undergoing a series of moults, gradually assume an adult form.

The subphylum includes so many different body plans that it is usually divided into six classes with numerous subclasses and orders. The crabs, lobsters, and shrimps (class Malacostraca, order Decapoda; *deka* = 10; *poda* = foot) number more than 10 thousand species. The vast majority of decapods are marine, but a few shrimps, crabs, and crayfishes occupy freshwater habitats. Some crabs also live in moist terrestrial habitats, where they scavenge dead vegetation, clearing the forest floor of debris.

All decapods exhibit extreme specialization of their appendages. In the American lobster, for example, each of the 19 pairs of appendages is different (see Figure 27.40c). Behind the antennae, mandibles, and maxillae, the thoracic segments have three pairs of maxillipeds, which shred food and pass it up to the mouth; a pair of large chelipeds (pinching claws); and four pairs of walking legs. The abdominal appendages include a pair specialized for sperm transfer (in males only); swimmerets for locomotion and for brooding eggs; and uropods, which, in combination with the telson (the tip of the abdomen), make a fan-shaped tail.

Representatives of several crustacean classes— fairy shrimps, amphipods, water fleas, ostracods, and copepods **(Figure 27.41)**—live as plankton in the upper waters of oceans and lakes. Most are only a few millimetres long but are present in huge numbers. They feed on microscopic algae or detritus and are themselves food for larger invertebrates, fishes, and some suspension-feeding marine mammals such as the baleen whales. Planktonic crustaceans are among the most abundant animals on Earth. The total biomass of a single species, *Euphausia superba*, is estimated at 500 million tonnes, more than the total mass of humans.

Adult barnacles (class Maxillopoda, subclass Cirripedia; *cirrus* = curl of hair; *poda* = foot) are sessile marine crustaceans that live within a strong, calcified, cup-shaped shell **(Figure 27.42)**. Their free-swimming larvae attach permanently to substrates—rocks, wooden pilings, the hulls of ships, the shells of molluscs, and

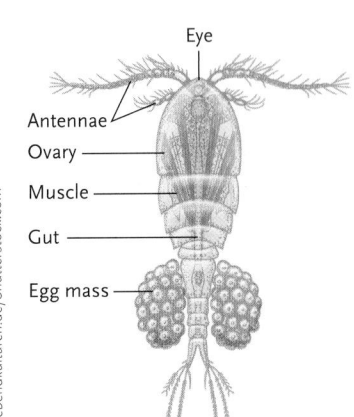

Figure 27.41

Copepods. Tiny crustaceans, such as this copepod (*Calanus* species on the left, *Cyclops* species on the right), occur by the billions in freshwater and marine plankton.

Eye
Antennae
Ovary
Muscle
Gut
Egg mass

Fleshy stalk
Ovary
Testis
Midgut
Digestive gland
Anus
Penis
Muscle (cut)
Mouth
Shell
Food-collecting legs

Figure 27.42

Barnacles. Gooseneck barnacles (*Lepas anatifera*) attach to the underside of floating debris. Like other barnacles, they open their shells and extend their feathery legs to collect particulate food from seawater.

even the skin of whales—and secrete the shell, which is a modified exoskeleton. To feed, barnacles open the shell and extend six pairs of feathery legs. The beating legs capture microscopic plankton and transfer it to the mouth. Unlike most crustaceans, barnacles are hermaphroditic.

Subphylum Myriapoda. The 3000 species of centipedes (class Chilopoda) and 10 thousand species of millipedes (class Diplopoda) are classified together in the subphylum Myriapoda (*murias* = 10 thousand *poda* = foot). Myriapods have two body regions: a head and a segmented trunk **(Figure 27.43)**. The head bears one pair of antennae, and the trunk bears one (centipedes) or two (millipedes) pairs of walking legs on most of its many segments. Myriapods are terrestrial, and many species live under rocks or dead leaves. Centipedes are fast and voracious predators; they generally feed on invertebrates, but some eat small vertebrates. Although most species are less than 10 cm long, some grow to 25 cm. The millipedes are slow but powerful herbivores or scavengers. The largest species attain a length of nearly 30 cm.

Subphylum Hexapoda. The subphylum Hexapoda (*hex* = six) includes the class Insecta, as well as some other smaller classes. In terms of sheer numbers and

a.

b.

Figure 27.43

Millipedes and centipedes. **(a)** Millipedes, such as *Spirobolus* species, feed on living and decaying vegetation. They have two pairs of walking legs on most segments. **(b)** Like all centipedes, this one, shown feeding on a mouse, is a voracious predator. Centipedes have one pair of walking legs per segment.

diversity, the approximately 1 million species of insects are the most successful animals on Earth, occupying virtually every terrestrial and aquatic habitat. They were among the first animals to colonize terrestrial habitats, where most species still live. The oldest Hexapod fossils date from the Devonian, about 400 Ma ago, and the first insect fossils appeared shortly after. Insects are generally small, ranging from 0.1 mm to 30 cm in length. The class is divided into about 30 orders **(Figure 27.44)**.

The insect body plan always includes a head, a thorax, and an abdomen **(Figure 27.45)**. The head is equipped with multiple mouthparts, a pair of compound eyes, and one pair of sensory antennae. The thorax has three pairs of walking legs and often one or two pairs of wings. Adult insects are the only invertebrates capable of flight. The origin of wings is uncertain. The traditional view holds that they are new structures arising as outgrowths of the body wall. However, on the basis of both fossil and molecular evidence, one of the foremost researchers in the field, Jarmila Kukalova-Peck of Carleton University in Ottawa, maintains that wings are derived from branches of a proximal (near the body) segment of the leg.

Insects exchange gases through a specialized **tracheal system** (see Chapter 49), a branching network of tubes that carries oxygen from small openings in the exoskeleton to individual cells throughout the body.

Insects excrete nitrogenous wastes through specialized **Malpighian tubules** (see Chapter 50) that transport wastes to the digestive system for disposal with the feces. These two organ systems also appear in some of the terrestrial chelicerates. Since these are not paraphyletic with hexapods, this is another example of convergent evolution.

Insect sensory systems are diverse and complex. Besides a pair of image-forming compound eyes, many insects have light-sensing ocelli on their heads. Many also have hairs, sensitive to touch, on their antennae, legs, and other regions of the body. Chemical receptors are particularly common on the legs and feet, allowing the identification of food. Many groups of insects have sound receptors to detect predators and potential mates. The familiar chirping of crickets, for example, is a mating call emitted by males that may repel other males and attract females. The beetles of the family Lampyridae emit light signals from their abdomens to attract mates (see Chapter 44).

As a group, insects use an enormous variety of materials as food, and their mouthparts may be modified to reflect the nature of the food source **(Figure 27.46)**. The basic plan is reflected in a plant feeder such as a locust or a generalized feeder such as a cockroach. The *labrum* is an anterior flaplike extension of the front of the head that covers the mouthparts and has sensory structures. The mouthparts themselves are modified

a. Silverfish (Thysanura, *Ctenolepisma longicaudata*) are primitive wingless insects.

b. Dragonflies, like the flame skimmer (Odonata, *Libellula saturata*), have aquatic larvae that are active predators; adults capture other insects in mid-air.

c. Male praying mantids (Mantodea, *Mantis religiosa*) are often eaten by the larger females during or immediately after mating.

d. This rhinoceros beetle (Coleoptera, *Dynastes granti*) is one of more than 250 thousand beetle species that have been described.

e. Fleas (Siphonoptera, *Hystrichopsylla dippiei*) have strong legs with an elastic ligament that allows these parasites to jump on and off their animal hosts.

f. Crane flies (Diptera, *Tipula* species) look like giant mosquitoes, but their mouthparts are not useful for biting other animals; the adults of most species live only a few days and do not feed at all.

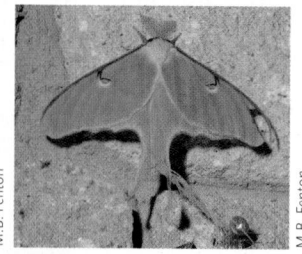

g. The luna moth (Lepidoptera, *Actias luna*), like other butterflies and moths, has wings that are covered with colourful microscopic scales.

h. Like many other ant species, fire ants (Hymenoptera, *Solenopsis invicta*) live in large cooperative colonies. Fire ants—named for their painful sting—were introduced into southeastern North America, where they are now serious pests.

Figure 27.44

Insect diversity. Insects are grouped into about 30 orders, 8 of which are illustrated here.

External anatomy of a grasshopper

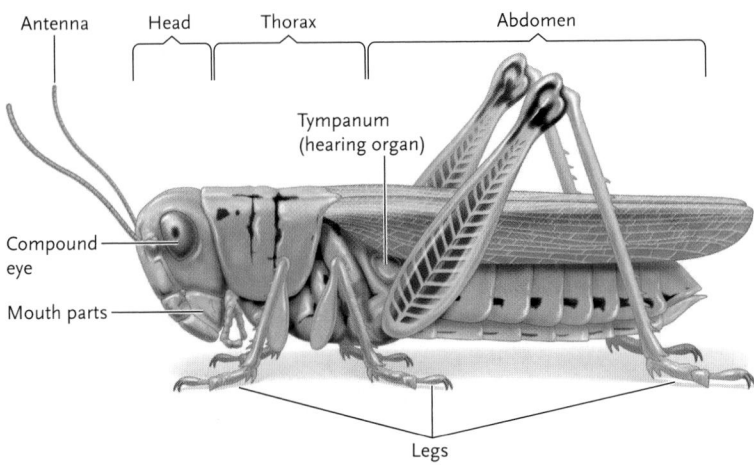

Internal anatomy of a female grasshopper

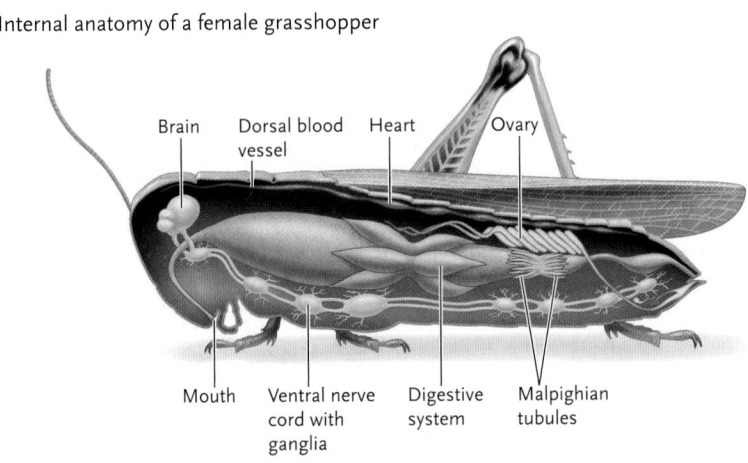

Figure 27.45
The insect body plan. Insects have a distinct head, thorax, and abdomen. Of all the internal organ systems, only the dorsal blood vessel, ventral nerve cord, and some muscles are strongly segmented.

appendages. The paired mandibles are chewing organs, and behind those are paired maxillae abundantly supplied with sense organs, particularly on its *palps* (jointed projections), which act to scoop the food. The most posterior is the *labium,* representing a fused

pair of appendages, which is well supplied with sensory structures and palps. This ancestral mandibulate pattern, with mouth parts representing three of the six segments that form the insect head, is modified in various ways to accommodate different modes of feeding. In some biting flies, such as mosquitoes, the mouthparts are piercing structures, with a narrow channel to suck up blood. In butterflies and moths, the mouthparts include a long proboscis to drink nectar. In houseflies, the mouthparts are adapted for sopping up food that has been moistened by its saliva.

Life on land requires internal fertilization (see Chapter 41). In insects, males may produce packets of sperm enclosed in spermatophores and insert them into the female ducts, or sperm transfer may be direct via a penis. Sperm are stored in the female until used to fertilize eggs at the time of egg laying. The eggs of most insects are covered with a waterproof shell before they are fertilized and have one or more minute pores to permit the entry of sperm.

Parthenogenesis occurs in a number of species. In aphids, not only are the females parthenogenetic at times when food plants are abundant, they also produce live young, and development is so telescoped that embryos within the mother already have embryos within their ovaries. This results in an enormously rapid increase in population when conditions are favourable. Under less favourable conditions, normal sexual reproduction occurs. In a few species, parthenogenesis is the only mode of reproduction and males are unknown.

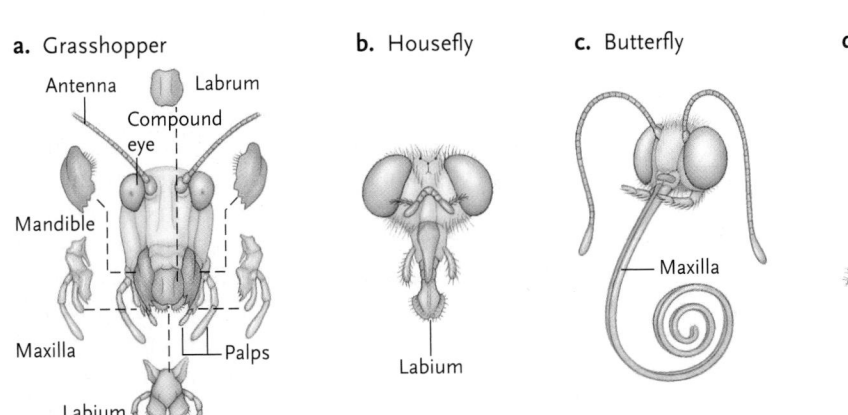

a. Grasshopper   b. Housefly   c. Butterfly   d. Mosquito

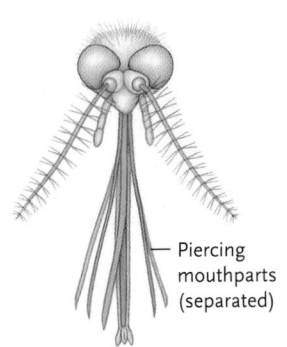

Figure 27.46
Specialized insect mouthparts. The **(a)** ancestral chewing mouthparts have been modified during evolution, allowing different insects to **(b)** sponge up food, **(c)** drink nectar, and **(d)** pierce skin to drink blood.

PEOPLE BEHIND BIOLOGY 27.2

V.B. Wigglesworth, *Cambridge University*

V.B. Wigglesworth was a British researcher who founded the subject of insect physiology. He had an active research career that spanned seven decades, from 1928 to 1991. He discovered the utility of the blood-sucking bug *Rhodnius prolixus* **(Figure 1)** for experimental work. Unfed, the insect remains in a state of suspended development. However, when it takes a blood meal, development to the next stage begins. Wigglesworth used this signal, together with clever surgical approaches **(Figure 2),** to establish the basic facts of the hormonal control of development in insects. Although he is perhaps best known for this work, he also established the basic facts of insect digestion, insect excretion, the operation of the Malpighian tubules, the operation of the tracheal system, and the properties of the cuticle. He used a keen sense of observation to identify appropriate experimental questions and devised and carried out clever experimental approaches to answer the questions he investigated.

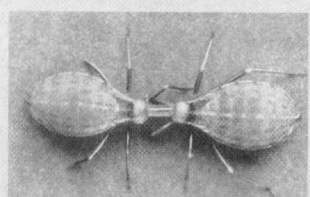

Plate II a opposite p. 45 of *The Physiology of Insect Metamorphosis* Issue 1 by V. B. Wigglesworth. Cambridge Univeresity Press, 1954. Reprinted with the permission of Cambridge University Press.

Figure 2
One of Wigglesworth's surgical procedures. The nymph on the right was decapitated within a day of feeding, before the hormones from the head governing moulting were secreted; therefore, it will not develop. The insect on the left was decapitated after the hormones were released and will develop normally. The two are joined so that their hemocoels are connected (a procedure called parabiosis). They will both initiate the formation of a new cuticle driven by hormones from the insect on the left.

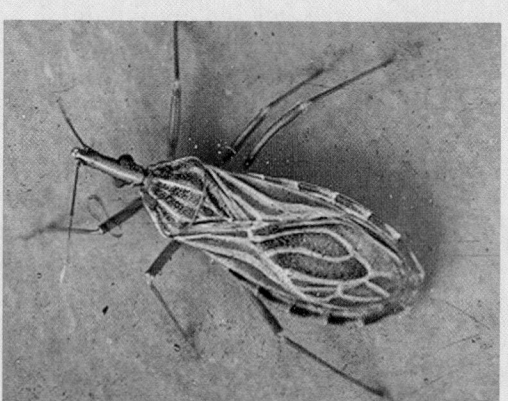

Ken Davey

Figure 1
The adult female of Rhodnius prolixus.

---

a. Incomplete metamorphosis

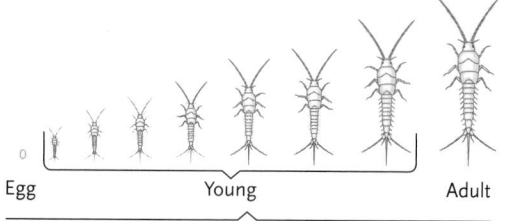

Egg   Young   Adult

Some wingless insects, like silverfish (order Thysanura), do not undergo a dramatic change in form as they grow.

b. Metamorphosis without a pupa

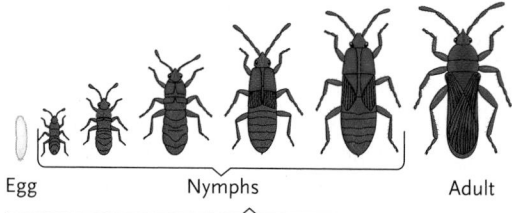

Egg   Nymphs   Adult

Some insects, such as the Order Hemiptera, undergo a metamorphosis that involves no major reorganization in form apart from the development of wings.

Figure 27.47
Patterns of postembryonic development in insects.

c. Metamorphosis with a pupa

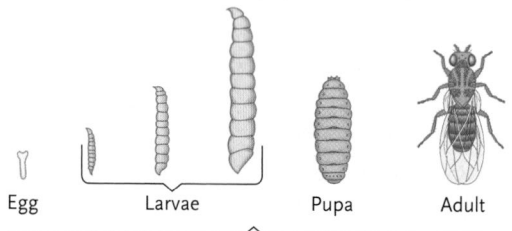

Egg   Larvae   Pupa   Adult

Fruit flies (order Diptera) and many other insects undergo a total reorganization of their internal and external anatomy when they pass through the pupal stage of the life cycle.

After it hatches from an egg, an insect passes through a series of developmental stages called **instars.** Several hormones control development and ecdysis, which marks the passage from one instar to the next. Insects exhibit one of three basic patterns of postembryonic development **(Figure 27.47).** Primitive, wingless species (order Thysanura) simply grow and shed their exoskeleton, undergoing only minimal changes in morphology. Early instars lack scales on their cuticle, and the appearance of scales corresponds

MOLECULE BEHIND BIOLOGY 27.3

Insect Juvenile Hormone

The juvenile hormone of insects is a family of hormones, each differing only slightly in structure **(Figure 1)**. Its existence was first demonstrated by the English researcher V.B. Wigglesworth (see "People behind Biology," Box 27.2) in 1948. Using surgical procedures, he showed that the corpus allatum, an endocrine organ just behind the brain, was the source of a factor governing metamorphosis. In the presence of the factor, the insect remained larval, and in its absence, metamorphosis occurred, leading to the adult insect. In the adult, the hormone governs egg production and other elements of reproduction. The precise structure of the first member of the family of molecules was not elucidated until 1967. Because juvenile hormone is an oil, it passes easily through the cuticle of insects **(Figure 2)**. This raised the possibility that mimics of the hormone might be useful as insecticides. Since the hormone has no obvious counterpart in other animals, it was argued that insecticides based on the hormone

Ken Davey

Figure 1
JH III, the most common of the family of molecules used as juvenile hormone in insects.

Figure 2
Juvenile hormone passes easily through the cuticle. In this experiment, Wigglesworth applied the appropriate concentration of the hormone to a localized area on the dorsal surface of a last-stage larva of Rhodnius prolixus *during the process of forming the adult cuticle. On the left, he applied the hormone to a single segment, and on the right, he applied it in the form of his initials. When the insects moulted to the adult, the new cuticle in the treated portions retained the characteristics of the larval cuticle, whereas the rest of the insect exhibited normal adult cuticle. This photo also shows that in some insects with "incomplete metamorphosis," the changes in morphology may be very great.*

should be safe. Some compounds have emerged as useful pesticides and have been particularly useful in controlling mosquito larvae in water bodies and fleas on pets.

---

to reproductive maturity. Moulting cycles may continue after reproductive maturity. Other species undergo what is often called **incomplete metamorphosis.** They hatch from the egg as a nymph, which lacks functional wings. In many species, such as grasshoppers (order Orthoptera), the nymphs resemble the adults. In other insects, such as dragonflies (order Odonata), the aquatic nymphs are morphologically very different from the adults. Even in insects following this developmental pattern, the adult form differs by more than the abrupt development of the wings. The nature or colour of the cuticle may differ (see "Molecule behind Biology," Box 27.3). The terminal segments are reorganized to produce the external genitalia. In general, however, the descendants of the cells present in the first instar produce these changes.

Most insects undergo **complete metamorphosis:** the larva that hatches from the egg differs greatly from the adult. Larvae and adults often occupy different habitats and consume different food. The larvae (caterpillars, grubs, or maggots) are often worm shaped, with chewing mouthparts. They grow and moult several times, retaining their larval morphology. Before

they transform into sexually mature adults, they spend a period of time as a sessile **pupa.** During this stage, most of the larval tissues are destroyed and replaced by groups of embryonic cells, called *disks,* that have been in place since hatching. Although these cells are not obviously differentiated, their developmental fate is determined. Thus, there are antennal disks, eye disks, wing disks, and so on. The process is fundamentally different from that in insects with incomplete metamorphosis. In the latter, existing cells are reprogrammed at the last moult to produce the adult form, whereas in insects with complete metamorphosis, entirely new cells, programmed during embryogenesis to produce adult tissues, are involved.

Moths, butterflies, beetles, wasps, and flies are examples of insects with complete metamorphosis. Their larval stages specialize in feeding and growth, whereas the adults are adapted for dispersal and reproduction. In some species, the adults never feed, relying on the energy stores accumulated during the larval stage. This mode of development has been highly successful. The four principal orders with a pupa—Lepidoptera, Coleoptera, Hymenoptera, and

Diptera—account for about two-thirds of all known species of animals.

The evolution of insects has been characterized by innovations in morphology, life cycle patterns, locomotion, feeding, and habitat use. Insects' well-developed nervous systems govern exceptionally complex patterns of behaviour, including parental care, a habit that reaches its zenith in the colonial social insects, the termites, ants, bees, and wasps (see "Polymorphic Development," p. 612). The factors that contribute to the insects' success also make them our most aggressive competitors. They destroy agricultural crops, stored food, wool, paper, and timber. They feed on blood from humans and domesticated animals, sometimes transmitting disease-causing pathogens such as malaria as they do so. Nevertheless, insects are essential members of terrestrial ecological communities. Many species pollinate flowering plants, including important crops. Many others attack or parasitize species that are harmful to human activities. Most insects are a primary source of food for other animals. Some make useful products, such as honey, shellac, beeswax, and silk, and many human cultures use them for food.

STUDY BREAK

1. What are the advantages of moulting in nematodes?
2. If an arthropod's rigid exoskeleton cannot be expanded, how does the animal grow?
3. How does the number of body regions differ among the four subphyla of living arthropods?
4. How do the life stages differ between insects that have incomplete metamorphosis and those that have complete metamorphosis?

Review

To access course materials such as Aplia and other companion resources, please visit www.NELSONbrain.com.

27.1 What Is an Animal?

- Animals are eukaryotic, multicellular organisms that are differentiated from plants by heterotrophy, motility, and direct contact between adjacent cells.
- Animals probably arose in the Precambrian from a hollow sphere of colonial flagellates that reorganized as a double-layered sac-within-a-sac.

27.2 Key Innovations in Animal Evolution

- Tissues, groupings of identical cells specialized to perform specific functions, are organized into two or three tissue layers: ectoderm; endoderm; and, in those animals with three layers, mesoderm. In some sponges, the specialized cells may be capable of dedifferentiation.
- Some animals exhibit radial symmetry; most exhibit bilateral symmetry. Bilaterally symmetrical animals have left and right sides, dorsal and ventral sides, and anterior and posterior ends.
- Acoelomate animals have no body cavity. Pseudocoelomate animals have a body cavity between the derivatives endoderm and mesoderm. Coelomate animals have a body cavity that is entirely lined by derivatives of mesoderm. The cavities are filled with fluid that separates and protects the organs and in some cases functions as a hydrostatic skeleton.
- Two lineages of animals differ in developmental patterns. Most protostomes exhibit spiral, determinate cleavage; the coelom (when present) is a schizocoelom; and the blastopore develops into the mouth. Deuterostomes have radial symmetry, indeterminate cleavage, and an enterocoelom, and their blastopore becomes the anus.
- The development of many protostomes includes a larval stage. This polymorphic development allows sessile animals to be distributed, permits parasitic forms to exist in widely different environments, and avoids competition between the young and the adults.
- Four animal phyla exhibit segmentation.

27.3 An Overview of Animal Phylogeny and Classification

- Sequence analyses of highly conserved structures such as rRNA, mitochondrial DNA, and DNA coding for specific proteins can be compared in various species. The closer the similarity, the more closely related the species are assumed to be. Phylogenetic trees based on such data have confirmed some relationships based on developmental and morphological data and challenged others.
- The Radiata includes animals with two tissue layers and radial symmetry, and the Bilateria includes animals with three tissue layers and bilateral symmetry.
- Bilateria is further subdivided into Protostomia and Deuterostomia. The phylogeny based on molecular evidence divides the Protostomia into the Lophotrochozoa and the Ecdysozoa.
- Molecular phylogeny suggests that ancestral protostomes had a coelom and that acoelomate and pseudocoelomate conditions were derived from the coelomate.
- Segmentation arose independently in three lineages: the annelids, the Onychophora/Arthropoda, and the Chordata.

27.4 Phylum Porifera

- Sponges (phylum Porifera) are asymmetrical animals, many with limited integration of cells in their bodies.

- The body of many sponges is a water-filtering system with incurrent pores, a spongocoel, and an osculum through which water exits the body. Flagellated choanocytes draw water into the body and capture particulate food.

27.5 Metazoans with Radial Symmetry

- The two major radiate phyla have two well-developed tissue layers with a gelatinous mesoglea between them. They lack organ systems but have well-developed nerve nets. All are aquatic or marine.

- The hydrozoans, jellyfishes, sea anemones, and corals (phylum Cnidaria) are predators that capture prey with tentacles and stinging nematocysts.

- The life cycles of cnidarians may include polyps, medusae, or both. Anthozoans lack a medusa stage, whereas in jellyfish (Scyphozoa and Cubozoa), medusae are prominent and hydroids may be absent; both are present in Hydrozoa.

- The small, translucent comb jellies (phylum Ctenophora) use long, sticky tentacles to capture particulate food. They are weak swimmers that use rows of cilia for locomotion.

27.6 Lophotrochozoan Protostomes

- The taxon Lophotrochozoa includes eight phyla that share either a characteristic type of larva or a specialized feeding structure.

- Flatworms (phylum Platyhelminthes) are either free living or parasitic. Free-living species have well-developed digestive, excretory, reproductive, and nervous systems. Parasitic flukes and tapeworms live within or upon animal hosts. They attach to hosts with suckers or hooks, and they produce numerous eggs. Some organ systems may be greatly reduced in parasitic species.

- The wheel animals (phylum Rotifera) are tiny and abundant inhabitants of freshwater and marine ecosystems. Movements of cilia in the corona control their locomotion and bring food to their mouths. Many are parthenogenetic.

- Three small phyla (Ectoprocta, Brachiopoda, and Phoronida) all use a lophophore to feed on particulate matter. Brachiopods live within a two-part shell; ectoprocts form flattened or branching colonies; and phoronids are small, usually tube-dwelling worms.

- The ribbon worms (phylum Nemertea) are elongate and often colourful animals with a proboscis housed in a unique structure, the rhynchocoel.

- Chitons, snails, clams, octopuses, and their relatives (phylum Mollusca) have fleshy bodies that are often enclosed in a hard shell. The molluscan body plan includes a head-foot, a visceral mass, and a mantle.

- Segmented worms (phylum Annelida) generally exhibit segmentation of the coelom and of the muscular, circulatory, excretory, respiratory, and nervous systems. Polychaetes have segmental appendages used in locomotion and gas exchange. Leeches have reduced segmentation.

27.7 Ecdysozoan Protostomes

- The taxon Ecdysozoa includes three phyla that periodically shed their cuticle.

- Roundworms (phylum Nematoda) feed on decaying organic matter or parasitize plants or animals. Locomotion depends on muscles contracting against a hydrostatic skeleton provided by a fluid-filled pseudocoel. Moulting is not essential for growth, but it permits changing the nature of the cuticle to accommodate different environments.

- The velvet worms (phylum Onychophora) have segmented bodies and unjointed legs. Some species bear live young, which develop in a uterus.

- The arthropods (phylum Arthropoda) are the most diverse animals on Earth. Their segmented bodies are often differentiated into distinct regions, and their jointed appendages are specialized for feeding, locomotion, or reproduction. They shed their firm, water-resistant exoskeleton to accommodate growth or to begin a new stage of the life cycle. Arthropods have an open circulatory system; numerous sense organs that provide input to a complex nervous system; and, in some groups, highly specialized respiratory and excretory systems.

- Arthropods are divided into five subphyla. The extinct trilobites (subphylum Trilobita), with three-lobed bodies and relatively undifferentiated appendages, were abundant in Paleozoic seas.

- Spiders, ticks, mites, scorpions, and horseshoe crabs (subphylum Chelicerata) have a cephalothorax and an abdomen; two pairs of appendages on the head serve in feeding.

- Lobsters, crabs, and their relatives (subphylum Crustacea) have a carapace that covers the cephalothorax, as well as highly modified appendages, including five pairs on the head.

- The centipedes and millipedes (subphylum Myriapoda) are largely terrestrial. They have a head and an elongate, segmented trunk.

- Insects and their relatives (subphylum Hexapoda) are also largely terrestrial. Insects have three body regions, three pairs of walking legs on the thorax, and a pair of antennae and three pairs of feeding appendages on the head.

- Most insects undergo metamorphosis. In incomplete metamorphosis, the cells of the nymph are reprogrammed to produce adult structures. In complete metamorphosis, an additional stage, the pupa, permits entirely new adult structures to replace larval cells.

Questions

Self-Test Questions

1. Which of the following characteristics is NOT typical of most animals?
   a. heterotrophic
   b. sessile
   c. radially symmetrical
   d. multicellular

2. Which term refers to a body cavity that separates the digestive system from the body wall but is NOT completely lined with mesoderm?
   a. schizocoelom
   b. mesentery
   c. peritoneum
   d. pseudocoelom

3. Which part of a mollusc secretes the shell?
   a. the visceral mass
   b. the trochophore
   c. the head-foot
   d. the mantle

4. Which of the following is NOT a result of polymorphic development?
   a. castes in social insects
   b. the alternation of generations in Cnidaria
   c. the pupal stage in insects
   d. the adult octopus

5. Ecdysis refers to a process in which
   a. bivalves use siphons to pass water across their gills
   b. arthropods and nematodes shed their cuticles
   c. cnidarians build skeletons of calcium carbonate
   d. squids escape from predators in a cloud of ink

6. The Burgess Shale contains fossils from the
   a. Precambrian   c. Devonian
   b. Pleistocene   d. Cambrian

7. Which of the following are NOT part of the cnidarian body plan?
   a. notochord   c. hydrozoa
   b. mantle   d. metazoa

8. Protostomes and deuterostomes differ markedly in which of the following?
   a. the pattern of embryological development
   b. the origin of the anus and the mouth
   c. the pattern of fertilization
   d. the presence of a mantle cavity

9. Which of the following does NOT belong to the Mollusca?
   a. tapeworms   c. limpets
   b. leeches   d. lobsters

10. In which of the following does the ventral solid nerve chord occur?
    a. Chordata   c. Platyhelminthes
    b. Cnidaria   d. Arthropoda

Questions for Discussion

1. Many invertebrate species are hermaphroditic. What selective advantages might this characteristic offer? In what kinds of environments might it be most useful?

2. In terms of numbers of species, insects are the dominant life form on Earth, but the individuals are also smaller in size than many other groups. What has contributed to their success, and why are they not larger?

3. The egg of the human parasite *A. lumbricoides* hatches in the small intestine. What experiments would you do to test whether this is a result of the egg shell being digested by the intestinal enzymes?

4. What is a parasite? In which groups do we find parasitic animals?

5. What role does the coelom play in the development of protostomes?

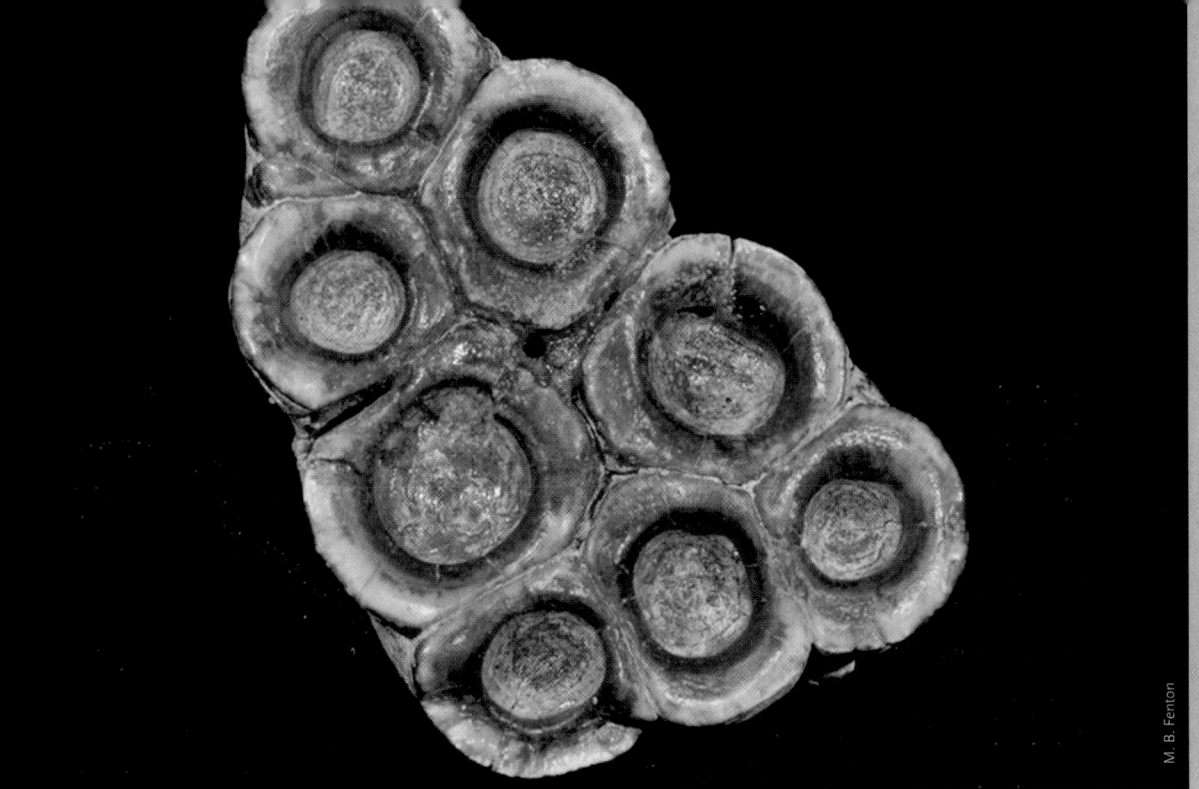

28

The molar tooth of an extinct mammal (Desmostylus) was used to crush and grind marine plants. Paleontologists discovered that Desmostylus was more walruslike than seacow (manatee)-like only when they found almost-complete skeletons.

Diversity of Animals 2: Deuterostomes: Vertebrates and Their Closest Relatives

WHY IT MATTERS

Based on molecular evidence, Xenoturbellida was identified as a new phylum in November 2006. This new phylum is closely related to the Chordata (traditional phylogeny in **Figure 28.1a, p. 634**), which includes vertebrates, among them *Homo sapiens*. Specifically, a phylogenetic analysis of 170 nuclear proteins and 13 mitochondrial proteins was used to derive the phylogeny that placed Xenoturbellida among the Deuterostomes (Figure 28.1b, p. 635). Look at some of the organisms **(Figure 28.2, p. 636)** arranged in the phylogeny.

Xenoturbellida (such as *Xenoturbella bocki;* see Figure 28.2f) was originally described in 1949. They are delicate, ciliated marine worms with simple body plans. They lack a gut with two openings, organized gonads, excretory structures, and a coelom. The nervous system is a diffuse net with no brain. Until 2006, there were more questions than answers about their phylogenetic position, and even to what phylum they should belong. At first, *X. bocki* was thought to be a turbellarian flatworm (Platyhelminthes), but it was also identified as a possible hemichordate or echinoderm (based on similarities in the nerve net). The details of *Xenoturbella*'s cilia are like those of hemichordates, which could indicate that it is really an acoelomorph flatworm.

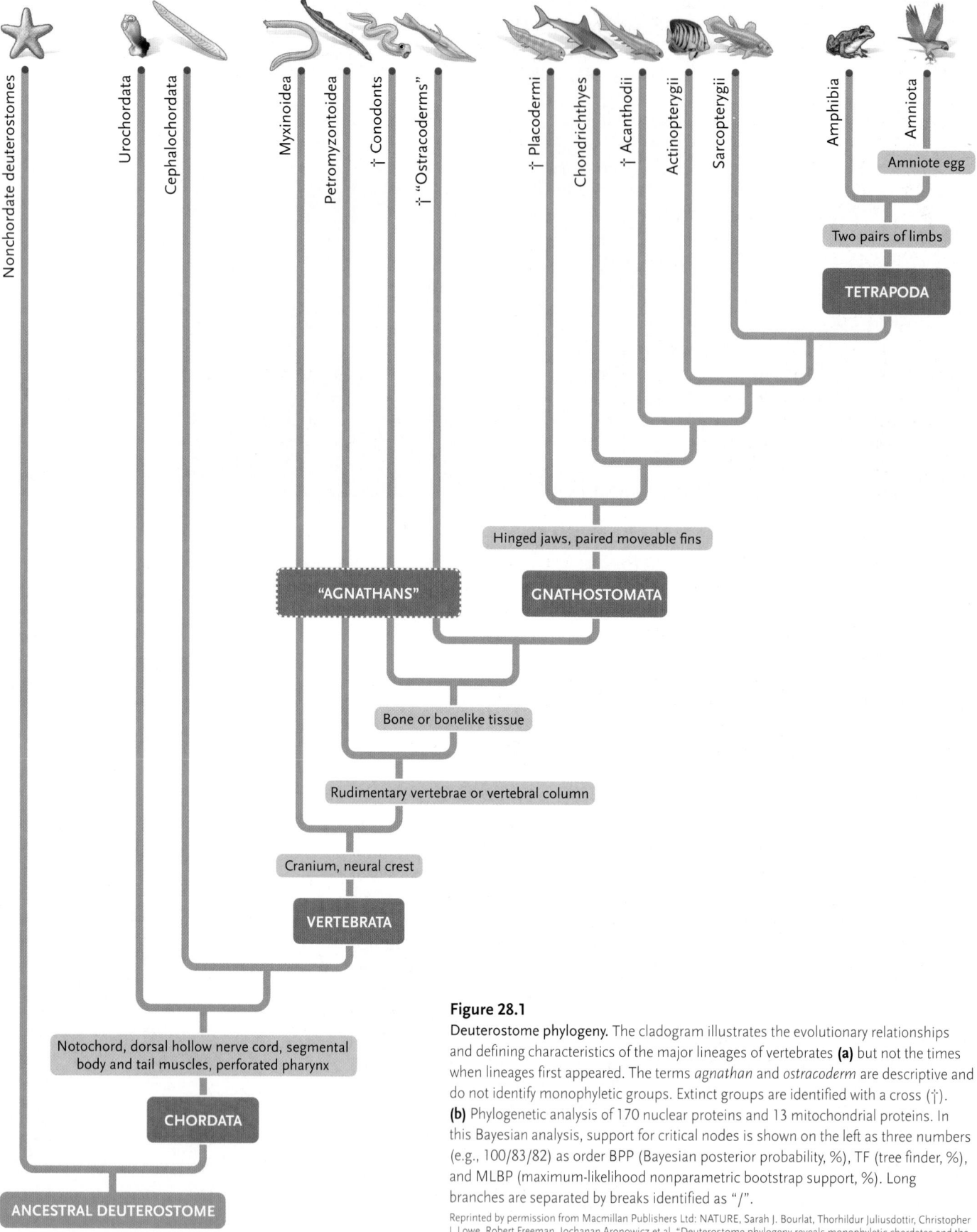

Figure 28.1

Deuterostome phylogeny. The cladogram illustrates the evolutionary relationships and defining characteristics of the major lineages of vertebrates **(a)** but not the times when lineages first appeared. The terms *agnathan* and *ostracoderm* are descriptive and do not identify monophyletic groups. Extinct groups are identified with a cross (†). **(b)** Phylogenetic analysis of 170 nuclear proteins and 13 mitochondrial proteins. In this Bayesian analysis, support for critical nodes is shown on the left as three numbers (e.g., 100/83/82) as order BPP (Bayesian posterior probability, %), TF (tree finder, %), and MLBP (maximum-likelihood nonparametric bootstrap support, %). Long branches are separated by breaks identified as "/".

Reprinted by permission from Macmillan Publishers Ltd: NATURE, Sarah J. Bourlat, Thorhildur Juliusdottir, Christopher J. Lowe, Robert Freeman, Jochanan Aronowicz et al. "Deuterostome phylogeny reveals monophyletic chordates and the new phylum Xenoturbellida," Vol. 444, 85–88, copyright 2006.

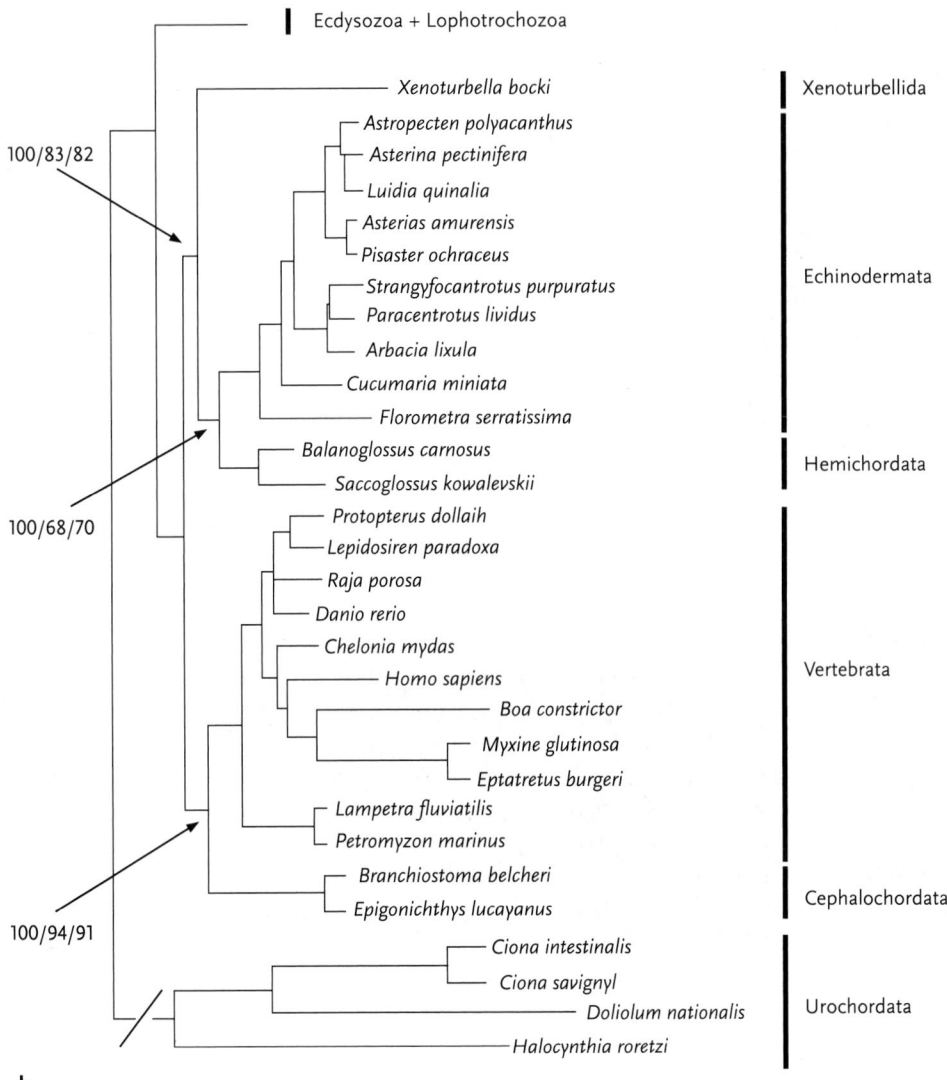

Ecdysozoa + Lophotrochozoa

100/83/82

100/68/70

100/94/91

Xenoturbella bocki — Xenoturbellida

Astropecten polyacanthus
Asterina pectinifera
Luidia quinalia
Asterias amurensis
Pisaster ochraceus
Strangyfocantrotus purpuratus
Paracentrotus lividus
Arbacia lixula
Cucumaria miniata
Florometra serratissima — Echinodermata

Balanoglossus carnosus
Saccoglossus kowalevskii — Hemichordata

Protopterus dollaih
Lepidosiren paradoxa
Raja porosa
Danio rerio
Chelonia mydas
Homo sapiens
Boa constrictor
Myxine glutinosa
Eptatretus burgeri
Lampetra fluviatilis
Petromyzon marinus — Vertebrata

Branchiostoma belcheri
Epigonichthys lucayanus — Cephalochordata

Ciona intestinalis
Ciona savignyl
Doliolum nationalis
Halocynthia roretzi — Urochordata

b.

In 1997, analysis of molecular phylogenetic data had been used to place *Xenoturbella* in the Mollusca, specifically among the bivalves. This arrangement was supported by the discovery of bivalvelike eggs and larvae within specimens of *Xenoturbella*. In 1997, it was easy to believe the molecular argument.

How can molecular data be challenged? In 1998, an alternative explanation was offered: mollusc genetic information appeared inside *Xenoturbella* because it eats molluscs. When the molluscan genetic information is ignored, *Xenoturbella* is clearly a deuterostome, most closely related to Echinodermata (see Figure 28.1b).

Thus, making correct choices about classification (see Chapter 20) means looking beyond appearance and may also require careful consideration of molecular data.

Although vertebrates are the best-known deuterostomes, echinoderms, hemichordates, and tunicates are also deuterostomes. In deuterostomes, the anus develops from the blastopore and the mouth arises as a separate opening, making them fundamentally different from protostomes. Deuterostome species show the same trophic diversity as protostomes, but with few species specialized as parasites. Vertebrates are classified as chordates, first known as small, filter-feeding animals. Vertebrates have an internal skeleton usually made of bone, partly accounting for their rich fossil record. Vertebrates include the largest animals known on Earth, the enormous plankton-feeding ichthyosaurs and whales.

28.1 Deuterostomes

Membership in the Deuterostomia (Greek, *deutero* = second; *stomia* = opening) is restricted to animals in which the anus develops from the blastopore and the mouth from a second opening. At first glance, deuterostome animals—such as echinoderms, chordates, and hemichordates, let alone *Xenoturbella*—are not obviously similar, reflecting modifications of their bodies that mask underlying developmental and genetic features.

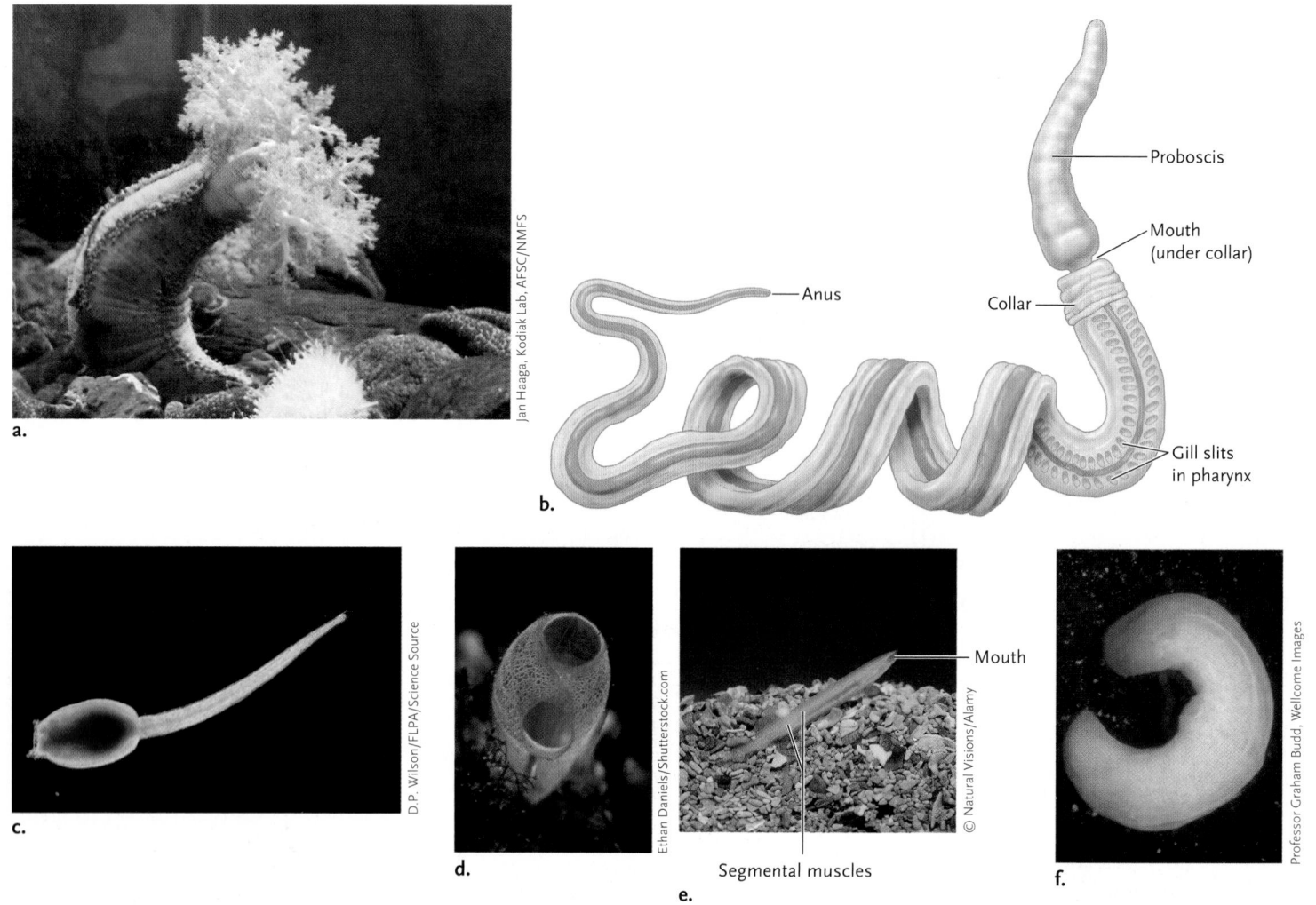

Figure 28.2

Deuterostomes. **(a)** Holothuroidea. A sea cucumber (*Cucumaraia miniata*) extends its tentacles, which are modified tube feet. **(b)** Phylum Hemichordata. Acorn worms draw food- and oxygen-laden water in through the mouth and expel it through gill slits in the anterior region of the trunk. **(c)** Urochordates. A tadpole-like tunicate larva will metamorphose into a sessile adult. **(d** and **e)** Cephalochordates. The unpigmented skin of an adult lancelet (*Brachiostoma* species) reveals its segmented body wall muscles. **(f)** A *Xenoturbella bocki* does not look very similar to any of the other animals illustrated here.

STUDY BREAK

Give one difference between protostomes and deuterostomes.

28.2 Phylum Echinodermata

The phylum Echinodermata (*echino* = spiny; *derm* = skin) includes 6500 species of sea stars, sea urchins, sea cucumbers, brittle stars, and sea lilies. These slow-moving or sessile bottom-dwelling animals are important herbivores and predators living in oceans from the shallow coastal waters to the depths. The phylum was diverse in the Paleozoic, but only a remnant of that fauna remains. Echinoderms vary in size from less than 1 cm in diameter to more than 50 cm long. Adult echinoderms develop from bilaterally symmetrical, free-swimming larvae. As the larvae develop, they assume a secondary radial symmetry, often organized around five rays or "arms" **(Figure 28.3)**. Many echinoderms have an oral surface, with the mouth facing the substrate, and an aboral surface facing in the opposite direction. Virtually all echinoderms have an internal skeleton made of calcium-stiffened ossicles that develop from mesoderm. In some groups, fused ossicles form a rigid container called a *test*. In most species with these features, spines or bumps project from the ossicles.

The internal anatomy of echinoderms is unique among animals (see Figure 28.3). They have a well-defined coelom and a complete digestive system (see Figure 28.3e) but no excretory or respiratory systems, and most have only a minimal circulatory system. In many, gases are exchanged and metabolic wastes eliminated through projections of the epidermis and peritoneum near the base of the spines. Given their radial symmetry, there is no head or central brain; the nervous system is organized around nerve cords that encircle the mouth and branch into the radii. Sensory cells are abundant in the skin.

a. Asteroidea: This sea star (*Fromia milleporella*) lives in the intertidal zone.

aquapix/Shutterstock.com

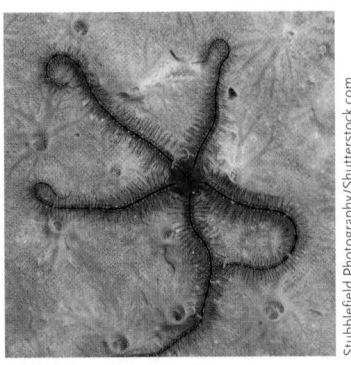

b. Ophiuroidea: A brittle star (*Ophiothrix swensonii*) perches on a coral branch.

Stubblefield Photography/Shutterstock.com

c. Echinoidea: A sea urchin (*Strongylocentrotus purpuratus*) grazes on algae.

M.B. Fenton

d. Crinoidea: A feather star (*Himerometra robustipinna*) feeds by catching small particles with its numerous tentacles.

Peter Scoones/Science Source

Figure 28.3

Echinoderm diversity. (a)–(d) Echinoderms exhibit secondary radial symmetry, usually organized as five rays around an oral–aboral axis. The coelom **(e)** is well developed in echinoderms, as illustrated by this cutaway diagram of a sea star. The water vascular system **(f)**, unique in the animal kingdom, operates the tube feet. Tube feet **(g)** are responsible for locomotion. Note the pedicillariae on the upper surface of the star's arm.

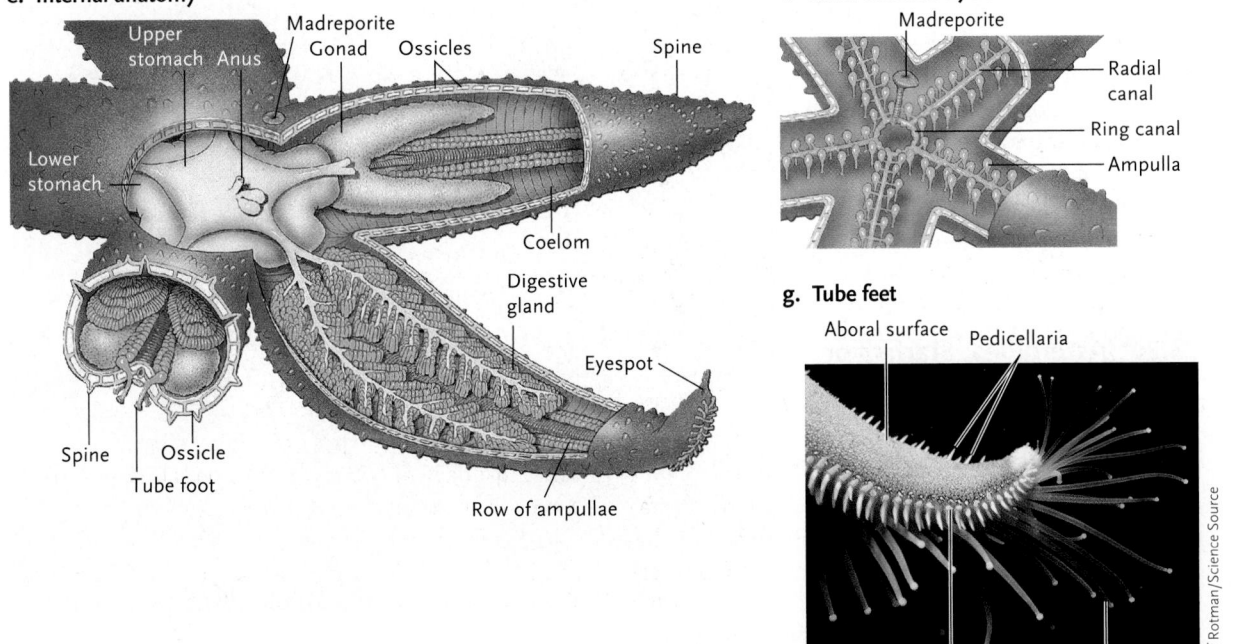

e. Internal anatomy

Upper stomach · Anus · Madreporite · Gonad · Ossicles · Spine · Lower stomach · Coelom · Digestive gland · Eyespot · Spine · Ossicle · Tube foot · Row of ampullae

f. Water vascular system

Madreporite · Radial canal · Ring canal · Ampulla

g. Tube feet

Aboral surface · Pedicellaria · Oral surface · Tube foot

Jeff Rotman/Science Source

Echinoderms move using a system of fluid-filled canals, the *water vascular system* (see Figure 28.3f). In a sea star, for example, water enters the system through the madreporite, a sievelike plate on the aboral surface. A short tube connects it to the *ring canal,* which surrounds the **esophagus**. The ring canal branches into five *radial canals* that extend into the arms. Each radial canal is connected to numerous *tube feet* that protrude through holes in the plates. Each tube foot has a mucus-covered, sucker like tip and a small muscular bulb, the *ampulla,* that lies inside the body. When an ampulla contracts, fluid is forced into the tube foot, causing it to lengthen and attach to the substrate (see Figure 28.3g). When the tube foot contracts, it pulls the animal along. As the tube foot shortens, water is forced back into the ampulla, and the tube foot releases its grip on the substrate. The tube foot can then take another step forward, reattaching to the substrate. Although each tube foot has limited strength, the coordinated action of hundreds or even thousands of them is so strong that they can hold an echinoderm to a substrate even against strong wave action.

Echinoderms have separate sexes, and most reproduce by releasing gametes into the water. Radial cleavage is so clearly apparent in the transparent eggs of some sea urchins that they are commonly used to demonstrate cleavage in introductory biology laboratories. A few echinoderms reproduce asexually by splitting in half and regenerating the missing parts. Other echinoderms regenerate body parts lost to predators. Four-day-old sand dollars (*Dendraster excentricus*) asexually clone themselves in response to the odour of fish (in mucus), apparently a defensive response.

Echinoderms are divided into six groups, the most recently described (1986) being the sea daisies (Concentricycloidea). These small, medusa-shaped animals occupy sunken, waterlogged wood in the deep sea. Sunken ships are often important habitats for these and other marine organisms. The five other groups, described below, are more diverse and better known.

28.2a Asteroidea: Starfish or Sea Stars

Sea stars live on rocky shorelines to depths of 10 000 m. Many are brightly coloured. The body consists of a central disk surrounded by 5–20 radiating "arms" (see Figure 28.3a), with the mouth centred on the oral surface. The ossicles of the endoskeleton are not fused, permitting flexibility of the arms and disk. **Pedicellariae** are small pincers at the base of short spines. They are used to remove debris that falls onto the animal's aboral surface (see Figure 28.3g). Many sea stars eat invertebrates and small fishes. Species that consume bivalve molluscs grasp the two valves with tube feet and slip their everted stomachs between the bivalve's shells **(Figure 28.4).** The stomach secretes digestive enzymes that dissolve the mollusc's tissues. Some sea stars are destructive predators of corals, endangering many reefs.

28.2b Ophiuroidea: Brittle Stars

The 2000 species of brittle stars and basket stars occupy roughly the same range of habitats as sea stars. Their bodies have a well-defined central disk and slender, elongated arms that are sometimes branched (see Figure 28.3b). Ophiuroids can crawl fairly swiftly across substrates by moving their arms in a coordinated fashion. As their common name implies, the arms are delicate and easily broken, an adaptation allowing them to escape from predators with only minor damage. Brittle stars feed on small prey, suspended plankton, or detritus that they extract from muddy deposits.

28.2c Echinoidea: Sea Urchins and Sand Dollars

The 950 species of sea urchins and sand dollars lack arms (see Figure 28.3c). Their ossicles are fused into solid tests that provide excellent protection but

Figure 28.4
Sea star feeding on a mussel. Even when the tide is out in Haida Gwaii, sea stars hunt mussels.

restrict flexibility. The test is spherical in sea urchins and flattened in sand dollars. These animals use tube feet in locomotion. Five rows of tube feet emerge through pores in the test. Most echinoids have movable spines, some with poison glands. A jab from some tropical species can cause a careless swimmer severe pain and inflammation. Echinoids graze on algae and other organisms that cling to surfaces. In the centre of an urchin's oral surface is a five-part nipping jaw that is controlled by powerful muscles. Some species damage kelp beds, disrupting the habitat of young lobsters and other crustaceans. Echinoid ovaries are a gourmet delicacy in many countries, making these animals a prized natural resource.

28.2d Holothuroidea: Sea Cucumbers

Sea cucumbers are elongated animals that lie on their sides on the ocean bottom (see Figure 28.2a, p. 636); they number about 1500 species. Although they have five rows of tube feet, their endoskeleton is reduced to widely separated microscopic plates. The body, which is elongated along the oral–aboral axis, is soft and fleshy, with a tough, leathery covering. Modified tube feet form a ring of tentacles around the mouth. The central disk and mouth point upward rather than toward the substrate. Some species secrete a mucus net that traps plankton or other food particles. The net and tentacles are inserted into the mouth, where the net and trapped food are ingested. Other species extract food from bottom sediments. Many sea cucumbers exchange gases through an extensively branched respiratory tree arising from the rectum, the part of the digestive system just inside the anus at the aboral end of the animal. A well-developed circulatory system distributes oxygen and nutrients to tissues throughout the body.

Sea cucumbers are actually home for a specialized symbiotic fish. *Carapus bermudensis,* the pearl fish, enters sea cucumbers' cloacal opening tail first. The cloaca is the chamber receiving urine, feces, and reproductive products. Pearl fish are members of a group that usually live in the tubes of other animals, including the cavities of bivalves. These fishes have elongated, thin bodies. They have lost pelvic fins and scales, and the anal opening has moved forward to a position under the head. This adaptation ensures that the fish defecates outside the body of the sea cucumber. These fishes use olfactory cues to find the "correct" host.

28.2e Crinoidea: Sea Lilies and Feather Stars

The 600 living species of sea lilies and feather stars are the surviving remnants of a diverse and abundant fauna 500 million years ago (mya) (see Figure 28.3d, p. 637). Most species occupy marine waters of medium depth. Between five and several hundred branched arms surround the disk that contains the mouth. New arms are added as a crinoid grows larger.

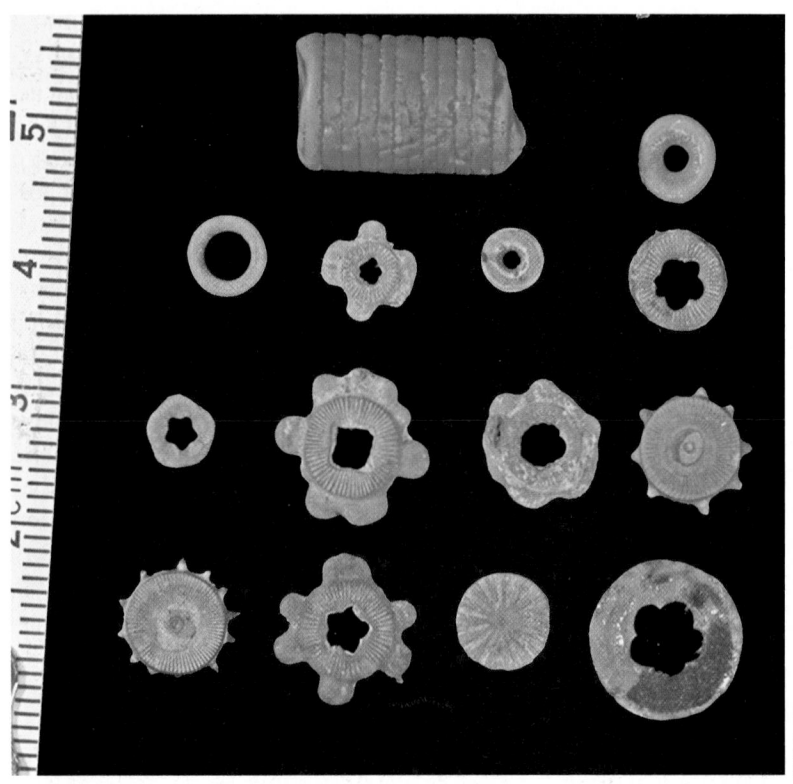

Figure 28.5

Fossil crinoid stems. Ossicles making up the stems of crinoids are commonly fossilized. The individual ossicles are from the Devonian of Ontario. The section of complete stem is *Encrinus liliiformis* from the Triassic of Germany. Scale is in millimetres.

The branches of the arms are covered with tiny, mucus-coated tube feet that trap suspended microscopic organisms. Sessile sea lilies have the central disk attached to a flexible stalk that can reach 1 m in length. By contrast, adult feather stars can swim or crawl weakly, attaching temporarily to substrates. The disks making up sea lily stalks, called ossicles, are common fossils in many deposits **(Figure 28.5)**.

STUDY BREAK

1. What are echinoderms? How do adult echinoderms develop?
2. Use a table to compare an echinoderm and a human by system: (a) digestive, (b) excretory, (c) respiratory, (d) circulatory, and (e) nervous.
3. Using a sea star as an example, describe how echinoderms move.

28.3 Phylum Hemichordata

The 80 species of **acorn worms** making up this phylum take their name from *hemi,* meaning half, and *chord,* referring to the phylum Chordata. Hemichordates have a stomochord that superficially resembles the notochord of chordates. Acorn worms are sedentary marine animals living in U-shaped tubes or burrows in coastal sand or mud. Their soft bodies range in length from

2 cm to 2 m and are organized into an anterior proboscis, a tentacled collar, and an elongated trunk (see Figure 28.2b, p. 636). They use the muscular, mucus-coated proboscis to construct burrows and trap food particles. Acorn worms also have pairs of gill slits in the pharynx, the part of the digestive system just posterior to the mouth. Beating cilia create a flow of water, which enters the pharynx through the mouth and exits through the gill slits. As water passes through, suspended food particles are trapped and shunted into the digestive system, and gases are exchanged across the partitions between gill slits. The dorsal nerve cord, coupled with feeding and respiration, reflects a close evolutionary relationship between hemichordates and chordates. Pterobranchia, or sea angels, are the other class of animals in this phylum. These uncommon marine animals are colonial and live in tubes. They superficially resemble some cnidarians.

STUDY BREAK

How do hemichordates feed?

28.4 Phylum Chordata

This phylum includes evolutionary lines of invertebrates, the Urochordata and Cephalochordata, as well as the more diverse line, the Vertebrata. A notochord, a dorsal hollow nerve cord, and gill slits (a perforated pharynx) are key morphological features distinguishing chordates from all other Deuterostomes. These features occur during at least some time in a chordate's life cycle. Chordates also have segmental muscles in the body wall and tail **(Figure 28.6)**. Collectively, these structures enable higher levels of activity and unique modes of aquatic locomotion, as well as more efficient feeding and oxygen acquisition.

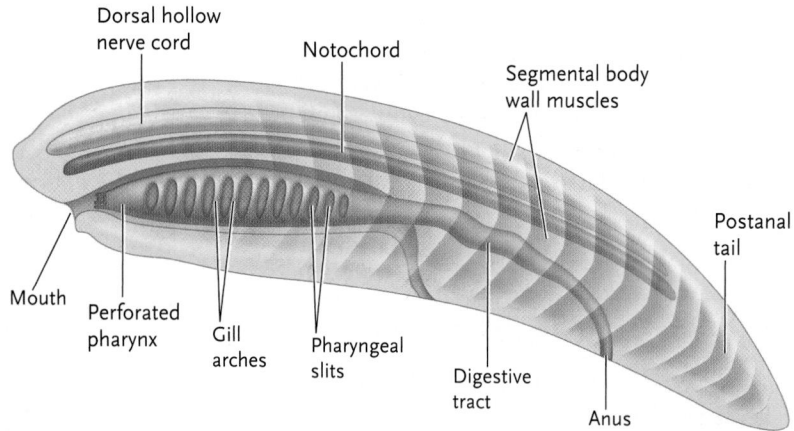

Figure 28.6

Diagnostic chordate characteristics. Chordates have a notochord, a dorsal hollow nerve cord, pharangeal (gill) slits (a perforated pharynx), and a muscular postanal tail with segmental body wall and tail muscles. Other basic features are shown as well, but they are not unique to chordates.

Early in chordate embryonic development, the **notochord** (*noto* = back; *chord* = string), a flexible rod, develops from mesoderm dorsal to the developing digestive system. The notochord is constructed of fluid-filled cells surrounded by tough connective tissue. It supports the embryo from head to tail. The notochord is the skeleton of invertebrate chordates, serving as an anchor for body wall muscles. When these muscles contract, the notochord bends but does not shorten. Waves of contractions pass down one side of the animal and then up the other, sweeping the body and tail back and forth in a smooth and continuous movement. Thus, the chordate body swings left and right during locomotion, propelling the animal forward. The chordate tail, which is posterior to the anus, provides most of the propulsion in some aquatic species. Segmentation allows each muscle block to contract independently. Unlike the bodies of annelids and other nonchordate invertebrates, the chordate body does not shorten when the animal is moving. Remnants of the notochord persist as gelatinous disks between the vertebrae of some adult vertebrates.

The central nervous system of chordates is a hollow nerve cord on the dorsal side of the embryo (see Chapter 42). Most nonchordate invertebrates have ventral, solid nerve cords. In vertebrates, an anterior enlargement of the nerve cord forms the brain. In invertebrates, an anterior concentration of nervous system tissue is a **ganglion** and may be referred to as a "brain."

Gill (pharyngeal) slits mean that the chordate pharynx is perforated. The pharynx is the part of the digestive system just behind the mouth. **Gill slits** are paired openings originating as exit holes for water that carried particulate food into the mouth, allowing chordates to gather food by filtration. Invertebrate chordates also collect oxygen and release carbon dioxide across the walls of the pharynx. In fishes, gill arches have evolved as supporting structures between the slits in the pharynx. Invertebrate chordates and fishes retain a perforated pharynx throughout their lives. In most air-breathing vertebrates, the slits are present only during embryonic development and in some larvae.

28.4a Subphylum Urochordata: Sea Squirts and Tunicates

The 2500 species of urochordates (*uro* = tail) float in surface waters or attach to substrates in shallow marine habitats. Sessile adults of many species secrete a gelatinous or leathery "tunic" around their bodies and squirt water through a siphon when disturbed. Adults can attain lengths of several centimetres (**Figure 28.7;** see also Figure 28.2c, p. 636). In the most common group of sea squirts (Ascidiacea), swimming larvae have notochords, dorsal hollow nerve cords, and gill slits, features lacking in the sessile adults. Larvae eventually attach to substrates and transform into sessile adults.

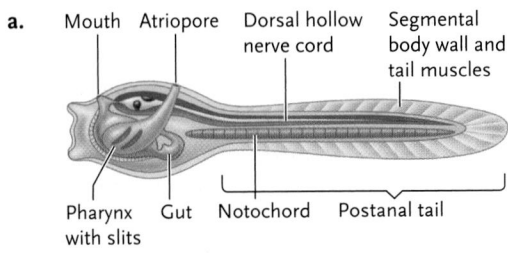

a. Mouth Atriopore Dorsal hollow nerve cord Segmental body wall and tail muscles

Pharynx with slits Gut Notochord Postanal tail

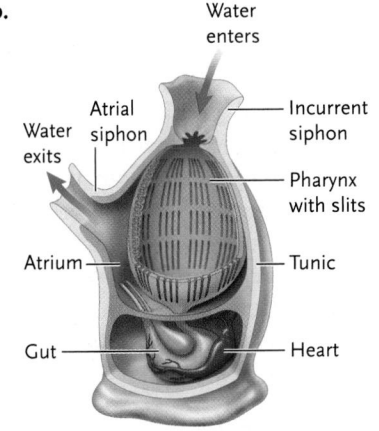

b. Water enters

Water exits Atrial siphon Incurrent siphon

Water exits Pharynx with slits

Atrium Tunic

Gut Heart

Figure 28.7

Diagrams of urochordates. The tadpolelike tunicate larva **(a)** metamorphoses into an adult, a sessile filter-feeder. **(b)** In the adult, the atriopore becomes the atrial siphon.

During metamorphosis, larvae lose most traces of the notochord, dorsal nerve cord, and tail, and their basketlike pharynx enlarges. In adults, beating cilia pull water into the pharynx through an incurrent siphon. A mucus net traps particulate food, which is carried with the mucus to the gut. Water passes through the gill slits, enters a chamber called the **atrium**, and is expelled through the **atrial siphon** along with digestive wastes and carbon dioxide. Oxygen is absorbed across the walls of the pharynx. In some urochordates, the larvae are neotenous, acquiring the ability to reproduce and remaining active throughout their life cycles.

28.4b Subphylum Cephalochordata: Lancelets

All 28 species of cephalochordates (*cephalo* = head) live in warm, shallow marine habitats, where they lie mostly buried in sand (see Figure 28.2d, e, p. 636).

Although generally sedentary, they have well-developed body wall muscles and a prominent notochord. Most species are included in the genus *Branchiostoma* (formerly *Amphioxus*). Lancelet bodies, which are 5 to 10 cm long, are pointed at both ends like the double-edged surgical tools for which they are named **(Figure 28.8)**. Adults have light receptors on the head as well as chemical sense organs on tentacles that grow from the **oral hood**. Lancelets use cilia to draw food-laden water through hundreds of pharyngeal slits; water flows into the atrium and is expelled through the **atriopore**. Most gas exchange occurs across the skin.

28.4c Subphylum Vertebrata: Vertebrates

Species in this subphylum have a distinct head making them craniate, and most have a **backbone (spine)** made up of individual bony vertebrae (see Chapters 20 and 42). This internal skeletal feature provides structural support for muscles and protects the nervous system and other organs. In addition, the internal skeleton and attached muscles allow most vertebrates to move rapidly. Vertebrates are the only animals that have bone, a connective tissue in which cells secrete the mineralized matrix that surrounds them (see Chapter 46). One vertebrate lineage, cartilaginous fishes (class Chondrichthyes), may have lost its bone over evolutionary time. These animals, mostly sharks and rays, have skeletons of cartilage, a dense, flexible connective tissue that can be a developmental precursor of bone (see Chapters 42 and 46).

At the anterior end of the vertebral column, the head is usually protected by a bony **cranium** or skull. The backbone surrounds and protects the dorsal nerve cord, and the bony cranium surrounds the brain. The cranium, vertebral column, ribs, and sternum (breastbone) make up the **axial skeleton.** Most vertebrates also have a **pectoral girdle** anteriorly and a **pelvic girdle** posteriorly that attach bones in the fins or limbs to the axial skeleton. The bones of the two girdles and the appendages constitute the appendicular skeleton.

Vertebrates have neural crest cells (see Chapter 42), a unique cell type distinct from endoderm, mesoderm, and ectoderm. Neural crest cells arise next to the developing nervous system but migrate throughout the body. Neural crest cells ultimately contribute to

Lancelet anatomy

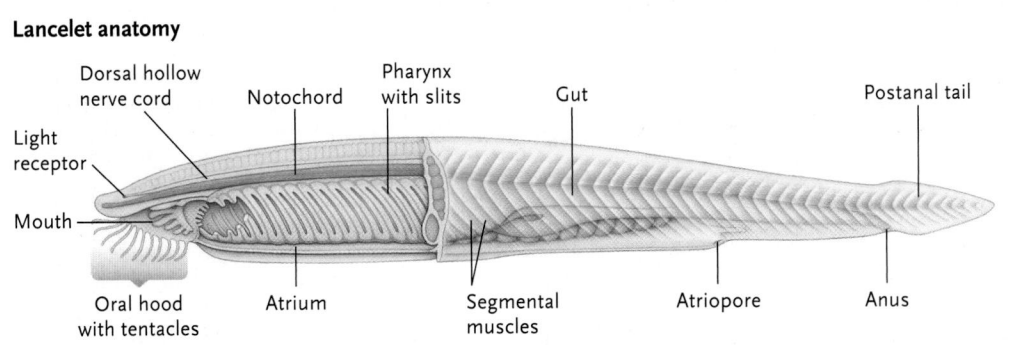

Dorsal hollow nerve cord Pharynx with slits
Light receptor Notochord Gut Postanal tail

Mouth

Oral hood with tentacles Atrium Segmental muscles Atriopore Anus

Figure 28.8

A drawing of the internal anatomy of an adult lancelet (*Branchiostoma*).

uniquely vertebrate structures such as parts of the cranium, teeth, sensory organs, **cranial nerves**, and the medulla (the interior part) of the adrenal glands.

The brains of vertebrates are larger and more complex than those of invertebrate chordates. Moreover, the vertebrate brain is divided into three regions, the forebrain, midbrain, and hindbrain, each governing distinct nervous system functions (see Chapter 44).

STUDY BREAK

1. List four morphological features distinguishing chordates from other deuterostomes.
2. Explain the purpose and structure of gill slits.
3. What are tunicates and lancelets? To which subphyla do they belong? What characteristics of swimming tunicate larvae are missing from the sessile adults?

28.5 The Origin and Diversification of Vertebrates

Biologists have used embryological, molecular, and fossil evidences to trace the origin of vertebrates and to chronicle the evolutionary diversification of the group to which humans belong. We suspect that vertebrates arose from a cephalochordate-like ancestor through duplication of genes that regulate development.

Vertebrates appear to be more closely related to cephalochordates than to urochordates (see Figure 28.1, p. 634). The change from cephalochordate-like creatures to vertebrates was marked by the emergence of neural crest cells, bone, and other vertebrate traits. Biologists hypothesize that an increase in the number of genes that control the expression of other genes (homeotic) may have facilitated the development of more complex anatomy. (For more about homeotic genes, see Chapter 42). When it comes to organization, there is no compelling reason to believe that "more complex" is superior to "simple."

Hox genes are homeotic genes that influence the three-dimensional shape of the animal and the locations of important structures such as eyes, wings, and legs, particularly along the head-to-tail axis of the body. *Hox* genes are arranged on chromosomes in a particular order, forming the *Hox* gene complex. Each gene in the complex governs the development of particular structures. Animal groups with the simplest structure, such as cnidarians, have two *Hox* genes. Those with more complex anatomy, such as insects, have 10. Chordates typically have up to 13 or 14. Lineages with many *Hox* genes generally have more complex anatomy than those with fewer *Hox* genes.

Molecular analyses reveal that the entire *Hox* gene complex was duplicated several times in the evolution of vertebrates, producing multiple copies of all the genes in the *Hox* complex **(Figure 28.9)**. The cephalochordate *Branchiostoma* has one *Hox* gene

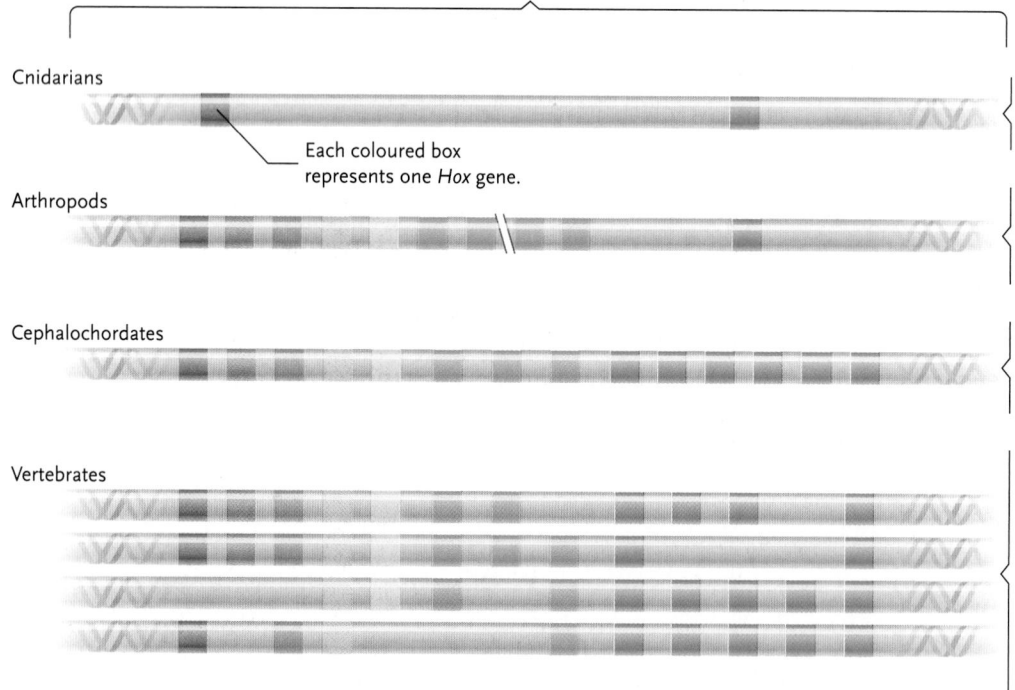

Each row of coloured boxes represents one *Hox* gene complex.

Cnidarians

Each coloured box represents one *Hox* gene.

Arthropods

Cephalochordates

Vertebrates

a. Invertebrates with simple anatomy, such as cnidarians, have a single *Hox* gene complex that includes just a few *Hox* genes.

b. Invertebrates with more complicated anatomy, such as arthropods, have a single *Hox* gene complex, but with a larger number of *Hox* genes.

c. Invertebrate chordates, such as cephalochordates, also have a single *Hox* gene complex, but with even more *Hox* genes than are found in nonchordate invertebrates.

d. Vertebrates, such as the laboratory mouse, have numerous *Hox* genes, arranged in two to seven *Hox* gene complexes. The additional *Hox* gene complexes are products of wholesale duplications of the ancestral *Hox* gene complex. The additional copies of *Hox* genes specify the development of uniquely vertebrate characteristics, such as the cranium, vertebral column, and neural crest cells.

Figure 28.9

Hox genes and the evolution of vertebrates. The *Hox* genes in different animals appear to be homologous, indicated here by their colour and position in the complex. Vertebrates have many more individual *Hox* genes than invertebrates, and the entire *Hox* gene complex was duplicated in the vertebrate lineage.

complex, whereas hagfish, the most ancestral living vertebrate, has two. All vertebrates with jaws have at least four sets of *Hox* genes, and some fishes have seven. Evolutionary biologists who study development hypothesize that the duplication of *Hox* genes and other tool-kit genes allowed the evolution of new structures. Although original copies of these genes maintained their ancestral functions, duplicate copies were available to assume *new* functions, leading to the development of novel structures such as the vertebral column and jaws. These changes coincided with the adaptive radiation of vertebrates.

The oldest known vertebrate fossils are from the early Cambrian (about 550 mya) in China. Both *Myllokunmingia* and *Haikouichthys* were fish-shaped animals about 3 cm long (**Figures 28.10** and **28.11**). In both species, the brain was surrounded by a cranium of fibrous connective tissue or cartilage. They also had segmental body wall muscles and fairly well-developed fins, but neither shows any evidence of bone.

The early vertebrates gave rise to numerous descendants (see Figure 28.1a, p. 634), which varied

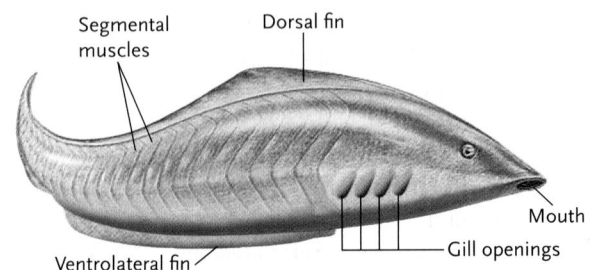

Figure 28.10

Cambrian agnathan, *Haikouichthys*, was more like a hagfish than a lamprey but generally similar to an ammocoetes larva of lampreys, living agnathans.

greatly in anatomy, physiology, and ecology. New feeding mechanisms and locomotor structures were correlated with their success. Today, vertebrates occupy nearly every habitat on Earth and eat virtually all other organisms. Biologists tend to identify vertebrates with four key morphological innovations: cranium, vertebrae, bone, and neural crest cells. We must remember that these structures did not evolve spontaneously.

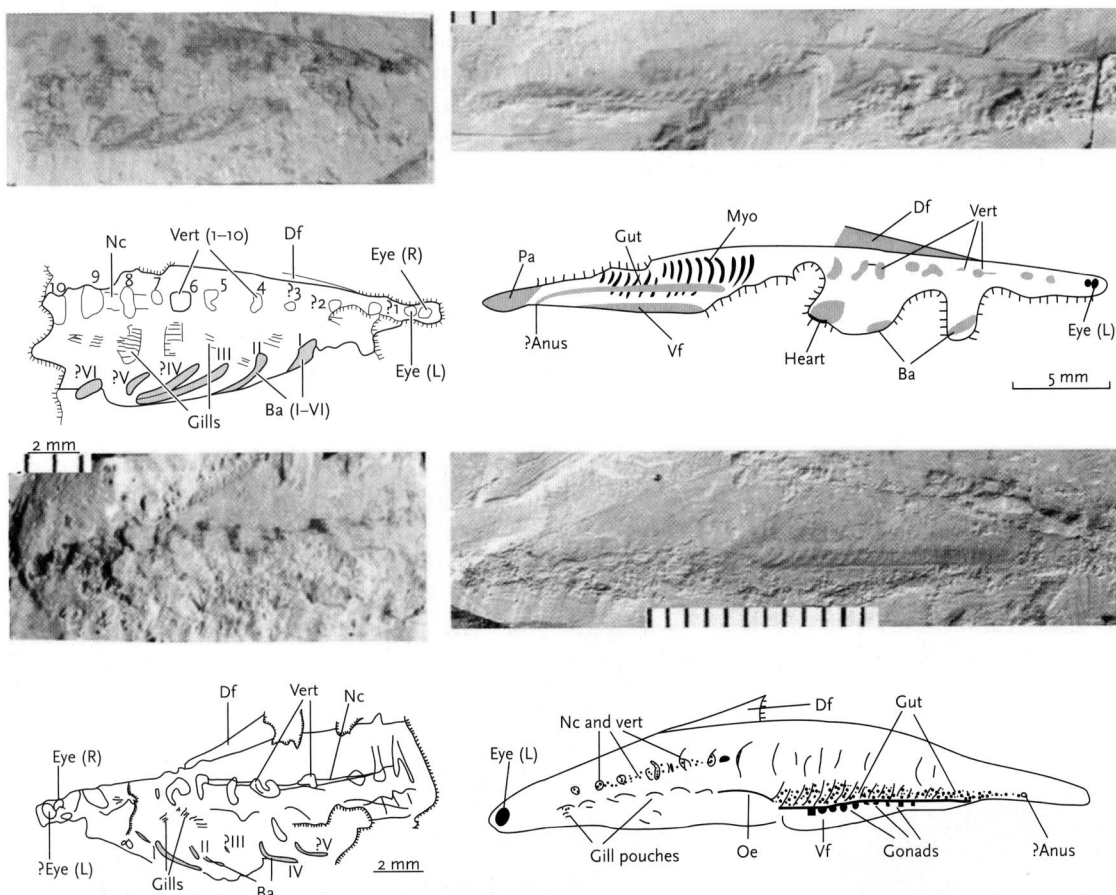

Figure 28.11

A diagram showing an early vertebrate. *Myllokunmingia* is one of the earliest vertebrates yet discovered. This species had no bones and was about 30 cm long. The labels identify features, some with abbreviations, including the following: Ap—anterior plates; Ba—branchial arches; Df—dorsal fin; Myo—myosepta; Nc—notochord; Nc and Vert—notochord with vertebral elements; Nos—nostril; Ns—nasal sacs; Oc—otic capsule; Oe—esophagus; Pa—postanal tail; Vert—vertebral elements; Vf—ventral fin fold; L—left; R—right.

Reprinted by permission from Macmillan Publishers Ltd: NATURE, "Head and backbone of the early Cambrian vertebrate Haikouichthys", Shu, D.-G., S. C. Morris, J. Han, Z-F. Zhang, K. Yasui, P. Janvaier, L. Chen, X-L. Zhang, J-N. Liu, and H-Q. Liu, vol. 421: 526–529, copyright 2003.

Important biological changes during the evolution of vertebrates included improved access to energy (food), which involved mobility and jaws, combined with effective aerobic metabolism (access to oxygen).

The earliest vertebrates lacked jaws (Agnatha, *a* = not; *gnath* = jawed), but Agnatha is not a monophyletic group. Although most became extinct by the end of the Paleozoic, two ancestral lineages, Myxinoidea (hagfishes) and Petromyzontoidea (lampreys), survive today. All other vertebrates have movable jaws and form the monophyletic lineage **Gnathostomata** (*gnath* = jawed; *stoma* = mouth). The first jawed fishes, the Acanthodii and Placodermi, are now extinct, but several other lineages of jawed fishes are still abundant. Included are Chondrichthyes, fishes with cartilaginous skeletons (sharks, skates, chimaeras), and Teleostei (actinopterygians and sarcopterygians), with bony endoskeletons. Although all jawless vertebrates and most jawed fishes are restricted to aquatic habitats, mudskippers (*Periophthalmus* species) and climbing perch (*Anabas* species) regularly venture onto land. Many fish have developed lunglike structures for breathing atmospheric oxygen, but most use gills to extract dissolved oxygen from water. Lungs may be an ancestral trait in vertebrates.

Gnathostomata also includes the monophyletic lineage **Tetrapoda** (*tetra* = four; *pod* = foot), most of which use four limbs for locomotion. Many tetrapods are amphibious, semiterrestrial, or terrestrial, although some, such as sea turtles and porpoises, have secondarily returned to aquatic habitats. Adult tetrapods generally use lungs to breathe atmospheric oxygen. Within the Tetrapoda, one lineage, the Amphibia (such as frogs and salamanders), typically needs standing water to complete its life cycle. Another lineage, the Amniota,

comprises animals with specialized eggs that can develop on land. Shortly after their appearance, amniotes diversified into three lineages, one ancestral to living mammals; another to living turtles; and a third to lizards, snakes, alligators, and birds.

STUDY BREAK

1. What is the function of a backbone?
2. What marked the change from a cephalochordate-like creature to a vertebrate?
3. What is a *Hox* gene, and how does it influence the diversity of vertebrates?

28.6 Agnathans: Hagfishes and Lampreys, Conodonts, and Ostracoderms

Lacking jaws, the earliest vertebrates used a muscular pharynx to suck water containing food particles into the mouth, and used gills both to acquire dissolved oxygen and to filter food from the water. The agnathans that flourished in the Paleozoic varied greatly in size and shape and possessed different combinations of vertebrate characters.

Lampreys and hagfishes, the two living groups of agnathans, have skeletons composed entirely of cartilage. Although as yet no fossilized lampreys or hagfishes have been found before the Devonian, the absence of bone in their living descendants suggests that they arose early in vertebrate history, before the evolution of bone. The first fossil lamprey is known from the Devonian of South Africa **(Figure 28.12)**.

a. Living jawless fishes

Hagfish

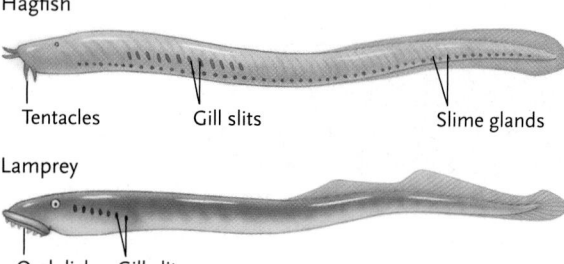

Tentacles   Gill slits   Slime glands

Lamprey

Oral disk   Gill slits

b. Mouth of a lamprey

© PureStock/Alamy

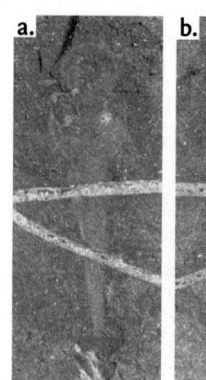

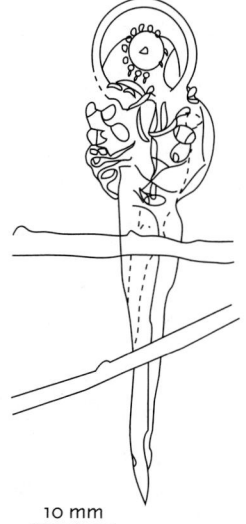

a.   b.

10 mm

Reprinted by permission from Macmillan Publishers Ltd: NATURE, "A lamprey from the Devonian period of South Africa," Robert W. Gess, Michael I. Coates and Bruce S. Rubidge, vol. 443, pp. 981–984, copyright 2006.

Figure 28.12

Living agnathans. Two groups of jawless fishes, the hagfishes and the lampreys **(a)**, are shown as diagrams with a photograph of a lamprey **(b)**. Also shown is the fossil and diagram of a Devonian lamprey from South Africa.

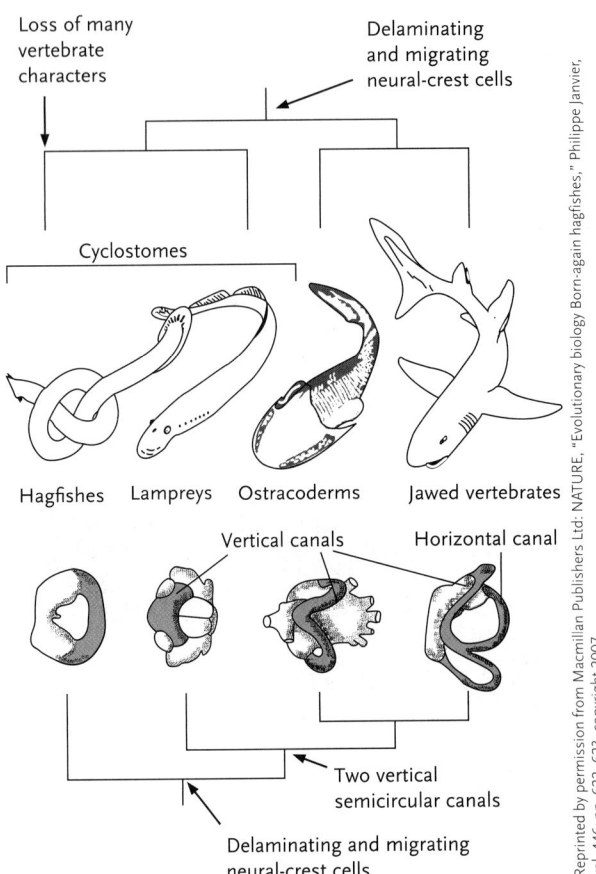

Reprinted by permission from Macmillan Publishers Ltd: NATURE, "Evolutionary biology Born-again hagfishes," Philippe Janvier, vol. 446, pp. 622–623, copyright 2007.

Figure 28.13

Two alternatives for hagfishes. The top tree implies that they are vertebrates that have lost features and the bottom one that they are a sister group of all other vertebrates. One striking difference is the presence of one semicircular canal in hagfishes and at least two in all other vertebrates. The truth remains elusive.

Hagfishes and lampreys have a well-developed notochord but no true vertebrae or paired fins. Their skin lacks scales. Individuals grow to a maximum length of about 1 m (see Figure 28.12). Two possible phylogenies for hagfishes and other vertebrates are presented **(Figure 28.13)**, but at this time, there are too few data to decide which is most likely to be correct.

The axial skeletons of the 60 living species of hagfishes include only a cranium and a notochord. No specialized structures surround the dorsal nerve cord. Hagfishes are marine scavengers that burrow in sediments on continental shelves. They feed on invertebrate prey and on dead or dying fishes. In response to predators, they secrete an immense quantity of sticky, noxious slime. When no longer threatened, a hagfish ties itself into a knot and wipes the slime from its body. The life cycle of a hagfish lacks a larval stage.

The 38 living species of lamprey have a more specialized axial skeleton than hagfishes. Their notochord is surrounded by dorsally pointing cartilage that partially covers the nerve cord, perhaps representing an early stage in the evolution of the vertebral column. About half of the living lamprey species are parasitic as adults and use the sucking disk around their mouths to attach to the bodies of fish (or other prey), rasp a hole in the host's body, and ingest body fluids. In most species, sexually mature adults migrate from the ocean or a lake to the headwaters of a stream, where they reproduce and then die. The filter-feeding **ammocoetes** larvae of lampreys resemble adult cephalochordates. They burrow into mud and develop for as long as seven years before metamorphosing and migrating to the sea or lake to live as adults.

Conodonts and ostracoderms were early jawless vertebrates with bony structures. Conodonts are mysterious bonelike fossils, mostly less than 1 mm long, occurring in oceanic rocks from the early Paleozoic through the early Mesozoic. Called **conodont** elements, these abundant fossils were originally described as supporting structures of marine algae or feeding structures of ancient invertebrates. Recent analyses of their mineral composition reveal that they were made of dentine, a bonelike component of vertebrate teeth. In the 1980s and 1990s, many questions about conodonts were answered by the discovery of fossils of intact conodont animals with these elements.

We now know that conodonts were elongate, soft bodied animals, 3–10 cm long. They had a notochord, a cranium, segmental body wall muscles, and large, movable eyes **(Figure 28.14a)**. The conodont elements at the front of the mouth were forward-pointing, hook-shaped structures (the original fossils) apparently used in the collection of food. Conodont elements in the pharynx were stouter, making them suitable for crushing food. Paleontologists now classify conodonts as vertebrates, the earliest ones with bonelike structures.

Ostracoderms (*ostrac* = shell; *derm* = skin) include an assortment of jawless fishes representing several evolutionary lines that lived from the

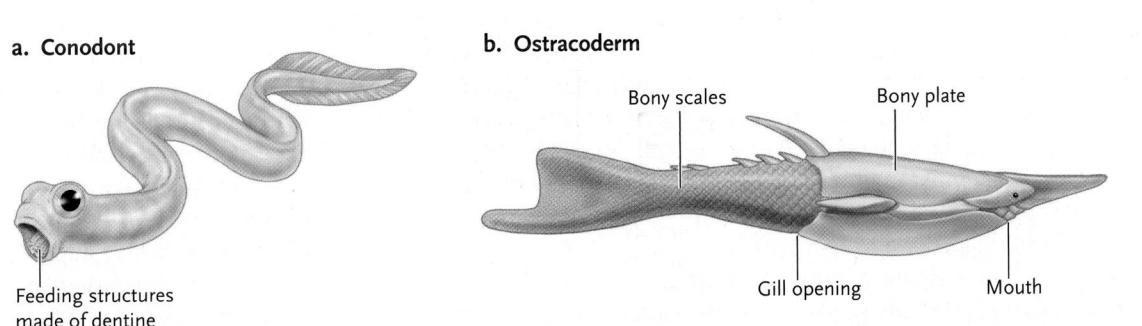

a. Conodont

Feeding structures made of dentine

b. Ostracoderm

Bony scales

Bony plate

Gill opening

Mouth

Figure 28.14

Extinct agnathans. (a) Conodonts were elaborate, soft-bodied animals with bonelike feeding structures in the mouth and pharynx.
(b) *Pteropsis*, an ostracoderm, had large bony plates on its head and small body scales on the rest of its body. It was about 6 cm long.

Figure 28.15
The evolution of jaws. In two early lineages of jawed fishes (Acanthodii and Placodermi), the upper jaw (**maxillae, premaxillae**) was firmly attached to the cranium, while the lower jaw moved up and down. This meant an inflexible mouth that simply snapped open and shut. Acanthodians and placoderms had bony internal skeletons.

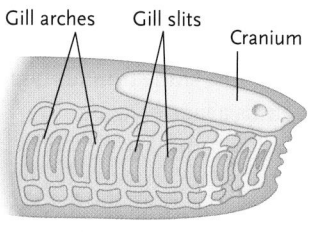

a. Jaws evolved from gill arches in the pharynx of jawless fishes.

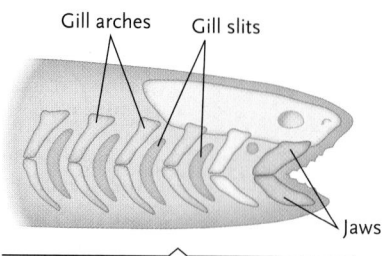

b. In early jawed fishes, the upper jaw was firmly attached to the cranium.

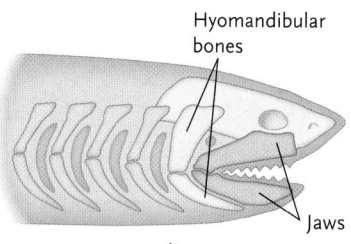

c. In later jawed fishes, the jaws were supported by the hyomandibular bones, which were derived from a second pair of gill arches.

Ordovician through the Devonian (Figure 28.14b, p. 645). Like their invertebrate chordate ancestors, ostracoderms probably used the pharynx to draw water with food particles into the mouth and used gills to filter food from water. The muscular pharynx was more efficient than that of agnathans, using currents generated by cilia. Greater flow rates allowed ostracoderms to collect food more rapidly and achieve larger body sizes. Although most ostracoderms were much smaller, some were 2 m long.

The skin of ostracoderms was heavily armoured with bony plates and scales. Although some had paired lateral extensions of their bony armour, they could not move them in the way living fishes move paired fins. Ostracoderms lacked a true vertebral column, but they had rudimentary support structures surrounding the nerve cord. Ostracoderms had other distinctly vertebrate-like characteristics. Their head shields indicate that their brains had the three regions (forebrain, midbrain, and hindbrain) typical of all later vertebrates (see Chapter 44).

STUDY BREAK

1. How did the earliest vertebrates feed without jaws?
2. Compare the hagfish and the lamprey based on body structure, feeding habits, and life cycles.

28.7 Jawed Fishes: Jaws Expanded the Feeding Opportunities for Vertebrates

The first gnathostomes were jawed fishes. Jaws meant that they could eat more than just filtered food particles and take larger food items with higher energy content. The renowned anatomist and paleontologist A.S. Romer (see "People behind Biology," Box 18.1, p. 422) described the evolution of jaws as "perhaps the greatest of all advances in vertebrate history." Hinged jaws allow vertebrates to grasp, kill, shred, and crush large

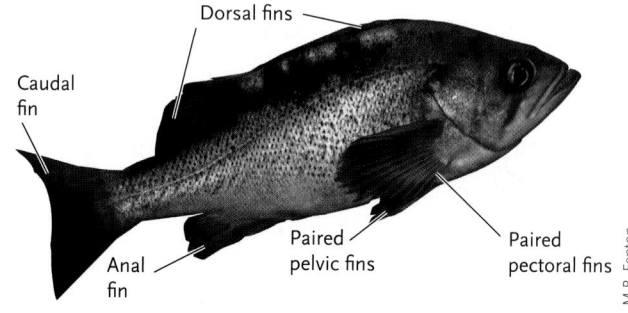

Figure 28.16
Fish fins. Most fishes have both paired and unpaired fins.

food items. Some species also use their jaws for defence, for grooming, to construct nests, and to transport young. Jaws may serve more than one purpose.

Embryological evidence suggests that jaws evolved from paired gill arches in the pharynx of a jawless ancestor (**Figure 28.15**). One pair of ancestral **gill arches** formed bones in the upper and lower jaws, whereas a second pair was transformed into the **hyomandibular bones** that braced the jaws against the cranium. Nerves and muscles of the ancestral suspension-feeding pharynx control the movement and actions of jaws. Jawed fishes also had fins, first appearing as folds of skin and movable spines that stabilized locomotion and deterred predators. Movable fins appeared independently in several lineages, and by the Devonian, most jawed fishes had unpaired (dorsal, anal, and caudal) and paired (pectoral and pelvic) fins (**Figure 28.16**).

28.7a Class Acanthodii

The spiny "sharks" (*acanth* = spine) persisted from the late Ordovician through the Permian. Most of these sharklike fishes were less than 20 cm long, with small, light scales, streamlined bodies, well-developed eyes, large jaws, and numerous teeth (**Figure 28.17a**). Although acanthodians were not true sharks, they were probably fast swimmers and efficient predators. Many of them lived in fresh water. Most had a row of ventral spines and fins with internal skeletal support on each side of the body. The anatomy of acanthodians suggests a close relationship to bony fishes of today.

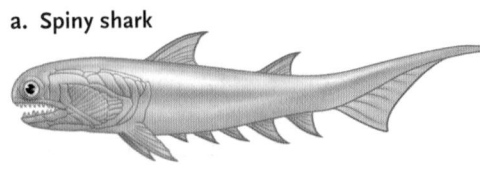

a. Spiny shark

b. Placoderm

c. Skull of *Dunkleosteus*

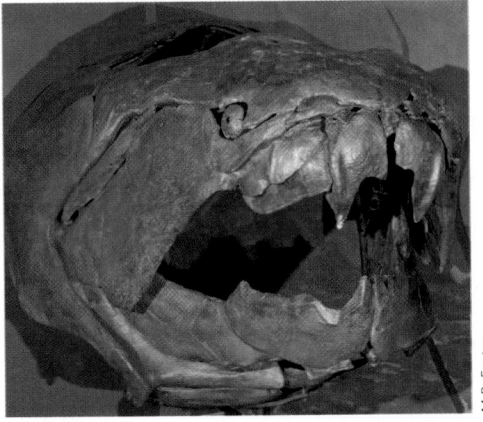

M.B. Fenton

Figure 28.17
Early gnathostomes. *Climatius*, an acanthodian (spiny shark) **(a)**, was small, about 8 cm long. The placoderm **(b)** *Dunkleosteus* was gigantic, growing to 10 m in length. Although some acanthodians had teeth, placoderms had only sharp cutting edges. The 3-m-long skull of a *Dunkleosteus* **(c)** demonstrates how impressive placoderms could be.

28.7b Class Placodermi

The placoderms (*plac* = plate; *derm* = skin) appeared in the Silurian and diversified in the Devonian and Carboniferous but left no direct descendants. Some, such as *Dunkleosteus* species (Figure 28.17b, c), reached lengths of 10 m. The bodies of placoderms were covered with large, heavy plates of bone anteriorly and smaller scales posteriorly. Their jaws had sharp cutting edges but no separate teeth, and their paired fins had internal skeletons and powerful muscles.

28.7c Class Chondrichthyes

The cartilaginous fishes (*chondr* = cartilage; *ichthy* = fish) are represented today by about 850 living species of sharks, skates and rays, and chimeras. As the name implies, their skeletons are entirely cartilaginous. However, the absence of bone is a derived trait because all earlier fishes had bony armour or bony endoskeletons. Most living chondrichthyans are grouped into two subclasses, the **Elasmobranchii** (skates, rays, and sharks; **Figure 28.18**) and the **Holocephali** (chimeras). Most are marine predators. With about 40 living species, holocephalians are the only cartilaginous fishes with an operculum (gill cover).

Skates and rays are dorsoventrally flattened (see Figure 28.18a) and swim by undulating their enlarged pectoral fins. Most are bottom dwellers that often lie partly buried in sand. They eat hard-shelled invertebrates (such as molluscs), which they crush with rows of flattened teeth (see Figure 28.18). The largest species, the manta ray (*Manta birostris*), measures 6 m across and eats plankton in the open ocean. Some rays

a. Manta ray

Masa Ushioda/age fotostock/Getty Images

b. Galapagos shark

c. Swell shark egg case

Photos.com

© BRUCE COLEMAN INC./Alamy

Figure 28.18
Chondricthyes.
(a) Skates and rays, such as the manta ray (*Manta birostris*), as well as **(b)** sharks, such as the Galapagos shark (*Carcharhinus galapagensis*), are grouped in the Elasmobranchii. The eggs of many sharks **(c)** include a large yolk that nourishes the developing embryo.

have electric organs that stun prey with shocks of as much as 200 volts. There are species of freshwater skates and rays in some rivers in the tropics; for example, in the Mekong River basin, some *Himantura chaophraya* are 2 m across.

Sharks (see Figure 28.18b, p. 647) are among the oceans' dominant predators. Flexible fins, lightweight skeletons, streamlined bodies, and the absence of heavy body armour allow most sharks to rapidly pursue prey. Their livers often contain **squalene**, an oil that is lighter than water, which increases their buoyancy. The great white shark (*Carcharodon carcharias*), the largest living predatory species of shark, can be 10 m long. At 18 m, the whale shark (*Rhincodon typus*) is the world's largest fish, and it eats only plankton. Sharks' teeth are designed for cutting. *Isisius plutodus,* the cookie-cutter shark, uses piercing teeth in its upper jaw to attach to its prey, biting with the lower jaw and its cutting teeth while rotating its body. The feeding process removes a disk of flesh from the prey. The combination of serrated teeth and flexible extensible jaws makes the effects of shark bites astonishing and frightening.

Elasmobranchs have remarkable adaptations for acquiring and processing food. Their teeth develop in whorls under the fleshy parts of the mouth. New teeth migrate forward as old, worn teeth break free **(Figure 28.19)**. In many sharks, the upper jaw is loosely attached to the cranium, and it swings down during feeding. As the jaws open, the mouth spreads wide, sucking in large, hard-to-digest chunks of prey, which are swallowed intact, allowing hurried eating. Although the elasmobranch digestive system is short, it includes a corkscrew-shaped **spiral valve**, which slows the passage of material and increases the surface area available for digestion and absorption.

Elasmobranchs also have well-developed sensory systems. In addition to vision and olfaction, they use **electroreceptors** to detect weak electric currents produced by other animals. Their **lateral line system**, a row of tiny sensors in canals along both sides of the body, detects vibrations in water (see Figure 45.5, p. 1102). They use urea as an osmolyte that makes their body fluids more concentrated than sea water. Freshwater skates have much lower concentrations of urea in their blood than their saltwater relatives do (for more about osmoregulation, see Chapter 50).

Chondrichthyans have evolved numerous reproductive specializations. Males have a pair of organs, the **claspers,** on the pelvic fins, which help transfer sperm into the female's reproductive tract. Fertilization occurs internally. In many species, females produce yolky eggs with tough leathery shells (see Figure 28.18c). Others retain the eggs within the oviduct until the young hatch. A few species nourish young in utero (see Chapters 41 and 42).

28.7d The Bony Fishes

In terms of diversity (numbers of species) and sheer numbers of individuals, fishes with bony endoskeletons (cranium, vertebral column with ribs, and bones supporting their movable fins) are the most successful of all vertebrates. The endoskeleton provides lightweight support compared with the bony armour of ostracoderms and placoderms, enhancing their locomotor efficiency. Some bony fishes have cartilaginous skeletons, but they are not chondrichthyans.

Bony fishes have numerous adaptations that increase swimming efficiency. The scales of most bony fishes are small, smooth, and lightweight, and their bodies are covered with a protective coat of mucus that retards bacterial growth and minimizes drag as water flows past the body.

Bony fishes first appeared in the Silurian and rapidly diversified into two lineages, Actinopterygii and Sarcopterygii. The ray-finned fishes (Actinopterygii; *acti* = ray; *ptery* = fin) have fins supported by thin and flexible bony rays, whereas the fleshy-finned fishes (Sarcopterygii; *sarco* = flesh) have fins supported by muscles and an internal bony skeleton. Ray-finned

Figure 28.19

Elasmobranch teeth. Barndoor skates (*Dipturus laevis*; see also Chapter 32) **(a)** are specialized for crushing hard prey, such as bivalve molluscs. Cookie-cutter sharks (*Isistius plutodus*) **(b)** have cutting teeth. In (a) and (b), the replacement pattern of the teeth (from back to front of the jaws) is obvious.

a.
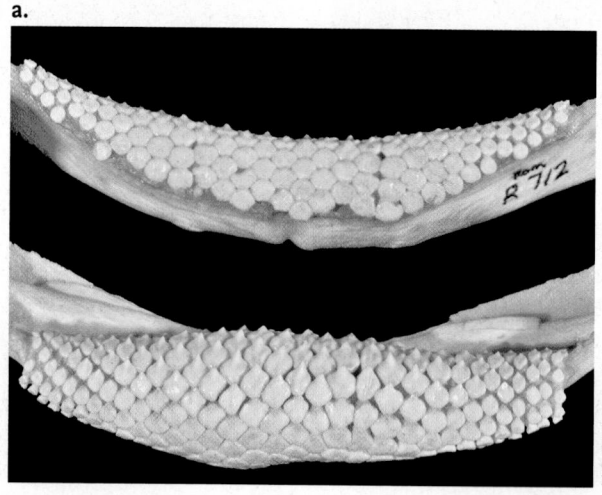

b.

M.B. Fenton

M.B. Fenton

a. Lake sturgeon

b. Long-nosed gar

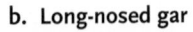

Figure 28.20

Ancestral actinopterygians (ray-finned bony fishes). Lake sturgeon (*Acipenser fulvescens*) **(a)**, and a long-nosed gar (*Lepidosteus sasteus*) **(b)**, are living representatives of early ray-finned fishes.

fishes are more diverse as measured by numbers of species and today vastly outnumber fleshy-finned fishes. The ~30 thousand living species of bony fishes occupy nearly every aquatic habitat and represent more than 95% of living fish species. Adults range from 1 cm to more than 6 m in length. In the Yangtze River basin, *Pseuphurus glodius,* the Chinese paddlefish, can weigh up to 500 kg.

Class Actinopterygii. Sturgeons **(Figure 28.20a)** and paddlefishes, the most ancestral members of this group, are characterized by mostly cartilaginous skeletons. These large fishes live in rivers and lakes of the Northern Hemisphere. Sturgeons eat detritus and invertebrates, whereas paddlefish eat plankton. Gars (Figure 28.20b) and bowfins are remnants of a more recent radiation. They occur in the eastern half of North America, where they eat fish and other prey. Gars are protected from predators by a heavy coat of bony scales.

The subclass Teleosteii represents the latest radiation of Actinopterygii, one that produced a wide range of body forms **(Figure 28.21, p. 650).** Teleosts have an internal skeleton made almost entirely of bone. On either side of the head, the **operculum**, a flap of the body wall, covers a chamber that houses the gills. Sensory systems (see Chapter 45) generally include large eyes, a lateral line system, sound receptors, chemoreceptive nostrils, and taste buds.

Variations in jaw structure allow different teleosts to consume plankton, macroalgae, invertebrates, or other vertebrates. Teleosts exhibit remarkable adaptations for feeding and locomotion. When some teleosts open their mouths, bones at the front of the jaws swing forward to create a circular opening. Folds of skin extend backward, forming a tube through which they suck food (see Figure 28.21f). Like Chondrichthyes, Actinopterygii exhibit great variation in tooth structure **(Figure 28.22, p. 651).** Species such as piranhas (*Sarrasalamus*) are notorious for their bites. Other species have teeth specialized for crushing hard prey, such as bivalve molluscs. Whereas the piranha's

teeth are on the premaxilla, maxilla, and mandible (as they are in mammals and many other vertebrates), the crushing teeth of ray-finned fishes often occur on the bones of the pharynx.

In many modern ray-finned fishes, a gas-filled **swim bladder** serves as a hydrostatic organ that increases buoyancy (see Figure 28.21a). The swim bladder is derived from an ancestral air-breathing lung that allowed early actinopterygians to gulp air, supplementing gill respiration in aquatic habitats, where dissolved oxygen concentration is low.

Many have symmetrical tail fins posterior to the vertebral column that provide power for locomotion. Their pectoral fins often lie high on the sides of the body, providing fine control over swimming. Some species use pectoral fins for acquiring food, for courtship, and for care of eggs and young. Some teleosts use pectoral fins for crawling on land (e.g., mudskippers, *Periophthalmus* species, and climbing perch, *Anabas* species) or gliding in the air (flying fish, family Exocoetidae).

Most marine species produce small eggs that hatch into larvae that live among the plankton. Eggs of freshwater teleosts are generally larger and hatch into tiny versions of the adults. Parents often care for their eggs and young, fanning oxygen-rich water over them, removing fungal growths, and protecting them from predators. Some freshwater species, such as guppies, give birth to live young (see "On the Road to Vivipary," p. 998).

Class Sarcopterygii. The two groups of fleshy-finned fishes—lobe-finned fishes and lungfishes—are represented by only eight living species **(Figure 28.23, p. 651).** Although lobe-finned fishes were once thought to have been extinct for 65 million years, a living coelacanth (*Latimeria chalumnae*) was discovered in 1938 near the Comoros Islands, off the southeastern coast of Africa. A population of these metre-long fishes live at depths of 70–600 m, feeding on other fishes and squid. Remarkably, a second population of coelacanths was discovered in 1998, 10 000 km east of the Comoros,

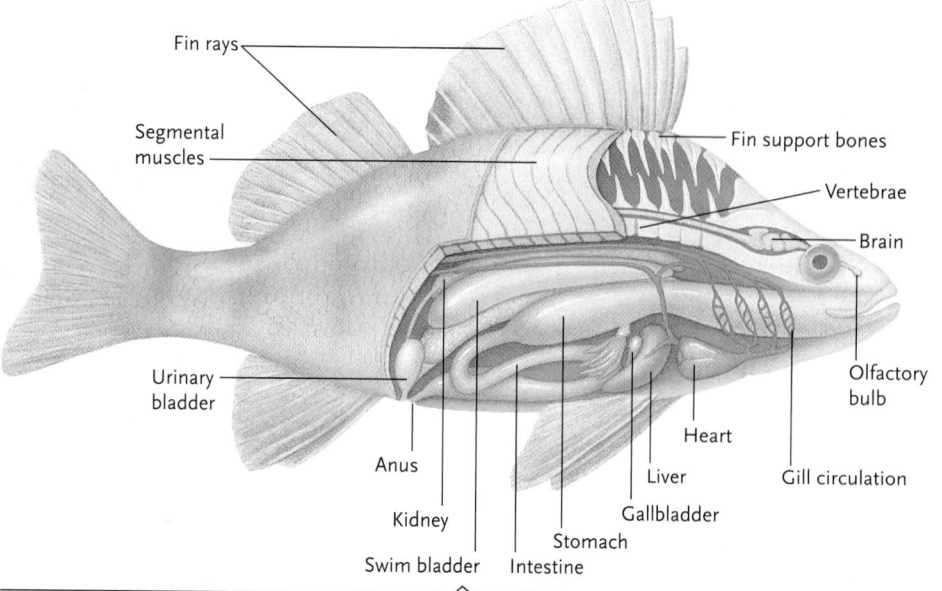

a. Teleost internal anatomy

Labels: Fin rays, Segmental muscles, Fin support bones, Vertebrae, Brain, Olfactory bulb, Gill circulation, Heart, Liver, Gallbladder, Intestine, Stomach, Swim bladder, Kidney, Anus, Urinary bladder

b. Sea horses, like the northern sea horse (*Hippocampus hudsonius*), use a prehensile tail to hold on to substrates; they are weak swimmers.

Digital Vision/Getty Images

c. The long, flexible body of a spotted moray eel (*Gymnothorax moringa*) can wiggle through the nooks and crannies of a reef.

Kit Kittle/Corbis

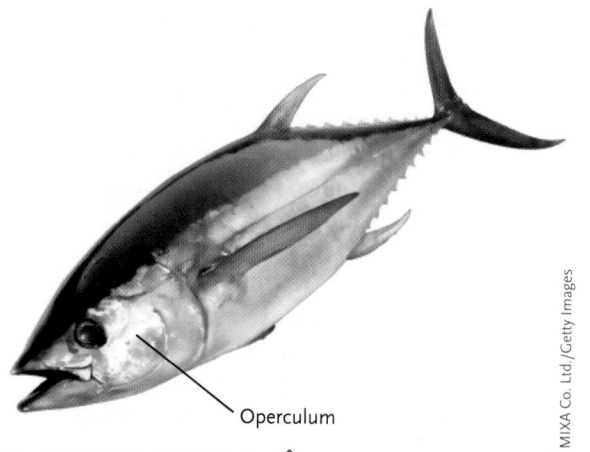

d. Flatfishes, like this European flounder (*Platichthys flesus*), lie on one side and leap at passing prey.

© blickwinkel/Alamy

Operculum

e. Open ocean predators, like the yellowfin tuna (*Thunnus albacares*), have strong, torpedo-shaped bodies and powerful caudal fins.

MIXA Co. Ltd./Getty Images

f. Kissing Gouramis (*Helostoma temmincki*) extend their jaws into a tube that sucks food into the mouth.

Arthur W. Ambler/Science Source

Figure 28.21

Teleost diversity. Although all teleosts (bony fish) share similar internal features, their diverse shapes adapt them to different diets and types of swimming.

a.

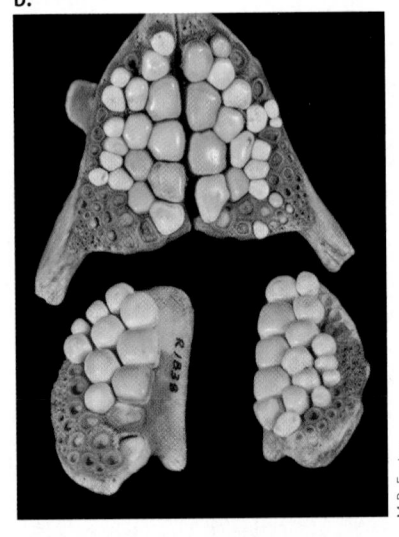

b.

Figure 28.22

Teleost teeth. Like Chrondricthyes, bony fishes have also developed cutting **(a)** and crushing **(b)** teeth. The cutting teeth are those of a piranha (*Sarrasalamus* species); the crushing teeth are from a black drum (*Pogones cromis*).

M.B. Fenton

M.B. Fenton

a. Coelacanth

b. Australian lungfish

AlessandroZocc/Shutterstock.com

Tom McHugh/Science Source

Figure 28.23

Sarcopterygians. The coelocanth (*Latimeria chalumnae*) **(a)** is now one of two living species of lobe-finned fishes. The Australian lungfish (*Neoceratodus forsteri*) **(b)** is one of six living lungfish species.

when a specimen was found in an Indonesian fish market. Analyses of the DNA of the Indonesian specimen indicated that it is a distinct species (*Latimeria menadoensis*).

Lungfishes have changed relatively little over the last 200 million years. Six living species are distributed on southern continents. Australian lungfishes live in rivers and pools, using their lungs to supplement gill respiration when dissolved oxygen concentration is low. South American and African species live in swamps and use their lungs for breathing during the annual dry season, which they spend encased in a mucus-lined burrow in the dry mud. When the rains begin, water fills the burrow and the fishes awaken from dormancy. During their periods of dormancy, these fishes excrete urea.

STUDY BREAK

1. What did the evolution of jaws mean for fish?
2. What anatomical and physiological characteristics make sharks dominant ocean predators?
3. What is the lateral line system? What does it do?

28.8 Early Tetrapods and Modern Amphibians

The fossil record suggests that tetrapods evolved in the late Devonian from a group of fleshy-finned fishes, the Osteolepiformes. Osteolepiformes and early tetrapods shared several derived characteristics, including dental and cranial features. Specifically, both had infoldings of tooth surfaces that probably increased the functional area of the tooth. They also shared shapes and positions of bones on the dorsal side of their crania and in their appendages.

Some problems of moving onto land were identified earlier. During dry periods in swampy, late Devonian habitats, drying pools may have forced osteolepiform ancestors to move overland to adjacent pools that still had water. During these excursions, the fish may have found that land plants, worms, and arthropods provided abundant food, and oxygen was more readily available in air than in water. Furthermore, there may well have been fewer terrestrial predators at that time, but this interpretation is open to question.

a. *Eusthenopteron*, an osteolepiform fish

b. *Ichthyostega*, an early tetrapod

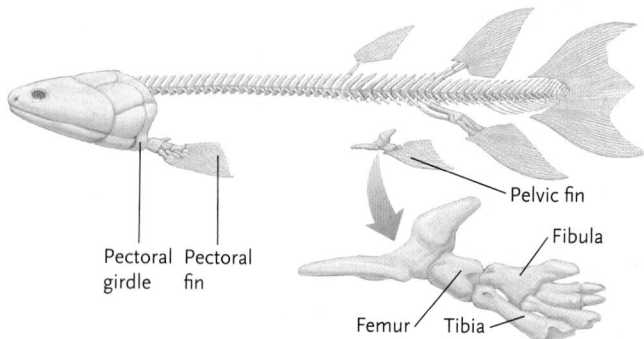

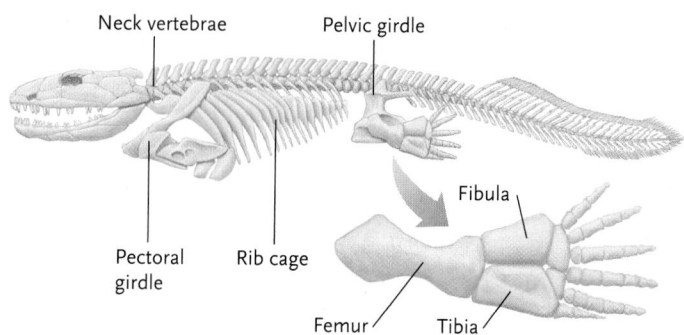

Figure 28.24

Evolution of tetrapod limbs. The limb skeleton of osteolepiform fishes such as **(a)** *Eusthenopteron* is homologous to that of early tetrapods, such as **(b)** *Ichthyostega*. Although *Ichthyostega* retained many fishlike characteristics, its pectoral girdle was completely freed from the cranium, and it had a heavy ribcage. Fossils of its forefoot have not yet been discovered.

Osteolepiformes **(Figure 28.24a)** usually had strong, stout fins that allowed them to crawl on mud. Of particular importance were crescent-shaped bones in their vertebral columns that provided strong intervertebral connections. Their nostrils led to sensory pits housing olfactory (odour) receptors (see Chapter 45). They almost certainly had lungs, allowing them to breathe atmospheric oxygen. Like living lungfishes, they could also have excreted urea or uric acid rather than ammonium, which is toxic.

The earliest tetrapod with nearly complete skeletal data is the semiterrestrial, metre-long *Ichthyostega* (Figure 28.24b). Compared with its fleshy-finned ancestors, *Ichthyostega* had a more robust vertebral column, sturdier limb girdles and appendages, a ribcage that protected its internal organs (including lungs), and a neck. Fishes lack necks because the pectoral girdle is fused to the cranium. In *Ichthyostega,* several vertebrae separated the pectoral girdle and the cranium, allowing the animal to move its head to scan the environment and capture food. *Ichthyostega* retained a fishlike lateral line system, caudal fin, and scaly body covering.

Life on land also required changes in sensory systems. In fishes, the body wall picks up sound vibrations and transfers them directly to sensory receptors. Sound waves are harder to detect in air. The appearance of a **tympanum** (ear drum) in early tetrapods apparently allowed them to detect vibrations in air associated with airborne sounds. The tympana are specialized membranes on either side of the head. The tympanum connects to the **stapes**, a bone homologous to the hyomandibula, which had supported the jaws of fishes (see Figure 20.14, Chapter 20). The stapes, in turn, transfers vibrations to the sensory cells of an inner ear.

28.8a Class Amphibia: Frogs and Toads, Salamanders, and Caecilians

Most of the 6000+ living species of amphibians (*amphi* = both; *bios* = life) are small, and their skeletons contain fewer bones than those of Paleozoic tetrapods such

as *Ichthyostega*. All living amphibians are carnivorous as adults, but the aquatic larvae of some are herbivores. Fossil amphibians, such as *Eryops* **(Figure 28.25),** were quite large and predatory.

The thin, scaleless skin of most living amphibians is well supplied with blood vessels and can be a major site of gas exchange. To operate in oxygen uptake, the skin must be moist and thin enough to bring blood into close contact with air. Having moist skin limits amphibians to moist habitats. Many species of living amphibians keep their skin surfaces moist, and some are lungless, but most use lungs in gaseous exchange. The evolution of lungs was accompanied by modifications of the heart and circulatory system that increase the efficiency with which oxygen is delivered to body tissues (see Chapter 42). Some adult anurans have a waxy coating on their skin, making them as waterproof as lizards **(Figure 28.26).**

The life cycles of many amphibians include larval and adult stages. In frogs, larvae (tadpoles) hatch from fertilized eggs and eventually metamorphose into adults (see Chapter 42). The larvae of most frog species are aquatic, but adults may live their lives in water (be aquatic), move between land and water (be amphibious), and live entirely on land (be terrestrial). Some salamanders are pedomorphic (see Chapter 17), which means that the larval stage attains sexual maturity without changing its form or moving to land. Some frogs and salamanders reproduce on land, omitting the larval stage altogether. In these species, tiny adults emerge directly from fully developed eggs. However, the eggs of terrestrial breeders dry out quickly unless they are laid in moist places.

Modern amphibians are represented by three lineages **(Figure 28.27, p. 654),** but the evolutionary origin of frogs, salamanders, and caecilians has remained unresolved. The 2008 description of a small fossil from the Lower Permian of Texas suggests that frogs and salamanders have a relatively close common ancestor, whereas caecilians are distantly related to them.

Figure 28.25
A fossil amphibian. This amphibian, *Eryops*, from the Texas Permian, was about 1.8 m long and was strikingly different from living amphibians.

M.B. Fenton

a.

M.B. Fenton

Figure 28.26
Waterproof frogs. **(a)** *Chiromantis xerampelina* from southern Africa and **(b)** *Phyllomedusa sauvagii* from South America make their skin waterproof with a waxy secretion. These frogs are as waterproof as chameleons. They also excrete uric acid to further conserve water.

b.

M.B. Fenton

Populations of practically all amphibians have declined rapidly in recent years. These declines are probably due to exposure to acid rain, high levels of ultraviolet B radiation, and fungal and parasitic infections. Another major factor in the decline of amphibians may be habitat splitting, the human-induced disconnection of habitats essential to the survival of amphibians. This aspect of **habitat fragmentation** (see "People behind Biology," Box 31.2, p. 770) can cause adult amphibians to move across inhospitable habitat (roads, power line rights-of-way) to reach breeding habitats.

Anura. The 3700 species of frogs and toads (*an* = not; *ura* = tail) have short, compact bodies, and the adults lack tails. Their elongated hind legs and webbed feet allow them to hop on land or to swim. A few species are adapted to dry habitats, encasing themselves in mucus cocoons to withstand periods of drought.

Urodela. The 400 species of salamanders (*uro* = tail; *del* = visible) have an elongated, tailed body and four legs. They walk by alternately contracting muscles on either side of the body, much the way fishes swim. Species in the most diverse group, the lungless salamanders, are fully terrestrial throughout their lives, using their skin and the lining of the throat for gas exchange.

a. A frog

b. A salamander

c. A caecelian

Figure 28.27

Living amphibians. Anurans **(a),** such as the northern leopard frog (*Rana pipiens*), have compact bodies and long hind legs. Urodeles **(b),** such as the red-spotted newt (*Notophthalmus viridescens*), have an elongate body and four legs. Caecilians **(c),** such as *Caecelia nigricans* from Colombia, are legless burrowers.

Gymnophonia. The 200 species of caecelians (*gymno* = naked; *ophioneos* = snakelike) are legless, burrowing animals with wormlike bodies. They occupy tropical habitats throughout the world. Unlike other extant amphibians, caecelians have small bony scales embedded in their skin. Fertilization is internal, and females give birth to live young. In some species, the mother's skin produces a milklike substance for the young, which use specialized teeth to collect it from the mother's body (see Chapter 42).

STUDY BREAK

1. Present four lines of evidence suggesting that tetrapods arose from Osteolepiformes.
2. Why was the development of the tympanum important to life on land?
3. What characteristics allow amphibians to use their skin as a major site of gas exchange?

28.9 The Origin and Mesozoic Radiations of Amniotes

The amniote lineage arose during the Carboniferous, a time when seed plants and insects began to invade terrestrial habitats, providing additional food and cover for early terrestrial vertebrates. Amniotes take their name from the **amnion,** a fluid-filled sac that surrounds the embryo during development (see Chapter 42). Although the fossil record includes many skeletal remains of early amniotes, it provides little direct information about soft body parts and physiology. Three key features of living amniotes allow life on dry land and liberate them from reliance on standing water. The changes involve being waterproof and producing waterproof eggs.

- First, skin is waterproof: keratin and lipids in the cells make skin relatively impermeable to water.

- Second, **amniote (amniotic) eggs** can survive and develop on dry land because they have four specialized membranes and a hard or leathery shell perforated by microscopic pores **(Figure 28.28).** Amniote eggs are resistant to desiccation. The membranes protect the developing embryo and facilitate gas exchange and excretion. The shell mediates the exchange of air and water between the egg and its environment. Developing amniote embryos can excrete uric acid, which is stored in the allantois of the embryo, which will later become the bladder. Generous supplies of **yolk** in the egg are the developing embryo's main energy source, whereas **albumin** supplies nutrients and water. There is no larval stage, and hatchling amniotes are miniature versions of the adult. Amniote eggs are the ancestral condition, but they

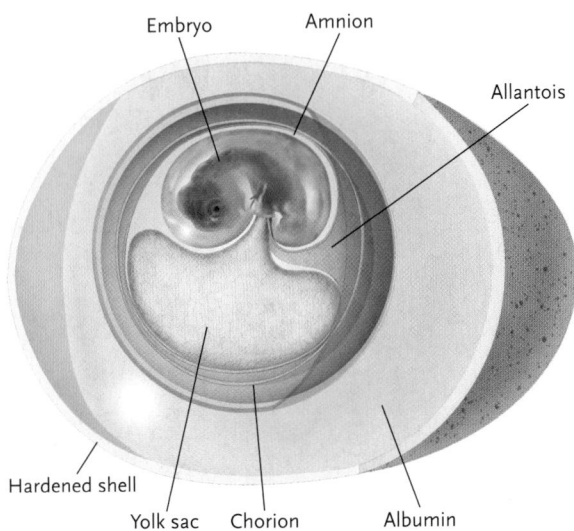

Embryo   Amnion

Allantois

Hardened shell

Yolk sac   Chorion   Albumin

Figure 28.28

The amniote egg. A water-retaining egg with four specialized membranes surrounded by a hard or leathery shell allowed amniotes and their descendants to reproduce in dry environments. The chorion surrounds the amnion which in turn surrounds the amniotic fluid.

are circumvented in most mammals (and some reptiles; see Chapter 18).

- Third, some amniotes produce urea and/or uric acid as a waste product of nitrogen metabolism (see Chapter 44). Although ammonia (NH_3^+) is less expensive (metabolically) to produce, it is toxic and must be flushed away with water. Urea is much less toxic than NH_3^+ and therefore easier to store and to void. Uric acid is even less toxic and, because it is insoluble, it can be stored or voided without risk while conserving water.

The abundance and diversity of fossils of amniotes indicate that they were extremely successful, quickly replacing many nonamniote species in terrestrial habitats. During the Carboniferous and Permian, amniotes produced three major radiations: synapsids, anapsids, and diapsids **(Figure 28.29, p. 656)**, distinguishable by the numbers of bony arches in the temporal region of the skull (in addition to the openings for the eyes (see **Figure 28.30, p. 657)**. The bony arches delimit fenestrae, openings in the skull that allow space for contraction (and expansion) of large and powerful jaw muscles.

Synapsids (Figure 28.30a), a group of small predators, were the first offshoot from ancestral amniotes. Synapsids (*syn* = with; *apsid* = connection) had one temporal arch on each side of the head. They emerged late in the Permian, and mammals are their living descendants.

Anapsida (Figure 28.30b), the second lineage (*an* = not), had no temporal arches and no spaces on the sides of the skull. Turtles are living representatives of this group.

Diapsida (Figure 28.30c, d; *di* = two) are the third lineage and included most Mesozoic amniotes. Diap-

sids had two temporal arches, and their descendants include the dinosaurs, as well as extant lizards and snakes, crocodilians, and birds. Arguably, birds are other examples of living diapsids.

28.9a Extinct Diapsids

Early diapsids differentiated into two lineages, **Archosauromorpha** (*archo* = ruler; *sauro* = lizard; *morph* = form) and **Lepidosauromorpha** (*lepi* = scale), which differed in many skeletal characteristics. Archosaurs (archosauromorphs), or "ruling reptiles," include crocodilians, pterosaurs, and dinosaurs. Crocodilians first appeared during the Triassic. They have bony armour and a laterally flattened tail, which is used to propel them through water. Pterosaurs, now extinct, were flying predators of the Jurassic and Cretaceous **(Figure 28.31, p. 657)**. The smallest were sparrow sized; the largest had wing spans of 11 m. Some evidence indicates that pterosaur wings attached to the side of their bodies at about the hips.

Two lineages of dinosaurs, "lizard-hipped" saurischians and "bird-hipped" ornithischians, proliferated in the Triassic and Jurassic **(Figure 28.32, p. 657)**. Saurischians included bipedal carnivores and quadrupedal herbivores. Some carnivorous saurischians **(Figure 28.33a, p. 658)** were swift runners, and some had short forelimbs (e.g., *Tyrannosaurus rex,* which was 12 m long and stood 6 m high; Figure 28.33b). One group of small carnivorous saurischians, the deinonychsaurs, is ancestral to birds (see Figure 28.33a).

By the Cretaceous, some herbivorous saurischians were gigantic, and many had long, flexible necks. *Apatosaurus* (previously known as *Brontosaurus*) was 25 m long and may have weighed 50 000 kg **(Figures 28.34 and 28.35, p. 658)**. The largely herbivorous ornithischian dinosaurs had large, chunky bodies. This lineage included armoured or plated dinosaurs (*Ankylosaurus* and *Stegosaurus*), duck-billed dinosaurs (*Hadrosaurus*), horned dinosaurs (*Styracosaurus*), and some with remarkably thick skulls (*Pachycephalosaurus*). Ornithischians were most abundant in the Jurassic and Cretaceous.

Lepidosaurs (Lepidosauromorpha) are the second major lineage of diapsids. This diverse group included both marine and terrestrial animals. Fossil lepidosaurs include champosaurs (see Figure 28.30d, p. 657), which were freshwater fish eaters, and the marine, fish-eating plesiosaurs, with long, paddlelike limbs they used like oars **(Figure 28.36, p. 658)**. Fossil lepidosaurs also included ichthyosaurs (see Figure 20.20, p. 464), porpoise like animals with laterally flattened tails. Like today's whales, ichthyosaurs were highly specialized for marine life and did not return to land to lay eggs. Indeed, it appears that ichthyosaurs, like today's whales, gave birth to live young. Squamates, the living lizards and snakes, are the third important group within this lineage. *Sphenodon,* the tuatara, is the last living genus of a once diverse group of lizard like squamates.

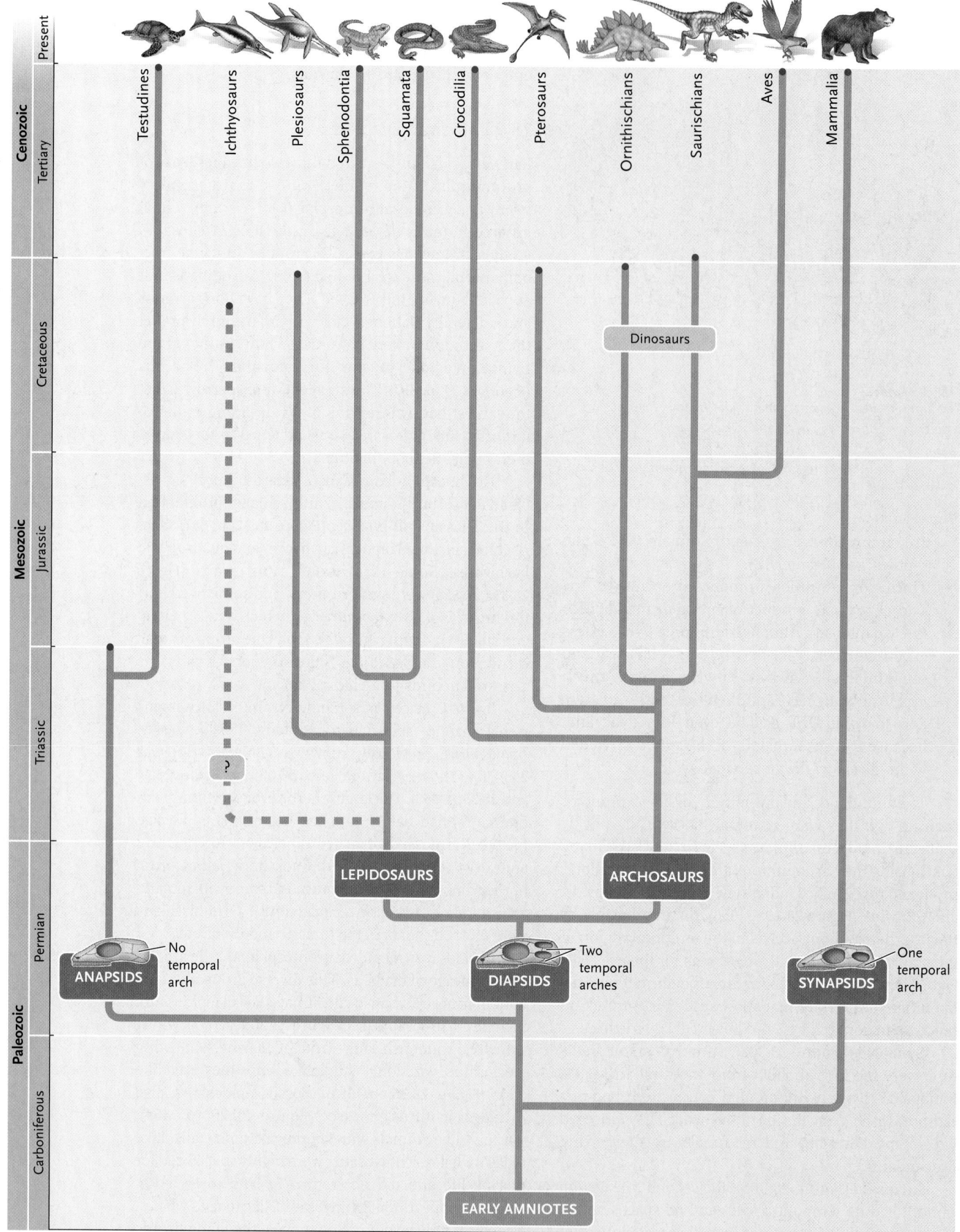

Figure 28.29

Amniote ancestry. The early amniotes gave rise to three lineages (anapsids, synapsids, and diapsids) and numerous descendants. The lineages are distinguished by the number of bony arches in the temporal region of the skull (indicated on the small icons).

a.

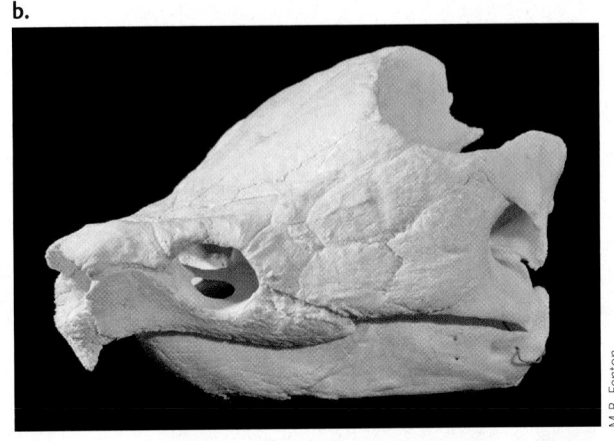

b.

Figure 28.30
Skulls of reptiles. The synapsid condition **(a),** shown by *Dimetrodon;* the anapsid condition **(b),** shown by a snapping turtle; and the diapsid conditions shown by *Camarosaurus* **(c)** and *Champsosaurus* **(d).**

c.

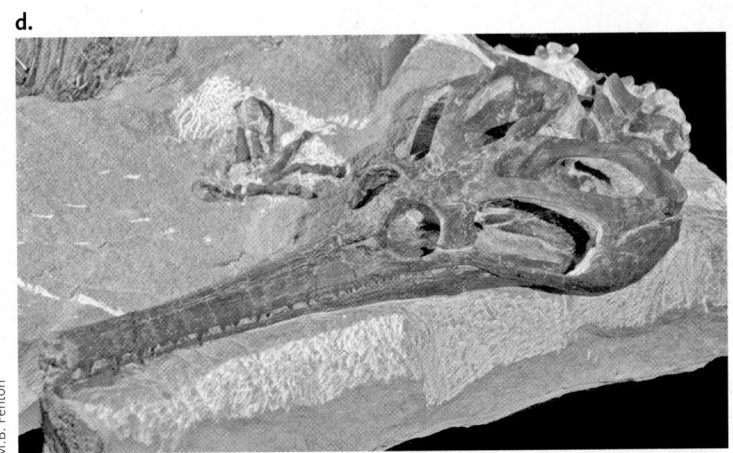

d.

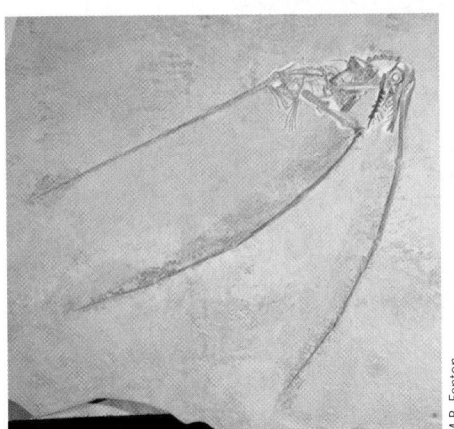

Figure 28.31
Rhamphorynchus meunsteri, a pterosaur with a wing span of about 1.7 m. Note the impressions of the wing membranes, the teeth, and the long tail. This species is known from the Upper Jurassic of Germany.

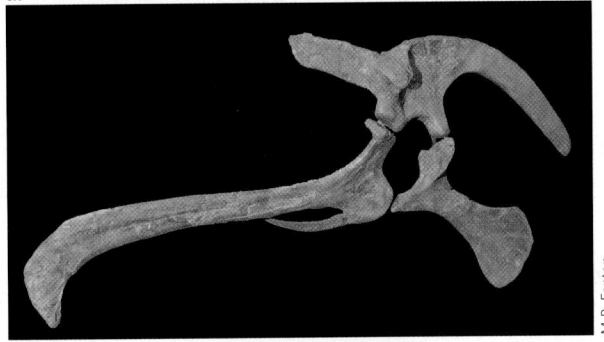

a.

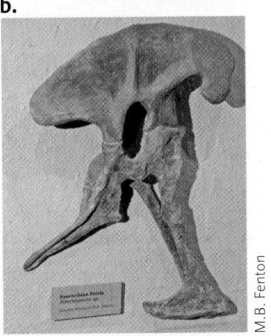

b.

Figure 28.32
Ornithischian **(a)** and saurischian **(b)** dinosaurs differed in their pelvic structures. The ornithischian is a hadrasaur (duck-billed dinosaur), the saurischian an Albertosaurus. In each case, the **acetabulum,** the socket receiving the head of the femur, is the large elliptical area in the middle.

Figure 28.33
Saurischian dinosaurs. Whereas *Ornitholestes hermanii* **(a)** stood less than 1 m at the shoulder, the fearsome *Tyrannosaurus rex* **(b)** was about 12 m long. Both were carnivores.

a.

b.

a. *Lambeosaurus*

b. *Stegoceras*

a.

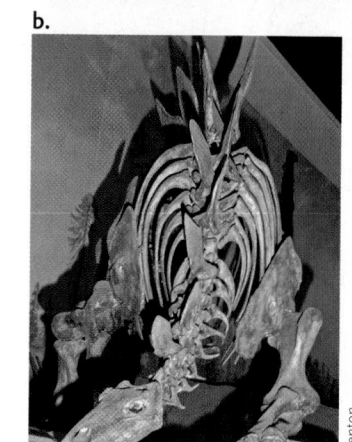

Figure 28.34
Ornithischian dinosaurs. These herbivores ranged in size from the 15-m-long **(a)** *Lambeosaurus lambei*, a smaller, thick-skulled **(b)** *Stegoceras*, to the 10-m-long **(c)** *Triceratops horridus*.

c. *Triceratops*

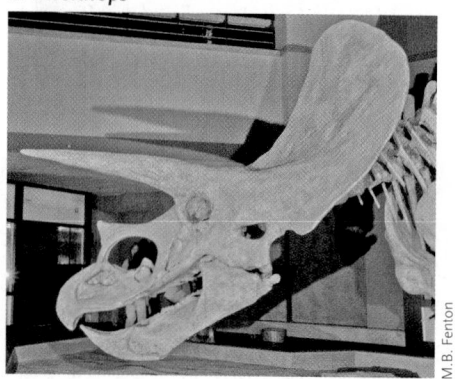

b.

Figure 28.35
Large, lumbering herbivores. Other ornithischian dinosaurs included the 18-m-long *Camarasaurus supremus* **(a)** and the 9-m-long *Stegosaurus armatus* **(b)**. The latter had distinctive plates along its back.

Figure 28.36
Paddles. Paddlelike forelimbs developed in sea turtles **(a)** and plesiosaurs **(b)**, *Trinacromerum bonneri*).

a.

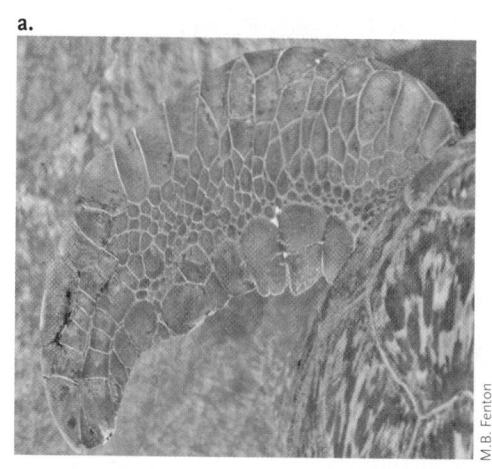

b.

a.

b.

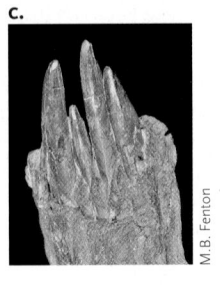
c.

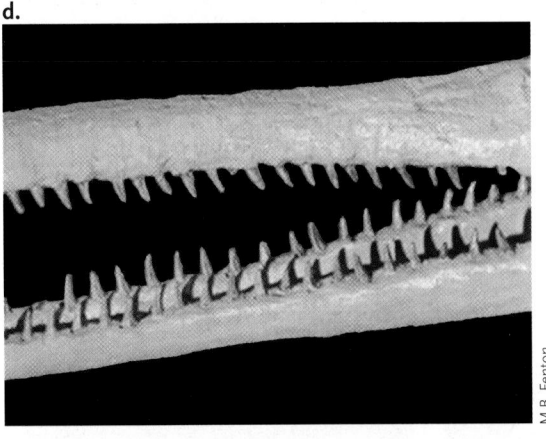
d.

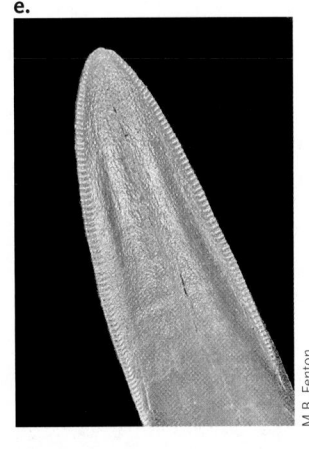
e.

f.

Figure 28.37
Reptile teeth. As usual, teeth reflect the dietary habits of vertebrates. Herbivorous dinosaurs, **(a)** *Diplodocus longus* and **(b)** hadrosaur, had teeth adapted for gathering plant material (a) and grinding it (b). They differ from those of a carnivorous dinosaur (*Daspletosaurus*) **(c)** or a fish-eating reptile such as a champhosaur **(d)**. A tooth of a *Tyrannosaurus rex* changed distinctly over its length. The biting part of the tooth **(e)** had enamel and serrated edges. There was no enamel on the part of the tooth located within the socket of the skull **(e, f)**.

The teeth of reptiles provide important clues about their diets **(Figure 28.37)** and show interesting parallels with the teeth of other vertebrates.

STUDY BREAK

1. Where do amniotes get their name? Why are amniote eggs resistant to desiccation?
2. What three key features liberate living amniotes from reliance on standing water?
3. Name and describe three major radiations of amniotes during the Carboniferous and Permian.

28.10 Subclass Testudinata: Turtles and Tortoises

The turtle body plan, largely defined by a bony, boxlike shell, has changed little since the group first appeared during the Triassic **(Figure 28.38).** A turtle's ribs are fused to the inside of the shell, and in contrast to other tetrapods, the pectoral and pelvic girdles lie within the ribcage. The shell is formed from large keratinized scales covering the bony plates.

The 250 living species occupy terrestrial, fresh water, and marine habitats. They range from 8 cm to 2 m in length. Turtles use a keratinized beak in feeding, whether they eat animal or plant material. When threatened, most species retract into their shells. Many species are now endangered because adults are hunted for meat

a. The turtle skeleton

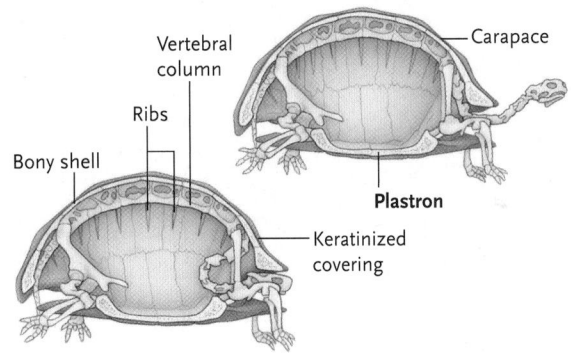

b. An aquatic turtle

Figure 28.38
Testudines. Most turtles **(a)** can withdraw their heads and legs into a bony shell. Aquatic turtles **(b)**, such as the eastern painted turtle (*Chrysmys picta*), often bask in the Sun to warm up. The sunlight may also help to eliminate parasites that cling to the turtle's skin.

and their eggs are eaten by humans and other predators. Young are often collected for the pet trade, and the beaches favoured as nesting sites by marine species are too often used as tourist attractions (see Chapter 32).

STUDY BREAK

Describe the body plan of a turtle.

28.11 Living Diapsids: Sphenodontids, Squamates, and Crocodilians

28.11a Infraclass Lepidosaura, Order Rhynchocephalia: The Tuatara

Sphenodon punctatus is one of two living species of sphenodontids (*sphen* = wedge; *dont* = tooth) or tuataras, a lineage that was diverse in the Mesozoic

(Figure 28.39a). These lizardlike animals are best known as tetrapods with a "third" or pineal eye, a reflection of earlier vertebrates such as lampreys with pineal eyes (see also photoreceptors in Chapter 1). They survive on a few islands off the coast of New Zealand. Adults are about 60 cm long. They live in dense colonies, where males and females defend small territories. They often share underground burrows with seabirds and eat invertebrates and small vertebrates. They are primarily nocturnal and maintain low body temperatures during periods of activity. Their survival is threatened by two introduced predators, cats and rats.

28.11b Infraclass Lepidosaura, Order Squamata: Lizards and Snakes

Lizards and snakes (Figure 28.39b) are covered by overlapping, keratinized scales (*squam* = scale) that protect against dehydration. Squamates periodically shed their skin while growing, much the way

a. Sphenodontia includes the tuatara (*Sphenodon punctatus*) and one other species.

b. Basilisk lizards (*Basiliscus basiliscus*) escape from predators by running across the surface of streams.

c. A western diamondback rattlesnake (*Crotalus atrox*) of the American southwest bares its fangs with which it injects a powerful toxin into prey.

Venom gland

Hollow fang

d. Crocodilia includes semiaquatic predators, like this resting African Nile crocodile (*Crocodylus niloticus*), that frequently bask in the Sun.

Figure 28.39
Living nonfeathered diapsids.

arthropods shed their exoskeletons (see Chapter 27). Most squamates regulate their body temperature behaviourally (see Chapter 50), so they are active only when weather conditions are favourable. They shuttle between sunny and shady places to warm up or cool down as needed.

Most of the 3700 lizard species are less than 15 cm long, but Komodo dragons (*Varanus komodoensis*) grow to nearly 3 m in length (see Figure 45.11, Chapter 45). Lizards occupy a wide range of habitats and are especially common in deserts and the tropics. One species (*Lacerta vivipara*) occurs within the Arctic Circle. Most lizards eat insects, although some consume leaves or meat.

The 2300 species of snakes evolved from a lineage of lizards that lost their legs over evolutionary time. Streamlined bodies make snakes efficient burrowers or climbers (Figure 28.39c). Many subterranean species are 10 or 15 cm long, whereas the giant constrictors may grow to 10 m. Unlike lizards, all snakes are predators that swallow prey whole. Compared with their lizard ancestors, snake skull bones are reduced in size and connected to each other by elastic ligaments. This gives snakes a remarkable capacity to stretch their mouths. Some snakes can swallow food items that are larger than their heads (see Chapter 48). Snakes also have well-developed sensory systems for detecting prey. The flicking tongue carries airborne molecules to sensory receptors in the roof of the mouth (see "Forked Tongues," p. 1118). Most snakes can detect vibrations on the ground, and some, like rattlesnakes, have heat-sensing organs (see Figure 45.26, Chapter 45). Many snakes kill by constriction, which suffocates prey, whereas other species produce venoms, toxins that immobilize, kill, and partially digest prey (see "Molecule behind Biology," Box 30.1, p. 718).

28.11c Infraclass Archosauria, Order Crocodylia: Crocodiles, Alligators, and Gavials

The 21 species of alligators and crocodiles, along with the birds, are the living remnants of the archosaurs (Figure 28.39d). Australian saltwater crocodiles (*Crocodylus porosus*) are the largest, growing to 7 m in length. Crocodilians are aquatic predators that eat other vertebrates. Striking anatomical adaptations distinguish them from living lepidosaurs, including a four-chambered heart that is homologous to the heart in birds, analogous to this structure in mammals. In some crocodilians, muscles that originate on the pubis insert on the liver and pericardium. When these muscles contract, the liver moves toward the tail, creating negative pressure in the chest cavity and drawing air in. This situation is analogous to the role of the diaphragm in mammals.

American alligators (*Alligator mississippiensis*) exhibit strong maternal behaviour, perhaps reflecting their relationship to birds. Females guard their nests ferociously and, after the young hatch, free their offspring from the nest. The young stay close to the mother for about a year, feeding on scraps that fall from her mouth and living under her watchful protection.

Many species of alligators and crocodiles are endangered because their habitats have been disrupted by human activities. They have been hunted for meat and leather and because larger individuals are predators of humans. There is hope, however, as some populations of *A. mississippiensis* have recovered in the wake of efforts to protect them. In Africa and Australia, crocodiles are farmed for their meat and skin.

In the past, crocodilians were more diverse in body form than they are today. *Dakosaurus andiniensis*, a Jurassic–Cretaceous marine crocodilian from western South America, differed dramatically from typical crocodilians **(Figure 28.40, p. 662)**.

STUDY BREAK

1. How do snakes kill their prey?
2. What features of crocodilians are homologous to those of birds? Which ones are analogous to those of mammals?

28.12 Aves: Birds

Birds (Aves; *avis* = bird) appeared in the Jurassic as descendants of carnivorous, bipedal dinosaurs (see Figure 20.18, Chapter 20). Birds belong to the archosaur lineage, and their evolutionary relationship to dinosaurs is evident in their skeletal anatomy and in the scales on their legs and feet. Powered flight gave birds access to new **adaptive zones**, likely contributing to their astounding evolutionary success **(Figure 28.41, p. 663)**. Some species of birds are flightless, and some of these are bipedal runners. Other birds are weak fliers.

Three skeletal features associated with flight in birds are the **keeled sternum** (breastbone), the **furculum** (wishbone), and the uncinate processes on the ribs **(Figure 28.42, p. 663)**. The keel on the sternum anchors the flight muscles (see Figure 28.41c); the furculum acts like a spring; and the uncinate processes, which effect overlap of adjoining ribs, give the ribcage strength and anchor intercostal muscles. In flightless species, the sternum often lacks a keel (see Figure 28.42), an exception being penguins that "fly" through the water. However, flightless species often have uncinate processes.

Birds' skeletons are light and strong (see Figure 28.41c). The skeleton of a 1.5 kg frigate bird (*Fregata magnificens*) weighs just 100 g, far less than the mass

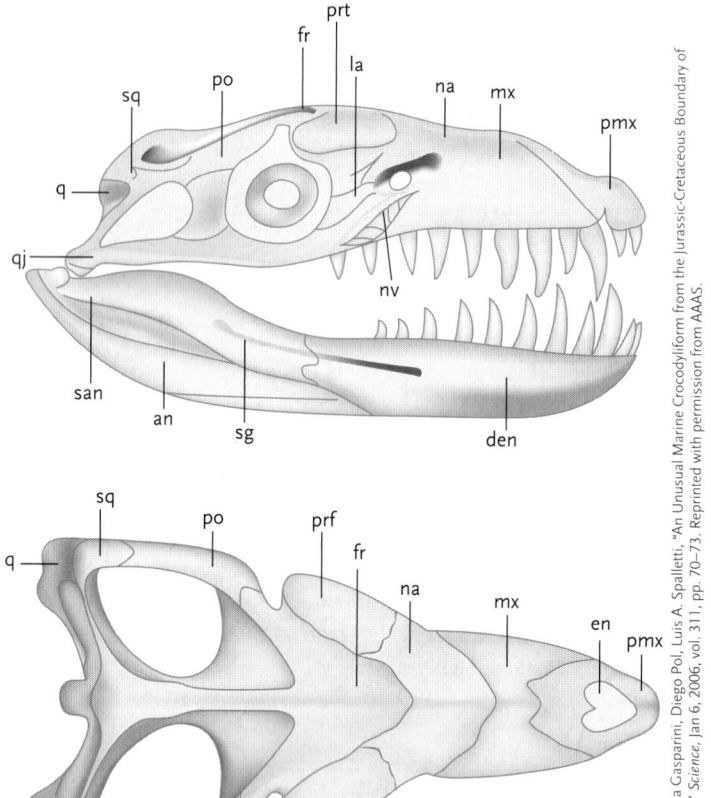

From Zulma Gasparini, Diego Pol, Luis A. Spalletti, "An Unusual Marine Crocodyliform from the Jurassic-Cretaceous Boundary of Patagonia," *Science*, Jan 6, 2006, vol. 311, pp. 70–73. Reprinted with permission from AAAS.

M.B. Fenton

Figure 28.40

Crocodilians. *Dakosaurus andinensis*, a crocodile from the Jurassic–Cretaceous boundary in Patagonia **(a)**, has a more robust skull and jaw than a more typical member of the group **(b)**, *Alligator mississipiensis*. The skull of *Dakosaurus* is more rounded than the wedge-shaped skull of *Alligator*. Bones are abbreviated: an—angular; den—dentary; en—external nares; eoc—exoccipital; fr—frontal; ic—internal carotid formane; la—lacrimal; na—nasal; nv—-neurovascular formina; pmx—premaxilla; po—postorbital; prf—prefrontal; pt—pterygoid; q—quadrate; qj—quadratojugal; san—surangular; sg—surangular groove; soc—supraoccipital; sq—squamosal.

of its feathers. Although the skeleton of a 20 g mammal weighs the same as that of a 20 g bird, the bird's bones are larger and lighter. Most birds have hollow limb bones with small supporting struts that criss-cross the internal cavities. Birds have reduced numbers of separate bony elements in the wings, skull, and vertebral column (especially the tail), so the skeleton is rigid. The bones associated with flight are generally large, and the wingbones are long (see Figure 28.41).

All extant birds **(Figure 28.43, p. 664)** use a keratinized bill for feeding rather than teeth, which are dense and heavy. Many species have a long, flexible neck that allows them to use their bills for feeding, grooming, nest building, and social interactions. Birds' soft internal organs are modified to reduce mass. Most birds lack a urinary bladder, so uric acid paste is eliminated with digestive wastes. Females have only one ovary and never carry more than one mature egg at a time. Eggs are laid as soon as they are shelled. Egg sizes give an indication of the range of size in birds **(Figure 28.44, p. 664)**. Birds range in size from a bee hummingbird (*Mellisuga helenae*) at 2 g to ostriches (*Struthio camelus*) at about 150 kg. The size spectrum is illustrated by a comparison of breast bones **(Figure 28.45, p. 664)**.

All birds have **feathers** (see Figure 28.41d), sturdy, lightweight structures derived from scales in the skin of their reptilian ancestors. Each feather has numerous barbs and barbules with tiny hooks and grooves that maintain the feathers' structures, even during vigorous activity. Flight feathers on the wings provide lift, whereas contour feathers streamline the surface of the body. Down feathers form an insulating cover close to the skin. Moulting replaces feathers once or twice each year. But not all animals with feathers are birds. Several extinct archosaurs had feathers, but these animals had none of the adaptations for flight.

Other adaptations for flight allow birds to harness the energy needed to power their flight muscles. Their metabolic rates are 8 to 10 times as high as those of comparably sized reptiles, allowing them to process energy-rich food rapidly. A complex and efficient respiratory system (see Chapter 49) and a four-chambered heart (see Chapter 40) enable them to consume and distribute oxygen efficiently. As a consequence of high rates of metabolic heat production, most birds maintain a high and constant body temperature (see Chapter 50).

Flying birds were abundant by the Cretaceous. Even in the Jurassic, *Archaeopteryx* had a furculum and was capable of at least limited flight. Until 2008, two main theories purported to explain the evolution of flight in birds. Proponents of the *top-down* theory argued that ancestral birds lived in trees and glided down from them in pursuit of insect prey. Gliding and access to prey are key elements of this theory. Proponents of the *bottom-up* theory proposed that a

a. Wing movements of an owl during flight

b. Skeletal system of birds

Skull
Radius
Ulna
Pectoral girdle
Humerus
Scapula
Furculum (wishbone)
Coracoid
Pelvic girdle
Keeled sternum

c. Pectoral girdle and flight muscles of bird in frontal view

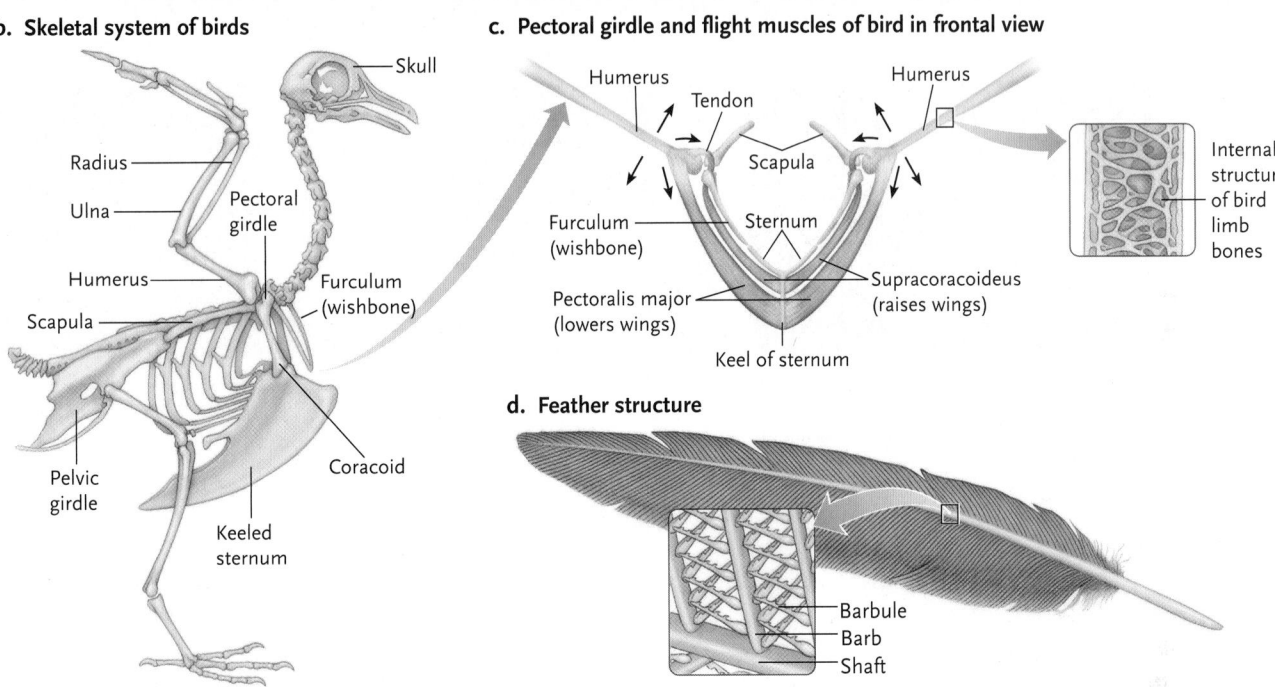

Humerus
Tendon
Humerus
Scapula
Furculum (wishbone)
Sternum
Pectoralis major (lowers wings)
Supracoracoideus (raises wings)
Keel of sternum
Internal structure of bird limb bones

d. Feather structure

Barbule
Barb
Shaft

Figure 28.41

Adaptations for flight in birds. The flapping movements **(a)** of a bird's wing provide thrust for forward momentum and lift to counteract gravity. The bird skeleton **(b)** includes a boxlike trunk, short tail, long neck, lightweight skull and beak, and well-developed limbs. In large birds, limb bones are hollow. Two sets of flight muscles **(c)** originate on the keeled sternum: one set raises the wings, whereas the other lowers them. Flexible feathers **(d)** form an airfoil on the wing surface.

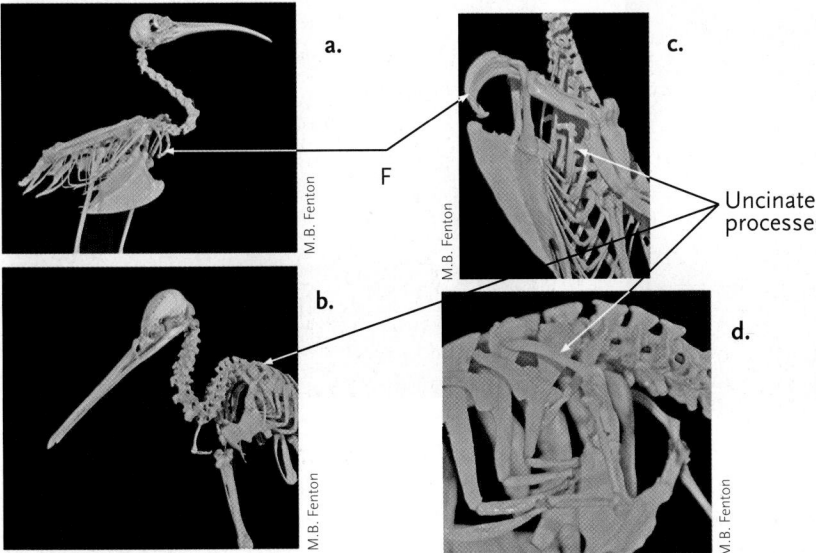

a.

b.

c.

d.

F

Uncinate processes

Figure 28.42

Bird skeletons and flight. Compared are the thoracic skeletons of a Hudsonian Curlew (*Numenius phaeopus*) **(a)**, a kiwi (*Apteryx australis*) **(b, d)**, and a penguin **(c)**. Note the wishbones (furcula F—plural of furculum), as well as keels on the sterna of the curlew and the penguin but not on the kiwi. Neither the penguin nor the kiwi can fly, but the penguin "flies" in water. The wings of the kiwi are drastically reduced **(d)**, and there is no furculum (wishbone), which is obvious in the penguin and the curlew. All three species have distinct uncinate processes on the ribs **(b, c, d)**.

Figure 28.43
Bird diversity.

Steve Oehlenschlager/Shutterstock.com

a. The Laysan albatross (Procellariiformes, *Phoebastria immutabilis*) has the long thin wings typical of birds that fly great distances.

Bildagentur Zoonar GmbH/Shutterstock.com

b. The roseate spoonbill (Ciconiiformes, *Ajaia ajaja*) uses its bill to strain food particles from water.

Colin Edwards Wildside/Shutterstock.com

c. The bald eagle (Falconiformes, *Haliaeetus leucocephalus*) uses its sharp bill and talons to capture and tear apart prey.

Wim Klomp/Foto Natura/Minden Pictures

d. A European nightjar (Caprimulgiformes, *Caprimulgus europaeus*) uses its wide mouth to capture flying insects.

Photos .com

e. A ruby-throated hummingbird. (Apodiformes, *Archilochus colubris*) hovers before a hibiscus blossom to drink nectar from the base of the flower.

Anatoliy Lukich/Shutterstock.com

f. The chestnut-backed chickadee (Passeriformes, *Parus rufescens*) uses its thin bill to probe for insects in dense vegetation.

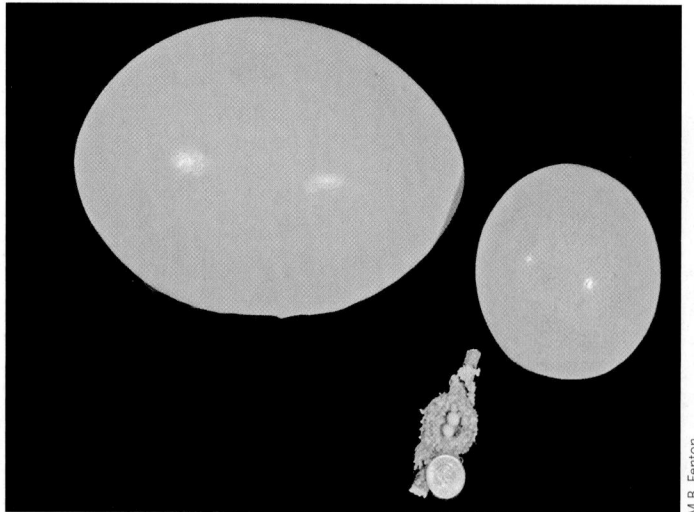

M.B. Fenton

Figure 28.44
Bird eggs. Bird eggs range in size from those of elephant birds (left, *Aepyornis* of Madagascar) to ostriches (*Struthio camelus,* right) and a hummingbird (bottom). The scale, a Canadian $2 coin, is 2.8 cm in diameter.

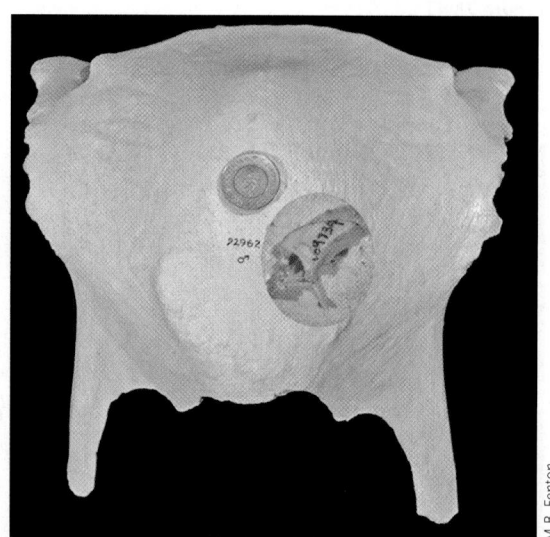

M.B. Fenton

Figure 28.45
Breastbones of birds. Compared are the unkeeled breastbone of an ostrich (*Struthio camelus*) and the keeled breastbone of a hummingbird (*Trochilus polytmus*). A Canadian $2 coin (2.8 cm in diameter) is shown for scale.

protobird was a runner (cursorial) and ran in pursuit of prey and jumped up to catch it.

In 2008, Kenneth P. Dial and two colleagues proposed the *ontogenic–transitional wing* (OTW) hypothesis to explain the evolution of flight in birds. They asserted that the transitional stages leading to the development of flight in modern birds corresponded to its evolutionary development. Key to the OTW theory is the observation that in developing from flightless hatchlings to flight-capable juveniles, individual birds move their protowings in the same ways as adults move fully developed wings. Dial and his colleagues noted that flap-running allows as yet flightless birds to move over obstacles. The OTW theory provides another look at the evolution of flight, and its predictions can be tested with fledglings of extant species. The combination of wings and bipedalism is central to the OTW hypothesis. Birds are bipedal, and pterosaurs may have been. Bats, however, are not bipedal, so the OTW hypothesis does not explain the evolution of flight in that group.

The first known radiation of birds produced the enantiornithines ("opposite" birds), the dominant birds of the Jurassic and Cretaceous. Ornithurines are modern birds **(Figure 28.46)**. Like dinosaurs, many mammals, and other organisms, the enantiornithines did not survive the extinctions that marked the end of the Cretaceous (see Chapter 32). Many enantiornithines flew, reflected by keeled sterna, furcula, and other "modern" skeletal features. Others, such as *Hesperornis*, were swimmers that used their feet for propulsion and, unlike penguins, had unkeeled sterna **(Figure 28.47, p. 666)**. Ornithurines include modern groups of wading birds and seabirds, first known from late Cretaceous rocks. Woodpeckers, perching birds, birds of prey, pigeons, swifts, the flightless ratites, penguins, and some other groups were all present by the end of the Oligocene. Birds continued to diversify through the Miocene.

The ~ 9000 living bird species show extraordinary ecological specializations built on the same body plan. Living birds are traditionally classified into nearly 30 orders. A bird's bill usually reflects its diet. Seed and nut eaters, such as finches and parrots, have deep, stout bills that crack hard shells. Carnivorous hawks and carrion-eating vultures have sharp beaks to rip flesh. Nectar-feeding hummingbirds and sunbirds have long slender bills to reach into flowers, although

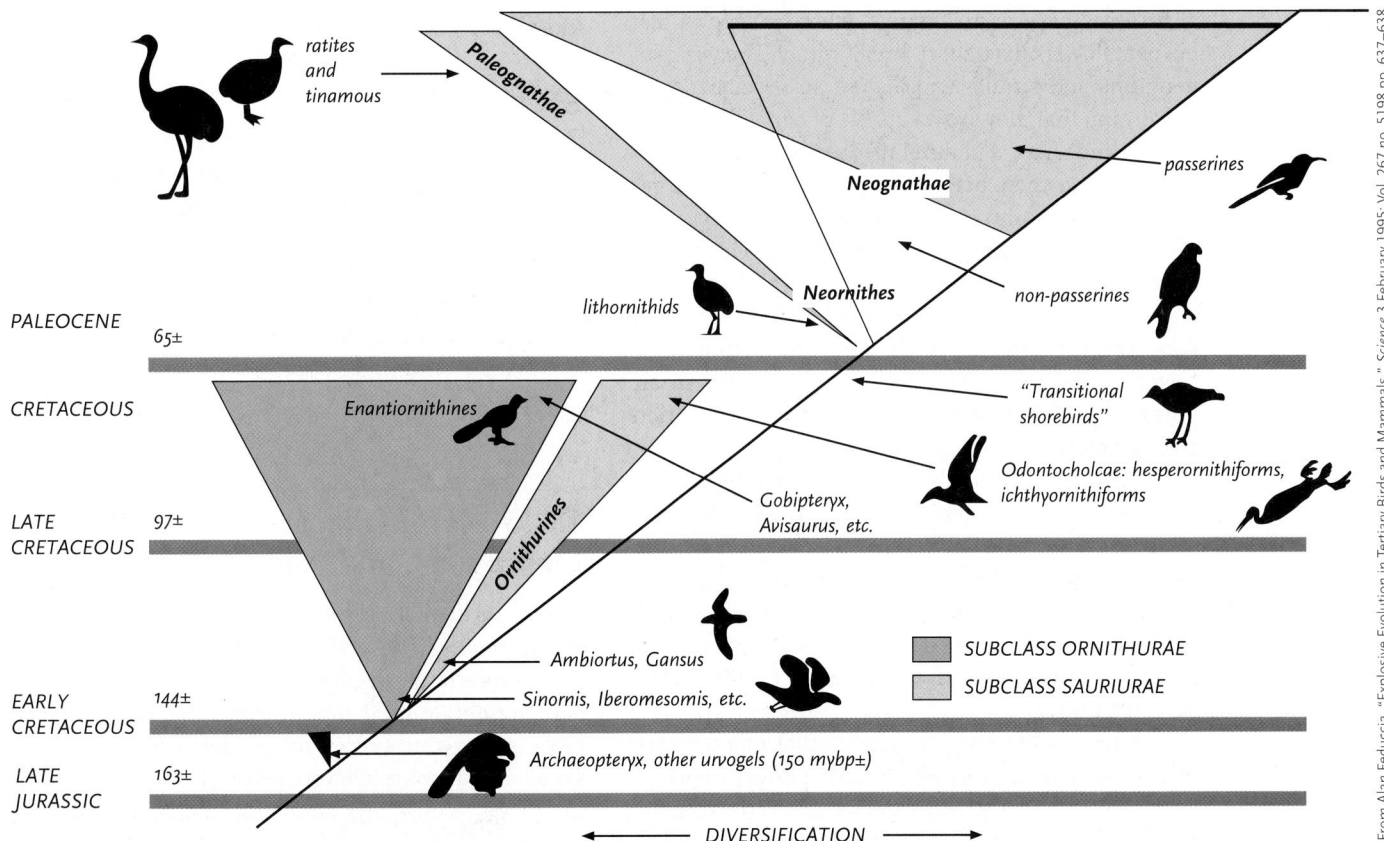

From Alan Feduccia, "Explosive Evolution in Tertiary Birds and Mammals," *Science* 3 February 1995: Vol. 267 no. 5198 pp. 637–638. Reprinted with permission from AAAS.

Figure 28.46

Evolution of birds. Enantiornithines, or opposite birds, were dominant in the Mesozoic but coexisted with ornithurine (more modern) birds in the early Cretaceous. The enantiornithines did not survive the extinctions at the end of the Cretaceous. By the Miocene, passerine birds became the dominant landbirds. Names of genera for some fossil birds make it easier to find out more about these animals.

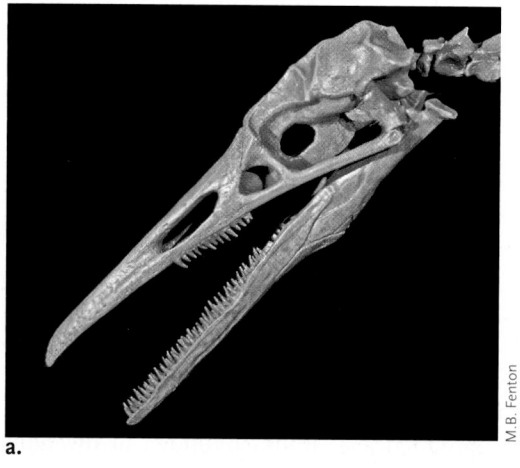
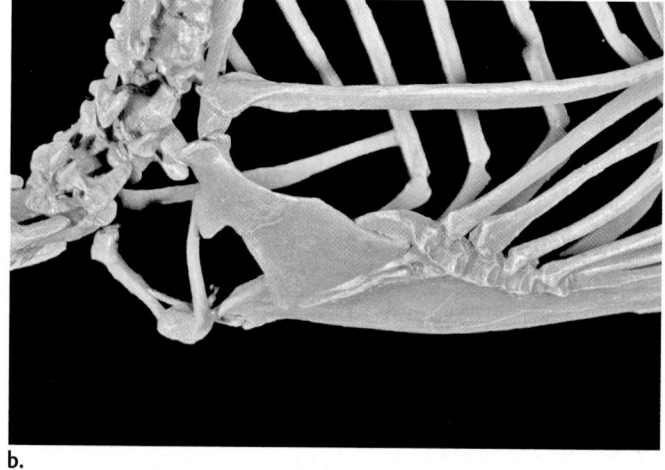

Figure 28.47

Skull **(a)** and sternum **(b)** of Hesperornis, a Cretaceous enantiornithine bird. Note the teeth, along with an unkeeled sternum and a furculum. This diving bird swam with its feet rather than its wings. The skull is 25 cm long.

many perching birds also have slender bills to feed on insects. The bills of ducks are modified to extract particulate matter from water.

Birds also differ in the structure of their feet and wings. Predators have large, strong talons (claws), whereas ducks and other swimming birds have webbed feet that serve as paddles. Long-distance fliers such as albatrosses have narrow wings, whereas species that hover at flowers have short, broad wings. The wings of penguins and similar species are so specialized for swimming that they are incapable of aerial flight.

All birds have well-developed sensory and nervous systems, and their brains are proportionately larger than those of comparably sized diapsids. Large eyes provide sharp vision, and most species also have good hearing, which nocturnal hunters such as owls use to locate prey. Vultures and some other species have a good sense of smell, which they use to find food. Migrating birds use polarized light, changes in air pressure, and Earth's magnetic field for orientation (see Chapter 45).

Many birds exhibit complex social behaviour, including courtship, territoriality, and parental care. Many species use vocalizations and visual displays to challenge other individuals or attract mates. Most raise their young in nests, using body heat to incubate eggs. The nest may be a simple depression on a gravel beach, a cup woven from twigs and grasses, or a feather-lined hole in a tree.

Many bird species make semiannual long-distance migrations (see Chapter 47). Golden plovers (*Pluvialis dominica*) and the godwit (*Limosa lipponica*) migrate over 20 000 km a year going to and from their summer and winter ranges. Migrations are a response to seasonal changes in climate. Birds travel toward the tropics as winter approaches. In spring, they return to high latitudes to breed and to use seasonally abundant food sources.

CONCEPT FIX Some people think that birds can fly because of air spaces between their cells. In reality, many of the bones of birds are laminated structures with hollows that reduce the density of their skeletons, but this is true even of flightless birds such as ostriches. Birds, bats, pterosaurs, and insects fly because they have wings and muscles to flap them in addition to other morphological and physiological specializations. ⬡

STUDY BREAK

1. What three skeletal features are associated with bird flight? Which ones are missing in flightless birds?
2. What adaptations make flight possible in birds and pterosaurs?
3. What characteristics maintain the structure of feathers and make them important to flight in birds?

28.13 Mammalia: Monotremes, Marsupials, and Placentals

Mammals are part of the synapsid lineage, the first of the amniotes to diversify. During the late Paleozoic, medium- to large-sized synapsids were the most abundant vertebrate predators in terrestrial habitats. Therapsids were one successful and persistent branch of synapsids. Therapsids were relatively mammal-like in their legs, skulls, jaws, and teeth and represented an early radiation of synapsids. By the end of the Triassic, the earliest mammals (most of them no bigger than a rat) had appeared. Several lineages of early mammals, such as multituberculates (see Chapter 32) and the lineage that includes the Mesozoic beaver (see Figures 18.6 and 18.7, Chapter 18), persisted and even flourished through much of the Mesozoic. These mammals coexisted with dinosaurs and other diapsids, as well as with the enantiornithine birds.

Paleontologists hypothesize that most Mesozoic mammals were nocturnal, perhaps to avoid diurnal predators and/or overheating. There are two living mammalian lineages **(Figure 28.48)**: the egg-laying Prototheria (or Monotremata) and the live-bearing Theria (marsupials and placentals).

Several features distinguish mammals from other vertebrates, but mammalian diversity makes it difficult to generalize absolutely about definitive characteristics. Living mammals are relatively easy to recognize. They are usually furry and have a diaphragm (a sheet of muscle separating the chest cavity from the viscera); most are **endothermic** (warm-blooded) and bear live young. In mammals, most blood leaves the heart through the **left aortic arch** (the main blood vessel leaving the heart; see Chapter 40). Mammals have two occipital condyles where the skull attaches to the neck, as well as a secondary palate (the plate of bones forming the roof of the mouth). They are **heterodont** and **diphyodont (Figure 28.49, p. 668)**. Heterodont means that different teeth are specialized for different jobs; diphyodont means that there are two generations of teeth (milk or deciduous teeth and adult teeth). But some mammals have no teeth, and others lay eggs. The secondary palate allows mammals to breathe while sucking, without releasing hold on the nipple—an essential part of nursing.

Endothermy means that mammals typically maintain an elevated and stable body temperature so that they can be active under different environmental conditions. They can do this because of their metabolic rates and insulation. Heterodont teeth make mammals more efficient at mechanically dealing with their food (chewing), reducing the lag between the time food is consumed and when the energy in it is available to the consumer. Heterodont teeth are correlated with improved jaw articulation, in mammals between the dentary (lower jaw) and squamosal (bone on the skull). The diaphragm means that mammals are reasonably efficient at breathing, and the circulatory system with a four-chambered heart makes them efficient at internal circulation of resources or collection of wastes. Milk is a rich food source, and by feeding it to their young, female mammals provide the best opportunity for growth and development. The **cortex** of the brain is central to information processing and learning. Mammals' brains are another key to their evolutionary success.

28.13a The Mammalian Radiation: Variations on Mammals

The egg-laying Prototheria (*proto* = first; *theri* = wild beast), also called Monotremata, and the live-bearing Theria are the two groups of living mammals. Among the Theria, the Metatheria (*meta* = between), also

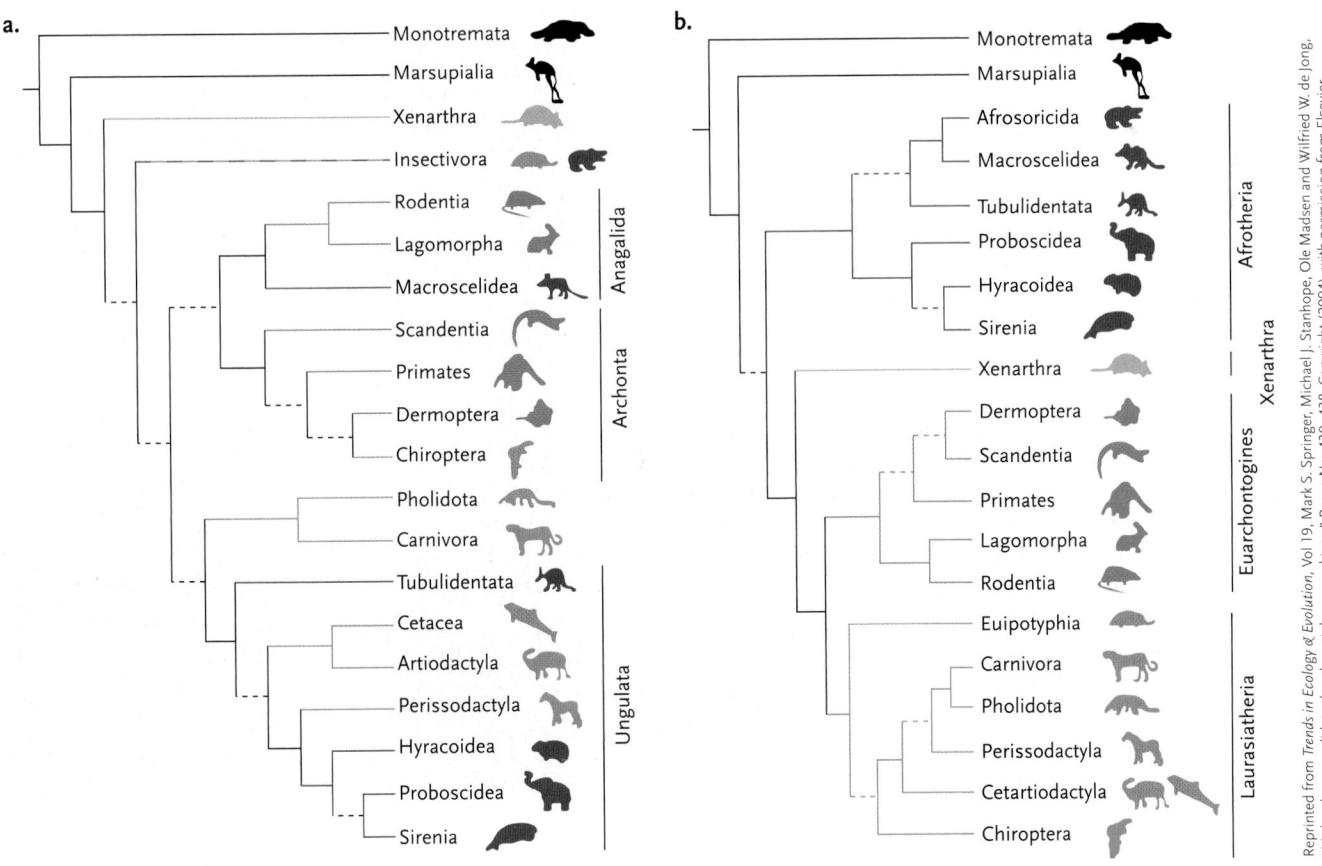

Figure 28.48
Modern mammals. Prevailing phylogenies of mammals derived from **(a)** morphological and **(b)** molecular data.

Reprinted from *Trends in Ecology & Evolution*, Vol 19, Mark S. Springer, Michael J. Stanhope, Ole Madsen and Wilfried W. de Jong, "Molecules consolidate the placental mammal tree," Pages No. 430–438, Copyright (2004), with permission from Elsevier.

Figure 28.49

Mammal teeth. In most mammals, the teeth are diphyodont, meaning that milk (deciduous) teeth are replaced by permanent teeth. The skull of a vampire bat (*Desmodus rotundus*) clearly shows four deciduous teeth (arrows), as well as permanent teeth **(a).** The teeth of mammals are also heterodont **(b),** meaning that different teeth are specialized to do different jobs. In this bear (*Ursus americana*), incisors (i), a canine (c), **premolars** (p), and molars (m) are obvious.

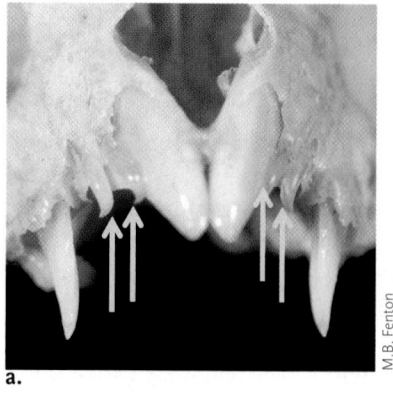

a.

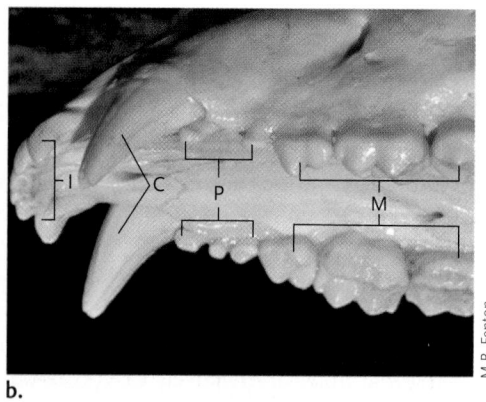

b.

called marsupials, and the Eutheria (*eu* = good), or placentals, differ in their reproductive adaptations.

Monotremata. The **monotremes** (*mono* = one; *trema* = perforation) are represented by three living species that occur only in the Australian region **(Figure 28.50).** Females lay leathery shelled eggs, and newly hatched young lap up milk secreted by modified sweat glands (mammary glands) on the mother's belly. The duck-billed platypus (*Ornithorhynchus anatinus*) lives in burrows along riverbanks and feeds on aquatic invertebrates. The two species of echidnas or spiny anteaters (*Tachyglossus aculeatus* and *Zaglossus bruijnii*) feed on ants or termites.

Marsupialia. Represented by 240 species, marsupials (*marsupion* = purse) (Metatheria) are characterized by short **gestation** periods. The young are briefly (as few as 8 to 10 days in some species and up to 30 days in others) nourished in the uterus via a placenta and are then born at an early stage of development. Newborns use their forelimbs to drag themselves from the vagina and across the mother's belly fur to her abdominal pouch, the marsupium, where they complete their development attached to a teat. Marsupials are prevalent among the native mammals of Australia and are also diverse in South America **(Figure 28.51).** One species, the opossum (*Didelphis virginiana*), occurs as far north as Canada. South America once had a diverse marsupial fauna, which declined after the Isthmus of Panama bridged the seaway between North and South America (see Chapter 19), allowing placental mammals to move southward.

Placental mammals (Eutheria) are represented by 4000 living species. They complete embryonic development in the mother's uterus, nourished through a **placenta** until they reach an advanced stage of development **(viviparous).** Some species, such as humans, are helpless at birth **(altricial),** but others, such as horses, are born with fur and are quickly mobile **(precocial).** Biologists divide the eutherians into about 18 orders, of which only 8 have more than 50 living species **(Figure 28.52).** Rodents (Rodentia) make up about 45%

a. Short-nosed echidna

b. Duck-billed platypus

Figure 28.50

Monotremes. The short-nosed echidna (*Tachyglossus aculeatus*) **(a),** is terrestrial. The duck-billed platypus (*Ornithorhynchus anatinus*) **(b),** raises its young in a streamside burrow.

Figure 28.51

Marsupials. **(a)** A kangaroo (*Macropus giganteus*) carries her "joey" in her pouch; **(b)** a male koala (*Phascolarctos cinereus*) naps; and **(c)** an opossum from Guyana (*Didelphis* species) emerges from its den after dark to feed.

of eutherian species, and bats (Chiroptera) make up another 22%. We belong to the primates, along with 169 other species, representing about 5% of the current mammalian diversity.

Some eutherians are obviously specialized for locomotion. Although whales and dolphins (order Cetacea) and manatees and dugongs (order Sirenia) are descended from terrestrial ancestors, they are aquatic (mainly marine) and can no longer function on land. By contrast, seals and walruses (order Carnivora) feed under water but rest and breed on land. Bats (order Chiroptera) use wings for powered flight.

a. The capybara (Rodentia, *Hydrochoerus hydrochaeris*), the largest rodent, feeds on vegetation in South American wetlands.

b. Most bats, like the Eastern small-footed bat (*Myotis leibii*), are nocturnal predators on insects.

c. Walruses (Carnivora, *Obodenus rosmarus*) feed primarily on marine invertebrates in frigid arctic waters.

d. The black rhinoceros (Perissodactyla, *Diceros bicornis*) feeds on grass in sub-Saharan Africa.

e. Arabian camels (Artiodactyla, *Camelus dromedarius*) use enlarged foot pads to cross hot desert sands.

Figure 28.52
Eutherian diversity.

Although early mammals appear to have been insectivorous, the diets of modern eutherians are diverse. Odd-toed ungulates (*ungula* = hoof) such as horses and rhinoceroses (order Perissodactyla), even-toed ungulates such as cows and camels (order Artiodactyla), and rabbits and hares (order Lagomorpha) all eat vegetation. Some of the vegetarians use fermentation to digest cellulose (see Chapter 48). **Carnivores** (order Carnivora) usually consume other animals, but some, such as the giant panda (*Ailuropoda melanoluca*), are vegetarians. Most bats eat insects, but some feed on flowers, fruit, or nectar, and some, the vampires, consume blood. Many whales and dolphins prey on fishes and other animals, but some eat plankton. Some groups, including rodents and primates, feed opportunistically on both plant and animal matter. Ants and termites are the preferred food of a variety of mammals, both prototherian and therian.

STUDY BREAK

1. How do monotremes differ from marsupials and placentals?
2. How do marsupials and placentals differ from one another?
3. What are four distinctive features of mammals?

28.14 Evolutionary Convergence and Mammalian Diversity: Tails to Teeth

In the discussion on the Mesozoic beaver (see Figures 18.6 and 18.7, Chapter 18), we learned that a dorsoventrally flattened tail occurred in a Mesozoic mammal and today occurs in a monotreme (duck-billed platypus) and in beavers (*Castor* species). Evolutionary convergences in design features such as these are common in mammals. Another good example is the development of protective spines (quills) from hairs. These occur in spiny anteaters (monotremes), porcupines (rodents), and hedgehogs and tenrecs (insectivores).

Another striking example of convergence among mammals is provided by the teeth and lumbar vertebrae of the Mesozoic *Fruitafossor windscheffeli* **(Figure 28.53)** and some living xenarthrans (armadillos and sloths). Like sloths and armadillos, *Fruitafossor* had round molars with open roots **(Figure 28.54).** Also like sloths and armadillos, *Fruitafossor* had processes in its lower back (**lumbar vertebrae**) known only from living xenarthrans. Although the teeth and vertebral structures converge between *Fruitafossor* and xenarthrans, *Fruitafossor* is not closely related to any living mammals. We do not know if *Fruitafossor* had other features of mammals such as mammary glands, a diaphragm, and vivipary.

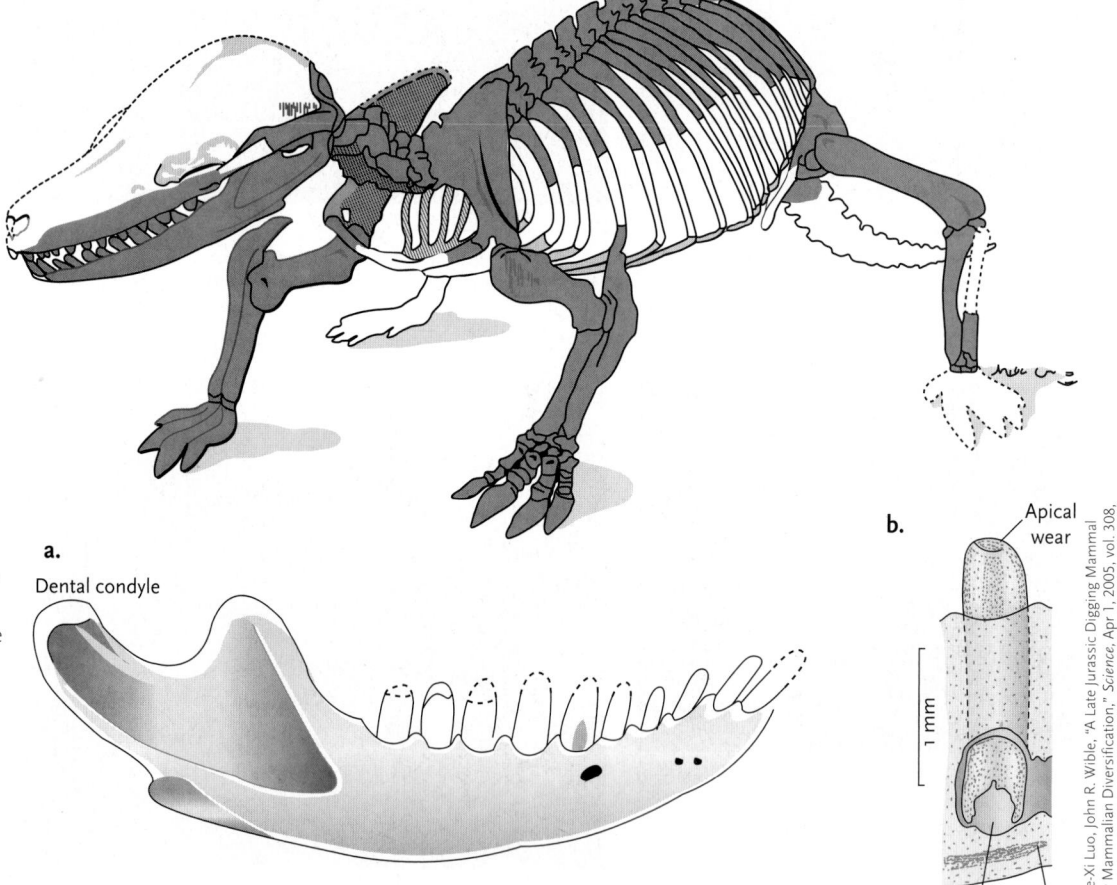

Figure 28.53
Fruitafossor, a mammal of the mid-Jurassic. Dark outlined bones are known from fossils; other bones are presumed. The *Fruitafossor* had round teeth with open roots **(a, b).**

a.
Dental condyle

3 mm

b.
Apical wear

1 mm

Open root-end mg

From Zhe-Xi Luo, John R. Wible, "A Late Jurassic Digging Mammal and Early Mammalian Diversification," *Science*, Apr 1, 2005, vol. 308, pp. 103–107. Reprinted with permission from AAAS.

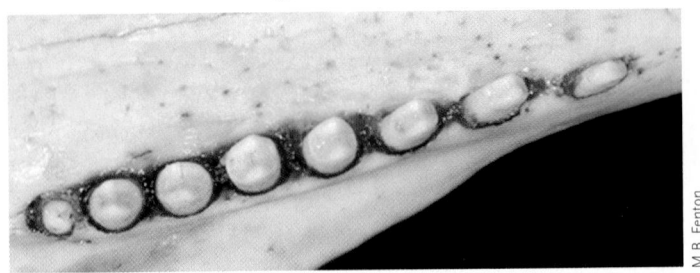

M.B. Fenton

Figure 28.54

Convergences in mammals. Like the *Fruitafossor* (Figure 28.53), round teeth with open roots are well known from edentate mammals like armadillos and tubulidentates, the aardvark (shown here).

But *Fruitafossor*'s bones, particularly its jaw joints and occipital condyles, make it a mammal.

28.14a Mammalian Teeth: Diversity of Form and Function

As in other vertebrates, mammals' teeth provide a good indication of diet. Some molars (cheek teeth) with W-shaped cusps cut and crush food **(Figure 28.55a, b)**, whereas others mainly crush (Figure 28.55c, d). Grinding teeth have appeared in a wide range of forms in mammals **(Figure 28.56)** and show considerable variation in the details of their design.

Animals such as the walrus **(Figure 28.57a, p. 672)** have tusks for digging and small flat molars for crushing the shells of bivalves (compare with Figures 28.19a, p. 648, and 28.22b, p. 651). The molars of *Desmostylus* have an artistic circular pattern (see photo on opening page of this chapter). These Miocene mammals were thought to have resembled Sirenia, the dugongs and manatees. Later discoveries revealed that they had massive limbs and, in body form, looked more

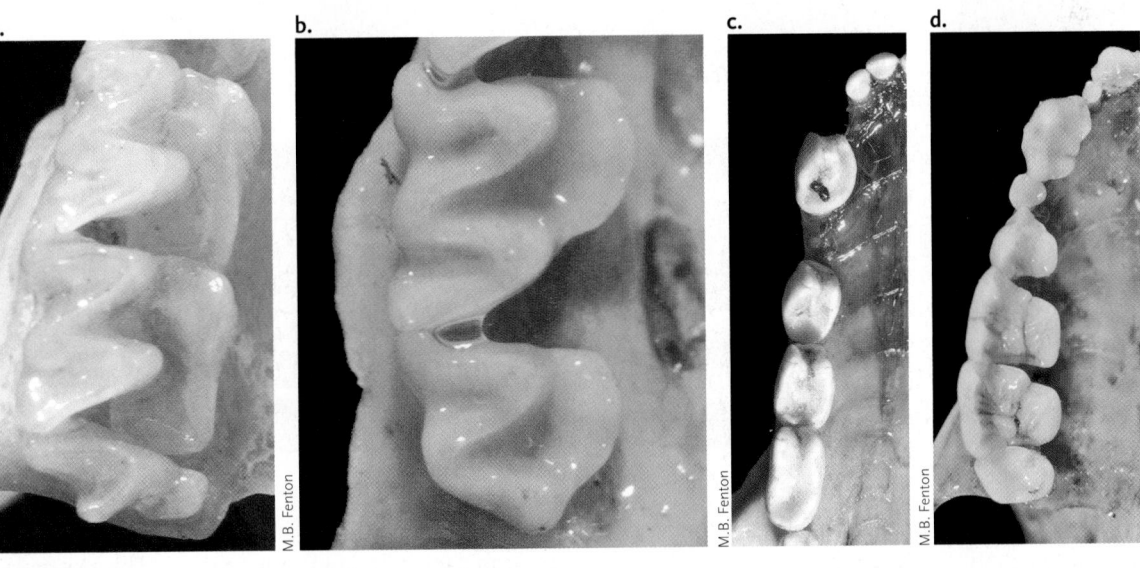

a.   b.   c.   d.

Figure 28.55

Molars that cut and crush **(a, b)**, and those that crush **(c, d)**. The molars (cheek teeth) of a mole **(a)**, an insectivorous bat **(b)**, an Old World fruit bat **(c)**, and a New World fruit bat **(d)**. The mole is *Condylura*, the insectivorous bat a *Taphozous*, the Old World fruit bat a *Pteropus*, the New World fruit bat a *Brachyphylla*.

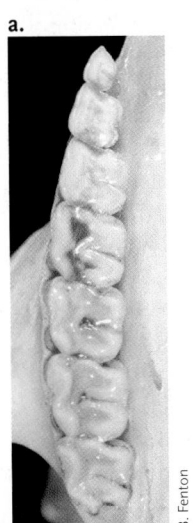
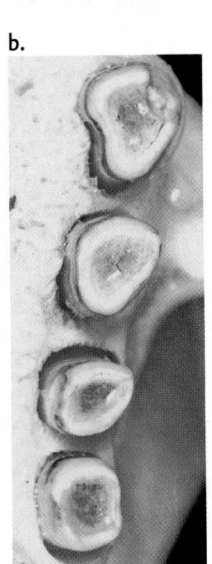
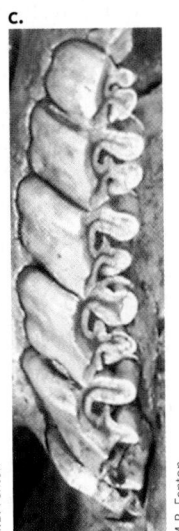
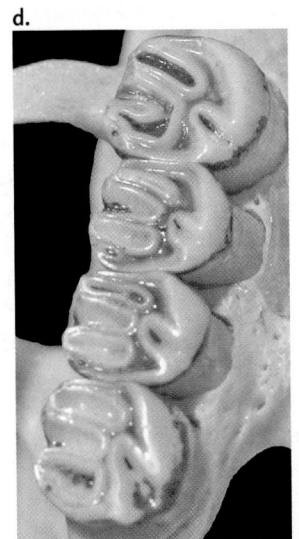

a.   b.   c.   d.

Figure 28.56

Molar teeth for crushing plant material (mostly leaves and stems). Look carefully at the details of the molars of a hyrax (*Heterohyrax brucei* from Africa) **(a)**, a three-toed sloth (*Bradypus tridactylus* from South America) **(b)**, a black rhino (*Diceros bicornis* from Africa) **(c)**, and a porcupine (*Erethizon dorsatum* from North America) **(d)**. The hyrax belongs to the order Hyracoidea, the sloth to the order Xenarthra, the rhino to the order Persissodactyla, and the porcupine to the order Rodentia.

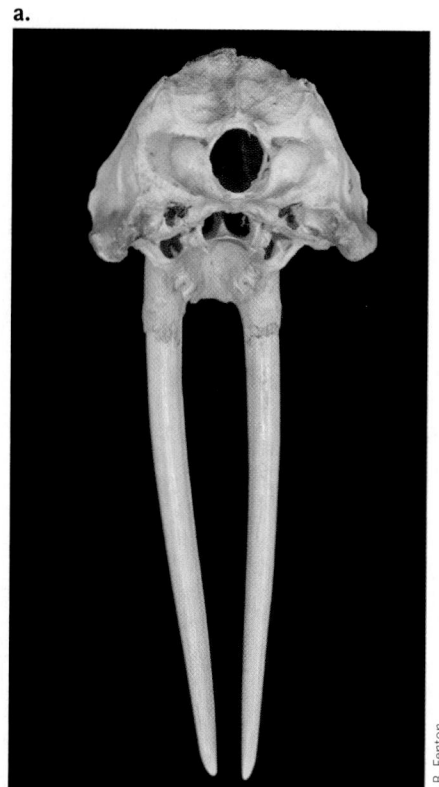

a.

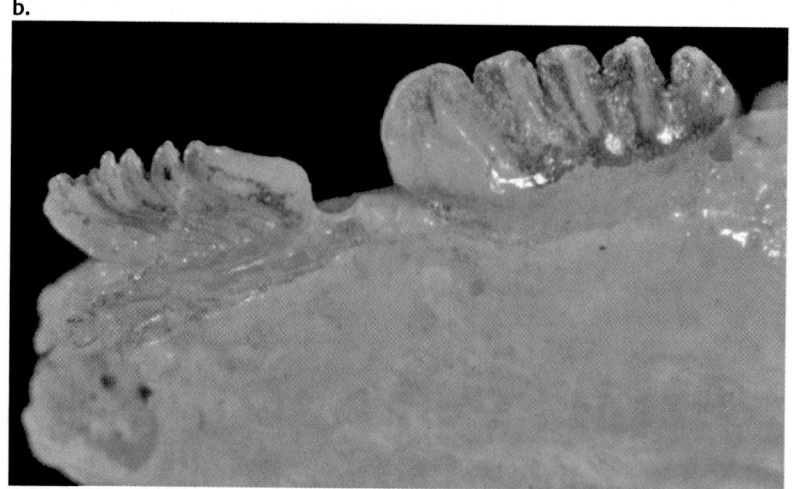

b.

M.B. Fenton

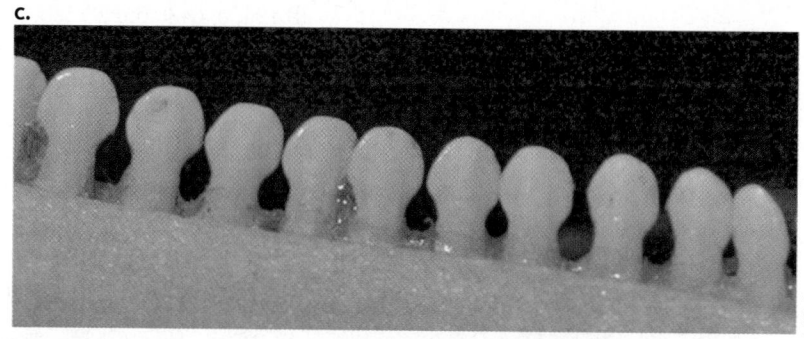

c.

M.B. Fenton

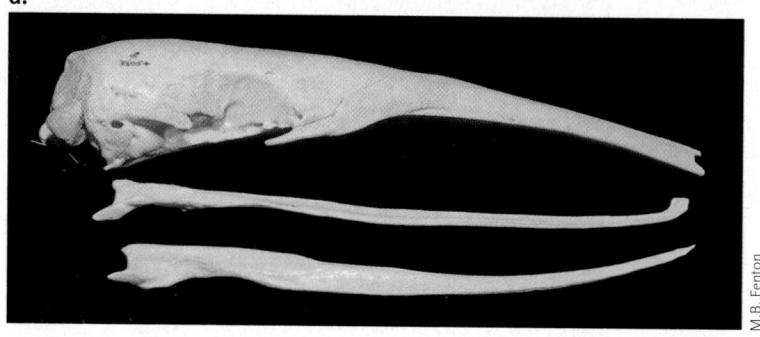

d.

M.B. Fenton

M.B. Fenton

Figure 28.57

Teeth (or no teeth) for different jobs. Whereas walruses (*Odobenus rosmarus*) **(a)** use tusks (which are canine teeth) for digging clams, which they crush with their molars, flying lemurs (*Cynocephalus*) **(b)** use comblike lower incisor teeth to comb their fur, and dolphins (*Tursiops*) **(c)** have rows of similar teeth (homodont) for grasping fish. The giant anteater (*Myrmecophaga*) **(d)** eats ants and termites and lacks teeth.

like modern hippos. They presumably used their teeth for crushing aquatic vegetation.

Mammals that eat mainly ants and termites (see Figure 28.57d) often lack teeth entirely. This way of life has been exploited by mammals across the tropics. The ant and termite eaters of the New World (armadillos—*Dasypus* spp—anteaters—*Myrmecophaga*) are different evolutionary lineages (see Figure 28.48, p. 667) than the African aardvark (*Orycteropus*, order Tubulidantata) or scaly anteater (*Manis*, order Pholidota). Australia has both monotreme and marsupial anteaters. The most astonishing anteater is the aardwolf, a variation on a hyena **(Figure 28.58)**; both of these species belong to the order Carnivora.

Whereas reptiles, amphibians, fish, and sharks can replace teeth many times (e.g., Figure 28.19, p. 648), mammals replace them only once. Teeth wear with age **(Figure 28.59)**. When the teeth are worn out, the animal can no longer feed itself properly and dies.

Elephants deal with this problem by having only four active molars in the jaw at any one time. The new molar grows in from the back (Figure 28.59), replacing the worn one. In rodents and some other mammals (and also in hydrasaur dinosaurs), molar (and for rodents and lagomorphs, incisor) teeth grow continuously. Here the teeth are curved so that pressure during biting is not directed at the points of growth (see Figures 18.7b, Chapter 18, and 47.4, Chapter 47).

STUDY BREAK

1. What features are found in most mammals and distinguish them from other vertebrates?
2. How is heterodont different from diphyodont?
3. Distinguish among monotremes, marsupials, and placentals.

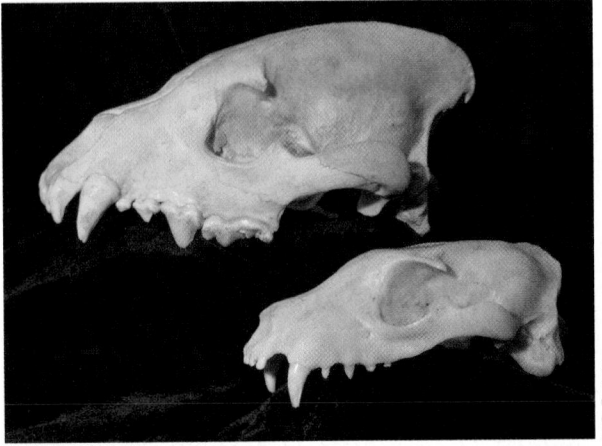

Figure 28.58
Divergence! Spotted hyena (top, *Crocuta crocuta*) and aardwolf (bottom, *Proteles capensis*) are in the same family (Hyaenidae). The spotted hyena is a carnivorous scavenger with massive teeth capable of cutting tendons and crushing bones. The aardwolf eats mainly ants and termites and has reduced teeth (and a differently shaped skull). The *Crocuta* skull is about 30 cm long. Both belong to the order Carnivora.

Figure 28.59
Tooth wear and replacement. Elephants (*Loxodonta africana*) have four functional molars in the mouth at any one time (one in each jaw quadrant). New molars push into the tooth row from the back.

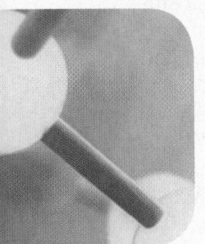

MOLECULE BEHIND BIOLOGY 28.1

Geckel

Geckos **(Figure 1)** can cling to vertical and even inverted smooth surfaces such as window glass. They do this using specialized keratinous setae (fine hairs) on their feet. There are spatulate extensions at the end of each seta. The other element in the sticking ability of geckos is an adhesive. But the remarkable feature of geckos is that the system allows rapid detachment of the foot. As many people have learned using Crazy Glue and its equivalents, getting stuck to something is easy; getting unstuck is a different story.

As remarkable as geckos, some mussels (see Chapter 27) secrete a specialized adhesive with a high concentration of catecholic amino acid 3,4-dihydroxyl-phenlalanine (DOPA). DOPA allows the mussels to cling firmly to wet surfaces. In contrast, the adhesive ability of geckos is diminished by full immersion in water.

Geckel is a new hybrid adhesive **(Figure 2)** combining the adhesive features of those used by geckos and mussels. Geckel is a thin layer of a synthetic polymer that retains its adhesive properties in dry and wet environments for more than 1000 contact cycles.

Work with geckos (probably *Rhoptropus biporosus*) from the Namib desert sheds light on the evolutionary background of their extraordinary clinging power. These small geckos weigh about 2 g and show great mobility on the variety of substrates they encounter—rough, undulant, and unpredictable, often providing few points of adhesion. The adhesive pads under the geckos' toes allow them to cling to the full spectrum of surfaces, and their ability to stick to glass is coincidental.

Figure 1
Geckos, such as this one, can walk on (stick to) glass.

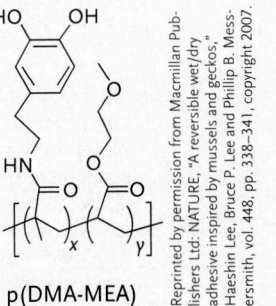

p(DMA-MEA)

Figure 2 *Geckel, a fabricated adhesive that mimics those produced and used by geckos and mussels.*

Reprinted by permission from Macmillan Publishers Ltd: NATURE, "A reversible wet/dry adhesive inspired by mussels and geckos," Haeshin Lee, Bruce P. Lee and Phillip B. Messersmith, vol. 448, pp. 338–341, copyright 2007.

Review

To access course materials such as Aplia and other companion resources, please visit www.NELSONbrain.com.

28.2 Phylum Echinodermata

- Echinodermata are slow-moving or sessile bottom-dwelling animals. They are important herbivores and predators that occur from shallow coastal waters to the oceans' depths. Adult echinoderms develop from bilaterally symmetrical, free-swimming larvae. Developing larvae assume a secondary radial symmetry, often organized around five rays or arms. The table below compares echinoderms and humans.

System	Echinoderms	Humans
digestive	complete system	complete system
excretory	not present	complete system
respiratory	not present	complete system
circulatory	minimal system	closed, complete system
nervous system	no head or central	cephalized system

- Water enters the fluid-filled canals of the water vascular system through the madreporite (a sievelike plate) on the aboral surface. A tube connects the madreporite to the ring canal, which surrounds the esophagus. The ring canal branches into radial canals that extend into each arm and is connected to numerous mucus-covered tube feet. When ampullae, small muscular bulbs in each tube foot, contract, they force water into the tube foot, causing it to lengthen and attach to a substrate. The tube foot contracts, pulling the animal along and pushing water back into the ampulla; this causes the tube foot to release the substrate. Echinoderms reproduce either sexually or asexually. Sexual reproduction is usually achieved by releasing gametes into the water. Asexual reproduction involves clonal budding and may be stimulated by the odour of a predator.

- Asteroidea (sea stars) consist of a central disk surrounded by 5 to 20 radiating arms. Small pincers at the base of pedicellariae (short spines) are used to remove debris that falls onto the animal's surface. The ossicles of their endoskeleton are not fused, permitting flexibility of the arms and disk. Most sea stars eat invertebrates and small fish.

- Ophiuroidea (brittle stars and basket stars) have a well-defined central disk and slender, elongated arms that are sometimes branched. They crawl swiftly across substrates by moving their arms in a coordinated fashion. They feed on small prey, suspended plankton, or detritus extracted from muddy deposits.

- Echinoidea (sea urchins and sand dollars) lack arms. Their ossicles are fused into tests that provide excellent protection but restrict flexibility. Echinoids use tube feet in locomotion. They graze on algae and other organisms that cling to marine surfaces.

- Holothuroidea (sea cucumbers) have a reduced endoskeleton consisting of widely separated microscopic plates. They have five rows of tube feet and a soft body elongated along the oral–aboral axis. Modified tube feet form a ring of tentacles around the mouth. Some species secrete a mucus net that traps plankton or other food particles. The net and tentacles are inserted into the mouth, where the food is ingested. Other species extract food from bottom sediments.

- Crinoidea (sea lilies and feather stars) have five to several hundred branched arms surrounding the disk containing the mouth. Branches of the arms are covered with mucus-coated tube feet that trap suspended microscopic organisms. Sessile sea lilies have a central disk attached to a flexible stalk that can reach a metre in length, whereas adult feather stars swim or crawl weakly.

28.3 Phylum Hemichordata

- Hemichordates (acorn worms) use a muscular, mucus-coated proboscis to construct burrows and trap food particles. Drawn in by beating cilia, water enters the pharynx and exits through the pharyngeal gill slits. As the water passes, suspended food is trapped and directed to the digestive system, while gases are exchanged across the partitions between gill slits.

28.4 Phylum Chordata

- A notochord, a dorsal hollow nerve cord, and gill slits distinguish chordates from all other deuterostomes. Gill slits are sets of paired openings in the pharynx. Water is drawn into the mouth, food is filtered out, and the water passes through the pharynx, where gas exchange takes place, and then out through the gill slits.

- Tunicates (sea squirts) belong to the subphylum Urochordata. They float in surface waters or attach to substrates in shallow waters. Many are sessile as adults and secrete a gelatinous or leathery "tunic" around their bodies. They squirt water through a siphon when disturbed. Some urochordates have larvae that resemble cephalochordates (or lancelets), and in a few species, these larvae are neotenous. Adult lancelets are mainly sedentary, lying partly buried in sand of shallow marine waters. They have well-developed body wall muscles and a prominent notochord. Adults have light receptors on the head and chemical sense organs on tentacles.

28.5 The Origin and Diversification of Vertebrates

- The internal skeleton of vertebrates provides structural support for muscles and protects the nervous system and other internal organs. The backbone surrounds and protects the dorsal nerve cord, and a bony cranium provides protection for the brain. The backbone acts as a place for muscle attachments, which allows quick movement.

- Homeotic (*Hox*) genes influence the three-dimensional shape of an animal and the locations of structures such as eyes, wings, and legs. *Hox* genes are arranged on chromosomes in a specific order to form the *Hox* gene complex. Each gene in the complex governs the development of particular structures. Species with simple anatomy have fewer *Hox* genes than more complex species, which have duplicated copies. These duplicate copies assumed new functions, directing the development of novel structures, such as the vertebral column and jaws.

- Vertebrates have four characteristic morphological innovations: cranium, vertebrae, bone, and neural crest cells.

28.6 Agnathans: Hagfishes and Lampreys, Conodonts, and Ostracoderms

- Agnatha are primitive vertebrates that use a muscular pharynx to suck water containing food particles into their mouths and gills to filter the food and perform gas exchange. Hagfishes and lampreys have a well-developed notochord but lack true vertebrae and paired fins. The hagfish skeleton is a cranium and a notochord. These marine scavengers feed on invertebrate prey and dead fish. They lack a larval stage. Lampreys have a more derived axial skeleton than hagfishes. Their notochord is surrounded by cartilage that partially covers the nerve cord, whereas hagfish

have no specialized structures surrounding the nerve cord. Some species of lampreys are parasitic as adults, attaching to a host. Ammocoetes, the larval stage of lampreys, resemble cephalochordates and may develop for up to seven years before metamorphosing into adults.

28.7 Jawed Fishes: Jaws Expanded the Feeding Opportunities for Vertebrates

- Jaws meant that fishes could feed on larger items of food with higher energy content. Jaws also function to defend against predators; groom; transport young; and grasp, kill, and shred food items.

- Flexible fins, lightweight skeletons, streamlined bodies, and an absence of heavy body armour allow sharks to pursue prey rapidly. Sharks and their relatives have squalene, an oily substance contained within the liver. Squalene is lighter than water and increases the animals' buoyancy.

- The lateral line system of elasmobranchs and other fishes consists of a row of tiny sensors in canals along both sides of the body. This system allows detection of vibrations in water, which can be used when hunting.

- In many bony fishes, a gas-filled swim bladder serves as a hydrostatic organ to increase buoyancy. Bony fish also have small, smooth, lightweight scales and bodies that are covered with a protective coat of mucus that retards bacterial growth and smoothes the flow of water past the body.

28.8 Early Tetrapods and Modern Amphibians

- Osteolepiformes (fleshy-finned fishes) and tetrapods had infoldings of their tooth surfaces. The shapes and positions of bones on the dorsum and side of their crania and in their appendages were similar. Osteolepiformes had strong fins enabling them to crawl on mud (making their move onto land) and possessed vertebral columns with crescent-shaped bones for support. Osteolepiformes had lungs allowing them to breathe atmospheric oxygen. They could excrete urea or uric acid rather than ammonium.

- The body wall of fish picks up sound vibrations and directly transfers them to sensory receptors. Sound waves are harder to detect in air. The development of a tympanum, or eardrum, allowed tetrapods to detect airborne vibrations and transfer them to the sensory cells of their inner ear.

- Amphibians have thin, scaleless skin, well supplied with blood vessels. Since some oxygen and carbon dioxide enters the body across a thin layer of water, most amphibians need moist skin, restricting them to aquatic or wet terrestrial habitats. Many amphibians need access to free-standing water to reproduce.

28.9 The Origin and Mesozoic Radiations of Amniotes

- Amniotes get their name from the amnion, a fluid-filled sac surrounding the embryo during development. Amniote eggs are resistant to desiccation because the developing embryos excrete uric acid that is stored in the allantois.

- Amniote eggs have four specialized membranes that protect the embryo and facilitate gas exchange and excretion. They also have a hard or leathery shell perforated by microscopic pores that mediates the exchange of air and water between the egg and its environment. Keratin and lipids are partly responsible for making the skin waterproof.

28.10 Subclass Testudinata: Turtles and Tortoises

- There have been three major radiations of amniotes: anapsids, synapsids, and diapsids, distinguishable by the numbers of bony arches in the temporal region of the skull. The bony arches allow space for contraction (and expansion) of large and powerful jaw muscles. Anapsids lacked temporal arches, synapsids have one pair of temporal arches, and diapsids have two pairs of arches.

- Surviving anapsids are turtles and tortoises. A turtle's body is defined by a bony, boxlike shell, which includes a dorsal carapace and a ventral plastron. Its ribs are fused to the inside of the carapace, and the pectoral and pelvic girdles lie within the ribcage. Large keratinized scales cover the bony plates that form the shell.

28.11 Living Diapsids: Sphenodontids, Squamates, and Crocodilians

- Diapsids evolved in two lines, lepidosaurs and archosaurs. Lepidosaurs include snakes and lizards and many extinct forms. Snakes and lizards use olfactory and vibrational cues to detect prey. Some even have thermal perception.

- Living archosaurs include crocodilians, animals with a four-chambered heart that is homologous to the heart in birds. Some crocodilians have muscles that originate on the pubis and insert on the liver. When these muscles contract, the liver moves toward the tail, creating negative pressure in the chest cavity. This situation is analogous to the role of the diaphragm in mammals.

28.12 Aves: Birds

- Birds' ability to fly reflects a keeled sternum (breastbone), a furculum (wishbone), and uncinate processes on the ribs. These main adaptations, coupled with lightweight, strong bones and feathers, contributed to the success of birds. Flightless birds often lack a keeled sternum. Most birds have hollow limb bones with small supporting struts that criss-cross the internal cavities. Birds have fewer separate bony

elements in the wings, skull, and vertebral column, so the skeleton is light and rigid. All modern birds have replaced dense and heavy teeth with a lightweight keratinized bill. Birds have much higher metabolic rates than comparably sized reptiles do, and they depend on energy-rich food. A complex and efficient respiratory system and a four-chambered heart enable them to consume and distribute oxygen efficiently. Other adaptations include modification of internal organs to reduce weight, elimination of a urinary bladder so that uric paste is eliminated with digestive wastes, and laying eggs as soon as they are shelled.

- Each feather has numerous barbs and barbules with tiny hooks and grooves that maintain the feathers' structure, even during vigorous activity. Flight feathers on the wings provide lift, whereas contour feathers streamline the surface of the body. Down feathers form an insulating cover close to the skin.

- The top-down theory suggests that ancestral birds lived in trees and glided down from those trees in pursuit of insect prey. The bottom-up theory proposes a cursorial ancestor that ran along in pursuit of prey and jumped up to catch it. The ontogenic–transitional wing (OTW) hypothesis is a third effort to explain the evolution of flight in birds. Its proponents suggest that flapping protowings gave the ancestors of birds greater mobility.

28.13 Mammalia: Monotremes, Marsupials, and Placentals

- Most living mammals are furry and endothermic (warm-blooded). Mammals usually bear live young and have a diaphragm, a left aortic arch leaving the heart, and two occipital condyles. Mammals also have a secondary palate and are heterodont (teeth specialized for different jobs) and diphyodont (two generations of teeth, milk or deciduous teeth and adult teeth). Heterodont teeth make mammals more efficient at mechanically dealing with their food (chewing), reducing the lag time between consumption of food and availability of the food's energy.

- Monotremes lay leathery shelled eggs. When newborns hatch, they lap up milk secreted by the mammary glands located on the mother's belly. Marsupials have short gestation periods of as few as 8 to 10 days. Young are born at an early stage of development and complete their development attached to a teat in the abdominal pouch (the marsupium) of their mother. Placental mammals complete embryonic development in the mother's uterus, nourished through a placenta until they reach a fairly advanced stage of development.

Questions

Self-Test Questions

1. Which phylum includes animals with a water vascular system?
   a. Echinodermata
   b. Hemichordata
   c. Chordata
   d. Arthropoda

2. Which of the following is NOT a characteristic of all chordates?
   a. a notochord
   b. a segmented nervous system
   c. a dorsal hollow nerve cord
   d. a perforated pharynx

3. Which group of vertebrates has adaptations allowing reproduction on land?
   a. agnathans
   b. gnathostomes
   c. amniotes
   d. ichthyosaurs

4. Which group of fishes has the most living species today?
   a. actinopterygians
   b. chondrichthyans
   c. acanthodians
   d. ostracoderms

5. Which is true about modern amphibians?
   a. They closely resemble their Paleozoic ancestors.
   b. They always occupy terrestrial habitats as adults.
   c. They never occupy terrestrial habitats as adults.
   d. They are generally larger than their Paleozoic ancestors.

6. Which one of the following key adaptations allows amniotes to occupy terrestrial habitats?
   a. the production of carbon dioxide as a metabolic waste product
   b. an unshelled egg protected by jellylike material
   c. a dry skin largely impermeable to water
   d. a lightweight skeleton with hollow bones

7. Which of the following characteristics are central to powered flight in birds?
   a. webbed feet, long legs, feathers
   b. efficient respiratory and excretory systems, flight muscles
   c. elongated forelimbs, keeled breast bone, flight muscles
   d. feathers, furculum, eyes

8. Which of the following characteristics did NOT contribute to the evolutionary success of mammals?
   a. extended parental care of young
   b. an erect posture and flexible hip and shoulder joints
   c. specializations of the teeth and jaws
   d. high metabolic rate and homeothermy

9. Why are Echinodermata and Chordata deuterostomes?
   a. because each has a notochord
   b. because neither has spiral cleavage
   c. because both have radial cleavage
   d. because both have a mouth

10. Why are lamprey eels and hagfish chordates?
   a. because both have a notochord, gill slits, and a dorsal hollow nerve chord
   b. because both have mouths
   c. because neither have paired fins
   d. because neither have jaws

Questions for Discussion

1. Most sharks and rays are predatory, but the largest species feed on plankton. Construct a hypothesis to explain this observation. How would you test your hypothesis?

2. What selection pressures did tetrapods face when they first ventured onto land? What characteristics allowed them to meet these pressures?

3. Use binoculars to observe several species of birds in different environments, such as lakes and forests. How are their beaks and feet adapted to their habitats and food habits?

4. Imagine that you unearthed the complete fossilized remains of a mammal. How would you determine its diet?

5. What evidence suggests that birds are really only specialized reptiles and should not be considered a distinct class (see also Chapter 19).

The Chemical, Physical, and Environmental Foundations of Biology

The Scientific Basis of Biology

The information contained in this textbook represents the culmination of hundreds of years of research involving a huge number of experiments carried out by countless scientists. The entire content of this book—every observation, experimental result, and generality—is the product of **biological research,** the collective effort of individuals who have worked to understand every aspect of the living world. This section describes how biologists working today pose and find answers to questions.

The Scientific Method

Beginning about 500 years ago in Europe, inquisitive people began to understand that direct observation is the most reliable and productive way to study natural phenomena. By the nineteenth century, researchers were using the **scientific method**—an investigative approach to acquiring knowledge in which scientists make observations about the natural world, develop working explanations about what they observe, and then test those explanations by collecting more information.

Application of the scientific method requires both curiosity and skepticism: successful scientists question the current state of our knowledge and challenge old concepts with new ideas and new observations. Explanations of natural phenomena must be backed up by objective evidence rooted in observation and measurement. Most important, scientists share their ideas and results by publishing their work.

Testing a Hypothesis Is Central to the Scientific Method

A **hypothesis** can be defined as a tentative explanation for an observation, phenomenon, or scientific problem that can be tested by further investigation. Scientific hypotheses have two fundamental elements. First, a hypothesis must be *testable*. That is, there must be some set of observations or experiments that can be undertaken to support the hypothesis. For example, you may be studying a gene in yeast that you find is activated when cells are placed under conditions of heat stress. You may hypothesize that the protein encoded by this gene is essential for the yeast to survive short-term exposure to high temperature. Using modern molecular techniques, you can test this hypothesis by inactivating the gene in a population of yeast cells and observing if there is a change in heat tolerance. Today, this hypothesis is easily testable. A scientist may have had a similar idea 30 years ago, but given the lack of molecular techniques, the hypothesis would not have been testable at that time.

The second key to a scientific hypothesis is that it must be *falsifiable*. That is, through observation or experimentation you must be able to show that the original hypothesis may not be correct. Getting back to the yeast analogy, it is very possible that through analysis you would find that inactivation of the gene does not change the ability of yeast cells to survive high temperatures.

Scientists test the predictions that come from hypotheses with experimental or observational tests that generate relevant data. And if data from just one study refute a scientific hypothesis (i.e., demonstrate that its predictions are incorrect), the scientist must modify the hypothesis and test it again or abandon it altogether.

No amount of data can prove beyond a doubt that a hypothesis is correct; there is always the chance that a contradictory example exists, and it is impossible to test every imaginable example. That is why scientists say that positive results are consistent with, support, or confirm a hypothesis.

Elements of the Scientific Method

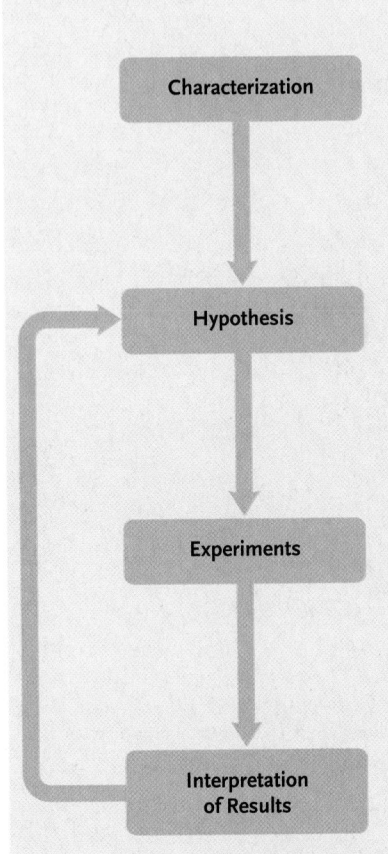

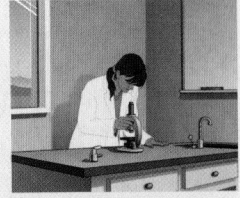

1 Before a new hypothesis is formulated, researchers today usually know a fair amount about the subject under study. This characterization comes from years of their own experiments as well as the published research of other scientists working in the same discipline.

2 Based on earlier findings, create a testable and falsifiable explanation (a hypothesis) of the information gathered. Hypotheses may be expressed in words or in mathematical equations.

3 Design and conduct a controlled experiment to test the predictions of the hypothesis, that is, what you would expect to observe if the hypothesis were correct. The experiment must be clearly defined so that it can be repeated by others.

4 Compare the results of the experiment with those predicted by the hypothesis. If the results do not match the predictions, the hypothesis is refuted, and it must be rejected or revised. If the prediction was correct, the hypothesis is confirmed. The data from one set of experiments are subsequently used to develop additional hypotheses to be tested.

An Example of Hypothesis Development and Testing

Consider this simple example of hypothesis development and testing. A friend gives you a plant that she grew on her windowsill. Under her care, the plant always flowered. You place the plant on your windowsill and water it regularly, but the plant never blooms. You know that your friend always gave fertilizer to the plant, and you wonder whether fertilizing the plant will make it flower. In other words, you create a hypothesis with a specific **prediction:** "This type of plant will flower if it receives fertilizer." This is a good hypothesis because it is not only testable but also falsifiable. To test the hypothesis, you would simply give the plant fertilizer. If it flowers, your hypothesis is confirmed. If it does not bloom, the data force you to reject or revise your hypothesis.

With all experiments it is important to include a **control**—a set of individuals that will not be subject to the treatment. To test this specific hypothesis, you need to compare plants that receive fertilizer (the experimental treatment) with plants grown without fertilizer (the control treatment). The presence or absence of fertilizer is the **experimental variable,** and in a controlled experiment, everything except the experimental variable—the flower pots, the soil, the amount of water, and exposure to sunlight—is kept the same between the treated and control individuals. This type of control ensures that any differences in flowering pattern observed between plants that receive the experimental treatment (fertilizer) and those that receive the control treatment (no fertilizer) can be attributed to the experimental variable.

Nearly all experiments in biology include **replicates,** multiple subjects that receive either the same experimental treatment or the same control treatment. Scientists use replicates in experiments because individuals typically vary in genetic makeup, size, health, or other characteristics—and because accidents may disrupt a few replicates. By exposing multiple subjects to both treatments, we can use a statistical test to compare the average result of the experimental treatment with the average result of the control treatment, giving us more confidence in the overall findings.

continued on next page

Question: Your friend fertilizes a plant that she grows on her windowsill, and it flowers. After she gives you the plant, you put it on your windowsill, but you do not give it any fertilizer and it does not flower. Will giving the plant fertilizer induce it to flower?

Friend added fertilizer.

You did not add fertilizer.

Experiment: Establish six replicates of an experimental treatment (identical plants grown with fertilizer) and six replicates of a control treatment (identical plants grown without fertilizer).

Experimental Treatment

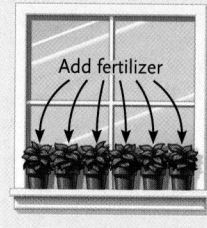
Add fertilizer

Control Treatment

No fertilizer

Possible Result 1: Neither experimental nor control plants flower.

Experimentals

Controls

Possible Result 2: Plants in the experimental group flower, but plants in the control group do not.

Experimentals

Controls

Conclusion: Fertilizer alone does not cause the plants to flower. Consider alternative hypotheses and conduct additional experiments, each testing a different experimental treatment, such as the amount of water or sunlight the plant receives or the temperature to which it is exposed.

Conclusion: The application of fertilizer induces flowering in this type of plant, confirming your original hypothesis. Pat yourself on the back and apply to graduate school in plant biology.

The Scientific Theory

When a hypothesis stands up to repeated experimental tests, it is gradually accepted as an accurate explanation of natural events. This acceptance may take many years, and it usually involves repeated experimental confirmations. When many different tests have consistently confirmed a hypothesis that addresses many broad questions, it may become regarded as a scientific **theory**—a scientifically acceptable, well-substantiated explanation of some aspect of the natural world. Most scientific theories are supported by exhaustive experimentation; thus, scientists usually regard them as established truths that are unlikely to be contradicted by future research.

In common usage, the word *theory* most often labels an idea as either speculative or downright suspect, as in the expression "It's only a theory." But when scientists talk about theories, they refer to concepts that have withstood the test of many experiments. Because of the difference between the scientific and common usage of the word *theory*, many people fail to appreciate the extensive evidence that supports scientific theories. For example, virtually every scientist accepts the theory of evolution as a fully supported scientific truth: all species change with time, new species are formed, and older species eventually die off. Although evolutionary biologists debate the details of how evolutionary processes bring about these changes, very few scientists doubt that the theory of evolution is essentially correct. Moreover, *no scientist who has tried to cast doubt on the theory of evolution has ever devised or conducted a study that disproves any part of it.* Unfortunately, the confusion between the scientific and common usage of the word *theory* has led, in part, to endless public debate about supposed faults and inadequacies in the theory of evolution.

Experimental versus Observational Science

In some scientific disciplines, the system under study may be too large or too complex to establish controlled experiments. In astronomy, for example, one cannot manipulate stars and galaxies as if they were potted plants. Astronomy is considered an observational science, as are research themes in ecology and evolutionary biology. Observational science relies on sophisticated statistical techniques to analyze detailed observational data in order to test hypotheses. The statistical tools provide a method for researchers to infer pattern and underlying cause from the collected data.

Many scientific disciplines rely on a combination of observational and experimental science. For example, ecology researchers studying global climate change often set up experiments that take place in the environment. These enable a certain level of control of variables under far more realistic conditions than would be possible in a laboratory. These so-called field experiments complement the analysis of observational data that may reflect changes to our climate that occurred hundreds of years ago.

Measurement and Scale

The SI system of Measurement

The International System of Units is the most widely used system of measurement in the world. Its abbreviation, *SI*, is from the French Système International d'Unités. It was adopted by the eleventh General Conference of Weights and Measures in 1960 and represents the latest modification of the metric system, which was first implemented by the French National Assembly in 1790.

The SI system uses seven base units, each of which measures or describes a different kind of physical quantity. Each unit is strictly defined, although the definitions have been modified (and made more accurate) over time. As an example, the metre was originally defined by the French Academy of Sciences as the length between two marks on a platinum–iridium bar that was designed to represent 1/10 000 000 of the distance from the equator to the North Pole through Paris. This definition was changed in 1983 by the International Bureau of Weights and Measures to become the distance travelled by light in absolute vacuum in 1/299 792 458 of a second.

The Seven Base Units of the SI System

Name	Symbol	Quantity
metre	m	length
kilogram	kg	mass
second	s	time
ampere	A	electric current
kelvin	K	temperature
mole	mol	amount of substance
candela	cd	luminous intensity

The SI system also uses a series of prefix names and prefix symbols to form the names and symbols of the decimal multiples of the base SI units. Note that the base unit for mass is the kilogram, not the gram. One kilogram equals 1000 g ($1 \text{ kg} = 10^3 \text{ g}$). This list has been extended several times: prefixes now range from yotta, at 10^{24} (one septillion), to yocto, at 10^{-24} (one septillionth).

Factor	Prefix	Symbol	Factor	Prefix	Symbol
10^{24}	yotta	Y	10^{-1}	deci	d
10^{21}	zetta	Z	10^{-2}	centi	c
10^{18}	exa	E	10^{-3}	milli	m
10^{15}	peta	P	10^{-6}	micro	μ
10^{12}	tera	T	10^{-9}	nano	n
10^{9}	giga	G	10^{-12}	pico	p
10^{6}	mega	M	10^{-15}	femto	f
10^{3}	kilo	k	10^{-18}	atto	a
10^{2}	hecto	h	10^{-21}	zepto	z
10^{1}	deca	da	10^{-24}	yocto	y

Derived SI Units

Several other units have been derived from combinations of the seven base units of measure. Three of the more common concern units of force (newton), pressure (pascal), and energy or heat (joule). The measurement of temperature in degrees Celsius is also considered a derived unit, even though one Celsius degree is the same size as one kelvin. However, $0°C = 273.16 \text{ K}$ (note that no degree symbol is used when expressing temperature in kelvins).

Name	Symbol	Quantity	Expression
newton	N	force	$m \cdot kg \cdot s^{-2}$
pascal	Pa	pressure	$N \cdot m^{-2}$
joule	J	energy and work	$N \cdot m$

Non-SI Units in Common Usage

A number of units not derived from the base SI units are accepted for use with SI units.

Name	Symbol	Value in SI Units
minute	min	60 s
hour	h	3600 s
day	d	86 400 s
litre	L	$1\ dm^3 = 10^{-3}\ m^3$
angstrom	Å	$10^{-10}\ m$
calorie, a measure of food energy*	cal	4.184 J
unified atomic mass unit or Dalton**	u or Da	$\sim\!1.66054 \times 10^{-24}\ kg$

*One food calorie = 1 Cal = 1000 cal
**Value determined experimentally to be 1/12 the mass of an unbound atom of carbon-12.

Scale in Biology

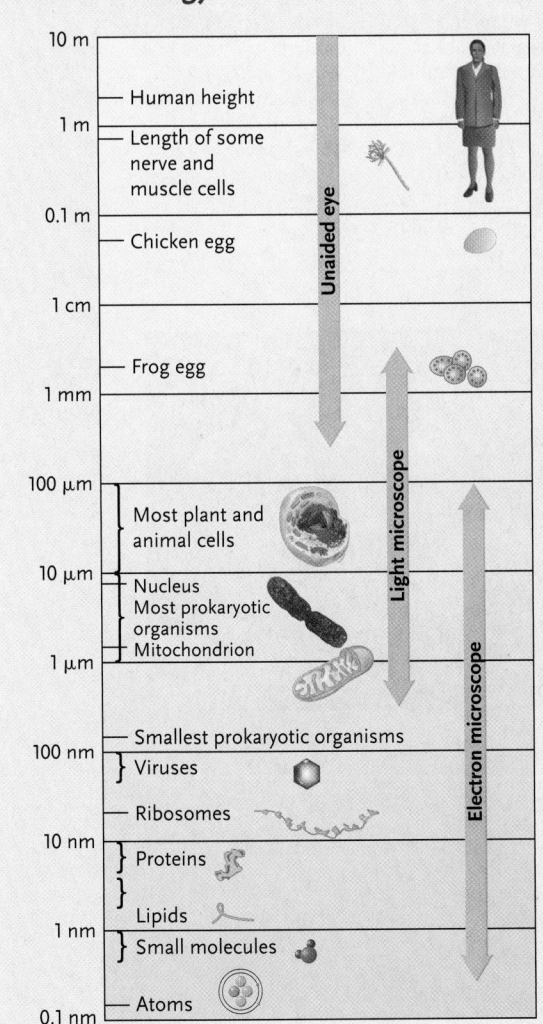

Why Everyone Should Use SI Units

In December 1998, NASA launched the Mars Climate Orbiter on a mission to study the Martian weather and climate. As it approached Mars, the spacecraft received instructions from flight control on Earth to fire thruster engines to enter into a proper orbit about 140 to 150 km above the Martian surface. However, as it approached the planet, a navigation error caused the spacecraft to descend into an orbit of only 57 km above the surface. The spacecraft was soon destroyed by the heat caused by atmospheric friction.

The review of the incident found that the root cause was a mix-up between the use of SI units and an older system of measure, imperial units (e.g., inches, feet, and pounds). More specifically, the software that was used to control the thruster engines of the spacecraft from the ground was written using the imperial unit of force, the pound-force, whereas onboard the spacecraft, information was interpreted in terms of newtons, the metric unit of force. Since 1 pound-force equals about 4.45 N, instructions from the ground were thus multiplied by 4.45.

The total cost of the mission was approximately $327 million.

The Organization of Matter

Any substance in the universe that has mass and occupies space is defined as **matter.** The fundamental scientific concepts that explain how matter is organized in biological systems are no different from those for nonliving forms of matter. Living organisms are built from the same chemical building blocks as nonliving systems and abide by the same fundamental laws of chemistry and physics. Because of this, a basic understanding of how all matter is organized is important for a complete picture of the structure and function of organisms.

Elements and Compounds

All matter is composed of elements. An **element** is a pure substance composed of only one type of atom. Ninety-two different elements occur naturally on Earth. Living organisms are composed of about 25 elements, with only 4 elements—carbon, hydrogen, oxygen, and nitrogen—accounting for more than 96% of the mass of an organism. Seven other elements—calcium, phosphorus, potassium, sulfur, sodium, chlorine, and magnesium—contribute most of the remaining 4%. The proportions by mass of different elements differ markedly in sea water, the human body, a fruit, and Earth's crust, as shown below.

A **compound** is a substance that contains two or more elements. For example, hydrogen and oxygen are the elements that make up the compound water (H_2O). The chemical and physical properties of compounds are typically distinct from those of their atoms or elements.

Sea water		Human		Pumpkin		Earth's crust	
Oxygen	88.3	Oxygen	65.0	Oxygen	85.0	Oxygen	46.6
Hydrogen	11.0	Carbon	18.5	Hydrogen	10.7	Silicon	27.7
Chlorine	1.9	Hydrogen	9.5	Carbon	3.3	Aluminium	8.1
Sodium	1.1	Nitrogen	3.3	Potassium	0.34	Iron	5.0
Magnesium	0.1	Calcium	2.0	Nitrogen	0.16	Calcium	3.6
Sulfur	0.09	Phosphorus	1.1	Phosphorus	0.05	Sodium	2.8
Potassium	0.04	Potassium	0.35	Calcium	0.02	Potassium	2.6
Calcium	0.04	Sulfur	0.25	Magnesium	0.01	Magnesium	2.1
Carbon	0.003	Sodium	0.15	Iron	0.008	Other elements	1.5
Silicon	0.0029	Chlorine	0.15	Sodium	0.001		
Nitrogen	0.0015	Magnesium	0.05	Zinc	0.0002		
Strontium	0.0008	Iron	0.004	Copper	0.0001		
		Iodine	0.0004				

Andriano/Shutterstock.com

©iStock.com/Hon Lau

The Atom

Elements are composed of **atoms**—the smallest units that retain the chemical and physical properties of an element. Any given element has only one type of atom identified by a standard one- or two-letter symbol. The element carbon is identified by the single letter C, which stands for both the carbon atom and the element.

Each atom consists of an atomic nucleus surrounded by one or more smaller, fast-moving particles called electrons. All atomic nuclei contain one or more positively charged particles called **protons.** The number of protons in the nucleus of each kind of atom is referred to as the **atomic number.** This number does not vary and thus specifically identifies the atom. The smallest atom, hydrogen, has a single proton in its nucleus, so its atomic number is 1. Carbon with six protons, nitrogen with seven protons, and oxygen with eight protons have atomic numbers of 6, 7, and 8, respectively.

With one exception, the nuclei of all atoms also contain uncharged particles called **neutrons,** which occur in variable numbers approximately equal to the number of protons. The single exception is the most common form of hydrogen, which has a nucleus that contains only a single proton. Atoms are assigned a **mass number** based on the total number of protons and neutrons in the atomic nucleus. Electrons are ignored in determinations of atomic mass because the mass of an electron is very small.

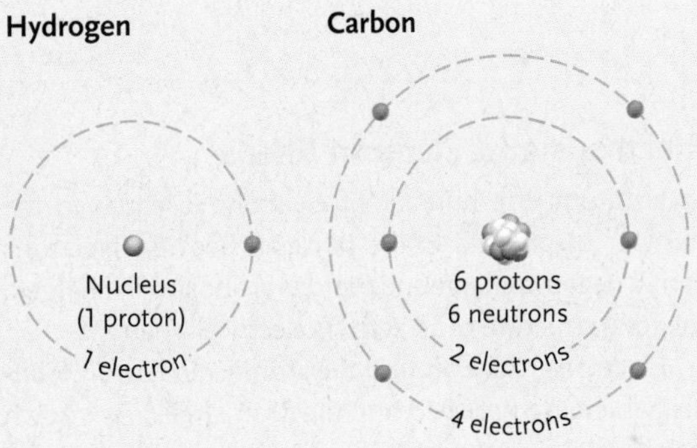

Atomic Number and Mass Number of the Most Common Elements in Living Organisms			
Element	Symbol	Atomic Number	Mass Number of the Most Common Form
Hydrogen	H	1	1
Carbon	C	6	12
Nitrogen	N	7	14
Oxygen	O	8	16
Sodium	Na	11	23
Magnesium	Mg	12	24
Phosphorus	P	15	31
Sulfur	S	16	32
Chlorine	Cl	17	35
Potassium	K	19	39
Calcium	Ca	20	40
Iron	Fe	26	56
Iodine	I	53	127

Hydrogen

Nucleus
(1 proton)
1 electron

Carbon

6 protons
6 neutrons
2 electrons
4 electrons

Isotopes

All atoms of a specific element have the same number of protons, but they may differ in the number of neutrons. These distinct forms of an element, where atoms have the same atomic number but different atomic masses, are called **isotopes.** The nuclei of some isotopes are unstable and break down, or *decay*, giving off particles of matter and energy that can be detected as radioactivity. The decay transforms the unstable, radioactive isotope—called a radioisotope—into an atom of another element. For example, the carbon isotope ^{14}C is unstable and undergoes radioactive decay in which one of its neutrons splits into a proton and an electron. The electron is ejected from the nucleus, but the proton is retained, giving a new total of seven protons and seven neutrons, which is characteristic of the most common form of nitrogen. Thus, the decay transforms the carbon atom into an atom of nitrogen.

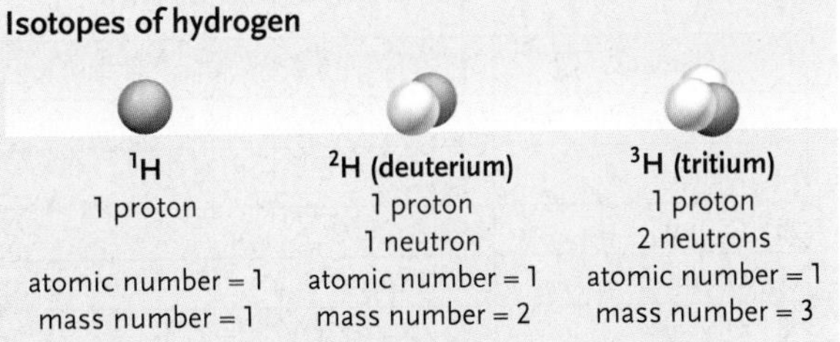

Isotopes of hydrogen

^{1}H
1 proton

atomic number = 1
mass number = 1

^{2}H (deuterium)
1 proton
1 neutron

atomic number = 1
mass number = 2

^{3}H (tritium)
1 proton
2 neutrons

atomic number = 1
mass number = 3

Isotopes of carbon

^{12}C
6 protons
6 neutrons
atomic number = 6
mass number = 12

^{13}C
6 protons
7 neutrons
atomic number = 6
mass number = 13

^{14}C
6 protons
8 neutrons
atomic number = 6
mass number = 14

Use of Radioisotopes

Radioactive decay occurs at a steady, clocklike rate. The length of time it takes for one-half of a sample of a radioisotope to decay is termed its **half-life.** Each type of radioisotope has a characteristic half-life. For example, carbon-14 decays with a fixed half-life of 5730 years, while uranium-238 has a half-life of 4.5 billion years. Because unstable isotopes decay at a fixed rate that is not affected by chemical reactions or environmental conditions such as temperature or pressure, they are used to estimate the age of organic material, rocks, and fossils. These radiometric techniques have been vital in dating animal remains and tracing evolutionary lineages.

A number of radioisotopes that have short half-lives are used in medical imaging and in the treatment of diseases. These isotopes include iodine-123 and thalium-201, which have half-lives of only 13.3 and 3.1 days, respectively.

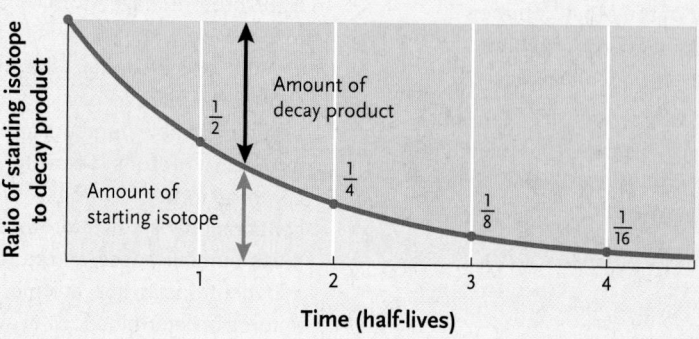

Electrons and Electron Shells

In an atom, the number of electrons is equal to the number of protons in the nucleus. Because electrons carry a negative charge and protons are positively charged, the total structure of an atom is electrically neutral.

Electrons move around the atomic nucleus in **orbitals,** which are grouped into **electron shells.** As shown below, the first shell (I) may be occupied by a maximum of two electrons. The second (II) and third (III) shells can hold a maximum of eight electrons each.

The fourth shell can hold 18 electrons (not all shown). Atoms with more than four electron shells are very rare in biological molecules.

The chemical behaviour of an atom depends primarily on the number of electrons in its outermost shell. This is referred to as the valence shell, which holds **valence electrons.** Atoms in which the valence shell is not completely filled with electrons tend to be chemically reactive; those with a completely filled valence shell are nonreactive, or inert.

For example, as shown on the left, hydrogen has a single, unpaired electron in its outermost and only electron shell and is highly reactive; helium has two valence electrons filling its single orbital and is unreactive or inert. Along with helium, neon and argon are also referred to as inert gases because their outer electron shell is full.

Because an unfilled electron shell is less stable than a filled one, atoms with an incomplete outer shell have a strong tendency to interact with other atoms in a way that causes them to either gain or lose enough electrons to achieve a completed outermost shell. All elements commonly found in living organisms have valence shells that are not completely filled with electrons (purple balls in figure at left). Because of this, these atoms readily participate in chemical reactions with other atoms.

Atomic number

	Element	I	II	III	IV
1	Hydrogen	○			
2	Helium	○○			
6	Carbon	○○	○○○○		
7	Nitrogen	○○	○○○○○		
8	Oxygen	○○	○○○○○○		
10	Neon	○○	○○○○○○○○		
11	Sodium	○○	○○○○○○○○	○	
12	Magnesium	○○	○○○○○○○○	○○	
15	Phosphorus	○○	○○○○○○○○	○○○○○	
16	Sulfur	○○	○○○○○○○○	○○○○○○	
17	Chlorine	○○	○○○○○○○○	○○○○○○○	
18	Argon	○○	○○○○○○○○	○○○○○○○○	
19	Potassium	○○	○○○○○○○○	○○○○○○○○	○
20	Calcium	○○	○○○○○○○○	○○○○○○○○	○○

Chemical Bonds

An atom with an incomplete valence shell has a strong tendency to interact with other atoms so that they have a completely filled valence shell. These interactions, called **chemical bonds,** are caused by closely associated atoms sharing or transferring electrons to complete the valence shell. Four types of chemical bonds are important in biological molecules: ionic bonds, covalent bonds, hydrogen bonds, and van der Waals forces. Because of their importance in hydrogen bonding, polar molecules are also discussed in this section.

Ionic Bonds

Ionic bonds form between atoms that gain or lose valence electrons completely. A sodium atom (Na) readily loses a single electron to achieve a full valence shell (see last section), and chlorine (Cl) readily gains an electron to do the same. After the transfer, the sodium atom, now with 11 protons and 10 electrons, carries a single positive charge. The chlorine atom, now with 17 protons and 18 electrons, carries a single negative charge. In this charged condition, the atoms are called **ions:** sodium, with a positive charge, is a **cation,** while chloride, with a negative charge, is an **anion.**

Ionic bonds are common among the forces that hold ions, atoms, and molecules together in living organisms because these bonds have three key features:

- They exert an attractive force over greater distances than any other chemical bond.
- Their attractive force extends in all directions.
- They vary in strength depending on the presence of other charged substances.

Ionic bond formation between sodium and chlorine

Crystals of sodium chloride (NaCl)

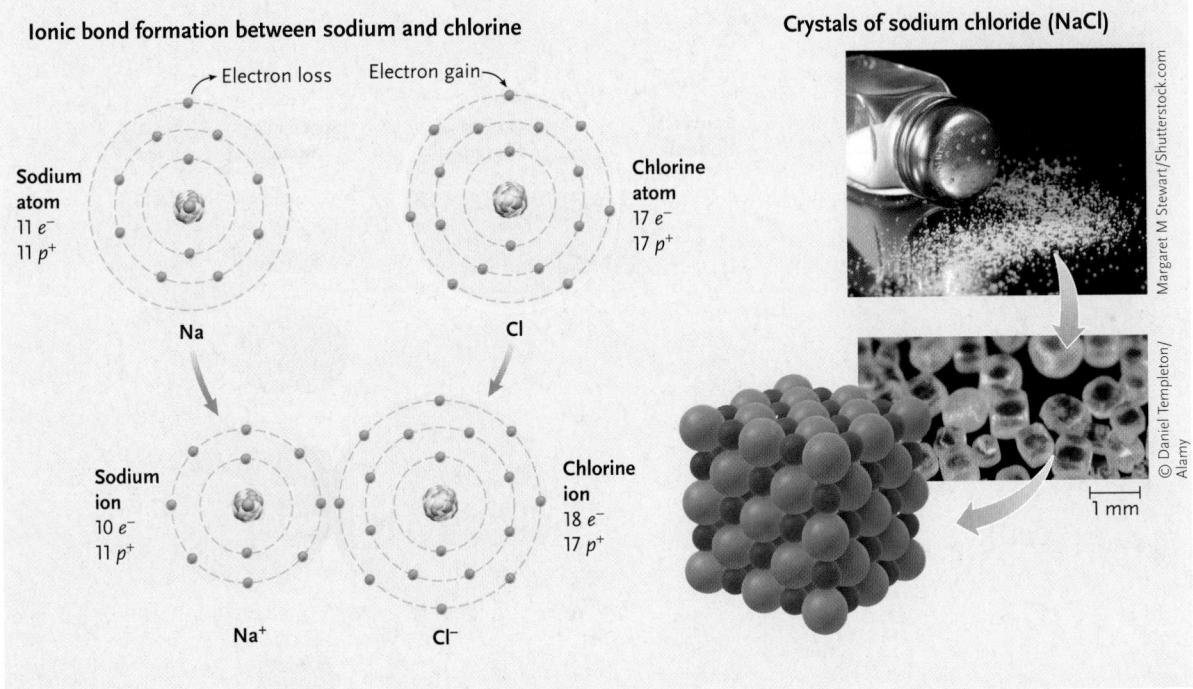

Electron loss

Electron gain

Sodium atom
11 e^-
11 p^+

Chlorine atom
17 e^-
17 p^+

Na

Cl

Sodium ion
10 e^-
11 p^+

Chlorine ion
18 e^-
17 p^+

Na$^+$

Cl$^-$

Margaret M Stewart/Shutterstock.com

© Daniel Templeton/Alamy

1 mm

Covalent Bonds

Covalent bonds form between two atoms when they share valence electrons. This is distinct from ionic bonds, where electrons are gained or lost from atoms. The term **molecule** refers to two or more atoms held together by covalent bonds. The formation of molecular hydrogen, H_2, by two hydrogen atoms is the simplest example of a covalent bond. If two hydrogen atoms collide, the single electron of each atom may join in a new, combined two-electron orbital that surrounds both nuclei. The two electrons fill the orbital; thus, the hydrogen atoms tend to remain linked stably together. The linkage formed by the shared orbital is a covalent bond.

A structural formula represents a covalent bond of a pair of shared electrons as a single line. For example, in H_2, the covalent bond that holds the molecule together is represented as H:H or H—H. Generally speaking, the covalent bonding capacity of an atom is equal to the number of valence shell electrons necessary to fill the shell: hydrogen, 1; oxygen, 2; nitrogen, 3; and carbon, 4.

As shown below, a single oxygen atom has six valence shell electrons, and two oxygen atoms form a single molecule.

Carbon, with four unpaired outer electrons, typically forms four covalent bonds to complete its outermost energy level. An example is methane, CH_4, the main component of natural gas.

Unlike ionic bonds, which extend their attractive force in all directions, the shared orbitals that form covalent bonds extend between atoms at discrete angles and directions, giving covalently bound molecules distinct, three-dimensional forms. For biological molecules such as proteins, which are held together primarily by covalent bonds, the three-dimensional form imparted by these bonds is critical to their functions.

The four covalent bonds formed by the carbon atom are fixed at an angle of 109.5° from each other, forming a tetrahedron. The tetrahedral arrangement of the bonds allows carbon atoms to link extensively to each other in chains and rings in both branched and unbranched form. Such structures form the backbones of an almost unlimited variety of molecules. Carbon can also form double bonds, in which atoms share two pairs of electrons, and triple bonds, in which atoms share three pairs of electrons.

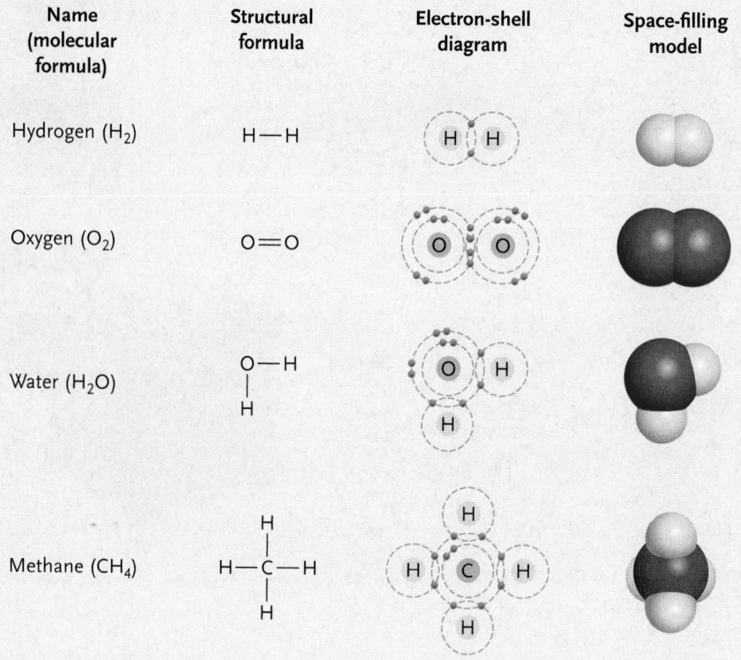

Name (molecular formula)	Structural formula	Electron-shell diagram	Space-filling model
Hydrogen (H_2)	H—H		
Oxygen (O_2)	O=O		
Water (H_2O)	O—H | H		
Methane (CH_4)	H—C—H (with H above and H below)		

Polarity and Hydrogen Bonding

All covalent bonds involve the sharing of valence electrons between two atoms. Yet the degree of electron sharing between the two atoms can differ widely. **Electronegativity** is the measure of an atom's attraction for the electrons it shares in a chemical bond with another atom. The more electronegative an atom is, the more strongly it attracts shared electrons. Oxygen is the most electronegative atom found in biological molecules, followed by nitrogen and sulfur. By comparison neither carbon nor hydrogen is considered electronegative.

The unequal sharing of electrons between two atoms that differ in their electronegativity results in a **polar covalent bond.** The atom that attracts the electrons more strongly carries a partial negative charge (denoted by the symbol d^-), and the atom deprived of electrons carries a partial positive charge (denoted by the symbol d^+). As shown below for a molecule of water, atoms carrying partial charges give the molecule partially positive and negative ends; this is referred to as polarity, and the molecule is termed **polar.**

Polar molecules attract and align themselves with other polar molecules and with charged ions and molecules. Polar molecules that associate readily with water because it is strongly polar are identified as **hydrophilic** (*hydro* = water; *philic* = preferring). Nonpolar substances that are excluded by water and other polar molecules are identified as **hydrophobic** (*phobic* = avoiding). Most common non-polar molecules consist primarily of C—H bonds (neither atom being electronegative).

Hydrogen atoms are made partially positive by sharing electrons unequally with oxygen, nitrogen, or sulfur. Because of this, the hydrogen atom may be attracted to other electronegative atoms that it is not directly bonded to (see figure below). his attractive force is the **hydrogen bond,** illustrated by a dotted line in structural diagrams of molecules. Hydrogen bonds may form between atoms in the same or different molecules.

Individual hydrogen bonds are about 1/20 the strength of a covalent bond. However, large biological molecules may offer many opportunities for hydrogen bonding, both within and between molecules. When numerous, hydrogen bonds are collectively strong and lend stability to the three-dimensional structure of molecules such as proteins.

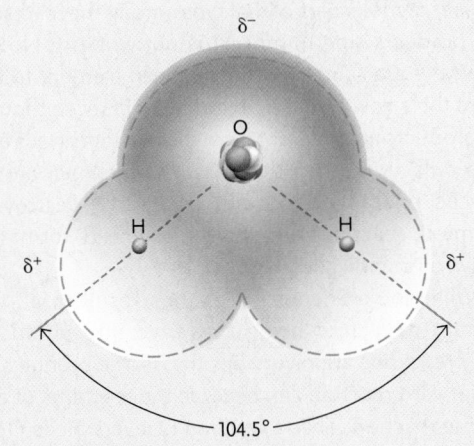

Hydrogen-bond donor		Hydrogen-bond acceptor
d^- d^+		d^-
—N—H	⋯⋯⋯⋯ Hydrogen bond	N—
—N—H	⋯⋯⋯⋯	O—
—O—H	⋯⋯⋯⋯	N—
—O—H	⋯⋯⋯⋯	O—

Van der Waals Forces

Van der Waals forces are even weaker than hydrogen bonds. These forces develop between nonpolar molecules or regions of molecules when, through their constant motion, electrons accumulate by chance in one part of a molecule or another. This process leads to zones of positive and negative charge, making the molecule polar. If they are oriented in the right way, the polar parts of the molecules are attracted electrically to one another and cause the molecules to stick together briefly. Although an individual bond formed with van der Waals forces is weak and transient, the formation of many bonds of this type can stabilize the shape of a large molecule, such as a protein.

A striking example of the collective power of van der Waals forces concerns the ability of geckos to cling to and walk up vertical smooth surfaces. The toes of the lizard are covered in millions of pads, each one forming a weak interaction—using van der Waals forces—with the molecules on the smooth surface.

Nathalie Speliers Ufermann/Shutterstock.com

Chemical Reactions

Chemical reactions occur when atoms or molecules interact to form new chemical bonds or break old ones. As a result of bond formation or breakage, atoms are added to or removed from molecules, or the linkages of atoms in molecules are rearranged. When any of these alterations occur, molecules change from one type to another, usually with different chemical and physical properties. In biological systems, chemical reactions are accelerated by *enzymes*, which are discussed in Chapter 4.

The atoms or molecules entering a chemical reaction are called the **reactants,** and those leaving a reaction are the **products.** A chemical reaction is written with an arrow showing the direction of the reaction; reactants are placed to the left of the arrow, and products are placed to the right. Both reactants and products are usually written in chemical shorthand as formulas.

For example, the overall reaction of photosynthesis, in which carbon dioxide and water are combined to produce sugars and oxygen (see Chapter 7), is written as follows:

$$6CO_2 \; + \; 6H_2O \; \rightarrow \; C_6H_{12}O_6 \; + \; 6O_2$$

Carbon dioxide   Water   A sugar   Molecular oxygen

The number in front of each formula indicates the number of molecules of that type among the reactants and products (the number 1 is not written). Notice that there are as many atoms of each element to the left of the arrow as there are to the right, even though the products are different from the reactants. This balance reflects the fact that in such reactions, atoms may be rearranged but not created or destroyed. Chemical reactions written in balanced form are known as **chemical equations.**

While some chemical reactions result in all the reactant molecules being converted into products, many reactions are reversible; that is, the products of the forward reaction can become the reactants of the reverse reaction. That a reaction is reversible is illustrated by using opposite-headed arrows. As an example, hydrogen and molecular nitrogen can react to produce ammonia, but ammonia can also break down to produce hydrogen and nitrogen:

$$3H_2 + N_2 \rightleftarrows 2NH_3$$

Water

All living organisms contain water, and many kinds of organisms live directly in water. Even those that live in dry environments contain water in all their structures—different organisms range from 50% to more than 95% water by mass. The water inside organisms is crucial for life: it is required for many important biochemical reactions and plays major roles in maintaining the shape and organization of cells and tissues.

Hydrogen Bonds and the Properties of Water

The properties of water molecules that make them so important to life depend to a great extent on their polar structure and their ability to link to each other by hydrogen bonds (See "Chemical Bonds").

Hydrogen bonds form readily between water molecules in both liquid water and ice. In liquid water, each water molecule establishes an average of 3.4 hydrogen bonds with its neighbours, forming an arrangement known as the **water lattice.** In liquid water, the hydrogen bonds that hold the lattice together constantly break and reform, allowing the water molecules to break loose from the lattice, slip past one another, and reform the lattice in new positions.

In ice, the water lattice is a rigid, crystalline structure in which each water molecule forms four hydrogen bonds with neighbouring molecules. The rigid ice lattice spaces the water molecules farther apart than the water lattice. Because of this greater spacing, water has the unusual property of being about 10% less dense when solid than when liquid. Imagine what Earth would be like if ice sank to the bottom, as most solids do.

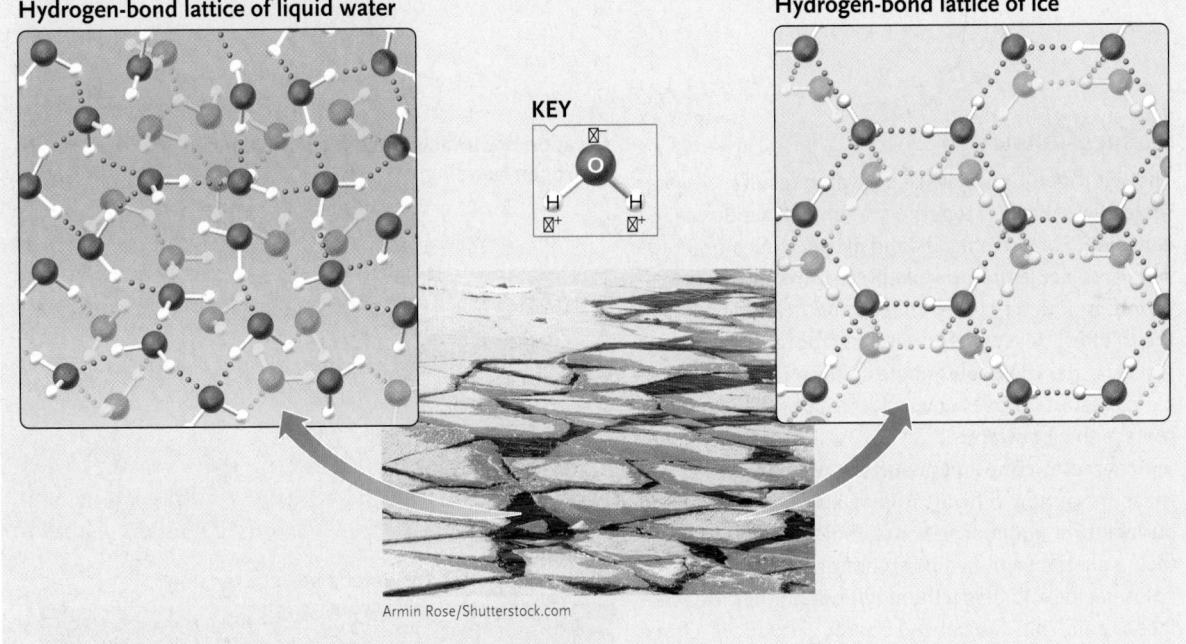

Hydrogen-bond lattice of liquid water

Hydrogen-bond lattice of ice

KEY

Armin Rose/Shutterstock.com

Specific Heat and Heat of Vaporization

The hydrogen-bond lattice of liquid water retards the escape of individual water molecules as the water is heated. As heat flows into water, much of it is absorbed in the breakage of hydrogen bonds. As a result, the temperature of water, reflected in the average motion of its molecules, increases relatively slowly as heat is added. This results in water having a high **specific heat,** defined as the amount of heat required to increase the temperature of a given quantity of water. For example, relatively high temperatures and the addition of considerable heat are required to break enough hydrogen bonds to make water boil. The high boiling point maintains water as a liquid over the wide temperature range of 0 to 100°C.

The unusual properties of water are more obvious if you compare it to H_2S, a molecule that has a similar molecular mass and structure. Compared to H_2O, H_2S boils at an astonishingly low temperature of −60°C.

The vast difference in boiling points between these two molecules is explained by oxygen being more electronegative than sulfur. This results in water being a more polar molecule and in turn being able to form a much stronger hydrogen bond lattice.

A large amount of heat, 586 cal/g, must be added to give water molecules enough energy of motion to break loose from liquid water and form a gas. This required heat, known as the **heat of vaporization,** allows humans and many other organisms to cool off when hot. In humans, water is released onto the surface of the skin by more than 2.5 million sweat glands; the heat energy absorbed by the water in sweat as the sweat evaporates cools the skin and the underlying blood vessels. The heat loss helps keep body temperature from increasing when environmental temperatures are high. Plants use a similar cooling mechanism as water evaporates from their leaves.

Surface Tension

The hydrogen-bond lattice of water results in water molecules staying together, a phenomenon called **cohesion.** For example, in land plants, cohesion holds water molecules in unbroken columns in the microscopic conducting tubes that extend from the roots to the highest leaves. As water evaporates from the leaves, water molecules in the columns, held together by cohesion, move upward through the tubes to replace the lost water.

Related to cohesion is **surface tension,** which is a measure of how difficult it is to stretch or break the surface of a liquid. The water molecules at surfaces facing air can form hydrogen bonds with water molecules beside and below them but not on the sides that face the air. This unbalanced bonding produces a force that places the surface water molecules under tension, making them more resistant to separation than the underlying water molecules. This force is strong enough to allow small insects such as water striders to walk on water.

Creation of surface tension by unbalanced hydrogen bonding

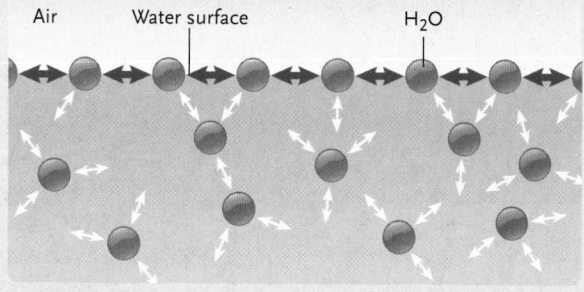

Spider supported by water's surface tension

Aqueous Solutions

Because water molecules are small and strongly polar, they readily surround other polar and charged molecules and ions. The surface coat, called a **hydration shell,** reduces the attraction between the molecules or ions and promotes their separation and entry into a **solution,** where they are suspended individually, surrounded by water molecules. Once in solution, the hydration shell prevents the polar molecules or ions from reassociating. In such an aqueous solution, water is called the **solvent,** and the molecules of a substance dissolved in water are called the *solute.*

Sodium chloride (salt) dissolves in water because water molecules quickly form hydration layers around the Na^+ and Cl^- ions in the salt crystals, reducing the attraction between the ions so much that they separate from the crystal and enter the surrounding water lattice as individual ions. In much the same way, hydration shells surround macromolecules such as nucleic acids and proteins, reducing their electrostatic interaction with other molecules.

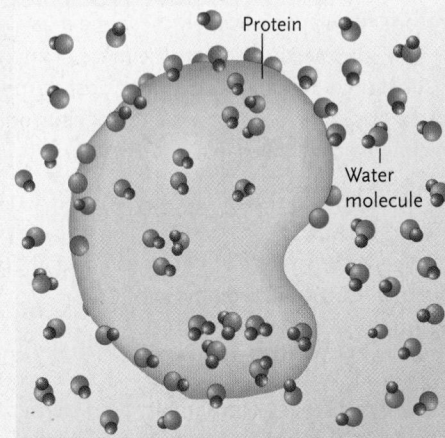

Calculating Solute Concentrations

In the cell, chemical reactions depend on solutes dissolved in aqueous solutions. To understand these reactions, you need to know the number of atoms and molecules involved. **Concentration** is the number of molecules or ions of a substance in a unit volume of space, such as 1 mL or 1 L. The number of molecules or ions in a unit volume cannot be counted directly, but it can be calculated indirectly by using the mass number of atoms as the starting point.

The mass number of an atom is equivalent to the number of protons and neutrons in its nucleus. From the mass number, and the fact that neutrons and protons have approximately the same mass (i.e., 1.66×10^{-24} g), you can calculate the mass of an atom of any substance. For an atom of the most common form of carbon, with six protons and six neutrons in its nucleus, the total mass is

$$12 \times (1.66 \times 10^{-24} \text{ g}) = 1.992 \times 10^{-23} \text{ g}$$

For an oxygen atom, with eight protons and eight neutrons in its nucleus, the total mass is

$$16 \times (1.66 \times 10^{-24} \text{ g}) = 2.656 \times 10^{-23} \text{ g}$$

Dividing the total mass of a sample of an element by the mass of a single atom gives the number of atoms in the sample. Suppose you have a carbon sample with a mass of 12 g—a mass in grams equal to the atom's mass number. (A mass in grams equal to the mass number is known as the atomic weight of an element.) Dividing 12 g by the mass of one carbon atom gives

$$\frac{12}{(10.992 \times 10^{-23} \text{g})} = 6.02 \times 10^{23} \text{ atoms}$$

If you divide the atomic weight of oxygen (16 g) by the mass of one oxygen atom, you get the same result:

$$\frac{16}{(2.656 \times 10^{-23} \text{g})} = 6.02 \times 10^{23} \text{ atoms}$$

In fact, dividing the atomic weight of any element by the mass of an atom of that element always produces the same number: 6.02×10^{23}. This number is called **Avogadro's number** after Amedeo Avogadro, the nineteenth-century Italian chemist who first discovered the relationship.

The same relationship holds for molecules. The **molecular weight** of any molecule is the mass in grams equal to the total mass number of its atoms. For NaCl, the total mass number is $23 + 35 = 58$ (a sodium atom has 11 protons and 12 neutrons, and a chlorine atom has 17 protons and 18 neutrons). The mass of an NaCl molecule is therefore

$$58 \times (1.66 \times 10^{-24} \text{ g}) = 9.628 \times 10^{-23} \text{ g}$$

Dividing the molecular weight of NaCl (58 g) by the mass of a single NaCl molecule gives

$$\frac{58}{(9.628 \times 10^{-23} \text{g})} = 6.02 \times 10^{23} \text{ atoms}$$

When concentrations are described, the atomic weight of an element or the molecular weight of a compound—the amount that contains 6.02×10^{23} atoms or molecules—is known as a **mole** (abbreviated *mol*). The number of moles of a substance dissolved in 1 L of solution is known as the **molarity** (abbreviated *M*) of the solution. This relationship is highly useful in chemistry and biology because we know that two solutions with the same volume and molarity but composed of different substances will contain the same number of molecules of the substances.

Dissociation of Water and pH

The most critical property of water that is unrelated to its hydrogen-bond lattice is its ability to separate or dissociate. This occurs when a hydrogen atom that is involved in a hydrogen bond between two water molecules moves from one molecule to the other. The proton (H^+) is what actually leaves; the electron is left behind. This proton switch results in the formation of a hydroxide ion (OH^-) and a hydronium ion (H_3O^+).

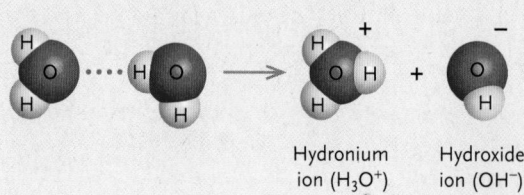

Hydronium
ion (H_3O^+)

Hydroxide
ion (OH^-)

It is convention to simply use H^+ (the hydrogen ion) to denote the hydronium ion. The proportion of water molecules that dissociate to release hydrogen and hydroxide ions is small. However, because of the dissociation, water always contains some H^+ and OH^- ions.

In pure water, the concentrations of H^+ and OH^- ions are equal. However, adding other substances may alter the relative concentrations of H^+ and OH^-, making them unequal. Some substances, called **acids,** are proton donors, which release hydrogen ions (and anions) when they are dissolved in water, effectively increasing the H^+ concentration. For example, hydrochloric acid (HCl) dissociates into H^+ and Cl^- when dissolved in water:

$$HCl \rightarrow H^+ + Cl^-$$

Other substances, called **bases,** are proton acceptors, which reduce the H^+ concentration of a solution. Most bases dissociate in water into hydroxide ions (OH^-) and cations. The hydroxide ion can act as a base by accepting a proton to produce water. For example, sodium hydroxide (NaOH) separates into Na^+ and OH^- ions when dissolved in water:

$$NaOH \rightarrow Na^+ + OH^-$$

The excess OH^- combines with H^+ to produce water,

$$OH^- + H^+ \rightarrow H_2O$$

thereby reducing the H^+ concentration. Basic solutions are also called *alkaline* solutions.

Other bases do not dissociate to produce hydroxide ions directly. For example, ammonia (NH_3), a poisonous gas, acts as a base when dissolved in water, directly accepting a proton from water, producing an ammonium ion, and releasing a hydroxide ion:

$$NH_3 + H_2O \rightarrow NH_4^+ + OH^-$$

The concentration of H^+ is measured on a numerical scale from 0 to 14, called the pH scale. Because the number of H^+ ions in solution increases exponentially as the acidity increases, the scale is based on logarithms of this number to make the values manageable:

$$pH = -\log_{10}[H^+]$$

In this formula, the brackets indicate concentration in moles per litre. The negative of the logarithm is used to give a positive number for the pH value. For example, in a water solution that is *neutral*—neither acidic nor basic—the concentration of *both* H^+ and OH^- ions is 1×10^{-7} M (0.000 000 1 M). The base 10 logarithm of 1×10^{-7} is -7. The negative of the logarithm -7 is 7. Acidic solutions have pH values less than 7, while basic solutions have pH values greater than 7. Each whole number on the pH scale represents a value 10 times or one-tenth the next number.

Hydrochloric acid (HCl)	Lemon juice, cola drinks, some acid rain		Black coffee		Urine (5.0–7.0)	[H^+] = [OH^-] Pure water	Egg white (8.0)	Phosphate detergents, bleach, antacids		Household ammonia (10.5–11.9)		Oven cleaner	
				Tomatoes	Bread								
0	**1**	**2**	**3**	**4**	**5**	**6**	**7**	**8**	**9**	**10**	**11**	**12**	**13**	**14**
	Gastric fluid (1.0–3.0)		Vinegar, wine, beer, oranges	Bananas	Typical rainwater	Milk (6.6)	Blood (7.3–7.5)	Seawater (7.8–8.3)		Soapy solutions		Hair remover		Sodium hydroxide (NaOH)

pH

Buffers Keep pH within Limits

Acidity is important to cells because even small changes, on the order of 0.1 or even 0.01 pH unit, can drastically affect biological reactions. In large part, a small change in pH can cause structural changes in proteins that can damage or destroy the proteins' function. Consequently, all living organisms have elaborate systems that control their internal acidity by regulating H^+ concentration near the neutral value of pH 7.

Living organisms control the internal pH of their cells with *buffers*—substances that compensate for pH changes by absorbing or releasing hydrogen ions. When hydrogen ions are released in excess by biological reactions, buffers combine with them and remove them from the solution; if the concentration of hydrogen ions decreases, buffers release H^+ to restore the balance. Most buffers are weak acids, weak bases, or combinations of these substances that dissociate reversibly in water solutions to release or absorb H^+ or OH^-. (Weak acids, such as acetic acid, or weak

bases, such as ammonia, release relatively few H^+ or OH^- ions in an aqueous solution, whereas strong acids or bases dissociate extensively. HCl is a strong acid; NaOH is a strong base.)

The buffering mechanism that maintains blood pH near neutral values is a good example. In humans and many other animals, blood pH is buffered by a chemical system based on carbonic acid (H_2CO_3), a weak acid. In water solutions, carbonic acid dissociates readily into bicarbonate ions (HCO_3^-) and H^+:

$$H_2CO_3 \rightarrow HCO_3^- + H^+$$

The reaction is reversible. If hydrogen ions are present in excess, the reaction is pushed to the left—the excess H^+ ions combine with bicarbonate ions to form H_2CO_3. If the H^+ concentration declines below normal levels, the reaction is pushed to the right—H_2CO_3 dissociates into HCO_3^- and H^+, restoring the H^+ concentration. The back-and-forth adjustments of the buffer system help keep human blood close to its normal pH of 7.4.

Carbon Compounds

Carbon Bonding

Compounds that contain carbon form the structures of living organisms and take part in all biological reactions as well as serving as energy sources. Collectively, molecules based on carbon are known as organic molecules. All other substances, that is, those without carbon atoms in their structures, are **inorganic molecules.** A few of the smallest carbon-containing molecules that occur in the environment as minerals or atmospheric gases, such as $CaCO_3$ and CO_2, are also considered inorganic molecules.

Carbon's central role in life's molecules arises from its bonding properties: it can assemble into an astounding variety of chain and ring structures that form the backbones of all biological molecules. This is because carbon has four unpaired outer electrons that it readily shares to complete its outer most energy level, forming four covalent bonds. With different combinations of single, double, and even triple bonds, an almost limitless array of molecules is possible. Carbon atoms bond covalently to each other and to other atoms, chiefly hydrogen, oxygen, nitrogen, and sulfur, in molecular structures that range in size from a few to thousands or even millions of atoms. Molecules consisting of carbon linked only to hydrogen atoms are called hydrocarbons (*hydro-* refers to hydrogen, not water). The simplest hydrocarbon, CH_4 (methane), consists of a single carbon atom bonded to four hydrogen atoms. Removing one hydrogen atom from methane leaves a methyl group, which occurs in many biological molecules:

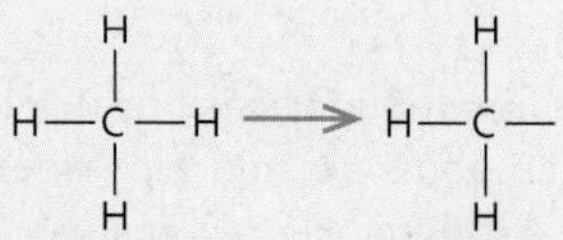

Methane   Methyl group

Now imagine bonding two methyl groups together. Removing a hydrogen atom from the maximum of four bonds, the number of hydrogen atoms in a molecule decreases as the resulting structure, ethane, produces an ethyl group:

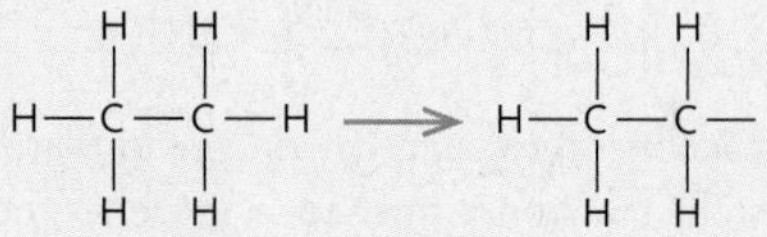

Ethane   Ethyl group

Repeating this process builds a linear hydrocarbon chain:

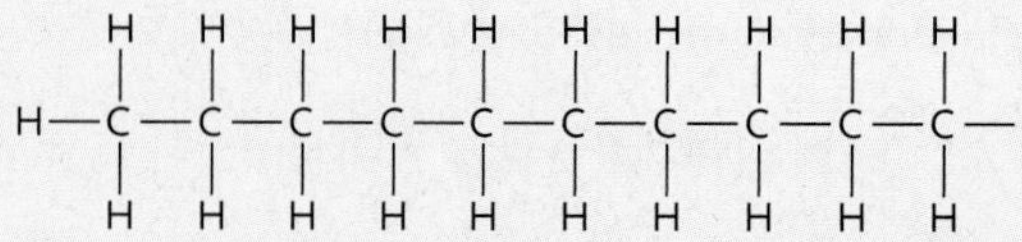

Branches can be added to produce a branched hydrocarbon chain:

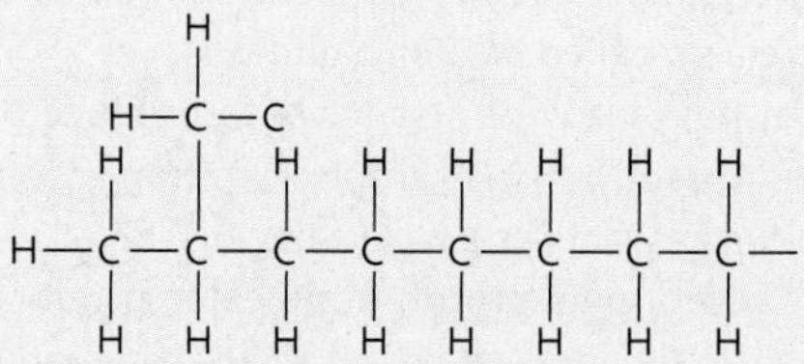

A chain can loop back on itself to form a ring. For example, cyclohexane, C_6H_{12}, has single covalent bonds between each pair of carbon atoms and two hydrogen atoms attached to each carbon atom:

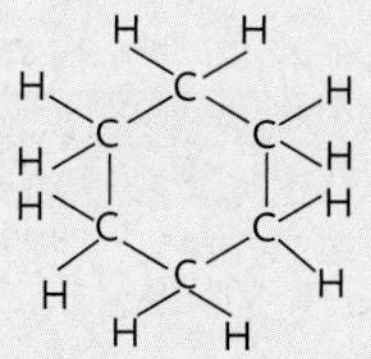

C_6H_{12}, cyclohexane

Hydrocarbons gain added complexity when neighbouring carbon atoms form double or triple bonds. Because each carbon atom can form a maximum of four bonds, the number of hydrogen atoms in a molecule decreases as the number of bonds between any two carbon atoms increases:

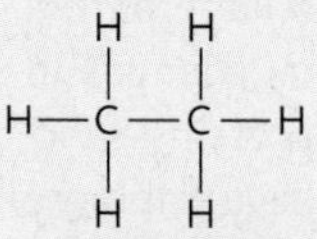

Single bonding: Double bonding: Triple bonding:
C_2H_6, ethane C_2H_4, ethene C_2H_2, ethyne
                 (ethylene)   (acetylene)

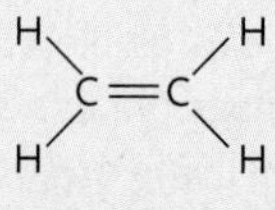

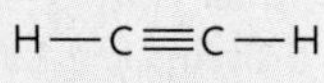

Double bonds between carbon atoms are also found in carbon rings:

Notice how each carbon in the ring above still maintains four covalent bonds.

To simplify things, the ring structure above is often simply depicted like this:

Many carbon rings can join together to produce larger molecules, as in the string of sugar molecules that makes up a polysaccharide chain:

There is almost no limit to the number of different hydrocarbon structures that carbon and hydrogen can form. However, the molecules of living systems typically contain other elements in addition to carbon and hydrogen. These other elements confer functional properties on organic molecules, producing the four major classes of organic molecules: *carbohydrates*, *lipids*, *proteins*, and *nucleic acids*.

Dehydration and Hydrolysis Reactions

In many of the reactions that involve functional groups, the components of a water molecule, —H and —OH, are removed from or added to the groups as they interact. When the components of a water molecule are *removed* during a reaction, usually as part of the assembly of a larger molecule from smaller subunits, the reaction is called a **dehydration synthesis reaction** or a condensation reaction. For example, this type of reaction occurs when individual sugar molecules combine to form a starch molecule. In **hydrolysis,** the reverse reaction, the components of a water molecule are *added* to functional groups as molecules are broken into smaller subunits. For example, the breakdown of a protein molecule into individual amino acids occurs by hydrolysis.

Dehydration synthesis reactions

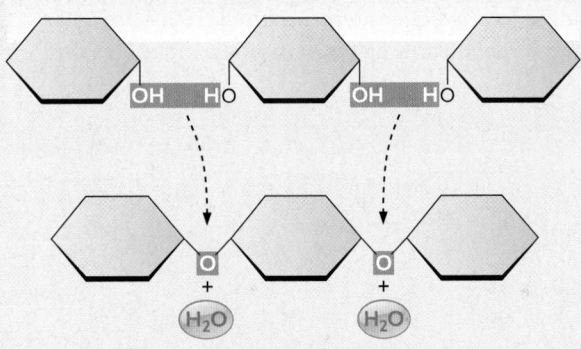

The components of a water molecule are removed as subunits join into a larger molecule.

Hydrolysis

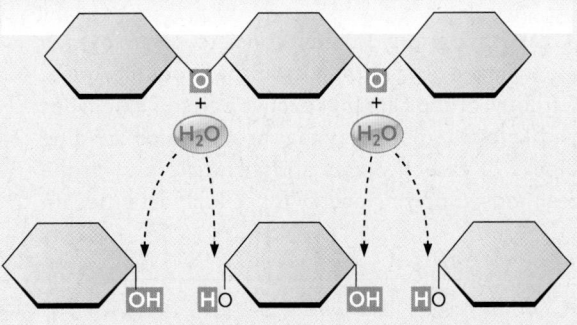

The components of a water molecule are added as molecules are split into smaller subunits.

Functional Groups

Carbohydrates, lipids, proteins, and nucleic acids are synthesized and degraded in living organisms through interactions between small, reactive groups of atoms attached to the organic molecules. The atoms in these reactive groups, called **functional groups,** occur in positions in which their covalent bonds are more readily broken or rearranged than the bonds in other parts of the molecules.

The functional groups that enter most frequently into biological reactions are the *hydroxyl, carbonyl, carboxyl, amino, phosphate,* and *sulfhydryl* groups. The unconnected covalent bonds written to the left of each structure link these functional groups to other atoms in biological molecules, usually carbon atoms. The symbol R is used to represent a chain of carbon atoms.

Common Functional Groups of Organic Molecules

Functional Group	Major Classes of Molecules	Example
Hydroxyl R—OH	Alcohols	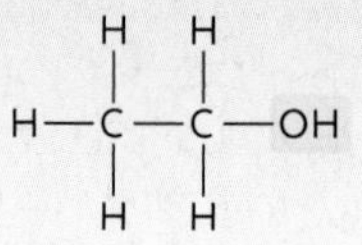 Ethyl alcohol (in alcoholic beverages)

A hydroxyl group (—OH) consists of an oxygen atom linked to a hydrogen atom. Hydroxyl groups are polar and confer polarity on the parts of the molecules that contain them. The presence of the hydroxyl group enables an alcohol to form linkages to other organic molecules through dehydration synthesis reactions.

Functional Group	Major Classes of Molecules	Example

Carbonyl 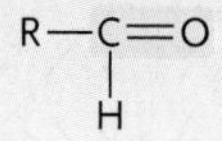
 	
R—C=O
  |
  H
	Aldehydes	
		Acetaldehyde

Carbonyl
 	
R—C=O
  |
  C
  |
	Ketones	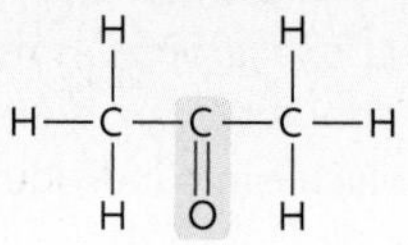
		Acetone (a solvent)

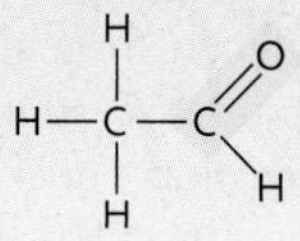

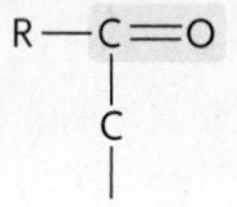

A carbonyl group (C=O) consists of an oxygen atom linked to a carbon atom by a double bond. Carbonyl groups are the reactive parts of aldehydes and ketones, molecules that act as major building blocks of carbohydrates and also take part in the reactions supplying energy for cellular activities. In an aldehyde, the carbonyl group is linked—along with a hydrogen atom—to a carbon atom at the end of a carbon chain, along with a hydrogen atom, as in acetaldehyde. In a ketone, the carbonyl group is linked to a carbon atom in the interior of a carbon chain, as in acetone.

Functional Group	Major Classes of Molecules	Example

Carboxyl
 	
R—COOH
or 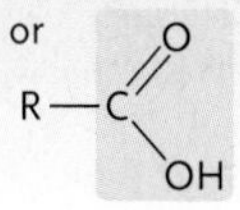
    O
    ||
R—C
    |
    OH
	Organic acids	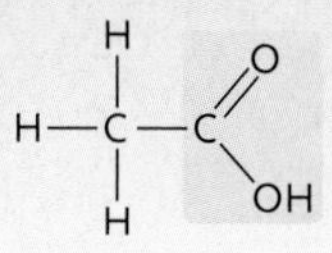
		Acetic acid (in vinegar)

A carboxyl group (—COOH) is formed by the combination of a carbonyl group and a hydroxyl group. The carboxyl group is the characteristic functional group of organic acids (also called carboxylic acids). The carboxyl group gives organic molecules acidic properties because its —OH group readily releases the hydrogen as a proton (H^+) in solution.

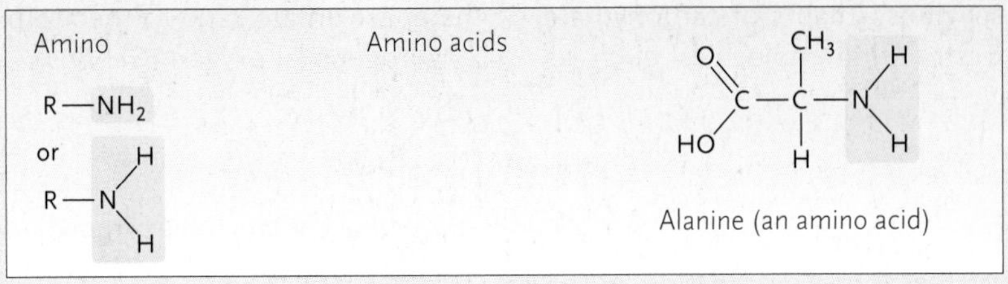

Amino   Amino acids

R—NH_2

or

R—$N\begin{smallmatrix}H\\\\H\end{smallmatrix}$

Alanine (an amino acid)

The amino group (—NH_2) consists of a nitrogen atom bonded on one side to two hydrogen atoms; in a molecule it is linked to an R group on the other side, as in the amino acid alanine and all other amino acids.

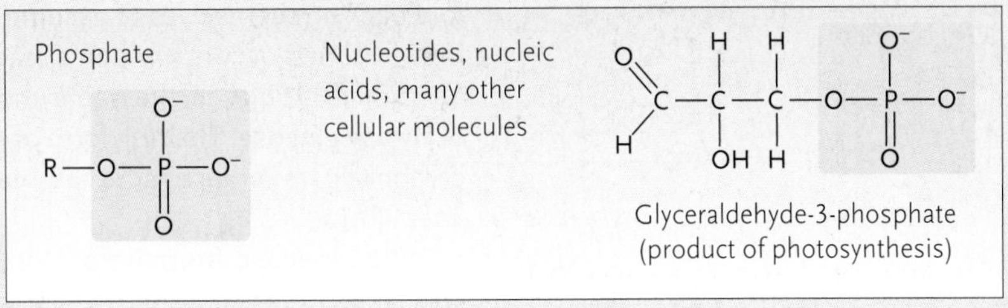

Phosphate   Nucleotides, nucleic acids, many other cellular molecules

Glyceraldehyde-3-phosphate
(product of photosynthesis)

The phosphate group (—OPO_3^{2-}) consists of a central phosphorus atom bonded to four oxygen atoms, as shown at left. Among the large biological molecules linked by phosphate groups is the nucleic acid DNA. Phosphate groups are added to or removed from biological molecules as part of reactions that conserve or release energy. In addition, they control biological activity—the activity of many proteins is turned on or off by the addition or removal of phosphate groups.

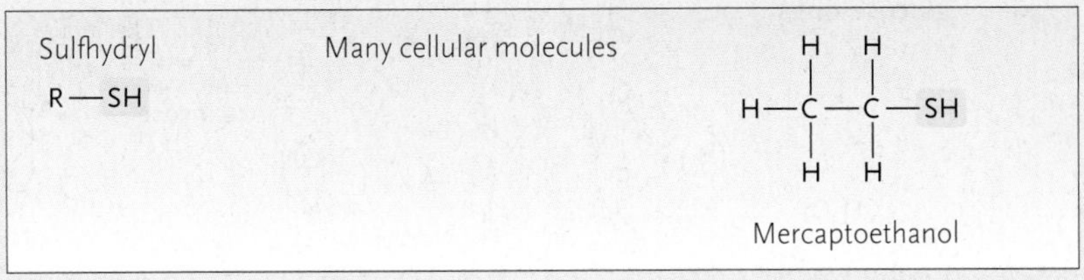

Sulfhydryl   Many cellular molecules

R—SH

Mercaptoethanol

In the sulfhydryl group (—SH), a sulfur atom is linked on one side to a hydrogen atom; in a molecule, the other side is linked to an R group. The sulfhydryl group is easily converted into a covalent linkage in which it loses its hydrogen atom as it binds. In many of these linking reactions, two sulfhydryl groups interact to form a disulfide linkage (—S—S—). In proteins, the disulfide bond contributes to tertiary structure.

Carbohydrates

Carbohydrates, the most abundant biological molecules, serve many functions. Together with fats, they act as the major fuel substances providing chemical energy for cellular activities. Chains of carbohydrate subunits also form structural molecules such as cellulose, one of the primary constituents of plant cell walls. Carbohydrates get their name because they contain carbon, hydrogen, and oxygen atoms, with the approximate ratio of the atoms being 1 carbon : 2 hydrogens : 1 oxygen (CH_2O).

Monosaccharides

Carbohydrates occur either as monosaccharides or as chains of monosaccharide units linked together. Monosaccharides are soluble in water, and most have a distinctly sweet taste. Of the monosaccharides, those that contain three carbons (*trioses*), five carbons (*pentoses*), and six carbons (*hexoses*) are most common in living organisms. All monosaccharides can occur in the linear form, where each carbon atom in the chain except one has both an —H and an —OH group attached to it.

Monosaccharides with five or more carbons can fold back on themselves to assume a ring form. Folding into a ring occurs through a reaction between two functional groups in the same monosaccharide, as occurs in glucose. The ring form of most five- and six-carbon sugars is much more common in cells than the linear form.

When glucose forms into a ring, two alternative arrangements are possible (α-glucose and β-glucose) that differ in the arrangements of the —OH group bound to the carbon at position 1. These two different forms of glucose are called isomers, which are discussed below.

Glyceraldehyde
(3 carbons;
a triose)

Ribose
(5 carbons;
a pentose)

Mannose
(6 carbons;
a hexose)

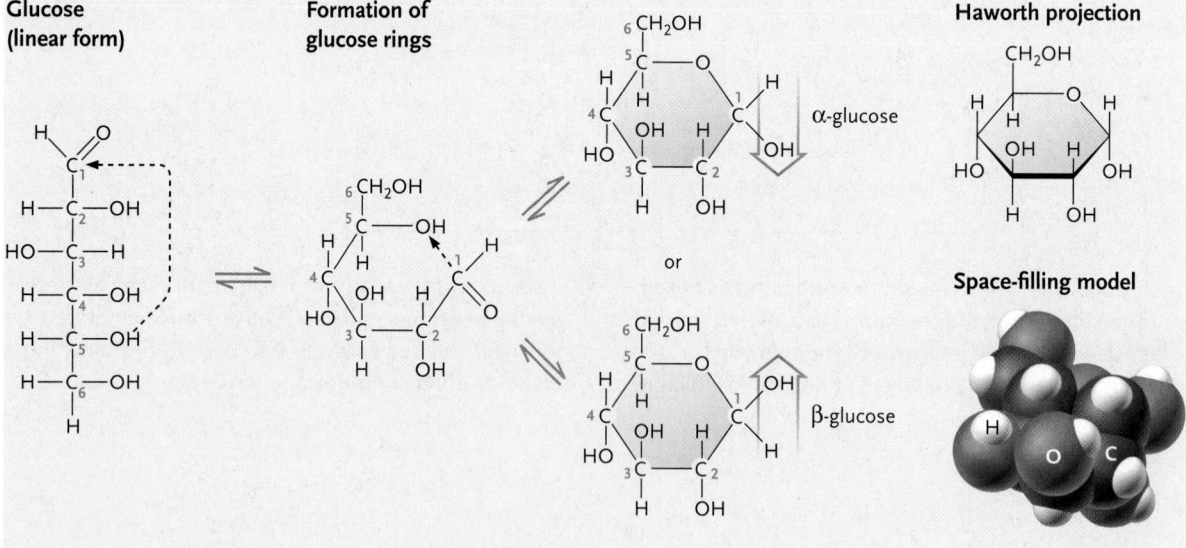

Glucose
(linear form)

Formation of
glucose rings

α-glucose

or

β-glucose

Haworth projection

Space-filling model

Isomers of the Monosaccharides

Typically, one or more of the carbon atoms in a monosaccharide links to four different atoms or chemical groups. Carbons linked in this way are called *asymmetrical* carbons; they have important effects on the structure of a monosaccharide because they can take either of two fixed positions with respect to other carbons in a carbon chain. For example, the middle carbon of the three-carbon sugar glyceraldehyde is asymmetrical because it shares electrons in covalent bonds with four different atoms or groups: —H, —OH, —CHO, and —CH_2OH. The —H and —OH groups can take either of two positions, with the —OH extending to either the left or the right of the carbon chain relative to the —CHO and —CH_2OH groups:

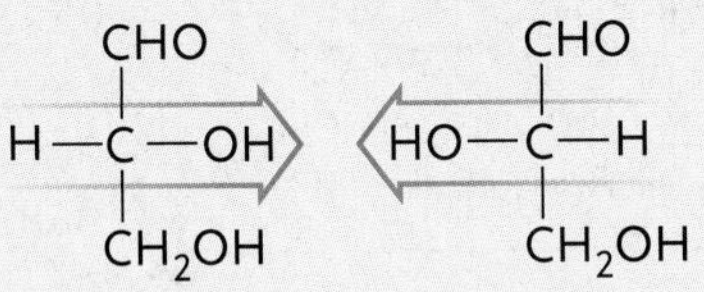

D-Glyceraldehyde   L-Glyceraldehyde

Note that the two forms of glyceraldehyde have the same chemical formula, $C_3H_6O_3$. The difference between the two forms is similar to the difference between your two hands. Although both hands have four fingers and a thumb, they are not identical; rather, they are mirror images of each other. That is, when you hold your right hand in front of a mirror, the reflection looks like your left hand, and vice versa.

Two or more molecules with the same chemical formula but different molecular structures are called isomers. Isomers that are mirror images of each other, like the two forms of glyceraldehyde, are called enantiomers, or optical isomers. One of the enantiomers—the one in which the hydroxyl group extends to the left in the view just shown—is called the l-form (*laevus* = left). The other enantiomer, in which the —OH extends to the right, is called the d-form (*dexter* = right). The difference between l- and d-enantiomers is critical to biological function. Typically, one of the two forms enters much more readily into cellular reactions; just as your left hand does not fit readily into a right-hand glove, enzymes (proteins that accelerate chemical reactions in living organisms) fit best to one of the two forms of an enantiomer. For example, most of the enzymes that catalyze the biochemical reactions of monosaccharides react more rapidly with the d-form, making this form much more common among cellular carbohydrates than the l-form. Many other kinds of biological molecules besides carbohydrates form enantiomers; an example is the amino acids.

In the ring form of many five- or six-carbon monosaccharides, including glucose, the carbon at the 1 position of the ring is asymmetrical because its four bonds link to different groups of atoms. This asymmetry allows monosaccharides such as glucose to exist as two different enantiomers. The glucose enantiomer with an —OH group pointing below the plane of the ring is known as *alpha-glucose*, or *a-glucose*; the enantiomer with an —OH group pointing above the plane of the ring is known as *beta-glucose*, or *b-glucose*. Other five- and six-carbon monosaccharide rings have similar α- and β-configurations.

The α- and β-rings of monosaccharides can give the polysaccharides assembled from them vastly different chemical properties. For example, starches, which are assembled from α-glucose units, are biologically reactive polysaccharides easily digested by animals; cellulose, which is assembled from β-glucose units, is relatively unreactive and, for most animals, completely indigestible.

Another form of isomerism is found in monosaccharides, as well as in other molecules. Two molecules with the same chemical formula but atoms that are arranged in different ways are called structural isomers. The sugars glucose and fructose are examples of structural isomers.

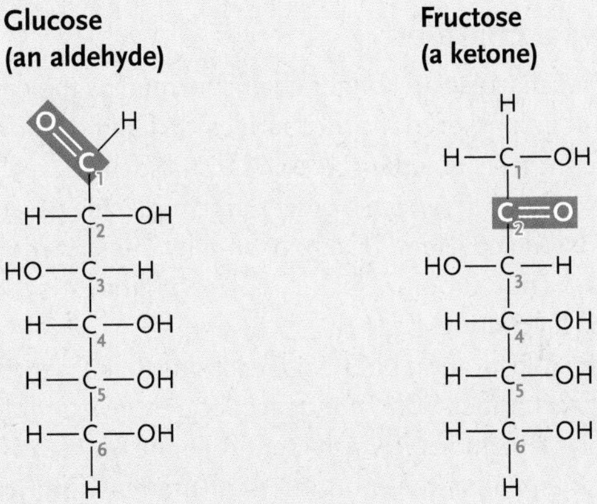

Glucose (an aldehyde)   **Fructose (a ketone)**

Disaccharides

Disaccharides are typically assembled from two monosaccharides linked by a dehydration synthesis reaction. For example, the disaccharide maltose is formed by the linkage of two α-glucose molecules with oxygen as a bridge between the number 1 carbon of the first glucose unit and the number 4 carbon of the second glucose unit. Bonds of this type, which commonly link monosaccharides into chains, are known as glycosidic bonds. A glycosidic bond between a 1 carbon and a 4 carbon is written in chemical shorthand as a 1 → 4 linkage. Linkages such as 1 → 2, 1 → 3, and 1 → 6 are also common in carbohydrate chains. The linkages are designated as α or β depending on the orientation of the —OH group at the 1 carbon that forms the bond. In maltose, the —OH group is in the α position. Therefore, the link between the two glucose subunits of maltose is written as an α (1 → 4) linkage. Maltose, sucrose, and lactose are common disaccharides.

Formation of maltose

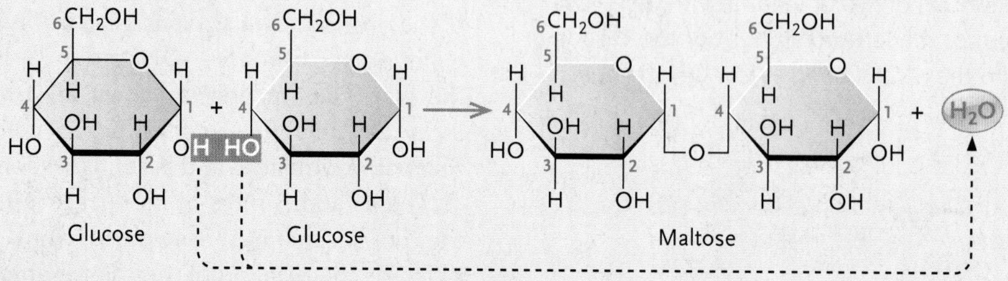

Sucrose

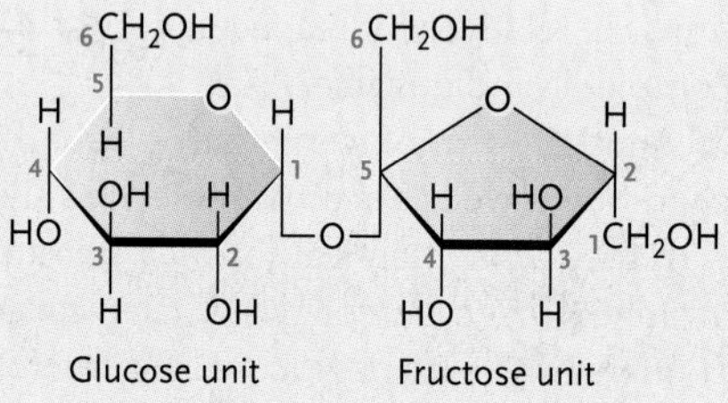

Glucose unit   Fructose unit

Lactose

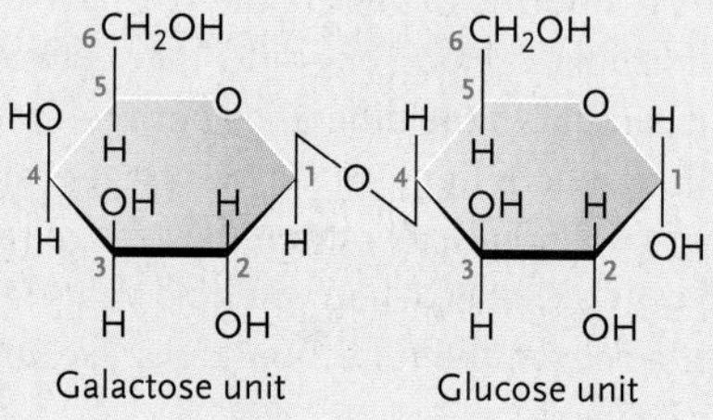

Galactose unit   Glucose unit

Polysaccharides

Polysaccharides are longer chains formed by the end-to-end linking of monosaccharides through dehydration synthesis reactions. A polysaccharide is a type of macromolecule, which is a very large molecule assembled by the covalent linkage of smaller subunit molecules. The subunit for a polysaccharide is the monosaccharide.

The dehydration synthesis reactions that assemble polysaccharides from monosaccharides are examples of polymerization, in which identical or nearly identical subunits, called the monomers of the reaction, join like links in a chain to form a larger molecule called a polymer. Linkage of a relatively small number of nonidentical subunits can create highly diverse and varied biological molecules. Many kinds of polymers are found in cells, not just polysaccharides. DNA is a primary example of a highly diverse polymer assembled from various sequences of only four different types of monomers.

The most common polysaccharides—the plant starches glycogen and cellulose—are all assembled from hundreds or thousands of glucose units. Other polysaccharides are built up from a variety of different sugar units. Polysaccharides may be linear, unbranched molecules, or they may contain one or more branches in which side chains of sugar units are attached to a main chain.

Amylose, formed from α-glucose units joined end to end in α (1 → 4) linkages. The coiled structures are induced by the bond angles in the α-linkages.

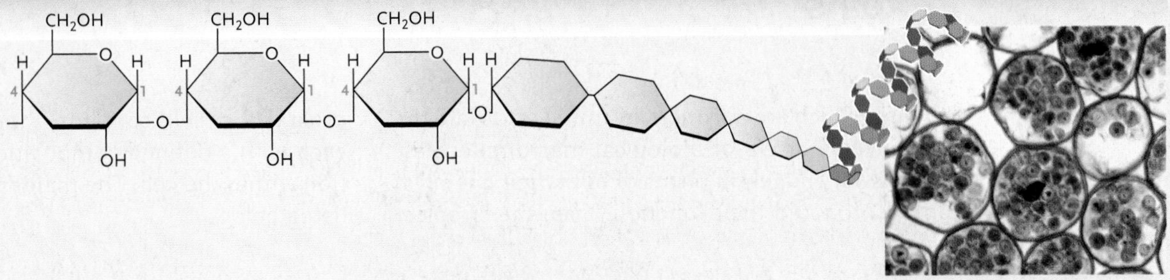

Amylose grains (purple) in plant root tissue

Glycogen, formed from glucose units joined in chains by α (1 → 4) linkages; side branches are linked to the chains by α (1 → 6) linkages (boxed in blue).

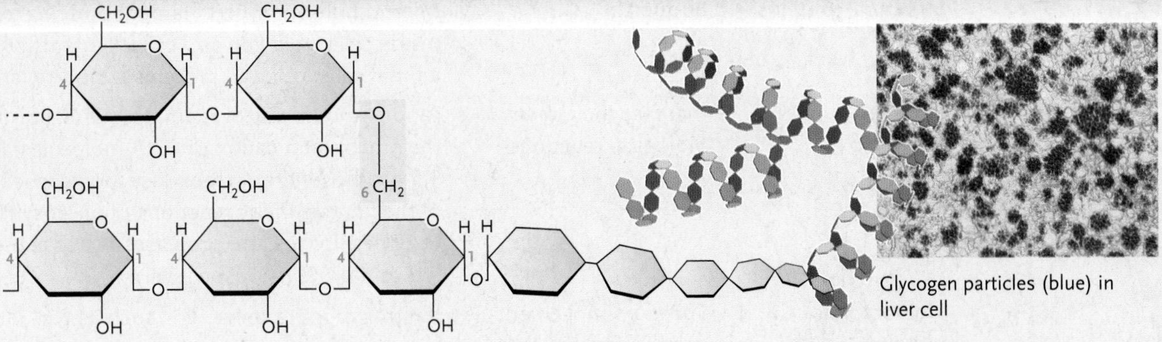

Glycogen particles (blue) in liver cell

Cellulose, formed from glucose units joined end to end by β (1 → 4) linkages. Hundreds to thousands of cellulose chains line up side by side, in an arrangement reinforced by hydrogen bonds between the chains, to form cellulose microfibrils in plant cells.

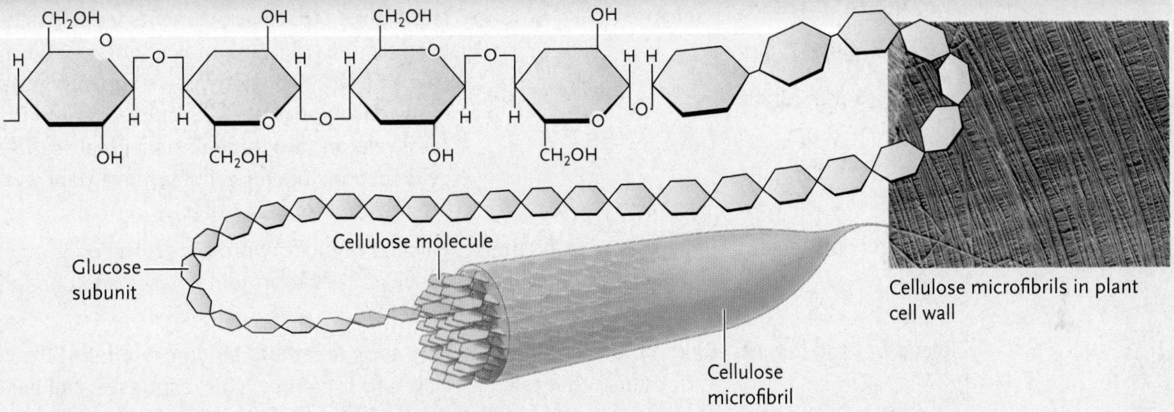

Glucose subunit
Cellulose molecule
Cellulose microfibril

Cellulose microfibrils in plant cell wall

Chitin, formed from β-linkages joining glucose units modified by the addition of nitrogen-containing groups. The external body armour of the tick is reinforced by chitin fibres.

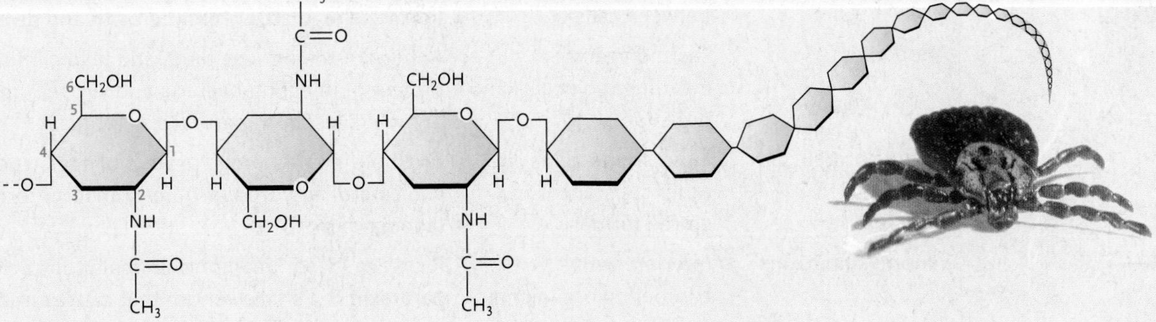

Proteins

Proteins, which are polymers of amino acids, are the most diverse group of biological macromolecules. Proteins vary hugely in terms of both their chemical composition and their function. Even the simplest prokaryotic cell contains thousands of proteins, each with a defined composition and specific function within the cell. The major protein functions are listed below.

Protein Type	Function	Examples
Structural proteins	Support	Microtubule and microfilament proteins form supporting fibres inside cells; collagen and other proteins surround and support animal cells; cell wall proteins support plant cells.
Enzymatic proteins	Increase the rate of biological reactions	Among thousands of examples, DNA polymerase increases the rate of duplication of DNA molecules; RuBP (ribulose 1,5-bisphosphate) carboxylase/oxygenase increases the rates of the first synthetic reactions of photosynthesis; the digestive enzymes lipases and proteases increase the rate breakdown of fats and proteins, respectively.
Membrane transport proteins	Speed up movement of substances across biological membranes	Ion transporters move ions such as Na^+, K^+, and Ca^{2+} across membranes; glucose transporters move glucose into cells; aquaporins allow water molecules to move across membranes.
Motile proteins	Produce cellular movements	Myosin acts on microfilaments (called thin filaments in muscle) to produce muscle movements; dynein acts on microtubules to produce the whipping movements of sperm tails, flagella, and cilia (the last two are whiplike appendages on the surfaces of many eukaryotic cells); kinesin acts on microtubules of the cytoskeleton (the three-dimensional scaffolding of eukaryotic cells responsible for cellular movement, cell division, and the organization of organelles).
Regulatory proteins	Promote or inhibit the activity of other cellular molecules	Nuclear regulatory proteins turn genes on or off to control the activity of DNA; protein kinases add phosphate groups to other proteins to modify their activity.
Receptor proteins	Bind molecules at cell surface or within cell; some trigger internal cellular responses	Hormone receptors bind hormones at the cell surface or within cells and trigger cellular responses; cellular adhesion molecules help hold cells together by binding molecules on other cells; LDL receptors bind cholesterol-containing particles to cell surfaces.
Hormones	Carry regulatory signals between cells	Insulin regulates sugar levels in the bloodstream; growth hormone regulates cellular growth and division.
Antibodies	Defend against invading molecules and organisms	Antibodies recognize, bind, and help eliminate essentially any protein of infecting bacteria and viruses, and many other types of molecules, both natural and artificial.
Storage proteins	Hold amino acids and other substances in stored form	Ovalbumin is a storage protein of eggs; apolipoproteins hold cholesterol in stored form for transport through the bloodstream.
Venoms and toxins	Interfere with competing organisms	Ricin is a castor bean protein that stops protein synthesis; bungarotoxin is a snake venom that causes muscle paralysis.

Amino Acids

All proteins are polymers of amino acids. The generalized structure of an amino acid has a central carbon atom attached to an amino group (—NH$_2$), a carboxyl group (—COOH), and a hydrogen atom:

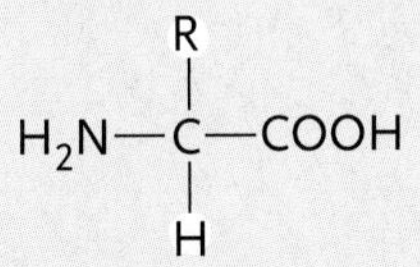

$$H_2N-\underset{\underset{H}{|}}{\overset{\overset{R}{|}}{C}}-COOH$$

The remaining bond of the central carbon is to 1 of 20 different side groups represented by the R. The R group, also called the side chain, ranges from a single hydrogen atom in the amino acid glycine to complex carbon chains or rings in some others. Differences in the side groups give the amino acids their individual properties. When discussing protein structure, amino acids are commonly referred to as amino acid residues or simply residues.

Proteins are synthesized from 20 different amino acids. These 20 are most commonly grouped according to the properties of their side chains. Here, the amino acids are shown in the ionic form common at the pH typical of a cell, 7.2.

Nonpolar amino acids

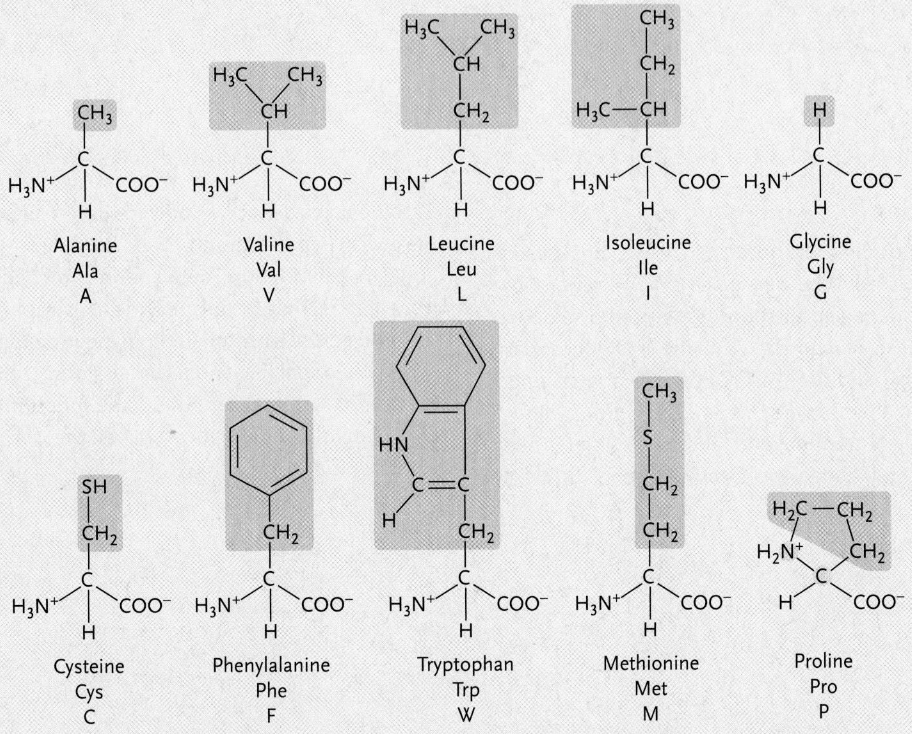

Alanine
Ala
A

Valine
Val
V

Leucine
Leu
L

Isoleucine
Ile
I

Glycine
Gly
G

Cysteine
Cys
C

Phenylalanine
Phe
F

Tryptophan
Trp
W

Methionine
Met
M

Proline
Pro
P

Uncharged polar amino acids

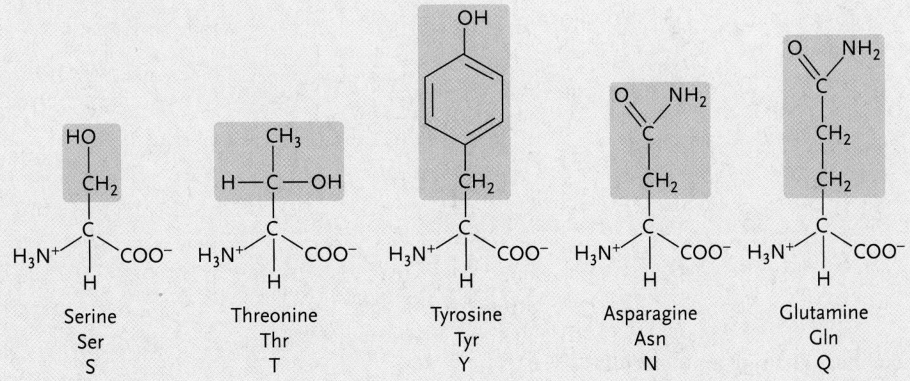

Serine
Ser
S

Threonine
Thr
T

Tyrosine
Tyr
Y

Asparagine
Asn
N

Glutamine
Gln
Q

continued on next page

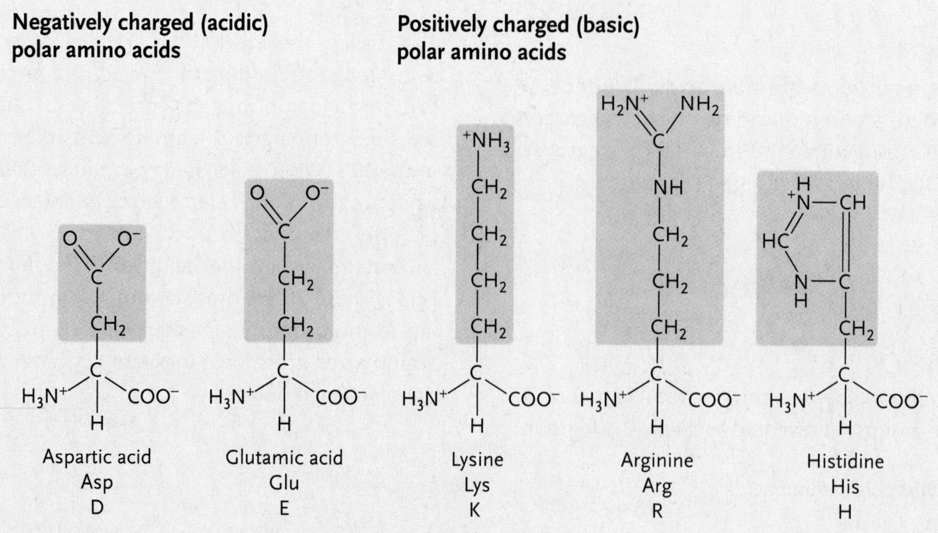

Negatively charged (acidic) polar amino acids

Positively charged (basic) polar amino acids

| Aspartic acid Asp D | Glutamic acid Glu E | Lysine Lys K | Arginine Arg R | Histidine His H |

Polypeptides

Covalent bonds link amino acids into chains called **polypeptides**. The link between each pair of amino acids in a polypeptide, a peptide bond, is formed by a dehydration synthesis reaction between the —NH_2 group of one amino acid and the —COOH group of a second. An amino acid chain always has an —NH_2 group at one end, called the N-terminal end, and a —COOH group at the other end, called the C-terminal end. In cells, amino acids are added only to the —COOH end of the growing peptide strand.

The distinction between a polypeptide and a protein is that a polypeptide is simply a string of amino acids. A protein is a polypeptide that has folded into the specific three-dimensional shape that is required for most proteins to be functional.

The following figure shows the formation of a peptide bond.

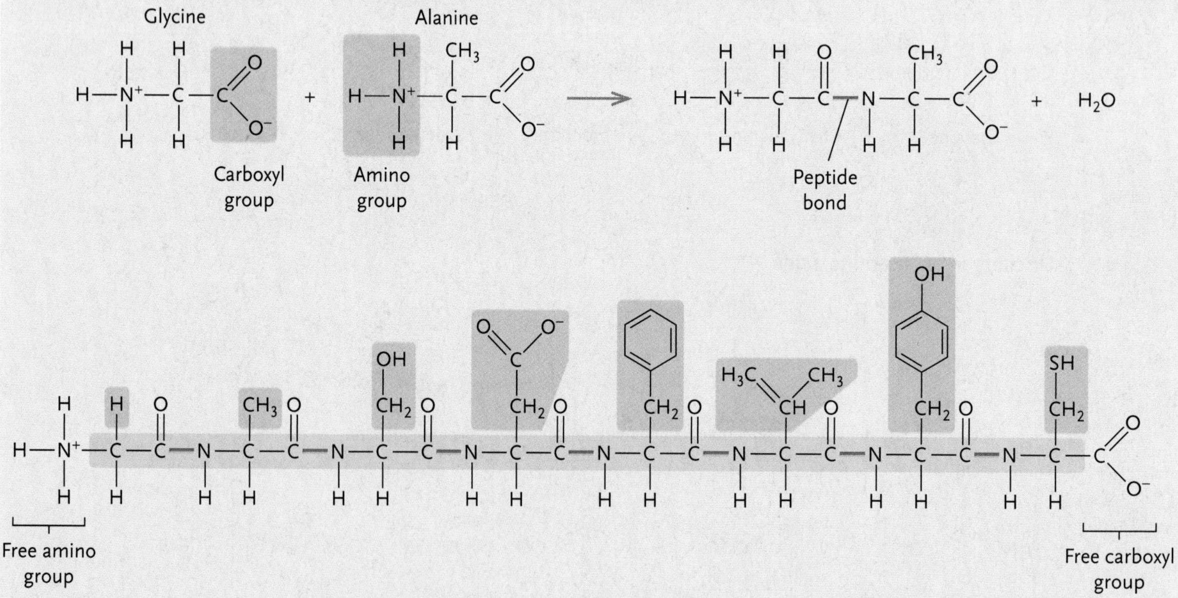

Glycine   Alanine

Carboxyl group   Amino group   Peptide bond   + H_2O

Free amino group   Free carboxyl group

A polypeptide—a linear chain of amino acids.

The backbone of the polypeptide is highlighted in the bottom figure above. The amino end of the polypeptide is called the N-terminus, while the carboxyl end is called the C-terminus.

The Four Levels of Protein Structure

Proteins have four potential levels of structure, with each level imparting different characteristics and degrees of structural complexity to the molecule. **Primary structure** is the particular and unique sequence of amino acids forming a polypeptide; **secondary structure** is produced by the twists and turns of the amino acid chain. Tertiary structure is the folding of the amino acid chain, with its secondary structures, into the overall three-dimensional shape of a protein. All proteins have primary, secondary, and tertiary structures. Quaternary structure, when present, refers to the arrangement of polypeptide chains in a protein that is formed from more than one chain. Each structural level depends upon the level before it.

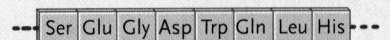

Primary structure: the sequence of amino acids in a protein

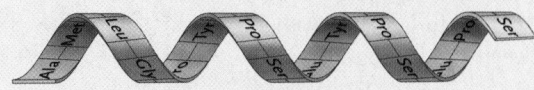

Secondary structure: regions of alpha helix, beta strand, or random coil in a polypeptide chain

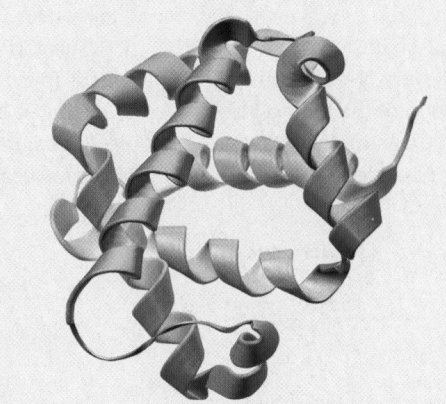

Tertiary structure: overall three-dimensional folding of a polypeptide chain

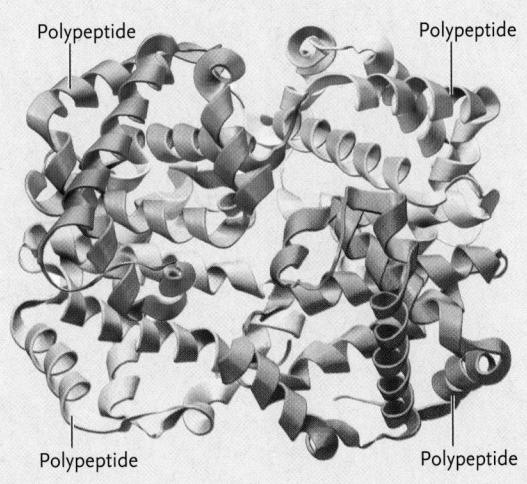

Polypeptide   Polypeptide

Polypeptide   Polypeptide

Quaternary structure: the arrangement of polypeptide chains in a protein that contains more than one chain

Primary Structure

The primary structure of a protein is simply its complete amino acid sequence. The primary sequence is determined by the nucleotide sequence of the coding region of the protein's corresponding gene.

H_3N^+ —| Phe | Val | Asn | Gln | His | Leu | Cys | Gly | Ser | His | Leu | Val | Glu | Ala | Leu | Tyr | Leu | Val | Cys | Gly | Glu | Arg | Gly | Phe | Phe | Tyr | Thr | Pro | Lys | Ala |— COO^-

Secondary Structure

The amino acid chain of a protein, rather than being stretched out in linear form, is folded into arrangements that form the protein's secondary structure. Secondary structure is based on hydrogen bonds between atoms of the backbone. More precisely, the hydrogen bonds form between the hydrogen atom attached to the nitrogen of the backbone and the oxygen attached to one of the carbon atoms of the backbone. Two highly regular secondary structures are the alpha helix and the beta sheet. A third, less regular arrangement, the random coil or loop, imparts flexibility to certain regions of the protein. Most proteins have segments of all three arrangements.

Experimental Research Figure

The α helix

A model of the α helix (left), a coil shape formed when hydrogen bonds form between every N—H group of the backbone and the C=O group of the amino acid four residues earlier. In protein diagrams (right), the α helix is depicted as a cylinder or barrel.

The β sheet

A β sheet is formed by side-by-side alignment of β strands (picture shows two strands). The sheet is formed by hydrogen bonds between atoms of each strand. In protein diagrams, the β strands are depicted by ribbons with arrowheads pointing toward the C-terminal.

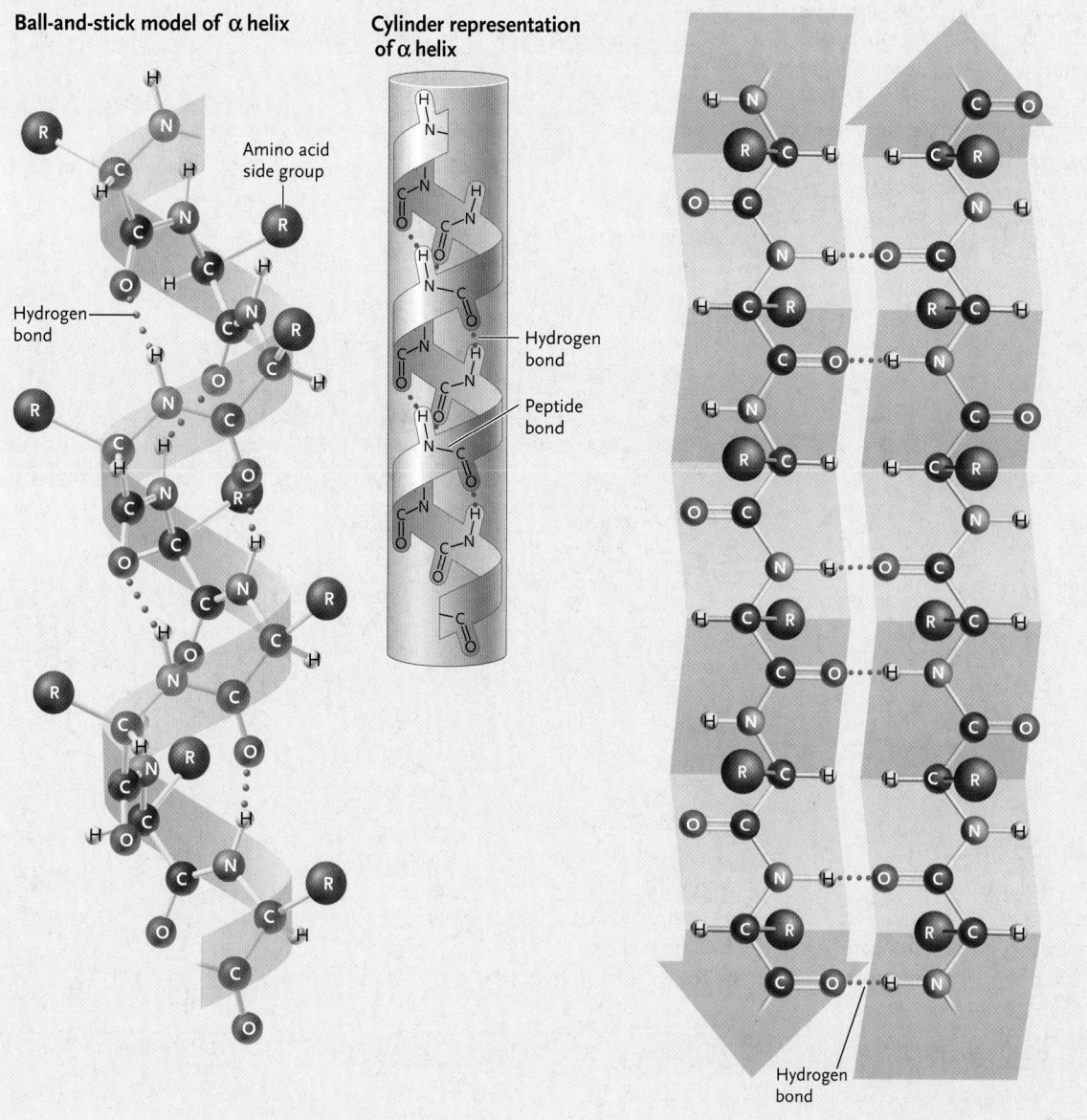

Ball-and-stick model of α helix

Cylinder representation of α helix

Tertiary Structure

The four major interactions between R groups that contribute to tertiary structure are shown below: (1) ionic bonds, (2) hydrogen bonds, (3) hydrophobic interactions, and (4) disulfide bridges. The tertiary structure of most proteins is flexible, allowing them to undergo limited alterations in three-dimensional shape known as conformational changes. These changes contribute to the function of many proteins, particularly enzymes, as well as other proteins involved in cellular movements or in the transport of substances across cell membranes.

Below are two representations of the three-dimensional structure of the enzyme lysozyme. In a ribbon diagram, α helices are shown as a cylinder, β strands are depicted as flat arrows, and random coils are shown as thin ropes. In a space-filling model, spheres represent different atoms. The sizes of the spheres and the intersphere distances are proportional to the actual dimensions. Atoms of different elements are represented by different colours. Disulfide bonds are shown in yellow.

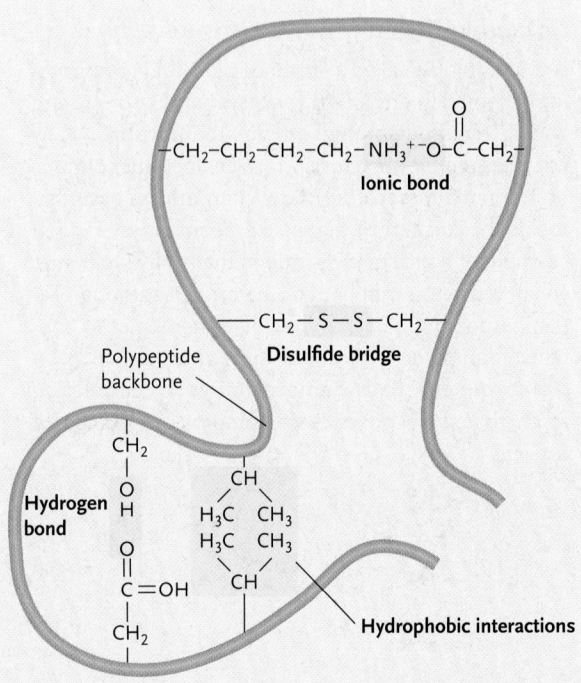

Lysozyme

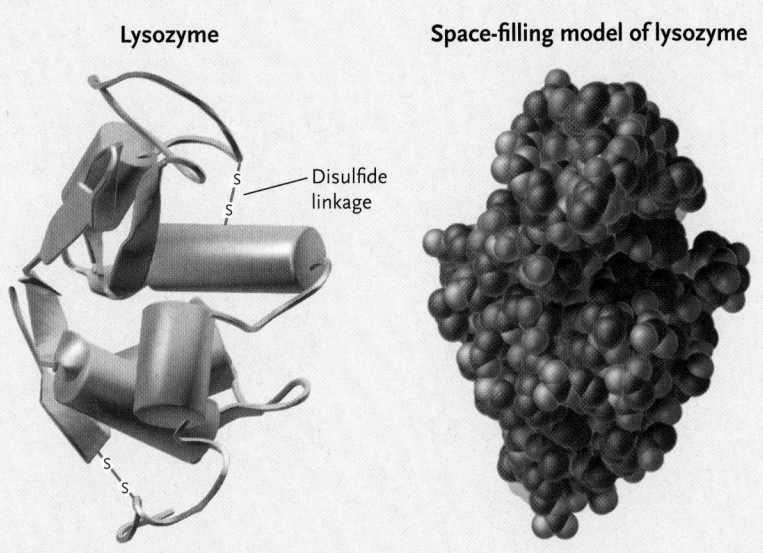

Space-filling model of lysozyme

Quaternary Structure

Some proteins consist of two or more polypeptides that come together to form a functional protein. An example of a protein that exhibits quaternary structure is collagen. The collagen molecule consists of three helical polypeptides that aggregate to form a triple-helix structure. Collagen is a major component of the connective tissue, is found exclusively in animals, and is the most abundant protein in mammals.

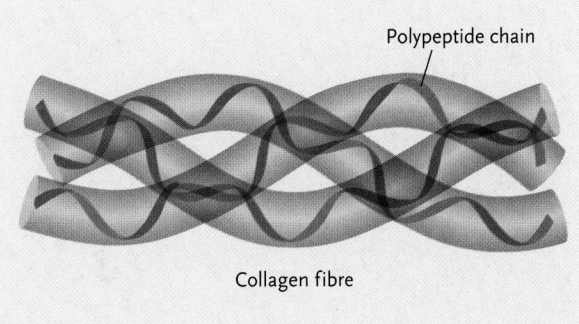

Polypeptide chain

Collagen fibre

Cofactors/Prosthetic Groups

A cofactor (also called a prosthetic group) is a nonprotein chemical compound that is bound to a protein and is required for the protein to function. Many enzymes require cofactors, which can be either organic or inorganic molecules. Many vitamins are essential to life because they act as key cofactors. A good example of a prosthetic group is the molecule heme, which is a key component of the oxygen-carrying protein hemoglobin. Each molecule of hemoglobin contains four heme molecules—one attached to each globin protein. Each heme contains a central iron atom that is responsible for binding molecules of oxygen.

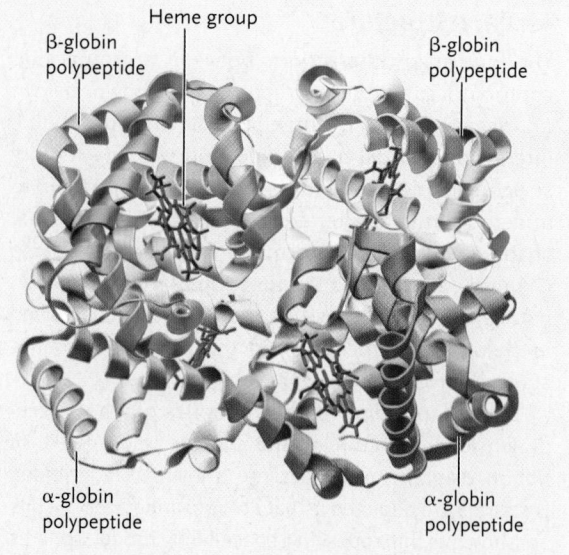

β-globin polypeptide

Heme group

β-globin polypeptide

α-globin polypeptide

α-globin polypeptide

Protein Domains

In many proteins, folding of the polypeptide(s) produces distinct, large structural subdivisions called domains. Often, one domain of a protein is connected to another by a segment of random coil. The hinge formed by the flexible random coil allows domains to move with respect to one another. That different domains of a protein are structurally distinct often reflects that they are functionally distinct as well.

Two domains in an enzyme that assembles DNA molecules

Domain a   Domain b

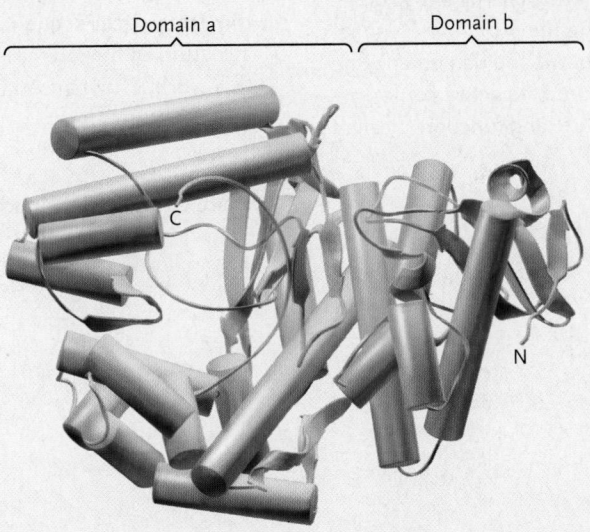

The same protein, showing the domain surfaces

Domain a   Domain b

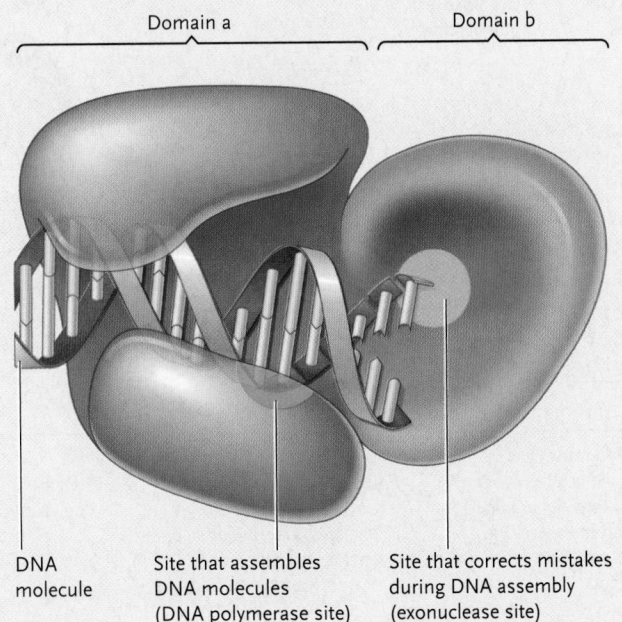

DNA molecule

Site that assembles DNA molecules (DNA polymerase site)

Site that corrects mistakes during DNA assembly (exonuclease site)

Protein Folding and Denaturation

A fundamental biochemical question is, "What determines how a protein will fold into the correct functional conformation?" The first insight came from a classic experiment published by Christian Anfinsen and Edgar Haber in 1962. The researchers studied ribonuclease, an enzyme that hydrolyzes RNA.

When they treated the enzyme chemically to break the disulfide linkages holding the protein in its functional state, the protein unfolded and had no enzyme activity. Unfolding a protein from its active conformation so that it loses its structure and function is called

denaturation. This most often involves the use of specific chemicals or heat.

When they removed the denaturing chemicals, the ribonuclease slowly regained full activity because the disulfide linkages reformed, enabling the protein to reassume its functional conformation. The reversal of denaturation is called **renaturation.**

The key conclusion from this experiment was that the amino acid sequence itself specifies the tertiary structure of a protein. Nothing else is required. For this work, Christian Anfinsen received a Nobel Prize in 1972.

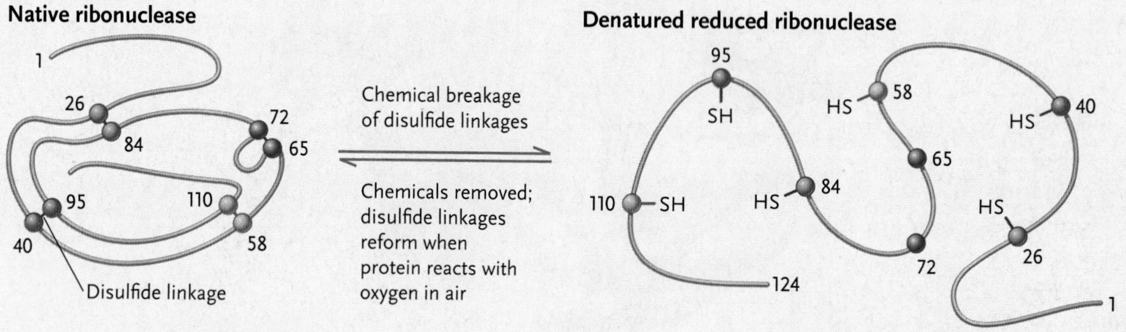

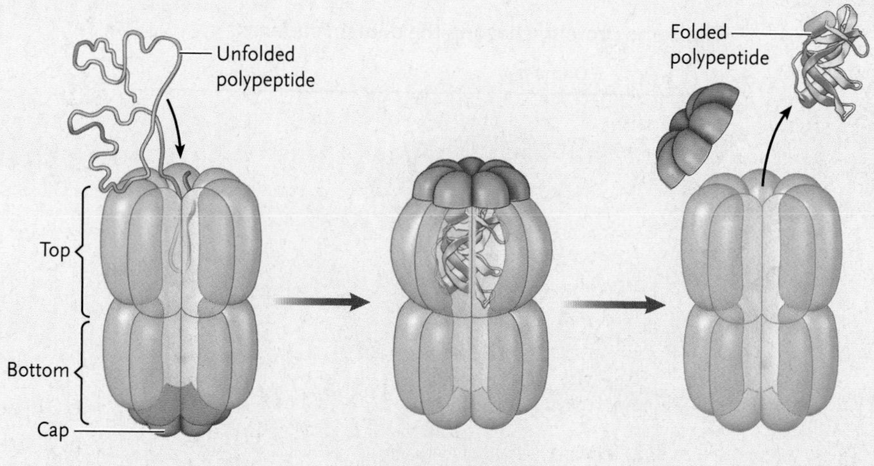

Native ribonuclease

Denatured reduced ribonuclease

Chemical breakage of disulfide linkages

Chemicals removed; disulfide linkages reform when protein reacts with oxygen in air

Disulfide linkage

Unfolded polypeptide

Folded polypeptide

Top

Bottom

Cap

1 An empty chaperonin molecule has the cap on the bottom. An unfolded polypeptide enters the chaperonin cylinder at the top.

2 The cap moves from the bottom to the top; the shape of the chaperonin changes, creating an enclosure that enables the polypeptide to fold.

3 The cap comes off, releasing the fully folded polypeptide.

Within the cell, the high density of newly synthesized proteins may impede the proper folding of individual proteins. For many proteins, correct folding is helped by a group of proteins called **chaperone proteins** or **chaperonins** (see figure above).

They function by temporarily binding with newly synthesized proteins, directing their conformation toward the correct tertiary structure, and inhibiting incorrect arrangements as the new proteins fold.

Nucleic Acids

Two types of nucleic acids exist: DNA and RNA. Deoxyribonucleic acid (DNA) stores the hereditary information in all eukaryotes, bacteria, and archaea. In all organisms, ribonucleic acid (RNA) carries out a diversity of functions.

RNA carries the instructions for assembling proteins from DNA to the site of protein synthesis, the ribosome, which is itself composed partially of RNA. Another type of RNA serves to bring amino acids to the ribosome for their assembly into proteins.

Nucleotides

All nucleic acids are polymers of nucleotides. A **nucleotide** consists of three parts linked by covalent bonds: (1) a nitrogenous base formed from rings of carbon and nitrogen atoms; (2) a five-carbon, ring-shaped sugar; and (3) one to three phosphate groups.

In nucleotides, the nitrogenous bases link covalently to a five-carbon sugar, either **deoxyribose** or **ribose.** The carbons of the two sugars are numbered with a prime symbol—1′, 2′, 3′, 4′, and 5′. The prime symbols are added to distinguish the carbons in the sugars from those in the nitrogenous bases, which are written without primes. The two sugars differ only in the chemical group bound to the 2′ carbon: deoxyribose has an —H at this position, and ribose has an —OH group.

The two types of nitrogenous bases are pyrimidines, with one carbon–nitrogen ring, and purines, with two rings. Three pyrimidine bases—uracil (U), thymine (T), and cytosine (C)—and two purine bases—adenine (A) and guanine (G)—form parts of nucleic acids in cells.

Overall structural plan of a nucleotide

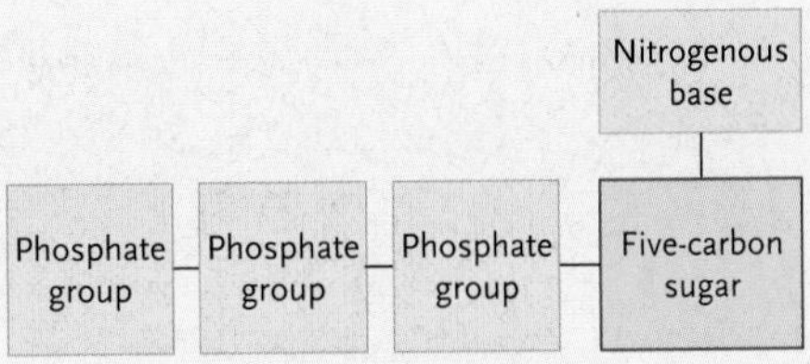

Chemical structures of nucleotides

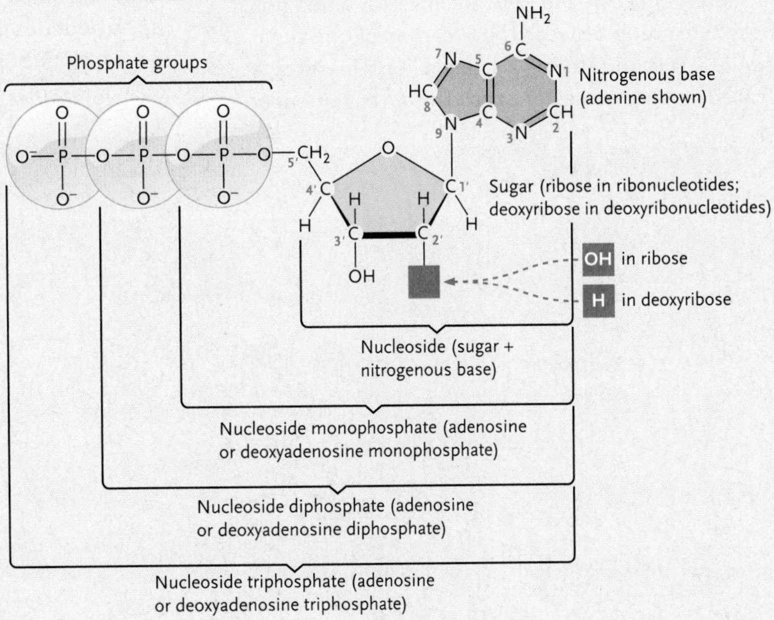

Pyrimidine and Purine Bases of Nucleic Acids

The figure shows the three single-ring pyrimidines and two double-ring purines that are the nitrogenous bases of nucleotides. The red arrows indicate where the bases link to ribose or deoxyribose sugars to form nucleotides.

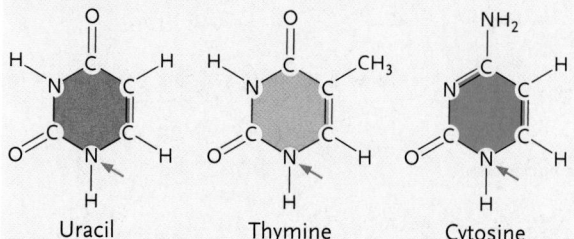

Pyrimidines

Uracil   Thymine   Cytosine

Purines

Adenine   Guanine

DNA and RNA Structure

Nucleotides in DNA and RNA are linked by a bridging phosphate group between the 5′ carbon of one sugar and the 3′ carbon of the next sugar in line. This linkage is called a phosphodiester bond. This arrangement of alternating sugar and phosphate groups forms the backbone of a nucleic acid. The nitrogenous bases of the nucleotides project from this backbone. Note that the nucleotide thymine (T) in DNA is not found in RNA; it is replaced by uracil (U).

DNA

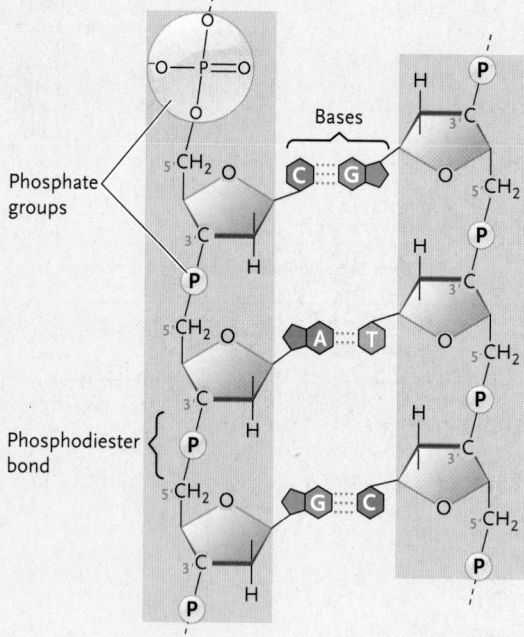

RNA

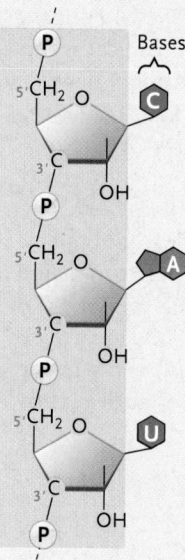

DNA Double Helix

In cells, DNA takes the form of a double helix: two nucleotide chains wrapped around each other in a spiral that resembles a twisted ladder. As shown below, the sides of the ladder are the sugar–phosphate backbones of the two chains, which twist around each other to form the double helix. The rungs of the ladder are the nitrogenous bases, which extend inward from the sugars toward the centre of the helix.

Each rung consists of a pair of nitrogenous bases held in a flat plane roughly perpendicular to the long axis of the helix. The two nucleotide chains of a DNA double helix are held together by hydrogen bonds between the base pairs. A DNA double-helix molecule is also referred to as double-stranded DNA. The space separating the sugar–phosphate backbones of a DNA double helix is just wide enough to accommodate a base pair that consists of one purine and one pyrimidine. Purine–purine base pairs are too wide and pyrimidine–pyrimidine pairs are too narrow to fit this space exactly. More specifically, of the possible purine–pyrimidine pairs, only two combinations, adenine with thymine and guanine with cytosine, can form stable hydrogen bonds so that the base pair fits precisely within the double helix. An adenine–thymine (A—T) pair forms two stabilizing hydrogen bonds; a guanine–cytosine (G—C) pair forms three.

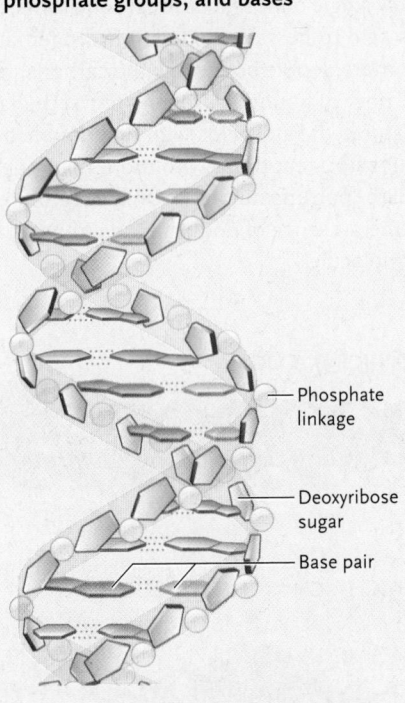

DNA double helix, showing arrangement of sugars, phosphate groups, and bases

Phosphate linkage
Deoxyribose sugar
Base pair

To deoxyribose
To deoxyribose
Adenine
Thymine

To deoxyribose
To deoxyribose
Guanine
Cytosine

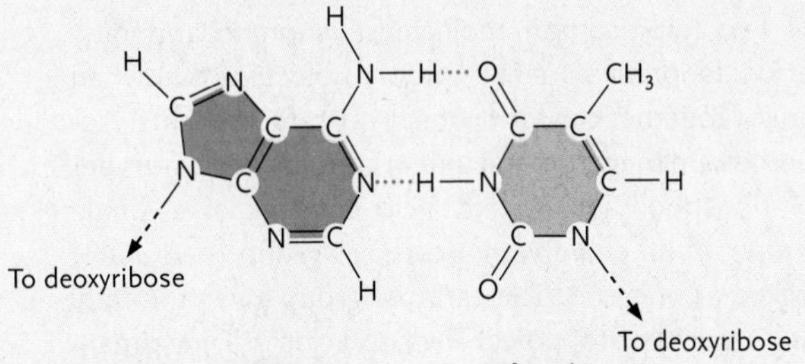

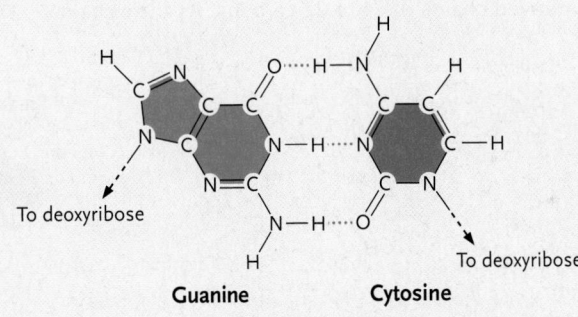

Lipids

Lipids are a diverse group of water-insoluble, primarily nonpolar biological molecules composed mostly of hydrogen and carbon (hydrocarbons). The term *lipid* is a catch-all word for a range of nonpolar molecules. They are not large enough to be considered true macromolecules and, unlike nucleic acids and proteins, are not considered polymers of defined monomeric subunits. As a result of their nonpolar character, lipids typically dissolve much more readily in nonpolar solvents, such as acetone and chloroform, than in water. Their insolubility in water underlies their ability to form cell membranes. In addition, some lipids are stored and used in cells as an energy source. Other lipids serve as hormones that regulate cellular activities. Lipids in living organisms can be grouped into one of three categories—fats, phospholipids, and steroids.

Isoprenes and Fatty Acids

The structural backbone of all lipids is derived from one of two hydrocarbon molecules: isoprene and fatty acids. Isoprenes are five-carbon molecules that when linked together can form long hydrocarbon chains. Isoprenes are the structural unit in steroids and a number of phospholipids. A fatty acid consists of a single hydrocarbon chain with a carboxyl group (—COOH) linked at one end. The carboxyl group gives the fatty acid its acidic properties. The fatty acids in living organisms contain four or more carbons in their hydrocarbon chain, with the most common forms having even-numbered chains of 14 to 22 carbons. As their chain length increases, fatty acids become progressively less water soluble and more solid.

If the hydrocarbon chain of a fatty acid binds the maximum possible number of hydrogen atoms, so that only single bonds link the carbon atoms, the fatty acid is said to be saturated with hydrogen atoms. If one or more double bonds link the carbons, reducing the number of bound hydrogen atoms, the fatty acid is unsaturated. Fatty acids with one double bond are monounsaturated; those with more than one double bond are polyunsaturated. Unlike saturated fatty acids, the presence of double bonds imparts a "kink" in the molecule.

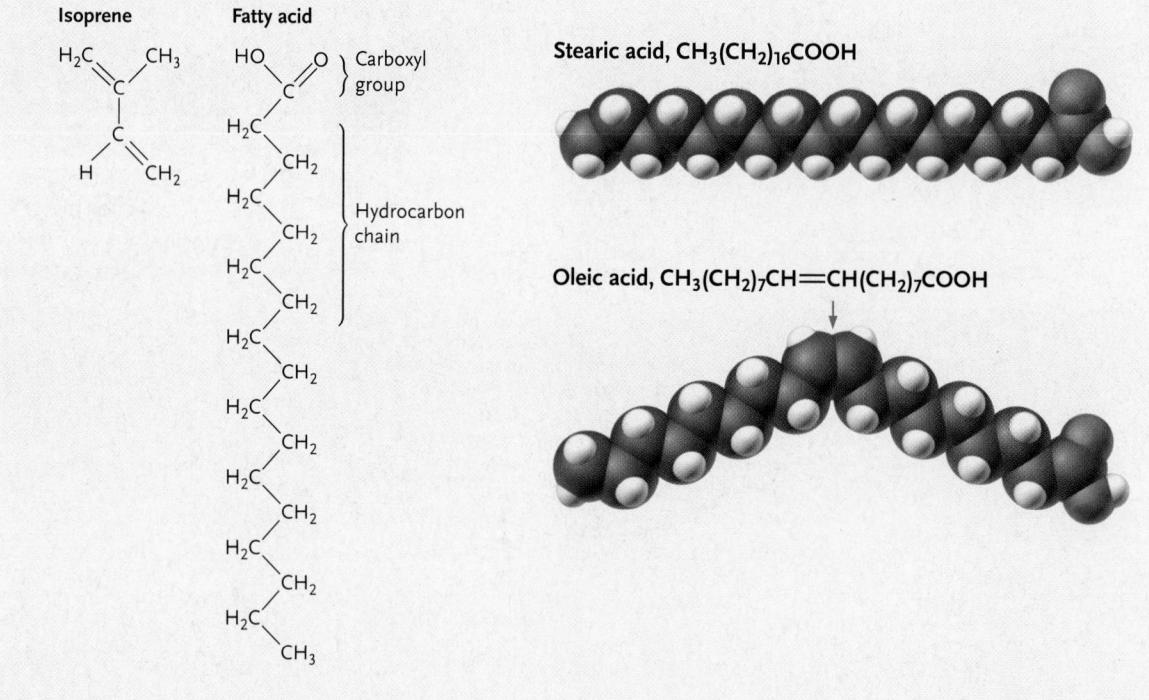

Stearic acid, $CH_3(CH_2)_{16}COOH$

Oleic acid, $CH_3(CH_2)_7CH{=}CH(CH_2)_7COOH$

Phospholipids

Phosphate-containing lipids, or phospholipids, are the primary lipids of cell membranes. In the most common phospholipids, glycerol forms the backbone for the molecule as in triglycerides, but only two of its binding sites are linked to fatty acids. The third site is linked to a polar phosphate group, which also binds to another polar unit. Thus, a phospholipid contains two hydrophobic fatty acids at one end, attached to a hydrophilic polar group, often called the head group. Molecules that contain both hydrophobic and hydrophilic regions are called amphipathic molecules.

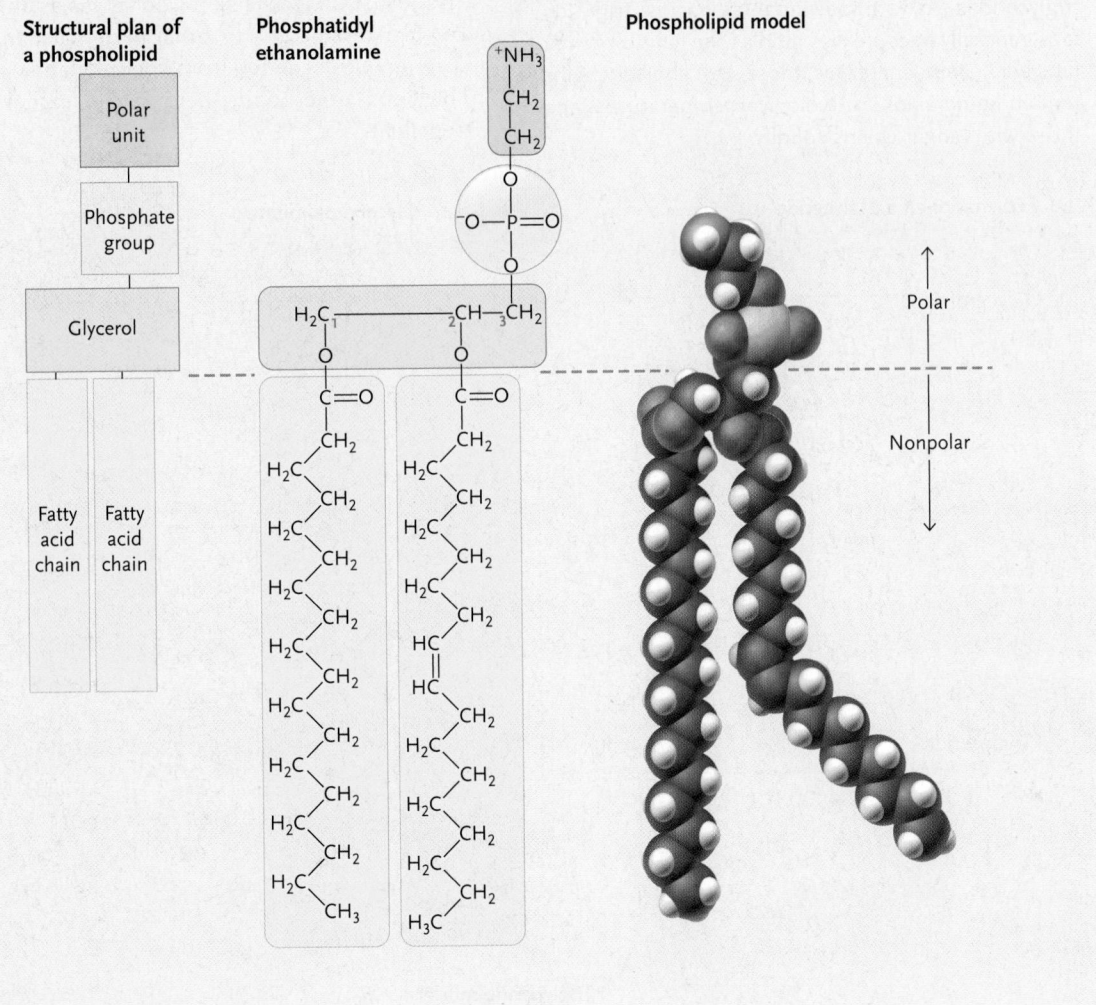

Fats

A fat consists of three fatty acid chains linked to a single molecule of glycerol. Because of this, fats are also often referred to as triacylglycerols or triglycerides. The three fatty acids linked to the glycerol may be different or the same. Different organisms usually have distinctive combinations of fatty acids in their triglycerides. As with individual fatty acids, triglycerides generally become less fluid as the length of their fatty acid chains increases; those with shorter chains remain liquid as oils at biological temperatures, and those with longer chains solidify.

Triglycerides are used widely as stored energy in animals. Gram for gram, they yield more than twice as much energy as carbohydrates. Therefore, fats are an excellent source of energy in the diet. Storing the equivalent amount of energy as carbohydrates rather than fats would add more than 45 kg to the mass of an average man or woman. A layer of fatty tissue just under the skin also serves as an insulating blanket in humans, other mammals, and birds. Triglycerides secreted from special glands in waterfowl and other birds help make feathers water repellent.

Formation of a triglyceride

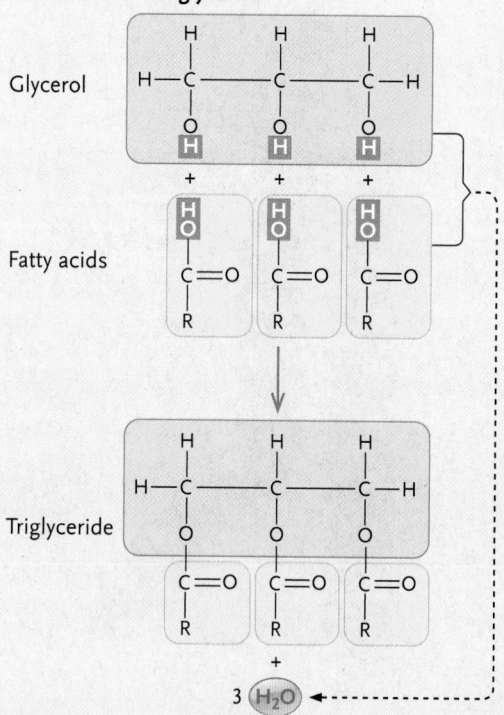

Glyceryl palmitate

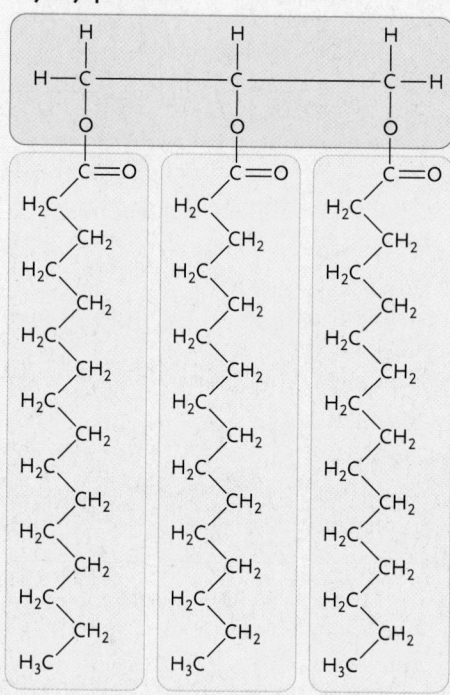

Triglyceride model

Steroids

Steroids are a group of lipids with structures based on a framework of four carbon rings that are derived from isoprene units. Small differences in the side groups attached to the rings distinguish one steroid from another. The most abundant steroids, the sterols, have a single polar —OH group linked to one end of the ring framework and a complex, nonpolar hydrocarbon chain at the other end. Although sterols are almost completely hydrophobic, the single hydroxyl group gives one end of the molecules a slightly polar, hydrophilic character. As a result, sterols also have dual solubility properties and, like phospholipids, tend to assume positions that satisfy these properties.

Cholesterol is an important component of the plasma membrane surrounding animal cells; similar sterols, called phytosterols, occur in plant cell membranes.

Arrangement of carbon rings in a steroid

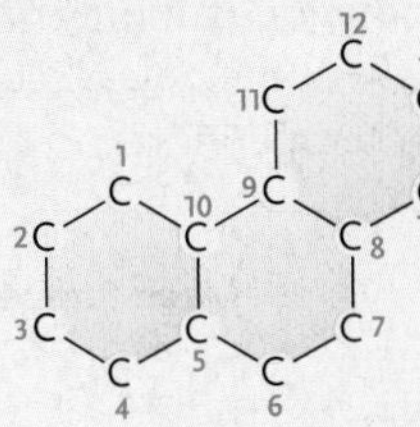

Cholesterol, a sterol

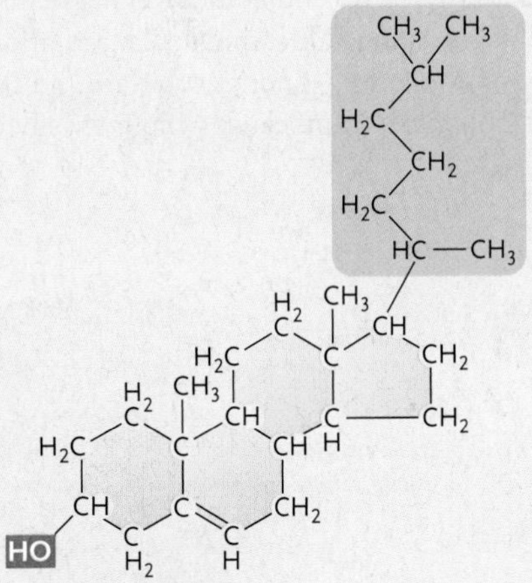

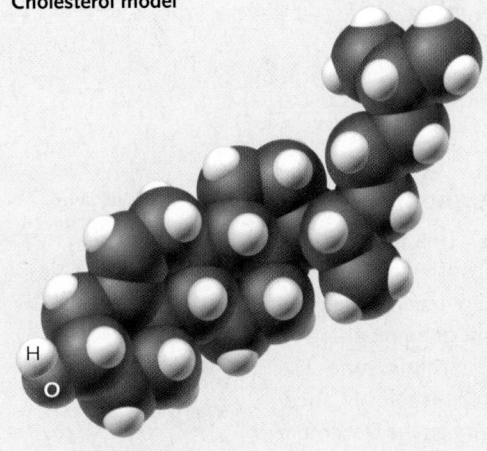

Cholesterol model

The Biosphere

The biosphere is the area occupied by life on Earth, from the depths of the ocean to the sky above. The various physical environments of Earth and their different abiotic factors, such as sunlight, temperature, humidity, wind speed, cloud cover, and rainfall, influence the evolution and diversity of organisms. These abiotic factors contribute to a region's climate, the weather conditions prevailing over an extended period of time. Climates vary on global, regional, and local scales and undergo seasonal changes almost everywhere.

Solar Radiation: Energy from the Sun

The global pattern of environmental diversity results from latitudinal variation in incoming solar radiation, Earth's rotation on its axis, and its orbit around the Sun. Earth's spherical shape causes the intensity of incoming solar radiation to vary from the equator to the poles. Solar radiation is more concentrated near the equator than it is at the poles, causing latitudinal variation in Earth's temperature.

Solar radiation

Near the poles, solar radiation travels a long distance through the atmosphere and strikes a large surface area.

Near the equator, solar radiation travels a short distance through the atmosphere and strikes a small surface area.

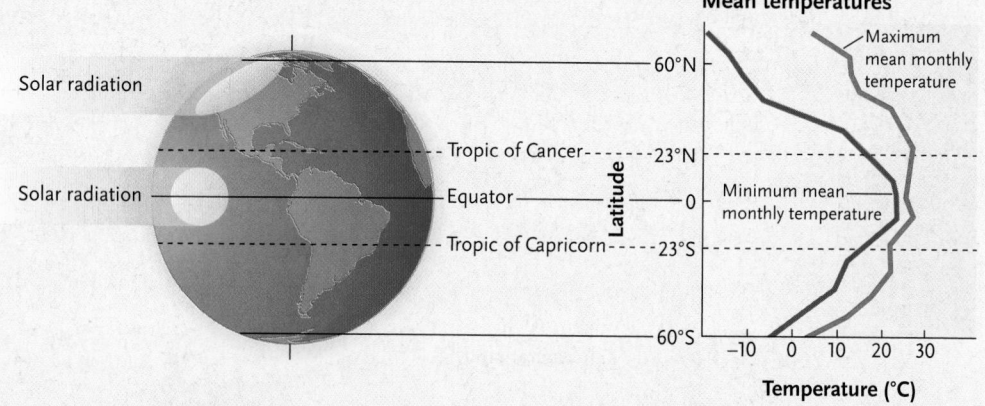

Mean temperatures

Maximum mean monthly temperature

Minimum mean monthly temperature

Seasonality: Weather throughout the Year

Earth is tilted on its axis by 23.5°. This tilt produces seasonal variation in the intensity of incoming solar radiation. The northern hemisphere receives its maximum illumination, and the southern hemisphere its minimum, on the June solstice (around June 21), when the Sun shines directly over the Tropic of Cancer (23.5° N latitude). The reverse is true on the December solstice (around December 21), when the Sun shines directly over the Tropic of Capricorn (23.5° S latitude). Twice each year, on the vernal and autumnal equinoxes (around March 21 and September 21, respectively), the Sun shines directly over the equator. Only the Tropics, the latitudes between the tropics of Cancer and Capricorn, ever receive intense solar radiation from directly overhead. Moreover, the tropics experience only small seasonal changes in temperature and day length, with high temperatures and day length of approximately 12 hours throughout the year.

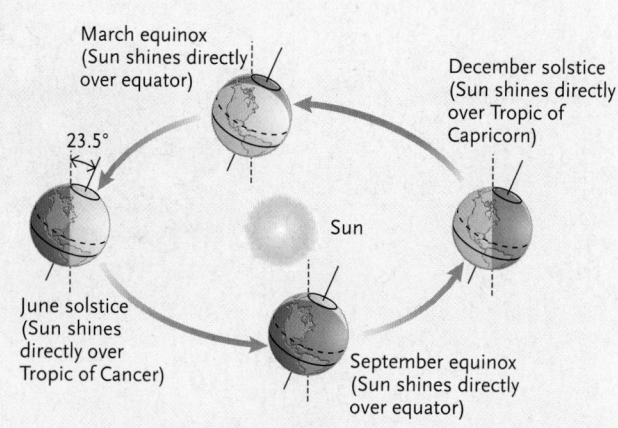

Seasonal variation in temperature and day length increases steadily toward the poles. Polar winters are long and cold, with periods of continuous darkness, and polar summers are short, with periods of continuous light.

Air Circulation: Wind Patterns

Sunlight warms air masses, causing them to expand, lose pressure, and rise in the atmosphere. The unequal heating of air at different latitudes initiates global air movements, producing three circulation cells in each hemisphere. Warm equatorial air masses rise to high altitude before spreading north and south. They eventually sink back to Earth at about 30° N and S latitude. At low altitude, some air masses flow back toward the equator, completing low-latitude circulation cells. Others flow toward the poles, rise at 60° latitude, and divide at high altitude. Some of this air flows toward the equator, completing the pair of middle-latitude circulation cells. The rest moves toward the poles, where it descends and flows toward the equator, forming the polar circulation cells.

The flow of air masses at low altitude creates winds near the planet's surface. But the planet's surface rotates beneath the atmosphere, moving rapidly near the equator, where Earth's diameter is greatest, and more slowly near the poles. Latitudinal variation in the speed of Earth's rotation deflects the movement of the rising and sinking air masses from a strictly north–south path into belts of easterly and westerly winds; this deflection is called the Coriolis effect. Winds near the equator are called the trade winds; those farther from the equator are the temperate westerlies and easterlies, named for their direction of flow.

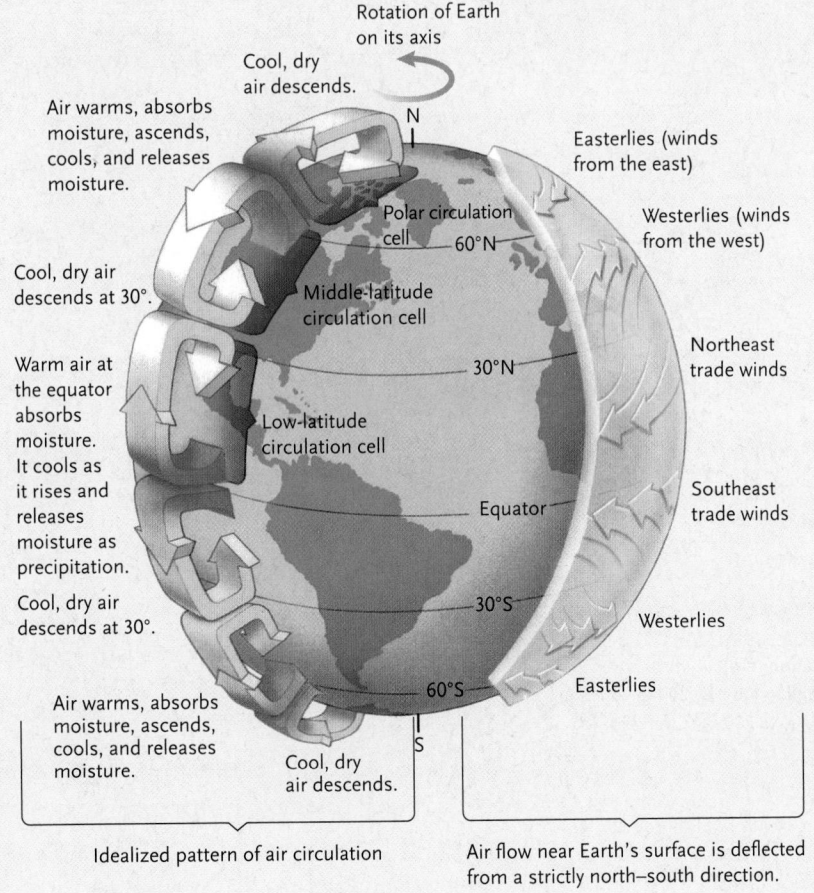

Rotation of Earth on its axis

Cool, dry air descends.

Air warms, absorbs moisture, ascends, cools, and releases moisture.

Cool, dry air descends at 30°.

Warm air at the equator absorbs moisture. It cools as it rises and releases moisture as precipitation.

Cool, dry air descends at 30°.

Air warms, absorbs moisture, ascends, cools, and releases moisture.

Cool, dry air descends.

Polar circulation cell

Middle-latitude circulation cell

Low-latitude circulation cell

60°N

30°N

Equator

30°S

60°S

N

S

Easterlies (winds from the east)

Westerlies (winds from the west)

Northeast trade winds

Southeast trade winds

Westerlies

Easterlies

Idealized pattern of air circulation

Air flow near Earth's surface is deflected from a strictly north–south direction.

Precipitation

Differences in solar radiation and global air circulation create latitudinal variations in rainfall. Warm air holds more water vapour than cool air does. As air near the equator heats up, it absorbs water, primarily from the oceans. However, the warm air masses expand as they rise, and their heat energy is distributed over a larger volume, causing their temperature to drop. A decrease in temperature without the actual loss of heat energy is called adiabatic cooling. After cooling adiabatically, the rising air masses release moisture as rain. Torrential rainfall is characteristic of warm equatorial regions, where rising, moisture-laden air masses cool as they reach high altitude.

As cool, dry air masses descend at 30° latitude, increased air pressure at low altitude compresses them, concentrating their heat energy, raising their temperature, and increasing their capacity to hold moisture. Descending air masses absorb water at these latitudes, which are typically dry. Some air masses continue moving poleward in the lower atmosphere. When they rise at 60° latitude, they cool adiabatically and release precipitation, creating moist habitats in the northern and southern temperate zones.

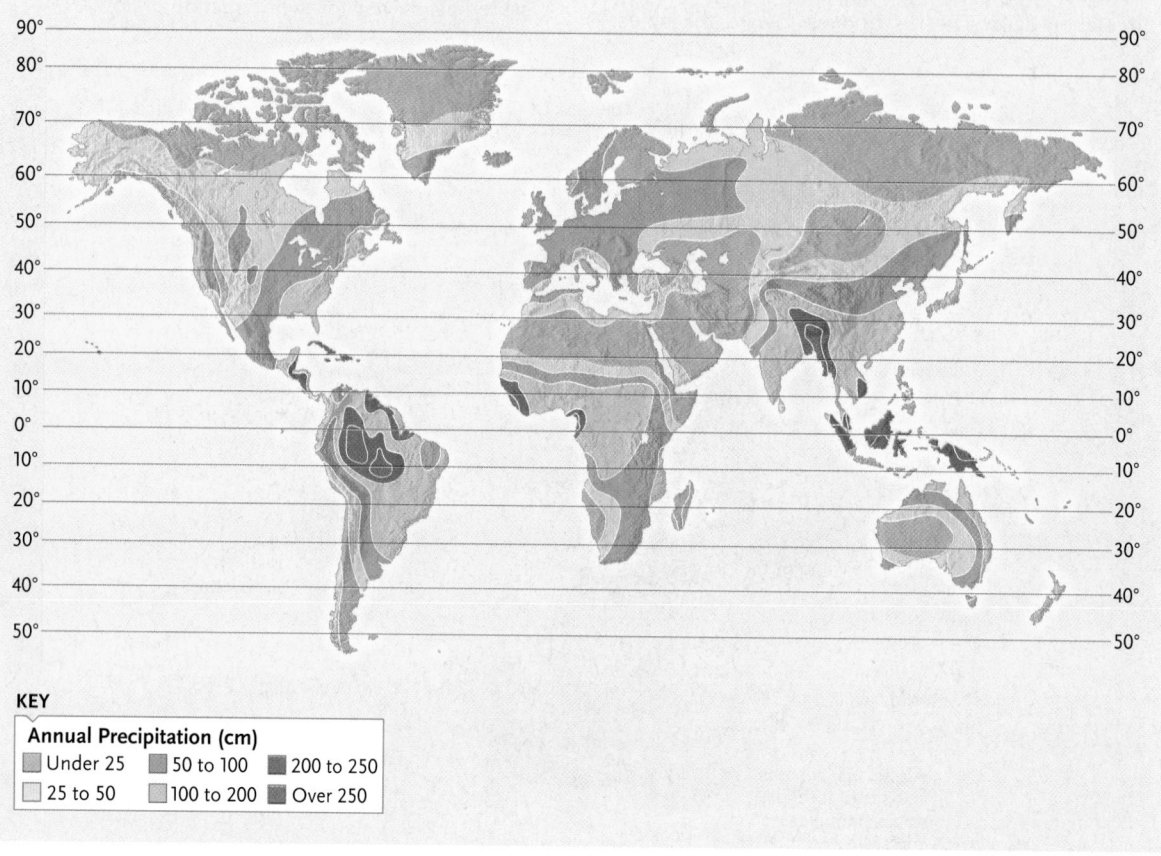

KEY

Annual Precipitation (cm)

- Under 25
- 25 to 50
- 50 to 100
- 100 to 200
- 200 to 250
- Over 250

Ocean Currents

Latitudinal variations in solar radiation also warm the oceans' surface water unevenly. Because the volume of water increases as it warms (5 decrease in density), sea level is about 8 cm higher at the equator than at the poles. The volume of water associated with this "slope" is enough to cause surface water to move in response to gravity. The trade winds and temperate westerlies also contribute to the mass flow of water at the ocean surface. Thus, surface water flows in the direction of prevailing winds, forming major currents. Earth's rotation, the positions of landmasses, and the shapes of ocean basins also influence the movements of these currents.

Oceanic circulation is generally clockwise in the northern hemisphere and counterclockwise in the southern hemisphere (see figure below). The trade winds push surface water toward the equator and westward until it contacts the eastern edge of a continent. Swift, narrow, and deep currents of warm, nutrient-poor water run toward the poles, parallel to the east coasts of continents. For example, the Gulf Stream flows northward along the east coast of North America, carrying warm water toward northwestern Europe. Cold water returns from the poles toward the equator in slow, broad, and shallow currents, such as the California Current, that parallel the west coasts of continents.

KEY

- Upwelling zone
- → Warm surface current
- → Cold surface current

Regional and Local Effects

Although global and seasonal patterns determine an area's climate, regional and local effects also influence abiotic conditions. Currents running along sea coasts exchange heat with air masses flowing above them, moderating the temperature over the nearby land. Breezes often blow from the sea toward the land during the day and in the opposite direction at night (see figure at right). These local effects sometimes override latitudinal variations in temperature. For example, the climate in London, England, is much milder than that in Winnipeg, even though Winnipeg is slightly farther south. London has a maritime climate, tempered by winds that cross the nearby North Atlantic Current, but Winnipeg's climate is continental, not moderated by the distant ocean.

Ocean currents also affect moisture conditions in coastal habitats. For example, the region off the southeast coast of Newfoundland known as the Grand Banks is one of the foggiest places on Earth. Here, as the warm Gulf Stream current meets the cold Labrador current, the air above the water cools and its water vapour condenses into heavy fog and rain.

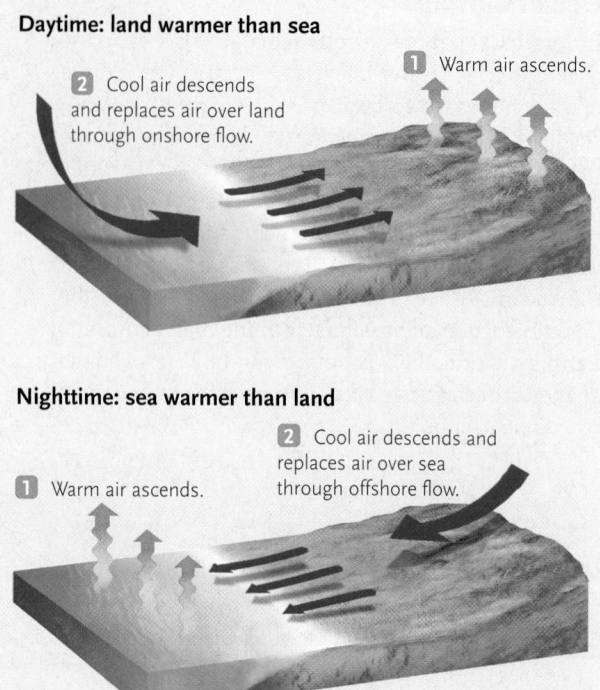

Daytime: land warmer than sea

2 Cool air descends and replaces air over land through onshore flow.

1 Warm air ascends.

Nighttime: sea warmer than land

2 Cool air descends and replaces air over sea through offshore flow.

1 Warm air ascends.

The Effects of Topography

Mountains, valleys, and other topographic features are a major influence on regional climates. In the northern hemisphere, south-facing slopes are warmer and drier than north-facing slopes because they receive more solar radiation. In addition, adiabatic cooling causes air temperature to decline 3 to 6°C for every 1000 m increase in elevation. Mountains also establish regional and local rainfall patterns. For example, warm air masses pick up moisture from the Pacific Ocean and then move inland toward the Rocky Mountains. As air rises to cross the mountains, it cools adiabatically and loses moisture, releasing heavy rainfall on the windward side (see below). After the now-dry air crosses the peaks, it descends and warms, absorbing moisture and forming a rain shadow. Habitats on the leeward side of mountains, such as the eastern slopes of the Rocky Mountains in Alberta, are typically drier than those on the windward side.

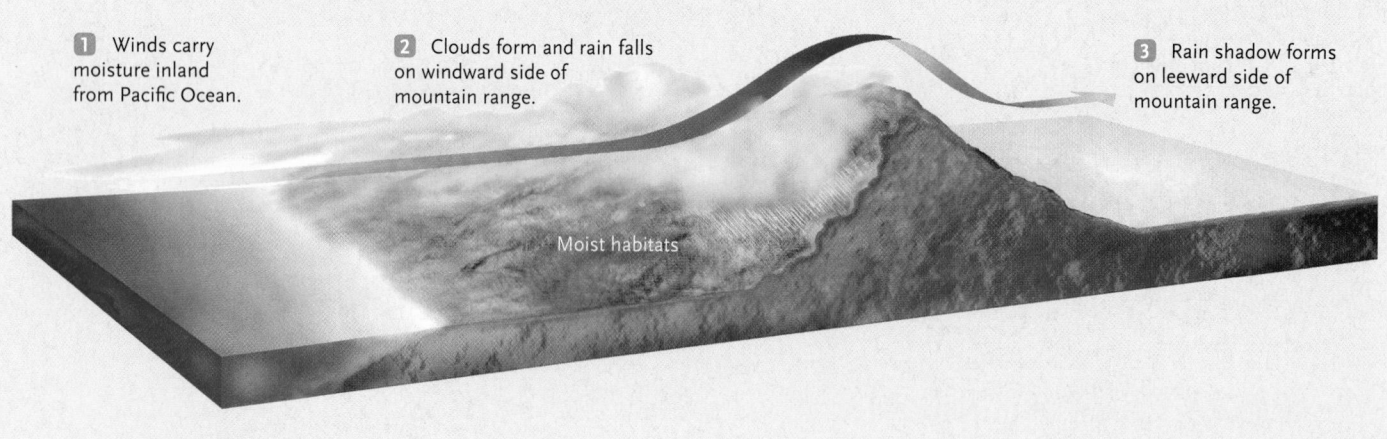

1 Winds carry moisture inland from Pacific Ocean.

2 Clouds form and rain falls on windward side of mountain range.

3 Rain shadow forms on leeward side of mountain range.

Moist habitats

Microclimate

Although climate influences the overall distributions of organisms, the abiotic conditions that immediately surround them, the microclimate, have the greatest effect on survival and reproduction. For example, a fallen log on the forest floor creates a microclimate in the underlying soil that is shadier, cooler, and moister than the surrounding soil, which is exposed to sun and wind. Many animals, including some insects, worms, salamanders, and snakes, occupy these sheltered sites and avoid the effects of prolonged exposure to the elements.

Aleksander Bolbot/Shutterstock.com

Biomes

Various climatic factors interact to create and regulate **biomes**—groups of ecosystems that share distinctive combinations of soils, vegetation, and animals. Fourteen different biomes have been defined (see below). Why is climate so important in defining biomes? Climatic factors, particularly temperature regimes and water availability, control the rate of photosynthesis by plants, which produce the organic molecules that provide the energy and carbon required by all other organisms in a biome.

In addition, climate influences the type of plants that make up the dominant vegetation of a biome through the selection pressures it creates: certain climatic regions favour certain adaptations and strategies. For example, in arid regions, the dominant plants have adaptations that store water or reduce water loss by evaporation, or that are metabolically active only in the wettest season, have an advantage over other plants. Biomes are often classified climatically (e.g., desert) or on the basis of the dominant vegetation (e.g., grassland, tropical rainforest).

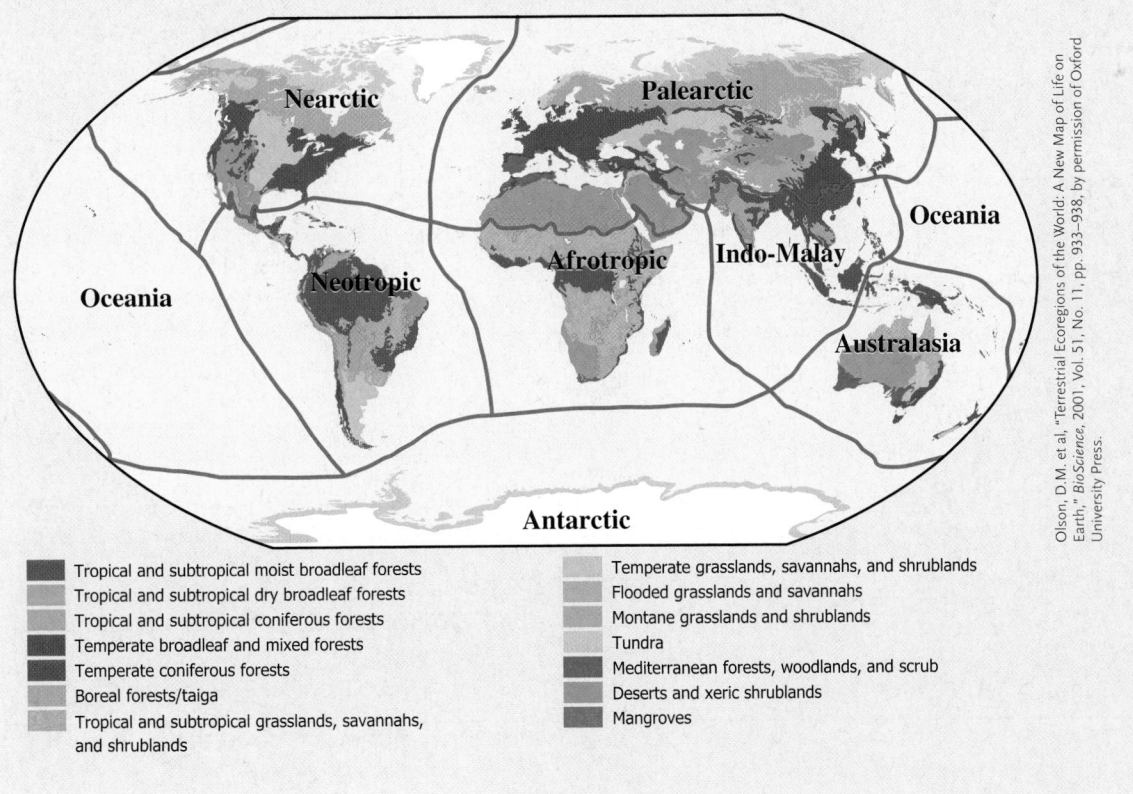

Olson, D.M. et al, "Terrestrial Ecoregions of the World: A New Map of Life on Earth," *BioScience*, 2001, Vol. 51, No. 11, pp. 933–938, by permission of Oxford University Press.

- Tropical and subtropical moist broadleaf forests
- Tropical and subtropical dry broadleaf forests
- Tropical and subtropical coniferous forests
- Temperate broadleaf and mixed forests
- Temperate coniferous forests
- Boreal forests/taiga
- Tropical and subtropical grasslands, savannahs, and shrublands
- Temperate grasslands, savannahs, and shrublands
- Flooded grasslands and savannahs
- Montane grasslands and shrublands
- Tundra
- Mediterranean forests, woodlands, and scrub
- Deserts and xeric shrublands
- Mangroves

History of Earth

Geological Time Scale and Major Evolutionary Events

The Geological Time Scale and Major Evolutionary Events

Eons (duration drawn to scale)	Eon	Era	Period	Epoch	Millions of Years Ago	Major Evolutionary Events
Cenozoic / Mesozoic / Paleozoic (Phanerozoic)	Phanerozoic	Cenozoic	Quaternary	Holocene		
					0.01	
				Pleistocene		Origin of humans; major glaciations
					1.7	
			Tertiary	Pliocene		Origin of apelike human ancestors
					5.2	
				Miocene		Angiosperms and mammals further diversify and dominate terrestrial habitats
					23	
				Oligocene		Divergence of primates; origin of apes
					33.4	
				Eocene		Angiosperms and insects diversify; modern orders of mammals differentiate
					55	
				Paleocene		Grasslands and deciduous woodlands spread; modern birds and mammals diversify; continents approach current positions
Proterozoic		Mesozoic	Cretaceous		65	Many lineages diversify: angiosperms, insects, marine invertebrates, fishes, dinosaurs; asteroid impact causes mass extinction at end of period, eliminating dinosaurs and many other groups
					144	
			Jurassic			Gymnosperms abundant in terrestrial habitats; first angiosperms; modern fishes diversify; dinosaurs diversify and dominate terrestrial habitats; frogs, salamanders, lizards, and birds appear; continents continue to separate
					206	
			Triassic			Predatory fishes and reptiles dominate oceans; gymnosperms dominate terrestrial habitats; radiation of dinosaurs; origin of mammals; Pangaea starts to break up; mass extinction at end of period
					251	

Eons (duration drawn to scale)	Eon	Era	Period	Epoch	Millions of Years Ago	Major Evolutionary Events
Archaean	Phanerozoic (continued)	Paleozoic	Permian			Insects, amphibians, and reptiles abundant and diverse in swamp forests; some reptiles colonize oceans; fishes colonize freshwater habitats; continents coalesce into Pangaea, causing glaciation and decline in sea level; mass extinction at end of period eliminates 85% of species
					290	
			Carboniferous			Vascular plants form large swamp forests; first seed plants and flying insects; amphibians diversify; first reptiles appear
					354	
			Devonian			Terrestrial vascular plants diversify; fungi and invertebrates colonize land; first insects appear; first amphibians colonize land; major glaciation at end of period causes mass extinction, mostly of marine life
					417	
			Silurian			Jawless fishes diversify; first jawed fishes; first vascular plants on land
					443	
			Ordovician			Major radiations of marine invertebrates and fishes; major glaciation at end of period causes mass extinction of marine life
					490	
			Cambrian			Diverse radiation of modern animal phyla (Cambrian explosion); simple marine communities
					543	
	Proterozoic					High concentration of oxygen in atmosphere; origin of aerobic metabolism; origin of eukaryotic cells; evolution and diversification of protists, fungi, soft-bodied animals
					2500	
	Archaean					Evolution of prokaryotes, including anaerobic bacteria and photosynthetic bacteria; oxygen starts to accumulate in atmosphere
					3800	
						Formation of Earth at start of era; Earth's crust, atmosphere, and oceans form; origin of life at end of era
					4600	

Model Research Organisms

Certain species or groups of organisms have become favourite subjects for laboratory and field studies because their characteristics make them relatively easy research subjects. In most cases, such **model organisms** became popular because they have rapid development, short life cycles, and small adult size. Thus, researchers can rear and house large numbers of them in the laboratory. Also, as fuller portraits of their genetics and other aspects of their biology emerge, their appeal as research subjects tends to grow because biologists have a better understanding of the biological context within which specific processes occur. Because the fundamental elements of biochemistry, development, and evolution are common to all organisms, research on these small and often simple model organisms provides insight into biological processes that operate in and among larger and more complex organisms.

As a cautionary note, you should also be aware that the very characteristics that make model organisms valuable for research may make them poor representatives of other organisms in that group. Thus, specific findings from *Drosophila* or *Caenorhabditis elegans* may not be generally applicable to other insects or nematodes, respectively. The use of model organisms only, to the exclusion of others, may obscure the richness of biological diversity.

Escherichia coli

We probably know more about *Escherichia coli* than any other organism. For example, microbiologists have deciphered the complete DNA sequence of the genome of a standard laboratory strain of *E. coli*, including the sequence of the approximately 4400 genes in its genome. The functions of about one-third of these genes are still unidentified; however, *E. coli* got its start in laboratory research because of the ease with which it can be grown in cultures. Because *E. coli* cells divide about every 20 minutes under optimal conditions, a clone of 1 billion cells can be grown in a matter of hours in only 10 mL of culture medium. The same amount of medium can accommodate as many as 10 billion cells before the growth rate begins to slow. *E. coli* strains can be grown in the laboratory with minimal equipment, requiring little more than culture vessels in an incubator held at 37°C.

The study of naturally occurring plasmids in *E. coli* and of enzymes that cut DNA at specific sequences eventually resulted in the development of recombinant DNA techniques—procedures to combine DNA from different sources. Today, *E. coli* is used extensively for creating such molecules and for amplifying (cloning) them once they are made. In essence, the biotechnology industry has its foundation in molecular genetics studies of *E. coli*.

Large-scale *E. coli* cultures are widely used as "factories" for the production of desired proteins. For example, the human insulin hormone, required for treatment of certain forms of diabetes, can be produced by *E. coli* factories.

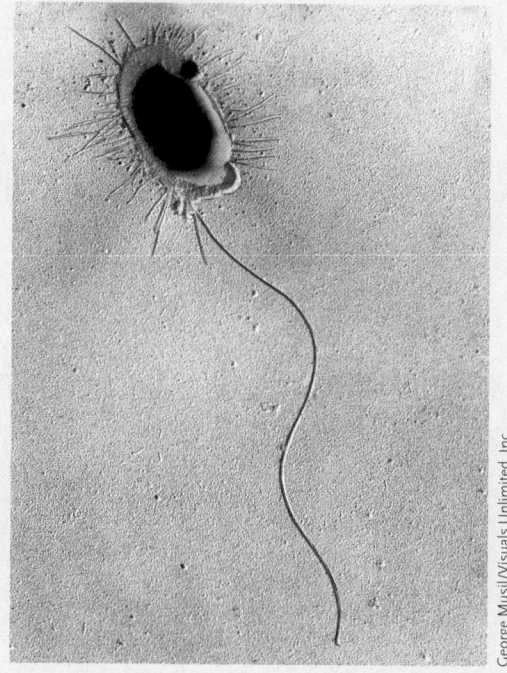

George Musil/Visuals Unlimited, Inc.

Saccharomyces cerevisiae

Commonly known as baker's yeast or brewer's yeast, *Saccharomyces cerevisiae* was probably the first microorganism to have been domesticated by humans—a beer-brewing vessel is basically a *Saccharomyces* culture. Favourite strains of baker's and brewer's yeasts have been kept in continuous cultures for centuries. The yeast has also been widely used in scientific research; its microscopic size and relatively short generation time make it easy and inexpensive to culture in large numbers in the laboratory.

The complete DNA sequence of *S. cerevisiae*, which includes more than 12 million base pairs that encode about 6000 genes, was the first eukaryotic genome to be determined. Plasmids, extrachromosomal segments of DNA, have been produced that are used to introduce genes into yeast cells. Using plasmids, researchers can experimentally alter any of the yeast genes to test their functions and can introduce genes or DNA samples from other organisms for testing or cloning. These genetic engineering studies have demonstrated that many mammalian genes can replace yeast genes when

introduced into the fungi, confirming their close relationships, even though mammals and fungi are separated by millions of years of evolution. *S. cerevisiae* has been so important to genetic studies in eukaryotes that it is often called the eukaryotic *E. coli*. Research with another yeast, *Schizosaccharomyces pombe,* has been similarly productive, particularly in studies of genes that control the cell cycle.

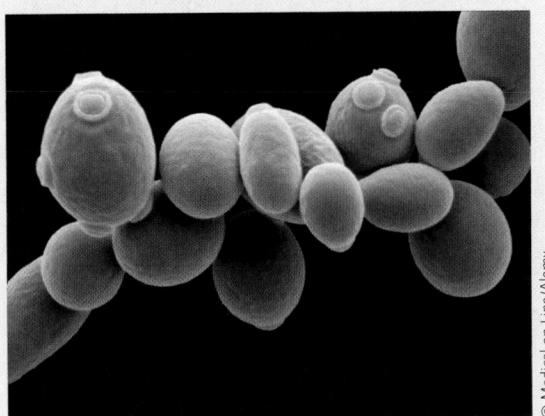

© Medical-on-Line/Alamy

Drosophila melanogaster

The pesky little fruit fly that appears seemingly from nowhere when rotting fruit or a fermented beverage is around is one of the mainstays of genetic research. It was first described in 1830 by C. F. Fallén, who named it *Drosophila*, meaning "dew lover." The species identifier became *melanogaster,* which means "black belly." The great geneticist Thomas Hunt Morgan began to culture *D. melanogaster* in 1909 in the famous "Fly Room" at Columbia University. Many important discoveries in genetics were made in the Fly Room, including sex-linked genes and sex linkage and the first chromosome map. The subsequent development of methods to induce mutations in *Drosophila* led, through studies of the mutants produced, to many other discoveries that collectively established or confirmed essentially all of the major principles and conclusions of eukaryotic genetics.

One reason for the success of *D. melanogaster* as a subject for genetics research is the ease of culturing it. It is usually grown at 25°C in small bottles stopped with a cotton or plastic foam wad and about one-third filled with a fermenting medium that contains water, cornmeal, agar, molasses, and yeast. The several hundred eggs laid by each adult female hatch rapidly and progress through larval and pupal stages to produce adult flies in about 10 days. These are ready to breed within 10 to 12 hours. Males and females can be identified easily with the unaided eye.

Many types of mutations produce morphological differences, such as changes in eye colour, wing shape, or the numbers and shapes of bristles, which can be

seen with the unaided eye or under a low-power binocular microscope. The salivary gland cells of the fly larvae have giant chromosomes that are so large that differences can be observed directly with a light microscope. The availability of a wide range of mutants and comprehensive linkage maps of each of its chromosomes, and the ability to manipulate genes readily by molecular techniques, made the fruit fly genome one of the first to be sequenced. The sequencing of *Drosophila*'s genome was completed in 2001; it has approximately 14 000 genes in its 165-million-base-pair genome. (A database of the *Drosophila* genome is available at http://flybase.org.) Importantly, the relationship between fruit fly and human genes is close, to the point that many human disease genes have counterparts in the fruit fly genome. This similarity enables the fly genes to be studied as models of human disease genes to better understand the functions of those genes and how alterations in them can lead to disease.

The analysis of fruit fly embryonic development has also contributed significantly to the understanding of development in humans. For example, experiments on mutants that affect fly development have provided insight into the genetic basis of many human birth defects. Before making a career as an environmentalist, David Suzuki studied temperature-sensitive neurological mutants at the University of British Columbia.

Roblan/Shutterstock.com

Caenorhabditis elegans

Researchers studying the tiny, free-living nematode *C. elegans* have made many advances in molecular genetics, animal development, and neurobiology. It is so popular as a model research organism that most workers simply refer to it as "the worm." Several attributes make *C. elegans* a model research organism. The adult is about 1 mm long and thrives on cultures of *E. coli* or other bacteria; thus, thousands can be raised in a culture dish. It completes its life cycle from egg to reproductive adult within three days at room temperature. Furthermore, stock cultures can be kept alive indefinitely by freezing them in liquid nitrogen or in an ultracold freezer ($-80°C$). Researchers can therefore store new mutants for later research without having to clean, feed, and maintain active cultures. Best of all, the worm is anatomically simple; an adult contains just 959 cells (excluding the gonads). Having a fixed cell number is relatively uncommon among animals, and developmental biologists have made good use of this trait. The eggs, juveniles, and adults of the worm are completely transparent, and researchers can observe cell divisions and cell movements in living animals with straightforward microscopy techniques. There is no need to kill, fix, and stain specimens for study. And virtually every cell in the worm's body is accessible for manipulation by laser microsurgery, microinjection, and similar approaches.

The genome of *C. elegans*, which was sequenced in 1998, is also simple, consisting of 100 million base pairs organized into roughly 17 000 genes on 6 pairs of chromosomes. The genome, which is about the same size as 1 human chromosome, specifies the amino acid sequences of about 10 000 protein molecules—far fewer than are found in more complex animals.

The knowledge gained from research on *C. elegans* is highly relevant to studies of larger and more complex organisms, including vertebrates. Recent research demonstrates some striking similarities among nematodes, fruit flies, and mice in the genetic control of development; in some of the proteins that govern important events such as cell death; and in the molecular signals used for cell-to-cell communication. Using a relatively simple model such as *C. elegans*, researchers can answer research questions more quickly and more efficiently than they could if they studied larger and more complex animals.

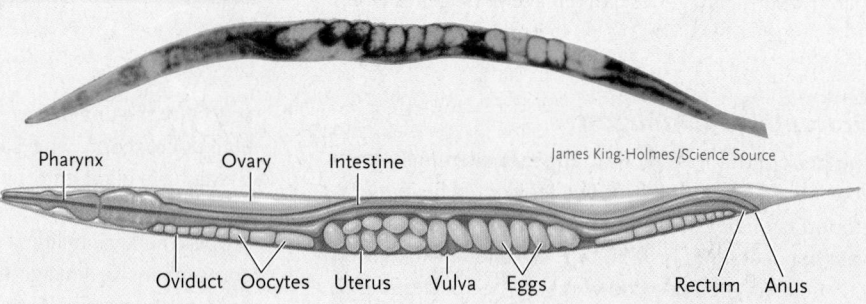

Pharynx · Ovary · Intestine

James King-Holmes/Science Source

Oviduct · Oocytes · Uterus · Vulva · Eggs · Rectum · Anus

Arabidopsis thaliana

For plant geneticists, the little white-flowered thale cress, *Arabidopsis thaliana*, has attributes that make it a prime subject for genetic research. A tiny member of the mustard family, *Arabidopsis* is revealing answers to some of the biggest questions in plant development and physiology. Each plant grows only a few centimetres tall, so little laboratory space is required to house a large population. As long as *Arabidopsis* is provided with damp soil containing basic nutrients, it grows easily and rapidly in artificial light. Seeds grow to mature plants in just over a month and then flower and reproduce themselves in another three to four weeks. This permits investigators to perform desired genetic crosses and obtain large numbers of offspring with known, desired genotypes with relative ease.

The *Arabidopsis* genome was the first complete plant genome to be sequenced. Researchers have identified approximately 28 000 genes arranged on 5 pairs of chromosomes. The genome contains relatively little repetitive DNA, so it is fairly easy to isolate *Arabidopsis* genes, which can then be cloned using genetic engineering techniques. Cloned genes are inserted into bacterial plasmids, and the recombinant plasmids are transferred to the bacterial species *Agrobacterium tumefaciens*, which readily infects *Arabidopsis* cells. Amplified by the bacteria, the genes and their protein products can be sequenced or studied in other ways. Typically, researchers use chemical mutagens or recombinant bacteria to introduce changes in the *Arabidopsis* genome.

© Custom Life Science Images/Alamy

Danio rerio

The zebrafish (*Danio rerio*) is a small (3 cm) freshwater fish that gets its name from the black and white stripes running along its body. Native to India, it has spread around the world as a favourite aquarium fish. Beginning about 30 years ago, it began to be used in scientific laboratories as a model vertebrate organism for studying the roles of genes in development. Its use is now so widespread that it has been dubbed the "vertebrate fruit fly."

The zebrafish brings many advantages as a model research organism. It can be maintained easily in an ordinary aquarium on a simple diet. Although its generation time is relatively long (3 months for the zebrafish compared with 6 weeks for the mouse), a female zebrafish produces about 200 offspring at a time, compared with an average of 10 for the mouse. Embryonic development of the zebrafish takes place in eggs released to the outside by the female. The embryos develop rapidly, taking only three days from egg laying to hatching. Best of all, the eggs and embryos are transparent, providing an open window that allows researchers to observe developmental stages directly, with little or no disturbance to the embryo. Observational conditions are so favourable that the origin and fate of each cell can be traced from the fertilized egg to the hatchling. Individual nerve cells can be traced, for example, as they grow and make connections in the brain, spinal cord, and peripheral body regions. Removing or transplanting cells and tissues is also relatively easy. Biochemical and molecular studies can be carried out by techniques ranging from the simple addition of reactants to the water surrounding the embryos to injection of chemicals into individual cells.

The advantages of working with the zebrafish have spurred efforts to investigate its genetics, with particular interest in genes that regulate embryonic development. This work has already identified mutants of more than 2000 genes, including more than 400 genes that influence development. Most of the mechanisms controlled by the developmental genes resemble their counterparts in humans and other mammals. Developmental and physiological studies have revealed functions of some zebrafish genes that were previously unknown for their mammalian equivalents.

David Dohnal/
Shutterstock.com

Mus musculus

The "wee, sleekit, cow'rin', tim'rous beastie," as the poet Robert Burns called the mouse (*Mus musculus*), has a much larger stature among scientists. The mouse and its cells have been used to great advantage as models for research on mammalian developmental genetics, immunology, and cancer. The availability of the mouse as a research tool enables scientists to carry out mammalian experiments that would not be practical or ethical with humans. Its small size makes the mouse relatively inexpensive and easy to maintain in the laboratory, and its short generation time, compared with most other mammals, allows genetic crosses to be carried out within a reasonable time span. Mice can be mated when they are 10 weeks old; in 18 to 22 days, the female gives birth to a litter of 5 to 10 offspring. A female may be rebred a little more than a day after giving birth.

Mice have a long and highly productive history as experimental animals. Gregor Mendel, the founder of genetics, is known to have kept mice as part of his studies. Toward the end of the nineteenth century, August Weissmann helped disprove an early evolutionary hypothesis, the inheritance of acquired characters, by cutting off the tails of mice for 22 successive generations and finding that it had no effect on tail length. The first example of a lethal allele was also found in mice, and pioneering experiments on the transplantation of tissues between individuals were conducted with mice. During the 1920s, Fred Griffith laid the groundwork for the research showing that DNA is the hereditary molecule in his work with pneumonia-causing bacteria in mice.

More recently, genetic experiments with mice have revealed more than 500 mutants that cause hereditary diseases, immunological defects, and cancer in mammals, including humans. The mouse has also been the mammal of choice for experiments that introduce and modify genes through genetic engineering. One of the most spectacular results of this research was the production of giant mice by introducing a human growth hormone gene into a line of dwarfed mice that were deficient for this hormone. Genetic engineering has also produced knockout mice (see "Knockouts: Genes and Behaviour," Chapter 47) in which a gene of interest is completely nonfunctional. The effects of this lack of function often help investigators determine the role of the normal form of the gene. Some knockout mice are defective in genes homologous to human genes that cause serious diseases, such as cystic

continued on next page

fibrosis, so researchers can study the disease in mice with the goal of developing cures or therapies.

The revelations in developmental genetics from studies with the mouse have been of great interest and importance in their own right. In 2002, the sequence of the mouse genome was reported. This sequence is enabling researchers to refine and expand their use of the mouse as a model organism for studies of mammalian biology and mammalian diseases. More and more, as we find that much of what applies to the mouse also applies to humans, the findings in mice have shed new light on human development and

opened pathways to the possible cures of human genetic diseases.

lostbear/Shutterstock.com

Anolis Lizards of the Caribbean

The lizard genus *Anolis* has been a model system for studies in ecology and evolutionary biology since the 1960s, when Ernest E. Williams of Harvard University's Museum of Comparative Zoology first began studying it. With more than 400 known species—and new ones being described all the time—*Anolis* is one of the most diverse vertebrate genera known. Most anoles are less than 10 cm long, not including the tail, and many occur at high densities, making it easy to collect a lot of data in a relatively short time. Male anoles defend territories, and their displays make them conspicuous even in dense forests.

Anolis species are widely distributed in South America and Central America, but nearly 40% occupy Caribbean islands. The number of species on an island is generally proportional to the island's size. Cuba, the largest island, has more than 50 species, whereas small islands have just one or two. Studies by Williams and others suggest that the anoles on some large islands are the products of independent adaptive radiations. Eight of the 10 *Anolis* species now found in Puerto Rico

probably evolved on that island from a common ancestor. Similarly, the seven *Anolis* species in Jamaica had a common ancestor, which was different from the ancestor of the Puerto Rican species. The anole faunas in Cuba and Hispaniola are the products of several independent radiations on each island. Williams discovered that these independent radiations had produced similar-looking species on different islands. He developed the concept of the *ecomorph*, a group of species that have similar morphological, behavioural, and ecological characteristics even though they are not closely related within the genus. Williams named the ecomorphs after the vegetation that they commonly used. For example, grass anoles are small, slender species that usually perch on low, thin vegetation. Trunk-ground anoles have chunky bodies and large heads, and they perch low on tree trunks, frequently jumping to the ground to feed. Although the grass anoles or the trunk-ground anoles on different islands are similar in many ways, they are not closely related to each other. Their resemblances are the products of convergent evolution.

A. krugi

Jason Patrick Ross/Shutterstock.com

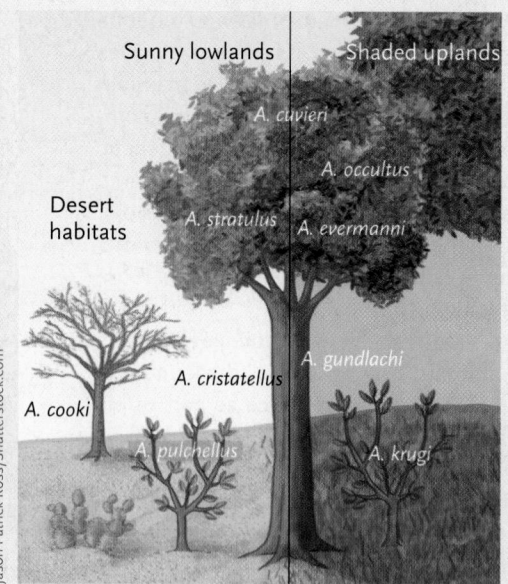

29

The large number of gulls (*Larus* species) is obvious at a landfill site near Thunder Bay, Ontario. The population of gulls reflects the local population of humans.

Population Ecology

WHY IT MATTERS

Controlling rabies in wildlife involves understanding many aspects of biology, from populations to epidemiology and behaviour. Rabies, from the Latin *rabere* (to rage or rave), affects the nervous system of terrestrial mammals. Caused by a Lyssavirus, rabies is usually spread by bites because the virus accumulates in the saliva of infected animals. Before 1885, when Louis Pasteur in France developed a vaccine for it, rabies was common in Europe, and many people died from it every year. In 2007, the World Health Organization estimated that worldwide, more than 50 000 people die annually from rabies, usually people in the developing world. Between 1980 and the end of 2000, 43 people in the United States and Canada died of rabies.

Animals with *furious* rabies become berserk, attacking anything and everything in their path, a behaviour that spreads the virus and helps ensure its survival. Animals with paralytic rabies (*dumb rabies*) suffer from increasing paralysis that progresses forward from the hindlimbs. Animals with either manifestation of rabies can spread the disease by biting when there is virus in their saliva. Paralysis of the throat muscles means that rabid animals cannot swallow the saliva they produce, so they appear to foam at the mouth.

Rabies is almost invariably fatal once an animal or a human shows clinical symptoms of the disease, so immunization of someone exposed to the disease should start as soon as possible after exposure. Since 1980, human diploid vaccines have been commonly available, raising the level of protection against rabies.

From the 1960s to the 1990s, a visit to almost any rural hospital in southern Ontario would have revealed at least one farmer receiving postexposure rabies shots. During that time, red foxes (*Vulpes vulpes*) were the main vector for rabies in Ontario, and cows (*Bos taurus*) exposed to rabies through fox bites in turn exposed farmers to the virus. Many farmers are accustomed to treating choking cows by reaching into the cow's gullet to clear an obstruction. A farmer dealing with a rabid cow could have been scratched and exposed to the virus, and then, after the cow died of rabies, the farmer would have received postexposure rabies shots. Rabies transmitted to cows from foxes posed a threat to human lives and was a significant drain on the economy through compensation paid to farmers whose cattle succumbed to the disease.

In 1967, 4-year-old Donna Featherstone of Richmond Hill, Ontario, died of rabies after being bitten by a stray cat. The resulting public outcry set the stage for a rabies eradication program in Ontario. Controlling fox rabies in southern Ontario was achieved by a combination of innovation and knowledge of basic biology. There were three phases: (a) developing an oral vaccine, (b) developing a means of vaccinating foxes, and (c) monitoring the impact of the program on the fox population.

First, two main baits for the oral vaccine were developed, and one, Evelyn, Rocketniki, Abelseth (ERA), was a modified live virus replicated in tissues of the mouth and throat. ERA successfully stimulated seroconversion in red foxes and vaccinated them against rabies. Second, foxes were vaccinated by eating ERA-containing baits scented with chicken. The baits were small, the size of restaurant packets of jam, and easy to distribute widely from low-flying aircraft, allowing vaccination of foxes across large areas of southern Ontario. Third, each bait contained tetracycline, a biomarker absorbed into the system of any mammal that ate the bait. Once in the body, some tetracycline penetrated the dentine of the animals' teeth, especially in younger individuals. Biologists sectioned and stained teeth from foxes taken by trappers. In the sections, bands of tetracycline in tooth rings **(Figure 29.1)** identified foxes that had taken baits, and biologists established that over 70% of red foxes had been vaccinated by this method.

Before the bait vaccination program, on average 211 cattle annually died of rabies in southern Ontario. The baiting program started in 1989, and by 1996, rabies in cattle dropped to an average of 11 cases a year and the levels of rabies in foxes in Ontario were dramatically reduced. The example demonstrates how problems in biology are solved by combined approaches, from population biology, behaviour, immunology, and epidemiology.

The purpose of this chapter is to introduce you to ecology in general and population ecology in particular. For more details about the incidence of a disease in a population, see Section 29.8.

Ecology is the study of the relationships among species and between species and the environments in which they occur. Studies of populations are fundamental to ecology and involve everything from the numbers of individuals to their age structure and patterns of reproduction. This connects with work on the life histories of species, often with a focus on

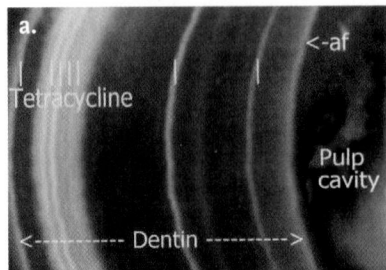
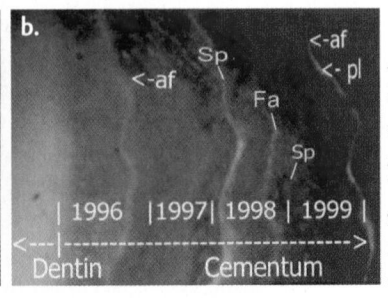

Reprinted from *Rabies*, Alan C. Jackson, "Rabies Control in Wildlife" by David H. Johnston, Rowland R. Tinline, Pages No. 445–471, Copyright 2003, with permission from Elsevier.

Figure 29.1

Tetracycline rings in carnivore teeth. Yellow fluorescent lines from ingestion of rabies baits with tetracycline as a biomarker. The sections are undecalcified, ultraviolet fluorescent × 100. **(a)** Coyote (*Canis latrans*) tooth with seven daily tetracycline lines from vaccine baits. **(b)** Canine tooth of a four-year-old raccoon (*Procyon lotor*) with yearly tetracycline lines in cementum. af = autofluorescent collagen; Fa = fall baits, 1998; pl = periodontal ligament; Sp = spring baits, 1998 and 1999.

reproduction and survival. Changes in the numbers of individuals of a species may lead to an explosion in numbers or to extinction, and both biotic and abiotic factors can influence changes in populations. The basics of population biology and ecology apply as much to our own species as they do to others.

29.1 The Science of Ecology

Ecology encompasses two related disciplines. In basic ecology, major research questions relate to the distribution and abundance of species and how they interact with each other and the physical environment. Using these data as a baseline, workers in **applied ecology** develop conservation plans and amelioration programs to limit, repair, and mitigate ecological damage caused by human activities (see also Chapter 32). Ecology has its roots in descriptive natural history dating back to the ancient Greeks. Modern ecology was born in 1870 when the German biologist Ernst Haeckel coined the term (from *oikos* = house). Contemporary researchers still gather descriptive information about ecological relationships, often as the starting point for other studies. Although ecological research is dominated by hypothetico-deductive approaches, initial inductive approaches allow biologists to generate appropriate hypotheses about how systems function. Research in ecology is often linked to work in genetics, physiology, anatomy, behaviour, paleontology, evolution, geology, geography, and environmental science. Many ecological phenomena, such as climate change, occur over huge areas and long time spans, so ecologists must devise ways to determine how environments influence organisms and how organisms change the environments in which they live. The responses of biological systems to climate change illustrate how the ecology of an organism (or group of organisms) reflects the impact(s) of a range of extrinsic and intrinsic factors on individuals. The range of points of impact includes physiological, reproductive, and energetic factors.

Ecology can be divided into four increasingly complex and inclusive levels of organization. First, in **organismal ecology**, researchers study organisms to determine the genetic, biochemical, physiological, morphological, and behavioural adaptations to the abiotic environment (see *The Purple Pages*). Second, in **population ecology**, researchers focus on groups of individuals of the same species that live together. Population ecologists study how the size and other characteristics of populations change in space and time. Third, in **community ecology**, biologists examine populations of different species that occur together in one area (are **sympatric**). Community ecologists study interactions between species, analyzing how predation, competition, and environmental disturbances influence a community's development, organization, and structure (see Chapter 31). Fourth, those studying **ecosystem ecology** explore how nutrients cycle and energy flows between the biotic components of an **ecological community** and the abiotic environment (see Chapter 31).

Ecologists can create hypotheses about ecological relationships and how they change through time or differ from place to place. Some formalize these ideas in mathematical models that express clearly defined, but hypothetical, relationships among important variables in a system. Manipulation of a model, usually with the help of a computer, can allow researchers to ask what would happen if some of the variables or their relationships changed. Thus, researchers can simulate natural events and large-scale experiments before investing time, energy, and money in fieldwork and laboratory work. Bear in mind that mathematical models are no better than the ideas and assumptions they embody, and useful models are constructed only after basic observations have defined the relevant variables.

Ecologists use field or laboratory studies to test predictions of their hypotheses about relationships among variables in systems. In controlled experiments, researchers compare data from an experimental treatment (involving manipulation of one or more variables) with data from a control (in which nothing is changed). In some cases, *natural experiments* can be conducted because of the patterns of distribution and/or behaviour of species. This has the advantage of allowing ecologists to test predictions about how systems are operating without manipulating variables. Two species of fish, cutthroat trout (*Oncorhynchus clarki*) and Dolly Varden char (*Salvelinus malma*), live in coastal lakes of British Columbia. Some lakes have either trout or char, but others contain both species. The natural distributions of these fishes allowed researchers to measure the effect of each species on the other. In lakes in which both species live, each restricts its activities to fewer areas and eats a smaller variety of prey than it does in lakes in which it occurs alone.

29.2 Population Characteristics

CONCEPT FIX Many people believe that the sizes of populations of animals and plants will increase until net demands for food (and other resources) exceed the supply. This crisis of carrying capacity leads to crashes in populations and even to extinction. Under natural conditions, however, interactions among individuals (of the same or different species) usually cause populations to stop growing well before they reach carrying capacity. In many populations, there are natural cycles of numbers. ⬡

Seven characteristics can be described for any population.

29.2a Geographic Range Is Determined by the Boundaries of Distribution

Populations have characteristics that transcend those of the individuals making up the populations. Every population has a **geographic range**, the overall spatial boundaries within which it lives. Geographic ranges vary enormously. A population of snails might inhabit a small tide pool, whereas a population of marine phytoplankton might occupy an area that is orders of magnitude as large. Every population also occupies a **habitat**, the specific environment in which it lives, as characterized by its biotic and abiotic features. Ecologists also measure other population characteristics, such as size, distribution in space, and age structure.

29.2b Population Density Is Based on the Number of Individuals per Unit Area

Population size is the number of individuals making up the population at a specified time (N_i). **Population density** is the number of individuals per unit area or per unit volume of habitat. Species with a large body size generally have lower population densities than those with a small body size **(Figure 29.2)**. Although population size and density are related measures, knowing a population's density provides more information about its relationship to the resources it uses. If a population of 200 oak trees occupies 1 hectare (ha; $10\,000$ m^2), the population density is $200 \times 10\,000$ m^{-2} or 1 tree per 50 m^2. But if 200 oaks are spread over 5 ha, the density is 1 tree per 250 m^2. Clearly, the second

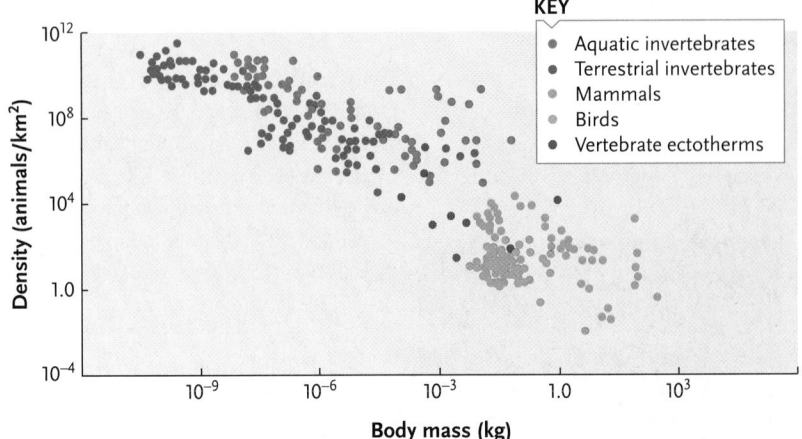

KEY
- Aquatic invertebrates
- Terrestrial invertebrates
- Mammals
- Birds
- Vertebrate ectotherms

Figure 29.2
Population density and body size. Population density generally declines with increasing body size among animal species. There are similar trends for other organisms.

population is less dense than the first, and its members will have greater access to sunlight, water, and other resources.

Ecologists measure population size and density to monitor and manage populations of endangered species, economically important species, and agricultural pests. For large-bodied species, a simple head count could provide accurate information. For example, ecologists survey the size and density of African elephant populations by flying over herds and counting individuals **(Figure 29.3)**. Researchers use a variation on that technique to estimate population size in tiny organisms that live at high population densities. To estimate the density of aquatic phytoplankton, for example, you might collect water samples of known volume from representative areas in a lake and count them by looking through a microscope. These data allow you to estimate population size and density based on the estimated volume of the entire lake. In other cases, researchers use the mark–release–recapture sampling technique (see "Capture–Recapture," p. 682). One ongoing challenge is measuring population size in organisms that are clones, for example, stands of poplar trees (*Populus* spp.).

29.2c Population Dispersion Is the Distribution of Individuals in Space

Populations can vary in their **dispersion**, the spatial distribution of individuals within the geographic range. Ecologists define three theoretical patterns of dispersion: *clumped, uniform,* and *random* **(Figure 29.4, p. 682)**.

Clumped dispersion (see Figure 29.4a) is common and occurs in three situations. First, suitable conditions are often patchily distributed. Certain pasture plants, for instance, may be clumped in small, scattered areas where cowpats had fallen for months, locally enriching the soil. Second, populations of some social animals (see Chapter 47) are clumped because mates are easy to locate within groups, and individuals may cooperate in rearing offspring, feeding, or defending themselves from predators. Third, populations can be clumped when species reproduce by asexual clones that remain attached to the parents.

Figure 29.3
Counting elephants. It is easy to think that large animals such as African elephants (*Loxodonta africana*) would be easy to count from the air **(a).** This may or may not be true, depending on vegetation. But it can be easy to overlook animals, particularly young ones **(b),** in the shade.

Capture–Recapture

Research Method Box

Ecologists use the mark–release–recapture technique to estimate the population size of mobile animals that live within a restricted geographic range. To do this, a sample of organisms (n_1) is captured, marked, and released. Ideally, the marks (or tags) are permanent and do not harm the tagged animal. Insects and reptiles are often marked with ink or paint, birds with rings (bands) on their legs, and mammals with ear tags or collars.

Later, a second sample (n_2) of the population is captured. In the second sample, the proportion of marked (n_2m) to unmarked individuals is used to estimate the total population (x) of the study area by solving the following equation for x:

$$n_1/x = n_{2m}/n_2$$

Assume that you capture a sample of 120 butterflies **(Figure 1)**, mark each one, and release them. A week later, you capture a sample of 150 butter-

Figure 1
This butterfly has been captured and marked before release in a capture–recapture experiment.

flies, 30 that you marked. Thus, you had marked 30 of 150, or 1 of every 5 butterflies, on your first field trip. Because you captured 120 individuals on that first excursion, you would estimate that the total population size is 120 × (150/30) = 600 butterflies.

The capture–recapture technique is based on several assumptions that are critical to its accuracy: (1) being marked has no effect on survival, (2) marked and unmarked animals mix randomly in the population, (3) no migration into or out of the population takes place during the estimating period, and (4) marked individuals are just as likely to be captured as unmarked individuals. (Sometimes animals become "trap shy" or "trap happy," a violation of the fourth assumption.)

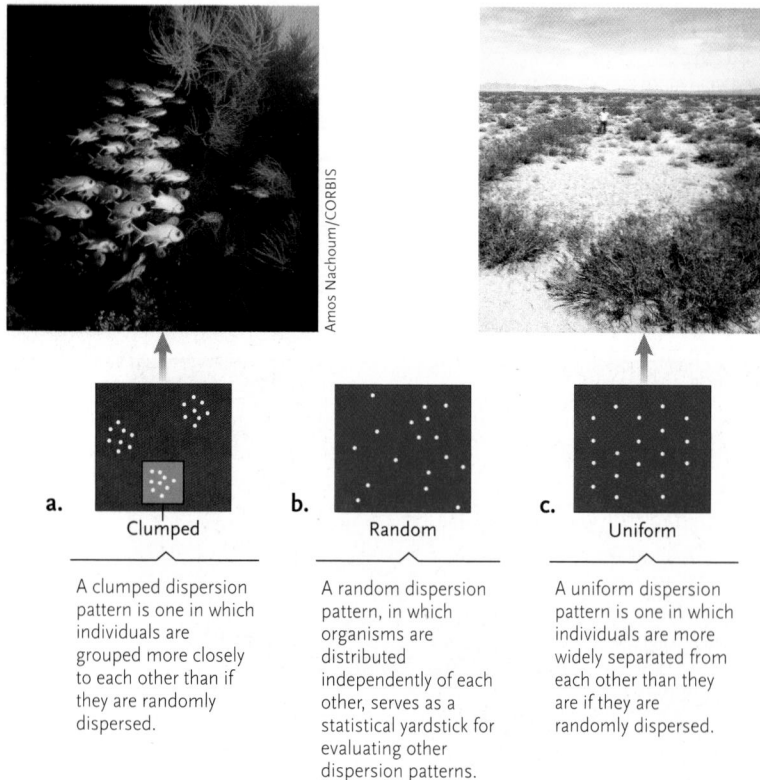

a. Clumped

A clumped dispersion pattern is one in which individuals are grouped more closely to each other than if they are randomly dispersed.

b. Random

A random dispersion pattern, in which organisms are distributed independently of each other, serves as a statistical yardstick for evaluating other dispersion patterns.

c. Uniform

A uniform dispersion pattern is one in which individuals are more widely separated from each other than they are if they are randomly dispersed.

Figure 29.4
Dispersion patterns. A clumped pattern **(a)** is evident in fish that live in social groups. A random pattern **(b)** of dispersion appears to be rare in nature, where it occurs in organisms that are neither attracted to nor repelled by conspecifics. Nearly uniform patterns **(c)** are demonstrated by creosote bushes (*Larrea tridentata*) near Death Valley, California.

Aspen trees and sea anemones reproduce this way and often occur in large aggregations (see Chapter 19). Clumping may also occur in species in which seeds, eggs, or larvae lack dispersal mechanisms and offspring grow and settle near their parents.

Uniform distributions can occur when individuals repel one another because resources are in short supply. Creosote bushes are uniformly distributed in the dry scrub deserts of the U.S. Southwest (see Figure 29.4c). Mature bushes deplete the surrounding soil of water and secrete toxic chemicals, making it impossible for seedlings to grow. This chemical warfare is called *allelopathy*. Moreover, seed-eating ants and rodents living at the bases of mature bushes eat any seeds that fall nearby. In these situations, the distributions of species of plants and animals can be uniform and interrelated. Territorial behaviour, the defence of an area and its resources, can also produce **uniform dispersion** in some species of animals, such as nests in colonies of colonial birds (see Chapter 47).

Random dispersion (see Figure 29.4b) occurs when environmental conditions do not vary much within a habitat, and individuals are neither attracted to nor repelled by others of their species (conspecifics). Ecologists use formal statistical definitions of *random* to establish a theoretical baseline for assessing the pattern of distribution. In cases of random dispersion, individuals are distributed unpredictably. Some spiders, burrowing clams, and rainforest trees exhibit random dispersion.

Black-Footed Ferret, *Mustela nigripes*

Black-footed ferrets **(Figure 1)** are crepuscular and nocturnal hunters of the prairie. Weighing 0.6 to 1.1 kg, these weasel relatives (family Mustelidae, order Carnivora) were once abundant in western North America, from Texas in the United States to Saskatchewan and Alberta in Canada. Like other mustelids, males are larger than females. In the wild, these predators probably fed mainly on prairie dogs (*Cynomys* species) and lived around prairie dog towns. Litters range in size from one to five. Females bear a single litter a year, and males and females are sexually mature at age one year.

By 1987, *M. nigripes* was probably extinct in the wild. The last known wild population was discovered near Meeteetse, Wyoming, in 1981. Seven animals from this population were captured and brought into captivity and served as the genetic founders for a captive breeding program. Over 4800 juvenile black-footed ferrets were produced by this program, and wildlife officials began to release captive-bred animals into suitable habitats.

At Shirley Basin, Wyoming, 228 captive-born black-footed ferrets were received between 1991 and 1994. By 1996, only 25 were observed in the wild and by 1997, only 5. This decline reflected the impact of diseases, specifically canine distemper and plague. In 1996, it seemed that the reintroductions would fail, and *M. nigripes* would again be extinct in the wild.

In 2003, however, 52 black-footed ferrets were observed in the field at Shirley Basin, and, since then, the population has increased significantly **(Figure 2)**. The increase reflects an intrinsic rate of increase (*r*) of 0.47, which reflects success in the first year of life and derives from a combination of survival and fertility.

There appears to be hope for the future of *M. nigripes*. It remains to be determined if the genetic bottleneck (see Chapter 18) that the species has endured will prove to be an important handicap to long-term survival.

Figure 1

Mustela nigripes, *the black-footed ferret. This critically endangered carnivore from the North American prairie shows evidence of a comeback.*

John E Marriott/All Canada Photos/Getty Images

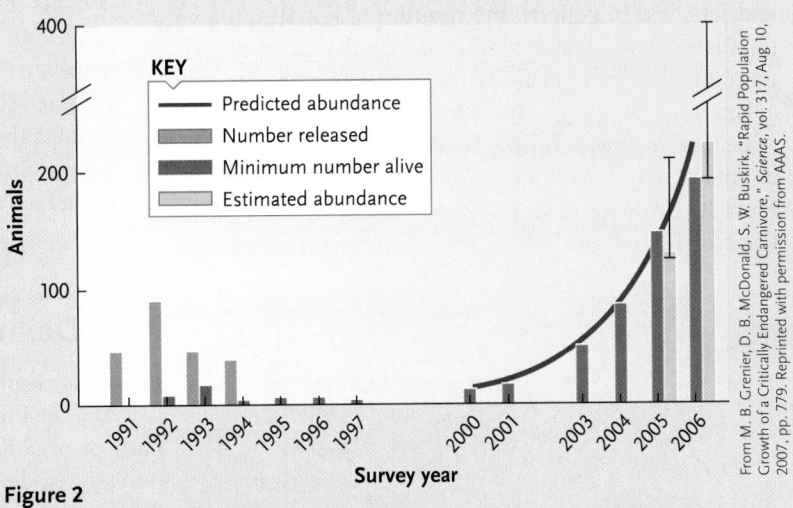

KEY
- Predicted abundance
- Number released
- Minimum number alive
- Estimated abundance

From M. B. Grenier, D. B. McDonald, S. W. Buskirk, "Rapid Population Growth of a Critically Endangered Carnivore," *Science*, vol. 317, Aug 10, 2007, pp. 779. Reprinted with permission from AAAS.

Figure 2

Population growth. Black-footed ferrets in Shirley Basin, Wyoming, have shown rapid population growth. The 95% confidence limits suggest a population of 192 to 401 in 2006.

Whether the spatial distribution of a population appears to be clumped, uniform, or random depends partly on the size of the organisms and of the study area. Oak seedlings may be randomly dispersed on a spatial scale of a few square metres, but over an entire mixed hardwood forest, they are clumped under the parent trees. Therefore, dispersion of a population depends partly on the researcher's scale of observation.

In addition, the dispersion of animal populations often varies through time in response to natural environmental rhythms. Few habitats provide a constant supply of resources throughout the year, and many animals move from one habitat to another on a seasonal cycle, reflecting the distribution of resources such as food. Tropical birds and mammals are often widely dispersed in deciduous forests during the wet season, when food is widely available. During the dry season, these species crowd into narrow *gallery forests* along watercourses, where evergreen trees provide food and shelter.

29.2d Age Structure Is the Numbers of Individuals of Different Ages

All populations have an **age structure**, a statistical description of the relative numbers of individuals in each age class (discussed further in Chapter 30). Individuals can be categorized roughly as prereproductive (younger than the age of sexual maturity), reproductive, or postreproductive (older than the maximum age of reproduction). A population's age structure reflects its recent growth history and predicts its future growth potential. Populations composed of many prereproductive individuals obviously grew rapidly in the recent past. These populations will continue to grow as young individuals mature and reproduce.

29.2e Generation Time Is the Average Time between Birth and Death

Another characteristic that influences a population's growth is its **generation time**, the average time between the birth of an organism and the birth of its offspring. Generation time is usually short in species that reach sexual maturity at a small body size **(Figure 29.5)**. Their populations often grow rapidly because of the speedy accumulation of reproductive individuals.

29.2f Sex Ratio: Females : Males

Populations of sexually reproducing organisms also vary in their **sex ratio**, the relative proportions of males and females. In general, the number of females in a population has a bigger impact on population growth than the number of males because only females actually produce offspring. Moreover, in many species, one male can mate with several females, and the number of males may have little effect on the population's reproductive output. In northern elephant seals (see Chapter 18), mature bulls fight for dominance on the beaches where the seals mate. Only a few males may ultimately inseminate a hundred or more females. Thus, the presence of other males in the group may have little effect on the size of future generations. In animals that form lifelong pair bonds, such as geese and swans, the numbers of males and females influence reproduction in the population.

29.2g The Proportion Reproducing Is the Incidence of Reproducing Individuals in a Population

Population ecologists try to determine the proportion of individuals in a population that are reproducing. This issue is particularly relevant to the conservation of any species in which individuals are rare or widely dispersed in the habitat (see Chapter 32).

STUDY BREAK

1. What is the difference between geographic range and habitat?
2. What are the three types of dispersion? What is the most common pattern found in nature? Why?
3. What is the common pattern of generation time among bacteria, protists, plants, and animals?

29.3 Demography

Populations grow larger through the birth of individuals and the **immigration** (movement into the population) of organisms from neighbouring populations. Conversely, death and **emigration** (movement out of the population) reduce population size. **Demography** is the statistical study of the processes that change a population's size and density through time.

Ecologists use demographic analysis to predict a population's growth. For human populations, these data help governments anticipate the need for social services such as schools, hospitals, and chronic care facilities. Demographic data allow conservation ecologists to develop plans to protect endangered species. Demographic data on northern spotted owls (*Strix occidentalis caurina*) helped convince the courts to restrict logging in the owl's primary habitat, the old-growth forests of the Pacific Northwest. Life tables and survivorship curves are among the tools ecologists use to analyze demographic data.

29.3a Life Tables Show the Number of Individuals in Each Age Group

Although every species has a characteristic life span, few individuals survive to the maximum age possible. Mortality results from starvation, disease, accidents,

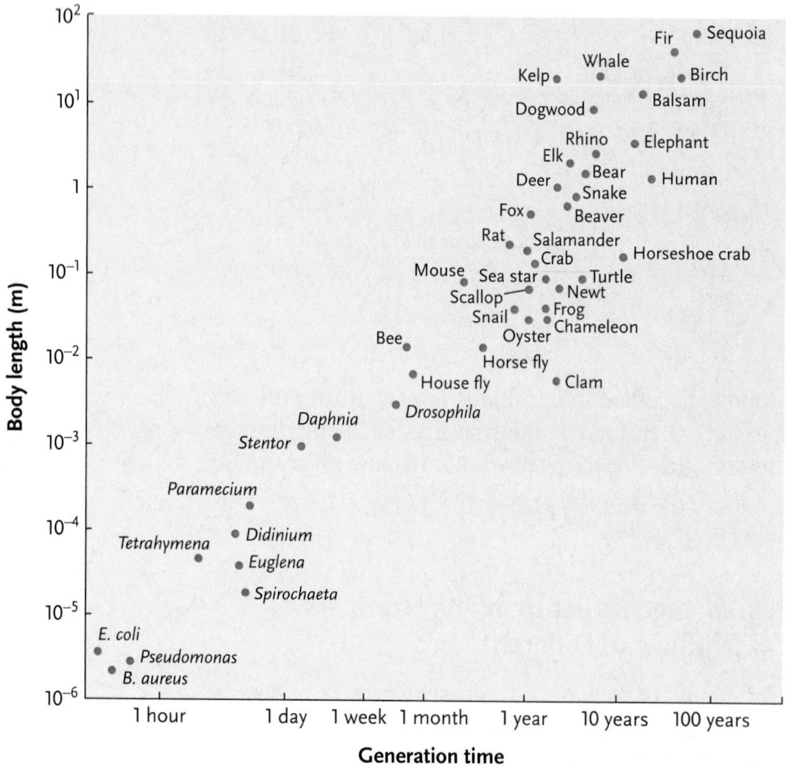

Figure 29.5

Generation time and body size. Generation time increases with body size among bacteria, protists, plants, and animals. The logarithmic scale on both axes compresses the data into a straight line.

predation, or inability to find a suitable habitat. Life insurance companies first developed techniques for measuring mortality rates (known as actuarial science), and ecologists adapted these approaches to the study of nonhuman populations.

A **life table** summarizes the demographic characteristics of a population **(Table 29.1)**. To collect life table data for short-lived organisms, demographers typically mark a **cohort**, a group of individuals of similar age, at birth and monitor their survival until all members of the cohort die. For organisms that live more than a few years, a researcher might sample the population for one or two years, recording the ages at which individuals die and then extrapolating these results over the species' life span. The approach to the timing of collection of data about reproduction and longevity will depend on the details of the species under study.

In any life table, life spans of organisms are divided into age intervals of appropriate length. For short-lived species, days, weeks, or months are useful, whereas for longer-lived species, years or groups of years will be better. Mortality can be expressed in two complementary ways. **Age-specific mortality** is the proportion of individuals alive at the start of an age interval that died during that age interval. Its more cheerful reflection, **age-specific survivorship**, is the proportion of individuals alive at the start of an age interval that survived until the start of the next age interval. Thus, for the data shown in Table 29.1, the age-specific mortality rate during the 3- to 6-month age interval is 195/722 = 0.270, and the age-specific survivorship rate is 527/722 = 0.730. For any age interval, the sum of age-specific mortality and age-specific survivorship must equal 1. Life tables also summarize the proportion of the cohort that survived to a particular age, a statistic identifying the probability that any randomly selected newborn will still be alive at that age. For the 3- to 6-month age interval in Table 29.1, this probability is 722/843 = 0.856.

Life tables also include data on **age-specific fecundity**, the average number of offspring produced by surviving females during each age interval. Table 29.1 shows that plants in the 3- to 6-month age interval produced an average of 300 seeds each. In some species, including humans, fecundity is highest in individuals of intermediate age. Younger individuals have not yet reached sexual maturity, and older individuals are past their reproductive prime. However, fecundity increases steadily with age in some plants and animals.

29.3b Survivorship Curves Graph the Timing of Deaths of Individuals in a Population

Survivorship data are depicted graphically in a **survivorship curve**, which displays the rate of survival for individuals over the species' average life span. Ecologists have identified three generalized survivorship curves (blue lines in **Figure 29.6, p. 686**), although most organisms exhibit survivorship patterns falling between these idealized patterns.

Type I curves reflect high survivorship until late in life (see Figure 29.6a, p. 686). They are typical of large animals that produce few young and provide them with extended care, which reduces juvenile mortality. Large mammals, such as Dall mountain sheep, produce only one or two offspring at a time and nurture them through their vulnerable first year. At that time, the young are better able to fend for themselves and are at

Table 29.1	Life Table for a Cohort of 843 Individuals of the Grass *Poa annua* (Annual Bluegrass)					
Age Interval (in months)	Number Alive at Start of Age Interval	Number Dying during Age Interval	Age-Specific Mortality Rate	Age-Specific Survivorship Rate	Proportion of Original Cohort Alive at Start of Age Interval	Age-Specific Fecundity (Seed Production)
0–3	843	121	0.144	0.856	1.000	0
3–6	722	195	0.270	0.730	0.856	300
6–9	527	211	0.400	0.600	0.625	620
9–12	316	172	0.544	0.456	0.375	430
12–15	144	90	0.625	0.375	0.171	210
15–18	54	39	0.722	0.278	0.064	60
18–21	15	12	0.800	0.200	0.018	30
21–24	3	3	1.000	0.000	0.004	10
24–	0	—	—	—	—	—

Source: *Population Ecology*, Begon, M., and M. Mortimer. Copyright © 1981 John Wiley and Sons. Reproduced with permission of Blackwell Publishing Ltd.

a. Dall mountain sheep (*Ovis dalli*)

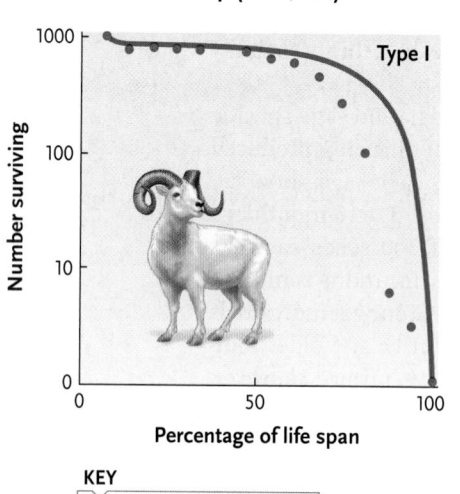

b. Five-lined skink (*Eumeces fasciatus*)

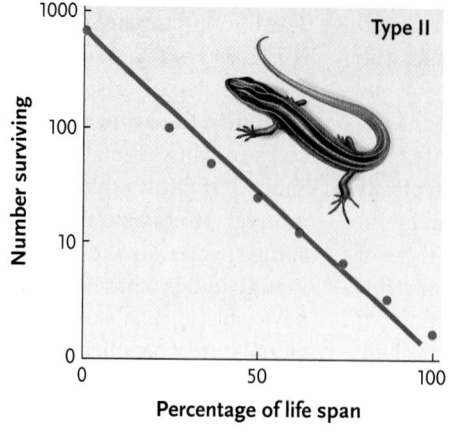

c. Perennial desert shrub (*Cleome droserifolia*)

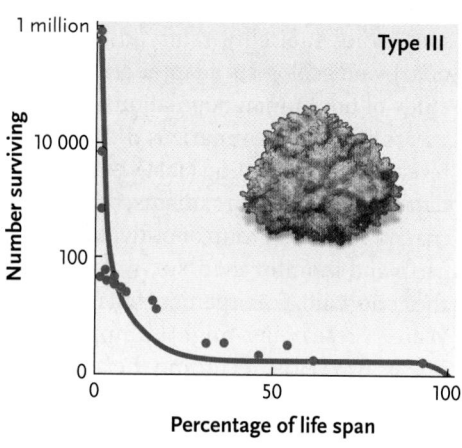

KEY

— Theoretical   • Data

Figure 29.6

Survivorship curves. The survivorship curves of many organisms (pink) roughly match one of three idealized patterns (blue).

lower risk for mortality (compared with younger animals). The picture of survivorship in mammals could change if one starts with the time of conception, as opposed to birth. The change would reflect problems of pregnancy (see Chapter 41) and health of mothers.

Type II curves reflect a relatively constant rate of mortality in all age classes, a pattern that produces steadily declining survivorship (see Figure 29.6b). Many lizards, such as the five-lined skink, as well as songbirds and small mammals, face a constant probability of mortality from predation, disease, and starvation and show a type II pattern.

Type III curves reflect high juvenile mortality, followed by a period of low mortality once offspring reach a critical age and size (see Figure 29.6c, in which the vertical scale is logarithmic). *Cleome droserifolia*, a desert shrub from the Middle East, experiences extraordinarily high mortality in its seed and seedling stages. Researchers estimate that for every 1 million seeds produced, fewer than 1000 germinate, and only about 40 individuals survive their first year. Once a plant becomes established, however, its likelihood of future survival is higher, and the survivorship curve flattens out. Many plants, insects, marine invertebrates, and fishes exhibit type III survivorship.

STUDY BREAK

1. What is the relationship between age-specific mortality and age-specific survivorship? If the age-specific mortality is 0.384, what is the age-specific survivorship?
2. What is age-specific fecundity?
3. Describe three survivorship curves. Which curve describes humans? Songbirds? Insects?

29.4 The Evolution of Life Histories

Analysis of life tables reveals how natural selection affects an organism's **life history**, which includes the lifetime patterns of growth, maturation, and reproduction. Ecologists study life histories to understand trade-offs in the allocation of resources to these three activities. The results of their research suggest that natural selection adjusts the allocation of resources to maximize an individual's number of surviving offspring.

Every organism is constrained by a finite **energy budget**, the total amount of energy it can accumulate and use to fuel its activities. An organism's energy budget is like a savings account. When the individual accumulates more energy than it needs, it makes deposits to this account, storing energy as starch, glycogen, or fat. When the individual expends more energy than it harvests, it makes withdrawals from its energy stores. But unlike a bank account, an organism's energy budget cannot be overdrawn, and no loans against future "earnings" are possible.

Just as humans find clever ways to finance their schemes, many organisms use different ways to mortgage their operations. Organisms that enter states of inactivity or dormancy can maximize the time over which they use stored energy. An extreme example is animals and plants that can survive freezing, an obvious strategy for conserving energy. Hibernation and estivation in animals are other examples (see Chapter 50). Hibernating animals use periods of reduced body temperature to survive prolonged periods of cold weather. Estivation is inactivity during prolonged periods of high temperatures. Specialized spores can be resistant to heat and desiccation. Migrating birds on long flights get energy by metabolizing fat as well as other body structures, such as muscle or digestive tissue. Organisms use the energy they harvest for three broadly defined functions: maintenance (the preservation of good physiological condition), growth, and reproduction. When an organism devotes energy to any one of these functions, the balance in its energy budget is reduced, leaving less energy for other functions.

A fish, a deciduous tree, and a mammal illustrate the dramatic variations existing in life history patterns. Larval coho salmon (*Oncorhynchus kisutch*) hatch in the headwaters of a stream, where they feed and grow for about a year before assuming their adult body form and swimming to the ocean. They remain at sea for a year or two, feeding voraciously and growing rapidly. Eventually, using a Sun compass and geomagnetic and chemical cues, salmon return to the rivers and streams where they hatched. The fishes swim upstream. Males prepare nests and try to attract females. Each female lays hundreds or thousands of relatively small eggs. After breeding, the body condition of males and females deteriorates, and they die.

Most deciduous trees in the temperate zone, such as oaks (genus *Quercus*), begin their lives as seeds (acorns) in late summer. The seeds remain metabolically inactive until the following spring or a later year. After germinating, seedling trees collect nutrients and energy and continue to grow throughout their lives. Once they achieve a critical size, they may produce thousands of acorns annually for many years. Thus, growth and reproduction occur simultaneously through much of the trees' life.

European red deer (*Cervus elaphus*) are born in spring, and the young remain with their mothers for an extended period, nursing and growing rapidly. After weaning, the young feed on their own. Female red deer begin to breed after reaching adult size in their third year, producing one or two offspring annually until they are about 16 years old, when they reach their maximum life span and die.

How can we summarize the similarities and differences in the life histories of these organisms? All three species harvest energy throughout their lives. Salmon and deciduous trees continue to grow until old age, whereas deer reach adult size fairly early in life. Salmon produce many offspring in a single reproductive episode, whereas deciduous trees and deer reproduce repeatedly. However, most trees produce thousands of seeds annually, whereas deer produce only one or two young each spring.

What factors have produced these variations in life history patterns? Life history traits, like all population characteristics, are modified by natural selection. Thus, organisms exhibit evolutionary adaptations that increase the fitness of individuals. Each species' life history is, in fact, a highly integrated "strategy" or suite of selection-driven adaptations.

In analyzing life histories, ecologists compare the number of offspring with the amount of care provided to each by the parents. They also determine the number of reproductive episodes in the organism's lifetime and the timing of first reproduction. Because these characteristics evolve together, a change in one trait is likely to influence others.

29.4a Fecundity versus Parental Care: Cutting Your Losses

If a female has a fixed amount of energy for reproduction, she can package that energy in various ways. A female duck with 1000 units of energy for reproduction might lay 10 eggs with 100 units of energy per egg. A salmon, which has higher fecundity, might lay 1000 eggs with 1 unit of energy in each. The amount of energy invested in each offspring before it is born is **passive parental care** provided by the female. Passive parental care is provided through yolk in an egg; endosperm in a seed; or, in mammals, nutrients that cross the placenta.

Many animals also provide **active parental care** to offspring after their birth. In general, species producing many offspring in a reproductive episode (e.g., the coho salmon) provide relatively little active parental care to each offspring. In fact, female coho salmon, each producing 2400 to 4500 eggs, die before their eggs even hatch. Conversely, species producing few offspring at a time (e.g., European red deer) provide much more care to each one. A red deer doe nurses its single fawn for up to eight months before weaning it.

29.4b How Often to Breed: Once or Repeatedly?

The number of reproductive episodes in an organism's life span is a second life history characteristic adjusted by natural selection. Some organisms, such as coho salmon, devote all of their stored energy to a single reproductive event. Any adult that survives the upstream migration is likely to leave some surviving offspring. Other species, such as deciduous trees and red deer, reproduce more than once. In contrast to salmon, individuals of these species devote only some of their energy budget to reproduction at any time, with the balance allocated to maintenance and growth. Moreover, in some plants, invertebrates, fishes, and reptiles, larger individuals produce more offspring than smaller ones. Thus, one advantage of using only part of the energy budget for reproduction is that continued growth may result in greater fecundity at a later age. However, if an organism does not survive until the next breeding season, the potential advantage of putting energy into maintenance and growth is lost.

29.4c Age at First Reproduction: When to Start Reproducing

Individuals that first reproduce at the earliest possible age may stand a good chance of leaving some surviving offspring. But the energy they use in reproduction is not available for maintenance and growth. Thus, early reproducers may be smaller and less healthy than individuals that delay reproduction in favour of other functions. Conversely, an individual that delays reproduction may increase its chance of survival and its future fecundity by becoming larger or more experienced. But there is always some chance that it will die before the next breeding season, leaving no offspring at all. Therefore, a finite energy budget and the risk of mortality establish a tradeoff in the timing of first reproduction. Mathematical models suggest that delayed reproduction will be favoured by natural selection if a sexually mature individual has a good chance of surviving to an older age, if organisms grow larger as they age, and

if larger organisms have higher fecundity. Early reproduction will be favoured if adult survival rates are low, if animals do not grow larger as they age, or if larger size does not increase fecundity. These characteristics apply more readily to some animals and plants than they do to others. Among animals, the features discussed above apply more readily to vertebrate than to invertebrate animals. Parasitic organisms may have quite different patterns of life history.

Life history characteristics vary from one species to another, and they can vary among populations of a single species. Predation differentially influences life history characteristics in natural populations of guppies (*Poecilia reticulata*) in Trinidad (see "Life Histories of Guppies").

Life Histories of Guppies

Some years ago, drenched with sweat and with fishnets in hand, two ecologists were engaged in fieldwork on the Caribbean island of Trinidad. They were after guppies (*Poecilia reticulata*), small fish most of us see in pet shops. In their native habitats, guppies bear live young in shallow mountain streams **(Figure 1),** and John Endler and David Reznick were studying the environmental variables influencing the evolution of their life history patterns.

Male guppies are easy to distinguish from females. Males stop growing at sexual maturity. They are smaller, and their scales have bright colours that serve as visual signals in intricate courtship displays. Females are drably coloured and continue to grow larger throughout their lives. In

the mountains of Trinidad, guppies live in different streams, even in different parts of the same stream. Two other species of fish eat guppies **(Figure 2).** In some streams, a small killifish (*Rivulus hartii*) preys on immature guppies but does not have much success with the larger adults. In other streams, a large pike–cichlid (*Crenicichla alta*) prefers mature guppies and rarely hunts small, immature ones.

Reznick and Endler found that the life history patterns of guppies vary among streams with different predators. In streams with pike–cichlids, male and female guppies mature faster and begin to reproduce at a smaller size and younger age than their counterparts in streams where killifish live. Female guppies from pike–cichlid streams reproduce more often, producing smaller and more numerous young. These differences allow guppies to avoid some predation. Those in pike–cichlid streams

begin to reproduce when they are smaller than the size preferred by that predator. Those from killifish streams grow quickly to a size that is too large to be consumed by killifish **(Figure 3).**

Although these life history differences were correlated with the

Male guppy (right) that shared a stream with pike–cichlids (below)

Male guppy (right) that shared a stream with killifish (below)

David Reznick/University of California, Riverside; Mark Smith/Science Source

David Reznick/University of California, Riverside; DEA/C. DANI/Contributor/ Getty Images

Figure 2

Male guppies from streams where pike–cichlids live (top) are smaller and more streamlined and have duller colours than those from streams where killifish live (bottom). The pike–cichlid prefers to eat large guppies, and the killifish feeds on small guppies. Guppies are shown approximately life sized; adult pike–cichlids grow to 16 cm in length, and adult killifish grow to 10 cm.

David Reznick/University of California, Riverside

Figure 1

David Reznick surveys a shallow stream in the mountains of Trinidad.

distributions of the two predatory fishes, they might result from some other, unknown differences between the streams. Endler and Reznick investigated this possibility with controlled laboratory experiments. They shipped groups of live guppies to California, where they bred guppies from each kind of stream for two generations. Both types of experimental populations were raised under identical conditions in the absence of predators. Even in the absence of predators, the two types of experimental populations retained their life history differences. These results provided evidence of a genetic (heritable) basis for the observed life history differences.

Endler and Reznick also examined the role of predators in the *evolution* of the size differences **(Figure 4)**. They raised guppies for many generations in the laboratory under three experimental conditions: some alone, some with killifish, and some with pike–cichlids. As predicted, the guppy lineage subjected to predation by killifish became larger at maturity. Individuals that were small at maturity were frequently eaten, and their reproduction was limited. The lineage raised with pike–cichlids showed a trend toward earlier maturity. Individuals that matured at a larger size faced a greater likelihood of being eaten before they had reproduced.

When they first visited Trinidad, Endler and Reznick had introduced guppies from a pike–cichlid stream to another stream that contained killifish but no pike–cichlids or guppies. There, 11 years later, guppy populations had changed. As the researchers predicted, the guppies became larger and reproduced more slowly, characteristics typical of natural guppy populations that live and die with killifish.

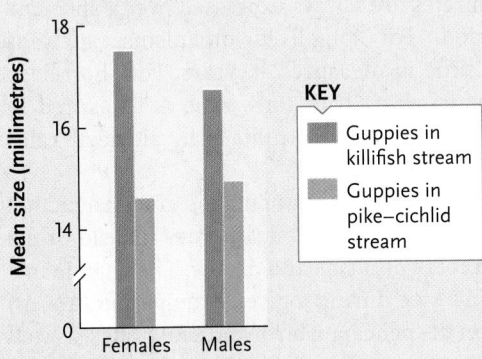

Figure 3
Guppies in streams occupied by pike–cichlids are smaller than those in streams occupied by killifish.

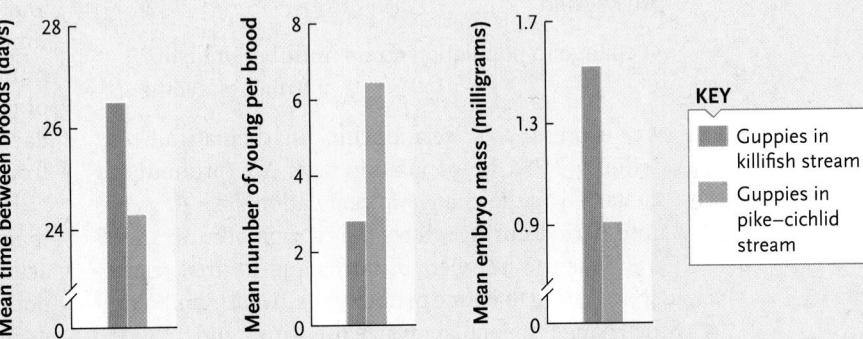

Figure 4
Female guppies from streams occupied by pike–cichlids reproduce more often (shorter time between broods) and produce more young per brood and smaller young (lower embryo mass) than females living in streams occupied by killifish.

STUDY BREAK

1. Organisms use energy for what three main operations?
2. Explain passive and active parental care in humans.
3. When would early reproduction be favoured?

29.5 Models of Population Growth

We now examine two mathematical models of population growth, exponential and logistic. **Exponential** models apply when populations experience unlimited growth. **Logistic** models apply when population growth is limited, often because available resources are finite. These simple models are tools that help ecologists refine their hypotheses, but neither provides entirely accurate predictions of population growth in nature. In the simplest versions of these models, ecologists define births as the production of offspring by any form of reproduction and ignore the effects of immigration and emigration.

29.5a Exponential Models: Populations Taking Off

Populations sometimes increase in size for a period of time with no apparent limits on their growth. In models of exponential growth, population size increases steadily by a constant ratio. Populations of bacteria and prokaryotes provide the most obvious examples, but multicellular organisms also sometimes exhibit exponential population growth.

Bacteria reproduce by binary fission. A parent cell divides in half, producing two daughter cells, and each can divide to produce two granddaughter cells. Generation time in a bacterial population is simply the time between successive cell divisions. If no bacteria in the population die, the population doubles in size each generation.

Bacterial populations grow quickly under ideal temperatures and with unlimited space and food. Consider a population of the human intestinal bacterium *Escherichia coli,* for which the generation time can be as short as 20 minutes. If we start with a population of one bacterium, the population doubles to two cells after one generation, to four cells after two generations, and to eight cells after three generations **(Figure 29.7)**. After only 8 hours (24 generations), the population will number almost 17 million. And after a single day (72 generations), the population will number nearly 5×10^{21} cells. Although other bacteria grow more slowly than *E. coli,* it is no wonder that pathogenic bacteria, such as those causing cholera or plague, can quickly overtake the defences of an infected animal.

When populations of multicellular organisms are large, they can grow exponentially, as we shall see below for our own species. In any event, over a given time period,

change in population size = number of births − number of deaths.

We express this relationship mathematically by defining N as the population size; ΔN (pronounced "delta N") as the change in population size; Δt as the time period during which the change occurs; and B and D as the numbers of births and deaths, respectively, during that time period. Thus, $\Delta N/\Delta t$ symbolizes the change in population size over time, and

$$\Delta N/\Delta t = B - D.$$

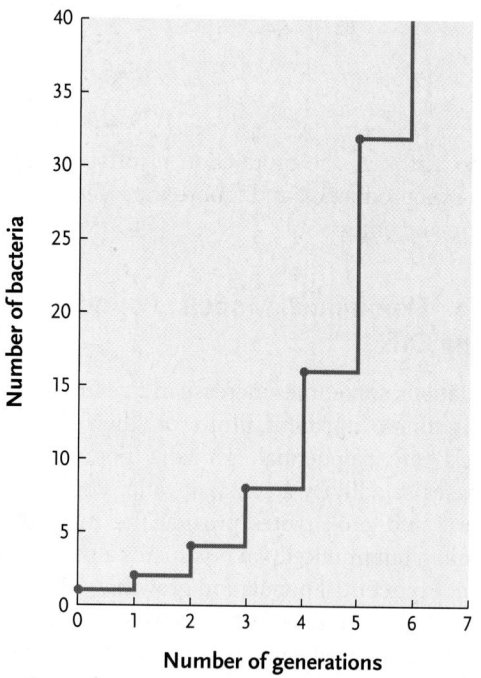

Figure 29.7

Bacterial population growth. If all members of a bacterial population divide simultaneously, a plot of population size over time forms a stair-stepped curve in which the steps get larger as the number of dividing cells increases.

The above equation applies to any population for which we know the exact numbers of births and deaths. Ecologists usually express births and deaths as per capita (per individual) rates, allowing them to apply the model to a population of any size. The per capita birth rate (b) is the number of births in the population during the specified time period divided by the population size: $b = (B/N)$. Similarly, the per capita death rate, d, is the number of deaths divided by the population size: $d = (D/N)$.

If in a population of 2000 field mice, 1000 mice are born and 200 mice die during 1 month, $b = 1000/2000 = 0.5$ births per individual per month, and $d = 200/2000 = 0.1$ deaths per individual per month. Of course, no mouse can give birth to half an offspring, and no individual can die one-tenth of a death. But these rates tell us the per capita birth and death rates *averaged over all mice in the population.* Per capita birth and death rates are always expressed over a specified time period. For long-lived organisms, such as humans, time is measured in years. For short-lived organisms, such as fruit flies, time is measured in days. We can calculate per capita birth and death rates from data in a life table.

Now we can revise the population growth equation to use per capita birth and death rates instead of the actual numbers of births and deaths. The change in a population's size during a given time period ($\Delta N/\Delta t$) depends on the per capita birth and death rates, as well as on the number of individuals in the population. Mathematically, we can write

$$\Delta N/\Delta t = B - D = bN - dN = (b - d)N$$

or, in the notation of calculus,

$$dN/dt = (b - d)N.$$

This equation describes the **exponential model of population growth.** (Note that in calculus, dN/dt is the notation for the population growth rate. The d in dN/dt is *not* the same d we use to symbolize the per capita death rate.)

The difference between the per capita birth rate and the per capita death rate, $b - d$, is the **per capita growth rate** of the population, symbolized by r. Like b and d, r is always expressed per individual per unit time. Using the per capita growth rate, r, in place of $b - d$, the exponential growth equation is written

$$dN/dt = rN.$$

If the birth rate exceeds the death rate, r has a positive value ($r > 0$), and the population is growing. In our example with field mice, r is $0.5 - 0.1 = 0.4$ mice per mouse per month. If, on the other hand, the birth rate is lower than the death rate, r has a negative value ($r < 0$), and the population is shrinking. In populations in which the birth rate equals the death rate, r is zero, and the population's size is not changing-a situation known as **zero population growth**, or ZPG. Even under ZPG, births and deaths still occur, but the numbers of births and deaths cancel each other out.

Populations will grow as long as the per capita growth rate is positive ($r > 0$). In our hypothetical population of field mice, we started with $N = 2000$ mice and calculated a per capita growth rate of 0.4 mice per individual per month. In the first month, the population grows by $0.4 \times 2000 = 800$ mice **(Figure 29.8)**. At the start of the second month, $N = 2800$ and r is still 0.4. Thus, in the second month, the population grows by $0.4 \times 2800 = 1120$ mice. Notice that even though r remains constant, the *increase* in population size grows each month because more individuals are reproducing. In less than two years, the mouse population will increase to more than one million! A graph of exponential population growth has a characteristic J shape, getting steeper through time. The population grows at an ever-increasing pace because the change in a population's size depends on the number of individuals in the population and its per capita growth rate.

Imagine a hypothetical population living in an ideal environment with unlimited food and shelter; no predators, parasites, or disease; and a comfortable abiotic environment. Under such circumstances (admittedly unrealistic), the per capita birth rate is very high; the per capita death rate is very low; and the per capita growth rate, r, is as high as it can be. This maximum per capita growth rate, symbolized r_{max}, is the population's **intrinsic rate of increase.** Under these ideal conditions, our exponential growth equation is

$$dN/dt = r_{max}N.$$

When populations grow at their intrinsic rate of increase, population size increases very rapidly. Across a wide variety of protists and animals, r_{max} varies inversely with generation time: species with a short generation time have higher intrinsic rates of increase than those with a long generation time **(Figure 29.9, p. 692)**.

The exponential model predicts unlimited population growth. But we know from even casual observations that population sizes of most species are somehow limited. We are not knee-deep in bacteria, rosebushes, or garter snakes. What factors limit the growth of populations? As a population gets larger, it uses more vital resources, perhaps leading to a shortage of resources. In this situation, individuals may have less energy available for maintenance and reproduction, causing decreases in per capita birth rates and increases in per capita death rates. Energy in food is not always equally available, and when an animal spends time handling food to eat it, the ratio of cost (handling) to benefit (energy in the food) diminishes, affecting return on investment. Such rate changes can affect a population's per capita growth rate, causing population growth to slow or stop.

29.5b Logistic Models: Populations and Carrying Capacity (K)

Environments provide enough resources to sustain only a finite population of any species. The maximum number of individuals that an environment can support indefinitely is termed its **carrying capacity,** symbolized as K. K is defined for each population. It is a property of the environment that can vary from one habitat to another and in a single habitat over time. The spring and summer flush of insects in temperate habitats supports large populations of insectivorous birds. But fewer insects are available in autumn and winter, causing a seasonal decline in K for birds, so autumnal migrations occur in birds seeking more food and less inclement weather. Other cycles are annual, such as variation in water levels in wetlands from year to year.

Month	Old Population Size		Net Monthly Increase		New Population Size
1	2 000	+	800	=	2 800
2	2 800	+	1 120	=	3 920
3	3 920	+	1 568	=	5 488
4	5 488	+	2 195	=	7 683
5	7 683	+	3 073	=	10 756
6	10 756	+	4 302	=	15 058
7	15 058	+	6 023	=	21 081
8	21 081	+	8 432	=	29 513
9	29 513	+	11 805	=	41 318
10	41 318	+	16 527	=	57 845
11	57 845	+	23 138	=	80 983
12	80 983	+	32 393	=	113 376
13	113 376	+	45 350	=	158 726
14	158 726	+	63 490	=	222 216
15	222 216	+	88 887	=	311 102
16	311 102	+	124 441	=	435 543
17	435 543	+	174 217	=	609 760
18	609 760	+	243 904	=	853 664
19	853 674	+	341 466	=	1 195 1340

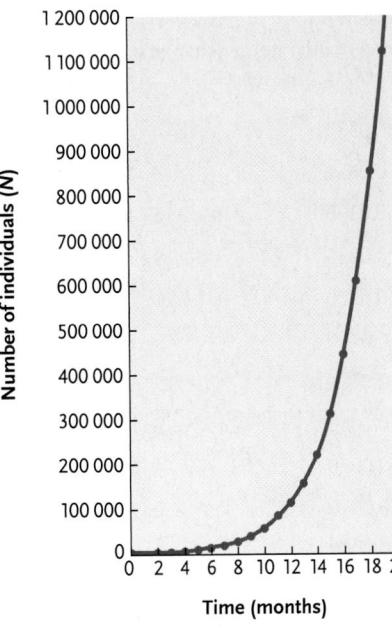

Figure 29.8
Exponential population growth. Exponential population growth produces a J-shaped curve when population size is plotted against time. Although the per capita growth rate (r) remains constant, the increase in population size gets larger every month because more individuals are reproducing.

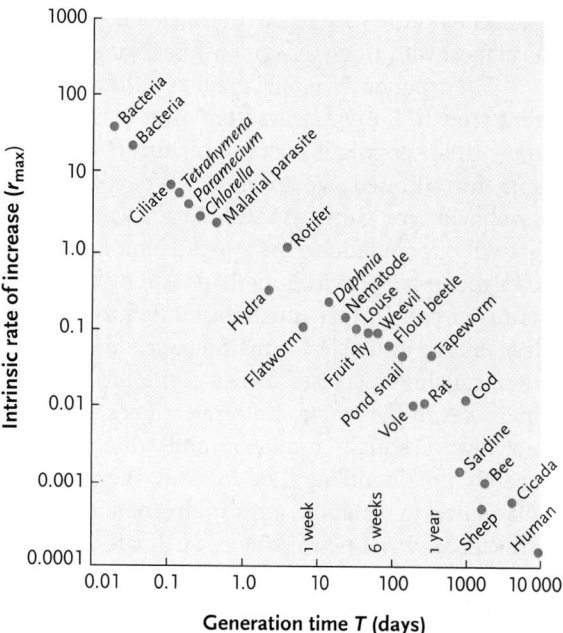

Figure 29.9

Generation time and r_{max}. The intrinsic rate of increase (r_{max}) is high for protists and animals with short generation times and low for those with long generation times.

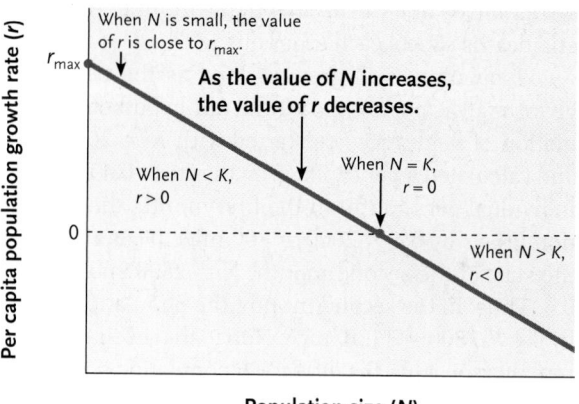

a. The predicted effect of N on r

When N is small, the value of r is close to r_{max}.

As the value of N increases, the value of r decreases.

When $N < K$, $r > 0$

When $N = K$, $r = 0$

When $N > K$, $r < 0$

Population size (N)

Per capita population growth rate (r)

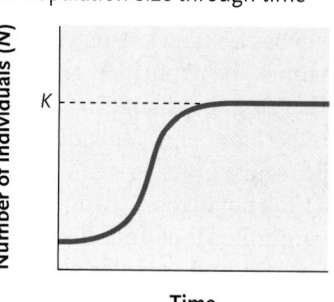

b. Population size through time

Number of individuals (N)

Time

Figure 29.10

The logistic model of population growth. The logistic model **(a)** assumes that the per capita population growth rate (r) decreases linearly as population size (N) increases. The logistic model also predicts that population size **(b)** increases quickly at first but then slowly approaches carrying capacity (K).

The **logistic model of population growth** assumes that a population's per capita growth rate, r, decreases as the population gets larger **(Figure 29.10)**. In other words, population growth slows as the population size approaches K. The mathematical expression $K - N$ tells us how many individuals can be added to a population before it reaches K. The expression $(K - N)/K$ indicates what percentage of the carrying capacity is still available.

To create the logistic model, we factor the impact of K into the exponential model by multiplying r_{max} by $(K - N)/K$ to reduce the per capita growth rate (r) from its maximum value (r_{max}) as N increases:

$$dN/dt = r_{max}N(K - N)/K.$$

The calculation of how r varies with population size is straightforward **(Table 29.2)**. In a very small population (N much smaller than K), plenty of resources are available; the value of $(K - N)/K$ is close to 1. Here the per capita growth rate (r) approaches the maximum possible (r_{max}). Under these conditions, population growth is close to exponential. If a population is large (N close to K), few additional resources are available. Now the value of $(K - N)/K$ is small, and the per capita growth rate (r) is very low. When the size of the population exactly equals K, $(K - N)/K$ becomes 0, as does the population growth rate, the situation defined as ZPG.

The logistic model of population growth predicts an S-shaped graph of population size over time, with the population slowly approaching K and remaining at that level **(Figure 29.11)**. According to this model, the population grows slowly when the population size is small because few individuals are reproducing. It also

Table 29.2	The Effect of N on r and ΔN^* in a Hypothetical Population Exhibiting Logistic Growth in which K equals 2000 and r_{max} is 0.04 per capita per year		
N (population size)	$(K - N)/K$ (% of K available)	$r = r_{max}(K - N/K)$ (per capita growth rate)	$\Delta N = rN$ (change in N)
50	0.975	0.0390	2
100	0.950	0.0380	4
250	0.875	0.0350	9
500	0.750	0.0300	15
750	0.625	0.0250	19
1000	0.500	0.0200	20
1250	0.375	0.0150	19
1500	0.250	0.0100	15
1750	0.125	0.0050	9
1900	0.050	0.0020	4
1950	0.025	0.0010	2
2000	0.000	0.0000	0

*ΔN rounded to the nearest whole number.

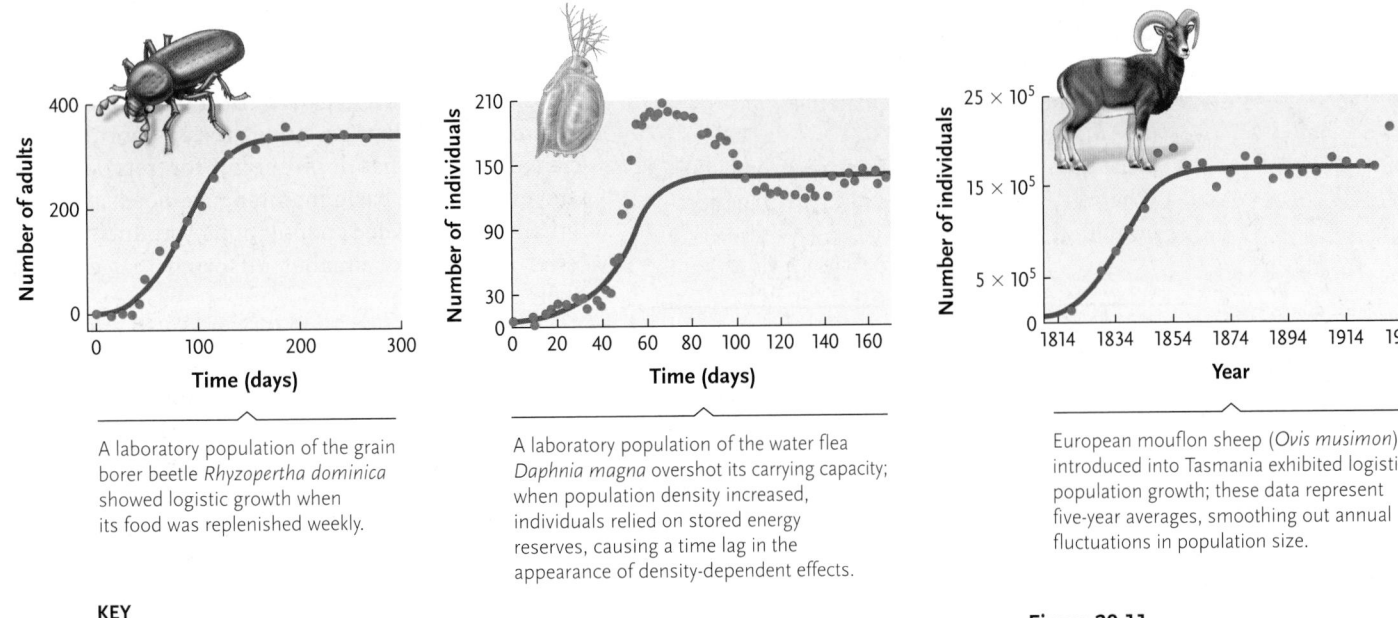

A laboratory population of the grain borer beetle *Rhyzopertha dominica* showed logistic growth when its food was replenished weekly.

A laboratory population of the water flea *Daphnia magna* overshot its carrying capacity; when population density increased, individuals relied on stored energy reserves, causing a time lag in the appearance of density-dependent effects.

European mouflon sheep (*Ovis musimon*) introduced into Tasmania exhibited logistic population growth; these data represent five-year averages, smoothing out annual fluctuations in population size.

KEY

— Theoretical   • Data

Figure 29.11
Examples of logistic population growth.

grows slowly when the population size is large because the per capita population growth rate is low. The population grows quickly (dN/dt is highest) at intermediate population sizes, when a sizable number of individuals are breeding and the per capita population growth rate (r) is still fairly high (see Table 29.2).

The logistic model assumes that vital resources become increasingly limited as a population grows. Thus, the model is a mathematical portrait of **intraspecific** (within species) **competition**, the dependence of two or more individuals in a population on the same limiting resource. For mobile animals, limiting resources could be food, water, nesting sites, and refuges from predators. For sessile species, space can be a limiting resource. For plants, sunlight, water, inorganic nutrients, and growing space can be limiting. The pattern of uniform dispersion described earlier often reflects intraspecific competition for limited resources.

In some very dense populations, accumulation of poisonous waste products may reduce survivorship and reproduction. Most natural populations live in open systems where wastes are consumed by other organisms or flushed away. But the build-up of toxic wastes is common in laboratory cultures of microorganisms. For example, yeast cells ferment sugar and produce ethanol as a waste product. Thus, the alcohol content of wine usually does not exceed 13% by volume, the ethanol concentration that poisons yeasts that are vital to the wine-making process.

How well do species conform to the predictions of the logistic model? In simple laboratory cultures, relatively small organisms, such as *Paramecium,* some crustaceans, and flour beetles, often show an S-shaped pattern of population growth (Figure 29.11 left,

middle). Moreover, large animals introduced into new environments sometimes exhibit a pattern of population growth that matches the predictions of the logistic model (Figure 29.11 right).

Nevertheless, some assumptions of the logistic model are unrealistic. For example, the model predicts that survivorship and fecundity respond immediately to changes in a population's density. Many organisms exhibit a delayed response (a **time lag**) because fecundity has been determined by resource availability at some time in the past. This may reflect conditions that prevailed when individuals were adding yolk to eggs or endosperm to seeds. Moreover, when food resources become scarce, individuals may use stored energy reserves to survive and reproduce. This delays the impact of crowding until stored reserves are depleted and means that population size may overshoot K (see Figure 29.11 middle). Deaths may then outnumber births, causing the population size to drop below K, at least temporarily. Time lags often cause a population to oscillate around K.

The assumption that the addition of new individuals to a population always decreases survivorship and fecundity is unrealistic. In small populations, modest population growth may not have much impact on survivorship and fecundity. In fact, most organisms probably require a minimum population density to survive and reproduce. Some plants flourish in small clumps that buffer them from physical stresses, whereas a single individual living in the open would suffer adverse effects. In some animal populations, a minimum population density is necessary for individuals to find mates. Determining the minimum viable population for a species is an important issue in conservation biology (see Chapter 32).

1. When do you use an exponential model rather than a logistic one?
2. Define the terms in the equation $dN/dt = (b - d)N$.
3. What does it mean when $r < 0$, $r > 0$, or $r = 0$? What is r_{max}, and how does it vary with generation time?

29.6 Population Regulation

What environmental factors influence population growth rates and control fluctuations in population size? Some factors affecting population size are **density dependent** because their influence increases or decreases with the density of the population. Intraspecific competition and predation are examples of density-dependent environmental factors. The logistic model includes the effects of density dependence in its assumption that per capita birth and death rates change with population density.

Numerous laboratory and field studies have shown that crowding (high population density) decreases individual growth rate, adult size, and survival of plants and animals **(Figure 29.12)**. Organisms living in extremely dense populations are unable to harvest enough resources. They grow slowly and tend to be small, weak, and less likely to survive.

Gardeners understand this relationship and thin their plants to achieve a density that maximizes the number of vigorous individuals that survive to be harvested.

Crowding has a negative effect on reproduction **(Figure 29.13)**. When resources are in short supply, each individual has less energy for reproduction after meeting its basic maintenance needs. Hence, females in crowded populations produce either fewer offspring or smaller offspring that are less likely to survive.

In some species, crowding stimulates developmental and behavioural changes that may influence the density of a population. Migratory locusts can develop into either solitary or migratory forms in the same population. Migratory individuals have longer wings and more body fat, characteristics that allow long-distance dispersal. High population density increases the frequency of the migratory form. Thus, many locusts move away from the area of high density **(Figure 29.14)**, reducing the size and thus the density of the original population.

Although these data about locusts confirm the assumptions of the logistic equation, they do not prove that natural populations are regulated by density-dependent factors. Experimental evidence is necessary to provide a convincing demonstration that an increase in population density causes population size to decrease, whereas a decrease in density causes it to increase.

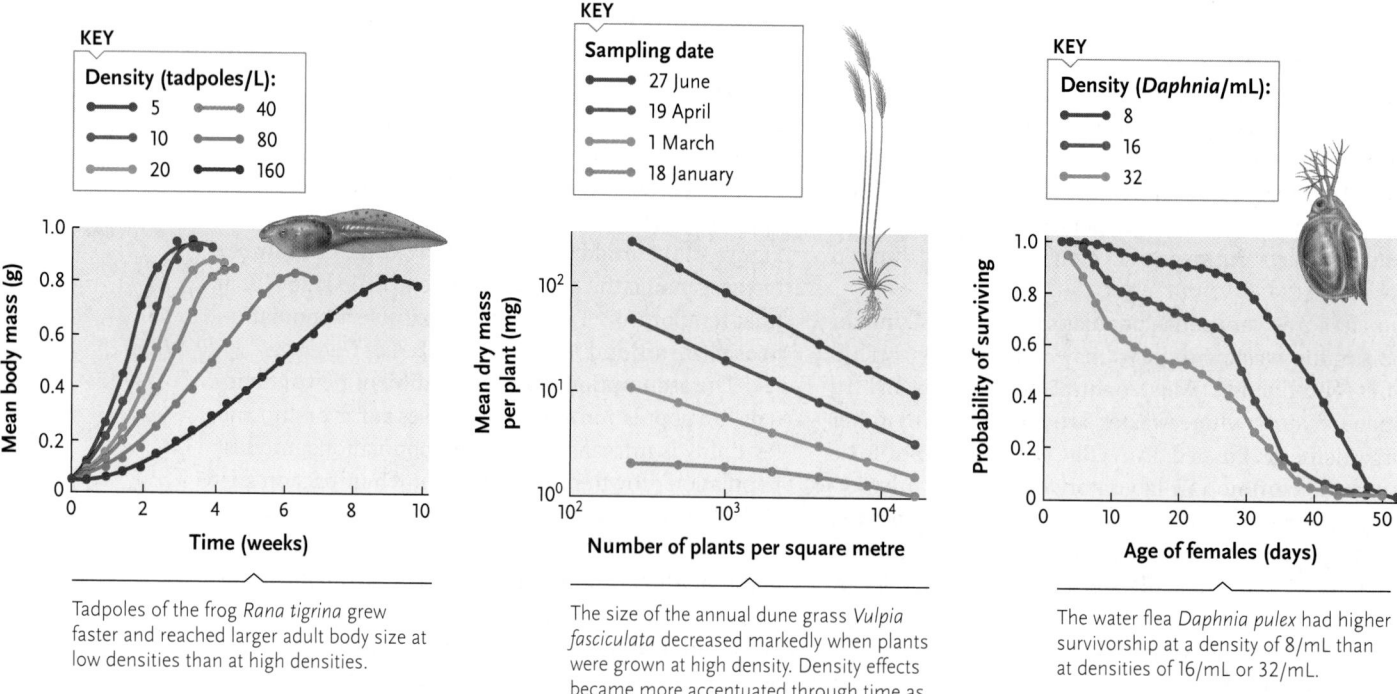

Tadpoles of the frog *Rana tigrina* grew faster and reached larger adult body size at low densities than at high densities.

The size of the annual dune grass *Vulpia fasciculata* decreased markedly when plants were grown at high density. Density effects became more accentuated through time as the plants grew larger (indicated by the progressively steeper slopes of the lines).

The water flea *Daphnia pulex* had higher survivorship at a density of 8/mL than at densities of 16/mL or 32/mL.

Figure 29.12
Effects of crowding on individual growth, size, and survival.

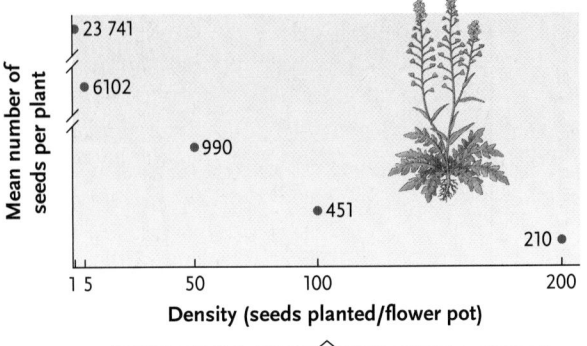

The number of seeds produced by shepherd's purse (*Capsella bursa-pastoris*) decreased dramatically with increasing density in experimental plots.

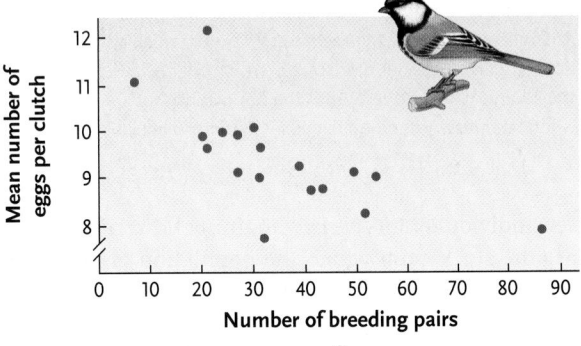

The mean number of eggs produced by the Great Tit (*Parus major*), a woodland bird, declined as the number of breeding pairs in Marley Wood increased.

Figure 29.13
Effects of crowding on fecundity.

Figure 29.14
A swarm of locusts. Migratory locusts (*Locusta migratoria*) moving across an African landscape can devour their own weight in plant material every day.

In the 1960s, Robert Eisenberg experimentally increased the numbers of aquatic snails (*Lymnaea elodes*) in some ponds, decreased them in others, and maintained natural densities in control ponds. Adult survivorship did not differ between experimental and control treatments. But there was a gradient in egg production from few eggs (snails in high-density ponds), to more (control density), to most (low density). Furthermore, survival rates of young snails declined as density increased. After four months, densities in the two experimental groups converged on those in the control, providing strong evidence of density-dependent population regulation.

At this stage, intraspecific competition appears to be the primary density-dependent factor regulating population size. Competition between populations of different species can also exert density-dependent effects on population growth (see Chapter 30). The Allee effect occurs when r begins to decline after N falls below some threshold. This is another example of a density-dependent regulator.

But predation can also cause density-dependent population regulation. As a particular prey species becomes more numerous, predators may consume more of it because it is easier to find and catch. Once a prey species exceeds some threshold density, predators may consume a larger percentage of its population, a density-dependent effect. On rocky shores in California, sea stars feed mainly on the most abundant invertebrate there. When one prey species becomes common, predators feed on it disproportionately, drastically reducing its numbers. Then they switch to now more abundant alternative prey.

Sometimes several density-dependent factors influence a population at the same time. On small islands in the West Indies, spiders are rare wherever lizards (*Ameiva festiva, Anolis carolinensis,* and *Anolis sagrei*) are abundant but common where the lizards are rare or absent. To test whether the presence of lizards limits the abundance of spiders, David Spiller and Tom Schoener built fences around plots on islands where these species occur. They eliminated lizards from experimental plots but left them in control plots. After two years, spider populations in some experimental plots were five times as dense as those in control plots, suggesting a strong impact of lizard populations on spider populations **(Figure 29.15, p. 696)**. In this situation, lizards had two density-dependent effects on spider populations. First, lizards ate spiders, and, second, they competed with them for food. Experimental evidence made it possible for biologists to better understand the situation.

Predation, parasitism, and disease can cause density-dependent regulation of plant and animal populations. Infectious microorganisms (e.g., rabies) spread quickly in a crowded population. In addition, if crowded individuals are weak or malnourished, they are more susceptible to infection and may die from diseases that healthy organisms would survive. Effects on survival can be direct or indirect.

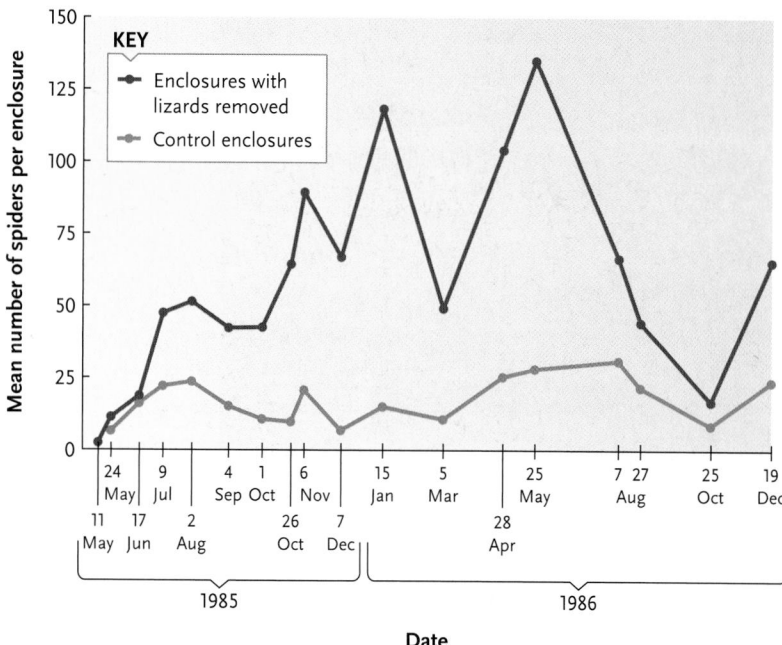

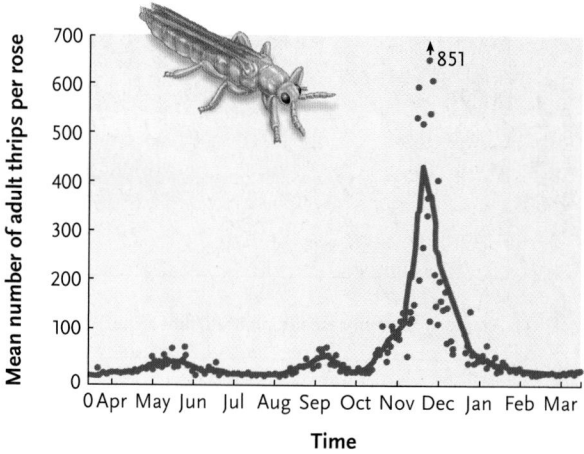

Figure 29.16

Booms and busts in a thrips population. Populations of the Australian insect *Thrips imaginis* grow exponentially when conditions are favourable during spring (which begins in September in the southern hemisphere). But the populations crash in summer when hot and dry conditions cause high mortality.

Experimental Research Figure 29.15 Populations of spiders (*Metepeira daytona*) on a small island in the Bahamas are influenced by the presence of lizards. Note how much higher the population densities of spiders are in the absence than in the presence of lizards.

29.6a Density-Independent Factors: Reducing Population in Spite of Density

Some populations are affected by **density-independent** factors that reduce population size regardless of its density. If an insect population is not physiologically adapted to high temperature, a sudden hot spell may kill 80% of them whether they number 100 or 100 000. Fires, earthquakes, storms, and other natural disturbances can contribute directly or indirectly to density-independent mortality. Because such factors do not cause a population to fluctuate around its *K*, these density-independent factors can reduce but do not regulate population size.

Density-independent factors have a particularly strong effect on populations of small-bodied species that cannot buffer themselves against environmental change. Their populations grow exponentially for a time, but shifts in climate or random events cause high mortality before populations reach a size at which density-dependent factors would regulate their numbers. When conditions improve, populations grow exponentially, at least until another density-independent factor causes them to crash again. A small Australian insect, a thrip (*Thrips imaginis*), eats the pollen and flowers of plants in the rose family. These thrips can be abundant enough to damage blooms. Populations of thrips grow exponentially in spring, when many flowers are available and the weather is warm and moist **(Figure 29.16).** But their populations crash predictably during summer because thrips do not tolerate hot and dry conditions. After the crash, a

few individuals survive in remaining flowers, and they are the stock from which the population grows exponentially the following spring.

29.6b Interactions between Density-Dependent and Density-Independent Factors: Sometimes Population Density Affects Mortality

Density-dependent factors can interact with density-independent factors and limit population growth. Food shortage caused by high population density (a density-dependent factor) may lead to malnourishment. Malnourished individuals may be more likely to succumb to the stress of extreme weather (a density-independent factor).

Populations can be affected by density-independent factors in a density-dependent manner. Some animals retreat into shelters to escape environmental stresses, such as floods or severe heat. If a population is small, most individuals can be accommodated in available refuges. But if a population is large (exceeds the capacity of shelters), only a proportion will find suitable shelter. The larger the population, the greater the percentage of individuals exposed to the stress(es). Thus, although the density-independent effects of weather limit populations of thrips, the availability of flowers in summer (a density-dependent factor) regulates the size of the starting populations of thrips the following spring. Hence, both density-dependent and density-independent factors influence the size of populations of thrips.

Other explanations focus on extrinsic control, such as the relationship between a cycling species and its food or predators. A dense population may exhaust its food supply, increasing mortality and decreasing reproduction. The die-off of large numbers of African elephants in Tsavo National Park in Kenya is an example of the impact of overpopulation. There elephants overgrazed vegetation in most of the park habitat. In 1970, the combination of overgrazing and a drought caused high mortality of elephants. The picture is not always clear, because experimental food supplementation does not always prevent decline in mammal populations. This suggests some level of intrinsic control.

29.6c Life History Characteristics: Evolution of Strategies for Population Growth

Even casual observation reveals tremendous variation in how rapidly population sizes change in different species. New weeds often appear in a vegetable garden overnight, whereas the number of oak trees in a forest may remain relatively stable for years. Why do only some species have the potential for explosive population growth? The answer lies in how natural selection has moulded life history strategies adapted to different ecological conditions. Some ecologists recognize two quite different life history patterns, **r-selected** species and **K-selected** species **(Table 29.3; Figure 29.17).**

On the face of it, r-selected species are adapted to rapidly changing environments, and many have at least some of the features outlined in Table 29.3. The success of an r-selected life history depends on flooding the environment with a *large quantity* of young because only some may be successful. Small body size means that compared with larger-bodied species, r-selected species lack physiological mechanisms to buffer them from environmental variation. Populations of r-selected species can be so reduced by changes in abiotic environmental factors (e.g., temperature or moisture) that they never grow large enough to reach K and face a shortage of limiting resources. In these cases, K cannot be estimated by researchers, and changes in population size are not accurately described by the logistic model of population growth. Although r-selected species appear to have poor tolerance of environmental change, they are said to be adapted to rapidly changing environments.

At the same time, K-selected species have at least some of the features outlined for them in Table 29.3. These organisms survive the early stages of life (type I or type II survivorship), and a low r_{max} means that their populations grow slowly. The success of a K-selected life history is linked to the production of a relatively small number of high-quality offspring that join an already well-established population. Generalizations about r-selected and K-selected species are misleading. We can recognize this by comparing two species of mammals.

Peromyscus maniculatus, deer mice, occur widely in North America. In southern Ontario, adults weigh 12 to 31 g, females produce average litters of four (range two to eight), and each can have four or five

litters a year. Females become sexually mature at age two months and breed in their first year. Occasionally, deer mice live to age three years in the wild. Throughout their extensive range in North America, *Myotis lucifugus*, little brown bats, weigh 7 to 12 g; females bear a single young per litter and have one litter per year. Females may breed a year after they are born, but many wait until they are two years old. In the wild, little brown bats can live over 30 years. Using these data, one small mammal (deer mouse) is an r-strategist, whereas another (little brown bat) is a K-strategist. To complicate matters, deer mice living in Kananaskis in the mountains near Calgary mature at one year and may have two litters per year, typically five young per litter. Compared to little brown bats, Kananaskis deer mice are r-strategists. Compared to Ontario deer mice, they are more like K-strategists.

Table 29.3 Characteristics of *r*-Selected and *K*-Selected Species

Characteristic	r-Selected Species	K-Selected Species
Maturation time	Short	Long
Life span	Short	Long
Mortality rate	Usually high	Usually low
Reproductive episodes	Usually one	Usually several
Time of first reproduction	Early	Late
Clutch or brood size	Usually large	Usually small
Size of offspring	Small	Large
Active parental care	Little or none	Often extensive
Population size	Fluctuating	Relatively stable
Tolerance of environmental change	Generally poor	Generally good

a. An r-selected species

© Nigel Cattlin/Alamy

b. A K-selected species

Photos.com

Figure 29.17

Life history differences. An r-selected species, **(a)** *Chenopodium quinoa*, matures in one growing season and produces many tiny seeds. Quinoa was a traditional food staple for indigenous people of North and South America. A K-selected species, **(b)** *Cocos nucifera*, a coconut palm, grows slowly and produces a few large seeds repeatedly during its long life.

Biologists may find the idea of *r*-strategists and *K*-strategists useful, but too often the idea means imposing some human view of the world on a natural system. *K*-strategists and *r*-strategists may be more like beauty, defined by the eye of the beholder. Elephants (*Loxodonta africana*, *Loxodonta cyclotis*, *Elephas maximus*) are big and meet all *K*-strategist criteria. Many insects are small but in all other respects meet the criteria considered typical of *K*-strategists because of their patterns of reproduction. Codfish (*Gadus morhua*) are big (compared to insects or bats) but meet most of the criteria used to identify *r*-strategists, such as their patterns of reproduction.

29.6d Population Cycles: Ups and Downs in Numbers of Individuals

Population densities of many insects, birds, and mammals in the northern hemisphere fluctuate between species-specific lows and highs in a multiyear cycle. Arctic populations of small rodents (*Lemmus lemmus*) vary in size over a 4-year cycle, whereas snowshoe hares (*Lepus americanus*), ruffed grouse (*Bonasa umbellus*), and lynx have 10-year cycles. Ecologists

documented these cyclic fluctuations more than a century ago, but none of the general hypotheses proposed to date explain cycles in all species. Availability and quality of food, abundance of predators, prevalence of disease-causing microorganisms, and variations in weather can influence population growth and declines. Furthermore, food supply and predators for a cycling population are themselves influenced by a population's size.

Theories of intrinsic control suggest that as an animal population grows, individuals undergo hormonal changes that increase aggressiveness, reduce reproduction, and foster dispersal. The dispersal phase of the cycle may be dramatic. When populations of Norway lemming (*Lemmus lemmus*), a rodent that lives in the Scandinavian Arctic, reach their peak density, aggressive interactions drive younger and weaker individuals to disperse. The dispersal of many thousands of lemmings during periods of population growth has sometimes been incorrectly portrayed in nature films as a suicidal mass migration.

Cycles in populations of predators could be induced by time lags between populations of predators and prey and vice versa **(Figure 29.18)**. The 10-year

Figure 29.18

The predator–prey model. Predator–prey interactions may contribute to density-dependent regulation of both populations. A mathematical model **(a)** predicts cycles in the numbers of predators and prey because of time lags in each species' responses to changes in the density of the other. (Predator population size is exaggerated in this graph.) **(b)** Canada lynx (*Lynx canadensis*) and snowshoe hare (*Lepus americanus*) were often described as a typical cyclic predator–prey interaction. The abundances of lynx (red line) and snowshoe hare (blue line) are based on counts of pelts trappers sold to the Hudson's Bay Company over a 90-year period. Recent research shows that population cycles in snowshoe hares are caused by complex interactions between the snowshoe hares, its food plants, and its predators.

Ed Cesar/Science Source

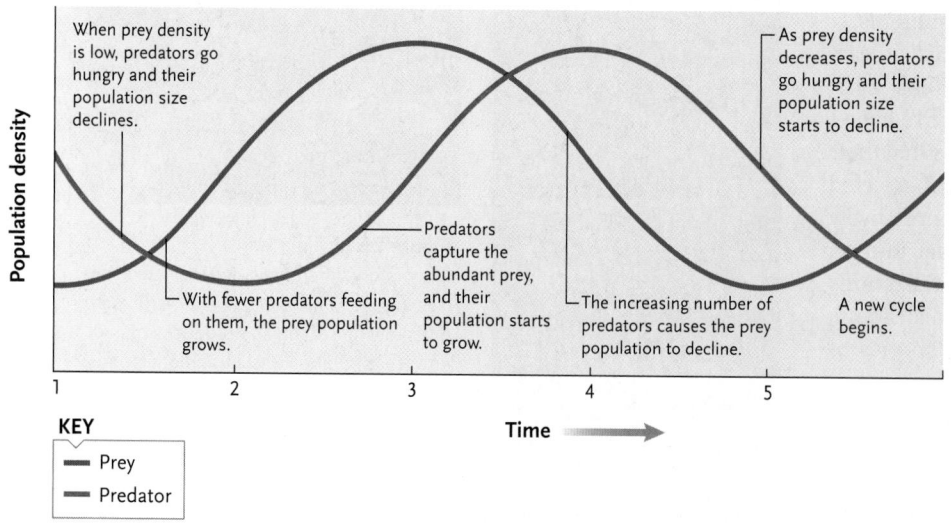

a. Predictions of a predator–prey model

When prey density is low, predators go hungry and their population size declines.

As prey density decreases, predators go hungry and their population size starts to decline.

With fewer predators feeding on them, the prey population grows.

Predators capture the abundant prey, and their population starts to grow.

The increasing number of predators causes the prey population to decline.

A new cycle begins.

Population density

Time

KEY
— Prey
— Predator

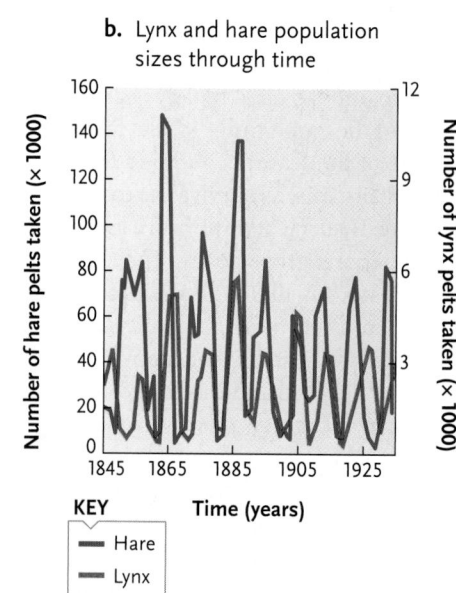

b. Lynx and hare population sizes through time

Number of hare pelts taken (× 1000)

Number of lynx pelts taken (× 1000)

Time (years)

KEY
— Hare
— Lynx

cycles of snowshoe hares and their feline predators, Canada lynx, were often cited as a classic example of such an interaction. But snowshoe hare populations can exhibit a 10-year fluctuation even on islands where lynx are absent. Thus, lynx are not solely responsible for population cycles in snowshoe hares. To further complicate matters, the database demonstrating fluctuations was often the numbers of pelts purchased by the Hudson's Bay Company. Here, fur price influenced the trapping effort and the numbers of animals harvested. This economic reality brought into question the relationship between the numbers of pelts and actual population densities of lynx and snowshoe hares.

Charles Krebs and his colleagues studied hare and lynx interactions with a large-scale, multiyear experiment in Kluane in the southern Yukon. Using fenced experimental areas, they could add food for snowshoe hares, exclude mammalian predators, or apply both experimental treatments while monitoring unmanipulated control plots. When mammalian predators were excluded, densities of snowshoe hares approximately doubled relative to controls. Where food was added, densities of snowshoe hares tripled relative to controls. In plots where food was added and predators were excluded, densities of snowshoe hares increased 11-fold compared with controls. Krebs and his colleagues concluded that neither food availability nor predation is solely responsible for population cycles in snowshoe hares. They postulated that complex interactions between snowshoe hares, their food plants, and their predators generate cyclic fluctuations in populations of snowshoe hares.

STUDY BREAK

1. What are density-dependent factors? Why do dense populations tend to decrease in size?
2. Define density-independent factors, and give some examples.
3. Describe two key differences between *r*-selected species and *K*-selected species.

29.7 Human Population Growth

How do human populations compare with those of other species? The worldwide human population was over 7 billion in 2014. Like many other species, humans live in somewhat isolated populations that vary in their demographic traits and access to resources. Although many of us live comfortably, at least a billion people are

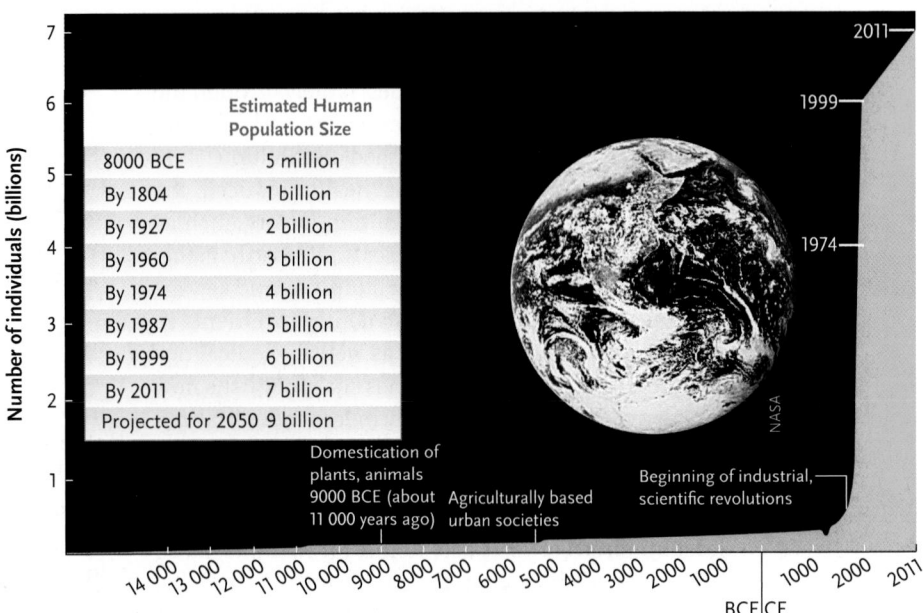

Figure 29.19

Human population growth. The worldwide human population grew slowly until 200 years ago, when it began to increase explosively. The dip in the mid-fourteenth century represents the death of 60 million Asians and Europeans from the bubonic plague. The table shows how long it took for the human population to add each billion people.

malnourished or starving, lack access to clean drinking water, and live without adequate shelter or health care.

For most of human history, our population grew slowly, reflecting the impact of a range of restraints. Over the past two centuries, the worldwide human population has grown exponentially **(Figure 29.19)**. Demographers identified three ways in which we have avoided the effects of density-dependent regulating factors.

First, humans have expanded their geographic range into virtually every terrestrial habitat, alleviating competition for space. Our early ancestors lived in tropical and subtropical grasslands, but by 40 thousand years ago, they had dispersed through much of the world. Their success resulted from their ability to solve ecological problems by building fires, assembling shelters, making clothing and tools, planning community hunts, and sharing information. Vital survival skills spread from generation to generation and from one population to another because language allowed communication of complex ideas and knowledge.

Second, we have increased *K* in habitats we occupy, isolating us, as a species, from restrictions associated with access to resources. This change began to occur about 11 thousand years ago, when populations in different parts of the world began to shift from hunting and gathering to agriculture (see Chapter 33). At that time, our ancestors cultivated wild grasses and other plants, diverted water to irrigate crops, and used domesticated animals for food and labour. Innovations such as these increased the availability of food, raising both *K* and rates of population growth. In the mid-eighteenth century, people harnessed the energy in fossil fuels, and industrialization began in western Europe and North America. Food supplies and *K* increased again, at least in industrialized countries, largely through the use of synthetic fertilizers, pesticides, and efficient methods of transportation and food distribution.

Third, advances in public health reduced the effects of critical population-limiting factors such as malnutrition, contagious diseases, and poor hygiene. Over the past 300 years, modern plumbing and sewage treatment, removal of garbage, and improvements in food handling and processing, as well as medical discoveries, have reduced death rates sharply. Births now greatly exceed deaths, especially in less industrialized countries, resulting in rapid population growth. Note, however, that problems of hygiene and access to fresh water and food had been solved in some societies at least hundreds of years ago. Rome, for example, had a population of about 1 million people by 2 CE, and this was supported by an excellent infrastructure for importing and distributing food, providing fresh water, and dealing with human wastes.

29.7a Age Structure and Economic Growth: Phases of Development

Where have our migrations and technological developments taken us? It took about 2.5 million years for the human population to reach 1 billion, 80 years to reach the second billion, and only 12 years to jump from 5 billion to 6 billion and another 12 years to reach 7 billion (see the inset table in Figure 29.19, p. 699). Rapid population growth now appears to be an inevitable consequence of our demographic structure and economic development.

29.7b Population Growth and Age Structure: Not All Populations Are the Same

In 2011, the worldwide annual growth rate for the human population averaged about 1.15% (*r* = 0.0115 new individuals per individual per year). Population experts expect that rate to decline, but even so, the human population will probably exceed 9 billion before 2050.

In 2000, population growth rates of individual nations varied widely, ranging from much less than 1% to more than 3% **(Figure 29.20a).** Industrialized countries of Western Europe have achieved nearly ZPG, but other countries, particularly those in Africa, Latin America, and Asia, will experience huge increases over the next 20 or 25 years (Figure 29.20b).

For all long-lived species, differences in age structure are a major determinant of differences in population growth rates **(Figure 29.21).** There are three basic patterns in the graphs in Figure 29.21. In the first, in countries with ZPG, there are approximately equal numbers of people of reproductive and prereproductive ages. The ZPG situation is exacerbated when reproductives have very few offspring, meaning that prereproductives may not even replace themselves in the population. Second, in countries with negative growth (without immigration), postreproductives outnumber reproductives, and these populations will not experience a growth spurt when today's children reach reproductive age. Third are countries with rapid growth, where reproductives vastly outnumber postreproductives.

Countries with rapid growth have a broad-based age structure (pattern three, above), with many youngsters

a. Mean annual population growth rates

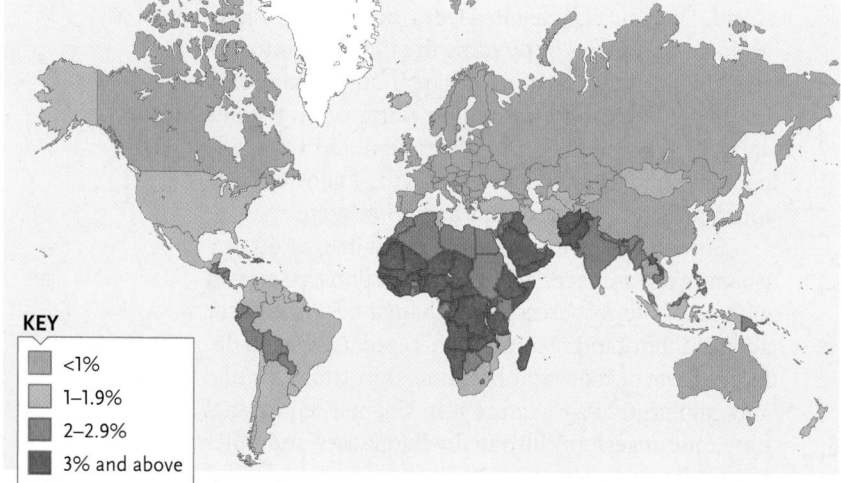

KEY
<1%
1–1.9%
2–2.9%
3% and above

b. Projected population sizes for 2025

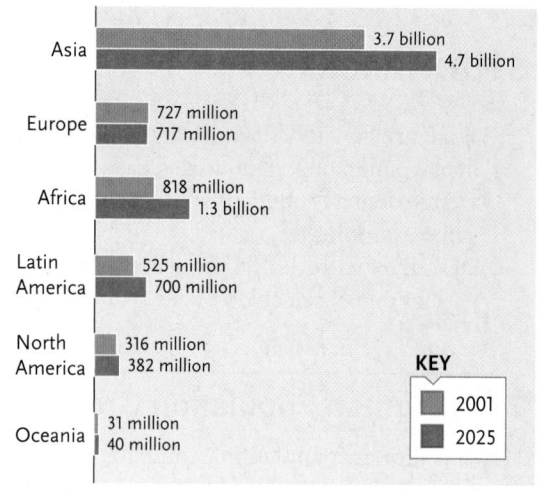

Asia — 3.7 billion / 4.7 billion
Europe — 727 million / 717 million
Africa — 818 million / 1.3 billion
Latin America — 525 million / 700 million
North America — 316 million / 382 million
Oceania — 31 million / 40 million

KEY
2001
2025

Figure 29.20

Local variation in human population growth rates. In 2001, **(a)** average annual population growth rates varied among countries and continents. In some regions **(b),** the population is projected to increase greatly by 2025 (red) compared with the population size in 2001 (orange). The population of Europe is likely to decline.

a. Hypothetical age distributions for populations with different growth rates

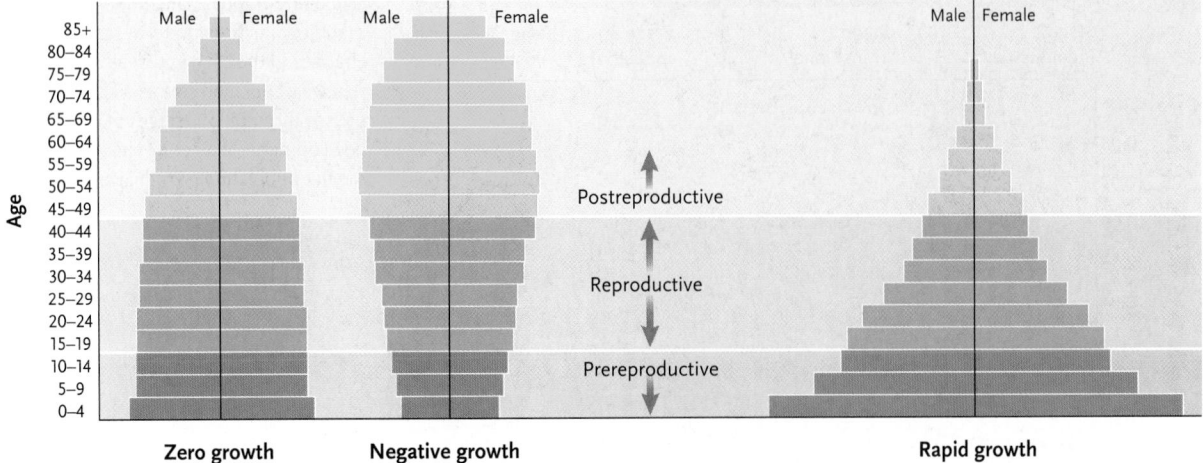

b. Age pyramids for the United States and Mexico in 2000

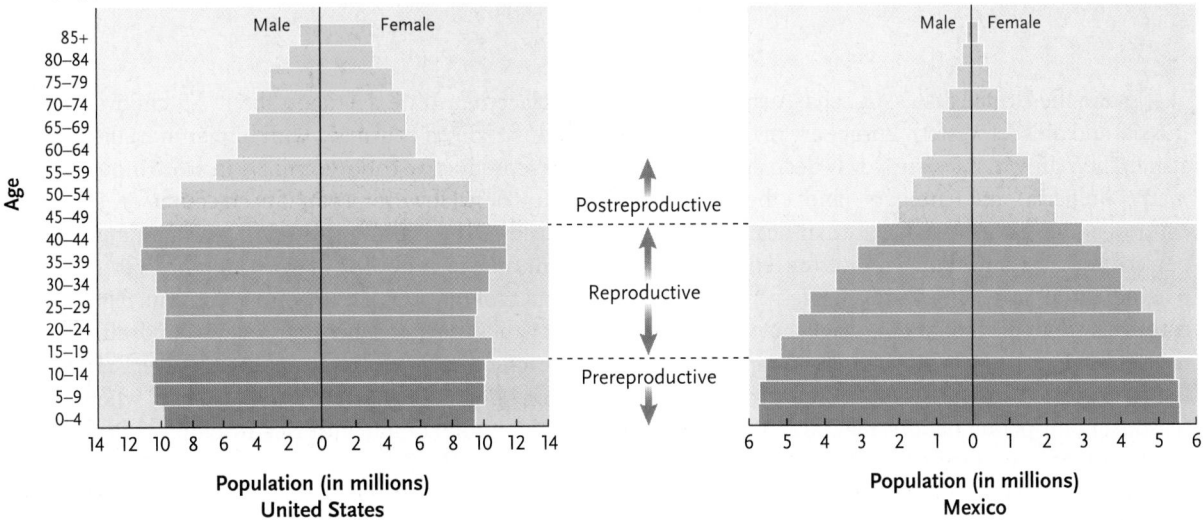

Figure 29.21

Age structure diagrams. Age structure diagrams **(a)** differ for countries with zero, negative, and rapid population growth rates. The width of each bar represents the proportion of the population in each age class. Age structure diagrams for the United States and Mexico **(b)** in 2000 (measured in millions of people) suggest that these countries will experience different growth rates.

born during the previous 15 years. Worldwide, more than one-third of the human population falls within this prereproductive base. This age class will soon reach sexual maturity. Even if each woman produces only two offspring, populations will continue to grow rapidly because so many individuals are reproducing. This situation can be described as a *population bomb*.

The age structures of the United States and Mexico differ, which has consequences for population growth in the two jurisdictions. Remember the potential importance of immigration and emigration when considering the longer-term impact of the population bomb.

29.7c Population Growth and Economic Development: Interconnections

The relationship between a country's population growth and its economic development can be depicted by the **demographic transition model (Figure 29.22,**

p. 702). This model describes historical changes in demographic patterns in the industrialized countries of western Europe. Today, we do not know if it accurately predicts the future for developing nations.

According to this model, during a country's preindustrial stage, birth and death rates are high, and the population grows slowly. Industrialization begins a *transitional* stage, when food production rises, and health care and sanitation improve. Death rates decline, resulting in increased rates of population growth. Later, as living conditions improve, birth rates decline, causing a drop in rates of population growth. When the industrial stage is in full swing, population growth slows dramatically. Now people move from countryside to cities, and urban couples often choose to accumulate material goods instead of having large families. ZPG is reached in the *postindustrial* stage. Eventually, the birth rate falls below the death rate, r falls below zero, and population size begins to decrease.

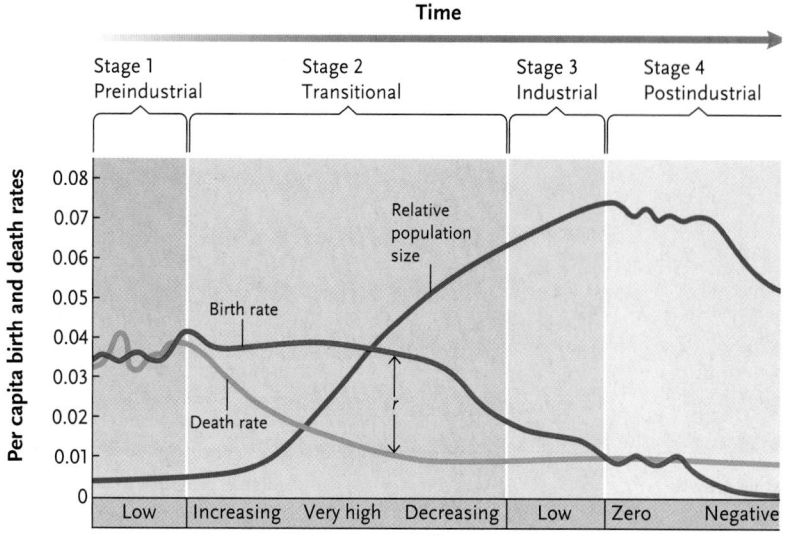

Figure 29.22
The demographic transition. The demographic transition model describes changes in the birth and death rates and relative population size as a country passes through four stages of economic development. The bottom bar describes the net population growth rate, *r*.

Today, the United States, Canada, Australia, Japan, Russia, and most of western Europe are in the industrial stage. Their growth rates are slowly decreasing. In Germany, Bulgaria, and Hungary (and other European countries), birth rates are lower than death rates, and populations are shrinking, indicating entry into the postindustrial stage. Kenya and other less industrialized countries are in the transitional stage, but they may not have enough skilled workers or enough capital to make the transition to an industrialized economy. For these reasons, many poorer nations may be stuck in the transitional stage. Developing countries experience rapid population increase because they experience declines in death rates associated with the transitional stage without the decreases in birth rates typical of industrial and postindustrial stages.

29.7d Controlling Reproductive Output: Planned Reproduction

Most governments realize that increased population size is now the major factor causing resource depletion, excessive pollution, and an overall decline in quality of life. The principles of population ecology demonstrate that slowing the rate of population growth and effecting an actual decline in population size can be achieved only by decreasing the birth rate or increasing the death rate. Increasing mortality is neither a rational nor a humane means of population control. Some governments use **family planning programs** in an attempt to lower birth rates. In other countries, any form of family planning is unlawful. This topic is discussed further in Chapter 32, where we will see that education of women is a vital undertaking.

To achieve ZPG, the average replacement rate should be just slightly higher than two children per couple. This is necessary because some female children die before reaching reproductive age. Today's replacement rate averages about 2.5 children in less industrialized countries with higher mortality rates in prereproductive cohorts and 2.1 in more industrialized countries. However, even if each couple on Earth produced only 2 children, the human population would continue to grow for at least another 60 years (the impact of the population bomb). Continued population growth is inevitable because today's children, who outnumber adults, will soon mature and reproduce. The worldwide population will stabilize only when the age distributions of all countries resemble that for countries with ZPG.

Family planning efforts encourage women to delay their first reproduction. Doing so reduces the average family size and slows population growth by increasing generation time (see Figure 29.9, p. 692). Imagine two populations in which each woman produces two offspring. In the first population, women begin reproducing at age 32 years, and in the second, they begin reproducing at age 16 years. We can begin with a cohort of newborn baby girls in each population. After 32 years, women in the first population will be giving birth to their first offspring, but women in the second population will be new grandmothers. After 64 years, women in the first population will be new grandmothers, but women in the second population will witness the birth of their first great-great grandchildren (if their daughters also bear their first children at age 16 years). Obviously, the first population will grow much more slowly than the second.

29.7e The Future: Where Are We Going?

Homo sapiens has arrived at a turning point in our cultural evolution and in our ecological relationship with Earth. Hard decisions await us, and we must make them soon. All species face limits to their population growth, and it is naive to assume that our unique abilities

Jack Millar, a professor of biology at Western University in London, and his students study the life histories of small mammals such as mice, voles, and wood rats. They do most of this work in the field, mainly at sites in the Kananaskis Valley in southwestern Alberta. The work involves trapping the small mammals and marking them so that they can recognize them later. Recaptures of known individuals allow the researchers to track the performances of individuals. Millar and his students have been following populations of deer mice for over 20 years, and their records have allowed them to ask basic questions about life history.

Most deer mice born in any year in the Kananaskis Valley are dead by the end of September of that year. Although a few females are able to bear two litters in a year, most do not. Compared with deer mice living in southwestern Ontario (see Section 29.6c), the mice in Kananaskis are barely hanging on. But Millar and his students were interested to learn what factors limit age at first reproduction and the ability of a female to breed more than once a year. Female mice given protein-rich diets (cat food) were sometimes able to breed in their first summer, suggesting a strong influence of food quality on life history traits.

With data on deer mice covering a span of more than 20 years, Millar was able to explore the possible effects of climate change on deer mice living in the Kananaskis Valley. Specifically, between 1985 and 2003, female deer mice typically conceived their first litters on May 2, and the first births occurred on May 26. There were no statistically significant changes in the timing of first births, although the average temperatures in early May had declined by about 2°C during this period. Spring breeding of the deer mice was not related to temperatures or snowfall. The decline in temperature had no effect on the mice's reproductive success. Changes in photoperiod appear to be responsible for initiating reproductive activity in the mice. Their access to protein did affect their reproductive output.

Discoveries about diseases associated with wildlife are a side benefit of Millar's endeavours. The work that Millar and his students have done provided a different view of the role of beavers in the spread of giardiasis, also known as "beaver fever." Giardiasis is caused by infections of a protozoan species in the genus *Giardia*. Humans can be infected if they drink water containing spores or trophozoites of *Giardia* species. Humans with giardiasis suffer from intestinal distress. As the name beaver fever implies, these aquatic rodents have been presumed to be the source of human infections. Using specimens, some provided by Millar and his students, P.M. Wallis and colleagues determined that 20 of 21 red-backed voles (*Clethrionomys gapperi*) were infected by *Giardia*—a much higher rate of infection than any other small mammals and also higher than beavers (2 of 50 infected).

Millar's work has demonstrated how long-term experimental research involving both observations and experiments can shed light on life history strategies.

---

exempt us from the laws of population growth. We have postponed the action of most factors that limit population growth, but no amount of invention and intervention can expand the ultimate limits set by resource depletion and a damaged environment. We now face two options for limiting human population growth: we can make a global effort to limit our population growth, or we can wait until the environment does it for us.

Return to Figure 29.19, p. 699 and observe that only the bubonic plague (also known as the Black Death) caused any deflection from the trajectory of the curve tracking growth in the human population. The plague appears to have been spread into Europe by the Mongols. The plague, long established in China, arrived in the Mongol summer capital of Shangdu in 1332. By 1351, the population of China had been reduced by 50 to 66%. By 1345, the plague had reached Feodosija in Ukraine (then Kaffa). Between 1340 and 1400, it is estimated that the population of Africa declined from 80 million to 68 million and the world population from 450 million to between 350 million and 375 million. These data do not include the Americas because the plague did not reach there until about 1600.

To put these percentages in context, consider human deaths associated with World War II. In this conflict, Great Britain lost less than 1% of its population, France about 1.5%, and Germany 9.1%. In Poland and Ukraine, where there was a postwar famine, 19% of the human populations there are said to have died.

These sobering figures remind us that we are animals, vulnerable to many of the factors that affect other species on Earth. Now, look back at the chapter-opening image and note the large numbers of gulls at a landfill site. The gulls and the landfill illustrate a fundamental point in population biology, namely, the ability of populations to reach large numbers and have large environmental impacts whether the species are gulls or people.

STUDY BREAK

1. In what three ways have humans avoided the effects of density-dependent regulation factors?
2. What is a population bomb?
3. What does family planning encourage women to do?

Progesterone

The advent of birth control pills **(Figure 1)** had a great impact on the behaviour of people. Women using birth control pills had more control over their fertility than others. Central to the development of an effective oral contraceptive was a change in the molecular structure of progesterone **(Figure 2a)**. Specifically, the addition of a CH_3 group (Figure 2b) meant that the new molecule, megestrol, had the same effect on a woman's reproductive system, but it was not quickly metabolized and remained in the system long enough to have the desired effect (suppressing ovulation). Similarly, slight modifications to the estradiol molecule turned it into ethinylestradiol **(Figure 3)**. Megestrol is an analogue of progesterone, and ethinylestradiol is an analogue of estradiol.

Today, biologists working in zoos use a variety of birth control methods to control the fertility of animals in their collections. For critically endangered species such as black-footed ferrets (see "Black-Footed Ferret, *Mustela nigripes*," p. 683), this means using information about cycles of fertility to maximize reproductive output.

For animals whose populations are growing at a rapid pace, birth control gives keepers the chance to control growth of the populations. The same principles apply to working with organisms in the wild, but getting African elephants to take their birth control pills has not proven to be easy.

Hormones and their -analogues are common in untreated municipal wastewaters. In some cases, male fish exposed to these wastewaters are becoming feminized. Specifically, some male fish produce vitellogenin mRNA and protein, substances normally associated with the maturation of oocytes in females. Males thus exposed produce early-stage eggs in their testes. This feminization occurs in the presence of estrogenic substances, including natural estrogen (17b-estradiol) and the synthetic estrogen 17a-ethinylestradiol.

Do a few feminized male fish in the population matter? Karen A. Kidd and six colleagues conducted a seven-year whole-lake experiment in northwestern Ontario (the Experimental Lakes Area). Male fathead minnows (*Pimephales promelas*) **(Figure 4)** chronically exposed to low levels (5–6 ng·L⁻¹) of estrogenic substances showed feminizing effects and the development of intersex males, whereas females had altered oogenesis. The situation led to the near-extinction of fathead minnows in the experimental lake.

Figure 1
Birth control pills, a selection of products.

a.

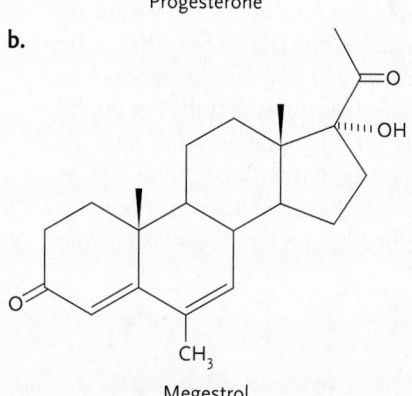

Progesterone

b.

Megestrol

Figure 2
Progesterone and the synthetic megestrol.

a.

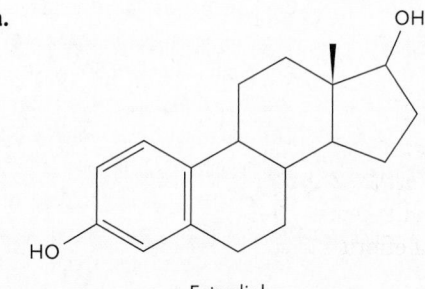

Estradiol

b.

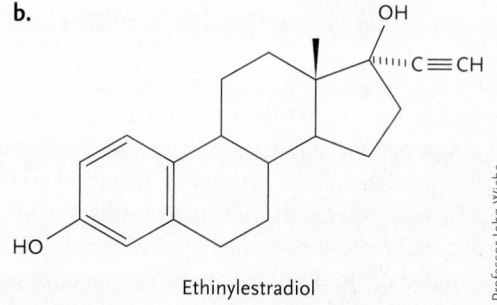

Ethinylestradiol

Figure 3
Estradiol and the synthetic ethinylestradiol.

Figure 4
Pimphales promelas, *the fathead minnow.*

29.8 Sisyphean Problems in Population Biology

According to the myth, Sisyphus revealed a secret of Zeus, the supreme Greek god, so in Hades, Sisyphus faced a life of eternal frustration because the boulder he had to push up a hill always rolled back down. The equivalent of this situation is a recurring theme in population biology. For example, on many oceanic islands, eradication of rats and cats is a Sisyphean problem because when either introduced pest is at levels of high populations, it can be easy to kill them with a relatively small investment per cat or per rat. Control or, better still, eradication of cats and rats may reduce the threat to native species (see also Chapter 32). When their populations diminish, the effort and cost per kill increases. In difficult financial times, it may be easy to relent on control measures and use the funds elsewhere. But as soon as the pressure is removed from the populations of rats and cats, the numbers increase, starting the cycle anew.

Sisyphean problems also occur in disease control. The prevalence rates of malaria in children (ages 2 to 10 years) in Zanzibar illustrates this problem (Figure 29.23). Zanzibar, an archipelago consisting of several islands off the east coast of Africa, has a population of 1.3 million people, compared to 1.1 billion in Africa. A recent study showed that the annual antimalarial efforts in Zanzibar prevent about 600 000 cases

of and about 3300 deaths from malaria. The cost is about US$1183 per death averted and US$34.50 per impact of reducing the incidence of the disease.

The data on the prevalence of the malarial parasite in children demonstrate the importance of maintaining antimalarial programs even in the face of success measured as fewer cases of malaria and fewer deaths from it. Arrival of people already infected with the malarial parasite in Zanzibar is part of the problem. The same situation applies to the control of measles, which, unlike malaria, can be controlled by vaccination. This problem is Sisyphean because as soon as efforts to control malaria are cut back, the incidence of the disease increases.

Infestations of bedbugs are yet another Sisyphean problem. The population biology roots of such problems lie in the patterns of population growth: the impact of r (intrinsic rate of increase), N (population size), and K (carrying capacity). Perhaps Sisyphus's challenge was modest compared to that posed by some problems in population biology!

STUDY BREAK

1. Why is it appropriate to consider malaria control as a Sisyphean problem?
2. Why is the malaria situation different between Zanzibar and continental Africa?

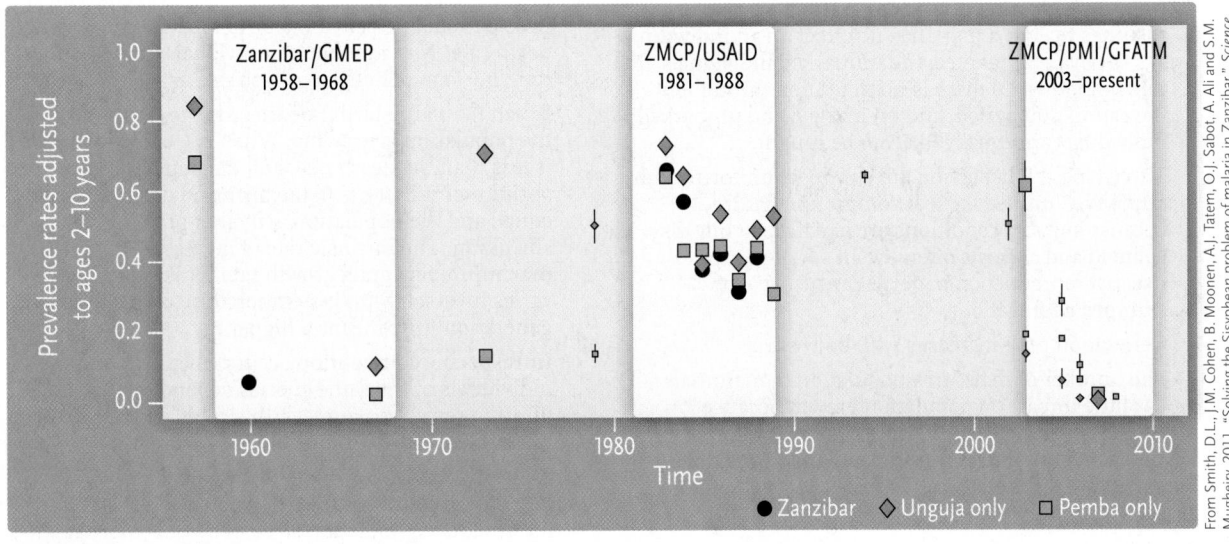

From Smith, D.L., J.M. Cohen, B. Moonen, A.J. Tatem, O.J. Sabot, A. Ali and S.M. Mugheiry, 2011. "Solving the Sisyphean problem of malaria in Zanzibar." *Science*, vol. 332: 1384–1385. Reprinted with permission from AAAS.

Figure 29.23

The incidence of malaria parasites in children between 1958 and 2010. During periods of concerted efforts to reduce or eliminate malaria, the incidence of parasites in children drops, only to rebound when control measures are stopped. Large symbols represent regional and national surveys; smaller symbols, smaller-scale surveys. GMEP was part of the Global Malaria Eradication Programme, ZMCP was the Zanzibar Malaria Control Programme assisted by USAID, and ZMCP/PMI/GFATM is the Zanzibar Malaria Control Programme assisted by the U.S. President's Malaria Initiative and the Global Fund to Fight AIDS, Tuberculosis, and Malaria. Here Zanzibar refers to all of the islands; Pemba and Unguja are specific islands included in the archipelago.

Review

To access course materials such as Aplia and other companion resources, please visit www.NELSONbrain.com.

29.1 The Science of Ecology

- Organismal ecology is the study of organisms to determine adaptations to the abiotic environment, including morphological, physiological, biochemical, behavioural, and genetic adaptations. Population ecologists document changes in size and other characteristics of populations of species over space and time. Community ecologists study sympatric populations, the interactions among them, and how these interactions affect the community's growth. Interactions may include predation and competition. Ecosystem ecologists study nutrient cycling and energy flow through the biotic and the abiotic environment.

- Mathematical models express hypotheses about ecological relationships and different variables, allowing researchers to manipulate the model and document resulting changes. In this way, researchers can simulate natural events before investing in lab work.

- Experimental and control treatments are necessary because they allow ecologists to separate cause and effect.

29.2 Population Characteristics

- Geographic range is the overall spatial boundary around a population. Individuals in the population often live in a specific habitat within the range.

- A lower population density means that individuals have greater access to resources such as sunlight and water. The capture−mark−recapture technique assumes that (1) a mark has no effect on an individual's survival, (2) marked and unmarked individuals mix randomly, (3) there is no migration throughout the estimation period, and (4) marked and unmarked individuals are equally likely to be caught.

- Three types of dispersion are clumped, uniform, and random. Clumped is most common in nature because suitable conditions are usually patchily distributed and animals often live in social groups. Asexual reproduction patterns can also lead to clumped aggregations.

- Generation time increases with body size.

- The number of males in a population of mammals has little impact on population growth because females bear the costs of reproduction (pregnancy and lactation), thus limiting population growth. Sea horses are different because males get pregnant.

29.3 Demography

- Age-specific mortality and age-specific survivorship deal with age intervals. In any one interval, age-specific mortality is the proportion of individuals that died during that time. Age-specific survivorship is the number surviving during the interval. The two values must sum to 1. For example, if age-specific survivorship is 0.616, then age-specific mortality is $1 - 0.616$ or 0.384.

- Age-specific fecundity is the average number of offspring produced by surviving females during each age interval.

- In a type I curve, high survivorship at a young age decreases rapidly later in life. Type I curves are common for large animals, including humans. In a type II curve, the relationship is linear because there is a constant rate of mortality across the life span. Songbirds fit in this category. A type III curve shows high mortality at a young age that stabilizes as individuals grow older and larger. Insects fall into this category.

29.4 The Evolution of Life Histories

- Maintenance, growth, and reproduction are the three main energy-consuming processes.

- Passive care occurs in animals that simply lay eggs and leave them or, in mammals, as nutrients cross the placenta from the mother to the developing baby. Active care involves nursing and other care provided after birth.

- Salmon have a short life span and devote a great deal of energy to reproduction. Deciduous trees may reproduce more than once and use only some energy in any reproductive event, balancing reproduction and growth.

- Early reproduction is favoured if adult survival rates are low or if, when animals age, they do not increase in size. In this case, fecundity does not increase with size.

29.5 Models of Population Growth

- An exponential model is used when a population has unlimited growth.

- dN/dt = change in a population's size during a given time period; b = per capita birth rate; d = per capita death rate; N = number of individuals in the population; $b - d$ = per capita growth rate = r.

- When $r > 0$, the birth rate exceeds the death rate, and the population is growing. When $r < 0$, the birth rate is less than the death rate, and the population is shrinking. When $r = 0$, the birth and death rates are equal, and the population is neither growing nor shrinking. The intrinsic rate of increase (r_{max}) is the maximum per capita growth rate. This value usually varies inversely with generation time, so a shorter generation time means a higher r_{max}.

- Intraspecific competition occurs when two or more individuals of the same species depend on the same limiting resource. For deer, this could include food, water, or refuge from predators.

- A logistic model has the following pattern: when the population growth is low, the population is small. At intermediate population sizes, growth is more rapid because more individuals breed and r is high. When population growth approaches K (carrying capacity), competition increases, r decreases, and the growth of the population is reduced.

29.6 Population Regulation

- Density-dependent factors include intraspecific competition and predation. At high density, fewer resources are available for individuals, which, in turn, use more energy in maintenance needs and less in reproduction. Offspring produced at higher population densities are often smaller in number or size and less likely to survive. At high population levels, adults may be smaller and weaker.

- Density-independent factors, such as fire, earthquakes, storms, floods, and other natural disturbances, reduce a population size regardless of density.

- *r*-selected species often have large numbers of small young, whereas *K*-selected species usually have small numbers of larger young. Other possibilities may include characteristics from Table 29.3.

- Extrinsic control includes interactions between individuals in a population and their food and predators. Once a food supply is exhausted, reproduction will decrease and mortality will increase. Intrinsic control can be hormonal changes within a population that cause increased aggressiveness, faster dispersal, and reduced reproduction. Aggression can cause weaker individuals to be forced to disperse to reduce the population density.

29.7 Human Population Growth

- Humans have avoided the effects of density-dependent regulation factors by expanding their geographic range into virtually every habitat, increasing *K* through agriculture, and reducing population-limiting factors resulting from poor hygiene, malnutrition, and contagious diseases.

- A population bomb is when many offspring are born in one time period, first forming the prereproductive base. At sexual maturity, populations can grow rapidly because of the large number of individuals in this cohort.

- The preindustrial stage is characterized by slow population growth, as birth and death rates are high. The transitional stage has better health care and sanitation, as well as increased food production. In the transitional stage, there is a decline in death rates, allowing population growth, but birth rates eventually decline as living conditions improve. In the industrial stage, there is slow population growth as family size decreases because couples choose to have fewer children and accumulate more material goods. In the postindustrial stage, the population size decreases as the birth rate falls below the death rate.

- Family planning encourages families to delay first reproduction, decreasing the size of the average family and, in turn, reducing the population size as generation time increases. Decisions about reproduction should involve couples.

29.8 Sisyphean Problems in Population Biology

- Sisyphean problems in biology range from the challenge of eradicating pests (such as introduced rats and cats) to the problem of controlling diseases such as malaria. In either case, when the incidence of the pests or the disease drops below some level and control operations stop, the populations of pests and the incidence of disease rebound.

Questions

Self-Test Questions

1. Ecologists sometimes use mathematical models to do which of the following tasks?
   a. simulate natural events before conducting detailed field studies
   b. make basic observations about ecological relationships in nature
   c. collect survivorship and fecundity data to construct life tables
   d. determine the geographic ranges of populations

2. Which term can be used to describe the number of individuals per unit area or volume of habitat?
   a. dispersion pattern
   b. density
   c. size
   d. age structure

3. Suppose that one day you caught and marked 90 butterflies in a population. A week later, you returned to the population and caught 80 butterflies, including 16 that had been marked previously. What is the size of the butterfly population?
   a. 170
   b. 450

   c. 154
   d. 186

4. What does a uniform dispersion pattern imply about the members of a population?
   a. They work together to escape from predators.
   b. They use resources that are patchily distributed.
   c. They may experience intraspecific competition for vital resources.
   d. They have no ecological interactions with each other.

5. What does the model of exponential population growth predict about the per capita population growth rate (*r*)?
   a. *r* does not change as a population gets larger.
   b. *r* gets larger as a population gets larger.
   c. *r* gets smaller as a population gets larger.
   d. *r* is always at its maximum level (r_{max}).

6. If a population of 1000 individuals experiences 452 births and 380 deaths in 1 year, what is the value of *r* for this population?
   a. 0.072/individual/year
   b. 0.452/individual/year
   c. 0.380/individual/year
   d. 0.820/individual/year

7. According to the logistic model of population growth, what happens to the absolute number of individuals by which a population grows during a given time period?
   a. It gets steadily larger as the population size increases.
   b. It gets steadily smaller as the population size increases.
   c. It remains constant as the population size increases.
   d. It is highest when the population is at an intermediate size.

8. Which example might reflect density-dependent regulation of population size?
   a. An exterminator uses a pesticide to eliminate carpenter ants from a home.
   b. Mosquitoes disappear from an area after the first frost.
   c. Northeast storms blow over and kill all willow trees along a lake.
   d. A clam population declines in numbers in a bay as the number of predatory herring gulls increases.

9. Which pattern is a *K*-selected species likely to exhibit?
   a. a type I survivorship curve and a short generation time
   b. a type II survivorship curve and a short generation time
   c. a type III survivorship curve and a short generation time
   d. a type I survivorship curve and a long generation time

10. Which one of the following is a reason that human populations have sidestepped factors that usually control population growth?
    a. Agriculture and industrialization have increased the carrying capacity for our species.
    b. The population growth rate (*r*) for the human population has always been small.
    c. The age structure of human populations has no impact on its population growth.
    d. Plagues have killed off large numbers of humans at certain times in the past.

Questions for Discussion

1. Do you expect to see a genetic bottleneck effect in *Mustela nigripes* populations in the wild in the future? If so, how long will they take to appear?

2. Design an income tax policy and social services plan that would encourage people to have either larger or smaller families.

3. Many city-dwellers have noted that the density of cockroaches in apartment kitchens appears to vary with the habits of the occupants. People who wrap food carefully and clean their kitchen frequently tend to have fewer arthropod roommates than those who leave food on kitchen counters and clean less often. Interpret these observations from the viewpoint of a population ecologist.

4. Why have bedbugs reemerged as all-too-common pests in many dwellings? How can they be controlled?

5. How could the use of composting by urban gardeners affect local populations of raccoons? Is there a "green" solution to the perceived problem posed by urban raccoons?

A hovering rubythroated hummingbird (*Archilochus colubris*) approaches a flower.

Population Interactions and Community Ecology

WHY IT MATTERS

The late Robert Whittaker, a well-known ecologist, referred to birds such as hummingbirds as "ornaments" of evolution. He made this observation based on the reality that although they were spectacular and distinctive, many of these plant-pollinating and seed-dispersing birds did not appear to contribute much to the overall productivity of ecosystems. This is an example of one important challenge facing biologists working to understand the nature and details of interactions among the species that coexist in and constitute communities. How can we assess detailed interactions among sympatric species?

If you watched hummingbirds (opening photograph) visiting flowers, it would be easy to believe that they subsisted on the nectar and perhaps pollen that are available from flowers. But if you watch the same birds bringing food to their nestlings, it would seem that they eat mainly insects. So which data set is more indicative of the role these "ornaments" play in the ecosystems in which they occur?

By examining and describing morphological specializations, biologists who study bats identified tropical and subtropical species apparently specialized for nectar feeding. Long snouts, small teeth **(Figure 30.1, p. 710)**, and extensible tongues **(Figure 30.2, p. 710)** are features of most of the bats that visit flowers and pollinate plants. Flowers pollinated mainly by bats are chiropterphilous, quite different from those pollinated by birds (ornithophilous). At flowers, the bats pump blood into their tongues, thus extending them into the flowers they visit. Papillae at the tips of their tongues soak up nectar. Flowers pollinated by insects and birds often have nectar guides, visual patterns that guide the pollinator to the nectar. The positioning of nectaries, the areas of nectar production and storage, ensures that the pollinator

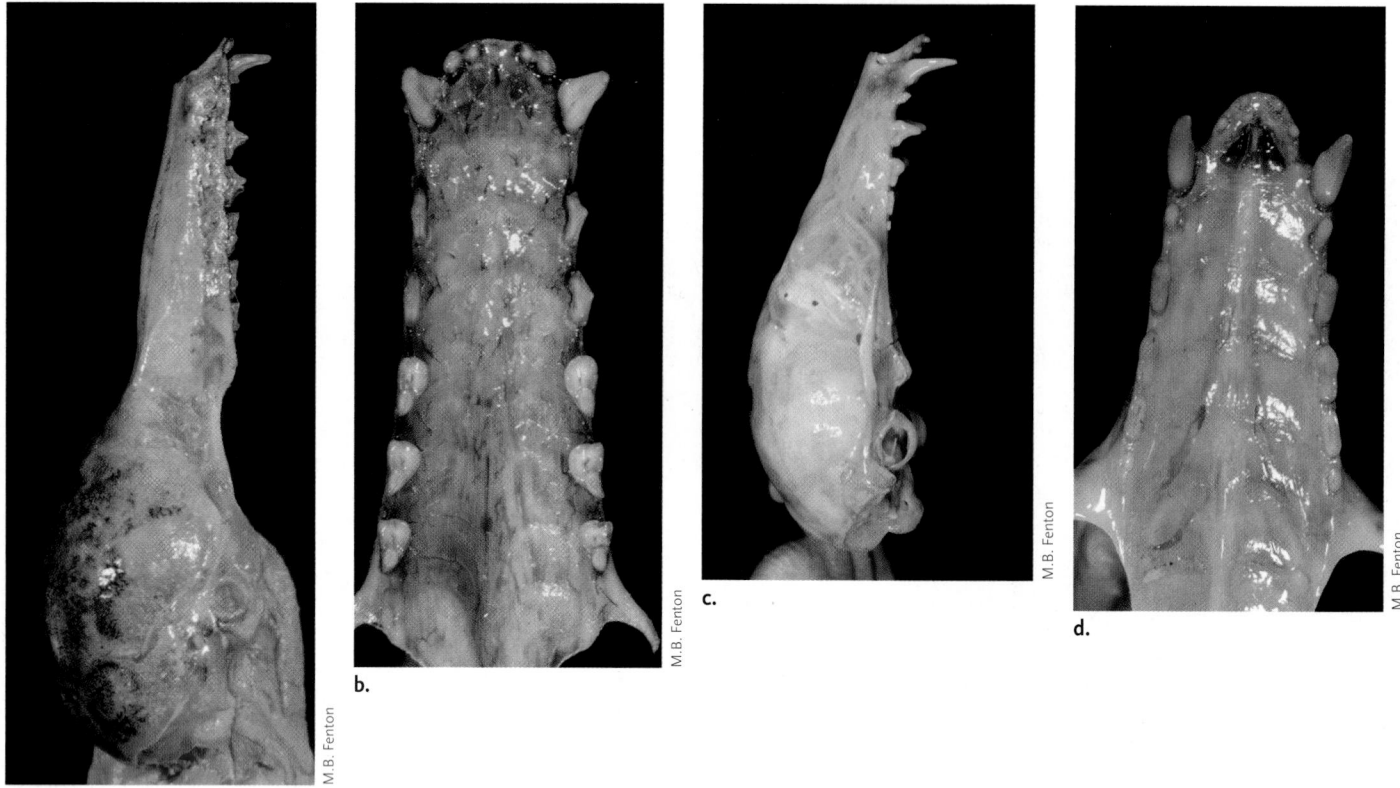

Figure 30.1

The skulls of nectar-feeding bats have long muzzles and relatively small teeth. This is true of a Mexican long-tongued bat (*Choeronycteris Mexicana*) **(a, b)** and a Woerman's bat (*Megaloglossus woermanii*) **(c, d).** (b) and (d) show a closer view of the teeth. The Mexican long-tongued bat is a New World leaf-nosed bat from the Neotropics. Woermna's bat is an Old World fruit bat from Africa. Flower-visiting has arisen independently in both lineages of bats.

Figure 30.2

Two exposures of a Pallas' long-tongued bat with tongue extended into a banana flower. Bananas are a chiropterophilous species.

gets coated with pollen in the process of feeding. Some chiropterophilous flowers have acoustic nectar guides that allow the hovering bat to position itself to achieve best access to the nectar.

In earlier studies, biologists had collected and analyzed the contents of the droppings of bats **(Figure 30.3).** This approach was at best tedious and provided only a general view of what the bats had eaten. DNA barcoding ("Barcode of Life," p. 730) changed the scene, perhaps indicating that the bats were not just ornaments in ecosystems.

The work of a team led by Canadian Beth (E. L.) Clare (currently in the Department of Biology and Chemistry at Queen Mary College in London, United Kingdom) overturned the conventional view of nectar-feeding bats. Their analysis revealed that Pallas' long-tongued bats frequently ate insects and fruit as well as nectar and pollen. The proof of this came from barcoding insect remains from the droppings of Pallas' long-tongued bat, observing the animals hunting insects, and monitoring their echolocation calls (see Chapter 47) and behaviour. The data set demonstrated that these bats actively hunted insects, and their echo-location calls made them relatively undetectable by many insects equipped with bat-detecting ears. The bats filled at least three trophic roles, eating insects, fruit, and nectar and pollen.

Earlier work with DNA barcode analysis of insect fragments in the droppings of insectivorous bats had revealed that six sympatric species in the same ecosystem in Jamaica showed relatively little overlap in diet, raising questions about the role of competition in

a.

b.

M.B. Fenton

M.B. Fenton

structuring this community. These results, combined with those about the diets of Pallas' long-tongued bats, give biologists a different view of the trophic structure of bat communities. Comparable analyses of the diets of other species, for example, fish and leeches, alerts us to the dynamic nature of communities and the diversity of interactions among the species that the communities comprise.

30.1 Interspecific Interactions

Interactions between species typically benefit or harm the organisms involved, although they may be neutral **(Table 30.1)**. Furthermore, where interactions with other species affect individuals' survival and reproduction, many of the relationships we witness today are the products of long-term evolutionary modification. Good examples range from predator–prey interactions to those associated with pollination or dispersal of seeds.

Interactions between species can change constantly, but remember that the interactions occur at the individual level. Some individuals of a species may be better adapted when another species exerts selection pressure on that species. This adaptation can, in turn, help these individuals exert selection pressure on the other species, which can exert selection pressure on the first species in the chain. The situation, known as **coevolution**, is defined as genetically based reciprocal adaptation in two or more interacting species. A good example is provided by the arms race between echolocating bats and insects with bat-detecting ears (see "Echolocation: Communication," Chapter 47).

Some coevolutionary relationships are straightforward. Ecologists describe the coevolutionary interactions between some predators and their prey as a race

Table 30.1	Population Interactions and Their Effects	
Interaction	Effects on Interacting Populations	
Predation	+/−	Predators gain nutrients and energy; prey are killed or injured.
Parasitism	+/−	Parasites gain nutrients and energy; hosts are injured or killed.
Herbivory	+/−	Herbivores gain nutrients and energy; plants are killed or injured.
Competition	−/−	Both competing populations lose access to some resources.
Commensalism	−/0	One population benefits; the other population is unaffected.
Mutualism	+/+	Both populations benefit.

in which each species evolves adaptations that temporarily allow it to outpace the other. When antelope populations suffer predation by cheetahs, natural selection fosters the evolution of faster antelopes. Faster cheetahs may be the result of this situation, and if their offspring are also fast, then antelopes will also become more fleet of foot. Other coevolved interactions provide benefits to both partners. Flower structures of different monkey-flower species have evolved characteristics that allow them to be visited by either bees or hummingbirds (see Figure 19.4, Chapter 19).

One can hypothesize a coevolutionary relationship between any two interacting species, but documenting the evolution of reciprocal adaptations is difficult. Coevolutionary interactions often involve more than two species, and most organisms experience complex interactions with numerous other species in their communities. Cheetahs take several prey species. Antelopes are prey for many species of predators, from cheetahs to lions, leopards, and hyenas, as well as some larger birds of prey. Not all predators use the same hunting strategy. Therefore, the simple portrayal of coevolution as taking place between two species rarely does justice to the complexity of these relationships.

STUDY BREAK

What is coevolution? Is it usually restricted to two species?

30.2 Getting Food

Because animals typically acquire nutrients and energy by consuming other organisms, **predation** (the interaction between predatory animals and the animal prey they consume) and **herbivory** (the interaction between herbivorous animals and the plants they eat) can be the most conspicuous relationships in ecological communities.

Both predators and **herbivores** have evolved characteristics allowing them to feed effectively. Carnivores use sensory systems to locate animal prey and specialized behaviours and anatomical structures to capture and consume it. Herbivores use sensory systems to identify preferred food or to avoid food that is toxic.

Rattlesnakes, such as species in the genus *Crotalus,* use heat sensors on pits in their faces (see Figure 45.26, Chapter 45) to detect warm-blooded prey. The snakes deliver venom through fangs (hollow teeth) by open-mouthed strikes on prey. After striking, the snakes wait for the venom to take effect and then use chemical sensors also on the roofs of their mouths to follow the scent trail left by the dying prey. The venom is produced in the snakes' salivary glands. It contains neurotoxins that paralyze prey and protease enzymes that begin to digest it. Elastic ligaments connecting the bones of the snakes' jaws (mandibles) to one another and the mandibles to the skull allow snakes to open their mouths very wide to swallow prey larger than their heads (see Figure 48.21, Chapter 48).

Herbivores have comparable adaptations for locating and processing their food plants. Insects use chemical sensors on their legs and mouthparts to identify edible plants and sharp mandibles or sucking mouthparts to consume plant tissues or sap. Herbivorous mammals have specialized teeth to harvest and grind tough vegetation (see Figure 28.57, Chapter 28). Herbivores, such as farmer ants (see Chapter 33), ruminants, and termites (see "Digesting Cellulose: Fermentation," Chapter 48), may also coopt other species to gain access to nutrients locked up in plant materials.

All animals select food from a variety of potential items. Some species, described as *specialists,* feed on one or just a few types of food. Among birds, Everglades Kites (*Rostrhamus sociabilis*) eat only apple snails (*Pomacea paludosa*). Koalas (see Figure 28.51, Chapter 28) eat the leaves of only a few of the many available species of *Eucalyptus*. Other species, described as generalists, have broader tastes. Crows (genus *Corvus*) take food ranging from grain to insects to carrion. Bears (genus *Ursus*) and pigs (genus *Sus*) are as omnivorous as humans.

How does an animal select its food? Why pizza rather than salad? Mathematical models, collectively described as **optimal foraging theory**, predict that an animal's diet is a compromise between the costs and benefits associated with different types of food. Assuming that animals try to maximize their energy intake at any meal, their diets should be determined by the ratio of costs to benefits: the costs of obtaining the food versus the benefits of consuming it. Costs are the time and energy it takes to pursue, capture, and consume a particular kind of food. Benefits are the energy provided by that food. A cougar (*Felis concolor*) will invest more time and energy hunting a mountain goat (*Oreamnos americanus*) than a jackrabbit (*Lepus townsendii*), but the payoff for the cat is a bigger meal. One important element in food choice is the relative abundance of prey. *Encounter rate* is usually influenced by population density and can influence a predator's diet. For the cougar, encounter rate determines the time between jackrabbits, and when they are abundant, they can be a more economical target than larger, scarcer prey.

Food abundance affects food choice. When prey are scarce, animals often take what they can get, settling for food that has a higher cost-to-benefit ratio. When food is abundant, they may specialize, selecting types that provide the largest energetic return. Bluegill sunfishes eat *Daphnia* and other small crustaceans. When crustacean density is high, these fishes hunt mostly large *Daphnia*, which provide more energy for their effort. When prey density is low, bluegills eat *Daphnia* of all sizes **(Figure 30.4)**.

Think of yourself at a buffet. The array of food can be impressive, if not overwhelming. But your state of hunger, the foods you like, the ones you do not like, and any to which you are allergic all influence your selection. You may also be influenced by choices made by others. In your feeding behaviour, you betray your animal heritage.

STUDY BREAK

1. How do predators differ from herbivores? How are they similar?
2. Is a koala a generalist or a specialist? What is the difference?
3. What does optimal foraging theory predict? Describe the costs and benefits central to this theory.

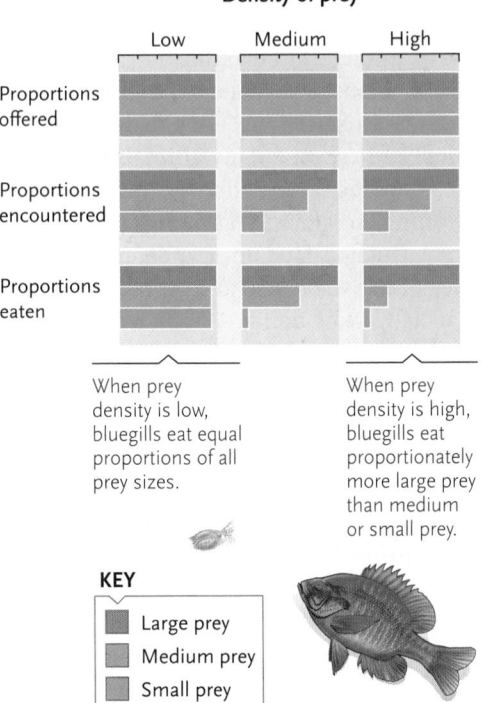

Density of prey

	Low	Medium	High
Proportions offered			
Proportions encountered			
Proportions eaten			

When prey density is low, bluegills eat equal proportions of all prey sizes.

When prey density is high, bluegills eat proportionately more large prey than medium or small prey.

KEY
- Large prey
- Medium prey
- Small prey

Figure 30.4

An experiment demonstrating that prey density affects predator food choice. Bluegill sunfishes (*Lepomis macrochirus*) were offered equal numbers of small, medium, and large prey (*Daphnia magna*) at three different total densities of prey. Because large prey are easy to find, the fishes encountered them more often, especially at the highest prey densities, than either medium-sized or small prey. The fishes' choice of prey varied with prey density, but they always chose the largest prey available.

30.3 Defence

CONCEPT FIX Some people believe that "natural" products (chemicals) are beneficial to us, while artificial ones are potentially harmful. In reality, many plants produce chemicals (natural products) that are dangerous and even deadly—to humans and to other animals and to some plants. Contact with the leaves (or stems, roots, flowers, or berries) of poison ivy may be enough to convince you that not all plant products are beneficial. If not, you can read about conine (see Figure 33.31, Chapter 33), an active ingredient in poison hemlock, the poison that killed Socrates. ⬡

Predation and herbivory negatively affect the species being eaten, so it is no surprise that animals and plants have evolved mechanisms to avoid being caught and eaten. Some plants use spines, thorns, and irritating hairs to protect themselves from herbivores. Plant tissues often contain poisonous chemicals that deter herbivores from feeding. When damaged, milkweed plants (family Asclepiadaceae) exude a milky,

irritating sap **(Figure 30.5)** that contains poisons that affect the heart (cardiac glycosides). Even small amounts of cardiac glycosides are toxic to the heart muscles of some vertebrates. Other plants have compounds that mimic the structure of insect hormones, disrupting the development of insects that eat them. Most of these poisonous compounds are volatile, giving plants their typical aromas. Some herbivores have developed the ability to recognize these odours and avoid toxic plants. Some plants increase their production of toxic compounds in response to herbivore feeding. Potato and tomato plants damaged by herbivores have higher levels of protease-inhibiting chemicals. These compounds prevent herbivores from digesting the proteins they have eaten, reducing the food value of these plants.

30.3a Be Too Big to Tackle

Size can be a defence. At one end of the spectrum, this means being too small to be considered food. At the other end, it means being so big that few, if any, predators can succeed in attacking and killing the prey. Today, elephants and some other large herbivores (megaherbivores) are species with few predators (other than humans). But 50 thousand years ago, there were larger predators (see Figure 17.1, Chapter 17), including one species of "lion" that was one-third larger than an African lion.

30.3b Eternal Vigilance: Always Be Alert

A first line of defence of many animals is avoiding detection. This often means not moving, but it also means keeping a sharp lookout for approaching predators and the danger they represent **(Figure 30.6, p. 714)**. Animals that live in groups benefit from the multitude of eyes and ears that can detect approaching danger, so the risk of predation can influence group size and social interactions.

Figure 30.5

Protective latex sap. Milky sap laced with cardiac glycosides oozes from a cut milkweed (*Asclepias* species) leaf. Milky sap does not always mean dangerous chemicals; for example, the sap of dandelions is benign.

Figure 30.6

Eternally vigilant. The sentry of a group of meerkats (*Suricata suricatta*).

30.3c Avoid Detection: Freeze—Movement Invites Discovery

Many animals are cryptic, camouflaged so that a predator does not distinguish them from the background. Patterns such as the stripes of a zebra (*Equus burchellii*) make the animal conspicuous at close range, but at a distance, patterns break up the outline, rendering the animals almost invisible. Many other animals look like something that is not edible. Some caterpillars look like bird droppings, whereas other insects look like thorns or sticks. Neither bird droppings nor thorns are usually eaten by insectivores.

30.3d Thwarting Attacks: Take Evasive Action

Animals resort to other defensive tactics once they have been discovered and recognized. Running away is a typical next line of defence. Taking refuge in a shelter and getting out of a predator's reach are an alternative. African pancake tortoises (*Malacochersus tornieri*) are flat, as the name implies. When threatened, they retreat into rocky crevices and puff themselves up with air, becoming so tightly wedged that predators cannot extract them.

If cornered by a predator, offence becomes the next line of defence. This can involve displays intended to startle or intimidate by making the prey appear large and/or ferocious. Such a display might dissuade a predator or confuse it long enough to allow the potential victim to escape. Many animals use direct attack in these situations, engaging whatever weapons they have (biting, scratching, stinging, etc.). Direct attacks are not usually a good primary defence because they involve getting very close to the predator, something prey usually avoid doing.

30.3e Spines and Armour: Be Dangerous or Impossible to Attack

Other organisms use active defence in the form of spines or thorns **(Figure 30.7)**. North American porcupines (genus *Erethizon*) release hairs modified into sharp, barbed quills that when stuck into a predator, cause severe pain and swelling. The spines detach easily from the porcupine, and the nose, lips, and tongue of an attacker are particularly vulnerable. There are records of leopards (*Panthera pardus*) being killed by porcupine spines. In these instances, the damage to the leopards' mouths, combined with infection, was probably the immediate cause of death. Many other mammals, from monotremes (spiny anteaters) to tenrecs (insectivores from Madagascar, *Tenrec* species and *Hemicentetes* species), hedgehogs (*Erinaceus* species), and porcupines in the Old World, use the same defence. So do some fishes and many plants.

Other organisms are armoured **(Figure 30.8)**. Examples include bivalve and gastropod molluscs, chambered nautiluses, arthropods such as horseshoe crabs (*Limulus* species), trilobites (see Chapter 27), fishes such as catfish (*Siluriformes*), reptiles (turtles; see Figure 28.38, Chapter 28), and mammals (armadillos, scaly anteaters). We know a great deal about extinct species that were armoured (see Chapter 20) because they often made good fossils.

30.3f Chemical Defence Ranges from Bad Taste to Deadly

Like plants that produce chemicals to repel herbivores, many animals make themselves chemically unattractive. At one level, this can be as simple as smelling or tasting bad. Have you ever had a dog or a cat that was sprayed by a skunk (*Mephitis mephitis*)? Many animals vomit and defecate on their attackers. Skunks and bombardier beetles escalate this strategy by producing and spraying a noxious chemical. Other animals go beyond spraying. Many species of cnidarians, annelids, arthropods, and chordates produce dangerous toxins and deliver them directly into their attackers. These toxins may be synthesized by the user (e.g., snake venom; see "Molecule behind Biology," Box 30.1, p. 718) or sequestered from other sources, often plants

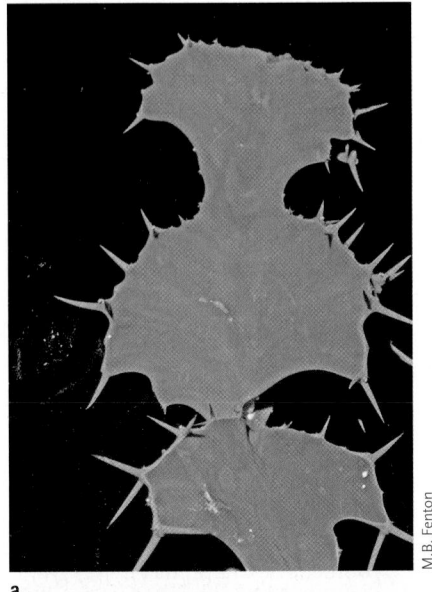

a.

b.

c.

d.

Figure 30.7
Defensive spines. Plants such as **(a)** the cowhorn euphorb (*Euphorbia grandicornis*) and **(b)** crown of thorns (*Euphorbia milli*) and animals such as **(c)** spiny anteaters (*Tachyglossus* species) and **(d)** porcupines (*Hystrix* species) use thorns or spines in defence. Pen shown for scale with quills in (d).

or other animals (see "Nematocysts," p. 720). Caterpillars of monarch butterflies are immune to the cardiac glycosides in the milkweed leaves they eat. They extract, concentrate, and store these chemicals, making the caterpillars themselves (and the adult butterflies) poisonous to potential predators. The concentrations of defensive chemicals may be higher in the animal than they were in its food. Cardiac glycosides persist through metamorphosis, making adult monarchs poisonous to vertebrate predators.

30.3g Warnings Are Danger Signals

Many animals that are noxious or dangerous are **aposematic:** they advertise their unpalatability with an appropriate display (**Figure 30.9;** see Chapter 18). Aposematic displays are designed to "teach" predators to avoid the signaller, reducing the chances of harm to would-be predators and prey. Predators that attack a brightly coloured bee or wasp and are stung learn to

Figure 30.8
Armour. Turtles and their allies (see Chapter 26) live inside shells. This leopard tortoise (*Geochelone pardalis*) is inspecting the remains of a conspecific. Armour does not guarantee survival.

Figure 30.9
Warning colours. This arrowhead frog gets its name from toxins in its skin that were used to poison arrowheads.

Figure 30.10

Mimicry. **(a)** Batesian mimics are harmless animals that mimic a dangerous one. The harmless drone fly (*Eristalis tenax*) is a Batesian mimic of the stinging honeybee (*Apis mellifera*). **(b)** Müllerian mimics are poisonous species that share a similar appearance. Two distantly related species of butterfly, *Heliconius erata* and *Heliconius melpomene*, have nearly indistinguishable patterns on their wings.

a. Batesian mimicry

Drone fly (*Eristalis tenax*), the mimic

Honeybee (*Apis mellifera*), the model

b. Müllerian mimicry

Heliconius erato

Heliconius melpomene

Heliconius melpomene

associate the aposematic pattern with the sting. Many predators quickly learn to avoid black-and-white skunks, yellow-banded wasps, or orange monarch butterflies because they associate the warning display with pain, illness, or severe indigestion.

But for every ploy there is a counterploy, and some predators eat mainly dangerous prey. Bee-eaters (family Meropidae) are birds that eat hymenopterans (bees and wasps). Some individual African lions specialize in porcupines, and animals such as hedgehogs (genus *Erinaceus*) seem able to eat almost anything and show no ill effects. Indeed, some hedgehogs first lick toads and then their own spines, anointing them with toad venom. Hedgehog spines treated with toad venom are more irritating (at least to people) than untreated ones, enhancing their defensive impact.

30.3h Mimicry Is Advertising, Whether True or False

If predators learn to recognize warning signals, it is no surprise that many harmless animals' defences are based on imitating (mimicking) species that are dangerous or distasteful. **Mimicry** occurs when one species evolves to resemble another **(Figure 30.10)**. **Batesian mimicry**, named for English naturalist Henry W. Bates, occurs when a palatable or harmless species (the **mimic**) resembles an unpalatable or poisonous one (the **model**). Any predator that eats the poisonous model and suffers accordingly will subsequently avoid other organisms that resemble it. However, the predator must survive the encounter. **Müllerian mimicry**, named for German zoologist Fritz Müller, involves two or more unpalatable

species looking the same, presumably to reinforce lessons learned by a predator that attacks any species in the mimicry complex.

For mimicry to work, the predator must learn (see Chapter 47) to recognize and then avoid the prey. The more deadly the toxin, the less likely an individual predator is to learn by its experience. In many cases, predators learn by watching the discomfort of a conspecific that has eaten or attacked an aposematic prey.

Plants often use toxins to protect themselves against herbivores. Is this also true of toxins in mushrooms (see Chapter 19)?

30.3i There Is No Perfect Defence

Helmets protect soldiers, skiers, motorcyclists, and cyclists, but not completely because no defence provides perfect protection. Some predators learn to circumvent defences. Many predators learn to deal with a diversity of prey species and a variety of defensive tactics. Orb web spiders confronting a captive in a web adjust their behaviour according to the prey. They treat moths differently from beetles, and they treat bees in yet another way. When threatened by a predator, headstand beetles raise their rear ends and spray a noxious chemical from a gland at the tip of the abdomen. This behaviour deters many would-be predators. But experienced grasshopper mice from western North America circumvent this defence. An experienced mouse grabs the beetle, averts its face (to avoid the spray), turns the beetle upside down so that the gland discharges into the ground, and eats the beetle from the head down **(Figure 30.11)**.

a. *Eleodes* **beetle**

b. Grasshopper mouse

Figure 30.11

Defence and learning. When confronted by a predator, **(a)** the headstand beetle (*Eleodes longicollis*) raises its abdomen and sprays a noxious chemical from its hind end. Experienced grasshopper mice (*Onychomys leucogaster*) **(b)** thwart the beetle's defence by grabbing it, turning it upside down, and eating it headfirst.

30.4 Competition

Different species using the same limiting resources experience **interspecific competition** (competition between species). Competing individuals may experience increased mortality and decreased reproduction, responses similar to the effects of intraspecific competition. Interspecific competition can reduce the size and population growth rate of one or more of the competing populations.

Community ecologists identify two main forms of interspecific competition. In **interference competition**, individuals of one species harm individuals of another species directly. Here animals may fight for access to resources, as when lions chase smaller predators, such as hyenas, jackals, and vultures, from their kills. Many plant species, including creosote bushes (see Figure 29.4, Chapter 29), release toxic chemicals into the soil, preventing other plants from growing nearby.

In **exploitative competition**, two or more populations use (*exploit*) the same limiting resource, and the presence of one species reduces resource availability for others. Exploitative competition need not involve snout-to-snout or root-to-root confrontations. In the deserts of the U.S. Southwest, many bird and ant species eat mainly seeds, and each seed-eating species may deplete the food supply available to others without necessarily encountering each other.

30.4a Competition and Niches: When Resources Are Limited

In the 1920s, the Russian mathematician Alfred J. Lotka and the Italian biologist Vito Volterra independently proposed a model of interspecific competition, modifying the logistic equation (see Chapter 29) to describe the effects of competition between two species. In their model, an increase in the size of one population reduces the population growth rate of the other.

In the 1930s, a Russian biologist, G. F. Gause, tested the model experimentally. He grew cultures of two *Paramecium* species (ciliate protozoans) under constant laboratory conditions, regularly renewing food and removing wastes. Both species feed on bacteria suspended in the culture medium. When grown alone, each species exhibited logistic growth. When grown together in the same dish, *Paramecium aurelia* persisted at high density, but *Paramecium caudatum* was almost eliminated **(Figure 30.12)**. These results inspired Gause to define the **competitive exclusion principle**. Populations of two or more species cannot coexist indefinitely if they rely on the same limiting resources and exploit them in the same way. One species inevitably harvests resources more efficiently; produces more offspring than the other; and, by its actions, negatively affects the other species.

Ecologists developed the concept of the **ecological niche** to visualize resource use and the potential for interspecific competition in nature. They define a population's niche by the resources it uses and the environmental conditions it requires over its lifetime. In this context, niche includes food, shelter, and nutrients, as well as nondepletable abiotic conditions such as light intensity and temperature. In theory, an almost infinite variety of conditions and resources could contribute to a population's niche. In practice, ecologists usually identify the critical resources for

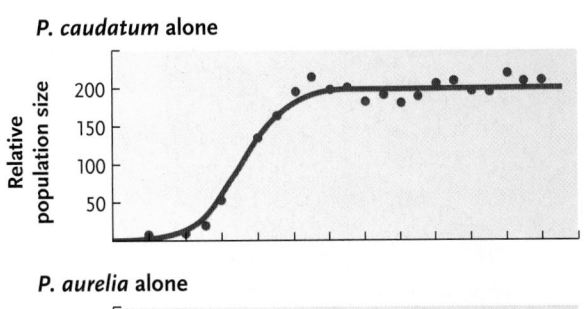

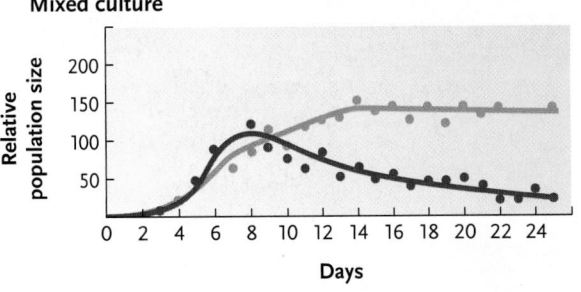

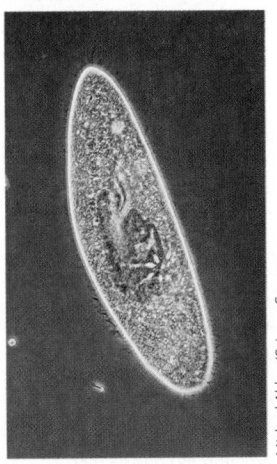

Paramecium caudatum

Paramecium aurelia

Experimental Research Figure 30.12 Gause's experiments on interspecific competition in *Paramecium*.

Taipoxin: Snake Presynaptic Phospholipase A$_2$ Neurotoxins

Snake venoms are typically a concoction of ingredients designed to immobilize and digest prey. Like those of the venom of nematocysts (see "Nematocysts," p. 720), the effects of snake venom can include symptoms associated with neurotoxins cardiotoxins, hemolytic actions, and digestion (necrosis) of tissues. Not all snakes (or other venomous animals) have the same venom.

Snake presynaptic phospholipase A$_2$ neurotoxins, or SPANs, have neurotoxic effects and work by blocking neuromuscular junctions. Phospholipase A$_2$ activity varies greatly among SPANs. Using mouse neuromuscular junction hemidiaphragm preparations and neurons in culture, M. Rigoni and seven colleagues explored the way in which SPANs work. They used SPANs from single-chain notexin (from *Notechis scutatus*, the eastern tiger snake); a two-subunit B-bungarotoxin (from *Bungarus multicinctus*, the many-banded krait); the three-subunit taipoxin (from *Oxyuranus scutellatus*, the taipan; **Figure 1**); and the five-unit textilotoxin (from *Pseudonaja textilis*, the eastern brown snake).

The results showed that administration of SPANs to neuromuscular junctions causes enlargement of the junctions and reduction in the contents of synaptic vesicles. SPANs also induce exocytosis of neurotransmitters. In other words, SPANs bind nerve terminals via receptors **(Figure 2)**; the

BMCL/Shutterstock.com

Figure 1

The taipan (Oxyuranus spp.) *is an extremely venomous snake from northern Australia and southern New Guinea.*

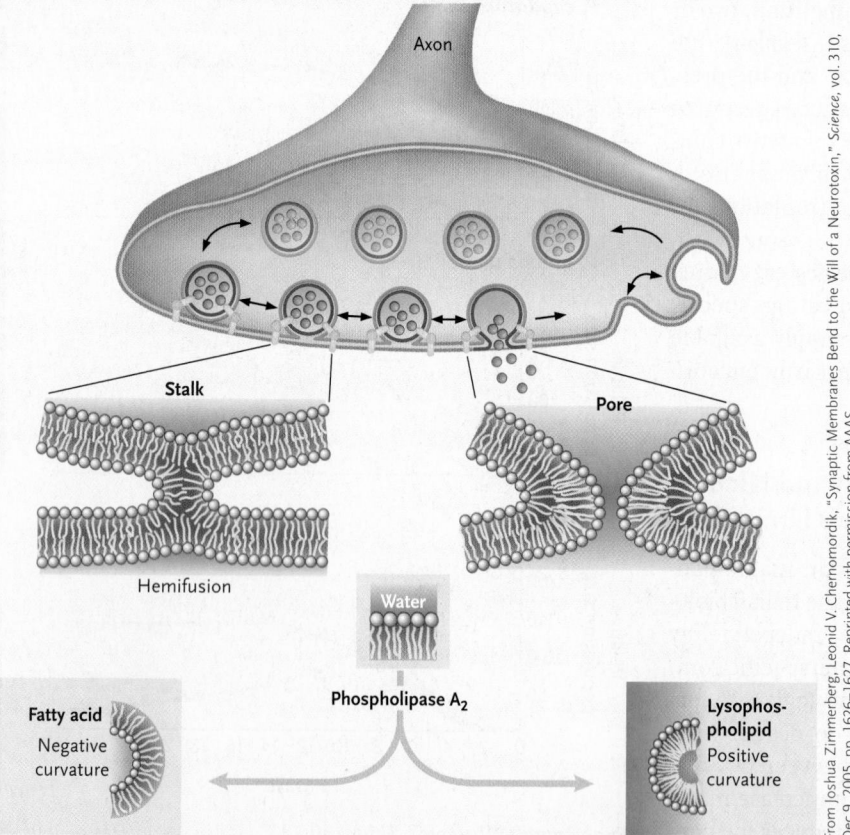

Figure 2

Taipoxin paralyzes the neuromuscular junction by causing membrane fusion between the plasma membranes (tan colour) of presynaptic ganglia and synaptic vesicles, forming a pore that allows mixing of lipids from inner (purple) and outer (green) leaflets. Although this is probably restricted by proteins (yellow ribbons) around the pore, phospholipase A$_2$ changes the curvature at the junction, resulting in the paralysis noted above.

From Joshua Zimmerberg, Leonid V. Chernomordik, "Synaptic Membranes Bend to the Will of a Neurotoxin," *Science*, vol. 310, Dec 9, 2005, pp. 1626–1627. Reprinted with permission from AAAS.

results indicate that venoms can be used to further our understanding of what happens at neuromuscular junctions.

Among extant lepidosauran reptiles (snakes and lizards), two lineages have venom delivery systems, advanced snakes and helodermatid (gila monster) lizards. The view that the evolution of venom systems is fundamental to the radiation of snakes has been supported by the prevalence of venom among snakes and its restriction to just two species of lizards. Using tools of molecular genetics, B. G. Fry and 13 colleagues explored the early evolution of venom systems in lizards and snakes. The ancestral condition, in venomous lizards, is lobed, venom-secreting glands on upper and lower jaws. Advanced snakes and two lizards have more derived venom systems. They have one pair of venom glands, either upper or lower glands. Analysis of venoms indicates that snakes, iguanians (monitor lizards), and anguimorphs form a single clade **(Figure 3)**, suggesting that venom is an ancestral trait in this evolutionary line of reptiles.

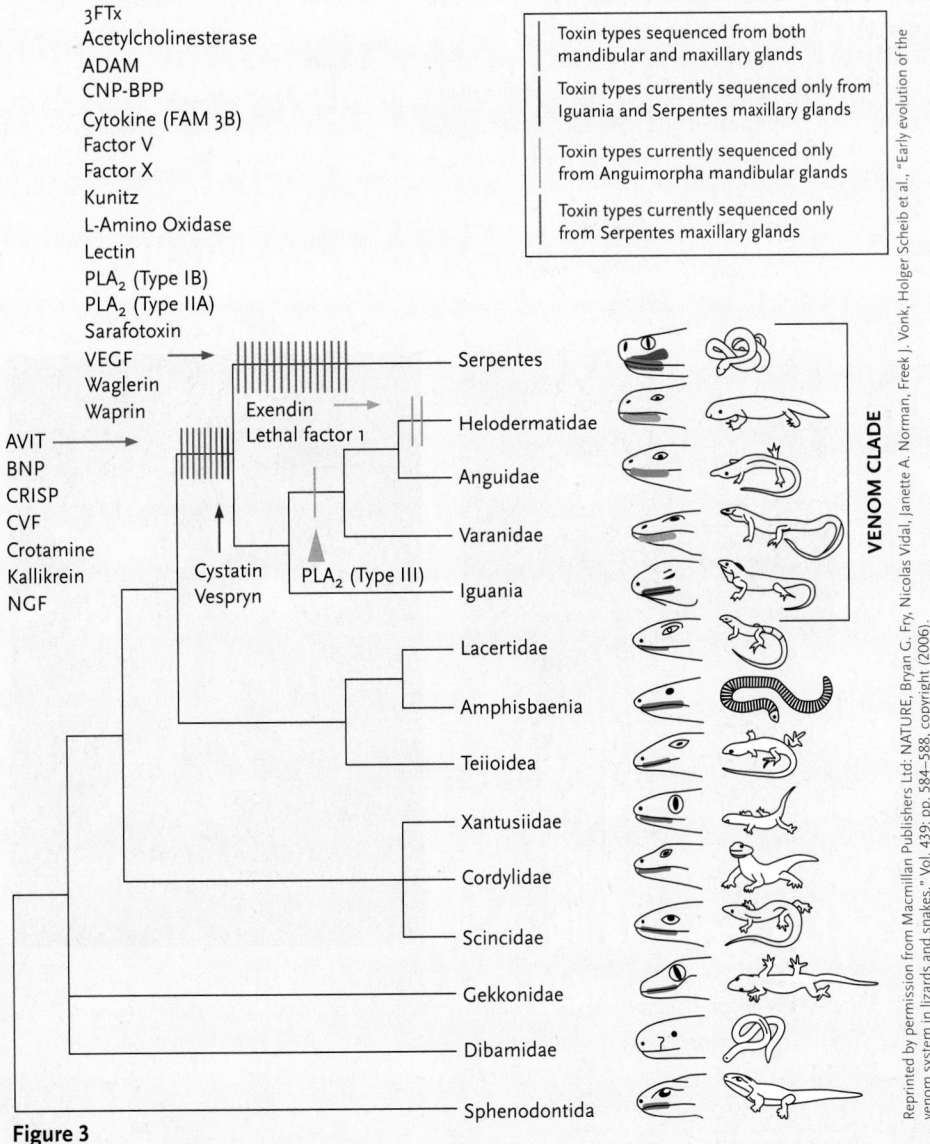

Reprinted by permission from Macmillan Publishers Ltd: NATURE, Bryan G. Fry, Nicolas Vidal, Janette A. Norman, Freek J. Vonk, Holger Scheib et al., "Early evolution of the venom system in lizards and snakes," Vol. 439: pp. 584–588, copyright (2006).

Figure 3
Snake and lizard phylogeny based on the appearance of venom. Shown here are the relative glandular development and appearance of toxin recruitment in squamate reptile phylogeny. Glands secreting mucus are blue, ancestral venom glands are red, and derived venom is orange. Elements in venom include the following: three-finger toxins (3FTx); a disintegrin and metalloproteinase (ADAM); C-type natriuretic peptide–bradykinin–potentiating peptide (CNP-BPP); cobra venom factor (CVF); nerve growth factor (NGF); and vascular endothelial growth factor (VEGF).

Nematocysts

Swimmers at ocean beaches in warmer parts of the world can be exposed to stings from the Portuguese man-o'-war. In the United States, at least three human deaths have been caused by exposure to the venom of its nematocysts **(Figure 1).** First aid for someone who has been stung includes (a) using sea water to flush away any tentacles still clinging to the victim (or picking them off if necessary), (b) applying ice or cold packs to the area of the sting(s) and leaving them in place for 5 to 15 minutes, (c) using an inhaled analgesic to reduce pain, and (d) seeking additional medical aid.

Beaches where Portuguese man-o'-war and other jellyfish may occur must be supervised by lifeguards who understand the danger. The inflatable bladders (sails; see **Figure 2**) make Portuguese man-o'-war easy to see in the water, and swimming is not permitted when they are near.

In 1968, lifeguards and others at a beach at Port Stephens, New South Wales, in Australia were surprised and concerned when they realized that some people had been stung by Portuguese man-o'-war when none of these animals had been spotted near the beach. The stinging animals turned out to be sea slugs, *Glaucus* species.

Since 1903, it had been known that sea slugs use nematocysts as a defence.

Glaucus atlanticus **(Figure 3)** feed on the cnidosacs that contain the nematocysts, preferentially selecting and storing those of Portuguese man-o'-war, which have two sizes of nematocysts. The sea slugs take the larger nematocysts that, when discharged, have the longest penetrants. It is likely that the same digestive processes other sea slugs use to extract chloroplasts can be used to extract nematocysts.

This situation demonstrates the versatility of defensive systems in animals.

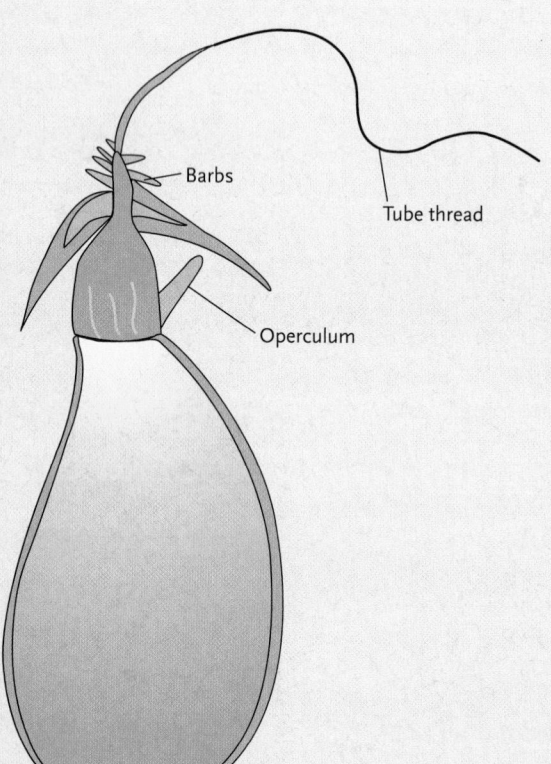

Figure 1

Nematocyst. Nematocysts are stinging cells occurring in animals in the phylum Cnidaria (see Chapter 27). Nematocysts of Physalia physalis (Portuguese man-o'-war; Figure 2) contain toxic proteins and at least six or seven enzymes that can be injurious. Unpurified nematocyst venoms have several effects, some of which can be lethal. The venoms can be neurotoxic, cardiotoxic, or myotoxic or cause lysis of red blood cells or mitochondria. Like other venoms, nematocyst venom from the Portuguese man-o'-war can interfere with the transport of Na^+ and Ca^{2+} ions.

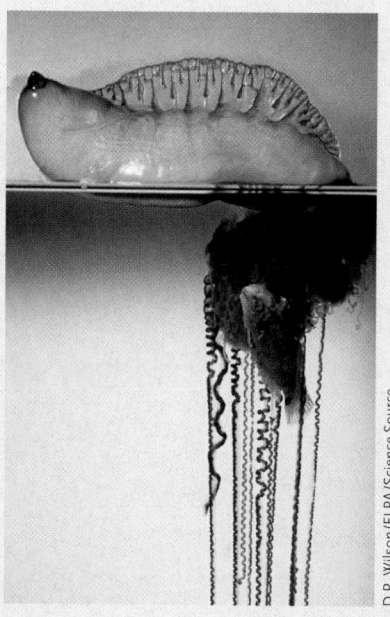

D.P. Wilson/FLPA/Science Source

Figure 2
Physalia physalis, *Portuguese man-o'-war.*

Dr. T.E. Thompson/Science Source

Figure 3
Glaucus atlanticus, *a sea slug that ingests nematocysts from* Physalia physalis.

which populations might compete. Sunlight, soil moisture, and inorganic nutrients are important resources for plants, so differences in leaf height and root depth, for example, can affect plants' access to these resources. Food type, food size, and nesting sites are important for animals. When several species coexist, they often use food and nest resources in different ways.

Ecologists distinguish the **fundamental niche** of a species, the range of conditions and resources it could tolerate and use, from its **realized niche**, the range of conditions and resources it actually uses in nature. Realized niches are smaller than fundamental niches, partly because all tolerable conditions are not always present in a habitat and partly because some resources are used by other species. We can visualize competition between two populations by plotting their fundamental and realized niches with respect to one or more resources **(Figure 30.13)**. If the fundamental niches of two populations overlap, they *might* compete in nature.

Observing that several species use the same resource does not demonstrate that competition occurs (or does not occur). All terrestrial animals consume oxygen but do not compete for oxygen because it is usually plentiful. Nevertheless, two general observations provide *indirect* evidence that interspecific competition may have important effects.

Resource partitioning occurs when several sympatric (living in the same place) species use different resources or the same resources in different ways. Although plants might compete for water and dissolved nutrients, they may avoid competition by partitioning these resources, collecting them from different depths in the soil **(Figure 30.14)**. This allows coexistence of different species.

Character displacement can be evident when comparing species that are sometimes sympatric and sometimes allopatric (living in different places). Allopatric populations of some animal species are morphologically similar and use similar resources, whereas sympatric populations are morphologically different and use different resources. Differences between sympatric species allow them to coexist without competing. Allen Keast studied honey-eaters (family Meliphagidae), a group of birds from Australia, to illustrate this situation. In mainland Australia, up to six species in the genus *Melithreptus* occur in some habitats. Just off the coast of Kangaroo Island, there are two species. When two species are sympatric, each feeds in a wider range of situations than when six species live in the same area, reflecting the use of broader niches. Behavioural and morphological differences are evident when species are compared between the different situations. Although well known for his work on birds, Keast also studied communities of fish. He spent most of his academic career at Queen's University in Kingston, Ontario.

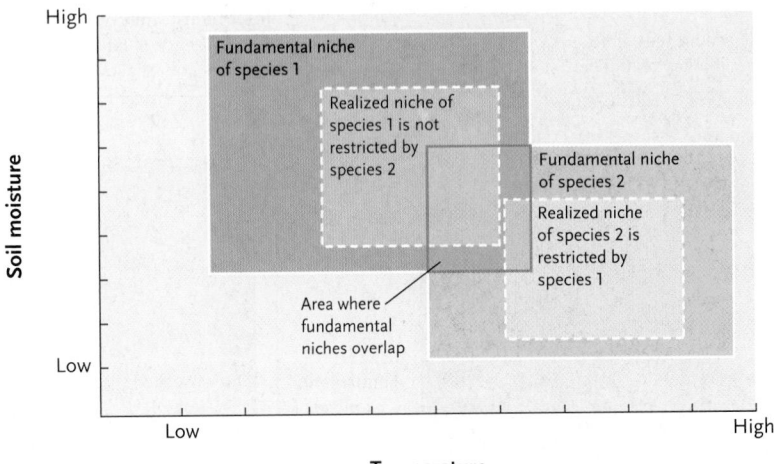

Figure 30.13

Fundamental versus realized niches. In this hypothetical example, both species 1 and species 2 can survive intermediate temperature conditions, as indicated by the shading where their fundamental niches overlap. Because species 1 actually occupies most of this overlap zone, its realized niche is not much affected by the presence of species 2. In contrast, the realized niche of species 2 is restricted by the presence of species 1, and species 2 occupies warmer and drier parts of the habitat.

Data on resource partitioning and character displacement suggest, but do not prove, that interspecific competition is an important selective force in nature. To demonstrate *conclusively* that interspecific competition limits natural populations, one must show that the presence of one population reduces the population size or density of its presumed competitor. In a classic field experiment, Joseph Connell examined

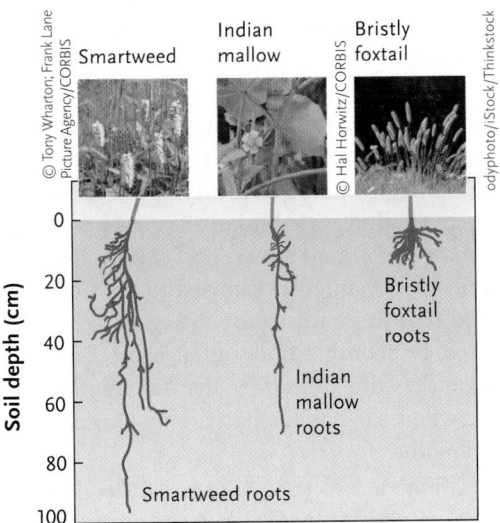

Figure 30.14

Resource partitioning. The root systems of three plant species that grow in abandoned fields partition water and nutrient resources in soil. Bristly foxtail grass (*Setaria faberi*) has a shallow root system, Indian mallow (*Abutilon theophrasti*) has a moderately deep taproot, and smartweed (*Polygonum pennsylvanicum*) has a deep taproot that branches at many depths.

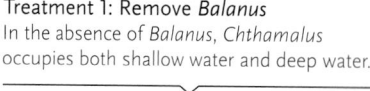

Realized niches before
experimental treatments.

Treatment 1: Remove *Balanus*
In the absence of *Balanus*, *Chthamalus*
occupies both shallow water and deep water.

Treatment 2: Remove *Chthamalus*
In the absence of *Chthamalus*, *Balanus*
still occupies only deep water.

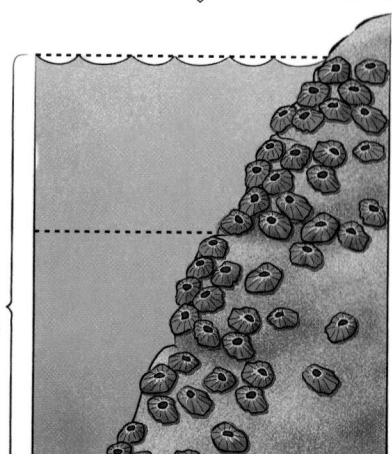

High tide

Chthamalus

Low tide

Balanus

Fundamental
niche of
Chthamalus

Fundamental
niche of
Balanus

Experimental Research Figure 30.15 Demonstration of competition between two species of barnacles.

competition between two barnacle species **(Figure 30.15).** Connell first observed the distributions of both species of barnacles in undisturbed habitats to establish a reference baseline. *Chthamalus stellatus* is generally found in shallow water on rocky coasts, where it is periodically exposed to air. *Balanus balanoides* typically lives in deeper water, where it is usually submerged.

In the absence of *Balanus* on rocks in deep water, larval *Chthamalus* colonized the area and produced a flourishing population of adults. *Balanus* physically displaced *Chthamalus* from these rocks. Thus, interference competition from *Balanus* prevents *Chthamalus* from occupying areas where it would otherwise live. Removal of *Chthamalus* from rocks in shallow water did not result in colonization by *Balanus*. *Balanus* apparently cannot live in habitats that are frequently exposed to air. Connell concluded that there was competition between the two species. But competition was asymmetrical because *Chthamalus* did not affect the distribution of *Balanus*, whereas *Balanus* had a substantial effect on *Chthamalus*.

30.4b Symbiosis: Close Associations

Symbiosis occurs when one species has a physically close ecological association with another (*sym* = together; *bio* = life; *sis* = process). Biologists define three types of symbiotic interactions: commensalism, mutualism, and parasitism (see Table 30.1, p. 711).

In **commensalism,** one species benefits from and the other is unaffected by the interactions. Commensalism

appears to be rare in nature because few species are unaffected by interactions with another. One possible example is the relationship between Cattle Egrets (*Bubulcus ibis*, birds in the heron family) and the large grazing mammals with which they associate **(Figure 30.16).** Cattle Egrets eat insects and other small animals that their commensal partners flush from grass. Feeding rates of Cattle Egrets are higher when they associate with large grazers than when they do not. The birds clearly benefit from this interaction, but the presence of birds has no apparent positive or negative impact on the mammals.

Figure 30.16

Commensalism. Cattle Egrets (*Bubulcus ibis*) feed on insects and other small animals flushed by the movements of large grazing animals such as African elephants (*Loxodonta africana*).

a. Flowering yucca plant

SeanPavonePhoto/Shutterstock.com

b. Female yucca moth

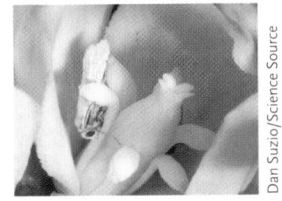

Dan Suzio/Science Source

A female yucca moth uses highly modified mouthparts to gather the sticky pollen and roll it into a ball. She carries the pollen to another flower, and after piercing its ovary wall, she lays her eggs. She then places the pollen ball into the opening of the stigma.

c. Yucca moth larva

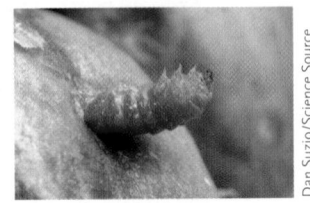

Dan Suzio/Science Source

When moth larvae hatch from the eggs, they eat some of the yucca seeds and gnaw their way out of the ovary to complete their life cycle. Enough seeds remain undamaged to produce a new generation of yuccas.

Figure 30.17

Mutualism between plants and animals. Several species of yucca plants (*Yucca* species) are each pollinated exclusively by one species of moth (*Tegeticula* species). The adult moth appears at the time of year when the plants are flowering. These species are so mutually interdependent that the larvae of each moth species can feed on only one species of yucca, and each yucca plant can be pollinated by only one species of moth. Extinction of the pollinator will usually lead to extinction of the plant. Most plant–animal mutualisms are less specific.

In **mutualism**, both partners benefit. Mutualism appears to be common and includes coevolved relationships between flowering plants and animal pollinators. Animals that feed on a plants' nectar or pollen carry the plants' gametes from one flower to another **(Figure 30.17)**. Similarly, animals that eat fruits disperse the seeds and "plant" them in piles of nutrient-rich feces. Mutualistic relationships between plants and animals do not require active cooperation, as each species simply exploits the other for its own benefit. Some associations between bacteria and plants are mutualistic. Perhaps the most important of these associations is between *Rhizobium* and leguminous plants such as peas, beans, and clover (see Chapter 37).

Mutualistic relationships between animal species are common. Cleaner fishes, small marine species, feed on parasites attached to the mouths, gills, and bodies of larger fishes **(Figure 30.18)**. Parasitized fishes hover motionless while cleaners remove their ectoparasites. The relationship is mutualistic because cleaner fishes get a meal, and larger fishes are relieved of parasites.

The relationship between the bull's horn acacia tree (*Acacia cornigera*) of Central America and a species of small ants (*Pseudomyrmex ferruginea*) is a highly coevolved mutualism **(Figure 30.19, p. 724)**. Each acacia is inhabited by an ant colony that lives in the tree's swollen thorns. Ants swarm out of the thorns to sting, and sometimes kill, herbivores that touch the tree. Ants also clip any vegetation that grows nearby. Acacia trees colonized by ants grow in a space free of herbivores and competitors, and occupied trees grow faster and produce more seeds than unoccupied trees. In return, the plants produce sugar-rich nectar consumed by adult ants and protein-rich structures that the ants feed to their larvae. Ecologists describe the coevolved mutualism between these species as *obligatory,* at least for the

ants, because they cannot subsist on any other food sources.

Many animals eat honey and sometimes also the bees that produce it. In Africa, Greater Honey-Guides (*Indicator indicator*) are birds that use a special guiding display to lead humans to beehives. In one tribe of Kenyans, the honey-gathering Borans use a special whistle to call *I. indicator*. Boran honey-gatherers that work with Greater Honey-Guides are much more efficient at finding beehives than those working alone. When the honey-gatherer goes to the hive and raids it to obtain honey, Greater Honey-Guides help themselves to bee larvae, left-over honey, and wax. Although *I. indicator* are said also to guide ratels (honey badgers, *Mellivora capensis*) to beehives, there are no firm data supporting this assumption.

Cleaner wrasse

Rand McMeins/Moment/ Getty Images

Figure 30.18

Mutualism between animal species. A large potato cod (*Epinephelus tukula*) from the Great Barrier Reef in Australia remains nearly motionless in the water while a striped cleaner wrasse (*Labroides dimidiatus*) carefully removes and eats ectoparasites attached to its lip. The potato cod is a predator; the striped cleaner wrasse is a potential prey. Here the mutualistic relationship supersedes the possible predator–prey interaction.

a.

b.

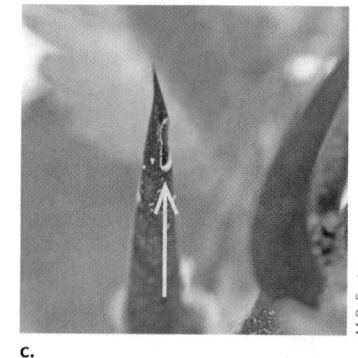

c.

d.

e.

Marie Read Natural History Photography

M.B. Fenton

Figure 30.19

Highly coevolved mutualisms. **(a)** Bull's horn acacia trees (*Acacia cornigera*) provide colonies for small ants (*Pseudomyrmex ferruginea*). In addition to providing homes (domatia—yellow arrow in **(b)**) in hollow thorns, the acacia also provides food for the ants (nectar—yellow arrow in **(c)**). In the same area in the New World tropics, a cowhorn orchid **(d)** is patrolled by ants (yellow arrow) that are also housed in **(c)** domatia (yellow arrow in **(e)**) on the plant.

In **parasitism**, one species—the parasite—uses another—the host—in a way that is harmful to the host. Parasite–host relationships are often considered to be specialized predator–prey relationships because one population of organisms feeds on another. But parasites rarely kill their hosts because a dead host is not a continuing source of nourishment.

Endoparasites, such as tapeworms, flukes, and roundworms, live *within* a host. Many endoparasites acquire their hosts passively when a host accidentally ingests the parasites' eggs or larvae. Endoparasites generally complete their life cycle in one or two host individuals. Ectoparasites, such as leeches, aphids, and mosquitoes, feed on the exterior of the host. Most animal ectoparasites have elaborate sensory and behavioural mechanisms, allowing them to locate specific hosts, and they feed on numerous host individuals during their lifetimes. Plants such as mistletoes (genus *Phoradendron*) live as ectoparasites on the trunks and branches of trees; their roots penetrate the host's xylem and extract water and nutrients. These differ from epiphytes, such as bromeliads or Spanish moss, that use the host only as a base. Other plants are root parasites, for example, *Conopholis americana*.

Not all parasites feed directly on a host's tissues. Some bird species are brood parasites, laying their eggs in the host's nest. It is quite common for female birds such as Canvasback Ducks, Brown-headed Cowbirds,

and Kirtland's Warblers (*Aythya valisineria*) to lay their eggs in the nests of conspecifics (members of the same species). Some species of songbirds often lay some eggs in the nests of others, a variation on hedging of genetic bets and on extra-pair copulations (see Chapters 41 and 47, respectively). Brood parasitism is the next level of escalation in this spectrum of parasitism. Brown-headed Cowbirds (*Molothrus ater*), like other brood parasites, always lay their eggs in the nest of other species, leaving it to the host parents to raise their young. This behaviour can have drastic repercussions for host species. Brown-headed Cowbirds, for instance, have played a large role in the near-extinction of Kirtland's Warblers (*Dendroica kirtlandii*).

The feeding habits of insects called parasitoids fall somewhere between true parasitism and predation. A female parasitoid lays her eggs in a larva or pupa of another insect species, and her young consume the tissues of the living host. But the parasitoid spends part of its life cycle as free living. It is the larval stage that usually kills the host. Because the hosts chosen by most parasitoids are highly specific, agricultural ecologists often try to use parasitoids to control populations of insect pests.

One of the most striking and perhaps startling example of symbioses is the rich biota of prokaryotes and Protozoa that inhabit our digestive tracts. This biota significantly expands our capacity for extracting nutrients and other important factors from the food we ingest. The producers of "probiotic" foods depend upon our being impressed by the importance of our symbionts.

STUDY BREAK

1. What is interspecific competition? What two types of interspecific competition have been identified by community ecologists?
2. Describe the competitive exclusion principle.
3. What is a species' ecological niche? How does a fundamental niche differ from a realized niche?

30.5 The Nature of Ecological Communities

The interactions that occur among species in an ecological community can be broadly categorized as antagonistic or mutually beneficial **(Figure 30.20)**. Trophic interactions are those associated with consumption (antagonistic)—one species eating another—the usual situation portrayed in food webs. Mutually beneficial interactions include, for example, those between flowering plants and their insect pollinators. Understanding community dynamics requires knowledge of the structure of community networks (e.g., food webs; see Chapter 31) as well as information about how structure influences the extinction or persistence of species. An overview of the dynamics of an ecological community is obtained through a combination of fieldwork and attendant statistical analysis of data to document ecosystem architecture. The second element, knowledge of the influence of architecture on species persistence, emerges from mathematical modelling.

To explore the nature of ecosystems, Elisa Thebault and Colin Fontaine examined pollination (mutualistic) and plant–herbivore (trophic) systems **(Figure 30.21, p. 726)**. They found that the structure of the network favouring ecosystem **stability** differs between trophic (herbivore) and mutually beneficial (pollination) networks. In pollination networks, the elements are highly connected and nested, promoting stability of communities. In herbivore networks, stability is greater in structures that are compartmentalized and weakly connected. The work identifies features that affect the stability of ecosystems, potentially informing those working to effect conservation at the system level.

Ecotones, the borders between communities, are sometimes wide transition zones. Ecotones are generally species rich because they include plants and animals from both neighbouring communities, as well as some species that thrive only under transitional conditions. Although ecotones are usually relatively broad, places where there is a discontinuity in a critical resource or important abiotic factor may have a sharp community boundary. Chemical differences between soils derived from serpentine rock and sandstone establish sharp boundaries between communities of native California wildflowers and introduced European grasses.

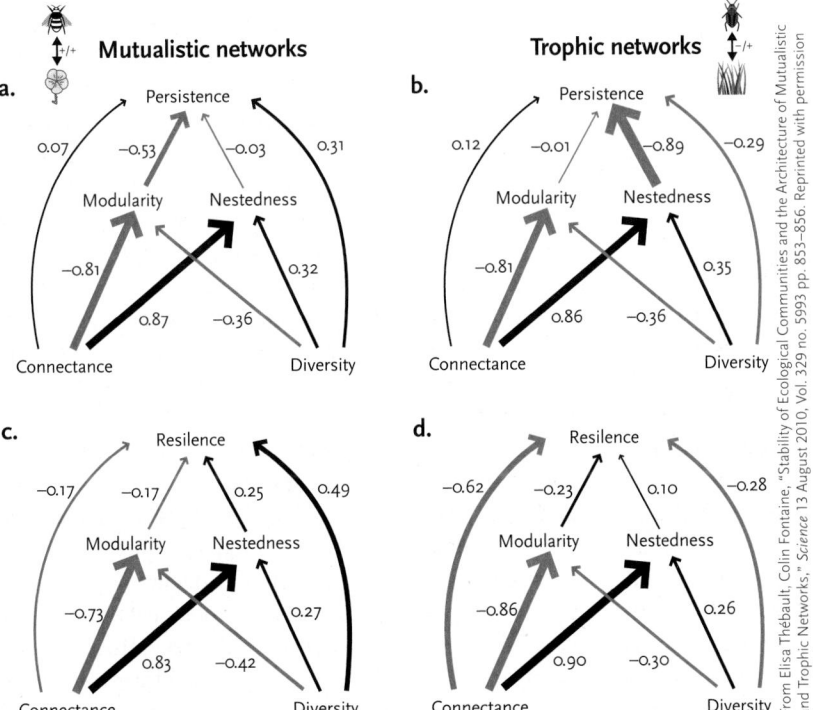

Figure 30.20

The persistence **(a, b)** and resilience **(c, d)** of pollinating and herbivore ecosystem patterns are summarized, revealing important differences. The thickness of arrows, scaled to standardized coefficients, illustrates the relative strength of the effects. Red identifies negative effects; black, positive ones. There is a further comparison of the effects of connectance and diversity, comparing direct and indirect effects, considering modularity and nestedness. The numbers in each diagram indicate the coefficients along the path.

a. Interactive hypothesis

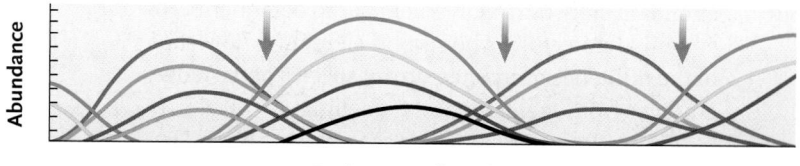

The interactive hypothesis predicts that species within communities exhibit similar distributions along environmental gradients (indicated by the close alignment of several curves over each section of the gradient) and that boundaries between communities (indicated by arrows) are sharp.

b. Individualistic hypothesis

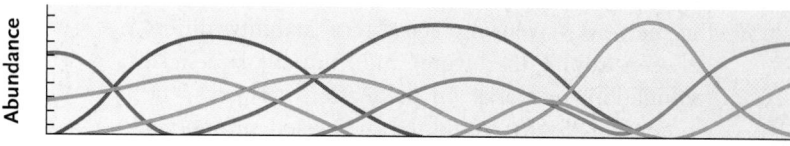

The individualistic hypothesis predicts that species distributions along the gradient are independent (indicated by the lack of alignment of the curves) and that sharp boundaries do not separate communities.

c. Siskiyou Mountains

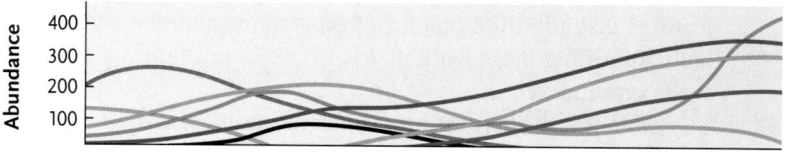

Most gradient analyses support the individualistic hypothesis, as illustrated by distributions of tree species along moisture gradients in Oregon's Siskiyou Mountains and Arizona's Santa Catalina Mountains.

d. Santa Catalina Mountains

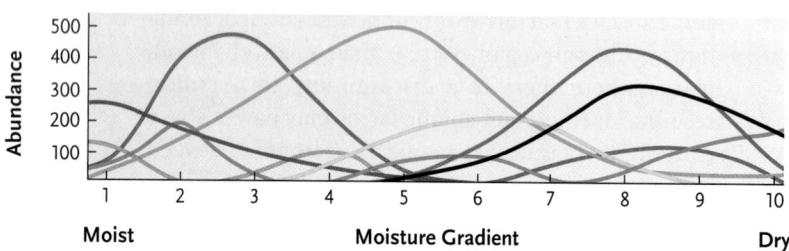

Figure 30.21
Two views of ecological communities.
Each graph line indicates a different species.

STUDY BREAK

1. Distinguish between antagonistic and mutualistic architectural structures in ecosystems.
2. Are ecotones generally species rich or species poor?

30.6 Community Characteristics

Growth forms (sizes and shapes) of plants vary markedly in different environments, so the appearances of plants can often be used to characterize communities. Warm, moist environments support complex vegetation with multiple vertical layers. Tropical forests include a canopy formed by the tallest trees, an understorey of shorter trees and shrubs, and a herb layer under openings in the canopy. Vinelike lianas and epiphytes grow on the trunks and branches of trees **(Figure 30.22)**. In contrast, physically harsh environments are occupied by low vegetation with simple structure. Trees on mountainsides buffeted by cold winds are short, and the plants below them cling to rocks and soil.

Other environments support growth forms between these extremes.

Communities differ greatly in **species richness**, the number of species that live within them. The harsh environment on a low desert island may support just a few species of microorganisms, fungi, algae, plants, and arthropods. In contrast, tropical forests that grow under milder physical conditions include many thousands of species. Ecologists have studied global patterns of species richness (see Chapter 32) for decades. Today, as human disturbance of natural communities has reached a crisis point, conservation biologists try to understand global patterns of species richness to determine which regions of Earth are most in need of preservation.

The relative abundances of species vary across communities. Some communities have one or two abundant species and a number of rare species. In others, the species are represented by more equal numbers of individuals. In a **temperate deciduous forest** in southern Quebec, red oak trees (*Quercus rubra*) and sugar maples (*Acer saccharum*) might together account for nearly 85% of the trees. A tropical forest in Costa Rica may have more than 200 tree species, each making up a small percentage of the total.

Some Perils of Mutualism

Living organisms offer many examples of mutualistic interactions in which one species (or group of species) shows varying levels of dependence on another or others. Mutualistic situations can place species on the edge of survival. Where one species depends entirely on another, the extinction of one must lead to change or the extinction of both (e.g., Dodos—see Chapter 32—and yucca plants and their moths). There are many other examples of close relationships, including a desert melon (*Cucumis humifructus*) that depends perhaps entirely on aardvarks (*Orycteropus afer*) for dispersal of its seeds. Aardvarks sniff out the underground melons, dig them up, and eat them to obtain water. When aardvarks bury their dung, they plant the melon's seeds and fertilize them. The survival of the melon depends on the aardvark but not vice versa.

Mutualistic interactions between species can be even more complex. In the African **savannah**, ants often live in mutualistic relationships with trees. In east Africa, whistling thorn acacia trees (*Acacia drepanolobium*) are host to four species of ants (see Figure 30.19, p. 724). One species of ant (*Crematogaster mimosae*) in particular depends on room (hollows in swollen thorns, called domatia) and board (carbohydrates secreted from extrafloral glands and the bases of leaves) provided by the trees. Another species of ant (*Crematogaster sjostedti*) also lives on the trees but usually nests in holes made by cerambycid beetles that burrow into and harm the trees.

The ants, particularly *C. mimosae*, attack animals that attempt to browse on the foliage or branches of *A. drepanolobium*. They deter many herbivores, from large mammals to wood-boring beetles (such as cerambycids). If large, browsing mammals are excluded from the area, *A. drepanolobium* produce fewer domatia and fewer carbohydrates for *C. mimosae*. The decline in this species of ant leads to higher damage by cerambycid beetles and increases in populations of *C. sjostedti*.

Many other plants also use ants as mercenaries (see Figure 30.19), and it is becoming clear that survival of these systems depends on the continued presence of participating species.

Figure 30.22

Layered forests. Tropical forests, such as one near the Mazaruni River in Guyana (South America), include a canopy of tall trees and an understorey of short trees and shrubs. Huge vines (lianas) climb through the trees, eventually reaching sunlight in the canopy. Epiphytic plants grow on trunks and branches, increasing the structural complexity of the habitat.

The factors underlying diversity and community structure can be expected to vary among groups of organisms, and the interactions between very different groups of organisms can have positive effects on both. Using an experimental mycorrhizal plant system (see Chapter 25), H. Maherali and J. N. Klironomos found that after one year, the species richness of mycorrhizal fungi correlated with higher plant productivity. In turn, the diversity and species richness of mycorrhizal fungi were highest when their starting community had more distinct evolutionary lineages. This example illustrates the importance of diversity and interactions.

30.6a Measuring Species Diversity and Evenness: Calculating Indices

The number of species is the simplest measure of diversity, so a forest with four tree species has higher **species diversity** than one with two tree species. But there can

Figure 30.23

Species diversity. In this hypothetical example, each of three samples of forest communities (A, B, and C) contains 50 trees. Indices allow biologists to express the diversity of species and evenness of numbers (see Table 30.2).

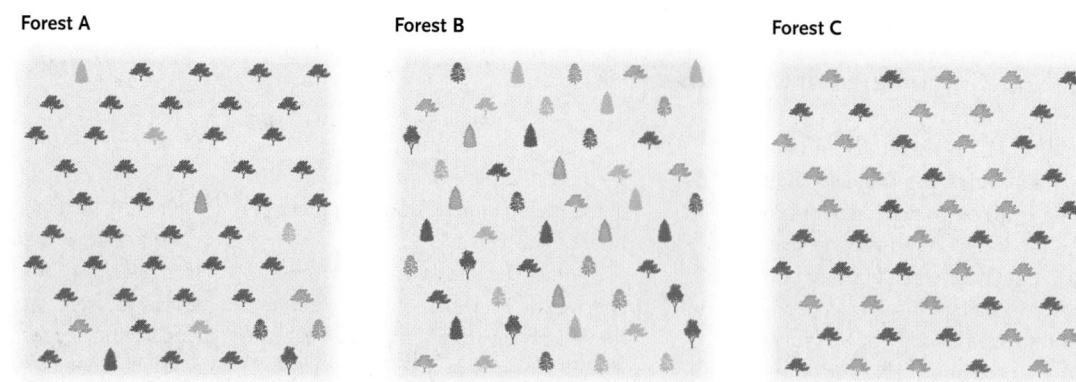

Forest A   Forest B   Forest C

be more to measuring diversity than just counting species. Biologists use indices of diversity to facilitate comparison of data sets documenting the numbers of species and of individuals. Shannon's index of diversity (H'), one commonly used measure, is calculated using the formula

$$H' = -\sum_{i=1}^{S} p_i \ln p_i$$

where S is the total number of species in the community (richness), p_i is the proportion of S made up by species I, and ln is the natural logarithm.

Another index, Shannon's evenness index (E_H), is calculated using the formula

$$E_H = \frac{H'}{\ln S}$$

where ln S is the natural logarithm of the number of species. Evenness is an indication of the mixture of species. Indices of diversity and evenness allow population biologists to objectively portray and compare the diversity of communities.

Use the two indices to compare the 3 forests of 50 trees each **(Figure 30.23)**. The number of species and number of individuals of each species in each forest are shown in **Table 30.2.** In Table 30.2, the values of H' and E_H indicate the diversity of the three hypothetical forests and the evenness of species representations. Lower values of H' and E_H suggest communities with few species (low H' values) or uneven distribution (low E_H values). Higher values of H' and E_H suggest a richer array of species with evenly distributed individuals.

Measures of diversity can be used to advantage. Ecologists refer to α diversity to represent the numbers of sympatric species in one community and β diversity to depict the numbers in a collection of communities. The number of herbivorous Lepidoptera species in one national park is α diversity, whereas β diversity is the number of species in the country in which the park is located. The trend to establish parks that cross international boundaries is a step toward recognizing the reality that political and biological boundaries can be quite different. Measures of diversity can be used directly in some conservation plans (see Chapter 32).

Table 30.2 — Shannon's Indices for Measuring Diversity and Evenness

Numbers of Individuals Per Species

	Forest A*	Forest B*	Forest C*
Species 1	39	5	25
Species 2	2	5	25
Species 3	2	5	0
Species 4	1	5	0
Species 5	1	5	0
Species 6	1	5	0
Species 7	1	5	0
Species 8	1	5	0
Species 9	1	5	0
Species 10	1	5	0
Shannon Indices			
H' diversity	0.6	2.3	0.7
E_h evenness	0.26	1.0	1.0

*Forests from Figure 30.23.

30.6b Trophic Interactions: Between Nourishment Levels

Every ecological community has trophic structure (*troph* = nourishment), comprising all plant–herbivore, predator–prey, host–parasite, and potential competitive interactions **(Figure 30.24).** We can visualize the trophic structure of a community as a hierarchy of **trophic levels,** defined by the feeding relationships among its species (see Figure 30.24a). Photosynthetic organisms are primary producers, the first trophic level. Primary producers are photoautotrophs (*auto* = self) because they capture sunlight and convert it into chemical energy that is used to make larger organic molecules that plants can use directly. Plants are the main primary producers in terrestrial communities. Multicellular algae and plants are the major primary producers in shallow freshwater and marine environments, whereas photosynthetic protists and cyanobacteria play that role in deep, open water.

All **consumers** in a community (animals, fungi, and diverse microorganisms) are heterotrophs

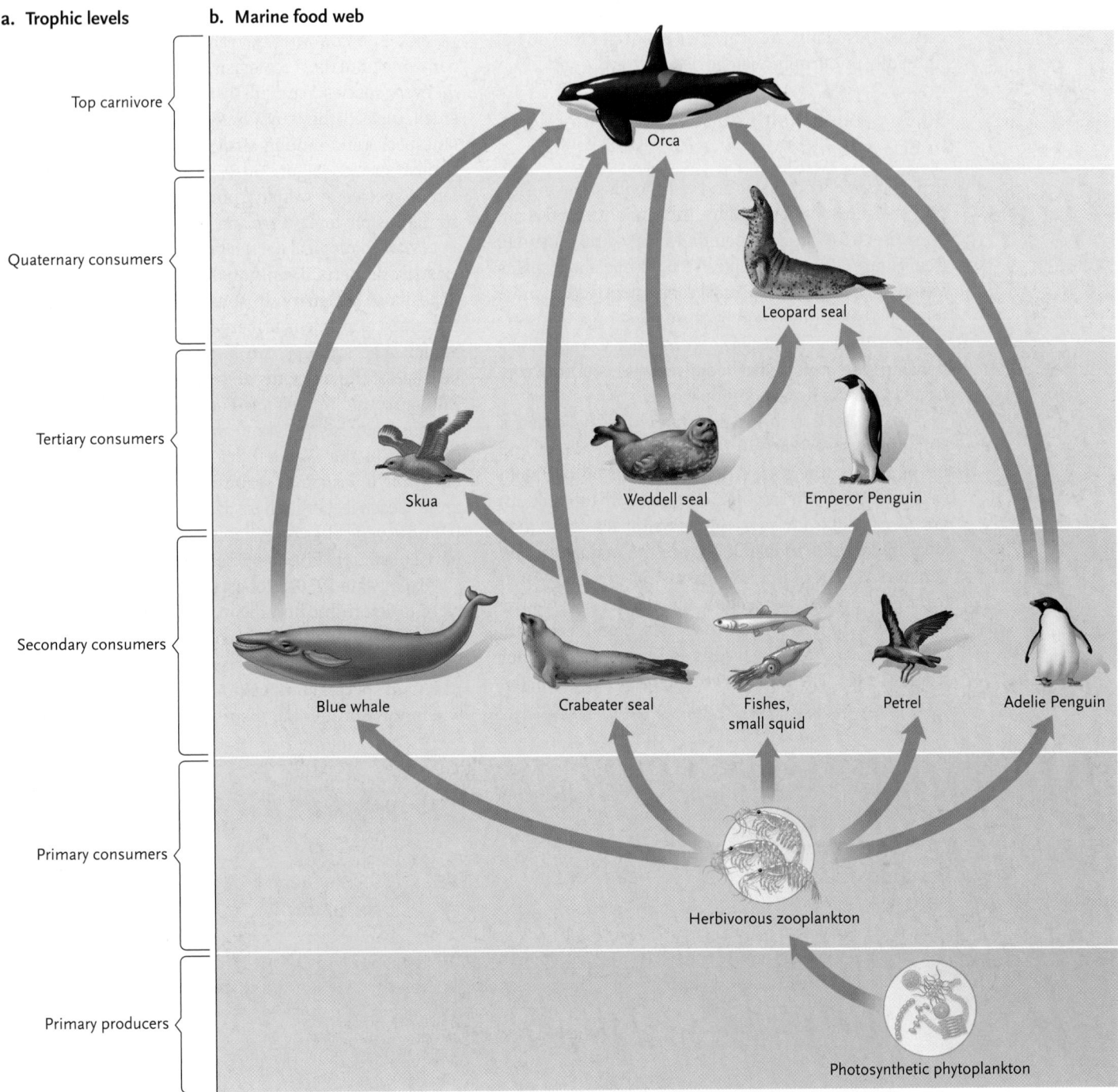

a. Trophic levels
b. Marine food web

Top carnivore — Orca

Quaternary consumers — Leopard seal

Tertiary consumers — Skua, Weddell seal, Emperor Penguin

Secondary consumers — Blue whale, Crabeater seal, Fishes, small squid, Petrel, Adelie Penguin

Primary consumers — Herbivorous zooplankton

Primary producers — Photosynthetic phytoplankton

Figure 30.24
The marine food web off the coast of Antarctica.

(*hetero* = other) because they acquire energy and nutrients by eating other organisms or their remains. Animals are consumers. Herbivores (primary consumers) feed directly on plants and form the second trophic level. Secondary consumers (mesopredators) eat herbivores and form the third trophic level. Animals that eat secondary consumers make up the fourth trophic level, the **tertiary consumers.** At one meal, animals that are omnivores (e.g., humans, pigs, and bears) can act as primary, secondary, and tertiary consumers.

Detritivores (scavengers) form a separate and distinct trophic level. These organisms extract energy from

organic detritus produced at other trophic levels. Detritivores include fungi, bacteria, and animals such as earthworms and vultures that ingest dead organisms, digestive wastes, and cast-off body parts such as leaves and exoskeletons. Decomposers, a type of detritivore, are small organisms, such as bacteria and fungi, that feed on dead or dying organic material. Detritivores and decomposers serve a critical ecological function because their activity reduces organic material to small inorganic molecules that producers can assimilate (see Chapter 22).

Although omnivores obviously do not fit exclusively into one trophic level, this can also be true of other organisms. Sea slugs that use chloroplasts or carnivorous plants are examples of species that do not fit readily into trophic categories.

30.6c Food Chains and Webs: Connections in Ecosystems

Ecologists use food chains and webs to illustrate the trophic structure of a community. Each link in a food chain is represented by an arrow pointing from food to consumer (see Figure 30.24b). Simple, straight-line food chains are rare in nature because most consumers feed on more than one type of food and because most organisms are eaten by more than one type of consumer. Complex relationships are portrayed as food webs—sets of interconnected food chains with multiple links.

In the food web for the waters off the coast of Antarctica (see Figure 30.24), primary producers and primary consumers are small organisms occurring in vast numbers. Microscopic diatoms (phytoplankton) are responsible for most photosynthesis, and small shrimplike krill (zooplankton) are the major primary consumers. These tiny organisms, in turn, are eaten by larger species such as fish and seabirds, as well as by suspension-feeding baleen whales. Some secondary consumers are eaten by birds and mammals at higher trophic levels. The top carnivore in this ecosystem, the orca, feeds on carnivorous birds and mammals.

Ideally, depictions of food webs would include all species in a community, from microorganisms to top consumer. But most ecologists simply cannot collect data on every species, particularly those that are rare or very small. Instead, they study links between the most important species and simplify analysis by grouping trophically similar species. Figure 30.24 categorizes the many different species of primary producers and primary consumers as phytoplankton and zooplankton, respectively.

Many biological *hot spots* (areas with many species) exist, from thermal vents on the floor of some oceans to deposits of bat guano in some caves. A more recently described example is icebergs drifting north from Antarctica. The icebergs can be hot spots of enrichment because of the nutrients and other materials they shed into surrounding waters. The water around two free-drifting icebergs ($0.1 \ km^2$ and $30.8 \ km^2$ in area) was sampled in the Weddell Sea. High concentrations of chlorophyll, krill, and seabirds extended about 3.7 km around each iceberg. These data, reported by K. L. Smith Jr. and seven colleagues, demonstrate that icebergs can have substantial effects on pelagic ecosystems.

In the late 1950s, Robert MacArthur analyzed food webs to determine how the many links between trophic levels may contribute to a community's stability. The stability of a community is defined as its ability to maintain **species composition** and relative abundances when environmental disturbances eliminate some

Barcode of Life

To calculate the diversity of organisms using an index such as the Shannon–Weaver one described in the text, you need to know how many different species are in your sample. The diversity of species can be overwhelming, so it is difficult to provide a confident estimate of how many species remain undescribed. For many groups of organisms, there may be very few authorities able to identify species and provide descriptions of "new" species, the ones not yet described and therefore nameless. The Barcode of Life Data Systems, based in Guelph, Ontario, offers one alternative to the challenge of knowing how many species are in the sample you have just acquired, or the origin of a mysterious mouse found in a shipment of frozen chickens from Thailand.

The Barcode of Life project depends upon variation in the mitochondrial cytochrome *c oxidase 1* (CO1) gene consisting of about 650 nucleotides. This genetic barcode is embedded in almost every cell and offers biologists a chance to identify a species even if they have only a small sample of feathers or fur, a leaf, a seed, or a caterpillar. Since identification of some species depends upon having a whole adult specimen, being able to make an identification from an egg, a larva, or a hair offers enormous potential. Identification of organisms with different life stages can be particularly challenging. Using morphology, it can be easy to identify a butterfly or a frog, but much more difficult to identify its caterpillar or its tadpole.

The Barcode of Life project is based on polymerase chain reaction (PCR) technology, which allowed biologists to process 100 samples every three hours. Subsequent advances in genomic technology have increased our capacity for efficient sequencing of DNA. The combination of this potential, an army of researchers collecting specimens, and global positioning satellite (GPS) technology to document locations means that the Barcode of Life project can deliver accurate (to 97.5%) identifications of specimens in a short time. Further developments could see biologists and naturalists armed with appropriately programmed handheld devices to obtain in-field identifications.

One important consequence of this project is that biologists will have a fighting chance to document more fully the diversity of life on Earth. On one hand, this means realizing that one species of the butterfly *Astraptes fulgerator* is actually ten species, or that what people had thought were several species is, in fact, one. Protecting species through CITES, the Convention on International Trade in Endangered Species, means being able to name them so that they can be placed on a protected list. The Barcode of Life project should allow a merchant to be sure that the ivory being sold in her shop is from an extinct mammoth, rather than a living species said to be endangered (Chapter 32). The same applies to food species in a market—is that fish really what the label says?

Everyone has experience with barcode operations because they are used in many retail outlets and therefore we all know that barcodes and readers do not always work. These limitations, as well as biological ones associated with genetics of different species, make some organisms more appropriate for Barcode of Life approaches than others.

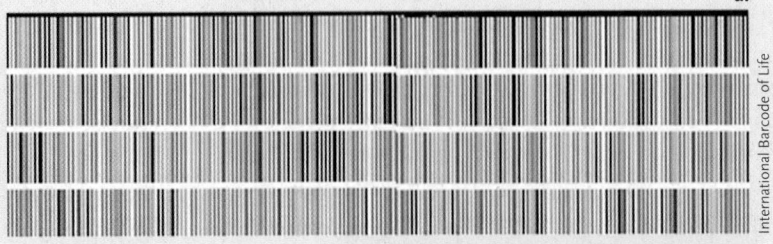

*Shown here is **(a)** the bar code of **(b)** a flying fringe-lipped bat* (Trachops cirrhosus).

species from the community. MacArthur hypothesized that in species-rich communities, where animals feed on many food sources, the absence of one or two species would have only minor effects on the structure and stability of the community as a whole. He proposed a connection between species diversity, food web complexity, and community stability.

Subsequent research has confirmed MacArthur's reasoning. The average number of links per species generally increases with increasing species richness. Comparative food web analysis reveals that the relative proportions of species at the highest, middle, and lowest trophic levels are reasonably constant across communities. In 92 communities, MacArthur found two or three prey species per predator species, regardless of species richness.

Interactions among species in most food webs can be complex, indirect, and hard to unravel. In contrast, rodents and ants living in desert communities of the U.S. Southwest potentially compete for seeds, their main food source. Plants that produce the seeds compete for water, nutrients, and space. Rodents generally prefer to eat large seeds, whereas ants prefer small seeds. Thus, feeding by rodents reduces the potential population sizes of plants that produce large seeds. As a result, the population sizes of plants that produce small seeds may increase, ultimately providing more food for ants (see Chapter 48). Compared with the Antarctic system described above (see Figure 30.24), this community is not particularly complex.

STUDY BREAK

1. Why are indices important for population biologists? What do Shannon's indices measure?
2. Differentiate between α and β diversity.
3. Are herbivores primary or secondary consumers? Which trophic level do they form? Where do omnivores belong?

30.7 Effects of Population Interactions on Community Structure

Observations of resource partitioning and character displacement suggested that some process had fostered differences in resource use among coexisting species, and competition provided the most straightforward explanation of these patterns.

Interspecific competition can cause local extinction of species or prevent new species from becoming established in a community, reducing its species richness. During the 1960s and early 1970s, ecologists emphasized competition as the primary factor structuring communities.

30.7a Competition: More Than One Species Competing for a Resource

To further explore the role of competition, ecologists undertook field experiments on competition in natural populations. The experiment on barnacles (see Figure 30.15, p. 722) is typical of this approach—the impact on one species' potential competitors of adding or removing another species changed patterns of distribution or population size. The picture that emerges from the results of these experiments is not clear, even to ecologists. In the early 1980s, Joseph Connell surveyed 527 published experiments on 215 species. He found that competition was demonstrated in roughly 40% of the experiments and more than 50% of species. At the same time, Thomas W. Schoener used different criteria to evaluate 164 experiments on approximately 400 species. He found that competition affected more than 75% of species.

It is not surprising that there is no single answer to the question about how competition works in and influences communities. Plant and vertebrate ecologists working with K-selected species generally believe that competition has a profound effect on species distributions and resource use. Insect and marine ecologists working with r-selected species argue that competition is not the major force governing community structure, pointing instead to predation or parasitism and physical disturbance. We know that even categorizing a species as r- or K-selected is open to discussion (see Chapter 29).

30.7b Feeding

Predators can influence the species richness and structure of communities by reducing the sizes of prey populations. On the rocky coast of British Columbia, different species that fill different trophic roles compete for attachment sites on rocks, a requirement for life on a wave-swept shore. Mussels are the strongest competitors for space, eliminating other species from the community (see "Effect of a Predator on the Species Richness of Its Prey"). At some sites, predatory sea stars preferentially eat mussels, reducing their numbers and creating space for other species to grow. Because the interaction between *Pisaster* and *Mytilus* affects other species as well, it qualifies as a strong interaction.

In the 1960s, Robert Paine used removal experiments to evaluate the effects of predation by *Pisaster* (see "Effect of a Predator on the Species Richness of Its Prey"). In predator-free experimental plots, mussels outcompeted barnacles, chitons, limpets, and other invertebrate herbivores, reducing species richness from 15 species to 8. In control plots containing predators, all 15 species persisted. Ecologists describe predators such as *Pisaster* as **keystone species**, defined as species with a greater effect on community structure than their numbers might suggest. Snowshoe hares (Chapter 29) are candidates to be keystone species in boreal forest ecosystems because they are prey for a range of predators. Pallas' long-tongued bats may emerge as keystone species because, as we have seen, they eat insects and fruit as well as nectar and pollen.

Herbivores also exert complex effects on communities. In the 1970s, Jane Lubchenco studied herbivory in a periwinkle snail, believed to be a keystone species on rocky shores in Massachusetts (see "The Complex Effects of a Herbivorous Snail on Algal Species Richness," p. 734). The features of plants and algae and the food preferences of animals that eat them can influence community structure.

STUDY BREAK

1. How does the importance of competition vary between K-selected and r-selected species?
2. Does predation or herbivory increase or decrease species richness? Explain.
3. What is a keystone species?

30.8 Effects of Disturbance on Community Characteristics

Recent research tends to support the individualistic view that many communities are not in equilibrium and that species composition changes frequently. Environmental disturbances such as storms, landslides, fires, floods, avalanches, and cold spells often eliminate some species and provide opportunities for others to become established. Frequent disturbances keep some ecological communities in a constant state of flux.

Physical disturbances are common in some environments. Lightning-induced fires commonly sweep through grasslands, powerful hurricanes often demolish patches of forest and coastal habitats, and waves wash over communities at the edge of the sea and sweep away organisms as well as landforms and other structures.

Joseph Connell and his colleagues conducted an ambitious long-term study of the effects of disturbance on coral reefs, shallow tropical marine habitats that are among the most species-rich communities on Earth. In some parts of the world, reefs are routinely battered by violent storms that wash corals off the substrate, creating bare patches in the reef. The scouring action of storms creates opportunities for coral larvae to settle on bare substrates and start new colonies.

From 1963 to 1992, Connell and his colleagues tracked the fate of the Heron Island Reef at the south end of Australia's Great Barrier Reef **(Figure 30.25)**. The inner flat and protected crests of the reef are sheltered from severe wave action during storms, whereas some pools and crests are routinely exposed to physical disturbance. Because corals live in colonies of variable size, the researchers monitored coral abundance by measuring the percentage of the substrate (i.e., the sea floor) that colonies covered. They revisited marked study plots at intervals, photographing and identifying individual coral colonies.

Effect of a Predator on the Species Richness of Its Prey

Experimental Research Box

Biologists used a predatory sea star (*Pisaster ochraceus*) to assess the influence a predator can have on species richness and relative abundance of prey **(Figure 1)**. *P. ochraceus* preferentially eats mussels (*Mytilus californicus*), one of the strongest competitors for space in rocky intertidal pools. Robert Paine removed *Pisaster* from caged experimental study plots, leaving control study plots undisturbed, and then monitored the species richness of *Pisaster*'s invertebrate prey over many years.

Paine documented an increase in mussel populations in the experimental plots as well as complex changes in the feeding relationships among species in the intertidal food web **(Figure 2)**. When he removed *P. ochraceus*, the top predator in this food web, he observed a rapid decrease in the species richness of invertebrates and algae. Species richness on control plots did not change over the course of the experiment.

Predation by *P. ochraceus* prevents mussels from outcompeting other invertebrates on rocky shores.

Figure 1

A predatory sea star (Pisaster ochraceus) *feeding on a mussel* (Mytilus californicus).

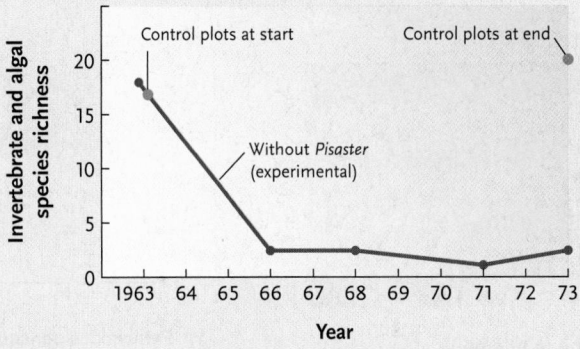

Figure 2

Changes in the species richness of invertebrates and algal species according to changes in populations of sea stars.

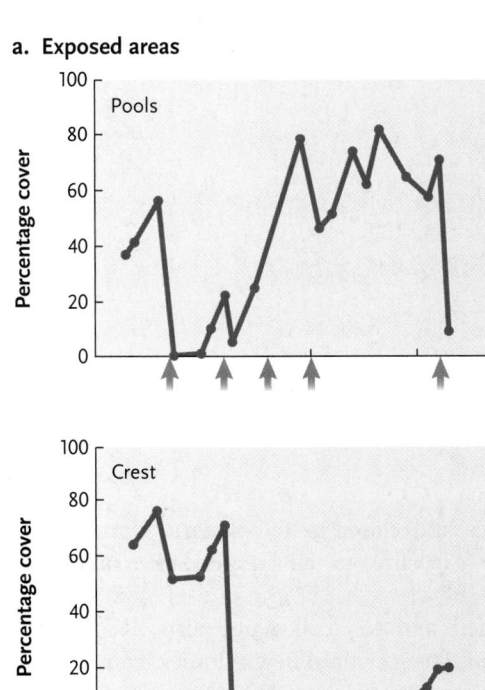

a. Exposed areas

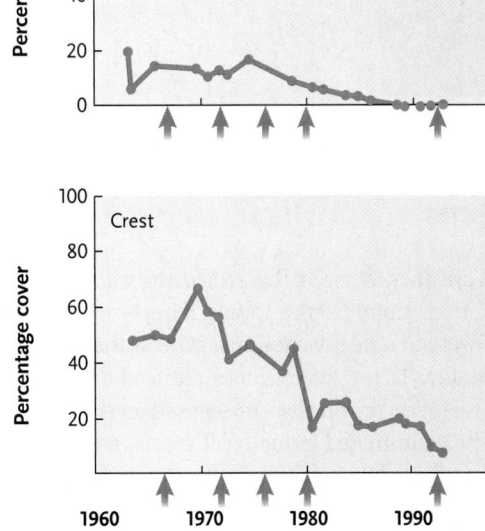

b. Sheltered areas

Figure 30.25

Major hydrodynamic disturbances to coral reefs, such as tsunamis and severe storms, have important impacts on coral reefs. Using oceanographic and engineering models, it is possible to predict the degree of dislodgement of benthic reef corals and, in this way, predict how coral shape and size indicate vulnerability to major disturbances. The use of these models is particularly important during times of climate change. The graphs show the effects of storms on corals. Five tropical cyclones (marked by grey arrows) damaged corals on the Heron Island Reef during a 30-year period. Storms reduced the percentage cover of corals in **(a)** exposed parts of the reef much more than in **(b)** sheltered parts of it. These data show that the 1970 event had the largest impact on some exposed and sheltered areas.

The Complex Effects of a Herbivorous Snail on Algal Species Richness

Experimental Research Box

Jane Lubchenco made enclosures that prevented periwinkle snails (*Littorina littorea*) from entering or leaving study plots in tide pools and on exposed rocks in rocky intertidal habitat **(Figure 1).** She then monitored the algal species composition in the plots, comparing them to the density of the periwinkles. In this way, she examined the influence of the periwinkles on the species richness of algae in intertidal communities.

The results varied dramatically between the study plots in tide pools and on exposed rocks. In tide pools, periwinkle snails preferentially ate *Enteromorpha*, the competitively dominant alga. At intermediate densities of *Enteromorpha*, the periwinkles remove some of these algae, allowing weakly competitive species to grow. The snails' grazing increases species richness. But grazing by periwinkles when *Enteromorpha* is at low or high densities reduces the species richness of algae in tide pools. On exposed rocks, where periwinkle snails rarely eat the competitively dominant alga *Chondrus*, feeding by snails reduces algal species richness **(Figure 2).**

Periwinkle snails (*Littorina littorea*)

William Warner/Shutterstock.com

Enteromorpha growing in tide pools

Wild Horizon/Contributor/UIG via Getty Images

Chondrus growing on exposed rocks

Ted Kinsman/Science Source

Figure 1

The distribution of periwinkle snails and two kinds of algae.

In tide pools

In tide pools, snails at low densities eat little algae and *Enteromorpha* competitively excludes other algal species, reducing species richness. At high snail densities, heavy feeding on all species reduces algal species richness. At intermediate snail densities, grazing eliminates some *Enteromorpha*, allowing other species to grow.

On exposed rocks

On exposed rocks, periwinkles don't eat much *Chondrus*, but they consume the tender, less successful competitors. Thus, feeding by periwinkles reinforces the competitive superiority of *Chondrus*: as periwinkle density increases, algal species richness declines.

Figure 2

Density of periwinkles versus algal species richness in tide pools and on exposed rocks.

Five major cyclones crossed the reef during the 30-year study period. Coral communities in exposed areas of the reef were in a nearly continual state of flux. In exposed pools, four of the five cyclones reduced the percentage of cover, often drastically. On exposed crests, the cyclone of 1972 eliminated virtually all corals, and subsequent storms slowed the recovery of these areas for more than 20 years. In contrast, corals in sheltered areas suffered much less storm damage. Nevertheless, their coverage also declined steadily during the study as a natural consequence of the corals' growth. As colonies grew taller and closer to the ocean's surface, their increased exposure to air resulted in substantial mortality.

Connell and his colleagues also documented *recruitment,* the growth of new colonies from settling larvae, in their study plots. They discovered that the rate at which new colonies developed was almost always higher in sheltered than in exposed areas. Recruitment rates were extremely variable, depending in part on the amount of space that storms or coral growth had made available.

This long-term study of coral reefs illustrates that frequent disturbances prevent some communities from reaching an equilibrium determined by interspecific interactions. Changes in the coral reef community at Heron Island result from the effects of external disturbances that remove coral colonies from the reef, as well as internal processes (growth and recruitment) that either eliminate colonies or establish new ones. In this community, growth and recruitment are slow processes and disturbances are frequent. Thus, the community never attains equilibrium, and moderate levels of disturbance can foster high species richness.

The **intermediate disturbance hypothesis**, proposed by Connell in 1978, suggests that species richness is greatest in communities experiencing fairly frequent disturbances of moderate intensity. Moderate disturbances create openings for *r*-selected species to arrive and join the community while allowing *K*-selected species to survive. Thus, communities that experience intermediate levels of disturbance contain a rich mixture of species. Where disturbances are severe and frequent, communities include only *r*-selected species that complete their life cycles between catastrophes. Where disturbances are mild and rare, communities are dominated by long-lived *K*-selected species that competitively exclude other species from the community.

Several studies in diverse habitats have confirmed the predictions of the intermediate disturbance hypothesis. Colin R. Townsend and his colleagues studied the effects of disturbance at 54 stream sites in the Taieri River system in New Zealand. Disturbance occurs in these communities when water flow from heavy rains moves rocks, soil, and sand in the streambed, disrupting animal habitats. Townsend and his colleagues measured how much the substrate moved in different streambeds to develop an index of the intensity of disturbance. Their results indicate that species richness is highest in areas that experience intermediate levels of disturbance **(Figure 30.26)**.

Some ecologists have suggested that species-rich communities recover from disturbances more readily than less diverse communities. In the United States, David Tilman and his colleagues conducted large-scale

experiments in midwestern grasslands. They examined relationships between species number and the ability of communities to recover from disturbance. Grassland plots with high species richness recover from drought faster than plots with fewer species.

STUDY BREAK

1. What did Connell's 30-year study of coral reefs illustrate about the ability of communities to reach a state of equilibrium?
2. What is the intermediate disturbance hypothesis? Describe one study that supports this hypothesis.
3. How does species richness affect the rate of recovery following a disturbance?

30.9 Succession

Ecosystems change over time in a process called **succession**, the change from one community type to another.

30.9a Primary Succession: The First Steps

Primary succession begins when organisms first colonize habitats without soil, such as those created by erupting volcanoes and retreating glaciers **(Figure 30.27, p. 736)**. Lichens are often among the very first colonists (see Chapter 25), deriving nutrients from rain and bare rock. They secrete mild acids that erode rock surfaces, initiating the slow development of soil, which is enriched by the organic material lichens produce. After lichens modify a site, mosses (see Chapter 26) colonize patches of soil and grow quickly.

As soil accumulates, hardy, opportunistic plants (grasses, ferns, and broad-leaved herbs) colonize the site from surrounding areas. Their roots break up rock, and when they die, their decaying remains enrich the soil. Detritivores and decomposers facilitate these processes. As the soil becomes deeper and richer, increased moisture and nutrients support bushes and, eventually, trees. Late successional stages are often dominated by *K*-selected species with woody trunks and branches that position leaves in sunlight and large root systems that acquire water and nutrients from soil.

In the classical view of ecological succession, long-lived species, which replace themselves over time, eventually dominate a community, and new species join it only rarely. This relatively stable, late successional stage is called a **climax community** because the dominant vegetation replaces itself and persists until an environmental disturbance eliminates it and allows other species to invade. Local climate and soil conditions, the surrounding communities where colonizing species originate, and chance events determine the species composition of climax communities. We now know that even climax communities change slowly in response to environmental fluctuations.

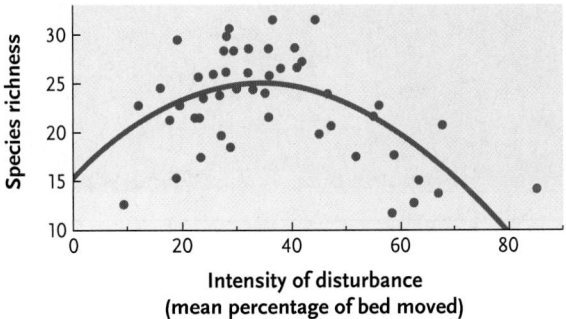

Figure 30.26

An observational study that supports the intermediate disturbance hypothesis. In the Taieri River system in New Zealand, species richness was highest in stream communities that experienced an intermediate level of disturbance.

1 The glacier has retreated about 8 m per year since 1794.

2 This site was covered with ice less than 10 years before this photo was taken. When a glacier retreats, a constant flow of melt water leaches minerals, especially nitrogen, from the newly exposed substrate.

3 Once lichens and mosses have established themselves, mountain avens (genus *Dryas*) grows on the nutrient-poor soil. This pioneer species benefits from the activity of mutualistic nitrogen-fixing bacteria, spreading rapidly over glacial till.

4 Within 20 years, shrubby willows (genus *Salix*), cottonwoods (genus *Populus*), and alders (genus *Alnus*) take hold in drainage channels. These species are also symbiotic with nitrogen-fixing microorganisms.

5 In time, young conifers, mostly hemlocks (genus *Tsuga*) and spruce (genus *Picea*), join the community.

6 As the years progress the smaller trees and shrubs are gradually replaced by larger trees.

Figure 30.27

Primary succession following glacial retreat. The retreat of glaciers at Glacier Bay, Alaska, has allowed ecologists to document primary succession on newly exposed rocks and soil.

30.9b Secondary Succession: Changes after Destruction

Secondary succession occurs after existing vegetation is destroyed or disrupted by an environmental disturbance, such as a fire, a storm, or human activity. The presence of soil makes disturbed sites ripe for colonization and may contain numerous seeds that germinate after disturbance. Early stages of secondary succession proceed rapidly, but later stages parallel those of primary succession.

30.9c Climax Communities: The Ultimate Ecosystems until Something Changes

Similar climax communities can arise from several different successional sequences. Hardwood forests can also develop in sites that were once ponds. During **aquatic succession**, debris from rivers and runoff accumulates in a pond, filling it to its margins. Ponds are first transformed into swamps, inhabited by plants adapted to a semisolid substrate. As larger plants get established, their high transpiration rates dry the soil, allowing other plant species to colonize. Given enough time, the site may become a meadow or forest in which an area of moist, low-lying ground is the only remnant of the original pond.

Because several characteristics of communities can change during succession, ecologists try to document how patterns change. First, because *r*-selected species are short lived and *K*-selected species are long lived, species composition changes rapidly in the early stages and more slowly in later stages of succession. Second, species richness increases rapidly during early stages because new species join the community faster than resident species become extinct. In later stages, species richness stabilizes or may even decline. Third, in terrestrial communities receiving sufficient rainfall, the maximum height and total mass of the vegetation increase steadily as large species replace small ones, creating the complex structure of the climax community.

Because plants influence the physical environment below them, the community itself increasingly moderates its **microclimate.** The shade cast by a forest canopy helps retain soil moisture and reduce temperature fluctuations. The trunks and canopy also reduce wind speed. In contrast, the short vegetation in an early successional stage does not effectively shelter the space below it.

Although ecologists usually describe succession in terms of vegetation, animals can show similar patterns. As the vegetation shifts, new resources become available, and animal species replace each other over time. Herbivorous insects, often with strict food preferences, undergo succession along with their food plants. And as herbivores change, so do their predators, parasites, and parasitoids. In old-field succession in eastern North America, different vegetation stages harbour a changing assortment of bird species **(Figure 30.28).**

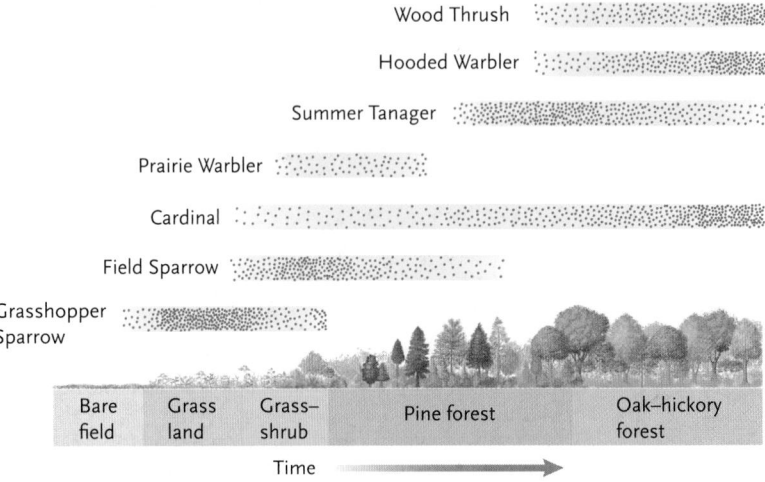

Figure 30.28

Succession in animals. Successional changes in bird species composition in an abandoned agricultural field in eastern North America parallel the changes in plant species composition. The residence times of several representative species are illustrated. The density of stippling inside each bar illustrates the density of each species through time.

Differences in dispersal abilities (see "Dispersal," p. 738), maturation rates, and life spans among species are partly responsible for ecological succession. Early successional stages harbour many *r*-selected species because they produce numerous small seeds that colonize open habitats and grow quickly. Mature successional stages are dominated by *K*-selected species because they are long lived. Nevertheless, coexisting populations inevitably affect one another. Although the role of population interactions in succession is generally acknowledged, ecologists debate the relative importance of processes that either facilitate or inhibit the turnover of species in a community.

30.9d Facilitation Hypothesis: One Species Makes Changes That Help Others

The **facilitation hypothesis** suggests that species modify the local environment in ways that make it less suitable for themselves but more suitable for colonization by species typical of the next successional stage. When lichens first colonize bare rock, they produce a small quantity of soil that is required by mosses and grasses that grow there later. According to this hypothesis, changes in species composition are both orderly and predictable because the presence of each stage facilitates the success of the next one. Facilitation is important in primary succession, but it may not be the best model of interactions that influence secondary succession.

30.9e Inhibition Hypothesis: One Species Negatively Affects Others

The **inhibition hypothesis** suggests that new species are prevented from occupying a community by species that are already present. According to this

Dispersal

Organisms often show astonishing dispersal abilities. In some cases, long-distance dispersal by plants in the Arctic is effected by the combination of strong winds and extensive expanses of ice and snow. The Svalbard Archipelago **(Figure 1)** is an interesting location for the study of plant dispersal. The islands were glaciated 20 thousand years ago, and it is likely that plants did not survive this condition. The fossil record indicates that plants have been present on Svalbard for fewer than 10 thousand years, although between 4000 and 9500 years ago, the climate was warmer there (by 1 to 2°C) than it is now.

Using DNA fingerprinting, I. G. Alsos and eight colleagues demonstrated that plant colonization of the Svalbard Archipelago has involved the arrival of plants from all possible adjacent regions **(Figure 2).** In eight of nine species, genetic evidence indicates multiple colonization events.

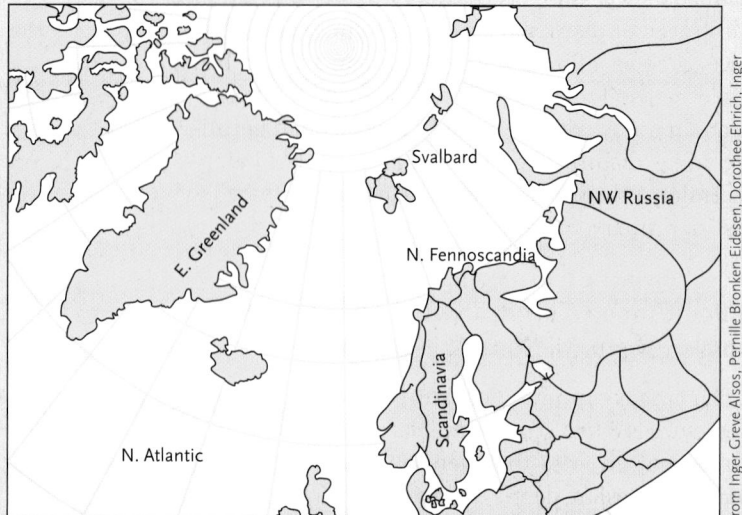

From Inger Greve Alsos, Pernille Bronken Eidesen, Dorothee Ehrich, Inger Skrede, Kristine Westergaard, Gro Hilde Jacobsen, Jon Y. Landvik, Pierre Taberlet, Christian Brochmann, "Frequent Long-Distance Plant Colonization in the Changing Arctic," *Science*, vol. 316, Jun 15, 2007, pp. 1606–1609. Reprinted with permission from AAAS.

Figure 1

The location of the Svalbard Archipelago.

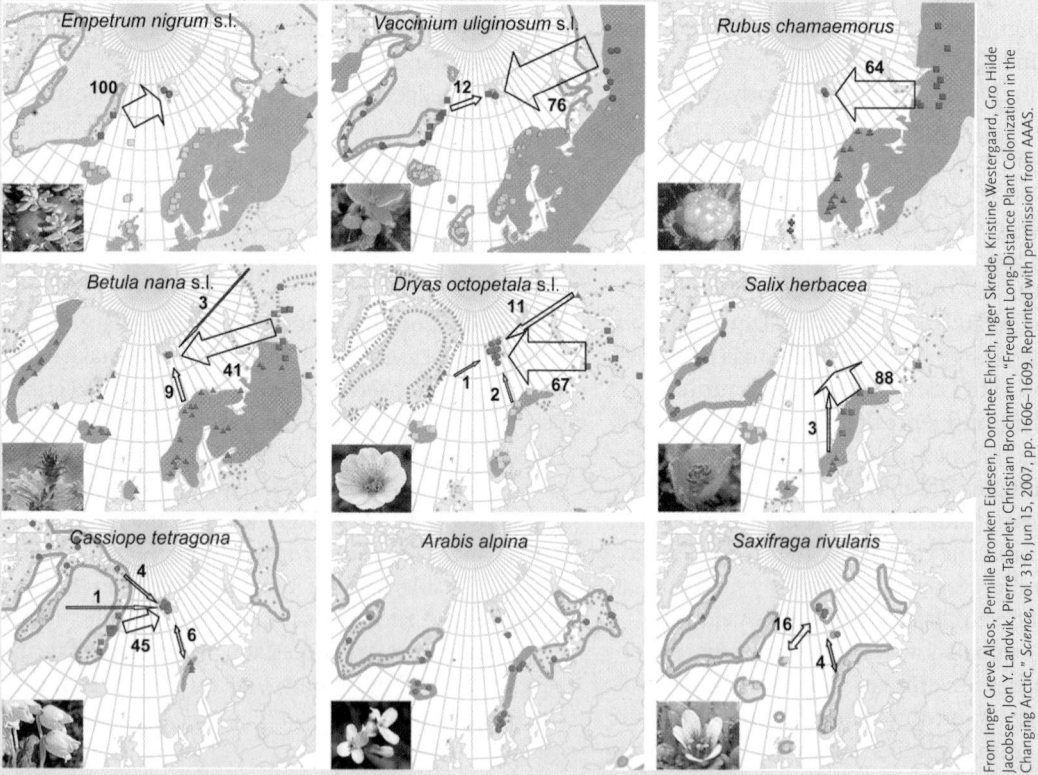

From Inger Greve Alsos, Pernille Bronken Eidesen, Dorothee Ehrich, Inger Skrede, Kristine Westergaard, Gro Hilde Jacobsen, Jon Y. Landvik, Pierre Taberlet, Christian Brochmann, "Frequent Long-Distance Plant Colonization in the Changing Arctic," *Science*, vol. 316, Jun 15, 2007, pp. 1606–1609. Reprinted with permission from AAAS.

Figure 2

Source regions for Svalbard plants. Shading shows the geographic distribution of nine species of plants, and dotted lines show the distributions of related species. The main genetic groups are represented by colours, although some populations (*) could not be assigned to a genetic group. Arrows identify **source populations,** *and the numbers indicate the percentage allocation by source region.*

Plants can obviously disperse without assistance from animals.

In other situations, plants disperse with the assistance of animals through pollination and seeds. Using *Prunus mahaleb*, the mahaleb cherry **(Figure 3)**, and genetic techniques, P. Jordano and two colleagues examined the role of birds and mammals in pollination and dispersing seeds. Small Passerine Birds dispersed seeds short distances (most less than 50 m) from the parent tree, whereas medium-sized birds (*Corvus corone and Turdus viscivorus*) usually dispersed seeds over longer distances (more than 110 m). Mammals (usually *Martes foina* and *Vulpes vulpes* but sometimes *Meles meles*) dispersed seeds about 500 m. The genetic work also indicated the extent of gene flow during pollination.

It is obvious that plants capable of self-fertilization or vegetative reproduction can be more effective colonists than those depending on outcrossing, especially with the help of animal pollinators.

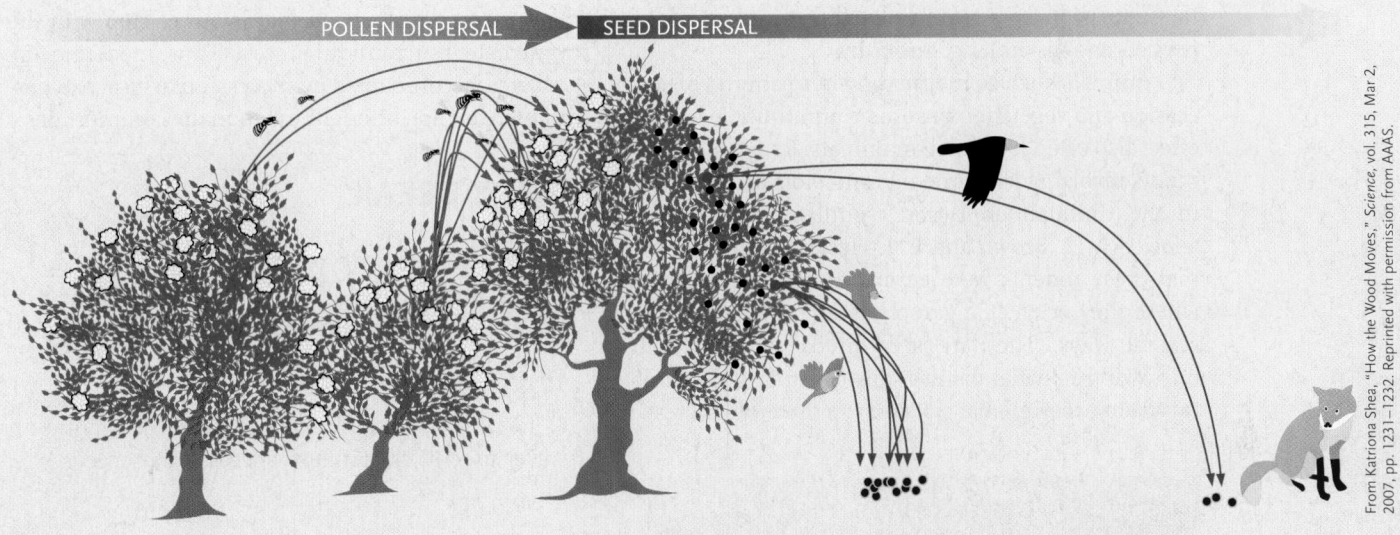

POLLEN DISPERSAL → **SEED DISPERSAL**

From Katriona Shea, "How the Wood Moves," *Science*, vol. 315, Mar 2, 2007, pp. 1231–1232. Reprinted with permission from AAAS.

Figure 3
The movement of pollen and seeds from mahaleb cherry trees. Gene flow occurs through pollination and seed dispersal (see Chapter 18).

hypothesis, succession is neither orderly nor predictable because each stage is dominated by the species that happened to have colonized the site first. Species replacements occur only when individuals of dominant species die of old age or when an environmental disturbance reduces their numbers. Eventually, long-lived species replace short-lived species, but the precise species composition of a mature community is open to question. Inhibition appears to play a role in some secondary successions. The interactions among early successional species in an old field are highly competitive. Horseweed inhibits the growth of asters that follow them in succession by shading aster seedlings and releasing toxic substances from their roots. Experimental removal of horseweed enhances the growth of asters, confirming the inhibitory effect.

30.9f Tolerance Hypothesis: Species Tolerate One Another

The **tolerance hypothesis** asserts that succession proceeds because competitively superior species replace competitively inferior ones. According to this model, early-stage species neither facilitate nor inhibit the growth of later-stage species. Instead, as more species arrive at a site and resources become limiting, competition eliminates species that cannot harvest scarce resources successfully. In the Piedmont region of North America, young hardwood trees are more tolerant of shade than are young pine trees, and hardwoods gradually replace pines during succession. Thus, the climax community includes only strong competitors. Tolerance may explain the species composition of many transitional and mature communities.

At most sites, succession probably results from some combination of facilitation, inhibition, and tolerance, coupled with interspecific differences in dispersal, growth, and maturation rates. Moreover, within a community, the patchiness of abiotic factors strongly influences plant distributions and species composition. In deciduous forests of eastern North America, maples (*Acer* species) predominate on wet, low-lying ground, but oaks (*Quercus* species) are more abundant at higher and drier sites. Thus, a mature deciduous forest is often a mosaic of species and not a uniform stand of trees.

Disturbance and density-independent factors play important roles, in some cases speeding successional change. Moose (*Alces alces*) prefer to feed on deciduous shrubs in northern forests. This disturbance accelerates the rate at which conifers replace deciduous shrubs. On Isle Royale in Lake Superior, however, grazing by moose strongly affects balsam fir (*Abies balsamea*), their preferred food there. The net effect is a severe reduction in conifers and an increase in deciduous shrubs. Disturbance can also inhibit successional change, establishing a **disturbance climax** or **disclimax community.** In many grassland communities, periodic fires and grazing by large mammals kill seedlings of trees that would otherwise become established. Thus, disturbance prevents the succession from grassland to forest, and grassland persists as a disclimax community.

Animals such as moose can alter patterns of succession and vegetation in some communities, but the effect also extends to small mammals. Removal experiments involving kangaroo rats and plots of shrubland in the Chihuahuan Desert (southeastern Arizona) allowed J. H. Brown and E. J. Heske to demonstrate that these rodents were keystones in some systems where they occur. Kangaroo rats affect the plants in several ways. They are seed predators, and their burrowing activities disturb soils. Excluding kangaroo rats from experimental plots led to a threefold increase in the density of tall perennials and annual grasses **(Figure 30.29)**, suggesting that by predation on seeds and burrowing, these rodents affected the vegetation in the experimental areas.

On a local scale, disturbances often destroy small patches of vegetation, returning them to an earlier successional stage. A hurricane, tornado, or avalanche may topple trees in a forest, creating small, sunny patches of open ground. Locally occurring *r*-selected species take advantage of newly available resources and quickly colonize the openings. These local patches then undergo succession that is out of step with the immediately surrounding forest. Thus, moderate disturbance, accompanied by succession in local patches, can increase species richness in many communities.

STUDY BREAK

1. What are the two types of succession? How do they differ?
2. What is a climax community? What determines the species composition of a climax community?
3. Identify and briefly describe the three hypotheses used to explain how succession proceeds.

30.10 Variations in Species Richness among Communities

Species richness often varies among communities according to a recognizable pattern. Two large-scale patterns of species richness—latitudinal trends and island patterns—have captured the attention of ecologists for more than a century.

30.10a Latitudinal Effects: From South to North

Ever since Darwin and Wallace travelled the globe (see Chapter 17), ecologists have recognized broad latitudinal trends in species richness. For many but not all plant and animal groups, species richness follows a latitudinal gradient, with the most species in the tropics and a steady decline in numbers toward the poles **(Figure 30.30).** Several general hypotheses may explain these striking patterns.

Some hypotheses propose historical explanations for the *origin* of high species richness in the tropics. The benign climate in tropical regions allows some tropical organisms to have more generations per year than their temperate counterparts. Small seasonal changes in temperature mean that tropical species may be less likely than temperate species to migrate from one habitat to another, reducing gene flow between geographically isolated populations (see Chapter 18). These factors may have fostered higher speciation rates

Figure 30.29
Predation and succession. Kangaroo rats (*Dipodomys*) were removed from the left side of the fence, which excluded them from the plot on the left. The top photograph was taken 5 years after the removal and the bottom one 13 years after. A large-seeded annual (after 5 years) and tall grasses are present in the *Dipodomys*-free plots.

From James H. Brown, Edward J. Heske, "Control of a Desert-Grassland Transition by a Keystone Rodent Guild," *Science*, vol. 250, Dec 21, 1990, pp. 1705–1707. Reprinted with permission from AAAS.

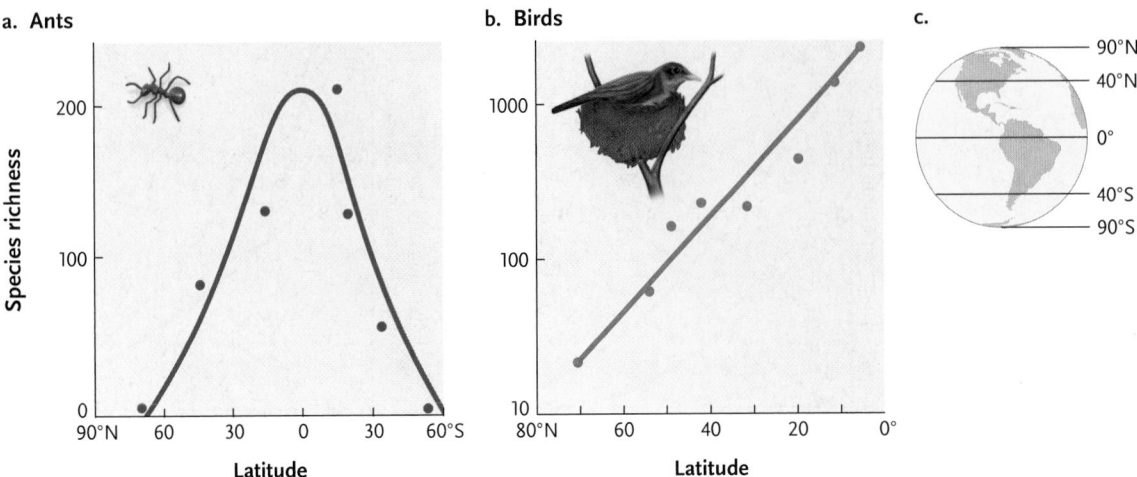

a. Ants

Species richness

200

100

0
90°N   60   30   0   30   60°S
Latitude

b. Birds

1000

100

10
80°N   60   40   20   0°
Latitude

c.

90°N
40°N
0°
40°S
90°S

Figure 30.30
Latitudinal trends in species richness. The species richness of many animals and plants varies with latitude **(c)** as illustrated here for **(a)** ants in North, Central, and South America and **(b)** birds in North and Central America. The species richness data used in (b) are based on records of where these birds breed.

in the tropics, accelerating the accumulation of species. Tropical communities may also have experienced severe disturbance less often than communities at higher latitudes, where periodic glaciations have caused repeated extinctions. Thus, new species may have accumulated in the tropics over longer periods of time.

Other hypotheses focus on ecological explanations for the *maintenance* of high species richness in the tropics. Some resources are more abundant, predictable, and diverse in tropical communities. Tropical regions experience more intense sunlight, warmer temperatures in most months, and higher annual rainfall than temperate and polar regions (see *The Purple Pages*). These factors provide a long and predictable growing season for the lush tropical vegetation, which supports a rich assemblage of herbivores, and through them many carnivores and parasites. Furthermore, the abundance, predictability, and year-round availability of resources allow some tropical animals to have specialized diets. Tropical forests support many species of fruit-eating bats and birds that could not survive in temperate forests where fruits are not available year-round.

Species richness may be a self-reinforcing phenomenon in tropical communities. Complex webs of population interactions and interdependency have coevolved in relatively stable and predictable tropical climates. Predator–prey, competitive, and symbiotic interactions may prevent individual species from dominating communities and reducing species richness.

30.10b Equilibrium Theory of Island Biogeography

In 1883, a volcanic eruption virtually obliterated the island of Krakatoa. Within 50 years, what was left of Krakatoa had been recolonized by plants and animals, providing biologists with a clear demonstration of the dispersal powers of many living species. The colonization of islands and the establishment of biological communities there have provided many natural experiments that have advanced our knowledge of ecology and populations. Islands are attractive sites for experiments

because although the species richness of communities may be stable over time, the species composition is often in flux as new species join a community and others drop out. In the 1960s, Robert MacArthur and Edward O. Wilson used islands as model systems to address the question of why communities vary in species richness. Islands provide natural laboratories for studying ecological phenomena, just as they do for evolution (see Chapter 17). Island communities can be small, with well-defined boundaries, and are isolated from surrounding communities.

MacArthur and Wilson developed the **equilibrium theory of island biogeography** to explain variations in species richness on islands of different size and different levels of isolation from other landmasses. They hypothesized that the number of species on any island was governed by give and take between two processes: the immigration of new species to an island and the extinction of species already there **(Figure 30.31, p. 742).**

According to their model, the mainland harbours a *species pool* from which species immigrate to offshore islands. Seeds and small arthropods are carried by wind or floating debris. Animals such as birds arrive under their own power. When only a few species are on an island, the rate at which new species immigrate to the island is high. But as more species inhabit the island over time, the immigration rate declines because fewer species in the mainland pool can still arrive on the island as *new* colonizers (see Chapter 17). Once some species arrive on an island, their populations grow and persist for variable lengths of time. Other immigrants die without reproducing. As the number of species on an island increases, the rate of species extinction also rises. Extinction rates increase over time partly because more species can go extinct there. In addition, as the number of species on the island increases, competition and predator–prey interactions can reduce the population sizes of some species and drive them to extinction.

According to MacArthur and Wilson's theory, an equilibrium between immigration and extinction determines the number of species that ultimately occupy an island (see Figure 30.30a). Once that equilibrium has

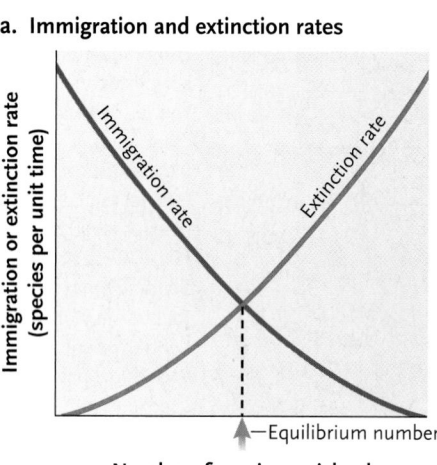

a. Immigration and extinction rates

Immigration or extinction rate (species per unit time)

Immigration rate

Extinction rate

↑—Equilibrium number

Number of species on island

The number of species on an island at equilibrium (indicated by the arrow) is determined by the rate at which new species immigrate and the rate at which species already on the island go extinct.

b. Effect of island size

Immigration or extinction rate (species per unit time)

Immigration rate (large island)

Immigration rate (small island)

Extinction rate (small island)

Extinction rate (large island)

Small island→↑ ↑-Large island

Number of species on island

Immigration rates are higher and extinction rates lower on large islands than on small islands. Thus, at equilibrium, large islands have more species.

c. Effect of distance from mainland

Immigration or extinction rate (species per unit time)

Immigration rate (near island)

Immigration rate (far island)

Extinction rate

Far island→↑ ↑-Near island

Number of species on island

Organisms leaving the mainland locate nearby islands more easily than distant islands, causing higher immigration rates on near islands. Thus, near islands support more species than far ones.

Figure 30.31
Predictions of the theory of island biogeography. The horizontal axes of the graphs are time.

been reached, the number of species remains relatively constant because one species already on the island becomes extinct in about the same time it takes a new one to arrive. The model does not specify which species immigrate or which ones already on the island become extinct. It simply predicts that the number of species on the island is in equilibrium, although species composition is not. The ongoing processes of immigration and extinction establish a constant turnover in the roster of species that live on any island.

The MacArthur–Wilson model also explains why some islands harbour more species than others. Large islands have higher immigration rates than small islands because they are larger targets for dispersing organisms. Moreover, large islands have lower extinction rates because they can support larger populations and provide a greater range of habitats and resources. At equilibrium, large islands have more species than small islands do (see Figure 30.31b). Islands near the mainland have higher immigration rates than distant islands because dispersing organisms are more likely to arrive at islands close to their point of departure. Distance does not affect extinction rates, so, at equilibrium, nearby islands have more species than distant islands (see Figure 30.31c).

The equilibrium theory's predictions about the effects of area and distance are generally supported by data on plants and animals **(Figure 30.32)**. Experimental work has verified some of the theory's basic assumptions. Amy Schoener found that more than 200 species of marine organisms colonized tiny artificial islands (plastic kitchen scrubbers) within 30 days after she placed them in a Bahamian lagoon. Her research also confirmed that immigration rate increases with island size. Daniel Simberloff and Edward O. Wilson exterminated insects on tiny islands in the Florida Keys and

monitored subsequent immigration and extinction (see "Experimenting with Islands," p. 744). Their research confirmed the equilibrium theory's predictions that an island's size and distance from the mainland influence how many species will occupy it.

The equilibrial view of species richness can also apply to mainland communities that exist as islands in a metaphorical sea of dissimilar habitat. Lakes are "islands" in a "sea" of dry land, and mountaintops are habitat "islands" in a "sea" of low terrain. Species richness in these communities is partly governed by the immigration of new species from distant sources and the extinction of species already present. As human activities disrupt environments across the globe, undisturbed sites function as islandlike refuges for threatened and endangered species. Conservation biologists apply the general lessons of MacArthur and Wilson's theory to the design of nature preserves (see Chapter 32).

The study of community ecology promises to keep biologists busy for some time to come.

STUDY BREAK

1. How does species richness change with increasing latitude?
2. In the island biogeography model proposed by MacArthur and Wilson, what processes govern the number of species on an island? What happens to the number of species once equilibrium is reached?
3. What effect do island size and distance from the mainland have on immigration and extinction of colonizing species?

PEOPLE BEHIND BIOLOGY 30.3

Bridget J. Stutchbury, *York University*

Bridget Stutchbury studies the behaviour and ecology of songbirds, working at sites in eastern North America (United States and Canada), as well as sites in the Neotropics. One aspect of her research is documenting the reproductive behaviour of birds. Although songbirds were thought to be monogamous over at least a breeding season, using genetic techniques, Stutchbury and others are discovering that both males and females often mate with a bird that is not their mate. This behaviour is called extrapair copulation if it is just mating or extrapair fertilization when young result from the matings.

Using radio tracking to follow individual birds combined with DNA fingerprinting, Stutchbury and her colleagues were able to look at the movement patterns of Acadian Flycatchers (*Empidonax virescens*) and determine how far males and females travelled to meet their extrapair partners. Males travelled 50 to 1500 m from their nests to meet partners.

Work with other species, such as Hooded Warblers (*Wilsonia citrina*), demonstrated that when these birds lived in small forest fragments, their mating behaviours were disturbed compared with the behaviours of those nesting in larger tracts of forest.

Overall, her research has demonstrated that whereas some songbirds in the eastern United States and eastern Canada depend on corridors connecting habitat fragments, other species cross open habitats to use different patches of forest.

In 2007, her book *Silence of the Songbirds* reported declines in numbers of migrating songbirds and raised concerns about their future.

a. Distance effect

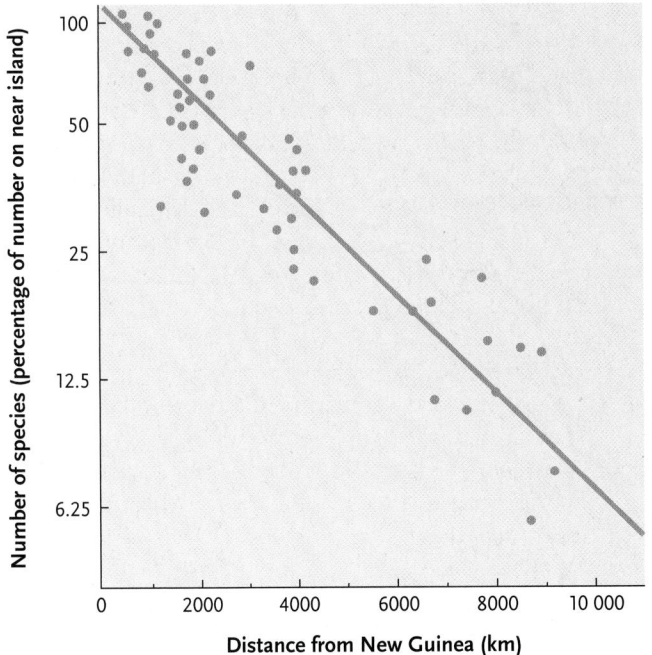

The number of lowland bird species on islands of the South Pacific declines with the islands' distance from the species source, the large island of New Guinea. Data in this graph were corrected for differences in the sizes of the islands. The number of bird species on each island is expressed as a percentage of the number of bird species on an island of equivalent size close to New Guinea.

b. Area effect

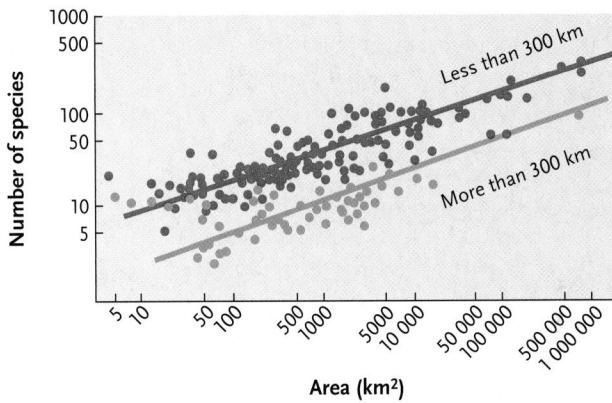

The number of bird species on tropical and subtropical islands throughout the world increases dramatically with island area. The data for islands near to a mainland source and islands far from a mainland source are presented separately to minimize the effect of distance. Notice that the distance effect reduces the number of bird species on islands that are more than 300 km from a mainland source.

Figure 30.32

Factors that influence bird species richness on islands. **(a)** Evidence that fewer bird species colonize islands that are distant from the mainland source. **(b)** Evidence that more bird species colonize large islands than small ones.

Experimenting with Islands

Shortly after Robert MacArthur and Edward O. Wilson published the equilibrium theory of island biogeography in the 1960s, Wilson and Daniel Simberloff, one of Wilson's graduate students at Harvard University, undertook an ambitious experiment in community ecology. Simberloff reasoned that the best way to test the theory's predictions was to monitor immigration and extinction on barren islands.

Simberloff and Wilson devised a system for removing all the animals from individual red mangrove trees in the Florida Keys. The trees, with canopies that spread from 11 to 18 m in diameter, grow in shallow water and are isolated from their neighbours. Thus, each tree is an island that harbours an arthropod community. The species pool on the Florida mainland includes about 1000 species of arthropods, but each mangrove island contains no more than 40 species at one time.

After cataloguing the species on each island, Simberloff and Wilson hired an extermination company to erect large tents over each mangrove island and fumigate them to eliminate all arthropods on them **(Figure 1).** The exterminators used methylbromide, a pesticide that does not harm trees or leave any residue. The tents were then removed.

Simberloff then monitored both the immigration of arthropods to the islands and the extinction of species that became established on them. He surveyed four islands regularly for two years and at intervals thereafter.

The results of this experiment confirm several predictions of MacArthur and Wilson's theory **(Figure 2).** Arthropods rapidly recolonized the islands, and within eight or nine months, the number of species living on each island had reached an equilibrium that was close to the original species number. The island nearest the mainland had more species than the most distant island. However, immigration and extinction were rapid, and Simberloff and Wilson

Figure 1
After cataloguing the arthropods, Simberloff and Wilson hired an extermination company to eliminate all living arthropods.

suspected that some species went extinct even before they had noted their presence. The researchers also discovered that three years after the experimental treatments, the species composition of the islands was still changing constantly and did not remotely resemble the species composition on the islands before they were defaunated.

Simberloff and Wilson's research was a landmark study in ecology because it tested the predictions of an important theory using a field experiment. Although such efforts are now almost routine in ecological studies, this project was one of the first to demonstrate that large-scale experimental manipulations of natural systems are feasible and that they often produce clear results.

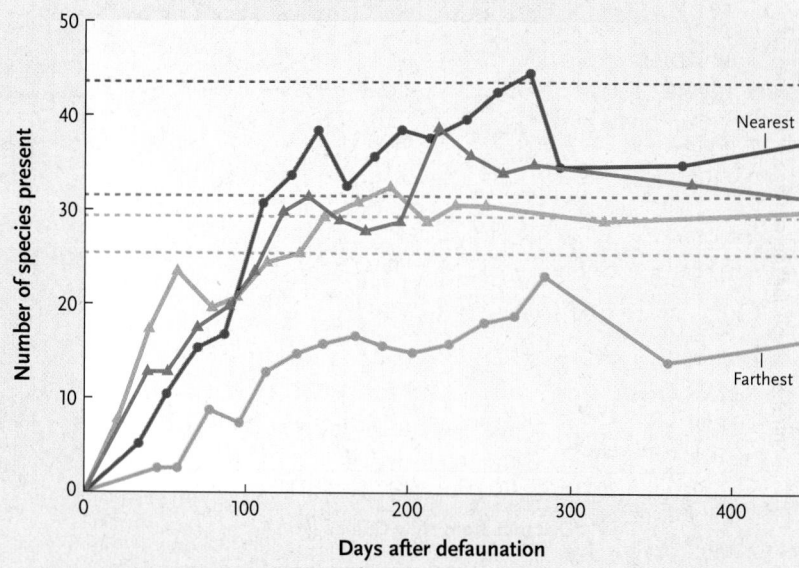

Figure 2
On three of four islands, species richness slowly returned to the predefaunation level (indicated by colour-coded dotted lines). The most distant island had not reached its predefaunation species richness after two years.

Review

30.1 Interspecific Interactions

- Coevolution involves genetically based, reciprocal adaptations in two or more interacting species. Coevolution is not restricted to two species but often involves complex interactions among several species in a community.

30.2 Getting Food

- Predators eat animal prey, whereas herbivores eat plants. Predators and herbivores are animals with characteristics allowing them to feed efficiently. Predation and herbivory are the most conspicuous relationships in ecological communities.

- A koala is a specialist because it eats the leaves of only a few of the available species of *Eucalyptus*. Specialists tend to eat only a few types of food, whereas generalists take a broader diet.

- Optimal foraging theory predicts that an animal's diet is a compromise between the costs and the benefits associated with different types of food. Costs include the time and energy it takes to pursue, capture, and consume a particular kind of food. Benefits are the energy that the food provides.

30.3 Defence

- Animals use seven types of defence: size, eternal vigilance, avoiding detection, counterattack, chemical defence, warnings, and mimicry. Below are examples of each.

- The large size of elephants means that they have few natural predators.

- Meerkats are constantly on the lookout for potential predators.

- Caterpillars that look like bird droppings are not recognized as edible.

- A bee or a scorpion has a sting, and porcupines and other mammals have spines.

- Skunks spray a noxious chemical at potential predators.

- Skunks have black and white coloration and monarch butterflies have orange and black coloration.

- The harmless drone fly mimics the coloration and behaviour of the stinging bee or wasp.

- Animals using chemical defences either synthesize the chemicals themselves or sequester them from other sources. This can include plants that the organism eats.

- Aposematic displays teach would-be predators to avoid the signaller.

- Batesian mimicry occurs when an edible or harmless species mimics an inedible or a poisonous one. In Müllerian mimicry, two or more unpalatable or poisonous species have a similar appearance.

30.4 Competition

- Intraspecific competition can occur between two different species. Two types of competition are interference and exploitation. The competitive exclusion principle states that two or more species cannot coexist indefinitely if both rely on the same limiting resources and exploit them in the same way. One species will be able to harvest the available resources better and eventually outcompete the other species.

- A population's ecological niche is defined as the resources it uses and the environmental conditions it requires over its lifetime. A fundamental niche, larger than a realized niche, includes all conditions and resources a population can tolerate. A realized niche is the range of conditions and resources that a population actually encounters in nature.

- Resource partitioning occurs when sympatric species use different resources or the same resources in different ways. Plants may position their root systems at different levels, avoiding competition for water and nutrients.

- Character displacement results in sympatric species that differ in morphology and use different resources even though they would not do so in allopatric situations. An example of character displacement is the honey-eaters of Australia.

30.5 The Nature of Ecological Communities

Type of Interaction	Effect on Species Involved	Example
Commensalism	One species benefits; the other is unaffected (+/0)	Egrets and the large grazers that flush insects out of grasses during feeding
Mutualism	Both species benefit (+/+)	Bull's horn acacia tree and a species of small ants
Parasitism	One species benefits (parasite); the other is harmed (host) (+/−)	Ectoparasites such as mosquitoes and leeches and their mammalian hosts

- Two hypotheses about ecological communities have been developed by ecologists. The interactive hypothesis predicts that mature communities are at equilibrium and, if disturbed, will return to the predisturbed state. The individualistic hypothesis predicts that communities do not achieve equilibrium but rather are in a steady state of flux in response to disturbance and environmental change.

- Ecotones are generally species rich because they contain species from both communities, as well as species that occur only in transition zones.

30.6 Community Characteristics

- Indices allow population biologists to objectively compare the diversity of communities. Shannon's indices provide a measure of diversity (H') and evenness (E_H).

- Alpha (α) diversity is the number of species living in a single community. Beta (β) diversity is the number of species living in a collection of communities.

- Herbivores are primary consumers and form the second trophic level. Omnivores can be primary, secondary, and tertiary consumers (second, third, and fourth trophic levels, respectively) in a single meal.
- Generally, communities that support complex food webs are more stable. The disappearance of one or even two species does not have a major impact on the food web and thus community structure.

30.7 Effects of Population Interactions on Community Structure

- Species distribution and resource use in *K*-selected species are profoundly affected by competition. However, competition seems to have little effect on the community structure of *r*-selected species.
- Predation and herbivory can increase and/or decrease species richness, depending on the circumstances. Species richness can increase if a predator eliminates a strong competitor, allowing other organisms to exploit the available resources, for example, predatory sea stars reducing populations of mussels. Species richness can decrease when a predator eats less abundant species, further reducing their numbers.
- A keystone species has a much greater effect on the community than its numbers might suggest. Only a few individuals can have a profound impact on community structure.

30.8 Effects of Disturbance on Community Characteristics

- A community may never attain equilibrium because of disturbances such as cyclones, mortality caused by internal processes, and the recruitment of new colonies.
- The intermediate disturbance hypothesis states that species richness is greatest in communities experiencing fairly frequent disturbances of moderate intensity. Data gathered about a river system in New Zealand revealed that areas with moderate disturbance (e.g., moved rocks, soil, and sand in the streambed) had the highest species diversity.
- Generally, communities with a higher species richness recover from disturbance much more quickly than those with a low species richness.

30.9 Succession

- Primary succession begins when organisms first colonize habitats without soil, whereas secondary succession occurs after existing vegetation is destroyed or disrupted by an environmental disturbance.
- A climax community is a late successional stage that can be found in both primary and secondary succession. Climax communities are dominated by a few species that replace themselves and persist until a disturbance eliminates them. Species composition of a climax community is determined by local climate and soil conditions, surrounding vegetation, and chance events.
- The facilitation hypothesis holds that species modify the environment in a way that makes it less suitable for themselves but more suitable for those species that follow them in succession. The inhibition hypothesis contends that species currently occupying a successional stage prevent new species from occupying the same community. The tolerance hypothesis holds that early-stage species neither facilitate nor inhibit the growth of new species. Instead, succession proceeds because new species are able to outcompete and replace early-stage species.

30.10 Variation in Species Richness among Communities

- Species richness generally decreases with increasing latitude.
- The numbers of species on an island is governed by immigration of new species and extinction of species already there. Once equilibrium between immigration and extinction is reached, the number of species on an island remains relatively constant. As one species goes extinct, it is replaced by a newly arrived immigrant species.
- Large islands have higher immigration rates and lower extinction rates than small islands. Islands near the mainland have higher immigration rates than distant islands. Distance does not affect extinction rates. As a result, at equilibrium, near islands have more species than far islands.

Questions

Self-Test Questions

1. According to optimal foraging theory, what do predators do?
   a. always eat the largest prey possible
   b. always eat the prey that are easiest to catch
   c. choose prey based on the costs of consuming it compared to the energy it provides
   d. eat plants when animal prey are scarce
   e. have coevolved mechanisms to overcome prey defences

2. What term refers to the use of the same limiting resource by two species?
   a. brood parasitism
   b. interference competition
   c. exploitative competition
   d. mutualism

3. What is the range of resources that a population of one species can possibly use called?
   a. its fundamental niche
   b. its realized niche
   c. resource partitioning
   d. its relative abundance

4. Differences in molar (tooth) structure of sympatric mammals may reflect which of the following?
   a. predation
   b. character displacement
   c. interference competition
   d. cryptic coloration

5. Bacteria that live in the human intestine assist human digestion and eat nutrients that the human consumes. Which term best describes this relationship?
   a. commensalism
   b. mutualism
   c. endoparasitism
   d. ectoparasitism

6. In the table below, the letters refer to four communities, and the numbers indicate how many individuals were recorded for each of five species. Which community has the highest species diversity?

	Species 1	Species 2	Species 3	Species 4	Species 5
a.	80	10	10	0	0
b.	25	25	25	25	0
c.		4	6	8	80
d.	20	20	20	20	20

7. Which sentence best describes a keystone species?
   a. It is usually a primary producer.
   b. It has a critically important role in determining the species composition of its community.
   c. It is always a predator.
   d. It usually exhibits aposematic coloration.

8. Species richness can be highest in communities with this type of disturbances.
   a. very frequent and severe
   b. very frequent and of moderate intensity
   c. very rare and severe
   d. of intermediate frequency and moderate intensity

9. Which term refers to a community's change in species composition from bare and lifeless rock to climax vegetation?
   a. competition
   b. secondary succession
   c. primary succession
   d. facilitation

10. What does the equilibrium theory of island biogeography predict about the number of species found on an island?
   a. It increases steadily until it equals the number in the mainland species pool.
   b. It is greater on large islands than on small ones.
   c. It is smaller on islands near the mainland than on distant islands.
   d. It is greater for islands near the equator than for islands near the poles.

Questions for Discussion

1. Many landscapes dominated by agricultural activities also have patches of forest of various sizes. What is the minimum amount of habitat required by different species? Focus on 10 species —5 animals and 5 plants. For each species, can you estimate the minimum viable population?

2. Using the terms and concepts introduced in this chapter, describe the interactions that humans have with 10 other species. Try to choose at least 8 species that we do not eat.

3. After reading about the two potential biases in the scientific literature on competition, describe how future studies of competition might avoid such biases.

4. What are the primary producers in a community of parasites?

5. What influence does agriculture have on population interactions and community development?

31

Among the fastest growing ecosystems in the world——shopping malls and residential areas sprawl in the north part of London, Ontario.

Ecosystems

WHY IT MATTERS

As shown in the chapter-opening photograph, urban ecosystems are the most rapidly growing habitat on the planet, replacing existing habitats at an astonishing rate, partly because of growth in populations and in economies. The system portrayed in the photograph is low-density housing with services (water and electricity), but in many parts of the world, housing expansions are high density, with few, if any, services. Apart from humans and our domesticated plant and animal species, which components of the original flora and fauna persist? Walk around your neighbourhood and check it out.

How does the urban ecosystem differ from what was there before? In what ways does it differ? Think of runoff from rain and snow, of the heat-absorbing and reflecting properties of buildings, concrete, and asphalt. What are the effects of gardeners and landscapers, however well meaning? The urban ecosystem offers people in general, and biologists in particular, many opportunities for research and study. We must determine what changes we can effect in the construction and design of neighbourhoods to maximize their compatibility with native organisms. How can we make urban neighbourhoods more useful to migrating songbirds?

Archaeological evidence indicates that urban sprawl occurred around 6000 years ago around the site of Tell Brak in what is now northern Syria. Then the "city" that stood at Tell Brak occupied about 55 ha when other contemporary settlements rarely exceeded 3 ha and the largest of its neighbours was just 15 ha. There is evidence of spatial separation between subcommunities at Tell Brak, where neighbourhoods were divided by walls and limited points of access.

Urban sprawl may not be new, but the current scale makes it a frontier for action to achieve conservation of biodiversity. The purpose of this chapter is to explore some aspects of ecosystems and introduce them as objects of biological study.

Ecosystems are often studied by following the movement of energy from one level to another. Photosynthetic organisms form the energetic basis for ecosystems, providing sources of food for other organisms (usually animals). Levels of biomass at different trophic levels (primary producers; primary, secondary, and tertiary consumers) generally reflect the movement of energy. The movement of markers such as DDT (see "Molecule behind Biology," Box 31.2, p. 758) through ecosystems provided an all-too-sobering view of the connectedness among the ecosystems of the globe.

31.1 Energy Flow and Ecosystem Energetics

Ecosystems receive a steady input of energy from an external source, usually the Sun. Energy flows through an ecosystem, but, as dictated by the laws of thermodynamics, much of that energy is lost without being used by organisms. In contrast, materials cycle between living and nonliving reservoirs, both locally and on a global scale. The flow of energy through and the cycling of materials around an ecosystem make resident organisms highly dependent on one another and on their physical surroundings.

Food webs define the pathways by which energy and nutrients move through an ecosystem's biotic components. In most ecosystems, nutrients and energy move simultaneously through a grazing food web and a detrital food web **(Figure 31.1).** The grazing food web includes the producer, herbivore, and secondary consumer trophic levels. The detrital food web includes detritivores and decomposers. Because detritivores and decomposers subsist on the remains and waste products of organisms at every trophic level, the two food webs are closely interconnected. Detritivores also contribute to the grazing food web when carnivores eat them.

All organisms in a particular trophic level are the same number of energy transfers from the ecosystem's ultimate energy source. Photosynthetic plants are one energy transfer removed from sunlight, herbivores

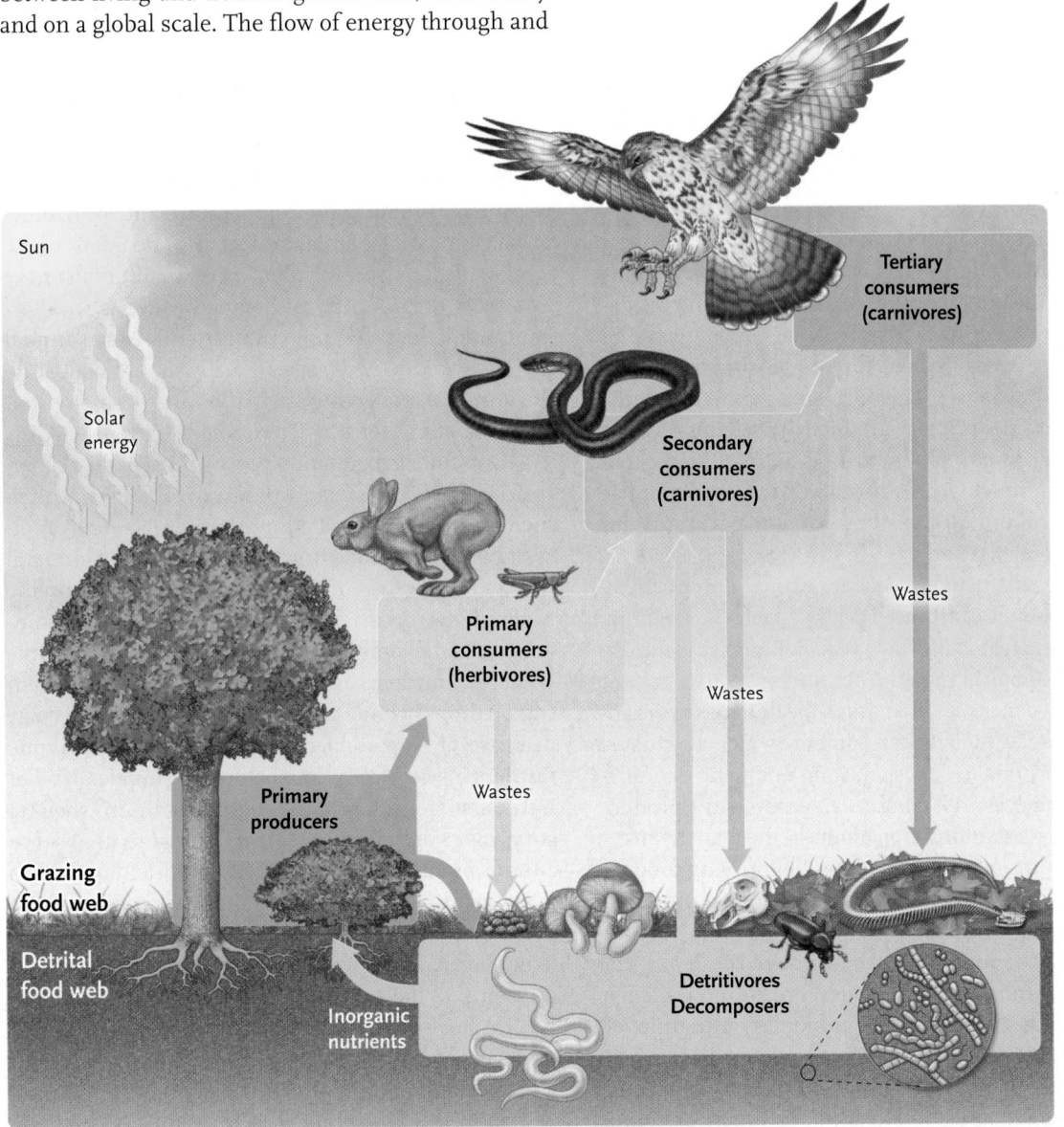

Figure 31.1
Grazing and detrital food webs. Energy and nutrients move through two parallel food webs in most ecosystems. The grazing food web includes producers, herbivores, and carnivores. The detrital food web includes detritivores and decomposers. Each box in this diagram represents many species, and each arrow represents many arrows.

(primary consumers) are two, secondary consumers are three, and tertiary consumers are four.

31.1a Primary Productivity Involves Fixing Carbon

Virtually all life on Earth depends on the input of solar energy. Every minute of every day, Earth's atmosphere intercepts roughly 80 kJ (kilojoules) of energy per square metre (see Chapter 1). About half of that energy is absorbed, scattered, or reflected by gases, dust, water vapour, and clouds before it reaches the planet's surface (see *The Purple Pages*). Most energy reaching the surface falls on bodies of water or bare ground, where it is absorbed as heat or reflected back into the atmosphere. Reflected energy warms the atmosphere. Only a small percentage contacts primary producers, and most of that energy evaporates water, driving transpiration in plants (see Chapter 7).

Ultimately, photosynthesis converts less than 1% of the solar energy arriving at Earth's surface into chemical energy. But primary producers still capture enough energy to produce an average of several kilograms of dry plant material per square metre per year. On a global scale, they produce more than 150 billion tonnes of new biological material annually. Some of the solar energy that producers convert into chemical energy is transferred to consumers at higher trophic levels.

The rate at which producers convert solar energy into chemical energy is an ecosystem's **gross primary productivity**. But, like other organisms, producers use energy for their own maintenance functions. After deducting energy used for these functions (see Chapter 6), whatever chemical energy remains is the ecosystem's **net primary productivity**. In most ecosystems, net primary productivity is 50 to 90% of gross primary productivity. In other words, producers use between 10 and 50% of the energy they capture for their own respiration.

Ecologists usually measure primary productivity in units of energy captured ($kJ \cdot m^{-2} \cdot year^{-1}$) or in units of biomass created ($kg \cdot m^{-2} \cdot year^{-1}$). **Biomass** is the dry mass of biological material per unit area or volume of habitat. (We measure biomass as the *dry* mass of organisms because their water content, which fluctuates with water uptake or loss, has no energetic or nutritional value.) Do not confuse an ecosystem's productivity with its **standing crop biomass**, the total dry mass of plants present at a given time. Net primary productivity is the *rate* at which the standing crop produces *new* biomass (see Chapter 7).

Energy captured by plants is stored in biological molecules, mostly carbohydrates, lipids, and proteins. Ecologists can convert units of biomass into units of energy or vice versa as long as they know how much carbohydrate, protein, and lipid a sample of biological material contains. For reference, 1 g of carbohydrate and

1 g of protein each contains about 17.5 kJ of energy. Thus, net primary productivity indexes the rate at which producers accumulate energy as well as the rate at which new biomass is added to an ecosystem. Ecologists measure changes in biomass to estimate productivity because it is far easier to measure biomass than energy content. New biomass takes several forms, including

- growth of existing producers,
- creation of new producers by reproduction, and
- storage of energy as carbohydrates.

Because herbivores eat all three forms of new biomass, net primary productivity also measures how much new energy is available for primary consumers.

The potential rate of photosynthesis in any ecosystem is proportional to the intensity and duration of sunlight, which varies geographically and seasonally (see Chapter 6, Chapter 7, and *The Purple Pages*). Sunlight is most intense and day length is least variable near the equator. In contrast, the intensity of sunlight is weakest and day length is most variable near the poles. This means that producers at the equator can photosynthesize for nearly 12 hours a day, every day of the year, whereas near the poles, photosynthesis is virtually impossible during the long, dark winter. In summer, however, photosynthesis occurs virtually around the clock.

Sunlight is not the only factor influencing the rate of primary productivity. Temperature and availability of water and nutrients also affect this rate. Many of the world's **deserts** receive plenty of sunshine but have low rates of productivity because water is in short supply and the soil is poor in nutrients. Mean annual primary productivity varies greatly on a global scale **(Figure 31.2)**, reflecting variations in these environmental factors (see *The Purple Pages*).

On a finer geographic scale, within a particular terrestrial ecosystem, mean annual net productivity often increases with the availability of water **(Figure 31.3)**. In systems with sufficient water, a shortage of mineral nutrients may be limiting. All plants need specific ratios of macronutrients and micronutrients for maintenance and photosynthesis (see Chapter 7). But plants withdraw nutrients from soil, and if nutrient concentration drops below a critical level, photosynthesis may decrease or stop altogether. In every ecosystem, one nutrient inevitably runs out before the supplies of other nutrients are exhausted. The element in shortest supply is called a **limiting nutrient** because its absence curtails productivity. Productivity in agricultural fields is subject to the same constraints as productivity in natural ecosystems. Farmers increase productivity by irrigating (adding water to) and fertilizing (adding nutrients to) their crops.

In freshwater and marine ecosystems, where water is always readily available, the depth of water and combined availability of sunlight and nutrients govern the rate of primary productivity. Productivity

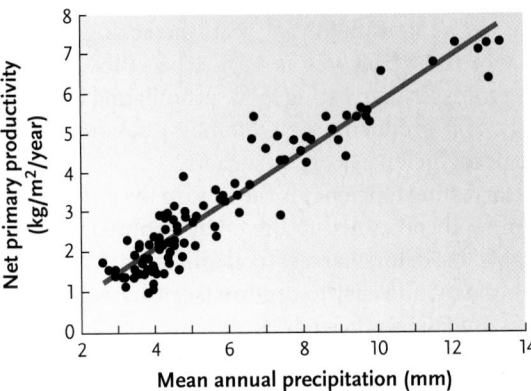

Figure 31.2
Global variation in primary productivity. Satellite data from 2002 provide a visual portrait of net primary productivity across Earth's surface. High-productivity regions on land are dark green; low-productivity regions are yellow. For aquatic environments, the highest productivity is red, down through orange, yellow, green, blue, and purple (lowest).

Figure 31.3
Water and net primary productivity. Mean annual precipitation at 100 sites in the Great Plains of North America. These data include only above-ground productivity.

Table 31.1	Standing Crop Biomass and Net Primary Productivity of Different Ecosystems	
Ecosystem	Mean Standing Crop Biomass (kg/m²)	Mean Net Primary Productivity (kg/m²/y¹)
Terrestrial Ecosystems		
Tropical rain forest	450	22.0
Tropical deciduous forest	350	16.0
Temperate rain forest	350	13.0
Temperate deciduous forest	300	12.0
Savannah	40	9.0
Boreal forest (**taiga**)	200	8.0
Woodland and shrubland	60	7.0
Agricultural land	10	6.5
Temperate grassland	16	6.0
Tundra and alpine tundra	6.0	1.4
Desert and thornwoods	7.0	0.9
Extreme desert, rock, sand, ice	0.2	0.03
Freshwater Ecosystems		
Swamp and marsh	150	20
Lake and stream	0.2	2.5
Marine Ecosystems		
Open ocean	0.03	1.3
Upwelling zones	0.2	5.0
Continental shelf	0.1	3.6
Kelp beds and reefs	20	25
Estuaries	10	15
World Total	**36**	**3.3**

Source: Based on Whittaker, R.H. 1975. Communities and Ecosystems. 2nd ed. Macmillan.

is high in near-shore ecosystems, where sunlight penetrates shallow, nutrient-rich waters. Kelp beds and coral reefs along temperate and tropical marine coastlines, respectively, are among the most productive ecosystems on Earth (**Table 31.1**; see also Figure 31.2). In contrast, productivity is low in the open waters of a large lake or ocean. There sunlight penetrates only the upper layers, and nutrients sink to the bottom; thus, the two requirements for photosynthesis—sunlight and nutrients—are available in different places.

Although ecosystems vary in their rates of primary productivity, these differences are not always proportional to variations in their standing crop biomass (see Table 31.1). For example, biomass amounts in temperate deciduous forests and **temperate grasslands** differ by a factor of 20, but the difference in their rates of net primary productivity is much smaller. Most biomass in trees is present in nonphotosynthetic tissues such as wood, so their ratio of productivity to biomass is low ($12\,\text{kg}\cdot\text{m}^{-2}/300\,\text{kg}\cdot\text{m}^{-2} = 0.04$). By contrast, grasslands do not accumulate much biomass because annual mortality, herbivores, and fires remove plant material

as it is produced, so their productivity to biomass ratio is much higher ($6.0\,\text{kg}\cdot\text{m}^{-2}/16\,\text{kg}\cdot\text{m}^{-2} = 0.375$).

Some ecosystems contribute more than others to overall net primary productivity (**Figure 31.4, p. 752**).

Ecosystems covering large areas make substantial total contributions, even if their productivity per unit area is low. Conversely, geographically restricted ecosystems make large contributions if their productivity is high. Open ocean and tropical rain forests contribute about equally to total global productivity, but for different reasons. Open oceans have low productivity, but they cover nearly two-thirds of Earth's surface. Tropical rain forests are highly productive but cover only a relatively small area.

Net primary productivity ultimately supports all consumers in grazing and detrital food webs. Consumers in the grazing food web eat some biomass at every trophic level except the highest. Uneaten biomass eventually dies and passes into detrital food webs. Moreover, consumers assimilate only a portion of the material they ingest, and unassimilated material passed as feces also supports detritivores and decomposers.

31.1b Secondary Productivity Involves Animals Eating Plants, Thus Moving Up the Trophic Scale

As energy is transferred from producers to consumers, some is stored in new consumer biomass, called **secondary productivity.** Nevertheless, two factors cause energy to be lost from the ecosystem every time it flows

from one trophic level to another. First, animals use much of the energy they assimilate for maintenance and locomotion rather than for production of new biomass. Second, as dictated by the second law of thermodynamics, no biochemical reaction is 100% efficient, so some of the chemical energy liberated by cellular respiration is converted to heat, which most organisms do not use.

31.1c Ecological Efficiency Is Measured by Use of Energy

Ecological efficiency is the ratio of net productivity at one trophic level to net productivity at the trophic level below. If plants in an ecosystem have a net primary productivity of $1.0 \, \text{kg} \cdot \text{m}^{-2} \cdot \text{year}^{-1}$ of new tissue, and the herbivores that eat these plants produce 0.1 kg of new tissue per square metre per year, the ecological efficiency of the herbivores is 10%. The efficiencies of three processes (harvesting food, assimilating ingested energy, and producing new biomass) determine the ecological efficiencies of consumers.

Harvesting efficiency is the ratio of the energy content of food consumed to the energy content of food available. Predators harvest food efficiently when prey are abundant and easy to capture (see Chapter 30).

Assimilation efficiency is the ratio of the energy absorbed from consumed food to the total energy content

Ecosystem	Percentage of Earth's Surface Covered	Percentage of Total Producer Biomass	Percentage of Total Net Primary Productivity
Open ocean	65.1	0.05	24.40
Continental shelf	5.2	0.01	5.60
Extreme desert, rock, sand, ice	4.7	0.03	0.04
Desert and thornwoods	3.5	0.71	0.94
Tropical rain forest	3.3	41.60	22.00
Savannah	2.9	3.30	7.94
Cultivated land	2.7	0.76	5.35
Boreal forest (taiga)	2.4	13.03	5.64
Temperate grassland	1.8	0.76	3.18
Woodland and shrubland	1.7	2.71	3.53
Tundra	1.6	0.27	0.65
Tropical deciduous forest	1.5	14.12	7.05
Temperate deciduous forest	1.4	11.41	4.94
Temperate rain forest	1.0	9.51	3.82
Swamp and marsh	0.4	1.62	2.35
Lake and stream	0.4	0.003	0.29
Estuaries	0.3	0.08	1.23
Algal beds and reefs	0.1	0.07	0.94

Figure 31.4

Biomass and net primary productivity. An ecosystem's percentage coverage of Earth's surface is not proportional to its contribution to total biomass of producers or its contribution to the total net primary productivity.

of the food. Because animal prey is relatively easy to digest, carnivores absorb between 60 and 90% of the energy in their food. Assimilation efficiency is lower for prey with indigestible parts such as bones or exoskeletons. Herbivores assimilate only 15 to 80% of the energy they consume because cellulose is not very digestible. Herbivores lacking cellulose-digesting systems are on the low end of the scale, whereas those that can digest cellulose are at the higher end.

Production efficiency is the ratio of the energy content of new tissue produced to the energy assimilated from food. Production efficiency varies with maintenance costs. Endothermic animals often use less than 10% of their assimilated energy for growth and reproduction because they use energy to generate body heat (see Chapter 50). Ectothermic animals channel more than 50% of their assimilated energy into new biomass.

The overall ecological efficiency of most organisms is 5 to 20%. As a rule of thumb, only about 10% of energy accumulated at one trophic level is converted into biomass at the next higher trophic level, as illustrated by energy transfers at Silver Springs, Florida **(Figure 31.5)**. Silver Springs is an ecosystem that has been studied for many years. Producers in the Silver Springs ecosystem convert 1.2% of the solar energy they intercept into chemical energy (represented by 86 986 kJ·m^{-2}·year^{-1} of gross primary productivity). However, plants use about two-thirds of this energy for respiration, leaving a net primary productivity, one-third of which is to be included in new plant biomass. All consumers in the grazing food web (on the right in Figure 31.5) ultimately depend on this energy source, which diminishes with each transfer between trophic levels. Energy is lost to respiration and export at each trophic level. In addition, organic

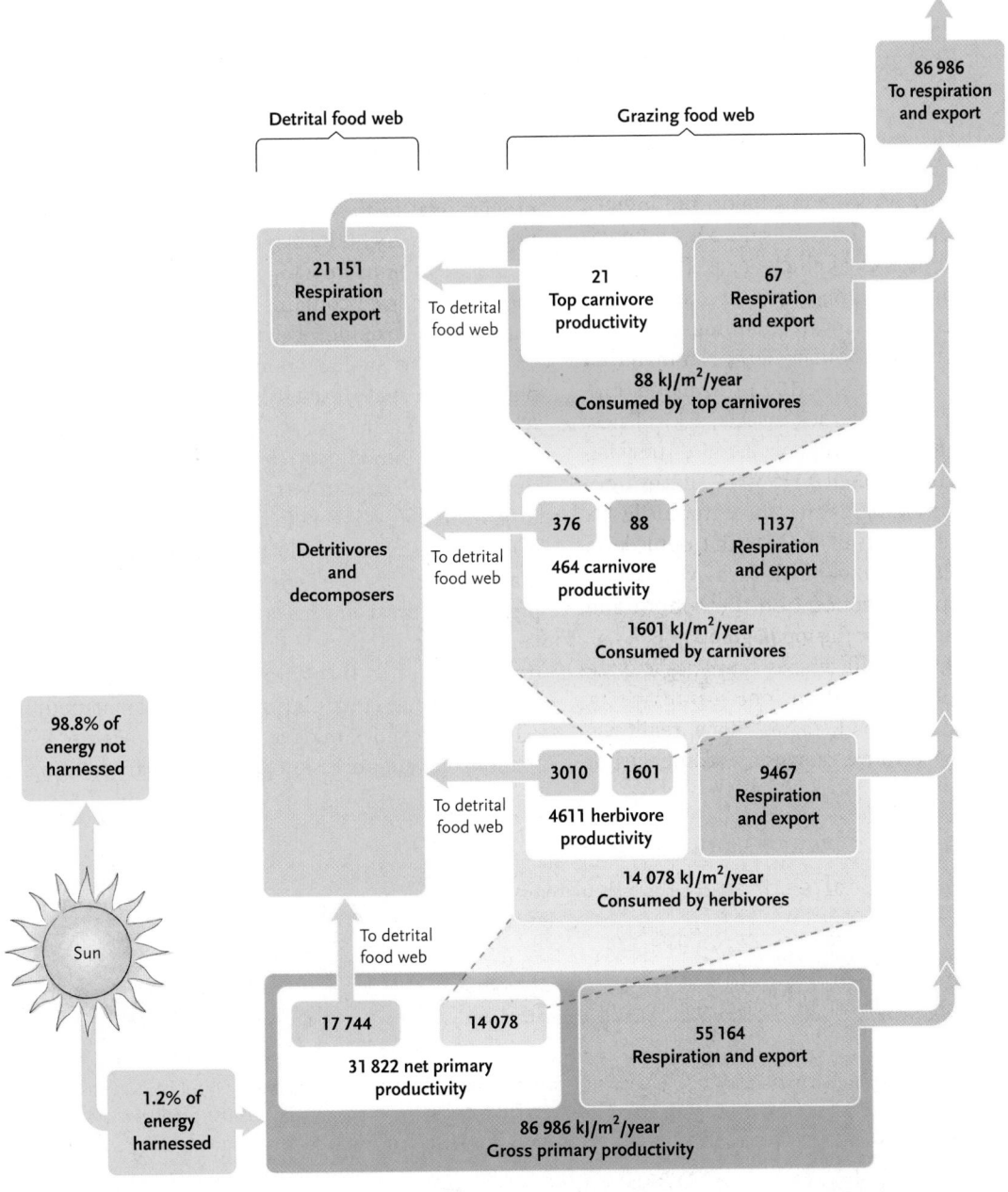

Observational Figure 31.5 Energy flow through the Silver Springs ecosystem.

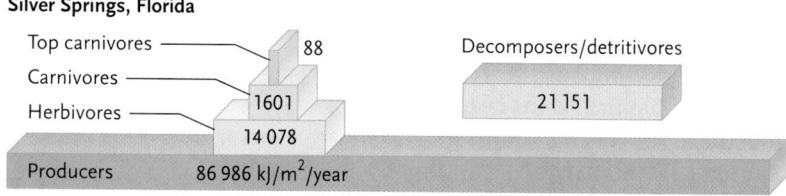

Figure 31.6

Pyramids of energy. The pyramid of energy for Silver Springs, Florida, shows that the amount of energy (kJ·m⁻²·year⁻¹) passing through each trophic level decreases as it moves up the food web.

wastes and uneaten biomass represent substantial energy that flows into the detrital food web (on the left in Figure 31.5). To determine the ecological efficiency of any trophic level, we divide its productivity by the productivity of the level below it. The ecological efficiency of midlevel carnivores at Silver Springs is 10.06%, $464 \text{ kJ·m}^{-2}\cdot\text{year}^{-1}/4611 \text{ kJ·m}^{-2}\cdot\text{year}^{-1}$.

31.1d Three Types of Pyramids Describe Ecosystems: Energy, Biomass, and Numbers

As energy works its way up a food web, energy losses are multiplied in successive energy transfers, greatly reducing the energy available to support the highest trophic levels (see Figure 31.5). Consider a hypothetical example in which ecological efficiency is 10% for all consumers. Assume that the plants in a small field annually produce new tissues containing 100 kJ of energy. Because only 10% of that energy is transferred to new herbivore biomass, the 100 kJ in plants produces 10 kJ of new herbivorous insects, 1 kJ of new songbirds that eat insects, and only 0.1 kJ of new falcons that eat songbirds. About 0.1% of the energy from primary productivity remains after three trophic levels of transfer. If the energy available to each trophic level is depicted graphically, the result is a **pyramid of energy**, with primary producers on the bottom and higher-level consumers on the top **(Figure 31.6)**.

The low ecological efficiencies that characterize most energy transfers illustrate one advantage of eating "lower on the food chain." This reality is reflected by major adaptive radiations of lineages of animals whose ancestors were secondary consumers when they switched to being primary consumers. Good examples of such radiations occur, for example, among insects, fish, dinosaurs, and mammals.

Even though humans digest and assimilate meat more efficiently than vegetables, we could feed more people if we all ate more primary producers directly instead of first passing them through another trophic level, such as cattle or chickens, to produce meat. Production of animal protein is costly because much of the energy fed to livestock is used for their own maintenance rather than production of new biomass. But despite the economic and health-related logic of a more vegetarian diet, changing our eating habits alone will not eliminate food shortages or the frequency of malnutrition. Many regions of Africa, Australia, North America, and South America support vegetation that is suitable only for grazing by large herbivores. These areas could not produce significant quantities of edible grains and vegetables without significant additions of water and fertilizer (see Chapter 33).

Inefficiency of energy transfer from one trophic level to the next has profound effects on ecosystem structure. Ecologists illustrate these effects in diagrams called **ecological pyramids.** Trophic levels are drawn as stacked blocks, with the size of each block proportional to the energy, biomass, or numbers of organisms present. Pyramids of energy typically have wide bases and narrow tops (see Figure 31.6) because each trophic level contains only about 10% as much energy as the trophic level below it.

Progressive reduction in productivity at higher trophic levels usually establishes a **pyramid of biomass (Figure 31.7).** The biomass at each trophic level is proportional to the amount of chemical energy temporarily stored there. Thus, in terrestrial ecosystems, the total mass of producers is generally greater than the total mass of herbivores, which is, in turn, greater than the total mass of predators (see Figure 31.7a). Populations of top predators, from killer whales to lions and crocodiles, contain too little biomass and energy to support another trophic level; thus, they have no nonhuman predators.

Freshwater and marine ecosystems sometimes exhibit inverted pyramids of biomass (see Figure 1.7b). In the open waters of a lake or ocean, primary consumers

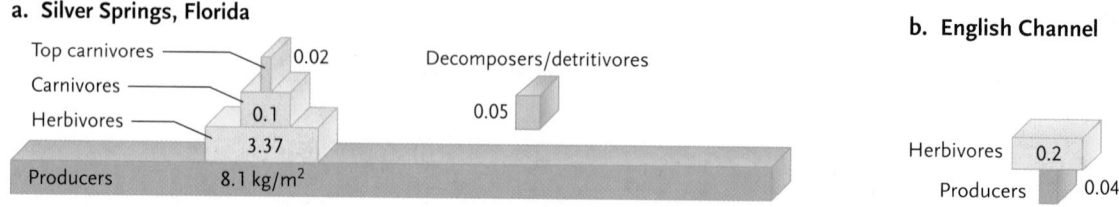

Figure 31.7

Pyramids of biomass. **(a)** The pyramid of standing crop biomass for Silver Springs is bottom heavy, as it is for most ecosystems. **(b)** Some marine ecosystems, such as that in the English Channel, have an inverted pyramid of biomass because producers are quickly eaten by primary consumers. Only the producer and herbivore trophic levels are illustrated here. The data for both pyramids are given in kilograms per square metre of dry biomass.

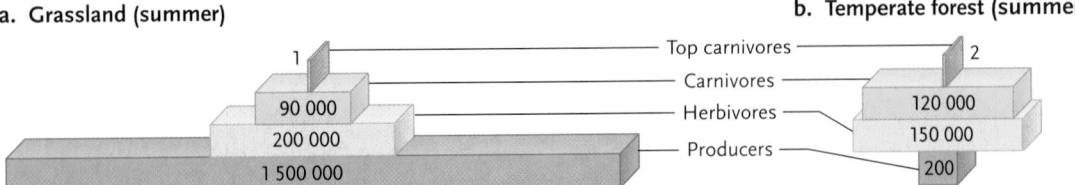

a. Grassland (summer)
b. Temperate forest (summer)

Top carnivores
Carnivores
Herbivores
Producers

1

90 000

200 000

1 500 000

2

120 000

150 000

200

Figure 31.8

Pyramids of numbers. **(a)** The pyramid of numbers (numbers of individuals per 1000 m²) for temperate grasslands is bottom heavy because individual producers are small and very numerous. **(b)** The pyramid of numbers for forests may have a narrow base because herbivorous insects usually outnumber the producers, many of which are large trees. Data for both pyramids were collected in summer. Detritivores and decomposers (soil animals and microorganisms) are not included because they are difficult to count.

(zooplankton) eat primary producers (phytoplankton) almost as soon as they are produced. As a result, the standing crop of primary consumers at any moment in time is actually larger than the standing crop of primary producers. Food webs in these ecosystems are stable because producers have exceptionally high **turnover rates.** In other words, producers divide and their populations grow so quickly that feeding by zooplankton does not endanger their populations or reduce the producers' productivity. However, on an annual basis, the *cumulative total* biomass of primary producers far outweighs that of primary consumers.

The reduction of energy and biomass affects population sizes of organisms at the top of a food web. Top predators can be relatively large animals, so the limited biomass present in the highest trophic levels is concentrated in relatively few animals **(Figure 31.8).** The extremely narrow top of this **pyramid of numbers** has grave implications for conservation biology (see Chapter 32). Top predators tend to be large animals with small population sizes. And because each individual must patrol a large area to find sufficient food, members of a population are often widely dispersed within their habitats. As a result, they are subject to genetic drift (see Chapter 18) and are highly sensitive to hunting, habitat destruction, and random events that can lead to extinction. Top predators may also suffer from the accumulation of poisonous materials that move through food webs (see the next section). Even predators that feed below the top trophic level often suffer the ill effects of human activities. Consumers sometimes regulate ecosystem processes.

Numerous abiotic factors, such as the intensity and duration of sunlight, rainfall, temperature, and the availability of nutrients, have significant effects on primary productivity. Primary productivity, in turn, profoundly affects the populations of herbivores and predators that feed on them. But what effect does feeding by these consumers have on primary productivity?

Consumers sometimes influence rates of primary productivity, especially in ecosystems with low species diversity and relatively few trophic levels. Food webs in lake ecosystems depend primarily on the productivity of phytoplankton **(Figure 31.9).** Phytoplankton are, in turn, eaten by herbivorous zooplankton, themselves

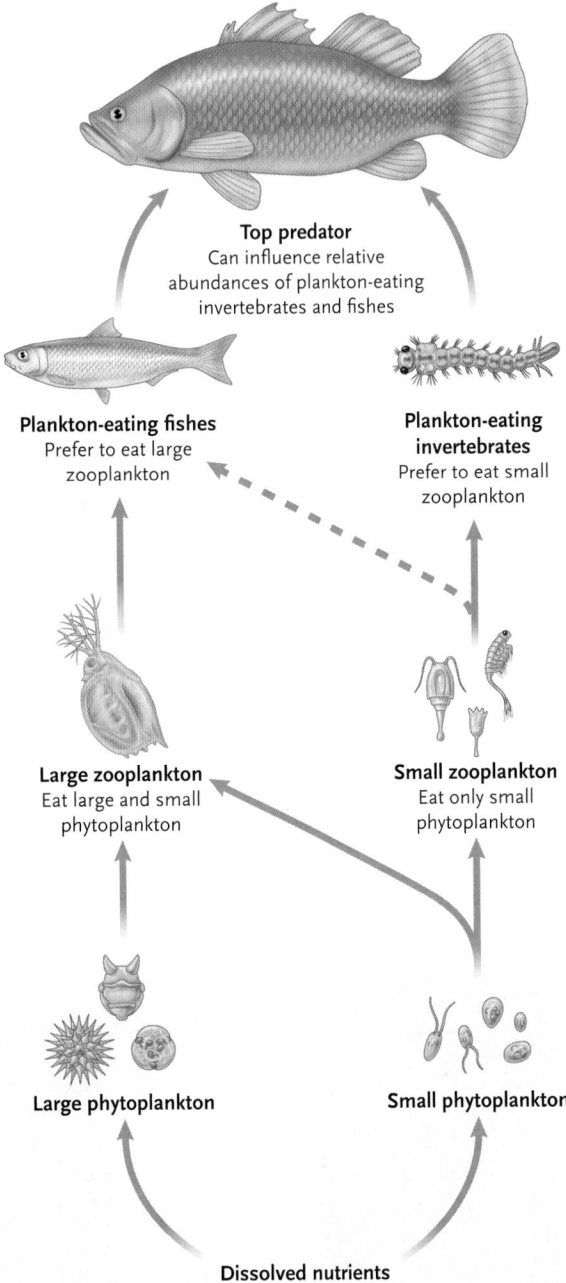

Top predator
Can influence relative abundances of plankton-eating invertebrates and fishes

Plankton-eating fishes
Prefer to eat large zooplankton

Plankton-eating invertebrates
Prefer to eat small zooplankton

Large zooplankton
Eat large and small phytoplankton

Small zooplankton
Eat only small phytoplankton

Large phytoplankton

Small phytoplankton

Dissolved nutrients

Figure 31.9

Consumer regulation of primary productivity. A simplified food web illustrates that lake ecosystems have relatively few trophic levels. The effects of feeding by top carnivores can cascade downward, exerting an indirect effect on the phytoplankton and thus on primary productivity.

Fishing Fleets at Loggerheads with Sea Turtles

Populations of loggerhead sea turtles (*Caretta caretta*) that nest on Western Pacific beaches in Australia and Japan have been in decline. Like other sea turtles, *C. caretta* hatch from eggs that females bury on sandy beaches. Immediately after hatching, the young turtles rush to the surf and the open ocean. Turtles mature at sea and return to their hatching beaches to lay eggs. Using mitochondrial DNA (mtDNA), Bruce Bowen and colleagues explored the situation sea turtles face.

The researchers took mtDNA samples from nesting populations in Australia and Japan, from populations of turtles feeding in Baja California, and from turtles drowned in fishing nets in the north Pacific. One 350-base-pair (bp) segment of mtDNA included sequence variations that are characteristic of different loggerhead populations. After samples were amplified by the polymerase chain reaction, sequencing revealed three major variants of mtDNA, which the researchers designated as sequences A, B, and C. The sequences were distributed among loggerhead turtles, as shown in **Table 1.**

The mtDNA of most *C. caretta* found in Baja California and in fishing nets in the north Pacific matched that of turtles from the Japanese nesting areas. These data support the idea that loggerhead turtles hatched in Japan make the 10 000-km-long migration across the North Pacific to Baja California. The data also indicate that a few turtles that hatched in Australia may follow the same migratory route.

This migration could be aided by the North Pacific Current, which moves from west to east, whereas the return trip from Baja to Japan could be made via the North Equatorial Current, which runs from east to west just north of the equator. Loggerhead turtles have been found in these currents, and further tests will reveal whether they have the mtDNA sequence characteristic of the individuals nesting in Japan and feeding in Baja California.

The nesting population of *C. caretta* in Japan is 2000 to 3000 females. It is uncertain if this population can survive the loss of thousands of offspring to fishing in the North Pacific. The number of female loggerhead turtles nesting in Australia has declined by 50 to 80% in the last decade, so the loss of only a few individuals in fishing nets could also have a drastic impact on this population. To save the loggerhead turtles, wildlife managers and international agencies must establish and enforce limits on the number of migrating individuals trapped and killed in the ocean fisheries.

Like other sea turtles, the loggerheads are severely impacted by fishing. As many as 4000 loggerheads drown in nets every year, and others are caught in longline fisheries. Adoption of a new fish hook **(Figure 1)** could reduce the turtle catch in longline fishing operations.

Table 1	Sources of Turtles by Nesting Grounds		
	Number of Turtles		
Location	Sequence A	Sequence B	Sequence C
Australian nesting areas	26	0	0
Japanese nesting areas	0	23	3
Baja California feeding grounds	2	19	5
North Pacific	1	28	5

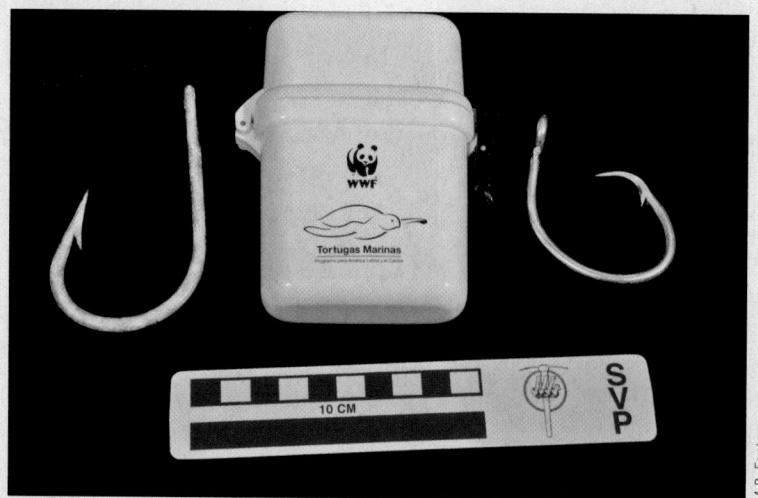

M.B. Fenton

Figure 1

Sea turtles and longlining. Conventional longline hooks (left) readily catch sea turtles, whereas the hook on the right does not. The World Wildlife Fund is promoting the use of hooks (right) that are friendly to sea turtles in an effort to reduce their losses to longline fishing.

consumed by predatory invertebrates and fishes. The top nonhuman carnivore in these food webs is usually a predatory fish.

Herbivorous zooplankton play a central role in regulation of lake ecosystems. Small zooplankton species consume only small phytoplankton. Thus, when *small* zooplankton are especially abundant, large phytoplankton escape predation and survive, and the lake's primary productivity is high. By contrast, large zooplankton are voracious, eating both small and large phytoplankton. When large zooplankton are especially abundant, they reduce the overall biomass of phytoplankton, lowering the ecosystem's primary productivity.

In this **trophic cascade**, predator–prey effects reverberate through population interactions at two or more trophic levels in an ecosystem. Feeding by plankton-eating invertebrates and fishes has a *direct* impact on herbivorous zooplankton populations and an *indirect* impact on phytoplankton populations (the ecosystem's primary producers). Invertebrate predators prefer small zooplankton. And when the invertebrates that eat small zooplankton are the dominant predators in the ecosystem, large zooplankton become more abundant; they consume many phytoplankton, causing a decrease in productivity. But plankton-eating fishes prefer to eat large zooplankton (see Figure 31.9, p. 755), so when they are abundant, small zooplankton become the dominant herbivores, leading large phytoplankton to become more numerous, which raises the lake's productivity.

Large predatory fishes may add an additional level of control to the system because they feed on and regulate the population sizes of plankton-eating invertebrates and fishes. Thus, the effects of feeding by the top predator can cascade downward through the food web, affecting the densities of plankton-eating invertebrates and fishes, herbivorous zooplankton, and phytoplankton. Research in Norway with brown trout (*Salmo trutta*), a top predator, and Arctic char (*Salvelinus alpinus*), the prey, demonstrated how culling prey can promote the recovery of top predators. In this case, Lake Takvatn was the scene of a large-scale experiment. Older, stunted prey species (*S. alpinus*) were removed. These fish had eaten small prey, so an increase in the availability of prey and recovery of the predator resulted. In this case, *S. trutta* was the top predator, and *S. alpinus*, an introduced species, was culled to rejuvenate the system. Another process of bioremediation, the addition of piscivorous fish to a lake, has also been successful in restoring ecosystem balance in other parts of the world.

31.1e Biological Magnification Is the Movement of Contaminants Up the Food Chain

DDT (a formerly popular insecticide; see "Molecule behind Biology," Box 31.1, p. 758) provided a clear demonstration of the interconnectedness of organisms.

Consumers accumulate DDT from all the organisms they eat in their lifetimes. Primary consumers, such as herbivorous insects, may ingest relatively small amounts of DDT, but a songbird that eats many of these insects accumulates all the collected DDT consumed by its prey. A predator such as a raptor, perhaps a Sharp-shinned Hawk (*Accipiter striatus*), that eats songbirds accumulates even more. Whether the food chain (web) is aquatic or terrestrial, the net effect on higher-level consumers is the same **(Figure 31.10)**.

Natural systems have provided many examples of **biological magnification.** In cities where DDT was used in an effort to control the spread of Dutch elm disease, songbirds died from DDT poisoning after eating insects that had been sprayed (whether or not they were involved in spreading the disease). In forests, DDT was used in an effort to control spruce budworm moths (*Choristoneura occidentalis*), and salmon died because runoff carried DDT into their streams and rivers, where their herbivorous prey consumed it.

Despite the ban on the use of DDT in the United States in 1973, in 1990, the California State Department of Health recommended closing a fishery off the coast of California because of DDT accumulating there. DDT discharged in industrial waste 20 years earlier was still moving through the ecosystem. The half-life of DDT in an organism's body fat is eight years.

Other contaminants emulate DDT. Mercury contamination is common in many parts of the world, often as a by-product of the pulp and paper industry. Minamata, the disease humans get from mercury

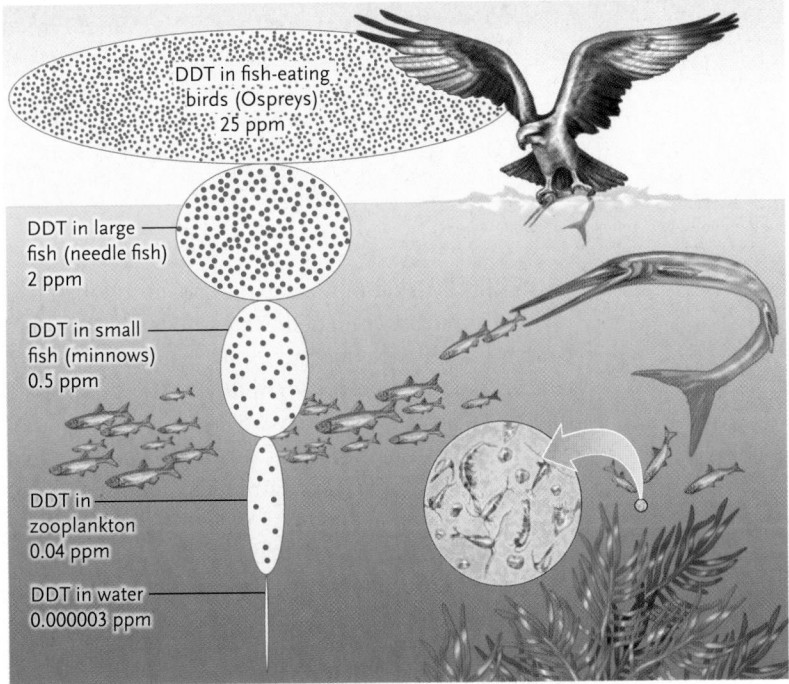

Figure 31.10

Biological magnification. In this marine food web in northeastern North America, DDT concentration (measured in parts per million, ppm) was magnified nearly 10 million times between zooplankton and the Osprey (*Pandion haliaetus*).

MOLECULE BEHIND BIOLOGY 31.1
DDT: Dichloro-Diphenyl-Trichloroethane

Figure 1
A molecule of DDT.

Originally formulated in 1873, DDT's potential as an insecticide was only recognized in 1939 by Paul Muller of Geigy Pharmaceutical in Switzerland. DDT **(Figure 1)**, the first of the chlorinated insecticides, was used extensively in some theatres of World War II, notably in Burma (now Myanmar) in 1944, when the Japanese forces were on the brink of moving into India. There Allied forces suffered from "three m's": mud, morale, and malaria. Meanwhile, in 1943 in southern Italy, DDT was instrumental in controlling populations of lice that plagued Canadian troops there. Widespread application of DDT in Burma reduced the incidence of malaria by killing mosquitoes, the vectors for the disease (see Chapter 27). After World War II, the use of DDT spread rapidly, and the World Health Organization (WHO) credited this molecule with saving 25 million

human lives (mainly through control of mosquitoes that carry malaria).

At first, DDT appeared to be an ideal insecticide. In addition to being inexpensive to produce, it had low toxicity to mammals (300 to 500 mg·kg^{-1} is the LD$_{50}$, the amount required to kill half of the target population). But many insects subsequently developed immunity to DDT, reducing its effectiveness.

DDT is chemically stable and soluble in fat, so instead of being metabolized by mammals, it is stored in their fat. The biological half-life of DDT is approximately eight years (it takes about eight years for a mammal to metabolize half the amount of DDT it has assimilated). DDT is released when fat is metabolized, so when mammals metabolize fat (for example, when humans go on a diet), they are exposed to higher concentrations of DDT in their blood. DDT also had dramatic effects on some birds, notably those higher up the food chain. Eggshell thinning was a consequence of exposure to DDT. Populations of birds such as Peregrine Falcons (*Falco peregrinus*) plummeted.

Since 1985, the use of DDT has been totally banned in Canada, and it is now banned in many other countries. But DDT is still produced in countries such as the United States and still used in countries where

malaria is a prominent problem because the ecological costs of DDT are considered secondary to the importance of controlling the mosquitoes. WHO estimates that every 30 seconds, a child dies of malaria. Approximately 40% of the world's population of humans is at risk of contracting malaria where they live, mainly in Africa. Malaria also remains a problem in tropical and subtropical Asia and Central and South America. People in southern Europe and the Middle East may also be at risk.

By the early 1970s, cetaceans in the waters around Antarctica had DDT in their body fat even though DDT had never been used there. The movement of DDT up the food chain and through food webs demonstrated the interconnections in biological systems. The movement of DDT also provides a graphic demonstration of the transfer of materials from one trophic level to another.

Removing DDT from the arsenal of products used to control insects has had other impacts. For example, there has been an upsurge recently in the number of houses and apartments infested by bedbugs (*Cimex lectularius*). DDT had been very effective in the control of bedbugs, but in its absence, populations of these insects have rebounded, renewing old challenges that our grandparents had experienced.

---

poisoning, is usually linked to the consumption of fish taken from contaminated watersheds. Eating fish contaminated with mercury can result in mercury concentrations in people's hair (0.9 to 94 mg·kg^{-1}) and in otters (*Lontra canadensis;* 0.49 to 54.37 mg·kg^{-1}). In southern Ontario, the hair of bats that eat insects that emerge from mercury-contaminated sediments contains concentrations up to 13 mg·kg^{-1}. Fish are obviously not essential to this chain of biomagnification.

Evidence for the impact of pesticides often comes from sources we might not have expected. An excellent example is the accumulation of Chimney Swift (*Chaetura pelagica*) droppings at the bottom of a chimney in Kingston, Ontario. As their name implies, Chimney Swifts often nest in chimneys, and analysis of samples from the accumulated droppings revealed how changes

in levels of insecticides coincided with changes in the birds' diets and, presumably, in populations of their prey. This research may help us to understand the reasons for declines in populations of aerial insectivorous birds (such as Chimney Swifts).

STUDY BREAK

1. What is net primary productivity? How does it differ from standing crop biomass? Are the pyramids useful?
2. Many deserts have low levels of productivity despite receiving a lot of sunlight. Why?
3. What are assimilation efficiency and production efficiency?

31.2 Nutrient Cycling in Ecosystems

The availability of nutrients is as important to ecosystem function as the input of energy. Photosynthesis requires carbon, hydrogen, and oxygen, which producers acquire from water and air. Primary producers also need nitrogen, phosphorus, and other minerals (see Chapter 48). A deficiency in any of these minerals can reduce primary productivity.

Earth is essentially a closed system with respect to matter, even though cosmic dust enters the atmosphere. Thus, unlike energy, for which there is a constant cosmic input, virtually all of the nutrients that will ever be available for biological systems are already present. Nutrient ions and molecules constantly circulate between the abiotic environment and living organisms in **biogeochemical cycles.** And, unlike energy, which flows through ecosystems and is gradually lost as heat, matter is conserved in biogeochemical cycles. Although there may be local shortages of specific nutrients, Earth's overall supplies of these chemical elements are never depleted or increased.

Nutrients take various forms as they pass through biogeochemical cycles. Materials such as carbon, nitrogen, and oxygen form gases that move through global *atmospheric cycles.* Geologic processes move other materials, such as phosphorus, through local *sedimentary cycles,* carrying them between dry land and the sea floor. Rocks, soil, water, and air are the reservoirs where mineral nutrients accumulate, sometimes for many years.

Ecologists use a **generalized compartment model** to describe nutrient cycling **(Figure 31.11).** Two criteria divide ecosystems into four compartments in which nutrients accumulate. First, nutrient molecules and ions are either *available* or *unavailable,* depending on whether they can be assimilated by organisms. Second, nutrients are present in either *organic* material, living or dead tissues of organisms, or *inorganic* material, such as rocks and soil. Minerals in dead leaves on the forest floor are in the available-organic compartment because they are in the remains of organisms that can be eaten by detritivores. Calcium ions in limestone rocks are in the unavailable-inorganic compartment because they are in a nonbiological form that producers cannot assimilate.

Nutrients move rapidly within and between the available compartments. Living organisms are in the available-organic compartment, and whenever heterotrophs consume food, they recycle nutrients within that reservoir (indicated by the circular arrow in the upper left of Figure 31.11). Producers acquire nutrients from the air, soil, and water of the available-inorganic compartment. Consumers acquire nutrients from the available-inorganic compartment when

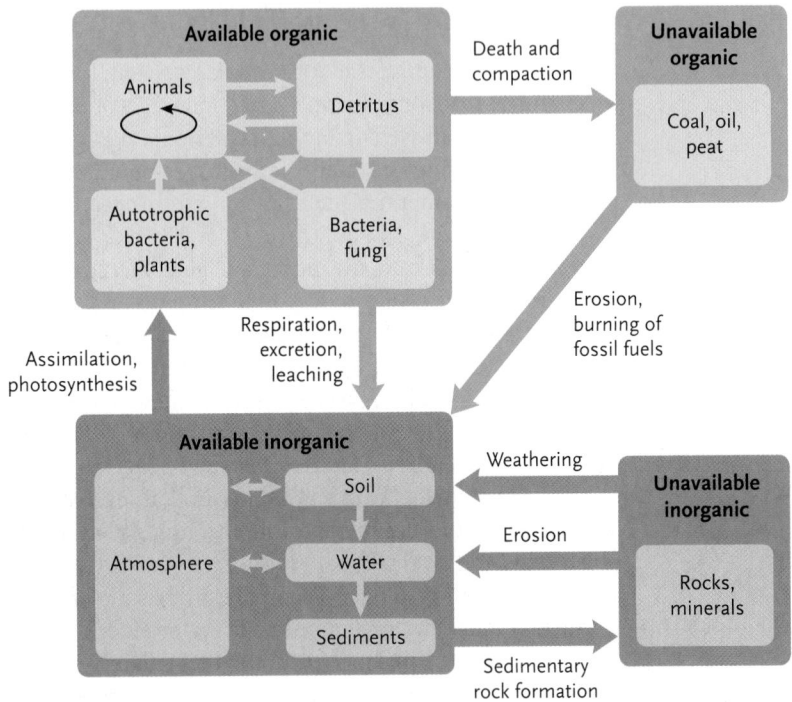

Figure 31.11

A generalized compartment model of nutrient cycling. Nutrients cycle through four major compartments within ecosystems. Processes that move nutrients from one compartment to another are indicated on the arrows. The circular arrow under "Animals" represents animal predation on other animals.

they drink water or absorb mineral ions through their integument. Several processes routinely transfer nutrients from organisms to the available-inorganic compartment. Respiration releases carbon dioxide, moving both carbon and oxygen from the available-organic compartment to the available-inorganic compartment.

By contrast, the exchange of materials into and out of the unavailable compartments is generally slow. Sedimentation, a long-term geologic process, converts ions and particles of the available-inorganic compartment into rocks of the unavailable-inorganic compartment. Materials are gradually returned to the available-inorganic compartment when rocks are uplifted and eroded or weathered. Similarly, over millions of years, the remains of organisms in the available-organic compartment were converted into the coal, oil, and peat of the unavailable-organic compartment.

Except for the input of solar energy, we have described energy flow and nutrient cycling as though ecosystems were closed systems. In reality, most ecosystems exchange energy and nutrients with neighbouring ecosystems. Rainfall carries nutrients into a forest ecosystem, and runoff carries nutrients from a forest into a lake or river. Ecologists have mapped biogeochemical cycles of important elements, often by using radioactively labelled molecules that they can follow in the environment.

31.2a Water Is the Staff of Life

Water is the universal intracellular solvent for biochemical reactions, but only a fraction of 1% of Earth's total water is present in biological systems at any time.

The cycling of water, the **hydrogeologic cycle**, is global, with water molecules moving from oceans into the atmosphere, to land, through freshwater ecosystems, and back to the oceans **(Figure 31.12)**. Solar energy causes water to evaporate from oceans, lakes, rivers, soil, and living organisms, entering the atmosphere as a vapour and remaining aloft as a gas, as droplets in clouds, or as ice crystals. Water falls as precipitation, mostly in the form of rain and snow. When precipitation falls on land, water flows across the surface or percolates to great depths in soil, eventually reentering the ocean reservoir through the flow of streams and rivers.

The hydrogeologic cycle maintains its global balance because the total amount of water entering the atmosphere is equal to the amount that falls as precipitation. Most water that enters the atmosphere evaporates from the oceans, which are the largest reservoir of water on the planet. A much smaller fraction evaporates from terrestrial ecosystems, and most of that is through transpiration by green plants.

Constant recirculation provides fresh water to terrestrial organisms and maintains freshwater ecosystems such as lakes and rivers. Water also serves as a transport medium that moves nutrients within and between ecosystems, as demonstrated in a series of classic experiments in the Hubbard Brook Experimental Forest (see "Studies of the Hubbard Brook Watershed").

31.2b Carbon Is the Backbone of Life

Carbon atoms provide the backbone of most biological molecules, and carbon compounds store the energy captured by photosynthesis (see Chapter 7). Carbon enters food webs when producers convert atmospheric carbon dioxide (CO_2) into carbohydrates. Heterotrophs acquire carbon by eating other organisms or detritus. Although carbon moves somewhat independently in sea and on land, a common atmospheric pool of CO_2 creates a global **carbon cycle** **(Figure 31.13, p. 762)**.

The largest reservoir of carbon is sedimentary rock, such as limestone. Rocks are in the unavailable-inorganic compartment, and they exchange carbon with living organisms at an exceedingly slow pace. Most *available* carbon is present as dissolved bicarbonate

a. The water cycle

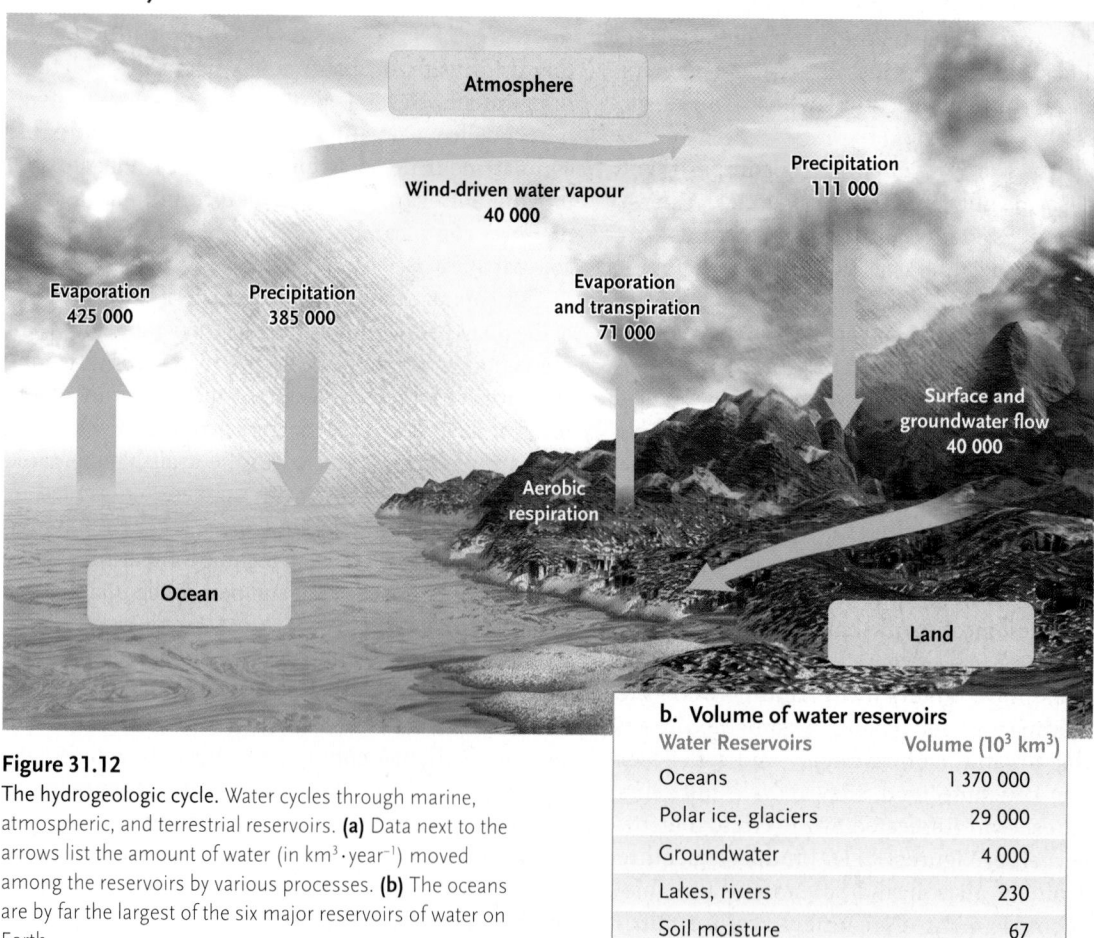

Figure 31.12

The hydrogeologic cycle. Water cycles through marine, atmospheric, and terrestrial reservoirs. **(a)** Data next to the arrows list the amount of water (in km³·year⁻¹) moved among the reservoirs by various processes. **(b)** The oceans are by far the largest of the six major reservoirs of water on Earth.

b. Volume of water reservoirs

Water Reservoirs	Volume (10^3 km³)
Oceans	1 370 000
Polar ice, glaciers	29 000
Groundwater	4 000
Lakes, rivers	230
Soil moisture	67
Atmosphere (water vapour)	14

Studies of the Hubbard Brook Watershed Box

Water flows downhill, so local topography affects the movement of dissolved nutrients in terrestrial ecosystems. A **watershed** is an area of land from which precipitation drains into a single stream or river. Each watershed represents a part of an ecosystem from which nutrients exit through a single outlet. When several streams join to form a river, the watershed drained by the river encompasses the smaller watersheds drained by the streams. The Mackenzie River watershed covers roughly 20% of Canada and includes the watersheds of the Peace and Athabasca rivers, as well as many other watersheds drained by smaller streams and rivers.

Watersheds are ideal for large-scale field experiments about nutrient flow in ecosystems because they are relatively self-contained units. Herbert Bormann and Gene Likens conducted a classic experiment on nutrients in watersheds in the 1960s. Bormann and Likens manipulated small watersheds of temperate deciduous forest in the Hubbard Brook Experimental Forest in the White Mountain National Forest of New Hampshire. They measured precipitation and nutrient input into the watersheds, the uptake of nutrients by vegetation, and the amount of nutrients leaving the watershed via streamflow. They monitored nutrients exported in streamflow by collecting water samples from V-shaped concrete weirs built into bedrock below the streams that drained the watersheds **(Figure 1).** Impermeable bedrock underlies the soil, preventing water from leaving the system by deep seepage.

Bormann and Likens collected several years of baseline data on six undisturbed watersheds. Then, in 1965 and 1966, they felled all of the trees in one small watershed and used herbicides to prevent regrowth. After

Gene E. Likens, from Gene E. Likens et al., *Ecology Monograph*, 40(1): 23–47, 1970

Figure 1

A weir used to measure the volume and nutrient content of water leaving a watershed by streamflow.

these manipulations, they monitored the output of nutrients in streams that drained experimental and control watersheds. They attributed differences in nutrient export between undisturbed watersheds (controls) and the clear-cut watershed (experimental treatment) to the effects of deforestation.

Bormann and Likens determined that vegetation absorbed substantial water and conserved nutrients in undisturbed watersheds. Plants used about 40% of the precipitation for transpiration. The rest contributed to runoff and groundwater. Control watersheds lost only about 8 to 10 kg of calcium per hectare each year, an amount replaced by erosion of bedrock and input from rain. Moreover, control watersheds actually accumulated about 2 kg of nitrogen per hectare per year and slightly smaller amounts of potassium.

The experimentally deforested watershed experienced a 40% annual increase in runoff, including a 300% increase during a four-month period in summer. Some mineral losses were similarly large. The net loss of calcium was 10 times as high **(Figure 2)** as in the control watersheds and of potassium was 21 times as high. Phosphorus losses did not increase because this mineral was apparently retained by the soil. The loss of nitrogen, however, was very large—120 kg·ha^{-1}·year^{-1}. The washing out of nitrogen meant that the stream draining the experimental watershed became choked with algae and cyanobacteria. The Hubbard Brook experiment demonstrated that deforestation increases flooding and decreases the fertility of ecosystems.

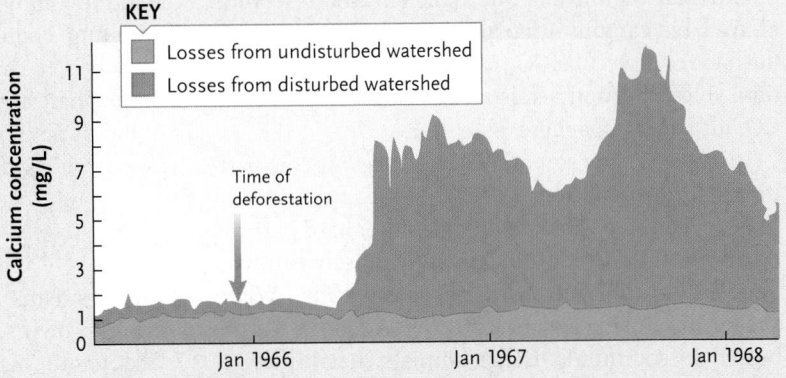

Figure 2

Calcium losses from the deforested watershed were much greater than those from controls. The arrow indicates the time of deforestation in early winter. Mineral losses did not increase until after the ground thawed the following spring. Increased runoff also caused large water losses from the watershed.

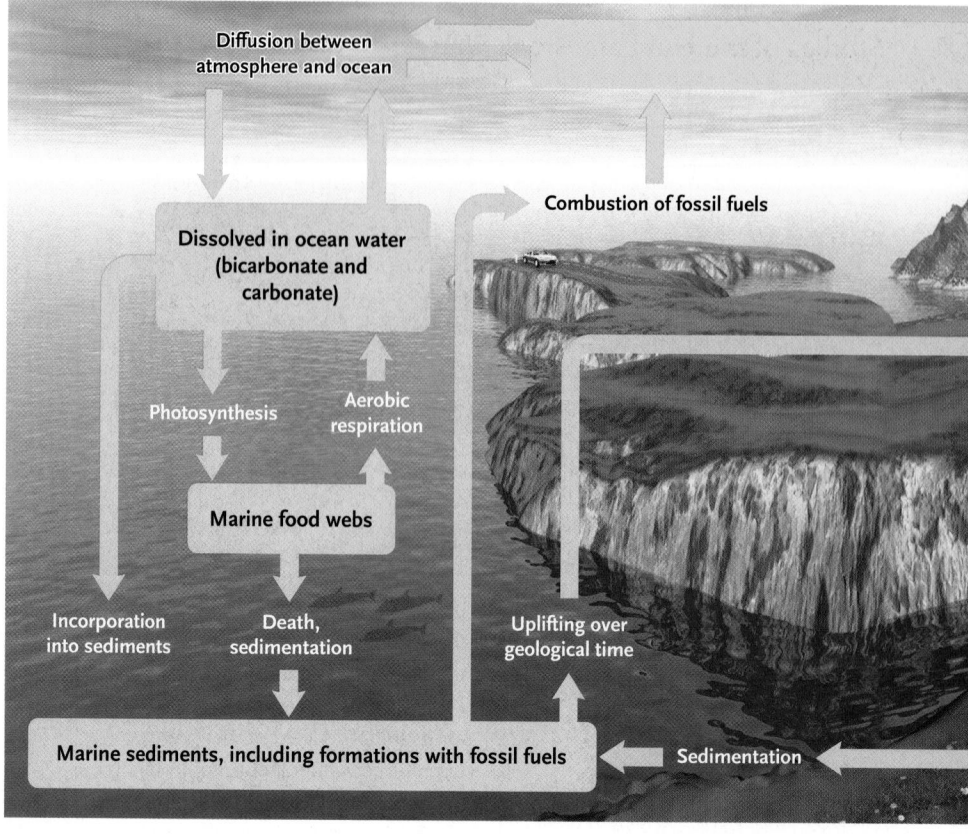

a. Amount of carbon in major reservoirs

Carbon Reservoirs	Mass (10^{12} g)
Sediments and rocks	770 000 000
Ocean (dissolved forms)	397 000
Soil	15 000
Atmosphere	7 500
Biomass on land	7 150

b. Annual global carbon movement between reservoirs

Direction of Movement	Mass (10^{12} kg)
From atmosphere to plants (carbon fixation)	1200
From atmosphere to ocean	1070
To atmosphere from ocean	1050
To atmosphere from plants	600
To atmosphere from soil	600
To atmosphere from burning fossil fuel	50
To atmosphere from burning plants	20
To ocean from runoff	4
Burial in ocean sediments	1

c. The global carbon cycle

Diffusion between atmosphere and ocean

Combustion of fossil fuels

Dissolved in ocean water (bicarbonate and carbonate)

Photosynthesis   Aerobic respiration

Marine food webs

Incorporation into sediments   Death, sedimentation   Uplifting over geological time

Marine sediments, including formations with fossil fuels   Sedimentation

Figure 31.13

The carbon cycle. Marine and terrestrial components of the global carbon cycle are linked through an atmospheric reservoir of carbon dioxide. **(a)** By far the largest amount of Earth's carbon is found in sediments and rocks. **(b)** Earth's atmosphere mediates most of the movement of carbon. **(c)** In this illustration of the carbon cycle, boxes identify major reservoirs and labels on arrows identify the processes that cause carbon to move between reservoirs.

ions (HCO_3^-) in the ocean. Soil, atmosphere, and plant biomass are significant, but much smaller, reservoirs of available carbon. Atmospheric carbon is mostly in the form of molecular CO_2, a product of aerobic respiration. Volcanic eruptions also release small quantities of CO_2 into the atmosphere.

Carbon atoms sometimes leave organic compartments for long periods of time. Some organisms in marine food webs build shells and other hard parts by incorporating dissolved carbon into calcium carbonate ($CaCO_3$) and other insoluble salts. When shelled organisms die, they sink to the bottom and are buried in sediments. Other animals, notably vertebrates, store calcium in bone. Insoluble carbon that accumulates as rock in deep sediments may remain buried for millions of years before tectonic uplifting brings it to the surface, where erosion and weathering dissolve sedimentary rocks and return carbon to an available form.

Carbon atoms are also transferred to the unavailable-organic compartment when soft-bodied organisms die and are buried in habitats where low oxygen concentration prevents decomposition. In the past, under suitable geologic conditions, these carbon-rich tissues were slowly converted to gas, petroleum, or coal, which we now use as fossil fuels. Human activities, especially burning fossil fuels, are transferring carbon into the atmosphere at an unnaturally high rate. The resulting change in the worldwide distribution of carbon is having profound consequences for Earth's atmosphere and climate, including a general warming of the climate and a rise in sea level (see "Disruption of the Carbon Cycle," p. 764).

31.2c Nitrogen Is a Limiting Element

All organisms require nitrogen to construct nucleic acids, proteins, and other biological molecules (see Chapter 48). Earth's atmosphere had a high nitrogen concentration long before life began. Today, a global **nitrogen cycle** moves this element between the huge atmospheric pool of gaseous molecular nitrogen (N_2) and several much smaller pools of nitrogen-containing compounds in soils, marine and freshwater ecosystems, and living organisms **(Figure 31.14)**.

Molecular nitrogen is abundant in the atmosphere, but triple covalent bonds bind its two atoms so tightly that most organisms cannot use it. Only certain microorganisms, volcanic action, and lightning can convert N_2 into ammonium (NH_4^+) and nitrate (NO_3^-) ions. This conversion is called **nitrogen fixation** (see Chapter 37). Once nitrogen is fixed,

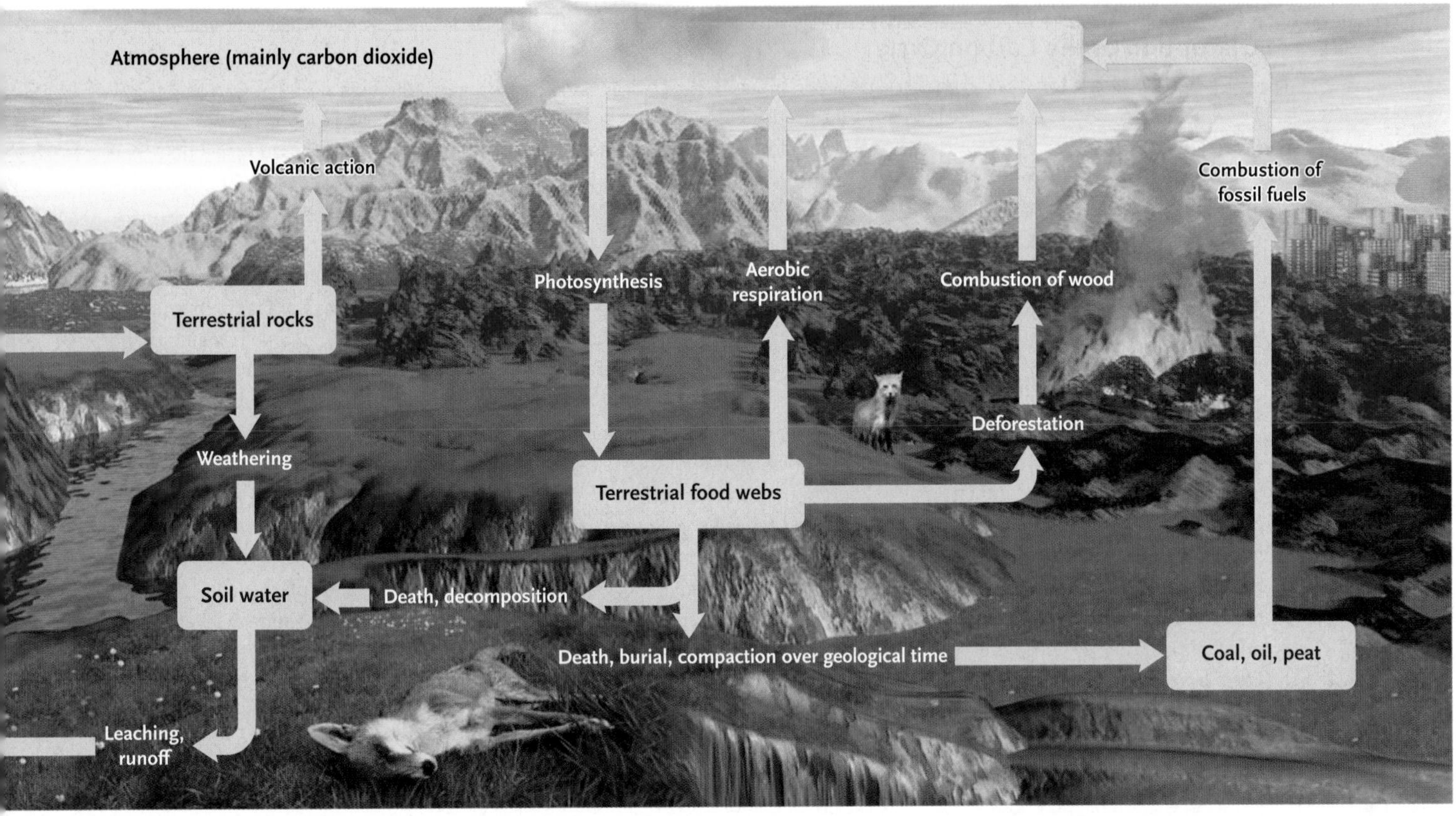

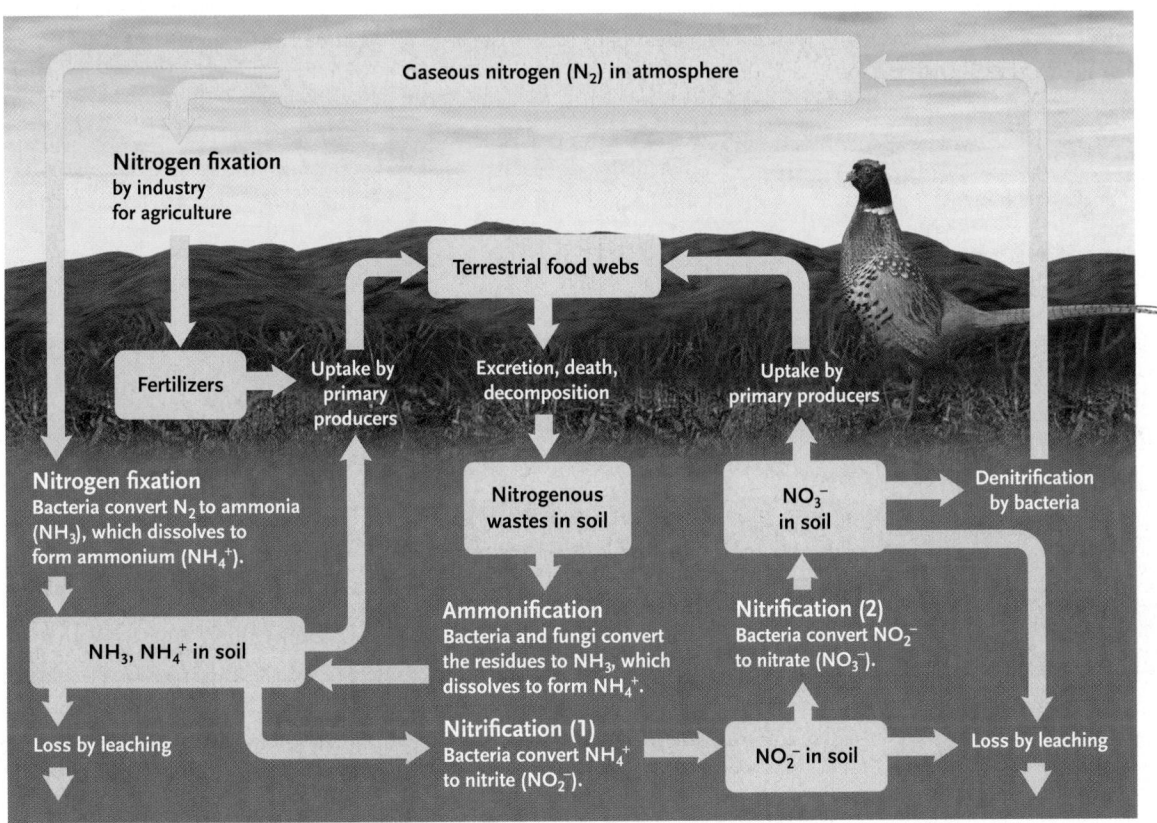

Atmosphere (mainly carbon dioxide)

Volcanic action

Photosynthesis   Aerobic respiration   Combustion of wood   Combustion of fossil fuels

Terrestrial rocks

Weathering

Deforestation

Soil water ← Death, decomposition ← Terrestrial food webs

Death, burial, compaction over geological time → Coal, oil, peat

Leaching, runoff

Gaseous nitrogen (N₂) in atmosphere

Nitrogen fixation by industry for agriculture

Terrestrial food webs

Fertilizers   Uptake by primary producers   Excretion, death, decomposition   Uptake by primary producers

NO_3^- in soil   Denitrification by bacteria

Nitrogen fixation
Bacteria convert N_2 to ammonia (NH_3), which dissolves to form ammonium (NH_4^+).

Nitrogenous wastes in soil

Nitrification (2)
Bacteria convert NO_2^- to nitrate (NO_3^-).

NH_3, NH_4^+ in soil

Ammonification
Bacteria and fungi convert the residues to NH_3, which dissolves to form NH_4^+.

Loss by leaching

Loss by leaching

Nitrification (1)
Bacteria convert NH_4^+ to nitrite (NO_2^-).

NO_2^- in soil

Figure 31.14

The nitrogen cycle in a terrestrial ecosystem. Nitrogen-fixing bacteria make molecular nitrogen available in terrestrial ecosystems. Other bacteria recycle nitrogen within the available-organic compartment through ammonification and two types of nitrification, converting organic wastes into ammonium ions and nitrates. Denitrification converts nitrate to molecular nitrogen, which returns to the atmosphere. Runoff carries nitrogen from terrestrial ecosystems into aquatic ecosystems, where it is recycled in freshwater and marine food webs.

Disruption of the Carbon Cycle

The concentrations of gases in the lower atmosphere have a profound effect on global temperature, in turn affecting global climate. Molecules of CO_2, water vapour, ozone, methane, nitrous oxide, and other compounds collectively act like a pane of glass in a greenhouse (hence the term *greenhouse gases*). They allow short wavelengths of visible light to reach Earth's surface while impeding the escape of longer, infrared wavelengths into space, trapping much of their energy as heat **(Figure 1).** Greenhouse gases foster the accumulation of heat in the lower atmosphere, a warming action known as the **greenhouse effect.** This natural process prevents Earth from being a cold and lifeless planet.

Data from air bubbles trapped in glacial ice indicate that atmospheric CO_2 concentrations have fluctuated widely over Earth's history **(Figure 2).** Since the late 1950s, scientists have measured atmospheric concentrations of CO_2 and other greenhouse gases at remote sampling sites such as the top of Mauna Loa in the Hawaiian Islands. These sites are free of local contamination and reflect average global conditions. Concentrations of greenhouse gases have increased steadily for as long as they have been monitored **(Figure 3).**

The graph for atmospheric CO_2 concentration has a regular zigzag pattern that follows the annual cycle of plant growth (see Figure 3). The concentration of CO_2 decreases during the summer because photosynthesis withdraws so much from the atmospheric available-inorganic pool. The concentration of CO_2 is higher during the winter when photosynthesis slows while aerobic respiration continues, returning carbon to the atmospheric available-inorganic pool. Whereas the zigs and zags in the data for CO_2 represent seasonal highs and lows, the midpoint of the annual peaks and troughs has increased steadily for 40 years. These data are evidence of a rapid buildup of atmospheric CO_2, representing a shift in the distribution of carbon in the major reservoirs on Earth. The best estimates suggest that CO_2 concentration has increased by 35% in the last 150 years and by more than 10% in the last 30 years.

The increase in the atmospheric concentration of CO_2 appears to result from combustion, whether we burn fossil fuels or wood. Today, humans burn more wood and fossil fuels than ever before. Vast tracts of tropical forests are being cleared and burned (see Chapter 32). To make matters worse, deforestation reduces the world's biomass of plants that assimilate CO_2 and help maintain the carbon cycle as it existed before human activities disrupted it.

The increase in the concentration of atmospheric CO_2 is alarming because plants with C_3 metabolism respond to increased CO_2 concentrations with increased growth rates. This is not true of C_4 plants (see Chapter 7). Thus, rising atmospheric levels of CO_2 will probably alter the relative abundances of many plant species, changing the composition and dynamics of their communities.

Simulation models suggest that increasing concentrations of any greenhouse gas may intensify the greenhouse effect, contributing to a trend of global warming. Should we be alarmed about the prospect of a warmer planet? Some models predict that the mean temperature of the lower atmosphere will rise by 4°C, enough to increase ocean surface temperatures. In some areas, such as

Figure 1
The greenhouse effect.

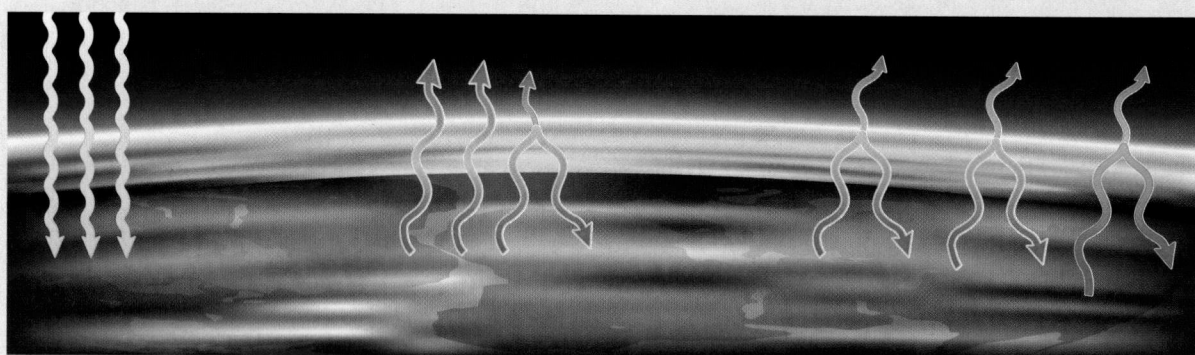

Sunlight penetrates the atmosphere and warms Earth's surface.

Earth's surface radiates heat (infrared wavelengths) to the atmosphere. Some heat escapes into space. Greenhouse gases and water vapour absorb some infrared energy and reradiate the rest of it back toward Earth.

When atmospheric concentrations of greenhouse gases increase, the atmosphere near Earth's surface traps more heat. The warming causes a positive feedback cycle in which rising ocean temperatures cause increased evaporation of water, which further enhances the greenhouse effect.

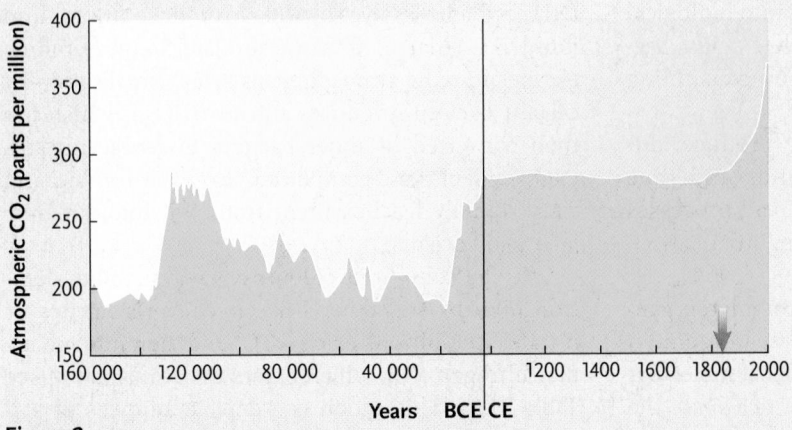

Figure 2

Carbon dioxide levels over time. The amount of atmospheric CO$_2$ has risen dramatically since about 1850 (arrow).

the Canadian Arctic and the Antarctic, warming has occurred much more rapidly than predicted or expected. Water expands when heated, and global sea levels could rise as much as 0.6 m just from this expansion. In addition, atmospheric temperature is rising fastest near the poles. Thus, global warming may also foster melting of glaciers and the Antarctic ice sheet, which might raise sea levels as much as 50 to 100 m, inundating low coastal regions. Waterfronts in Vancouver, Los Angeles, Hong Kong, Durban, Rio de Janeiro, Sydney, New York, and London would be submerged. So would agricultural lands in India, China, and Bangladesh, where much of the world's rice is grown. Moreover, global warming could disturb regional patterns of precipitation and temperature. Areas that now produce much of the world's grains would become arid scrub or deserts, and the now-forested areas to their north would become dry grasslands.

Many scientists believe that atmospheric levels of greenhouse gases will continue to increase at least until the middle of the twenty-first century and that global temperature may rise by several degrees. At the Earth Summit in 1992, leaders of the industrialized countries agreed to try to stabilize CO$_2$ emissions by the end of the twentieth century. We have already missed that target, and some countries, including the United States (then the largest producer of greenhouse gases), have now forsaken that goal as too costly. Stabilizing emissions at current levels will not reverse the damage already done, nor will it stop the trend toward global warming. We should begin preparing for the consequences of global warming now. We might increase reforestation efforts because a large tract of forest can withdraw significant amounts of CO$_2$ from the atmosphere. We might also step up genetic engineering studies to develop heat-resistant and drought-resistant crop plants, which may provide crucial food reserves in regions of climate change.

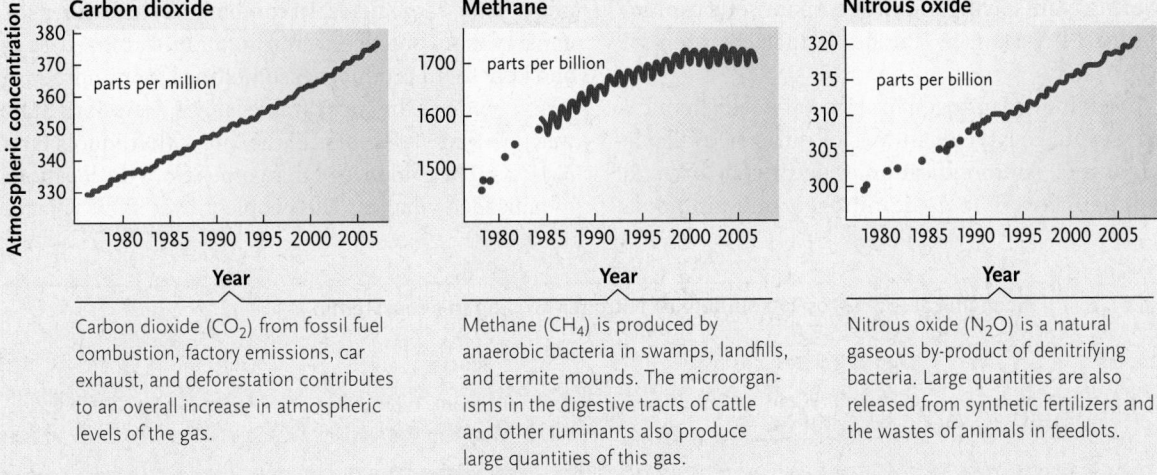

Carbon dioxide (CO$_2$) from fossil fuel combustion, factory emissions, car exhaust, and deforestation contributes to an overall increase in atmospheric levels of the gas.

Methane (CH$_4$) is produced by anaerobic bacteria in swamps, landfills, and termite mounds. The microorganisms in the digestive tracts of cattle and other ruminants also produce large quantities of this gas.

Nitrous oxide (N$_2$O) is a natural gaseous by-product of denitrifying bacteria. Large quantities are also released from synthetic fertilizers and the wastes of animals in feedlots.

Figure 3

Increases in atmospheric concentrations of three greenhouse gases, mid-1970s through 2004. The data were collected at a remote monitoring station in Australia (Cape Grim, Tasmania) and compiled by scientists at the Commonwealth Scientific and Industrial Research Organization, an agency of the Australian government.

primary producers can incorporate it into biological molecules such as proteins and nucleic acids. Secondary consumers obtain nitrogen by consuming these molecules.

Several biochemical processes produce different nitrogen-containing compounds and thus move nitrogen through ecosystems. These processes are nitrogen fixation, ammonification, nitrification, and denitrification **(Table 31.2)**.

In nitrogen fixation, several kinds of microorganisms convert molecular nitrogen (N_2) to ammonium ions (NH_4^+). Certain bacteria, which collect molecular nitrogen from the air between soil particles, are the major nitrogen fixers in terrestrial ecosystems (see Table 31.2). The cyanobacteria partners in some lichens (see Chapter 25) also fix molecular nitrogen. Other cyanobacteria are important nitrogen fixers in aquatic ecosystems, whereas the water fern (genus *Azolla*) plays that role in rice paddies. Collectively, these organisms fix an astounding 200 million tonnes of nitrogen each year. Plants and other primary producers assimilate and use this nitrogen in the biosynthesis of amino acids, proteins, and nucleic acids, which then circulate through food webs.

Some plants, including legumes (such as beans and clover), alders (*Alnus* species), and some members of the rose family (Rosaceae), are mutualists with nitrogen-fixing bacteria. These plants acquire nitrogen from soils much more readily than plants that lack such mutualists. Although these plants have the competitive edge in nitrogen-poor soil, nonmutualistic species often displace them in nitrogen-rich soil. In an interesting twist on the usual predator–prey relationships, several species of flowering plants living in nitrogen-poor soils capture and digest insects (see "Pitcher Plant Ecosystems," pp. 767–768).

In addition to nitrogen fixation, other biochemical processes make large quantities of nitrogen available to producers. **Ammonification** of detritus by bacteria and fungi converts organic nitrogen into ammonia (NH_3), which dissolves in water to produce ammonium ions (NH_4^+) that plants can assimilate. Some ammonia escapes into the atmosphere as a gas. **Nitrification** by certain bacteria produces nitrites (NO_2^-), which are then converted by other bacteria to usable nitrates (NO_3^-). All of these compounds are water soluble, and water rapidly leaches them from soil into streams, lakes, and oceans.

Under conditions of low oxygen availability, **denitrification** by still other bacteria converts nitrites or nitrates into nitrous oxide (N_2O) and then into molecular nitrogen (N_2), which enters the atmosphere (see Table 31.2). This action can deplete supplies of soil nitrogen in waterlogged or otherwise poorly aerated environments, such as bogs and swamps.

In 1909, Fritz Haber developed a process for fixing nitrogen, and with the help of Carl Bosch, the process was commercialized for fertilizer production. The Haber–Bosch process has altered Earth's nitrogen cycles and is said to be responsible for the existence of 40% of the people on Earth. Before the implementation of the Haber–Bosch process, the amount of nitrogen available for life was limited by the rates at which N_2 was fixed by bacteria or generated by lightning strikes. Today, spreading fertilizers rich in nitrogen is the basis for most of agriculture's productivity. This practice has quadrupled some yields over the past 50 years (see Chapter 33). Of all nutrients required for primary production, nitrogen is often the least abundant. Agriculture routinely depletes soil nitrogen, which is removed from fields through the harvesting of plants that have accumulated nitrogen in their tissues. Soil erosion and leaching remove more. Traditionally, farmers rotated their crops, alternately planting legumes and other crops in the same fields. In combination with other soil conservation practices, crop rotation stabilized soils and kept them productive, sometimes for hundreds of years. Some of the most arable land in New York State was farmed by members of the Mohawk Iroquois First Nations. The evidence of this comes from the locations of palisaded villages. The people moved their villages

Table 31.2 | **Biochemical Processes That Influence Nitrogen Cycling in Ecosystems**

Process	Organisms Responsible	Products	Outcome
Nitrogen fixation	Bacteria: *Rhizobium, Azotobacter, Frankia* Cyanobacteria: *Anabaena, Nostoc*	Ammonia (NH_3), ammonium ions (NH_4^+)	Assimilated by primary producers
Ammonification of organic detritus	Soil bacteria and fungi	Ammonia (NH_3), ammonium ions (NH_4^+)	Assimilated by primary producers
Nitrification			
(1) Oxidation of NH_3	Bacteria: *Nitrosomonas, Nitrococcus*	Nitrite (NO_2^-)	Used by nitrifying bacteria
(2) Oxidation of NO_2^-	Bacteria: *Nitrobacter*	Nitrate (NO_3^-)	Assimilated by primary producers
Denitrification of NO_3^-	Soil bacteria	Nitrous oxide (N_2O), molecular nitrogen (N_2)	Released to atmosphere

Pitcher Plant Ecosystems

Pitcher plants have modified leaves (pitchers) that act as pitfall traps for drowning and digesting insect prey. Pitchers have developed in at least five different evolutionary lines of vascular plants (see Chapter 35). Throughout much of North America, pitcher plants (the provincial flower of Newfoundland and Labrador; **Figure 1**) are common in bogs. *Sarracenia purpurea*, like other carnivorous plants, obtain much of their nitrogen from the insects they capture.

The captured arthropod prey, mainly ants and flies, is the base of a food web inside the pitchers. These are shredded and partly consumed by larvae of midges (*Metriocnemus knabi*) and sarcophagid flies (*Fletcherimyia fletcheri*; **Figure 2**). A subweb of bacteria and protozoa processes shredded prey, which are eaten by filter-feeding rotifers (*Habrotrocha rosa*; **Figure 3, p. 768**) and mites (*Sarraceniopus gibsonii*). Mosquito larvae (*Wyeomyia smithii*) eat the bacteria, protozoa, and rotifers, whereas the larger sarcophagid fly larvae eat the rotifers and smaller mosquito larvae. Populations of bacteria, protozoa, and rotifers grow much more rapidly than populations of mosquito or midge larvae, making the system sustainable.

Pitchers are essential to the life cycles of two species of insects whose larvae live in them. A mosquito and a midge coexist in the same pitchers, and their populations are limited by

M.B. Fenton

Figure 1
Sarracenia purpurea, *a pitcher plant. The flower on a long stalk extends above the pitchers. One pitcher is shown in the photo on the right.*

M.B. Fenton

the availability of insect carcasses. In any pitcher, growth in populations of the midge larvae is not affected by increases in the numbers of mosquito larvae. But populations of mosquito larvae increase as populations of midge larvae increase (see Figure 2).

The situation is an example of processing-chain commensalism because the action of one species creates opportunities for another. In this case, midge larvae feed on the

hard parts of insect carcasses and break them up in the process. Mosquito larvae are filter feeders, consuming particles derived from the decaying matter. The feeding of the midges generates additional food for the mosquito larvae. Although the populations of midge and mosquito larvae can be large in any pitcher, only a single sarcophagid fly larva occurs in any pitcher. *F. fletcheri* is a *K*-strategist (see Chapter 29) and gives birth to

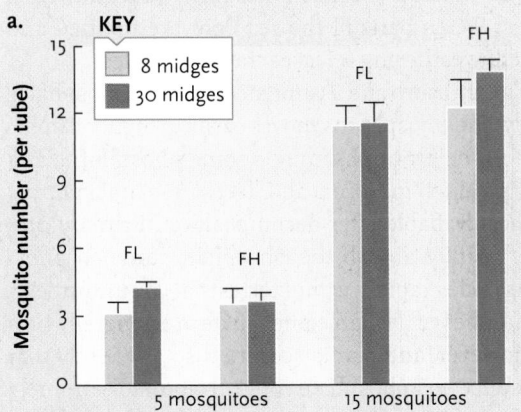

a.

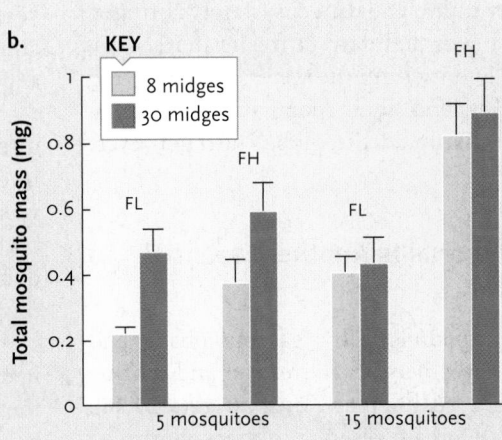

b.

Figure 2
Midge and mosquito larvae in pitchers.
(a) The density and (b) total dry mass of mosquito larvae are the same whether the population of midges is low (8 midges) or high (30 midges). FH = high food availability; FL = low food availability. Error bars show standard errors of the mean.

(Continued)

Pitcher Plant Ecosystems (*Continued*)

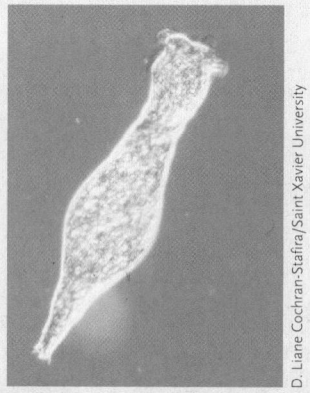

Figure 3
A bdelloid rotifer, Habrotrocha rosa, *from a* Sarracenia purpurea *pitcher.*

a.

b.

Courtesy of the Biodiversity Institute of Ontario

Courtesy of Daniel Handfield

Figure 4
Moths whose caterpillars eat Sarracenia purpurea. *The caterpillars of* **(a)** Exyra fax *and* **(b)** Papaipema appassionata *feed on pitcher plants, either (a) the lining of pitchers or (b) the rhizomes.*

larvae. If you place more than one *F. fletcheri* larva in a pitcher, a fight ensues. The larger larva either wins or leaves the pitcher to pupate in the sphagnum around it.

These insects do not appear to compete with their hosts, the pitcher plants. The abundance of rotifers living in the pitchers of *S. purpurea* is negatively associated with the presence of midge and mosquito larvae (which eat the rotifers). Rotifers are detritivores, and their excretory products (NO_3^-N, NH_4OH, P) account for a major portion of the N acquired by the plants from their insect prey.

Two species of moths also exploit *S. purpurea* (**Figure 4**). *Exyra fax* and *Papaipema appassionata* do not live in the pitchers. *Exyra fax* caterpillars eat the interior surface of the pitcher chambers, whereas *P. appassionata* caterpillars consume the rhizomes. Although predation by *E. fax* caterpillars does not kill the plants, predation by *P. appassionata* does. To what trophic level does one assign moths whose caterpillars are herbivores feeding on primary producers that eat insects?

and farming operations every 10 to 20 years, changing fields repeatedly over hundreds of years.

The production of synthetic fertilizers is expensive, using fossil fuels as both raw material and an energy source. Fertilizer becomes increasingly costly as supplies of fossil fuels dwindle. Furthermore, rain and runoff leach excess fertilizer from agricultural fields and carry it into aquatic ecosystems. Nitrogen has become a major pollutant of freshwater ecosystems, artificially enriching the waters and allowing producers to expand their populations. Human activities have disrupted the global nitrogen cycle (**Figure 31.15**).

31.2d Phosphorus Is Another Essential Element

Phosphorus compounds lack a gaseous phase, and this element moves between terrestrial and marine ecosystems in a sedimentary cycle (**Figure 31.16**). Earth's crust is the main reservoir of phosphorus, as it is for other minerals, such as calcium and potassium, that also undergo sedimentary cycles.

Phosphorus is present in terrestrial rocks in the form of phosphates (PO_4^{3-}). In the **phosphorus cycle**, weathering and erosion add phosphate ions to soil and carry them into streams and rivers, which eventually transport them to the ocean. Once there, some phosphorus enters marine food webs, but most of it precipitates out of solution and accumulates for millions of years as insoluble deposits, mainly on continental shelves. When parts of the sea floor are uplifted and exposed, weathering releases the phosphates.

Plants absorb and assimilate dissolved phosphates directly, and phosphorus moves easily to higher trophic levels. All heterotrophs excrete some phosphorus as a waste product in urine and feces; the phosphorus becomes available after decomposition. Primary producers readily absorb the phosphate ions, so phosphorus cycles rapidly *within* terrestrial communities.

Supplies of available phosphate are generally limited, however, and plants acquire it so efficiently that they reduce soil phosphate concentration to extremely low levels. Thus, like nitrogen, phosphorus is a common ingredient in agricultural fertilizers, and excess phosphates are pollutants of freshwater

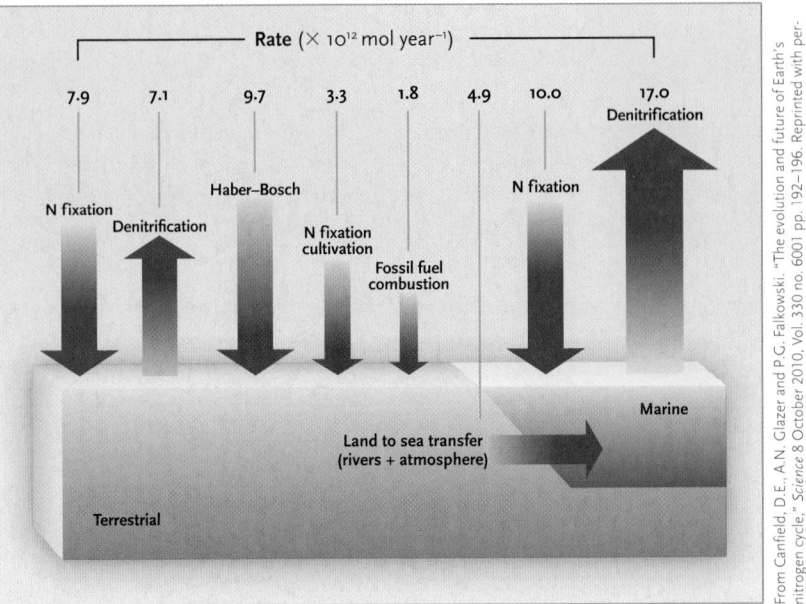

Figure 31.15
Modern global nitrogen flux depends upon the efficiency of transfer of N between reservoirs. Thickness of arrows indicates relative size of flux. Anthropogenic inputs are shown as dark brown arrows.

From Canfield, D.E., A.N. Glazer and P.G. Falkowski: "The evolution and future of Earth's nitrogen cycle," *Science* 8 October 2010, Vol. 330 no. 6001 pp. 192–196. Reprinted with permission from AAAS.

ecosystems. A particularly good example is Lake Erie, a Great Lake that was heavily affected by accumulations of phosphorus. The example here is more convincing because the problem has largely been resolved over the years.

For many years, phosphate for fertilizers was obtained from guano (the droppings of seabirds that consume phosphorus-rich food), which was mined on small islands that hosted seabird colonies, for example, in Polynesia and Micronesia. We now obtain most phosphate for fertilizer from phosphate rock mined in places such as Saskatchewan that have abundant marine deposits.

STUDY BREAK

1. How is balance maintained in the hydrogeologic cycle?
2. How do consumers obtain carbon?
3. What is the role of cyanobacteria in the nitrogen cycle? Why is their role important?

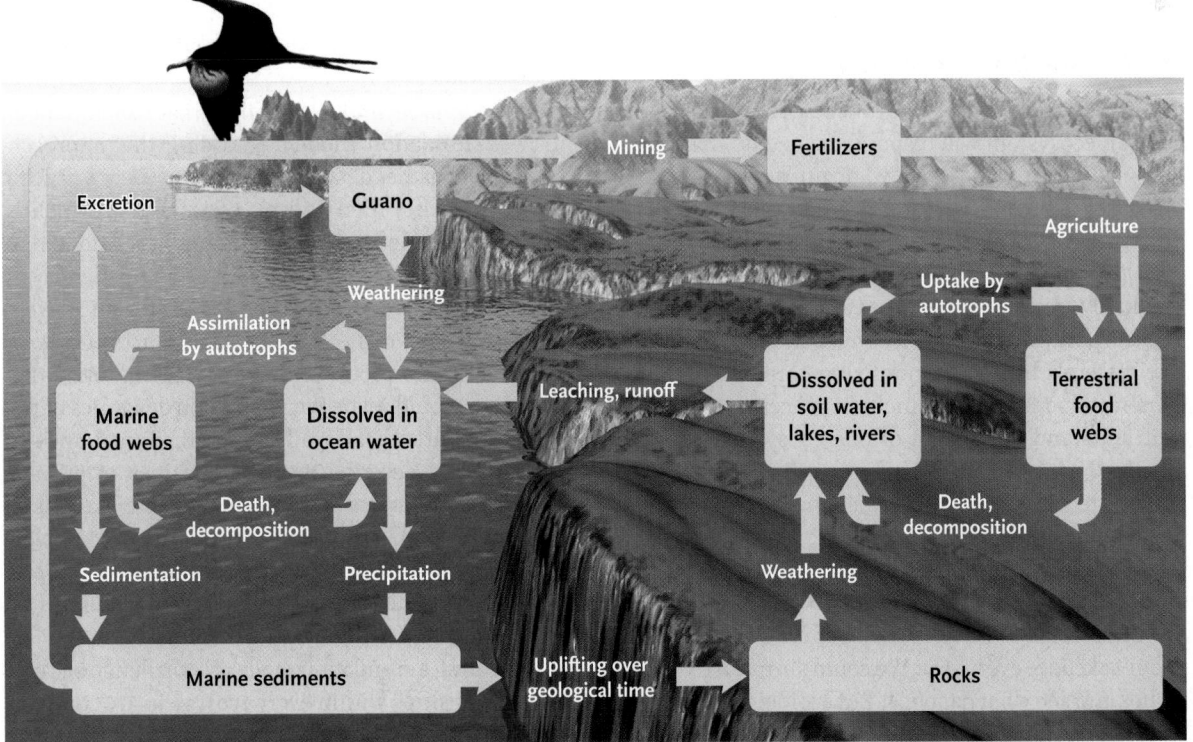

Figure 31.16
The phosphorus cycle. Phosphorus becomes available to biological systems when wind and rainfall dissolve phosphates in rocks and carry them into adjacent soil and freshwater ecosystems. Runoff carries dissolved phosphorus into marine ecosystems, where it precipitates out of solution and is incorporated into marine sediments.

Lenore Fahrig, *Carleton University*

The fragmentation of habitats is a ubiquitous effect of human activity on landscapes. In many parts of the world, land areas that used to be continuous forest are now large expanses of agricultural or urban landscapes dotted with small fragments of forest **(Figure 1).** Lenore Fahrig examines the impact of landscape structure on the abundance, distribution, and persistence of organisms.

In her research, Fahrig uses a variety of organisms, from beetles to plants and birds. She considers habitats and the impacts of roads and fence lines. She and her students try to identify the habitat features associated with the persistence of species after fragmentation and the role of connectivity between fragments in the persistence of populations in the fragments.

Using a combination of theoretical work and fieldwork, she has assessed the responses of species in different trophic roles to the fragmentation of habitat. Her work demonstrates that

not all species respond in the same way and that some benefit from fragmentation.

The connections between theoretical work and reality emerge

clearly from her research, and the implications for conservation of biodiversity (see Chapter 32) are clear.

Figure 1
An aerial view of farmland in southwestern Ontario illustrates isolated patches of forest (woodlots) and bands of woodland (riparian) along the edges of a creek. The woodlots are varied in their size and shape and in the degree of their isolation or connection to other woodlots.

31.3 Ecosystem Modelling

Ecologists use modelling to make predictions about how an ecosystem will respond to specific changes in physical factors, energy flow, or nutrient availability. Analyses of energy flow and nutrient cycling allow us to create a *conceptual model* of how ecosystems function **(Figure 31.17).** Energy that enters ecosystems is gradually dissipated as it flows through a food web. By contrast, nutrients are conserved and recycled among the system's living and nonliving components. This general model does not include processes that carry nutrients and energy out of one ecosystem and into another.

More important, the model ignores the nuts-and-bolts details of exactly how specific ecosystems function. Although it is a useful tool, a conceptual model does not really help us predict what would happen, say, if we harvested 10 million tonnes of introduced salmon from Lake Erie every year. We could simply harvest the fishes and see what happens. But ecologists prefer less

intrusive approaches to studying the potential effects of disturbances.

One approach to predicting "what would happen if..." is **simulation modelling.** Using this approach, researchers gather detailed information about a specific ecosystem. They then derive a series of mathematical equations that define its most important relationships. One set of equations might describe how nutrient availability limits productivity at various trophic levels. Another might relate the population growth of zooplankton to the productivity of phytoplankton. Other equations would relate the population dynamics of primary carnivores to the availability of their food, and still others would describe how the densities of primary carnivores influence reproduction in populations at both lower and higher trophic levels. Thus, a complete simulation model is a set of interlocking equations that collectively predict how changes in one feature of an ecosystem might influence other features.

Creating a simulation model is a challenge because the relationships within every ecosystem are complex.

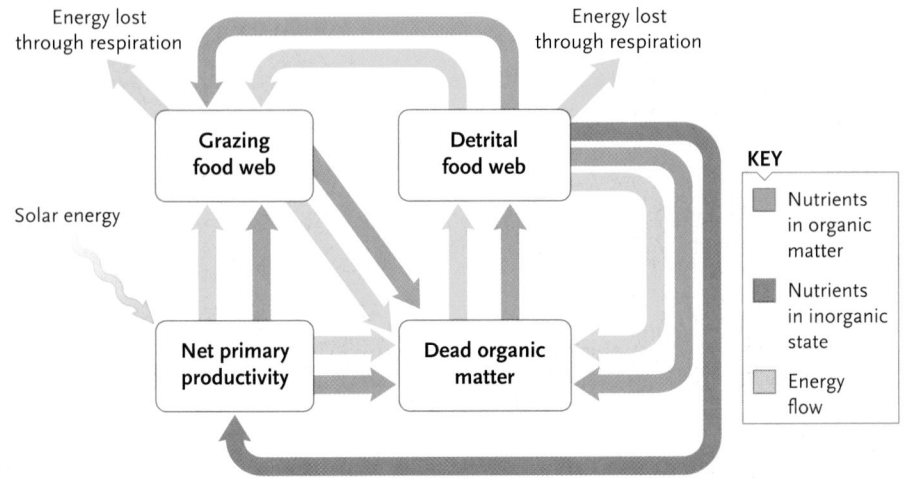

Energy lost through respiration

Energy lost through respiration

Grazing food web

Detrital food web

Solar energy

Net primary productivity

Dead organic matter

KEY

Nutrients in organic matter

Nutrients in inorganic state

Energy flow

Figure 31.17

A conceptual ecosystem model. A simple conceptual model of an ecosystem illustrates how energy flows through the system and is lost from both detrital and grazing food webs. Nutrients are recycled and conserved.

First, you must identify the important species, estimate their population sizes, and measure the average energy and nutrient content of each. Next, you must describe the food webs in which they participate, measure the quantity of food each species consumes, and estimate the productivity of each population. And, for the sake of completeness, you must determine the ecosystem's energy and nutrient gains and losses caused by erosion, weathering, precipitation, and runoff. You must repeat these measurements seasonally to identify annual variation in these factors. Finally, you might repeat the measurements over several years to determine the effects of year-to-year variation in climate and chance events.

After collecting these data, you must write equations that quantify the relationships in the ecosystem, including information about how temperature and other abiotic factors influence the ecology of each species. Having completed that job, you would begin to predict, for example, possibly in great detail, the effects of adding 1000 new housing units to an area of native prairie or boreal forest. Of course, you must refine the model whenever new data become available.

Some ecologists devote their professional lives to studying ecosystem processes and creating simulation models. The long-term initiative at the Hubbard Brook Forest provides a good example (see "Studies of the Hubbard Brook Watershed"). As we attempt to understand larger and more complex ecosystems (and as we create larger and more complex environmental problems), modelling becomes an increasingly important tool. If a model is based on well-defined ecological relationships and good empirical data, it can allow us to make accurate predictions about ecosystem changes without the need for costly and environmentally damaging experiments. But, like all ideas in science, a model is only as good as its assumptions, and models must constantly be adjusted to incorporate new ideas and recently discovered facts.

STUDY BREAK

1. Briefly describe the process of simulation modelling.
2. Why is simulation modelling necessary?

31.4 Scale, Ecosystems, Species

As we have seen, the complex interactions between and among species combine with abiotic and biotic factors to produce even more complex situations. Several questions emerge from this situation: What determines which species occur in an ecosystem? What controls the size of the populations of species in an ecosystem? How do species in an ecosystem interact? What effect does scale have on the situation?

Ecosystems span scales from millimetres to kilometres. Consider the microorganisms in a biofilm of water compared to the species, some of them microorganisms, in the water contained in the pitcher of a pitcher plant. Furthermore, the community of organisms may vary among pitchers on one plant. Like the pitcher-based community, terrestrial organisms living on an island may be relatively isolated. Consider the differences among islands, such as the British Isles, the Hawaiian Islands, or the Galapagos Islands, with respect to the combination of size (area), degree of isolation (distance from mainland), and range of habitats.

Variations in the scale of interactions **(Figure 31.18, p. 772)** help to put the nature of ecosystems in context.

Compare Figure 31.18 to Figure 3 in "Dispersal," Box 30.7, Chapter 30, about pollen and seed dispersal. More obvious in the latter is the influence of mobility on patterns of dispersal and connections. Large animals disperse seeds farther than small ones. Animals that can fly have greater potential as dispersers of seeds and pollen than those that walk or run. Data on

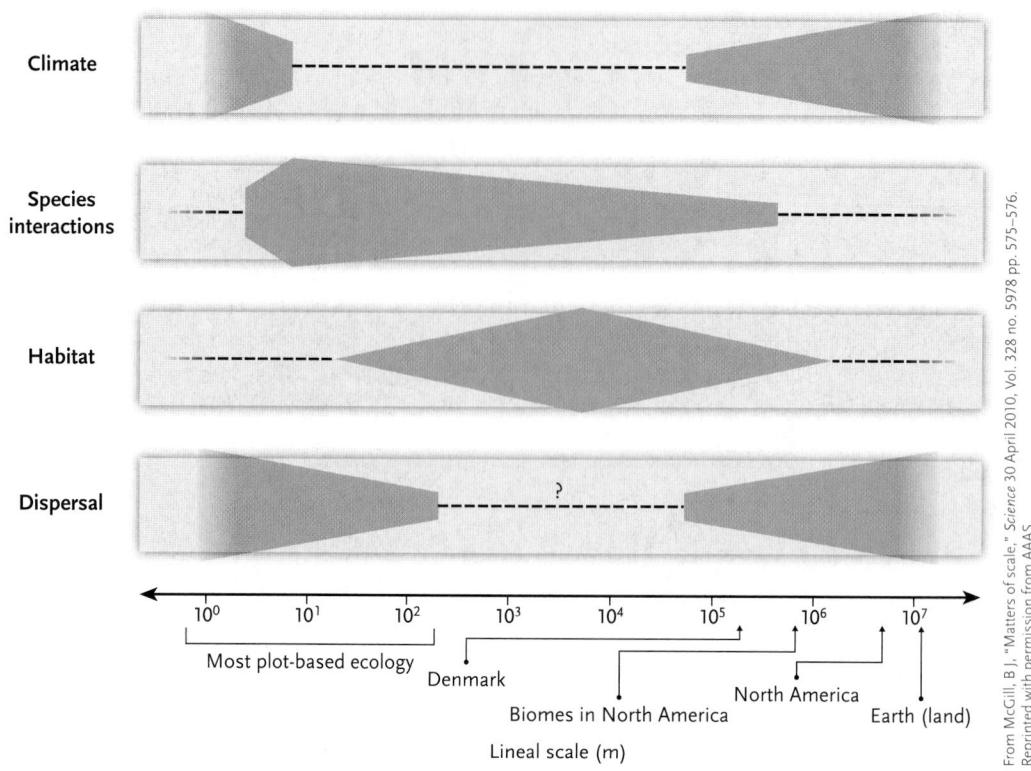

Figure 31.18

Climate, species interactions, habitat, and dispersal are four main factors controlling the distribution of species (vertical axis). Note the variation in scale (horizontal axis) across which these factors can act.

Climate

Species interactions

Habitat

Dispersal

10^0 10^1 10^2 10^3 10^4 10^5 10^6 10^7

Most plot-based ecology

Denmark

Biomes in North America

North America

Earth (land)

Lineal scale (m)

From McGill, B J., "Matters of scale," *Science* 30 April 2010, Vol. 328 no. 5978 pp. 575–576. Reprinted with permission from AAAS.

the distribution and habitat associations of terrestrial birds in Denmark reveal how species in the same genus and those filling similar niches have more influence on the patterns of distribution of one another than less similar species.

Studies of salmon along the northwest coast of North America (Great Bear Rainforest in British Columbia and sites around Bristol Bay in Alaska) reveal how species in the genus *Oncorhynchus* can influence the plant communities bordering the streams in which they spawn. Nutrients from salmon enter these communities when salmon die after spawning, or when they are taken and eaten by predators such as bears. Healthy populations of salmon affect the nutrient loading in terrestrial plants along the rivers and streams. The density of salmon and the characteristics of the watershed (steep versus shallow banks) influence the situation. Nutrient input from salmon leads to an increase in plants such as salmonberry, associated with nutrient-rich soils. Lower input from salmon is associated with plants associated with nutrient-poor soils (e.g., blueberries). Increases in nutrient-rich soils coincide with decreases in plant diversity **(Figure 31.19)**.

The above data from salmon at sites along 50 watersheds in British Columbia demonstrate the local impact that species in one genus can have. Work from sites in Alaska shows how the inherent diversity

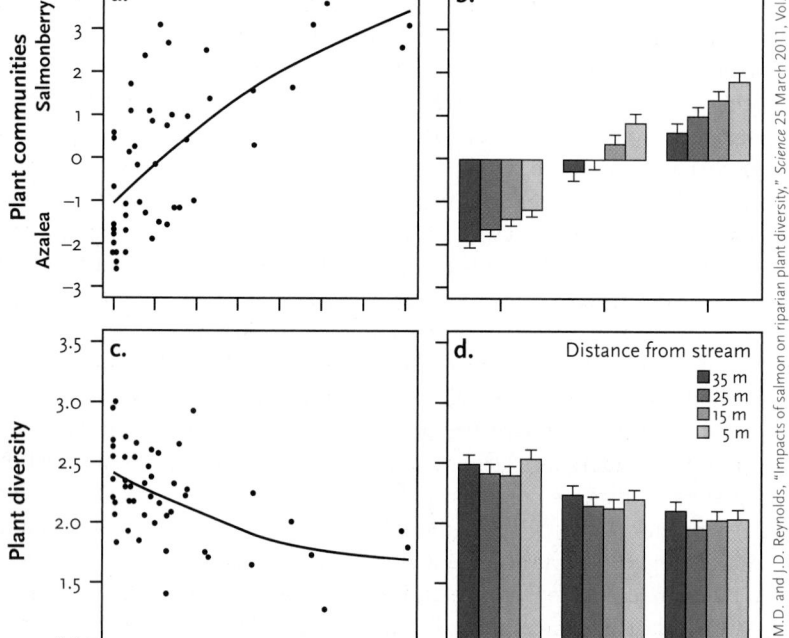

From Hocking, M.D. and J.D. Reynolds, "Impacts of salmon on riparian plant diversity," *Science* 25 March 2011, Vol. 331 no. 6024 pp. 1609–1612. Reprinted with permission from AAAS.

Figure 31.19

The influence of stream level density of spawning salmon (horizontal axis) on the structure of the community of understorey plants **(a, b)**, and on the diversity of plants **(c, d)** (vertical axis). Note the impact of distance from the stream.

of populations of sockeye salmon (*Oncorhynchus nerka*) is vital to the survival of the species and, by extension, the ecosystem. The data from sockeye salmon show how damping variance in the population provides stability. One example of variance is the timing of returns of salmon to the streams in which they hatched. The diversity is part of the *portfolio effect,* named because it is analogous to the impact of asset diversity on the stability of financial portfolios.

The diversity inherent in several hundred discrete watershed-based populations of sockeye is less than half the diversity that would occur if the sockeye were a single homogeneous population. The diversity also makes the sockeye more resilient to pressures of fishing. Studies of food webs in the watersheds provide further evidence of diversity and extend the portfolio effect.

In short, work with salmon advances our knowledge of the fundamental nature of ecosystems and helps us to appreciate the importance of maintaining biodiversity, setting the stage for Chapter 32.

STUDY BREAK

1. Why do ecosystems range so greatly in scale?
2. How do salmon have such a great influence on ecosystems around the streams in which they spawn?

31.5 Biomes and Ecozones

Biomes and ecozones are clusters of ecosystems that can occur anywhere in the world where the environmental conditions are appropriate. Biomes are **biota** (groups of organisms) characteristic of specific climatic and topographic situations (e.g., rain forest versus montane rain forest). Ecozones, or extensive ecological regions, are generally reflected by their dominant biotas or groupings of species. Like biomes, the occurrence of ecozones generally reflects climatic and topographic conditions. Both biomes and ecozones are categories of habitats that encompass a variety of ecosystems. **Table 31.3, p. 774** and **Figure 31.20, p. 776** illustrate the ecozones of Canada.

The view from the ground provides a better impression of ecozones. For example, continental shelf biome waters **(Figure 31.21a, p. 776)** are oceanic waters above the continental shelf within the Pacific Maritime ecozone. The coastal rain forest in Figure 31.21a occurs in the Pacific Maritime Biome and the Pacific Maritime ecozone. In Figure 31.21b, Lake Erie is a lentic biome, and the adjacent terrestrial biome is in the Mixedwood Biome, which here consists of a patchwork of habitat including forest (woodlots) and an open agroecosystem.

For instance, although we often think of rain forests as tropical, they also occur in temperate areas or anywhere in the world where there are forests and high levels of rainfall. For any biome, similarity of structure (e.g., plains or mountains) and dominant life forms (e.g., trees or grasses) are the key features, whether the focus is rain forest or desert. The range of terrestrial biomes, in addition to freshwater and marine biomes, in Canada illustrates their diversity. Two other Canadian biomes are shown in **Figure 31.22a and b, p. 777.**

Biomes reflect the climatic and biotic qualities that currently prevail, as well as some aspects of the history of the areas where they occur. Biomes in most of Canada resulted from changes that occurred over the last 12 thousand to 8 thousand years with the disappearance of continental ice sheets. The fossil record from sites in southern Canada and the adjacent United States reveals that at the peak of glaciation, organisms that currently occur in Arctic biomes occurred much farther south.

After settlement by Europeans, extensive clearing of forests (mixedwood plains) in what is now southwestern Ontario meant that over about 100 years biomes changed substantially to more prairielike conditions. More recently, changes in agricultural practices (e.g., time of ploughing) and land use (taking land out of agricultural production) mean that many areas cleared by 1900 CE are now reforested. Before the arrival of European settlers, some First Nations peoples also cleared forests to grow crops, but apparently not as extensively as the new arrivals. Changes in habitat are also reflected by changes in the distribution of species. For example, birds of open grasslands, such as Eastern Meadowlarks (*Sturnella magna*) and Bobolinks (*Dolichonyx oryzivorus*), became widespread in southern Ontario. Populations of some grassland bird species are now in decline in southern Ontario, reflecting changes in land use and habitat.

Turbines associated with commercial wind energy projects are prominent features of three of the terrestrial biomes shown in Figures 31.21 and 31.22. Wind farms are another example of technical modifications of existing habitats. The size and density of turbines may be less disruptive of natural systems than agricultural operations. However, both birds and bats are killed by turbines, and there are allegations that wind farms have a negative impact on human health. The longer-term impact of turbines on biomes and ecozones remains to be determined.

31.5a Cities Are Urban Techno Ecosystems

Cities are the fastest-growing ecosystems in the world **(Figure 31.23, p. 777).** In many countries, more people now live in cities (urban) than in the countryside (rural). By the year 2 CE, Rome was the first western city to reach a population of one million people, reflecting an extensive infrastructure for delivering food and services and removing wastes. In size, human populations in cities range from many with tens of

Table 31.3 | **Characteristics of Some of the Terrestrial Ecozones of Canada**

Each of the terrestrial ecozones is distinct in terms of its geomorphology, climate, and prevailing ecological communities. MAT is the mean annual temperature; MST is the mean summer temperature.

Ecozone	Region	Geomorphology	Climate	Soil	Prevailing Vegetation
Northern Arctic	Northern Quebec; mainland and eastern islands of Nunavut	Rocky lowlands; much exposed bedrock and glacial debris	High-Arctic; long cold winters, short cool summers; MAT −17 to −11°C; MST −2 to 4°C; precipitation 10–20 cm/yr	Regosol; permafrost throughout; active layer 30–50 cm; generally moist soil	High-Arctic tundra (Arctic desert or semidesert); dominant cover is lichens, mosses, and low-growing vascular plants
Arctic Cordillera	Northern Labrador, eastern Arctic islands of Nunavut	Mountains and rocky uplands, glacial ice fields; much exposed bedrock and glacial debris	High-Arctic; long cold winters, short cool summers; MAT −20 to −6°C; MST −2 to 6°C; precipitation 10–60 cm/yr	Regosol; permafrost throughout; active layer 20–40 cm; generally moist soil	75% of terrain is rocks or ice; elsewhere high-Arctic tundra of lichens, mosses, and low-growing vascular plants
Southern Arctic	Northern Quebec and across mainland Nunavut and northern Northwest Territories	Extensive rolling terrain and lowlands; much exposed bedrock and glacial debris	Low-Arctic; long cold winters, short cool–warm summers; MAT −11 to −7°C; MST 4 to 6°C; precipitation 20–40 cm/yr	Regosol; permafrost throughout (except under large lakes and rivers); active layer 50–70 cm; generally moist soil	Low-Arctic tundra, with more continuous cover than in high Arctic; low-shrub heath and graminoid meadows
Taiga Plains	Western Northwest Territories, northern British Columbia, northwestern Alberta; taiga watershed of Mackenzie River	Rolling plains and uplands; postglacial sediment and debris abundant; lakes and wetlands common	Subarctic; long cold winters, short warm summers; MAT −10 to −1°C; MST 7 to 14°C; precipitation 20–50 cm/yr	Podsol; permafrost discontinuous; active layer > 80 cm; generally moist soil	Open boreal forest (taiga) of relatively short, well-spaced, slow-growing trees, mostly spruce and pine with aspen, poplar, and birch; periodic wildfires
Taiga Shield	Central Quebec, Labrador, southeastern Northwest Territories, northern Saskatchewan and Manitoba	Rolling terrain on quartzitic shield bedrock; much exposed bedrock and glacial debris; lakes and wetlands common	Boreal continental; long cold winters, short warm summers; MAT −8 to 0°C; MST 6 to 11°C; precipitation 20–50 cm/yr	Thin podsol, permafrost discontinuous; active layer > 80 cm; generally moist soil	Open boreal forest (taiga) of relatively short, well-spaced, slow-growing trees, mostly spruce and pine with aspen, poplar, and birch; also open tundralike areas of low shrubs; periodic wildfires
Boreal Shield	Newfoundland and southern Labrador, southern Quebec, northern Ontario, central Manitoba, northern Saskatchewan	Rolling terrain on quartzitic shield bedrock; much exposed bedrock and glacial debris; lakes and wetlands common	Boreal continental; long cold winters, short warm summers; MAT −4°C in continental areas to 5.5°C in maritime Newfoundland; MST 11 to 15°C; precipitation 10–50 cm/yr in continental and 90–160 cm/yr in maritime	Podsol; generally moist soil	Closed boreal forest, mostly of spruce, pine, fir with aspen, poplar, and birch; periodic wildfires
Atlantic Maritime	New Brunswick, Nova Scotia, Prince Edward Island, adjacent Gaspé of Quebec	Rolling terrain on various bedrock, from quartzitic to sedimentary; abundant glacial debris; lakes and wetlands common	Temperate coastal to continental; cold winters, long warm summers; MAT 4 to 7°C; MST 13 to 16°C; precipitation 90–150 cm/yr	Complex soils, from podsol to brunisol; generally moist soil	Mixed-species forests of temperate trees, ranging from angiosperm dominated to coniferous dominated
Mixedwood Plains	Southern Quebec and Ontario within Great Lakes–St. Lawrence valley	Gently rolling terrain over sedimentary, often limestone bedrock	Temperate continental; cold winters, long hot summers; MAT 5 to 8°C; MST 16 to 18°C; precipitation 70–100 cm/yr	Deep, base-rich (high-calcium) brunisol, especially on postglacial lakebed parent materials; generally moist soil	Mixedwood forest, mostly angiosperm dominated; most of the natural cover is converted to agriculture and urban uses

Ecozone	Region	Geomorphology	Climate	Soil	Prevailing Vegetation
Boreal Plains	Central Manitoba and Saskatchewan to northern Alberta and northeastern British Columbia	Rolling terrain over moraine and flatter areas of postglacial lake sediment; abundant wetlands	Boreal continental; long cold winters, short hot summers; MAT −2 to 2°C; MST 13 to 16°C; precipitation 30–63 cm/yr	Deep podsol to brunisol; generally moist soil	Mostly conifer-dominated forest, with angiosperm dominated in the south
Prairies	Southern and central Manitoba, Saskatchewan, Alberta	Rolling terrain over moraine and flatter areas of postglacial lake sediment; abundant ponds and wetlands	Temperate continental; cold winters, long hot summers; MAT 2 to 4°C; MST 14 to 16°C; precipitation 25–70 cm/yr	Chernozem; soil dry to moist	Prairie dominated by grasses and forbs, ranging from tallgrass to mixedgrass to shortgrass types; most of the natural cover is converted to agriculture and urban uses
Taiga Cordillera	Western Northwest Territories and northern Yukon Territory	Steep to rolling terrain of northern Rocky Mountains and foothills; streams and rivers, fewer lakes	Subarctic coastal to continental; long cold winters, short warm summers; MAT −10 to −5°C; MST 7 to 10°C; precipitation 30–70 cm/yr	Podsol; generally moist soil	Because of altitudinal range, vegetation ranges from alpine tundra to subarctic boreal forest of spruce and birch
Boreal Cordillera	Northern British Columbia and southern Yukon Territory	Rugged mountainous terrain with foothills and deep wide valleys; streams and rivers, fewer lakes	Boreal continental; long cold winters, short warm summers; MAT 1 to 6°C; MST 10 to 12°C; precipitation < 30 cm/yr in rain shadow to > 150 cm/yr of orographic precipitation (i.e., increased by mountainous terrain that forces moist air masses to rise in altitude, cool, and precipitate their water content)	Podsol; permafrost discontinuous in north, with active layer > 80 cm; generally moist soil	Because of altitudinal range, vegetation ranges from alpine tundra to montane forest of spruce and aspen; open forest and grasslands in southern areas
Pacific Maritime	Coastal British Columbia	Rugged mountainous terrain with foothills and narrow coastal lowlands; streams and rivers, fewer lakes	Temperate coastal; short cool winters, long warm summers; MAT 5 to 9°C; MST 10 to 16°C; precipitation 60 cm/yr in dry Gulf islands to 400 cm/yr if orographic precipitation; generally 150–300 cm/yr	Podsol; moist soil	Because of altitudinal range and variable rainfall, vegetation ranges from alpine tundra to open dry forest to old-growth mixed-species conifer rain forest
Montane Cordillera	Southwestern Alberta and southern British Columbia	Rugged mountainous terrain with foothills; streams and rivers, fewer lakes	Temperate continental; short cold winters, long warm summers; MAT 1 to 8°C; MST 11 to 17°C; precipitation 30 cm/yr in rain shadow to 120 cm/yr if orographic precipitation	Podsol; drier soil	Because of altitudinal range and variable rainfall, vegetation ranges from alpine tundra to open grassland–forest to closed mesic forest
Hudson Plains	Northwestern Quebec, northern Ontario, northeastern Manitoba	Lowlands of postglacial James and Hudson Bays; surface waters abundant	Boreal coastal to continental; long cold winters, short cool summers; MAT −4 to −2°C; MST 11 to 12°C; precipitation 40–80 cm/yr	Regosol and podsol; permafrost discontinuous, with active layer > 80 cm; generally moist soil	Coastal areas have salt marsh, then tundra farther inland, and then open boreal coniferous forest

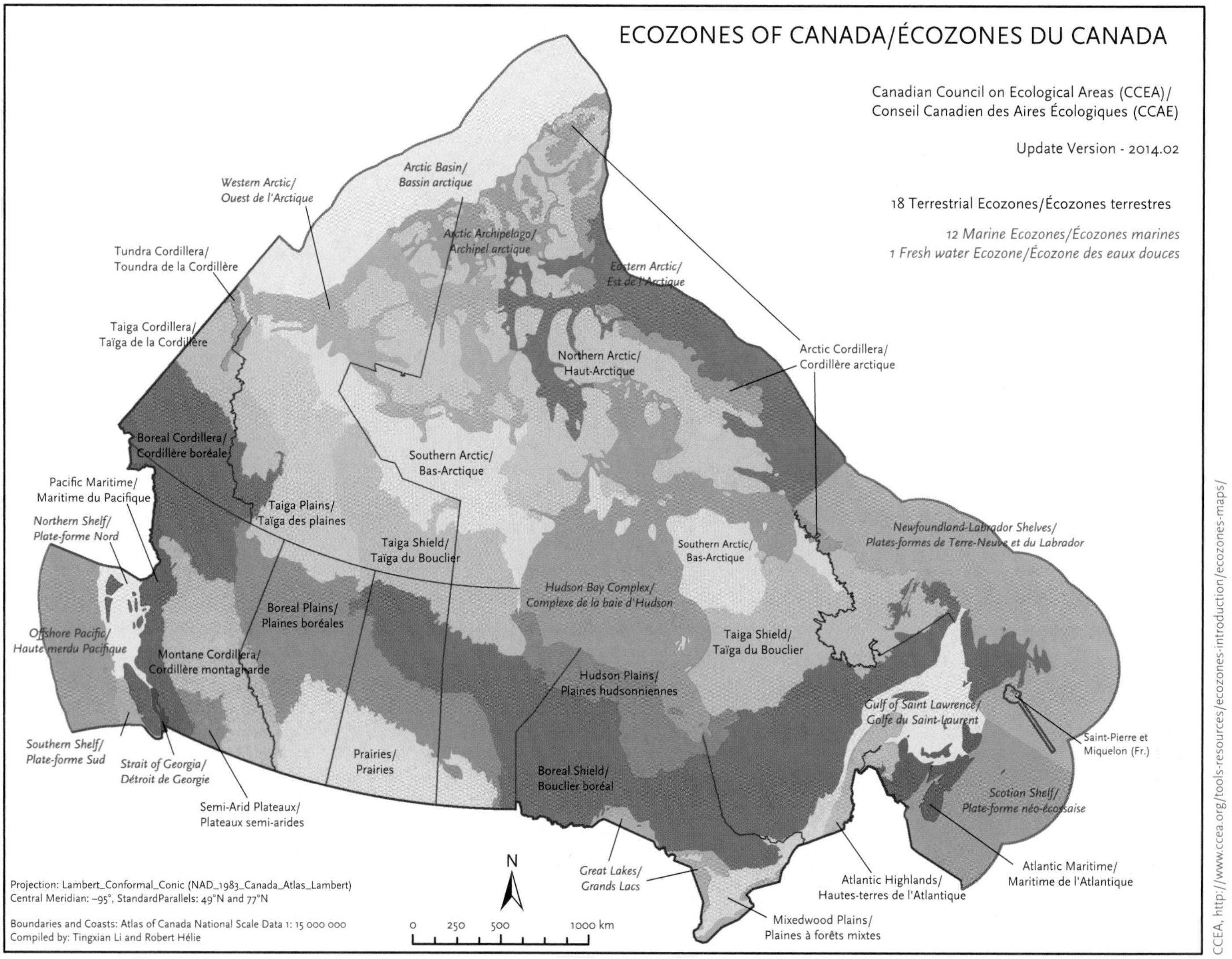

ECOZONES OF CANADA/ÉCOZONES DU CANADA

Canadian Council on Ecological Areas (CCEA)/
Conseil Canadien des Aires Écologiques (CCAE)

Update Version - 2014.02

18 Terrestrial Ecozones/Écozones terrestres

12 Marine Ecozones/Écozones marines
1 Fresh water Ecozone/Écozone des eaux douces

Western Arctic/
Ouest de l'Arctique

Arctic Basin/
Bassin arctique

Arctic Archipelago/
Archipel arctique

Eastern Arctic/
Est de l'Arctique

Tundra Cordillera/
Toundra de la Cordillère

Northern Arctic/
Haut-Arctique

Arctic Cordillera/
Cordillère arctique

Taiga Cordillera/
Taïga de la Cordillère

Boreal Cordillera/
Cordillère boréale

Southern Arctic/
Bas-Arctique

Pacific Maritime/
Maritime du Pacifique

Taiga Plains/
Taïga des plaines

Newfoundland-Labrador Shelves/
Plates-formes de Terre-Neuve et du Labrador

Northern Shelf/
Plate-forme Nord

Taiga Shield/
Taïga du Bouclier

Southern Arctic/
Bas-Arctique

Offshore Pacific/
Haute mer du Pacifique

Boreal Plains/
Plaines boréales

Hudson Bay Complex/
Complexe de la baie d'Hudson

Taiga Shield/
Taïga du Bouclier

Montane Cordillera/
Cordillère montagnarde

Hudson Plains/
Plaines hudsonniennes

Gulf of Saint Lawrence/
Golfe du Saint-Laurent

Saint-Pierre et
Miquelon (Fr.)

Southern Shelf/
Plate-forme Sud

Strait of Georgia/
Détroit de Georgie

Prairies/
Prairies

Boreal Shield/
Bouclier boréal

Scotian Shelf/
Plate-forme néo-écossaise

Semi-Arid Plateaux/
Plateaux semi-arides

Atlantic Maritime/
Maritime de l'Atlantique

Great Lakes/
Grands Lacs

Atlantic Highlands/
Hautes-terres de l'Atlantique

Mixedwood Plains/
Plaines à forêts mixtes

N

Projection: Lambert_Conformal_Conic (NAD_1983_Canada_Atlas_Lambert)
Central Meridian: –95°, StandardParallels: 49°N and 77°N

Boundaries and Coasts: Atlas of Canada National Scale Data 1: 15 000 000
Compiled by: Tingxian Li and Robert Hélie

0   250   500   1000 km

Figure 31.20
Distribution of the terrestrial ecozones and marine and freshwater ecozones of Canada.

a.

b.

M.B. Fenton

M.B. Fenton

Figure 31.21
Four Canadian examples of distinct ecozones are shown in these two pictures. In **(a)**, the Pacific ecozone, inhabited by humpback whales (*Megaeptera novae-angliae*), along with the Pacific maritime biome in the background. The setting is Haida Gwaii off the west coast of British Columbia. In **(b)**, along the north shore of Lake Erie, what had been mixedwood plains is now a mosaic of open agroecosystems and woodlots.

a.

b.

Figure 31.22

(a) A boreal shield ecozone east of Thunder Bay in Northern Ontario. **(b)** A view of prairie near Head-Smashed-In Buffalo Jump in south-western Alberta. Turbines of commercial wind energy projects are present in both (a) and (b).

thousands of citizens, to Tokyo with over 35 million. Like other biomes, cities are dynamic and changing.

In a 2013 model, Luís Bettencourt identified four simple assumptions that could be used to advantage in understanding the dynamics of cities. First, cities have the capacity for mixing populations of citizens that live there. This means that citizens can afford to fully explore the city and use it to advantage. Second, infrastructure develops gradually and incrementally, accommodating the expanding population. The main-stay of this assumption is a city's network of roads. Third, G (the product of gross domestic product and road volume per capita) is mainly independent of N (population size). G reflects an increasing demand by

cities on the mental and physical efforts of their citizens. This also involves communication networks. Fourth, socioeconomic outputs are proportional to local social interactions. In other words, cities are concentrations of social interactions, not just of people. Unlike biological systems that appear to minimize dissipation of energy, cities are systems in which energy dissipation is maximized. This reflects ongoing processes such as transportation, as well as heating and cooling.

Cities share many features with more biological ecosystems, and thinking of them in this way may teach us more about these different kinds of systems and how they operate.

Figure 31.23

A view of an urban techno ecosystem, specifically the campus of Western University as seen in 2007. Note that expanses of woodland may be more extensive than those in agroecosystems (Figure 31.21b).

Review

To access course materials such as Aplia and other companion resources, please visit www.NELSONbrain.com.

31.1 Energy Flow and Ecosystem Energetics

- Net primary productivity is the chemical energy remaining in a system after energy has been used by producers to complete life processes and cellular respiration. Net primary productivity differs from standing crop biomass in that net primary productivity is a measure of energy, whereas standing crop biomass is a measure of dry mass.

- Other factors affect primary productivity, such as water and access to nutrients.

- Assimilation efficiency refers to energy absorbed from eating compared with the total energy in the food. Production efficiency is the energy content of new tissue material compared with the energy absorbed from food intake.

- Some energy is lost during transfer by consumption. The process becomes less efficient as the number of transfers increases, meaning that less energy is transferred to the final consumer.

- Biological magnification occurs when material (e.g., DDT) present in small amounts in a producer or low-level organism is consumed by another organism, transferring the material to the predator. DDT accumulates with each successive transfer. Top predators exhibit the highest concentrations of contaminants such as DDT.

31.2 Nutrient Cycling in Ecosystems

- The amount of water that leaves Earth and enters the atmosphere through evaporation is equal to the amount of water reaching Earth by precipitation.

- Producers transform atmospheric carbon (CO_2) into carbohydrates. Consumers then eat the producers and take in the carbohydrates.

- Cyanobacteria can fix nitrogen, which is crucial because although atmospheric nitrogen levels are high, this nitrogen is not accessible to plants or animals. Atmospheric nitrogen must be converted or fixed into a usable form such as ammonium or nitrate.

- Phosphate ions are carried to bodies of water through weathering and erosion, where most of the phosphate precipitates. Eventually, weathering releases phosphates, which are then directly absorbed by plants.

31.3 Ecosystem Modelling

- Modelling involves collecting data about an ecosystem and deriving mathematical equations about the relationships in the ecosystem. Data collected over different seasons and different years can be used to simulate the effects of a disruption on various levels of the ecosystem in question.

- Simulation modelling helps us understand and predict the impact of influences on certain ecosystems without actually conducting an experiment. Altering anything in an ecosystem without knowledge of its possible effects can be devastating on many or all levels.

31.4 Scale, Ecosystems, Species

- Biotic and abiotic factors interact in ecosystems, and these can occur across scales from millimetres to kilometres.

- At one end, the microorganisms in a biofilm of water provide an example of ecological interactions, as do the organisms living in the water in a pitcher plant.

- Islands provide another example of a range of scales and interactions, from the Galápagos Islands as a group to individual islands.

- Plants disperse using different mechanisms, from wind to the activities of animals.

- In the west coast rain forest of North America, salmon (*Oncorhynchus*) influence plant communities along the rivers and streams in which they spawn.

31.5 Biomes and Ecozones

- *Biota* refers to species, so the biota of an area is the species that occur there.

- Biomes are defined by their biota, perhaps as a list of species.

- *Ecozones* are extensive ecological regions (typically consisting of more than one biome).

- *Biome* and *ecozone* are terms for identifying categories of habitats, often connecting the species to the prevailing climatic conditions.

Questions

Self-Test Questions

1. Which of the following events moves energy and material from a detrital food web into a grazing food web?
   a. an earthworm eating dead leaves on the forest floor
   b. a robin catching and eating an earthworm
   c. a crow eating a dead robin
   d. a bacterium decomposing the feces of an earthworm

2. What is the definition of the total dry mass of plant material in a forest?
   a. a measure of the forest's gross primary productivity
   b. a measure of the forest's net primary productivity
   c. a measure of the forest's standing crop biomass
   d. a measure of the forest's ecological efficiency

3. Which ecosystem has the highest rate of net primary productivity?
   a. open ocean
   b. temperate deciduous forest
   c. tropical rainforest
   d. agricultural land

4. Endothermic animals exhibit a lower ecological efficiency than ectothermic animals for which reason?
   a. Endotherms are less successful hunters than ectotherms.
   b. Endotherms eat more plant material than ectotherms.
   c. Endotherms are larger than ectotherms.
   d. Endotherms use more of their energy to maintain body temperature than ectotherms.

5. What determines the amount of energy available at the highest trophic level in an ecosystem?
   a. only the gross primary productivity of the ecosystem
   b. only the net primary productivity of the ecosystem
   c. the net primary productivity and the ecological efficiencies of herbivores
   d. the net primary productivity and the ecological efficiencies at all lower trophic levels

6. Which pyramid is inverted in some freshwater and marine ecosystems exhibit?
   a. biomass
   b. energy
   c. numbers
   d. ecological efficiency

7. Which process moves nutrients from the available-organic compartment to the available-inorganic compartment?
   a. respiration
   b. assimilation
   c. sedimentation
   d. photosynthesis

8. Which of the following materials has a sedimentary cycle?
   a. oxygen
   b. nitrogen
   c. phosphorus
   d. carbon

9. Which statement is supported by the results of studies at the Hubbard Brook Experimental Forest?
   a. Most energy captured by primary producers is lost before reaching the highest trophic level in an ecosystem.
   b. Deforested watersheds experience a more significant decrease in runoff than undisturbed watersheds.
   c. Deforested watersheds lose more calcium and nitrogen in runoff than undisturbed watersheds.
   d. Nutrients generally move through biogeochemical cycles very quickly.

10. Biological magnification describes which phenomenon?
    a. Certain materials become increasingly concentrated in the tissues of animals at higher trophic levels.
    b. Certain materials become most concentrated in the tissues of animals at the lowest trophic levels.
    c. Certain materials accumulate only in the tissues of tertiary consumers.
    d. Certain materials accumulate only in the tissues of detritivores.

Questions for Discussion

1. Identify 12 ecosystem changes associated with hydroelectric power projects. Consider upstream and downstream changes as well as those associated with transmission of generated power. How does preparing your answers draw on information presented in this chapter?

2. A lake near your home becomes overgrown with algae and pondweeds a few months after a new housing development is built nearby. What kind of data would you collect to determine whether the housing development might be responsible for the changes in the lake?

3. Some politicians question whether recent increases in atmospheric temperature result from our release of greenhouse gases into the atmosphere. They argue that atmospheric temperature has fluctuated widely over Earth's history, and the changing temperature is just part of a historical trend. What information would allow you to refute or confirm their hypothesis? From another perspective, describe the pros and cons of reducing greenhouse gases as soon as possible versus taking a "wait-and-see" approach to this question.

4. What are the ecological consequences of converting crops from food to biofuels?

5. Look at the birds in your neighbourhood. How many have been introduced? How does their behaviour differ from those of native species of birds living in the same immediate areas?

32

A leopard photographed in the wild in South Africa.

Conservation of Biodiversity

WHY IT MATTERS

Achieving preservation of Earth's biodiversity is one of the most pressing challenges facing our species today. Numbers of species are a simple indicator of **biodiversity**, perhaps the most apparent and easy to grasp. But as we have seen, many species of organisms remain undescribed and unnamed. Without names and descriptions, how can we recognize or count them? As we shall see shortly, being unnamed means being unprotected. The Barcode of Life project (see "Barcode of Life," Chapter 30) is one promising effort to better catalogue biodiversity by identifying and allowing us to name its components.

Research on ecosystems shows repeatedly how the numbers of species are associated with stability and productivity, and, as we have seen (Chapter 30), antagonistic and mutualistic systems differ in community composition. The natural order (association between productivity and biodiversity), however, does not coincide with the productivity that our own species must achieve to feed our ever-expanding populations. Creating and maintaining agricultural monocultures is a way for us to maximize food production and efficiency of harvest. In many areas, this approach leads to the disappearance of family-operated farms. Is this progress justified by efforts to increase efficiency and yield? Humans also use genetically modified organisms to increase productivity and marketability, as well as other features, such as shelf life and portability. All too often, increased agricultural productivity is achieved by the use of more fertilizer, water, and energy. Does agriculture have to be the enemy of biodiversity?

Some people have connected humans' attitude to Earth and its riches with religious teachings. In 1967, Lynn White Jr., a professor of medieval history, explored the historical roots of our ecological crisis. He focused on the Christian view of creation, the importance of science, and the separation of humans from their environmental roots. A dualism between humans and nature had emerged in some Christian societies more than in others. Inherent in these societies was the prevailing idea that it is God's will that humans exploit nature for their own ends. White

Figure 32.1
St. Francis of Assisi.

nominated St. Francis of Assisi **(Figure 32.1)** as the patron saint of ecologists. He said that appreciating the virtue of humility was key to understanding the teachings of St. Francis. His point was that as soon as an animal or a plant (or a meadow, lake, or grove of trees) has its own place in nature (in God's eyes), then it can become as important as we believe we are.

The onus is on us as citizens of the planet to conserve biodiversity, whether the focus is species or habitats. One of the main problems we must overcome is the attitude of many humans, as reviewed by White. Today a common reflection of this attitude is that being able to do something (afford to, have the means to) is justification enough for doing it—whether the project involves making space for a shopping mall by draining a wetland or cutting down the trees in a woodlot.

If we as a species can recognize the importance of biodiversity and accept that the world is not ours to do with as we please, what is the best route to protecting and conserving biodiversity? Should we focus on species? On genetic diversity? On ecosystems? How should we blend these approaches to achieve the best support for the endeavour? How can we engage people in this important activity and perhaps move them away from a human-centric view of the world?

As we shall see, at almost every turn there are examples of human activities driving other species to extinction. The motivations for human actions range from

little more than greed to the daily effort to survive. The purpose of this chapter is to introduce you to a range of situations and examples associated with the reduction of biodiversity by extinction and the threat of extinction. We also consider steps that can be taken to protect biodiversity, including some successes and some failures.

Although extinction has been a recurring phenomenon in the history of life on Earth, the current extinction of species as a result of human activities is of grave concern to biologists and others concerned about our future. Changes wrought by humans include introduction of alien species and overharvesting of natural resources. These changes often have unexpected impacts on other species. Adoption of standardized criteria for identifying species at risk is an important part of conservation. Efforts to conserve biodiversity are focused at the species and/or habitat levels. In the final analysis, many of them depend upon taking steps to control the population of our own species and the related consumption of resources and habitat. We also need to change our attitudes and recognize that we are but one of millions of species on the planet and that, ultimately, our survival will depend upon theirs.

32.1 Extinction

Extinction is part of the process of evolution. Given that life has been on Earth for about 3 billion years, today there are more extinct than living species. Occasionally, the fossil record demonstrates a continuum in time from one species to another, sometimes blurring the boundaries between taxa (see Chapter 19). In this case, one could argue that the original species in a series lives on in its descendants. For example, the discovery that the genome of *Homo sapiens sapiens* contains some genes from *Homo sapiens neanderthalensis* leaves open the question about the distinctness of the two taxa. Although data for fossil species usually do not permit us to assess the levels of gene flow between populations, the Neanderthals provide an interesting exception. The difficulties inherent in applying the species concept to fossil material are familiar to paleontologists but less so to biologists.

Species and lineages have been going extinct since life first appeared. We should expect species to disappear at some low rate, the **background extinction rate;** as environments change, poorly adapted organisms do not survive or reproduce. In all likelihood, more than 99.9% of the species that have ever lived are now extinct. David Raup has suggested that, on average, as many as 10% of species go extinct every million years and more than 50% go extinct every 100 million years. Thus, the history of life has been characterized by an ongoing turnover of species.

The fossil record indicates that extinction rates rose well above the background rate at least five times in Earth's history. These events are referred to as mass extinctions. One extinction occurred at the end of the Ordovician and the beginning of the Devonian, the next at the end of the Devonian, then the end of the Permian, the end of the Triassic, and the end of the Cretaceous. The Permian extinction was the most severe, and more than 85% of the species alive at that time disappeared forever. This extinction was the end of the trilobites, many amphibians, and the trees of the coal swamp forests. During the last mass extinction, at the end of the Cretaceous, half of the species on Earth, including most dinosaurs, disappeared. A sixth mass extinction, potentially the largest of all, is occurring now as a result of human degradation of the environment.

Different factors were responsible for the five **mass extinctions.** Some were probably caused by tectonic activity and associated changes in climate. For example, the Ordovician extinction occurred after Gondwana moved toward the South Pole, triggering a glaciation that cooled the world's climate and lowered sea levels. The Permian extinction coincided with a major glaciation and a decline in sea level induced by the formation of Pangaea (see Chapter 20).

Many researchers believe that an asteroid impact caused the Cretaceous mass extinction. The resulting dust cloud may have blocked the sunlight necessary for photosynthesis, setting up a chain reaction of extinctions that began with microscopic marine organisms. Geologic evidence supports this hypothesis. Rocks dating to the end of the Cretaceous period (65 mya) contain a highly concentrated layer of iridium, a metal that is rare on Earth but common in asteroids. The impact from an iridium-laden asteroid only 10 km in diameter could have caused an explosion—equivalent to a billion tonnes of TNT—that scattered iridium dust around the world. Geologists have identified the submarine Chicxulub crater, 180 km in diameter, off Mexico's Yucatán peninsula as the likely site of the impact.

Although scientists agree that an asteroid struck Earth at that time, many question its precise relationship to the mass extinction. Dinosaurs had begun their decline at least 8 million years earlier, but many persisted for at least 40 thousand years after the impact. Moreover, other groups of organisms did not suddenly disappear, as one would expect after a global calamity. The Cretaceous extinction took place over tens of thousands of years. Furthermore, some organisms survived periods of extinction, such as ginkgo trees (*Ginkgo biloba*), horseshoe crabs (*Limulus polyphemus*), and coelocanths (*Latimeria chalumnae*).

Even today we cannot blame the extinctions of most species on the activities of humans. But our increasing technological capability and prowess coincide with a burgeoning population of people. This situation is exacerbated by the philosophical view that humans are disconnected from nature. Thus, we are becoming better and better at destroying the biota of the planet. Taking action requires identifying root causes and then trying to make changes that will alleviate the problems.

First, we consider extinctions not linked to humans, and then we review examples of situations in which our actions have either directly or indirectly led to the extinction of species. The fact that extinction is integral to the process of evolution is hardly justification or rationalization for our driving so many species there. Put another way, invoking "survival of the fittest" may not be adequate justification for eradicating other species.

32.1a Dinosaurs: A Most Notable Extinction

Why do species go extinct? There could be as many theories as there are extinct species! The disappearance of the dinosaurs is one of the best-known extinction events in Earth's history. At the end of the Cretaceous about 65.5 mya, the dinosaurs disappeared. The ancestors of dinosaurs had appeared in the Triassic, and the group underwent extensive adaptive radiation reflected in body size, lifestyle, and distribution. Although people think of large and spectacular carnivorous dinosaurs such as *Tyrannosaurus rex* or the huge herbivore *Apatosaurus* (previously known as *Brontosaurus*), in reality, many species of dinosaurs were small and delicate.

Evidence from deposits in Alberta suggests that the carnivorous dinosaur *Albertosaurus* **(Figure 32.2)** showed age-specific mortality and high juvenile survival **(Figure 32.3a)**. Indeed, the survivorship curves (see Chapter 29) for *Albertosaurus* resemble those for

Figure 32.2
Mounted skeleton of *Albertosaurus* on display in the Royal Tyrrell Museum, Drumheller, Alberta. This late Cretaceous carnivore is abundant in fossil beds in Alberta and elsewhere.

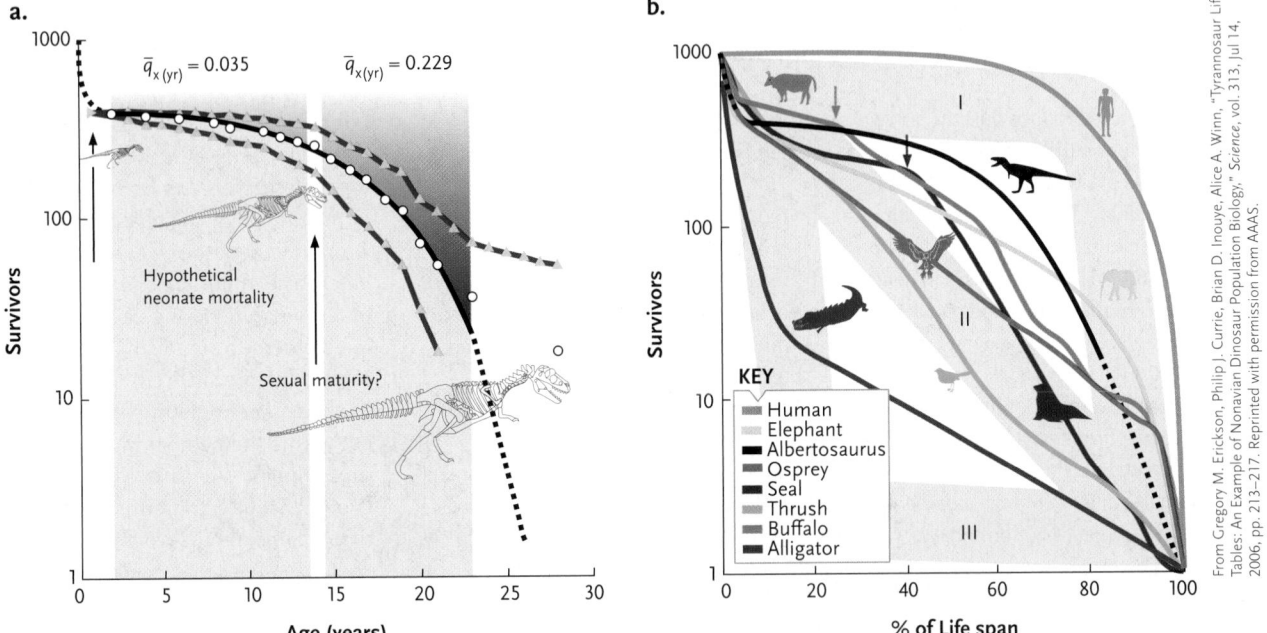

From Gregory M. Erickson, Philip J. Currie, Brian D. Inouye, Alice A. Winn, "Tyrannosaur Life Tables: An Example of Nonavian Dinosaur Population Biology," *Science*, vol. 313, Jul 14, 2006, pp. 213–217. Reprinted with permission from AAAS.

Figure 32.3

(a) Survivorship curve for a hypothetical cohort of 1000 *Albertosaurus* with presumed neonatal mortality of 60%. **(b)** Survivorship of *Albertosaurus* compared with that of other animals, including humans from developed countries, short-lived birds, mammals, and lizards, as well as crocodilians and some captive mammals.

humans (Figure 32.3b). The data do not provide any indication of a flaw that predisposed dinosaurs to extinction. This is sobering, given the similarity between some aspects of dinosaur population biology and those of our own. The prevailing view today is that the disappearance of the dinosaurs is linked to the impact of an asteroid. Many other theories have been proposed to explain extinction, but the fact that birds and mammals and many other groups of organisms showed widespread extinctions at the end of the Cretaceous implies a pervasive catastrophic event.

32.1b Multituberculates Are a Mammalian Example of Extinction

There is more to extinction than dinosaurs. Competition (see Chapter 30) has been proposed as a mechanism that can lead to extinction. Among mammals, the adaptive radiation of rodents (order Rodentia; **Figure 32.4a**) in the early Oligocene coincides with the disappearance of multituberculates (order Multituberculata; Figure 32.4b). As a group, multituberculates were prominent and persisted for 100 million years (compared with 150 million years for dinosaurs), making them the most successful mammals to date.

Multituberculates ranged from small (~20 g) to medium (5 to 10 kg) in size and exhibited both terrestrial and arboreal lifestyles. We can only speculate what happened to them, and why they became extinct. The widespread success of rodents almost worldwide could lend credibility to competition as the reason for

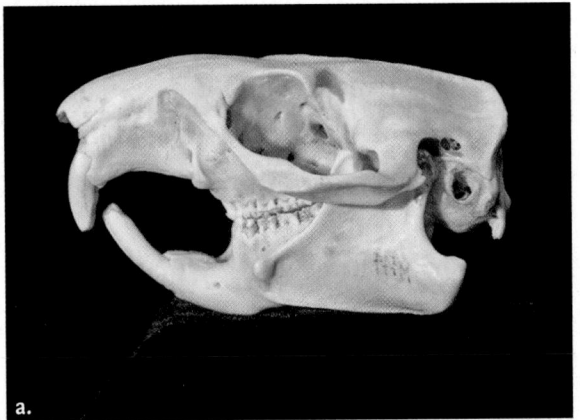

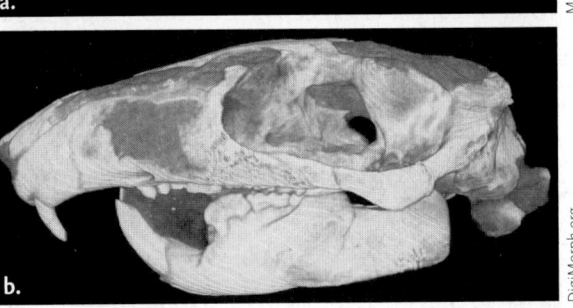

Figure 32.4

(a) The skull of a groundhog (*Marmota monax*), a North American rodent, compared with **(b)** the skull of a multituberculate (*Kryptobaatar dashzevegi*) from the Mongolian late Cretaceous.

the multituberculates' demise. However, the fossil record does not tell us what rodents and multituberculates competed for: food? nest sites?

STUDY BREAK

1. What is extinction?
2. What were multituberculates?

Sex Determination and Global Warming

Failure to reproduce puts the survival of a species on the edge, so anything that interferes with reproduction can be threatening. Genetic recombination is a fundamental benefit of sexual reproduction, enabling increases in genetic diversity and elimination of deleterious mutants. Effective sexual reproduction means having male and female systems, sometimes in one individual (hermaphrodites) and perhaps more often in different individuals. Males and females differ in many fundamental ways—genetically, hormonally, physiologically, and anatomically.

In humans and many other animals, gender is determined by genotype, with males having an X and a Y chromosome and females having two X chromosomes. The reverse is true in many other animals, for example birds. But in many reptiles gender is determined environmentally. Eggs incubated at some temperatures develop into males; when incubated at other temperatures, they produce females.

In 2008, D. A. Warner and R. Shine reported the results of experiments done with jacky dragons (*Amphibolurus muricatus*), an Australian lizard in which gender is determined by temperature. Eggs incubated at 23 to 26°C or 30 to 33°C produce females; those incubated from 27 to 29°C produce males. Warner and Shine tested the hypothesis that temperature-dependent sex determination ensured production of females when they had an advantage and males when the advantage was to them. Using a combination of temperature and hormonal manipulations, Warner and Shine could produce males or females at any temperature. They analyzed paternity to assess the reproductive output of these males and observed eggs laid and hatched to document these females' reproductive output.

In female jacky dragons, larger body sizes occur at higher temperatures, and larger females have higher fecundity than smaller ones. Higher temperatures also correlate with larger body size in males. However, males hatched from eggs incubated between 27 and 29°C sired more offspring than those hatched from eggs incubated at lower or higher temperatures.

Change in climate, such as global warming, could put species with temperature-dependent sex determination at risk by effectively eliminating males or females from the population. Eggs incubated at the wrong temperatures would fail to hatch. The importance of variation in temperature during development in ectothermic organisms could explain the prevalence of genotypic-dependent sex determination in euthermic (homeothermic) viviparous animals. Viviparous or ovoviviparous ectotherms (fish, amphibians, reptiles, other animals) could also rely on temperature-dependent gender determination, provided that their developing young experience an appropriate range of temperatures.

Recent work from Mexico reveals that since 1975, 12% of local populations of lizards have disappeared, likewise 4% of worldwide local populations. Like other ectotherms, lizards have a narrow thermal range in which they thrive, and climate change has altered the thermal niches available to them. This and temperature-dependent gender determination put lizards in double jeopardy.

32.2 The Impact of Humans

When it comes to extinctions, we know most about those resulting from our activities, usually because these records are relatively recent and accessible. If you recently visited Mauritius, you might have noticed that the few remaining Mauritian calvaria trees are slowly dying of old age. Their passing will mark the extinction of this species, which has occurred even though the trees continued to bloom and produce seeds. The key to the pending extinction of *Sideroxylon majus* is the earlier extinction of Dodos. To germinate, seeds of Mauritian calvaria trees had to pass through the Dodo's digestive tract. The Dodo (**Figure 32.5**) was a medium-sized flightless bird that lived on the island of Mauritius. When European sailors first visited the island, they used Dodos as a source of fresh meat. Then, as the island was settled, the birds were exposed to introduced predators (cats, dogs, rats) and an expanding human population. Dodos vanished by 1690.

Species confined to islands often have small populations and are unaccustomed to terrestrial predators, making them

Figure 32.5
(a) A reconstruction of a *Raphus cucullatus*, the Dodo, an extinct flightless bird from **(b)** Mauritius.

Raphus cucullatus by Roelandt Savery, 1626

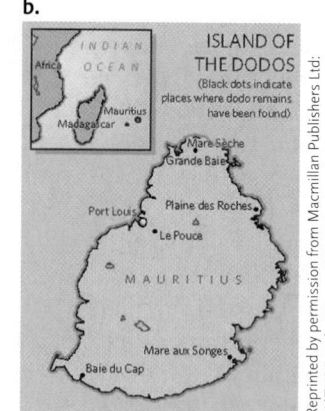

Reprinted by permission from Macmillan Publishers Ltd: NATURE, Vol. 443: pp. 138–140, "Ornithology: Digging for dodo" by Henry Nicholls, copyright (2006).

vulnerable to extinction. The fossil and subfossil records show that many species of birds disappeared from islands in the South Pacific as Polynesians arrived there from the west. This occurred from Tonga to Easter Island and beyond **(Figure 32.6)**. The Galápagos, only discovered by people in 1535, was sheltered from the wave of human-induced extinctions. On Easter Island, **endemic species** of sea birds and other species disappeared soon after people settled there. These examples demonstrate that humans do not have to be industrial or "high tech" to effect extinctions.

Meanwhile, in the North Atlantic, people hunted *Pinguinus impennis,* the Great Auk, to extinction. However, land birds with large distributions and huge populations have also disappeared, such as *Ectopistes migratorius,* the Passenger Pigeon, in eastern North America. Large-scale harvesting of these birds, combined with their low reproductive rate (clutch size: one egg), made the birds vulnerable in spite of their enormous populations. Animals that produce one young per year and suffer "normal" mortality must live at least 10 years to replace themselves in the population (see Chapter 29).

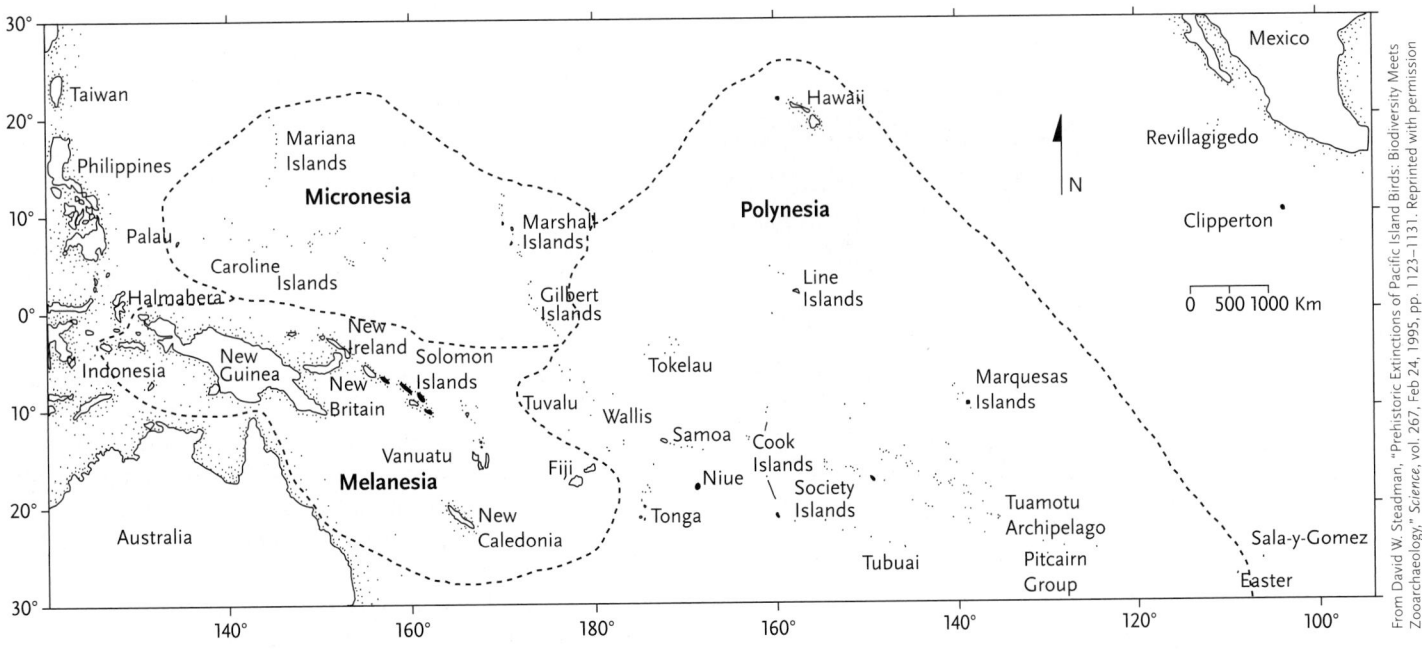

From David W. Steadman, "Prehistoric Extinctions of Pacific Island Birds: Biodiversity Meets Zooarchaeology," *Science,* vol. 267, Feb 24, 1995, pp. 1123–1131. Reprinted with permission from AAAS.

Figure 32.6

Islands in the South Pacific where the arrival of Polynesians coincided with the extinction of many island species of birds.

1. When did the five mass extinctions occur? Which was the most severe? Which extinction affected the dinosaurs?
2. What caused these extinctions?
3. Why are island animals and plants particularly susceptible to extinction due to human impacts?

CONCEPT FIX It is easy to believe that we must focus conservation efforts on large, charismatic animals and plants because these can be the poster images of conservation. We now realize that it is often more important to protect ecosystems, recognizing that the many components of ecosystems play a vital role in maintaining biodiversity, including the large and charismatic members of ecological communities. ⬡

32.3 Introduced and Invasive Species

Humans cause extinction through hunting and by the introduction of other species. House cats, *Felis domesticus,* are among the worst introductions people have made. Anecdotal records suggest that in 1894, one house cat (named Tibbles) exterminated an entire population of flightless wrens **(Figure 32.7)** on Stephen's Island, a 2.6 km² island off the north shore of New Zealand. Fossils indicate that the wrens had occurred widely in New Zealand. This record stands for one individual, Tibbles, taking out the remaining approximately 10 pairs and exterminating the species.

Figure 32.7

Stephen's Island Wren, *Xenicus lyalli.* This species was exterminated by one cat.

Figure 32.8

(a) Zebra mussels, *Dreissena polymorpha,* were introduced to the Great Lakes in North America, where they have spread rapidly. **(b)** They are directly responsible for the declines in eastern pond mussels (*Lampsilis radiata*), a local mussel species.

It should be obvious that moving species from one part of the world to another, whether done willfully or by accident, can have calamitous impacts. The invaders, once arrived and established, may outcompete resident species, laying waste to species and ecosystems. The list of introduced organisms is very long and includes many domesticated or commensal species of animals and plants. The arrival of zebra mussels **(Figure 32.8a)** in the Great Lakes is the main reason for the decline of the now endangered eastern pond mussels (Figure 32.8b). The immigrant mussels outcompeted and overgrew the native ones, reducing their range and populations to levels that resulted in eastern pond mussels being recommended for listing as endangered in Canada in 2007.

Meanwhile, in parts of the British Isles, flatworms (*Arthurdendyus triangulatus;* **Figure 32.9, p. 788;** see also Chapters 27 and 48) introduced from New Zealand are deadly predators of earthworms. Since their arrival in garden pots, the flatworms have thrived and spread rapidly, coinciding with the demise of earthworms. Ironically, although we often think of gardeners as individuals in touch with nature, their propensity to introduce exotic species may not be compatible with conservation. Earthworms themselves have been introduced widely to places around the world.

Some organisms move about in ballast water. Since about 1880, ships have regularly used water for ballast. In the early 1990s, a survey of ballast water in 159 cargo ships in Coos Bay, Oregon, revealed 367 taxa

MOLECULE BEHIND BIOLOGY 32.3

The 2,4-D Molecule and Resiliency

Resiliency is one of the most impressive features of life at the species and/or ecosystem levels. In one respect, this feature complicates the challenges of conserving biodiversity because introduced species can be so invasive, reflecting their adaptability. Humans first identified 2,4-D (2,4-dichlorophenoxyacetic acid; **Figure 1**) in 1942, and from 1944, it was marketed as a herbicide more effective against broad-leaved plants than against grasses. Technically, 2,4-D is a hormone absorbed by the plant and translocated to the growing points of roots and shoots. 2,4-D kills weeds by inhibiting growth. The global market for 2,4-D is probably more than U.S.$300 million, and it is mainly used to control broad-leaved weeds in cereal crops. According to the World Health Organization (WHO), 2,4-D is a "moderately hazardous pesticide" known to affect a variety of animals (e.g., dogs but not rats). Curiously, it turns out that other animals may use 2,4-D for their own ends.

In 1971, Thomas Eisner and colleagues reported that a grasshopper (*Romalea microptera;* **Figure 2**) produced a froth of chemicals **(Figure 3)** for protection against ants. One of the main ingredients in the froth was 2,5-dichlorophenol, apparently derived from 2,4-D. This is an astonishing demonstration of adaptability that can underlie resiliency.

Resiliency and the recuperative powers of ecosystems are demonstrated by stories of "lost cities," for example, structures built by Maya in Central America, being found in a jungle. Archaeological evidence reveals that in some habitats, these buildings and pyramids were overgrown by the rain forest in about 100 years. The Great Zimbabwe Ruins in southern Africa were overgrown by savannah woodland in a period of 100 to 200 years and only latterly "discovered" by European explorers.

Figure 1
2,4-dichlorophenoxyacetic acid, 2,4-D.

Figure 2
Romalea microptera, a grasshopper that uses an ant repellent with a 2,4-D derivative.

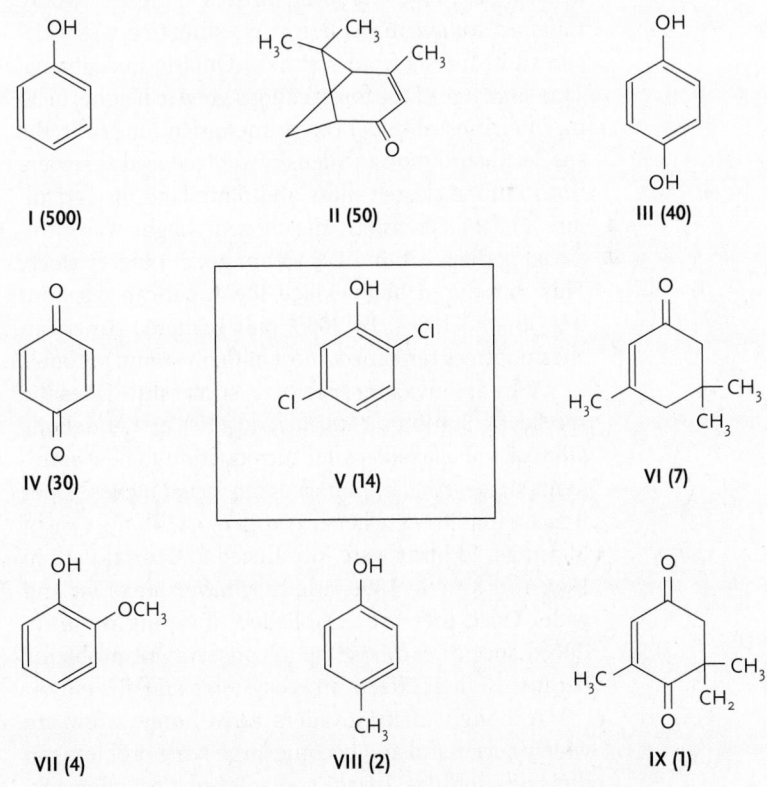

Figure 3
Active ingredients in the defensive froth of the grasshopper, Romalea microptera. *2,5-dichlorophenol (boxed) is apparently derived from 2,4-D.*

Figure 32.9

This earthworm-eating planarian (*Arthurdendyus triangulatus*) was introduced to the British Isles from New Zealand. It has had a devastating effect on local populations of earthworms.

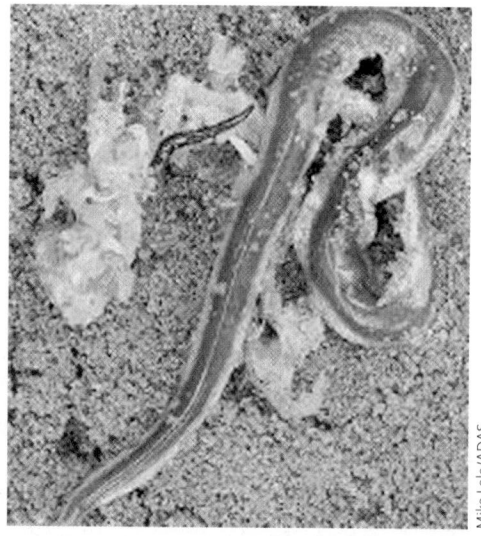

Mike Lole/ADAS

representing 16 animal and 3 protist phyla, as well as 3 plant divisions. The samples included all major and most minor phyla. Organisms in the ballast water included carnivores, herbivores, omnivores, deposit feeders, scavengers, suspension feeders, primary producers, and parasites. Ballast water is taken on in one port and discharged in another, providing many species with almost open access to waters around the world.

Meanwhile, introduced diseases (and the organisms that cause them) have decimated, if not obliterated, resident species. When Europeans arrived in the New World, *Castanea dentata,* the American chestnut tree, was widespread in forests from southern Ontario to Alabama. This large tree of the forest canopy grew to heights of 30 m. Often most abundant on prime agricultural soils, the species' distribution and density were reduced as settlers from Europe cleared more and more land for agriculture. *Endothia parasitica,* the chestnut blight, was introduced perhaps around 1904 from Asian nursery stock. This introduced blight killed the American chestnut trees by the 1930s. By 2000, only scattered American chestnut trees remained, most of them stump sprouts.

Why are invading species so successful? Does the spread of Starlings (*Sturnus vulgaris*) or dandelions (*Taraxacum officinale*) after introduction to new continents suggest that they moved into vacant niches? Does it mean that they are better competitors? In the case of Starlings, 13 birds were introduced to Central Park in New York City in 1890, and they have spread far and wide. Once they are established, invading or introduced species can pose huge conservation problems because of their effects on ecosystems and diversity.

Although many invaders arrive, only a few are widely successful and become large-scale problems in their new settings. Invading plants are most often successful in nutrient-rich habitats, where they can achieve high growth rates, early reproduction, and maximal production of offspring. What happens in resource-poor settings? In the past, conventional wisdom has suggested that low-resource settings could be reservoirs for native species that could outcompete invaders.

However, an experimental examination of the responses of native and introduced species to challenging conditions revealed that invasive plant species almost always fared better (**Figure 32.10**). Resource use efficiency (RUE), calculated by measuring carbon assimilation per unit of resource, provides an indicator of success. Many invasive species, such as ferns, C_3 and C_4 grasses, herbs, shrubs, and trees, were more successful in low-resource systems than native species were.

This research was conducted in Hawaii, an excellent place for studying invasive species because so many are there. Among the invaders were *Bromus tectorum* (cheatgrass), *Heracleum mantegazzianum* (cartwheel flower or giant hogweed), and *Pinus radiata* (Monterey pine). Humans have introduced these plants for gardening (cheatgrass and cartwheel flower) or commercial timber production (Monterey pine). The data demonstrate that attempting to restore ecosystems and exclude invading species by reducing resource availability does not succeed because of the efficiency with which some species use resources.

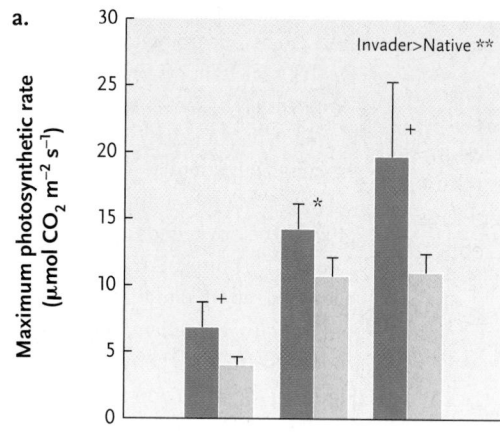

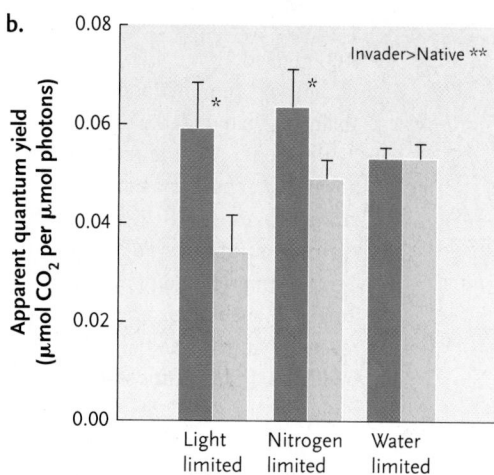

Figure 32.10

(a) Photosynthetic rates (RUE) and **(b)** light-use efficiency of invasive plant species (blue bars) make them more competitive than native ones (yellow bars). The plants were from three different habitats in Hawaii.

In the graphs, + denotes $P < 0.01$, while * denotes $P < 0.05$.

** Indicates that in both instances, invaders are significantly more efficient than native species

a.

b.

32.4 How We Got/Get There

Lamentably, we know that humans can exterminate species that are populous and widespread as well as ones that have small populations and occur in a small area.

32.4a Poaching Caused the Demise of the Black Rhinoceros

It is estimated that 60 thousand black rhinos (*Diceros bicornis*) lived in the wild in Africa in 1960 (**Figure 32.11a**). This large (1.5 m at the shoulder, 1400 kg) browsing mammal was widespread in sub-Saharan Africa (Figure 32.11b). Adult males and females have two distinctive "horns" (**Figure 32.12a;** see also Figure 32.11a), actually formed from hair. Rhinos use the horns to protect themselves and their young from predators and other rhinos. By 1981, the populations in the wild had been reduced to between 10 thousand and 15 thousand, and again reduced to about 3500 by 1987. Today only a few individuals survive in some protected areas in Africa. In less than 30 years, the species was almost exterminated in the wild.

In 1960, black rhinos were one of the "big five" on the list of big game for which hunters made safaris to Africa to shoot as trophies. Others on the list were the African lion, African elephant, Cape buffalo (*Syncerus caffer*), and leopard. Safari hunters then paid large sums of money to go to Africa and obtain licences to kill trophy specimens of each of the big five. But this hunting pressure, which has since stopped, did not lead to the extermination of black rhinos.

Figure 32.11

(a) Black rhinos (*Diceros bicornis*) were widespread and common in Africa in 1960 (orange area on the map in (b)). **(b)** Today their range (dark spots in orange areas) is much reduced, reflecting diminished populations. Note the oxpecker (*Buphagus africanus*) sitting on the rhino.

People have long used the horns of all species of rhino in different ways. In China, bowls made from rhino horn (Figure 32.12b) were believed to have magical properties in that they could remove or neutralize poisons. Travelling nobles were served wine in their own rhino horn bowls to minimize the chances of their being poisoned. In India and some other areas from India to Korea, powdered rhino horn was used as a fever suppressant. Contrary to popular belief, rhino

a.

b.

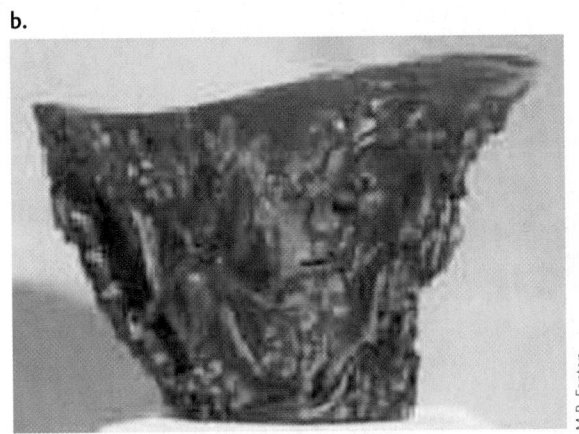

c.

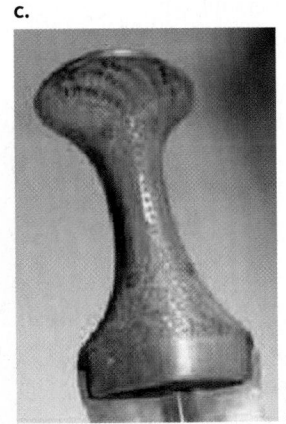

Figure 32.12
(a) A horn from a black rhino in Zimbabwe is shown with **(b)** a rhino horn bowl from China and **(c)** a jambiya with a rhino horn handle.

Figure 32.13

A Kalashnikov assault rifle (an AK), a weapon widely used in the poaching of animals in many parts of the world.

horn does not appear to have been used as an aphrodisiac, an early version of Viagra.

A tradition among some people of the Arabian Peninsula is the carrying of a jambiya or ceremonial dagger. Jambiyas with rhino horn handles (Figure 32.12c) were highly prized. In 1973, when the price of oil jumped from U.S.$4 to U.S.$12 a barrel, the ensuing "energy crisis" meant a larger market for jambiyas because more people could afford them. Increased cash flow and easy access to military weapons such as Kalashnikov assault rifles **(Figure 32.13)** provided an incentive and a means to kill rhinos. The epidemic of poaching started in northern Kenya and spread southward throughout the continent. Thus, poaching for their horns led to the catastrophic reduction in the populations of black rhinos. The large population of rhinos that had long survived in the presence of predators, including *Homo sapiens,* was not protected from extermination. In 1984, going for a walk at night around the headquarters of Mana Pools National Park in Zimbabwe almost always meant meeting a black rhino. By 1987, the rhinos were very scarce, and by 1990 they did not exist in the area.

The demise of black rhinos can only be attributed to human greed.

32.4b White-Nose Syndrome and Bats: A Different Challenge

In March 2006, at sites near Albany, New York, biologists counting bats hibernating in caves and abandoned mines were shocked to find thousands of dead bats where they had expected thousands of live ones. The bats had died from White-Nose Syndrome (WNS), which is caused by a cold-loving fungus (*Pseudogymnoascus destructans* (formerly *Geomyces destructans*)) that interrupted their rhythm of hibernation. Some infected bats were easy to recognize by the white funguslike

structures around their nostrils **(Figure 32.14)**. To survive hibernation, bats minimize the number of times they arouse from torpor because of the metabolic cost of waking up (raising the body temperature from 2–5°C to over 35°C). At most Canadian hibernation sites bats normally go about 90 days between arousals because each arousal costs them energy that they could use in 60 days of hibernation. WNS causes them to arouse much more often and exhaust their stores of body fat in January or February, well before spring and the re-emergence of insect prey.

The initial focal area for WNS in North America was specific sites around Albany. By March 2010, WNS had spread to underground hibernation sites in Ontario and Quebec, as well as to many other sites in the United States. In the intervening years WNS continued to spread. In June 2014, in Canada, WNS had not been found west of Wawa in Ontario and had not been reported from sites in Newfoundland and Labrador. But WNS has continued to spread west and south in the United States. The spread of WNS surely reflects the movements and behaviour of the bats.

At known hibernation sites, populations of Little Brown Bats (*Myotis lucifugus*) declined by over 95%. Professor Craig Willis from the University of Winnipeg and his colleagues demonstrated that the strain of the fungus causing WNS in North American bats originated in Europe. It was presumably inadvertently transferred to the sites near Albany by cave explorers or bat biologists. The European strain did not cause WNS in bats there, just as the North American strain did not cause WNS in bat species from North America.

Can the populations of Little Brown Bats recover? Probably not, because like most bats, Little Browns live in the slow life history lane (see Chapter 29). They reproduce slowly (a single young per year) and like most species of bats of the temperate regions, up to 60% of young do not survive their first year. The combination of low reproductive output and low survival of first year translates into low potential for increase of their populations.

There are 19 species of bats in Canada, not all of which are exposed to WNS because not all of them hibernate in underground sites (usually caves and abandoned mines). So, perhaps WNS does not mean the end of our bats, but likely the loss of more than half of the species. WNS is a stark example of how a widespread and abundant species can become endangered.

32.4c The Bay Scallop Is Affected by Overfishing of Sharks

Populations of organisms we harvest for food often show marked declines. The annual harvest of bivalve molluscs has been a local fishery in Chesapeake Bay in the United States and elsewhere along the eastern seaboard for hundreds of years. In 1999, populations of

Figure 32.14

Three Little Brown Myotis, one (middle) showing characteristic signs of infection by the fungus *Pseudogymnoascus destructans* (formerly *Geomyces destructans*) that causes white-nose syndrome (WNS) in bats that hibernate underground in the United States and Canada.

Figure 32.15

A handful of bay scallops (*Argopecten irradians*).

Figure 32.17

A cownose ray (Rhinoptera bonasus).

bay scallops (*Argopecten irradians;* **Figures 32.15** and **32.16**), a main target of the fishery, were very low. The immediate reason for the low populations was the impact of predation by skates and rays that feed heavily on bivalve molluscs. Skates and rays are tertiary consumers and in turn are eaten by larger elasmobranchs, specifically various species of sharks.

Among tertiary consumers, the cownose ray **(Figure 32.17)** showed a marked increase in population. Evidence from surveys on the U.S. Atlantic coast estimates an order-of-magnitude increase in populations of cownose rays, and the total population of 14 species of rays and skates exceeds 40 million. So the decline in scallop (and other bivalve) populations can be explained by the increase in predation by tertiary consumers, especially skates and rays.

The picture becomes clearer when the population data for the local great sharks are added to the mix **(Figure 32.18, p. 792).** Prolonged and intensive fishing of 12 species of sharks accounts for a 35-year decline in their populations (see Figure 32.18, top row). The sharks have been taken primarily for their fins and meat. In some parts of the world, shark fins sell for around U.S.$700 per kilogram and are used to make shark fin soup.

The data demonstrate how a century-old scallop fishery was effectively destroyed because of predation by tertiary consumers, whose populations, in turn, had been enhanced (see Figure 32.18, middle row) by the removal of top predators, the great sharks. The data illustrate a cascading ecological effect and demonstrate the potential long-term harm that our species can do to ecosystems and the species inhabiting them. The demise of bay scallops and other bivalves can be attributed to the impact of large-scale harvesting of marine resources. The late Ransome Myers and his colleagues documented this cascade of effects.

The examples above are merely samples from a long list of species. Evidence of declines of populations of native species can be found almost everywhere. Whether the root cause is overharvesting, introduced species, or destruction of habitat, species from whales to songbirds are threatened by human activity. What can we do about it?

STUDY BREAK

1. What is the significance of rhino horn in the conservation of these animals?
2. What connects the decline in the scallop fishery to sharks?
3. What is White-Nose Syndrome?

32.5 Protecting Species

The widespread recognition of trademarks such as the World Wildlife Fund (WWF) panda demonstrates how associating a cause with an icon can be very successful. It is not surprising that many conservation efforts began with a focus on one species—such as giant pandas (*Ailuropoda melanoleuca*), polar bears (*Ursus maritimus*), or redwood trees (*Sequoia sempervirens*).

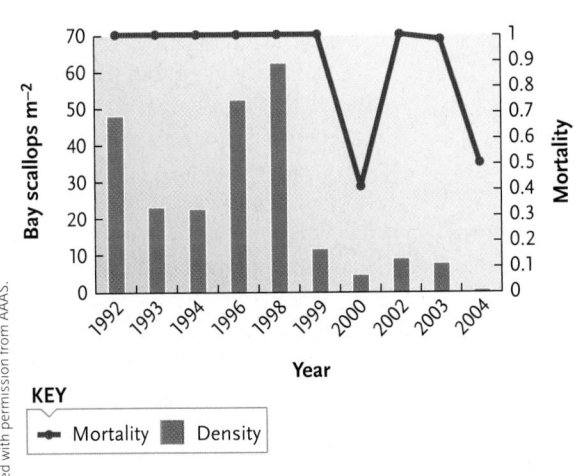

KEY

— Mortality ■ Density

Figure 32.16

Numbers of bay scallops off the east coast of the United States.

From Ransom A. Myers, Julia K. Baum, Travis D. Shepherd, Sean P. Powers, Charles H. Peterson. "Cascading Effects of the Loss of Apex Predatory Sharks from a Coastal Ocean," *Science*, vol. 315, Mar 30, 2007, pp. 1846–1850. Reprinted with permission from AAAS.

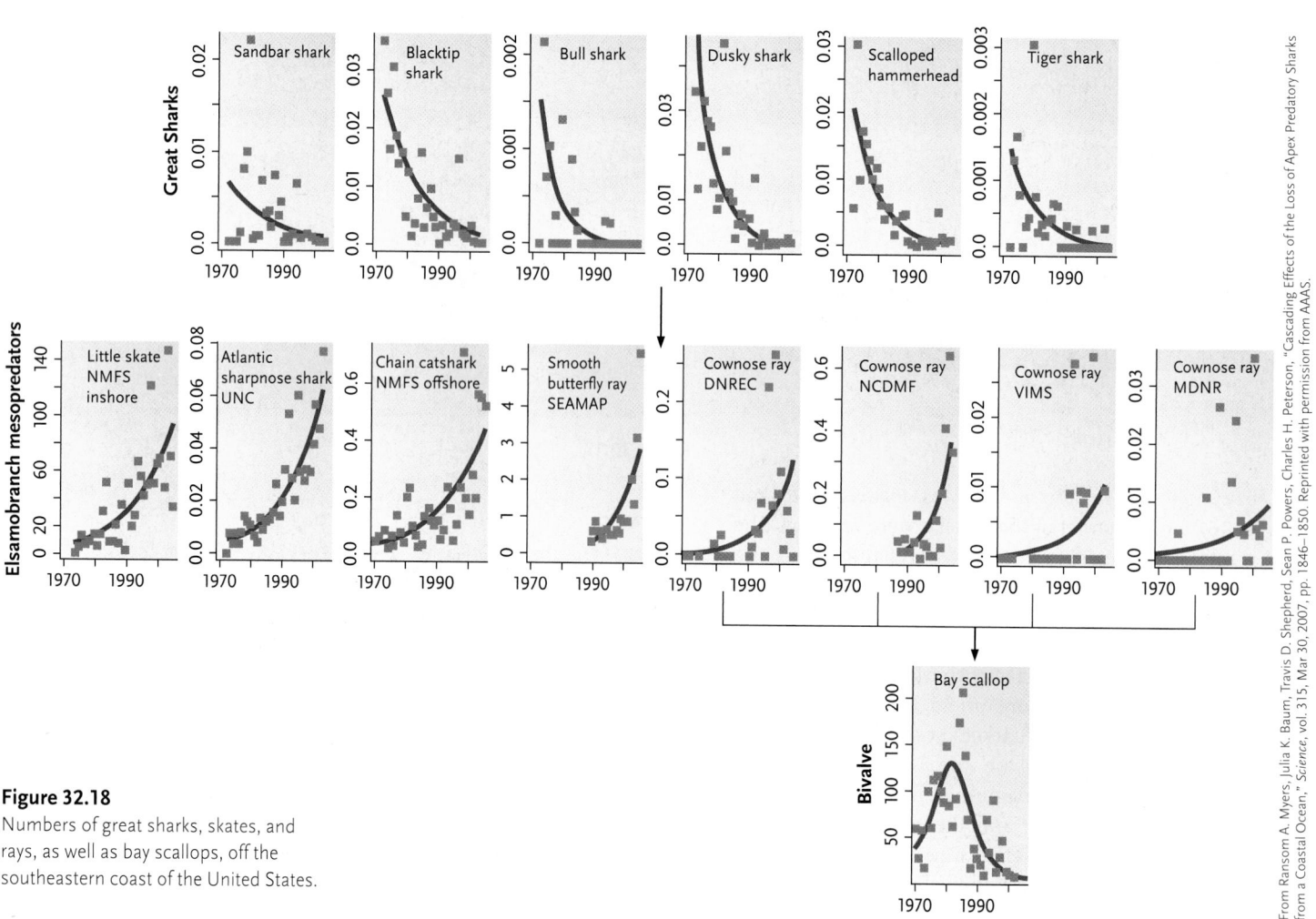

From Ransom A. Myers, Julia K. Baum, Travis D. Shepherd, Sean P. Powers, Charles H. Peterson, "Cascading Effects of the Loss of Apex Predatory Sharks from a Coastal Ocean," *Science*, vol. 315, Mar 30, 2007, pp. 1846–1850. Reprinted with permission from AAAS.

Figure 32.18

Numbers of great sharks, skates, and rays, as well as bay scallops, off the southeastern coast of the United States.

Hunting: Threat or Salvation?

We saw earlier (see Chapter 19) how the Linnaean system of nomenclature is used to name species. Once a species has a name, however acquired, it may benefit from protection under CITES, the Convention on International Trade in Endangered Species of Wild Fauna and Flora. But will data-based decisions about what counts as endangered be consistent and predictable? The answer is "yes" and "no." The example of black rhinos showed one situation in which protection under CITES did not work. There are others.

Also in Africa, the leopard (see the chapter-opening photograph) was accorded protection under CITES. The passing of the Endangered Species Act (ESA) in the United States (1972) precipitated an interesting situation: it

obliged Americans to "obey" the listing of leopards on CITES Appendix 1, which banned the importation of leopard skins, including those shot on safari hunts. The rationale for the listing was the belief that leopards were endangered and their survival was threatened by hunting.

There were quick, negative responses to the ban on importing leopard skins into the United States from two different groups. First were the hunting and related associations and lobbies whose members were anxious to be able to bring home trophies. Second, leaders and governments in many African countries that benefited from the hunts objected to the ban because safaris were (and still are) an important source of foreign

exchange. In many of these countries, "safari hunting areas" were set aside to accommodate visitors, and these large tracts of land also protected populations of nongame species and appropriate habitat.

What do the data show? Leopards are 40 to 80 kg, solitary cats that hunt by stealth. They are widespread in Africa but have been little studied. The estimate is that there are more than 700 000 leopards in the wild in Africa, with resident populations in all but very small countries with high human population densities. In 2000, Zimbabwe alone had a population of more than 16 000 leopards in the wild. The 1969 safari harvest of 6100 leopards throughout Africa and the export of their skins were not a threat to the population in

Zimbabwe, let alone to leopards in the whole continent.

Ecologists studied the population of leopards in the Matetsi Safari Area in Zimbabwe. Before 1974, the 4300 km² area was a cattle ranch whose operators made strong efforts to eradicate leopards to protect their livestock. After conversion to a hunting area, people on the first safaris rarely succeeded in shooting leopards. By 1984, the leopard population in the Matetsi Safari Area was 800 to 1000, and in 1988, the annual safari quota there was 3.6% (12 to 28 leopards). When leopards shot in the mid-1980s were compared with those taken in the 1970s, no change in leopard size was found. But by 1986, the average age of leopards taken as trophies was 5.4 years, compared with 3.2 years from the earlier period. These data show that leopards can persist even when subjected to heavy hunting pressure. On average, leopards live longer in a safari hunting regime than when they are being hunted in the context of predator control operations. Other evidence suggests that populations of leopards persist even in urban areas—trapping evidence suggests that resident leopards live in Nairobi, the capital of Kenya.

Leopards are an interesting example of human responses to conservation. Hunting or some other form of harvesting is not necessarily a threat to the survival of some species. Indeed, some harvesting may be critical to the livelihood of some people and can advance efforts to protect some species. But decisions about harvesting made in one part of the world can influence what happens elsewhere.

Today there are quotas for the numbers of leopards that can be harvested in different countries in Africa. Safari hunters must obtain licences to take trophies, and skins exported must be accompanied by paperwork showing that the harvest was legal. The documentation allows a citizen, for example, of Canada or of a European Union country, to import a leopard skin. This was not possible in the United States in the 1970s, but it is now. In Africa, local farmers are permitted to kill "problem" animals that threaten their livestock or themselves and their families and may be supported in this by government officials.

Key elements in the success of harvesting include having data about the population of organisms, the rates of reproduction, and the rates of harvest. Enforcement of quotas is essential if this approach is to succeed. Legal harvest quotas do not require people who object to hunting to be hunters. Trophy hunting is not the exclusive preserve of countries in Africa. On April 3, 2007, *The Globe and Mail* reported that the economy of Nunavut received C$2.9 million from polar bear **(Figure 1)** hunting. Hunters can pay U.S.$20 000 for a polar bear hunt.

In 1992, saola **(Figure 2)** made the news as one of the first "new" species of large mammals to be discovered in recent times. These goatlike animals live in a restricted area of Vietnam, where they have been and are hunted by local people. Saolas are rare, and little is known about them. There are no quotas for the local hunters, and it is not practical to enforce a ban on their harvest. In reality, we probably lack critical information about the biology of many species of wildlife today. However, once they have names, they have a chance of being protected.

Figure 1
Polar bear, Ursus maritimus.

Figure 2
(a) Saola (Pseudoryx nghetinhensis) *and (b) its distribution.*

From Richard Stone, "The Saola's Last Stand," *Science,* vol. 314, Dec. 1, 2006, pp. 1380–1383. Reprinted with permission from AAAS.

The lure of conservation movements that focus on charismatic species is very strong. But charismatic organisms may not need protection, whereas some species that are unattractive, dangerous, or mundane are in desperate need of our assistance. Unfortunately, mundane, ugly, and dangerous (to us) species are unlikely to serve as a call to arms (or to attract financial support). Worldwide, the WWF panda is one of the most recognized logos, whether or not pandas are in the neighbourhood.

A critical first step toward conservation is the development and adoption of objective, data-based criteria for assessing the risk posed to different species. This process has been developed on several fronts around the world. The criteria and assessment procedures perfected by the International Union for Conservation of Nature (IUCN) are used widely. There are many records of success, but there also are many examples of species with which and situations in which we have failed. Making arguments based on data does not guarantee success. Using a data-based approach, some species emerge as being in need of protection, but others do not. Being rare or unusual, by itself, will not warrant protection. The species concept and the Linnaean system of nomenclature (see Chapter 19) are fundamental to conservation.

In Canada, recommendations about the conservation status of species involve the Committee on the Status of Endangered Wildlife in Canada (COSEWIC). The definition of wildlife includes plants and animals. Like IUCN, COSEWIC recognizes six categories for assessing species at risk:

- *Extinct* wildlife species no longer exist.
- *Extirpated* species no longer exist in one location in the wild but occur elsewhere.
- *Endangered* species face imminent extirpation or extinction.
- *Threatened* species are likely to become endangered if limiting factors are not reversed.
- *Special concern* species may become threatened or endangered because of a combination of biological characteristics and identified threats.
- *Data deficient* is a category used when available information is insufficient either to resolve a wildlife species' eligibility of assessment or to permit an assessment of its risk of extinction.

A seventh category—*not at risk*—is used to identify species not at risk of extinction under current circumstances.

COSEWIC members vote on the appropriate conservation category for each species whose status they review. The members consider the area of occupancy, which is an indication of the range of a species and the availability of suitable habitat. They take into consideration population information, including trends in the numbers of organisms, correcting for species that show extreme fluctuations in numbers from year to year. They consider the demographics of the species and the variability in the habitat where the species occurs. Generation time is also considered, along with specific habitat features that may be essential for the species' survival. Data on population size, particularly the numbers of reproducing adults, are important, as well as risks to the species' survival.

In a biological context, the criteria used by COSEWIC (and similar agencies elsewhere) are familiar to population biologists (see Chapter 29). The data describe the numbers of individuals in the population, fecundity, mortality, and the intrinsic rate of increase. Carrying capacity is also important, as is the area (range) over which the species occurs. These criteria are designed to promote data-based decisions about the conservation status of species.

STUDY BREAK

1. What is IUCN? What role does it play in conservation?
2. What criteria would identify a species as endangered? Give an example.

32.6 Protecting What?

Before data are used to address questions of species-at-risk status, conservation biologists must decide about eligibility. The conservation jargon for this is "designatable unit." Are the organisms "real" species? Are they subspecies? Are they distinct populations? Are they really Canadian? Do they regularly occur in Canada or perhaps turn up here by accident? If the species does not breed here, is the habitat they use in Canada essential to their survival? Most species of wildlife in Canada occur close to the border with the United States, and many species widespread in the United States just make it into Canada. In some cases, a distinct population is treated as a designatable unit. Distinct populations may be recognized by their geographic distribution and/or their genetic structure.

Questions about what units are designatable hearken back to the definition of species (see Chapter 19). Off the west coast of Canada, striking differences in behaviour can be used to distinguish between two "kinds" of killer whales. The *resident* killer whales eat mainly fish and often echolocate. The *transient* killer whales eat mainly marine mammals and rarely produce echolocation signals. Furthermore, repeated sightings of recognizable individual whales indicate that different groups of these animals live in different areas along the coast (**Figure 32.19**).

a.

b.

c.

M.B. Fenton

Figure 32.19
Three views of a killer whale (*Orcinus orca*). **(a)** A captive animal in Vancouver, **(b)** a wild orca swimming off the Queen Charlotte Islands, and **(c)** a Haida representation.

In reviewing the conservation status of killer whales, COSEWIC recognized different designatable units based on behaviour and geography **(Figure 32.20, p. 796).** The different units faced different threats to their survival.

Questions about what to protect often reflect different realities of biology. Migrating birds may be blown off course and end up in southern Ontario instead of their usual habitat much farther south. Marine birds or mammals may feed in Canadian

An Endangered Species

Banff Springs snails, *Physella johnsoni* **(Figure 1),** live and eat algae in five hot springs on Sulphur Mountain in Banff National Park, Alberta. Not very long ago, Banff Springs snails were found in nine springs. In 1996, the total population of snails was about 5000. Water temperatures in the springs occupied by the snails range from 26 to 48°C, but temperatures less than 44°C seem best for them. Their very limited occurrence makes them vulnerable to extinction (COSEWIC, 2000).

Humans appear to be the main threat to the survival of Banff Springs snails. By discarding unsightly (to humans) accumulations of algae from pools, people have killed some snails that were in the algal mats. Changes in the patterns of water circulation may subject some snails to high temperatures that could be lethal. Well-wishers

that throw copper coins into the pools may have harmed snails because of contamination arising from the interaction of copper with sulphurous water in the springs.

Other impacts of people are not clear, but one threat is entertaining to contemplate—people "skinny-dipping" in the pools are thought to threaten the snails. Some skinny-dippers have been caught and charged. Bathers in the pools, clad or unclad, may have crushed snails while getting into or out of the water. Bathers doused in sunscreen or insect repellents may have

introduced chemicals into the snails' habitat and further reduced their populations.

Banff Springs snails are neither charismatic nor prominent, but the data-based approach to decision making has provided the basis for identifying them as endangered.

Degner, M. & L, used with permission of Parks Canada

Figure 1
Banff Springs snail, Physella johnsoni.

CHAPTER 32 CONSERVATION OF BIODIVERSITY

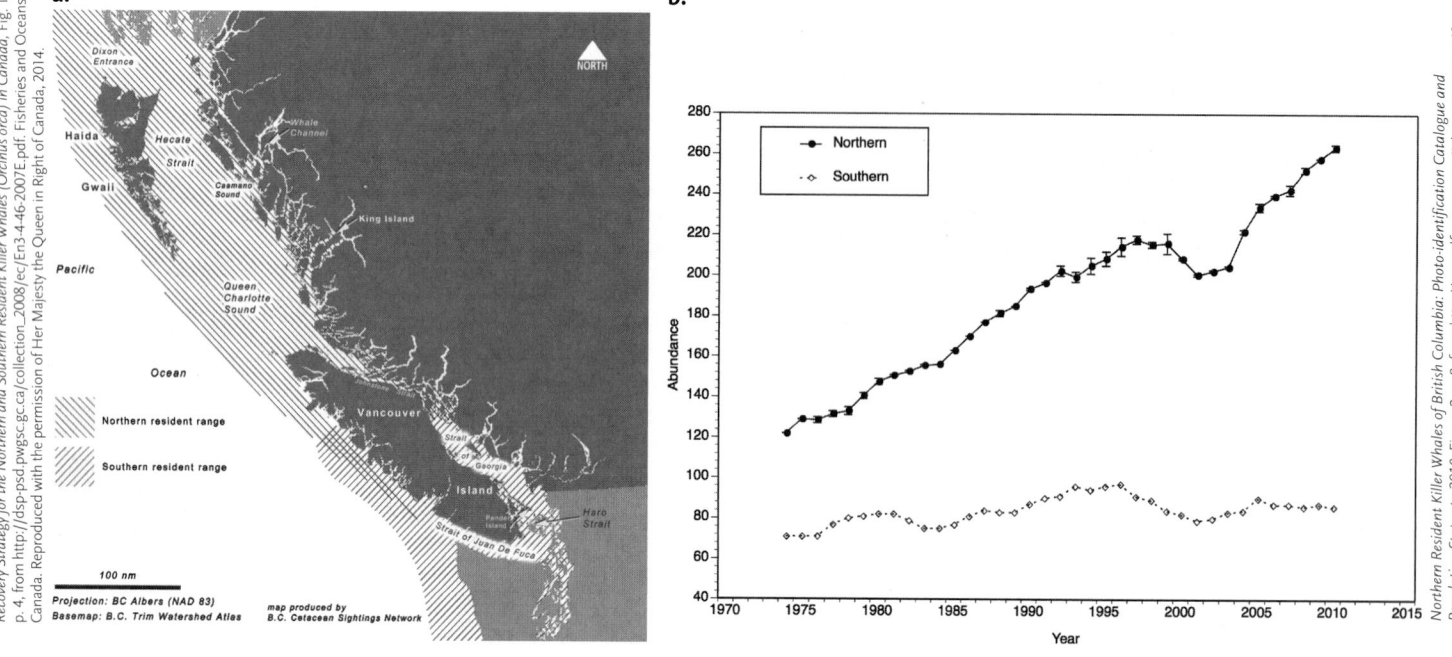

Figure 32.20

(a) The population distribution (designatable units) of killer whales off the coast of British Columbia and **(b)** estimates of population sizes of northern and southern resident killer whales (British Columbia).

waters but breed elsewhere. Many organisms commonly hitchhike, using ocean vessels, aircraft, or automobiles as vehicles of dispersal. But some hitchhikers, for example, some snails, travel with birds, making the association and the dispersal more "natural."

People can be quick to try to protect species they consider to be important or distinctive. In 2003, the Ontario Ministry of Natural Resources reported four to six white-coloured moose (*Alces alces*) among the approximately 1900 moose in two wildlife management areas near Foleyet in northeastern Ontario. Should white-coloured moose be protected? There was local support for protecting the moose, animals that have cultural and spiritual significance for First Nations communities. White moose have been reported from other places in northern Ontario, Newfoundland and Labrador, and elsewhere. Although the population of white moose is small and widespread, there is no evidence that they are a designatable unit. In Canada, they have not been accorded special protection.

There is protection at the international level. CITES, the Convention on International Trade in Endangered Species of Wild Fauna and Flora, plays a pivotal role in protecting species. International trade in wildlife is a leading threat to conserving biodiversity because in addition to directly affecting local populations of threatened species, it can also spread infectious diseases and promote the spread of invasive species. Membership in CITES includes 180 countries, and CITES tries to regulate trade in almost

36 000 species. Basic, accurate, and reliable biological data about species are essential for informing decisions about which species should be protected. Yet, decisions about what species are protected by CITES are political and not necessarily uniformly acclaimed. Between 2014 and 2016, the annual budget of the secretariat of CITES averages U.S.$6.2 million, coming from donations. Budget restrictions influence the effectiveness of CITES at the secretariat level by affecting the capacity for detailed collection and analysis of basic data.

At local levels, however, unmonitored trade in wildlife continues to occur openly, often in clear violation of CITES. Orchids are a clear example. In a period when CITES records showed 20 cases of orchids listed by CITES coming into Thailand from four neighbouring countries (Lao People's Democratic Republic, Myanmar, Cambodia, and Vietnam), 168 cases were observed in local markets. At one market in the Mekong Delta, one trader can sell more orchids in one day than reported by CITES over a period of nine years!

STUDY BREAK

1. Give an example of a designatable unit that is a species.
2. Give an example of a designatable unit that is a population.
3. Is designatable unit synonymous with species?

Who Gets Protection?

Being recognized as rare and considered to be endangered does not necessarily translate into protection.

Because the Endangered Species Act (ESA) in the United States does not protect hybrids, this can affect the conservation of, for example, the Florida panther **(Figure 1),** a subspecies of cougar. Cougars, also known as panthers, used to occur widely in North, South, and Central America. Although still widespread in some areas, the current range of cougars in most of the United States and Canada is much less than it was when Columbus arrived in the New World in 1492. Florida panthers, a small population recognized as a subspecies, occur mainly in the Florida keys. Florida panthers were protected under the ESA.

Using techniques of molecular genetics, biologists determined that Florida panthers carried the genes of cougars from South America. This situation probably arose when panthers originally caught in South America were brought to the United States as zoo animals or for display in circuses or animal shows. Some of

these animals escaped and interbred with local Florida panthers. Florida panthers with genes from South American cougars are technically hybrids and are therefore not protected by the ESA.

There are many other examples of situations in which genetic tools allow clearer delineation of boundaries between populations (designatable units) and species. In some cases, however, removal of protection from other "species" because of their genetic status can lead to their extinction. *Ammodramus maritimus nigrescens,* or the Dusky Seaside Sparrow, was previously considered to be a distinct form living in Florida. When genetic evidence showed that these darker animals were not genetically distinct, they lost their protected status and have virtually disappeared.

Other species, such as roundnosed grenadier, have suffered calamitous declines in population. These codlike fish **(Figure 2)** were taken in large numbers after cod populations had declined **(Figure 3).** The species was on the verge of extinction even before much was known about it. We do know that round-nosed grenadiers are late to mature, and their populations are slow to recover.

Although round-nosed grenadiers and at least four other species meet the IUCN criteria for listing as endangered, these fish have not been

Figure 2
Coryphaenoides rupestris, a round-nosed grenadier.

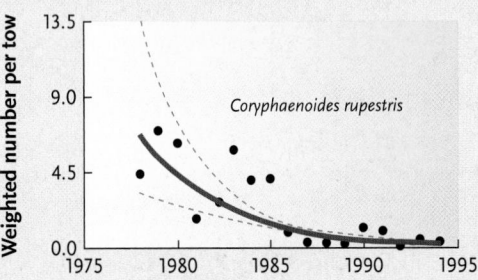

Figure 3
Captures of round-nosed grenadiers.

and are not protected. Fisheries and Oceans Canada has not supported a move to protect round-nosed grenadiers. Changing fishing practices to avoid catching the few remaining round-nosed grenadiers is not economically feasible when other species are still being caught in sufficient numbers to justify a continued fishery. The situation differs only from that facing the Little Brown Bat in that the round-nosed grenadiers have been the targets of an active fishery.

We have seen that the hunt for polar bears can bring significant income to the economy of Nunavut, and the same is true of other jurisdictions within the bear's range.

(Continued)

Figure 1
A Florida panther (Felis concolor coryi).

Who Gets Protection? (*Continued*)

There are distinct populations of polar bears **(Table 1)** within Canada's jurisdiction. The occurrence of bears in political jurisdictions including Canada, the United States, Russia, Iceland, Denmark (Greenland), Norway, Finland, and Sweden makes it more difficult to protect them. The apparent vulnerability of the bears to climate change and their value as trophies may combine to hasten their demise.

Table 1 — **Status of Canadian Polar Bear Populations (January 1997)**

Population	% Females in Harvest	Number	Sustainable Annual Kill	Mean Annual Kill	Environ. Concern	Status	Quality of Estimate	Degree of Bias	Age of Estimate	Harvest/ Capture Data
Western Hudson Bay	31	1200	54	44	None	S[a]	Good	None	Current	Good (>15 yr)
Southern Hudson Bay	35	1000	43	45	None	S[a]	Fair	Moderate	Old	Fair (5–10 yr)
Foxe Basin	38	2300	91	118	None	S[a]	Good	None	Current	Good (>15 yr)
Lancaster Sound	25	1700	77	81	None	S[a]	Fair	None	Current	Good (>15 yr)
Baffin Bay	35	2200	94	122	None	D?[b]	Fair	None	Current	Fair (>15 yr)
Norwegian Bay	30	100	4	4	None	S[a]	Fair	None	Current	Good (>15 yr)
Kane Basin	37	200	8	6	None	S	Fair	None	Current	Fair (>15 yr)
Queen Elizabeth	–	(200?)	9?	0	Possible	S?[b]	None	–	–	–
Davis Strait	36	1400	58	57	None	S?[b]	Fair	Moderate	Outdated	Good (>15 yr)
Gulf of Boothia	42	900	32	37	None	S[a]	Poor	Moderate	Outdated	Good (>15 yr)
M'Clintock Channel	33	700	32	25	None	S[a]	Poor	Moderate	Outdated	Good (>15 yr)
Viscount Melville Sound	0	230	4	0	None	I	Good	None	Current	Good (>15 yr)
Northern Beaufort Sea	43	1200	42	29	None	S	Good	None	Recent	Good (>15 yr)
Southern Beaufort Sea	36	1800	75	56	None	S	Good	Moderate	Recent	Good (>15 yr)

[1]D = decreasing; I = increasing; S = stationary; ? = indicated trend uncertain.

[a]Population is managed with a flexible quota system in which overharvesting in a given year results in a fully compensatory reduction to the following year's quota.

[b]See text at link below, "Population Size and Trend," for discussion.

Source: *COSEWIC Assessment and Update Status Report on the Polar Bear Ursus maritimus in Canada,* 2002. http://www.sararegistry.gc.ca/document/default_e.cfm?documentID=248. © Her Majesty The Queen in Right of Canada, Environment Canada, 2014. Reproduced with the permission of the Minister of Public Works and Government Services Canada.

32.7 The Downside of Being Rare

Whether the commodity is coins, stamps, antiques, or endangered species, as soon as something is rare enough, there is a market for it. This "get them while they last" attitude is exemplified by trade in *Leucopsar rothschildi*, Bali Starlings **(Figure 32.21)**. This bird, another island species, faces immediate extinction, but it is in high demand as an exotic pet. In 1982, when there were fewer than 150 individuals in the wild, 35 were for sale as pets, 19 in Singapore, and 16 in Bali.

Rare species may also be in demand for use of their body parts in traditional medicine. One stark example is the swim bladders of *Bahaba taipingensis,* the Chinese bahaba. At a time when fewer than 6 individuals are caught each year, more than 100 boats are trying to catch them. The swim bladders are used in traditional medicine. They are worth at least seven times their weight in gold. Shark fins are even more valuable. These are extreme examples similar to the earlier story about rhino horns and jambiyas.

Before criticizing and condemning the users or consumers of jambiyas or Chinese bahaba swim bladders, think about the overall impact of our lifestyle on other species of animals and plants. Of particular note is an insatiable demand for energy. Are sport utility vehicles necessary? Personal watercraft? Snowmobiles? All-terrain vehicles? The list goes on. Is a Canadian as justified in buying a large SUV as a North Yemenese a jambiya with a rhino horn handle? Once again, might (the ability or capacity to do something) may not be right.

Figure 32.21
A Bali Starling (*Leucopsar rothschildi*).

STUDY BREAK

1. What is the value of the International Union for Conservation of Nature (IUCN)?
2. Why do species description and formal naming affect the Convention on International Trade in Endangered Species of Wild Fauna and Flora (CITES)?
3. What is the difference between an extinct species and an extirpated species? Give an example of each.

32.8 Protecting Habitat

It is obvious from many of the examples above that protecting species has not been entirely successful as a conservation strategy. As a species, we are much better at killing than we are at conserving. Whether this is direct or indirect, the end result can be the same. It is also clear that destruction of habitat is an effective way to remove a species. For example, populations of mosquitoes can be limited by denying them places to lay their eggs. This is a common theme in public education programs designed to reduce the incidence of West Nile virus (or other mosquito-borne diseases).

Is protection of habitat an effective strategy? The answer can be "yes," particularly for species that are not motile. Many species of plants have specific habitat requirements. From trees to shrubs, forbs, ferns, and mosses, we know that we can protect species by protecting habitat. Furthermore, protecting large tracts of habitat can also protect large, mobile species. Rainforests, whether tropical or temperate, are examples of habitats that can be flagships for protection and conservation. They are also considered by many to be storehouses of wealth associated with biodiversity, from building materials to compounds of pharmacological value.

The case of the black rhino demonstrated how a species targeted for harvesting can be driven to the brink of extinction even when it is protected (or lives in national parks or game reserves). *Panax quinquefolius,* American ginseng, is another target species, now endangered in Canada because of harvesting. The species used to grow wild from southwestern Quebec and southern Ontario south to Louisiana and Georgia. This 20- to 70-cm-tall perennial is long lived in rich, moist, mature, sugar maple–dominated woods. Although the species has been listed on Appendix II of COSEWIC since 1973, populations have continued to decline. In 2000, there were 22 viable populations in Ontario and Quebec, but none were secure. Black rhinos and ginseng were common about 50 years ago, but by 2008, both demonstrated the risks of being rare and expensive. They are also examples of the need for immediate on-the-ground enforcement of regulations and laws protecting species and habitats.

Protecting habitats can be most challenging in areas with larger human populations. All of the viable populations of American ginseng in Ontario and Quebec were close to roads, making the plants vulnerable to anyone who knew about them and wished to take advantage of the economic opportunity they presented. *Sorex bendirii,* the Pacific water shrew, is another example of a species whose future in Canada is threatened by expanding human populations and the associated value of real estate **(Figure 32.22).**

In British Columbia, expanding human population and the wine industry in the southern Okanagan

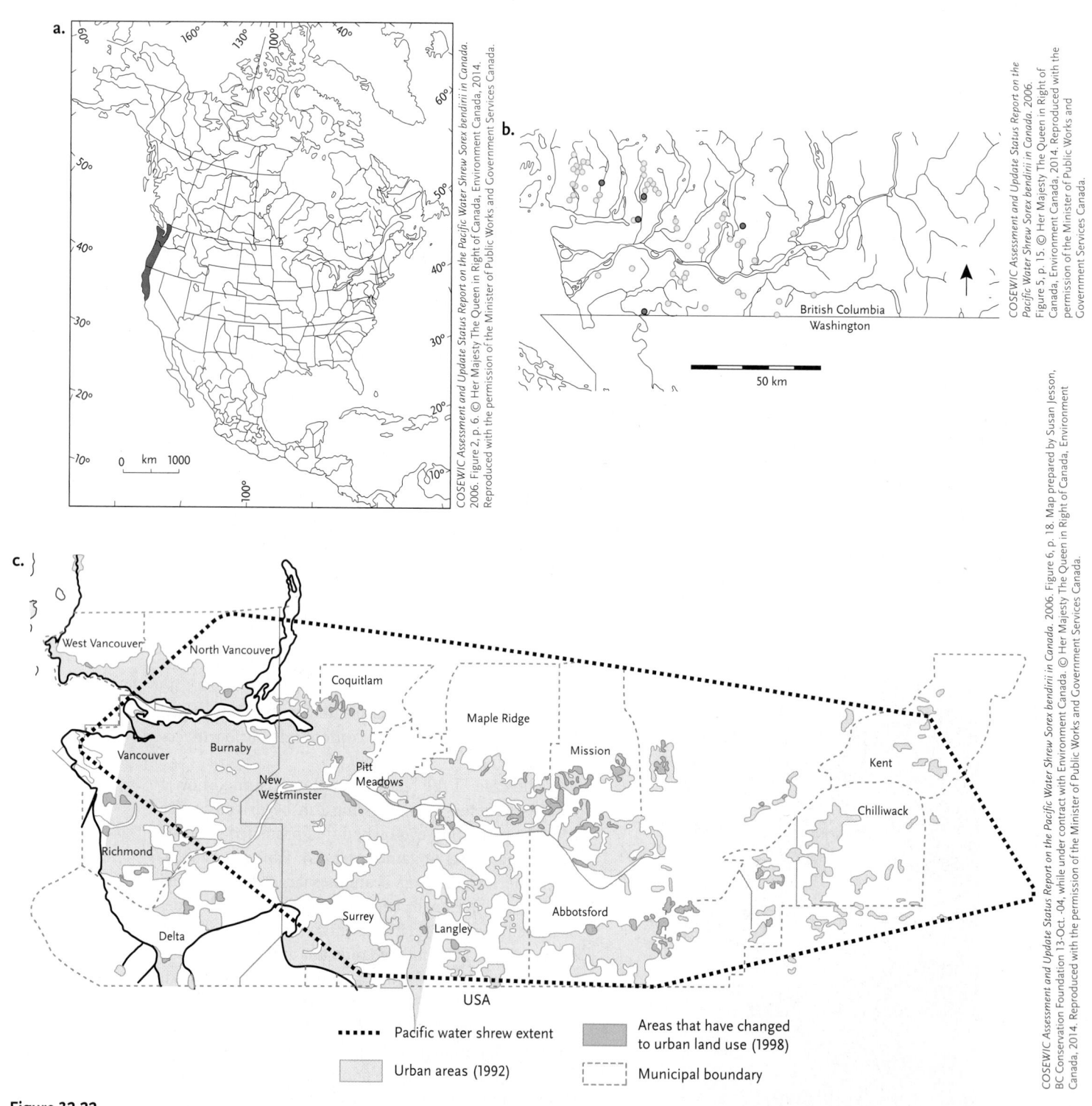

Figure 32.22

(a) The distribution of *Sorex bendiri,* the Pacific water shrew. **(b)** Lower Fraser Valley locations where it was found (solid circles) or not found (open circles). **(c)** For comparison, the same area is shown with changes in the availability of urban lands in 1992 and 1998. Data on the map are from 2004.

Baseline Thematic Mapping Present Land Use Mapping at 1:220 000.

Figure 32.23
Antelope bush, *Purshia tridentata*, showing **(a)** the bush and **(b)** a cross-section of the stem. These woody shrubs have long life spans, and the ecosystem they typify is home to a variety of species.

Valley have combined to dramatically reduce a local ecosystem dominated by antelope bush **(Figure 32.23)**. The antelope bush system, one of the most endangered ecosystems in Canada, is home to a number of species of plants and animals whose future is now threatened by the demise of the habitat they require. The boom in real estate for people looking for retirement properties, more than just the density of human populations, is a key factor in this situation. Meanwhile, in southern Ontario, the demand for real estate to accommodate the expanding housing and business market is reducing both the available natural habitats and farmland.

STUDY BREAK

1. Give examples of how protecting a habitat can work.
2. How do market forces influence the abundance of species?

32.9 Effecting Conservation

Today we face many challenges when trying to protect biodiversity. Too many of the immediate threats are the direct or indirect consequences of human activities. We must protect species by acting at levels ranging from species to populations and habitats.

32.9a Human Population Is a Root Problem in Conservation

One fundamental root cause of declining biodiversity is the human population and the energy and habitat consumed in trying to feed, house, and protect our flourishing species. Visit the website http://www.ined.fr/en/everything_about_population/world_population_me/ and use it to determine the estimated human population in the year you were born and then for the years in which your parents and grandparents were born. Even when many people are killed, the momentum of our population increase does not slow down. The December 2004 tsunami killed approximately 250 000 people, at a time when the world population was estimated at 6 billion. By comparison, the 1883 explosion of Krakatoa (and resulting tsunamis) is thought to have killed 35 000 people when the global human population was about 1.5 billion. If these estimates are correct, $4.1 \times 10^{-3}\%$ of the human population at the time was killed by the 2004 tsunami and $2.3 \times 10^{-3}\%$ by the explosion of Krakatoa. Neither calamity caused the human population growth curve (see Chapter 29) to waver.

If human population growth continues at the same rate as it is growing now, it will double in 40 years. However, studies show that our population is not growing as quickly as it did during much of the twentieth century. The United Nations Development Program (UNDP) has released data on human fertility (the total number of births per woman) for 162 countries **(Table 32.1, p. 802)**. Compared with 1970–75, 152 countries had lower human fertility in 2000–05, 3 countries showed increases in fertility, and 7 showed no change.

Concerned about the global population and its effect on Earth, world leaders adopted the United Nations Millennium Development Goals in 2000, committing their nations to achieving the following goals by 2015:

- ending poverty and hunger,
- universal education,

Table 32.1	Variations in Fertility Rate (Total Births per Woman): A Sample of UNDP Data for 162 Countries		
Country	Human Development Index (HDI) Rank	1970–75	2000–05
Norway	2	2.2	1.8
Canada	4	2.0	1.5
United States	12	2.0	2.0
Portugal	29	2.7	1.5
Brazil	70	4.7	2.3
China	81	4.9	1.7
Indonesia	107	5.2	2.4
India	128	5.4	3.1

- gender equality,
- child health,
- maternal health,
- combatting HIV/AIDS,
- environmental sustainability, and
- global partnerships.

These goals can be achieved only if reproduction is controlled (see Chapter 29). Go to the United Nations Millennium Goals website at http://www.un.org/millenniumgoals/bkgd.shtml to see how we are faring. In 1994, the United Nations held the International Conference on Population and Development (ICPD), which set a target for global investment in family planning. By 2004, the amount spent had fallen to 13% of this target. Consequently, family planning information and devices (usually for fertility control) are not readily available in many of the lowest-income countries. In 1950, Sri Lanka and Afghanistan had the same population. Sri Lanka began strong efforts to make family planning available in culturally acceptable ways. This did not happen in Afghanistan. By 2050, Afghanistan will have four times as many people as Sri Lanka. The solution centres around controlling the fertility of women, but more particularly on giving them the power to control their own fertility in culturally acceptable ways. As seen in Chapter 29, the growth potential of a population is determined by the numbers of females of reproductive age. Why females? Because females are the limiting step in reproduction—the ones who produce the eggs or young.

32.9b Signs of Stress Show Up on Systems and on Species

People's demand for food, water, and energy puts thousands of other species at risk. We do not have to look far to see examples of species and ecosystems under stress (see Chapter 31). For example, we are losing birds. We know this because for years, bird-watchers and ornithologists have counted them and monitored their behaviour and activity. Locally, birds are affected by changes in habitat availability as cities and towns and their suburbs expand into adjoining land. Birds also lose habitat when agricultural operations expand to increase productivity. Birds that make annual migrations from temperate areas of the world to tropical and subtropical ones must survive the changes that accumulate across their entire circuit of habitats, each one essential to their survival.

Avian influenza (also called bird flu) is a looming crisis for humans, one that appears to involve birds as central players. The issue here is another one involving basic biology, namely the outcome when a disease-causing organism jumps from one species (host) to another. Bird flu could have as much to do with our insatiable demand for poultry as food as it does with birds. In 2006, 12 billion chickens were farmed in China. Worldwide, poultry farms housed over 100 billion broiler chickens. Raising organisms at very high densities (see Chapter 29) provides an ideal setting for the spread of disease. Humans have responded to the threat of bird flu by wholesale slaughter of fowl, raising concerns about the roles played by migrating birds, and trying to develop a vaccine that will protect humans from bird flu. All involve basic biology.

Drylands are arid, semiarid, and subhumid areas where precipitation is scarce and more or less unpredictable. In drylands, the combination of high temperatures, low relative humidities, and abundant solar radiation means high potential evapotranspiration. Drylands cover approximately 41% of Earth's land surface and are home to about 38% of the human population. Drylands are not just a problem of deserts but cover large expanses, for example, of Canada's prairie provinces. However, between 10 and 20% of the drylands are subject to some form of severe land degradation, directly affecting the lives of at least 250 million people. Climate change, combined with increasing pressure on water resources for these people, their crops, and their animals, compounds the problems that confront them. Competition for limited resources, such as water, can generate local and international strife.

We have seen that complexity is an important and pervasive feature of ecosystems. Biodiversity is intimately associated with complexity, and disruption of this complexity often translates into reduced biodiversity and decay of ecosystems. Ironically, many social and economic systems that humans have developed are also subject to disruption by stress. This places the onus on our species to develop sustainable operations, whether in the area of agriculture, resource use and exploitation, or conservation.

1. How is reproductive effort different between males and females in birds and in mammals?
2. List the United Nations Millennium Development Goals.
3. How are drylands at risk?

32.10 Taking Action

It is easy to believe that nothing can change, that as individuals we have no power. Yet we can also think of things that have changed dramatically in a relatively short time. Two good examples are the abolition of slavery and the emancipation of women, proving humans' capacity for effecting change. On a more local level, the acceptance of the use of tobacco in public has declined remarkably in the last 20 years—in Canada and elsewhere. We also have seen the abolition of capital punishment and much more ready access to abortion in Canada.

But none of these changes is universal. In the daily news we find stories about people living in virtual slavery, of people executed in public, of women with few or no rights in their home countries. To complicate the matter, not everyone agrees that the changes listed above are for the better.

Effecting changes in our approach to conservation means identifying the root causes for the erosion of biodiversity and the things that are impediments to conservation. This means starting by changing our own lifestyles, including the food we eat and our use of energy. We must be wary of simple, and often misleading, solutions and avoid blaming someone else as a way of self-exoneration. We must respect the rights of others; use education and training to become informed; and learn to be objective, to examine and evaluate data or evidence. The outpouring of support for victims of the 2004 tsunami demonstrated that humans have great empathy for their fellows, and we need to extend this concern to the other species with whom we share the planet.

We have seen that action is needed at the species and the habitat level, and there is a propensity to focus more on species. But in the human view, all species are not equal. The 2006 IUCN list of threatened species shows that whereas 20% of the described species of mammals were listed as threatened, only 0.07% of the insect species received this level of attention. Other interesting numbers from this table are 12% of described species of birds listed as threatened, 4% of fish species, 3.5% of dicotyledonous plants, and 0.006% of species of mushrooms. In Canada, the same situation prevails, with mammals and birds dominating the list of threatened species, with other taxa receiving less attention. Do these data about threatened species mean that mammals are more vulnerable than insects? That we care more about mammals than about insects? Or does it mean that there are more "experts" to offer opinions and data about mammals than about insects? Are the possibilities mutually exclusive?

Biology can be at the centre of the movement to achieve conservation of biodiversity while being part of our efforts to achieve sustainable use of the resources we need as a species **(Figure 32.24)**. Conservation begins at home when we modify our lifestyles and become active on any front, from protecting local habitat and species to protecting charismatic species elsewhere. To better appreciate the situation, try to answer the questions posed in **Figure 32.25**. Elephants are an excellent example of how the objectivity that can be inherent in data is vulnerable to emotional responses.

Figure 32.24
Eat yourself out of house and home—like this African elephant (*Loxodonta africana*) trekking across the shore to Lake Kariba.

M.B. Fenton

a.

b.

c.

Figure 32.25
To understand some of the dilemmas facing conservationists, use the Internet to explore the situation of African elephants. **(a)** How many species are there? What are the populations in the wild? What products from elephants do we use **(b)** and **(c)**? Are elephants endangered? How can they be protected? What are the main threats to their survival?

Review

32.1 Extinction

- A species is said to be extinct when there are no living representatives known on Earth. Conservation organizations usually say that a species is extinct when it has not been seen or recorded for 50 years.

- Mass extinctions occurred at the end of the Ordovician and the beginning of the Devonian, at the end of the Devonian, at the end of the Permian, at the end of the Triassic, and at the end of the Cretaceous. The Permian extinction was the most severe, and more than 85% of the species alive at that time disappeared forever, including the trilobites, many amphibians, and the trees of the coal swamp forests. Dinosaurs did not survive the extinction that occurred at the end of the Cretaceous.

- The extinction at the end of the Cretaceous is believed to have been caused by an asteroid impact. Dust clouds resulting from the impact blocked the sunlight necessary for photosynthesis, setting up a chain reaction of extinctions that began with microscopic marine organisms and finished with dinosaurs (as well as many birds and mammals).

- Measured by time on Earth, multituberculates were the most successful mammals.

32.2 The Impact of Humans

- Species (particularly flightless birds) that are confined to islands often have small populations and are unaccustomed to introduced terrestrial predators (such as cats, dogs, rats), making them vulnerable to extinction when human populations settle and expand.

- The demise of the calvaria trees (*Sideroxylum majus*) on the island of Mauritius will occur even though the trees continue to bloom and produce seeds. The extinction of the tree is linked to the earlier extinction of Dodos, since *S. majus* seeds had to pass through the Dodo's digestive tract to germinate.

32.3 Introduced and Invasive Species

- Stephen's Island wrens were flightless and unaccustomed to predators. The population on the island was small, and it was easy for Tibbles the cat to catch and kill the remaining 20 birds.

- Since about 1880, ships have regularly used ballast water. A survey of ballast water in 159 ships in Coos

Bay, Oregon, revealed 367 species of organisms representing 19 animal phyla and 3 plant divisions. When ships empty their ballast, the organisms in it are introduced to the system where the ship is anchored.

- RUE, resource use efficiency, is measured as carbon assimilation per unit resource. Many invasive plant species in Hawaii are more efficient than native species. This means that conserving native biodiversity in the face of invasive and introduced organisms is a pervasive problem.

32.4 How We Got/Get There

- Horns of rhinos, particularly black rhinos (*Diceros bicornis*), have been used to make handles for ornamental daggers (jambiyas) in some parts of the Arabian peninsula. Increasing oil prices in the early 1970s increased the demand for jambiyas. The main source of rhino horn was from poaching rhinos in Africa.

- Since March 2006, White-Nose Syndrome has killed millions of Little Brown Bats in the northeastern United States and adjacent Canada.

- The demise of the bay scallops occurred because of an increase in the populations of skates and rays that are predators of bay scallops. The increase in skates and rays was attributed to the decline in populations of their predators, sharks. The sharks were extensively fished for their fins. Large-scale harvesting of the scallops for human consumption also compounded the impacts and contributed to their demise.

32.5 Protecting Species and 32.6 Protecting What?

- The International Union for Conservation of Nature (IUCN) has established objective criteria identifying species that are at risk. Extinct means the species no longer exists, extirpated means the species is locally extinct, endangered means the species is facing imminent extirpation or extinction, threatened means the species is likely to become endangered if limiting factors are not reversed, and special concern means the species may become threatened or endangered because of biological characteristics and identified threats. The criteria take into account data on populations, their patterns of distribution, and their population status.

- The Convention on International Trade in Endangered Species of Wild Fauna and Flora (CITES) attempts to prohibit international trade in endangered species. Newly described and as yet undescribed (and therefore unnamed) species are not protected because they have no legal identity.

- Species such as passenger pigeons or Dodos that have been exterminated are extinct. Extirpated species are locally extinct. Black-footed ferrets (see "Black-Footed Ferret, *Mustela nigripes*," Chapter 29) have been extirpated in Canada but still occur in the United States.

- In Canada, recommendations about the conservation status of species involve the Committee on the Status

of Endangered Wildlife in Canada (COSEWIC). COSEWIC members vote on the appropriate conservation category for each species whose status they review and use IUCN criteria to assess the status of species.

32.7 The Downside of Being Rare

- An animal on the list of endangered species is more likely to become a commodity in high demand because it has become rare. The Bali Starling is an example.

32.8 Protecting Habitat

- Protecting habitat can be particularly difficult.

- Networks of protected areas can influence conservation.

32.9 Effecting Conservation

- In 1994, the International Conference on Population and Development (ICPD) outlined a plan for investing in family planning.

- The United Nations Millennium Development Goals of 2000 are ending poverty and hunger, universal education, gender equality, child and maternal health, combatting HIV/AIDS, environmental sustainability, and global partnerships.

- Drylands cover 41% of Earth's land surface, and 10 to 20% of drylands are subject to severe land degradation, affecting, in 2008, the lives of at least 250 million people.

- Climate change and increasing pressure on water supplies negatively affect drylands.

- The case of leopards (*Panthera pardus*) demonstrates how some species persist even in the face of considerable hunting pressure.

- Targeted hunting—selection of trophy or spectacular specimens—can be less threatening to a species' survival than disease or eradication programs (bounties on predators such as wolves). Extensive killing, even of species with large populations, can drive them to the brink of extinction. Black rhinos are a telling example.

- The overgrowth of Mayan cities or ruins in Africa demonstrates the resiliency of ecosystems. A grasshopper's use of 2,4-D to synthesize an ant repellent demonstrates the resiliency of individuals.

- Hybrids are not protected by the U.S. Endangered Species Act, putting species such as Florida panthers at risk because their populations have been genetically contaminated.

32.10 Taking Action

- African elephants epitomize some of the challenges involved in taking action to protect a species and its habitats.

- This species also illustrates the potential importance of hunting and poaching.

Questions

Self-Test Questions

1. Extinction is a natural part of the process of speciation. What do some estimates suggest is the percentage of the species that have ever lived that are now extinct?
   a. more than 20%
   b. more than 30%
   c. more than 50%
   d. more than 80%

2. Some researchers use evidence from a variety of sources to support the suggestion that an asteroid striking Earth in this period largely explains the extinction of the dinosaurs.
   a. Ordovician
   b. Triassic
   c. Cretaceous
   d. Pleistocene

3. If our species first appeared 200 000 years before present, the multituberculates survived this many times as long as we have to date.
   a. 50
   b. 100
   c. 500
   d. 1000

4. Hunting by people is largely responsible for the extinction of which of the following?
   a. multituberculates, Dodos, and passenger pigeons
   b. black-footed ferrets and giant auks
   c. passenger pigeons, giant auks, and Dodos
   d. black rhinos, Bali starlings, and ginseng

5. The ballast water of ships is responsible for the spread of which species?
   a. *Arthurdendyus triangulatus*
   b. *Dreissena polymorpha*
   c. *Rattus norvegicus*
   d. *Lampsilis radiata*

6. In Hawaii, high resource use efficiency (RUE), measured as carbon use, partly explains the success of these invading species.
   a. ferns
   b. C_3 and C_4 grasses
   c. flatworms
   d. Both b and c are correct.

7. Which tertiary consumers have experienced increases in populations, which may explain the demise of scallops off the southeastern coast of the United States?
   a. skates and rays
   b. sharks
   c. killer whales
   d. pelagic seabirds

8. Species such as black-footed ferrets (*Mustela nigripes*) no longer occur in Canada but still live in the United States. Which term describes their status?
   a. extinct
   b. extirpated
   c. highly endangered
   d. not at risk

9. CITES is designed to stop international trade in which of the following species, among others?
   a. passenger pigeons
   b. black rhinos
   c. Canada geese
   d. leopards

10. Differences in government support for family planning explain the differences in the growth of human populations in which two countries?
    a. Great Britain and France
    b. Afghanistan and Sri Lanka
    c. Mexico and Germany
    d. India and South Africa

Questions for Discussion

1. Should gardeners and farmers be exempt from rules concerning the introduction of foreign species? Why or why not?

2. In situations where the behaviour of one endangered species threatens the survival of another (or others), how should authorities proceed?

3. What species are "rare" on your campus? What is a good working definition of rare? What steps can you take to protect rare species?

4. How will the cost of food influence efforts to conserve species at risk?

5. How does the biological definition of species influence our efforts to conserve species?

33

Sunflowers. Originally from the New World, sunflowers (*Helianthus annuus*) are grown as a source of oil. In terms of harvest and area under cultivation, in 1998, sunflowers ranked twelfth in importance among domesticated plants in the world. Domesticated sunflowers often hybridize with local wild species, creating a challenge for those concerned about biodiversity.

Putting Selection to Work

WHY IT MATTERS

In 1960, an estimated 1.8 billion people in the world (60% of the population) did not receive enough food every day to sustain themselves fully over the longer period—they were hungry. This number was reduced to 1.1 billion (17%) in 2000. Even though the world population had grown by 3 billion in the intervening period, about 700 million fewer people were hungry in 2000.

Worldwide in 2000, subsistence farmers accounted for about 66% of the hungry people. The reduction in the numbers of hungry people can be tied to changes in agriculture that have increased yields. Specifically, the combination of new genetic strains, better fertilizers, better irrigation, more effective pest control, and more efficient harvesting and processing means more productivity. One indication of this change is provided by data about corn yields. In Iowa in 1935, corn yields were about 1600 kg·ha^{-1} compared with about 10 700 kg·ha^{-1} in 2000. Changes in crop yield are part of the "green revolution." Agriculture in general and the green revolution in particular have allowed humans to continue to redefine one element of carrying capacity (see Chapter 29): the amount of food available to our populations.

But agricultural improvements are not enough. Climate also influences crop yield. In 2006 in southwestern Ontario (~42° N in Canada), the corn yield was about 10 000 kg·ha^{-1}, whereas in Zimbabwe (~18° S), on commercial farms, it was about 5500 to 6600 kg·ha^{-1}, compared with about 500 to 1000 kg·ha^{-1} on communal lands where farming was low tech. Irrigation also influences yield: in Zimbabwe, irrigated cornfields produce 8500 to 10 000 kg·ha^{-1}, much more than nonirrigated commercial farms.

Although increases in crop yield and a reduced incidence of malnutrition and starvation sound like good news, in 2011, hunger still claimed the lives of about 8500 children a day. Worldwide, one child in three is underweight and malnourished. Ironically, at the same time in some developed countries, obesity in children reached almost epidemic proportions.

In addition to the *biology* of domestication and increasing crop yields, the *security* of food sources plays a vital role in the survival of tens of thousands of people. In this context, security refers to social and political factors and stability. In 2011, changes in the political structure of several countries in North Africa and the Middle East may have been as attributable to rising costs of food and their impact on average people as to any other single factor. Availability of fresh water is inextricably associated with agriculture and the global food supply.

Changes in diet may have been fundamental to the origin and adaptive radiation of species in the genus *Homo*. The use of fire and tools influenced our ancestors' abilities to obtain food. Diets rich in "brain food," such as many aquatic animals, as well as in starches **(Figure 33.1)** could have heralded important changes in our ancestors.

Our species has turned natural selection to its advantage by domesticating other species, selectively breeding strains with traits that provide us with more food or other desirable commodities. Domestication emerged from cultivation of wild forms, probably in association with a more sedentary lifestyle, and has occurred in different parts of the world, from central Asia to the Far East, from the Middle East to the New World and southeast Asia. We depend on the productivity of many domesticated food plants to sustain our populations, and on domesticated animals for labour, hides, and food. Domestication provides repeated examples of how our species has benefited from selection, and unravelling the history of the process depends heavily on genetic tools.

33.1 Domestication

The purpose of this chapter is to explore how humans have used selection to put biodiversity to work. Biologists and anthropologists believe that our ancestors originally gathered plants and hunted animals in the wild for use as food, building products, or fuel (see "Molecule behind Biology," Box 33.1, p. 811). From gathering, our ancestors progressed to cultivating plants, a process involving the systematic sowing of wild plant seeds. Over time, cultivation improved when people provided more care to their crops and eventually involved repetitive cycles of sowing, collecting, and sowing wild stock **(Figure 33.2). Domestication** is more than just taming. It occurs when people selectively breed individuals of other species (plants and animals) to increase the desirable characteristics in the progeny (e.g., in plants: yield, taste, colour, shelf life). This marked the birth of agriculture. The progression from gathering to cultivation to domestication of plants occurred independently at several locations around the world. The beginning of the Neolithic Period is often defined by the domestication of other species, and this period started at different times in different parts of the world.

But agriculture is not the exclusive domain of humans. Recall that about 50 mya, ants of the tribe Attini were the first to manipulate other species (fungi) to increase food availability (see Chapter 25). Today at least 200 species of ants in this tribe are obligate farmers. These early farmers have lost their own digestive enzymes and rely on fungal enzymes to digest the food for them. The ant farmers propagate their fungal crops asexually, with each colony working with one species. Therefore, any single species of farmer ant may propagate several different species of fungi. These ants have been involved in at least five domestication

Figure 33.1

A diagrammatic presentation of the progression in diet and brain size across 4 million years of hominin history.

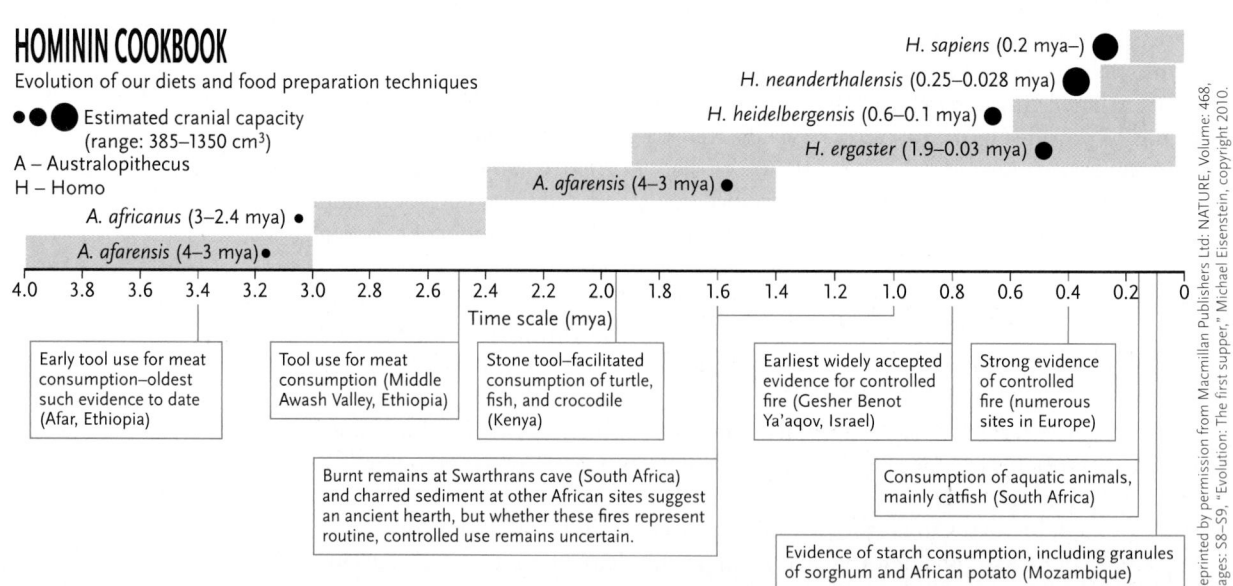

Reprinted by permission from Macmillan Publishers Ltd: NATURE, Volume: 468, Pages: S8–S9, "Evolution: The first supper," Michael Eisenstein, copyright 2010.

Figure 33.2
The way in which garlic (*Allium sativum*) is grown influences the size and development of the bulbs. From left to right: one domesticated, two cultivated, and two wild garlic bulbs.

M.B. Fenton

events, and there are a number of interesting parallels between these ants and people.

The list of species that humans have domesticated is long. It includes many land plants (~250 species), some yeasts, and terrestrial animals from insects to birds and mammals (~44 species). Biogeographic and genetic evidence shows that domestication of some species by humans occurred in different places and at different times. Domestication was not a one-time (or one-location) event and appears to have arisen independently in 8 to 10 environmentally and biotically diverse areas in the world.

As intriguing as which species were domesticated is the fact that very few available species of animals, plants, and fungi were domesticable. At the same time, humans were exploiting and continue to exploit many species without actually domesticating them.

33.1a When and Where Did Domestication Take Place?

Once they were domesticated, many domesticated plants and animals became widely used, becoming staple foods carried with humans as they moved to occupy many of the land areas on the planet. Data provided by the tools of molecular genetics (see Chapter 17) have made it easier to determine where and when domestication events took place. In the past, archaeologists had to try to recognize the remains of domesticated species and distinguish them from wild species. This was often impossible because individual bones or pieces of plant did not always provide a clear indication

of domestication. In 1973, radiocarbon dates suggested that the first dogs (*Canis familiaris*) were domesticated by 9500 years B.P. (before present), based on remains found in England and elsewhere in Europe. In 2002, mitochondrial DNA (mtDNA) evidence suggested an East Asian origin of domestication of dogs dating from 15 thousand years B.P. But pictures based on genetic evidence can also change. In 2003, morphological and genetic evidence suggested a southeast Asian origin of domesticated pigs (*Sus scrofa*), whereas in 2005, new genetic data indicated multiple origins of domestication of pigs across Eurasia **(Figure 33.3, p. 810).**

Worldwide, domestication of aquatic species has lagged behind that of terrestrial ones. Although there are about 180 species of domesticated freshwater animals, about 250 species of marine animals, and about 19 species of marine plants, all were domesticated in the last 1000 years, and most in the last 100 years **(Figure 33.4, p. 810).**

33.1b How Long Did Domestication Take? Archaeological Evidence

The time it takes to progress from harvesting tended wild crops to cultivating them and then to domesticating them varies with species and situation. When there are clear morphological or chemical differences between cultivated and domesticated stocks, determining the place and time of domestication is possible. Wheat is an example of such a morphological change. Wheat and other cereal crops are grasses that disperse their seed explosively by shattering. The process of

Figure 33.3

Origins of domestication of pigs.
Mitochondrial DNA obtained from
pigs indicates 14 clusters of related
lineages, each identified by a different
colour. The geographic relationships
are shown with the phylogeny. Pigs
were domesticated in numerous
centres.

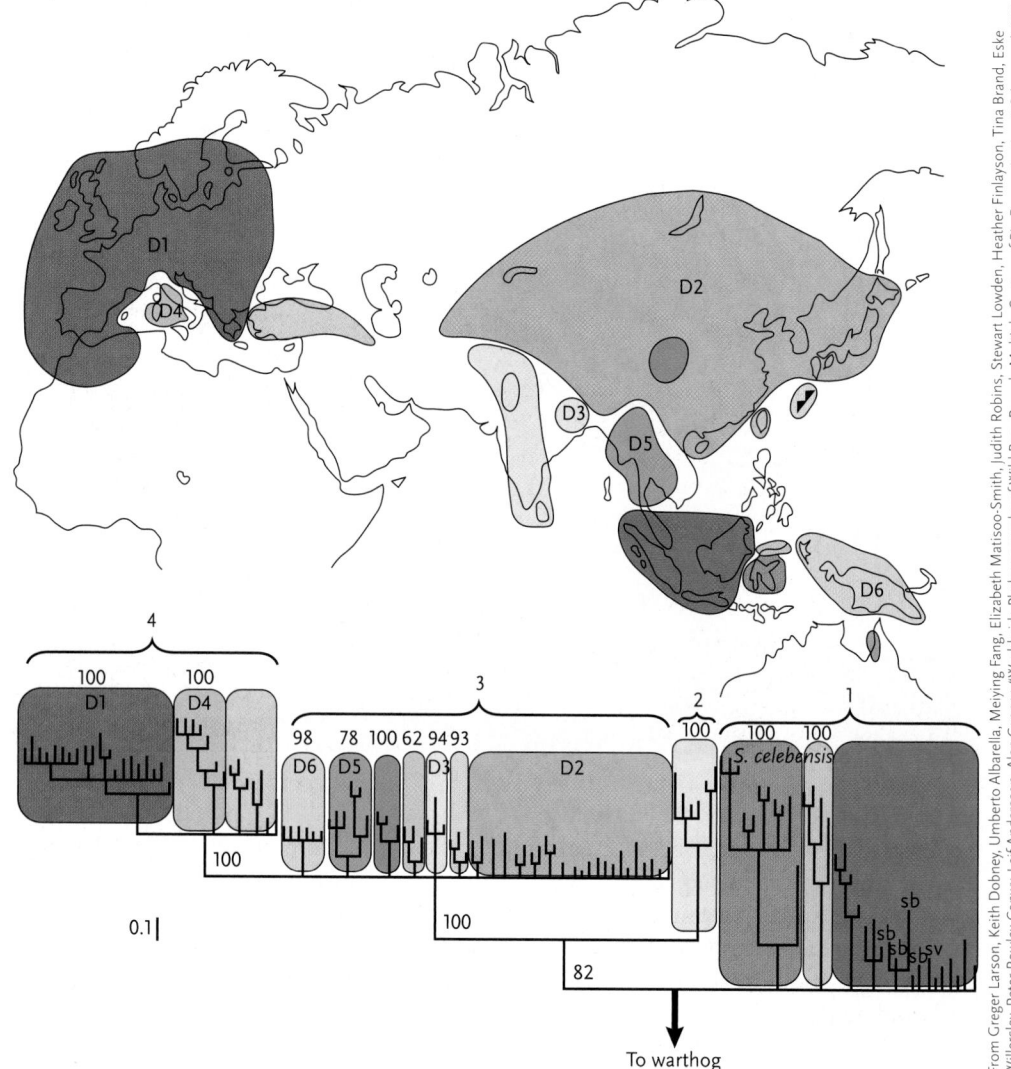

From Greger Larson, Keith Dobney, Umberto Albarella, Meiying Fang, Elizabeth Matisoo-Smith, Judith Robins, Stewart Lowden, Heather Finlayson, Tina Brand, Eske Willerslev, Peter Rowley-Conwy, Leif Andersson, Alan Cooper, "Worldwide Phylogeography of Wild Boar Reveals Multiple Centers of Pig Domestication," *Science*, vol. 307, Mar 11, 2005, pp. 1618–1621. Reprinted with permission from AAAS.

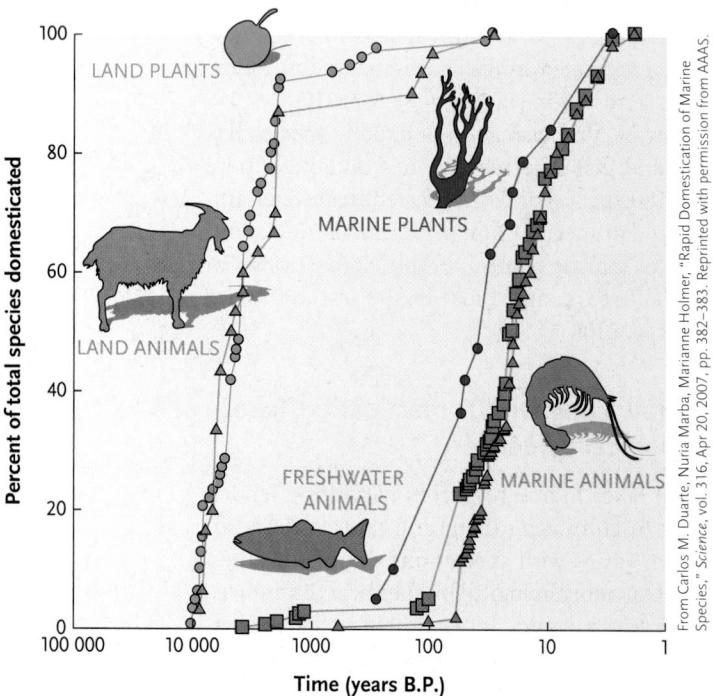

From Carlos M. Duarte, Nuria Marba, Marianne Holmer, "Rapid Domestication of Marine Species," *Science*, vol. 316, Apr 20, 2007, pp. 382–383. Reprinted with permission from AAAS.

Figure 33.4

Most land species were domesticated much earlier than most aquatic ones.

domestication of wheat and cereal crops meant developing stocks that do not shatter (indehiscence) from stocks that shattered (dehiscence) **(Figure 33.5, p. 811)**. Ripe indehiscent grains are easily gathered (harvested) compared with dehiscent ones that naturally scatter. Indehiscence in wheat results from a naturally occurring mutation.

But, in addition to selecting indehiscent stock, the early farms also had to select for plants whose seeds did not go through a period of dormancy. This change would allow repeated sowing of crops when conditions were appropriate. Material recovered from sites in northeastern Syria and Turkey has been radiocarbon dated and shows that wild varieties of wheat were cultivated for at least 1000 years before domestication **(Figure 33.6)**. When sexual reproduction is involved in the breeding process, the time to domestication is partly determined by life cycle, so finding stock that can self-fertilize can accelerate the domestication process.

When organisms reproduce asexually, domestication may occur more rapidly. Common figs (*Ficus*

Salicylic Acid

Figure 1
The molecular structure of salicylic acid.

Salicylic acid
2-OH-C$_6$C$_4$CO$_2$H

The precursor of the main active ingredient in aspirin is salicylic acid, which is obtained from the bark of willow trees (*Salix* species). Over 2500 years B.P., Chinese medical practitioners used an extract of willow bark to relieve pain and fever. The same kinds of extracts were used in medicine as practised in Greece and in Assyria. In Iceland 500 years ago, willow bark extracts were used to treat the symptoms of colds and headaches. Willow extract was widely used among First Nations people in North America, who commonly used it to stanch bleeding. They also used the supple willow twigs in other applications, from snares for catching mammals to nets for catching fish. This is not an example of domestication, but it demonstrates how the spread of traditional knowledge about plants and their products among peoples is pervasive. For more information about salicylic acid, see Chapter 38, p. 932.

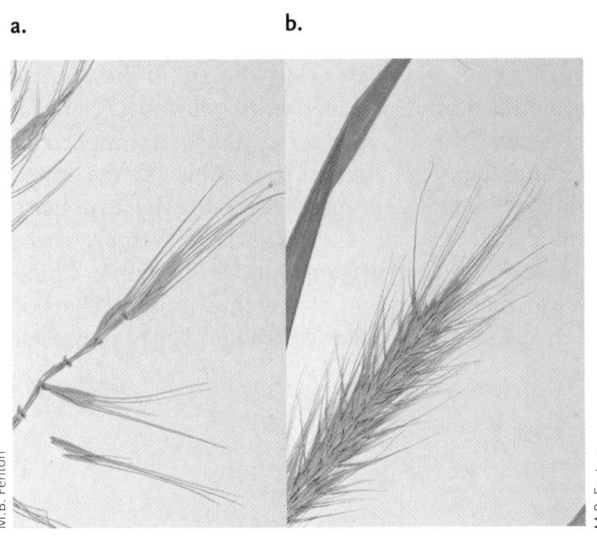

M.B. Fenton

Figure 33.5
Some plants readily shed ripe seeds from the inflorescence; these plants show dehiscence. Indehiscence is the propensity to hold seeds. These two herbarium specimens, **(a)** bottle brush grass (*Elymus hystrix*) and **(b)** riverbank wild rye (*Elymus riparius*), illustrate dehiscence and indehiscence, respectively.

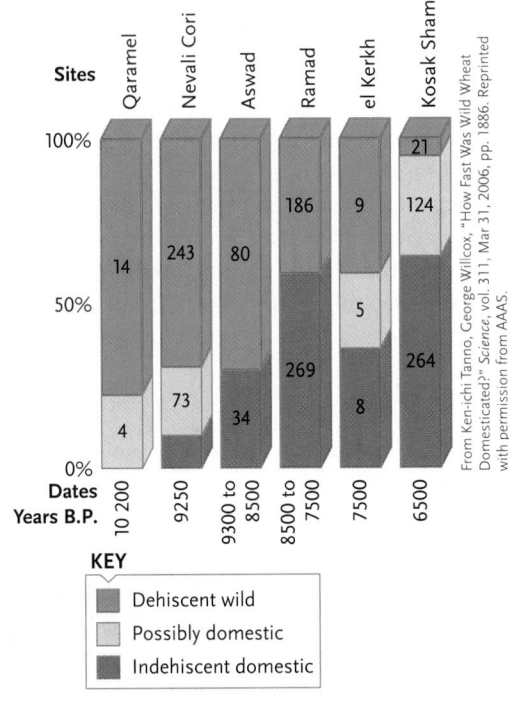

From Ken-ichi Tanno, George Willcox, "How Fast Was Wild Wheat Domesticated?" *Science*, vol. 311, Mar 31, 2006, pp. 1886. Reprinted with permission from AAAS.

KEY
- Dehiscent wild
- Possibly domestic
- Indehiscent domestic

Figure 33.6
Timing of domestication of wheat. Data from archaeological digs at six locations in the Middle East demonstrate the transition from wild (dehiscent) to domesticated (indehiscent) wheat from 10 200 to 6500 years B.P.

carica var. *domestica*) are gynodioecious and provide an example of more rapid domestication. In parthenocarpic female figs, ovaries develop without pollination and fertilization. Parthenocarpic figs can be propagated by cutting branches, sticking them in the ground, and waiting for them to grow into trees. When figs reproduce sexually, symbiotic fig wasps (*Blastophaga psenes;* **Figure 33.7**) serve as pollinators. The absence of access holes for wasps in fossil figs allows biologists to recognize parthenocarpic figs and date early incidences of fig domestication. At one site in the lower Jordan Valley (Middle East), parthenocarpic figs date to between 10 500 and 11 400 years B.P., perhaps preceding the domestication of cereal crops by about 1000 years.

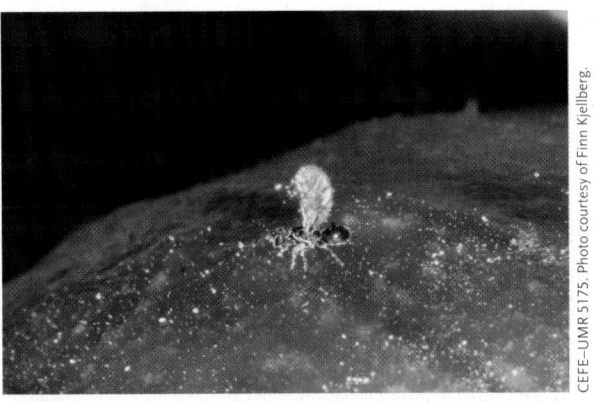

CEFE–UMR 5175. Photo courtesy of Finn Kjellberg.

Figure 33.7
Female flowers on some fig trees (*Ficus carica*) are fertilized by symbiotic wasps, *Blastophaga psenes*.

Figure 33.8

In Ghana (West Africa), fire is still used in slash and burn to clear the underbrush from an area where forest trees have been felled.

33.1c In Which Habitats Did Domestication Occur?

The transition from nomadic hunters and gatherers to people living more localized lives in more permanent dwellings appears to have been a prelude to cultivation and domestication. These changes meant that people would have been available to care for their "crops," whether grown from seeds or from parthenocarpic plants, and whether cultivated or domesticated.

Although controlled burning has been documented at many sites in the last 10 thousand years, there is evidence of it 50 thousand to 55 thousand years ago at sites near Mossel Bay in South Africa, and it still occurs today **(Figure 33.8)**. Archaeological evidence indicates that some humans were increasingly using some plant resources and using local burning to increase productivity. The increased use of plant resources and fire occurred during a period of harsh environmental conditions. These changes in human behaviour coincided with the appearance of more sophisticated tools, the use of marine organisms as food, and the first use of ochre for decoration. These modifications suggest differences in human behaviour that may have assisted the emergence of domestication.

Were changes in habitat associated with domestication? Evidence from pollen shows that from 7500 years B.P. in the lower Yangtze region of China, people used fire to clear alders, which are small woody bushes (*Alnus* species). This element in the process of domestication is called *niche construction* or *ecosystem engineering*, modifying the environment and setting the stage for domestication. At the lower Yangtze Neolithic site, people used fire first to prepare and then to maintain sites in lowland swamps, where they cultivated rice **(Figure 33.9)**. This region in China was a major centre of rice domestication. The evidence suggests that rice cultivation began in coastal

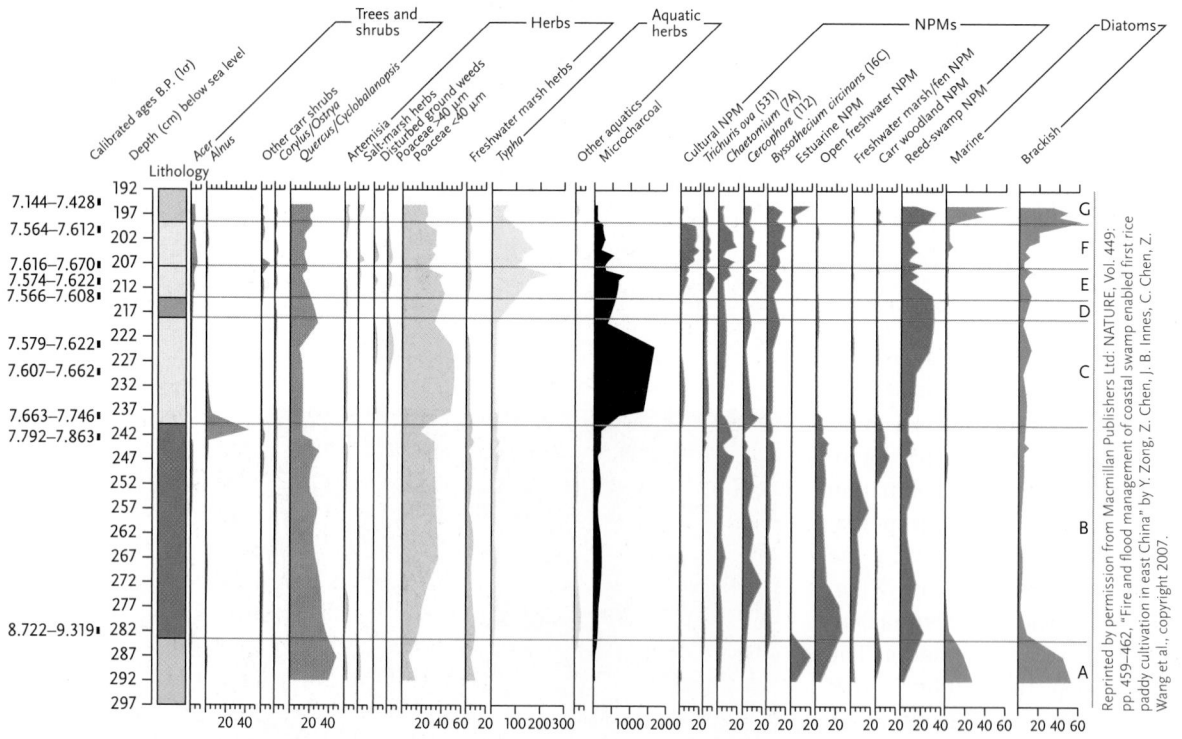

Figure 33.9

The climatic setting for rice domestication at Kuahugiao in China. Shown here are generalized stratigraphic units and their associated pollen (shown in colours) and microscopic charcoal (black) that indicate conditions of climate and habitat. The data support the use of fire to establish favourable conditions for growing rice. NPMs are nonpollen microfossils that provide paleoecological data. The increases in charcoal, grasses (Poaceae), and reed-swamp microfossils in phases C, E, and F (right edge) indicate the use of fire to establish favourable conditions for growing rice.

Reprinted by permission from Macmillan Publishers Ltd: NATURE, Vol. 449: pp. 459–462, "Fire and flood management of coastal swamp enabled first rice paddy cultivation in east China" by Y. Zong, Z. Chen, J. B. Innes, C. Chen, Z. Wang et al., copyright 2007.

wetlands in an ecosystem vulnerable to coastal change. This system was very fertile and productive and used for at least 200 years before the land was inundated by sea water.

Meanwhile, at sites on the coast of Peru, occupied between 3800 and 3500 years B.P., people ate marine organisms as the main animal food, combined with cultivated plants (squashes, *Cucurbita* species; beans, *Phaseolus lunatus* and *Phaseolus vulgaris;* peppers, *Capsicum* species; jicama, *Pachyrhizus tuberosus*) and wild plants (guava, *Psidium guajava;* lacuma, *Lucuma bifera;* and pacay, *Inga feuillei*). Cotton was an important crop used for making fishing tackle and clothing. The findings from these Peruvian sites and many other sites around the world suggest a progression toward domestication, including the range of foods consumed, the development of more sophisticated tools, and the use of materials from plants and animals as tools, as well as in food. Domestication, therefore, involved the spread among peoples of the practice of using plants and animals to advantage.

33.1d Abu Hureyra on the Euphrates Is an Example of a Setting for Domestication

The prehistoric settlement of Abu Hureyra (a recent photograph is shown in **Figure 33.10**) on the south side of the Euphrates River (35° 52 N, 38° 24 E) about 130 km from Aleppo (a modern Syrian city) also illustrates progression toward domestication. The first habitations that we know of in Abu Hureyra date from about 12 thousand years ago. Its population was estimated at 100 to 200 people who lived in semi-subterranean pit dwellings clustered together on a low promontory overlooking the river. By 7000 to 9400 years ago, 4000 to 6000 people lived at the same site, now in multiroomed family dwellings made of mud and brick. This settlement was built over the remains of the earlier one.

People living at Abu Hureyra about 12 thousand years B.P. ate the fruits and seeds of over 100 species of local plants as well as local animals such as gazelles. Many of the plants and animals appear to have come from the adjoining oak-dominated park woodland. It appears to have been a time of plentiful food. The situation changed, however, and by 9400 years ago, the climate was cooler and drier, and the people relied more on cultivated plants and less on wild ones. By this time, there was little evidence of use of plants from the oak-dominated parkland, which by then was at least 14 km from the settlement. These changes were evident in pollen records and in plant and animal remains associated with the dwellings. The climate change likely triggered the start of cultivation of foods that could serve as caloric staples. Despite the changing climate and the focus on fewer food staples, the human population at the site dramatically increased.

33.2 Why Some Organisms Were Domesticated

We can surmise, perhaps accurately, that securing a sustainable food supply provided an initial motivation for cultivation and domestication. It is certainly true that cultivated plants such as beans, squash, corn, rice, and cereal grains all help feed many, many people worldwide. People eat different parts of plants, from flowers and fruits to seeds, leaves, stems, roots, and tubers. Plants may be a source of energy (calories), or their products may be used to enhance flavours, to control and repel pests, or as medicines. Still others, such as the bottle gourd (*Lagenaria siceraria*), are used as containers. Domesticated cereal grains have been derived from variants (local varieties, sometimes known as breeds, cultivars, or landraces) with four important features: (a) nonshattering, (b) large seeds, (c) self-compatibility, and (d) no required dormancy. This suite of characters makes them valuable because they can be readily fertilized, harvested, and planted whenever conditions permit. Domesticated animals provide food, but many are also used as a source of labour. The following are examples of four very different domesticated species and how people use them.

33.2a Cattle

Cattle were among the first of the large herd mammals to be domesticated, at least 9000 years B.P. One theory proposes that the domesticators of cattle were sedentary farmers, not nomadic hunters. Some anthropologists maintain that a religious motivation was behind the domestication of cattle because the curve of their horns resembled the crescent of the Moon and hence the mother-goddess. Imposing horns were particularly prominent in some male *Bos primigenius* (called *urus*), the apparent Pleistocene ancestor of domesticated cattle. Whatever the original impetus, today there are two basic stocks of cattle (Figure 33.10), the humped *Bos indicus* and the humpless *Bos taurus*. Cattle provide us with labour, milk, meat, hides, and blood, and in some societies, they are symbols of wealth. At the root of the domestication of cattle are some biological realities: they are relatively docile animals that live in herds and can be useful in many ways.

Figure 33.10

Today there are two varieties of domesticated cattle: **(a)** the humped *Bos indicus* and **(b)** the humpless *Bos taurus*.

33.2b Honeybees

Domestic honeybees provide us with honey and pollination services. Steps to domestication of honeybees included changes in their behaviour compatible with large population size in hives. Features of honeybees that make them suitable for domestication are their colonial and food-storing behaviours, unlike other species of bees that are solitary and do not store honey. The changes could have involved hygiene, aggression, and foraging. Although everyone recognizes honey as a product of bees, the service provided by bees is often overlooked. In 2000 in the United States, it is estimated that bees contributed about U.S.$14.5 billion through their role as pollinators. Plants such as alfalfa, apples, almonds, onions, broccoli, and sunflowers are exclusively pollinated by insects, usually more than 90% by honeybees. Many beekeepers earn significant income by moving their bees from location to location, thus providing a mobile pollinator service for farmers. Declines in populations of honeybees have serious economic implications throughout the world, but many conservationists are also concerned about the impact of populations of honeybees on native bee species.

33.2c Cotton

At least four species of cotton **(Figure 33.11)** have been domesticated: two diploid species from the Old World (*Gossypium arboreum, Gossypium herbaceum*) and two tetraploids from the New World (*Gossypium hirsutum* and *Gossypium barbadense*). The domestication events appear to have been independent, and one site on the Mexican gulf coast of Tabasco shows evidence of people growing cotton by 4400 years B.P. Cotton seeds were a source of oil, whereas fibre was and is still used in applications ranging from clothing to implements such as ropes and nets.

Figure 33.11

Cotton, *Gossypium herbaceum*, showing flowers and cotton bolls.

33.2d Yeast

Strains of the yeast *Saccharomyces cerevisiae* have been used by people in bread-making beginning at least 6000 years B.P. Evidence of this is in archaeological finds in Egypt, indicating the presence of bakeries and breweries, two yeast-based operations. Analysis of 12 DNA microsatellites obtained from 651 strains of *S. cerevisiae* collected at 56 locations around the world revealed 575 distinct genotypes. Yeasts associated with bread were intermediate between wild types and those used in making beer and wine, whereas those used in the production of rice wine and sake were more similar to those used for beer. About 28% of the genetic variation in yeast genotypes was associated with geographic location. The basal group of these 12 DNA microsatellites was samples from Lebanon, suggesting a Mesopotamian origin and a spread of yeast types along the Danube River and around the Mediterranean. Different strains of yeast have different capacities for maltose fermentation. Commercial bakers' yeast strains are more effective at maltose fermentation than nonindustrial strains. Domesticated yeast makes important contributions to providing humans with food and drink and supports lucrative industries.

33.2e Rice

Rice (*Oryza sativa*) is one of the world's most important food crops, and its domestication depended on the change from dehiscence (shattering) to indehiscence (nonshattering). Domesticated rice is derived from two wild species: *Oryza rufipogon* and *Oryza indica*. In 2006, Changbao Li and his colleagues reported that three quantitative trait loci (QTL) in F_1 hybrids between these two species were responsible for a reduction of grain shattering (dehiscence) in rice. Specifically, *sh3*, *sh4*, and *sh8* were involved, with *sh4* explaining 64% of the phenotypic variance. In the wild species, *sh4* was

dominant and caused the shattering. The genetic changes in *sh3*, *sh4*, and *sh8* affected normal development of the abscission layer, explaining the change to indehiscence **(Figure 33.12)**. We can now better understand the genetics of changes associated with the domestication of rice.

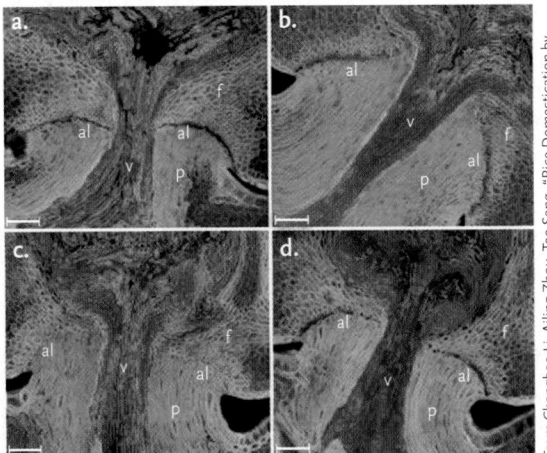

From Changbao Li, Ailing Zhou, Tao Sang, "Rice Domestication by Reducing Shattering," *Science*, vol. 311, Mar 31, 2006, pp. 1936–1939. Reprinted with permission from AAAS.

Figure 33.12

Rice dehiscence and indehiscence. Under a fluorescence microscope, a longitudinal section of the junction between the rice flower and its pedicel shows a complete abscission layer (al in **(a)**) and an incomplete one (al in **(b)**). In these figures, f = flower side; p = pedicel side; and v = vascular bundles. *Oryza nivara* is shown in **(a)**, *Oryza sativa* in **(b)**, *Oryza sativa japonica* in **(c)**, and transformed *O. sativa japonica* in **(d)**.

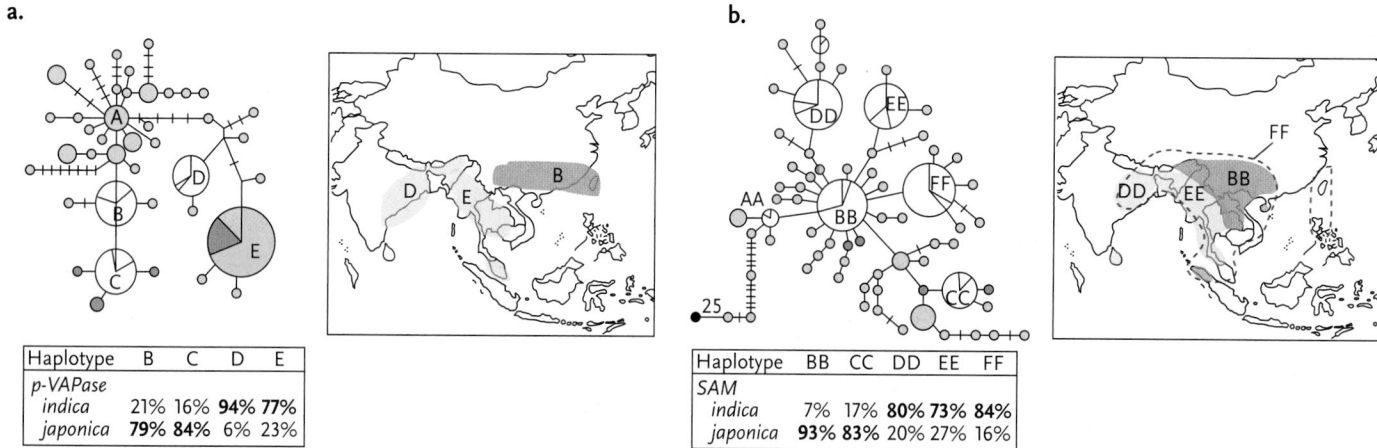

a.

Haplotype	B	C	D	E
p-VAPase				
indica	21%	16%	**94%**	**77%**
japonica	**79%**	**84%**	6%	23%

b.

Haplotype	BB	CC	DD	EE	FF
SAM					
indica	7%	17%	**80%**	**73%**	**84%**
japonica	**93%**	**83%**	20%	27%	16%

Figure 33.13

Geography of rice domestication. Genetic haplotype information allows identification of major domestication regions for rice. The haplotype network of the neutral nuclear p-VAPase region is shown in **(a)**, compared with **(b)**, the haplotype of the functional nuclear SAM region. In this diagram, orange represents *Oryza japonica;* blue, *Oryza indica* (Aus cultivar); and yellow, *O. indica.* Lines joining haplotypes represent mutational steps.

Londo, J. Y-C Chiang, K-H Hung, T-Y Chiang, and B. Schaal. (2006) "Phylogeography of Asian wild rice, Oryza rufipogon reveals multiple independent domestications of cultivated rice, Oryza sativa," PNAS, 103: 9578–9583. Copyright 2006 National Academy of Sciences, U.S.A.

These changes occur in the two major varieties of domesticated rice, *Oryza sativa japonica* and *Oryza sativa indica.* Jason Londo and his colleagues used DNA sequence variation to demonstrate multiple independent domestication events. Although *O. sativa indica* was probably domesticated in eastern India, Myanmar, and Thailand south of the Himalayan mountains, *O. sativa japonica* was domesticated in southern China **(Figure 33.13).**

33.2f Wheat

Other major domestications appear to stem from single domestication events, including wheat, barley **(Figure 33.14),** and corn. DNA fingerprinting was used by Manfred Heun and his colleagues to identify the Karacadağ mountains **(Figure 33.15)** in today's Turkey as *the* site of domestication of einkorn wheat. Einkorn wheat is derived from the wild *Triticum monococcum boeoticum.* Einkorn wheat is selfing and nonshattering and has a firm stalk, heavier seeds,

and denser seed masses than its progenitor (see Figure 19.16, Chapter 19), all contributing to the ease and efficiency of harvest. The genetic changes between einkorn wheat and the wild stock from which it was derived were relatively minor, and domestication probably occurred in a short period of time.

Archaeological evidence reveals that einkorn wheat spread rapidly throughout the immediate area and beyond. The same single domestication event and rapid spread of the new crop also appear to be true of barley **(Figure 33.16b);** *Hordeum vulgare* derived from *Hordeum spontaneum).* More recent wheat domestication is discussed in "Marquis Wheat," p. 819.

33.2g Lentils

Lentils (*Lens culinaris* from *Lens orientalis;* Figure 33.16a) were more difficult to domesticate than wheat or rice. Wild lentils are small plants, producing, on

Figure 33.14

A barley field near London, Ontario, illustrates the consistency of a monoculture, one plant dominating an area, in this case for artificial reasons.

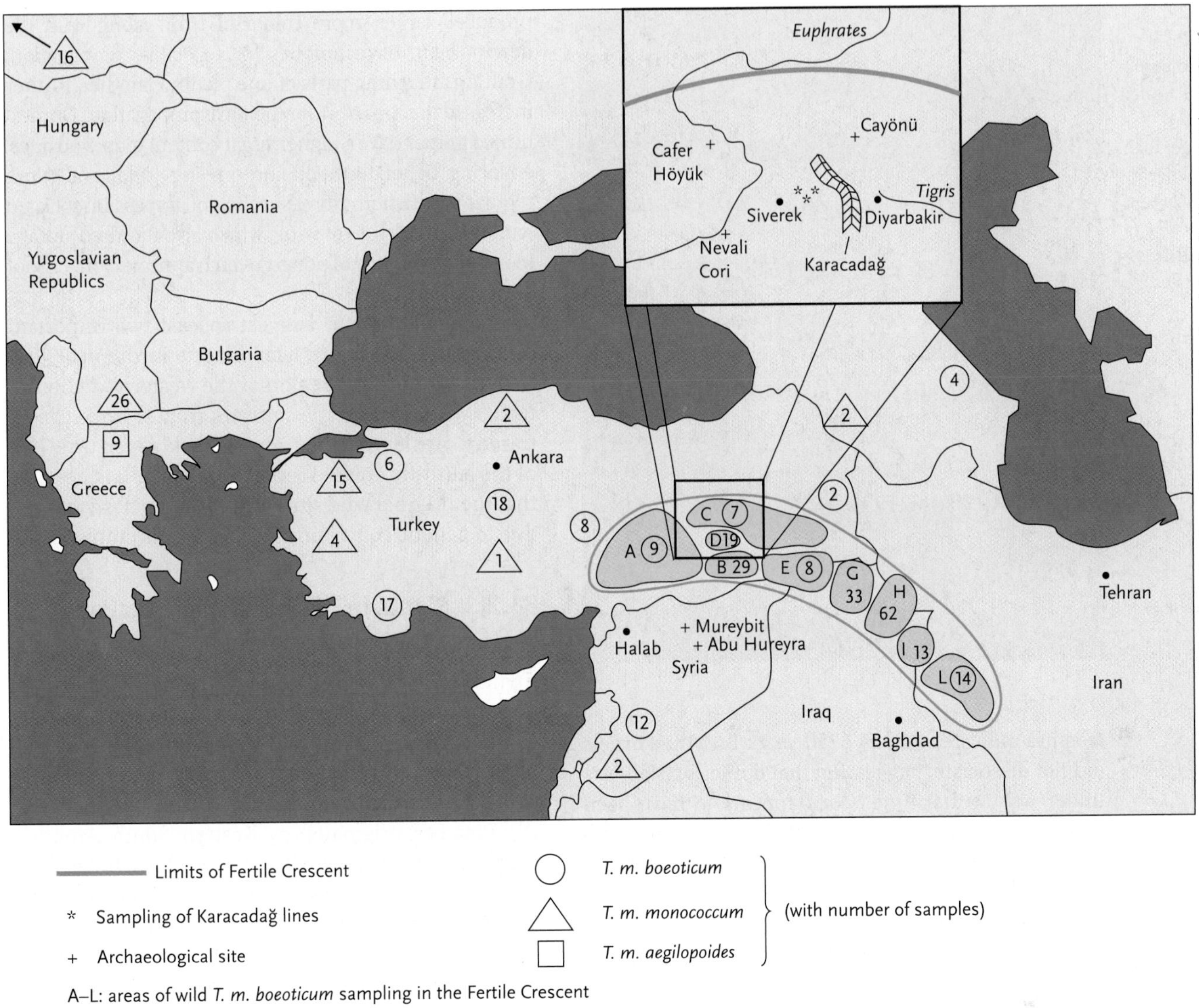

From Manfred Heun, Ralf Schafer-Pregl, Dieter Klawan, Renato Castagna, Monica Accerbi, Basilio Borghi, Francesco Salamini, "Site of Einkorn Wheat Domestication Identified by DNA Fingerprinting," *Science*, vol. 278, Nov 14, 1997, pp. 1312–1314. Reprinted with permission from AAAS.

——— Limits of Fertile Crescent

*   Sampling of Karacadağ lines

+   Archaeological site

○   *T. m. boeoticum*

△   *T. m. monococcum*   } (with number of samples)

□   *T. m. aegilopoides*

A–L: areas of wild *T. m. boeoticum* sampling in the Fertile Crescent

Figure 33.15

Site of wheat domestication. A phylogenetic analysis based on allelic frequency at 288 amplified fragment length polymorphism marker loci revealed that the progenitor of *T. m. boeoticum* came from what is now southeastern Turkey.

M.B. Fenton

Figure 33.16

Two important domesticated crops are **(a)** lentils (*Lens culinaris*) and **(b)** barley (*Hordeum vulgare*). In each case, individual grains are about 6 mm long.

average, 10 seeds per plant. Furthermore, these seeds go through a period of programmed dormancy. The combination of low yield and dormancy means that relatively few seeds germinate. Lentils could have been

cultivated only after a dormancy-free mutant had appeared. Archaeological evidence suggests that the first stages of lentil domestication (loss of dormancy) occurred in what is now southeastern Turkey and northern Syria, suggesting a single initial domestication event. The second phase was selection for strains that produced large numbers of seeds. This change may have occurred some distance (hundreds or even thousands of kilometres) from the original sites of domestication.

33.2h Corn

Corn (*Zea mays*), like wheat and barley, appears to have been domesticated at one location. Unlike wheat or barley, domestication of corn required drastic changes from the ancestral teosinte (*Zea mays parviglumis*; **Figure 33.17, p. 818**). Corn kernels do not dissociate from the cob, presumably reflecting changes associated with domestication (and analogous to indehiscence). Analysis of some of the most ancient inflorescences of

Figure 33.17
Teosinte and
domesticated corn.
Ears of domesticated
corn (*Zea mays*,
right) are larger than
those of one species
of its wild relative
teosinte (*Zea diplo-
perennis*, left). Cross-
ing domesticated
corn and teosinte
produces intermedi-
ate forms (centre).

John Doebley

teosinte indicates that by 6250 years B.P. the kernels did not dissociate, suggesting that domestication was under way at that time. Corn appears to have been domesticated in the Central Balsas Valley of south central Mexico or in the Mexican State of Puebla at altitudes of 1000 to 1500 m. The deposits that contained the undissociated teosinte kernels also had the remains of squash (*Cucurbita pepo*). Although squash appears to have been domesticated in the same area, remains from a cave in Puebla and Oxa States in Mexico suggest that the people there were cultivating it by about 4000 years before corn. These data again demonstrate that some people were responsible for multiple domestication events. Corn, beans, and squash, the "three sisters" of early farmers in the New World, together provided different foods. The beans climbed on the corn stalks and provided nitrogen fixation services (see Chapter 31), while the squash leaves shaded out the weeds.

33.2i Grapes

The Eurasian grape (*Vitis vinifera sylvestris*), a dioecious plant, is widespread from the Atlantic coasts of Europe to the western Himalayas, and its fruits were often eaten by Paleolithic hunter–gatherers. Domestication of *V. vinifera* involved the selection of hermaphroditic genotypes that produced larger, more colourful fruit, along with the development of techniques for vegetative propagation. Grafting, attaching parts of one plant to another, further increased the capacity for vegetative propagation. Domesticated grapes have a higher sugar content than wild ones, ensuring better fermentation, greater yield, and more regular production. Domestication of grapes is associated with the production of wine, which also required storage in containers made of pottery, which appeared only about 10 500 years B.P.

Genetic analyses suggest at least two important origins of grape domestication: one in the near East and the other in the region of the western Mediterranean. Many wine-grape cultivars from Europe can be traced to western Mediterranean stock, as can over 70% of the cultivars on the Iberian Peninsula. It is possible that the original wild stock of *V. vinifera* has vanished due to genetic contamination by various cultivars.

33.2j Plants in the Family Solonaceae

Solanaceous plants include species such as deadly nightshade (*Atropa belladonna*) that produce virulent poisons, as well as food species (tomatoes, *Solanum lycopersicum*; potatoes, *Solanum tuberosum*; and eggplant, *Solanum melongena*) that are staples of many meals worldwide **(Figure 33.18)**.

Potatoes originated in western South America, specifically in areas of what is now Chile. Today, local varieties (also known as breeds, cultivars, or landraces) of potatoes are adapted to the local conditions where they grow. These landraces form the *Solanum brevicaule* complex. Genotypes from multilocus amplified

M.B. Fenton

Figure 33.18
Solonaceae. Domesticated species include **(a)** capsicum peppers, **(b)** eggplant, **(c)** potato, and **(d)** tomato.

Marquis Wheat

Different strains of domesticated crops, such as wheat and corn, have different features that make them better suited to some areas (climates) than to others. But wheat crops provide both flour for bread and straw for thatching and animal bedding. Whereas Red Fife wheat matures in 130 days and produces 3270 kg·ha^{-1}, Hard Red Calcutta wheat matures in 110 days but yields only 1240 kg·ha^{-1}. In an effort to find a wheat variety that would grow well and produce a good yield in the Canadian Prairies, Sir Charles E. Saunders made extensive crosses and developed Marquis wheat by 1906. To do this involved selective breeding: a planned program of hybridization of different wheat varieties, rigid selection of the best available material, preliminary and final evaluations of the results from replicated trials, and extensive testing of the new varieties. In the 1880s, Dr. William Saunders (Sir Charles's father) had introduced and tested many strains of wheat, often from Russia and India.

Crosses that led to the emergence of Marquis wheat were mainly focused on Hard Red Calcutta and Red Fife. The products of the crosses were tested at stations near Agassiz, British Columbia; Indian Head, Saskatchewan; and Brandon, Manitoba. At the latter two locations, these two varieties differed by three weeks in reaching maturity.

Sir Charles is famous for the "chewing test" he used to test the products of the crosses. He observed that chewing allowed him to determine the elasticity of the gummy substance produced (gluten). Other tests included baking bread with flour ground from the different strains to ensure that the product was satisfactory.

In 1906, Marquis wheat emerged as the product of a cross between a Hard Red Calcutta female and a Red Fife male. The kernels were dark red and hard, medium in size, and short. Heads were medium in length and bearded. Marquis ripens a few days before Red Fife and produces flour that is strong and of good colour. Tested at Brandon in 1908, Marquis wheat was the earliest to ripen and yielded 4336 kg·ha^{-1}—the best among the strains compared. Note that in Ottawa, Marquis wheat was not as productive, yielding only 2522 kg·ha^{-1}, reflecting the effect of climate and conditions on yield.

---

fragment length polymorphisms were determined from 261 wild and 98 landrace samples. The resulting data suggest that potatoes (*S. tuberosum*) arose from one domestication event. Today, wild potatoes still occur in Chile, along with eight cultivar groups of potatoes—some diploid and some triploid, others tetraploid or pentaploid (see Chapter 19). They vary noticeably in leaf shape, floral patterns, and tuber colour. Domesticated potatoes have been selected for short stolons, large tubers, and various colours of tubers from white to yellow to black. After being introduced to Europe, potatoes became an important food staple, supplanting traditional crops such as grains, and millions of people were adversely affected when a blight caused widespread failure of the potato crop. The blight was caused by the oomycete protist *Phytophthora infestans*. The "potato famine" in Ireland had huge social repercussions: many people died, and others emigrated to Canada, the United States, and Australia.

Tomatoes (*Solanum lycopersicum*) were domesticated from plants that grew in South and Central America, but there is considerable debate about when they were domesticated. Two modern forms, wild cherry tomatoes and currant tomatoes, were recently domesticated from stock native to eastern Mexico. Eaten raw or cooked, tomatoes come in different sizes, shapes, and colours. Different strains grow well in a variety of situations, meaning that tomatoes can be grown in many different climate zones. When grown in greenhouses, tomatoes are often pollinated by resident bumblebees, although many cultivars are self-pollinating.

Eggplants include three closely related cultivated species: *Solanum melongena,* the brinjal eggplant or aubergine; *Solanum aethiopicum,* the scarlet eggplant; and *Solanum macrocarpum,* the gboma eggplant. All cultivated species are native to the Old World, with *S. macrocarpum* and *S. aethiopicum* having been domesticated in Africa. The origin of the brinjal eggplant is less certain, but it may have originated in Africa and been domesticated in India and southeast China. During the Arab conquests, it spread from there to the Mediterranean and today is cultivated around the world. Brinjal eggplants and tomatoes are autogamous diploids.

Chili peppers, *Capsicum* species, are another member of the Solonaceae that originated in the New World. Known as producers of capsaicin ("Molecule behind Biology," Box 45.2, Chapter 45), chili peppers are often used to spice food. Cultivation of *Capsicum* species was well advanced and widespread in the Americas by 6000 years B.P. Then, as now, they were used as condiments.

People use other members of this family as the source of hallucinogenic compounds. Notable examples are tobacco and species in the genus *Datura.* Jimson weed or locoweed (*Datura stramonium*) contains strong poisons, including belladonna alkaloids, atropine, and scopalamine. It grows in many parts of the world and has often been used as a hallucinogenic drug because one active ingredient interferes with neurotransmitters (see Chapter 44) and can induce violent hallucinations. The name "locoweed" is a useful, important warning.

Stepping outside the Solonaceae to the family Brassicaceae, species in the genus *Brassica* also include many varieties seen on dinner tables worldwide (see "Domesticated Plants in the Genus *Brassica*," p. 823).

33.2k Squash

In the family Cucurbitaceae, at least five species of squash (*Cucurbita;* **Figure 33.19**) were domesticated in the Americas before European settlers arrived, some of them at least 10 thousand years B.P. Squash, beans, and corn were the "three sisters," staple foods farmed by many First Nations peoples in the New World (see Section 33.2h). Genetic data obtained from an intron region of the mitochondrial *nad1* gene suggest that at least six independent domestication events occurred. *Cucurbita argyrosperma* appears to have been domesticated from *Cucurbita sororia,* a wild Mexican gourd that grew in the same general area of Mexico as teosinte. *Cucurbita moschata* was probably domesticated somewhere in lowland South America and *Cucurbita maxima* in the humid lowlands of Bolivia from *Cucurbita andreana*. The *Cucurbita pepo* complex seems to be derived from at least two domestication events, one in eastern North America and one in northeastern Mexico.

Many people are familiar with *C. pepo* as the pumpkin, but the species also includes summer squashes and zucchinis. *Cucurbita maxima* is the Hubbard and other winter squashes, which also include some *C. pepo* and *C. moshata*. The diversity of these cultivars is astonishing. The intraspecific variations provide another example of the difficulty of applying the species concept to the diversity of life (see Chapter 19).

Figure 33.19
Many cultivars of squash, *Cucurbita* species, are New World domesticates, some first domesticated in southwestern Mexico at least 10 thousand years B.P.

33.2l Dogs

Behaviour provides clues to important aspects about the domestication of dogs. Researchers assessed the abilities of dogs, chimps, and wolves to read human signals indicating the location of food. Even young puppies with little human contact were more skillful at these social cognition skills than chimpanzees and wolves. Interspecific communication appears to have been strongly selected for during the domestication of dogs, building on the evolutionary history of social skills associated with cooperative hunting inherited from their wolf ancestors.

Dogs were among the first animals to have been domesticated, presumably to help people with hunting. Genetic, behavioural, and morphological evidence indicates that dogs were derived from wolves. The earliest morphological evidence suggests domestication of dogs by 14 500 years B.P. whereas mtDNA data suggest a date of 15 000 years B.P. Genetic data suggest an East Asian origin for dogs. Other mtDNA data from specimens in Latin American and Alaska indicate that dogs crossed into North America via the Bering Land Bridge, with people producing a group (clade) of dogs unique to the New World. The mtDNA data imply that either European colonists or native aboriginals actively prevented dogs that they brought with them from interbreeding with dogs already present in the New World.

33.2m *Salmo salar,* Atlantic Salmon

Atlantic salmon naturally occurs around the North Atlantic (locations in North America, Greenland, and Europe), but intensive fishing has reduced its natural stocks to the brink of extinction in some areas. However, Atlantic salmon have also been introduced to many sites around the world, from Jordan and Greece to Australia, New Zealand, Chile, Argentina, Brazil, and the Falkland Islands. There are both landlocked natural and introduced freshwater populations of Atlantic salmon. This species has been a traditional target of subsistence, sport, and commercial fishing and more recently the focus of aquaculture operations. The farmed fish have been selectively bred; thus, they are domesticated. Farmed fish are larger and more aggressive than those from wild stock, and they mature later.

Aquaculture operations have proved to be very lucrative, leading to a proliferation of these facilities in many areas. But some aquacultural operations have negative impacts. For example, escaped fish are thought to interbreed with local species (on the west coast of North America), threatening their genetic survival. This threat may be reduced by using sterile triploid Atlantic salmon for aquaculture. Sterile triploids can be mass produced, making it relatively feasible to use them in many areas. But concerns about the productivity and survival of triploid fish compared with diploid individuals have slowed the spread of their use.

Aquaculture can bring other problems. Sea lice, such as *Lepeophtheirus salmonis*, are parasitic copepod crustaceans that can cause serious problems for salmon aquacultural facilities. Recurrent sea lice infestations of aquacultured populations have spread to and decimated some wild salmon populations. Infestations by sea lice originating from cultured salmon have also caused a 99% collapse of some wild populations of pink salmon (*Oncorhynchus gorbuscha*) in coastal British Columbia.

The scale of aquaculture operations involving Atlantic salmon is astounding. In 2006, in Nova Scotia alone, 35 thousand tonnes of Atlantic salmon were produced by aquaculture. Atlantic salmon also dominate farmed stock in British Columbia. The scale of production has wide environmental, social, and economic implications for human nutrition and employment.

33.2n Some Organisms Were Domesticated for More Than One Use

Some animals and plants provide more than one crop. Cotton, as noted above, provides oilseed and fibre. Cattle are sources of meat, milk, blood, hides, and labour. Sheep provide wool, milk, meat, and hides. Sheepskins with the wool attached are used to make clothing.

33.2o Some Organisms Were Cultivated but Not Domesticated

People cultivate many species that have not been domesticated because there is no evidence of selective breeding. Mushrooms are examples because they have been cultivated and used as a source of protein for several thousand years without selective breeding of specific lines (= domestication).

Ostriches (*Struthio camelus*) are ranched (= cultivated) for their meat, hides, feathers, and eggs, but they are not domesticated. Crocodiles (*Crocodylus* species) are also ranched for their hides and meat. Oysters (*Pinctada fucata*) and other species of molluscs have been cultivated for hundreds of years mainly for pearls **(Figure 33.20)**.

Figure 33.20
Pearls for sale in the Pearl Market in Beijing, China.

Other animals, such as *Python regius* (ball pythons), are bred and sold to snake fanciers. Breeders may select individuals with specific traits in their breeding programs, technically making the animals domesticated because the definition does not speak to use.

STUDY BREAK

1. Distinguish between cultivation and domestication. Give examples.
2. How has domestication affected K, the carrying capacity?
3. Use two of the examples in this section to compare the timing, location, and path of domestication. Be sure to explain how the domesticated organisms are used.
4. How did the domestication of rice, wheat, barley, and lentils differ from that of squash and potatoes?
5. Use an example to show how domestication of animals depends on their behaviour.

33.3 Yields

In his book *The Upside of Down: Catastrophe, Creativity and the Renewal of Civilization*, Thomas Homer-Dixon calculated the amount of energy it would have taken to build the Coliseum in Rome. He estimated that it would have taken 44 billion kilocalories of energy: 34 billion for oxen and 10 billion for human workers (assuming that both worked 220 days a year for 5 years and that the humans received $12\,500$ kJ·day^{-1}). The Roman workers would have eaten grain, mainly wheat, as well as legumes, vegetables, wine, and a little meat. The oxen would have been fed hay, mainly alfalfa, as well as legumes, millet, clover, tree foliage, and wheat chaff. Records from the time indicate a yield of wheat of about 1160 kg·ha^{-1} and alfalfa of about 2600 kg·ha^{-1}. Wheat delivers 1.0×10^7 kJ·ha^{-1} and alfalfa 1.6×10^7 kJ·ha^{-1}. Growing wheat would have required 58 days of slave labour per hectare per year. If farmers had had to pay labourers, the cost of production would have been higher.

Based on these data, Homer-Dixon calculated that building the Coliseum would have required the wheat grown on 19.8 km^2 of land and the alfalfa on 35.2 km^2. At its peak around 1 and 2 CE the population of the city of Rome was 1 million, and that number would have required the food produced on 8800 km^2 of land, equivalent to the area of Lebanon today. Much of the wheat that fed Rome came from North Africa, as well as Sicily and other parts of Italy.

On average, one adult human needs 8300 kJ·day^{-1} or 3.0×10^6 kJ·yr^{-1} to maintain a stable body mass. This assumes a much lower level of exercise and

Figure 33.21
The field marked with the * is 20 ha and, if planted in wheat, should yield 13 200 kg in southwestern Ontario. Note that the area is also being used to harvest wind energy.

M.B. Fenton

physical exertion than the Roman worker in Homer-Dixon's calculations. If people were to meet their caloric demands from wheat alone, and if wheat delivers about 8700 kJ·kg⁻¹, at 8300 kJ·day⁻¹, each person would need to consume about 350 kg of wheat a year. In 2007 in southwestern Ontario, a wheat yield of 6600 kg·ha⁻¹ meant that 1 ha of land would support 18.8 people for a year. A city of 50 thousand, eating only wheat, would need the wheat produced on 2660 ha (**Figures 33.21** and **33.22**).

The data in **Table 33.1**, demonstrate that it takes energy input to generate energy, in this case food. So variations in crop yields noted at the beginning of this chapter have huge repercussions. From a farmer's perspective, the difference in farm income would be substantial if the yield of corn were 1600 kg·ha⁻¹ versus

10 000 kg·ha⁻¹ (Table 33.1). But, even without considering the costs of transportation (from farm to market), the data illustrate the impact of the cost of fuel. Spraying, planting, and trucking all require diesel fuel, so any change in the price of this commodity influences the costs of farm operations.

Terrain influences the level of technology that can be used in farming practices. Small terraced plots (**Figure 33.23, p. 824**) must be worked by hand or with

Table 33.1

Balancing Cost and Yield, a Farm in Southwestern Ontario (Costs as of Autumn, 2007)		
Crop/Material/Process	Yield kg·ha⁻¹	C$·ha⁻¹
Corn	10 000	1650
Soy beans	3 500	1300
Wheat	6 600	900
Costs		
Seed		150
Fertilizer		125–225
Spray		50–100
Planting		38
Labour		40–190
Combine		100
Trucking		25
Crop insurance		20
Field rental		325

a.

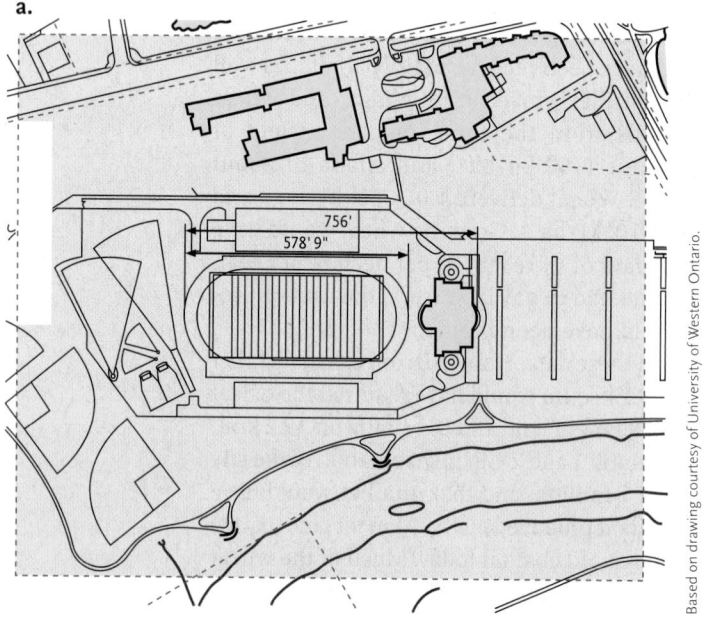

Based on drawing courtesy of University of Western Ontario.

b.

M.B. Fenton

Figure 33.22
Envisioning 20 ha. Two views of the same area: a drawing (a) in which the coloured area represents 20 ha and an aerial photograph (b) showing the same football stadium on the campus of the Western University, in London, Ontario.

Domesticated Plants in the Genus *Brassica*

Some plants have provided humans with an embarrassment of riches. Imagine plants in one genus (*Brassica*) whose flowers, roots, stems, leaves, and seeds are all important food crops **(Figure 1)**. Foods from these vegetables are high in vitamin C and soluble fibre. They contain a rich mixture of nutrients, including some (diindolyl-methane, sulforaphane, and selenium) thought to have anticancer effects.

A sample of an all-*Brassica* meal could include broccoli, cauliflower, Brussels sprouts, kale, and cabbage, all variations of one species, *Brassica olera-cea*. You could add rutabaga (*Brassica napus*) along with turnip and rapini (*Brassica rapa*) to the menu and use canola oil (see "People behind Biology," Box 33.2) and mustard (*B. rapa*, *Brassica carinata*, *v elongata*, *Brassica juncea*, *Brassica nigra*, *Brassica ruprestris*) to dress the parts of the meal presented as salad.

Nuclear restriction fragment length polymorphisms (RFLPs) obtained from 10 different *Brassica rapa*, 9 cultivated types of *B. oleracea*, and 6 other species in *Brassica* and related genera suggest two basic evolutionary pathways **(Figure 2)** for diploid species: one that gave rise to *Brassica fruticulosa*, *B. nigra*, and

Figure 1

Brassicas in our diets. *Shown here are cultivars of brassicas that commonly appear in people's diets, including **(a)** radishes, **(b)** Brussels sprouts, **(c)** cauliflower, **(d)** broccoli, **(e)** turnip, **(f)** rutabaga, **(g)** mustard, **(h)** cabbage, and **(i)** canola oil.*

Sinapis arvensis (*Brassica adpressa* is a close relative), and the other to *B. oleracea* and *B. rapa* **(Figure 3)**. *Raphanus sativus* and *Eruca sativa* appear to be intermediate between the two lineages (see Figure 2). Europe and East Asia appear to have been centres of domestication for *Brassica* species. The related *Arabidopsis* is an important experimental tool used in understanding the genetics and selection of desirable traits in *Brassica*.

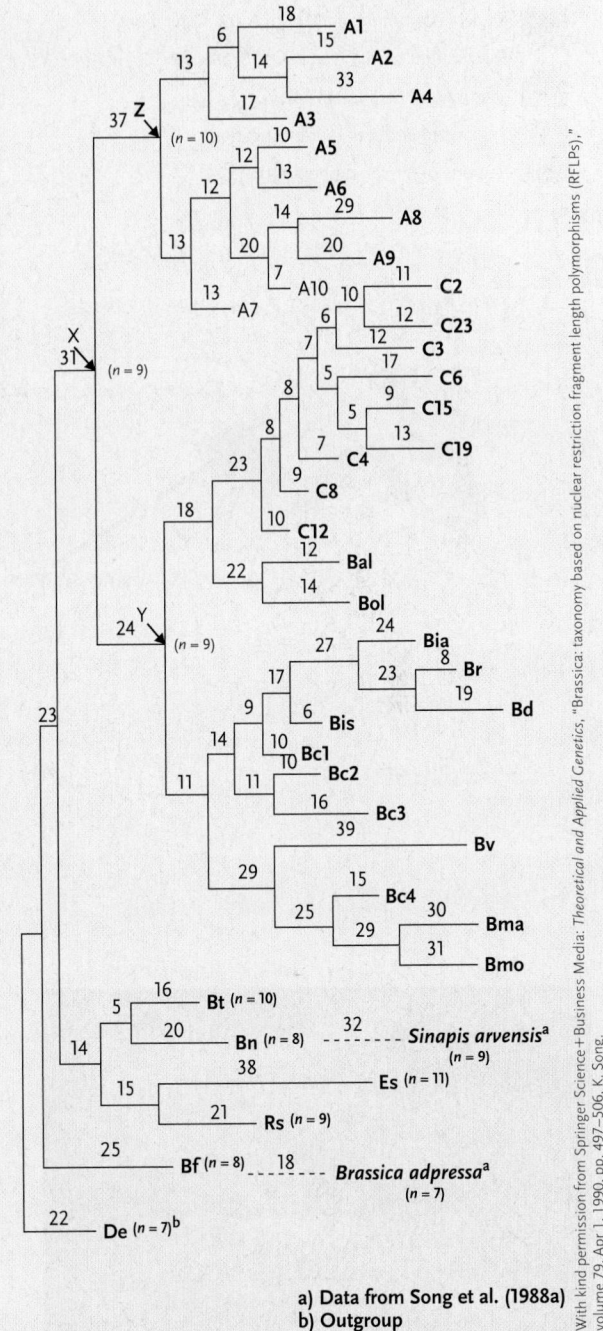

a) Data from Song et al. (1988a)
b) Outgroup

With kind permission from Springer Science+Business Media: *Theoretical and Applied Genetics*, "Brassica: taxonomy based on nuclear restriction fragment length polymorphisms (RFLPs)," volume 79, Apr 1, 1990, pp. 497–506, K. Song.

Figure 2

The shortest phylogenetic tree showing the relationships between different species of Bras-sica, including cultivars and wild species. A1–A5 B. rapa *cultivars:* 1 = *flowering pak choi;* 2 = *pak choi;* 3 = B. narinosa; 4 = *Chinese cabbage;* 5 = *turnip;* A6–A10 B. rapa *wild:* Bal = B. alboglabra; Bc 1–4 = B. cretica; Bd = B. drepanensis; Bia = B. incana; Bis = B. isularis Bma = B. macrocarpa; Bmo = B. montana; Bol = B. oleracea; Br = B Rupestris; BBv = B. villosa; C2–C23 B. oleracea *cultivars:* 2 = *broccoli;* 3 = *broccoli (packman);* 4 = *cabbage;* 8 = *Portuguese tree kale;* 12 = *Chinese kale;* 15 = *kohlrabi;* 19 = *borecole;* 23 = *cauliflower;* Bf = B. fruticulosa; Bn = B. nigra; Bt = B. tournefortii; De = Diplotaxis erucoides; Es = Eruca sativa; Rs = Raphanus sativus.

(Continued)

the help of animals. Large expanses of relatively flat land can be worked effectively and efficiently with machinery **(Figure 33.24, p. 824)**, increasing the energy consumption associated with farming but also the yield. Modern agriculture relies on energy from petro-leum products, so increases in the price of oil can undermine food production, especially if crops origi-nally grown for food are used in the production of biofuels.

STUDY BREAK

1. How must crop yield be balanced against the cost of achieving it? Does this apply to a kitchen garden (as opposed to a functioning farm)?
2. What factors were fundamental in allowing the city of Rome to have a population of over 1 million people in 2 CE ?

Domesticated Plants in the Genus *Brassica* (Continued)

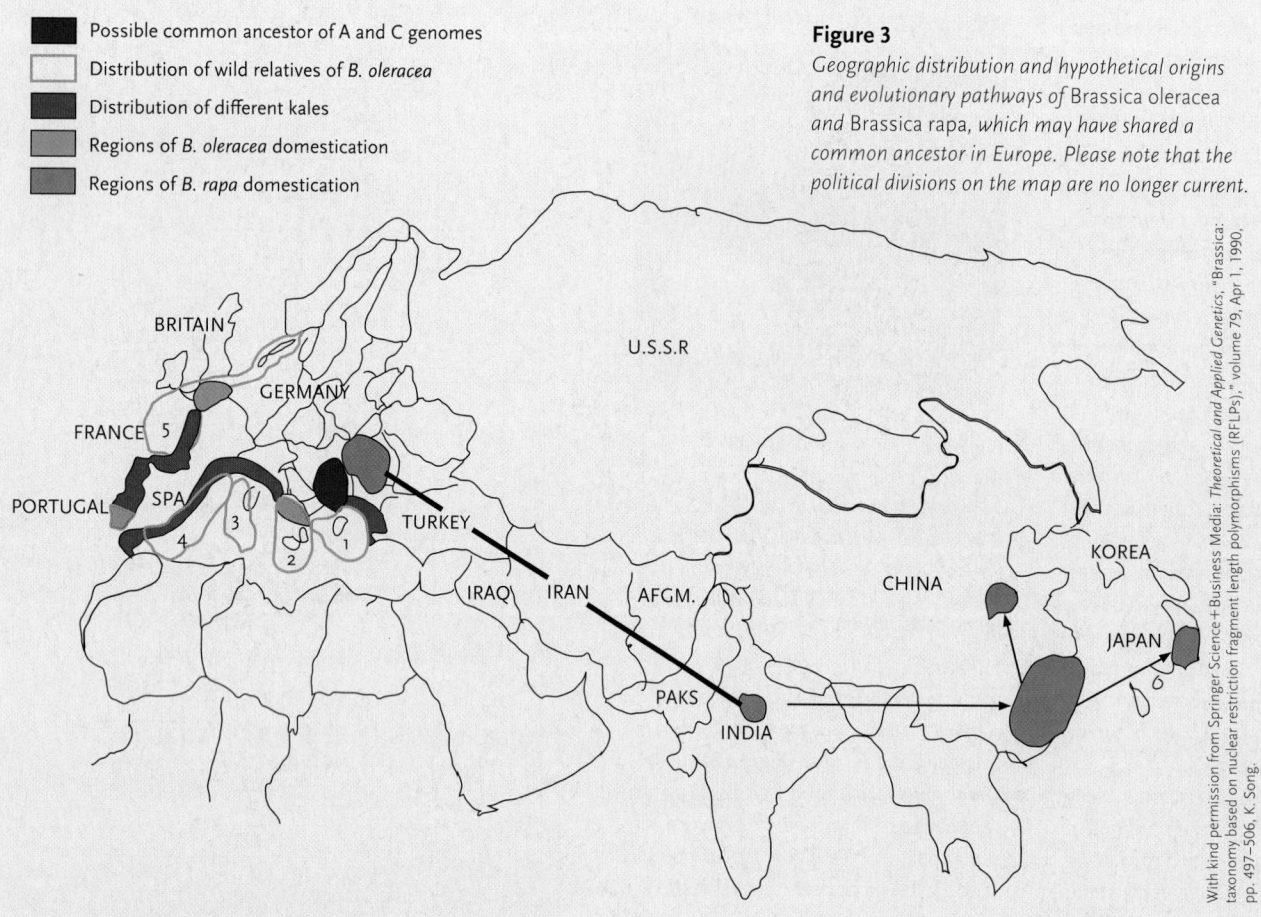

Legend:
- Possible common ancestor of A and C genomes
- Distribution of wild relatives of *B. oleracea*
- Distribution of different kales
- Regions of *B. oleracea* domestication
- Regions of *B. rapa* domestication

Figure 3

Geographic distribution and hypothetical origins and evolutionary pathways of Brassica oleracea and Brassica rapa, which may have shared a common ancestor in Europe. Please note that the political divisions on the map are no longer current.

With kind permission from Springer Science+Business Media: *Theoretical and Applied Genetics*, "Brassica: taxonomy based on nuclear restriction fragment length polymorphisms (RFLPs)," volume 79, Apr 1, 1990, pp. 497–506, K. Song.

Figure 33.23
A series of rock walls creates terraced areas for growing crops near Beijing in China. The terraces hold soil, but the setting precludes extensive use of mechanized farming equipment.

Figure 33.24
Large expanses of relatively flat land lend themselves to mechanized farming, allowing more uniform conditions and crops.

33.4 Complications

As anyone who has ever gardened or worked a farm knows well, there is more to growing crops than putting seeds or small plants in the ground and then harvesting the crops.

33.4a Fertilizer, Water, Yield, and Pests Are Factors in the Care of Crops

In the course of operating an experimental farm in the Negev Desert in southern Israel **(Figure 33.25),** researchers established several basic truths. By providing 20 m^3 of manure (sheep and goat) and 600 kg of ammonium sulfate per hectare, they could obtain good yields: 4800 kg·ha^{-1} of barley (where less-tended crops yielded 400 to 600 kg·ha^{-1}) and 4400 kg·ha^{-1} (nanasit strain) or 2700 kg·ha^{-1} (Florence strain) of wheat. They were able to produce 750 kg·ha^{-1} of carrots and 650 kg·ha^{-1} of onions. Achieving these yields required cultivation of the soil, irrigation, and dealing with a variety of pests. Their farm became an oasis of green that attracted hares, gazelles, porcupines, desert partridges, and a host of insects, meaning that control of pest species had to be routine.

Irrigation was a key to good crops, and the experiment had been designed to test the prediction that by collecting runoff water and storing it in cisterns, the people there could farm in an area with little and highly seasonal rainfall. In 2006–07, the 60 mm total rainfall in the area occurred between November 20 and April 16. Water collected as runoff and stored in a 1400 m^3 cistern could last the farm (people, animals, and crops) over two years.

33.4b Cats Are Sometimes Workers

The need to control rodent pests in areas where grain is stored **(Figure 33.26)** might be one factor explaining the domestication of cats, *Felis silvestris catus*. Carlos A. Driscoll and colleagues examined short tandem repeat (STR) and mtDNA data from 979 cats and wild progenitors to examine relationships among them. The evidence suggested at least five founder populations, including the European wildcat (*Felis silvestris silvestris*), near Eastern wildcat (*Felis silvestris lybica*), central Asian wildcat (*Felis silvestris ornata*), southern African wildcat (*Felis silvestris cafra*), and Chinese desert cat (*Felis silvestris bieti*). Each of these populations represents a distinct subspecies. Cats were thought to have been domesticated in the Near East, and their descendants were transported across the world with assistance from humans **(Figure 33.27 , p. 826).** Driscoll and his colleagues proposed that the domestication of cats coincided with the development of agriculture in different locations.

33.4c Crops Can Become Contaminated

People living in rural parts of Bosnia, Bulgaria, Croatia, Romania, and Serbia exhibit a high incidence of a devastating renal disease termed *endemic Balkan nephropathy* (EN). People afflicted with EN progress

Figure 33.25
An experimental farm plot near Avdot in the Negev Desert in southern Israel. The green areas are irrigated with water stored in an underground cistern. The experimental farm was established on an ancient farm site. Other fields previously under cultivation are shown. By collecting runoff during and after rainfall, farmers have stored water for their families and crops for hundreds of years.

Figure 33.26
Harvested crops can be stored in different ways. Near Tien in China, farmers hang collections of corn cobs after harvest. This approach to crop storage suggests a dearth of local birds and rodents that might consume the corn.

Figure 33.27

The origins of domestic cats. **(a)** Genetic assessment of 979 cats (*Felis silvestris catus*) based on short tandem repeats (STR) and mtDNA identifies different contributors to cat genotypes. **(b)** The accompanying phenogram of 851 domesticated and wild cats illustrates the relationships between domestic cats and wild genetic contributors.

From Carlos A. Driscoll, Marilyn Menotti-Raymond, Alfred L. Roca, Karsten Hupe, Warren E. Johnson, Eli Geffen, Eric H. Harley, Miguel Delibes, Dominique Pontier, Andrew C. Kitchener, Nobuyuki Yamaguchi, Stephen J. O'Brien, David W. Macdonald, "The Near Eastern Origin of Cat Domestication," *Science*, vol. 317, Jul 27, 2007, pp. 519–523. Reprinted with permission from AAAS.

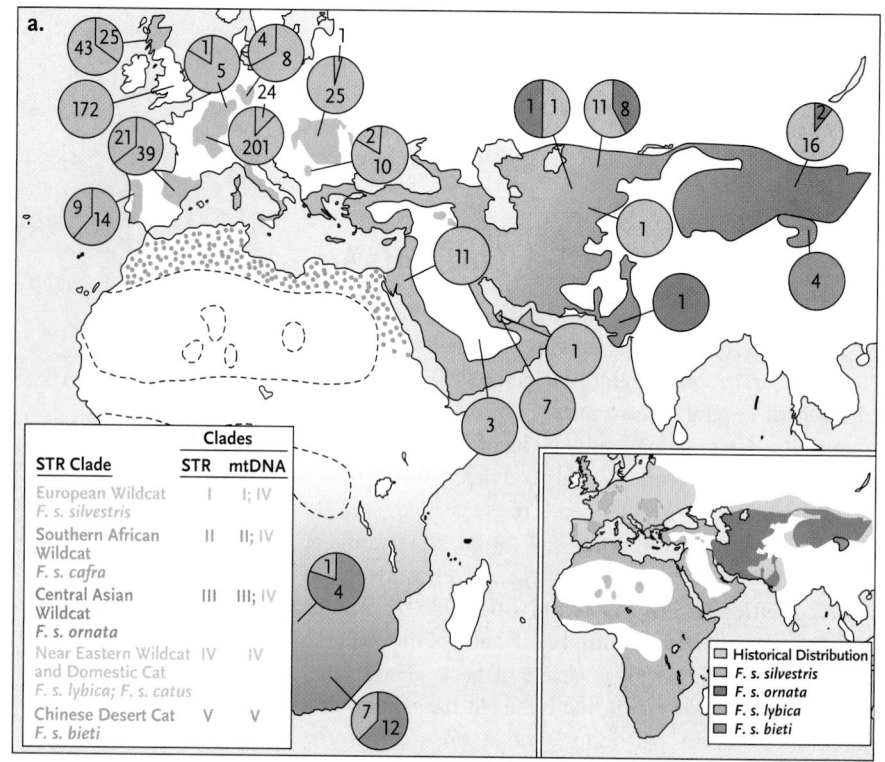

STR Clade	Clades	
	STR	mtDNA
European Wildcat *F. s. silvestris*	I	I; IV
Southern African Wildcat *F. s. cafra*	II	II; IV
Central Asian Wildcat *F. s. ornata*	III	III; IV
Near Eastern Wildcat and Domestic Cat *F. s. lybica*; *F. s. catus*	IV	IV
Chinese Desert Cat *F. s. bieti*	V	V

Historical Distribution
F. s. silvestris
F. s. ornata
F. s. lybica
F. s. bieti

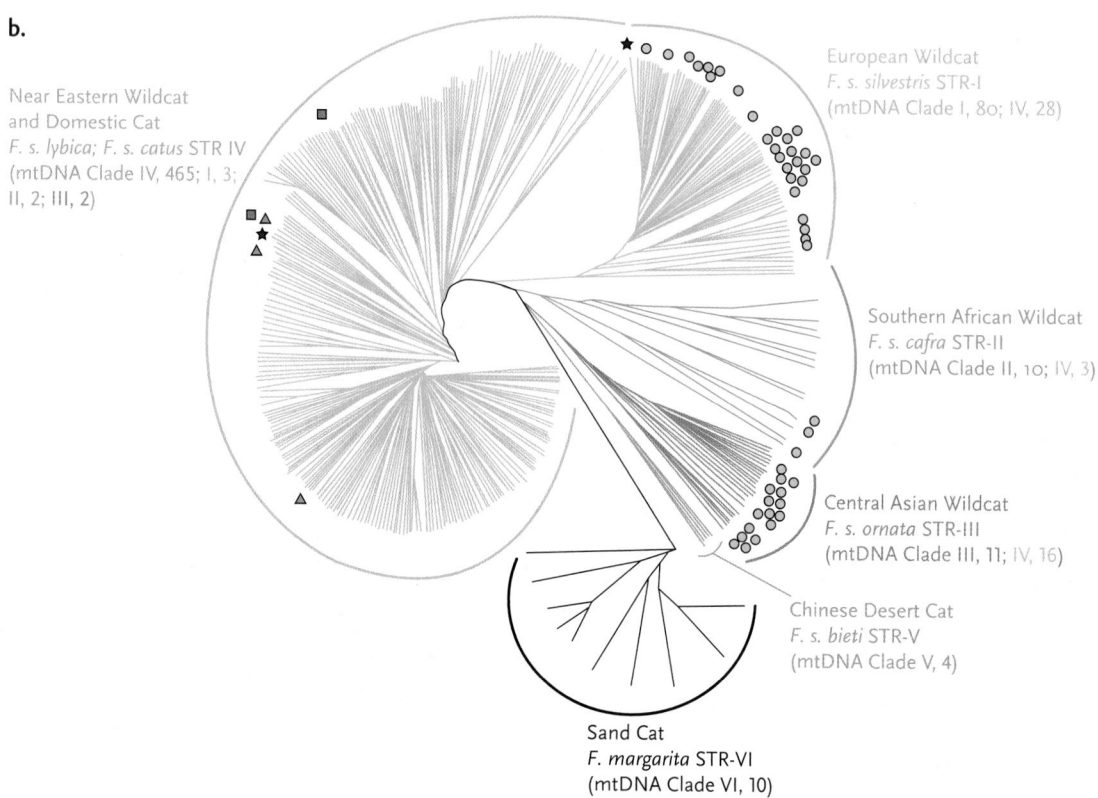

Near Eastern Wildcat and Domestic Cat *F. s. lybica*; *F. s. catus* STR IV (mtDNA Clade IV, 465; I, 3; II, 2; III, 2)

European Wildcat *F. s. silvestris* STR-I (mtDNA Clade I, 80; IV, 28)

Southern African Wildcat *F. s. cafra* STR-II (mtDNA Clade II, 10; IV, 3)

Central Asian Wildcat *F. s. ornata* STR-III (mtDNA Clade III, 11; IV, 16)

Chinese Desert Cat *F. s. bieti* STR-V (mtDNA Clade V, 4)

Sand Cat *F. margarita* STR-VI (mtDNA Clade VI, 10)

from chronic renal failure to a high incidence of cancer of the upper urinary tract. EN and its associated cancer can be related to chronic dietary poisoning by aristolochic acid. The source of the poisoning is contamination of grain crops with the plant *Aristolochia clematitis*. A clue to this situation came from horses that developed renal failure after being fed hay contaminated with *A. clematitis*. The presence of weeds or other contaminants in crops **(Figure 33.28)** poses an important challenge to

farmers, one that goes beyond how the productivity of the crop is affected by weeds' needs for water and nutrients.

Ironically, the medicinal virtues of extracts of *A. clematitis* are extolled on some websites selling homeopathic remedies. A first step in solving the mystery of EN came from case studies of some Belgian women who had developed renal problems after taking extracts of *A. clematitis* as part of a weight loss program.

Figure 33.28
Weeds, such as the milkweeds (*Asclepias* species) growing in this barley field in southwestern Ontario, can pose a problem at harvesting. If the weeds are toxic, they must be extracted from the harvested crop to ensure the safety of animals and people that consume the barley.

Plants that look the same often produce quite different chemicals. Poison hemlock, for example, superficially resembles Queen Anne's lace (*Daucus carota*), a common weed. It can also be confused with fennel (*Foeniculum vulgare*) and parsley (*Petroselinum crispum*), two herbs often used in cooking.

The speed with which changes can occur in biological systems can be astonishing, exemplified by the assimilation of new foods into our own diets whether the product is natural or synthetic. This is not a feature unique to humans. Work on ant farmers revealed that one species (*Cyphomyrmex rimosus*) introduced to Florida quickly acquired a crop cultivated by another species of ant indigenous there (*Cyphomyrmex minutus*).

STUDY BREAK

1. What factors influence crop yield?
2. How does crop yield vary?
3. What are some of the problems associated with crops?

33.5 Chemicals, Good and Bad

Plants are often treasured as much for the chemicals they produce as for their use as food, and until the time of Linnaeus (Chapter 19), plants were mainly classified by their medicinal properties. The chef who adds rosemary (*Rosmarinus officinalis;* **Figure 33.29**) to a dish as it is cooking knows that the flavour will enhance the final product. Other plant products are used as medicines. Phenolic compounds are responsible for the distinctive flavours of coffee, cinnamon, cloves, and nutmeg, which add flavour and aroma to food. Some of these compounds, such as caffeine, can be addictive.

The secondary compounds of plants may add taste to our food but can also be lethal. Consider how the plants use these compounds. Ginsenosides (obtained from ginseng, *Panax quinquefolius*), for example, appear to be used by ginseng plants as fungicides, whereas humans take them to stimulate the immune system. Other plant products are toxic. Conine, produced by poison hemlock (*Conium maculatum;* not to be confused with the hemlock tree), is an active ingredient used by the plant to defend against herbivores (**Figure 33.30**). It was the hemlock used to kill Socrates.

Figure 33.29
Herbs such as rosemary (*Rosmarinus officinalis*) are used to add flavour during cooking.

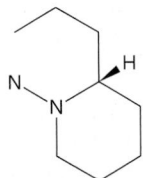

Figure 33.30
Conine, the poison from poison hemlock.

Also in Florida, the grasshopper *Romalea microptera* uses a mix of chemicals to repel ants (see "Molecule behind Biology," Box 32.3, Chapter 32). Soon after people began to use 2,4-D (2,4-dichlorophenoxyacetic acid) to kill weeds, the grasshopper added 2,5- dichlorophenol to its ant-repellent mixture. This situation is analogous to humans cultivating and domesticating new crop species.

The pharmacological potential of plant products for treating and preventing human ailments has not been lost on three groups of people: those interested in conserving biodiversity, those concerned about alleviating human suffering, and those anxious to make money by selling biopharmaceuticals to others.

STUDY BREAK

1. Why do humans use 2,4-D? What other uses could it have?
2. Why are spices important?
3. What is poison hemlock?

33.6 Molecular Farming

CONCEPT FIX Some people believe that genetically modified organisms pose a serious risk to those who consume them. The reality is that genetically modified food organisms, such as rice, are not dangerous and may be beneficial (e.g., golden rice with enriched Vitamin A). The other reality is that some genetically modified organisms pose a risk to wild stock of the same species. ⬡

Many native peoples in the New World used tobacco (*Nicotiana tabacum*) as a traditional medicine to ease the pain of childbirth, stave off hunger, and treat various ailments **(Figure 33.31)**. Tobacco was dried and smoked in ceremonies; used in poultices; brewed as a tea; and used as an emetic, an expectorant, and a laxative. By 1540, tobacco reached Europe, where it was said to cure illnesses ranging from epilepsy to plague. People have also used nicotine, an alkaloid in tobacco, in the same way as tobacco plants do, as an insecticide. The recreational use of tobacco has given the plant a bad name, even though other *Nicotiana* species are commonly used in gardens for their aromatic white or pink flowers.

Tobacco, however, is undergoing a rebirth as a positive contributor. *Molecular farming* involves the use of plants as producers of specific proteins useful as human medications. Plants can be genetically modified to produce large amounts of these proteins at low cost. Concerns about using plants as molecular farms reflect the possibility of biologically active products entering ecosystems and dispersing through food webs, via pollen or seeds, all of which could negatively affect existing crops. Enter tobacco, a nonfood plant that is harvested before flowering. Tobacco is not cold hardy (in Canada), but it is easy to genetically modify and is highly productive

Figure 33.31
A tobacco field in southwestern Ontario.

(40-day production cycle). Using appropriate DNA technology (see Chapter 15), T-DNA containing the human *IL-10* gene was inserted into the tobacco genome by connecting it to a plant promoter and an *Agrobacterium* terminator. In this way, tobacco plants are modified to produce interleukin-10 (*IL-10*), which can be used to treat irritable bowel disease in humans. Interleukin is a cytokine involved in the regulation of inflammatory diseases, reducing the production of necrosis factor by tumours. The IL-10 was found not to enter the soil in which the tobacco was grown or the aphids or other insects that fed on the tobacco plants. The vast amount of agricultural land that was used to grow tobacco for recreational use can now be used for molecular farming. This change to molecular farming of tobacco could produce many different useful compounds.

STUDY BREAK

1. How are plant products used in addition to their role as food?
2. What is molecular farming?
3. What conservation risks are associated with molecular farming? With agriculture in general?

Feeding People and Keeping Them Healthy

We have seen how humans' ability to harness nature has allowed us to reduce the number of hungry people in the world. Nevertheless, in 2008, converting food crops to biofuels in the interest of being "green" meant less food. This development coincided with climate change, which further reduced the availability of food worldwide. This impact was amplified by increases in the prices of food crops, in turn making food less accessible to many poor people. This brings us back to the price of oil, the energy used in agriculture to produce food, and the cost of food.

We use plants for more than food. In 2008, over 1 million people died from malaria, caused by a blood parasite (see Chapter 27). Some 200 million people worldwide suffer from malaria. Most (over 75%) of the deaths were of children under 5 years of age and living in Africa. For years, alkaloids (quinine, quinidine, cichonidine, and cinchonine) extracted from the bark of four species of tree in the genus *Cinchona* have been used to treat agues and periodic fevers. Cinchona alkaloids have been particularly effective against malaria, but some strains of malaria are resistant to quinines and other treatments, suggesting that they are not the solution to malaria.

Enter qinghao, also known as huang hua hao, an extract from a ubiquitous shrub, *Artemisia annua*. Extracts of qinghao (artemisinin) have long been used in traditional Chinese medicine. In 1971, Chinese scientists discovered its effectiveness in treating malaria and

reported it to the world in 1979. In 2008, artemisinin combination treatments became first-line drugs for some forms of malaria (uncomplicated falciparum malaria), but they were not available worldwide. Using improved agriculture techniques—selection of high-yielding hybrids, microbial production, and synthetic peroxidases—could lower prices, increase the availability of artemisinin, and save many lives.

In 2007, about one-third of the annual U.S.$30 billion worldwide investment in agriculture research was aimed at solving the problems of agriculture in developing countries—home to about 80% of the global population. This investment is less than 3% of the amount that countries of the Organization for Economic Co-operation and Development (OECD) spend to subsidize their own agricultural production. To ensure continued reduction of human hunger, OECD countries must invest more to solve the agriculture problems in the developing world. One important first step in this process is recognizing the energy input required to grow enough food to feed the human population (see Section 33.3).

We know that our agricultural prowess can be used to increase production of food and medicines such as artemisinin. Golden rice provides another example. In 1984, the World Health Organization estimated that 250 thousand to 500 thousand children a year had diets lacking in vitamin A. This deficiency damaged their retinas and corneas, so many of them went blind and half of them died.

By the early 2000s, Ingo Potrykus and Peter Byer had used genetic engineering to splice two daffodil genes and a bacterial gene into the rice genome. The genetically modified rice, **golden rice,** was golden in colour and produced precursors to vitamin A in its endosperm. Golden rice offered a way to address the results of vitamin A deficiency. Growing golden rice was field tested in Louisiana in 2004 and 2005.

In 2008, it appeared that even though golden rice offered the ability to prevent the suffering arising from vitamin A deficiency, it was unlikely that any would be planted before 2012. The delay reflects a combination of widespread public suspicion about genetically modified organisms (GMOs) and the high cost of obtaining approval to use GMOs. The costs associated with getting approval mean that only large companies with large budgets are likely to succeed in getting GMO products approved. The controversy persists today.

Moving more people away from "the edge" means investing more in finding solutions to agricultural operations in the developing world to ensure that discoveries such as artemisinin and golden rice are used to advantage. This requires policies and investments that support small-scale operations, probably subsidizing the costs of energy and water necessary to grow enough food. Our ancestors, who domesticated crops such as wheat and corn, lentils and squash, and potatoes and rice, changed the world for us. We should ensure that their legacy lives on.

33.7 The Future

It is tempting to believe that the diversity of life will continue to provide humans with solutions to many of their problems in the world: food for the hungry, poisons to selectively control pests, and biopharmaceuticals to cure diseases. Molecular farming potentially allows new approaches to solving old problems. Although we can continue to domesticate other species, effectively taking evolution in directions that suit

us, we must remember that when new forms require higher investments in energy and fertilizer, costs can outweigh benefits.

In 2008, global increases in the price of food were partly due to our using traditional food crops (e.g., corn) to produce ethanol to fuel vehicles. In 2007, the energy return on investment (EROI) for biofuel (ethanol, or food for internal combustion engines) from corn was about 1. This means that every litre of ethanol produced from corn consumes about a litre of petroleum fuel.

Other crops, for example, poplar trees, have much better EROIs. Although biofuels are potentially low-carbon-emitting energy sources, the means of production dramatically influences their "greenness." In many cases, production of biofuels generates a *biofuel carbon debt*. In Brazil, Southeast Asia, and the United States, converting native habitat (rain forest, peatlands, savannah, and grasslands) to produce biofuels generates 17 to 420 times as much CO_2 as the habitats normally produce. When these costs are taken into account, biofuels are not feasible environmentally friendly alternatives.

Calculating EROI and tracking it over time helps put energy use in perspective. In Rome in 1 CE, the EROI for wheat was 12:1, and for alfalfa, it was 27:1. In 2007, the EROI value for gasoline was about 17:1, whereas in the 1930s, it was about 100:1. The values of EROI speak to the sustainability of a process, so Canadians must be concerned that the EROI on the tar sands is less than 4:1 without taking into consideration the water consumed by the process.

The development of resistance to toxins, whether of bacteria to antibiotics or of insect pests to insecticides or of weeds to herbicides, demonstrates that evolution works both ways. Perfecting genetic strains of crops protected by resistance to a pathogen or pest and using them exclusively can make the crops vulnerable to pathogens or pests that are resistant to the defence(s). The potato famine, discussed in Section 33.2j, was exacerbated by the lack of genetic diversity of the potatoes used in Europe: they were vulnerable to blight.

Over evolutionary history, individuals able to exploit other species had an advantage over those that did not, whether within a society or between societies. The same principles apply to our own species. Our advantages of exploiting other species include increased access to food (quantity and quality), labour, materials for constructing things, and chemicals for treating disorders or controlling pests. These advantages were amplified through domestication, which meant increasing control over the other species, leading to several net effects. One effect was achieving larger populations of humans because of better access to food and/or protection from disease. Ironically, living closer to more animals also exposes us to diseases such as avian flu, swine flu, and smallpox. Another factor was probably the increase in available time for the development and perfection of new tools and techniques and the emergence of groups of people in society who did not contribute directly by gathering or processing food. Such people could have contributed to society through their talents as artisans, soldiers, or even politicians.

The range of possibilities seems endless, particularly with the advent of the ability to directly modify genotypes and thus phenotypes. Perhaps we should be glad that the Attine ants have domesticated only fungi.

STUDY BREAK

What is artemisinin? What is golden rice? Why are they important?

Review

aplia

To access course materials such as Aplia and other companion resources, please visit www.NELSONbrain.com.

33.1 Domestication

- The green revolution is credited with reducing the number of hungry people in the world by increasing agricultural productivity. Increased productivity reflects the use of improved strains of crops, increased applications of fertilizers, and more extensive irrigation. Higher productivity (crop yields) also reflects more dependence on mechanized (fossil fuel–based) farming operations.

- There are three stages in the exploitation of biodiversity for human benefit. The initial stage involves hunting and gathering, collecting organisms in the wild. The second stage is cultivation or caring for organisms under progressively controlled conditions. The third stage, domestication, involves selective breeding to enhance desired characteristics and features. The domestication process can apply to animals; plants; or other organisms, such as yeast and fungi.

- Domestication of asexually reproducing organisms does not involve selective breeding, which provides more control, allows shorter generation times, or speeds up the process of selection.

- Sexually reproducing organisms that are successfully domesticated are usually capable of self-fertilization and/or readily propagated by grafting.

- Living in the same sites year-round allowed people to tend and protect their crops. Increased time in one place would also have facilitated the process of selective breeding.

- At least 200 species of ants in the tribe Attini are obligate farmers. Other ants also tend seed gardens.

- The first evidence of domestication appears to be figs by about 12 000 years B.P. Cultivation may have been practised for 1000 years before domestication. Many crops had been domesticated by 6000 years B.P.

33.2 Why Some Organisms Were Domesticated

- Yeast appears to have been domesticated in Egypt by 6000 years B.P., when it was used in making bread and beer. Cotton may have been domesticated in at least four sites, two in the Old World and two in the New World. Cotton is used as a source of oil seed and fibre, and some domestication had occurred by 4400 years B.P.

- The change from dehiscence (shattering) to indehiscence (nonshattering) was critical in the domestication of plants, such as grasses, whose seeds were the crop to be harvested.

- Squash and potatoes provide carbohydrates from fruits or tubers; the seeds are not the target of domestication.

- The domestication of dogs appears to have been based, in part, on their social behaviour, including their ability to communicate with people. Some experiments show that dogs are better at reading communication signals from people than other animals, such as chimps.

- Some animals are farmed for meat and hides (and for feathers and eggs), but there is no evidence of selective breeding.

33.3 Yields

- The level of farming intensity, the strains of crops, the application of fertilizers, and irrigation affect crop yields. Terrain can influence the level of mechanization.

- The data on corn and wheat yields clearly demonstrate variation over time and location.

33.4 Complications

- Rich patches of food reflect higher productivity but can attract pests.

33.5 Chemicals, Good and Bad

- Various plant chemicals are used as stimulants, medicines, hallucinogens, pesticides, and flavourings.

33.6 Molecular Farming

- Molecular farming is the use of plants to produce various useful proteins in large amounts at lower cost. Molecular farming may prove to be an inexpensive way to manufacture medicines and other substances.

33.7 The Future

- Energy return on investment (EROI) is an important concept whether the topic is crops or oil. When the cost of seeing a crop from planting to harvest exceeds the return, there is little point in continuing to use the crop. The same should be true for harvesting energy, for example, oil from the tar sands.

Questions

Self-Test Questions

1. According to available evidence, when did humans first domesticate other organisms?
   a. 6000 years ago
   b. 8000 years ago
   c. 12 000 years ago
   d. 20 000 years ago

2. The process of moving from cultivation to domestication involved dehiscence (shattering) to indehiscence (nonshattering) in the following crops.
   a. rice
   b. wheat
   c. squash
   d. Both a and b are correct.

3. Parthenocarpy was important in the domestication of which species?
   a. dogs
   b. figs
   c. lentils
   d. cotton

4. At Abu Hureyra, which of the following did settlers domesticate?
   a. olives
   b. some grains
   c. cattle
   d. hot peppers

5. Which group includes only domesticated organisms?
   a. yeast, mushrooms, pigs, and oysters
   b. honeybees, yeast, rice, and ostriches
   c. cattle, pigs, cats, and dogs
   d. lentils, mushrooms, yeast, and crocodiles

6. Plants in the family Solonaceae include which of the following?
   a. tomatoes, potatoes, and eggplant
   b. potatoes, hot peppers, and deadly nightshade
   c. corn, tobacco, and hot peppers
   d. Both a and b are correct.

7. If a person needs $8300\ kJ \cdot day^{-1}$, and wheat yields $6638\ kg \cdot ha^{-1}$, how much land would be needed to produce enough wheat for 1 year for a city of 20 thousand people who eat only wheat?
   a. 155 ha
   b. 1055 ha
   c. 2055 ha
   d. 5055 ha

8. Endemic Balkan nephropathy is an example of a disorder arising when people eat crops that have been subjected to the following.
   a. irrigation with polluted water
   b. contamination with *Aristolochia clematitis*
   c. contamination with *Rosmarinus officinalis*
   d. contamination with *Asclepias exultata*

9. *Nicotiana tabacum* is being used to produce which of the following medications?
   a. nicotine
   b. interleukin
   c. acetylcholine
   d. conine

10. EROI, energy return on investment, suggests that alfalfa in Roman times was this many times as efficient at energy production as the tar sands in Alberta in 2008.
    a. 4
    b. 6
    c. 10
    d. 50

Questions for Discussion

1. When is domestication complete? At what point is a domesticated stock a separate species? Is domestication ever complete?

2. Why have so few aquatic species been domesticated? Why has there been a recent increase in the numbers of domesticated aquatic organisms?

3. How do domesticated populations threaten native species? Should this be a concern for conservation biologists? What can be done to minimize this threat to native species?

4. Why is the price of oil so important in food production?

5. How can we ensure continued supplies of water for food production?

SYSTEMS AND PROCESSES

Larva of a monarch butterfly (*Danaus plexippus*) feeding on milkweed (*Asclepias* sp.).

In the next three units, we focus on the key characteristics of systems and processes in plants and animals, and make several comparisons between land plants and terrestrial animals. The emphasis on plants and animals reflects the approach emphasized in the text and is not meant to downplay the importance of other kingdoms of life.

Plants and animals share many features: their bodies are composed of eukaryotic cells that form tissues and organs. Why then are plants and animals so different? Part of the answer lies in how long ago plants and animals arose from their last common ancestor. The ancestors of modern-day plants and animals are thought to have diverged about 1.6 billion years ago (bya), when both groups consisted of unicellular organisms. The unicellular organism that was the common ancestor of both plants and animals had already incorporated the endosymbiont that would become mitochondria, which is why *both* plants and animals have mitochondria (see Figure 2.17, Chapter 2). However, while plants and animals both inherited mitochondria, the lineage that gave rise to plants also incorporated another endosymbiont, a photosynthetic cyanobacterium that over evolutionary time would become the chloroplast. From this group of photosynthetic organisms, one specific lineage, a type of multicellular green algae, gave rise to land plants. From these ancestors, land plants also inherited cellulose cell walls and apical meristems, cells at the tips of shoots and roots that are analogous to the stem cells in animals that generate specialized tissues.

The earliest known fossils of multicellular land plants and animals date from about 570 million years ago (mya), telling us that multicellularity evolved independently in plants and animals. This divergence of plants and animals before either group was multicellular in organization explains major differences in how their cells interact to form tissues and organs. For example, the development of an embryo with differentiated tissues occurs by very different processes in animals and plants. In animals, embryo development involves migrations of cells from one location to another. In contrast, the walls around plant cells prevent them from moving about within an embryo. Instead, changes in plant form rely on changes in the plane of cell division. These are controlled by the position of the mitotic spindle, which in turn alters the position and orientation of cell walls in daughter cells.

These fundamental differences between plants and animals reflect two very different solutions to the challenges of life on land. If we generalize about "typical" plants and "typical" animals, we get a clearer picture of these two different strategies. Animals obtain both energy and carbon from the food they eat. Their food sources tend to have fairly high concentrations of nutrients. Almost all terrestrial animals are motile (some aquatic animals are stationary) and can move from one place to another in search of food, water, or a mate. They can also flee from predators or move away from unfavourable conditions. Animal bodies must be fairly compact to facilitate movement.

In contrast, most plants are "self-feeders" that need sunlight, carbon dioxide (available in air), and water (available in soil). In addition, plants require other nutrients usually obtained from soil. These nutrients are usually patchily distributed in the soil and often available only at low concentrations. Therefore, plants gather diffuse nutrients from both air and soil. How best to capture these diffuse nutrients? Large surface areas are important, both above and below ground, and thus plant bodies are not compact but spreading and branched in form (the term for this form is *dendritic*, which literally means "treelike"). To visualize this dendritic growth, think about how the branches of a deciduous tree look in the spring before they have leafed out (see Figure 34.1 in Chapter 34). The root system of the tree is also dendritic, branching and spreading below ground. Thus, the evolutionary response to the challenges posed by life on land has resulted in a plant body consisting of two closely linked but quite different components—a photosynthetic *shoot system* extending upward into the air **(Figure 1)** and a nonphotosynthetic *root system* extending downward into the soil.

Obviously, a plant cannot just pick up this extensive root system and move around in search of better conditions. Instead, plants are fixed in place (sessile). This means that they search for nutrients and water, find mates, and defend themselves from predators—everything that animals have to do—while fixed in place. Adaptation to meet these challenges without the ability to move to another location has resulted in specialized body plans with highly specialized tissues and functions. As you read through the chapters in Units 9, 10, and 11, think about how differences in animal and plant growth, development, reproduction, and other processes relate to the fundamental difference between being motile—obtaining the materials they need to build bodily structures from the food they eat—and being sessile—obtaining the primary material they need to build body structures (carbon) from the air and sunlight they capture.

Figure 1

Dendritic growth shown by the above-ground portion of a tree.

costas anton dumitrescu/Shutterstock.com

Adrian Jones, IAN Image Library (www.ian.umces.edu/imagelibrary)

Isabella M. Gioia

Banyan tree (*Ficus* sp.), one type of "strangler fig" with close-up view of strangler fig roots on the trunk of a host plant (inset photo).

Organization of the Plant Body

WHY IT MATTERS

What is the largest plant in the world? The answer depends on how we define largest—is it the tallest? The one with the greatest mass? If we define largest as the plant with the biggest canopy (stem and branches), then the winner of the contest is the banyan tree (see photograph above). The banyan is one of several kinds of figs (*Ficus* species) that are known as strangler figs due to their aggressive growth habit. The seeds of strangler figs, dispersed by birds, are often deposited high up on the branches of other tree species in tropical rain forests. The seeds germinate in the bark of their host tree and send thin roots down the trunk of the host plant to the ground. Once the roots enter the ground, they grow and thicken quickly. Roots that cross over each other fuse together, trapping the host plant's trunk in a cage of roots that eventually fuse into a more or less solid mass. Meanwhile, the stem of the fig climbs upward, twining itself around the host's stem, and soon overtops the host, putting out many thick leaves that shade the host plant's leaves. The strangler fig can now outcompete the host plant for sunlight and for water and nutrients from the soil. The network of roots that surrounds the host's trunk prevents further lateral growth of the host, which eventually starves to death. The host trunk rots away, leaving a hollow cylinder of roots that forms the main trunk of the fig tree. Some of these fig species continue to send down roots from their branches. When these aerial roots reach the ground, they become additional trunks to help support the canopy, which can become massive. In this way, a single fig tree and its numerous, interconnected trunks can spread out over a very large area. The largest banyan tree in the world has a canopy that is 420 m in diameter! This aggressive growth strategy is a definite advantage in rain forests, where competition for light under the dense upper canopy is fierce.

Even though strangler figs are unusual in that their seeds germinate in the bark of another plant rather than in the soil, fig seedlings develop into mature plants via the same processes as other plants. As you saw in Chapter 26, plants were able to successfully colonize diverse land habitats only as adaptations in form and function helped them solve problems posed by the terrestrial environment. These evolutionary adaptations included

- a *shoot system* that helps support leaves and other body parts in air,
- a *root system* that anchors the plant in soil and provides access to soil nutrients and water,
- tissues for internal transport of water and nutrients, and
- specializations for preventing water loss.

What structures make up the root and shoot systems of a plant? How do the different parts of a plant develop? How do plants grow in height (upward for shoots and downward for roots)? How do some plants, such as strangler figs, become woody? Starting in this chapter and continuing through the next three chapters, we investigate these questions and explore the structure and functioning of plants—their morphology, anatomy, and physiology.

A plant's *morphology* is its external form, such as the shape of its leaves, and its *anatomy* is the structure and arrangement of its internal parts. Plant *physiology* refers to the mechanisms by which the plant's body functions in its environment. Our focus in this chapter is on the angiosperms, or flowering plants, which are the most successful plants on Earth in terms of distribution and sheer numbers of species.

34.1 Plant Structure and Growth: An Overview

In this chapter, we focus on the key characteristics of plant structure and growth and make several comparisons between land plants and terrestrial animals. We could compare plants with many other organisms since plants and animals are just two of the many kingdoms of life, but we tend to be most familiar with animals. It is obvious that plants are very different from animals, but why are they so different? We can think of plants and animals as representatives of two very different solutions to the challenges of life on land. If we generalize about "typical" plants and "typical" animals, we can get a clearer picture of these two different strategies. Animals are chemoheterotrophs: they obtain both energy and carbon from the food they eat. Their food sources tend to have fairly high concentrations of nutrients. All terrestrial animals are motile (some aquatic animals are stationary) and can move from one place to another in search of food, water, or a mate. They can also flee from predators or move away from unfavourable conditions. Animal bodies, then, need to be fairly compact to facilitate moving around.

In contrast, most plants are photosynthetic autotrophs—"self-feeders"—that need sunlight, carbon dioxide (available in air), and water (available in soil). In addition, plants require other nutrients that are usually available only in soil; these nutrients are usually

patchily distributed in the soil and often available only at low concentrations. Thus, unlike animals, plants have to gather diffuse nutrients from both air and soil. How best to capture these diffuse nutrients? A large surface area is important, both above ground and below ground, so plant bodies are not compact but spreading and branched in form (the term for this form is *dendritic*, which literally means "treelike"). To visualize this dendritic growth, think about how the branches of an aspen or another poplar tree look in the spring before they have leafed out **(Figure 34.1)**. The root system of the tree is also dendritic, branching and spreading below ground. Thus, the evolutionary response to the challenges posed by life on land has resulted in a plant body consisting of two closely linked but quite different components—a photosynthetic *shoot system* extending upward into the air and a nonphotosynthetic *root system* extending downward into the soil **(Figure 34.2)**.

Obviously, a plant cannot just pick up this extensive root system and move around in search of better conditions. Instead, plants are fixed in place (sessile), and they must therefore search for nutrients and water, find mates, and defend themselves from predators—everything that animals have to do—while fixed in one place. As you read through this chapter, think about how plant morphology and growth relate to being sessile photoautotrophs.

34.1a Cells of All Plant Tissues Share Some General Features

Both root and shoot systems consist of various **organs**—body structures that contain two or more types of tissues and have a definite form and function.

Figure 34.1
Dendritic growth shown by the above-ground portion of a tree.

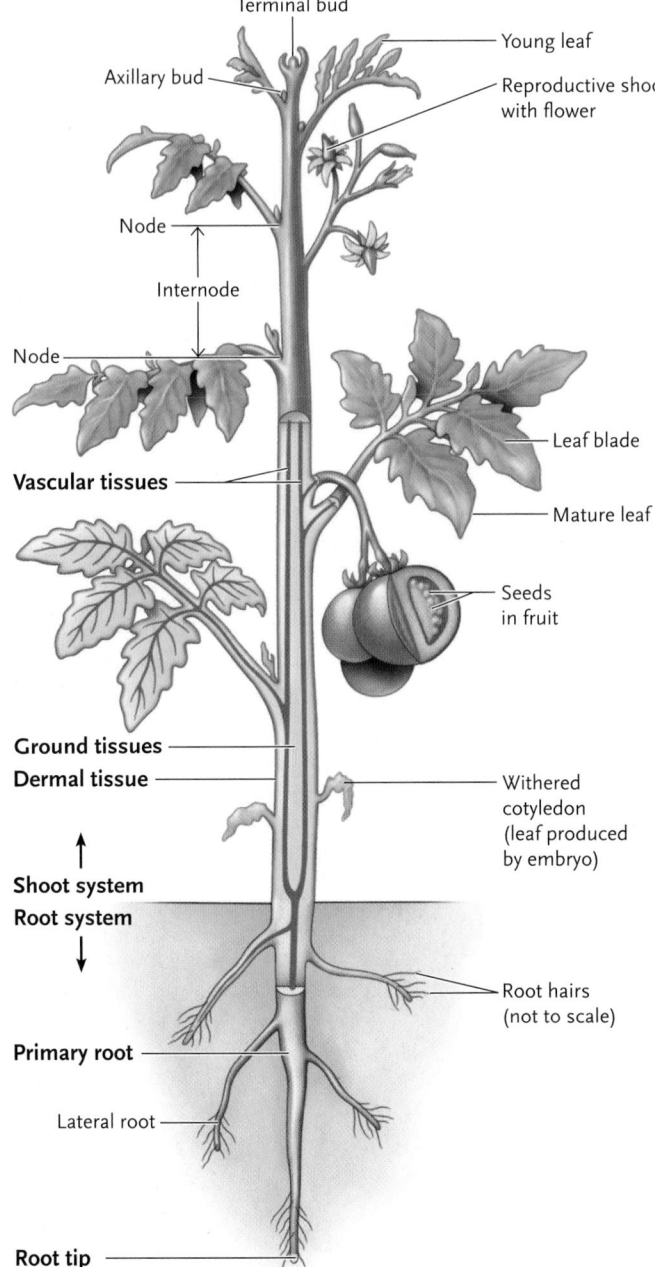

Figure 34.2

Basic body plan for the tomato plant *Solanum lycopersicum*, a typical angiosperm. Vascular tissues (purple) conduct water, dissolved minerals, and organic substances. They thread through ground tissues, which make up most of the plant body. Dermal tissues (epidermis, in this case) cover the surfaces of the root and shoot systems.

plant tissues, the cells have a **primary cell wall** surrounding the plasma membrane and cell contents (cytoplasm and organelles). These cell walls are the "skeleton" of a plant, serving to support the plant's body, as the skeleton for an animal does. A primary cell wall is made largely of microfibrils of **cellulose**, a polymer of glucose, embedded in a matrix of other polysaccharides **(Figure 34.3).** Cellulose is the most abundant polysaccharide on Earth and is currently being investigated as a source of biofuel (see "Molecule behind Biology," Box 34.1, p. 838). The combination of cellulose fibrils and other polysaccharides gives the cell wall strength and flexibility. Primary cell walls also contain various proteins. Some of these are structural proteins that contribute to the wall's strength, whereas others are enzymes that catalyze the formation and modification of the cell wall. **⬤ CONCEPT FIX** Don't think of the cell wall as a solid barrier, like a cement wall, but rather as a semipermeable "mesh" or filter, which allows some molecules (e.g., water) to pass through into the cell. As well, cytoplasmic connections between adjacent cells, called **plasmodesmata** (singular, *plasmodesma*), allow solutes such as amino acids and sugars to move from one cell to the next. The space between the primary cell walls of adjacent cells is filled with a polysaccharide layer called the middle lamella. ⬢

As a young plant grows, different types of cells deposit additional cellulose and other materials inside the primary wall, forming a strong **secondary cell wall.** Secondary walls often contain **lignin,** a complex water-insoluble polymer (see Chapter 25), which makes cell walls very strong, rigid, and impermeable to water. As we learned in Chapter 26, the evolution of large vascular plants became possible only after biochemical pathways producing lignin evolved, through modification of existing pathways, allowing a plant to produce lignified cells that both provided support and conducted water through the

Plant organs include leaves, stems, and roots. A **tissue** is a group of cells and intercellular substances that function together in one or more specialized tasks.

Plant cells share some features with animal cells but differ in that they typically have a cell wall, a large vacuole, and, in many cells, chloroplasts. Chloroplasts function in photosynthesis and are discussed in more detail in Chapter 7. The vacuole may occupy most of the volume in a mature plant cell and plays an important role in cell elongation and maintenance of rigid tissues. Vacuoles may also act as storage compartments. In all

Primary Plant Cell Wall

- Cellulose
- Structural protein
- Hemicellulose
- Pectin

Figure 34.3
Structure of a plant cell wall.

Cellulose

Cellulose **(Figure 1)** is the most abundant organic compound on Earth, thanks to its presence in every cell wall of every plant. We have many uses for cellulose already: in paper, clothing, insulation, and a variety of industrial uses. But could we also use this abundant substance as a source of fuel? Driven by a desire for more green sources of energy and the high price of oil, people have been looking at various sources of biofuel—fuel from plants.

One such biofuel, ethanol, is currently produced primarily from corn. Corn kernels are mostly starch (and water), which is easily broken down into sugars that are then fermented to produce ethanol. Like starch, cellulose is also a polymer of glucose, so it, too, can be broken down to sugars that can be fermented. However, it is much more difficult to do this with cellulose. Why? The linkages between glucose monomers in cellulose are different from those in starch (see *The Purple Pages*). This seemingly minor difference makes a very big difference in

not only the characteristics of the resulting polymers—cellulose is linear, whereas starch is coiled—but also in how difficult it is for decomposers to break the bonds between the monomers. Starch is readily decomposed by many organisms, including humans, but cellulose can be broken down only by a few organisms, mostly fungi and prokaryotic organisms. Moreover, cellulose may be protected by lignin in some plant tissues, making the cellulose even harder to break down. Consequently, converting cellulose into liquid fuel is currently a very complex, difficult, and energy-consuming process.

But there are issues related to the use of corn for ethanol. World food prices rose 75% between 2002 and 2008, partly because of rising biofuel demand; diverting corn from food to fuel use may be contributing to food shortages in some countries. There is also concern about the long-term effects on soil fertility of cultivating corn.

An advantage of converting cellulose to fuel is that all parts of the plants, not just the starch- and sugar-rich

parts, could be converted to fuel. In Canada and the United States, a very promising source of cellulose for biofuel is switchgrass (*Panicum virgatum*), a native grass of the tallgrass prairies. It grows very quickly, is able to grow on marginal land that is unsuitable for crop production, and can withstand both drought and flooding. Switchgrass is perennial, so it does not need to be replanted every year. Although the problems related to converting cellulose from switchgrass to liquid fuel (ethanol) remain, we could avoid these problems if we focused on other forms of fuel. For example, in Canada, the biggest contributor to greenhouse gases is heating, not transportation. Researchers have found that switchgrass stems can be dried and compressed into fuel pellets that can be burned. Used in this way, switchgrass produced a whopping 540 times the amount of energy as was needed to grow, harvest, and process it. This yield is seven times as much energy per hectare as corn yields. Switchgrass pellets could also be used to generate electricity.

Cellulose, formed from glucose units joined end to end by β(1→4) linkages. Hundreds to thousands of cellulose chains line up side by side, in an arrangement reinforced by hydrogen bonds between the chains, to form cellulose microfibrils in plant cells.

Glucose subunit

Cellulose molecule

Cellulose microfibril

Cellulose microfibrils in a plant cell wall.

Biophoto Associates/Science Source

Figure 1
Cellulose, the major component of plant cell walls.

plant body. Lignin is also very resistant to decomposition, so its presence in cell walls makes the cell more resistant to attack by microbes (see Chapter 25).

As in animals, all of a plant's cells have the same genes in their nuclei. So how do specialized cells such as xylem arise in plants? As each cell matures and *differentiates* (becomes specialized for a particular function), specific genes are activated. For the most

part, fully differentiated animal cells perform their functions while alive, but some types of plant cells die after differentiating, and their cytoplasm disappears. The walls that remain, however, serve key functions, particularly in xylem.

Most plant cells have a much more flexible differentiation than do animal cells. In general, once an animal cell has differentiated, it cannot easily dedifferentiate or "turn back" into an unspecialized cell (this is why it has proved so difficult to clone animals). Almost any plant cell, even one that has become specialized, can dedifferentiate and divide to produce an entire plant (obviously, cells in which the cytoplasm has been lost, such as xylem cells, are not able to dedifferentiate). This ability of almost any cell to give rise to all other parts of a plant is known as **totipotency.** You can see totipotency in action if you take a cutting of a shoot and place it in water: in a few days, roots will form on the bottom of the stem. Cloning of plants is very easy, something that many plants do all the time as a means of reproduction and many gardeners use as a means of propagation. What are the advantages of totipotency? It allows plants to heal wounds and, as mentioned above, is also one means of asexual reproduction; for example, in many plants (such as raspberries), if a branch or stem comes into contact with the soil for long enough, roots will develop at the point where the stem touches the ground, forming a new plant.

34.1b Shoot and Root Systems Perform Different but Integrated Functions

A flowering plant's **shoot system** typically consists of stems, leaves, buds, and—during part of the plant's life cycle—reproductive organs known as flowers (see Figure 34.2, p. 837). A stem with its attached leaves and buds is a *vegetative* (nonreproductive) shoot; a bud eventually gives rise to an extension of the shoot or to a new, branching shoot. A *reproductive* shoot produces flowers, which later develop fruits containing seeds.

The shoot system is highly adapted for photosynthesis. Leaves greatly increase a plant's surface area and thus its exposure to light. Stems are frameworks for upright growth, which favourably positions leaves for light exposure and flowers for pollination. Some parts of the shoot system also store carbohydrates manufactured during photosynthesis. Many plants can change the orientation of their leaves to maximize light absorption or, in arid habitats, to prevent overheating.

The **root system** usually grows below ground. It anchors the plant and supports its upright parts. It also absorbs water and dissolved minerals from soil and stores carbohydrates. Adaptations in the structure and function of plant cells and tissues were an integral part of the evolution of shoots and roots, for example, the development of vascular tissues specialized to serve as internal pipelines that conduct water, minerals, and organic substances throughout the plant. The root hairs sketched in Figure 34.2 are surface cells specialized for absorbing water and nutrients from soil.

34.1c Meristems Produce New Tissues throughout a Plant's Life

Most animals grow to a certain size, and then their growth slows dramatically or stops. This pattern is called **determinate growth.** In contrast, plants can grow throughout their lives, a pattern called **indeterminate growth.** Individual plant parts, such as leaves, flowers, and fruits, exhibit determinate growth, but every plant also has self-perpetuating embryonic tissue, called meristem (*merizein* = to divide), at the tips of shoots and roots. Under the influence of plant hormones, these **meristems** produce new tissues more or less continuously while the plant is alive.

Why do plants have indeterminate growth? A capacity for indeterminate growth gives plants a great deal of flexibility—or what biologists often call *plasticity*—in their possible responses to changes in environmental factors such as light, temperature, water, and nutrients. This plasticity has major adaptive benefits for an organism that cannot move about, as most animals can. For example, if external factors (such as a houseplant's owner) change the direction of incoming light for photosynthesis, stems can "shift gears" and grow in that direction. These and other plant movements, called tropisms, are a major topic of Chapter 37.

Remember, too, that nutrients are patchily distributed and diffuse in soil. Indeterminate growth allows a root system to extend and grow out of regions in which nutrients have been depleted and forage for patches with more nutrients; if plant root systems were determinate, plants would soon exhaust local nutrient supplies and be unable to forage for more.

As you know, animals grow mainly by mitosis, which increases the number of body cells. Plants, however, grow by two mechanisms—an increase in the number of cells by mitotic cell division in the meristems *and* an increase in the size of individual cells. In regions adjacent to the meristems in the tips of shoots and roots, the daughter cells rapidly increase in size—especially in length—for some time after they are produced. In contrast, when animal cells divide mitotically, the daughter cells are usually roughly the same size as the parent cell.

34.1d Meristems Are Responsible for Growth in Both Height and Girth

Some plants have only one kind of meristem, whereas others have two **(Figure 34.4, p. 840).** All plants have **apical meristems,** clusters of self-perpetuating tissue at the tips of their buds, stems, and roots (see Figure 34.4a). Tissues that develop from apical meristems are called **primary tissues** and make up the **primary plant body.** Growth of the primary plant body is called **primary growth.**

Some plants—herbaceous plants such as grasses, for example—have only primary growth, which occurs at the tips of roots and shoots. Others have **secondary growth** as well as primary growth. Secondary growth originates at cylinders of tissue called **lateral meristems** and increases the diameter of older roots and stems (see Figure 34.4b). Tissues that develop from lateral meristems are called **secondary tissues.** Woody plants, such as trees and shrubs, including the strangler fig discussed earlier, all have secondary tissues.

a. Plants increase in length by cell divisions in apical meristems and by elongation of the daughter cells derived from the apical meristems.

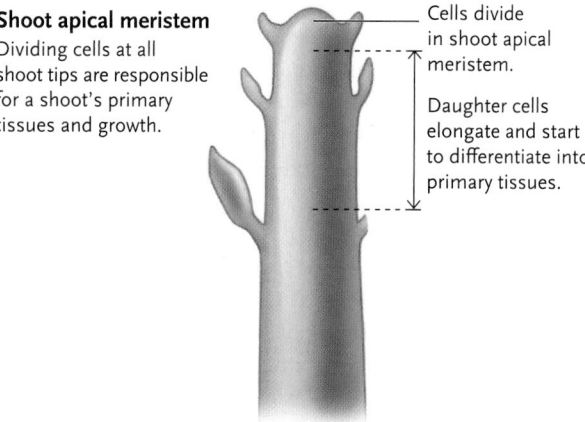

Shoot apical meristem
Dividing cells at all shoot tips are responsible for a shoot's primary tissues and growth.

Cells divide in shoot apical meristem.

Daughter cells elongate and start to differentiate into primary tissues.

Root apical meristem
Dividing cells at root tips behind the root caps are responsible for a root's primary tissues and growth.

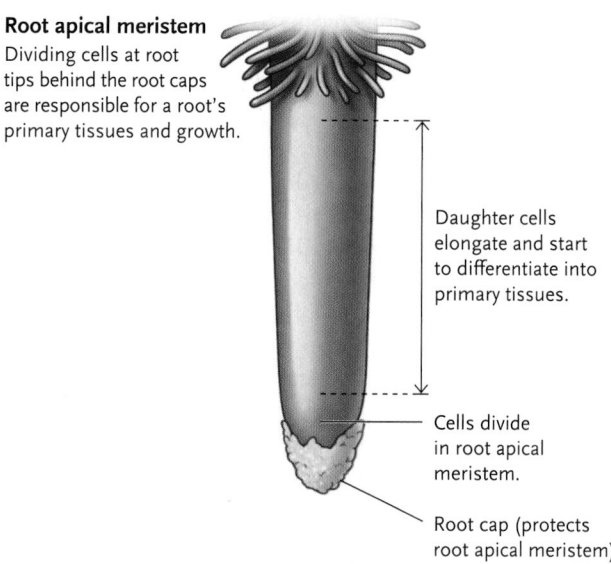

Daughter cells elongate and start to differentiate into primary tissues.

Cells divide in root apical meristem.

Root cap (protects root apical meristem)

b. The stems of some plants increase in girth by way of cell divisions in lateral meristems: the vascular cambium and cork cambium.

Vascular cambium   Cork cambium

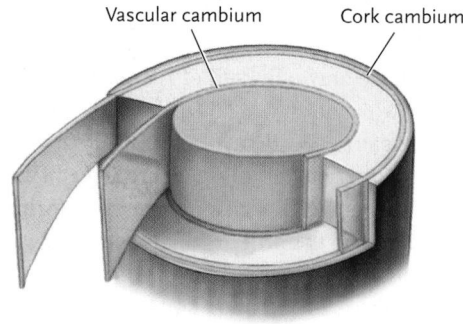

Figure 34.4
Approximate locations of types of meristems that are responsible for increases in the length and diameter of the shoots and roots of a vascular plant.

Primary and secondary growth can go on simultaneously in a single plant, with primary growth increasing the length of shoots and roots and secondary growth adding girth to these organs. Each

spring, for example, a poplar tree undergoes primary growth at each of its root and shoot tips, and secondary growth increases the diameter of its older, woody parts. Plant hormones govern these growth processes and other key events described in Chapter 38.

34.1e Monocots and Eudicots Are the Two General Structural Forms of Flowering Plants

Several broad categories of body architecture arose as flowering plants evolved, with the two major categories being the **monocot** and **eudicot** lineages. Grasses, lilies, cattails, corn, and rice are examples of monocots. Eudicots include nearly all familiar angiosperm trees and shrubs, as well as many nonwoody (herbaceous) plants. Examples are poplars, willows, oaks, cacti, roses, poppies, sunflowers, and garden beans and peas.

Monocots and eudicots get their names from the number of *cotyledons*, the leaves produced by the embryo, sometimes called seed leaves (see Chapter 26). Monocot seeds have one cotyledon and eudicot seeds have two. Although monocots and eudicots have similar types of tissues, their body structures differ in distinctive ways **(Table 34.1)**. As we discuss the morphology of flowering plants, we refer frequently to these structural differences.

34.1f Flowering Plants Can Be Grouped According to Type of Growth and Life Span

As you learned above, we can distinguish between flowering plants depending on whether they are herbaceous or woody plants and whether they are monocots or eudicots. We can also distinguish plants by life span. **Annuals** are herbaceous plants in which the life cycle is completed in one growing season. With minimal or no secondary growth, annuals typically have only apical meristems. Examples are tomatoes (a eudicot) and corn (a monocot). **Biennials** such as carrots complete their life cycle in two growing seasons, and limited secondary growth occurs in some species. In the first season, roots, stems, and leaves form; in its second year of growth, the plant flowers, forms seeds, and dies. In **perennials**, vegetative growth and reproduction continue year after year. Many perennials, such as trees, shrubs, and some vines, have secondary tissues, although others, such as irises and daffodils, do not.

STUDY BREAK

1. Explain how plant cell secondary walls differ from primary walls.
2. Compare the components and functions of a land plant's shoot and root systems.
3. Explain what meristem tissue is, and name and describe the functions of apical and lateral meristems.

Table 34.1	Eudicots and Monocots Compared	
Character	Eudicots	Monocots

Cotyledons — Cotyledons, First leaf, Seed coat, Cotyledon, First leaf, Seed coat, Primary root

Floral parts — Usually four or five floral parts (or multiples of four or five) · Usually three floral parts (or multiples of three)

Leaf veins — Leaf veins usually in a netlike array · Leaf veins usually running parallel to one another

Location of vascular bundles — Vascular bundles organized as a ring · Vascular bundles distributed throughout ground tissue

Root system — Usually a main taproot with smaller lateral roots · Usually a branching fibrous root system

34.2 The Three Plant Tissue Systems

As in animals, plant organs are composed of tissue systems. Each tissue system includes several types of tissue, and each tissue is made up of cells with specializations for different functions **(Table 34.2, p. 842).** *Simple* tissues have only one type of cell. Other tissues are *complex*, with organized arrays of two or more types of cells. **Figure 34.5, p. 842,** will help you interpret images of plant tissues, beginning with the tissues in a transverse section of a stem, shown in **Figure 34.6, p. 842.**

Unlike animals, which have a wide range of tissues, plant organs are composed of just three tissue systems. The **ground tissue system**, which makes up most of the plant body, functions in metabolism (including photosynthesis), storage, and support. The **vascular tissue system** consists of xylem and phloem, which transport water and nutrients throughout the plant. The cylinders of vascular tissue are embedded in ground tissue. The **dermal tissue system** is a skinlike protective covering for the plant body. As shown in Figure 34.2, these tissues generally have a radial arrangement in the plant body, with dermal tissues surrounding ground tissues, in which vascular tissues are embedded. Below we discuss the key features of each tissue system.

34.2a Ground Tissues Are All Structurally Simple but Exhibit Important Differences

Plants have three types of ground tissue systems, each with a distinct structure and function—*parenchyma, collenchyma,* and *sclerenchyma* **(Figure 34.7, p. 842).** Each type is structurally simple, being composed mainly of one kind of cell. In a very real sense, the cells in ground tissues are the "worker bees" of plants, carrying out photosynthesis, storing carbohydrates, providing mechanical support for the plant body, and performing other basic functions. Each kind of cell has a distinctive wall structure, and some have variations in the cytoplasmic contents as well.

Parenchyma: Soft Primary Tissues. Parenchyma (*para* = around; *chein* = fill in or pour) makes up the bulk of the primary growth of roots, stems, leaves, flowers, and fruits. Most parenchyma cells have only a thin primary wall and so are pliable and permeable to water. Often the cells are spherical or many sided, as in Figure 34.7a. Parenchyma cells typically have air spaces between them, especially in leaves (see Section 34.3). Stems and leaves in aquatic plants often have very large air spaces between parenchyma cells, which facilitate the movement of oxygen to submerged parts of the plant and help the leaves float upward toward the light.

Table 34.2 | **Summary of Flowering Plant Tissues and Their Components**

Tissue System	Name of Tissue	Cell Types in Tissue	Tissue Function
Ground tissue	Parenchyma	Parenchyma cells	Photosynthesis, respiration, storage, secretion
	Collenchyma	Collenchyma cells	Flexible strength for growing plant parts
	Sclerenchyma	Fibres or sclereids	Rigid support, deterring herbivores
Vascular tissue	Xylem	Conducting cells (tracheids, vessel members), parenchyma cells, sclerenchyma cells	Transport of water and dissolved minerals
	Phloem	Conducting cells (sieve tube members), parenchyma cells, sclerenchyma cells	Sugar transport
Dermal tissue	Epidermis	Undifferentiated cells, guard cells, other specialized cells	Control of gas exchange, water loss, protection
	Periderm	Cork, cork cambium, phelloderm	Protection

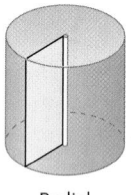 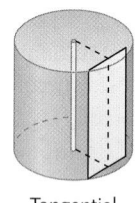 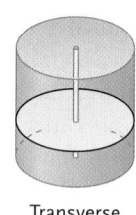

Radial   Tangential   Transverse

Figure 34.5

Terms that identify how tissue specimens are cut from a plant. Along the radius of a stem or root, longitudinal cuts give radial sections. Cuts at right angles to a root or stem radius give tangential sections. Cuts perpendicular to the long axis of a stem or root give transverse sections (cross-sections).

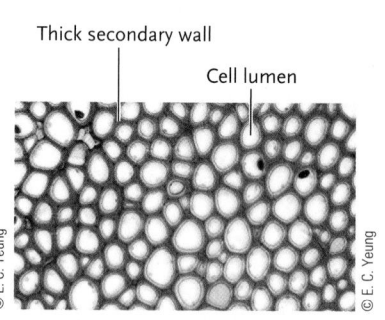

Figure 34.6

Locations of ground, vascular, and dermal tissues in one kind of plant stem, transverse section. Ground tissues are simple tissues, whereas vascular and dermal tissues are complex, containing various types of specialized cells.

Figure 34.7

Examples of ground tissues from the stem of a pepper plant (*Capsicum*) **(a, b)** and a sunflower plant (*Helianthus annuus*) **(c).**

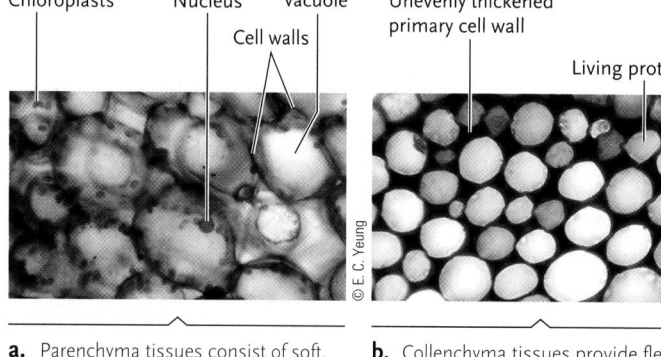

a. Parenchyma tissues consist of soft, living cells specialized for storage, other functions.

b. Collenchyma tissues provide flexible support.

c. Sclerenchyma tissues provide rigid support and protection.

Parenchyma cells may be specialized for tasks as varied as storage, secretion, and photosynthesis. For example, the photosynthetic cells of leaves are parenchyma cells. In many plant species, modified parenchyma cells are specialized for short-distance transport of solutes. Such cells are common in tissues in which water and solutes must be rapidly moved from cell to cell. Parenchyma cells usually remain alive and metabolically active when mature.

Collenchyma: Flexible Support. The "strings" in celery are examples of the flexible ground tissue called **collenchyma** (*kolla* = glue; see Figure 34.7b), which helps strengthen plant parts that are still elongating.

Collenchyma cells are typically elongated, and, collectively, they often form strands under the dermal tissue of growing shoot regions and leaf stalks.

The primary walls of collenchyma cells are built of alternating layers of cellulose and pectin and are unevenly thickened. These walls can stretch as the cell enlarges, making them very suitable for flexible support of young, growing organs. Mature collenchyma cells are alive and metabolically active, and they continue to synthesize primary wall layers as the plant grows.

Sclerenchyma: Rigid Support and Protection. Mature plant parts gain additional mechanical support and protection from **sclerenchyma** (*skleros* = hard), cells with thick, lignified secondary walls (see Figure 34.7c). Some regions of the cell wall lack secondary wall material, forming a *pit* where the cell wall is more porous than elsewhere. Water can flow from one sclerenchyma cell to another through these pits. After lignification occurs, sclerenchyma cells die because their cytoplasm can no longer exchange gases, nutrients, and other materials with the environment. The walls, however, remain to provide protection and support.

The two types of sclerenchyma cells—*sclereids* and *fibres*—differ in their shape and arrangement. **Sclereids** tend to be short and are often branched **(Figure 34.8a)**; they sometimes aggregate into protective sheets, forming the hard casings of a coconut shell or a peach pit, for example. Sclereids can also be scattered in tissue; cube-shaped sclereids dispersed in the flesh of a pear give it its gritty texture (Figure 34.8b). **Fibres** are long, tapered cells (Figure 34.8c) that resist stretching but are more pliable than sclereids. Fibres often occur in bundles in stems and leaves, strengthening and supporting these tissues. We use plant fibres to manufacture rope, paper, and cloth. Linen, for example, is made of fibres extracted from the stems of flax plants (*Linum usitatissimum*).

34.2b Vascular Tissues Are Specialized for Conducting Fluids

Vascular tissue systems consist of specialized conducting cells, parenchyma cells, and fibres. *Xylem* and *phloem,* the two kinds of vascular tissues in flowering plants, are organized into cylinders of interconnected cells that extend throughout the plant.

Xylem: Transporting Water and Minerals. **Xylem** (*xylon* = wood) conducts water and dissolved minerals absorbed from the soil upward from a plant's roots to the shoot. The evolution of xylem cells was a key adaptation allowing plants to make the transition to life on land (see Chapter 26). The two types of conducting cells, *tracheids* and *vessel members,* develop thick, lignified secondary cell walls and die at maturity, forming pipelines for water and minerals.

a.

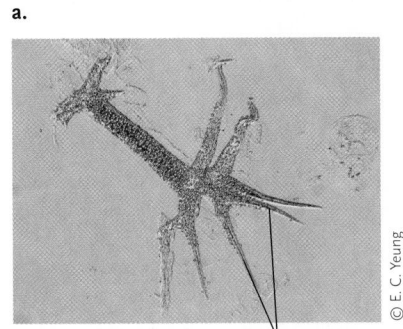

© E. C. Yeung

Branches of astrosclereid cell

b.

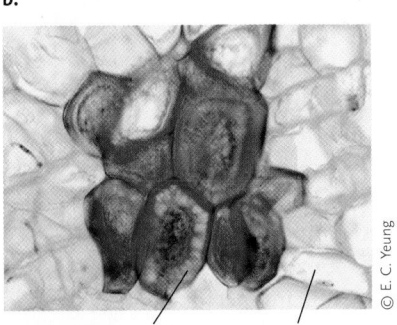

© E. C. Yeung

Thick secondary wall   Parenchyma
of sclereid cell   cell

c.

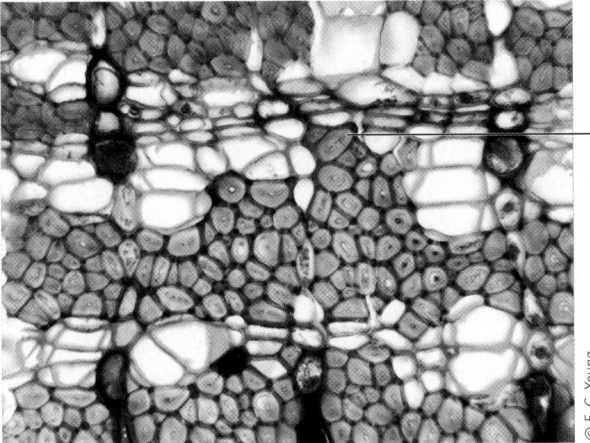

© E. C. Yeung

— Fibre with thick
secondary wall

Figure 34.8
Examples of sclerenchyma cells. **(a)** Astrosclereid, a radiately branched type of sclereid. This cell was isolated from the cells surrounding it by maceration of the tissue. **(b)** Cross-section of stone cells with thick lignified walls in the flesh of a pear (*Pyrus*). Sclereids are distinguished from the surrounding cells by their thick secondary cell walls (stained red in the figure). **(c)** Cross-section of phloem fibres from the stem of a linden tree (*Tilia*). Note that fibres occur in clusters and, like sclereids, are distinguished from the surrounding cells by their thick secondary cell walls (stained red in the figure).

a. Tracheids, tangential section

b. A vessel member

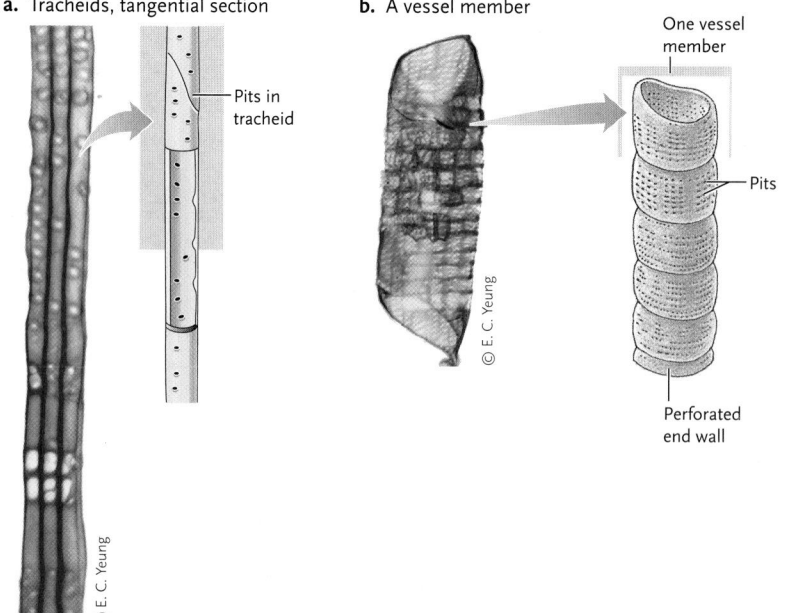

One vessel member

Pits in tracheid

© E. C. Yeung

Pits

© E. C. Yeung

Perforated end wall

© E. C. Yeung

Figure 34.9

Representative tracheids and vessel members from woody stems, elements in xylem that conduct water and dissolved mineral salts through the body of a vascular plant. These images show longitudinal views of **(a)** tracheids from pine (*Pinus*) and **(b)** a vessel member from oak (*Quercus*).

Tracheids are elongated cells, with tapered, overlapping ends **(Figure 34.9a)**. As in sclerenchyma, water can move from cell to cell through pits. Usually, a pit in one cell is opposite a pit of an adjacent cell, so water flows laterally from tracheid to tracheid.

Vessel members (or vessel elements) are shorter and wider cells than tracheids and are joined end to end in tubelike columns called vessels (Figure 34.9b). **Vessels** are typically several centimetres long, and may even be metres long in some vines and trees. Like tracheids, vessel members have pits. However, they have another adaptation that greatly enhances water flow. As vessel members mature, enzymes break down portions of their end walls, producing perforations. Some vessel members have a single, large perforation, so that the end is completely open (see Figure 34.9b). Others have a cluster of small, round perforations, or ladderlike bars, extending across the open end. Water moves more efficiently through vessels than tracheids due to their greater diameter and perforated ends.

Fossil evidence shows that the forerunners of modern vascular plants relied solely on tracheids for water transport, and today ferns and most gymnosperms still have only tracheids. Nearly all angiosperms and a few gymnosperms and seedless vascular plants have *both* tracheids and vessel members, however, which confers an adaptive advantage. Flowing water sometimes contains air bubbles, which are a potentially lethal threat to the plant. Water can flow rapidly through vessel members that

are linked end to end, but the open channel cannot prevent air bubbles from forming and possibly blocking the flow through the whole vessel. By contrast, even though water moves more slowly in tracheids, the pits are impermeable to air bubbles, and a bubble that forms in one tracheid stays there; water continues to move between other tracheids. This ability to filter out air bubbles might also make xylem an effective and inexpensive way to provide clean drinking water in developing countries (see "People behind Biology," Box 34.2).

Phloem: Transporting Sugars and Other Solutes. The vascular tissue **phloem** (*phloios* = tree bark) transports the sugars made in photosynthesis and other organic molecules throughout the plant body. The main conducting cells of phloem are **sieve tube members (Figure 34.10)**, which are connected end to

a. Sieve tube members

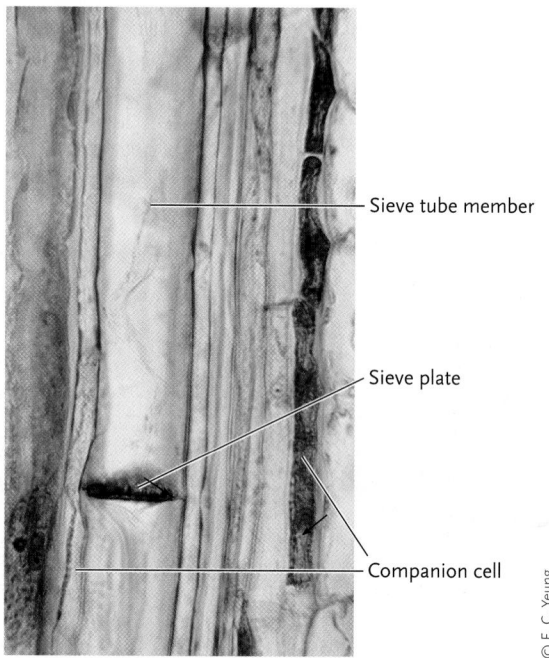

Sieve tube member

Sieve plate

Companion cell

© E. C. Yeung

b. Sieve plate

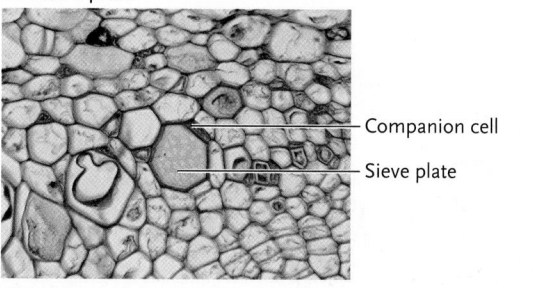

Companion cell

Sieve plate

© E. C. Yeung

Figure 34.10

Structure of sieve tube members. **(a)** Micrograph showing sieve tube members of cucumber (*Cucumis*) in longitudinal section. Long tubes of sieve tube members conduct sugars and other organic compounds. **(b)** Sieve plate in a cell in phloem of cucumber (*Cucumis*), cross-section.

Rohit Karnik, *Massachusetts Institute of Technology*

Clean, safe drinking water that is free of chemical contaminants and biological pathogens is a scarce resource for millions of people around the world. The World Health Organization (WHO) reports that 1.6 million people—mostly children under the age of five—die every year from diarrhoeal diseases and hundreds of millions of others are infected by waterborne pathogens. The WHO and many other agencies and individuals are working on approaches that can provide safe drinking water, but a major challenge is to develop methods that are both effective and inexpensive. For example, boiling water for long enough to kill biological contaminants requires a large amount of fuel, which may also be scarce or expensive. Membrane-filtration systems often require pumps to overcome low flow rates, and existing membranes are expensive and tend to clog easily. A team of researchers from the Massachusetts Institute of Technology (MIT), led by Rohit Karnit, looked to nature for inspiration for alternative membrane materials and discovered that xylem is an effective natural filter, removing bacteria and chemical particles.

As explained in Section 34.2b, xylem cells transport water efficiently but can also filter out small air bubbles as water flows from one xylem cell to another through pits (small pores) **(Figure 1a)**. Pit diameter ranges from a few nanometres to a few hundred nanometres, depending on plant species; Karnik and his team thus hypothesized that this same filtration process would remove waterborne bacteria and protists, which are larger than most pits.

The researchers constructed filters from white pine branches by removing the bark and inserting short sections of wood into plastic tubing (Figure 1b). Preliminary tests showed that water poured into the tubing flowed readily

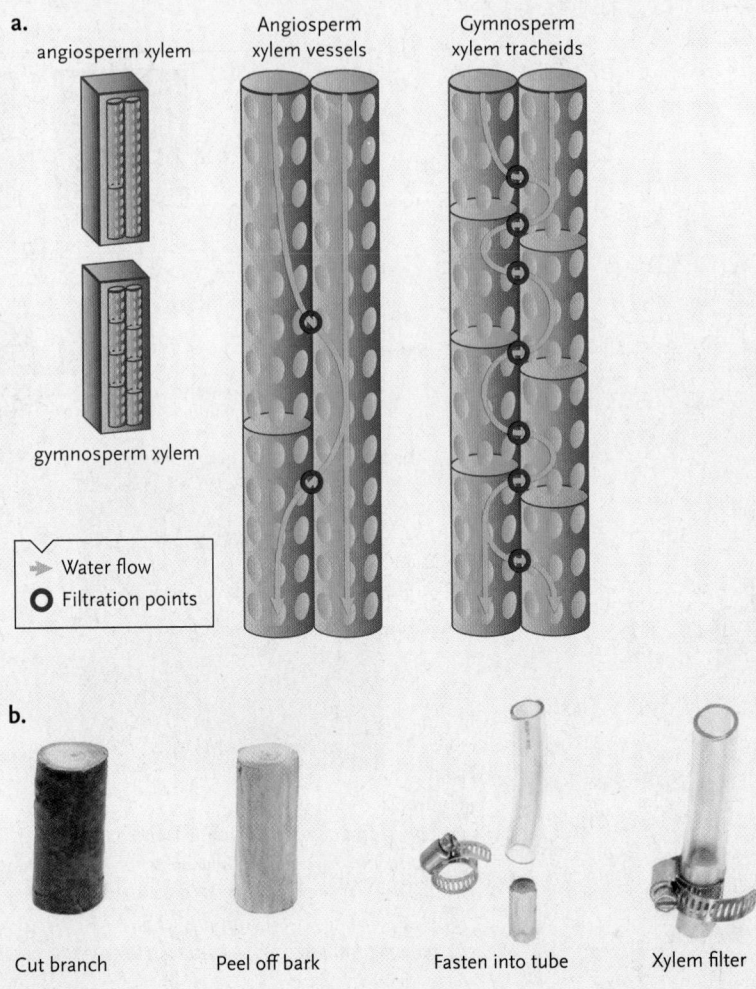

Figure 1

(a) *Structure of xylem vessels in flowering plants and tracheids in conifers. The longer length of the vessels provides pathways that can bypass filtration through pit membranes.* **(b)** *Preparation of xylem filters.*

Boutilier MSH, Lee J, Chambers V, Venkatesh V, Karnik R (2014) Water Filtration Using Plant Xylem. PLoS ONE 9(2): e89934. doi:10.1371/journal.pone.0089934. © 2014 Boutilier et al.

through the xylem filter under gravitational pressure alone, without the need for a pump. Even though the filter cross-sectional area was quite small (approximately 1 cm²), the flow rate was about 4 L per day, which is sufficient to meet the water needs of an adult. When water contaminated with *E. coli* bacteria was filtered through the xylem, more than 99.9% of the bacterial cells were removed from the water **(Figure 2, p. 846)**. Plant xylem thus has the potential to be used for simple, effective water filters that are inexpensive enough to be discarded after use—and disposal would be environmentally friendly as well. The white pine used in this experiment has pits of about 100 nm diameter, which is small enough to remove most bacteria but not viruses. Karnik's research group is now investigating the use of other woody plants to determine if other species can filter out smaller particles.

(Continued)

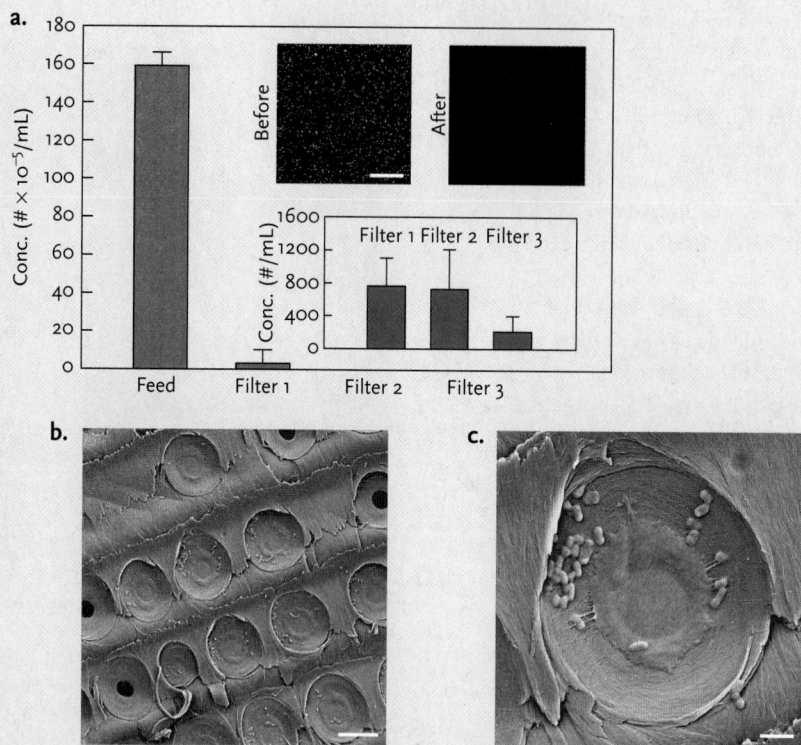

Figure 2

(a) *Concentration of fluorescently labelled bacterial cells in feed (unfiltered) and filtered solutions. The inset graphs show bacterial concentrations in the filtered solutions on a smaller scale. The inset images show fluorescence images of the feed (before) and filtered solutions (after) (scale bar = 200 μm).* *(b), (c)* *Scanning electron micrograph images showing bacteria accumulated on pits after filtration (scale bars = 10 μm and 2 μm, respectively).*

Boutilier MSH, Lee J, Chambers V, Venkatesh V, Karnik R (2014) Water Filtration Using Plant Xylem. PLoS ONE 9(2): e89934. doi:10.1371/journal.pone.0089934. © 2014 Boutilier et al.

end to form a **sieve tube.** As the name implies, the end walls of sieve tubes, called sieve plates, contain numerous pores. In flowering plants, phloem tissue often contains fibres and sclereids in addition to conducting cells; these sclerenchyma cells strengthen stems.

Immature sieve tube members contain the usual plant organelles. Over time, however, the cell nucleus and internal membranes in plastids break down, mitochondria shrink, and the cytoplasm is reduced to a thin layer lining the interior surface of the cell wall. Even without a nucleus, the cell lives up to several years in most plants and much longer in some trees.

In many flowering plants, specialized parenchyma cells known as **companion cells** are connected to mature sieve tube members by plasmodesmata. Unlike sieve tube members, companion cells retain

their nuclei when mature. Companion cells assist sieve tube members with both the uptake of sugars and the unloading of sugars in tissues engaged in food storage or growth. They may also help regulate the metabolism of mature sieve tube members. We will take a deeper look at the functions of xylem and phloem cells in Chapter 35.

34.2c The Dermal Tissue System Protects Plant Surfaces

A complex tissue called **epidermis** covers the primary plant body in a single continuous layer (Figure 34.6, p. 842) or sometimes in multiple layers of tightly packed cells. The external surface of epidermal cell walls is coated with waxes that are embedded in cutin, a network of chemically linked fats. Epidermal cells

a. Leaf epidermis

Cuticle   Epidermal cell

Parenchyma cell inside leaf

Jubal Harshaw/Shutterstock.com

b. Leaf surface

Cuticle-coated cell
of lower epidermis   Guard cells

One stoma

Dr. Jeremy Burgess/Science Source

c. Root hairs

Root   Root hair

Photographer: Michael Clayton, University of
Wisconsin Plant Teaching Collection, http://
botit.botany.wisc.edu

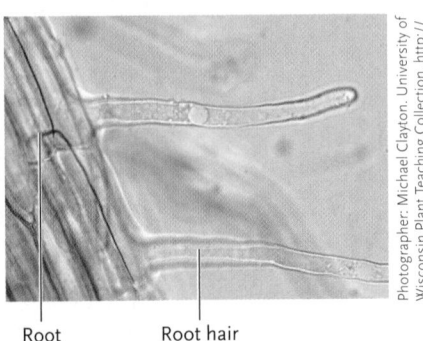

Figure 34.11
Structure and examples of epidermal tissue. **(a)** Cross-section of leaf epidermis from a bush lily (*Clivia miniata*). **(b)** Scanning electron micrograph of a leaf surface, showing cuticle-covered epidermal cells and stomata. **(c)** Root hairs, an epidermal specialization.

secrete this coating, or **cuticle** (see **Figure 34.11a**), which resists water loss and helps protect against attacks by microbes. A cuticle coats all plant parts except the very tips of the shoot and the most absorptive parts of roots; other root regions have an extremely thin cuticle.

Most epidermal cells are relatively unspecialized, but some are modified in ways that represent important adaptations for plants. Young stems, leaves, flower parts, and even some roots have pairs of crescent-shaped **guard cells** (see Figure 34.11b). Unlike other cells of the epidermis, guard cells contain chloroplasts and so can carry out photosynthesis. The pore between a pair of guard cells is called a **stoma** (plural, *stomata*). Water vapour, carbon dioxide, and oxygen cross the epidermis through the stomata. Guard cells regulate opening and closing of stomata via mechanisms we consider in Chapter 35.

Other epidermal specializations are the single-celled or multicellular outgrowths collectively called **trichomes**, which give the stems or leaves of some plants a hairy appearance. Some trichomes exude sugars that attract insect pollinators. Leaf trichomes of *Urtica,* the stinging nettle, provide protection by injecting an irritating toxin into the skin of animals that brush against the plant or try to eat it (see Figure 34.17d, p. 851). **Root hairs**, extensions of the outer wall of root epidermal cells (Figure 34.11c), are also trichomes. Root hairs absorb much of a plant's water and minerals from the soil.

The epidermal cells of flower petals (which are modified leaves) synthesize pigments that are partly responsible for a blossom's colours.

STUDY BREAK

1. Describe the defining features, cellular components, and functions of the ground tissue system.
2. What are the functions of xylem and phloem?
3. What are the cellular components and functions of the dermal tissue system?

34.3 Primary Shoot Systems

A young flowering plant's shoot system consists of the main stem, leaves, and buds, as well as flowers and fruits. Chapter 36 looks more closely at flowers and fruits; here we focus on the growth and organization of stems, buds, and leaves of the primary shoot system.

34.3a Stems Are Adapted to Provide Support, Routes for Vascular Tissues, Storage, and New Growth

Stems are structurally adapted for four main functions:

- Stems provide mechanical support, generally along a vertical (upright) axis, for body parts involved in growth, photosynthesis, and reproduction. These parts include meristematic tissues, leaves, and flowers.
- Stems have the vascular tissues (xylem and phloem), which transport products of photosynthesis, water and dissolved minerals, hormones, and other substances throughout the plant.
- Stems are often modified to store water and food.
- Buds and specific stem regions contain meristematic tissue that gives rise to new cells of the shoot.

The Modular Organization of a Stem. A plant stem develops in a pattern that divides the stem into modules, each consisting of a *node* and an *internode.* A **node** is a place on the stem where one or more leaves are attached; the region between two nodes is thus an **internode**. New primary growth occurs in buds—a **terminal bud** at the apex of the main shoot, and **axillary buds**, which produce branches (lateral shoots) at the point where leaves meet the stem. Meristematic tissue in buds gives rise to leaves, flowers, or both **(Figure 34.12, p. 848)**.

🔧 **CONCEPT FIX** Many people think that flowering plants grow from the base of their stems, as if they pushed upward from the soil surface. But this is not true: shoot growth occurs from the apical meristem, not the base of the stem. ⬡

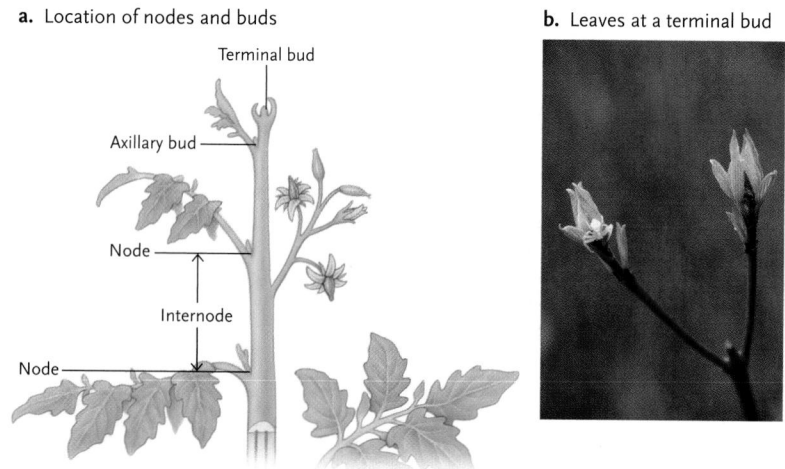

a. Location of nodes and buds

Terminal bud

Axillary bud

Node

Internode

Node

b. Leaves at a terminal bud

altrendo nature/Altrendo/Getty Images

Figure 34.12

Modular structure of a stem. **(a)** The arrangement of nodes and buds on a plant stem. **(b)** Formation of leaves at a terminal bud of a dogwood (genus *Cornus*).

readily after grazing or mowing because the meristem is not removed.

Terminal buds release a hormone (auxin) that inhibits the growth of nearby **lateral buds**, a phenomenon called **apical dominance**. Gardeners who want a bushier plant can stimulate axillary bud growth by periodically cutting off the terminal bud. The flow of hormone signals then dwindles to a level low enough that lateral buds begin to grow. In nature, apical dominance is an adaptation that directs the plant's resources into growing up toward the light (see Chapter 37).

Primary Growth and Structure of a Stem. Primary growth, the cell divisions and enlargement that produce the primary plant body, begins in the shoot and root apical meristems. The sequence of events is shown for a eudicot shoot in **Figure 34.13**.

The shoot apical meristem is a dome-shaped mass of cells at the tip of shoots, surrounded by developing leaves. When a cell of this meristem divides, one of its daughter cells remains part of the meristem, whereas the other begins to differentiate to follow a particular developmental path.

The differentiating cells give rise to three **primary meristems:** *protoderm, procambium,* and *ground meristem* (see Figure 34.13a). These primary meristems are relatively unspecialized tissues with cells that differentiate, in turn, into specialized cells and tissues. In eudicots, the primary meristems are also responsible for elongation of the plant body.

The **protoderm**, a meristem that gives rise to the stem's epidermis, is the outermost layer of the shoot tip, as shown in Figure 34.13a. Inward from the protoderm

In eudicots, most growth in a stem's length occurs directly below the apical meristem as internode cells divide and elongate. Internode cells nearest the apex are most active, so the most visible new growth occurs at the ends of stems. So why isn't the growth of grasses stopped when you mow your lawn or when cattle graze on them? In grasses and some other monocots, the upper cells of an internode stop dividing as the internode elongates, and cell divisions are limited to a meristematic region at the base of the internode. The stems of bamboo and other grasses elongate as the internodes are "pushed up" by the growth of such meristems. This adaptation allows grasses to grow back

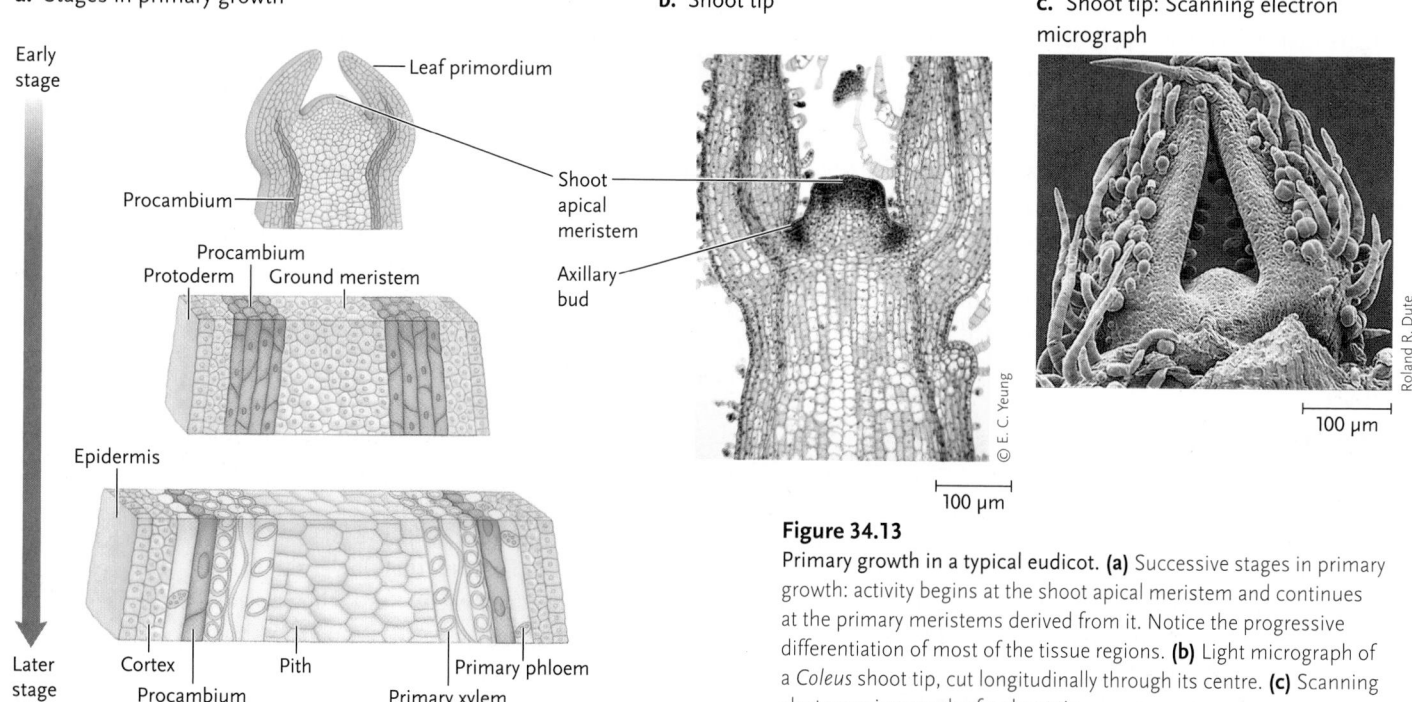

a. Stages in primary growth

Early stage

Leaf primordium

Procambium

Procambium
Protoderm   Ground meristem

Epidermis

Later stage   Cortex   Pith   Primary phloem
   Procambium   Primary xylem

b. Shoot tip

Shoot apical meristem

Axillary bud

© E. C. Yeung

100 μm

c. Shoot tip: Scanning electron micrograph

Roland R. Dute

100 μm

Figure 34.13

Primary growth in a typical eudicot. **(a)** Successive stages in primary growth: activity begins at the shoot apical meristem and continues at the primary meristems derived from it. Notice the progressive differentiation of most of the tissue regions. **(b)** Light micrograph of a *Coleus* shoot tip, cut longitudinally through its centre. **(c)** Scanning electron micrograph of a shoot tip.

is the **ground meristem**, which will give rise to ground tissue. **Procambium**, which produces the primary vascular tissues, is sandwiched between ground meristem layers. Procambial cells are long and thin, and their spatial orientation foreshadows the function of the tissues they produce. In most plants, inner procambial cells give rise to xylem and outer procambial cells to phloem. In plants with secondary growth, a thin region of procambium between the primary xylem and phloem remains undifferentiated. Later it will give rise to a lateral meristem.

The developing vascular tissues become organized into **vascular bundles**, cylinders of primary xylem and phloem that are sometimes wrapped in sclerenchyma. Eudicot stems have vascular bundles arranged in a circle that separates the ground tissue in the centre of the stem (the **pith**) from the ground tissue under the epidermis (the cortex) **(Figure 34.14a)**. Both cortex and pith consist mainly of parenchyma; in some plant species, the pith parenchyma stores starch reserves. Monocot stems also have vascular bundles, but these are scattered throughout the ground tissue, so distinct pith and cortical regions do not form (Figure 34.14b). In some monocots, including bamboo, the pith breaks down, leaving the stem with a hollow core. The hollow stems of certain hard-walled bamboo species are used to make bamboo flutes.

As leaves and buds develop along a stem, some vascular bundles in the stem branch off into these tissues. The arrangement of vascular bundles in a plant ultimately depends on the number of branch points to leaves and buds and on the number and distribution of leaves.

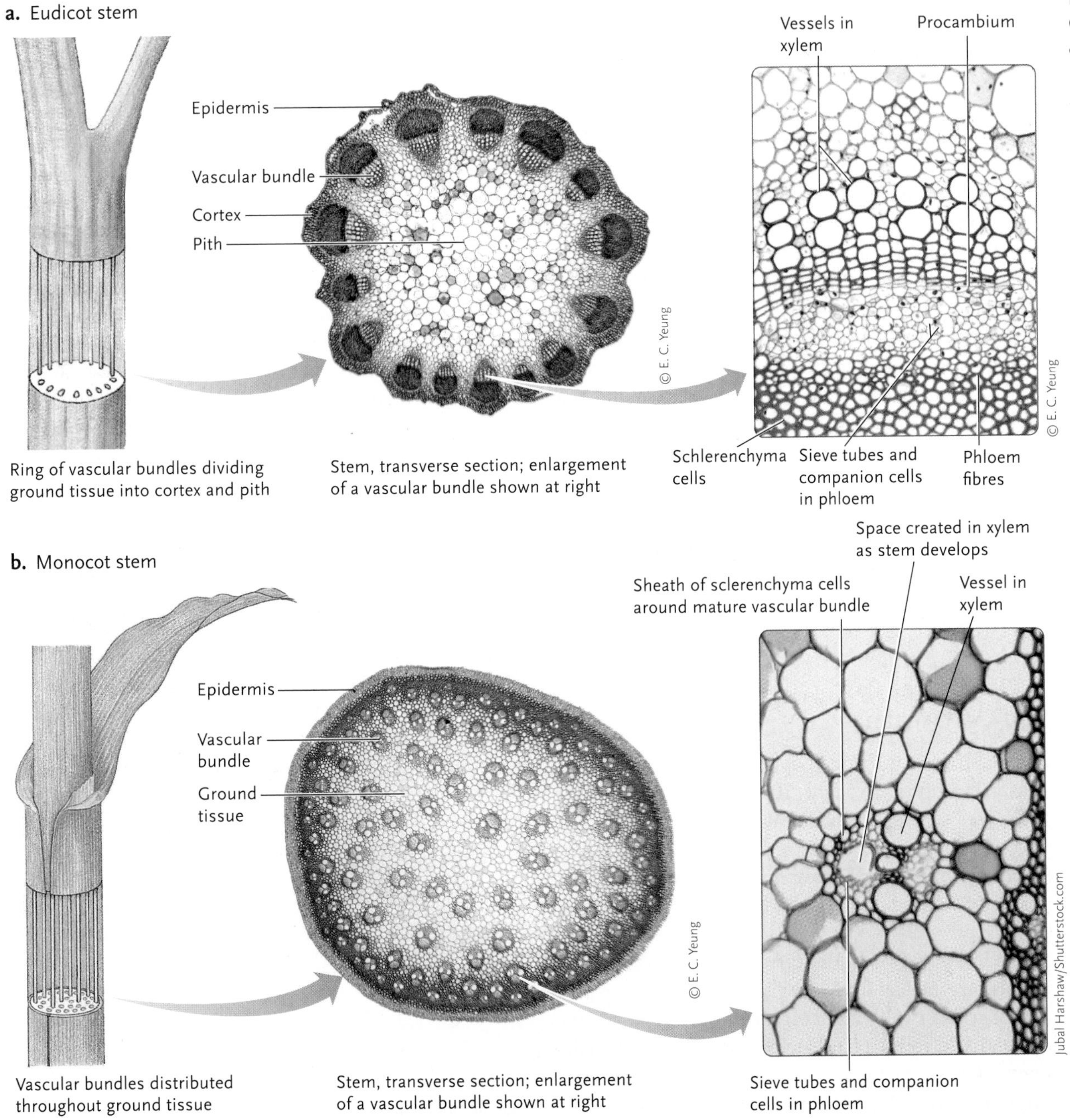

a. Eudicot stem

Epidermis

Vascular bundle

Cortex

Pith

Ring of vascular bundles dividing ground tissue into cortex and pith

Stem, transverse section; enlargement of a vascular bundle shown at right

© E. C. Yeung

Vessels in xylem   Procambium

Schlerenchyma cells   Sieve tubes and companion cells in phloem   Phloem fibres

© E. C. Yeung

b. Monocot stem

Epidermis

Vascular bundle

Ground tissue

Vascular bundles distributed throughout ground tissue

Stem, transverse section; enlargement of a vascular bundle shown at right

© E. C. Yeung

Sheath of sclerenchyma cells around mature vascular bundle

Space created in xylem as stem develops

Vessel in xylem

Sieve tubes and companion cells in phloem

Jubal Harshaw/Shutterstock.com

Figure 34.14
Organization of cells and tissues inside the stem of a eudicot and a monocot. **(a)** Part of a stem from sunflower (*Helianthus*), a eudicot. In many species of eudicots and conifers, the vascular bundles develop in a more or less ring-like array in the ground tissue system, as shown here. The enlarged photo at right is of a vascular bundle of a sunflower (*Helianthus*). **(b)** Part of a stem from corn (*Zea mays*), a monocot. In most monocots and some herbaceous eudicots, vascular bundles are scattered through the ground tissue, as shown here.

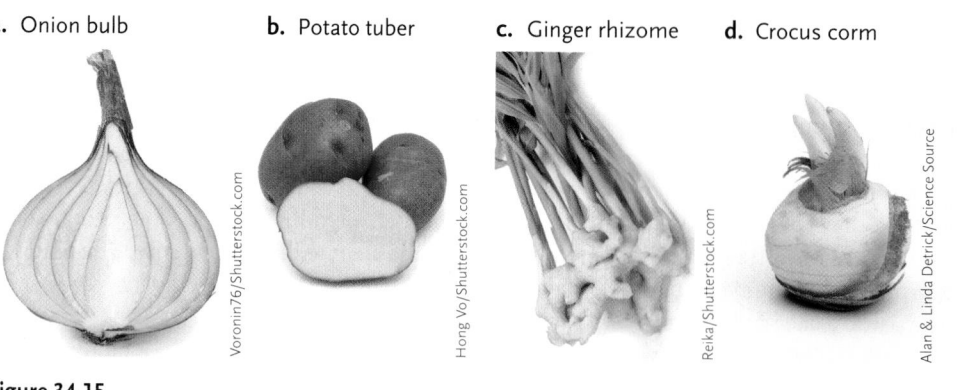

a. Onion bulb   **b.** Potato tuber   **c.** Ginger rhizome   **d.** Crocus corm   **e.** Strawberry stolons

Figure 34.15

A selection of modified stems. **(a)** The fleshy bulbs of onions (*Allium cepa*) are modified shoots in which the plant stores starch. **(b)** A potato (*Solanum tuberosum*), a tuber. **(c)** Ginger "root," the pungent, starchy rhizome of the ginger plant (*Zingiber officinale*). **(d)** Crocus plants (genus *Crocus*) typically grow from a corm. **(e)** A strawberry plant (*Fragaria ananassa*) and stolons.

Stem Modifications. Evolution has produced a range of stem specializations, including structures modified for reproduction, food storage, or both **(Figure 34.15)**. An onion or a garlic head is a *bulb,* a modified shoot that consists of a bud with fleshy leaves. *Tubers* are stem regions enlarged by the presence of starch-storing parenchyma cells; the potato is an example of a plant that forms tubers. The "eyes" of a potato are buds at nodes of the modified stem. Many grasses, such as quackgrass (*Elymus repens*), and some weeds are difficult to eradicate because they have *rhizomes*—long underground stems that can extend as much as 50 cm deep into the soil and rapidly produce new shoots when existing ones are pulled out. The pungent, starchy "root" of ginger is also a rhizome. Crocuses and some other ornamental plants develop elongated, fleshy underground stems called *corms,* another starch-storage adaptation. Tubers, rhizomes, and corms all have meristematic tissue at nodes from which new plants can be propagated—a vegetative (asexual) reproductive mode. Other plants, including strawberries (*Fragaria* spp.), reproduce vegetatively via slender stems called *stolons,* which grow along the soil surface. New plants arise at nodes along the stolon.

34.3b Leaves Carry Out Photosynthesis and Gas Exchange

Each spring, a mature maple tree heralds the new season by unfurling roughly 100 thousand leaves. Some other tree species produce leaves by the millions. For these and most other plants, leaves are the main organs of photosynthesis and gas exchange (the movement of carbon dioxide and oxygen into and out of the leaf).

Leaf Morphology and Anatomy. In both eudicots and monocots, the leaf **blade** provides a large surface area for absorbing sunlight and carbon dioxide **(Figure 34.16)**. Leaves of flowering plants are generally oriented on the stem axis so that they can capture the maximum amount of sunlight; the stems and leaves of some plants

change position to follow the Sun's movement during the day (this phenomenon is described in Chapter 38).

Many eudicot leaves, such as those of maples, have a broad, flat blade attached to the stem by a stalklike **petiole** (see Figure 34.16); the celery stalks that we eat are petioles. Petioles hold leaves away from the stem and help prevent individual leaves from shading one another. In many plant species, petioles allow leaves to move in the breeze—think about trembling aspen (*Populus tremuloides*) leaves rustling in a breeze— enhancing air circulation around leaves, thus replenishing the supply of carbon dioxide for photosynthesis. In most monocot leaves, such as those of grass and corn, the blade is longer and narrower and its base simply forms a sheath around the stem (see Figure 34.16).

As with other plant parts, however, the adaptation of land plants to different environments has produced tremendous variety in leaf morphology. For instance, spiny margins on the leaves of the carnivorous Venus flytrap (*Dionaea muscipula*) **(Figure 34.17a)** prevent the escape of insects that become trapped when the seemingly hinged leaves snap shut around them—a movement that takes only about a tenth of a second. Leaves or parts of leaves may also be modified into tendrils, like those of the sweet pea (Figure 34.17b). Cactus leaves are modified as spines, while leaves of other plants have trichomes that take the form of hairs or hooks (Figure 34.17c) that help defend against grazing by herbivores. In some plants, these trichomes

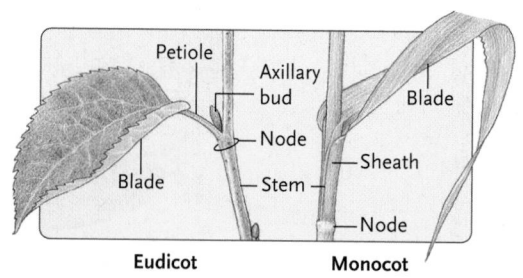

Figure 34.16

Leaf forms. Common forms of eudicot and monocot leaves.

a. Interlocking spines of Venus flytrap leaves

©iStock.com/Ryan Poling

b. Tendrils of a sweet pea

Maxine Adcock/Science Source

50 μm

c. Hairs and glandular structures on a tomato leaf

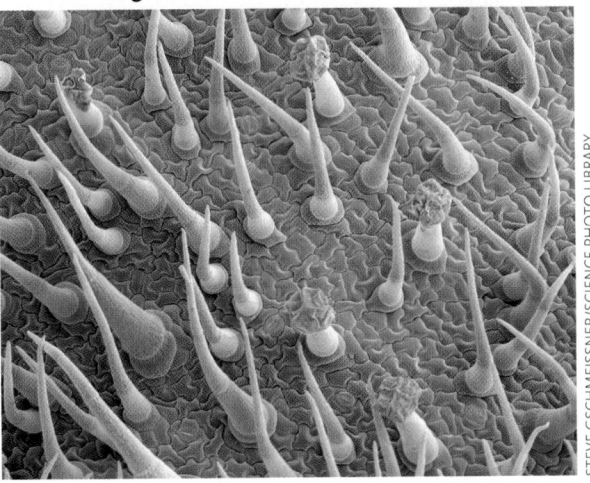

STEVE GSCHMEISSNER/SCIENCE PHOTO LIBRARY

d. Stinging hairs on nettle leaf

Nigel Cattlin/Visuals Unlimited, Inc.

Figure 34.17

A few adaptations of leaves. **(a)** Margins on the leaves of the Venus flytrap (*Dionaea muscipula*) are modified into long, interlocking spines. **(b)** The tendrils of a sweet pea (*Lathyrus odoratus*) help to support the climbing plant's stem. **(c)** Specializations on a tomato leaf include hooklike hairs and lobed glandular structures that release an insect-deterring chemical. **(d)** Stinging hairs (trichomes) on a leaf of *Urtica* (stinging nettle) defend the plant against herbivores.

seem more aggressive than defensive: the leaves and stem of stinging nettle (*Urtica dioica*) are covered with stinging trichomes (Figure 34.17d). When a herbivore or unlucky passerby brushes against a leaf, the tips of the trichomes break off, converting each trichome into hypodermic needles that inject the stinging chemicals in the base of the trichome into the victim's skin.

Leaf Primary Growth and Internal Structure. As the shoot apical meristem divides, it produces a series of bumps on its sides, the **leaf primordia**, which give rise to leaves (see Figure 34.13a, p. 848). As the plant grows and the internodes elongate, the leaves that form from leaf primordia become spaced at intervals along the length of the stem or its branches.

A leaf is typically composed of several layers **(Figure 34.18, p. 852)**. Uppermost is the epidermis, with cuticle covering its outer surface. Just beneath the epidermis is **mesophyll** (*mesos* = middle; *phyllon* = leaf), ground tissue composed of loosely packed parenchyma cells that contain chloroplasts. In the leaves of many plants, especially eudicots, the mesophyll is differentiated into palisade mesophyll and spongy mesophyll. *Palisade mesophyll* cells contain more chloroplasts than the spongy mesophyll cells and are arranged in

compact columns with smaller air spaces between them, typically toward the upper leaf surface. *Spongy mesophyll*, which tends to be located toward the underside of a leaf, consists of irregularly arranged cells with a conspicuous network of air spaces—between 15% and 50% of the leaf's volume—that give this layer a spongy appearance. What is the role of these air spaces? They enhance the uptake of carbon dioxide and release of oxygen during photosynthesis. Mesophyll may also contain collenchyma and sclerenchyma cells, which support the photosynthetic cells.

Below the mesophyll is another cuticle-covered epidermal layer. Except in grasses and a few other plants, this layer contains most of the stomata through which water vapour exits the leaf and gas exchange occurs. For example, the upper surface of an apple leaf has no stomata, whereas a square centimetre of the lower surface has more than 20 thousand. A square centimetre of the upper epidermis of a tomato leaf has about 1200 stomata, whereas the same area of the lower epidermis has 13 thousand. Why are more stomata located on the underside of the leaf? This positioning protects stomata from direct exposure to sunlight, thus limiting water loss by evaporation through stomatal openings.

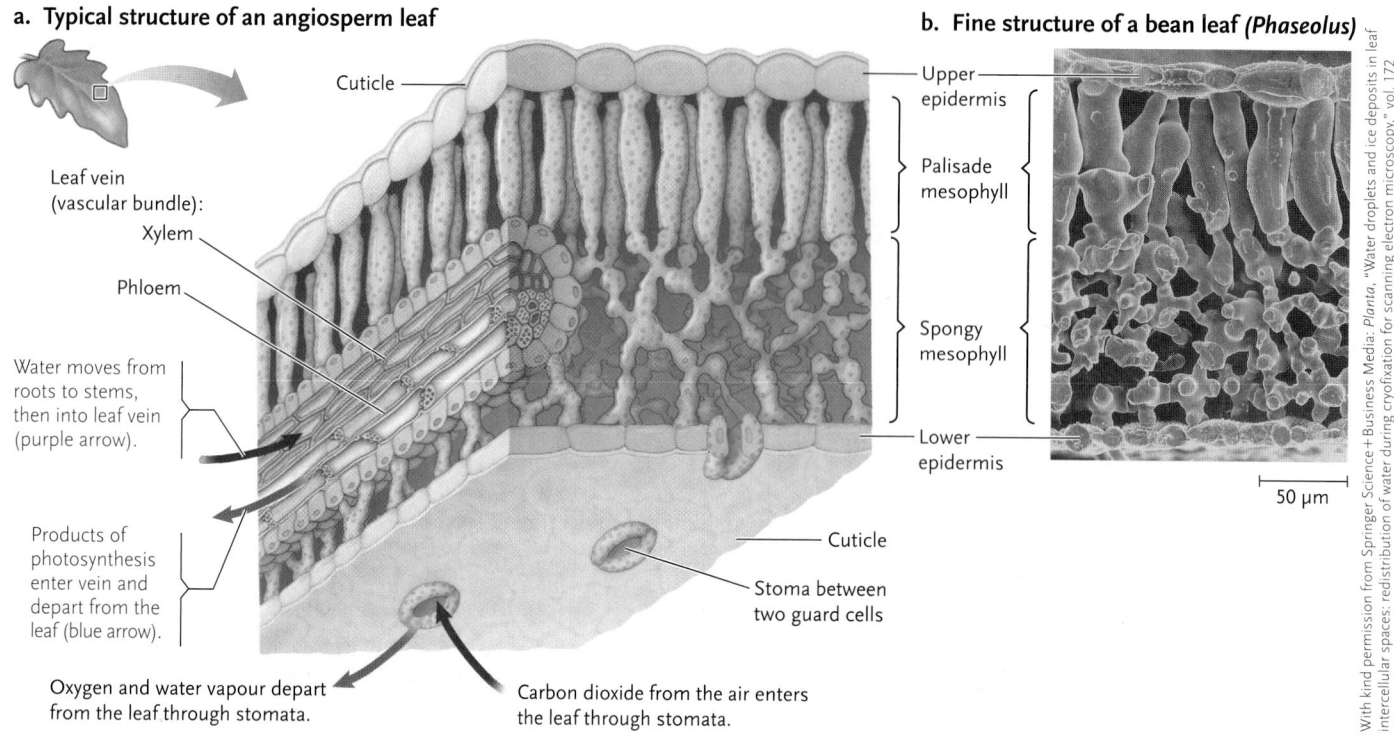

a. Typical structure of an angiosperm leaf

Cuticle

Leaf vein
(vascular bundle):

Xylem

Phloem

Water moves from
roots to stems,
then into leaf vein
(purple arrow).

Products of
photosynthesis
enter vein and
depart from the
leaf (blue arrow).

Oxygen and water vapour depart
from the leaf through stomata.

Carbon dioxide from the air enters
the leaf through stomata.

b. Fine structure of a bean leaf *(Phaseolus)*

Upper
epidermis

Palisade
mesophyll

Spongy
mesophyll

Lower
epidermis

Cuticle

Stoma between
two guard cells

50 μm

With kind permission from Springer Science+Business Media: *Planta*, "Water droplets and ice deposits in leaf intercellular spaces: redistribution of water during cryofixation for scanning electron microscopy," vol. 172 (1):20–37, C.E. Jeffree, et al, 1987.

Figure 34.18

Internal structure of a leaf. **(a)** Diagram of a typical leaf structure for many kinds of flowering plants. See Figure 34.11b, p. 847, for a scanning electron micrograph of stomata. **(b)** Scanning electron micrograph of tissue from the leaf of a kidney bean plant (*Phaseolus*), transverse section. Notice the compact organization of epidermal cells.

Vascular bundles form a lacy network of **veins** throughout the leaf. Eudicot leaves typically have a branching vein pattern; in monocot leaves, veins tend to run in parallel along the length of the leaf (see Table 34.1, p. 841).

In temperate regions, most leaves are temporary structures. In deciduous species such as birches and maples, hormonal signals cause the leaves to drop from the stem as the days shorten in autumn. Other temperate plants, such as most conifers, also drop their leaves (which are modified into needles in conifers), but they appear "evergreen" because the leaves may persist for several years and do not all drop at the same time.

STUDY BREAK

1. Describe the functions of stems and stem structures.
2. Explain how primary growth occurs in stems. What are the three primary meristems that form and to what tissues does each give rise?
3. Compare the arrangement of the three primary tissues in monocot and dicot stems.
4. Explain the general function of leaves and how leaf anatomy supports this role in eudicots and monocots.
5. Describe the steps in the primary growth of a leaf and the structures that result from the process.

34.4 Root Systems

Plants cannot move around to find water and nutrients when they have depleted supplies in their immediate soil neighbourhood, so they must be able to forage for new supplies. Once these supplies are found, the plant must absorb enough water and dissolved minerals to sustain growth and routine cellular maintenance. These tasks can require a tremendous root surface area, at least part of which is regularly replaced. In one study, rye plants (*Secale cereale*) that had been growing for only four months were measured. One plant's root system had a surface area of more than 700 m²—about 130 times the surface area of its shoot system!

In addition to taking up water and nutrients, roots store nutrients produced by photosynthesis, some of which is used by root cells and some transported later to cells of the shoot. As the root system penetrates downward and spreads out, it also anchors the above-ground parts.

34.4a Taproot and Fibrous Root Systems Are Specialized for Particular Functions

Most eudicots have a **taproot system**—a single main root, or taproot, that is adapted for storage, plus smaller branching roots called **lateral roots (Figure 34.19a)**. As the main root grows downward, the diameter of the upper, older part of the root increases, and the lateral roots emerge along the length of its older, differentiated

a. Taproot system

b. Fibrous root system

c. Adventitious roots

jirapong/Shutterstock.com

Figure 34.19

Types of roots. **(a)** Taproot system of a California poppy (*Eschscholzia californica*). **(b)** Fibrous root system of a grass plant. **(c)** Example of adventitious roots, the numerous prop roots of red mangrove trees (*Rhizophora*).

regions. The youngest lateral roots are near the root tip. Carrots and dandelions have a taproot system, as do pines and many other conifers. A pine's taproot system can penetrate 6 m or more into the soil.

Grasses and many other monocots develop a **fibrous root system** in which several main roots branch to form a dense mass of smaller roots (Figure 34.19b). Fibrous root systems are adapted to absorb water and nutrients from the upper layers of soil and tend to spread out laterally from the base of the stem. Fibrous roots are important ecologically because dense root networks help hold topsoil in place and prevent erosion. During the 1930s, overgrazing by livestock and intensive farming in the prairie provinces of Canada and the U.S. Midwest destroyed hundreds of thousands of acres of native prairie grasses, contributing to soil erosion on a massive scale. Swirling clouds of soil particles prompted journalists to name the area the Dust Bowl and gave this decade the name the "Dirty Thirties."

In some plants, **adventitious roots** arise from the stem of the young plant. **Adventitious** refers to any structure arising at an unusual location, such as roots that grow from stems or leaves. Adventitious roots of Virginia creeper (*Parthenocissus quinquefolia*) and some other climbing plants produce a gluelike substance that allows them to cling to vertical surfaces. The *prop roots* of a corn plant are adventitious roots that develop from the shoot node nearest the soil surface; they support the plant and absorb water and nutrients. Mangroves and other trees that grow in marshy habitats often have huge prop roots, which develop from branches and from the main stem (Figure 34.19c).

34.4b Root Structure Is Specialized for Underground Growth

Like shoots, roots have distinct anatomical parts, each with a specific function. In most plants, primary growth of roots begins when an embryonic root emerges from a germinating seed and its apical meristem becomes active. **Figure 34.20, p. 854,** shows the structure of a root tip. Notice that the root apical meristem terminates in a dome-shaped cell mass, the **root cap.** The meristem produces the cap, which, in turn, surrounds and protects the meristem as the root elongates through the soil. Certain cells in the cap respond to gravity, guiding the root tip downward. Cap cells also secrete a polysaccharide-rich substance that lubricates the tip and eases the growing root's passage through the soil. Outer root cap cells are continually abraded off and replaced by new cells at the cap's base.

Zones of Primary Growth in Roots. In most plants, root growth is a continuous process that only stops if environmental conditions become unfavourable for growth (e.g., drought).

The root apical meristem and the actively dividing cells behind it form the **zone of cell division.** As in the stem, cells of the apical meristem divide to produce cells that remain as part of the meristem and other cells that differentiate into the three primary meristems. The arrangement of the primary meristems is

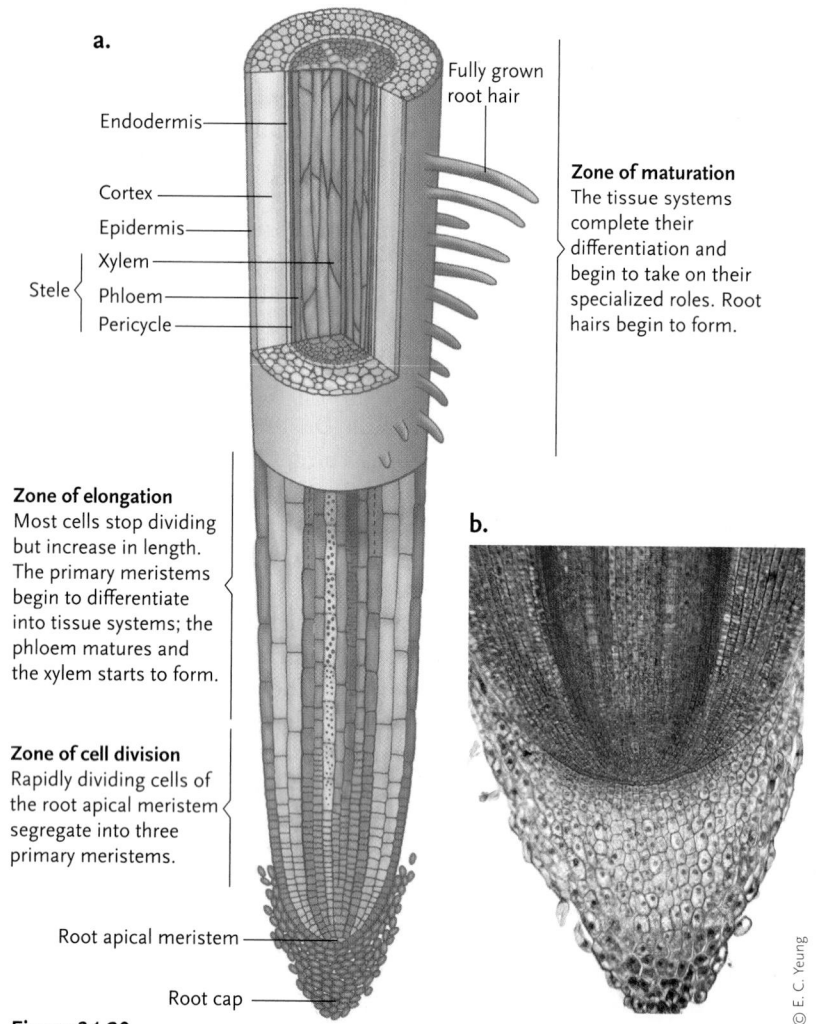

a.

Endodermis

Cortex

Epidermis

Xylem

Phloem

Pericycle

Stele

Fully grown root hair

Zone of maturation
The tissue systems complete their differentiation and begin to take on their specialized roles. Root hairs begin to form.

b.

Zone of elongation
Most cells stop dividing but increase in length. The primary meristems begin to differentiate into tissue systems; the phloem matures and the xylem starts to form.

Zone of cell division
Rapidly dividing cells of the root apical meristem segregate into three primary meristems.

Root apical meristem

Root cap

© E. C. Yeung

Figure 34.20

Tissues and zones of primary growth in a root tip. **(a)** Generalized root tip, longitudinal section. **(b)** Micrograph of a corn (*Zea mays*) root tip, longitudinal section.

different in the root than in the shoot: cells in the centre of the root tip become the procambium, those just outside the procambium become ground meristem, and those on the periphery of the apical meristem become protoderm.

The zone of cell division merges into the **zone of elongation.** Most of the increase in a root's length comes from this region, where cells become longer as their vacuoles fill with water. This *hydraulic* elongation pushes the root cap and apical meristem through the soil by as much as several centimetres a day.

Above the zone of elongation, cells do not increase in length, but they may differentiate further and take on specialized roles in the **zone of maturation.** For example, epidermal cells in this zone give rise to root hairs, and the procambium, ground meristem, and protoderm complete their differentiation in this region.

Tissues of the Root System. Together with the primary growth of the shoot, primary root growth produces a unified system of vascular pipelines extending from root tip to shoot tip. The root procambium produces cells

that mature into the root's xylem and phloem **(Figure 34.21).** Ground meristem gives rise to the root's cortex, its ground tissue of starch-storing parenchyma cells that surround the stele. In eudicots, the stele runs through the centre of the root (see Figure 34.21a). In corn and some other monocots, the vascular cylinder (**stele**) forms a ring that divides the ground tissue into cortex and pith (see Figure 34.21b).

The root cortex often contains air spaces that allow oxygen to reach all of the living root cells. In many flowering plants, the outer root cortex cells give rise to an **exodermis,** a thin band of cells that, among other functions, may limit water losses from roots and help regulate the absorption of ions. The innermost layer of the root cortex is the **endodermis,** a thin, selectively permeable barrier that helps control the movement of water and dissolved minerals into the stele. We look in more detail at the roles of exodermis and endodermis in Chapter 35.

The outermost part of the stele, between the endodermis and the phloem, is the **pericycle,** consisting of one or more layers of parenchyma cells that have retained the ability to function as meristem. The pericycle initiates the formation of lateral roots **(Figure 34.22)** in response to chemical growth regulators. These lateral roots grow out through the cortex and epidermis, producing enzymes that help break down the intervening cells. The distribution and frequency of lateral root formation partly control the overall shape of the root system and the extent of the soil area it can penetrate.

The outer surface of some cells in the developing root epidermis become elongated into root hairs (see Figure 34.20). Root hairs can be more than a centimetre long and can form in less than a day. Collectively, the thousands or millions of them on a plant's roots greatly increase the plant's absorptive surface. But it is not just the increased surface area provided by root hairs that increases nutrient uptake: each hair is a slender tube with thin walls made sticky on their surface by a coating of pectin. Soil particles tend to adhere to the walls, providing an intimate association between the hair and the surrounding earth, thus facilitating the uptake of water molecules and mineral ions from soil. When plants are transplanted, rough handling can tear off much of this fragile absorptive surface. Unable to take up enough water and minerals, the transplant may die before new root hairs can form.

STUDY BREAK

1. Compare the two general types of root systems.
2. Describe the zones of primary growth in roots.
3. Describe the various tissues that arise in a root system and their functions.
4. Compare the arrangement of the three primary tissues in a dicot root to the arrangement in a monocot root and in a dicot stem.

a. Eudicot root

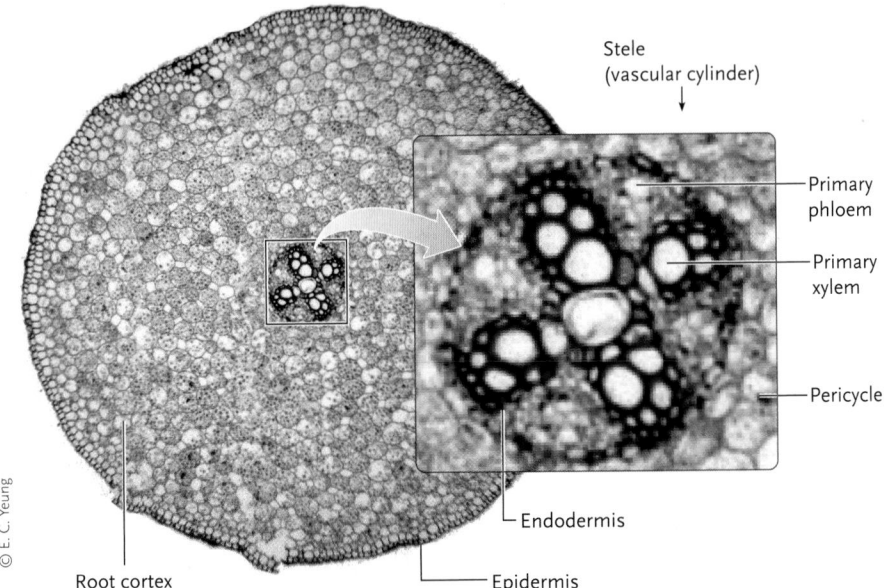

Stele
(vascular cylinder)
↓

Primary phloem

Primary xylem

Pericycle

Endodermis

Epidermis

Root cortex

© E. C. Yeung

b. Monocot root

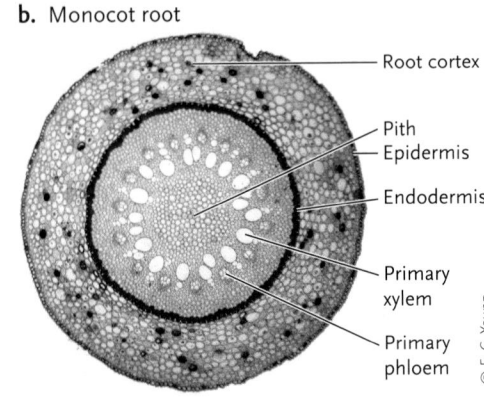

Root cortex

Pith
Epidermis

Endodermis

Primary xylem

Primary phloem

© E. C. Yeung

Figure 34.21
Stele structure in eudicot and monocot roots compared.
(a) A young root of the buttercup *Ranunculus*, a eudicot.
The close-up shows details of the stele. **(b)** Root of
Smilax, a monocot. Notice how the vascular cylinder
divides the ground tissue into cortex and pith. Both
roots are shown in transverse section.

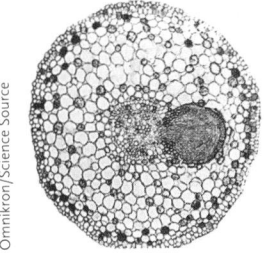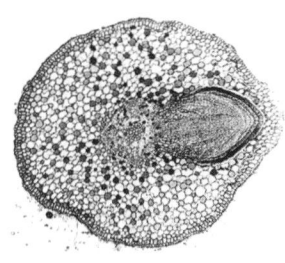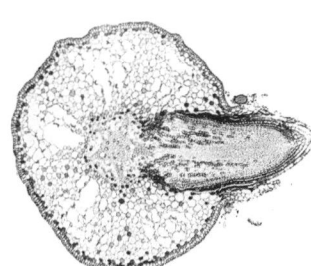

Omnikron/Science Source

Figure 34.22
Micrographs showing the formation of a lateral
root from the pericycle of a willow tree (*Salix*). These
micrographs show transverse sections.

34.5 Secondary Growth

All plants undergo primary growth of the root and
stem. In addition, some plants have secondary growth
processes that add girth to roots and stems over two or
more growing seasons.

CONCEPT FIX Many people think that only woody
plants have secondary growth. In reality, many herbaceous plants also have secondary growth; for example,
carrot roots are mostly secondary tissues. And if you've
ever tried to dig up a dandelion, you'll know how large
and tough their taproots are, which is due to secondary
growth in these roots. ⬡

In plant species that have secondary growth, older
stems and roots become more massive and woody
through the activity of two types of lateral meristems
called *cambia* (singular, *cambium*). One of these meristems, the **vascular cambium**, produces secondary
xylem and phloem. The other, called the **cork cambium**, produces **cork**, a secondary tissue that replaces
the original epidermis of the plant. In contrast to the
cells of the apical meristems, the cells of the lateral
meristems divide perpendicular to the stem's longitudinal axis, so their descendants add girth to the stem
instead of length.

34.5a Vascular Cambium Gives Rise to Secondary Growth in Stems

Recall that after the stem of a woody plant completes its
primary growth, each vascular bundle contains a layer of
undifferentiated cells between the primary xylem and
the primary phloem. These cells, along with parenchyma
cells between the bundles, eventually give rise to a continuous cylinder of vascular cambium that surrounds the
xylem and pith of the stem **(Figure 34.23, p. 856).** Secondary growth takes place as the cells of the vascular
cambium divide. Division of the vascular cambium produces secondary xylem toward the inside of the stem and
secondary phloem toward the outside of the stem.

With time, the mass of secondary xylem inside the
ring of vascular cambium increases, forming the hard
tissue known as **wood.** Outside the vascular cambium,
secondary phloem cells are also added each year
(Figure 34.24, p. 856). (The primary phloem cells, which
have thin walls, are destroyed as they are pushed outward by secondary growth.) As a stem increases in
diameter, the growing mass of new tissue eventually
causes the cortex, and the epidermis beyond it, to rupture. Such breaks in the outer protective "skin" of the
plant are potentially harmful as they would allow easy
entrance for pathogens. The cork cambium—produced

Figure 34.23

Secondary and primary growth compared. In a woody plant, primary growth resumes each spring at the terminal and lateral buds. Secondary growth resumes at the vascular cambium inside the stem.

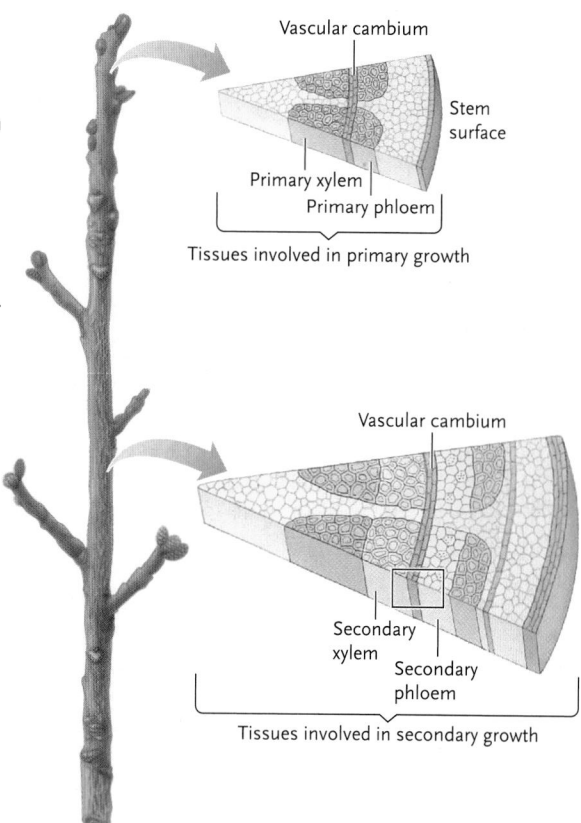

Vascular cambium

Stem surface

Primary xylem

Primary phloem

Tissues involved in primary growth

Vascular cambium

Secondary xylem

Secondary phloem

Tissues involved in secondary growth

Bark encompasses all the tissues outside the vascular cambium; it thus includes the secondary phloem, and the periderm (see Figure 34.25). Girdling a tree by removing a strip of bark around the trunk is lethal because it destroys the secondary phloem layer, so nutrients from photosynthesis in leaves cannot reach the tree's roots. Cork for use in flooring and as bottle stoppers is harvested from the thick outer bark of the cork oak, *Quercus suber* **(Figure 34.26)**. Cork can be harvested from these trees once they are 25 years old and can be sustainably harvested every 9 to 12 years thereafter. Some trees can yield about 1 tonne of cork over the course of their lives!

How do the vascular cambium and other living tissues in a secondary stem obtain oxygen, given that the bark can be very thick on some trees? In some regions of the stem, the cork cambium divides very actively, forming tissue with abundant air spaces (*lenticels*). Lenticels allow the exchange of oxygen and carbon dioxide between the living tissues and the outside air.

As a tree ages, changes also unfold in the appearance and function of the wood itself. In the centre of its older stems and roots is **heartwood**, dry tissue that no longer transports water and solutes and is a storage depot for some defensive compounds. In time, these substances—including resins, oils, gums, and tannins—clog and fill in the oldest xylem pipelines. Typically, they darken the heartwood, strengthen it, and make it more aromatic and resistant to decay. **Sapwood** is secondary growth located between heartwood and the vascular cambium. Compared with heartwood, it is wet and not as strong (see Figure 34.25).

In temperate climates, trees produce secondary xylem seasonally, with larger-diameter cells produced in spring, when water is generally abundant, and smaller-diameter cells in summer, when less water is available to be transported. The resulting "spring wood" and

early in the stem's secondary development by meristem cells in the cortex or epidermis—replaces the lost epidermis with cork cells. The walls of cork cells contain lignin and thick layers of **suberin**, a waxy substance that is impermeable to water and gases. Cork cells are dead at maturity. Cork is produced toward the outside of the stem; like the vascular cambium, the cork cambium also produces cells toward the inside of the stem, called phelloderm. Together, the cork, cork cambium, and phelloderm make up the **periderm (Figure 34.25)**.

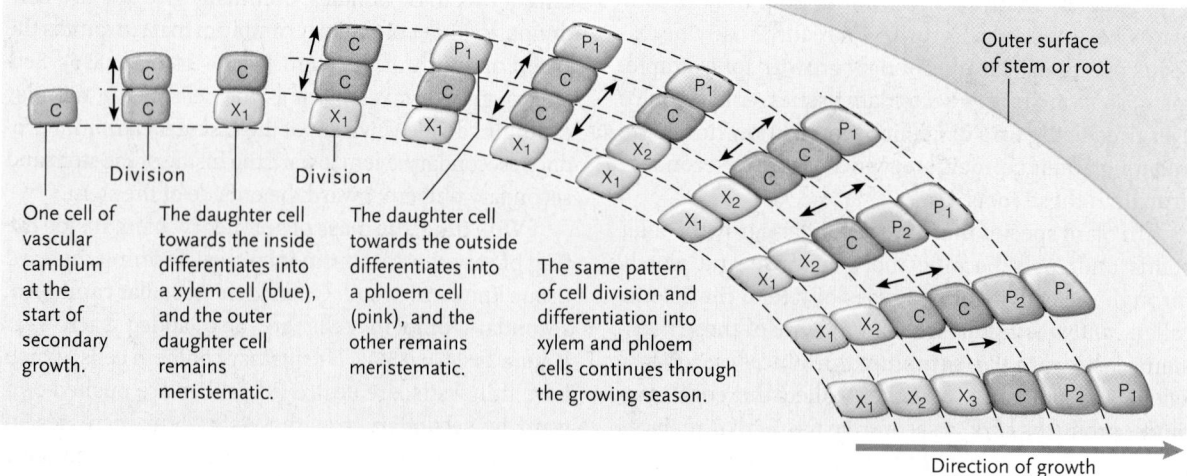

One cell of vascular cambium at the start of secondary growth.

Division

The daughter cell towards the inside differentiates into a xylem cell (blue), and the outer daughter cell remains meristematic.

Division

The daughter cell towards the outside differentiates into a phloem cell (pink), and the other remains meristematic.

The same pattern of cell division and differentiation into xylem and phloem cells continues through the growing season.

Outer surface of stem or root

Direction of growth

Figure 34.24

Relationship between the vascular cambium and its derivative cells (secondary xylem and phloem). The drawing shows stem growth through successive seasons. Notice how the ongoing divisions displace the cambial cells, moving them steadily outward even as the core of xylem increases the stem or root thickness.

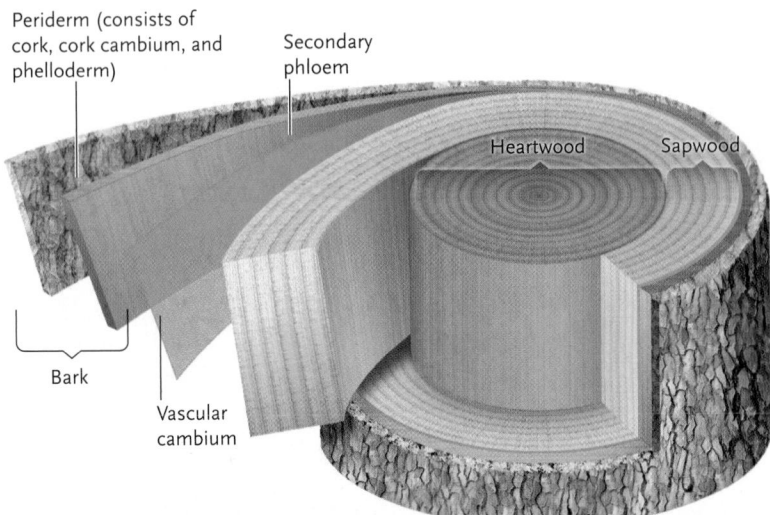

Periderm (consists of cork, cork cambium, and phelloderm)

Secondary phloem

Heartwood

Sapwood

Bark

Vascular cambium

Figure 34.25
Structure of a woody stem showing extensive secondary growth. Heartwood, the mature tree's core, has no living cells. Sapwood, the cylindrical zone of xylem between the heartwood and the vascular cambium, contains some living parenchyma cells among the nonliving vessels and tracheids. Everything outside the vascular cambium is bark. Everything inside it is wood.

© adrian davies/Alamy

Figure 34.26
Cork oak (*Quercus suber*) that has recently had part of its bark harvested.

"summer wood" reflect light differently, and it is possible to identify them as alternating light and dark bands. The alternating bands represent annual growth layers known as "growth rings" **(Figure 34.27, p. 858)**. The age of a tree can be determined by counting the growth rings.

Growth rings also provide information on past climates: the wider spaced the rings, the more growth a tree was able to put on in one year, so the better the conditions (i.e., warmer and wetter). Dendroclimatologists use tree rings and other biological information to reconstruct past environments. This line of research is making significant contributions to our understanding of how the global climate has changed over time (see "People behind Biology," Box 34.3, p. 858).

34.5b Secondary Growth Can Also Occur in Roots

The roots of grasses, palms, and other monocots are almost always produced by primary growth alone, but in some plants, secondary growth also occurs in roots, although it is different from that in stems. In a root, the vascular cambium arises in part from a procambium layer between the xylem and phloem **(Figure 34.28, p. 860, step 1)** and in part from the pericycle (step 2), eventually forming a complete cylinder (step 3). The vascular cambium functions in roots as it does in stems, producing secondary xylem toward the inside and secondary phloem toward the outside. As secondary xylem accumulates, older roots can become extremely thick and woody. Their ongoing secondary growth is powerful enough to break through concrete sidewalks and even dislodge the foundations of homes.

The pericycle also produces cork cambium in roots. In many woody eudicots and in all gymnosperms, most of the root epidermis and cortex falls away, and the surface consists entirely of tissue produced by the cork cambium (see Figure 34.28, step 4).

34.5c Secondary Growth Is an Adaptive Response

Like all living organisms, plants compete for resources, and woody stems and roots confer some advantages. Plants with taller stems or wider canopies that defy the pull of gravity can intercept more of the light energy from the Sun. With a greater energy supply for photosynthesis, they have the metabolic means to

Doug Larson, *University of Guelph*

You might think that the biggest trees would also be the oldest, but this isn't necessarily the case. Some big trees are very old—the giant sequoias of California, for example—but sometimes the oldest trees are slow-growing survivors of marginal habitats. In Canada, the oldest trees east of the Rocky Mountains are white cedars (*Thuja occidentalis*) growing in "vertical forests" on the cliffs of the Niagara Escarpment (**Figure 1**). The escarpment runs through southwestern Ontario from near Niagara Falls north to the Bruce Peninsula that juts out into Lake Huron.

These ancient trees were discovered in the 1980s by Doug Larson, a biologist at the University of Guelph, Ontario, and his graduate students. Larson was interested in how environmental gradients influence a plant community and had decided that the cliffs of the Niagara Escarpment would be an interesting gradient to study. As part of describing the forest community on the cliffs, Larson and his students sampled the trees to determine their age. This step sounds simple, but because the trees grow on steep cliffs, sampling them meant dangling over the edge of a cliff in harness and helmet using rock-climbing skills. The trees were stunted,

Figure 1

White cedar (Thuja occidentalis) *growing on a cliff of the Niagara Escarpment, Ontario.*

twisted, and very small, ranging from a few centimetres to a few metres in height and less than 25 cm in diameter. Thin cores were taken from living trees using a core borer. Rings are usually very obvious in both slices and cores (see Figure 34.27), but when Larson looked at the cores in the lab, he couldn't see any rings. Only after the cores were polished with sandpaper and checked under a microscope could the rings be counted. To the team's amazement, the tree was 350 years old! Larson did more extensive sampling of

the cliff forests, in collaboration with Peter Kelly, a dendrochronologist (a biologist who uses tree-ring data to date past events), and found that these trees were indeed ancient, with some living specimens that were more than 1300 years old (they germinated in about 690 CE). Larson had discovered an ancient forest that had survived for centuries in one of the most heavily populated parts of Canada. Other ancient cliff forests—which are, literally, life on the edge—have since been discovered elsewhere in the world.

a.

Primary growth, some secondary growth | Secondary growth

Year 1 | 2 | 3

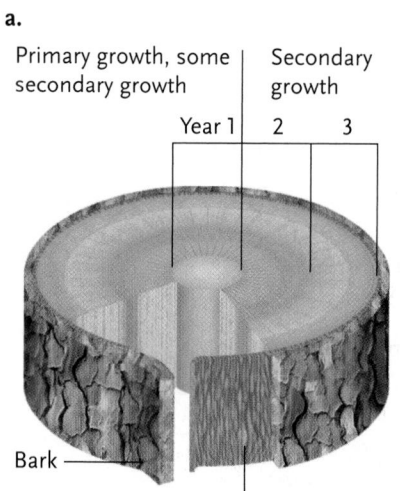

Bark

Vascular cambium

b.

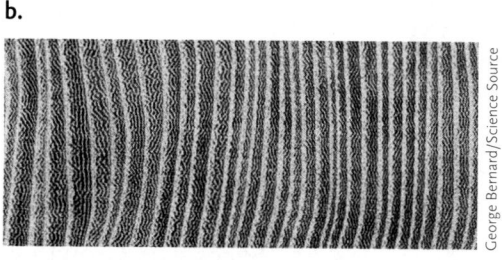

Figure 34.27

Secondary growth and tree ring formation. **(a)** A radial cut through a woody stem (shown here in cross or transverse section) that has three annual rings. **(b)** Tree rings in an elm (*Ulmas*). Each ring corresponds to one growing season. Differences in the widths of tree rings correspond to shifts in climate, including the availability of water.

Not all plants shed their leaves at the end of the growing season; evergreen conifers retain their leaves for several years. The record holder for longest-lived leaves is also a gymnosperm, and its leaves can be over 500 years old. This gymnosperm, *Welwitschia mirabilis*, is one of the weirdest plants in the world, and grows only in the Namib Desert. The Namib Desert stretches along the Atlantic coast of southwestern Africa and receives an average of less than 10 mm of rain per year. In some years, there is no rain at all. Another source of moisture in the Namib Desert is the coastal fogs that form at night as the cold, moist ocean air meets the hot air rising off the desert. The fogs disperse in the early morning, leaving a crucial source of water for the few organisms that live in the desert. *Welwitschia* is one of the plants that can survive life in the Namib.

At first glance, *Welwitschia* doesn't really look like a plant **(Figure 1)** but rather like a pile of old leaves. The leaves really are the dominant feature of this plant. Like all seed plants, *Welwitschia* has roots, a stem, and leaves. Its woody stem is very short, growing only about 50 cm high. Its roots can extend to great depths in the sandy soil. The plant generally has only

Figure 1
Welwitschia mirabilis *growing in the Namib desert of Africa.*

one pair of leaves, which are evergreen and shaped like broad, flat ribbons. Unlike other plants, *Welwitschia* never sheds these leaves, nor does it produce more leaves. The original leaves just continue to grow for the entire life of the plant—which is about 500 years, on average, with the oldest specimens being about 2000 years old. The leaves can be several metres long, although they sometimes split along their length, and the ends get tattered and torn by the wind. Wouldn't such large leaves be a disadvantage in a desert? Many desert plants have very reduced leaves—cacti, for example, have leaves reduced to spines. *Welwitschia's* large leaves are beneficial: they shade the soil around the stem of the plant, keeping that region of soil much cooler than the surrounding area, which can reach temperatures of up to 65°C. The leaves also absorb moisture from the nighttime fogs. This bizarre plant is a "living fossil," a survivor from the Jurassic period, when gymnosperms dominated the Earth.

increase their root and shoot systems and thus are better able to acquire resources—and ultimately to reproduce successfully.

In every stage of a plant's growth cycle, growth maintains a balance between the shoot system and the root system. Leaves and other photosynthetic parts of the shoot must supply root cells with enough sugars to support their metabolism, and roots must provide the shoot structures with water and minerals. As long as a plant is growing, this balance is maintained, even as the complexity of the root and shoot systems increases. This happens in all plants, whether they live only a few months or—like some bristlecone pines—for 6000 years.

In this chapter, we've established the basic anatomy and morphology of the plant body, with an emphasis on angiosperms. In Chapter 38, we will look at how plants control the patterns of growth and development described in this chapter. First, however, we need to consider how water and nutrients are transported in the plant body (see Chapter 35) and how angiosperm seedlings are formed (see Chapter 36).

STUDY BREAK

1. Define secondary growth, explaining how it is similar to and different from primary growth, and where it typically occurs in plants.
2. Explain how the vascular cambium produces secondary xylem and phloem. How does division of cambial cells maintain the vascular cambium?
3. What is cork and how is it formed? What are the differences between cork, bark, periderm, and wood?
4. Compare secondary growth in stems and in roots.

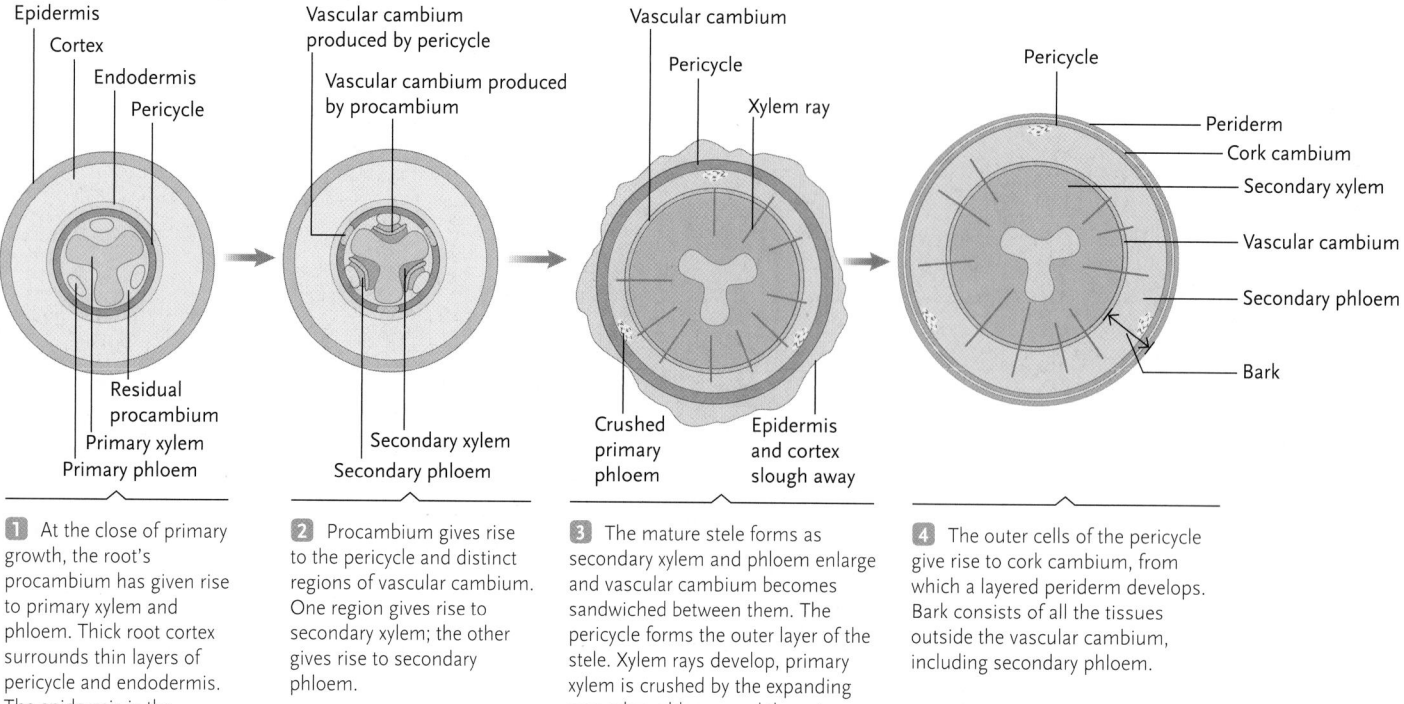

Epidermis
Cortex
Endodermis
Pericycle

Residual
procambium
Primary xylem
Primary phloem

1 At the close of primary growth, the root's procambium has given rise to primary xylem and phloem. Thick root cortex surrounds thin layers of pericycle and endodermis. The epidermis is the outermost layer of the root.

Vascular cambium produced by pericycle
Vascular cambium produced by procambium

Secondary xylem
Secondary phloem

2 Procambium gives rise to the pericycle and distinct regions of vascular cambium. One region gives rise to secondary xylem; the other gives rise to secondary phloem.

Vascular cambium
Pericycle
Xylem ray

Crushed primary phloem
Epidermis and cortex slough away

3 The mature stele forms as secondary xylem and phloem enlarge and vascular cambium becomes sandwiched between them. The pericycle forms the outer layer of the stele. Xylem rays develop, primary xylem is crushed by the expanding secondary phloem, and the epidermis and underlying root cortex begin to slough away.

Pericycle
Periderm
Cork cambium
Secondary xylem
Vascular cambium
Secondary phloem
Bark

4 The outer cells of the pericycle give rise to cork cambium, from which a layered periderm develops. Bark consists of all the tissues outside the vascular cambium, including secondary phloem.

Figure 34.28
Secondary growth in the root of one type of woody plant.

Review

aplia™

To access course materials such as Aplia and other companion resources, please visit www.NELSONbrain.com.

34.1 Plant Structure and Growth: An Overview

- Differences between the structures and growth of plants and animals reflect their modes of nutrition.
- The plant body of an angiosperm consists of an above-ground shoot system with stems, leaves, and flowers and an underground root system (Figure 34.2).
- Meristems give rise to the plant body and are responsible for a plant's lifelong growth. Each meristem cell produces two daughter cells, one of which remains part of the meristem, whereas the other differentiates into a cell of one of the three primary tissues (protoderm, ground tissue, or procambium).
- Primary growth of roots and shoots originates at apical meristems at root and shoot tips (Figure 34.4). Some plants have lateral meristems that produce secondary growth and increase the diameter of stems and roots.
- The two major classes of flowering plants (angiosperms) are monocots and eudicots (Table 34.1); angiosperms can also be differentiated based on pattern of growth (annuals versus perennials, woody versus herbaceous).

34.2 The Three Plant Tissue Systems

- All plant cells have primary cell walls composed primarily of cellulose (Figure 34.3). In some cells, secondary walls are laid down inside the primary walls. Maturing

cells become specialized for specific functions, with some functions accomplished by the walls of dead cells.

- Plants have three tissue systems. Ground tissues make up most of the plant body, vascular tissues serve in transport, and dermal tissue forms a protective cover (Figure 34.6).
- Of the three types of ground tissues, parenchyma is active in photosynthesis, storage, and other tasks, whereas collenchyma and sclerenchyma provide mechanical support (Figure 34.7).
- Xylem and phloem are the plant vascular tissues. Xylem conducts water and dissolved minerals taken up from the soil and consists of conducting cells called tracheids and vessel members (Figure 34.9). Phloem, which conducts the products of photosynthesis from the leaves to the rest of the plant, contains living cells (sieve tube members) joined end to end in sieve tubes (Figure 34.10).
- The dermal tissue, epidermis, is coated with a waxy cuticle that restricts water loss (Figure 34.11a). Water vapour and other gases enter and leave the plant through pores called stomata, which are flanked by specialized epidermal cells called guard cells (Figure 34.11b). Epidermal specializations also include trichomes, such as root hairs (Figure 34.11c).

34.3 Primary Shoot Systems

- The primary shoot system consists of the main stem, leaves, and buds, plus any attached flowers and fruits. Stems provide mechanical support, house vascular tissues, and may store food and fluid.

- Stems are organized into modular segments. Nodes are points where leaves and buds are attached, and internodes are the regions between nodes (Figure 34.12). The terminal bud at a shoot tip consists of shoot apical meristem. Lateral buds occur at intervals along the stem. Meristem tissue in buds gives rise to leaves, flowers, or both.
- Derivatives of the apical meristem produce three primary meristems: protoderm produces the stem's epidermis, procambium gives rise to primary xylem and phloem, and ground meristem differentiates into ground tissue (Figure 34.13).
- Vascular tissues are organized into vascular bundles, with phloem outside the xylem in each bundle (Figure 34.14).
- Monocot and eudicot leaves have blades of different forms (Figure 34.16), all providing a large surface area for absorbing sunlight and carbon dioxide. Leaf modifications are adaptive responses to environmental selection pressures. Leaf characteristics such as shape or arrangement may change over the life cycle of a long-lived plant.

34.4 Root Systems

- Roots absorb water and dissolved minerals and conduct them to aerial plant parts; they anchor and sometimes support the plant and often store food. Root morphologies include taproot systems, fibrous root systems, and adventitious roots.
- During primary growth of a root, the primary meristem and actively dividing cells make up the zone of cell division, which merges into the zone of elongation (Figure 34.20). Past the zone of elongation, cells may differentiate and perform specialized roles in the zone of cell maturation.
- A root's vascular tissues (xylem and phloem) are usually arranged as a central stele (Figure 34.21). Parenchyma tissue around the stele forms the root cortex. The root endodermis also wraps around the stele. Inside it is the pericycle, containing parenchyma that can function as meristem. It gives rise to root primordia from which lateral roots emerge. Root hairs from the epidermis greatly increase the surface available for absorbing water and solutes.

34.5 Secondary Growth

- In plants with secondary growth, older stems and roots become more massive and woody via the activity of vascular cambium and cork cambium.
- Vascular cambium produces secondary phloem toward the outside and secondary xylem toward the inside of the stem (Figure 34.24).
- Cork cambium gives rise to cork toward the outside of the stem, which replaces epidermis lost when stems increase in diameter, and to phelloderm toward the inside of the stem. These three tissues together make up the periderm.
- Bark consists of all tissues outside the vascular cambium (secondary phloem and periderm (Figure 34.25)).
- In root secondary growth, a thin layer of procambium cells between the xylem and phloem differentiates into vascular cambium (Figure 34.28). The pericycle produces root cork cambium.

Questions

Self-Test Questions

1. Plants are said to have a dendritic pattern of growth. What does a "dendritic growth pattern" mean, specifically?
   a. Both primary and secondary growth occur.
   b. The plant continues to grow throughout its lifetime.
   c. Both above- and below-ground parts are highly branched.
   d. A plant can lose individual roots, leaves, etc., and still continue to function.

2. What is the specific advantage of a dendritic growth pattern?
   a. Roots and shoots can continuous grow into new spaces.
   b. The plant will be more stable (e.g., less likely to tip over).
   c. The plant can survive loss of part of a root or shoot system.
   d. It provides a large surface area for uptake of nutrients and light.

3. Which of the following is the correct pairing of a tissue with its components?
   a. sclerenchyma : tracheids
   b. parenchyma : sclereids
   c. epidermis : companion cells
   d. phloem : sieve tube members

4. In which type(s) of plants are tracheids found?
   a. in all land plants
   b. in all vascular plants
   c. only in seed plants
   d. only in angiosperms

5. Which of the following is absent in a eudicot leaf?
   a. pericycle
   b. vascular bundles
   c. spongy mesophyll
   d. palisade mesophyll

6. A student leaves a carrot in her refrigerator. Three weeks later, she notices slender white fibres growing from its surface. They are not a fungus. Which of the following are they?
   a. adventitious roots
   b. lateral roots on a taproot
   c. root hairs on a lateral root
   d. root hairs on a fibrous root

7. Which of the following tissues is formed by primary growth in plants?
   a. the pith
   b. the cork
   c. the periderm
   d. the heartwood

8. How many rings of vascular cambium would be found in a tree that is six years old?
   a. 0
   b. 1
   c. 3
   d. 6

9. To what is the driving force that pushes a root through the soil primarily due?
   a. continuous cell division in the root cap at the tip of the root
   b. continuous cell division of the apical meristem just behind the root cap
   c. elongation of the cells behind the root apical meristem
   d. maturation of cells and formation of root hairs after cells elongate

10. Which of the following statements about growth in plants is correct?
    a. Stems, but not roots, have secondary growth.
    b. Primary growth, but not secondary growth, occurs at meristems.
    c. Primary growth, but not secondary growth, occurs in all seed plants.
    d. Trees and other woody plants have secondary growth but not primary growth.

Questions for Discussion

1. While camping in a national park, you notice a "Do Not Litter" sign nailed to the trunk of a mature fir tree about 2 m off the ground. When you return five years later, will the sign be at the same height, or will the tree's growth have raised it higher? Explain your answer.

2. African violets and some other flowering plants are propagated commercially using leaf cuttings. A leaf detached from a parent plant is placed in a growth medium. In time, adventitious shoots and roots develop from the leaf blade, producing a new plant. Are all cells in the original leaf tissue equally likely to give rise to the new structures? If not, which ones are most likely to have done so? What property of the cells makes this propagation method possible?

3. You completely remove the bark from around a section of a tree. You notice that the leaves retain their normal appearance for several weeks, but the tree eventually dies. What tissue(s) did you completely remove when you stripped off the bark? What tissue(s) were left functional?

4. Your yard is infested with a weed called quackgrass (*Agropyron repens*), a monocot. In order to find out how best to get rid of this weed, you decide to read more about it. Your book on Canadian weeds says that these plants spread via lateral roots, but the "Wonderful World of Weeds" webpage says the plants spread via rhizomes (underground stems). In order to determine whether the plant spreads by lateral roots or by rhizomes, you do a cross-section of one of these structures. Explain what you would expect to see in the cross-section if this structure were a rhizome and how this would differ from what you would see if the structure were a lateral root.

5. Where are the meristematic tissues of a woody plant? What tissues are formed by each meristem?

Giant Sitka Spruce (*Picea sitchensis*), such as these trees growing in the coastal forests of Vancouver Island, can grow to be more than 70 m tall and can live to be several hundred years old. Such extremely tall trees exemplify the ability of plants to move water and solutes from roots to shoots over amazingly long distances.

© Chris Cheadle/Alamy.

35

Transport in Plants

WHY IT MATTERS

Conifer trees growing in the coastal rain forests of British Columbia take life to extremes. Many of these giant trees can live for more than 400 years, and like those shown above, they can grow very tall. In fact, Vancouver Island is home to two of the tallest trees on Earth: the Red Creek fir is the tallest Douglas fir (*Pseudotsuga menziesii*) in the world, measuring 73 m high and over 13 m in circumference. Larger still, the tallest spruce tree on Earth is the Carmanah Giant, a Sitka spruce (*Picea sitchensis*) tree that soars 95 m above the forest floor. Such massive plants consume thousands of litres of water each day to survive. And that water—with its cargo of dissolved nutrients—must be transported the great distances between roots and leaves.

At first, movement of fluids and solutes 70 m or more from a tree's roots to its leafy crown may seem to challenge the laws of physics. If you wanted to raise water that high above ground in a pipe, you would need a powerful mechanical pump at the base and substantial energy to move the water against the pull of gravity. You also require a pump—your heart—to move fluid over a vertical distance of (usually) less than 2 m in your body. Yet a giant tree has no pump, so how does it move water from its roots to its leaves? In addition to moving water, plants must also move the sugars produced in their leaves to the rest of the plant—how do they manage to do this without a pump to circulate fluids? As you'll learn in this chapter, plants are able to move large volumes of fluid over great distances by harnessing the cumulative effects of seemingly weak interactions such as cohesion, the effects of solutes on the tendency of water to move across membranes, and the tremendous tension that can be created by the evaporation of water from plant leaves. Overall, plant transport mechanisms solve a fundamental biological problem—the need to acquire materials from the environment and distribute them throughout the plant body.

Our discussion begins with a brief review of the principles of water and solute movement in plants, a topic introduced in Chapter 5. Then we examine how these principles apply to the movement of water and solutes into and throughout a plant's vascular system.

35.1 Principles of Water and Solute Movement in Plants

Plants require large volumes of water to maintain the turgor pressure necessary to support their tissues and drive cell expansion, as well as for cooling by transpiration. Water is also required for photosynthesis and other metabolic processes. Plants must move water and nutrients long distances throughout their bodies in response to changing demands for those substances **(Figure 35.1a),** which involves their specialized transport systems—the vascular tissues called xylem and phloem (Figure 35.1b). In long-distance transport, water and dissolved minerals travel in the xylem from roots to shoots and leaves, and the products of photosynthesis

move via the phloem from the leaves and stems into roots and other structures **(Figure 35.2).** However, in plants, as in all organisms, the movement of water and solutes begins at the level of individual cells and relies on short-distance transport mechanisms that move substances across membranes. Examples of short-distance transport are water and soluble minerals entering roots by crossing the cell membranes of root hairs (Figure 35.1c), or sugars in the phloem crossing plasma membranes into metabolically active cells. Ultimately, the movement of materials throughout a plant results from the integrated activities of individual cells, tissues, and organs.

We will look at long-distance transport processes in the xylem and phloem later in this chapter. First, we will focus on short-distance mechanisms that move water and solutes into and out of specific cells in roots, leaves, and stems. Keep in mind that although the plant cell wall is very strong, it is also porous, meaning that it is not a barrier to the flow of water and solutes to and from the semipermeable cell membrane. Water and solutes can also move from one cell to another

Figure 35.1

Overview of transport routes in plants. **(a)** Water and minerals are transported upward in xylem from roots to leaves, where some of the water evaporates into the air. Sugars are transported in phloem from sites where they are made or stored to sites where they are needed. **(b)** Vascular tissues are involved in long-distance transport. Long-distance transport of water and minerals throughout the plant occurs in the xylem. Long-distance transport of carbohydrates, hormones, and other compounds from leaves and stems into other tissues occurs in the phloem. **(c)** Short-distance transport of water and minerals into and between cells of the root. Short-distance transport also includes movement of water, minerals, sugars, and other materials through and between cells of other tissues, and into and out of xylem and phloem.

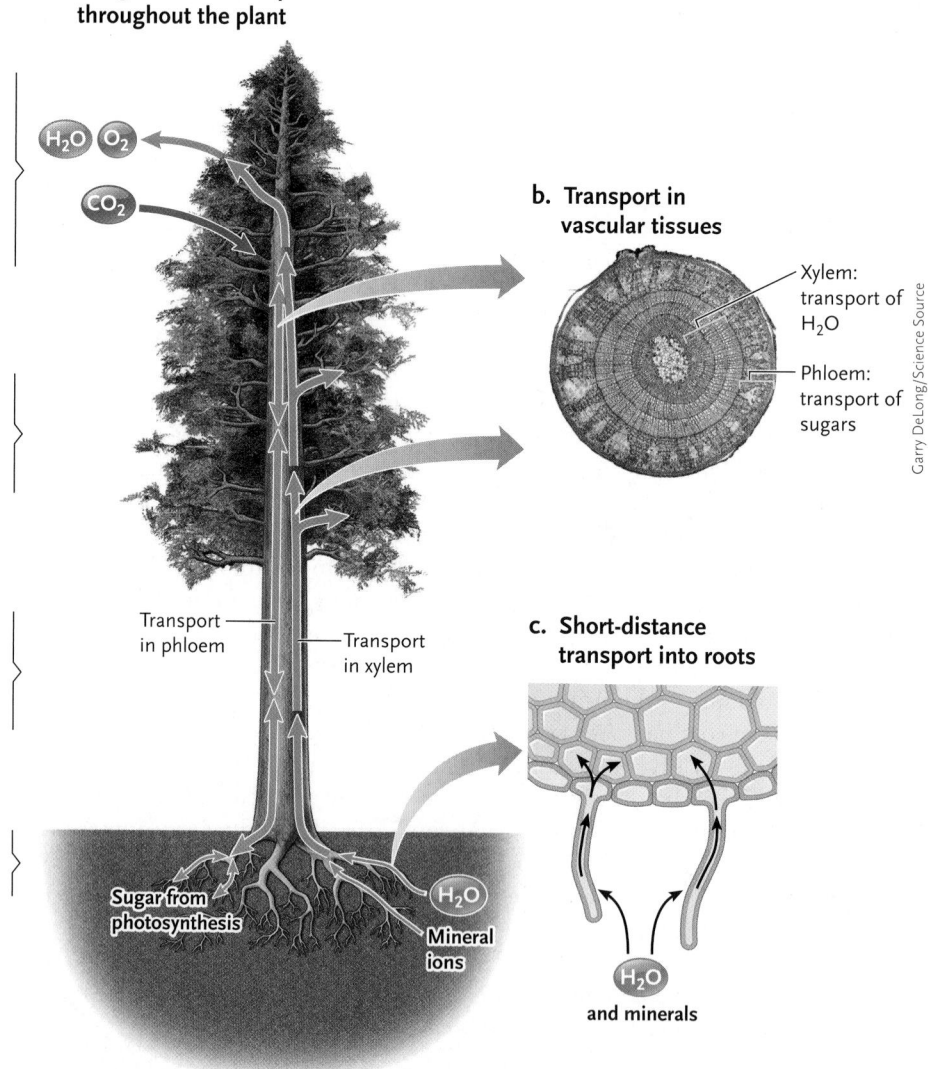

a. Long-distance transport throughout the plant

Sugars produced in photosynthesis are loaded and unloaded into and out of the phloem. Water vapour and oxygen are released from plant leaves.

Vascular tissue distributes substances throughout the plant, sometimes over great distances.

Water and mineral ions travel from root hairs into xylem vessels by passing through or between cells.

Water and solutes from soil enter plant roots through the root hairs.

Transport in phloem

Transport in xylem

Sugar from photosynthesis

H_2O

Mineral ions

H_2O O_2

CO_2

b. Transport in vascular tissues

Xylem: transport of H_2O

Phloem: transport of sugars

Garry DeLong/Science Source

c. Short-distance transport into roots

H_2O

and minerals

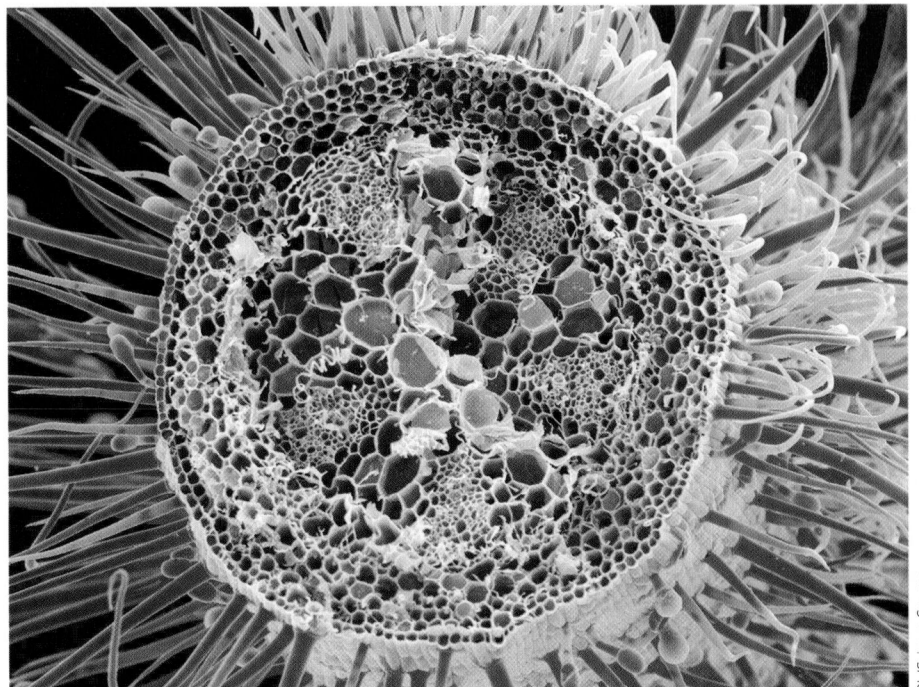

Figure 35.2
Cross-section of the stem of geranium (*Pelargonium*). This false-colour image shows large-diameter xylem vessels, which carry water and minerals, as whitish cells in the centre and radiating out from the centre. Phloem cells, which transport sugars and other organic molecules, are shown in pale green and are located between the "arms" of xylem. Layers of parenchyma cells (pink) surround the vascular tissue; epidermal cells form the outermost layer.

SPL/Science Source

through cytoplasmic connections (plasmodesmata) between adjacent cells (see Chapter 34).

35.1a Short-Distance Transport Mechanisms Move Molecules across Plant Cell Membranes

Recall from Chapter 5 that there are two general mechanisms for transporting molecules across a plasma membrane, passive transport and active transport. In **passive transport**, a substance moves with its concentration gradient or, if the substance is an ion, with its electrochemical gradient (see Figure 5.12), which means that cells do not need to expend energy. Passive transport includes both simple diffusion (transport of nonpolar molecules and small polar molecules that can readily diffuse across the lipid portion of a membrane) and facilitated diffusion (transport of polar and charged molecules that move across the membrane via transport proteins). Like facilitated diffusion, **active transport** moves ions and large molecules across membranes via transport proteins, but because these solutes are being moved against their concentration gradient, cells must expend energy. The energy for active transport can be provided either by ATP hydrolysis (for primary active transport) or by harnessing the energy in a concentration gradient (for secondary active transport).

Our focus in this section is on understanding the movement of water across cell membranes, as this is one of the most important aspects of plant physiology. You'll recall from Chapter 5 that individual cells gain and lose water by **osmosis**, the passive transport of water across a selectively permeable membrane either by simple diffusion or by facilitated diffusion through water-conducting channel proteins called **aquaporins**, which allow rapid movement of water across a membrane. Osmosis occurs in response to solute concentration gradients, a pressure gradient, or both (see Chapter 5). The combined influence of solute concentration and pressure on the movement of water is captured by the parameter of **water potential**. Water potential refers to the potential energy of water or its tendency to move from one place to another.

35.1b The Relationship between Osmosis and Water Potential

Water potential is symbolized by the Greek letter psi (ψ_w) and is measured in units of pressure (**megapascals**; MPa). Water potential is a relative value that is defined in reference to pure water at atmospheric pressure (such as water in an open container), which has a ψ_w value of 0 MPa. Two factors that determine water potential in living plants are the presence of solutes and physical pressure.

The movement of water across a membrane is strongly influenced by the concentration of solutes on either side of the membrane. The effect of dissolved solutes on water potential is called *solute potential*, symbolized by ψ_s. When solutes are added to water, they disrupt some of the hydrogen bonding between water molecules, and the polar water molecules then interact with the solutes, forming a hydration shell that surrounds the solute molecules (see *The Purple Pages*). Water molecules in a hydration shell are constrained from moving, and thus addition of solutes decreases the free energy of the water in the solution. Thus, water potential is *lower* in a solution with more solutes than in pure water; pure water has a ψ_s of 0 MPa, while solutions containing solutes will have ψ_s values less than zero. The relationship between water potential and solute potential is vital to understanding transport in plants because water moves by osmosis from regions where water potential is higher (closer to zero or less negative) to regions where it is lower (farther from zero or more negative), as shown in **Figure 35.3.** Solutes are usually more concentrated inside plant cells than in the fluid surrounding them. This means that the water potential is higher outside plant cells than inside them, so water tends to enter the cells by osmosis; this is the process by which soil water is drawn into a plant's roots.

Pressure can also change how water moves. We can investigate how changes in pressure influence the movement of water by considering a simple U-tube experiment in which the solutions on either side of the U-tube are separated by a membrane that is permeable to water but not to solutes. We can change the movement of water between the two sides of the U-tube by pushing or pulling on a plunger inserted in one side of the tube. If we push on the plunger, we will exert a positive pressure on the water, giving it more potential energy than the water in the other side of the U-tube, causing water to flow to that side. In contrast, if we pull

on the plunger, we put the water under negative pressure (or tension), reducing its potential energy, thus effectively pulling water from the other side of the U-tube to the side under tension. Pressure potential is symbolized as ψ_p and can be positive (greater than 0 MPa) if water is under positive pressure, negative (less than 0 MPa) if water is under tension, or equal to 0 MPa (at atmospheric pressure).

Together, solute potential and pressure potential determine a cell's water potential; the equation to express this relationship is $\psi_w = \psi_s + \psi_p$. In all cases, water will move from a solution of higher (less negative) potential to a solution of lower (more negative) potential.

35.1c Osmosis in Plant Cells Creates Turgor Pressure, Which Is Necessary for Plant Support

The movement of water in cells is more complex than in a simple U-tube, of course. Most of the volume of a mature plant cell is occupied by a large **central vacuole**, which is surrounded by a vacuolar membrane **(tonoplast)** and contains a dilute solution of sugars, proteins, other organic molecules, and salts. The cell

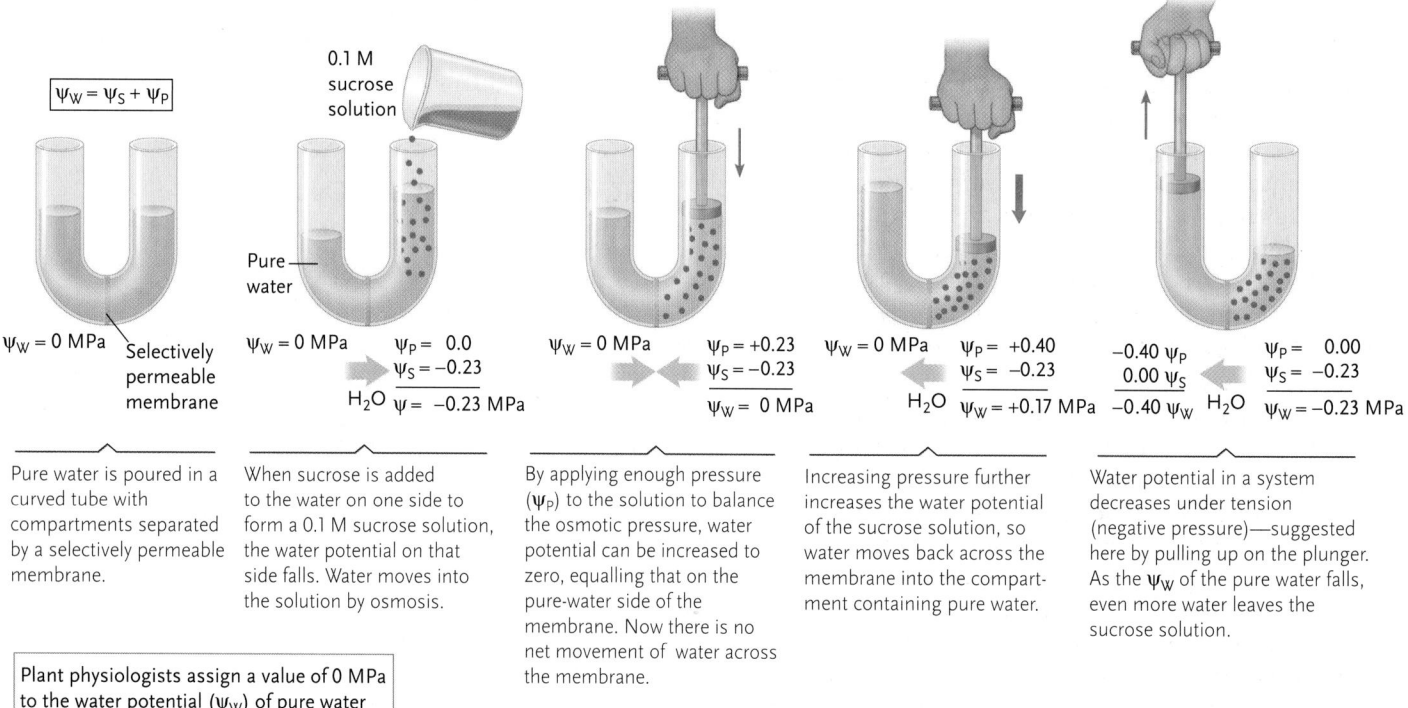

Pure water is poured in a curved tube with compartments separated by a selectively permeable membrane.

When sucrose is added to the water on one side to form a 0.1 M sucrose solution, the water potential on that side falls. Water moves into the solution by osmosis.

By applying enough pressure (ψ_p) to the solution to balance the osmotic pressure, water potential can be increased to zero, equalling that on the pure-water side of the membrane. Now there is no net movement of water across the membrane.

Increasing pressure further increases the water potential of the sucrose solution, so water moves back across the membrane into the compartment containing pure water.

Water potential in a system decreases under tension (negative pressure)—suggested here by pulling up on the plunger. As the ψ_w of the pure water falls, even more water leaves the sucrose solution.

Plant physiologists assign a value of 0 MPa to the water potential (ψ_w) of pure water in an open container under normal atmospheric pressure and temperature.

Figure 35.3

The relationship between osmosis and water potential (ψ). As discussed in more detail in the text, water potential in plant cells is determined by physical pressure (*p*) and the concentration of solutes (*s*). If the water potential is higher on one side of a membrane that is permeable to water and not to solutes, water will cross the membrane to the side with lower water potential. This diagram shows pure water on one side of a selectively permeable membrane and a simple sucrose solution on the other side. In an organism, however, the selectively permeable membranes of cells are rarely, if ever, in contact with pure water.

cytoplasm is confined to a thin layer between the tonoplast and the plasma membrane. Many solutes that enter a plant cell are actively transported from the cytoplasm into the central vacuole through ion channels in the tonoplast. As the solutes accumulate in the vacuole, water follows by osmosis. Recall from Chapter 5 that when animal cells are placed in a hypotonic solution, they may swell to the point of bursting. In plants, bursting is prevented by the cell wall, which resists the further inward movement of water. The pressure of the water-filled vacuole and cytoplasm against the wall keeps the cell firm or **turgid**, so we refer to the pressure of the cytoplasm on the wall as **turgor pressure**. Turgor pressure develops as a result of osmosis and increases until it is high enough to prevent more water from entering a cell by osmosis. Turgor pressure in many plant cells is in the range of 0.6 MPa, which is more than three times the pressure in a car tire!

The water mechanics we have been discussing have major implications for land plants, which obtain water and mineral nutrients from the soil surrounding their roots. As long as the ψ_w of soil is higher than that of the root epidermal cells, water will follow the ψ_w gradient and flow into root cells, making them turgid. However, if the soil around the roots dries out, the soil water potential becomes more negative than that of root epidermal cells and water no longer flows into the roots from the soil. The turgor pressure inside plant cells falls, and the protoplasts shrink away from the cell walls. This disruption in water uptake from the soil can lead to the drooping of leaves and stems called **wilting**, which occurs when turgor pressure in the cells of leaves and stems drops to very low levels. We can mimic these changes in water potential by placing plant cells in solutions of different water potentials **(Figure 35.4)**.

STUDY BREAK

1. How is facilitated diffusion similar to active transport? How are the two processes different?
2. What are the two components of water potential in plant cells?
3. Explain how adding more dissolved solutes to the cytoplasm of a plant cell would affect the cell's water potential.
4. What is turgor pressure? How is it generated and why is it important in plant cells?

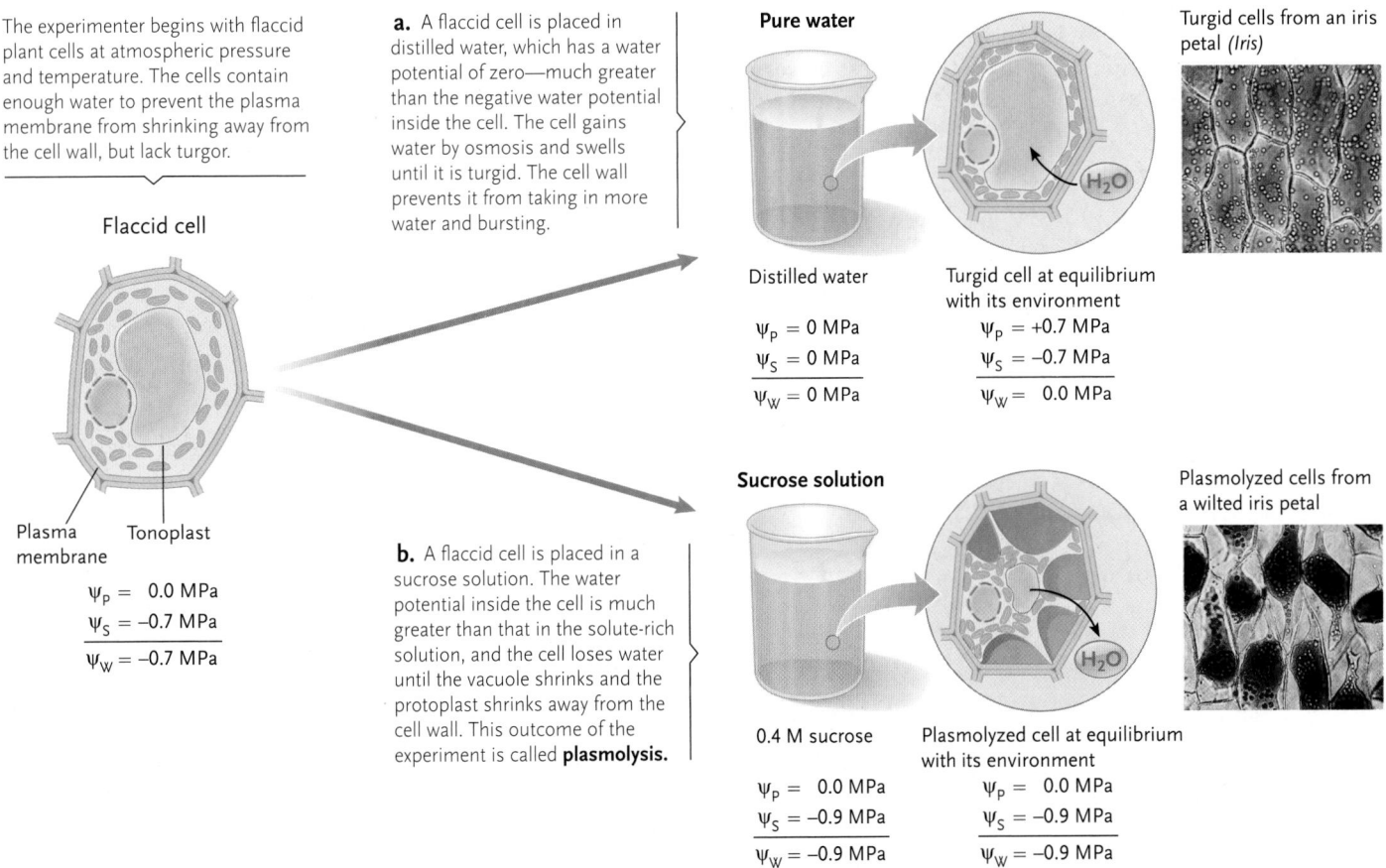

Figure 35.4

An experiment to test the effects of different osmotic environments on plant cells. Notice that in both **(a)** and **(b)**, the final condition is the same: the water potentials of the plant cell and its environment become equal. (Micrographs: © Claude Nuridsany and Marie Perennou/Science Photo Library/Photo Researchers, Inc.)

35.2 Uptake and Transport of Water and Solutes by Roots

As mentioned above, land plants obtain water and mineral nutrients from the soil surrounding their roots. In this section, we will investigate how water and minerals enter roots and how they move across the root to the xylem that will transport them to the rest of the plant.

35.2a Water Travels across the Root to the Root Xylem by Two Pathways

Soil water and the minerals dissolved in it always enter a root through the root epidermis but can then travel by two different paths across the root to the xylem, either through interconnected cytoplasm of living cells or through cell walls and intercellular spaces **(Figure 35.5)**. The living cells make up the **symplast** (*sym* = with, together; *plast* refers to the cytoplasm) and are interconnected by plasmodesmata, allowing water to flow from the cytoplasm of one cell to the next via the **symplastic pathway**. In contrast, the continuous network of cell walls and spaces between cells makes up the nonliving areas of the root or the **apoplast** (*apo* = away from). Water moves through the **apoplastic pathway** as it flows through these nonliving spaces without crossing plasma membranes.

When water enters a root **(Figure 35.6a)**, some diffuses across the plasma membranes of epidermal cells, entering the symplast, but most moves into the apoplast, flowing through cell walls and intercellular spaces. This apoplastic water (and any solutes dissolved in it) travels rapidly inward until it encounters the **endodermis**, the innermost layer of the cortex. Cells in the root cortex generally have air spaces between them (which helps aerate the tissue), but endodermal cells are tightly packed (Figure 35.6b). Each endodermal cell also has a ribbonlike **Casparian strip** in its radial and transverse walls, positioned somewhat like a ribbon of packing tape around a rectangular package (Figure 35.6c). The Casparian strip is impregnated with **suberin**, a waxy substance that is impermeable to water and blocks the apoplastic movement of water at the endodermis. What happens when apoplastic water and solutes reach the Casparian strip? They are forced to detour from the apoplast, moving across the plasma membrane of endodermal cells and into the symplastic pathway, which is not blocked by the Casparian strips. Water and solutes then pass through plasmodesmata to cells in the outer layer of the stele.

What is the benefit of blocking the apoplastic pathway so that water and solutes can't just flow automatically into the xylem? Although water molecules can easily cross an endodermal cell's plasma membrane, the semipermeable membrane allows only certain solutes to cross. Undesirable solutes may be barred, whereas desirable ones may move into the cell by facilitated diffusion or active transport. Conversely, the endodermis prevents needed substances in the xylem from leaking out, back into the root cortex. Thus, the Casparian strips allow the endodermis to control which substances enter and leave a plant's vascular tissue. The roots of most flowering plants also have a second layer of cells with Casparian strips just inside the root epidermis. This layer, the exodermis, functions like the endodermis.

35.2b Roots Take Up Ions by Active Transport

As described above, mineral ions dissolved in soil water also enter roots through the epidermis. Some enter the apoplast along with water, but most ions important for plant nutrition tend to be much more concentrated in roots than in the surrounding soil, so they cannot follow a concentration gradient into root epidermal cells. Instead, the epidermal cells actively transport ions inward by means of membrane-bound protein transporters. Ions can enter the symplast immediately and travel to the xylem via the symplastic pathway or they can move inward, following the apoplastic pathway until they reach the Casparian strip of the endodermis. In short, mechanisms that control which solutes will be absorbed by

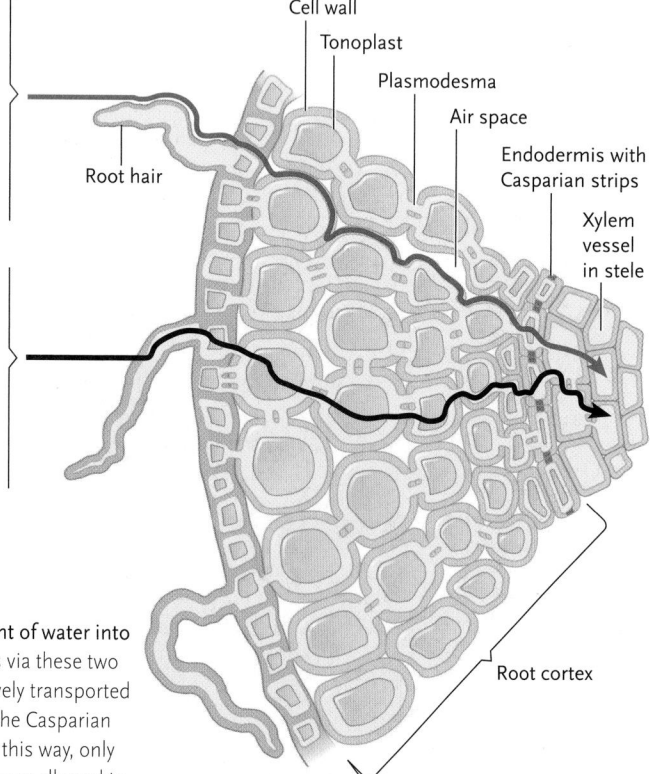

In the **apoplastic pathway** (red), water moves through nonliving regions—the continuous network of adjoining cell walls and tissue air spaces. However, when it reaches the endodermis, it must pass through the cytoplasm of endodermal cells.

In the **symplastic pathway** (black), water passes into and through living cells. After being taken up into root hairs, water diffuses through the cytoplasm and passes from one living cell to the next through plasmodesmata.

Figure 35.5
Pathways for the movement of water into roots. Ions also enter roots via these two pathways but must be actively transported into cells when they reach the Casparian strip of the endodermis. In this way, only certain solutes in soil water are allowed to enter the stele.

Cell wall
Tonoplast
Plasmodesma
Air space
Endodermis with Casparian strips
Xylem vessel in stele
Root hair
Root cortex
Epidermis

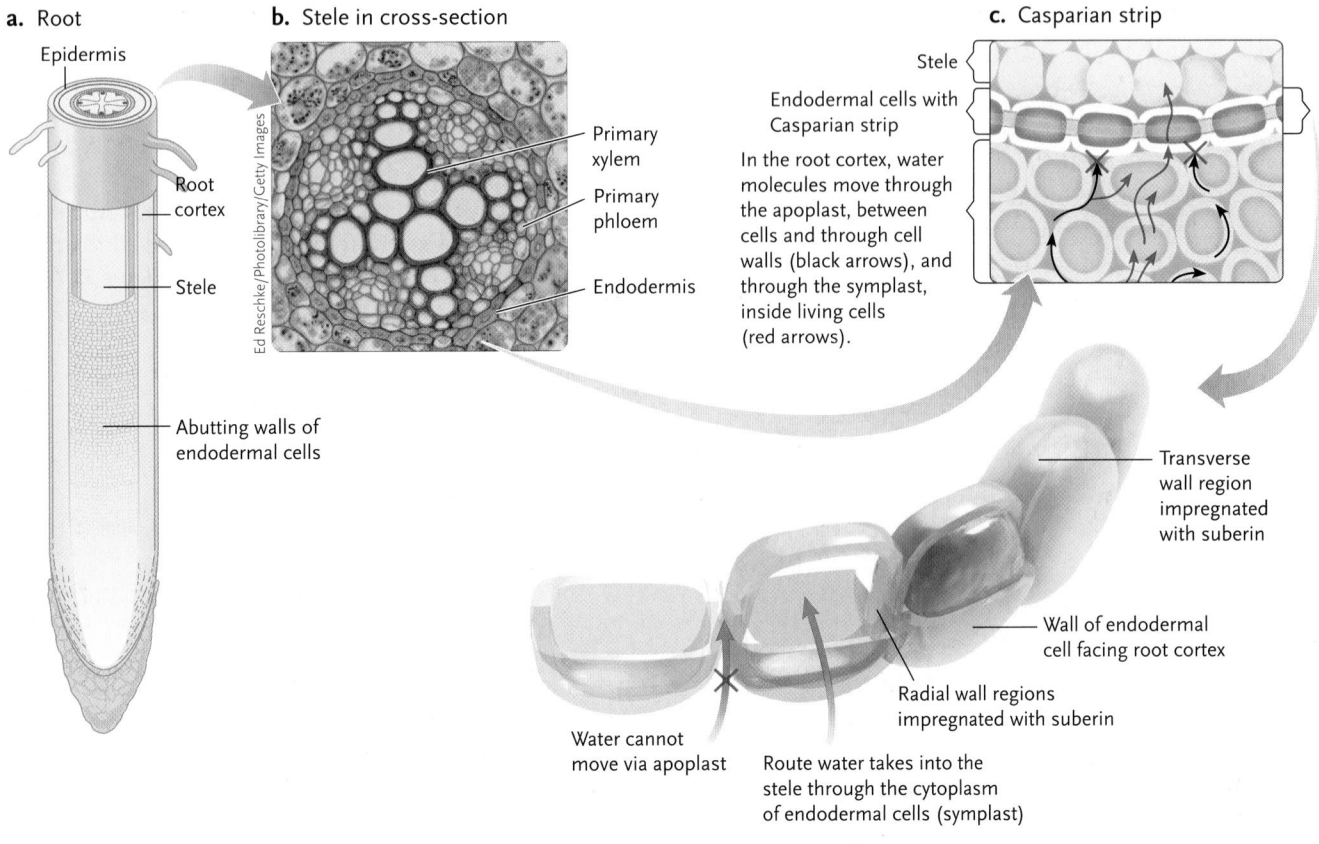

a. Root

Epidermis

Root cortex

Stele

Abutting walls of endodermal cells

Ed Reschke/Photolibrary/Getty Images

b. Stele in cross-section

Primary xylem

Primary phloem

Endodermis

c. Casparian strip

Stele

Endodermal cells with Casparian strip

In the root cortex, water molecules move through the apoplast, between cells and through cell walls (black arrows), and through the symplast, inside living cells (red arrows).

Transverse wall region impregnated with suberin

Wall of endodermal cell facing root cortex

Radial wall regions impregnated with suberin

Water cannot move via apoplast

Route water takes into the stele through the cytoplasm of endodermal cells (symplast)

Figure 35.6
Location and function of Casparian strips in roots.

Waxy, water-impervious Casparian strip (gold) in abutting walls of endodermal cells that control water and nutrient uptake.

root cells ultimately determine which solutes will be distributed through the plant.

Once an ion reaches the stele, it diffuses from cell to cell until it is "loaded" into the xylem. Experiments to determine whether the loading is passive (by diffusion) or active have been inconclusive, so the details of this final step are not entirely clear. Because the xylem's conducting elements are not living, water and ions in effect reenter the apoplastic pathway when they reach the xylem. Once in the xylem, water and mineral ions can move laterally to and from tissues or travel upward in the conducting elements. Minerals are distributed to living cells and taken up by active transport. The following section examines how this long-distance transport occurs.

STUDY BREAK

1. Explain two key differences in how the apoplastic and symplastic pathways direct substances laterally in roots.
2. Where in a root is the Casparian strip, and what is its function?
3. How does an ion enter a root hair and then move to the xylem?

35.3 Long-Distance Transport of Water and Minerals in the Xylem

We return now to the question that opened this chapter: How does the solution of water and minerals called xylem sap move—70 m or more in the tallest trees—from roots to stems and then into leaves? Inside a plant's tubelike vascular tissues, large amounts of water travel by **bulk flow**—the mass movement of molecules in response to a difference in pressure between two locations, like water in a closed plumbing system gushing from an open faucet. By the same principle, the dilute solution of water and ions that flows in the xylem, called **xylem sap**, moves by bulk flow from roots to shoots. However, because mature xylem cells are dead, they cannot expend energy to move water into and through the plant shoot. Instead, the driving force for the upward movement of xylem sap from root to shoot is the evaporation of water from leaves and other above-ground parts of land plants **(transpiration)**. How is evaporation of water sufficiently powerful to pull water more than 70 m up through a tree? Transpiration is able to pull xylem sap upward through the plant body because of the cohesion of water molecules and the tension created by the evaporation of water from plant surfaces, as we will discuss in this section.

35.3a The Properties of Water Play a Key Role in Its Transport

The Purple Pages review several biologically important properties of water, two of which are important in understanding water movement in the xylem. First, water molecules are strongly *cohesive*: they tend to form hydrogen bonds with one another. Second, water molecules are *adhesive*: they form hydrogen bonds with molecules of other substances, including the carbohydrates in plant cell walls. Together, water's cohesive and adhesive forces pull water molecules into exceedingly small spaces, such as crevices in cell walls or narrow tubes such as those making up xylem in roots, stems, and leaves. In 1914, plant physiologist Henry Dixon explained the ascent of sap in terms of the relationship between transpiration and water's properties. His theory of xylem transport is now called the **cohesion–tension theory of water transport (Figure 35.7)**.

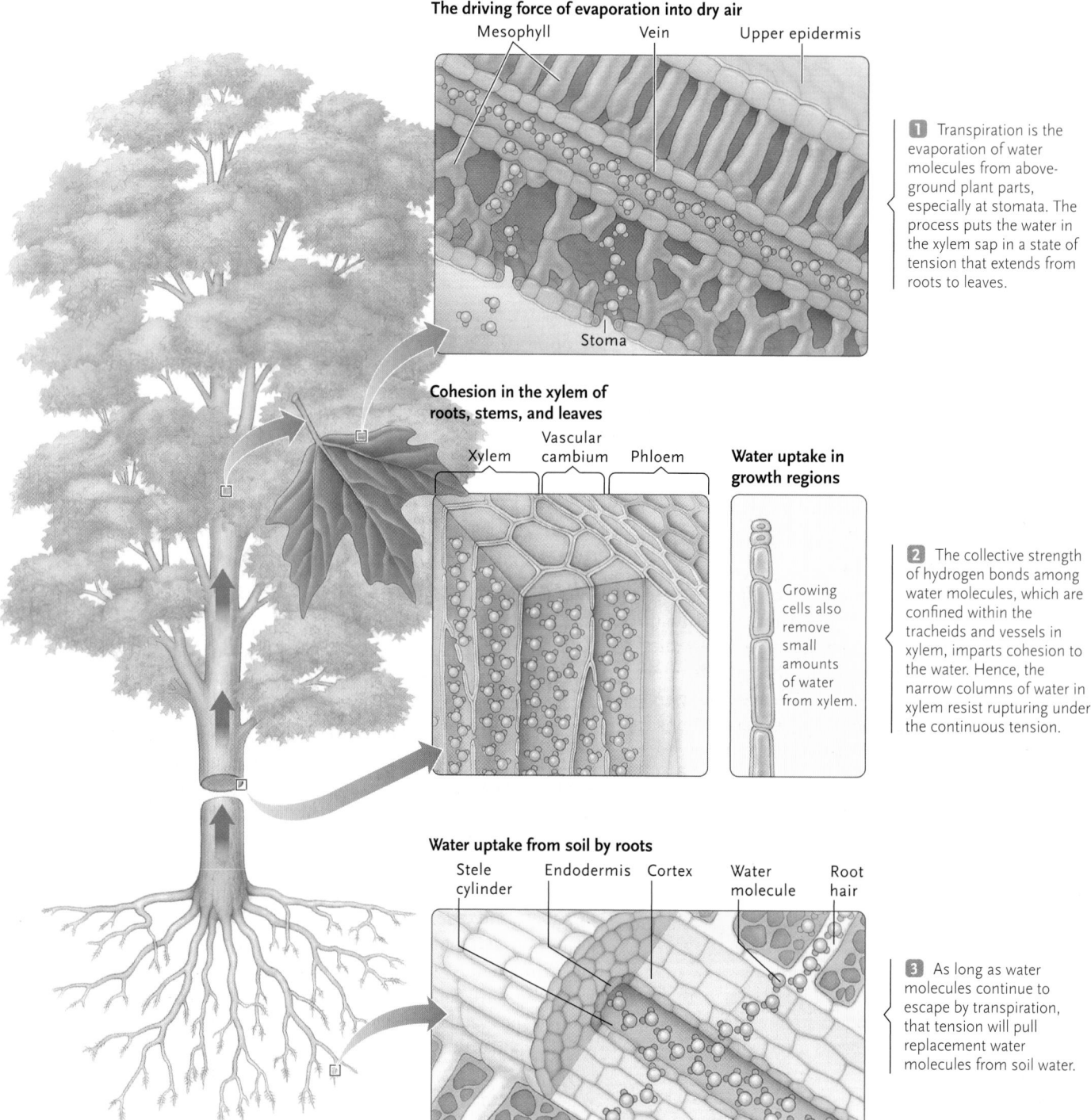

The driving force of evaporation into dry air

Mesophyll · Vein · Upper epidermis

Stoma

1 Transpiration is the evaporation of water molecules from above-ground plant parts, especially at stomata. The process puts the water in the xylem sap in a state of tension that extends from roots to leaves.

Cohesion in the xylem of roots, stems, and leaves

Xylem · Vascular cambium · Phloem

Water uptake in growth regions

Growing cells also remove small amounts of water from xylem.

2 The collective strength of hydrogen bonds among water molecules, which are confined within the tracheids and vessels in xylem, imparts cohesion to the water. Hence, the narrow columns of water in xylem resist rupturing under the continuous tension.

Water uptake from soil by roots

Stele cylinder · Endodermis · Cortex · Water molecule · Root hair

3 As long as water molecules continue to escape by transpiration, that tension will pull replacement water molecules from soil water.

Figure 35.7

Cohesion–tension mechanism of water transport. Transpiration, the evaporation of water from shoot parts, creates tension on the water in xylem sap. This tension, which extends from root to leaf, pulls upward columns of water molecules that are hydrogen-bonded to one another.

According to the cohesion–tension theory, water transport begins as water evaporates from the walls of mesophyll cells inside leaves and into the intercellular spaces. This water vapour escapes by transpiration through open stomata, the pores in the leaf surface. As water molecules exit the leaf, they are replaced by others from the cytoplasm of mesophyll cells. The water lost from mesophyll cells gradually reduces their water potential below the water potential in the leaf xylem. Now, water from the xylem in the leaf veins follows the gradient into cells, replacing the water lost in transpiration.

In the xylem, water molecules are confined in narrow, tubular xylem cells. The water molecules form a long chain, like a string of weak magnets, held together by hydrogen bonds between individual molecules. When a water molecule moves out of a leaf vein into the mesophyll, its hydrogen bonds with the next molecule in line stretch but don't break. The stretching creates *tension*—a negative pressure gradient—in the column. Adhesion of the water column to xylem vessel walls adds to the tension. Under continuous tension from above, the entire column of water molecules in xylem is pulled upward, similar to how water is pulled up through a drinking straw. This tension in the xylem is transmitted distally in roots due to the presence of the Casparian strips in the root epidermis, thus forming a continuous water potential gradient from the soil through the plant to the air surrounding the plant's leaves.

Transpiration continues regardless of whether the water lost from leaves is replenished by water rapidly taken up from the soil. Wilting is visible evidence that the water potential gradient between soil and a plant's stem and leaves has shifted. As soil dries out, the remaining water molecules in the soil are held ever more tightly by the soil particles; this tension pulls the water molecules closer to the soil particles and reduces the water potential in the soil surrounding the roots, causing the roots to take up water more slowly. However, because the water that evaporates from the plant's leaves is no longer being fully replaced, the leaves wilt as turgor pressure drops. Reducing the water potential in soil by adding solutes such as those that make up fertilizers can cause the same wilting effect. When the water potential in the soil finally equals that in leaf cells, a gradient no longer exists and the movement of water from the soil into roots and up to the leaves comes to a halt.

35.3b Leaf Anatomy Contributes to Cohesion–Tension Forces

Leaf anatomy is key to the processes that move water upward in plants. To begin with, as much as two-thirds of a leaf's volume consists of air spaces, meaning that there is a large internal surface area for evaporation of water. Leaves may also have thousands to millions of stomata through which water vapour escapes. Both of these factors increase transpiration. Also, every square centimetre of a leaf contains thousands of tiny xylem veins, so most leaf cells lie within half a millimetre of a vein. This close proximity supplies water to cells and the spaces between them, from which the water can readily evaporate.

As water evaporates from a leaf, the water film on the surface of mesophyll cells becomes thinner, meaning that the water molecules in the film are held more tightly to the surface of the cell walls. The decreased water potential on the surface of the cells results in increased tension, pulling water from the mesophyll cells and ultimately from the leaf veins **(Figure 35.8)**. This tension is multiplied many times over in all of the leaves and xylem veins of a plant. It increases further as the plant's metabolically active cells take up xylem sap.

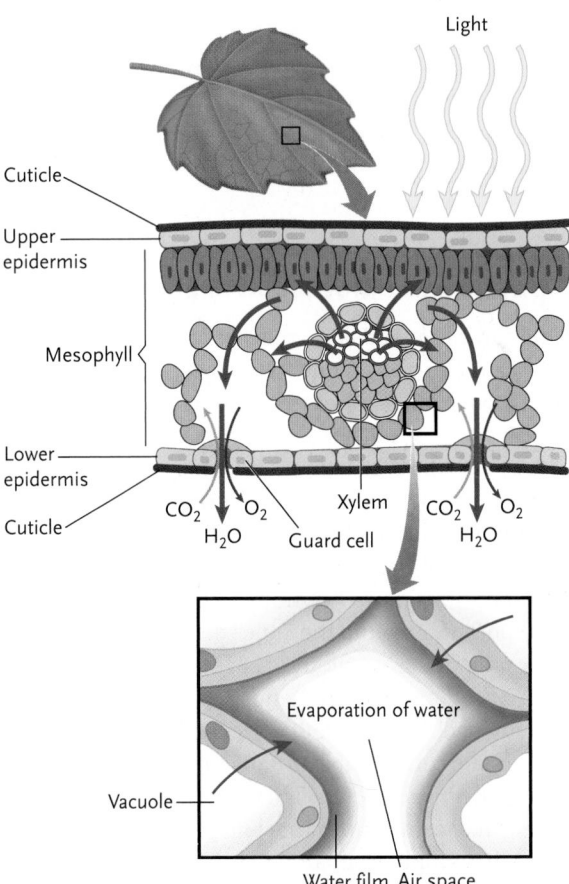

Figure 35.8

Water evaporating from a leaf causes increasing tension on the remaining water inside the leaf, eventually pulling water up out of the xylem. As water evaporates from the water-lined air spaces in a leaf, the film of water on the mesophyll cells around the air space becomes thinner and thinner. The water molecules in this film are under increasing tension, causing a decrease in water potential in the film, which pulls water out of mesophyll cells and, ultimately, out of the xylem.

35.3c In the Tallest Trees, the Cohesion–Tension Mechanism May Reach Its Physical Limit

Numerous experiments have tested the cohesion–tension theory, and thus far, all of the data strongly support it. For example, the theory predicts that xylem sap will begin to move upward at the top of a tree early in the day when water starts to evaporate from leaves. Experiments with several different tree species have confirmed that this is the case. The experiments also showed that sap transport peaks at midday, when evaporation is greatest, and then tapers off in the evening, as transpiration slows down.

Other experiments have probed the relationship between xylem transport and tree height. One team of researchers studied eight of the tallest living redwoods, including one that towers nearly 113 m above the forest floor. When the scientists measured the maximum tension exerted in the xylem sap in twigs at the tops of the trees, they discovered that it approached the theoretical limit at which the bonds between water molecules in a column of water in a conifer's xylem will break. Based on this finding and other evidence, the team predicted that the maximum height for a healthy coast redwood tree is 122 to 130 m, so it is possible that the tallest redwood alive today may grow taller still!

35.3d Root Pressure Contributes to Upward Water Movement in Some Plants

The cohesion–tension mechanism accounts for upward water movement in tall trees. In some shorter nonwoody plants, however—grasses, for instance—a positive pressure can develop in roots and force xylem sap upward. This **root pressure** occurs under conditions that reduce transpiration, such as high humidity or low light. In fact, the mechanism that produces root pressure often operates at night, when transpiration slows or stops. At this time, active transport of ions into the stele sets up a water potential gradient across the root. Because the Casparian strip of the endodermis tends to prevent ions from moving back into the root cortex, the water potential in the stele can be substantially lower than the water potential in the cortex. This water potential difference can become quite large and can cause the movement of enough water and dissolved solutes into the stele to produce a relatively high positive pressure in the xylem. Although this root pressure is not sufficient to force water to the top of a very tall plant, it can force water out of leaf openings in some smaller plant species, in a process called **guttation (Figure 35.9)**. Pushed up and out of vein endings by root pressure, tiny droplets of xylem fluid that look like dew in the early morning emerge from modified stomata at the margins of leaves.

Figure 35.9

Guttation caused by root pressure. The drops of water appear at the endings of xylem veins along the leaf edges of a strawberry plant (*Fragaria*).

Scott Camazine/Science Source

35.3e Stomatal Movements Regulate the Loss of Water by Transpiration

Three environmental conditions have major effects on the rate of transpiration: relative humidity, air temperature, and air movement. The most important is relative humidity, which is a measure of the amount of water vapour in air. The less water vapour in the air, the lower the water potential of the air and the more water that will evaporate from leaves. Increasing air temperatures at the leaf surface also increase the rate of transpiration: the amount of water lost can double with each 10°C rise in air temperature. Air movement at the leaf surface carries water vapour away from the leaf surface and so makes a steeper water potential gradient. Together these factors explain why on extremely hot, dry, windy days the leaves of certain plants must completely replace their water each hour.

Even when conditions are not so extreme, more than 90% of the water moving into a leaf can be lost through transpiration. Of the remaining water, about 2% is used in photosynthesis and other metabolic activities. Obviously, it is crucial for plants to be able to control transpiration: if water loss from leaves exceeds water uptake by roots, the plant will not have enough water to carry out normal metabolic functions and may even wilt and die.

Plants have evolved strategies to limit and regulate water loss. The cuticle-covered epidermis of

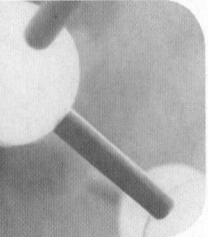

Quebecol

Quebecol is a phenolic compound produced when maple sap is boiled to form maple syrup.

Maple sap is rich in sucrose but also contains amino acids, organic acids, minerals, and numerous phenolic compounds that play many roles in plants (e.g., defence against herbivores). Sap is about 98% water and so must be boiled down to produce syrup—it takes about 40 L of sap to make 1 L of syrup. As sap is transformed into syrup, the heating process concentrates the sugars and phenolics but also triggers reactions among the components of sap to form new phenolic compounds, such as quebecol **(Figure 1)**. Not only do these phenolics make maple syrup taste good—they might also be beneficial to human health. Research has shown that extracts of maple syrup do have antioxidant and antimutagenic properties. And, recently, quebecol has been found to stop the proliferation of human cancer cells *in vitro*.

Contrary to what you might think, the sap used to make maple syrup is not the sugar-rich phloem sap but rather the xylem sap. In this chapter, we have emphasized that xylem sap consists of water and the mineral nutrients dissolved in it, so how does the sap end up containing so much sugar? And what causes the sap to flow out when holes are cut in the tree? Despite the long history and importance of maple syrup, the answers to these questions have only recently been worked out, and some details are still not known.

In the spring, holes are bored in the trunks of sugar maples, and the sap that is forced out of the hole under pressure is collected **(Figure 2)**. Sap collection is only possible in spring because the thaw–freeze cycles of warm days followed by cooler nights are required for sap to be forced out under pressure. How does the sap rise up the trunk of the tree in spring? The cohesion–tension mechanism cannot be responsible, as the trees do not yet have leaves, so there is no transpiration to pull xylem sap upward. Nor is root pressure responsible because it does not occur in maple trees (although it does in birch trees, the xylem sap of which is also used to make syrup). Instead, the key lies in the combination of warm days and cool nights: as temperatures rise during the day, the living cells of the xylem (xylem parenchyma cells known as ray cells) produce CO_2 gas in respiration; this gas accumulates in the spaces between the cells during the day. The xylem sap also contains CO_2 gas, which expands as temperatures rise. The accumulation and expansion of CO_2 cause positive pressure to build up in the xylem, forcing sap out of the

holes bored in the trunk. At night, as temperatures drop below freezing, the CO_2 gas condenses and contracts. Some of the sap also freezes, compressing the gas dissolved in it. These changes reduce water potential in the xylem such that the sap is under tension, not pressure, which pulls water up from the roots. When temperatures rise the next morning, the upward flow of sap begins again.

The sugars in the sap come from carbohydrates that accumulated in the xylem parenchyma cells during the previous growing season and were stored as starch over winter. In the spring, the starch is hydrolyzed to sucrose, which moves into the sap flowing through the conducting cells of the xylem.

Maple syrup has been a part of life in Canada for centuries: the Aboriginal peoples of what is now eastern Canada and the northeastern United States harvested sap from sugar maple trees (*Acer saccharum*) each spring and taught the process to European colonists. Today, Canada supplies 85% of the maple syrup in the world, with the remaining 15% produced by the United States.

Figure 1

The chemical structure of quebecol.
This phenolic compound is produced as maple sap is boiled to produce syrup.

Ed (Edgar181)

Norman Pogson/Shutterstock.com
GoodMood Photo/Shutterstock.com

Figure 2

(a) *Collection of sugar maple xylem sap in spring.* **(b)** *Maple sap is boiled down to produce darker, thicker maple syrup.*

leaves and stems reduces the rate of water loss from above-ground plant parts, and plants can reduce transpiration by closing their stomata. Although both of these adaptations reduce water loss, they also limit the rate at which CO_2 for photosynthesis can diffuse into the leaf, so plants have to balance their need to conserve water against their need to fix carbon. This *transpiration–photosynthesis compromise* involves the regulation of transpiration and gas exchange by opening and closing stomata as environmental conditions change.

Opening and Closing of Stomata. Two guard cells flank each stoma ("mouth"; plural, *stomata*) **(Figure 35.10).** The inner cell walls of these guard cells are thicker and less elastic than the outer walls, and the cell walls are reinforced by cellulose microfibrils that wrap around the walls in a radial pattern, similar to the steel belts in an automobile's radial tires (Figure 35.10). These features play important roles in the regulation of stomatal opening by guard cells.

Stomata open when guard cells accumulate ions in their cytoplasm and water follows by osmosis; the resulting swelling of the guard cells causes the pair of cells to pull apart, due to the radial arrangement of cellulose microfibrils in the walls (Figure 35.10c, d), thus increasing the aperture of the stoma. How do guard cells accumulate ions, and which ions are involved? The key ion is potassium (K^+): the K^+ concentration in turgid guard cells can be four to eight times that in flaccid (limp) guard cells **(Figure 35.11).** Accumulation of K^+ ions inside guard cells is driven by proton pumps (H^+-ATPase pumps) in the plasma membrane of guard cells, which use the energy of ATP hydrolysis to pump protons out of the guard cell cytoplasm, thereby creating an electrochemical gradient across the guard cell membranes. Guard cells use this electrochemical gradient to drive the uptake of K^+ through ion channels into the cytoplasm. To maintain the charge difference across the plasma membrane, the accumulation of positively charged K^+ ions inside the cell is balanced by uptake of Cl^- and other anions though anion channels in the plasma membrane. Most of these anions and the K^+ ions are transported into the vacuole rather than remaining in the cytoplasm, causing the water that flows into the guard cell by osmosis to move into the vacuole. Thus, the increase in the size of the guard cell is due mostly to increases in the volume of the vacuole (Figure 35.11).

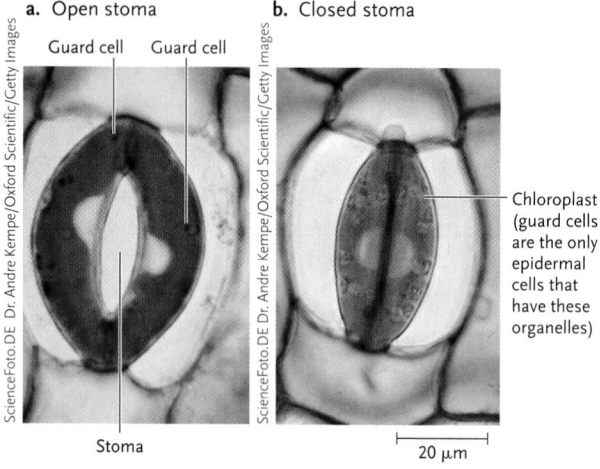

a. Open stoma   **b.** Closed stoma

Guard cell   Guard cell

Chloroplast (guard cells are the only epidermal cells that have these organelles)

Stoma

20 μm

c. Cells turgid/stoma open   **d.** Cells flaccid/stoma closed

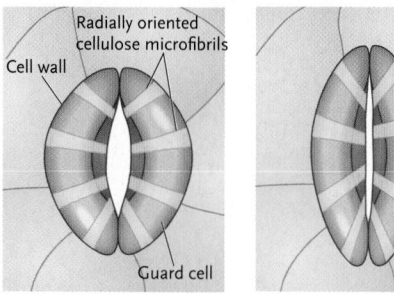

Radially oriented cellulose microfibrils

Cell wall

Guard cell

Figure 35.10

Guard cells and stomatal action. **(a)** An open stoma in the needle-like leaf of the rock needlebush (*Hakea gibbosa*). The osmotic flow of water into the guard cells increases the turgor pressure inside the cells, causing them to move apart, thus opening the stoma. **(b)** A closed stoma. The osmotic flow of water out of guard cells reduces their turgor pressure, causing them to collapse against each other and close the stoma. **(c)** Arrangement of cellulose microfibrils in guard cells around open stoma. **(d)** Orientation of cellulose microfibrils in guard cells when stoma is closed.

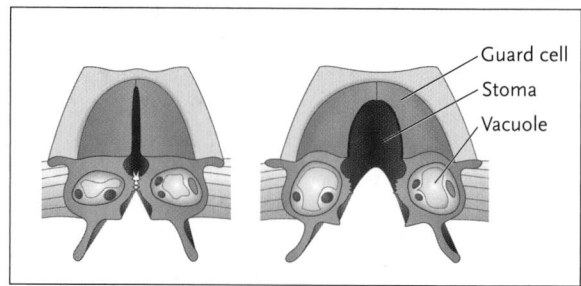

Guard cell

Stoma

Vacuole

	Closed stomata	Open stomata
Stomatal aperture	0 μm	8 μm
Guard cell turgor	1.0 MPa	4.5 MPa
K^+ content	0.3 pmol	2.5 pmol

Figure 35.11

Evidence for potassium accumulation in stomatal guard cells undergoing expansion. Strips from the leaf epidermis of a day-flower (*Commelina communis*) were immersed in a solution containing a stain that binds preferentially with potassium ions. In leaf samples with closed stomata, most of the potassium was concentrated in epidermal cells adjacent to the guard cells. In leaf samples with open stomata, most of the potassium was concentrated in the guard cells. Three-dimensional representation of guard cells when stoma is closed (left) and open (right). Data in the table indicate that changes in K^+ ion content of guard cells cause an influx of water into the cells, increasing their volume (primarily the vacuoles). The increased volume causes the guard cells to bend and change shape, opening the stoma. (From Roelfsema and Hedrich 2005. New Phytologist 167: 665–691).

During stomatal closure, anions are released from the guard cells by anion channels in the plasma membrane. This loss of anions reduces the charge difference across the guard cell plasma membrane, triggering the efflux of K^+ ions from the guard cells, and water follows by osmosis. When the water content of the guard cells drops, so does turgor pressure. The guard cells collapse against each other, closing the stoma (Figure 35.11).

In most plants, stomata are open in daylight and closed at night, indicating that guard cells clearly respond to light. But experiments have shown that guard cells can respond to a number of other environmental and internal signals, any of which can induce the ion flows that open and close stomata. These signals include humidity, CO_2 concentration in the air spaces inside leaves, the amount of water available to the plant, time of day, and the presence of microbes.

35.3f In Dry Climates, Plants Exhibit Various Adaptations for Conserving Water

Many plants have evolutionary adaptations that conserve water, including modifications in leaf structure or physiology (**Figure 35.12**). The stomata of oleanders, for example, lie at the bottom of pitlike invaginations (see Figure 35.12a, b). These sunken stomata are less exposed to drying breezes, so that water evaporates from the leaf much more slowly.

The leaves of *xerophytes*—plants adapted to hot, dry environments in which water stress can be severe—often have a thickened cuticle that gives them a leathery feel and provides enhanced protection against evaporative water loss. In other xerophytes, such as cacti, leaves are reduced to thin spines, which reduces the surface area from which water is lost by transpiration; photosynthesis occurs in the stems, which are thickened, leaflike pads (see Figure 35.11c). These structural alterations reduce the surface area for transpiration.

One intriguing variation on water conservation mechanisms occurs in CAM plants, including cacti, orchids, and most succulents. As discussed in Chapter 7, **Crassulacean acid metabolism** (CAM) is a biochemical variation of photosynthesis that was discovered in a member of the family Crassulaceae (the plant family to which a common houseplant, the jade plant, belongs). CAM plants generally have fewer stomata than other types of plants, and unlike most plants, their stomata are closed during the day. At

a. Oleanders

©iStock.com/Jublub

b. Oleander leaf

Cuticle
} Multilayer epidermis

Recessed stoma

Dr. Keith Wheeler/Science Source

c. Spines (modified leaves) on a cactus stem

©iStock.com/Joseph Justice

d. CAM plant

Steve Mann/Shutterstock.com

Figure 35.12

Some adaptations that enable plants to survive water stress. **(a)** Oleanders (*Nerium oleander*) are adapted to arid conditions. **(b)** As shown in the micrograph, oleander leaves have recessed stomata on their lower surface and a multilayer epidermis covered by a thick cuticle on the upper surface. **(c)** Like many other cacti, the leaves of the Graham dog cactus (*Opuntia grahamii*) are modified into spines that protrude from the underlying stem. Transpiration and photosynthesis occur in the green stems, such as the oval stem in this photograph. **(d)** *Sedum*, a CAM plant, in which the stomata open only at night.

night, when temperatures are cooler and the relative humidity is higher, CAM plants open their stomata to take up carbon dioxide, which is converted to malate, an organic acid. In the daytime, the CO_2 is liberated from malate and diffuses into chloroplasts, so photosynthesis takes place even though a CAM plant's stomata are closed. This adaptation prevents heavy evaporative water losses during the heat of the day.

STUDY BREAK

1. Explain the key steps in the cohesion–tension mechanism of water transport in a plant.
2. How and when do stomata open and close? In what ways is their functioning important to a plant's ability to manage water loss?

PEOPLE BEHIND BIOLOGY 35.2
Uwe Hacke, *University of Alberta*

In trees, the tension (or negative pressure) in the xylem can reach values of −10 MPa. Why don't xylem cells implode under such tensions? This is one of the questions that Uwe Hacke of the University of Alberta has explored in his research, and he has found that, surprisingly, the density of wood appears to be a crucial factor in the ability of plants to resist tension. We tend to think of wood density as playing an important role in supporting woody plants against external stresses such as gravity, high winds, and heavy snow cover. However, when Hacke and his colleagues investigated wood density in 48 species of woody angiosperms and gymnosperms from a range of habitats, they found the highest wood densities not in the tallest trees or the trees subject to high winds and snow at the tree line in mountains, but rather in low-elevation species adapted to very dry habitats, such as small desert shrubs **(Figure 1)**. Stronger wood may, therefore, be necessary to withstand the very negative xylem pressures

common to plants growing in such dry environments. The ability of these plants to transport water under these negative pressures comes at a cost, as denser wood is more expensive in terms of the carbon and energy required to build it. Conifer xylem is cheaper to build than is angiosperm xylem because of its simpler structure (it is composed of tracheids rather than vessels; see Chapter 34), giving

conifers an advantage in arid and cold environments where water availability limits photosynthesis. Hacke's research suggests that the simpler xylem of conifers may give these plants not only a competitive advantage over angiosperms in some habitats but also a strategy for successful long-term growth, perhaps explaining why both the tallest and oldest trees on Earth are conifers.

Figure 1
Juniperus osteosperma (*Utah Juniper*).

35.4 Transport of Organic Substances in the Phloem

We have explored how plants move water and the mineral nutrients dissolved in it through their bodies via their xylem. Plants have another long-distance transport system, the phloem, which carries huge amounts of carbohydrates; lesser but vital amounts of amino acids, fatty acids, and other organic compounds; and still other essential substances, such as hormones and signalling molecules. Unlike the xylem's unidirectional upward flow, the phloem transports substances throughout the plant to wherever they are used or stored. Organic compounds and water in the sieve tubes of phloem are under pressure and driven by concentration gradients.

35.4a Organic Compounds Are Stored and Transported in Different Forms

One problem in transporting the carbohydrates, proteins, and other organic compounds in the phloem is that most of these molecules are too large to cross the

cell membranes and leave the cells in which they are made. They may also be too insoluble in water to be transported to other regions of the plant body. Consequently, in leaves and other plant parts, specific reactions convert organic compounds to transportable forms. For example, hydrolysis of starch liberates glucose units, which combine with fructose to form sucrose, the main form in which sugars are transported through the phloem of most plants. Proteins are broken down into amino acids, and lipids are converted into fatty acids. These forms are better able to cross cell membranes by passive or active mechanisms.

35.4b Organic Solutes Move by Translocation

In plants, the long-distance transport of substances such as sucrose is called **translocation**. We have the best understanding of how translocation occurs in flowering plants. In these plants, the phloem is composed of sieve tube member cells, which are joined together by their perforated end walls (sieve plates) to form interconnected sieve tubes (see Figure 34.10). Recall from Chapter 34 that as sieve tube members mature, they lose many of their organelles, including nuclei. At maturity, like xylem cells, sieve tubes form a pipeline for transport, but unlike tracheids and vessels, sieve tube cells are alive and contain cytoplasm. The large pores of the perforated end walls allow water and organic compounds, collectively called **phloem sap**, to flow rapidly through the sieve tubes, another example of a structural adaptation that suits a particular function. But sieve tubes do not move phloem sap by themselves; the companion cells connected via plasmodesmata to sieve tubes and to surrounding parenchyma cells are also necessary for transport of phloem sap.

"Ant feeding on honeydew" by Jmalik (talk). Original uploader was Jmalik at en.wikipedia - Transferred from en.wikipedia (Original text: I (Jmalik (talk)) created this work entirely by myself). Licensed under Creative Commons Attribution-Share Alike 3.0 via Wikimedia Commons - http://commons.wikimedia.org/wiki/File:Ant_feeding_on_honeydew .JPG#mediaviewer/File:Ant_feeding_on_honeydew.JPG

Figure 35.13
Aphid releasing honeydew.

35.4c Phloem Sap Moves from Source to Sink under Pressure

Over the decades, plant physiologists have proposed several mechanisms of translocation, but it was the tiny aphid, an insect that annoys gardeners, that helped demonstrate that organic compounds flow under high pressure in the phloem—the pressure can be as high as five times the pressure in an automobile tire! An aphid attacks plant leaves and stems, forcing its needlelike stylet (a mouthpart) into sieve tubes, with incredible precision, to obtain the dissolved sugars and other nutrients inside. When an aphid feeds on phloem sap, this pressure forces the fluid through the aphid's gut and (minus nutrients absorbed) out its anus as "honeydew" **(Figure 35.13)**. If you park your car under a tree being attacked by aphids, it might get spattered with sticky honeydew droplets, thanks to the high fluid pressure in the tree's phloem.

In flowering plants, sucrose-laden phloem sap flows from a starting location, called the *source*, to another site, the *sink*, along gradients of decreasing solute concentration and pressure. A **source** is any region of the plant where organic substances are being loaded into the phloem's sieve tube system; actively photosynthesizing leaves, for example, are sources **(Figure 35.14)**. A **sink** is any region where organic substances are being unloaded from the sieve tube system and used or stored; a developing fruit is a sink.

The same plant organ can act as a source or a sink at different times during development. For example, think about a tulip bulb: in spring, food stored in the bulb is mobilized for transport upward to growing plant parts, but after the plants bloom, the bulb becomes a sink, as sugars manufactured in the tulip plant's leaves are moved into it for storage. Leaves, roots, and fruits generally start out as sinks, only to become sources when the season changes or the plant enters a new developmental phase. In general, sinks receive organic compounds from sources closest to them. Hence, the lower leaves on a rose bush may

supply sucrose to roots, whereas leaves farther up the shoot supply the rapidly growing shoot tip.

What causes sucrose and other solutes produced in leaf mesophyll to flow from a source to a sink? All of the evidence to date indicates that solute movement through the sieve tubes is driven by a pressure gradient between a source and a sink: as sugars and other macromolecules are loaded into the phloem at a source, the resulting pressure pushes these solutes by bulk flow toward a sink, where they are unloaded, resulting in lower pressure. **Figure 35.15, p. 878,** summarizes this **pressure-flow mechanism.**

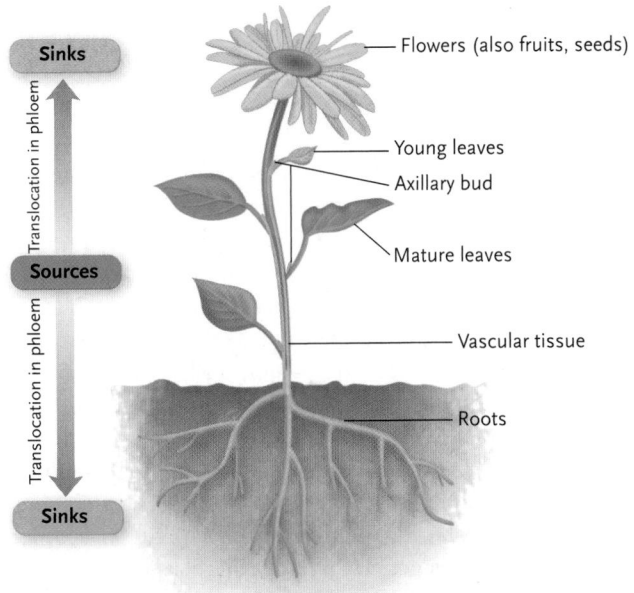

Figure 35.14
Sources and sinks in a plant. Mature leaves of this plant act as sources for sugars and other organic substances that are transported via the phloem to sinks, regions of the plants that have a high demand for these substances.

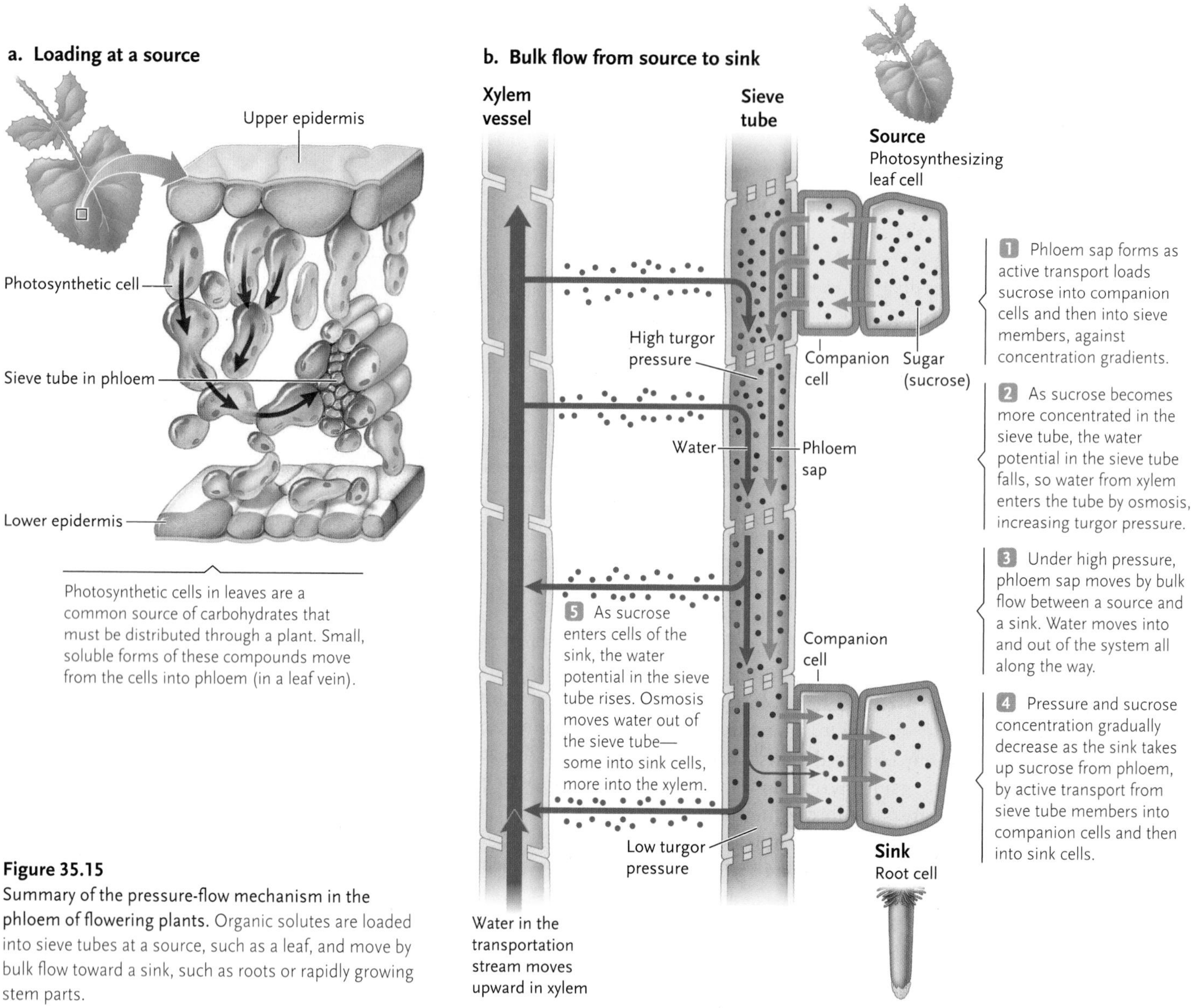

a. Loading at a source

Upper epidermis

Photosynthetic cell

Sieve tube in phloem

Lower epidermis

Photosynthetic cells in leaves are a common source of carbohydrates that must be distributed through a plant. Small, soluble forms of these compounds move from the cells into phloem (in a leaf vein).

Figure 35.15

Summary of the pressure-flow mechanism in the phloem of flowering plants. Organic solutes are loaded into sieve tubes at a source, such as a leaf, and move by bulk flow toward a sink, such as roots or rapidly growing stem parts.

b. Bulk flow from source to sink

Xylem vessel

Sieve tube

Source
Photosynthesizing leaf cell

High turgor pressure

Companion cell · Sugar (sucrose)

Water · Phloem sap

5 As sucrose enters cells of the sink, the water potential in the sieve tube rises. Osmosis moves water out of the sieve tube— some into sink cells, more into the xylem.

Companion cell

Low turgor pressure

Sink
Root cell

Water in the transportation stream moves upward in xylem

1 Phloem sap forms as active transport loads sucrose into companion cells and then into sieve members, against concentration gradients.

2 As sucrose becomes more concentrated in the sieve tube, the water potential in the sieve tube falls, so water from xylem enters the tube by osmosis, increasing turgor pressure.

3 Under high pressure, phloem sap moves by bulk flow between a source and a sink. Water moves into and out of the system all along the way.

4 Pressure and sucrose concentration gradually decrease as the sink takes up sucrose from phloem, by active transport from sieve tube members into companion cells and then into sink cells.

Most substances carried in phloem are loaded into sieve tube members by active transport (Figure 35.15a). For example, the sucrose formed in mesophyll cells of leaves is actively pumped into companion cells by an H^+/sucrose symport, where H^+ ions move into the cell through the same carrier that takes up the sugar molecules. From the companion cells, most sucrose crosses into the living sieve tube members through plasmodesmata.

In some plants, companion cells become modified into **transfer cells**, which facilitate the short-distance transport of organic solutes from one cell to another. Transfer cells generally form when large amounts of solutes must be loaded or unloaded into the phloem, and they shunt substances through plasmodesmata to sieve tube members. How do transfer cells facilitate solute transport? As a transfer cell is forming, parts of its cell wall grow inward like pleats **(Figure 35.16)**. The underlying plasma membrane, packed with trans-

port proteins, enfolds each pleat in the cell wall, thus increasing the surface area across which solutes can be taken up. Xylem parenchyma transfer cells also enhance transport between living cells in the xylem, and they occur in glandlike tissues that secrete nectar. Transfer cells occur in species from every taxonomic group in the plant kingdom, as well as in fungi and algae.

When sucrose is loaded into sieve tubes of small leaf veins, its concentration rises inside the tubes. The increased solute concentration causes a decrease in water potential inside the sieve tube members, and water flows into the cells by osmosis. As water enters the sieve tubes, turgor pressure in the tubes increases, and the sucrose-rich fluid moves by bulk flow into the increasingly larger sieve tubes of larger veins. Eventually, the fluid is pushed out of the leaf into the stem and toward a sink (Figure 35.15). Solute unloading is mostly symplastic in sink cells. When sucrose is unloaded into

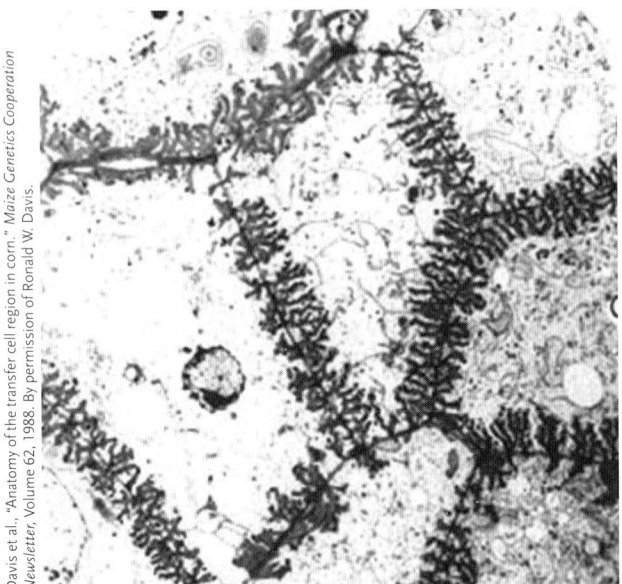

Davis et al., "Anatomy of the transfer cell region in corn." *Maize Genetics Cooperation Newsletter*, Volume 62, 1988. By permission of Ronald W. Davis.

Figure 35.16

Transmission electron micrograph showing transfer cells in corn (*Zea mays*) in cross-section. Both the cell wall and the underlying plasma membrane of these transfer cells are highly folded, increasing the surface area for solute transfer.

the sink, the amount of solutes in the sieve tube member decreases, resulting in an increase in the water potential and the movement of water out of the sieve tube by osmosis. Some of this water "follows the solutes" into the sink cells, but most of it enters the xylem (Figure 35.15b, step 5).

Sieve tubes are mostly passive conduits for translocation. The system works because companion cells supply most of the energy that loads sucrose and

other solutes at the source, and because solutes are removed at their sinks. As sucrose enters a sink, for example, its concentration in the sieve tubes decreases, and water moves out of the cell, causing the pressure potential to decrease. Thus, for sucrose and other solutes transported in the phloem, there is always a gradient of concentration from source to sink, and a pressure gradient that keeps the solute moving along.

As noted previously, phloem sap moving through a plant carries a wide variety of substances, including amino acids, organic acids, agricultural chemicals, and organic nitrogen compounds and mineral ions that are removed from dying leaves and stored for reuse in root tissue. The phloem also transports hormones and other signal molecules such as RNA from one part of a plant to another and so plays a much greater role in the plant than just a pipeline to move sucrose: it acts as an "information superhighway."

The transport functions of xylem and phloem are closely integrated with plant reproduction, development of embryos, and the hormone-based regulation of plant growth. We will explore these topics in Chapter 38.

STUDY BREAK

1. What is the difference between translocation and transpiration?
2. Using sucrose as your example, summarize how a substance moves from a source into sieve tubes and is then unloaded at a sink. What is this mechanism called, and why?

Review

aplia™

To access course materials such as Aplia and other companion resources, please visit www.NELSONbrain.com.

35.1 Principles of Water and Solute Movement in Plants

- Plants have mechanisms for moving water and solutes (1) into and out of cells, (2) laterally from cell to cell, and (3) over long distances from the root to the shoot and vice versa (Figure 35.1).

- Both passive and active transport mechanisms move substances across plant cell membranes.

- Water crosses plant cell membranes by osmosis, which is driven by differences in water potential (ψ_w) between a cell and its surroundings. Water tends to move from regions where water potential is higher to regions where it is lower (Figure 35.3).

- Water potential has two components, pressure potential and solute potential. Water potential is measured in megapascals (MPa).

- Water and solutes also move across the tonoplast between the cell's central vacuole and the cytoplasm (Figure 35.4). Water in the central vacuole is vital for maintaining turgor pressure inside a plant cell.

35.2 Uptake and Transport of Water and Solutes by Roots

- Water and mineral ions entering roots travel laterally through the root cortex to the root xylem, following one or both of two pathways: the apoplastic pathway and the symplastic pathway (Figure 35.5).

- In the apoplastic pathway, water diffuses into roots in the cell walls of root epidermal cells and moves across the cortex to the endodermis via cell walls and intercellular spaces. By contrast, water and solutes absorbed by roots can flow from the cytoplasm of one cortical cell to the next via plasmodesmata.

- The innermost layer of the cortex, the endodermis, has cell walls with Casparian strips that block apoplastic movement, forcing all water and solutes into the symplast (Figure 35.6). This ensures that all water and solutes pass through a plasma membrane in order to enter the stele, allowing the plant to regulate the ions that pass into the vascular tissue.

35.3 Long-Distance Transport of Water and Minerals in the Xylem

- Transpiration, the evaporation of water from leaves and shoots, creates tension (negative pressure) on the water in the xylem. Cohesion between water molecules creates a continuous column of water molecules that is pulled by the tension created as water exits a plant's leaves (Figure 35.7). Thus, the negative pressure generated in the shoot drives bulk flow of xylem sap.
- In some plants, notably herbaceous species, positive pressure sometimes develops in roots when transpiration is low, contributing to the upward movement of xylem sap.
- Transpiration and carbon dioxide uptake occur mostly through stomata. Environmental factors such as relative humidity, air temperature, and air movement at the leaf surface affect the transpiration rate.
- Most plants lose water and take up carbon dioxide during the day, when stomata are open. At night, when stomata are closed, plants conserve water and the inward movement of carbon dioxide falls.
- Stomata open when activation of proton pumps (H^+-ATPase pumps) in the plasma membrane of guard cells creates an electrochemical gradient that drives the accumulation of K^+ ions in guard cell cytoplasm. Water flows into the guard cells by osmosis, increasing turgor pressure, which causes the guard cells to swell and draw apart, opening the stomata (Figure 35.10). Other factors, such as light and water stress, also influence the opening and closing of stomata.

35.4 Transport of Organic Substances in the Phloem

- Phloem sap carries much more than just sugars: it also transports amino acids, organic acids, organic nitrogen compounds, hormones, and other signal molecules.
- In flowering plants, phloem sap is translocated in sieve tube members. Differences in pressure between source and sink regions drive the flow. Sources include mature leaves; sinks include growing tissues and storage regions (such as the tubers of a potato) (Figure 35.14).
- In leaves, the sugar sucrose is actively transported into companion cells adjacent to sieve tube members and then loaded into the sieve tubes through plasmodesmata (Figure 35.15).
- As the sucrose concentration increases in the sieve tubes, water potential decreases. The resulting influx of water causes pressure to build up inside the sieve tubes, so the sucrose-laden fluid flows in bulk toward the sink, where sucrose and water are unloaded and distributed among surrounding cells and tissues.

Questions

Self-Test Questions

1. Which of the following statements is the best description of turgor pressure?
   a. It is the equivalent of water potential.
   b. It is the movement of water into a cell by osmosis.
   c. It is the driving force for osmotic movement of water (ψ).
   d. It is the pressure exerted by fluid inside a plant cell against the cell wall.

2. Which of the following statements is the best description of water potential?
   a. It is the driving force for the osmotic movement of water into plant cells.
   b. It is less negative in a solution that has more solute molecules relative to water molecules.
   c. It is a measure of the combined effects of a solution's pressure potential and its solute potential.
   d. It is a measure of the physical pressure required to halt osmotic water movement across a membrane.

3. What will happen if you place a plant cell with ψ_s of −0.4 MPa and ψ_p of 0.2 MPa in a chamber filled with pure water that is pressurized with 0.5 MPa?
   a. Water will flow into the cell.
   b. Water will flow out of the cell.
   c. The cell will be crushed.
   d. The cell will explode.

4. A plant cell with a solute potential of −0.65 MPa maintains a constant volume when bathed in a solution in an open container that has a solute potential of −0.3 MPa. What can you conclude from this information?
   a. This cell has a pressure potential of +0.65 MPa.
   b. This cell has a water potential of −0.65 MPa.
   c. This cell has a pressure potential of +0.35 MPa.
   d. This cell has a water potential of 0 MPa.

5. Which of the following changes would not contribute to water uptake by a plant cell in solution?
   a. an increase in the water potential of the surrounding solution
   b. adding more solutes to the cell's cytoplasm
   c. a decrease in the water potential of the cytoplasm
   d. an increase in tension on the surrounding solution

6. Which of the following statements about water movement in roots is incorrect?
   a. The symplast is a network of connected living cells.
   b. Water can enter the xylem without entering the symplast.
   c. Water can enter the xylem without entering the apoplast.
   d. Casparian strips prevent water from moving from one endodermal cell to another.
   e. The endodermis is a layer in the root cortex.

7. An indoor gardener leaving for vacation completely wraps a potted plant with clear plastic. Temperature and light are left at low intensities. Which of the following correctly describes the effect of this strategy on the plant?
   a. Photosynthesis will stop.
   b. Transpiration will be reduced.
   c. Guard cells will shrink and stomata open.
   d. Evaporation from leaf mesophyll cells will be increased.

8. Which of the following processes would result in opening of stomata?
   a. K^+ flows out of guard cells.
   b. CO_2 concentration inside the leaf increases.
   c. Turgor pressure in the guard cells increases.
   d. The H^+-ATPase pumps in the guard cell membranes stop pumping.

9. Which of the following factors contributes to the movement of water up a plant stem?
   a. active transport of water into the root hairs
   b. absorption of raindrops on a leaf's epidermis
   c. cohesion of water molecules in the stem and leaf xylem
   d. higher (less negative) water potential in the leaf's mesophyll layer than in the xylem

10. Which of the following processes occurs during the translocation of sucrose-rich phloem sap?
    a. Companion cells pump sucrose into sieve tube members.
    b. The sap flows toward a source as pressure builds up at a sink.
    c. Sucrose diffuses into companion cells, while H^+ simultaneously leaves the cells by a different route.
    d. Companion cells use energy to load solutes at a source and the solutes then follow their concentration gradients to sinks.

Questions for Discussion

1. Many popular houseplants are actually native to tropical rainforests. Among other characteristics, these plants have extraordinarily broad, flat leaves, some so large that indigenous people use them as umbrellas. What environmental conditions might make a broad leaf adaptive in tropical regions, and why?

2. Insects such as aphids that prey on plants by feeding on phloem sap generally attack only young shoot parts. Other than the relative ease of piercing less mature tissues, suggest a reason why it may be more adaptive for these animals to focus their feeding effort on younger leaves and stems.

3. So-called systemic insecticides are often mixed with water and applied to the soil in which a plant grows. The chemicals are effective against sucking insects no matter which plant tissue the insects attack, but often don't work as well against chewing insects. Propose a reason for this difference.

4. Concerns about global warming and the greenhouse effect centre on rising levels of greenhouse gases, including atmospheric carbon dioxide. Plants use CO_2 for photosynthesis, and laboratory studies suggest that increased CO_2 levels could cause a rise in photosynthetic activity. However, as one biologist noted, "What plants do in environmental chambers may not happen in nature, where there are many other interacting variables." Strictly from the standpoint of physiological effects, what are some possible ramifications of a rapid doubling of atmospheric CO_2 on plants in temperate environments? In arid environments?

36

The reproductive structures of an ornamental poppy (*Papaver rhoeas*). Male reproductive structures, which produce pollen, surround the female reproductive structure, which produces eggs and is the site of fertilization and seed development.

Reproduction and Development in Flowering Plants

WHY IT MATTERS

What kinds of plants do we rely on most for our food? Think about the food that you eat in a given day: bread, pasta, rice, potatoes, salad greens, fruit—all of these foods and more come from just one group of land plants, the angiosperms (flowering plants). Although we do eat the vegetative parts of many angiosperms, we rely particularly heavily on the seeds and fruits produced by these plants. Worldwide, the top 10 crop plants are angiosperms; for example, consider the top three crop plants in the world: rice, maize, and wheat **(Figure 36.1).** Millions of people rely either directly on the seeds and/or fruit of these plants or on products made from these parts (e.g., flour). As in other flowering plants, the fruits and seeds of rice, wheat, and corn result from sexual reproduction. Angiosperms have elaborate reproductive systems—housed in flowers—that produce and protect gametes and developing embryos. The flowers of many species also serve as invitations to animal pollinators.

While some angiosperms (including the top three crop plants mentioned above) are wind pollinated, many others rely on animals to carry pollen from one plant to another to complete sexual reproduction. Because of the heavy reliance of humans on animal-pollinated species such as fruit trees, when populations of pollinators are threatened or decline, seed and fruit production is also put at risk.

This dire situation is now facing many angiosperms in North America that depend on honeybees for their pollination. About one-third of North American plants, including many important crop plants, rely on honeybees for pollination **(Figure 36.2).** In North America, honeybees pollinate more than $16 billion worth of almonds, cucumbers, berries, apples, and canola. These honeybees are not native

a.

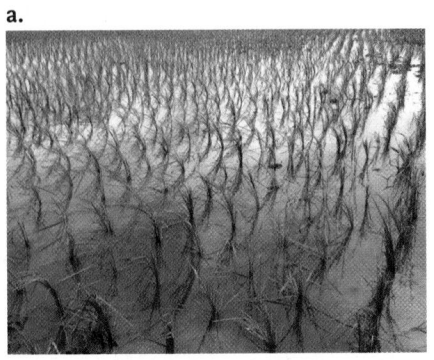

b.

Kashfia Rahman/StockXchng

Isa Fernandez Fernandez/Shutterstock.com

c.

Photos.com

Figure 36.1
The world's three most important crop plants (from a human perspective): **(a)** rice (*Oryza sativa*) plants in a rice paddy, **(b)** maize (*Zea mays*) plants, and **(c)** wheat (*Triticum aestivum*) plants.

to North America but were introduced (along with many crop plants that they pollinate) about 400 years ago from Europe. These introduced bees displaced most native bees by the 1920s, and we now depend on them to pollinate many crops. In 2004, some beekeepers started to report wide-scale disappearance of colonies: hives were virtually abandoned, containing only a few larvae, sometimes a queen, and a lot of honey—but no adults. By 2006, the scale of these disappearances was large enough that they hit the news and were termed *Colony Collapse Disorder* (CCD).

The cause of CCD is still a mystery; explanations put forward include various pathogens and pests, long-distance transport of beehives for crop pollination, the effects of genetically modified plants, drought, and even interference from cellphone towers (this last hypothesis is not supported by any evidence). Perhaps several factors are acting together to create a "perfect storm" that results in the collapse of honeybee populations. If CCD spreads, populations of both cultivated and native angiosperms that rely on honeybees to complete their life cycles could be at risk.

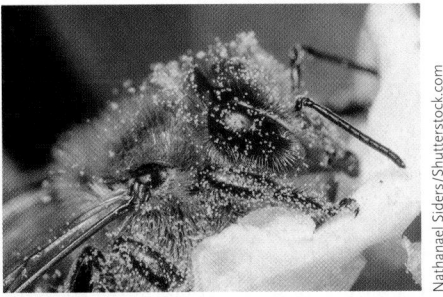

Nathanael Siders/Shutterstock.com

Figure 36.2
Honeybee covered with pollen.

In this chapter, we first investigate how sexual reproduction occurs in flowering plants. We then compare sexual reproduction with asexual reproduction, which occurs in many angiosperms under certain circumstances to produce clones that are genetically identical to their parents. Whether formed by sexual or asexual reproduction, once a new individual begins to grow, finely regulated gene interactions guide the development of flowers and other plant parts. Using methods of molecular biology and a variety of model organisms, plant biologists are beginning to understand some of the mechanisms by which these developmental pathways unfold; we conclude the chapter by looking at some of these mechanisms.

36.1 Overview of Flowering Plant Reproduction

In plants, as in animals, sexual reproduction occurs when male and female haploid gametes unite to create a diploid zygote, which then embarks on a developmental course of mitotic cell divisions, cell enlargement, and **cell differentiation**. In flowering plants, subsequent steps result in distinctive haploid and diploid forms of an individual.

36.1a Diploid and Haploid Generations Arise in the Angiosperm Life Cycle

An angiosperm zygote develops into an embryo enclosed within a seed. In a seed, early versions of the basic plant tissue systems are already in place, so the embryo is already a **sporophyte**—a term that refers to the diploid, spore-producing body of a plant (see Chapter 26). What we see when we look at a flowering plant, such as a wild rose (*Rosa acicularis*), is the sporophyte **(Figure 36.3, p. 884).**

At some point during one or more seasons of an angiosperm sporophyte's growth and development, one or more of its vegetative shoots undergo changes in structure and function and become reproductive shoots that will give rise to a flower or an **inflorescence** (a group of flowers on the same floral shoot). Within the sexual organs of the flower, certain cells divide by meiosis. Unlike in animals, however, meiosis in plants does not produce gametes. Instead, meiosis

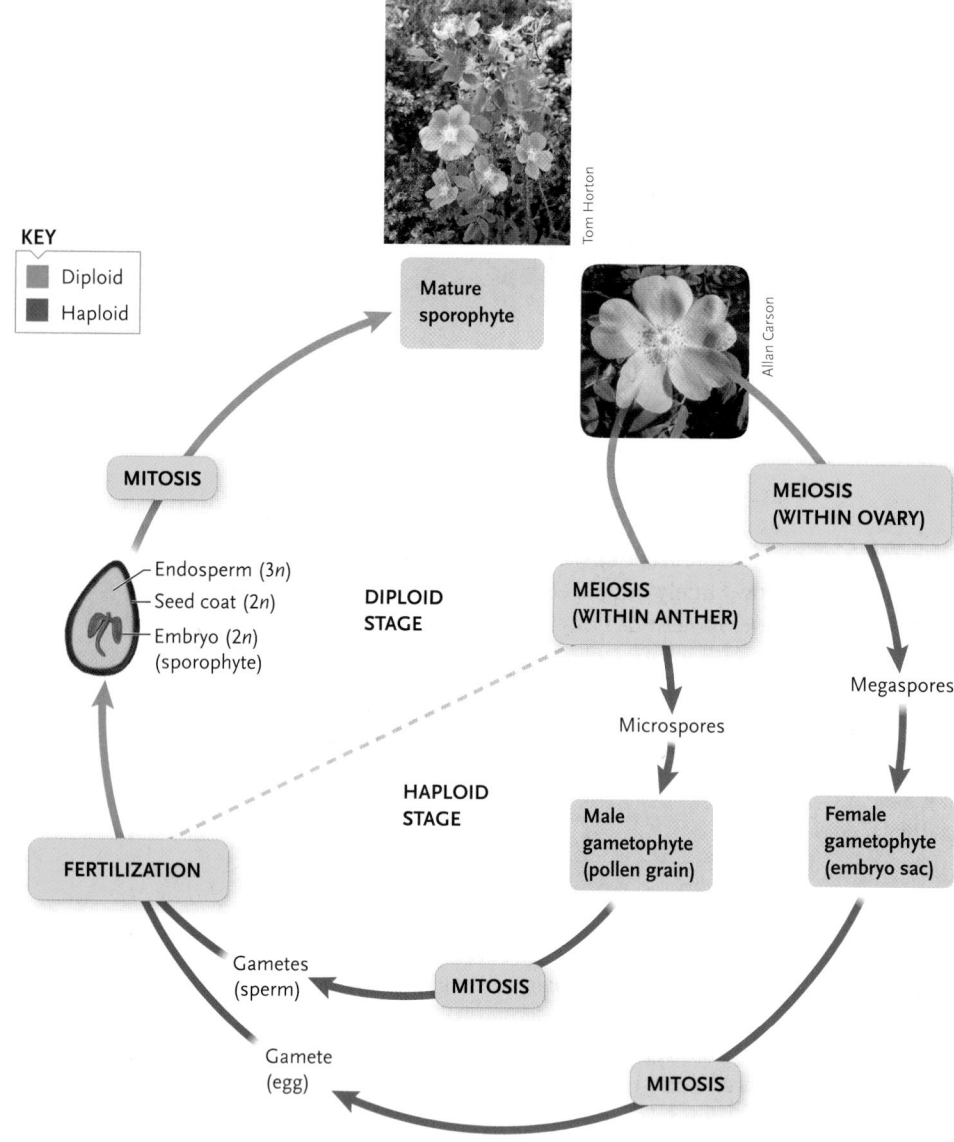

KEY
- Diploid
- Haploid

Mature sporophyte

Tom Horton

Allan Carson

MITOSIS

MEIOSIS (WITHIN OVARY)

MEIOSIS (WITHIN ANTHER)

Endosperm (3*n*)
Seed coat (2*n*)
Embryo (2*n*) (sporophyte)

DIPLOID STAGE

Megaspores

Microspores

HAPLOID STAGE

FERTILIZATION

Male gametophyte (pollen grain)

Female gametophyte (embryo sac)

Gametes (sperm)

MITOSIS

Gamete (egg)

MITOSIS

Figure 36.3

Overview of the flowering plant life cycle, using the wild rose (*Rosa acicularis*) as an example. This type of reproductive cycle, alternation of generations, has a haploid phase in which multicellular but reduced gametophytes produce gametes, which fuse to form a zygote. This zygote develops into a multicellular embryo within a seed and then into a mature sporophyte. Meiotic divisions in the flower of the sporophyte produce spores, which give rise to new gametophytes.

nourished by the gametophyte. In ferns, which are seedless vascular plants, the gametophyte is much smaller than the sporophyte and, while it is free living for much of its life and nourishes itself by photosynthesis, it does not live as long as the sporophyte generation. In angiosperms and other seed plants, gametophytes are so reduced in size that they are retained *inside* sporophyte tissue for all or part of their lives. The female gametophyte of a flowering plant usually consists of only seven cells, which are embedded in floral tissues, as you will read shortly. Male gametophytes are released into the environment as pollen grains and are so small that they are measured in micrometres (μm). The pollen grain matures when it reaches floral tissue, producing a pollen tube that grows through floral tissue to the egg, carrying sperm with it. When the pollen tube enters the ovules inside the ovary, the sperm are released, resulting in fertilization and production of a new generation of seeds.

Sporophytes may also reproduce asexually. For instance, strawberry (*Fragaria* species) plants send out horizontal stems or stolons, and new roots and shoots develop at each node along the stems. Short underground stems of onions and lilies put out buds that grow into new plants. In summer and fall, quackgrass (*Elymus repens*) produces new plants at nodes along its subterranean rhizomes. Asexual reproduction can also be induced artificially. Whole orchards of genetically identical fruit trees are grown from the cuttings or buds of a single parent tree.

We turn now to our consideration of sexual reproduction in angiosperms, beginning with the crucial step in which flowers develop.

gives rise to haploid **spores**, walled cells that develop by mitosis into multicellular haploid **gametophytes.** The gametophytes produce haploid gametes, again by mitosis. Male gametophytes produce sperm, and female gametophytes produce eggs. This division of a life cycle into a diploid, spore-producing generation and a haploid, gamete-producing one is called **alternation of generations** (a phenomenon described more fully in Chapter 26).

In virtually all plants, the gametophyte and the sporophyte are strikingly different from one another in both function and structure. As you learned in Chapter 26, in mosses and other bryophytes, the gametophyte is longer lived than the sporophyte; the sporophyte grows out of the gametophyte and remains attached to and

STUDY BREAK

1. What are the two alternating generations of plants? How do these two life phases differ in structure and function?
2. How does meiosis in plants differ from meiosis in animals? How is it similar?
3. How do the gametophytes of nonseed plants compare to those of seed plants?

36.2 Flower Structure and Formation of Gametes

Flowering marks a developmental shift for an angiosperm. What triggers the formation of flowers? Biochemical signals—triggered in part by environmental cues such as day length and temperature—travel to the apical meristem of a shoot, as you will see in Chapter 38, and set in motion changes in the activity of cells there. Instead of continuing vegetative growth, the shoot is modified into a floral shoot that will give rise to floral organs.

36.2a A Flower Consists of Both Sterile and Fertile Parts

A flower develops from the end of the floral shoot, called the **receptacle.** Flowers consist of four concentric circles (*whorls*) of organs, all of which are modified leaves; **Figure 36.4** shows a typical flower. The two outer whorls consist of sterile (nonfertile), vegetative organs. The outermost whorl (whorl 1) is made up of leaflike **sepals.** Sepals are usually green and enclose all of the other parts early in the flower's development, as in an unopened rosebud. The next whorl is made up of **petals,** the showy parts of flowers. Petals have distinctive colours, patterning, and shapes, which play important roles in promoting pollen dispersal by wind in wind-pollinated species and in attracting bees and other animal pollinators in animal-pollinated species.

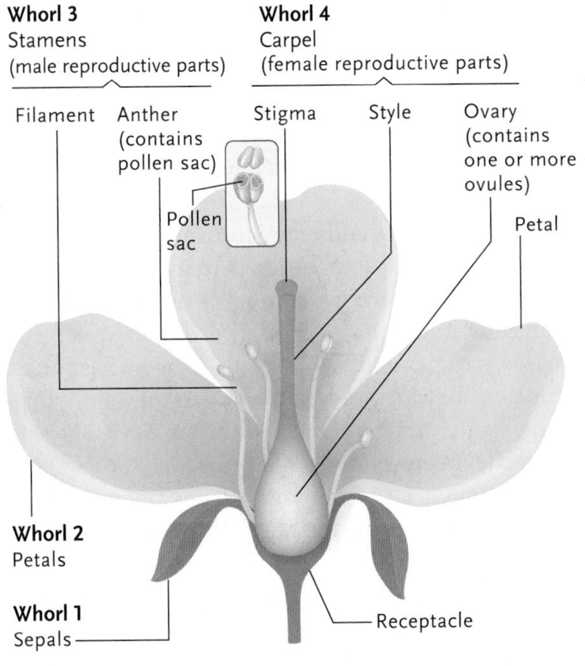

Glands that produce nectar, a sugary liquid that attracts animal pollinators, are often located at the base of petals.

A flower's two inner whorls are the reproductive organs. Inside the petals are the **stamens** (whorl 3), in which male gametophytes form. In almost all living flowering plant species, a stamen consists of a slender **filament** (stalk) capped by an **anther.** Each anther is composed of four **pollen sacs,** in which pollen develops.

The innermost whorl (whorl 4) consists of one or more **carpels,** in which female gametophytes form. The lower part of a carpel is the **ovary.** Inside the ovary are one or more **ovules,** in which an egg develops and fertilization takes place. A seed is a mature ovule. In many flowers that have more than one carpel, the carpels fuse into a single, common ovary containing multiple ovules and, after fertilization, multiple seeds. Typically, the carpel's slender **style** widens at its upper end, terminating in the **stigma,** a receptive surface for pollen. Fused carpels may share a single stigma and style, or each may retain separate ones. The name angiosperm (seed vessel) refers to the carpel.

Not all plants have all four whorls: many plants have flowers that have both male and female sexual organs, such as the flower shown in Figure 36.4, but other plants' flowers lack stamens or carpels. These **imperfect flowers** are further divided according to whether individual plants produce both sexual types of flowers or only one. In **monoecious** (= "one house") species, such as corn (*Zea mays*), each plant has some "male" flowers with only stamens and some "female" flowers with only carpels. In **dioecious** (= "two houses") species, such as willows (*Salix* species), a given plant produces flowers with only stamens or carpels **(Figure 36.5, p. 886).** In addition, some flowers lack the showy petals of the typical flower in Figure 36.4 or may have highly modified petals; these modifications often relate to the ability of a plant to attract specific pollinators (see "Molecule behind Biology," Box 36.2, p. 890). With this basic angiosperm reproductive anatomy in mind, we now turn to the processes that produce male and female gametes.

36.2b Pollen Grains Arise from Microspores in Anthers

Most of a flowering plant's reproductive life cycle, from production of spores to production of a mature seed, takes place within its flowers. **Figure 36.6, p. 887,** shows this cycle as it unfolds in a flower with both stamens and carpels. The spores that give rise to male gametophytes are produced in anthers (see Figure 36.6, left). The pollen sacs inside each anther are microsporangia and contain diploid microsporocytes (also called *microspore mother cells*); each microsporocyte produces four small haploid **microspores** by meiosis. Inside the spore wall, each microspore divides again, this time by mitosis. The result is a haploid male gametophyte—a

Whorl 3
Stamens
(male reproductive parts)

Filament   Anther
(contains
pollen sac)

Pollen
sac

Whorl 4
Carpel
(female reproductive parts)

Stigma   Style   Ovary
(contains
one or more
ovules)

Petal

Whorl 2
Petals

Whorl 1
Sepals

Receptacle

Figure 36.4
Structure of a generalized flower, with the four whorls indicated. The anthers of the stamens produce haploid pollen. The stigma of the carpel receives pollen, and ovules inside the ovary contain haploid eggs.

Figure 36.5
Examples of monoecious and dioecious plants. **(a)** Corn (*Zea mays*) has separate male (the tassels at the top of the plants) and female (the "ears" of the corn) flowers on the same plant. **(b)** Willows (*Salix* species) have separate female (top photo) and male (bottom photo) plants.

a.

©iStock.com/Justin Voight

b.

Martin Fowler/Shutterstock.com

Kiev.Victor/Shutterstock.com

pollen grain. A pollen grain consists of two cells: a larger vegetative (or tube) cell that will later produce a pollen tube and a smaller **generative cell** that will later divide to produce two gametes or sperm. In many angiosperms, the male gametophyte is in this two-celled stage when it is released from the anther; in others, division of the generative cell to produce two sperm occurs before the pollen grain is released.

At maturity, a male gametophyte consists of three cells: two sperm cells and the **tube cell**. When pollen lands on a stigma, it germinates and forms a pollen tube, which grows through the tissues of the carpel and carries the sperm to the ovule.

The walls of pollen grains are tough enough to protect the male gametophyte during the somewhat precarious journey from anther to stigma. These walls are so distinctive that the family to which a plant belongs can usually be identified from pollen alone—based on the size and wall sculpturing of the grains, as well as the number of pores in the wall **(Figure 36.7, p. 888)**. Because they withstand decay, pollen grains fossilize well and can provide revealing clues about the evolution of seed plants, as well as help biologists reconstruct ancient plant communities and determine how climates have changed over time.

36.2c Eggs and Other Cells of Female Gametophytes Arise from Megaspores in Ovaries

The ovary of a flower contains one or more ovules (see Figure 36.6, right), which will develop into seeds after fertilization if all goes well. Only one ovule forms in the carpel of some flowers, such as the cherry. Dozens, hundreds, or thousands may form in the carpels of other flowers, such as those of a bell pepper plant (*Capsicum annuum*). Each ovule consists of a stalk bearing a **nucellus** (the inner part of an ovule, in which the embryo sac develops; equivalent to a megasporangium) enveloped in additional layers of sporophyte tissue called **integuments**, with an opening (the **micropyle**) at one end.

Formation of the female gametophyte varies among plant species, but we will consider only the most commonly observed pattern of development. Inside the ovule, a diploid megasporocyte (also called a *megaspore mother cell*) divides by meiosis, forming four haploid **megaspores**. In most plants, three of these megaspores disintegrate. The remaining megaspore enlarges and develops into the female gametophyte in a sequence of steps tracked in Figure 36.6.

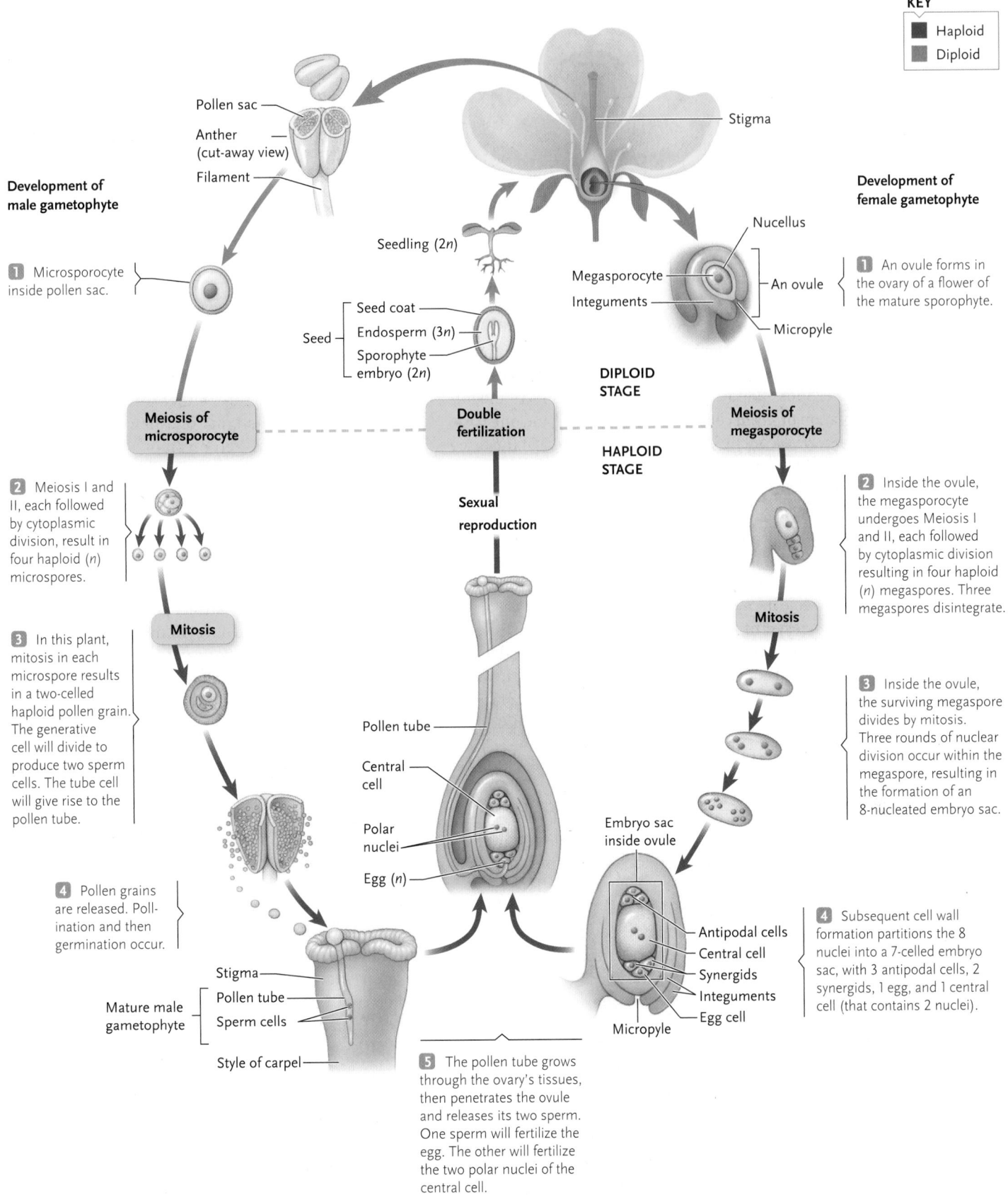

KEY
■ Haploid
■ Diploid

Pollen sac

Anther (cut-away view)

Filament

Stigma

Development of male gametophyte

Development of female gametophyte

Nucellus

Seedling (2n)

Megasporocyte

Integuments

An ovule

Micropyle

1 An ovule forms in the ovary of a flower of the mature sporophyte.

1 Microsporocyte inside pollen sac.

Seed coat

Seed — Endosperm (3n)

Sporophyte embryo (2n)

DIPLOID STAGE

Meiosis of microsporocyte

Double fertilization

Meiosis of megasporocyte

HAPLOID STAGE

2 Meiosis I and II, each followed by cytoplasmic division, result in four haploid (n) microspores.

Sexual reproduction

2 Inside the ovule, the megasporocyte undergoes Meiosis I and II, each followed by cytoplasmic division resulting in four haploid (n) megaspores. Three megaspores disintegrate.

Mitosis

Mitosis

3 In this plant, mitosis in each microspore results in a two-celled haploid pollen grain. The generative cell will divide to produce two sperm cells. The tube cell will give rise to the pollen tube.

3 Inside the ovule, the surviving megaspore divides by mitosis. Three rounds of nuclear division occur within the megaspore, resulting in the formation of an 8-nucleated embryo sac.

Pollen tube

Central cell

Polar nuclei

Egg (n)

Embryo sac inside ovule

4 Pollen grains are released. Pollination and then germination occur.

Antipodal cells

Central cell

Synergids

Integuments

Egg cell

Micropyle

4 Subsequent cell wall formation partitions the 8 nuclei into a 7-celled embryo sac, with 3 antipodal cells, 2 synergids, 1 egg, and 1 central cell (that contains 2 nuclei).

Stigma

Pollen tube

Mature male gametophyte — Sperm cells

Style of carpel

5 The pollen tube grows through the ovary's tissues, then penetrates the ovule and releases its two sperm. One sperm will fertilize the egg. The other will fertilize the two polar nuclei of the central cell.

Figure 36.6

Sexual reproduction in a generalized angiosperm. Pollen grains develop in pollen sacs (microsporangia) within the anthers. An embryo sac forms inside each ovule within an ovary, and an egg forms within the embryo sac. Pollen grains are released and deposited on the stigma and then germinate to produce a pollen tube that carries the two sperm to the ovary, where double fertilization occurs. An embryo sporophyte and nutritive endosperm develop and become encased in a seed coat.

CHAPTER 36 REPRODUCTION AND DEVELOPMENT IN FLOWERING PLANTS |

Figure 36.7

Examples of pollen grain diversity. Scanning electron micrographs of pollen grains from **(a)** a grass, **(b)** chickweed (*Stellaria*), and **(c)** ragweed (*Ambrosia*) plants.

a.
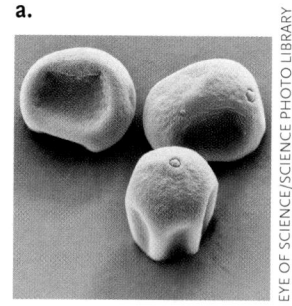
EYE OF SCIENCE/SCIENCE PHOTO LIBRARY

b.
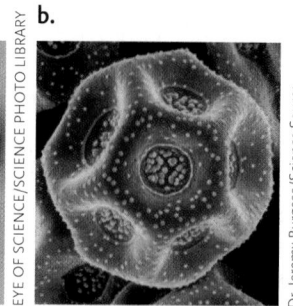
Dr. Jeremy Burgess/Science Source

c.

daniel mathys/Stock/Thinkstock

First, three rounds of mitosis occur *without* cytoplasmic division; these divisions produce a single cell with eight nuclei, four at each pole of the cell. Next, one nucleus in each group migrates to the centre of the cell; these two **polar nuclei** ("polar" because they migrate from opposite ends of the cell) may fuse or remain separate. Subsequent cell wall formation results in the formation of seven cells, three at either pole and one large *central cell* containing the two polar nuclei. Of the three cells that form a cluster near the micropyle, one is an **egg cell** that may eventually be fertilized; the other two, called *synergids,* play a role in fertilization (as discussed in "People behind Biology," Box 36.1). At the other end of the embryo sac are three cells called antipodals; their function is unknown, although in some plants they may play a role in nutrient transfer from maternal tissues to the endosperm. The eventual result of all of these events is an **embryo sac** containing seven cells and eight nuclei. This embryo sac is the female gametophyte.

As the male and female gametophytes complete their maturation, the stage is set for fertilization and the development of a new individual.

STUDY BREAK

1. What is the biological role of flowers, and what fundamental physiological change must occur before an angiosperm can produce a flower?
2. Explain the steps leading to the formation of a mature male gametophyte, beginning with microsporocytes in a flower's anthers. Which structures are diploid, and which are haploid? What is the difference between an immature male gametophyte and a mature male gametophyte?
3. Trace the development of a female gametophyte, beginning with the megasporocyte in an ovule of a flower's ovary. Which structures are diploid, and which are haploid?

36.3 Pollination, Fertilization, and Germination

CONCEPT FIX Most of us are familiar with the concept of pollination: for example, most people know that animals such as birds and bees are involved in moving pollen from one flower to another. But there is also a lot of confusion about the process: many people think pollination is the same as fertilization, and that it only happens in flowering plants. In reality, pollination occurs in all seed plants; it is the transfer of pollen to a female sexual structure, either a female cone of a gymnosperm or the stigma of a flower in an angiosperm. Gymnosperms rely on air or water currents to transport pollen, whereas pollen transfer in angiosperms can involve air or wind, but also various animals, including birds, bats, and insects. (In Chapter 26, we discussed the complex relationship between some flowering plants and their animal pollinators; see also "Molecule behind Biology," Box 36.2, p. 890, for more on how flowers attract pollinators.) Pollination and fertilization are not synonymous: pollination is only the first step in a series of events that can lead to *fertilization,* the fusion of an egg and sperm inside an ovule, but because there are many other steps before successful fertilization occurs, pollination does not guarantee that fertilization will happen. A fertilized ovule matures into a seed, and the zygote that results from fertilization matures into an embryo, a young sporophyte. When the seed *germinates,* or sprouts, the sporophyte begins to grow. ⬡

36.3a Pollination Requires Compatible Pollen and Female Tissues

Even after pollen reaches a stigma, in most cases, pollination and fertilization can take place only if the pollen and stigma are compatible. For example, if pollen from one species lands on a stigma belonging to a different species, chemical incompatibilities usually prevent pollen tubes from developing.

Even when the pollen and stigma are from the same species, pollination may not lead to fertilization unless the pollen and stigma belong to genetically distinct individuals. For instance, when pollen from a given plant lands on that plant's own stigma, a pollen tube may begin to develop but stop before reaching the embryo sac. How is self-pollination detected and blocked? **Self-incompatibility** is a biochemical recognition and rejection process that prevents **self-fertilization**, and it apparently results from interactions between proteins encoded by *S* (self) genes.

Tetsuya Higashiyama, *Nagoya University, Japan*

How do pollen tubes find their way through the style to the egg? This question has intrigued plant biologists for almost 150 years, since researchers first observed pollen tubes growing from pollen grains isolated from a plant and placed in a sugar medium. Biologists suggested that ovules produced some chemical that would guide the tube toward the egg. What part of the ovule produced the attractant, and what exactly was the attractant? Tetsuya Higashiyama of Nagoya University in Japan was fascinated by this question not only out of curiosity but also because of the importance of plant reproduction in the crop plants upon which humans depend. To investigate how pollen tubes are guided to eggs, Higashiyama chose to work with an unusual plant, *Torenia fournieri*

(commonly known as bluewings or wishbone flower; **Figure 1**).

In this plant, the embryo sac protrudes out of the ovule **(Figure 2),** making direct observations of interactions between pollen tubes and embryo sacs possible. For example, when ovules of this plant were moved by a micromanipulator, pollen tubes were able to track the movement of the synergids, as shown in Figure 2. In a series of experiments using a laser, Higashiyama and his colleagues removed various cells of the embryo sac and then observed how pollen tube movement was affected; only when the synergids were removed did pollen tube attraction stop, revealing that in this plant species, the synergids are the source of the chemical attractant guiding the

pollen tube. The race was on to identify the attractant, with several research labs around the world all working to solve the problem. Again using laser technology, Higashiyama's research group was the first to positively identify the attractants: two polypeptides that are produced by and secreted onto the surface of the synergids. While synergids in all angiosperms likely produce attractants to guide pollen tubes, Higashiyama's research shows that different plant species rely on different molecules: pollen tubes produced by one species were not attracted to ovules of another related species. This specificity is another way, in addition to those described in Section 36.3, that a plant can control its fertilization.

Figure 1
Flower of Torenia fourniera.

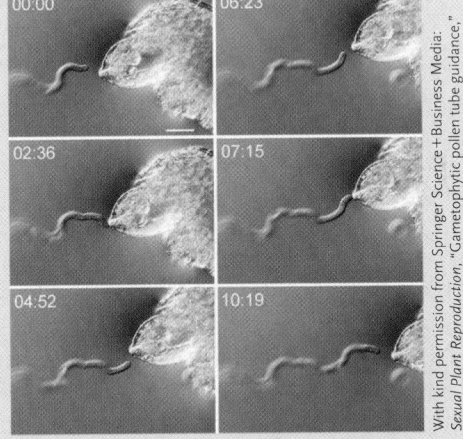

Figure 2
Pollen tube elongation toward an embryo sac of Torenia fourniera. *An ovule attracting a pollen tube in an in vitro system was manipulated with a microma-nipulator; note the micropylar end (where the synergids and egg are located) of the embryo sac projecting from the ovule. This series of sequential images (time is indicated as minutes:seconds) shows how the pollen tube is able to follow the embryo sac as the ovule is moved. Scale bar = 30 μm. Modified from Higashiyama and Hamamura (2008).*

Research has shown that *S* genes usually have multiple alleles—in some species, there may be hundreds—and a common type of incompatibility occurs when pollen and stigma carry an identical *S* allele. The result is a biochemical signal that prevents proper formation of the pollen tube **(Figure 36.8, p. 891).** For example, studies on plants of the mustard family (*Brassicaceae,* which includes canola) have revealed that pollen contacting an incompatible stigma produces a protein that prevents the stigma

Octenol

How do orchids trick insects into thinking that their flowers are actually mushrooms? Not all flowers resemble the "typical" flower shown in Figure 36.4. Consider the flower shown in **Figure 1a:** this is *Dracula chestertonii*, an orchid. Orchid flowers are often extensively modified versions of the typical flower shown below, and as you can see, flowers of this species of *Dracula* are modified to resemble a mushroom—the pattern on the lower lip mimics the gills on the underside of a mushroom cap. And the mimicry doesn't end there: the flower produces a very "mushroomy" scent. The chemical that creates this mushroomy smell is octenol (an alcohol; Figure 1b), which is also produced by fungi and gives them their characteristic odour. Together, the scent and appearance of *Dracula* flowers attract female flies that normally lay their eggs in mushrooms. The flies are so convinced that the flower is really a mushroom that they deposit their eggs in the "mushroom cap" petal; in the process, they pick up pollen, which they will then carry to the next *Dracula* flower that tricks them in the same way. But isn't it harmful for the flower to attract insects that will lay eggs in it? Unfortunately for the fly, its larvae can't develop in the floral tissue, so it has not only been deceived into helping the flower but also wasted its eggs in the process.

a.

Andreas Philipp

b.

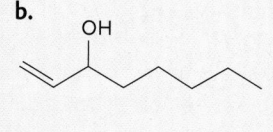

Figure 1
(a) Dracula chestertonii *flower.*
(b) *Structure of octenol.*

---

from hydrating the relatively dry pollen grain, an essential step if the pollen tube is to grow. Researchers have discovered a wide range of self-incompatibility responses. In some plants, when incompatible pollen contacts a stigma, a pollen tube grows normally, but a hormonal response soon causes the flower to drop off the plant, preventing fertilization.

Why is it desirable for plants not to pollinate themselves? Self-incompatibility prevents inbreeding and promotes genetic variation, which is the raw material for natural selection and adaptation. Even so, many flowering plants do self-pollinate, either partly or exclusively, because that mode, too, has benefits in some circumstances. (Mendel's peas are a classic example.) For instance, *selfing* may help preserve adaptive traits in a population. It also reduces or eliminates a plant's reliance on wind, water, or animals for pollination and thus ensures that seeds will form when conditions for cross-pollination are unfavourable, as when pollinators or potential mates are scarce.

36.3b Double Fertilization Results in the Formation of Embryos and Endosperm

If a pollen grain lands on a compatible stigma, it absorbs moisture and germinates a pollen tube, which burrows through the stigma and style toward an ovule. How does the pollen tube locate the embryo sac? This question is discussed in "People behind Biology," Box 36.1, p. 889. Before or during these events, the pollen grain's haploid generative cell divides by mitosis, forming two haploid sperm. When the pollen tube reaches the ovule, it enters through the micropyle, and an opening forms in its tip. By this time, one synergid has begun to die (an example of programmed cell death), and the two sperm are released into the disintegrating cell's cytoplasm. Experiments suggest that elements of the synergid's cytoskeleton guide the sperm onward, one to the egg cell and the other to the central cell.

Next, a remarkable sequence of events occurs called **double fertilization**, which has been observed only in flowering plants and (in a somewhat different version) in the gnetophyte *Ephedra* (see Chapter 26). Typically, one sperm fuses with the egg to form a diploid ($2n$) zygote. The other sperm fuses with the central cell, forming a cell with a triploid ($3n$) nucleus. This $3n$ nucleus (the primary endosperm nucleus) divides repeatedly along with its surrounding cytoplasm to form a tissue called **endosperm** (= inside the seed). The endosperm nourishes the embryo and, in monocots, the seedling, until its leaves form and photosynthesis has begun.

Embryo-nourishing endosperm forms only in flowering plants, and its evolution coincided with a reduction in the size of the female gametophyte. In other land plants, such as gymnosperms and ferns, the gametophyte itself contains enough stored food to nourish the embryonic sporophytes. Endosperm offers an advantage over female gametophyte tissue as a

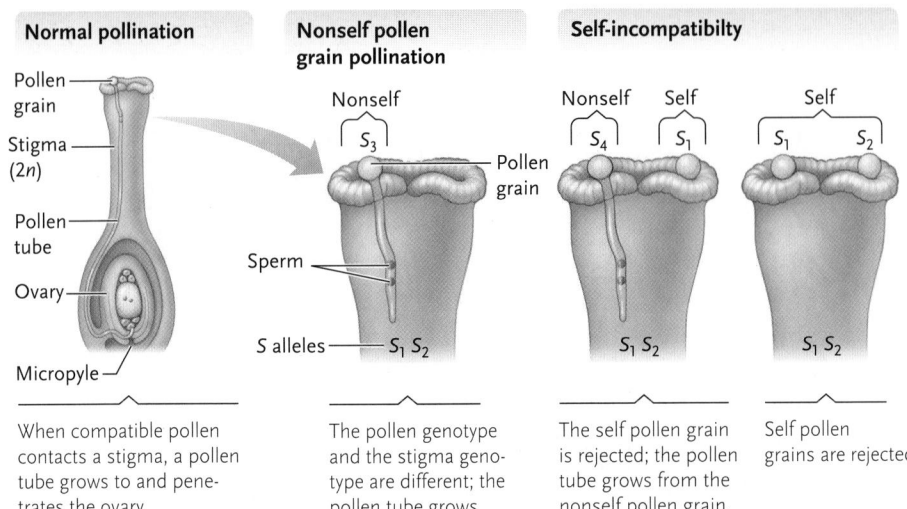

Figure 36.8

Self-incompatibility. When a pollen grain has an *S* allele that matches one in the stigma (which is diploid), the result is a biochemical response that prevents fertilization—in this illustration, by preventing the growth of a pollen tube.

Normal pollination

Pollen grain

Stigma (2*n*)

Pollen tube

Ovary

Micropyle

When compatible pollen contacts a stigma, a pollen tube grows to and penetrates the ovary.

Nonself pollen grain pollination

Nonself
S_3

Pollen grain

Sperm

S alleles — S_1 S_2

The pollen genotype and the stigma genotype are different; the pollen tube grows.

Self-incompatibilty

Nonself   Self
S_4 S_1

S_1 S_2

The self pollen grain is rejected; the pollen tube grows from the nonself pollen grain.

Self
S_1 S_2

S_1 S_2

Self pollen grains are rejected.

nutrient source for embryos because its development is tied to that of the embryo: if no embryo forms, the plant does not commit resources to endosperm. In gymnosperms, resources are committed to female gametophyte tissue even if no embryo forms. And if an angiosperm embryo is aborted, which can happen if environmental conditions become unfavourable for embryo development (e.g., in the case of drought), endosperm development also ceases, saving the plant energy and resources.

36.3c After Fertilization, Ovaries Develop into Fruits That Protect Seeds and Aid Seed Dispersal

Most angiosperm seeds are housed inside fruits, which provide protection for the developing seeds and often aid in seed dispersal. Contrary to popular assumption, the fruit does not provide any nutrients to the developing seeds. A **fruit** is a mature or ripened ovary. Usually, fruits begin to develop after ovules are fertilized, but some plants can produce fruit without fertilization, via a process called parthenocarpy. Parthenocarpy occurs naturally in some plants, and can be induced in others, to produce seedless fruits such as bananas, pineapples, and watermelons. The fruit wall (**pericarp**) develops from the ovary wall and can have several layers. Hormones in pollen grains provide the initial stimulus that turns on the genetic machinery leading to fruit development; additional signals come from hormones produced by the developing seeds.

Fruits are extremely diverse, and biologists classify them into types based on combinations of structural features. A key feature is whether the pericarp is fleshy (as in a peach) or dry (as in a hazelnut). A fruit is also classified according to the number of ovaries or flowers from which it develops. **Simple fruits**, such as peaches (*Prunus persica*) and tomatoes (*Solanum lycopersicum*), develop from a single ovary, and in many of them, at least one layer of the pericarp is fleshy and juicy

(Figure 36.9, p. 892). Other simple fruits, including grains and nuts, have a thin, dry pericarp, which may be fused to the seed coat. The garden pea (*Pisum sativum*) is a simple fruit, the peas being the seeds and the surrounding pod the pericarp. **Aggregate fruits** are formed from several ovaries in a single flower. Examples are raspberries (*Rubus* species) and strawberries, which develop from clusters of ovaries. Strawberries also qualify as accessory fruits, in which floral parts in addition to the ovary become incorporated as the fruit develops. Anatomically, the fleshy part of a strawberry is an expanded receptacle (the end of the floral shoot) and the strawberry fruits are the tiny, dry nubbins (called *achenes*) you see embedded in the fleshy tissue of each berry. **Multiple fruits** develop from several ovaries in multiple flowers. For example, a pineapple (*Ananas* species) is a multiple fruit that develops from the enlarged ovaries of several flowers clustered together in an inflorescence. Figure 36.9 shows examples of some different types of fruits.

Fruits have two functions: they protect seeds, and they aid seed dispersal in specific environments. For example, the shell of a sunflower seed is a pericarp that protects the seeds within. A pea pod is a pericarp that in nature splits open to disperse the seeds (peas) inside. Maple fruits have winglike extensions for dispersal (see Figure 36.9e). When the fruit drops, the wings cause it to spin sideways and can carry it away on a breeze. This aerodynamic property propels maple seeds to new locations, where they will not have to compete with the parent tree for water and minerals. Fruits may also have hooks, spines, hairs, or sticky surfaces that allow them to be carried to new locations when they adhere to feathers or fur of animals, or the clothing of people that brush against them. Fleshy fruits such as blueberries and cherries are nutritious food for many animals, and their seeds are adapted for surviving digestive enzymes in the animal gut. The seeds are distributed away from the parent plant in the animal's feces.

Figure 36.9

Fruits. (a) Peach (*Prunus persica*), a fleshy simple fruit. **(b)** Raspberry (*Rubus*), an aggregate fruit. **(c)** Strawberry (*Fragaria ananassa*), an accessory fruit that is also an aggregate fruit. **(d)** Pineapple (*Ananas comosus*), a multiple fruit. **(e)** Winged fruits of maple (*Acer*).

a. Peach (*Prunus*), a simple fruit

b. Raspberry (*Rubus*), an aggregate fruit

c. Strawberry (*Fragaria*), an accessory fruit

Fleshy pericarp

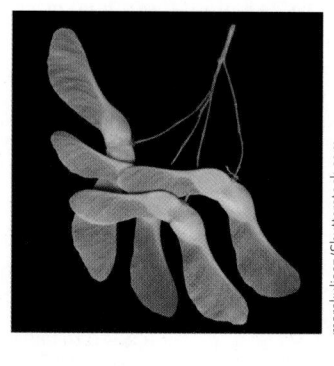

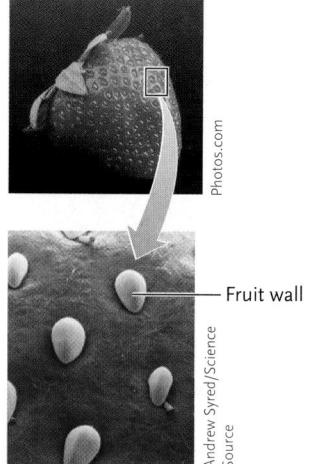

Fruit wall

d. Pineapple (*Ananus comosus*), a multiple fruit

e. Maple (*Acer*) fruit

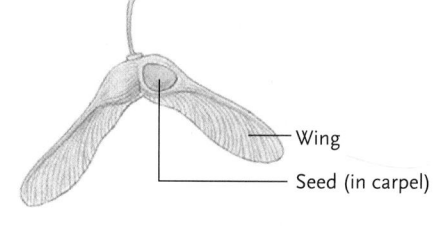

One of many individual fruits

Wing

Seed (in carpel)

36.3d The Embryonic Sporophyte Develops inside a Seed

When the zygote first forms, it starts to develop and elongate even before mitosis begins. Most of the organelles in the zygote, including the nucleus, become situated in the top half of the cell, whereas a vacuole takes up most of the lower half. The first round of mitosis divides the zygote into an upper *apical cell* and a lower *basal cell* **(Figure 36.10).** The apical cell then gives rise to the multicellular embryo, although most descendants of the basal cell form a simple row of cells, the **suspensor,** which transfers nutrients from the parent sporophyte to the embryo (see Figure 36.10).

The first apical cell divisions produce a globe-shaped structure attached to the suspensor. As they continue to grow, eudicot embryos become heart-shaped (see Figure 36.10); each lobe of the "heart" is a developing **cotyledon** (seed leaf), which provides nutrients for growing tissues. By the time the ovule is mature—that is, a fully developed seed—it has become encased by a protective **seed coat.** Inside the seed, the sheltered embryo has a lengthwise axis with

a root apical meristem at one end and a shoot apical meristem at the other.

In some eudicots, such as castor bean, endosperm is maintained as a tissue outside the embryo. In the seeds of these eudicots, the cotyledons form an interface between the rest of the embryo and the endosperm; they produce enzymes that digest the endosperm and transfer the liberated nutrients to the seedling. In other eudicot seeds, the cotyledons absorb much of the nutrient-storing endosperm and become plump and fleshy. For instance, the mature seeds of a sunflower (*Helianthus annuus*) have no endosperm at all. Monocots have one large cotyledon that acts like pea seed cotyledons; that is, it is an interface between the endosperm and the embryo, transferring nutrients to the embryo.

Figure 36.11a and **b** illustrates the structure of the seeds of two eudicots, the kidney bean (*Phaseolus vulgaris*) and the castor bean (*Ricinus communis*). The kidney bean has broad, fleshy cotyledons, whereas the castor bean has much thinner ones, but in other ways, the embryos are quite similar. The **radicle,** or embryonic root, is located near the micropyle, where the pollen tube entered the ovule before fertilization. The

Figure 36.10
Stages in the development of a eudicot embryo.

radicle attaches to the cotyledon at a region of cells called the **hypocotyl** (= below the cotyledons). Beyond the hypocotyl is the **epicotyl** (= above the cotyledons), which has the shoot apical meristem at its tip and which often bears a cluster of tiny foliage leaves, the **plumule**. At germination, when the root and shoot first elongate and emerge from the seed, the cotyledons are positioned at the first stem node, with the epicotyl above them and the hypocotyl below them.

The embryos of monocots such as corn differ structurally from those of eudicots in several ways (Figure 36.11c). In addition to having only one very large cotyledon, monocots also have protective tissues that shield the root and shoot apical meristems. The shoot apical meristem and first leaves are covered by a **coleoptile**, a sheath of cells that protects them during upward growth through the soil. A similar covering, the **coleorhiza**, sheathes the radicle until it breaks out of the seed coat and enters the soil as the primary root. The actual embryo of a corn plant is buried deep within the corn "kernel," which is technically called a *grain*. Most of the moist interior of a fresh corn grain is endosperm; the single cotyledon forms a plump mass that absorbs nutrients from the endosperm.

a. Kidney bean (*Phaseolus vulgaris*)

c. Corn (*Zea mays*)

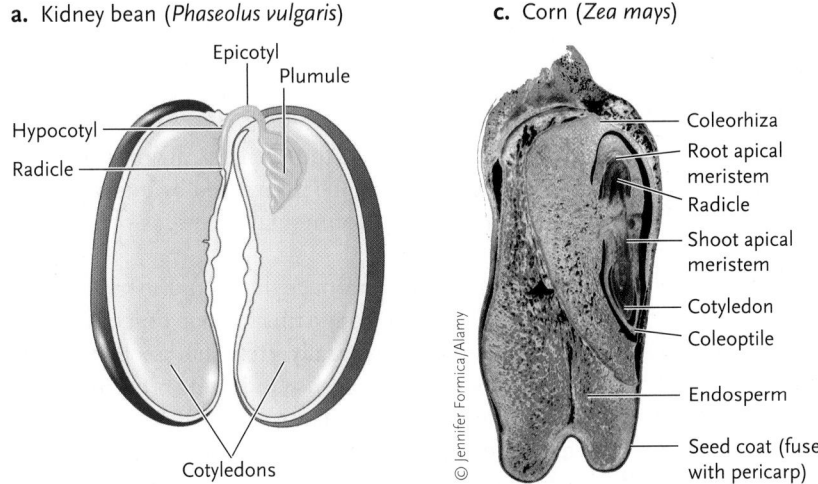

b. Castor bean (*Ricinus communis*)

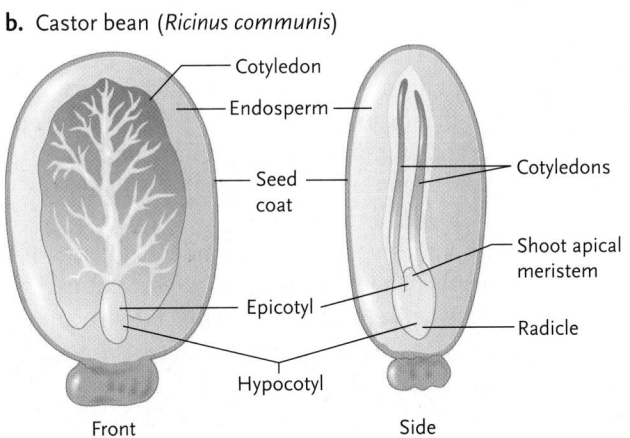

Figure 36.11
The structure of eudicot and monocot seeds. Eudicot seeds have two cotyledons, which store food absorbed from the endosperm, but the timing of this function varies in different species. **(a)** The cotyledons of a kidney bean (*Phaseolus vulgaris*) take up nutrients from endosperm while the seed develops, becoming plump and fleshy. **(b)** In the castor bean (*Ricinus communis*), the endosperm is thick and the cotyledons are thin until the seed germinates, when the cotyledons begin to take up endosperm nutrients. The drawing on the right gives a side view of the embryo. **(c)** A kernel of corn (*Zea mays*), a representative monocot seed, shown here in longitudinal section. Monocot seeds have a single cotyledon, which develops into a shield-shaped cotyledon that absorbs nutrients from endosperm.

36.3e Seed Germination Continues the Life Cycle

A mature seed is essentially dehydrated. Why is being dehydrated important? It allows the seed to stay in a state of "suspended animation." On average, only about 10% of a seed's mass is water—too little for **cell expansion** or metabolism. After a seed is dispersed and germinates, the embryo inside it becomes hydrated and resumes growth. Ideally, a seed germinates when external conditions favour the survival of the embryo and growth of the new sporophyte. This timing is important because once germination is underway, the embryo loses the protection of the seed coat and other structures that surround it. Overall, the amount of soil moisture and oxygen, the temperature, the day length, and other environmental factors influence when germination takes place.

In some species, the life cycle may include a period of seed **dormancy** (*dormire* = to sleep), in which biological activity is suspended. Plant biologists have described a striking array of variations in the conditions required for dormant seeds to germinate. For instance, seeds may require minimum periods of daylight or darkness, repeated soaking, mechanical abrasion, exposure to certain enzymes, the high heat of a fire, or a freeze–thaw cycle before they finally break dormancy. In some desert plants, hormones in the seed coat inhibit growth of a seedling until heavy rains flush the hormones away. This adaptation prevents seeds from germinating unless there is enough water in the soil to support the growth of the plant through the flowering and seed production stages before the soil dries once again. Many desert plants—and plants in harsh environments such as alpine tundra—cycle from germination to growth, flowering, and seed development in the space of a few weeks, and their offspring remain dormant as seeds until conditions once again favour germination and growth. Many seeds will not germinate until they have passed through the gut of an animal: their seed coats contain germination-inhibiting substances that are broken down by the acids and enzymes of an animal's digestive tract, allowing the seeds to germinate after they are deposited in the animal's feces.

How long can a seed remain viable? The seeds of some species appear to remain viable for amazing lengths of time: for example, 1000-year-old lotus seeds (*Nelumbo lutea*) discovered in a dry lakebed have germinated without difficulty. The record for the oldest seed to germinate is a 2000-year-old date palm seed that was germinated in 2005; as of 2010, the seedling was about 2 m tall.

Germination begins with **imbibition**, in which water moves into the seed, attracted to hydrophilic groups of stored proteins. As water enters, the seed swells, rupturing the seed coat, and the radicle begins its downward growth into the soil. Within this general framework, however, there are many variations among plants.

Once the seed coat splits, water and oxygen move more easily into the seed. Metabolism switches into high gear as cells divide and elongate to produce the seedling. Enzymes that were synthesized before dormancy become active; other enzymes are produced as the genes encoding them begin to be expressed. Among other roles, the increased gene activity and enzyme production mobilize the seed's food reserves in cotyledons or endosperm. Nutrients released by the enzymes sustain the rapidly developing seedling until its root and shoot systems are established.

The events of seed germination have been studied extensively in cereal grains, which are monocots. As a hydrating seed imbibes water, the embryo produces *gibberellin,* a hormone that stimulates the production of enzymes. Some of these enzymes digest components of endosperm cell walls; others digest the proteins, nucleic acids, and starch of the endosperm, releasing nutrient molecules for use by cells of the young root and shoot. Although it is clear that nutrient reserves are also mobilized by metabolic activity in eudicots and in gymnosperms, the details of the process are not well understood.

Inside a germinating seed, embryonic root cells are generally the first to divide and elongate, producing a radicle. When the radicle emerges from the seed coat as the primary root, germination is complete. **Figure 36.12** and **Figure 36.13, p. 896,** depict the stages of early development in a kidney bean, a eudicot, and in corn, a monocot. As the young plant grows, its development continues to be influenced by interactions of hormones and environmental factors, as you will read in Chapter 38.

Many plants produce large numbers of seeds because, in nature, only a tiny fraction of seeds survive, germinate, and eventually grow into another mature plant. Also, flowers, seeds, and fruits represent major investments of plant resources. Asexual reproduction, discussed next, is a more "economical" means by which many plants can propagate themselves.

STUDY BREAK

1. Explain the sequence of events in a flowering plant that begins with the formation of a pollen tube and culminates with the formation of a diploid zygote and the $3n$ cell that will give rise to endosperm in a seed.
2. Early angiosperm embryos undergo a series of general changes as a seed matures. Summarize this sequence, and then describe the structural differences that develop in the seeds of monocots and eudicots.

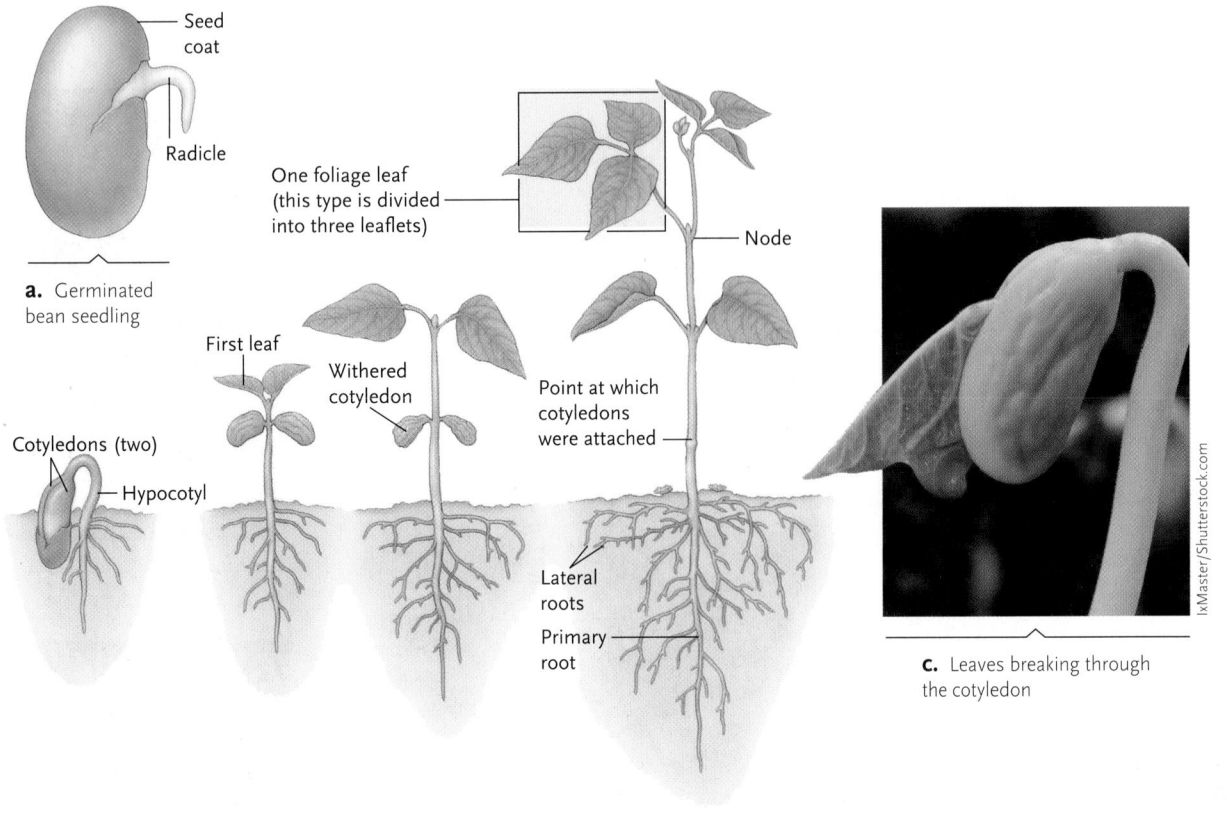

a. Germinated bean seedling

b. Food-storing cotyledons are lifted above the soil surface when cells of the hypocotyl elongate. The hypocotyl becomes hook shaped and forces a channel through the soil as it grows. At the soil surface, the hook straightens in response to light. For several days, cells of the cotyledons carry out photosynthesis; then the cotyledons wither and drop off. Photosynthesis is taken over by the first leaves that develop along the stem and later by foliage leaves.

c. Leaves breaking through the cotyledon

Figure 36.12
Stages in the development of a representative eudicot, the kidney bean (*Phaseolus vulgaris*).

36.4 Asexual Reproduction in Flowering Plants

As noted in Chapter 34, nodes in the stolons of strawberries and the rhizomes of quackgrass can each give rise to new individuals. So can *suckers* that sprout from the roots of raspberry bushes and *eyes* in the tubers of potatoes. All of these examples involve asexual or **vegetative reproduction** from a nonreproductive plant part, usually a bit of meristematic tissue in a bud on the root or stem. All of them produce offspring that are clones of the parent. Vegetative reproduction relies on an intriguing property of plants—namely, that many fully differentiated plant cells are **totipotent** (= all powerful); that is, they have the genetic potential to develop into a whole, fully functional plant, as discussed in Chapter 34. Under appropriate conditions, a totipotent cell can *dedifferentiate*: it returns to an unspecialized embryonic state, and the genetic program that guides the development of a new individual is turned on.

Some animal stem cells are also totipotent (those in the first stage of embryo development); cells from later stages of animal development are *pluripotent* (cannot grow into a whole organism but can become many different kinds of cells) or *multipotent* (can only become certain kinds of cells). In contrast, most plant cells are totipotent regardless of the stage of development.

36.4a Vegetative Reproduction Is Common in Nature

Various plant species have developed different mechanisms for reproducing asexually. In the type of vegetative reproduction called **fragmentation**, cells in a piece of the parent plant dedifferentiate and then regenerate missing plant parts. Many gardeners have discovered to their frustration that a piece of dandelion root left in the soil can rapidly grow into a new dandelion plant in this way.

When a leaf falls or is torn away from a jade plant (*Crassula* species), a new plant can develop from meristematic tissue in the detached leaf adjacent to the wound surface. In the mother of thousands plant, *Kalanchoe daigremontiana*, meristematic tissue in notches along the leaf margin gives rise to tiny plantlets **(Figure 36.14, p. 896)** that eventually fall to the ground, where they can sprout roots and grow to maturity.

Some flowering plants, including Kentucky bluegrass (*Poa pratensis*), a common lawn grass in Canada,

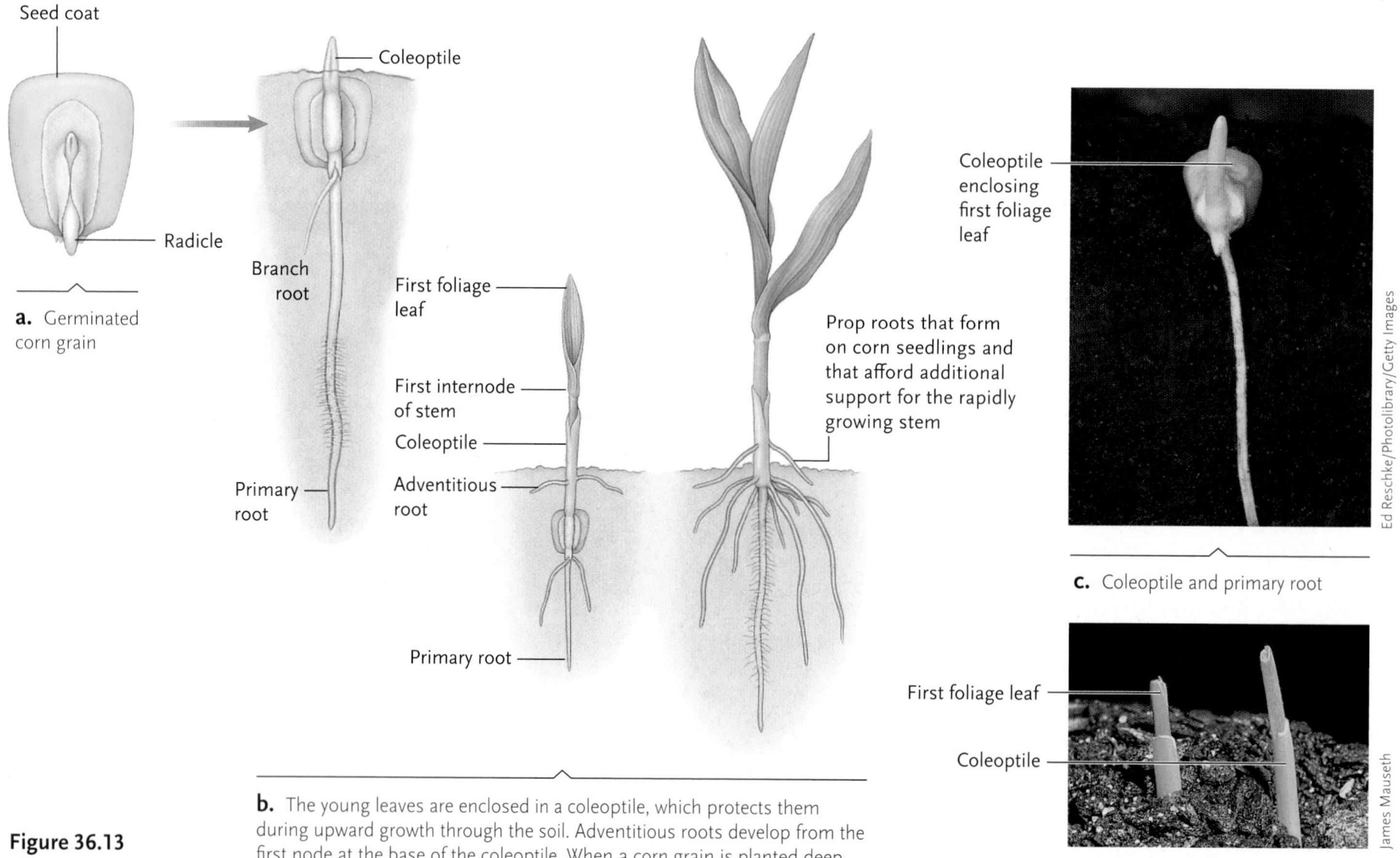

Seed coat

a. Germinated corn grain

Radicle

Coleoptile

Branch root

Primary root

First foliage leaf

First internode of stem

Coleoptile

Adventitious root

Primary root

Prop roots that form on corn seedlings and that afford additional support for the rapidly growing stem

Coleoptile enclosing first foliage leaf

c. Coleoptile and primary root

First foliage leaf

Coleoptile

d. Coleoptile and first foliage leaf of two seedlings breaking through the soil surface

Figure 36.13

Stages in the development of a representative monocot, the corn plant (*Zea mays*).

b. The young leaves are enclosed in a coleoptile, which protects them during upward growth through the soil. Adventitious roots develop from the first node at the base of the coleoptile. When a corn grain is planted deep, the first internode elongates, separating the primary and adventitious roots. When a grain is planted close to the soil surface, light inhibits elongation of the first internode and the primary and adventitious roots look as if they originate in the same region of the stem.

can reproduce asexually through a mechanism called **apomixis**. Typically, a diploid embryo develops from an unfertilized egg or from diploid cells in the ovule tissue around the embryo sac. The resulting seed is said to contain a **somatic embryo**, which is genetically identical to the parent.

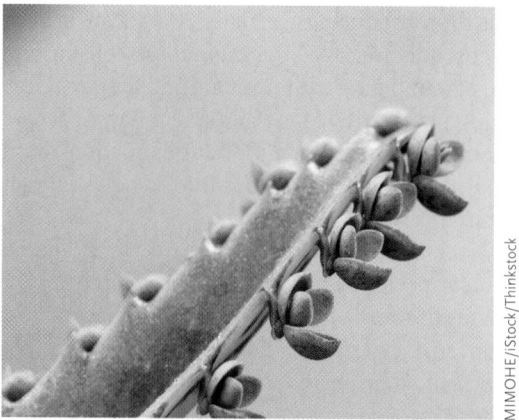

Figure 36.14

Kalanchoe daigremontiana, the mother of thousands plant. Each tiny plant growing from the leaf margin can become a new, independent adult plant.

In native plant species, most types of asexual reproduction result in offspring located near the parent. These clonal populations lack the variability provided by sexual reproduction, variation that enhances the odds for survival when environmental conditions change. Yet asexual reproduction offers an advantage in some situations. It usually requires less energy than producing complex reproductive structures such as seeds and showy flowers to attract pollinators. Moreover, clones are likely to be well suited to the environment in which the parent grows.

For centuries, gardeners and farmers have used asexual plant propagation to grow particular crops and trees and some ornamental plants. They routinely use *cuttings,* pieces of stems or leaves, to generate new plants; placed in water or moist soil, a cutting may sprout roots within days or a few weeks. Vegetative propagation can also be used to grow plants from single cells. Rose bushes and fruit trees from nurseries, and commercially important fruits and vegetables such as Bartlett pears, McIntosh apples, Thompson seedless grapes, and asparagus come from plants produced vegetatively in tissue culture conditions that cause their cells to dedifferentiate to an embryonic stage.

36.4b Plants Can Be Propagated Asexually Using Tissue Culture

Researchers have taken advantage of the totipotency of plant cells to develop plant tissue culture techniques, which allow them to produce clones of plants with desirable traits or to generate entire plants from single cells that have been genetically modified, among other goals. Plant tissue culture is simple in its general outlines **(Figure 36.15)**. Pieces of tissue are excised from a plant and grown in a nutrient medium. The procedure disrupts normal interactions between cells in the tissue, and the cells dedifferentiate and form an unorganized cell mass called a **callus**. When cultured with nutrients and growth hormones, some cells of the callus regain totipotency and develop into plantlets with roots and shoots.

Plant tissue culture is the foundation for a new field of research dealing with *somatic embryogenesis* in plants. Single cells derived from a callus generated from shoot meristem are placed in a medium containing nutrients and hormones that promote cell differentiation. With some species, totipotent cells in the sample eventually give rise to diploid somatic embryos that can be packaged with nutrients and hormones in artificial "seeds." Endowed with the same traits as their parent, crop plants grown from somatic embryos are genetically uniform.

Regardless of how it comes into being, a young sporophyte changes significantly as it begins the developmental journey toward maturity, when it will be capable of reproducing. Next we explore what researchers are learning about these developmental changes.

STUDY BREAK

1. Describe three modes of asexual reproduction that occur in flowering plants.
2. What is the major disadvantage of asexual reproduction (relative to sexual reproduction)? Are there any advantages? If so, what are they?
3. What is totipotency? How does it differ from pluripotency?
4. How do methods of tissue culture exploit the totipotency of plant cells?

36.5 Early Development of Plant Form and Function

As you learned in Chapter 34, one difference between plants and animals is that plant organs, such as leaves and flowers, may arise from meristems throughout an

---

1 Pieces of tissue are excised and cultured in a nutrient medium, under strictly controlled environmental conditions.

2 Within a few days, cells in the excised tissue dedifferentiate and form an unorganized tissue mass called a callus.

3 Individual callus cells can be separated out and cultured in a medium containing growth hormones.

4 These totipotent cells eventually give rise to plantlets with roots and shoots.

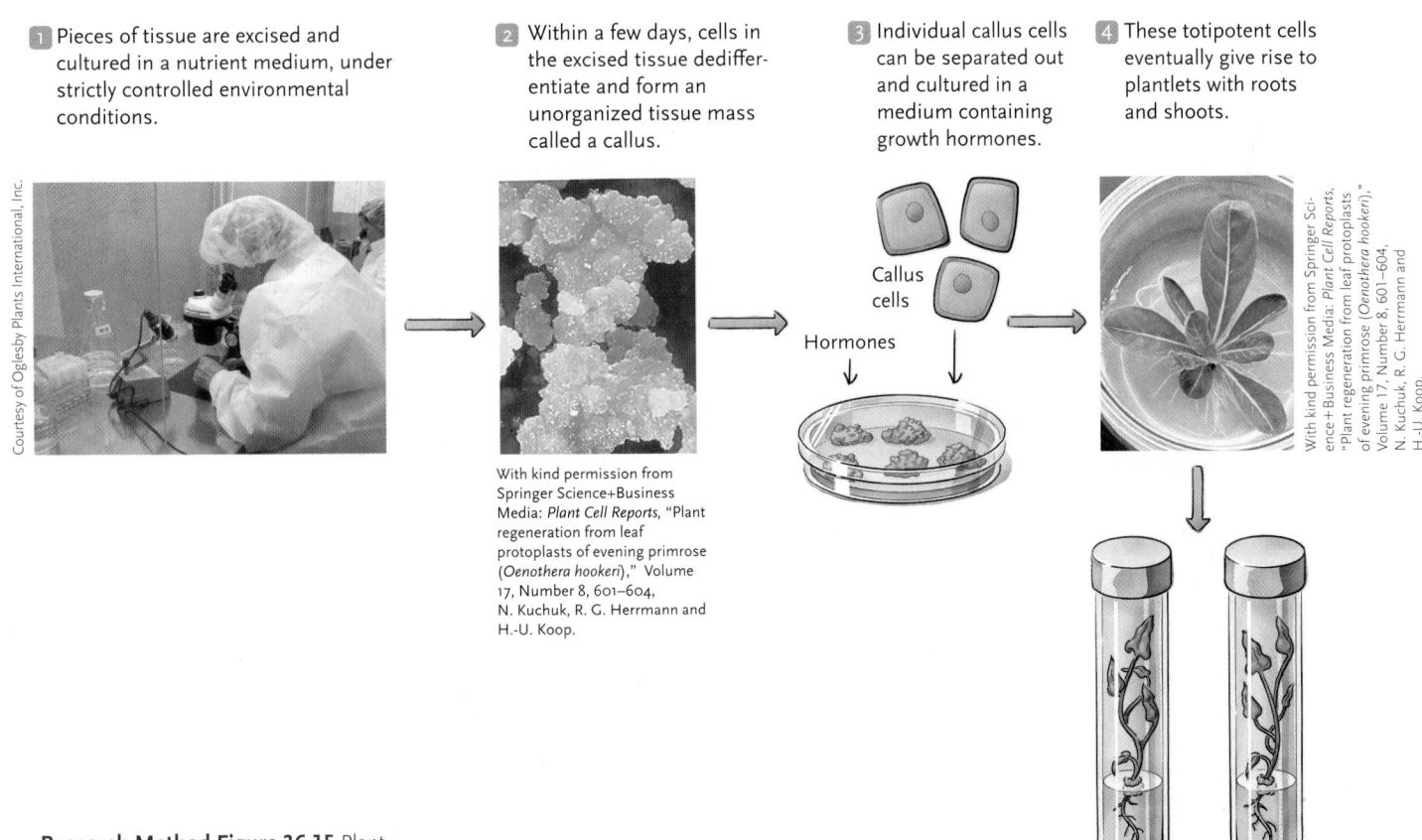

Courtesy of Oglesby Plants International, Inc.

With kind permission from Springer Science+Business Media: *Plant Cell Reports*, "Plant regeneration from leaf protoplasts of evening primrose (*Oenothera hookeri*)," Volume 17, Number 8, 601–604, N. Kuchuk, R. G. Herrmann and H.-U. Koop.

Callus cells

Hormones

With kind permission from Springer Science+Business Media: *Plant Cell Reports*, "Plant regeneration from leaf protoplasts of evening primrose (*Oenothera hookeri*)," Volume 17, Number 8, 601–604, N. Kuchuk, R. G. Herrmann and H.-U. Koop.

Research Method Figure 36.15 Plant tissue culture protocol.

individual's life, sometimes over a period of thousands of years. Accordingly, in plants, the biological role of embryonic development is not to generate the tissues and organs of the adult but to establish a basic body plan—the root–shoot axis and the radial, "outside-to-inside" organization of epidermal, ground, and vascular tissues (see Chapter 34)—and the precursors of the primary meristems. Although they may sound simple, these fundamentals and the stages beyond them all require an intricately orchestrated sequence of molecular events that plant scientists are defining through sophisticated experimentation.

One of the most fruitful experimental approaches has been the study of plants with natural or induced gene mutations that block or otherwise affect steps in development and thus lend insight into the developmental roles of the normal, wild-type versions of these abnormal genes. Although researchers work with various species to probe the genetic underpinnings of early plant development, the thale cress (*Arabidopsis thaliana*) has become an important model organism for plant genetic research.

The entire *Arabidopsis* genome has been sequenced, providing a powerful molecular database for determining how various genes contribute to shaping the plant body. Experimenters' ability to trace the expression of specific genes has shed considerable light on how the root–shoot axis is set and how the three basic plant tissue systems arise.

36.5a Within Hours, an Early Plant Embryo's Basic Body Plan Is Established

What determines which part of the embryo will be the root and which part will be the shoot? Studies done with *Arabidopsis* have revealed that the first, asymmetrical division of the zygote, in which the apical cell receives the majority of the zygote's cytoplasm, whereas the basal cell receives the zygote's large vacuole and less cytoplasm, results in the two daughter cells

receiving very different mixes of mRNAs. This means that the apical and basal cells will produce very different proteins; of particular interest are the different *transcription factors* (proteins that regulate transcription) that will be produced in the two daughter cells. With different transcription factors triggering the expression of different genes in the apical and basal cells, distinct biochemical pathways unfold in the two cells, which set in motion steps leading to the differentiation of root and shoot systems.

As development proceeds, cells at different sites become specialized in prescribed ways as a particular set of genes is expressed in each type of cell—a process known as *differentiation*. Differentiated cells, in turn, are the foundation of specialized tissues and organs. We are starting to unravel how plants regulate differentiation, but much of this topic is beyond the scope of this chapter.

In nature, genes that govern plant development switch on or off in response to changing environmental conditions. Their signals determine the course of a plant's vegetative growth throughout its life. In many perennials, new leaves begin to develop inside buds in autumn and then become dormant until the following spring, when external conditions favour further growth. Environmental cues stimulate the gene-guided production of hormones that travel through the plant in xylem and phloem, triggering renewed leaf growth and expansion. Leaves and other shoot parts also age, wither, and fall away from the plant as hormonal signals change. The far-reaching effects of plant hormones on growth and development are the subject of Chapter 38.

STUDY BREAK

1. What does the term *differentiation* mean?
2. What event sets the stage for differentiation of a plant into root and shoot systems?

Review

36.1 Overview of Flowering Plant Reproduction

- In most flowering plant life cycles, a multicellular diploid stage, the sporophyte (spore-producing plant), alternates with a multicellular haploid stage, the gametophyte (gamete-producing plant) (Figure 36.3). The sporophyte develops roots; stems; leaves; and, at some point, flowers. The separation of a life cycle into diploid and haploid stages is called alternation of generations.

36.2 Flower Structure and Formation of Gametes

- A flower develops at the tip of a floral shoot. It can have up to four whorls supported by the receptacle. The outermost two whorls consist of the sepals and petals. The third whorl consists of stamens, and carpels make up the innermost whorl (Figure 36.4).

- Flowers can contain both stamens and carpels, or they may contain only male or only female sex organs. Monoecious species have separate male and female flowers on the same plant; in dioecious species, the male and female flowers are on different plants (Figure 36.5).

- The anther of each stamen contains sacs where pollen grains develop. If compatible pollen lands on the stigma, the receptive surface of the carpel, it produces a pollen tube that grows down the style to the ovary, where ovules are formed (Figure 36.6). Eggs are produced by female gametophytes inside ovules.
- In pollen sacs, meiosis produces haploid microspores. Mitosis inside each microspore produces a pollen grain, an immature male gametophyte that consists of a generative cell and a vegetative (or tube) cell. The generative cell later divides to produce two sperm cells, the male gametes of flowering plants. The tube cell produces the pollen tube (Figure 36.7).
- In the ovule, four haploid megaspores form following meiosis; usually all but one disintegrate. The remaining megaspore undergoes mitosis three times without cytokinesis, producing eight nuclei in a single large cell. Two of these, called polar nuclei, migrate to the centre of the cell. When cytokinesis occurs, cell walls form around the nuclei, with the two polar nuclei enclosed in a single wall. The result is the seven-celled embryo sac, one cell of which is the haploid egg. The cell with two polar nuclei will help give rise to endosperm.

36.3 Pollination, Fertilization, and Germination

- Upon pollination, the pollen grain resumes growth. A pollen tube develops from the vegetative (tube) cell; mitosis of the male gametophyte's generative cell produces two sperm cells.
- In double fertilization, one sperm fuses with one egg nucleus to form a diploid ($2n$) zygote. The other sperm and the two polar nuclei of the remaining cell also fuse, forming a triploid ($3n$) nucleus that divides to produce endosperm in the seed.
- After the endosperm forms, the ovule expands, and the embryonic sporophyte develops. A mature ovule is a seed and is encased by a protective seed coat. Inside the seed, the embryo has a lengthwise axis with a root apical meristem at one end and a shoot apical meristem at the other.

- Eudicot embryos have two cotyledons. The embryonic shoot consists of an upper epicotyl and a lower hypocotyl; also present is an embryonic root, the radicle. The single cotyledon of a monocot absorbs nutrients from endosperm. The root and shoot apical meristems of a monocot embryo are protected by a coleoptile over the shoot tip and a coleorhiza over the radicle.
- A fruit is a matured or ripened ovary. Fruits protect seeds and disperse them by animals, wind, or water.
- Fruits are simple, aggregate, or multiple, depending on the number of flowers or ovaries from which they develop (Figure 36.9). Fruits also vary in the characteristics of their pericarp, which surrounds the seed.
- The seeds of most plants remain dormant until external conditions—moisture, oxygen, temperature, number of daylight hours, and other aspects—favour the survival of the embryo and the development of a new sporophyte.

36.4 Asexual Reproduction in Flowering Plants

- Many flowering plants also reproduce asexually, as when new plants arise by mitotic divisions at nodes or buds along modified stems of the parent plant. New plants may also arise by vegetative propagation, either natural or induced.
- Tissue culture methods for developing new plants from a parent plant's somatic (nonreproductive) cells include somatic embryogenesis (Figure 36.15).

36.5 Early Development of Plant Form and Function

- In plants that reproduce sexually, development starts at fertilization. In a sequence of gene-guided processes, a new embryo acquires its root–shoot axis, and cells in different regions begin to differentiate, becoming specialized for particular functions.

Questions

Self-Test Questions

1. In angiosperms, an egg cell is produced in which structure and by which process?
   a. embryo sac; meiosis
   b. embryo sac; mitosis
   c. central cell; mitosis
   d. female gametophyte; meiosis

2. What cells are produced when the microspore of an angiosperm divides?
   a. two generative cells
   b. two vegetative (tube) cells
   c. a generative cell and a tube cell
   d. a sperm cell, a generative cell, and a tube cell

3. Which statement best describes double fertilization?
   a. Two sperm simultaneously fertilize two eggs in two separate ovules.
   b. One sperm fertilizes the egg, and a second sperm fertilizes the polar nuclei.

   c. One microspore becomes a pollen grain; the other microspore becomes a sperm-producing cell.
   d. One sperm fertilizes the egg, and a second sperm fertilizes a synergid, forming endosperm.

4. Which of the following happens in angiosperm reproduction?
   a. Ovaries become fruit.
   b. Megaspores become eggs.
   c. The tube cell produces two sperm cells.
   d. Motile sperm swim down a pollen tube.

5. What is the term for the transfer of a pollen grain from an anther to a stigma?
   a. fertilization
   b. pollination
   c. germination
   d. sexual reproduction

6. Which of the following best describes a seed?
   a. endosperm
   b. mature ovary
   c. mature ovule
   d. mature megaspore

7. How have fruit contributed to the success of angiosperms?
   a. They nourish the plant that makes them.
   b. They attract insects to the pollen inside.
   c. They nourish the developing seedling.
   d. They facilitate seed dispersal by animals.

8. What does a primary root develop from?
   a. a radicle
   b. an epicotyl
   c. a hypocotyl
   d. a coleoptile

9. Which of the following accurately describe why a seed coat ruptures?
   a. emergence of the cotyledons
   b. activation of enzymes in the endosperm
   c. the entry of water, by imbibition, into the seed
   d. a sudden increase in cell division in cotyledons

10. Which of the following is a correct description of the offspring produced by vegetative reproduction in angiosperms?
    a. haploid
    b. tetraploid
    c. genetically identical
    d. genetically inferior to those produced by sexual reproduction

Questions for Discussion

1. Explain the differences among a carpel, an ovary, and an ovule, using a labelled diagram. Be sure to clarify the physical relationship among them (which is enclosed by which), and identify which specific structure is responsible for producing the female gametophyte. To which generation (gametophyte or sporophyte) does each of these structures belong?

2. Using labelled diagrams, explain the relationship among the following structures: (a) a megasporangium, a megaspore, and a female gametophyte of an angiosperm, and (b) a microsporangium, a microspore, and a male gametophyte of an angiosperm.

3. A germinating pea seedling kept in a dark closet will actually grow several centimetres high before dying. What nutrient supply sustains this growth?

4. A plant physiologist has succeeded in cloning a gene for pest resistance into petunia cells. How can she use tissue culture to propagate a large number of petunia plants with the gene?

5. In grocery stores, displays of fruits and vegetables are separated according to typical uses for these plant foods. For instance, bell peppers, cucumbers, tomatoes, and eggplants are in the vegetable section, whereas apples, pears, and peaches are displayed with other fruits. How does this practice relate to the biological definition of a fruit?

Lush azaleas (*Rhododendron*) and a stately Southern live oak (*Quercus virginiana*) draped with the unusual flowering plant called Spanish moss (*Tillandsia usneoides*). The roots of shrubs, trees, and most other plants take up water and minerals from soil, but Spanish moss is an epiphyte—it lives independently on other plants and obtains nutrients by way of absorptive hairs on its leaves and stems.

Plant Nutrition

WHY IT MATTERS

Tropical rain forests are remarkable for many reasons, but for biologists the key one may be that they are among the most biologically diverse ecosystems on Earth. In addition to containing countless thousands of species of animals, fungi, protists, and prokaryotes, these lush domains are dense with broadleaved evergreen trees; sinuous vines; and other vegetation. With rain a near-daily event, it may not seem surprising that the trees' foliage is a deep, luxuriant green (**Figure 37.1**).

Yet tropical rain forests are demanding places for plants to survive, in large part because the soil is chronically deficient in nutrients necessary for plant metabolism. This nutrient scarcity is a direct outcome of the incessant rain and the high acidity of tropical rain forest soil. There is ample moisture in the upper layer of soil, but in acidic soil, minerals vital to plant metabolism, such as potassium, calcium, magnesium, and phosphorus, are subject to **leaching**—being washed into deeper soil levels that are not as accessible to plant roots. In addition, in the warm, moist environment of a tropical rain forest, bacteria and fungi speedily decompose fallen leaves and other organic remains. Just as rapidly, established trees and vines take up any nutrients these decomposers have released, leaving few or none to enrich the soil. As falling rain dissolves some atmospheric CO_2, it creates carbonic acid—a type of "acid rain"—which exacerbates the leaching problem even more.

Such poor soil and the near perpetual twilight at the forest floor make it extremely difficult for small shrubs and herbaceous plants to survive there. Nearly all such plants climb upward as vines, using the tree trunks for mechanical support, or they live attached to the upper branches of taller species, where they can absorb needed minerals from falling dust or from the surfaces of other plants. These intricate adaptations to a particular environment allow the plants to secure energy and raw materials and to use both for growth and development.

Figure 37.1
A lush tropical rain forest
in Southeast Asia.

InnervisionArt/Shutterstock.com

Tropical rain forests are not unique in posing nutritional challenges for plants. In fact, plants rarely have ready access to a full complement of necessary resources. In a rain forest, the carbon, hydrogen, and oxygen that plants need for photosynthesis are relatively easy to come by: plants there usually get enough carbon from the CO_2 in air, and their roots can take up enough water to gain the necessary hydrogen and oxygen. But soils in other environments are frequently dry, making water a limited resource, and almost nowhere in nature do soils hold lavish amounts of dissolved minerals such as nitrogen, calcium, and others that are vital for a plant's survival. In response to the challenge of obtaining nutrients, plants have evolved the range of structural and physiological adaptations that we consider in this chapter.

37.1 Plant Nutritional Requirements

No organism grows normally when deprived of a chemical element essential for its metabolism. In the latter half of the nineteenth century, plant physiologists exploited rapid advances in chemistry to probe both the chemical composition of plants and the essential nutrients plants need to survive. In recent times, researchers have brought to bear sophisticated methods to expand our understanding of the range of plant nutrients, including those required only in trace amounts.

37.1a Plants Require Macronutrients and Micronutrients for Their Metabolism

By weight, the tissues of most plants are more than 90% water. Early researchers could obtain a rough idea of the composition of a plant's dry weight by burning the plant and then analyzing the ash. This method typically yielded a long list of elements, but the results were flawed. Chemical reactions during burning can dissipate quantities of some important elements, such as nitrogen. Also, plants take up a variety of ions that they do not use; depending on the minerals present in the soil where a plant grows, a plant's tissues can contain nonnutritive elements such as gold, lead, arsenic, and uranium.

Studying Plant Nutrition Using Hydroponics. In 1860, German plant physiologist Julius von Sachs pioneered an experimental method for identifying the minerals absorbed into plant tissues that are essential for plant growth. Sachs carefully measured amounts of compounds containing specific minerals and mixed them in different combinations with pure water. He then grew plants in the solutions, a method now called **hydroponic culture** (*hydro* = water; *ponos* = work). By eliminating one element at a time and observing the results, Sachs deduced a list of six essential plant nutrients, in descending order of the amount required: nitrogen, potassium, calcium, magnesium, phosphorus, and sulfur.

Sachs's innovative research paved the way for decades of increasingly sophisticated studies of plant nutrition, and the eventual identification of many more essential plant nutrients. In the spirit of his work, one basic experimental method involves growing a plant in a solution containing a complete spectrum of known and possible essential nutrients **(Figure 37.2a)**. The healthy plant is then transferred to a solution that is identical, except that it lacks one element having an unknown nutritional role (Figure 37.2b). Abnormal growth of the plant in this solution is evidence that the missing element is essential. If the plant grows normally, the missing element may not be essential; however, only further experimentation can confirm this hypothesis.

In a typical, modern hydroponic apparatus, the nutrient solution is refreshed regularly, and air is bubbled into it to supply oxygen to the roots. Without sufficient oxygen for respiration, the plants' roots do not absorb nutrients efficiently. (The same effect occurs in poorly aerated soil.) Variations of this technique are used on a commercial scale to grow some vegetables, such as lettuce and tomatoes.

Essential Macronutrients and Micronutrients. Hydroponics research has revealed that plants generally require 17 essential elements **(Table 37.1, p. 904).** By definition, an **essential element** is necessary for normal growth and reproduction, cannot be functionally replaced by a different element, and has one or more roles in plant metabolism. With enough sunlight and the 17 essential elements, plants can synthesize all the compounds they need.

Nine of the essential elements are **macronutrients**, meaning that plants incorporate relatively large amounts of them into their tissues. Three of these elements—carbon, hydrogen, and oxygen—account for about 96% of a plant's dry mass. Together, these three elements are the key components of lipids and of carbohydrates such as cellulose; with the addition of nitrogen, they form the basic building blocks of proteins and nucleic acids. Plants also use phosphorus in constructing nucleic acids, ATP, and phospholipids, and they use potassium for functions ranging from enzyme activation to mechanisms that control the opening and closing of stomata. Rounding out the list of macronutrients are calcium, sulfur, and magnesium. All macronutrients except carbon, hydrogen, and oxygen are classified as minerals, which chemists usually define as elements or compounds with a crystalline structure that are formed by geological processes. Minerals are available to plants through the soil as ions dissolved in water, and most minerals that serve as nutrients in plants are derived from the weathering of rocks and inorganic particles in the Earth's crust.

The other elements essential to plants are also minerals, and are classed as **micronutrients** because plants require them only in trace amounts. Nevertheless, they are just as vital as macronutrients to a plant's health and survival. For example, 5 t of potatoes contains roughly the amount of copper in a single copper-plated penny—yet without it, potato plants are sickly and do not produce normal tubers.

PURPOSE: In studies of plant nutritional requirements, using hydroponic culture allows a researcher to manipulate and precisely define the types and amounts of specific nutrients that are available to test plants.

PROTOCOL: In a typical hydroponic apparatus, many plants are grown in a single solution containing pure water and a defined mix of mineral nutrients. The solution is replaced or refreshed as needed and is aerated with a bubbling system.

a. Basic components of a hydroponic apparatus

b. Procedure for identifying elements essential for proper plant nutrition

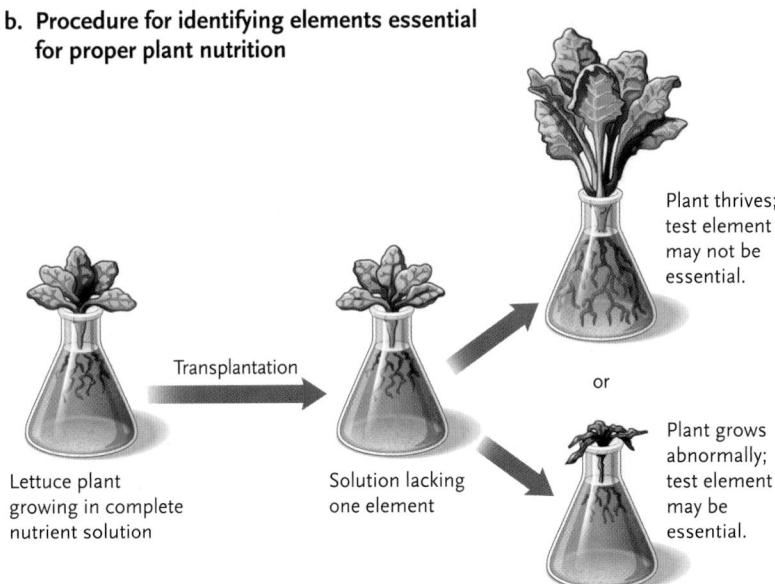

A "complete" solution contains all the known and suspected essential plant nutrients. An "incomplete" solution contains all but one of the same nutrients, in the same amounts. For experiments, researchers first grow plants in a complete solution, then transplant some of the plants to an incomplete solution.

INTERPRETING THE RESULTS: Normal growth of test plants suggests that the missing nutrient is not essential, whereas abnormal growth is evidence that the missing nutrient may be essential.

Research Method Figure 37.2 Hydroponic Culture.

Chlorine, generally present in soil in its anionic form Cl^- (chloride), was identified as a micronutrient nearly a century after Sachs's experiments. Chloride functions in some reactions of photosynthesis and (along with K^+) in the opening and closing of stomata, among other roles. The researchers who discovered its importance in plant nutrition performed hydroponic culture experiments in a California laboratory near the Pacific Ocean, where the air, like coastal air everywhere, contains

Table 37.1

Essential Plant Nutrients and Their Functions

Element	Commonly Absorbed Forms	Some Known Functions	Some Deficiency Symptoms
Macronutrients			
Carbon*	CO_2	Raw materials for photosynthesis	Rarely deficient
Hydrogen*	H_2O		No symptoms; available from water
Oxygen*	O_2, H_2O, CO_2		No symptoms; available from water and CO_2
Nitrogen	NO_3^-, NH_4^+	Component of proteins, nucleic acids, coenzymes, chlorophylls	Stunted growth; light-green newer leaves; older leaves yellow and die (chlorosis)
Phosphorus	$H_2PO_4^-$, HPO_4^{2-}	Component of nucleic acids, phospholipids, ATP, several coenzymes	Purplish veins; stunted growth; fewer seeds, fruits
Potassium	K^+	Activation of enzymes; key role in maintaining water–solute balance and so influences osmosis	Reduced growth; curled, mottled, or spotted older leaves; burned leaf edges; weakened plant
Calcium	Ca^{2+}	Roles in formation and maintenance of cell walls and in membrane permeability; enzyme cofactor	Leaves deformed; terminal buds die; poor root growth
Sulfur	SO_4^{2-}	Component of most proteins, coenzyme A	Light-green or yellowed leaves; reduced growth
Magnesium	Mg^{2+}	Component of chlorophyll; activation of enzymes	Chlorosis; drooping leaves
Micronutrients			
Chlorine	Cl^-	Role in root and shoot growth and in photosynthesis	Wilting; chlorosis; some leaves die (deficiency not seen in nature)
Iron	Fe^{2+}, Fe^{3+}	Roles in chlorophyll synthesis, electron transport; component of cytochrome	Chlorosis; yellow and green striping in grasses
Boron	H^3BO^3	Roles in germination, flowering, fruiting, cell division, nitrogen metabolism	Terminal buds, lateral branches die; leaves thicken, curl, and become brittle
Manganese	Mn^{2+}	Role in chlorophyll synthesis; coenzyme action	Dark veins, but leaves whiten and fall off
Zinc	Zn^{2+}	Role in formation of auxin, chloroplasts, and starch; enzyme component	Chlorosis; mottled or bronzed leaves; abnormal roots
Copper	Cu^+, Cu^{2+}	Component of several enzymes	Chlorosis; dead spots in leaves; stunted growth
Molybdenum	MoO_4^{2-}	Component of enzyme used in nitrogen metabolism	Pale green, rolled or cupped leaves
Nickel	Ni^{2+}	Component of enzyme required to break down urea generated during nitrogen metabolism	Dead spots on leaf tips (deficiency not seen in nature)

*Carbon, hydrogen, and oxygen are the nonmineral plant nutrients. All others are minerals.

sodium chloride. The investigators found that their test plants could obtain tiny but sufficient quantities of chloride from the air, as well as from sweat (which also contains NaCl) on the researchers' own hands. Great care had to be taken to exclude chlorine from the test plants' growing environment to prove that it was essential.

In some cases, plant seeds contain enough of certain trace minerals to sustain the adult plant. For example, nickel (Ni^{2+}) is a component of urease, the enzyme required to hydrolyze urea. Urea is a toxic by-product of the breakdown of nitrogenous compounds, and it will kill cells if it accumulates. In the late 1980s, investigators found that barley seeds contain enough nickel to sustain two complete generations of barley plants. Plants grown in the absence of nickel did not begin to show signs of nickel deficiency until the third generation.

Besides the 17 essential elements, some species of plants may require additional micronutrients. Experiments suggest that many, perhaps most, plants adapted to hot, dry conditions require sodium; many

plants that photosynthesize by the C_4 pathway (see Chapter 7) appear to be in this group. A few plant species require selenium, which is also an essential micronutrient for animals. Horsetails (*Equisetum*) require silicon, and some grasses (such as wheat) may also need it. Scientists continue to discover additional micronutrients for specific plant groups.

Both micronutrients and macronutrients play vital roles in plant metabolism. Many function as cofactors or coenzymes in protein synthesis, starch synthesis, photosynthesis, and aerobic respiration. Some also have a role in creating solute concentration gradients across plasma membranes, which are responsible for the osmotic movement of water.

37.1b Nutrient Deficiencies Cause Abnormalities in Plant Structure and Function

Plants differ in the quantity of each nutrient they require—the amount of an essential element that is adequate for one plant species may be insufficient for another. Lettuce and other leafy plants require more nitrogen and magnesium than do other plant types, for example, and alfalfa requires significantly more potassium than do lawn grasses. An adequate amount of an essential element for one plant may even be harmful to another. For example, the amount of boron required for normal growth of sugar beets is toxic for soybeans. For these reasons, the nutrient content of soils is an important factor in determining which plants will grow well in a given location.

Plants that are deficient in one or more of the essential elements develop characteristic symptoms, (Table 37.1 lists some observable symptoms of nutrient deficiencies.) The symptoms give some indication of the metabolic roles the missing elements play. Deficiency symptoms typically include stunted growth, abnormal leaf color, dead spots on leaves, or abnormally formed stems **(Figure 37.3).** For instance, iron is a component of the cytochromes on which the cellular electron transfer system depends, and it plays a role in reactions that synthesize chlorophyll. Iron deficiency causes **chlorosis**, a yellowing of plant tissues that results from a lack of chlorophyll (see Figure 37.3b). Because ionic iron (Fe^{3+}) is relatively insoluble in water, gardeners often fertilize plants with a soluble iron compound called chelated iron to stave off or cure chlorosis. Similarly, because magnesium is a necessary component of chlorophyll, a plant deficient in this element has fewer chloroplasts than normal in its leaves and other photosynthetic parts. It appears paler green than normal, and its growth is stunted because of reduced photosynthesis (see Figure 37.3c).

Plants that lack adequate nitrogen may also become chlorotic (see Figure 37.3d), with older leaves yellowing first because the nitrogen is preferentially shunted to younger, actively growing plant parts. This adaptation is not surprising, given nitrogen's central role in the synthesis of amino acids, chlorophylls, and other compounds vital to plant metabolism. With some other mineral deficiencies, young leaves are the first to show symptoms. These kinds of observations underscore the point that plants use different nutrients in specific, often metabolically complex ways.

Soils are more likely to be deficient in nitrogen, phosphorus, potassium, or some other essential mineral than to contain too much, and farmers and gardeners typically add nutrients to suit the types of plants they wish to cultivate. They may observe the deficiency symptoms of plants grown in their locale or have soil tested in a laboratory and then choose a fertilizer with the appropriate balance of nutrients to

a. Plant grown using a complete growth solution

b. Iron deficiency

c. Magnesium deficiency

d. Nitrogen deficiency

e. Phosphorus deficiency

f. Potassium deficiency

g. Sulfur deficiency

h. Calcium deficiency

Photos by E. Epstein, University of California, Davis

Research Method Figure 37.3 Leaves and stems of tomato plants showing visual symptoms of seven different mineral deficiencies. The plants were grown in the laboratory, where the experimenter could control which nutrients were available.

compensate for the deficiencies. Packages of commercial fertilizers use a numerical shorthand (for example, 15-30-15) to indicate the percentages of nitrogen, phosphorus, and potassium they contain.

STUDY BREAK

1. What are the two main categories of the essential elements plants need? Give several examples of each.
2. Do all plants require the same basic nutrients in the same amounts? Explain.

37.2 Soil

Soil anchors plant roots and is the main source of the inorganic nutrients plants require. It is also the source of water for most plants and of oxygen for respiration in root cells. The physical texture of soil is a factor in whether root systems have access to sufficient water and dissolved oxygen. Together, physical and chemical properties of soils have a major impact on the ability of plants to grow, survive, and reproduce in particular habitats.

37.2a The Components of a Soil and the Size of the Particles Determine Its Properties

Soil is a complex mix of mineral particles, chemical compounds, ions, decomposing organic matter, air, water, and assorted living organisms. Most soils develop from the physical or chemical weathering of rock (which also liberates mineral ions). The different kinds of soil particles range in size from sand (2.0–0.02 mm) to silt (0.02–0.002 mm) and clay (diameter less than 0.002 mm). These mineral particles are usually mixed with various organic components, including **humus**—decomposing parts of plants and animals, animal droppings, and other organic matter. Dry humus has a loose, crumbly texture. It can absorb a great deal of water, contributing to the capacity of soil to hold water. Organic molecules in humus are reservoirs of nutrients, including nitrogen, phosphorus, and sulfur, that are vital to living plants.

The relative proportions of the different sizes of mineral particles give soil its basic texture—gritty if the soil is largely sand, smooth if silt predominates, and dense and heavy if clay is the major component. A soil's texture in turn helps determine the number and volume of pores—air spaces—that it contains. The relative amounts of sand, silt, and clay determine whether a soil is sticky when wet, with few air spaces (mostly clay), or dries quickly and may wash or blow away (mostly sand). Clay soils are more than 30% clay, whereas sandy soils contain less than 20% clay or silt.

The piles of bagged humus for sale at garden centres each spring reflect the fact that the amount of humus in a soil also affects plant growth. Its plentiful organic material feeds decomposers, whose metabolic activities in turn release minerals that plant roots can take up, but that is not its only value in soil. Humus helps retain soil water and, with its loose texture, helps aerate soil as well. Well-aerated soils containing roughly equal proportions of humus, sand, silt, and clay are **loams**, and they are the soils in which most plants do best.

37.2b In Turn, Plants and Other Organisms Influence Soil Features

A square metre of fertile soil contains trillions of bacteria, hundreds of millions of fungi, and several million nematodes, plus an array of other worms and insects. It also contains dead plant roots, leaves, and other parts. Bacteria and fungi decompose this and other organic matter on and in the soil, and burrowing creatures such as earthworms aerate the soil. The roots and other tissues of plants may also play a key role in shaping the characteristics and composition of soil, including the abundance of soil-dwelling organisms.

Experiments document these soil-shaping activities. For example, Edward Ayres and his colleagues at Colorado State University's Natural Resource Ecology Laboratory studied soil properties in Colorado's San Juan Mountains, where stands of trembling aspen (*Populus tremuloides*), lodgepole pine (*Pinus contorta*), or Engelmann spruce (*Picea engelmannii*) live in close proximity. P. tremuloides trees have a more open growth form than pines and spruce trees do, all their leaves drop each year in autumn, and previous research had shown that P. tremuloides leaf litter has about twice the nitrogen content of the other two species. With these facts in mind, the Ayres team hypothesized that in their four study areas, such species-specific characteristics would influence the physical, chemical, and biological properties of the soil. The data they gathered supported parts of their hypothesis and also raised questions. For example, they found that in all study areas, the soil in which the aspens grew was significantly warmer—a difference that the team attributed to increased sunlight reaching the ground through the relatively open aspen canopies. The soil littered with aspen leaves also contained more nitrate—a form of nitrogen that plant roots can readily take up—than the nearby soil where the lignin-rich needlelike leaves of pines and spruces accumulated. The study was not designed to attempt a comprehensive analysis of the diversity of the soil's bacterial, fungal, and microscopic animal communities, but the researchers did document markedly different arrays of soil-dwelling organisms associated with the aspens, pines, and spruces in each study area. Clearly, we have a lot more to learn about the intricate interactions between plants, soils, and communities of soil organisms.

As soils develop naturally, they tend to take on a characteristic vertical profile, with a series of layers or **horizons (Figure 37.4)**. Each horizon has a distinct texture and composition that varies with soil type. The top layer of surface litter—organic matter like twigs and leaves, animal dung, and fungi—is accordingly called the *O horizon*. **Topsoil**, the most fertile layer, occurs just below and forms the *A horizon*. This fairly loose layer may be less than a centimetre deep on steep slopes to more than a metre deep in grasslands. It consists of humus mixed with mineral particles and is where the roots of most herbaceous plants are located. Below the topsoil is the **subsoil** or *B horizon*, a layer of larger soil particles containing relatively little organic matter. Mineral ions, including those that serve as nutrients in plants, tend to accumulate in the *B horizon*, and mature tree roots generally extend down into this layer. Under it is the *C horizon*, a layer of mineral particles and rock fragments that extends down to bedrock.

Regions where the topsoil is naturally deep and rich in humus are ideal for agriculture. Prime examples are the vast former grasslands of the North American Midwest and Ukraine now converted to fields of corn, wheat, soybeans, and other crops. Modern cultivation practices that sharply reduce erosion have helped maintain this soil resource. In arid regions, chronic, sparse rainfall generally correlates with low natural soil humus, and agriculture is only possible with intensive irrigation and soil management. Nor can agriculture flourish for long on land cleared of a tropical rain forest, because of the soil leaching and lack of nutrients described in the chapter introduction.

O horizon
Fallen leaves and other organic material littering the surface of mineral soil

A horizon
Topsoil, which contains some percentage of decomposed organic material and which is of variable depth; here it extends about 30 cm below the soil surface

B horizon
Subsoil; larger soil particles than the A horizon, not much organic material, but greater accumulation of minerals; here it extends about 60 cm below the A horizon

C horizon
No organic material, but partially weathered fragments and grains of rock from which soil forms; extends to underlying bedrock

Bedrock

© William Ferguson

Figure 37.4
Soil horizons in a grassland.

37.2c The Characteristics of Soil Affect Root–Soil Interactions

Roots are superbly adapted to penetrate soil and extract needed nutrients from it, but they are also quite sensitive to variations in the properties of soil. In the following section, we consider some adaptations plants have evolved in many otherwise inhospitable soil environments. First, however, we consider the general ways in which soil composition influences the ability of plant roots to obtain water and minerals.

Water Availability. As water flows into and through soil, gravity pulls much of the water down through the spaces between soil particles into deeper soil layers. This available water is part of the **soil solution (Figure 37.5),** a combination of water and dissolved substances that coats soil particles and partially fills pore spaces. The solution develops through ionic interactions between water

molecules and soil particles. Clay particles and the organic components in soil (especially proteins) often bear negatively charged ions on their surfaces. The negative charges attract the polar water molecules, which form hydrogen bonds with the soil particles (see *The Purple Pages*).

Unless a soil is irrigated, the amount of water in the soil solution depends largely on the amount and

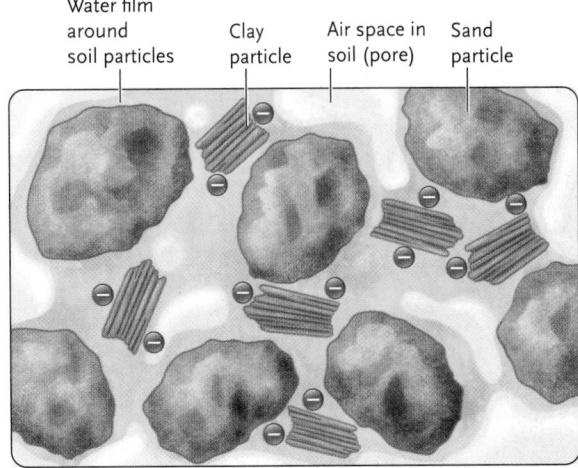

Water film around soil particles   Clay particle   Air space in soil (pore)   Sand particle

Figure 37.5
Location of the soil solution. Negatively charged ions on the surfaces of soil particles attract water molecules, which coat the particles and fill spaces between them (blue). Hydrogen bonds between water and soil components counteract the pull of gravity and help hold some water in the soil spaces.

pattern of precipitation (rain or snow) in a region. How much of this water is actually available to plants depends on the soil's composition—the size of the air spaces in which water can accumulate and the proportions of water-attracting particles of clay and organic matter. By volume, soil is about one-half solid particles and one-half air space.

The type and size of the particles in a given soil has a major effect on how well plants will grow there. Sand particles are small, and sandy soil has relatively large air spaces, so water drains rapidly below the top two soil horizons, where most plant roots are located. Soils rich in clay or humus often hold quite a bit of water, but in the case of clay, ample water is not necessarily an advantage for plants. Whereas a humus-rich soil contains lots of air spaces, the closely layered particles in clay allow few air spaces—and what spaces there are tend to hold tightly the water that enters them. The lack of air spaces in clay soils also severely limits supplies of oxygen available to roots for cellular respiration, and the plant's metabolic activity suffers. Thus, few plants can flourish in clay soils, even when water content is high. (Overwatered houseplants die because their roots are similarly "smothered" by water.) Plants do not fare much better in drier clay-rich soils, because roots cannot extract the existing water and cannot easily penetrate the densely packed clay. These characteristics explain why good agricultural soils tend to be sandy or silty loams, which contain a mix of humus and coarse and fine particles.

As you learned in Chapter 34, root hairs are specialized extensions of root epidermal cells; they directly contact the soil solution and allow roots to absorb water

(and dissolved ions). And as you saw in Section 35.2, differences in water potential govern the osmotic movement of water into plant roots. The soil solution usually contains fewer dissolved solutes than does the water in the cells of plant roots. Accordingly, water tends to move from wet soil, where the water potential is higher, into roots, where the water potential is lower. The water potential in clay soils is significantly lower than in other soil types, even when clay is relatively wet. As roots extract water from clay soil, the water potential may fall below that in a plant's roots, making it impossible for water to diffuse into the roots. Water still continues to evaporate from leaves and to be used in photosynthesis, however, so the plant eventually wilts. Plants that survive in deserts or in salty soils have adaptations that permit their roots to absorb water even when osmotic conditions in soil do not favour water movement into the plant.

Mineral Availability. Some mineral nutrients enter plant roots as cations (positively charged ions) and some as anions (negatively charged ions). Although both cations and anions may be present in soil solutions, they are not equally available to plants.

Cations such as magnesium (Mg^{2+}), calcium (Ca^{2+}), and potassium (K^+) cannot easily enter roots because they are attracted by the net negative charges on the surfaces of soil particles. To varying degrees, they become reversibly bound to negative ions on the surfaces. Attraction in this form is called *adsorption*. Roots do acquire cations, however, through **cation exchange**. In this mechanism, one cation, usually H^+, replaces a soil cation **(Figure 37.6a)**. The protons (H^+)

a. Adsorption of cations to a clay particle

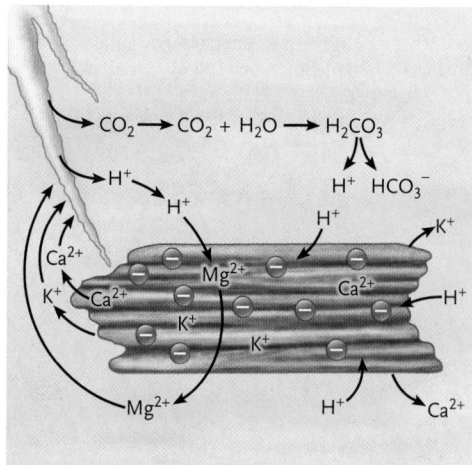

b. Adding gypsum to the soil

Wasu Watcharadachaphong/Shutterstock.com

Figure 37.6

(a) Cation exchange on the surface of a clay particle. When cations come into contact with the negatively charged surface of the particle, they become adsorbed. As one type of cation, such as H^+, becomes adsorbed, other ions are liberated and can be taken up by plant roots. **(b)** A farmer in rural Thailand sprinkling gypsum on a plowed field. Gypsum, a complex compound of calcium, sulfur, and other elements, may alter the chemical characteristics of soil in various ways, including increasing the availability to plant roots of calcium and sulfur, reducing acidity, and counteracting excess sodium.

come from two main sources. Respiring root cells release carbon dioxide, which dissolves in the soil solution, yielding carbonic acid (H_2CO_3). Subsequent reactions ionize H_2CO_3 to produce bicarbonate (HCO_3^-) and H^+. Reactions involving organic acids inside roots also produce H^+, which is excreted. As H^+ enters the soil solution, it displaces adsorbed mineral cations attached to clay and humus, freeing them to move into roots. Other types of cations may also participate in this type of exchange, as shown in Figure 37.6a.

By contrast, anions in the soil solution, such as nitrate (NO_3^-), sulfate (SO_4^{2-}), and phosphate (PO_4^{3-}), are only weakly bound to soil particles, and so they generally move fairly freely into root hairs. However, because they are so weakly bound compared with cations, anions are more subject to loss from soil by leaching.

The pH of soil affects the availability of some mineral ions. Soil pH is a function of the balance between cation exchange and other processes that raise or lower the concentration of H^+ in soil. As noted earlier, in areas that receive heavy rainfall, soils tend to become acidic (that is, they have a pH of less than 7). This acidification occurs in part because moisture promotes the rapid decay of organic material in humus; as the material decomposes, it releases its organic acids. **Acid precipitation**, which results from the release of sulfur and nitrogen oxides into the air (in large measure from the burning of fossil fuels and industrial emissions), also contributes to soil acidification. By contrast, the soil in arid regions, where precipitation is low, is often alkaline (the pH is greater than 7).

Although most plants are not directly sensitive to soil pH, chemical reactions in very acid (pH < 5.5) or very alkaline (pH > 9.5) soils can have a major impact on whether plant roots take up various mineral cations. For example, experiments have showed that in the presence of OH– in alkaline soil, calcium and phosphate ions react to form insoluble calcium phosphates. The phosphate captured in these compounds is as unavailable to roots as if it were completely absent from the soil.

For a soil to sustain plant life over long periods, the mineral ions that plants take up must be replenished naturally or artificially (see Figure 37.6b). Over the long run, some mineral nutrients enter the soil from the ongoing weathering of rocks and smaller bits of minerals. In the shorter run, minerals, carbon, and some other nutrients are returned to the soil by the decomposition of organisms and their parts or wastes. Airborne compounds, such as sulfur in volcanic and industrial emissions, may enter soil when they dissolve in rain and fall to Earth. Minerals, including compounds of nitrogen and phosphorus, may also enter soil in fertilizers.

Although the use of commercial fertilizers maintains high crop yields, agricultural chemicals do not add humus to the soil. Their use can also cause serious pollution problems, as when nitrogen-rich runoff from agricultural fields promotes the severe overgrowth of algae in lakes and bays. In many parts of the world, industrial pollutants such as cadmium, lead, and mercury are increasingly grave soil contaminants.

STUDY BREAK

1. Why is humus an important component of fertile soil?
2. How does the composition of a soil affect a plant's ability to take up water?
3. What factors affect a plant's ability to absorb minerals from the soil?

37.3 Root Adaptations for Obtaining and Absorbing Nutrients

Soil managed for agriculture can be plowed, precisely irrigated, and chemically adjusted to provide air, water, and nutrients in optimal quantities for a particular crop. By contrast, in natural habitats, wide variations in soil minerals, humus, pH, the presence of other organisms, and other factors influence the availability of essential elements. Although adequate carbon, hydrogen, and oxygen are typically available from the air and soil water, other essential elements that must be obtained from soil may not be as abundant. In particular, nitrogen, phosphorus, and potassium are often relatively scarce. The evolutionary solutions to these challenges include an array of adaptations in the structure and functioning of plant roots.

37.3a Root Systems Allow Plants to Locate and Absorb Essential Nutrients

Sessile organisms such as plants must locate nutrients in their immediate environment, and for plants the adaptive solution to this problem is an extensive root system. Roots make up 20% to 50% of the dry weight of many plants, and even more in species growing where water or nutrients are especially scarce, such as Arctic tundra. As long as a plant lives, its root system continues to grow, branching out through the surrounding soil. Roots do not necessarily grow deeper as a root system branches out, however. In arid regions, a shallow-but-broad root system may be better positioned to take up water from occasional rains that may never penetrate below the first few inches of soil.

You may recall from Section 35.2 that roots take up ions in the regions just behind the root tips, where root hairs are present. These diminutive absorptive structures, shown in Figure 34.11c, are a major adaptation for the uptake of mineral ions and water. Over successive growing seasons, long-lived plants such as trees can develop millions, even billions, of root tips, each one a potential absorption site. In a plant such as a mature red oak (*Quercus rubra*), which has a vast root system, the total number of root hairs is astronomical. Even in young plants, root hairs greatly increase the root surface area available for absorbing water and ions. Recently, a team led by Chinese researcher Keke Yi uncovered a genetic "master switch" that regulates the growth of root hairs. Experiments with *Arabidopsis thaliana* plants revealed that the activity of a transcription factor called RSL4 (for root-hair defective 6-like 4) activates downstream genes that promote the growth of long root hairs, while the absence of RSL4 stunts root hair growth. Apparently, expression of the *rsl4* gene is modulated by external cues, including soil phosphate levels and signals from the plant hormone auxin that you will read more about in Chapter 38.

Chapter 35 also mentioned another plant adaptation for gaining access to mineral ions—ion-specific transport proteins in plant cell membranes by which the cells selectively absorb ions from soil. For example, from studies of plants such as *Arabidopsis thaliana*, a weed that has become a key model organism for plant research, we know that transport channels for potassium ions (K$^+$) are embedded in the cell membranes of root cortical cells. Such ion transporters absorb more or less of a particular ion depending on chemical conditions in the surrounding soil.

37.3b The Discovery That Roots Also Secrete Substances into Soil Expands Our Understanding of Plant Adaptations for Obtaining Nutrients

In addition to acquiring substances from the surrounding soil, roots of various plant species also release into soil a long list of organic compounds, including carbohydrates, amino acids, various organic and fatty acids, and enzymes and other proteins. Today experiments are revealing the details of how such "root exudates" may improve a plant's access to particular nutrients. For example, a study headed by Corey D. Broeckling of Colorado State University showed that the roots of *A. thaliana* and *Medicago trunculata* (a legume commonly called barrel medic) secrete organic compounds that help determine which species of soil fungi can thrive near the roots. As you will read shortly, such fungi are partners in symbiotic relationships that help nourish many plant species. Similarly, Eric Paterson and his coworkers at the University of Aberdeen in Scotland found that roots of barley plants (*Hordeum vulgare*) exposed to above-normal nitrogen increased their release of organic substances that promote the growth of soil organisms that convert nitrogen into a chemical form plant roots can absorb. Other compounds released by roots enhance the uptake of phosphate.

37.3c Nutrients Move into and through the Plant Body by Several Routes

Plants obtain carbon, hydrogen, and oxygen from the air, but most mineral ions enter plant roots passively along with the water in which they are dissolved. Some enter root cells immediately. Others travel in solution *between* cells—in the apoplast—until they meet the endodermis sheathing the root's stele (see Figure 35.5). At the endodermis, the ions are actively transported into the endodermal cells and then into the xylem for transport throughout the plant. Inside cells, most mineral ions enter vacuoles or remain in the cytoplasm, where they are immediately available for metabolic reactions.

Some nutrients, such as nitrogen-containing ions, move in phloem from site to site in the plant, as dictated by growth and seasonal needs. In plants that shed their leaves in autumn, before the leaves age and fall, significant amounts of nitrogen, phosphorus, potassium, and magnesium move out of them and into twigs and branches. This evolutionary adaptation conserves the nutrients, which will be used in new growth the next season. Likewise, in late summer, mineral ions move to the roots and lower stem tissues of perennial range grasses that typically die back during the winter. These activities are regulated by hormonal signals, which are the topic of Chapter 38.

Given the essential role of roots in a plant's nutritional survival, it is not surprising that research is uncovering an ever-growing list of root adaptations for exploiting and enhancing nutrient resources in soil. We now take a closer look at two of these adaptations—the associations called *mycorrhizae* and interactions with nitrogen-fixing microorganisms.

37.3d Mycorrhizae and Nitrogen-Fixing Bacteria: Adaptive Interactions That Increase Plant Access to Scarce Nutrients

The course of plant evolution has resulted in vitally important symbioses that enhance access to nutrients. **Mycorrhizae** are crucial symbiotic associations between a fungus and the roots of a plant (see Section 25.3b). They promote the uptake of water and nutritionally vital ions—especially phosphate—in most species of plants. As shown in Figure 25.22, the fungal partner in the association often grows as a network of hyphal filaments around and beyond the plant's roots. Collectively, the hyphae provide a tremendous surface area for absorbing ions from a large volume of soil. As with plant roots, transport proteins shepherd ions into hyphae. Experiments have verified that hyphal transport proteins are encoded by the DNA of the fungus, not that of the plant. Some of the plant's sugars and nitrogenous compounds nourish the fungus, and as the root grows, it uses some of the minerals that the fungus has secured. In other types of mycorrhizae, the fungus actually lives inside cells of the root **cortex.** Orchids, for example, depend on this type of mutualistic association. And, as you will see shortly, some other plants gain access to adequate nitrogen by way of mutually beneficial associations with bacteria.

37.3e Plants Depend on Bacteria for an Adequate Supply of Usable Nitrogen

It might seem that plants live surrounded by nitrogen. For example, nitrogen steadily enters the soil in organic compounds released when dead organisms and animal wastes decompose. Dried blood is about 12% nitrogen by weight—although the nitrogen is bound up in complex organic molecules such as amino acids and proteins. Air contains plenty of gaseous nitrogen—this N_2 is almost 80% by volume—but plants cannot extract it because they lack the enzyme necessary to break apart the three covalent bonds in each N_2 molecule ($N \equiv N$). Plants can absorb atmospheric nitrogen that reaches the soil in the form of nitrate, NO_3^-, and ammonium ion, NO_4^+, and experiments show that roots of at least some plant species directly take up amino acids. Even so, lack of nitrogen is the single most common limit to plant growth because there is usually not nearly enough nitrogen available in these forms to meet plants' ongoing needs.

Instead, the main natural processes that replenish soil nitrogen and convert it to an absorbable form are carried out by bacteria. These processes, which we'll now consider, are part of the *nitrogen cycle*, the global movement of nitrogen in its various chemical forms from the environment to organisms and back to the environment, which is described in Chapter 31.

Production and Assimilation of Ammonium and Nitrate. The incorporation of atmospheric nitrogen into compounds that plants can take up is called **nitrogen fixation. Figure 37.7** summarizes the basic steps. Metabolic pathways of *nitrogen-fixing bacteria* living in the soil or in mutualistic association with plant roots add hydrogen to atmospheric N_2, producing two molecules of NH_3 (ammonia) and one H_2 for each N_2 molecule. The process requires a substantial input of ATP and is catalyzed by the enzyme nitrogenase. In a final step, H_2O and NH_3 react, forming NO_4^+ (ammonium) and OH^-.

Another bacterial process, called **ammonification,** also produces NH_4^+ when soil bacteria known as ammonifying bacteria break down decaying organic matter. In this way, nitrogen already incorporated into plants and other organisms is recycled.

Although plants use NH_4^+ to synthesize organic compounds, most plants absorb nitrogen in the form of nitrate, NO_3^-. Nitrate is produced in soil by **nitrification,** in which NH_4^+ is oxidized to NO_3^-. Soils generally teem with *nitrifying bacteria*, which carry out this process. Because of ongoing nitrification, nitrate is far more abundant than ammonium in most soils. Usually, the only soils from which plant roots take up ammonium directly are highly acidic, such as in bogs, where the low pH is toxic to nitrifying bacteria.

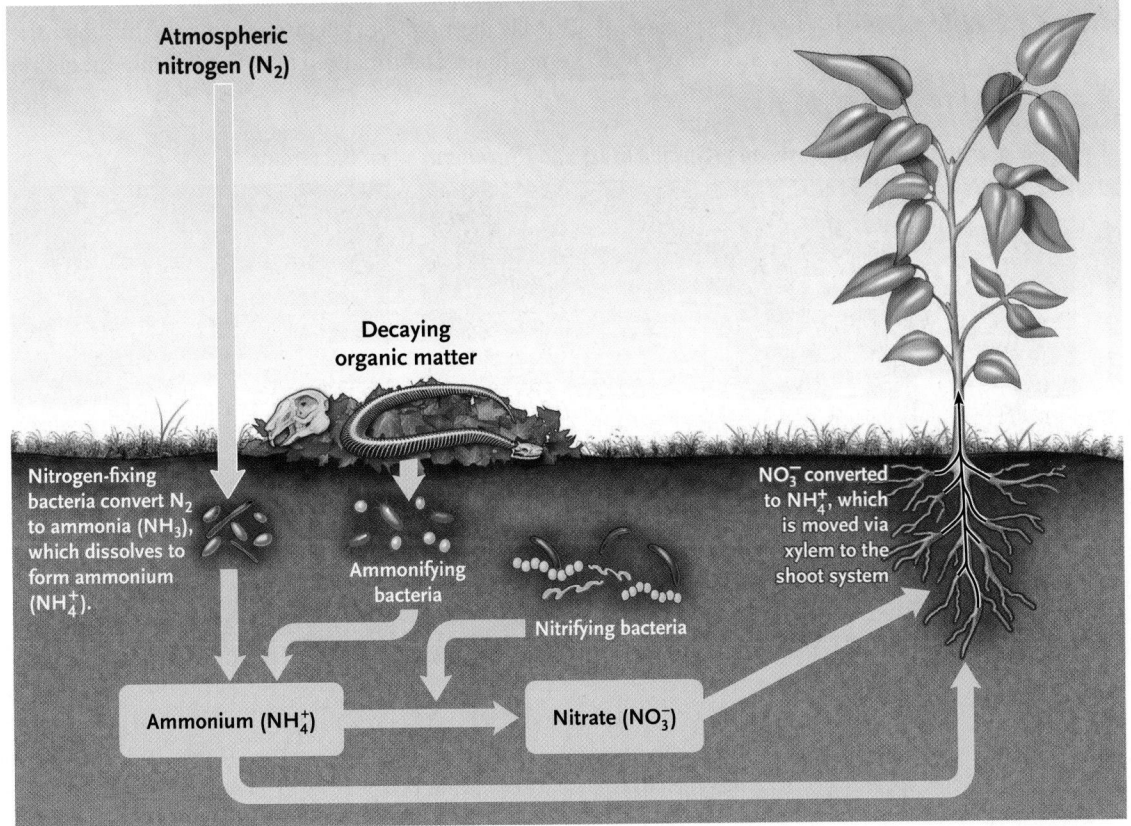

Figure 37.7

How plants obtain nitrogen from soil. Many commercial nitrogen fertilizers are in the chemical form of nitrate, which plant roots readily take up, or in the form of ammonium, which nitrifying bacteria convert to nitrate.

Atmospheric nitrogen (N_2)

Decaying organic matter

Nitrogen-fixing bacteria convert N_2 to ammonia (NH_3), which dissolves to form ammonium (NH_4^+).

Ammonifying bacteria

Nitrifying bacteria

NO_3^- converted to NH_4^+, which is moved via xylem to the shoot system

Ammonium (NH_4^+)

Nitrate (NO_3^-)

Nitrogen Assimilation. Once inside root cells, absorbed NO_3^- is converted by a multistep process back to NH_4^+. In this form, nitrogen is rapidly used to synthesize organic molecules, mainly amino acids. These molecules pass into the xylem, which transports them throughout the plant. In some plants, the nitrogen-rich precursors for needed substances travel in xylem to leaves, where different organic molecules are synthesized. These molecules travel to other plant cells in the phloem.

Nitrogen Fixation in Plant–Bacteria Associations. Although some nitrogen-fixing bacteria live free in the soil, by far the largest percentage of nitrogen is fixed by species of *Rhizobium* and *Bradyrhizobium*, which form mutualistic associations with the roots of plants in the legume family. The host plant supplies organic molecules that the bacteria use for cellular respiration, and the bacteria supply NH_4^+ that the plant uses to produce proteins and other nitrogenous molecules. In legumes—a large family that includes peas, beans, clover, alfalfa, and alders, among others—the nitrogen-fixing bacteria reside in **root nodules**, localized swellings on roots **(Figure 37.8)**. Farmers may exploit root nodules to increase soil nitrogen by rotating crops (for example, planting soybeans and corn in alternating years). When the legume crop is harvested, the root nodules and other tissues remaining in the soil enrich its nitrogen content.

For a plant, an association with nitrogen-fixing bacteria offers the selective advantage of a steady source of absorbable nitrogen. Decades of research have revealed the details of how this remarkable relationship unfolds. Usually, a single species of nitrogen-fixing bacteria colonizes a single legume species, drawn to the plant's roots by chemical attractants—primarily compounds called flavonoids—that the roots secrete. Through a sequence of exchanged molecular signals, bacteria are able to penetrate a root hair and form a colony inside the root cortex.

An association between a soybean plant (*Glycine max*) and a bacterium (*Bradyrhizobium japonicum*) illustrates the process. In response to a specific flavonoid released by soybean roots, bacterial genes called *nod* genes (for *nodule*) begin to be expressed **(Figure 37.9a)**. Products of the *nod* gene cause the tip of the root hair to curl toward the bacteria and trigger the release of bacterial enzymes that break down the root hair cell wall (Figure 37.9b). As bacteria enter the cell and multiply, the plasma membrane forms a tube called an **infection thread** that extends into the root cortex, allowing the bacteria to invade cortex cells (Figure 37.9c). The enclosed bacteria, now called **bacteroids**, enlarge and become immobile. Stimulated by still other *nod* gene products, cells of the root cortex begin to divide. This region of proliferating cortex cells forms the root nodule. Typically, each cell in a root nodule contains several thousand bacteroids; the plant takes up some of the nitrogen fixed by the bacteroids, and the bacteroids use some compounds produced by the plant.

Inside bacteroids, N_2 is reduced to NH_4^+ (ammonium) using ATP produced by cellular respiration. The process is catalyzed by nitrogenase. Ammonium is highly toxic to cells if it accumulates, however. Thus, NH_4^+ is moved out of bacteroids into the surrounding nodule cells immediately and converted to other compounds, such as the amino acids glutamine and asparagine.

One factor encoded by the bacterial *nod* genes stimulates plant nodule cells to produce a protein called **leghemoglobin** ("legume hemoglobin"). Like the hemoglobin of animal red blood cells, leghemoglobin

a. Root nodules

Root nodule

Dan Guravich/Science Source

b. Field experiment with soybeans (*Glycine max*) and *Rhizobium*

From Takuji Ohyama, FNCA Biofertilizer Manual, 2006

c. Bacteroids

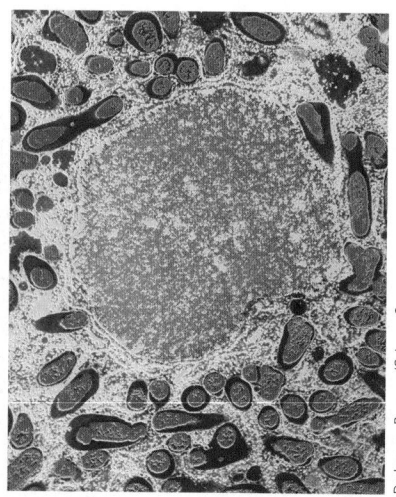

Dr. Jeremy Burgess/Science Source

Figure 37.8

The beneficial effect of root nodules. **(a)** Root nodules on a soybean plant (*Glycine max*). **(b)** Soybean plants growing in nitrogen-poor soil. The plants on the right were inoculated with *Rhizobium* cells and developed root nodules. **(c)** False-colour transmission electron micrograph showing membrane-bound bacteroids (red) in a root nodule cell. Membranes that enclose the bacteroids appear blue. The large yellow-green structure is the cell's nucleus.

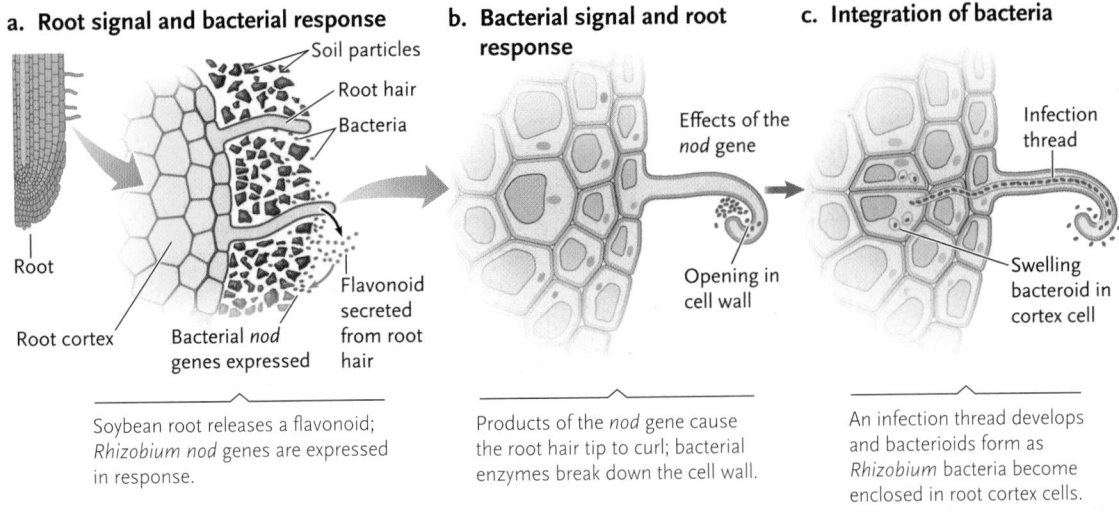

a. Root signal and bacterial response

Soil particles
Root hair
Bacteria
Root
Root cortex
Bacterial *nod* genes expressed
Flavonoid secreted from root hair

Soybean root releases a flavonoid; *Rhizobium nod* genes are expressed in response.

b. Bacterial signal and root response

Effects of the *nod* gene
Opening in cell wall

Products of the *nod* gene cause the root hair tip to curl; bacterial enzymes break down the cell wall.

c. Integration of bacteria

Infection thread
Swelling bacteroid in cortex cell

An infection thread develops and bacterioids form as *Rhizobium* bacteria become enclosed in root cortex cells.

Research Method Figure 37.9 Root nodule formation in legumes. Legume root nodules typically form as a result of a mutualistic association with the nitrogen-fixing bacteria *Rhizobium* and *Bradyrhizobium*. The association develops as the products of bacterial *nod* genes allow bacteria to infect root cortex cells and proliferate there.

contains a reddish, iron-containing heme group that binds oxygen. Its color gives root nodules a pinkish cast (see Figure 37.8a). Leghemoglobin picks up oxygen at the cell surface and shuttles it inward to the bacteroids. This method of oxygen delivery is vital, because nitrogenase, the enzyme responsible for nitrogen fixation, is irreversibly inhibited by excess O_2. Leghemoglobin delivers just enough oxygen to maintain bacteroid respiration without shutting down the action of nitrogenase.

37.3f Some Plants Obtain Nutrients in Unusual Ways

The Venus flytrap, the cobra lily, and various species of sundews are members of a curious group of plants that obtain nitrogen and other nutrients by trapping and digesting animals. A few species of tropical pitcher plants even capture and digest mice and small rats. Such "carnivorous" ("meat-eating") plants have become adapted to survive in nutrient-deficient, and especially nitrogen-deficient, environments such as boggy and sandy areas through elaborate mechanisms for extracellular digestion and absorption. The cobra lily (*Darlingtonia californica*; **Figure 37.10a, p. 914**) is a good example. Its leaves form a "pitcher" that is partly filled with digestive enzymes. Insects lured in by attractive odours often wander deeper into the pitcher, encountering downward-pointing leaf hairs that have a slick, waxy coating and speed the insect's descent into the pool of enzymes. The plant then absorbs monomers released as the animal tissues are digested.

Dodders (Figure 37.10b) and thousands of other species of flowering plants are parasites that obtain some or all of their nutrients from the tissues of other plants. Parasitic species develop *haustorial roots* that

penetrate deep into the host plant and tap into its vascular tissues. Although some parasitic plants, like mistletoe, contain chlorophyll and thus can photosynthesize, dodders and other nonphotosynthesizers rob the host of sugars as well as water and minerals.

The snow plant (*Sarcodes sanguinea*) shows a variation on this theme. As its deep red color suggests (Figure 37.10c), it lacks chlorophyll, but it doesn't have haustorial roots. Instead, the snow plant's roots take up nutrients from mycorrhizae they "share" with the roots of nearby conifers.

Epiphytes, such as the tropical orchid pictured in Figure 37.10d, are not parasitic even though they grow on other plants. Some trap falling debris and rainwater among their leaves, whereas their roots (including mycorrhizae, in the case of the orchid) invade the moist leaf litter and absorb nutrients from it as the litter decomposes. In temperate forests, many mosses and lichens are epiphytes.

These and other strategies plants have evolved for obtaining nutrients and water are only part of the survival equation, however. Plants use nutrients not only for growth and maintenance, but also, of course, for building structures such as pollen, flowers, and seeds used in reproduction—our topic in Chapter 36.

STUDY BREAK

1. What is a mycorrhiza, and why are mycorrhizal associations so vital to many plants?
2. Distinguish between nitrogen fixation, ammonification, and nitrification.
3. Summarize the mechanism by which associations with bacteria supply nitrogen to plants such as legumes.

a. Cobra lily (*Darlingtonia californica*)

b. Dodder (*Cuscuta*)

c. Snow plant (*Sarcodes sanguinea*)

d. Lady-of-the-night orchid (*Brassavola flagellaris*)

Figure 37.10

Some plants with unusual adaptations for obtaining nutrients. **(a)** Cobra lily (*Darlingtonia californica*), a carnivorous plant. The patterns formed by light shining through the plant's pitcherlike leaves are thought to confuse insects that have entered the pitcher, making an exit more difficult. **(b)** A parasitic dodder, one of the more than 150 *Cuscuta* species. Dodders have slender yellow-orange stems that twine around the host plant before producing haustorial roots that absorb nutrients and water from the host's xylem and phloem. **(c)** Snow plant (*Sarcodes sanguinea*), which pops up in the deep humus of shady conifer forests after snow has melted in spring. This species lacks chlorophyll and does not photosynthesize. Instead its roots intertwine with hyphae of soil fungi that also form associations with the roots of nearby conifers. Radiocarbon studies have shown that the fungi take up sugars and other nutrients from the trees and pass a portion of this food to the snow plant. **(d)** The lady-of-the-night orchid (*Brassavola nodosa*), a tropical epiphyte.

REVIEW

 To access course materials such as Aplia and other companion resources, please visit www.NELSONbrain.com.

37.1 Plant Nutritional Requirements

- Plants require 17 essential nutrients (Table 37.1). With enough sunlight and these nutrients, plants can synthesize all the compounds they require to survive.

- Nine essential elements are macronutrients, required in relatively large amounts. Of these, carbon, hydrogen, oxygen, and nitrogen are the main building blocks in the synthesis of carbohydrates, lipids, proteins, and nucleic acids. Macronutrients dissolved in the soil solution are nitrogen, potassium, calcium, magnesium, phosphorus, and sulfur.

- Plants require essential micronutrients in much smaller amounts. Known micronutrients are chlorine, iron, boron, manganese, zinc, copper, molybdenum, and nickel.

- Each plant species requires specific amounts of specific nutrients. Typical deficiency symptoms are stunted growth, yellowing or other abnormal changes in leaf colour, dead spots on leaves, or abnormally formed stems (Figure 37.3).

- Most mineral ions enter plant roots dissolved in water. Inside cells, most mineral ions enter vacuoles or the cell cytoplasm, where they are available for metabolism. Some elements, such as nitrogen and potassium, are mobile—they can move from site to site in phloem as the plant grows.

37.2 Soil

- Soil consists of sand, silt, and clay particles, usually held together by humus and other organic components. Humus absorbs considerable water and contributes to the water-holding capacity of soil.

- The relative proportions of various soil mineral particles and humus give soil its basic texture and structure. The best agricultural soils are loams that contain clay, sand, silt, and humus in roughly equal proportions. Topsoil is the most fertile soil layer (Figure 37.4).

- Soil particles are thinly coated by the soil solution, a mixture of water and solutes (Figure 37.5). Root hairs and other root epidermal cells absorb water and solutes from this solution.

- The amount of water available to plant roots depends mainly on the relative proportions of different soil components. Water moves quickly through sandy soils, whereas soils rich in clay and humus tend to hold water.

- Cations are adsorbed on the negatively charged surfaces of soil particles, potentially limiting their uptake by roots. Cation exchange, in which mineral cations are replaced by H^+, helps make these nutrients available to plants (Figure 37.6). Anions are more weakly bound to soil particles; they move more readily into root hairs but are also more apt to leach out of topsoil. In nature, the soil solution surrounding plant roots generally contains only tiny amounts of essential mineral ions.

- Plants influence the physical, chemical, and biological features of soil, as when dead plant parts decompose and substances they contain enter the soil.

37.3 Root Adaptations for Obtaining and Absorbing Nutrients

- Numerous adaptations help plants solve the problems of obtaining and absorbing nutrients. Roots penetrate the soil toward nutrients and water; huge numbers of root hairs increase the root's absorptive surface. Ion-specific transporters in root cortical cells adjust the plant's uptake of particular ions. Mycorrhizal associations between fungi and plant roots enhance the absorption of nutrients, notably phosphorus.

- Nitrogen is usually the scarcest nutrient in soil, and nitrogen-fixing bacteria produce much of the usable soil nitrogen. Nitrogen fixation reduces atmospheric N_2 to NH_4^+ (ammonium) in a reaction that requires nitrogenase as a catalyst. Nitrifying bacteria rapidly convert NH_4^+ to nitrate, the form in which the roots of most plants absorb nitrogen (Figure 37.7).

- In legumes and a few other species, nitrogen-fixing bacteria reside in root nodules in a mutualistic association (Figure 37.8).

- Bacteria (bacteroids) enclosed in a root nodule reduce N_2 to NH_4^+ (Figure 37.9). The toxic NH_4^+ is moved out of the bacteroids and converted to nitrogen-rich, nontoxic compounds such as amino acids. In plants that do not form root nodules, nitrate absorbed by roots is reduced to ammonium, which is then converted to nontoxic forms.

- In many plant species, root cells synthesize amino acids and other organic nitrogenous compounds, and these molecules are transported in phloem throughout the plant. In some plants, the nitrogen-rich precursors travel in xylem to leaves, where different organic molecules are synthesized. Those molecules move to other cells in phloem.

- A few plant species have evolved alternative mechanisms for obtaining some or all of their nutrients (Figure 37.10). Carnivorous plants have structures that physically trap insects or other small animals and produce solutions of enzymes that digest the animal tissues, releasing absorbable nutrients.

- Some plant species parasitize other plants. The parasite may or may not contain chlorophyll and carry out photosynthesis; species that do not photosynthesize obtain all of their nutrition from the host. Epiphytes grow on other plants but obtain nutrients independently.

Questions

Self-Test Questions

1. Which best applies to a micronutrient?
   a. It makes up 96% of the plant's dry mass.
   b. It cannot be replaced artificially.
   c. It is early on the periodic table compared with macronutrients.
   d. It is required in large amounts during sunlight hours.
   e. It is an essential element.

2. Nutrient runoff from fertilizing lush lawns often causes "algal blooms" in nearby lakes, making swimming impossible. Which of the following fertilizer components most likely caused the blooms?
   a. iron, magnesium, and nitrogen
   b. nitrogen, phosphorus, and sulfur
   c. nitrogen, potassium, and phosphorus
   d. selenium, magnesium, and potassium
   e. nitrogen, magnesium, and nickel

3. Which of the following are NOT among the ideal soil conditions for growing crops?
   a. extremely large air spaces
   b. sandy or silty loam
   c. blend of sand and clay
   d. less than 5% humus
   e. thick topsoil

4. Which of the following processes contributes to the uptake of mineral ions by plant roots?
   a. chlorosis
   b. osmosis
   c. cation exchange
   d. anion leaching
   e. growth of root hairs

5. Which of the following does NOT influence soil pH?
   a. rainfall
   b. hydroponic growth
   c. release of sulfur and nitrogen oxides into the air
   d. decomposition of organisms
   e. weathering of rock

6. Which of the following is a common process that makes usable nitrogen available to plants?
   a. nitrogen-fixing bacteria synthesizing nitrate
   b. ammonifying bacteria using ammonium to produce nitrate
   c. nitrifying bacteria converting NH_4^+ to NO_3^-
   d. the direct absorption of NH_4^+ by root hairs
   e. the absorption of atmospheric N_2 into the xylem

7. The *nod* genes in the bacteria in soybean nodules allow the bacteria to fix nitrogen. Which of the following, if any, is NOT a step in this process?
   a. The products of *nod* genes cause cells of the root cortex to divide and become the root nodule in which bacteroids fix nitrogen for the plant.
   b. In the cortex cells, bacteria enlarge and become immobile, forming bacteroids.
   c. Bacteria enter the root hair cell and multiply, causing the cell plasma membrane to form an infection thread that extends into the root cortex.
   d. Roots release flavonoid, which turns on the expression of bacterial *nod* genes. Products of *nod* genes cause the tip of the root hair to curl toward the bacteria.
   e. Root hairs trigger release of bacterial enzymes that break down root hair cell walls.

8. Being "carnivorous" is a plant adaptation to obtain which of the following?
   a. oxygen
   b. phosphorus
   c. potassium
   d. nitrogen
   e. carbon

9. Haustorial roots are characteristic of plants that are which of the following?
   a. parasites
   b. epiphytes
   c. nitrate fixers
   d. leghemoglobin users
   e. carnivorous

10. Identify the correct match of a nutrient with its function.
    a. chlorine: component of several enzymes
    b. potassium: component of nucleic acids
    c. phosphorus: component of most proteins
    d. manganese: role in shoot and root growth
    e. calcium: maintenance of cell walls and membrane permeability

Questions for Discussion

1. If you want to study factors that affect plant nutrition in nature, what are the advantages and disadvantages of using a hydroponic culture method?

2. Gardeners often add a humus-rich "soil conditioner" to garden plots before they plant. Adding the conditioner helps aerate the soil, and the decomposing organic materials in humus provide nutrients. If the plot is for annual plants, it often must be reconditioned year after year, even though the gardener faithfully pulls weeds, fertilizes seedlings, applies chemicals to curtail disease-causing soil microbes, and immediately tosses out the mature plants (along with any plant debris) when they have finished bearing. Suggest some reasons why reconditioning is necessary in this scenario, and some strategies that could help limit the need for it.

3. One effect of acid rain is to dissolve rock, liberating minerals into soil. Accordingly, can a case be made that acid rain confers environmental benefits as well as doing harm? What are some other factors, especially with regard to plant adaptations for gaining nutrients, that bear on this question?

4. Using Table 37.1 as a guide, describe some of the known roles of nitrogen, phosphorus, and potassium in plant function. What are some of the signs that a plant suffers a deficiency in these elements?

Sunflower plants (*Helianthus*) with flower heads that orient toward the Sun's rays—an example of a plant response to shifting light levels in the environment.

Plant Signals and Responses to the Environment

WHY IT MATTERS

Larrea tridentata—creosote bush—won't win many botanical beauty contests, but it is one tough plant. Native to desert areas of the southwestern United States, this shrubby eudicot can withstand droughts of two years or more **(Figure 38.1)**. The species may well be the most drought-tolerant perennial in North America, and individual plants may survive hundreds of years, physiologically "hunkering down" during multiyear dry spells as their fleshy leaves shrivel and reduce photosynthesis to minimal levels. A thick, waxy cuticle helps slow water loss, but the leaves also tend to retain water due to their high internal water potential—a feature related to the plant's adaptations for defence against predation. The leaf tissue is infused with resinous compounds and other chemicals that deter nearly all herbivores as well as pathogenic fungi and most insect predators. Wet leaves exude a tarry odor that gives creosote bush its common name.

With adequate rainfall, *L. tridentata* produces small yellow flowers and feathery white fruits packed with seeds (Figure 38.1, inset). However, for a variety of reasons, relatively few *L. tridentata* seeds germinate successfully. Selection pressure related to this high seed mortality may explain another striking *L. tridentata* trait—the formation of large clonal populations by vegetative reproduction. After eight or nine decades, a plant's interior branches may begin to die, eventually leaving what appear to be clusters of separate, widely spaced plants. Carbon dating indicates that some living clusters of these genetically identical L. tridentata plants are descended from single seeds that germinated 9 to 11 thousand years ago—qualifying them as some of the oldest living organisms on Earth.

The drought tolerance, chemical defences, and reproductive flexibility of the creosote bush are prime examples of adaptations plants have evolved for responding

38

Figure 38.1

Creosote bush, *Larrea tridentata*, growing in the Mojave Desert. Inset: Flowers, fruits, and the resin-impregnated leaves of *L. tridentata*.

to both biotic (living) and abiotic (nonliving) shifts and stresses in their environment. Such adaptations are the focus of this chapter, starting with the chemical signalling of hormones that regulate many, if not most, aspects of plant growth and development. Next we survey responses that help protect plants against predation or allow them to adjust to abiotic factors that are powerful forces in shaping the physical and chemical environments in which plants live. Collectively these adaptations are the evolutionary solution to a key plant "problem"—the need to respond to changes in the environment despite being rooted in one place (sessile).

38.1 Introduction to Plant Hormones

A **hormone** (*hormon* = to stimulate) is a signalling molecule that regulates or helps coordinate some aspect of growth, metabolism, or development. Plant hormones may also be called *phytohormones*, especially in the scientific literature. At a fundamental level, hormones control the plant life cycle. They serve as triggers for seed germination and govern the development of a plant's body form; the shift from a vegetative growth phase to a reproductive phase or vice versa; and the timed death of flowers, leaves, and other parts. Beyond these basic roles, hormones mediate changes in the structure and functioning of plant parts in response to external biotic factors such as predation, and abiotic factors such as the availability of light, moisture, and soil nutrients and effects of air currents, gravity, and physical contact with other objects. Some plant hormones are transported from the tissue that produces them to another plant part, whereas others exert their effects in the tissue where they are synthesized. Often, hormonal effects involve changes in gene expression, although sometimes other mechanisms are at work.

All plant hormones are rather small organic molecules that are active in extremely low concentrations. Hormones that have effects outside the tissue where

they are produced typically diffuse to their target site(s) or travel to the site via vascular tissues. Plant hormones vary greatly in their effects, although each one affects a given tissue in a particular way. For instance, some stimulate one or more facets of growth or development, while others have an inhibiting influence. A given hormone can also have different effects in different plant tissues, and the effects can differ depending on a target tissue's stage of development. Adding to the complexity, many physiological responses result from the interaction of two or more hormones. Biologists recognize at least seven major classes of plant hormones **(Table 38.1)**: gibberellins, auxins, cytokinins, ethylene, brassinosteroids, abscisic acid (ABA), and jasmonates. Recent discoveries have added other hormonelike signalling agents to this list (see Table 38.1).

38.1a Plant Hormones Exert Their Effects via Signal Transduction Pathways

In general, the target cells for a particular hormone have receptors that can bind the hormone and cellular pathways that are activated in response. This mechanism unfolds in three basic steps introduced in Chapter 5. We'll briefly review these steps here and consider current scientific understanding of the signal transduction pathways by which specific hormones exert their effects.

Recall from Chapter 5 that in the first step of a signal transduction pathway, a target cell receptor receives the signal, which may be a molecule or an environmental cue such as sunlight. Next, the signal is transduced—that is, its "message" is changed into a form that can trigger the cellular response. In the third step, the transduced signal causes the cellular response **(Figure 38.2, p. 920).** Some transduced signals activate or turn off genes and so alter protein synthesis; others set in motion events that modify existing cell proteins. Various plant hormones and growth factors bind to receptors at the target cell's plasma membrane. Others cross the plasma membrane and bind to receptors inside the cell. These receptors may be located on the endoplasmic reticulum (ER), in the cytoplasm, or in the nucleus. In many cases, hormone binding causes the receptor to change shape. Regardless, binding of a hormone or growth factor triggers a complex pathway that leads to the cell response—the opening of ion channels, activation of transport proteins, or some other event. Only cells with the appropriate receptor can respond to a particular signalling molecule. For example, certain cells in developing seeds and maturing fruits have receptors for the "ripening hormone" ethylene, but cells in stems generally do not.

We can think of plant hormones and other signalling molecules as external "first messengers" that deliver the initial physiological signal to a target cell. Often, binding of the signal molecule triggers the synthesis of internal second messengers. These go-between molecules diffuse

Table 38.1   Major Plant Hormones and Signalling Molecules

Hormone/Signalling Compound	Where Synthesized	Tissues Affected	Effects
Auxins	Apical meristems, developing leaves and embryos	Growing tissues, buds, roots, leaves, fruits, vascular tissues	Promote growth and elongation of stems; promote formation of lateral roots and dormancy in lateral buds; promote fruit development; inhibit leaf abscission; orient plants with respect to light, gravity
Gibberellins	Root and shoot tips, young leaves, developing embryos	Stems, developing seeds	Promote cell divisions and growth and elongation of stems; promote seed germination and bolting
Cytokinins	Mainly in root tips	Shoot apical meristems, leaves, buds	Promote cell division; inhibit senescence of leaves; coordinate growth of roots and shoots (with auxin)
Ethylene	Shoot tips, roots, leaf nodes, flowers, fruits	Seeds, buds, seedlings, mature leaves, flowers, fruits	Regulates elongation and division of cells in seedling stems, roots; in mature plants regulates senescence and abscission of leaves, flowers, and fruits
Brassinosteroids	Young seeds; shoots and leaves	Mainly shoot tips, developing embryos	Stimulate cell division and elongation, differentiation of vascular tissue
Abscisic acid	Leaves, chloroplasts, possibly roots in drying soils	Mainly shoot tips, developing embryos, leaves (stomata)	Stimulates cell division and elongation, differentiation of vascular tissue, opening/closing of stomata
Jasmonates	Roots, seeds, probably other tissues	Various tissues, including damaged ones	In defence responses, promote transcription of genes encoding protease inhibitors; possible role in plant responses to nutrient deficiencies
Oligosaccharins	Cell walls	Damaged tissues; possibly active in most plant cells	Promote synthesis of phytoalexins in injured plants; may also have a role in regulating growth
Systemin	Damaged tissues	Damaged tissues	To date known only in tomato; roles in defence, including triggering jasmonate-induced chemical defences
Salicylic acid	Damaged tissues	Many plant parts	Triggers synthesis of pathogenesis-related (PR) proteins, other general defences

rapidly through the cytoplasm and provide the main chemical signal that alters cell functioning.

38.1b Second Messenger Systems Enhance the Plant Cell's Response to a Hormone Signal

Second messengers are usually synthesized in the sequence of chemical reactions that convert an external signal into internal cell activity. For many years the details of plant second messenger systems were sketchy and hotly debated. Fairly early on, calcium ions were found to play a second messenger role in some hormonal responses. Recent experimental evidence indicates that the hormonal signal from auxins is often conveyed by cAMP (cyclic adenosine monophosphate), a major second messenger in cells of animals and other organisms. **Inositol triphosphate (IP$_3$),** a second messenger in plants, fungi, and animals, is involved in the reactions that close plant stomata in response to a signal from abscisic acid, ABA. Other research suggests that a molecule called cGMP (cyclic guanosine monophosphate) serves as a second messenger for auxins, ABA, and a few other plant hormones.

In addition to the basic signal transduction pathways described here, other routes may exist that are unique to plant cells. Light is the driving force for photosynthesis,

and researchers are investigating the possibility that plants have evolved other unique light-related biochemical pathways as well. For instance, experiments are extending our knowledge of how plant cells respond to blue light, which, as you will read, triggers some photoperiod responses such as the opening and closing of stomata. Other exciting research suggests that it is probably common for signal response pathways in plant cells to include many steps in which different types of proteins and other molecules are mobilized, not unlike the steps in animal cells.

Auxins were the first plant hormones to be identified, and we start with them as we consider each major class of plant hormones and discuss some newly discovered signalling molecules as well.

38.1c Auxins Promote Growth

Auxins are synthesized mainly in the shoot apical meristem and young stems and leaves. They are crucial to plant growth. Among other effects, auxins are essential for the normal progression of the cell cycle, and they stimulate the elongation of cells in growing stems and coleoptiles. Auxins also mediate growth responses to light and gravity. Indoleacetic acid (IAA) is the most important natural auxin. Botanists often use the general term "auxin" to refer to IAA, a practice we follow here.

Experimental Research Figure 38.2 Overview of signal transduction pathways in plant cells. The three stages of a signal response pathway are reception of the signal, transduction into a form the cell can recognize, and the cell's response. Proteins or second messenger molecules may be the intermediaries in the transduction stage of a signal response pathway (compare Figure 38.6, p. 924).

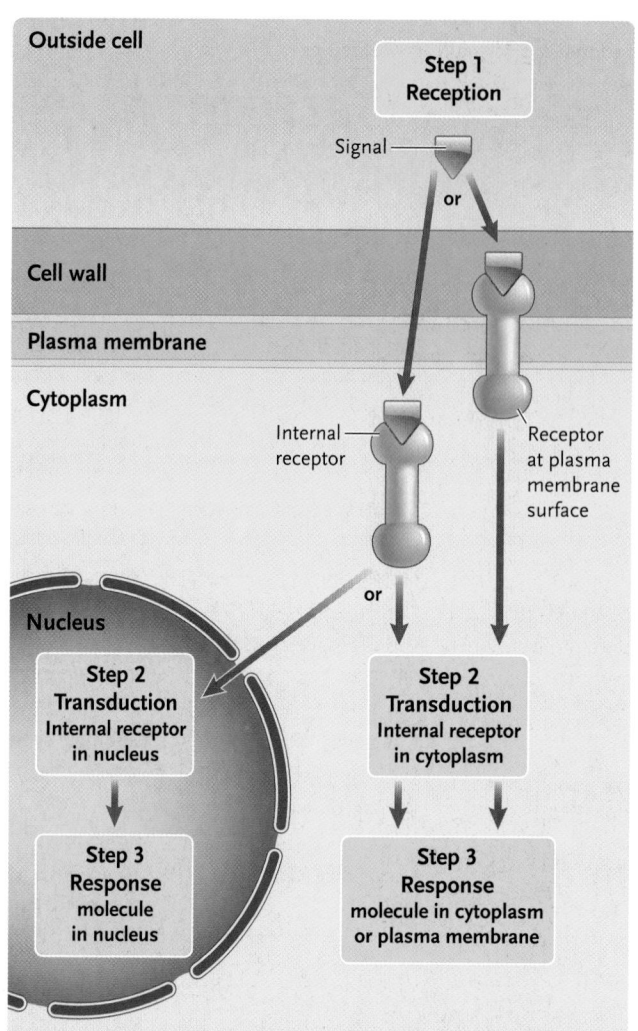

Outside cell

Step 1 Reception

Signal

or

Cell wall

Plasma membrane

Cytoplasm

Internal receptor

Receptor at plasma membrane surface

or

Nucleus

Step 2 Transduction Internal receptor in nucleus

Step 2 Transduction Internal receptor in cytoplasm

Step 3 Response molecule in nucleus

Step 3 Response molecule in cytoplasm or plasma membrane

1 An arriving signal binds and activates a receptor in the plasma membrane, cytoplasm, ER, or nucleus. In most cases, the receptor changes shape, which triggers the transduction pathway inside the cell.

2 As the transduction pathway unfolds, receptor activation leads to activation of one or more proteins. This activation step may set in motion a cascade of protein phosphorylation, or it may mobilize second messenger molecules.

3 Phosphorylated proteins or second messenger molecules trigger the cell response, such as a change in ion flow into or out of the cell, a shift in translation of mRNA, or altering gene transcription in the nucleus.

Experiments That Led to the Discovery of Auxins. The path to the discovery of auxins began in the late nineteenth century in the library of Charles Darwin's home in the English countryside. Among his interests, Darwin was fascinated by plant **tropisms**—movements such as the bending of a houseplant toward light. This growth response, triggered by exposure to a directional light source, is an example of a **phototropism**.

Working with his son Francis, Darwin explored phototropisms by germinating seeds of two species of grasses, oats (*Avena sativa*) and canary grass (*Phalaris canariensis*), in pots on the sill of a sunny window. Recall from Chapter 34 that the shoot apical meristem and plumule of grass seedlings are sheathed by a coleoptile—a protective structure that is extremely sensitive to light. Darwin did not know this detail, but he observed that as the emerging shoots grew, within a few days they bent toward the light. He hypothesized that the tip of the shoot detected light and communicated that information to the coleoptile. Darwin and his son tested this idea in several ways **(Figure 38.3, p. 922)** and concluded that when seedlings are illuminated from the side, "some influence is transmitted from the upper to the lower part, causing them to bend."

The Darwins' observations spawned decades of studies that illustrate how scientific understanding typically advances step by step, as one set of experimental findings stimulates new research. First, scientists in Denmark and Poland showed that the bending of a shoot toward a light source was caused by something that could move through agar (a jellylike culture material derived from certain red algae) but not through a sheet of the mineral mica. This finding prompted experiments establishing that indeed the stimulus was a chemical produced in the shoot tip. Soon afterward, in 1926, experiments by the Dutch plant physiologist Frits Went confirmed that the growth-promoting chemical diffuses downward from the shoot tip to the stem below **(Figure 38.4, p. 923).** Using oat seeds, Went first sliced the tips from young shoots that had been grown under normal light conditions. He then placed the tips on agar blocks and left them there long enough for diffusible substances to move into the agar. Meanwhile, the decapitated stems stopped growing, but growth quickly resumed in seedlings that Went "capped" with

When we think of Charles Darwin, we think of course of his theory of evolution by natural selection, which is the foundation of modern biology. But Darwin's contribution to biology is much more than just his work on evolution, key though it is: he was a very knowledgeable and creative naturalist who carried out experiments on several other important biological questions. He published a detailed analysis of how earthworms improve the soil ("The Formation of Vegetable Mould through the Action of Worms") and wrote books on several botanical topics, among them plants that eat animals (*Insectivorous Plants*), pollination and fertilization systems (*Fertilisation in Orchids and the Effects of Self- and Cross-Fertilisation*), and the tendency of plants to grow toward sunlight (*The Power of Movement in Plants*). Experiments on this last topic, carried out with his son Francis, are still cited today for their role in helping us understand how plants respond to light.

Darwin's study. Darwin undertook most of his life's work in this room at Down House. He hesitated to discard old papers and specimens, believing that he would find a use for them as soon as they were carried away in the garbage.

William Perlman/Star Ledger/Corbis

---

the agar blocks (see Figure 38.4a). Clearly, a growth-promoting substance in the excised shoot tips had diffused into the agar, and from there into the seedling stems. Went also attached an agar block to one side of a decapitated shoot tip; when the shoot began growing again, it bent away from the agar (see Figure 38.4b). Importantly, Went performed his experiments in total darkness, to avoid any "contamination" of his results by the possible effects of light.

Went did not determine the mechanism—differential elongation of cells on the shaded side of a shoot—by which the growth promoter controlled phototropism. However, he did develop a test that correlated specific amounts of the substance, later named auxin (*auxein* = to increase), with particular growth effects. This careful groundwork culminated several years later when other researchers identified auxin as indoleacetic acid (IAA).

Effects of Auxin. Auxin is one of the first chemical signals to help shape the plant body. When the zygote first divides, forming an embryo that consists of a basal cell and an apical cell, auxin exported by the apical cell to the basal cell helps guide the development of the various features of the embryonic shoot. It plays a key role in when and where leaf primordia form in the apical meristem. As the embryo develops, the leaf primordium of the young shoot becomes the main source of IAA; a secondary signal stimulates the primary growth of the stem and root as long as the embryonic plant is underground. Once an elongating shoot breaks through the soil surface, its tip is exposed to sunlight, and the first leaves unfurl and begin photosynthesis. Shortly thereafter, the leaf tip stops producing IAA and that task is assumed first by cells at the leaf edges and then by cells at base of the young leaf. Even so, as described in Section 38.3, IAA continues to influence a plant's responses to light and plays a role in its growth responses to gravity as well. IAA also stimulates cell division in the vascular cambium and promotes the formation of secondary xylem, as well as the formation of new root apical meristems, including lateral meristems. Not all of auxin's effects promote growth, however. IAA also maintains apical dominance, which inhibits growth of lateral meristems on shoots and restricts the formation of branches. Hence, auxin is a signal that the shoot apical meristem is present and active.

Commercial orchardists spray synthetic IAA on fruit trees because it promotes uniform flowering, helps set the fruit, and also helps prevent fruit from dropping off the plant prematurely. Various synthetic auxins are used as **herbicides** (generally, weed killers), essentially stimulating a target plant to "grow itself to death." The most widely used herbicide in the world is the synthetic auxin 2,4-D (2,4-dichlorophenoxyacetic acid). This chemical kills broadleaf eudicots but spares monocots such as grasses. It is used extensively to prevent broadleaf weeds from growing in fields of cereal crops such as corn (which are monocots).

QUESTION: Why does a plant stem bend toward the light?

Experiment 1: The Darwins observed that the first shoot of an emerging grass seedling, which is sheathed by a coleoptile, bends toward sunlight shining through a window. They removed the shoot tip from a grass seedling and illuminated one side of the seedling.

Original observation

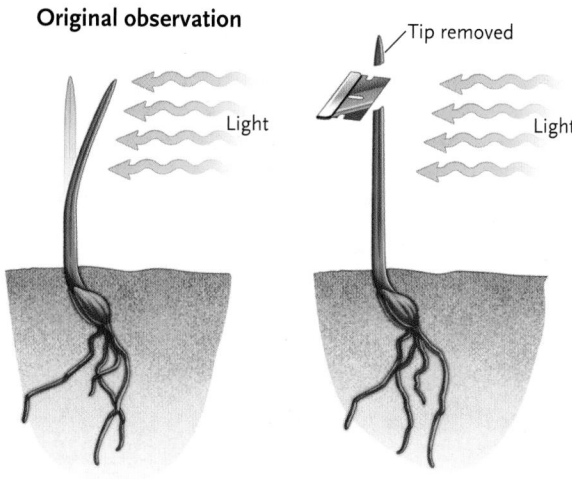

RESULT: The seedling neither grew nor bent.

Experiment 2: The Darwins divided seedlings into two groups. They covered the shoot tips of one group with an opaque cap and the shoot tips of the other group with a translucent cap. All of the seedlings were illuminated from the same side.

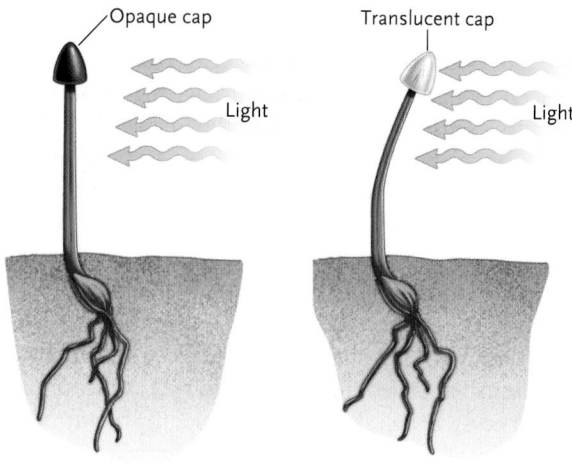

RESULT: The seedlings with opaque caps grew but did not bend. Those with translucent caps both grew *and* bent toward the light.

CONCLUSION: When seedlings are illuminated from one side, an unknown factor transmitted from a seedling's tip to the tissue below causes it to bend toward the light.

Experimental Research Figure 38.3 The Darwins' experiments on phototropism.

Based on C. R. Darwin. 1880. *The Power of Movement in Plants.* London: John Murray.

Auxin Transport. To exert its far-reaching effects on plant tissues, auxin must travel away from its main synthesis sites in shoot meristems and young leaves. Although IAA moves through plant tissues slowly—roughly 1 cm/h—this rate is 10 times as fast as could be explained by simple diffusion. How, then, is auxin transported?

Researchers adapted the agar block method pioneered by Went to trace the direction and rate of auxin movements in different kinds of tissues. A team led by Winslow Briggs at Stanford University determined that the shaded side of a shoot tip contains more IAA than the illuminated side. Hypothesizing that light causes IAA to move laterally from the illuminated to the shaded side of a shoot tip, the team then inserted a vertical barrier (a thin slice of mica) between the shaded and illuminated sides of a shoot tip. IAA could not cross the barrier, and when the shoot tip was illuminated it did not bend. In addition, the concentrations of IAA in the two sides of the shoot tip remained about the same. When the barrier was shortened so that the separated sides of the tip again touched, the IAA concentration in the shaded area increased significantly, and the tip did bend. The study confirmed that IAA initially moves laterally in the shoot tip, from the illuminated side to the shaded side, where it triggers the elongation of cells and curving of the tip toward light. Subsequent research showed that IAA then moves downward in a shoot by way of a top-to-bottom mechanism called **polar transport**. That is, IAA in a coleoptile or shoot tip travels from the apex of the tissue to its base, such as from the tip of a developing leaf to the stem. **Figure 38.5** outlines the experimental method that demonstrated polar transport. When IAA reaches roots, it moves toward the root tip.

In a stem, parenchyma cells next to vascular bundles apparently transport IAA. The hormone moves through and between cells, apparently travelling by polar transport. Figure 38.5 diagrams one widely accepted model for this process. As you can see, the IAA enters at one end by diffusing passively through cell walls, driven by concentration and electrochemical gradients produced by H^+ pumps in the plasma membrane. The hormone exits at the opposite end by active transport across the plasma membrane.

Although auxin is typically not found in xylem, there is increasing evidence that it sometimes moves from parenchyma cells into phloem and travels rapidly through plants. As this work continues, researchers will undoubtedly gain a clearer understanding of how plants distribute this crucial hormone to their growing parts.

Insights into Auxin Signal Transduction. Only in recent years have researchers begun to understand how target cells detect auxins and how subsequent transduction steps unfold. Experiments suggest that auxin binds different receptors, depending on the nature of the ensuing response. Responses that require a change in gene expression rely on a family of proteins often called simply TIR1, after the original protein discovered. Research in several laboratories confirmed that when IAA in a cell enters the nucleus and binds to TIR1, binding removes an existing

a. The procedure showing that IAA promotes elongation of cells below the shoot tip

b. The procedure showing that cells in contact with IAA grow faster than those farther away

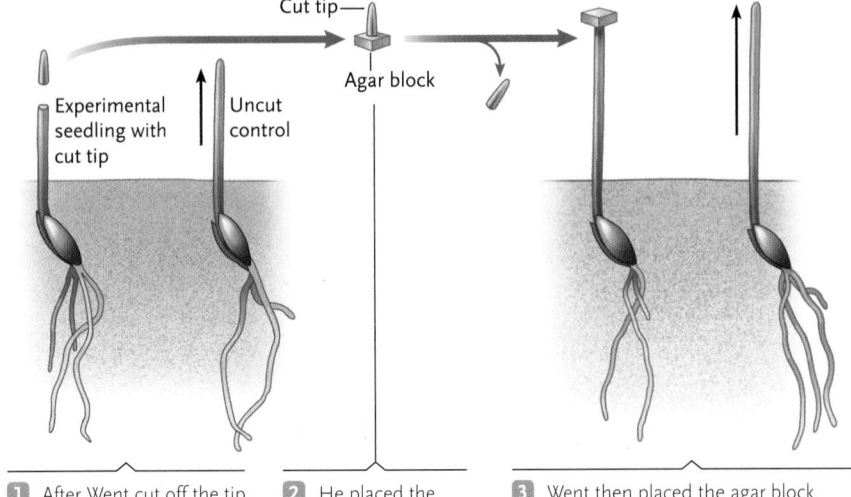
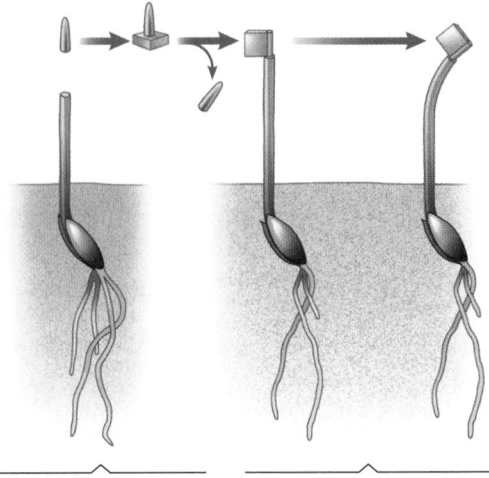

1 After Went cut off the tip of an oat seedling, the shoot stopped elongating, while a control seedling with an intact tip continued to grow.

2 He placed the excised tip on an agar block for 1–4 hours. During that time, IAA diffused into the agar block from the cut tip.

3 Went then placed the agar block containing auxin on another detipped oat shoot, and the shoot resumed elongation, growing about as rapidly as in a control seedling with an intact shoot tip.

1 Went removed the tip of a seedling and placed it on an agar block.

2 He placed the agar block containing auxin on one side of the shoot tip. Auxin moved into the shoot tip on that side, causing it to bend away from the hormone.

Figure 38.4

Two experiments by Frits Went demonstrating the effect of IAA on an oat coleoptile. Went carried out the experiments in darkness to prevent effects of light from skewing the results.

inhibition of the transcription of particular genes. As a result, the previously repressed genes are turned on and the cell's activity changes in some way. In responses that don't require a change in gene expression (such as swelling of the protoplast during cell expansion) auxin binds to a receptor called ABP1 (for auxin binding protein 1) at the outer surface of the plasma membrane. ABP1 binding triggers additional transduction steps within the cell, although the steps are not well understood. **Figure 38.6, p. 924,** sketches these pathways as researchers currently envision them.

Possible Mechanisms of IAA Action. Ever since auxins were discovered, researchers have sought to understand how IAA stimulates plant cells to elongate. As you may recall from Section 36.5, in an elongating plant cell the cellulose meshwork of the cell wall is first loosened and then stretched by turgor pressure. Several hormones, and auxin especially, apparently increase the plasticity (stretching) of the cell wall. Two major hypotheses have sought to explain this effect, and both may be correct.

Plant cell walls grow much faster in an acidic environment—that is, when the pH is less than 7. The **acid-growth hypothesis** proposes that auxin causes cells to secrete acid (H^+) into the cell wall by stimulating

Figure 38.5

The polar transport of auxin in plant shoots. Studies of IAA transport in plants have demonstrated that the hormone moves only in one direction, from the shoot tip downward to plant parts below. This diagram shows one model for this polar auxin transport. In this model, a plasma membrane H^+ pump maintains gradients of pH and electrical charge across the membrane, moving H^+ out of the cell using energy from ATP hydrolysis. Following the gradients, at the basal pole of a cell IAA diffuses through the transport proteins into the cell wall, then (as IAAH) into the next cell in line.

Figure 38.6

Model for auxin signal transduction pathways.

a. Fast response to IAA

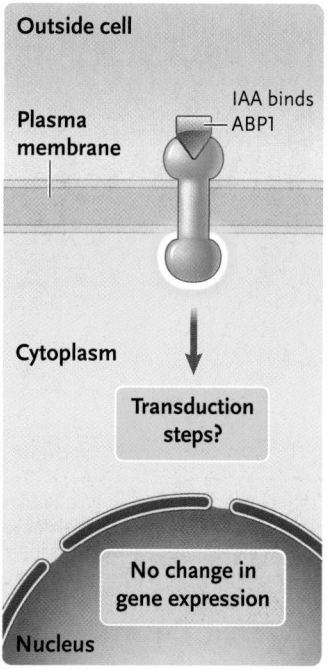

Outside cell

Plasma membrane

IAA binds ABP1

Cytoplasm

Transduction steps?

No change in gene expression

Nucleus

b. IAA responses requiring change in gene expression

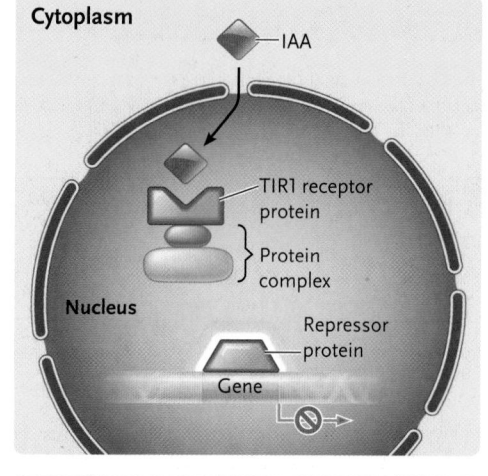

Cytoplasm

IAA

TIR1 receptor protein

Protein complex

Nucleus

Repressor protein

Gene

1 IAA moves from the cytoplasm into the nucleus, where it can bind to TIR1 linked to a protein complex.

Until IAA binds TIR1, a repressor protein prevents transcription of the gene responsible for the cellular response to IAA.

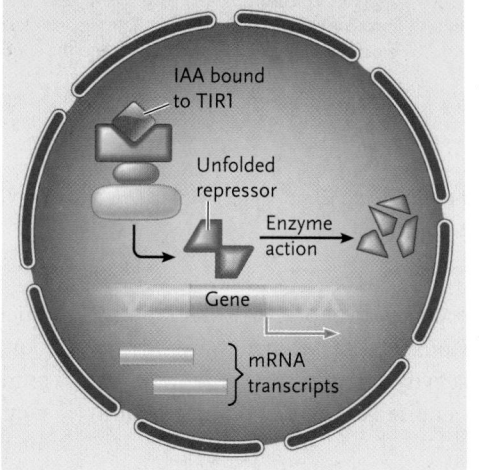

IAA bound to TIR1

Unfolded repressor

Enzyme action

Gene

mRNA transcripts

2 When IAA binds to TIR1, the protein complex is activated. It unfolds the repressor and the gene can be transcribed. Enzymes degrade the unfolded repressor and the released amino acids are recycled.

the plasma membrane H⁺ pumps to move protons from the cell interior into the cell wall; the increased acidity activates proteins called *expansins*, which penetrate the cell wall and disrupt bonds between cellulose microfibrils in the wall **(Figure 38.7)**. Activation of the plasma membrane H⁺ pump also produces a membrane potential that pulls K⁺ and other cations into the cell; the resulting osmotic gradient draws water into the cell, increasing turgor pressure and helping to stretch the "loosened" cell walls. (Experiments have shown that all of these effects occur shortly after ABP1 binds auxin.)

It is also possible that IAA triggers the expression of genes encoding enzymes that play roles in the synthesis of new wall components. Plant cells exposed to IAA do not show increased growth if they are treated with a chemical that inhibits protein synthesis. On the other hand, certain mRNAs rapidly increase in concentration within 10 to 20 min after stem sections are treated with auxin. Expression of mRNAs in response to auxin is likely regulated by the activity of microRNAs, as described in Chapter 14. It is not yet known exactly which proteins these mRNAs encode.

38.1d Gibberellins Also Stimulate Growth, Including the Elongation of Stems

Gibberellins stimulate various aspects of plant growth. Collectively they make up the largest class of plant hormones, with more than 130 recognized chemical variations. Gibberellins have been isolated from fungi and from flowering plants, including eudicots and some monocots. They may also exist in other plant groups. Their effects include triggering stem elongation and prompting seeds and buds to break dormancy. Research on barley embryos showed that a gibberellin provides signals during germination that lead to the enzymatic breakdown of endosperm, releasing nutrients that nourish the developing seedling (see Section 36.3).

Perhaps most apparent to humans is the ability of gibberellins to promote the lengthening of plant stems by stimulating both cell division and cell elongation. Synthesized in shoot and root tips and young leaves, gibberellins, like auxins, modify the properties of plant cell walls in ways that promote expansion (although the gibberellin mechanism does not involve

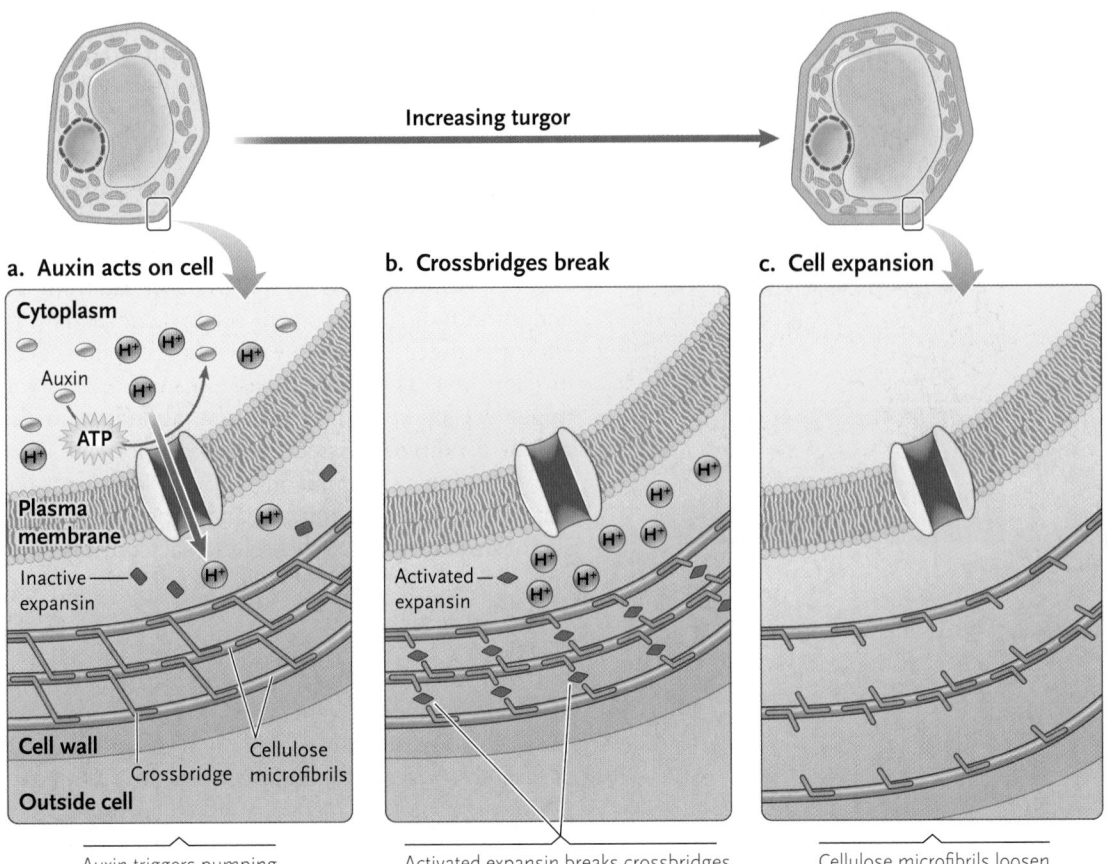

Figure 38.7
How auxin may regulate expansion of plant cells. According to the acid-growth hypothesis, plant cells secrete acid (H⁺) when auxin stimulates the plasma membrane H⁺ pumps to move protons into the cell wall; the increased acidity activates enzymes called expansins, which disrupt bonds between cellulose microfibrils in the wall. As a result, the wall becomes extensible and the cell can expand.

Increasing turgor

a. Auxin acts on cell

Cytoplasm

Auxin

ATP

Plasma membrane

Inactive expansin

Cell wall

Crossbridge   Cellulose microfibrils

Outside cell

Auxin triggers pumping of H⁺ into the cell wall.

b. Crossbridges break

Activated expansin

Activated expansin breaks crossbridges between cellulose microfibrils.

c. Cell expansion

Cellulose microfibrils loosen.

---

acidification of the cell wall). It may be that the two hormones both affect expansins, or are functionally linked in some other way. Experiments show that the general signal transduction pathway for gibberellins is essentially identical to that for auxins—the target cell's activity is altered when a gene repressor in the cell nucleus is inactivated and the gene begins to be expressed (see Figure 38.6).

In most plant species analyzed to date, the main controller of stem elongation is the gibberellin called GA1. Normally, GA1 is synthesized in small amounts in young leaves and transported throughout the plant in the phloem. When GA1 synthesis goes awry, the outcome is a dramatic change in the plant's stature. For example, experiments with a dwarf variety of peas (*Pisum sativum*) and some other species show that these plants and their taller relatives differ at a single gene locus. Normal plants make an enzyme required for gibberellin synthesis; dwarf plants of the same species lack the enzyme, and their internodes barely elongate at all.

Another stark demonstration of the effect gibberellins can have on internode growth is **bolting**, growth of a floral stalk in plants that form vegetative rosettes, such as cabbages (*Brassica oleracea*), iceberg lettuce (*Lactuca sativa*), and tarweeds such as the silversword (*Argyroxiphium sandwicense*). In a rosette plant, stem internodes are so short that the leaves appear to arise from a single node. When these plants flower, however, the stem elongates rapidly and flowers develop on the new stem parts **(Figure 38.8, p. 926)**. In nature, external cues such as increasing day length or warming after a cold snap stimulate gibberellin synthesis, and bolting occurs soon afterward. This observation supports the hypothesis that in rosette plants and possibly some others, gibberellins switch on internode lengthening when environmental conditions favour a shift from vegetative growth to reproductive growth.

Beyond the effects just mentioned, in monoecious species (having flowers of both sexual types on the same plant), applications of a gibberellin seem to encourage proportionately more "male" flowers to develop. As a result, there may be more pollen available to pollinate "female" flowers and, eventually, more fruit produced.

38.1e Cytokinins Enhance Growth and Retard Aging

Cytokinins play a major role in stimulating cell division (hence the name, which refers to cytokinesis), and many of their effects appear to come about as they interact with auxins. They were first discovered during experiments designed to define the nutrient media required for plant tissue culture. Researchers found that in addition to a carbon source such as sucrose or

Figure 38.8

Bolting. This Haleakala silversword, one of 28 endangered tarweed species that evolved in the isolation of the Hawaiian Islands, grows only along the upper fringe of Maui's volcanic Haleakala Crater. The plants may grow in a compact rosette form for more than half a century before several weeks of bolting produce a tall floral shoot, as shown here. After flowering, the plant dies.

Steven Maltby/Shutterstock.com

glucose, minerals, and certain vitamins, cells in culture also required two other substances. One was auxin, which promoted the elongation of plant cells but did not stimulate the cells to divide. The other substance could be coconut milk, which is actually liquid

endosperm, or it could be DNA that had been degraded into smaller molecules by boiling. When either was added to a culture medium along with an auxin, the cultured cells would begin dividing and grow normally.

We now know that the active ingredients in both boiled DNA and endosperm are cytokinins, which have a chemical structure similar to that of the nucleic acid base adenine. The most abundant natural cytokinin is zeatin, so-called because it was first isolated from the endosperm of young corn seeds (*Zea mays*). In endosperm, zeatin probably promotes the burst of cell division that takes place as a fruit matures. As you might expect, cytokinins are also abundant in the rapidly dividing meristem tissues of root and shoot tips. Cytokinins occur not only in flowering plants but also in many conifers, mosses, and ferns. They are also synthesized by many soil-dwelling bacteria and fungi and may be crucial to the growth of mycorrhizae, which help nourish thousands of plant species (see Section 37.3). Conversely, *Agrobacterium* and other microbes that cause plant tumours carry genes that regulate the production of cytokinins.

Cytokinins are synthesized mainly in root tips and apparently travel through the plant in xylem sap. Besides promoting cell division generally, they stimulate the growth of lateral buds and have other developmental and metabolic effects. For example, cytokinins promote expansion of young leaves (as leaf cells expand), cause chloroplasts to mature, and retard leaf aging. In concert with auxin they also coordinate the growth of roots and shoots. Investigators culturing tobacco tissues found that the relative amounts of auxin and a cytokinin strongly influenced not only growth, but also development, as illustrated in **Figure 38.9.**

Natural cytokinins can prolong the life of stored vegetables. Similar synthetic compounds are already widely used to prolong the shelf life of lettuces and mushrooms and to keep cut flowers fresh.

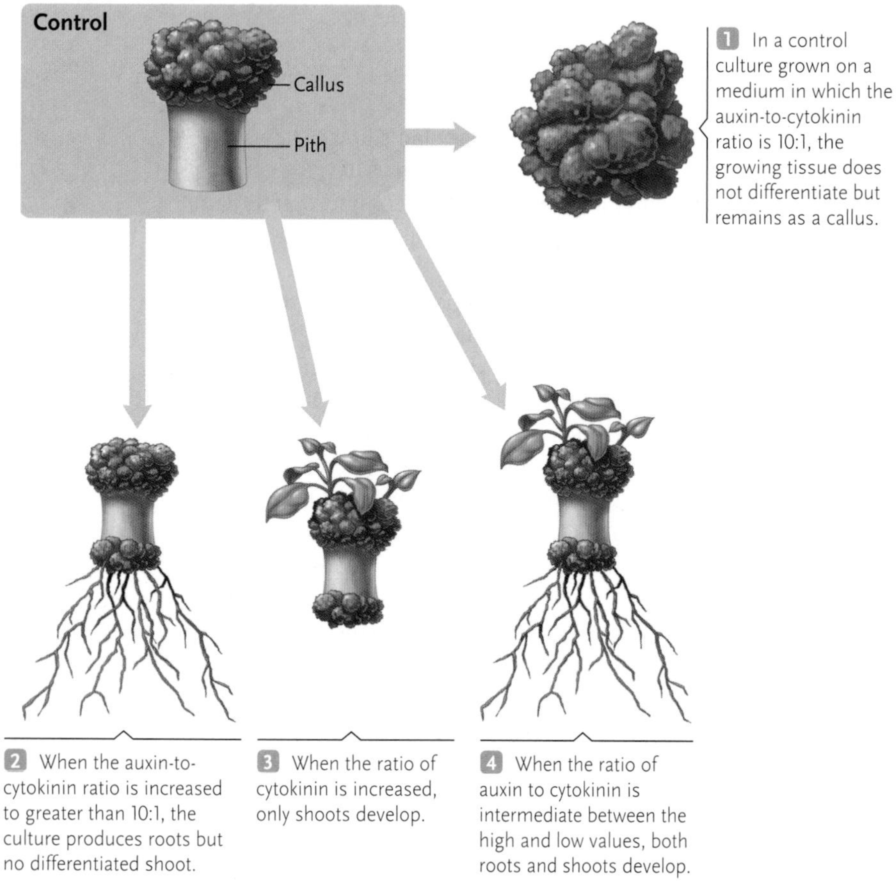

38.1f Ethylene Regulates Plant Aging and Some Other Responses

Most parts of a plant can produce **ethylene,** which is present in fruits, flowers, seeds, leaves, and roots and helps regulate a wide variety of plant physiological responses. Ethylene is also an unusual hormone, in part because it is structurally simple (see Table 38.1, p. 918) and in part because it is a gas at normal temperature and pressure.

Ethylene and Plant Senescence. The aging of plant parts is termed **senescence.** Senescence is a closely controlled process of deterioration that leads to the death of plant

---

Control

Callus

Pith

1 In a control culture grown on a medium in which the auxin-to-cytokinin ratio is 10:1, the growing tissue does not differentiate but remains as a callus.

2 When the auxin-to-cytokinin ratio is increased to greater than 10:1, the culture produces roots but no differentiated shoot.

3 When the ratio of cytokinin is increased, only shoots develop.

4 When the ratio of auxin to cytokinin is intermediate between the high and low values, both roots and shoots develop.

Figure 38.9

Interaction of auxin and cytokinin in the development of plant roots and shoots. Using callus tissue cultured from tobacco pith (spongy parenchyma from a stem), Folke Skoog and Carlos Miller demonstrated how normal development of tobacco plants requires the proportional interaction of auxin and cytokinin.

cells. In autumn the leaves of deciduous trees senesce, often turning yellow or red as chlorophyll and proteins break down, allowing other pigments to become more noticeable. Ethylene triggers the expression of genes leading to the synthesis of chlorophyllases and proteases, enzymes that launch the breakdown process. In many plants, senescence is associated with **abscission**, the dropping of flowers, fruits, and leaves in response to environmental signals. In this process, ethylene apparently stimulates the activity of enzymes that digest cell walls in an abscission zone—a localized region at the base of the petiole. The petiole detaches from the stem at that point **(Figure 38.10)**.

Senescence appears to require a range of cues. For some species, the funnelling of nutrients into reproductive parts may help to trigger senescence of leaves, stems, and roots. When the drain of nutrients is halted by removing each newly emerging flower or seed pod, a plant's leaves and stems stay green and vigorous much longer. Gardeners routinely remove flower buds from many plants to maintain vegetative growth. Evidence suggests that other cues are important, too. For instance, when a cocklebur is induced to flower under winterlike conditions, its leaves turn yellow regardless of whether the nutrient-demanding young flowers are left on or pinched off. It is as if a "death signal" forms that leads to flowering and senescence when there are fewer hours of daylight (typical of winter days). This observation underscores the general theme that many plant responses to the environment involve the interaction of multiple molecular signals.

Fruit Ripening: A Form of Senescence. Although the precise mechanisms are not well understood, ripening begins when a fruit starts to synthesize ethylene. The ripening process may involve the conversion of starch or organic acids to sugars, the softening of cell walls, or the rupturing of the cell membrane and loss of cell fluid. The same kinds of events occur in wounded plant tissues, which also synthesize ethylene.

Ethylene from an outside source can stimulate senescence responses, including ripening, when it binds to specific protein receptors on plant cells. The ancient Chinese observed that they could induce picked fruit to ripen faster by burning incense; later, it was found that the incense smoke contains ethylene. Today ethylene gas is widely used to ripen tomatoes, pineapples, bananas, honeydew melons, mangoes, papayas, and other fruit that have been picked and shipped while still green. Ripening fruit itself gives off ethylene, which is why placing a ripe banana in a closed sack of unripe peaches (or some other green fruit) can cause the fruit to ripen. Oranges and other citrus fruits may be exposed to ethylene to brighten their rind. Conversely, limiting fruit exposure to ethylene can delay ripening. Apples will keep for months without rotting if they are exposed to a chemical that inhibits ethylene production or if they are stored in an environment that inhibits the hormone's effects—including low atmospheric pressure and a high concentration of CO_2, which may bind ethylene receptors.

38.1g Brassinosteroids Regulate Plant Growth Responses

The dozens of steroid hormones classed as **brassinosteroids** all appear to be vital for normal growth in plants, for they stimulate cell division and elongation in a wide range of plant cell types. Confirmed as plant hormones in the 1980s, brassinosteroids are now the subject of intense research on their sources and effects. Although brassinosteroids have been detected in a wide variety of plant tissues and organs, the highest concentrations are found in shoot tips and in developing seeds and embryos—all examples of young, actively developing parts. In laboratory studies, the hormones have different effects depending on the tissue where they are active. They have promoted cell elongation, differentiation of vascular tissue, and elongation of a pollen tube after a flower is pollinated. By contrast, they inhibit the elongation of roots. First isolated from pollen of a plant in the mustard family, *Brassica napus* (a type of canola), brassinosteroids seem to regulate the expression of genes associated with a plant's growth responses to light. This role was underscored by the outcomes of experiments using

Biophoto Associates/Science Source. Colorization by: Mary Martin

Abscission zone at base of leaf where it joins the stem

Figure 38.10
Abscission zone in a maple (*Acer*). This longitudinal section is through the base of the petiole of a leaf.

mutant *Arabidopsis* plants that were homozygous for a defective gene called *bri1* (for brassinosteroid-insensitive receptor). The results provided convincing evidence that brassinosteroids mediate growth responses to light.

38.1h Abscisic Acid Suppresses Growth and Influences Responses to Environmental Stress

The hormone **abscisic acid** (ABA) has a variety of effects, many of which represent evolutionary adaptations to environmental challenges. Plants apparently synthesize ABA from carotenoid pigments inside plastids in leaves and possibly other plant parts. Several ABA receptors have been identified, and in general, we can group ABA effects into changes in gene expression that result in long-term inhibition of growth, and rapid, short-term physiological changes that are responses to immediate stresses, such as a lack of water, in a plant's surroundings. As its name suggests, at one time ABA was thought to play a central role in abscission. As already described, however, we now know that ethylene is the major abscission trigger.

Suppressing Growth in Buds and Seeds. Operating as a counterpoint to growth-stimulating hormones like gibberellins, ABA inhibits growth in response to environmental cues, such as seasonal changes in temperature and light. This growth suppression can last for many months or even years. For example, one of ABA's major growth-inhibiting effects is apparent in perennial plants, in which the hormone promotes dormancy in leaf buds—an important adaptive advantage in places where winter cold can damage young leaves. If ABA is applied to a growing leaf bud, the bud's normal development stops, and instead protective *bud scales*—modified, nonphotosynthetic leaves that are small, dry, and tough—form around the apical meristem and insulate it from the elements **(Figure 38.11).** After the scales develop, most cell metabolic activity shuts down and the leaf bud becomes dormant.

Figure 38.11

Bud scales, here on the bud of a perennial cornflower (*Centaurea montana*).

Dominator/Shutterstock.com

In some plants that produce fleshy fruits, such as apples and cherries, ABA is associated with the dormancy of seeds as well. As the seed develops, ABA accumulates in the seed coat, and the embryo does not germinate even if it becomes hydrated. The build-up of ABA in developing seeds does more than simply inhibit development, however. As early development draws to a close, ABA stimulates the transcription of certain genes, and large amounts of their protein products are synthesized. These proteins are thought to store nitrogen and other nutrients that the embryo will use when it eventually does germinate. Before such a seed can germinate, it will usually require a long period of cool, wet conditions, which stimulate the breakdown of ABA. Commercial growers often apply ABA and related growth inhibitors to plants slated to be shipped to plant nurseries. Dormant plants suffer less shipping damage, and the effects of the inhibitors can be reversed by applying a gibberellin.

Responses to Environmental Stress. ABA also triggers plant responses to various environmental stresses, including cold snaps, high soil salinity, and drought. A great deal of research has focused on how ABA influences plant responses to a lack of water. When a plant is water stressed, ABA helps prevent excessive water loss by stimulating stomata to close. As described in Chapter 35, flowering plants depend heavily on the proper functioning of stomata. When a lack of water leads to wilting, mesophyll cells in wilted leaves rapidly synthesize and secrete ABA. The hormone diffuses to guard cells, where an ABA receptor binds it. Binding stimulates the release of K^+ and water from the guard cells, and within minutes the stomata close.

ABA's role in stomatal closure—triggered by water stress or some other environmental cue—begins when the hormone activates a receptor in the plant cell plasma membrane. Experiments have shown that this binding launches a complex sequence of events that transduce the hormone signal. In a first step, binding activates G proteins that in turn activate phospholipase C. This enzyme then stimulates the synthesis of second messengers such as inositol triphosphate (IP_3). The second messenger diffuses through the cytoplasm and binds with calcium channels in structures such as the ER and tonoplast. Those channels then open, releasing calcium ions that activate protein kinases in the cytoplasm. In turn, these enzymes activate their target proteins by phosphorylating them. It's likely that the cellular response to ABA also involves cleaving phosphates needed for the phosphorylation step. Experiments have shown that an *Arabidopsis* mutant unable to respond to ABA lacks an enzyme that removes phosphate groups from certain proteins.

As described in Section 5.7, the original hormone signal is greatly amplified by a cascade of activated protein kinases, each one of which can activate a large number of target proteins.

38.1i Jasmonates and Oligosaccharins Regulate Growth and Function in Defence

In recent years, studies of plant growth and development have helped define the roles—or revealed the existence—of several other hormonelike compounds in plants. Like the well-established plant hormones just described, these substances are organic molecules, and only tiny amounts are required to alter some aspect of a plant's functioning. Some have long been known to exist in plants, but the extent of their signalling roles has only recently become better understood. This group includes **jasmonates** (JA), a family of about 20 compounds derived from fatty acids. Experiments with *Arabidopsis* and other plants have revealed numerous genes that respond to JA, including genes that help regulate root growth and seed germination. JA also appears to help plants "manage" stresses caused by deficiencies of certain nutrients (such as K$^+$). The JA family is best known, however, as part of the plant arsenal to limit damage by pathogens and predators, the topic of the following section.

Some other substances are also drawing keen interest from plant scientists, but because their signalling roles are still poorly understood, they are not widely accepted as confirmed plant hormones. A case in point involves the complex carbohydrates that are structural elements in the cell walls of plants and some fungi. Several years ago, researchers observed that in some plants, some of these oligosaccharides could serve as signalling molecules. Such compounds were named **oligosaccharins**, and one of their known roles is to defend the plant against pathogens. In addition, oligosaccharins have been proposed as growth regulators that adjust the growth and differentiation of plant cells, possibly by modulating the influences of growth-promoting hormones such as auxin. At this writing, researchers in many laboratories are pursuing a deeper understanding of this curious subset of plant signalling molecules.

STUDY BREAK

1. Which plant hormones promote growth, and which inhibit it?
2. Give examples of how some hormones have both promoting and inhibiting effects on growth in different parts of the plant at different times of the life cycle.
3. Summarize the various ways that chemical signals reaching plant cells are converted to changes in cell functioning.

38.2 Plant Chemical Defences

Plants do not have immune systems like those that have evolved in animals (the subject of Chapter 51), but higher plants have evolved an array of means for coping with biotic stressors in the environment. Over the millennia, virtually constant exposure to predation by herbivores and the onslaught of pathogens have resulted in a striking array of chemical defences that ward off or reduce damage to plant tissues from infectious bacteria, fungi, worms, or plant-eating insects **(Table 38.2, p. 930)**. You will discover in this section that, as with the defensive strategies of animals, plant defences include both general responses to any type of attack and specific responses to particular threats. Some get under way almost as soon as an attack begins, whereas others help promote the plant's long-term survival. And more often than not, multiple chemicals interact as the response unfolds.

38.2a Jasmonates and Other Compounds Interact in a General Response to Wounds

When an insect begins feeding on a leaf or some other plant part, the plant may respond to the resulting wound by launching what in effect is a cascade of chemical responses. These complex signalling pathways often rely on interactions among jasmonates, ethylene, or some other plant hormone. As the pathway unfolds, it triggers expression of genes leading to chemical and physical defences at the wound site. For example, in some plants, jasmonate induces a response leading to the synthesis of protease inhibitors, which disrupt an insect's capacity to digest proteins in the plant tissue. The protein deficiency in turn hampers the insect's growth and functioning.

A plant's capacity to recognize and respond to the physical damage of a wound apparently has been subject to strong selection pressure during plant evolution. When a plant is wounded experimentally, numerous defensive chemicals can soon be detected in its tissues. One of these, **salicylic acid**, or **SA** (a compound similar to aspirin, which is acetylsalicylic acid), seems to have multiple roles in plant defences, including interacting with jasmonates (see "Molecule behind Biology," p. 932).

Researchers are regularly discovering new variations of hormone-induced wound responses in plants. For example, experiments have elucidated some of the steps in an unusual pathway that thus far is known only in tomato (*Solanum lycopersicum*) and a few other plant species. As diagrammed in **Figure 38.12, p. 931,** the wounded plant rapidly synthesizes **systemin**, the first peptide hormone to be discovered in plants. (Various animal hormones are peptides, a topic covered in Chapter 43.) Systemin enters the phloem and is transported throughout the plant. Although various details of the signalling pathway have yet to be worked out, when receptive cells bind systemin, their plasma membranes release a lipid that is the chemical precursor of jasmonate. Next, jasmonate is synthesized, and it in turn sets in motion the expression of genes that encode protease inhibitors, which protect the plant

Table 38.2	Summary of Plant Chemical Defences	
Type of Defence	**Effects**
General Defences	
Jasmonate (JA) responses to wounds/injury by pathogens; pathways often include other hormones, such as ethylene	Synthesis of defensive chemicals such as protease inhibitors
Hypersensitive response to infectious pathogens (e.g., fungi, bacteria)	Physically isolates infection site by surrounding it with dead cells
PR (pathogenesis-related) proteins	Enzymes, other proteins that degrade cell walls of pathogens
Salicylic acid (SA)	Mobilizes during other responses and independently; induces the synthesis of PR proteins; operates in systemic acquired resistance
Systemin (in tomato)	Triggers JA response
Secondary Metabolites	
Phytoalexins	Antibiotic
Oligosaccharins	Trigger synthesis of phytoalexins
Systemic acquired resistance (SAR)	Long-lasting protection against some pathogens; components include SA and PR proteins that accumulate in healthy tissues
Specific Defences	
Gene-for-gene recognition of chemical features of specific pathogens (by binding with receptors coded by R genes)	Triggers defensive response (e.g., hypersensitive response, PR proteins) against pathogens
Other	
Heat-shock responses (encoded by heat-shock genes)	Synthesizes of chaperone proteins that reversibly bind other plant proteins and prevent denaturing caused by heat stress
"Antifreeze" proteins	In some species, stabilize cell proteins under freezing conditions

against attack, even in parts remote from the original wound.

38.2b The Hypersensitive Response and PR Proteins Are Other General Defences

Often, a plant that becomes infected by pathogenic bacteria or fungi counters the attack by way of a **hypersensitive response**—a defence that physically cordons off an infection site by surrounding it with dead cells. Initially, cells near the site respond by producing a burst of highly reactive oxygen-containing compounds (such as hydrogen peroxide, H_2O_2) that can break down nucleic acids, inactivate enzymes, or have other toxic effects on cells. The burst is catalyzed by enzymes in the plant cell's plasma membrane. It may begin the process of killing cells close to the attack site, and, as the response advances, programmed cell death may also come into play. In short order, the "sacrificed" dead cells wall off the infected area from the rest of the plant. Thus denied an ongoing supply of nutrients, the invading pathogen dies. A common sign of a successful hypersensitive response is a dead spot surrounded by healthy tissue **(Figure 38.13)**.

While the hypersensitive response is under way, salicylic acid triggers other defensive responses by an infected plant. One of its effects is to induce the synthesis of **pathogenesis-related proteins**, or **PR**

proteins. Some PR proteins are hydrolytic enzymes that break down components of a pathogen's cell wall. Examples are chitinases, which dismantle the chitin in the cell walls of fungi and so kill the cells. In some cases, plant cell receptors also detect the presence of fragments of the disintegrating wall and set in motion additional defence responses.

38.2c Secondary Metabolites Defend against Pathogens and Herbivores

Many plants counter bacteria and fungi by making **phytoalexins**, biochemicals of various types that function as antibiotics. When an infectious agent breaches a plant part, genes encoding phytoalexins begin to be transcribed in the affected tissue. For instance, when a fungus invades plant tissues, the enzymes it secretes may trigger the release of oligosaccharins. In addition to their roles as growth regulators (described in Section 38.1), these substances can also promote the production of phytoalexins, which have toxic effects on a variety of fungi. Plant tissues may also synthesize phytoalexins in response to attacks by viruses.

Phytoalexins are among many *secondary metabolites* produced by plants. Such substances are termed "secondary" because they are not routinely synthesized in all plant cells as part of basic metabolism. A wide range of plant species deploy secondary metabolites as defences

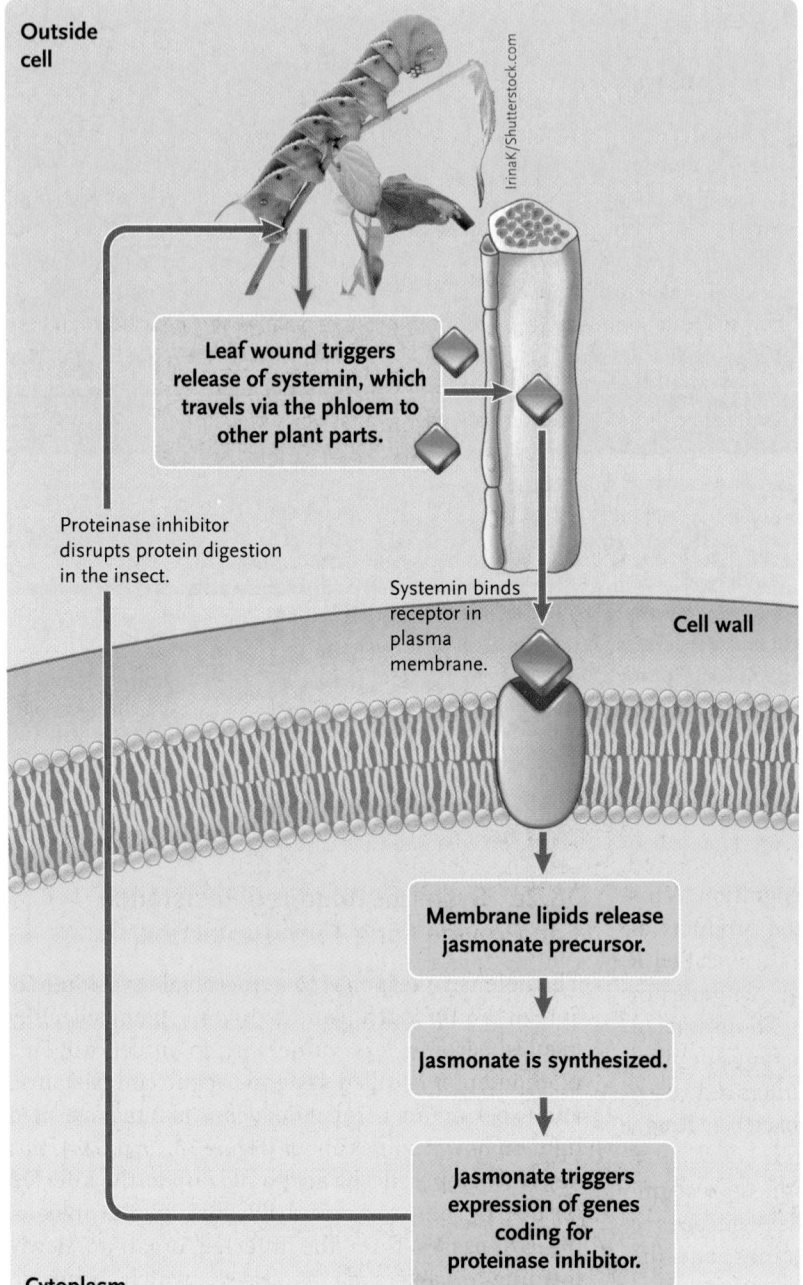

Figure 38.12

The systemin response to wounding. When a plant is wounded, it responds by releasing the protein hormone systemin. Transported through the phloem to other plant parts, in receptive cells systemin sets in motion a sequence of reactions that lead to the expression of genes encoding protease inhibitors—substances that can seriously disrupt an insect predator's capacity to digest protein.

Within the diagram:

Outside cell

Leaf wound triggers release of systemin, which travels via the phloem to other plant parts.

Proteinase inhibitor disrupts protein digestion in the insect.

Systemin binds receptor in plasma membrane.

Cell wall

Membrane lipids release jasmonate precursor.

Jasmonate is synthesized.

Jasmonate triggers expression of genes coding for proteinase inhibitor.

Cytoplasm

Figure 38.13
Evidence of the hypersensitive response. The dead brown spots on these leaves of a strawberry plant (*Fragaria*) are sites where a pathogen invaded, triggering the defensive destruction of the surrounding cells.

against feeding herbivores. Examples are alkaloids such as caffeine, cocaine, and the poison strychnine (in seeds of the *nux vomica* tree, *Strychnos nux-vomica*); tannins such as those in oak acorns; and various terpenes. The terpene family includes insect-repelling substances in cotton and the resins of conifers and the creosote bush, and essential oils produced by sage and basil plants. Because these terpenes are volatile—they easily diffuse out of the plant into the surrounding air—they can also provide indirect defence to a plant. Released from the wounds created by a munching insect, they attract other insects that prey on the herbivore.

38.2d Gene-for-Gene Recognition Allows Rapid Responses to Specific Threats

One of the most interesting questions with respect to plant defences is how plants first sense that an attack is under way. In some instances plants can apparently detect an attack by a specific predator through a

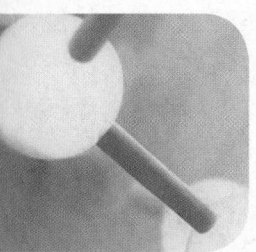

Salicylic Acid

In plants, salicylic acid **(Figure 1)** functions in systemic acquired resistance, in which a damaged plant part signals to other parts, triggering defence responses in those tissues. We don't yet understand how this response works. Salicylic acid has also been the basis for one of the most widely consumed medicines, aspirin. As far back as 500 BCE, the Greek physician Hippocrates noted that chewing the bark of *Salix alba* (white willow) could relieve fevers and pain. The aboriginal people of North American boiled *Salix* bark as remedies for aches and fevers. In the 1820s, the active ingredient, salicin, was isolated from willow bark and also from leaves of meadowsweet

(*Spiraea ulmaria*). In the human body, salicin is broken down to salicylic acid, so that acid was soon synthesized commercially and sold as medicine. However, salicylic acid is very bitter tasting and causes stomach irritation (because it reduces the stomach wall secretions that protect the stomach lining from acid produced in digestion), so a modified version, acetylsalicylic acid (ASA), was developed that avoided these problems. This form of ASA was trademarked as "Aspirin," with "a" standing for acetyl, and "spir" for *Spiraea*, the source of the salicin used to make salicylic acid **(Figure 2)**. Why don't we just take salicin isolated from plants instead of going through

the industrial process of making aspirin? It is safer to take a commercial product so that you know exactly how much of the active ingredient you are ingesting; the dosage of salicin in plant tissue is unpredictable, varying with conditions such as whether the plant has ramped up its production in response to an attack by a predator.

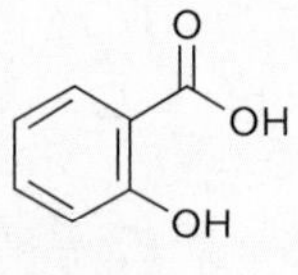

Figure 1
Structure of salicylic acid.

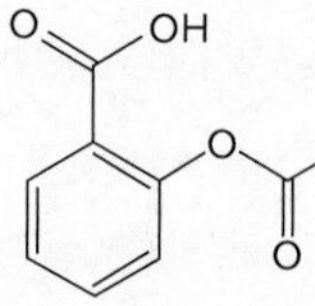

Figure 2
Structure of aspirin.

---

mechanism called **gene-for-gene recognition**. This term refers to a matchup between the products of dominant alleles of two types of genes: a so-called ***R* gene** (for "resistance") in a plant, and an ***Avr* gene** (for "avirulence") in a particular pathogen. Thousands of *R* genes have been identified in a wide range of plant species. Dominant *R* alleles confer enhanced resistance to plant pathogens including bacteria, fungi, and nematode worms that attack roots.

The basic mechanism of gene-for-gene recognition is simple: the dominant *R* allele encodes a receptor in plant cell plasma membranes, and the dominant pathogen *Avr* allele encodes a molecule that can bind the receptor. "Avirulence" implies that the pathogen becomes "not virulent" because binding of its *Avr* gene product triggers an immediate defence response in the plant. Trigger molecules run the gamut from proteins to lipids to carbohydrates that have been secreted by the pathogen or released from its surface **(Figure 38.14)**. Experiments have demonstrated a rapid-fire sequence of early biochemical changes that follow binding of the *Avr*-encoded molecule; these include changes in ion concentrations inside and outside plant cells and the production of biologically active oxygen compounds, heralding the hypersensitive response. In fact, of the instances of gene-for-gene recognition plant scientists have observed thus far, most trigger the hypersensitive response and the ensuing synthesis of PR proteins, with their antibiotic effects.

38.2e Systemic Acquired Resistance Can Provide Long-Term Protection

The defensive response to a microbial invasion may spread throughout a plant, so that the plant's healthy tissues become less vulnerable to infection. This phenomenon is called **systemic acquired resistance**, and experiments using *Arabidopsis* plants have shed light on how it comes about **(Figure 38.15, p. 934)**. In a key early step, salicylic acid builds up in the affected tissues. By some route, probably through the phloem, the SA passes from the infected organ to newly forming organs such as leaves, which begin to synthesize PR proteins—again, providing the plant with a "home-grown" antimicrobial arsenal. How does the SA exert this effect? It seems that when enough SA accumulates in a plant cell's cytoplasm, a regulatory protein called NPR-1 (for *n*onexpressor of *p*athogenesis-related genes) moves from the cytoplasm into the cell nucleus. There it interacts with factors that promote the transcription of genes encoding PR proteins.

In addition to synthesizing SA that will be transported to other tissues by a plant's vascular system, the damaged leaf also synthesizes a chemically similar compound, methyl salicylate. This substance is volatile, and researchers speculate that it may serve as an airborne "harm" signal, promoting defence responses in the plant that synthesized it and possibly in nearby plants as well.

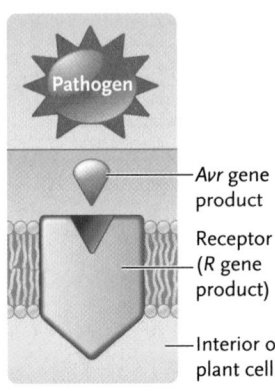

Required precondition
A plant has a dominant *R* gene encoding a receptor that can bind the product of a specific pathogen-dominant *Avr* gene.

Avr gene product

Receptor (*R* gene product)

Interior of plant cell

Figure 38.14

Model of how gene-for-gene resistance may operate. For resistance to develop, the plant must have a dominant *R* gene and the pathogen must have a corresponding dominant *Avr* gene. Products of such "matching" genes can interact physically, rather like the lock-and-key mechanism of an enzyme and its substrate. Most *R* genes encode receptors at the plasma membranes of plant cells. As diagrammed in step 1, when one of these receptors binds an *Avr* gene's product, the initial result may be changes in the movements of specific ions into or out of the cell and the activation of membrane enzymes that catalyze the formation of highly reactive oxygen-containing molecules. Such events help launch other signalling pathways that lead to a variety of defensive responses, including the hypersensitive response (step 2).

1 When the *R*-encoded receptor binds its matching *Avr* product, the binding triggers signalling pathways, leading to various defence responses in the plant.

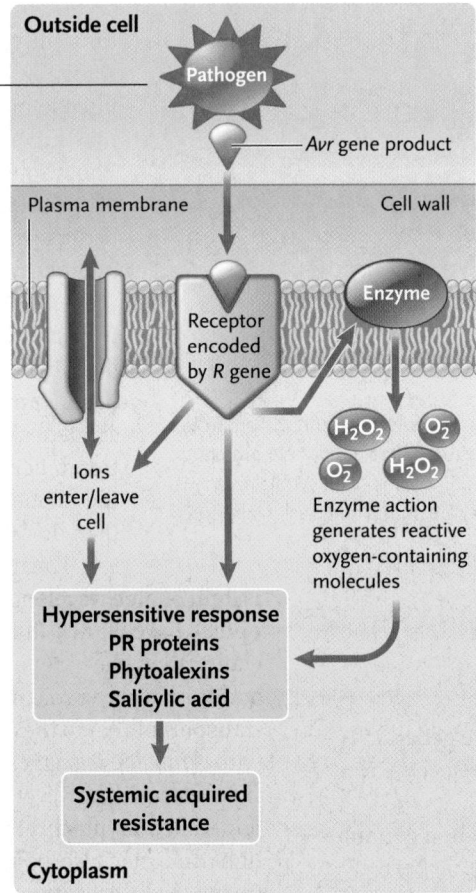

Outside cell

Pathogen

Avr gene product

Plasma membrane

Cell wall

Receptor encoded by *R* gene

Enzyme

H_2O_2 O_2^-
O_2^- H_2O_2

2 Fluxes of ions and enzyme activity at the plasma membrane contribute to the hypersensitive response. Soon PR proteins, phytoalexins, and salicyic acid (SA) are synthesized. The PR proteins and phytoalexins combat pathogens directly. SA promotes systemic acquired resistance.

Ions enter/leave cell

Enzyme action generates reactive oxygen-containing molecules

**Hypersensitive response
PR proteins
Phytoalexins
Salicylic acid**

Systemic acquired resistance

Cytoplasm

38.2f Extremes of Heat and Cold Also Elicit Protective Chemical Responses

Plant cells also contain **heat-shock proteins (HSPs)**, a type of chaperone protein found in cells of many species. In general, HSPs bind and stabilize other proteins, including enzymes, which might otherwise stop functioning if they were to become denatured by rising temperature. Plant cells may rapidly synthesize HSPs in response to various stimuli, including a sudden temperature rise. For example, experiments with cells and seedlings of soybean (*Glycine max*) showed that when the temperature rose 10 to 15°C, in less than 5 min smRNA transcripts coding for as many as 50 different HSPs were present in cells. When the temperature returns to a normal range, HSPs release bound proteins, which can then resume their usual functions. Further studies have revealed that HSPs help protect plant cells subjected to other environmental stresses as well, including drought, salinity, and cold.

Like extreme heat, freezing can also be lethal to plants. If ice crystals form in cells, they can literally tear the cell apart. In many cold-resistant species, dormancy (discussed in Section 38.4) is the long-term strategy for dealing with cold, but in the short term, such as an unseasonable cold snap, some species also undergo a rapid shift in gene expression that equips cold-stressed cells with so-called antifreeze proteins. Like HSPs, these molecules are thought to help maintain the structural integrity of other cell proteins.

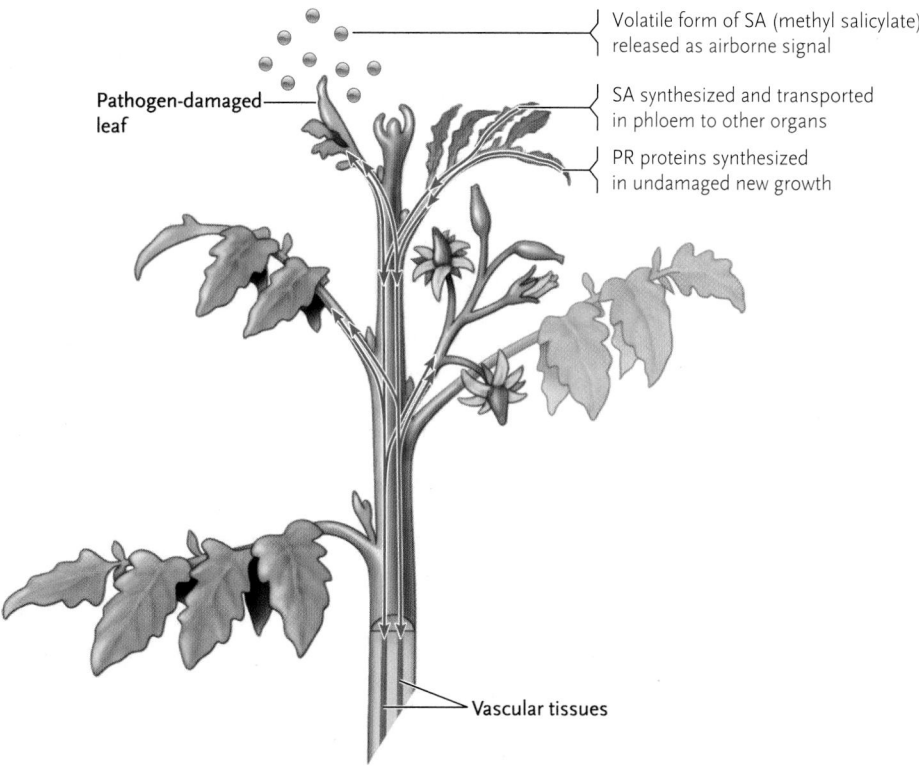

Pathogen-damaged leaf

Volatile form of SA (methyl salicylate) released as airborne signal

SA synthesized and transported in phloem to other organs

PR proteins synthesized in undamaged new growth

Vascular tissues

Figure 38.15

A proposed mechanism for systemic acquired resistance. When a plant successfully fends off a pathogen, the defensive chemical SA is transported in the phloem to other plant parts, where it may help protect against another attack by stimulating the synthesis of PR proteins. In addition, the plant synthesizes and releases a slightly different, more volatile form of SA called methyl salicylate. It may serve as an airborne signal to other parts of the plant as well as to neighbouring plants.

STUDY BREAK

1. Which plant chemical defences are general responses to attack, and which are specific to a particular pathogen?
2. Why is salicylic acid considered to be a general systemic response to damage?
3. How is the hypersensitive response integrated with other chemical defences?

38.3 Plant Movements

Although a plant cannot move from place to place as external conditions change, plants do alter the orientation of their body parts in response to environmental stimuli. As noted earlier in the chapter, growth toward or away from a unidirectional stimulus, such as light or gravity, is called a tropism. Tropic movement involves permanent changes in the plant body because cells in particular areas or organs grow differentially in response to the stimulus. Plant physiologists do not fully understand how tropisms occur, but they are fascinating examples of the complex abilities of plants to adjust to their environment. This section will

also touch on two other kinds of movements—developmental responses to physical contact, and changes in the position of plant parts that are not related to the location of the stimulus.

38.3a Phototropisms Are Responses to Light

Light is a key abiotic stimulus for many kinds of organisms. Phototropisms, which we have already discussed in the section on auxins, are growth responses to a directional light source. As the Darwins discovered, if light is more intense on one side of a stem, the stem may curve toward the light **(Figure 38.16a).** Phototropic movements are extremely adaptive for photosynthesizing organisms because they help maximize the exposure of photosynthetic tissues to sunlight.

How do auxins influence phototropic movements? In a coleoptile that is illuminated from one side, IAA moves by polar transport into the cells on the shaded side (Figure 38.16b–d). Phototropic bending occurs because cells on the shaded side elongate more rapidly than do cells on the illuminated side.

The main stimulus for phototropism is light of blue wavelengths. Experiments on corn coleoptiles have shown that a large, yellow pigment molecule called phototropin can absorb blue wavelengths, and it may play a role in stimulating the initial lateral transport of IAA to the dark side of a shoot tip. Studies with *Arabidopsis* suggest there is more than one blue light receptor, however. One is a light-absorbing protein called **cryptochrome,** which is sensitive to blue light and may also be an important early step in the various light-based growth responses. As you will read later, cryptochrome appears to have a role in other plant responses to light as well.

38.3b Gravitropism Orients Plant Parts to the Pull of Gravity

Plants show growth responses to Earth's gravitational pull, a phenomenon called **gravitropism.** After a seed germinates, the primary root curves down, toward the "pull" (positive gravitropism), and the shoot curves up (negative gravitropism).

Several hypotheses seek to explain how plants respond to gravity. The most widely accepted hypothesis proposes that plants detect gravity much as animals do—that is, particles called **statoliths** in certain cells move in the direction gravity pulls them. In the semicircular canals of human ears, calcium carbonate crystals serve as statoliths; in most plants the statoliths

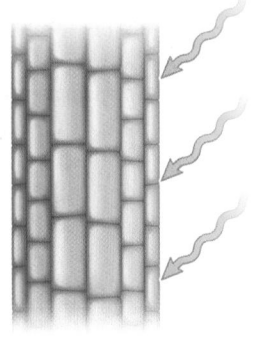

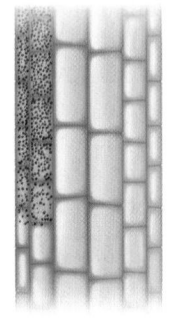

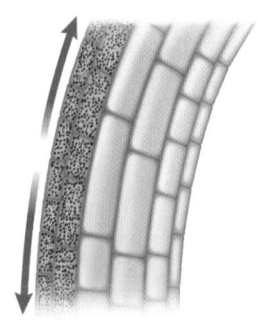

a. Seedlings bend toward light.

b. Rays from the Sun strike one side of a shoot tip.

c. Auxin (red) diffuses down from the shoot tip to cells on its shaded side.

d. The auxin-stimulated cells elongate more quickly, causing the seedling to bend.

Figure 38.16

Phototropism in seedlings. **(a)** Tomato seedling grown in darkness; its right side was illuminated for a few hours before it was photographed. **(b–d)** Hormone-mediated differences in the rates of cell elongation bring about the bending toward light. (Auxin is shown in red.)

are amyloplasts, modified plastids that contain starch grains. In eudicot angiosperm stems, amyloplasts are often present in one or two layers of cells just outside the vascular bundles. In monocots such as cereal grasses, amyloplasts are located in a region of tissue near the base of the leaf sheath. In roots, amyloplasts occur in the root cap. If the spatial orientation of a plant cell is shifted experimentally, its amyloplasts sink through the cytoplasm until they come to rest at the bottom of the cell **(Figure 38.17).**

How do amyloplast movements translate into an altered growth response? The full explanation appears to be complex, and there is evidence that somewhat different mechanisms operate in stems and in roots. In stems, the sinking of amyloplasts may provide a mechanical stimulus that triggers a gene-guided redistribution of IAA. For example, when a potted sunflower seedling is turned on its side in a dark room, within 15 to 20 min cell elongation decreases markedly on the upper side of the growing horizontal stem, but

increases on the lower side. With the adjusted growth pattern, the stem curves upward, even in the absence of light. Using different types of tests, researchers have been able to document the shifting of IAA from the top to the bottom side of the stem. The changing auxin gradient correlates with the altered pattern of cell elongation.

In roots, a high concentration of auxin has the opposite effect—it inhibits cell elongation. If a root is placed on its side, amyloplasts in the root cap accumulate near the side wall that is now the bottom side of the cap. This shift stimulates cell elongation in the opposite wall, and within a few hours the root once again curves downward. In root tips of many plants, however, especially eudicots, researchers could not detect a change in IAA concentration that correlates with the changing position of amyloplasts. Eventually experiments on gravitropism in soybean (*Glycine max*) root tips suggested that the IAA signal is transduced by way of a signal cascade. The hormone induces the accumulation of

a. Root oriented vertically

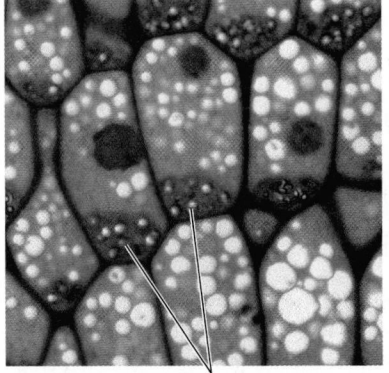

Statoliths

b. Root oriented horizontally

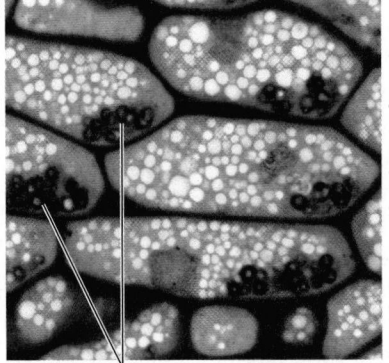

Statoliths

Figure 38.17

Evidence that supports the statolith hypothesis. When a corn root was laid on its side, amyloplasts—statoliths—in cells from the root cap settled to the bottom of the cells within 5 to 10 minutes. Statoliths may be part of a gravity-sensing mechanism that redistributes auxin through a root tip.

nitric oxide (NO) at the downward side of the root tip, where NO in turn appears to induce cGMP. This second messenger then delivers the original IAA signal. The sequence inhibits the elongation of cells on the tip's downward side, so the root curves downward.

Along with IAA, calcium ions (Ca^{2+}) appear to play a major role in gravitropism. For example, if Ca^{2+} is added to an otherwise untreated agar block that is then placed on one side of a root cap, the root will bend toward the block. In this way, experimenters have been able to manipulate the direction of growth so that the elongating root forms a loop. Similarly, if an actively bending root is deprived of Ca^{2+}, the gravitropic response abruptly stops. By contrast, the negative gravitropic response of a shoot tip is inhibited when the tissue is exposed to excess calcium.

Just how Ca^{2+} interacts with IAA in gravitropic responses is unknown. One hypothesis posits that calcium functions as an activator. Calcium binds to a small protein called *calmodulin*, activating it in the process. Activated calmodulin in turn can activate a variety of key cell enzymes in many organisms, both plants and animals. One possibility is that calcium-activated calmodulin stimulates cell membrane pumps that enhance the flow of both IAA and calcium through a gravity-stimulated plant tissue.

Some of the most active research in plant biology focuses on the intricate mechanisms of gravitropism. For example, there is increasing evidence that in many plants, cells in different regions of stem tissue differ in their sensitivity to IAA, and that gravitropism is linked in some fundamental way to these differences. In a few plants, including some cultivated varieties of corn and radish, the direction of the gravitropic response by a seedling's primary root is influenced by light. Clearly there is much more to be learned.

38.3c Thigmotropism and Thigmomorphogenesis Are Responses to Physical Contact

Varieties of peas, grapes, and some other plants demonstrate **thigmotropism** (*thigma* = touch), which is growth in response to contact with a solid object. Thigmotropic plants typically have long, slender stems and cannot grow upright without physical support. They often have *tendrils*, modified stems or leaves that can rapidly curl around a fencepost or the sturdier stem of a neighbouring plant. If one side of a grapevine stem grows against a trellis, for example, specialized epidermal cells on that side of the stem tendril shorten whereas cells on the other side of the tendril rapidly elongate. Within minutes the tendril starts to curl around the trellis, forming tight coils that provide strong support for the vine stem. **Figure 38.18** shows thigmotropic twisting in the passion flower (*Passiflora*).

Figure 38.18
Thigmotropism in a passion flower (*Passiflora*) tendril, which is twisted around a support.

Auxin and ethylene may be involved in thigmotropism, but most details of the mechanism remain elusive.

The rubbing and bending of stems caused by frequent strong winds, rainstorms, grazing animals, and even farm machinery can inhibit the overall growth of plants and can alter their growth patterns. In this phenomenon, called **thigmomorphogenesis**, a stem stops elongating and instead adds girth when it is regularly subjected to mechanical stress. Merely shaking some plants daily for a brief period will inhibit their upward growth, but although such plants may be shorter, their thickened stems will be stronger. Thigmomorphogenesis helps explain why plants growing outdoors are often shorter, have somewhat thicker stems, and are not as easily blown over as plants of the same species grown indoors. Trees growing near the snowline of windswept mountains show an altered growth pattern that reflects this response to wind stress.

Research on the cellular mechanisms of thigmomorphogenesis has begun to yield tantalizing clues. In one study, investigators repeatedly sprayed *Arabidopsis* plants with water and imposed other mechanical stresses and then sampled tissues from the stressed plants. The samples contained as much as double the usual amount of mRNA for at least four genes, which had been activated by the stress. The mRNAs encoded calmodulin and several other proteins that may have roles in altering *Arabidopsis* growth responses. The test plants were also short, generally reaching only half the height of unstressed controls.

38.3d Nastic Movements Are Nondirectional

Tropisms are responses to directional stimuli, such as light striking one side of a shoot tip, but many plants also exhibit **nastic movements** (*nastos* = pressed close together)—reversible responses to nondirectional stimuli, such as mechanical pressure or humidity. We see nastic movements in leaves, leaflets, and even flowers. For instance, certain plants exhibit nastic sleep movements, holding their leaves (or flower petals) in roughly horizontal positions during the day but folding them closer to the stem at night **(Figure 38.19)**. Tulip flowers "go to sleep" in this way.

Many nastic movements are temporary and result from changes in cell turgor. For example, the daily opening and closing of stomata in response to changing light levels are nastic movements, as is the traplike closing of the lobed leaves of the Venus fly-trap when an insect brushes against hairlike sensory structures on the leaves. The leaves of *Mimosa pudica*, the sensitive plant, also close in a nastic response to mechanical pressure. Each *Mimosa* leaf is divided into pairs of leaflets **(Figure 38.20a)**. Touching even one leaflet at the leaf tip triggers a chain reaction in which each pair of leaflets closes up within seconds (Figure 38.20b).

In many turgor-driven nastic movements, water moves into and out of the cells in **pulvini** (*pulvinus* = cushion), thickened pads of tissue at the base of a leaf or petiole. Stomatal movements depend on changing concentrations of ions within guard cells, and pulvinar cells drive nastic leaf movements in *Mimosa* and numerous other plants by the same mechanism (Figure 38.20c).

How is the original stimulus transferred from cells in one part of a leaf to cells elsewhere? The answer lies in the polarity of charge across cell plasma

Figure 38.19
Nastic sleep movements in a bloodroot plant (*Sanguinaria canadensis*).

membranes (see Section 5.5). Touching a *Mimosa* leaflet triggers an **action potential**—a brief reversal in the polarity of the membrane charge. When an action potential occurs at the plasma membrane of a pulvinar cell, the change in polarity causes potassium ion (K^+) channels to open, and ions flow out of the cell, setting up an osmotic gradient that draws water out as well. As water leaves by osmosis, turgor pressure falls, pulvinar cells become flaccid, and the leaflets move together. Later, when the process is reversed, the pulvinar cells regain turgor and the leaflets spread apart. Action potentials travel between parenchyma cells in the pulvini via plasmodesmata at the rate of about 2 cm/s. Animal nerves conduct

a. Undisturbed plant

b. Plant response to touch

Figure 38.20
Nastic movements in leaflets of *Mimosa pudica*, the sensitive plant. **(a)** In an undisturbed plant the leaflets are open. If a leaflet near the leaf tip is touched, changes in turgor pressure in pulvini at the base cause the leaf to fold closed **(b, c)**. The diagram sketches this folding movement in cross-section. Other leaflets close in sequence as action potentials transmit the stimulus along the leaf.

c. Leaf folding mechanism

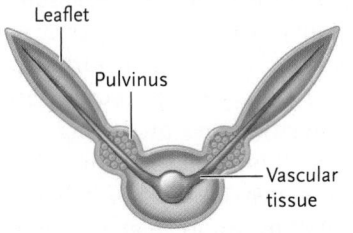

Leaflet

Pulvinus

Vascular tissue

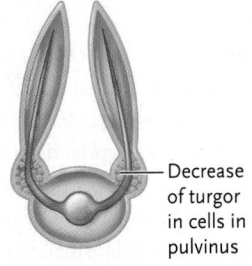

Decrease of turgor in cells in pulvinus

similar changes in membrane polarity along their plasma membranes (see Chapter 44). These changes in polarity, which are also called action potentials, occur much more rapidly—at velocities between 1 and 100 m/s.

Stimuli other than touch can also trigger action potentials leading to nastic movements. Cotton, soybean, sunflower, and some other plants display *solar tracking*, nastic movements in which leaf blades are oriented toward the east in the morning and then steadily change their position during the day, following the Sun across the sky. Such movements maximize the amount of time that leaf blades are perpendicular to the Sun, which is the angle at which photosynthesis is most efficient.

STUDY BREAK

1. What is the direct stimulus for phototropism? For gravitropism?
2. Explain how nastic movements differ from tropic movements.

38.4 Plant Biological Clocks

Like all eukaryotic organisms, plants have internal time-measuring mechanisms called **biological clocks** that adapt the organism to recurring environmental changes. In plants, biological clocks help adjust both daily and seasonal activities.

38.4a Circadian Rhythms Are Based on 24-Hour Cycles

Some plant activities occur regularly in cycles of about 24 hours, even when environmental conditions remain constant. These are **circadian rhythms** (*circa* = around, *dies* = day). In Chapter 34, we noted that stomata open and close on a daily cycle, even when plants are kept in total darkness. Nastic sleep movements, described earlier, are another example of a circadian rhythm. Even when a plant that exhibits such movements is kept in constant light or darkness for a few days, it folds its leaves into the "sleep" position at roughly 24-hour intervals. In some way, the plant measures time without sunrise (light) and sunset (darkness). Such experiments demonstrate that internal controls, rather than external cues, largely govern circadian rhythms.

Circadian rhythms and other activities regulated by a biological clock help ensure that plants of a single species do the same thing, such as flowering, at the same time. For instance, flowers of the aptly named four-o'clock plant (*Mirabilis jalapa*) open predictably every 24 hours—in nature, in the late afternoon. Such coordination can be crucial for successful pollination.

Although some circadian rhythms can proceed without direct stimulus from light, many biological clock mechanisms are influenced by the relative lengths of day and night.

38.4b Photoperiodism Involves Seasonal Changes in the Relative Length of Night and Day

Obviously, environmental conditions in a 24-hour period are not the same in summer as they are in winter. In North America, for instance, winter temperatures are cooler and winter day length is shorter. Experimenting with tobacco and soybean plants in the early 1900s, two American botanists, Wightman Garner and Henry Allard, elucidated a phenomenon they called **photoperiodism**, in which plants respond to changes in the relative lengths of light and dark periods in their environment during each 24-hour period. Through photoperiodism, the biological clocks of plants (and animals) make seasonal adjustments in their patterns of growth, development, and reproduction.

In plants, we now know that a family of blue-green pigments collectively called **phytochrome** often serves as a switching mechanism in the photoperiodic response, signalling the plant to make seasonal changes. Plants synthesize phytochrome in an inactive form, P_r, which absorbs light of shorter wavelengths (about 660 nm) at the "red" end of the spectrum (see Figure 1.3). Sunlight contains relatively more red light than far-red light, which has a longer wavelength (about 730 nm). During daylight hours when red wavelengths dominate, P_r absorbs red light. Absorption of red light triggers the conversion of phytochrome to an active form designated P_{fr}, which absorbs light of far-red wavelengths. At sunset, at night, or even in shade, where far-red wavelengths predominate, P_{fr} reverts to P_r **(Figure 38.21)**.

In nature a high concentration of P_{fr} "tells" a plant that it is exposed to sunlight. The plant's ability to sense sunlight is vital given that over time sunlight provides favourable conditions for leaf growth, photosynthesis, and flowering. The exact mechanism of this crucial transfer of environmental information is still not fully understood. Phytochrome activation may stimulate plant cells to take up Ca^{2+} ions, or it may induce certain plant organelles to release them. Either way, when free calcium ions combine with calcium-binding proteins (such as calmodulin), they may initiate at least some responses to light. Botanists suspect that P_{fr} controls the types of enzymes being produced in particular cells—and different enzymes are required for seed germination; stem elongation and branching; leaf expansion; and the formation of flowers, fruits, and seeds. When plants adapted to full sunlight are grown in darkness, they put more resources into

a. Interconversion of phytochrome

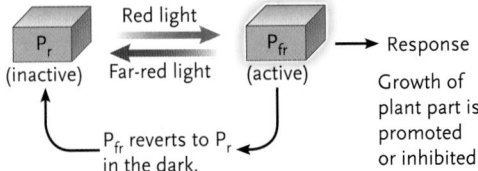

Red light

P_r (inactive) → P_{fr} (active) → Response

Far-red light

Growth of plant part is promoted or inhibited.

P_{fr} reverts to P_r in the dark.

b. Changing absorption spectra

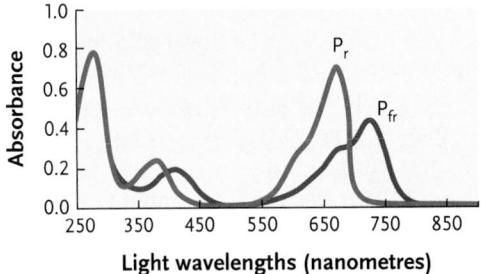

Absorbance

P_r

P_{fr}

Light wavelengths (nanometres)

Figure 38.21
The phytochrome switching mechanism, which can promote or inhibit growth of different plant parts. **(a)** Interconversion of phytochrome from the active form (P_{fr}) to the inactive form (P_r). **(b)** The absorption spectra associated with the interconversion of P_r and P_{fr}.

G. R. 'Dick' Roberts/NSIL/Visuals Unlimited, Inc.

Figure 38.22
Effects of the absence of light on young bean plants (Phaseolus). The seedling on the left was grown in darkness for several days. Its leaves are yellow because it could form carotenoids but not chlorophyll in darkness. It also has a larger stem, smaller leaves, and a smaller root system than the seedling on the right, which was grown in the light.

stem elongation and less into leaf expansion or stem branching **(Figure 38.22)**.

Cryptochrome—which, recall, is sensitive to blue light and appears to influence light-related growth responses—also interacts with phytochromes in producing circadian responses. Researchers have recently discovered that cryptochrome occurs not only in plants but also in animals such as fruit flies and mice. Does it act as a circadian photoreceptor in both kingdoms? Only further study will provide the answer.

38.4c Cycles of Light and Dark Often Influence Flowering

Photoperiodism is especially apparent in the flowering process. Like other plant responses, flowering is often keyed to changes in day length through the year and to the resulting changes in environmental conditions. Corn, soybeans, peas, and other annual plants begin flowering after only a few months of growth. Roses and other perennials typically flower every year or after several years of vegetative growth. Carrots, cabbages, and other biennials typically produce roots, stems, and leaves the first growing season, die back to soil level in autumn, then grow a new flower-forming stem the second season.

In the late 1930s, Karl Hamner and James Bonner grew cocklebur plants (*Xanthium strumarium*) in chambers in which the researchers could carefully control environmental conditions, including photoperiod. And they made an unexpected discovery: flowering occurred only when the test plants were exposed to a single night of 8.5 hours of uninterrupted

darkness. The length of the "day" in the growth chamber did not matter, but if light interrupted the dark period for even a minute or two, the plant would not flower at all. Subsequent research confirmed that for most angiosperms, it is the length of darkness, not light, that controls flowering.

Kinds of Flowering Responses. The photoperiodic responses of flowering plants are so predictable that botanists have long used them to categorize plants **(Figure 38.23)**. The categories, which refer to day length, reflect the fact that scientists recognized the phenomenon of photoperiodic flowering responses long before they understood that darkness, not light,

Flowers

Jan Zeevaart

Figure 38.23
Effect of day length on spinach (*Spinacia oleracea*), a long-day plant. The plant on the left was not exposed to light long enough to trigger flowering. The plant on the right did flower after light exposure that mimicked the long days of spring.

was the cue. **Long-day plants**, such as irises, daffodils, and corn, usually flower in spring when dark periods become shorter and day length becomes longer than some critical value—usually 9 to 16 hours. **Short-day plants**, including cockleburs, chrysanthemums, and potatoes, flower in late summer or early autumn when dark periods become longer and day length becomes shorter than some critical value. **Intermediate-day plants,** such as sugarcane, flower only when day length falls in between the values for long-day and short-day plants. **Day-neutral plants,** such as dandelions and roses, flower whenever they become mature enough to do so, without regard to photoperiod.

Experiments demonstrate what happens when plants are grown under the "wrong" photoperiod regimes. For instance, spinach, a long-day plant, flowers and produces seeds only if it is exposed to no more than 10 hours of darkness each day for 2 weeks (see Figure 38.23). **Figure 38.24** illustrates the results of an experiment to test the responses of short-day and long-day plants to night length. In this experiment, bearded iris plants (*Iris* species), which are long-day plants, and chrysanthemums, which are short-day plants, were exposed to a range of light conditions. In each case, when the researchers interrupted a critical dark period with a pulse of red light, the light reset the plants' clocks. The experiment provided clear evidence that short-day plants flower only when nights are longer than a critical value—and long-day plants flower only when nights are shorter than a critical value.

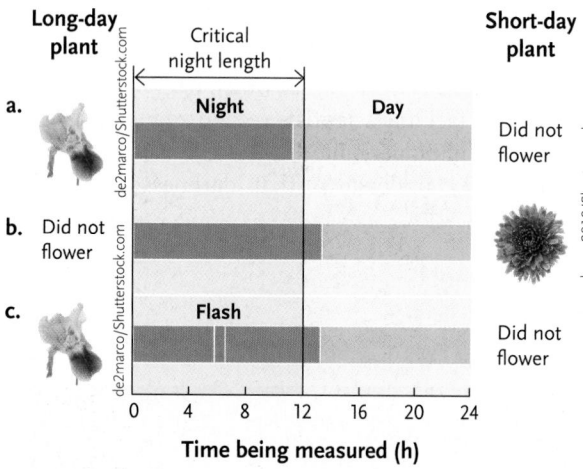

Figure 38.24

Experiments showing that short-day and long-day plants flower by measuring night length. Each horizontal bar signifies 24 hours. Blue bars represent night, and yellow bars day. **(a)** Long-day plants such as bearded irises flower when the night is shorter than a critical length, whereas **(b)** short-day plants such as chrysanthemums flower when the night is longer than a critical value. **(c)** When an intense red flash interrupts a long night, both kinds of plants respond as if it were a short night; the irises flowered but the chrysanthemums did not.

Chemical Signals for Flowering. When photoperiod conditions are right, what sort of chemical message stimulates a plant to develop flowers? In the 1930s botanists began postulating the existence of "florigen," a hypothetical hormone that served as the flowering signal. In a somewhat frustrating scientific quest, researchers spent the rest of the twentieth century seeking this substance in vain. Recently, however, molecular studies using *Arabidopsis* plants have defined a sequence of steps that may collectively provide the internal stimulus for flowering. Here again, we see one of the recurring themes in plant development—major developmental changes guided by several interacting genes.

Figure 38.25 traces the steps of the proposed flowering signal. To begin with, a gene called *CONSTANS* is expressed in a plant's leaves in tune with the daily light/dark cycle, with expression peaking at dusk (step 1). The gene encodes a regulatory protein called CO (not to be confused with carbon monoxide). As days lengthen in spring, the concentration of CO rises in leaves, and as a result a second gene is activated (step 2). The product of this gene, a regulatory protein called FT (for flowering locus T), travels in the phloem to shoot tips (step 3). Once there, FT interacts with a second regulatory protein (step 4) that is synthesized only in shoot apical meristems (step 5). The encounter apparently sparks the development of a flower by promoting the expression of floral organ identity genes in the meristem tissue (see Section 36.5). Key experiments that uncovered this pathway all relied on analysis of DNA microarrays, a technique introduced in Section 16.3 and featured in "Using DNA Microarray Analysis to Track Down 'Florigen,'" p. 942.

Vernalization and Flowering. Flowering is more than a response to changing night length. Temperatures also change with the seasons in most parts of the world, and they too influence flowering. For instance, unless buds of some biennials and perennials are exposed to low winter temperatures, flowers do not form on stems in spring. Low-temperature stimulation of flowering is called **vernalization** ("making springlike").

In 1915 the plant physiologist Gustav Gassner demonstrated that it was possible to influence the flowering of cereal plants by controlling the temperature of seeds while they were germinating. In one case, he maintained germinating seeds of winter rye (*Secale cereale*) at just above freezing (1°C) before planting them. In nature, winter rye seeds in soil germinate during the winter, giving rise to a plant that flowers months later, in summer. Plants grown from Gassner's test seeds, however, flowered the same summer even when the seeds were planted in the late spring. Home gardeners can induce flowering of daffodils and tulips

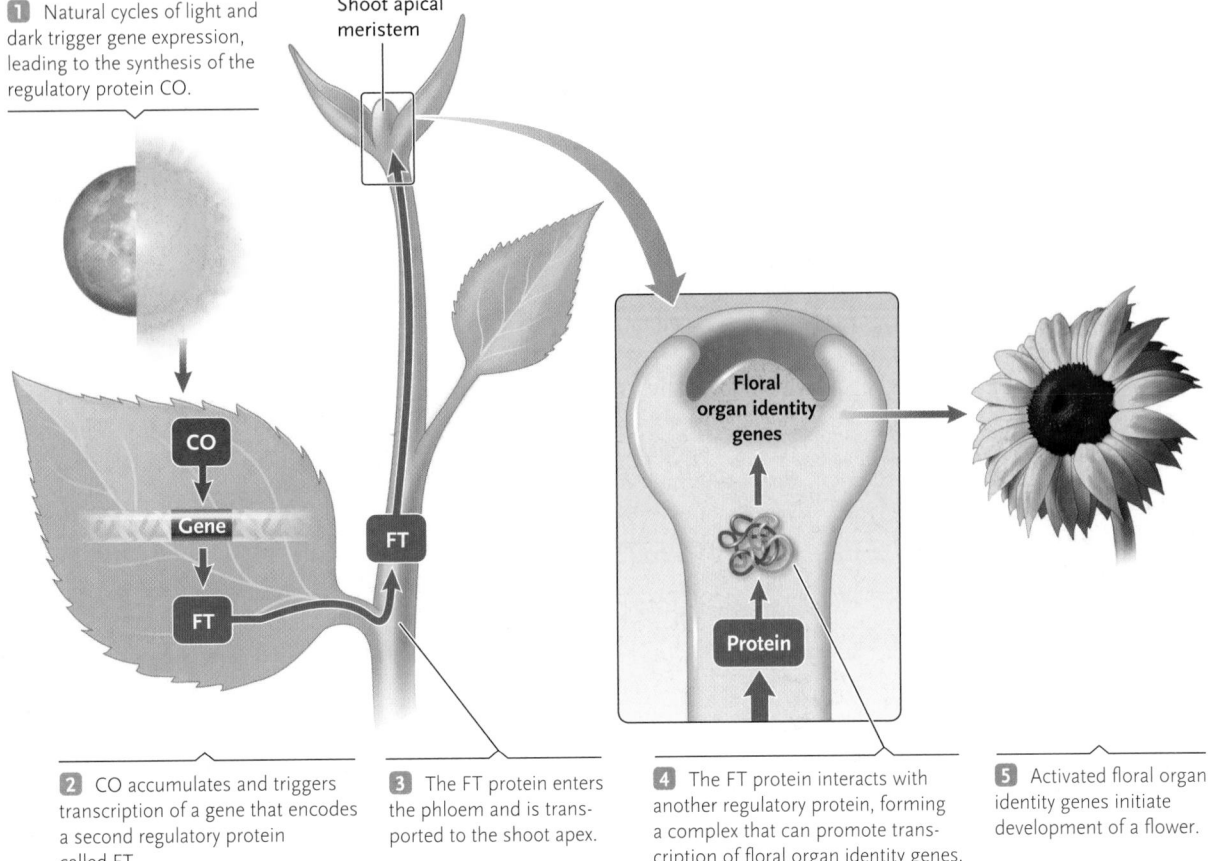

1 Natural cycles of light and dark trigger gene expression, leading to the synthesis of the regulatory protein CO.

Shoot apical meristem

Floral organ identity genes

CO

Gene

FT

FT

Protein

2 CO accumulates and triggers transcription of a gene that encodes a second regulatory protein called FT.

3 The FT protein enters the phloem and is transported to the shoot apex.

4 The FT protein interacts with another regulatory protein, forming a complex that can promote transcription of floral organ identity genes.

5 Activated floral organ identity genes initiate development of a flower.

Figure 38.25

Proposed pathway for the flowering signal. The pathway starts as shifting cycles of light and dark trigger expression of the *CONSTANS* gene. As described in the text, this step is the first in a sequence that leads to the activation of floral organ identity genes in the shoot apical meristem. When these genes are expressed, a flower develops.

by putting the bulbs (technically, *corms*) in a freezer for several weeks before early spring planting. Commercial growers use vernalization to induce millions of plants, such as Easter lilies, to flower just in time for seasonal sales.

38.4d Dormancy Is an Adaptation to Seasonal Changes or Stress

As autumn approaches and days grow shorter, growth slows or stops in many plants even if temperatures are still moderate, the sky is bright, and water is plentiful. When a perennial or biennial plant stops growing under conditions that seem (to us) quite suitable for growth, it has entered a state of **dormancy.** Ordinarily, its buds will not resume growth until early spring.

Short days and long nights—conditions typical of winter—are strong cues for dormancy. In one experiment, in which a short period of red light interrupted the long dark period for Douglas firs, the plants responded as if nights were shorter and days were longer; they continued to grow taller **(Figure 38.26, p. 942).** Conversion of Pr to Pfr by red light during the

dark period prevented dormancy. In nature, buds may enter dormancy because less Pfr can form when day length shortens in late summer. Other environmental cues are at work also. Cold nights, dry soil, and a deficiency of nitrogen apparently also promote dormancy.

The requirement for multiple dormancy cues has adaptive value. For example, if temperature were the only cue, plants might flower and seeds might germinate in warm autumn weather—only to be killed by winter frost.

A dormancy-breaking process is at work between fall and spring. Depending on the species, breaking dormancy probably involves gibberellins and abscisic acid, and it requires exposure to low winter temperatures for specific periods **(Figure 38.27, p. 943).** The temperature needed to break dormancy varies greatly among species. For example, the Delicious variety of apples grown in Utah requires 1230 hours near 6°C; apricots grown there require only 720 hours at that temperature. Generally, trees growing in the southern United States or in Italy require less cold exposure than those growing in Canada or in Sweden.

Using DNA Microarray Analysis to Track Down "Florigen"

The more plant scientists learn about plant genomes, the more they are relying on DNA microarray assays to elucidate the activity of plant genes.

Recall from Section 16.3 that a DNA microarray, also called a DNA chip, allows an investigator to explore questions such as how the expression of a particular gene differs in different types of cells. This procedure can be manipulated to reveal the relative amounts of expression of more than one of a cell's genes.

Philip A. Wigge and his colleagues used this method to learn more about the signalling pathway that causes a plant's apical meristem to give rise to flowers. Previous research had established that in leaves, lengthening spring days coincided with rising concentrations of CO, a regulatory protein encoded by the *CONSTANS* gene. But what did CO regulate? Working with *Arabidopsis thaliana*, Wigge's group was able to narrow down the field to four genes, and using microarray analysis of DNA from leaf cells they pinpointed one called FT (for flowering locus T). The researchers found that in leaves, CO causes strong expression of FT: when enough CO is present, FT mRNA is rapidly transcribed and then enters the phloem. (The transport of mRNA in phloem is not unusual.) By contrast, when they tested CO's effects in shoot apex cells, they found that it triggers far less gene expression there. Clearly, CO was not directly triggering the development of flowers. However, FT mRNA moves in the phloem to the shoot apex, where it is translated into protein. Was that protein the direct flowering signal? Other studies had implicated a regulatory protein called FD, which microarray analysis had shown was expressed *only*—but very strongly—in the shoot apex.

To sort out this final piece of the puzzle, the Wigge team examined flowering responses in normal *A. thaliana* plants as well as in mutants having a normal FT protein but a defective *fd*, and vice versa. Flowering was abnormal in both types of mutants, possibly because the mutated "partner" suppressed some aspect of the functioning of the normal protein. On the other hand, in wild-type plants, which had a functioning FD protein, expression of FT triggered a marked increase in the expression of the floral organ gene *APETALA1* (**Figure 1**). These results have two major implications. First, they support the hypothesis that FT and FD interact in a normal flowering response. Second, the study suggests that FT, the CO-induced signal from leaves, conveys the environmental signal that it is time for a plant to flower. In that sense, FT may be the long-sought "florigen." However, only by interacting with FD does FT "know" where to deliver its flowering signal—in the apical meristems of shoots.

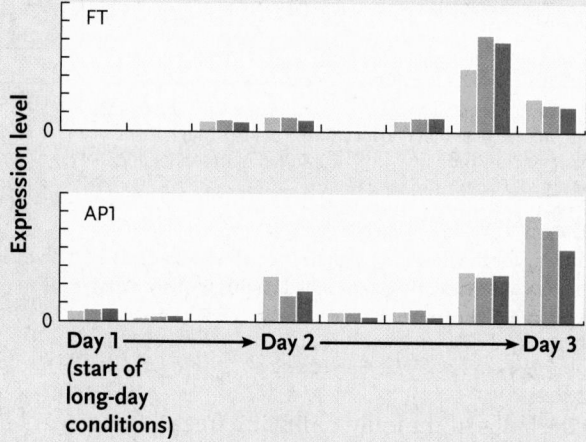

Figure 1

Effect of the FT protein on expression of the *APETALA1 (AP1)* floral organ identity gene. In nature, *Arabidopsis thaliana* is a long-day plant, and the experiment was carried out under long-day (i.e., short-night) conditions. Three groups of replicates shown here in yellow, orange, and red, respectively, were monitored for both *AP1* and FT. After a brief delay, the expression of *AP1* closely tracked the appearance of the FT regulatory protein, which had been activated by its interaction with the FD protein.

Figure 38.26

Effect of the relative length of day and night on the growth of Douglas firs (*Pseudotsuga menziesii*). The young tree at the left was exposed to alternating periods of 12 hours of light followed by 12 hours of darkness for a year; its buds became dormant because day length was too short. The tree at the right was exposed to a cycle of 20 hours of light and 4 hours of darkness; its buds remained active and growth continued. The middle plant was exposed each day to 12 hours of light and 11 hours of darkness, with a 1-hour light in the middle of the dark period. This light interruption of an otherwise long dark period also prevented buds from going dormant.

Potted plant grown inside a greenhouse did not flower.   Branch exposed to cold outside air flowered.

Eric Welzel/Fox Hill Nursery, Freeport, Maine

Figure 38.27

Effect of cold temperature on dormant buds of a lilac (*Syringa vulgaris*). In this experiment, a lilac plant was grown in winter inside a warm greenhouse with one branch extending through a hole to the outdoors. Only the buds on the branch exposed to low outside temperatures resumed growth in spring. This experiment suggests that low-temperature effects are localized in plants.

Review

aplia™

To access course materials such as Aplia and other companion resources, please visit www.NELSONbrain.com.

38.1 Introduction to Plant Hormones

- Hormones and environmental stimuli alter the behaviour of target cells, which have receptors to which signal molecules can bind. By means of a response pathway that transduces the hormone signal, a signal can induce changes in the cell's shape or internal structure or influence its metabolism or the transport of substances across the plasma membrane (Figures 38.1 and 38.2).

- Some plant hormones and growth factors may bind to receptors at the target cell's plasma membrane, changing the receptor's shape. This binding often triggers the release of internal second messengers that diffuse through the cytoplasm and provide a chemical signal that alters gene expression.

- Second messengers usually act by way of a reaction sequence that amplifies the cell's response to a signal. The sequence activates a series of proteins, including G proteins and enzymes that stimulate the synthesis of second messengers (such as IP_3) that bind ion channels on endoplasmic reticulum (ER). Binding releases calcium ions, which enter the cytoplasm and activate protein kinases, enzymes that activate specific proteins that produce the cell response.

- At least seven classes of hormones govern flowering plant development, including germination, growth, flowering, fruit set, and senescence (Table 38.1).

- Auxins, mainly IAA, promote elongation of cells in the coleoptile and stem, among other effects (Figures 38.3–38.7).

- Gibberellins promote stem elongation and help seeds and buds break dormancy (Figure 38.8).

- Cytokinins stimulate cell division, promote leaf expansion, and retard leaf aging (Figure 38.9).

- Ethylene promotes senescence, as well as fruit ripening and abscission (Figure 38.10).

- Brassinosteroids stimulate cell division and elongation.

- Abscisic acid (ABA) promotes stomatal closure and may trigger seed and bud dormancy (Figure 38.11).

- Jasmonates regulate growth and have roles in defence.

38.2 Plant Chemical Defences

- Plants have diverse chemical defences that limit damage from bacteria, fungi, worms, or plant-eating insects (Figure 38.12).

- The hypersensitive response isolates an infection site by surrounding it with dead cells (Figure 38.13).

- Oligosaccharins can trigger the synthesis of phytoalexins, secondary metabolites that function as antibiotics.

- Gene-for-gene recognition enables a plant to chemically recognize a specific pathogen and mount a defence (Figure 38.14).

- Systemic acquired resistance provides long-term protection against some pathogens. Salicylic acid (SA) has a central role in this response (Figure 38.15).

- Heat-shock proteins (HSPs) can reversibly bind enzymes and other proteins in plant cells and prevent them from denaturing when the plant is under certain types of stress.

- Some plants can synthesize "antifreeze" proteins that stabilize cell proteins when cells are threatened with freezing.

38.3 Plant Movements

- Plants adjust their growth patterns in response to environmental rhythms and unique environmental circumstances. These responses include tropisms.

- Phototropisms, mainly stimulated by blue light, are growth responses to a directional light source. (Figure 38.16).
- Gravitropism is a growth response to Earth's gravitational pull. Stems exhibit negative gravitropism (growing upward), whereas roots show positive gravitropism (Figure 38.17).
- Some plants or plant parts demonstrate thigmotropism, growth in response to contact with a solid object (Figure 38.18). Mechanical stress can cause thigmomorphogenesis, which causes the stem to add girth.
- Some plant species show nastic leaf movements in response to certain environmental cues. Changes in fluid pressure in cells of a pulvinus, a pad of tissue at the base of a leaf or petiole, cause the movements (Figures 38.19 and 38.20).

38.4 Plant Biological Clocks

- Plants have biological clocks, internal time-measuring mechanisms with a biochemical basis. Environmental cues can "reset" the clocks, enabling plants to make seasonal adjustments in growth, development, and reproduction.
- In photoperiodism, plants respond to a change in the relative length of daylight and darkness in a 24-hour period. A switching mechanism involving the pigment phytochrome promotes or inhibits germination, growth, and flowering and fruiting (Figure 38.21).
- Long-day plants flower when day length is long relative to night. Short-day plants flower when day length is relatively short, and intermediate-day plants flower when day length falls in between the values for long-day and short-day plants. Flowering of day-neutral plants is not regulated by light. In vernalization, a period of low temperature stimulates flowering (Figures 38.22–38.24).
- The direct trigger for flowering may begin in leaves, when the regulatory protein CO triggers the expression of the FT gene. The resulting mRNA transcripts move in phloem to apical meristems, where translation of the mRNAs yields a second regulatory protein, which in turn interacts with a third. This final interaction activates genes that encode the development of flower parts (Figure 38.25).
- Senescence is the sum of processes leading to the death of a plant or plant structure.
- Dormancy is a state in which a perennial or biennial stops growing even though conditions appear to be suitable for continued growth (Figures 38.26 and 38.27).

Questions

Self-Test Questions

1. Which of the following plant hormones does NOT stimulate cell division?
   a. auxins
   b. cytokinins
   c. ethylene
   d. gibberellins
   e. abscisic acid

2. Which is the correct pairing of a plant hormone and its function?
   a. salicylic acid: triggers synthesis of general defence proteins
   b. brassinosteroids: promote responses to environmental stress
   c. cytokinins: stimulate stomata to close in water-stressed plants
   d. gibberellins: slow seed germination
   e. ethylene: promotes formation of lateral roots

3. Which of the following is a characteristic of auxin (IAA) transport?
   a. IAA moves by polar transport from the base of a tissue to its apex.
   b. IAA moves laterally from a shaded to an illuminated side of a plant.
   c. IAA enters a plant cell in the form of IAAH, an uncharged molecule that can diffuse across cell membranes.
   d. IAA exits one cell and enters the next by means of transporter proteins clustered at both the apical and basal ends of the cells.
   e. All of the above are characteristics of auxin transport in different types of cells.

4. Hanging wire fruit baskets have many holes or open spaces. Which of the following is the major advantage of these spaces?
   a. They prevent gibberellins from causing bolting or the formation of rosettes on the fruit.
   b. They allow the evaporation of ethylene and thus slow ripening of the fruit.
   c. They allow oxygen in the air to stimulate the production of ethylene, which hastens the abscission of fruits.
   d. They allow oxygen to stimulate brassinosteroids, which hasten the maturation of seeds in/on the fruits.
   e. They allow carbon dioxide in the air to stimulate the production of cytokinins, which promotes mitosis in the fruit tissue and hastens ripening.

5. Which of the following is NOT an example of a plant chemical defence?
   a. ABA inhibits leaves from budding if conditions favour attacks by sap-sucking insects.
   b. Jasmonate activates plant genes encoding protease inhibitors that prevent insects from digesting plant proteins.
   c. Acting against fungal infections, the hypersensitive response allows plants to produce highly reactive oxygen compounds that kill selected tissue, thus forming a dead tissue barrier that walls off the infected area from healthy tissues.
   d. Chitinase, a PR hydrolytic protein produced by plants, breaks down chitin in the cell walls of fungi and thus halts the fungal infection.
   e. Attack by fungi or viruses triggers the release of oligosaccharins, which in turn stimulate the production of phytoalexins having antibiotic properties.

6. Which of the following statements about plant responses to the environment is true?
   a. The heat-shock response induces a sudden halt to cellular metabolism when an insect begins feeding on plant tissue.
   b. In gravitropism, amyloplasts sink to the bottom of cells in a plant stem, causing the redistribution of IAA.
   c. The curling of tendrils around a twig is an example of thigmotropism.
   d. Phototropism results when IAA moves first laterally and then downward in a shoot tip when one side of the tip is exposed to light.
   e. Nastic movements, such as the sudden closing of the leaves of a Venus flytrap, are examples of a plant's ability to respond to specific directional stimuli.

7. In nature the poinsettia, a plant native to Mexico, blooms only in or around December. What does this pattern suggest?
   a. The long daily period of darkness (short day) in December stimulates the flowering.
   b. Vernalization stimulates the flowering.
   c. The plant is dormant for the rest of the year.
   d. Phytochrome is not affecting the poinsettia flowering cycle.
   e. A circadian rhythm is in effect.

8. Which of the following steps is NOT part of the sequence that is thought to trigger flowering?
   a. Cycles of light and dark stimulate the expression of the *CONSTANS* gene in a plant's leaves.
   b. CO proteins accumulate in the leaves and trigger expression of a second regulatory gene.
   c. mRNA transcribed during expression of a second regulatory gene moves via the phloem to the shoot apical meristem.
   d. Interactions among regulatory proteins promote the expression of floral organ identity genes in meristem tissue.
   e. CO proteins in the floral meristem interact with florigen, a so-called flowering hormone, which provides the final stimulus for expression of floral organ identity genes.

9. Damage from an infectious bacterium, fungus, or worm may trigger a plant defensive response when the pathogen or a substance it produces binds to which of the following?
   a. a receptor encoded by the plant's *avirulence* (*Avr*) gene
   b. an *R* gene in the plant cell nucleus
   c. a receptor encoded by a dominant *R* gene
   d. PR proteins embedded in the plant cell plasma membrane
   e. salicylic acid molecules released from the besieged plant cell

10. What happens in the sequence that unfolds after molecules of a hormone such as ABA bind to receptors at the surface of a target plant cell?
   a. First messenger molecules in the cytoplasm are mobilized, and then G proteins carry the signal to second messengers such as protein kinases, which alter the activity of cell proteins such as IP3.
   b. Binding activates G proteins, which in turn activate second messengers such as IP3; subsequent steps are thought to involve activation of genes that encode protein kinases.
   c. Binding activates phospholipase C, which in turn activates G proteins, which then activate molecules of IP3, a step that leads to the synthesis of protein kinases.
   d. Binding stimulates G proteins to activate protein kinases, which then bind calcium channels in the ER; the flux of calcium ions activates second messenger molecules that alter the activity of cell proteins or enter the cell nucleus and alter the expression of target genes.
   e. Binding activates G proteins, which in turn activate phospholipase C; this substance then stimulates the synthesis of second messenger molecules, the second messengers bind calcium channels in the cell's ER, and finally protein kinases alter the activity of proteins by phosphorylating them.

Questions for Discussion

1. You work for a plant nursery and are asked to design a special horticultural regimen for a particular flowering plant. The plant is native to northern Spain, and in the wild it grows a few long, slender stems that produce flowers each July. Your boss wants the nursery plants to be shorter, with thicker stems and more branches, and she wants them to bloom in early December in time for holiday sales. Outline your detailed plan for altering the plant's growth and reproductive characteristics to meet these specifications.

2. Synthetic auxins such as 2,4-D can be weed killers because they cause an abnormal growth burst that kills the plant within a few days. Suggest reasons why such rapid growth might be lethal to a plant.

3. In some plant species, an endodermis is present in both stems and roots. In experiments, the shoots of mutant plants lacking differentiated endodermis in their root and shoot tissue do not respond normally to gravity, but roots of such plants do respond normally. Explain this finding, based on your reading in this chapter.

39

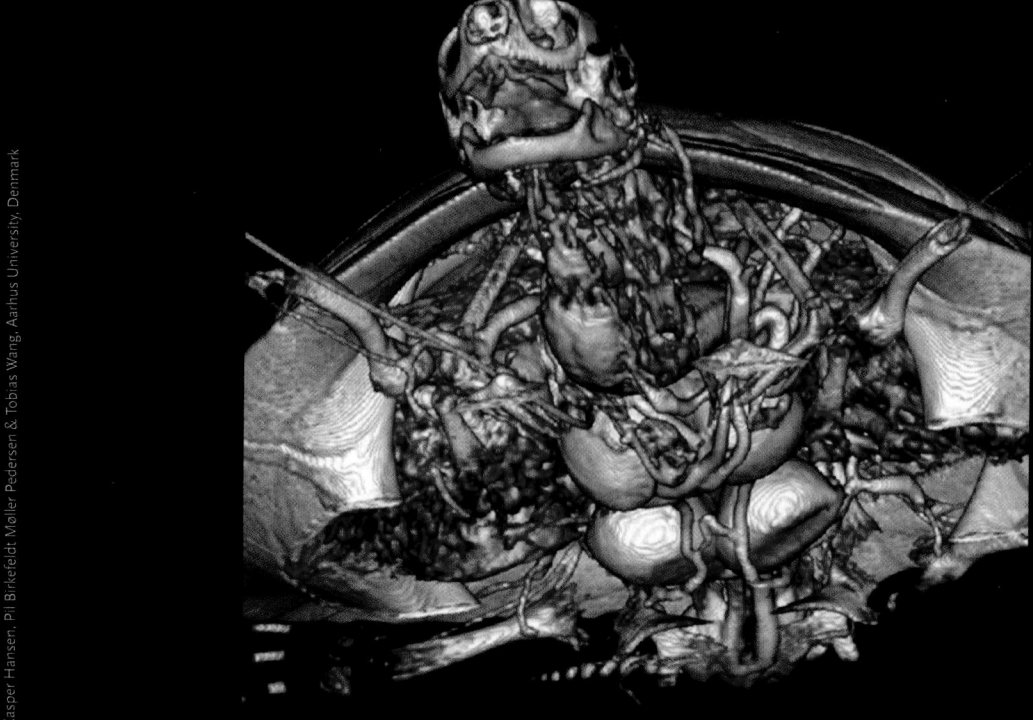

Magnetic resonance image (MRI) of a turtle. Various organs can be seen in the scan: the whitish skeleton throughout the body, the brain within the skull, and the lungs (dark) in the chest.

Introduction to Animal Organization and Physiology

WHY IT MATTERS

When winter sets in across Canada, different species of animals begin to prepare for the cold in different ways. Many species begin to slow body processes and prepare their cells to withstand freezing as a way of surviving until the spring. Others undergo changes that will allow them to remain fully active throughout the cold snowy months. Yet other species cannot tolerate the cold and migrate to warmer climates for the winter **(Figure 39.1)**. To fully understand the factors that determine species tolerances and distributions requires a detailed knowledge of how species are built and how their various parts work.

In multicellular organisms, cells differentiate to form different tissues. In animals, tissues combine in different ways to form organs and organ systems. Each organ system has a unique function. This specialization allows multicellular organisms to establish an internal environment around their cells that can be maintained relatively independent of the external environment through a process called **homeostasis** (*homeo* = the same; *stasis* = standing or stopping). The processes and activities responsible for homeostasis are called **homeostatic mechanisms.** Being able to maintain an internal environment relatively independent of changes in the external environment greatly broadens the geographic range over which animals can live actively. Understanding biodiversity and species distribution ultimately requires an understanding of the form and function of the organs and organ systems within each species, which in turn requires an understanding of the specialized cells and tissues that compose them. **Anatomy** is the study of the structures of organisms, and **physiology** is the study of the functions of the cells, tissues, organs, and organ systems.

Figure 39.1

Ways of surviving the winter include avoidance (migratory birds), cold adaptation (small mammals), and freeze tolerance (many insects).

In this chapter, we begin by examining the organization of individual cells into tissues, organs, and organ systems, the major body structures that carry out animal activities. Our discussion continues by examining the coordination of the processes and activities of organ systems that accomplish homeostasis. The other 12 chapters in this unit discuss the individual organ systems that carry out major body functions such as digestion, movement, and reproduction. Although we emphasize vertebrates throughout the unit (including humans), we also make comparisons with invertebrates, to keep the structural and functional diversity of the animal kingdom in perspective and to understand the evolution of the structures and processes involved.

39.1 Organization of the Animal Body

39.1a In Animals, Specialized Cells Are Organized into Tissues, Tissues into Organs, Organs into Organ Systems, and Organ Systems into Organisms

The individual cells of multicellular animals have the same requirements as cells of any kind, including single-celled organisms. They must be surrounded by an aqueous solution that contains ions and molecules required by the cells, including complex organic molecules that can be used as energy sources. The concentrations of these molecules and ions must be balanced to keep cells from shrinking or swelling excessively due to osmotic water movement. Most animal cells also require oxygen to serve as the final acceptor for electrons removed in oxidative reactions. Animal cells must be able to release waste molecules and other by-products of their activities, such as carbon dioxide and nitrogenous wastes, to their environment. The physical conditions of the cellular environment, such as temperature, must also remain within tolerable limits.

The evolution of multicellularity (see Section 3.5) made it possible for organisms to create an *internal fluid environment* that supplies all the needs of individual cells, including nutrient supply, waste removal, and osmotic balance. By regulating this internal environment despite changes in the external environment, multicellular organisms can occupy diverse habitats, including dry terrestrial habitats that would be lethal to single cells. While multicellular organisms can become very large, their individual cells remain small enough to exchange ions and molecules easily with the internal fluid.

The evolution of multicellularity also allowed specialized groups of cells to differentiate and take on specific life functions, with each group of cells concentrating primarily on a single activity. Specialization greatly increases the efficiency by which animals carry out these functions. In most animals, these specialized groups of cells are organized into tissues, the tissues into organs, and the organs into organ systems **(Figure 39.2, p. 948)**. A **tissue** is a group of cells with the same structure and function, working together as a unit to carry out one or more activities. Four types of tissue are shown in Figure 39.2: epithelial, connective, muscle, and nervous tissue. An **organ** integrates two or more different tissues into a structure that carries out a specific function. The eye, liver, and stomach are examples of organs. An **organ system** coordinates the activities of two or more organs to carry out a major body function such as digestion, excretion, or

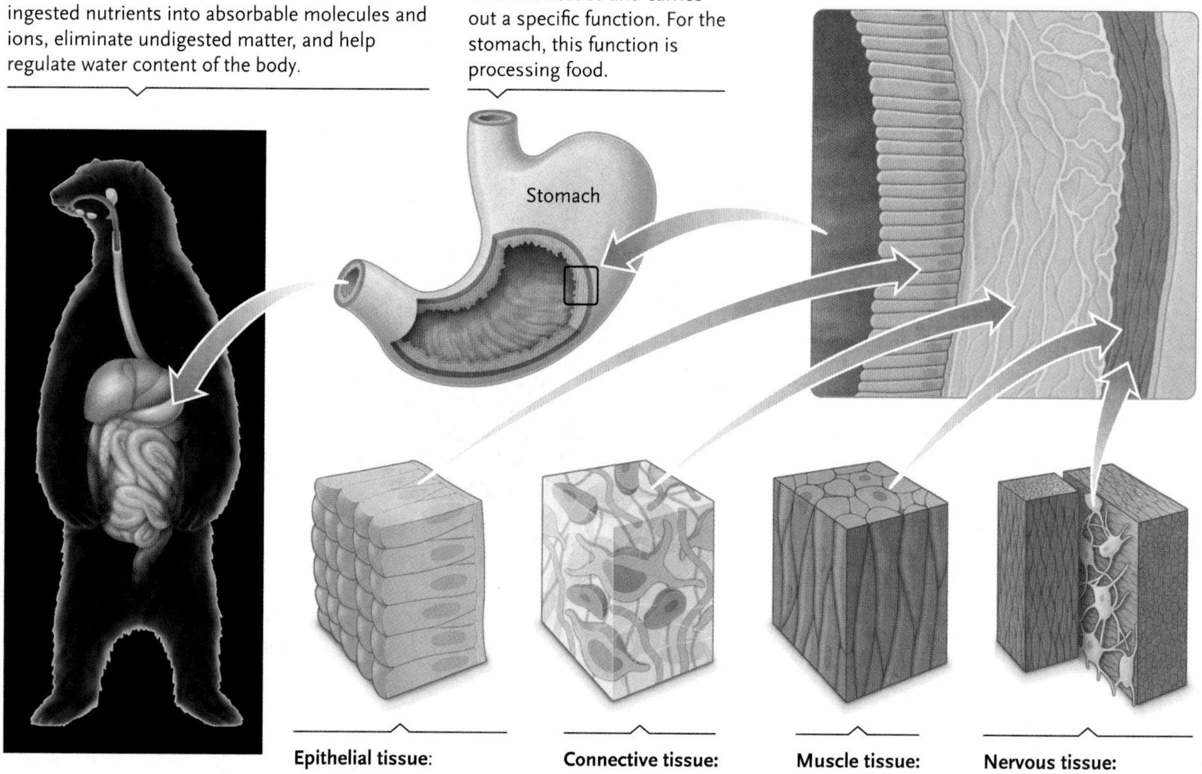

Figure 39.2

Organization of animal cells into tissues, organs, and organ systems as exemplified here by the digestive system.

Organ system:
A set of organs that interacts to carry out a major body function. The digestive system coordinates the activities of organs, including the mouth, esophagus, stomach, small and large intestines, liver, pancreas, rectum, and anus, to convert ingested nutrients into absorbable molecules and ions, eliminate undigested matter, and help regulate water content of the body.

Organ:
Body structure that integrates different tissues and carries out a specific function. For the stomach, this function is processing food.

Stomach

Epithelial tissue:
Protection, transport, secretion, and absorption of nutrients released by digestion of food

Connective tissue:
Structural support

Muscle tissue:
Movement

Nervous tissue:
Communication, coordination, and control

reproduction. The organ system carrying out digestion, for example, coordinates the activities of organs, including the mouth, stomach, pancreas, liver, and small and large intestines. Some organs contribute functions to more than one organ system. For instance, the pancreas forms part of the endocrine system as well as the digestive system.

STUDY BREAK

1. What are the differences among tissues, organs, and organ systems?
2. Why must multicellular organisms have organs and organ systems?

39.2 Animal Tissues

Although the most complex animals may contain hundreds of distinct cell types, all can be classified into one of only four basic tissue groups: *epithelial, connective, muscle,* and *nervous* (see Figure 39.2). Each tissue type is assembled from individual cells. The properties of these cells determine the structure and, therefore, the

function of the tissue. More specifically, the structure and function of a tissue depend on the structure and organization of the cytoskeleton within the cell, the type and organization of the extracellular matrix surrounding the cell, and the junctions holding the cells together. The **extracellular matrix** (ECM) is a nonliving material secreted by cells consisting of a variety of proteins and glycoproteins. The ECM provides support and shape for tissues and organs. The cell walls of plants and the cuticle of arthropods are examples of specialized ECM.

Junctions of various kinds link cells into tissues (see **Figure 39.3**). *Anchoring junctions* form buttonlike spots or belts that weld cells together. They are most abundant in tissues subject to stretching, such as skin and heart muscle. *Tight junctions* seal the spaces between cells, keeping molecules and even ions from leaking between cells. For example, tight junctions in the tissue lining the urinary bladder prevent waste molecules and ions from leaking out of the bladder into other body tissues. *Gap junctions* are open channels between cells in the same tissue, allowing ions and small molecules to flow freely from one cell to another. For example, gap junctions between muscle cells help muscle tissue to function as a unit.

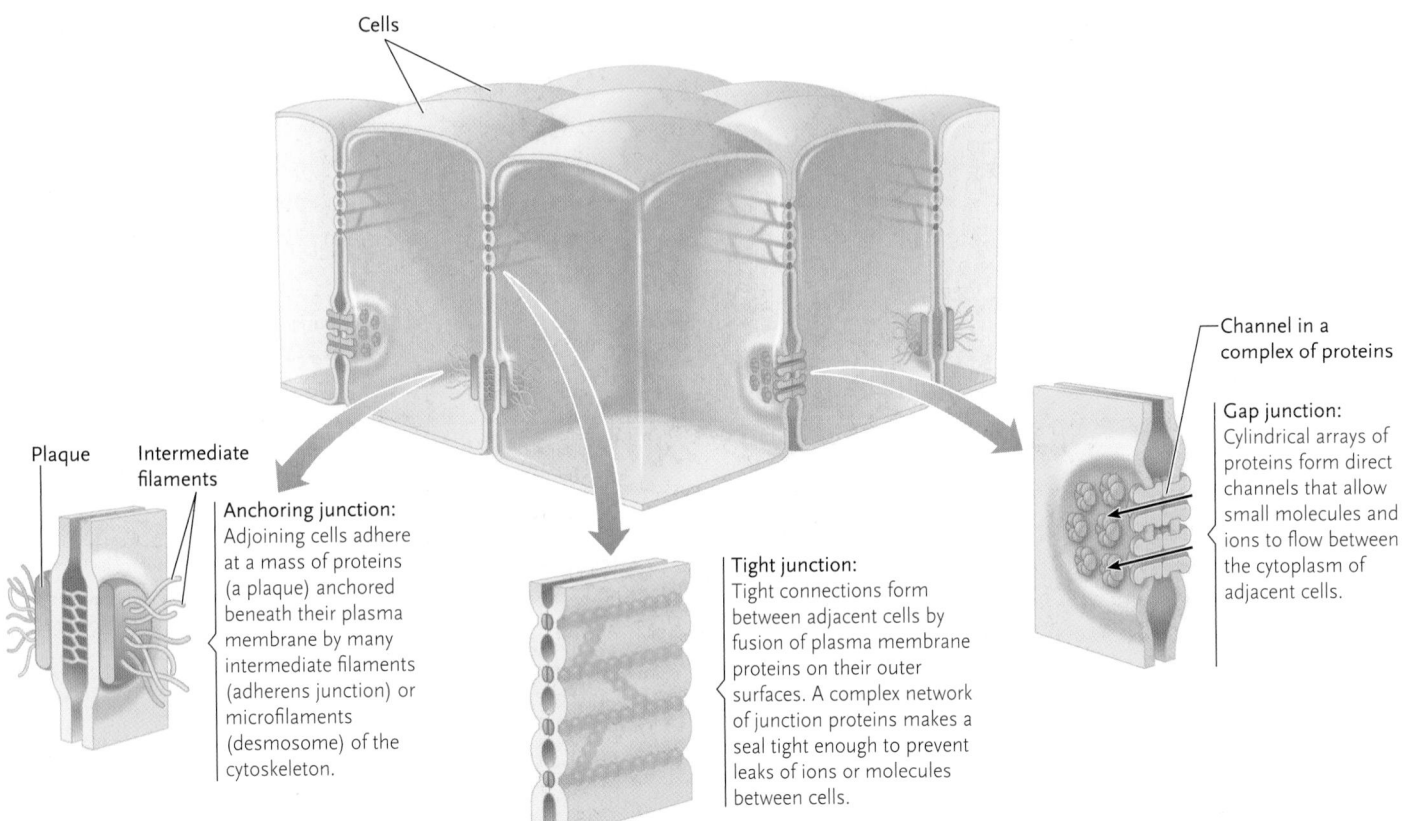

Figure 39.3

Anchoring junctions, tight junctions, and gap junctions, which connect cells in animal tissues. Anchoring junctions reinforce the cell-to-cell connections made by cell adhesion molecules, tight junctions seal the spaces between cells, and gap junctions create direct channels of communication between animal cells.

39.2a Epithelial Tissue Forms Protective, Secretory, and Absorptive Coverings and Linings of Body Structures

Epithelial tissue (*epi* = over; *thele* = covering) consists of sheetlike layers of cells that are usually joined tightly together, with little ECM material between them **(Figure 39.4, p. 950).** Also called *epithelia* (singular, *epithelium*), these tissues cover body surfaces and the surfaces of internal organs and line cavities and ducts within the body. They protect body surfaces from invasion by bacteria and viruses and secrete or absorb substances. They often have other roles. For example, the epithelium covering a fish's gill structures serves as a barrier to bacteria and viruses, while at the same time serving as an exchange site for oxygen, carbon dioxide, and ions with the aqueous environment. The epithelium of the external surface of arthropods secretes the tough cuticle that in addition to acting as a barrier to the environment, functions as their skeleton. Some epithelial cells in the epidermis of vertebrates also contain a network of fibrous proteins (keratin) that form such protective structures as scales, nails, claws, hooves, and horns, as well as hair and feathers.

Some epithelia, such as those lining the capillaries of the circulatory system, act as filters, allowing ions and small molecules to leak from the blood into surrounding tissues while barring the passage of blood cells and large molecules such as proteins.

Because epithelia form coverings and linings, they have a free (or outer) surface and an inner surface. The outer **apical surface** may be exposed to water, air, or fluids within the body. In internal cavities and ducts, the apical surface is often covered with *cilia*, which beat like oars to move fluids through the cavity or duct. The epithelium lining the oviducts in mammals, for example, is covered with cilia that generate fluid currents to move eggs from the ovaries to the uterus. In free-living flatworms, the ventral epithelium of the animal is frequently ciliated, allowing the worm to glide over surfaces. In some epithelia, including the lining of the small intestine, the free surface is crowded with *microvilli,* fingerlike extensions of the plasma membrane that increase the area available for secretion or absorption.

The inner, **basal surface** of an epithelium adheres to a layer of ECM secreted by the epithelial cells called the **basal lamina** (see Figure 39.4). In many cases, such as the intestinal epithelium of vertebrates, a further layer of fibres is secreted by underlying connective tissue, but this is lacking in most invertebrate epithelia. The entire assemblage is the **basement membrane.**

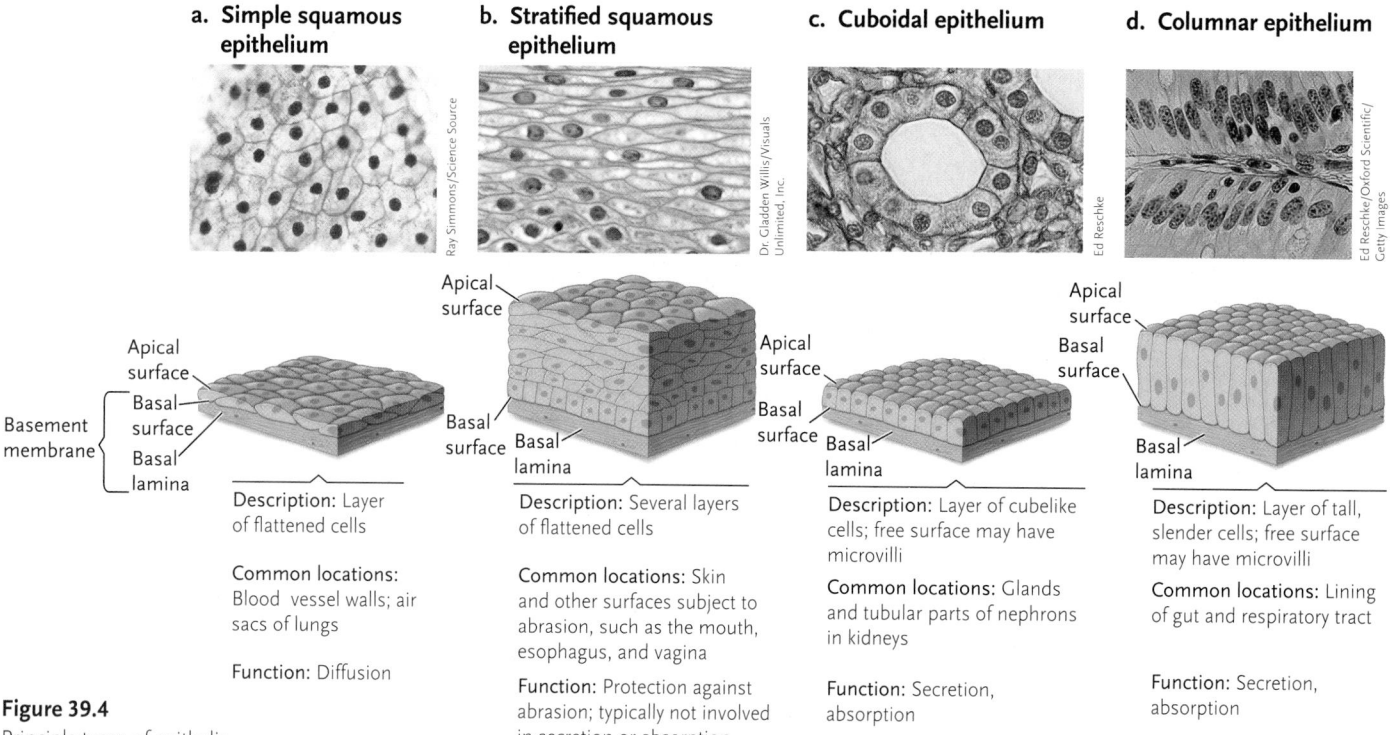

a. Simple squamous epithelium

Ray Simmons/Science Source

Description: Layer of flattened cells

Common locations: Blood vessel walls; air sacs of lungs

Function: Diffusion

b. Stratified squamous epithelium

Dr. Gladden Willis/Visuals Unlimited, Inc.

Description: Several layers of flattened cells

Common locations: Skin and other surfaces subject to abrasion, such as the mouth, esophagus, and vagina

Function: Protection against abrasion; typically not involved in secretion or absorption

c. Cuboidal epithelium

Ed Reschke

Description: Layer of cubelike cells; free surface may have microvilli

Common locations: Glands and tubular parts of nephrons in kidneys

Function: Secretion, absorption

d. Columnar epithelium

Ed Reschke/Oxford Scientific/ Getty Images

Description: Layer of tall, slender cells; free surface may have microvilli

Common locations: Lining of gut and respiratory tract

Function: Secretion, absorption

Figure 39.4

Principle types of epithelia.

Types of Epithelia. Epithelia are classified as *simple*—formed by a single layer of cells—or *stratified*—formed by multiple cell layers (see Figure 39.4). The shapes of cells within an epithelium may be *squamous* (mosaic, flattened, and spread out), *cuboidal* (shaped roughly like dice or cubes), or *columnar* (elongated, with the long axis perpendicular to the epithelial layer). Four principle types of epithelia are found in the body (see Figure 39.4).

The cells of some epithelia, such as those forming the skin and the lining of the intestine, divide constantly to replace worn and dying cells. New cells are produced through division of stem cells in the basal (lowest) layer of the skin. *Stem cells* are undifferentiated (unspecialized) cells in the tissue that divide to produce more stem cells as well as cells that differentiate (i.e., become specialized into one of the many cell types of the body). Stem cells are found in both adult organisms and embryos. Besides the skin, adult stem cells are found in tissues of the brain, bone marrow, blood vessels, skeletal muscle, and liver ("People behind Biology," Box 39.1, describes the discovery of stem cells in the brain). Stem cells are important for development in many invertebrates. In some cases, the stem cells may already be programmed for a specific cell type, as in the eye or wing disks of insect pupae, whereas in others, the stem cells may be totipotent, as in flatworms (Chapter 27). A totipotent cell has the capacity to form an entire organism.

Glands Formed by Epithelia. Epithelia typically contain or give rise to cells that are specialized for secretion. Some of these secretory cells are scattered among nonsecretory cells within the epithelium. Others form structures called **glands**, which are derived from pockets of epithelium during embryonic development.

Some glands, called **exocrine glands** (*exo* = external; *crine* = secretion), remain connected to the epithelium by a duct, which empties their secretion at the epithelial surface. Exocrine secretions include mucus, saliva, digestive enzymes, sweat, earwax, oils, milk, and venom (**Figure 39.5a** shows an exocrine gland in the skin of a poisonous tree frog). Other glands, called **endocrine glands**, may not be composed of epithelial cells. They have no ducts but secrete their products, hormones, into the interstitial fluid to be picked up by the blood for circulation to the organs and tissues of the body (Figure 39.5b). The endocrine glands are considered in detail in Chapter 43.

Some glands contain both exocrine and endocrine elements. For example, some cells of the pancreas, form an exocrine gland that secretes pancreatic juice through a duct into the small intestine, where it plays an important role in food digestion (see Chapter 48), while different cells of the pancreas serve an endocrine function by secreting the hormones insulin and glucagon into the bloodstream to help regulate glucose levels in the blood (see Chapter 43).

Some epithelial cells, particularly in the epidermis of vertebrates, contain a network of fibres of keratin, a family of tough proteins. Keratin forms the scales of fish and reptiles (including the shells of turtles); the feathers of birds; and the hair, claws, hooves, horns, and fingernails of mammals.

Photos.com

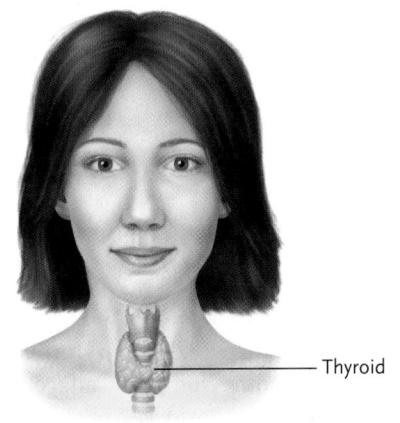

— Thyroid

Figure 39.5

Exocrine and endocrine glands. **(a)** The poison secreted by the blue poison frog (*Dendrobates azureus*) is one of the most lethal glandular secretions known. **(b)** The hormones secreted by the thyroid gland are vital for growth, development, maturation, and metabolism in all vertebrates.

Pore → Secretory product

— Epithelium

Exocrine gland cell (mucous gland)   Exocrine gland cell (poison gland)

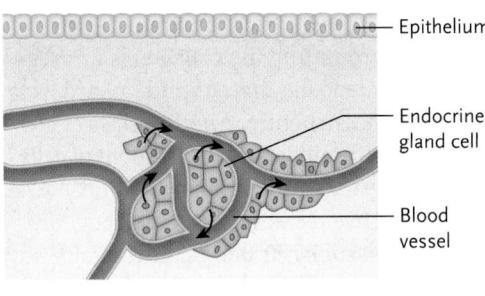

— Epithelium

— Endocrine gland cell

— Blood vessel

a. Examples of exocrine glands: The mucus- and poison-secreting glands in the skin of a blue poison frog

b. Example of an endocrine gland: The thyroid gland, which secretes hormones that regulate the rate of metabolism and other body functions

39.2b Connective Tissue Supports Other Body Tissues

Most animal body structures contain one or more types of **connective tissue.** Connective tissues support other body tissues, transmit mechanical and other forces, and in some cases act as filters. They consist of cells that form networks or layers in and around body structures and that are separated by nonliving material, specifically the ECM secreted by the cells of the tissue. Many forms of connective tissue have more nonliving ECM material (both by weight and by volume) than living cellular material.

The mechanical properties of a connective tissue depend on the type and quantity of its ECM. The consistency of the ECM ranges from fluid (as in blood and lymph), through soft and firm gels (as in tendons), to hard and crystalline (as in bone). In most connective tissues, the ECM consists primarily of the fibrous glycoprotein **collagen** embedded in a network of proteoglycans—glycoproteins that are very rich in carbohydrates. Collagen is the most abundant family

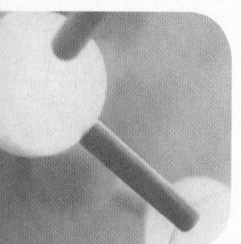

MOLECULE BEHIND BIOLOGY 39.2

Resilin: An Insect Rubber with Amazing Properties

While Torkel Weis-Fogh, a young Danish comparative physiologist, was conducting his groundbreaking studies on insect flight in the Zoological Laboratory in Cambridge, United Kingdom, in 1960, he discovered patches of highly elastic cuticle in the wing joints of locusts. Subsequently, the same elastic protein proved to be important to other insect movements. For example, it is involved in the movements of the membrane that produce the song of cicadas. Fleas are able to jump very large distances because muscle contractions store energy in pads of resilin in the hind legs **(Figure 1).** The sudden release of this energy, in less than a millisecond, propels the jump with an instantaneous acceleration greater than that of the space shuttle.

Resilin belongs to the same family of proteins as elastin, an elastic protein in many animals, but it is restricted to insects and a few crustacea. Resilin's elastic properties have proven to be astonishing. It can be stretched to four times its length without breaking (elastin only manages two). It is 97% efficient, so that when energy stored in it by stretching is released, only 3% is released as heat. It survives huge numbers of cycles of stretch and relaxation. The membrane producing the song of cicadas vibrates several thousand times per second. Like other elastic proteins, resilin is composed of coiled protein molecules cross-linked to one another, and stretching involves uncoiling. In 2005, a gene coding for *Drosophila* resilin was cloned in *E. coli* by a team led by Christopher Elvin of the Commonwealth Scientific and Industrial Research Organisation in Australia. That lab is now able to produce significant quantities of resilin. Ultimately, this insect rubber should find application, for example, as replacements for spinal disks in humans or as artificial blood vessels.

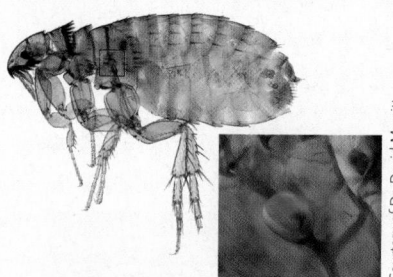

Figure 1

The pad of resilin (blue) in the coxa (part of the thigh) of the hind leg of a flea.

of proteins in animals. More than 25 different forms have been described, and some form of collagen occurs in all Metazoa, including Porifera (sponges). The collagen molecule is thus an ancient one that has been modified during evolution. In bone, the glycoprotein network surrounding the collagen is impregnated with mineral deposits that produce a hard, yet still somewhat elastic, structure. Another class of glycoproteins, **fibronectin,** aids in the attachment of cells to the ECM and helps hold the cells in position.

In some connective tissues, another rubbery protein, **elastin,** adds elasticity to the ECM. It is able to return to its original shape after being stretched, bent, or compressed. Elastin fibres, for example, help the skin return to its original shape when pulled or stretched and give the lungs the elasticity required for their alternating inflation and deflation. Resilin is a protein related to elastin that occurs only in insects and some Crustacea. It is the most elastic material known and is the basis for the jumping of fleas and locusts (see "Molecule behind Biology").

Vertebrates have six major types of connective tissue: *loose connective tissue, fibrous connective tissue, cartilage, bone, adipose tissue,* and **blood.** Each type has a characteristic function correlated with its structure **(Figure 39.6).**

Loose Connective Tissue. Loose connective tissue consists of sparsely distributed cells surrounded by a more or less open network of collagen and other glycoprotein fibres (see Figure 39.6a). The cells, called **fibroblasts,** secrete most of the collagen and other proteins in this connective tissue.

In vertebrates, loose connective tissues support epithelia and form a corsetlike band around blood vessels, nerves, and some internal organs; they also reinforce deeper layers of the skin. Sheets of loose connective tissue, covered on both surfaces with epithelial cells, form the **mesenteries,** which hold the abdominal organs in place and provide lubricated, smooth surfaces that prevent chafing or abrasion between adjacent structures as the body moves. In insects, and perhaps some other invertebrates, the loose connective tissues suspending organs and providing support for epithelia are the products of specialized cells circulating in the blood.

Fibrous Connective Tissue. In **fibrous connective tissue,** fibroblasts are sparsely distributed among dense masses of collagen and elastin fibres that are lined up in highly ordered, parallel bundles (see Figure 39.6b). The parallel arrangement produces maximum tensile strength and elasticity. Examples include **tendons,** which attach muscles to bones, and **ligaments,** which connect bones to each other at a joint. The cornea of the eye is a transparent fibrous connective tissue formed from highly ordered collagen molecules.

In some invertebrates, fibrous connective tissue provides shape to the animal, as in many sponges (see Chapter 27) and echinoderms. In sea cucumbers

a. Loose connective tissue

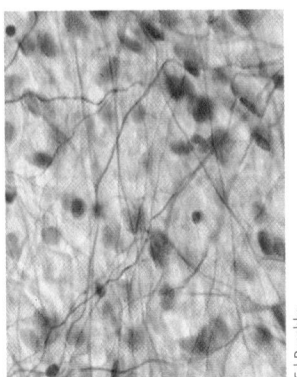

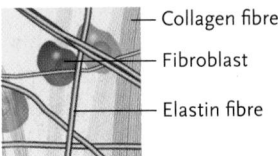

— Collagen fibre
— Fibroblast
— Elastin fibre

Description: Fibroblasts and other cells surrounded by collagen and elastin fibres forming a glycoprotein matrix

Common locations: Under the skin and most epithelia

Function: Support, elasticity, diffusion

b. Fibrous connective tissue

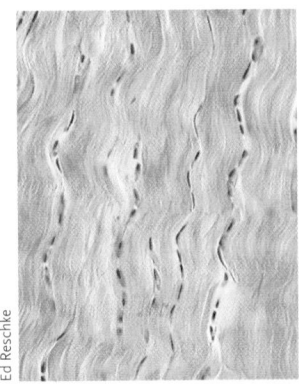

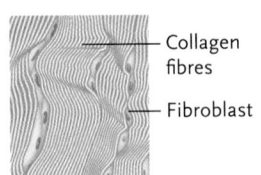

— Collagen fibres
— Fibroblast

Description: Long rows of fibroblasts surrounded by collagen and elastin fibres in parallel bundles with a dense ECM

Common locations: Tendons, ligaments

Function: Strength, elasticity

c. Cartilage

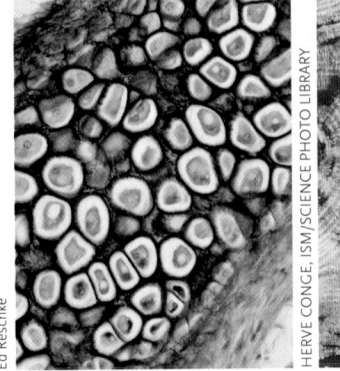

HERVE CONGE, ISM/SCIENCE PHOTO LIBRARY

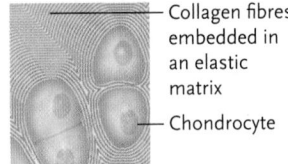

— Collagen fibres embedded in an elastic matrix
— Chondrocyte

Description: Chondrocytes embedded in a pliable, solid matrix of collagen and chondroitin sulfate

Common locations: Ends of long bones, nose, parts of airways, skeleton of vertebrate embryos

Function: Support, flexibility, low-friction surface for joint movement

d. Bone tissue

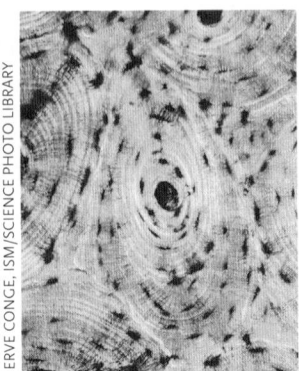

— Fine canals
— Central canal containing blood vessel
— Osteocytes

Description: Osteocytes in a matrix of collagen and glycoproteins hardened with hydroxyapatite

Common locations: Bones of vertebrate skeleton

Function: Movement, support, protection

Figure 39.6
The six major types of connective tissues in vertebrates.

e. Adipose tissue

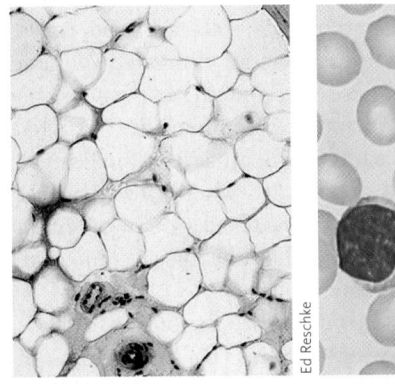

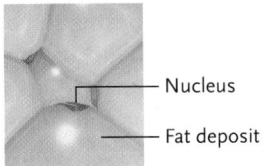

— Nucleus
— Fat deposit

Description: Large, tightly packed adipocytes with little ECM

Common locations: Under skin; around heart, kidneys

Function: Energy reserves, insulation, padding

f. Blood

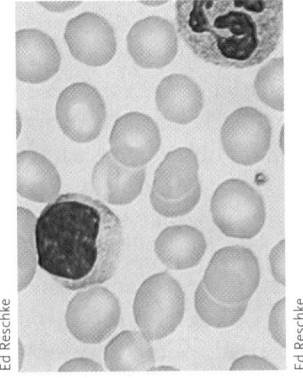

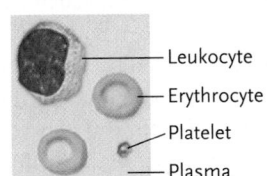

— Leukocyte
— Erythrocyte
— Platelet
— Plasma

Description: Leukocytes, erythrocytes, and platelets suspended in a plasma matrix

Common locations: Circulatory system

Function: Transport of substances

(see Chapter 28), the rigidity of the connective tissue can be changed quickly by the animal, resulting in a loss or change of shape. This acts as an escape response.

Cartilage. **Cartilage** consists of sparsely distributed cells called **chondrocytes**, surrounded by networks of collagen fibres embedded in a tough but elastic matrix of the glycoprotein *chondroitin sulfate* (see Figure 39.6c). Elastin is also present in some forms of cartilage.

The elasticity of cartilage allows it to resist compression and stay resilient, like a piece of rubber. Bending your ear or pushing the tip of your nose, which are supported by cores of cartilage, will give you a good idea of the flexible nature of this tissue. Cartilage also supports the larynx, trachea, and smaller air passages in the lungs. It forms the disks cushioning the vertebrae in the spinal column and the smooth, slippery capsules around the ends of bones in joints such as the hip and knee. Cartilage also serves as a precursor to bone during embryonic development; in sharks and rays and their relatives, almost the entire skeleton remains as cartilage in adults.

Bone. The densest form of connective tissue, **bone**, forms the skeleton, which supports the body, protects softer body structures such as the brain, and contributes to body movements by forming levers for muscle to pull on.

Mature bone consists primarily of cells called **osteocytes** (*osteon* = bone) embedded in an ECM containing collagen fibres and glycoproteins impregnated with *hydroxyapatite*, a calcium–phosphate mineral (see Figure 39.6d, p. 953). The collagen fibres give bone tensile strength and elasticity; the hydroxyapatite resists compression and allows bones to support body weight. Cells called **osteoblasts** (*blast* = bud or sprout) produce the collagen and mineral of bone—as much as 85% of the weight of bone is mineral deposits. Osteocytes, in fact, are osteoblasts that have become trapped and surrounded by the bone materials they themselves produce. **Osteoclasts** (*clast* = break) remove the minerals and recycle them through the bloodstream. Bone is not a stable tissue; it is reshaped continuously by the bone-building osteoblasts and the bone-degrading osteoclasts.

Although bones appear superficially to be solid, they are actually porous structures consisting of a system of microscopic spaces and canals. The structural unit of bone is the **osteon**. It consists of a minute central canal surrounded by osteocytes embedded in concentric layers of mineral matter (see Figure 39.6d). A blood vessel and extensions of nerve cells run through the central canal, which is connected to the spaces containing cells by very fine, radiating canals filled with interstitial fluid. The blood vessels supply nutrients to the cells with which the bone is built, and the nerve cells innervate the blood vessels as well as the connective tissue (periosteum) surrounding the bone.

Adipose Tissue. The connective tissue called **adipose tissue** mostly contains large, densely clustered cells called *adipocytes* that are specialized for fat storage (see Figure 39.6e). It has little ECM. Adipose tissue also cushions the body and, in mammals, forms an especially important insulating layer under the skin.

The animal body stores limited amounts of carbohydrates, primarily in muscle and liver cells. Unlike plants, in animals excess carbohydrates are converted into the fats stored in adipocytes. The storage of chemical energy as fats offers animals a weight advantage. For example, the average human would weigh about 45 kg more if the same amount of chemical energy were stored as carbohydrates instead of fats. Adipose tissue is richly supplied with blood vessels, which move fats or their components to and from adipocytes.

In invertebrates, fat storage may occur in a variety of tissues. Insects have a fat body, an organ that functions both for storage and as an important structure for metabolism, much like the vertebrate liver.

Blood. Blood is considered to be a connective tissue because the fluid portion is essentially a fluid form of ECM. Blood functions as the principal transport vehicle to carry nutrients, oxygen (in most animals), and hormones to the tissues and to remove metabolic wastes for transport to the organs specialized for waste removal. It is also frequently involved in defence against disease (Chapter 51) and may be important in wound healing.

Vertebrates have two basic types of cells suspended in a straw-coloured fluid, the plasma (see Figure 39.6f). Erythrocytes (*erythros* = red), or red blood cells, contain hemoglobin, a protein to which O_2 binds; these are specialized for O_2 transport. Several types of leukocytes (*leukos* = white), or white blood cells, protect the body against foreign elements such as viruses and bacteria. These are considered in Chapter 40. Vertebrate blood also contains platelets (often called thrombocytes), which are membrane-bound fragments of specialized leukocytes. They play an essential role in the formation of blood clots to heal wounds.

In invertebrates, the oxygen-carrying pigments, such as hemoglobin, may be present either within special cells, as in some annelids, or free in the plasma. Blood cells in insects are also known to take part in wound healing and protection against foreign bodies.

39.2c Muscle Tissue Produces Movement

Muscle tissue consists of cells that have the ability to contract (shorten). The contractions, which depend on the interaction of two proteins—*actin* and *myosin*—move body limbs and other structures, pump the blood, and produce a squeezing pressure in organs such as the intestine and uterus. Three types of muscle tissue, *skeletal, cardiac,* and *smooth,* produce body movements in vertebrates **(Figure 39.7)**.

Skeletal Muscle. **Skeletal muscle** is so called because most muscles of this type are attached by tendons to the skeleton. Skeletal muscle cells are also called **muscle fibres** because each is an elongated cylinder (see Figure 39.7a). These cells contain many nuclei and are packed with actin and myosin molecules arranged in highly ordered, parallel units that give the tissue a banded or striated appearance when viewed under a microscope. Muscle fibres packed side by side into parallel bundles surrounded by sheaths of connective tissue form many body muscles.

Skeletal muscle contracts in response to signals carried by the nervous system. The contractions of skeletal muscles, which are characteristically rapid and powerful, move body parts and maintain posture. The contractions also release heat as a by-product of cellular metabolism. This heat helps mammals, birds, and some other vertebrates maintain their body temperatures, particularly through the process of shivering when environmental temperatures fall. (Skeletal muscle is discussed further in Chapter 46.)

Cardiac Muscle. **Cardiac muscle** is the contractile tissue of the heart (see Figure 39.7b). Cardiac muscle has a striated appearance because it contains actin and myosin molecules arranged like those in skeletal muscle. However, cardiac muscle cells are short and

a. Skeletal muscle

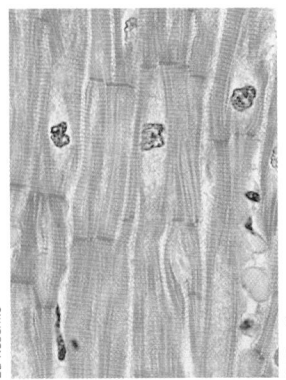

b. Cardiac muscle

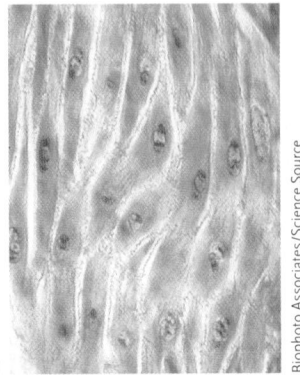

c. Smooth muscle

Ed Reschke

Ed Reschke

Biophoto Associates/Science Source

Figure 39.7
Structure of skeletal, cardiac, and smooth muscle.

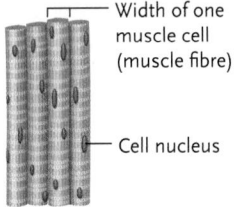

- Width of one muscle cell (muscle fibre)
- Cell nucleus

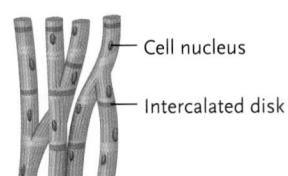

- Cell nucleus
- Intercalated disk

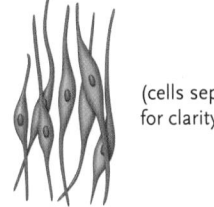

(cells separated for clarity)

Description: Bundles of long, cylindrical, striated, contractile cells called muscle fibres

Typical location: Attached to bones of skeleton

Function: Locomotion, movement of body parts

Description: Cylindrical, striated cells that have specialized end junctions

Location: Wall of heart

Function: Pumping of blood within circulatory system

Description: Contractile cells with tapered ends

Typical location: Wall of internal organs, such as stomach

Function: Movement of internal organs

branched, with each cell connecting to several neighbouring cells; the joining point between two such cells is called an *intercalated disk*. Cardiac muscle cells thus form an interlinked network, which is stabilized by anchoring junctions and gap junctions. The gap junctions transmit electrical signals through the movement of ions that make cardiac muscles contract as a unit. In this network, because of the gap junctions when one cell contracts, all cells in the heart contract, and because each cell is connected tightly to many of its neighbours, heart muscle contracts in all directions, with the cells pulling on one another to produce a squeezing or pumping action rather than a lengthwise, unidirectional contraction as seen in skeletal muscle.

Smooth Muscle. **Smooth muscle** is found in the walls of tubes and cavities in the body, including blood vessels, the stomach and intestine, the bladder, and the uterus. Smooth muscle cells are relatively small and spindle shaped (pointed at both ends), and their actin and myosin molecules are arranged in a loose network rather than in bundles (see Figure 39.7c). This loose network makes the cells appear smooth rather than striated when viewed under a microscope. Smooth muscle cells are connected by gap junctions and enclosed in a mesh of connective tissue. As in cardiac muscle, the gap junctions transmit signals that make smooth muscles contract as a unit, typically producing a squeezing motion. Although smooth

muscle contracts more slowly than skeletal and cardiac muscles do, its contractions can be maintained at steady levels for a much longer time. These contractions move and mix the stomach and intestinal contents, constrict blood vessels, and push the infant out of the uterus during childbirth.

Invertebrate Muscle. In general, most invertebrates have striated muscles throughout, even muscles involved with structures such as the heart, intestine, and reproductive ducts. The striated muscle of invertebrates can't be subdivided into skeletal and cardiac, as in vertebrates, although different types of striation patterns do occur. In insects, the striated muscles that control the movements of some of the viscera, such as the ovaries and parts of the digestive system, are frequently branched and interconnected to form a lattice. Smooth muscle has been reported in some invertebrates only.

39.2d Nervous Tissue Receives, Integrates, and Transmits Information

Nervous tissue contains cells called **neurons** (also called *nerve cells*) that serve as lines of communication and control between body parts. Billions of neurons are packed into the human brain; others have extremely long processes that extend throughout the body. Nervous tissue also contains **glial cells** (*glia* = glue), which physically support and provide nutrients to

Figure 39.8
Neurons and their structure. The micrograph shows a network of motor neurons, which relay signals from the brain or spinal cord to muscles and glands.

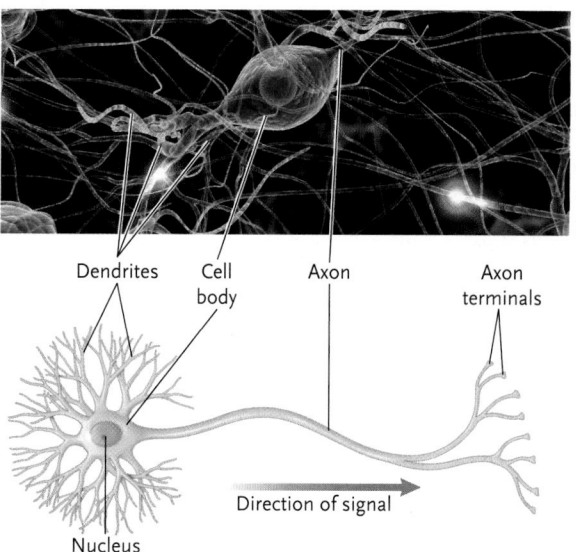

Dendrites   Cell body   Axon   Axon terminals

Direction of signal

Nucleus

neurons, provide electrical insulation between them, and scavenge cellular debris and foreign matter.

A neuron consists of a *cell body*, which houses the nucleus and organelles, and two types of cell extensions, dendrites and axons **(Figure 39.8)**. *Dendrites* receive chemical signals from other neurons or from body cells of other types and convert them into an electrical signal that is transmitted to the cell body of the receiving neuron. Dendrites are usually highly branched. *Axons* conduct electrical signals away from the cell body to the axon terminals, or endings. At their terminals, most axons convert the electrical signal to a chemical signal that stimulates a response in nearby muscle cells, gland cells, or other neurons (direct electrical connections are discussed further in Chapter 40). Axons are usually unbranched except at their terminals. Depending on the type of neuron and its location in the body, its axon may extend from a few micrometres or millimetres to more than a metre (the cell body of a neuron innervating the foot of a giraffe is in its

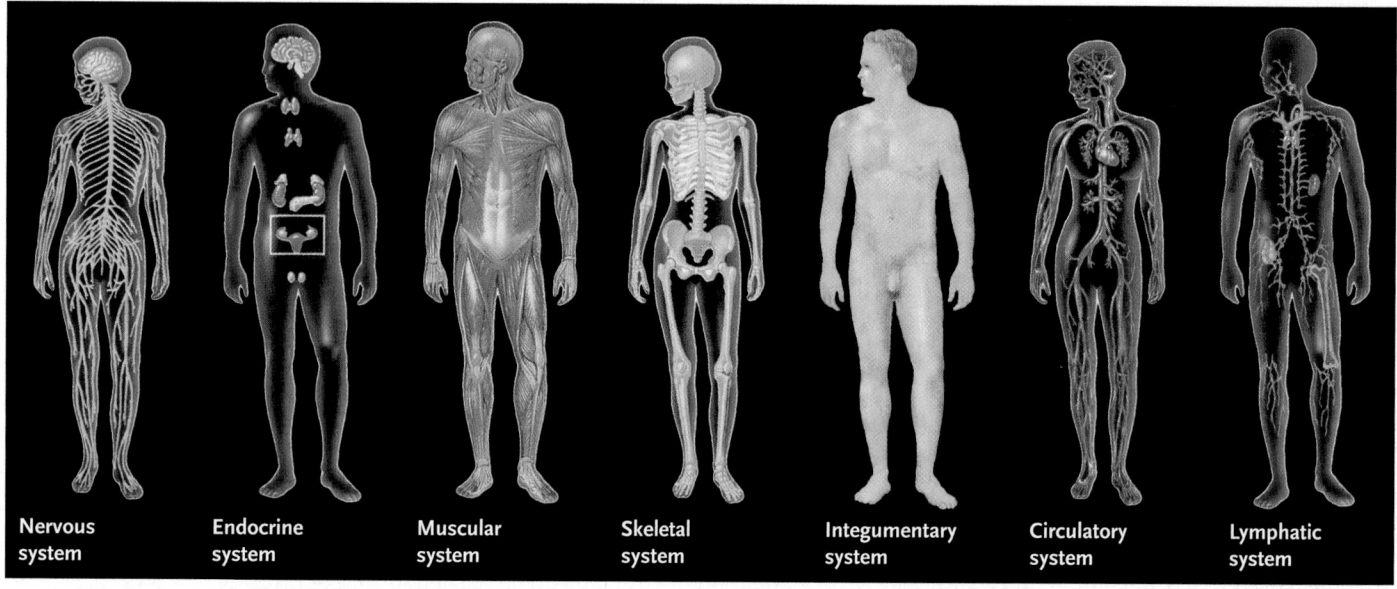

Nervous system	Endocrine system	Muscular system	Skeletal system	Integumentary system	Circulatory system	Lymphatic system
Main organs: Brain, spinal cord, peripheral nerves, sensory organs	**Main organs:** Pituitary, thyroid, adrenal, pancreas, and other hormone-secreting glands	**Main organs:** Skeletal, cardiac, and smooth muscle	**Main organs:** Bones, tendons, ligaments, cartilage	**Main organs:** Skin, sweat glands, hair, nails	**Main organs:** Heart, blood vessels, blood	**Main organs:** Lymph nodes, lymph ducts, spleen, thymus
Main functions: Principal regulatory system; monitors changes in internal and external environments and formulates compensatory responses; coordinates body activities. Nervous systems are present in all metazoans except sponges.	**Main functions:** Regulates and coordinates body activities through secretion of hormones. Endocrine systems are also present in most metazoans.	**Main functions:** Moves body parts; helps run bodily functions; generates heat. Specialized muscle cells do not appear in evolution until triploblastic animals.	**Main functions:** Supports and protects body parts; provides leverage for body movements. An internal skeleton composed of bone and/or cartilage occurs only in the vertebrates. Similar functions in invertebrates are carried out by an external skeleton or by internal hydrostatic pressure.	**Main functions:** Covers external body surfaces and protects against injury and infection; helps regulate water content and body temperature. All Metazoa except sponges have an integument of some sort.	**Main functions:** Distributes water, nutrients, oxygen, hormones, and other substances throughout the body and carries away carbon dioxide and other metabolic wastes; helps stabilize internal temperature and pH. Specialized circulatory systems occur in all vertebrates and in the annelids, molluscs, and arthropods.	**Main functions:** Returns excess fluid to the blood; defends the body against invading viruses, bacteria, fungi, and other pathogens as part of the immune system. Invertebrates do not have a specialized lymphatic system.

spinal cord). (Neurons and their organization in body structures are discussed further in Chapter 44.)

All four major tissue types—epithelial, connective, muscle, and nervous—combine to form the organs and organ systems of animals. The next section depicts the major organs and organ systems of vertebrates and outlines their main tasks.

STUDY BREAK

1. Embryonic stem cells can differentiate into any tissue. How many major tissue types are there in vertebrates, and what are the differences between them in terms of their structure and function?
2. Distinguish between exocrine and endocrine glands. What do they have in common? What makes them different?

39.3 Coordination of Tissues in Organs and Organ Systems

39.3a Organs and Organ Systems Function Together to Enable an Animal to Survive

In the tissues, organs, and organ systems of an animal, each cell engages in the basic metabolic activities that ensure its own survival and performs one or more functions of the system to which it belongs. All vertebrates have 12 major organ systems, which are summarized in **Figure 39.9**. Most invertebrates have the same systems but do not have a separate system of lymphatic ducts.

The functions of all these organ systems are coordinated and integrated to collectively accomplish a series of tasks that are vital to all animals, whether a flatworm, a salmon, a moose, or a human. These tasks include

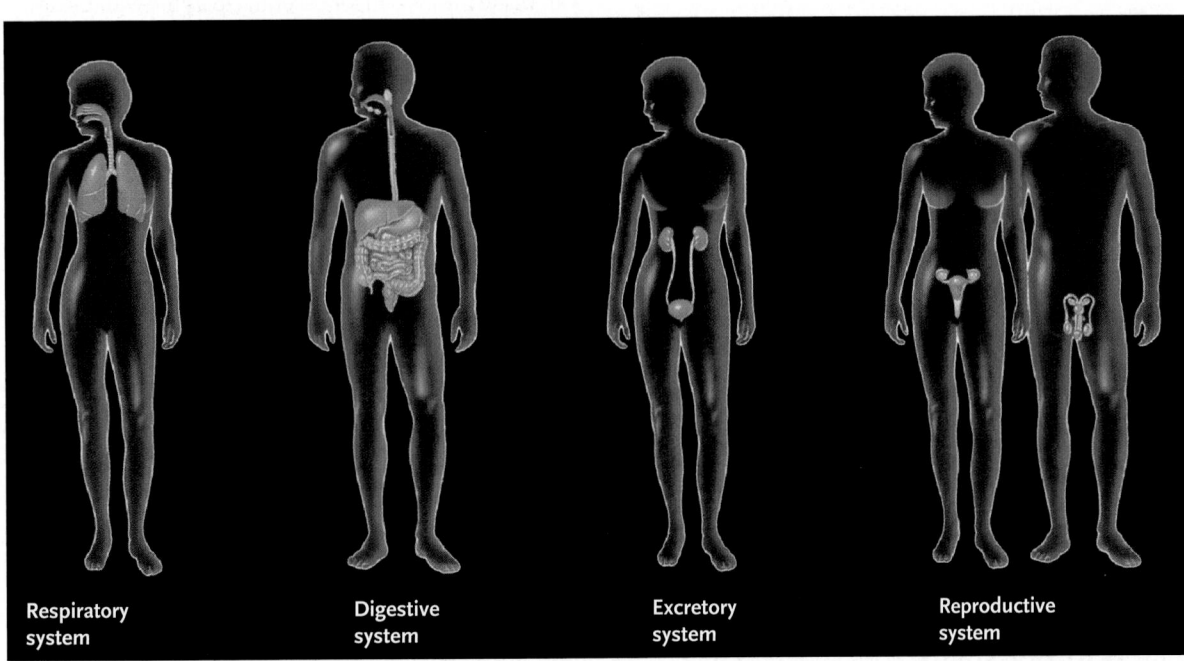

Figure 39.9
Organ systems of the human body. The immune system, which is primarily a cellular system, is not shown. The functions performed by these organ systems are performed by all animals, although different organs and systems may be involved.

From Starr/Taggart, *Biology: The Unity and Diversity of Life*, 8E. © 1998 Cengage Learning.

Respiratory system

Main organs:
Lungs, diaphragm, trachea, and other airways

Main functions:
Exchanges gases with the environment, including uptake of oxygen and release of carbon dioxide. Fish have a respiratory system that involves gills. Some form of specialized respiratory system occurs in most invertebrates.

Digestive system

Main organs:
Pharynx, esophagus, stomach, intestines, liver, pancreas, rectum, anus

Main functions:
Converts ingested matter into molecules and ions that can be absorbed into the body; eliminates undigested matter; helps regulate water content. Most metazoans, with the exception of some parasitic forms, have a digestive system.

Excretory system

Main organs:
Kidneys, bladder, ureter, urethra

Main functions:
Removes and eliminates excess water, ions, and metabolic wastes from body; helps regulate internal osmotic balance and pH. All animals perform these functions. All vertebrates have kidneys, and most invertebrates have specialized excretory organs and systems.

Reproductive system

Main organs:
Female: ovaries, oviducts, uterus, vagina, mammary glands
Male: testes, sperm ducts, accessory glands, penis

Main functions:
Maintains the sexual characteristics and passes on genes to the next generation. Most triploblastic animals have specialized reproductive organs and systems.

1. acquiring nutrients and other required substances, such as oxygen; coordinating their processing; distributing them throughout the body; and disposing of wastes;
2. synthesizing the protein, carbohydrate, lipid, and nucleic acid molecules required for body structure and function;
3. sensing and responding to changes in the environment, such as temperature, pH, and ion concentrations;
4. protecting the body against injury or attack from other animals and from viruses, bacteria, and other disease-causing agents; and
5. reproducing and, in many instances, nourishing and protecting offspring through their early growth and development.

Together these tasks maintain homeostasis, preserving the internal environment required for survival of the body. Homeostasis is the topic of the next section.

CONCEPT FIX While each organ system has a major role to perform for the body (e.g., the circulatory system moves essential materials around the body and removes wastes from tissues, and the integumentary system provides mechanical protection), *it is not true that most organ systems serve only one role.* Many have minor supporting roles in other functions, and none of these systems work in isolation. ⬢

STUDY BREAK

1. What are the major functions of each of the 12 organ systems? What are the key organs in each?
2. What are the major organ systems in a duck? In a shark? In an insect? In an earthworm?

39.4 Homeostasis

To live, cells of all organisms must take in nutrients and O_2 from the external environment and eliminate wastes such as CO_2 and nitrogenous wastes to the external environment. A single-celled organism such as an amoeba is in direct contact with the external environment. Although most cells of a multicellular animal are isolated from direct contact with the external environment, these cells have the same needs for nutrient and O_2 input and waste elimination. These needs are met by the specialized tissues of the different organ systems that establish an internal environment in the form of the **extracellular fluid** (ECF) **(Figure 39.10).** The ECF has two components:

- **plasma,** the fluid portion of the blood, and
- **interstitial fluid** (*inter* = between; *stitial* = that which stands), the fluid that surrounds the cells.

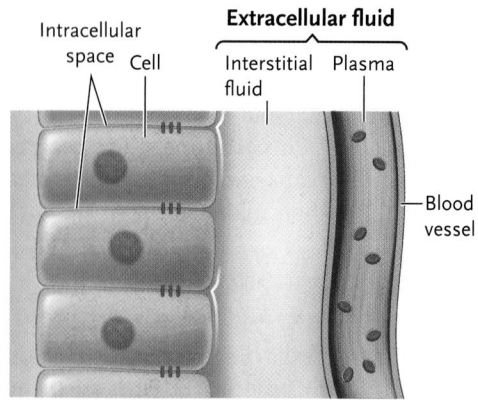

Figure 39.10
Nature of the extracellular fluid (ECF).

The ECF connects all cells to the external environment. Thus, no matter where a cell is within the body, it can make the exchanges essential to its life with the interstitial fluid. Particular organ systems enable these exchanges between the external and internal environments. The digestive system processes incoming food and transfers absorbed nutrients into the plasma of the blood. The nutrients reach all parts of the body by the action of the circulatory system, along with O_2 that enters the blood by the action of the respiratory system. The nutrients and O_2 in the plasma reach the interstitial fluid through the capillaries, and, from there, they enter the cells as needed. Waste moves in the opposite direction: from the cells into the interstitial fluid and then into the plasma. The respiratory system handles the removal of CO_2, and the excretory system handles the metabolic wastes.

For optimal function of these systems, the composition and state of the ECF must be maintained within a narrow range so that cells have available the necessary nutrients and O_2 and wastes can be eliminated. Further, other aspects of the internal environment that are important for cellular (and therefore organismal) life, such as temperature, must also be regulated within a tolerable range.

Animals fall into two major categories in this regard: **regulators** maintain factors of the internal environment in a relatively constant state, and **conformers** have internal environments that match the external environment. Any given animal may regulate some factors and conform to others. For instance, animals such as fishes, reptiles, and insects are thermoconformers. Their body temperatures match that of the external environment. They are also osmoregulators, meaning that they maintain the ionic composition of their ECF relatively constant regardless of the composition of the external environment. Homeostasis is the process by which animals regulate their internal environment to maintain a relatively stable state. Homeostasis is a dynamic process, in which internal adjustments are made continuously to compensate for changes in the internal or external environment.

39.4a Many Factors of the Internal Environment Are Homeostatically Regulated

Factors of the internal environment that are regulated homeostatically include the following:

1. *Nutrient concentration.* Energy production by cells requires a constant supply of nutrient molecules. The energy generated by catabolizing the nutrients is used for basic cellular processes and any specialized activities of the cell.
2. *Concentration of O_2.* Cellular respiration (see Chapter 6), the process that generates energy from catabolic reactions, requires a constant supply of O_2 for optimal productivity.
3. *Concentration of CO_2.* The CO_2 produced by the catabolic reactions of cellular respiration must be removed as waste or else the ECF would become increasingly acidic.
4. *Concentration of waste chemicals.* Particular biochemical reactions in the cell generate products that would be toxic to the cell if not removed as waste.
5. *Concentration of water and NaCl.* The relative concentrations of water and NaCl in the ECF affect how much water enters or leaves a cell and hence the cell's volume. These concentrations must be regulated to maintain a cell volume that is optimal for function; swollen or shrunken cells are typically functionally impaired.
6. *pH.* Changes in the pH of the ECF can adversely affect enzymatic activities within cells, as well as the functions of all other functional proteins (ion channels, receptors, etc.).
7. *Volume and pressure of the plasma.* Both the volume and the pressure in blood vessels must be maintained at adequate levels to distribute the fluid throughout the body. This circulation is vitally important for supplying cells with their needs and removing wastes.
8. *Temperature.* Body cells function optimally within a specified temperature range. Outside that range, chemical reactions change their rates and may be completely inhibited. If cells become too cold, the rates of enzymatic reactions decrease too much, and, if cells become too hot, structural and enzymatic proteins can be denatured and become inactive.

39.4b Homeostasis Is Accomplished by Negative Feedback Control Systems

The primary mechanism of homeostasis is **negative feedback**, in which a stimulus resulting from a change in the external or internal environment triggers a response that compensates for the environmental change **(Figure 39.11)**. The components of a negative feedback control system are as follows:

- A **stimulus** is an environmental change (external or internal) that triggers a response.
- A **sensor** is a tissue or organ that detects the environmental change (such as external temperature, or the internal concentration of a molecule such as glucose).

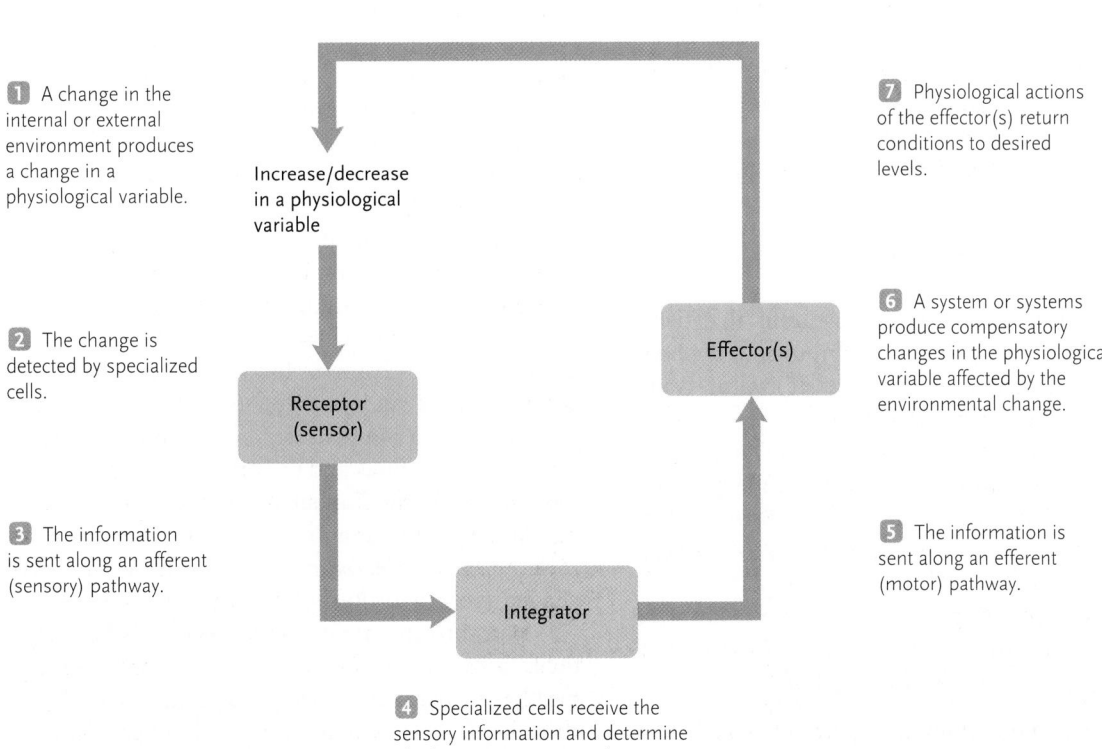

1. A change in the internal or external environment produces a change in a physiological variable.

Increase/decrease in a physiological variable

2. The change is detected by specialized cells.

Receptor (sensor)

3. The information is sent along an afferent (sensory) pathway.

4. Specialized cells receive the sensory information and determine whether action is required.

Integrator

5. The information is sent along an efferent (motor) pathway.

6. A system or systems produce compensatory changes in the physiological variable affected by the environmental change.

Effector(s)

7. Physiological actions of the effector(s) return conditions to desired levels.

Figure 39.11
Components of a negative feedback mechanism maintaining homeostasis. The integrator coordinates a response by comparing the level of an environmental condition with a set point that indicates where the level should be.

- An **integrator** is a control centre that compares the detected environmental change with a **set point**, the level at which the condition controlled by the pathway is to be maintained. In most animals, the integrator is part of the central nervous system or the endocrine system.
- An **effector** is a system, activated by the integrator, that returns the condition to the set point if it has strayed away. Effectors may include parts of any body tissue or organ.

The Thermostat as a Negative Feedback Mechanism. The concept of negative feedback may be most familiar in systems designed by human engineers. The thermostat maintaining temperature at a chosen level in a house provides an example. In the thermostat, the following are the components of the negative feedback system:

- The stimulus is a change in room temperature of a few degrees (up or down) from the temperature set on the thermostat.
- The sensor is the thermometer within the thermostat that measures the temperature.
- The integrator is the electrical circuit in the thermostat that activates the effectors.
- The effector returns the room temperature to the set point. If the temperature has fallen below the set point, the effector is the furnace, which adds heat to the house until the temperature rises to the set point. If the temperature has risen above the set point, the furnace is turned off and an air conditioner may be turned on to remove the heat more quickly from the room until the temperature falls to the set point.

Thermostats and Negative Feedback Mechanisms in Animals. Mammals and birds also have a homeostatic mechanism that maintains body temperature within a relatively narrow range around a set point. The components of the negative feedback system are as follows:

- The stimulus is a change in body temperature beyond normal levels.
- The sensors are groups of neurons that detect changes in the temperature throughout the body.
- The integrator is the temperature control centre in a region of the brain called the *hypothalamus* that compares the changes in temperature of the brain and the rest of the body with the set point. For most mammals, including humans, the set point has a relatively narrow range centred at about 37°C.
- The effectors are physiological or behavioural responses that function to return the body temperature to desired levels (the set point).

All birds and mammals regulate their body temperature around a relatively narrow set point. If the temperature falls below the lower limit, the hypothalamus activates effectors that constrict the blood vessels in the skin. The reduction in blood flow means that less heat is conducted from the blood through the skin to the environment; in short, heat loss from the skin is reduced. Small muscles may be activated to cause the fur or feathers to stand up, increasing the thickness of the animal's insulation. Other effectors may induce shivering, a physical mechanism to generate body heat. Some animals are capable of nonshivering heat production using specialized fat cells (brown fat). Also, integrating neurons in the brain, stimulated by signals from the hypothalamus, initiate behavioural responses such as moving to a warmer area (see Chapter 50 for details).

Conversely, if the blood temperature rises above the set point, the hypothalamus triggers effectors that dilate the blood vessels in the skin, increasing blood flow to the skin and heat loss from it. Other effectors cause fur and feathers to flatten, reducing the insulation layer. Yet other effectors can induce sweating on bare patches of skin, which cools the skin and the blood flowing through it as the sweat evaporates. And again, through integrating neurons in the brain, animals may consciously sense being overheated, which may be counteracted by moving to cooler locations, or taking a dip in a pool of water. **Figure 39.12,** illustrates how a dog responds to activity at high environmental temperatures.

CONCEPT FIX It is commonly believed that *all physiological variables that are homeostatically regulated are always held more or less constant.* This is not always the case. For instance, sometimes the temperature set point changes, and the negative feedback mechanisms then operate to maintain body temperature at the new set point. Thus, when animals become infected by certain viruses and bacteria, the temperature set point increases to a higher level, producing a fever to help overcome the infection. Once the infection is combatted, the set point is readjusted down again to its normal level. ⬡

Whereas mammals and birds regulate their internal body temperature within a narrow range around a set point, certain other vertebrates regulate over a broader range. These vertebrates use other negative feedback mechanisms for their temperature regulation. Snakes and lizards, for example, respond behaviourally to compensate for variations in environmental temperatures and use other, less precise negative feedback mechanisms for their temperature regulation. They may absorb heat by basking on sunny rocks in the cool early morning and move to cooler, shaded spots in the heat of the afternoon.

Many insects employ similar mechanisms to raise their body temperature. Some caterpillars group together, increasing their body temperatures by a degree or two and shortening the time of development by as much as three days. Flight requires energy, and the

7 Physiological actions of the effector(s) lead to cooling and return body temperature to normal levels.

1 Activity on a hot, dry day leads to a rise in body temperature.

6 Panting is produced by respiratory muscles to produce evaporative heat loss from the respiratory passages.

Increase in skin and body temperature

2 The change is detected by temperature-sensitive cells in the skin and the hypothalamus.

Blood vessels dilate, conducting metabolically generated heat to evaporative surfaces (lungs, throat, mouth, tongue, feet).

Effector(s)

Receptor (sensor)

Salivary glands secrete fluid to increase evaporation from the tongue, mouth, and throat.

3 The information is sent along an afferent (sensory) pathway.

5 The information is sent along an efferent (motor) pathway.

Integrator

4 Neurons in the hypothalamus receive the sensory information and determine whether action is required.

Ermolaev Alexander/Shutterstock.com

Figure 39.12
Homeostatic mechanisms maintaining the body temperature of a husky when environmental temperatures are high.

39.4c Animals Also Have Positive Feedback Mechanisms That Do Not Result in Homeostasis

Under certain circumstances, animals respond to a change in internal or external environmental condition by a **positive feedback** mechanism that intensifies or adds to the change. Such mechanisms, with some exceptions, do not result in homeostasis. They operate when the animal is responding to life-threatening conditions (an attack, for instance), or as part of reproductive processes, and produce sudden explosive events.

The birth process in mammals is a prime example. During human childbirth, initial contractions of the uterus push the head of the fetus against the **cervix**, the opening of the uterus into the vagina. The pushing causes the cervix to stretch. Sensors that detect the stretching signal the hypothalamus to release a hormone, oxytocin, from the pituitary gland. Oxytocin increases the uterine contractions, intensifying the squeezing pressure on the fetus and further stretching the cervix. The stretching results in more oxytocin

flight muscles operate best at higher temperatures. Some insects bask in the Sun to warm the muscles. Many, such as dragonflies, bumblebees, butterflies, and moths, contract the flight muscles rapidly in a process similar to shivering in order to warm them. This is particularly important in moths that fly at night, when the environmental temperature is lower. Honeybees form masses in the winter and maintain their temperature by contracting the wing muscles. Once insects are in flight, however, the energy production is so high that they must dissipate the heat produced. In bees, the most important method is evaporative cooling by regurgitation of some of the intestinal contents onto the mouthparts, a process equivalent to panting in vertebrates.

release and stronger uterine contractions, repeating the positive feedback circuit and increasing the squeezing pressure until the fetus is pushed entirely out of the uterus.

Because positive feedback mechanisms such as the one triggering childbirth do not result in homeostasis, most occur less commonly than negative feedback in animals. Others, such as the nerve action potential that will be discussed in Chapter 44, are very regular events.

Review

aplia™

To access course materials such as Aplia and other companion resources, please visit www.NELSONbrain.com.

39.1 Organization of the Animal Body

- Multicellularity permits organisms to maintain an internal environment, allowing them to exploit a greater variety of external environments. It allows organisms to become larger but requires cells to differentiate and become specialized to perform specific functions. Since not all cells perform all functions, the extracellular fluid (ECF) serves to ensure that the net results of the specialized cells are relayed to every single cell in the organism (Figure 39.10).

- In most animals, cells are specialized and organized into tissues, tissues into organs, and organs into organ systems (Figure 39.2). A tissue is a group of cells with the same structure and function, working as a unit to carry out one or more activities. An organ is an assembly of tissues integrated into a structure that carries out a specific function. An organ system is a group of organs that carry out related steps in a major physiological process.

39.2 Animal Tissues

- Animal tissues are classified as epithelial, connective, muscle, or nervous. The properties of the cells of these tissues determine the structures and functions of the tissues.

- Various kinds of junctions link cells in a tissue (Figure 39.3). Anchoring junctions "weld" cells together. Tight junctions seal the cells into a leakproof layer. Gap junctions form direct avenues of communication between the cytoplasm of adjacent cells in the same tissue.

- Epithelial tissue consists of sheetlike layers of cells that cover body surfaces and the surfaces of internal organs and line cavities and ducts within the body.

- Exocrine glands are secretory structures derived from epithelia. Exocrine glands are connected by a duct that empties onto the epithelial surface. Endocrine glands are ductless. Not all endocrine glands are derived from epithelia (Figure 39.5).

- Connective tissue consists of cell networks or layers and a prominent extracellular matrix (ECM) of dead material that separates the cells. It supports other body tissues and transmits mechanical and other forces.

- Loose connective tissue consists of sparsely distributed fibroblasts surrounded by an open network of collagen and other glycoproteins. It supports epithelia and organs of the body and forms a covering around blood vessels, nerves, and some internal organs.

- Fibrous connective tissue contains sparsely distributed fibroblasts in a matrix of densely packed, parallel bundles of collagen and elastin fibres. It forms high–tensile strength structures such as tendons and ligaments.

- Cartilage consists of sparsely distributed chondrocytes surrounded by a network of collagen fibres embedded in a tough but highly elastic matrix of branched glycoproteins. Cartilage provides support, flexibility, and a low-friction surface for joint movement.

- In bone, osteocytes are embedded in a collagen matrix hardened by mineral deposits. Osteoblasts secrete collagen and minerals for the ECM; osteoclasts remove the minerals and recycle them into the bloodstream.

- Adipose tissue consists of cells specialized for fat storage. It also cushions the body and provides an insulating layer under the skin.

- Blood in most animals consists of a fluid matrix, the plasma, in which cells may be suspended. In vertebrates, the erythrocytes carry oxygen to body cells and the leukocytes produce antibodies and initiate the immune response against disease-causing agents.

- Muscle tissue contains cells that have the ability to contract forcibly. Skeletal muscle, containing long cells called muscle fibres, moves body parts and maintains posture.

- Cardiac muscle, which contains short contractile cells with a branched structure, forms the heart.

- Smooth muscle consists of spindle-shaped contractile cells that form layers surrounding body cavities and ducts.

- Nervous tissue contains neurons and glial cells. Neurons communicate information between body parts in the form of electrical and chemical signals. Glial cells support the neurons or provide electrical insulation between them.

39.3 Coordination of Tissues in Organs and Organ Systems

- Organs and organ systems are coordinated to carry out vital tasks, including maintenance of internal body conditions; nutrient acquisition, processing, and distribution; waste disposal; molecular synthesis; environmental sensing and response; protection against injury and disease; and reproduction.

- In all vertebrates, the major organ systems that accomplish these tasks are the nervous, endocrine, muscular, skeletal, integumentary, circulatory, lymphatic, immune, respiratory, digestive, excretory, and reproductive systems (Figure 39.9). Many invertebrates also have these organ systems, with the exception of a lymphatic system.

39.4 Homeostasis

- Homeostasis is the process by which animals maintain their internal environment at conditions their cells can tolerate. It is a dynamic state in which internal adjustments are made continuously to compensate for environmental (external or internal) changes.

- Homeostasis is accomplished by negative feedback mechanisms that include a sensor, which detects a change in an external or internal condition; an integrator, which compares the detected change with a set point; and an effector, which returns the condition to the set point if it has varied (Figure 39.11).

- Animals also have positive feedback mechanisms, in which a change in an internal or external condition triggers a response that intensifies the change and typically does not result in homeostasis.

Questions

Self-Test Questions

Any number of answers from a to e may be correct.

1. Which structure has the highest level of organization (i.e., contains the other structures)?
   a. the liver
   b. the epithelium
   c. mitochondria
   d. the hepatic (liver) cell

2. What is a muscle, such as the biceps muscle, composed of?
   a. similar tissues
   b. different tissues
   c. similar cells
   d. different organs
   e. similar organs

3. Which tissue is a constant source of adult stem cells in a mammal?
   a. kidneys
   b. pancreas
   c. basal lamina
   d. heart muscle
   e. bone marrow

4. The bones of an elderly woman break more easily than those of a younger person. Which cell type would you surmise diminishes in activity with aging?
   a. osteocyte
   b. osteoblast
   c. osteoclast
   d. fibroblast
   e. chondrocyte

5. Where is interstitial fluid found in organisms?
   a. in the cytoplasm
   b. in the cell membrane
   c. in the cell walls
   d. between cells

6. What type of junction allows ions and molecules to flow between cells by way of channels?
   a. tight junction
   b. gap junction
   c. anchoring junction

7. What type of cell can a chemical signal be passed to after leaving a neuron?
   a. gland cell
   b. muscle cell
   c. another neuron
   d. all of the above

8. Which of the following is NOT a homeostatic response?
   a. The basketball players are dripping sweat at halftime.
   b. The pupils in the eyes constrict when looking at a light.
   c. The brain is damaged when a fever rises above 40.5°C.
   d. Slower breathing in sleep changes carbon dioxide and oxygen levels in the blood, which affect blood pH.
   e. In a contest, a student eats an entire chocolate cake in 10 minutes. Due to hormonal secretions, his blood glucose level does not change dramatically.

9. A decrease in body temperature causes the pituitary gland to release a hormone that stimulates the release of thyroxine from the thyroid gland. Thyroxine increases metabolism, generating heat. As the body temperature increases, the release of the pituitary hormone decreases and less thyroxine is released. What is this an example of?
   a. integration
   b. osmolarity
   c. positive feedback
   d. negative feedback
   e. environmental sensing

10. Which system coordinates other organ systems?
    a. the skeletal system
    b. the muscular system
    c. the nervous system
    d. the reproductive system
    e. the endocrine system

Questions for Discussion

1. Astronauts lose bone mass during space travel. Why do you think this happens? To test your hypothesis, can you devise an experiment that does not involve space travel?

2. There are at least 25 known collagens. What information would you need to have, and how would you use that information to propose a hypothesis that explains the way that evolution has acted to produce so many versions of the same molecule?

3. Positive feedback mechanisms are rarer in animals than negative feedback mechanisms. Why do you think this is so?

4. Name four important functions of epithelial tissue. Provide an example of one tissue for each function.

5. Why was the discovery of neural stem cells so important?

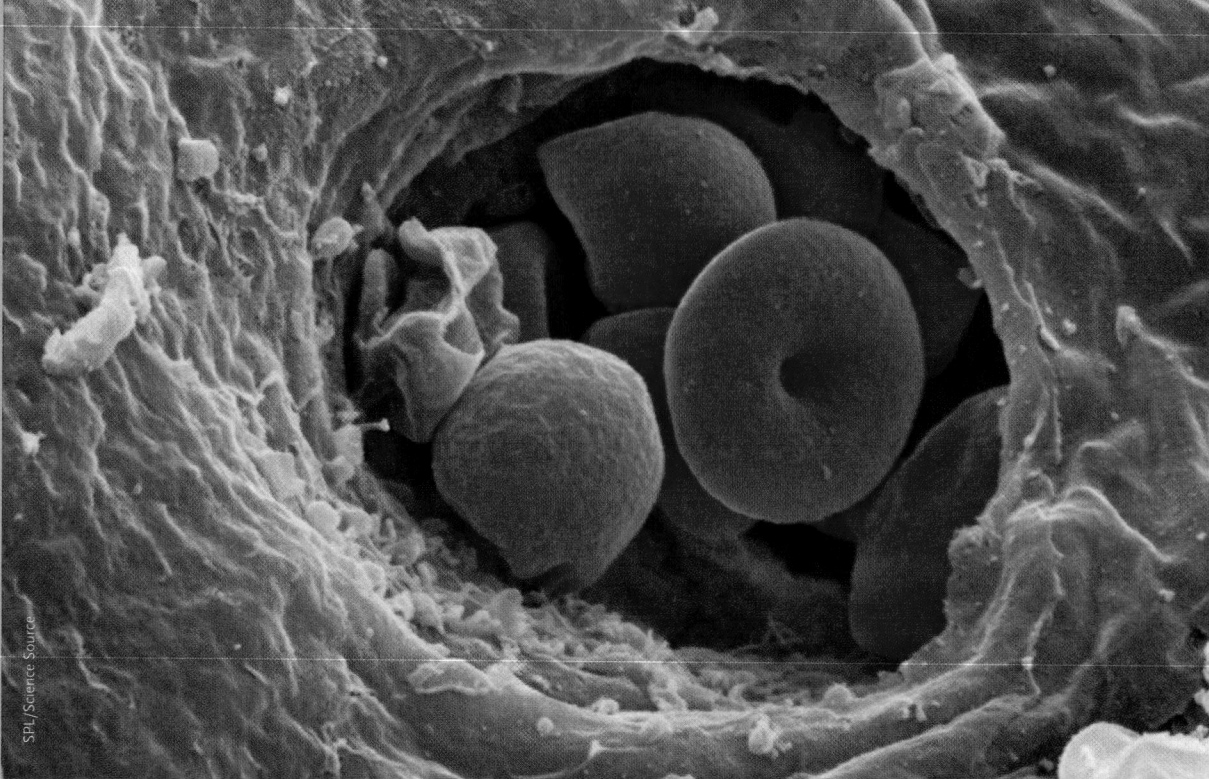

Confocal image of red blood cells in a small arteriole in muscle. In 1628, William Harvey published the first account of the circulation of the blood with a clear account of the action of the heart and the movement of blood around the body in blood vessels such as this, forming a continuous circuit.

40

Transport in Animals: The Circulatory System

WHY IT MATTERS

Jimmie the bulldog stood on the stage of a demonstration laboratory at a meeting of the Royal Society in London in 1909, with one front paw and one rear paw in laboratory jars containing salt water **(Figure 40.1)**. Wires leading from the jars were connected to a galvanometer, a device that can detect electrical currents.

Jimmie's master, Dr. Augustus Waller, a physician at St. Mary's Hospital, was relating his experiments in the emerging field of *electrophysiology*. Among other discoveries, Waller found that his apparatus detected the electrical currents produced each time the dog's heart beat.

Waller had originally experimented on himself. He already knew that the heart produces an electrical current as it beats; other scientists had discovered this by attaching electrodes directly to the heart of experimental animals. Looking for a painless alternative to that procedure, Waller reasoned that because the human body can conduct electricity, his arms and legs might conduct the currents generated by the heart if they were connected to a galvanometer. Accordingly, Waller set up two metal pans containing salt water and connected wires from the pans to a galvanometer. He put his bare left foot in one pan and his right hand in the other. The technique worked; the indicator on the galvanometer jumped each time his heart beat. And it worked with Jimmie, too.

a. Jimmie the bulldog

b. Electrocardiogram

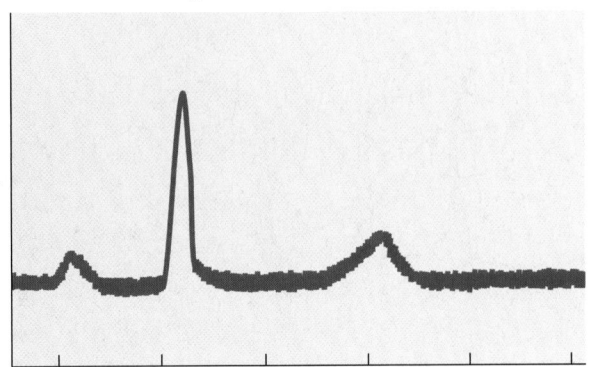

From A. D. Waller, Physiology, *The Servant of Medicine*, Hitchcock Lectures, University of London Press, 1910.

The first electrocardiograms. **(a)** Jimmie the bulldog standing in laboratory jars containing salt water, with wires leading to a galvanometer that recorded the electrical currents produced by his heartbeat. **(b)** One of Waller's early electrocardiograms.

Waller also invented a method for recording the changes in current, which became the first electrocardiogram (ECG). He constructed a galvanometer by placing a column of mercury in a fine glass tube, with a conducting salt solution layered above the mercury. Changes in the current passing through the tube caused corresponding changes in the surface tension of the mercury, which produced movements that could be detected by reflecting a beam of light from the mercury surface. By placing a moving photographic plate behind the mercury tube, Waller could record the movements of the reflected light on the plate (Figure 40.1b shows one of his records). These were the first ECGs.

As discussed in Chapter 39, with the evolution of multicellularity came the evolution of specialized tissues with specific functions. In this chapter, we discuss how the circulatory system is designed to allow these specialized tissues to exchange important molecules, and often cells, between tissues. Examples of transported molecules are oxygen (O_2), nutrients, hormones, and wastes.

The **circulatory system** consists of a fluid and a pump (usually a heart) and vessels for moving the fluid. It also consists of an accessory system, a lymphatic system consisting of its own vessels and organs, that balances the distribution of fluid between the blood and the extracellular fluid (ECF) surrounding tissues and participates in the body's defences against invading disease organisms.

40.1 Animal Circulatory Systems: An Introduction

Protostomes with simple body plans, including sponges, flatworms, and nematodes, function with no specialized circulatory system. Nearly all of these animals are aquatic or, like parasitic flatworms, live surrounded by the body fluids or intestinal contents of a host animal. Their bodies are structured as thin sheets of cells that lie close to the fluids of the surrounding environment **(Figure 40.2a, p. 966)**. The products of digestion diffuse among the cells via the interstitial fluids, and O_2 and CO_2 are exchanged with the medium through the surface of the animal. Nematodes have a fluid-filled body cavity but no special mechanism for circulating that fluid.

40.1a Animal Circulatory Systems Share Basic Elements

In larger and more complex animals, most cells lie in cell layers too deep within the body to exchange substances directly with the environment via diffusion. Instead, the animals have a circulatory system, composed of tissues and organs, that conducts O_2, CO_2, nutrients, and the products of metabolism among the cells and tissues. The circulatory system connects specialized regions of the animal where substances are exchanged with the external environment. For example, oxygen is absorbed from the environment in the gills or lungs of many animals and is carried by the blood to all parts of the body; CO_2 released from body cells is carried by the blood to the lungs or gills, where it is released to the environment. Soluble wastes are conducted from body cells to the kidneys or other excretory organs, which remove wastes from circulation and excrete them into the environment.

Animal circulatory systems carrying out these roles share certain basic features:

- A specialized fluid medium, usually containing at least some cells, carries nutrients from the digestive system, the products of metabolism, and soluble wastes. With the conspicuous exception of the insects, it also transports O_2 and CO_2.
- The fluid is usually contained in tubular vessels that distribute it to the various organs. Animal circulatory systems take one of two forms, either *open* or *closed*. In an **open circulatory system**, vessels leaving the heart release fluid, usually termed **hemolymph,** directly into body spaces or into sinuses surrounding organs (Figure 40.2b; **Figure 40.3a, p. 967**). Thus, the organs and tissues are directly bathed in the hemolymph. The hemolymph reenters the heart through valves in the heart wall

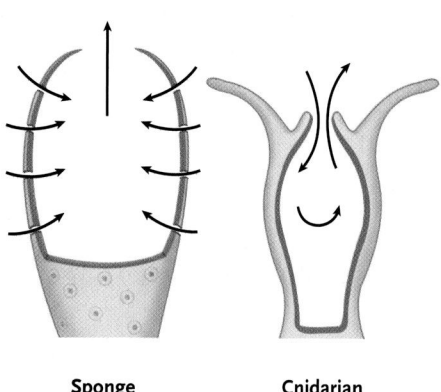

Sponge   **Cnidarian**

a. Circulation of external fluid through an open body cavity

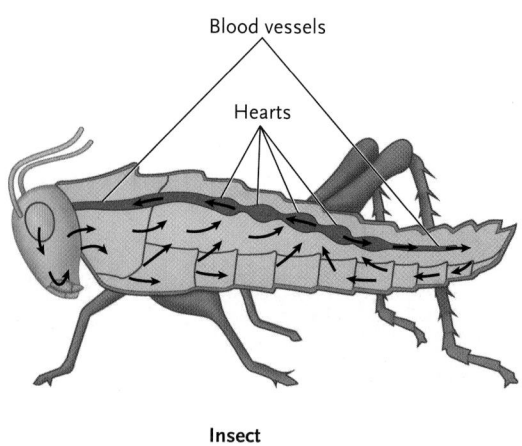

Blood vessels

Hearts

Insect

b. Circulation of internal fluid through an open circulatory system

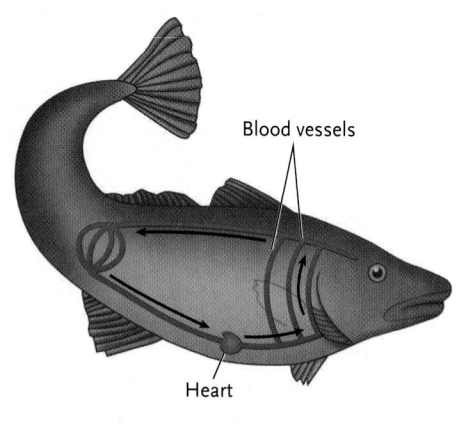

Blood vessels

Heart

Fish

c. Circulation of internal fluid through a closed circulatory system

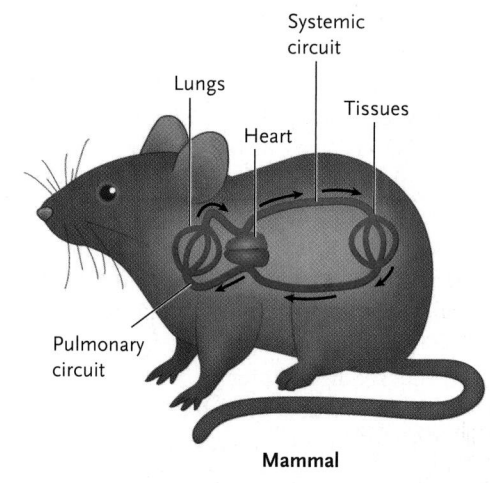

Systemic circuit

Lungs

Heart

Tissues

Pulmonary circuit

Mammal

d. Circulation of internal fluid through a closed circulatory system with a double circuit

Figure 40.2

The general plan of the circulatory system in different animals. **(a)** Multicellular organisms such as sponges and cnidarians use the external medium to transport molecules between cells. **(b)** Most invertebrates circulate an internal fluid through an open circulatory system. **(c)** Some invertebrates and all vertebrates move an internal fluid through a closed circulatory system in which blood is separated from the interstitial fluid. **(d)** In birds and mammals, two separate circuits serve the lungs and all other tissues of the body.

that close each time the heart pumps, thereby maintaining a unidirectional flow. In a **closed circulatory system**, the blood is confined to blood vessels and is distinct from the interstitial fluid (Figure 40.2c, d; Figure 40.3b). Substances are exchanged between the blood and the interstitial fluid and then between the interstitial fluid and cells.

- A muscular heart pumps the fluid through the circulatory system. Words associated with the heart often include *cardio,* from *kardia,* Greek for heart.

40.1b Most Invertebrates Have Open Circulatory Systems

Among the protostomes, arthropods and most molluscs have open circulatory systems with one or more muscular hearts (Figure 40.2b; **Figure 40.4**). The blood

is not conveyed directly to all cells by tubes but is ejected from the open ends of the blood vessels and directly bathes the body tissues. In an open system, most of the fluid pressure generated by the heart dissipates when the blood is released from vessels into body spaces. Although the pressure remains low, the rate at which the blood circulates can be increased by an increase in the rate of beating of the heart. Because of one-way valves in the heart and vessels, the beating heart draws hemolymph in from the tissue spaces as they expand and pushes it out through the vessels as the heart contracts. In highly active invertebrates, such as flying insects, the heart rate may rise to three or more times the resting rate, increasing the circulation of the hemolymph among the tissues. In addition, most insects have accessory hearts associated with the wings and each leg, which experience the same

a. Open circulatory system: no distinction between hemolymph and interstitial fluid

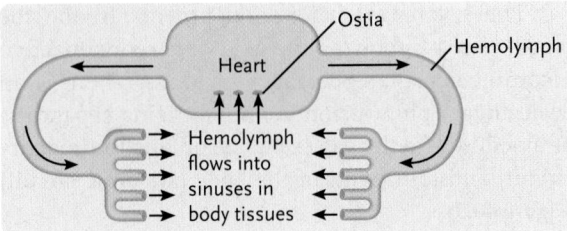

b. Closed circulatory system: blood separated from interstitial fluid

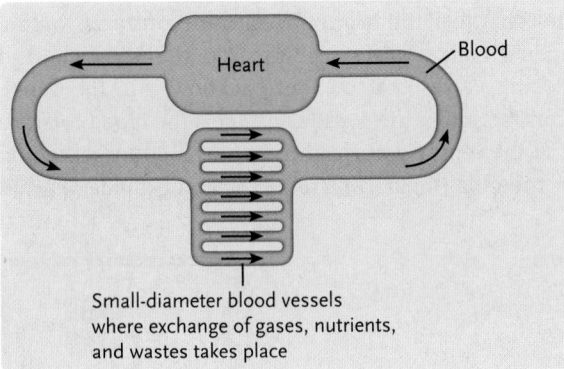

Figure 40.3

(a) Open circulatory system: hemolymph bathes the organs and body tissues. **(b)** Closed circulatory system: blood is confined in tubes that lie among the cells of all tissues.

increases in rate during periods of high metabolic activity. In insects and molluscs, heart rate is controlled in some cases by nerves but largely by a variety of amine and peptide hormones (see Chapter 44).

40.1c Some Invertebrates and All Vertebrates Have Closed Circulatory Systems

Annelids, cephalopod molluscs such as squids and octopuses, most deuterostome invertebrates, and all vertebrates have closed circulatory systems (Figure 40.2c, Figure 40.4). In these systems, vessels called **arteries** conduct blood away from the heart at relatively high pressure. From the arteries, the blood eventually enters highly branched networks of microscopic, thin-walled vessels called **capillaries** that are well adapted for diffusion of substances. Nutrients and wastes are exchanged between the blood and body tissues as the blood moves through the capillaries. The blood then flows at relatively low pressure from the capillaries to larger vessels, the **veins**, which carry the blood back to the heart. Typically, the blood is maintained at a higher pressure and moves more rapidly through the body in closed systems than in open systems. In many animals, closed systems allow precise control of the distribution and rate of blood flow to

Figure 40.4

Evolutionary trends in animal circulatory systems.

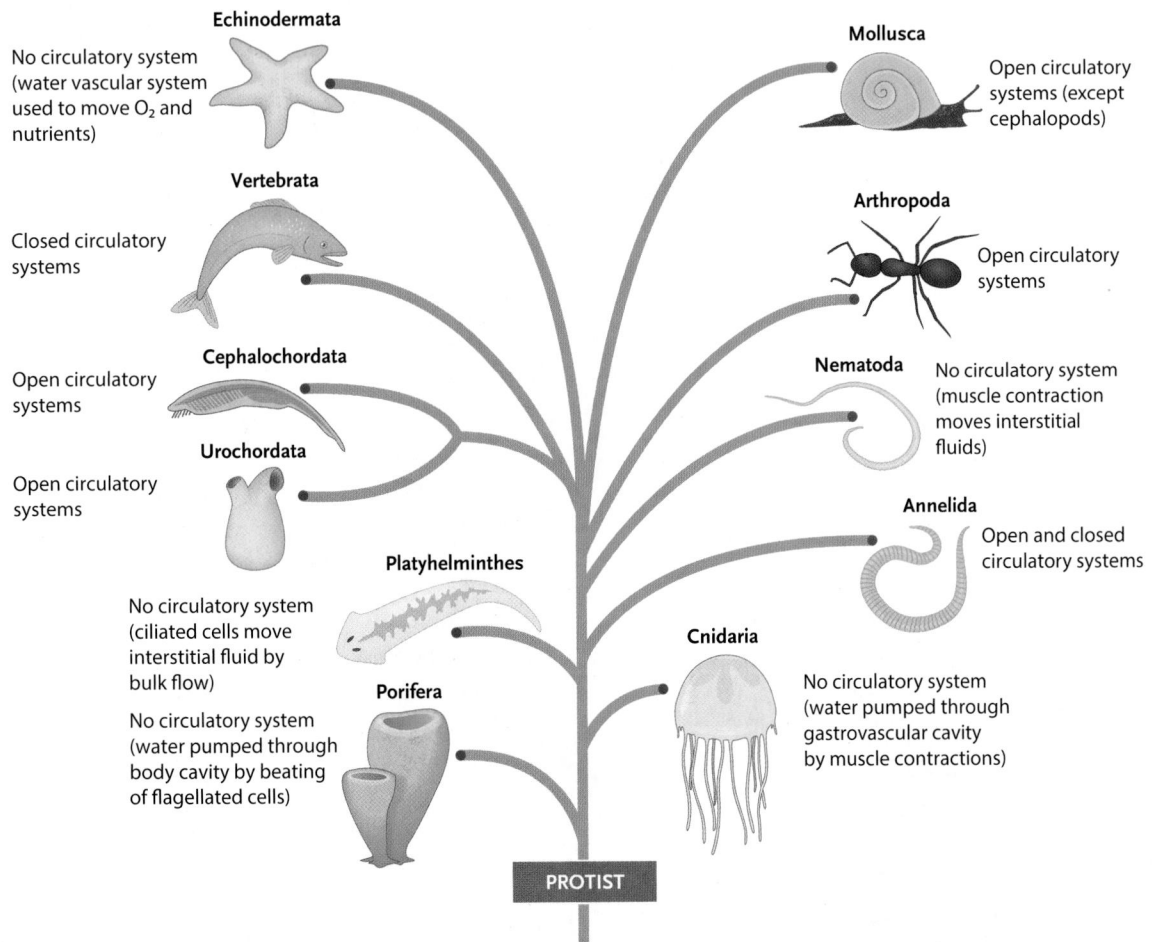

different body regions by means of muscles that contract or relax to adjust the diameter of the blood vessels. Contraction leads to reduced blood vessel diameter, while relaxation increases the blood vessel diameter (see Section 40.4).

40.1d Vertebrate Circulatory Systems Have Evolved from Single to Double Blood Circuits

A comparison of the different vertebrate groups reveals several evolutionary trends that accompanied the invasion of terrestrial habitats. Among the most striking are the changes that occurred in the major vessels of the body and the heart. These changes converted the single-circuit system of sharks and bony fish, in which the gills are in the same circuit as the rest of the blood vessels, to a double-circuit system in which the circulation to the lungs parallels the circulation to the rest of the body (Figure 40.2d, p. 966).

There were two major developments. In one, the blood vessels supplying the gills were reorganized to accommodate the appearance of lungs. There is an evolutionary progression to an increasing separation of blood flow to the gas exchange organs (**pulmonary circuit**) and to the rest of the body (**systemic circuit**) **(Figure 40.5)**.

The second involved developments in the structure of the heart. In a shark or bony fish **(Figure 40.6a)**, venous, deoxygenated blood from the tissues enters the first chamber, the atrium. The atrium contracts, forcing open flaplike valves leading into the ventricle and closing valves that prevent backflow into the veins. Contraction of the ventricle propels the blood forward into the ventral aorta leading to the blood vessels going to the gills (the aortic arches); the gill capillaries; and

Figure 40.5
Evolution of the circulatory system. The single circulation of most fishes gives rise to the double circulation of tetrapods.

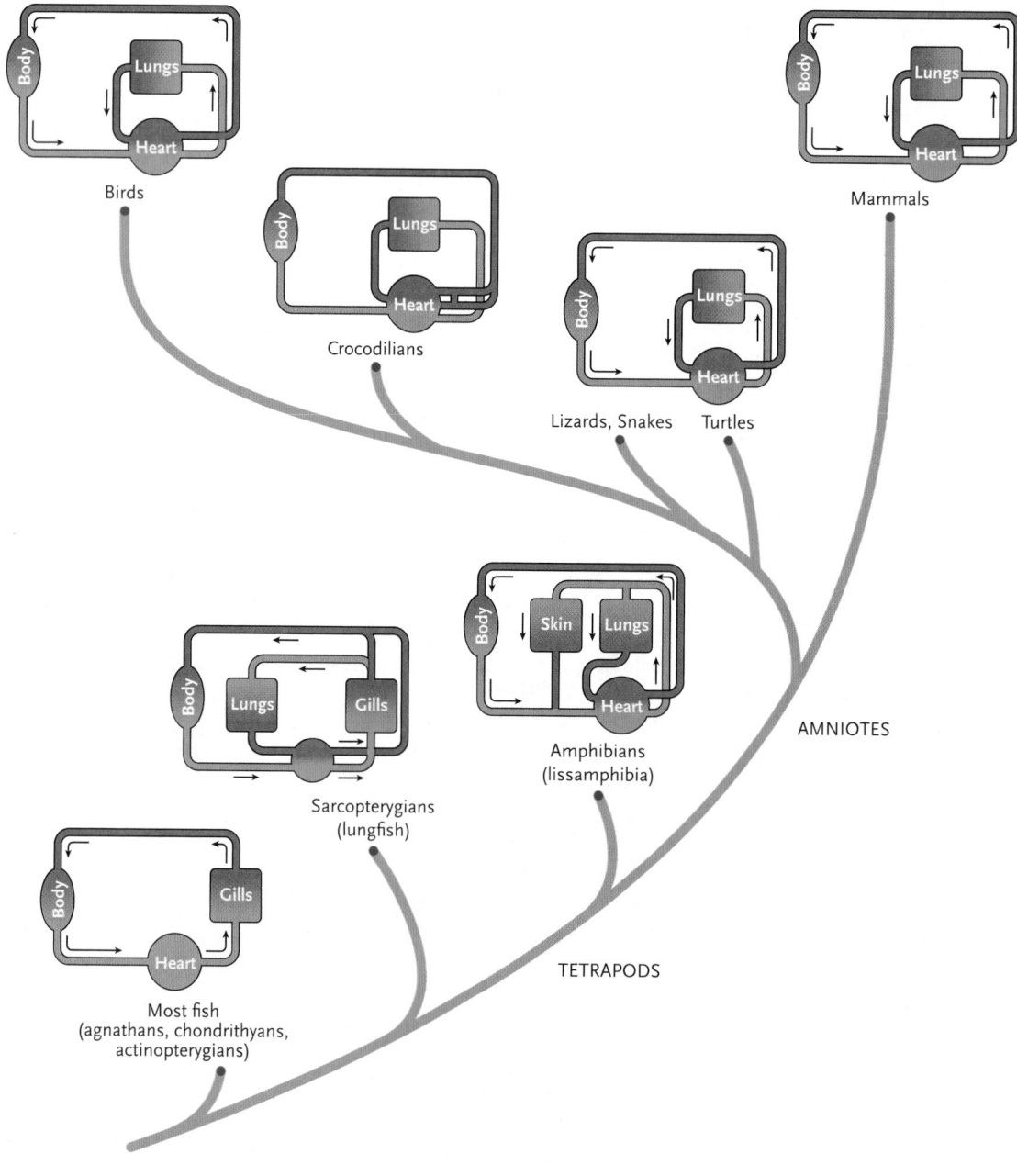

968 | UNIT TEN SYSTEMS AND PROCESSES: ANIMALS

NEL

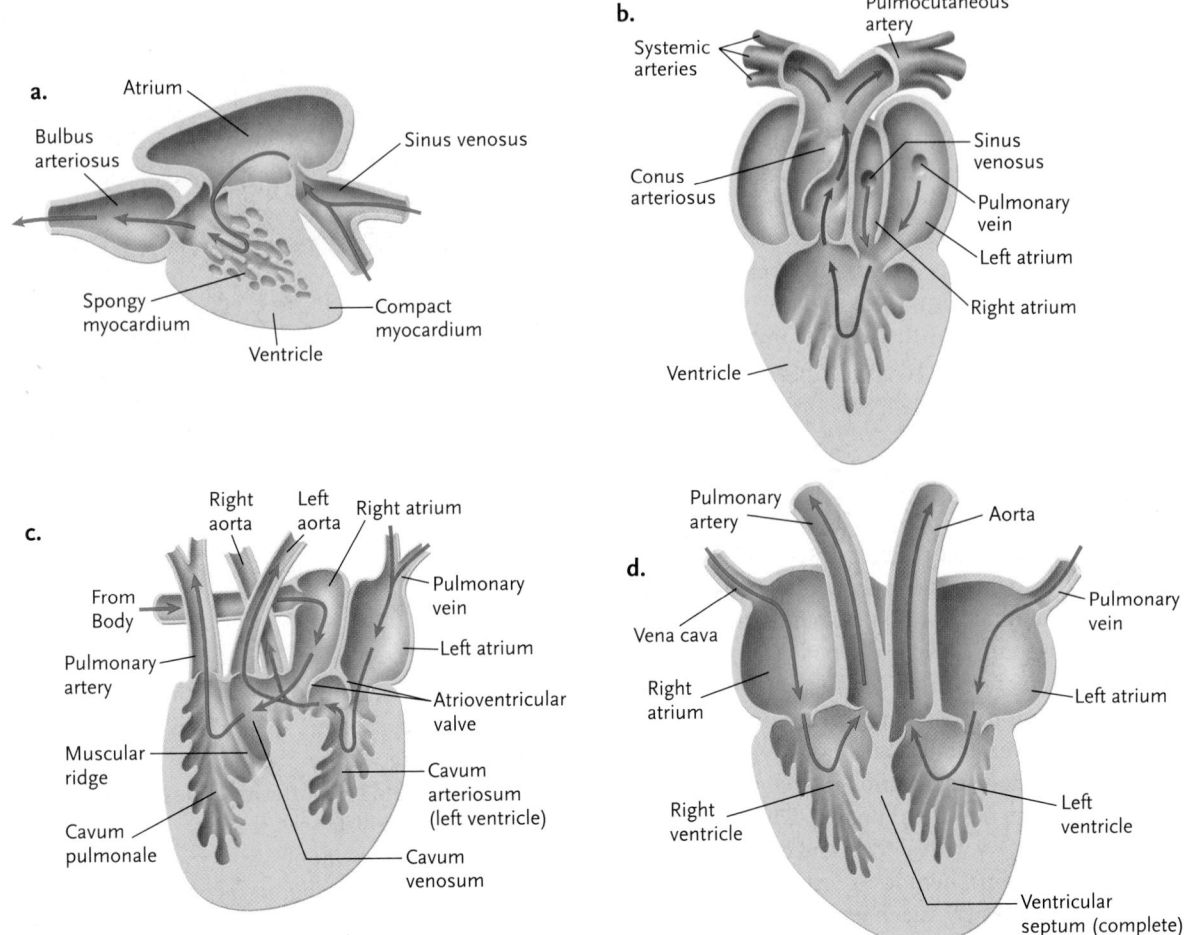

Figure 40.6

Major steps in the evolution of the heart leading to separate circulation for the lungs. **(a)** The two-chambered heart in a shark or a bony fish, **(b)** two atria in an amphibian, **(c)** partial division of the ventricle in modern reptiles, and **(d)** complete separation into two ventricles in crocodilians, birds, and mammals.

the dorsal aorta, which carries blood to the tissues. In amphibians, the atrium is divided, with one side (left atrium) receiving oxygenated blood from the lungs and the other side (right atrium) receiving deoxygenated blood from the body and skin (Figure 40.6b). In the single ventricle, some separation of the two streams is achieved by the spongy nature of the ventricle, which prevents open mixing, and by a flaplike structure in the vessel leaving the heart that can direct blood into the arteries going to the skin and lungs or into the arteries going to the rest of the body. In modern reptiles, such as lizards and snakes, the ventricle is partially divided (Figure 40.6c).

CONCEPT FIX *The partially divided ventricle of most reptiles is often referred to as incompletely divided, with the inference that this is a phylogenetic artifact and less than ideal.* The cardiovascular systems of reptiles, however, are extraordinarily flexible and no less adaptive for their lifestyles than the systems found in birds and mammals. Evolution of the cardiovascular system does not represent a progressive improvement in design but rather different adaptive alternatives for meeting the different demands that different lifestyles place on the circulatory system. In the case of the reptiles, the amount of mixing of blood returning from the two atria that occurs in the ventricles can be carefully regulated. ⬢

Full separation of the blood supply to the lungs occurs in mammals, birds, and crocodilians (alligators and crocodiles share ancestry with birds) by complete division of the ventricle (Figure 40.6d). There are thus two separate circuits. One circuit delivers oxygenated blood from the lungs into the left atrium, which then propels it into the left ventricle. The contraction of the left ventricle sends blood to the body circulation via the carotid arteries to the head and the dorsal aorta, which supplies the remainder of the body. In the second circuit, deoxygenated blood from the body and head enters the right atrium and, via the right ventricle, is propelled to the lungs in the pulmonary artery.

This progressive separation of the body and lung circulation (illustrated in Figure 40.5), and the accompanying changes in the architecture of the heart (illustrated in Figure 40.6), demonstrate that although evolution happens by changes in existing structures, changes in one set of structures (the aortic arches) are correlated with changes in other structures (the heart).

1. What are the differences between open and closed circulatory systems? Has the closed system evolved only once? What are the advantages of a closed system over an open one?
2. Which vertebrates have a separate pulmonary circulation? Compare and contrast the circulatory systems of a fish, a frog, and a beaver and describe the changes in the structure of the heart that accompany them. What might be the advantages or disadvantages of the differences?

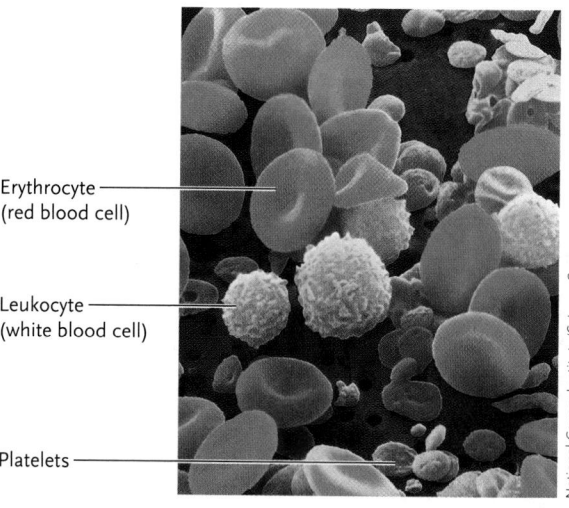

Erythrocyte (red blood cell)

Leukocyte (white blood cell)

Platelets

National Cancer Institute/Science Source

Figure 40.7

Typical components of human blood. The colourized scanning electron micrograph shows the three major cellular components.

40.2 Blood and Its Components

In both vertebrates and invertebrates, blood is a complex connective tissue that may contain blood cells suspended in a liquid called the *plasma*. Although the blood of all vertebrates contains blood cells, the blood of some invertebrates may consist exclusively of plasma with few or no suspended cells, as in the Nematoda. In other invertebrates, such as the arthropods, blood cells or hemocytes of various recognizable types may occur in large numbers (up to 275 000 per µL in crickets) that can vary with activity and developmental stage. Whereas some hemocytes circulate with the hemolymph, others may attach temporarily to various tissues. These hemocytes can be mobilized rapidly and enter the circulation, for example, to take part in wound healing or in defence against disease and parasites. In addition to transporting nutrients, dissolved gases, and metabolic wastes, blood helps stabilize the internal pH and salt composition of body fluids and serves as a highway for cells of the immune system and the antibodies produced by some of these cells.

In vertebrates, blood also helps regulate body temperature by transferring heat between warmer and cooler body regions and between the body and the external environment (see Chapter 50). The total blood volume of most vertebrates is 5 to 8% of body mass (about 4 to 5 L in an average-sized adult human). The *plasma*, a clear, straw-coloured fluid, is about 45% to 55% of the volume of blood in most vertebrates (55% in human males and 58% in human females). Suspended in the plasma are three main types of blood cells, *erythrocytes, leukocytes,* and *platelets,* which account for the remainder of the blood volume. The typical components of human blood are shown in **Figure 40.7.**

40.2a Plasma Is an Aqueous Solution of Proteins, Ions, Nutrient Molecules, and Gases

Plasma is complex, and its composition varies depending on many factors. Its average composition in humans is given in **Table 40.1.** The plasma proteins of vertebrates fall into three classes: the albumins, the globulins, and fibrinogen. The **albumins**, the most abundant proteins of the plasma, are important for osmotic balance and pH buffering. They also transport a wide variety of substances through the circulatory system, including hormones and metabolic wastes. Because of their similar chemical composition to hormones, many therapeutic drugs are designed to be transported this way. The **globulins** transport lipids (including cholesterol) and fat-soluble vitamins; a specialized subgroup of globulins, the **immunoglobulins**, includes antibodies and other molecules that contribute to the immune response. Some globulins are also enzymes. **Fibrinogen** plays a central role in the clotting mechanism of the blood.

The ions of the plasma include Na^+, K^+, Ca^{2+}, Cl^-, and HCO_3^- (bicarbonate). The Na^+ and Cl^- ions are the most abundant and are present in concentrations similar to those of seawater, reflecting evolutionary ancestry. Some of the ions, particularly the bicarbonate ion, help maintain arterial blood at its characteristic pH (see Chapter 50).

40.2b Erythrocytes Are the Oxygen Carriers of Vertebrate Blood

Erythrocytes, or red blood cells, carry O_2 from the lungs to body tissues. Each microlitre of human blood normally contains about 5 million erythrocytes, which are small, flattened, and disclike. They measure about 7 µm in diameter and 2 µm in thickness. Microtubules of the cytoskeleton (see Chapter 2) are arranged beneath the surface of the cell so that they are *biconcave*—thinner in the middle than at the edges (see Figure 40.7). The proteins of the cytoskeleton that determine their shape also give them the flexibility to squeeze through narrow capillaries.

Table 40.1   The Composition of Human Blood

The sketch of the test tube shows what happens when you centrifuge a blood sample. The blood separates into three layers: a thick layer of straw-coloured plasma on top, a thin layer containing leukocytes and platelets, and a thick layer of erythrocytes. The table shows the relative amounts and functions of the various components of blood.

Plasma Portion (55–58% of total volume)

Components	Relative Amounts	Functions
1. Water	91–92% of plasma volume	Solvent
2. Plasma proteins (albumin, globulins, fibrinogen, etc.)	7–8%	Defence, clotting, lipid transport, roles in ECF volume, and so on
3. Ions, sugars, lipids, amino acids, hormones, vitamins, dissolved gases, urea and uric acid (metabolic wastes)	1–2%	Roles in ECF volume, pH, eliminating waste products, and so on

Cellular Portion (42–45% of total volume):

	Relative Amounts	Functions
1. Erythrocytes (red blood cells)	4 800 000–5 400 000 per microlitre	Transport oxygen, carbon dioxide
2. Leukocytes (white blood cells)		
Neutrophils	3000–6750	Phagocytosis during inflammation
Lymphocytes	1000–2700	Immune response
Monocytes/macrophages	150–720	Phagocytosis in all defence responses
Eosinophils	100–360	Defence against parasitic worms
Basophils	25–90	Secretion of substances for inflammatory response and for fat removal from blood
3. Platelets	250 000–300 000	Roles in clotting

Plasma

Leukocytes and platelets

Packed cell volume, or hematocrit

Erythrocytes

Like all blood cells, erythrocytes arise from stem cells (see Chapter 39) in the red bone marrow. As they mature, mammalian erythrocytes lose their nucleus, cytoplasmic organelles, and ribosomes. Because they are no longer capable of synthesizing new proteins, this limits their metabolic capabilities and their life span (as short as 35 days in a chicken but 120 days in a human). The remaining cytoplasm contains enzymes, which carry out glycolysis, and large quantities of *hemoglobin,* the O_2-carrying protein of the blood. The erythrocytes of nearly all other vertebrates retain a nucleus.

Hemoglobin, the molecule that gives erythrocytes, and thus blood, their red colour, consists of four polypeptides, each linked to a nonprotein *heme* group (see "Molecule behind Biology," Box 40.1, p. 972) that contains an iron atom in its centre. The iron atom binds O_2 molecules as the blood circulates through the lungs and releases the O_2 as the blood flows through other body tissues. This is described in more detail in Chapter 49. It is the oxygenation of the iron that gives arterial blood its bright red colour (and red blood cells their name), just as the oxygenation of iron in nature gives rust its colour. When hemoglobin releases its oxygen, the venous blood changes to a much darker colour.

Some 2 to 3 *million* erythrocytes are produced in the average human each second. The life span of an erythrocyte in the circulatory system is about 120 days.

At the end of their useful life, erythrocytes are engulfed and destroyed by *macrophages* (*macro* = big; *phagein* = to eat), a type of large leukocyte, in the spleen, liver, and bone marrow.

A negative feedback mechanism keyed to the blood's O_2 content stabilizes the number of erythrocytes in blood. If the O_2 content drops below the normal level, the kidneys synthesize **erythropoietin**, a peptide hormone that stimulates stem cells in bone marrow to increase erythrocyte production. Erythropoietin is also secreted after blood loss and when mammals move to higher altitudes. As new red blood cells enter the bloodstream, the O_2-carrying capacity of the blood rises. If the O_2 content of the blood rises above normal levels, erythropoietin production falls and red blood cell production drops. Erythropoietin has been used in "blood doping" by some athletes to improve their performance.

40.2c   Leukocytes Provide the Body's Front Line of Defence against Disease

Leukocytes eliminate dead and dying cells from the body, remove cellular debris, and provide the body's first line of defence against invading organisms. They are called white blood cells because they are colourless, in contrast to the red blood cells. Because leukocytes retain their nuclei, cytoplasmic organelles, and ribosomes, they are fully functional cells.

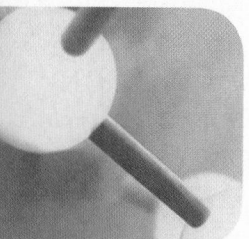

MOLECULE BEHIND BIOLOGY 40.1

Heme

Heme is a member of a family of chemicals called porphyrins **(Figure 1).** Among the properties of porphyrins is their ability to bind metals. In heme, the metal bound is iron. The complex attracts O_2, which binds to the iron. Heme is a cofactor of several important animal proteins and, as such, is an important carrier of O_2 in biological systems. Because these protein molecules are coloured, they are often referred to as oxygen-carrying pigments.

Hemoglobin is a protein that includes four heme molecules. In the lungs, the higher concentration of O_2 loads the hemoglobin in the erythrocytes with O_2 (oxidizing the iron and turning the blood bright red), and in the tissues, the low concentration leads to unloading (deoxidizing the iron and turning the blood a darker colour). The process of unloading the O_2 is assisted by higher concentrations of CO_2 in the tissues. Slight changes in the structure of the hemoglobin molecule lead to changes in the ability of the protein to bind oxygen. Animals that live in oxygen-poor environments possess hemoglobins with stronger binding affinities for oxygen than those that live in oxygen-rich environments. The mammalian fetus, for example, lives in a relatively oxygen-poor environment and has a special embryonic hemoglobin with a very strong binding affinity.

Myoglobin in muscle, neuroglobin in nervous tissue, and cytoglobin in all tissues are proteins that bind oxygen with an affinity higher than that of hemoglobin. They facilitate diffusion of oxygen through tissues, scavenge nitric oxide or reactive oxygen species, or serve a protective function by increasing oxygen availability during oxidative stress. The oxygen that is bound to myoglobin also serves as a reservoir that can be called on during periods of high metabolic demand resulting from increased muscle activity.

a.

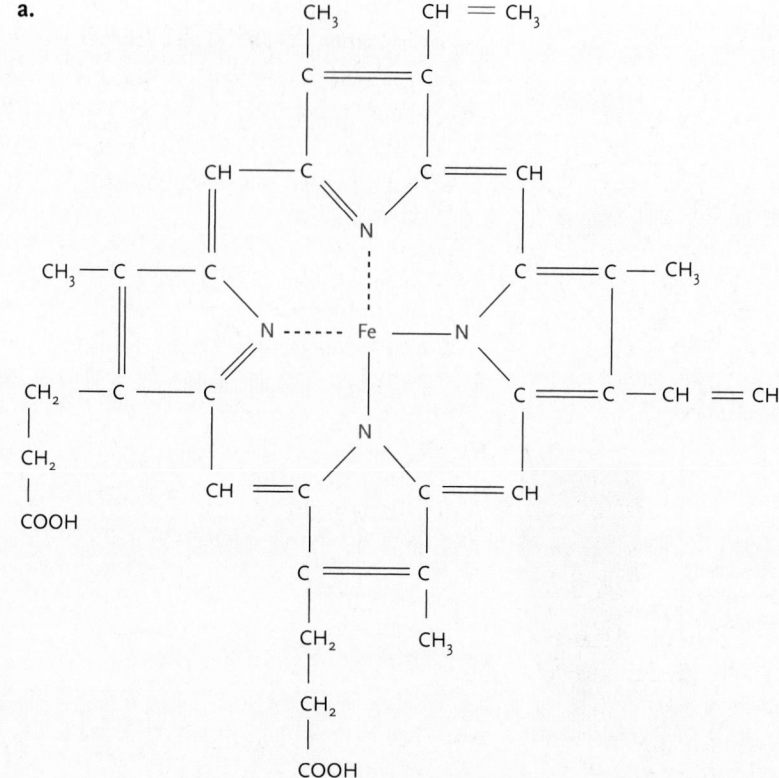

b.

Photos.com

c.

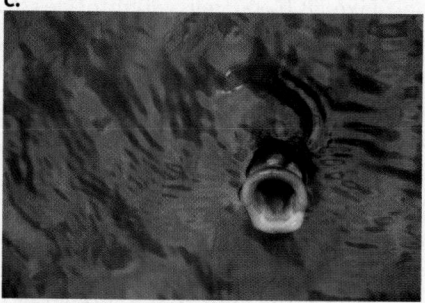

Photos.com

d.

Kul Bhatia/Science Source

Figure 1

(a) The heme molecule. Animals with heme-containing hemoglobins that readily bind oxygen include those that live in oxygen-poor environments such as *(b)* the llama at altitude; *(c)* fish, such as the carp, in hypoxic waters; and *(d)* mammalian fetuses such as that of an elephant shown here.

Cytochrome oxidase, the terminal enzyme in the complex that transfers electrons to molecular O_2 (see Chapter 6), also contains a heme molecule.

Other oxygen-carrying pigments, such as hemocyanin, which occurs in molluscs and some arthropods, and hemerythrin, found in a variety of invertebrates, also function as O_2 carriers. They do not, however, contain heme as the functional group.

Like red blood cells, leukocytes arise from the division of stem cells in red bone marrow. As they mature, they are released into the bloodstream, from which they enter body tissues in large numbers. Some types of leukocytes are capable of continued division in the blood and body tissues. The specific types of leukocytes and their functions in the immune reaction are discussed in Chapter 51.

40.2d Platelets Induce Blood Clots That Seal Breaks in the Circulatory System

Blood **platelets** are oval or rounded cell fragments, 2 to 4 μm in diameter, each enclosed in its own plasma membrane. They are produced in red bone marrow by the division of stem cells. Platelets contain enzymes and other factors that take part in blood clotting. When blood vessels are damaged, collagen fibres in the extracellular matrix are exposed to the leaking blood. Platelets in the blood stick to the collagen fibres and release signalling molecules that induce additional platelets to stick to them. The process continues, forming a plug that helps seal off the damaged site. As the plug forms, the platelets release other factors that convert the soluble plasma protein, fibrinogen, into long, insoluble threads of **fibrin.** Cross-links between the fibrin threads form a meshlike network that traps blood cells and platelets and further seals the damaged area **(Figure 40.8).** The entire mass is a blood clot.

STUDY BREAK

1. Why is blood considered a tissue? What are its three main cellular components, and what are their functions?
2. Hemoglobin occurs in the blood of many animals, but in many invertebrates, it occurs free in the plasma. What advantages can you think of that might have led to its incorporation in erythrocytes during the evolution of vertebrates? The nonvertebrate chordates do not have hemoglobin in their blood, but they do have cells in their blood that have a role in defence, suggesting that erythrocytes may have originated from such cells. What evidence could be used to examine this hypothesis?

40.3 The Heart

The vertebrate heart is composed of cardiac muscle cells (see Chapters 39 and 46). In mammals, we have seen (Figures 40.5, p. 968 and 40.6, p. 969) that the heart is a four-chambered pump, with two atria (singular, atrium) and two ventricles **(Figure 40.9).** The atria pump blood into the ventricles, and then powerful contractions of the ventricles push the blood at relatively

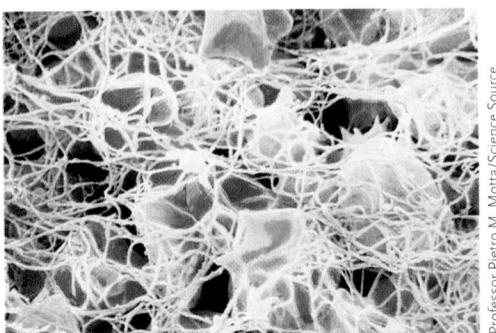

Figure 40.8
Red blood cells caught in a meshlike network of fibrin threads during formation of a blood clot.

Professor Pietro M. Motta/Science Source

high pressure into arteries leaving the heart. This arterial pressure is responsible for blood circulation. Valves between the atria and the ventricles, and between the ventricles and the arteries leaving the heart, keep the blood from flowing backward.

The mammalian heart pumps the blood through two completely separate circuits of blood vessels: the systemic circuit and the pulmonary circuit **(Figure 40.10, p. 974).** The right atrium (toward the right side of the body) receives blood returning from the entire body, except for the lungs. The *superior vena cava* conveys blood returning from the head and forelimbs, and the *inferior vena cava* conveys blood returning from the abdominal organs and hindlimbs. This blood is depleted of O_2 and has a high CO_2 content. The right atrium pumps the blood into the right ventricle, which contracts to push the blood into the *pulmonary arteries* leading to the lungs. In the capillaries of the lungs, the blood releases CO_2 and picks up O_2. The oxygenated

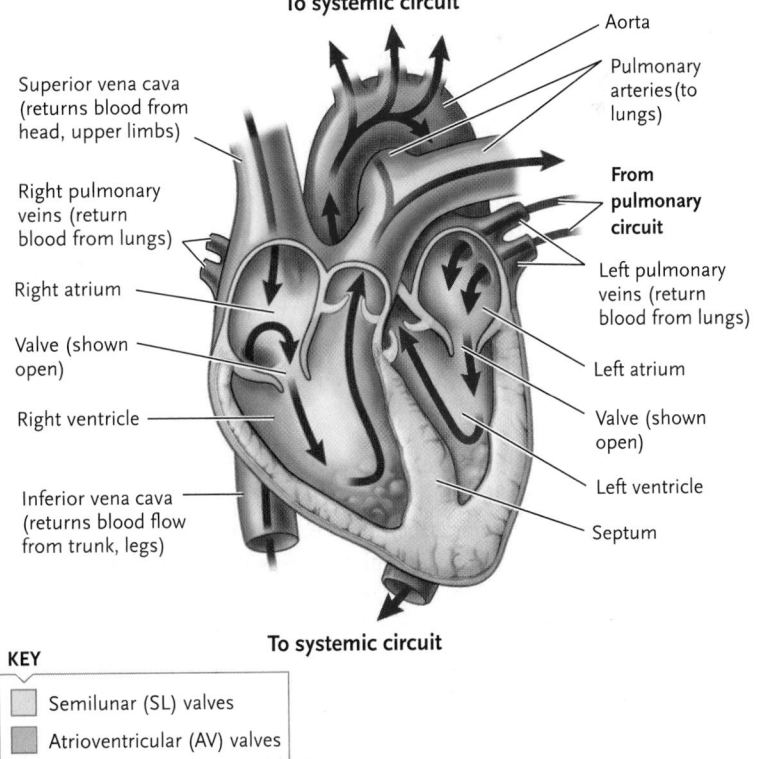

To systemic circuit

Aorta

Pulmonary arteries (to lungs)

From pulmonary circuit

Superior vena cava (returns blood from head, upper limbs)

Right pulmonary veins (return blood from lungs)

Right atrium

Valve (shown open)

Right ventricle

Inferior vena cava (returns blood flow from trunk, legs)

Left pulmonary veins (return blood from lungs)

Left atrium

Valve (shown open)

Left ventricle

Septum

To systemic circuit

KEY

☐ Semilunar (SL) valves

▨ Atrioventricular (AV) valves

Figure 40.9
Cutaway view of a mammalian heart showing its internal organization.

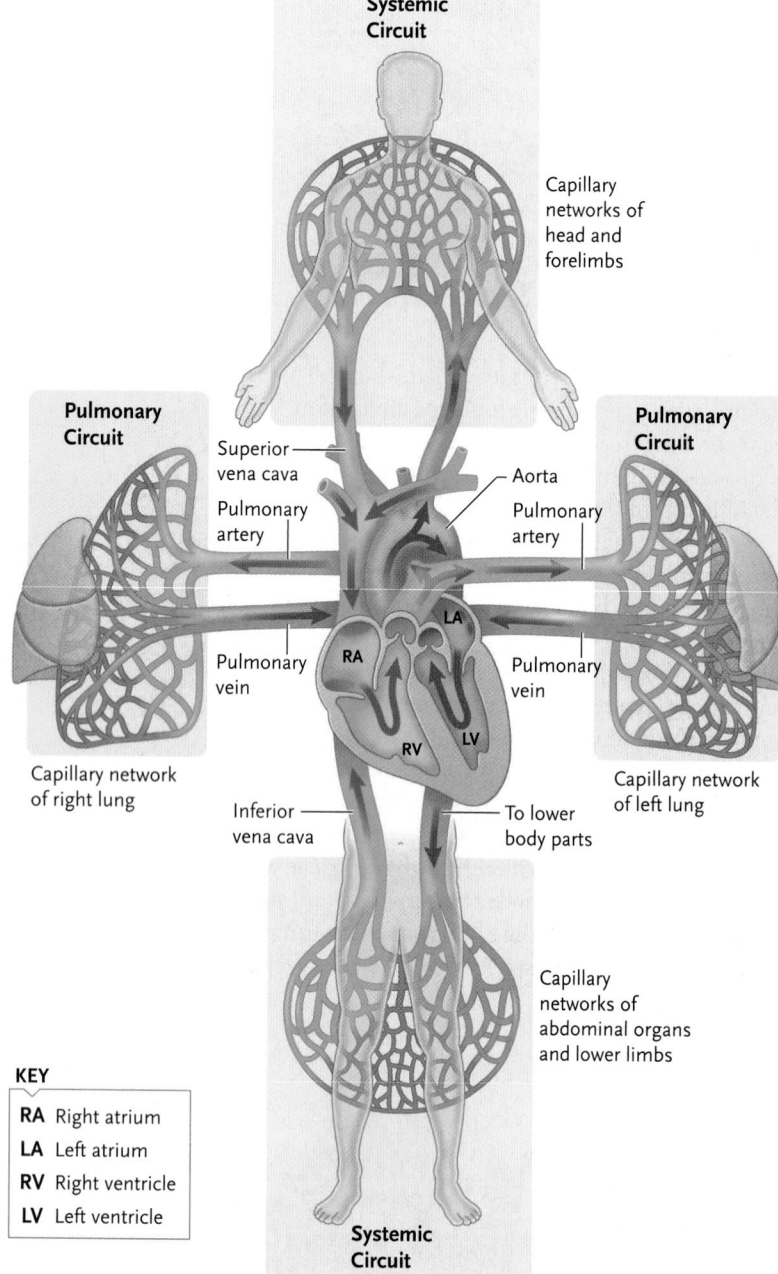

Systemic Circuit

Capillary networks of head and forelimbs

Pulmonary Circuit

Superior vena cava

Pulmonary artery

Pulmonary vein

RA

RV

Aorta

Pulmonary artery

LA

LV

Pulmonary Circuit

Pulmonary vein

Capillary network of left lung

Capillary network of right lung

Inferior vena cava

To lower body parts

Capillary networks of abdominal organs and lower limbs

KEY

RA	Right atrium
LA	Left atrium
RV	Right ventricle
LV	Left ventricle

Systemic Circuit

Figure 40.10

The pulmonary and systemic circuits of a typical mammal (a human). The right half of the heart pumps blood into the pulmonary circuit, and the left half of the heart pumps blood into the systemic circuit.

blood completes this pulmonary circuit by returning to the heart in *pulmonary veins.*

Blood returning from the pulmonary circuit enters the left atrium, which pumps it into the left ventricle. This ventricle, the most thick walled and powerful chamber, contracts to send the oxygenated blood into a large artery, the **aorta**, which branches into arteries leading to all body regions except the lungs.

The arteries divide into smaller and smaller arteries and then into capillary networks, in which the blood releases O_2 and picks up CO_2. The O_2-depleted blood collects in veins, which complete the systemic

circuit. The blood from the veins enters the right atrium. The amount of blood pumped by the two halves of the heart is normally balanced so that neither side pumps more than the other.

The heart also has its own circulation, called the *coronary circulation.* Two small *coronary arteries* branch off the aorta and then branch extensively over the heart, leading to dense capillary beds that serve the cardiac muscle cells. The blood from the capillary networks collects into veins that empty into the right atrium. If a coronary artery becomes blocked, the muscle cells it supplies can die and the person can suffer a heart attack (see "People behind Biology," Box 40.2).

40.3a The Heartbeat Is Produced by a Cycle of Contraction and Relaxation of the Atria and Ventricles

Average heart rates vary among mammals (and among vertebrates generally), depending on body size and the overall level of metabolic activity. An adult human heart beats 72 times each minute, on average, with each beat lasting about 0.8 second. The heart rate of a trained endurance athlete is typically much lower. In infants and young children the heart beats 120 to 160 times each minute. The heart of a flying bat may beat 1200 times a minute, whereas that of an elephant beats only 30 times a minute. **Systole** is the period of contraction and emptying of the heart, and **diastole** is the period of relaxation and filling of the heart between contractions. The systole–diastole sequence of the heart is called the **cardiac cycle (Figure 40.11)**. The following discussion goes through one cardiac cycle.

Starting when both atria and ventricles are relaxed in diastole, the atria begin to fill with blood (step 1 in Figure 40.11). At this point, the **atrioventricular (AV) valves** between each atrium and ventricle and the **semilunar (SL) valves** between the ventricles and the aorta and pulmonary arteries are closed. As the atria fill, the pressure pushes open the AV valves and begins to fill the relaxed ventricles (step 2). When the ventricles are about 80% full, the atria contract and completely fill the ventricles with blood (step 3). Although there are no valves where the veins open into the atria, the atrial contraction compresses the openings, sealing them so that little backflow occurs into the veins.

As the ventricular muscles begin to contract, rising pressure in the ventricular chambers forces the AV valves shut (step 4). As they continue to contract, the pressure in the ventricular chambers rises above that in the arteries leading away from the heart, forcing open the SL valves. Blood now rushes from the ventricles into the aorta and the pulmonary arteries (step 5).

Completion of the contraction squeezes about two-thirds of the blood in the ventricles into the arteries. Now the ventricles relax, lowering pressure in the ventricular chambers below that in the arteries.

Lorrie Kirshenbaum, *University of Manitoba*

Cardiac muscle cells do not divide after birth; the growth of the heart is the result of an increase in the size of the muscle cells. When the cells are damaged in a heart attack, they cannot repair themselves, nor can they be replaced.

Lorrie Kirshenbaum, Canada Research Chair in Molecular Cardiology at the University of Manitoba, explores the molecular events controlling the growth and death of cardiac muscle. *BNIP3* is one of a family of genes that initiate cell death. In particular, it initiates cardiac cell death when there is a lack of oxygen.

Kirshenbaum's lab has been active in describing the pathways that activate the *BNIP3* gene and the intracellular pathways that the protein product of the gene uses to kill the cells. Kirshenbaum is exploring how this information might be used to prevent the activation of the gene or its effects and thus prevent the death of cardiac cells when deprived of oxygen. He is also studying whether the gene could be activated to kill cancer cells in other tissues.

A different approach involves exploring ways to replace damaged heart cells. Cardiac muscle cells, which

do not divide, have the normal cell division pathway blocked. Kirshenbaum's lab is using growth factors, delivered as genes in viruses, to activate the cell division pathway. However, it turns out that adult myocytes that have the cell division pathway turned on enter a pathway leading to death. To get around this difficulty, Kirshenbaum also delivers growth factors that block the self-destructive pathway together with the growth factors that stimulate cell division. This is a promising alternative to using stem cells to replace damaged heart cells.

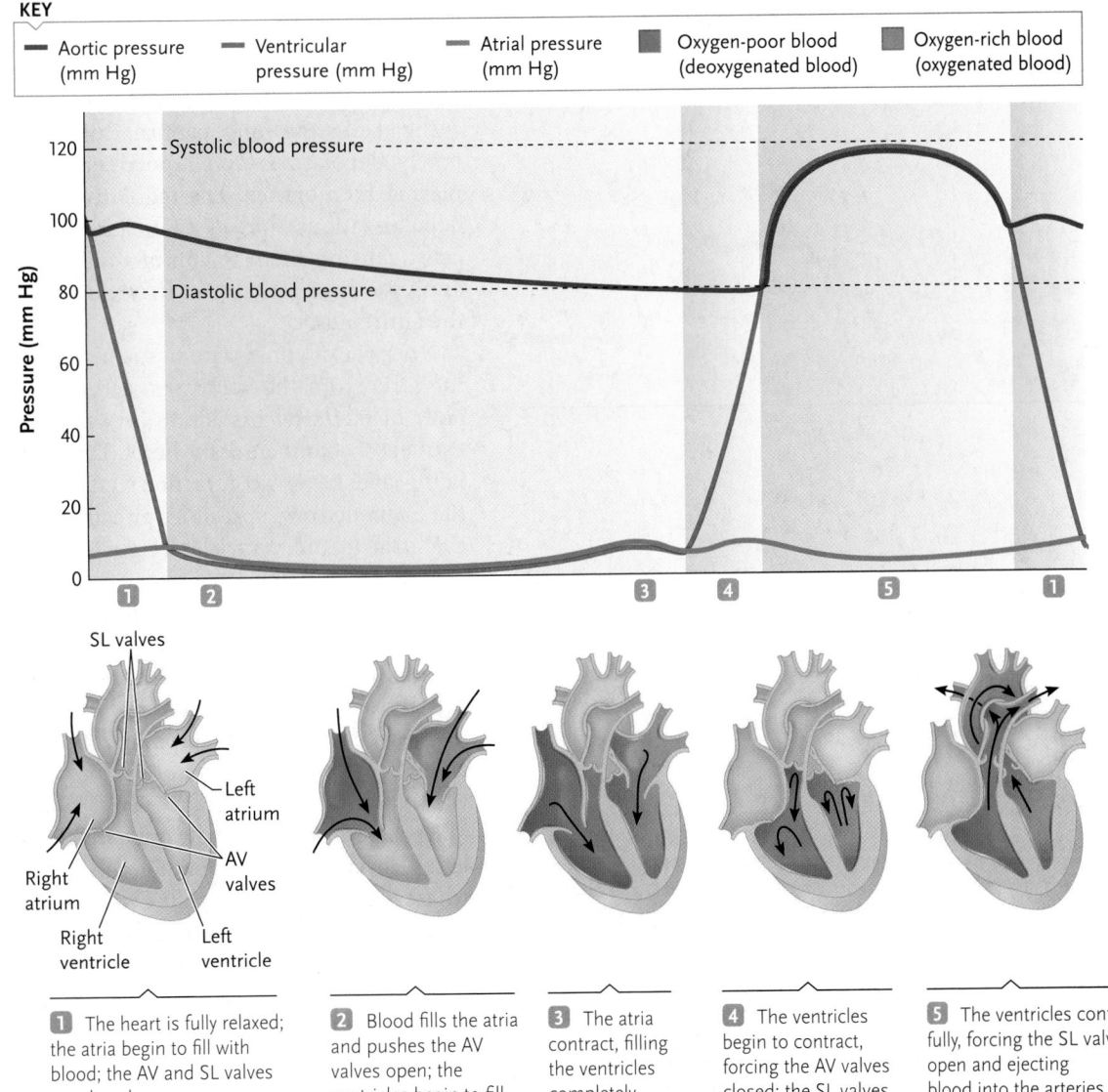

KEY

| — Aortic pressure (mm Hg) | — Ventricular pressure (mm Hg) | — Atrial pressure (mm Hg) | ■ Oxygen-poor blood (deoxygenated blood) | ■ Oxygen-rich blood (oxygenated blood) |

Figure 40.11

The cardiac cycle.

1 The heart is fully relaxed; the atria begin to fill with blood; the AV and SL valves are closed.

2 Blood fills the atria and pushes the AV valves open; the ventricles begin to fill.

3 The atria contract, filling the ventricles completely.

4 The ventricles begin to contract, forcing the AV valves closed; the SL valves remain closed.

5 The ventricles contract fully, forcing the SL valves open and ejecting blood into the arteries.

This reversal of the pressure gradient reverses the direction of blood flow in the regions of the SL valves, causing them to close. For about half a second, both the atria and the ventricles remain in diastole and blood flows into the atria and ventricles. Then the blood-filled atria contract, and the cycle repeats.

In most vertebrates at rest, each minute a ventricle pumps roughly an amount equivalent to the entire volume of blood in the body; that is, blood leaving the heart takes roughly one minute to complete a single circulation. At maximum rate and strength, the hearts of most vertebrates pump about 10 times the resting amount. The human heart is only capable of about half this increase.

40.3b The Cardiac Cycle Is Initiated within the Heart

Contraction of cardiac muscle cells is triggered by action potentials that spread across the muscle cell membranes. Recall that cardiac muscle cells are connected to one another by gap junctions (see Chapters 39 and 46). As a result, electrical activity can pass rapidly from one cell to its neighbours. Crustaceans, such as crabs and lobsters, have **neurogenic hearts,** that is, hearts that beat under the control of signals from the nervous system. Each contraction is initiated by signals from a cardiac ganglion located in the heart, and the heart will continue to beat and respond to some environmental signals in isolation from the central nervous system, as long as the ganglion is intact. Other animals, including all insects and all vertebrates, have **myogenic hearts,** which maintain their contraction rhythm with no requirement for signals from the nervous system. Isolated cardiac myocytes contract rhythmically when grown in a suitable medium. Both neurogenic and myogenic hearts can also be influenced by signals from the central nervous system and by hormones.

The rate and timing of the contraction of individual cardiac muscle cells in a mammalian myogenic heart are coordinated by a region of the heart called the **sinoatrial node (SA node).** The SA node consists of **pacemaker cells,** which are specialized cardiac muscle cells in the upper wall of the right atrium **(Figure 40.12,** step 1). Ion channels in these cells open in a cyclic, self-sustaining pattern that alternately depolarizes and repolarizes their plasma membranes. The regularly timed depolarizations spread to neighbouring cells, causing them to contract and initiating waves of contraction that travel over the entire heart.

A layer of connective tissue separates the atria from the ventricles, acting as a layer of electrical insulation between the two sets of chambers of the heart. The insulating layer keeps a contraction signal from the SA node from spreading directly from the atria to the ventricles (Figure 40.12, step 2). Instead, the atrial wave of contraction excites cells of the **atrioventricular (AV) node,** located in the heart wall between the right atrium and the right ventricle, just above the insulating layer of connective tissue. The signal produced travels from the AV node to the bottom of the heart via *Purkinje fibres* (step 3). These fibres follow a path downward, through the insulating layer, to the bottom of the heart, where they branch through the walls of the ventricles. The signal carried by the Purkinje fibres induces a wave of contraction that begins at the bottom of the heart and spreads from cell to cell upward, squeezing the blood from the ventricles into the aorta and pulmonary arteries (step 4). The transmission of a signal from the AV node to the

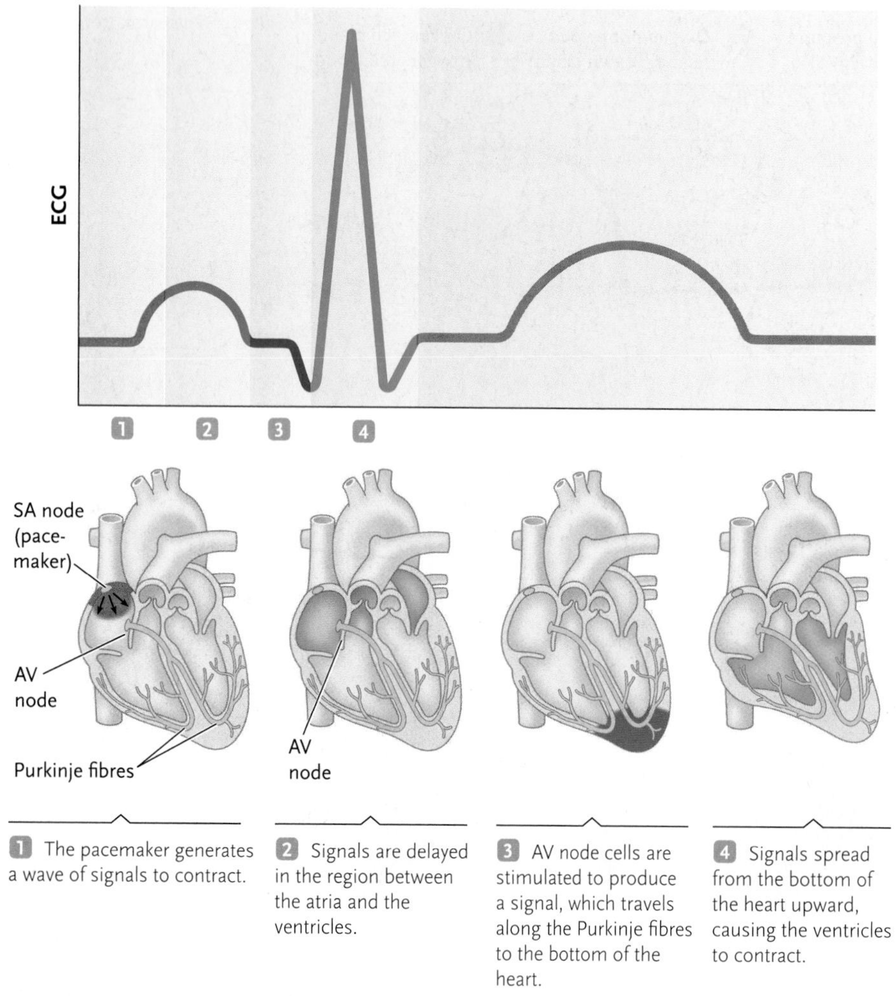

1 The pacemaker generates a wave of signals to contract.

2 Signals are delayed in the region between the atria and the ventricles.

3 AV node cells are stimulated to produce a signal, which travels along the Purkinje fibres to the bottom of the heart.

4 Signals spread from the bottom of the heart upward, causing the ventricles to contract.

Figure 40.12

The electrical control of the cardiac cycle. The bottom part of the figure shows how a signal originating at the SA node leads to ventricular contraction. The top part of the figure shows the electrical activity for each of the stages as seen in an ECG. The colours in the hearts show the location of the signal at each step and correspond to the colours in the ECG.

ventricles takes about 0.1 second; this delay gives the atria time to finish their contraction before the ventricles contract.

As Augustus Waller found in experiments with Jimmie the bulldog, the electrical signals passing through the heart can be detected by attaching electrodes to different points on the surface of the body. The signals change in a regular pattern corresponding to the electrical signals that trigger the cardiac cycle, producing what is known as an **electrocardiogram** (**ECG**; also EKG, from the German *Elektrokardiogramm*). The highlighted region of the ECG above each stage of the cardiac cycle in Figure 40.12 indicates the electrical activity measured in those stages.

40.3c Arterial Blood Pressure Cycles between a High Systolic and a Low Diastolic Pressure

The pressure exerted by a fluid in a confined space is called *hydrostatic pressure*. That is, fluid in a container exerts some pressure on the wall of the container. Blood vessels are essentially tubular containers that are part of a closed system filled with fluid. Hence, the blood in vessels exerts hydrostatic pressure against the walls of the vessels. *Blood pressure* is the measurement of that hydrostatic pressure on the walls of the arteries as the heart pumps blood through the body. Blood pressure is determined by the force and amount of blood pumped by the heart and the size and stiffness of the arteries. In any animal, blood pressure changes in response to activity, temperature, body position, behaviour, time of day, and diet.

As the ventricles contract, a surge of high-pressure blood moves outward through the arteries leading from the heart. This peak of high pressure, called the *systolic blood pressure,* can be felt as a *pulse* by pressing a finger against an artery that lies near the skin, such as the arteries of the neck or the artery that runs along the inside of the wrist. Between ventricular contractions, the arterial blood pressure reaches a low point called the *diastolic blood pressure.* In healthy humans at rest, the systolic pressure, measured in the large artery in the forearm, is equivalent to between 90 and 120 mm of mercury (mm Hg) and the diastolic to between 60 and 80 mm Hg. These pressures are illustrated in Figure 40.11, p. 975.

The blood pressure in the systemic and pulmonary circuits is highest in the arteries leaving the heart and drops as the blood passes from the arteries into the capillaries. By the time the blood returns to the heart, its pressure has dropped to 2 to 5 mm Hg, with no differentiation between systolic and diastolic pressures. The reduction in pressure occurs because the blood encounters resistance as it moves through the vessels, primarily due to the friction created when blood cells and plasma proteins move over each other and over vessel walls.

STUDY BREAK

1. Distinguish between systolic and diastolic blood pressure. Why are these two values different? Why doesn't pressure fall to zero during diastole when the heart is not contracting?
2. What is the function of the sinoatrial node? What is the function of the AV node? How do their roles differ?

40.4 Blood Vessels of the Circulatory System

Both the systemic and the pulmonary circuit consist of a continuum of different blood vessel types that begin and end at the heart (**Figure 40.13**). From the heart, large arteries carry blood and branch into progressively smaller arteries, delivering blood to the various parts of the body. When a small artery reaches the organ it supplies, it branches into yet smaller vessels, the **arterioles**. Within the organ, arterioles branch into capillaries, the smallest vessels of the circulatory system. Capillaries form a network in the organ, where they exchange substances between the blood and the surrounding interstitial fluid. Capillaries rejoin to form small **venules,** which merge into the small veins that leave the organ. The small veins progressively join to form larger veins that eventually become the large veins that enter the heart.

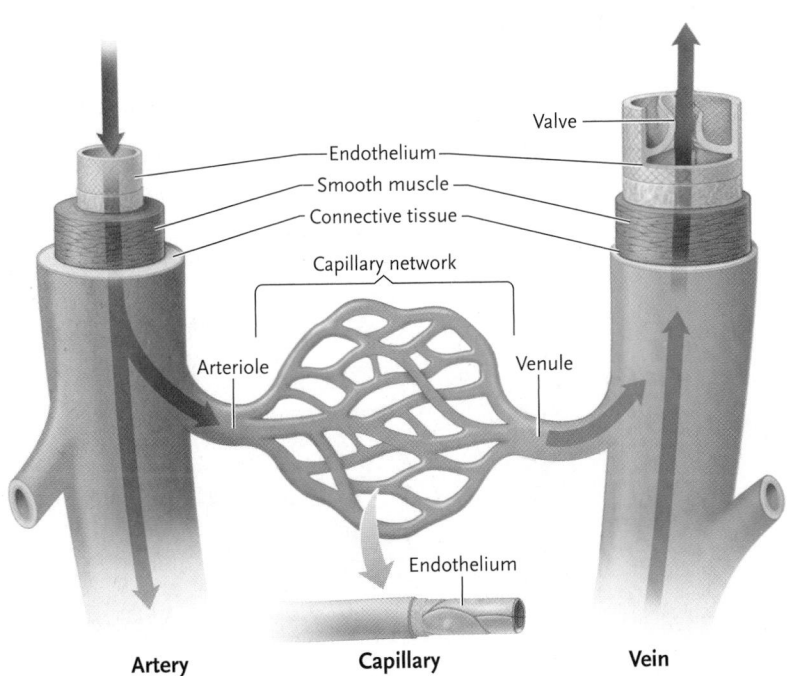

Figure 40.13

The structure of arteries, capillaries, and veins, and their relationship in blood circuits.

40.4a Arteries Transport Blood Rapidly to the Tissues and Serve as a Pressure Reservoir

Arteries have relatively large diameters and therefore provide little resistance to blood flow. They are structurally adapted to the relatively high pressure of the blood passing through them. The walls of arteries consist of three major tissue layers (see Figure 40.13):

- an outer layer of connective tissue containing collagen fibres mixed with fibres of the protein elastin, making the vessel elastic and giving it the ability to recoil;
- a relatively thick middle layer of vascular smooth muscle cells also mixed with elastin fibres; and
- an inner layer of flattened cells only one cell in thickness, forming an endothelium.

In addition to being conduits for blood travelling to the tissues, arteries also act as a pressure reservoir for blood movement when the heart is relaxing. When contraction of the ventricles pumps blood into the arteries, a greater volume of blood enters the arteries than leaves them to flow into the smaller vessels downstream because of the higher resistance to blood flow in these smaller vessels. Arteries accommodate the excess volume of blood because of their elastic walls, which allow the arteries to expand in diameter. When the heart relaxes and blood is no longer being pumped into the arteries, the arterial walls recoil passively back to their original state and the pressure in them falls from systolic to diastolic levels. The longer the interval between heartbeats, the more the diastolic pressure falls. The recoil pushes the excess blood from the arteries into the smaller downstream vessels. As a result, blood flow to tissues is continuous during systole and diastole even though the heart is relaxing and not contracting during diastole.

40.4b Capillaries Are the Sites of Exchange between the Blood and Interstitial Fluid

Capillaries thread through nearly every tissue in the body and are arranged in networks bringing them within 0.01 mm of most body cells. In humans, they are estimated to have a surface area of about 2600 km² for the exchange of gases, nutrients, and wastes with the interstitial fluid. Capillary walls consist of a single layer of endothelial cells, resting on a thin basement membrane. They do not contain smooth muscle (see Chapter 39).

Control of Blood Flow through Capillaries. Blood flow through capillary networks is controlled by contraction of smooth muscle in the arterioles that feed them **(Figure 40.14).** In addition to the normal layer of smooth muscle, some arterioles have circular rings of smooth muscle at the entrance to the capillary bed, called *precapillary sphincter muscles*. When the arteriole and sphincter smooth muscles are relaxed, blood flows readily through the arterioles and capillary networks. When they are contracted, blood flow through the arterioles and capillary networks is limited. By varying the contraction of the arteriole and sphincter smooth muscles, the rate of flow through the capillary networks of individual organs can be adjusted. For example, during exercise, the flow of blood through the capillary networks of the intestines is decreased while that through the muscles of the legs is increased.

The Velocity of Blood Flow through Capillaries. Although their total surface area is astoundingly large, the diameter of individual capillaries is so small that

a. Relaxed

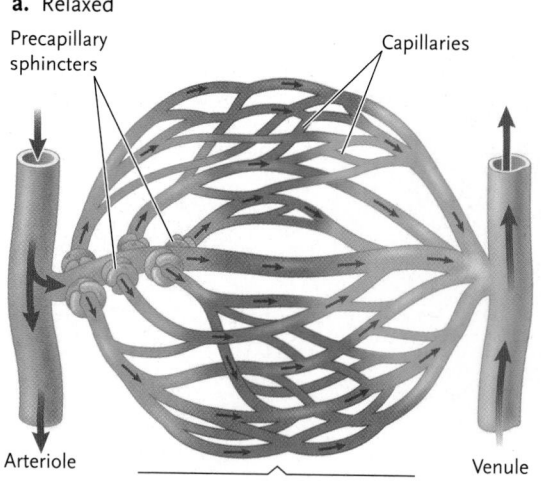

b. Contracted

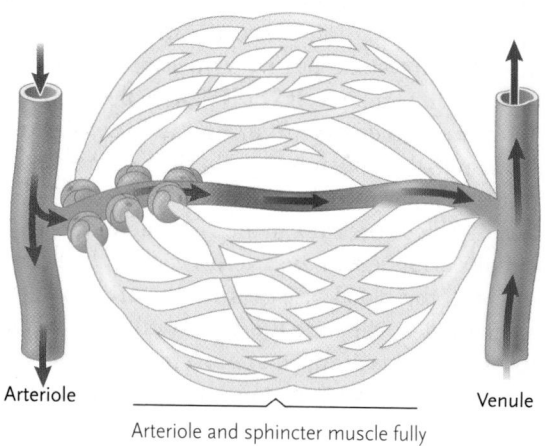

Figure 40.14

Control of blood flow through capillary networks. **(a)** Maximal blood flow when arteriole and sphincter muscles are fully relaxed. **(b)** Minimal blood flow when the arteriole and sphincter muscles are fully contracted.

a.

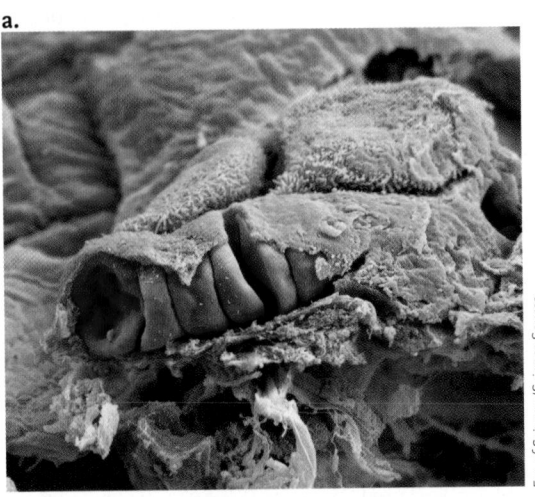

Eye of Science/Science Source

b.

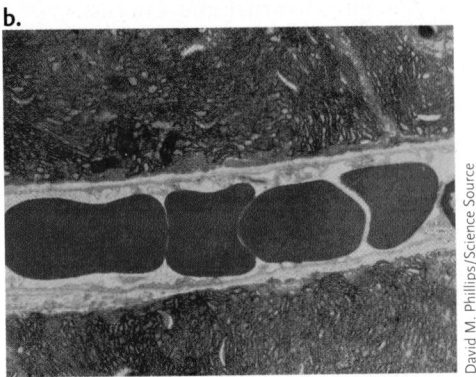

David M. Phillips/Science Source

c.

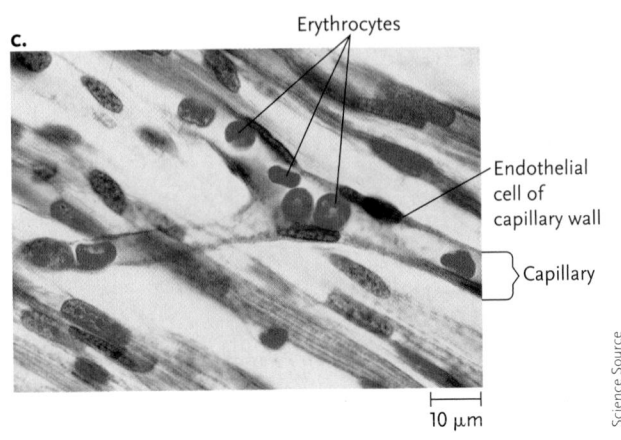

Erythrocytes

Endothelial cell of capillary wall

Capillary

10 μm

Science Source

Figure 40.15

Erythrocytes moving through a capillary that is just wide enough to admit the cells in single file as seen by **(a)** confocal microscopy, **(b)** cross-section and staining using light microscopy, and **(c)** sagittal section through muscle fibres.

red blood cells must squeeze through most of them in single file **(Figure 40.15)**. As a result, each capillary presents a high resistance to blood flow. In addition, there are so many billions of capillaries in the networks that their combined diameter is about 1300 times the cross-sectional area of the aorta. As a result of the resistance and the vastly increased diameter of the combined tubes, blood slows considerably as it moves through capillaries, maximizing opportunities for exchange between the blood and the interstitial fluids. As they leave the tissues, capillaries rejoin to form venules and veins. Veins have a total cross-sectional area that is smaller than the total cross-sectional area of the capillaries, so the velocity of flow increases as blood returns to the heart.

The Exchange of Substances across Capillary Walls. In most body tissues, narrow spaces between the capillary endothelial cells allow water, ions, and small molecules such as glucose to pass freely between blood and interstitial fluid. Erythrocytes, platelets, and most plasma proteins are too large to pass between the cells and are retained inside capillaries, except for molecules that are transported through epithelial cells by specific carriers. Leukocytes, however, can squeeze actively between the cells and pass from the blood to the interstitial fluid.

There are exceptions to these general properties. In the brain, endothelial cells are tightly sealed together, preventing all molecules and ions from passing between them. The tight seals set up the *blood–brain barrier* (Chapter 45). This limits the exchange between capillaries and brain tissues to molecules and ions that are specifically transported through the capillary endothelial cells. At the other extreme are capillaries in the liver, in which the spaces between endothelial cells are wide enough to admit most plasma proteins (most plasma proteins are synthesized in the liver), and in the small intestines, where wide spaces between capillary endothelial cells allow many nutrient molecules to pass into the bloodstream. In bone marrow and other sites of erythrocyte production, spaces are large enough to admit red blood cells.

40.4c Venules and Veins Serve as Blood Reservoirs in Addition to Conduits to the Heart

Because of the high resistance in the capillaries, the pressure remaining in the blood is low as it enters the venules and veins. Accordingly, the walls of venules and veins are thinner than those of arteries and contain little elastin. Many veins have flaps of connective tissue that extend inward from their walls. These flaps form

Figure 40.16

How skeletal muscle contraction and the valves inside veins help move blood toward the heart.

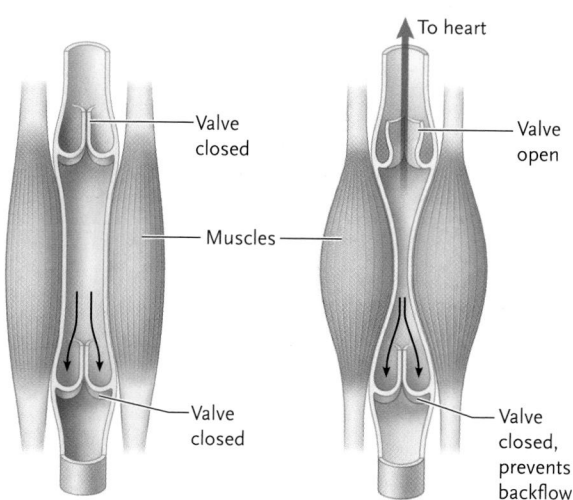

one-way valves that keep blood flowing toward the heart (see Figure 40.13, p. 977).

Because of the differences in their structure, the relatively thin walls of venules and veins can expand over a relatively wide range without developing much recoil. This allows them to act as blood reservoirs as well as conduits. At times, venules and veins may contain from 60 to 80% of the total blood volume of the body. The stored volume is adjusted by skeletal muscle contraction and the valves, in response to metabolic conditions and signals carried by hormones and neurotransmitters.

Although blood pressure in the venous system is relatively low, several mechanisms assist the movement of blood back to the heart. The contraction of skeletal muscles compresses nearby veins, increasing their internal pressure **(Figure 40.16)**. The one-way valves in the veins, especially numerous in the larger veins of the limbs, keep the blood from flowing backward when the muscles relax. Respiratory movements also force blood from the abdomen toward the chest cavity.

STUDY BREAK

1. Why must arterial walls be thick and elastic?
2. During strenuous exercise, more capillary beds open up to receive blood flow. Where does the extra blood that it takes to fill them come from?
3. How do skeletal muscles assist in the circulation of blood?

40.5 Maintaining Blood Flow and Pressure

Arterial blood pressure is the principal force moving blood to the tissues. Blood pressure must be regulated carefully so that the brain and other tissues receive adequate blood flow, but it must not be so high that the heart is overburdened, risking damage to blood vessels. The three main regulators of blood pressure are

- *cardiac output*, the amount of blood pumped by the left and right ventricles combined, which is a product of the rate at which the heart beats (*heart rate*) and the amount of blood pumped with each beat (*stroke volume*);
- the degree of constriction of the blood vessels (primarily the arterioles); and
- the total blood volume.

The autonomic nervous system and the endocrine system interact to coordinate the mechanisms controlling these factors. The system counteracts the effects of constantly changing internal and external conditions, such as movement from rest to physical activity or ending a period of fasting by eating a large meal. In humans, for example, moderate physical activity results in an increase in blood flow to the heart itself by 360%, to the muscles of the skin by 370% (increases loss of heat), and to the skeletal muscles by 1060%. Flow is decreased to the digestive tract and liver by 60%, to the kidneys by 40%, and to the bone and most other tissues by 30%. Only the blood flow to the brain remains unchanged. These changes are the result of changes in cardiac output together with adjustments to the muscles in the arterioles supplying the various organs.

40.5a Cardiac Output Is Controlled by Regulating the Rate and Strength of the Heartbeat

Regulation of the strength and rate of the heartbeat starts at **stretch receptors** called *baroreceptors* (a type of mechanoreceptor; see Chapter 45), located in the walls of blood vessels. The baroreceptors in the cardiac muscle, aorta, and carotid arteries (which supply blood to the brain) are the most crucial. By detecting the amount of stretch of the vessel walls, baroreceptors constantly provide information about blood pressure, sending signals to the medulla within the brain stem. In response, the brain stem sends signals to the heart (primarily the SA node) and muscles of the blood vessels via the autonomic nervous system (see Chapter 44). The sympathetic system, using norepinephrine, stimulates the heart, whereas the parasympathetic system uses acetylcholine to slow the rate. These signals adjust the rate and force of the heartbeat: the heart beats more slowly and contracts less forcefully (pumping less blood with each heartbeat) when arterial pressure is above normal levels, and it beats more rapidly and contracts more forcefully (pumping more blood with each heartbeat) when arterial pressure is below normal levels.

The O_2 content of the blood, detected by chemoreceptors in the aorta and carotid arteries, also influences cardiac output. If O_2 concentration falls below normal levels, the brain stem integrates this information with the baroreceptor signals and issues signals that increase the rate and force of the heartbeat. Too much O_2 in the blood has the opposite effect, reducing cardiac output.

40.5b Hormones Regulate Both Cardiac Output and Arteriole Diameter

Hormones secreted by several glands contribute to the regulation of blood pressure and flow. For example, as part of the stress response, the adrenal medulla reinforces the action of the sympathetic nervous system by secreting epinephrine and norepinephrine into the bloodstream (see Chapter 43). Epinephrine in particular raises cardiac output by increasing the strength and rate of the heartbeat. It also stimulates vasoconstriction (decrease in diameter) of arterioles in some parts of the body, including the skin, gut, and kidneys, and induces vasodilation (increase in diameter) of arterioles that deliver blood to the heart, skeletal muscles, and lungs. Thus, epinephrine increases blood flow to the structures essential for dealing with stress and reduces blood flow to those structures that are not necessary at that time.

40.5c Local Controls Also Regulate Arteriole Diameter

Several automated mechanisms also operate locally to increase the flow of blood to body regions engaged in increased metabolic activity. Repeated contraction of the muscles of the legs during the escape response of an antelope produces low O_2 and high CO_2 concentrations in the legs because of the increased oxidation of glucose and other fuels. This increases vasodilation of the arterioles and hence the blood supply serving the muscles. At least part of the vasodilation is caused by nitric oxide (NO) produced by arterial endothelial cells. NO is broken down quickly after its release, ensuring that its effects are local.

CONCEPT FIX The interaction between blood pressure, cardiac output, and blood vessel resistance is complex. *People often believe that heart rate is the key regulated variable in the circulatory system.* This is because heart rate changes, but the changes maintain constant blood pressure, which is the true controlled variable. The reason that a constant blood pressure is so important is that each capillary bed is capable of controlling its own blood flow (as just described). The job of the heart is to ensure that the total blood flow, that is, the cardiac output, is equal to the local demands of all the tissues. The way this occurs is by monitoring blood pressure and altering cardiac output to keep it constant (increasing cardiac output when blood pressure falls and dropping cardiac output when it increases). Since the local tissues alter their flow by constricting and dilating blood vessels, which alters resistance and hence blood pressure, if the heart can maintain blood pressure constant, then the total flow will equal all the local demands. ⬡

"Life on the Edge," Box 40.3, p. 982, discusses aquatic animals and their specialized circulatory systems, which allow deep and prolonged dives by the careful regulation of blood flow and blood pressure.

STUDY BREAK

1. What are the three factors by which blood pressure and flow are controlled? Why is it so important to control blood pressure?
2. How do O_2 and CO_2 concentrations affect local blood flow?
3. How do changes in local blood flow alter blood pressure?
4. What are baroreceptors, and how do they affect circulation?
5. In some people, the pressure of the blood pooling in the legs leads to a condition called *varicose veins,* in which the veins stand out like swollen, purple knots. Explain why this might happen and why veins closer to the leg surface are more susceptible to the condition than those in deeper leg tissues.

40.6 The Lymphatic System

Under normal conditions, a little more fluid from the blood plasma in the capillaries enters the interstitial fluid than is reabsorbed into the plasma. The **lymphatic system** is an extensive network of vessels that collects excess interstitial fluid and returns it to the venous blood **(Figure 40.17a, p. 983).** Interstitial fluid picked up by the lymphatic system is called **lymph.** This system also collects fats that have been absorbed from the small intestine and delivers them to the blood circulation. The lymphatic system is also a key component of the immune system (see Chapter 51).

40.6a Vessels of the Lymphatic System Extend throughout Most of the Body

Vessels of the lymphatic system collect lymph and transport it to *lymph ducts,* which empty into the veins of the circulatory system. *Lymph capillaries,* the smallest vessels of the system, are distributed throughout the body, intermixed intimately with the capillaries of the circulatory system. Although they are several times as large in diameter as the blood capillaries, the walls of lymph capillaries also consist of a single layer of endothelial cells resting on a basement membrane. Interstitial fluid becomes lymph when it enters the lymph capillaries at sites in their walls where the endothelial cells overlap, forming a flap that is forced open by the higher pressure of the interstitial

Taking a Long Dive—How Do They Do It?

Reptiles, birds, and mammals evolved as land animals, but many species (e.g., snakes, turtles, penguins, loons, otters, seals, and whales; see **Figure 1**) have taken up life in freshwater or marine environments and feed under water. This may involve prolonged dives, sometimes to considerable depths and for extended times. Among the champion divers are marine turtles, emperor penguins, elephant seals, and some species of whales. For example, sperm whales, *Physeter macrocephalus*, grow up to 18 m in length and weigh up to 40 tonnes. They feed on schools of deep-water squid and other animals that live near the bottom of the ocean. The whales can dive to depths of 2 km (but typically less than 1 km (1000 m)) and remain submerged for as long as 90 minutes. Most species, however, do not dive to great depths and remain submerged for a relatively brief period. The time they spend submerged, however, still exceeds the abilities of land dwellers. The supply of oxygen to the muscles, the brain, and some other tissues needs to be maintained throughout the dive, which raises the question, "How do these animals remain submerged for so long (particularly in the case of the champion divers!) while swimming actively?"

To begin with, most diving species possess increased oxygen stores. Diving mammals often have large blood volumes with more erythrocytes per unit of blood than other animals, and these erythrocytes are larger. The muscles of many divers also contain far more myoglobin (see "Molecule behind Biology," p. 972) than terrestrial mammals. Weddell seals can store five times as much oxygen in the blood as humans can and sperm whales store four times as much oxygen in muscle as humans do. The lungs of most divers are not large and do not store much oxygen. This is largely because these animals either exhale before they dive (reducing buoyancy) or because the lungs are severely compressed by the increase in pressure at depth and often collapse.

For many species, diving is a brief event, for feeding or avoiding predation. Other species spend their entire life underwater, returning to the surface only to breathe. In infrequent divers, and in all species during prolonged dives, metabolic rate is reduced (often assisted by reducing body temperature), and the *diving reflex* is invoked. All birds and mammals (including humans) possess a diving reflex, a defensive reflex that inhibits breathing while an animal is underwater and produces changes in blood flow designed to conserve oxygen for the tissues that need it most. The diving reflex is particularly well developed in marine mammals. It involves a reduction in metabolic rate (reducing the need for oxygen), and a redirection of blood flow, by adjustment of arteriole muscles, away from skin, skeletal muscles, and digestive organs and toward the heart and brain. During these dives, heart rate slows, sometimes by as much as 90%. This *bradycardia* is required to prevent blood pressure from rising due to the profound vasoconstriction of the blood vessels in all of the nonessential organs. The extent of this diving bradycardia depends on dive duration in voluntary dives so that short dives involve little or no bradycardia, while long dives involve a profound bradycardia.

When the animals return to the surface, the air in the lungs expands and circulation returns to the surface condition. CO_2 accumulated during the dive is exhaled during the familiar "blow," and the animal breathes on the surface to restore depleted oxygen and excrete CO_2 before diving again.

a.

b.

c.

d.

Figure 1
*Different species exhibit different degrees of commitment to an aquatic lifestyle. The beaver **(a)** dives briefly for food, while sea lions **(b)** alternate between periods on land and periods at sea. Dolphins **(c)** and killer whales **(d)** are fully aquatic.*

a.

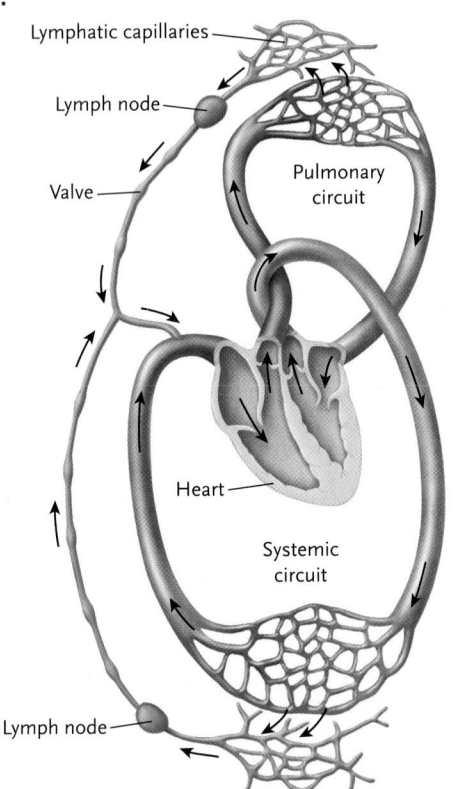

Lymphatic capillaries
Lymph node
Valve
Pulmonary circuit
Heart
Systemic circuit
Lymph node

b.

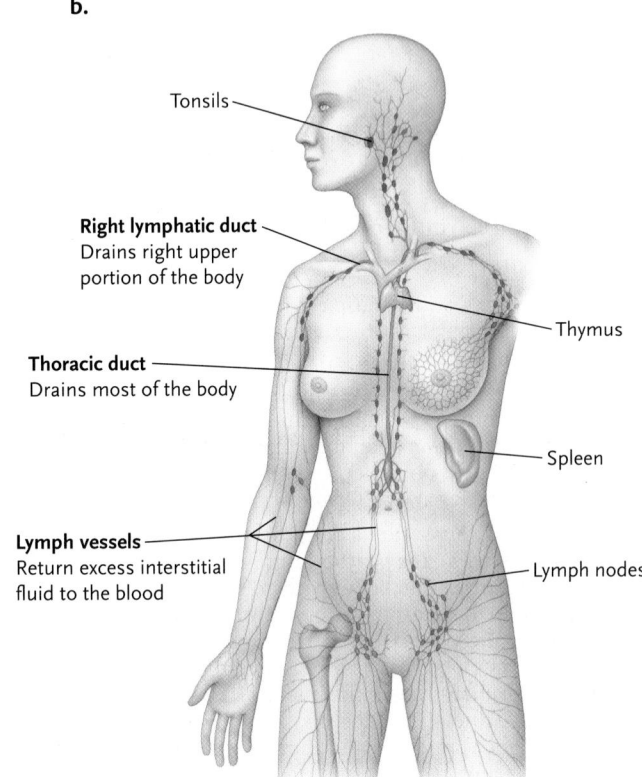

Tonsils

Right lymphatic duct
Drains right upper portion of the body

Thymus

Thoracic duct
Drains most of the body

Spleen

Lymph vessels
Return excess interstitial fluid to the blood

Lymph nodes

Figure 40.17

The human lymphatic system. (a) The lymphatic system is an extensive network of vessels that collect excess interstitial fluid from the tissues and return it to the lowest-pressure component of the systemic circulation, the venous system outside the heart. **(b)** The tissues and organs of the lymphatic system include the lymph nodes, patches in the small intestine, the spleen, the thymus, the tonsils and the appendix.

fluid. The openings are wide enough to admit all components of the interstitial fluid, including bacteria, damaged cells, cellular debris, and **lymphocytes.**

Lymph capillaries merge into *lymph vessels,* which contain one-way valves that prevent the lymph from flowing backward. Lymph vessels lead to the thoracic duct and the right lymphatic duct (see Figure 40.17b), which empty the lymph into a vein beneath the clavicles (collarbones).

Breathing movements and movements of skeletal muscles adjacent to lymph vessels help move lymph through the vessels, just as they help move the blood through veins. Over a day, the human lymphatic system returns about 3 to 4 L of fluid to the bloodstream. In fishes, amphibians, and reptiles, the lymphatic vessels have lymphatic hearts, regions of the ducts equipped with striated muscle that propels the lymph through the vessels.

40.6b Lymphoid Tissues and Organs Act as Filters and Participate in the Immune Response

Tissues and organs of the lymphatic system include the *lymph nodes,* the *spleen,* the **thymus,** and the *tonsils.* They play primary roles in filtering viruses, bacteria,

damaged cells, and cellular debris from the lymph and bloodstream and in defending the body against infection and cancer. Lymphoid tissue also occurs in other regions of the body, particularly the digestive tract, where patches of lymph cells can be found beneath the epithelium of the intestine, in the colon, and in the appendix.

Lymph nodes are small, bean-shaped organs spaced along the lymph vessels and clustered along the sides of the neck, in the armpits and groin, and in the centre of the abdomen and chest cavity (see Figure 40.17b). Spaces in nodes contain macrophages, a type of leukocyte that engulfs and destroys cellular debris and infecting bacteria and viruses in the lymph. The lymph nodes also contain other leukocytes, which produce antibodies that aid in the destruction of invading pathogens (see Chapter 51).

STUDY BREAK

1. Why do we need a lymphatic system? What are its three functions?
2. If this is a system, what are its main organs?

Review

40.1 Animal Circulatory Systems: An Introduction

Those invertebrates with a simple internal structure (sponges, cnidarians, flatworms, and nematodes) have no specialized circulatory systems.

- Animals with circulatory systems have a muscular heart that pumps a specialized fluid, such as blood, from one body region to another through tubular vessels. The blood carries O_2 and nutrients to body tissues and carries away CO_2 and wastes.

- Animal circulatory systems are either open or closed. In an open system, the heart pumps hemolymph into vessels that empty into body spaces. The hemolymph is returned to the heart in various ways that may or may not involve vessels. In a closed system, the blood is confined in blood vessels throughout the body and does not mix directly with the interstitial fluid (Figures 40.2 and 40.3). Closed systems circulate the blood at higher pressures and allow more rapid distribution of O_2 and nutrients and clearance of CO_2 and wastes (Figures 40.2 and 40.3).

- In the protostome invertebrates, open circulatory systems occur in arthropods and most molluscs, whereas closed circulatory systems occur in annelids and cephalopod molluscs (squids and octopuses).

- In vertebrates, the circulatory system has evolved from a heart with a single series of chambers, pumping blood through a single circuit to the gills and then on to the brain and other organs, to a completely separated double heart in birds and mammals that pumps blood through separate pulmonary and systemic circuits (Figure 40.5).

40.2 Blood and Its Components

- Mammalian blood is a fluid connective tissue consisting of erythrocytes, leukocytes, and platelets suspended in a fluid matrix, the plasma.

- Plasma contains water, ions, dissolved gases such as O_2 and CO_2, glucose, amino acids, lipids, vitamins, hormones, and plasma proteins. The plasma proteins include albumins, which transport substances through the blood or act as enzymes; globulins, which transport lipids and include antibodies; and fibrinogen, which takes part in the clotting reaction (Table 40.1).

- Erythrocytes contain hemoglobin, which transports O_2 between the lungs and all body regions. Erythrocytes also transport some CO_2 from interstitial fluid to the lungs and contribute to the reactions maintaining blood pH. They are formed by division of stem cells in the red bone marrow.

- Leukocytes engulf cellular debris and dead and diseased cells, and they defend the body against infecting pathogens. Leukocytes are produced by division of stem cells in the red bone marrow and by division of existing leukocytes.

- Platelets are functional cell fragments that trigger clotting reactions at sites of damage to the circulatory system.

40.3 The Heart

- The mammalian heart is a four-chambered pump. Two atria at the anterior of the heart pump the blood into two ventricles at the posterior of the heart, which pump blood into two separate pulmonary and systemic circuits of blood vessels (Figure 40.9).

- In both circuits, the blood leaves the heart in large arteries, which branch into smaller arteries, the arterioles. The arterioles deliver the blood to capillary networks, where substances are exchanged between the blood and the interstitial fluid. Blood is collected from the capillaries in small veins, the venules, which join into larger veins that return the blood to the heart (Figure 40.10).

- Contraction of the ventricles pushes blood into the arteries at a peak pressure, the systolic pressure. Between contractions, the blood pressure in the arteries falls to a minimum pressure, the diastolic pressure. The systole–diastole sequence is the cardiac cycle (Figure 40.11).

- Contraction of the atria and ventricles is initiated by signals from the sinoatrial (SA) node (pacemaker) of the heart. The signals move over the atria and then activate cells in the atrioventricular (AV) node. From there, the signals move along Purkinje fibres to trigger contraction of the ventricles, starting at the bottom of the heart and moving upward (Figure 40.12).

40.4 Blood Vessels of the Circulatory System

- The walls of arteries consist of an inner endothelial layer, a middle layer of smooth muscle, and an outer layer of elastic fibres. By expanding during systole and rebounding during diastole, the elastic arteries ensure that blood flow is continuous even though ejection of blood from the heart is not (Figure 40.13).

- Capillary walls consist of a single layer of endothelial cells and their basement membrane. Blood flow through capillaries is controlled by contraction and relaxation of the smooth muscles of arterioles and precapillary sphincters.

- In the capillary networks, the rate of blood flow is considerably slower than in arteries and veins. This maximizes the time for exchange of substances between blood and tissues. Two major mechanisms drive the exchange of substances: diffusion along concentration gradients and bulk flow.

- Venules and veins have thinner walls than arteries, allowing the vessels to expand and contract over a wide range. As a result, they act as both blood reservoirs and conduits.

- The return of blood to the heart is aided by pressure exerted on the veins when surrounding skeletal muscles contract and by respiratory movements, which force blood from the abdomen toward the chest cavity. One-way valves in the veins prevent the blood from flowing backward.

40.5 Maintaining Blood Flow and Pressure

- Local control of blood flow responds primarily to O_2 and CO_2 concentrations in tissues. Low O_2 and high

CO_2 concentrations cause dilation of arteriole walls, increasing the arteriole diameter and blood flow. High O_2 and low CO_2 concentrations have the opposite effects. Nitric oxide released by arterial endothelial cells acts locally to increase arteriole diameter and blood flow (Figure 40.14).

- These changes in arteriole diameter lead to changes in blood pressure. By regulating blood pressure, the total flow from the heart, the cardiac output, will match the needs of all the individual organs.

- Blood pressure and flow are regulated by controlling cardiac output, the degree of constriction of blood vessels (primarily arterioles), and the total blood volume. The autonomic nervous system and the endocrine system interact to coordinate these mechanisms.

- Regulation of cardiac output starts with baroreceptors, which detect blood pressure changes in the large arteries and veins and send signals to the medulla of the brain stem. In response, the brain stem sends signals via the autonomic nervous system that alter the rate and force of the heartbeat.

- Hormones secreted by several glands contribute to the regulation of blood pressure and flow. Epinephrine increases blood pressure by increasing the strength and rate of the heartbeat and stimulating vasoconstriction of arterioles in some areas of the body.

40.6 The Lymphatic System

- The lymphatic system is an extensive network of vessels that collect excess interstitial fluid, which becomes lymph, and return it to the venous blood. The system also collects fats absorbed from the small intestine and delivers them to the blood circulation, and it is a key component of the immune system (Figure 40.17).

- The tissues and organs of the lymphatic system include the lymph nodes, the spleen, the thymus, and the tonsils. They remove viruses, bacteria, damaged cells, and cellular debris from the lymph and bloodstream and defend the body against infection and cancer.

Questions

Self-Test Questions

Any number of answers from a to e may be correct.

1. Functions of the circulatory system include
   a. transport of oxygen
   b. nutrient transport
   c. transport of waste products
   d. hormone circulation
   e. transport of carbon dioxide

2. Which circulatory system best describes the given animal or animals?
   a. Amphibians and reptiles use a two-chambered heart to separate oxygenated and deoxygenated blood.
   b. Fishes have a single-chambered heart with an atrium that pumps blood through gills for oxygen exchange.
   c. Squids and octopuses have open circulatory systems with ventricles that pump blood away from the heart.
   d. Birds and mammals pump blood to separate pulmonary and systemic systems from two separate ventricles in a four-chambered heart.
   e. Amphibians have the most oxygenated blood in the pulmocutaneous (leading to and from the lungs and skin) circuit and the most deoxygenated blood in the systemic circuit.

3. Which of the following is a characteristic of blood circulation through or to the mammalian heart?
   a. The superior vena cava conveys blood to the head.
   b. The inferior vena cava conveys blood to the right atrium.
   c. The pulmonary veins convey blood into the left ventricle.
   d. The pulmonary arteries convey blood from the lungs to the left atrium.
   e. The aorta branches into two coronary arteries that convey blood from the heart muscle.

4. In general, the path of blood in the systemic circulation in a vertebrate occurs in the following order:
   a. veins, venules, capillaries, arterioles, arteries
   b. arterioles, capillaries, arteries, venules, veins
   c. venules, veins, capillaries, arteries, arterioles
   d. arteries, arterioles, capillaries, venules, veins

5. Pulmonary arteries carry blood that is
   a. low in O_2
   b. low in CO_2
   c. high in O_2
   d. high in CO_2
   e. on its way to the lungs

6. Which of the following is correct?
   a. Vertebrate hearts are neurogenic.
   b. Erythropoietin is secreted by red blood cells.
   c. Red blood cells in vertebrates contain no nuclei.
   d. Carotid arteries supply the brain with oxygenated blood.
   e. The mammalian aorta is derived from aortic arch VI on the left side.

7. Which of the following is a characteristic of veins and venules?
   a. thick walls
   b. large muscle mass in walls
   c. a large quantity of elastin in the walls
   d. low blood volume compared with arteries
   e. one-way valves to prevent backflow of blood

8. Functions of the blood include
   a. gas transport
   b. defence
   c. clotting
   d. buffering pH

9. Systolic pressure
   a. occurs during the relaxation phase of the heart cycle
   b. is the blood pressure at the peak of the contraction phase of the heart cycle
   c. is generated by the atrium
   d. is generated by the ventricle
   e. is measured in the pulmonary veins

10. Which of the following actions would increase cardiac output?
    a. Baroreceptors in the brain signal the sympathetic nerves.
    b. The brain stem signals the baroreceptors, causing the heart to beat faster.
    c. Chemoreceptors, stimulated by excessive blood oxygen, increase the rate of the heartbeat.
    d. The adrenal medulla and sympathetic nervous system secrete epinephrine and norepinephrine.
    e. The autonomic nervous system responds to low oxygen detected by chemoreceptors and decreases the force of the heartbeat.

Questions for Discussion

1. What is the adaptive significance of a closed circulatory system?

2. *Aplastic anemia* develops when certain drugs or radiation destroy red bone marrow, including the stem cells that give rise to erythrocytes, leukocytes, and platelets. Predict some symptoms a person with aplastic anemia would be likely to develop. Include at least one symptom related to each type of blood cell.

3. Hemoglobin occurs in the blood of many animals, but in many invertebrates, it occurs free in the plasma. What advantages can you think of that might have led to its incorporation in erythrocytes during the evolution of vertebrates? The nonvertebrate chordates do not have hemoglobin in their blood, but they do have cells in their blood that have a role in defence, suggesting that erythrocytes may have originated from such cells. What evidence could be used to examine this hypothesis?

4. Discuss possible adaptations that may occur in animals that allow them to be well suited for life at high altitude—or for life as a diver.

5. In some people, the pressure of the blood pooling in the legs leads to a condition called *varicose veins,* in which the veins stand out like swollen, purple knots. Explain why this might happen and why veins closer to the leg surface are more susceptible to the condition than those in deeper leg tissues.

Dolphins on a coral reef.

Reproduction in Animals

WHY IT MATTERS

Over a few nights each year along the Great Barrier Reef off the east coast of Australia, many species of reef-building corals synchronously spawn, releasing eggs and sperm into the water. Timing is important for the corals because they rely on external fertilization. Male and female gametes must be released at the same time to maximize the chances of meeting. The circadian clocks of the corals control the reproductive event, which is synchronized to the lunar cycle. The mass spawning occurs over several nights after the full moon. Spawning is triggered by changes in lunar irradiance intensity. Many species of corals release eggs and sperm at the same time, but chemicals in the egg coatings and in the sperm head interact to ensure fertilization of eggs only by sperm of the same species.

Bottle-nosed dolphins, on the other hand, have a breeding season during which males compete for females. Mating involves copulation that ensures that sperm and eggs from the same species meet in a protected site inside the body of the female. The fertilized egg is nourished inside the uterus of the female for 12 months on average and then the newborn is nursed by the female for a further 18 to 20 months. These are two very different strategies for allocating resources for reproduction—mass spawning with no parental care and selective reproduction with a large investment in parental care.

MOLECULE BEHIND BIOLOGY 41.1

Cryptochromes

Acropora millepora is one coral involved in the mass reproductive event described in "Why It Matters." It is a sensitivity to blue light (see Chapter 1) that entrains the circadian clock controlling its reproduction. The light sensors are proteins called cryptochromes (CRYs) (blue, ultraviolet-A receptors).

The prefix *crypto-* was used to describe these receptors because the identity of the pigments was unknown for some time. As early as 1881, Charles Darwin had reported that the growth of plants toward the Sun (heliotropism) could be eliminated by filtering blue wavelengths (380 to 500 nm) from the light reaching the plant. The proteins involved in this process, however, were first only described in higher animals (vertebrates and insects), and related but different proteins were not shown to be involved in similar processes in plants and eubacteria until 2007.

Cryptochromes are DNA photolyase-like receptor proteins **(Figure 1a)**.

DNA photolyases are molecules involved in DNA repair. These molecules are flavoproteins. We now know that the absorption spectrum of flavins is similar to the action spectrum of blue light and that the photoreceptors that regulate the circadian clocks in animals and plants depend on the genes for these proteins in organisms ranging from corals (Figure 1b) to *Arabidopsis* (a small plant related to mustard) and from *Drosophila* (Figure 1c) to mammals (Figure 1d).

Figure 1
(a) *DNA photolyase.*
(b) *Coral*—Acropora millepora.
(c) *Fruit* fly—Drosophila melanogaster.
(d) *Arctic Ground Squirrel*—Spermophilus parryii.

a.

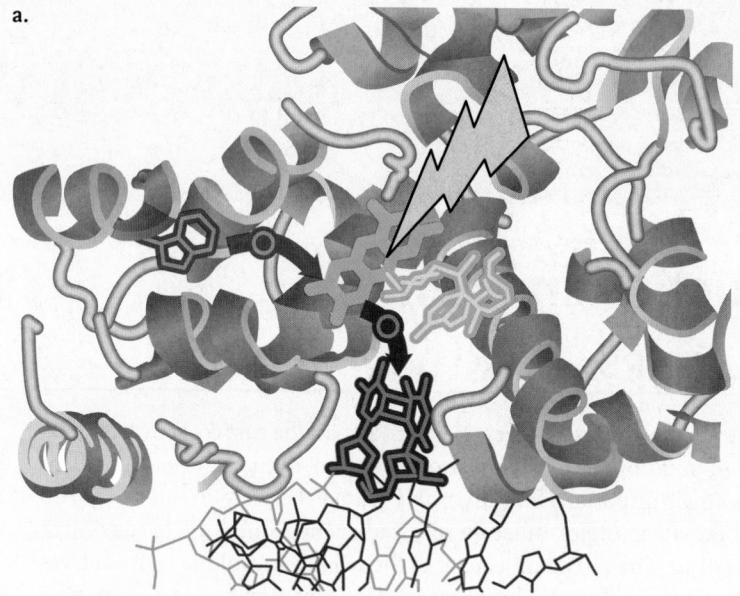

b.

Tobias Bernhard/Oxford Scientific/Getty Images

c.

THOMAS DEERINCK, NCMIR/Science Photo Library/Getty Images

d.

Photos.com

As in plants, bacteria, fungi, and all other life, reproduction ensures the transfer of genes (as allowed by natural selection). There is a strong drive to reproduce, and considerable time and resources are devoted to reproductive processes. For animals that reproduce by eggs and sperm, the adaptations are as diverse as the number of species on Earth. This diversity allows individuals of the same species to find each other and unite eggs and sperm. Within the diversity, however, are underlying patterns that are shared by all animals. Both the underlying patterns and the diversity of animal reproduction are the subjects of this chapter. We also discuss the development of eggs and sperm and the union of egg and sperm that begins the development of a new individual. The next chapter continues with the events of development after eggs and sperm have united.

41.1 The Drive to Reproduce

The success of any species is contingent on its reproductive success. Accordingly, organisms have a strong drive to reproduce and go to considerable lengths to ensure that their genes are represented in future generations. All energy that does not go into growth, development, and maintenance goes into reproduction. This chapter outlines various strategies taken by different species to ensure that this energy is spent wisely. The strategies employed are diverse, reflecting differences in habitat and lifestyle, but all are designed to ensure that the species survives. As we go through the chapter, keep in mind that energy spent on any one activity is energy not available for other purposes, and imagine the selection pressures that have led to the diversity we see.

41.2 Asexual and Sexual Reproduction

Reproduction is the means of passing on an individual's genes to a new generation, making it the most vital function of living organisms. In **asexual reproduction**, a single individual gives rise to offspring with no genetic input from another individual. In **sexual reproduction**, male and female parents produce zygotes (fertilized eggs) through the union of egg and sperm.

CONCEPT FIX A common misconception is that sexual reproduction is always a better **reproductive strategy** than asexual reproduction. As you will see in the next sections, asexual reproduction gives rise to genetic uniformity of offspring, which can be advantageous in stable, uniform environments. Sexual reproduction, on the other hand, gives rise to genetic diversity of offspring, which is advantageous in unstable, changing environments. Thus, each may be a superior strategy depending on environmental conditions. ⬢

41.2a Asexual Reproduction: Reproduction without Recombination Is Advantageous in Stable Environments

Many aquatic invertebrates and some terrestrial annelids and insects reproduce asexually. This mode of reproduction is much less common among vertebrates. In asexual reproduction, from one to many cells of a parent's body develop directly into a new individual. Cells involved in asexual reproduction in animals are usually produced by mitosis and sometimes by meiosis. When cells involved in asexual reproduction are produced by mitosis, the resulting offspring are genetically identical to one another and to the parent (clonal reproduction).

Genetic uniformity of offspring can be advantageous in stable, uniform environments. In these cases, successful individuals with the "best" combinations of genes perpetuate the most competitive genotypes through asexual reproduction. Individuals do not have to expend energy to produce gametes or find a mate. Asexual reproduction is also advantageous to individuals living in sparsely settled populations or to sessile (immobile) animals.

In animals, asexual reproduction involving mitosis occurs by three basic mechanisms: *fission, budding,* and *fragmentation.* In **fission**, the parent splits into two or more offspring of approximately equal size. Some species of planarians (Platyhelminthes) reproduce asexually by fission, dividing transversely or longitudinally **(Figure 41.1a, p. 990)**. In **budding**, a new individual grows and develops while attached to the parent. Sponges, tunicates, and some cnidarians reproduce asexually by budding, and offspring may break free from the parent or remain attached to form a *colony.* In the cnidarian *Hydra,* an offspring buds and grows from one side of the parent's body and then detaches to become a separate individual (Figure 41.1b). In corals, buds often remain attached when their growth is complete, forming colonies of thousands of interconnected individuals. In **fragmentation**, pieces separate from a parent's body and develop (regenerate) into new individuals. Many species of cnidarians, flatworms, annelids, and some echinoderms can reproduce by fragmentation (Figure 41.1c).

Some animals produce offspring by **parthenogenesis** (*parthenos* = virgin; *genesis* = birth), which is the growth and development of an unfertilized egg. Offspring produced by parthenogenesis may be haploid or diploid, depending on the species (see Chapter 9). Because the egg from which a parthenogenetic offspring is produced derives from meiosis in the female parent, the offspring are not genetically identical to the parent or to each other. (We describe below how chromosome segregation and genetic recombination during meiosis produce gametes with gene combinations different from the parent.) In some species, the offspring are all female, whereas in other species, they

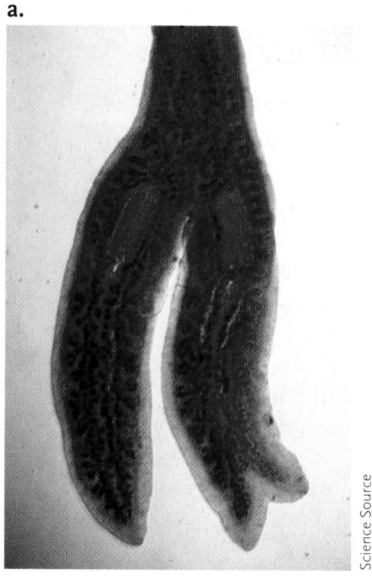

a.

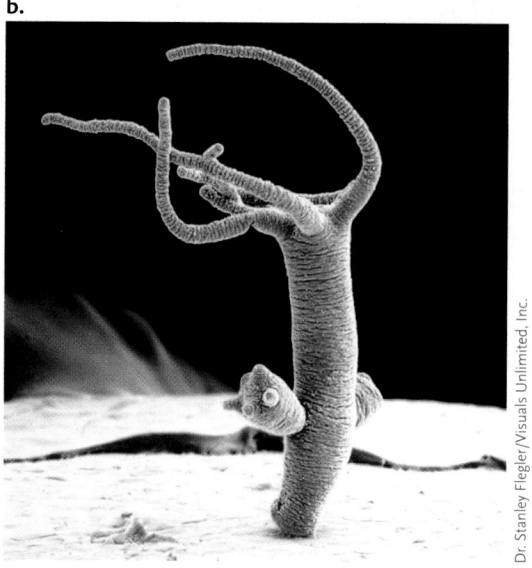

b.

c.

Science Source

Dr. Stanley Flegler/Visuals Unlimited, Inc.

© Wolfgang Pölzer/Alamy

Figure 41.1
Asexual reproduction by **(a)** fission in *Planaria* species, **(b)** budding in *Hydra* species, and **(c)** fragmentation in *Echinoderm* species.

are all males. All whip-tail lizard (*Cnemidophorus* spp.; **Figure 41.2**) species consist of females produced by parthenogenesis. These females still go through the motions of mating with each other.

Parthenogenesis occurs in some invertebrates, including certain aphids, water fleas, bees, and crustaceans. In bees, haploid drones (males) are produced parthenogenetically from unfertilized eggs produced by reproductive females (queens). New queens and sterile workers develop from fertilized eggs. Parthenogenesis also occurs in some vertebrates (e.g., certain fishes, salamanders, amphibians, lizards, and turkeys).

41.2b Sexual Reproduction: Reproduction with Recombination Is Advantageous in Changing Environments

Animals reproduce sexually through the union of **sperm** (motile gamete) and **eggs** (nonmotile gametes), both produced by **meiosis**. The overriding advantage of sexual reproduction is the generation of genetic diversity among offspring. Genetic diversity of offspring is advantageous in unstable, changing environments. Genetic diversity increases the chances that at least some offspring will be better adapted to altered

Karl H. Switak/Science Source

Figure 41.2
A whip-tail lizard (*Cnemidophorus deppei*), in which all individuals are female.

conditions and be more likely to survive and reproduce. Genetic recombination and independent assortment of chromosomes are two mechanisms of meiosis that give rise to genetic diversity in eggs and sperm (see Chapter 8). Genetic recombination mixes the alleles of parents into new combinations within chromosomes. Independent assortment randomly combines maternal and paternal chromosomes in the gamete nuclei. Additional variability is generated at fertilization when eggs and sperm from genetically different individuals fuse together at random to initiate the development of new individuals. Adding to the effects of genetic recombination and independent assortment, random mutations in DNA are the ultimate source of variability for both sexual and asexual reproduction.

Sexual reproduction can be a disadvantage because of the costs in energy and raw materials associated with producing gametes and finding mates. Finding mates can expose animals to predation and may conflict with the need to find food and shelter and care for existing offspring.

STUDY BREAK

1. What are the advantages of sexual reproduction over asexual reproduction?
2. Which would be the better strategy (asexual versus sexual reproduction) for an organism to use if it were living in a stable versus an unstable environment? Explain your answer in detail.

41.3 Mechanisms of Sexual Reproduction

The cellular mechanisms of sexual reproduction include **gametogenesis**, the formation of male and female gametes, and **fertilization**, the union of gametes that initiates development of a new individual. Mating is the pairing of a male and a female for sexual reproduction.

41.3a Gametogenesis: Production of Gametes

Germ cells are cell lines, set aside early in embryonic development in most animals, that can undergo meiosis to produce haploid cells called gametes. They are distinct from the **somatic cells** that compose the rest of the body, which are only ever capable of undergoing mitosis. During development, germ cells collect in gonads, the specialized gamete-producing organs: **testes** (singular, *testis*) in males and **ovaries** in females. Germ cells initially undergo mitotic division to produce large numbers of **spermatogonia** (singular, *spermatogonium*) in males and **oogonia** (singular, *oogonium*) in females. It is these cells that then undergo meiosis to produce gametes **(Figure 41.3)**. In some animals, germ cells also give rise to families of cells that assist gamete development.

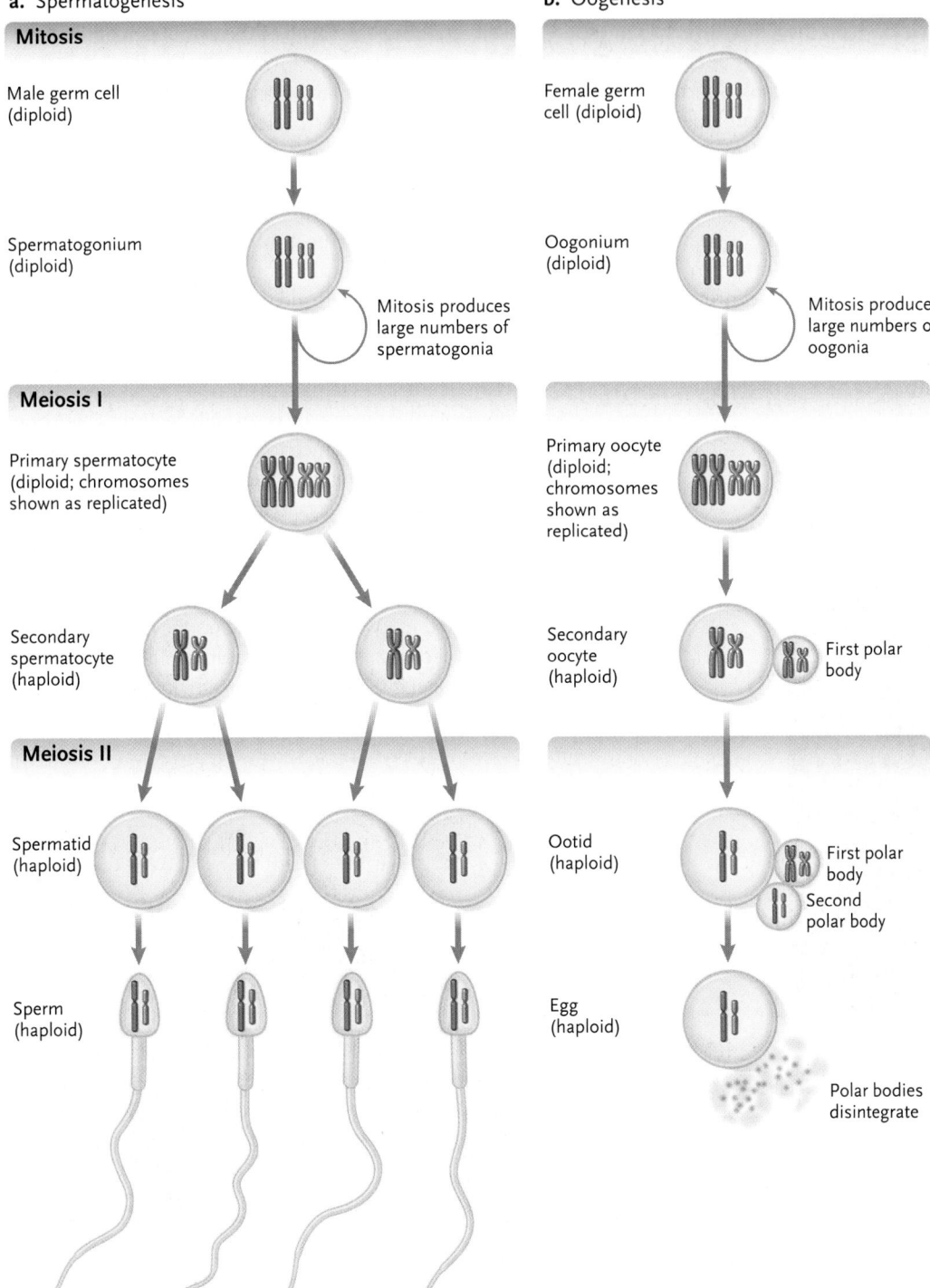

Figure 41.3
The mitotic and meiotic divisions that produce eggs and sperm from germ cells. **(a)** Spermatogenesis. **(b)** Oogenesis. The first **polar body** may or may not divide, depending on the species, so that either two or three polar bodies may be present at the end of meiosis. Two are shown in this diagram.

Meiosis reduces the number of chromosomes from diploid to haploid. Thus, while somatic cells have two copies of each chromosome, gametes have one. Fertilization, the fusion of a haploid sperm and a haploid egg, restores the diploid condition and produces a **zygote** or fertilized egg, the first cell of a new individual. The zygote will then divide by mitosis to produce a new adult within which only one cell line will retain the ability to undergo meiosis: the germ cells that produce gametes.

Spermatogenesis produces haploid cells specialized to deliver their nuclei to eggs produced by members of the same species. Two meiotic divisions produce four haploid spermatids (see Figure 41.3a, p. 991) that each develop into a mature sperm (**spermatozoa**; singular, *spermatozoon* = sperm) (see **Figure 41.4**) or eggs (**ova**; singular, *ovum*). Sperm are specialized to move toward, contact, and penetrate eggs. During sperm maturation, most of the cytoplasm is lost. Mitochondria are concentrated in the cytoplasm around the base of the flagellum. These mitochondria produce ATP, the energy source for beating of the flagellum. At the opposite end of the sperm, the **acrosome** is a specialized secretory vesicle forming a cap over the nucleus. The acrosome contains enzymes and other proteins that help the sperm attach to and penetrate the surface coatings of an egg of the same species.

Although sperm are usually smaller than eggs, they show considerable range in size, even in related species; for example, they are from 4.5 to 16.5 µm long among *Drosophila* species. The size of individual sperm influences the speed at which they swim (longer sperm move more quickly than shorter sperm). But producing longer sperm may reduce total sperm production and limit a male's reproductive success because sperm must be numerous to maximize the chances of fertilization. In mammals, variations in sperm size, volume of ejaculate, and sperm density often reflect mating behaviour (see Chapter 47).

Oogenesis produces ova. Only one of the cell products of meiosis develops into a functional egg, with that cell retaining almost all of the cytoplasm of the parent. The other products form polar bodies (see Figure 41.3b). Unequal cytoplasmic divisions concentrate nutrients and other molecules required for development in the egg. In most species, polar bodies eventually disintegrate and do not contribute to fertilization or embryonic development.

The oocytes of most animals do not complete meiosis until the time of fertilization. At this time, the number of oocytes that complete meiosis may vary. In mammals, for example, oocytes stop developing at the end of the first meiotic prophase within a few weeks after a female is born. The oocytes remain in the ovary at this stage of development until the female is sexually mature. Then, one to several oocytes advance to the metaphase of the second meiotic division and are released from the ovary at intervals ranging from days to months, or at certain seasons, depending on the species. In humans, some oocytes may remain in prophase of the first meiotic division for perhaps 50 years. In other animals, such as the corals described in "Why It Matters," on the other hand, thousands of eggs are released simultaneously once a year.

The egg **(Figure 41.5)** typically has specialized features, including stored nutrients required for at least

a. Human sperm

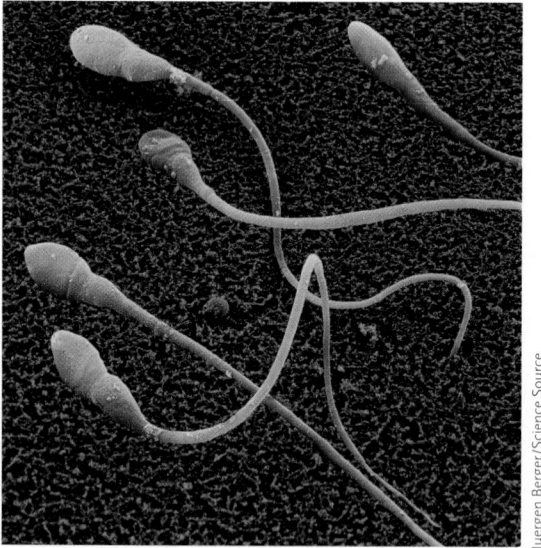

b. Sperm structure

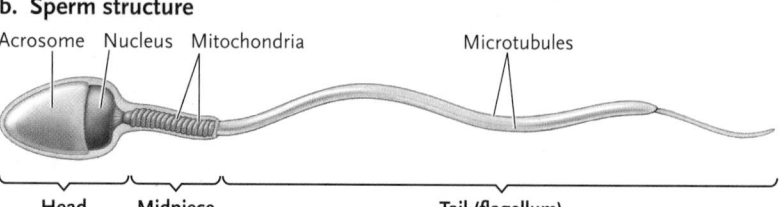

Acrosome Nucleus Mitochondria   Microtubules

Head   Midpiece   Tail (flagellum)

Figure 41.4
Spermatozoa. **(a)** Photomicrograph of human sperm and **(b)** the structure of a sperm.

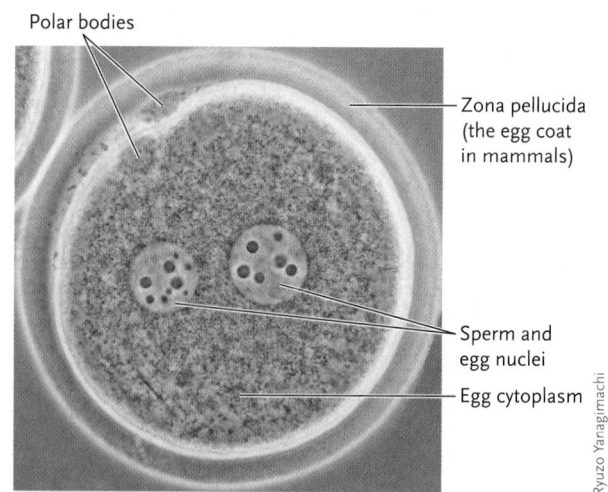

Polar bodies

Zona pellucida (the egg coat in mammals)

Sperm and egg nuclei

Egg cytoplasm

Figure 41.5
A mature hamster egg that has been fertilized. The sperm and egg nuclei are about to fuse together.

the early stages of embryonic development, as well as one or more kinds of coatings that protect the egg from mechanical injury and infection. In some species, egg coats protect the embryo immediately after fertilization and prevent penetration by more than one sperm.

Egg coats are surface layers added during oocyte development or fertilization. The **vitelline coat** (the **zona pellucida** in mammals; see Figure 41.5) is a gel-like matrix of proteins, glycoproteins, and/or polysaccharides lying immediately outside the plasma membrane of the egg cell. Eggs that are laid or deposited outside the body of the female usually have extra layers to protect them and prevent desiccation. Insect eggs have additional outer protein coats forming a hard, water-impermeable layer that prevents desiccation. In amphibians and some echinoderms, egg jelly forms the outer coat protecting the egg from desiccation. In birds, reptiles, and monotremes (see Chapter 28), egg white, a thick solution of proteins, surrounds the vitelline coat. Outside the white is the *shell* of the egg, which is flexible and leathery in reptiles and mineralized and brittle in birds. Both egg white and shell are added while the egg moves along the **oviduct,** the tube connecting the ovary to the outside of the body. For the egg to be fertilized, it must encounter sperm before these layers are added to the egg.

In the mammalian ovary, the egg is surrounded by **follicle cells** during its development. Follicle cells grow from ovarian tissue and nourish the developing egg. They also make up part of the zona pellucida while the egg is in the ovary and remain as a protective layer after it is released.

Mature eggs can be the largest cells in an animal (see Figure 28.44, Chapter 28). Mammalian eggs are microscopic, with few stored nutrients, because the embryo develops inside the mother and is supplied with nutrients by her body. The eggs of birds are huge because they contain all of the nutrients required for complete embryonic development. The bird egg includes the *yolk,* which contains the nutrients for the developing egg; the ovum or egg cell; and the white. Regardless of size, cytoplasm makes up most of the volume of an animal egg, and the nucleus of the egg is usually microscopic.

41.3b Fertilization: Union of Egg and Sperm

Eggs and sperm are delivered from the ovaries and testes to the site of fertilization by oviducts (females) and sperm ducts (males). In many species, external accessory sex organs participate in the delivery of gametes. The basic design of vertebrate and invertebrate reproductive systems **(Figure 41.6)** is similar. Nonmotile eggs move through oviducts on currents generated by the beating of cilia that line the oviducts or by contractions of the oviducts or the body wall. Sperm are ejaculated but are motile and then swim on their own.

a. Insect (fruit fly)

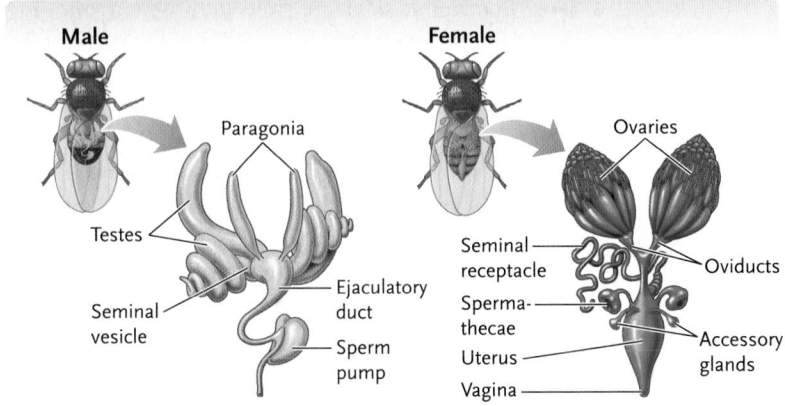

b. Amphibian (frog)

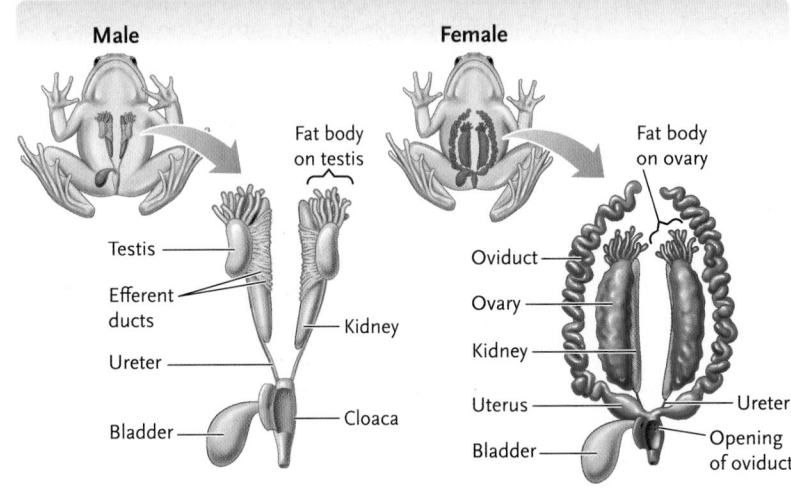

c. Mammal (cat)

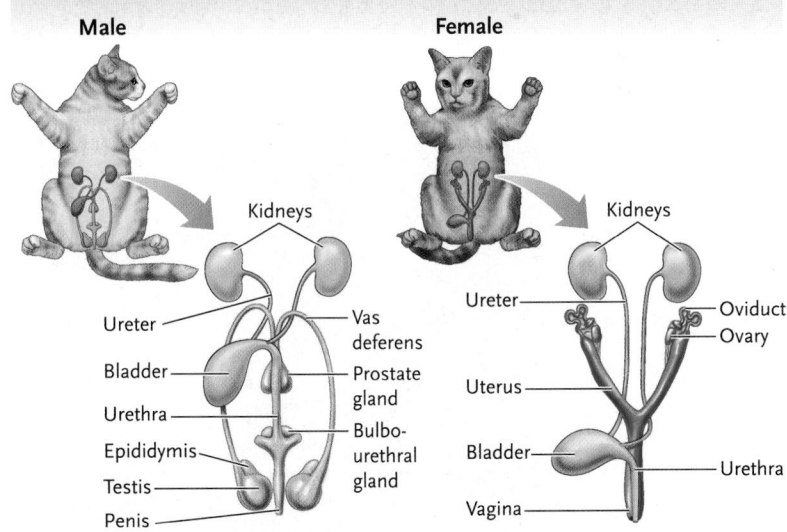

Figure 41.6

Some reproductive systems. **(a)** An insect, *Drosophila* (fruit fly). **(b)** An amphibian, a frog. **(c)** A mammal, a cat. Female systems are shown in blue and male systems in yellow.

Both eggs and sperm are small, fragile, and prone to desiccation. For fertilization to be external (outside the body of either parent), it must be in a watery medium. Otherwise it must be internal in a watery fluid inside a female's body. **External fertilization** occurs in most aquatic invertebrates, bony fishes, and amphibians. Sperm and eggs are shed into the surrounding water. Sperm swim until they collide with an egg of the same species. A disadvantage of external fertilization in an aquatic environment is the possibility of dispersion of the gametes before fertilization can occur. The process is helped by synchronization of the release of eggs and sperm and by the enormous numbers of gametes released (see "Why It Matters"). In other cases, it is helped by releasing the gametes into a protected nest. In animals such as sea urchins and amphibians, sperm are attracted to eggs by diffusible attractant molecules released by the egg.

Another means of increasing the odds of fertilization is by the juxtaposition of genital openings between the male and female during the release of gametes. Most amphibians, even terrestrial species such as toads, mate in an aquatic environment. Frogs typically mate by a reflex response called *amplexus,* in which the male clasps the female tightly around the body with his forelimbs **(Figure 41.7)**. Amplexus stimulates the female to shed a mass of eggs into the water through the *cloaca.* The cloaca is the cavity into which intestinal, urinary, and genital tracts empty in reptiles, birds, amphibians, and many fishes. As the eggs are released, they are fertilized by sperm released by the male.

Internal fertilization is widespread in terrestrial animals. Internal fertilization occurs in invertebrates such as annelids, some arthropods, and some molluscs and in vertebrates from fishes and salamanders to reptiles, birds, and mammals. In internal fertilization, sperm are released by the male close to or inside the entrance to the female's reproductive tract. Sperm swim through fluids in the reproductive tract until one reaches and fertilizes an egg. In some species, molecules released by the egg attract the sperm. Internal fertilization involves copulation, which occurs when a male's accessory sex organ (e.g., a penis) is inserted into a female's accessory sex organ (e.g., a vagina). Internal fertilization makes terrestrial life possible because the female's body provides the aquatic medium required for fertilization without the danger of gametes drying when exposed to the air. Effecting internal fertilization means close contact between individuals.

Male sharks and rays use a pair of modified pelvic fins as accessory sex organs that channel sperm directly into the female's cloaca. Male reptiles, birds, and mammals also use accessory sex organs to place sperm directly inside the reproductive tract of females, where fertilization takes place. In reptiles and birds that lay their eggs, sperm must fertilize eggs as they are released from the ovary and travel through the oviducts, before the shell is added. In mammals, the penis delivers sperm into the female's vagina, which is a specialized structure for reproduction (but see "Hormones and External Genitalia," p. 1030). Fertilization takes place when a sperm meets an egg in the oviducts.

Once a sperm touches the outer surface of an egg of the same species **(Figure 41.8a)**, receptor proteins in the sperm plasma membrane bind the sperm to the vitelline coat or zona pellucida (Figure 41.8b, step 4). In most animals, only a conspecific (same species) sperm is recognized and binds to the egg surface. Species recognition between sperm and eggs is particularly important in animals using external fertilization because water surrounding the egg may contain sperm from many different species. This aspect is less important in species using internal fertilization, where structural adaptations and behavioural patterns usually limit sperm transfer from males to females of the same species (see "Variations in Internal Fertilization," p. 996).

After initial attachment of sperm to egg, the events of fertilization proceed in rapid succession (see Figure 41.8b). The acrosome of the sperm releases its contents, including enzymes that dissolve a path through the egg coats. The sperm, with its tail still beating, follows the path until its plasma membrane touches and fuses with the egg's plasma membrane (step 5). Fusion introduces the sperm nucleus into the egg cytoplasm and activates the egg to complete meiosis and begin development.

Figure 41.7
A male leopard frog (*Rana pipiens*) clasping a female during a mating embrace known as amplexus. The tight squeeze by the male frog stimulates the female to release her eggs, which will stream from her body embedded in a mass of egg jelly. Sperm released by the male fertilize the eggs as they pass from the female.

41.3c Polyspermy: Keeping Sperm Out of the Fertilized Egg

Protection against polyspermy (more than one sperm fertilizing an egg) is widespread in the animal kingdom. Two mechanisms help prevent polyspermy: a **fast block** that works within seconds of fertilization and a **slow block** that works in minutes. In invertebrate species such as the sea urchin, fusion of egg and sperm

a. Sperm adhering to egg

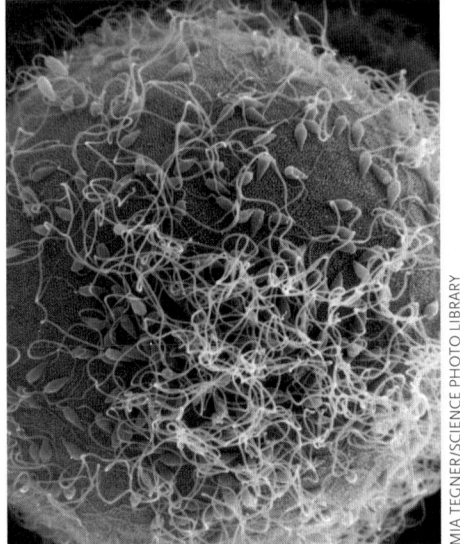

MIA TEGNER/SCIENCE PHOTO LIBRARY

Figure 41.8

Fertilization. **(a)** Sperm adhering to the surface coat of a sea urchin egg. Of the many sperm that may initially adhere to the outer surface of the egg, usually only one accomplishes fertilization. **(b)** Steps of fertilization in a sea urchin.

b. Steps in fertilization

1 A sperm contacts the jelly layer of the egg.

2 The acrosomal reaction begins: Enzymes contained in the acrosome are released and dissolve a path through the jelly layer.

3 Proteins in its plasma membrane bind the sperm to the vitelline coat.

4 The sperm lyses a hole in the vitelline coat. The sperm and egg plasma membranes fuse.

5 Membrane depolarization produces the fast block to polyspermy.

6 The sperm nucleus and centriole enter the egg. The sperm nucleus then fuses with the egg nucleus.

7 The fusion of egg and sperm triggers the release of Ca^{2+} ions, which trigger the cortical reaction, the fusion of secretory **cortical granules** with the egg's plasma membrane. The enzymes of the granules released to the outside alter the egg coats, producing the slow block to polyspermy.

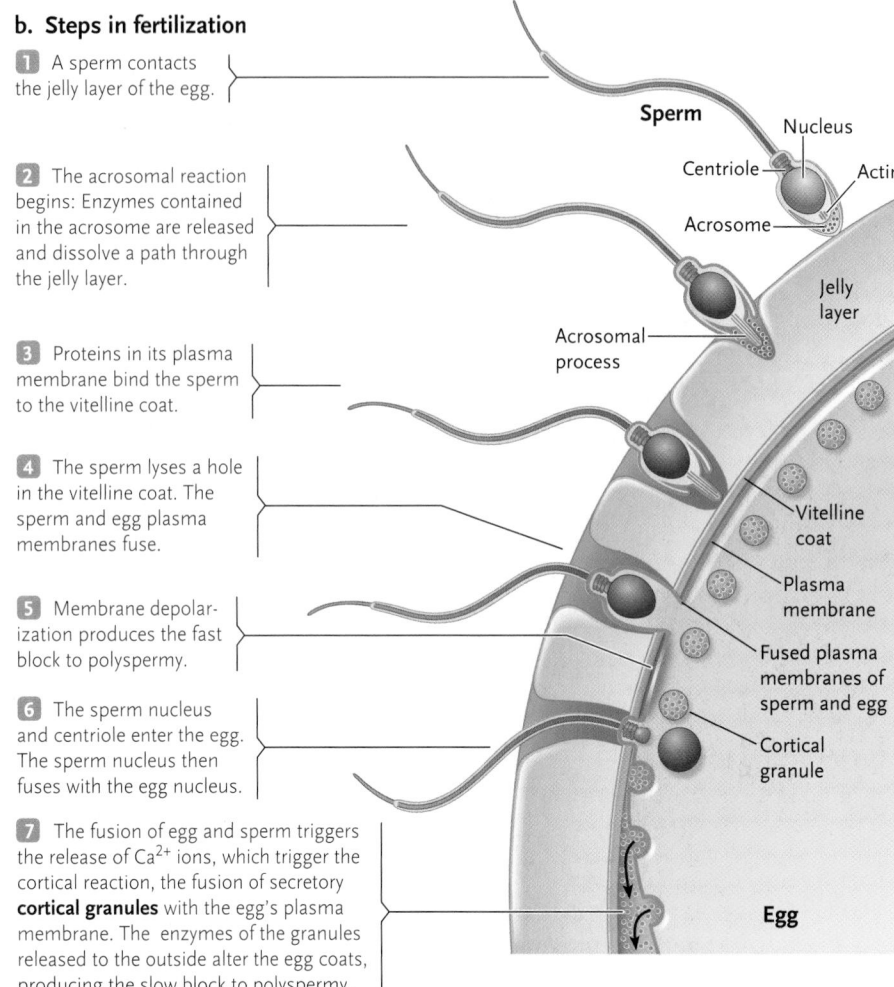

opens ion channels in the egg's plasma membrane, spreading a wave of electrical depolarization over the egg surface (much like a nerve impulse travelling along a neuron). Depolarization alters the egg plasma membrane so that it cannot fuse with any additional sperm, eliminating the possibility of more than one set of paternal chromosomes entering the egg. This fast block occurs within a few seconds of fertilization.

The fast block depends on a change in the egg's membrane potential from negative to positive. It is not established when the membrane potential of a sea urchin egg is experimentally kept at a negative value. In this case, additional sperm fuse with the plasma membrane. Fertilization was entirely blocked if the membrane was kept positive before sperm contact.

In vertebrates, the wave of membrane depolarization following sperm–egg fusion is not as pronounced as it is in sea urchins and does not prevent additional sperm from fusing with the egg. However, additional sperm nuclei that enter the egg cytoplasm usually break down and disappear, so only the first sperm nucleus to enter fuses with the egg nucleus.

In both invertebrates and vertebrates, fusion of egg and sperm also triggers the release of stored calcium (Ca^{2+}) ions from the endoplasmic reticulum (ER) into the cytosol. Ca^{2+} ions cause cortical granules to fuse with the egg's plasma membrane and release their contents to the outside (see Figure 41.8b, step 7). Enzymes released from cortical granules alter the egg coats within minutes of fertilization, so no further sperm can attach to or penetrate the egg. This process is the slow block to polyspermy.

The importance of Ca^{2+} to cortical granule release has been demonstrated experimentally. Granules are released in unfertilized eggs if Ca^{2+} is added experimentally to the cytoplasm. Conversely, if chemicals that bind Ca^{2+} are added to the cytoplasm of unfertilized eggs, the concentration of Ca^{2+} cannot rise and cortical granule release does not occur after fertilization.

After the sperm nucleus enters the egg cytoplasm, microtubules move the sperm and egg nuclei together in the egg cytoplasm until they fuse. The chromosomes of egg and sperm nuclei then assemble together and enter mitosis. The subsequent, highly programmed events of embryonic development convert the fertilized egg into an individual capable of independent existence.

Variations in Internal Fertilization: Penetrating Sex Is Bedbug Success

Internal fertilization benefits animals in several ways. The placement of sperm in the female's reproductive system maximizes the chances of fertilization, giving both males and females more control over which sperm fertilize which eggs. Specializations of the male and female genitalia increase the precision with which sperm is delivered to the egg, and a combination of behavioural and structural (often referred to as "lock-and-key") arrangements minimizes the chances of mistaken identity (cross-species matings).

Invertebrate animals such as bedbugs, insects in the family Cimicidae, have well-developed genitalia but practise *traumatic insemination*. This involves the male bedbug (e.g., *Cimex lectularius*) piercing the female's abdominal wall with his external genitalia and delivering sperm directly into her coelom (body cavity). The male always pierces the female's body at the spermalege, a special site where the intromittant organ penetrates easily and the sperm are deposited **(Figure 1)**.

Females are wounded during traumatic insemination, and the spermalege appears to be an anatomical counterstrategy to minimize the damage associated with this form of mating. The spermalege also reduces the chances of infection by pathogens transferred in the mating process.

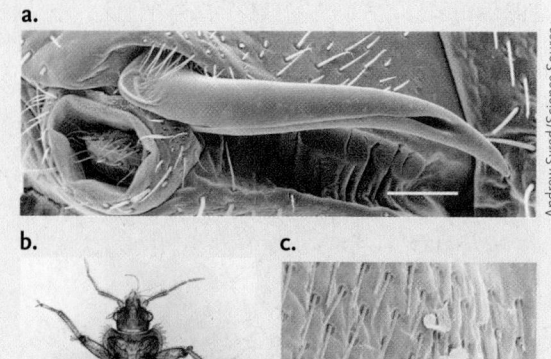
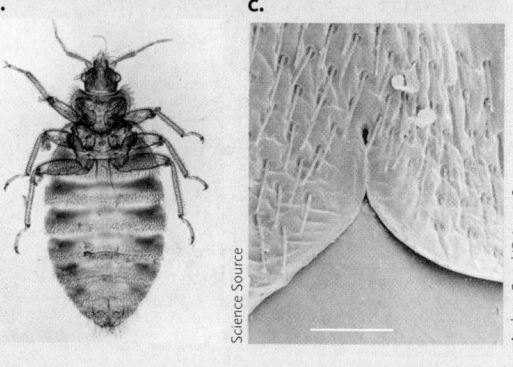

Figure 1
(a) The intromittant organ (penis) of Cimex lectularius *(scale bar 0.1 μm) is shown with* **(b)** *a ventral view of the whole animal and* **(c)** *the ectospermalege, an incurving in one of the female's sternites that guides the male's intromittant organ (scale bar 1.5 μm).*

The paternal chromosomes, the microtubule organizing centre, and one or two centrioles (see Chapter 8) are the only components of sperm to survive in the egg. Therefore, almost all cytoplasmic structures of the embryo and of the new individual are maternal in origin. The centrioles of the new individual are normally of paternal origin.

41.3d Patterns of Development: Moving from Zygote to Complete Organism

In animals with internal fertilization, three major types of support for embryonic development have evolved: *oviparity*, meaning egg laying; *viviparity*, meaning giving birth to live offspring; and *ovoviviparity*, meaning giving birth to live offspring that first hatch internally from eggs.

- **Oviparous** animals (*ovum* = egg, *parere* = to bring forth, to bear) lay eggs that contain the nutrients needed for development of the embryo outside the mother's body. Examples are insects, spiders, most reptiles, and birds. The only oviparous mammals are the *monotremes:* the echidnas and the duck-billed platypus (*Ornithorhynchus anatinus*), both of which are native to Australia.
- **Ovoviviparous** animals retain fertilized eggs within the body, and the embryo develops using the nutrients provided by the egg. When development is complete, the eggs hatch inside the mother and the offspring are released to the exterior. Ovoviviparity is seen in some fishes, lizards, and amphibians; many snakes; and many invertebrates.
- **Viviparous** animals (*vivus* = alive) retain the embryo within the mother's body and nourish it during at least early embryo development. All mammals except the monotremes are viviparous. Viviparity is also seen in all other vertebrate groups except for crocodilians, turtles, and birds. An exceptionally well preserved Devonian fossil placoderm (*Materpiscis attenboroughi*) reveals that viviparity is an ancient trait in vertebrates. In viviparous animals, development of the embryo takes place in a specialized portion of the female reproductive tract, the **uterus** (*womb*). Various structural adaptations are present in different species to enhance nutrient, waste, and gas exchange between the developing embryo and the mother. Perhaps the most advanced are seen among the mammals. One group, called the *placental mammals* or *eutherians,* has a specialized temporary structure, the **placenta,** which connects the embryo to the uterus. The placenta facilitates the transfer of nutrients from the blood of the mother to the embryo and the movement of wastes in the opposite direction. Humans are placental mammals. Another group of mammals, the *marsupials* or *metatherians,* were originally called nonplacental mammals because of a belief that they lacked a placenta. In fact, they do have a placenta, but it derives from a different

tissue than that of eutherians and does not connect the embryo and the uterus. Instead, it provides nutrients to the embryo from an attached membranous sac containing yolk for only the early stages of its development. In many metatherians, the embryo is then born at an early stage and crawls over the mother's fur to reach the **marsupium**, an abdominal pouch within which it attaches to nipples and continue its development **(Figure 41.9).** Kangaroos, koalas, wombats, and opossums are marsupials.

41.3e Hermaphroditism: Producing Eggs and Sperm in One Individual

Hermaphroditic (from *Hermes* + *Aphrodite,* a Greek god and goddess) individuals can produce both eggs and sperm. **Hermaphroditism** is more common among sponges (Porifera), cnidaria, flatworms (Platyhelminthes), earthworms, land snails, and some other invertebrates than it is in vertebrates.

Simultaneous hermaphrodites are individuals that develop functional ovaries and testes at the same time. Earthworms, as shown in **Figure 41.10,** are a good example of **simultaneous hermaphroditism.** The only known vertebrate simultaneous hermaphrodites are hamlets (genus *Hypoplectrus*), a group of predatory sea basses **(Figure 41.11).** Most simultaneous hermaphroditic individuals do not fertilize themselves. Self-fertilization is prevented by anatomical barriers that preclude introduction of sperm into the hermaphrodite's own body or by mechanisms that cause eggs and sperm to mature at different times. The prevention of self-fertilization maintains the genetic variability of sexual reproduction.

Sequential hermaphrodites are individuals that change from one sex to the other. **Sequential hermaphroditism** occurs in many invertebrates (e.g., some crustaceans) and some ectothermic vertebrates. Well-known

a. Mating earthworms

Erni/Shutterstock.com

b. Sex organs

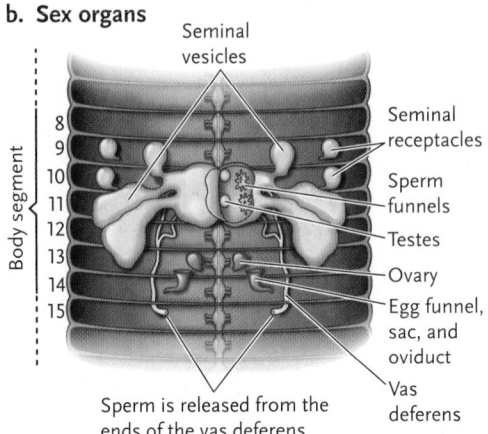

Figure 41.10
Simultaneous hermaphroditism in the earthworm.
(a) Copulation by a mating pair of earthworms, in which each individual releases sperm that fertilizes the eggs in its partner. **(b)** The sex organs in the earthworm.

examples are the genus *Amphiprion,* in which, in some species, the initial sex is male (as in clownfish), whereas in others, it is female. In still other species, individuals may waver between sexes with no discernable order, the sex of any individual being determined by the ratio of sexes in its immediate community (many species of gobies).

Kenneth M. Highfill/Science Source

Figure 41.9
Developing offspring of a marsupial mammal, an opossum (*Didelphis virginiana*), attached to a nipple in the marsupium (pouch) of its mother.

Andrew J. Martinez/Science Source

Figure 41.11
Hypoplectrus gummigutta, a predatory sea bass that is a simultaneous hermaphrodite.

On the Road to Vivipary

Guppies, familiar aquarium fish (see "Life Histories of Guppies," Chapter 29), and other related fish in the genus *Poeciliopsis* exhibit the full range of reproductive patterns described in Section 41.3d. Some species are oviparous. Females in other species retain eggs in their bodies after fertilization and give birth to live young with no further provisioning (ovoviviparous), whereas the females of still other species develop a *follicular pseudoplacenta* that functions much like the placenta of a mouse (or a human) (viviparous). The level of maternal investment (matrotrophic index) ranges from none in the oviparous species to extensive (ovoviviparous and viviparous) **(Figure 1)**. A phylogeny of *Poeciliopsis* and other species **(Figure 2)** shows repeated evolution of placenta-like structures in these fish. All species thrive and are successful; the differences in strategy employed reflect differences in habitat and life history strategies (see Section 29.4, Chapter 29).

Vivipary appears in different evolutionary lines from elasmobranchs (sharks, rays, and holocephalians; see Chapter 28) to mammals. The young of various rays and tiger sharks

(*Galeocerdo cuvier*) develop inside the mother and are nourished by her, but there is no evidence of a placenta. Species such as grey nurse sharks (*Ginglymostoma cirratum*), mako sharks (*Isurus oxyrinchus*), porbeagle

sharks (*Lamna nasus*), and thresher sharks (*Alopias* species) are oophagous. This means that while pregnant, the females of these species of sharks continue to produce large numbers of small eggs that are consumed by the

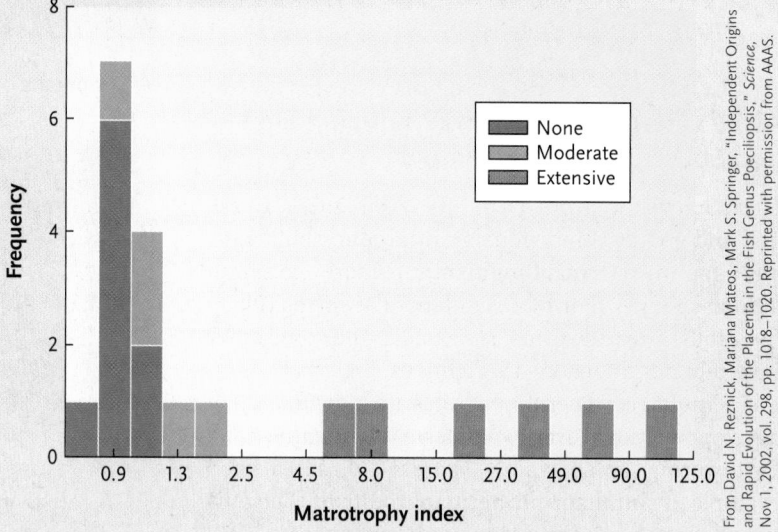

From David N. Reznick, Mariana Mateos, Mark S. Springer, "Independent Origins and Rapid Evolution of the Placenta in the Fish Genus Poeciliopsis," *Science*, Nov 1, 2002, vol. 298, pp. 1018–1020. Reprinted with permission from AAAS.

Figure 1

A matrotrophy index (MI) results when values are assigned to the levels of investment that females make in their young. Matrotrophy indices range from low (ovipary) to high (some ovovivipary and vivipary), and there are intermediate conditions. This figure illustrates the range of MI indices for species in the genus Poeciliopsis. Low MI values are the most common (highest frequency values), but there are significant differences among species. In this figure, species with the same colour do not differ significantly in MI values.

STUDY BREAK

1. When does meiosis occur in animal life cycles? How does this compare with the situation in plants?
2. Why do the eggs of animals differ so much in size? What are the implications for development?
3. Why do many species produce vast amounts of sperm but only a few eggs?
4. What are the costs and benefits of internal fertilization? What are the costs and benefits of external fertilization?

41.4 Sexual Reproduction in Mammals

Reproductively, humans are typical eutherian (placental) mammals. Males and females each have a pair of gonads (testes or ovaries). As in other vertebrates, gonads serve a dual function, producing gametes and secreting hormones responsible for sexual development and mating behaviour (see Chapters 42, 43, and 47).

41.4a Females Produce Eggs, Get Pregnant, and Lactate

Human females have a pair of ovaries suspended in the abdominal cavity. An oviduct leads from each ovary to the uterus, which is hollow, with walls that contain smooth muscle. The uterus is lined by the endometrium, formed by layers of connective tissue with embedded glands and richly supplied with blood vessels. If an egg is fertilized and begins development, it must implant in the endometrium to continue developing. The lower end of the uterus, the cervix, opens into a muscular canal, the vagina, which leads to the exterior. Sperm enter the female reproductive tract via the vagina, and at birth, the baby passes from the uterus through the vagina to the outside.

developing embryos. At least one species of fossil holocephalan chondrichthyian fish (*Delphyodontos dacriformes*) appears to have been oophagous.

Although mammals could be considered the archetypical viviparous animals, not all living mammals are viviparous. Furthermore, living therian mammals (marsupials and placentals) do not all have the same kind of placenta. We have no data about whether early mammals such as *Castorocauda* (see Figures 18.6 and 18.7, Chapter 18) were viviparous. At least 50 gene loci regulate the development of the placenta in placental mammals such as mice and rats. Included are gene families that produce protein hormones and hemoglobin. Some of these genes and gene families are adapted to fetal development.

By retaining developing young in their bodies, adults (usually females) better protect them from predators; provide an appropriate environment for development; and, in many cases, ensure an adequate food supply. As a recurring theme in animals, viviparity is of little value in determining phylogeny (evolutionary relationships). We explore this subject in more detail when we consider development in Chapter 42.

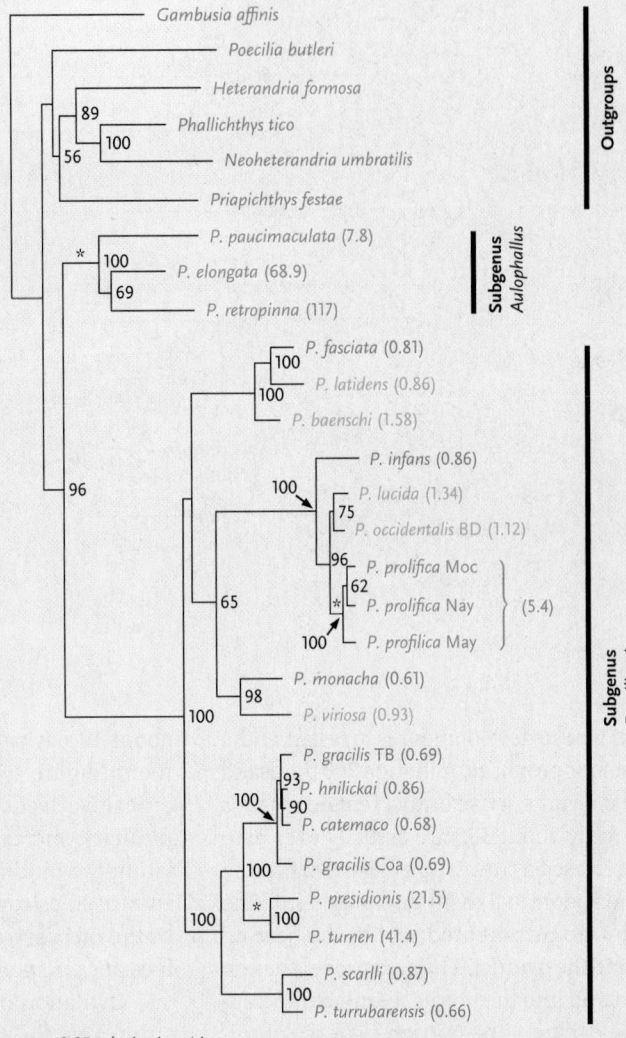

Figure 2

Matrotrophy index (MI) values are shown for species in a phylogeny (phylogram) that illustrates proposed evolutionary relationships among species in the genus Poeciliopsis and other species.

This phylogeny demonstrates that different evolutionary lineages of guppy-like fish have independently achieved different levels of association between mothers and their developing young. The colour codes are as per the MI indices in Figure 1.

From David N. Reznick, Mariana Mateos, Mark S. Springer, "Independent Origins and Rapid Evolution of the Placenta in the Fish Genus Poeciliopsis," *Science*, Nov 1, 2002, vol. 298, pp. 1018—1020. Reprinted with permission from AAAS.

The **vulva**, the external female sex organs (genitalia), surround the opening of the vagina **(Figure 41.12, p. 1000).** Two folds of tissue, the **labia minora**, run from front to rear on either side of the opening. Labia minora are partially covered by **labia majora**, a pair of fleshy, fat-padded folds that also run from front to rear on either side of the vagina. At the anterior end of the vulva, the labia minora join to partly cover the **clitoris**, a bulb-like erectile organ with the same embryonic origins as the penis. Two **greater vestibular glands** open near the entrance to the vagina and secrete a mucus-rich fluid that lubricates the vulva. The urethra that conducts urine from the bladder to the outside opens between the clitoris and the vaginal opening. Most nerve endings associated with erotic sensations are concentrated in the clitoris and the labia minora and around the opening of the vagina. When a human female is born, a thin flap of tissue, the **hymen**, partially covers the opening of the vagina. This membrane, if it has not already been ruptured by physical exercise or other disturbances, is broken during the first sexual intercourse.

In most vertebrates, **ovulation**, the release of the egg from the ovary, usually occurs during a well-defined mating season, the time of year when males and females are *fertile*—physiologically and behaviourally ready to reproduce. The timing of mating seasons is usually under the general control of day length (photoperiod), with some adjustment for local weather. This pattern ensures that the young are born at a time of year when food is plentiful. Humans, however, do not show any evidence of a mating season. Mating and fertilization can occur at any time of the year. Furthermore, ovulation appears to be cryptic in humans, meaning that women do not know when they are ovulating, nor do their partners.

Reproduction in human females is under neuro-endocrine control, involving complex interactions between the hypothalamus, pituitary, ovaries, and uterus. The **ovarian cycle** occurs from puberty to menopause and involves the events in the ovaries leading to the release of a mature egg approximately every 28 days. In human females, at birth, each ovary contains about

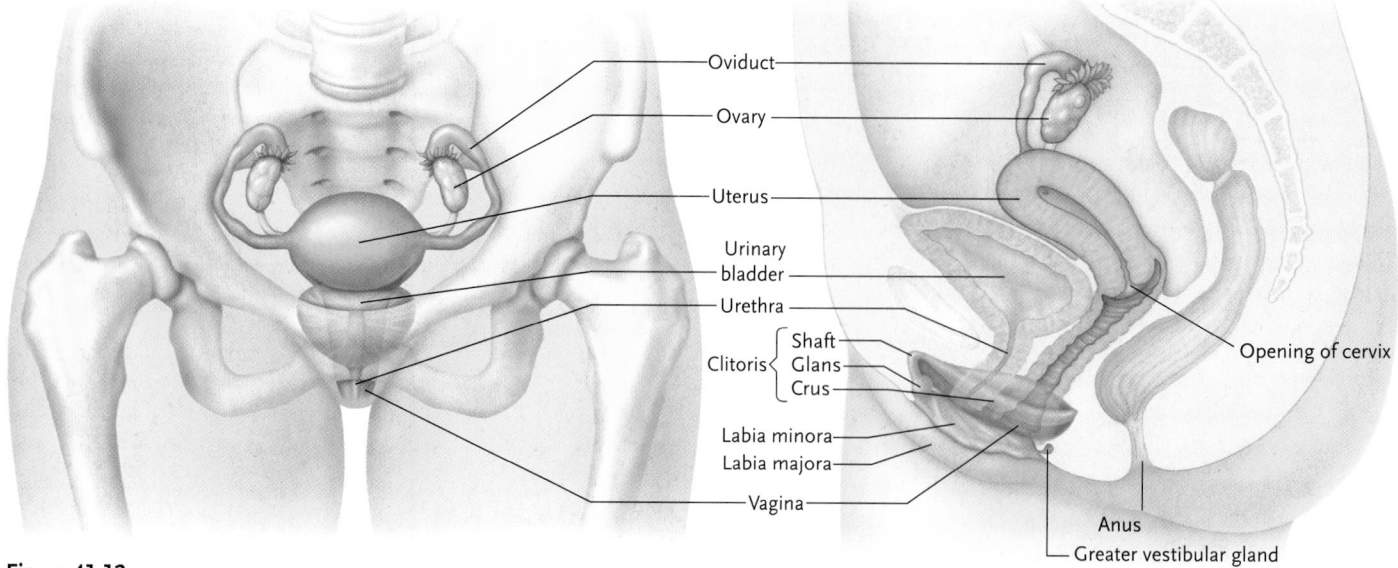

Figure 41.12
The reproductive organs of a human female.

1 million oocytes whose development is arrested at the end of the first meiotic prophase. Although 200 thousand to 380 thousand oocytes survive until a female reaches sexual maturity, only about 380 are actually ever ovulated. These are released as immature eggs, usually one per cycle, into the abdominal cavity and pulled into the nearby oviduct by the current produced by the beating of the cilia that line the oviduct. The cilia propel the egg along the oviduct and into the uterus. Fertilization usually occurs in the oviduct. The ovarian cycle is coordinated with the **uterine cycle** or **menstrual cycle** (*menses* = month), events in the uterus that prepare it for implantation of the egg if fertilization occurs.

The beginning of the ovarian cycle **(Figure 41.13, p. 1002)** is stimulated by the release of gonadotropin-releasing hormone (GnRH) by the hypothalamus. GnRH stimulates the pituitary to release follicle-stimulating hormone (FSH) and luteinizing hormone (LH) into the bloodstream **(Figure 41.14a, p. 1003).** FSH stimulates 6 to 20 oocytes in the ovaries to begin meiosis. As oocytes develop, they become surrounded by cells that form a **follicle** (the ovum and follicle cells) (see Figure 41.13, step 1 and Figure 41.14a, day 2). During this phase, the follicle grows and develops and, at its largest size, becomes filled with fluid and may be 12 to 15 mm in diameter. Usually, only one follicle develops to maturity with release of the egg (secondary oocyte) by ovulation. Multiple births can result if two or more follicles develop and their eggs ovulate in one cycle.

As the follicle enlarges, FSH and LH interact to stimulate estrogen (female sex hormone, primarily estradiol) secretion by follicular cells. Initially, estrogens are secreted in low amounts and have a negative feedback effect on the pituitary, inhibiting secretion of FSH. As a result, FSH secretion declines briefly. But estrogen secretion increases steadily, and its level peaks

about 12 days after the beginning of follicle development (Figure 41.14c, day 12). High estrogen level has a positive feedback effect on the hypothalamus and pituitary, increasing secretion of GnRH and stimulating the pituitary to release a burst of FSH and LH. Increased estrogen levels convert the mucus secreted by the uterus to a thin and watery consistency, making it easier for sperm to swim through the uterus.

Ovulation occurs after the burst in LH secretion stimulates the follicle cells to release enzymes that digest away the wall of the follicle, causing it to rupture and release the egg (see Figure 41.13, step 5). LH also causes the follicle cells remaining at the surface of the ovary to grow into the **corpus luteum,** an enlarged, yellowish structure (*corpus* = body; *luteum* = yellow), initiating the luteal phase. The corpus luteum (see Figure 41.13, step 6) acts as an endocrine gland that secretes estrogens, as well as large quantities of progesterone, a second female sex hormone, and **inhibin,** another hormone. Progesterone stimulates growth of the uterine lining and inhibits contractions of the uterus. Progesterone and inhibin have a negative feedback effect on the hypothalamus and pituitary. Progesterone inhibits secretion of GnRH and, in turn, secretion of FSH and LH by the pituitary. Inhibin specifically inhibits FSH secretion. The fall in FSH and LH levels diminishes the signal for follicular growth, and no new follicles begin to grow in the ovary.

If fertilization does not occur, the corpus luteum gradually shrinks, perhaps because of the low levels of LH. About 10 days after ovulation, the shrinkage has inhibited secretion of estrogen, progesterone, and inhibin. In the absence of progesterone, *menstruation* begins. As progesterone and inhibin levels decrease, FSH and LH secretion is no longer inhibited, and a new monthly cycle begins.

Delaying Reproduction

Many animals have a distinct reproductive season. Its timing is often triggered by changes in photoperiod that herald the changing seasons of the year or by lunar events (e.g., corals). Other animals are more opportunistic. Sea turtles and seals are two examples of animals that separate the acts of mating from the timing of egg laying or birth. Both sea turtles and seals must go ashore to lay eggs or give birth to their young, but for the rest of the year, the animals lead largely pelagic lives. In sea turtles **(Figure 1),** males gather off the beaches where females come to lay their eggs, and mate with females en route to the beach. But males and females also mate during chance meetings in the open ocean. Females can store sperm for up to five years so that they can be ready to lay eggs and fertilize them (from their supply of stored sperm) at any time.

Many species of seals also disassociate the act of mating from, in their case, birth (also known as parturition). Males defend territories on beaches where females haul out to give birth. Females undergo a postpartum estrus, so they are ready to mate (fertile) immediately after giving birth. The seals mate, the egg is fertilized, and the zygote is formed, but implantation is delayed for several months, and the young are born a year later. The time between mating and birth is considerably longer than the gestation period (the time needed for growth and development of the fetus). This approach to reproduction maximizes the chances of males and females finding mates, yet still giving birth at opportune times.

Bats in the families Rhinolophidae and Vespertilionidae separate the acts of copulation and ovulation. The

Figure 2
A pair of little brown bats, Myotis lucifugus, *mate in an abandoned mine in southern Ontario. The male is on the female's back.*

gestation periods in these species are about 60 days. In temperate regions, species in both families mate in late summer and early autumn **(Figure 2).** Females store the sperm in their uteri and then enter hibernation. Ovulation and fertilization occur when females leave hibernation in the spring, and the young are born when spring is well advanced.

Fertilization followed by delays in development or implantation can allow males and females more control over mate choice through interactions among sperm or between sperm and the females' reproductive tract. Delays achieved by sperm storage and postponement of ovulation raise possibilities of competition between sperm or other mechanisms for selecting the sperm that fertilizes the egg.

Figure 1
A pair of green turtles (Chelonia mydas) *mate in the waters of Tortuguero in Costa Rica.*

The uterine (menstrual) cycle includes the changes in the uterus over one ovarian cycle. The hormones that control the ovarian cycle also control the menstrual cycle (Figure 41.14d), physiologically connecting the two processes. Day 0 of the monthly cycle is the beginning of follicular development in the ovary (Figure 41.14b), and in the uterus, menstrual flow begins.

Menstrual flow results from the breakdown of the endometrium, which releases blood and tissue breakdown products from the uterus to the outside through the vagina. When the flow ceases (at day 4 to 5 of the cycle), the proliferation phase begins as the endometrium begins to grow again. As the endometrium gradually thickens, oocytes in both ovaries begin to develop further, eventually leading to ovulation (usually a single egg from one ovary) at about 14 days into the cycle. The uterine lining continues to grow for another 14 days after ovulation. This is the

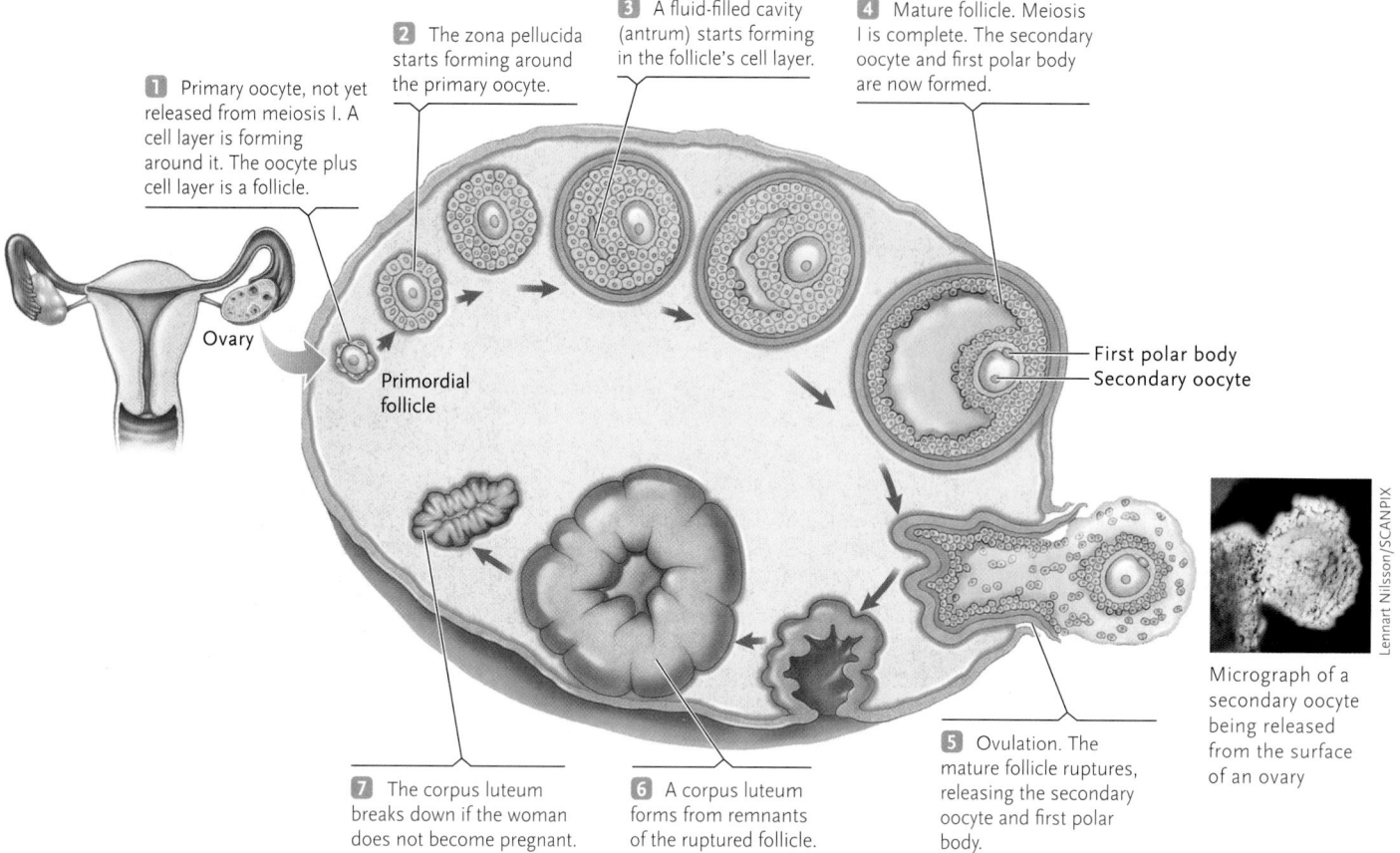

1 Primary oocyte, not yet released from meiosis I. A cell layer is forming around it. The oocyte plus cell layer is a follicle.

2 The zona pellucida starts forming around the primary oocyte.

3 A fluid-filled cavity (antrum) starts forming in the follicle's cell layer.

4 Mature follicle. Meiosis I is complete. The secondary oocyte and first polar body are now formed.

Ovary

Primordial follicle

First polar body
Secondary oocyte

7 The corpus luteum breaks down if the woman does not become pregnant.

6 A corpus luteum forms from remnants of the ruptured follicle.

5 Ovulation. The mature follicle ruptures, releasing the secondary oocyte and first polar body.

Lennart Nilsson/SCANPIX

Micrograph of a secondary oocyte being released from the surface of an ovary

Figure 41.13

The growth of a follicle, ovulation, and the formation of the corpus luteum in a human ovary.

secretory phase. At that time, if fertilization has not taken place, the absence of progesterone results in contraction of the arteries supplying blood to the uterine lining, shutting down the blood supply and causing the lining to disintegrate. The menstrual flow begins. Contractions of the uterus, no longer inhibited by progesterone, help expel the debris. **Prostaglandins** released by the degenerating endometrium add to uterine contractions, making them severe enough to be felt as the pain of cramps and sometimes producing other effects, such as nausea, vomiting, and headaches.

Menstruation occurs only in human females and our closest primate relatives, gorillas and chimpanzees. In other mammals, the uterine lining is completely reabsorbed if a fertilized egg does not implant during the period of reproductive activity. The uterine cycle in these mammals is called the *estrous* cycle, and females are said to be *in estrus* when fertile.

41.4b Males Produce and Deliver Sperm

Organs that produce and deliver sperm make up the male reproductive system **(Figure 41.15, p. 1004)**. Human males have a pair of testes (singular, *testis*), suspended in a baglike **scrotum.** Keeping the testes at cooler temperatures than the body core provides an

optimal environment for sperm development. Some land mammals, such as elephants and monotremes with relatively low body temperatures, have internal (cryptic) testes carried within the body. Marine mammals such as whales and dolphins also have internal testes despite relatively high body temperatures. In these animals, countercurrent exchange between cool blood flowing from the tail flukes to the testis cools them enough to allow the production of fertile sperm. In many mammals (e.g., grey squirrels, *Sciurus carolinensis*), the testes descend into the scrotum only during the mating season. Otherwise, they are cryptic, kept in the body captivity, where temperatures are too warm to produce fertile sperm.

In human males, each testicle is packed with about 125 metres of **seminiferous tubules,** in which sperm proceed through all stages of spermatogenesis **(Figure 41.16, p. 1005).** The entire process, from spermatogonium to sperm, takes 9 to 10 weeks, and the testes produce about 130 million fertile sperm each day.

Sertoli cells are supportive cells that completely surround the developing spermatocytes in the seminiferous tubules. Sertoli cells supply nutrients to the spermatocytes and seal them off from the body's blood supply. **Leydig cells,** located in the tissue surrounding the developing spermatocytes, produce the male sex hormones (**androgens**), particularly **testosterone.**

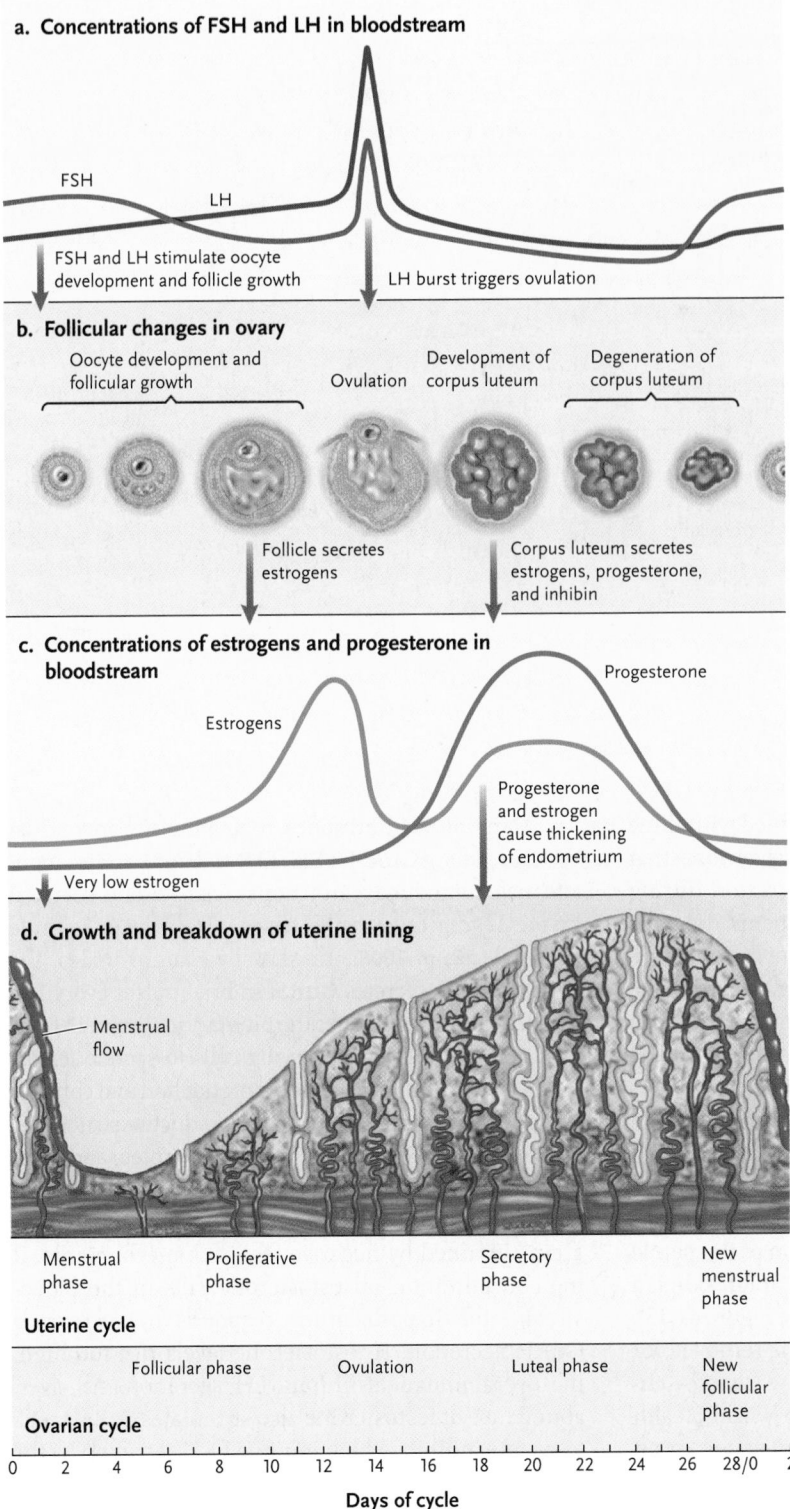

a. Concentrations of FSH and LH in bloodstream

FSH

LH

FSH and LH stimulate oocyte
development and follicle growth

LH burst triggers ovulation

b. Follicular changes in ovary

Oocyte development and
follicular growth

Ovulation

Development of
corpus luteum

Degeneration of
corpus luteum

Follicle secretes
estrogens

Corpus luteum secretes
estrogens, progesterone,
and inhibin

c. Concentrations of estrogens and progesterone in bloodstream

Progesterone

Estrogens

Progesterone
and estrogen
cause thickening
of endometrium

Very low estrogen

d. Growth and breakdown of uterine lining

Menstrual
flow

| Menstrual phase | Proliferative phase | Secretory phase | New menstrual phase |

Uterine cycle

| Follicular phase | Ovulation | Luteal phase | New follicular phase |

Ovarian cycle

0 2 4 6 8 10 12 14 16 18 20 22 24 26 28/0 2

Days of cycle

Figure 41.14
The ovarian and uterine (menstrual) cycles of a human female. The days of the monthly cycle are given in the scale at the bottom of the diagram. **(a)** The changing concentrations of FSH and LH in the bloodstream, triggered by GnRH secretion by the hypothalamus. **(b)** The cycle of follicle development, ovulation, and formation of the corpus luteum in the ovary. **(c)** The concentrations of estrogens and progesterone in the bloodstream. **(d)** The growth and breakdown of the uterine lining.

Mature sperm flow from seminiferous tubules into the **epididymis,** a coiled storage tubule attached to the surface of each testis. Rhythmic muscular contractions of the epididymis move sperm into a thick-walled,

muscular tube, the **vas deferens** (plural, *vasa deferentia*), which extends through the abdominal cavity. Just below the bladder, the vasa deferentia join the urethra. During ejaculation, muscular contractions force the sperm into the urethra and out of the penis. At this time, the sperm are activated and become motile when they come into contact with alkaline secretions added to the ejaculated fluid by accessory glands.

About 150 to 350 million sperm are released in a single ejaculation. **Semen,** the ejaculate, is a mixture of sperm and the secretions of several accessory glands. In humans, about two-thirds of the volume is produced by a pair of **seminal vesicles** that secrete seminal fluid, a thick, viscous liquid, into the vasa deferentia near the point where they join with the urethra. Seminal fluid contains prostaglandins that, when ejaculated into the female, trigger contractions of the female reproductive tract that help move the sperm into and through the uterus.

The **prostate gland,** which surrounds the region where the vasa deferentia empty into the urethra, adds a thin, milky fluid to the semen. The alkaline prostate secretion makes up about one-third of the volume of semen, raising its pH (and that of the vagina) to about pH 6, the level of acidity best tolerated by sperm. This pH level also fosters sperm motility. As part of the prostate secretion, a fast-acting enzyme converts the semen to a thick gel at ejaculation. The thickened consistency helps keep the semen from draining from the vagina when the penis is withdrawn. A second, slower-acting enzyme in the prostate secretion gradually breaks down the semen clot and releases the sperm to swim freely in the female reproductive tract.

Finally, a pair of **bulbourethral glands** secretes a clear, mucus-rich fluid into the urethra before and during ejaculation. This fluid lubricates the tip of the penis and neutralizes the acidity of any residual urine in the urethra. In total, the secretions of the accessory glands make up more than 95% of the volume of semen; less than 5% is sperm.

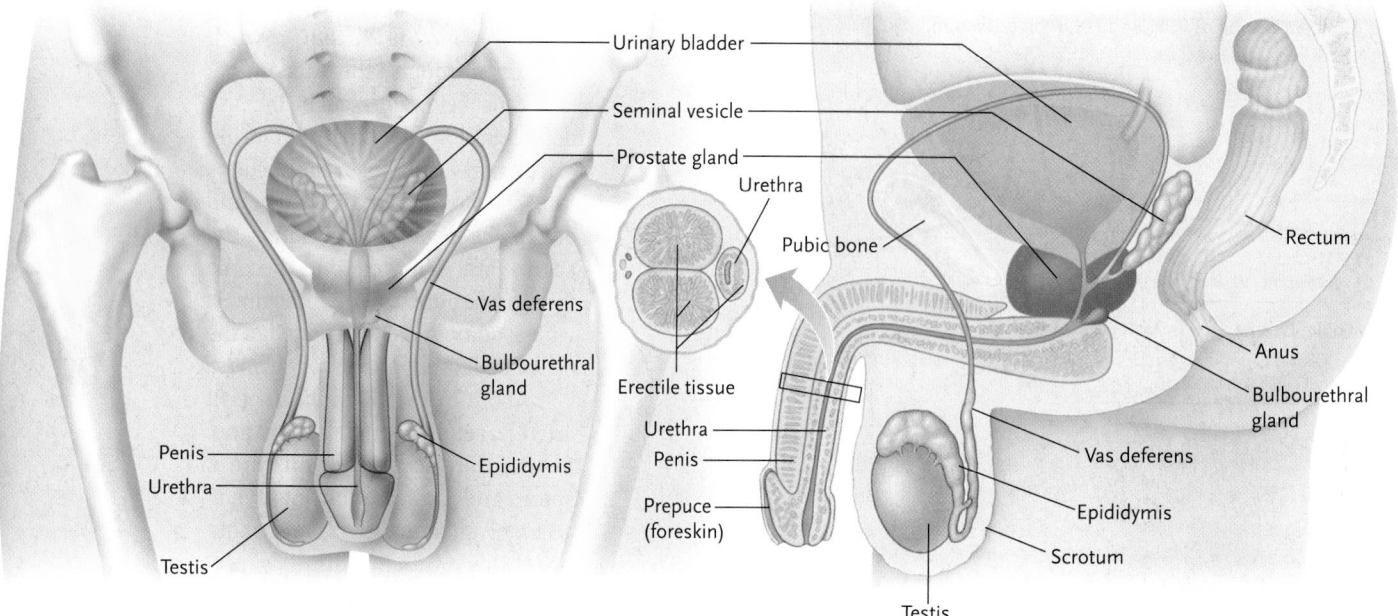

Figure 41.15
The reproductive organs of a human male.

Most of the interior of the penis is filled with three cylinders of spongelike tissue (corpora cavernosa) that become filled with blood and cause erection during sexual arousal. Although the human penis depends solely on engorgement of spongy tissue for erection, the males of many mammals, including bats, rodents, carnivores, and most other primates, have a baculum or penis bone **(Figure 41.17)** that helps maintain the penis in an erect state. The presence of bacula in a species usually coincides with the presence of a baubellum (clitoris bone) in females.

The penis ends in the **glans**, a soft, caplike structure. Most nerve endings producing erotic sensations are crowded into the glans and the region of the penile shaft just behind the glans. The **prepuce** or **foreskin** is a loose fold of skin that covers the glans (see Figure 41.15). In many human cultures, the foreskin is removed for hygienic, religious, or other ritualistic reasons by **circumcision** (= around cut). In 2007, the World Health Organization stated that male circumcision is an important strategy to prevent heterosexually acquired HIV infection in males. Female circumcision, the removal of the labia minora and the clitoris, is often called female genital mutilation or FGM. FGM is practised in various countries in western, eastern, and north-eastern Africa and in parts of Asia and the Middle East. The World Health Organization estimates that 100 to 140 million women and girls around the world have experienced the procedure. It is now considered a human rights violation carried out to reduce libido and control women's sexuality. The United Nations holds an International Day of Zero Tolerance to Female Genital Mutilation each February 6.

Many of the hormones regulating the menstrual cycle, including GnRH, FSH, LH, and inhibin, also regulate male reproductive functions. Testosterone, secreted by the Leydig cells in the testes, also plays a key role **(Figure 41.18, p. 1006).** In sexually mature males, the hypothalamus secretes GnRH in brief pulses every 1 to 2 hours. GnRH stimulates the pituitary to secrete LH and FSH. LH stimulates the Leydig cells to secrete testosterone, which stimulates sperm production and controls the growth and function of male reproductive structures. FSH stimulates Sertoli cells to secrete a protein and other molecules required for spermatogenesis.

Concentrations of male reproductive hormones are maintained by negative feedback mechanisms. If the concentration of testosterone falls in the bloodstream, the hypothalamus responds by increasing GnRH secretion. If testosterone levels rise too high, the overabundance inhibits LH secretion. An overabundance of testosterone also stimulates Sertoli cells to secrete inhibin, which inhibits FSH secretion by the pituitary. As a result, testosterone secretion by the Leydig cells drops off, returning the concentration to optimal levels in the bloodstream.

When the male is sexually aroused, sphincter muscles controlling the flow of blood to the spongy erectile tissue of the penis relax, allowing the tissue to become engorged with blood (the penis is a hydrostatic skeleton structure; see Chapter 46). As the spongy tissue swells, it maintains the pressure by compressing and almost shutting off the veins draining blood from the penis. The engorgement produces an erection in which the penis lengthens, stiffens, and enlarges. During continued sexual arousal, lubricating fluid secreted by the

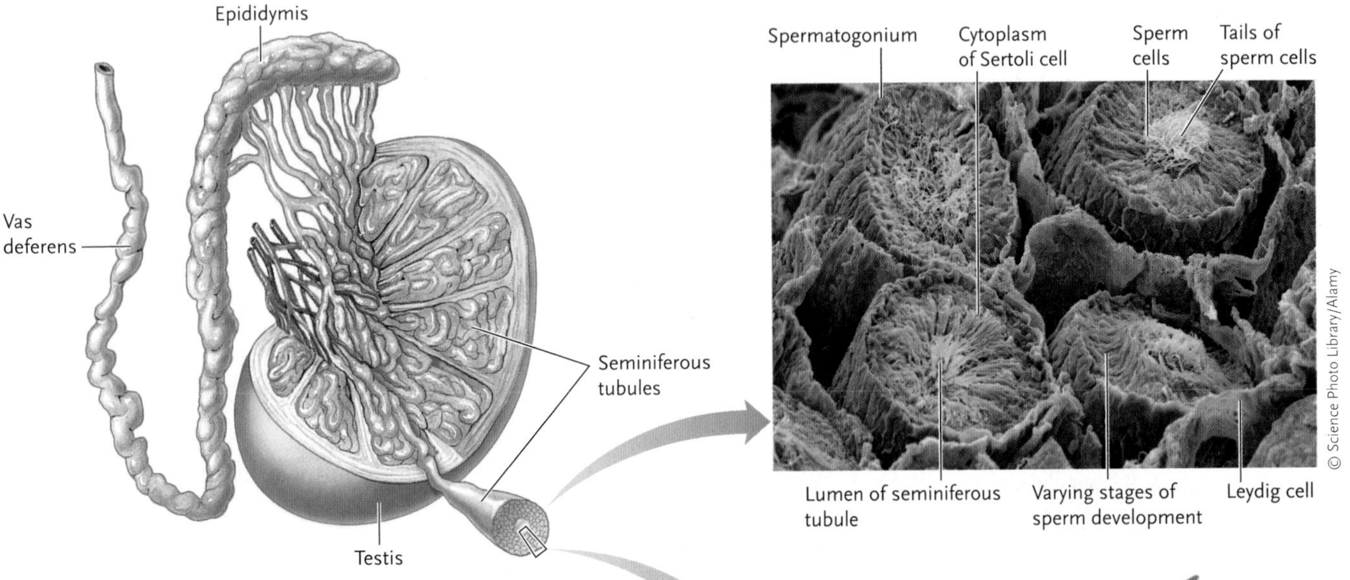

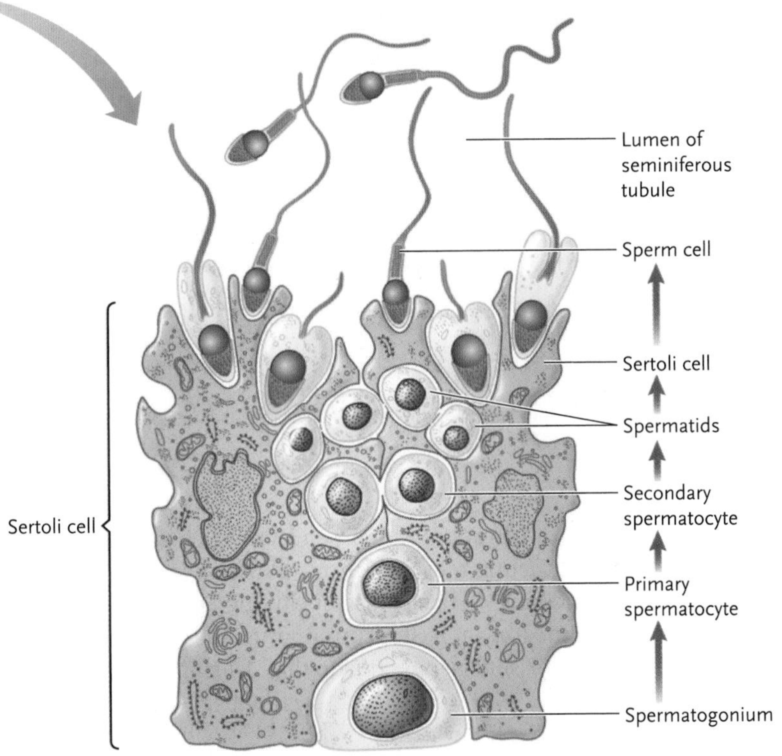

Figure 41.16

The structure of seminiferous tubules and the stages of spermatogenesis. Spermatogonia are located nearest the outer wall and mature sperm cells nearest the tubule lumen. Sertoli cells completely surround the developing spermatocytes and protect them from attack by the immune system.

Epididymis

Vas deferens

Seminiferous tubules

Testis

Spermatogonium

Cytoplasm of Sertoli cell

Sperm cells

Tails of sperm cells

© Science Photo Library/Alamy

Lumen of seminiferous tubule

Varying stages of sperm development

Leydig cell

Lumen of seminiferous tubule

Sperm cell

Sertoli cell

Spermatids

Secondary spermatocyte

Primary spermatocyte

Spermatogonium

Sertoli cell

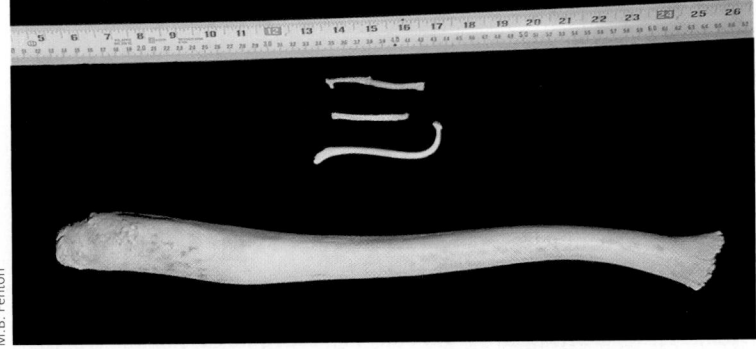

Figure 41.17

The bacula of a wolverine (*Gulo luscus*), a red fox (*Vulpes fulva*), a raccoon (*Procyon lotor*), and a walrus (*Odobenus rosmarus*) (top to bottom).

M.B. Fenton

bulbourethral glands may be released from the tip of the penis.

Female sexual arousal results in enlargement and erection of the clitoris, in a process analogous to erection of the penis. The labia minora become engorged with blood and swell in size, and lubricating fluid is secreted onto the surfaces of the vulva by the vestibular glands. In addition to these changes, the nipples become erect by contraction of smooth muscle cells, and the breasts swell in size due to engorgement with blood.

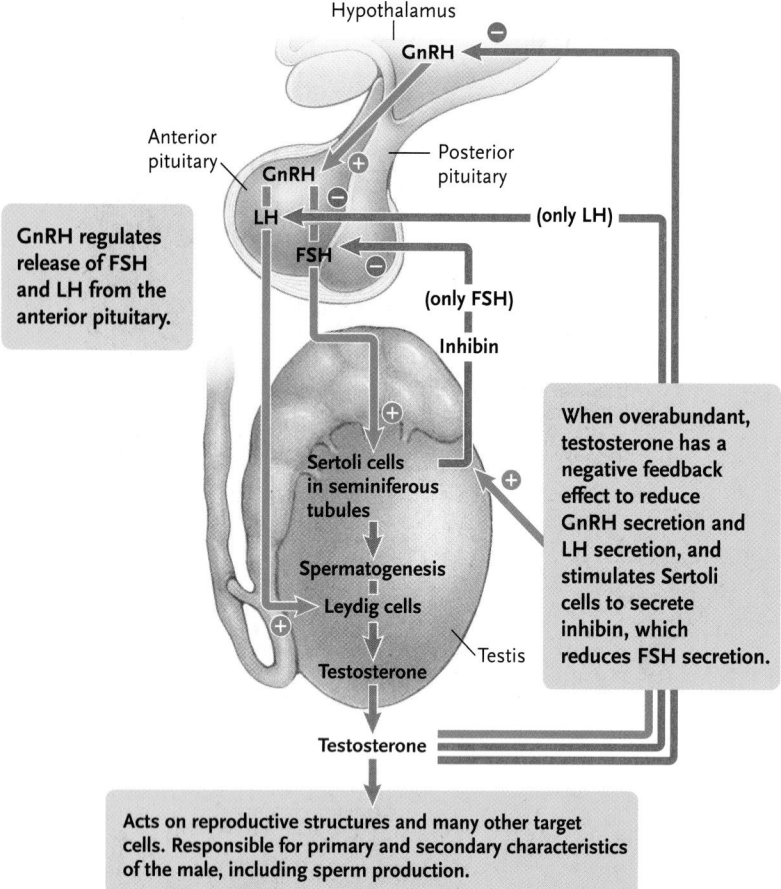

Figure 41.18

Hormonal regulation of reproduction in the male and the negative feedback systems controlling hormone levels.

Insertion of the penis into the vagina and the thrusting movements of copulation lead to the reflex actions of ejaculation, including spasmodic contractions of muscles surrounding the vasa deferentia, accessory glands, and urethra. During ejaculation, the sphincter muscles controlling the exit from the bladder close tightly, preventing urine from mixing with the ejaculate. Ejaculation is usually accompanied by *orgasm,* a sensation of intense physical pleasure that is the peak (climax) of excitement for sexual intercourse, followed by feelings of relaxation and gratification.

The motions of copulation stretch the vagina and stimulate the clitoris, sometimes inducing orgasm in females. Vaginal stretching also stimulates the hypothalamus to secrete oxytocin, which induces contractions of the uterus. The contractions keep the sperm in suspension and aid their movement through the reproductive tract. Uterine contractions are also induced by the prostaglandins in the semen.

Sperm reach the site of fertilization in the oviducts within 30 minutes of being ejaculated. Of the millions of sperm released in a single ejaculation, only a few hundred actually reach the oviducts. After orgasm, the penis, clitoris, and labia minora gradually return to their unstimulated size. Females can experience additional orgasms within minutes or even seconds of a first orgasm, but most males enter a *refractory period* lasting 15 minutes or longer before they can regain an erection and have another orgasm.

41.4c Fertilization of Human Eggs: Producing Zygotes

A human egg can be fertilized only during its passage through the third of the oviduct nearest the ovary. If the egg is not fertilized during the 12 to 24 hours that it is in this location, it disintegrates and dies. Sperm do not swim randomly for a chance encounter with the egg. Rather, they first swim up the cervical canal to reach the oviduct and then are propelled up the oviduct by contractions of the oviduct's smooth muscles. There is evidence that eggs release chemical attractant molecules that the sperm recognize, causing them to swim directly toward the egg.

PEOPLE BEHIND BIOLOGY 41.2

John P. Wiebe, *Western University*

What happens to hormones after they are produced? John P. Wiebe, professor emeritus in the Department of Biology at Western University in London, Ontario, and his graduate students study two steroid hormones derived from progesterone. Specifically, 5α-pregnane-3,20-dione (5αP) and 3α-hydroxy-4-pregen-20-one (3αP) are metabolites of progesterone. 5αP is cancer promoting, stimulating cells to proliferate (tumour growth) and detach (metastasis). 3αP is cancer inhibiting because it suppresses cell proliferation and metastasis.

Wiebe and his colleagues have demonstrated that tumorigenic cells produce higher levels of 5αP and lower levels of 3αP. When cells become tumorigenic, there are strong increases in the mRNA expression of the enzyme that catalyzes conversion of progesterone to 5αP, specifically 5α-reductase. This increase is paralleled by a decrease in 3α-hydroxysteroidoxidoreductase, which catalyzes conversion to 3αP.

It is clear that this work has potential in the control and treatment of breast cancer. It also alerts us to the impact of exposure to hormones, whether in the food we eat or in the water we drink (see "Molecule behind Biology," Box 29.2, Chapter 29).

CONCEPT FIX Although 150 to 350 million sperm are released in a single ejaculation by a human male, an ovum can only be fertilized by one sperm. It is a misconception that several sperm can enter a single ovum. ⬡

Sperm must first penetrate the layer of follicle cells surrounding the egg, aided by enzymes in the sperm plasma membrane **(Figure 41.19)**. Then the sperm adhere to receptor molecules on the surface of the zona pellucida. This contact triggers the **acrosome reaction**, in which enzymes contained in the acrosome are released from the sperm onto the zona pellucida, where they digest a path to the plasma membrane of the egg. As soon as the first sperm cell reaches the egg, sperm and egg plasma membranes fuse, and the sperm cell is engulfed by the cytoplasm of the egg. Although only one sperm fertilizes the egg, the combined release of acrosomal enzymes from many sperm greatly increases the chance that a complete channel will be opened through the zona pellucida. This is in part why a low sperm count is often a source of male infertility. Low sperm counts can be caused by infection, heat,

a. Sperm attached to zona pellucida

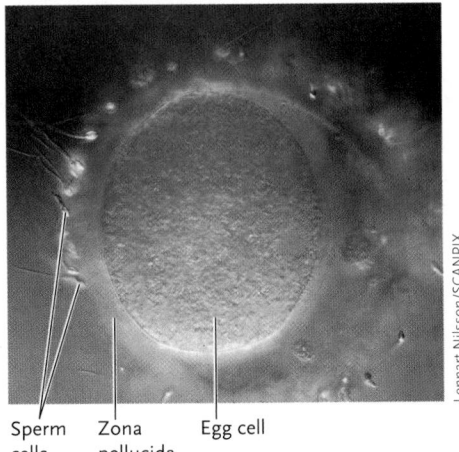

Lennart Nilsson/SCANPIX

Sperm cells   Zona pellucida   Egg cell

b. Early steps in fertilization in mammals

Fertilization

Oviduct

Uterus

Ovary   Ovulation

Opening of cervix

Vagina

Sperm enter vagina

1 The fertilizing sperm penetrates the layer of follicle cells and binds to receptors on the zona pellucida (receptors not shown).

2 The binding of sperm to receptors triggers the acrosome reaction in which hydrolytic enzymes in the acrosome are released onto the zona pellucida.

3 The acrosomal enzymes digest the zona pellucida, creating a pathway to the plasma membrane of the egg cell. When the sperm reaches the egg cell, the plasma membranes of the two cells fuse.

4 The sperm nucleus enters the egg cytoplasm.

5 The sperm stimulates release of Ca^{2+} stored in the egg, which, in turn, triggers the cortical reaction, leading to the slow block in polyspermy.

Follicle cells

Zona pellucida

Sperm plasma membrane

Acrosomal vesicle

Egg plasma membrane

Cortical granules

Egg cytoplasm

Sperm basal body

Sperm nucleus

Figure 41.19

Fertilization in mammals. **(a)** Sperm attached to the zona pellucida of a human egg cell. **(b)** Early steps in the fertilization process.

frequent ejaculation, smoking, and excess alcohol consumption.

Membrane fusion activates the egg. The sperm that enters the egg releases nitric oxide that stimulates the release of stored Ca^{2+} in the egg. Ca^{2+} triggers cortical granule release to the outside of the egg. Enzymes from the cortical granules cross-link molecules in the zona pellucida, hardening it and sealing the channels opened by acrosomal enzymes. The enzymes also destroy receptors that bind sperm to the surface of the zona pellucida. As a result, no further sperm can bind to the zona pellucida or reach the plasma membrane of the egg. The Ca^{2+} also triggers the completion of meiosis of the egg. The sperm and egg nuclei then fuse, and the cell is now considered a zygote. Mitotic divisions of the zygote soon initiate embryonic development.

The first cell divisions of embryonic development take place while the fertilized egg is still in the oviduct. About seven days after ovulation, the embryo passes from the oviduct and implants in the uterine lining. During and after implantation, cells associated with the embryo secrete **human chorionic gonadotropin (hCG)**, a hormone that keeps the corpus luteum in the ovary from breaking down. Excess hCG is excreted in the urine; its presence in urine or blood provides the basis of pregnancy tests.

Continued activity of the corpus luteum keeps estrogen and progesterone secretion at high levels, maintaining the uterine lining and preventing menstruation. The high progesterone level also thickens the mucus secreted by the uterus, forming a plug that seals the opening of the cervix from the vagina. The plug keeps bacteria, viruses, and sperm cells from further copulations from entering the uterus.

About 10 weeks after implantation, the placenta takes over the secretion of progesterone, hCG secretion drops off, and the corpus luteum regresses. However, the corpus luteum continues to secrete the hormone *relaxin,* which inhibits contraction of the uterus until near the time of birth.

STUDY BREAK

1. How does the reproductive pattern of human females differ from that of other mammals? What is the significance of this? (See also Chapters 28, 30, and 33.)
2. Given all of the changes in hormones associated with the ovarian and uterine cycles, which hormones can be used as good predictors of ovulation?
3. Compare and contrast Figures 41.8 and 41.19 (steps of fertilization in sea urchin and mammal). How would you determine whether this was a case of convergent evolution versus the retention of an ancestral trait?

41.5 Controlling Reproduction

Knowledge about the details of reproduction can allow us to control fertility. In some cases, this means increasing the chances of reproducing, whereas in other cases, it means minimizing them. In human society, pregnancy can be a blessing or a disaster, depending on the situation. Statistics describing the effectiveness of different means of limiting human reproductive output **(Table 41.1)** illustrate our progress in the area of family planning. Knowledge about the timing of ovulation, for example (see Figure 41.14, p. 1003), can provide the means to maximize or minimize the chances of pregnancy.

Biologists working to conserve biodiversity often attempt to control reproduction. When a species is on the brink of extinction, the goal is to maximize reproductive output. Techniques can range from the use of foster parents to raise young to using reproductive technologies such as *in vitro* fertilization and implantation of embryos. For example, biologists working with black-footed ferrets (see "Black-Footed Ferret, *Mustela nigripes,*" Chapter 29), which are highly endangered, strive to maximize reproductive output to increase the population.

Table 41.1	Pregnancy Rates for Birth Control Methods	
Method	Lowest Expected Rate of Pregnancy[a]	Typical-Use Rate of Pregnancy[b]
Rhythm method	1–9%	25%
Withdrawal	4%	19%
Condom (male)	3%	14%
Condom (female)	5%	21%
Diaphragm and spermicidal jelly	6%	20%
Vasectomy (male sterilization)	0.1%	0.15%
Tubal ligation (female sterilization)	0.5%	0.5%
Contraceptive pill (combination estrogen–progestin)	0.1%	5%
Contraceptive pill (progestin only)	0.5%	5%
Implant (progestin)	0.09%	0.09%
Intrauterine device (IUD) (copper T)	0.6%	0.8%

[a]Rate of pregnancy when the birth control method was used correctly every time.
[b]Rate of pregnancy when the method was used typically, meaning that it may not have always been used correctly every time.

Source: U.S. Food and Drug Administration, http://www.fda.gov/fdac/features/1997/conceptbl.html. Data reported in 1997 for effectiveness of methods in a one-year period.

Similarly, when an increasing population of one species threatens to overwhelm other species or an ecosystem, the goal is to prevent reproduction. Biologists faced with growing populations of African elephants try to reduce reproductive output using techniques ranging from the application of contraceptives to females (see "Molecule behind Biology," Box 29.2, Chapter 29) to culling (killing) individuals in the population. Culling usually targets females because they produce young (see Chapter 29). The same principles of controlling reproductive output have been central to humans' domestication of other organisms (see Chapter 33).

Review

To access course materials such as Aplia and other companion resources, please visit www.NELSONbrain.com.

41.1 The Drive to Reproduce

- The drive to reproduce is strong; thus all energy that does not go into growth, development, and maintenance goes into reproduction. The strategies employed are diverse, reflecting differences in habitat and lifestyle.

41.2 Asexual and Sexual Reproduction

- One important advantage of asexual reproduction over sexual reproduction is the exact production of successful genotypes in organisms living in stable environments. Asexual reproduction occurs by mitosis through fission, budding, or fragmentation, or by meiosis through parthenogenesis.

- One important advantage of sexual reproduction over asexual reproduction is the generation of genetic diversity in offspring. Genetic recombination and independent assortment of chromosomes during meiosis give rise to diversity and reduce vulnerability to deleterious effects carried on recessive alleles.

41.3 Mechanisms of Sexual Reproduction

- Meiosis in animals occurs only during the production of gametes. Spermatogenesis produces sperm, while oogenesis produces eggs (oocytes). In plants and fungi, meiosis is not always used to produce gametes.

- The amount of yolk in eggs varies enormously, reflecting the time of incubation and development, i.e., the time required until the animal can obtain nutrients from other sources. Development in eggs with large amounts of yolk often involves a small embryonic disk floating on top of the yolk (e.g., bird) as opposed to development around the yolk (e.g., frog).

- External fertilization occurs in most aquatic organisms. Various strategies exist to enhance the odds of sperm and eggs from members of the same species uniting.

- Internal fertilization can reduce the amounts of sperm necessary to achieve fertilization. Internal fertilization can also increase the certainty of mate choice where only the sperm of a male that mated come into contact with the egg(s) of a female. It requires more energy, however, leaving less for production of eggs and sperm.

- Polyspermy, more than one sperm entering an egg, is prevented in two ways. The fast block depends on a change in the egg's membrane potential from negative to positive, whereas the slow block involves Ca^{2+} ions that cause cortical granules to fuse with the egg's plasma membrane and release their enzyme contents to the outside.

- Viviparous animals give birth to live young. Before birth, the mother provides the developing embryo with food and oxygen and removes its metabolic wastes. Usually, but not always, this occurs in the body of the female. Ovoviviparous animals may also give birth to live young, but here the eggs develop and grow inside the mother using yolk as the source of energy. Oviparous animals lay eggs that develop outside the mother's body.

41.4 Sexual Reproduction in Mammals

- Most mammals show a well-defined mating season, the time when copulation, fertilization, and development take place. The onset of the mating season is often triggered by changes in photoperiod (day length). All events are timed so that the young are born at the time of year when food is most available for growth and development. Mammalian females typically display behavioural and physiological signs of fertility (estrus). Humans, on the other hand, do not have a mating season, and both sexes are receptive to mating at any time.

- Reproduction is under neuroendocrine control. In the female this consists of two cycles, the ovarian cycle and the uterine cycle, both of which are highly integrated. The ovarian cycle begins when GnRH released by the hypothalamus stimulates the pituitary to release leutinizing hormone (LH) and follicle-stimulating hormone (FSH). This ultimately leads to ovulation. Subsequent steps depend upon whether fertilization occurs or not.

- Cryptic testes (e.g., elephants and whales) are housed inside the body. In many mammals, testes are housed in a scrotum outside the body because sperm develop best at temperatures below the core temperature of a mammal. A baculum is a penis bone, a feature of many mammals but not humans. The baubellum (clitoris bone) is the equivalent in females.

- Alkaline prostate secretions raise the pH of semen and the vagina to about pH 6, activating sperm motility. Prostaglandins in seminal fluid trigger contractions of the female reproductive tract that help move the sperm into and through the uterus.

- When enzymes contained within the acrosome of a sperm are released into the zona pellucida, they digest a path to the plasma membrane of the egg.

This process is initiated by contact between the sperm and receptor molecules on the surface of the zona pellucida and adherence of the sperm and the molecules. When the sperm meets the egg cell, the plasma membranes of the two cells fuse and the sperm nucleus enters the egg cytoplasm, where it ultimately fuses with the nucleus of the egg.

41.5 Controlling Reproduction

- Controlling (promoting or reducing) reproductive output is important for our species in social and political contexts. It is also important in conservation.

Questions

Self-Test Questions

Any number of answers from a to e may be correct.

1. Which action can be involved in asexual reproduction?
   a. fission
   b. budding
   c. copulation
   d. parthenogenesis

2. A sea star is cut into two separate pieces and each develops into a complete sea star. What is this an example of?
   a. external fertilization
   b. fission
   c. budding
   d. fragmentation

3. Germ cells give rise to which of the following?
   a. eggs
   b. somatic cells
   c. sperm
   d. Only a and b are correct.

4. In which group is internal fertilization rarely seen?
   a. in platyhelminthes
   b. in bedbugs
   c. in mammals
   d. in frogs

5. In slow blocks, more than one sperm is prevented from fertilizing an egg by changes in which of the following?
   a. in Ca^{2+} ions
   b. in Cl^- ions
   c. in Na^+ ions
   d. in cortical granules

6. Ovulation in women is signalled by a rise in which of the following hormones?
   a. relaxin
   b. estrogen
   c. testosterone
   d. luteinizing hormone

7. Which mammal has cryptic testes?
   a. humans
   b. elephants
   c. dogs
   d. bulls

8. Which action is involved in the reproductive cycle of some bats and turtles?
   a. delayed fertilization
   b. delayed implantation
   c. delayed development
   d. postpartum estrus

9. What colour of light are cryptochromes sensitive to?
   a. red light
   b. white light
   c. blue light
   d. green light

10. Which animals engage in the mating behaviour termed amplexus?
   a. birds
   b. frogs and toads
   c. salmon
   d. mammals

Questions for Discussion

1. How do plants exploit animals to effect pollination and dispersal of seeds? Are there examples of animals exploiting plants to achieve reproduction?

2. How does variation in the pattern of fertilization (internal versus external) differ among the animal phyla? Do these variations indicate different ancestral conditions? How do these patterns differ between plants and animals?

3. What methods do zoos and botanical gardens use to control reproduction of captive organisms? Is this appropriate?

4. What steps could conservation biologists take to increase reproductive output of rare and endangered species?

5. Explain why species recognition is important for animals using external fertilization.

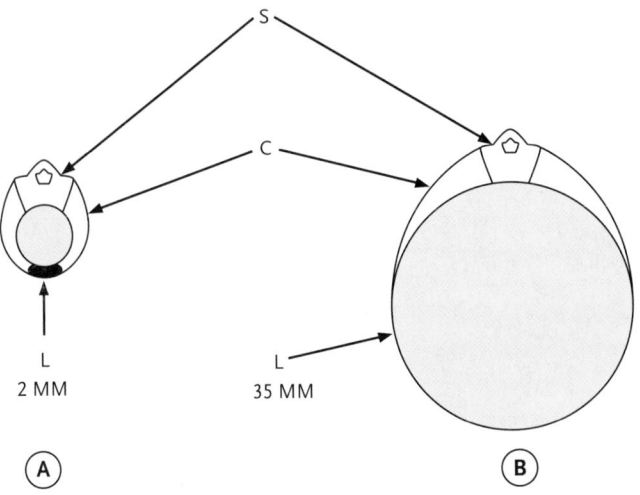

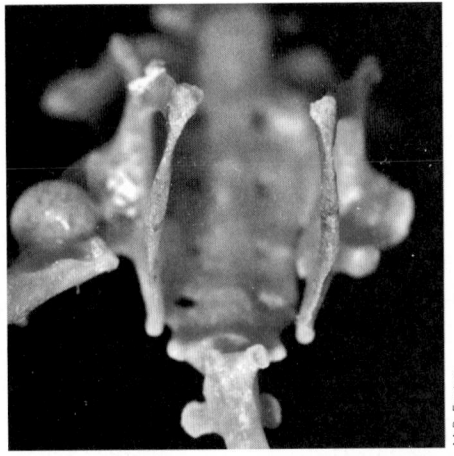

M.B. Fenton

A diagrammatic view of the bony birth canal of (A) a nonpregnant Brazilian free-tailed bat (*Tadarida brasiliensis*) and (B) a female in the process of giving birth. The sacrum (S), coxal bone (C), and interpubic ligament (L) are shown. The photograph (right) shows the ventral view of the pelvis of another nonpregnant free-tailed bat. There is no connection between the pubic bones, making it possible for the animal to achieve the expansion of the birth canal required during birth.

Animal Development

WHY IT MATTERS

The size of the mother and of the young can be a challenging aspect of the process of giving birth (parturition). Bats are an extreme example when it comes to offspring size. Typically, a female bat bears a pup that is 25 to 30% of her normal body mass—the equivalent of a 60-kg woman bearing a 15- to 20-kg baby. The birth canal of placental mammals (such as bats and people) passes between the two halves of the pelvic girdle. Not surprisingly, the pelvis of adult female placental mammals differs from that of males.

The magnitude of the challenge of parturition to a female Brazilian free-tailed bat is shown above. In a nonpregnant female, the bony birth canal is 2 mm in diameter, but it expands to 35 mm in diameter during birth. In these bats, birth takes about 90 seconds.

The elasticity of the interpubic ligament is the key to the birth process in female mammals that give birth to large young. This ligament has an abundance of elastic fibres, which intermingle with collagen fibres at the ligament's core. The situation in bats appears to be the same as in other mammals: elastic fibres stretch, whereas collagen fibres slide in relation to each other. The hormone relaxin plays a fundamental role in this process, promoting the stretchability of interpubic ligaments at the time of parturition (see "Molecule behind Biology," Box 42.1).

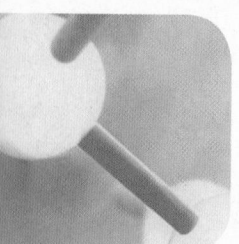

MOLECULE BEHIND BIOLOGY 42.1

Relaxin

The hormone relaxin **(Figure 1)** is a polypeptide produced by the ovaries during pregnancy. In humans, relaxin occurs at higher levels earlier in pregnancy than closer to parturition. Relaxin promotes angiogenesis, the growth of new blood vessels, and influences the interface between the uterus and the placenta. Relaxin inhibits muscular contractions of the uterus that could terminate pregnancy and stimulates the growth of glands that produce milk in breast tissue.

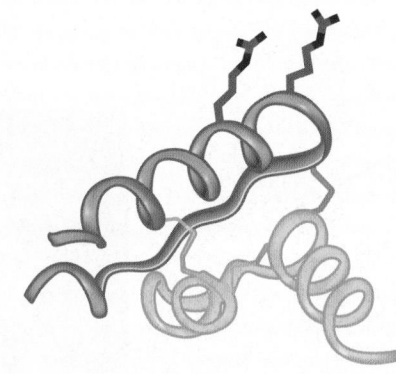

Figure 1
Relaxin. Near parturition, relaxin causes relaxation of the pubic ligaments and softens and enlarges the cervical opening.

The process of development takes an animal from the zygote (fertilized egg) to the complete adult stage of the life cycle. The surroundings of the developing animal (embryo or larva) can influence the process, although the mechanisms and the major patterns of development are common among multicellular animals. Genetic controls underlie the cellular and molecular processes involved in animal development and include apoptosis, or programmed death of cells. Exploring the development processes in a range of animals illustrates both diversity and underlying principles.

care has several different stages (see "On the Road to Vivipary," Chapter 41). Although we associate vivipary with mammals, many species of fish are mouth-breeders, keeping eggs and, for a time, developing young in their mouths. Other fish, such as sea horses and pipefish (family Syngnathidae, order Gasterosteiformes; **Figure 42.1**), keep eggs and developing young in specialized incubation areas, called brood pouches, located on the tail or trunk of the male. "Pregnancy" in male sea horses represents an increase in parental investment. It also allows males to be confident about the paternity of the young they raise.

42.1 Housing and Fuelling Developing Young

Some animal parents invest significant energy in housing and feeding their developing young. This is one aspect of the genetically selfish drive to ensure that their genes are represented in future generations.

42.1a Housing Provides a Place in Which the Embryo Can Develop

There is a recurring tendency across phyla for parents to put eggs and developing young in situations that minimize their exposure to predators and parasites while maximizing favourable conditions for growth and development. Many species of birds use nests to house their eggs and unfledged young. Parents of other species, such as some species of scorpions, frogs, and insects, carry their young with them, often on their backs. This allows the parent (parents) to avoid or actively deter would-be predators.

An escalation in **parental investment** is moving eggs and young inside the parent's body (vivipary and ovovivipary; see Chapter 41). This approach to parental

Figure 42.1
A male sea horse gives birth.

Some amphibians also show high levels of parental care. In Australia, female frogs, *Rheobatrachus silus,* use their stomachs as brood pouches. While the young are developing, they secrete prostaglandin E$_2$, which inhibits the secretion of gastric acid in the stomach and saves the developing young from being digested. On Mount Nimba in West Africa, female toads (*Nectophrynoides occidentalis*) harbour developing young in their uterus, where the young feed on uterine secretions in the absence of a placenta. The gestation period for these toads is nine months, and newborns are 7 to 8 mm long and weigh 30 to 60 mg. Retention of developing embryos in the oviducts has evolved independently in each of the three living groups of Amphibia: Anura, Urodela, and Gymnophiona (see Chapter 28).

42.1b Feeding Aids and Encourages Developing Young

Almost everyone has seen pictures of parent birds feeding their young (see Figures 47.2 and 47.4, respectively). In many species, both males and females deliver food to the nestlings. Some fruit-eating adult birds feed insects to their young because a higher-protein diet promotes rapid growth of the young. Producing high-quality food is the next level of parental investment, and "milk" is a prime example.

The term *milk* is usually applied to secretions of the **mammary glands** of mammals, and it is the quintessentially mammalian food. However, other animals also make milk. Female cockroaches **(Figure 42.2)** house developing embryos in a brood sac and give birth to them as first-instar (first-stage) larvae. The brood sac is an infolding of a ventral intersegmental membrane, and its epithelium produces the milk, a blend of water-soluble proteins encoded by a multigene family.

Both male and female discus fish **(Figure 42.3)** feed their hatchling young (known as *fry*) skin secretions,

Figure 42.3
Symphysodon discus, a fish that produces milk to feed its young.

M.B. Fenton

the first and only food eaten by the fry. This fish milk appears as a slight mucus coating on the adults' bodies, particularly above the lateral lines. Skin feeding also occurs in caecilians (an amphibian). The skin in brooding females **(Figure 42.4** and **Figure 42.5, p. 1014)** is transformed to provide a rich supply of nutrients, and the young have specialized teeth for peeling and eating the outer layer of their mother's skin **(Figure 42.6, p. 1014).** In some other caecilians, young develop in the uterus and feed on the lining of the oviduct. Milk has also been reported in birds, where the crop milk of pigeons (*Columba livia domestica*) is fed to young (known as squabs) from hatching to about age 19 days. In some cases, pigeon lactation continues to day 28. Pigeon crop milk is composed mainly of proteins and lipids and is highly nutritious. Female mammals have not cornered the milk market, either. There are records of male mammals lactating, the most notable being

Figure 42.2
Diploptera punctata, a cockroach that produces live young.
Starting from left, an adult female and male, egg, last-stage fetus, larval instars.

Joseph G. Kunkel

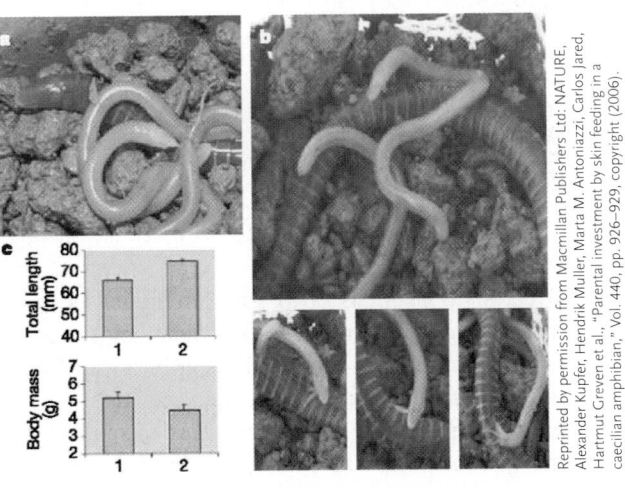

Figure 42.4
A female caecilian (*Boulengerula taitanus*) feeding her young skin secretions.

Reprinted by permission from Macmillan Publishers Ltd: NATURE, Alexander Kupfer, Hendrik Muller, Marta M. Antoniazzi, Carlos Jared, Hartmut Greven et al., "Parental investment by skin feeding in a caecilian amphibian," Vol. 440, pp. 926–929, copyright (2006).

Reprinted by permission from Macmillan Publishers Ltd: NATURE, Alexander Kupfer, Hendrik Muller, Marta M. Antoniazzi, Carlos Jared, Hartmut Greven et al., "Parental investment by skin feeding in a caecilian amphibian," Vol. 440, pp. 926–929, copyright (2006).

Figure 42.5

Details of the skin of **(a)** a nonbrooding and **(b)** a brooding female caecilian.

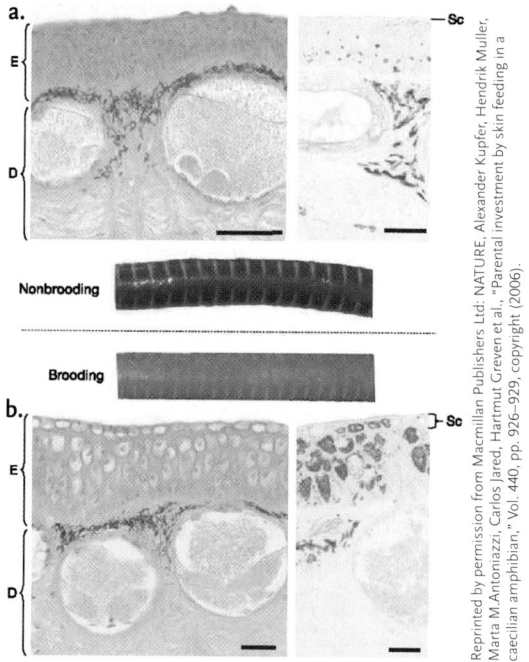

Figure 42.6

Scanning electron micrographs of the specialized teeth of young caecilians.

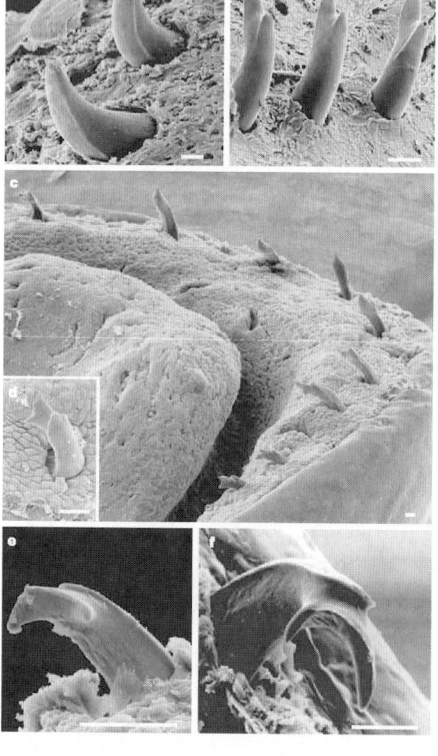

Reprinted by permission from Macmillan Publishers Ltd: NATURE, Alexander Kupfer, Hendrik Muller, Marta M. Antoniazzi, Carlos Jared, Hartmut Greven et al., "Parental investment by skin feeding in a caecilian amphibian," Vol. 440, pp. 926–929, copyright (2006).

Dyacopterus spadiceus, a fruit bat from Indonesia. Some male *D. spadiceus* produce milk, although not as much as females, but the behavioural significance of male lactation in this species remains unknown.

Perhaps female tsetse flies (*Glossina* spp.) are the most astonishing in the matter of caring for developing young. Females ovulate, releasing a single egg that enters the functional equivalent of a mammalian uterus and is fertilized. A milk gland associated with the uterus provides the developing embryo with food, and eventually the female gives birth to a larva that weighs more than she does. The larva then burrows into the ground and pupates. An adult fly later emerges from the pupa.

STUDY BREAK

1. What is relaxin? What role does it play?
2. Where can embryos develop in a parent?
3. What is milk? Which animals produce it?

42.2 Mechanisms of Embryonic Development

When a sperm fertilizes an egg, a zygote is produced. At this point, embryonic development begins, ultimately producing a free-living individual. All of the instructions required for development are packed into the zygote. Mitotic divisions of the zygote are the beginning of developmental activity (see Chapter 8).

Information that directs the initiation of development is stored in two locations in the zygote. The nucleus houses the DNA derived from egg and sperm nuclei. This DNA directs development as individual genes are activated or turned off in a regulated and ordered manner. The balance of the information is stored in the zygote's cytoplasm.

Because sperm contribute essentially no cytoplasm to the zygote, the zygote's cytoplasm is maternal in origin. The mRNA and proteins stored in the egg cytoplasm are known as **cytoplasmic determinants**, which direct the first stages of animal development before genes become active. Depending on the animal group, control of early development by cytoplasmic determinants may be limited to the first few divisions of the zygote (e.g., in mammals), or it may last until the actual tissues of the embryo are formed (e.g., in most invertebrates).

The zygote's cytoplasm also contains ribosomes and other cytoplasmic components required for protein synthesis and early divisions of embryonic cells. Zygote cytoplasm contains the tubulin molecules required to form spindles for early cell divisions, as well as mitochondria and nutrients stored in granules in the yolk and in lipid droplets. In many animals, zygotes contain pigments that colour the egg or regions of it.

Yolk contains nutrients. In the eggs of typical insects, reptiles, and birds, large amounts of yolk supply all of the nutrients for development of the embryo. In contrast, the eggs of placental mammals contain very little yolk, which is used only to support the earliest stages of development.

Depending on the species, yolk may be concentrated at one end or in the centre or distributed evenly throughout the egg. Yolk distribution influences the rate and location of cell division during early embryonic development. Typically, cell division proceeds more slowly in the region of the egg containing the yolk. In the large, yolky eggs of birds and reptiles, cell division takes place only in a small, yolk-free patch at the egg's surface.

Unequal distribution of yolk and other components in the egg is termed **polarity.** In most species, the egg's nucleus is located toward one end, called the **animal pole.** The animal pole typically gives rise to surface structures and the anterior end of the embryo. The opposite end of the egg, the **vegetal pole,** typically gives rise to internal structures such as the gut, along with the posterior end of the embryo. When yolk is unequally distributed in the egg cytoplasm, it is usually concentrated in the vegetal half of the egg. Egg polarity plays a role in setting the three body axes of bilaterally symmetrical animals, namely the anterior–posterior axis, the dorsal–ventral (back-front) axis, and the left–right axis **(Figure 42.7).**

42.2a Cleavage and Gastrulation: Zygote to Multicellular Embryo

Soon after fertilization, the zygote begins a series of mitotic cleavage divisions in which cycles of DNA replication and division occur without the production of new cytoplasm. Thus, the cytoplasm of the zygote is partitioned into successively smaller cells without increasing the size or mass of the embryo **(Figure 42.8).** In the frog *Xenopus laevis,* 12 cleavage divisions produce an embryo of about 4000 cells that collectively occupy about the same volume and mass as the original zygote.

Cleavage is the first of three major developmental stages that, with modifications, are common to the

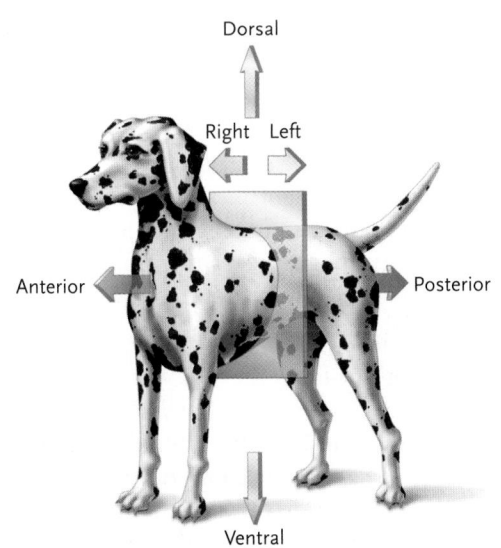

Figure 42.7
Body axes: anterior–posterior, dorsal–ventral, and left–right.

a. Fertilized egg

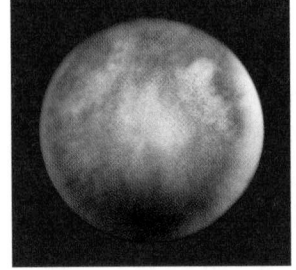

b. Two-cell stage

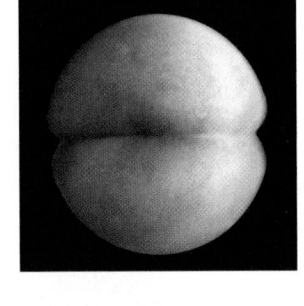

c. Four-cell stage

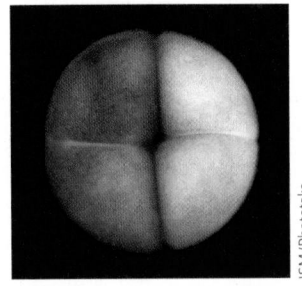

d. Eight-cell stage

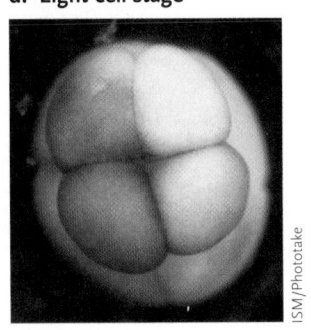

Figure 42.8
The first three cleavage divisions of a frog embryo, which convert the fertilized egg into the eight-cell stage. Note that the cleavage divisions cut the volume of the fertilized egg into successively smaller cells.

early development of most animals. **Gastrulation,** the stage following cleavage, produces an embryo with three distinct primary tissue layers. **Organogenesis** follows gastrulation and gives rise to the development of major organ systems. At the end of organogenesis, the embryo has the body organization characteristic of its species. Cell division, cell movements, and cell rearrangements occur during gastrulation and in organogenesis. **Figure 42.9, p. 1016,** shows these stages as part of the life cycle of a frog.

In frogs, cleavage divisions form two different structures in succession. The **morula** (= mulberry) is a solid ball or layer of cells. As cleavage divisions continue, the ball or layer hollows out to form the **blastula** (*blast* = bud or offshoot; *ula* = small), the second structure, in which cells, now called **blastomeres** (*mere* = part or division), enclose a fluid-filled cavity, the **blastocoel** (*coel* = hollow).

After cleavage is complete, cells of the blastula migrate and divide to produce the **gastrula** (*gaster* = gut or belly). Gastrulation, a morphogenetic process, dramatically rearranges the cells of the blastula into the three **primary cell layers** of the embryo: **ectoderm,** the outer layer (*ecto* = outside; *derm* = skin); **endoderm,** the inner layer (*endo* = inside); and **mesoderm** (*meso* = middle), the middle layer between ectoderm and endoderm. Gastrulation establishes the body pattern. Each tissue and organ of the adult animal originates from one of the three primary cell layers of the gastrula **(Table 42.1, p. 1016).** Cell movements also

Figure 42.9
Stages of animal development shown in a frog.

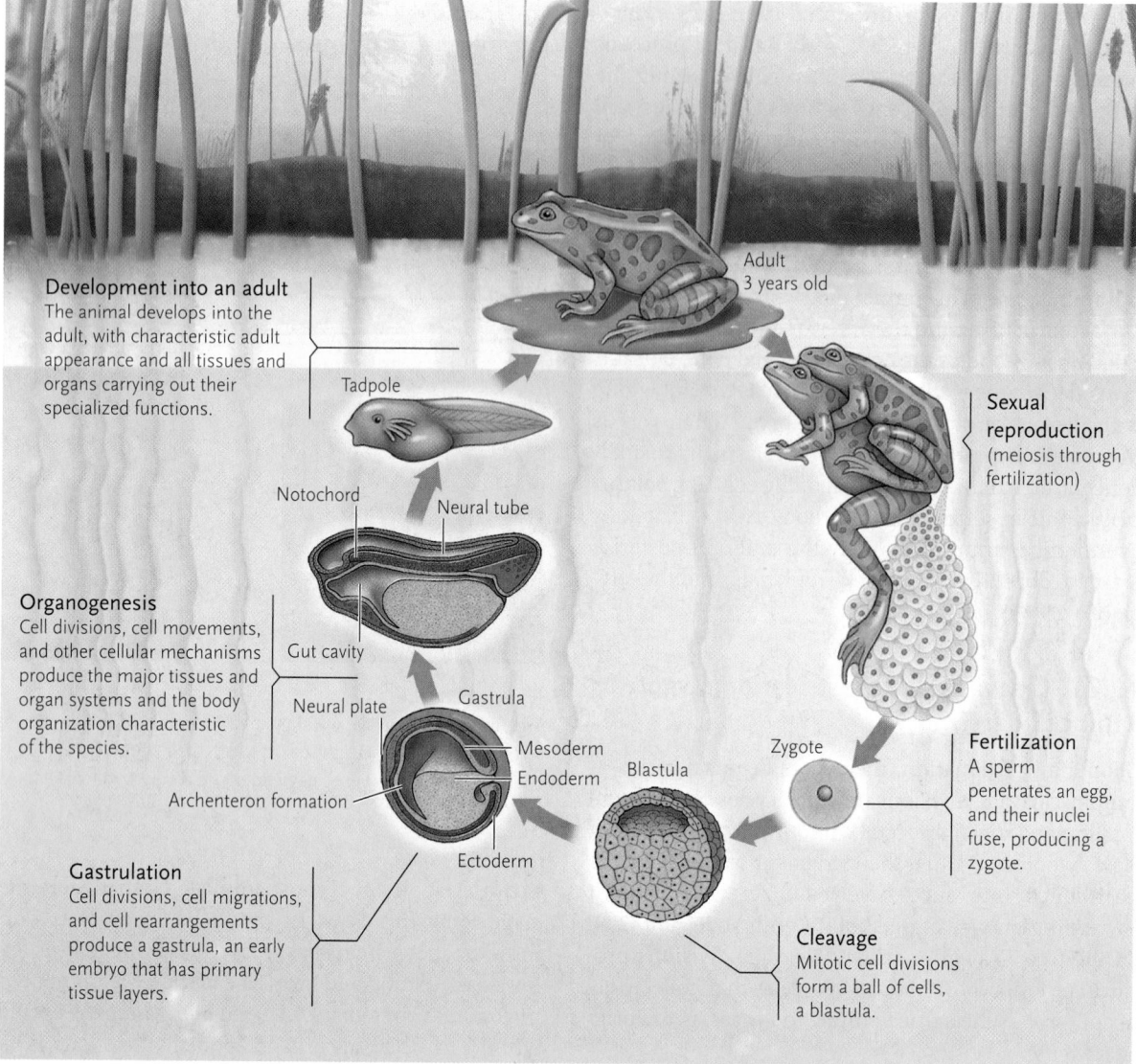

Adult
3 years old

Development into an adult
The animal develops into the adult, with characteristic adult appearance and all tissues and organs carrying out their specialized functions.

Tadpole

Sexual reproduction
(meiosis through fertilization)

Notochord · Neural tube

Organogenesis
Cell divisions, cell movements, and other cellular mechanisms produce the major tissues and organ systems and the body organization characteristic of the species.

Gut cavity

Neural plate · Gastrula

Mesoderm
Endoderm · Blastula · Zygote

Fertilization
A sperm penetrates an egg, and their nuclei fuse, producing a zygote.

Archenteron formation

Ectoderm

Gastrulation
Cell divisions, cell migrations, and cell rearrangements produce a gastrula, an early embryo that has primary tissue layers.

Cleavage
Mitotic cell divisions form a ball of cells, a blastula.

contribute to the formation of the **archenteron** (*arch* = beginning; *enteron* = intestine or gut), a new cavity within the embryo that is lined with endoderm.

As the blastula develops into the gastrula, embryonic cells begin to differentiate, becoming recognizably different in biochemistry, structure, and function. The developmental potential of each cell becomes more limited than that of the zygote from which it restriction of developmental potential resulted because cells lost all of their genes except those required for the structure and function of the cell type they would become. However, differentiating cells each contain the complete genome of the organism, but each type of cell has a different program of gene expression.

Although development in all animals is accomplished by mechanisms under genetic control, the mechanisms are influenced to some extent by environmental factors such as temperature. The six mechanisms are as follows:

- Mitotic cells divide.
- Cells move.
- **Selective cell adhesions** occur in which cells make and break specific connections to other cells or to the extracellular matrix (ECM).
- In **induction**, one group of cells (inducing cells) causes or influences another nearby group of cells (responding cells) to follow a particular developmental

Table 42.1	Origins of Adult Tissues and Organs in the Three Primary Tissue Layers
Primary Tissue Layer	**Adult Tissues and Organs**
Ectoderm	Skin and its elaborations, including hair, feathers, scales, and nails; nervous system, including brain, spinal cord, and peripheral nerves; lens, retina, and cornea of eye; lining of mouth and anus; sweat glands, mammary glands, adrenal medulla, and tooth enamel
Mesoderm	Muscles; most of skeletal system, including bones and cartilage; circulatory system, including heart, blood vessels, and blood cells; internal reproductive organs; kidneys and outer walls of digestive tract
Endoderm	Lining of digestive tract, liver, pancreas, lining of respiratory tract, thyroid gland, lining of urethra, and urinary bladder

pathway. The key to induction is that only certain cells can respond to the signal from the inducing cells. Induction typically involves signal transduction events (see Chapter 43). Some induction events are triggered by direct cell–cell contact involving interaction between a membrane-embedded protein on the inducing cell and a receptor protein on the responding cell's surface. Others are triggered by a signal molecule released by the inducing cell that interacts with a receptor on the responding cell (e.g., paracrine regulation; see Chapter 43).

- **Determination** sets the developmental fate of a cell. Before determination, a cell has the potential to become any cell type of the adult. Afterward, the cell is committed to becoming a particular cell type. Typically, determination results from induction, although in some cases, it results from the asymmetric segregation of cellular determinants.
- **Differentiation** follows determination and involves the establishment of a cell-specific developmental program in the cells. Differentiation results in cell types with clearly defined structures and functions. These features are derived from specific patterns of gene expression in cells.

STUDY BREAK

1. What is yolk? What role does it play?
2. What mechanisms are involved in animal development from the zygote?

42.3 Major Patterns of Cleavage and Gastrulation

42.3a Sea Urchin

Cleavage divisions proceed at approximately the same rate in all regions of a sea urchin embryo (**Figure 42.10, step 1**), reflecting uniform distribution of yolk in the egg. These divisions continue until a blastula containing about a thousand cells is formed (step 2).

Gastrulation begins at the vegetal pole of the blastula. Through induction, some cells in the middle of the vegetal pole become elongated and cylindrical, causing the region to flatten and thicken. Then some cells (primary mesenchyme; *mesen* = middle; *chyme* = juice) break loose and migrate into the blastocoel (step 3), making and breaking adhesions until eventually they attach along the ventral sides of the blastocoel. These cells form the future mesoderm (see Figure 42.10, step 7), which give rise to skeletal elements of the embryo. Next, the flattened vegetal pole of the blastula invaginates, pushing gradually into the interior (steps 4 and 5). The cells that invaginate will become endoderm cells. The inward movement, much like

Figure 42.10
Cleavage and gastrulation in the sea urchin.

Cleavage

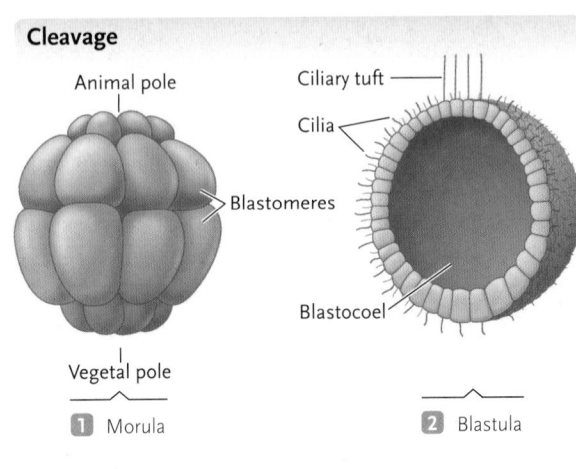

1 Morula

2 Blastula

Gastrulation

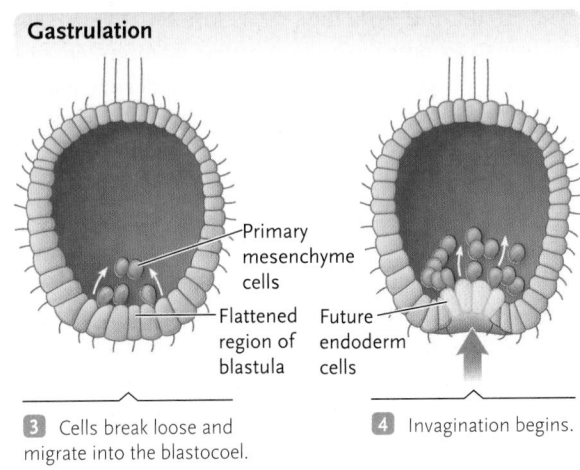

3 Cells break loose and migrate into the blastocoel.

4 Invagination begins.

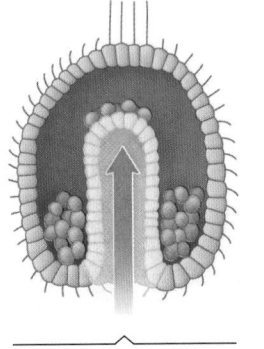

5 Invagination continues.

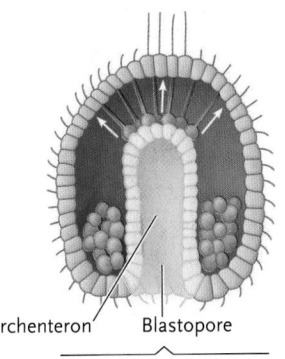

6 The archenteron forms; cells of invagination stretch across the blastocoel and adhere to the ectoderm.

KEY

- Ectoderm
- Mesoderm
- Endoderm

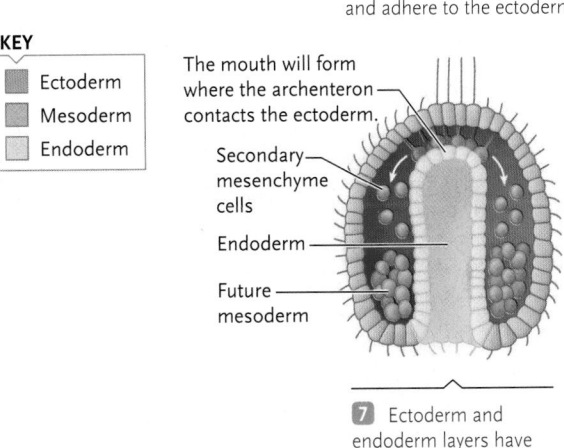

7 Ectoderm and endoderm layers have formed; mesoderm cells are between them.

pushing in the side of a hollow rubber ball, generates the archenteron, a new cavity that opens through the blastopore.

As the archenteron forms, extensions of cells of the invaginated layer stretch across the blastocoel and contact the inside of the ectoderm (Figure 42.10, p. 1017, step 6). These extensions make tight adhesions and then contract, pulling the invaginated cell layer inward with them, eliminating most of the blastocoel.

Now the embryo has two complete cell layers. The outer layer—the original blastula surface—forms embryonic ectoderm. Cells of the second, inner layer are derived from the archenteron and become endoderm. Mesodermal cells, which begin to form a third layer, are derived from the primary mesenchyme cells and from *secondary mesenchyme* cells that migrated into the space between the ectoderm and endoderm (step 7). After the formation of the three primary cell layers, cells begin to differentiate based on synthesis of different proteins in each layer.

As ectoderm, mesoderm, and endoderm layers develop, the embryo lengthens into an ellipsoidal shape, with the blastopore marking the posterior end of the embryo. From here on, further cell divisions, combined with cell movements, selective cell adhesions, induction, and differentiation, lead to differentiation of organ systems. In sea urchins and other deuterostomes, the blastopore forms the anus, and the mouth will form at the opposite, anterior end of the gut.

42.3b Amphibians

In the eggs of amphibians such as frogs, yolk is concentrated in the vegetal half, giving it a pale colour. The animal half is darkly coloured because of a layer of pigment granules just below the surface. A sperm normally fertilizes the egg in the animal half (**Figure 42.11,** step 1). After fertilization, the pigmented layer of cytoplasm rotates toward the site of sperm entry, exposing a crescent-shaped region of underlying cytoplasm at the side opposite the point of sperm entry (step 2). This region, the **grey crescent**, establishes the dorsal–ventral

axis of the embryo and marks the future dorsal side of the animal.

Normally, the first cleavage division runs perpendicular to the long axis of the grey crescent and divides the crescent equally between resulting cells (step 3). If the first two blastomeres are experimentally divided so that one does not receive grey crescent material, and the two cells are separated, the blastomere without grey crescent material divides but ends up in a disordered mass that stops developing. The blastomere receiving grey crescent material produces a normal embryo. Cytoplasmic material localized in the grey crescent is essential to normal development in frog embryos.

As cleavage of the frog embryo continues, cell divisions proceed more rapidly in the animal half, producing smaller and more numerous cells there than in the yolky vegetal half. By the time cleavage has produced an embryo with 15 thousand cells, the animal half has hollowed out, forming the blastula (**Figure 42.12,** step 1, and **Figure 42.13a**).

Gastrulation begins when cells from the animal pole begin to migrate across the embryo surface to reach the region derived from the grey crescent. This site is marked by a crescent-shaped depression rotated 90° clockwise and called the **dorsal lip of the blastopore** (see Figure 42.12, step 2, and Figure 42.13b). These cells invaginate, changing shape and pushing inward from the surface to produce the depression. The depression eventually forms a complete circle (the blastopore) after further inward movement of additional cells (see Figure 42.12, step 3, and Figure 42.13c).

By **involution**, cells migrate into the blastopore, and the pigmented cell layer of the animal half expands to cover the entire embryo surface (see Figure 42.12, step 4, and Figure 42.13c). Cells of the vegetal half are enclosed by this cell migration, becoming visible on the outside as a yolk plug in the blastopore. The blastopore gives rise to the anus.

Continuing involution moves cells into the interior and upward (see Figure 42.12, steps 3 and 4), forming two layers that line the inside top half of the embryo. Dorsal mesoderm (shown in red) is

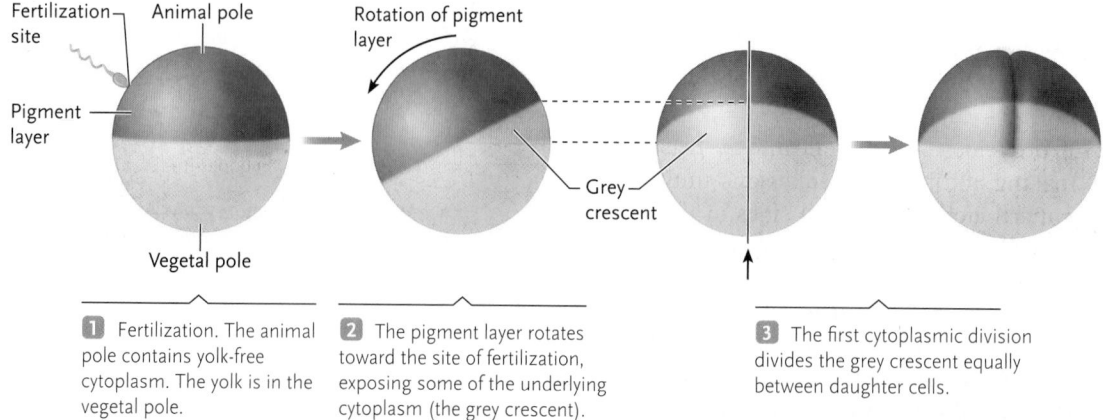

Figure 42.11

Rotation of the pigment layer and development of the grey crescent after fertilization in a frog egg. The grey crescent marks the site where gastrulation of the embryo will begin.

Fertilization site | Animal pole | Rotation of pigment layer

Pigment layer

Vegetal pole

Grey crescent

1 Fertilization. The animal pole contains yolk-free cytoplasm. The yolk is in the vegetal pole.

2 The pigment layer rotates toward the site of fertilization, exposing some of the underlying cytoplasm (the grey crescent).

3 The first cytoplasmic division divides the grey crescent equally between daughter cells.

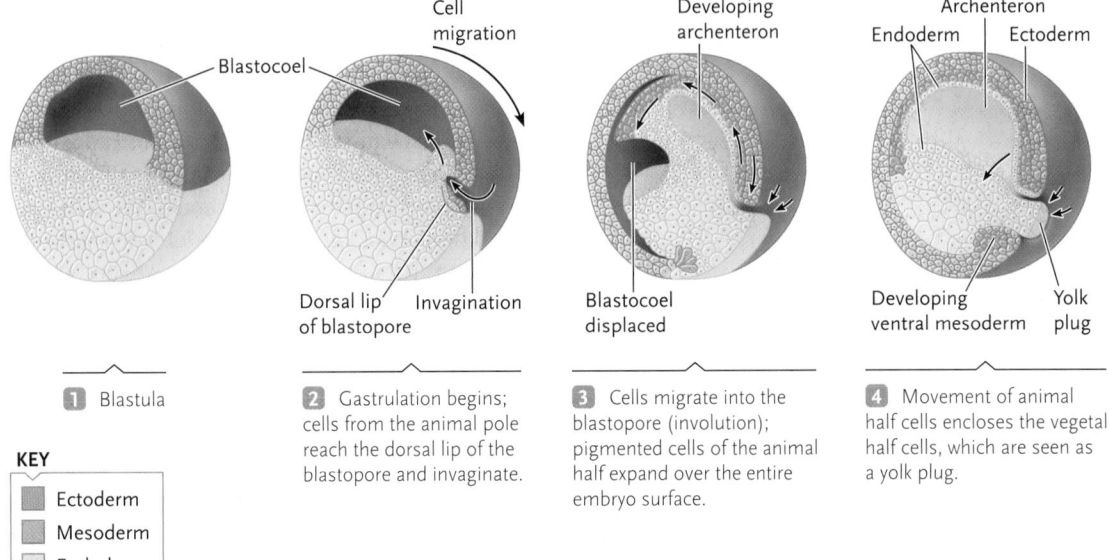

Cell migration

Blastocoel

Developing archenteron

Archenteron
Endoderm — Ectoderm

Figure 42.12
Gastrulation in a frog embryo. Yolk cells are shown in paler yellow.

Dorsal lip — Invagination
of blastopore

Blastocoel displaced

Developing — Yolk
ventral mesoderm — plug

1 Blastula

2 Gastrulation begins; cells from the animal pole reach the dorsal lip of the blastopore and invaginate.

3 Cells migrate into the blastopore (involution); pigmented cells of the animal half expand over the entire embryo surface.

4 Movement of animal half cells encloses the vegetal half cells, which are seen as a yolk plug.

KEY

Ectoderm
Mesoderm
Endoderm

the uppermost of these induced layers. Beneath it is the endoderm (shown in yellow), containing cells originating from both the outer surface of the embryo and the yolky interior. Ectoderm (shown in blue) forms from pigmented cells remaining at the surface of the embryo. Induction of the ventral mesoderm begins near the vegetal pole.

As mesoderm and endoderm form, the depression created by inward cell movements gradually deepens and extends inward as the archenteron (see Figure 42.12, steps 3 and 4), displacing the blastocoel. Cells of the three primary cell layers continue to increase in number by further migrations and divisions as development proceeds.

The major induction centre during frog gastrulation is the dorsal lip of the blastopore. If cells are removed from the dorsal lip and transplanted elsewhere in the egg, they form a second blastopore (and a second embryo).

42.3c Birds

Gastrulation in amniotes (see Chapter 28) such as birds and reptiles is modified by the distribution of the yolk, which occupies almost the entire volume of the egg. A thin layer of cytoplasm at the egg's surface gives rise to primary tissues of the embryo. Although mammalian eggs have relatively little yolk, gastrulation in them follows a similar pattern.

Early cleavage divisions in birds produce the **blastodisc**, a thin layer of cells at the yolk's surface (**Figure 42.14, p. 1020**, step 1). The complete blastodisc is a layer with about 20 thousand cells. Cells of the blastodisc then separate into two layers, the **epiblast** (top layer) and the **hypoblast** (bottom layer). The blastocoel is the flattened cavity between them (step 2).

Gastrulation begins as cells in the epiblast stream toward the midline of the blastodisc, thickening it in this region. The thickened layer (or **primitive streak**) is first

a. Blastula

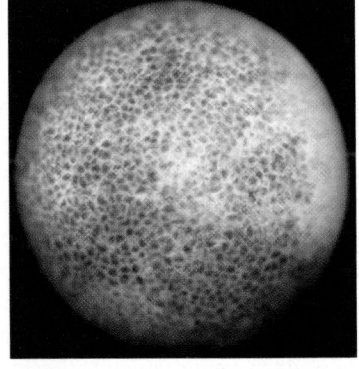

b. Early gastrulation

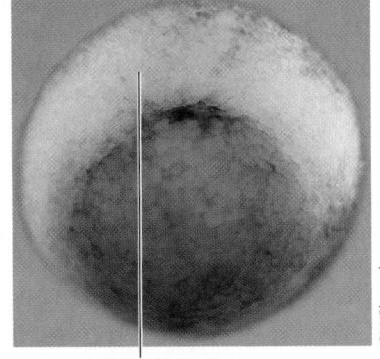

Dorsal lip
of blastopore

c. Late gastrulation

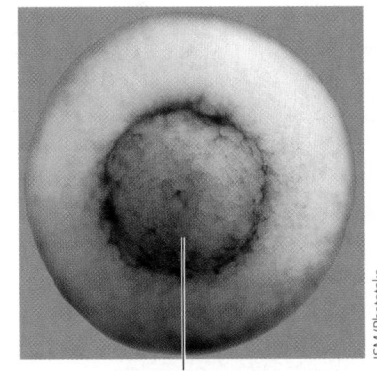

Blastopore
with yolk plug
in centre

Figure 42.13
Photomicrographs of a frog embryo.
(a) Blastula.
(b) Early gastrulation and the formation of the dorsal lip of the blastopore. **(c)** Late gastrulation, showing the completed blastopore, closed by the yolk plug.

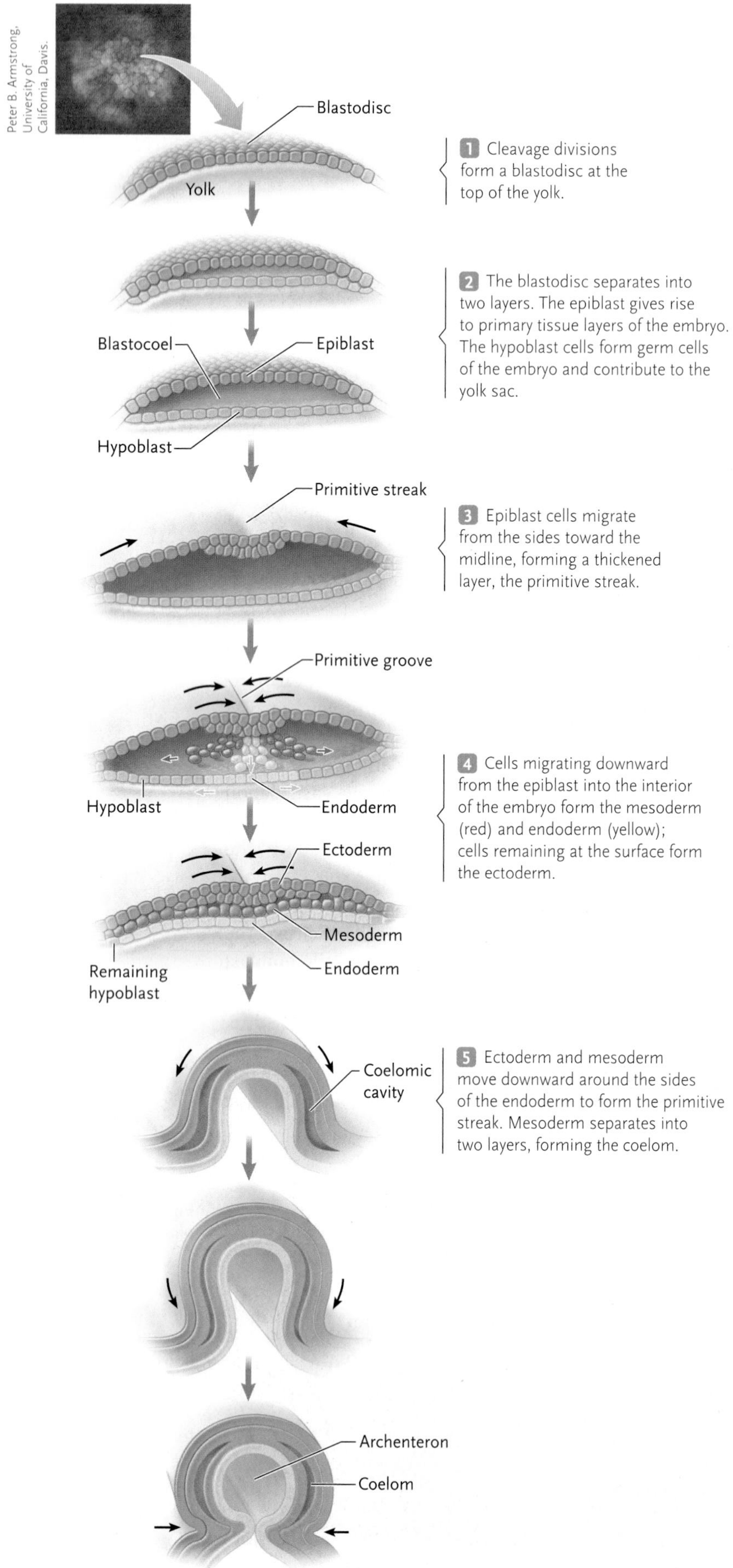

Peter B. Armstrong, University of California, Davis.

1 Cleavage divisions form a blastodisc at the top of the yolk.

Blastodisc

Yolk

2 The blastodisc separates into two layers. The epiblast gives rise to primary tissue layers of the embryo. The hypoblast cells form germ cells of the embryo and contribute to the yolk sac.

Blastocoel

Epiblast

Hypoblast

Primitive streak

3 Epiblast cells migrate from the sides toward the midline, forming a thickened layer, the primitive streak.

Primitive groove

Hypoblast

Endoderm

4 Cells migrating downward from the epiblast into the interior of the embryo form the mesoderm (red) and endoderm (yellow); cells remaining at the surface form the ectoderm.

Ectoderm

Mesoderm

Remaining hypoblast

Endoderm

Coelomic cavity

5 Ectoderm and mesoderm move downward around the sides of the endoderm to form the primitive streak. Mesoderm separates into two layers, forming the coelom.

Archenteron

Coelom

Figure 42.14
Gastrulation in a bird embryo.

evident in the posterior end of the embryo and extends toward the anterior end as more cells of the epiblast (see Figure 42.14) move into it (step 3). A thickening at the anterior end of the primitive streak (the primitive knot) is the functional equivalent of the amphibian dorsal lip of the blastopore. The primitive streak initially marks the future posterior end of the embryo, and by the time it has elongated fully, it has established the left and right sides of the embryo. The streak forms on what will become the dorsal side of the embryo, with the ventral side below.

As the primitive streak forms, its midline sinks, forming the **primitive groove**, a conduit for migrating cells to move into the blastocoel. Epiblasts are the first cells to migrate through the primitive groove (see Figure 42.14, step 4) and produce the endoderm. Mesoderm is formed from cells migrating laterally between the epiblast and the endoderm. Epiblast cells remaining at the surface of the blastodisc form ectoderm (step 4).

In the bird embryo, all primary tissue layers arise from the epiblast. Only a few of the hypoblast cells—near the posterior end of the embryo—contribute directly to the embryo. These form **germ cells** that later migrate to developing gonads, founding cell lines leading to eggs and sperm (see Chapter 8).

Initially, ectoderm, mesoderm, and endoderm are located in three more or less horizontal layers in the chick embryo. During gastrulation, the endoderm pushes upward along its midline, and its left and right sides fold downward, forming a tube that is oriented parallel to the primitive streak (see Figure 42.14, step 5). The archenteron is the central cavity of the tube, the primitive gut. Mesoderm separates into two layers, forming the coelom, a fluid-filled body cavity lined with mesoderm. These movements complete the formation of the gastrula.

42.3d Extra-embryonic Membranes: Amnion, Chorion, Allantois

Each primary tissue layer of a bird embryo extends outside the embryo to form **extra-embryonic membranes (Figure 42.15)** that conduct nutrients from yolk to embryo, exchange gases with the environment outside the egg, or store metabolic wastes removed from the embryo. The **yolk sac** is an extension of mesoderm and endoderm enclosing the yolk. Although the yolk sac remains connected to the gut of the embryo by a stalk, yolk does not directly enter the embryo by this route. Rather, it is absorbed by blood in vessels in the membrane, which then transport the nutrients to the embryo.

The **chorion**, produced from ectoderm and mesoderm, completely surrounds the embryo and yolk sac and lines the inside of the shell. The chorion exchanges oxygen and carbon dioxide with the environment through the egg's shell. The **amnion** closes over the embryo to form the amniotic cavity. Cells of the amnion secrete amniotic fluid into the cavity, which bathes the embryo and provides an aquatic environment in which it can develop. Reptilian and mammalian embryos are

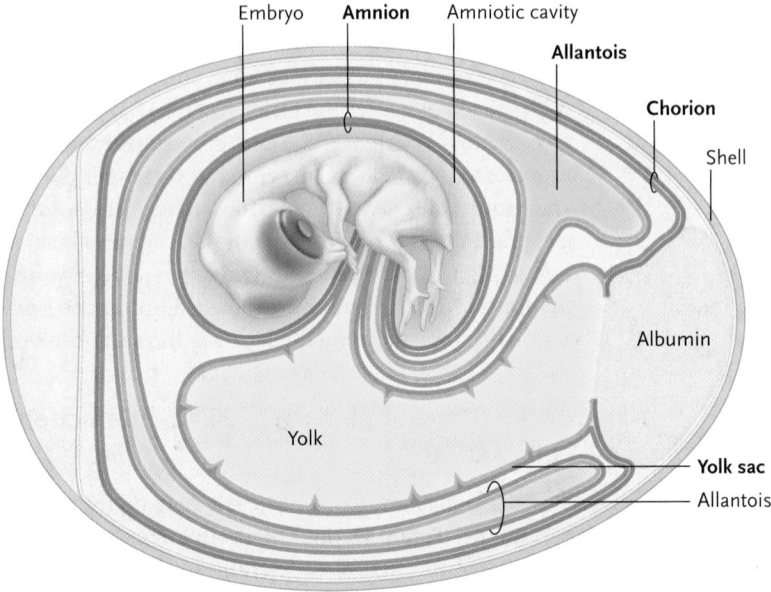

Embryo   **Amnion**   Amniotic cavity
Allantois
Chorion
Shell
Albumin
Yolk
Yolk sac
Allantois

Figure 42.15
The four extra-embryonic membranes in a bird embryo (in bold).

also surrounded by an amnion and amniotic fluid. Providing the embryo with an aquatic environment is presumed to have been a key factor in the evolution of fully terrestrial vertebrates, the **Amniota**.

The **allantoic membrane** forms from mesoderm and endoderm that have bulged outward from the gut. It encloses the allantois, a sac that closely lines the chorion and fills much of the space between the chorion and the yolk sac. The **allantois** stores nitrogenous wastes (primarily uric acid) removed from the embryo. The part of the allantoic membrane lining the chorion forms a rich bed of capillaries connected to the embryo by arteries and veins. This circulatory system delivers carbon dioxide to the chorion and picks up the oxygen that is absorbed through the shell and chorion. At hatching, part of the allantoic membrane becomes the lining of the bladder.

STUDY BREAK

1. What are the main developmental differences between protostomes and deuterostomes?
2. What are important differences between typical patterns of development in birds and in mammals?
3. What are the features of an amniote egg? What is its evolutionary significance?

42.4 Organogenesis: Gastrulation to Adult Body Structures

Following gastrulation, organogenesis gives rise to the body organization characteristic of the species. Organogenesis involves the same mechanisms used in gastrulation, namely cell division, cell movements, selective cell adhesion, induction, and differentiation. Organogenesis also involves an additional mechanism, **apoptosis**, in which certain cells are programmed to die (see Chapter 8). To illustrate how cellular mechanisms of development interact in organogenesis, we follow the formation of major organ systems in the bird embryo. Then we describe the generation of one organ, the eye, which follows a pathway typical of eye development in all vertebrates.

42.4a Ectoderm and the Nervous System: Neural Tube and Neural Crest Cells

In vertebrates, organogenesis begins with **neurulation**, which is the development of nervous tissue from ectoderm. As a preliminary to neurulation, cells of the mesoderm form the notochord, a solid rod of tissue extending the length of the embryo under the dorsal ectoderm. Notochord cells carry out a major induction, causing the overlying ectoderm to form the **neural plate**, a thickened and flattened longitudinal band of cells (**Figure 42.16, p. 1022**, steps 1 and 2). The neural plate does not form if the notochord is removed.

Once induced, the neural plate sinks downward along its midline (steps 2 and 3), creating a deep longitudinal groove and ridges (neural crests) that rise along the sides of the neural plate. The neural tube forms when the neural crests move together and close over the centre of the groove along the length of the developing embryo (steps 4 and 5). The neural tube then pinches off from the overlying ectoderm, which closes over the tube (step 6). The central nervous system, including the brain and spinal cord, develops directly from the neural tube.

During formation of the neural tube, **neural crest** cells (see Figure 42.16) migrate into the mesoderm and follow specific routes to reach distant points in the developing embryo, where they contribute to the formation of a variety of organ systems. Some cells develop into cranial nerves in the head, whereas others contribute to the bones of the inner ear and skull, cartilage of facial structures, and teeth. Still others form ganglia of the autonomic nervous system, peripheral nerves leading from the spinal cord to body structures, and nerves of the developing gut. Neural crest cells also move to the skin, where they form pigment cells, and to the adrenal glands, where they form the medulla of the kidney. The migration of neural crest cells contributes to development in all vertebrates (see "People behind Biology," Box 42.2, p. 1023).

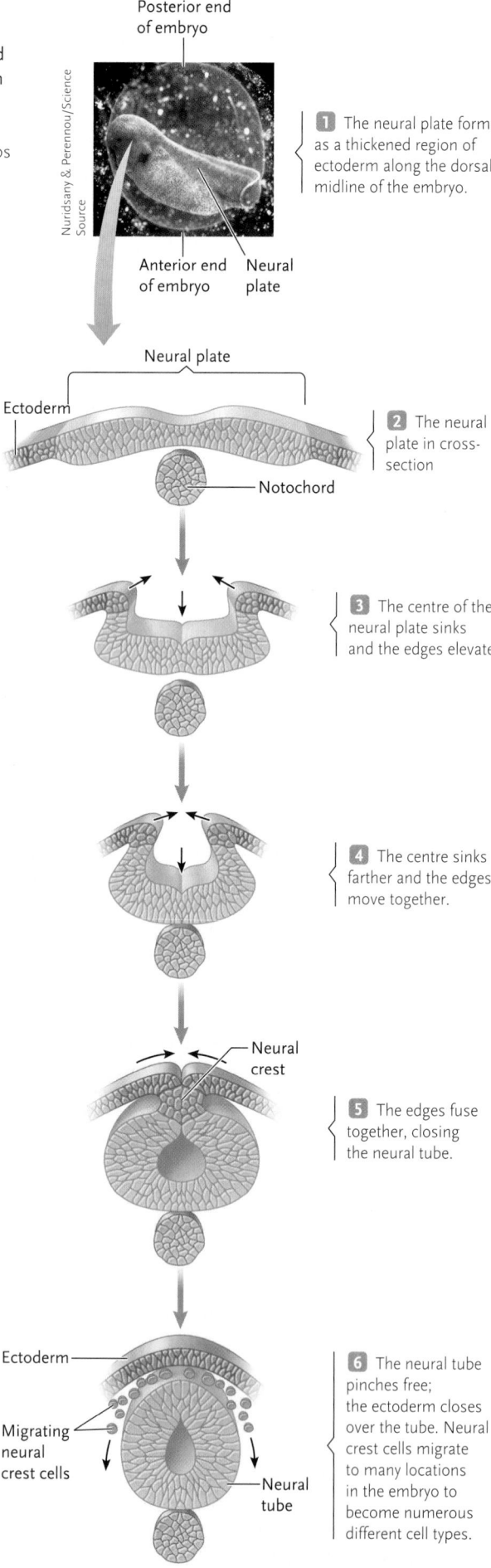

Figure 42.16
Development of the neural tube and neural crest cells in vertebrates. Photo is of an embryo; drawings show steps in a bird embryo.

Posterior end of embryo

Nuridsany & Perennou/Science Source

Anterior end of embryo   Neural plate

1 The neural plate forms as a thickened region of ectoderm along the dorsal midline of the embryo.

Neural plate

Ectoderm

2 The neural plate in cross-section

Notochord

3 The centre of the neural plate sinks and the edges elevate.

4 The centre sinks farther and the edges move together.

Neural crest

5 The edges fuse together, closing the neural tube.

Ectoderm

Migrating neural crest cells

Neural tube

6 The neural tube pinches free; the ectoderm closes over the tube. Neural crest cells migrate to many locations in the embryo to become numerous different cell types.

Other structures differentiate in the embryo while the neural tube is forming. On each side of the notochord, mesoderm separates into **somites**, blocks of cells spaced one after the other **(Figure 42.17)**. Somites give rise to the vertebral column, ribs, repeating sets of muscles associated with the ribs and vertebral column, and limb muscles. Mesoderm outside the somites extends around the primitive gut (lateral mesoderm in Figure 42.17) and splits into two layers, one covering the surface of the gut and the other lining the body wall. The space between the layers is the adult coelom.

42.4b Development of the Eye: Interactions between Cells

Eyes develop by the same basic five-step pathway in all vertebrates **(Figure 42.18)**. The brain forms at the anterior end of the neural tube from a cluster of hollow vesicles that swell outward from the neural tube (see Figure 42.18, step 1). Each of the two optic vesicles develops into an eye. Figure 42.18 depicts the development of a frog eye; the development of the rest of the brain is not shown.

The optic vesicles grow outward until they contact the overlying ectoderm, inducing a series of developmental responses in both tissues. The optic cup, a

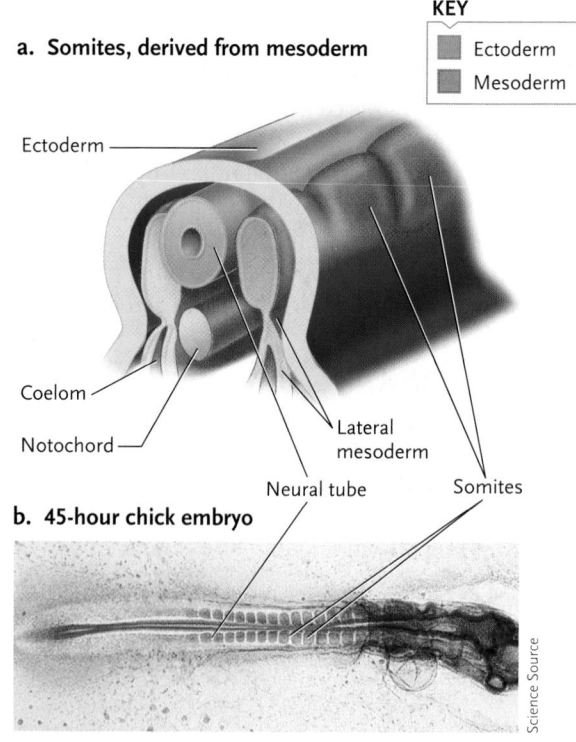

a. Somites, derived from mesoderm

KEY
■ Ectoderm
■ Mesoderm

Ectoderm

Coelom

Notochord

Lateral mesoderm

Neural tube

Somites

b. 45-hour chick embryo

Science Source

Figure 42.17
Later development of the mesoderm. **(a)** Somites develop into segmented structures such as the vertebrae, the ribs, and the musculature between the ribs. The lateral mesoderm gives rise to other structures, such as the heart and blood vessels and the linings of internal body cavities. **(b)** The somites in a 45-hour chick embryo.

double-walled structure, forms when the outer surface of the optic vesicle thickens and flattens at the region of contact and then pushes inward. The optic cup ultimately becomes the retina. The lens forms from the lens placode, a disclike swelling that arises when the optic cup induces thickening of overlying ectoderm (step 2). The centre of the lens placode sinks inward toward the optic cup, and its edges eventually fuse together, forming the lens vesicle, a ball of cells (step 3).

The developing lens cells begin to synthesize crystallin, a fibrous protein that collects into clear, glassy deposits. Lens cells finally lose their nuclei and form the elastic, crystal-clear lens.

As the lens develops, it contacts the overlying ectoderm that has closed over it. In response, the ectoderm cells form the cornea by losing their pigment granules and becoming clear. Eventually, the developing cornea joins with the edges of the optic cup to complete the primary structure of the eye (step 4). Other cells contribute to accessory structures of the eye. Mesoderm and neural crest cells contribute to reinforcing tissues in the wall of the eye and the muscles that move the

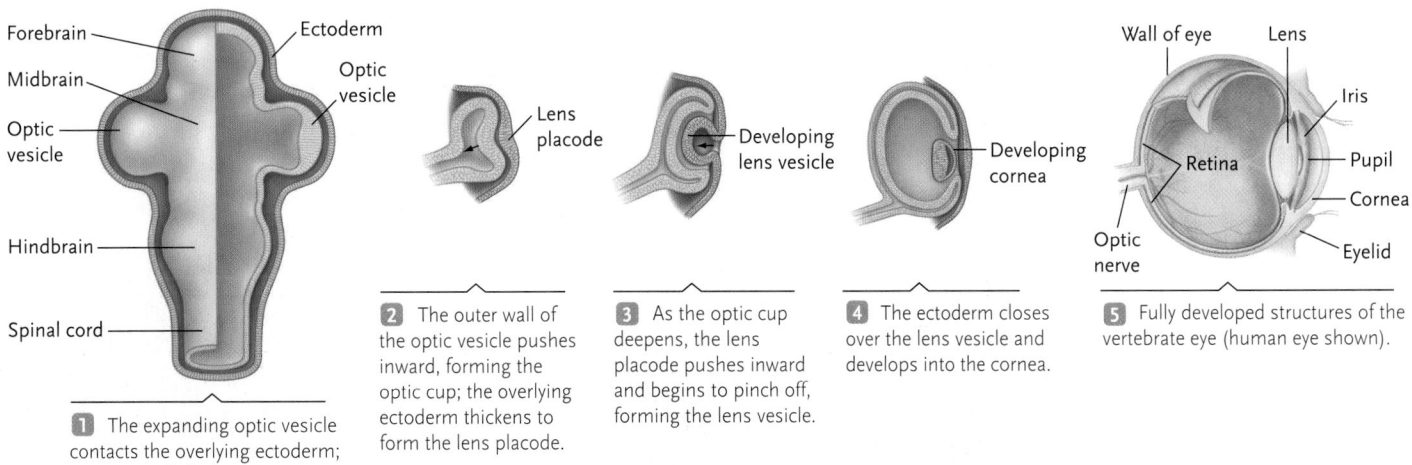

1 The expanding optic vesicle contacts the overlying ectoderm; its outer wall thickens.

2 The outer wall of the optic vesicle pushes inward, forming the optic cup; the overlying ectoderm thickens to form the lens placode.

3 As the optic cup deepens, the lens placode pushes inward and begins to pinch off, forming the lens vesicle.

4 The ectoderm closes over the lens vesicle and develops into the cornea.

5 Fully developed structures of the vertebrate eye (human eye shown).

Figure 42.18
Stages in the development of the vertebrate eye from the optic vesicle of the brain and the overlying ectoderm.

eye. Figure 42.18, page 1023, step 5, shows a fully developed vertebrate eye.

Initial induction by optic vesicles is necessary for the development of the eye. If an optic vesicle is removed before lens formation, ectoderm fails to develop into the lens placode and vesicle. Moreover, placing a removed optic vesicle under the ectoderm in other regions of the head causes a lens to form in the new location. If ectoderm over an optic vesicle is removed and ectoderm from elsewhere in the embryo is grafted in its place, a normal lens develops in the grafted ectoderm. This occurs even though in its former location it would not differentiate into lens tissue.

Eye development also demonstrates differentiation. Ectoderm cells induced to form the lens synthesize crystallin. In other locations, ectoderm cells synthesize mainly keratin, a different protein. Keratin is a component of surface structures such as skin, hair, feathers, scales, and horns. In response to induction by the optic vesicle, genes of ectoderm cells coding for crystallin are activated, but genes coding for keratin are not.

42.4c Apoptosis: Programmed Cell Death

Induction and differentiation build complex, specialized organs from three fundamental tissue types. Apoptosis (see Chapter 8), programmed cell death, complements these processes by removing tissues needed during development but not present in the fully formed organ. Apoptosis plays an important role in the development of both invertebrates and vertebrates. The development of wings and hind feet in short-tailed fruit bats (*Carollia perspicillata*) **(Figure 42.19)** demonstrates how apoptosis in the hind limbs leads to "normal" mammalian feet, but wings on the forelimbs.

STUDY BREAK

1. What processes and embryonic layers are involved in the development of the eye?
2. How does the neural tube develop?

Figure 42.19

Bone morphogenetic proteins (Bmps) trigger apoptosis of interdigital mesenchyme. In short-tailed fruit bats (*Carollia perspicillata*), apoptosis occurs in the hind feet where there are no antagonists to Bmps. Apoptosis does not occur in the forelimbs of the bats because antagonists to Bmps inhibit apoptosis. Compared here are expressions of Bmps inhibitors in the forelimbs and hind feet of the bats. Roman numerals indicate digit numbers.

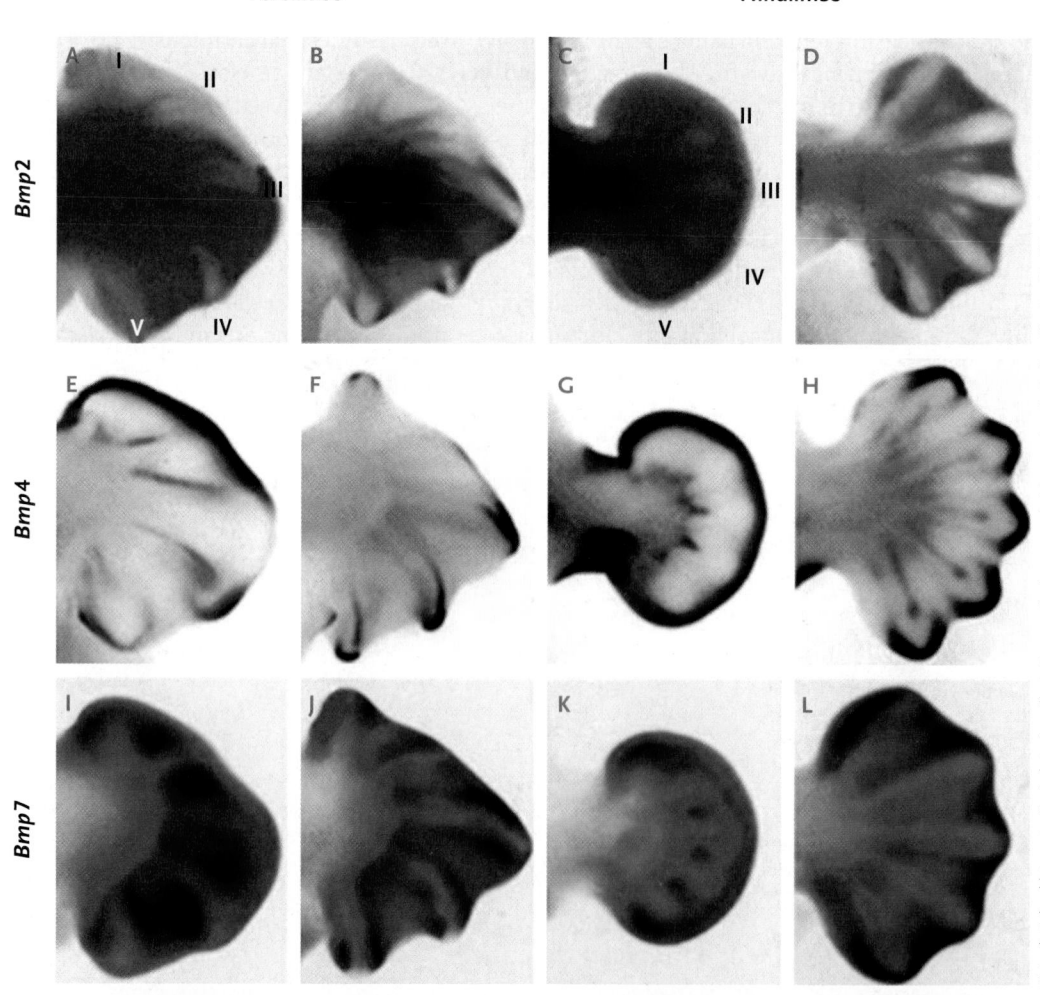

"Interdigital webbing retention in bat wings illustrates genetic changes underlying amniote limb diversification," by Weatherbee, Behringer, Rasweiler, and Niswander. *PNAS* October 10, 2006 vol. 103 no. 41, 15103–15107. Copyright 2006 National Academy of Sciences, U.S.A.

42.5 Embryonic Development of Humans and Other Mammals

The embryonic development of humans is representative of placental mammals. In the uterus, the embryo is nourished by the placenta, which supplies oxygen and nutrients to the embryo and carries carbon dioxide and nitrogenous wastes away from it.

Pregnancy or gestation, the period of mammalian *in utero* development, varies among species. Larger mammals bearing larger young tend to have longer gestation periods. From fertilization to birth, pregnancy lasts about 600 days in elephants, about 365 days in blue whales, and 21 days in hamsters.

In humans, gestation takes an average of 266 days, about 38 weeks. Because the date of fertilization can be difficult to establish, human gestation is usually calculated from the beginning of the menstrual cycle in which fertilization took place. The nine-month period is divided into three **trimesters**, each three months long.

Major developmental events in human gestation—cleavage, gastrulation, and organogenesis—take place during the first trimester. By week 4, the embryo's heart is beating, and by the end of week 8, the major organs and organ systems have formed. From this point until birth, the developing human is called a **fetus**. Only 5 cm long by the end of the first trimester, the fetus grows during the second and third trimesters to an average length of 50 cm and an average mass of 3.5 kg.

Cleavage occurs during the passage of the developing embryo down the fallopian tube and while it is still enclosed in the zona pellucida, the original coat of the egg **(Figure 42.20)**.

By day 4, the morula, a ball of 16 to 32 cells, has been produced. By the time the endometrium (uterine lining) is ready for implantation (about seven days after ovulation), the morula has reached the uterus and, through further cell divisions and differentiation, has become a blastocyst. The **blastocyst** is a hollow ball of about 120 cells in a single layer, with a fluid-filled cavity, the blastocoel, which has a dense mass of cells localized on one side. This **inner cell mass** will become the embryo itself, whereas the outer single layer of cells of the blastocyst, the **trophoblast**, will become tissues that support development of the embryo in the uterus.

When ready to implant, the blastocyst breaks out of the zona pellucida and sticks to the endometrium on its inner cell mass side **(Figure 42.21a, p. 1026)**. Implantation begins when the trophoblast cells that overlie the inner cell mass secrete proteases that digest pathways between endometrial cells. Dividing trophoblast cells fill in the digested spaces, appearing as fingerlike projections into the endometrium. These cells continue to digest nutrient-rich endometrial cells, producing a hole in the endometrium for the blastocyst and releasing nutrients for the developing embryo after it has

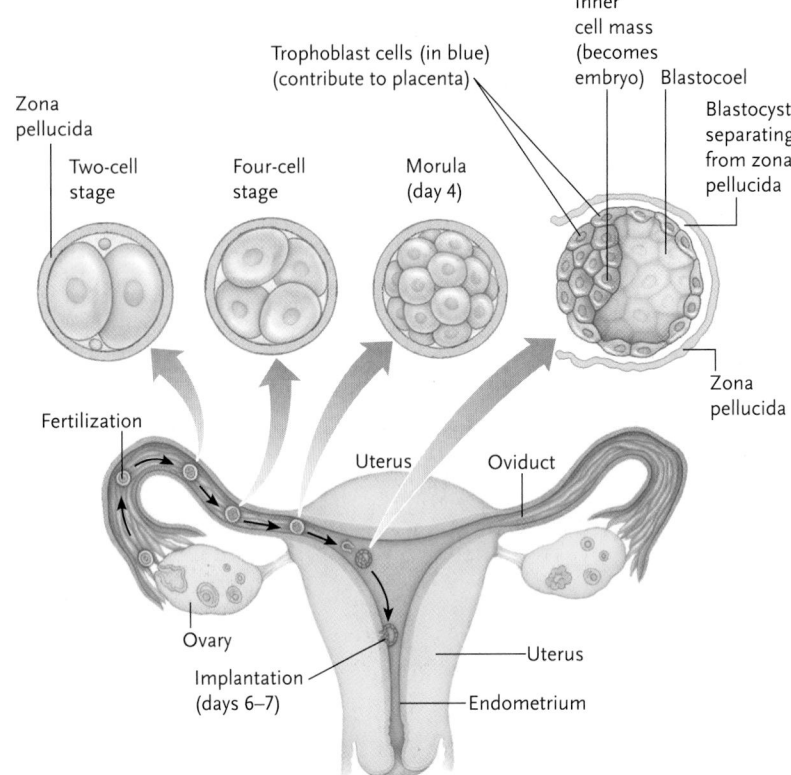

Figure 42.20
Early stages in the development of the human embryo.

consumed the small amount of yolk contained in egg cytoplasm. While the blastocyst burrows into the endometrium, the inner cell mass separates into the embryonic disc, which consists of two distinct cell layers (see Figure 42.21a). The epiblast, the layer farther from the blastocoel, gives rise to the embryo proper. The hypoblast, the layer nearer the blastocoel, generates part of the extra-embryonic membranes. When implantation is complete, the blastocyst has completely burrowed into the endometrium and is covered by a layer of endometrial cells (Figure 42.21b).

Gastrulation proceeds as in birds (see Figure 42.14, p. 1020), with the formation of a primitive streak in the epiblast. Soon after the inner cell mass separates into epiblast and hypoblast, a layer of cells separates from the epiblast along its top margin (see Figure 42.21b). The amniotic cavity is the fluid-filled space created by the separation. The layer of cells forming its roof becomes the amnion, which expands until it completely surrounds the embryo, suspending it in amniotic fluid.

Also as in birds, the hypoblast develops into the yolk sac. In mammals, the mesoderm of the yolk sac gives rise to the blood vessels in the embryonic portion of the placenta. The allantois stores nitrogenous wastes in birds, but it is a small, vestigial sac in human embryos because most nitrogenous wastes are transferred across the placenta to the mother via blood vessels in the placenta and the umbilical cord.

a. Days 6–7

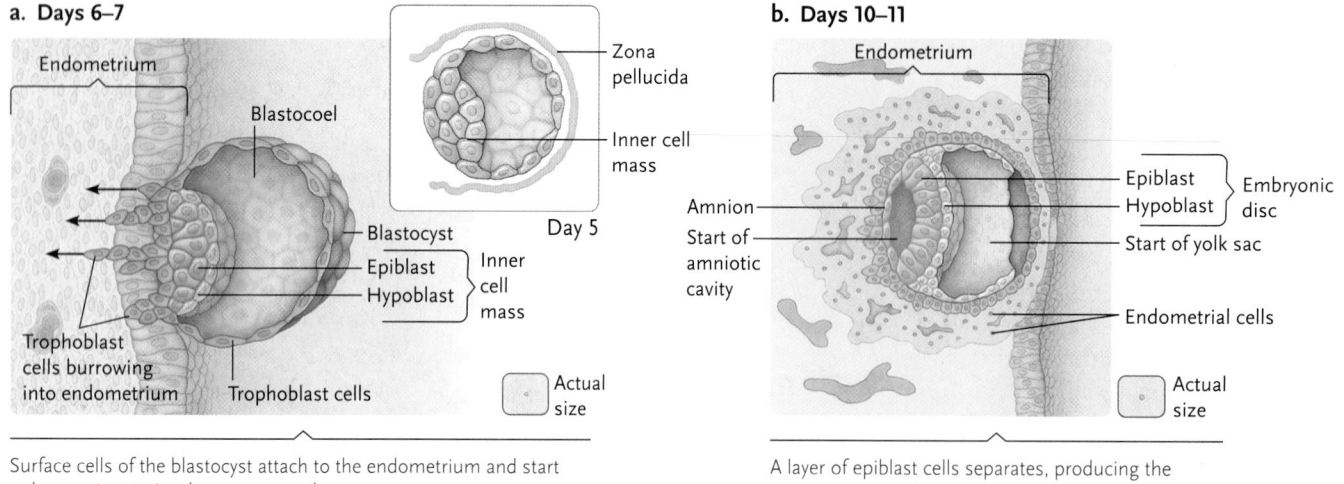

Endometrium
Blastocoel
Zona pellucida
Inner cell mass
Day 5
Blastocyst
Epiblast
Hypoblast
Inner cell mass
Trophoblast cells burrowing into endometrium
Trophoblast cells
Actual size

Surface cells of the blastocyst attach to the endometrium and start to burrow into it. Implantation is under way.

b. Days 10–11

Endometrium
Amnion
Start of amniotic cavity
Epiblast
Hypoblast
Embryonic disc
Start of yolk sac
Endometrial cells
Actual size

A layer of epiblast cells separates, producing the amniotic cavity. The cells above the cavity become the amnion, which eventually surrounds the embryo. The hypoblast begins to form around the yolk sac.

c. Day 12

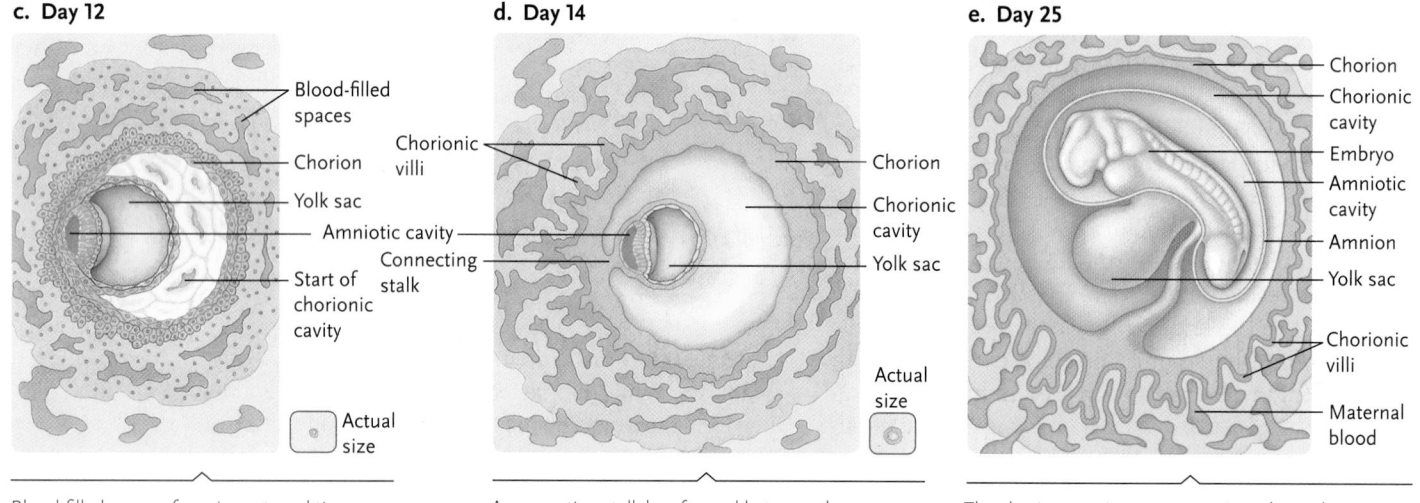

Blood-filled spaces
Chorion
Yolk sac
Amniotic cavity
Start of chorionic cavity
Actual size

Blood-filled spaces form in maternal tissue. The chorion forms, derived from trophoblast cells, and encloses the chorionic cavity.

d. Day 14

Chorionic villi
Chorion
Chorionic cavity
Connecting stalk
Yolk sac
Actual size

A connecting stalk has formed between the embryonic disc and chorion. Chorionic villi, which will be features of a placenta, start to form.

e. Day 25

Chorion
Chorionic cavity
Embryo
Amniotic cavity
Amnion
Yolk sac
Chorionic villi
Maternal blood

The chorion continues to grow into the endometrium, producing the chorionic villi. The chorion growth stimulates blood vessels of the endometrium to grow into the maternal circulation of the placenta.

f. Day 45

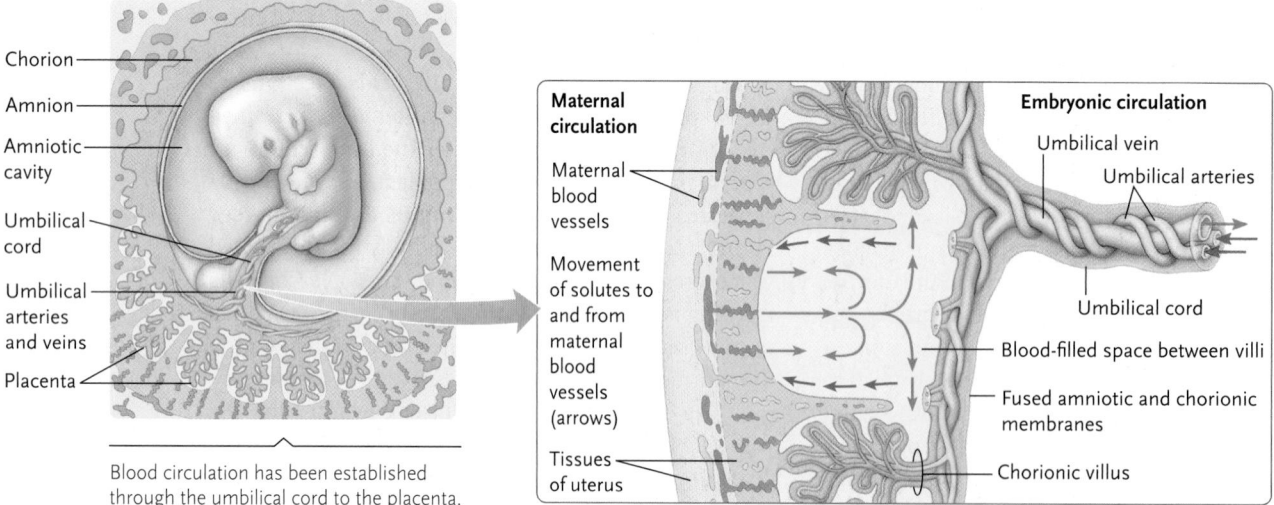

Chorion
Amnion
Amniotic cavity
Umbilical cord
Umbilical arteries and veins
Placenta

Blood circulation has been established through the umbilical cord to the placenta.

Maternal circulation
Maternal blood vessels
Movement of solutes to and from maternal blood vessels (arrows)
Tissues of uterus

Embryonic circulation
Umbilical vein
Umbilical arteries
Umbilical cord
Blood-filled space between villi
Fused amniotic and chorionic membranes
Chorionic villus

Figure 42.21
Implantation of a human blastocyst in the endometrium of the uterus and the establishment of the placenta.

While the amnion is expanding around the embryo, blood-filled spaces form in maternal tissue, and trophoblast cells grow rapidly around both the embryo and the amnion to form the chorion (Figure 42.21c), the membrane that forms most of the embryonic portion of the placenta. Next, a connecting stalk forms between the embryonic disc and the chorion, which begins to grow into the endometrium as finger-like extensions called **chorionic villi** (singular, *villus*) (Figure 42.21d). Chorionic villi increase the surface area of the chorion. The placenta forms in the area where these villi grow into the endometrium. As the chorion develops, mesodermal cells of the yolk sac grow into it and form a rich network of blood vessels, the embryonic circulation of the placenta. At the same time, the expanding chorion stimulates the blood vessels of the endometrium to grow into the maternal circulation of the placenta (Figure 42.21e).

Within the placenta of humans, apes, monkeys, and rodents, the maternal circulation opens into spaces where maternal blood directly bathes capillaries coming to the placenta from the embryo (Figure 42.21f). (Other mammals have different types of placentas.) Embryonic circulation remains closed so that the embryonic blood and the maternal blood do not mix directly. This isolation prevents the mother from developing an immune reaction against cells of the embryo, which may be recognized as foreign. Eventually, the placenta and its blood circulation grow to cover about a quarter of the inner surface of the enlarged uterus and reach the size of a dinner plate.

When the amnion forms (see Figure 42.21e), the embryo remains connected to the developing placenta through the **umbilicus**, a cord of tissue. Blood vessels in the umbilical cord conduct blood between the embryo and the placenta (see Figure 42.21f, inset). Within the placenta, nutrients and oxygen pass from the mother's circulation into the circulation of the embryo. Besides nutrients and oxygen, many other substances taken in by the mother, such as alcohol, caffeine, drugs, pesticide residues, and toxins in smog and cigarette smoke, can pass from mother to embryo. Carbon dioxide and nitrogenous wastes pass from the embryo to the mother and are disposed of by the mother's lungs and kidneys.

Cells from the embryonic portion of the placenta or from the amniotic fluid are derived from the embryo. To test for the presence of genetic diseases such as cystic fibrosis or Down syndrome, these cells can be obtained by chorionic villus sampling or by amniocentesis (*centesis* = puncture, referring to the use of a needle, which is pushed through the abdominal wall to obtain fluid from the amniotic cavity). Chorionic villus sampling can be carried out as early as the eighth week of pregnancy, compared with 14 weeks for amniocentesis.

42.5a Birth: The Fetus Leaves the Mother

By the end of its fourth week, a human embryo is 3 to 5 mm long, 250 to 500 times the size of the zygote **(Figure 42.22a)**. It has a tail and gill arches, embryonic

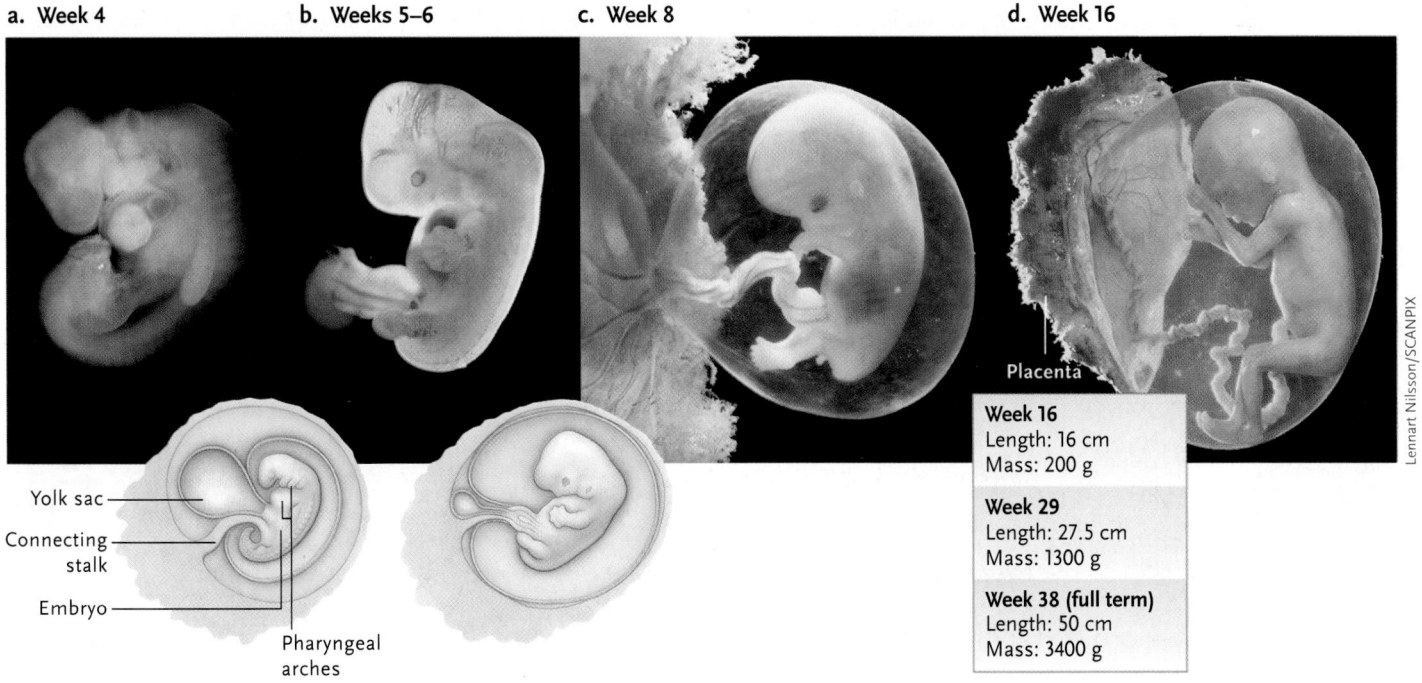

a. Week 4   b. Weeks 5–6   c. Week 8   d. Week 16

Yolk sac
Connecting stalk
Embryo
Pharyngeal arches

Placenta

Week 16
Length: 16 cm
Mass: 200 g

Week 29
Length: 27.5 cm
Mass: 1300 g

Week 38 (full term)
Length: 50 cm
Mass: 3400 g

Lennart Nilsson/SCANPIX

Figure 42.22
The human embryo at various stages of development, beginning at week 4. The chorion has been moved aside to reveal the embryo in the amnion at week 8 and week 16. By week 16, movements begin as nerves make functional connections with the forming muscles.

features of all vertebrates (see Chapter 28). Gill arches contribute to the formation of the face, neck, mouth, nasal cavities, larynx, and pharynx. After five to six weeks, most of the tail has disappeared, and the embryo begins to be a recognizable human form (Figure 42.22b). At eight weeks, the embryo, now a fetus, is about 2.5 cm long (Figure 42.22c). Its organ systems have formed, and its limbs, with fingers or toes at their ends, have developed.

After about 38 weeks, fetal growth comes to a close, the cervix of the uterus softens, and the fetus typically turns so that its head is downward, pressed against the cervix. At this time, a steep rise in levels of estrogen secreted by the placenta cause uterine cells to express the gene for the receptor of the hormone *oxytocin* (secreted by the pituitary). Receptors become inserted into the plasma membranes of uterine cells. Oxytocin binds to its receptors, triggering contractions of smooth muscle cells of the uterine wall, beginning the rhythmic contractions of labour. These contractions mark the beginning of the three steps culminating in birth or **parturition** (*parturire* = to be in labour).

Contractions push the fetus against the cervix and stretch its walls (**Figure 42.23,** step 1). In response, stretch receptors in the walls send nerve signals to the hypothalamus, which responds by stimulating the pituitary to secrete more oxytocin. Oxytocin stimulates more forceful contractions of the uterus, pressing the fetus more strongly against the cervix and further stretching its walls. The positive feedback cycle continues, steadily increasing the strength of the uterine contractions.

As the contractions force the head of the fetus through the cervix (step 2), the amniotic membrane bursts, releasing the amniotic fluid. Usually after 12 to 15 hours from the onset of uterine contractions, the head passes entirely through the cervix. Once the head is through, the rest of the body follows quickly and the entire fetus is forced out through the vagina, still connected to the placenta by the umbilical cord (step 3).

After the baby takes its first breath, the umbilical cord is cut and tied off by the birth attendant. Uterine contractions continue expelling the placenta and any remnants of the umbilical cord and embryonic membranes as the afterbirth, usually within 15 to 60 minutes after the infant's birth. The short length of umbilical cord still attached to the infant dries and shrivels within a few days. Eventually, it separates entirely and leaves a scar, the umbilicus or navel, to mark its former site of attachment during embryonic development. Immediately after birth, some mammals (e.g., *Gazella* species) can stand and are soon able to run. The newborns of these precocial species contrast with those of altricial species, which are immobile and helpless for some considerable time after birth. The same terms, *precocial* and *altricial*, also apply to other animals.

Many people believe that all mammals give birth to live young. In reality, mammals such as the duck-billed platypus and spiny anteater lay eggs. These mammals are the living representatives of an ancient group of mammals, the monotremes. Although they lay eggs, female platypuses and spiny anteaters feed their young milk like other mammals do.

42.5b Milk: Food for the Young

Before birth, estrogen and progesterone secreted by the placenta stimulate the growth of the mammary glands in the mother's breasts. But high levels of

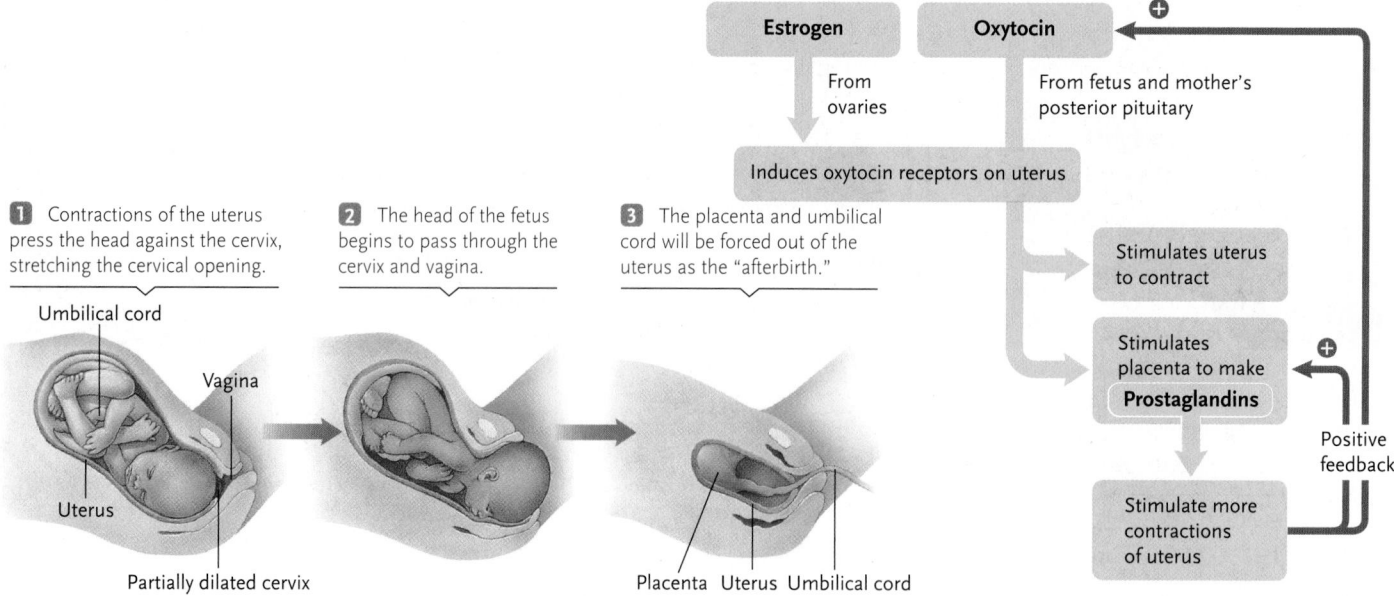

1 Contractions of the uterus press the head against the cervix, stretching the cervical opening.

2 The head of the fetus begins to pass through the cervix and vagina.

3 The placenta and umbilical cord will be forced out of the uterus as the "afterbirth."

Umbilical cord
Vagina
Uterus
Partially dilated cervix

Placenta Uterus Umbilical cord

Estrogen
Oxytocin
⊕
From ovaries
From fetus and mother's posterior pituitary
Induces oxytocin receptors on uterus
Stimulates uterus to contract
Stimulates placenta to make **Prostaglandins**
⊕
Positive feedback
Stimulate more contractions of uterus

Figure 42.23
Birth of the fetus. Hormonal events of birth are at the top, and physical events are at the bottom.

these hormones prevent mammary glands from responding to **prolactin**, the hormone secreted by the pituitary that stimulates the glands to produce milk. After birth and the release of the placenta, levels of estrogen and progesterone in the mother's bloodstream fall steeply, and the breasts begin to produce milk (stimulated by prolactin) and secrete it (stimulated by oxytocin).

Continued milk secretion depends on whether the infant suckles. Stimulation of the nipples sends nerve impulses to the hypothalamus, which responds by signalling the pituitary to release a burst of prolactin and oxytocin. Hormonal stimulation of milk production and secretion continues as long as the infant is breast-fed.

42.5c Gonadal Development: The Gender of the Fetus

Gonads and their ducts begin to develop in the fetus during week 4 of gestation. Until week 7, male and female embryos have the same set of internal structures derived from mesoderm, including a pair of gonads **(Figure 42.24a)**. Each gonad is associated with two primitive ducts, the **Wolffian duct** and the **Müllerian duct**, that lead to a cloaca. These internal structures are bipotential because they can develop into either male or female sexual organs.

The presence or absence of a Y chromosome determines whether the internal structures develop into male or female sexual organs. In a fetus with an XY combination of sex chromosomes, *SRY* (the sex-determining region of the Y), a single gene on the Y chromosome, becomes active in week 7. The protein encoded by *SRY* induces a molecular switch that causes primitive gonads to develop into testes. Fetal testes secrete two hormones, testosterone and the **anti-Müllerian hormone** (*AMH*). Testosterone stimulates development of Wolffian ducts into a male reproductive tract, including the epididymis, vas deferens, and seminal vesicles (Figure 42.24b). *AMH* causes the Müllerian ducts to degenerate and disappear. Testosterone also stimulates development of male genitalia.

In a fetus with XX chromosomes, no *SRY* protein is produced. The primitive gonads, under the influence of estrogens and progesterone secreted by the placenta, develop into ovaries. Müllerian ducts develop into oviducts, the uterus, and part of the vagina, and the Wolffian ducts degenerate and disappear (Figure 42.24c). Female sex hormones also stimulate the development of the external female genitalia (see "Hormones and External Genitalia," p. 1030).

42.5d Further Development

Once fetal development is over, the newborn animal follows a prescribed course of further growth and

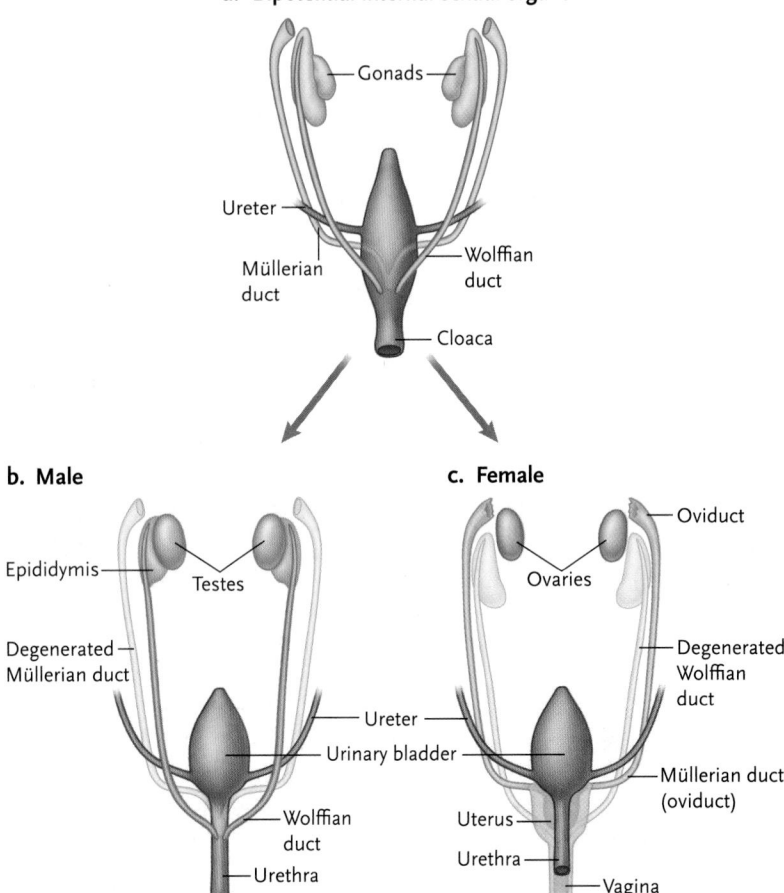

Figure 42.24

Development of the internal sexual organs of males and females from common bipotential origins.

development leading to the mature adult. In humans, internal and external sexual organs mature and secondary sexual characteristics appear at puberty. Similar changes occur in most mammals.

There are many examples among different animal groups of developmental changes that take place after hatching or birth. In some cases, offspring hatch in forms distinctly different in structure from the adult. Examples among invertebrates are insects such as *Drosophila* and butterflies, in which eggs hatch to produce larvae that undergo metamorphosis into adults. Some frogs hatch as tadpoles, which undergo metamorphosis to produce adults.

STUDY BREAK

1. Define the terms *blastocyst*, *chorionic villi*, and *parturition*.
2. What is a placenta? Name some animals with placentas.
3. What is *SRY*? What role does it play?

Hormones and External Genitalia

Figure 1
Crocuta crocuta, *a spotted hyena.*

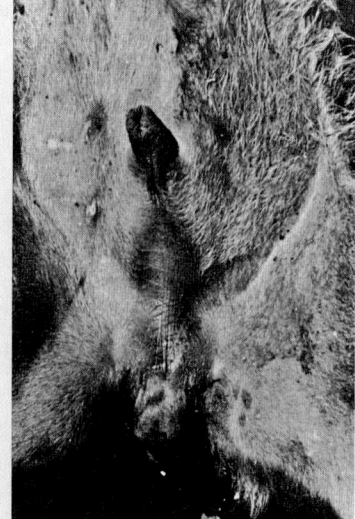

Racey, P.A. and J.S. Skinner, "Endocrine aspects of sexual mimicry in spotted hyaenas, Crocuta crocuta," *Journal of Zoology*, 187:315–326. Copyright © 2009, John Wiley and Sons.

Figure 2
In spotted hyenas (C. crocuta), the external genitalia of a nulliparous female (left) resembles that of a male (right). Note the peniform clitoris and pseudoscrotum on the female.

Many Africans believe that spotted hyenas (*Crocuta crocuta*, **Figure 1**) are hermaphroditic because the females have external genitalia resembling those of the males. Specifically, the clitoris is peniform, and there is a pseudoscrotum **(Figure 2)**. The combined effect means that anyone looking at a spotted hyena easily confuses males and females. Confusion turns to puzzlement when the apparent male is obviously lactating, nursing young. Spotted hyenas are not hermaphroditic. Males mate with females, and fertilized eggs develop into fetuses that are born as in other placental mammals.

How do females come to look like males? In mammals, the neutral condition for genitalia is the arrangement typical of females. The derived condition is what we associate with males. Genetically, female embryos exposed to testosterone during gestation develop malelike genitalia. Blood samples obtained from pregnant female spotted hyenas had relatively high levels of circulating testosterone, 5α-dihydrotestosterone, and androstenedione, albeit not as high as in males. Male and female fetuses experience the same levels of maternal androgens; therefore, females have malelike genitalia. What selective advantage would female spotted hyenas gain from looking like males? Spotted hyenas are social animals that live in clans. As adults, females are larger and more aggressive than males. During greeting ceremonies, when members of a clan meet after a separation, individuals sniff one another's genitals. During these encounters, males erect their penises and females their peniform clitorises. Females appear to dominate spotted hyena societies even though their levels of testosterone and 5α-dihydrotestosterone are lower than in adult males. This does not support the proposal that females are more aggressive because of levels of circulating male hormones. Spotted hyena cubs are precocial and aggressive to the point of siblicide, perhaps reflecting hormone levels at birth. Masculinization of female genitalia could be a side effect of selection for aggressive neonates.

Masculinization of female genitalia occurs in some other members of the order Carnivora, for example, the fossa (*Cryptoprocta ferox*) from Madagascar **(Figure 3).** Transient masculinization of young female (but not adult female) fossas could allow them to avoid sexual harassment by adult males and to escape from aggression by adult females.

a.
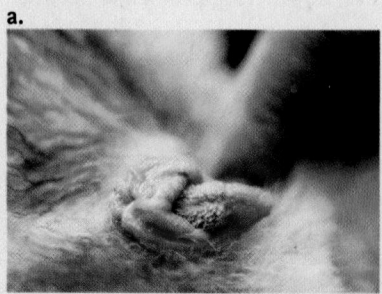

b.
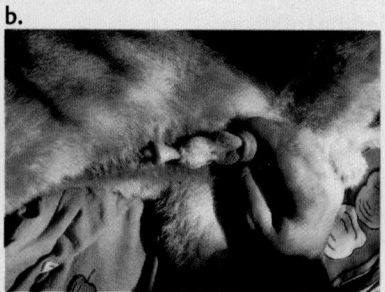

c.
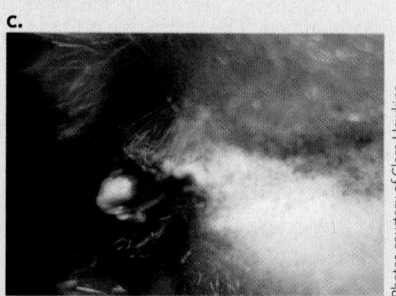

Photos courtesy of Clare Hawkins

Figure 3
In their external genitalia, specifically their clitorises, (a) subadult female fossas (Cryptoprocta ferox) resemble (b) males, whereas (c) adult females do not.

M.B. Fenton

42.6 Cellular Basis of Development

Orientation and rate of mitotic cell division have special significance in the development of the shape, size, and location of organ systems of the embryo. Regulation of the orientation and rate occurs at all stages of development.

Orientation of cell division refers to the angles at which daughter cells are added to older cells as development proceeds. Orientation is determined by the location of the furrow separating the cytoplasm after mitotic division of the nucleus (see Chapter 8). The furrow forms in alignment with the spindle midpoint so that when the spindle is centrally positioned in the cell, the furrow leads to symmetrical division of the cell. When the spindle is displaced to one end of the cell, the furrow leads to asymmetrical division into a smaller cell and a larger cell. Little is known about how spindle positioning is regulated.

The rate of cell division primarily reflects the time spent in the G_1 period of interphase (see Chapter 8). Once DNA replication begins, the rest of the cell cycle takes the same time in all cells of a species. As an embryo develops and cells differentiate, the time spent in interphase increases and varies in different cell types. Therefore, different cell types proliferate at various rates as they differentiate, giving rise to tissues and organs with different cell numbers. When fully differentiated, some cells remain fixed in interphase and stop replicating DNA or dividing. Nerve cells in the mammalian brain and spinal cord stop dividing once the nervous system is fully formed. Ultimately, the rate of cell division is under genetic control.

Frog egg cleavage provides examples of how both changes in orientation and rate of mitotic division affect development. The first two cleavages start at the animal pole and extend to the vegetal pole, producing four equal blastomeres (see Figure 42.8, p. 1015). The third cleavage occurs equatorially, but because of yolk in the vegetal region of the embryo, this cleavage furrow forms toward the animal pole and produces an eight-cell embryo with four small blastomeres near the animal pole and four large blastomeres in the vegetal region. Blastomeres in the animal region of the embryo proceed to divide rapidly, whereas blastomeres in the vegetal region divide more slowly because division is inhibited by yolk. The resulting morula consists of an animal region with many small cells and a vegetal region with relatively few but larger blastomeres.

42.6a Microtubules and Microfilaments: Movements of Cells

Embryonic cells undergo changes in shape that generate movements, such as the infolding of surface layers to produce endoderm or mesoderm. Entire cells also move during embryonic growth, both singly and in groups. Movements are also produced by changes

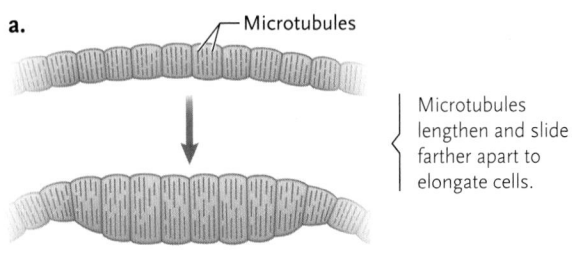

a.

Microtubules lengthen and slide farther apart to elongate cells.

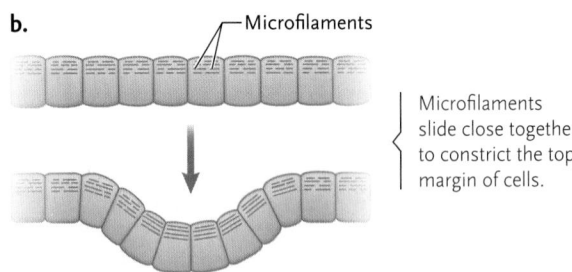

b.

Microfilaments slide close together to constrict the top margin of cells.

Figure 42.25
The roles of **(a)** microtubules and **(b)** microfilaments in the changes in cell shape that produce developmental movements.

in rates of growth or by breakdowns of microtubules and microfilaments. Changes in both cell shape and cell movement play important roles in cleavage, gastrulation, and organogenesis.

42.6b Change in Cell Shape: Adjusting to New Roles

Changes in cell shape typically result from reorganization of the cytoskeleton. During development of the neural plate in frogs, the ectoderm flattens and thickens and cells in the ectoderm layer change from cubelike to columnar in shape **(Figure 42.25a)**.

Sinking of the neural plate downward along its midline reflects changes in cell shape from columnar to wedgelike (Figure 42.25b). As one end of each cell narrows, the entire cell layer invaginates (is forced inward). How does this occur? Each wedge-shaped cell contains a group of microfilaments arranged in a circle at the top. Microfilaments slide over each other, tightening the ring like a drawstring and narrowing the top of the cell. If an experimenter adds cytochalasin, a chemical that interferes with microfilament assembly, to the cells, the microfilament circle disperses, and invagination does not occur.

42.6c Movements of Whole Cells to New Positions in the Embryo

Cell movements during gastrulation and long-distance migrations of neural crest cells are striking examples of movements of whole cells during embryonic development. These movements involve coordinated activity by microtubules and microfilaments. The typical

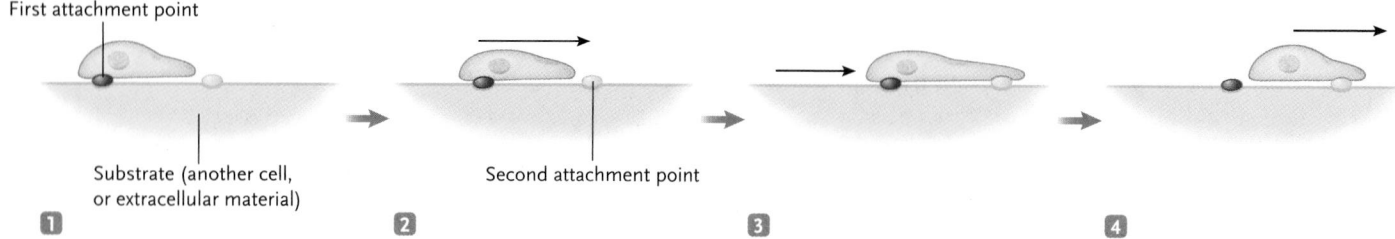

First attachment point

Substrate (another cell,
or extracellular material)

Second attachment point

1 **2** **3** **4**

Figure 42.26
The cycle of attachment, stretching, and contraction by which a cell moves over other cells or extracellular materials in embryos.

pattern of movement is a repeating cycle of extension, anchoring, and contracting. First, a cell attaches to the substrate (**Figure 42.26,** step 1), and then it moves forward by elongating from the point of attachment (step 2). The cell then makes a new attachment at the advancing tip (step 3) and contracts until the rearmost attachment breaks (step 4). The front attachment now serves as the base for another movement.

How do the cells know where to go? Typically, cells migrate over the surfaces of stationary cells in one of the embryo's layers. In many developmental systems, migrating cells follow tracks formed by molecules of the ECM, secreted by cells along the route over which they travel. Fibronectin, an important track molecule, is a fibrous, elongated protein of the ECM. Migrating cells recognize and adhere to fibronectin, and, in response, internal changes in cells trigger their movement in a direction based on the alignment of fibronectin molecules.

Some migrating cells follow concentration gradients rather than molecular tracks. Gradients are created by diffusion of molecules (often proteins) released by cells in one part of an embryo. Cells with receptors for the diffusing molecule follow the gradient toward or away from the source.

Selective cell adhesion—the ability of an embryonic cell to make and break specific connections to other cells—is closely related to cell movement. As development proceeds, many cells break initial adhesions, move, and form new adhesions in different locations. Final cell adhesions hold the embryo in its correct shape and form. Junctions of various kinds, including tight, anchoring, and gap junctions, reinforce final adhesions. Selective cell adhesions were first demonstrated in a classic experiment by Johannes Holtfreter and P. L. Townes (see "Making Cell Connections").

Many cell surface proteins are responsible for selective cell adhesions, including **cell adhesion molecules** (CAMs) and **cadherins (calcium-dependent adhesion molecules)**. Cadherins require calcium ions to set up adhesions—hence their name. As cells develop, different types of CAMs or cadherins appear or disappear from their surfaces as they make and break cell adhesions. The changes reflect alterations in gene activity, often in response to molecular signals arriving from other cells. During early development in

a chick, cells of the ectoderm are held together by E- and N-cadherins. As the neural plate appears, cells destined to form the neural tube lose their cadherins, and N-CAMs appear on their surfaces. As these surface molecules appear, the neural tube cells break loose from the ectoderm and adhere to each other to form the neural tube. If N-cadherin is added experimentally to ectoderm cells, the neural tube stays anchored and never separates.

42.6d Induction: Interactions between Cells

Induction is the process in which a group of inducing cells causes or influences a nearby group of responding cells to follow a particular developmental pathway. Induction is the major process responsible for determination, in which the developmental fate of a cell is set. Induction occurs through the combination of signal molecules with surface receptors on the responding cells. The signal molecules may be located on the surface of the inducing cells or released by inducing cells. The surface receptors are activated by binding the signal molecules. In the activated form, signal molecules trigger internal response pathways that produce the developmental changes (see Chapter 14). Responses often include changes in gene activity.

In the 1920s, Hans Spemann and Hilde Mangold conducted the first experiments identifying induction in embryos. They found that if the dorsal lip of a newt embryo was removed and grafted into a different position of another embryo, cells moving inward from the dorsal lip induced a neural plate, a neural tube, and eventually an entire embryo at the new location (see "Spemann and Mangold's Experiment Demonstrating Induction in Embryos," p. 1034). They proposed that the dorsal lip is an organizer, acting on other cells to alter the course of development. This action is now known as induction. Spemann received the Nobel Prize in 1935 for his research. (Mangold had died in a tragic accident in 1924 when her kitchen gasoline heater exploded, the year their research paper was published. She would likely have also received the Nobel Prize, but it has only once been awarded posthumously.)

Spemann and Mangold's findings touched off a search for inducing molecules that must pass from inducing cells to responding cells. In 1992, researchers

Making Cell Connections: Selective Adhesion Properties of Cells

1. Holtfreter and Townes separated ectoderm, mesoderm, and endoderm tissue from amphibian embryos soon after the neural tube had formed. They used embryos from amphibian species that had cells of different colours and sizes, so they could follow under the microscope where each cell type ended up. (The colours shown here are for illustrative purposes only.)

2. The researchers placed the tissues individually in alkaline solutions, which caused the tissues to break down into single cells.

3. Holtfreter and Townes then combined suspensions of single cells in various ways. Shown here are ectoderm + mesoderm and ectoderm + mesoderm + endoderm. When the pH was returned to neutrality, the cells formed aggregates. Through a microscope, the researchers followed what happened to the aggregates on agar-filled Petri dishes.

RESULT: In time, the reaggregated cells sorted themselves with respect to cell type; that is, instead of the cell types remaining mixed, each cell type became separated spatially. That is, in the ectoderm + mesoderm mixture, the ectoderm moved to the periphery of the aggregate, surrounding mesoderm cells in the centre. In no case did the two cell types remain randomly mixed. The ectoderm + mesoderm + endoderm aggregate showed further that cell sorting in the aggregates generated cell positions reflecting the positions of the cell types in the embryo. That is, the endoderm cells separated from the ectoderm and mesoderm cells and became surrounded by them. In the end, the ectoderm cells were located on the periphery, the endoderm cells were internal, and the mesoderm cells were between the other two cell types.

Amphibian embryos of different species

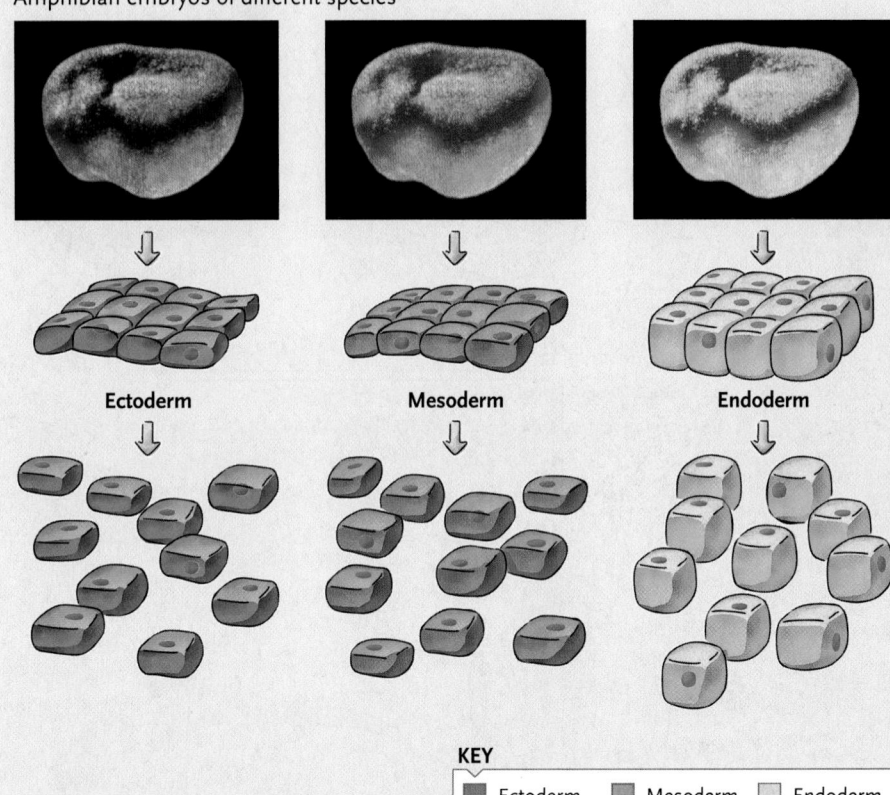

Ectoderm   Mesoderm   Endoderm

KEY
■ Ectoderm   ■ Mesoderm   □ Endoderm

Ectoderm + Mesoderm

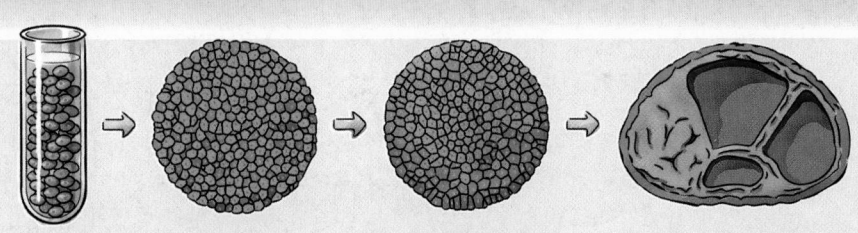

Ectoderm + Mesoderm + Endoderm

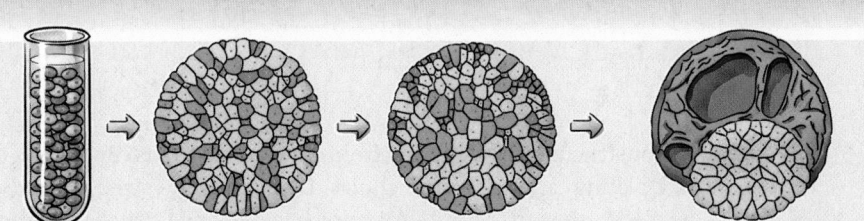

CONCLUSION: Holtfreter interpreted the results to mean that cells have selective affinity for each other; that is, cells have selective adhesion properties. Specifically, he proposed that ectoderm cells have positive affinity for mesoderm cells but negative affinity for endoderm cells, whereas mesoderm cells have positive affinity for both ectoderm cells and endoderm cells. In modern terms, these properties result from cell surface molecules that give cells specific adhesion properties.

Spemann and Mangold's Experiment Demonstrating Induction in Embryos

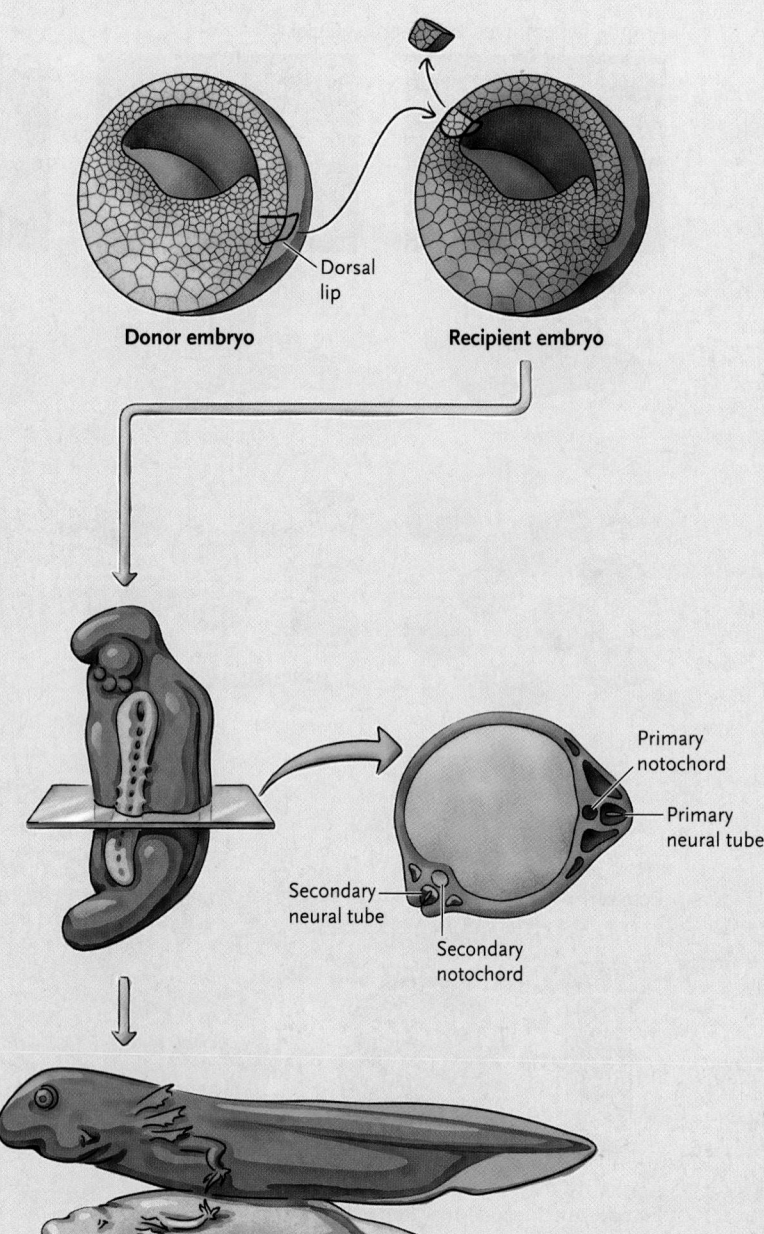

Donor embryo

Recipient embryo

Dorsal lip

Primary notochord

Primary neural tube

Secondary neural tube

Secondary notochord

EXPERIMENT: To investigate the process of induction, Hans Spemann and Hilde Mangold performed transplantation experiments with newt embryos, the results of which demonstrated that specific induction of development occurs in the embryos. The researchers removed the dorsal lip of the blastopore from one newt embryo and grafted it onto a different position—the ventral side—of another embryo. The two embryos were from different newt species that differed in pigmentation, allowing them to follow the fate of the tissue easily. The embryo with the transplant was allowed to develop.

CONCLUSION: The grafted dorsal lip of the blastopore induced a second gastrulation and subsequent development in the ventral region of the recipient embryo. The result demonstrated the ability of particular cells to induce the development of other cells.

constructed a DNA library from *Xenopus* gastrulas. By isolating and cloning cellular DNA in gene-sized pieces, they made mRNA transcripts of cloned genes and injected them into early *Xenopus* embryos in which the inducing ability of mesoderm had been destroyed by exposure to ultraviolet light. Some injected mRNAs, translated into proteins in the embryos, were able to induce formation of a neural plate and tube, leading to a normal embryo. More than 10 proteins that act as inducing molecules have been identified in the *Xenopus* system.

Differentiation produces specialized cells without the loss of genes. By this process, cells that have committed to a particular developmental fate by

the determination process (see Section 42.2) develop into specialized cell types with distinct structures and functions. As part of differentiation, cells concentrate on the production of molecules characteristic of the specific types. For example, 80 to 90% of the total protein synthesized by lens cells is crystallin.

Research into differentiation confirmed that as cells specialize, they retain all of the genes of the original egg cell. Except in rare instances, differentiation does not occur through selective gene loss. Robert Briggs and Thomas King, and later John B. Gurdon, used ultraviolet light to destroy the nucleus of a fertilized frog egg. A micropipette was then used to transfer a nucleus from a fully differentiated tissue, intestinal

a. Founder cells

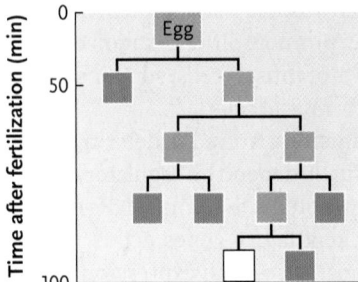

b. Cell lineage for intestinal cells

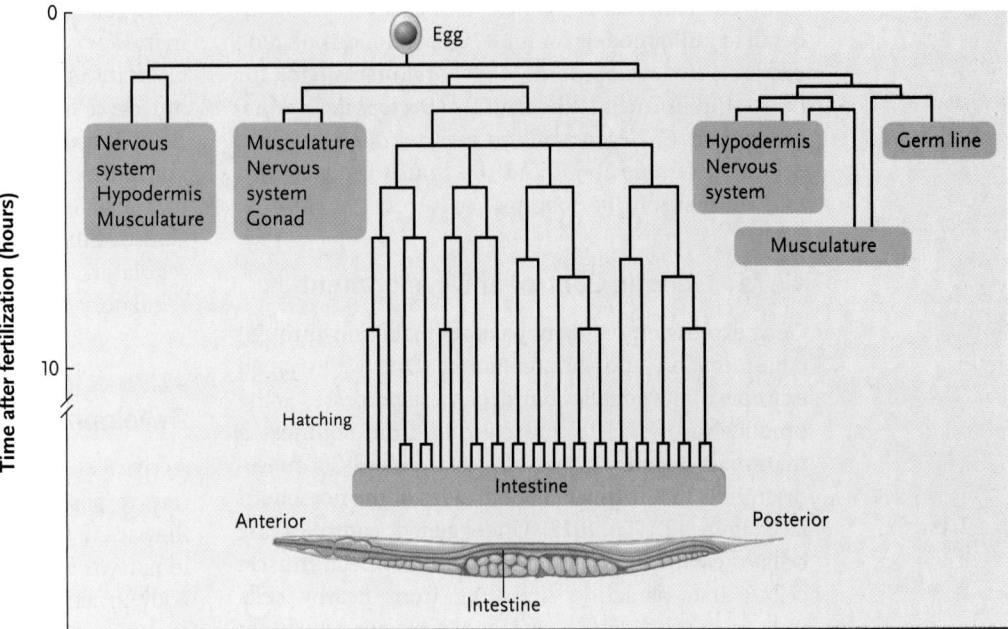

Figure 42.27

Cell lineages of *C. elegans*. **(a)** The founder cells (blue) produced in early cell divisions from which all adult somatic cells are produced. The white cell gives rise to germ-line cells. **(b)** The cell lineage for cells that form the intestine. The detailed lineages for the other parts of the adult are not shown.

epithelium, to the enucleated egg. Some eggs receiving transplanted nuclei subsequently developed into normal tadpoles and adult frogs. This outcome is possible only if the differentiated intestinal cells retained the full complement of genes. This conclusion was extended to mammals in 1997 when Ian Wilmut and his colleagues successfully cloned a sheep (named Dolly) starting with an adult cell nucleus.

From the early days of studying development, embryologists focused on describing not only how embryos form and develop but also exactly how adult tissues and organs are produced from embryonic cells. An important goal of embryology was to trace cell lineages from embryo to adult. For most organisms, it is not possible to trace lineages at the individual cell level, primarily because of the complexity of the developmental process and the opacity of embryos. Experimenters developed **fate mapping**, mapping adult or larval structures onto the region of the embryo from which each structure developed. Fate mapping is done by following the development of living embryos under the microscope, either by using species in which the embryo is transparent or by marking cells so they can be followed. Cells may be marked with vital dyes (that do not kill cells), fluorescent dyes, or radioactive labels. Fate maps have been produced for *Xenopus,* the chick, and *Drosophila.*

In most cases, a fate map is not detailed enough to show how particular cells in the embryo gave rise to cells of the adult. The exception is the fate map of the nematode *Caenorhabditis elegans,* an organism with a fixed, reproducible developmental pattern. *C. elegans*

has a transparent body, and scientists have mapped the fate (traced the **cell lineage**) of every cell as the zygote divides and the resulting embryo differentiates into a 959-cell adult hermaphrodite or 1031-cell adult male **(Figure 42.27)**. All somatic cells of the adult can be traced from five somatic founder cells produced during early development. Knowing the cell lineages of *C. elegans* has been a valuable tool for research into the genetic and molecular control of development because mutants affecting development can be easily visualized.

STUDY BREAK

1. What role do microtubules and microfilaments play in development?
2. What happens during induction? Where does it occur?
3. How has work with *Caenorhabditis elegans* advanced our knowledge of developmental biology? What about *Xenopus*?

42.7 Genetic and Molecular Control of Development

Developmental biologists are interested in identifying and characterizing the genes involved in development and defining how gene products regulate and bring about elaborate events. One productive research approach has been to isolate mutants that affect developmental processes and then identify the genes involved, clone these genes, and analyze them in detail to build models for molecular functions of gene products in development. Model organisms used for these studies include the fruit fly (*Drosophila melanogaster*) and *C. elegans* among invertebrates, and the zebrafish (*Danio rerio*) and the house mouse (*Mus musculus*) among vertebrates.

42.7a Genetic Control of Development

Gene expression regulates changes that occur through determination and differentiation. One well-studied example of the genetic control of these processes is the production of skeletal muscle cells from somites in mammals **(Figure 42.28)**. Somites are blocks of mesoderm cells that form along both sides of the notochord (see Figure 42.17, p. 1022). Under genetic control, some cells of each somite differentiate into skeletal muscle cells. First, paracrine signalling from nearby cells induces somite cells to express the master regulatory gene, *myoD*. The product of *myoD* is the transcription factor MyoD. By turning on specific muscle-determining genes, the action of MyoD brings about determination of these cells, converting them to undifferentiated muscle cells called **myoblasts**. Among genes that MyoD regulates are *myogenin* and *MEF* genes. Both are regulatory genes, expressing transcription factors in myoblasts

that turn on yet another set of genes. The products of these genes, including myosin (a major protein involved in muscle contraction), promote differentiation of myoblasts into specific types of muscle cells, such as skeletal muscle cells or cardiac muscle cells.

Molecular mechanisms involved in determination and differentiation usually depend on regulatory genes that encode regulatory proteins that control the expression of other genes. Regulatory genes act as master regulators, and, in most cases, the expression of the regulatory genes is controlled by induction.

42.7b Gene Control of Pattern Formation: Developing a Body

As part of the signals guiding differentiation, cells receive positional information that tells them where they are in the embryo. Positional information is vital to **pattern formation**, the arrangement of organs and body structures in their proper three-dimensional relationships. Positional information is laid down primarily as concentration gradients of regulatory molecules produced by genetic control. In most cases, gradients of several different regulatory molecules interact to tell a cell, or a cell nucleus, where it is in the embryo. Genetic control of pattern formation is well documented in *Drosophila melanogaster*. Developmental principles discovered in *D. melanogaster* also apply to many other animal species, including humans.

42.7c Embryogenesis in *Drosophila*: Fruit Fly Model

Production of an adult fruit fly from a fertilized egg occurs in a sequence of genetically controlled development events. Following fertilization, division of the nucleus begins by mitosis. This produces a multinucleate blastoderm because the cytoplasm does not divide in the early embryo (cytokinesis does not occur) **(Figure 42.29)**. At the tenth nuclear division, the nuclei migrate to the periphery of the embryo, where, three divisions later, the 6000 or so nuclei are organized into separate cells. At this stage, the embryo is a cellular blastoderm, corresponding to the late blastula stage in the animals discussed above. Ten hours after fertilization, the cellular blastoderm develops into a segmented embryo (an embryo with distinct segments). About 24 hours after fertilization, the egg hatches into a larva that will undergo three moults before becoming a pupa. The adult fly emerges after metamorphosis 10 to 12 days after fertilization. The colours in Figure 42.29 illustrate how segments of the embryo can be mapped to the segments of the adult fly.

The study of developmental mutants has provided important information about *Drosophila* development. Three researchers performed key, pioneering research with developmental mutants: Edward B. Lewis, Christiane Nüsslein-Volhard, and

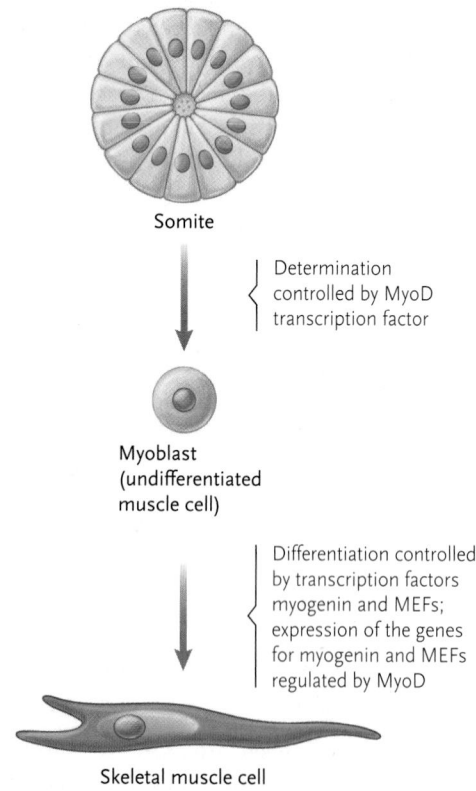

Somite

Determination controlled by MyoD transcription factor

Myoblast (undifferentiated muscle cell)

Differentiation controlled by transcription factors myogenin and MEFs; expression of the genes for myogenin and MEFs regulated by MyoD

Skeletal muscle cell

Figure 42.28

The genetic control of determination and differentiation involved in mammalian skeletal muscle cell formation.

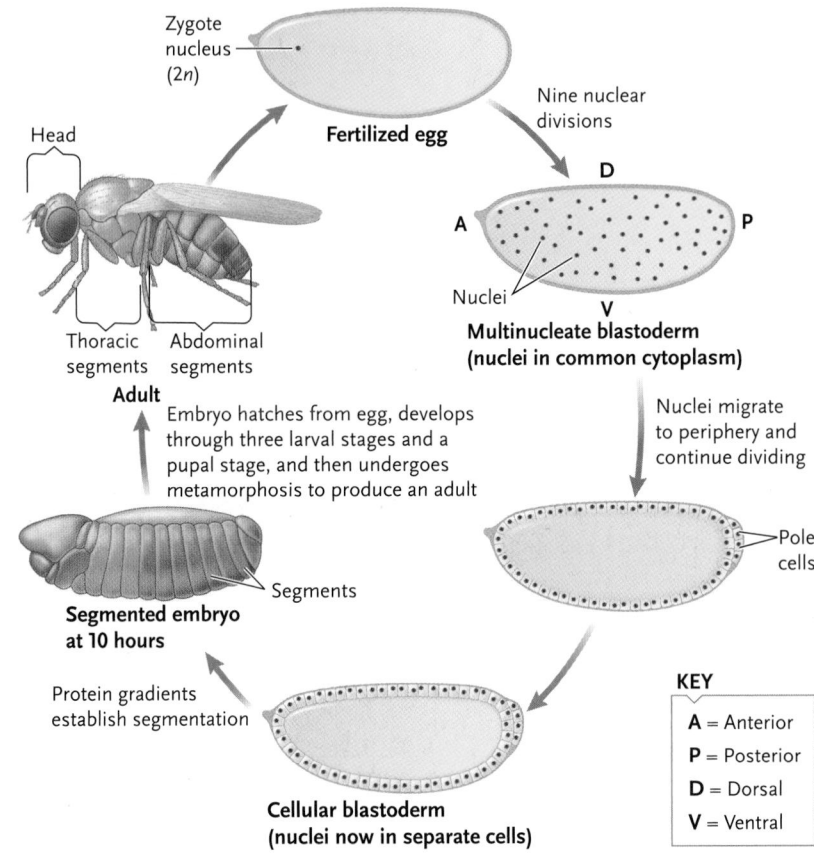

Figure 42.29

Embryogenesis in *Drosophila* and the relationship between segments of the embryo and segments of the adult.

Eric Wieschaus. The three shared a Nobel Prize in 1995 "for their discoveries about the genetic control of early embryonic development."

Nüsslein-Volhard and Wieschaus studied early embryogenesis. They searched for *every* gene required for early pattern formation in the embryo by looking for recessive, embryonic lethal mutations. When homozygous, these mutations result in embryo death during development. By determining the stage at which the embryo died and how development was disrupted, they gained insight into the role of the particular genes in embryogenesis.

Lewis studied mutants that changed the fates of cells in particular embryonic regions, producing structures in the adult that were normally produced by other regions. His work was the foundation of research identifying master regulatory genes that control the development of body regions in a wide range of organisms.

42.7d Maternal-Effect Genes and Segmentation Genes: Segmenting the Body

A number of genes control the establishment of the embryo's body plan. These genes regulate the expression of other genes. Two classes, maternal-effect genes and segmentation genes, work sequentially **(Figure 42.30, p. 1038)**. Many **maternal-effect genes** are expressed by the mother during oogenesis. These genes control egg polarity and thus embryo polarity. Some of these genes control formation of embryonic anterior structures, whereas others control formation of posterior structures. Still others control formation of the terminal end.

The *bicoid* gene is the key maternal-effect gene responsible for development of the head and thorax. This gene is transcribed in the mother during oogenesis, and the resulting mRNAs are deposited in the egg, localizing near the anterior pole **(Figure 42.31, p. 1038)**. After fertilization, translation of mRNAs produces BICOID protein, which diffuses through the zygote to form a gradient with concentration highest at the anterior end, fading to none at the posterior end. BICOID is a transcription factor that activates some genes and represses others along the anterior–posterior axis of the embryo. Embryos with mutations in the *bicoid* gene lack thoracic structures but have posterior structures at each end. In normal embryos, the *bicoid* gene is a master regulator gene controlling the expression of genes for the development of anterior structures (head and thorax).

The activities of products of other maternal-effect genes in gradients in the embryo are also involved in axis formation. The *nanos* gene is the key maternal-effect gene for the posterior structures. When the *nanos* gene is mutated, embryos lack abdominal segments.

Once the axis of the embryo is set, expression of at least 24 **segmentation genes** progressively subdivides the embryo into regions, determining the segments of the embryo and the adult (see Figure 42.30). Gradients of BICOID and other proteins encoded by maternal-effect genes regulate expression of the embryo's segmentation genes differentially. So each segmentation gene is expressed at a particular time and in a particular location during embryogenesis.

Three sets of segmentation genes are regulated in a cascade of gene activations. **Gap genes**, such as *hunchback* and *tailless*, are the first to be expressed. These genes are activated based on their positions in the maternally directed anterior–posterior axis of the zygote by reacting to the concentrations of BICOID and other proteins. Products of gap genes control subdivision of the embryo into several broad regions along

Figure 42.30
Maternal-effect genes and segmentation genes and their role in *Drosophila* embryogenesis.

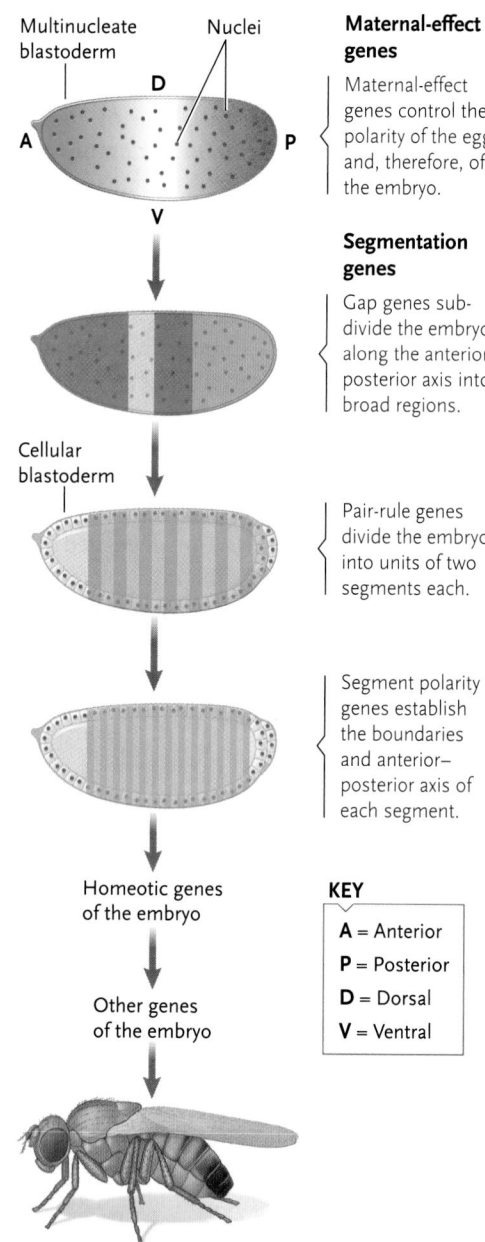

Multinucleate blastoderm   Nuclei

Cellular blastoderm

Homeotic genes of the embryo

Other genes of the embryo

Normal adult *Drosophila*

Maternal-effect genes

Maternal-effect genes control the polarity of the egg and, therefore, of the embryo.

Segmentation genes

Gap genes subdivide the embryo along the anterior–posterior axis into broad regions.

Pair-rule genes divide the embryo into units of two segments each.

Segment polarity genes establish the boundaries and anterior–posterior axis of each segment.

KEY

A = Anterior
P = Posterior
D = Dorsal
V = Ventral

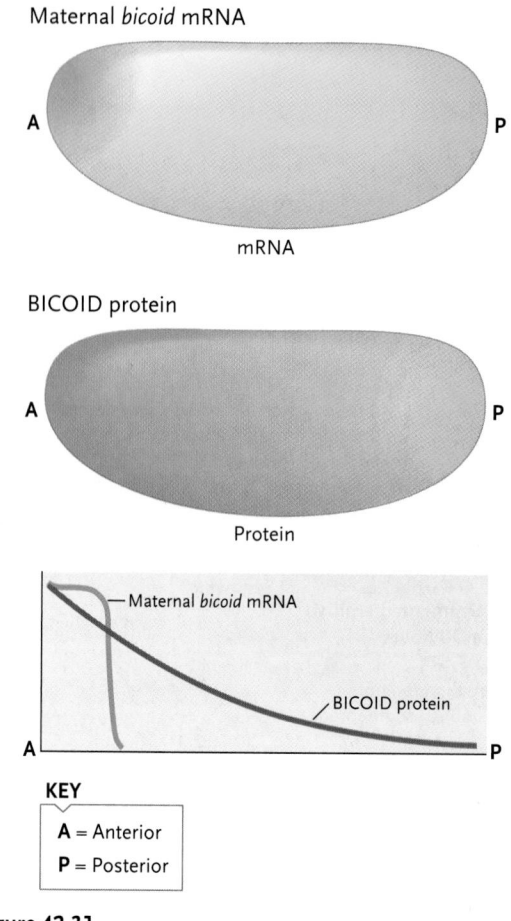

Maternal *bicoid* mRNA

mRNA

BICOID protein

Protein

Maternal *bicoid* mRNA

BICOID protein

KEY

A = Anterior
P = Posterior

Figure 42.31
Gradients of *bicoid* mRNA and BICOID protein in the *Drosophila* egg.

the anterior–posterior axis. Mutations in gap genes result in the loss of one or more body segments in the embryo **(Figure 42.32a)**.

Products of gap genes are transcription factors that activate **pair-rule genes**, such as *even-skipped* and *fushi tarazu*. The products of pair-rule genes divide the

Figure 42.32
Examples of mutations in the different types of segmentation genes of *Drosophila*. Orange highlights indicate wild-type segments that are mutated. **(a)** Gap gene mutants lack one or more segments. **(b)** Pair-rule gene mutants are missing every other segment. **(c)** Segment polarity genes have segments with one part missing and the other part duplicated as a mirror image.

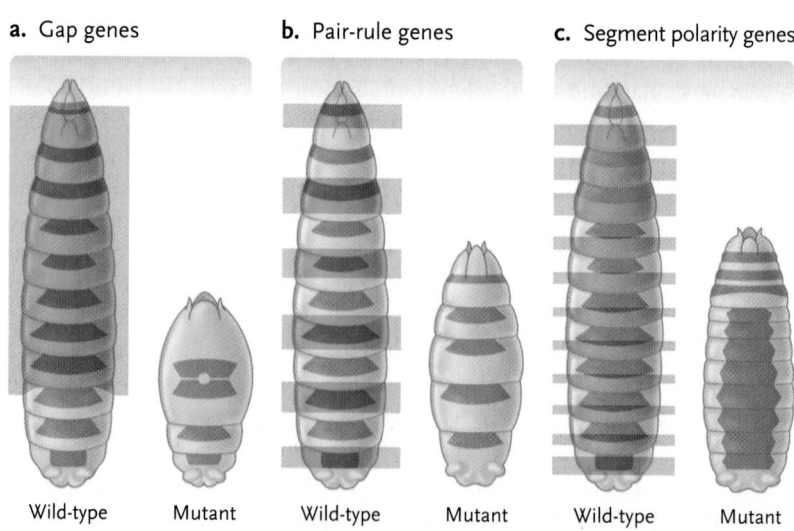

a. Gap genes   **b.** Pair-rule genes   **c.** Segment polarity genes

Wild-type   Mutant   Wild-type   Mutant   Wild-type   Mutant

embryo into units of two segments each. Mutations in pair-rule genes lead to the deletion of every other segment of the embryo (Figure 42.32b).

Expression of **segment polarity genes**, *engrailed* and *gooseberry*, is regulated by products of pair-rule genes. The actions of the products of segment polarity genes set boundaries and the anterior–posterior axis of each segment in the embryo. Mutations in segment polarity genes produce segments in which one part is missing and the other part is duplicated as a mirror image (Figure 42.32c). The products of segment polarity genes are transcription factors and other molecules that regulate other genes involved in laying down the pattern of the embryo.

42.7e Homeotic Genes: Structure and Determining Outcomes

Once the segmentation pattern has been set, **homeotic** (structure-determining) **genes** of the embryo specify what each segment will become after metamorphosis. In normal flies, homeotic genes are master regulatory genes controlling development of structures such as eyes, antennae, legs, and wings on particular segments **(Figure 42.33)**. Mutations of these genes allowed researchers to discover the role of homeotic genes. In the *Antennapedia* mutant fly, legs develop instead of antennae (see Figure 42.33, right).

How do homeotic genes regulate development? Homeotic genes encode transcription factors that regulate expression of genes responsible for the development of adult structures. Each homeotic gene has a common region called a homeobox that is key to its function. A homeobox corresponds to an amino acid section of the encoded transcription factor called the **homeodomain**. The homeodomain of each protein binds to a region in the promoters of the genes whose transcription it regulates.

Eight homeobox (*Hox*) genes in *Drosophila* are organized along a chromosome in the same order as

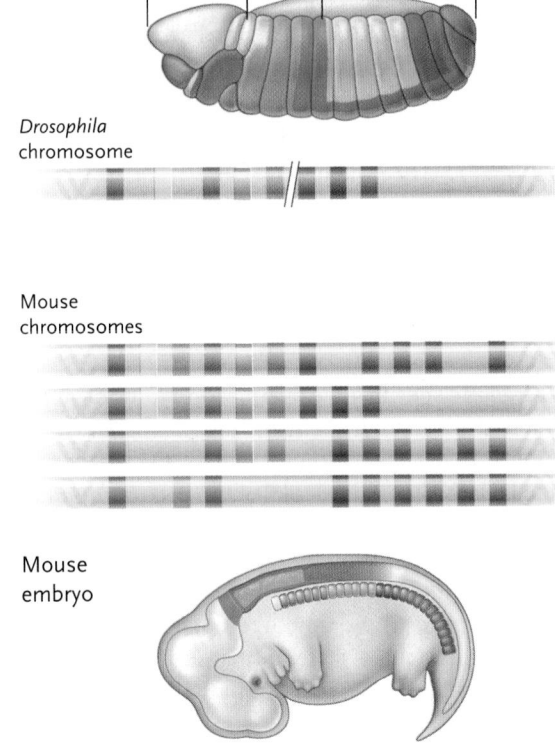

Figure 42.34
The *Hox* genes of the fruit fly and the corresponding regions of the embryo they affect. The mouse has four sets of *Hox* genes on four different chromosomes. Their relationship to the fruit fly genes is shown by the colours.

they are expressed along the anterior–posterior body axis **(Figure 42.34)**. The discovery of *Hox* genes in *Drosophila* led to a search for equivalent genes in other organisms. *Hox* genes are present in all major animal phyla, where they control the development of segments or regions of the body and are arranged in order in the genome. Homeobox sequences in *Hox* genes are highly conserved, indicating common function in the wide range of animals in which they occur. Homeobox sequences of mammals are the same as or similar to those of the fruit fly (see Figure 42.34). Homeotic genes are also found in plants, where they affect flower development. Homeobox genes have been identified and analyzed in *Arabidopsis* (see Chapter 37).

42.7f Cell-Death Genes: Apoptosis

Apoptosis plays a role in the breakdown of a tadpole's tail and in many other patterns of development in vertebrates and invertebrates. In humans, developing

Normal

Antennapedia mutant

Figure 42.33
Antennapedia, a homeotic mutant of *Drosophila*, in which legs develop in place of antennae.

fingers and toes are initially connected by tissue, forming paddle-shaped structures. Later in development, cells of this tissue die by apoptosis, resulting in separated fingers and toes. Like many other mammals, kittens and puppies are born with their eyes sealed shut by an unbroken layer of skin. Just after birth, cells die in a thin line across the middle of each eyelid, freeing the eyelids and allowing them to open. During pupation from caterpillar to butterfly, many tissues of the larva break down by apoptosis to be replaced by newly formed adult tissues.

Apoptosis results from gene activation in response to molecular signals from receptors on the surfaces of marked cells. In effect, the signals are death notices, delivered at a specific time during embryonic development. In the nematode *C. elegans*, division of the zygote produces 1090 cells. Of these, exactly 131 die at prescribed times to produce a total of 959 cells in the adult hermaphrodite.

In *C. elegans*, a death signal molecule that binds to a receptor in the plasma membrane of the target cell results in apoptosis. When the receptor is activated, it leads to activation of proteins that kill the cell. The killing proteins remain inactive in the absence of the death signal.

In the absence of a death signal, the membrane receptor is inactive **(Figure 42.35a)**. This allows CED-9, a protein associated with the outer mitochondrial membrane (encoded by the *ced-9* cell-death gene), to inhibit CED-4 (encoded by the *ced-4* gene) and CED-3 (encoded by the *ced-3* gene). These two proteins are needed to turn on the cell-death program. Cells with the *ced-9* gene expressed and its product CED-9 active normally survive in the adult nematode. When a death signal binds to and activates a receptor, the resulting events are typical of signal transduction pathways (see Figure 42.35b). Now the activated receptor leads to inactivation of CED-9. In the absence of CED-9, CED-4 is activated, which, in turn, activates CED-3. Activated CED-3 triggers a cascade of reactions, including activation of proteases and nucleases that degrade cell structures and chromosomes.

Studies of mutants have helped us understand the role of cell-death genes in *C. elegans*. In mutants lacking normal *ced-3* or *ced-4* genes, the 131 marked cells fail to die, producing a disorganized embryo. In the nervous system, the 103 cells that normally die by apoptosis live to form extra neurons in mutants. These extra neurons are inserted at random in the embryo, leading to a disorganized, nonfunctional nervous system.

a. No death signal　　　　　　　　**b. Death signal**

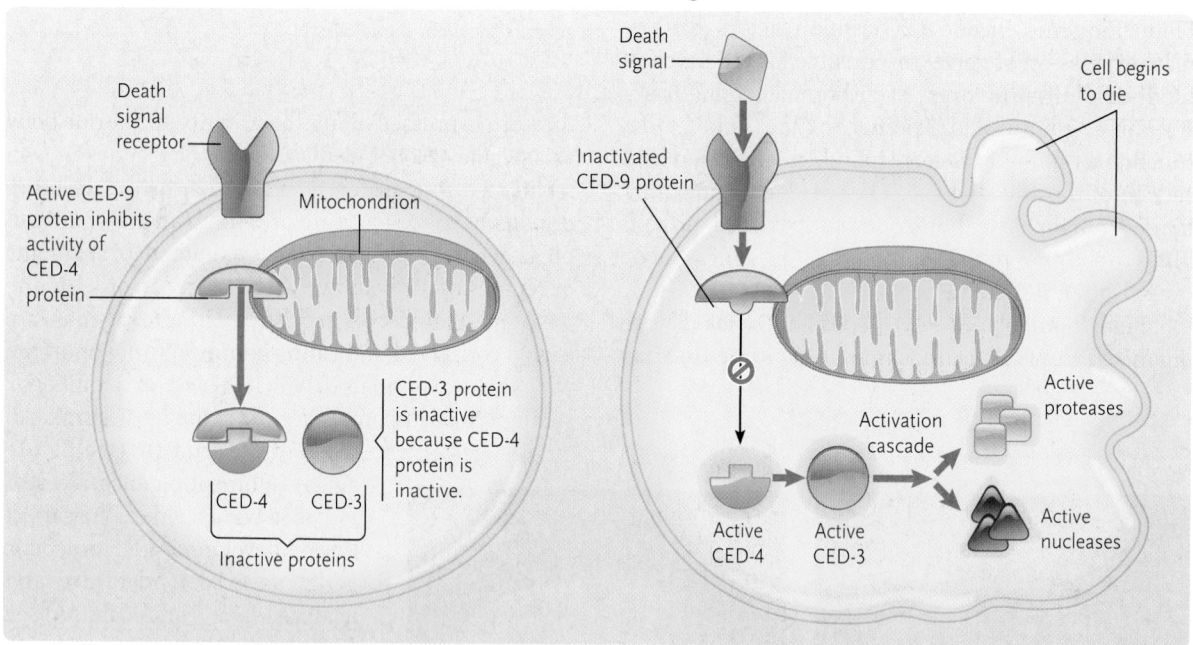

Apoptosis is inhibited as long as CED-9 protein is active; cell remains alive.

When a death signal binds to the death signal receptor, it activates the receptor, which leads to inactivation of CED-9 protein. As a result, CED-4 protein is no longer inhibited and becomes active, activating CED-3 protein. Active CED-3 triggers a cascade of activations producing active proteases and nucleases, which cause the changes seen in apoptotic cells and eventually lead to cell death.

Figure 42.35

The molecular basis of apoptosis in *C. elegans*. **(a)** In the absence of a death signal, no apoptosis occurs. **(b)** In the presence of a death signal, activation of CED-4 and CED-3 proteins triggers a pathway that leads to the cell's death.

Genes related to *ced-3* and *ced-4* occur in all animals tested for their presence. In humans and other mammals, the *caspase-9* gene is equivalent to *ced-3*. The *caspase-9* gene, which encodes a protease that degrades cell structures, is activated in cells that form the webbing between the fingers and toes in a human embryo, causing it to break down. The equivalent of *ced-4* is the *Apaf* gene (for *apoptotic protease-activating factor*). Mammalian cells are saved from death by the *Bcl* family of genes, which are the equivalent of *ced-9* in *C. elegans*. The genes are so closely related that they retain their effects if they are exchanged between *C. elegans* and human cells.

42.7g Epigenetics

Each cell in a multicellular organism has identical DNA, so how does the animal end up with so many different cell phenotypes—why does your hair look different from your fingernails? **Epigenetics** (*epi* = above) refers to nongenetic cellular memory that reflects developmental and environmental conditions. Note that epigenetic modifications alter the expression of DNA but not its sequence. Epigenetic changes are not mutations. Epigenetic features should be self-perpetuating and reversible, because they literally cause genes to be switched on and off. These changes may involve modifications to histones, proteins around which DNA wraps and that can be involved in down-regulating or transcription of genes. Methylation tightens packing of genes around histones, affecting down-regulating and inhibiting transcription, so demethylation can play an important role in epigenesis. Acetylation loosens packing and encourages transcription and translation. Without these epigenetic modifications, all genes would be functional in every cell, severely inhibiting or preventing cell differentiation.

During gametogenesis and embryogenesis, many metazoan animals show genome-wide epigenetic reprogramming involving DNA methylation and histone modification. In mammals, epigenetic processes occur at two stages during development, first in primordial germ cells and second in the zygote from just after fertilization to just before implantation. These changes can involve erasure of DNA methylation throughout the genome in the zygote and in primordial germ cells.

Epigenetic changes account for different patterns of activation of the X chromosome in mammals. Each female embryo receives two X chromosomes (one maternal, the other paternal), but often only one is activated. The details of activation vary among mammals, differing considerably among mice, rabbits, and humans. The situation also varies among cell lines within a developing embryo, sometimes differing between the inner cell mass of the blastocyst and the epiblast. In mice, the paternal X is inhibited early in embryogenesis by imprinted (parental influence on the genome is genetic imprinting) expression of regulatory RNA. In embryonic humans and rabbits, both X chromosomes are activated in many cells. It is clear that epigenetic effects can account for striking differences among cells and in whole-organism phenotypes. An epigenetic effect associated with the X chromosome causes tortoiseshell colouring in female cats, so only female cats can be true tortoiseshell in colour.

The impact of such changes is also very obvious in some insects. In a hive of honeybees (*Apis mellifera*), the queen (diploid) and workers (haploid) have the same genome even though they differ behaviourally and reproductively. These postmitotic differences result from the relative amounts of royal jelly consumed during the larval stage acting in concert with different patterns of DNA methylation. A comparison of the brains of worker and queen bees revealed that over 550 genes showed significant differences in methylation, apparently partly caused by modifications to sites for gene splicing. Once again, epigenetic factors influence phenotype.

Epigenetic factors may also play a role in hybrid sterility (see Chapter 19), providing another example of postzygotic isolation. Specifically, in mice, the gene *Prdm9* (responsible for protein coding) activates proteins essential for meiosis by methylation of histone H3 at lysine 4.

Now biologists better appreciate the many implications of epigenetics across the spectrum, from developmental and medical settings to the processes of speciation.

STUDY BREAK

1. Give two examples of genetic control of development.
2. What are maternal-effect genes?
3. What are homeotic genes, and what do they do?

Review

 To access course materials such as Aplia and other companion resources, please visit www.NELSONbrain.com.

42.1 Housing and Fuelling Developing Young

- In mammals, relaxin is a polypeptide hormone produced by the ovaries during pregnancy. Relaxin inhibits muscular contractions of the uterus and promotes the growth of glands that produce milk. As the time for parturition approaches, relaxin causes relaxation of the pubic ligaments and softens and enlarges the opening to the cervix.

- Viviparous animal embryos often develop in the uterus (or a uterus-like structure in the reproductive tract). Ovoviviparous animal embryos may develop in the oviduct, the stomach (at least one species of frog), the mouth (several species of fish), or a brood pouch (other fish, such as sea horses). Oviparous animals lay eggs, and the embryos develop outside the body.

- Strictly defined, milk is produced by female mammals to feed their young. Milklike substances (milk analogues) are produced by many other animals: the crop milk of pigeons; secretions of the skin in some fish and caecilians; or the uterine milk of a variety of animals, from some cockroaches to elasmobranchs.

- Yolk is food housed within the egg and used to support the growth and development of the embryo. The eggs of birds, insects, reptiles, and many other animals have large deposits of yolk that support the complete development of the young. Other animals have varying amounts of yolk corresponding to shorter periods of in-egg development.

42.2 Mechanisms of Embryonic Development

- The process of progressing from a zygote to a complete organism involves mitotic cell divisions, movements of cells, selective cell adhesions, induction, determination, and differentiation.

- The zygote divides by cleavage, without increasing the overall size or mass of the embryo. It forms a morula, or sphere of blastomeres (cells), which forms the blastula, a hollow ball of cells in a single layer enclosing the blastocoel, a fluid-filled cavity. The blastula invaginates to form the gastrula, a hollow ball of cells with an opening, the blastopore.

42.3 Major Patterns of Cleavage and Gastrulation

- Protostomes show determinant development in the progression from egg to animal. From the first divisions, blastomeres develop into specific tissues and organs. Deuterostomes show indeterminant development. Blastomeres produced by the first several divisions remain totipotent, capable of developing into a complete organism.

- Differences in the patterns of development of birds and typical mammals are functions of ovipary (egg-laying) versus vivipary (bearing live young). Mammalian eggs have little yolk, and the developing embryo depends on its mother for food and oxygen, as well as the collection and removal of wastes. The eggs of birds have large amounts of yolk, and the complete development of the embryo takes place independent of the mother's body. Bird embryo development begins with the formation of the blastodisc, a layer of cells on the surface of the yolk.

42.4 Organogenesis: Gastrulation to Adult Body Structures

- In the amniote egg, the amnion (an extra-embryonic membrane) encloses the developing embryo in a pool of amniotic fluid. The chorion, another extra-embryonic membrane, is produced from ectoderm and mesoderm and lines the inside of the egg shell. It is the site of oxygen exchange for the developing embryo. The allantois is enclosed by the allantoic membrane, which forms from mesoderm and endoderm. The allantois is the storage site for metabolic end products such as uric acid. The amniote egg allows development outside of water and was a fundamental breakthrough in the evolution of fully terrestrial vertebrates.

- In vertebrates, the eye forms when two sides of the anterior end of the neural tube (optic vesicles) swell outward until they contact the ectoderm. The optic cup, a double-walled structure, forms when the outer surface of the optic vesicle thickens and flattens at the region of contact and then pushes inward. The optic cup becomes the retina. The lens forms from the lens placode, a swelling arising when the optic cup induces thickening of the overlying ectoderm. Developing lens cells synthesize crystallin, a fibrous protein that collects into glassy deposits. Lens cells lose their nuclei and form the lens.

42.5 Embryonic Development of Humans and Other Mammals

- The chorion of the mammalian embryo forms chorionic villi, fingerlike extensions into the endometrium that increase the surface area of the chorion where the placenta will form.

- The placenta is the interface between the developing embryo and its mother. There are a variety of placental structures in mammals, and placenta-like structures (analogues) occur in fishes and insects. There is a rich diversity of placentas in guppylike fishes, demonstrating repeated evolution of this type of structure.

- *SRY* is the sex-determining region of the Y chromosome. *SRY* encodes a protein that induces a molecular switch that causes the primitive gonads to develop into testes. Fetuses with XX chromosomes do not produce *SRY*, and under the influence of estrogens and progesterone, the primitive gonads develop into ovaries. High levels of circulating testosterone in a pregnant female cause masculinization of the genitalia in some mammals.

- In many animals, a larval stage is intermediate between embryo and adult. Larval stages occur in some species that produce eggs with small amounts of yolk. Larvae are often strikingly different from the adults and may be the feeding and/or dispersal stage of the species.

42.6 Cellular Basis of Development

- Microtubules and microfilaments produce movements of whole cells and changes in cell shape.
- Induction is a process by which a group of cells causes or influences changes in a nearby group of cells, leading them to follow a particular pathway. Induction works through interactions between signal molecules from inducing cells and surface receptors on responding cells. When signal molecules bind to surface receptors, the responding cells are activated or inactivated.

42.7 Genetic and Molecular Control of Development

- Work with the nematode worm *Caenorhabditis elegans* has allowed biologists to follow development patterns and trace the fate of every cell produced from the zygote. *C. elegans* is transparent, making it even more appropriate for this kind of work.

- Skeletal muscles in mammals develop from somites, blocks of mesoderm along either side of the notochord. Paracrine signalling by nearby cells induces somite cells to express *myoD,* which turns on specific muscle-determining genes, converting them to myoblasts, undifferentiated muscle cells.
- Maternal-effect genes expressed by the mother during oogenesis control egg polarity and thus embryo polarity. The *bicoid* gene is a key maternal-effect gene responsible for the development of the head and thorax. Segmentation genes subdivide the embryo into regions, determining the segmentation in the body plan.
- Homeotic genes determine structure. In *Drosophila,* they are master regulatory genes controlling the development of body parts such as eyes, antennae, legs, and wings. In each homeotic gene is a common region called a homeobox. Homeobox-containing genes are called *Hox* genes.
- Apoptosis is programmed cell death, a process central to the development of some parts of the body. The process results from gene activation in response to molecular signals from receptors on the surfaces of marked cells.
- Epigenetics refers to nongenetic, cellular memory reflecting developmental and environmental conditions.

Questions

Self-Test Questions

1. Vivipary occurs in which class?
   a. class Osteichthyes
   b. class Amphibia
   c. class Chondrichthyes
   d. All of the above are correct.

2. Where does gastrulation occur?
   a. only in amniotes
   b. only in echinoderms
   c. only in amphibians
   d. in all metazoa

3. Large amounts of yolk occur in the eggs of all of which of the following?
   a. bony fish
   b. birds
   c. mammals
   d. None of the above.

4. In vertebrates, the blastopore becomes which of the following?
   a. the mouth
   b. the nostrils
   c. the anus
   d. the auditory meatus

5. In the development of which of the following does apoptosis occur?
   a. mammals
   b. reptiles
   c. bony fish
   d. all of the above

6. In which of the following do *Pax-6* genes control the development of eyes?
   a. in chordates
   b. in arthropods
   c. in cnidarians
   d. all of the above

7. Milk produced by mammary glands is a characteristic of which of the following?
   a. discus fish
   b. pigeons
   c. caecilians
   d. mammals

8. Induction results from interactions between cells. In the development of which of the following does it occur?
   a. tissues
   b. embryos
   c. nervous systems
   d. all of the above

9. Segmentation of developing embryos is controlled by which of the following?
   a. *Pax-6* genes
   b. gap genes
   c. homeotic genes
   d. b and c

10. In which of the following is relaxin is important in the process of birth?
    a. sharks
    b. mammals
    c. birds
    d. cockroaches

Questions for Discussion

1. How could apoptosis be used in the treatment of cancer?

2. How does the process of development differ between fraternal and identical twins? In humans, what is the incidence of fraternal twins with two fathers? How does this compare with other species of mammals?

3. How does the pattern of development of compound eyes differ from that of the eyes of vertebrates and molluscs?

4. What is the role of ectoderm in the development of the nervous system?

5. What is the main advantage of vivipary?

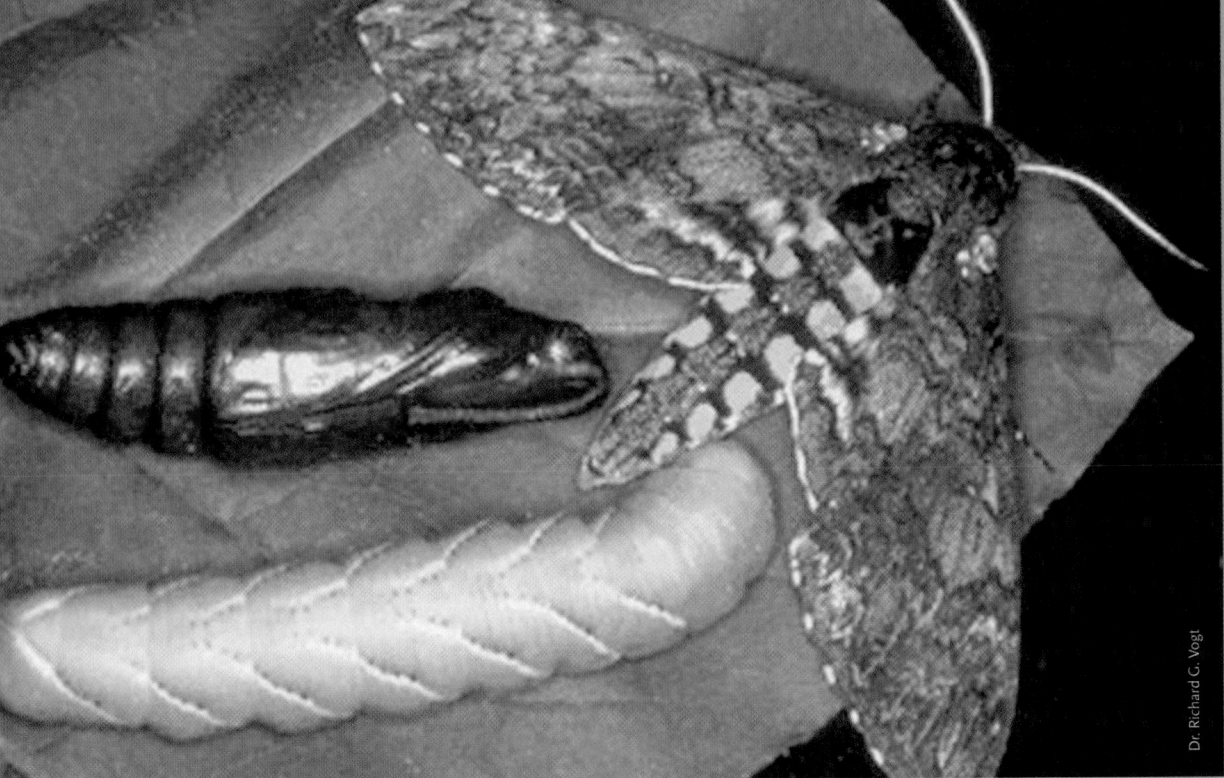

The larva, pupa, and adult moth of the tobacco hornworm, *Manduca sexta*, a model insect that has been used in exploring the hormonal control of metamorphosis.

43

Control of Animal Processes: Endocrine Control

WHY IT MATTERS

The larva, or caterpillar, of the tobacco hornworm, *Manduca sexta*, having reached its critical weight, stops feeding, drops to the ground, and burrows into the soil, where it moults into the pupal stage. Within the pupa, nearly all of the old larval tissues are destroyed and replaced by the tissues of the adult moth, which have been waiting in embryonic form for the signal to develop. When the moth is formed, it wriggles, still enclosed in the pupal cuticle, to the surface of the soil. It begins the behaviour leading to rupture of the pupal cuticle and its emergence as the adult moth, a soft animal with still rumpled wings. Once emergence is complete, it inflates its body and expands its wings. Only then does the cuticle harden. Moths are nocturnal, and the female, feeding on the nectar of several species of flowers, completes the development of her eggs, begun in the pupal stage, and releases a pheromone that will attract males. After mating, the female takes flight and searches for a suitable host plant in the family Solanaceae, where she lays the 100 or so eggs that she carries. The eggs hatch into larvae and the cycle begins again.

This carefully timed sequence of developmental and behavioural events is orchestrated by several hormones released in response to internal and external environmental cues. This marvel of communication between the environment and the cells, tissues, and organs of animals involving interactions between the nervous system and endocrine structures is a feature of everyday life in even the simplest of organisms.

As discussed in Chapter 39, multicellular organisms have specialized tissues and organs that allow them to function in a highly efficient manner. This requires communication between the highly differentiated cells in different organ systems leading to the effective coordination of their functions. Along with the central nervous system (CNS), hormones of the endocrine system provide the communication that coordinates the activities of multicellular life. These two systems are structurally, chemically, and functionally related, but they control different types of activities. It is impossible to understand the functioning of any animal in the absence of knowledge of the endocrine and nervous systems.

Hormones (*horme* = to excite) are secreted by cells of the **endocrine system** (*endo* = within; *krinein* = separate). In general, the endocrine system controls activities that involve slower, longer-acting responses of multiple tissues or organs. Some of these responses may be relatively quick (less than a minute) and directed, as in the stimulation of milk secretion by the suckling of an infant (see discussion of oxytocin, Section 43.3b), but many also involve transcription and translation of DNA, leading to sustained long-lasting responses (hours, weeks, months, or even years), incorporating the activities of many tissues and organs. For more complex animals such as insects and mammals, a galaxy of hormones regulates a host of functions, from the concentration of salt and glucose in body fluids to body growth and sexual maturation.

The nervous system (Chapter 44), on the other hand, acts through high-speed electrical signals to enable an organism to react rapidly to changes in its internal or external environment. The nervous system also activates or inhibits highly specific localized targets. Ultimately, the nervous system also controls many aspects of the endocrine system.

The mechanisms and functions of the endocrine system are the subjects of this chapter, while those of the nervous system are the subjects of Chapters 44 and 45.

43.1 Hormones and Their Secretion

Cells signal other cells in several ways. Local regulators are used to communicate between neighbouring cells, while hormones and neurotransmitters (see Chapter 44) are used to communicate with distant organs. Our focus in this chapter is on hormones and local regulators, while the following chapter examines the role of neurotransmitters.

43.1a The Endocrine System Includes Four Major Types of Cell Signalling

Four types of cell signalling occur in the endocrine system. In *autocrine regulation,* a local regulator acts on the same cells that release it **(Figure 43.1a)**. This is a common mechanism used by cells to either reduce or increase their sensitivity to other stimuli. In *paracrine regulation,* a cell releases a signalling molecule that

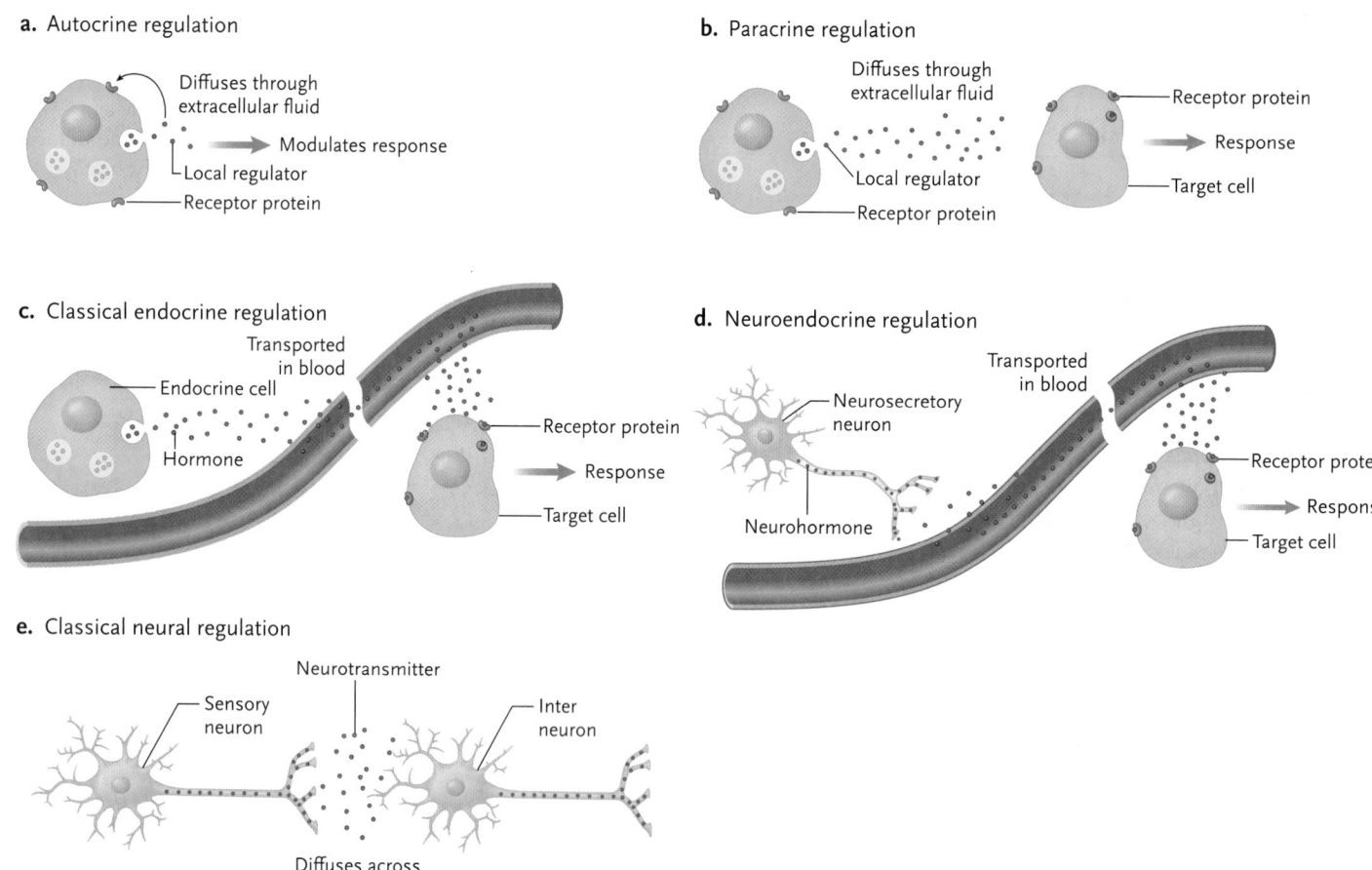

a. Autocrine regulation

Diffuses through extracellular fluid
Modulates response
Local regulator
Receptor protein

b. Paracrine regulation

Diffuses through extracellular fluid
Receptor protein
Response
Local regulator
Target cell
Receptor protein

c. Classical endocrine regulation

Transported in blood
Endocrine cell
Receptor protein
Response
Hormone
Target cell

d. Neuroendocrine regulation

Transported in blood
Neurosecretory neuron
Receptor protein
Response
Neurohormone
Target cell

e. Classical neural regulation

Neurotransmitter
Sensory neuron
Inter neuron
Diffuses across synapse

Figure 43.1
The major types of cell signalling in the endocrine and nervous systems.

diffuses through the extracellular fluid and acts on nearby cells (Figure 43.1b). In both of these instances, regulation is *local* rather than at a distance. Many of the growth factors that regulate cell division and differentiation act in both an autocrine and a paracrine fashion. In *classical endocrine regulation,* hormones are secreted into the blood or extracellular fluid by the cells of ductless secretory organs called **endocrine glands** (Figure 43.1c). Hormones are circulated throughout the body in the blood or other body fluids, and, as a result, most body cells are constantly exposed to a wide variety of hormones. Only the *target cells* of a hormone, those with *receptor proteins* (Chapter 5) recognizing and binding that hormone, respond to it. Hormones are cleared from the body at a steady rate by enzymatic breakdown in their target cells or blood or organs such as the liver or kidneys, and the breakdown products are excreted.

▶ **CONCEPT FIX** *Exocrine glands,* such as the sweat and salivary glands, release their secretions into ducts that lead outside the body or into the cavities of the digestive tract (see Chapter 39) **(Figure 43.2a).** Secretions from the stomach, liver, pancreas, gall bladder, intestine, and so on, that are released into the digestive tract are not endocrine secretions (Figure 43.2b) but, rather, are exocrine, since the **lumen** of the digestive tract is technically outside the body. The digestive tract extends from the mouth to the anus, and unless substances are absorbed across cell membranes along the digestive tract, they never really enter the body. As a result, secretions that enter the digestive tract via ducts from glands along its length are exocrine in nature. ⬤

In *neuroendocrine regulation,* specialized **neurosecretory neurons** respond to and conduct electrical signals, but rather than synapsing with target cells, they release a neurohormone into the circulation when appropriately stimulated (Figure 43.1d). The hormone is produced in the cell body and packaged in membrane-bound vesicles that are transported along the axon to the release sites. The neurohormone is usually distributed in blood or other body fluids and elicits a response in target cells some distance away that have receptors for the neurohormone. For instance, the peptide vasopressin secreted by the pituitary gland in the brain circulates in the blood and acts on the kidney, reducing the water excreted in the urine, and on muscles of blood vessels, increasing blood pressure. It is also released directly into the brain to cause a myriad of social effects, such as pair bonding in some mammals. In contrast, in *neural regulation,* **neurons** synapse directly with target cells, releasing neurotransmitters into a synapse (Figure 43.1e) or neuromuscular junction (see Chapters 44 and 46, respectively, for details). Note that both neurohormones and neurotransmitters are secreted by neurons. Neurohormones are distinguished from neurotransmitters in that neurohormones affect distant target cells, whereas neurotransmitters affect adjacent cells.

However, both neurohormones and neurotransmitters function in the same way—they cause cellular responses by interacting with specific receptors on target cells.

a. Exocrine gland

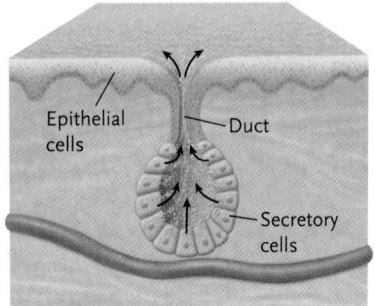

b. Endocrine gland

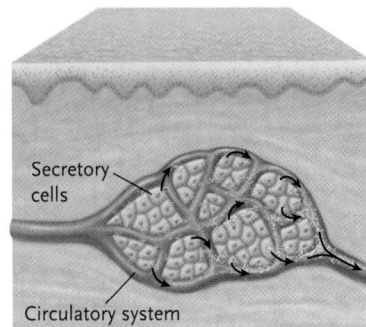

Figure 43.2

The structure of exocrine and endocrine glands. Exocrine glands **(a)** secrete chemicals into ducts that lead to the surface of the body or the digestive tract, while endocrine glands **(b)** secrete hormones directly into body fluids, especially the circulatory system.

43.1b Hormones and Local Regulators Can Be Grouped into Four Classes Based on Their Chemical Structures

More than 60 hormones and local regulators have been identified in humans. Many human hormones are either identical or very similar in structure and function to those in other animals, but other vertebrates, as well as invertebrates, may have hormones not found in humans. Most of these chemicals can be grouped into four molecular classes: amine, peptide, steroid, and fatty acid–derived molecules.

Amine hormones are involved in classical endocrine signalling and neuroendocrine signalling. Most amine hormones are based on tyrosine. With one major exception, they are hydrophilic molecules, which diffuse readily into the blood and extracellular fluids. On reaching a target cell, they bind to receptors at the cell surface. The amine hormones include dopamine; epinephrine; norepinephrine; and, in protostomes, octopamine, which are all neurotransmitters released by some neurons (see Chapter 44). The exception is the thyroid hormones secreted by the thyroid gland. These hormones, based on a pair of tyrosines, enter the cell by receptor-mediated endocytosis. Inside the cell, one form of the hormone binds to nuclear receptors in the same way as described for steroids below. Thyroid hormones also act via membrane receptors not only on the surface of the cells but also on mitochondrial membranes.

Peptide hormones consist of amino acid chains, ranging in length from as few as 3 amino acids to more than 200. Some have carbohydrate groups attached. They are involved in classical endocrine signalling and neuroendocrine signalling. Mostly hydrophilic hormones, peptide hormones are released into the blood or extracellular fluid by exocytosis when cytoplasmic vesicles containing the hormones fuse with the plasma membrane. One large group of peptide hormones, the **growth factors,** regulates the division and differentiation of many cell types in the body. Many growth factors act in both a paracrine and an autocrine manner, as well as in classical endocrine signalling. Because they can switch cell division on or off, growth factors are an important focus of cancer research.

Steroid hormones are involved in classical endocrine signalling. All are hydrophobic molecules derived from cholesterol and are sparingly soluble in water. They combine with hydrophilic carrier proteins to form water-soluble complexes that diffuse easily in blood or other fluids. On contacting a cell, the hormone is released from its carrier protein, passes through the plasma membrane of the target cell (a process that is sometimes mediated by receptors), and binds to internal receptors in the nucleus or cytoplasm. Steroid hormones include aldosterone; cortisol; the vertebrate sex hormones; and ecdysone, the hormone that governs the formation of new cuticles in ecdyzoan protostomes. Steroid hormones may vary little in structure but produce very different effects. Testosterone and estradiol, two major sex hormones responsible for the development of mammalian male and female characteristics, respectively, differ only in the presence or absence of a methyl group. Steroids can also act via membrane receptors, controlling cellular events such as apoptosis and cell proliferation and more complex events such as behaviour.

Fatty acids represent a very specialized category of hormones. In arthropods and possibly annelids, hormones derived from farnesoic acid include the juvenile hormones that govern metamorphosis and reproduction (see "Molecule behind Biology," Box 27.3, Chapter 27). Prostaglandins and their relatives are important local regulators derived from arachidonic acid. They are involved in paracrine and autocrine regulation in all animals. First discovered in semen, they enhance the transport of sperm through the female reproductive tract by increasing the contractions of muscle cells in both vertebrates and insects. In at least some insects, prostaglandins act as endocrines: they are synthesized in the sperm storage organs of mated females and initiate egg laying by acting on the oviducts and possibly the nervous system.

43.1c Many Hormones Are Regulated by Feedback Pathways

The secretion of many hormones is regulated by feedback pathways, some of which operate partially or completely independently of neuronal controls. Most pathways are controlled by negative feedback, in which a product of the pathway inhibits an earlier step in the pathway. In vertebrates, secretion by the thyroid gland is regulated by a negative feedback loop **(Figure 43.3)**. Neurosecretory neurons in the hypothalamus secrete thyroid-releasing hormone (TRH) into a vein connecting the hypothalamus to the pituitary (neuroendocrine regulation). In response, the pituitary releases thyroid-stimulating hormone (TSH) into the blood, which stimulates the thyroid gland to release thyroid hormones (classical endocrine regulation). As the thyroid hormone

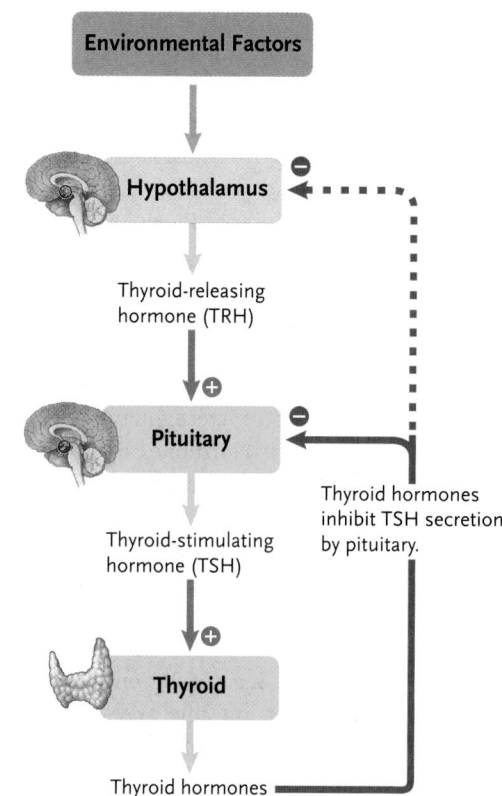

Figure 43.3
A negative feedback loop regulating secretion of thyroid hormones. In this case, when the concentration of thyroid hormones in the blood increases, the hormones inhibit earlier steps in the pathway (indicated by the red arrows and negative sign) to maintain their levels more or less constant. This feedback control can be exclusively hormonal (yellow boxes) but may also involve neurosecretory pathways coordinating hormone levels to environmental factors (blue box). Green arrows produce stimulation while red arrows produce inhibition.

concentration in the blood increases, it begins to inhibit TRH secretion by the hypothalamus. In turn, TSH and secretion of the thyroid hormones are reduced. This feedback control can be exclusively hormonal (yellow boxes) but may also involve neurosecretory pathways coordinating hormone levels to environmental factors (blue box).

43.1d Body Processes Are Regulated by Coordinated Hormone Secretion

Although we mostly discuss individual hormones in the remainder of the chapter, most body processes are affected by more than one hormone. The blood concentrations of glucose; fatty acids; and ions such as Ca^{2+}, K^+, and Na^+ are regulated by the coordinated activities of several hormones secreted by different glands. Similarly, body processes such as oxidative metabolism, digestion, growth, sexual development, and reactions to stress are all controlled by multiple hormones.

In many of these systems, negative feedback loops adjust the levels of secretion of hormones that act in antagonistic (opposing) ways, creating a balance in their effects that maintains body homeostasis (see Chapter 50). Consider the regulation of fuel molecules such as glucose, fatty acids, and amino acids in the blood. We usually eat three meals a day and fast to some extent between meals. During these periods of eating and fasting, five hormone systems act in a coordinated fashion to keep the fuel levels in balance: (1) gastrin and ghrelin secreted by the stomach and secretin from the intestine; (2) insulin and glucagon, secreted by the pancreas; (3) growth hormone, secreted by the anterior pituitary; (4) epinephrine–norepinephrine, released by the sympathetic nervous system and the adrenal medulla; and (5) glucocorticoid hormones, released by the adrenal cortex.

The entire system of hormones regulating fuel metabolism resembles the fail-safe mechanisms designed by engineers in which redundancy, overlapping controls, feedback loops, and multiple safety valves ensure that vital functions are maintained at appropriate levels in the face of changing and even extreme circumstances.

STUDY BREAK

1. What are the functions of the endocrine and nervous systems? How are they the same, and how do they differ? What is the significance of this?
2. What are the major types of cell signalling that occur in the endocrine system? How do they work?

43.2 Mechanisms of Hormone Action

Hormones control cell functions by binding to receptor molecules on or in their target cells. Small quantities of hormones can typically produce profound effects in cells and body functions due to **amplification.** In amplification, an activated receptor activates many proteins, which then activate an even larger number of proteins for the next step in the cellular pathway, and so on in each subsequent step (see Chapter 5). It has been estimated that, by amplification, a single molecule of epinephrine acting on a liver cell will liberate over 100 molecules of glucose from stored glycogen.

43.2a The Secreted Hormone May Not Be in an Active Form

Many hormones are secreted in an inactive or less active form (a *prohormone*) and converted by target cells or enzymes in the blood or other tissues to the active form. The best-known example is thyroxine, discussed below. Many other hormones are subject to similar processes. **Ecdysone,** a steroid governing the formation of new cuticle in insects, is converted to the much more active functional hormone 20-OH ecdysone by the addition in the target cells of a single hydroxyl group. Peptide hormones are commonly synthesized as prohormones

that undergo post-translational conversion to the active forms in the source cell. In some cases, however, further conversion occurs once the hormone has been secreted. **Angiotensin** is a hormone that governs blood pressure in humans. It is secreted by the liver as angiotensinogen. An inactive form of angiotensin is cleaved from angiotensinogen by an enzyme. This inactive form is converted to the active hormone by angiotensin-converting enzyme (ACE). ACE inhibitors are often prescribed for control of high blood pressure.

43.2b Hydrophilic Hormones Bind to Surface Receptors, Activating Protein Kinases Inside Cells

Hormones that bind to receptor molecules in the plasma membrane produce their responses through signal transduction pathways. In brief, when a surface receptor binds a hormone, it transmits a signal through the plasma membrane. Within the cell, the signal is transduced, changing into a form that causes the cellular response **(Figure 43.4a, p. 1050).** Typically, the reactions of signal transduction pathways involve protein kinases, which are enzymes that add phosphate groups to proteins. Adding a phosphate group to a protein may activate or inhibit it, depending on the protein and the reaction. The particular response produced by a hormone depends on the kinds of protein kinases activated in the cell and the types of target proteins they phosphorylate (Chapter 8). The receptor molecule may be a tyrosine kinase molecule, a receptor with a built-in protein kinase on the cytoplasmic side of the receptor itself, or it may be a **G protein–coupled receptor** that secondarily activates protein kinases within the cell. These hormones generally act on functional proteins already present in the cell, such as enzymes, ion channels, and transport proteins, although the signal transduction pathway may not stop at the cytoplasm: many growth factors and some peptide hormones ultimately affect events in the nucleus. Although action via membrane receptors is characteristic of and most extensively studied in peptide and amine hormones, many steroid and fatty acid hormones also exert some of their actions in this way.

The peptide hormone glucagon illustrates the mechanisms triggered by surface receptors. When glucagon binds to surface receptors on liver cells, it triggers a series of steps leading to the phosphorylation and activation of the enzyme governing the breakdown of glycogen stored in those cells into glucose.

43.2c Hydrophobic Hormones Bind to Receptors Inside Cells, Activating or Inhibiting Genetic Regulatory Proteins

After passing through the plasma membrane, the hydrophobic steroid and thyroid hormones bind to internal receptors in the nucleus or cytoplasm (Figure 43.4b). Binding of the hormone activates the receptor, which

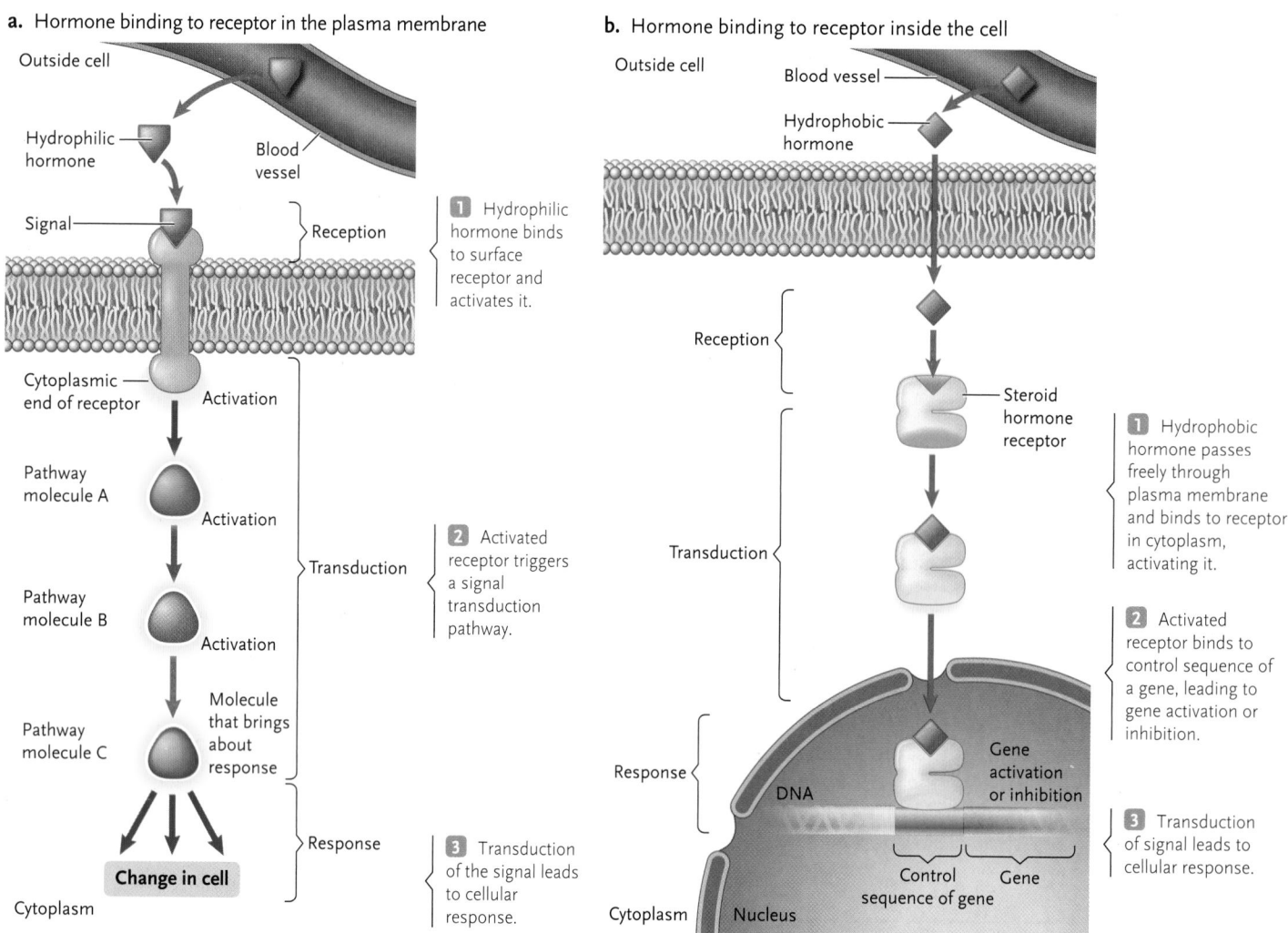

a. Hormone binding to receptor in the plasma membrane

Outside cell

Hydrophilic hormone

Blood vessel

Signal

Reception — 1 Hydrophilic hormone binds to surface receptor and activates it.

Cytoplasmic end of receptor

Activation

Pathway molecule A

Activation

Transduction — 2 Activated receptor triggers a signal transduction pathway.

Pathway molecule B

Activation

Pathway molecule C

Molecule that brings about response

Change in cell

Response — 3 Transduction of the signal leads to cellular response.

Cytoplasm

b. Hormone binding to receptor inside the cell

Outside cell

Blood vessel

Hydrophobic hormone

Reception

Steroid hormone receptor — 1 Hydrophobic hormone passes freely through plasma membrane and binds to receptor in cytoplasm, activating it.

Transduction — 2 Activated receptor binds to control sequence of a gene, leading to gene activation or inhibition.

Response

DNA

Gene activation or inhibition

Control sequence of gene

Gene — 3 Transduction of signal leads to cellular response.

Cytoplasm   Nucleus

Figure 43.4

The reaction pathways activated by hormones that bind to receptor proteins in plasma membrane **(a)** or inside cells **(b)**. In both mechanisms, the signal—the binding of the hormone to its receptor—is transduced to produce the cellular response.

then binds to a control sequence of specific genes. Depending on the gene, binding the control sequence either activates or inhibits its transcription, leading to changes in protein synthesis that accomplish the cellular response. The characteristics of the response depend on the specific genes controlled by the activated receptors and on the presence of other proteins that modify the activity of the receptor.

One of the actions of the steroid hormone aldosterone illustrates the mechanisms triggered by internal receptors. If blood pressure falls below optimal levels, aldosterone is secreted by the adrenal glands. The hormone circulates throughout the body in the blood but affects only cells (mostly in the kidney but also in sweat glands and the colon) that contain the aldosterone receptor in their cytoplasm (step 1 in Figure 43.4b). When activated by aldosterone, the receptor binds to the control sequence of a gene (step 2 in Figure 43.4b), leading to the synthesis of proteins that increase **reabsorption** of Na^+ by the kidney cells (step 3 in Figure 43.4b). The resulting increase in Na^+ concentration in

body fluids increases water retention and, with it, blood volume. The increase in blood volume returns (increases) blood pressure back to normal, removing the stimulus for aldosterone secretion from the adrenal glands.

43.2d Target Cells May Respond to More Than One Hormone, and Different Target Cells May Respond Differently to the Same Hormone

A single target cell may have receptors for several hormones and respond differently to each hormone. Vertebrate liver cells have receptors for the pancreatic hormones insulin and glucagon. Insulin increases glucose uptake and conversion to glycogen, which decreases blood glucose levels, whereas glucagon stimulates the breakdown of glycogen into glucose, which increases blood glucose levels.

Conversely, particular hormones interact with different types of receptors in or on a range of target cells. Different responses are then triggered in each target cell type because the receptors trigger different transduction pathways. For example, the amine hormone epinephrine prepares the body for handling stress (including dangerous situations) and physical activity. In mammals, epinephrine can bind to three different plasma membrane–embedded receptors: α, β_1, and β_2 receptors. When epinephrine binds to α receptors on smooth muscle cells, such as those of the blood vessels, it triggers a response pathway that causes the cells to constrict, cutting off circulation to peripheral organs. When epinephrine binds to β_1 receptors on heart muscle cells, the contraction rate of the cells increases, which, in turn, enhances blood supply. When epinephrine binds to β_2 receptors on liver cells, it stimulates the breakdown of glycogen to glucose, which is released from the cell. The overall effect of these and a number of other responses to epinephrine secretion is to supply energy to the major muscles responsible for locomotion, preparing the animal for stress or for physical activity. Similar tissue-specific diversification of responses is known for many hormones.

Moreover, the response to a hormone may differ in different animals. For example, melatonin, an amine derived from tryptophan, is important in regulating daily and annual cycles in most animals. However, it also plays a role in regulating the salt gland of marine birds. Thyroxine promotes metamorphosis in amphibians but inhibits metamorphosis in cyclostomes. The same hormone may have different functions at different stages in the life of an animal. The juvenile hormone of insects acts to maintain insects in a larval state but also controls reproduction in the adult.

In summary, the mechanisms by which hormones work have four major features:

1. Only the cells that contain surface or internal receptors for a particular hormone respond to that hormone.
2. Once bound by their receptors, hormones may produce a response that involves stimulation or inhibition of cellular processes through the specific types of internal molecules activated by the hormone action.
3. Because of the amplification that occurs through both the surface and internal receptor mechanisms, hormones are effective in very small concentrations.
4. The response to a hormone differs among target organs.

In the next two sections, we discuss the major endocrine cells and glands of vertebrates. The locations of these cells and glands in mammals (including humans) and their functions are summarized in **Figure 43.5** and **Table 43.1, p. 1052.**

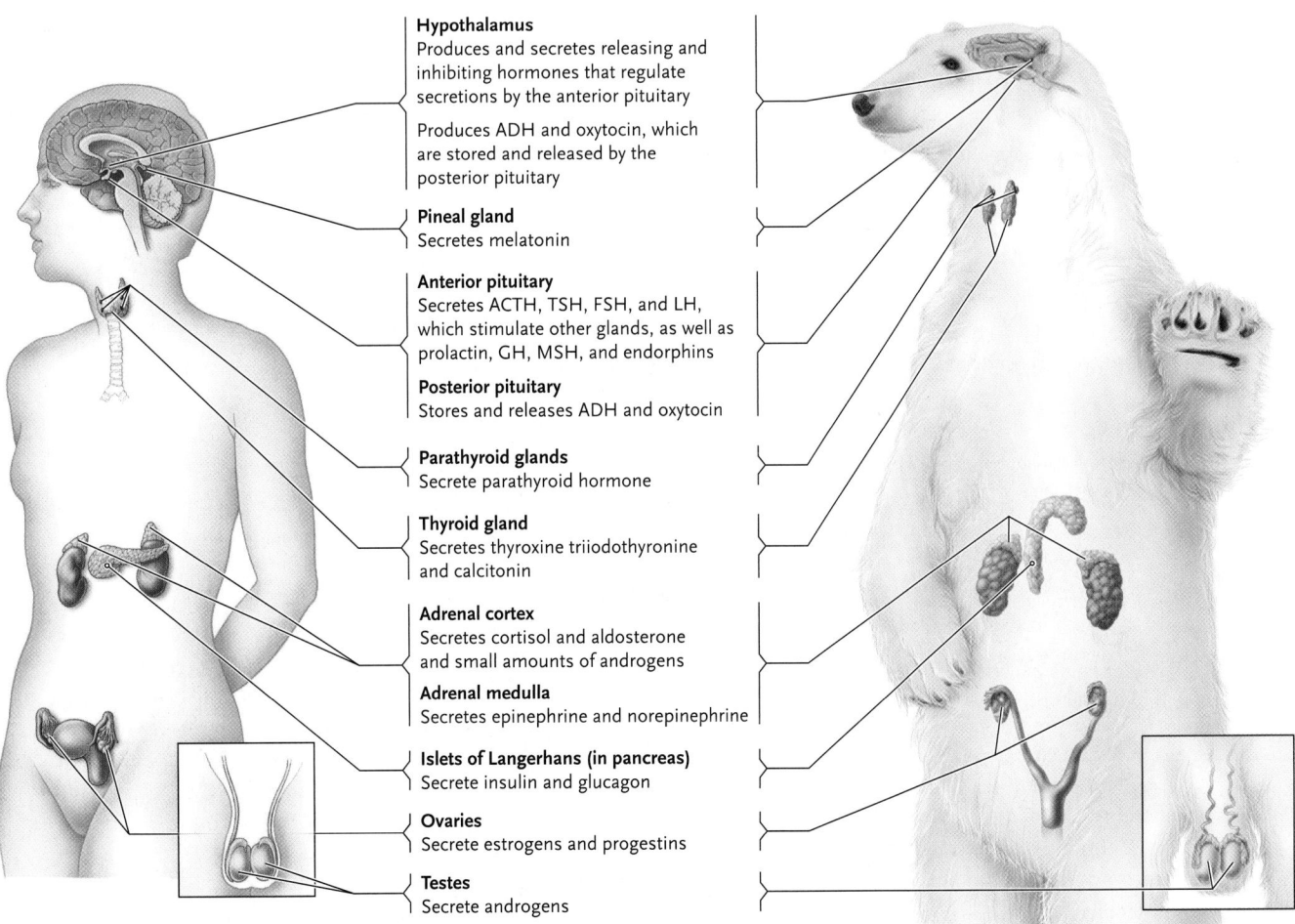

Hypothalamus
Produces and secretes releasing and inhibiting hormones that regulate secretions by the anterior pituitary

Produces ADH and oxytocin, which are stored and released by the posterior pituitary

Pineal gland
Secretes melatonin

Anterior pituitary
Secretes ACTH, TSH, FSH, and LH, which stimulate other glands, as well as prolactin, GH, MSH, and endorphins

Posterior pituitary
Stores and releases ADH and oxytocin

Parathyroid glands
Secrete parathyroid hormone

Thyroid gland
Secretes thyroxine triiodothyronine and calcitonin

Adrenal cortex
Secretes cortisol and aldosterone and small amounts of androgens

Adrenal medulla
Secretes epinephrine and norepinephrine

Islets of Langerhans (in pancreas)
Secrete insulin and glucagon

Ovaries
Secrete estrogens and progestins

Testes
Secrete androgens

Figure 43.5
This diagram shows the major endocrine glands in a mammal's body, using a human and a bear as models.

Table 43.1   The Major Human Endocrine Glands and Hormones

Secretory Tissue or Gland	Hormones	Molecular Class	Target Tissue	Principal Actions
Hypothalamus	Releasing and inhibiting hormones	Peptide	Anterior pituitary	Regulate secretion of anterior pituitary hormones
Anterior pituitary	Thyroid-stimulating hormone (TSH)	Peptide	Thyroid gland	Stimulates secretion of thyroid hormones and growth of thyroid gland
	Adrenocorticotropic hormone (ACTH)	Peptide	Adrenal cortex	Stimulates secretion of glucocorticoids by adrenal cortex
	Follicle-stimulating hormone (FSH)	Peptide	Ovaries in females, testes in males	Stimulates egg growth and development and secretion of sex hormones in females; stimulates sperm production in males
	Luteinizing hormone (LH)	Peptide	Ovaries in females, testes in males	Regulates ovulation in females and secretion of sex hormones in males
	Prolactin (PRL)	Peptide	Mammary glands	Stimulates breast development and milk secretion
	Growth hormone (GH)	Peptide	Bone, soft tissue	Stimulates growth of bones and soft tissues; helps control metabolism of glucose and other fuel molecules
	Melanocyte-stimulating hormone (MSH)	Peptide	Melanocytes in skin of some vertebrates	Promotes darkening of the skin
	Endorphins	Peptide	Pain pathways of PNS	Inhibit perception of pain
Posterior pituitary	Antidiuretic hormone (ADH)	Peptide	Kidneys	Raises blood volume and pressure by increasing water reabsorption in kidneys
	Oxytocin	Peptide	Uterus, mammary glands	Promotes uterine contractions; stimulates milk ejection from breasts
Thyroid gland	Calcitonin	Peptide	Bone	Lowers calcium concentration in blood
	Thyroxine and triiodothyronine	Amine	Most cells	Increase metabolic rate; essential for normal body growth
Parathyroid glands	Parathyroid hormone (PTH)	Peptide	Bone, kidneys, intestine	Raises calcium concentration in blood; stimulates vitamin D activation
Adrenal medulla	Epinephrine and norepinephrine	Amine	Sympathetic receptor sites throughout body	Reinforce sympathetic nervous system; contribute to responses to stress
Adrenal cortex	Aldosterone (mineralocorticoid)	Steroid	Kidney tubules	Helps control body's salt–water balance by increasing Na^+ reabsorption and K^+ excretion in kidneys
	Cortisol (glucocorticoid)	Steroid	Most body cells, particularly muscle, liver, and adipose cells	Increases blood glucose by promoting breakdown of proteins and fats
Testes	Androgens, such as testosterone*	Steroid	Various tissues	Control male reproductive system development and maintenance; most androgens are made by the testes
	Oxytocin	Peptide	Uterus	Promotes uterine contractions when seminal fluid is ejaculated into vagina during sexual intercourse
Ovaries	Estrogens, such as estradiol**	Steroid	Breasts, uterus, other tissues	Stimulate maturation of sex organs at puberty and development of secondary sexual characteristics
	Progestins, such as progesterone**	Steroid	Uterus	Prepare and maintain uterus for implantation of fertilized eggs and the growth and development of embryos

*Small amounts secreted by ovaries and adrenal cortex.
**Small amounts secreted by testes.

Secretory Tissue or Gland	Hormones	Molecular Class	Target Tissue	Principal Actions
Pancreas (islets of Langerhans)	Glucagon (alpha cells)	Peptide	Liver cells	Raises glucose concentration in blood; promotes release of glucose from glycogen stores and production from noncarbohydrates
	Insulin (beta cells)	Peptide	Most cells	Lowers glucose concentration in blood; promotes storage of glucose, fatty acids, and amino acids
Pineal gland	Melatonin	Amine	Brain, anterior pituitary, reproductive organs, immune system, possibly others	Helps synchronize body's biological clock with day length; may inhibit gonadotropins and initiation of puberty
Many cell types	Growth factors	Peptide	Most cells	Regulate cell division and differentiation
	Prostaglandins	Fatty acid	Various tissues	Have many diverse roles

In addition to these major endocrine organs, important hormones are also secreted by organs that have other primary functions, including the kidney, heart, liver, and intestine. In particular, the digestive system is the source of several peptide hormones, many of which are also produced elsewhere. It is the only known source for peptides such as gastrin, secretin, and ghrelin, which coordinate the digestive secretions of the gut and its associated glands and send signals associated with hunger to the brain. The gut is increasingly recognized as an important endocrine organ in many animals. Among vertebrates, it is a more important source for circulating levels of melatonin than the pineal body with which that hormone is traditionally associated. In insects, several peptide hormones, such as proctolin, produced by neuroendocrine cells in the CNS are also produced by cells in the intestine.

STUDY BREAK

1. What are the four major features of the mechanism by which hormones work?
2. Explain how one type of target cell could respond to different hormones, and how the same hormone could produce different effects in different cells.
3. Explain how a small amount of hormone can produce very large responses.

43.3 The Hypothalamus and Pituitary

Hormones of vertebrates work in coordination with the nervous system. The action of several hormones is closely coordinated by the hypothalamus–pituitary complex.

The hypothalamus is a region of the brain located in the floor of the cerebrum (see Chapter 44). The **pituitary,** consisting mostly of two fused lobes, is suspended just below the hypothalamus by a slender stalk

of tissue that contains both neurons and blood vessels **(Figure 43.6, p. 1054).** The **posterior pituitary** contains axons and endings of neurosecretory neurons that originate in the hypothalamus. The **anterior pituitary** contains nonneuronal endocrine cells that form a distinct gland. The two lobes are separate in structure and embryonic origins.

43.3a Under Regulatory Control by the Hypothalamus, the Anterior Pituitary Secretes Eight Hormones

The secretions of the anterior pituitary are under the control of peptide neurohormones, called **releasing hormones (RHs)** and **inhibiting hormones (IHs),** produced by the hypothalamus. These neurohormones are carried in the blood to the anterior pituitary in a *portal vein,* a special vein that connects the capillaries of the two glands. The portal vein provides a critical link between the brain and the endocrine system, ensuring that most of the blood reaching the anterior pituitary first passes through the hypothalamus.

RHs and IHs are **tropic hormones** (*tropic* means "stimulating," not to be confused with *trophic,* which means "nourishing") that regulate hormone secretion by another endocrine gland, in this case, the anterior

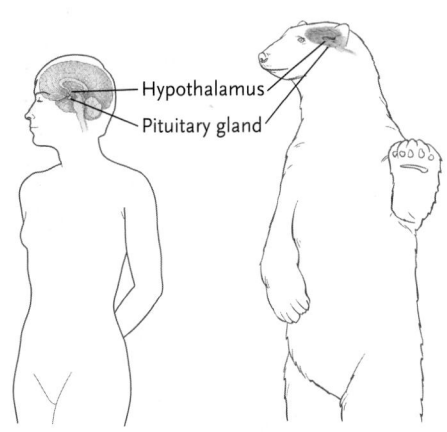

Hypothalamus
Pituitary gland

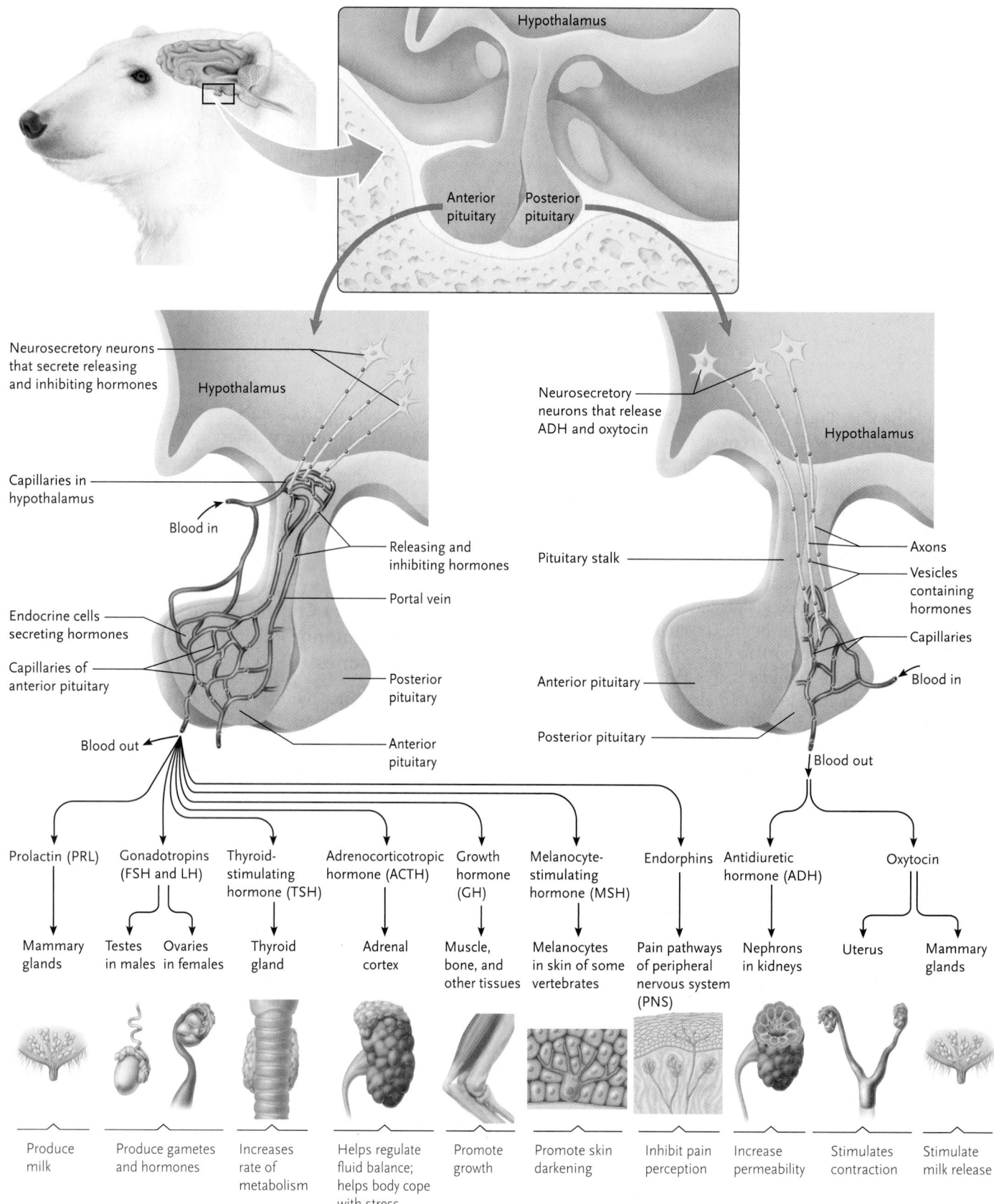

Figure 43.6

The hypothalamus and pituitary. Hormones secreted by the anterior and posterior pituitary are controlled by neurohormones released in the hypothalamus.

pituitary. The hormones of the anterior pituitary, in turn, control many other endocrine glands of the body and some body processes directly.

Under the control of the hypothalamus, the anterior pituitary secretes six major hormones into the bloodstream (see Figure 43.6): prolactin, growth hormone, thyroid-stimulating hormone, adrenocorticotropic hormone, follicle-stimulating hormone, and luteinizing hormone. **Prolactin (PRL)**, a *nontropic hormone* (a hormone that does not regulate hormone secretion by another endocrine gland), influences reproductive activities and parental care in vertebrates. In mammals, PRL stimulates development of the secretory cells of **mammary glands** during late pregnancy and milk synthesis after birth. Stimulation of the mammary glands and the nipples, as occurs during suckling, leads to PRL release. PRL occurs in nonmammalian vertebrates, where it has a variety of functions. In fish, for example, it is among the hormones controlling water balance. In all vertebrates, it has a role in promoting both maternal and paternal behaviour.

Growth hormone (GH) stimulates cell division, protein synthesis, and bone growth in children and adolescents, thereby causing body growth. GH also stimulates protein synthesis and cell division in adults. For these actions, GH acts as a tropic hormone by binding to target tissues, mostly liver cells, causing them to release **insulin-like growth factor (IGF)**, a peptide that directly stimulates growth processes. GH also acts as a nontropic hormone to control a number of major metabolic processes in mammals of all ages, including the conversion of glycogen to glucose and fats to fatty acids as a means of regulating their levels in the blood. GH also stimulates body cells to take up fatty acids and amino acids and limits the rate at which muscle cells take up glucose. These actions help maintain the availability of glucose and fatty acids to tissues and organs between feedings; this is particularly important for the brain. In humans, deficiencies in GH secretion during childhood produce *pituitary dwarfs,* who remain small in stature **(Figure 43.7)**. Overproduction of GH during childhood or adolescence, often due to a tumour of the anterior pituitary, produces *pituitary giants,* who may grow to above 2 m in height.

Many of the other hormones secreted by the anterior pituitary are tropic hormones that control endocrine glands elsewhere in the body. **Thyroid-stimulating hormone (TSH)** stimulates the thyroid gland to grow in size and secrete thyroid hormones. **Adrenocorticotropic hormone (ACTH)** triggers hormone secretion by cells in the adrenal cortex. **Follicle-stimulating hormone (FSH)** affects egg development in females and sperm production in males. It also has a tropic effect by stimulating the secretion of sex hormones in female mammals. **Luteinizing hormone (LH)** regulates part of the menstrual cycle in human females and the secretion of sex hormones in males. FSH and LH are grouped together as **gonadotropins** because they

Figure 43.7
The results of overproduction and underproduction of growth hormone by the anterior pituitary. The man on the left is of normal height. The man in the centre is a pituitary giant, whose pituitary produced excess GH during childhood and adolescence. The man on the right is a pituitary dwarf, whose pituitary produced too little GH.

regulate the activity of the gonads (ovaries and testes). The roles of the gonadotropins and sex hormones in the reproductive cycle are described in Chapter 41.

Melanocyte-stimulating hormone (MSH) and **endorphins** are nontropic hormones secreted by the anterior pituitary. MSH is named because of its effect in some vertebrates on melanocytes, skin cells that contain the black pigment melanin. An increase in secretion of MSH produces a marked darkening of the skin of fishes, amphibians, and reptiles. The darkening is produced by a dispersal of melanin in melanocytes so that it covers a greater area. In humans, an increase in MSH secretion also causes skin darkening, although the effect is by no means as obvious as in the other vertebrates mentioned. MSH secretion increases in pregnant women. Combined with the effects of increased estrogens, MSH results in increased skin pigmentation. The effects are reversed after the birth of the child.

Endorphins, nontropic peptide hormones produced by the hypothalamus and pituitary, are also released by the intermediate lobe of the pituitary. In the **peripheral nervous system (PNS)**, endorphins act as neurotransmitters in pathways that control pain, thereby inhibiting the perception of pain. Hence, endorphins are often called natural painkillers.

43.3b The Posterior Pituitary Secretes Two Hormones into the Body Circulation

The neurosecretory neurons in the posterior pituitary secrete two nontropic peptide hormones, antidiuretic hormone and oxytocin, directly into the body circulation (see Figure 43.6, p. 1054).

Antidiuretic hormone (**ADH**, also known as vasopressin) stimulates kidney cells to absorb more water from urine, thereby increasing the volume of the blood. The hormone is released when sensory receptor cells of the hypothalamus detect an increase in the blood's Na^+ concentration during periods of dehydration or after a salty meal. Ethyl alcohol and caffeine inhibit ADH secretion, explaining in part why alcoholic drinks and coffee increase the volume of urine excreted. Nicotine and emotional stress, in contrast, stimulate ADH secretion and water retention. After severe stress is relieved, the return to normal ADH secretion often makes a trip to the bathroom among our most pressing needs. The hypothalamus also releases a flood of ADH when an injury results in heavy blood loss or some other event triggers a severe drop in blood pressure. ADH helps maintain blood pressure by reducing water loss and by causing small blood vessels in some tissues to constrict.

Hormones with structure and action similar to those of ADH are also secreted in fishes, amphibians, reptiles, and birds. In amphibians, these ADH-like hormones increase the amount of water entering the body through the skin and from the urinary bladder.

Oxytocin stimulates the ejection of milk from the mammary glands of a nursing mother. Stimulation of the nipples in suckling sends neuronal signals to the hypothalamus and leads to the release of oxytocin from the posterior pituitary. The released oxytocin stimulates more oxytocin secretion by a positive feedback mechanism. Oxytocin causes the smooth muscle cells surrounding the mammary glands to contract, forcibly expelling the milk through the nipples. The entire cycle, from the onset of suckling to milk ejection, takes less than a minute in mammals. Oxytocin also plays a key role in childbirth (see Chapter 41).

In males, oxytocin is secreted into the seminal fluid by the testes. When the seminal fluid is ejaculated into the vagina during sexual intercourse, the hormone stimulates contractions of the uterus that aid movement of sperm through the female reproductive tract.

STUDY BREAK

1. Distinguish between tropic and nontropic hormones. What is the significance of having different types?
2. Distinguish between the anterior and the posterior pituitary. How is the release of hormones from each of these controlled?
3. If oxytocin stimulates more oxytocin release, what terminates milk ejection?

43.4 Other Major Endocrine Glands of Vertebrates

In addition to the hypothalamus and pituitary, the body has seven major endocrine glands or tissues, many of them regulated by the hypothalamus–pituitary connection. Included are the thyroid gland, parathyroid glands, adrenal medulla, adrenal cortex, gonads, pancreas, and pineal gland (shown in Figure 43.5, p. 1053, and summarized in Table 43.1, p. 1052).

43.4a The Thyroid Hormones Stimulate Metabolism, Development, and Maturation

The **thyroid gland** is located in the front of the throat in all terrestrial vertebrates, including humans, and is shaped like a bowtie. It also secretes the same hormones in all vertebrates. The thyroid hormones have an extraordinarily wide range of effects. The primary thyroid hormone, **thyroxine**, is known as T_4 because it contains four iodine atoms. The thyroid also secretes smaller amounts of a closely related hormone, **triiodothyronine** or T_3, which contains three iodine atoms. A supply of iodine in the diet is necessary for production of these hormones. Normally, their concentrations are kept at finely balanced levels in the blood by negative feedback loops such as the loop described in Figure 43.3, p. 1048. Most of the circulating hormone is bound to a transport protein, thyroglobulin, and only the free hormone is available to enter cells.

Once inside a cell, the T_4 is deiodinated, forming T_3. T_3 is the form that combines with internal receptors in the nucleus. Binding of T_3 to receptors alters gene expression, which brings about many of the hormone's effects.

Thyroid hormones are vital to growth, development, maturation, and metabolism in all vertebrates. They interact with GH for their effects on growth and development. Thyroid hormones also increase the sensitivity of many body cells to the effects of epinephrine and norepinephrine, hormones released by the adrenal medulla as part of the *fight-or-flight response* (discussed further below).

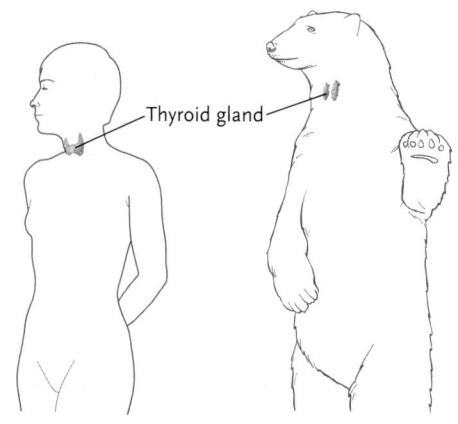

Thyroid gland

Insufficient iodine in the diet can cause *goitre,* enlargement of the thyroid. Without iodine, the thyroid cannot make T_3 and T_4 in response to stimulation by TSH. Because the thyroid hormone concentration remains low in the blood, TSH continues to be secreted, and the thyroid grows in size. Dietary iodine deficiency has been eliminated in developed regions of the world by the addition of iodine to table salt.

In amphibians, rising concentrations of thyroid hormones trigger **metamorphosis,** or a change in body form from tadpole to adult **(Figure 43.8).** Teleost fish undergo a form of metamorphosis during their early development, and the transformation from a hatchling *larval* form to a juvenile form is also triggered by rising concentrations of thyroid hormones. Curiously, however, the opposite is true in the agnathan lamprey. Its metamorphosis is triggered by decreasing concentrations of T_4. Thyroid hormones also contribute to seasonal moulting, leading to changes in the plumage of birds and coat colour in mammals.

The thyroid also has specialized cells that secrete **calcitonin,** a peptide originally discovered in fish by Harold Copp, working at the University of British Columbia. The hormone lowers the level of Ca^{2+} in the blood by inhibiting the ongoing dissolution of calcium from bone. Calcitonin secretion is stimulated when Ca^{2+} levels in blood rise above the normal range and inhibited when Ca^{2+} levels fall below the normal range.

Although the specialized cells of the thyroid are the principal source, calcitonin is also synthesized in the lung and intestine. In nonmammalian vertebrates, a separate gland, the ultimobrachial gland, produces calcitonin.

43.4b The Parathyroid Glands Regulate Ca^{2+} Levels in the Blood

The **parathyroid glands** occur only in tetrapod vertebrates (amphibians, reptiles, birds, and mammals). Each is a spherical structure about the size of a pea. Mammals have four parathyroids located on the posterior surface of the thyroid gland, two on each side. The single hormone they produce, a nontropic hormone called **parathyroid hormone (PTH),** is secreted in response to a fall in blood Ca^{2+} levels. PTH stimulates bone cells to dissolve the mineral matter of bone tissues, releasing both calcium and phosphate ions into the blood. The released Ca^{2+} is available for enzyme activation, conduction of nerve signals across synapses, muscle contraction, blood clotting, and other uses. How blood Ca^{2+} levels control PTH and calcitonin secretion is shown in **Figure 43.9, p. 1058.**

PTH also stimulates enzymes in the kidneys that convert **vitamin D,** a steroidlike molecule, into its fully active form in the body. The activated vitamin D increases the absorption of Ca^{2+} and phosphates from ingested food by promoting the synthesis of a calcium-binding protein in the intestine. It also increases the release of Ca^{2+} from bone in response to PTH.

PTH underproduction causes Ca^{2+} concentration to fall steadily in the blood, disturbing nerve and muscle function—the muscles twitch and contract uncontrollably, and convulsions and cramps occur. Without treatment, the condition is usually fatal because the severe muscular contractions, particularly of the muscles of the abdomen and thorax, interfere with breathing. Overproduction of PTH results in the loss of so much calcium from the bones that they become thin and fragile. At the same time, the elevated Ca^{2+} concentration in the blood causes calcium deposits to form in soft tissues, especially in the lungs, arteries, and kidneys (where the deposits form kidney stones).

Although fish do not have a parathyroid gland, they produce PTH, and PTH receptors are known to

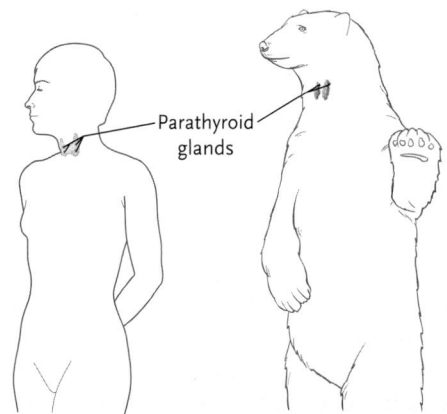

Figure 43.8

Metamorphosis of a tadpole into an adult frog, under the control of thyroid hormones. As part of the metamorphosis, changes in the gene activity lead to a change from an aquatic to a terrestrial habitat. TRH: thyroid-releasing hormone; TSH: thyroid-stimulating hormone.

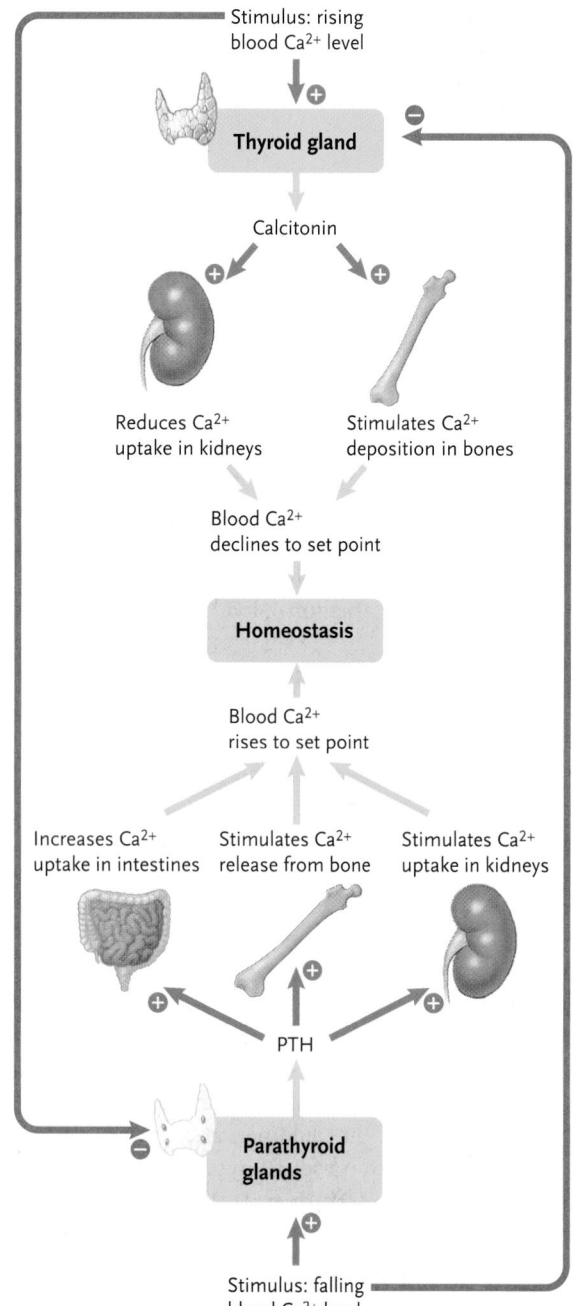

Stimulus: rising
blood Ca^{2+} level

Thyroid gland

Calcitonin

Reduces Ca^{2+}
uptake in kidneys

Stimulates Ca^{2+}
deposition in bones

Blood Ca^{2+}
declines to set point

Homeostasis

Blood Ca^{2+}
rises to set point

Increases Ca^{2+}
uptake in intestines

Stimulates Ca^{2+}
release from bone

Stimulates Ca^{2+}
uptake in kidneys

PTH

**Parathyroid
glands**

Stimulus: falling
blood Ca^{2+} level

Figure 43.9
Negative feedback control of PTH and calcitonin secretion by
blood Ca^{2+} levels.

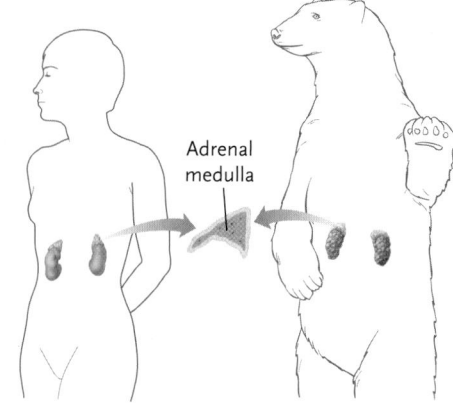

Adrenal
medulla

be present in fish, but the origin of the hormone and
its precise function remain uncertain.

43.4c The Adrenal Medulla Releases Two
Fight-or-Flight Hormones

The adrenal glands (*ad* = next to; *renes* = kidneys) of
mammals have two distinct regions. The central region,
the **adrenal medulla,** contains highly modified neurosecretory neurons that have lost their axons and dendrites. The tissue surrounding it, the **adrenal cortex,**
contains nonneural endocrine cells. The two regions
secrete hormones with entirely different functions.

Nonmammalian vertebrates have glands equivalent to
the adrenal medulla and adrenal cortex of mammals,
but the two parts are separate entities. Most of the hormones produced by these glands have essentially the
same functions in all vertebrates. The only major exception is aldosterone, which is secreted by the adrenal
cortex or its equivalent only in tetrapod vertebrates.

In most species, the adrenal medulla secretes two
nontropic amine hormones, **epinephrine** and **norepinephrine,** which are **catecholamines,** chemicals
derived from tyrosine that can act as hormones or neurotransmitters. They bind to receptors in the plasma
membranes of their target cells. Norepinephrine is
also released as a neurotransmitter by neurons of the
sympathetic nervous system.

Epinephrine and norepinephrine reinforce the
action of the sympathetic nervous system and are
secreted when the body encounters stresses such as
emotional excitement, danger (fight-or-flight situations),
anger, fear, infections, injury, and even midterm and
final exams. Epinephrine in particular prepares the body
for handling stress or physical activity. The heart rate
increases. Glycogen and fats break down, releasing glucose and fatty acids into the blood as fuel molecules. In
the heart, skeletal muscles, and lungs, the blood vessels
dilate to increase blood flow. Elsewhere in the body, the
blood vessels constrict, raising blood pressure, reducing
blood flow to the intestine and kidneys, and inhibiting
smooth muscle contractions, which reduces water loss
and slows down the digestive system. Airways in the
lungs also dilate, helping to increase the flow of air.

The effects of norepinephrine on heart rate, blood
pressure, and blood flow to the heart muscle are similar to those of epinephrine. However, in contrast to
epinephrine, norepinephrine causes blood vessels in
skeletal muscles to constrict. This contrary effect is
largely cancelled out because epinephrine is secreted
in much greater quantities.

43.4d The Adrenal Cortex Secretes Two
Groups of Steroid Hormones

The adrenal cortex of mammals secretes two major classes
of steroid hormones: **glucocorticoids** help maintain the

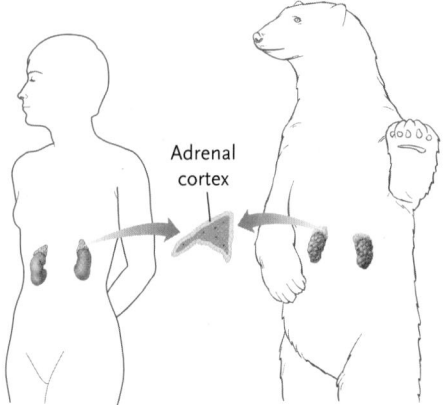

Adrenal
cortex

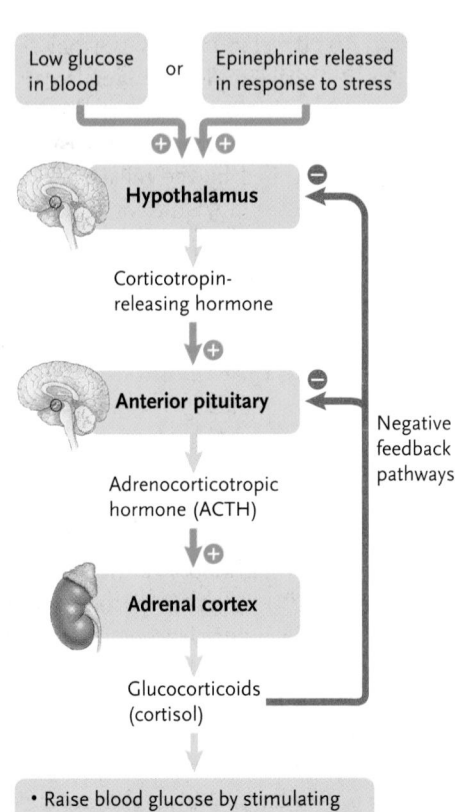

Figure 43.10

Pathways linking secretion of glucocorticoids to low blood sugar and epinephrine secretion in response to stress.

blood concentration of glucose and other fuel molecules, and **mineralocorticoids** regulate the levels of Na$^+$ and K$^+$ ions in the blood and extracellular fluid.

The Glucocorticoids. The glucocorticoids help maintain glucose levels in the blood by three major mechanisms: (1) stimulating the synthesis of glucose from noncarbohydrate sources such as fats and proteins; (2) reducing glucose uptake by body cells except those in the CNS; and (3) promoting the breakdown of fats and proteins, which releases fatty acids and amino acids into the blood as alternative fuels when glucose supplies are low. The favouring of glucose uptake in the CNS keeps the brain well supplied with glucose between meals and during periods of extended fasting. **Cortisol** is the major glucocorticoid secreted by the adrenal cortex.

Secretion of glucocorticoids is ultimately under the control of the hypothalamus **(Figure 43.10)**. Low glucose concentrations in the blood, or elevated levels of epinephrine secreted by the adrenal medulla in response to stress, are detected in the hypothalamus, leading to secretion of the tropic hormone ACTH by the anterior pituitary. ACTH promotes the secretion of glucocorticoids by the adrenal cortex.

Glucocorticoids also have anti-inflammatory properties; consequently, they are used to treat conditions such as arthritis and dermatitis. They also suppress the immune system and are used in the treatment of autoimmune diseases such as rheumatoid arthritis.

The Mineralocorticoids. In tetrapods, the mineralocorticoids, primarily **aldosterone**, increase the amount of Na$^+$ reabsorbed from the urine in the kidneys and absorbed from foods in the intestine. They also reduce the amount of Na$^+$ secreted by salivary and sweat glands and increase the rate of K$^+$ excretion by the kidneys. The net effect is to keep Na$^+$ and K$^+$ balanced at the levels required for normal cellular functions, including those of the nervous system. Secretion of aldosterone is linked tightly to blood volume and indirectly to blood pressure. The adrenal cortex also secretes small amounts of androgens, steroid sex hormones responsible for maintenance of male characteristics, which are synthesized primarily by the gonads.

43.4e The Gonadal Sex Hormones Regulate the Development of Reproductive Systems, Sexual Characteristics, and Mating Behaviour

The **gonads**, the testes and ovaries, are the primary source of sex hormones in vertebrates. The steroid hormones they produce, the **androgens**, **estrogens**, and **progestins**, have similar functions in regulating the development of male and female reproductive systems, sexual characteristics, and mating behaviour. Both males and females produce all three types of hormones, but in different proportions. Androgen production is predominant in males, whereas estrogen and progestin production is predominant in females. An outline of the actions of these hormones is presented here; a more complete picture was given in Chapter 41.

The **testes** (singular, *testis*) of male vertebrates secrete androgens, steroid hormones that stimulate and control the development and maintenance of male reproductive systems. The principal androgen is testosterone, the male sex hormone. In young adult males, a jump in **testosterone** levels stimulates puberty and the development of secondary sexual characteristics, including the growth of facial and body hair, muscle development, changes in vocal cord morphology, and

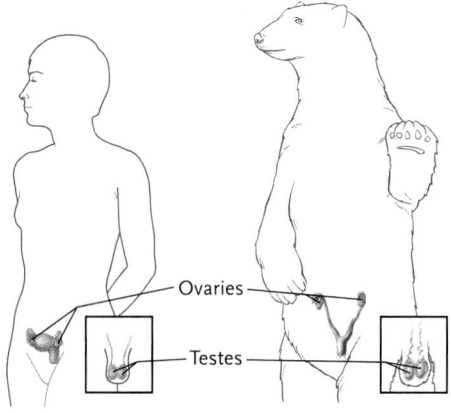

development of normal sex drive. The synthesis and secretion of testosterone by cells in the testes are controlled by the release of LH from the anterior pituitary, which, in turn, is controlled by **gonadotropin-releasing hormone (GnRH)**, a tropic hormone secreted by the hypothalamus.

Androgens are natural types of **anabolic steroids**, hormones that stimulate muscle development. Natural and synthetic anabolic steroids have been in the news over the years because of their use by bodybuilders and other athletes from sports in which muscular strength is important.

The **ovaries** of females produce estrogens, steroid hormones that stimulate and control the development and maintenance of female reproductive systems. The principal estrogen is **estradiol,** which stimulates maturation of sex organs at puberty and the development of secondary sexual characteristics. Ovaries also produce progestins, principally **progesterone,** the steroid hormone that prepares and maintains the uterus for implantation of a fertilized egg and the subsequent growth and development of an embryo. The synthesis and secretion of progesterone by cells in the ovaries are controlled by the release of FSH from the anterior pituitary, which, in turn, is controlled by the same GnRH as in males.

43.4f The Pancreatic Islets of Langerhans Hormones Regulate Glucose Metabolism

Most of the **pancreas**, a relatively large gland located just behind the stomach, forms an exocrine gland that secretes digestive enzymes into the small intestine (see Chapter 48). About 2% of the cells in the pancreas are endocrine cells that form the **islets of Langerhans.** Found in all vertebrates, the islets secrete the peptide hormones insulin and glucagon into the bloodstream.

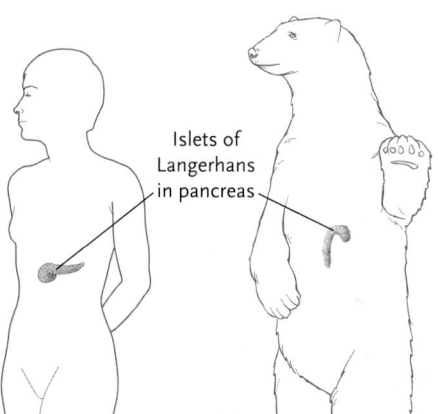

Insulin and glucagon regulate the metabolism of fuel substances in the body. **Insulin** (see "Molecule behind Biology," Box 43.1) is secreted by *beta cells* in the islets. It acts mainly on nonworking skeletal muscles, liver cells, and adipose tissue (fat). Brain cells do not require insulin for glucose uptake. Insulin lowers blood glucose, fatty acid, and amino acid levels and promotes their storage. The actions of insulin include stimulation of glucose transport into cells, glycogen synthesis from glucose, uptake of fatty acids by adipose tissue cells, fat synthesis from fatty acids, and protein synthesis from amino acids. Insulin inhibits glycogen degradation to glucose, fat degradation to fatty acids, and protein degradation to amino acids.

Glucagon, secreted by *alpha cells* in the islets, has effects opposite to those of insulin: it stimulates glycogen, fat, and protein degradation. Glucagon also results in amino acids and other noncarbohydrates being used for glucose synthesis; this aspect of glucagon function operates during fasting. Negative feedback mechanisms keyed to the concentration of glucose in the blood control secretion of both insulin and glucagon to maintain glucose homeostasis **(Figure 43.11)**.

Diabetes mellitus, a disease that afflicts more than 2 million people in Canada, results from problems with insulin production or action. The three classic diabetes symptoms are frequent urination, increased thirst (and consequently increased fluid intake), and increased appetite. Frequent urination occurs because the ability of body cells to take up glucose is impaired in diabetics, leading to abnormally high glucose concentration in the blood. Excretion of the excess glucose in the urine requires water to carry it, which causes increased fluid loss and frequent trips to the bathroom. The need to replace the excreted water causes increased thirst. Increased appetite comes about because cells have low glucose levels as a result of the insulin defect; therefore, proteins and fats are broken down as energy sources. Food intake is necessary to offset the negative energy balance, or weight loss will occur. Two of these classic symptoms gave the disease its name: diabetes is derived from a Greek word meaning "siphon," referring to the frequent urination, and mellitus, a Latin word meaning "sweetened with honey," refers to the sweet taste of a diabetic's urine. (Before modern blood or urine tests were developed, physicians tasted a patient's urine to detect the disease.)

The disease occurs in two major forms, called *type 1* and *type 2.* Type 1 diabetes, which occurs in about 10% of diabetics, results from insufficient insulin secretion by the pancreas. This type of diabetes is usually caused by an autoimmune reaction that destroys pancreatic beta cells. To survive, type 1 diabetics must receive regular insulin injections (typically, a genetically engineered human insulin called Humulin); careful dieting and exercise also have beneficial effects, because active skeletal muscles do not require insulin to take up and utilize glucose.

In type 2 diabetes, insulin is usually secreted at or above normal levels, but the target cells of affected people have significantly reduced responsiveness to the hormone compared with the cells of normal people. About 90% of patients in the developed world with type 2 diabetes are obese. A genetic

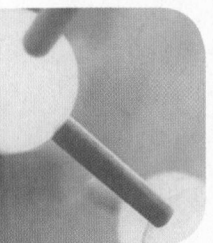

Insulin: More than Diabetes

Insulin was discovered in 1922 by J. R. Banting and his colleagues, J. J. R. McLeod, Charles Best, and James Collip, working at the University of Toronto. Banting and McLeod were awarded the Nobel Prize in 1923. (They were disturbed that their colleagues had not been recognized and shared the prize with them, Banting with Best, and McLeod with Collip.) In 1955, Frederick Sanger of Cambridge University worked out insulin's complete amino acid sequence (the first protein to be fully sequenced) and was awarded a Nobel Prize in 1958 for this work.

Insulin is a very large peptide of 51 amino acids. Molecular studies show that the insulin gene is present in all vertebrates but is absent from invertebrates. Insulin is a member of a family of genes. Two other members of that family encode two structurally related peptides, insulin-like growth factor (IGF) I and II. Although they are similar in structure to insulin, they have different but structurally related receptors, and their cellular action is different. They act as growth factors, regulating cell and tissue growth. IGFs are widely distributed in animals, protists, bacteria, and fungi. In molluscs, insects, and nematodes, these insulin-like peptides (ILPs) are neurohormones, expressed in neurosecretory cells in the brain. Surgical removal of these cells and their reimplantation demonstrate that they control growth, like the IGFs in vertebrates. The impact of these studies on the origin and evolution of the insulin gene is not yet clear.

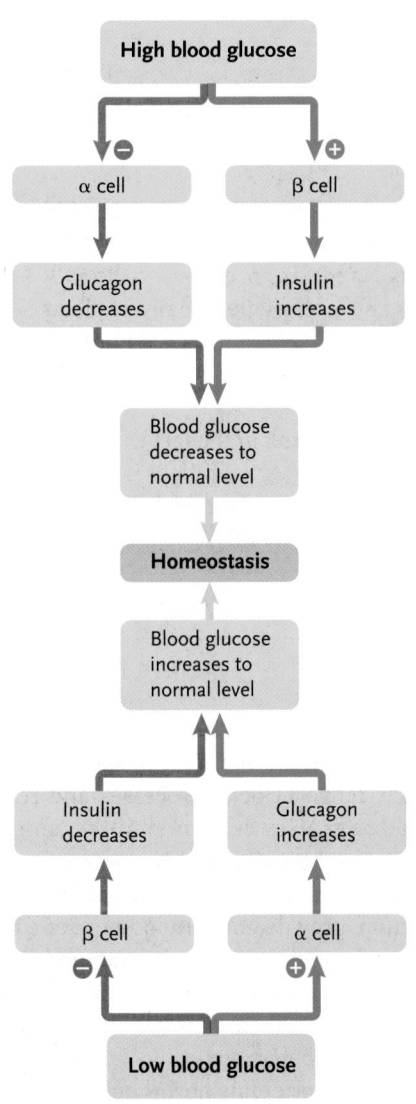

Figure 43.11

The action of insulin and glucagon in maintaining the concentration of blood glucose at an optimal level.

predisposition can also be a factor. Most affected people can lead a normal life by controlling their diet and weight, exercising, and taking drugs that enhance insulin action or secretion.

Diabetes has long-term effects on the body. The body's cells, unable to utilize glucose as an energy source, start breaking down proteins and fats to generate energy. The protein breakdown weakens blood vessels throughout the body, particularly in the arms and legs and in critical regions such as the kidneys and retina of the eye. The circulation becomes so poor that tissues degenerate in the arms, legs, and feet. At advanced stages of the disease, bleeding in the retina causes blindness. The breakdown of circulation in the kidneys can lead to kidney failure. In addition, in type 1 diabetes, acidic products of fat breakdown (**ketones**) are produced in abnormally high quantities and accumulate in the blood. The resulting lowering of blood pH can disrupt heart and brain function, leading to coma and death if the disease is untreated.

43.4g The Pineal Gland Regulates Some Biological Rhythms

The **pineal gland** is found at different locations in the brains of vertebrates. In most vertebrates, it is on the surface of the brain just under the skull and is directly sensitive to light. For the most part in birds and mammals, the gland is exclusively endocrine and no longer involved in photoreception. The pineal gland regulates some biological rhythms.

The earliest vertebrates had a third, light-sensitive eye at the top of the head, and *Sphenodon* and some lizards have an eyelike structure in this location. In most vertebrates, the third eye became modified into

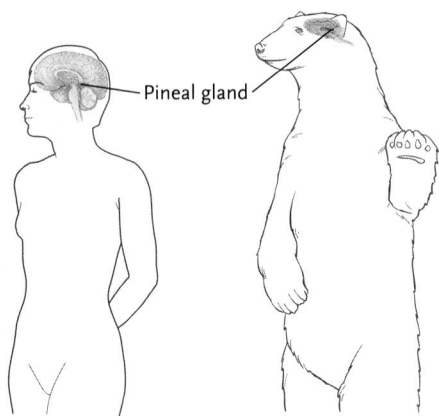

Pineal gland

a pineal gland, which in many groups retains some degree of photosensitivity. In mammals, it is too deeply buried in the brain to be affected directly by light; nonetheless, specialized photoreceptors in the eyes make connections to the pineal gland.

The pineal gland secretes the amine hormone **melatonin,** derived from tryptophan, which helps maintain daily biorhythms. Secretion of melatonin is regulated by an inhibitory pathway. Light hitting the eyes generates signals that inhibit melatonin secretion; consequently, the hormone is secreted most actively during periods of darkness. Melatonin targets a part of the hypothalamus called the *suprachiasmatic nucleus,* which is the primary structure coordinating body activity to a daily cycle. The nightly release of melatonin may help synchronize the biological clock with daily cycles of light and darkness. The physical and mental discomfort associated with jet lag may reflect the time required for melatonin secretion to reset a traveller's daily biological clock to match the period of daylight in a new time zone.

Melatonin occurs throughout the animal kingdom, as well as in many plants and fungi. In some fishes, amphibians, and reptiles, melatonin and other hormones produce changes in skin colour through their effects on *melanophores,* the pigment-containing cells of the skin. Skin colour may vary with the season, the animal's breeding status, or the colour of the background. In invertebrates, it is known to be important in the control of diurnal (daily) rhythms.

STUDY BREAK

1. What are the hormones controlling Ca^{2+} levels in the blood of vertebrates, and how do they control Ca^{2+} levels? Why is it important to control Ca^{2+} levels?
2. Distinguish between the adrenal medulla and the adrenal cortex. What hormones do they secrete, and what are their functions?
3. How are levels of glucose in the blood maintained? Which tissues do not require insulin to regulate glucose uptake? Why is this important?
4. How do you explain the fact that both epinephrine and norepinephrine produce the same effect in some cells but different effects in other cells?

43.5 Endocrine Systems in Invertebrates

Some invertebrates have fewer hormones, regulating a narrower range of body processes and responses, than vertebrates. However, in even the simplest animals, such as the cnidarian *Hydra,* hormones produced by neurosecretory neurons control the reproduction, growth, and development of some body features. In annelids, arthropods, and molluscs, endocrine cells and glands produce hormones that regulate development, reproduction, water balance, heart rate, sugar levels, and behaviour.

The known vertebrate hormones, particularly the peptides, also occur in a wide range of organisms. Thus, insulin-like hormones can be found in most invertebrates, and receptors are known from insects

The notion of *stress* may be familiar to all as a vague concept. It was originated by Hans Selye, working with rats at the University of Montreal, as the "General Adaptation Syndrome," in which what he called "nocuous agents" ranging from physical damage to psychological events led to the activation of the fight-or-flight response. This was characterized at the time as the activation of the hypothalamus–pituitary pathway leading to the release of a number of hormones, particularly those from the adrenal glands, which lead to increased heart rate and blood pressure. They also lead to release of glucose and fatty acids into the blood and increased blood flow to heart, lungs and skeletal muscle. This prepares the body for handling stress or physical activity. They reduce blood flow to organs not involved in the fight-or-flight response. If stress persists, however, continued release of these hormones causes cessation of growth; and, in severe cases, tissue damage and, ultimately, death.

Although we are accustomed to thinking of stress as a human or mammalian phenomenon, it has emerged as a response in a wide range of animals. Any unaccustomed sensory input (whether from external or internal events) leads to release of many hormones, and if the input continues, this can result in pathology or death. Stress, with the release of the corticosteroid hormones, is well studied in most vertebrates. Even in cockroaches, forced activity or forced inactivity will cause the release of many neurohormones, and the insect will die.

and nematodes. The protist *Tetrahymena* binds and exhibits responses to insulin and T_4. Whereas some hormones, such as the peptide proctolin and the insect juvenile hormones, do not occur in vertebrates, many of the growing number of peptide hormones identified in invertebrates have structural homologues in the vertebrates, although their functions may be different. Some peptides controlling diuresis in insects are structural homologues of vertebrate corticotropin-releasing factor. Other diuretic peptides in insects are related to calcitonin. The larva of the tapeworm *Spirometra mansonoides* has developed the capacity to secrete vertebrate growth hormone so that its host rat grows larger.

The endocrinology of the more complex invertebrates, formerly thought to be relatively simple, is emerging as very complex: about 200 bioactive peptides have thus far been described in insects, which are probably more closely governed by hormones than any other animals. The development of the eggs and the egg-laying behaviour of insects are controlled by more than a dozen hormones.

43.5a Hormones Regulate Development in Insects

Among the best known invertebrate hormonal systems is the one governing growth and development in insects **(Figure 43.12, p. 1064)**. As insects grow, they undergo a series of moults during which a new cuticle is laid down beneath the old cuticle and the old cuticle is shed (see Chapter 27). The signal to the epidermal cells to begin the process is provided by a steroid hormone, ecdysone, from the prothoracic glands. The prothoracic glands are stimulated to secrete ecdysone by a tropic peptide, prothoracicotropic hormone (PTTH), produced in neuroendocrine cells in the brain and released from the corpus cardiacum. The corpus cardiacum secretes several other hormones and contains both the nerve endings of neurons in the brain and neuroendocrine cells that lack axons and dendrites.

The corpus allatum is an endocrine gland that secretes **juvenile hormone**, a fatty acid derivative (see "Molecule behind Biology," Box 27.3, Chapter 27). Juvenile hormone controls metamorphosis: when it is present, the insect remains larval. In its absence, the next moult is metamorphic, producing a pupa and then an adult in those insects with a pupal stage or proceeding directly to the adult in those lacking a pupal stage. In the adult of most insects, the corpus allatum becomes active once more, secreting juvenile hormone and stimulating a number of reproductive processes, especially egg development. The secretion of juvenile hormone by the corpus allatum is controlled by both inhibitory and stimulatory tropic peptides from the brain.

The intricate process of shedding the old cuticle involves complex behaviours that are controlled by the interaction of up to five neurohormones, and the hardening of the new cuticle requires a sixth.

Hormones that control moulting have also been detected in crustaceans, including lobsters, crabs, and crayfish. During the period between moults, **moult-inhibiting hormone (MIH)**, a peptide neurohormone secreted by cells in the eyestalks (extensions of the brain leading to the eyes), inhibits ecdysone secretion. The first step in the moulting process is the inhibition of MIH secretion. Ecdysone secretion increases, and the processes leading to the

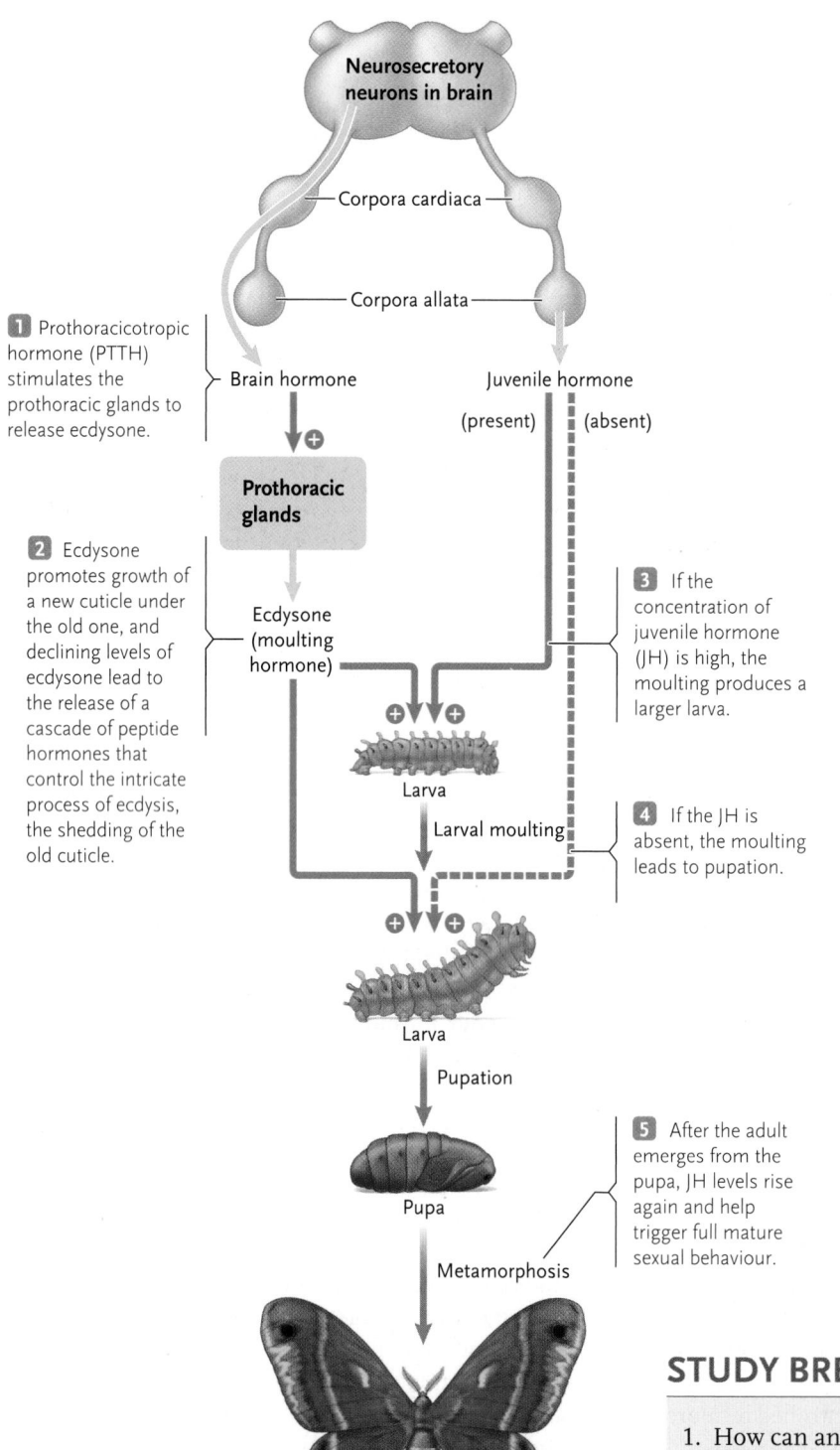

Figure 43.12

The roles of **brain hormone**, ecdysone, and juvenile hormone in the development of a silkworm moth.

① Prothoracicotropic hormone (PTTH) stimulates the prothoracic glands to release ecdysone.

② Ecdysone promotes growth of a new cuticle under the old one, and declining levels of ecdysone lead to the release of a cascade of peptide hormones that control the intricate process of ecdysis, the shedding of the old cuticle.

③ If the concentration of juvenile hormone (JH) is high, the moulting produces a larger larva.

④ If the JH is absent, the moulting leads to pupation.

⑤ After the adult emerges from the pupa, JH levels rise again and help trigger full mature sexual behaviour.

Neurosecretory neurons in brain

Corpora cardiaca

Corpora allata

Brain hormone

Juvenile hormone

(present) (absent)

Prothoracic glands

Ecdysone (moulting hormone)

Larva

Larval moulting

Larva

Pupation

Pupa

Metamorphosis

Adult

STUDY BREAK

1. How can an invertebrate surrounded by a rigid exoskeleton grow? Describe the process of moulting and growth in insects, and describe how it is regulated.
2. How are the pituitary of vertebrates and the corpus cardiacum of invertebrates similar? How do they differ?
3. The occurrence in many invertebrates of peptide hormones that closely resemble those in vertebrates is striking. What are the possible explanations for this in evolutionary terms? What research would you do to help you choose among the possibilities?

replacement of the exoskeleton are initiated. As in insects, metamorphosis and reproduction are governed by a hormone different from but structurally related to JH.

Review

43.1 Hormones and Their Secretion

- Hormones are molecules secreted by cells of the endocrine system that control the activities of cells elsewhere in the body. The cells that respond to a hormone are its target cells. This contrasts with the nervous system, which controls specific target cells in close proximity to its endings.

- The endocrine system includes four major types of cell signalling: classical endocrine signalling, in which endocrine glands secrete hormones; neuroendocrine signalling, in which neurosecretory neurons release neurohormones into the circulation; paracrine regulation, in which cells release local regulators that diffuse through the extracellular fluid to regulate nearby cells; and autocrine regulation, in which cells release local regulators that regulate the same cells that produced them.

- Most hormones and local regulators fall into one of four molecular classes: amines, peptides, steroids, and fatty acids.

- Neurosecretory neurons secrete hormones under direct control of the CNS.

- Many hormones are controlled by negative feedback mechanisms in which a hormone inhibits the reactions that synthesize or release it when its concentration rises in the body.

43.2 Mechanisms of Hormone Action

- Many hormones undergo modification after release that renders them more active.

- Only cells that have receptors for the hormone can respond to the hormone. Cells may respond by stimulation or inhibition of a process. Because of amplification involved in receptor mechanisms, hormones are present in body fluids at low concentrations. The response to a hormone may differ among cells and tissues.

- Hormones may bind to receptor proteins in the plasma membrane. When a receptor binds a hormone, its cytoplasmic end is activated, triggering a series of cytoplasmic reactions that may include the activation of protein kinases.

- Hydrophobic hormones bind to receptors in the cytoplasm or nucleus, activating them so that they can bind to the control sequences of specific genes in the cell nucleus. Binding to the control sequence either stimulates or inhibits transcription of the target gene, leading to changes in protein synthesis. They may also bind to membrane receptors.

- The major endocrine cells and glands of vertebrates are the hypothalamus, pituitary, thyroid gland, parathyroid gland, adrenal medulla, adrenal cortex, testes, ovaries, islets of Langerhans of the pancreas, and pineal gland.

43.3 The Hypothalamus and Pituitary

- The hypothalamus and pituitary together regulate many other endocrine cells and glands in the body.

- The posterior pituitary contains the terminals of neurosecretory cells in the hypothalamus.

- The anterior pituitary contains endocrine cells derived from nonnervous tissue.

- The hypothalamus produces tropic hormones (releasing hormones and inhibiting hormones) that control the secretion of eight hormones by the anterior pituitary. Prolactin (PRL), a nontropic hormone, regulates mammary gland development and milk secretion in mammals. As a nontropic hormone, growth hormone (GH) stimulates body growth in children and adolescents, and as a tropic hormone, it stimulates liver cells to make insulin-like growth factor (IGF), which stimulates growth processes. Melanocyte-stimulating hormone (MSH) controls reversible skin darkening. Endorphins can reduce pain. The four other hormones are tropic hormones: thyroid-stimulating hormone (TSH) stimulates secretion by the thyroid gland, adrenocorticotropic hormone (ACTH) regulates hormone secretion by the adrenal cortex, follicle-stimulating hormone (FSH) controls egg development and the secretion of sex hormones by the ovaries in female mammals and the production of sperm cells in males, and luteinizing hormone (LH) regulates part of the menstrual cycle in human females and the secretion of sex hormones in human males.

- Antidiuretic hormone (ADH) and oxytocin are secreted by the posterior pituitary. ADH regulates body water balance. In female mammals, oxytocin stimulates the contraction of smooth muscle in the uterus as part of childbirth and triggers milk release from the mammary glands during suckling of the young.

- The intermediate lobe of the pituitary produces the nontropic hormones MSH and endorphins. MSH secretion in some vertebrates produces a darkening of the skin. Endorphins are neurotransmitters that affect pain pathways in the peripheral nervous system, inhibiting the perception of pain.

43.4 Other Major Endocrine Glands of Vertebrates

- The thyroid gland secretes the thyroid hormones and, in mammals, calcitonin. In mammals, the thyroid hormones stimulate the oxidation of carbohydrates and lipids and coordinate with growth hormone to stimulate body growth and development. Calcitonin lowers the Ca^{2+} level in the blood by inhibiting the release of Ca^{2+} from bone. In many vertebrates, thyroid hormones control metamorphosis.

- The parathyroid gland secretes parathyroid hormone (PTH), which stimulates bone cells to release Ca^{2+} into the blood. PTH also stimulates the activation of vitamin D, which promotes Ca^{2+} absorption into the blood from the small intestine.

- The adrenal medulla secretes epinephrine and norepinephrine, which reinforce the sympathetic nervous system in responding to stress. The adrenal cortex secretes glucocorticoids and mineralocorticoids. Glucocorticoids help maintain glucose at normal levels in the blood; mineralocorticoids regulate Na^+ balance and extracellular fluid volume. The adrenal cortex also secretes small amounts of androgens.

- The gonadal sex hormones—androgens, estrogen, and progestins—regulate the development of reproductive systems, sexual characteristics, and mating behaviour. Both sexes secrete all three, but males primarily produce androgens (secreted by the testes), and females primarily produce estrogens and progestins (secreted by the ovaries).

- The islets of Langerhans of the pancreas secrete insulin and glucagon. Insulin lowers the concentration of glucose in the blood by stimulating glucose uptake by cells, glycogen synthesis from glucose, uptake of fatty acids by adipose tissue cells, fat synthesis from fatty acids, and protein synthesis from amino acids; it inhibits the conversion of noncarbohydrate molecules into glucose. Glucagon raises blood glucose by stimulating glycogen, fat, and protein degradation. The balance of insulin and glucagon regulates the concentration of fuel substances in the blood.

- The pineal gland secretes melatonin, which interacts with the hypothalamus to set the body's daily rhythms.

43.5 Endocrine Systems in Invertebrates

- Even the simplest protostomes use hormones to coordinate growth and reproduction.

- Many of the hormones that occur in vertebrates also occur in invertebrates, although their function may be different.

- Three major hormones—prothoracicotropic hormone (PTTH) from the brain, ecdysone from prothoracic glands, and juvenile hormone (JH) from the corpus allatum—control moulting and metamorphosis in insects. PTTH is a tropic hormone that stimulates the secretion of ecdysone, which initiates and maintains the secretion of the new cuticle. If JH is present, metamorphosis is suppressed, and in its absence, metamorphosis proceeds. The activity of the corpus allatum is governed by stimulatory and inhibitory tropic hormones from the brain. The shedding of the old cuticle and the hardening of the new cuticle are governed by a cascade of neuropeptides.

- JH controls reproduction in the adult insect.

- Similar hormones that control moulting and reproduction are also present in crustaceans, but the secretion of ecdysone is under the control of an inhibitory neurohormone.

Questions

Self-Test Questions

Any number of answers from a to e may be correct.

1. Which of the following is a chemical released from an epithelial cell in a gland that enters the blood to affect the activity of another cell some distance away?
   a. hormone
   b. neurohormone
   c. pheromone
   d. bile

2. Control of hormone secretion by a negative feedback mechanism generally includes which of the following?
   a. no change in hormone secretion
   b. decrease in hormone secretion
   c. change that maintains homeostasis
   d. increase in hormone secretion

3. When the concentration of thyroid hormone in the blood increases, which of the following does it do?
   a. It activates a positive feedback loop.
   b. It stimulates the pituitary to secrete TSH.
   c. It stimulates the pituitary to secrete TRH.
   d. It stimulates a secretion by the hypothalamus.
   e. It inhibits TRH secretion by the hypothalamus.

4. Which molecules directly regulate blood levels of calcium?
   a. PTH made by the pituitary
   b. vitamin D activated in the liver
   c. calcitonin secreted by specialized thyroid cells
   d. insulin synthesized by the alpha cells of the pancreas
   e. prolactin synthesized by the intermediate lobe of the pituitary

5. What is proctolin?
   a. a peptide secreted by insects
   b. a steroid acting on nematodes
   c. a hormone that is secreted by the intestine of mammals
   d. a hormone that acts on smooth muscle in invertebrates
   e. a hormone that governs egg development in invertebrates

6. Which function in mammals is NOT controlled by a hormone from the anterior pituitary?
   a. metabolic rate
   b. egg production
   c. milk production
   d. growth of muscle
   e. contraction of uterine muscles

7. During an acute stress response, what do catecholamines cause?
   a. an increase in heart rate
   b. an increase in blood pressure
   c. the breakdown of glycogen and fatty acids
   d. dilation of airways

8. Increased skeletal growth results directly and/or indirectly from the activity of which of the following?
   a. aldosterone
   b. growth hormone
   c. hormones produced by the hypothalamus
   d. progestins
   e. epinephrine

9. Diabetes mellitus is a condition that can be described by which of the following?
   a. lack of insulin
   b. lack of glucagon
   c. lack of antidiuretic hormone (ADH)
   d. lack of cortisol
   e. none of the above

10. Which event occurs when blood glucose rises in healthy humans?
    a. Target cells decrease their insulin receptors.
    b. Glucagon uses amino acids as an energy source.
    c. The beta cells of the pancreas increase insulin production.
    d. The alpha cells of the pancreas increase glucagon secretion.
    e. The pituitary secretes a tropic hormone controlling the pancreas.

Questions for Discussion

1. The occurrence in many invertebrates of peptide hormones that closely resemble those in vertebrates is striking. What are the possible explanations for this in evolutionary terms? What research would you do to help you choose among the possibilities?

2. Stress is commonly regarded as a diseaselike condition that can lead to death. Stresslike phenomena are widely spread in other taxa. If it is pathological, why has evolution not eliminated it? What are the advantages that have led to its retention?

3. Juvenile hormone in insects controls very different functions in the larva than in the adult. Propose different ways that this can occur, and design experiments to test these hypothesis.

4. An increasing number of sports figures have been charged and found guilty of using performance-enhancing drugs such as steroids, growth hormone, insulin like growth factors, erythropoietin, and other drugs. What are the benefits of each of these hormones for performance enhancement, and what are the potential side effects of each on the athlete?

5. Although this is less common, hormones are also involved in some positive feedback cycles. Identify one such case, and explain it in detail. Include a statement illustrating why this is beneficial.

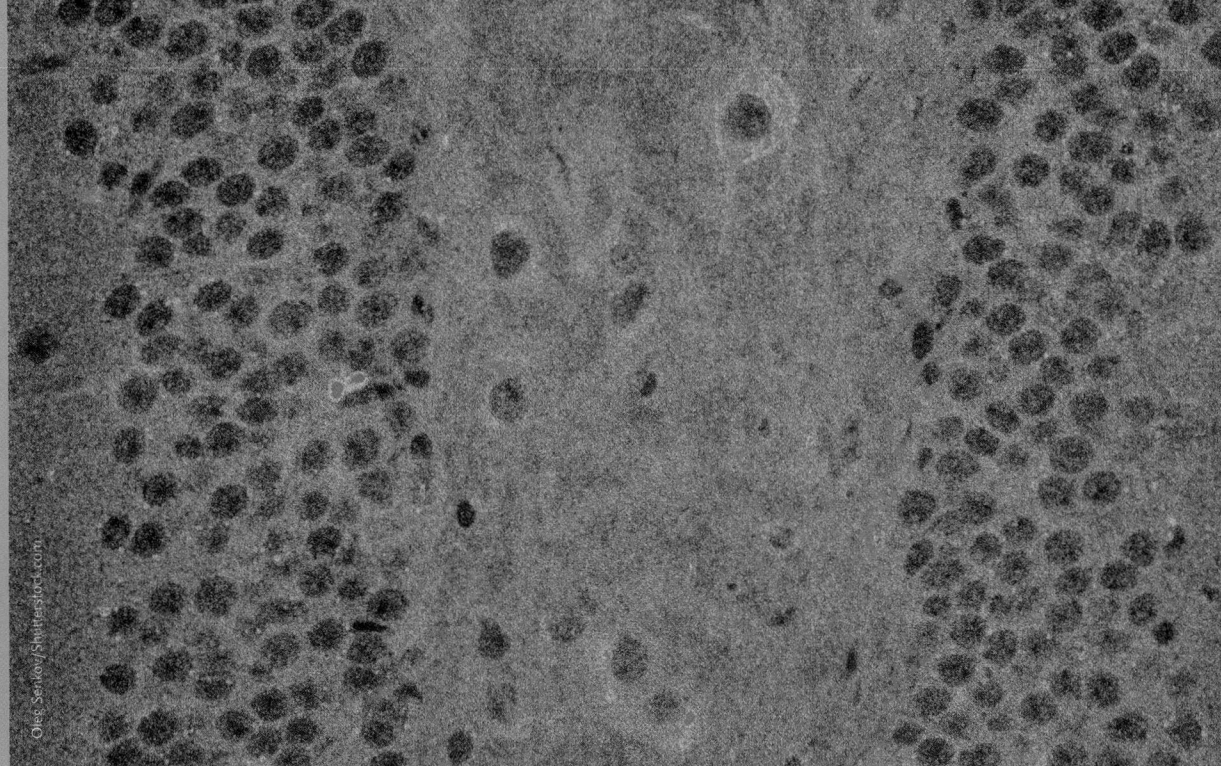

Neurons in the hippocampus of a mammalian brain. Cell bodies are stained in green.

Control of Animal Processes: Neural Control

WHY IT MATTERS

On a warm evening in early summer, the twilight in a garden in Montreal is punctuated by brief bursts of light from the abdomen of a flying male of the beetle *Photuris versicolor,* the firefly. The flashes of light have a specific duration and come at specific intervals, constituting a code unique to that species. These visible mating calls are answered by a female perched on the vegetation below **(Figure 44.1),** which emits flashes with the same code. The male flies toward her. This photonic conversation continues until the male lands on the vegetation and mates with the female, which then ceases flashing. A day or two later, the mated female has begun to make eggs and again responds to flashes from males flying overhead. But now she responds to and mimics codes of flashing from males of other species of firefly. A male, lured to her by her mimicry of the flashing code for his species, lands and expects to mate but becomes prey and provides nutrition, enabling her to enhance egg production.

This behaviour requires the complex interaction of cells within the nervous systems of the flies. The brain or **central nervous system (CNS)** of the male sends the appropriate rhythmic signals (the code) to the light-producing organ in his abdomen. It also sends signals to the muscles controlling the wings so that he can fly toward the female. The female's eyes detect the light flashes, and her nervous system processes the information, causing her brain to send the appropriate signals to her own light-producing organ so that she responds with the appropriate code. All of this can happen in under a second. Mating is a complex behaviour involving coordinated movements not only of the genital apparatus but also of the other appendages. This act of mating turns off the flashing response of the female and signals the endocrine system of the female to release the hormones involved in egg production. A chemical transferred by the male in his semen acts on the brain of the female so that it no longer responds to the code for her species and causes her to mimic the

Figure 44.1
Photinus species female with abdomen flashing.

codes of other species. Sensing of the environment leading to rapid, specific communication between cells in multicellular organisms is the domain of the nervous system and the topic of this chapter.

One of the consequences of multicellularity and tissue differentiation, discussed in Chapter 39, is that cells must communicate over long distances. There are several ways this can occur, but the fastest is via transmission by neurons. Most multicellular animals have a nervous system. Hundreds to billions of neurons (depending on the animal) compose the nervous system and work together carrying electrical messages that control every aspect of the body. Messages are carried in the form of action potentials, and the rate of conduction and strength of each message are affected by various physical and chemical factors. Neurons are connected to one another and other body tissues via chemical or electrical synapses. The detection and flow of information from the environment and between the individuals, and the instantaneous analysis and processing of that information to produce specific behaviour, are astounding, even in relatively simple animals. In this chapter, we first examine the properties of the cells that make up the nervous system that are responsible for receiving, transmitting, and analyzing the information. These functions result from the activities of only two major cell types: *neurons* and *glial cells*. In most animals, these cells are organized into complex networks called *nervous systems*. In the peripheral nervous system (PNS), the long slender projections of neurons (*axons*) are bundled into cablelike projections called **nerves** that provide a common pathway between different structures and the CNS. In the CNS, networks are organized into *ganglia* and *brains*.

44.1 Neurons and Their Organization in Nervous Systems: An Overview

Communication between cells in an animal by **neural signalling** involves the flow of information at two levels: within a single neuronal cell or **neuron** and between neurons within networks or circuits. In most animals, the four components of neural signalling are *reception, transmission, integration,* and *response.*

- **Reception** is the detection of a stimulus.
- **Integration** is the sorting and interpretation of sensory inputs or neural messages and determination of the appropriate response(s).
- **Transmission** is the sending of a message along a neuron to another neuron or to a muscle or gland.
- **Response** is the output or action resulting from the integration of neural messages. For a *P. versicolor* male flying at dusk, for example, sensors in the eye (see Chapter 45) detect flashes of light, and this information is transmitted to the brain, where it is integrated with internal information to determine whether or not to send outputs along nerves controlling the flight muscles commanding them to fly toward the female.

44.1a Neurons and Neural Circuits Are Specialized for the Reception, Integration, and Transmission of Informational Signals

Neurons vary widely in shape and size. All have an enlarged cell body and two types of extensions or processes: dendrites and axons **(Figure 44.2, p. 1070)**.

- **Reception:** The **dendrites**, and often the **cell body**, receive signals that they integrate and transmit toward a specialized part of the neuron called the spike initiation zone. The dendrites are generally highly branched, forming a treelike outgrowth at one end of the neuron (*dendros* = tree). Dendrites and cell bodies conduct graded electrical signals produced by ions flowing down electrochemical gradients through channels in the plasma membrane.
- **Integration:** The spike initiation zone is the first site along the neuron capable of generating an action potential. If the magnitude of the signal arriving from the dendrites and cell body is large enough, an action potential is initiated (see Section 44.2c).
- **Transmission:** Axons (*axon* = axis) conduct signals away from the spike initiation zone to another neuron or an effector. Neurons typically have a single axon that arises from a junction with the cell body called an **axon hillock**. The axon has branches at its tip that end as small, buttonlike swellings called **axon terminals**. The more terminals contacting a neuron, the greater its capacity to integrate incoming information.
- **Response:** This is the output or action resulting from the generation of the action potential. It could be the stimulation of another neuron, the release of a hormone from a gland, or the contraction of a muscle.

Figure 44.2

Neural signals: steps in processing of information in single neurons. There is considerable variation in the size and shape of neurons, but most of them are divided into distinct regions that serve in receiving signals, integrating them, conducting them, and transmitting them to other cells.

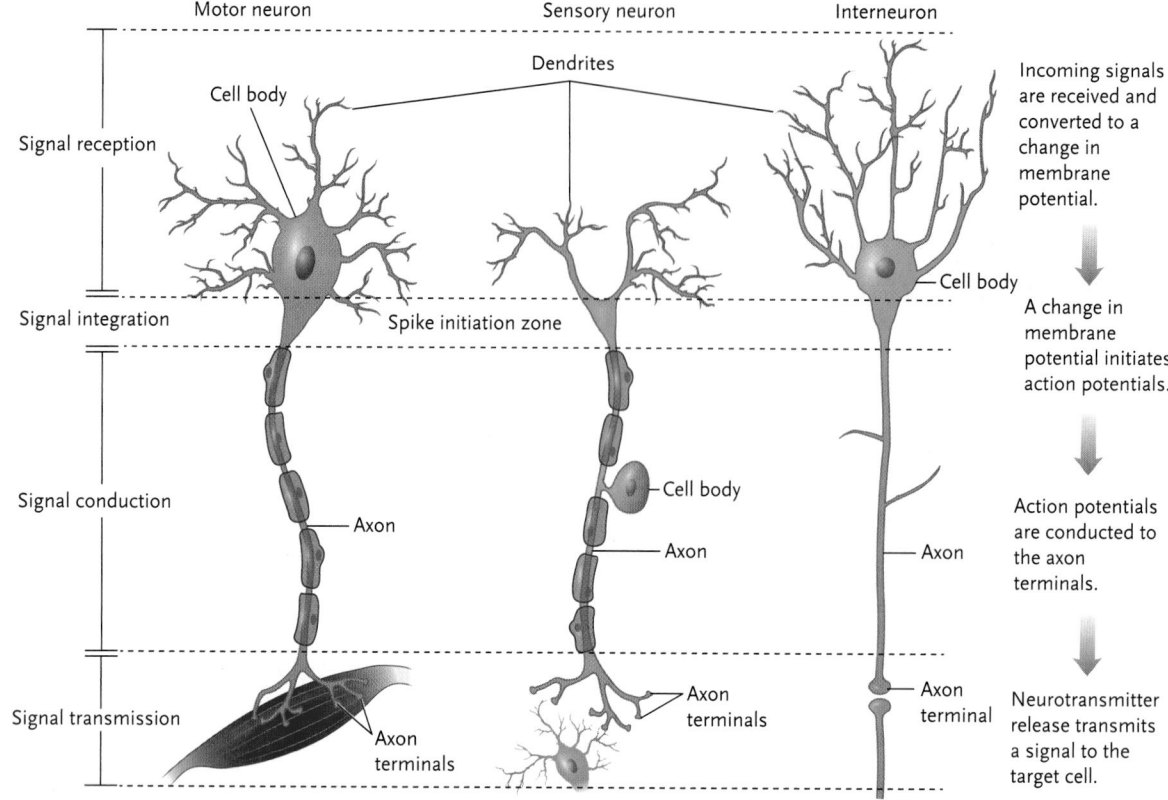

The cell body, which contains the nucleus and the majority of cell organelles, synthesizes most of the proteins, carbohydrates, and lipids of the neuron.

Connections between the axon terminals of one neuron and the dendrites or cell body of a second neuron link neurons into **neuronal circuits**. These circuits involve the same four components of neural signalling but at a higher level.

- **Reception:** Afferent neurons (also called **sensory neurons**) transmit stimuli collected by sensory receptors on those neurons to interneurons.
- **Integration:** Interneurons integrate the information to formulate an appropriate response. In humans and some other primates, 99% of all neurons are interneurons. Interneurons may receive input from several axons and may, in turn, connect to other interneurons and several efferent neurons. In this way, circuits combine into networks that interconnect the parts of the nervous system.
- **Transmission:** Efferent neurons carry the signals initiating a response away from the interneuron networks to the **effectors.**
- **Response:** This is carried out by muscles and glands. Efferent neurons that carry signals to skeletal muscle are called **motor neurons.**

In vertebrates, the afferent (sensory) neurons and efferent neurons collectively form the PNS. The interneurons form the brain and spinal cord, which make up the CNS. As depicted in **Figure 44.3**, afferent (carrying toward) information is ultimately transmitted to the CNS, where efferent (carrying away) information is initiated. The nervous systems of most invertebrates are also composed of central and peripheral divisions.

44.1b Neurons Are Supported Structurally and Functionally by Glial Cells

Glial cells are nonneuronal cells that provide nutrition and support to neurons. One type, called **astrocytes** because they are star-shaped **(Figure 44.4)**, were formerly thought to play only a supporting role in the CNS by maintaining ion concentrations in the interstitial fluid surrounding the neurons. More recently, however, scientists have realized that in vertebrates and some invertebrates, astrocytes communicate with neurons and may influence their activity (see next Concept Fix).

Two other types of glial cells, **oligodendrocytes** in the CNS and **Schwann cells** in the PNS, form tightly wrapped layers of plasma membrane, called myelin sheaths, around axons **(Figure 44.5, p. 1072)**. These myelin sheaths act as electrical insulators due to the membrane's high lipid content. The gaps between Schwann cells, called **nodes of Ranvier**, expose the axon membrane directly to extracellular fluids. This arrangement of insulated stretches of the axon punctuated by gaps speeds the rate at which electrical impulses move along the axons they protect.

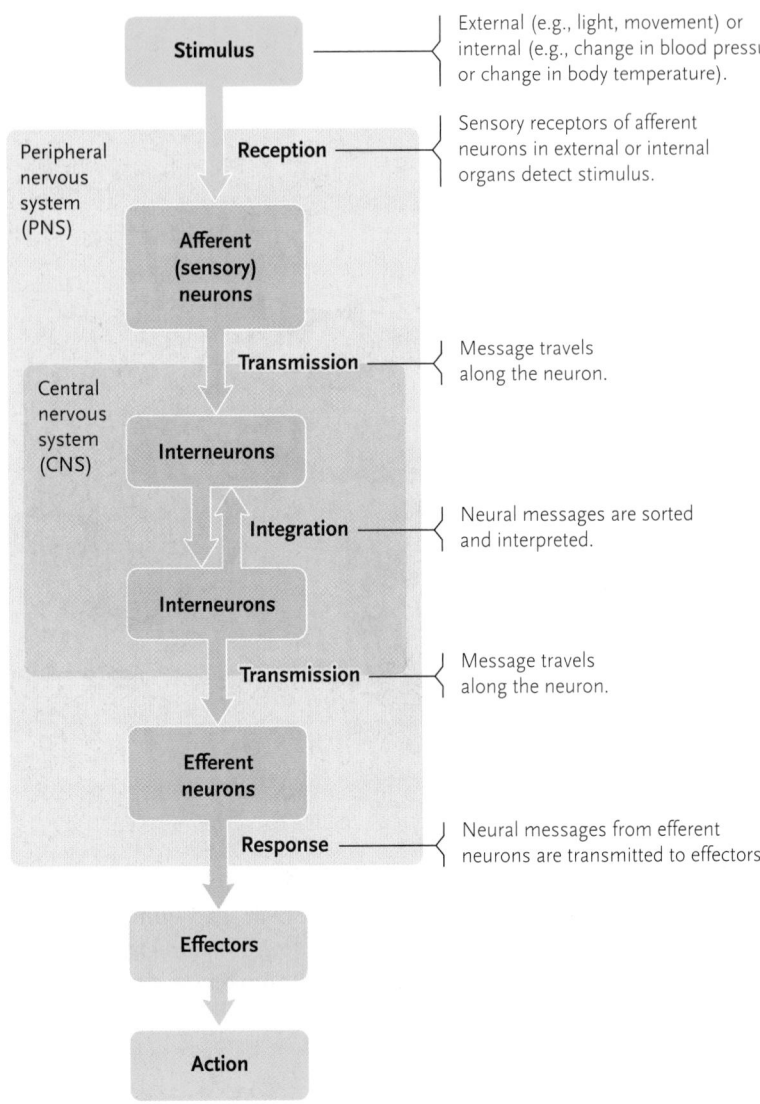

Figure 44.3

Neural signalling: the information-processing steps in neural circuits.

Stimulus — External (e.g., light, movement) or internal (e.g., change in blood pressure or change in body temperature).

Peripheral nervous system (PNS)

Reception — Sensory receptors of afferent neurons in external or internal organs detect stimulus.

Afferent (sensory) neurons

Transmission — Message travels along the neuron.

Central nervous system (CNS)

Interneurons

Integration — Neural messages are sorted and interpreted.

Interneurons

Transmission — Message travels along the neuron.

Efferent neurons

Response — Neural messages from efferent neurons are transmitted to effectors.

Effectors

Action

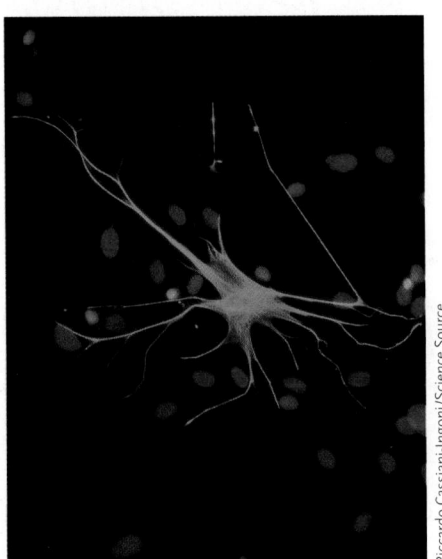

Figure 44.4

An astrocyte.

Riccardo Cassiani-Ingoni/Science Source

Unlike most neurons, glial cells retain the capacity to divide throughout the life of the animal. This capacity allows glial tissues to replace damaged or dead cells but also makes them the source of almost all brain tumours, which are produced when regulation of glial cell division is lost.

CONCEPT FIX That glial cells only support and provide nutrition to neurons no longer appears to be an accurate assessment of their roles. Some glial cells regulate the clearance of neurotransmitters from the synaptic cleft and release factors such as ATP, which modulate presynaptic function. During early embryogenesis, other glial cells also direct the migration of neurons and produce molecules that modify the growth of axons and dendrites. In the past, glial cells were not believed to have chemical synapses or to release neurotransmitters. They were regarded as a "glue" in the nervous system, as their name implies. More recently it has been shown that the only notable differences between neurons and glial cells are that the latter lack axons and dendrites and they lack the ability to generate action potentials. Attention is now being paid to their roles in the formation and modulation of synapses and in the repair of neurons after injury. ⬡

CONCEPT FIX Neurons are sometimes called nerve cells, but this is technically incorrect. Many neurons, such as those within the CNS, do not form nerves, and nerves usually contain nonneuronal cells, such as Schwann cells, that coat the axons in myelin. ⬡

44.1c Neurons Function as Circuits and Communicate via Synapses

A **synapse** (*synapsis* = juncture) is a site where a neuron makes a communicating connection with either another neuron or an effector such as a muscle fibre or gland. On one side of the synapse is the axon terminal of a **presynaptic cell**, the neuron that transmits the signal. On the other side is the dendrite or cell body of a **postsynaptic cell**, the neuron or the surface of an effector that receives the signal. Communication across a synapse may occur by the direct flow of an electrical signal or by means of a

Figure 44.5
Myelinated neurons have axons wrapped in Schwann cells that coat the axon in a myelin sheath, which acts as an electrical insulator.

As many as 300 overlapping layers of the Schwann cell plasma membrane wind around an axon like a jelly roll.

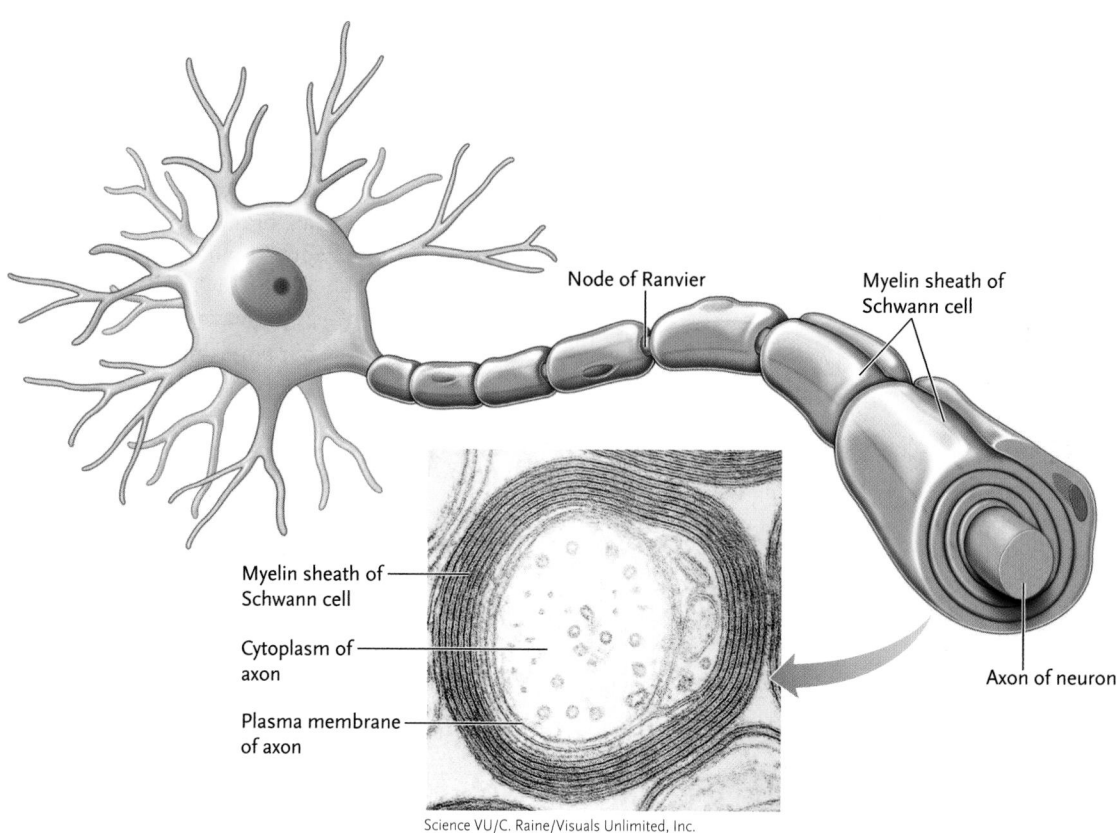

Node of Ranvier

Myelin sheath of Schwann cell

Myelin sheath of Schwann cell

Cytoplasm of axon

Plasma membrane of axon

Axon of neuron

Science VU/C. Raine/Visuals Unlimited, Inc.

neurotransmitter, a chemical released by an axon terminal at a synapse.

In **electrical synapses**, the plasma membranes of the presynaptic and postsynaptic cells are in direct contact **(Figure 44.6a)**. When an electrical impulse arrives at the axon terminal, gap junctions (see Chapter 39) allow ions to flow directly between the two cells, leading to unbroken transmission of

a. Electrical synapse

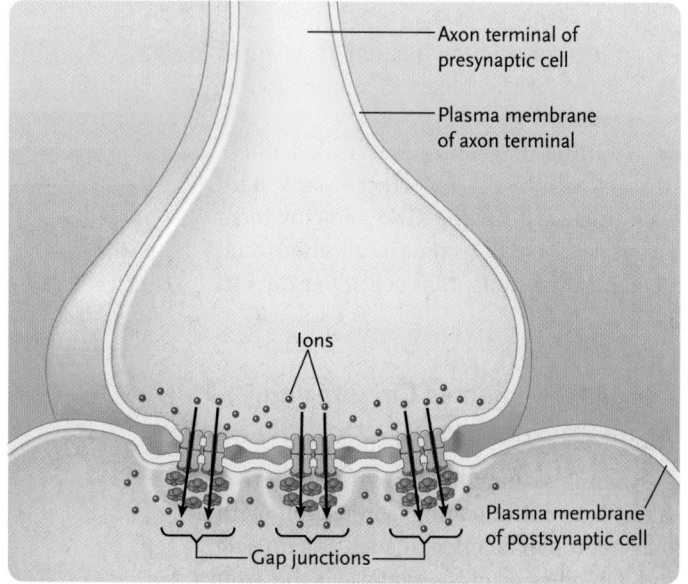

Axon terminal of presynaptic cell

Plasma membrane of axon terminal

Ions

Gap junctions

Plasma membrane of postsynaptic cell

In an electrical synapse, the plasma membranes of the presynaptic and post-synaptic cells make direct contact. Ions flow through gap junctions that connect the two membranes, allowing impulses to pass directly to the postsynaptic cell.

b. Chemical synapse

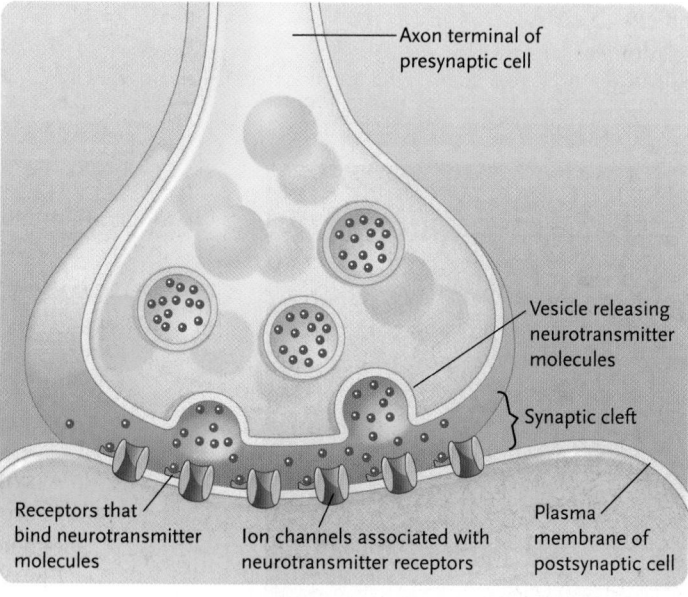

Axon terminal of presynaptic cell

Vesicle releasing neurotransmitter molecules

Synaptic cleft

Receptors that bind neurotransmitter molecules

Ion channels associated with neurotransmitter receptors

Plasma membrane of postsynaptic cell

In a chemical synapse, the plasma membranes of the presynaptic and post-synaptic cells are separated by a narrow synaptic cleft. Neurotransmitter molecules diffuse across the cleft and bind to receptors in the plasma membrane of the postsynaptic cell. The binding opens channels to ion flow that may generate an impulse in the postsynaptic cell.

Figure 44.6
The two types of synapses by which neurons communicate with other neurons or effectors.

the electrical signal. Electrical synapses are useful for two types of functions:

- They allow for very rapid transmission. They were first discovered in the nervous system of crayfish, where they are involved in the rapid movements needed for escape from predators.
- They allow for synchronous activity in a group of neurons. For example, the neurons controlling the secretion of hormones from the hypothalamus of mammals are connected by electrical synapses, thus ensuring a coordinated burst of secretion of some hormones.

The vast majority of vertebrate neurons, however, communicate by means of neurotransmitters (Figure 44.6b). In these **chemical synapses,** the plasma membranes of the presynaptic and postsynaptic cells are separated by a narrow gap, about 25 nm wide, called the **synaptic cleft.** When an electrical impulse arrives at an axon terminal, it causes the release of a neurotransmitter into the synaptic cleft. The neurotransmitter diffuses across the synaptic cleft and binds to a receptor in the plasma membrane of the postsynaptic cell. If enough neurotransmitter molecules bind to these receptors, an action potential will be generated at the spike initiation zone of the postsynaptic cell that will then travel along its axon to reach a synapse with the next neuron or effector in the circuit. Many factors determine whether a new electrical impulse will be generated in the postsynaptic cell, including the levels of neurotransmitters that excite the cell and neurotransmitters that inhibit that cell rather than stimulate it. The balance of stimulatory and inhibitory effects in chemical synapses contributes to the integration of incoming information in a receiving neuron.

STUDY BREAK

1. Describe *reception, integration, transmission,* and *response* at the level of a single neuron as well as in a complex neural circuit.
2. What are the differences between an electrical synapse and a chemical synapse? What are the advantages of each?

44.2 Signal Initiation by Neurons

44.2a All Cells Have a Resting Membrane Potential

As you learned in Chapter 5, plasma membranes are *selectively* permeable in that they allow some molecules but not others to move across the membrane through protein channels embedded in the phospholipid bilayer. Many of these molecules are charged

(remember that **anions** are ions that carry a positive charge and cations are ions that carry a negative charge; see *The Purple Pages*). All animal cells have more negatively charged molecules (anions, amino acids, nucleic acids, etc.) inside the cell than outside the cell. The separation of positive and negative charges across the plasma membrane produces an electrical gradient or potential difference across the plasma membrane called a **membrane potential.** All cells exhibit a resting potential and are said to be *polarized.*

In all cells at rest, there are more K^+ ions inside the cell, with a net tendency to diffuse out due to their concentration gradient, but a net tendency to remain inside the cell due to their electrical charge (like charges repel and unlike charges attract). These don't quite balance out, and thus there is a small net tendency (i.e., a small net **electrochemical gradient**) for K^+ to diffuse out of the cell through various channels. There are more Na^+ ions outside the cell than inside, and both electrical and concentration gradients (i.e., electrochemical gradient) favour movement into the cell. The plasma membrane, however, is not normally very permeable to Na^+, and so only a small amount leaks in. Plasma membrane–embedded Na^+/K^+ active transport pumps use energy from ATP hydrolysis to simultaneously pump three Na^+ out of the cell for every two K^+ pumped in, maintaining a steady state and a steady resting membrane potential. For most cells, this membrane potential is very stable. The distribution of ions inside and outside an axon that produces the membrane potential is shown in **Figure 44.7, p. 1074.**

Like all cells, neurons also have a voltage difference between the inside and the outside of the cell. However, neurons and muscle cells are *excitable.* In response to electrical, chemical, mechanical, and certain other types of stimuli, their membrane potential changes rapidly and transiently. Excitability, produced by a sudden flow of ions across the plasma membrane, is the basis for nerve impulse generation. It also gives neurons the ability to store, recall, and distribute information.

The membrane of a neuron that is not conducting an impulse exhibits a steady negative membrane potential called the **resting potential** because the neuron is at rest. The resting potential has been measured at about -70 mV in isolated neurons. The change in membrane potential that occurs when a neuron is excited is the action potential, which will be discussed shortly. As we will see in the following discussion, the action potential results from the opening and closing of *voltage-gated ion channels* for Na^+ and K^+.

CONCEPT FIX The statement that the diffusion of ions is due to differences in concentration between two sites is not always true. While ions diffuse along concentration gradients, this is not the only factor that

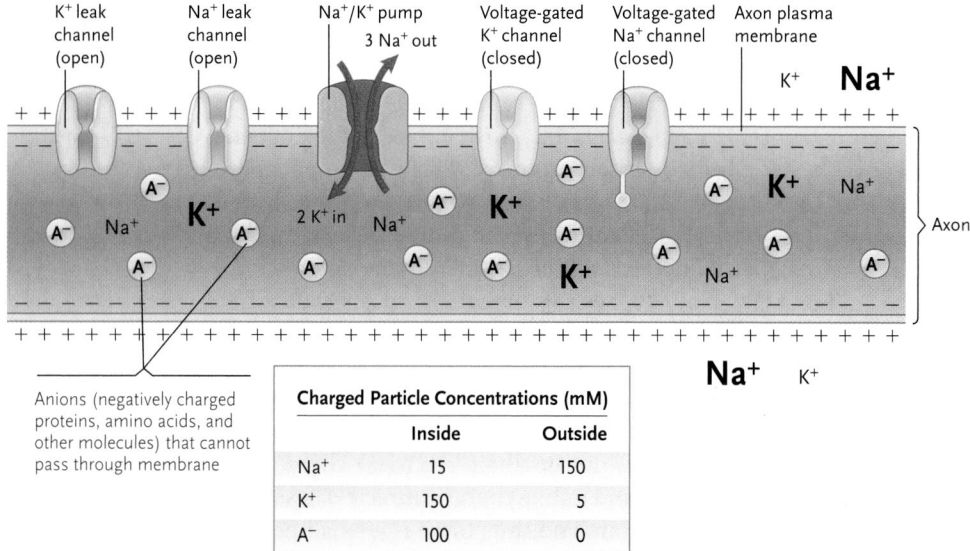

K⁺ leak channel (open) · **Na⁺ leak channel (open)** · **Na⁺/K⁺ pump 3 Na⁺ out** · **Voltage-gated K⁺ channel (closed)** · **Voltage-gated Na⁺ channel (closed)** · **Axon plasma membrane**

Anions (negatively charged proteins, amino acids, and other molecules) that cannot pass through membrane

Charged Particle Concentrations (mM)		
	Inside	Outside
Na⁺	15	150
K⁺	150	5
A⁻	100	0

Figure 44.7

The distribution of ions inside and outside an axon that produces the resting potential, -70 mV. A⁻ incorporates the distribution of Cl⁻ and other negatively charged anions, amino acids, and nucleic acids. Na⁺ and K⁺ diffuse along their electrochemical gradients through *leak* channels, and the Na⁺/K⁺ pump returns them back again. The voltage-gated ion channels only open when the membrane potential changes.

determines the direction of the net diffusion of ions. Ions really diffuse along energy gradients, and the energy gradient for any specific ion is also determined by electrical charge (for charged molecules), temperature, and pressure. Diffusion is only due to concentration gradients alone if all other factors are equal, something that is not common in living cells. ⬡

44.2b Graded Potentials Can Occur in All Cells Due to Changes in Membrane Permeability to Ions and Can Vary in Magnitude

The permeability of most membranes to ions is not constant but changes as various channels in the membrane open and close. If, as a result of such events, positively charged ions (such as Na⁺) enter the cell or negatively charged ions leave the cell, the charge across the membrane will become less negative and the membrane will be less polarized, or **depolarized (Figure 44.8)**. If, on the other hand, positively charged ions (such as K⁺) leave the cell or negatively charged ions (such as Cl⁻) enter, the membrane will become more polarized, or **hyperpolarized**. The more ions that cross the membrane, the greater the depolarization or hyperpolarization; that is, these events are graded. As the ions are restored to their initial levels, the membrane is repolarized and the

cell membrane returns to its resting membrane potential. These changes in membrane potential due to changes in membrane permeability to ions are called graded potentials. Because the ions that enter through any channel will disperse within the cell, graded potentials will radiate over the cell membrane, decreasing in magnitude as they move farther away from the open ion channel. They can occur in any cell and are not confined to nerve cells. In neurons, graded potentials are part of the integration that takes place in dendrites and cell bodies.

44.2c An Action Potential Is Not Graded but Is a Rapid, Reversible Event

Although graded potentials cannot be transmitted over long distances, action potentials can. Only certain parts of a neuron can generate action potentials (as a rule, dendrites and cell bodies cannot), and the site where action potentials are initiated in the neuron is referred to as the spike initiation zone (or trigger zone; see Figure 44.2, p. 1070). If graded potentials spreading over the dendrites and cell body are sufficient to depolarize the membrane at the site of the spike initiation zone to a level known as the **threshold potential**, about -50 to -55 mV in isolated neurons, an action potential occurs at this site. In less than 1 ms ($=$ millisecond), the inside of the plasma membrane at this site becomes positive because of an influx of positive ions across the cell membrane, momentarily reaching a value of $+30$ mV or more **(Figure 44.9)**. The membrane potential at

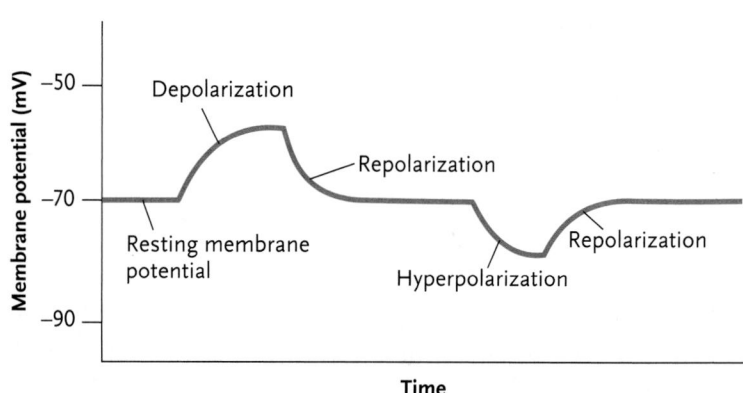

Figure 44.8

Changes in membrane potential in a cell due to changes in ion permeability. The resting membrane potential in this cell is -70 mV. During depolarization, the membrane potential becomes less negative. During hyperpolarization, the membrane potential becomes more negative. During repolarization, the membrane potential returns to its resting level.

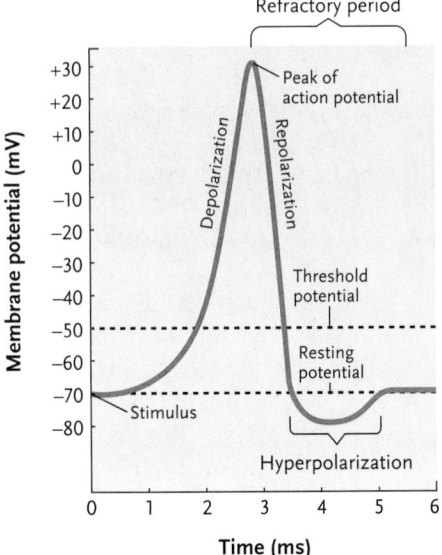

Figure 44.9
Changes in membrane potential during an action potential.

as the membrane potential changes (see **Figure 44.10, p. 1076**). Voltage-gated Na⁺ channels have two gates, an *activation gate* and an *inactivation gate*, whereas voltage-gated K⁺ channels have one gate, an *activation gate*.

Figure 44.10 shows how the two voltage-gated ion channels operate when generating an action potential. When the membrane is at its resting potential, the activation gates of both the Na⁺ and the K⁺ channel are closed. A depolarizing stimulus, such as a neurotransmitter substance, produces a graded depolarization that spreads to the spike initiation zone and raises the membrane potential to the threshold. This change in membrane charge pulls the activation gate of the voltage-gated Na⁺ channels open, allowing Na⁺ ions to flow into the axon along their electrochemical gradient. Once above the threshold, the more the membrane depolarizes, the more Na⁺ channels open, causing a rapid inward flow of positive charges that raises the membrane potential to the peak of the action potential. As the action potential peaks, the change in charge at the plasma membrane causes the inactivation gates of the Na⁺ channels to close (resembling putting a stopper in a sink), which stops the inward flow of Na⁺. The refractory period now begins.

At the same time, the activation gates of the K⁺ channels begin to open, allowing K⁺ ions to flow rapidly outward in response to their electrochemical gradient (the cell is now positive on the inside and negative on the outside so that both concentration and electrical gradients favour the movement of K⁺ out of the cell). The movement of K⁺ ions contributes to the refractory period and compensates for the inward movement of Na⁺ ions, returning the membrane to the resting potential. As the resting potential is re-established, the activation gates of the K⁺ channels close, as do those of the Na⁺ channels, and the inactivation gates of the Na⁺ channels open. These events end the refractory period and ready the membrane for another action potential. (The opening of the inactivation gates and the resetting of the activation gates on the NA⁺ and K⁺ channels end the refractory period and allow another action potential to be generated.) The opening and closing of the gates are the result of interactions between charge on the plasma membrane and charge on the gates themselves, either pulling the gates toward the membrane and opening them or pushing the gates away and closing them; this is the basis of the operation of all voltage-gated channels.

In some neurons, closure of the voltage-gated K⁺ channels lags, and K⁺ continues to flow outward for a brief time after the membrane returns to the resting potential. This excess outward flow causes the hyperpolarization shown in Figure 44.9 and Figure 44.10 (step 6), in which the membrane potential dips briefly below the resting potential.

At the end of an action potential, the membrane potential has returned to its resting state, but the ion distribution has been changed slightly. That is, some

this site then falls, in many cases becoming hyperpolarized and dropping to about −80 mV before rising again to the resting potential. The entire change, from initiation of the action potential to the return to the resting potential, takes less than 5 ms in the fastest neurons. Action potentials take the same basic form in neurons of all types, although there may be differences in the values of the resting potential and the peak of the action potential and in the time required to return to the resting potential.

An action potential is produced only if the depolarization arriving at the spike initiation zone is strong enough to cause the membrane potential to reach threshold. Furthermore, once triggered, the changes in membrane potential take place independently of the strength of the stimulus. No matter how strong the stimulus, once threshold is reached, the action potential will always be the same size. This is referred to as the **all-or-nothing principle.**

Beginning at the peak of an action potential, the membrane enters a **refractory period** of a few milliseconds, during which the threshold required for generation of an action potential is much higher than normal. The refractory period lasts until the membrane has stabilized at the resting potential. As we will see, the refractory period keeps impulses travelling in a one-way direction in neurons.

44.2d The Action Potential Is Produced by Ion Movements through the Plasma Membrane

The action potential is produced by movements of Na⁺ and K⁺ through the plasma membrane that are controlled by specific **voltage-gated ion channels,** membrane-embedded proteins that open and close

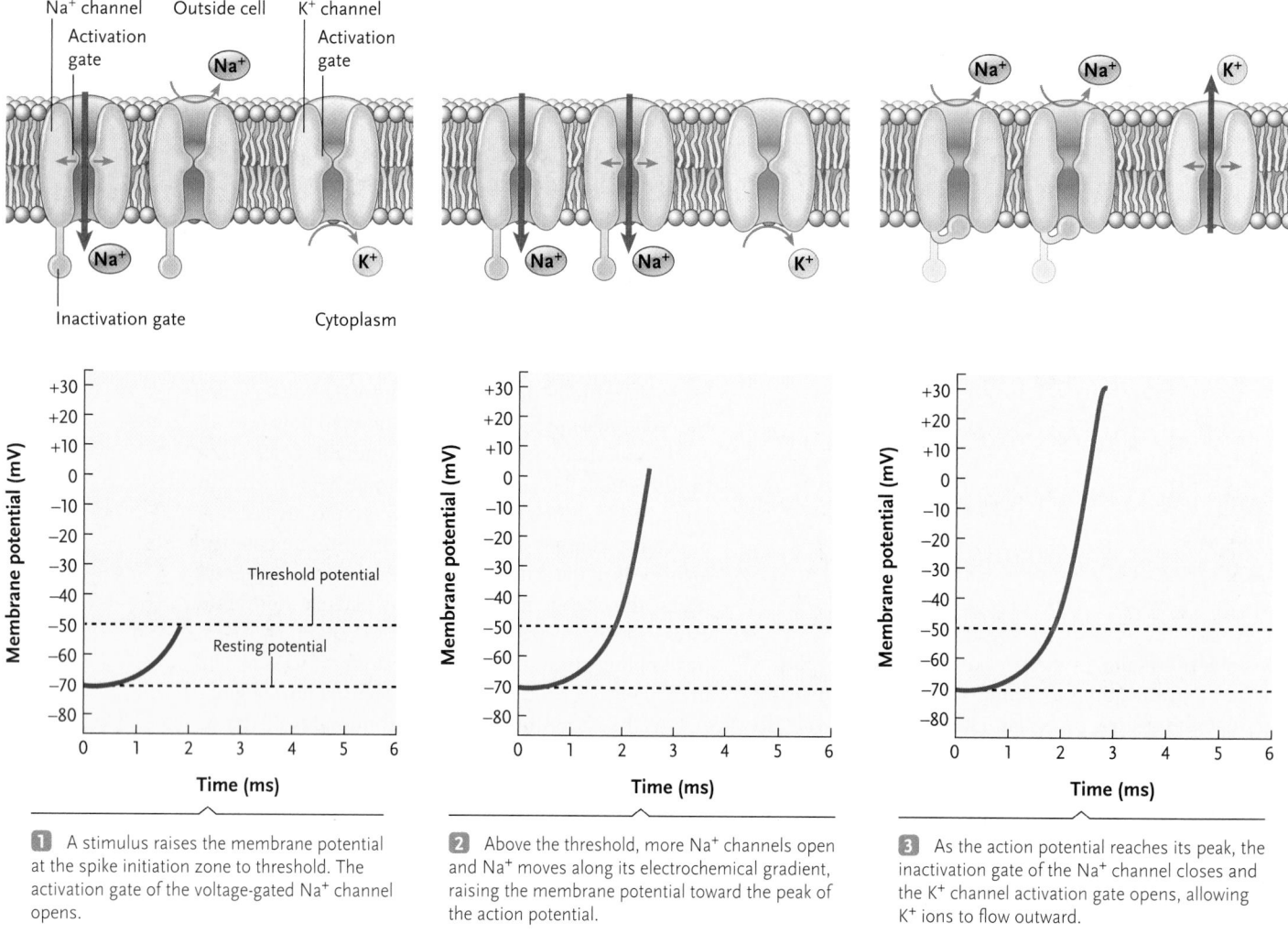

1 A stimulus raises the membrane potential at the spike initiation zone to threshold. The activation gate of the voltage-gated Na⁺ channel opens.

2 Above the threshold, more Na⁺ channels open and Na⁺ moves along its electrochemical gradient, raising the membrane potential toward the peak of the action potential.

3 As the action potential reaches its peak, the inactivation gate of the Na⁺ channel closes and the K⁺ channel activation gate opens, allowing K⁺ ions to flow outward.

Figure 44.10
Changes in voltage-gated Na^+ and K^+ channels that produce the action potential.

Na^+ has entered the cell and some K^+ has left the cell. Actually, relatively few of the total number of Na^+ and K^+ ions change locations during an action potential (ions only flow for 5 ms). Hence, additional action potentials can occur without the need to completely correct the altered ion distribution. In the long term, the Na^+/K^+ active transport pumps restore the Na^+ and K^+ to their original locations.

44.2e Sensory Receptors Are the First Step in Transmitting Information to the Nervous System

Sensory systems begin with **sensory receptors (transducers)** that detect sensory information, convert it to neural activity, and pass the information along neurons to the CNS. Sensory receptors are formed by the dendrites of afferent neurons or by specialized receptor cells **(Figure 44.11, p. 1078)**. Receptors collect information about the internal and external environments of organisms. In organisms with a developed head region (cephalized), many receptors for external

stimuli are located there so that the organism can collect information about where it is going. Receptors associated with eyes, ears, skin, and other surface organs detect stimuli from the external environment. Sensory receptors associated with internal organs detect stimuli arising in the body interior.

Sensory transduction occurs when stimuli cause changes in membrane potentials in the sensory receptors. This is usually achieved by changes in rates at which channels conduct positive ions (Na^+, K^+, or Ca^{2+}) across the plasma membrane. Stimuli may be in the form of light, heat, sound waves, mechanical stress, or chemicals **(Figure 44.12, p. 1078)**. The change in membrane potential may generate one or more action potentials that travel along the axon of an afferent neuron to reach interneuron networks of the CNS. These interneurons integrate the action potentials, and the brain formulates a compensating response, that is, a response appropriate for the stimulus. In animals with complex nervous systems, interneuron networks may produce an awareness of a stimulus in the form of a conscious sensation or perception.

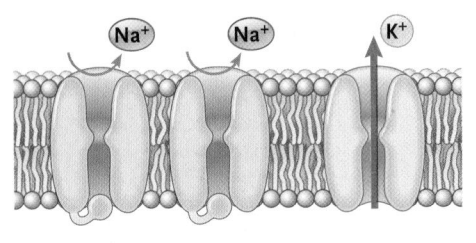
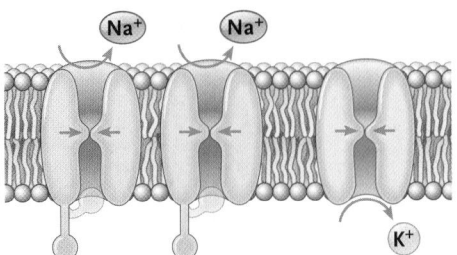
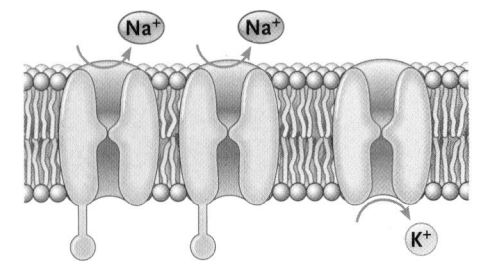

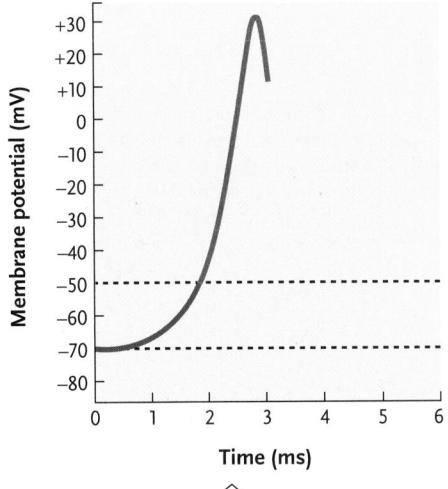

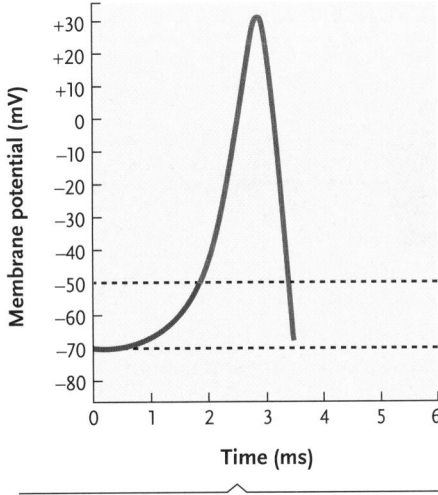

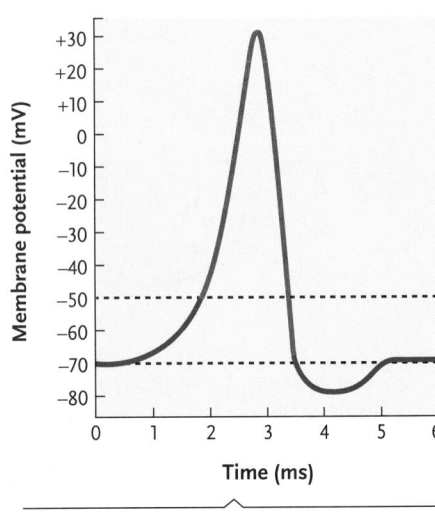

4 The outward flow of K^+ along its electrochemical gradient causes the membrane potential to begin to fall.

5 As the membrane potential reaches the resting value, the activation gate of the Na^+ channel closes and the inactivation gate opens. The K^+ activation gate also closes.

6 Closure of the K^+ activation gate stabilizes the membrane potential at the resting value.

44.2f Basic Types of Receptors: What Can Animals Sense?

Many sensory receptors are positioned individually in body tissues. Others are part of complex sensory organs, such as the eyes or ears, specialized for reception of physical or chemical stimuli (see Chapter 45). Receptors, particularly for external information, usually occur in pairs, providing the opportunity for the animal to localize the stimulus. Eyes, ears, and antennae are examples of paired sensory organs. There are exceptions. *Opabinia* species from the Burgess Shale (see Chapter 27) had five eyes, and some species of praying mantis have only one ear. Some spiders have rows of simple eyes.

Sensory receptors are classified into five major types, based on the type of stimulus that each detects:

- **Mechanoreceptors** detect mechanical energy when it deforms membranes (Figure 44.12, p. 1078). Changes in pressure, body position, or acceleration are detected by mechanoreceptors, for instance. The auditory receptors in the ears are examples of mechanoreceptors.
- **Photoreceptors** detect the energy of light. In vertebrates, photoreceptors are mostly located in the retina of the eye.
- **Chemoreceptors** detect specific molecules or chemical conditions such as acidity. Taste buds on the tongue are examples of chemoreceptors.

- **Thermoreceptors** detect the flow of heat energy. Receptors of this type are located in the skin, where they detect changes in the temperature of the body surface.
- **Nociceptors** detect tissue damage or noxious chemicals; their activity registers as pain. Pain receptors are located in the skin and in some internal organs.

Some animals also have receptors that detect electrical or magnetic fields. Traditionally, humans are said to have five senses: vision, hearing, taste, smell, and touch. In reality, we can detect many more types of environmental stimuli. The traditional list should also include external heat; internal temperature; gravity; acceleration; the positions of muscles and joints; body balance; internal pH; and the internal concentrations of substances such as oxygen, carbon dioxide, salts, and glucose. Note, however, that these are all detected by one of the five classes of receptors listed above.

CONCEPT FIX Not all parts of neurons can generate action potentials. In most, the dendrites and cell bodies can only generate graded potentials. As a result, some integration takes place at every synapse in a neural circuit. Higher-level integration takes place in the CNS, where interneurons are recruited specifically to integrate information coming in from many different sites in the body. ⬡

a. Sensory receptor consisting of free nerve endings—dendrites of an afferent neuron

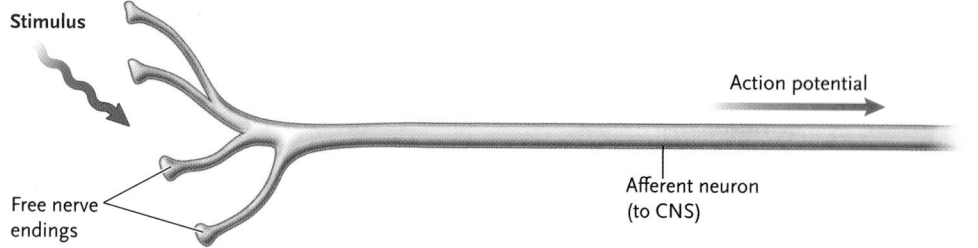

In sensory receptors consisting of the dendrites of afferent neurons, a stimulus causes a change in membrane potential that generates action potentials in the axon of the neuron. Examples are pain receptors and some mechanoreceptors.

b. Sense organ—sensory receptor involving nerve endings of an afferent neuron enclosed in a specialized structure

In sensory receptors involving nerve endings enclosed in a specialized structure, a stimulus affecting the structure triggers an action potential in the afferent neuron. Some mechanoreceptors are of this type.

c. Sensory receptor formed by a cell that synapses with an afferent neuron

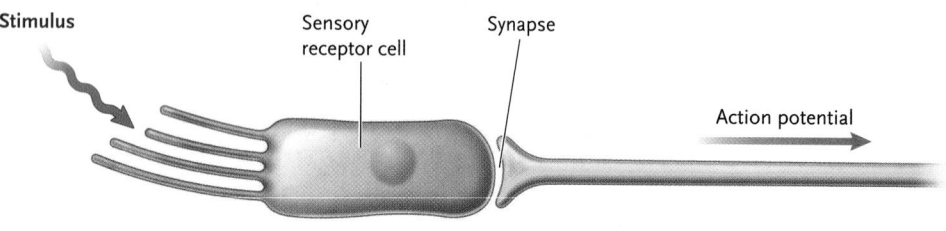

In sensory receptors consisting of separate cells, a stimulus causes a change in membrane potential that releases a neurotransmitter from the cell. The neurotransmitter triggers an action potential in the axon of an afferent neuron to which the sensory receptor cell is synapsed. Examples are photoreceptors, chemoreceptors, and some mechanoreceptors.

Figure 44.11

Sensory receptors, formed **(a)** by the dendrites of an afferent neuron, **(b)** by nerve endings enclosed in specialized structures, or **(c)** by a separate cell or structure that communicates with an afferent neuron via a neurotransmitter.

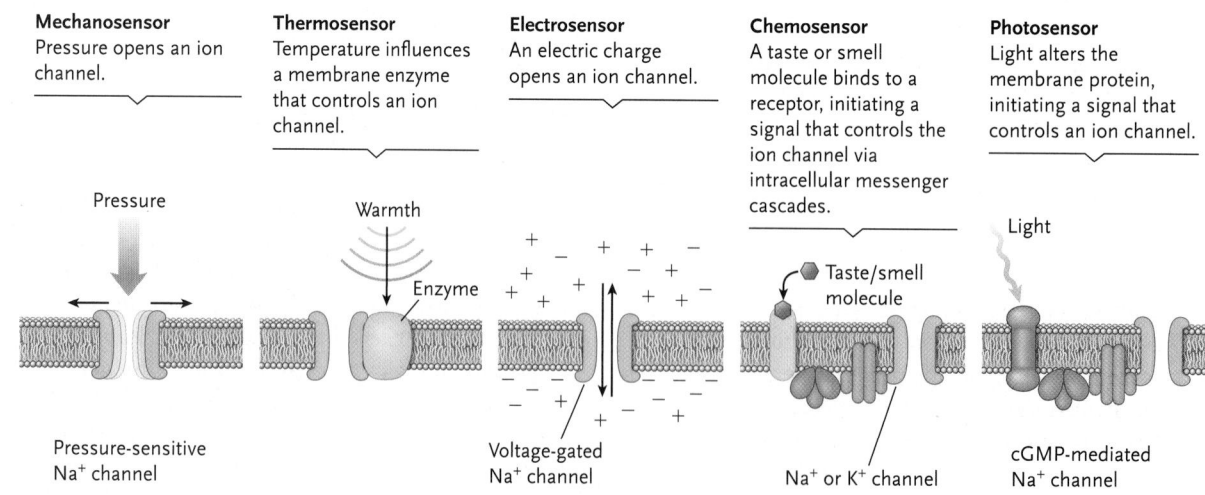

Figure 44.12

Sensory cell membrane proteins respond to stimuli. Sensory stimuli modify receptor proteins in the membranes of sensors, which in turn modify ion channels. The receptors in mechanoreceptors, thermosensors, and electrosensors are themselves ion channels. In chemosensors and photosensors, activated receptor proteins initiate biochemical cascades that eventually open or close ion channels.

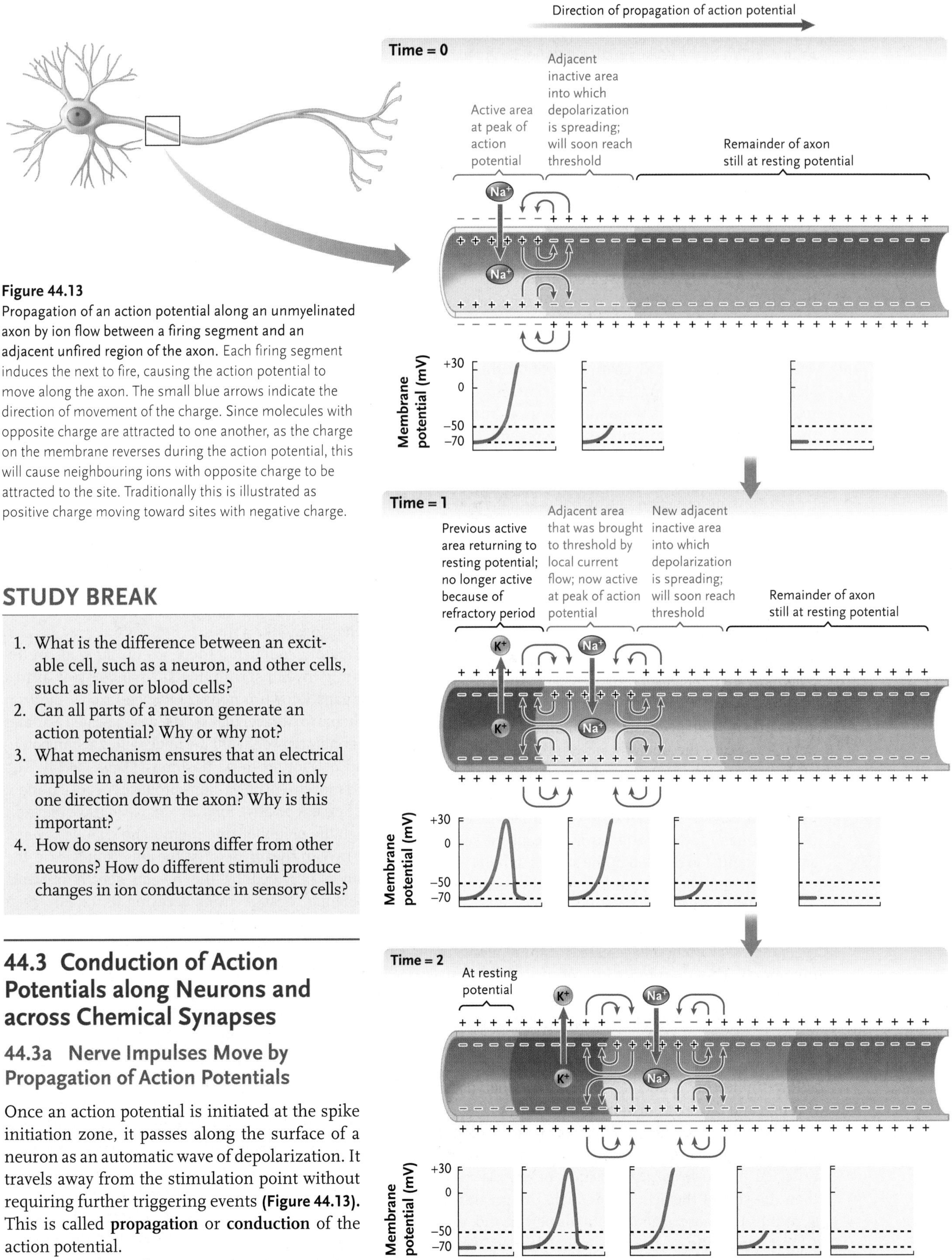

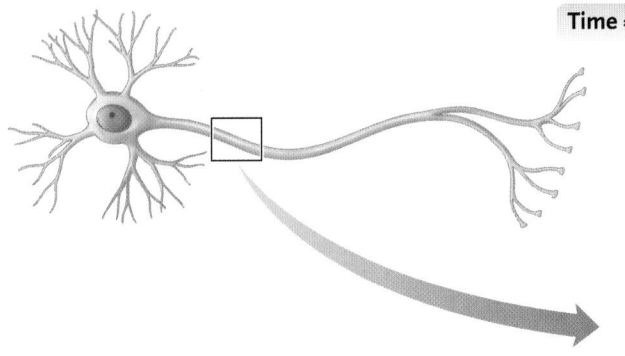

Time = 0

Active area at peak of action potential

Adjacent inactive area into which depolarization is spreading; will soon reach threshold

Remainder of axon still at resting potential

Na⁺

Na⁺

Membrane potential (mV): +30, 0, −50, −70

Time = 1

Previous active area returning to resting potential; no longer active because of refractory period

Adjacent area that was brought to threshold by local current flow; now active at peak of action potential

New adjacent inactive area into which depolarization is spreading; will soon reach threshold

Remainder of axon still at resting potential

K⁺

Na⁺

K⁺

Na⁺

Membrane potential (mV): +30, 0, −50, −70

Time = 2

At resting potential

K⁺

Na⁺

K⁺

Na⁺

Membrane potential (mV): +30, 0, −50, −70

Figure 44.13

Propagation of an action potential along an unmyelinated axon by ion flow between a firing segment and an adjacent unfired region of the axon. Each firing segment induces the next to fire, causing the action potential to move along the axon. The small blue arrows indicate the direction of movement of the charge. Since molecules with opposite charge are attracted to one another, as the charge on the membrane reverses during the action potential, this will cause neighbouring ions with opposite charge to be attracted to the site. Traditionally this is illustrated as positive charge moving toward sites with negative charge.

STUDY BREAK

1. What is the difference between an excitable cell, such as a neuron, and other cells, such as liver or blood cells?
2. Can all parts of a neuron generate an action potential? Why or why not?
3. What mechanism ensures that an electrical impulse in a neuron is conducted in only one direction down the axon? Why is this important?
4. How do sensory neurons differ from other neurons? How do different stimuli produce changes in ion conductance in sensory cells?

44.3 Conduction of Action Potentials along Neurons and across Chemical Synapses

44.3a Nerve Impulses Move by Propagation of Action Potentials

Once an action potential is initiated at the spike initiation zone, it passes along the surface of a neuron as an automatic wave of depolarization. It travels away from the stimulation point without requiring further triggering events (**Figure 44.13**). This is called **propagation** or **conduction** of the action potential.

In a segment of an axon generating an action potential, the outside of the membrane becomes temporarily negative and the inside positive. Because opposites attract, as the membrane potential reverses in the region of the spike initiation zone, local current flow occurs between the area undergoing an action potential and the adjacent downstream inactive area, both inside and outside the membrane (see arrows, Figure 44.13). This current flow (flow of charge, like the flow of electrical current in a wire) makes nearby regions of the axon membrane less positive on the outside and more positive on the inside; in other words, the membrane of these adjacent regions depolarizes.

In regions of the neuron capable of propagating action potentials, voltage-gated channels occur throughout the membrane. The local depolarization spreading from the spike initiation zone is large enough to push the membrane potential of the neighbouring voltage-gated Na^+ and K^+ channels past the threshold, starting an action potential in the downstream adjacent region. In this way, each segment of the axon stimulates the next segment to fire, and the action potential moves rapidly along the axon as a nerve impulse.

The refractory period keeps an action potential from reversing direction at any point along an axon; only the region in front of the action potential can fire. The refractory period begins when the inactivation gate closes on the voltage-gated Na^+ channel; this site on the membrane is not capable of generating another action potential until this gate reopens and the activation gates on the Na^+ and K^+ channels are reset. That is, once they are opened to their activated state, the upstream voltage-gated ion channels need time to reset to their original positions before they can open again. Therefore, only downstream voltage-gated ion channels are able to open, ensuring the one-way movement of the action potential along the axon toward the axon terminals. By the time the refractory period ends in a membrane segment that has just fired an action potential, the action potential has moved too far away to cause a second action potential to develop in the same segment.

The magnitude of an action potential stays the same as it travels along an axon, even where the axon branches at its tips. This is because the action potential is being regenerated at each voltage-gated channel along the membrane. Thus, the propagation of an action potential resembles a burning fuse, which burns with the same intensity along its length and along any branches, once it is lit at one end. Unlike a fuse, however, an axon can fire another action potential of the same intensity within a few milliseconds after an action potential passes through and the refractory period is complete.

The all-or-nothing principle of action potential generation means that the intensity of a stimulus is reflected in the *frequency* of action potentials rather than the size of the action potential. The greater the stimulus, the more action potentials per second, up to a limit depending on the axon type. For most neuron types, the limit lies between 10 and 100 action potentials per second.

44.3b Saltatory Conduction Increases Propagation Rate in Small-Diameter Axons

In the propagation pattern shown in Figure 44.13, an action potential is regenerated at every voltage-gated channel along the length of the axon. The rate of conduction increases with the rate of local current flow, which increases with the diameter of the axon. Some specialized axons with very large diameters occur in invertebrates such as lobsters, earthworms, and squids, as well as a few marine fishes. Giant axons typically carry signals that produce an escape or withdrawal response, such as the sudden flexing of the tail (abdomen) in lobsters that propels the animal backward. The largest known axons, 1.7 mm in diameter, occur in fanworms (Phylum Annelida, Class Polychaeta; see Figure 27.30, Chapter 27, Fanworms are a group of which the feather duster worm is a member). The signals they carry contract a muscle that retracts the fanworm's body into a protective tube when the animal is threatened. The giant axons of the squid were used in the early experiments that led to the current conceptual model of how axons work.

Although large-diameter axons can conduct impulses as rapidly as 25 m/s, they take up a great deal of space. In the jawed vertebrates, insulating axons with myelin sheaths (Figure 44.5, p. 1072) produces **saltatory conduction** (*saltere* = to leap), allowing action potentials to "hop" rapidly along axons instead of burning slowly like a fuse.

Saltatory conduction depends on the gaps in the insulating myelin sheath that surrounds many axons. These gaps, known as nodes of Ranvier, expose the axon membrane to extracellular fluids. Voltage-gated Na^+ and K^+ channels crowded into the nodes allow action potentials to develop at these positions **(Figure 44.14).** The inward movement of Na^+ ions produces depolarization, but the excess positive ions are unable to leave the axon through the neighbouring membrane regions covered by the myelin sheath. Because the ions cannot leave, the insulation provided by the myelin sheath prevents the potential change from decaying as rapidly as it spreads along the neuron. Instead, the local current spreads much farther. As long as the next node is close enough that the local current does not decay below threshold, it will cause depolarization, inducing an action potential at that node. As this mechanism repeats, the action potential jumps rapidly along the axon from node to node. Saltatory conduction proceeds at rates up to 130 m/s, whereas an unmyelinated axon of the same diameter conducts action potentials at about 1 m/s.

Saltatory conduction allows thousands to millions of fast-transmitting axons to be packed into a relatively small diameter. For example, in humans, the 3-mm-diameter optic nerve leading from the eye to the brain is packed with more than a million axons. If those axons were unmyelinated, each would have to be about 100 times as thick to conduct impulses at the same velocity, producing an optic nerve about 300 mm in diameter.

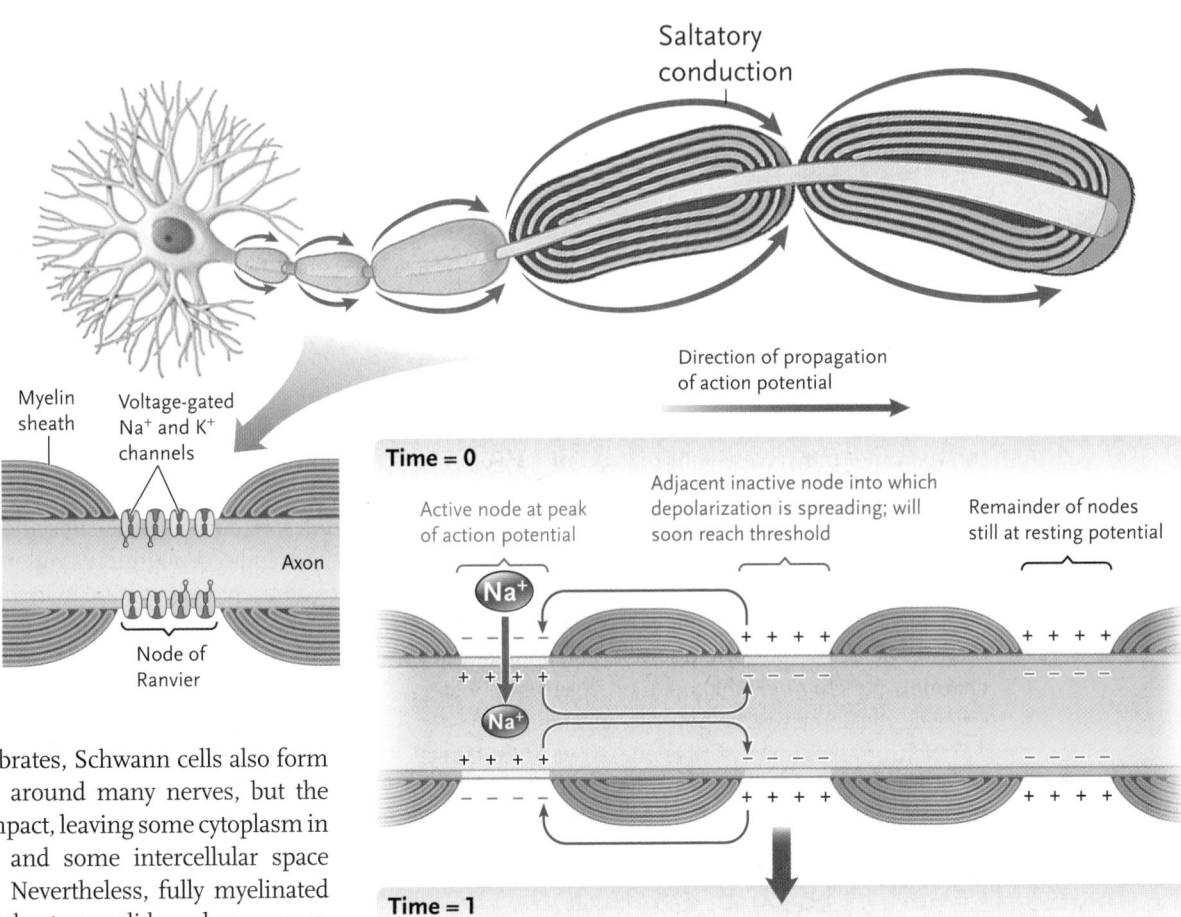

Saltatory conduction

Myelin sheath

Voltage-gated Na⁺ and K⁺ channels

Axon

Node of Ranvier

Direction of propagation of action potential

Time = 0

Active node at peak of action potential

Adjacent inactive node into which depolarization is spreading; will soon reach threshold

Remainder of nodes still at resting potential

Time = 1

Previous active node returning to resting potential; no longer active because of refractory period

Adjacent node that was brought to threshold by local current flow now active at peak of action potential

New adjacent inactive node into which depolarization is spreading; will soon reach threshold

Time = 2

At resting potential

Previous active node returning to resting potential; no longer active because of refractory period

Adjacent node that was brought to threshold by local current flow now active at peak of action potential

Among invertebrates, Schwann cells also form concentric coatings around many nerves, but the layers are not so compact, leaving some cytoplasm in the Schwann cells and some intercellular space between the layers. Nevertheless, fully myelinated fibres occur in oligochaete annelids and some crustaceans, complete with gaps to permit saltatory transmission. The occurrence of myelin in some protostome invertebrates and its absence from the lower vertebrates suggests that this important mechanism has evolved more than once, presenting another example of convergent evolution. The embryonic origin of Schwann cells is different in vertebrates and invertebrates, confirming the independent evolution of myelination in vertebrates and invertebrates.

The disease *multiple sclerosis* (*sclero* = hard) underscores the importance of myelin sheaths to the operation of the vertebrate nervous system. In this disease, myelin is attacked by the immune system and is progressively lost from axons and replaced by hardened scar tissue. The changes block or slow the transmission of action potentials, producing numbness, muscular weakness, faulty coordination of movements, and paralysis that worsens as the disease progresses. Although clear genetic

Figure 44.14

Saltatory conduction of the action potential by a myelinated axon. The action potential jumps from node to node, greatly increasing the speed at which it travels along the axon. The small blue arrows indicate the direction of movement of the charge. Since molecules with opposite charge are attracted to one another, as the charge on the membrane reverses during the action potential, this will cause neighbouring ions with opposite charge to be attracted to the site. Traditionally this is illustrated as positive charge moving toward sites with negative charge.

factors are involved, the environment also plays a role: the incidence of the disease increases with the distance from the equator. The incidence in Canada, 2.4 people per 1000 population, is one of the highest in the world.

44.3c Nerve Impulses Are Conducted across Chemical Synapses by Neurotransmitters

Action potentials are transmitted directly across electrical synapses (Figure 44.6, p. 1072), but they cannot jump across the synaptic cleft in a chemical synapse. Instead, the arrival of an action potential causes neurotransmitter molecules synthesized in the cell body of the neuron to be released across the plasma membrane of the axon terminal, called the **presynaptic membrane (Figure 44.15).** The neurotransmitter diffuses across the cleft and alters ion conduction by activating *ligand-gated ion channels* in the **postsynaptic membrane**, the plasma membrane of the postsynaptic cell. **Ligand-gated ion channels** are channels that open or close when a specific chemical, the ligand, binds to the channel.

Neurotransmitters work in one of two ways. **Direct neurotransmitters** (Figure 44.15) bind directly to a ligand-gated ion channel in the postsynaptic membrane, which opens or closes the channel gate and alters the flow of a specific ion or ions in the postsynaptic cell. The time between arrival of an action potential at an axon terminal and alteration of the membrane potential in the postsynaptic cell may be as little as 0.2 ms.

Indirect neurotransmitters work more slowly (on the order of hundreds of milliseconds). They act as *first messengers,* binding to G-protein-coupled receptors in the postsynaptic membrane, which activates the receptors and triggers the generation of a *second messenger* such as cyclic AMP or other processes. The cascade of second-messenger reactions opens or closes ion-conducting channels in the postsynaptic membrane. Indirect neurotransmitters typically have effects that may last for minutes or hours. Some substances can act as either direct or indirect neurotransmitters, depending on the types of receptors they bind to in the receiving cell.

Not all of the chemicals released at nerve terminals directly stimulate the postsynaptic neuron to fire. Some may inhibit the neuron from firing, whereas others may enhance the action of other transmitters. Actions that modify the effects of other transmitters may not be confined to the single synaptic cleft but may act to coordinate

Figure 44.15
Structure and function of chemical synapses.

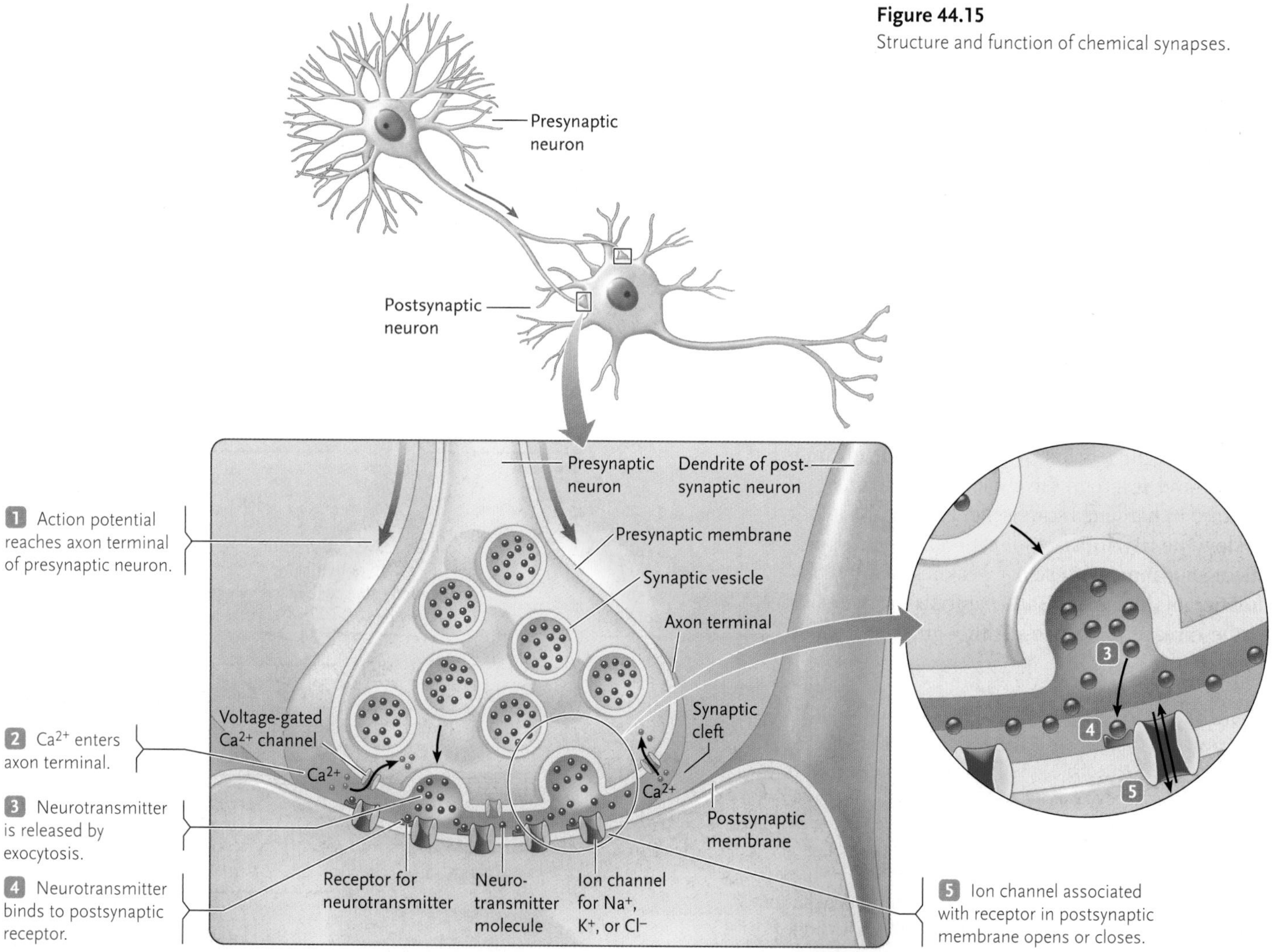

1 Action potential reaches axon terminal of presynaptic neuron.

2 Ca²⁺ enters axon terminal.

3 Neurotransmitter is released by exocytosis.

4 Neurotransmitter binds to postsynaptic receptor.

Presynaptic neuron

Postsynaptic neuron

Presynaptic neuron
Dendrite of post-synaptic neuron
Presynaptic membrane
Synaptic vesicle
Axon terminal
Synaptic cleft
Postsynaptic membrane

Voltage-gated Ca²⁺ channel
Ca²⁺
Ca²⁺

Receptor for neurotransmitter
Neurotransmitter molecule
Ion channel for Na⁺, K⁺, or Cl⁻

5 Ion channel associated with receptor in postsynaptic membrane opens or closes.

groups of neurons. These transmitters are sometimes called neuromodulators.

The time required for the release, diffusion, and binding of neurotransmitters across chemical synapses delays transmission compared with the almost instantaneous transmission of impulses across electrical synapses. However, communication through chemical synapses allows postsynaptic neurons to integrate inputs from many presynaptic axons at the same time. Some neurotransmitters have stimulatory effects, whereas others have inhibitory effects. All of the information received at a postsynaptic membrane is integrated to produce a response that consists of the receptor neuron firing with a particular frequency.

44.3d Neurotransmitters Are Released by Exocytosis

Neurotransmitters are stored in secretory vesicles, called synaptic vesicles, in the cytoplasm of an axon terminal. The arrival of an action potential at the terminal releases the neurotransmitters by *exocytosis:* the vesicles fuse with the presynaptic membrane and release the neurotransmitter molecules into the synaptic cleft (Figure 44.15).

The release of synaptic vesicles depends on voltage-gated Ca^{2+} channels in the plasma membrane of an axon terminal (see Figure 44.15). Ca^{2+} ions are constantly pumped out of all animal cells by an active transport protein in the plasma membrane, keeping their concentration higher outside than inside. As an action potential arrives, the change in membrane potential opens the Ca^{2+} channel gates in the axon terminal, allowing Ca^{2+} to flow back into the cytoplasm. The rise in Ca^{2+} concentration triggers a protein in the membrane of the synaptic vesicle that allows the vesicle to fuse with the plasma membrane, releasing neurotransmitter molecules into the synaptic cleft.

Each action potential arriving at a synapse typically causes approximately the same number of synaptic vesicles to release their neurotransmitter molecules. For example, arrival of an action potential at one type of synapse causes about 300 synaptic vesicles to release a neurotransmitter called acetylcholine. Each vesicle contains about 10 000 molecules of the neurotransmitter, giving a total of some 3 million acetylcholine molecules released into the synaptic cleft by each arriving action potential.

When a stimulus is no longer present, action potentials are no longer generated. When action potentials stop arriving at the axon terminal, the voltage-gated Ca^{2+} channels in the axon terminal close, and the Ca^{2+} in the axon cytoplasm is quickly pumped to the outside. The drop in cytoplasmic Ca^{2+} stops vesicles from fusing with the presynaptic membrane, and no further neurotransmitter molecules are released. Any free neurotransmitter molecules remaining in the cleft are broken down by enzymes in the cleft or else reuptake occurs, meaning that the neurotransmitter molecules

are pumped back into the axon terminals or into glial cells by active transport. Transmission of impulses across the synaptic cleft ceases within milliseconds after action potentials stop arriving at the axon terminal.

44.3e Most Neurotransmitters Alter Flow through Na^+ or K^+ Channels

Most neurotransmitters work by opening or closing membrane-embedded ligand-gated ion channels that conduct Na^+ or K^+ across the postsynaptic membrane, although some regulate chloride ions (Cl^-). The resulting ion flow may stimulate or inhibit the generation of action potentials by the postsynaptic cell. If Na^+ channels are opened, the inward Na^+ flow brings the membrane potential of the postsynaptic cell toward the threshold (the membrane becomes depolarized). If K^+ channels are opened, the outward flow of K^+ has the opposite effect (the membrane becomes hyperpolarized). The combined effects of the various stimulatory and inhibitory neurotransmitters at all of the chemical synapses of a postsynaptic neuron or muscle cell determine whether the postsynaptic cell triggers an action potential (see Section 44.4).

44.3f Many Different Molecules Act as Neurotransmitters

Nearly 100 different substances are known or suspected to be neurotransmitters. Most of them are relatively small molecules that diffuse rapidly across the synaptic cleft. Some axon terminals release only one type of neurotransmitter, whereas others release several types. Depending on the type of receptor to which it binds, the same neurotransmitter may stimulate or inhibit the generation of action potentials in the postsynaptic cell. **Table 44.1, p. 1084,** lists some examples of neurotransmitters, the types of molecules they represent, their sites and type of action, and some drugs and other molecules that affect neurotransmission. (Note that this list is not complete, listing only the most common transmitter substances.)

STUDY BREAK

1. How does the presence of a myelin sheath affect the conduction of impulses in neurons? What type of cell forms the myelin sheath?
2. Describe the steps from the arrival of an action potential at an axon terminal to the release of a neurotransmitter.
3. Describe how a direct neurotransmitter in a presynaptic neuron controls ion flow in a postsynaptic neuron.
4. Why aren't synaptic vesicles being released all the time? Why are they only released when an action potential arrives? How is this controlled?

Table 44.1   Examples of Neurotransmitters and Drugs That Affect Neurotransmission

Neurotransmitter	Site(s) of Action	Type of Action	Drugs and Other Molecules That Affect Neurotransmission
Acetylcholine	Between some neurons of CNS and at neuromuscular junctions in PNS; acetylcholine-releasing neurons in the brain degenerate in people with Alzheimer disease, in which memory, speech, and perceptual abilities decline	Mostly excitatory; inhibitory at some sites	• Curare—blocks release; blocks muscle contraction and produces paralysis, potentially leading to death • Atropine—blocks receptors; used to relax iris muscles to dilate pupils for eye exam • Nicotine—activates receptors, causing increased attention and decreased stress and irritability
Monoamines (biogenic amines)			
Norepinephrine	CNS interneurons involved in diverse brain and body functions, such as memory, mood, sensory perception, muscle movements, maintenance of blood pressure and sleep; also released into body circulation as a hormone	Excitatory or inhibitory	• Amphetamines—stimulate release of norepinephrine and dopamine and block their reuptake; cause increased attention and focus and decreased fatigue • Methylphenidate (Ritalin)—increases release; used to treat attention deficit hyperactivity disorder (ADHD) • Certain antidepressants—prevent reuptake; used to treat depression and obsessive-compulsive disorder
Dopamine	CNS interneurons involved in many pathways similar to norepinephrine, but especially involved in reinforcements and addiction; not released into circulation	Mostly excitatory	• Amphetamines—stimulate release of norepinephrine and dopamine and block their reuptake; cause increased attention and focus and decreased fatigue • Cocaine—stimulates release of norepinephrine and dopamine and blocks dopamine reuptake; produces intense euphoria followed by depression
Serotonin	CNS interneurons in a number of pathways, including those regulating appetite, reproductive behaviour, muscular movement, sleep, and emotional states such as anxiety	Inhibitory or modulatory	• Fluoxetine (Prozac), sertraline (Zoloft), paroxetine (Paxil)—block serotonin reuptake; used to treat depression and obsessive-compulsive disorder
Amino Acids			
Glutamate	Many CNS pathways, including those involved in vital brain functions such as memory and learning	Excitatory	• Phencyclidine (PCP or angel dust)—blocks receptor; causes feelings of power and confusion; can cause violent behaviour
Gamma-aminobutryic acid (GABA)	Many CNS pathways; often acts in same circuits as glutamate	Inhibitory (main inhibitory neurotransmitter in mammalian CNS)	• Alcohol (ethanol)—stimulates GABA neurotransmission, increases dopamine neurotransmission, and inhibits glutamate neurotransmission; causes relaxation, a sense of euphoria, and sleepiness • Some antianxiety/sedative drugs such as diazepam (Valium), alprazolam (Xanax), and flunitrazepam (Rohypnol, the "date rape" drug)—bind to GABA receptors and increase GABA neurotransmission; result in sedation, sleepiness, and possibly amnesia • Tetanus toxin—released by bacterium *Clostridium tetani*; blocks GABA release in synapses that control muscle contraction, causing muscles to contract forcibly (once respiratory muscles are affected, victim dies quickly)
Purines			
ATP	Released simultaneously with many neurotransmitters; may have independent actions, an area under active investigation	Modulates action of other neurotransmitters	

Table 44.1

Table 44.1	Examples of Neurotransmitters and Drugs That Affect Neurotransmission (*Continued*)		
Neurotransmitter	Site(s) of Action	Type of Action	Drugs and Other Molecules That Affect Neurotransmission
Neuropeptides			
Endorphins ("endogenous morphines")	Most act on CNS and PNS as well as on effectors such as muscle, reducing pain and, in some cases, also inducing euphoria; released during periods of pleasurable experience (such as eating or sexual intercourse), or physical stress (such as childbirth or extended physical exercise)	Inhibitory—modulate pain response	
Enkephalins (subclass of endorphins)	Active only in CNS	Inhibitory—modulate pain response	
Substance P	Released by special, unmyelinated sensory neurons in spinal cord, which produce rapid conduction for resulting signals; increases neural signals associated with intense, persistent, or severe pain. Endorphins are antagonistic to substance P, reducing the perception of pain.	Excitatory	
Gaseous Neurotransmitters			
Nitrous oxide (NO)	Diffuses across cell membranes in PNS rather than being released at synapses; relaxes smooth muscles in walls of blood vessels, causing vessels to dilate, which increases blood flow; contributes to many nervous system functions such as learning, sensory responses, and muscle movements	Modulatory	· Sildenafil citrate (Viagra, an impotency drug)—aids erection by inhibiting enzyme that normally reduces effects of NO in vascular beds of penis

44.4 Integration of Incoming Signals by Neurons

Most neurons receive a multitude of stimulatory and inhibitory signals carried by both direct and indirect neurotransmitters. These signals are integrated by the postsynaptic neuron into a response that reflects their combined effects. The integration depends primarily on the patterns, number, types, and activity of the synapses that the postsynaptic neuron receives from presynaptic neurons. Inputs from other sources, such as indirect neurotransmitters and other signal molecules, can modify the integration. The response of the postsynaptic neuron is elucidated by the frequency of action potentials it generates.

44.4a Integration at Chemical Synapses Occurs by Summation

As mentioned earlier, depending on the type of receptor to which it binds, a neurotransmitter may stimulate or inhibit the generation of action potentials in the postsynaptic neuron. If a neurotransmitter opens a ligand-gated Na^+ channel, Na^+ enters the cell, causing a depolarization. This change in membrane potential pushes the neuron closer to threshold; that is, it is excitatory and is called an **excitatory postsynaptic potential**, or **EPSP**. On the other hand, if a neurotransmitter opens a ligand-gated ion channel that allows Cl^- to flow into the cell and K^+ to flow out, hyperpolarization occurs. This change in membrane potential pushes the neuron farther from threshold; that is, it is inhibitory and is called an **inhibitory postsynaptic potential**, or **IPSP**. EPSPs and IPSPs are **graded potentials**, in which the membrane moves up or down in potential without necessarily triggering an action potential. There are no refractory periods for EPSPs and IPSPs.

A neuron typically has hundreds to thousands of chemical synapses formed by axon terminals of presynaptic neurons contacting its dendrites and cell body **(Figure 44.16, p. 1086)**. The events that occur at each synapse produce either an EPSP or an IPSP in that postsynaptic neuron. But how is an action potential

Botulinum Toxin: Potent Toxin, Valuable Therapy, and "Good-Looking" Poison

Botulinum toxin is the product of the bacterium *Clostridium botulinum*, a common organism that thrives in anaerobic environments, such as preserved foods. The toxin is a protein that is destroyed by heat, but if it is ingested, it is extraordinarily toxic—1 000 000 times as toxic as strychnine. The protein is produced as a large molecule that is cleaved into a toxin and a protease. The toxin acts at acetylcholine-mediated synapses (nerve–muscle junctions in vertebrates) by promoting the entry of the protease into the neuron at the synapse, where the enzyme attacks one of the molecules essential for the release of acetylcholine. As a result this prevents muscles from contracting. The effect is long lasting: up to several months. Recovery from the effect has been hypothesized to involve the sprouting of new axon terminals.

These properties have been used since 1989 to treat a number of disorders that involve muscle spasms, such as strabismus (crossed eyes), by injecting minute quantities of the A serotype directly into the muscles. Ophthalmologists using the toxin in this way noted that frown lines around the eyes disappeared as a side effect, and the Botox cosmetic industry was born **(Figure 1)**. Other uses for this toxin now include the treatment of migraine headaches.

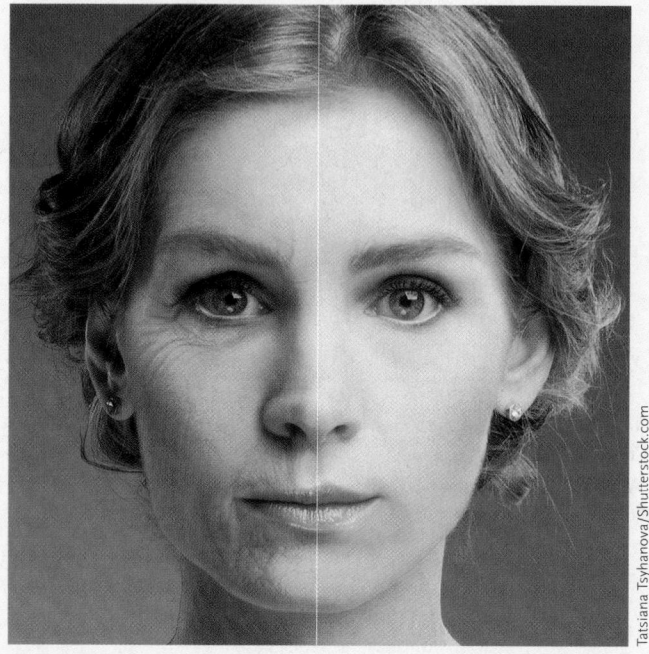

Figure 1

Botox is now used in plastic surgery to relax muscles and remove stress lines as shown in this figure that compares before (left) and after (right) images of a patient receiving such treatment.

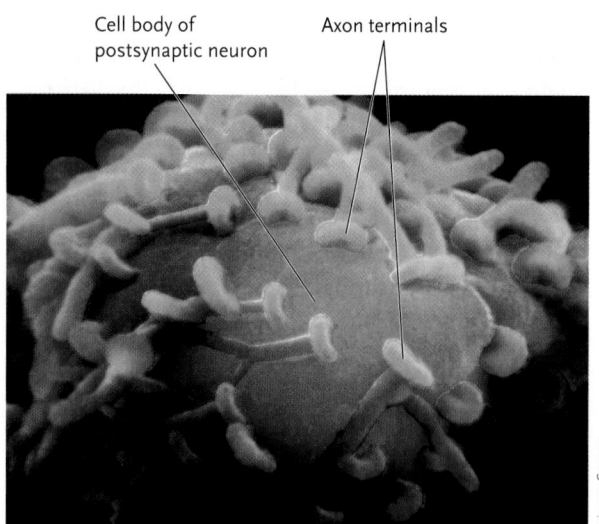

Figure 44.16

The multiple chemical synapses relaying signals to a neuron. The drying process used to prepare the neuron for electron microscopy has toppled the axon terminals and pulled them away from the neuron's surface.

produced if a single EPSP is not sufficient to push the postsynaptic neuron to threshold? The answer involves the summation of all the inputs received through all the chemical synapses formed by presynaptic neurons. At any given time, some or many of the presynaptic neurons may be firing, producing EPSPs and/or IPSPs in the postsynaptic neuron. The sum of all the EPSPs and IPSPs at a given time determines the total potential in the postsynaptic neuron and, therefore, how that neuron responds. **Figure 44.17,** shows, in a greatly simplified way, the effects of EPSPs and IPSPs on membrane potential and how the summation of inputs brings a postsynaptic neuron to threshold.

The postsynaptic neuron in Figure 44.17 has three neurons, N1 to N3, forming synapses with it. Suppose that the axon of N1 releases a neurotransmitter that produces an EPSP in the postsynaptic cell (see Figure 44.17a). The membrane depolarizes, but not enough to reach threshold. If N1 input causes a new EPSP after the first EPSP has died down, it will be of the same magnitude as the first EPSP, so no progression toward

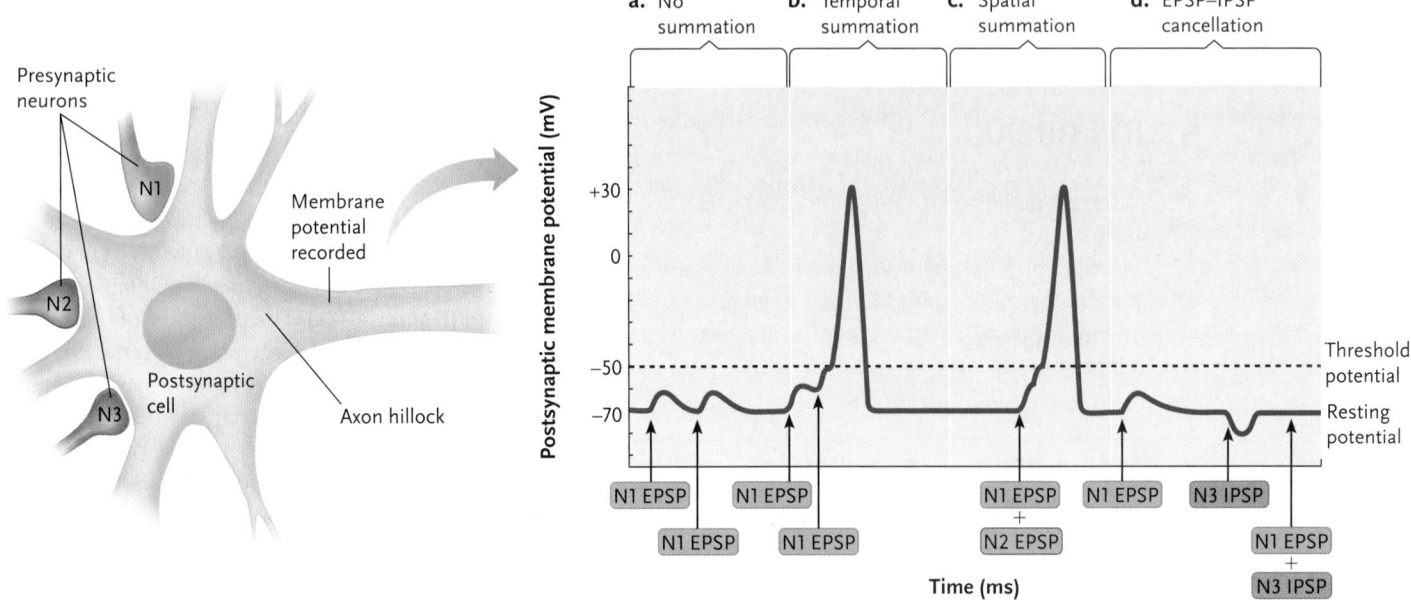

a. No summation: The axon of Ex1 releases a neurotransmitter, which produces an EPSP in the postsynaptic cell. The membrane depolarizes, but not enough to reach threshold. If Ex1 input causes a new EPSP after the first EPSP has died down, it will be of the same magnitude as the first EPSP and no progression toward threshold has taken place—no summation has occurred.

b. Temporal summation: If instead, Ex1 input causes a new EPSP before the first EPSP has died down, the second EPSP will sum with the first and a greater depolarization will have taken place. This summation of two (or more) EPSPs produced by successive firing of a single presynaptic neuron over a short period of time is temporal summation. If the total depolarization achieved in this way reaches threshold, an action potential is produced in the postsynaptic neuron.

c. Spatial summation: The postsynaptic neuron may be brought to threshold by spatial summation, the summation of EPSPs produced by the simultaneous firing of two different excitatory presynaptic neurons, such as Ex1 and Ex2.

d. Summation resulting in cancellation: EPSPs and IPSPs can sum to cancel each other out. In the example, firing of the excitatory presynaptic neuron Ex1 alone produces an EPSP, firing of presynaptic inhibitory neuron In3 alone produces an IPSP, while firing of Ex1 and In3 simultaneously produces no change in the membrane potential.

Figure 44.17
Summation of EPSPs and IPSPs by a postsynaptic neuron.

threshold happens because no summation has occurred. If, instead, N1 input causes a new EPSP before the first EPSP has died down, the second EPSP will sum with the first, leading to a greater depolarization (see Figure 44.17b). This summation of several EPSPs at a common site produced by successive firing of a single presynaptic neuron over a short period of time is called **temporal summation**. If the total depolarization achieved in this way spreads to the axon hillock or spike initiation zone and brings it to threshold, an action potential will be produced in the postsynaptic neuron.

The postsynaptic cell may also be brought to threshold by **spatial summation**, the summation of EPSPs produced by the firing of different presynaptic neurons at different sites on the postsynaptic cell, such as N1 and N2 (see Figure 44.17c). Lastly, EPSPs and IPSPs can cancel each other out. In the example shown in Figure 44.17d, firing of N1 alone produces an EPSP, and firing of N3 alone produces an IPSP, whereas the simultaneous firing of N1 and N3 produces no change in the membrane potential.

The key summation point for EPSPs and IPSPs is the axon hillock or spike initiation zone of the postsynaptic neuron. EPSPs and IPSPs spread over the membrane of the dendrites and cell body as graded potentials summing or cancelling each other as they meet. If the net change in membrane potential is sufficient to bring the spike initiation zone (the first site along the neuron where voltage-gated Na$^+$ and K$^+$ channels occur) to threshold, an action potential will result. This action potential will then be conducted or propagated along the neuron.

44.4b The Patterns of Synaptic Connections Contribute to Integration

The total number of connections made by a neuron may be very large. Some single interneurons in the human brain, for example, form as many as 100 thousand synapses with other neurons. The synapses are not absolutely fixed; they can change through modification, addition, or removal of synaptic connections, or even entire neurons, as animals mature and experience changes in their environments. The combined activities of all the neurons in an animal provide the flow of information on which the integrated functioning of increasingly complex organisms depends. In the remainder of this chapter, we explore the ways that neurons are organized into nervous systems in the various major groups of animals.

STUDY BREAK

1. Differentiate between spatial and temporal summation.
2. Describe the steps from the repeated release of an excitatory neurotransmitter at a synapse to the initiation of an action potential at an axon hillock.

44.5 Evolutionary Trends in Neural Integration: Networks, Ganglia, and Brains

The nervous systems of all animals are designed to effectively sense environmental changes (internal and external), integrate this information, and produce appropriate responses. The organization of the nervous systems of the different groups of invertebrate and vertebrate animals reflects differences in lifestyle and habitat. Most nervous systems contain sensory (afferent) pathways that collect information, integrating centres (where decisions are made), and motor (afferent) pathways that produce responses (see Section 44.1a and Figure 44.3, p. 1071).

44.5a Cnidarians Have Nerve Nets

Cnidarians (including jellyfish and sea anemones) are radially symmetrical protostomes with body parts arranged around a central axis. Their nervous systems are composed of **nerve nets**, loose meshes of neurons that extend over the entire organism just beneath the epithelium **(Figure 44.18a)**. The processes of their neurons do not fit the normal description of dendrites and axons. Instead, all processes can conduct action potentials, and instead of having synapses at their terminals, there are synapses wherever the processes of neurons cross processes of other neurons. Moreover, both processes involved in a synapse may produce transmitters and have receptors for transmitters. When part of the animal is stimulated, impulses are conducted through the nerve net in all directions from the point of stimulation. Although there is no cluster of neurons that plays the coordinating role of a brain, nerve cells may be more concentrated in some regions. In the cnidarian hydra, they are more concentrated around the oral opening, allowing better coordination of tentacles for feeding. In the cnidarian jellyfish, which swim by rhythmic contraction of their bells, neurons are denser in a ring around the bell, in the

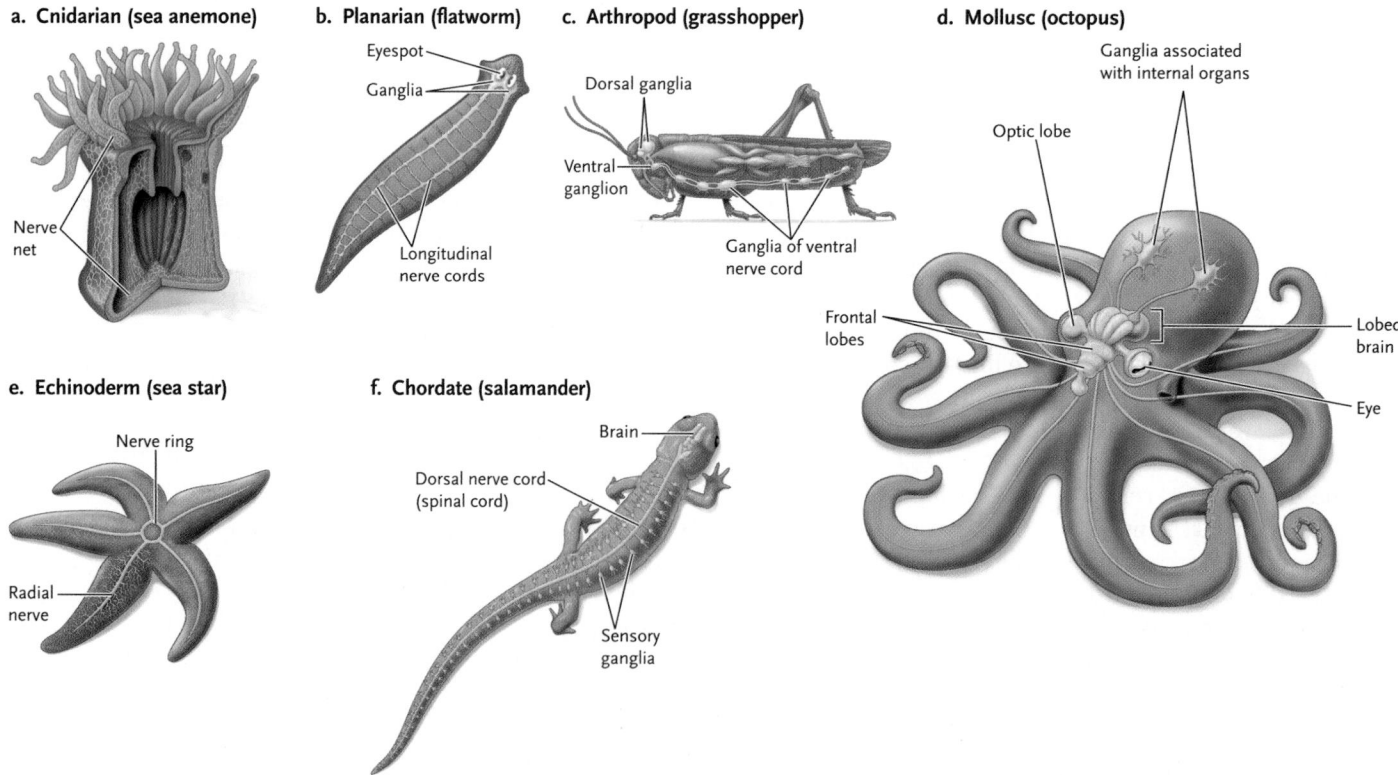

Figure 44.18
Invertebrate and vertebrate nervous systems compared, showing increasing cephalization. The exception to the trend is the echinoderms. The diagrams are not drawn to the same scale.

same area as the contractile cells that produce the swimming movements.

44.5b Other Invertebrates Have Cephalized Nervous Systems

More complex invertebrates have neurons with clearly defined axons and dendrites and more specialized functions. Some neurons are concentrated into functional clusters called ganglia (singular, *ganglion*) **(Figure 44.19).** This anatomical localization of interconnections allows rapid integration of sensory information and more complex reactions to that information. We saw in Chapter 27 that segmental development has occurred twice in the evolution of the protostomes, which had profound effects on the development of the nervous system. In these organisms, each segment has a separate pair of ganglia, joined by interconnections, both between members of a pair within a segment and between the pairs of ganglia in the segments in front and behind.

Another key evolutionary development in invertebrates is a trend toward *cephalization,* the formation of a distinct head region. This head region usually contains major sensory structures, reflecting the tendency for the head of the animal, as it moves through its environment, to encounter new stimuli first. The formation of a distinct head is the result of the fusion of anterior segments and the paired ganglia associated with these segments to form a **brain.** This concentration of neurons at a site close to the sensory structures reduces the transmission times for processing incoming signals and for integrating signals to produce appropriate responses. One or more solid **nerve cords**—bundles of nerves—extend from the central ganglia to the rest of the body; they are connected to smaller nerves. Another evolutionary trend is toward bilateral symmetry of the body and the nervous system, in which body parts are mirror images on left and right sides. These trends toward cephalization and bilateral symmetry are well illustrated in flatworms, arthropods, and molluscs.

In flatworms, a small brain consisting of a pair of ganglia at the anterior end is connected by two or more longitudinal nerve cords to nerve nets in the rest of the body (Figure 44.18b). The brain integrates inputs from sensory receptors, including a pair of anterior eyespots with receptors that respond to light. The brain and longitudinal nerve cords constitute the flatworm's CNS, the simplest one known, while the nerves from the CNS to the rest of the body constitute the PNS (see Section 44.5d).

Arthropods, such as insects, have a head region that contains a brain, consisting of dorsal and ventral pairs of ganglia, and major sensory structures, usually eyes and antennae (Figure 44.18c). The brain exerts centralized control over the remainder of the animal. A ventral nerve enlarges into a pair of ganglia in each body segment. In arthropods with fused body segments, as in the thorax of insects, the ganglia are also fused into masses forming secondary control centres.

Although different in basic plan from the arthropod system, the nervous systems of molluscs (such as clams, snails, and octopuses) also rely on neurons clustered into paired ganglia connected by major nerves. Different molluscs have varying degrees of cephalization, with cephalopods having the most pronounced cephalization of any invertebrate group. In the head of an octopus, for example, a cluster of ganglia fuses into a complex, lobed brain with clearly defined sensory and motor regions. Paired nerves link different lobes with muscles and sensory receptors, including prominent optic lobes linked by nerves to large, complex eyes (Figure 44.18d). Octopuses are capable of rapid movement to hunt prey and to escape from predators, behaviours that rely on rapid, sophisticated processing of sensory information.

The echinoderms, including sea stars, are an exception to this trend. Like cnidarians, this group is also radially symmetrical. These animals lack a *cephalized brain* and instead have a series of ganglia connected by a nerve ring that surrounds the centrally located mouth. Neurons are organized into radial nerves (Figure 44.18e). If the nerve serving an arm is cut, the arm can still move in response to stimuli, but not in coordination with the other arms. Echinoderms are descended from bilaterally symmetrical ancestors, and in many species the larvae are bilaterally

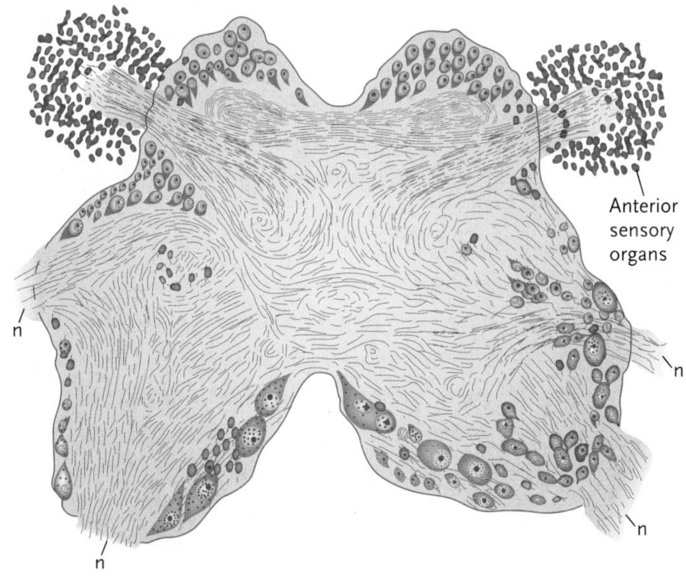

Figure 44.19
A section of a typical invertebrate cerebral ganglion. The cell bodies (blue) are located on the periphery, with a mass of axons and dendrites forming the **neuropile** in the centre. Nerves (n) bring sensory information into the ganglion and carry motor information outward.

Anterior sensory organs

symmetrical and develop radial symmetry during metamorphosis into adults.

44.5c Vertebrates Have Complex Nervous Systems

In vertebrates, the CNS consists of the brain and spinal cord, and the PNS consists of all the nerves and ganglia that connect the brain and spinal cord to the rest of the body (Figure 44.18f). All vertebrate nervous systems are highly cephalized, with major concentrations of neurons in a brain located in the head. In contrast to invertebrate nervous systems, which have solid nerve cords located ventrally, the brain and nerve cord of vertebrates are hollow, fluid-filled structures located dorsally. The head contains specialized sensory organs, which are connected directly to the brain by nerves. Compared with those in invertebrates, the ganglia are greatly reduced in mass and functional activity (except in the gut, which contains extensive interneuron networks), while the brains are greatly enlarged.

The structure of the vertebrate nervous system reflects its pattern of development. The nervous system of a vertebrate embryo begins as a hollow **neural tube** (discussed more in Chapter 42), the anterior end of which develops into the brain and the rest into the **spinal cord**. The cavity of the neural tube becomes the fluid-filled **ventricles** of the brain and the **central canal** through the spinal cord. Adjacent tissues give rise to nerves that connect the brain and spinal cord with all the body regions. Just as in the invertebrates, the brain and longitudinal nerve cord (now the spinal cord) constitute the CNS, while the nerves connecting the CNS to the rest of the body constitute the PNS.

44.5d The Vertebrate Central Nervous System and Its Functions

Early in development, the anterior part of the neural tube enlarges into three distinct regions: the **forebrain**, **midbrain**, and **hindbrain (Figure 44.20).** A little later, the embryonic hindbrain subdivides into the *metencephalon* and *myelencephalon,* the midbrain develops into the *mesencephalon,* and the forebrain subdivides into the *telencephalon* and *diencephalon.*

Figure 44.20

Development of the human brain from the anterior end of an embryo's neural tube.

Regions in 4-week embryo	Regions in 5-week embryo	Regions in adult	Functions in adult
Forebrain	Telencephalon	Telencephalon (cerebrum)	Higher functions, such as thought, action, and communication
	Diencephalon	Thalamus	Receives sensory input and relays it to regions of the cerebral cortex
		Hypothalamus	Centre for homeostatic control of internal environment
Midbrain	Mesencephalon	Midbrain	Coordinates involuntary reactions and relays signals to telencephalon
Hindbrain	Metencephalon	Cerebellum	Integrates signals for muscle movement
		Pons	Centre for information flow between cerebellum and telencephalon
	Myelencephalon	Medulla oblongata	Controls many involuntary tasks

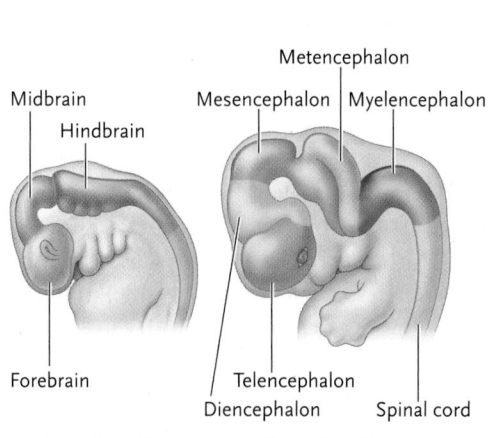

4-week embryo   **5-week embryo**

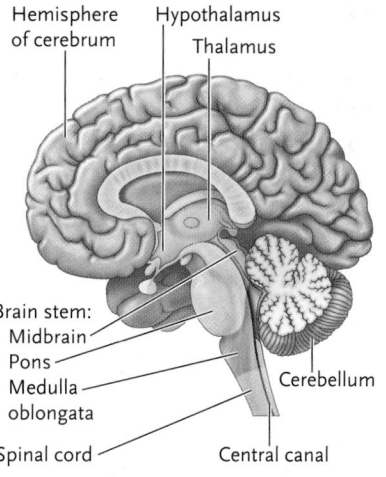

Adult brain regions

The myelencephalon gives rise to the *medulla oblongata* (commonly shortened to medulla), which controls many vital involuntary tasks, such as respiration and blood circulation. The metencephalon, associated with the developing ear (when present) and balance organs gives rise to the *cerebellum*, which integrates sensory signals from the eyes, ears, and muscle spindles, and the *pons*, a major traffic centre for information passing between the cerebellum and the higher integrating centres of the adult telencepahlon. The midbrain, or mesencephalon, receives input from the eyes and from the ears and coordinates reflex responses (involuntary reactions) to visual and auditory (hearing) input. It also relays this information to the telencephalon for further processing. The diencephalon gives rise to the *thalamus,* a centre that receives sensory input and relays it to the regions of the cerebral cortex concerned with motor responses to such input, and the *hypothalamus,* the primary centre for homeostatic control over the internal environment. The embryonic telencephalon develops into the adult *cerebrum.* The cerebrum controls higher functions such as thought, memory, language, and emotions as well as voluntary movements.

The general pattern of brain development underwent major modification in the evolution of various groups of animals **(Figure 44.21).** In sharks, the cerebrum is relatively small, but the olfactory bulbs are prominent, testifying to the importance of olfaction in these very successful predators. Frogs are hunters that rely on vision, so the optic lobes of the mesencephalon are prominent, whereas the olfactory bulbs are less so. Birds also rely on vision for feeding and navigation, and their optic lobes reflect that.

One of the major trends in the evolution of the brain, however, is the increasing prominence of the cerebrum. Beginning with reptiles, it increased in size relative to the rest of the brain. In mammals, convolutions or folds appeared, increasing the amount of brain material in a particular volume. As well, the total mass of the brain relative to the size of the animal increased, permitting animals to undertake more complex tasks. The mass of bird and mammal brains is about 15 times that of other taxa when corrected for the size of the animal. With their advanced locomotor and navigational skills, birds and mammals also exhibit an increase in the cerebellum, a major coordinating centre for automatic activities.

44.5e The Vertebrate Peripheral Nervous System and Its Functional Parts

The PNS can be divided into two main systems. The afferent system of the PNS includes all the neurons that transmit sensory information from receptors to the CNS. The efferent system consists of the axons of neurons that carry signals to the muscles and glands acting as effectors. The efferent system is further divided into somatic and autonomic systems **(Figure 44.22, p. 1092).**

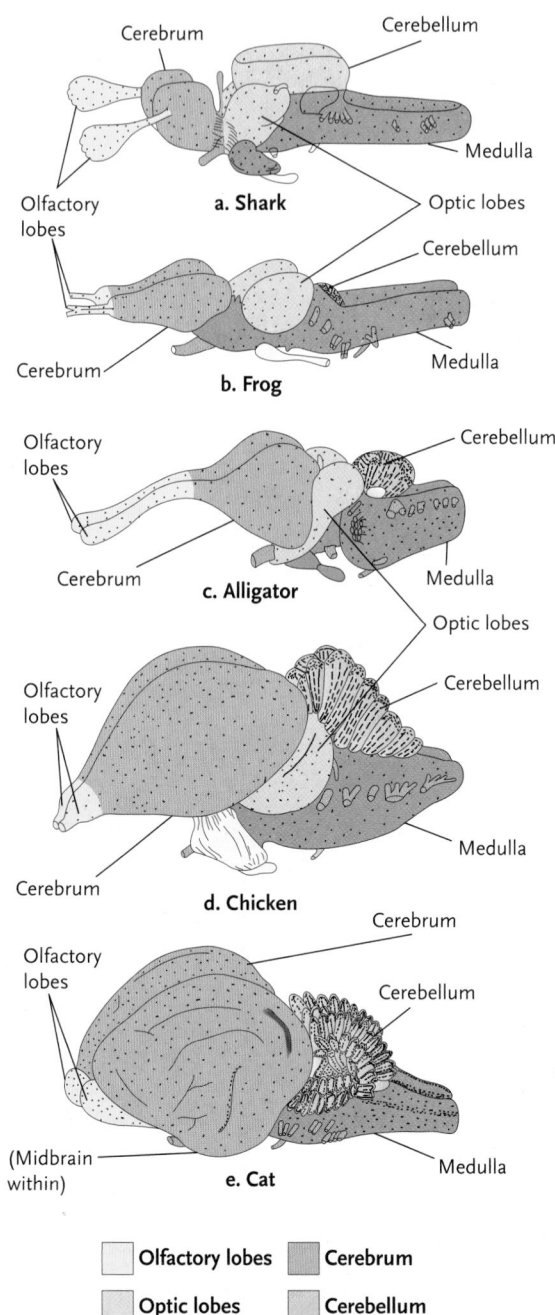

Figure 44.21
A comparison of brain structures in five different groups of vertebrates, illustrating the evolutionary trends described in the text.

The Somatic System Controls the Contraction of Skeletal Muscles. The **somatic nervous system** controls body movements that are primarily conscious and voluntary. Its efferent neurons, called motor neurons, carry signals from the CNS to the skeletal muscles. The dendrites and cell bodies of motor neurons are located in the spinal cord. Their axons extend from the spinal cord, emanating from between the vertebrae, to the skeletal muscles they control. The cell bodies for the sensory nerves (the afferent neurons) are located outside the spinal cord within the dorsal root, forming the dorsal root ganglion.

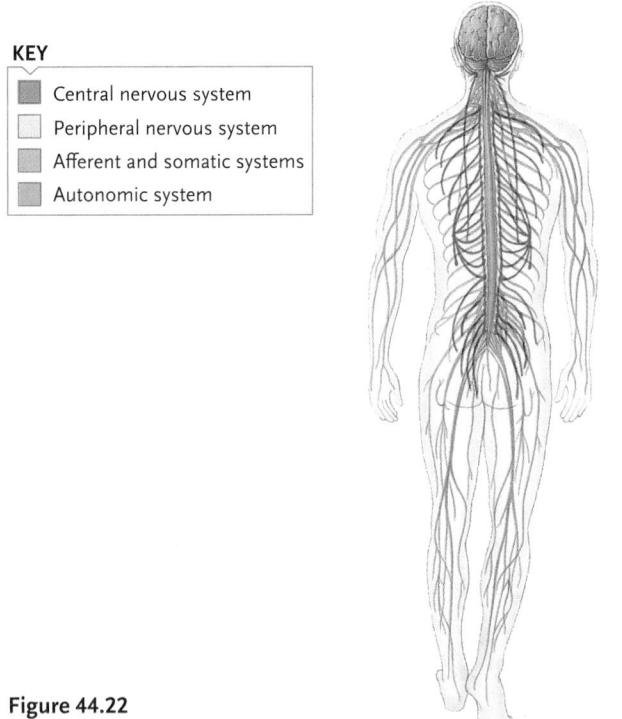

KEY

- **Central nervous system**
- **Peripheral nervous system**
- **Afferent and somatic systems**
- **Autonomic system**

Figure 44.22

The CNS and PNS and their subsystems.

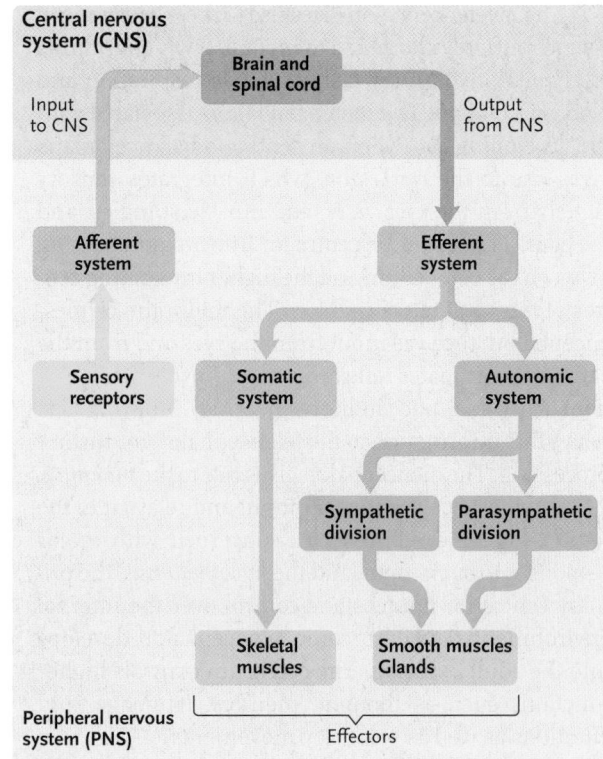

Central nervous system (CNS)

- Input to CNS
- **Brain and spinal cord**
- Output from CNS
- **Afferent system**
- **Efferent system**
- **Sensory receptors**
- **Somatic system**
- **Autonomic system**
- **Sympathetic division**
- **Parasympathetic division**
- **Skeletal muscles**
- **Smooth muscles Glands**

Peripheral nervous system (PNS)

Effectors

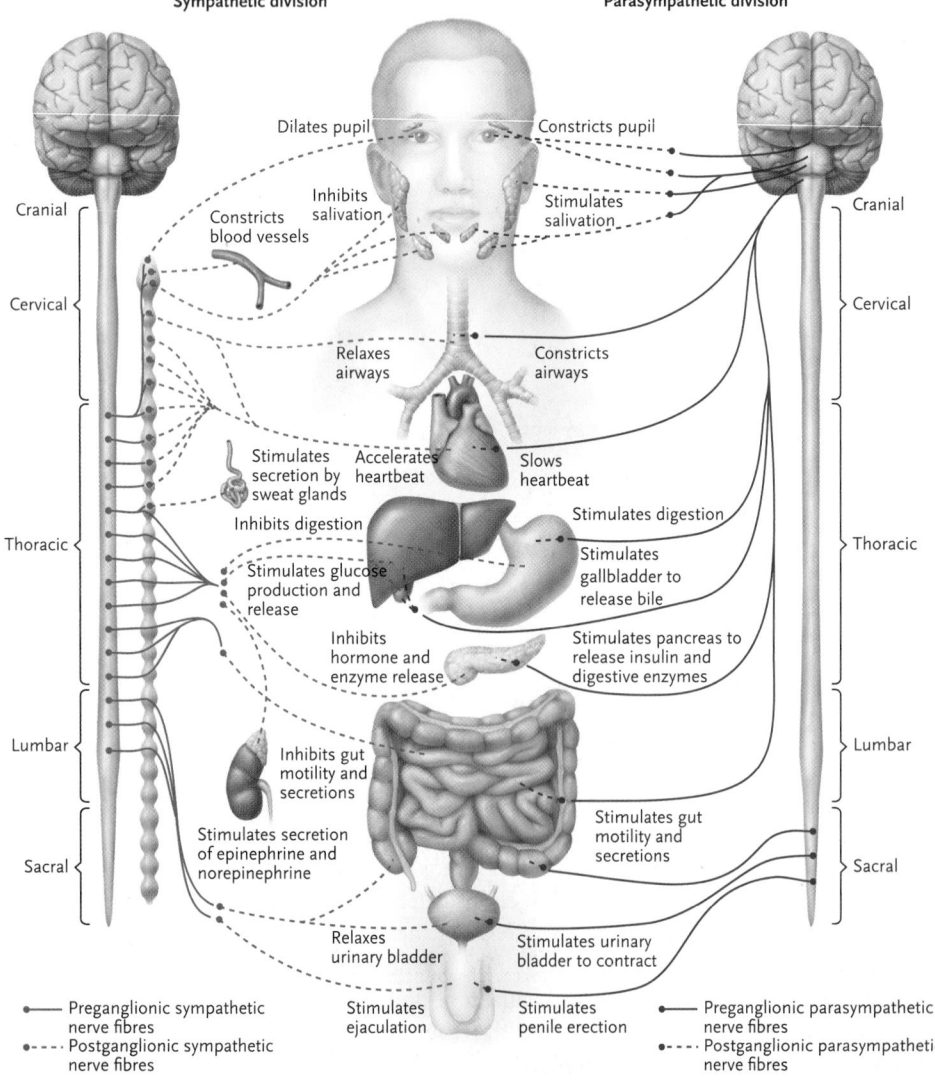

Sympathetic division

Parasympathetic division

- Dilates pupil — Constricts pupil
- Inhibits salivation — Stimulates salivation
- Constricts blood vessels
- Cranial
- Cervical
- Relaxes airways — Constricts airways
- Stimulates secretion by sweat glands — Accelerates heartbeat — Slows heartbeat
- Inhibits digestion — Stimulates digestion
- Thoracic
- Stimulates glucose production and release — Stimulates gallbladder to release bile
- Inhibits hormone and enzyme release — Stimulates pancreas to release insulin and digestive enzymes
- Lumbar
- Inhibits gut motility and secretions — Stimulates gut motility and secretions
- Stimulates secretion of epinephrine and norepinephrine
- Sacral
- Relaxes urinary bladder — Stimulates urinary bladder to contract
- Stimulates ejaculation — Stimulates penile erection

- •— Preganglionic sympathetic nerve fibres
- •-- Postganglionic sympathetic nerve fibres
- •— Preganglionic parasympathetic nerve fibres
- •-- Postganglionic parasympathetic nerve fibres

Cranial, Cervical, Thoracic, Lumbar, Sacral

Although the somatic system is primarily under conscious, voluntary control, some contractions of skeletal muscles are unconscious and involuntary. These include reflexes, shivering, and the constant muscle contractions that maintain body posture and balance.

The Autonomic System Is Divided into Sympathetic and Parasympathetic Pathways. The **autonomic nervous system** controls largely involuntary processes such as digestion, secretion by sweat glands, circulation of the blood, many functions of the reproductive and excretory systems, and contraction of smooth muscles in all parts of the body. It is organized into *sympathetic* and *parasympathetic* divisions, which are always active and have opposing effects on the organs that they affect, thereby enabling precise control **(Figure 44.23).** For example, in the circulatory system, sympathetic neurons stimulate the force and rate of the heartbeat, and

Figure 44.23

Effects of sympathetic and parasympathetic divisions on organ and gland function. Only one side of each division is shown; both are duplicated on the left and right sides of the body.

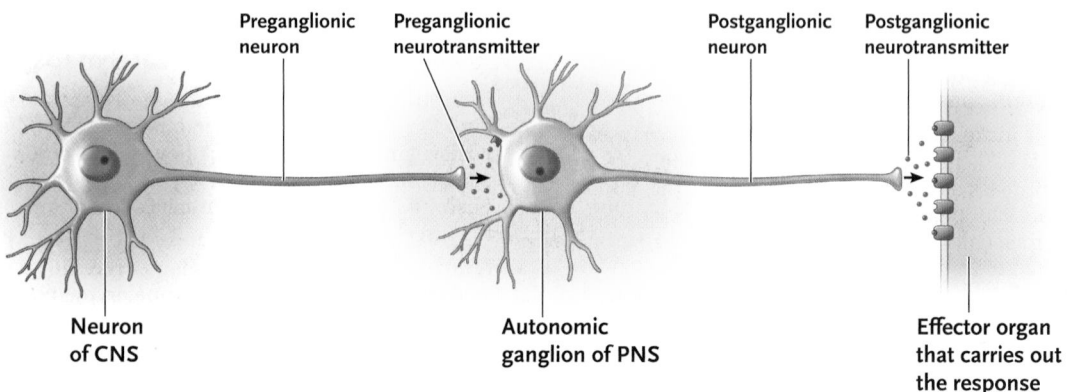

Figure 44.24
An autonomic nervous system pathway.

Preganglionic neuron

Preganglionic neurotransmitter

Postganglionic neuron

Postganglionic neurotransmitter

Neuron of CNS

Autonomic ganglion of PNS

Effector organ that carries out the response

parasympathetic neurons inhibit these activities. In the digestive system, sympathetic neurons inhibit the smooth muscle contractions that move materials through the small intestine, whereas parasympathetic neurons stimulate the same activities. These opposing effects control involuntary body functions precisely.

The pathways of the autonomic nervous system include two neurons **(Figure 44.24).** The first neuron has its dendrites and cell body in the CNS, and its axon extends to a ganglion outside the CNS. There it synapses with the dendrites and cell body of the second neuron in the pathway. The axon of the second neuron extends from the ganglion to the effector carrying out the response.

The **sympathetic division** predominates in situations involving stress, danger, excitement, or strenuous physical activity. Signals from the sympathetic division increase the force and rate of the heartbeat, raise the blood pressure by constricting selected blood vessels, dilate air passages in the lungs, induce sweating, and open the pupils wide. Activities that are less important in an emergency, such as digestion, are suppressed by the sympathetic system. The ganglia between the first and second neurons in each sympathetic pathway occur as a chain of segmental ganglia just outside (ventral to) the vertebral column in the thoracic and abdominal regions (see Figure 44.23).

The **parasympathetic division**, in contrast, predominates during quiet, low-stress situations, such as relaxation. Under its influence, the effects of the sympathetic division, such as rapid heartbeat and elevated blood pressure, are reduced, and maintenance activities such as digestion predominate. The parasympathetic pathways originate either directly from the brain or from the sacral region of the spinal cord. The ganglia between the first and second neurons in the parasympathetic pathways are associated with the target organs (see Figure 44.23).

STUDY BREAK

1. Distinguish between a neuron, a nerve net, nerves, and nerve cords.
2. What is cephalization and why does it occur?
3. What is the significance of ganglia, brains, and nerve cords?
4. How do differences in brain structure reflect differences in lifestyle?
5. What two systems make up the PNS, and what do they generally control?
6. When an elk attempts to evade an attacking pack of wolves, what division of its autonomic nervous system dominates? What effects might result?

Review

aplia

To access course materials such as Aplia and other companion resources, please visit www.NELSONbrain.com.

44.1 Neurons and Their Organization in Nervous Systems: An Overview

- Neurons are excitable cells specialized for the rapid transmission of signals. They have dendrites and cell bodies, which receive and integrate information, and axons, which generate and conduct signals away from the cell body to another neuron or an effector.

- The nervous system of an animal (1) receives information about conditions in the internal and external environment, (2) integrates the information to formulate an appropriate response, (3) conducts the message along neurons, and (4) transmits the signal to effector organs to produce a response.

- Afferent neurons conduct information from sensory receptors to interneurons, which integrate the information into a response. The response signals are passed to efferent neurons, which activate the effectors carrying out the response.

- Glial cells provide structural and functional support to neurons. They help maintain the balance of ions surrounding neurons and form insulating layers around the axons.

- Neurons make connections by two types of synapses, electrical and chemical. In an electrical synapse, impulses pass directly from the sending to the receiving cell. In a chemical synapse, neurotransmitter molecules released by the presynaptic neuron diffuse across a narrow synaptic cleft and bind to receptors in the plasma membrane of the postsynaptic cell. Binding of the neurotransmitters may generate an electrical impulse in the postsynaptic cell.

44.2 Signal Initiation by Neurons

- The membrane potential of a cell results from the unequal distribution of positive and negative charges on either side of the membrane. This establishes a potential difference, the resting potential, across the membrane.

- The resting potential is maintained by an active transport pump that maintains the concentration gradients of Na^+ ions (higher outside) and K^+ ions (higher inside).

- Not all parts of a neuron can generate an action potential; only those parts that possess voltage-gated Na^+ and K^+ channels can.

- An action potential is generated when a stimulus pushes the resting potential at the spike initiation zone to the threshold value at which voltage-gated Na^+ channels open in the plasma membrane. The inward flow of Na^+ changes the membrane potential abruptly from a negative to a positive peak, which opens the voltage-gated K^+ channels. The potential falls to the resting value again as the gated K^+ channels allow this ion to flow out.

- Action potentials move along an axon as the ion flows generated in one location on the axon depolarize the potential in the adjacent location.

- Action potentials are prevented from reversing direction by a brief refractory period (a few milliseconds), during which a sector of membrane that has just generated an action potential cannot be stimulated to produce another. During this period, the action potential has moved too far away for its electrical disturbances to cause the preceding sector to depolarize again.

- In myelinated axons, ions can flow across the plasma membrane only at the nodes of Ranvier, where the insulating myelin sheath is interrupted.

- The intensity of a stimulus is reflected in the frequency of action potentials.

- Receptors in the sensory system collect information (i.e., stimuli) from internal and external sensors (transducers) and convert (transduce) the information into neural activity. Dendrites of an afferent neuron pick up the stimuli. The axon of the afferent neuron conveys the stimulus to the CNS, providing the organisms with sensory data used to influence behaviour and homeostasis.

- Mechanoreceptors detect mechanical energy (pressure), photoreceptors detect the energy of light, chemoreceptors detect specific molecules or chemical conditions, thermoreceptors detect the flow of heat energy, and nociceptors detect tissue damage or noxious chemicals.

44.3 Conduction of Action Potentials along Neurons and across Chemical Synapses

- Neurotransmitters released into the synaptic cleft bind to receptors in the plasma membrane of the postsynaptic cell, altering the flow of ions across the plasma membrane of the postsynaptic cell and pushing its membrane potential toward or away from the threshold potential.

- A direct neurotransmitter binds to a receptor associated with a ligand-gated ion channel in the postsynaptic membrane; the binding opens or closes the channel.

- An indirect neurotransmitter works as a first messenger, binding to a receptor in the postsynaptic membrane and triggering generation of a second messenger, which leads to the opening or closing of a gated channel.

- Neurotransmitters are released from synaptic vesicles into the synaptic cleft by exocytosis, which is triggered by entry of Ca^{2+} ions into the cytoplasm of the axon terminal through voltage-gated Ca^{2+} channels opened by the arrival of an action potential.

- Neurotransmitter release stops when action potentials cease arriving at the axon terminal. Neurotransmitters remaining in the synaptic cleft are broken down by enzymes or taken up by the axon terminal or glial cells.

- Types of neurotransmitters include acetylcholine, amino acids, biogenic amines, neuropeptides, and gases such as NO and CO.

44.4 Integration of Incoming Signals by Neurons

- Integration of incoming information by a neuron determines the frequency at which the receiving neuron sends out action potentials.

- Neurons carry out integration by summing excitatory postsynaptic potentials (EPSPs) and inhibitory postsynaptic potentials (IPSPs). The summation may occur over time (temporal) or from different neurons at the same time (spatial). This summation pushes the membrane potential of the receiving cell toward or away from the threshold for an action potential.

44.5 Evolutionary Trends in Neural Integration: Networks, Ganglia, and Brains

- Identifiable nerves first appear in radially symmetrical animals as nerve nets composed of single neurons that are more concentrated around the mouth and that may have localized groupings of cell bodies that control particular functions.

- The development of bilateral symmetry resulted in the concentration of parts of the nerve net into several longitudinal nerve cords composed of several axons, with paired ventral cords becoming increasingly dominant.

- Ganglia, local concentrations of nerve cell bodies permitting enhanced coordination of sensory and motor functions, first appeared in the flatworms, with the anterior ganglia prominent and acting as a brain. Protostome ganglia have the cell bodies at the periphery and the neuropile of axons and dendrites in the interior.

- Molluscs have well-developed nervous systems with paired ventral nerve cords and a brain consisting of several fused ganglia that permits advanced behaviour in the cephalopods.
- In the development of the vertebrate brain, the hindbrain subdivides into the myelencephalon, which becomes the medulla oblongata, or brain stem, responsible for many involuntary functions, and the metencephalon, associated with hearing and balance. The cerebellum, a major processing centre for balance and navigation, is an outgrowth of the metencephalon. The midbrain, or mesencephalon, coordinates hearing and vision. The forebrain subdivides into the diencephalon, which gives rise to the optic nerves, and the telencephalon, which is responsible for olfaction and gives rise to the cerebrum, the major processing centre of the brain.
- During evolution, some parts of the brain become more prominent, depending on the lifestyle of the animal. Sharks have large olfactory centres, whereas frogs have larger optic centres.
- The evolution of the brain also involves an increase in its mass relative to the body mass of the animal and,

in particular, an increase in the mass of the cerebrum relative to the rest of the brain. Birds and mammals also exhibit an increased prominence of the cerebellum.
- Afferent neurons in the PNS conduct signals to the CNS, and signals from the CNS go via efferent neurons to the muscles and glands that carry out responses.
- The somatic system of the PNS controls the skeletal muscles that produce voluntary body movements, as well as involuntary muscle contractions that maintain balance, posture, and muscle tone.
- The autonomic system of the PNS controls involuntary functions such as heart rate and blood pressure, glandular secretion, and smooth muscle contraction.
- The autonomic system is organized into sympathetic and parasympathetic divisions that balance and fine-tune involuntary body functions. The sympathetic system predominates in situations involving stress, danger, or strenuous activities, whereas the parasympathetic system predominates during quiet, low-stress situations.

Questions

Self-Test Questions

Any number of answers from a to e may be correct.

1. What does neural signalling involve?
   a. differentiation
   b. integration
   c. response
   d. transmission
   e. reception

2. The nodes of Ranvier
   a. speed up conduction of electrical impulses
   b. make up a myelin sheath
   c. expose the axon
   d. are gaps between adjacent Schwann cells

3. Electrical synapses
   a. occur where the cytoplasm of pre- and postsynaptic cells are in direct contact
   b. have narrow gaps called clefts between cells
   c. are electrical impulses that flow between cells through gap junctions
   d. involve the release of neurotransmitters

4. What does the resting potential result from?
   a. differential permeability of membranes that results in accumulation of more proteins and other anions carrying negative charge inside the cell
   b. an opening of voltage-gated sodium channels
   c. higher concentrations of Na^+ and K^+ inside and outside the cells created by the Na^+/K^+ pump
   d. an opening of ligand-gated sodium channels

5. Which statement best describes how nerve signals travel?
   a. The axons of oligodendrocytes transmit nerve impulses to the dendrites of astrocytes.
   b. An axon of a motor neuron receives the signal, its cell body transmits the signal to a sensory neuron's dendrite, and the signal is sent to the target.

   c. A dendrite of a sensory neuron receives the signal, its cell body transmits the signal to a motor neuron's axon, and the signal is sent to the target.
   d. Efferent neurons conduct nerve impulses toward the cell body of sensory neurons, which send them on to interneurons and, ultimately, to afferent motor neurons.
   e. A dendrite of a sensory neuron receives a signal, the cell's axon transmits the signal to an interneuron, and then the signal is transmitted to dendrites of a motor neuron and sent via its axon to the target.

6. A synapse could be the site where which of the following takes place?
   a. Postsynaptic neurons transmit a signal across a cleft to a presynaptic neuron.
   b. The axons of a presynaptic neuron directly contact the dendrites of a postsynaptic neuron.
   c. The neurotransmitters released by an axon travel across a gap and are picked up by receptors on a muscle cell.
   d. An electrical impulse arrives at the end of a dendrite, causing ions to flow onto axons of presynaptic neurons.
   e. An on–off switch stimulates an electrical impulse in a presynaptic cell to stimulate, not inhibit, other presynaptic cells.

7. Nerve impulses do which of the following?
   a. always travel in both directions along an axon
   b. sometimes travel in both directions along an axon
   c. travel in only one direction along an axon
   d. always travel with the same intensity along the length of an axon
   e. diminish in intensity as they travel along an axon

8. Which phrase describes spatial summation?
   a. the change in membrane potential of a postsynaptic cell brought on by the firing of different presynaptic neurons
   b. the change in membrane potential of a postsynaptic cell brought on by successive firing of a single presynaptic neuron over a short period of time
   c. the total of the membrane potentials that occur along an axon as the electrical impulse is transmitted

9. Ganglia first became enlarged and fused into a lobed brain in the evolution of which of the following?
   a. vertebrates
   b. annelids
   c. flatworms
   d. cephalopods
   e. mammals

10. The metencephalon is the origin of which of the following?
   a. the spinal cord
   b. the cerebellum
   c. the mesencephalon
   d. the medulla oblongata
   e. the cerebrum

Questions for Discussion

1. The mechanism for the propagation of the action potential along an axon was worked out using the giant axons of squids. How confident should we be that this model applies to vertebrates? Is there an evolutionary link between the nerves of vertebrates and those of molluscs, or did the mechanism arise twice?

2. How did evolution of chemical synapses make higher brain function possible?

3. Describe the fundamental ways in which different sensory modalities (chemical, thermal, mechanical, electrical, light) lead to changes in the membrane potential of sensory cells. How are they similar? How are they different?

4. Discuss the selective pressure for cephalization in animals. Would there be a difference between those with sessile or motile life histories?

5. Many types of local anaesthetic compounds are used to reduce pain sensations during medical procedures. Identify different types of local anaesthetics, and explain how they work to reduce pain sensations.

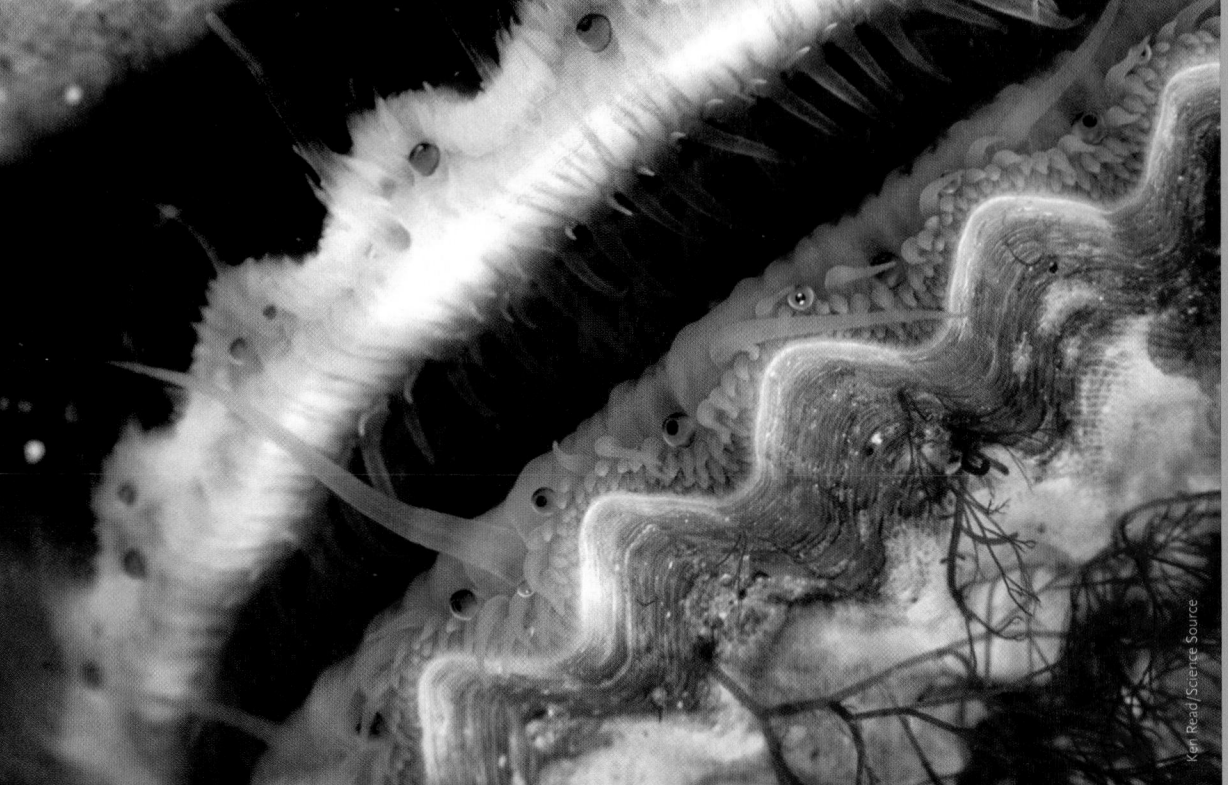

Close-up view of a blue-eyed scallop (*Argopecten irradians*) showing some of the many blue eyes that detect light and movement, allowing it to escape from predators such as sea stars and crabs.

Ken Read/Science Source

Control of Animal Processes: Neural Integration

WHY IT MATTERS

An insectivorous bat leaves its cave to look for food. As it flies the bat emits a steady stream of ultrasonic clicking noises. Receptors in the bat's ears detect echoes of the clicks bouncing off objects in the environment and send signals to the brain, where they are integrated into a sound map that the animal uses to navigate through the complex environment. This ability, called **echolocation**, is so keenly developed that a bat can detect and avoid a thin wire in the dark.

Besides recognizing obstacles, the bat's auditory system is sharply tuned to the distinctive pattern of echoes from the fluttering wings of its favourite food, a moth. Although the slow-flying moth seems doomed to become a meal for the foraging bat, natural selection has provided some species of moths with an astoundingly sensitive auditory sense as well. On each side of its abdomen is an "ear," a thin membrane that resonates at the frequencies of the clicks emitted by the bat. The moth's ears register the clicks while the bat is still about 30 m away and initiate a response that turns its flight path directly away from the source of the clicks.

In spite of the moth's evasive turns, if the bat approaches within about 6 m of it, echoes from the moth begin to register in the bat's auditory system, and the bat increases the frequency of its clicks, enabling it to pinpoint the moth's position **(Figure 45.1, p. 1098).**

Figure 45.1

A greater horseshoe bat (*Rhinolophus ferrumequinum*) hunting a moth. The bat uses its sensory systems to pursue prey, and the moth uses its sensory systems to avoid capture.

The moth has not exhausted its evasive tactics, however. As the bat closes in, the increased frequency of the clicks sets off another programmed response that alters the moth's flight into sudden loops and turns, ending with a closed-wing, vertical fall toward the ground. After dropping a few feet, the moth resumes its fluttering flight and may or may not be detected again by the bat.

Echolocation is not confined to bats. Porpoises and dolphins use echolocation to locate food fishes in murky waters, and whales use echolocation to keep track of the sea bottom and rocky obstacles. Bird species, such as the Oilbird and the Cave Swiftlet, use echolocation to avoid obstacles and find their nests in dark caves.

Natural selection has produced highly adaptive sensory receptors and neural circuits to integrate and process their input in all animals. These systems, the subject of this chapter, provide animals with a steady stream of information about their internal and external environments, leading to appropriate responses. By integrating sensory information in the central nervous system (CNS), animals respond in ways that enable them to survive and reproduce.

The focus of this chapter is on neural integration. While integration occurs at every synapse in the nervous system through the processes of spatial and temporal summation, as described in the previous chapter (see Section 44.4), integration occurs on a larger scale in two areas: at sensory structures, where information is first transduced into electrical signals (see Sections 44.2e and 44.2f), and in the CNS (central ganglia and brains), where information from multiple sources is collected and interpreted (see Section 44.5). We begin this chapter with a survey of animal sensory systems. Sensory receptors are involved in neural signalling via the same four basic components as described in the previous chapter: reception, transduction, integration, and response. At the level of the single receptor cell, this ultimately involves the opening or closing of ion channels that alter the membrane potential of the cell. Depending on the nature of the stimulus, the manner in which it leads to changes in channel configuration varies. In highly specialized sensory organs, different anatomical features have evolved to amplify, filter, or modulate the stimulus. These are secondary sensory structures such as the ear or the eye; the primary structures are the sensory neurons themselves. The characteristics of several of these are examined for individual receptor types in the first part of the chapter. This sensory information is conducted by peripheral nerves to the CNS, where complex neural processing occurs, integrating information from many sources to produce memory and appropriate responses. This is described for the vertebrate CNS (and, in particular, for the human nervous system, for which the most is known) in the second part of this chapter.

45.1 Overview of Sensory Integration

45.1a The Strength of a Stimulus Is Encoded in Several Ways

Sensory pathways begin at a sensory receptor and proceed by afferent neurons to the CNS. Each type of receptor conveys information to a specific part of the CNS. Action potentials arising in the retina of the eye, for instance, travel along the optic nerves to the visual cortex, where they are interpreted by the brain as differences in pattern, colour, and intensity of light.

The frequency of action potentials that the stimulus generates in the afferent neuron (number per unit time) can indicate the intensity and extent of the stimulus. Stronger stimuli cause more action potentials than weaker ones (see Chapter 44). A light touch to the hand, for example, causes action potentials to flow at low frequencies along the axons leading to the primary somatosensory area of the cerebral cortex. As the pressure increases, the number of action potentials per second rises in proportion. In the brain, the increase is interpreted as greater pressure on the hand.

The number of afferent neurons sending action potentials in response to a stimulus can also convey information about the intensity and extent of a stimulus. The more sensory receptors that are activated, the more axons carry information to the brain. A light touch activates a relatively small number of receptors in a small area near the surface of the finger. But as the pressure increases, the resulting indentation of the finger's surface increases in area and depth, activating more receptors. In the appropriate somatosensory area of the brain, the larger number of axons carrying action potentials is interpreted as an increase in pressure spread over a greater area of the finger.

45.1b Many Receptors Adapt When Stimuli Remain Constant

In many sensory systems, the effect of a stimulus is reduced if it continues at a constant level. This reduction is called **sensory adaptation** (do not confuse this with adaptation used in the context of evolution). Some receptors adapt quickly and broadly; other receptors adapt only slightly. For example, when you walk outside on a pleasantly sunny day, it may seem exceptionally bright for a minute or two but the sensation passes and normal vision returns. In contrast, receptors detecting painful stimuli show little or no adaptation. Pain signals a danger

to some part of the body, and the signal is maintained until a response by the animal compensates for the stimulus causing the pain. Being able to adapt to light intensity, on the other hand, allows the visual system to retain its sensitivity whether on a moonlit night or in the middle of a sunny afternoon (remember that a neuron can only increase its action potential frequency so much).

Sensory adaptation also increases the sensitivity of receptor systems to *changes* in environmental stimuli. For some stimuli, these can be more important to survival than keeping track of constant environmental factors. For example, when something touches your skin you are initially aware of the touch and pressure. Within a few minutes, however, the sensation lessens or is lost, even though the pressure remains the same. The loss reflects adaptation of mechanoreceptors in your skin. If the intensity of the stimulus changes, the mechanoreceptors again become active. There are some sensations that organisms must always be aware of and others where receiving constant reminders are of less value than being able to perceive changes in the level of a stimulus.

45.1c Perception of a Stimulus Results from Interpretation of Sensory Input

Perception is the conscious awareness of our external and internal environments derived from the processing of sensory input. That is, the action potentials from sensory receptors are the signals the brain uses to generate an interpretation—the perception—of the external and internal environments. There are multiple aspects of this. In the first instance, an organism's perception of the world depends on the types of receptors it possesses. Our human perception of the world is significantly different from the perceptions of other organisms. For instance, many animals can detect much higher sound frequencies and many insects can "see" ultraviolet light, making their perception of the same environment different than ours (see Figure 1.28, Chapter 1). Then, during processing, different forms of sensory input are given unique characteristics. Action potentials arriving in the sensory cortex from different receptors on the tongue may create a sweet or a tart sensation. These are a result of the processing. All taste receptors generate action potentials, but the action potentials arriving from different receptors give rise to different sensations. Finally, input is further processed by the cerebral cortex, including comparison of the particular input with other incoming sensory input and with memories of similar situations. Each individual (perhaps with the exception of identical twins) has slightly different perceptions of the world.

Next, we will examine several individual receptor types (mechanoreceptors, photoreceptors, chemoreceptors, thermoreceptors, electroreceptors, and magnetoreceptors; see Chapter 44) and their characteristics. The focus will be on receptors that sense the external environment, but keep in mind that there is also a rich variety of receptors that sense the internal environment; many of these will be discussed briefly in subsequent chapters dealing with the regulation of the internal environment.

STUDY BREAK

1. What are the different roles of primary and secondary sensory structures? What do the secondary structures contribute?
2. How is the strength of a sensory stimulus conveyed to the brain?
3. Why is it important that some receptors allow the effect of a stimulus to be reduced, whereas other receptors do not?

45.2 Mechanoreceptors and the Tactile and Spatial Senses

Mechanoreceptors detect mechanical stimuli such as touch and pressure. The mechanical force of a stimulus creates tension in the plasma membrane of a receptor cell, which causes ion channels to open. Ion flow changes the membrane potential of the receptors and generates action potentials in afferent neurons leading from the receptors to the CNS. Sensory information from these receptors informs the brain of the body's contact with objects in the environment, providing information on the movement, position, and balance of body parts, and underlies the sense of hearing. Mechanoreceptors are involved in sensing the external environment (touch and pressure), the internal environment (position and movement of the limbs, blood pressure, lung inflation), and the relationship between the two environments (balance and equilibrium related to position in space). As we will see, while the mechanism of transduction remains constant at the level of the receptor cell, there is a variety of ways in which the mechanical event leads to the deformation of the receptor cell.

45.2a Receptors for Touch and Pressure Occur throughout the Body

In vertebrates, mechanoreceptors that detect touch and pressure are embedded in the skin and other surface tissues, in skeletal muscles, in the walls of blood vessels, and in internal organs. In humans, touch receptors in the skin are concentrated in greatest numbers in the fingertips, lips, and tip of the tongue, giving these regions the greatest sensitivity to mechanical stimuli. In other areas, such as the skin of the back, arms, and legs, the receptors are more widely spaced.

You can compare the spacing of receptors by pressing two toothpicks lightly against a fingertip and then against the skin of your arm or leg. On your fingertip, two toothpicks separated by ~1 mm can be discerned as two separate points. On your arm or leg, the two toothpicks must be nearly 5 cm apart to be distinguished as two separate points. If they are closer together, they feel like only one point.

Skin contains several types of touch and pressure receptors (**Figure 45.2**). Some are free nerve endings: dendrites of afferent neurons with no specialized structures surrounding them. Free nerve endings wrapped around hair follicles respond when the hair is bent, making you instantly aware of a spider exploring your arm or leg as it brushes against the hairs. Other mechanoreceptors, such as Pacinian corpuscles, have structures surrounding the nerve endings that contribute to reception of stimuli.

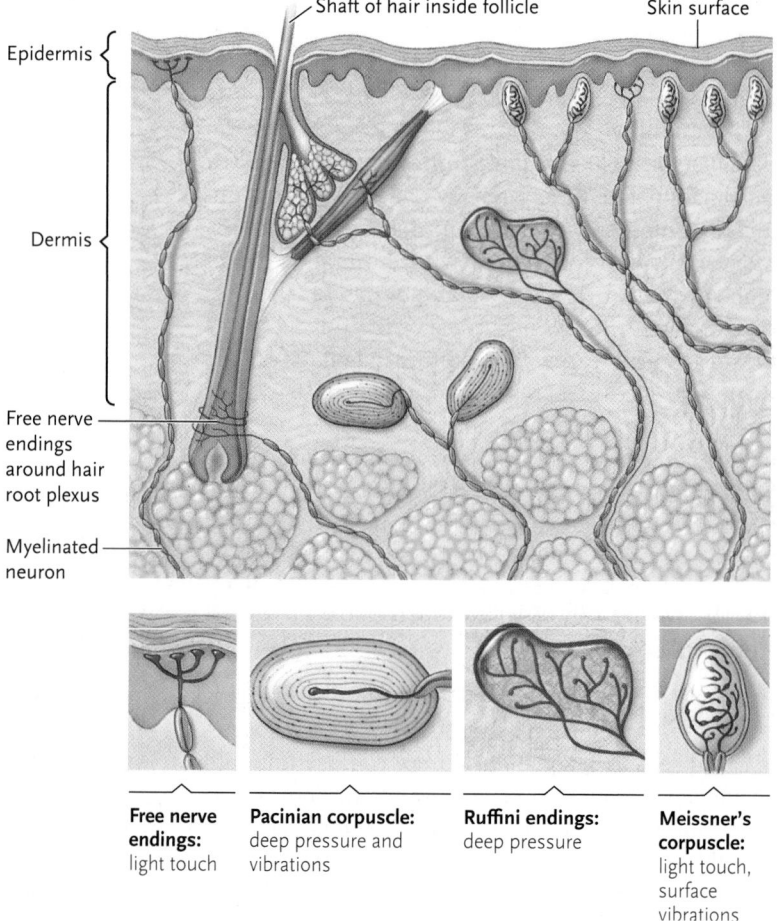

Free nerve endings: light touch

Pacinian corpuscle: deep pressure and vibrations

Ruffini endings: deep pressure

Meissner's corpuscle: light touch, surface vibrations

Figure 45.2

In human skin, four types of mechanoreceptors detect tactile stimulation.

45.2b Proprioceptors Provide Information about Movements and the Position of the Body

Proprioceptors are mechanoreceptors (*proprius* = one's own) that detect stimuli that are used by the CNS to maintain body balance and equilibrium and to monitor changes in the position of the head and limbs. The activity of proprioceptors allows you to touch the tip of your nose with your eyes closed or to precisely reach and scratch an itch on your back.

Statocysts (*statos* = standing; *kystis* = bag) are proprioceptors in aquatic invertebrates such as jellyfishes, some gastropods, and some arthropods. Most statocysts are fluid-filled chambers enclosing one or more movable stonelike bodies called **statoliths**. The chamber walls contain **sensory hair cells (Figure 45.3)**. In lobsters (*Homarus americanus*), statoliths are sand grains stuck together by mucus. When the animal moves, the statoliths lag behind the movement, bending the sensory hairs and triggering action potentials in afferent neurons. Thus, statocysts signal the brain about the body's position and orientation with respect to gravity. If you replace the sand grain statoliths with iron filings, you can use a magnet and emulate the lobster's response to the pull of gravity. In plants, statoliths control the direction of growth (see Chapter 38).

Information about self-motion is particularly important for flying animals and must be quickly (almost instantaneously) available. Typical insects have two pairs of wings, but species in the order Diptera (flies) have one. The second pair are reduced and persist as *halteres* **(Figure 45.4)**. Halteres are club-shaped and oscillate at the wing-beat frequency. They transduce information about pitch (oscillation around a horizontal axis perpendicular to the direction of movement), roll (sway on the axis parallel to the direction of movement), and yaw (oscillation about a vertical axis) movements to the CNS. Coriolis (gyroscopic) forces cause the halteres to deviate in their plane of motion. Hawk moths, with two pairs of wings, use mechanosensors on their antennae to mediate flight control. Mechanical input is essential for flight stability in moths. In bats, small hairs on the ventral surfaces of the wings are important for complex flight manoeuvres. Birds must also have mechanoreceptors associated with wings and flight, and, presumably, pterosaurs did as well.

The **lateral line system** in fishes and some aquatic amphibians uses mechanoreceptors to detect vibrations and currents in the water **(Figure 45.5, p. 1102)**. Fishes have *neuromasts,* mechanoreceptors that provide information about the fish's orientation with respect to gravity, as well as its swimming velocity. In some fishes, neuromasts are exposed on the body surface; in others, they are recessed in water-filled canals with porelike openings to the outside (see Figure 45.5). Sensory hairs are clustered at the base of each dome-shaped neuromast hair cell. One surface of the hair cell

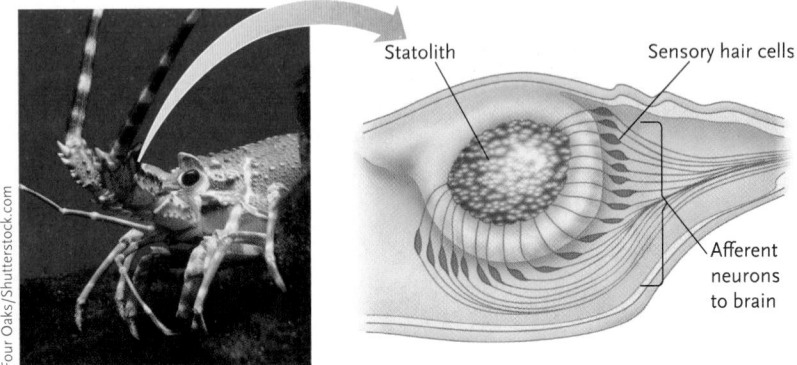

Figure 45.3

A statocyst, an organ of equilibrium in invertebrates, is located at the base of the antenna of a lobster. The statoliths inside are usually formed from fused grains of sand (calcium carbonate) in lobsters.

is covered with **stereocilia**, microvilli or cell processes reinforced by bundles of microfilaments. Stereocilia extend into a gelatinous structure, the **cupula** (*cupule* = little cup), which moves with pressure changes in the surrounding water. Movement of the cupula bends the stereocilia, causing depolarization of the hair cell's plasma membrane and release of neurotransmitter molecules that generate action potentials in associated afferent neurons.

Vibrations detected by the lateral line enable fishes to avoid obstacles, orient in a current, and monitor the presence of other moving objects in the water. The system is also responsible for the ability of schools of fish to move in unison, turning and diving in what appears to be a perfectly synchronized aquatic ballet. In actuality, the movement of each fish creates a pressure wave in the water that is detected by the lateral line systems of other fishes in the school. Schooling fishes can still swim in unison even if blinded, but if the nerves leading from the lateral line system to the brain are severed, the ability to school is lost.

Figure 45.4

Halteres, vestigial hind wings of flies, transduce information about pitch, roll, and yaw during flight. The fly shown here is a *crane fly*. Note the normal wings in the photo and the small vestigial wings below them. The head of the fly is at the top of the photo and the tail extends out of the photo at the bottom.

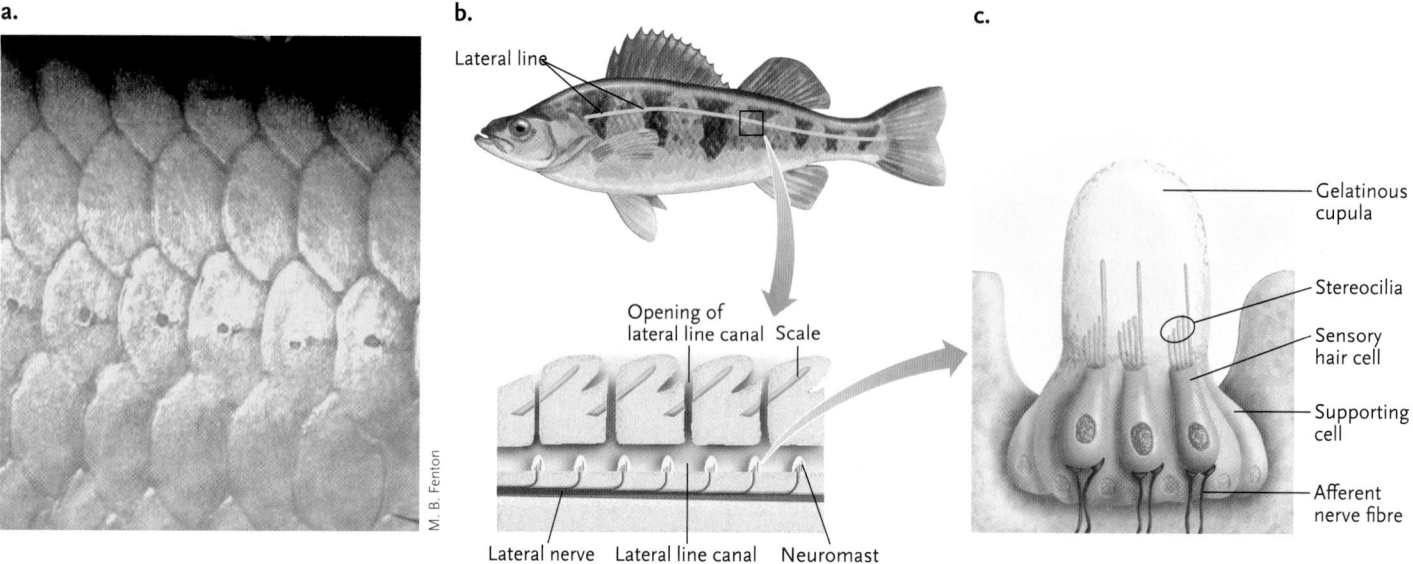

a.

b.

Lateral line

Opening of
lateral line canal   Scale

Lateral nerve   Lateral line canal   Neuromast

M. B. Fenton

c.

Gelatinous
cupula

Stereocilia

Sensory
hair cell

Supporting
cell

Afferent
nerve fibre

Figure 45.5

The lateral line system of fishes. **(a)** Pores along the lateral line of an arrowhana (*Scleropages* species). **(b)** Neuromasts are the sensory receptors in the lateral line system. **(c)** Neuromasts have a gelatinous cupula that is pushed and pulled by vibrations and currents transmitted through the lateral line canal. As the cupula moves, the stereocilia of the sensory hair cells are bent, generating action potentials in afferent neurons that lead to the brain.

45.2c The Vestibular Apparatus of Vertebrates Provides a Sense of Balance and Orientation

The inner ear of most terrestrial vertebrates has two specialized sensory structures, the *vestibular apparatus* and the *cochlea*. The **vestibular apparatus** is responsible for perceiving the position and motion of the head and is essential for maintaining equilibrium and for coordinating head and body movements. The cochlea is used in hearing (see Section 45.3).

The vestibular apparatus **(Figure 45.6)** consists of three **semicircular canals** and two chambers, the **utricle** and the **saccule**, filled with a fluid called *endolymph*. The semicircular canals are positioned at angles corresponding to the three planes of space. They detect rotational (spinning) motions. Each canal has an *ampulla*, a swelling at its base that is topped with sensory hair cells embedded in a cupula similar to that found in lateral line systems. Cupulas protrude into the endolymph of the canals. When the body or head rotates horizontally, vertically, or diagonally, endolymph in the semicircular canal corresponding to that direction lags behind, pulling the cupula with it. Displacement of the cupula bends the sensory hair cells and generates action potentials in afferent neurons that make synapses with the hair cells.

CONCEPT FIX It is commonly thought that the semicircular canals are responsible for maintaining body balance. In fact, the semicircular canals only detect body movement and send this information to the CNS, including to the cerebellum. It is the job of the CNS to interpret this information and produce the responses that maintain balance. Clearly, receiving this information

is essential for maintenance of balance, but it is only the first step in the process. ●

The utricle and saccule provide information about the position of the head with respect to gravity (up versus down), as well as changes in the rate of linear movement of the body. The utricle and saccule are oriented approximately 30° to each other, and each contains sensory hair cells with stereocilia. The hair cells are covered with a gelatinous *otolithic membrane* (which is similar to a cupula) in which **otoliths**, small crystals of calcium carbonate (*oto* = ear; *lithos* = stone), are embedded; the function of otoliths is analogous to that of statoliths of invertebrates (see Figure 45.3).

When a tetrapod is standing in its normal posture, the sensory hairs in the utricle are oriented vertically and those in the saccule are oriented horizontally. When the head is tilted in any other direction or when there is a change in the linear motion of the body, the otolithic membrane of the utricle moves and bends the sensory hairs. Depending on the direction of movement, the hair cells release more or less neurotransmitter, and the brain integrates the signals it receives and generates a perception of the movement. In humans, the saccule responds to the tilting of the head away from the horizontal (such as in diving) and to a change in movement up and down (such as in jumping). The utricle and saccule adapt quickly to the body's motion, decreasing their response when there is no change in the rate and direction of movement.

Senses of up and down vary among animals, suggesting differences in how data from the utricle and saccule are interpreted. The "normal" posture of upside-down catfish (*Synodontis nigriventris*), many bats (Chiroptera), and sloths (genera *Choloepus* and

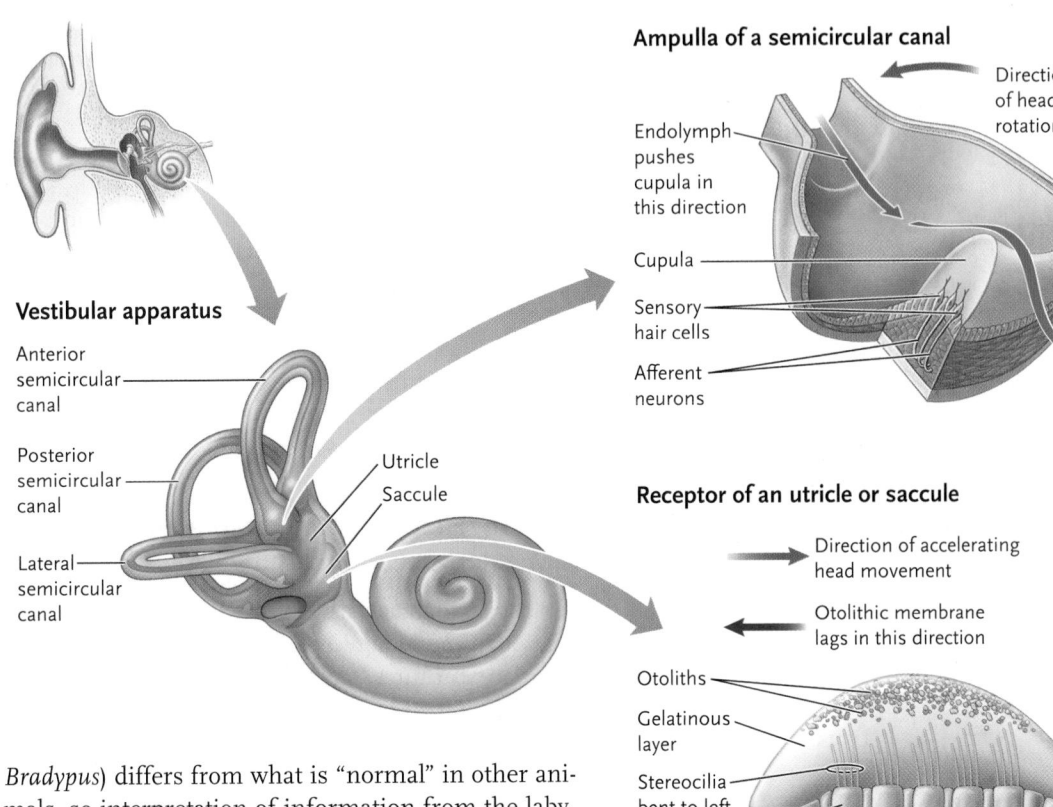

Vestibular apparatus

Anterior
semicircular
canal

Posterior
semicircular
canal

Lateral
semicircular
canal

Utricle

Saccule

Ampulla of a semicircular canal

Direction
of head
rotation

Endolymph
pushes
cupula in
this direction

Cupula

Sensory
hair cells

Afferent
neurons

Figure 45.6
The vestibular apparatus (and bony labyrinth) of the human ear. The ampulla at the base of each semicircular canal detects rotational movement of the head and body. The otolith-containing receptors in the utricle and saccule detect accelerating and decelerating movements and the position of the head relative to gravity.

Receptor of an utricle or saccule

Direction of accelerating
head movement

Otolithic membrane
lags in this direction

Otoliths

Gelatinous
layer

Stereocilia
bent to left

Sensory
hair cells

Synapse

Afferent
neurons

Bradypus) differs from what is "normal" in other animals, so interpretation of information from the labyrinth must differ as well. As bipeds, the normal posture of humans is also aberrant, so interpretation of postural information must differ here too.

The bony labyrinths of mammals vary considerably **(Figure 45.7)**, reflecting lifestyle. Agile, arboreal mammals have semicircular canals with large radii. The large radii make the vestibular system extremely sensitive to changes in body position. Such sensitivity is not compatible with the lifestyles of cetaceans because they frequently make fast body rotations. Cetaceans typically have semicircular canals with small radii (see Figure 45.7).

45.2d Stretch Receptors in Vertebrates Keep Track of Tension on Muscles

Stretch receptors are proprioceptors in the muscles and tendons of vertebrates that detect the position and movement, for example, of the limbs. Stretch receptors in muscles are **muscle spindles**, bundles of small, specialized muscle cells wrapped with the dendrites of afferent neurons and enclosed in connective tissue **(Figure 45.8, p. 1104)**. When the muscle stretches, the spindle stretches too, stimulating the dendrites and triggering the production of action potentials. The strength of the response of stretch receptors to stimulation depends on how much and how fast the muscle is stretched. Proprioceptors of tendons, called **Golgi tendon organs**, are dendrites that branch within the fibrous connective tissue of the tendon (shown in Figure 45.8). These nerve endings measure stretch and compression of the tendon as muscles contract and move limbs.

Proprioceptors allow the CNS to monitor the body's position and help keep the body in balance. They allow

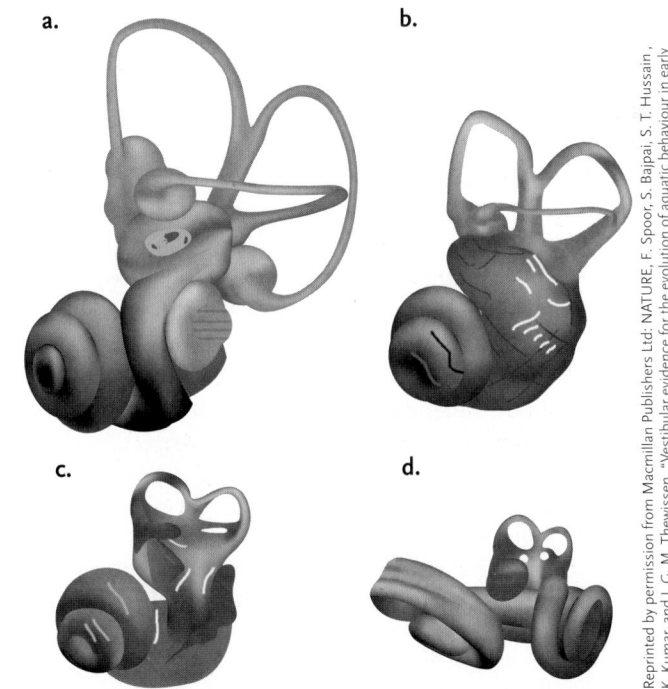

a.

b.

c.

d.

Reprinted by permission from Macmillan Publishers Ltd: NATURE, F. Spoor, S. Bajpai, S. T. Hussain, K. Kumar, and J. G. M. Thewissen, "Vestibular evidence for the evolution of aquatic behaviour in early cetaceans," Vol. 417, pp. 163–166, copyright (2002).

Figure 45.7
Bony labyrinths of four species of mammals. Lateral left view of the labyrinth organs of *Galago moholi* **(a)**, *Ichthyolestes pinfoldi* **(b)**, *Indocetus ramani* **(c)**, and *Tursiops truncatus* **(d)**. *Ichthyolestes pinfoldi* and *I. ramani* are fossil cetaceans (whales), *T. truncatus* is an extant species of dolphin, and *G. moholi* is an arboreal primate.

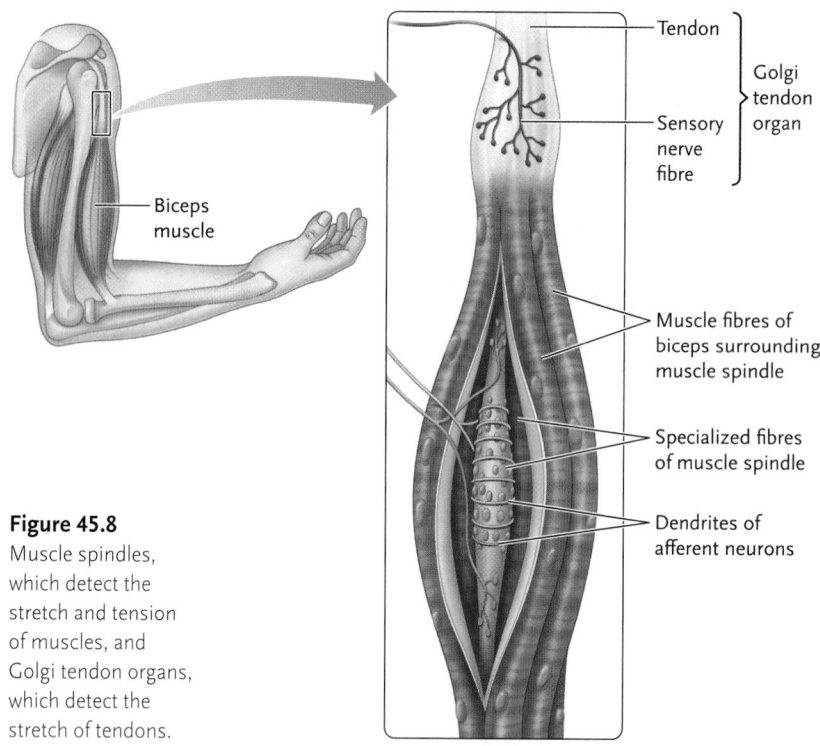

Figure 45.8
Muscle spindles, which detect the stretch and tension of muscles, and Golgi tendon organs, which detect the stretch of tendons.

Labels on figure: Tendon; Golgi tendon organ; Sensory nerve fibre; Biceps muscle; Muscle fibres of biceps surrounding muscle spindle; Specialized fibres of muscle spindle; Dendrites of afferent neurons

muscles to apply constant force under a constant load and to adjust almost instantly as the load changes. When you hold a cup while someone fills it with coffee, the muscle spindles in your biceps muscle detect the additional stretch as the cup becomes heavier. Signals from the spindles allow you to compensate for the additional weight by increasing the contraction of the muscle, keeping your arm level with no conscious effort on your part. Proprioceptors are typically slow to adapt, so the body's position and balance are constantly monitored.

STUDY BREAK

1. If two fruit flies landed on you, one on your face and the other on your leg, which one would you be most likely to detect by touch receptors? Why?
2. What is the purpose of the vestibular apparatus in the inner ear of most vertebrates?
3. On what does the strength of the response of stretch receptors depend?

45.3 Mechanoreceptors and Hearing

45.3a Sound

Sounds are vibrations that travel as waves produced by the alternating compression and decompression of air or water. The loudness, or *intensity*, of a sound depends on the amplitude (height) of the wave. It is measured in decibels (dB). Whether the *pitch* of a sound—a musical tone, for example—is a high note or a low note depends on the

frequency of the waves, measured in hertz (Hz; cycles per second). The more cycles per second, the higher the pitch. Some animals, such as the bat in the introduction to this chapter, can hear sounds well above 100 000 Hz. Humans can hear sounds between about 20 and 20 000 Hz, which is why we cannot hear the bat's sonar clicks. Although sound waves travel through air at about 340 m·s⁻¹ at sea level, individual air molecules transmitting the waves move back and forth over only a short distance as the wave passes. Water is denser than air, so sounds move approximately three times as fast under water. Infrasounds, frequencies below the range of human hearing (20 Hz), are used by African elephants to communicate.

45.3b Hearing in Invertebrates Involves Mechanoreceptors and Ears

Most invertebrates detect sound and other vibrations through mechanoreceptors in their skin or on other surface structures. An earthworm, for example, quickly retracts into its burrow at the smallest vibration of the surrounding earth, even though it has no specialized structures serving as ears. Cephalopods (squids and octopuses) have a system of mechanoreceptors on their head and tentacles, similar to the lateral line of fishes. These mechanoreceptors detect vibrations in the surrounding water. In many insects and other arthropods, hairs or bristles act as sensory receptors, vibrating in response to sound waves, often at particular frequencies.

Insects such as grasshoppers and crickets have ears, complex auditory organs on each side of the abdomen or on the first pair of walking legs **(Figure 45.9)**, whereas in moths, these "ears" have been found on the

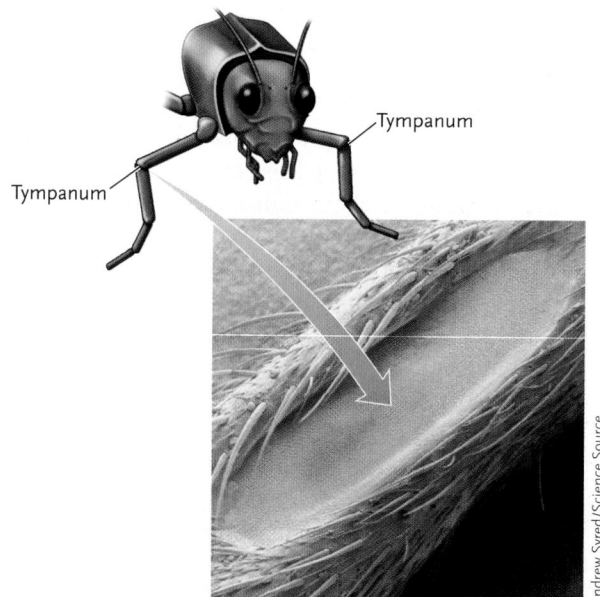

Andrew Syred/Science Source

Figure 45.9
The tympanum or eardrum of a cricket, located on the front walking legs.

head (mouthpart), thorax, and abdomen. These "ears" consist of a thinned region of the insect's exoskeleton forming a **tympanum** (= drum) over a hollow chamber. Sounds reaching the tympanum cause it to vibrate. Mechanoreceptors connected to the tympanum translate vibrations into nerve impulses. Some insect ears respond to sounds only at certain frequencies, such as the pitch of a cricket's song.

45.3c Hearing in Vertebrates Involves Auditory Systems

The auditory structures of terrestrial vertebrates transduce vibrations in air (sound) to sensory hair cells that respond by triggering action potentials. The auditory system of humans is typical for mammals **(Figure 45.10)**. The **pinna** (**outer ear**; *pinna* = wing or leaf) concentrates and focuses sound waves. Some animals have pinnae; others lack them **(Figure 45.11, p. 1106)**. Sound waves enter the auditory canal and strike a thin sheet of tissue (tympanic membrane or eardrum) and start it vibrating.

Vibrations in the tympanic membrane generate vibrations in the auditory ossicles located in the **middle ear**, which is an air-filled cavity. Mammals have three auditory ossicles, the **malleus** (hammer), **incus** (anvil), and **stapes** (stirrup). The manubrium of the malleus sits immediately behind the eardrum, and the eardrum's

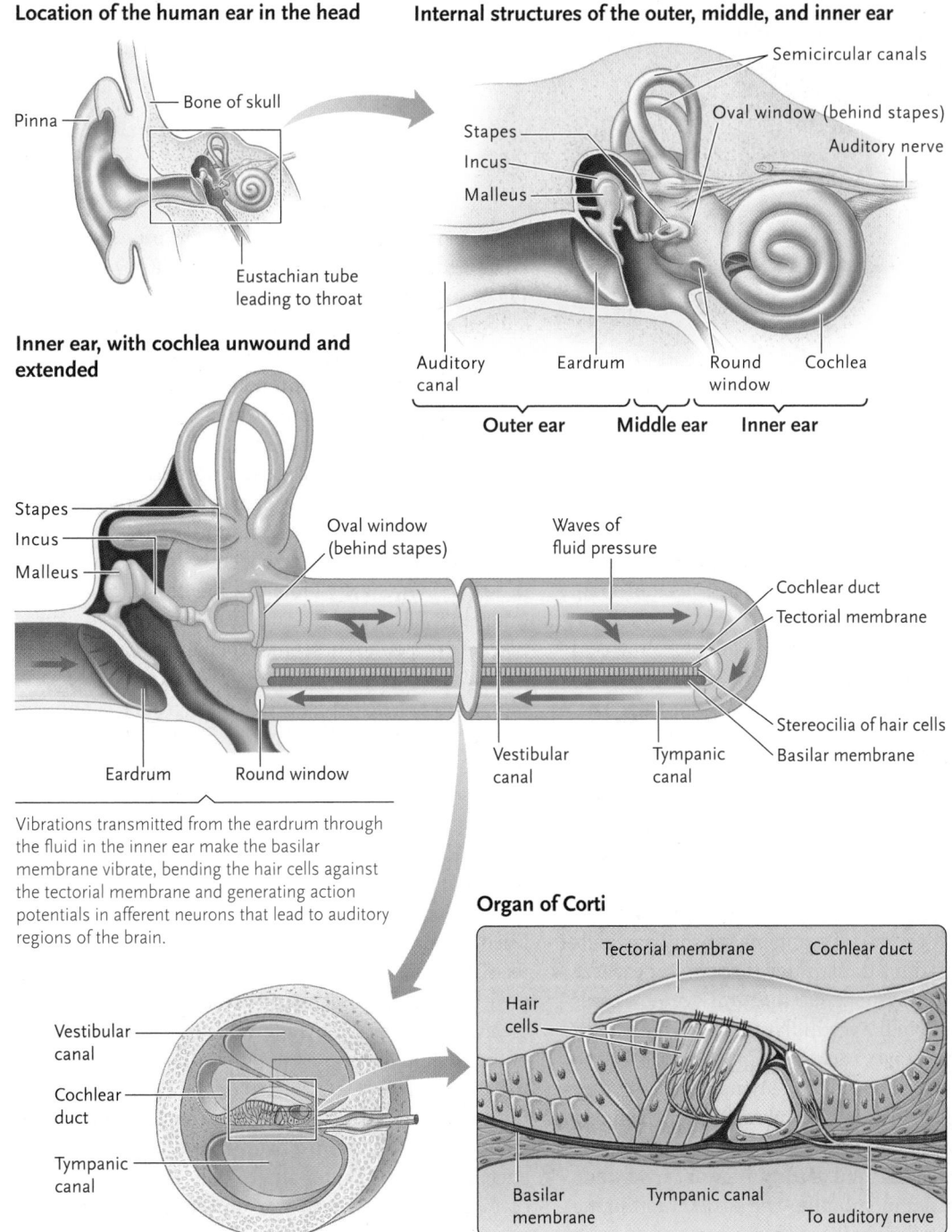

Figure 45.10
Structures of the human ear.

Location of the human ear in the head

Pinna
Bone of skull
Eustachian tube leading to throat

Internal structures of the outer, middle, and inner ear

Semicircular canals
Oval window (behind stapes)
Auditory nerve
Stapes
Incus
Malleus
Auditory canal
Eardrum
Round window
Cochlea

Outer ear   Middle ear   Inner ear

Inner ear, with cochlea unwound and extended

Stapes
Incus
Malleus
Oval window (behind stapes)
Waves of fluid pressure
Cochlear duct
Tectorial membrane
Stereocilia of hair cells
Basilar membrane
Eardrum
Round window
Vestibular canal
Tympanic canal

Vibrations transmitted from the eardrum through the fluid in the inner ear make the basilar membrane vibrate, bending the hair cells against the tectorial membrane and generating action potentials in afferent neurons that lead to auditory regions of the brain.

Vestibular canal
Cochlear duct
Tympanic canal

Organ of Corti

Tectorial membrane
Cochlear duct
Hair cells
Basilar membrane
Tympanic canal
To auditory nerve

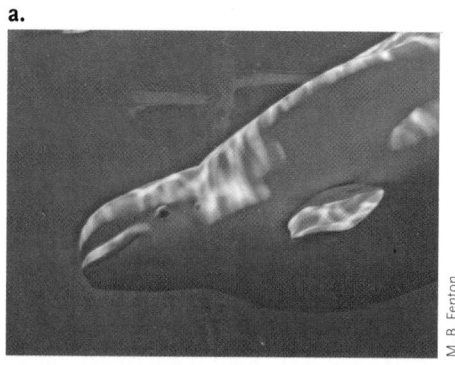

a.

b.

c.

d.

Figure 45.11

Pinnae (external ears) are lacking in mammals such as **(a)** the beluga (*Delphinapterus leucas*), **(b)** birds (*Struthio camelus*), and **(c)** reptiles (*Varanus komodoensis*), but large and conspicuous in **(d)** a bat (*Otonycteris hemprichii*).

vibrations are conducted from the malleus to the incus and the stapes. The stapes abuts the inner ear at the **oval window,** elastic membrane where vibrations in bone are converted to vibrations in the fluid in the vestibular canal. Between the eardrum and the oval window, sounds are amplified at least 20 times.

The **inner ear** contains several fluid-filled compartments, the vestibular apparatus (see Section 45.2), and the **cochlea,** a spiral tube (*kochlias* = snail). In humans, the cochlea twists through about 2.5 turns (if straightened, it would be about 3.5 cm long in an adult). The spiralling of the cochlea appears to make it more sensitive to lower-frequency sounds. Thin membranes divide the cochlea into three longitudinal chambers, the *vestibular canal* at the top, the *cochlear duct* in the middle, and the *tympanic canal* at the bottom (see Figure 45.10). The vestibular canal and the tympanic canal join at the outer tip of the cochlea, so the fluid they contain is continuous. The **organ of Corti** lies within the cochlear duct. It contains sensory hair cells that detect vibrations transmitted to the inner ear (see Figure 45.10). Vibrations of the oval window pass through the fluid in the vestibular canal, make the turn at the end, and travel back through the fluid in the tympanic canal. At the end of the tympanic canal, they are transmitted to the **round window,** a thin membrane that faces the middle ear.

Vibrations in the fluid of the inner ear cause vibrations in the **basilar membrane.** The basilar membrane forms part of the floor of the cochlear duct and anchors the sensory hair cells in the organ of Corti. The stereocilia of these cells are embedded in the *tectorial membrane* extending the length of the cochlear canal. Vibrations of the basilar membrane cause the hair cells to bend, stimulating them to release a neurotransmitter that triggers action potentials in afferent neurons leading from the inner ear.

The basilar membrane is narrowest near the oval window and gradually widens toward the outer end of the cochlear duct. High-frequency vibrations produced by high-pitched sounds vibrate the basilar membrane most strongly near its narrow end, whereas vibrations of lower frequency vibrate the membrane nearer the outer end. Thus, each frequency of sound waves causes hair cells in a different segment of the basilar membrane to initiate action potentials. More than 15 thousand hair cells are distributed in small groups along the basilar membrane. Each group of hairs is connected by synapses to afferent neurons, the axons of which are bundled together in the *auditory nerve,* a cranial nerve leading to the thalamus. From there, the signals are routed to specific regions in the auditory centre of the temporal lobe.

The **eustachian tube,** a duct leading from the air-filled middle ear to the throat (see Figure 45.10), protects the eardrum from damage caused by changes in environmental atmospheric pressure. As we swallow or yawn, the tube opens, allowing air to flow into or out of the middle ear, equalizing pressure on both sides of the eardrum. When swelling or congestion prevents the tube from admitting air, we complain of having stopped-up ears because we sense a pressure difference between the outer and middle ear caused by the eardrum bulging inward or outward; this interferes with the transmission of sounds.

John Ford, *Department of Fisheries and Oceans and the University of British Columbia*

In 1973, Mike Biggs, working at the Pacific Biological Station in Nanaimo, British Columbia, revolutionized field studies of killer whales when he realized that the natural markings on the bodies of the whales could be used to identify individuals. By photographing the dorsal fin and grey "saddle" patch at the base of the fin, along with the unique pattern of nicks and scars, he developed a technique of photoidentification that allowed his group to begin to examine the natural history and population ecology of this charismatic species in detail **(Figure 1)**.

In 1977, John Ford began to study the underwater vocalizations of the whales with respect to their behaviour and social structure. Killer whales, like other toothed whales, produce a wide variety of sounds that serve different purposes. Rapid series of clicks are used for echolocation and navigation (like the bats mentioned at the start of this chapter), while other sounds resembling whistles, squeals,

squawks, and screams are used for social communication. A large percentage of these sounds are distinctly different in different groups of whales. These vocal variations, known as dialects, have allowed researchers to distinguish family groups. A group's dialect appears to be learned by each calf by mimicking its mother, and thus each pod of killer whales can be readily identified using sound analyzers. Dialects appear to be used by the whales as acoustic indicators of group identity and membership. As a result, they have been used to gain insight into the

social history of populations. Pods with similar dialects belong to a clan, a continuous lineage descended from a common ancestral pod. As pods grew in size over time, they gradually split into new pods and their common dialect drifted apart. Pods with similar dialects probably split in the recent past, while others with fewer similarities are likely more distantly related. Different clans have no dialect features in common and probably have very ancient links. Thus, vocal dialects of resident killer whale pods provide information on how communities have evolved in the past **(Figure 2)**.

Figure 1
A pod of killer whales foraging off the coast of British Columbia.

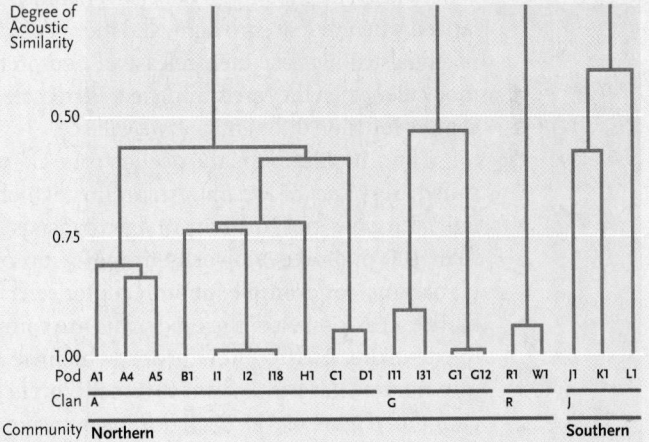

Figure 2
This diagram shows all 19 resident pods of killer whales organized according to the degree of similarity in their call repertoire. Pods with almost identical dialects are linked with a high index of acoustic similarity. Pods that have related dialects belong to the same clan. The four resident clans have no acoustic similarity and are not linked. The family tree shown for each clan reflects its historical genealogy.
Reprinted with permission from *Killer Whales*, 2nd edition by John Ford, Kenneth Balcomb and Graeme Ellis © University of British Columbia Press, 2000. All rights reserved by the Publisher.

STUDY BREAK

1. How do most invertebrates detect sound? Give an example.
2. Explain in detail how a human detects sound. Is this fundamentally different from the way an invertebrate hears?
3. Why are the echolocation calls of many bats inaudible to humans despite their high intensity? If humans can't hear bats, how did they discover that bats could emit calls and hear in this frequency range?

45.4 Photoreceptors and Vision

The great majority of animals have receptors that can detect and respond to light. As animals evolved and became more complex, the complexity of their visual sensory receptors increased, leading to the highly developed eyes of cephalopods and vertebrates.

45.4a Vision Involves Detection and Perception of Radiant Energy

Photoreceptors detect light at particular wavelengths, and centres in a brain or central ganglion integrate signals arriving from the receptors into a perception of light. All animals use forms of a single lipidlike pigment, *retinal* (synthesized from vitamin A), in photoreceptors to absorb light energy. The simplest eyes are capable only of distinguishing light from dark; the most complex eyes distinguish shapes and colours and focus an accurate image of objects being viewed onto a layer of photoreceptors.

45.4b Invertebrate Eyes Take Many Forms

Some invertebrates, such as earthworms, do not have visual organs; instead, photoreceptors in their skin allow them to sense and respond to light. Earthworms respond negatively to light, as you can easily discover by shining a flashlight on an earthworm outside its burrow at night.

The eyes of other invertebrates are diverse, ranging from collections of photoreceptors with no lens and no image-forming capability to eyes remarkably like those of vertebrates. The photoreceptors of invertebrates are depolarized when they absorb light, and they generate action potentials or increase their release of neurotransmitter molecules when they are stimulated. Vertebrate photoreceptors function differently, as we will see.

The simplest eye is the **ocellus** (plural, *ocelli;* also called an *eyespot* or *eyecup*). An ocellus, which detects light but does not form an image, consists of fewer than 100 photoreceptor cells lining a cup or pit. In planarians, for example, photoreceptor cells in a cuplike depression below the epidermis are connected to the dendrites of afferent neurons, which are bundled into nerves that travel from the ocelli to the cerebral ganglion **(Figure 45.12)**. Each ocellus is covered on one side by a layer of pigment cells that blocks most of the light rays arriving from the opposite side of the animal. As a result, most of the light received by the pigment cells enters the ocellus from the side that it faces. Through integration of information transmitted to the cerebral ganglion from the eyecups, planarians orient themselves so that the amount of light falling on the two ocelli is equal and diminishes as they swim. This reaction carries them directly away from the source of the light. Similar ocelli are found in a variety of animals, including insects, arthropods, and molluscs.

Two main types of image-forming eyes have evolved in invertebrates: compound eyes and single-lens eyes. The **compound eye** of insects, crustaceans, and a few annelids and molluscs contains hundreds to thousands of faceted visual units called **ommatidia** (*omma* = eye) fitted closely together **(Figure 45.13)**. Each ommatidium samples a small part of the visual field. In insects, light entering an ommatidium is focused by a transparent **cornea** and a *crystalline cone* (just below the cornea) onto a bundle of photoreceptor cells. Microvilli of these cells interdigitate like the fingers of clasped hands, forming a central axis rich in rhodopsin, a retinal-containing **photopigment** (light-absorbing pigment). Absorption of light by rhodopsin causes action potentials to be generated in afferent neurons connected to the base of the ommatidium. From these signals, the brain receives a mosaic image of the world. Because even the slightest motion is detected simultaneously by many ommatidia, compound eyes are extraordinarily adept at detecting movement—a lesson soon learned by fly-swatting humans.

The **single-lens eye** of cephalopods **(Figure 45.14)** resembles a vertebrate eye in that both types operate like a camera. In the cephalopod eye, light enters through the transparent cornea; a **lens** concentrates the light; and a layer of photoreceptors at the back of the eye, the **retina**, records the image. Behind the cornea is the **iris**, which surrounds the **pupil**, the opening through which light enters the eye. Muscles in the iris adjust the size of the pupil to vary the amount of light entering the eye. When the light is bright, circular muscles in the iris contract, shrinking the size of the pupil and reducing the amount of light that enters. In dim light, radial muscles contract and enlarge the pupil, increasing the amount of light that enters the eye. Muscles move the lens forward and back with respect to the retina to focus the image. This is an example of **accommodation**, a process by which the lens changes to enable the eye to focus on objects at different distances. A neural network lies under the retina, meaning that light rays do not have to pass through the neurons to reach the photoreceptors. The vertebrate eye has the opposite arrangement. This and other differences in structure and function indicate that cephalopod and vertebrate eyes evolved independently but are remarkably similar.

Figure 45.12

The ocellus of a planarian flatworm, and the arrangement of pigment cells on which its orientation response is based.

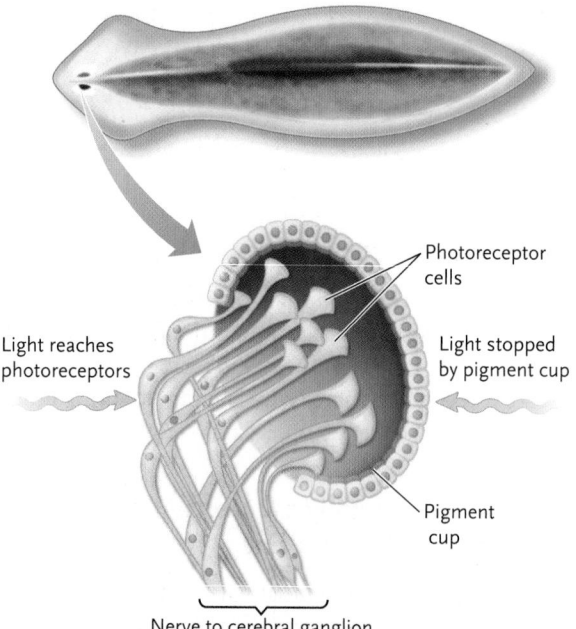

Photoreceptor cells

Light reaches photoreceptors

Light stopped by pigment cup

Pigment cup

Nerve to cerebral ganglion

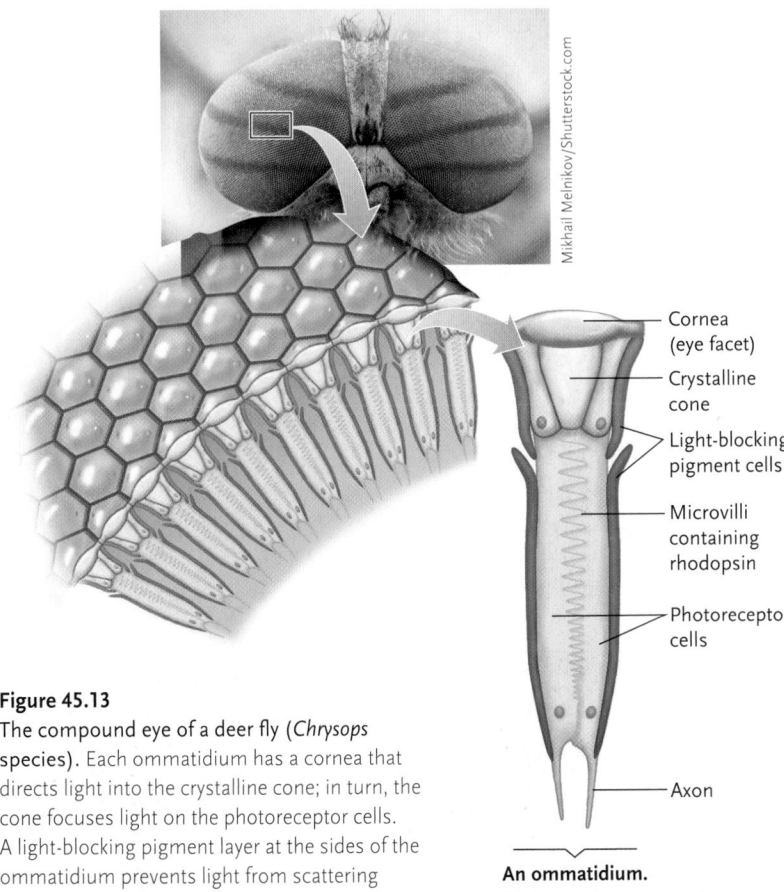

Figure 45.13

The compound eye of a deer fly (*Chrysops* species). Each ommatidium has a cornea that directs light into the crystalline cone; in turn, the cone focuses light on the photoreceptor cells. A light-blocking pigment layer at the sides of the ommatidium prevents light from scattering laterally in the compound eye.

Cornea (eye facet)

Crystalline cone

Light-blocking pigment cells

Microvilli containing rhodopsin

Photoreceptor cells

Axon

An ommatidium.
The unit of a compound eye

45.4c Vertebrate Eyes Have a Complex Structure

The human eye **(Figure 45.15, p. 1110)** has similar structures—cornea, iris, pupil, lens, and retina—to those of the cephalopod eye just described. Light entering the eye through the cornea passes through the iris and then the lens. The lens focuses an image on the retina, and the axons of afferent neurons originating in the retina converge to form the optic nerve leading from the eye to the brain. A clear fluid called the **aqueous humour** fills the space between the cornea and the lens. This fluid carries nutrients to the lens and cornea, which do not contain any blood vessels. The main chamber of the eye, located between the lens and the retina, is filled with the jellylike **vitreous humour** (*vitrum* = glass). The outer wall of the eye contains a tough layer of connective tissue (the *sclera*). Inside it is a darkly pigmented layer (the *choroid*) that prevents light from entering except through the pupil. It also contains the blood vessels nourishing the retina.

Two types of photoreceptors, rods and cones, occur in the retina along with layers of neurons that carry out an initial integration of visual information before it is sent to the brain. The **rods** are specialized for detection of light at low intensities; the **cones** are specialized for detection of different wavelengths

(colours). Accommodation does not occur by forward and backward movement of the lens, as described for cephalopods. Rather, the lens of most terrestrial vertebrates is focused by changing its shape. The lens is held in place by fine ligaments that anchor it to a surrounding layer of connective tissue and muscle, the **ciliary body**. These ligaments keep the lens under tension when the ciliary muscle is relaxed. The tension flattens the lens, which is soft and flexible, and focuses light from distant objects on the retina **(Figure 45.16a, p. 1111).** When the ciliary muscles contract, they relieve the tension of the ligaments, allowing the lens to assume a more spherical shape and focusing light from nearby objects on the retina (Figure 45.16b).

45.4d The Retina of Mammals and Birds Contains Rods and Cones and a Complex Network of Neurons

The retina of a human eye contains about 120 million rods and 6 million cones organized into a densely packed single layer. Neural networks of the retina are layered on top of the photoreceptor cells, so that light rays focused by the lens on the retina must pass through the neurons before reaching the photoreceptors. The light must also pass through a layer of fine blood vessels covering the surface of the retina.

In mammals and birds with eyes specialized for daytime vision, cones are concentrated in and around a small region of the retina, the **fovea** (see Figure 45.15). The image focused by the lens is centred on the fovea, which is circular and less than a millimetre in diameter in humans. The rods are spread over the remainder of

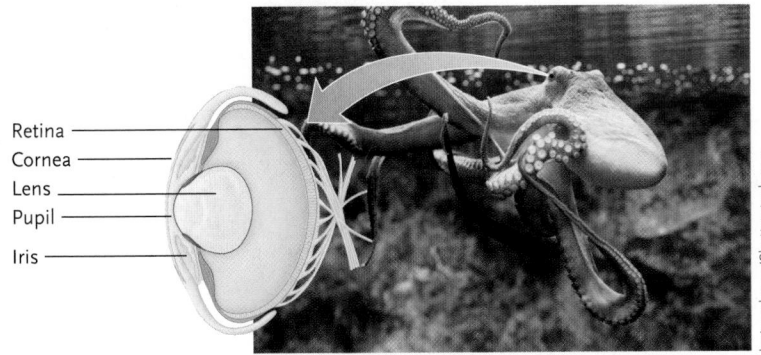

Retina
Cornea
Lens
Pupil
Iris

Figure 45.14
The eye of an octopus, a cephalopod mollusc.

Seeing with your Feet: Using the Genome as an Investigative Tool

It is one thing to demonstrate that an animal responds behaviourally to external stimuli, but it can be very challenging to determine how the stimuli were detected. For example, there is evidence of animals such as garden toads (*Bufo bufo*) changing their behaviour in advance of an earthquake, but we do not know what cues trigger the response.

For some time, it has been clear that echinoderms such as sea urchins respond to changing light conditions, but nobody had found photoreceptors in these animals (see Chapter 1 for a discussion of the significance of light and light sensing). The publication of the genome of purple urchins (*Strongylocentrotus purpuratus*) **(Figure 1)** provided biologists with a means of investigating photoreception in these animals.

Specifically, data in the genome showed that sea urchins possess several genes that code for a widely occurring eye protein, opsin. Discovering this, the researchers designed antibodies against different opsin proteins and performed *in situ* hybridization (see Chapters 15 and 16 for discussions of DNA technologies and genomics). They found that the urchins possess *Sp-opsin4* and *Sp-pax6*, two proteins that regulate phototaxis. This approach also allowed them to visualize where the photoreceptor cells were located—in the urchin's tube feet. The sea urchin photoreceptive cells are microvillar r-opsin, previously known only from protostomes. Since tube feet are found all over the body of the sea urchin, it appears that the entire adult sea urchin acts as a huge compound eye!

There are many other mysteries about the sensory world of animals. We know that many animals show a magnetic sense, but in most cases we do not know the details of the receptor.

 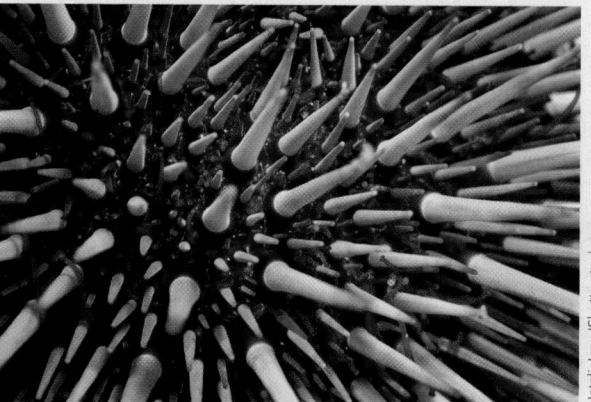

Figure 1

(a) *The purple sea urchin* Strongylocentrotus purpuratus. *(b)* *Close-up showing the tube feet between the spines. The photoreceptors are located at the tips of the tube feet.*

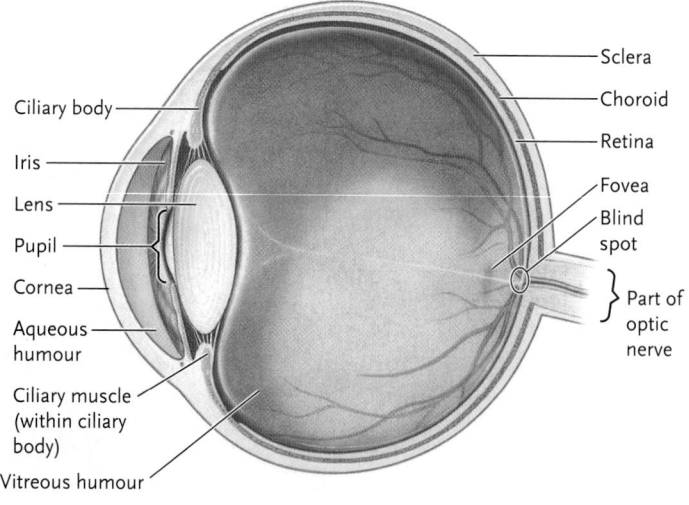

Figure 45.15
Structures of the human eye.

the retina. We can see distinctly only the image focused on the fovea; the surrounding image is what we term *peripheral vision.* Mammals and birds with eyes specialized for night vision have retinas containing mostly rods and lacking a defined fovea. Some fishes and many reptiles have cones generally distributed throughout their retina and very few rods.

The rods of mammals are much more sensitive than the cones to low-intensity light; in fact, they can respond to a single photon of light. This is why, in dim light, we can detect objects better by looking slightly to the side of the object. This action directs the image away from the cones in the fovea to the highly light-sensitive rods in surrounding regions of the retina.

a. Focusing on distant object

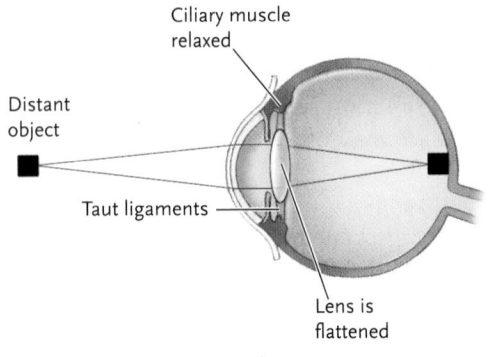

Distant object

Ciliary muscle relaxed

Taut ligaments

Lens is flattened

When the eye focuses on a distant object, the ciliary muscles relax, allowing the ligaments that support the lens to tighten. The tightened ligaments flatten the lens, bringing the distant object into focus on the retina.

b. Focusing on near object

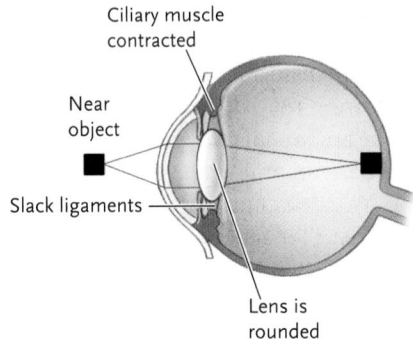

Ciliary muscle contracted

Near object

Slack ligaments

Lens is rounded

When the eye focuses on a near object, the ciliary muscles contract, loosening the ligaments and allowing the lens to become rounder. The rounded lens focuses a near object on the retina.

Figure 45.16
Accommodation in terrestrial vertebrates occurs when the lens changes shape to focus on distant **(a)** and near **(b)** objects.

Sensory Transduction by Rods and Cones: Converting Signals to Electrical Impulses. A photoreceptor cell has three parts:

- an outer segment, consisting of stacked, flattened, membranous discs;
- an inner segment, where the cell's metabolic activities occur; and

- the synaptic terminal, where neurotransmitter molecules are stored and released **(Figure 45.17a).**

The light-absorbing pigment of rods and cones, retinal, is bonded covalently to **opsins** to produce photopigments. The photopigments are embedded in the membranous discs of the photoreceptors' outer segments (Figure 45.17b). The retinal–opsin photopigment in rods is **rhodopsin.**

a. Structure of cones and rods

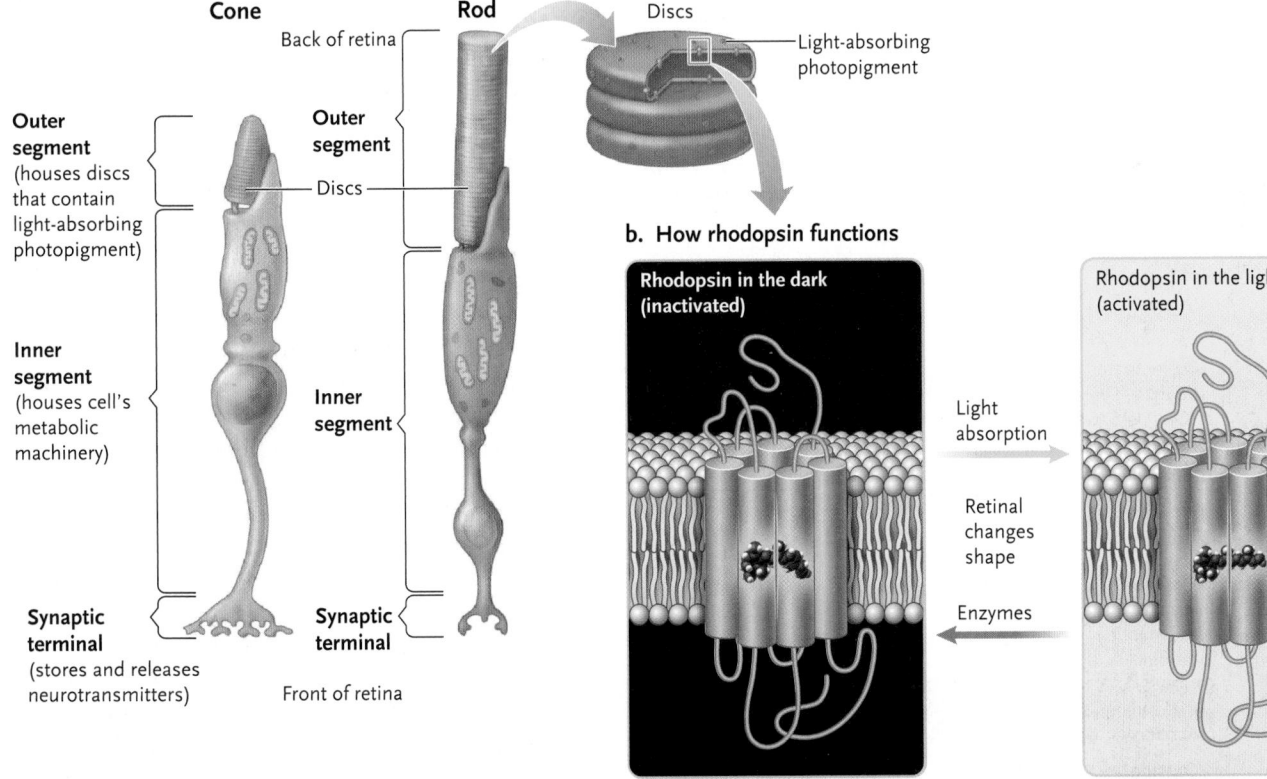

Figure 45.17
Photoreceptors. Structure of cones and rods **(a),** the photoreceptors of all mammals, and the location of photopigments in stacked, membranous discs. The photopigment rhodopsin (found in rods) **(b),** which consists of the opsin protein retinal. In response to light, the retinal changes from a bent to a straight structure.

In the dark, the retinal segment of unstimulated rhodopsin is *cis*-retinal, an inactive form (see Figure 45.17b), and rods steadily release the neurotransmitter glutamate. When rhodopsin absorbs a photon of light, retinal converts to *trans*-retinal, the active form, and the rods *decrease* the amount of glutamate they release.

Rhodopsin is a membrane-embedded G protein–coupled receptor. An extracellular signal received by a G protein–coupled receptor activates the receptor, triggering a signal transduction pathway within the cell and generating a cellular response. Here, activated rhodopsin triggers a signal transduction pathway that leads to the closure of Na$^+$ channels in the plasma membrane **(Figure 45.18)**. Closure of the channels hyperpolarizes the photoreceptor's membrane, decreasing neurotransmitter release. The response is graded because as light absorption by photopigment molecules increases, the amount of neurotransmitter released is reduced proportionately. If light absorption decreases, neurotransmitter release by the photoreceptor increases proportionately. Transduction in rods works in the opposite way from most sensory receptors in which a stimulus increases neurotransmitter release.

Visual Processing in the Retina: Events at the Back of the Eye. In the vertebrate retina, the two types of photoreceptors are linked to a network of neurons that carry out initial integration and processing of visual information. The retina of mammals has four types of neurons **(Figure 45.19)**. There is a layer of **bipolar cells** just in front of the rods and cones. These neurons synapse with rods or cones at one end and with **ganglion cells**, a layer of neurons, at the other end. The axons of ganglion cells extend over the retina and collect at the back of the eyeball to form the optic nerve, which transmits action potentials to the brain. The point where the optic nerve exits the eye lacks photoreceptors. This *blind spot* can be several millimetres in diameter in humans. **Horizontal cells** connect photoreceptor cells, whereas **amacrine cells** connect bipolar and ganglion cells.

In the dark, the steady release of glutamate from rods and cones depolarizes some postsynaptic bipolar cells and hyperpolarizes others. In the light, the decrease in neurotransmitter release from rods and cones results in the depolarized bipolar cells becoming hyperpolarized and the hyperpolarized bipolar cells becoming depolarized.

Signals from the rods and cones may move vertically or laterally in the retina. Signals move vertically from the photoreceptors to bipolar cells and then to ganglion cells. Whereas the human retina has over 120 million photoreceptors, it has only about 1 million ganglion cells. This disparity is explained by the fact that each ganglion cell receives signals from a clearly defined set of photoreceptors constituting the *receptive field* for that cell. Therefore, stimulating numerous photoreceptors in a ganglion cell's receptive field results in only a single message to the brain from that cell. Receptive fields are typically circular and are of different sizes. Smaller receptive fields result in sharper images because they send more precise information to the brain about the location in the retina where the light was received.

Lateral movement of signals from a rod or cone proceeds to a horizontal cell and continues to bipolar cells with which the horizontal cell makes inhibitory connections. To understand this, consider a spot of light falling on the retina. Photoreceptors detect the light and send a signal to bipolar cells and horizontal cells. Horizontal cells inhibit more-distant bipolar cells that are outside the spot of light, causing the light spot to appear lighter and its surrounding dark area to appear darker. This type of visual processing is called **lateral inhibition** and serves both to sharpen the edges of objects and to enhance contrast in an image.

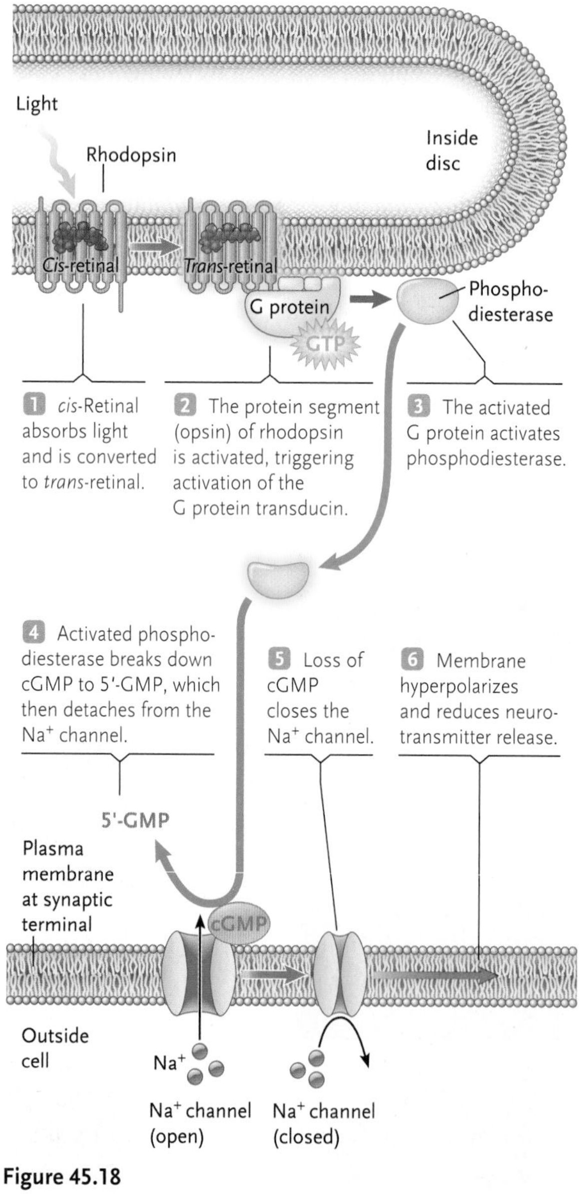

Figure 45.18

The signal transduction pathway that closes Na$^+$ channels in photoreceptor plasma membranes when rhodopsin absorbs light.

Light

Rhodopsin

Inside disc

Cis-retinal   *Trans*-retinal

G protein

GTP

Phosphodiesterase

1 *cis*-Retinal absorbs light and is converted to *trans*-retinal.

2 The protein segment (opsin) of rhodopsin is activated, triggering activation of the G protein transducin.

3 The activated G protein activates phosphodiesterase.

4 Activated phosphodiesterase breaks down cGMP to 5'-GMP, which then detaches from the Na$^+$ channel.

5 Loss of cGMP closes the Na$^+$ channel.

6 Membrane hyperpolarizes and reduces neurotransmitter release.

5'-GMP

Plasma membrane at synaptic terminal

cGMP

Outside cell

Na$^+$

Na$^+$ channel (open)   Na$^+$ channel (closed)

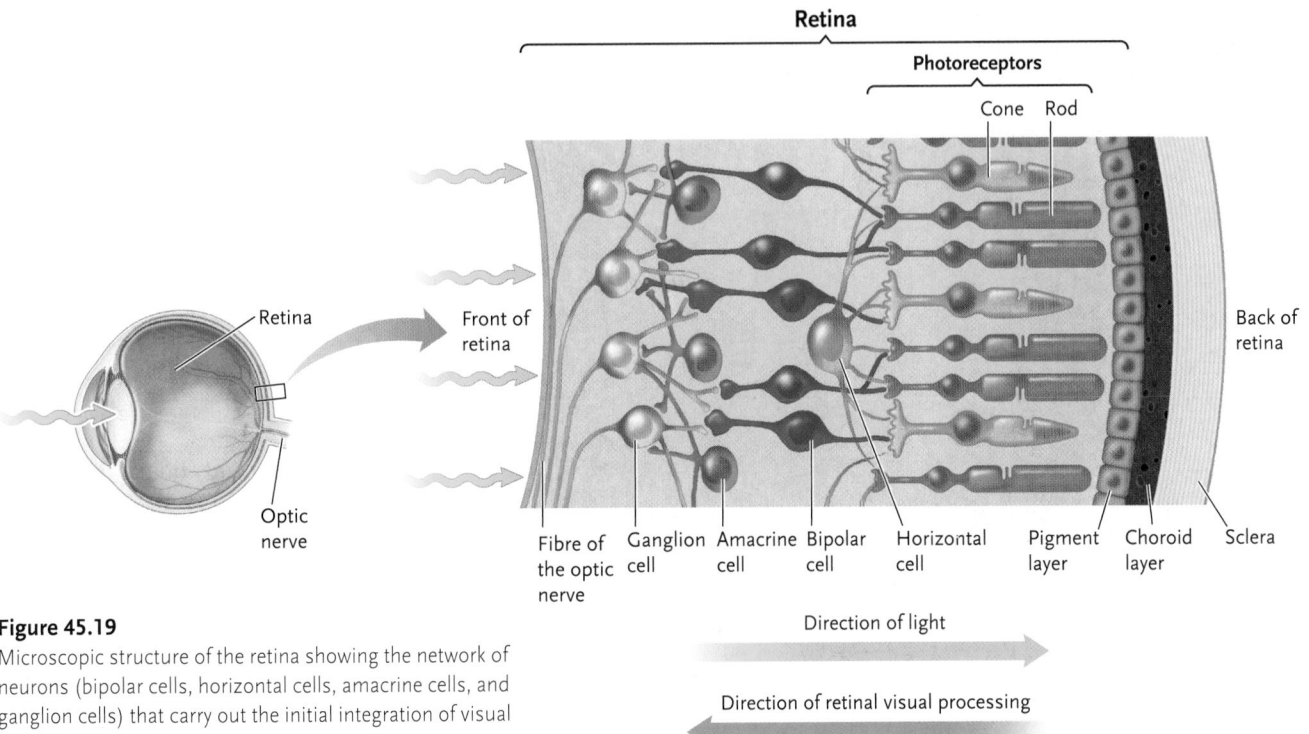

Retina

Photoreceptors

Cone　Rod

Front of retina

Back of retina

Fibre of the optic nerve

Ganglion cell

Amacrine cell

Bipolar cell

Horizontal cell

Pigment layer

Choroid layer

Sclera

Direction of light

Direction of retinal visual processing

Retina

Optic nerve

Figure 45.19

Microscopic structure of the retina showing the network of neurons (bipolar cells, horizontal cells, amacrine cells, and ganglion cells) that carry out the initial integration of visual information.

45.4e　Three Kinds of Opsin Pigments Underlie Colour Vision

Many invertebrates and some species in each class of vertebrates have colour vision, which depends on cones in the retina. Most mammals have two types of cones, whereas humans and other primates have three types. Each human or primate cone cell contains one of three **photopsins** in which retinal is combined with different opsins. The three photopsins absorb light over different, but overlapping, wavelength ranges, with peak absorptions at 445 nm (blue light), 534 nm (green light), and 570 nm (red light). The farther a wavelength is from the peak colour absorbed, the less strongly the cone responds. Having more types of cones translates into better colour vision. Birds have four photopsins and are able to distinguish shades of colour that humans cannot.

Overlapping wavelength ranges for the three photoreceptors mean that light at any visible wavelength stimulates at least two of three types of cones. Maximal absorption by each type of cone at a different wavelength leads to differential stimulation of different types of cones. These differences are relayed to the visual centres of the brain, where they are integrated into the perception of a colour corresponding to the particular wavelength absorbed. Light stimulating all three receptor types equally is seen as white.

Colour-blindness results from inherited defects in opsin proteins of one or more of the three types of cones. For example, people with a mutation preventing cones from making a functional form of red-absorbing opsin see orange, yellow, and red as the same grey or greenish colour.

45.4f　The Visual Cortex Generates Images in the Brain

Just behind the eyes, the optic nerves converge before entering the base of the brain. A portion of each optic nerve crosses over to the opposite side, forming the **optic chiasm** (*chiasma* = crossing place). Most axons enter the **lateral geniculate nuclei** in the thalamus, where they synapse with interneurons leading to the visual cortex **(Figure 45.20, p. 1114)**.

Because of the optic chiasm, the left half of the image seen by both eyes is transmitted to the visual cortex in the right cerebral hemisphere, and the right half of the image is transmitted to the left cerebral hemisphere. The right hemisphere thus sees objects to the left of the centre of vision, and the left hemisphere sees objects to the right of the centre of vision. Communication between the right and left hemispheres integrates this information into a perception of the entire visual field seen by the two eyes.

If you look at a nearby object with one eye and then the other, you will notice that the point of view is slightly different. Integration of the visual field by the brain creates a single picture with a sense of distance and depth. The greater the difference between the images seen by the two eyes, the closer the object appears to the viewer.

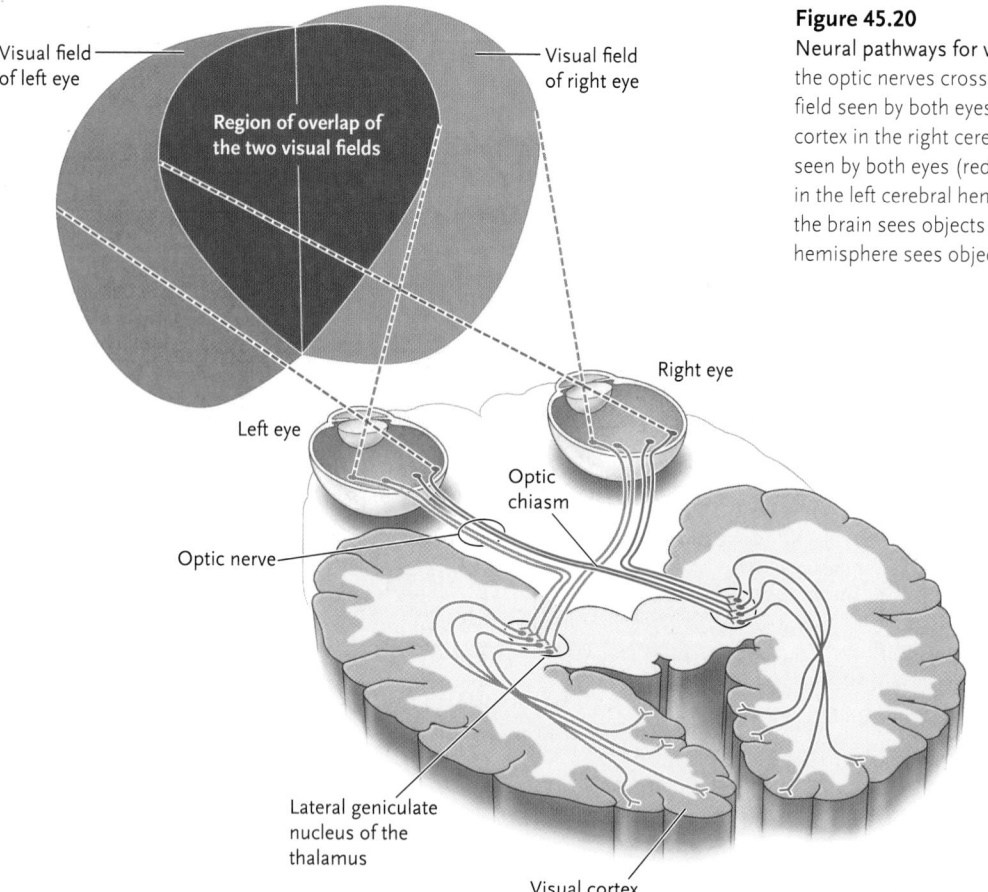

Visual field of left eye — Region of overlap of the two visual fields — Visual field of right eye

Right eye

Left eye

Optic chiasm

Optic nerve

Lateral geniculate nucleus of the thalamus

Visual cortex

Figure 45.20

Neural pathways for vision. Because half of the axons carried by the optic nerves cross over in the optic chiasma, the left half of the field seen by both eyes (green segment) is transmitted to the visual cortex in the right cerebral hemisphere. The right half of the field seen by both eyes (red segment) is transmitted to the visual cortex in the left cerebral hemisphere. As a result, the right hemisphere of the brain sees objects to the left of the centre of vision, and the left hemisphere sees objects to the right of the centre of vision.

different positions. They also correct for curvature of the water droplet's trajectory. Other fish also spit at aerial prey, and some birds hunt fish from above the water's surface; both deal with the problems of refraction from a different standpoint.

In humans, the two optic nerves together contain more than a million axons, more than all other afferent neurons of the body put together. Almost one-third of the grey matter of the cerebral cortex is devoted to visual information. These numbers give some idea of the complexity of the information integrated into the visual image formed by the brain and the importance of visual information for everyday activity.

Archerfish **(Figure 45.21)** live in fresh water and knock flying or resting insects onto the water's surface with spit droplets. The fish then catch and eat the insects. During the spitting attacks, the fishes' eyes are below the surface of the water, posing a potentially serious problem because of refraction, the deflection of rays of light at the air–water interface. Some evidence suggests that archerfish spit from directly under the prey, but further observations show that this is not always true. Archerfish correctly set their spitting angle to compensate for the refraction they experience at

STUDY BREAK

1. What is the "simplest" eye? Why is it an eye, and how does it differ from image-forming eyes?
2. What causes colour-blindness?
3. Why are compound eyes so adept at detecting motion?
4. What is accommodation?

Figure 45.21

An archerfish, *Toxotes chatareus*, projecting spit at an insect prey.

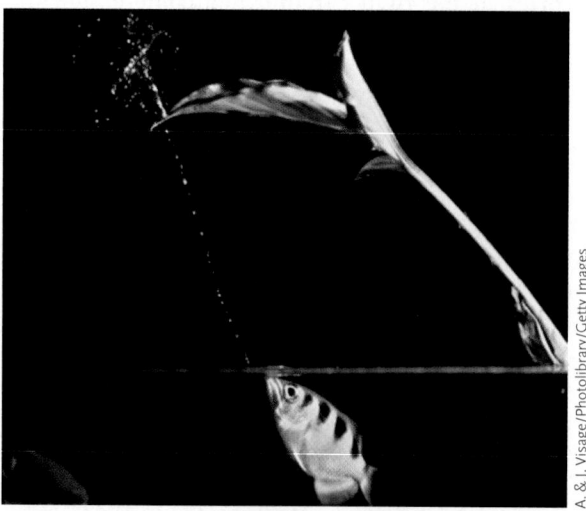

A. & J. Visage/Photolibrary/Getty Images

45.5 Chemoreceptors

Chemoreceptors provide information about taste (gustation) and smell (olfaction), as well as measures of intrinsic levels of molecules such as oxygen, carbon dioxide, and hydrogen ions. All chemoreceptors probably work through membrane receptor proteins that are stimulated when they bind with specific molecules in their environment (internal or external) and generate action potentials in afferent nerves leading to the CNS. In this section, we only discuss sensing of external stimuli through taste and smell.

45.5a Invertebrate Animals Experience a Rich World of Odours

In many invertebrates, the same receptors serve for sensing smell and taste. These receptors may be concentrated around the mouth or distributed over the body surface. The cnidarian *Hydra* has chemoreceptors around its mouth that respond to glutathione, a chemical released from prey organisms ensnared in the cnidarian's tentacles. Stimulation of chemoreceptors by glutathione causes the tentacles to retract, resulting in ingestion of the prey. In contrast, earthworms have taste and smell receptors distributed over the entire body surface.

Some terrestrial invertebrates have clearly differentiated receptors for taste and smell. In insects, taste receptors occur inside hollow sensory bristles called *sensilla* (singular, *sensillum*), usually located on the antennae, mouthparts, or feet **(Figure 45.22)**. Pores in the sensilla admit molecules from potential food to the chemoreceptors, which are specialized to detect sugars, salts, amino acids, or other chemicals. Many female insects have chemoreceptors on their ovipositors, allowing them to lay their eggs on food appropriate for the larvae when they hatch.

Olfactory receptors detect airborne molecules such as pheromones, the chemicals used in communication by both animals and plants (see Chapter 47). Insects are excellent examples of animals that make extensive use of pheromones. Female insects use pheromones to attract males, and vice versa. Olfactory receptors in the bristles on the antennae of male silkworm moths (*Bombyx mori*) **(Figure 45.23, p. 1116)** bind a pheromone released by female conspecifics. If an antenna from a silkworm moth is connected to electrodes at its base and tip, action potentials from olfactory receptors can be detected at pheromone concentrations as low as one attractant molecule per 10^{17} air molecules! The male moth responds by flying rapidly when as few as 40 of the 20 thousand receptor cells on an antenna have been stimulated by pheromone molecules. Ants, bees, and wasps may use odour to recognize conspecifics, to identify members of the same hive or nest, or to alert nestmates to danger.

45.5b Vertebrate Animals: More Variations on the Sense of Taste and Smell

Taste involves the detection of potential food molecules in objects touched by a receptor. Smell involves the detection of airborne molecules. Taste and smell receptors have hairlike extensions that contain proteins that bind environmental molecules. Hairs of taste receptors are derived from microvilli and contain microfilaments. Hairs of smell receptors are derived from cilia and contain microtubules. Information from taste receptors is typically processed in the parietal lobes of the brain, whereas information from smell receptors

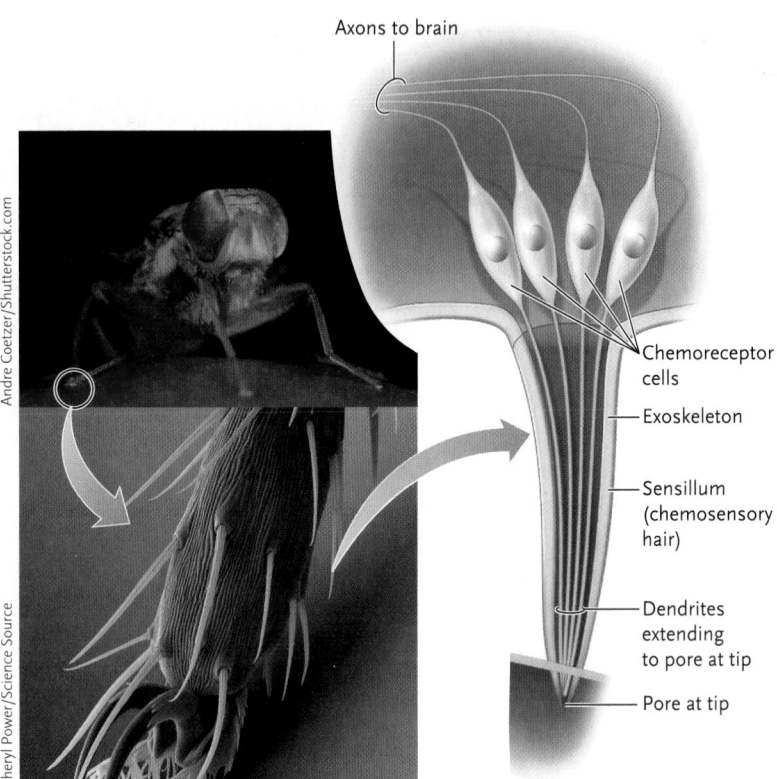

Figure 45.22
Taste receptors on the foot of a fruit fly, *Drosophila*.

Axons to brain

Chemoreceptor cells

Exoskeleton

Sensillum (chemosensory hair)

Dendrites extending to pore at tip

Pore at tip

is processed in olfactory bulbs and the temporal lobes of the brain.

Taste. Taste receptors of most vertebrates form part of a structure called a taste bud, a small, pear-shaped capsule with a pore at the top opening to the exterior **(Figure 45.24, p. 1116)**. Sensory hairs of taste receptors pass through the pore of a taste bud and project to the exterior. The opposite end of the receptor cells synapses with dendrites of an afferent neuron.

Taste receptors of aquatic vertebrates (e.g., fishes and amphibian tadpoles) are generally found throughout the oral cavity but in some species may be found distributed all over the body surface. In terrestrial vertebrates, they are concentrated in the mouth. Humans have about 10 thousand taste buds, each 30 to 40 μm in diameter, scattered over the tongue, roof of the mouth, and throat. Those on the tongue are embedded in outgrowths called *papillae* (*papula* = pimple), which give the surface of the tongue its rough or furry texture. Taste receptors on the human tongue respond to five basic tastes: sweet, sour, salty, bitter, and umami (savoury). Some receptors for umami respond to the amino acid glutamate (familiar as monosodium glutamate or MSG).

Signals from taste receptors are relayed to the thalamus. From there, some signals lead to gustatory centres in the cerebral cortex, which integrate them into the perception of taste. Others lead to the brain

Figure 45.23

Figure 45.23

The brushlike antennae of a male silkworm moth. Fine sensory bristles containing olfactory receptor cells cover the filaments of the antennae.

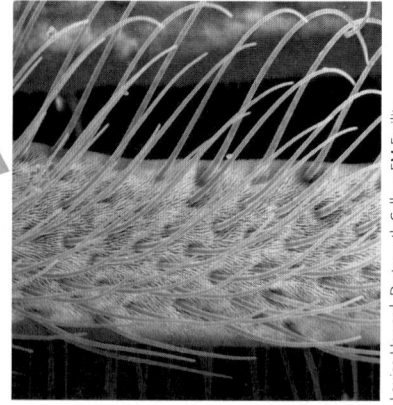

25 μm

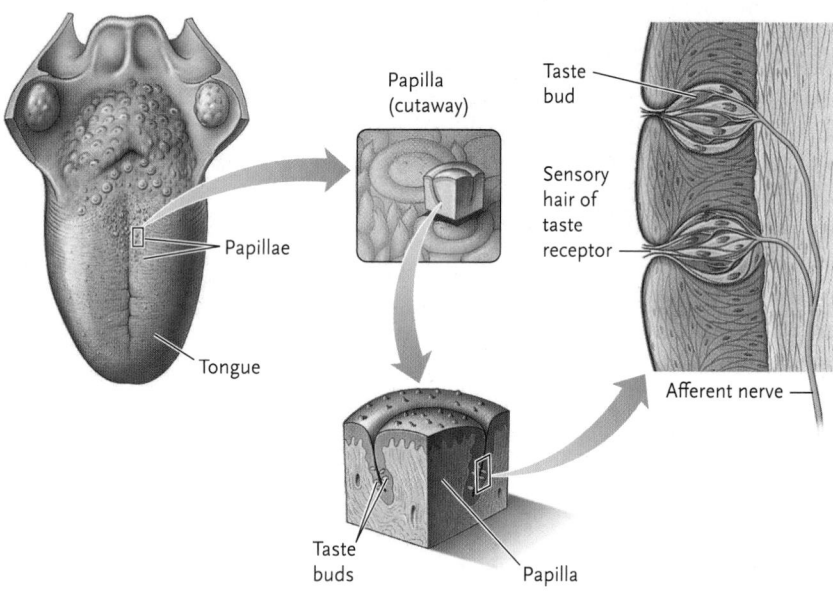

Papilla (cutaway)

Papillae

Tongue

Taste buds

Papilla

Taste bud

Sensory hair of taste receptor

Afferent nerve

Figure 45.24

Taste receptors in the human tongue. The receptors occur in microscopic taste buds that line the sides of the furry papillae.

stem and limbic system, which link tastes to involuntary visceral and emotional responses. Through brain stem and limbic connections, a pleasant taste may lead to salivation, secretion of digestive juices, sensations of pleasure, and sexual arousal, whereas an unpleasant taste may produce revulsion, nausea, and vomiting.

Smell. For water-dwelling vertebrates (e.g., fishes and amphibian tadpoles), the olfactory system detects chemicals present in the surrounding water. These receptors are found inside nasal sacs that open to the water through nares but that are blind ending and not used for breathing. For air-breathing vertebrates, the olfactory system primarily detects volatile (airborne) chemicals by receptors located in the nasal cavities. Bloodhounds have more than 200 million olfactory receptors in patches of olfactory epithelium in the upper nasal passages; humans have about 5 million olfactory receptors. On one end, each olfactory receptor cell has 10 to 20 sensory hairs projecting into a layer of mucus covering the olfactory area in the nose. To be detected, airborne molecules must dissolve in the watery mucus solution. At the other end, the olfactory receptor cells synapse with interneurons in the olfactory bulbs. Olfactory receptors are the only receptor cells that make direct connections with brain interneurons rather than via afferent neurons. It has commonly been believed that, at least in mammals, the olfactory epithelium does not detect odorants in water. However, star-nosed moles (*Condylura cristata*) and water shrews (*Sorex palustris*) exhale bubbles while diving. They re-inhale the bubbles that equilibrate with the water around them and, in this way, obtain airborne olfactory cues from the water **(Figure 45.25).**

From the olfactory bulbs, nerves conduct signals directly to the olfactory centres of the cerebral cortex. This is the only sense that is not relayed through the thalamus in the brain (see Section 45.9f). Here they are integrated into the perception of tantalizing or unpleasant odours from a rose to a rotten egg. Most odour perceptions arise from combinations of different olfactory receptors. About 1000 different human genes give rise to an equivalent number of olfactory receptor types, each specific for a different class of chemicals. Recent experiments demonstrate that rats smell in stereo, accurately localizing odours in one or two sniffs. They could do so only with bilateral sampling. Some neurons in the olfactory bulb respond differently to stimuli from the left than from the right. Furthermore, some receptors in the olfactory cortex of mammals fire only upon stimulation by

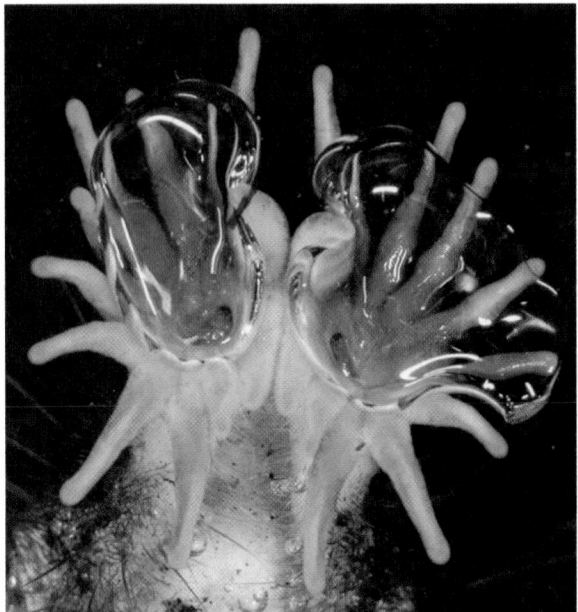

Figure 45.25

Star-nosed moles (*Condylura cristata*) have papillae around their noses. Shown here, the papillae capture bubbles of air, allowing the submerged mole to smell airborne odours.

combinations of odorants, perhaps explaining why mixes of odours are perceived as novel by humans.

As in taste, other connections from the olfactory bulbs lead to the limbic system and brain stem, where the signals elicit emotional and visceral responses similar to those caused by pleasant and unpleasant tastes. Olfaction contributes to the sense of taste because vaporized molecules from foods are conducted from the throat to the olfactory receptors in the nasal cavities. This is the reason why anything that dulls your sense of smell, such as a head cold or holding your nose, diminishes the apparent flavour of food.

Many mammals use odours as a means of communication. Individuals of the same family or colony are identified by their odour; odours are also used to attract mates and to mark territories and trails. Dogs, for example, use their urine to mark home territories with identifying odours. Humans use the fragrances of perfumes and colognes as artificial sexual attractants.

STUDY BREAK

1. What is the difference between taste and smell? What are the similarities?
2. Distinguish among receptor proteins, receptor cells, and receptor organs.
3. For terrestrial vertebrates, describe the pathway by which a signal generated by taste receptors leads to a response.

45.6 Thermoreceptors and Nociceptors

Thermoreceptors detect changes in the surrounding temperature. Nociceptors respond to stimuli that may potentially damage the surrounding tissues. Both types of receptors consist of free nerve endings formed by the dendrites of afferent neurons, with no specialized receptor structures surrounding them.

45.6a Thermoreception Is Heat Detection

Most animals have thermoreceptors. Invertebrates such as mosquitoes and ticks use thermoreceptors to locate warm-blooded prey. Some vertebrates, notably snakes such as rattlesnakes and pythons, also use thermoreceptors to detect the body heat of warm-blooded prey animals. These receptors are located in the pits of some pit vipers **(Figure 45.26)**, whereas those of pythons and boas may not have an opening to the surface. Vampire bats have infrared receptors on their noseleafs **(Figure 45.27, p. 1119)**, allowing them to detect places where blood (their food) flows close to the skin.

In vertebrates, distinct thermoreceptors respond to heat and cold. Researchers have shown that three members of the *transient receptor potential* (TRP)-gated

Pit organs

Figure 45.26

The pit organs of an albino western diamondback rattlesnake (*Crotalus atrox*) are located in depressions on both sides of the head below the eyes. These thermoreceptors detect infrared radiation emitted by warm-blooded prey such as mice and kangaroo rats.

Forked Tongues

The forked tongues of serpents (see Figure 1) are deeply embedded in the world's religious iconography, often as a representation of deceit and malevolence. Aristotle thought that the forked tongue could double taste sensations. Hodierna proposed that the forked tongue allowed snakes to pick dirt out of both nostrils simultaneously.

Snake tongues are involved in chemoreception and serve as the delivery mechanism for paired sensors in Jacobson's organs (vomeronasal organs) on the roofs of snakes' mouths. Jacobson's organs connect with the oral cavity through two small openings in the palate (vomeronasal fenestrae).

However, forked tongues are not restricted to snakes. Lepidosaurian reptiles (see Chapter 28) show considerable variation in tongue structure **(Figure 1).** Forked tongues allow snakes and lizards to follow the pheromone trails of prey and conspecifics. The forked tongue specifically allows animals to use tropotaxis, which means simultaneously sampling chemical stimuli at two points. In some snakes, varanid lizards, and teiid lizards, the distance between the tongue tips exceeds the width of the head. Forked tongues have evolved at least twice and perhaps as many as four times in lepidosaurian reptiles **(Figure 2).**

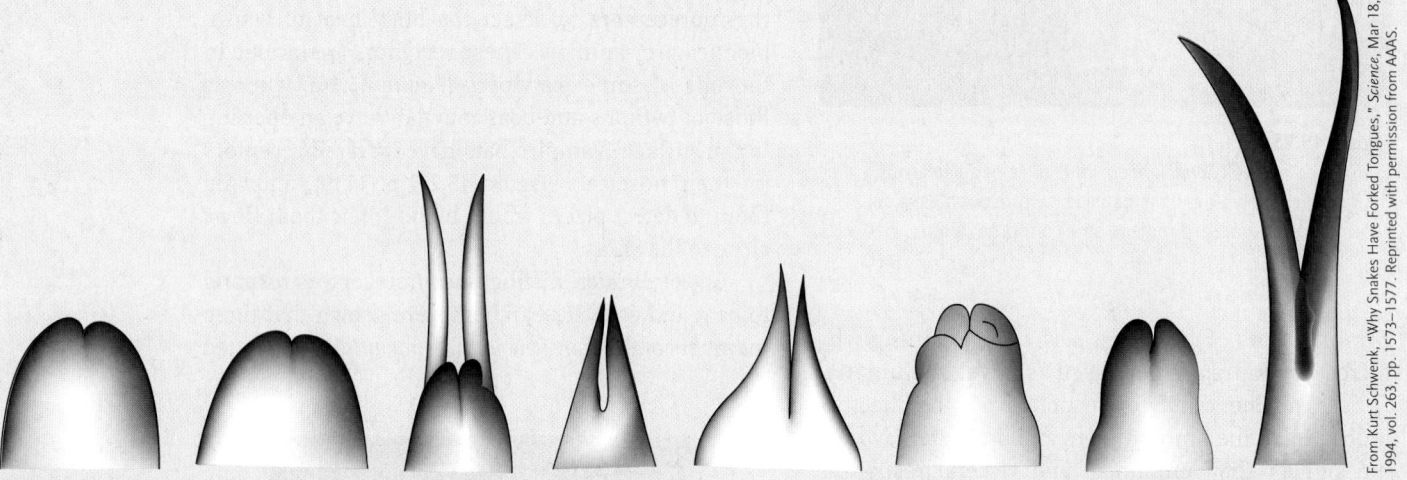

From Kurt Schwenk, "Why Snakes Have Forked Tongues," *Science,* Mar 18, 1994, vol. 263. pp. 1573–1577. Reprinted with permission from AAAS.

Figure 1

Tongue tips vary in squamate reptiles, from simple notches to deep forks. Shown here, from left to right, are the tongues of Sceloporus (Iguania), Coleonyx (Gekkonidae), Cnemidophorus (Teiidae), Lacerta (Lacertidae), Bipes (Amphisbaenia), Scincella (Scincidae), Abronia (Anguidae), and Varanus (Varanidae). Most snake tongues look like that of the Varanus.

From Kurt Schwenk, "Why Snakes Have Forked Tongues," *Science,* Mar 18, 1994, vol. 263, pp. 1573–1577. Reprinted with permission from AAAS.

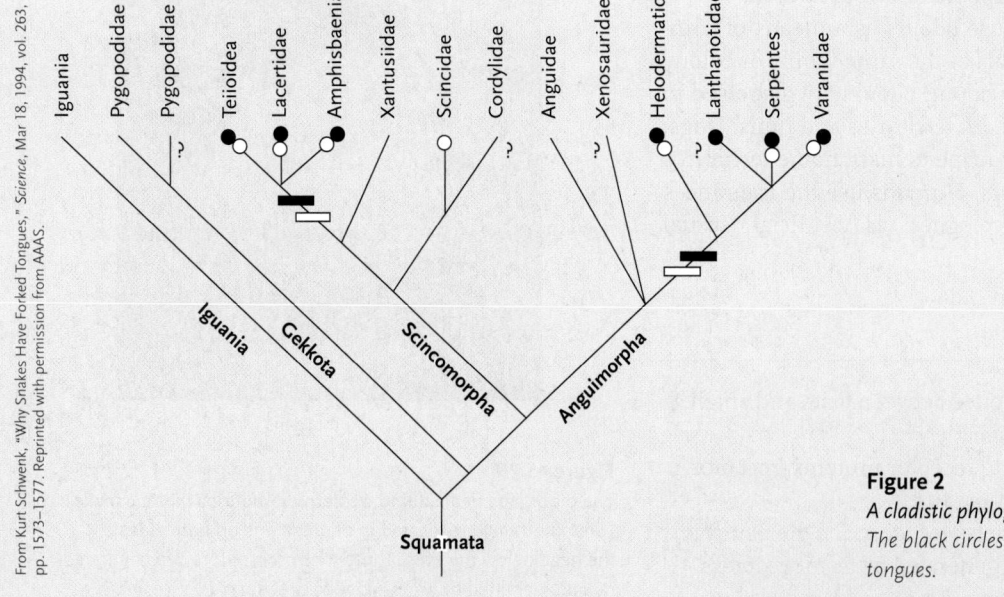

Figure 2

A cladistic phylogeny of Squamata. The black circles identify taxa with forked tongues.

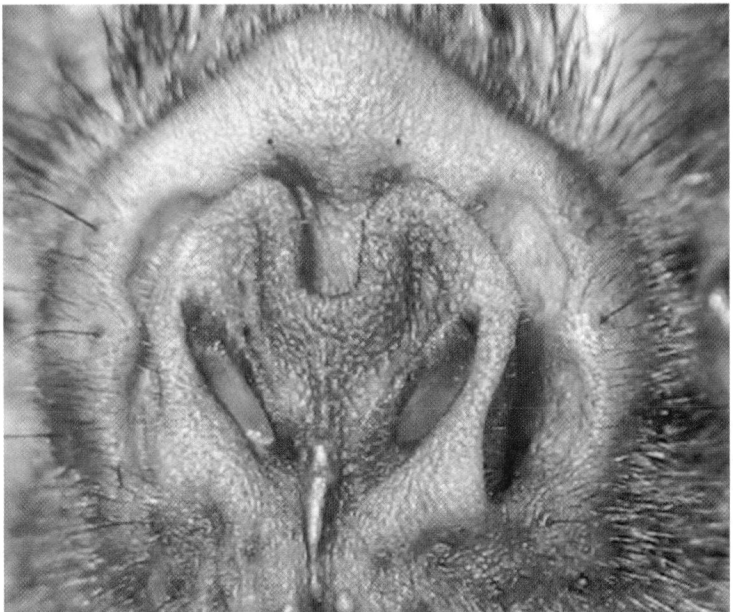

Figure 45.27

The noseleaf on the face of a vampire bat (*Desmodus rotundus*) houses an infrared detector, allowing the bat to find places where blood flows close to the skin. The bat then uses razor-sharp teeth to remove a divot of skin and anticlotting chemicals in its saliva to allow it to get a blood meal.

Ca²⁺ channel family act as heat receptors. One responds when the temperature reaches 33°C and another responds above 43°C, at which point, heat starts to be painful. Both receptors are believed to be involved in thermoregulation. The third receptor responds at 52°C and above, in this case producing a pain response rather than being involved in thermoregulation.

Two cold receptors are known in mammals. One responds between 8 and 28°C and is thought to be involved in thermoregulation. The second responds to temperatures below 8°C and appears to be associated with pain rather than thermoregulation. The molecular mechanisms controlling the opening and closing of heat and cold receptor chemical channels are not currently known.

Some neurons in the hypothalamus of mammals function as thermoreceptors, sensing changes in brain temperature and receiving afferent thermal information. They are highly sensitive to shifts from the normal body temperature and trigger involuntary responses such as sweating, panting, or shivering, which restore normal body temperature.

45.6b Nociceptors Detect Pain

Signals from nociceptors in mammals and possibly other vertebrates detect damaging stimuli that are interpreted by the brain as pain. Pain is a protective mechanism. In humans, pain prompts us to do something immediately to remove or decrease the damaging stimulus. Pain often elicits a reflex response, such as withdrawing the hand from a hot stove, that proceeds before we are consciously aware of the sensation.

Mechanical damage, such as a cut, pin prick, or blow to the body, and temperature extremes can cause pain. Some nociceptors are specific for a particular type of damaging stimulus, whereas others respond to more than one kind. Axons that transmit pain are part of the somatic system of the peripheral nervous system (PNS) (see Section 45.9). They synapse with interneurons in the grey matter of the spinal cord and activate neural pathways to the CNS by releasing the neurotransmitter glutamate or substance P (see Chapter 44). Glutamate-releasing axons produce sharp, prickling sensations that can be localized to a specific body part, such as the pain of stepping on a tack. Substance P–releasing axons produce dull, burning, or aching sensations that are not easily localized, such as the pain of tissue damage when you stub your toe.

As part of their protective function, pain receptors adapt very little, if at all (see Section 45.1b). Some pain receptors gradually intensify the rate at which they send out action potentials if the stimulus continues at a constant level. The CNS also has a pain-suppressing system. In response to stimuli, such as exercise, hypnosis, and stress, the brain releases *endorphins* (see Table 44.1, p. 1084), natural pain-killers that bind to membrane receptors on substance P neurons, reducing the amount of neurotransmitter released.

Nociceptors contribute to the taste of some spicy foods, particularly those containing capsaicin, the active ingredient in hot peppers (see "Molecule behind Biology," Box 45.2). Researchers who study pain use capsaicin to identify nociceptors. To some, the burning sensation from capsaicin is addictive because in its presence, nociceptors in the mouth, nose, and throat immediately transmit pain messages to the brain. The brain responds by releasing endorphins that act as a painkiller and create temporary euphoria.

STUDY BREAK

1. What do thermoreceptors and nociceptors have in common?
2. Why is it important that pain receptors not adapt?

MOLECULE BEHIND BIOLOGY 45.2

Capsaicin

Biting into a jalapeño pepper (a variety of *Capsicum annuum*) can produce a burning pain in your mouth strong enough to bring tears to your eyes. This painfully hot sensation is due primarily to capsaicin, a chemical that probably evolved in pepper plants as a defence against foraging animals. The defence is obviously ineffective for the humans who relish peppers and other foods containing capsaicin (such as hot sauce).

David Julius and his coworkers revealed the molecular basis for detection of capsaicin by nociceptors. They designed their experiments to test the hypothesis that the responding nociceptors have a cell surface receptor that binds capsaicin. Binding the chemical opens a membrane channel in the receptor that admits cations and initiates action potentials interpreted by the brain as pain **(Figure 1)**.

The Julius team isolated the total complement of messenger RNAs from nociceptors that respond to capsaicin

and made complementary DNA (cDNA) clones of the mRNAs (see Section 15.1). The cDNAs contained thousands of different sequences that encode proteins made in the nociceptors. The team transferred the cDNAs individually into embryonic kidney cells (which do not normally respond to capsaicin), and the transformed cells were screened with capsaicin to identify which cells took in calcium ions. These would be the cells that had received a cDNA encoding a capsaicin receptor. Messenger RNA transcribed from the identified cDNA clone was injected into both frog oocytes and cultured mammalian cells. Both oocytes and cultured cells responded to capsaicin by admitting calcium ions, confirming that the researchers had found the capsaicin receptor cDNA.

Among the effects noted when the receptor was introduced into oocytes was a response to heat. Increasing the temperature of the solution surrounding the oocytes from 22°C to about

48°C produced a strong calcium inflow. In short, by binding to the nociceptor, the capsaicin molecule produces the same sensation that excessive heat would, explaining the feeling that your mouth is on fire when you eat a hot pepper. As far as your nociceptors and CNS are concerned, it *is* on fire.

The ion channel involved in producing the flow of cations has subsequently been shown to be a member of the superfamily of TRP ion channels, and as such is now referred to as TRPV1. A number of different TRP ion channels have been shown to be sensitive to different ranges of temperature and probably are responsible for our range of temperature sensation.

Chili peppers were domesticated in different parts of the New World (from Chile to the Caribbean) by 6000 years B.P., and their use in cooking has spread throughout the world (see Chapter 33).

a.

b.

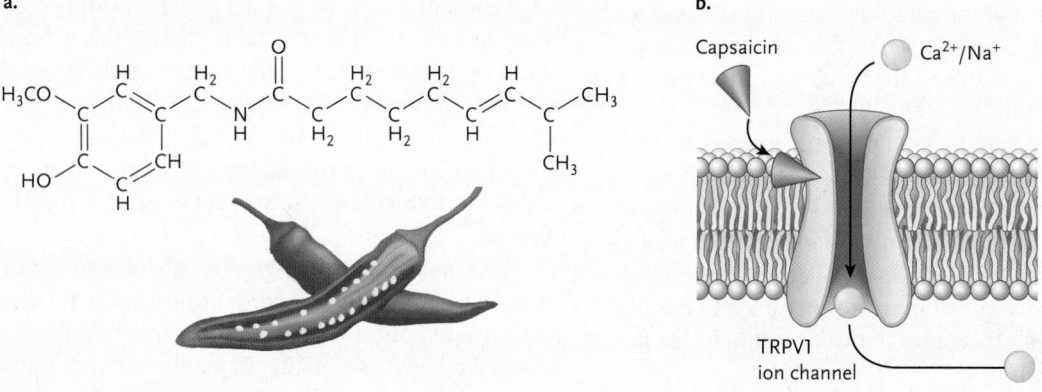

Figure 1

(a) A capsaicin molecule. **(b)** A TRPV1 channel allows cations to flow across the membrane of a heat-sensitive neuron in response to capsaicin, depolarizing the cell and generating action potentials.

45.7 Electroreceptors and Magnetoreceptors

Some animals gain information about their environment by sensing electrical or magnetic fields. In so doing, they directly sense stimuli that humans can detect only with scientific instruments.

45.7a Electroreception

Electroreception is an ancient trait in vertebrates. Although it was lost in ancestral bony fishes, it persists today in many sharks and has reappeared in some bony fishes and some amphibians. Mammals such as the star-nosed mole and duck-billed platypus detect electric fields with specialized **electroreceptors.**

Electroreceptors depolarize in an electric field, and the plasma membrane of an electroreceptor cell generates action potentials. The electrical stimuli detected by the receptors are used in different ways. Electrical information can be used to locate prey; to negotiate a way around obstacles in muddy water; or, by some fishes, to communicate. Some electroreception systems are passive, detecting electric fields in the environment, not the animal's own electric currents. Passive systems are used mainly to find prey. Sharks and rays use electroreceptors to locate prey buried under sand by detecting electrical currents generated by the prey's heartbeat or by the muscle contractions moving water over the gills.

45.7b Electric Fishes

Fishes in the orders Mormyriformes (elephant fish from Africa) and Gymnotiformes (knifefish from South America; **Figure 45.28a**) emit and receive low-voltage electrical signals, using them in prey location (electrolocation) and intraspecific communication. Some electric fishes can produce discharges of several hundred volts (e.g., *Electrophorus electricus*, the electric eel, Figure 45.28b, and *Malapterurus electricus*, the electric catfish) that stun or kill prey. The voltage discharged by an electric eel is high enough to stun, but not kill, a human.

45.7c Magnetoreception and Navigation

Just as humans can use a magnetic compass to navigate, some animals use magnetic compasses in long-distance navigation. **Magnetoreceptors** allow animals to detect and use Earth's magnetic field as a source of directional information. The list includes butterflies, beluga whales, sea turtles (see "Magnetic Sense in Sea Turtles," p. 1122), homing pigeons, and foraging honeybees (*Apis mellifera*).

The pattern of Earth's magnetic field differs from region to region yet remains almost constant over time, largely unaffected by changing weather or day and night. Animals with magnetic receptors can reliably monitor their location. Although little is known about the receptors that detect magnetic fields, they may depend on the fact that moving a conductor, such as an electroreceptor cell, through a magnetic field generates an electric current. Some magnetoreceptors may depend on the effect of Earth's magnetic field on the mineral *magnetite,* which is found in the bones or teeth of many vertebrates, including humans, and in insects, such as in the abdomen of the honeybee and the heads and abdomens of certain ants.

Animals such as homing pigeons (*Columbia livia*), famous for their ability to find their way back to their nests even when released far from home, navigate by detecting their position with reference to both Earth's magnetic field and the Sun. Magnetite is located in the beaks of these birds, which is where magnetoreception likely occurs. Big brown bats (*Eptesicus fuscus*) also have a magnetic sense that influences their navigational abilities.

Figure 45.28
Two electric fishes from South America. *Eigenmannia eigenmannia* **(a)** is a weakly electric fish that uses electrolocation, whereas *Electrophorus electricus* **(b)**, the electric eel, stuns prey with an electric discharge.

STUDY BREAK

1. How do animals use electrical information?
2. What are the advantages of a magnetic navigational system?

45.8 Overview of Central Neural Integration

The sensory pathways just described conduct information by afferent neurons to the CNS. Each type of receptor conveys information to a specific part of the CNS. Action potentials arising in the retina of the eye, for instance, travel along the optic nerves to the visual cortex, where they are interpreted by the brain as differences in pattern, colour, and intensity of light. In the next part of the chapter, we describe the roles of the major areas of the CNS where complex neural

Magnetic Sense in Sea Turtles

Experimental Research Box

To determine if loggerhead sea turtles (*Caretta caretta*) use a magnetoreceptor system for orientation, Kenneth Lohmann and colleagues tested the responses of hatchling turtles to magnetic fields. They placed each turtle hatchling they tested in a harness and tethered it to a swivelling electronic system in the centre of a circular pool of water **(Figure 1a)**. The pool was surrounded by a large coil system, allowing the researchers to reverse the direction of the magnetic field (Figure 1b). The direction the turtle swam was recorded by the tracking system and relayed to a computer.

The turtles swam under two experimental conditions: half of them in Earth's magnetic field and the other half in a reversed magnetic field. Turtle hatchlings tested in Earth's magnetic field swam, on average, in an east-to-northeast direction, mimicking the direction they follow normally when migrating at sea. The hatchlings tested in the reversed magnetic field swam, on average, in a direction 180° opposite that of the hatchlings swimming in Earth's magnetic field.

The results indicate that loggerhead sea turtle hatchlings can detect Earth's magnetic field and use it to help them orient their migration. Their direction of migration, east to northeast, matches the inclination of Earth's magnetic field in the Atlantic Ocean where they migrate (see Figure 1c).

a.

Kenneth Lohmann/University of North Carolina

b.

c.

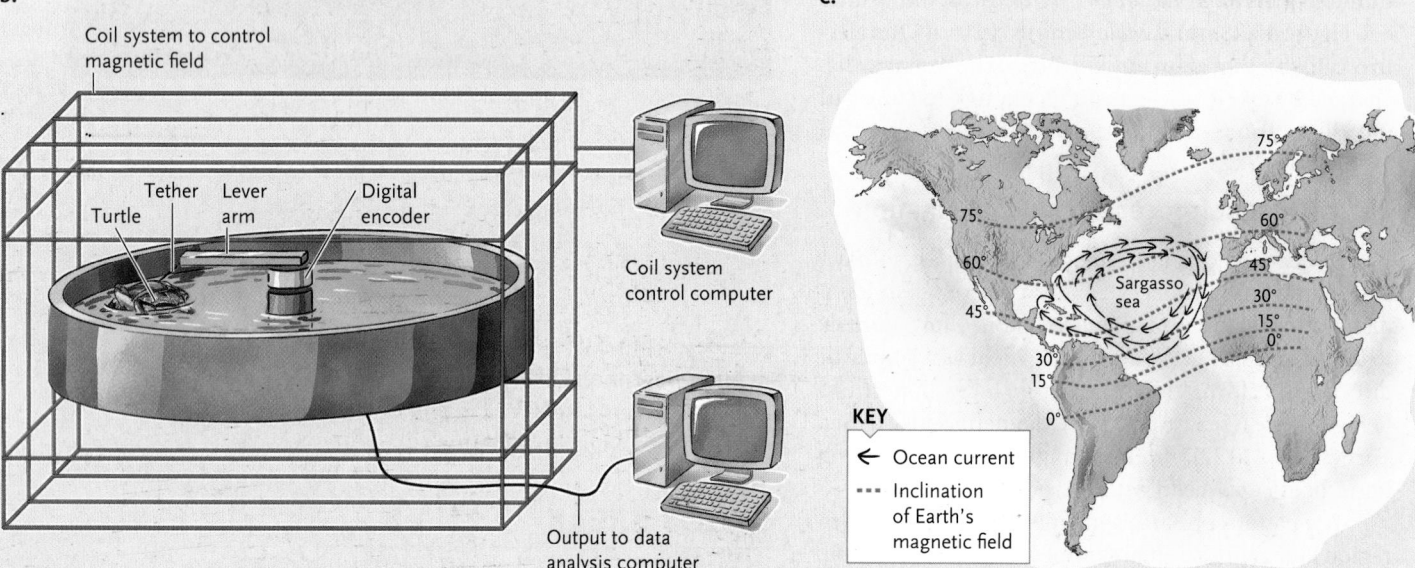

Figure 1
Harnessed hatchling loggerhead sea turtles (a) were tested in a circular pool in which the magnetic field could be altered (b). (c) Hatchlings swimming in the normal magnetic field of Earth swam in the directions they would travel at sea on migration.

---

processing occurs. This is described for the vertebrate CNS (and, in particular, for the human nervous system, about which the most is known).

CONCEPT FIX It is commonly thought that our perception of the environment arises from the sensory cells themselves. Perception, however, is the consequence of the processing of sensory information by the CNS. Action potentials arriving in the sensory cortex from different receptors on the tongue may create a sweet or a tart sensation. These are a result of the processing. All taste receptors generate action potentials, but the action potentials arriving from different receptors give rise to different sensations. ⬡

45.9 The Central Nervous System and Its Functions

The CNS consists of the brain and spinal cord. It manages body activities by integrating incoming sensory information from the PNS into compensating responses.

45.9a The Central Nervous System Is Protected by the Meninges and by Cerebrospinal Fluid

The brain and spinal cord are surrounded and protected by three layers of connective tissue, the **meninges** (*meninga* = membrane), and by the **cerebrospinal fluid**, which circulates through the central canal of the spinal cord, through the ventricles of the brain, and between two of the meninges. The fluid cushions the brain and spinal cord from jarring movements and impacts, nourishes the CNS, and protects the CNS from toxic substances.

45.9b The Spinal Cord Relays Signals between the Peripheral Nervous System and the Brain and Controls Reflexes

The spinal cord, which extends dorsally from the base of the brain, carries impulses between the brain and the PNS and also contains interneuron circuits that control motor reflexes. In cross-section, the spinal cord has a butterfly-shaped core of **grey matter**, consisting of nerve cell bodies and dendrites. This is surrounded by **white matter**, consisting of axons, many of them surrounded by myelin sheaths (**Figure 45.29**, step 3).

Pairs of **spinal nerves** connect with the spinal cord at spaces between the vertebrae.

The afferent (incoming) axons entering the spinal cord synapse with interneurons in the grey matter, which send axons upward through the white matter of the spinal cord to the brain. Conversely, axons from interneurons of the brain pass downward through the white matter of the cord and synapse with the dendrites and cell bodies of efferent neurons in the grey matter of the cord. The axons of these efferent (outgoing) neurons exit the spinal cord through the spinal nerves.

The grey matter of the spinal cord also contains interneurons of the pathways involved in **reflexes**, programmed movements that take place without conscious effort, such as the sudden withdrawal of a hand from a hot surface (shown in Figure 45.29). When your hand touches a hot surface, the heat stimulates an afferent neuron, which makes connections with at least two interneurons in the spinal cord. One of these interneurons stimulates an efferent neuron, causing the *flexor* muscle of the arm to contract. This bends the arm and withdraws the hand almost instantly from the hot surface. The other interneuron synapses with an efferent neuron connected to an *extensor* muscle, relaxing it so that the flexor can move more

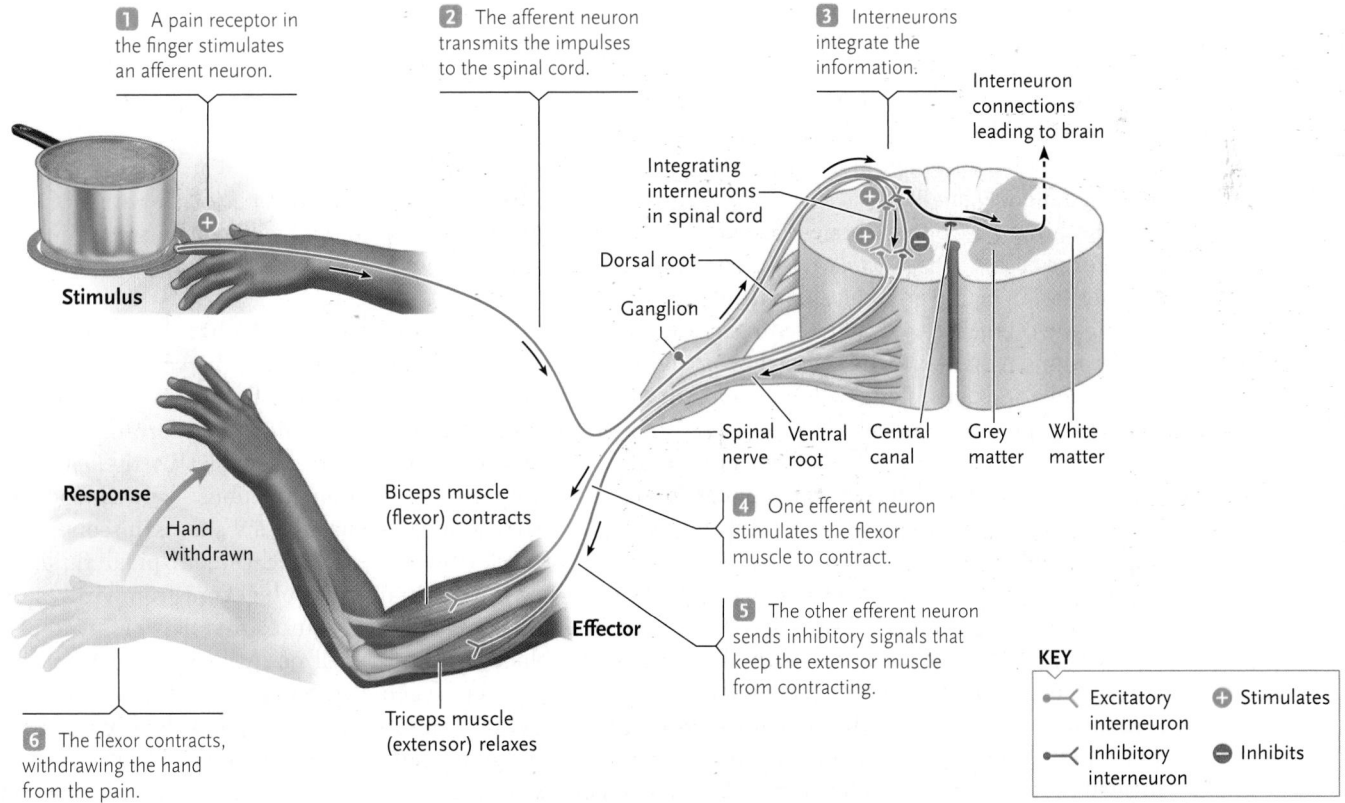

1 A pain receptor in the finger stimulates an afferent neuron.

2 The afferent neuron transmits the impulses to the spinal cord.

3 Interneurons integrate the information.

Interneuron connections leading to brain

Integrating interneurons in spinal cord

Dorsal root

Ganglion

Stimulus

Spinal nerve · Ventral root · Central canal · Grey matter · White matter

Response

Hand withdrawn

Biceps muscle (flexor) contracts

4 One efferent neuron stimulates the flexor muscle to contract.

Effector

5 The other efferent neuron sends inhibitory signals that keep the extensor muscle from contracting.

6 The flexor contracts, withdrawing the hand from the pain.

Triceps muscle (extensor) relaxes

KEY

Excitatory interneuron   ⊕ Stimulates

Inhibitory interneuron   ⊖ Inhibits

Figure 45.29

Organization of the spinal cord and the withdrawal reflex. The withdrawal reflex is an example of a relatively simple neuron circuit that integrates incoming information to produce an appropriate response. The reflex movement produced by this circuit is so rapid that the hand is withdrawn before the brain recognizes the sensation of pain.

quickly. Interneurons connected to the reflex circuits also send signals to the brain, making you aware of the stimulus causing the reflex. You know from experience that when a reflex movement withdraws your hand from a hot surface or other damaging stimulus, you feel the pain shortly *after* the hand is withdrawn. This is the extra time required for impulses to travel from the neurons of the reflex to the brain.

45.9c The Brain Integrates Sensory Information and Formulates Compensating Responses

The brain is the major centre that receives, integrates, stores, and retrieves information. Its interneuron networks generate responses that provide the basis for our voluntary movements, consciousness, behaviour, emotions, learning, reasoning, language, and **memory**, among many other complex activities.

Major Brain Structures. You learned earlier that the three major divisions of the embryonic brain—the forebrain, the midbrain, and the hindbrain—give rise to the structures of the adult brain. Like the spinal cord, each brain structure contains both grey matter and white matter (the myelination of the axons gives white matter its colour) and is surrounded by meninges and circulating cerebrospinal fluid **(Figure 45.30)**.

The hindbrain of vertebrates develops into the *medulla oblongata* (the *medulla*) (see Figure 44.20, Chapter 44). In many vertebrates, a mass of fibres connecting the cerebellum to higher centres in the brain is so prominent that it is identified as the *pons* (bridge). The medulla and pons, along with the midbrain, form a stalklike structure known as the **brain stem**, which connects the forebrain with the spinal cord. All but 2 of the 12 pairs of cranial nerves also originate from the brain stem. The *cerebellum,* with its deeply folded surface, is an outgrowth of the pons.

The forebrain, which makes up most of the mass of the brain in mammals, forms the *telencephalon (cerebrum)*. Its surface layer, the **cerebral cortex**, is a thin layer of grey matter in which numerous unmyelinated neurons are found. The cerebrum, which is divided into right and left *cerebral hemispheres,* is corrugated by fissures and folds that increase the surface area of the cerebral cortex (see Figure 45.30). This structure reflects two of the evolutionary tendencies in the brain of mammals: the corrugation of the hemispheres and the development of a layer of grey matter

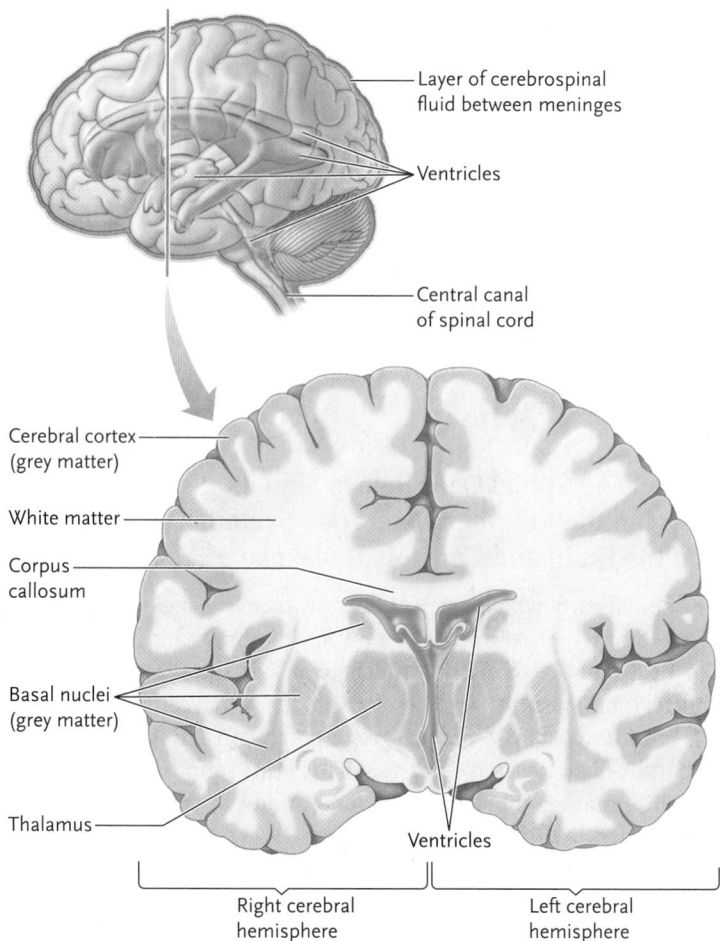

Figure 45.30

The mammalian brain, illustrating the distribution of grey matter, and the locations of the ventricles (in blue) with their connection to the central canal of the spinal cord.

on the periphery. The *basal nuclei,* consisting of several regions of grey matter (cell bodies), are located deep within the white matter.

The Blood–Brain Barrier. Unlike the epithelial cells that form capillary walls elsewhere in the body, which allow small molecules and ions to pass freely from the blood to surrounding fluids, those forming capillaries in the brain are sealed together by tight junctions (Chapter 39). The tight junctions set up a **blood–brain barrier** that prevents most substances dissolved in the blood from entering the cerebrospinal fluid, protecting the brain and spinal cord from viruses, bacteria, and toxic substances that may circulate in the blood. A few types of nonpolar molecules and ions, such as oxygen, carbon dioxide, alcohol, and anaesthetics, can move directly across the lipid bilayer of the epithelial cell membranes by diffusion. A few other substances are moved across the plasma membrane by highly selective transport proteins. The most significant of these transported molecules is glucose, an important source of metabolic energy for the cells of the brain.

45.9d The Brain Stem Regulates Many Vital Housekeeping Functions of the Body

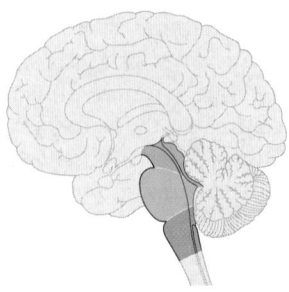

Physicians and scientists have learned much about the functions of various brain regions by studying animals as well as patients with brain damage from stroke, infection, tumours, and mechanical disturbances. Techniques such as *functional magnetic resonance imaging* (*fMRI*) and *positron emission tomography* (*PET*) allow researchers to identify the normal functions of specific brain regions in noninvasive ways. The instruments record a subject's brain activity during various mental and physical tasks by detecting increases in blood flow or metabolic activity in specific regions **(Figure 45.31).**

From such analyses, we know that grey-matter centres in the brain stem control many vital body functions without conscious involvement or control by the cerebrum. Among these functions are the heart and respiration rates, blood pressure, constriction and dilation of blood vessels, coughing, and reflex activities of the digestive system such as vomiting. These functions are so vital to life that damage to the brain stem has serious and often lethal consequences.

45.9e The Cerebellum Integrates Sensory Inputs to Coordinate Body Movements

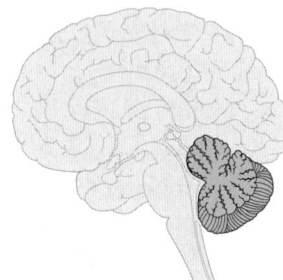

The cerebellum is separate in structure and function from the brain stem. Through its extensive connections with other parts of the brain, the **cerebellum** receives sensory input originating from receptors in muscles and joints; from balance receptors in the inner ear; and from the receptors of touch, vision, and hearing. These signals convey information about how the body trunk and limbs are positioned, the degree to which different muscles are contracted or relaxed, and the direction in which the body or limbs are moving. The cerebellum integrates these sensory signals and compares them with signals from the cerebrum that control voluntary body movements. Outputs from the cerebellum to the cerebrum, brain stem, and spinal cord modify and fine-tune the movements to keep the body in balance and

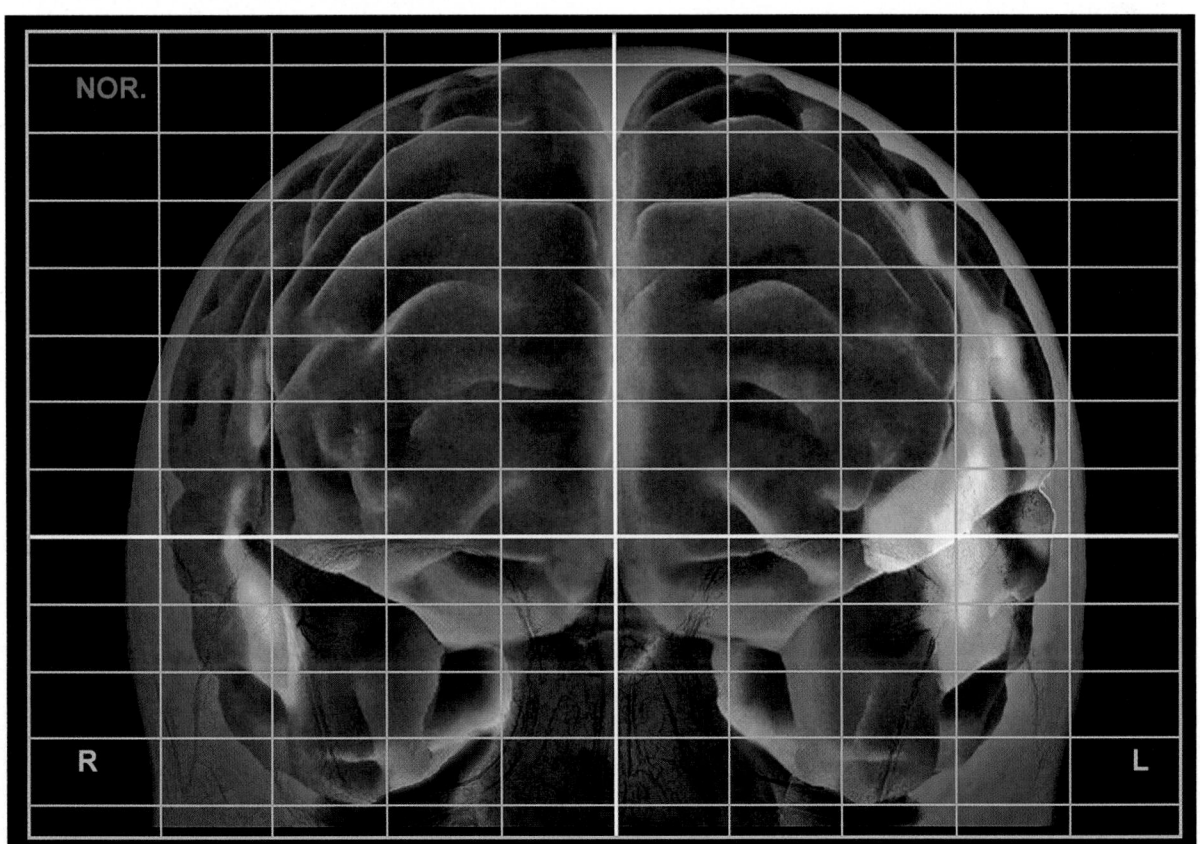

Figure 45.31

Activity in the human brain while reading aloud. The image combines an fMRI of a male brain with a PET scan, which shows the blood circulation increases in the language, hearing, and vision areas of the brain, especially in the left hemisphere.

directed toward targeted positions in space. The cerebellum is particularly important in flying birds, which make greater use of three-dimensional space, and, like the mammalian cerebellum, it has a folded structure, increasing its relative size.

45.9f The Basal Nuclei, the Thalamus, and the Hypothalamus Control a Variety of Functions

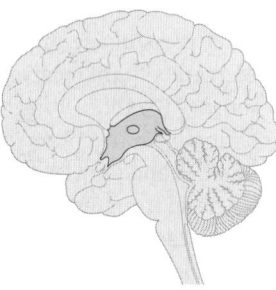

Grey-matter centres derived from the embryonic telencephalon include the thalamus, hypothalamus, basal nuclei, and limbic system (see Figure 45.30, p. 1124). They contribute to the control and integration of voluntary movements, body temperature, glandular secretions, osmotic balance of the blood and extracellular fluids, wakefulness, and the emotions, among other functions. Some of the grey-matter centres route information to and from the cerebral cortex and between the forebrain, brain stem, and cerebellum.

The **thalamus** (see Figure 45.30) forms a major switchboard that receives sensory information and relays it to the appropriate regions of the cerebral cortex. It also plays a role in alerting the cerebral cortex to full wakefulness or in inducing drowsiness or sleep.

The **hypothalamus** is a relatively small conical area found in all vertebrates. It contains centres that regulate basic homeostatic functions of the body. Some centres set and maintain body temperature by triggering reactions such as shivering and sweating. Others constantly monitor the osmotic balance of the blood by testing its composition of ions and other substances. If departures from normal levels are detected, the hypothalamus triggers responses such as thirst or changes in urine output that restore the osmotic and fluid balance. The hypothalamus is an important part of the endocrine system (see Chapter 43). It produces some of the hormones released by the pituitary and governs the release of other pituitary hormones.

The centres of the hypothalamus that detect blood composition and temperature are directly exposed to the bloodstream: they are the only parts of the brain *not* protected by the blood–brain barrier. Parts of the hypothalamus also coordinate responses triggered by the autonomic system (see Section 44.5), making it an important link in such activities as control of the heartbeat, contraction of smooth muscle cells in the digestive system, and glandular secretion. Some regions of the hypothalamus establish a biological clock that sets up daily metabolic rhythms, such as the regular changes in body temperature, metabolic rate, and sleep state that occur on a daily cycle.

The **basal nuclei** are grey-matter centres that surround the thalamus on both sides of the brain (see Figure 45.30). They moderate voluntary movements directed by motor centres in the cerebrum and can be recognized in all amniotes. Damage to the basal nuclei can affect the planning and fine-tuning of movements, leading to stiff, rigid motions of the limbs and unwanted or misdirected motor activity, such as tremors of the hands and inability to start or stop intended movements at the intended place and time. Parkinson's disease, in which affected individuals exhibit all of these symptoms, results from degeneration of centres in and near the basal nuclei.

Parts of the thalamus, hypothalamus, and basal nuclei, along with other nearby grey-matter centres—the amygdala, **hippocampus**, and olfactory bulbs—form a functional network called the **limbic system** (*limbus* = belt), sometimes called our emotional brain. The **amygdala** works as a switchboard, routing information about experiences that have an emotional component through the limbic system. The **olfactory bulbs** relay inputs from odour receptors to both the cerebral cortex and the limbic system. The olfactory connection to the limbic system may explain why certain odours can evoke particular, sometimes startlingly powerful, emotional responses.

The limbic system controls emotional behaviour and influences the basic body functions regulated by the hypothalamus and brain stem. Stimulation of different parts of the limbic system produces anger, anxiety, fear, satisfaction, pleasure, or sexual arousal. Connections between the limbic system and other brain regions bring about emotional responses such as smiling, blushing, or laughing.

45.9g The Cerebral Cortex Carries Out All Higher Brain Functions

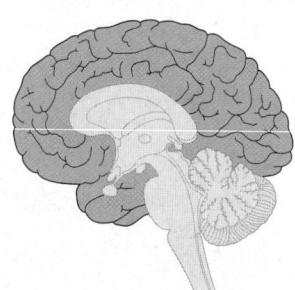

Over the course of evolution, the surface area of the cerebral cortex increased by continuously folding in on itself, thereby expanding the structure into sophisticated information-encoding and processing centres. Primates have cerebral cortices with the largest number of convolutions. In humans, each cerebral hemisphere is divided by surface folds into *frontal, parietal, temporal,* and *occipital* lobes **(Figure 45.32)**. Uniquely in mammals,

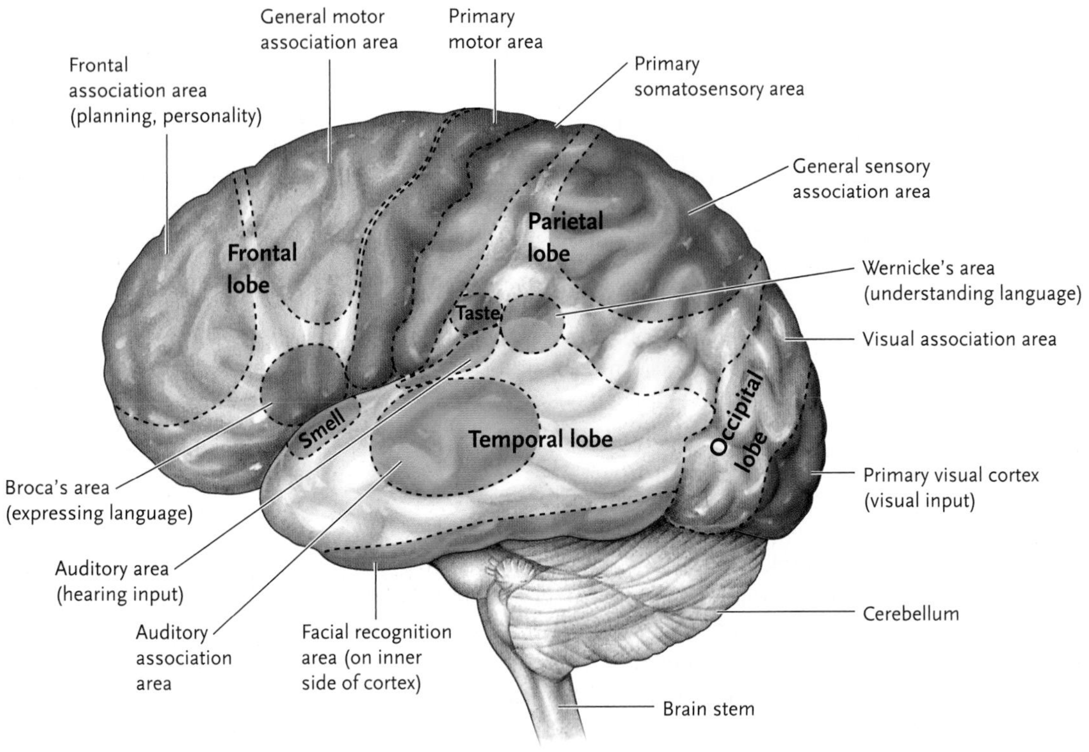

Figure 45.32
The lobes of the cerebrum of the human brain, showing major regions and association areas of the cerebral cortex.

the top layer of the cerebral hemispheres is organized into six layers of neurons called the *neocortex* (*neo* = new; these layers are the newest part of the cerebral cortex in an evolutionary sense).

The two cerebral hemispheres can function separately, and each has its own communication lines internally and with the rest of the CNS and the body. The left cerebral hemisphere responds primarily to sensory signals from, and controls movements in, the right side of the body. The right hemisphere has the same relationships to the left side of the body. The opposite connection and control reflect the fact that the nerves carrying afferent and efferent signals cross from left to right within the spinal cord or brain stem. Thick axon bundles, forming a structure called the **corpus callosum**, connect the two cerebral hemispheres and coordinate their functions.

Sensory Regions of the Cerebral Cortex. Areas that receive and integrate sensory information are distributed over the cerebral cortex. In each hemisphere, the **primary somatosensory area,** which registers information on touch, pain, temperature, and pressure, runs in a band across the parietal lobes of the brain (see Figure 45.32). Experimental stimulation of this band in one hemisphere causes prickling or tingling sensations in specific parts on the opposite side of the body, beginning with the toes at the top of each hemisphere and running through the legs, trunk, arms, and hands, to the head (**Figure 45.33, p. 1128** and "People behind Biology," Box 45.3, p. 1128).

Other sensory regions of the cerebral cortex have been identified with hearing, vision, smell, and taste (see Figure 45.33). Regions of the temporal lobes on both sides of the brain receive auditory inputs from the ears, whereas inputs from the eyes are processed in the primary visual cortex in both occipital lobes. Olfactory input from the nose is processed in the olfactory lobes, located on the ventral side of the temporal lobes. Regions in the parietal lobes receive inputs from taste receptors on the tongue and other locations in the mouth.

Motor Regions of the Cerebral Cortex. The **primary motor area** of the cerebral cortex runs in a band just in front of the primary somatosensory area (see Figure 45.33). Experimental stimulation of points along this band in one hemisphere causes movement of specific body parts on the opposite side of the body, corresponding generally to the parts registering in the primary somatosensory area at the same level (see Figure 45.33). Other areas that integrate and refine motor control are located nearby.

In both the primary somatosensory and motor areas, some body parts, such as the lips and fingers, are represented by large regions, and others, such as the arms and legs, are represented by relatively small regions. As shown in Figure 45.33, the relative sizes produce a distorted image of the human body that is quite different from the actual body proportions. The differences are reflected in the precision of touch and movement in structures such as the lips, tongue, and fingers.

Born in the United States at the end of the nineteenth century, Wilder Penfield graduated in literature from Princeton and won a Rhodes Scholarship to Oxford. He graduated as an M.D. from Johns Hopkins and worked in neurosurgery at Columbia. Attracted to McGill University in Montreal in 1928, he realized his dream by establishing the Montreal Neurological Institute, where scientists and clinicians could work together. It was here that he established the "Montreal Procedure" for the treatment of epilepsy. With patients under local anaesthesia and fully conscious, he exposed the entire cerebrum and stimulated various parts of the brain while the patients described their sensations. In this way, he could identify the area of the brain responsible for the epileptic seizures and, if feasible, remove or destroy it. In the course of this work, he was able to identify those areas of the cerebral cortex that related to particular parts of the body and developed the well-known map shown in Figure 45.33. This was a remarkably courageous procedure at the time and required his patients to trust him absolutely.

Association Areas. The sensory and motor areas of the cerebral cortex are surrounded by **association areas** (see Figure 45.32) that integrate information from the sensory areas, formulate responses, and pass them on to the primary motor area. Two of the most important association areas are *Wernicke's area* and *Broca's area,* which function in spoken and written language. They are usually present on only one side of the brain, in the left hemisphere in 97% of the human population. Comprehension of spoken and written language depends on Wernicke's area, which coordinates inputs from the visual, auditory, and general sensory association areas. Interneuron connections lead from Wernicke's area to Broca's area, which puts together the motor program for coordination of the lips, tongue, jaws, and other structures producing the sounds of speech and passes the program to the primary motor area. The brain scan images in Figure 45.31, p. 1125,

Figure 45.33
The primary somatosensory and motor areas of the cerebrum. The distorted images of the human body show the relative areas of the sensory and motor cortex devoted to different body regions.

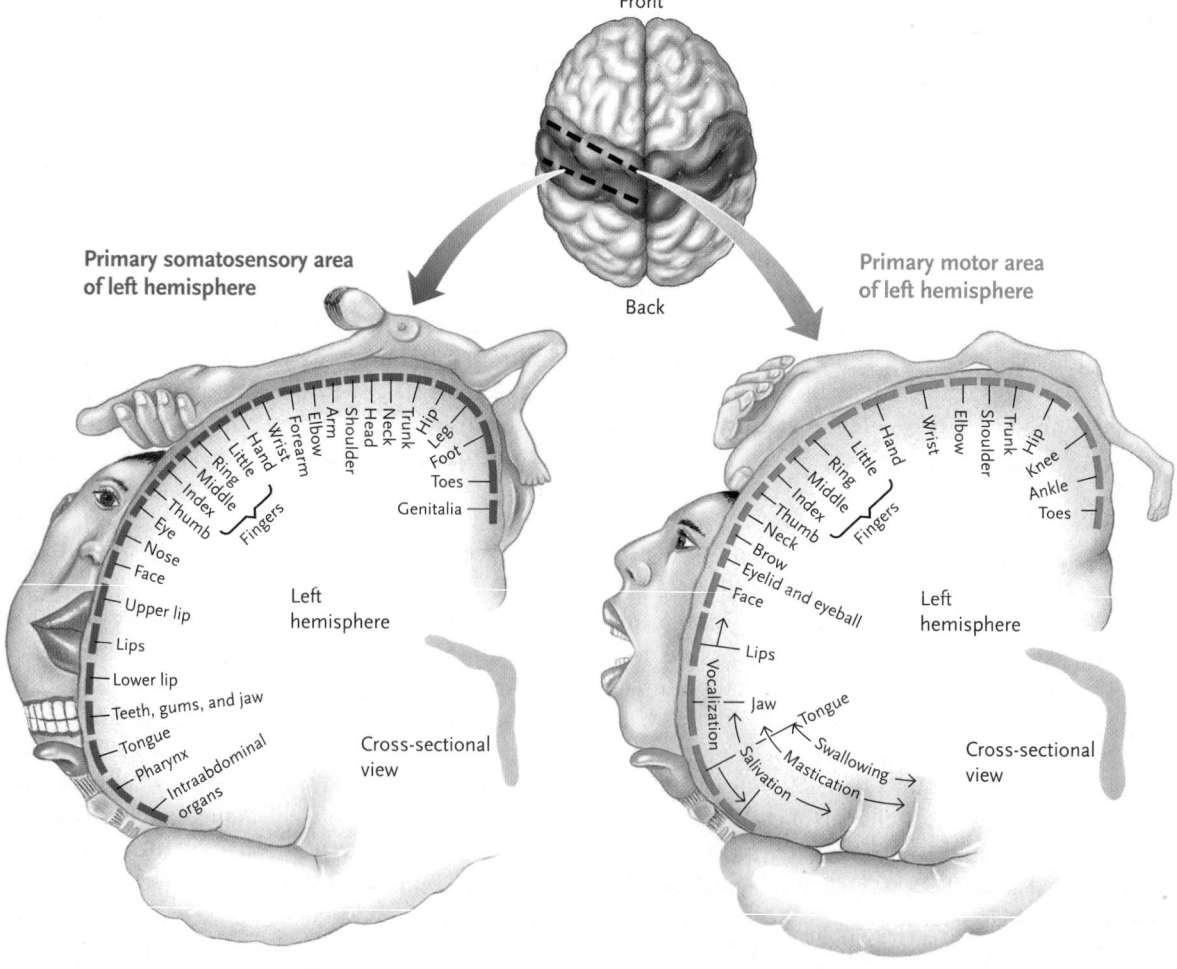

dramatically illustrate how these brain regions participate as a person performs different linguistic tasks.

45.9h Some Higher Functions Are Distributed in Both Cerebral Hemispheres, While Others Are Concentrated in One Hemisphere

Most of the other higher functions of the human brain, such as abstract thought and reasoning; spatial recognition; mathematical, musical, and artistic ability; and the associations forming the basis of personality, involve the coordinated participation of many regions of the cerebral cortex. Some of these regions are equally distributed in both cerebral hemispheres, and some are more concentrated in one hemisphere.

Among the functions more or less equally distributed between the two hemispheres is the ability to recognize faces. Consciousness, the sense of time, and recognizing emotions also seem to be distributed in both hemispheres.

Typically, some brain functions are more localized in one of the two hemispheres, a phenomenon called **lateralization**. Studies of people with split hemispheres and surveys of brain activity by PET and fMRI have confirmed that for the vast majority of people, the left hemisphere specializes in spoken and written language, abstract reasoning, and precise mathematical calculations. The right hemisphere specializes in nonverbal conceptualizing; intuitive thinking; musical and artistic abilities; and spatial recognition functions, such as fitting pieces into a puzzle. The right hemisphere also handles mathematical estimates and approximations that can be made by visual or spatial representations of numbers. Thus, the left hemisphere in most people is verbal and mathematical, and the right hemisphere is intuitive, spatial, artistic, and musical.

STUDY BREAK

1. What is the blood–brain barrier and what is its function?
2. What is the difference between white matter and grey matter?
3. What is the function of the brain stem?
4. Distinguish the functions of the cerebellum from those of the cerebral cortex.

45.10 Memory, Learning, and Consciousness

We set memory, learning, and consciousness apart from the other CNS functions because they appear to involve coordination of structures from the brain stem to the cerebral cortex. **Memory** is the storage and retrieval of a thought or a sensory or motor experience. **Learning** involves a change in the response to a stimulus, based on information or experiences stored in memory. **Consciousness** is not easily defined. In a narrow sense, it involves awareness, a state of alertness to our surroundings. But there is a broader and deeper meaning that involves awareness of ourselves, our identity, and an understanding of the significance and likely consequences of events that we experience. Later in this section, we deal with sleep as a decrease in awareness.

45.10a Memory Takes Two Forms: Short Term and Long Term

Psychology research and our everyday experience indicate that humans have at least two types of memory. **Short-term memory** stores information for seconds, minutes, or at most an hour or so. **Long-term memory** stores information from days to years or even for life. Short-term memory, but not long-term memory, is usually erased if a person experiences a disruption such as a sudden fright, a blow, a surprise, or an electrical shock. For example, a person knocked unconscious by an accident typically cannot recall the accident itself or the events just before it, but longstanding memories are not usually disturbed.

To explain these differences, investigators propose that short-term memories depend on transient changes in neurons that can be erased relatively easily, such as changes in the membrane potential of interneurons caused by excitatory and inhibitory postsynaptic potentials (EPSPs and IPSPs) and the action of indirect neurotransmitters that lead to reversible changes in ion transport. By contrast, storage of long-term memory is considered to involve more or less permanent molecular, biochemical, or structural changes in interneurons, which establish signal pathways that cannot be switched off easily.

All memories probably register initially in short-term form. They are then either erased and lost or committed to long-term form. The intensity or vividness of an experience, the attention focused on an event, emotional involvement, and the degree of repetition may all contribute to the conversion from short-term to long-term memory.

The storage pathway typically starts with an input at the somatosensory cortex that then flows to the amygdala, which relays information to the limbic system, and to the hippocampus, which sends information to the frontal lobes, a major site of long-term memory storage. People with injuries to the hippocampus cannot remember information for more than a few minutes; long-term memory is limited to information stored before the injury occurred. Squirrels hoard food for the winter in a number of caches and can locate these by remembering the location from landmarks rather than by tracking a smell. Each autumn, the hippocampus of a squirrel increases in size by about 15%.

How are neurons and neuron pathways permanently altered to create long-term memory? One change that has been much studied is **long-term potentiation**: a long-lasting increase in the strength of synaptic connections in activated neural pathways following brief periods of repeated stimulation. The synapses become increasingly sensitive over time, so that a constant level of presynaptic stimulation is converted into a larger postsynaptic output that can last hours, weeks, months, or years. Other changes consistently noted as part of long-term memory include more or less permanent alterations in the number and the area of synaptic connections between neurons, in the number and branches of dendrites, and in gene transcription and protein synthesis in interneurons. Experiments on both vertebrates and invertebrates demonstrate that long-term memory depends on protein synthesis. For example, goldfish were trained to avoid an electrical shock by swimming to one end of an aquarium when a light was turned on. The fish could remember the training for about a month under normal conditions, but if they were exposed to a protein synthesis inhibitor while being trained, they forgot the training within a day.

45.10b Learning Involves Combining Past and Present Experiences to Modify Responses

As with memory, most animals appear to be capable of learning to some degree. Learning involves three sequential mechanisms: (1) storing memories, (2) scanning memories when a stimulus is encountered, and (3) modifying the response to the stimulus in accordance with the information stored as memory.

One of the simplest forms of learning is an increased responsiveness to mild stimuli after experiencing a strong stimulus, often called **sensitization**. The process was nicely illustrated by Eric Kandel of Columbia University and his associates in experiments with a shell-less marine snail, the Pacific sea hare, *Aplysia californica*. The first time the researchers administered a single sharp tap to the siphon (the structure that admits water to the gills), the slug retracted its gills by a reflex movement. However, at the next touch, whether hard or gentle, the siphon retracted much more quickly and vigorously. Sensitization in *Aplysia* has been shown to involve changes in synapses. The synapses become more reactive because more of the neurotransmitter serotonin is released by each action potential. The cephalopod molluscs, such as the octopus, are capable of much more complex learning: they can distinguish and remember shapes and textures using only their tentacles.

45.10c Consciousness Involves Different States of Awareness

Most animals that have been investigated, including some invertebrates, experience a daily rhythm of activity and inactivity. The inactive period, sleep, is essential to normal functioning. Sleep deprivation leads to disruption of a number of functions, including memory and learning, and, if prolonged, can be fatal. During sleep, there is some degree of awareness since external stimuli such as sound or internal stimuli such as a full bladder can interrupt sleep.

In humans and other mammals, sleep is accompanied by changes in the electrical activity of the cerebrum as detected by electrodes applied to the scalp during an *electroencephalogram*. The waking state is characterized by rapid, irregular *beta waves* **(Figure 45.34)**. As the eyes close and you become fully relaxed, these give way to slower, more regular *alpha waves*. As you become more drowsy, these are replaced by slower *theta waves*. Full sleep is characterized by even slower *delta waves*. The heart rate falls, and the muscles are relaxed.

During full sleep, the brain returns at intervals to periods of beta waves, during which the heart rate increases, the muscles may twitch, and the eyes move rapidly behind the closed lids, giving these periods the name **rapid eye movement**, or **REM, sleep**. This brief period of about 10 to 15 minutes occurs every 90 minutes or so in healthy adults. Sleepers do most of their dreaming during REM sleep, and most individuals awakened from REM sleep report they were experiencing vivid dreams.

This pattern of alternating periods of greater or lesser cerebral activity is also characteristic of bird sleep, although birds may sleep on one side of the brain while the other remains fully alert as protection against predators. This is also true of some marine mammals. Reptiles, with their less-developed cerebrum, experience alternating patterns of activity in the amygdala.

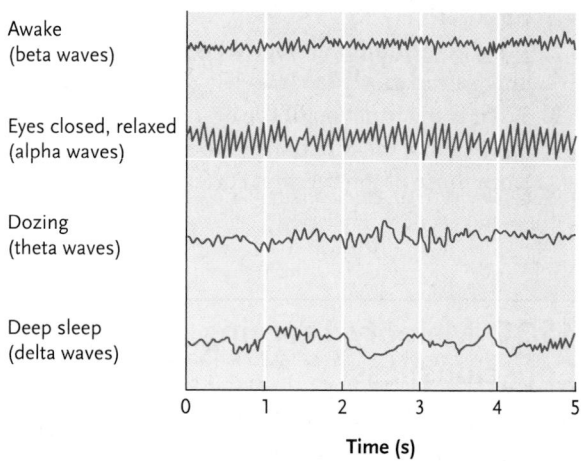

Figure 45.34
Brain waves characteristic of various states of consciousness.

Although we know that sleep is essential, we do not understand the physiological basis for these effects. What do these patterns of neural activity in the brain reflect, and why are they necessary? The fruit fly *Drosophila melanogaster* exhibits cycles of sleep. At night, it feeds and then seeks out an isolated place and becomes inactive for about eight hours. Interrupting the sleep interferes with memory and learning in the flies. Many labs are now using the flies as models to identify genes involved in the sleep process.

STUDY BREAK

Using long-term potentiation as an example, describe how neurons and neuron pathways can be altered to create memory.

Review

 aplia™

To access course materials such as Aplia and other companion resources, please visit www.NELSONbrain.com.

45.1 Overview of Sensory Integration

- Receptors in the sensory system collect information (i.e., stimuli) from internal and external sensors (transducers) and convert (transduce) the information into neural activity. Dendrites of an afferent neuron pick up the stimuli. The axon of the afferent neuron conveys the stimulus to the central nervous system (CNS), providing the organisms with sensory data used to influence behaviour and homeostasis.

- Mechanoreceptors detect mechanical energy (pressure), photoreceptors detect the energy of light, chemoreceptors detect specific molecules or chemical conditions, thermoreceptors detect the flow of heat energy, and nociceptors detect tissue damage or noxious chemicals.

- Some receptors allow the effect of a stimulus to be reduced over time. This adaptation also allows receptors to retain their sensitivity over a broad range of stimulus intensities.

45.2 Mechanoreceptors and the Tactile and Spatial Senses

- You would more likely detect a fruit fly walking on your face than on your leg because touch receptors are closer together on a human's face than on a human's leg.

- The vestibular apparatus and cochlea are specialized sensory structures. The vestibular apparatus is responsible for maintaining equilibrium and coordinating head and body movements. The cochlea is used in hearing.

- The strength of the responses of stretch receptors depends on how much and how fast the muscle is stretched.

45.3 Mechanoreceptors and Hearing

- Most invertebrates detect sound through mechanoreceptors in their skin or other surface structures. An example are ears in the common cricket. Crickets detect sound using tympana (ears) on each side of the abdomen or on the first pair of walking legs. The ears are areas of thin exoskeleton.

- Sound waves cause vibrations of the tympanum, which are converted into nerve impulses that travel along the auditory nerve to the CNS.

- Vertebrates use ears to hear sounds. Some vertebrates have pinnae (outer ears) that collect sounds (vibrations) and channel them down the auditory canal to the tympanum (eardrum). There, vibrations in air are converted (transduced) into vibrations of the membrane making up the tympanum. These vibrations are amplified by vibrations of the malleus, incus, and stapes (auditory ossicles) and conveyed to the oval window, where they are converted to vibrations in the fluid of the coiled cochlea. Vibrations in the fluid inside the cochlea cause vibrations of the basilar membrane, which are detected by cilia and converted to nerve impulses. The nerve impulses move down the auditory nerve to the brain, where they are processed and interpreted.

- The echolocation calls of bats range from about 8 kHz to over 200 kHz. Many are inaudible to humans because they are ultrasonic, above the range of human hearing. Infrasounds (< 20 Hz), used by elephants and whales, are below the range of human hearing.

45.4 Photoreceptors and Vision

- The ocellus, the simplest "eye," lacks a lens and is therefore not image forming. Ocelli are light receptors that often allow animals to detect differences in the brightness of light. Image-forming eyes (compound eyes and single-lens eyes) are photoreceptors too, but they have lenses that allow light to be focused on the retina, the layer with photosensitive cells.

- Colour-blindness is a result of inherited defects in opsin proteins of one or more of the three types of cones. Genes controlling colour vision are located on the X chromosome. Therefore, human males have only one set of genes controlling colour vision, whereas females have two sets. Colour-blindness is relatively common in men and relatively rare in women.

- Compound eyes are composed of ommatidia, many individual visual units. Each ommatidium samples a small part of the visual field, and many ommatidia provide the animal with an image that is a mosaic of many individual views. Motion is detected by many ommatidia at once, giving compound eyes special sensitivity to motion.

- Accommodation is the movement of the lens to focus the image on the retina. In cephalopods, muscles move the lens forward and back. In terrestrial vertebrates, muscles change the shape of the lens.

- Rods and cones, the photoreceptor cells in the retina, consist of an outer segment of stacked, flattened membranous discs with photopigments; an inner segment for cellular metabolic activities; and the synaptic terminal for storage and release of neurotransmitter.

45.5 Chemoreceptors

- Taste is the detection of potential food molecules *touched* by a receptor, whereas smell is the detection of *airborne* particles and molecules. Information from taste receptors is processed in the parietal lobes. Information from smell is processed in the olfactory bulb and temporal lobes. Both taste and smell receptors have hairlike extensions that bind molecules.

- The five basic tastes are sweet, sour, bitter, salty, and umami (savoury). Some lead to the gustatory centres of the cerebral cortex, whereas others are linked to the brain stem and limbic system, producing visceral and emotional responses, including physiological responses such as salivation and secretion of gastric juices. The same responses may occur in response to smells.

45.6 Thermoreceptors and Nociceptors

- Thermoreceptors and nociceptors consist of free nerve endings formed by the dendrites of afferent neurons. No specialized receptor structures surround them. All are members of the *transient receptor potential* (TRP)-gated Ca^{2+} channel family. There are three types of heat receptors; one responds to temperatures above 33°C, one to temperatures above 43°C, and the third to temperatures of 52°C and above. The first two are involved in thermoregulation, whereas the last elicits a strong pain response. There are two types of cold receptors; one responds to temperatures between 8 and 28°C and one to temperatures below 8°C. The first is involved in thermoregulation and the last elicits a strong pain response.

- The pain response (nociception) is a protective mechanism. These receptors do not adapt; otherwise, organisms would not withdraw from a prolonged painful stimulus, increasing the level of damage associated with the pain.

45.7 Electroreceptors and Magnetoreceptors

- Electric field information can be used to detect prey, for example, a shark's passive system. Some fishes generate electric signals to detect obstacles and prey and to communicate.

- The pattern of the magnetic field of Earth varies from region to region. Animals with a magnetic compass can detect the magnetic field and use the information in navigation (as people use compasses). Earth's magnetic field remains constant over time and so is reliable from year to year. Many animals have magnetic receptors. Other animals use Sun compasses.

45.8 Overview of Central Neural Integration

- In vertebrates, the CNS consists of a large brain located in the head and a hollow spinal cord, and the peripheral nervous system (PNS) consists of all the nerves and ganglia connecting the CNS to the rest of the body.

45.9 The Central Nervous System and Its Functions

- The CNS consists of the brain and spinal cord. The spinal cord carries signals between the brain and the PNS. Its neuron circuits also control reflex muscular movements and some autonomic reflexes.

- The adult derivatives of the hindbrain—the pons, medulla oblongata, and cerebellum—together with the relatively reduced midbrain, form the brain stem, which connects the telencephalon with the spinal cord.

- The telencephalon (cerebrum) is divided into right and left cerebral hemispheres, which are connected by a thick band of nerve fibres, the corpus callosum. The cerebral cortex, the surface of the cerebrum, is formed by grey matter. Other collections of grey matter, such as the thalamus, hypothalamus, and basal nuclei, lie at deeper layers of the telencephalon.

- Cerebrospinal fluid provides nutrients to and cushions the CNS. A blood–brain barrier set up by tight junctions between the cells of the capillary walls in the CNS allows only selected substances to enter the cerebrospinal fluid.

- Grey-matter centres in the pons and medulla control involuntary functions such as heart rate, blood pressure, respiration rate, and digestion. Centres in the midbrain coordinate responses to visual and auditory sensory inputs.

- The cerebellum integrates sensory inputs on the positions of muscles and joints, along with visual and auditory information, to coordinate body movements.

- Certain grey-matter centres of the telencephalon control a number of functions. The thalamus receives, filters, and relays sensory and motor information to and from regions of the cerebral cortex. The hypothalamus, the only part of the brain not protected by the blood–brain barrier, regulates basic homeostatic functions of the body and contributes to the endocrine control of body functions. The basal nuclei affect the planning and fine-tuning of body movements.

- The limbic system includes parts of the thalamus, hypothalamus, and basal nuclei, as well as the amygdala and hippocampus. It controls emotional behaviour and influences the basic body functions controlled by the hypothalamus and brain stem.

- The primary somatosensory areas of the cerebral cortex register incoming information on touch, pain, temperature, and pressure from all parts of the body. The temporal lobes receive input from the ears, the primary visual cortex from the eyes, the olfactory lobes from the nose, and the parietal lobes from taste receptors in the mouth. In general, the right cerebral hemisphere receives sensory information from the left side of the body, and vice versa.

- The primary motor areas of the cerebrum control voluntary movements of skeletal muscles in the body.

- The association areas integrate sensory information and formulate responses that are passed on to the primary motor areas. Importantly, Wernicke's area integrates visual, auditory, and other sensory information into the comprehension of language, whereas Broca's area coordinates movements of the lips, tongue, jaws, and other structures to produce the sounds of speech.

- Some functions, such as long-term memory and consciousness, are equally distributed between the two cerebral hemispheres. In contrast, the left hemisphere in most people specializes in spoken and written language, abstract reasoning, and precise mathematical calculations. The right hemisphere specializes in nonverbal conceptualizing, mathematical estimation, intuitive thinking, spatial recognition, and artistic and musical abilities.

45.10 Memory, Learning, and Consciousness

- Memory is the storage and retrieval of a sensory or motor experience, or a thought. Short-term memory involves temporary storage of information, probably resulting from changes in the membrane potential of interneurons, whereas long-term memory is essentially permanent, involving molecular, biochemical, or structural changes in interneurons.

- Learning involves modification of a response through comparisons made with information or experiences that are stored in memory.

- Consciousness is the awareness of ourselves, our identity, and our surroundings. It varies through states from full alertness to sleep.

- Sleep is characterized by alternations in patterns of electrical activity in the cerebrum between slow, relatively regular delta waves, characteristic of deep sleep, and brief periods of rapid, irregular beta waves signalling REM sleep.

Questions

Self-Test Questions

Any number of answers from a to e may be correct.

1. Which change would be involved in sensory adaptation?
   a. a reduction in the effect of stimuli
   b. the loss of eyes in cave-dwelling fish
   c. the development of an acute sense of smell
   d. the development of ears in insects preyed upon by bats

2. Which of the following is an example of proprioceptors?
   a. eyes
   b. statocysts
   c. ears
   d. halteres
   e. b and d

3. Which structure is involved in the vestibular system of vertebrates?
   a. retina
   b. utricle
   c. Golgi apparatus
   d. semicircular canals
   e. c and d

4. Which of the following is vital to the development of vision?
   a. vitamin E
   b. rhodopsin
   c. vitamin A
   d. fatty acids
   e. chlorophyll

5. When does accommodation occur in the eyes of cephalopods?
   a. when light is focused on the fovea
   b. when the shape of the lens is changed
   c. when the retina moves toward the lens
   d. when the retina moves away from the lens
   e. when the lens moves toward or away from the retina

6. Thermoreceptors are widespread in
   a. birds
   b. mammals
   c. pit vipers
   d. earthworms
   e. vampire bats

7. Which is the principle integration centre of homeostatic regulation and leads to the release of hormones?
   a. the cerebellum
   b. the thalamus
   c. the cerebrum
   d. the hypothalamus
   e. the association area

8. The regulation of blood pressure is primarily under the control of the
   a. somatic nervous system
   b. autonomic nervous system
   c. parasympathetic nervous system
   d. sympathetic nervous system

9. What part of the brain coordinates muscular activity?
   a. the cerebrum
   b. the cerebellum
   c. the myencephalon
   d. the medulla oblongata

10. Which of the following are nociceptors sensitive to?
    a. pheromones
    b. pain
    c. touch
    d. light
    e. vibration

Questions for Discussion

1. What is an eye? What are the key elements in the definition? Can robots have eyes?

2. Which are better at evoking memories in humans: visual or olfactory stimuli? What is the evidence supporting either point of view? Why would one kind of stimulus be more effective than the other?

3. Find examples of redundancy in the sensory systems of animals. What are the advantages of redundant systems? What are the disadvantages?

4. In some individuals, a genetic mutation causes them to perceive colours when listening to music. This condition is called synaesthesia. In terms of the sensory systems, explain how this can occur.

5. The cerebellum is relatively large in fishes. It is much larger in free-swimming fishes than in bottom-dwelling species. Why might this be so?

A successful strike by a great blue heron (*Ardea herodias*). All voluntary movements of animals occur as a result of contractions and relaxations of skeletal muscles. When stimulated by the nervous system, actin filaments in the muscles slide over myosin filaments to cause muscle contractions.

Muscles, Skeletons, and Body Movements

WHY IT MATTERS

On an early summer morning throughout the lakes of much of Canada one can hear the haunting calls of loons **(Figure 46.1)**. These charismatic birds are excellent divers, specialized for catching fish. They may dive as deep as 60 m while foraging. They are also excellent fliers, capable of long-distance migrations. During diving, they "fly" underwater using their feet for propulsion. To aid in this, their legs are positioned near the rear of their bodies. While ideal for diving, this is not ideal for walking. As a result, these birds are clumsy on land. All of these activities, vocalization, diving, and flying, require the coordinated activity of skeletal muscles. In the case of vocalization, the muscles act on the lungs to force air over the vocal cords at the entrance to the respiratory system, while for locomotion they act on bones of the limbs to produce movement of feet and wings. Not visible but also involved in all of these activities are the actions of the heart and blood vessels, which deliver oxygen to the skeletal muscles to support their work. These too depend on muscles—cardiac muscle in the case of the heart and smooth muscle in the case of the blood vessels. These muscles act to produce the forces needed to propel the nutrient-containing blood throughout the body.

Muscle tissue is another excitable tissue. Like the cells of the nervous system, cells of the muscular system, called **muscle fibres**, can generate and propagate action potentials. These cells, however, also contain a special group of proteins, contractile proteins, that are capable of shortening when excited. Thus, action potentials generated by muscle fibres lead to muscle contraction. The speed and strength of contraction vary as a result of the manner in which individual cells are constructed and excited. The action produced by the contraction, from the churning of the stomach

Figure 46.1
A loon (*Gavia sp.*).

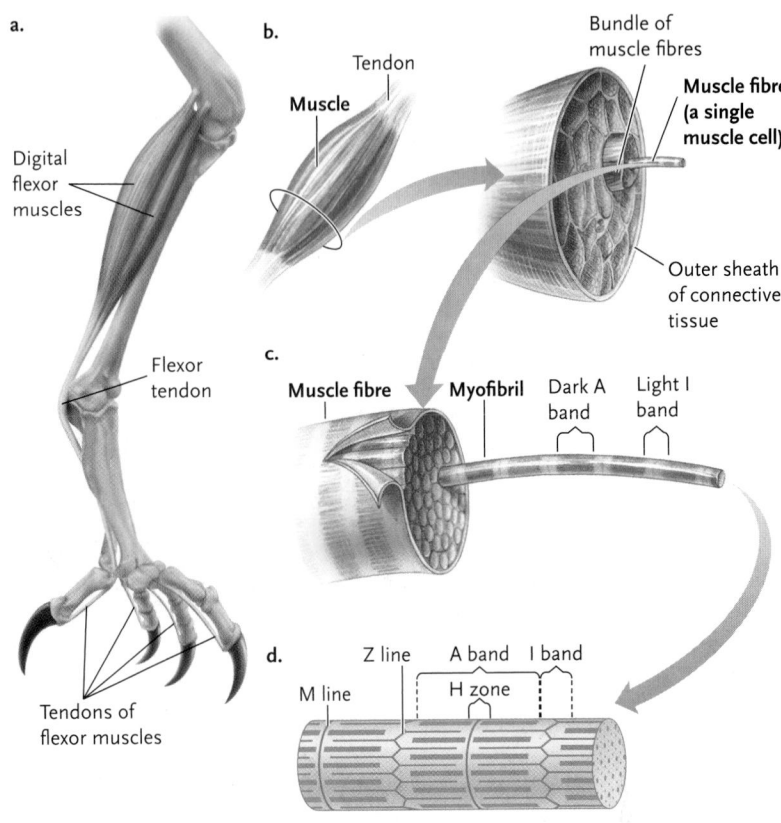

Figure 46.2
Skeletal muscle structure. Skeletal muscles are only attached at their ends to bones by tendons. Muscles are composed of bundles of cells called muscle fibres; within each muscle fibre are longitudinal bundles of contractile proteins called myofibrils.

to the flapping of wings, will be a function of the structures to which the muscles are attached. In this chapter, we describe the structure and function of skeletal muscles, the skeletal systems found in invertebrates and vertebrates, and the methods by which skeletal muscles bring about movement.

46.1 Vertebrate Skeletal Muscle: Structure and Function

Most skeletal muscles in vertebrates are attached at both ends, across a joint, to bones of the skeleton. Some, such as those that move the lips, are attached to other muscles or connective tissues under the skin. Skeletal muscles are attached to bones by cords of connective tissue called *tendons* (see Chapter 39 and **Figure 46.2a**). Tendons vary in length from a few millimetres to over a metre long (such as those in the forelimbs of a giraffe). Depending on its points of attachment, contraction of a single skeletal muscle may extend or bend body parts or may rotate one body part with respect to another. Most vertebrates (humans included) have more than 600 skeletal muscles, ranging in size from the small muscles that move the eyelids to the large muscles that move the legs. Skeletal muscles are controlled by the somatic nervous system (see Chapter 44).

46.1a The Striated Appearance of Skeletal Muscle Fibres Results from a Highly Organized Internal Structure

A skeletal muscle consists of bundles of elongated, cylindrical cells called muscle fibres, which are 10 to 100 μm in diameter and run the entire length of the muscle (Figure 46.2b). These fibres are formed by the fusion of cells, called myoblasts, generated during embryonic development in response to the appropriate signalling mechanisms. Reflecting this multicellular developmental origin, a single muscle fibre contains multiple nuclei. Some very small muscles, such as some of the muscles of the face, contain only a few hundred muscle fibres; others, such as the larger leg muscles, contain hundreds of thousands. In both cases, the muscle fibres are held in parallel bundles by sheaths of connective tissue that surround them in the muscle (Figure 46.2c) and merge with the tendons that connect muscles to bones or other structures. Muscle fibres are richly supplied with nutrients and oxygen by an extensive network of blood vessels that penetrates the muscle tissue.

Muscle fibres are packed with **myofibrils**, cylindrical bundles of contractile proteins about 1 mm in diameter that run lengthwise inside the cells (Figure 46.2d). Each myofibril **(Figure 46.3a, p. 1136)** consists of a regular arrangement of **thick filaments** (13–18 nm in diameter) and **thin filaments** (5–8 nm in diameter) (Figure 46.3 b and c). The thick and thin filaments alternate with one another in a stacked set.

The arrangement of thick and thin filaments forms a pattern of alternating dark bands and light bands, giving skeletal muscle a striated appearance under the microscope (see Figure 46.2c). As a result, skeletal muscle is often called striated muscle. The dark bands, called *A bands,* consist of stacked thick filaments along with the parts of thin filaments that overlap both ends (see Figure 46.3). The lighter-appearing

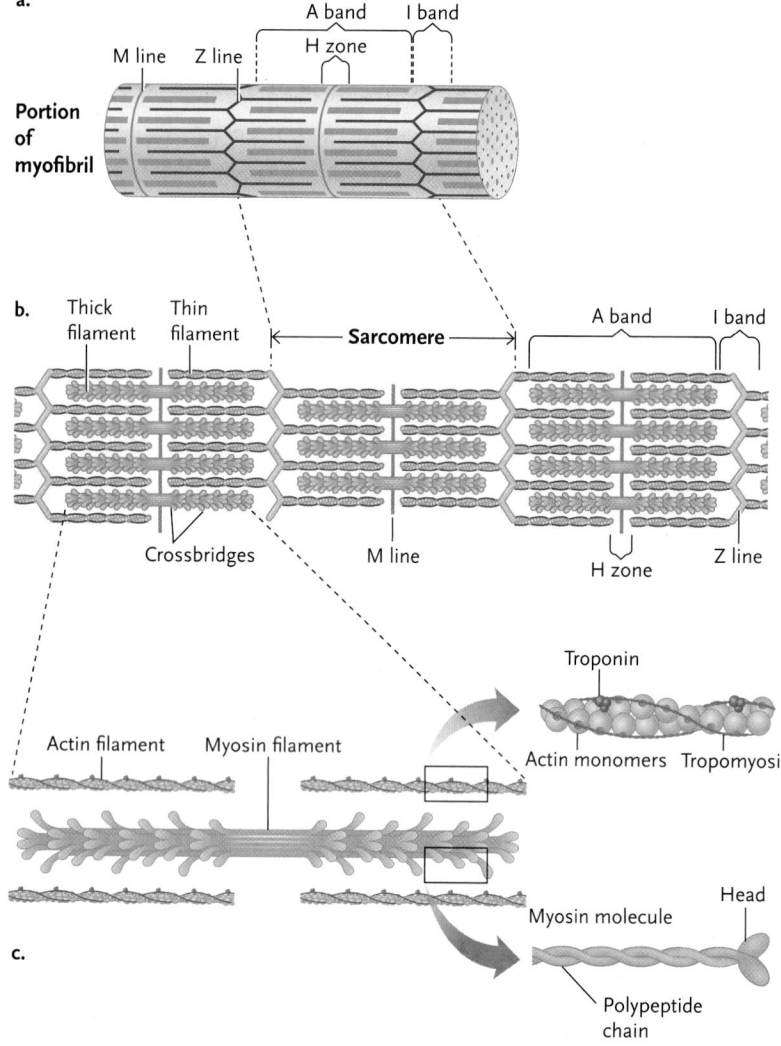

a.

Portion of myofibril

M line Z line

A band I band

H zone

b.

Thick filament Thin filament

Sarcomere

A band I band

Crossbridges M line H zone Z line

Troponin

Actin filament Myosin filament

Actin monomers Tropomyosin

Head

Myosin molecule

c.

Polypeptide chain

Figure 46.3

The unit of contraction within a myofibril, the sarcomere, consists of overlapping myosin thick filaments and actin thin filaments. The myosin molecules in the thick filaments each consists of two subunits organized into a head and a double-helical tail. The actin subunits in the thin filaments form twisted, double helices, with tropomyosin molecules arranged head to tail in the groove of the helix and troponin bound to the tropomyosin at intervals along the thin filaments.

is bent toward the adjacent thin filament to form a *crossbridge*. In vertebrates, each thick filament contains some 200 to 300 myosin molecules and forms as many crossbridges. The thin filaments consist mostly of two linear chains of actin molecules twisted into a double helix, which creates a groove running the length of the molecule. Bound to the actin are *tropomyosin* and *troponin* proteins. Tropomyosin molecules are elongated fibrous proteins that are organized end to end next to the groove of the actin double helix. Troponin is a three-subunit globular protein that binds to tropomyosin at intervals along the thin filaments. While these proteins are not involved directly in muscle shortening, they play key roles in regulating muscle shortening (as described below).

At each junction of an A band and an I band, the plasma membrane folds in to form a **T (transverse) tubule (Figure 46.4),** which extends deep within the muscle fibre and contains extracellular fluid vital for conducting electrical signals. The tubules come into close proximity with the intracellular endoplasmic reticulum, which in muscle cells takes a specialized form that wraps around the myofibrils and is referred to as the sarcoplasmic reticulum. The association between the T tubules and the sarcoplasmic reticulum is critical for coordinating the contraction of all of the individual myofibrils in a muscle fibre.

An axon of an efferent neuron leads to each muscle fibre. The axon terminal makes a single, broad synapse with a muscle fibre called a **neuromuscular junction** (see Figure 46.4). The neuromuscular junction, T tubules, and sarcoplasmic reticulum are key components in the pathway for stimulating skeletal muscle contraction by neural signals—which starts with action potentials travelling down the efferent neuron—as described next.

46.1b During Muscle Contraction, Thin Filaments on Sarcomeres Slide over Thick Filaments

The precise control of body motions depends on an equally precise control of muscle contraction by a signalling pathway that carries information from nerves to muscle fibres. An action potential arriving at the neuromuscular junction leads to an increase in the concentration of Ca^{2+} in the cytosol of the muscle fibre. The increase in Ca^{2+} triggers a process in which the thin filaments on each side of a sarcomere slide over the thick filaments toward the centre of the A band, which brings the Z lines closer together, shortening the sarcomeres and contracting the muscle (see Figure 46.5). This *sliding filament mechanism* of muscle contraction depends on dynamic interactions between actin and myosin proteins in the two filament types. That is, the myosin crossbridges make and break contact with actin and pull

middle region of an A band, which contains only thick filaments, is the *H zone.* In the centre of the H zone is a disk of proteins called the *M line,* which holds the stack of thick filaments together. The light bands, called *I bands,* consist of the parts of the thin filaments not in the A band. In the centre of each I band is a thin *Z line,* a disk to which the thin filaments are anchored. The region between two adjacent Z lines is a **sarcomere** (*sarco* = flesh; *meros* = segment). Sarcomeres are the basic units of contraction in a myofibril and are repeated along the entire length of each myofibril.

The thick filaments are parallel bundles of myosin molecules; each myosin molecule consists of two protein subunits that together form a *head* connected to a long double helix forming a *tail* (Figure 46.3). The head

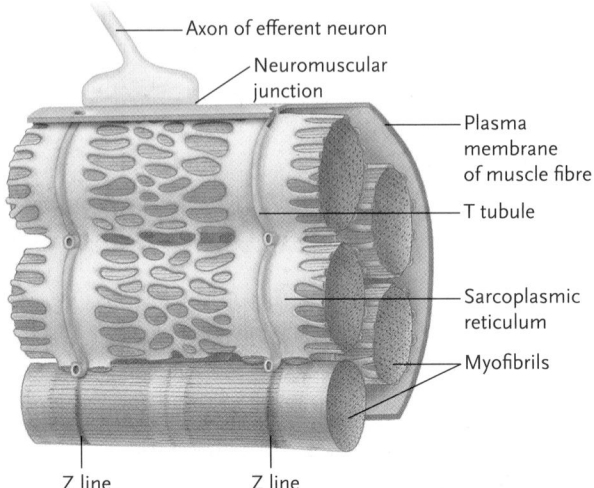

Figure 46.4

Components in the pathway for the stimulation of skeletal muscle contraction by neural signals. T (transverse) tubules are infoldings of the plasma membrane into the muscle fibre originating at each Z line in a sarcomere. The sarcoplasmic reticulum encircles the sarcomeres, and segments of it end in close proximity to the T tubules.

the thin filaments over the thick filaments—the action is similar to the walking, rowing, or ratcheting process described for kinesin in Chapter 2 (Figure 2.20). A model for muscle contraction is shown in **Figure 46.5.** The precise details of this process are described below.

Conduction of an Action Potential into a Muscle Fibre. Like neurons, skeletal muscle fibres are *excitable,* meaning that the electrical potential of their plasma membrane can change in response to a stimulus. When an action potential arrives at the neuromuscular junction, the axon terminal releases a neurotransmitter, *acetylcholine,* which triggers an action potential in the muscle fibre (see **Figure 46.6a, p. 1138,** step 1). The action potential travels in all directions over the muscle fibre's surface membrane and penetrates into the interior of the fibre along the walls of the T tubules, which are extensions of the plasma membrane.

Release of Calcium into the Cytosol of the Muscle Fibre. In the absence of a stimulus, the Ca^{2+} concentration is kept low in the cytosol by active transport proteins that continuously pump Ca^{2+} out of the cytosol and into the sarcoplasmic reticulum. When an action potential reaches the end of a T tubule, it opens ion channels in the sarcoplasmic reticulum that allow Ca^{2+} to flow out into the cytosol (see Figure 46.6a, step 2).

When Ca^{2+} flows into the cytosol, the troponin molecules of the thin filament bind the calcium and undergo a **conformational change** that causes the tropomyosin fibres to slip into the grooves of

QUESTION: What is the mechanism of muscle contraction?

EXPERIMENT: By 1954, researchers had established the locations and arrangements of actin and myosin in striated muscle. In that year, two independent teams—Andrew Huxley and Ralph Niedergerke of the University of Cambridge, United Kingdom, and Hugh Huxley and Jean Hanson of the Massachusetts Institute of Technology—used high-resolution light microscopy techniques to study how the actin and myosin arrangements changed during muscle contraction.

RESULTS: Their micrographs provided important evidence for the sliding of filaments. The stages of muscle contraction they outlined are shown in the figure.

a. Relaxed sarcomere

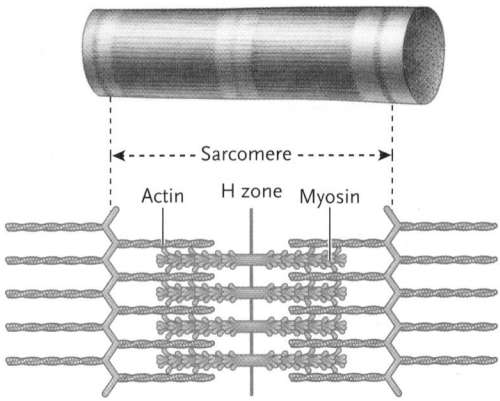

b. Contracted sarcomere

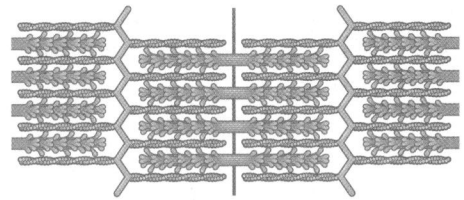

The figure illustrates a muscle fibre in a completely relaxed state **(a)**, and **(b)** as completely contracted. Note how the H zone, I band, and A band relate to one another at each stage. As the muscle fibre contracted, the key observations the researchers made were that: (1) the I band and the H zone each decrease in length in proportion to the shortening of the sarcomere, and (2) the A band remains constant in length.

CONCLUSION: The light microscopy evidence supported a model—the sliding filament model—in which muscle shortening (sarcomere shortening) results from increased overlap of thick and thin filaments, not from any change in length of those filaments. This model was revolutionary in the field of muscle physiology, as models for muscle contraction at the time all proposed folding or coiling of the protein molecules in the filaments.

Figure 46.5

Shortening of sarcomeres by the sliding filament mechanism, in which the thin filaments are pulled over the thick filaments toward the M line in the middle of the A band.

A. F. Huxley and R. Niedergerke. 1954. Structural changes in muscle during contraction. *Nature* 173:971–973; H. Huxley and J. Hanson. 1954. Changes in the cross-striations of muscle during contraction and stretch and their structural interpretation. *Nature* 173:973–976.

the actin double helix. The slippage uncovers the actin's binding sites for the myosin crossbridges (see Figure 46.6, step 3), and the myosin heads bind to the actin (step 4).

The Crossbridge Cycle. The crossbridge cycle that causes sarcomere shortening is driven by chemical and structural changes occurring solely within the myosin

a.

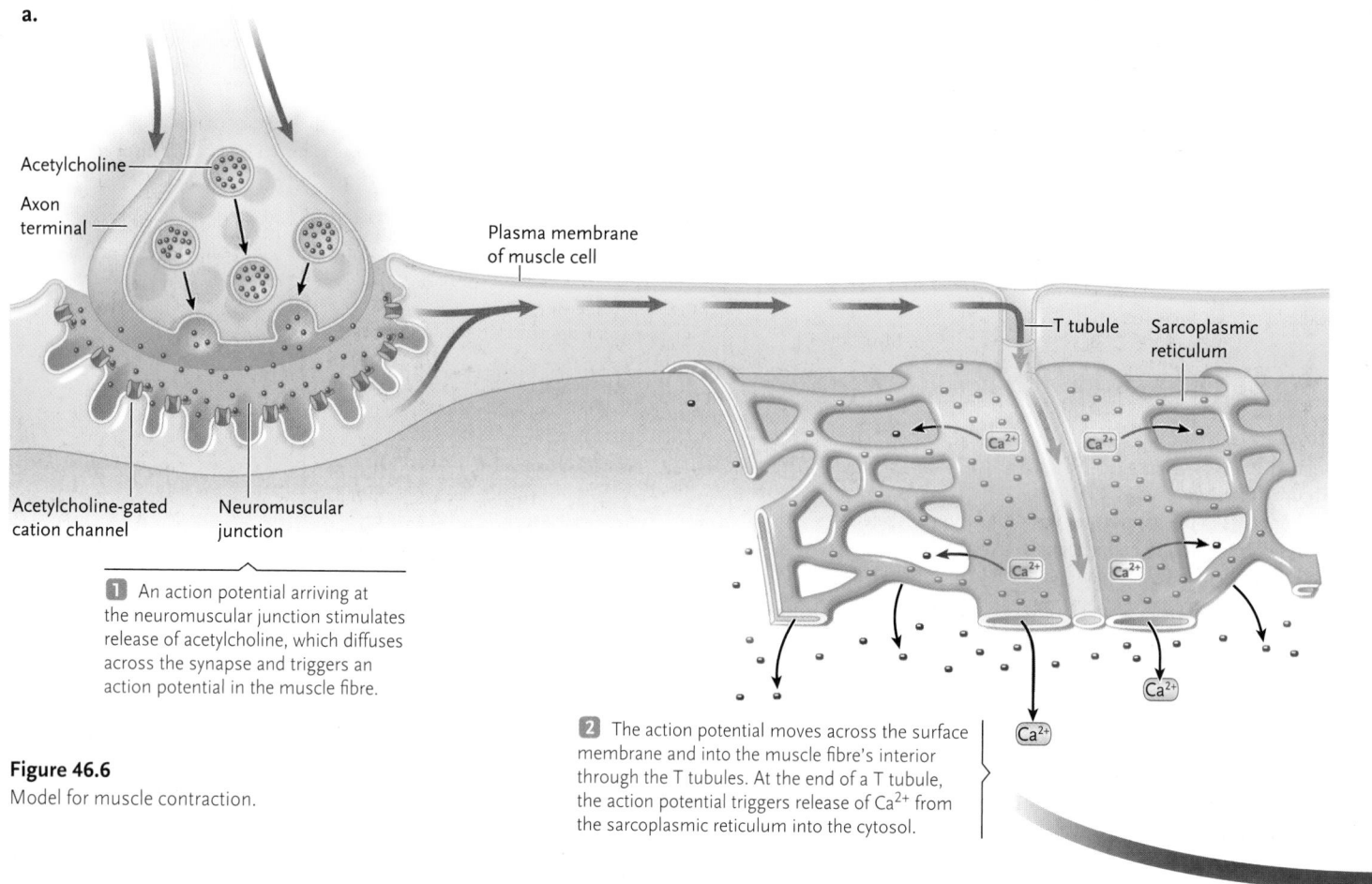

Acetylcholine

Axon terminal

Plasma membrane of muscle cell

T tubule   Sarcoplasmic reticulum

Ca²⁺ Ca²⁺

Ca²⁺ Ca²⁺

Ca²⁺

Acetylcholine-gated cation channel

Neuromuscular junction

Ca²⁺

1 An action potential arriving at the neuromuscular junction stimulates release of acetylcholine, which diffuses across the synapse and triggers an action potential in the muscle fibre.

Figure 46.6
Model for muscle contraction.

2 The action potential moves across the surface membrane and into the muscle fibre's interior through the T tubules. At the end of a T tubule, the action potential triggers release of Ca²⁺ from the sarcoplasmic reticulum into the cytosol.

head. It is essential to remember that every time a bond is made or broken, there are conformational or structural changes in the shapes of the proteins involved and in their energy states.

During the period of muscle relaxation, ATP bound to the myosin head is hydrolyzed, and the ADP, phosphate, and energy released are all stored in the myosin head. The binding of myosin to actin creates a crossbridge between the two molecules and results in a change in the conformation of the myosin head that triggers the release of the ADP and phosphate. It also releases the stored energy, which snaps the head of the myosin back toward the tail, producing the power stroke (motor) that pulls the thin filament over the thick filament (Figure 46.6, step 5). If no ATP is available, the myosin remains tightly bound to actin in this position. If ATP is available, it binds to an ATP-binding site on the myosin head. When the ATP binds, the myosin changes conformation and loses its affinity for actin and the crossbridge is released (step 6). Myosin is an ATPase, and when the crossbridge is released, the ATPase is activated and ATP is hydrolyzed. This causes the myosin crossbridge to bend away from the tail (step 7) and bind to a newly exposed myosin crossbridge binding site farther along the actin molecule, and the cycle repeats (starting at step 4). Crossbridge cycles based on actin

and myosin power movements in all living organisms, from cytoplasmic streaming in plant cells and amoebas to muscle contractions in animals. (Go to www.biologyedl2e.nelson.com to access animations illustrating crossbridge cycling.)

CONCEPT FIX For a long time, it was believed that ATP remains bound to the myosin head in relaxed muscle. It is now believed that while the binding of ATP to the myosin head causes the bond between myosin and actin to break, the ATP is quickly hydrolyzed to ADP and phosphate. This causes a change in the shape of the myosin, returning the head to its initial position (the position found in a relaxed muscle). The ADP, phosphate, and energy, however, are all stored in the myosin head. At the start of the next contraction, the binding of the myosin head to an actin molecule causes a conformational change in the myosin head, releasing the ADP, phosphate, and energy. The energy is then used to create the working stroke of the crossbridge cycle. ⬡

Although we have focused on the events taking place between a single myosin head and an actin filament in this discussion, within a sarcomere, each thick filament is surrounded by six thin filaments, and each thin filament is surrounded by three thick filaments **(Figure 46.7, p. 1140)**. While the force produced by a single myosin crossbridge is comparatively small, it is multiplied by the hundreds of crossbridges formed between a single thick filament and the thin filaments that surround it, and by the billions of thin filaments sliding in a contracting sarcomere. The force, multiplied further by the many sarcomeres and myofibrils in a muscle fibre, is transmitted to the plasma membrane of a muscle fibre by the attachment of myofibrils to elements of the cytoskeleton. From the plasma membrane, it is transmitted to bones and other body parts by the connective tissue sheaths surrounding the muscle fibres and by the tendons.

From Contraction to Relaxation. As long as action potentials continue to arrive at the neuromuscular junction, Ca^{2+} is released and remains in high concentration in the cytosol. As long as ATP is available, the crossbridge cycle continues to run, shortening the sarcomeres and contracting the muscle fibres. When action potentials stop, excitation of the T tubules ceases, and the Ca^{2+} release channels in the sarcoplasmic reticulum close. The active transport pumps quickly remove the remaining Ca^{2+} from the cytosol. In response, troponin releases its Ca^{2+} and the tropomyosin fibres are pulled back to cover the myosin-binding sites in the thin filaments. The crossbridge cycle stops, and contraction of the muscle fibre ceases.

When an animal dies, ATP levels decline. As a result, Ca^{2+} is no longer actively sequestered in the sarcoplasmic reticulum but enters the cytosol. This allows the myosin to bind actin, but with no ATP available, these bonds remain firmly attached and muscles become locked in *rigor mortis*.

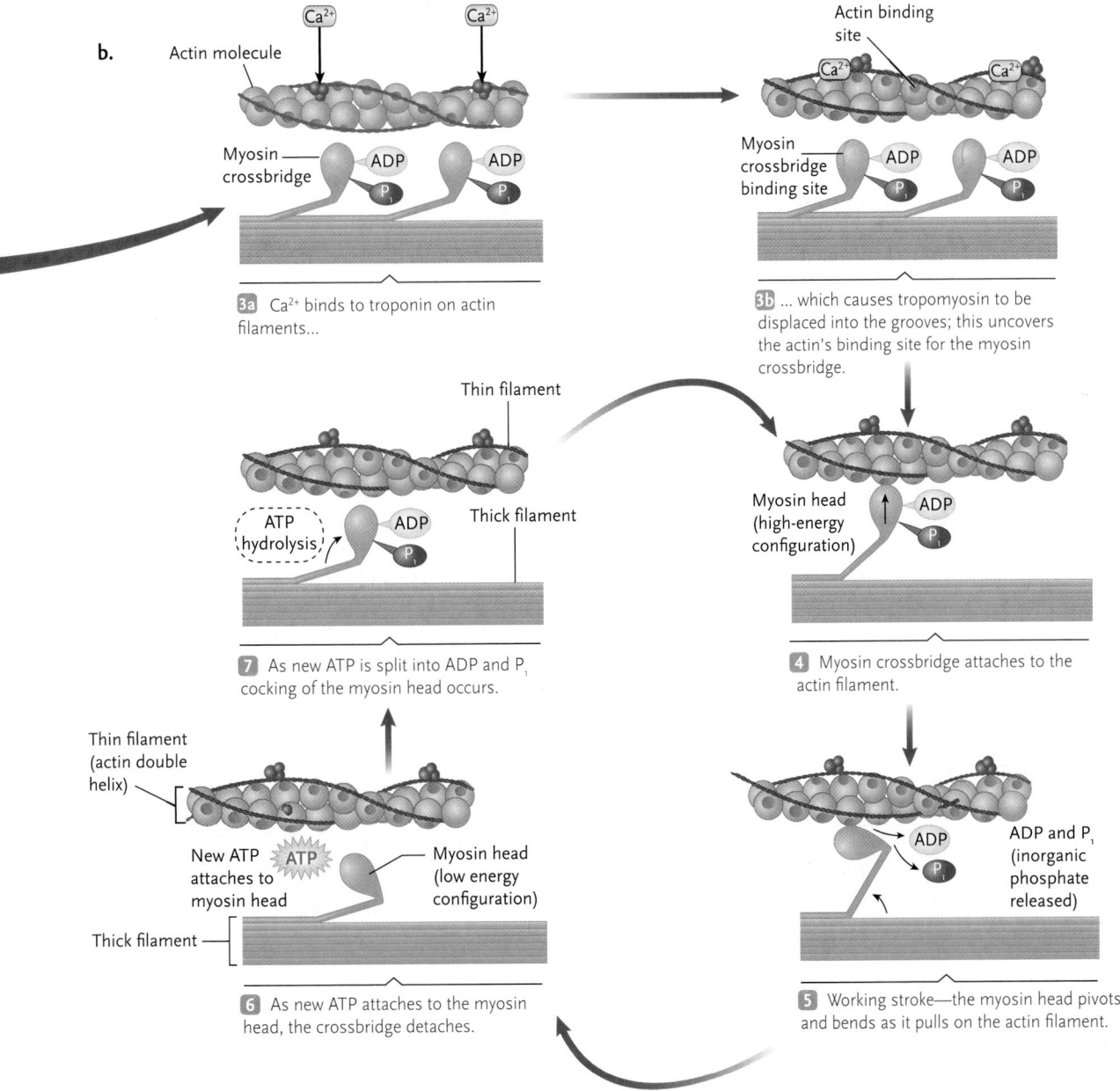

b.

3a Ca^{2+} binds to troponin on actin filaments...

3b ... which causes tropomyosin to be displaced into the grooves; this uncovers the actin's binding site for the myosin crossbridge.

7 As new ATP is split into ADP and P_i cocking of the myosin head occurs.

4 Myosin crossbridge attaches to the actin filament.

6 As new ATP attaches to the myosin head, the crossbridge detaches.

5 Working stroke—the myosin head pivots and bends as it pulls on the actin filament.

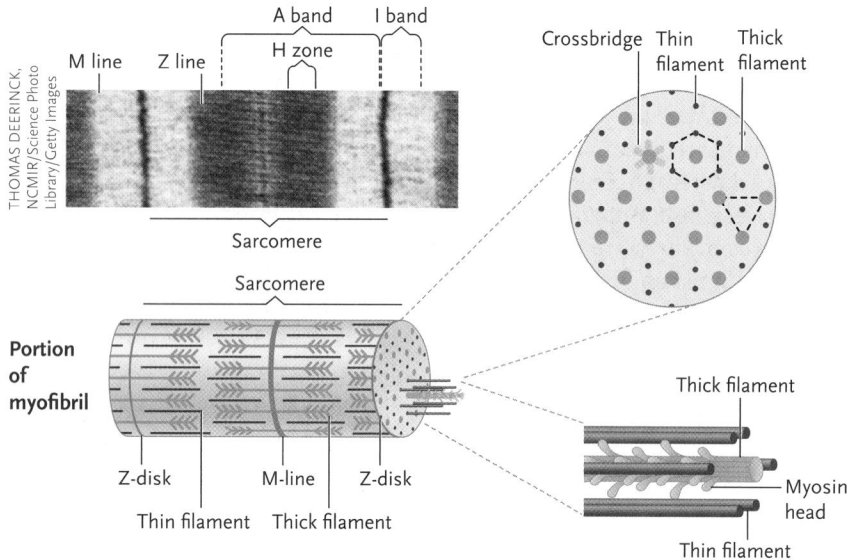

M line Z line A band I band

H zone

Sarcomere

Crossbridge Thin filament Thick filament

THOMAS DEERINCK, NCMIR/Science Photo Library/Getty Images

Portion of myofibril

Sarcomere

Z-disk M-line Z-disk

Thin filament Thick filament

Thick filament

Myosin head

Thin filament

Figure 46.7

Three-dimensional arrangement of thick and thin filaments within a sarcomere.

peaks as the action potential runs its course through the T tubules and the Ca^{2+} channels begin to close. Tension then decreases as the Ca^{2+} ions are pumped back into the sarcoplasmic reticulum, falling to zero about 50 ms after the peak.

If a muscle fibre is restimulated after it has relaxed completely, a new twitch identical to the first is generated (see Figure 46.8a). However, if a muscle fibre is restimulated before it has relaxed completely, the second twitch is added to the first, producing what is called *twitch summation,* or *temporal summation,* which is basically a summed, stronger contraction (Figure 46.8b). And if action potentials arrive so rapidly (about 25 ms apart) that the fibre cannot relax between stimuli, the Ca^{2+} channels remain open continuously and twitch summation produces a peak level of continuous contraction called **tetanus** (Figure 46.8c). Contractile activity will then decrease if either the stimuli cease or the muscle fatigues.

Tetanus is an essential part of muscle fibre function. Even body movements that require relatively little effort, such as standing still but in balance, involve tetanic contractions of some muscle fibres.

46.1c The Response of a Muscle Fibre to Action Potentials Ranges from Twitches to Tetanus

A single action potential arriving at a neuromuscular junction usually causes a single, weak contraction of a muscle fibre called a **muscle twitch (Figure 46.8a).** After a muscle twitch begins, the tension of the muscle fibre increases in magnitude for about 30 to 40 ms and then

46.1d Skeletal Muscle Control Is Divided among Motor Units

The control of muscle contraction extends beyond the simple ability to turn the crossbridge cycle on and off. We can adjust a handshake from a gentle squeeze to a

Figure 46.8

The relationship of the tension produced in a muscle fibre to the frequency of action potentials.

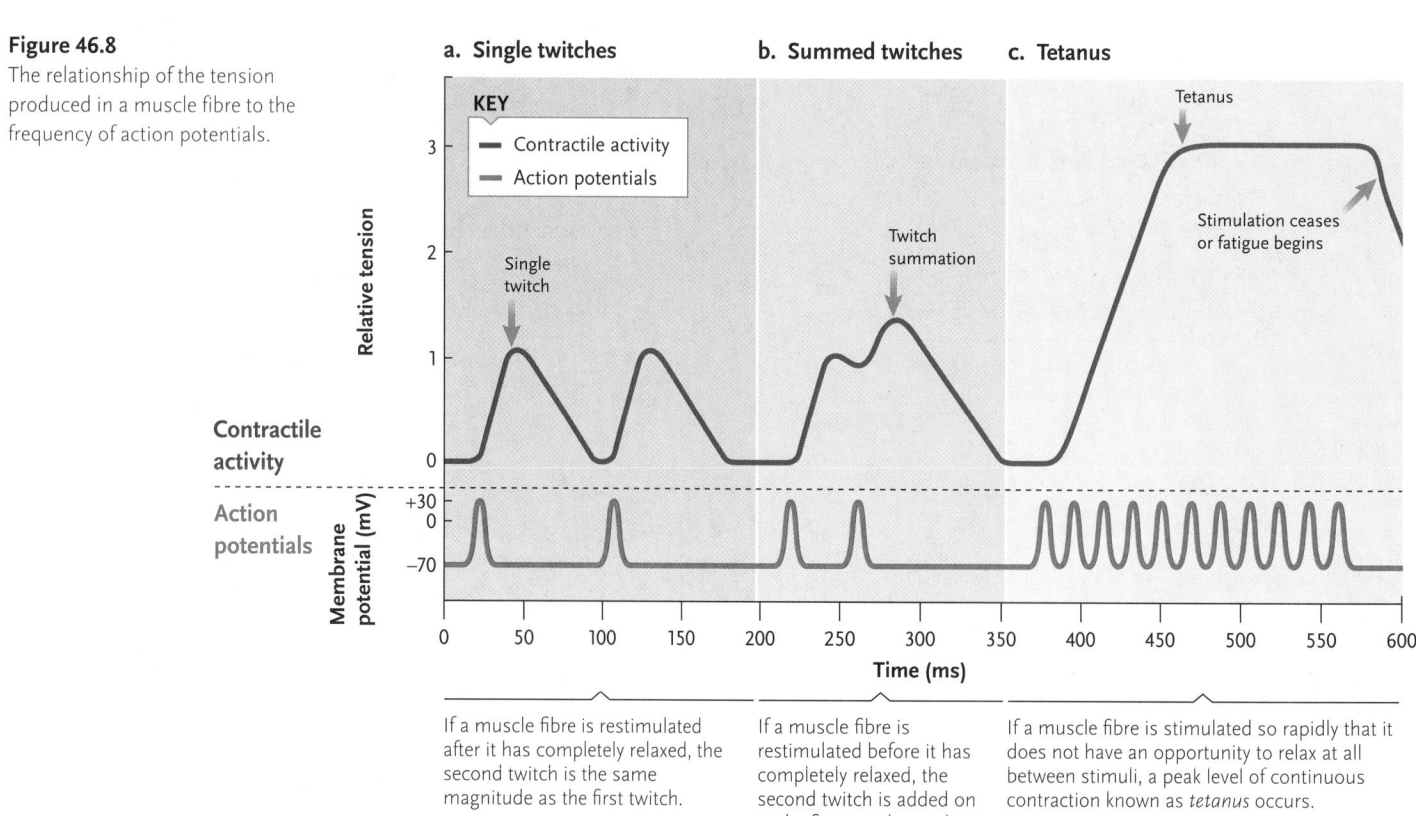

a. Single twitches **b. Summed twitches** **c. Tetanus**

KEY
— Contractile activity
— Action potentials

Relative tension

Single twitch

Twitch summation

Tetanus

Stimulation ceases or fatigue begins

Contractile activity

Action potentials

Membrane potential (mV)

Time (ms)

If a muscle fibre is restimulated after it has completely relaxed, the second twitch is the same magnitude as the first twitch.

If a muscle fibre is restimulated before it has completely relaxed, the second twitch is added on to the first twitch, resulting in summation.

If a muscle fibre is stimulated so rapidly that it does not have an opportunity to relax at all between stimuli, a peak level of continuous contraction known as *tetanus* occurs.

strong grasp or exactly balance a feather or a dumbbell in the hand. How are entire muscles controlled in this way? The answer lies in activation of the muscle fibres in blocks called **motor units.**

The muscle fibres in each motor unit are controlled by branches of the axon of a single efferent neuron **(Figure 46.9).** As a result, all of those fibres contract each time the neuron fires an action potential. When a motor unit contracts, its force is distributed throughout the entire muscle because the fibres are dispersed throughout the muscle rather than being concentrated in one segment, as shown in the figure.

For a delicate movement, only a few efferent neurons carry action potentials to a muscle, and only a few motor units contract. For more powerful movements, more efferent neurons carry action potentials, and more motor units contract. This is called *spatial summation,* the summing together of the activities of many motor units.

Muscles that can be precisely and delicately controlled, such as those moving the fingers in monkeys and humans, have many motor units in a small area,

with only a few muscle fibres—about 10 or so—in each unit. Muscles that produce grosser body movements, such as those moving the legs, have fewer motor units in the same volume of muscle but thousands of muscle fibres in each unit. In the calf muscle that raises the heel, for example, most motor units contain nearly 2000 muscle fibres. Other skeletal muscles fall between these extremes, with an average of about 200 muscle fibres per motor unit.

The maximum force any muscle can produce comes when there is maximal temporal and spatial summation; all motor units are tetanically stimulated.

46.1e Muscle Fibres Differ in Their Rate of Contraction and Susceptibility to Fatigue

Muscle fibres differ in their rate of contraction and resistance to fatigue and thus can be classified as slow, fast aerobic, and fast anaerobic muscle fibres. Their properties are summarized in **Table 46.1.** The proportions of the three types of muscle fibres tailor the contractile characteristics of each muscle to suit its function within the body.

Slow muscle fibres contract relatively slowly, and the intensity of contraction is low because their myosin crossbridges hydrolyze ATP relatively slowly. They can remain contracted for relatively long periods without fatiguing. Slow muscle fibres typically contain many mitochondria and make most of their ATP by oxidative phosphorylation (aerobic respiration). They have a low capacity to make ATP by anaerobic glycolysis. They also contain high concentrations of the oxygen-storing protein **myoglobin**, which greatly enhances their oxygen supplies. Myoglobin is closely related to hemoglobin, the oxygen-carrying protein of red blood cells (see Chapter 40). Myoglobin gives slow muscle fibres,

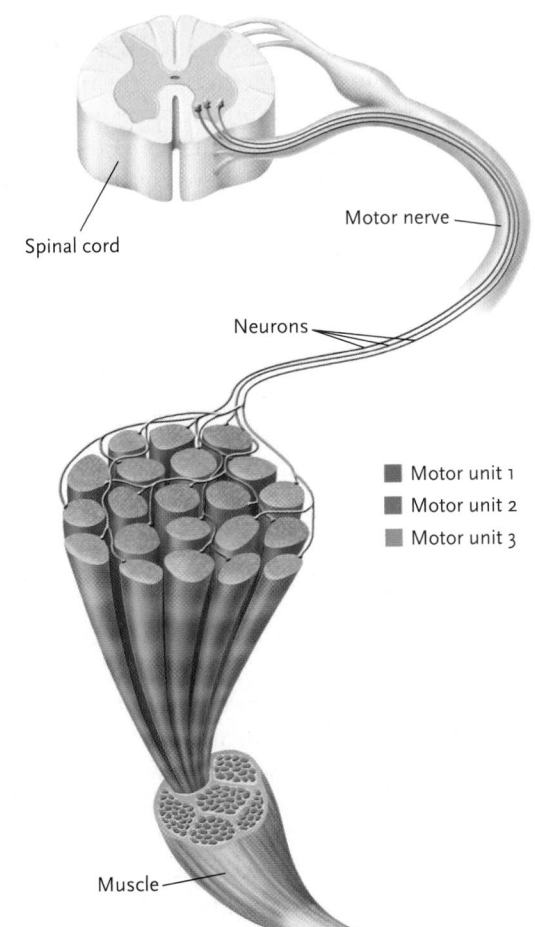

Figure 46.9

Motor units in vertebrate skeletal muscles. Each motor unit consists of groups of muscle fibres activated by branches of a single efferent (motor) neuron.

Spinal cord

Motor nerve

Neurons

■ Motor unit 1
■ Motor unit 2
■ Motor unit 3

Muscle

Table 46.1	Characteristics of Slow and Fast Muscle Fibres in Skeletal Muscle		
		Fibre Type	
Property	Slow	Fast Aerobic	Fast Anaerobic
Contraction speed	Slow	Fast	Fast
Contraction intensity	Low	Intermediate	High
Fatigue resistance	High	Intermediate	Low
Myosin–ATPase activity	Low	High	High
Oxidative phosphorylation capacity	High	High	Low
Enzymes for anaerobic glycolysis	Low	Intermediate	High
Mitochondria	Many	Many	Few
Myoglobin content	High	High	Low
Fibre colour	Red	Red	White
Glycogen content	Low	Intermediate	High

such as those in the legs of ground birds such as quail, chickens, and ostriches, a deep red colour. In sharks and bony fishes, strips of slow muscles concentrated in a band on either side of the body are used for slow, continuous swimming and maintaining body position.

Fast muscle fibres contract relatively quickly and powerfully because their myosin crossbridges hydrolyze ATP faster than those of slow muscle fibres. *Fast aerobic fibres* have abundant mitochondria, a rich blood supply, and a high concentration of myoglobin, which makes them red in colour. They have a high capacity for making ATP by oxidative phosphorylation, and an intermediate capacity for making ATP by anaerobic glycolysis. They fatigue more quickly than slow fibres, but not as quickly as fast anaerobic fibres. Fast aerobic muscle fibres are abundant in the flight muscles of migrating birds such as ducks and geese.

Fast anaerobic fibres typically contain high concentrations of glycogen, relatively few mitochondria, and a more limited blood supply than fast aerobic fibres. They generate ATP mostly by anaerobic respiration (glycolysis) and have a low capacity to produce ATP by oxidative respiration. Fast anaerobic fibres produce especially rapid and powerful contractions but are more susceptible to fatigue. Because their myoglobin supply is limited and they contain few mitochondria, they are pale in colour. Some ground birds have flight muscles consisting almost entirely of fast anaerobic muscle fibres. These muscles can produce a short burst of intensive contractions, allowing the bird to escape a predator, but they cannot produce sustained flight. Most muscles of lampreys, sharks, fishes, amphibians, and reptiles also contain fast anaerobic muscle fibres, allowing the animals to move quickly to capture prey and avoid danger.

The muscles of most animals are mixed and contain different proportions of slow and fast muscle fibres, depending on their functions. Muscles specialized for prolonged, slow contractions, such as the postural muscles of the back, have a high proportion of slow fibres and are a deep red colour. The muscles of the forearm that move the fingers have a higher proportion of fast fibres and are a paler red than the back muscles. These muscles can contract rapidly and powerfully, but they fatigue much more rapidly than the back muscles. It is worth noting that all muscle fibres in a single motor unit are of the same fibre type.

CONCEPT FIX It is commonly believed that lactic acid is only produced by muscles under anaerobic conditions and is the source of muscle fatigue. There are several myths associated with this.

To begin with, muscles do not produce lactic acid but rather they produce lactate. As a result, lactate production does not cause the acidosis associated with strenuous exercise. The hydrogen ions that cause the acidosis are produced from the breakdown of ATP to ADP + phosphate. Under aerobic conditions, when there is no acidosis, the protons produced by the breakdown of ATP are consumed by the mitochondria during ATP production. Under anaerobic conditions, however, the mitochondria are incapable of consuming all of the hydrogen ions produced from ATP breakdown, so hydrogen ions accumulate and acidosis ensues.

Secondly, some muscle cells produce lactate even under resting conditions when there is ample oxygen available. The lactate can remain in the cell for energy or leave the cell and travel to the heart and to active and inactive muscles to be used as a fuel (many cells prefer lactate as a fuel source over glucose). It will also travel to the liver, where it is converted back into glucose.

Finally, it has recently been shown that the accumulation of hydrogen ions (acidosis) may actually counteract the effects of muscle fatigue. Strenuous muscle contraction also precipitates a variety of other disturbances to cell homeostasis, including perturbations to energy charge and ion balances (particularly K^+) that lead to loss of tetanic force. Whether K^+ imbalance always, sometimes, or hardly ever leads to fatigue during exercise is uncertain. In muscles studied *in isolation,* however, hydrogen ion accumulation offered a degree of protection against fatigue produced by extracellular K^+ accumulation.

Thus, increased lactate production coincides with cellular acidosis but is not the cause of the metabolic acidosis. If muscle did not produce lactate and acidosis, muscle fatigue would occur more quickly and exercise performance would be severely impaired. ⬡

46.1f Invertebrates Move Using a Variety of Striated Muscles

Invertebrates also have muscle cells in which actin-based thin filaments and myosin-based thick filaments produce movements by the same sliding mechanism as in vertebrates. In most invertebrates (annelids, molluscs, echinoderms, nematodes, and arthropods), the actin and myosin fibrils are arranged in sarcomeres, forming striated muscle. In general, this is the dominant muscle type for these invertebrates and functions not only in locomotion but also in movements of the viscera, such as the gut and heart. In some invertebrates, such as Cnidaria and flatworms, muscle cells lacking striations may occur. In the muscles that close the shells of clams and other bivalves (see Chapter 27), smooth muscle cells are present among the striated muscle cells.

In invertebrates, an entire muscle is typically controlled by one or a few motor neurons. Nevertheless, invertebrate muscles are capable of finely graded contractions because individual neurons make large numbers of synapses with the muscle cells. In arthropods, the muscles may receive up to three types of innervation: fast, slow, and inhibitory. All muscles receive fast innervation, in which release of a neuromuscular transmitter produces a twitch. Some also receive slow innervation, in which a graded response results from increased action potentials. As action potentials arrive more frequently, more Ca^{2+} is released into the cells, and they contract more strongly. In addition, there may be inhibitory nerves that prevent the release of Ca^{2+}. The excitatory transmitter for both fast and slow nerves is glutamate, and the inhibitory transmitter is GABA (see Chapter 44).

The muscles responsible for the movement of the wings in insects are highly specialized striated muscles called fibrillar muscles. They possess a large number of gigantic mitochondria, in some cases about the size of a vertebrate red blood cell, so that the energetic demands of flight can be met. The frequency of wing beat of many flies, bees, and wasps is very high, up to 600 beats per second in mosquitoes. How is this achieved without tetanus being induced? The answer is that each contraction does not require new activation by another action potential. Flight muscles are stretch activated and occur in antagonistic pairs **(Figure 46.10)**. When one muscle of the pair contracts, the other is stretched back to its relaxed length. The stretching of one activates the other and vice versa. Nerve impulses arrive only about three times per second, keeping the muscles activated. The frequency with which they contract is determined by the elastic properties of the whole system.

STUDY BREAK

1. Compare thick and thin muscle filaments. How does their structure determine their function?
2. What is the role of the sarcoplasmic reticulum in muscle contraction?
3. What are the three types of muscle fibres, and how do they differ?
4. Greyhounds have muscles specialized for extreme speed over short distances. Sled dogs have muscles specialized for endurance over long distances. What kinds of muscle characteristics would each have?

46.2 Skeletal Systems

Animal skeletal systems provide physical support for the body and protection for the soft tissues. They also act as a framework against which muscles work to move parts of the body or the entire organism. Three main types of skeletons are found in both invertebrates and vertebrates: hydrostatic skeletons, exoskeletons, and endoskeletons.

46.2a A Hydrostatic Skeleton Consists of Muscles and Fluid

A **hydrostatic skeleton** (*hydro* = water; *statikos* = causing to stand) is a structure consisting of muscles and fluid

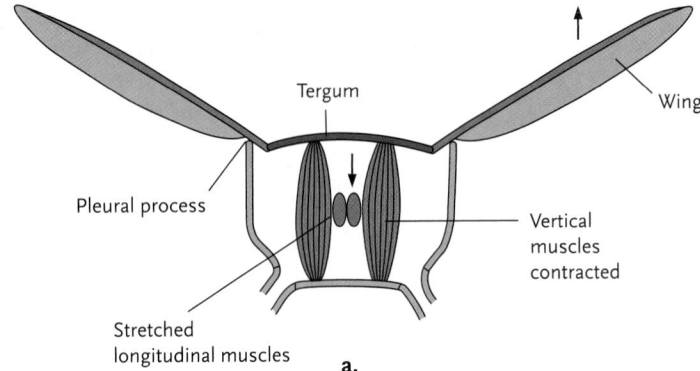

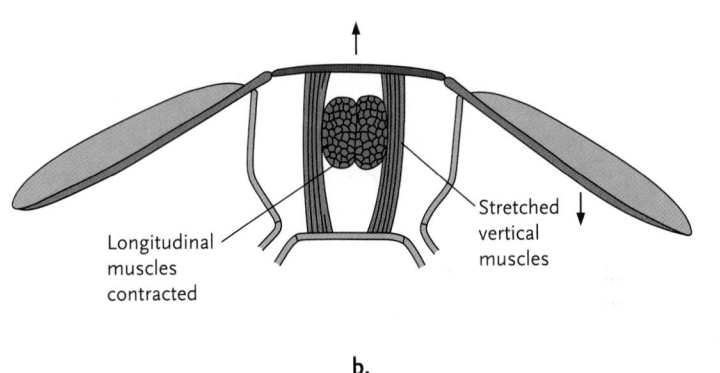

Figure 46.10

The muscles in a flying insect. When the vertical flight muscles contract, they pull on the cuticle forming the top of the segment, elevating the wings. The segment is constructed so that this action also elongates the cuticle of the thorax from front to rear (at right angles to the plane of the page), extending the longitudinal muscles. This stimulates the activated longitudinal muscles to contract, pushing up the tergum, elevating the wings, and elongating the vertical muscles. Because the cuticle is elastic, the whole system vibrates, producing very rapid wing beats.

that, by themselves, provide support for the animal or part of the animal; no rigid support, such as bone, is involved. A hydrostatic skeleton consists of a body compartment or compartments filled with water or body fluids, which are incompressible liquids. When the muscular walls of the compartment contract, they pressurize the contained fluid. If muscles in one part of the compartment are contracted while muscles in another part are relaxed, the pressurized fluid will move to the relaxed part of the compartment, distending it. In short, the contractions and relaxations of the muscles surrounding the compartments change the shape of the animal.

Hydrostatic skeletons are the primary support systems of cnidarians, flatworms, roundworms, and annelids. In all of these animals, compartments containing fluids under pressure make the body semirigid and provide a mechanical support on which muscles act. For example, sea anemones have a hydrostatic skeleton consisting of several fluid-filled body cavities. The body wall contains longitudinal and circular muscles that work against that skeleton. Between meals, longitudinal muscles are contracted (shortened), whereas the circular ones are relaxed, and the animal looks short and squat **(Figure 46.11a)**. It lengthens into its upright feeding position by contracting the circular muscles and relaxing the longitudinal ones (Figure 46.11b). In flatworms, roundworms, and annelids, striated muscles in the body wall act on the hydrostatic skeleton to produce creeping, burrowing, or swimming movements. Among these animals, annelids have the most highly developed musculoskeletal systems, with an outer layer of circular muscles surrounding the body, and an inner layer of longitudinal muscles **(Figure 46.12)**. Contractions of the circular muscles reduce the diameter of the body and increase the length; contractions of the longitudinal muscles shorten the body and increase its diameter. Because the coelom and musculature are divided into segments, expansion and contraction can be localized to individual segments. Annelids move along a surface or burrow by means of alternating waves of contraction of the two muscle layers that pass along the body, working against the fluid-filled body compartments of the hydrostatic skeleton.

Some structures of echinoderms are supported by hydrostatic skeletons. The tube feet of sea stars and sea urchins, for example, have muscular walls enclosing the fluid of the water vascular system (see Chapter 28).

a. Resting position   b. Feeding position

Figure 46.11
Sea anemones in (a) the resting and (b) the feeding positions. In (a), longitudinal muscles in the body wall are contracted, and circular muscles are relaxed. In (b), the longitudinal muscles are relaxed, and the circular muscles are contracted. Both sets of muscles work against a hydrostatic skeleton.

In vertebrates, the erectile tissue of the penis is a fluid-filled hydrostatic skeletal structure, although many mammals other than humans also possess a penis bone, the *os penis* or baculum (see Figure 46.14d, p. 1147, and Figure 41.17, page 1149).

Hydrostatic movement may not involve a fluid but may simply depend on the incompressibility of muscles themselves. Although the muscles in the structure may contract, the total body of muscles remains at a constant volume. Our tongues and lips are capable of a range of movements, but no skeletal element supports the movement. The elephant's trunk can lift a large log but can also pick up objects of a few millimetres. How is this achieved? The

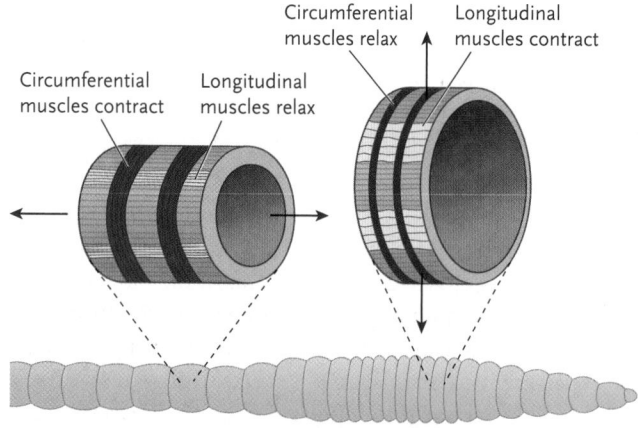

Circumferential muscles contract   Longitudinal muscles relax

Circumferential muscles relax   Longitudinal muscles contract

Figure 46.12
Movement of an earthworm, showing how muscles in the body wall act on its hydrostatic skeleton. Contraction of the circular muscles reduces body diameter and increases body length, whereas contraction of the longitudinal muscles decreases body length and increases body diameter.

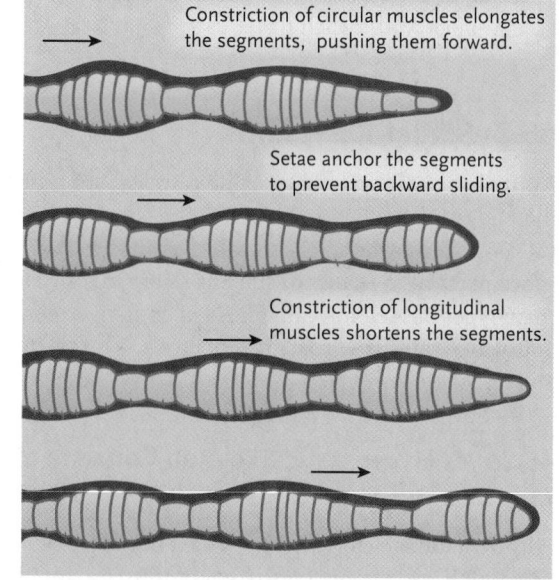

Constriction of circular muscles elongates the segments, pushing them forward.

Setae anchor the segments to prevent backward sliding.

Constriction of longitudinal muscles shortens the segments.

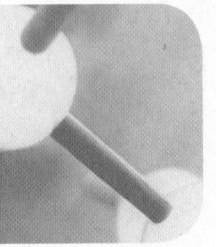

MOLECULE BEHIND BIOLOGY 46.1

Chitin

Experimental Research Box

Chitin is a polymer composed of many repeating units of N-acetyl-D-glucosamine **(Figure 1).** It is a strong, insoluble material that is the principal component of the cuticle of arthropods, where it is crosslinked to proteins. When arthropods moult, the proteins are resorbed, and the cast cuticle is made up almost entirely of chitin. Given the dominance of arthropods in Earth's biomass, it is not surprising that, after cellulose, chitin is the most abundant polymer on Earth, with an annual production of about 10 billion tonnes. Chitin also occurs in molluscs, the eggshells of nematodes, and the cell walls of fungi and algae.

Chitin is degraded by bacteria, particularly members of the genus *Vibrio*. Because chitin does not occur in vertebrates, chitin synthesis is a target of some successful insecticides.

Chitin and its closely related compound chitosan, produced by heating chitin in a strongly alkaline solution, are widely used in a variety of fields. Chitosan was found to have healing properties and has been used for medical sutures and as a support for growing skin over severe wounds. The compounds are also said to confer protection against disease in plants, and seeds treated with chitosan have gained some acceptance.

Chitin stimulates the growth of soil bacteria that secrete material toxic to nematodes that attack plants; compounds for this purpose are under development. Because the compounds have a strong positive charge, they have the potential to bind negatively charged compounds. Chitosans have been used in water purification plants for many years. More recently, chitosans, because of their ability to bind fats, have been promoted by the natural health products industry as a means to prevent fat absorption, leading to weight loss, but these claims are not supported by good scientific evidence.

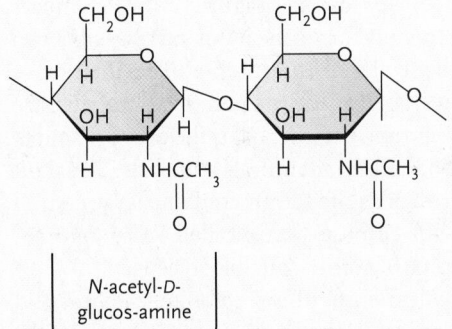

Figure 1

Chitin is a polymer made up of many repeating units of N-acetyl-D-glucosamine and is the principal component of the exoskeletons of arthropods.

trunk is basically an extension of the nose and upper lip. It has no skeleton but consists of an enormous number of muscle units attached to the skin or to one another. The muscle mass remains at a constant volume, and contractions of local muscle groups result in movement.

46.2b An Exoskeleton Is a Rigid External Body Covering

An **exoskeleton** (*exo* = outside) is a rigid external body covering, such as a shell that provides support. In an exoskeleton, the force of muscle contraction is applied against that covering. An exoskeleton also protects delicate internal tissues.

Many molluscs, such as clams and oysters, have an exoskeleton consisting of a hard calcium carbonate shell secreted by glands in the mantle. Arthropods, such as insects, spiders, and crustaceans, have an external skeleton in the form of a chitinous cuticle (see "Molecule behind Biology," Box 46.1), secreted by the underlying epidermis, that covers the outside surfaces of the animals. Like a suit of armour, the arthropod exoskeleton has movable joints, flexed and extended by muscles. Most muscles attach directly to the cuticle by extensions of the myofibrils and extend from the inside surface of one section of the cuticle to the inside surface of another section. Since the sections are separated by flexible cuticle, contraction results in movement about the joint **(Figure 46.13)**. The exoskeleton protects against dehydration, serves as armour against predators, and provides the levers against which muscles work. In many flying insects, elastic flexing of the exoskeleton contributes to the movements of the wings (see Figure 46.10).

In vertebrates, the shell of a turtle or tortoise is an exoskeletal structure.

46.2c An Endoskeleton Consists of Supportive Internal Body Structures Such as Bones

An **endoskeleton** (*endon* = within) consists of internal body structures, such as bones, that provide support. In an endoskeleton, the force of contraction is applied against those structures. The endoskeleton is the primary skeletal system of vertebrates. Most vertebrates have an endoskeleton arranged in two structural groups **(Figure 46.14)**. The **axial skeleton**, which includes the skull, vertebral column, sternum, and ribcage, forms the central part of the structure (shaded in red in Figure 46.14a), defines the long axis of the vertebrate body, provides sites for muscle attachment, and supports most of the weight. The **appendicular skeleton** (shaded in green) includes the shoulder, hip, leg, and arm bones and provides the levers that are used to produce locomotion. Four mammalian skeletons are included in Figure 46.14, illustrating how skeletons are adapted for particular lifestyles. The human skeleton reflects our upright movement, using only our hindfeet, while those of the monkey, lemur, and raccoon reflect arboreal (tree-living), gliding, and terrestrial lifestyles. Like exoskeletons, endoskeletons also protect delicate internal tissues such as the brain and respiratory organs.

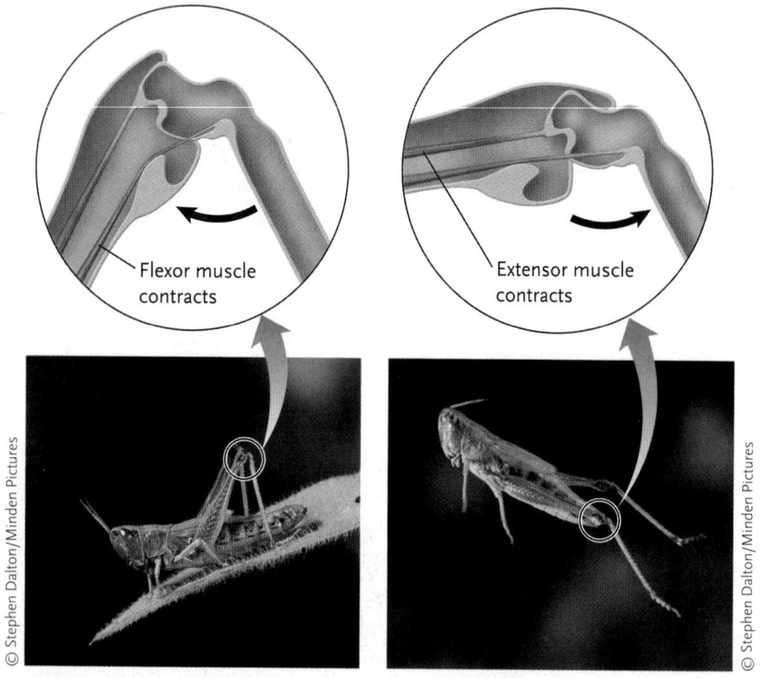

46.2d Bones of the Vertebrate Endoskeleton Are Organs with Several Functions

The vertebrate endoskeleton supports and maintains the overall shape of the body and protects internal organs. Bones are complex organs built up from multiple tissues, including bone tissue, with cells of several kinds; blood vessels; nerves; and, in some, stores of adipose tissue. Bone tissue is distributed between dense, compact bone regions, which have essentially no spaces other than the microscopic canals of the osteons (see Chapter 39), and spongy bone regions, which may open into larger spaces (see Figure 46.14). Compact bone tissue generally forms the outer surfaces of bones and spongy bone tissue the interior. The interior of some flat bones, such as the hip bones and the ribs, are filled with *red marrow,* a tissue that is the primary source of new red blood cells in mammals and birds. The shaft of long bones such as the femur contains large central canal filled with adipose tissue called *yellow marrow,* which is a source of some white blood cells.

Throughout the life of a vertebrate, calcium and phosphate ions are constantly deposited and withdrawn from bones. Hormonal controls maintain the concentration of Ca^{2+} ions at optimal levels in the blood and extracellular fluids (see Chapter 43), ensuring that calcium is available for proper functioning of the nervous system, muscular system, and other physiological processes.

Figure 46.13

Muscles are attached to the inside surfaces of the exoskeleton in a typical insect leg, such as those of the grasshopper shown here.

STUDY BREAK

1. Describe the similarities and differences between hydrostatic skeletons, exoskeletons, and endoskeletons. Give examples of each and describe which animal groups each can be found in.
2. Explain how the endoskeleton differs in humans compared with monkeys, lemurs, and raccoons (Figure 46.14), and, as best as you can, explain why these differences occur.
3. Astronauts in zero gravity tend to lose bone mass, particularly in the hips, the long bones of the legs, and the vertebrae. Instruments that measure bone loss cannot be taken on space missions. Can you think of experimental procedures that might be used on Earth to explore the phenomenon?

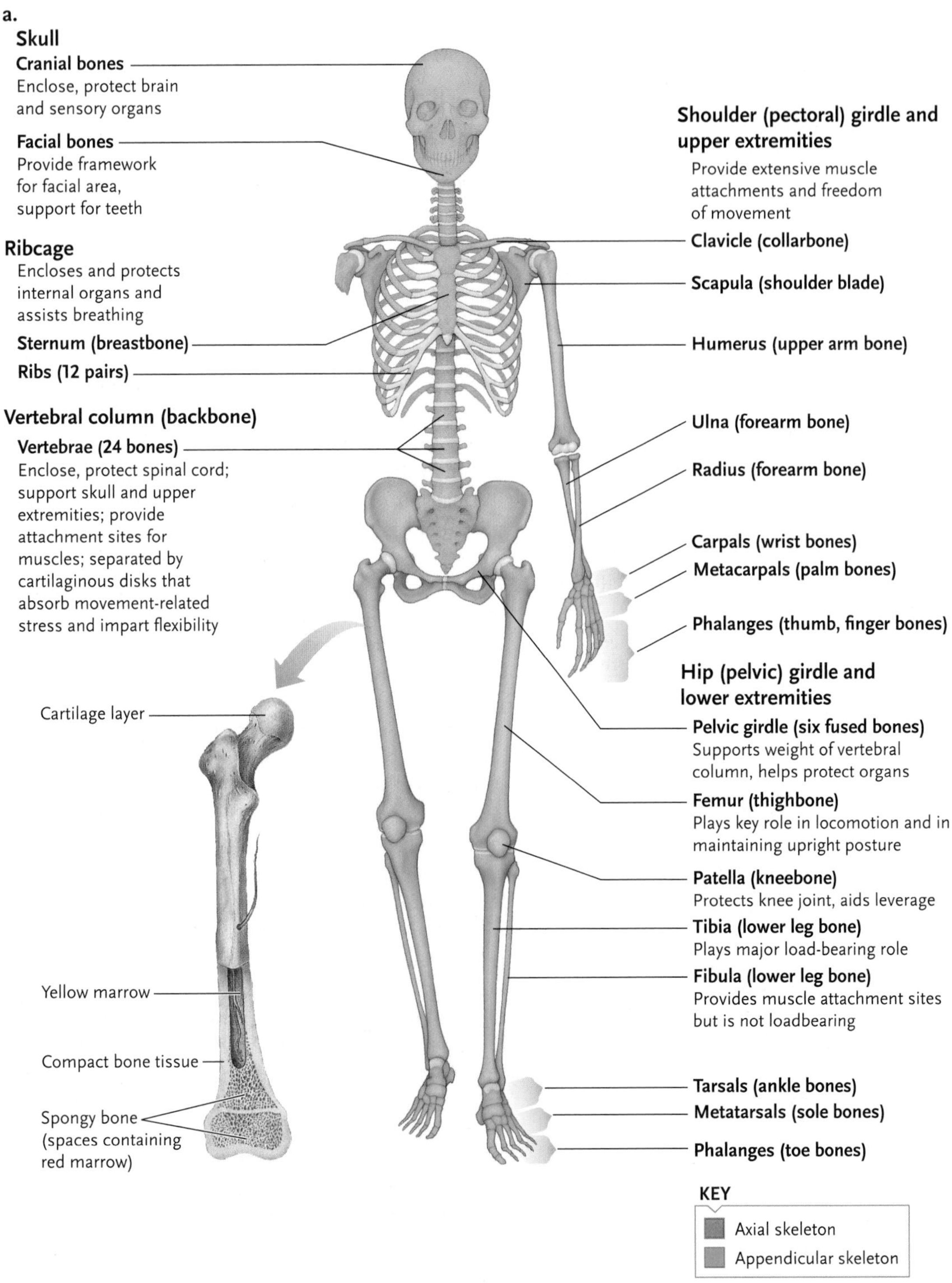

a.

Skull

Cranial bones
Enclose, protect brain and sensory organs

Facial bones
Provide framework for facial area, support for teeth

Ribcage
Encloses and protects internal organs and assists breathing

Sternum (breastbone)

Ribs (12 pairs)

Vertebral column (backbone)

Vertebrae (24 bones)
Enclose, protect spinal cord; support skull and upper extremities; provide attachment sites for muscles; separated by cartilaginous disks that absorb movement-related stress and impart flexibility

Cartilage layer

Yellow marrow

Compact bone tissue

Spongy bone (spaces containing red marrow)

Shoulder (pectoral) girdle and upper extremities
Provide extensive muscle attachments and freedom of movement

Clavicle (collarbone)

Scapula (shoulder blade)

Humerus (upper arm bone)

Ulna (forearm bone)

Radius (forearm bone)

Carpals (wrist bones)

Metacarpals (palm bones)

Phalanges (thumb, finger bones)

Hip (pelvic) girdle and lower extremities

Pelvic girdle (six fused bones)
Supports weight of vertebral column, helps protect organs

Femur (thighbone)
Plays key role in locomotion and in maintaining upright posture

Patella (kneebone)
Protects knee joint, aids leverage

Tibia (lower leg bone)
Plays major load-bearing role

Fibula (lower leg bone)
Provides muscle attachment sites but is not loadbearing

Tarsals (ankle bones)

Metatarsals (sole bones)

Phalanges (toe bones)

KEY

Axial skeleton

Appendicular skeleton

Figure 46.14
Mammalian skeletons.
(a) Major bones in the human. Inset shows the structure of the femur (thigh bone), with the location of red and yellow marrow. Internal spaces lighten the bone's density. At the joints a cartilage layer forms a smooth slippery cushion between bones. Compare the general features of this skeleton with those shown in (b), (c), and (d). Note general and specific resemblances among the skeletons of the four mammals. **(b)** The new world monkey (family Cebidae) lives in trees. **(c)** The gliding lemur (family Cynocephalidae) is also arboreal, but a glider. **(d)** The raccoon (family Procyonidae) is terrestrial. Note the differences in skull shape, limb lengths, and feet. The raccoon is a male, reflected in the conspicuous baculum or penis bone (see also Figure 41.17, Chapter 41).

b.

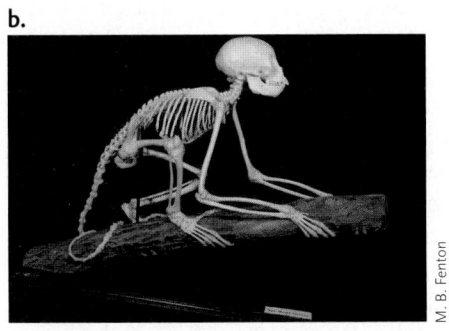

M. B. Fenton

c.

M. B. Fenton

d.

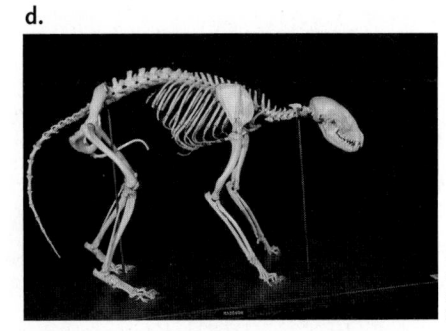

M. B. Fenton

CHAPTER 46 MUSCLES, SKELETONS, AND BODY MOVEMENTS

46.3 Vertebrate Movement: The Interactions between Muscles and Bones

The skeletal system acts as a framework against which muscles work to move parts of the body or the entire organism. In this section, the muscle–bone interactions that are responsible for the movement of vertebrates are described.

46.3a Joints of the Vertebrate Endoskeleton Allow Bones to Move and Rotate

The bones of the vertebrate skeleton are connected by joints, many of them movable. The most movable joints, including those of the shoulders, elbows, wrists, fingers, knees, ankles, and toes, are *synovial joints*, consisting of the ends of two bones enclosed by a fluid-filled capsule of connective tissue **(Figure 46.15a).** Within the joint, the ends of the bones are covered by a smooth layer of cartilage and lubricated by synovial fluid, which makes the bones slide easily as the joint moves. Synovial joints are held together by straps of connective tissue called *ligaments,* which extend across the joints outside the capsule (Figure 46.15b). The ligaments

restrict the motion of the joint and help prevent it from buckling or twisting under heavy loads.

In other, less movable joints, called *cartilaginous joints,* the ends of bones are covered with layers of cartilage but have no fluid-filled capsule surrounding them. Fibrous connective tissue covers and connects the bones of these joints, which occur between the vertebrae and some rib bones.

In still other joints, called *fibrous joints,* stiff fibres of connective tissue join the bones and allow little or no movement. Fibrous joints occur between the bones of the skull and hold the teeth in their sockets.

The bones connected by movable joints work like levers. A lever is a rigid structure that can move around a pivot point known as a *fulcrum* **(Figure 46.16a).** The most common type of lever system in the body—exemplified by the elbow joint—has the fulcrum at one end, the load at the opposite end, and the force applied at a point between the ends. Levers differ with respect to where the muscle is attached along the lever and, hence, where the force is applied (Figure 46.16b). If the muscle is inserted (attached) near the fulcrum, the muscle favours speed, as a short contraction will move the lever a large distance in a short period of time (Figure 46.16c, left). If the muscle is inserted more distally to the fulcrum, it favours strength. The same amount of contraction will move the forearm a shorter distance in the same period of time (i.e., it moves more slowly), but it will lift more weight (Figure 46.16c, right). Levers also differ with respect to where the fulcrum is located along the lever. In this case, if the fulcrum is located more distally, it favours speed, while if it is located more proximally, it favours strength (Figure 46.16d). A good example of the difference in design for speed versus strength is seen by comparing the forelimbs of burrowing and running species. Burrowing mammals have short, very stout forelimbs with muscles arranged so that contraction provides relatively little movement but great force **(Figure 46.17a).** Running species, on the other hand, have long, slender forelimbs with muscles arranged so that contraction provides rapid motion but less force (Figure 46.17b).

Most of the bones of vertebrate skeletons are moved by muscles arranged in **antagonistic pairs:** *extensor muscles* extend the joint, meaning increasing the angle between the two bones, whereas *flexor muscles* do the opposite. (Antagonistic muscles are also used in invertebrates for movement of body parts—e.g., the limbs of insects and arthropods.) In humans, one such pair is formed by the biceps brachii muscle at the front of the upper arm and the triceps brachii muscle at the back of the upper arm **(Figure 46.18).** When the biceps muscle contracts, the bone of the lower arm is bent (flexed) around the elbow joint, and the triceps muscle is passively stretched (see Figure 46.18a); when the triceps muscle contracts, the lower arm is straightened (extended), and the biceps muscle is passively stretched (see Figure 46.18b).

Figure 46.15
A synovial joint.
(a) Cross-section of a typical synovial joint.
(b) Ligaments reinforcing the knee joint.

a. Synovial joint cross-section

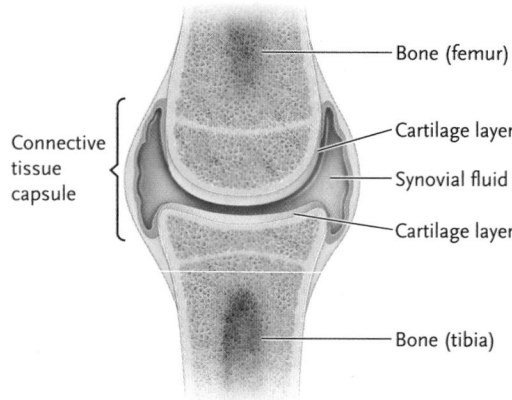

Bone (femur)
Connective tissue capsule
Cartilage layer
Synovial fluid
Cartilage layer
Bone (tibia)

b. Knee joint ligaments

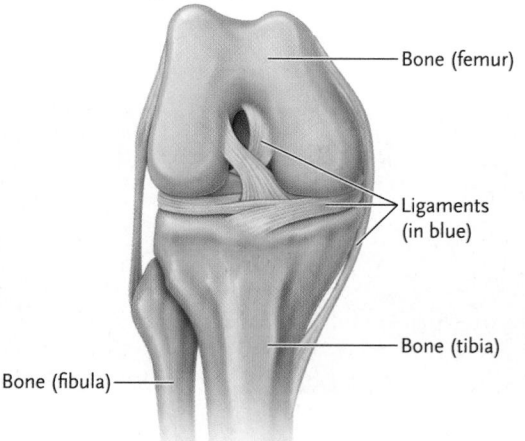

Bone (femur)
Ligaments (in blue)
Bone (tibia)
Bone (fibula)

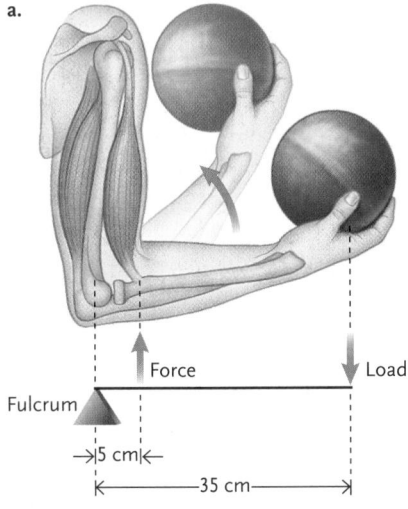

a.

Force   Load

Fulcrum

|← 5 cm →|

|←——— 35 cm ———→|

Figure 46.16
A body lever. **(a)** The lever formed by the bones of the forearm. The fulcrum (the hinge or joint) is at one end of the lever, the load is placed on the opposite end, and the force is exerted at a point on the lever between the fulcrum and the load. **(b)** Muscles can be attached proximal or distal to the fulcrum. **(c)** Proximal insertion favours speed, while distal insertion favours strength. **(d)** The fulcrum can also vary in position, favouring either speed or strength.

b.

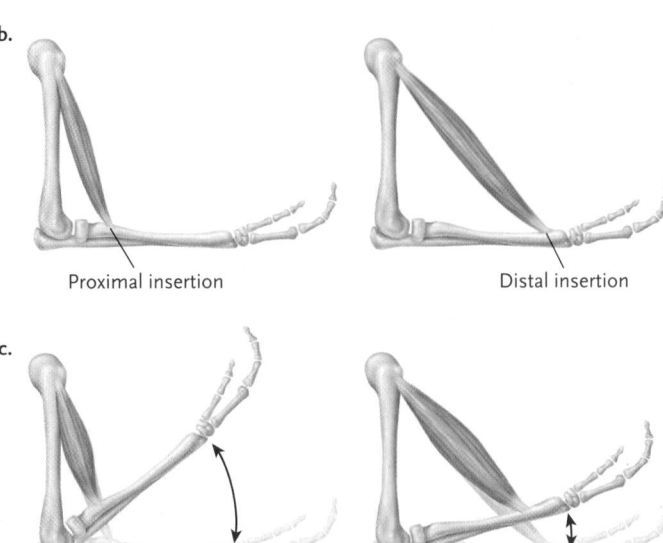

Proximal insertion   Distal insertion

c.

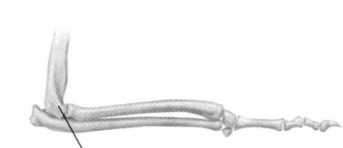
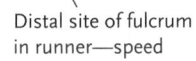
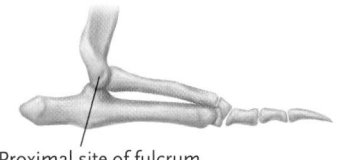

Proximal insertion of muscle—speed   Distal insertion of muscle—strength

d.

Distal site of fulcrum   Proximal site of fulcrum
in runner—speed   in digger—strength

46.3b Vertebrates Have Muscle–Bone Interactions Optimized for Specific Movements

Vertebrates differ widely in the relative size and shapes of individual bones, the patterns by which muscles connect to the bones, the sites of the joints along the bones, and the length and mechanical advantage of the levers produced by these connections. All of these differences reflect the lifestyles of the animals as reflected in the need for strength and speed of movement. Through evolutionary pressures to produce optimal interactions for specific movements, we see many examples of both conserved traits and convergent evolution.

a.   b.

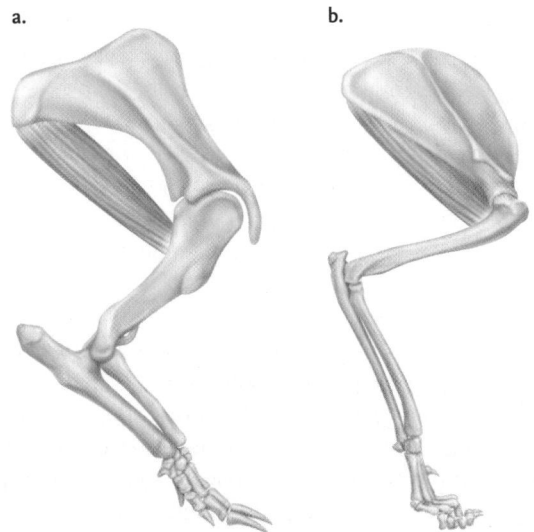

Figure 46.17
Design for strength versus speed in forelimbs can be seen by comparing **(a)** an armadillo with **(b)** a cat. Both forelimbs are drawn to be the same overall length. Note the differences in the size and mass of the bones, the points of insertion of the muscles, and the position of the joints along the bones.

a. When the biceps muscle contracts and raises the forearm, its antagonistic partner, the triceps muscle, relaxes.

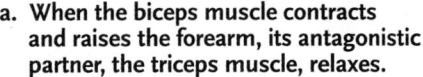

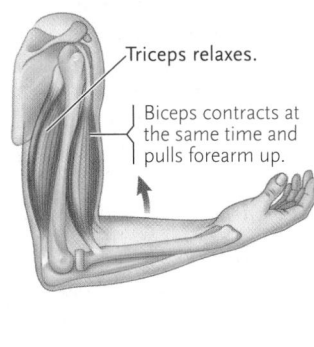

Triceps relaxes.

Biceps contracts at the same time and pulls forearm up.

b. When the triceps muscle contracts and extends the forearm, the biceps muscle relaxes.

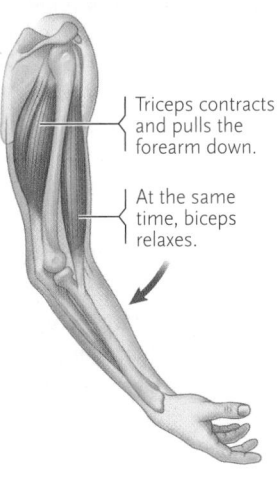

Triceps contracts and pulls the forearm down.

At the same time, biceps relaxes.

Figure 46.18
The arrangement of skeletal muscles in antagonistic pairs. **(a)** When the biceps muscle contracts and raises the forearm, its antagonistic partner, the triceps muscle, relaxes. **(b)** When the triceps muscle contracts and extends the forearm, the biceps muscle relaxes.

The act of walking is both unconscious and deceptively simple when viewed from a human perspective, involving two legs. But in insects, in which walking on six legs usually involves two legs on one side and one leg on the other being lifted, while the remaining legs form a triangular support, and in nonhuman mammals, in which the order of movement of the legs may differ for various speeds of locomotion, the real complexity of walking becomes more obvious.

Keir G. Pearson of the University of Alberta became interested in this challenging problem while he was a Rhodes Scholar at Oxford University, where he demonstrated the complexity of the patterns of nervous activity controlling a single muscle. He has continued that interest for more than 40 years, progressing to an analysis of walking in insects and vertebrates (usually cats). The techniques he has used are challenging, involving the recording of electrical activity in the leg muscles and the ganglia controlling those muscles. They have revealed that walking in insects is driven by a pattern of rhythmic activity, the central pattern generator (CPG) in the neurons of the ganglia associated with the limbs, influenced by information flowing from proprioceptors (see Chapter 37) in the limbs and by environmental information processed through the brain. These results have influenced the construction of walking robots and physiotherapy for humans with spinal cord damage.

Pearson recently turned his attention to the way in which visual information influences walking. A simple and revealing experiment involved cats **(Figure 1).** A cat approaching a small barrier steps over the barrier, first with the forelegs and then with the hindlegs. To a less inquiring mind, that might seem simple enough. But a scientist such as Pearson notes that when the hindlegs step over the barrier, the barrier is no longer in the visual field. The CPG in some sense remembers the position of the barrier. Pearson then asked the question: What if there is a delay after the forefeet step over? By offering food, he was able to interrupt the cat before its hindlegs had stepped over the barrier. It turns out that the hindlegs remember the position of the barrier for at least 10 minutes, even if the barrier is removed after the front feet have stepped over it. This is far longer than the memory of the image in the eye. These clear and simple results demonstrate the way in which information processed unconsciously through the brain influences "automatic" events.

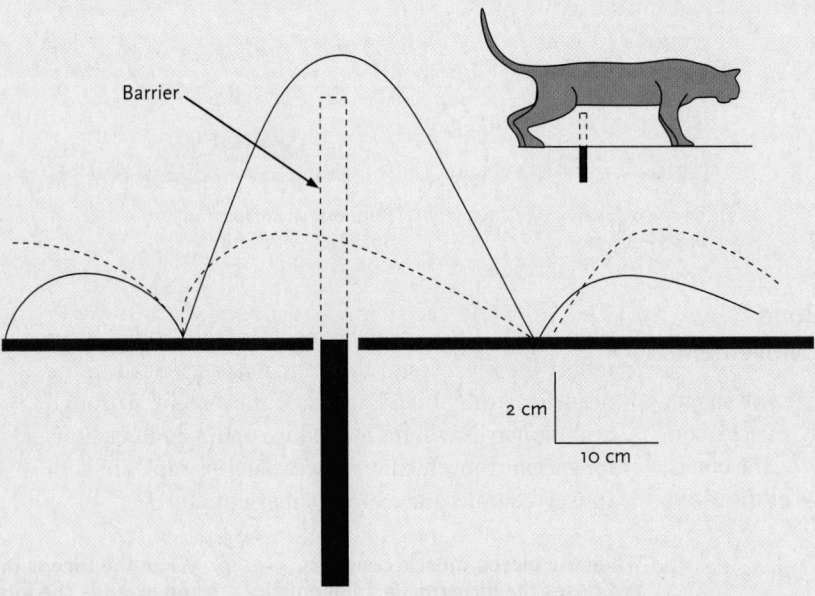

Figure 1

Experiment demonstrating that the locomotor CPG in the cat has a memory. The cat stepped over the barrier with the front feet but was then delayed by offering it food. The barrier, which it could no longer see, was removed. The solid line indicates the path taken by the hindfoot when walking resumed up to 10 minutes later. The dotted line indicates the path taken by the hindfoot when the cat does not encounter a barrier.

An example of an optimal strategy that has been conserved is seen in the undulatory movements of many vertebrate species. The segmental muscles of fish are attached to the vertebrae, ribs, and skin. They are efficient in propelling fish through the water with side-to-side movements. With the appearance of limbs and the movement onto land, this form of locomotion was combined with leg movement in most amphibians and reptiles but reappears as the dominant form of locomotion in snakes **(Figure 46.19).**

There are many examples of convergent evolution. Sea turtles, for instance, have paddlelike forelimbs that they use in swimming, very much like those of penguins and sea lions **(Figure 46.20).**

The development of flight, represented by birds and, among the mammals, bats, is accompanied by some common modifications **(Figure 46.21).** In both groups, but particularly in birds, the sternum is greatly enlarged to provide attachment for the powerful pectoral muscles. The bones of the forelimbs are greatly modified to produce wings. While this is another example of convergent evolution, note that the way in which each wing has formed is different. In bats, the bones of the hand are extended to provide a framework

a.

b.

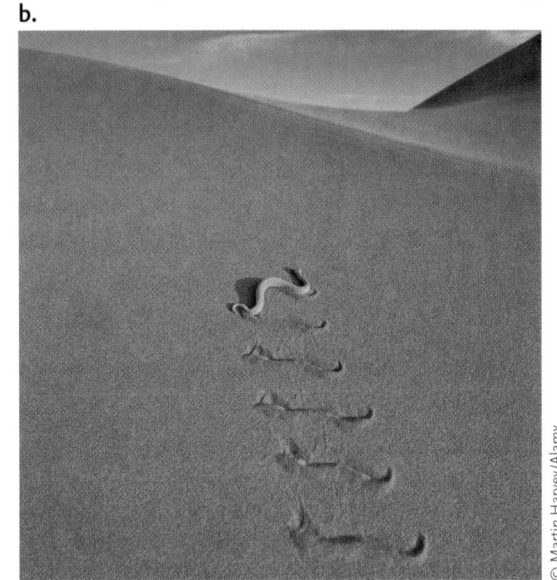

Figure 46.19
Almost all fish, amphibians, and reptiles depend on lateral undulations of the body wall for locomotion, but this is best seen in long, limbless species such as eels **(a)** and snakes **(b)**.

a.

b.

c.

Figure 46.20
Examples of convergent evolution. **(a)** A sea turtle. Note that the forelegs are winglike paddles that enhance swimming. **(b)** A penguin swimming, using modified forelimbs (wings). **(c)** A sea lion swimming using modified forelimbs for manoeuvring. Note the resemblance in shape of the forelimbs in all three species.

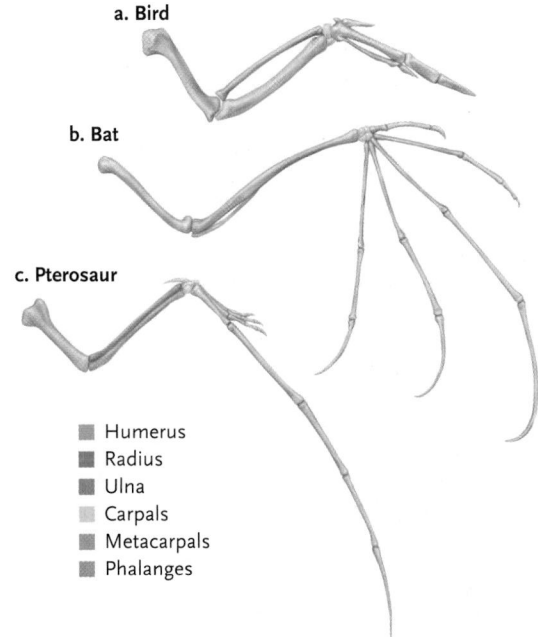

a. Bird

b. Bat

c. Pterosaur

- ■ Humerus
- ■ Radius
- ■ Ulna
- ☐ Carpals
- ■ Metacarpals
- ■ Phalanges

Figure 46.21
The bones of the wing of **(a)** a bird, **(b)** a bat, and **(c)** a pterosaur. Note the convergent evolution presented here in the structures.

for the sheets of skin that form the wing. In birds, the bones of the hand are fused and support the feathers that make up the wing. The bones generally contain many cavities or spaces, making the body lighter.

Also note that in both examples of convergent evolution presented here, the structures (wings in bats and birds; forelimb paddles in turtles, penguins, and sea lions) of the forelimb have diverged from that of other members of their classes. The bodies of animals are full of many similar examples, and these are the topic of comparative, functional anatomy.

STUDY BREAK

1. Distinguish synovial joints, cartilaginous joints, and fibrous joints.
2. Distinguish between ligaments and tendons.
3. What differences are there in the design of musculoskeletal systems for speed versus strength?
4. What are antagonistic muscle pairs? Give an example describing how they differ in the bone movements they produce.

Review

 To access course materials such as Aplia and other companion resources, please visit www.NELSONbrain.com.

46.1 Vertebrate Skeletal Muscle: Structure and Function

- Skeletal muscles move the joints of the body. They are formed from long, cylindrical cells called muscle fibres, which are packed with myofibrils, which are contractile elements consisting of myosin thick filaments and actin thin filaments. The two types of filaments are arranged in an overlapping pattern of contractile units called sarcomeres.

- Infoldings of the plasma membrane of the muscle fibre form T tubules. The sarcomeres are encircled by the sarcoplasmic reticulum, a system of vesicles with segments separated from T tubules by small gaps.

- In the sliding filament mechanism of muscle contraction, the simultaneous sliding of thin filaments on each side of sarcomeres over the thick filaments shortens the sarcomeres and the muscle fibres, producing the force that contracts the muscle.

- The sliding motion of thin and thick filaments is produced in response to an action potential arriving at the neuromuscular junction. The action potential causes the release of acetylcholine, which triggers an action potential in the muscle fibre that spreads over its plasma membrane and stimulates the sarcoplasmic reticulum to release Ca^{2+} into the cytosol. The Ca^{2+} combines with troponin, inducing a conformational change that moves tropomyosin away from the myosin-binding sites on thin filaments. Exposure of the sites allows myosin crossbridges to bind and initiate the crossbridge cycle in which the myosin heads of thick filaments attach to a thin filament, pull, and release in cyclic reactions powered by ATP hydrolysis.

- When action potentials stop, Ca^{2+} is pumped back into the sarcoplasmic reticulum, leading to Ca^{2+} release from troponin, which allows tropomyosin to cover the myosin-binding sites on the thin filaments, thereby stopping the crossbridge cycle.

- A single action potential arriving at a neuromuscular junction causes a muscle twitch. Restimulation of a muscle fibre before it has relaxed completely causes a second twitch, which is added to the first, causing a summed, stronger contraction. Rapid arrival of action potentials causes the twitches to sum to a peak level of contraction called tetanus (temporal summation). Normally, muscles contract in a tetanic mode.

- Vertebrate muscle fibres occur in three types. Slow muscle fibres contract relatively slowly but do not fatigue rapidly. Fast aerobic fibres contract relatively quickly and powerfully and fatigue more quickly than slow fibres. Fast anaerobic fibres can contract more rapidly and powerfully than fast aerobic fibres, but fatigue more rapidly. The fibres differ in their number of mitochondria and capacity to produce ATP.

- Skeletal muscles are divided into motor units, consisting of a group of muscle fibres activated by branches of a single motor neuron. The total force produced by a skeletal muscle is determined by the number of motor units that are activated (spatial summation).

- Most invertebrate muscles contain thin and thick filaments arranged in sarcomeres, and contract by the same sliding filament mechanism that operates in vertebrates. In arthropods, fast twitches and slower, graded contractions result from differences in innervation. Insect flight muscle is specialized to contract at high frequency.

46.2 Skeletal Systems

- A hydrostatic skeleton is a structure consisting of a muscle-surrounded compartment or compartments filled with fluid under pressure. Contraction and relaxation of the muscles change the shape of the animal.

- With an exoskeleton, a rigid external covering provides support for the body. The force of muscle contraction is applied against the covering. An exoskeleton can also protect delicate internal tissues.

- With an endoskeleton, the body is supported by rigid structures within the body, such as bones. The force of muscle contraction is applied against these structures. Endoskeletons also protect delicate internal tissues. In vertebrates, the endoskeleton is the primary skeletal system. The vertebrate axial skeleton consists of the skull, vertebral column, sternum, and ribcage, whereas the appendicular skeleton includes the shoulder bones, the forelimbs, the hip bones, and the hindlimbs.

- Bone tissue is distributed between compact bone, with no spaces except the microscopic canals of the osteons, and spongy bone tissue, which has spaces filled by red or yellow marrow.

- Calcium and phosphate ions are constantly exchanged between the blood and bone tissues. The turnover keeps the Ca^{2+} concentration balanced at optimal levels in body fluids.

46.3 Vertebrate Movement: The Interactions between Muscles and Bones

- The bones of a skeleton are connected by joints. A synovial joint, the most movable type, consists of a fluid-filled capsule surrounding the ends of the bones forming the joint. A cartilaginous joint, which is less movable, has smooth layers of cartilage between the bones, with no surrounding capsule. The bones of a fibrous joint are joined by connective tissue fibres that allow little or no movement.

- The bones moved by skeletal muscles act as levers, with a joint at one end forming the fulcrum of the lever, the load at the opposite end, and the force applied by attachment of a muscle at a point between the ends.

- At a joint, an agonist muscle, perhaps assisted by other muscles, causes movement. Most skeletal muscles are arranged in antagonistic pairs, in which the members of a pair pull a bone in opposite directions. When one member of the pair contracts, the other member relaxes and is stretched.

- Vertebrates have a variety of patterns in which muscles connect to bones, giving different properties to the levers produced. These properties are specialized for the activities of the animal.

Questions

Self-Test Questions

Any number of answers from a to e may be correct.

1. Which of the following is true for vertebrate skeletal muscle?
   a. may bend but not extend body parts
   b. is attached to bone by means of ligaments
   c. is usually attached at each end to the same bone
   d. may rotate one body part with respect to another
   e. is found in the walls of blood vessels and intestines

2. Which term refers to the connective tissue that joins bones together on either side of a joint?
   a. periosteum
   b. stratum corneum
   c. tendons
   d. ligaments
   e. meninges

3. Which of the following occurs in a resting muscle fibre?
   a. Sarcomeres are regions between two H zones.
   b. Z lines are adjacent to H zones, which attach thick filaments.
   c. I bands are composed of the same thick filaments seen in the A bands.
   d. Disks of M line proteins called the A band separate the thick filaments.
   e. Dark A bands contain overlapping thick and thin filaments with a central thin H zone composed only of thick filaments.

4. Which statement describes the sliding filament contractile mechanism?
   a. It lengthens the sarcomere to separate the I regions.
   b. It is inhibited by the influx of Ca^{2+} into the muscle fibre cytosol.
   c. It uses myosin crossbridges to stimulate delivery of Ca^{2+} to the muscle fibre.
   d. It depends on the isolation of actin and myosin until a contraction is completed.
   e. It causes thick and thin filaments to slide toward the centre of the A band, bringing the Z lines closer together.

5. During the crossbridge cycle, ATP does which of the following?
   a. causes the detachment of myosin from actin
   b. causes the bending of myosin crossbridges
   c. causes rigor mortis
   d. binds to troponin

6. Where is the Ca^{2+} that is directly involved in muscle contraction stored?
   a. in the T tubules
   b. in the cytosol
   c. in the sarcoplasmic reticulum
   d. in the extracellular fluid

7. Which is true for fast anaerobic muscle fibres?
   a. contain a high concentration of glycogen
   b. have few mitochondria
   c. have extensive vascularization
   d. are red in colour

8. In skeletal muscles, how is the force of contraction adjusted?
   a. by controlling the recruitment of motor units
   b. by controlling the speed of contraction
   c. by contracting antagonistic muscle pairs
   d. by tetanic contractions

9. Which of the following is NOT an example of a hydrostatic skeletal structure?
   a. the penis of mammals
   b. the body wall of annelids
   c. the tube feet of sea urchins
   d. the body wall of cnidarians
   e. the body wall of a grasshopper

10. Endoskeletons
    a. cannot be found in molluscs and echinoderms
    b. are composed of appendicular structures that form the skull
    c. are composed of the arms and legs, which are part of the axial skeleton
    d. differ from exoskeletons in that endoskeletons do not support the external body
    e. protect internal organs and provide structures against which the force of muscle contraction can work

Questions for Discussion

1. Not all snakes move by undulatory movement. Some use concertina movement for locomotion. What is this and how does it differ from undulatory movement? Why is it necessary?

2. If you were a leading researcher in a pharmaceutical company interested in controlling bone loss, what cells and processes would you target?

3. What are muscle cramps and what is their most likely cause?

4. Based on your knowledge of muscle contraction, explain what causes rigor mortis (stiffening of the body following death).

5. Phosphocreatine is a high-energy store for muscles. Why is energy stored in this fashion rather than as ATP? In your answer consider speed of availability and also refer back to Chapter 4 and the discussion of end-product feedback.

6. For a gram of tissue, muscle can be fast or strong but not both. Why? In your answer, consider what the need for mitochondria and sarcoplasmic reticulum do to the numbers of myofibrils contained in that gram of tissue and what the consequences of this would be.

47

M.B. Fenton

A Little Brown Bat (*Myotis lucifugus*) flies through an abandoned mine. Mouth open, the animal produces echolocation calls that, in this setting, allow it to orient through the underground space.

Animal Behaviour

WHY IT MATTERS

When it comes to food, many animals quickly learn to take advantage of new opportunities and show great versatility in behaviour from hunting to planning. Here are five examples.

During the Vietnam War (1959–1975), tigers (*Panthera tigris*) learned to associate the sound of gunfire with an opportunity to eat. The tigers' behaviour meant that some wounded soldiers waiting for treatment received a different kind of attention than what they expected. During World War II, wolves (*Canis lupus*) showed the same behaviour in some areas of Poland. A food reward is a strong reinforcer of behaviour.

In the 1970s, Kim McCleneghan and Jack Ames were studying sea otters (*Enhydra lutris*) in California waters. These otters dive and collect food (sea urchins, *Pisaster brevispinus,* and clams, *Saxidomus nuttalli*) from the bottom and bring their catch to the surface to eat it. The observers were surprised to see some otters resurfacing with empty beverage cans. These otters would lie on their backs in the ocean swells, take a can, bite it open, and, in some cases, remove and eat something before discarding the can. Some cans appeared to be empty and were discarded after opening. The biologists collected their own beverage cans and discovered that many harboured young octopods (*Octopus* species). Populations of these cephalopods are limited by the number of shelters available. Young octopods were exploiting new opportunities for shelter, and the sea otters, in turn, were taking advantage of the molluscs' behaviour.

Meanwhile, in savannah woodlands in Senegal (West Africa), Jill Pruetz and Paco Bertolani observed chimpanzees (*Pan troglodytes*) hunting bushbabies (*Galago senegalensis*). The fact that the chimps were not vegetarians was no surprise because they had been reported using grass stalks to fish for termites and working in gangs to hunt young baboons (*Papio ursinus*). The discovery that savannah chimps in Senegal used "spears" to impale bushbabies hidden in tree hollows extended the repertoire of chimps. Pruetz and Bertolani watched chimps modifying branches they had broken off by biting to sharpen them before using them against bushbabies. The chimps that Pruetz and Bertolani studied appeared to plan their hunts in advance.

Other experiments have revealed how Western Scrub Jays (*Aphelocoma californica*) cache food in preparation for the next day's breakfast. Proving that animals plan ahead means that the experiments have to demonstrate that the animal executes a novel action or combination of actions and anticipates an emotional state different from the one at the time of planning. These two conditions rule out behaviours associated with migration and hibernation or those associated with meeting an immediate need for food.

In foraging behaviour, animals exhibit an array of opportunism and adaptation that we often believe is the exclusive domain of *Homo sapiens*. The purpose of this chapter is to introduce you to the topic of **animal behaviour**.

The behaviour of animals is an excellent overall indication of their diversity. Behaviour reflects a rich blend of genetic and environmental control, as well as a combination of instinct and learning. Hormones exert a strong influence over an animal's behaviour, and the structure of the nervous system and, by extension, the sensory system, often stimulate and mediate behaviour. Some animals are territorial, defending all or part of their home range, usually in association with reproduction. Migration involves movements to and from different areas, usually between seasons, and implies the use of navigational cues. Animals show rich repertoires of behaviour around mating and reproduction. A few species live in large groups, and others exhibit complex social behaviour.

47.1 Genes, Environment, and Behaviour

Learning, as demonstrated by the foraging animals introduced above, illustrates how some behaviour patterns are acquired rather than inherited. But animal behaviourists had long debated whether animals are born with the ability to perform most behaviours completely or whether experience is necessary to shape their actions. Today, the emerging picture is that no behaviour is determined entirely by genetics or entirely by environmental factors. Rather, behaviours develop through complex gene–environment interactions.

Why do adult male White-crowned Sparrows sing a song that no other species sings **(Figure 47.1)**? They could have an innate (inborn) ability to produce their particular song, an ability so reliable that young males sing the "right" song the first time they try. According to this hypothesis, their distinctive song would be an **instinctive behaviour**, one genetically or developmentally "programmed" that appears in complete and functional form the first time it is used. An alternative hypothesis is that they acquire the song as a result of certain experiences, such as hearing the songs of adult male White-crowned Sparrows that live nearby. If so, this species' distinctive song might be an example of a **learned behaviour**, one that depends on having a particular kind of experience during development.

Figure 47.1

Songbirds and their songs. Sound spectrograms (visual representations of sound graphed as frequency versus time) illustrate differences in the songs of the White-crowned Sparrow (*Zonotrichia leucophrys*), Song Sparrow (*Melospiza melodia*), and Swamp Sparrow (*Melospiza georgiana*).

How can we determine which of these two hypotheses is correct? If the White-crowned Sparrow's song is instinctive, isolated male nestlings that have never heard other members of their species should be able to sing their species' song when they mature. If the learning hypothesis is correct, young birds deprived of certain essential experiences should not sing "properly" when they become adults.

Peter Marler tested these two hypotheses. He took newly hatched White-crowned Sparrows from nests in the wild and reared them individually in soundproof cages in his laboratory. Some of the chicks heard recordings of a male White-crowned Sparrow's song when they were 10 to 50 days of age, whereas others did not. Juvenile males in both groups first started to vocalize at about 150 days of age. For many days, the birds produced whistles and twitters that only vaguely resembled the songs of adults. Gradually, the young males that had listened to tapes of their species' song began to sing better and better approximations of that song. At about 200 days of age, these males were right on target, producing a song that was nearly indistinguishable from the one they had heard months before. Captive-raised males that had not heard recordings of White-crowned Sparrow songs never sang anything close to the songs typical of wild males.

These results show that learning is essential for a young male White-crowned Sparrow to acquire the full song of its species. Although birds isolated as nestlings sang instinctively, they needed the acoustical experience of listening to their species' song early in life if they were to reproduce it months later. These data allow us to reject the hypothesis that White-crowned Sparrows hatch from their eggs with the ability to produce the "right" song. Their species-specific song, and perhaps the songs of many other songbirds, include both instinctive and learned components.

Early researchers generally classified behaviours as either instinctive or learned, but we now know that most behaviours include both instinctive and learned components. Nevertheless, some behaviours have a stronger instinctive component than others.

STUDY BREAK

1. How can the study of bird song lend itself to understanding the influence of genes on behaviour?
2. How is behaviour learned?

47.2 Instinct

Instinctive behaviours can presumably be performed without the benefit of previous experience. They can be grouped into functional categories, such as feeding, defence, mating, and parental care. We assume that they have a strong genetic basis and that natural selection has preserved them as adaptive behaviours.

Many instinctive behaviours are highly stereotyped. When an animal is triggered by a specific cue, it performs the same response over and over in almost exactly the same way. These **fixed action patterns** are triggered by **sign stimuli.** Very young Herring Gull chicks use a begging response **(Figure 47.2a),** a fixed action pattern, to secure food from their parents. Begging chicks peck at the red spot on the parent's bill, and the tactile stimulus serves as a sign stimulus inducing the adult to regurgitate food from its crop. Baby gulls eat the chunks of fish, clams, or other food that have been regurgitated for them. We know that the spot on the parent's bill releases the begging response of the young gull because the same response is triggered by an artificial bill that looks only vaguely like an adult bill, provided that it has a dark contrasting spot near the tip (Figure 47.2b). Simple cues can activate fixed action patterns.

Human infants often respond innately to the facial expressions of adults **(Figure 47.3).** Researchers can

a.

Marie Read Natural History Photography

b.

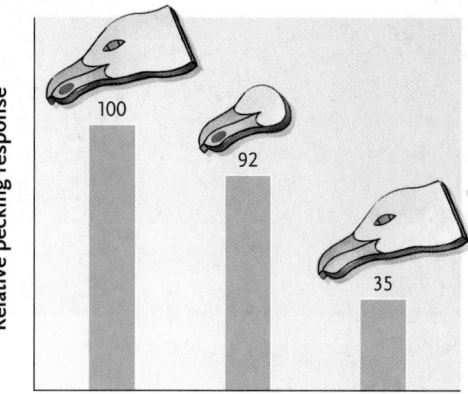

Model presented

Experimental Research Figure 47.2 (a) A Herring Gull (*Larus argentatus*) chick begs its parent for food. **(b)** Nestling Herring Gulls also begged when presented with various models of an adult gull. In the experiment, nestlings pecked at a model with a red spot on the lower jaw almost as often as they did to a real gull. A model lacking the jaw spot elicited much fewer begging pecks from the nestling.

Figure 47.3

Instinctive responses in humans. The smiling face of an adult is a sign stimulus that triggers smiling in very young infants.

Figure 47.4

This European Cuckoo (*Cuculus canorus*) is a brood parasite that stimulates food delivery by its foster parent, a Hedge Sparrow (*Prunella modularis*). The cuckoo elicits food delivery by displaying exaggerated versions of the sign stimuli used by the host offspring. The exaggerated stimuli are *releasers*, initiating the appropriate behaviour from a parent with food.

trigger smiling in even very young babies simply by moving a mask toward the infant, as long as the mask has two simple, diagrammatic eyes. Clearly, the infant, like a nestling Herring Gull, is not reacting to every feature of a face but rather to simple cues that function as sign stimuli releasing a fixed behavioural response.

Natural selection has moulded the behaviour of some parasitic species to exploit the relationship between sign stimuli and fixed action patterns for their own benefit. In effect, they have broken another species' code. Birds that are brood parasites lay their eggs in the nests of other species. When the brood parasite's egg hatches, the alien nestling mimics sign stimuli ordinarily exhibited by their hosts' own chicks. The parasitic chick begs for food by opening its mouth, bobbing its head, and calling more vigorously than the host's chicks. These exaggerated behaviours elicit feeding by the foster parents, and the young brood parasite often receives more food than the hosts' own young **(Figure 47.4)**.

Although instinctive behaviours are often performed completely the first time an animal responds to a stimulus, they can be modified by an individual's experiences. The fixed action patterns of a young Herring Gull change over time. Although the youngster initially begs by pecking at almost anything remotely similar to an adult gull's bill, it eventually learns to recognize the distinctive visual and vocal features associated with its parents. The chick uses this information to become increasingly selective about which stimuli elicit its begging behaviour. During their early performances, instinctive behaviours can be modified in response to particular experiences.

Behavioural differences between individuals may reflect genetic differences because performance of instinctive behaviours does not depend on previous experience. Stevan Arnold studied innate responses of captive newborn garter snakes to olfactory stimuli

provided by potential food items they had never before encountered. Arnold measured the snakes' responses to cotton swabs that had been dipped in a smelly extract of banana slug, a shell-less mollusc. Young snakes born to a mother captured in coastal California, where adult garter snakes regularly eat banana slugs, almost always began tongue-flicking at slug-scented cotton swabs **(Figure 47.5)**. Newborn snakes whose parents came from central California, where banana slugs do

a. Banana slug

b. Adult coastal garter snake eating a banana slug

c. Newborn garter snake "smelling" slug extract

Figure 47.5

Genetic control of food preference. **(a)** Banana slugs (*Ariolimax columbianus*) are a preferred food of **(b)** an adult garter snake (*Thamnophis elegans*) from coastal California. **(c)** A newborn snake from a coastal population of garter snakes flicks its tongue at a cotton swab drenched with tissue fluids from a banana slug.

not occur, rarely tongue-flicked at the swabs. Although the coastal and inland snakes belong to the same species, their instinctive responses to banana slug chemicals differed markedly.

Arnold then tested whether newborn snakes would eat bite-sized chunks of slug. After a brief flick of the tongue, 85% of newborn snakes from a coastal population routinely struck at the piece of slug and swallowed it even though they had no previous experience with this food. Even when no other food was available, only 17% of newborn snakes from the inland population consistently tongue-flicked at or ate pieces of slug. Arnold hypothesized that coastal and inland garter snakes have different alleles at one or more gene loci controlling their odour-detection mechanisms and leading to differences in their behaviour. Arnold crossbred coastal and inland snakes. If genetic differences contribute to the food preferences of snakes from the two populations,

hybrid offspring receiving genetic information from each parent should behave in an intermediate fashion. The results of the experiment confirmed the prediction. When presented with bite-sized chunks of slug, 29% of the newborn snakes of mixed parentage ate them every time.

Many other experiments have confirmed that genetic differences between individuals can translate into behavioural differences between them (see "Knockouts: Genes and Behaviour"). Bear in mind, however, that single genes do not control complex behaviour patterns directly. Rather, the alleles determine the kinds of enzymes that cells can produce, influencing biochemical pathways involved in the development of an animal's nervous system. The resulting neurological differences can translate into a behavioural difference between individuals that have certain alleles and those that do not.

Knockouts: Genes and Behaviour

Almost all eukaryotic organisms share a series of developmental interactions called the *wingless/Wnt* pathway. The name comes from the original discovery of the pathway in the fruit fly *Drosophila melanogaster*, in which mutant genes of the pathway cause alterations in the wings and other segmental structures. Recently, three genes closely related to *disheveled*, one of the genes of the *Drosophila wingless/Wnt* pathway, were isolated from and identified in mice. No functions have yet been identified for the proteins encoded in the three mouse *disheveled* genes, but they are highly active in both embryos and adults. Their function must be important, but what could it be?

Nardos Lijam and his coworkers sought an answer to this question by developing a line of mice that totally lacked one of the *disheveled* genes, *Dvl1* in genetic shorthand. First, they constructed an artificial copy of the *Dvl1* gene with the central section scrambled so that no functional proteins could be made from its encoded directions. Next, they introduced the artificial gene into embryonic mouse cells. Cells that successfully incorporated the gene

were injected into very early mouse embryos. Some mice grown from these embryos were heterozygotes, with one normal copy of the *Dvl1* gene and one nonfunctional copy. Interbreeding of the heterozygotes produced some individuals that carried two copies of the altered *Dvl1* gene and no normal copies. Individuals lacking the normal gene are called *knockout* mice for the missing gene.

Surprisingly, knockout mice grew to maturity with no apparent morphological defects in any tissue examined, including the brain. Their motor skills, sensitivity to pain, cognition, and memory all appeared to be normal. However, their social behaviour was different. In cages with normal mice, the knockouts failed to take part in the common activities of mouse social groups: social grooming, tail pulling, mounting, and sniffing. Although normal mice build nests and sleep in huddled groups, knockouts tended to sleep alone, without constructing full nests from cage materials. Mice heterozygous for the *Dvl1* gene (those with one normal and one altered copy of the gene) behaved normally in all of these social activities.

The knockout mice also jumped around wildly in response to an abrupt, startling sound, whereas the response of normal mice was less extreme. A neural circuit of the brain inhibits the startle response of normal mice, so the reaction of knockout mice suggested that this inhibitory circuit was probably altered. Humans with schizophrenia, obsessive-compulsive disorders, Huntington disease, and some other brain dysfunctions also show an intensified startle reflex similar to that of the *Dvl1* knockout mice.

The researchers' analysis revealed that the *Dvl1* gene modifies developmental pathways affecting complex social behaviour in mice and probably in other mammals. It is one of the first genes identified that affects mammalian behaviour. The similarity in startle reflex intensity between the knockout mice and humans with neurological or psychiatric disorders suggests that mutations in the *Dvl* genes and the *wingless* developmental pathway may underlie some human mental illnesses. If so, further studies of the *Dvl* genes may give us clues to the molecular basis of these diseases and a possible means to their cure.

1. What are the differences between instinctive and learned behaviours?
2. What are fixed action patterns and sign stimuli?
3. How did experiments with garter snakes demonstrate the influence of genetics on behaviour?

47.3 Learning

Unlike instinctive behaviours, learned behaviours are not performed accurately or completely the first time an animal responds to a specific stimulus. They change in response to environmental stimuli that an individual experiences as it develops. Behavioural scientists generally define learning as a process in which experiences change an animal's behavioural responses. Different types of learning occur under different environmental circumstances.

Imprinting occurs when animals learn the identity of a caretaker or the key features of a suitable mate during a **critical period**, a stage of development early in life. Newly hatched geese imprint on their mother's appearance and identity, staying near her for months. When they reach sexual maturity, young geese try to mate with other geese exhibiting the visual and behavioural stimuli on which they had imprinted as youngsters. When Konrad Lorenz, a founder of **ethology** (the study of animal behaviour), tended a group of newly hatched Greylag Geese, they imprinted on him rather than on an adult of their own species **(Figure 47.6)**. Male geese not only followed Lorenz, but at sexual maturity, they also courted humans.

Other forms of learning can occur throughout an animal's lifetime. Ivan Pavlov, a Russian physiologist, demonstrated **classical conditioning** in experiments with dogs. Like many other animals, dogs developed a mental association between two phenomena that are usually unrelated. Dogs typically salivate when they eat. Food is an *unconditioned stimulus* because the dogs instinctively respond to it and do not need to learn to salivate when presented with food. Pavlov rang a bell just before offering food to dogs. After about 30 trials in which dogs received food immediately after the bell rang, the dogs associated the bell with feeding time and drooled profusely whenever it rang, even when no food was forthcoming. The bell had become a *conditioned stimulus,* one that elicited a particular learned response. In classical conditioning, an animal learns to respond to a conditioned stimulus (e.g., the bell) when it precedes an unconditioned stimulus (e.g., food) that normally triggers the response (e.g., salivation). If your pet cat becomes exceptionally friendly whenever it hears the sound of a can opener, its behaviour is the result of classical conditioning.

Operant conditioning, trial-and-error learning, is another form of associative learning. Here animals learn to link a voluntary activity, an **operant,** with its favourable consequences, a **reinforcement.** A laboratory rat will explore a new cage randomly. If the cage is equipped with a bar that releases food when it is pressed, the rat eventually leans on the bar by accident (the operant) and immediately receives a morsel of food (the reinforcement). After a few such experiences, a hungry rat learns to press the bar in its cage more frequently, provided that the bar-pressing behaviour is followed by access to food. Laboratory rats have also learned to press bars to turn off disturbing stimuli, such as bright lights.

Insight learning occurs when an animal can abruptly learn to solve problems without apparent trial-and-error attempts at the solution. Captive chimpanzees solved a novel problem: how to get bananas hung far out of reach. The chimps studied the situation and then stacked several boxes, stood on them, and used a stick to knock the fruit to the floor.

Habituation occurs when animals lose their responsiveness to frequent stimuli not quickly followed by the usual reinforcement. Habituation can save the animal the time and energy of responding to stimuli that are no longer important. Sea hares (*Aplysia* species) are shell-less molluscs that typically retract their gills when touched on the side. Gill retraction helps protect sea hares from approaching predators. But a sea hare stops retracting its gills when it is touched repeatedly over a short period of time with no harmful consequences.

Learning occurs across different time frames, from habituation (above) to lifelong changes in behaviour. Cross-fostering experiments with two songbirds, Blue Tits (*Cyanistes caeruleus*) and Great Tits (*Parus major*), revealed that early learning is essential to the realization of ecological niches. In this work, Tore Slagsvold and Karen Wiebe transferred fertilized eggs from the nests of Great Tits to those of Blue Tits and vice versa. Compared with their genetic parents, the fostered young shifted their feeding niches, and the shift was lifelong. The changes in foraging behaviour were greater for fostered Great Tits, the species with more specialized foraging behaviour.

Figure 47.6

Imprinting. Having imprinted on him shortly after hatching, young Greylag Geese (*Anser anser*) frequently joined Konrad Lorenz for a swim.

47.4 Neurophysiology and Behaviour

Research in neuroscience has shown that all behavioural responses, whether mostly instinctive or mostly learned, depend on an elaborate physiological foundation provided by the biochemistry and structure of the nerve cells. Nerve cells that regulate an innate response and make it possible for an animal to learn something are products of a complex developmental process. Here genetic information and environmental contributions are intertwined. Although the anatomical and physiological basis for some behaviours is present at birth, an individual's experiences alter the cells of its nervous system in ways that produce particular patterns of behaviour.

Marler's experiments helped explain the physiological underpinnings of singing behaviour in male White-crowned Sparrows. If acoustical experience shapes singing, a sparrow chick's brain must be able to acquire and store information present in the songs of other males. Then, months later, when the young male starts to sing, its nervous system must have special features enabling the bird to match its vocal output to the stored memory of the song that it heard earlier. Eventually, when it achieves a good match, the sparrow's brain must "lock" on the now complete song and continue to produce it when the bird sings.

Additional experiments demonstrated that when young birds did not hear a taped song during their critical period (10–50 days of age), they never produced the full song of their species, even if they heard it later in life. In addition, young birds that heard recordings of *other* bird species' songs during the critical period never generated replicas of those songs as they matured. These and other findings suggested that certain nerve cells in the young male's brain are influenced only by appropriate stimuli, in this case, acoustical signals from individuals of its own species, and only during the critical period. Neuroscientists have identified nuclei, clusters of nerve cells, that make song learning and song production possible.

Every behavioural trait appears to have its own neural basis. Another songbird, a male Zebra Finch **(Figure 47.7)**, can discriminate between the songs of strangers and those of established neighbours. These finches live in **territories** (see Section 47.8), plots of land defended by individual males or breeding pairs. Defence of the territory ensures that the residents have exclusive access to food and other necessary resources.

Figure 47.7
Zebra Finches (*Taeniopygia guttata*) are native to Indonesia. They have played an important role in studies of the physiological basis of song learning. The male has the striped throat.

Zebra Finches' ability to discriminate between the songs of neighbours and those of strangers involves a nucleus in the forebrain. Cells in this nucleus fire frequently the first time the Zebra Finch hears the song of a new conspecific. As the song is played again and again, the cells of this nucleus cease to respond, indicating that the bird has become habituated to a now familiar song. The same bird still reacts to the songs of strangers. Neurophysiological networks that make this selective learning possible enable male Zebra Finches to behave differently toward familiar neighbours, which they largely ignore, than they do to unfamiliar singers, which they attack and drive away.

Molecular and cellular techniques have been used to identify the role of genes in learning. When a bird is exposed to relevant acoustical stimuli, such as songs of potential rivals, certain genes are quickly "turned on" within neurons in the song-controlling nuclei of the bird's brain. When a Zebra Finch hears the elements of its species' song, a gene called *zenk* rapidly becomes active in the brain, producing an enzyme that changes the structure and function of neurons. The ZENK enzyme programs nerve cells of the bird's brain to anticipate key acoustical events of potential biological importance. When they occur, these events trigger additional changes in the bird's brain, affecting its actions. In this way, a territory owner learns to ignore (= habituate to) a singing neighbour with which it shares an established territorial boundary. The same bird retains the ability to detect and respond to new intruding conspecifics because they are a real threat to its continued control of its territory.

STUDY BREAK

1. What role does the ZENK enzyme play?
2. How do nerve connections influence behaviour?

47.5 Hormones and Behaviour

Hormones are chemical signals that can trigger the performance of specific behaviours. Hormones often work by regulating the development of neurons and neural networks or by stimulating cells within endocrine organs to release chemical signals.

How did the neurons in an adult Zebra Finch acquire the remarkable capacity to change in response to specific stimuli? In Zebra Finches, only males produce courtship songs. Very early in its life, certain cells in the brain of a male songbird produce estrogen, which affects target neurons in the higher vocal centre, an area of the developing brain. Estrogen leads to a complex series of biochemical changes resulting in the production of more nerve cells in the parts of the brain that regulate singing. Brains of developing females do not produce estrogen. In the absence of estrogen, the number of neurons in the higher vocal centre of females *declines* over time **(Figure 47.8)**. If young female Zebra Finches are given estrogen, they produce more nerve cells in the higher vocal centre and are capable of singing. Specific stimuli, such as the songs of familiar or unfamiliar males, can alter the genetic activity of the nerve cells that control the behaviour of adult birds.

Just as estrogen influences the development of singing ability in Zebra Finches, other hormones mediate the development of the nervous system in other species. A change in the concentration of a certain hormone can be the physiological trigger that induces important changes in an animal's behaviour as it matures.

As they age, worker honeybees perform different tasks. Nurse bees, typically less than 15-day-old adults, tend to care for larvae and maintain the hive. Forager bees, typically more than 15-day-old adults, often make foraging excursions from the hive to collect food, nectar, and pollen **(Figure 47.9)**. These behavioural changes are induced by rising concentrations of juvenile hormone (see Chapter 43) released by a gland near the bee's brain. Despite its name, circulating levels of juvenile hormone actually increase as a honeybee ages.

Juvenile hormone may exert its effect on bee behaviour by stimulating genes in certain brain cells to produce proteins that affect nervous system function. Octopamine, for example, stimulates neural transmissions and reinforces memories. Octopamine is concentrated in the antennal lobes, parts of the bee's brain that contribute to the analysis of chemical scents in the external environment. Octopamine is found at higher concentrations in older, foraging bees that have higher levels of juvenile hormone. When extra juvenile hormone is experimentally administered to bees, their production of octopamine increases. Increased octopamine levels in the antennal lobes may help a foraging

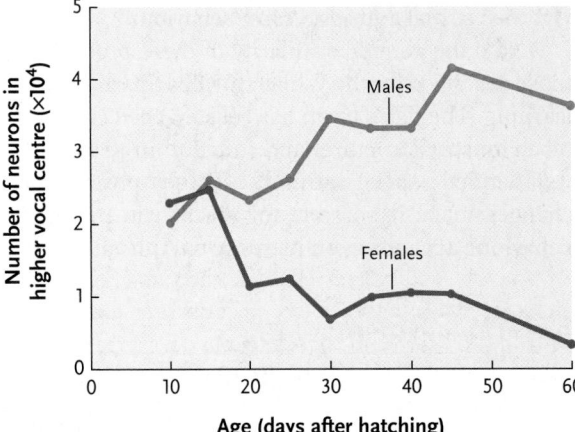

Figure 47.8
Hormonally induced changes in brain structure. The brains of young male Zebra Finches secrete estrogen, which stimulates production of additional neurons in the higher vocal centre. Lacking estrogen, young female Zebra Finches have fewer neurons in this region of the brain.

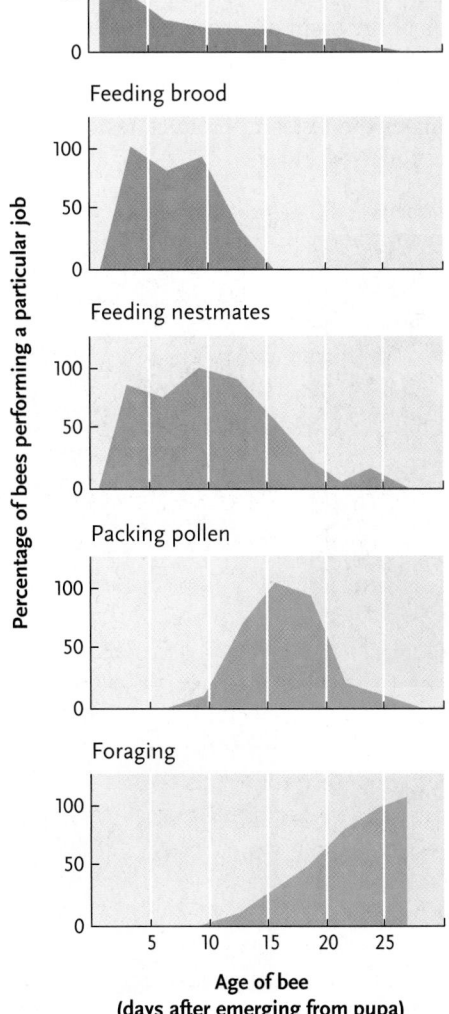

Figure 47.9
Age and task specialization in honeybee (*Apis mellifera*) workers. Newly emerged adult bees (nurses) typically clean cells and feed the brood, whereas older workers (foragers) leave the hive to forage for food.

bee home in on the odours of flowers from which it can collect nectar and pollen.

The honeybee example illustrates how genes and hormones interact in the development of behaviour. Genes code for the production of hormones that become part of the intracellular environment of assorted target cells. Hormones then directly or indirectly change genetic activity and enzymatic biochemistry in their targets. When the target cells are neurons, changes in biochemistry translate into changes in the animal's behaviour.

The African cichlid fish illustrates how hormones regulate reproductive behaviour. Some adult males maintain nesting territories on the bottom of Lake Tanganyika in East Africa **(Figure 47.10)**. Territory holders are relatively brightly coloured and exhibit elaborate behavioural displays to attract egg-laden females. These males defend their real estate aggressively against neighbouring territory holders and against incursions by males without territories of their own. Nonterritorial males (called *drifters*) are much less colourful and aggressive and do not control a patch of suitable nesting habitat. They make no effort to court females.

Differences in levels of GnRH (gonadotropin-releasing hormone; see Chapter 43) cause behavioural differences between the two types of males. In the hypothalamus of the brain of territorial males, large, biochemically active cells produce GnRH. The same cells in the brains of drifters are small and inactive. GnRH stimulates the testes to produce testosterone

and sperm. When circulating sex hormones are carried to the brain of the fish, they modulate the activity of nerve cells that regulate sexual and aggressive behaviour. In the absence of GnRH, male fish do not court females or attack other males.

What causes the differences in the neuronal and hormonal physiology of the two types of male fish? Russell Fernald and his students manipulated the territorial status of males. Some territorial males were changed into nonterritorial males and vice versa, whereas the territorial status of other males was left unchanged as a control. Four weeks after the changes, Fernald and his students compared experimental and control fishes. They considered coloration and behaviour, as well as the size of the GnRH-producing cells in the brains. Territorial males that had been changed to nonterritorial males quickly lost their bright colours and stopped being combative. Moreover, their GnRH-producing cells were smaller than those of the territory-holding controls. Conversely, males that gained a territory in the experiment quickly developed bright colours and displayed aggressive behaviours toward other males. GnRH-producing cells in their brains were larger than those of fish that had maintained their status as non-territory-holding controls.

This example shows that what is happening inside a fish affects its environmental situation—its success or failure at gaining and holding a territory. Fish can detect and store information about their aggressive interactions. Neurons that process this information transmit their input to the hypothalamus, where it affects the size of cells producing GnRH, in turn dictating the hormonal state of the male. A decrease in GnRH production can turn a feisty territorial male into a subdued drifter. Drifters bide their time and build energy reserves for a future attempt at defeating a weaker male and taking over his territory. If successful in regaining territorial status, the male's GnRH levels will increase again. The once peaceful male reverts to vigorous sexual and aggressive behaviour.

Note the general similarity of these processes to those described for the White-crowned Sparrow's song learning. The fish's brain has cells that can change its biochemistry, structure, and function in response to well-defined social stimuli. These physiological changes make it possible for the fish to modify its behaviour depending on its social circumstances.

African cichlid fish (*Haplochromis burtoni*)

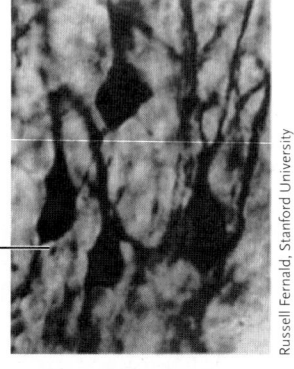

Experimental Research Figure 47.10

Photograph compares a nonterritorial male *Haplochromis burtoni* (top) with a territorial one. Photomicrographs at right compare gonadotropin-releasing hormone (GnRH) cells in the corresponding nonterritorial and territorial males.

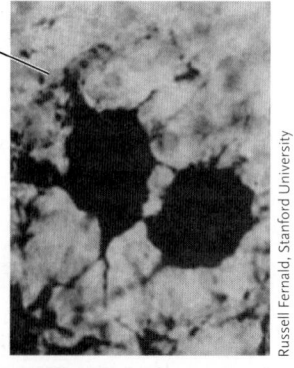

STUDY BREAK

1. How has research on White-crowned Sparrows and Zebra Finches advanced our knowledge of the impact of neurobiology on behaviour?
2. How does juvenile hormone affect the behaviour of adult bees?
3. How does GnRH affect the behaviour of fish?

47.6 Neural Anatomy and Behaviour

Some specific behaviours are produced by anatomical structures in an animal's nervous system. The nervous systems of many animal species allow them to respond rapidly to key stimuli. Sensory systems are often structured to acquire a disproportionately large amount of information about the stimuli that are most important to survival and reproductive success.

Important information acquired by the senses can be relayed directly to motor neurons, for example, providing prey animals with behaviour that can save them from a predator's attack. Insects such as crickets that fly mainly at night avoid predatory birds that typically fly by day; however, flying at night exposes them to attacks by insectivorous bats.

Insectivorous bats hunting at night use echolocation to detect and track flying prey (see Chapter 45). The echolocation calls of bats hunting flying insects are usually intense, with a sound pressure level of about 130 dB at 10 cm, making the calls stronger than the sound of a smoke detector alarm. The bats' calls can cover frequencies from ~10 kHz to >200 kHz, well beyond the 20 kHz upper limit of human hearing. By comparing its calls with the echoes from its calls, the bat uses echolocation to detect, assess, and track its flying prey. However, a bat's echolocation calls give crickets (and other prey; see Chapter 45) warning of their approach (see "Echolocation: Communication," p. 1164).

With ears on their front legs, black field crickets hear bat echolocation calls (see Figure 45.9, Chapter 45), and the anatomical structure of the cricket's nervous system produces a behavioural response that takes the cricket out of harm's way. Sensory neurons connected to the ears fire in response to the bat's calls, and the information is immediately translated into evasive action. When a bat attacks from the cricket's right side, the right ear receives a stronger stimulation than the left ear. The cricket's nervous system relays incoming messages from the *right* ear to the motor neurons controlling the *left* hindleg. Sufficient stimulation on the right side induces firing by motor neurons for the left hindleg, causing the leg to jerk up. This, in turn, blocks the movement of the left hindwing and reduces the flight power generated on the left side of the cricket's body. These changes cause the flying cricket to swerve sharply to the left and lose altitude, effectively diving down and away from the approaching bat **(Figure 47.11)**.

The structure and neural connections of sensory systems allow some animals to distinguish potentially life-threatening situations from more mundane stimuli. Fiddler crabs live and feed on mud flats, where they dig burrows that provide safe refuge from predators such as crab-hunting shorebirds. By distinguishing between predatory gulls and other fiddler crabs, a crab can use its burrow to best advantage and does not dash

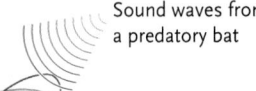

A flying cricket usually holds its hind legs close to the body so that they don't get in the way of its wings.

When a cricket hears the ultrasonic call of a bat coming from its left side, it automatically lifts its right hind leg.

The raised leg interrupts the right wing's movement, causing the insect to swerve down to the right and away from the approaching predator.

Figure 47.11
A neural mechanism for escape behaviour in the black field cricket (*Teleogryllus oceanicus*).
Mike May, "Aerial Defense Tactics of Flying Insects," *AMERICAN SCIENTIST* 79: 316–328.

for cover whenever anything moves in its field of vision.

Fiddler crabs have long-stalked eyes held above their **carapaces** and perpendicular to the ground **(Figure 47.12)**. John Layne wondered whether a crab might use a divided field of view to distinguish dangerous predators from fellow crabs. An approaching large gull would stimulate receptors on the upper part of the eye, whereas another crab's movements would be slightly below the midpoint of the eyes. A split field of view would allow the crab to distinguish between the two kinds of stimuli. To achieve this, receptors above and below the retinal equator must relay signals to different groups of neurons, effectively wiring the crab's nervous system to distinguish for a split field of view. If this were the case, stimulation of receptors above the midline of the eye would activate neurons controlling an escape response, triggering a dash for the burrow. A moving stimulus at or below eye level would stimulate a different response. Responses to other crabs are likely to be gender dependent.

To explore this, Layne placed crabs one at a time in a glass jar on an elevated platform. He presented a black square to each crab from two different heights.

Figure 47.12
A fiddler crab, *Uca pugilator*.

Echolocation: Communication

In echolocation, the echolocator stores the outgoing signal in its brain for comparison with returning echoes. The difference between what the animal "says" and what it hears is the data used in echolocation. However, when an echolocating bat (see the photo at the beginning of the chapter) or dolphin produces echolocation signals, the signals can also be heard by other animals.

When the bat or dolphin is foraging, potential prey (certain insects for the bat; certain fish for the dolphin) hear the signals and flee from the sound source (= negative phonotaxis) in an effort to evade the approaching predator. When the bat is close (strong echolocation signals), moths with ears dive to the ground or go into erratic flight to evade the bats. Moths with ears sensitive to bat echolocation calls avoid bat attacks 40% of the time. Insects lacking bat detectors are caught at much higher rates, sometimes >90% of the time. Acoustic warfare between bats and insects entertains biologists, involving measures and countermeasures by both predator and prey.

The same echolocation calls that alert potential prey are also available to any other animals within earshot, provided that their ears are sensitive to the frequencies in the signals. Little Brown Bats may use feeding buzzes (signals associated with attacks on prey) to locate concentrations of prey.

Spotted bats (*Euderma maculatum*) either approach a calling conspecific, apparently to chase it away, or turn and leave the area. Resident killer whales (*Orcinus orca*) in the Pacific Ocean off the west coast of Canada typically use echolocation to detect, track, and locate the salmon they eat. Transient killer whales in the same area feed mainly on marine mammals. These killer whales rarely echolocate. Local marine mammals, such as seals, quickly leave the water when they hear killer whales approaching.

The study of echolocation is a rich source of information about signals, signal design, hearing systems, and behaviour.

Sometimes the stimulus circled the jar above the crab's eyes; sometimes it circled below them. Stimuli activating the upper part of the retina induced escape behaviour, whereas those below the retinal equator were usually ignored (**Figure 47.13**). Specific nervous system connections between a fiddler crab's eyes and brain provide appropriate responses to different specific stimuli.

The match between the structure of an animal's nervous system and the real-world challenges it faces extends beyond the ability to avoid predators. Star-nosed moles live in wet tunnels in North American marshlands and spend almost all of their lives in complete darkness. Like nocturnal insect-eating bats, star-nosed moles must find food without the benefit of visual cues. Like the bats, its receptor–perceptual system enables it to feed effectively. A star-nosed mole eats mainly earthworms it locates with its nose, but not by smell. As the mole proceeds down its tunnel, 22 fingerlike tentacles on its nose sweep the area directly ahead of it. Each tentacle is covered with thousands of Eimer's organs (touch receptors; **Figure 47.14**). Sensory nerve terminals in Eimer's organs generate complex and detailed patterns of signals about the objects they contact. These messages are relayed by neurons to the cortex of the mole's brain, much of which is devoted to the analysis of information received from the nose's touch receptors.

The structural basis of the mole's sensory analysis is reflected by the amount of brain tissue responding to signals from its nose. The mole's brain contains many more cells decoding input from Eimer's glands than the combined input received from all other parts of the animal's body (see Figure 47.14b). Moreover, the brain does not treat inputs from all 22 of the mole's "nose fingers" equally. Instead, the brain devotes more cells to input from tentacles closest to the mouth. Fewer cells analyze messages from those farther away. Processing tactile information by star-nosed moles is related to the importance of finding food in dark, underground tunnels. Moreover, the extra attention given to signals from certain tentacles helps the star-nosed mole locate prey that are close to its mouth, in turn allowing it to feed more efficiently (see Figure 45.25, Chapter 45).

Animals' nervous systems do not offer neutral and complete pictures of the environment. Instead, the

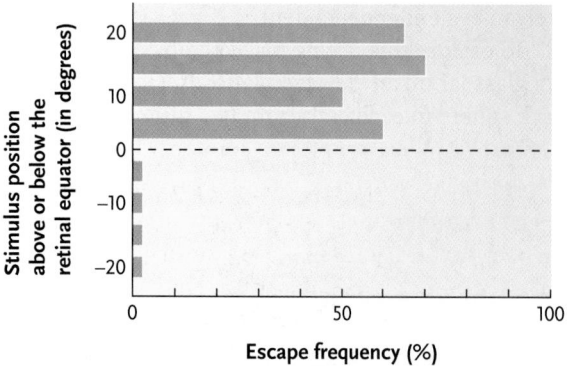

Experimental Research Figure 47.13 Stimuli that activated the upper part of the retinas of *Uca pugilator* elicited escape behaviour much more often than those activating the lower retinas.

a. Sensory organs on the tentacle of a star-nosed mole

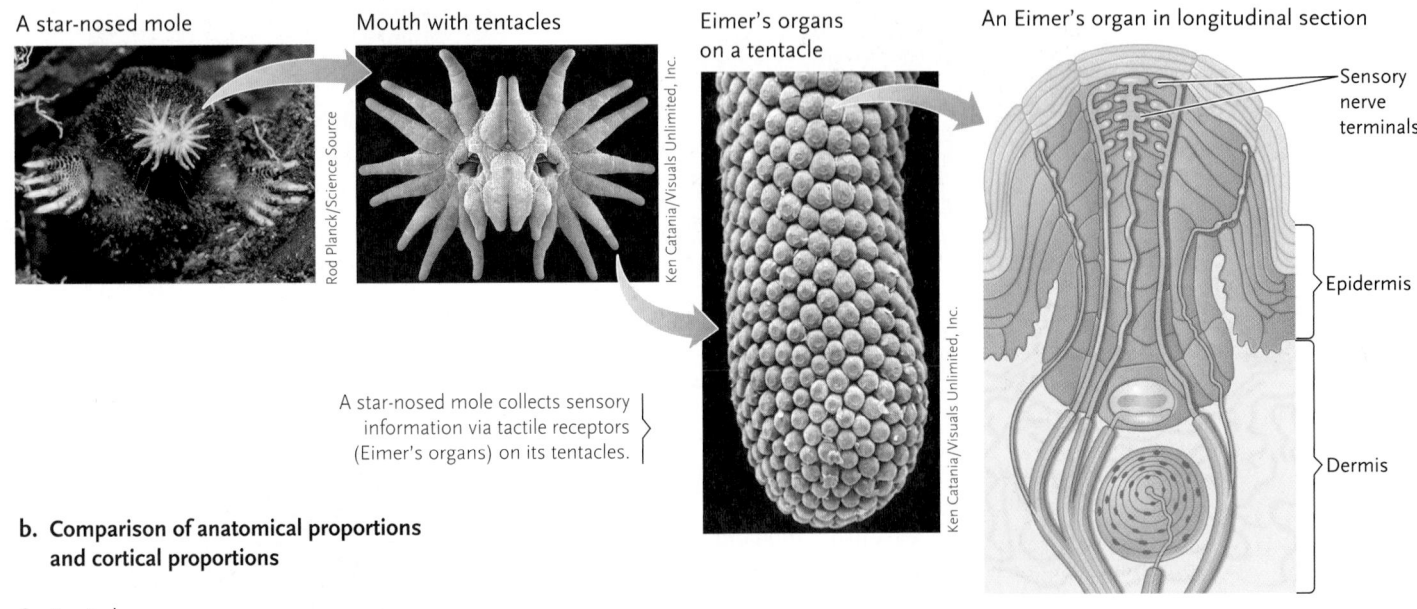

A star-nosed mole

Mouth with tentacles

Eimer's organs on a tentacle

An Eimer's organ in longitudinal section

Rod Planck/Science Source

Ken Catania/Visuals Unlimited, Inc.

Ken Catania/Visuals Unlimited, Inc.

Sensory nerve terminals

Epidermis

Dermis

A star-nosed mole collects sensory information via tactile receptors (Eimer's organs) on its tentacles.

b. Comparison of anatomical proportions and cortical proportions

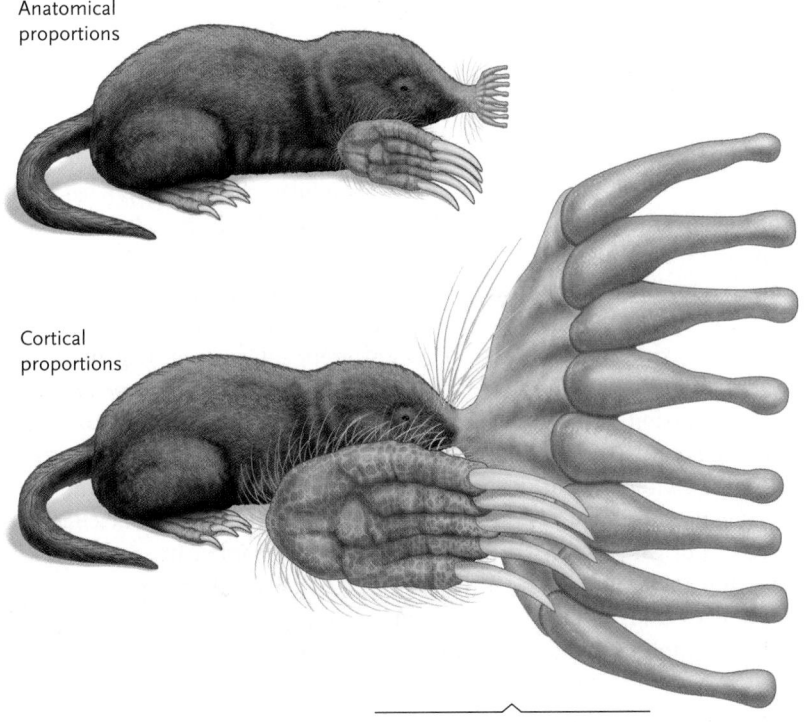

Anatomical proportions

Cortical proportions

Most of the mole's cerebral cortex is devoted to the tentacles and front, digging feet.

Figure 47.14

The collection and analysis of sensory information by the star-nosed mole (*Condylura cristata*). **(a)** The mole's nose has 22 fleshy tentacles covered with cylindrical tactile receptors called Eimer's organs. Each Eimer's organ contains sensory nerve terminals. **(b)** The mole's cerebral cortex devotes far more space and neurons to analysis of input from the tentacles than from elsewhere on the body. These drawings compare the relative amounts of sensory information coming from different parts of a mole's body.

pictures are distorted, but the unbalanced perceptions of the world are advantageous because certain types of information are far more important than others for the animals' survival and reproductive success.

STUDY BREAK

1. How do crickets hear the echolocation calls of bats?
2. How does what they see influence the behaviour of fiddler crabs?

47.7 Communication

In animal communication, one individual produces a signal that is received by another, changing the behaviour of one or both individuals in a way that benefits the signaller and/or the signal receiver. The signaller is the individual transmitting information (the signal), and the signal receiver **(Figure 47.15, p. 1166)** is the one receiving the signal. Some animals have broken the signal codes of others and exploited them to their advantage. Some people who study animal communication consider that only signals intended to communicate

Figure 47.15
Song birds, such as this Grey Vireo (*Vireo vincinior*), use songs to advertise their presence to unmated females and to other males.

Figure 47.16
Visual display. A male peacock spider (*Maratus volans;* Salticidae) approaches a female **(a)** and then displays to her **(b)**.

should be called communication (think back to problems with the definition of species or genes).

Animals use a variety of sensory modalities when producing signals, including acoustical, chemical, electrical, vibrational, and visual. Some signals combine modalities. Sometimes the animal itself is a signal; in other situations, the animal's excretory or eliminated products are signals.

Bird songs are acoustical signals heard by the signal receivers. The song of a male Whippoorwill (*Caprimulgus vociferus*) advertises his presence to females and may help him secure a mate. The same song is heard by other males, who recognize it as a territorial display. After the eggs have been laid, the same song is heard by the young developing in the eggs. Other birds, such as male Club-winged Manakins (e.g., *Machaeropterus deliciosus*), use sounds produced by feather stridulations as their acoustic courting signal. Sounds are used as signals by many other animals, such as insects and rattlesnakes. Pacific herring (*Clupea pallasii*) communicate with conspecifics through the noise generated with little bursts of gas (known colloquially as "farts") passed from the anus.

A striped skunk's (*Mephitis mephitis*) black and white stripes constitute a **visual signal**. Other examples are humans' facial expressions and body language. These visual signals are available to anyone viewing them. Visual signals can be enhanced by morphological features, such as the erectile crest of a Royal Kingbird (*Tyrannus melancholicus*), or semaphore flags used by people. In darkness, some animals use bioluminescent signals (e.g., Figure 48.18, Chapter 48). In many animals, visual signals are *ritualized*—they have become exaggerated and stereotyped, enhancing their function as signals **(Figure 47.16)**.

Many species produce chemical signals, well known to anyone who has walked a dog. Pheromones are distinctive volatile chemicals released in minute amounts to influence the behaviour of conspecifics. The body of a worker ant contains a battery of glands, each releasing a different pheromone **(Figure 47.17)**. One set of pheromones recruits fellow workers to battle colony invaders, whereas another stimulates workers to collect food that has been discovered outside the colony. Pheromones are used by some animals to attract mates. Female silkworm moths (*Bombyx mori*) produce the pheromone bombykol (see "Molecule behind Biology," Box 47.1). A single molecule of bombykol can generate a message in specialized receptors on the antennae of any male silkworm moth that is

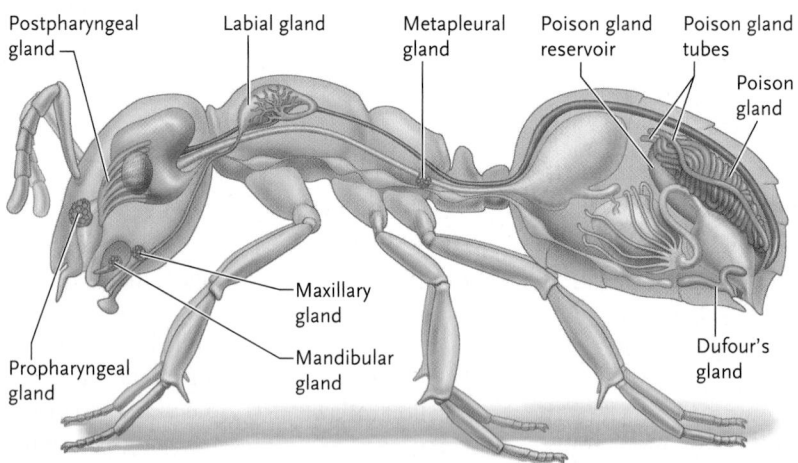

Figure 47.17
Chemical signals. An ant's body contains a host of pheromone-producing glands, each of which manufactures and releases its own volatile chemical or chemicals.

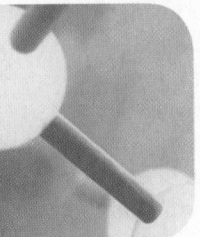

MOLECULE BEHIND BIOLOGY 47.1

Bombykol

Male silkworm moths (*Bombyx mori*) respond to the pheromone bombykol **(Figure 1)** produced and released by females. Bombykol is a pheromone designed to function in communication. Animals use pheromones to bring males and females together. Male *B. mori* detect bombykol using specialized receptors on their antennae (see Chapter 45).

Not surprisingly, predators exploit the powerful attractiveness of pheromones to lure prey. Female bolas spiders (*Mastophora cornigera* and other species in the genus) use a sticky ball of web impregnated with a chemical that mimics the odour of sex pheromones secreted by female moths. Male moths respond to the lure of these odours, approach the pheromone-soaked web, and are captured by the spiders.

Bolas spiders do not prey on just one species of moth. Adult female

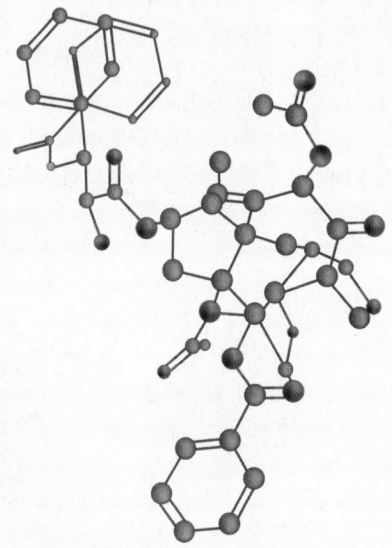

Figure 1
Bombykol.

M. cornigera produce three sex pheromone compounds, (Z)-9-tetradecenyl acetate, (Z)-9-tetradecenal, and (Z)-11-hexadecenal. These

pheromones attract several different species of moths. Moth attractants produced by *Mastophora hutchinsoni* are effective on several moths that use quite different pheromones. *Mastophora hutchinsoni* adjust the production of pheromone mimic to match the times of maximum activity by *Tetanolita mynesalis* (smoky tetanolita) and *Lacinipolia renigera* (bristly cutworm). The pheromone blend for the early-flying *L. renigera* interferes with attraction of the late-flying *T. mynesalis*, so the spider adjusts the blend of pheromone it uses in its lure. This spider lures the early-flying moth with one blend and the late-flying one with another.

---

downwind (see more about the influence of genes on behaviour in "Knockouts: Genes and Behaviour," p. 1158). Chemicals used as signals are often exploited by predators.

In many species, touch conveys important messages from a signaller to a receiver. **Tactile signals** can operate only over very short distances, but for social animals living in close company, they play a significant role in the development of friendly bonds between individuals **(Figure 47.18)**.

Some freshwater fish species, especially those that occupy murky tropical rivers where visual signals cannot be seen, use weak **electrical signalling** to communicate (see Figure 45.28, Chapter 45). These fish have electric organs that can release charges of variable intensity, duration, and frequency, allowing substantial modulation of the message that a signaller sends. Among the New World knifefish (order Gymnotiformes), including the electric eel (Figure 45.28b, Chapter 45), electrical discharges can signal threats, submission, or a readiness to breed.

Animals often use several channels of communication simultaneously. Karl von Frisch demonstrated that the famous dance of the honeybee involves tactile, acoustical, and chemical modes **(Figure 47.19, p. 1168)**. When a foraging honeybee discovers a source of pollen or nectar, it returns to its colony. There, in the darkness

M.B. Fenton

Figure 47.18
Tactile signals. Grooming by Hyacinth Macaws (*Anodorhynchus hyacinthinus*) removes ectoparasites and dirt from feathers. The close physical contact promotes friendly relationships between groomer and groomee.

of the hive, it performs a dance on the vertical surface of the honeycomb. The dancer moves in a circle, attracting a crowd of workers. Some workers follow and maintain physical contact with the dancer. The dance delivers information about the food source, its quality, and the distance and direction observers will need to fly to locate it.

a. Round dance

b. Waggle dance

c. Coding direction in the waggle dance

When the bee moves straight down the comb, other bees fly to the source directly away from the Sun.

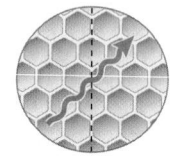

When the bee moves 45° to the right of vertical, other bees fly at a 45° angle to the right of the Sun.

When the bee moves straight up the comb, other bees fly straight toward the Sun.

Figure 47.19

Dance communication by honeybees (*Apis mellifera*). Foraging honeybees transmit information about the location and quality of a food source by dancing on a vertical honeycomb. **(a)** If the food source is close to the hive, the forager performs a *round dance*. **(b)** When food is farther from the hive, the honeybee performs a *waggle dance*. **(c)** The dancing bee indicates the direction to the distant food source by the angle of the waggle run.

When the food source is less than 75 m from the hive, the bee performs a *round dance* (see Figure 47.19a). Here the bee moves in tight circles, swinging its abdomen back and forth. Bees surrounding the dancer produce a brief acoustical signal that stimulates the dancer to regurgitate a sample of the food it discovered. The regurgitated sample serves as a chemical cue for other workers that search for the food.

When the food source is farther away, the forager performs the *waggle dance:* a half-circle in one direction, then a straight line while waggling its abdomen, and then a half-circle in the other direction (see Figure 47.19b). With each waggle, the dancer produces a brief buzzing sound. The angle of the waggle run relative to the vertical honeycomb indicates the direction of the food source relative to the position of the Sun (see Figure 47.19c). The duration of the waggles and buzzes carries information about distance to the food. The more time spent waggling and buzzing, the farther the food is from the hive.

Signal receivers often respond to communication from signallers in predictable ways. A male White-crowned Sparrow generally avoids entering a neighbouring territory simply because it hears the song of the resident male. Similarly, young male baboons and mandrills often retreat without a fight when they see an older male's visual threat display **(Figure 47.20)**, even

with the loss of a chance to mate with a female. Why do these receivers behave in ways that appear to be beneficial to their rivals but not to themselves?

Explaining behavioural interactions often means considering how an animal's actions affect its reproductive output. The retreating White-crowned Sparrow avoids wasting time and energy on a battle it is likely to lose. By retreating, the would-be intruder minimizes the chances of being injured or killed by a resident male. Moreover, ousting the current resident might be more tiring and risky than finding a suitable unoccupied breeding site. Resident males usually win physical contests, and intruders typically succeed in gaining a territory from a resident only after a prolonged series of exhausting clashes. Observations of territorial species, such as birds, lizards, frogs, fish, and insects, generally support these predictions.

Applying a similar argument to competition among male mandrills, we can predict that smaller or younger males will concede females to threatening older rivals without fighting. The signal receiver retreats after receiving the threat because he judges that he would not win—a male mandrill's canine teeth are not just for show. Evolutionary analyses suggest that the signaller and signal receiver benefit from the exchange of signals, for example, by allowing two animals to avoid a physical altercation that might result in injury.

In winter, Common Ravens (*Corvus corax*) may emit a strange "yell" call when they find the carcass of a deer. The loud yell attracts a crowd of hungry ravens.

EBFoto/Shutterstock.com

Figure 47.20

Threat display. Exposed canines epitomize the threat display of a dominant male mandrill (*Mandrillus sphinx*), which is used to drive away rival males.

The calling behaviour puzzled Bernd Heinrich, who noted that when paired, territory-holding adult ravens found a carcass, they fed quietly and did not yell. Yells are produced by young, wandering ravens that happen on a carcass in another bird's territory. The yells attracted other ravens, which collectively overwhelmed the residents' efforts to defend the carcass and their territory. Wanderers used yells to exploit the food supply, whereas residents just ate. Heinrich concluded that the reproductive benefit of resident ravens was enhanced by uninterrupted feeding. Wandering ravens succeeded in their trespassing only when they attracted others.

47.7a Language Consists of Syntax and Symbols

Although language is communication, not all communication is language. Many people believe that language is the exclusive domain of humans, but the distinction is not clear. The round and waggle dances of honeybees contain both syntax (the order in which information is presented) and symbols (a display that represents something else) and are considered by many to meet the criteria for language. Furthermore, by blackening the dancer's ocelli (see Chapter 45), James L. Gould was able to get a dancing bee to lie to other bees. When there is a light in the hive, the dancer orients the waggle dance to the light as if it were the Sun. Dancers with blackened ocelli do not see the light as do other bees—thus, the "lie."

Vervet monkeys have a repertoire of signals to alert conspecifics to different predators. Vervet monkeys use one signal for snakes, another for leopards, and still another for raptors, and they show different predator-specific defensive behaviours. Chickadees (*Parus atricapillus*) also use different alarm calls to alert others to approaching danger. Captive, trained chimpanzees and gorillas (*Gorilla gorilla*) have been reported to use American Sign Language (ASL).

In the area of communication, humans are not as distinct from other animals as some people would like to believe. To appreciate redundancy in animal communication, observe the body language and facial expressions of someone talking on a telephone. The eloquence of these signals is not conveyed to the signal receiver at the other end of the phone!

STUDY BREAK

1. What sensory modalities do animals use in communication?
2. How do "yells" influence the behaviour of ravens? Explain.
3. What is the meaning and importance of syntax and symbols in signalling?

47.8 Space

The geographic range of many animal species includes a mosaic of habitat types. The breeding ranges of White-crowned Sparrows can encompass forests, meadows, housing developments, and city dumps. Other animals have a limited range; for example, a Kirtland's Warbler (*Dendroica kirtlandii*) is found only in young jackpine forests. An animal's choice of habitat is critically important because the habitat provides food, shelter, nesting sites, and the other organisms with which it interacts. If an animal chooses a habitat that does not provide appropriate resources, it will not survive and reproduce.

On a large spatial scale, animals almost certainly use multiple criteria to select the habitats they occupy, but no research has yet established any general principles about how animals make these choices. When a migrating bird arrives at its breeding range, it probably cues on large-scale geographic features, such as a pond or a patch of large trees. If the bird does not find the food or nesting resources it needs, or if other individuals have already occupied the space and perhaps depleted those resources, it may move to another habitat patch.

On a very fine spatial scale, basic responses to physical factors enable some animals to find suitable habitats. **Kinesis** (*kine* = movement; *es* = inward) is a change in the rate of movement or the frequency of turning movements in response to environmental stimuli. Wood lice (terrestrial crustaceans in the order Isopoda) typically live under rocks and logs or in other damp places. Although these arthropods are not attracted to moisture *per se*, when a wood louse encounters dry soil, it scrambles around, turning frequently. When it reaches a patch of moist soil, it moves much less. This kinesis results in wood lice accumulating in moist habitats. Wood lice exposed to dry soil quickly dehydrate and die, so those that move to moister habitats are more likely to survive.

A **taxis** (= ordered movement) is a response directed either toward or away from a specific stimulus. Cockroaches (order Blattodea) exhibit negative phototaxis, meaning that they actively avoid light and seek darkness. Negative phototaxis makes cockroaches less vulnerable to predators that use vision to find their food.

Biologists generally assume that habitat selection is adaptive and has been shaped by natural selection. Some animals instinctively select habitats where they are well camouflaged and less detectable by predators. Predators would discover and eliminate individuals that did not select a matching background, along with any alleles responsible for the mismatch. Many insects have inherited preferences for the plants they eat as larvae (e.g., caterpillars). Adults often lay their eggs only on appropriate food plants, effectively selecting the habitats where their offspring will live and feed.

Vertebrates sometimes exhibit innate preferences, as demonstrated by two closely related species of European birds, Blue Tits (*Cyanistes caeruleus*) and Coal Tits (*Parus ater*). Adult Blue Tits forage mainly in oak trees and Coal Tits in pines. When researchers reared the young of both species in cages without any vegetation and then offered them a choice between oak branches and pine branches, Coal Tits immediately gravitated toward pines and Blue Tits toward oaks, suggesting an innate preference **(Figure 47.21)**. Each species feeds most efficiently in the tree species it prefers.

Habitat preferences can also be moulded by experiences early in life. Tadpoles of red-legged frogs (*Rana aurora*) usually live in aquatic habitats cluttered with sticks, strands of algae, and plant stems. In the laboratory, these tadpoles prefer striped backgrounds to plain ones. In contrast, tadpoles of the closely related cascade frog (*Rana cascadae*) live over gravel bottoms and prefer plain substrates over striped ones. These habitat preferences do not appear when red-legged frogs are reared over plain substrates and cascade frogs over striped substrates and are later given a choice of substrate.

47.8a Home Range and Territory Are Occupied and Defended Areas, Respectively

Space is an important resource for animals. Although many animals are motile, moving about in space, others are sessile. Sessile species such as barnacles

Figure 47.22
Pronghorn antelopes, *Antilocapra americana*.

(see Chapter 27) anchor themselves to the substrate but are motile as larvae. Barnacles that live on whales or the hulls of ships are sessile but mobile because of the substrate they selected. Motile animals have a home range, the space they regularly traverse during their lives. Home ranges or parts of home ranges become territories when they are defended. In species such as pronghorn antelopes **(Figure 47.22)**, some males hold territories, but others do not. Females are not usually territorial. There is a direct connection between territory quality and male reproductive success. Male pronghorn antelopes defending the "best" territories (those with the best food resources) attract the most females, offering the male the most opportunities to mate with the most females.

Male Jarrow's spiny lizards **(Figure 47.23)** are normally territorial during the autumnal mating season, when they have elevated levels of testosterone in their blood. Catherine Marler and Michael Moore implanted small doses of testosterone or a placebo under the skins of experimental animals during the nonmating season (June and July). Testosterone-enhanced males were more active and displayed more often than control males. Experimental males spent less time feeding, even though they used about 30% more energy per day than control males. In one seven-week period,

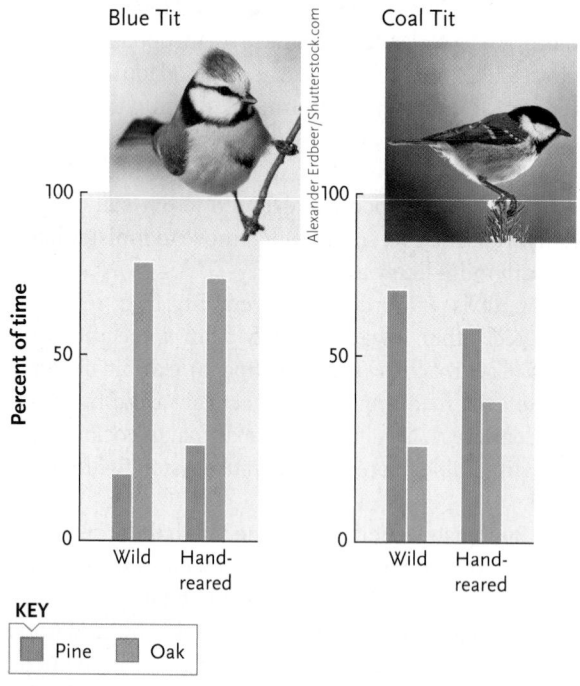

KEY

Pine   Oak

Figure 47.21
Habitat selection by birds. Wild Blue Tits (*Cyanistes caeruleus*) show a strong preference for oak trees; Coal Tits (*Parus ater*) show a strong preference for pines. Hand-reared birds raised in a vegetation-free environment showed identical but slightly weaker responses.

Figure 47.23
Jarrow's spiny lizard, *Sceloporus jarrovi*.

testosterone-enhanced males suffered significantly higher mortality than placebo males. It can be expensive to be territorial.

Territorial defence is always a costly activity. Patrolling territory borders, performing displays hundreds of times per day, and chasing intruders take time and energy. Moreover, territorial displays increase an animal's likelihood of being injured or detected and captured by a predator.

But territorial behaviour has its benefits, such as access to females. Territorial surgeonfish (*Acanthurus lineatus*) living in coral reefs around American Samoa may engage in as many as 1900 chases per day, defending their small territories from incursions by other algae-eating fish. Territorial surgeonfish eat five times as much food as non-territory-holders because they have more exclusive access to the food in their territories. It also costs the territory holders more to patrol and defend their realm.

STUDY BREAK

1. Define *kinesis* and *taxis*.
2. What is the difference between a home range and a territory?
3. How are home ranges and territories different from a species' range?

47.9 Migration

CONCEPT FIX Our knowledge of bird migration makes it easy to believe that migrating animals travel in groups. It is now clear that many animals migrate alone, and we have evidence of this from a variety of animals, including birds, bats, fur seals, and sea turtles. The reality, however, is that we know very little about the migrations and migratory behaviour of most animals.

Many animal species make a seasonal **migration**, travelling from the area where they were born or hatched to a distant and initially unfamiliar destination. The migration is complete when they later return to their natal site. The Arctic Tern, a seabird, makes an annual round-trip migration of 40 000 km **(Figure 47.24)**. Other vertebrate species, such as grey whales (*Eschrichtius robustus*) and salmon (*Salmo* species), undertake long and predictable journeys. The same is true of arthropods such as spiny lobsters that form long lines and move seasonally between coral reefs and the open ocean floor **(Figure 47.25, p. 1172)**.

Moving animals use various mechanisms to find their way during migration. There are three categories of way-finding mechanisms: **piloting**, **compass orientation**, and **navigation**. Many species probably use some combination of these mechanisms to guide their movements.

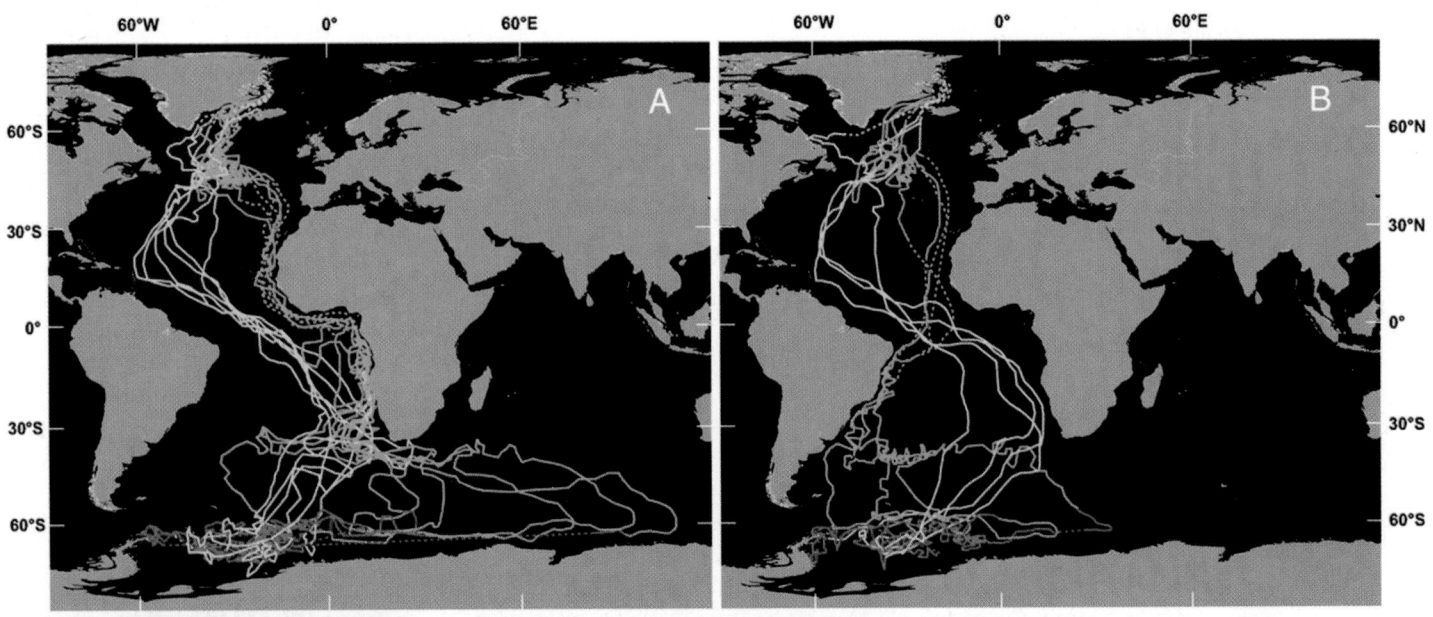

"Tracking of Arctic terns *Sterna paradisaea* reveals longest animal migration." Carsten Egevang, Iain J. Stenhouse, Richard A. Phillips, Aevar Petersen, James W. Fox, and Janet R. D. Silk. PNAS February 2, 2010 vol. 107 no. 5 2078–2081.

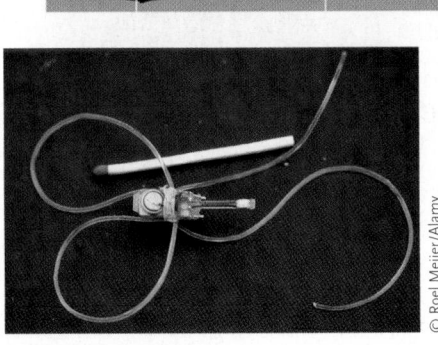

Figure 47.24
Geolocators provided details of the movements of 11 Arctic Terns tagged at breeding colonies in Greenland (10) and Iceland (1). Two migration routes emerged, one along the coast of Africa (A), the other along the coast of South America (B). Left inset, a small geolocator; right an Arctic Tern.

47.9a Piloting Is Finding the Way

Piloting is the simplest way-finding mechanism, involving the use of familiar landmarks to guide a journey. Grey whales migrate from Alaska to Baja California and back using visual cues provided by the Pacific coastline of North America. When it is time to breed and lay eggs, Pacific salmon use olfactory cues to pilot their way from the ocean back to the stream in which they hatched.

Animals that do not migrate also use specific landmarks to identify their nest site or places where they have stored food. Female digger wasps (*Philanthus triangulum*) nest in soil. In 1938, Niko Tinbergen showed that after foraging flights, these wasps used visual landmarks to find their nests **(Figure 47.26)**. While the female wasp was in the nest, Tinbergen arranged pinecones in a circle around it. As she left, the wasp flew around the area, apparently noting nearby landmarks. Tinbergen then moved the circle of pinecones a short distance away. Each time the female returned, she searched for her nest within the pinecone circle. She never once found her nest unless the pinecones were returned to their original position. Later, Tinbergen rearranged the pinecones into a triangle after females left their nests and added a ring of stones nearby. The returning females looked for their nest in the stone circle. Tinbergen concluded that digger wasps respond to the general outline or geometry of landmarks around their nests and not to the specific objects making up the landmarks.

47.9b Compass: Which Way Is North?

Animals using compass orientation move in a particular direction, often over a specific distance or for a prescribed length of time. Some day-flying migratory

Figure 47.25

Migrating arthropods. Spiny lobsters (*Panulirus argus*) make seasonal migrations between coral reefs and the open ocean floor. As many as 50 individuals march in single file for several days.

Wasp's flight pattern on leaving nest

Wasp's return, looking for nest

Nest

Experimental Research Figure 47.26 Female digger wasps find their nest. A ring of pinecones serves as a landmark for a female digger wasp (*Philanthus triangulum*). By moving landmarks, Nikko Tinbergen demonstrated the role they serve in the wasp's orientation behaviour.

Mark Conlin/Oxford Scientific/Getty Images

birds orient themselves using the Sun's position in the sky in conjunction with an internal biological clock (see Chapter 1). The internal clock allows the bird to use the Sun as a compass, compensating for changes in its position through the day. The clock may also allow some birds to estimate how far they have travelled since beginning their journey. Other animals, including birds, mammals, reptiles, amphibians, fish, crustaceans, and insects, use Earth's magnetic field as a compass. This requires detection of weak magnetic fields (~50 µT—microteslas). In 2008, some biologists suggested that magnetically sensitive free radical reactions could be the basis of a magnetic sense, but the search for the transducer(s) (see Chapter 45) continues.

47.9c Stars Are Used for Celestial Navigation

Some birds that migrate at night determine their direction by using the positions of stars. The Indigo Bunting flies about 3500 km from the northeastern United States to the Caribbean or Central America each fall and makes the return journey each spring. Stephen Emlen demonstrated that Indigo Buntings direct their migration using celestial cues **(Figure 47.27)**. Emlen confined individual buntings in cone-shaped test cages whose sides were lined with blotting paper. He placed inkpads on the cage bottoms and kept the cages in an outdoor enclosure so that the birds had a full view of the night sky. Whenever a bird made a directed

movement, its inky footprints indicated the direction in which it was trying to move. On clear nights in fall, the footprints pointed to the south, but in spring, they pointed north. On cloudy nights, when the buntings could not see the stars, Emlen recorded that their footprints were evenly distributed in all directions. The data indicated that the compass of Indigo Buntings required a view of the stars.

47.9d Navigation Is a Complex Challenge

Navigation is the most complex way-finding mechanism. It occurs when an animal moves toward a specific destination, using both a compass and a *mental map* of where it is in relation to the destination. Hikers in unfamiliar surroundings routinely use navigation to find their way home. They use a map to determine their current position and the necessary direction of movement and a compass to orient themselves in that direction. Scientists have documented true navigation in a few animal species, notably the Homing Pigeon (*Columba livia*). These birds can navigate to their home coops from any direction, probably using the Sun's position as their compass and olfactory cues as their map.

47.9e Animals Migrate for Several Reasons

Migrations by White-crowned Sparrows and many other species are triggered by changes in day length. Shortening

Indigo Bunting

Mark Tegges/Shutterstock.com

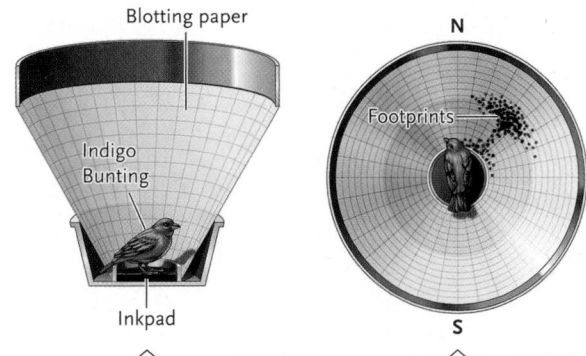

Side (left) and overhead (right) views of the test cage with blotting paper on the sides and an inkpad on the bottom.

Experimental Research Figure 47.27
Orientation by Indigo Buntings. The footprints of inked Indigo Buntings (*Passerina cyanea*) demonstrated migrating birds' responses to celestial cues.

Based on S. T. Emlen. 1967. "Migratory orientation of the indigo bunting, *Passerina cyanea*. Part I: Evidence for use of celestial cues." *The Auk* 84: 309–342.

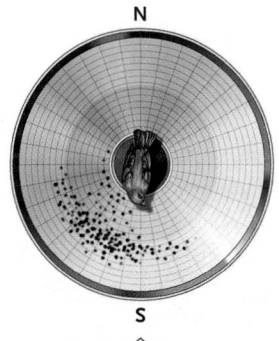

In autumn, the bunting footprints indicated that they were trying to fly south.

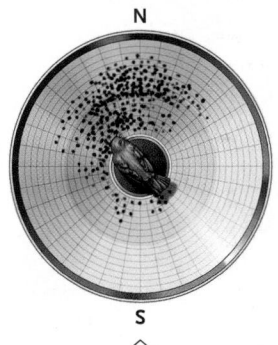

In spring, the bunting footprints indicated that they were trying to fly north.

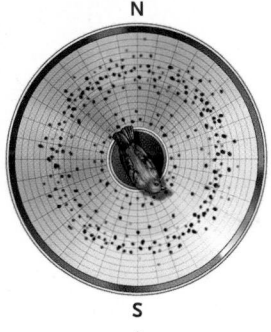

On cloudy nights, when buntings could not see the stars, their footprints indicated a random pattern of movement.

day length indicates approaching fall and winter; lengthening day length indicates spring. Day length changes the anterior pituitary of the bird's brain to generate a series of hormonal changes. In response, birds feed heavily and accumulate the fat reserves necessary to fuel their long journey. Sparrows also become increasingly restless at night until, one evening, they begin their nocturnal migration. Their ability to adopt and maintain a southerly orientation in autumn (and a northerly one in spring) rests in part on their capacity to use the positions of stars to provide directional information.

Migratory behaviour entails obvious costs, such as the time and energy devoted to the journey and the risk of death from exhaustion or predator attack. Migratory behaviour is not universal—many animals never migrate, spending their lives in one location. Why do some species migrate? What ecological pressures give migrating individuals higher fitness than individuals that do not migrate? Remember that many species of terrestrial animals migrate, such as wildebeest and caribou.

For migratory birds, seasonal changes in food supply are the most widely accepted hypothesis to explain migratory behaviour. Insects can be abundant in higher-latitude (greater than 50°N or S) habitats during the warm spring and summer, providing excellent resources for birds to raise offspring. As summer wanes and fall and winter approach, insects all but disappear. Bird species that remain in temperate habitats over winter eat mainly seeds and dormant insects. When it is winter at higher latitudes, energy supplies are more predictably available in the tropical grounds used by overwintering migratory birds.

Two-way migratory journeys may provide other benefits. Avoiding the northern winter is probably adaptive because endotherms must increase their metabolic rates just to stay warm in cold climates (see Chapter 50). Moreover, summer days are longer at high latitudes than they are in the tropics (see *The Purple Pages*), giving adult birds more time to feed and rear a brood.

Seasonal changes in food supply also underlie the migration of monarch butterflies that eat milkweed leaves as caterpillars and milkweed nectar as adults **(Figure 47.28a)**. In eastern North America,

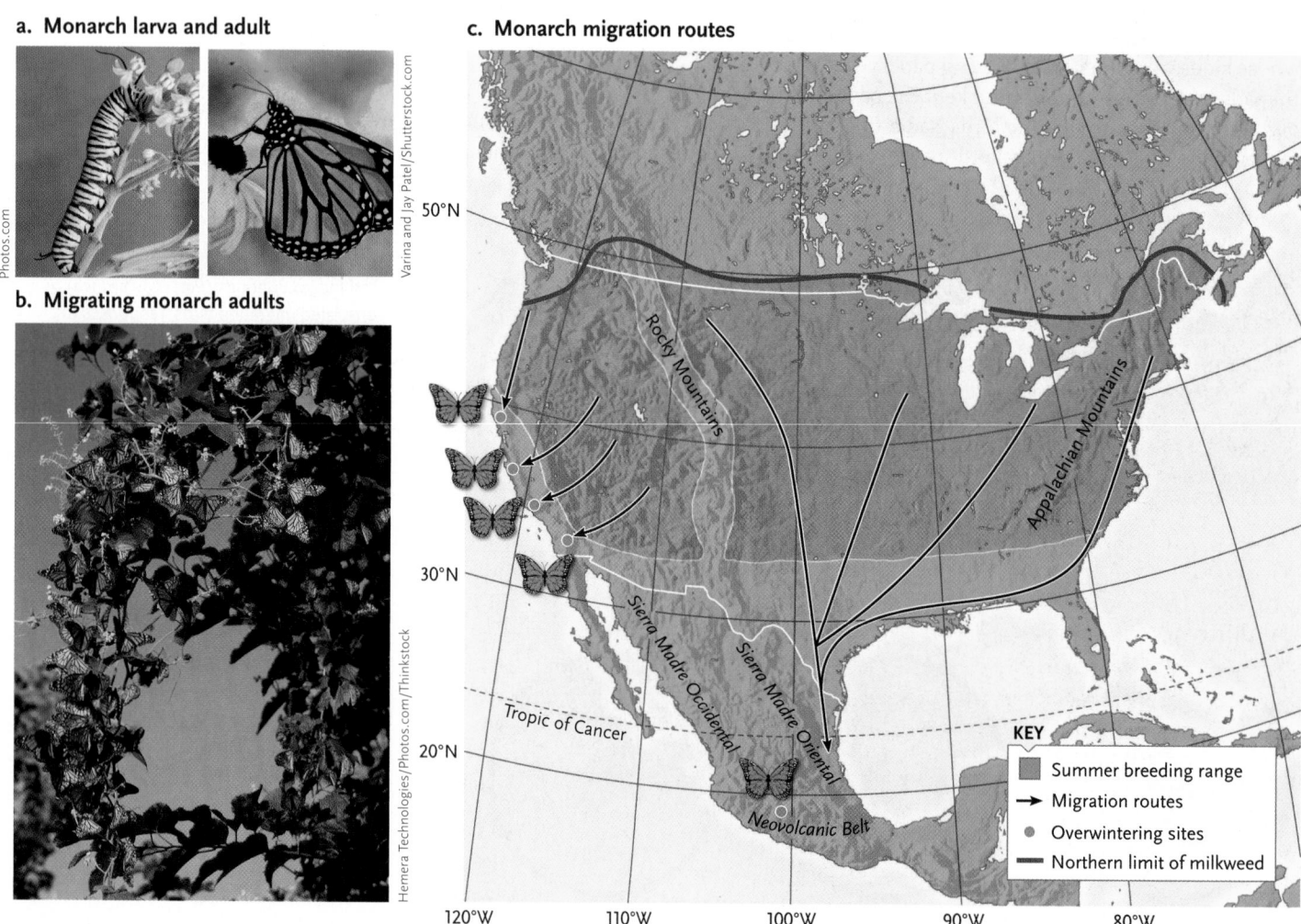

a. Monarch larva and adult

b. Migrating monarch adults

c. Monarch migration routes

KEY
- Summer breeding range
- → Migration routes
- • Overwintering sites
- ▬ Northern limit of milkweed

Figure 47.28

Migrating monarch butterflies. **(a)** Monarch butterflies (*Danaus plexippus*) eat milkweed plants as caterpillars. **(b)** When milkweed plants in their breeding range die back at the end of summer, monarchs migrate south. The following spring, after passing the winter in a semidormant state, they migrate north. **(c)** Monarchs that live and breed east of the Rocky Mountains migrate to Mexico. Those living west of the Rocky Mountains overwinter in coastal California.

milkweed plants grow only during spring and summer. Many adult monarchs head south in late summer, when the plants begin to die. Some migrate as much as 4000 km from eastern and central North America to central Mexico, where they cluster in spectacular numbers (Figure 47.28b and c), apparently using olfactory cues to find preferred resting places. Unlike migrant birds, these insects do not feed while at their overwintering grounds. Instead, their metabolic rate decreases in the cool mountain air, and the butterflies become inactive for months, conserving precious energy reserves. When spring arrives, the butterflies become active again and begin the return migration to northern breeding habitats. The northward migration is slow, however, and many individuals stop along the way to feed and lay eggs. These offspring, and their offspring, continue the northward migration through the summer. Some descendants of these migrants eventually reach Canada for a final round of breeding. The summer's last generation then returns south to the spot where their ancestors, two to five generations removed, spent the previous winter.

For other animals, migration to breeding grounds may provide the special conditions necessary for reproduction. Grey whales migrate south, where females give birth to their young in quiet, shallow lagoons where predators are rare and warm water temperatures are more conducive to the growth of their calves.

STUDY BREAK

1. Define migration. Give examples of migratory animals, including some not mentioned in the text. Do any humans migrate?
2. How do migrating animals find their way? Distinguish between navigation and compass orientation.

47.10 Mates as Resources

Mating systems have evolved to maximize reproductive success, partly in response to the amount of parental care that offspring require and partly in response to other aspects of a species' ecology. **Monogamy** describes the situation in which a male and a female form a pair bond for a mating season or, in some cases, for the individuals' reproductive lives. **Polygamy** occurs when one male has active pair bonds with more than one female (**polygyny**) or one female has active pair bonds with more than one male (**polyandry**). **Promiscuity** occurs when males and females have no pair bonds beyond the time it takes to mate. In polygyny, males often contribute nothing to reproduction but sperm; in polyandry, females nothing but eggs. The details will

vary according to the physiology of reproduction. In viviparous animals, the animals that get pregnant (usually females) may bear the costs of housing and feeding developing young.

When young require a great deal of care that both parents can provide, monogamy often prevails. Songbirds, such as the White-crowned Sparrow **(Figure 47.29)**, are altricial (naked and helpless) when they hatch. They beg for food, and both parents can bring it to them. Males and females achieve higher rates of reproduction when both parents are actively involved with raising young. In mammals, the situation is different because females provide the food (milk). Monogamy occurs in species in which males indirectly feed the young by bringing food to the mother.

If males have high-quality territories, the females living there may be able to raise young on their own. These males may be polygynous (mate with several females). The male's role is that of sperm donor and protector of the space rather than that of an active parent to all of his young. In birds such as Red-winged Blackbirds (*Agelaius phoenecius*), some males hold large, resource-filled territories that support several females. These males will be attractive to females even if a female (or females) already lives on the territory. Polygyny is prevalent among mammals because, compared with males, females make a much larger investment in raising young (through egg development and care of the young).

Promiscuous mating systems occur when females are only with males long enough to receive sperm and

Figure 47.29

Reproductive success. Parental care is just one of the many behaviours required for successful reproduction in White-crowned Sparrows and in many other animal species. The number of surviving nestlings will determine the reproductive success of their parents and the representation of their genes in the next generation.

© Dr. Edgar T. Jones/VIREO

Figure 47.30

Lekking behaviour. Male Sage Grouse (*Centrocercus urophasianus*) use their ornamental feathers in visual courtship displays performed at a lek. There each male has his own small territory. The smaller brown females observe the performing males before picking a mate.

there is no pair bond. These males make no contribution to raising young. Sage Grouse **(Figure 47.30)** and hammer-headed bats (*Hypsignathus monstrosus*) are examples of this approach. Both species form **leks**, congregations of displaying males, where females come only to mate. There are more details about Sage Grouse below.

STUDY BREAK

What do the terms *monogamy, polygamy,* and *promiscuity* mean?

47.11 Sexual Selection

Given the drive to reproduce (see Chapter 41), competition for access to mates coupled with mate choice sets the stage for sexual selection. **Sexual dimorphism,** in which one gender is larger or more colourful than the other, can be an outcome of sexual selection. When males compete for females, males are often larger than females and may have ornaments and weapons, such as horns and antlers, useful for attracting females and for butting, stabbing, or intimidating rival males. Displays of adornments or weapons can simultaneously warn off other males and attract the attention of females. Peacocks strut in front of female peahens while spreading a gigantic fan of tail feathers, which they shake, rattle, and roll.

Why should females choose males with exaggerated structures conspicuously displayed? A male's large size, bright feathers, or large horns might indicate that he is particularly healthy. His appearance could indicate that he can harvest resources efficiently

or simply that he has managed to survive to an advanced age. The features are, in effect, signals of male quality, and if they reflect a male's genetic makeup, he is likely to fertilize a female's eggs with sperm containing successful alleles. Large showy males may hold large, rich territories. Females that choose these males can gain access to the resources their territories contain.

The degree to which females *actively* choose genetically superior mates varies among species. In northern elephant seals, female choice is more or less passive. Large numbers of females gather on beaches to give birth to their pups before becoming sexually receptive again (see "Delaying Reproduction," Chapter 41). Males locate clusters of females and fight to keep other males away. Males that win have exceptional reproductive success because they mate with many females, but only after engaging in violent and relentless combat with rival males. In this mating system, the females struggle during a male's attempts to mate with them. A female's struggles attract other males, who try to interrupt the attempted mating. Only the largest and most powerful males are not interrupted in their copulations, and they inseminate the most females. These attributes may be associated with alleles that will increase their offspring's chances of living long enough to reproduce.

In other species, females exercise more active mate choice, mating only after inspecting several potential partners. Among birds, active female mate choice is most apparent at leks, display grounds where each male holds a small territory from which it courts attentive females. The male is the only resource on the territory. Male Sage Grouse in western North America gather in open areas among stands of sagebrush. Each male defends a few square metres, where it struts in circles while emitting booming calls and showing off its elegant tail feathers and big neck pouches (see Figure 47.30). Females wander among displaying males, presumably observing the males' visual and acoustical displays. Eventually, each female selects one mate from among the dozens of males that are present. Females repeatedly favour males that come to the lek daily, defend their small area vigorously, and display more frequently than the average lek participant. Males preferred by females sustain their territorial defence and high display rate over long periods, abilities that may correlate with other useful genetic traits. Ultimately, the male holding the "best" position in the lek mates with the most females.

The results of experiments with peafowl suggest that the top Peacocks (*Pavo cristatus*) supply advantageous alleles to their offspring. In nature, peahens prefer males whose tails have many ornamental eyespots **(Figure 47.31)**. In an experiment on captive birds, some peahens were mated to peacocks with highly attractive tails, but others were paired with males

Figure 47.31

Sexual selection for ornamentation. The attractiveness of a peacock to peahens depends in part on the number of eyespots in his extraordinary tail. The offspring of males with elaborate tails are more successful than the offspring of males with plainer tails.

whose tails were less impressive. The offspring of both groups were reared under uniform conditions for several months and then released into an English woodland. After three months on their own, the offspring of fathers with impressive tails survived better and weighed significantly more than did those whose fathers had less attractive tails. The evidence demonstrates that a peahen's mate choice influences her offspring's chances of survival.

According to the handicap hypothesis, females select males that are successful—the ones with ornate structures. These structures may impede their locomotion, and their elaborate displays may attract the attention of predators. Females select ornate males because they have survived *despite* carrying such a handicap. Successful alleles responsible for the ornamental handicap are passed to the female's offspring.

STUDY BREAK

What are the distinguishing features of a lek?

47.12 Social Behaviour

Social behaviour, the interactions that animals have with other members of their species, has profound effects on an individual's reproductive success. Some animals are solitary, getting together only briefly to mate (e.g., house flies and leopards). Others spend most of their lives in small family groups (e.g., gorillas). Still others live in groups with thousands of relatives (e.g., termites and honeybees). Some species, such as caribou and humans, live in large social units composed primarily of nonrelatives. In many species, the level of social interactions varies seasonally, usually reflecting the timing of reproduction, which, in turn, is influenced by changes in day length.

47.12a African Lions May Commit Infanticide

African lions (*Panthera leo*) usually live in prides, one adult male with several females and their young. Males typically sire the young born to the females in their pride, achieving a high reproductive output. Females benefit from the support of the others in the group, which includes caring for young and cooperating in foraging. Female lions living in prides wean more young per litter than those living alone. The females in a pride are often genetically related, and their estrus cycles are usually synchronized. Male lions are bigger (~200 kg) than females (~150 kg), and males fight vigorously for the position of pride male. Males protect their females from incursions by other males.

When a new male takes over a pride, he kills all nursing young, bringing the females into estrus. At first, this infanticide seems counterproductive. However, it benefits the male because it increases the chances of his succeeding at reproducing. Were he to wait until the females had raised their dependent young, his reproductive contributions could be delayed for some time, perhaps as much as a year or more. Furthermore, in the intervening period he could lose the opportunity to sire any young at all.

Females are not large enough to protect their young from the male. If a female takes her nursing young and leaves the pride, her efficiency as a hunter declines, and she is less able to protect her young. Her reproductive success plummets. Females can be more productive (measured by output of young) when they are part of a pride.

But why live in a group in the first place? By hunting together, lions are more efficient foragers than when they hunt alone; therefore, they raise more young. Perhaps more important is the threat posed by spotted hyenas, which live in large groups (clans). Although individually smaller (~60 kg), when spotted hyenas outnumber lions, they can chase lions from their kills. Furthermore, many of the lion's main prey also live in groups, and group defences affect lions' hunting success.

The situation in lions exemplifies some biological realities. Males and females do not have the same strategies when it comes to reproduction. Understanding behaviour means considering genetic relatedness and production of offspring, as well as the setting in which the animals live.

47.12b Group Living Has Its Costs and Benefits

Social Behaviours. Ecological factors have a large impact on the reproductive benefits and costs of social living. Groups of cooperating predators frequently capture prey more effectively than they would on their own. White Pelicans (*Pelecanus erythrorhynchos*) often encircle a school of fish before attacking, so being part of a group provides a better yield to individuals than working alone. On the other hand, prey subject to intense predation may benefit from group defence. This can mean more pairs of watchful eyes or ears to detect an approaching danger. It may also translate into multiple lures so that when a predator attacks, it is more difficult to focus on an individual. When you are part of a group that is attacked, it may be someone other than you that is captured, diluting the risk to any one group member.

When attacked by wolves, adult muskoxen form a circle around the young, so attackers are always confronted by horns and hooves **(Figure 47.32)**. Insects such as Australian sawfly caterpillars also show cooperative defensive behaviour **(Figure 47.33)**. When predators disturb the caterpillars, all group members rear up, writhe about, and regurgitate sticky, pungent oils. The caterpillars collect the oils from the eucalyptus leaves they eat. The oils do not harm the caterpillars but are toxic and repellent to birds.

Living in groups can also be expensive. One cost can be increased competition for food. When thousands of royal penguins crowd together in huge colonies **(Figure 47.34),** the pressure on local food supplies is great, increasing the risk of starvation. Communal living may facilitate the spread of contagious diseases and parasites. Nestlings in large colonies of Cliff Swallows (*Petrochelidon pyrrhonota*) are often stunted in

growth because the nests swarm with blood-feeding, bedbuglike parasites, *Oeciacus vicarius* **(Figure 47.35)**. The parasites move readily from nest to nest in crowded conditions. Some social animals learn to recognize and avoid diseased group members. Caribbean spiny lobsters live in groups but avoid conspecifics infected by a lethal virus (PaV1). It is no surprise that most animals live alone.

Group living brings both costs and benefits. Not all animals that live in groups are *social,* a term implying some organization of the group. The 10 million Brazilian free-tailed bats emerging from a cave roost near San Antonio, Texas, are no more a social group than the dozens of people leaving a high-rise apartment or university residence. Within the aggregation, there may be social units, but the aggregation itself is not necessarily a social unit.

Social animals usually live in groups characterized by some form of structure. Some individuals may dominate others (a **dominance hierarchy**), manifested in access to resources. Dominant (alpha or α) individuals get priority access to food (or mates or sleeping sites). In some situations, only dominant individuals (a male and a female) reproduce. Dominance hierarchies may be absolute, such as when the same individual always has priority access to any resource. In relative dominance hierarchies, an individual's status depends on the circumstance. One individual may dominate at a food source, while another may dominate in access to mates.

George Holton/Science Source

Figure 47.32
Muskoxen.

Auscape/UIG/Universal Images Group/Getty Images

Figure 47.33
Cooperative defensive behaviour. Australian sawfly (*Perga dorsalis*) caterpillars clump together on tree branches. They regurgitate yellow blobs of sticky aromatic fluid that repels birds. The accumulation of regurgitate from a group of caterpillars is an effective defence.

Figure 47.34
Colonial living. Royal Penguins (*Eudyptes schlegeli*) on Macquarie Island between New Zealand and Antarctica experience benefits and costs from living together in huge groups.

Dominance brings its costs. In animals such as wild dogs (*Lycaeon pictus*) and grey wolves, dominant animals must constantly defend their status. Dominants often have high levels of cortisol and other stress-related hormones in their blood (see Chapter 43) compared with subordinates. Elevated cortisol levels may induce high blood pressure, the disruption of sugar metabolism, and other pathological conditions.

Subordinance brings its benefits. Subordinate group members, like all members of the group, gain protection from predators. They may also gain experience by helping dominant individuals raise young. Over time, subordinate individuals can rise in a dominance hierarchy and avoid some of the side effects of dominance. Many social animals cannot survive on their own **(Figure 47.36, p. 1180).**

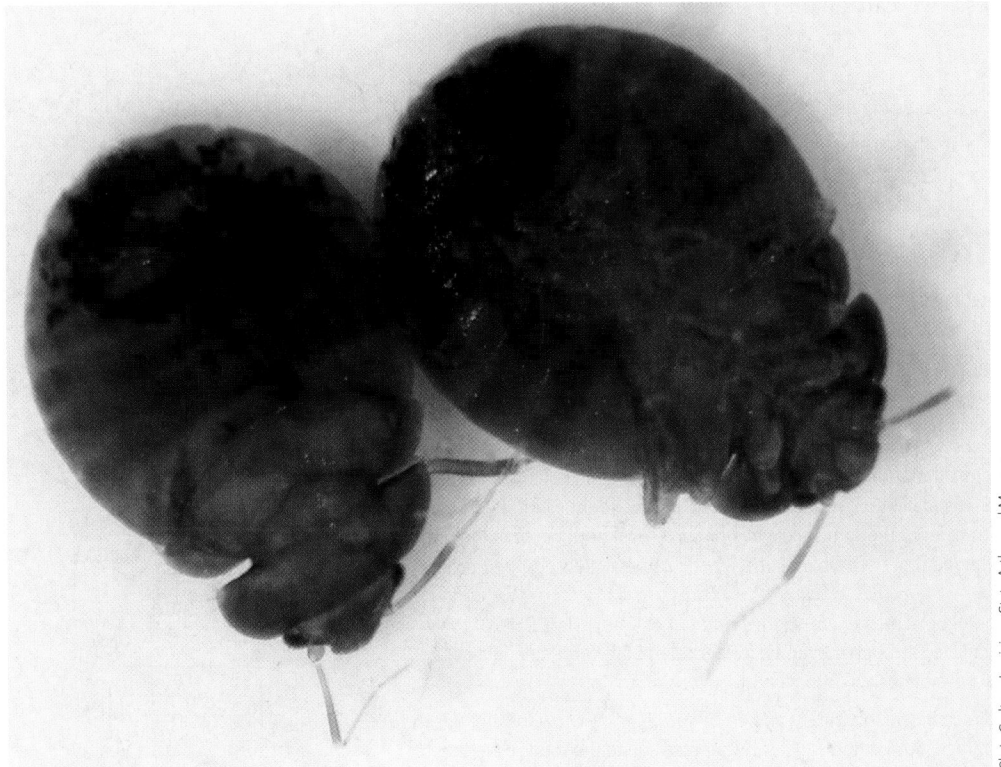

Figure 47.35
Dorsal and ventral views of a bedbug that specializes on birds (*Ornithocoris pallidus*).

Figure 47.36

When attacked by a leopard (*Panthera pardus*), a solitary baboon (*Papio anubis*) is defiant but unlikely to survive.

STUDY BREAK

1. What is infanticide? Why does it occur?
2. Give some examples of the advantages and disadvantages of group living.

47.13 Kin Selection and Altruism

Behavioural ecologist William D. Hamilton recognized that helping genetic relatives effectively propagates the helper's genes because family members share alleles inherited from their ancestors. By calculating the degree of relatedness, we can quantify the average percentage of alleles shared by relatives **(Figure 47.37)**. Half-siblings, by definition, share one genetic parent, so they share, on average, 0.25 of their alleles by inheritance from their shared parent. Their degree of relatedness is 0.25. Full siblings share both parents' share, 0.25 of their alleles through the mother and 0.25 through the father, for a total, on average, of 0.25 + 0.25 = 0.5 of their alleles. The degree of relatedness between a nephew or niece and an aunt or uncle is 0.25 and between first cousins is 0.125. Individuals should be more likely to help close relatives because increasing a close relative's fitness means that the individual is helping to propagate some of its own alleles. This is **kin selection.**

A male grey wolf helps his parents rear four pups to adulthood, pups that would have died without the extra assistance he provided. The pups are his younger full siblings, sharing 0.5 of his genes, so, on average, the helper has created "by proxy" two (0.50 × 4 = 2) copies of any allele they shared. However, the costs of his helping must be measured against this indirect reproductive success. If he had found a mate, sired offspring, and raised two of them, each would have carried half of his alleles, preserving only one (0.50 × 2 = 1) copy of a given allele. In this situation, reproducing on his own would have produced fewer copies of his alleles in the next generation than helping to raise his siblings. Sibling helpers have been

Research Method Figure 47.37 Calculating degrees of relatedness.

PURPOSE: The kin selection hypothesis suggests that the extent of altruistic behaviour exhibited by one individual to another is directly proportional to the percentage of alleles they share. The hypothesis therefore predicts that individuals are more likely to help close relatives because by increasing a close relative's fitness, the individual is helping to propagate some of its own alleles. Researchers calculate the degree of relatedness between individuals to test this prediction.

PROTOCOL: To calculate the degree of relatedness between any two individuals, we first draw a family tree that shows all of the genetic links between them. The alleles of a parent are shuffled by recombination and independent assortment in the gametes they produce, so we can calculate only the average percentage of a parent's alleles that offspring are likely to share.

We start by considering half siblings, those who share only one genetic parent. Each sibling receives half of its alleles from its mother. Because a parent has only two alleles at each gene locus, the probability of sibling A getting a particular allele from its mother is 0.5 (decimal notation for 50%). Similarly, the probability of sibling B getting the same allele from its mother is also 0.5. Statistically, the probability that two independent events—in this case, the transfer of an allele to sibling A and the transfer of the same allele to sibling B—will both occur is the product of their separate probabilities. Thus, the likelihood that both siblings receive the same allele from their mother is 0.5 × 0.5 = 25.

Now consider two full siblings, who share the same genetic mother and father. They share 25% of their alleles through the mother plus 25% of their alleles through the father, for a total of 50% (half their alleles). In other words, the degree of relatedness for full siblings is 0.50.

INTERPRETING THE RESULTS: Each link drawn between a parent and an offspring or between full siblings indicates that those two individuals share, on average, 50% of their alleles. We can calculate the total relatedness between any two individuals by multiplying out the probabilities across all of the links between them. Thus, the degree of r.

Half siblings

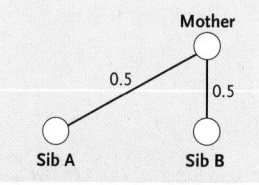

Relatedness = (0.5)(0.5) = 0.25

Full siblings

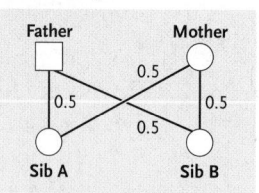

Relatedness
Through mother = (0.5)(0.5) = 0.25
Through father = (0.5)(0.5) = 0.25
Total relatedness = 0.25 + 0.25 = 0.5

First cousins

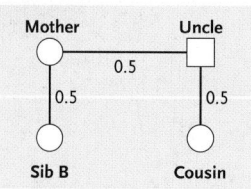

Relatedness = (0.5)(0.5)(0.5) = 0.125

documented in many species of birds and mammals. The phenomenon is especially common among animals in which inexperienced parents are not very successful at reproducing offspring on their own. By helping, they gain experience and realize some genetic benefit.

Altruism involves doing something that enhances the situation of another individual, but Hamilton's kin selection theory demonstrates why parental behaviour (or helping parents raise siblings) is genetically selfish, not altruistic. Therefore, the behaviour of the wolf mentioned above is not altruistic. Robert Trivers proposed that individuals will help nonrelatives if they are likely to return the favour in the future. Trivers called this **reciprocal altruism** because each member of the partnership can potentially benefit from the relationship. Trivers hypothesized that reciprocal altruism would be favoured by natural selection as long as individuals that do *not* reciprocate (cheaters) are denied future aid.

Among the many features of social animals, the evolution of cooperative behaviour can be one of the most challenging to understand. Why has cooperative behaviour arisen in populations of animals? How does it arise? And how is it maintained in populations?

John M. McNamara and three colleagues wrote about the coevolution of choosiness and cooperation. Using modelling and simulation experiments, these authors examined the consequences arising in situations in which one individual's cooperativeness influences the decisions about actions by other individuals toward group members. They postulated a situation of *competitive altruism* in which individuals actually compete with one another to be more cooperative.

The results of their analysis suggest that longer-lived species are more likely to develop cooperative behaviour than shorter-lived ones. This is important because the model does not require intermediate situations involving negotiation behaviour. The model helps us understand the appearance of cooperative behaviour in all animals, including in *Homo sapiens.*

Reciprocal altruism is one element of cooperative behaviour, and dolphins may be reciprocal altruists. Many species of dolphins are long lived and social, living in groups. Dolphins and other cetaceans show many forms of aid-giving behaviour, from attending injured group members to assisting with difficult births. They also use group behaviour to protect themselves from the attacks of sharks. Richard Connor and Kenneth Norris proposed that the persistent threat of attacks by sharks and the perils of living in the ocean combined to provide dolphins with many opportunities to help one another, or even members of other species. Connor and Norris did not have specific details of genetic relationships among group members, but they proposed that dolphins are reciprocal altruists.

STUDY BREAK

1. What is the main argument in Hamilton's kin selection theory?
2. Imagine that four of your first cousins, two siblings, and two half-siblings are about to fall from a cliff and die. You have the option of taking their place. In terms of kin selection, which is more beneficial to you, your life or the life of your genetic relatives?
3. Which of the following behaviours is altruistic: parental care, mate selection, courtship feeding, self-defence, and/or helping nonrelatives? Explain your choices.

47.14 Eusocial Animals

Hamilton's insights led to the prediction that self-sacrificing behaviour should be directed to kin. Evidence from many species of animals, particularly bees, ants, termites, and wasps, overwhelmingly supports this prediction. In a colony of **eusocial** insects, thousands of genetically related individuals, most of them sterile workers, live and work together for the reproductive benefit of a single queen and her mate(s). The workers may even die in defence of their colonies.

How did this social behaviour evolve, and why does it persist over time? A colony of honeybees may contain 30 to 50 thousand related individuals, but only the queen bee is fertile. All of the workers are her daughters **(Figure 47.38).** The queen's role in the colony is to reproduce. The workers perform all of the other tasks in maintaining the hive, from feeding the queen and her larvae to constructing new honeycomb and foraging for nectar and pollen. They also transfer food to one another (trophallaxis) and sometimes guard the entrance to the

a. Queen with sterile workers

b. Workers sharing food and passing pheromones

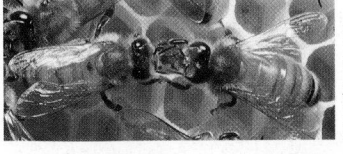

Figure 47.38

(a) In a hive of honeybees, a court of sterile workers (daughters) surround their mother (the queen). **(b)** Worker bees routinely share food (trophallaxis) and transfer pheromones to one another.

The drone (male parent) has only one set of chromosomes (symbolized by a red circle), which he contributes to the genome of every female worker. Thus, the workers are related to each other by 50% through their male parent.

Drone
♂

Queen
♀

The queen (female parent) has two sets of chromosomes (symbolized by a green triangle and a blue square). She contributes half of her alleles to each female offspring (either a triangle or a square) in this simplified presentation.

Workers that receive different alleles from the queen share no genetic relationship through their female parent. Thus, they are related to each other by 50% (the alleles inherited from their male parent).

Workers that receive the same alleles from the queen are related to each other by 50% through their male parent plus an additional 50% through their female parent.

Figure 47.39

Haplodiploidy. The genetic system of eusocial insects produces full siblings with exceptionally high degrees of relatedness. Although this simplified model ignores recombination between the queen's two sets of chromosomes, it demonstrates how half of the workers are related to each other by 50% and half are related to each other by 100%. On average, the relatedness between workers is 75%.

hive. Some pay the ultimate sacrifice when they sting intruders because stinging tears open the bee's abdomen, leaving the stinger and the poison sac behind in the intruder's skin and killing the bee.

In bees and other eusocial insects, sex is determined genetically through **haplodiploidy (Figure 47.39)**. Female bees are diploid because they receive a set of chromosomes from each parent. Male bees (drones), however, are haploid because they hatch from unfertilized eggs. All of the sperm carried by a drone will be genetically identical because he has just one set of chromosomes. When a queen bee mates with just one male, all of her worker offspring will inherit exactly the same set of alleles from their male parent, ensuring at least a 50% degree of relatedness among them. Like other diploid organisms, workers are related to each other by an average of 25% through their female parent. Adding these two components of relatedness, workers are related to each other by an average of 75%, a higher degree of

LIFE ON THE EDGE 47.2

Advantage: Social Behaviour

The McNamara et al. model of competitive altruism helps explain the evolution of blood-sharing behaviour in vampire bats. Vampire bats are the only euthermic blood-feeders, and like many other bats, they are long lived in the wild (recorded to at least 19 years of age). The three living species of vampire bats, Common Vampire Bat (*Desmodus rotundus*), White-Winged Vampire Bat (*Diaemus youngi*), and Hairy-Legged Vampire Bat (*Diphylla ecaudata*), all practise food sharing. An individual unsuccessful in foraging can return to its roost and beg blood from a successful forager among its roost mates. The donor bat regurgitates some of its blood meal to the recipient. G. S. Wilkinson's work with Common Vampire Bats demonstrated that individuals roost with both genetic relatives and nonrelatives. Familiarity, not relatedness, was the key to food sharing by these bats.

The selection process for the behaviour can be placed in context by evidence about a bat's success. Adult Common Vampire Bats are typically unsuccessful in obtaining blood one night per month. An adult can survive two days (daytime periods) without feeding but not three. This means that on any night in any month in a colony of 30 adult vampire bats, one individual will benefit from the cooperativeness of a roost mate.

Even more important, young bats may be unsuccessful three or four times a week. Blood-feeding bats thus live on the edge of survival and likely depend on a network of cooperation by roost mates. The social network demonstrated for Common Vampire Bats probably applies to white-winged and hairy-legged species as well. The network is based on cooperation and may be the key to being a successful euthermic blood-feeder.

relatedness than they would have to any offspring they would have produced had they been fertile.

The high degree of relatedness among workers in some colonies of eusocial insects may explain their exceptional level of cooperation. When Hamilton first worked out this explanation of eusocial behaviour, he suggested that workers devote their lives to caring for their siblings (the queen's other offspring) because a few of those siblings, those carrying 75% of the workers' alleles, may become future queens and produce enormous numbers of offspring themselves.

Naked mole rats are eusocial mammals with non-breeding workers. In East Africa, these small, almost hairless animals live in underground colonies of 70 to 80 individuals. Like eusocial insects, naked mole rats share an exceptionally high proportion of alleles (see "Naked Mole Rats," p. 1185).

Animals living in groups, whether they are aggregations or social units, may be at greater risk of inbreeding than those living alone. Dispersal is a mechanism that can reduce the chances of incestuous matings and inbreeding. Although spotted hyenas live in clans, the males tend to disperse from their natal units, minimizing the risk of inbreeding. Using microsatellite profiling, O. P. Höner and colleagues showed that a female preferred mates that had been born into or immigrated into the clan after she was born.

STUDY BREAK

What is haplodiploidy? How does it relate to Hamilton's prediction about self-sacrificing behaviour?

47.15 Human Social Behaviour

Humans and chimpanzees share 96% of their genomes. Compared with humans, both chimpanzees and bonobos (*Pan paniscus*) live in relatively unstructured social groups. The brains of *Homo sapiens* are approximately three times as large as those of great apes such as the chimps and bonobos. Brain tissue has a very high metabolic rate, so growing and operating a large brain imposes significant costs.

The cultural intelligence hypothesis proposes that large brain size in humans reflects cognitive skill sets absent from great apes. Large brains allow humans to perform many cognitive tasks more rapidly and efficiently than other species with smaller brains. The tasks include those associated with memory, learning time, long-range planning, and complexity of interindividual interactions.

To test this, Esther Herrmann and her colleagues administered a large battery of cognitive tests to chimpanzees, orangutans (*Pongo pygmaeus*), and two-and-a-half-year-old human children. The children in the experiment were preschool and preliteracy. Although the children, chimpanzees, and orangutans had similar cognitive skills for dealing with the physical world, the children had more sophisticated cognitive skills for dealing with the social world **(Figure 47.40)**. The data support the hypothesis that cultural intelligence is an important way to distinguish humans from their closest living relatives.

The ultimatum game is an economic decision-making tool for assessing the responses of individuals to opportunities and the behaviour of others. Responses allow researchers to distinguish between players on the basis of sensitivity and sense of fairness. Keith Jensen and his colleagues used the ultimatum game to compare humans and chimpanzees. Two anonymous individuals can play a round of this game. One, the proposer, is offered a sum of money (or a food reward) and can decide whether to share it with the other, the responder. The responder can accept or reject the proposer's offer. If the responder accepts the offer, then both receive their share of the reward. If the responder rejects the offer, then neither gets any reward. The economic model predicts that the proposer will offer the responder the minimum award.

When humans and chimps play the ultimatum game, their behaviour differs **(Figure 47.41, p. 1184)**. Chimps are rational maximizers because proposers

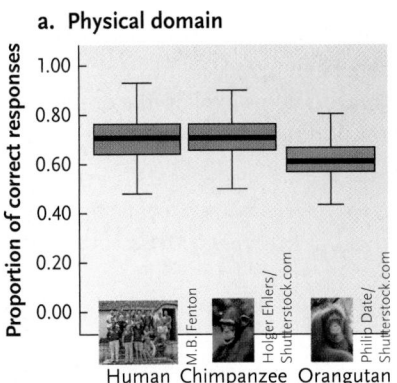

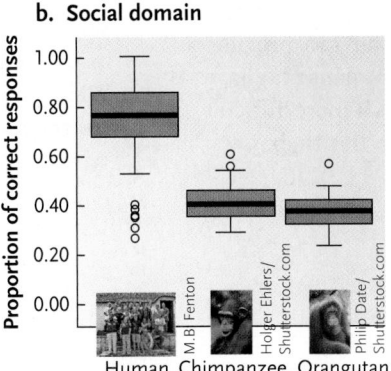

Figure 47.40

Humans, chimpanzees, and orangutans. Box plots showing the proportion of correct responses to survey questions in the physical and the social domains. In the social domain, outlying data points (circles) were at least 1.5 times the interquartile distances (shown by the error bars).

From Esther Herrmann, Josep Call, Maria Victoria Hernandez-Lloreda, Brian Hare, Michael Tomasello, "Humans Have Evolved Specialized Skills of Social Cognition: The Cultural Intelligence Hypothesis," *Science*, vol. 317, Sep 7, 2007, pp. 1360–1366. Reprinted with permission from AAAS.

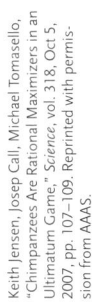

Keith Jensen, Josep Call, Michael Tomasello, "Chimpanzees Are Rational Maximizers in an Ultimatum Game," *Science*, vol. 318, Oct 5, 2007, pp. 107–109. Reprinted with permission from AAAS.

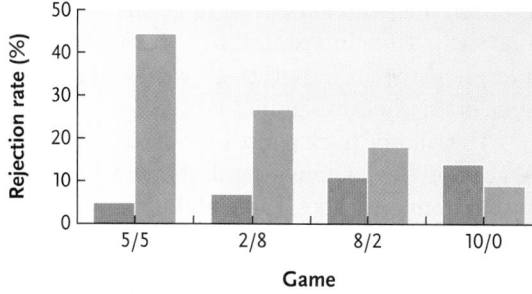

Figure 47.41

The ultimatum game. Data from chimpanzees (orange bars) and humans (green bars) show rejection rates (percentage of offers) indicating fundamental differences in the way that humans and chimps approach issues of fairness. The chimps are rational maximizers, whereas the humans are not.

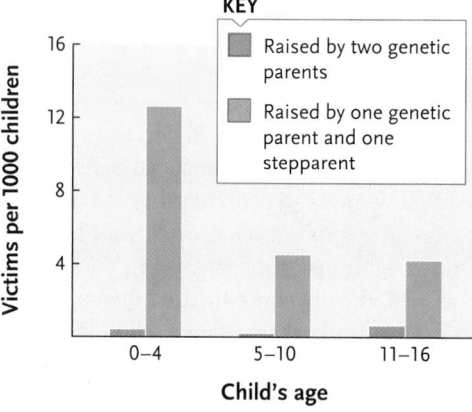

Observational Research Figure 47.42 Children raised by one genetic parent and one stepparent were 40 times as likely to suffer criminal abuse at home as children living with two genetic parents.

typically offer 40 to 50% of the reward, and responders typically reject offers of less than 20%. They follow the economic model and show little sensitivity to fairness or the interests of others. Humans are not rational maximizers because they are sensitive to fairness and the interests of others (see Figure 47.41). Together, the cultural intelligence hypothesis and the results of the ultimatum game suggest social differences between humans and their closest relatives. These findings support the views of people who believe that humans are not animals.

In other ways, humans behave like other animals. In the area of reproduction and genetic selfishness, some humans show little difference from their mammalian cousins. Kin selection predicts that humans (and other animals) that are genetic relatives will benefit from assisting the members of their family. What happens when there is no close genetic tie between parents and children?

Margo Wilson and Martin Daly wondered if child abuse might be more common in families with stepparents who are not genetically related to all of the children in their care. They examined data on criminal child abuse within families, made available by the police department of a Canadian city. They found that the chance that a young child would be subject to criminal abuse was 40 times as high when children lived with one stepparent and one genetic parent as with children living with both genetic parents **(Figure 47.42).**

This example illustrates the insights that an evolutionary analysis of human behaviour can provide. Wilson and Daly made the point that humans may have some genetic characteristic that makes it more difficult to invest in children they know are not their own,

particularly if they also care for their own genetic children. They did not excuse child abusers or claim that abusive stepparenting is acceptable. These results are not just academic. Most stepparents cope well with the difficulties of their role, but a few do not. Knowing the familial circumstances under which child abuse is more likely to occur may allow us to provide social assistance that could prevent some children from being abused in the future.

In recent years, the application of evolutionary thinking to human behaviour has produced research on many kinds of questions. Some questions are interesting or even profound: Why do some tightly knit ethnic groups discourage intermarriage with members of other groups? At other times, the issues may seem frivolous: Why do men often find women with certain physical characteristics attractive? Although evolutionary hypotheses about the adaptive value of behaviour can be tested, helping us understand why we behave as we do, the hypotheses should never be used to justify behaviour that is harmful to other individuals. Understanding why we get along or fail to get along with each other and the ability to make moral judgments about our behaviour are uniquely human characteristics that set us apart from other animals.

STUDY BREAK

1. What is the ultimatum game? How does it help us understand behaviour?
2. What genetic reason helps explain the domestic risks to foster children and stepchildren?

Naked Mole Rats

Naked mole rats are sightless and essentially hairless burrowing mammals **(Figure 1)** that live in mazes of subterranean tunnels in parts of Ethiopia, Somalia, and Kenya. Colonies of naked mole rats may number from 25 to several hundred individuals. In each colony, a single "queen" and one to three males are the breeders. All of the others, males and females, are nonbreeding workers that, like worker bees, ants, and termites in insect colonies, do all of the labour, including digging and defending the tunnels and caring for the queen and her mates. H. Kern Reeve and his colleagues set out to determine if close kinship could explain the behaviour of worker naked mole rats. They used molecular techniques resembling DNA finger-printing analysis (see Chapter 15) to obtain data about relatedness. The technique depends on a group of

Figure 1
Naked mole rats (Heterocephalus glaber) *live in colonies containing many workers that are effectively sterile.*

repeated DNA sequences that vary to a greater or lesser extent among individuals (e.g., they are polymor-phic). No two individuals (except identical twins) are likely to have exactly the same combination of sequences. Brothers and sisters with the same parents have the most closely related sequences, and differences increase as genetic relationships become more distant. Reeve and his colleagues captured mole rats living in four colonies in Kenya. Individuals from the same colony were placed together in a system of artificial tunnels. Samples of the entire DNA complement were extracted from individuals that died naturally in the artificial colonies. The extracted DNA was then "probed" with radioactively labelled DNA sequences that paired with and marked the three distinct groups of polymorphic sequences in the mole rat DNA (see Chapter 15).

Naked mole rat sequences were then fragmented by treatment with a restriction endonuclease. This procedure produced a group of fragments that, reflecting the varia-tions in polymorphic sequences, is unique for each individual. As a final experimental step, the fragments for each individual were separated into a pattern of bands by gel electrophore-sis. The pattern of bands, different for each individual, is the DNA fingerprint.

Reeve and his colleagues compared the DNA fingerprint of each mole rat with those of other members of the same and other colonies. In the comparisons, bands that were the same in two individuals were scored as *hits*. The number of hits was then analyzed to assign relatedness by noting which individuals shared the greatest number of bands.

Individuals in the same mole rat colony were found to be closely related. They shared an unusually high number of bands, higher than human siblings and approaching the kin similarity of identical twins. The number of bands shared between individuals of different colonies was significantly lower but still higher than that noted between unrelated individuals of other vertebrate species. Close relatedness of even separate colonies may be due to similar selection pressures or to recent common ancestry among colonies in the same geographic region.

In naked mole rats, close genetic relatedness among individuals in a colony could explain the altruistic behaviour of workers. The persistence of the social organization reflects its importance to the survival of individ-ual naked mole rats, rather like the situation in lion prides.

PEOPLE BEHIND BIOLOGY 47.3
R. F. (Griff) Ewer, *University of Ghana*

Experimental Research Box

Born Rosalie Griffith, Griff Ewer was an outstanding leader in the study of animal behaviour. She married Denis William (Jakes) Ewer, and with their two children, they moved to the University of Natal in 1946 and from there to Rhodes University. In 1963, the family moved to the University of Ghana. To the unsuspecting person meeting her, the fact that she wore her hair short, dressed in slacks and a suit jacket, and smoked a pipe meant that some visitors were at first confused. On one paleontological expedition, she used the newspaper in which meat had been wrapped as a tablecloth. Her young daughter started to read the paper, drawing Ewer's attention to the fact that she was "in the tablecloth"— a story about the findings of a "woman paleontologist." Later, when she met the reporter who had written the story, she proceeded to express her annoyance about being identified as a woman paleontologist. As she and her daughter walked away and were out of earshot, her daughter observed that she must have been annoyed with the man. Ewer said that she was not annoyed and was then told that her

(Continued)

foot must have been. Ewer observed that from a child's perspective, her foot stamping was more obvious than it had been to the reporter!

Ewer had an insatiable curiosity about animal behaviour. In her classic book *Ethology of Mammals*, published in 1968, she encouraged would-be students of animal behaviour to take animals into their homes and live with them. Animals had free run of her house. She observed that if there was something about their behaviour that you did not understand, the animal would patiently demonstrate the behaviour time and time again. A visitor to her house was usually quickly scent-marked by the resident mongooses. The book was dedicated to her pet meerkats. *Ethology of Mammals* stands as a classic book, a tribute to Ewer and her contributions.

Review

To access course materials such as Aplia and other companion resources, please visit www.NELSONbrain.com.

47.1 Genes, Environment, and Behaviour

- Instinctive behaviours are genetically or developmentally programmed. They appear in complete and functional form the first time they are used. Examples are eating, defence, mating, and parental care. Learned behaviours depend on having a particular kind of experience during development. Examples are language, mobility, and foraging. Marler's work with White-crowned Sparrows demonstrated some aspects of learned and instinctive behaviours.

47.2 Instinct

- Fixed action patterns are triggered by specific cues (sign stimuli). Fixed action patterns are repeated over and over in almost exactly the same way. The begging behaviour of Herring Gull chicks is a good example.

- Garter snakes from some areas of California often eat slugs, whereas those from other areas do not. Snakes that regularly eat slugs flick their tongues in response to the odour of slugs. Cross-breeding snakes from populations that eat slugs with those that do not demonstrates that the response to slugs was partly under genetic control.

- The feeding situations described in "Why It Matters" demonstrate how animals learn to adjust their foraging behaviour according to the availability of prey. The examples also demonstrate how some animals plan their meals ahead. The examples demonstrate the flexibility of animal behaviour.

47.3 Learning

- Research with White-crowned Sparrows and Zebra Finches demonstrated how the brains of these birds are involved in learning and matching song outputs. The work has also revealed the neurophysiological networks involved in selective learning, including how individual birds learn to ignore (habituate to) familiar signals.

47.4 Neurophysiology and Behaviour

- Studies of bird brains have demonstrated how specific enzymes are turned on to activate different patterns of behaviour.

- Changes in the level of factors controlling synapses can clearly influence animal behaviour.

47.5 Hormones and Behaviour

- In honeybees, juvenile hormone stimulates genes in certain brain cells to produce proteins that affect functions of the nervous system. Octopamine is a product that stimulates neural transmissions and reinforces memories—it occurs in higher concentrations in older bees. This example shows how hormones and genes interact to affect behaviour.

- Gonadotropin-releasing hormone (GnRH) causes differences in the behaviour of fish such as *Haplochromis burtoni*. Levels of GnRH are useful predictors of territorial behaviour.

47.6 Neural Anatomy and Behaviour

- Some crickets use ears (tympanic membranes) on their front legs to detect the echolocation calls of bats. Differential stimulation of left and right ears allows crickets to rapidly change their flight behaviour in response to approaching bats.

- Fiddler crabs (*Uca pugilator*) have compound eyes that give them a split field of view, which allows them to distinguish between the movements of other crabs and the approach of potential predators.

47.7 Communication

- Animals use at least five classes of signals in communication: acoustic signals such as songs used to attract a mate or warn intruders; visual signals such as the bioluminescent display of fireflies; chemical signals such as pheromones like bombykol, produced by female silkworm moths and predatory spiders; tactile signals such as grooming in group-living primates; and electrical signals such as the pulses produced by knifefish or elephant fish.

- The "yells" of ravens that have found a patch of food attract others. This behaviour allows nonresidents to gang up on resident ravens and obtain food that would otherwise not be available to them.

- Syntax (or grammar) and symbols are characteristic of language. Humans use words to represent objects (chair, table, dog), and the order of words (syntax) affects meaning (she was bitten by a dog versus she bit a dog). Dancing honeybees use symbols and syntax, and animals such as vervet monkeys use different calls to represent different predators.

47.8 Space

- Kinesis is a change in the rate of movement or the frequency of turning movements in response to environmental stimuli. A taxis is a response directed either toward or away from a specific stimulus.

- An animal's home range is the area it regularly uses, typically to move from where it sleeps to where it feeds. A territory is space that is defended to allow exclusive use by an individual or group of individuals. Home ranges and territories are features of individuals, whereas a range is a feature of a species (a population of individuals).

47.9 Migration

- Migration is a seasonal movement to and from an area. Migrations can be lengthy or short. Some birds, bats, insects, whales, and caribou migrate between different areas according to the season. Some nomadic populations of humans (and some retired people) move to and from habitats, also according to the season. Migrations are usually triggered by some combination of changes in day length and weather.

- Animals may find their way by piloting, landmarks, compass, or celestial (stars and the Sun) cues in navigation. Compass orientation involves movement in a specific direction as indicated by an external source such as the Sun, the stars, or Earth's magnetic field. Animals continue to move in the same direction until they have completed their journey. Navigation is more complicated. Although still using compass orientation, navigation also requires the use of a mental map, some independent indication of the animal's location. The mental map enables the animal to determine its relative position to the target location, and the compass provides the direction back to the target location.

47.10 Mates as Resources

- In many species, the behaviour of individuals of one sex is determined by the distribution of the other sex. In polygynous species, the distribution of males is influenced by the distribution of females, which are defended by the males. In other cases, albeit less commonly, males are defended by females. Be careful to distinguish situations in which the individuals are the resource that is defended as opposed to access to food or roosts (for example).

47.11 Sexual Selection

- Monogamy describes the situation when a male and a female form a pair bond and do not mate with others.

- Polygamy occurs when a male (polygyny) or a female (polyandry) mates with multiple partners. Polygamy usually involves pair bonds between the mating individuals. In promiscuity, there are no pair bonds, and males and females may mate with multiple partners.

- Lek mating systems are promiscuous. Males congregate in a display area (an arena), where they are visited by females. Females mate with the most attractive male (the one with the best display area). Males are the only resource at the display site. In leks, females typically mate with one male and males with multiple females.

47.12 Social Behaviour

- Infanticide is the killing of conspecific young. Male African lions practise infanticide when they take over a pride (group of females) because stopping nursing brings females back into heat, allowing the male to increase his direct fitness.

- Group living can provide significant advantages to animals, such as easier-to-locate food resources either through cooperative hunting or simply an increased searching capacity (many eyes), ease in finding a mate, more efficient raising of young, more effective defence against predators, and increased vigilance. Group living can also present disadvantages, including increased competition for food, increased spread of parasites and disease, and increased conspicuousness to predators.

47.13 Kin Selection and Altruism

- Hamilton's kin selection theory states that helping genetic relatives effectively propagates the helper's genes because family members share alleles inherited from their ancestors. Individuals should be more likely to help close relatives because increasing a close relative's fitness means that the individual is helping to propagate some of its own alleles.

- The degree of relatedness between four first cousins, two full siblings, and two half-siblings is calculated as follows:

$$= (4 \times \text{first cousin relatedness}) +$$
$$(2 \times \text{sibling relatedness}) +$$
$$(2 \times \text{half-sibling relatedness})$$
$$= (4 \times 0.125) + (2 \times 0.5) + (2 \times 0.25)$$
$$= 0.5 + 1 + 0.5$$
$$= 2$$

- Helping only nonrelatives is altruistic because each of the other examples involves genetically selfish behaviour (= getting the actor's genes into the next generation).

47.14 Eusocial Animals

- Haplodiploidy occurs when the females in a colony are diploid and the males are haploid. Females receive a set of chromosomes from each parent, but males (drones) hatch from unfertilized eggs (one set of chromosomes). This affects the level of genetic relationship among females. Worker bees hatching from eggs laid by one queen bee could all have the same father if the queen mated with only one male. In this scenario, the workers will share 75% of their genes. We can use Hamilton's kinship theory to understand the behaviour of eusocial animals.

47.15 Human Behaviour

- In the ultimatum game, players have the opportunity to demonstrate their sense of fairness and their sensitivity to others. When people and chimps play the ultimatum game, the results demonstrate a fundamental difference in their levels of social behaviour.

Questions

Self-Test Questions

1. As concluded by Peter Marler, under what circumstances can White-crowned Sparrows learn their species' song?
   a. after receiving hormone treatments
   b. during a critical period of their development
   c. under natural conditions
   d. from their genetic father

2. In cichlid fish, which of the following occurs with high levels of the hormone GnRH?
   a. Females are receptive to male attention.
   b. Males are sexually aggressive but not territorial.
   c. A male defends its territory.
   d. Males lose their bright colours.

3. Sensory bias in the nervous system of a cricket ensures that ultrasound perceived on one side of the body causes which of the following?
   a. a movement in a leg on the same side of the body
   b. a movement in a leg on the opposite side of the body
   c. the cricket to respond with a vocalization
   d. the cricket to fly toward the sound

4. In the brain of a star-nosed mole, more cells decode this.
   a. tactile information from its feet than from all other parts of its body
   b. tactile information from the tentacles on its nose than from all other parts of its body
   c. tactile information from its mouth than from all other parts of its body
   d. visual information from the bottom part of its visual field than the top part

5. Which of the following statements about animal migration is true?
   a. Piloting animals use the position of the Sun to acquire information about their direction of travel.
   b. Animals migrating by compass orientation use mental maps of their position in space.
   c. Navigating animals use familiar landmarks to guide their journey.
   d. Navigating animals use a compass and a mental map of their position to reach a destination.

6. The squashing of an ant on a picnic blanket often attracts many other ants to its "funeral." What kind of signal does squashing of the ant likely produce?
   a. a visual signal
   b. an acoustical signal
   c. a chemical signal
   d. a tactile signal

7. Compared with males, the females of many animal species do which of the following?
   a. compete for mates
   b. choose mates that are well camouflaged in their habitats
   c. choose to mate with many partners
   d. choose their mates carefully

8. The following is true about cooperative behaviour (not eusocial behaviour).
   a. It is exhibited only by animals that live in groups with close relatives.
   b. It evolved because group living provides benefits to individuals in the group.
   c. It is never observed in insects and other invertebrate animals.
   d. It can only be explained by the hypothesis of kin selection.

9. Altruism is a behaviour that does the following.
   a. It advances the welfare of the entire species.
   b. It increases the number of offspring an individual produces.
   c. It can indirectly spread the altruist's alleles.
   d. It can only evolve in animals with a haplodiploid genetic system.

10. Naked mole rats are like eusocial insects for the following reason.
    a. They live underground.
    b. They show evidence of behavioural castes.
    c. Females mate with more than one male.
    d. Some females are "queens."

Questions for Discussion

1. When can communication behaviour be called *language*? What is language?

2. Using an example from your own experience, explain why habituation to a frequent stimulus might be beneficial. Describe an example in which habituation might be harmful or even dangerous.

3. Is learning always superior to instinctive behaviour? If you think so, why do so many animals react instinctively to certain stimuli? Are there environmental circumstances in which being able to respond "correctly" the first time would have a big payoff?

4. What effects might climate change have on animal species that undertake seasonal migrations?

5. Develop three evolutionary hypotheses to explain why male birds are likely to involve themselves in caring for their young.

Soil provides nutrients for plants that provide nutrients for herbivores (such as deer) that provide nutrients for carnivores (such as the grey wolf).

Animal Nutrition

WHY IT MATTERS

While many animal herbivores obtain nutrients directly from the plants they eat, two different groups of sea slugs have evolved ways of harnessing the ability of plants to convert the Sun's energy into sugars and other nutrients in very different ways. They use plants or plastids from plant cells in "solar panels" to obtain nutrients more directly. Some herbivorous nudibranchs (sacoglossans) have evolved branches of their gut that ramify throughout the body wall. These *cerata* greatly extend the surface area of the animals and contain plastids, the photosynthesizing "factories" extracted from the algae on which they feed. The cerata act as solar panels, rotating and exposing the plastids to sunlight for photosynthesis, and in return the plastids provide the sea slugs with valuable nutrients they need. In *Elysia chlorotica,* one of the solar-powered sea slugs found in Canadian waters **(Figure 48.1, p. 1190)**, the plastids are chloroplasts, but some sacoglossans feed on red and brown algae and keep the plastids from these algae alive as well. Other, carnivorous, nudibranchs have evolved ways of harvesting entire single-celled plants (zooxanthellae) living symbiotically in animals that the nudibranchs ate. They then keep the zooxanthellae alive in their bodies as their own symbionts. Clearly, this form of symbiosis has evolved many times within the nudibranchs, with examples in many quite unrelated families and orders.

The subject of this chapter is animal **nutrition**—a topic that includes what nutrients are, and the processes by which nutrients are obtained and absorbed into body cells and fluids. We begin with a discussion of the basic categories of nutrients. We then discuss **ingestion** in animals, the feeding methods used to take food into the digestive cavity, and the process of **digestion**, which is the splitting of carbohydrates, proteins, lipids, and nucleic acids in foods into chemical subunits small enough to be absorbed into an animal's body fluids and cells. The chapter also presents the main structural and functional features of digestive systems, with special emphasis

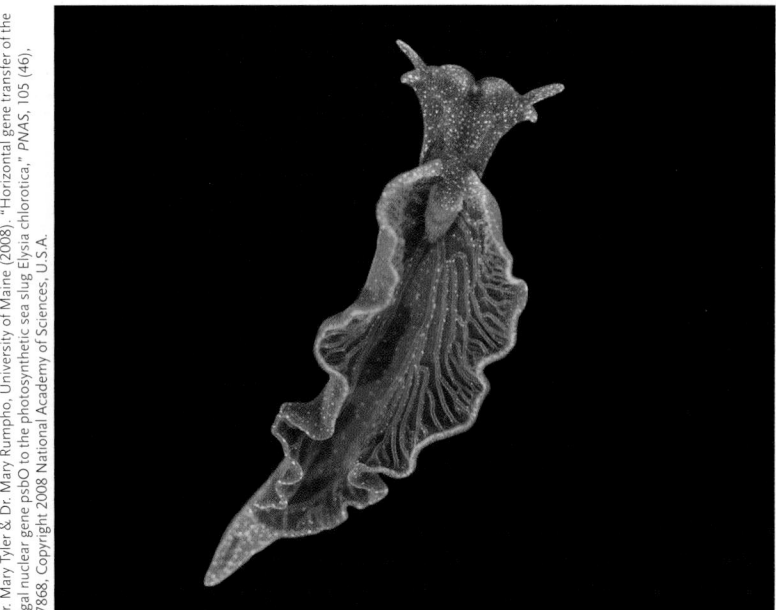

Dr. Mary Tyler & Dr. Mary Rumpho, University of Maine (2008). "Horizontal gene transfer of the algal nuclear gene psbO to the photosynthetic sea slug Elysia chlorotica." PNAS, 105 (46), 17868. Copyright 2008 National Academy of Sciences, U.S.A.

Figure 48.1

Elysia chlorotica, a solar-powered sea slug found in Canadian waters.

on vertebrates. The adaptations animals use to obtain and digest food are among their most strongly defining anatomical and functional characteristics.

48.1 Nutrients: Essential Materials

All organisms require nutrients to grow, maintain essential life functions, and reproduce. Some of these nutrients are used as building blocks to make structures, but most are used as sources of energy that is released and harnessed by carefully breaking chemical bonds (see Chapter 2). No organism can survive when deprived of the chemical elements essential for its metabolism. In the latter half of the nineteenth century, physiologists exploited rapid advances in chemistry to explore the chemical composition of the **essential nutrients** animals need to survive. They discovered that some elements essential for growth and maintenance were needed in relatively large amounts (**macronutrients**), while others were needed in only small trace amounts (**micronutrients**). They also discovered that only some of these nutrients were essential in the food of animals but that many could either be replaced with, or converted from, another element.

48.1a Essential Elements for Animals

Animals require a diet of organic molecules as a source of energy and nutrients that they cannot make for themselves. Animals can be classified according to the sources of organic molecules they use for these purposes. **Primary consumers** eat plants, whereas **secondary consumers** primarily eat other animals. "Primarily" is an appropriate modifier because many

herbivores sometimes eat animal matter (e.g., insects on the plants), and secondary consumers often eat plant material (e.g., a cat eating grass). Animals that regularly take food from different trophic levels are omnivores.

Organic molecules are the basis of two of the most fundamental processes of life, namely as fuels for oxidative reactions supplying energy and as building blocks for making complex biological molecules.

Fuel. Animals must acquire enough fuel in their diets to cover their basic costs of operation. Carbohydrates and fats are primary organic fuel molecules used in cellular respiration (see Chapter 6). Undernourished animals suffer from inadequate intake of organic fuels or abnormal assimilation of these fuels. **Undernutrition** is commonly referred to as **malnutrition**, a condition resulting from an improper diet. **Overnutrition** is caused by excessive intake of specific nutrients and is another form of malnutrition (see "The Puzzling Biology of Weight Control," p. 1210).

An undernourished animal can be starving for one or more nutrients or just be eating fewer calories than needed for daily activities. Animals with chronic undernutrition lose weight because they use molecules of their own bodies as fuels. In times when food is abundant, some animals accumulate stores of fat for use in lean times. Birds on long migratory flights metabolize fat, as well as other tissues, to meet their energy needs. Relatively short-term use of an animal's own proteins as fuels leads to the wasting of muscles and other tissues and cannot be sustained over long periods of time.

Building Blocks. Organic molecules serve as building blocks for carbohydrates, lipids, proteins, and nucleic acids. Animals can synthesize many of the organic molecules that they do not obtain directly in their diet by converting one type of building block into another. But there are some amino acids and fatty acids that most animals cannot synthesize and must obtain from organic molecules in their food. Lack of these essential amino acids and **essential fatty acids** can have serious consequences. Protein synthesis cannot continue unless all 20 amino acids are present. In the absence of essential amino acids in the diet, an animal would have to break down its own proteins to obtain the necessary building blocks for new protein synthesis.

Vitamins and Coenzymes. Animals must also ingest **vitamins**, organic molecules that are required in small quantities. Many animals cannot synthesize these for themselves. Many vitamins are coenzymes, nonprotein organic subunits associated with enzymes that assist in enzymatic catalysis (see Chapter 4). Individual species differ in the vitamins, essential amino acids, and fatty acids they require in their diets. Various species also have different dietary requirements for inorganic elements such as calcium, iron, and magnesium. Required inorganic elements are known collectively as essential nutrients. Essential nutrients include amino acids, fatty acids, vitamins, and minerals, but the precise list varies from species to species, even from individual to individual.

Elemental macronutrients and micronutrients are essential in the animal diet (see **Table 48.1**). Humans require macronutrients in amounts ranging from 50 mg to more than 1 g per day and micronutrients (**trace elements**) such as zinc in small amounts, some less than 1 mg per day. All of the minerals, although

Table 48.1 | **Essential Elements and Their Functions in Animals**

Element	Commonly Absorbed Forms	Macronutrient or Micronutrient and Some Known Functions	Some Deficiency Symptoms	Sources for Humans
Calcium	Ca^{2+}	**Macronutrient** Bone and tooth formation, blood clotting, action of nerves and muscles	Stunted growth, diminished bone mass (osteoporosis in humans)	Dairy products, leafy green vegetables, legumes, whole grains, nuts
Chlorine	Cl^-	**Macronutrient** Formation of HCl (in stomach), contributes to acid–base balance, neural function, water balance	Muscle cramps, impaired growth, poor appetite	Table salt, meat, eggs, dairy products
Chromium	Cr	**Macronutrient** Roles in carbohydrate metabolism	Impaired responses to insulin (increased risk of type 2 diabetes mellitus in humans)	Meat, liver, cheese, whole grains, brewer's yeast, peanuts
Cobalt	Co	**Macronutrient** Constituent of vitamin B_{12} (required for normal red blood cell maturation)	Same as for vitamin B_{12} (see Table 48.2, p. 1193)	Meat, liver, fish, milk
Copper	Cu^+, Cu^{2+}	**Micronutrient** Component of several enzymes. Used in synthesis of melanin, hemoglobin, and in some electron transport chain components in mitochondria	Anemia, changes in bone and blood vessels	Nuts, legumes, seafood, drinking water, whole grains
Fluorine	F	**Macronutrient** Bone and tooth maintenance	Tooth decay	Fluoridated water, seafood, tea
Iodine	I	**Macronutrient** Thyroid hormone formation	Goitre (enlarged thyroid); metabolic disorders	Marine fish, shellfish, iodized salt
Iron	Fe^{2+}, Fe^{3+}	**Macronutrient** Component of hemoglobin and myoglobin	Iron-deficiency anemia	Liver, whole grains, green leafy vegetables, legumes, nuts, eggs, lean meat, molasses, dried fruit, shellfish
Magnesium	Mg^{2+}	**Macronutrient** Activation of enzymes, roles in functioning of nerves and muscles	Weak and sore muscles, impaired neural function	Whole grains, green vegetables, legumes, nuts, dairy products
Manganese	Mn^{2+}	**Macronutrient** Role is activation of enzymes, coenzyme action; plays a role in synthesis of urea and fatty acids	Abnormal bone and cartilage	Whole grains, nuts, legumes, many fruits
Molybdenum	MoO_4^{2-}	**Macronutrient** Components of some enzymes	Impaired nitrogen excretion	Dairy products, whole grains, green vegetables, legumes
Nitrogen	NO_3^-, NH_4^+	**Macronutrient** Component of proteins, nucleic acids, coenzymes		

(Continued)

Table 48.1		Essential Elements and Their Functions in Animals (*Continued*)		
Element	Commonly Absorbed Forms	Macronutrient or Micronutrient and Some Known Functions	Some Deficiency Symptoms	Sources for Humans
Phosphorus	$H_2PO_4^-$, HPO_4^{2+}	**Macronutrient** Component of nucleic acids, phospholipids, ATP, several coenzymes component of bones	Muscular weakness, loss of minerals from bone	Whole grains, legumes, poultry, red meat, dairy products
Potassium	K^-	**Macronutrient** Activation of enzymes, key role in maintaining water-solute balance and so influences osmosis involved in actions of nerves and muscles	Muscular weakness	Meat, many fruits and vegetables
Selenium	Se	**Macronutrient** Constituent of several enzymes, antioxidant	Muscle pain	Meat, seafood, cereal grains, poultry, garlic
Sodium	Na^+	**Macronutrient** Acid–base balance, roles in functioning of nerves and muscles	Muscle cramps	Table salt, dairy products, meats, eggs
Sulfur	SO_4^{2-}	**Macronutrient** Component of most proteins, coenzyme A	Same symptoms as those associated with protein deficiencies	Meat, eggs, dairy products
Zinc	Zn^{2+}	**Micronutrient** Component of digestive enzymes, transcription factors; role in normal growth, wound healing, sperm formation, and smell	Impaired growth, scaly skin, impaired immune function	Whole grains, legumes, nuts, meats, seafood

listed as elements, are ingested by animals as compounds or as ions in solution.

48.1b Essential Elements for Humans

Adult humans require eight essential amino acids: lysine, tryptophan, phenylalanine, threonine, valine, methionine, leucine, and isoleucine. Infants and young children also require histidine. Proteins in fish, meat, egg whites, milk, and cheese supply all of the essential amino acids as long as they are eaten in adequate quantities. In contrast, the proteins of many plants are deficient in one or more of the amino acids essential to humans. Corn contains inadequate amounts of lysine, and beans contain little methionine. Vegetarians, especially vegans, whose diet includes no animal-derived nutrients, must choose their foods carefully to obtain all of the essential amino acids **(Figure 48.2)**.

Protein deficiency occurs when essential amino acids are not part of the diet. Consequently, many enzymes and other proteins cannot be synthesized in sufficient quantities. Protein deficiency is most damaging to the young because of their need for proteins for normal development and growth. Even mild protein starvation during pregnancy or for some months after birth can retard a child's growth and have negative effects on mental and physical development.

Two fatty acids, linoleic acid and linolenic acid, are essential because they are used in the synthesis of

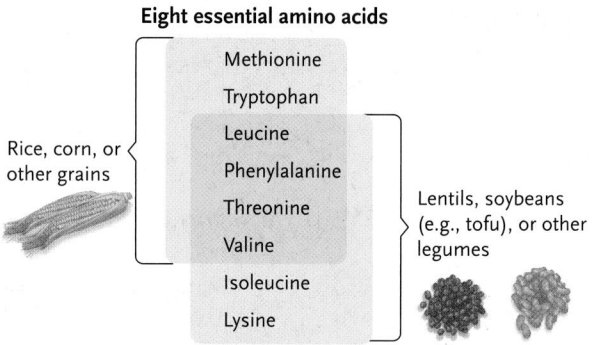

Eight essential amino acids

Rice, corn, or other grains

Methionine
Tryptophan
Leucine
Phenylalanine
Threonine
Valine
Isoleucine
Lysine

Lentils, soybeans (e.g., tofu), or other legumes

Figure 48.2
Obtaining essential amino acids in a human vegetarian diet.

phospholipids that form parts of biological membranes and certain hormones. Because almost all foods contain these fatty acids, most people have no problem obtaining them. However, people on a low-fat diet deficient in linoleic acid and linolenic acid are at serious risk for developing coronary heart disease. There is an inverse correlation between the concentration of these essential fatty acids in the diet and the incidence of coronary heart disease. Thus, Hindu vegetarians from India that eat mainly low-fat grains and legumes have a higher rate of coronary heart disease than rates in the United States and Europe, where dietary fat content is higher.

Humans require 13 known vitamins in their diet **(Table 48.2)**. Many metabolic reactions depend on

Table 48.2

Table 48.2 — Vitamins: Sources, Functions, and Effects of Deficiencies in Humans

Vitamin	Common Sources	Main Functions	Effects of Chronic Deficiency
Fat-Soluble Vitamins			
A (retinol)	Yellow fruits, yellow or green leafy vegetables; also in fortified milk, egg yolk, fish liver	Used in synthesis of visual pigments, bone, teeth; maintains epithelial tissues	Dry, scaly skin; lowered resistance to infections; night blindness
D (calciferol)	Fish liver oils, egg yolk, fortified milk; manufactured when body exposed to sunshine	Promotes bone growth and mineralization; enhances calcium absorption from gut	Bone deformities (rickets) in children; bone softening in adults
E (tocopherol)	Whole grains, leafy green vegetables, vegetable oils	Antioxidant; helps maintain cell membrane and red blood cells	Lysis of red blood cells; nerve damage
K (napthoquinone)	Intestinal bacteria; also in green leafy vegetables, cabbage	Promotes synthesis of blood-clotting protein by liver	Abnormal blood clotting, severe bleeding (hemorrhaging)
Water-Soluble Vitamins			
B_1 (thiamine)	Whole grains, green leafy vegetables, legumes, lean meats, eggs, nuts	Connective tissue formation; folate utilization; coenzyme forming part of enzyme in oxidative reactions	Beriberi; water retention in tissues; tingling sensations; heart changes; poor coordination
B_2 (riboflavin)	Whole grains, poultry, fish, egg white, milk, lean meat	Coenzyme	Skin lesions
Niacin	Green leafy vegetables, potatoes, peanuts, poultry, fish, pork, beef	Coenzyme of oxidative phosphorylation	Sensitivity to light; contributes to pellagra (damage to skin, gut, nervous system, etc.)
B_6 (pyridoxine)	Spinach, whole grains, tomatoes, potatoes, meats	Coenzyme in amino acid and fatty acid metabolism	Skin, muscle, and nerve damage
Pantothenic acid	In many foods (meats, yeast, egg yolk especially)	Coenzyme in carbohydrate and fat oxidation; fatty acid and steroid synthesis	Fatigue; tingling in hands; headaches; nausea
Folic acid	Dark green vegetables, whole grains, yeast, lean meats; intestinal bacteria produce some folate	Coenzyme in nucleic acid and amino acid metabolism; promotes red blood cell formation	Anemia; inflamed tongue; diarrhea; impaired growth; mental disorders; neural tube defects and low birth weight in newborns
B_{12} (cobalamin)	Poultry, fish, eggs, red meat, dairy foods (not butter)	Coenzyme in nucleic acid metabolism; necessary for red blood cell formation	Pernicious anemia; impaired nerve function
Biotin	Legumes, egg yolk; colon bacteria produce some	Coenzyme in fat and glycogen formation and amino acid metabolism	Scaly skin (dermatitis); sore tongue; brittle hair; depression; weakness
C (ascorbic acid)	Fruits and vegetables, especially citrus, berries, cantaloupe, cabbage, broccoli, green pepper	Vital for collagen synthesis; antioxidant	Scurvy; delayed wound healing; impaired immunity

vitamins, and the absence of one vitamin can affect the functions of the others. Vitamins fall into two classes: **water-soluble** (hydrophilic) **vitamins** and **fat-soluble** (hydrophobic) **vitamins**. The body stores excess fat-soluble vitamins in adipose tissues (fat), but any amount of water-soluble vitamins above daily nutritional requirements is passed in urine. Thus, meeting the daily minimum requirements of water-soluble vitamins is critical. The body can tap its stores of fat-soluble vitamins to meet daily requirements; however, these stores can be quickly depleted, and prolonged deficiencies of the fat-soluble vitamins may also become critical to health.

Humans can synthesize vitamin D (calciferol) through the action of ultraviolet light on lipids in the skin. People who are not exposed to enough sunlight to make sufficient quantities of the vitamin must rely on dietary sources. Although we cannot make vitamin K, much of what we require is supplied through the metabolic activity of bacteria living in our large intestine. Vitamin K deficiency is rare in healthy people. Vitamin K plays a role in blood clotting, so individuals with vitamin K deficiency will bruise easily and show increased blood-clotting times. Vitamin K deficiency can be caused in people on long-term antibiotic therapy because the antibiotics kill intestinal bacteria.

Other mammals have basically the same vitamin requirements as humans, with some differences. Most mammals can synthesize vitamin C, but not primates, guinea pigs, and fruit bats. To date, no mammals are known to synthesize B vitamins, but ruminants such as cattle and deer are supplied with these vitamins by microorganisms living in the digestive tract.

CONCEPT FIX It is a commonly held myth that eating fat makes you fat. The truth is that consuming more calories of anything (fat, carbohydrate, or protein) than you burn will lead to weight gain. It is also true that fat is a more concentrated form of calories than other substrates. Thus, gram for gram, you obtain more calories from fat than from other substrates. For humans, 20 to 30% of daily caloric intake should come from fats. Remember that we do need the essential fatty acids. The opposite is also true: gram for gram, animals can store more energy in fat than in any other way. This is why, no matter how we obtain our excess calories, they will end up being deposited as fat. ◗

STUDY BREAK

1. What is the difference between a macronutrient and a micronutrient?
2. What elements are essential for humans?
3. Why is iron essential for animals? What are the symptoms of iron deficiency in animals?
4. What do vitamins do for animals? Why are they necessary in their diets?

48.2 Obtaining and Absorbing Nutrients

48.2a Animals Eat to Acquire Nutrients

There are four basic feeding methods in animals, reflecting the physical states of the organic molecules they eat: fluid feeders, suspension feeders, deposit feeders, and bulk feeders **(Figure 48.3)**.

Fluid feeders ingest liquids containing organic molecules in solution. Among invertebrates, aphids, mosquitoes, leeches, butterflies, and spiders are fluid feeders. Among vertebrates, lamprey eels, hummingbirds, nectar-feeding bats, and vampire bats are examples of fluid feeders (see Figure 48.3a). Many fluid feeders have mouthparts specialized for reaching the source of their nourishment. Mosquitoes, bedbugs, and aphids have needlelike mouthparts that pierce body surfaces. Nectar-feeding butterflies, birds, and bats have long tongues that can extend deep within flowers. Some fluid feeders use enzymes or other chemicals to liquefy their food or to keep it liquid during feeding. Spiders inject digestive enzymes that liquefy tissues inside their victim, providing a nutrient soup they can ingest. The saliva

a. Fluid feeder

b. Suspension feeder

Baleen

c. Deposit feeder

d. Bulk feeder

Figure 48.3

General feeding methods in animals. (a) Hummingbirds, which eat nectar, are an example of a fluid feeder. **(b)** The northern right whale (*Balaena glacialis*), which gulps tonnes of water and filters out plankton, is an example of a suspension feeder. **(c)** The fiddler crab (*Uca* species), which sifts edible material from detritus, is an example of a deposit feeder. **(d)** This python, which can open its mouth very wide and take in large objects, such as a gazelle, is an example of a bulk feeder.

of mosquitoes, leeches, and vampire bats includes chemicals that keep blood from clotting.

Suspension feeders ingest small organisms suspended in water, such as bacteria, protozoa, algae, and small crustaceans, or fragments of these organisms. Suspension feeders include aquatic invertebrates, such as clams, mussels, and barnacles, and vertebrates, such as many species of fishes, as well as some birds, pterosaurs, and whales (see Figure 48.3b). Suspension feeders strain (filter) food particles suspended in water through a body structure covered with sticky mucus or through a filtering network of bristles, hairs, or other body parts. Trapped particles are funnelled into the animal's mouth, and the water is pushed out. Bits of organic matter are trapped by the gills of bivalves such as clams and oysters, and plankton is filtered from water by the sievelike fringes of horny fibre, called baleen, hanging in the mouths of baleen whales (see Figure 48.3b).

Deposit feeders pick up or scrape particles of organic matter from solid material they live in or on. Earthworms are deposit feeders that eat their way through soil, taking the soil into their mouth and digesting and absorbing any organic material it contains. Some burrowing molluscs and tube-dwelling polychaete worms use body appendages to gather organic deposits from the sand or mud around them. Mucus on the appendages traps the organic material, and cilia move it to the mouth. The fiddler crab (*Uca* species) is a deposit feeder (see Figure 48.3c) with front claws differing dramatically in size. The small claw picks up sediment and moves it to the mouth, where the contents are sifted (the large claw is used in signalling). The edible parts of the sediment are ingested, and the rest is put back on the sediment as a small ball. The feeding-related movement of the small claw over the larger claw looks as if the crab is playing the large claw like a fiddle, giving the crab its name.

Bulk feeders consume sizeable food items whole or in large chunks. Most mammals eat this way, as do reptiles, most birds and fishes, and adult amphibians. Depending on the animal, adaptations for bulk feeding include teeth for tearing or chewing, as well as claws and beaks for holding large food items. Some bulk feeders have flexible jaws, allowing them to ingest objects that are larger in diameter than their head (see Figure 48.3d).

STUDY BREAK

1. Describe the four basic feeding types in animals. Give examples of species that use each.
2. Can a single species employ more than one feeding type? How would you categorize yourself?

48.3 Digestive Processes in Animals

Most invertebrates and all vertebrates have a tubelike digestive system with two openings, a mouth for ingesting food and an anus for eliminating unused material. In these animals, contents move in one direction along the tube. The lumen of a digestive tube (also known as a gut, alimentary canal, digestive tract, or gastrointestinal tract) is external to all body tissues.

Mechanical and chemical digestive processes break food into its component parts, eventually breaking molecules into molecular subunits that can be absorbed into body fluids and transported to and moved into cells. Mechanical breakdown often involves grinding, sometimes with teeth or sometimes in a muscular **gizzard**. Chemical breakdown occurs by **enzymatic hydrolysis**, in which chemical bonds are broken by the addition of H^+ and OH^-, the components of water (see *The Purple Pages*). Specific enzymes speed these reactions: *amylases* catalyze the hydrolysis of starches, *lipases* break down fats and other lipids, *proteases* hydrolyze proteins, and *nucleases* digest nucleic acids. Enzymatic hydrolysis of food molecules may take place inside or outside the body cells, depending on the animal.

48.3a Intracellular Digestion

In intracellular digestion, primarily in sponges and some cnidarians, cells take in food particles by endocytosis. Inside the cell, endocytic vesicles containing food particles fuse with a lysosome, a vesicle containing hydrolytic enzymes. The molecular subunits produced by the hydrolysis pass from the vesicle to the cytosol. Any undigested material remaining in the vesicle is released to the outside of the cell by exocytosis.

In sponges, water-containing particles of organic matter and microorganisms enter the body through pores in the body wall (see Figure 27.8, Chapter 27). In the body cavity, individual *choanocytes* (collar cells) lining the body wall trap food particles take them in by endocytosis, and transport them to amoeboid cells, where intracellular digestion takes place.

48.3b Extracellular Digestion

Extracellular digestion takes place in a pouch or tube enclosed within the body but outside the body cells—the digestive tract. Epithelial cells lining the pouch or tube secrete enzymes that digest the food. Processing food in this specialized compartment prevents self-digestion of the body tissues of the animal itself. Extracellular digestion, which occurs in most invertebrates and all vertebrates, greatly expands the range of available food sources by allowing animals to deal with much larger food items than those that can be engulfed by single cells. Extracellular digestion also allows animals to eat large batches of food that can be stored and digested while the animal continues other activities.

Some animals, including flatworms and cnidarians such as hydras, corals, and sea anemones, have a saclike digestive system with a single opening that serves as both the entrance for food and the exit for undigested material. In some of these animals, such as the flatworm *Dugesia* (Figure 48.4), the digestive cavity is called a gastrovascular cavity because it circulates and digests food. Food is brought to the mouth by a protrusible pharynx (a throat that can be everted) and then enters the gastrovascular cavity, where glands in the cavity wall secrete enzymes that begin the digestive process. Cells lining the cavity then take up the partially digested material by endocytosis and complete digestion intracellularly. Undigested matter is released to the outside through the pharynx and mouth.

48.3c Gastrointestinal Tracts

In most animals with a digestive tube, digestion occurs in five successive steps, each taking place in a specialized region of the tube. The tube is a biological disassembly line, with food entering at one end and leftovers leaving from the other. Five main processes occur from ingestion of food to expulsion of wastes:

1. *Mechanical processing.* Chewing, grinding, and tearing food chunks into smaller pieces makes them easier to move through the tract and increases the surface area exposed to digestive enzymes.
2. *Secretion of enzymes and other digestive aids.* Enzymes and other substances that aid the process of digestion, such as acids, emulsifiers, and lubricating mucus, are released into the tube.
3. *Enzymatic hydrolysis.* Food molecules are broken down through enzyme-catalyzed reactions into absorbable molecular subunits.
4. *Absorption.* The molecular subunits are absorbed from the digestive contents into body fluids and cells.

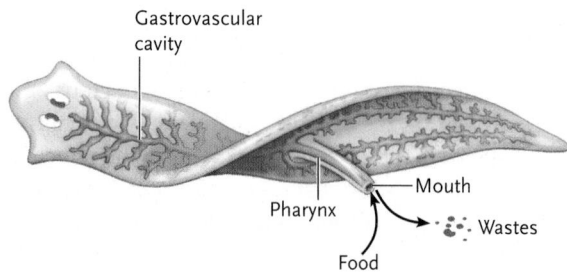

Figure 48.4
The digestive system of a flatworm (*Dugesia* species). The gastrovascular cavity (in blue) is a blind sac with one opening to the exterior through which food is ingested and wastes are expelled.

5. *Elimination.* Undigested materials are expelled through the anus.

Material being digested is pushed through the digestive tube by peristalsis, muscular contractions of its walls. During its progress through the tube, the digestive contents may be stored temporarily at one or more locations. Storage allows animals to take in larger quantities of food than they can process immediately, so feedings can be spaced in time rather than continuous.

Digestion in an Annelid. The earthworm (*Lumbricus* species, **Figure 48.5a**) is a deposit feeder that ingests a great deal of material, only some of which is edible. As it burrows, it pushes soil particles into its mouth. The particles pass from the mouth, through the esophagus, and into the crop (an enlargement of the digestive tube), where contents are stored and mixed with lubricating mucus. This mixture enters the muscular gizzard, where muscular contractions and abrasion by sand grains grind the food mixture into fine particles. The pulverized mixture then enters a long intestine, where organic matter is hydrolyzed by enzymes secreted into the digestive tube. As muscular contractions of the intestinal wall move the mixture along, cells lining the intestine absorb the molecular subunits produced by digestion. The absorptive surface of the intestine is increased by folds of the wall called *typhlosoles*. At the end of the intestine, the undigested residue is expelled through the anus.

Digestion in an Insect. Herbivorous insects such as grasshoppers (Figure 48.5b) are more selective in what they ingest. When eating, grasshoppers tear leaves and other plant parts into small particles with their mandibles, the hard external mouthparts. From the mouth, food particles pass through the pharynx, where salivary secretions moisten the mixture before it enters the esophagus and passes into the crop and begins the process of chemical digestion. From the crop, the food mass enters the muscular gizzard, where it is ground into smaller pieces. Food particles then enter the stomach, where food is stored and digestion continues. In gastric ceca (saclike outgrowths of the stomach; *cecum* = blind), enzymes hydrolyze food, and the products of digestion are absorbed through the walls of the ceca. Undigested food moves into the intestine for further digestion and absorption. At the end of the intestine, water is absorbed from undigested matter and the remnants (frass) are expelled through the anus. The digestive systems of other arthropods are similar to the insect system.

Digestion in a Bird. A pigeon (Figure 48.5c) is also selective in what it ingests. A pigeon picks up seeds with its bill and uses its tongue to move them into its mouth, where they are moistened by mucus-filled saliva and swallowed whole. Seeds pass through the pharynx and into the esophagus. Birds such as parrots

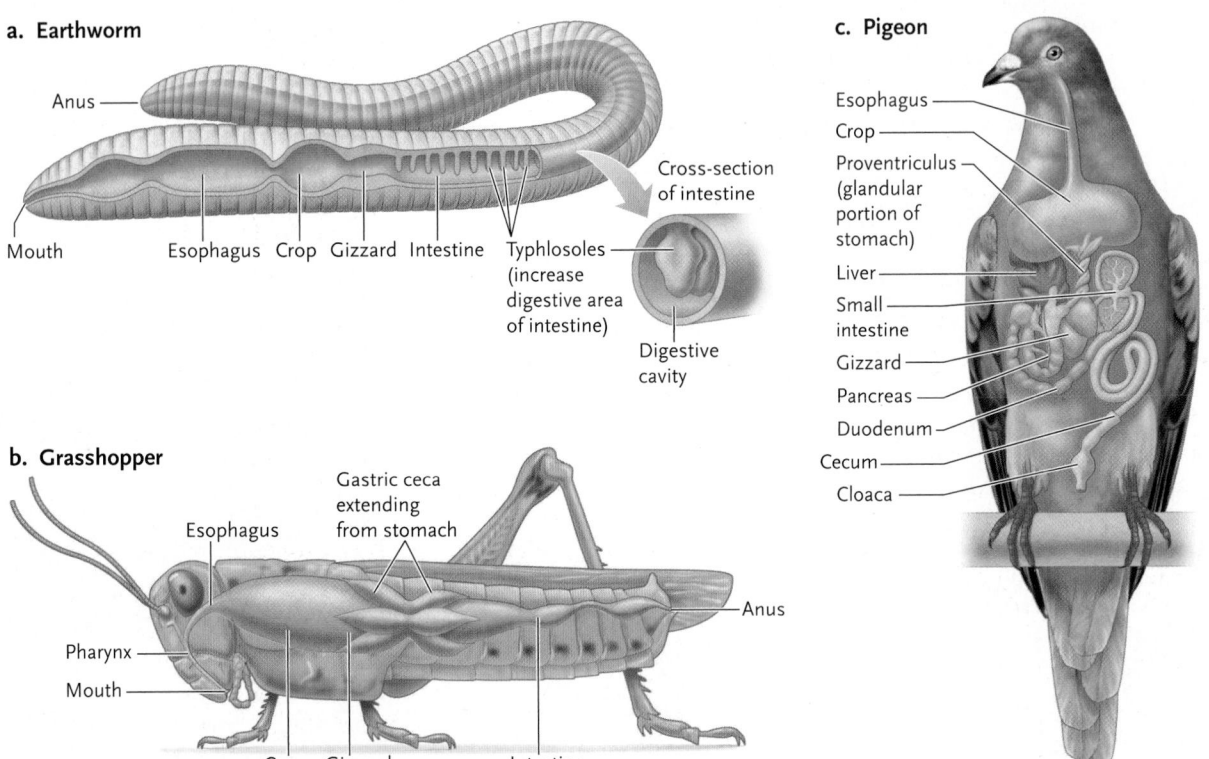

a. Earthworm

Anus

Mouth   Esophagus   Crop   Gizzard   Intestine   Typhlosoles (increase digestive area of intestine)

Cross-section of intestine

Digestive cavity

b. Grasshopper

Gastric ceca extending from stomach

Esophagus

Anus

Pharynx

Mouth

Crop   Gizzard   Intestine

c. Pigeon

Esophagus
Crop
Proventriculus (glandular portion of stomach)
Liver
Small intestine
Gizzard
Pancreas
Duodenum
Cecum
Cloaca

Figure 48.5
The digestive systems of **(a)** an annelid (earthworm), **(b)** an insect (grasshopper), and **(c)** a bird (pigeon).

or cardinals use their bills to crack open seeds, but ingestion occurs as it does in pigeons. The anterior end of the esophagus is tubelike, but it opens into a crop, where food can be stored. The food moves to the proventriculus, the anterior glandular portion of the stomach that secretes digestive enzymes and acids. The food then passes to the gizzard, where muscular action grinds the seeds into fine particles, aided by ingested bits of sand and rock. Food particles then enter the intestine, where secretions from the liver (**bile**) and pancreas (digestive enzymes) are added. Molecular subunits produced by enzymatic digestion are absorbed as the mixture passes along the intestine, and the undigested residues are expelled through the anus, which opens to the cloaca (a common chamber that receives the contents from the digestive tract, the kidneys, and the reproductive tract). Structures such as the mouth, pharynx, esophagus, stomach, intestine, liver, and pancreas occur in almost all vertebrates.

STUDY BREAK

1. What is the difference between intracellular and extracellular digestion? Name a species where each is found.
2. Describe a general digestive tract. What are the five processes that occur in one?
3. What is a gizzard? What does it do? Which animals have one?

48.4 Digestion in Mammals

The mammalian digestive system consists of a series of specialized regions that include the mouth, pharynx, esophagus, stomach, small and large intestines, rectum, and anus, which perform the five steps listed above **(Figure 48.6, p. 1198)**. Each region is under the control of the nervous and endocrine systems. The system allows mammals to obtain fuel molecules, which it breaks down to obtain energy as well as a wide range of nutrients, including the molecular building blocks of carbohydrates, lipids, proteins, and nucleic acids. If the diet is adequate, the digestive system also absorbs the essential nutrients (the amino acids, fatty acids, vitamins, and minerals that cannot be synthesized within our bodies). In all mammals, differences in diet are reflected by the structure of the digestive tract **(Figure 48.7, p. 1199)**. A carnivore's diet is relatively easy to digest; therefore, it does not require as long an intestine as a herbivore does. Plant matter, especially the cell walls, is particularly difficult to digest—hence the longer intestine. The rabbit cecum houses symbiotic, plant-digesting microorganisms that help extract nutrients from plants (see "Digesting Cellulose: Fermentation," p. 1200).

48.4a Gut Layers

The wall of the gut in mammals and other vertebrates contains four major layers, each with

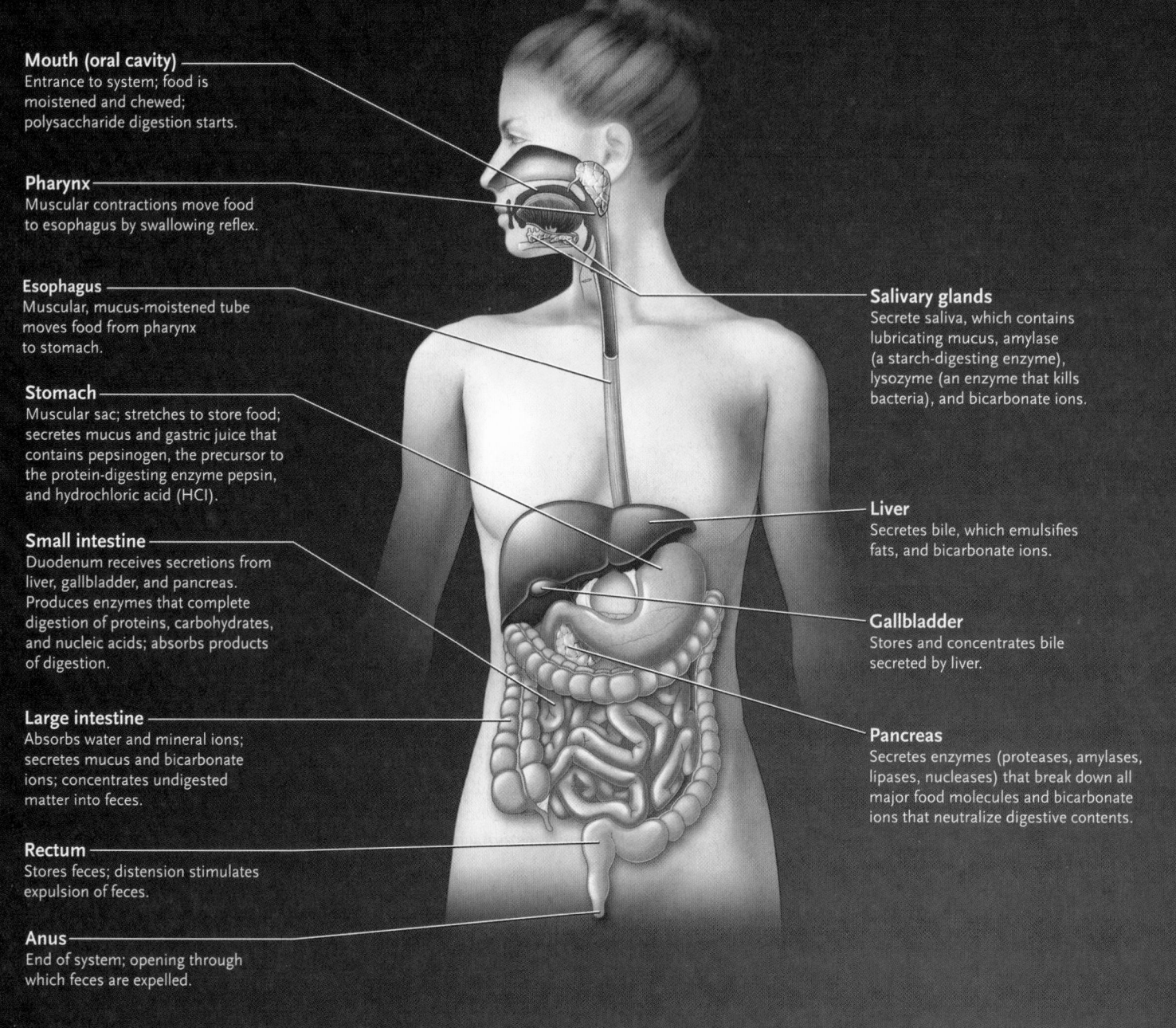

Mouth (oral cavity)
Entrance to system; food is moistened and chewed; polysaccharide digestion starts.

Pharynx
Muscular contractions move food to esophagus by swallowing reflex.

Esophagus
Muscular, mucus-moistened tube moves food from pharynx to stomach.

Stomach
Muscular sac; stretches to store food; secretes mucus and gastric juice that contains pepsinogen, the precursor to the protein-digesting enzyme pepsin, and hydrochloric acid (HCl).

Small intestine
Duodenum receives secretions from liver, gallbladder, and pancreas. Produces enzymes that complete digestion of proteins, carbohydrates, and nucleic acids; absorbs products of digestion.

Large intestine
Absorbs water and mineral ions; secretes mucus and bicarbonate ions; concentrates undigested matter into feces.

Rectum
Stores feces; distension stimulates expulsion of feces.

Anus
End of system; opening through which feces are expelled.

Salivary glands
Secrete saliva, which contains lubricating mucus, amylase (a starch-digesting enzyme), lysozyme (an enzyme that kills bacteria), and bicarbonate ions.

Liver
Secretes bile, which emulsifies fats, and bicarbonate ions.

Gallbladder
Stores and concentrates bile secreted by liver.

Pancreas
Secretes enzymes (proteases, amylases, lipases, nucleases) that break down all major food molecules and bicarbonate ions that neutralize digestive contents.

Figure 48.6
The human digestive system.

specialized functions **(Figure 48.8),** from the inner surface outward:

1. The **mucosa** contains epithelial and glandular cells and lines the inside of the gut. Epithelial cells absorb digested nutrients and seal off the digestive contents from body fluids. The glandular cells secrete enzymes, aids to digestion such as lubricating mucus, and substances that adjust the pH of the digestive contents.

2. The **submucosa** is a thick layer of elastic connective tissue containing neuron networks and blood and lymph vessels. Neuron networks provide local control of digestive activity and carry signals between the gut

and the central nervous system. Lymph vessels carry absorbed lipids to other parts of the body.

3. In most regions of the gut, the **muscularis** is formed by two smooth muscle layers, a *circular layer* that constricts the diameter of the gut when it contracts and a *longitudinal layer* that shortens and widens the gut. The stomach also has an *oblique layer* running diagonally around its wall. Peristalsis occurs when circular and longitudinal muscle layers of the muscularis coordinate their activities to push the digestive contents through the gut **(Figure 48.9, p. 1201).** In **peristalsis,** the circular muscle layer contracts in a wave that passes along the gut, constricting the gut and

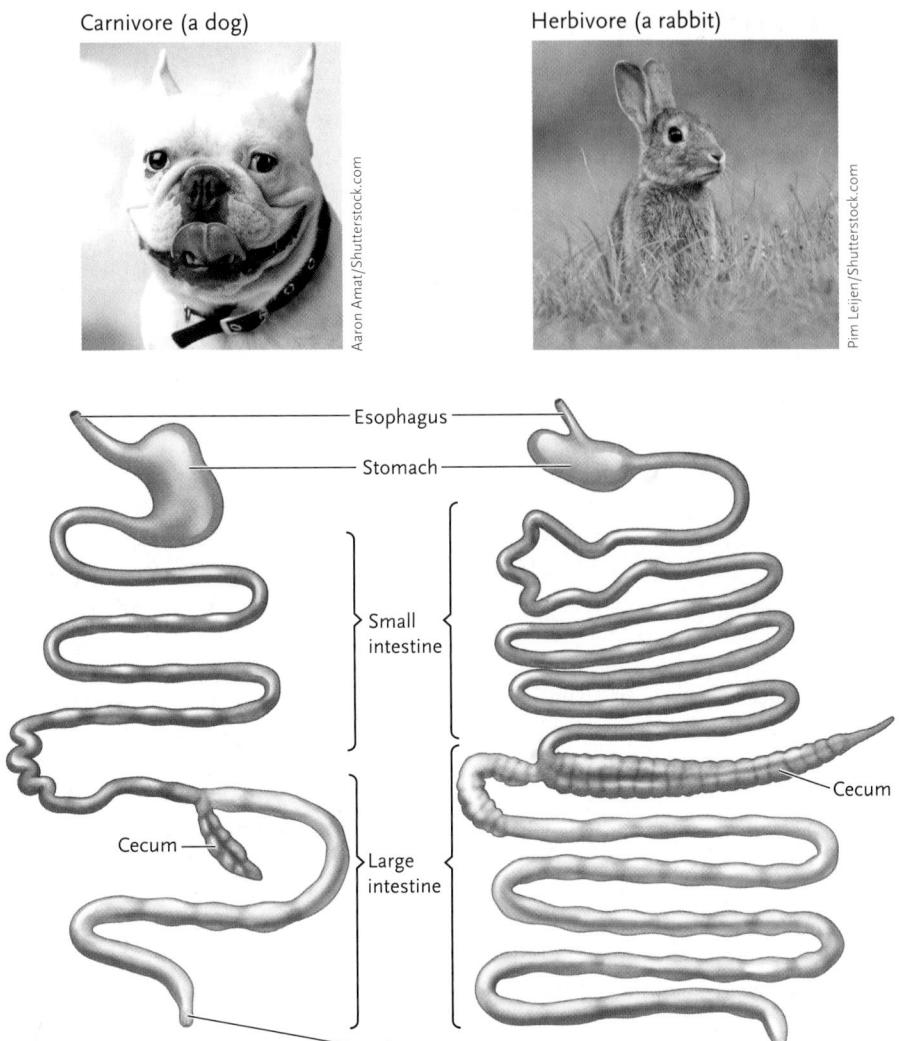

Carnivore (a dog)

Aaron Amat/Shutterstock.com

Herbivore (a rabbit)

Pim Leijen/Shutterstock.com

Esophagus

Stomach

Small intestine

Cecum

Large intestine

Cecum

Anus

Figure 48.7
Comparison of the lengths of the digestive systems of a carnivore (*Canis familiaris*) and a herbivore (*Oryctolagus cuniculus*). Note differences in length and the development of the ceca.

pushing the digestive contents onward. Just in front of the advancing constriction, the longitudinal layer contracts, shortening and expanding the tube and making space for the contents to advance.

4. The **serosa** is the outermost gut layer. It consists of connective tissue that secretes an aqueous,

slippery fluid that lubricates areas between the digestive organs and other organs, reducing friction between them as they move together as a result of muscle movement. The serosa is continuous with the mesentery along much of the length of the digestive system.

Figure 48.8
Layers of the gut wall in vertebrates as seen in the stomach wall.

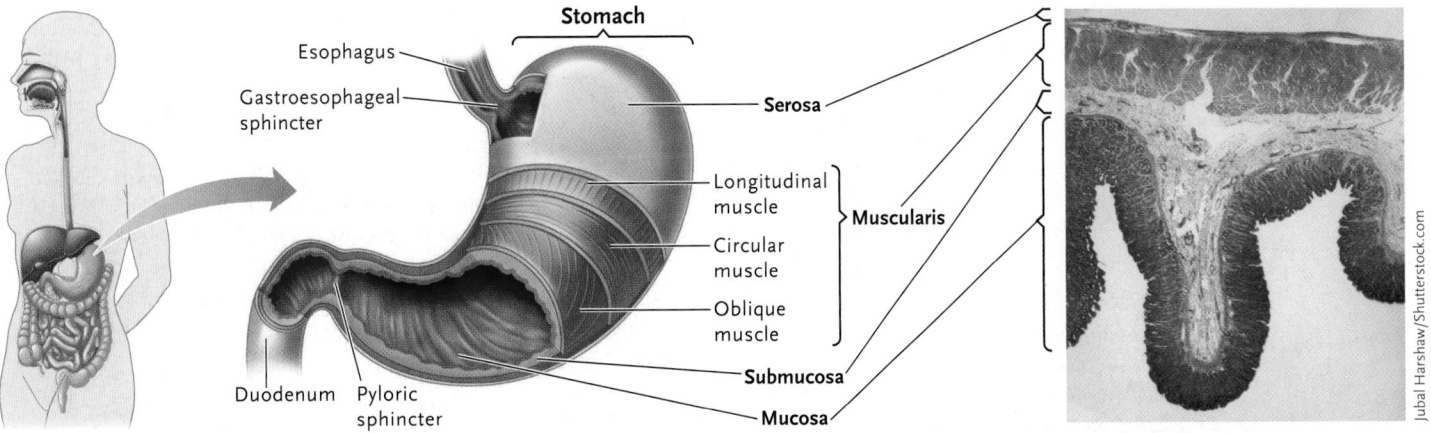

Stomach

Esophagus

Gastroesophageal sphincter

Serosa

Longitudinal muscle

Muscularis

Circular muscle

Oblique muscle

Submucosa

Duodenum   Pyloric sphincter

Mucosa

Jubal Harshaw/Shutterstock.com

Digesting Cellulose: Fermentation

The digestive tracts of mammals vary in length (see Figure 48.7). Most animals cannot digest cellulose because they lack cellulase, which hydrolyzes cellulose into glucose subunits. Many herbivorous animals (primary consumers) use the hydrolytic capabilities of microorganisms that do produce cellulase. In this way, bacteria, protists, and fungi help other animals to digest plant material.

Herbivores using microorganisms house these symbionts in specialized structures along the alimentary canal. These structures occur in the esophagus, stomach, or ceca, depending on the species. Ruminant mammals (Bovidae, Cervidae) and termites are well-known examples of animals that use symbionts to digest cellulose.

Ruminants use their teeth to crop and chew plant material. They swallow the masticated material, moving it to a complex, four-chambered rumen **(Figure 1).** The first three chambers of the rumen are derived from the esophagus, whereas the fourth, the abomasum, is the stomach. Swallowed food material arrives in the reticulum and then moves to the rumen. Then ruminants chew their cuds, regurgitating material from the reticulum and rumen, rechewing it, and macerating it into smaller fragments before swallowing it again. This exposes more surface area to microbial enzymes, giving them more time to act.

Fermentation by the microorganisms occurs in the reticulum and in the rumen. Oxygen levels in the chambers are too low to support mitochondrial reactions (see Chapter 7). Matter digested and liquefied by microorganisms moves to the *omasum,* where water is absorbed from the mass. In the *abomasum* (the ruminant's true stomach), acids and pepsin are added to the food mass, killing the microorganisms and starting the process of "typical" vertebrate digestion. As the food mass moves to the small intestine, dead microorganisms, themselves a rich source of proteins, vitamins, and other nutrients, are digested and absorbed along with other hydrolyzable molecules.

Fermentation generates products such as alcohols and amino acids that are used as nutrients. It also produces volatile fatty acids that move from the rumen to the blood and are used as sources of carbon and energy. Microorganisms use 40 to 60% of the food protein produced by fermentation, and, in turn, their bodies are protein for the host. Methane, another product, collects in the fermentation chambers, so ruminants belch the gas in huge quantities. One cow can release more than 400 L of methane per day. Cattle are estimated to contribute 20% of the methane polluting our atmosphere. A 500 kg cow with a 70 L rumen produces about 60 L of saliva a day and ingests 40 L of water. Fermentation takes time: the leaves eaten by a cow take about 55 hours to move through its digestive system.

Many other mammals have esophageal or gastric chambers that house plant-digesting symbiotic microorganisms. Biologists had long thought that the weight of the fermentation chamber made digestion by fermentation inaccessible to birds. But at least one species of bird, the South American Hoatzin, uses fermentation **(Figure 2).** Freshly caught Hoatzins smell like cattle dung, perhaps giving a clue to their use of fermentation. Hoatzins eat young leaves that are fermented in the enlarged forestomach. This fermentation centre takes up some space occupied by flight muscles in "normal" birds. Hoatzins are weak fliers, probably because of reduction in the mass of their flight muscles to accommodate the enlarged forestomach.

Michael S. Nolan/age fotostock/Getty Images

Chewing, swallowing, regurgitation, rechewing, and reswallowing of food through esophagus

II. Rumen

I. Reticulum

III. Omasum

IV. Abomasum (true stomach)

To small intestine

Figure 1
Ruminants, such as the pronghorn antelope (Antilocapra americana), *have a four-chambered stomach system for digesting plant material (cellulose) by fermentation.* Some other mammalian herbivores use the same approach, but others, such as the rabbit (Oryctolagus cuniculus), achieve fermentation in other parts of the digestive tract (see Figure 48.7).

Figure 2

The Hoatzin (Opisthocomus hoazin) *uses foregut fermentation to digest cellulose to meet a high percentage of its energy requirements.* The contents of the crop and lower esophagus account for over 15% of adult mass. Deep ridges in the lining of the crop increase its surface area.

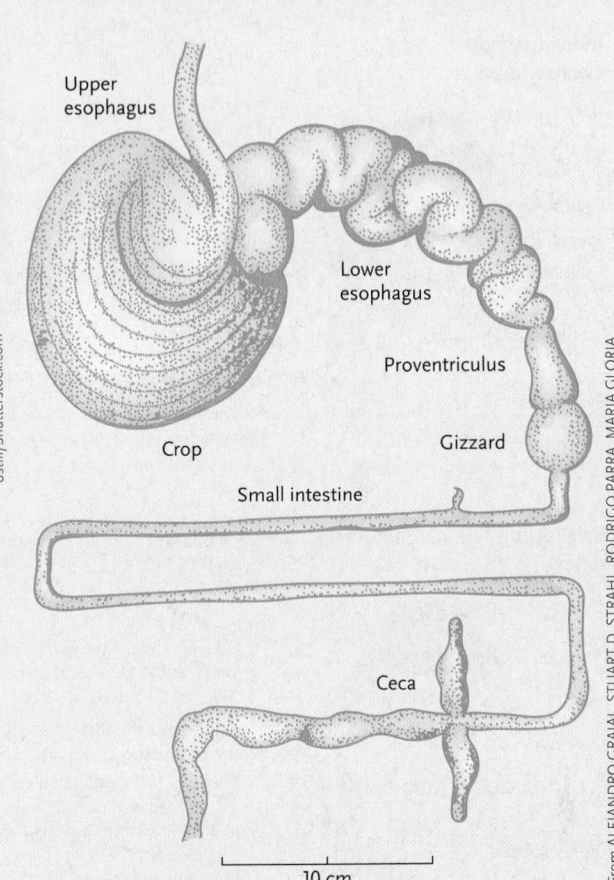

Upper esophagus

Lower esophagus

Proventriculus

Crop

Gizzard

Small intestine

Ceca

10 cm

From ALEJANDRO GRAJAL, STUART D. STRAHL, RODRIGO PARRA, MARIA GLORIA DOMINGUEZ, ALFREDO NEHER, "Foregut Fermentation in the Hoatzin, a Neotropical Leaf-Eating Bird," *Science,* vol. 245, Sep 15, 1989, pp. 1236–1238. Reprinted with permission from AAAS.

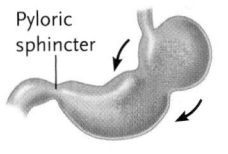 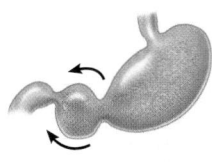 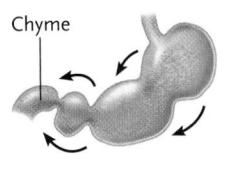

Pyloric sphincter

Chyme

1 The circular layer of the muscularis contracts in a wave, constricting the gut and pushing the digestive contents onward.

2 The longitudinal layer contracts, shortening and expanding the gut and making space for the contents to advance.

3 Partially processed food (chyme) enters the small intestine.

Figure 48.9
The waves of peristaltic contractions moving food through the stomach.

Mesenteries, thin tissues attached to the stomach and intestines, suspend the digestive system from the inner wall of the abdominal cavity.

Sphincters are powerful rings of smooth muscle that form valves between major regions of the digestive tract. By contracting and relaxing, the sphincters control the passage of the digestive contents from one region to another and through the anus. The adult human digestive tract in its normal living contracted state is about 4.5 m long. It is about twice as long when fully extended, as in a cadaver (all muscles relaxed).

48.4b Down the Tube

Food begins its travel through the gastrointestinal tract in the mouth, where the teeth cut, tear, and crush food items. During chewing, three pairs of **salivary glands** secrete saliva through ducts that open on the inside of the cheeks and under the tongue. Saliva, which is more than 99% water, moistens the food and, as we have seen, begins digestion with **salivary amylase** (see "Molecule behind Biology," Box 48.3). Saliva also contains mucus to lubricate the food mass, and bicarbonate ions (HCO_3^-) to neutralize acids in the food and keep the pH of the mouth between 6.5 and 7.5, the optimal range for salivary amylase to function. Saliva also contains a *lysozyme,* an enzyme that kills bacteria by breaking open their cell walls.

After a suitable period of chewing, the food mass, called a **bolus**, is pushed by the tongue to the back of the mouth, where touch receptors detect the pressure and trigger the *swallowing reflex* **(Figure 48.10, p. 1202).** This reflex is an involuntary action produced by contractions of muscles in the walls of the pharynx that direct food into the esophagus. Peristaltic contractions of the esophagus, aided by mucus secreted by the esophagus, propel the bolus toward the stomach. The passage down

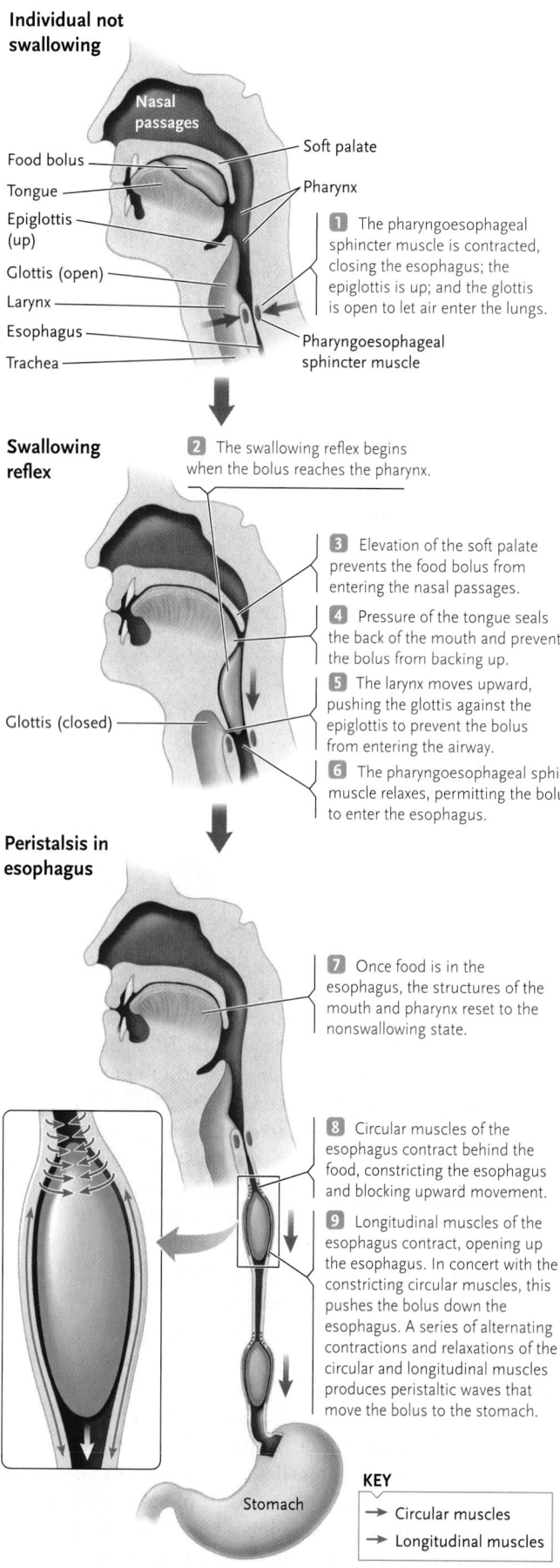

Individual not swallowing

- Nasal passages
- Food bolus
- Tongue
- Epiglottis (up)
- Glottis (open)
- Larynx
- Esophagus
- Trachea
- Soft palate
- Pharynx
- Pharyngoesophageal sphincter muscle

1 The pharyngoesophageal sphincter muscle is contracted, closing the esophagus; the epiglottis is up; and the glottis is open to let air enter the lungs.

Swallowing reflex

2 The swallowing reflex begins when the bolus reaches the pharynx.

3 Elevation of the soft palate prevents the food bolus from entering the nasal passages.

4 Pressure of the tongue seals the back of the mouth and prevents the bolus from backing up.

5 The larynx moves upward, pushing the glottis against the epiglottis to prevent the bolus from entering the airway.

- Glottis (closed)

6 The pharyngoesophageal sphincter muscle relaxes, permitting the bolus to enter the esophagus.

Peristalsis in esophagus

7 Once food is in the esophagus, the structures of the mouth and pharynx reset to the nonswallowing state.

8 Circular muscles of the esophagus contract behind the food, constricting the esophagus and blocking upward movement.

9 Longitudinal muscles of the esophagus contract, opening up the esophagus. In concert with the constricting circular muscles, this pushes the bolus down the esophagus. A series of alternating contractions and relaxations of the circular and longitudinal muscles produces peristaltic waves that move the bolus to the stomach.

- Stomach

KEY
→ Circular muscles
→ Longitudinal muscles

Figure 48.10
The swallowing reflex.

the esophagus stimulates the gastroesophageal sphincter at the junction between the esophagus and the stomach to open and admit the bolus into the stomach. After the bolus enters the stomach, the sphincter closes tightly. If the closure is imperfect, the acidic stomach contents can enter the esophagus, in humans causing *acid reflux* or heartburn.

Mammals consciously initiate the swallowing reflex, but once it has begun, they cannot stop it because whereas the muscles of the pharynx and upper esophagus are skeletal muscles under voluntary control, the muscles below are smooth muscles under involuntary control.

Involuntary movements of the tongue and soft palate at the back of the mouth prevent food from backing into the mouth or nasal cavities. The glottis (space between vocal cords) and the **epiglottis**, a flaplike valve, prevent entry of food into the trachea.

48.4c Stomach

The stomach is a muscular, elastic sac that stores food and adds secretions, furthering digestion. The stomach lining, the mucosa, is covered with tiny *gastric pits*, entrances to millions of *gastric glands*. These glands extend deep into the stomach wall and contain cells that secrete some of the products needed to digest food. Entry of food into the stomach activates stretch receptors in its wall. Signals from stretch receptors stimulate the secretion of **gastric juice (Figure 48.11)**, which contains the digestive enzyme pepsin, hydrochloric acid (HCl), and lubricating mucus. The stomach secretes about 2 L of gastric juice each day.

Pepsin begins the digestion of proteins by creating breaks in polypeptide chains. Pepsin is secreted in the form of an inactive precursor molecule, pepsinogen, by cells called *chief cells*. Pepsinogen is converted to pepsin by the highly acidic conditions of the stomach. Once produced, pepsin itself can catalyze the reaction, converting more pepsinogen to pepsin. The activation of pepsin illustrates a common theme in digestion. Powerful hydrolytic enzymes such as pepsin would be dangerous to the cells that secrete them. However, enzymes are synthesized as inactive precursors and not converted into active form until exposed to the digestive contents.

Parietal cells in the gastric pits secrete H^+ and Cl^-, which combine to form HCl in the lumen of the stomach. The HCl lowers the pH of the digestive contents to pH $\leq$ 2, the level at which pepsin reaches optimal activity. To put this pH in perspective, lemon juice is pH 2.4, and sulfuric acid or battery acid is about pH 1. The acidity of the stomach helps break up food particles and causes proteins in the stomach contents to unfold, exposing their peptide linkages to hydrolysis by pepsin. The acid also kills most bacteria that reach the stomach and stops the action of salivary amylase. Some nectar-feeding bats that digest pollen drink their own urine to make their stomach more acid.

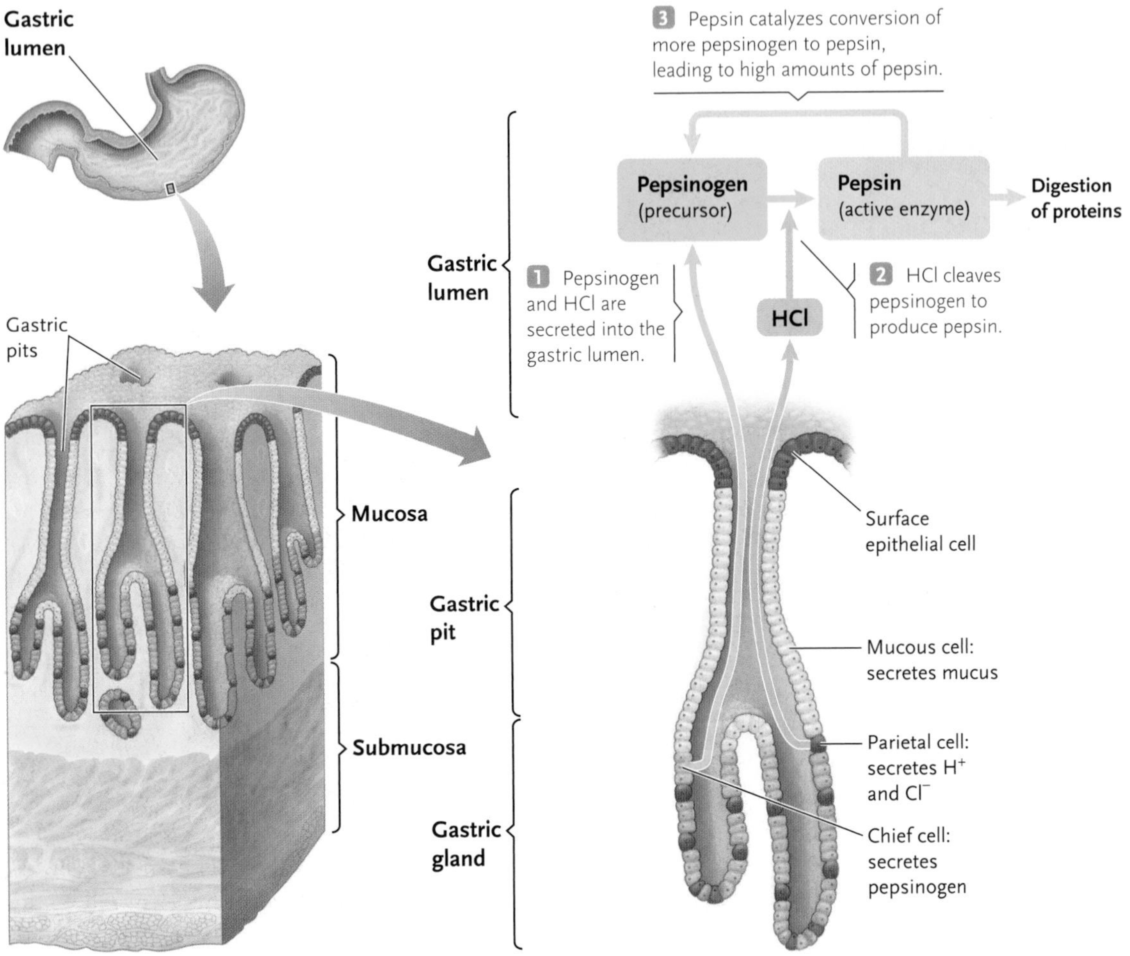

Gastric lumen

Gastric pits

Mucosa

Gastric pit

Submucosa

Gastric gland

3 Pepsin catalyzes conversion of more pepsinogen to pepsin, leading to high amounts of pepsin.

Figure 48.11
Cells that secrete mucus, pepsin, and HCl in the stomach lining.

Gastric lumen

Pepsinogen (precursor) → **Pepsin** (active enzyme) → **Digestion of proteins**

1 Pepsinogen and HCl are secreted into the gastric lumen.

HCl

2 HCl cleaves pepsinogen to produce pepsin.

Surface epithelial cell

Mucous cell: secretes mucus

Parietal cell: secretes H^+ and Cl^-

Chief cell: secretes pepsinogen

A thick coating of alkaline mucus is secreted by *mucus cells* and protects the stomach lining from attack by pepsin and HCl. Behind the mucus barrier, tight junctions between cells prevent gastric juice from seeping into the stomach wall. Even so, there is some breakdown of the stomach lining. The damage is normally repaired by rapid division of mucosal cells, replacing the entire stomach lining about every three days. Most bacteria cannot survive the highly acid environment of the stomach, but one, *Helicobacter pylori,* thrives there. Ulcers result when *H. pylori* breaks down the mucus barrier and exposes the stomach wall to attack by HCl and pepsin (see "People behind Biology," Box 22.2, Chapter 22).

Contractions of the stomach walls continually mix and churn the contents. Peristaltic contractions move the digestive contents toward the *pyloric sphincter* (*pylorus* = gatekeeper) at the junction between the stomach and the small intestine. The arrival of a strong stomach contraction relaxes and opens the valve briefly, releasing a pulse of the stomach contents, **chyme**, into the small intestine.

Feedback controls regulate the rate of gastric emptying, matching it to the rate of digestion, so that food is not moved along faster than it can be chemically processed. In particular, chyme with high fat content and high acidity stimulates the secretion of hormones

by cells in the mucosal layer of the duodenum. These hormones slow the process of stomach emptying. Fat is digested in the lumen of the small intestine more slowly than other nutrients, so further emptying of the stomach is prevented until fat processing has been completed in the small intestine. A fatty meal, such as a greasy pizza, feels heavy in the stomach because it sits there so much longer than a less fatty meal. Highly acidic chyme must be neutralized by bicarbonate in the small intestine. Unneutralized stomach acid inactivates digestive enzymes secreted in the small intestine and inhibits further emptying of the stomach until it is neutralized.

48.4d Small Intestine

The small intestine completes digestion and begins the absorption of nutrients. Nutrients are not absorbed in the mouth, pharynx, or esophagus. Substances such as alcohol, aspirin, caffeine, and water are absorbed in the stomach, but most absorption occurs in the small intestine, where digestion is completed. The small intestine is smaller in diameter than the large intestine. The lining of the small intestine is folded into ridges densely covered by microscopic, fingerlike extensions, the intestinal villi (singular, *villus*). In addition, the epithelial cells covering the villi

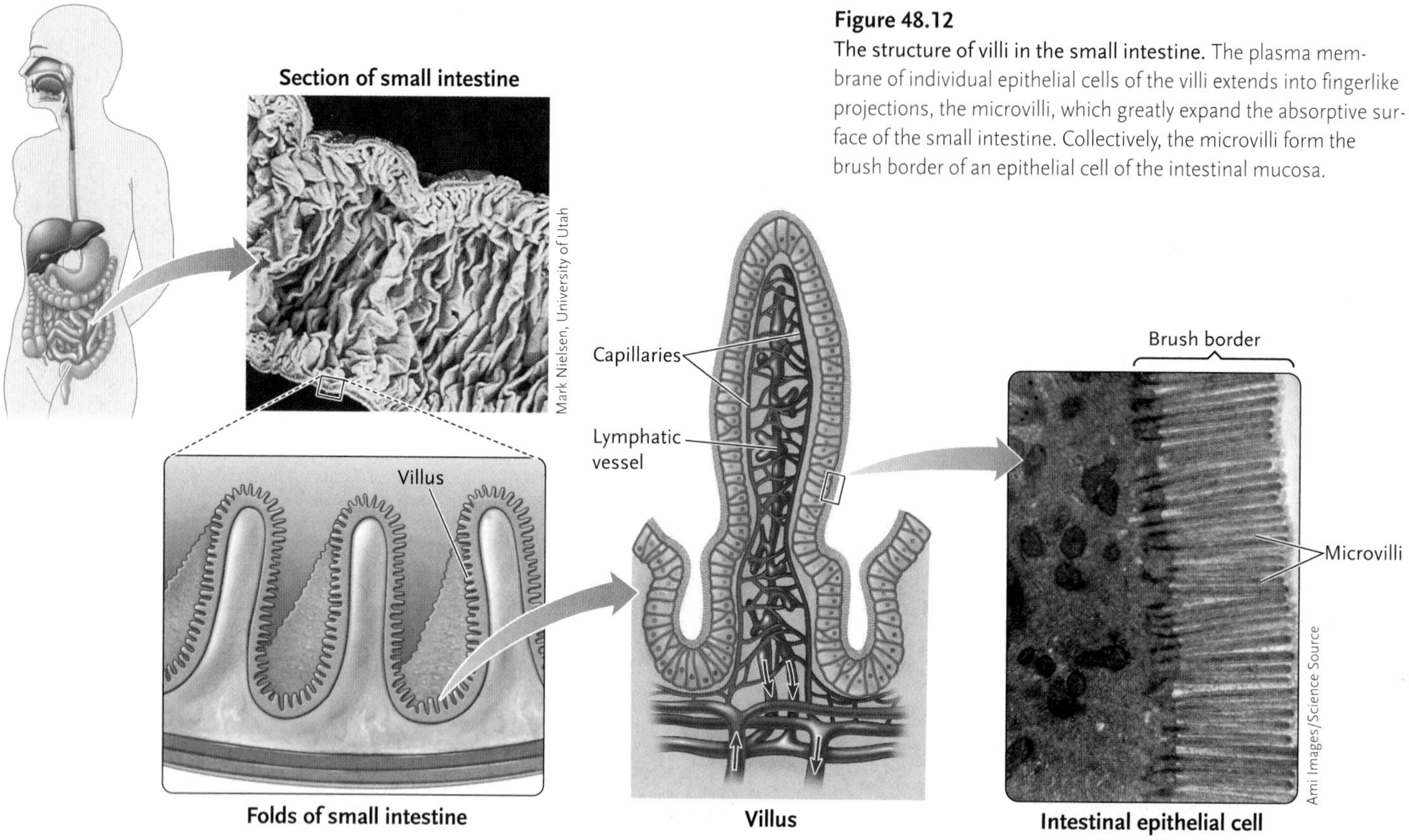

Section of small intestine

Mark Nielsen, University of Utah

Figure 48.12
The structure of villi in the small intestine. The plasma membrane of individual epithelial cells of the villi extends into fingerlike projections, the microvilli, which greatly expand the absorptive surface of the small intestine. Collectively, the microvilli form the brush border of an epithelial cell of the intestinal mucosa.

Villus

Capillaries

Lymphatic vessel

Brush border

Microvilli

Ami Images/Science Source

Folds of small intestine

Villus

Intestinal epithelial cell

themselves have fingerlike extensions, the microvilli. The microvilli are so fine that the surface of each epithelial cell looks like a brush, and hence the fingerlike projections of the plasma membrane are referred to as a *brush border* **(Figure 48.12)**. The intestinal villi and microvilli in humans increase the absorptive

surface area of the small intestine to 300 m², about the size of a doubles tennis court.

Digestion in the small intestine depends on enzymes and other substances secreted by the intestine itself and by the pancreas and liver. Secretions from the pancreas and liver enter a common duct that empties into the lumen of the **duodenum**, a 20-cm-long segment of the small intestine **(Figure 48.13)**.

About 95% of the volume of material leaving the stomach is absorbed as water and nutrients as digestive contents travel along the small intestine. Movement of the contents from the duodenum to the end of the small intestine takes three to five hours. By the time the digestive contents reach the large intestine, almost all nutrients have been hydrolyzed and absorbed.

In humans, the pancreas is an elongated, flattened gland located between the stomach and the duodenum (Figure 48.13; see also Figure 48.6, p. 1198). Exocrine cells in the pancreas secrete bicarbonate ions (HCO_3^-) and pancreatic enzymes into ducts that empty into the lumen of the duodenum. The bicarbonate ions neutralize the acid in chyme, bringing the digestive contents to a slightly alkaline pH. Alkaline pH allows optimal activity of the enzymes secreted by the pancreas, including proteases, an amylase, nucleases, and lipases. All of these enzymes act in the lumen of the small intestine. Like pepsin, the proteases released by the pancreas are secreted in an inactive precursor form and are activated by contact with the digestive contents. The enzyme mixture includes trypsin, which hydrolyzes bonds within polypeptide chains, and

Figure 48.13
The ducts delivering bile and pancreatic juice to the duodenum of the small intestine.

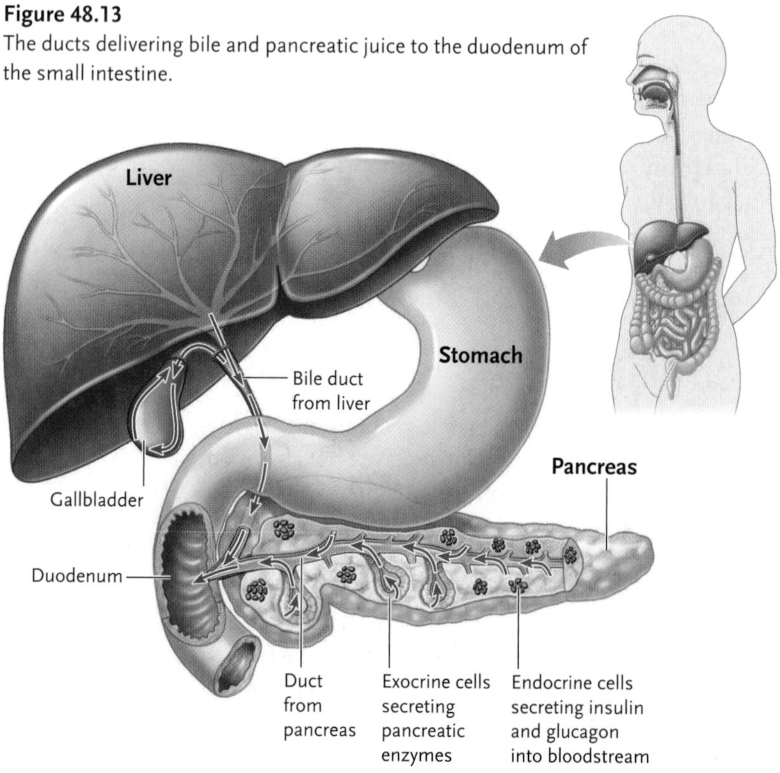

Liver

Stomach

Bile duct from liver

Gallbladder

Pancreas

Duodenum

Duct from pancreas

Exocrine cells secreting pancreatic enzymes

Endocrine cells secreting insulin and glucagon into bloodstream

carboxypeptidase, which cuts amino acids from polypeptide chains one at a time.

The liver secretes bicarbonate ions and bile, a mixture of substances including bile salts, cholesterol, and bilirubin. Bile salts are derivatives of cholesterol and amino acids that aid fat digestion through their detergent action. They form a hydrophilic coating around fats and other lipids, allowing the churning motions of the small intestine to emulsify fats. During emulsification, fats are broken down into tiny droplets called micelles, much the same effect as mixing oil and vinegar in a salad dressing. Lipase, a pancreatic enzyme, can then hydrolyze fats in the micelles to produce monoglycerides and free fatty acids. Bilirubin, a waste product derived from worn-out red blood cells, is yellow and gives the bile its colour. Bacterial enzymes in the intestines modify the pigment, resulting in the characteristic brown colour of feces.

The liver secretes bile continuously. Between meals, when no digestion is occurring, bile is stored in the **gallbladder**, where it is concentrated by the removal of water. After a meal, entry of chyme into the small intestine stimulates the gallbladder to release the stored bile into the small intestine.

Microvilli on the villi of the small intestine secrete water and mucus into the intestinal contents. They also carry out intracellular digestion by transporting products of earlier digestion, including disaccharides, peptides, and nucleotides, across their plasma membranes and producing enzymes to complete hydrolysis of these nutrients. Different disaccharidases break maltose, lactose, and sucrose into individual monosaccharides. Two proteases complete protein digestion: an aminopeptidase cuts amino acids from the end of a polypeptide, and a dipeptidase splits dipeptides into individual amino acids. Nucleases and other enzymes complete digestion of nucleic acids into five-carbon sugars and nitrogenous bases **(Figure 48.14).**

Water-soluble products of digestion enter the intestinal mucosal cells by active transport or facilitated diffusion **(Figure 48.15a, p. 1206)**, and water follows by osmosis. The nutrients are then transported from the mucosal cells into the extracellular fluids, from where they enter the bloodstream in the capillary networks of the submucosa. The absorption of fatty acids, monoglycerides, fat-soluble vitamins, and cholesterol and other products of lipid breakdown by lipase occurs with the assistance of the micelles formed by bile salts (Figure 48.15b). When a micelle contacts the plasma membrane of a mucosal cell, the hydrophobic molecules within the droplet penetrate the membrane and enter the cytoplasm.

In mucosal cells, fatty acids and monoglycerides are combined into fats (triglycerides) and packaged into **chylomicrons**, small droplets covered by a protein coat. Cholesterol absorbed in the small intestine is also packed into the chylomicrons. The protein coat of the chylomicrons provides a hydrophilic surface that keeps

	Carbohydrates	Proteins	Fats	Nucleic acids
Mouth	Polysaccharides ↓ **Salivary amylase** Smaller polysaccharides, disaccharides			
Stomach		Proteins ↓ **Pepsin** Peptides		
Lumen of small intestine	Polysaccharides ↓ **Pancreatic amylase** Disaccharides	Proteins ↓ **Trypsin, chymotrypsin** Peptides Large peptides ↓ **Carboxypeptidase** Amino acids	Triglycerides and other lipids ↓ **Lipase** Fatty acids, monoglycerides	DNA, RNA ↓ **Pancreatic nucleases** Nucleotides
Epithelial cells (brush border) of small intestine	Disaccharides (maltose, sucrose, lactose) ↓ **Disaccharidases** Monosaccharides (e.g., glucose)	Large peptides Dipeptides ↓ **Amino peptidase** ↓ **Dipeptidase** Amino acids Amino acids		Nucleotides ↓ **Nucleotidases, nucleosidases, phosphatases** Nitrogenous bases, five-carbon sugars, and phosphates

Figure 48.14
Enzymatic digestion of carbohydrates, proteins, fats, and nucleic acids in the human digestive system.

a. Absorption of water-soluble products of digestion by intestinal mucosal cells

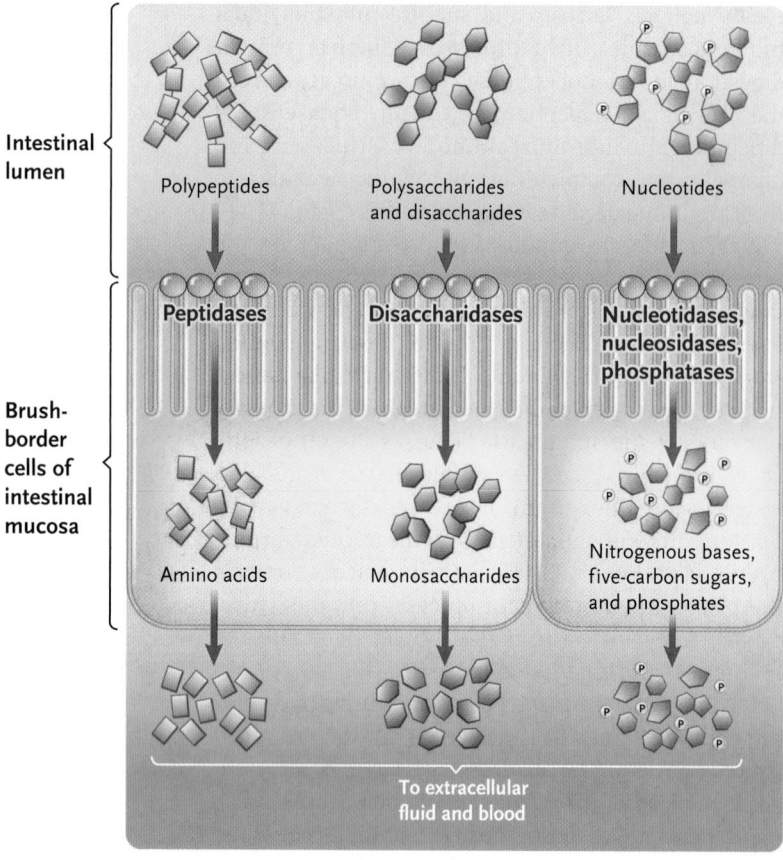

Intestinal lumen

Polypeptides

Polysaccharides and disaccharides

Nucleotides

Brush-border cells of intestinal mucosa

Peptidases

Disaccharidases

Nucleotidases, nucleosidases, phosphatases

Amino acids

Monosaccharides

Nitrogenous bases, five-carbon sugars, and phosphates

To extracellular fluid and blood

Water-soluble molecules are broken into absorbable subunits at brush borders of mucosal cells and transported inside; the subunits are transported on the other side to extracellular fluid and blood.

b. Absorption of fat-soluble products of digestion by intestinal mucosal cells

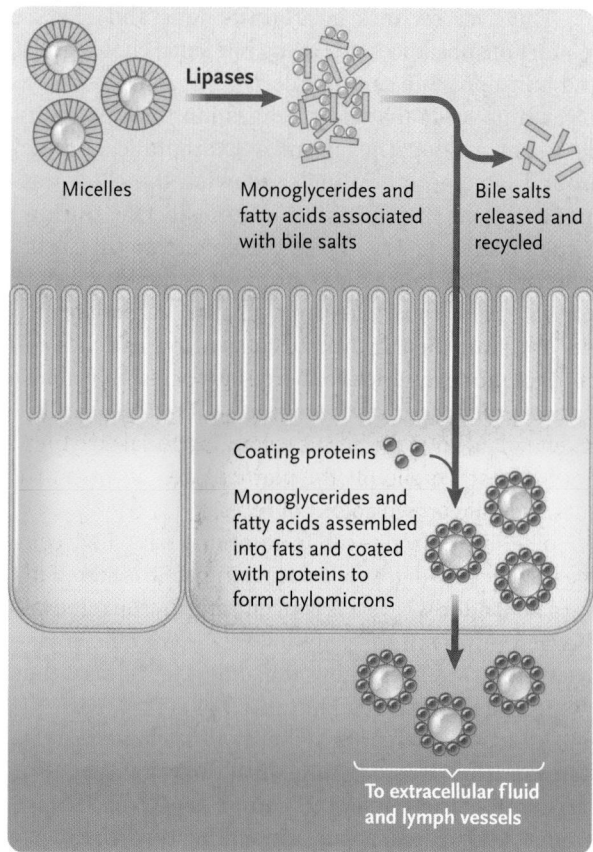

Lipases

Micelles

Monoglycerides and fatty acids associated with bile salts

Bile salts released and recycled

Coating proteins

Monoglycerides and fatty acids assembled into fats and coated with proteins to form chylomicrons

To extracellular fluid and lymph vessels

Micelles (fats coated with bile salts) are digested to monoglycerides and fatty acids, which penetrate into cells and are assembled into fats. The fats are coated with proteins to form chylomicrons, which are released by exocytosis to extracellular fluids, where they are picked up by lymph vessels.

Figure 48.15
Absorption of digestive products by the epithelial cells of the intestinal mucosa.

the droplets suspended in the cytosol. After travelling across the mucosal cells, the chylomicrons are secreted into the interstitial fluid of the submucosa, where they are taken up by lymph vessels. Eventually, they are transferred by lymph into the blood circulation.

Many nutrients absorbed by the small intestine are processed by the liver. Capillaries absorbing nutrient molecules in the small intestine collect into veins that join to form the hepatic portal vein, a larger blood vessel that leads to capillary networks in the liver. In the liver, some nutrients leave the bloodstream and enter liver cells for chemical processing. Among the reactions taking place in the liver is the combination of excess glucose units into glycogen that is stored in liver cells. This reaction reduces the glucose concentration in the blood exiting the liver to about 0.1%. If the glucose concentration in the blood entering the liver falls below 0.1% between meals, the reaction reverses. The reversal adds glucose to return the blood concentration to the 0.1% level before it exits the liver.

The liver also synthesizes lipoproteins that transport cholesterol and fats in the bloodstream, detoxifies ethyl alcohol and other toxic molecules, and inactivates

steroid hormones and many types of drugs. As a result of the liver's activities, the blood leaving it has a markedly different concentration of nutrients than the blood carried into the liver by the **hepatic portal vein**. From the liver, blood goes to the heart and is then pumped to deliver nutrients to all parts of the body.

48.4e Large Intestine

From the small intestine, the contents move on to the large intestine, or **colon**. A sphincter at the junction between the small and large intestines controls the passage of material and prevents backward movement of contents. The inner surface of the large intestine is relatively smooth and contains no villi.

The large intestine has several distinct regions. At the junction with the small intestine, a part of the large intestine forms the **cecum**, a blind pouch. A fingerlike sac, the **appendix**, extends from the cecum. The cecum merges with the colon, which forms an inverted U, finally connecting with the **rectum**, the terminal part of the large intestine.

The large intestine secretes mucus and bicarbonate ions and absorbs water and other ions, primarily

STUDY BREAK

1. Describe the four layers of the gut, and outline their distinctive features.
2. What is the role of saliva?
3. What is peristalsis? What does it do? Is it unique to mammals?
4. Why is the stomach so acidic?
5. What are the roles of the small and large intestine? Which comes first along the digestive tract?
6. Why are feces brown?

Na^+ and Cl^-. The absorption of water condenses and compacts the digestive contents into solid masses, the **feces**. Normally, fecal matter reaching the rectum contains less than 200 mL of the fluid that enters the digestive tract each day. Animals suffering from diarrhea produce liquid fecal matter. Diarrhea is a higher-than-normal rate of movement of materials through the small intestine, which does not leave adequate time for absorption of water. Diarrhea can be caused by infection, emotional stress, or irritation of the small intestine wall.

As many as 500 species of bacteria make up 30 to 50% of the dry matter of feces in humans and other vertebrates. Most of these bacteria live as essentially permanent residents in the large intestine. *Escherichia coli* is the most common in humans and other mammals. Intestinal bacteria metabolize sugars and other nutrients remaining in the digestive residue. They produce useful fatty acids and vitamins (such as vitamin K, the B vitamins, folic acid, and biotin), some of which are absorbed in the large intestine. Bacterial activity in the large intestines produces large quantities of gas (*flatus*), primarily CO_2, methane, and hydrogen sulfide. Most of the gas is absorbed through the intestinal mucosa, and the rest is expelled through the anus in the process of *flatulence*. The amount and composition of flatus depend on the type of food ingested and the particular population of bacteria present in the large intestine. Foods such as beans contain carbohydrates that humans cannot digest. These carbohydrates can be metabolized by gas-producing intestinal bacteria, however, explaining the connection between beans and flatulence.

Feces entering the rectum stretch its walls, at some point triggering a *defecation reflex* that opens the *anal sphincter* and expels the feces through the anus. Because the anal sphincter contains rings of voluntary skeletal muscle as well as involuntary smooth muscle, animals can resist the defecation reflex by voluntarily tightening the striated muscle ring—but only for a short period before the involuntary reflex wins out.

48.5 Regulation of the Digestive Processes

The digestive processes are regulated and coordinated largely by automated controls, most of which originate in the neuron networks of the submucosa of the digestive tract itself. Other controls, particularly those regulating appetite and oxidative metabolism, originate in the brain, in control centres that form part of the hypothalamus.

Movement of food through the digestive system is controlled by receptors in and hormones secreted by various parts of the system **(Figure 48.16)**. Control starts with the mouth, where the presence of food activates receptors that increase the rate of salivary secretion by as much as tenfold over the resting state.

Swallowed food expands the stomach and sets off signals from stretch receptors in the stomach walls. Chemoreceptors in the stomach respond to the presence of food molecules, particularly proteins. Signals from these receptors are integrated in neuron networks in the stomach and the autonomic nervous system to produce several reflex responses. One response is an increase in the rate and strength of

Figure 48.16
Control of digestion by receptors and hormones in the digestive system.

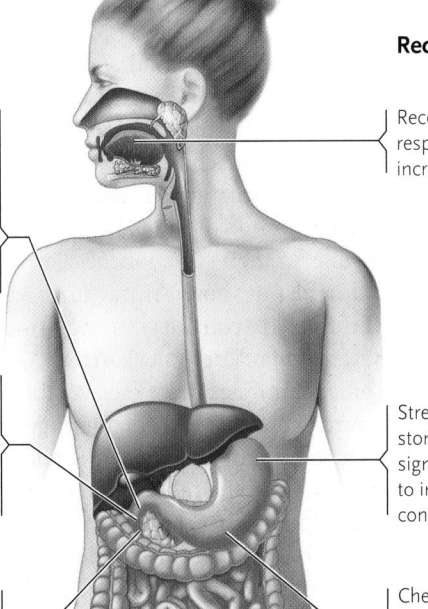

Hormonal controls

Acidic chyme stimulates release into the bloodstream of the hormone **secretin** from glandular cells in the small intestine. Secretin inhibits gastric emptying and gastric secretion and stimulates HCO_3^- secretion into the duodenum.

Fat (mostly) in chyme stimulates release of the hormone **cholecystokinin (CCK)**. CCK inhibits gastric activity and stimulates secretion of pancreatic enzymes.

A meal entering the digestive tract stimulates **GIP** (glucose-dependent insulinotropic peptide) secretion, which triggers **insulin** release. Insulin stimulates the uptake and storage of glucose from the digested food.

Receptor controls

Receptors in the mouth respond to food by increasing salivary secretion.

Stretch receptors in the stomach respond to food, signalling neuron networks to increase stomach contractions.

Chemoreceptors in the stomach respond to food, signalling neuron networks to stimulate the stomach to secrete the hormone **gastrin**, which in turn stimulates the stomach to secrete HCl and pepsinogen.

When the Canadian Government conducted its poll in 2004 to determine who was *The Greatest Canadian*, Sir Frederick Banting was voted fourth place. His claim to fame was isolating and purifying insulin from the islets of Langerhans in the pancreas, and demonstrating that the administration of these insulin extracts could control diabetes. The road to this discovery is a classic example of how intuition and determination pay off. In 1921, Banting, who was a general practitioner and surgeon in London, Ontario, and who had never done any research, approached J.J.R. MacLeod, a Professor at the University of Toronto, with an idea. At the time it was thought that insulin, a protein hormone, was produced in the islets of Langerhans in the pancreas and was necessary for controlling the metabolism of sugar. The pancreas also secretes proteolytic enzymes; however, attempts to extract insulin from ground-up pancreas failed since these proteolytic enzymes were also released and digested the insulin.

Based on his reading, Banting had an idea of how this could be circumvented. Since highly trained physiologists had been unable to solve this dilemma for over 30 years, MacLeod was doubtful of success but, as he was about to go on vacation, against his better judgment, he provided Banting with an assistant, Charles Best, and the facilities and opportunity to pursue his idea. Within 10 weeks, Banting and Best had prepared the first crude extract of insulin and successfully used it to save the life of a dying dog. By the following year, the purification and commercial preparation of insulin was saving the lives of dying diabetic patients. In 1923, Banting and MacLeod received the Nobel Prize for their work. Banting shared his prize with his assistant, Charles Best.

As a tribute to Banting, and all the people that have lost their lives to diabetes, a *flame of hope* was lit in Sir Frederick Banting Square in London, Ontario, in 1989, where it will remain lit until there is a cure for diabetes **(Figure 1).**

Figure 1
Flame of Hope at Banting House in London, Ontario.

stomach contractions. Another is secretion of a hormone, *gastrin,* into the blood leaving the stomach. After travelling through the circulatory system, gastrin returns to the stomach, where it stimulates the secretion of HCl and pepsinogen. These molecules are used in the digestion of the protein in the food that was responsible for their secretion. Gastrin also stimulates stomach and intestinal contractions, which serve to keep the digestive contents moving through the digestive system when a new meal arrives.

Three hormones secreted into the lumen of the duodenum participate in regulating digestive processes. When chyme is emptied into the duodenum, its acidic nature stimulates the release of the hormone *secretin.* Secretin inhibits further gastric emptying to prevent more acid from entering the duodenum until the newly arrived chyme is neutralized. Secretin also inhibits gastric secretion to reduce acid production in the stomach and stimulates HCO_3^- secretion into the lumen of the duodenum to neutralize the acid. If the acid is not neutralized, the duodenal wall can be damaged.

Fat and, to a lesser extent, protein in the chyme entering the duodenum stimulate the release of the hormone *cholecystokinin* (CCK). CCK inhibits gastric activity, allowing time for nutrients in the duodenum to be digested and absorbed. CCK also stimulates the secretion of pancreatic enzymes to digest macromolecules in chyme.

The hormone glucose-dependent insulinotropic peptide (GIP) acts primarily to stimulate insulin release into the blood by the pancreas. Insulin changes the metabolic state of the body after a meal is ingested so that new nutrients, particularly glucose, are used and stored. Glucose in the duodenum increases GIP secretion, triggering the release of insulin (see "People behind Biology," Box 48.1).

Two interneuron centres in the hypothalamus work in opposition to control appetite and oxidative metabolism. One centre stimulates appetite and reduces oxidative metabolism; the other stimulates the release of α-melanocyte-stimulating hormone (α-MSH), a peptide hormone inhibiting appetite. Leptin (*leptos* = thin), a peptide hormone, is a major link between these two pathways. Leptin was discovered in mice by Jeffrey Friedman and his coworkers. Fat-storing cells secrete leptin when deposition of fat

LIFE ON THE EDGE 48.2

Fuelling Hovering Flight

Hovering flight is extremely expensive in terms of fuel consumption, whether the hoverer is a Harrier Jump Jet, a helicopter, or a hummingbird. In a hovering hummingbird, more than 90% of the animal's metabolic rate (overall energy consumption) is accounted for by the flight muscles. Hovering allows hummingbirds and some nectar-feeding bats to feed at flowers and tank up with nectar, which is essentially sugar-water. How do these animals pay the costs of hovering?

Some humans eat a lot of sugar before exercising, but only 25 to 30% of the energy they burn comes from the sugar ingested just before or during exercise. It is a mistake to think that eating a bar of chocolate as you run will immediately increase the energy available to you.

In contrast, hummingbirds fuel around 95% of the cost of hovering from the sugar they are ingesting as they hover. High levels of sucrase activity in the birds' intestines translate into rapid hydrolysis of sucrose, explaining the rapid mobilization of this fuel. Nectar produced by the flowers visited by hummingbirds is high in sucrose.

What happens with flower-visiting bats? The Pallas' long-tongued bat (*Glossophaga soricina*, family Phyllostomidae) looks like a nocturnal version of a hummingbird. These bats hover in front of flowers and use long, extensible tongues to extract nectar. Kenneth Welch and two colleagues used measures of oxygen consumption to indirectly measure metabolism and the ratios of $^{13}C/^{12}C$ in exhaled CO_2 to identify sources of energy for hovering Pallas' long-tongued bats. Their measurements indicated that these bats mobilized about 78% of the energy needed for hovering from sucrose ingested during hovering. Sucrase levels in the intestines of the bats are about half those recorded for hummingbirds, probably accounting for the difference in immediate access to fuel. These results suggest convergent evolution between flower-visiting bats and birds. It remains to be determined if the flower-visiting bats of the Old World tropics (family Pteropodidae) have evolved the same adaptations.

The nectar content of flowers pollinated by hummingbirds is about 60% sucrose, whereas that of flowers pollinated by bats is 20% sucrose. These differences may make the bats less efficient at immediately covering the costs of hovering than the hummingbirds. Hovering to extract high-energy food (nectar) amounts to living on the edge, and animals that do so are highly adapted to pay the costs of this specialized flight behaviour (see "The Puzzling Biology of Weight Control," p. 1210).

---

increases in the body. Leptin travels in the bloodstream and binds to receptors in both centres in the hypothalamus. Binding stimulates the centre that reduces appetite and inhibits the centre that stimulates appetite. Leptin also binds to receptors on body cells, triggering reactions that oxidize fatty acids rather than converting them to fats. When fat storage is reduced, leptin secretion drops off, and signals from other pathways activate the appetite-stimulating centre in the hypothalamus and turn off the appetite-inhibiting centre. These controls closely match the activity of the digestive system to the amount and types of foods ingested and coordinate appetite and oxidative metabolism with the body's needs for stored fats (see "The Puzzling Biology of Weight Control," p. 1210, and "Molecule behind Biology," Box 48.3, p. 1212).

STUDY BREAK

1. What is the role of the stretch receptors in the stomach?
2. What are the roles of gastrin, secretin, and cholecystokinin?
3. What is leptin? What role does it play in digestion?

48.6 Variations in Obtaining Nutrients

While the four feeding methods described in Section 48.2 apply in general to all animals, there are many intriguing variations on each theme. Here is a sampling.

48.6a Gutless Animals

Most molluscs have a prominent alimentary tract, from mouth to stomach to intestine to anus, whether the animal is a bivalve, a gastropod, or a cephalopod. But several species in the bivalve genus *Solemya* have smaller guts. At least one benthic species (*Solemya borealis*) from the northeastern Pacific is gutless, lacking any evidence of a digestive tract. This burrow-dwelling species lives in areas rich in nutrients and appears to use secretions from the pedal gland to effect extra-organism digestion. Ctenidial lamellae on the gills are probably used to absorb dissolved organic molecules. The lamellae are well serviced by circulating blood, and cilia clean the sediment from them. Gutlessness is also known from species in the phylum Pogonophora (beardworms), as well as in tapeworms (Platyhelminthes, Cestoda).

The Puzzling Biology of Weight Control

Experimental Research Box

Obesity can lead to elevated blood pressure, heart disease, stroke, diabetes, and other ailments. Obese humans are 20% or more heavier than an optimal body weight. The body mass index (BMI), a measure of an individual's body fat, is a standard way to estimate obesity: BMI = weight in kilograms ÷ (height in metres)^2.

People with a BMI between 18.5 and 24.9 have a normal weight, whereas those with a BMI of 25 to 29.9 are considered overweight. People with a BMI of 30.0 or more are considered obese. People with a BMI greater than 27 have a moderately increased risk for developing type 2 diabetes, high blood pressure, and heart disease. Those with a BMI greater than 30 have a greatly increased risk for these conditions.

Conventional wisdom asserts that eating less is the way to lose weight. The relationship between eating and excessive body weight, however, is much more complex. One complicating factor is the probable existence of a genetically determined, homeostatic set point for body weight. If our body weight varies from the set point, compensating mechanisms adjust metabolism and eating behaviour to return body weight to the set point. Thus, if we diet and eat fewer calories, compensating mechanisms reduce the number of calories we use and make the diet less effective. Research shows that the metabolic activity of most body cells in a person on a crash diet decreases by about 15%. Each person has a different set point: people of the same height who eat the same number of daily calories vary widely in body weight. Some remain thin, whereas others grow fatter every day on the same amount of food.

If a genetically determined set point governs human body weight, then why is the incidence of obesity increasing in the population? Some researchers speculate that our set points are changing toward greater deposition of fat because, on average, we are less physically active and have greater access to food, especially fatty foods and sugars. The average male Mennonite farmer takes more then 20 000 steps a day. Wear a pedometer and see how your activity compares. Our evolutionary history may also be a factor. Humans evolved under conditions in which food was occasionally scarce, so our built-in physiological mechanisms may actually favour raising the set point and thus storing extra nutrients when food is available. This would provide some protection against starvation if food became unavailable.

The search for factors governing the set point and general concern over growing obesity have sparked intensive research into the genetic mechanisms that might control fat deposition and maintain body weight. In this area, Jeffrey Friedman discovered leptin, and Louis Tartaglia and coworkers found leptin receptors. Leptin is a circulating hormone derived from adipocytes (fat cells). It informs the brain about energy stores. Mice with mutant forms of the genes encoding leptin or leptin receptors become morbidly fat.

The discovery of leptin seemed to offer a "magic bullet" to control human obesity: administer leptin to people and they will lose weight. Further studies revealed, however, that the genetics of weight control are considerably more complex in humans than they are in mice. In addition to genes controlling the production of leptin and the formation of receptors for it, in humans, at least four other genes are involved in appetite control and weight gain, complicating the effects of leptin compared with the situation in mice. In trials with humans, obese patients with mutant leptin genes benefited from leptin injections, but some with normal leptin genes unexpectedly gained weight. Furthermore, none of the obese patients with normal leptin genes had any deficiency in leptin production. Obese individuals produced more leptin than people of normal weight.

Inconclusive results of leptin trials have turned attention to the development of other drugs to control obesity. PYY (pancreatic polypeptide YY), a recently discovered hormone, stimulates the appetite-suppressive centre in the hypothalamus and inhibits the appetite-stimulating centre. Trial injections of PYY in mice, rats, and humans have led to a significant decrease in appetite and eating.

Perhaps one of the best examples of weight control in nature is that of some of the species of hibernating mammals. These animals spend the summer months eating and putting on weight in the form of fat as energy stores for the long winter when food is not available. In winter they go into hibernation, greatly reducing their metabolic demands for energy and living off the stored fat. It has been shown in some studies that animals maintained under summer conditions all year are preprogrammed to lose this stored weight. Thus, during the winter months when they should have been hibernating, they ate less and slowly lost weight at the same rate as the animals that were hibernating without access to food. Thus both hibernating and nonhibernating animals emerged in the spring weighing roughly the same.

48.6b A Termite-Eating Flatworm: Unusual Lifestyles

At night in a garden in Harare, Zimbabwe, flatworms, *Microplana termitophaga*, gather around the vents of termite mounds where they harvest termites, *Odontotermes transvaalensis* (Figure 48.17). *Microplana termitophaga* are most numerous at termite mounds one to two hours after dawn, with peaks of activity after a rain. When hunting termites, *M. termitophaga* attach the posterior third of their bodies to the ground at the entrance to the termite mound, leaving the anterior end of the body mobile. The planarians appear to visually detect termites at ranges of 5 mm. Their eyes consist of groups of photoreceptors (see Figure 1.12, Chapter 1), and although not image forming, their eyes allow them to detect movement.

The flatworm uses its head, apparently with mucus and perhaps suction, to capture a termite, which is subdued and held against the pharyngeal region. In this position, *M. termitophaga* digests its prey, ingests the juices produced by digestion, and then expels the indigestible remains. In 135 captures, *M. termitophaga* took mainly worker termites rather than soldiers. On average, a flatworm ate a termite in 5 minutes 58 seconds. In 3 hours 14 minutes, one *M. termitophaga* ate 13 termites.

This pattern of feeding appears to be typical of predatory planarians. They may release poisonous secretions (to immobilize prey), adhesive mucus (to hold it), and copious amounts of digestive fluid from their extensible pharynges. Other species of flatworms eat earthworms (see Chapter 32), and some make group attacks on giant African land snails (*Achatina* species).

Figure 48.17
Microplana termitophaga hunting termites at a termite mound in Harare, Zimbabwe. One planarian (centre) has caught a termite. The termites include both soldiers (large head and jaws) and workers. Go to http://www.youtube
.com/watch?v=ZpXSz8byR_s
to see live footage of the flatworms foraging on termites.

David Shale/npl/Minden Pictures

Figure 48.18
An anglerfish with a bioluminescent lure.

48.6c Fish Predation: More Than You Imagined

Invisible in the inky darkness, a deep-sea anglerfish (*Chaenophryne longiceps;* order Lophiiformes) lies in wait for prey, its gaping mouth lined with sharp teeth. Just above the mouth dangles a glowing lure suspended from a fishing rod–like structure that is a spine of the fish's dorsal fin (Figure 48.18). The lure resembles a tiny fish, wiggling back and forth. The glow is produced by bioluminescent bacteria that live symbiotically in the fish's lure.

Attracted to the lure, a hapless fish comes within range. The oral cavity of the anglerfish expands suddenly, and powerful suction draws the prey into the gaping mouth. The angler's backward-angling teeth keep the prey from escaping, and it is swallowed. The strike takes 6 ms, among the fastest of any known fish. Contractions of throat muscles send the prey to the anglerfish's stomach, which can expand to accommodate a meal as large as the fish itself. The anglerfish now digests, sits, and waits. Some cephalopods, some other fish, and the siphonophore *Erenna* species are examples of other animals using bioluminescent lures.

Other predatory fish hunt from ambush. Moray eels, such as *Muraena retifera,* lie in wait in holes. Their eel-like shape makes it easy for them to fit into small openings. Like the anglerfish, moray eels can swallow items (fish and cephalopods) larger than their heads. But although moray eels lack the effective suction mechanisms of anglerfish, they have two sets of jaws (Figure 48.19, p. 1213). Like other gnathostomes, the moray eel's mouth is bordered by upper (maxilla and premaxilla) and lower (mandibular) jaws bearing teeth. Unlike other gnathostomes, moray eels also have pharyngeal jaws. In the two upper pharyngeal jaws, pharyngobranchial bones bear teeth and connect to the lower pharyngeal jaws, which also bear teeth. Upper and lower

Salivary Amylase—What Makes Food Taste Sweet

Salivary amylase **(Figure 1)**, an enzyme produced by the saliva glands in mammals, initiates digestion of starch during chewing and can break down significant amounts of starch even before the food is swallowed. Although most digestion in mammals occurs beyond the esophagus, salivary amylase presents an interesting exception. There are several benefits to this. Taste buds sense the levels of sugar and starch and give the animal an indication of the quality of the food it is eating. By hydrolyzing starch into the disaccharide maltose, salivary amylase can have an immediate impact on levels of blood sugar. The same impact on blood sugar is not achieved when people eat (and carefully chew) low-starch foods such as apples. Immediate oral access to energy can be important. Furthermore, in the stomach and intestines, salivary amylase augments the activity of pancreatic amylase and may further buffer the body against the impact of digestive disorders.

In humans, the salivary amylase gene (AMY1) varies in the numbers of copies according to the amount of starch in the diet. More copies of AMY1 coincide with higher levels of salivary amylase in saliva **(Figure 2)**. This is true both within a population as well as between populations. Humans living in agricultural societies and those in hunter–gatherer societies in arid environments have high-starch diets. Hunter–gatherers in rain forest and circumarctic habitats, as well as some pastoralists, have low-starch diets.

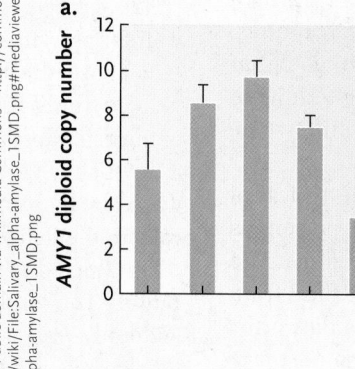
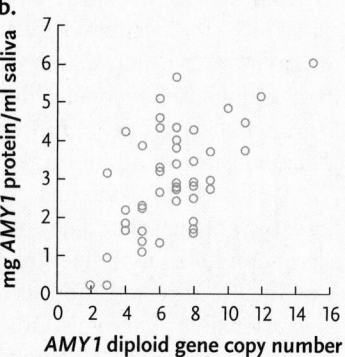

"Salivary alpha-amylase 1SMD" by Own work—From PDB entry 1SMD. Licensed under Public domain via Wikimedia Commons—http://commons.wikimedia.org/wiki/File:Salivary_alpha-amylase_1SMD.png#mediaviewer/File:Salivary_alpha-amylase_1SMD.png

Figure 1
Salivary amylase molecule.

Figure 2
Salivary amylase. (a) Example of variation in copies of AMY1 gene in European American individuals. (b) More copies of the gene AMY1 correlate with higher levels of salivary amylase and more starch in the diet.

Reprinted by permission from Macmillan Publishers Ltd: NATURE GENETICS, Vol. 39: pp. 1256–1260, "Diet and the evolution of human amylase gene copy number variation" by Paul George H Perry, Nathaniel J Dominy, Katrina G Claw, Arthur S Lee, Heike Fiegler et al., copyright (2007).

---

pharyngeal jaws are connected by the epibranchial bones. The jaw arrangement of moray eels allows them to grab and transport (swallow) prey in a system similar to the mechanisms in snakes. The combination of hunting from ambush and quick swallowing of prey makes moray eels efficient predators.

Vandellia cirrhosa, the dreaded candiru **(Figure 48.20),** is a specialized catfish that takes a completely different approach to feeding. Normally living as a gill parasite on larger fish, this small and slender predator lodges itself in the gills of a fish. There, like a leech, mosquito, tick, or lamprey eel, candirus drink the host's blood. To understand why the candiru is "dreaded," consider other aspects of its behaviour. Candirus locate hosts by swimming into currents, especially those bearing metabolic end-products. Freshwater fish often pass metabolic end-products out across their gills. The behaviour can also lead the candiru into the urethra of an animal urinating while submerged in the water. In South American waters inhabited by candirus, local humans avoid urinating when swimming or bathing and may wear protective covers cut from coconuts to foil invasive candirus.

48.6d Egg-Eating Snakes: Caviar of a Different Sort

Eggs well provisioned with yolk may be the ultimate food, rich in proteins and carbohydrates—everything

a.

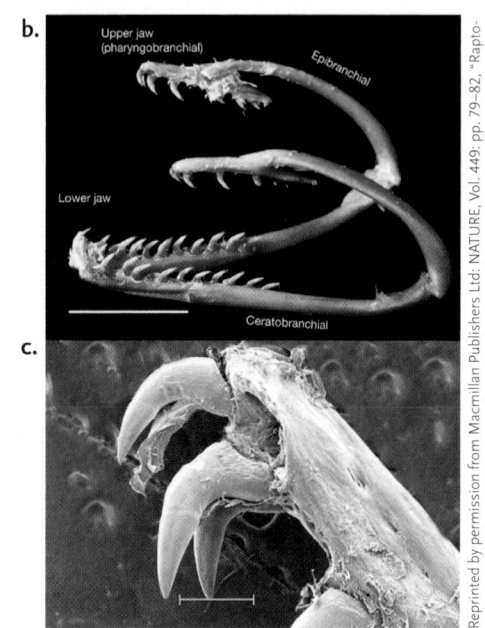
b.

Upper jaw
(pharyngobranchial)
Epibranchial
Lower jaw
Ceratobranchial

c.

Figure 48.19
(a) A moray eel with its mouth open does not reveal its pharyngeal jaws, the second part of its bite. **(b)** The skeleton and teeth of the pharyngeal jaws. **(c)** A scanning electron micrograph view of the upper pharyngeal recurved teeth. Scale bars 1 cm (b) and 500 μm (c).

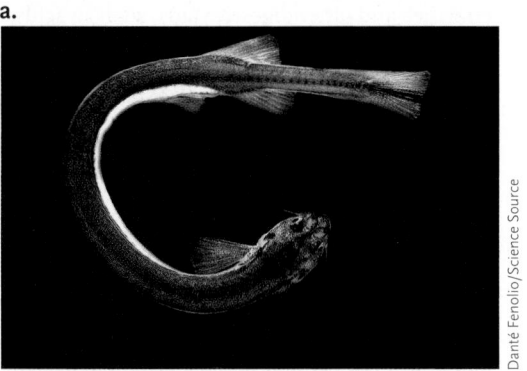
a.

Figure 48.20
The candiru **(a)**, *Vandellia cirrhosa*, is a catfish specialized as a gill parasite. It is attracted to metabolic wastes and has been known to enter the urethra of animals urinating while standing in the water **(b)**.

b.

a predator could want. Mammals such as mongooses break ostrich eggs by throwing rocks against them. Egyptian vultures (*Neophron percnopterus*) break these large eggs by dropping sticks or stones on them. But egg-eating snakes cannot throw stones or sticks.

Dasypeltis scabra is one of the best studied species of egg-eating snakes. Widespread in Africa, these snakes find birds' nests and help themselves to eggs. Like many other snakes, the jaws of *D. scabra* are loosely connected with elastic tendons and ligaments, allowing them to open their mouths extremely wide **(Figure 48.21)**. In a bird's nest, the snake can push an egg against the rim of the nest and swallow it whole. A wide gape is important, but so are an extensible epiglottis and a ribbed trachea, which together allow the snake to breathe while it swallows an egg.

Figure 48.21
Many species of snakes feed on eggs. Their jaws are loosely connected with elastic tendons allowing them to open wide. They break the egg open by squeezing it against ventral hypophyses (projections from the spine) that protrude down against the digestive tract.

After the egg has passed out of the mouth and into the esophagus, the snake makes a coil in front of the swallowed egg. By moving the coil backward along its body, the snake moves the egg down its digestive tract. You can watch the egg move down the snake, but then, with a cracking noise, the outline of the egg disappears. The snake has pushed the egg against ventral hypophyses, anteriorly pointed extensions of specialized vertebrae that protrude into the lumen of the gastrointestinal tract. *Dasypeltis scabra* has a built-in egg-cracker.

STUDY BREAK

Using any source of information you choose, describe what you consider is a species with an unusual way of obtaining nutrients.

Review

To access course materials such as Aplia and other companion resources, please visit www.NELSONbrain.com.

48.1 Nutrients: Essential Materials

- Living organisms require a variety of materials to survive. Use Table 48.1 to review nutrients essential to animals.

- Malnutrition can manifest itself as undernutrition or overnutrition, involving inadequate intake of organic fuels or abnormal ingestion of fuels, respectively. Undernutrition is commonly referred to as malnutrition and may involve ingestion of too little energy to fuel daily activities or failure to ingest essential nutrients. Malnutrition kills thousands of people annually (see Chapter 33). Overnutrition can result in excessive gain in body mass and other health problems.

- Animals such as birds on long migratory flights may mobilize energy from their own bodies during prolonged periods of not feeding.

- In animals, vitamins are essential for different metabolic operations.

48.2 Obtaining and Absorbing Nutrients

- Deposit feeders such as some burrowing molluscs and tube-dwelling polychaete worms (see Chapter 27) pick up or scrape particles of organic matter from their surroundings. Arthropods such as fiddler crabs sift food (organic matter) from sediments.

48.3 Digestive Processes in Animals

- The gizzard is a muscular structure that grinds food into small particles. In annelids such as earthworms, the gizzard uses sand to increase abrasive action. Birds and reptiles often pick up stones for the gizzard to achieve the same effect. Insects often have gizzards.

- Sponge choanocytes (collar cells) trap food particles and take them in by endocytosis. Amoeboid cells transfer food throughout the sponge.

- The diet of *Dugesia* is not typical of free-living flatworms (see Chapter 27). Whereas *Dugesia* mainly feed on detritus, other flatworms are predatory, taking earthworms, snails, or termites.

- Along the gastrointestinal tract, food is first mechanically processed, breaking it into smaller pieces. Then enzymes and other digestive aids, such as acids and mucus, are secreted and commence chemical breakdown of food. Enzymatic hydrolysis breaks food molecules into absorbable molecular subunits. This involves reactions catalyzed by enzymes. Then the food molecules are absorbed from the gastrointestinal tract. Finally, undigested materials are expelled through the anus.

- The Hoatzin is a flying bird known to use fermentation to digest cellulose. Just-caught Hoatzins often smell like cow dung. Their digestive systems include a fermentation centre that uses space normally occupied by flight muscles. Hoatzins are not strong fliers. Other animals, from termites to some mammals, use fermentation to digest cellulose.

48.4 Digestion in Mammals

- While living birds lack teeth and use the gizzard to break food into finer particles, many (but not all) mammals use teeth to mechanically break up food.

- Mammals such as humans require eight essential amino acids: lysine, tryptophan, phenylalanine, threonine, valine, methionine, leucine, and isoleucine.

- The layers in the gut of a mammal are the mucosa, submucosa, muscularis, and serosa. Each plays a different role in the operation of the gastrointestinal tract. Sphincter muscles control the movement of material through the gastrointestinal tract.

- Peristalsis is rhythmic contractions of circular bands of smooth muscle. Peristalsis constricts the gut and moves food along the gastrointestinal tract. Peristalsis usually occurs in waves.

- Gastric juices are produced in gastric glands that line the stomach. Production and release of gastric juices are stimulated by output of stretch receptors. The 2 L of gastric juices secreted daily by the average human includes pepsin, hydrochloric acid, and lubricating mucus.

- The liver secretes bile salts, cholesterol, and bilirubin. Bile salts are derivates of cholesterol and amino acids and aid the digestion of fat. They operate by forming a hydrophilic coating around fats and other lipids, allowing the churning of the intestine to emulsify fats. Bilirubin is a waste product derived from worn-out erythrocytes.

- As many as 500 species of bacteria may be living in the gastrointestinal tract of mammals. Bacterial activity is partly responsible for flatus in the gastrointestinal tract: the production of CO_2, methane, and hydrogen sulfide. Bacteria may make up 20 to 50% of the dry matter in feces.

48.5 Regulation of the Digestive Processes

- Leptin is a peptide hormone that links two inter-neuron centres in the hypothalamus. One centre stimulates appetite and reduces oxidative metabolism; the other stimulates the release of

- α-melanocyte-stimulating hormone (α-MSH), a peptide hormone inhibiting appetite. Fat-storing cells secrete leptin when deposition of fat increases in the body. Leptin moves through the bloodstream and binds to receptors in both centres in the hypothalamus. Binding stimulates the centre that reduces appetite and inhibits the centre that increases appetite.

48.6 Variations in Obtaining Nutrients

- Tapeworms and some species of molluscs lack digestive tracts.

- Some moray eels have two functional sets of jaws, each armed with teeth. In addition to their "normal" jaws, moray eels have pharyngeal jaws that make it easier for them to seize and then swallow prey. The mechanism works like a ratchet and is reminiscent of similar systems in some snakes.

- Some egg-eating snakes swallow birds' eggs whole. They use hypophyses—ventral, anterior-pointing processes from some vertebrae—as egg-crackers.

Questions

Self-Test Questions

1. Vitamins are which of the following?
   a. coenzymes
   b. fatty acids
   c. organic molecules
   d. amino acids
   e. carbohydrates

2. Required molecules that animals cannot synthesize are called which of the following?
   a. nutrients
   b. essential nutrients
   c. enzymes
   d. proteins
   e. carbohydrates

3. Which of the following is NOT an essential nutrient in humans?
   a. vitamin B
   b. calcium
   c. glycogen
   d. linoleic acid
   e. vitamin K

4. In which type of animal is the crop part of the digestive tract?
   a. birds
   b. reptiles
   c. earthworms
   d. a and c

5. How many essential amino acids do humans require?
   a. 2
   b. 5
   c. 8
   d. 10

6. The role of the liver in digestion is which of the following?
   a. Synthesize aminopeptidase and dipeptidase to digest polypeptides.
   b. Synthesize lipase to form free fatty acids.
   c. Secrete trypsin to break the bonds in polypeptides.
   d. Secrete bile and bicarbonate ions to help emulsify fats.
   e. Store bile between meals.

7. The order of successive steps in digestion is which of the following?
   a. Absorption follows enzymatic hydrolysis.
   b. Secretion of enzymes follows absorption of digestive material.
   c. Mechanical processing follows enzyme secretion.
   d. Mechanical processing follows enzymatic hydrolysis.
   e. Enzymatic hydrolysis precedes secretion of digestive aids.

8. Cellulose is digested by fermentation in which of the following animals?
   a. humans
   b. ruminants
   c. the Hoatzin
   d. b and c

9. Which set of animals are among those that eat blood and body fluids?
   a. leeches, vampire bats, and candirus
   b. tapeworms, flukes, and mosquitoes
   c. blackflies, lamprey eels, and bedbugs
   d. wolves, cats, and vampire bats
   e. a and c

10. Which of the following best describes regulation of digestion?
    a. GIP inhibits insulin release from the pancreas.
    b. Gastrin stimulates pancreatic secretion of HCl and pepsinogen.
    c. Secretin stimulates gastric emptying into the duodenum.
    d. CCK stimulates gastric activity to activate the duodenum.
    e. Leptin binds different hypothalamic receptors to stimulate or inhibit appetite.

Questions for Discussion

1. How does secondment of chloroplasts benefit animals (see Chapter 24)? In what animals does it occur? Do these animals have genetic control over the chloroplasts? If so, how do they acquire this control?

2. Although humans evolved as omnivores, some eat only meat, others are vegetarian, and many (perhaps most) eat a combination of plant and animal material. What should a balanced diet include?

3. What is the advantage of a tubelike digestive system over a saclike digestive system?

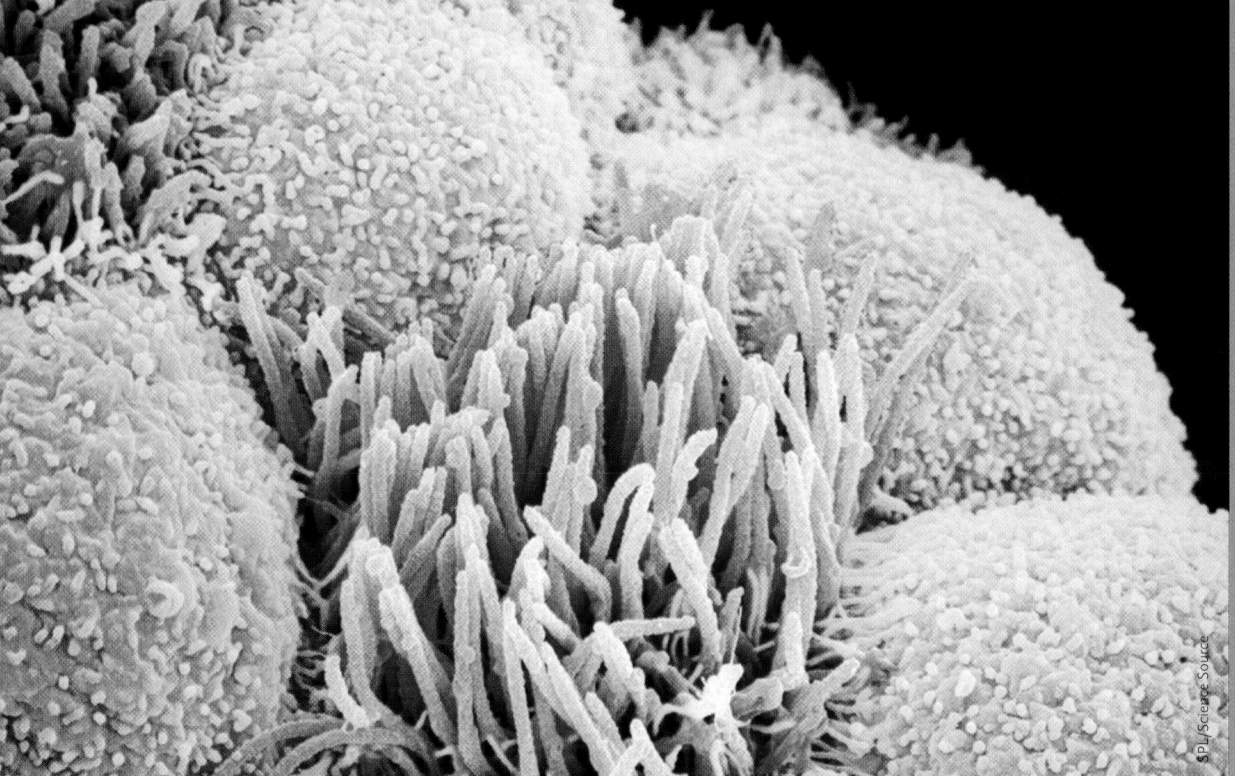

49

Lining of the trachea (windpipe) shown in a colourized scanning electron micrograph, with mucus-secreting cells (white) and epithelial cells with cilia (pink). The trachea is positioned between the larynx and the lungs, providing a conduit for air entering and leaving the body. The mucus traps foreign particles in the air, and the cilia move the trapped particles up and out of the respiratory tract, filtering the air and protecting the delicate alveoli of the lung where gas exchange takes place.

Gas Exchange: The Respiratory System

WHY IT MATTERS

In Africa, a huge swarm of the desert locust, *Schistocerca gregaria* **(Figure 49.1, p. 1218)**, composed of millions of individuals, takes off, looking for food (a swarm may consume the equivalent of food for 2500 people in a single day). Driven by the wind, the swarm normally descends in the evening to take off again in the morning but may remain airborne for more than 24 hours. A swarm may continue to fly during the day for several days until the wind delivers it to food. The wings of each locust, which weighs about 2.5 g, beat about 20 times per second. Their relatively large flight muscles require very large amounts of oxygen to provide energy that is derived from the metabolism of fat stores. Insects in flight have the highest O_2 consumption per gram ever recorded for animals.

Like these locusts, all organisms with active metabolism need to exchange gases with their surroundings. The mitochondria of eukaryotic cells need a constant supply of O_2, which is required as the terminal electron acceptor of respiratory electron transport (see Chapter 6). In addition, respiration also produces CO_2, which needs to be rapidly removed from animal cells because high cellular CO_2 is a narcotic poison that damages nerve function. High cellular CO_2 also produces changes in cellular pH, which in turn alter the activity levels of all functional proteins (enzymes, receptors, transport proteins, etc.).

In this chapter, we introduce the physical laws that are the basis for gas exchange and describe how evolution has produced a range of adaptations that maximize the rate of gas exchange both into and out of the tissues of animals living in different

Figure 49.1
(a) Migratory locust. **(b)** A small portion of a locust swarm in Mauritania in 2004.

environments. For single-celled organisms, exchange is simply across the plasma membrane. For large, multicellular organisms, however, exchange generally entails a series of steps known as the gas transport cascade. These steps (shown for an air-breathing vertebrate in **Figure 49.2**) transport oxygen and carbon dioxide between the environment and chloroplasts or mitochondria that are too far from the body surface to exchange gases by simple diffusion alone.

49.1 General Principles of Gas Exchange

Air normally contains about 78% nitrogen (N_2), 21% oxygen (O_2), and less than 1% carbon dioxide (CO_2) and other gases. This percentage composition of air remains constant from sea level up into the atmosphere. As you climb a mountain, however, the air gets thinner; there are fewer molecules of O_2 (and any other gas) in the environment. The density (thickness or thinness) of the air in the atmosphere is measured as the atmospheric pressure, which is greater the closer you are to sea level and falls with increasing altitude. The unit of measurement is often millimetres of mercury (mm Hg). At sea level, the atmospheric pressure is 760 mm Hg: the pressure generated by a column of air descending from the atmosphere to the surface of Earth is sufficient to support a vertical column of mercury 760 mm high. This pressure is the sum of the pressures of all the gases in a mixture. This is Dalton's law of partial pressures. The individual pressure exerted by each gas within a mixture of gases such as air is defined as its **partial pressure**. For any one gas, the partial pressure is calculated by multiplying the fractional composition of that gas by the atmospheric pressure. Given that O_2 is 0.21 of air, the partial pressure of O_2, abbreviated PO_2, at sea level is 0.21 × 760 mm Hg or 160 mm Hg.

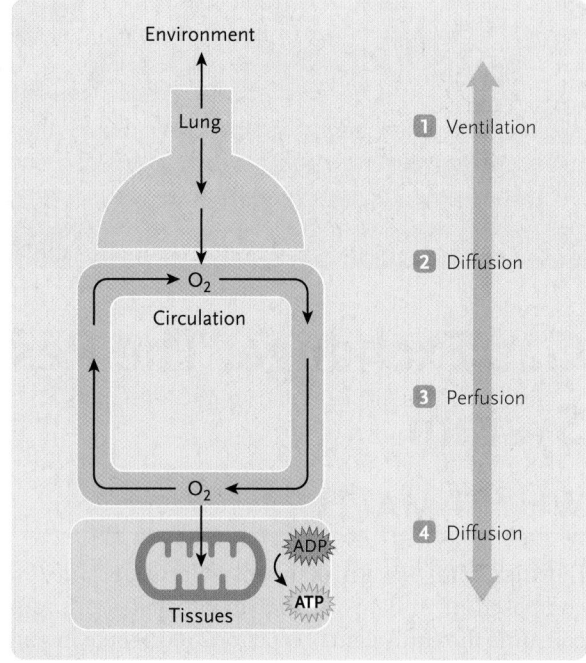

Figure 49.2
The steps in the cascade for the transport of oxygen between the environment and the mitochondria in an air-breathing vertebrate are shown here. There are four steps: (1) ventilation to move gas into and out of the lungs, (2) diffusion of oxygen into the blood, (3) perfusion or transport of blood by the heart to the tissues, and (4) diffusion of oxygen from the blood in the capillaries in the tissues into the mitochondria in the cells. The steps for the transport of CO_2 from the cells to the environment are the reverse of this.

CONCEPT FIX It is a myth that dissolved gases in aqueous solutions always diffuse from areas of high to areas of low concentration. Gases are not equally soluble in different fluids or in the same fluid at different temperatures. For instance, oxygen is far more soluble in lipid than it is in water, and it is more soluble in cold fluids than in warm ones. Thus, when oxygen is in equilibrium between two different solutions, the concentrations in each may be very different if the solubility of the gas in each solution is different, but their partial pressures will be the same. Just as a solute will move by simple diffusion from an area of high concentration to an area of low concentration, a gas will move down a partial pressure gradient from a region of high partial pressure to an area of low partial pressure. ⬢

49.1a Fick's Equation of Diffusion

While gases diffuse between two sites due to differences in partial pressure, the rate (amount per unit time) at which a gas will diffuse depends on a set of factors, only one of which is the difference in partial pressure between two regions. Fick's equation, which is important to understanding the diffusion of gases, recognizes the importance of these other factors. Fick's equation can be stated in a number of ways, but for the diffusion of gas across a membrane or other surface, it can be stated as

$$Q = \frac{DA \times \Delta P}{L}$$

where the variables are defined as follows:

- Q is the rate of diffusion between the two sides of the membrane.
- D is the diffusion coefficient for the gas involved. This is simply a factor specific to the gas molecule that recognizes its size and vibrational activity, the medium (gas, solid, liquid) in which the diffusion occurs, and the temperature.
- A is the area across which diffusion takes place.
- ΔP is the difference in partial pressures of the gases at the two locations.
- L is the path length or distance between the two locations (the thickness of the membrane).

It can be seen from this equation that anything that increases the diffusion coefficient, increases the surface area of the exchange surface, enhances the partial pressure gradient, or reduces the diffusion path will speed the rate of diffusion. We will briefly discuss some of these factors next.

49.1b Diffusion Coefficients Vary with the Medium; Water and Air Have Advantages and Disadvantages

An important variable in Fick's equation is the diffusion coefficient, D. This factor varies with the gas, the medium, and the temperature. The difference between D in gas and D in water is large: for O_2 at 20°C, the value in air is approximately 10 000 times that in water. Thus, the rate of diffusion of O_2 in air is 10 000 times as fast as in water. In addition, for the same volume, there is approximately $\frac{1}{30}$ the amount of O_2 in water as

in air at 15°C for the same partial pressure. These two factors require animals that obtain O_2 from water to pass a vastly greater volume of water over the respiratory surface in order to be exposed to the same volume of O_2 as an animal that obtains O_2 from air. Moreover, the density of water is about 1000 times that of air, and its viscosity is about 50 times that of air. Therefore, it takes significantly more energy to move water than air over a respiratory surface.

In addition, temperature and solutes affect the O_2 content of water. That is, as either the temperature or the amount of solute increases, the amount of gas that can dissolve in water decreases. Therefore, with respect to obtaining O_2, aquatic animals that live in warm water are at a disadvantage compared with those that live in cold water. And because solutes (such as sodium chloride) are higher in sea water than in fresh water, animals living in a marine environment are at a disadvantage compared to those living in a freshwater environment.

There are advantages to breathing water, however. Since CO_2 is roughly 20 times as soluble in water as is O_2, CO_2 excretion is easily achieved as a result of the ventilation required to obtain oxygen in aquatic organisms. Another advantage is that the exchange surface is always moist.

Thus, aquatic animals face challenges in obtaining O_2 from water compared with terrestrial animals but have a much easier time eliminating CO_2.

For air-breathing organisms, the relatively high O_2 content, low density, and low viscosity of air greatly reduce the energy required to ventilate the respiratory surface to obtain O_2. There are disadvantages to breathing air, however. CO_2 is not eliminated as easily in air-breathing organisms, so the excretion of CO_2 becomes the major drive to breathe. A second major disadvantage of air is that it constantly evaporates water from the respiratory surface unless it is saturated with water vapour. Therefore, except in an environment with 100% humidity, animals lose water by evaporation during breathing and must replace the water to keep the respiratory surface from drying.

49.1c Some Adaptations Increase the Surface Available for Gas Exchange in Organisms

In animals, gas exchange occurs across a **respiratory surface**, which may consist of each individual cell of the organism, the general external surface of the organism, or highly specialized exchange surfaces restricted to specialized areas of the body **(Figure 49.3, p. 1220)**. The first two strategies are only effective if all cells of the body are close to the body surface (Figure 49.3a). It is the evolution of larger specialized respiratory surfaces and/or some means of transporting gases to and from the surfaces of cells within the organism that has permitted the development of larger and more complex

a. Extended body surface: flatworm

©iStock.com/Piero Malaer

b. External gills: mudpuppy

Jack Dermid/Visuals Unlimited, Inc.

c. Lungs: human

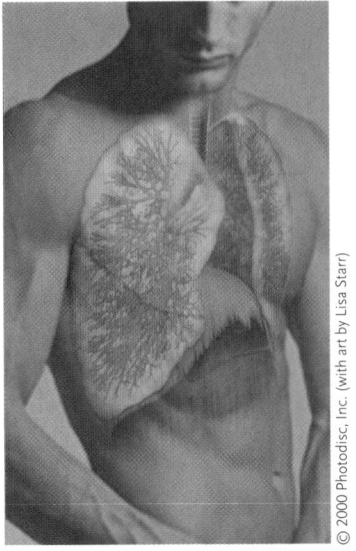

© 2000 Photodisc, Inc. (with art by Lisa Starr)

Figure 49.3
Adaptations increasing the area of the respiratory surface. **(a)** The flattened and elongated body surface of a flatworm. **(b)** The highly branched, feathery structure of the external gills in an amphibian, the mudpuppy (*Necturus*). **(c)** The many branches and pockets expanding the respiratory surface in the human lung.

organisms. In aquatic organisms, these specialized surfaces usually take the form of gills as outward extensions of the body surface (Figure 49.3b). In terrestrial organisms, they take the form of inward, moist, protected surfaces such as lungs (Figure 49.3c). The specialized respiratory surfaces are often very large, increasing A in Fick's equation. The total area of the lungs in humans is about 100 m². In addition, the cells that make up the specialized respiratory layer are thin, squamous epithelium, decreasing L.

49.1d Ventilation and Perfusion Are Adaptations That Increase Gas Exchange across a Surface

Although all gas exchange occurs by diffusion, two adaptations help most animals maintain the difference in concentration between gases outside and inside the respiratory surface. In Fick's equation, they maximize the value of ΔP, maintaining a steep gradient of the partial pressure of the gas across the respiratory surface and enhancing the rate of diffusion. One is **ventilation**, the flow of the respiratory medium (air or water, depending on the animal) over the respiratory surface. The second is **perfusion**, the flow of blood or other body fluids on the internal side of the respiratory surface (Figure 49.2).

Ventilation. As they respire, animals remove O_2 from the respiratory medium and replace it with CO_2. Without ventilation, the concentration of O_2 would fall in the respiratory medium close to the respiratory surface, and the concentration of CO_2 would rise, gradually reducing the value of ΔP in Fick's equation for both gases and reducing the rate of diffusion to below that necessary to sustain life. Examples of ventilation are the one-way flow of water over the gills in fishes

and many other aquatic animals and the in-and-out flow of air in the lungs of most vertebrates and in the tracheal system of insects at rest.

Perfusion. The rate at which blood or other fluids are replaced on the internal side of the respiratory surface similarly helps keep ΔP at an acceptable level. The circulatory system in animals that have one brings blood to the internal side of the respiratory surface, transporting CO_2 (often in the form of bicarbonate) from all cells of the body. At the surface, CO_2 is released into the medium, and a fresh supply of O_2 is picked up. Insects, as we will see, do not use blood to transport these gases.

The evolution of muscular pumps to ventilate specialized gas exchange surfaces and hearts to transport gases between the environment and the tissues throughout the bodies of animals was essential for the evolution of large multicellular organisms.

STUDY BREAK

1. What are the variables in Fick's equation that affect the rate of diffusion across a membrane?
2. What are the advantages and disadvantages of air versus water as a respiratory medium?

49.2 Adaptations for Gas Exchange

For the gases to pass across an epithelium and enter the cells of plants and animals, they must be dissolved and in solution. For aquatic organisms, for which the **respiratory medium** is water, that is already the case. For terrestrial organisms, the respiratory medium is air, and, thus, the respiratory epithelium must be covered by a thin film of fluid. In animals, the most highly evolved respiratory exchange surfaces are correspondingly large, thin, moist, and delicate and are usually highly protected to prevent damage and desiccation (see Figure 49.3c).

49.2a Gas Can Be Exchanged by Simple Diffusion

Relying on diffusion alone for gas exchange limits both the size and to some degree the shape of the organism. The importance of these factors is made obvious by a consideration of the surface to volume ratio. Bacteria, with a volume of about 10^{-18} m^3 and a surface area of about 6×10^{-12} m^2, have a surface area to volume ratio of 6 000 000:1. They can clearly rely on diffusion alone for gas exchange because the surface area is large with respect to the volume, and the distance that the gases must diffuse is relatively small. The same is true of protists, with a surface area to volume ratio of about 60 000:1. Among multicellular organisms, however, an increase in size can be accommodated only if the distance over which diffusion must occur is minimized. Gas exchange by diffusion can only occur if the organisms are thin and flat.

Flatworms (see Figure 49.3a) represent an example of a multicellular organism that relies on simple diffusion for gas exchange. Most free-living flatworms are small, but they may range up to 10 cm or more in length, and parasitic forms such as tapeworms may be as long as 3 m or more (see Chapter 7). But all are thin, so L in Fick's equation is minimized and A is maximized.

49.2b Insects Use a Tracheal System for Gas Exchange

Insects breathe air by a unique respiratory system consisting of air-conducting tubes called tracheae (*trachea* = windpipe) **(Figure 49.4)**. The tracheae are invaginations of the outer epidermis of the animal and as such consist of the epithelial cells and the cuticle secreted by those cells. They are lined with a thin layer of the same cuticle as the exoskeleton and are reinforced by rings of cuticle. They lead from the body surface and branch repeatedly. With each branching, the diameter of the tracheae is reduced, and the tracheae ultimately end as tracheoles less than 1 μm in diameter. Every cell except the hemocytes in an insect's body makes contact with at least one tracheole. In the case of large, metabolically active cells, such as the flight muscles (see "Why It Matters"), tracheoles may penetrate the cell via invaginations of the cell membrane. Tracheoles are dead-end tubes with very small tips filled with fluid that are in contact with cells of the body. Air is transported by the tracheal system to those tips, and gas exchange occurs directly across the very thin cuticle and epithelium of the tracheoles and the plasma membranes of the body cells. At places within the body, the tracheae may expand into internal air sacs that act as reservoirs to increase the volume of air in the system.

Air enters and leaves the tracheal system at openings in the insect's chitinous exoskeleton called **spiracles** (*spiraculum* = airhole). The spiracles are located in a row on either side of the thorax and abdomen, typically one pair per body segment. Each spiracle incorporates a muscle that allows the spiracles to open and close. For many insects at rest, the

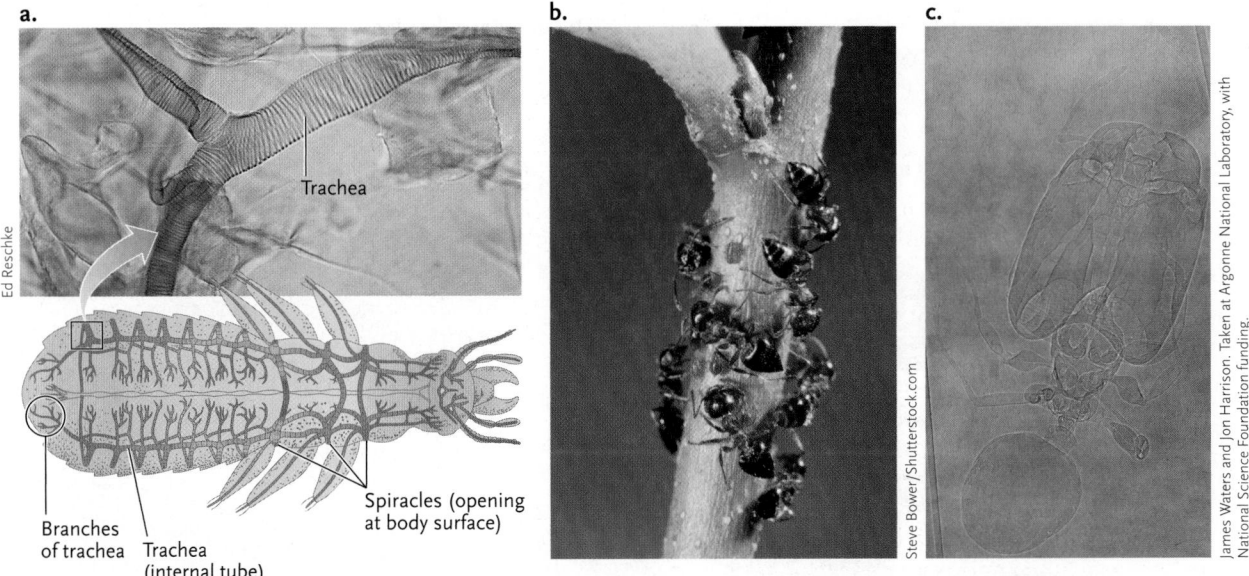

a.
Trachea

Branches of trachea   Trachea (internal tube)   Spiracles (opening at body surface)

Ed Reschke

b.
Steve Bower/Shutterstock.com

c.
James Waters and Jon Harrison. Taken at Argonne National Laboratory, with National Science Foundation funding.

Figure 49.4

(a) The tracheal system of insects. The photograph shows the chitinous rings that reinforce the tracheae, keeping them from collapsing. The tracheal system terminates in many tracheolar end cells that have branches with a diameter of less than 1 μm.
(b) A photograph of the ant, *Pheidole tepicana*. **(c)** An x-ray synchrotron image of the ant. If you look carefully you can see the trachea running throughout the body delivering oxygen directly from the atmosphere to the tissues.

spiracles are minimally open, allowing a very small current of air to enter. O_2 is consumed by the tissues, and CO_2 is taken up by the bicarbonate buffering system, resulting in a small negative pressure inside the tracheal system. The small inward current of air flowing through the reduced opening of the spiracle prevents water vapour from escaping. As the bicarbonate buffering system becomes saturated, free CO_2 builds up inside the insect, causing the spiracles to open briefly, allowing CO_2 (and water vapour) to escape. In periods of greater activity, as in flight, this mechanism is replaced by one in which alternating compression and expansion of the thorax by the flight muscles also pumps air through the tracheal system. The spiracles open and close in synchrony with this rhythm.

49.2c Some Animals Have External and Others Have Internal Gills

Gills are respiratory surfaces that are branched and folded evaginations (outward extensions) of the body. They increase the area over which diffusion can take place. **External gills (Figure 49.5a)** extend out from the body and do not have protective coverings. They occur in some molluscs, some annelids, the larvae of some aquatic insects, the larvae of some fishes, and the larvae of amphibians. **Internal gills (Figure 49.5b, c, d)** are located within chambers of the body. This not only provides protection for delicate structures but also allows currents of water to be directed over the gills.

Most crustaceans, molluscs, sharks, and bony fishes have internal gills. Some invertebrates, such as clams and oysters, use beating cilia to circulate water over their internal gills (Figure 49.5b). Others, such as the cuttlefish, use contractions of the muscular mantle to pump water over their gills (see Figure 49.5c). In adult bony fishes, the gills extend into a chamber covered by gill flaps or *opercula* (singular, *operculum* = little lid) on either side of the head. The operculum serves as part of a one-way pumping system that ventilates the gills (Figure 49.5d).

49.2d Many Animals with Internal Gills Use Countercurrent Flow to Maximize Gas Exchange

Sharks, fishes, and some Crustacea take advantage of one-way flow of water over the gills to maximize the amounts of O_2 and CO_2 exchanged with water. In this mechanism, called **countercurrent exchange**, the water flowing over the gills moves in a direction opposite to the flow of blood under the respiratory surface.

Figure 49.6 illustrates countercurrent exchange in the uptake of O_2. At the point where fully oxygenated water first passes over a gill filament in countercurrent flow, the blood flowing beneath it in the opposite direction is also almost fully oxygenated. However, the water still contains O_2 at a higher concentration than the blood, and the gas diffuses from the water into the blood, raising the concentration of O_2 in the blood almost to the level of the fully oxygenated water. At the opposite end of the filament, much of the O_2 has been removed from the water, but the blood flowing under the filament, which has just arrived from body tissues

Figure 49.5
External and internal gills. **(a)** The external gills of a nudibranch (*Flabellina iodinea*). **(b)** The internal gills in a clam. **(c)** The internal gills of a cuttlefish. **(d)** The internal gills of a bony fish. Water enters through the mouth and passes over the filaments of the gills before exiting through an opening at the edges of the flaplike protective covering, the operculum.

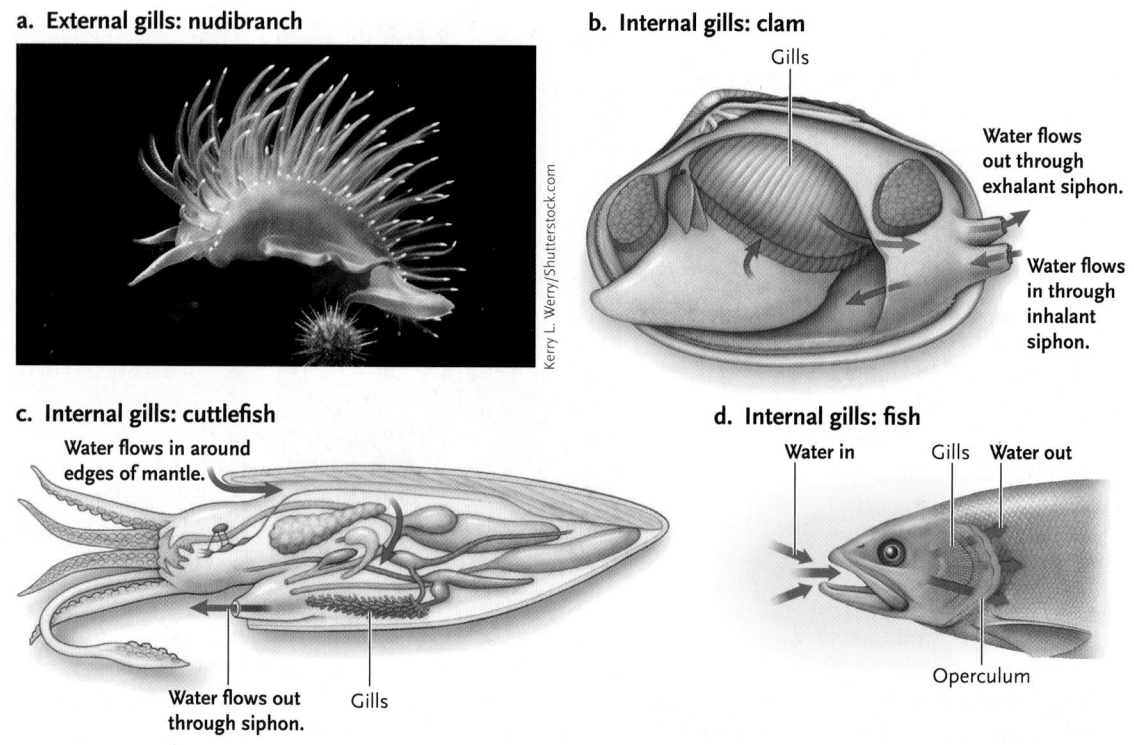

a. External gills: nudibranch

Kerry L. Werry/Shutterstock.com

b. Internal gills: clam

Gills

Water flows out through exhalant siphon.

Water flows in through inhalant siphon.

c. Internal gills: cuttlefish

Water flows in around edges of mantle.

Water flows out through siphon. Gills

d. Internal gills: fish

Water in   Gills   Water out

Operculum

and is fully deoxygenated, contains even less O_2. As a result, O_2 also diffuses from the water to the blood at this end of the filament. All along the gill filament, the same relationship exists, so that at any point, the water is more highly oxygenated than the blood. ΔP is maximized, and O_2 diffuses at a high rate from the water and into the blood across the respiratory surface.

The overall effect of countercurrent exchange is the removal of 80% to 90% of the O_2 content of water as it flows over the gills. In comparison, by breathing in and out and constantly reversing the direction of air flow, mammals manage to remove only about 25% of the O_2 content of air. Efficient removal of O_2 from water is important because of the much lower O_2 content of water compared with air (see Section 49.1b).

Countercurrent exchange is a remarkably efficient method of exchange and has been exploited frequently in biological systems. Other examples include heat exchange in birds and mammals (see Section 50.7) as well as exchange of ions and water in vertebrate kidneys (see Section 50.3).

49.2e Lungs Allow Animals to Live in Completely Terrestrial Environments

Lungs are one of the primary adaptations that allowed vertebrates to fully invade terrestrial environments. Many researchers believe that the bony fish evolved from a freshwater ancestor that had both fins and lungs. The lungs arose as invaginations of the upper digestive tract. Two lines evolved from this ancestor. In one line, the lung lost its connection to the digestive system and became the swim bladder, which controls buoyancy in the modern teleost fishes. The other line (Sarcopterygii), represented by only a few living species, retained the lung, enabling them to survive in O_2-poor water or in periods when pools dried up. This line gave rise to the tetrapod vertebrates. In these fish, air is obtained by **positive pressure breathing**, a gulping or swallowing motion that forces air into the lungs (see Chapter 28).

The lungs of mature amphibians such as frogs and salamanders are also thin-walled sacs with relatively little folding or pocketing. Amphibians also fill their lungs by positive pressure breathing **(Figure 49.7, p. 1224)**. In most adult amphibians, a breathing cycle begins with expansion of the buccal (mouth) cavity with the nostrils open and the entrance into the lungs constricted by the glottis. This draws fresh air into the mouth. Next, the glottis opens, and gas from the lungs enters the buccal cavity, where it mixes with the fresh air to varying degrees as it exits via the mouth and nares, which remain open. The nares and mouth then close, and buccal compression forces buccal gas into the lungs. The glottis then closes, and any excess gas left in the buccal cavity is expelled through the nares or mouth at the end of the buccal compression phase. Rhythmic motions of the floor of the mouth with the

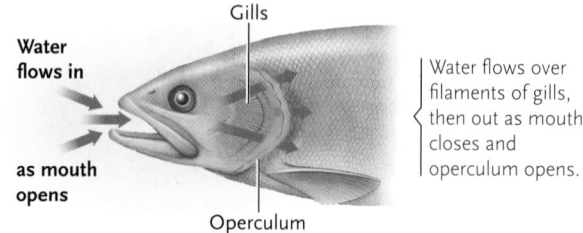

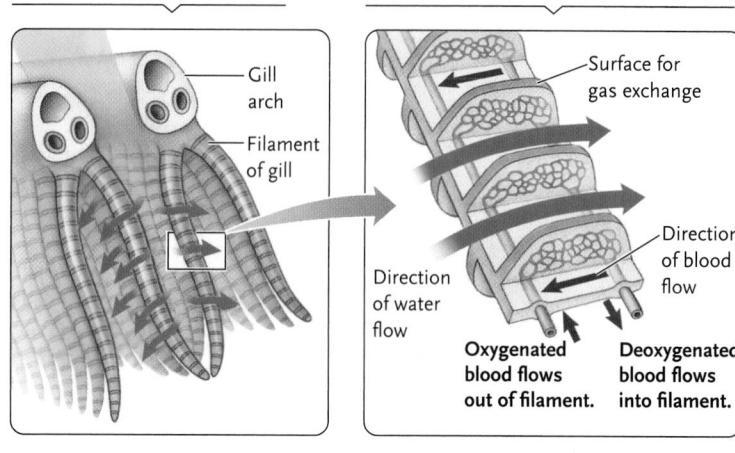

a. The flow of water around the gill filaments

b. Countercurrent flow in fish gills, in which the blood and water move in opposite directions

c. In countercurrent exchange, blood leaving the capillaries has the same O_2 content as fully oxygenated water entering the gills.

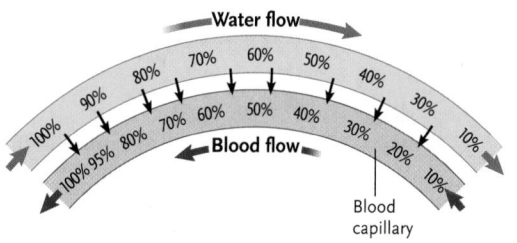

Figure 49.6

Ventilation and countercurrent exchange in bony fishes. **(a)** Water flows around the gill filaments. **(b)** Water and blood flow in opposite directions through the gill filaments. **(c)** Countercurrent exchange: oxygen from the water diffuses into the blood, raising its oxygen content. The percentages indicate the degree of oxygenation of water (blue) and blood (red).

nostrils open ensure that the buccal cavity contains fresh air for the beginning of the next cycle. The efficiency of the system is increased because much of the CO_2 is lost through the skin. (Remember from Chapter 40 that frogs have a pulmonary-cutaneous circulation.)

In reptiles, birds, and mammals, the lungs become more folded, with many pockets, increasing the surface for gas exchange. Mammalian lungs consist of millions of tiny air pockets, the **alveoli** (singular, *alveolus*), each surrounded by dense capillary networks. Reptiles, birds, and mammals fill their lungs by **negative pressure breathing**, in which muscular contractions expand the chest and lungs, lowering the pressure of the air in the lungs and causing air to be pulled inward. The muscles involved in doing this are

Figure 49.7
Positive pressure breathing in an amphibian (frog).

  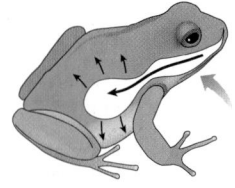 

1 The frog lowers the floor of the mouth and inhales through its nostrils.

2 Air in the lungs is exhaled when the glottis opens due to elastic recoil of the lungs and body wall.

3 The frog closes its nostrils and elevates the floor of the mouth, forcing air into the lungs.

4 Rhythmic movements flush the mouth cavity with fresh air for the next cycle.

a. Lungs and air sacs of a bird

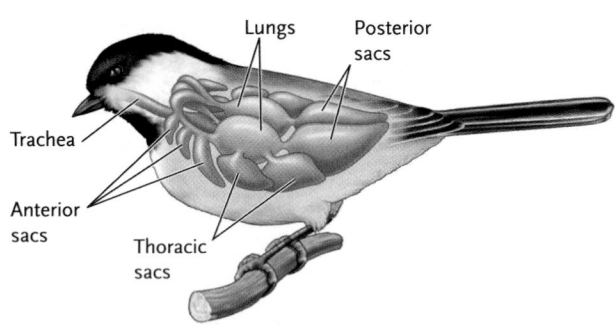

b. Crosscurrent exchange

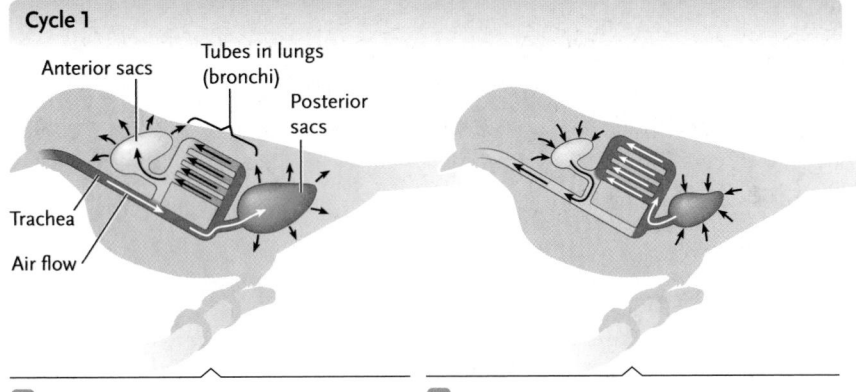

Cycle 1

1 During the first inhalation, most of the oxygen flows directly to the posterior air sacs. The anterior air sacs also expand but do not receive any of the newly inhaled oxygen.

2 During the following exhalation, both anterior and posterior air sacs contract. Oxygen from the posterior sacs flows into the gas-exchanging tubes (bronchi) of the lungs.

Cycle 2

1 During the next inhalation, air from the lung (now deoxygenated) moves into the anterior air sacs.

2 In the second exhalation, air from anterior sacs is expelled to the outside through the trachea.

Figure 49.8
Crosscurrent exchange in bird lungs. **(a)** Unlike mammalian lungs, bird lungs do not expand and contract. Changes in pressure in the expandable air sacs move air in and out. **(b)** Air flows in one direction through the tubes of the lungs; blood flows across this direction in the surrounding capillary network. Two cycles of inhalation and exhalation are needed to move a specific volume of air through the bird respiratory system.

largely those of the ribcage but can be assisted by other muscles. In crocodilians, for example, the contraction of a muscle connecting the liver to the pelvis pulls the liver back, causing the lungs to expand, while compression of the abdomen pushes the liver forward, forcing gases out of the lungs. The mechanism in mammals is described in detail in the next section.

In birds, a crosscurrent exchange system makes their lungs the most efficient vertebrate lungs **(Figure 49.8)**. In addition to paired lungs, birds have up to nine pairs of air sacs that branch off the respiratory tract. The air sacs, which collectively contain several times as much air as the lungs, are not respiratory surfaces. Unlike other vertebrate lungs, bird lungs are rigid and do not expand or contract. The air sacs do, however, and they set up a pathway that allows air to flow in one direction through the lungs rather than in and out, as in other vertebrates. As illustrated in Figure 49.8, two cycles of inhalation and exhalation are needed to move a specific volume of air through the bird respiratory system. Within the lungs, air always flows from back to front through an array of fine, parallel tubes that are surrounded by a capillary network. The blood flows in a direction across that of the air flow, setting up a crosscurrent exchange. The crosscurrent exchange allows bird lungs to extract more of the O_2 from the air than the lungs of mammals do, but less than with the countercurrent exchange system seen in fish **(Figure 49.9)**.

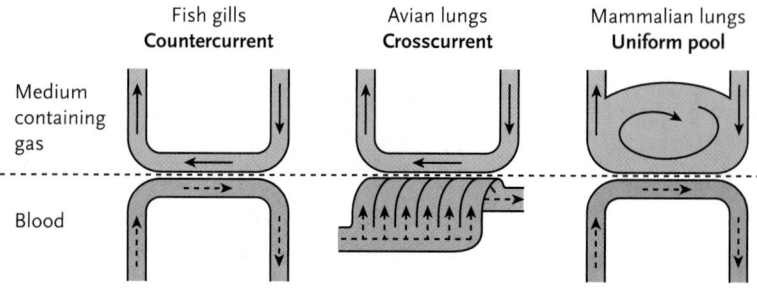

Figure 49.9

The three patterns of gas transfer encountered in vertebrate gas exchange organs are countercurrent, crosscurrent, and uniform (mixed) pool.

STUDY BREAK

1. What variable in Fick's equation is most affected by the countercurrent mechanism in gas exchange in teleost fish?
2. What is the difference between positive pressure breathing and negative pressure breathing?

49.3 The Mammalian Respiratory System

All mammals have a pair of lungs and a diaphragm in the chest cavity that plays an important role in negative pressure breathing. Rapid ventilation of the respiratory surface and perfusion by blood flow through dense capillary networks maximizes gas exchange.

49.3a The Airways Leading from the Exterior to the Lungs Filter, Moisten, and Warm the Entering Air

The human respiratory system is typical for a terrestrial mammal **(Figure 49.10, p. 1226)**. Air enters and leaves the respiratory system through the nostrils and mouth. Hairs in the nostrils and mucus covering the surface of the airways filter out and trap dust and other large particles. Inhaled air is moistened and warmed as it moves through the mouth and nasal passages.

Next, air moves into the throat or **pharynx**, which forms a common pathway for air entering the **larynx** or voice box and food entering the esophagus, which leads to the stomach. The airway through the larynx is open except during swallowing.

From the larynx, air moves into the **trachea**, which branches into two airways, the **bronchi** (singular, *bronchus*). The bronchi lead to the two elastic, cone-shaped lungs, one on each side of the chest cavity. Inside the lungs, the bronchi narrow and branch repeatedly, becoming progressively narrower and more numerous. The terminal airways, the **bronchioles**, lead into cup-shaped pockets, the alveoli (shown in Figure 49.10 insets).

Each of the 150 million alveoli in each lung is surrounded by a dense network of capillaries. By the time inhaled air reaches the alveoli, it has been moistened to the saturation point and brought to body temperature. The many alveoli provide an enormous area for gas exchange. If the alveoli of an adult human were flattened out in a single layer, they would cover an area approaching 100 m², about the size of a tennis court! The epithelium of the alveoli is composed of very thin squamous cells. In terms of Fick's law, A is very large and L is minimized.

The trachea and larger bronchi are nonmuscular tubes encircled by rings of cartilage that prevent the tubes from compressing (recall the analogous but not homologous arrangement in the tracheae of insects). The largest of the rings, which reinforces the larynx, stands out at the front of the throat as the Adam's apple, which is more prominent in males. The walls of the smaller bronchi and the bronchioles contain smooth muscle cells that contract or relax to control the diameter of these passages and with it the amount of air flowing to and from the alveoli.

The epithelium lining each bronchus contains cilia and mucus-secreting cells (see opening photo to this chapter). Bacteria and airborne particles such as dust and pollen are trapped in the mucus (see "Molecule behind Biology," Box 49.1, p. 1227) and then moved upward and into the throat by the beating of the cilia lining the airways. Infection-fighting macrophages (see Chapter 51) also patrol the respiratory epithelium.

49.3b Contractions of the Diaphragm and Muscles between the Ribs Ventilate the Lungs

The lungs are located in the ribcage above the *diaphragm*, a dome-shaped sheet of skeletal muscle separating the chest cavity from the abdominal cavity. The lungs are covered by a double layer of epithelial tissue called the **pleura**. The inner pleural layer is attached to the surface of the lungs, and the outer layer is attached to the surface of the chest cavity. A narrow space between the inner and outer layers is filled with slippery fluid, which allows the lungs to move within the chest cavity without rubbing or abrasion as they expand and contract. This fluid also creates a surface tension between the chest wall and the outer covering of the lung. As a result, the lungs cling to the chest wall (like any two wet surfaces) and expand and contract passively as the volume of the chest cavity changes. If air is introduced into this space, due to either rupture of the lung or a penetrating injury to the chest wall (pneumothorax), the lung on that side will collapse.

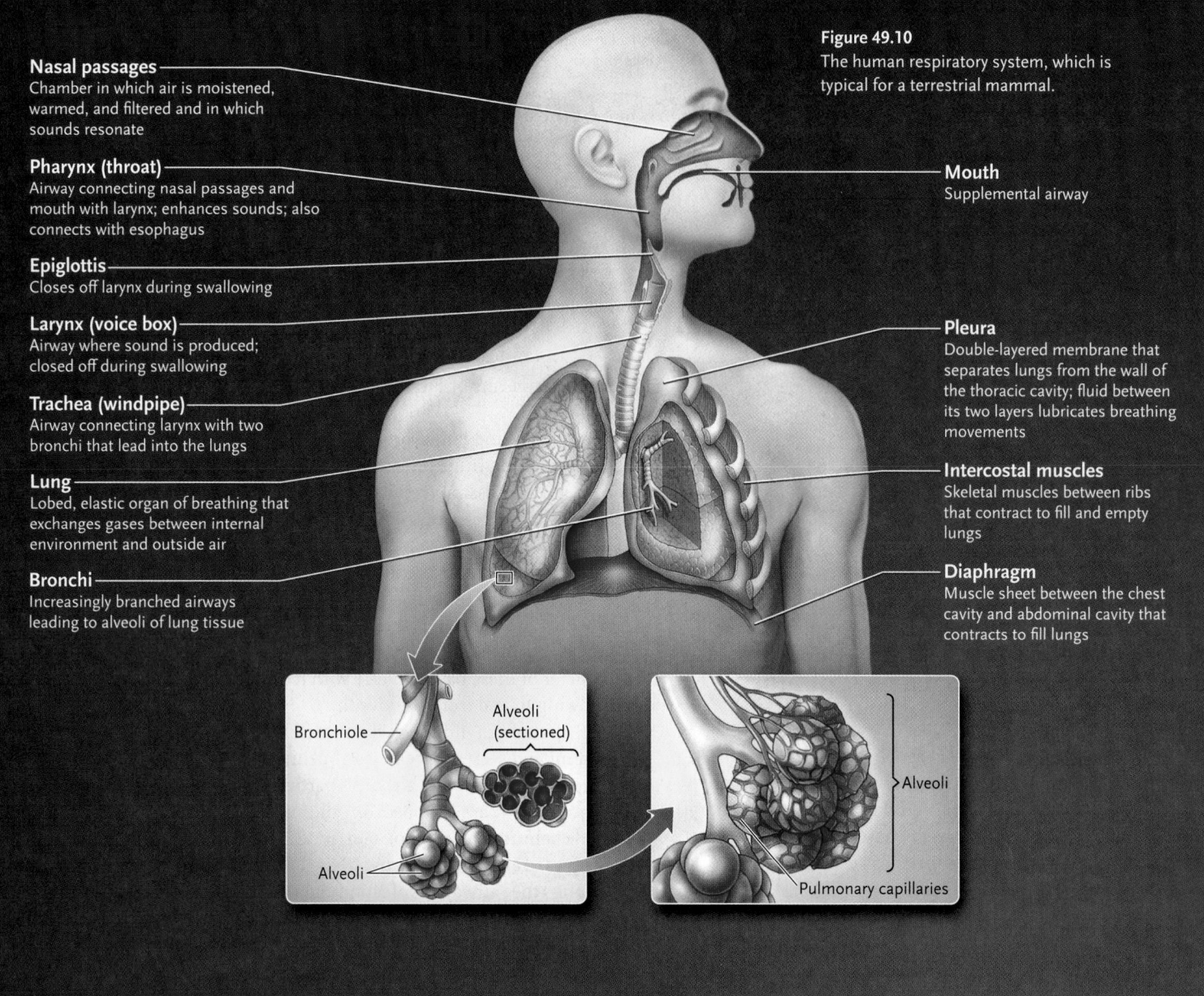

Nasal passages
Chamber in which air is moistened, warmed, and filtered and in which sounds resonate

Pharynx (throat)
Airway connecting nasal passages and mouth with larynx; enhances sounds; also connects with esophagus

Epiglottis
Closes off larynx during swallowing

Larynx (voice box)
Airway where sound is produced; closed off during swallowing

Trachea (windpipe)
Airway connecting larynx with two bronchi that lead into the lungs

Lung
Lobed, elastic organ of breathing that exchanges gases between internal environment and outside air

Bronchi
Increasingly branched airways leading to alveoli of lung tissue

Mouth
Supplemental airway

Pleura
Double-layered membrane that separates lungs from the wall of the thoracic cavity; fluid between its two layers lubricates breathing movements

Intercostal muscles
Skeletal muscles between ribs that contract to fill and empty lungs

Diaphragm
Muscle sheet between the chest cavity and abdominal cavity that contracts to fill lungs

Bronchiole

Alveoli (sectioned)

Alveoli

Alveoli

Pulmonary capillaries

Contraction of the diaphragm and the intercostal muscles between the ribs brings air into the lungs by a negative pressure mechanism. As an inhalation begins, the diaphragm contracts and flattens, and one set of muscles between the ribs, the external intercostal muscles, contracts, pulling the ribs upward and outward **(Figure 49.11, p. 1228)**. These movements expand the chest cavity and lungs, lowering the air pressure in the lungs below that of the atmosphere. As a result, air is drawn into the lungs, expanding and filling them.

The expansion of the lungs is much like filling two rubber balloons. Like balloons, the lungs are elastic and resist stretching as they are filled. And also like balloons, the stretching stores energy that can be released to expel air from the lungs. During an exhalation by a person at rest, the diaphragm and muscles between the ribs relax, and the elastic recoil of the lungs expels the air.

When physical activity increases the body's demand for O_2, contractions of other muscles help expel the air by forcefully reducing the volume of the chest cavity. That is, the abdominal wall muscles contract, which increases abdominal pressure. That pressure exerts an upward-directed force on the diaphragm, which is pushed upward. In addition, internal intercostal muscles contract, pulling the chest wall inward and downward, causing it to flatten. As a result, the dimensions of the chest cavity decrease.

49.3c The Volume of Inhaled and Exhaled Air Varies over Wide Ranges

The volume of air entering and leaving the lungs during inhalation and exhalation is called the **tidal volume.** In a person at rest, the tidal volume amounts to about 500 mL. As physical activity increases, the tidal volume

Mucin: Sticky Lubricant

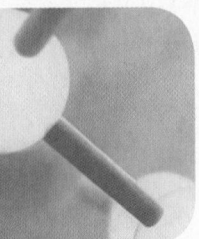

The airways leading to the lungs have a surface coating of mucus secreted by specialized epithelial cells **(Figure 1a).** The sticky mucus traps dust particles, bacteria, and other foreign bodies and is swept upward by the action of cilia to be swallowed into the digestive tract or expelled by the act of blowing your nose or spitting. Mucus is composed of a rodlike protein, mucin, that is very heavily glycosylated (the addition of sugars) after translation (see Section 15.3) and that forms giant polymers up to 10 million Da (daltons, a *unit* used for mass on a molecular scale). The very dense coating of sugar provides a lot of water-holding capacity. This gives mucin its slippery character, which makes it useful as a lubricant.

Mucins constitute a family of proteins, and at least 19 genes are known for humans. Three of these are expressed in airway epithelium. Mucins are also important in the mucus that coats the intestinal epithelium, where it serves as a lubricant and protects the epithelium from gastric acids and enzymes. Mucus also lines the reproductive tract. Mucins are important constituents of saliva, keeping the membranes of the oral cavity moist and lubricating the food as it is chewed (Figure 1b). Tears contain mucins, continuously washing the eye free of dust (Figure 1c). All of these mucins are produced continuously, often in large quantities.

Other mucins, the *tethered* or membrane-associated mucins, are attached to cell membranes, where they serve a variety of functions. In addition to the mucins in tears, mucins are also attached to the membranes of the surface of the eye, a slippery surface over which tears can move more easily. The function of the several membrane-associated mucins is far from clear. They may have a role in preventing infection or in cell-to-cell attachment, or, in some cases, they may be part of a cell-signalling mechanism.

Mucins are ancient molecules and occur in most metazoan phyla as well as in protists. In insects, a mucin gene is expressed in the intestine, and mucins are known from nematodes. In molluscs, mucins are found in the matrix that contains calcite to form the shell.

a.

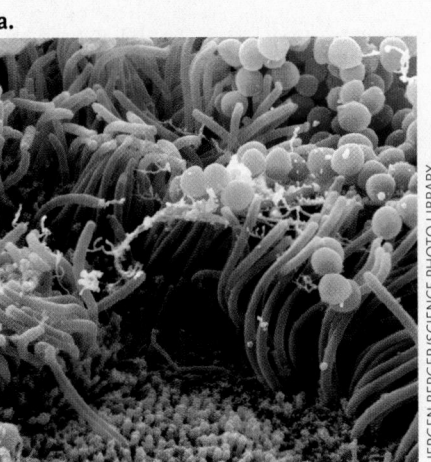

JUERGEN BERGER/SCIENCE PHOTO LIBRARY

b.

Michael Fitzsimmons/iStock/Thinkstock

c.

carlo dapino/Shutterstock.com

Figure 1
Mucins act as lubricants on many membranes. (a) Staphylococcus aureus *bacteria (yellow) on human nasal epithelial cells adhering to mucus (blue) on the hairlike cilia that protrude from the epithelial cells. Mucins are also common in (b) saliva and (c) tears.*

increases to match the body's demands for O_2; at maximal levels, the tidal volume reaches about 3400 mL in females and 4800 mL in males. This maximum tidal volume is called the **vital capacity** of an individual.

Even after the most forceful exhalation, about 1200 mL of air remains in the lungs in males and about 1000 mL in females; this is the **residual volume** of the lungs. In fact, the lungs cannot be deflated completely because small airways collapse during forced exhalation, blocking further outflow of air. Because air cannot be removed from the lungs completely, some gas exchange can always occur between blood flowing through the lungs and the air in the alveoli.

49.3d Ventilation Is Initiated and Controlled by Centres in the Brain

The respiratory movements are controlled by centres in the medulla and pons, part of the brain stem (see Chapter 45). Nerve signals from these centres to the muscles involved in breathing can vary the intake of air from as little as 5 to 6 L per minute to as much as 150 L per minute (for very brief periods). These centres integrate information about O_2 and CO_2 in the blood from O_2 and CO_2 receptors. These receptors are located in special sense organs (the carotid bodies) in the

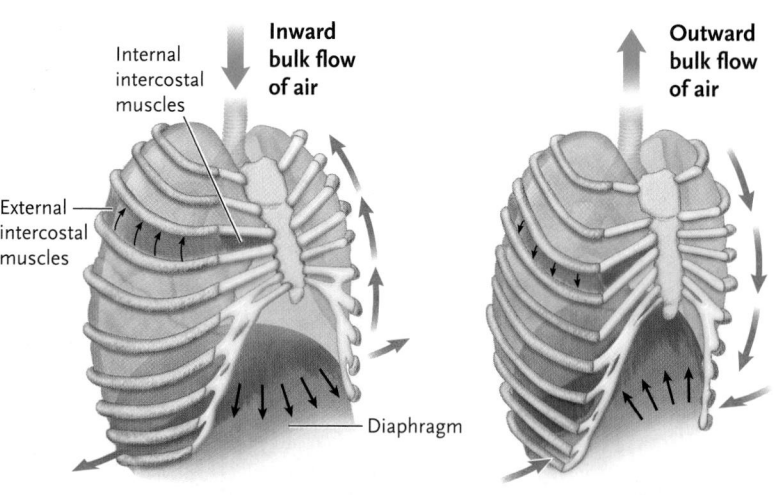

Internal intercostal muscles

Inward bulk flow of air

External intercostal muscles

Diaphragm

Outward bulk flow of air

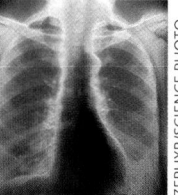

Inhalation.
The diaphragm contracts and moves down. The external intercostal muscles contract and lift the rib cage upward and outward. The lung volume expands.

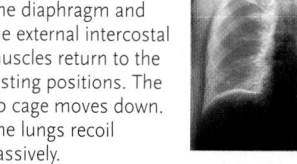

ZEPHYR/SCIENCE PHOTO LIBRARY

Exhalation during breathing or rest.
The diaphragm and the external intercostal muscles return to the resting positions. The rib cage moves down. The lungs recoil passively.

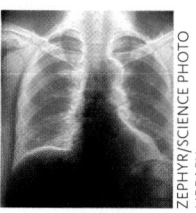

ZEPHYR/SCIENCE PHOTO LIBRARY

Figure 49.11
The respiratory movements of humans during breathing at rest. The movements of the ribcage and diaphragm fill and empty the lungs. Inhalation is powered by contractions of the external intercostal muscles and diaphragm, and exhalation is passive. During exercise or other activities characterized by deeper and more rapid breathing, contractions of the internal intercostal muscles and the abdominal muscles add force to exhalation. The X-ray images show how the volume of the lungs increases during inhalation and decreases during exhalation.

carotid arteries, which supply the brain, and in the aorta (the **aortic body**), which supplies blood to the rest of the body. These receptors are more sensitive to changes in CO_2: the P_{O_2} must drop below about 100 mm Hg before their activity increases significantly. The medulla integrates this information with information coming from its own receptors, which monitor the pH of the cerebrospinal fluid. The pH of this fluid is determined mostly by the CO_2 concentration in the blood. (Remember that the pH decreases as CO_2 levels increase.) In general, the CO_2 level is most closely monitored. The O_2 receptors act as a backup system, which comes into play only when blood O_2 concentration falls to critically low levels. The level of CO_2 in the blood and body fluids is much more closely monitored and has a much greater effect on breathing than the O_2 level (see Section 49.1). This reflects the fact that air-breathing vertebrates have far more trouble eliminating CO_2 than obtaining O_2 under normal conditions, along with the fact that small changes in pH due to changes in the levels of CO_2 throughout the body profoundly affect the activity of all functional proteins, such as enzymes.

STUDY BREAK

1. Compare and contrast the ways in which bird and mammalian lungs are designed to enhance gas exchange in terms of the components of Fick's equation.
2. What is tidal volume, and why is called by this name?

49.4 Mechanisms of Gas Transport

In this section, we consider the means by which gases are transported between the respiratory exchange surfaces and other body tissues. Recall that O_2 is not very soluble in fluids and so transport within the body is aided in most animals by pigment molecules that can bind and transport larger quantities of O_2. Hemoglobin is the vertebrate respiratory pigment.

At both the respiratory exchange surface and body tissues, gas exchange occurs when the gas diffuses from an area of higher partial pressure to an area of lower partial pressure. At the sites of gas exchange with the environment, the P_{O_2} in the environment is higher than the P_{O_2} in deoxygenated blood entering the network of capillaries in the gills or lungs **(Figure 49.12)**. As a result, O_2 readily diffuses into the plasma solution in the capillaries.

49.4a Hemoglobin Greatly Increases the O_2-Carrying Capacity of the Blood

In vertebrates, after entering the plasma, O_2 diffuses into erythrocytes, where it combines with hemoglobin. The combination with hemoglobin removes O_2 from the plasma, lowering the P_{O_2} of the plasma and increasing ΔP between alveolar air and the blood. This increases the rate of diffusion of O_2 across the alveoli and into the plasma.

Recall from Chapter 40 that a mammalian hemoglobin molecule has four heme groups, each containing an iron atom that can combine reversibly with an O_2 molecule. A hemoglobin molecule can therefore

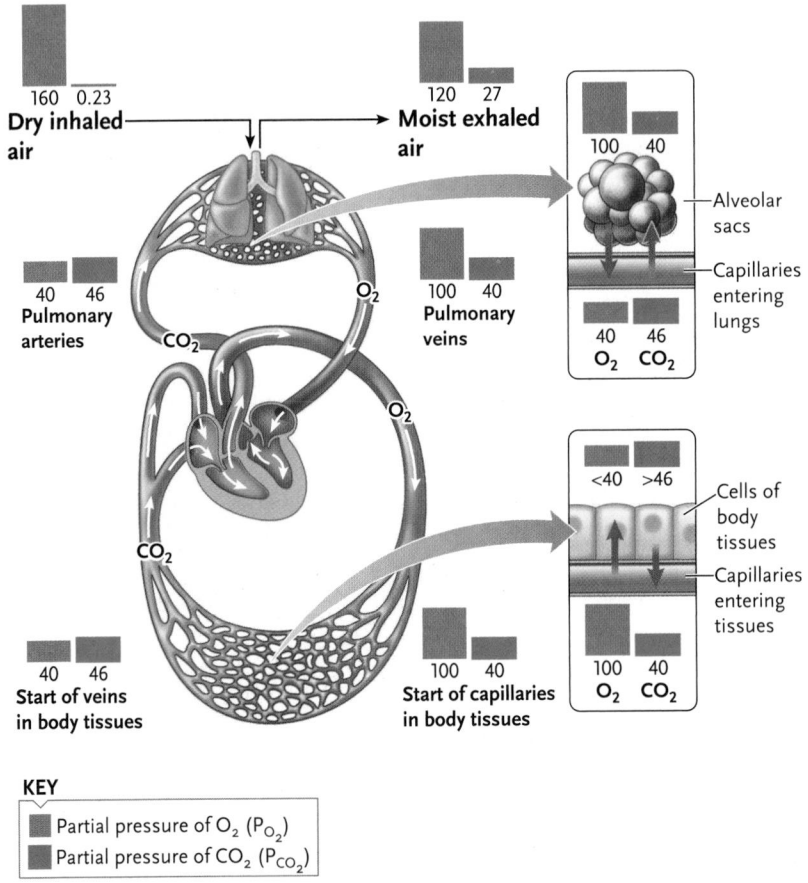

160 0.23
Dry inhaled air

Moist exhaled air
120 27

100 40
- Alveolar sacs
- Capillaries entering lungs

40 46
Pulmonary arteries

O_2

CO_2

100 40
Pulmonary veins

40 46
O_2 CO_2

O_2

<40 >46
- Cells of body tissues
- Capillaries entering tissues

CO_2

40 46
Start of veins in body tissues

100 40
Start of capillaries in body tissues

100 40
O_2 CO_2

100 40
O_2 CO_2

KEY

- ▇ Partial pressure of O_2 (P_{O_2})
- ▇ Partial pressure of CO_2 (P_{CO_2})

Figure 49.12
The partial pressures of O_2 (pink) and CO_2 (blue) in various locations in the body.

potentially bind a total of four molecules of O_2. The combination of O_2 with hemoglobin allows blood to carry about 60 times as much O_2 (about 200 mL per litre) as it could if the O_2 simply dissolved in the plasma (about 3 mL per litre). About 98.5% of the O_2 in blood is carried by hemoglobin, and about 1.5% is carried in solution in the blood plasma.

The reversible combination of hemoglobin with O_2 is related to the P_{O_2} in a pattern shown by the *hemoglobin–O$_2$ equilibrium curve* in **Figure 49.13.** (The curve is generated by measuring the amount of hemoglobin saturated at a given P_{O_2}.) In air-breathing vertebrates, the curve is not linear but S-shaped, with a plateau region. As the P_{O_2} of the plasma increases, hemoglobin binds to the O_2 until every hemoglobin molecule is bound to four molecules of O_2. Once the hemoglobin is fully saturated, further increases in P_{O_2} lead to only a small extra amount of O_2 going into solution—the hemoglobin can hold no more. Note that over the

a. Hemoglobin saturation level in lungs

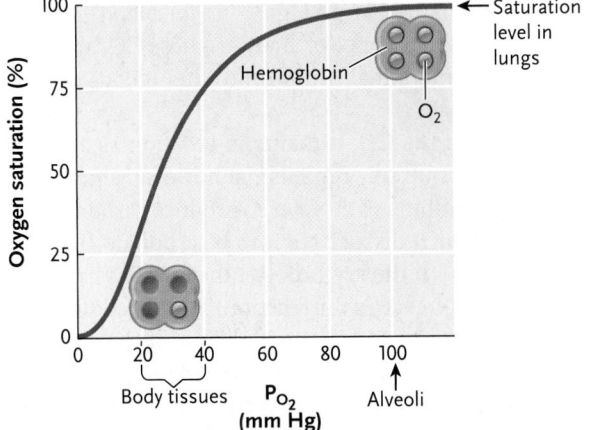

In the alveoli, in which the P_{O_2} is about 100 mm Hg and the pH is 7.4, most hemoglobin molecules are 100% saturated, meaning that almost all have bound four O_2 molecules.

Figure 49.13
Hemoglobin–O_2 equilibrium curves, which show the degree to which hemoglobin is saturated with O_2 at increasing P_{O_2}.

b. Hemoglobin saturation range in body tissues

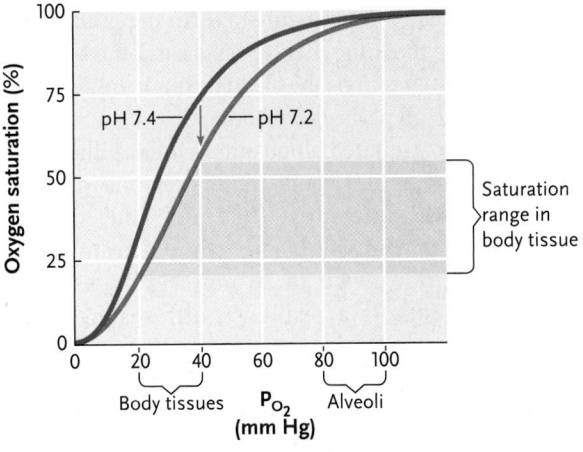

In the capillaries of body tissues, where the P_{O_2} varies between about 20 and 40 mm Hg, depending on the level of metabolic activity, and the pH is about 7.2, hemoglobin can hold less O_2. As a result, most hemoglobin molecules release two or three of their O_2 molecules to become between 25% and 50% saturated. Note that the drop in pH to 7.2 (red line) in active body tissues reduces the amount of O_2 hemoglobin can hold as compared with pH 7.4. The reduction in binding affinity at lower pH increases the amount of O_2 released in active tissues.

Enhancing Performance: Blood Doping

Because red blood cells (RBCs) greatly increase the ability of blood to transport oxygen, enhancing the number of RBCs in the bloodstream can enhance athletic performance. Initially, blood doping was achieved by the transfusion of blood, either from other individuals (homologous transfusion) or withdrawn earlier from the athlete and stored until just before a competition (autologous transfusion). Both types of transfusion can be dangerous for a host of reasons, including improperly stored blood, risks of communication of infectious diseases, possibility of a transfusion reaction, and risk of infection. More recently, blood doping has been achieved by the use of the hormone erythropoietin (EPO). EPO is a naturally occurring growth factor that stimulates the formation of new RBCs. Injections of EPO can increase RBC counts for more than six months, and its use has become widespread in endurance sports. Again, it is not without risk since if the RBC count becomes too high, the blood becomes more viscous (thicker and heavier), increasing the work that must be done by the heart.

Becky Scott, now a retired Canadian cross-country skier, was originally awarded a bronze medal at the 2002 Salt Lake City Olympic Games for the 5 km pursuit. Scott's performance was all the more impressive given the nature of her competition. Her medal was later upgraded to a silver, then gold after both of the other medallists tested positive for darbepoetin, a pharmacological version of EPO that stimulates erythropoiesis.

---

steep part of the curve between 0 and 60 mm Hg (the range found in the capillaries throughout the bodies of most vertebrates), small changes in PO_2 result in large changes in the amount of O_2 bound to hemoglobin.

Because the P_{O_2} in alveolar air is about 100 mm Hg, most of the hemoglobin molecules in the blood leaving the alveolar networks are fully saturated, meaning that most of the hemoglobin molecules have bound four O_2 molecules (see Figure 49.13a). The P_{O_2} of the O_2 in solution in the blood plasma has risen to approximately the same level as in the alveolar air, about 100 mm Hg. This blood will also change colour, reflecting the bright red colour of oxygenated hemoglobin compared with the darker red colour of deoxygenated hemoglobin.

The oxygenated blood exiting from the alveoli collects in venules, which merge to form the pulmonary veins leaving the lungs. These veins carry the blood to the heart, which pumps the blood through the systemic circulation to all parts of the body.

As the oxygenated blood enters the capillary networks of body tissues, it encounters regions in which the P_{O_2} in the interstitial fluid and body cells is lower than that in the blood, ranging from about 40 mm Hg downward to 20 mm Hg or less (see Figure 49.13b). As a result, O_2 diffuses from the blood plasma into the interstitial fluid and from the fluid into body cells. As O_2 diffuses from the blood plasma into body tissues, it is replaced by O_2 released from hemoglobin.

Several factors contribute to the release of O_2 from hemoglobin, including increased acidity (lower pH) in active tissues. The acidity increases because oxidative reactions release CO_2, which combines with water to form carbonic acid (H_2CO_3). The lowered pH reduces the affinity of hemoglobin for O_2, which is released and used in cellular respiration.

The net diffusion of O_2 from blood to body cells continues until the blood leaves the capillary networks in the body tissues, at which point much of the O_2 has been removed from hemoglobin. The blood, now with a P_{O_2} of 40 mm Hg or less, returns in veins to the heart, which pumps it through the pulmonary arteries to the lungs for oxygenation.

49.4b Carbon Dioxide Diffuses down Partial Pressure Gradients from Body Tissues and into the Blood and Alveolar Air

The CO_2 produced by cellular oxidation diffuses from active cells into the interstitial fluid, where it reaches a partial pressure of about 46 mm Hg. Because this P_{CO_2} is higher than the 40 mm Hg P_{CO_2} in the blood entering the capillary networks of body tissues, CO_2 diffuses from the interstitial fluid into the blood plasma **(Figure 49.14a)**.

Some of the CO_2 remains in solution as a gas in the plasma. In many organisms, however, significant amounts combine with water to produce carbonic acid (H_2CO_3), which dissociates into bicarbonate (HCO_3^-) and H^+ ions. In the erythrocyte, the enzyme carbonic anhydrase accelerates the reaction. This reaction maintains a maximal concentration gradient of CO_2 between the cells and the blood and is a means of temporarily storing the gas in a harmless form until it can be transported to the respiratory surface of the animal for release once more as a gas.

Most of the H^+ ions produced by the **dissociation** of carbonic acid combine with hemoglobin or with proteins in the blood. This combination, by removing excess H^+ from the blood solution, *buffers* the pH of the blood, helping to maintain it at the set point appropriate for the species, usually about 7.4.

a. Body tissues

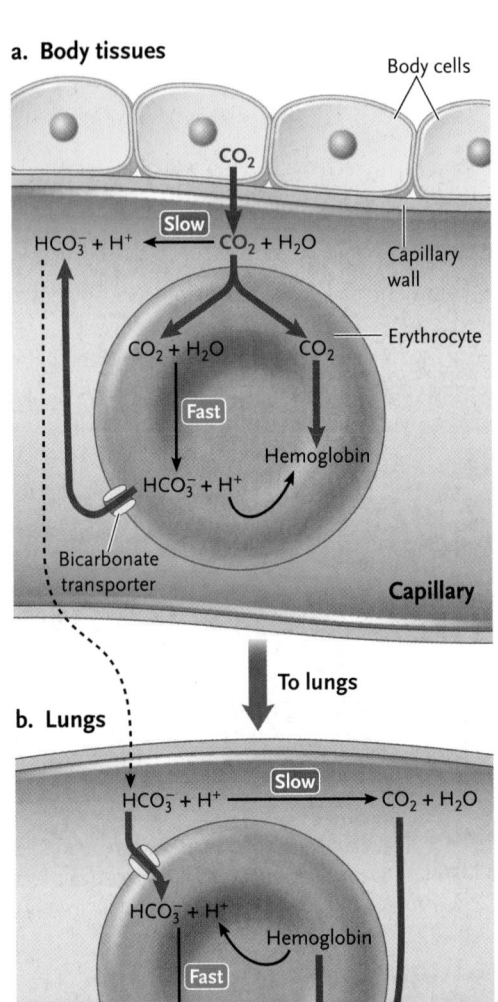

In body tissues, some of the CO_2 released into the blood combines with water in the blood plasma to form HCO_3^- and H^+. However, most of the CO_2 diffuses into erythrocytes, where some combines directly with hemoglobin and some combines with water to form HCO_3^- and H^+. The H^+ formed by this reaction combines with hemoglobin; the HCO_3^- is transported out of erythrocytes to add to the HCO_3^- in the blood plasma.

In the lungs, the reactions are reversed. Some of the HCO_3^- in the blood plasma combines with H^+ to form CO_2 and water. However, most of the HCO_3^- is transported into erythrocytes, where it combines with H^+ released from hemoglobin to form CO_2 and water. CO_2 is released from hemoglobin. The CO_2 diffuses from the erythrocytes and, with the CO_2 in the blood plasma, diffuses from the blood into the alveolar air.

Figure 49.14

The reactions occurring during the transfer of CO_2 from body tissues to alveolar air.

resulting in changes in breathing. The combination of solution in the plasma, conversion to bicarbonate, and combination with hemoglobin operate to maximize the ΔP of the gaseous CO_2 so that the rate of diffusion from the interstitial fluid into the blood is optimal.

The blood leaving the capillary networks of body tissues is collected in venules and veins and returned to the heart, which pumps it through the pulmonary arteries into the lungs. As the blood enters the capillary networks surrounding the alveoli, the entire process of CO_2 uptake is reversed (Figure 49.14b). The P_{CO_2} in the blood, now about 46 mm Hg, is higher than the P_{CO_2} in the alveolar air, about 40 mm Hg (shown in Figure 49.12, p. 1229). As a result, CO_2 diffuses from the blood and into the air. The diminishing CO_2 concentrations in the plasma, along with the lower pH encountered in the lungs, promote the release of CO_2 from hemoglobin. As CO_2 diffuses away, bicarbonate ions in the blood combine with H^+ ions, forming carbonic acid molecules that break down into water and additional CO_2. This CO_2 adds to the quantities diffusing from the blood into the alveolar air. By the time the blood leaves the capillary networks in the lungs, its P_{CO_2} has been reduced to the same level as that of the alveolar air, about 40 mm Hg.

CONCEPT FIX It is a common misconception that in mammals (including humans), exhaled air is depleted of oxygen. In actual fact, for most humans, exhaled air still contains 16% O_2 (inspired air contains 21%). For many animals under resting conditions, a very significant reserve of oxygen remains in the lung at the end of a breath. The exhaled air also contains 4% CO_2, however, and for air-breathing terrestrial vertebrates, it is the need to eliminate CO_2 that is the primary drive to breathe.

Most of the H^+ ions produced by the dissociation of carbonic acid combine with hemoglobin or with proteins in the plasma, so that the pH is maintained. Note, however, that if CO_2 levels are high, pH will fall,

STUDY BREAK

1. Why is hemoglobin so important for O_2 transport in the blood?
2. Why is the pH of the blood different in the tissues and in the lungs? What is the effect of the change in pH between these two sites for O_2 binding to hemoglobin?

PEOPLE BEHIND BIOLOGY 49.2

Peter Hochachka, *University of British Columbia*

Dr. Peter Hochachka (1937–2002), of the University of British Columbia, spent his career exploring the various biochemical mechanisms that allow animals to exploit extreme environments. Among his interests were the metabolic characteristics of animals in environments low in O_2. He brought Sherpas from the Himalaya and Quechuas from the high Andes as volunteer research subjects to his lab and used positron emission tomography (PET) and magnetic resonance spectroscopy, two techniques that permit the noninvasive characterization of metabolic activity, as well as magnetic resonance imaging (MRI), to understand their metabolism. He and his collaborators showed that the brains of the volunteers living at high altitudes metabolized O_2 at lower rates. Their hearts relied more on glucose as a fuel, an arrangement that produces more work per O_2 molecule than the greater reliance on fatty acids

characteristic of the hearts of people living nearer sea level. Thus, although the increased O_2-binding capacity of hemoglobin is important in animals that live at high altitudes, it is only one of a suite of genetic adaptations to living at high altitudes.

The discovery by Hochachka and others that there is a strong genetic component in humans and other animals and that this genetic component is not limited to the hemoglobin gene raises interesting questions. Which genes govern the improved performance at high altitudes? Did these genetic adaptations permit the astonishing feat of Rheinhold Messner, who, together with Peter Habeler in 1978, was the first human to climb to the pinnacle of Mount Everest without the use of supplementary O_2? Several others have matched this achievement, and, indeed, Messner repeated his feat alone in 1980. Messner was born and

raised in the mountainous Tyrol region of northern Italy. Although this in no way diminishes this extraordinary accomplishment, he may have had some genetic predisposition that permitted his survival, however agonizing, at an altitude where the atmospheric pressure is reduced from 760 mm Hg at sea level to 250 mm Hg and the P_{O_2} is only 53 mm Hg **(Figure 1)**.

David Evison/Shutterstock.com

Figure 1
Mount Everest. The barometric pressure at the summit is less than 250 mm Hg and the P_{O_2} is roughly 53 mm Hg, one-third of that found at sea level.

LIFE ON THE EDGE 49.3

Prospering in Thin Air

Experimental Research Box

With increasing altitude, atmospheric pressure decreases, and with it, the P_{O_2} also decreases. At an elevation of 5000 m, the atmospheric pressure is about half that at sea level, and the P_{O_2} is thus 380×0.21 or 80 mm Hg, about half that at sea level. This reduces the ΔP between the alveolar air and the blood, and, in turn, the supply of O_2 to the tissues is reduced. Humans who normally live at or near sea level and move to higher elevations above about 2500 m experience fatigue, dizziness, and nausea until their systems produce additional erythrocytes, a physiological response to the stress of reduced O_2.

However, some animals live at high altitudes, such as the llama (*Lama glama*) from the Andes at about 4000 m, or the bar-headed goose (*Anser indicus*), which migrates over the Himalayan mountains, sometimes at elevations in excess of 7000 m. One of the factors that permits these animals to exploit what is a marginal environment for other animals is a genetic difference in the hemoglobin molecule that produces a higher affinity for O_2. The hemoglobin in these animals shifts the hemoglobin–oxygen dissociation curve in Figure 49.13, p. 1229, to the left so that the hemoglobin is closer to saturation at lower P_{O_2} levels.

Hemoglobin is particularly polymorphic: the gene has a number of alleles, and the alleles present in the animals that can live at very high altitudes produce the appropriate forms of hemoglobin. This is well illustrated by the deer mouse, *Peromyscus maniculatus*, which occupies an extreme range of altitudes from below sea level in Death Valley to above 4300 m in the Sierra Nevada mountains. The populations of deer mice at higher altitudes have alleles of the hemoglobin genes with higher affinities for O_2 than the alleles of mice at low altitudes.

Review

 To access course materials such as Aplia and other companion resources, please visit www.NELSONbrain.com.

49.1 General Principles of Gas Exchange

- The percentage of a gas in a mixture of gases times the atmospheric pressure yields the partial pressure.
- Fick's equation describes the rate of diffusion across a biological membrane. The rate of diffusion is proportional to the product of the area over which diffusion occurs and the difference in partial pressures of the gas on either side of a membrane and is inversely proportional to the distance over which diffusion must occur.
- Gas exchange by simple diffusion is limited to small or flattened organisms.
- In larger animals, respiratory surfaces are increased, and the difference in partial pressures across a membrane is optimized by ventilation and perfusion.
- Water and air, as respiratory media, have different advantages and challenges. Water contains less oxygen, the rate of diffusion is greatly reduced, and its density requires greater energy for ventilation.

49.2 Adaptations for Gas Exchange

- Insects have a tracheal system that brings air directly to every cell.
- Gills are evaginations of the body surface. Water moves over gills by the beating of cilia or by muscular pumping.
- Water moves over the gills of sharks, bony fishes, and some arthropods, allowing countercurrent exchange to maximize the difference in partial pressure across the respiratory surface.
- Lungs are invaginated body surfaces with greatly expanded respiratory surfaces. They may be ventilated by positive pressure breathing, in which air is forced into the lungs, or by negative pressure breathing. Negative pressure breathing relies on muscles that alternately increase and decrease the pressure within the body cavity.

49.3 The Mammalian Respiratory System

- Air enters and leaves the lungs via the nostrils and mouth leading to the trachea, which branch into two bronchi. The bronchi branch many times into bronchioles leading to alveoli, which are surrounded by blood capillaries.
- Mammals rely on a negative pressure mechanism. The tidal volume is the volume of air moved in and out of the lungs during normal breathing. The vital capacity is the volume that can be moved in and out by breathing as deeply as possible. The residual volume is the volume remaining in the lungs after exhaling as much as possible.
- Breathing is controlled by centres in the brain reacting to sensors for CO_2 and O_2 in the carotid arteries. The concentration of CO_2 has the greatest influence.

49.4 Mechanisms of Gas Transport

- The site of gas exchange in mammals is at the alveolar surface. The P_{CO_2} in the alveolar air is greater than that in the blood, causing O_2 to diffuse into the blood and enter the erythrocytes, where it is bound by hemoglobin, thus maximizing the difference in partial pressure of that gas between the air and the blood.
- In the tissues, the reverse is true, and O_2 leaves the blood for the tissues.
- The P_{CO_2} is higher in the tissues than in the blood; CO_2 leaves the tissues and dissolves in the plasma and enters the erythrocytes, where most of it is converted into H^+ and HCO_3^- and released back into the plasma. This causes a slight drop in pH, promoting the release of O_2 from hemoglobin. The remaining CO_2 combines with hemoglobin. At the alveolar surface, the P_{CO_2} in the alveolar cells and the air in the lungs is lower than that in the blood, the HCO_3^- releases its CO_2, and CO_2 flows down the gradient.

Questions

Self-Test Questions

1. Which statement is correct concerning Fick's law of diffusion?
   a. The larger the partial pressure difference, the lower the rate of diffusion.
   b. The greater the surface area for exchange, the greater the rate of diffusion.
   c. The thinner the barrier, the higher the rate of diffusion.
   d. The lower the temperature, the higher the rate of diffusion.

2. Which statement is NOT true?
   a. The partial pressure of O_2 increases with altitude.
   b. Gases are in aqueous solution at respiratory surfaces.
   c. The solubility of O_2 decreases with an increase in temperature.
   d. Countercurrent flow optimizes the difference in partial pressures of a gas across a respiratory surface.
   e. The rate of diffusion of a gas increases with an increase in the difference of partial pressures on either side of a membrane.

3. Which of the following is a disadvantage associated with breathing water?
   a. low oxygen solubility
   b. dense respiratory medium
   c. must breathe a greater amount of the medium
   d. high oxygen content

4. Which of the following characterizes tracheal systems?
   a. positive pressure breathing
   b. closed tubes that circulate gases
   c. CO_2 sensors in the segmental ganglia
   d. the transport of respiratory gases directly to every cell
   e. uncontrolled diffusion of gases between the atmosphere and the tissues

5. Which of the following statements is correct?
   a. CO_2 receptors in the carotid body influence mammalian breathing.
   b. The concentration of O_2 in water rises with increasing temperature.
   c. Birds use a one-way flow of air through their lungs for gas exchange.
   d. O_2 receptors in the medulla have the greatest influence on mammalian breathing.
   e. An advantage of breathing in air (rather than water) is the reduced energy required to move the gases over the respiratory surface.

6. Which is true about countercurrent exchange?
   a. It is used by bony fish.
   b. It maximizes gas exchange by maintaining a diffusion gradient.
   c. It is described as the respiratory medium moving in the same direction as the blood flow.
   d. It is described as the respiratory medium moving in the opposite direction of the blood flow.

7. Suppose Canadian Olympic speed skating champion Christine Nesbitt is finishing her last lap. What occurs at this time?
   a. Her tidal volume is at vital capacity.
   b. Positive pressure brings air into her lungs.
   c. Her residual volume momentarily reaches zero.
   d. Her lungs undergo an elastic recoil when she inhales.
   e. Her diaphragm and rib muscles contract when she exhales.

8. The majority of CO_2 in the blood is which of the following?
   a. bound to hemoglobin
   b. transported as HCO_3^-
   c. dissolved as CO_2 in the plasma
   d. dissolved as CO_2 in the red blood cells

9. You and your pet dog live at sea level. The atmospheric Po_2 is 150 mm Hg. Your dog's arterial Po_2 is 100 mm Hg, and his tissue Po_2 is 10 mm Hg. What would you expect your dog to do?
   a. die
   b. accumulate CO_2
   c. have a serious but nonlethal O_2 deficit
   d. become dizzy from too much O_2
   e. function normally

10. What does the hemoglobin–O_2 dissociation curve do?
    a. shows a shift to the left when pH rises
    b. shows a lack of dependence on CO_2 levels
    c. reflects about 50% dissociation in the alveoli
    d. shows that hemoglobin holds less O_2 when the pH rises
    e. explains how hemoglobin can bind more O_2 in the lungs and release it at the tissues where the pH is lower

Questions for Discussion

1. The ability to live at high elevations appears to have genetic components beyond the properties of the blood in many animals. What experiments can you devise to explore this possibility? (Hint: Some animals have high- and low-elevation populations.)

2. Hospital patients frequently have a small ring on the end of a finger that shines a red light from the pad of tissue to a detector on the fingernail. What do you think this apparatus measures, and why is it important to measure it continuously? What factors could change the value of this measurement?

3. The control of the spiracular opening in insects is assumed to be important in avoiding water loss during gas exchange. Suggest an experiment to test this hypothesis. How does this mechanism differ from that controlling the opening of the stomata in plants?

4. You have two patients, one with chronic obstructive pulmonary disease and one with reduced blood flow to the lungs. Which one will benefit the most from application of supplemental oxygen, and why?

5. For an animal of your choice, use Fick's equation of diffusion to explain how it has become adapted for its lifestyle.

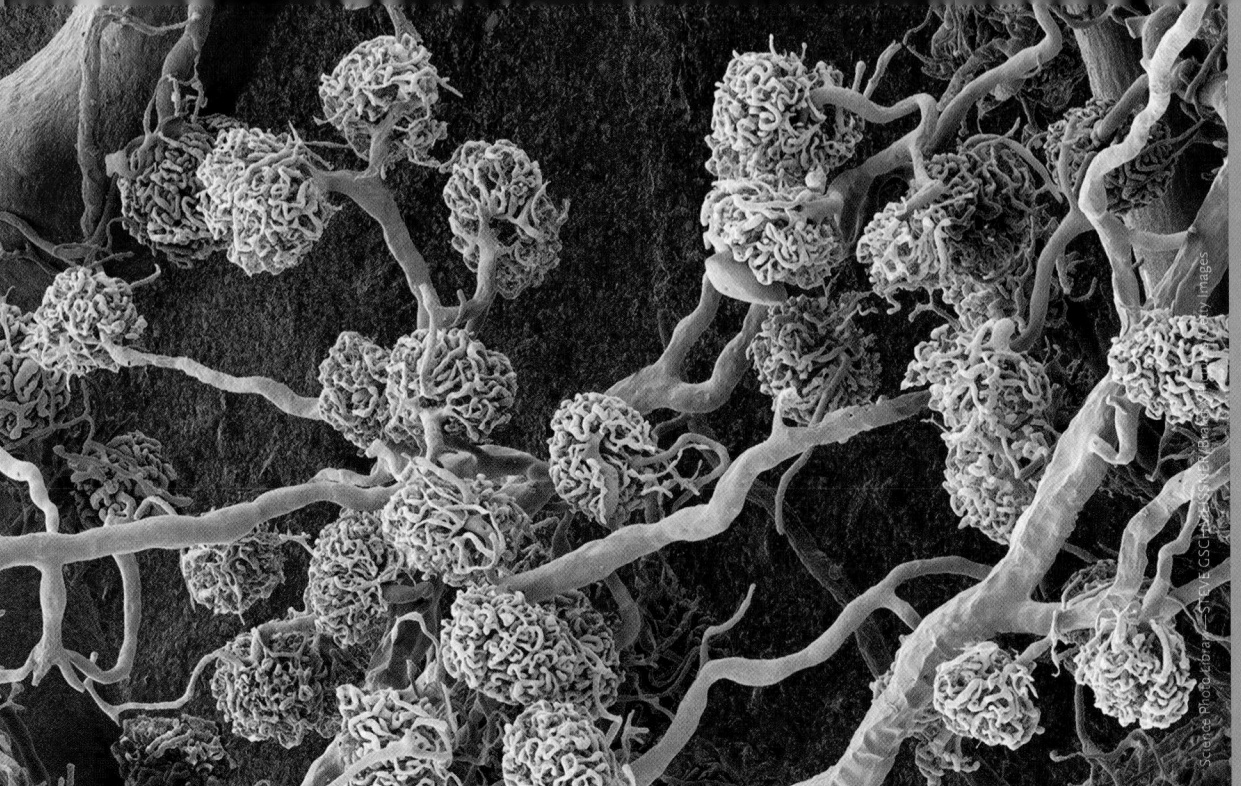

50

Nephrons in a human kidney (colourized scanning electron micrograph). Nephrons are the specialized tubules in kidneys that filter the blood to conserve nutrients and water, balance salts in the body, and concentrate wastes for excretion from the body.

Regulating the Internal Environment

WHY IT MATTERS

In the Miramichi River of New Brunswick, an Atlantic salmon (*Salmo salar*) has spent two or three years growing from an egg to a fish about 10 to 25 cm in length. In the spring, as day length increases and the water temperature begins to rise, it undergoes a number of physiological, morphological, and behavioural changes. It loses some of its mottled coloration, becoming silvery in appearance, and joins other similar fish in migrating downstream. It may pause at the estuary of the river for a day or two, but then it abandons its freshwater environment and enters the sea, where it will remain for two or more years, feeding and growing to maturity. Eventually, it will return to the Miramichi to spawn **(Figure 50.1, p. 1236).**

When the salmon leaves the fresh water of the river and enters the salt water of the North Atlantic Ocean, it moves from an environment in which the total concentration of solutes is about 0.1% to one with a concentration of about 3.5%. Its own body fluids contain about 1% solutes. Since water and many salts move freely across the gills of fish, which have a large surface area for gas exchange (see Chapter 49), the salmon moves, over the course of a few days, from a situation where it is constantly losing salts and taking up water from the environment to one in which the opposite occurs; it constantly takes up salts and loses water to its environment. Since all the cells of the body need to retain a relatively constant level of ions and water (neither shrinking nor swelling), dramatic changes must occur in the physiological processes that correct for these effects and maintain ion and water balance in body fluids. These physiological changes accompany its morphological changes and allow it to prosper in both environments.

All organisms may be subject to short-term fluctuations in their external environment that present challenges for them to maintain not only the integrity of their

Figure 50.1

Adult Atlantic salmon, ready to return to the river of their birth. The map and photo show the estuary of the Miramichi River in New Brunswick.

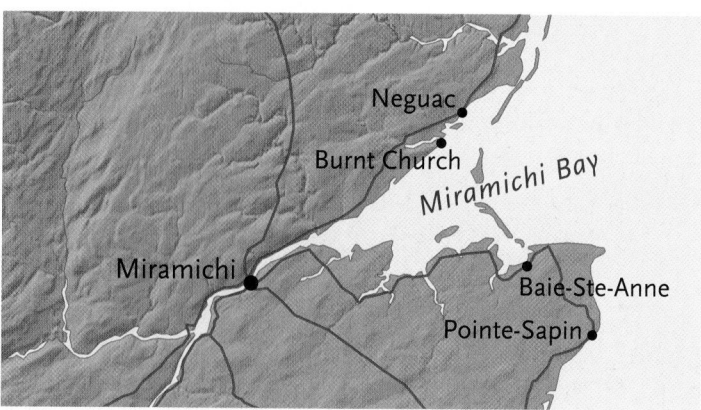

internal ions and fluids but also the functioning of all the systems that sustain life. Many are subject to seasonal changes that can also be challenging. The maintenance of a steady internal environment is called **homeostasis** (introduced in Chapter 39). In previous chapters, we have discussed why and how blood pressure (Chapter 40), blood glucose levels (Chapter 43), and blood oxygen levels (Chapter 49) are maintained relatively constant. In this chapter, we discuss the homeostatic control of ion, water, and temperature balance.

During evolution, a variety of physiological and behavioural mechanisms have appeared that have permitted organisms to exploit environments that may be highly variable. In the background are those mechanisms that accompanied the emergence from living in water to living on land. There are some important differences among marine, freshwater, and terrestrial environments. Organisms living in marine and terrestrial environments require mechanisms to obtain and conserve water; those in freshwater environments do not. On the other hand, obtaining salts and ions is relatively easy in marine environments but more difficult in freshwater and terrestrial ones. Finally, terrestrial environments are subject to much greater variation in temperature than either aquatic environment. The temperature of aquatic environments is seldom greater than about 25°C, and the lower limit is a little above −2°C (salt water freezes at about this temperature). By contrast, terrestrial organisms in the northern temperate zone of Canada, for example, encounter temperatures that may range between approximately 40 and −40°C. In this chapter, we explore the ways in which animals have adapted to overcome these challenges.

50.1 Introduction to Osmoregulation and Excretion

Living cells contain water, are surrounded by water, and constantly exchange water with their environment. The water of the external environment directly surrounds the cells of the simplest animals. For more complex animals, an aqueous extracellular fluid surrounds the cells and is separated from the external environment by a body covering. In animals with a circulatory system, the extracellular fluid includes both the blood and the interstitial fluid immediately surrounding the cells (see Chapter 39).

In this section, we review the mechanisms cells use to exchange water and solutes with the surrounding fluid through *osmosis*. We also look at how animals harness osmosis to regulate their internal *water balance*, the equilibrium between the inward and outward flow of water.

50.1a Osmosis Is Passive Diffusion

In osmosis (see Chapter 5), water molecules move across a selectively permeable membrane (one that lets water through but excludes most solutes) from a region of lesser solute concentration to a region of greater solute concentration. Selective permeability is a key factor in osmosis because it helps maintain differences in solute concentration on either side of biological membranes. Proteins are among the most important solutes in establishing the conditions producing osmosis. The passive transport of water occurs constantly in living cells (it will accompany any net ion transport), developing forces that lead to cell swelling or shrinking.

The total solute concentration of a solution may be measured in one of two ways. **Osmolarity** is measured in *osmoles*—the number of solute molecules and ions (in moles)—per litre of solution. **Osmolality** is measured in osmoles per kilogram of solution. Osmolality is used more frequently to measure the osmotic concentration of a solution, partly because the volume of water changes with temperature while the mass does not.

Because the total solute concentration in the body fluids of most animals is less than 1 osmole, osmolality is usually expressed in thousandths of an osmole, or *milliosmoles* (mOsm) per kilogram. As shown in **Figure 50.2,** the osmolality of body fluids in terrestrial vertebrates (including humans) is about 300 mOsm/kg. It is similar in freshwater organisms; in a goldfish, a freshwater teleost, it is about 290 mOsm/kg, while in freshwater invertebrates it is about 225 mOsm/kg. By contrast, sharks and many marine invertebrates, such as lobsters, have osmolalities close to that of sea water, about 1000 mOsm/kg, Interestingly, the osmolality in marine teleosts (bony fish such as the flounder) is only about 330 mOsm/kg. The relatively low osmotic concentration in marine teleosts reflects their evolutionary history. Early marine teleosts invaded fresh water and prospered there. During the extensive radiation of the group in fresh water, many have reinvaded the marine habitat.

A solution of higher osmolality on one side of a selectively permeable membrane is said to be *hyperosmotic* to a solution of lower osmolality on the other side, and a solution of lower osmolality is said to be *hypoosmotic* to a solution of higher osmolality. If the solutions on either side of a membrane have the same osmotic concentrations, they are *isoosmotic*. Water moves across the membrane between solutions that differ in osmolality, whereas when two solutions are isoosmotic, no *net* water movement occurs, although water exchanges from one side to the other. (Compare this to the discussion of hypotonic, hypertonic, and isotonic in Section 5.4d, Chapter 5.)

The principles that determine osmotic concentration also determine the freezing point, boiling point, and vapour pressure of a solution. Increasing solute concentration (increasing osmolality) reduces the freezing point, increases the boiling point, and lowers the vapour pressure of a solution.

CONCEPT FIX It is a misconception that once a cell is killed, diffusion and osmosis stop. Diffusion and osmosis are passive processes and will occur as long as the cell membrane is intact, providing a selectively permeable barrier. Since ion transporters are no longer working, the ability of the cell to correct for passive movement will be gone, and, of course, with time, the cell membrane will break down. ⬡

50.1b Animals Use Different Approaches to Regulate Osmosis

Because even small differences in osmotic concentration can cause cells to swell or shrink, animals must keep their cellular and extracellular fluids isoosmotic. In some animals, called **osmoconformers**, the osmotic concentrations of the cellular and extracellular solutions simply match that of the environment.

Many marine invertebrates are osmoconformers: when placed in dilute sea water, the osmotic concentration of their body fluids decreases, and their weight increases as a result of the osmotic influx of water. Other animals, called **osmoregulators,** use control mechanisms to keep the osmolality of cellular and extracellular fluids constant (i.e., homeostatically controlled) but at levels that may differ from the osmolality of the surroundings. Most freshwater and terrestrial invertebrates, and almost all vertebrates, are osmoregulators. It is important to recognize that the various solutes contributing to the osmotic concentration may be at different concentrations inside the cell, in the extracellular fluids, and in the environment.

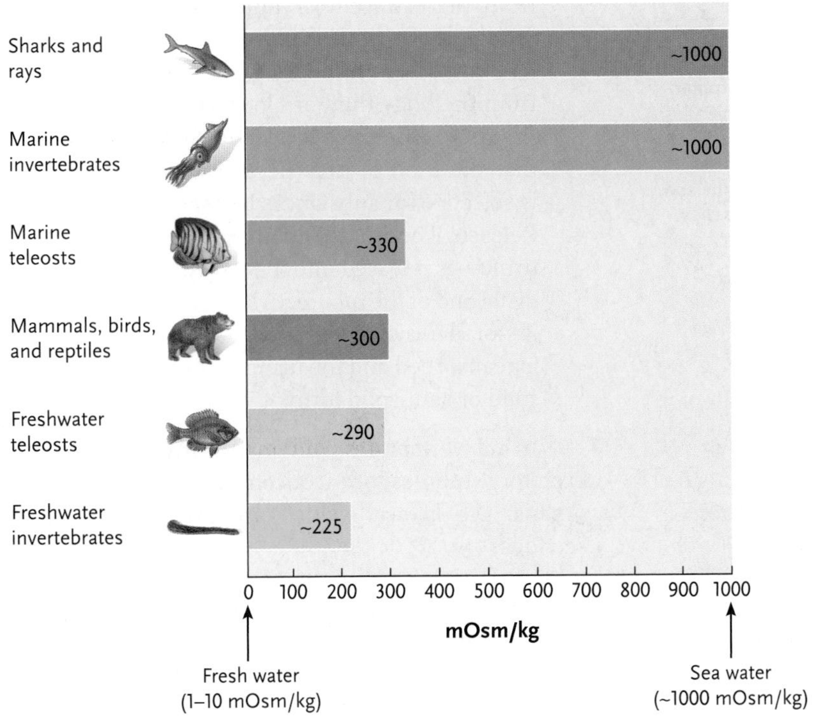

	mOsm/kg
Sharks and rays	~1000
Marine invertebrates	~1000
Marine teleosts	~330
Mammals, birds, and reptiles	~300
Freshwater teleosts	~290
Freshwater invertebrates	~225

Fresh water (1–10 mOsm/kg)

Sea water (~1000 mOsm/kg)

Figure 50.2
Osmolality of body fluids in some animal groups.

50.1c Excretion Is Closely Tied to Osmoregulation

Cells must control their ionic and pH balance as well as their osmotic concentration. This may require the removal of certain ions from cells and body fluids and their release into the environment. Animals excrete excess H^+ ions to keep the pH of body fluids near the neutral levels required by cells for survival. They also excrete toxic products of metabolism, such as nitrogenous (nitrogen-containing) compounds resulting from the breakdown of proteins and nucleic acids, and breakdown products of poisons and toxins. Excretion of ions and metabolic products is accompanied by water excretion because water serves as a solvent for those molecules. Animals that take in large amounts of water may also excrete water to maintain osmolality.

50.1d Microscopic Tubules Form the Basis of Excretion in Most Animals

Except in the simplest animals, minute tubular structures carry out osmoregulation and excretion (**Figure 50.3**). The tubules are immersed in body fluids at one end (called the *proximal end* of the tubules) and open directly or indirectly to the body exterior at the other end (called the *distal end* of the tubules). The tubules are formed from a **transport epithelium**—a layer of cells with specialized transport proteins in their plasma membranes. The transport proteins move specific molecules and ions into and out of the tubule by either active or passive transport, depending on the particular substance and its concentration gradient.

Typically, the tubules function in a four-step process:

1. **Filtration. Filtration** is the nonselective movement of water and a number of solutes—ions and small molecules, but not large molecules such as proteins—into the proximal end of the tubules through spaces between cells. In animals with an open circulatory system, the water and solutes come from body fluids, with movement into the tubules driven by the higher pressure of the body fluids compared with the fluid inside the tubule. In animals with a closed circulatory system, such as humans, the water and solutes come from the blood in capillaries that surround the tubules, with the movement into the tubules similarly driven by hydrostatic pressure. (Open and closed circulatory systems are described in Section 40.1, Chapter 40.)

2. **Reabsorption.** In reabsorption, some molecules (e.g., glucose and amino acids) and ions are transported by the transport epithelium back into the extracellular fluid and eventually into the blood (in animals with closed circulatory systems) as the filtered solution moves through the excretory tubule.

3. **Secretion.** Secretion is a selective process in which specific small molecules and ions are transported from the extracellular fluid and blood into the tubules. Secretion is the second and more important route for eliminating particular substances from the body fluid or blood, filtration being the first. The difference between the two processes is that filtration is nonselective whereas secretion is selective for substances transported.

4. **Release.** The fluid containing waste materials—urine—is released into the environment from the distal end of the tubule. In some animals, the fluid is stored in a bladder; in others, much of the water is reabsorbed and the urine is concentrated into a solid or semisolid form.

In all vertebrates and many invertebrates, the excretory tubules are concentrated in specialized organs, the *kidneys,* which are discussed in later sections.

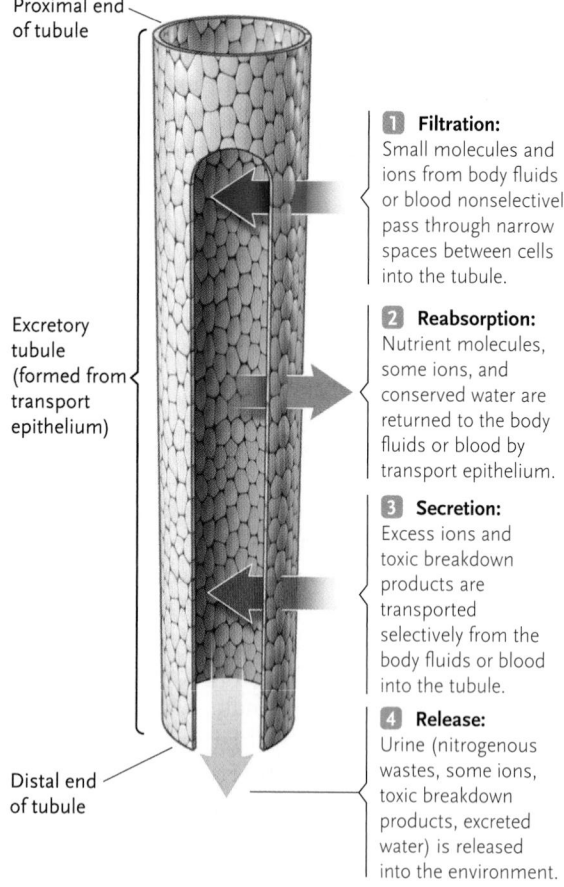

Proximal end of tubule

Excretory tubule (formed from transport epithelium)

1 Filtration: Small molecules and ions from body fluids or blood nonselectively pass through narrow spaces between cells into the tubule.

2 Reabsorption: Nutrient molecules, some ions, and conserved water are returned to the body fluids or blood by transport epithelium.

3 Secretion: Excess ions and toxic breakdown products are transported selectively from the body fluids or blood into the tubule.

4 Release: Urine (nitrogenous wastes, some ions, toxic breakdown products, excreted water) is released into the environment.

Distal end of tubule

Figure 50.3
Common structures and operations of the tubules carrying out osmoregulation and excretion in animals. The tubules are typically formed from a single layer of cells with transport functions.

50.1e Animals Excrete Nitrogen Compounds as Metabolic Wastes

The metabolism of ingested food is a source of both energy and molecules for the biosynthetic activities of an animal. Importantly, metabolism of ingested food produces *metabolic water,* which is used in chemical reactions as well as being involved in physiological processes such as the excretion of wastes.

The proteins, amino acids, and nucleic acids in food are continually broken down as part of digestion (see Chapter 48) and by the constant turnover and replacement of these molecules in body cells. The nitrogenous products of this breakdown are excreted by most animals as *ammonia, urea,* or *uric acid* or a combination of these substances **(Figure 50.4).** The particular molecule or combination of molecules depends on a balance among toxicity, water conservation, and energy requirements.

Ammonia. Ammonia (NH_3) results from the metabolism of amino acids and proteins and is highly toxic: it can be safely transported and excreted from the body only in dilute solutions. Those animals with a plentiful supply of water, such as aquatic or marine invertebrates, teleost fish, and larval amphibians, excrete ammonia as their primary nitrogenous waste. Other animals detoxify ammonia by converting it to urea or uric acid.

Urea. All mammals, most amphibians, some reptiles, some marine fishes, and some terrestrial invertebrates combine ammonia with HCO_3^- and convert the product in a series of steps to *urea,* a soluble substance that is less toxic than ammonia. Although producing urea requires more energy than forming ammonia, excreting urea instead of ammonia requires much less water.

Uric Acid. Water is conserved further in some animals, including many terrestrial invertebrates, reptiles, and birds, by the formation of uric acid instead of ammonia or urea. Uric acid is nontoxic, but its great advantage is its low solubility. During the concentration of the urine in the final stages of its formation, the uric acid precipitates as crystals that can be expelled with minimal water. (The white substance in bird droppings is uric acid.) The embryos of reptiles and birds, which develop within leathery or hard-shelled eggs that are impermeable to liquids, also conserve water by forming uric acid, which is stored as a waste product inside the shell. Similarly, the pupae of insects store uric acid in the rectum.

Many animals have the capacity to form all three products of nitrogen metabolism. Mammalian urine, for example, contains small amounts of uric acid, although urea predominates. Some tree frogs, such as *Phyllomedusa sauvagii,* have uric acid as their principal excretory product. This has enabled them to exploit the woodlands of South America, where the dry season is extremely arid. Conversely, the American cockroach, *Periplaneta americana,* an insect that normally lives in damp environments, uses ammonia as its primary excretory product and stores uric acid in special cells during periods when water is less available.

Sharks and rays maintain their osmotic concentration near that of their marine environment by retaining urea in their tissues and blood.

In the following sections, we look at the specifics of osmoregulation and excretion in different animal groups, beginning with the invertebrates.

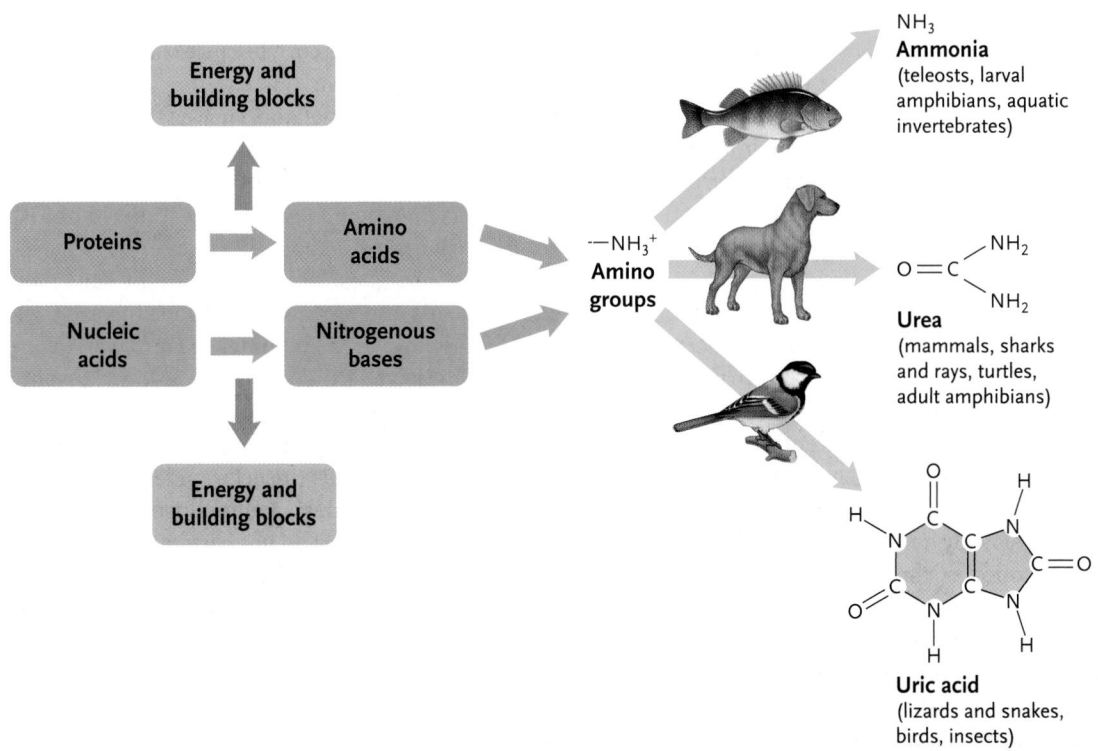

Figure 50.4
Nitrogenous wastes excreted by different animal groups. Although humans and other mammals primarily excrete urea, they also excrete small amounts of ammonia and uric acid.

50.2 Osmoregulation and Excretion in Invertebrates

Both osmoconformers and osmoregulators occur among the invertebrates, and most carry out excretion by specialized excretory structures.

50.2a Invertebrates Can Be Osmoconformers or Osmoregulators

Many marine invertebrates (sponges, cnidarians, some molluscs, and echinoderms) are osmoconformers. If placed in dilute solutions of sea water, they increase in weight because of the entry of water. They release nitrogenous wastes, usually in the form of ammonia, directly from body cells to the surrounding sea water. The cells of these animals do not normally swell or shrink because the osmotic concentrations of their intracellular and extracellular fluids and the surrounding sea water are the same, about 1000 mOsm/kg. Although they do not expend energy to maintain their osmolality, osmoconformers do expend energy to keep some ions, such as Na^+, at concentrations different from the concentration in sea water.

In general, the invertebrates that spend their entire lives in the open sea, where the environment is osmotically stable, have very little capacity for osmoregulation. Thus, many marine molluscs, such as squid and octopus, are osmoconformers, as are most marine arthropods, such as lobsters.

Other marine invertebrates are more diverse in their responses to variations in the osmotic concentration of the environment. Invertebrates living in the intertidal zone or at the mouths of tidal rivers experience regular changes in the osmotic concentration of their environment. Some marine annelids and arthropods that live in such environments are capable of short-term osmoregulation, slowing or delaying the changes in osmotic concentration of their body fluids that result from dilution of sea water by the outflow from rivers **(Figure 50.5)**. Some animals that live in the intertidal zone use behavioural responses, such as closing their shells (mussels and clams) or retreating into burrows (some annelids) to avoid desiccation when the tide is out.

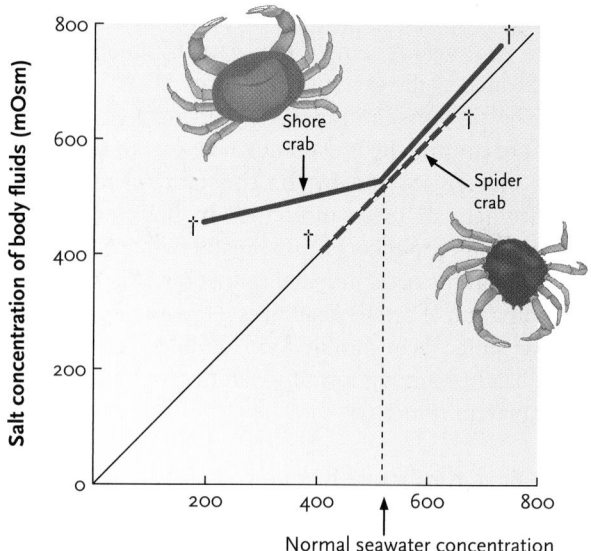

Figure 50.5

The concentration of the body fluids of two crabs when immersed in sea water of different concentrations. The spider crab lives in the sea, where it is not exposed to variation in the concentration of the water. The concentration of its fluids follows the 45° line that marks equivalency between the seawater concentration and the body fluid concentration. By contrast, the shore crab, which lives in the intertidal zone and in estuaries, can regulate its body fluids to some degree. The cross at the end of each line indicates the concentration at which each crab dies. Note that osmotic concentrations were not measured in this experiment.

By contrast, all freshwater and terrestrial invertebrates are osmoregulators. Those that live in freshwater environments are faced with a potential influx of water, diluting their body fluids. Although these invertebrates can live in more varied habitats than osmoconformers can, it comes at a cost, because osmoregulation is energetically expensive.

Freshwater osmoregulators such as flatworms and mussels are hyperosmotic relative to their environment, which causes water to move constantly from the surroundings into their bodies. This excess water must be excreted, at a considerable cost in energy, to maintain homeostasis. These animals obtain the salts they need from foods and by actively transporting salt ions from the water into their bodies (even fresh water contains some dissolved salts). This active ion transport occurs through the body surface or gills.

Terrestrial osmoregulators include annelids (earthworms), arthropods (insects, spiders and mites, millipedes, and centipedes), and molluscs (land snails and slugs). Although they do not have to excrete water entering by osmosis, they must constantly replace water lost from their bodies by evaporation and excretion. Most obtain water from their food, and some drink water. Like their freshwater relatives, these invertebrates must obtain salts from their surroundings, usually in their foods.

50.2b Specialized Excretory Tubules Participate in Osmoregulation

Most invertebrates (except marine osmoconformers) use specialized tubular structures to carry out excretion. These include *protonephridia* in flatworms and larval molluscs, *metanephridia* in annelids and most adult molluscs, and *Malpighian tubules* in insects and other arthropods. In protonephridia, the excretory tubules are open only at one end. Body fluids do not enter protonephridia directly. An *ultrafiltrate* enters the tubule through narrow extracellular spaces that permit only small molecules to enter and exclude larger molecules such as proteins. Metanephridia, by contrast, are open at both ends. They are characteristic of animals with coeloms, and the coelomic fluid is already an ultrafiltrate of the blood in the closed circulatory system.

Protonephridia. The flatworm *Dugesia* is an example of the simplest form of invertebrate excretory tubule, the **protonephridium** (*proto* = before; *nephros* = kidney). In *Dugesia,* two branching networks of protonephridia run the length of the body **(Figure 50.6)**. The cell at the blind end of each tubule has a bundle of cilia on its inner surface. The synchronous beating of the cilia resembles the flickering of a flame, so these cells are called *flame cells*. The cilia help draw a filtrate of body fluids through very small spaces between the cell membranes of the flame cell and those of the adjacent tubule cell and propel the filtrate along the tubule. As the fluids pass along the tubule, some molecules and ions are reabsorbed, whereas others are secreted into the tubules. The urine resulting from this filtration system is released through pores that connect the network of protonephridia to the body surface. The principal nitrogenous excretory product is ammonia. Although some ammonia passes out in the urine, most of it passes through the body wall.

Metanephridia. Animals with metanephridia have coelomic cavities (see Chapter 27) that are separate from the circulatory system. The fluid in the coelom is a filtrate of the hemolymph or blood. Coelomic fluid enters the proximal end of the excretory tubule, and ions and other solutes are reabsorbed or secreted as the fluid moves along the tubule. In annelids, the metanephridium **(Figure 50.7)** is a segmental structure. The proximal ends of a pair of metanephridia are located in each body segment, one on each side of the animal. A funnel-like opening surrounded by cilia admits coelomic fluid. Each tubule of the pair extends into the following segment, where it bends and folds into a convoluted arrangement surrounded by a network of blood vessels. Reabsorption and secretion of specific molecules and ions take place in the convoluted section. Urine from the distal end of the tubule collects in a saclike storage organ, the *bladder,* from where it is released through a pore in the surface of the segment.

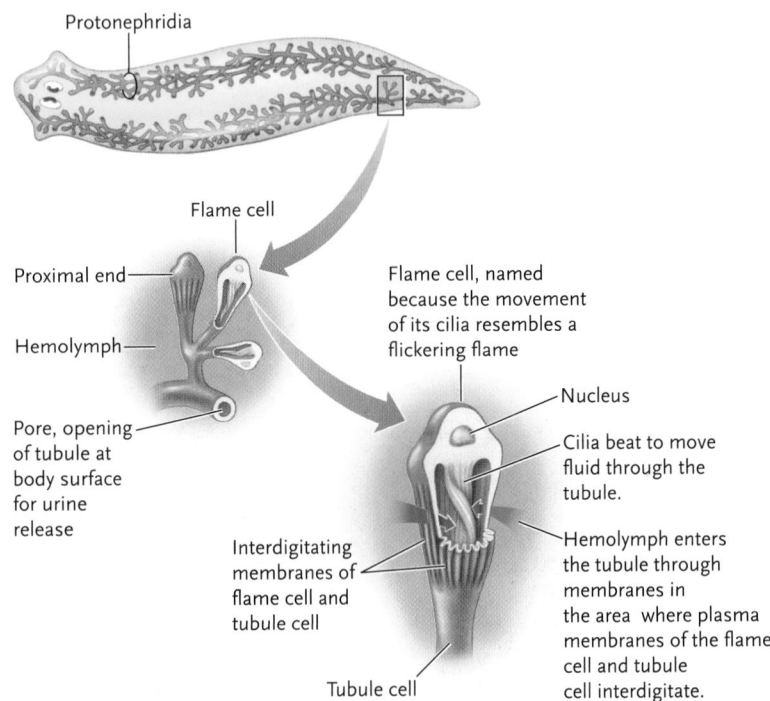

Figure 50.6
The protonephridia of the planarian *Dugesia,* showing the flame cells.

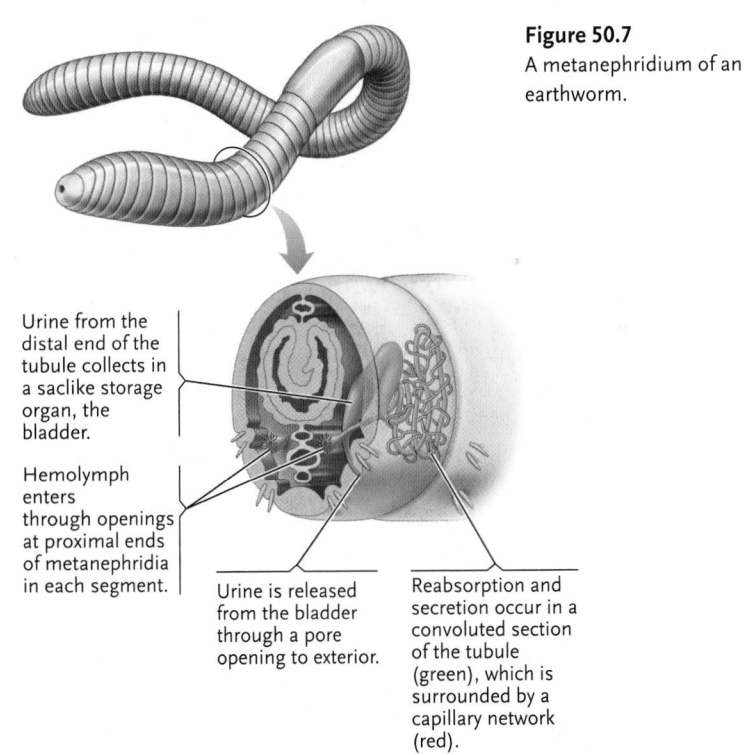

Figure 50.7
A metanephridium of an earthworm.

Malpighian Tubules. The excretory tubules of insects, the *Malpighian tubules,* have a closed proximal end that is immersed in the hemolymph **(Figure 50.8, p. 1242)**. The distal ends of the tubules empty into the gut. The fluid in the tubules results primarily from secretion, although in some insects, an ultrafiltrate of the hemolymph may enter the upper part of the tubule through

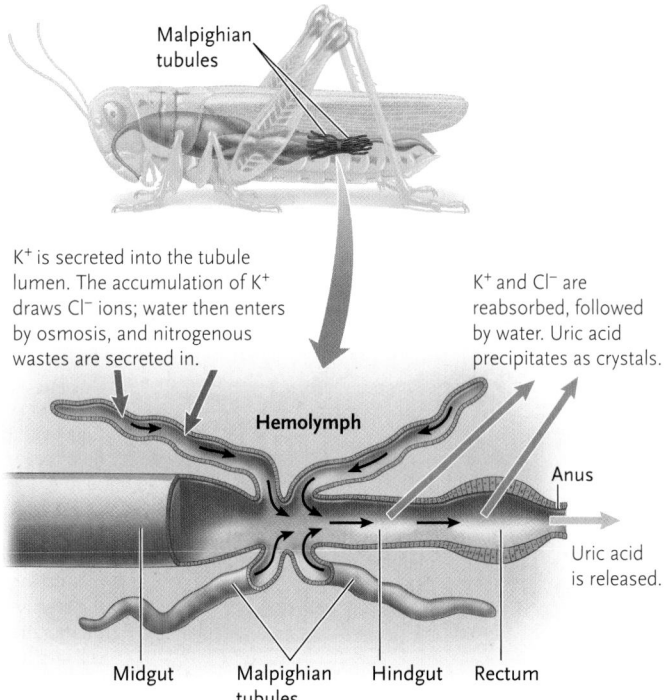

K+ is secreted into the tubule lumen. The accumulation of K+ draws Cl⁻ ions; water then enters by osmosis, and nitrogenous wastes are secreted in.

K+ and Cl⁻ are reabsorbed, followed by water. Uric acid precipitates as crystals.

Hemolymph

Anus

Uric acid is released.

Midgut

Malpighian tubules

Hindgut

Rectum

Figure 50.8
Excretion through Malpighian tubules in a grasshopper.

extracellular spaces. In particular, uric acid and several ions, including Na+ and K+, are actively secreted into the tubules. As the concentration of these substances rises, water moves osmotically from the hemolymph into the tubule. The fluid then passes into the hindgut (intestine and rectum) of the insect as dilute urine. Cells in the hindgut wall actively reabsorb most of the Na+ and K+ back into the hemolymph, and water follows by osmosis. The uric acid left in the gut precipitates into crystals, which mix with the undigested matter in the rectum and are released with the feces. This arrangement is important in conserving water.

Besides these specialized tubular structures, the body wall is also important in osmoregulation in many invertebrates. Researchers have cut parasitic cod worms and made sausagelike sacs by removing the intestine and closing the cut ends with ligatures. The sacs are capable of maintaining the internal osmotic concentration in environments of different osmotic concentrations.

STUDY BREAK

1. How does a protonephridium differ from a metanephridium? In what ways are they similar? In which animal groups are each of these found?
2. What is the excretory product of most insects? How does it get into the urine?
3. Some insects feed on plants, such as tobacco, that contain poisons. Can you devise an experiment to test whether such poisons are eliminated by Malpighian tubules?

LIFE ON THE EDGE 50.1

Life without Water

Some organisms live in temporary aquatic environments, consisting of ponds that dry up completely during periods of prolonged drought. One survival strategy involves complete desiccation of the animal, leading to anhydrobiosis (life without water). For example, the aquatic larvae of a midge (a small dipteran), *Polypedilum vanderplanki* (**Figure 1**), inhabit pools in Africa that can dry up completely for long periods. The larvae construct nests of mud, but within these nests, their water content is almost completely eliminated, and signs of life are absent. These desiccated larvae can withstand exposure to temperatures as low as −270°C and as high as +106°C. Immersed in water, the larvae recover within less than an hour, even after as long as 17 years of life without water. The precise mechanisms are not fully understood, but the animals accumulate high concentrations of the disaccharide trehalose as they enter the anhydrobiotic state. This sugar is thought to form a glasslike structure that prevents the formation of crystals that would damage the cells. Similar mechanisms are known in yeasts and other microorganisms.

Figure 1
The larva of Polypedilum vanderplanki *(a) in its active state and (b) during desiccation, showing the accumulation of trehalose. Optical and FTIR imaging data for a slowly dehydrated larva and a quickly dehydrated larva. Warm colours indicate higher intensity, i.e., larger amounts of trehalose. (Scale bar: 500 µm.)*

a.

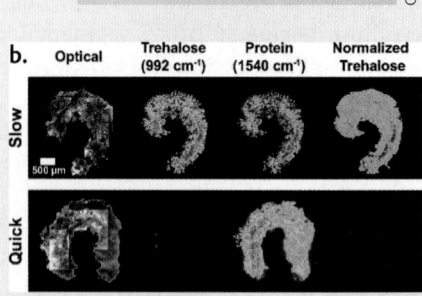

Courtesy of Takashi Okuda

"Vitrification is essential for anhydrobiosis in an African chironomid, *Polypedilum vanderplanki*." Minoru Sakurai et al. *PNAS* April 1, 2008 vol. 105 no. 13 5093–5098. Copyright 2008 National Academy of Sciences, U.S.A.

50.3 Osmoregulation and Excretion in Nonmammalian Vertebrates

In all vertebrates, specialized excretory tubules contribute to osmoregulation and excretion. The excretory tubules, called **nephrons,** are located in a specialized organ, the kidney. In all nonmammalian vertebrates, the kidneys produce a urine that is either hypoosmotic (dilute) or isoosmotic to body fluids, with the exception of some birds, which can produce urine that is weakly hyperosmotic. Mammals, on the other hand, can produce a very concentrated urine (up to 25 times the solute concentration of body fluids). This ability is unique to mammals and is discussed in the next section. The particular adaptations that maintain osmolality and water balance among the nonmammalian vertebrates vary depending on whether retention of water or of salts is the major issue.

50.3a Marine Fishes Conserve Water and Excrete Salts

Marine teleosts live in sea water, which is strongly hyperosmotic to their body fluids. As a result, they continually lose water to their environment by osmosis and must replace it by continuous drinking. The kidneys of marine teleosts play little role in regulating salt in their body fluids because they cannot produce hyperosmotic urine that would both remove salt and conserve water. Instead, excess Na^+, K^+, and Cl^- ions are eliminated from the body by specialized cells in the gills, called *chloride cells,* which actively transport Cl^- into the surrounding sea water; the Na^+ and K^+ ions are also actively transported to maintain electrical neutrality **(Figure 50.9a).** Divalent ions in the ingested sea water, such as Ca^{2+} and Mg^{2+}, are removed by the kidneys in an isoosmotic urine. On balance, a marine teleost is able to retain most of the water it drinks and eliminate most of the salt, allowing its tissue fluids to remain hypoosmotic to the surrounding water without producing hyperosmotic urine. The kidneys play little role in the removal of nitrogenous wastes; these are released from the gills, primarily as ammonia, by simple diffusion.

Sharks and rays have a different adaptation to sea water—the osmolality of their body fluids is maintained close to that of sea water by retaining high levels of urea in body fluids, along with another nitrogenous waste, *trimethylamine oxide.* Elasmobranchs (see Chapter 28) may have concentrations of urea as high as 1300 mg per 100 mL of blood. The match in osmolality keeps sharks and rays from losing water to the surrounding sea by osmosis, and they do not have to drink sea water continuous to maintain their water balance. Excess salts ingested with food are excreted in the kidney and by specialized secretory cells in a *rectal salt gland* located near the anal opening. The importance of urea as an osmolyte is illustrated by those species of

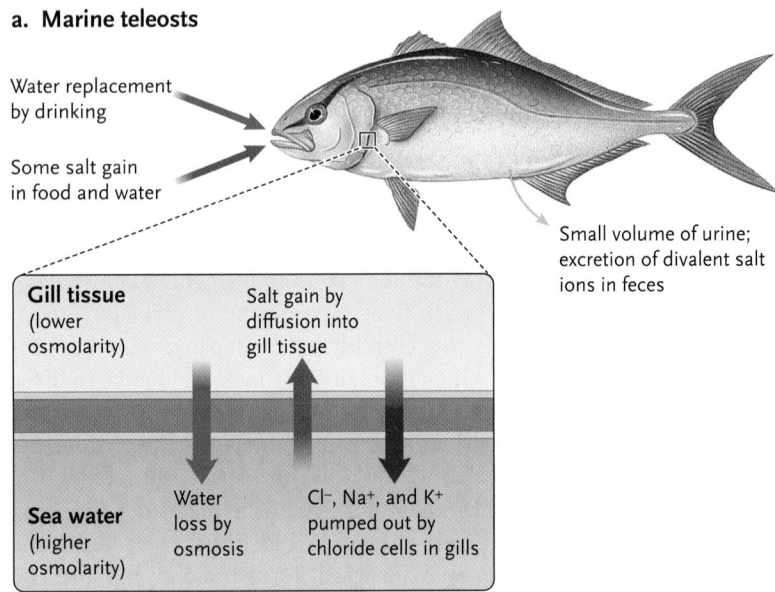

a. Marine teleosts

Water replacement by drinking

Some salt gain in food and water

Small volume of urine; excretion of divalent salt ions in feces

Gill tissue (lower osmolarity)

Salt gain by diffusion into gill tissue

Sea water (higher osmolarity)

Water loss by osmosis

Cl^-, Na^+, and K^+ pumped out by chloride cells in gills

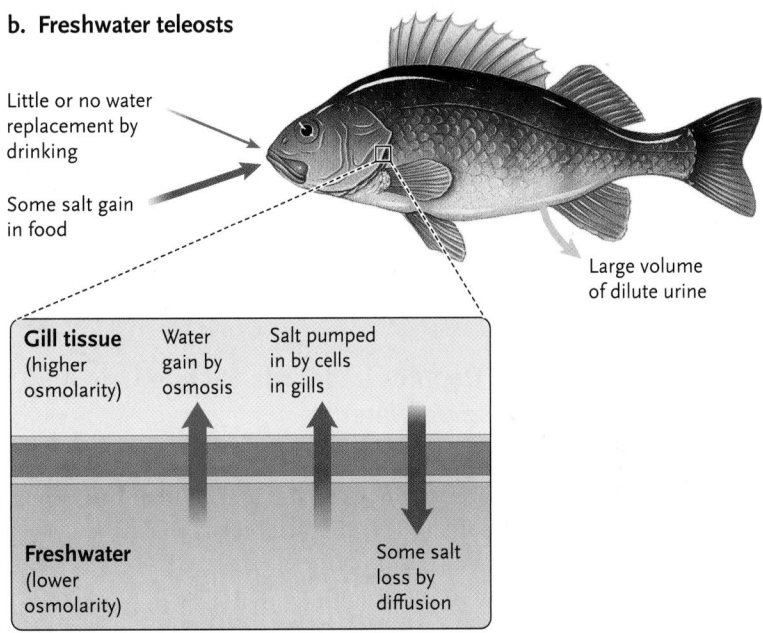

b. Freshwater teleosts

Little or no water replacement by drinking

Some salt gain in food

Large volume of dilute urine

Gill tissue (higher osmolarity)

Water gain by osmosis

Salt pumped in by cells in gills

Freshwater (lower osmolarity)

Some salt loss by diffusion

Figure 50.9
The mechanisms balancing the water and salt content of **(a)** marine teleosts and **(b)** freshwater teleosts.

stingrays that inhabit fresh water. In such species, the concentration of urea is reduced to about 2 to 3 mg per 100 mL of blood.

50.3b Freshwater Fishes and Amphibians Excrete Water and Conserve Salts

The body fluids of freshwater fishes and aquatic amphibians (no amphibians live in sea water, although the crab-eating frog (*Fejervarya cancrivora*) lives in mangrove swamps in Southeast Asia and can tolerate saltwater conditions) are hyperosmotic to the surrounding water, which usually ranges from about 1 to 10 mOsm/kg.

Water therefore moves osmotically into their tissues. Such animals rarely drink, and they excrete large volumes of dilute urine to get rid of excess water (Figure 50.9b). In freshwater fishes, salt ions lost with the urine are replaced by salt in foods and by active transport of Na^+, K^+, and Cl^- into the body by the gills. Aquatic amphibians obtain salt in the diet and by active transport across the skin from the surrounding water. Nitrogenous wastes are excreted from the gills as ammonia in both freshwater fishes and aquatic amphibians.

Terrestrial amphibians must conserve both water and salt, which is obtained primarily in foods. In these animals, the kidneys secrete salt into the urine, causing water to enter the urine by osmosis. In the bladder, the salt is reclaimed by active transport and returned to body fluids. The water remains in the bladder, making the urine very dilute; during times of drought, the water can be reabsorbed. Terrestrial amphibians also have behavioural adaptations that help minimize water loss, such as seeking shaded, moist environments and remaining inactive during the day. Larval amphibians, which are completely aquatic, excrete nitrogenous wastes from their gills as ammonia.

Most adult amphibians excrete nitrogenous wastes through their kidneys as urea. The leaf frog, *Phyllomedusa sauvagii* **(Figure 50.10),** however, produces uric acid as the principal nitrogenous waste. In addition, it secretes a waxy substance from glands in its skin and uses its legs to smear this over the entire surface, thereby minimizing water loss.

50.3c Reptiles and Birds Excrete Uric Acid to Conserve Water

Terrestrial reptiles and most birds conserve water by excreting nitrogenous wastes in the form of an almost water-free paste of uric acid crystals. Further water conservation occurs as the epithelial cells of the cloaca, the common exit for the digestive and excretory

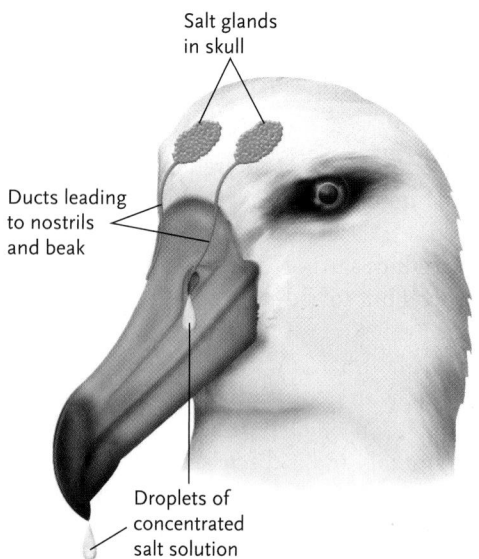

Figure 50.11
Salt glands in a bird living on a seacoast.

systems, absorb water from feces and urine before those wastes are eliminated. This arrangement is similar to the strategy used by insects, described earlier. In reptiles, the scales covering the skin allow almost no water to escape through the body surface.

Reptiles, such as crocodilians, sea snakes, and sea turtles, and birds, such as seagulls, penguins, and pelicans, that live in or around sea water take in large quantities of salt with their food and rarely or never drink fresh water. These animals typically excrete excess salt through specialized *salt glands* located in the head **(Figure 50.11)** that remove salts from the blood by active transport. The salts are secreted to the environment as a water solution in which salts are two to three times as concentrated as in body fluids. The secretion exits through the nostrils of birds and lizards, through the mouth of marine snakes, and as salty tears from the eye sockets of sea turtles and crocodilians.

STUDY BREAK

1. What are the osmoregulatory problems faced by marine and freshwater teleosts, and how are they solved?
2. What excretory strategy is used by birds and reptiles to conserve water?
3. How do marine birds and reptiles excrete excess salts?

Figure 50.10
Phyllomedusa sauvagii is a tree frog that prospers in the dry woodlands of South America. Among its many adaptations to a dry environment are the production of uric acid and the secretion from skin glands of a waterproofing waxy material.

Martin Fowler/Shutterstock.com

50.4 Osmoregulation and Excretion in Mammals

Since water moves by osmosis from areas of low solute concentration to areas of high solute concentration, it would seem impossible to produce a hyperosmotic urine, one in which water is reabsorbed into the body, leaving high concentrations of salt behind in the urine. We next describe of the structure and function of the mammalian kidney and the unique way in which "the impossible" has been achieved.

50.4a The Kidneys, Ureters, Bladder, and Urethra Constitute the Urinary System

Mammals have a pair of kidneys, located on each side of the vertebral column at the dorsal side of the abdominal cavity **(Figure 50.12)**. Internally, the mammalian kidney is divided into an outer **renal cortex** surrounding a central region, the **renal medulla.**

A **renal artery** carries blood to each kidney, where metabolic wastes and excess water and ions are excreted in the urine by the action of the nephrons. The blood is routed away from the kidney by the **renal vein**. The urine is produced in individual nephrons and is then processed further in **collecting ducts** that drain into a central cavity in the kidney called the **renal pelvis.**

From the renal pelvis, the urine flows through a tube called the **ureter** to the **urinary bladder,** a storage sac located outside the kidneys. Urine leaves the bladder through another tube, the **urethra,** which opens to the outside. Two sphincter muscles control the flow of urine from the bladder to the urethra. In human females, the opening of the urethra is just in

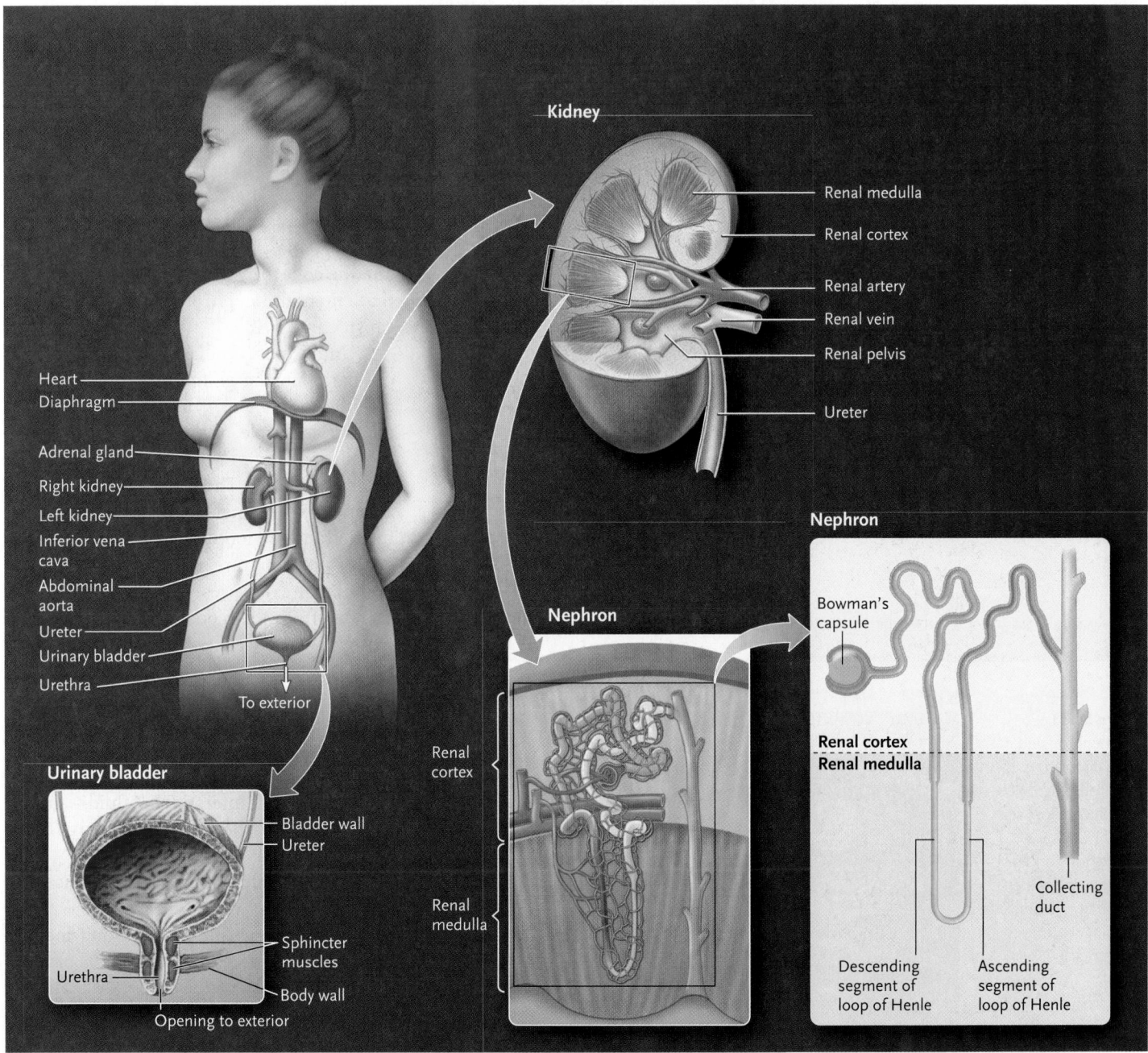

Figure 50.12
Human kidneys and urinary system in a female.

front of the vagina; in males, the urethra opens at the tip of the penis. The two kidneys and two ureters, the urinary bladder, and the urethra constitute the mammalian urinary system.

50.4b Regions of Nephrons Have Specialized Functions

Mammalian nephrons are differentiated into regions that perform successive steps in excretion. At its proximal end, a nephron forms **Bowman's capsule**, an infolded region that cups around a ball of blood capillaries called the **glomerulus (Figure 50.13)**. The capsule and glomerulus are located in the renal cortex. Filtration takes place as fluids are forced into Bowman's capsule from the capillaries of the glomerulus.

Following Bowman's capsule, the nephron forms a **proximal convoluted tubule** in the renal cortex, which descends into the medulla in a U-shaped bend called the **loop of Henle** and then ascends again to form a **distal convoluted tubule.** The distal tubule drains the urine into a branching system of collecting ducts that lead to the renal pelvis. As many as eight nephrons may drain into a single branch of a collecting duct. The combined activities of the proximal convoluted tubule, the loop of Henle, the distal convoluted tubule, and the collecting duct convert the filtrate that enters the nephron at the Bowman's capsule into urine.

Unlike most capillaries in the body, the capillaries in the glomerulus do not lead directly to venules. Instead, they form another arteriole that branches into a second capillary network called the **peritubular capillaries.** These capillaries thread around the proximal and distal convoluted tubules and the loop of Henle. Some molecules and ions are reabsorbed into the peritubular capillaries, whereas others are secreted from the blood into the nephron. However, because the capillaries and the tubules are not in physical contact due to the interstitial fluid between them, this transfer is not direct. Instead, the molecules or ions leave the tubule by passing through the one-cell-thick endothelial wall, diffuse through the interstitial fluid, and then pass into the capillary through its endothelial wall.

Each human kidney has more than a million nephrons. Of these, about 20% (the *juxtamedullary nephrons*) have long loops that descend deeply into the medulla of the kidney. The remaining 80% (the *cortical nephrons*) have shorter loops, most of which are located entirely in the cortex, and the remainder of which extend only partway into the medulla.

a.

Nephron

Renal cortex

Proximal convoluted tubule

Efferent arteriole

Afferent arteriole

Artery (branch of renal artery)

Glomerulus

Distal convoluted tubule

Bowman's capsule

Collecting duct

Vein (drains ultimately into renal vein)

Ascending segment of loop of Henle

Descending segment of loop of Henle

Peritubular capillaries

Renal medulla

To renal pelvis

b.

Figure 50.13

(a) A nephron and its blood circulation. For a detailed description of the blood flow, see text. **(b)** Glomeruli, ball-like tufts of capillaries in the nephrons of a kidney (colourized scanning electron micrograph).

Science Photo Library—STEVE GSCHMEISSNER/Brand X Pictures/ Getty Images

50.4c Nephrons and Other Kidney Structures Produce Hyperosmotic Urine

In mammals, urine is hyperosmotic to body fluids. Except for a few aquatic bird species, all other vertebrates produce urine that is hypoosmotic to body fluids or is at best isoosmotic. Production of hyperosmotic urine, a water-conserving adaptation, arises due to both structural (anatomical) and functional (physiological) features. Three features interact to conserve nutrients and water, balance salts, and concentrate wastes for excretion from the body:

- the structural arrangement of the loop of Henle, which descends into the medulla and returns to the cortex;
- differences in the permeability of successive regions of the nephron to water and ions, established by a specific group of membrane transport proteins in each region; and
- a gradient in the concentration of molecules and ions in the interstitial fluid of the kidney, established by these structural and functional processes, which increases gradually from the renal cortex to the deepest levels of the renal medulla.

Researchers determined the transport activities of specific regions of nephrons by dissecting them out of an animal and experimentally manipulating them *in vitro*. They placed segments in different buffered solutions and passed solutions containing various components of filtrates through the segment. By labelling specific molecules or ions radioactively, the scientists followed the movements of molecules in the solution surrounding the nephron segment or in the filtrate.

50.4d Filtration in Bowman's Capsule Begins the Process of Excretion

The mechanisms of excretion **(Figure 50.14,** and summarized in **Table 50.1, p. 1248)** begin in Bowman's capsule. The cells forming the walls of the capillaries of the glomerulus within Bowman's capsule, and the cells of the capsule itself, are loosely connected, leaving spaces just wide enough to allow water; ions; small nutrient molecules, such as glucose and amino acids; and nitrogenous waste molecules, primarily urea, to pass into the lumen of the capsule. The higher pressure of the blood drives fluid containing these molecules and ions from the capillaries of the glomerulus into the capsule. A thin net of connective tissue between the capillary and Bowman's capsule epithelia contributes to the filtering process. Blood cells and plasma proteins are too large to pass and are retained inside the capillaries. The fluid entering the capsule is an ultrafiltrate of the blood.

Note that this ultrafiltrate contains many molecules (nutrients and ions) that the body does not want to lose. The fundamental process that occurs in the

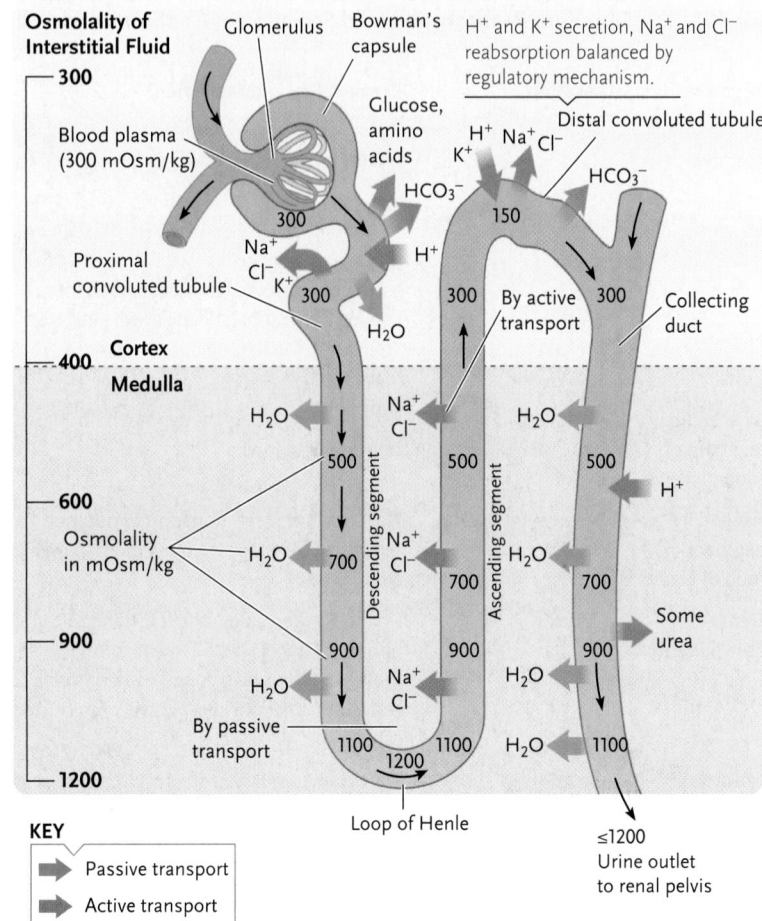

Figure 50.14

The movement of ions, water, and other molecules to and from nephrons and collecting tubules in the mammalian kidney. Nephrons in birds work in a similar fashion. The numbers are osmolality values in milliosmoles per kilogram.

nephron consists of eliminating all small molecules and then taking back what is important, rather than actively secreting waste molecules and toxic substances. The advantage of producing urine in this manner is that animals have not had to evolve transport proteins for foreign and unwanted substances. They pass directly into the filtrate and are not reabsorbed.

Two factors help maintain the pressure driving fluid into Bowman's capsule. First, the diameters of the arteriole delivering blood to the glomerulus (called the **afferent arteriole**) and of the capillaries of the glomerulus itself are larger than those of arterioles and capillaries elsewhere in the body. The larger diameters maintain blood pressure by presenting less resistance to blood flow. Second, the diameter of the arteriole that receives blood from the glomerulus (called the **efferent arteriole**) is smaller than the diameter of the afferent arteriole, producing a damming effect that backs up the blood in the glomerulus and helps keep the pressure high.

In humans, Bowman's capsules collectively filter about 180 L of fluid each day, from a daily total of 1400 L of blood that passes through the kidneys. The human

Table 50.1		Filtration, Reabsorption, and Secretion in Nephrons and Collecting Ducts			
Segment	Location	Permeability and Movement	Osmolality of Filtrate and Urine	Result of Passage
Bowman's capsule	Cortex	Water, ions, small nutrients, and nitrogenous wastes move through spaces between epithelia	300 mOsm/kg, same as surrounding interstitial fluid	Water and small substances, but not proteins, forced into nephron
Proximal convoluted tubule	Cortex	Na^+ and K^+ actively reabsorbed, Cl^- follows; water leaves through aquaporins; H^+ actively secreted; HCO_3^- actively reabsorbed into plasma of peritubular capillaries; glucose, amino acids, and other nutrients actively reabsorbed	300 mOsm/kg	67% of ions, 65% of water, 50% of urea, and all nutrients return to interstitial fluid; pH maintained
Descending segment of loop of Henle	Cortex into medulla	Water leaves through aquaporins; no movement of ions or urea	From 300 mOsm/kg at top to 1200 mOsm/kg at bottom of loop	Water drawn into interstitial fluid
Ascending segment of loop of Henle	Medulla into cortex	Na^+ and Cl^- actively transported out; no movement of water; no movement of urea	From 1200 mOsm/kg at bottom to 150 mOsm/kg at top of loop	Ions pumped into interstitial fluid, creating osmotic gradient
Distal convoluted tubule	Cortex	K^+ and Na^+ secreted via active transport into urine; Na^+ and Cl^- actively reabsorbed; water moves out of urine through aquaporins; HCO_3^- actively reabsorbed into plasma of peritubular capillaries	From 150 mOsm/kg at beginning to 300 mOsm/kg at junction with collecting duct	Ions balanced, pH balanced
Collecting ducts	Cortex through medulla, empties into renal pelvis	Water moves out via aquaporins; some active secretion of ions; some urea leaves at bottom of duct	From 300 to 1200 mOsm/kg at junction with renal pelvis	More water and some urea moves to interstitial fluid; some H^+ added to urine

MOLECULE BEHIND BIOLOGY 50.2

Aquaporins: Facilitating Osmotic Water Transport

Aquaporins are membrane proteins that form channels through which water can diffuse more rapidly than it would otherwise do. They are widely distributed, occurring in organisms from bacteria and yeast to mammals. In humans, at least 10 different aquaporins are known. The very narrow pores (Figure 1) permit water molecules to pass in single file, but because the molecules forming the channel are charged, other molecules of similar dimensions, such as H_3O^+, are excluded. The water moves in either direction in response to osmotic gradients: a single channel can permit as many as 3 billion molecules to cross the membrane per second.

Aquaporins are important in the functioning of mammalian kidneys. For example, one aquaporin,

aquaporin-2, resides on the membranes of vesicles within the cells of the collecting ducts. If the osmotic concentration of the body fluids increases, antidiuretic hormone from the pituitary gland causes the aquaporins from the vesicles to be inserted into the membrane of the cells of the collecting ducts. The presence of more aquaporins in the membrane greatly increases the rate of osmotic reabsorption of water. The urine becomes more concentrated, and the osmotic concentration of body fluids is reduced.

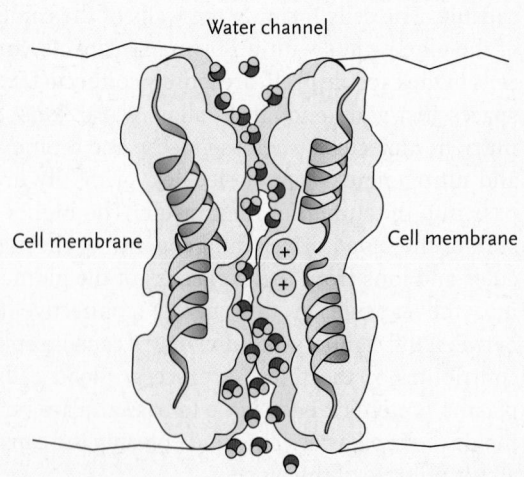

Figure 1
An aquaporin channel.

body contains only about 2.75 L of blood plasma, meaning that the kidneys filter a fluid volume equivalent to 65 times the volume of the blood plasma each day. On average, more than 99% of the filtrate, mostly water, is reabsorbed in the nephrons, leaving about 1.5 L to be excreted daily as urine.

50.4e Reabsorption and Secretion Occur in the Nephron

The fluid filtered into Bowman's capsule contains water, other small molecules, and ions at the same concentrations as the blood plasma. By the time the fluid reaches the distal end of the collecting duct, reabsorption out of and secretion into the tubules and collecting duct have markedly altered the concentrations of all components of the filtrate.

The Proximal Convoluted Tubule. This is the segment of the nephron where the valuable nutrients and ions that the body needs to retain are reabsorbed. Na^+/K^+ pumps in the epithelium of the proximal convoluted tubule move Na^+ and K^+ from the filtrate into the interstitial fluid surrounding the tubule (see Figure 50.14, p. 1247). The movement of positive charges sets up a voltage gradient that causes Cl^- ions to move out of the tubule with the positive ions. Specific active transport proteins move essentially all the glucose, amino acids, and other nutrient molecules out of the filtrate into the interstitial fluid, making the filtrate hypoosmotic to the interstitial fluid surrounding the tubule. As a result, water moves from the tubule into the interstitial fluid by osmosis. The osmotic movement is aided by *aquaporins,* proteins that form passages for water molecules in the transport epithelium of the tubule cells (see "Molecule behind Biology," Box 50.2). At this point, all of these substances that were filtered out of the blood into the filtrate in the tubule have been transported out of the filtrate into the interstitial fluid. The nutrients and water that entered the interstitial fluid then move into the capillaries of the peritubular network.

Some substances are secreted from the interstitial fluid into the tubule, primarily H^+ ions by active transport and the products of detoxified poisons by passive secretion (detoxification takes place in the liver). Small amounts of ammonia are also secreted into the tubule. The secretion of H^+ ions into the filtrate helps balance the acidity constantly generated in the body by metabolic reactions. H^+ secretion is coupled with HCO_3^- reabsorption from the filtrate in the tubule to the plasma in the peritubular capillaries.

In all, the proximal convoluted tubule reabsorbs about 67% of the Na^+, K^+, and Cl^- ions; 65% of the water; 50% of the urea; and essentially all the glucose, amino acids, and other nutrient molecules from the filtrate. The ions, nutrients, and water reabsorbed by the tubule are transported into the interstitial fluid and then into capillaries of the peritubular network.

Although 50% of the urea is reabsorbed, the constant flow of filtrate through the tubules and the excretion of the remaining urea in the urine keep the concentration of nitrogenous wastes low in body fluids.

The proximal convoluted tubule has structural specializations that fit its function. The epithelial cells that make up its walls are carpeted on their inner surface by a brush border of microvilli. These microvilli greatly increase the surface area available for reabsorption and secretion.

The Descending Segment of the Loop of Henle. The filtrate flows from the proximal convoluted tubule into the descending segment of the loop of Henle. As this tubule segment descends, it passes through regions of increasingly high solute concentrations in the interstitial fluid of the medulla (shown in Figure 50.14). (The generation of this concentration gradient is described later.) As a result, water moves out of the tubule by osmosis as the fluid travels through the descending segment.

The descending segment has aquaporins, which allow the rapid transport of water, but it has no other transport proteins. The outward movement of water concentrates the molecules and ions inside the tubule, gradually increasing the osmolality of the fluid to a peak of about 1200 mOsm/kg at the bottom of the loop. This is the same as the osmolality of the interstitial fluid at the bottom of the medulla.

The Ascending Segment of the Loop of Henle. The fluid then moves into the ascending segment of the loop of Henle, where Na^+ and Cl^- are actively transported into the interstitial fluid. The ascending segment has membrane proteins that transport salt ions, but it lacks aquaporins. Because water is trapped in the ascending segment, the osmolality of the urine is progressively reduced moving up this segment as salt ions, primarily Na^+ and Cl^-, are pumped out of the tubule.

The active transport of salt ions from the tubule into the interstitial fluid establishes the concentration gradient of the medulla: high near the renal pelvis and low near the renal cortex. The energy required to transport NaCl from the ascending segment makes the kidneys one of the major ATP-consuming organs of the body.

By the time the fluid reaches the cortex at the top of the ascending loop, its osmolality has dropped to about 150 mOsm/kg. During the travel of fluid around the entire loop of Henle, water, nutrients, and ions have been conserved and returned to the peritubular capillaries, and the total volume of the filtrate in the nephron has been greatly reduced. Urea and other nitrogenous wastes have been concentrated in the filtrate. Little secretion into the tubule occurs in either the descending or the ascending segment of the loop of Henle.

The Distal Convoluted Tubule. The transport epithelium of the distal convoluted tubule removes additional water from the filtrate in the tubule and works to balance the salt and bicarbonate concentrations of the filtrate against body fluids. In response to hormones triggered by changes in the body's salt concentrations, varying amounts of K^+ and H^+ ions are secreted into the filtrate, and varying amounts of Na^+ and Cl^- ions are reabsorbed. Bicarbonate ions are reabsorbed from the filtrate as in the proximal tubule.

In total, more ions move outward than inward in the distal tubule, and, as a consequence, water moves out of the tubule by osmosis through aquaporins. The amounts of urea and other nitrogenous wastes remain the same. By the time the filtrate, now urine, enters the collecting ducts at the end of the nephron, its osmolality is about 300 mOsm/kg.

The Collecting Ducts. The collecting ducts concentrate the urine. These ducts, which are permeable to water but not to salt ions, descend downward from the cortex through the medulla of the kidney. As the ducts descend, they travel through the gradient of increasing solute concentration in the medulla. This increase makes water move osmotically out of the ducts and greatly increases the concentration of the urine, which can become as high as 1200 mOsm/kg at the bottom of the medulla. Near the bottom of the medulla, the walls of the collecting ducts contain passive urea transporters that allow a portion of this nitrogenous waste to pass from the duct into the interstitial fluid. This urea adds significantly to the concentration gradient of solutes in the medulla.

In addition to these mechanisms, H^+ ions are actively secreted into the fluid by the same mechanism as in the proximal and distal convoluted tubules. The balance of the H^+ and bicarbonate ions established in the urine, interstitial fluid, and blood, achieved by secretion of H^+ into the urine by the nephrons and collecting ducts, is important for regulating the pH of blood and body fluids. The kidneys thus provide a safety valve if the acidity of body fluids rises beyond levels that can be controlled by the blood's buffer system (see Chapter 40).

At its maximum value of 1200 mOsm/kg, reached when water conservation is at its maximum, the urine reaching the bottom of the collecting ducts is about four times as concentrated as body fluids. But it can also be as low as 50 to 70 mOsm/kg when very dilute urine is produced in response to conditions such as excessive water intake.

The high osmolality of the interstitial fluid toward the bottom of the medulla would damage the medulla cells if they were not protected against osmotic water loss. The protection comes from high concentrations of otherwise inert organic molecules called *osmolytes* in these cells. The osmolytes, of which the most important is a sugar-alcohol called *sorbitol*, raise the osmolality of the cells to match that of the surrounding interstitial fluid. Urine flows from the end of the collecting ducts into the renal pelvis and then through the ureters into the urinary bladder, where it is stored. From the bladder, urine exits through the urethra to the outside.

50.4f Terrestrial Mammals Have Water-Conserving Adaptations

Terrestrial mammals have other adaptations that complement the water-conserving activities of the kidneys. One is the location of the lungs deep inside the body, which reduces water loss by evaporation during breathing (see Chapter 49). Another is a body covering of keratinized skin. Skin is so impermeable that it almost eliminates water loss by evaporation, except for the controlled loss through evaporation of sweat in mammals with sweat glands.

Among mammals, water-conserving adaptations reach their greatest efficiency in desert rodents such as the kangaroo rat (Figure 50.15). The proportion of nephrons with long loops extending deep into the kidney medulla of kangaroo rats is very high, allowing them to excrete urine that is 20 times as concentrated as body fluids. Further, most of the water in the feces

	Kangaroo Rat	Human
Water gain (millilitres)		
From ingesting food	6.0	850
From drinking liquids	0.0	1400
By metabolism	54.0	350
	60.0	2600
Water loss (millilitres)		
In urine	13.5	1500
In feces	2.6	200
By evaporation	43.9	900
	60.0	2600

Figure 50.15

A comparison of the sources of water for a human and a kangaroo rat (*Dipodomys species*). Water conservation in the kangaroo rat is so efficient that the animal never has to drink water.

is absorbed in the large intestine and rectum. Lacking sweat glands, they lose little water by evaporation from the body surface. Much of the moisture in their breath is condensed and recycled by specialized passages in the nasal cavities. They stay in burrows during the daytime and come out to feed only at night.

About 90% of the kangaroo rat's daily water supply is generated from oxidative reactions in its cells. (Humans, in contrast, can make up only about 12% of their daily water needs from this source.) The remaining 10% of the kangaroo rat's water comes from its food. These structural and behavioural adaptations are so effective that a kangaroo rat can survive in the desert without ever drinking water.

Marine mammals, including whales, seals, and manatees, eat foods that are high in salt content. They are able to survive the high salt intake because they produce urine that is more concentrated than sea water. As a result, they are easily able to excrete all the excess salt they ingest in their diet.

The adaptations described in this section allow animals to maintain the concentration of body fluids at levels that keep cells from swelling or shrinking and permit excretion of toxic wastes. An equally important challenge is maintaining an internal temperature that allows the organ systems to function with maximum efficiency. We look at these processes in the next section.

STUDY BREAK

1. What can mammalian kidneys do that the kidneys of other vertebrates cannot?
2. Where does active transport of ions occur in the nephron?
3. What is the major event in the ascending segment of the loop of Henle? Why is this important?
4. What is the major event in the collecting duct?
5. Compare how marine birds and marine mammals deal with the problem of excreting the excess salt they take in with their food.

50.5 Introduction to Thermoregulation

Environmental temperatures vary enormously across Earth's surface. Temperatures in deserts in Australia, Africa, and the United States may reach 50°C, whereas temperatures across the Canadian prairies and the north can fall to −50°C in winter, and some locations in Antarctica experience −80°C. There are also seasonal variations. A single location in the boreal forest of Canada might experience temperatures as low as −40°C in the winter and as high as 35°C in the summer. However, animal cells can function only within a temperature range from about 0 to 45°C. Not far below 0°C, the lipid bilayer of a biological membrane changes from a fluid to a frozen gel, which disrupts vital cell functions. Without protective measures, ice crystals will destroy the cell's organelles. At the other extreme, as temperatures approach 45°C, the kinetic motions of molecules become so great that most proteins and nucleic acids unfold from their functional form. Either condition leads quickly to cell death. Animals, therefore, usually maintain internal body temperatures somewhere within the 0 to 45°C limits, and most species can only operate over restricted portions of this range. As a consequence, most animals regulate their body temperatures to remain within their operable limits.

Temperature regulation (**thermoregulation**) is based on negative feedback pathways in which temperature receptors called *thermoreceptors* (see Chapter 45) monitor body temperature and integrate this information by comparing it to a temperature *set point* (see Section 39.4b, Chapter 39, for a description of the set point). Differences from the set point trigger physiological and behavioural responses that return the temperature to the set point. The responses triggered by negative feedback mechanisms (see Chapter 39) involve adjustments in the rate of heat-generating oxidative reactions within the body, coupled with adjustments in the rate of heat gain or loss at the body surface. The particular adaptations that accomplish these responses vary widely among species, however. And although body temperature is closely regulated around a set point in all endotherms, the set point itself may vary over the course of a day and between seasons (see Chapter 39).

In this section, we describe the structures, mechanisms, and behavioural adaptations that enable animals to regulate their temperature.

50.5a Thermoregulation Allows Animals to Reach Optimal Physiological Performance

Within the 0 to 45°C range of tolerable internal temperatures, an animal's *organismal performance* varies greatly. Organismal performance is a term that describes the rate and efficiency of an animal's biochemical, physiological, and whole-body processes. For instance, the speed at which a tropical fish can swim (one measure of organismal performance) is low when the animal's body temperature is cold, rises smoothly with body temperature until it levels to a fairly broad plateau, and then drops off dramatically with further increases in body temperature **(Figure 50.16a, p. 1252)**. The range of temperatures that provides optimal organismal performance varies from one species to another and may also vary within a species as a function of season (Figure 50.16b). Similar patterns of temperature dependence are observed for numerous other body functions.

a. Maximum running speed of a lizard at various body temperatures

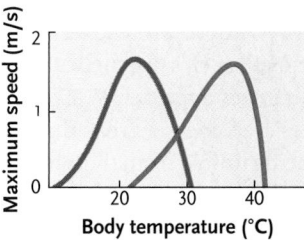

b. Maximum swimming speed of a fish at various body temperatures in winter and summer

c. Range of optimal physiological performance

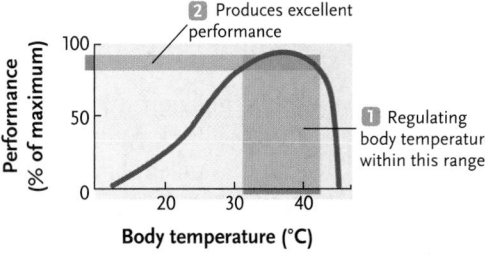

2 Produces excellent performance

1 Regulating body temperature within this range

Figure 50.16

Body temperature and organismal performance. **(a)** The maximum running speed of a lizard changes dramatically with body temperature. **(b)** This relationship varies relative to the environment, as shown here for a temperate species of fish between seasons (winter = blue; summer = red). **(c)** An animal's other behavioural and physiological processes respond to temperature changes in similar ways. The advantage of regulating body temperature within the range indicated by the bar on the horizontal axis is a high level of organismal performance, indicated by the bar on the vertical axis.

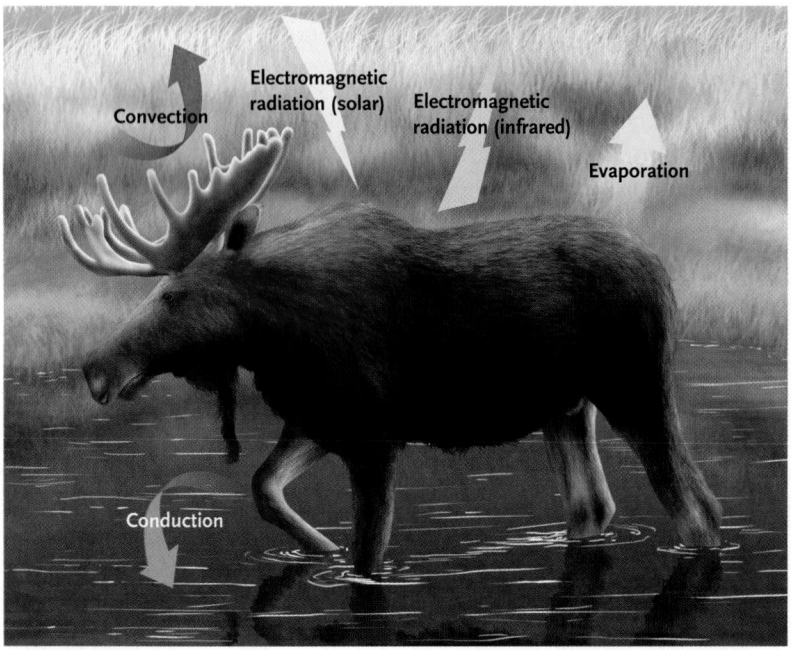

Electromagnetic radiation (solar)

Convection

Electromagnetic radiation (infrared)

Evaporation

Conduction

Figure 50.17

Heat flows into and out of a moose on a hot, sunny day. Unlike conduction, convection, and evaporation, which take place through the kinetic movement of molecules, electromagnetic radiation (infrared) is transmitted through space as waves of energy.

Animals that maintain their body temperature within the fairly narrow temperature range that corresponds to their optimal temperature can move quickly, digest food efficiently, and carry out necessary activities and processes rapidly and effectively (as shown in Figure 50.16c). Thus, in addition to keeping body temperatures within tolerable limits, thermoregulation allows animals to maintain an optimal level of organismal performance.

50.5b Animals Exchange Heat with Their Environment

As part of thermoregulation, animals exchange heat with their environment. Virtually all heat exchange occurs at surfaces where the body meets the external environment. As with all physical bodies, heat flows into animals if they are cooler than their surroundings and flows outward if they are warmer. This heat exchange occurs by four mechanisms: *conduction, convection, radiation,* and *evaporation* **(Figure 50.17)**.

Conduction is the flow of heat between atoms or molecules in direct contact. An animal loses heat by conduction when it contacts a cooler object and gains heat when it contacts an object that is warmer. **Convection** is the transfer of heat from a body to a fluid (air or water) that passes over its surface. The movement maximizes heat transfer by replacing fluid that has absorbed or released heat with fluid at the original temperature. **Radiation** is the transfer of heat energy as electromagnetic radiation. Any object warmer than absolute zero ($-273°C$) radiates heat; as the object's temperature rises, the amount of heat it loses as radiation increases as well. Animals also gain heat through radiation, particularly by absorbing radiation from the Sun. **Evaporation** is heat transfer through the energy required to change a liquid to a gas.

CONCEPT FIX Animals do not lose heat directly by sweating. Evaporation of water from a surface, however, is an efficient way to transfer heat; when the water in sweat evaporates from the body surface, the body cools down because heat is being transferred to the evaporated water in the surrounding air. In hot, humid regions where the air is saturated with water vapour, sweat does not evaporate and no heat can be lost in this manner. Thus, it is not the sweating that dissipates heat but the evaporation of the sweat. ⬡

All animals gain or lose heat by a combination of these four mechanisms. A moose struggling with the heat on a sunny summer day (Figure 50.17) loses heat by the evaporation of sweat from the skin and from the surface of the lungs, by convection as air flows over the skin, by conduction from the legs to the pond it is standing in, and by outward infrared radiation. It gains heat from internal biochemical reactions (especially oxidations), and by absorbing infrared and solar radiation. To maintain a constant body temperature, the heat gained and lost through these pathways must balance.

50.5c Animals Can Be Ectothermic or Endothermic

Different animals use one of two major strategies to balance heat gain and loss. Animals that obtain heat primarily from the external environment are known as **ectotherms** (*ecto* = outside); those obtaining most of their heat from internal physiological sources are called **endotherms** (*endo* = inside). All ectotherms generate at least some heat from internal reactions, however, and endotherms can obtain heat from the environment under some circumstances.

Most invertebrates, fishes, amphibians, and reptiles are ectotherms. Although these animals are popularly described as cold-blooded, the body temperature of some, such as an active lizard, may be as high as or higher than ours on a sunny day. Ectotherms regulate body temperature by controlling the rate of heat exchange with the environment. Through behavioural and physiological mechanisms, they adjust body temperature toward a level that allows optimal physiological performance. However, most ectotherms are unable to maintain optimal body temperature when the temperature of their surroundings departs too far from that optimum, particularly when environmental temperatures fall. As a result, the body temperatures of ectotherms fluctuate with environmental temperatures, and ectotherms are typically less active when it is cold.

The endotherms—birds and mammals—keep their bodies at an optimal temperature by regulating two processes: (1) the amount of heat generated by internal oxidative reactions and (2) the amount of heat exchanged with the environment. Because endotherms use internal heat sources to maintain body temperature at optimal levels, they can remain active over a broader range of environmental temperatures than ectotherms. However, endotherms require a nearly constant supply of energy to maintain their body temperatures. And because that energy is provided by food, endotherms must typically consume much more food than ectotherms of equivalent size. Some fishes, sea turtles, and invertebrates are also capable of generating significant amounts of internal heat.

The difference between ectotherms and endotherms is reflected in their metabolic responses to environmental temperature **(Figure 50.18)**. The metabolic rate of a resting mouse *increases* steadily as the environmental temperature falls from 25 to 10°C. This increase reflects the fact that to maintain a constant body temperature in a colder environment, endotherms must process progressively more food and generate more heat to compensate for their increased rate of heat loss.

By contrast, the metabolic rate of a resting lizard typically *decreases* steadily over the same temperature range. Because ectotherms do not maintain a constant body temperature, their biochemical and

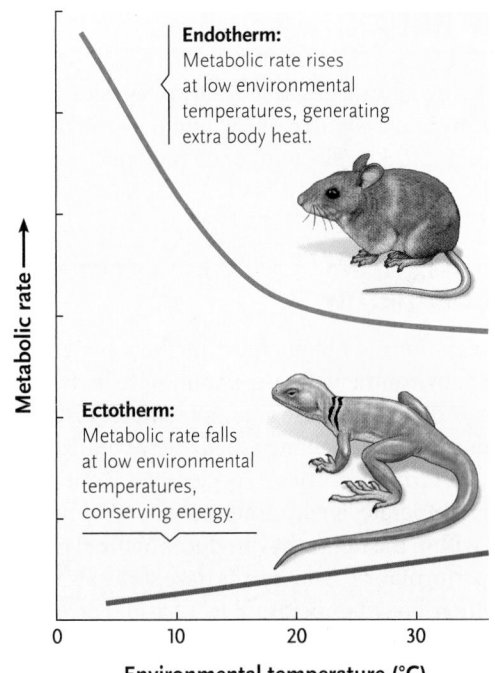

Endotherm:
Metabolic rate rises at low environmental temperatures, generating extra body heat.

Ectotherm:
Metabolic rate falls at low environmental temperatures, conserving energy.

Metabolic rate →

Environmental temperature (°C)

Figure 50.18
Metabolic responses of ectotherms and endotherms to cooling environmental temperatures. At any temperature, the metabolic rates of endotherms are always higher than those of ectotherms of comparable size.

physiological functions, including oxidative reactions, slow down as environmental and body temperatures decrease. Thus, an ectotherm consumes less food and requires less energy when it is cold than when it is warm.

Ectothermy and endothermy represent different strategies for coping with the variations in environmental temperature that all animals encounter; neither strategy is inherently superior to the other. Endotherms can remain fully active over a wide temperature range. Cold weather does not prevent them from foraging, mating, or escaping from predators, but it does increase their energy and food needs—and to satisfy their need for food, they may not have the option of staying curled up safely in a warm burrow. Ectotherms do not have the capacity to be active when environmental temperatures drop too low; they move sluggishly and are unable to capture food or escape from predators. However, because their metabolic rates are lower under such circumstances, so are their food needs, and they do not have to actively look for food and expose themselves to danger to the extent that endotherms do.

Having laid the ground rules of heat transfer and weighed the relative advantages and disadvantages of ectothermy and endothermy, we now begin a more detailed examination of how individual animals actually regulate their body temperatures within these overall strategies.

What are the advantages and disadvantages of ectothermy and endothermy? Would these be different in tropical versus temperate regions?

50.6 Ectothermy

Because ectotherms obtain most of their body heat from their environment, they generally have body temperatures very similar to the ambient temperature that surrounds them. To change body temperature to improve performance, these species live in or seek warm or temperate environments, where temperatures fall within the range that produces optimal physiological performance. Ectotherms have some ability to regulate their body temperature by physiological, but mainly behavioural, means, and those with a greater ability to thermoregulate generally occupy more varied habitats.

50.6a Ectotherms Are Found in All Invertebrate Groups

Aquatic invertebrates are limited thermoregulators. Their body temperature closely follows the temperature of their surroundings. Intertidal marine invertebrates, however, that are routinely exposed to air, use behavioural responses to regulate body temperature. For example, a South American intertidal mollusc, *Echinolittorina peruviana,* is longer than it is wide. Researchers in Chile have shown that this animal orients itself as a means of thermoregulation. On sunny summer days, it faces the Sun, offering a smaller surface area for the Sun's rays. On overcast summer days, or during the winter, it orients itself with its side, which has the larger surface area, toward the Sun's rays.

Invertebrates living in terrestrial habitats regulate their body temperatures more closely. Many also use behavioural responses, such as moving between shaded and sunny regions, to regulate body temperature. Some winged arthropods, including bees, moths, butterflies, and dragonflies, use a combination of behavioural and heat-generating physiological mechanisms for thermoregulation. In cool weather, these animals warm up before taking flight by rapidly vibrating the large flight muscles in the thorax, in a mechanism similar to shivering in humans. The tobacco hawkmoth (*Manduca sexta*) vibrates its flight muscles until its thoracic temperature reaches about 36°C before flying. During flight, metabolic heat generated by the flight muscles sustains the elevated thoracic temperature, so much so that a flying sphinx moth produces more heat per gram of body weight than many mammals. Honeybees (*Apis mellifera*) form masses in the hive in winter and use the heat generated by vibrating their flight muscles to maintain temperature inside the hive. Even in a Manitoba winter, with external temperatures below −20°C, the temperature in the mass of bees is normally about 30°C, and the bees may continue to raise offspring, using food stored in the hive.

50.6b Most Fishes, Amphibians, and Reptiles Are Ectotherms

Vertebrate ectotherms (most fishes, amphibians, and reptiles) also vary widely in their ability to thermoregulate. Most aquatic species have a more limited thermoregulatory capacity than that found among terrestrial species, particularly the reptiles. Some fishes, however, are highly capable thermoregulators.

Fishes. The body temperatures of most fishes remain within one or two degrees of their aquatic environment. However, many fishes use behavioural mechanisms to keep body temperatures at levels that allow good physiological performance. Many freshwater species perform better at lower temperatures and may use opportunities provided by the thermal stratification of lakes and ponds to sustain optimal performance in the summer. They remain in deep, cool water during hot summer days, moving to the shallows to feed only during early morning and late evening when air and water temperatures are lower.

Some cold-water marine teleosts (such as tunas and mackerels) and some sharks (such as the great white), on the other hand, use endothermy in their aerobic swimming muscles to maintain muscle temperature as much as 10 to 12°C warmer than their surroundings to improve performance. These animals have in common the fact that they move over long distances, swimming continuously. The action of the muscles generates heat that permits the muscles and other organs to operate more efficiently. Much of this heat generated in other fish is lost at the gill–water interface. However, a *countercurrent heat exchanger* system between the arteries and the veins in the swimming muscles of these specialized fishes minimizes this loss (see Chapter 49 for a description of countercurrent exchange). The anatomical details of the heat exchanger vary. In principle, however, the venules containing warm blood from the muscles form a network with arterioles containing cold blood coming from the gills. The heat from the venules is transferred to the arterioles and returned to the muscles. This transfer not only increases the temperature of the exercising muscle but reduces the temperature of the blood from the muscles returning to the heart on its way to the gills, and, thus, heat loss to the environment is minimized.

Amphibians and Reptiles. The body temperature of most amphibians also closely matches the environmental temperature. The tadpoles of foothill yellow-legged frogs (*Rana boylii*) regulate their body temperature to some degree by changing their location in ponds and lakes to take advantage of temperature differences between deep and shallow water or between sunny and shaded regions. Some terrestrial amphibians bask in the Sun to raise their body temperature and seek shade to lower body temperature. However, basking can be dangerous to amphibians because they lose water rapidly through their permeable skin. We have already noted that the leaf frog *Phyllomedusa sauvagii,* which often basks in sunlight, avoids this problem by coating itself with waterproofing lipids secreted by glands in its skin.

Thermoregulation is more pronounced among terrestrial reptiles. Some lizard species can maintain temperatures that are nearly as constant as those of endotherms **(Figure 50.19)**. For small lizards, the most common behavioural thermoregulatory mechanism is moving between sunny (warmer) and shady (cooler) regions. In the desert, lizards and other reptiles retreat into burrows during the hottest part of summer days. Some, such as the desert iguana (*Dipsosaurus dorsalis*), lose excess heat by *panting*—rapidly moving air in and out of the airways. The air movement increases heat loss by evaporation of water from the respiratory tract.

Lizards also frequently adjust their posture to foster heat exchange with the environment and control the angle of their body relative to the rays of the Sun. Horned lizards (genus *Phrynosoma*) often warm up by flattening themselves against warm, sunlit rocks to maximize their rate of heat gain by

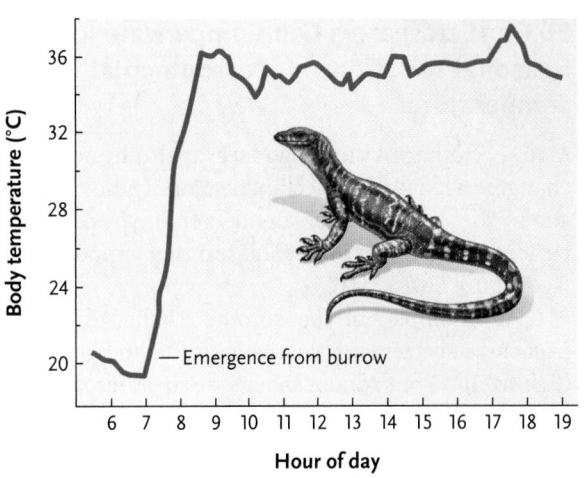

Figure 50.19
An example of excellent thermo-regulation in ecto-therms. The body temperature of the Australian lizard *Varanus varius* rises quickly after the animal emerges from its burrow and remains relatively stable throughout the day.

conduction from the rock and radiation from the Sun. Snakes and lizards can often be found on large rocks and on roads on chilly nights, taking advantage of the heat retained by the stone or concrete. *Agama savignyi,* a lizard that lives in the Negev Desert in Israel, cools off at midday by climbing into shady bushes, moving away from the hot sand, and catching a cooling breeze.

Researchers have demonstrated experimentally that several lizard species couple physiological responses to behavioural mechanisms of thermoregulation. When a Galápagos marine iguana (*Amblyrhynchus cristatus*) is exposed to heat from infrared radiation, blood flow increases in the heated regions of the skin. The blood absorbs heat rapidly and carries it to critical organs in the core of the body. Conversely, when an area of skin is experimentally cooled, blood flow to it is restricted, thereby preventing the loss of heat to the external environment.

PEOPLE BEHIND BIOLOGY 50.3

Ken Storey, *Carleton University*

Ken Storey, who leads a busy laboratory at Carleton University in Ottawa, explores the biochemical changes associated with hibernation and estivation in a wide range of animals, including both invertebrates and vertebrates. The wood frog, *Rana sylvatica*, is of particular interest because it spends the winter in a frozen state under the leaves on the forest floor. These frozen frogs have no heartbeat, breathing, or brain activity.

Storey's research has shown that the wood frog's tolerance of freezing involves the liberation of glucose from glycogen stores in the liver and its accumulation at extremely high concentrations within the cells, resulting in a slurry of small ice crystals and glucose. This suggests a suspension of the function of insulin, which normally controls glucose concentrations (see "Life on the Edge," Box 50.4, page 1261). Although

it may be obvious that the frogs freeze from the outside in and that heart and brain function will be the last to be suspended, it is also true that the frogs thaw from the inside out. The coordination of the events governing freezing and thawing involves a number of signal cascades, which Storey's lab has characterized. Storey is currently exploring gene expression during the process of freezing and thawing.

50.6c Ectotherms Can Compensate for Seasonal Variations in Environmental Temperature

Many ectotherms undergo seasonal physiological changes called **thermal acclimatization**. These changes allow the animals to attain good physiological performance at both winter and summer temperatures (see Figure 50.16, p. 1252).

For example, in the summer, bullhead catfish (*Ameiurus* species) can survive water temperatures as high as 36°C but cannot tolerate temperatures below 8°C. In the winter, however, the bullhead cannot survive water temperatures above 28°C but can tolerate temperatures near 0°C. Scientists have hypothesized that the production of different versions of some enzymes (perhaps encoded by different genes or produced as a result of alternative splicing) with optimal activity at cooler or warmer temperatures underlies such acclimatization.

Another acclimatizing change involves the phospholipids of biological membranes. Membrane phospholipids have higher proportions of double bonds in carp living in colder environments than in carp living in warmer environments. The higher proportion of double bonds makes it harder for the membrane to freeze. A higher proportion of cholesterol also protects membranes from freezing.

STUDY BREAK

1. Describe mechanisms an ectothermic animal can use to regulate its temperature.
2. What is thermal acclimatization? Why is it important?

50.7 Endothermy

Endotherms (mostly birds and mammals) have the most elaborate and extensive thermoregulatory adaptations of all animals. Set points (the core body temperatures that endotherms maintain homeostatically) vary with species and lie between about 39 and 42°C in birds and 32 and 39°C in mammals. We have already noted that the range of environmental temperatures that different organisms encounter is very great. A single species may encounter seasonal variations in environmental temperatures ranging over 70°C or more between winter and summer, yet their body temperatures do not vary.

We begin by describing the basic feedback mechanisms that maintain body temperature in this group.

50.7a The Hypothalamus Integrates Information from Thermoreceptors

Thermoreceptors are found in various locations in the bodies of endotherms, including the **integument** (skin), spinal cord, and hypothalamus. Two types of thermoreceptors occur in skin (see Chapters 44 and 45). In mammals, *warm receptors* send signals to the hypothalamus as the skin temperature rises above 30°C and reach maximum activity when the temperature rises above 40°C. Another type, the *cold receptor,* sends signals when skin temperature falls below about 35°C and reaches maximum activity at 25°C. By contrast, the highly sensitive thermoreceptors in the hypothalamus itself produce signals when the blood temperature shifts from the set point by as little as 0.01°C.

Signals from the thermoreceptors are integrated in the hypothalamus and other regions of the brain to bring about compensating physiological and behavioural responses (**Figure 50.20;** see also Figure 39.12, Chapter 39). The responses keep body temperature close to the set point, which varies normally in most mammals between 35 and 39°C for the head and trunk. The appendages may vary more widely in temperature. In very cold weather, for example, the legs, the feet, and especially the ears and nose are typically lower in temperature than the body core.

The hypothalamus was identified as a major thermoreceptor and response integrator in mammals by experiments on animals in which various regions of the brain were heated or cooled with a temperature probe. Within the brain, cooling and warming only the hypothalamus produced thermoregulatory responses such as shivering and panting. Later experiments revealed a similar response if regions of the spinal cord were cooled, indicating that thermoreceptors also occur in this location. The hypothalamus is also a major thermoreceptor and response integrator in fishes and reptiles. In birds, thermoreceptors in the spinal cord appear to be more significant in thermoregulation.

50.7b The Skin Controls Heat Transfer with the Environment

Besides its defensive role against infection, the skin of birds and mammals is an organ of heat transfer. It is a very large surface, in direct contact with the environment, across which heat can be transferred readily.

The outermost living tissue of human skin, the **epidermis**, consists of cells that divide and grow rapidly **(Figure 50.21),** becoming packed with fibres of a highly insoluble protein, *keratin*. When fully formed, the epidermal cells die and become compacted into a tough, impermeable layer.

Below the epidermis lies the **dermis**, a tissue layer packed with connective tissue fibres, such as collagen, that resist the compression, tearing, or puncture of the skin.

The dermis also contains thermoreceptors and a dense network of arterioles, capillaries, and venules. The arterioles delivering blood to the capillary networks of the skin constrict or dilate to control blood

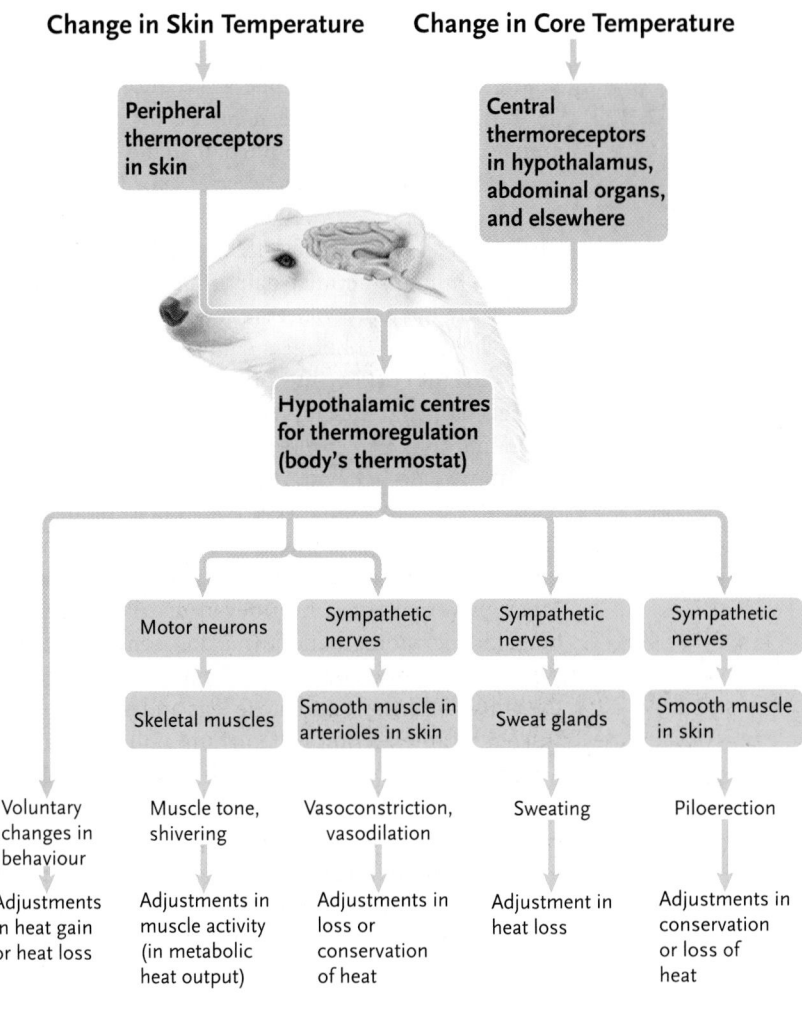

Change in Skin Temperature → Peripheral thermoreceptors in skin

Change in Core Temperature → Central thermoreceptors in hypothalamus, abdominal organs, and elsewhere

↓

Hypothalamic centres for thermoregulation (body's thermostat)

Motor neurons | Sympathetic nerves | Sympathetic nerves | Sympathetic nerves

Skeletal muscles | Smooth muscle in arterioles in skin | Sweat glands | Smooth muscle in skin

Voluntary changes in behaviour | Muscle tone, shivering | Vasoconstriction, vasodilation | Sweating | Piloerection

Adjustments in heat gain or heat loss | Adjustments in muscle activity (in metabolic heat output) | Adjustments in loss or conservation of heat | Adjustment in heat loss | Adjustments in conservation or loss of heat

Figure 50.20

The physiological and behavioural responses of birds and mammals to changes in skin and core temperature (see also Figure 39.12, Chapter 39).

flow and with it the amount of heat transferred from the body core to the surface. This is why our skin flushes pink on hot days, when the vessels dilate to dump heat, and turns white on cold days, when they constrict to prevent heat loss. Sweat glands and hair and feather follicles are also embedded in the dermis.

The innermost layer of the skin, the **hypodermis**, contains larger blood vessels and additional reinforcing connective tissue. The hypodermis also contains an insulating layer of fatty tissue below the dermal capillary network, which ensures that heat flows between the body core and the surface primarily through the blood. The insulating layer is thickest in mammals that live in cold environments, such as whales, seals, walruses, and polar bears, in which it is known as *blubber*.

50.7c The Body Reduces Heat Loss When Core Temperature Falls below the Set Point

When thermoreceptors signal a fall in core temperature below the set point, the hypothalamus triggers compensating responses by sending signals through the autonomic nervous system (see Chapter 44). Among the immediate responses is constriction of the arterioles in the skin (vasoconstriction), which reduces the flow of blood to the skin's capillary networks. The reduced flow cuts down the amount of heat delivered to the skin and therefore lost from the body surface. The reduction in flow is most pronounced in the skin covering the extremities, where blood flow may be reduced by as much as 99%

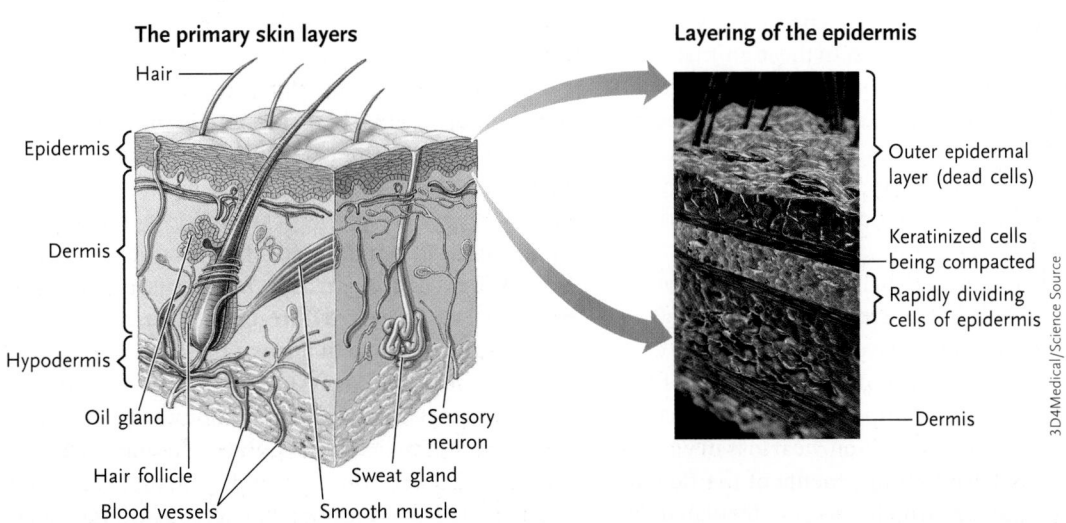

The primary skin layers

Hair
Epidermis
Dermis
Hypodermis
Oil gland
Hair follicle
Blood vessels
Sensory neuron
Sweat gland
Smooth muscle

Layering of the epidermis

Outer epidermal layer (dead cells)
Keratinized cells being compacted
Rapidly dividing cells of epidermis
Dermis

3D4Medical/Science Source

Figure 50.21
The structure of human skin.

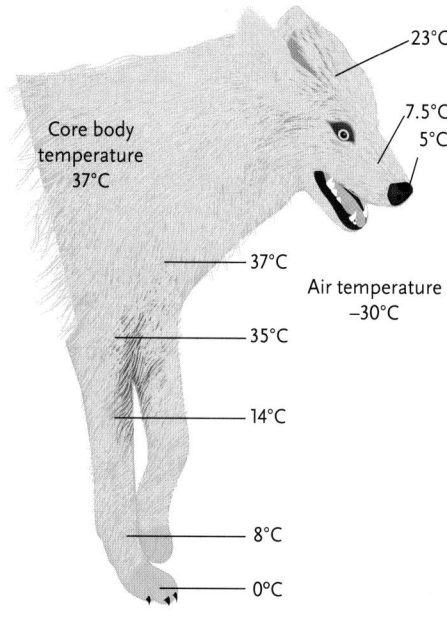

Core body
temperature
37°C

23°C

7.5°C
5°C

Air temperature
−30°C

37°C

35°C

14°C

8°C

0°C

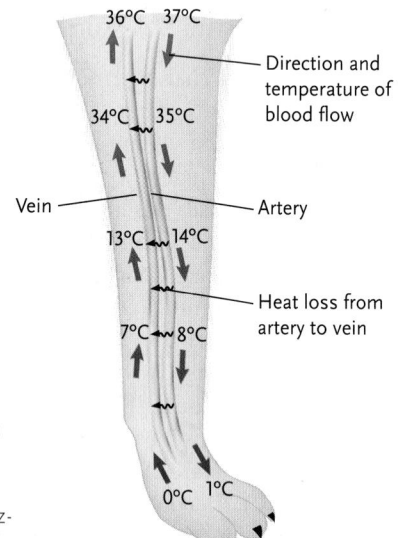

36°C   37°C

Direction and
temperature of
blood flow

34°C   35°C

Vein

Artery

13°C   14°C

Heat loss from
artery to vein

7°C   8°C

0°C   1°C

Figure 50.22
Countercurrent circulation in the leg of an Arctic wolf. The vein and artery are parallel and close together so that heat from the warm blood in the artery is transferred to the cold blood returning from the foot, minimizing heat loss through the foot.

when core temperature falls. This is particularly important for animals in polar regions. In these animals, the veins and arteries to the legs may form a simple countercurrent heat exchange system in which cold blood returning from the foot takes heat from the arterial blood entering the foot **(Figure 50.22)**. This minimizes heat loss from the foot while maintaining a nutritive flow of blood to the extremity.

In marine mammals such as whales and seals, heat loss is regulated by adjustments in the blood flow through the thick blubber layer to the skin. In cold water, blood flow is minimized by constriction of the vessels, making the skin temperature close to that of the surrounding water. In addition, heat loss in whales and seals is controlled by adjustments of the flow of blood to the flippers, which are not insulated by

blubber and act as heat radiators. When heat must be conserved to maintain core temperature at the set point, blood flow to the flippers is reduced.

Another immediate response is contraction of the smooth muscles that erect the hair shafts in mammals and feather shafts in birds (*piloerection*). This traps air in pockets over the skin, reducing convective heat loss. The response is minimally effective in animals such as pigs and humans because hair is sparse on most parts of the body, but it produces the goose bumps we experience when the weather gets chilly. However, in mammals with fur coats or in birds, erection of the hair or feather shafts significantly increases the thickness of the insulating layer that covers the skin, trapping more air.

Immediate behavioural responses triggered by a reduction in skin temperature also help reduce heat loss from the body. Mammals may reduce heat loss by moving to a warmer location or curling into a ball. Many mammals have an uneven distribution of fur that aids thermoregulation. In a dog, for example, the fur is thickest over the back and sides of the body and the tail and thinnest under the legs and over the belly. In cold weather, a dog will curl up, pull in its limbs, wrap its tail around its body, and bury its nose in its tail so that only body surfaces insulated by thick fur are exposed to the air **(Figure 50.23a)**. Many birds and mammals also huddle together to conserve heat. We have all seen puppies huddled together to keep warm; birds such as penguins also keep warm by huddling (Figure 50.23b).

If these immediate responses do not return body temperature to the set point, the hypothalamus triggers further responses, most notably the rhythmic tremors of skeletal muscle we know as shivering. The heat released by the muscle contractions and the oxidative reactions powering them can raise the total heat production of the body substantially. At the same time, the hypothalamus triggers secretion of *epinephrine* (from the adrenal medulla) and *thyroid hormone* (see Chapter 43), both of which increase heat production by stimulating the oxidation of fats and other fuels. The generation of heat by oxidative mechanisms in non-muscle tissue throughout the body is termed **nonshivering thermogenesis**.

In the young of many mammals (including newborn human infants), the most intense heat generation is by nonshivering thermogenesis that takes place in a specialized **brown adipose tissue** (also called brown fat), which can produce heat rapidly. Heat is generated by a mechanism that uncouples electron transport from ATP production in mitochondria (see Chapter 6); the heat is transferred throughout the body by the blood. Animals that hibernate or are active in cold regions contain larger amounts of brown adipose tissue. In most mammals, brown adipose tissue is concentrated between the shoulders in the back and around the neck where it preferentially warms the heart and the blood leaving the heart in arterial blood.

a.

b.

d.

c.

Jens Ottoson/Shutterstock.com

© david tipling/Alamy

Photos.com

Sumikophoto/Shutterstock.com

Figure 50.23
Structural and behavioural adaptations controlling heat transfer at the body surface. **(a)** A husky (*Canis lupus familiaris*) conserving heat by curling up with the limbs under the body and the tail around the nose. **(b)** Penguins huddle together to conserve heat. **(c)** An elephant cools off by spraying itself with water. **(d)** A jackrabbit (*Lepus californicus*) dissipating heat from its ears on a hot summer day. Notice the dilated blood vessels in its large ears. Both the large surface area of the ears and the extensive network of blood vessels promote the dissipation of heat by convection and radiation.

In human newborns, this tissue accounts for about 5% of body weight. The tissue normally shrinks during late childhood and is absent or nearly so in most adults. However, if exposure to cold is ongoing, the tissue remains. Some Japanese and Korean divers who harvest shellfish in frigid waters and Finns who work outside during the winter have significant amounts of brown adipose tissue.

If none of these responses succeed in raising body temperature to the set point, the result is **hypothermia**, a condition in which the core temperature falls below normal for a prolonged period. In humans, a drop in core temperature of only a few degrees affects brain function and leads to confusion; progressive or continued hypothermia can lead to coma and death.

50.7d The Body Increases Heat Loss When Core Temperature Rises above the Set Point

When the core temperature rises above the set point, the hypothalamus sends signals through the autonomic system that trigger responses that lower body temperature. As an immediate response, the signals relax smooth muscles of arterioles in the skin (vasodilation), increasing blood flow and with it the heat lost from the body surface. In marine mammals in warmer water, blood flow to the skin above the blubber increases, allowing excess heat to be lost from the body surface. In addition, when a whale generates excessive internal heat through the muscular activity of swimming, the flow of blood from the body core to the flippers increases.

The smooth muscles to feathers and fur in birds and mammals in hot climates also relax, and these layers are flattened, removing trapped air and decreasing their insulating capability. In addition, in mammals with sweat glands, such as antelopes, cows, humans, and horses, signals from the hypothalamus trigger the secretion of sweat, which absorbs heat as it evaporates from the surface of the skin. Some endotherms, including dogs (which have sweat glands only on their feet) and many birds (which have no sweat glands), use panting as a major way to release heat.

These physiological changes are reinforced by behavioural responses such as seeking shade or a cool burrow, plunging into cold water, wallowing in mud, or taking a cold drink. Elephants take up water in their trunks and spray it over their bodies to cool off in hot weather (Figure 50.23c); dogs spread their limbs, turn on their side or back, and expose the relatively bare skin of the belly, which acts as a heat radiator. In hot weather, many birds fly with their legs extended so that heat flows from their legs into the passing air. Similarly, penguins expose featherless patches of skin under their wings to cool off on days when the weather is too warm. Jackrabbits (Figure 50.23d) and elephants dissipate heat from their large ears, which are richly supplied with blood vessels. In times of significant heat stress, kangaroos and rats spread saliva on their fur to increase heat loss by evaporation; some bats coat their fur with both saliva and urine.

When the heat gain of the body is too great to be counteracted by these responses, **hyperthermia** results. An increase of only a few degrees above normal for a prolonged period is enough to disrupt vital biochemical reactions and damage brain cells. Most adult humans become unconscious if their body temperature reaches 41°C and die if it goes above 43°C for more than a few minutes.

As with ectotherms, many mammals also undergo thermal acclimatization with seasonal temperature changes. Although in many cases a change in day length appears to be the actual trigger, the development of a thick fur coat in winter, which is shed in summer, enables them to adapt to seasonal temperatures. Some arctic and subarctic mammals develop a thicker layer of insulating fat in winter.

50.7e Birds and Mammals Have Daily and Seasonal Rhythms

The temperature set point in many birds and mammals varies in a regular cycle during the day. In some, the daily variations are relatively small. In others, larger variations are correlated with daily or seasonal temperature changes. These rhythms are a response to day length rather than temperature change.

Humans are among the endotherms for which daily variations in the temperature set point are small. Normally, human core temperature varies from a minimum of about 35.5°C in the morning to a maximum of about 37.7°C in the evening. Women also show a monthly variation keyed to the menstrual cycle, with temperatures rising about 0.5°C from the time of ovulation until menstruation begins. The physiological significance of these variations is unknown.

Camels undergo a daily variation of as much as 7°C in set point temperature. During the day, a camel's set point gradually resets upward, an adaptation that allows its body to absorb a large amount of heat. The heat absorption conserves water that would otherwise be lost by evaporation to keep the body at a lower set point. At night, when the desert is cooler, the thermostat resets again, allowing the body temperature to cool several degrees, releasing the excess heat absorbed during the day.

When the environmental temperature is cool, having a lowered temperature set point greatly reduces the energy required to maintain body temperature. In many animals, the lowered set point is accompanied by reductions in metabolic, nervous, and physical activity (including slower respiration and heartbeat), producing a sleeplike state known as **torpor.**

Entry into **daily torpor,** a period of inactivity keyed to variations in daily temperature, is typical of many small mammals and birds. These animals typically expend more

energy per unit of body weight to keep warm than larger animals because the ratio of body surface area to volume increases as body size decreases; that is, the ratio of surface area across which they lose heat increases while the volume of cells producing heat decreases. Hummingbirds feed actively during the daytime, when their set point is close to 40°C. During the cool of night, however, the set point drops to as low as 13°C. This allows the birds to conserve enough energy to survive overnight when they are unable to feed and would otherwise be unable to obtain enough fuel to produce heat; they would literally starve to death. Some nocturnal animals, including bats and small rodents, such as the deer mouse, become torpid in cool locations during daylight hours when they do not actively feed. At night, their temperature set point rises and they become fully active **(Figure 50.24).**

Many animals enter a prolonged state of torpor tied to the seasons, triggered in most cases by a change in day length that signals the transition between summer and winter. The importance of day length has been demonstrated by laboratory experiments in which animals have been induced to enter seasonal torpor by changing the period of artificial light to match the winter or summer day length.

Many mammals during winter enter a more extreme state of metabolic suppression called **hibernation** (*hiberna* = winter), greatly reducing metabolic expenditures when food is unobtainable. Hibernators must store large quantities of fats to serve as energy reserves. The drop in body temperature during hibernation varies with the mammal. In some, such as hedgehogs, groundhogs, and squirrels, body temperature falls to within 1°C of environmental temperatures down to just above freezing. In some species of hedgehogs, body temperature falls from about 38°C in the summer to as low as 5 to 6°C during winter hibernation. The Arctic ground squirrel's body supercools (goes to a below-freezing, unfrozen state) during hibernation, with its body temperature dropping to about −3°C.

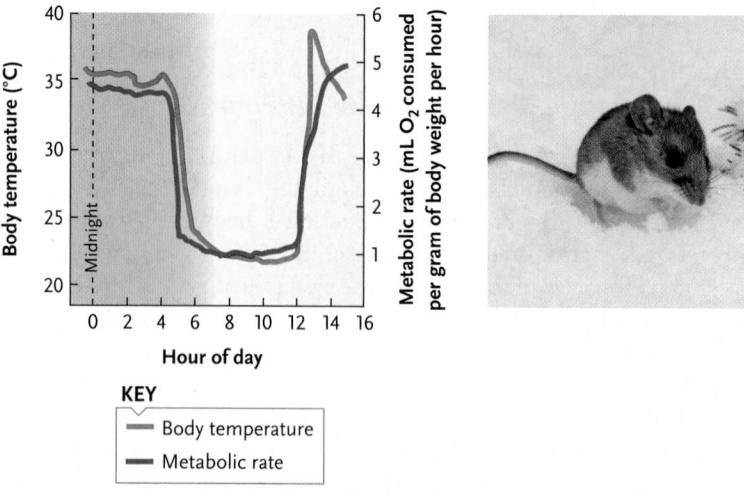

Figure 50.24

Cycle of daily torpor in a deer mouse (*Peromyscus maniculatus*).

Hibernating mammals may experience brief periods of arousal during the course of the winter.

In larger mammals, such as bears, the degree of metabolic suppression and fall in body temperature is less pronounced. The core temperature of bears drops only 5 to 10°C. Although sluggish, hibernating bears will waken readily if disturbed. They also waken normally from time to time, as when females wake to give birth during the winter season.

Some ectotherms, including amphibians and reptiles living in northern latitudes, also become torpid during winter. The Antarctic codfish, *Notothenia cordiceps*, spends the summer feeding on phytoplankton. In the winter, however, phytoplankton are reduced as a result of the low levels of light. The fish enter a state similar to hibernation. They remain relatively immobile in refuges, and their metabolic rate drops by about one-third. Because the temperature of the water remains constant over the seasons, the fish clearly have the capacity to control their metabolic rate independent of temperature.

Some mammals enter seasonal torpor during summer, called **estivation** (*aestivalis* = of summer), when environmental temperatures are high and water is scarce. Some ground squirrels remain inactive in the cooler temperatures of their burrows during extreme summer heat. Many ectotherms, among them land snails, lungfishes, many toads and frogs, and some desert-living lizards, weather such climates by digging into the soil and entering a state of estivation that lasts throughout the hot, dry season.

STUDY BREAK

1. What can birds and mammals do to dissipate heat on very hot days? How would this be affected by high humidity?
2. Mammals that live in the desert are often nocturnal. What advantages are there in terms of thermoregulation?

LIFE ON THE EDGE 50.4

Surviving Freezing

For birds and mammals, freezing temperatures lead to the formation of ice crystals that disrupt cell membranes and kill cells. Many ectotherms, however, encounter temperatures in winter that are well below the freezing point of their body fluids and yet their cells do not die. The tiny second-stage caterpillar of the spruce budworm (*Choristoneura fumiferana*) spends the winter at the tips of spruce trees in Canada, where temperatures may reach −40°C **(Figure 1a, b)**. Their ability to survive depends on the production of an antifreeze protein. Antifreeze proteins also occur in marine fishes that occupy habitats where the water temperature may be below the freezing point of their body fluids, and they occur in some amphibians **(Figure 1c)**, such as the wood frog (*Rana sylvatica*, see **Figure 1d** and "People behind Biology," Box 50.3). Research led by Peter Davies at Queen's University has shown that these proteins bond with forming ice crystals, limiting their growth, so although the water freezes, the crystals remain small enough that cell structure is not disrupted.

The molecular structure of the antifreeze proteins is diverse, and at least four types are known. This diversity appears even within a single species: work in Garth Fletcher's laboratory at Memorial University in St. John's, Newfoundland, has shown that the protein found in plasma of several species of marine teleosts is different from that in the skin cells or gill cells. This diversity suggests that genes for the various types of antifreeze proteins in fish arose independently during the relatively recent cooling of the polar oceans.

William M. Ciesla, Forest Health Management International, Bugwood.org

Jerald E. Dewey, USDA Forest Service, Bugwood.org

Zureks, CC-BY-SA-3.0

Courtesy of Kenneth Storey, Carleton University

Figure 1
(a) *The effects of spruce budworm on stands of western Douglas fir.* **(b)** *Spruce budworm* (Choristoneura fumiferana). **(c)** *The icefish* Trematomus bernacchi. **(d)** *Frozen specimens of the wood frog* Rana sylvatica.

Review

To access course materials such as Aplia and other companion resources, please visit www.NELSONbrain.com.

50.1 Introduction to Osmoregulation and Excretion

- Osmotic concentration of a solution is measured as osmolality in milliosmoles per kilogram (mOsm/kg) of solute. Both molecules and ions contribute to osmolality. Water moves osmotically from a solution of lower osmolality to one of higher osmolality. When comparing two solutions of different osmolality, the solution of higher osmolality is hyperosmotic and the solution of lower osmolality is hypoosmotic. Solutions of the same osmolality are isoosmotic.

- Osmoregulation describes the mechanisms that keep the osmolality of intracellular and extracellular fluids isotonic. Osmoregulators keep the osmolality of body fluids different from that of the environment. Osmoconformers allow the osmolality of their body fluids to match that of the environment.

- Osmoregulation is closely related to excretion because molecules and ions must be removed from the body to maintain the isotonicity of cells and extracellular fluids. Excretion also removes excess water, nitrogenous wastes, excess acid, and toxic molecules from the body.

- In most animals, tubules formed from a transport epithelium carry out the combined processes of osmoregulation and excretion. Extracellular fluids are filtered into the proximal end of the tubules; as the fluid moves through the tubules, some ions and molecules are reabsorbed from the fluid, and others are secreted into the fluid. Water is added or removed from the fluid to maintain the animal's water balance. The processed fluid is released to the exterior of the animal as urine.

- Nitrogenous wastes from the metabolism of amino acids and nucleic acids are excreted as ammonia, urea, or uric acid or as a combination of these substances.

50.2 Osmoregulation and Excretion in Invertebrates

- Most marine invertebrates are osmoconformers, whereas the invertebrates living in freshwater or terrestrial environments are osmoregulators.

- Because the body fluids of marine osmoconformers are isoosmotic to sea water, they expend little or no energy on maintaining water balance. Body fluids in osmoregulators living in freshwater and terrestrial environments are hyperosmotic to their surroundings. They must therefore expend energy to excrete water moving into their cells by osmosis and to obtain the salts required to maintain osmolality.

- Most invertebrates have specialized excretory tubules such as protonephridia, metanephridia, or Malpighian tubules that eliminate nitrogenous wastes and may assist in osmoregulation. A few groups eliminate wastes by diffusion.

- Protonephridia are blind tubes that produce an ultrafiltrate of the body fluids and adjust its ionic composition by secretion and reabsorption. Metanephridia are open tubes that take in coelomic fluid and adjust the ionic concentration.

- Malpighian tubules are blind tubes, and ions, uric acid, and other compounds enter by secretion.

50.3 Osmoregulation and Excretion in Nonmammalian Vertebrates

- Marine teleosts must continuously drink sea water to replace body water lost by osmosis to their hyperosmotic environment. The Na^+, K^+, and Cl^- in the ingested sea water are excreted from the gills. Nitrogenous wastes are excreted by the gills as ammonia.

- Sharks use urea and trimethylamine oxide as osmolytes to maintain their body fluids isoosmotic with sea water. As a result, sharks and rays do not lose water by osmosis and do not drink sea water. Excess salts ingested in foods are excreted in the kidney and by a rectal salt gland.

- Body fluids of freshwater fishes and amphibians are hyperosmotic to their environment, and these animals must excrete the excess water that enters by osmosis. Body salts are obtained from food and, in fishes, by active transport through the gills. Nitrogenous wastes are excreted from the gills of fish and larval amphibians as ammonia and through the kidneys of adult amphibians as urea.

- Most reptiles and birds conserve water by secreting nitrogenous wastes as uric acid. Water is also absorbed from the urine and the feces in the cloaca.

- Marine birds and reptiles secrete excess salts through a gland in the head.

50.4 Osmoregulation and Excretion in Mammals

- In mammals and other vertebrates, excretory tubules are concentrated in a specialized excretory organ, the kidney. Each of the pair of mammalian kidneys is divided into an outer renal cortex surrounding a central renal medulla. The mammalian excretory tubule, the nephron, has a proximal end at which filtration takes place, a middle region in which reabsorption and secretion occur, and a distal end that releases urine.

- The interstitial fluid reabsorbs ions, water, and other molecules from the nephron. A network of capillaries surrounding the nephron absorbs these materials. The urine leaving individual nephrons is processed further in collecting ducts and then pools in the renal pelvis. From there it flows through the ureter to the urinary bladder and through the urethra from the bladder to the exterior of the animal.

- At its proximal end, the mammalian nephron forms a cuplike Bowman's capsule around a cluster of capillaries, the glomerulus. A filtrate consisting of water, other small molecules, and ions is forced from the glomerulus into Bowman's capsule, from which it travels through the nephron and drains into the collecting ducts and renal pelvis.

- The proximal convoluted tubule of the nephron secretes H^+ into the filtrate and actively reabsorbs Na^+, K^+, HCO_3^-, and nutrients such as glucose and amino acids.
- In the descending segment of the loop of Henle, water is reabsorbed by osmosis.
- In the ascending segment of the loop of Henle, Na^+ and Cl^- move from the tubule by active transport. This establishes the osmotic gradient in the medulla of the kidney that is essential for water conservation.
- In the distal convoluted tubule, regulatory mechanisms operate to balance the concentrations of H^+ and salts between the urine and the interstitial fluid surrounding the nephron. K^+ and Na^+ move by active transport.
- In the collecting ducts, additional H^+ is secreted into the urine and water is reabsorbed; some urea is also reabsorbed at the bottom of the ducts.

50.5 Introduction to Thermoregulation

- Animals maintain body temperature at a level that provides optimal physiological performance. Heat flows between animals and their environment by conduction, convection, radiation, and evaporation.
- Ectothermic animals obtain heat energy primarily from the environment, and endothermic animals obtain heat energy primarily from internal reactions.

50.6 Ectothermy

- Ectotherms, including all invertebrates and amphibians, reptiles, and most fishes among the vertebrates, control body temperature by regulating heat exchange with the environment. Their thermoregulatory responses may be physiological or behavioural. For most ectotherms, the ability to thermoregulate is limited, and body temperature does not differ widely from environmental temperature.
- Many animals undergo thermal acclimatization, a change in the limits of tolerable temperatures as the environment alternates between warm and cool seasons or between day and night. The acclimatization may involve structural changes and/or metabolic alterations.

50.7 Endothermy

- Endotherms, mostly birds and mammals, maintain body temperature over a narrow range by balancing internal heat production against heat loss from the body surface.
- Internal heat production is controlled by negative feedback pathways triggered by thermoreceptors in the skin, hypothalamus, and spinal cord.
- Signals from the receptors are integrated in the hypothalamus to bring about compensating responses by activating the autonomic nervous system, the endocrine system, and the motor nerves. These responses return the core temperature to a set point when deviations occur.
- When body temperature falls below a set point, responses include an increase in heat-generating metabolic reactions and a reduction of blood flow to the body surface. Behavioural responses also reduce heat loss.
- When body temperature rises above the set point, blood flow to the skin increases and sweating is induced in mammals with sweat glands. Mammals with no or few sweat glands and birds, which have no sweat glands, release heat by panting. Behavioural responses also contribute to heat loss.
- The skin of endotherms is water impermeable, reducing heat loss by direct evaporation of body fluids. Mammals with sweat glands regulate evaporative heat loss by releasing moisture to the skin surface when body temperature rises. The blood vessels of the skin regulate heat loss by constricting or dilating. A layer of insulating fatty tissue under the vessels limits losses to the heat carried by the blood. The hair of mammals and feathers of birds also insulate the skin. Erection of the hair or feathers reduces heat loss by thickening the insulating layer.
- The temperature set point in many birds and mammals varies in daily and seasonal patterns. During cooler conditions, a lowered set point is accompanied by torpor: a reduction in metabolic, nervous, and physical activity. Seasonal torpor includes winter hibernation and summer estivation.
- Some animals, such as certain cold-water marine fishes, exhibit a form of endothermy in which part, but not all, of their core is maintained at a temperature significantly higher than the surrounding environment.

Questions

Self-Test Questions

1. In terms of osmoregulation, why are most sharks like freshwater fish?
   a. They are osmoregulators.
   b. They are osmoconformers.
   c. The salt in the animals is more concentrated than the water in which they live.
   d. The salt in the animals is less concentrated than the water in which they live.

2. Which structure can carry out filtration and/or excretion?
   a. ciliated metanephridia in insects
   b. a nephron and a bladder in insects
   c. the hindgut of earthworms
   d. protonephridia containing flame cells in flatworms

3. If an animal secretes uric acid crystals, in which environment could you conclude that the animal most likely lives?
   a. fresh water
   b. marine
   c. terrestrial and very dry
   d. terrestrial and very wet

4. Which statement describes what happens in the kidneys?
   a. In the descending loop of Henle, water leaves the tubule.
   b. In the collecting duct, water enters the tubule via aquaporins.
   c. The fluid that enters Bowman's capsule is an ultrafiltrate of the blood.
   d. In the ascending loop of Henle, Na^+ and K^+ enter the tubule by simple diffusion.

5. Substances that are NOT reabsorbed will do the following.
   a. be excreted from the body in urine
   b. remain in the nephron
   c. leave the kidney through the renal vein
   d. return to the glomerulus

6. What is unique to endotherms?
   a. torpor
   b. thermal acclimatization
   c. thermoregulation by the hypothalamus
   d. a body temperature that does not change

7. Boa constrictors wrap themselves around their eggs and shiver. Which term refers to this warming mechanism?
   a. evaporation
   b. conduction
   c. convection
   d. radiation

8. Thermal acclimatization can be described as follows.
   a. It involves regulation of body temperature in response to information provided by a thermoreceptor.
   b. It is the change in an animal's physiology that accompanies seasonal changes.
   c. It occurs when core body temperature is below normal for an extended period of time.
   d. It occurs when core body temperature is above normal for an extended period of time.

9. If the set point for temperature regulation were increased, which would you expect to occur?
   a. decreased epinephrine
   b. shivering
   c. vasodilation
   d. both a and c

10. Which statement best exemplifies ectotherms?
    a. All invertebrates are ectotherms.
    b. Food demand decreases when environmental temperatures decrease.
    c. The metabolic rate increases as the environmental temperature decreases.
    d. Body temperature remains constant when environmental temperatures change.

Questions for Discussion

1. Compare and contrast the challenges for salt and water balance of animals living in marine, freshwater, and terrestrial environments.

2. Describe how the structure of aquaporins allows water to pass through a membrane but not protons.

3. Hockey players are often advised to consume sports drinks containing salt before, during, and after a game. Why do you think this is?

4. Mammals that live in the desert are often nocturnal. What advantages are there in terms of thermoregulation?

5. The internal temperature of a reptile, such as a crocodile, varies with its developmental stage. Adult crocodiles tend to have a higher internal temperature that is less subject to change due to environmental variation in temperature. Why do you think this is?

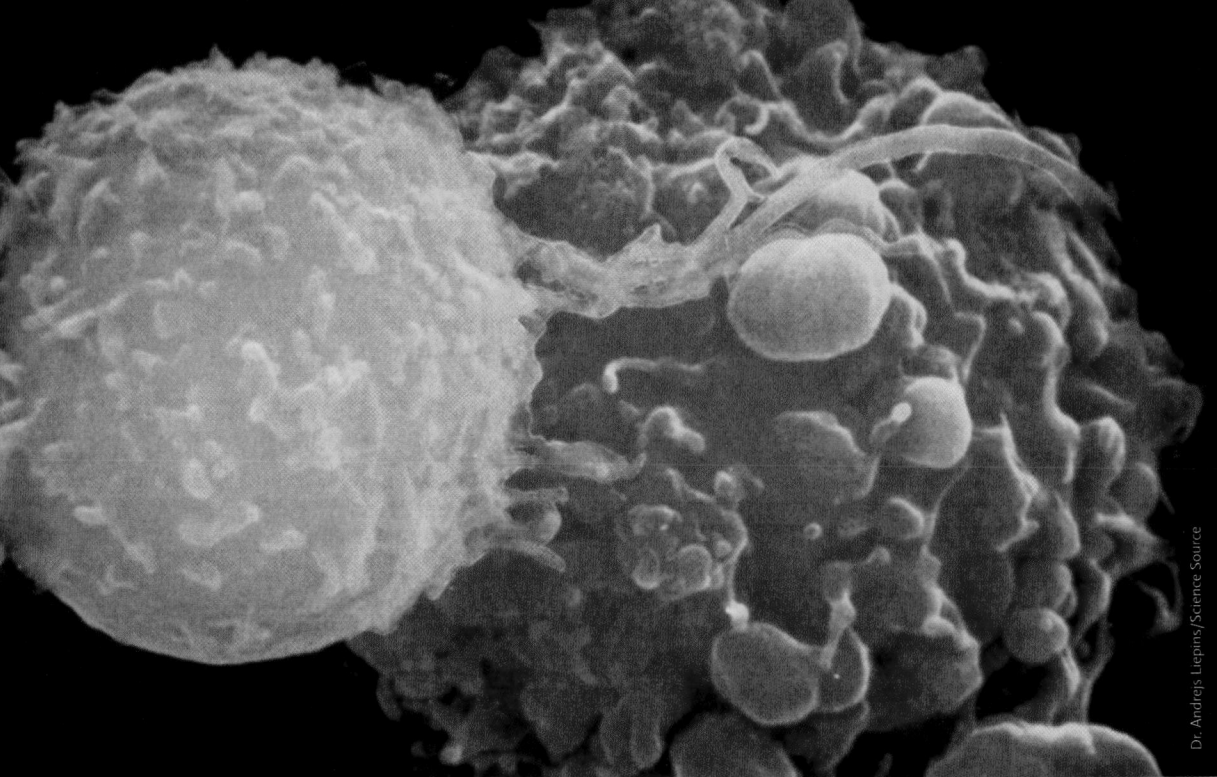

Death of a cancer cell. A cytotoxic T cell (orange) induces a cancer cell (mauve) to undergo apoptosis (programmed cell death). Cytotoxic T cells are part of the body's immune response system and are programmed to seek out, attach to, and kill cancer cells and pathogen-infected host cells.

Dr. Andrejs Liepins/Science Source

Defences against Disease

WHY IT MATTERS

Have you ever wondered whether you suffer more often from diseases than your friends? Have you ever wondered why some people never appear to get sick? Why do so many old people and newborns suffer from common diseases that never seem to bother teenagers? If everyone is equally susceptible to a disease, should we all suffer equally? How and why are we different with respect to our ability to fight off infections? Diseases have plagued all organisms, including humans, for billions of years. All animals, even insects, starfish, worms, and organisms too small to see with the naked eye, can suffer from some sort of disease. In humans, acquired immune deficiency syndrome (AIDS), first identified in the early 1980s, now affects about 40 million people worldwide and continues to spread. Malaria affects around 500 million people. Similar diseases in other organisms, such as avian influenza and avian malaria, may significantly reduce bird populations, and insect populations may be killed by outbreaks of specific viruses. Overall, bacteria, viruses, and parasites can all cause disease in their hosts. However, immune systems are able to regulate or eliminate the majority of pathogens that cause disease. We combat the other pathogens by developing effective drug treatments that eliminate the disease-causing organism or by vaccines that provide protection from infection.

The development of vaccines began with efforts to control smallpox, a dangerous and disfiguring viral disease that once infected millions of people worldwide, killing more than one-third of its victims. As early as the twelfth century, healthy individuals in China sought out people who were recovering from mild smallpox infections, ground up scabs from their lesions, and inhaled the powder or pushed it into their skin. Variations on this treatment were effective in protecting many people against smallpox infection.

In 1796, an English country doctor, Edward Jenner, used a more scientific approach. He knew that milkmaids never got smallpox if they had contracted cowpox, a similar but mild disease of cows that can be transmitted to humans. Jenner decided to see if a deliberate infection with cowpox would protect humans from smallpox. He scratched material from a cowpox sore into a boy's arm. Six weeks later, after the cowpox infection had subsided, he scratched fluid from human smallpox sores into the boy's skin. (Jenner's use of the boy as an experimental subject would now be considered highly unethical.) Remarkably, the boy remained free from smallpox. Jenner carried out additional, carefully documented case studies with other patients with the same results. His technique became the basis for worldwide **vaccination** (*vacca* = cow) against smallpox. With improved vaccines, smallpox has now been eradicated from the human population.

Vaccination takes advantage of the **immune system** (*immunis* = exempt), the natural protection that is our main defence against infectious disease. This chapter focuses on the roles of the different components of the immune system that deal with infection. We describe aspects of immune systems in different organisms, from insects to humans, to examine their similarities and differences.

CONCEPT FIX Some people think that *disease* and *parasites* are synonymous. A disease and a parasite are not the same thing. A parasite or pathogen is an organism that may infect us, whereas a disease is the manifestation of symptoms we show (sneezing, diarrhea, swollen liver, fever). Many individuals in a group may become infected with the same pathogen or parasite. Some of us will have genetic factors that prevent the foreign organism from developing, multiplying, and causing disease, while others will come down with full-blown symptoms. Sometimes parasites can develop and be transmitted from one individual to another without the infected person knowing they have an infection. These asymptomatic carriers can be important in maintaining an outbreak of disease as it is difficult to identify them as infected individuals and then treat them to eliminate the problem. ⬣

51.1 Three Lines of Defence against Invasion

Every organism is constantly exposed to *pathogens* and *parasites,* potentially disease-causing organisms such as viruses, bacteria, fungi, protists, and larger parasites such as trematodes, cestodes, and nematodes. Humans and other animals have three lines of defence against these threats. The first line of defence involves physical barriers that prevent the entry of pathogens. The second is the *innate immune system*—inherited mechanisms that protect the body from pathogens in a nonspecific way. The third is the *adaptive immune system,* found only in vertebrates, which involves inherited mechanisms that lead to the synthesis of molecules such as antibodies that target pathogens in a specific

way. Reaction to an infection takes minutes in the case of the innate immune system versus several days for the adaptive immune system. But a hallmark of the adaptive response includes its memory of previous infections and a much more rapid response to the same pathogen in subsequent infections.

51.1a The Epithelium Is a Barrier to Infection

An organism's first line of defence is the body surface—the skin covering the body exterior, the cuticle of an arthropod, the outer layer of a plant, and the epithelial surfaces covering internal body cavities and ducts, such as the lungs and intestinal tract. The body surface forms a barrier of tight junctions between epithelial cells that keeps most parasites and pathogens (as well as toxic substances) from entering the body.

In the respiratory tract, ciliated cells constantly sweep the mucus with its trapped bacteria and other foreign matter into the throat, where it is coughed out or swallowed. Many of the body cavities lined by mucus membranes have environments that are hostile to pathogens. For example, the strongly acidic environment inside the stomach kills most ingested bacteria and destroys many viruses, including those trapped in swallowed mucus from the respiratory tract. Most of the pathogens that survive the stomach acid are destroyed by the digestive enzymes and bile secreted into the small intestine. There are, however, some parasites and pathogens that are resistant to stomach acid and digestive enzymes and establish in the stomach or intestinal tract, where they may cause disease. Reproductive tracts may be acidic or basic, preventing many pathogens from surviving there, and many epithelial tissues secrete proteins such as defensins or lysozymes that are lethal to many bacteria.

51.1b Immune Systems Protect the Body

The body's second line of defence is a series of generalized internal chemical, physical, and cellular reactions that attack pathogens that have breached the first line. These defences include inflammation, which creates internal conditions that inhibit or kill many pathogens, and specialized cells that engulf or kill pathogens or infected body cells. These initial nonspecific responses to pathogens are components of the **innate immune system**.

The recognition of the parasite/pathogen as nonself is the first essential step in initiating any immune response. The innate immune response relies on germ line–encoded receptors that recognize highly conserved molecular patterns (i.e., found on many similar organisms) on the surface of pathogens but not found on host cells (see Section 43.1c). **Innate immunity** provides an immediate, *nonspecific* response; that is, it targets any invading pathogen and has no memory of prior exposure to that specific pathogen.

Invertebrates, fungi, and plants rely solely on innate immune responses, whereas vertebrates use the innate immune system in conjunction with the adaptive response for a more powerful overall response. This most complex line of defence, found only in vertebrates, is called **adaptive** (or **acquired**) **immunity.** Adaptive immunity is *specific:* it recognizes individual pathogens and mounts an attack that directly neutralizes or eliminates them. It is stimulated and shaped by the presence of a specific pathogen or foreign molecule in the body. This mechanism, which takes several days to become protective, is triggered by specific molecules on pathogens that are recognized as being foreign to the body. The body retains a memory of the first exposure to a foreign molecule, enabling it to respond more quickly if the same pathogen is encountered again in the future. This is the basis of the vaccination programs we will discuss later.

Innate immunity and adaptive immunity together constitute the immune system, and the defensive reactions of the system are termed the **immune response.** Functionally, all components of the immune system interconnect and communicate at the chemical and molecular levels. The immune system is the product of long-term coevolutionary interactions between pathogens and their hosts. Over millions of years, the mechanisms by which pathogens attack and invade have become more efficient, but the defences of organisms against the invaders have kept pace.

51.1c Organisms Can Recognize Pathogens

The basic tenet of immunity is that an organism can recognize a pathogen as being different from the host. This is often termed recognition of *nonself* and is the essential first step before any immune response can be initiated. Different organisms do this in different ways. Most organisms recognize unique *pathogen-associated molecular patterns* (PAMPs) found on many microbial organisms, using host molecules called *pattern recognition receptors* (PRRs). Common PAMPs include carbohydrates, glycoproteins, lipids, and nucleic acids, and are essential to the existence of the organisms. Two classic examples of PAMPs are the bacterial cell wall components lipoteichoic acid, found on Gram-positive bacteria, and lipopolysaccharide, found on Gram-negative bacteria. Major invertebrate PRRs include the Gram-negative bacteria binding protein and β 1,3-glucan recognition proteins that recognize and bind to common molecules found on many groups of pathogenic organisms rather than recognizing each pathogen species individually. For example, lipopolysaccharide is found on the outer surface of all Gram-negative bacteria and serves as a general PAMP to which a PRR can bind. Once specific PRRs are activated by the presence of the PAMP, signalling cascades are initiated that activate various components of the innate or acquired immune responses.

Plants respond to pathogens in a similar manner. They have two branches of immune responses: a system that uses transmembrane PRRs that respond to the PAMPs described for invertebrates and an intracellular response that uses protein products encoded by *R* genes that respond to infection and systemic signals from nearby infected cells. Damage to the plant cells by pathogens is believed to be the signal that activates the expression of plant R proteins.

Different PAMPs may activate the same or different signalling pathways and elicit different responses. This has been studied best in the fruit fly, *Drosophila melanogaster*. Infection of *Drosophila* with fungi and bacteria activates the Toll, JAK-STAT, and immune deficiency signalling (IMD) pathways that result in nuclear factor (NF)-κB-like transcription factors being translocated to the nucleus, activating many components of the innate immune response. Thus, recognizing the parasite/pathogen activates a complex chain of events that ultimately produces an effective immune response and often a complete elimination of the microorganism. Many similar pathways and molecules are found in humans. Researchers who discovered the Toll-receptors in *Drosophila,* which allowed for the discovery of Toll-like receptors in humans, received the Nobel Prize in Medicine and Physiology in 2011.

CONCEPT FIX The implication in this chapter is that all associations between viruses, bacteria, fungi, and parasites with vertebrate or invertebrate hosts have a negative impact on the hosts. Negative impacts are caused by many organisms, such as those that cause human disease. In other cases, however, these associations are beneficial to one or both organisms. Symbiotic relationships (two organisms living together) include phoresis, in which one organism uses another to be transported to new areas; **mutualism**, often an obligate relationship in which both symbionts benefit, such as the relationship between termites and their intestinal protozoans, or cows and their symbiotic bacteria that help digest cellulose, or the gut microflora in humans that help prevent pathogen invasion and produce nutrients we cannot obtain from our diet; and **commensalism**, where one symbiont benefits with no effect on the other, such as epiphytic plants or ciliates on crustaceans. Not all associations between different organisms cause problems. The immune system of each organism must be able to recognize and differentiate between friend and foe and learn not to attack and eliminate beneficial symbionts. ⬡

STUDY BREAK

1. What features of epithelial surfaces protect against pathogens?
2. What are the key differences between innate immunity and adaptive immunity?
3. How are pathogens recognized as nonself?

51.2 Nonspecific Defences: Innate Immunity

Invertebrates have only an innate response, and the PAMP–PRR interactions activate the signalling pathways described above. These activate processes such as **phagocytosis** (the internalization and destruction of particulate matter) of small pathogens by hemocytes (blood cells) and the coagulation of the hemolymph (invertebrate blood). Larger pathogens may be encapsulated by hemocytes and covered in a melanin-like material that kills them **(melanotic encapsulation)**. This may be helped by the release of reactive intermediates of nitrogen and oxygen. The third component of the innate response involves the production of small **antimicrobial peptides** that kill pathogens not eliminated by the other responses. Some intracellular pathogens, such as viruses, are inaccessible to these immune factors. These often are eliminated through apoptosis, a form of programmed cell death (see Section 51.2a). This coordinated, multifaceted, and integrated approach eliminates potential pathogens, preventing them from harming or killing the host.

The study of invertebrate immunology began when a Russian scientist poked a rose thorn into the body of a starfish (see "People behind Biology," Box 51.2, p. 1287) and watched how the hemocytes surrounded the thorn **(Figure 51.1)**. The innate immune responses of invertebrates, now a rapidly developing field of research, have allowed us to understand how the innate system works in both invertebrates and vertebrates, as well as what molecules are activated in response to different stimuli. Plants respond in a similar manner. They can wall off cells that are infected, undergo cell death to prevent pathogen development, and produce several small antimicrobial peptides that kill microbial pathogens.

Vertebrates use similar types of specific host cell surface receptors that recognize the various PAMPs found on microbial pathogens. Some receptors activate signalling pathways that bring about the secretion of lethal antimicrobial peptides that kill the pathogen. Other receptors trigger the host cell to engulf the pathogen, as was described for phagocytosis in invertebrates, but in vertebrates, this may also initiate an inflammation response and activate the soluble receptors of the *complement system* described below. The innate responses that recognize and initiate immune pathways in plants, invertebrates, and vertebrates are strikingly similar.

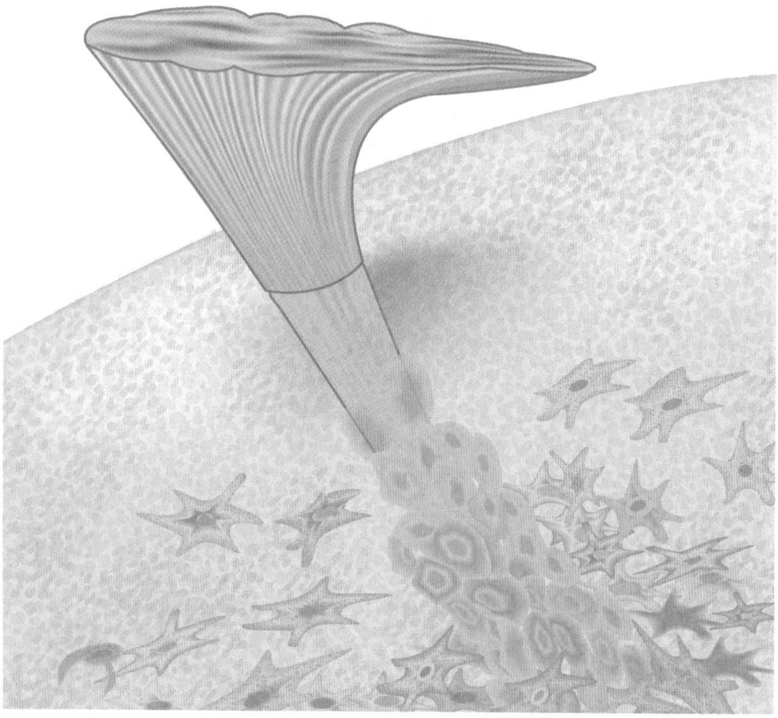

Antimicrobial Peptides. All epithelial surfaces, namely skin; the lining of the gastrointestinal tract; the lining of the nasal passages, gills, and lungs; and the lining of the genitourinary tracts, are protected by antimicrobial peptides, such as the *defensins* (see "Molecule behind Biology"). These epithelial cells secrete defensins upon attack by a microbial pathogen. The defensins attack the plasma membranes of the pathogens, eventually disrupting them and thereby killing the cells. In particular, defensins play a significant role in the innate immunity of the intestinal tracts of vertebrates and invertebrates. Antimicrobial peptides such as the defensins are highly conserved in plants, invertebrates, vertebrates, and even single-celled organisms, indicating their important role in immunity throughout evolution. Other antimicrobial peptides are found only in specific groups, suggesting a more specialized role that has maintained their existence.

Inflammation. A tissue's rapid response to injury, including infection by most pathogens, involves **inflammation** (*inflammare* = to set on fire): the heat, pain, redness, and swelling that occur at the site of an infection.

Several interconnecting mechanisms initiate inflammation **(Figure 51.2, p. 1270)**. Consider bacteria entering a tissue as a result of a wound. **Monocytes** (a type of leukocyte or white blood cell) enter the damaged tissue from the bloodstream through the endothelial wall of the capillary. Once in the damaged tissue, the monocytes differentiate into **macrophages** (= big eaters), which are phagocytes that are usually the first to recognize pathogens at the cellular level. (**Table 51.1, p. 1270**, lists the major types of leukocytes, including macrophages; see also **Figure 51.3, p. 1271**) Cell surface receptors on the macrophages recognize and bind to

Figure 51.1
Innate immune responses demonstrated by the role of starfish blood cells responding to a thorn that has penetrated the cuticle.

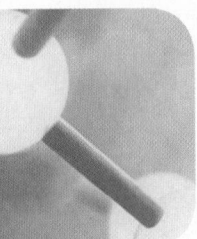

MOLECULE BEHIND BIOLOGY 51.1

Defensins

Defensins are ubiquitous cationic molecules used in the defences of essentially all organisms. Defensins have been isolated from several orders of the higher insects, such as the Diptera (flies) and Coleoptera (beetles), and from ancient insects, such as the Odonata (dragonflies). Functional analogues have been isolated and characterized from amoebas, nematodes, scorpions, molluscs, mammals (including humans), and plants. In some organisms, these molecules are secreted into the body cavity, whereas in others, they are intracellular and are released only at a wound site.

This strong conservation suggests that this molecule is of ancient origin and has been maintained throughout evolution due to its importance in limiting the growth of microbial pathogens. Defensins are composed of a series of structures: an N-terminal loop; an α-helix; and twisted, antiparallel β-sheets **(Figure 1)**. The three-dimensional shape of defensins is stabilized by the presence of three disulfide bridges. Defensins are active against many Gram-positive bacteria, some Gram-negative bacteria, and some fungi. The lethality occurs as the one region of the peptide binds to the outer surface of the bacteria, allowing other regions to form pores in the microbial membranes, causing a permeabilization that causes a loss of cytoplasmic potassium, a depolarization of the inner membrane, reduced amounts of cytoplasmic ATP, and a reduction in respiration. This can occur in single-celled organisms, in the body cavity of an insect, in plant tissues, or in the white blood cells of a vertebrate. Defensins represent one family of molecules conserved throughout all taxa for the same function and are truly universal immune molecules.

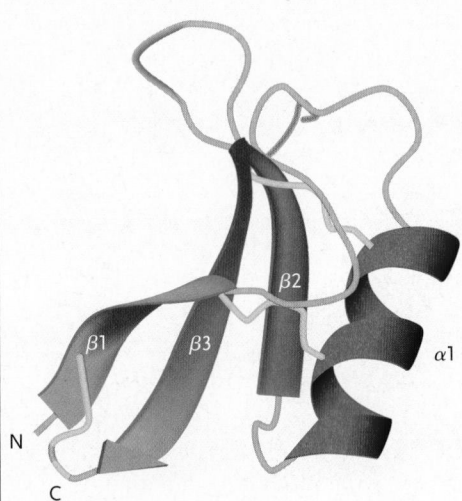

Figure 1
Computer-generated model of a defensin molecule showing the arrangement of the coiled α-helix (red) and the β-sheets (blue) that are held together by disulfide bridges (yellow).

---

surface molecules on the pathogen, activating the macrophage to phagocytize (engulf) the pathogen (see Figure 51.2, step 1). There may not be enough macrophages present at the site of infection to eliminate all of the pathogens. Activated macrophages also secrete **cytokines**. These are small proteins released by cells that affect the behaviour of other cells and may signal, activate, and recruit more immune cells to increase the system's response to the pathogen.

The death of cells caused by the pathogen at the infection site activates cells that are dispersed throughout the connective tissue, called **mast cells**, which then release histamine (see Figure 51.2, step 2, p. 1270). This histamine, along with the cytokines from activated macrophages, dilates local blood vessels around the infection site and increases their permeability. This increases blood flow and leakage of fluid from the vessels into body tissues (step 3). The response initiated by cytokines directly causes the heat, redness, and swelling of inflammation.

Cytokines also make the endothelial cells of the blood vessel wall stickier, causing circulating **neutrophils** (another type of phagocytic leukocyte) to attach to them in massive numbers. From there, the neutrophils are attracted to the infection site by **chemokines**, proteins also secreted by activated macrophages (see Figure 51.2, step 4). To get to the infection site, the neutrophils pass between endothelial cells of the blood vessel wall. Neutrophils may also be attracted directly to the pathogen by molecules released from the pathogens themselves. Like macrophages, neutrophils have cell surface receptors that enable them to recognize and engulf pathogens (step 5).

Once a macrophage or neutrophil has engulfed the pathogen, it uses a variety of mechanisms to destroy it. These mechanisms include the secretion of enzymes and defensins located in lysosomes and the production of toxic compounds. The harshness of these attacks usually kills the neutrophils as well, whereas macrophages usually survive to continue their pathogen-scavenging activities. Dead and dying neutrophils, in fact, are a major component of the pus formed at infection sites. The pain of inflammation is caused by the migration of macrophages and neutrophils to the infection site and their activities there.

Some parasitic worms are too large to be engulfed by macrophages or neutrophils. In that case, macrophages, neutrophils, and **eosinophils** (another type of leukocyte) cluster around the worm and secrete lysosomal enzymes in amounts usually sufficient to kill the parasite.

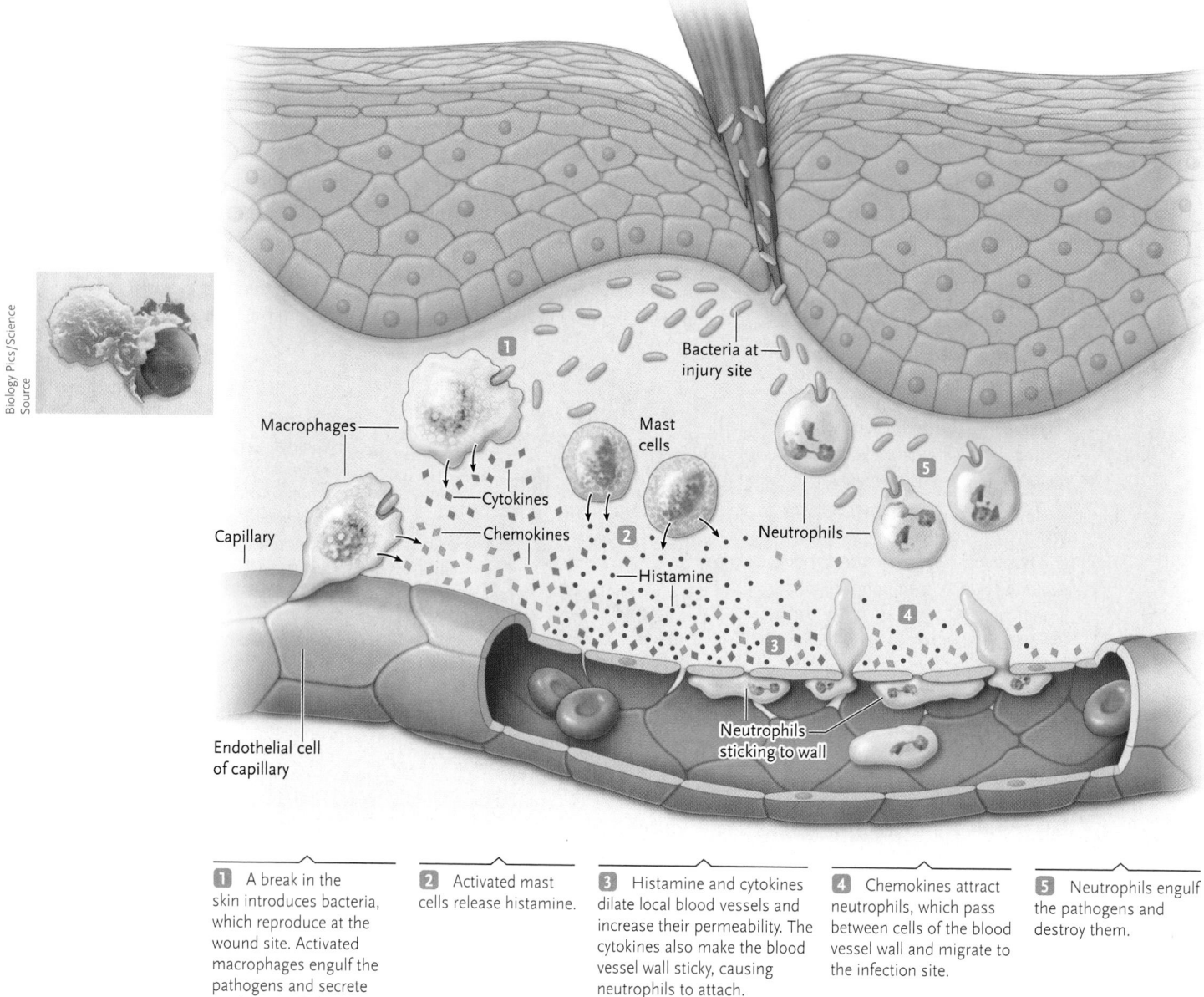

Biology Pics/Science Source

1 A break in the skin introduces bacteria, which reproduce at the wound site. Activated macrophages engulf the pathogens and secrete cytokines and chemokines.

2 Activated mast cells release histamine.

3 Histamine and cytokines dilate local blood vessels and increase their permeability. The cytokines also make the blood vessel wall sticky, causing neutrophils to attach.

4 Chemokines attract neutrophils, which pass between cells of the blood vessel wall and migrate to the infection site.

5 Neutrophils engulf the pathogens and destroy them.

Figure 51.2
The steps producing inflammation. The colourized micrograph on the left shows a macrophage engulfing a yeast cell.

Table 51.1	Major Types of Leukocytes and Their Functions
Type of Leukocyte	Function
Monocyte	Differentiates into a macrophage when released from blood into damaged tissue
Macrophage	Phagocyte that engulfs infected cells, pathogens, and cellular debris in damaged tissues; helps activate lymphocytes in carrying out immune response
Neutrophil	Phagocyte that engulfs pathogens and tissue debris in damaged tissues
Eosinophil	Secretes substances that kill eukaryotic parasites such as worms
Lymphocyte	Main subtypes involved in innate and adaptive immunity: natural killer (NK) cells, B cells, plasma cells, helper T cells, and cytotoxic T cells. NK cells function as part of innate immunity to kill virus-infected cells and some cancerous cells of the host. The other cell types function as part of adaptive immunity: they produce antibodies; destroy infected and cancerous body cells; and stimulate macrophages and other leukocyte types to engulf infected cells, pathogens, and cellular debris.
Basophil	Responds to IgE antibodies in an allergy response by secreting histamine, which stimulates inflammation

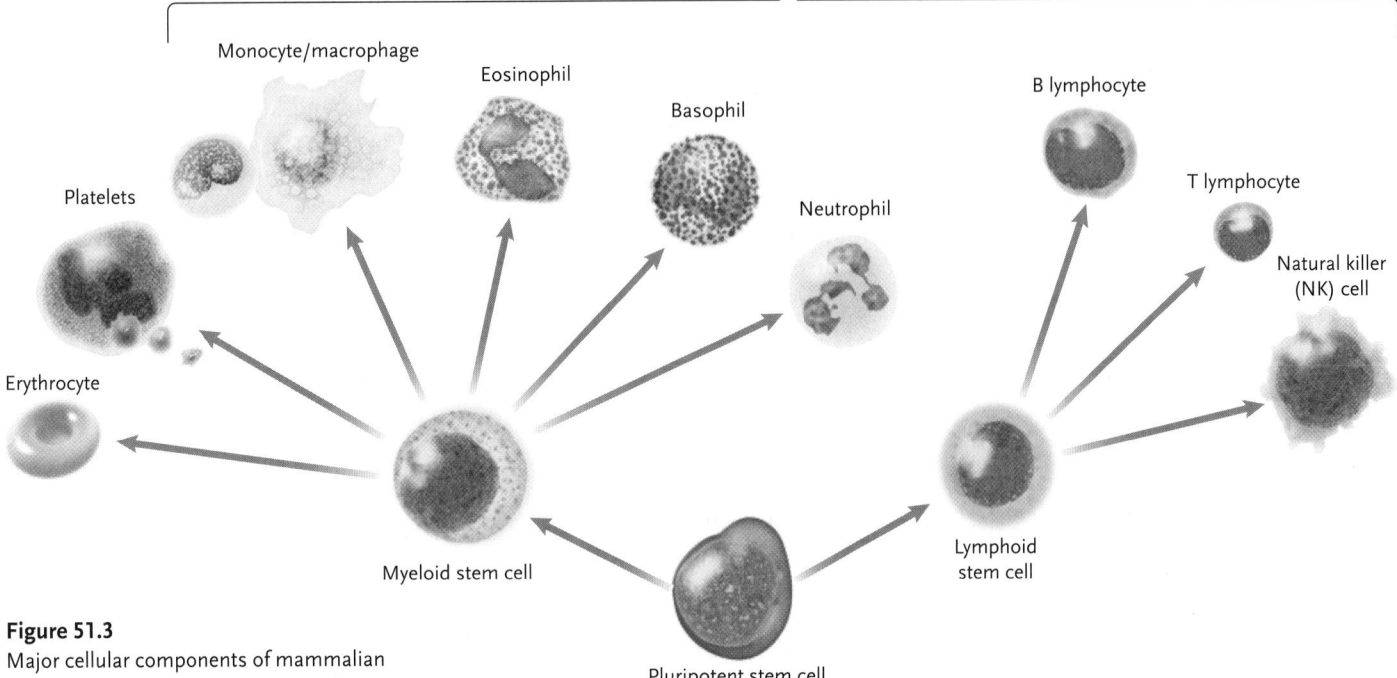

Figure 51.3
Major cellular components of mammalian blood and their origins from stem cells.

Labels in figure: Leukocytes; Monocyte/macrophage; Eosinophil; Basophil; B lymphocyte; T lymphocyte; Neutrophil; Natural killer (NK) cell; Platelets; Erythrocyte; Myeloid stem cell; Lymphoid stem cell; Pluripotent stem cell

Cells involved in inflammation provide an important connection to the adaptive immune response. Specifically, macrophages and dendritic cells (DCs: described below) function as antigen-presenting cells (APCs), which direct the outcome of the adaptive response toward pathogens and cytokines produced during inflammation. When DCs are recruited to an inflamed area, they encounter and recognize pathogen-specific PAMPs via their PRRs (see Section 51.1c). These interactions distinguish between intracellular and extracellular pathogens, and the DC responds by expressing different combinations of costimulatory molecules and cytokines, depending on the type of pathogen. When these APCs migrate to the lymph nodes, they initiate an adaptive immune response that is most appropriate for targeting the type of pathogen that was encountered. For example, signals induced by PRRs within the cell help initiate adaptive responses for killing intracellular pathogens such as viruses, whereas signals detected at the external cell surface trigger responses appropriate for killing and clearing extracellular agents.

The Complement System. Another nonspecific defence mechanism activated by invading pathogens is the **complement system,** a group of more than 30 interacting soluble plasma proteins that circulate in the blood and interstitial fluid. Normally inactive, the proteins are activated when they recognize molecules on the surfaces of pathogens. Activated complement proteins participate in a cascade of reactions on pathogen surfaces, producing large numbers of different complement proteins, some of which assemble into **membrane attack complexes.** These complexes insert into the plasma membrane of many types of bacterial cells and create pores that allow ions and small molecules to pass readily through the membrane. As a result, the bacteria can no longer maintain osmotic balance, and they swell and lyse. For other types of bacterial cells, the cascade of reactions coats the pathogen with fragments of the complement proteins. Cell surface receptors on phagocytes then recognize these fragments and engulf and destroy the pathogen.

Several activated proteins in the complement cascade also act individually to enhance the inflammatory response. For example, some of the proteins stimulate mast cells to enhance histamine release, whereas others increase the blood vessel permeability.

51.2a Three Main Strategies Are Used to Combat Pathogenic Viruses

Specific molecules on pathogens such as bacteria are key to initiating innate immune responses. The innate immune system, however, often cannot distinguish between surface molecules of viral pathogens and host cells or cannot enter host cells to eliminate intracellular pathogens. The host must, therefore, use other strategies to provide some immediate protection against these infections until the adaptive immune system, which can discriminate between pathogen and host proteins, is effective. Three main strategies in this regard involve RNA interference, interferon, and **natural killer (NK) cells.**

RNA Interference. *RNA interference (RNAi)* is a cellular mechanism that is triggered by double-stranded (ds) RNA molecules. The dsRNA interferes with the ability of a cell to transcribe specific genes. Because dsRNA is a natural part of the life cycle of many viruses, the use of RNAi can inhibit the dsRNA found in many viruses and eliminate the infection.

Similarly, the virus can interfere with the host immune machinery, including RNAi pathways. Thus, the activities within a cell, and the success of a pathogen such as a virus, depend on the interplay between pro- and anti-infection responses. The only well-established antiviral mechanisms reported in insects such as the fruit fly, *D. melanogaster,* are RNAi and apoptosis. To combat this host response, many viruses contain genes that encode RNAi suppressors or inhibitors of apoptosis that help them survive.

Interferon. Viral dsRNA may also cause the infected host cell to produce two cytokines, interferon-α and interferon-β. **Interferons** can be produced by most cells of the body. Interferons act both on the infected cell that produced them, an autocrine effect, and on neighbouring uninfected cells, a paracrine effect (see Chapter 43). Interferons bind to cell surface receptors, triggering a signal transduction pathway that changes the gene expression pattern of the cells. Key changes include the activation of a ribonuclease enzyme that degrades most cellular RNA and the inactivation of a key protein required for protein synthesis, thereby inhibiting most protein synthesis in the cell. These effects on RNA and protein synthesis inhibit pathogen replication, while putting the cell in a weakened state from which it can often recover.

Apoptosis. Apoptosis, or programmed cell death, is a process inherent to all eukaryotic cells that has been highly conserved through evolution. It represents an intrinsic form of cell death that is tightly regulated by a variety of internal and external cellular signals to avoid killing healthy, productive cells. During development, apoptosis is required to sculpt tissues, remove old and dying cells, and eliminate embryonic cells with damaged DNA. Throughout life, apoptosis is also used as an immune response against intracellular pathogens and parasites. These organisms often trigger abnormal cellular activity, which induces intrinsic apoptotic pathways, activating initiator and effector caspases that dismantle the cell's structure. Once effector caspases have been activated, the cell is destined to die. Apoptotic cells fragment into membrane-bound apoptotic bodies that are readily phagocytosed and digested by macrophages or by neighbouring cells without generating an inflammatory response. If the pathogen is recognized as non-self and the apoptotic response is initiated in time, both the infected cell and the pathogen are eliminated, and no disease is seen. However, as discussed in Section 51.6, some pathogens have developed mechanisms to inactivate this response.

Natural Killer Cells. Cells that have been infected with a virus must be destroyed. That is the role of *natural*

killer (NK) cells. NK cells are a type of *lymphocyte,* a leukocyte that carries out most of its activities in the tissues and organs of the blood and lymphatic circulatory systems (see Figure 40.17, Chapter 40). NK cells circulate in the blood and kill target host cells—not only cells that are infected with virus but also some cells that have become cancerous.

NK cells can be activated by cell surface receptors or by interferons secreted by virus-infected cells. NK cells are not phagocytes; instead, they secrete granules containing *perforin,* a protein that creates pores in the target cell's membrane. Unregulated diffusion of ions and molecules through the pores causes osmotic imbalance, swelling, and rupture of the infected cell. NK cells also kill target cells indirectly through the secretion of *proteases* (protein-degrading enzymes) that pass through the pores. The proteases trigger apoptosis (see Section 8.5h, Chapter 8). That is, the proteases activate other enzymes that cause the degradation of DNA, which, in turn, induces pathways leading to the cell's death.

How does an NK cell distinguish a target cell from a normal cell? The surfaces of most vertebrate cells contain particular *major histocompatibility complex* (*MHC*) *proteins.* You will learn about the role of these proteins in adaptive immunity in the next section. NK cells monitor the level of MHC proteins and respond differently depending on their level. An appropriately high level, as occurs in normal cells, inhibits the killing activity of NK cells. Because intracellular pathogens often inhibit the synthesis of MHC proteins in the cells they infect, these cells are recognized by NK cells. Cancer cells also have low or, in some cases, no MHC proteins on their surfaces, which makes them a target for destruction by NK cells.

STUDY BREAK

1. What are the usual characteristics of the inflammatory response?
2. What processes specifically cause each characteristic of the inflammatory response?
3. What is the complement system?
4. Why does combatting viral pathogens require a different response by the innate immune system than combatting bacterial pathogens?
5. What are the four main strategies a host uses to protect against viral infections?

51.3 Specific Defences: Adaptive Immunity

Adaptive immunity is a defence mechanism that recognizes specific molecules as being foreign and clears these molecules from the body. The foreign or abnormal molecules that are recognized may be free, as in the case of toxins, or found on the surface of a virus or cell, including pathogenic bacteria, cancer cells, virus- or pathogen-infected cells, pollen, and cells of transplanted organs. Adaptive immunity develops specifically in response to the presence of foreign molecules and therefore takes several days to become effective. This time delay to mount an adaptive response would be a significant problem in eliminating pathogens were it not for the innate immune system, which combats the invading pathogens in its nonspecific way within minutes after they enter the body.

There are two key distinctions between innate and adaptive immunity:

- innate immunity is nonspecific, whereas adaptive immunity is specific, and
- innate immunity retains no memory of exposure to the pathogen, whereas adaptive immunity retains a memory of the foreign molecule that triggered the response, enabling a rapid, more powerful response if that pathogen is encountered again.

51.3a Antigens Can Be Cleared by B Cells or T Cells

A foreign molecule that triggers an adaptive immunity response is called an **antigen** (= *anti*body *gen*erator). Antigens are macromolecules; most are large proteins (including glycoproteins and lipoproteins) or polysaccharides (including lipopolysaccharides). Some types of nucleic acids can also act as antigens, as can various large, artificially synthesized molecules.

Antigens may be *exogenous,* meaning that they enter the body from the environment, or *endogenous,* meaning that they are generated within the body. Exogenous antigens include antigens on pathogens introduced beneath the skin, antigens in vaccinations, and inhaled and ingested macromolecules such as toxins. Endogenous antigens include proteins encoded by viruses that have infected cells and abnormal proteins produced by mutated genes, such as those produced in cancer cells.

Antigens are recognized in the body by two types of lymphocytes, B cells and T cells. **B cells** differentiate from stem cells in the bone marrow (see Chapter 39). It is easy to remember this as "B for bone." However, the "B" actually refers to the *bursa of Fabricius,* a lymphatic organ found only in birds, where B cells were first discovered. After their differentiation, B cells are released into the blood and circulate throughout the body in the blood and lymphatic circulatory systems. Hematopoietic stem cells in the bone marrow migrate through the circulatory system to the thymus, where they differentiate into **T cells.** The T cells enter the circulation and peripheral lymphoid tissues, including the thymus (the "T" in "T cell" refers to the thymus).

The role of lymphocytes in adaptive immunity was demonstrated by experiments in which all of the leukocytes in mice were killed by irradiation. These mice were unable to develop adaptive immunity. Injecting lymphocytes from normal mice into the irradiated mice restored the response; other body cells extracted from normal mice and injected could not restore the response. (For more on the use of mice as an experimental organism in biology, see "Research Organisms: The Mighty Mouse and the Lowly Fruit Fly")

There are two types of adaptive immune responses: **antibody-mediated immunity** (also called *humoral immunity*) and **cell-mediated immunity.** The steps involved in the adaptive immune response are similar for antibody-mediated immunity and cell-mediated immunity:

1. **Lymphocyte encounter.** The lymphocytes encounter, recognize, and bind to an antigen.
2. **Lymphocyte activation.** The lymphocytes are activated by binding to the antigen and proliferate by cell division to produce large numbers of clones.
3. **Antigen clearance.** The activated lymphocytes are responsible for clearing the antigen from the body.
4. **Development of immunological memory.** Some of the activated lymphocytes differentiate into **memory cells,** which circulate in the blood and lymph, ready to initiate a rapid immune response on subsequent exposure to the same antigen.

These steps are explained in more detail in the following discussions of antibody-mediated immunity and cell-mediated immunity.

51.3b Immunity Can Be Mediated by Antibodies or Antibody-Mediated Immunity

An adaptive immune response begins as soon as an antigen is encountered in the body and is recognized as foreign.

Antigen Encounter and Recognition by Lymphocytes. Exogenous antigens are encountered by lymphocytes in the blood or lymphatic systems. As already mentioned, the two key lymphocytes that recognize antigens are B cells and T cells. Each B cell and each T cell is specific for a particular antigen, meaning that the cell can bind to only one particular molecular structure. The binding is so specific because the plasma membrane of each B cell and T cell is studded with thousands of identical receptors for the antigen; in B cells, they are called **B-cell receptors (BCRs),** and in T cells, they are called **T-cell receptors (TCRs) (Figure 51.4, p. 1275).** The populations of B cells and T cells contain cells capable of recognizing any antigen, and each antigen can be recognized by multiple cells. For example, each of us has about 10 trillion B cells that collectively have about 100 million different kinds of BCRs. And all of these cells are present *before* the body has encountered the antigens.

Experimental Research: The Mighty Mouse and the Lowly Fruit Fly

The house mouse (*Mus musculus*) **(Figure 1a)** and its cells have been used as models for research on mammalian developmental genetics, immunology, and cancer and have enabled scientists to carry out experiments that would not be practical or ethical with humans. Mice are small, are easy to maintain in the laboratory, and have been used extensively as experimental animals. Gregor Mendel, the founder of genetics, kept mice as part of his studies. More recently, mouse genetic experiments have revealed more than 500 mutants that cause hereditary diseases, immunological defects, and cancer in mammals, including humans. The mouse has also been the model used to introduce and modify genes through genetic engineering, producing giant mice by introducing a human growth hormone gene, or "knockout" mice, in which a gene of interest is rendered nonfunctional (see Chapter 15, and "Knockouts: Genes and Behaviour," Chapter 47) to determine its normal function. By knocking out mouse genes that are homologous to human genes involved in diseases such as cystic fibrosis, researchers can study human diseases in these model organisms, opening pathways to cure human genetic diseases. In 2002, the sequence of the mouse genome was published, enabling researchers to refine and expand their use of the mouse as a model organism for studies of mammalian biology and mammalian diseases.

Similarly, the fruit fly (*Drosophila melanogaster*) (Figure 1b) has become a major organism to study basic genetics, aspects of gene regulation, developmental biology, and especially the role of innate immunity against pathogens. This insect can be raised quickly, cheaply, and in massive numbers. *Drosophila* has been instrumental to our understanding of dorsal–ventral patterning during development and has been one of the major organisms used to identify pathways of immune signalling. Because insects do not have an adaptive immune system, the innate immune system can be studied by itself without the interaction with components of the adaptive system. Toll receptors, important in immune signalling, were first found in *Drosophila* and subsequently used to identify similar molecules in vertebrates called toll-like receptors. Mutant lines lacking specific functional genes have been generated in *Drosophila* to study the roles of these genes in all organisms. Because the signalling pathways of the innate immune pathways are highly conserved, information learned on how fruit flies recognize and eliminate pathogens can be transferred to similar studies in other organisms. The genome of *Drosophila* was completed in 2000, allowing for comparisons among and between the genomes of vertebrates and invertebrates.

The mighty mouse and the common fruit fly have provided researchers with amazing amounts of information to understand how our bodies work, how similar genes function in different groups of animals, and how we can apply what we learn about one animal to another.

a.

lostbear/Shutterstock.com

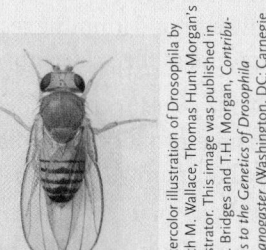

b.

Watercolor illustration of Drosophila by Edith M. Wallace, Thomas Hunt Morgan's illustrator. This image was published in C.B. Bridges and T.H. Morgan, *Contributions to the Genetics of Drosophila melanogaster* (Washington, DC: Carnegie Institution; 1919), CIW publication #278.

Figure 1
Two common research organisms: a mouse and a fruit fly.

The binding between antigen and receptor is an interaction between two molecules that fit together like an enzyme and its substrate. A given BCR or TCR typically does not bind to the whole antigen molecule but to small regions of it called **epitopes** or *antigenic determinants*. Therefore, several different B cells and T cells may bind to the population of a particular antigen they encounter.

BCRs and TCRs are encoded by different genes and thus have different structures (see "The Generation of Antibody Diversity"). When the BCR on a naive B cell matches a detected antigen, it is activated and may differentiate into a plasma cell that proliferates and secretes antibodies that recognize the same antigen (see Figure 51.4a).

As you will learn in more detail, an antibody molecule is a protein consisting of four polypeptide chains. At one end is a region that embeds in a plasma membrane, whereas at the other end are two identical *antigen-binding sites,* regions that bind to a specific antigen. TCRs are simpler than BCRs, consisting of a protein made up of two different polypeptides (see Figure 51.4b). Like BCRs, TCRs have an antigen-binding site at one end and a membrane-embedded region at the other end.

Antibodies. Antibodies are the core molecules of antibody-mediated immunity. Antibodies are large, complex proteins that belong to a class of proteins known as *immunoglobulins* (Ig). Each antibody molecule consists of four polypeptide chains: two identical **light chains** and two identical **heavy chains** about twice or more the size of the light chain (see Figure 51.4a). The chains are held together in the complete protein by disulfide (–S–S–) linkages and fold into a Y-shaped structure. The bonds between the two arms of the Y form a hinge that allows the arms to flex independently of one another.

a. B-cell receptor (BCR)

Identical antigen-binding sites

Identical light chains

V

V

C

Light chain

C

S-S

S-S

C

C

S-S

C

C

V

V

C

Disulfide linkage

Identical heavy chains

Heavy chain

KEY

V = variable region

C = constant region

Plasma membrane

Transmembrane domains

b. T-cell receptor (TCR)

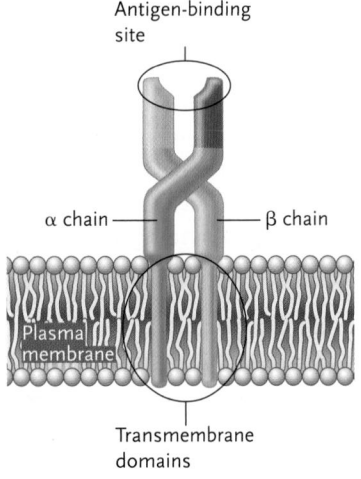

Antigen-binding site

α chain

β chain

Plasma membrane

Transmembrane domains

Figure 51.4
(a) Antigen-binding receptor on a B cell and the arrangement of light and heavy polypeptide chains in the antibody molecule. As shown, two sites, one at the tip of each arm of the Y, bind the same antigen. **(b)** Antigen-binding receptor on a T cell.

The Generation of Antibody Diversity

The human genome has approximately 20 thousand to 25 thousand genes, far fewer than necessary to encode 100 million different antibodies if two genes encoded one antibody, one gene for the heavy chain and one for the light chain. The great diversity in antigen-binding capability of these receptors is generated in a different way from one gene per chain. During B-cell differentiation, the DNA segments that encode parts of the light and heavy chains undergo three rearrangements. The genes for the two different subunits of the T-cell receptor undergo similar rearrangements. The Nobel Prize for Medicine or Physiology in 1987 was awarded to Dr. Susumu Tonegawa for discovering the genetic basis for the generation of antibody diversity.

The light chain expressed by an undifferentiated B cell is encoded by three types of DNA segments, and one of each type is needed to make a complete, functional light-chain gene. In humans, about 40 different V segments encode most of the variable regions of the chain, 5 different J (joining) segments encode the rest of the variable region, and only 1 copy of the segment makes up the constant (C) part of the chain. Thus, a complete light chain comprises one V segment, adjacent to one J segment, adjacent to the C region, which is the same for all light chains regardless of V or J segment usage (see Figure 51.4).

During B-cell differentiation, a DNA rearrangement occurs in which one random V segment and one random J segment join with the C segment to form a functional light-chain gene. During this assembly, there is a deletion of DNA between the V and J segments, and the positions at which the DNA breaks and rejoins in the V- and J-joining reaction occur randomly over a distance of several nucleotides, which

adds greatly to the variability of the final gene assembly. The DNA between the J segment and the C segment becomes an intron in the final assembled gene. Transcription of this newly assembled gene produces a typical pre-mRNA molecule (see Chapter 13). The introns are removed during the production of the mRNA by RNA processing. Translation of the mRNA produces the light chain with both the variable and the constant regions.

The assembly of functional heavy-chain genes occurs similarly. However, whereas light-chain genes have one C segment, heavy-chain genes have five types of C segments, each of which encodes one of the constant regions of IgM, IgD, IgG, IgE, and IgA. The inclusion of one of the five C-segment types in the functional heavy-chain gene therefore specifies the class of antibody that will be made by the B cell.

Each polypeptide chain of an antibody molecule has a *constant region* and a *variable region*. Each antibody type has the same amino acid sequence in the constant region of the heavy chain and likewise for the constant region of the light chain. The variable regions of both the heavy and the light chains, by contrast, have different amino acid sequences for each antibody molecule in a population. Structurally, the variable regions are the top halves of the polypeptides in the arms of the Y-shaped molecule. The three-dimensional folding of the heavy chain and light chain variable regions of each arm creates the antigen-binding site. The antigen-binding site is identical on both arms of the same antibody molecule because both ends of the Y have the same amino acid sequences in their variable regions. However, the antigen-binding sites are different from antibody molecule to antibody molecule (produced by different B-cell clones) because of the amino acid differences in the variable regions of the two chain types.

The constant regions of the heavy chains in the tail part of the Y-shaped structure determine the *class* of the antibody, that is, its location and function. Humans have five different classes of antibodies: IgM, IgG, IgA, IgE, and IgD **(Table 51.2)**.

IgM antibodies are the first antibodies produced in the early stages of an antibody-mediated response after BCRs are activated and B cells differentiate into plasma cells. When they bind an antigen, IgM antibodies activate the complement system and stimulate the phagocytic activity of macrophages.

IgG antibodies circulate in the highest concentration in the blood and lymphatic system, where they also stimulate phagocytosis and activate the complement system when it binds an antigen. IgG is produced in large amounts when the body is exposed a second time to the same antigen and can cross the placenta to provide protection to the fetus.

IgA is found mainly in body secretions such as saliva, tears, breast milk, and the mucus coating of body cavities such as the lungs, digestive tract, and vagina. In these locations, the antibodies bind to surface groups on pathogens and block their attachment to body surfaces. Breast milk transfers IgA antibodies, and thus immunity, to a nursing infant.

IgE is secreted by plasma cells of the skin and the tissues lining the gastrointestinal tract and respiratory tract. IgE binds to basophils and mast cells, where it mediates many allergic responses, such as hay fever, asthma, and hives. When its specific antigen binds to IgE, the basophils or mast cells release histamine, which triggers an inflammatory response. IgE also contributes to mechanisms that combat infection by parasitic worms.

IgD occurs with IgM as a receptor on the surfaces of B cells; its function is not well understood but may be involved in B-cell activation.

T-Cell Activation. We now follow the development of an antibody-mediated immune response by linking the recognition of an antigen by lymphocytes, the

Table 51.2		**Five Classes of Antibodies**	
Class	Structure	Location	Functions
IgM	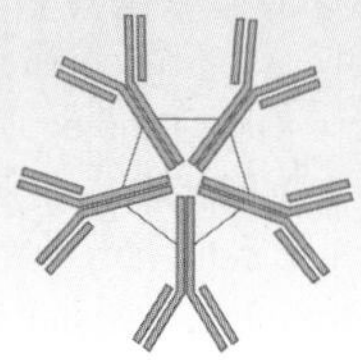	Surfaces of unstimulated B cells; free in circulation	First antibody to be secreted by B cells in primary response; when bound to antigen, promotes agglutination reaction, activates complement system, and stimulates phagocytic activity of macrophages
IgG	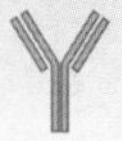	Blood and lymphatic circulation	Most abundant antibody in primary and secondary responses; crosses placenta, conferring passive immunity to fetus; stimulates phagocytosis and activates complement system
IgA	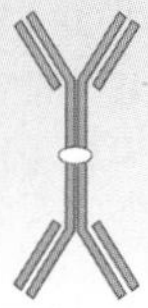	Body secretions such as tears, breast milk, saliva, and mucus	Blocks attachment of pathogens to mucus membranes; confers passive immunity for breast-fed infants
IgE	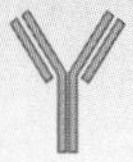	Skin and tissues lining gastrointestinal and respiratory tracts (secreted by plasma cells)	Stimulates mast cells and basophils to release histamine; triggers allergic responses
IgD	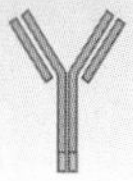	Surface of unstimulated B cells	Membrane receptor for mature B cells; probably important in B-cell activation (clonal selection)

Antibody-mediated immune response: T-cell activation

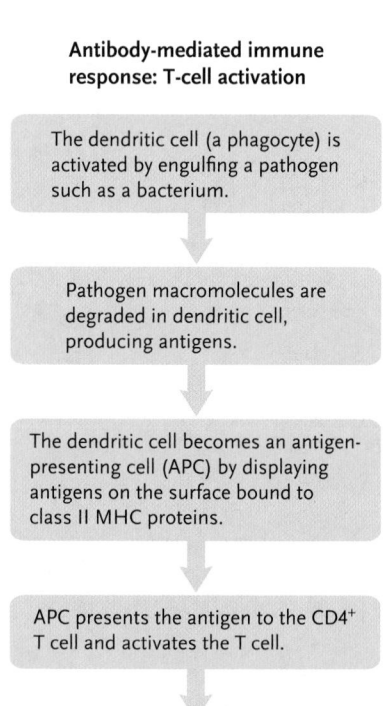

The dendritic cell (a phagocyte) is activated by engulfing a pathogen such as a bacterium.

↓

Pathogen macromolecules are degraded in dendritic cell, producing antigens.

↓

The dendritic cell becomes an antigen-presenting cell (APC) by displaying antigens on the surface bound to class II MHC proteins.

↓

APC presents the antigen to the CD4+ T cell and activates the T cell.

↓

The CD4+ T cell proliferates to produce a clone of cells.

↓

Clonal cells differentiate into helper T cells, which aid in effecting the specific immune response to the antigen.

Figure 51.5

An outline of T-cell activation in antibody-mediated immunity.

activation of lymphocytes by antigen binding, and the production of antibodies. Typically, the pathway begins when a type of T cell becomes activated and follows the steps outlined in **Figure 51.5,** which determine the fate of pathogenic bacteria that have been introduced under the skin. Circulating viruses in the blood follow the same pathway.

First, a type of phagocyte called a **dendritic cell** engulfs a bacterium in the infected tissue by phagocytosis (**Figure 51.6,** step 1). Dendritic cells are so named because they have many surface projections resembling the dendrites of neurons. They have the same origin as leukocytes and recognize a bacterium as foreign by the same recognition mechanism used by macrophages in the innate immune system. In essence, the dendritic cell connects the innate and adaptive immune systems by detecting pathogens and then translating the information so as to direct an appropriate adaptive response.

Engulfing a bacterium activates the dendritic cell; the cell now migrates to a nearby lymph node. Within the dendritic cell, the endocytic vesicle containing the bacterium fuses

with a lysosome. In the lysosome, the bacterium's proteins are degraded into short peptides, which are antigens (see Figure 51.6, step 2). The antigens bind to **class II major histocompatibility complex (MHC)** proteins (step 3), and the interacting molecules then migrate to the cell surface, where the antigen is displayed (step 4). These steps in the lymph node, which are recapped in **Figure 51.7, p. 1278,** step 1, have converted the cell into an **antigen-presenting cell (APC)**, ready to present the antigen to T cells in the next step of antibody-mediated immunity.

MHC proteins are named for the large cluster of genes encoding them, called the **major histocompatibility complex.** The complex spans 4 million base pairs and contains 128 genes. Many of these genes play important roles in the immune system. Each individual of each vertebrate species has a unique combination of MHC proteins on almost all body cells, meaning that no two individuals of a species, except identical twins, are likely to have exactly the same MHC proteins on their cells. There are two classes of MHC proteins, class I and class II, which have different functions in adaptive immunity, as we will see.

The key function of an APC is to present the antigen to a lymphocyte. In the antibody-mediated immune response, the APC presents the antigen, bound to a class II MHC protein, to a type of T cell in the lymphatic system called a **CD4+ T cell** because it has receptors named CD4 on its surface. A specific CD4+ T cell, which has a TCR with an antigen-binding site that recognizes the antigen, binds to the antigen on the APC (see Figure 51.7, step 2). The CD4 receptor on the T cell helps link the two cells together by binding to the MHC II.

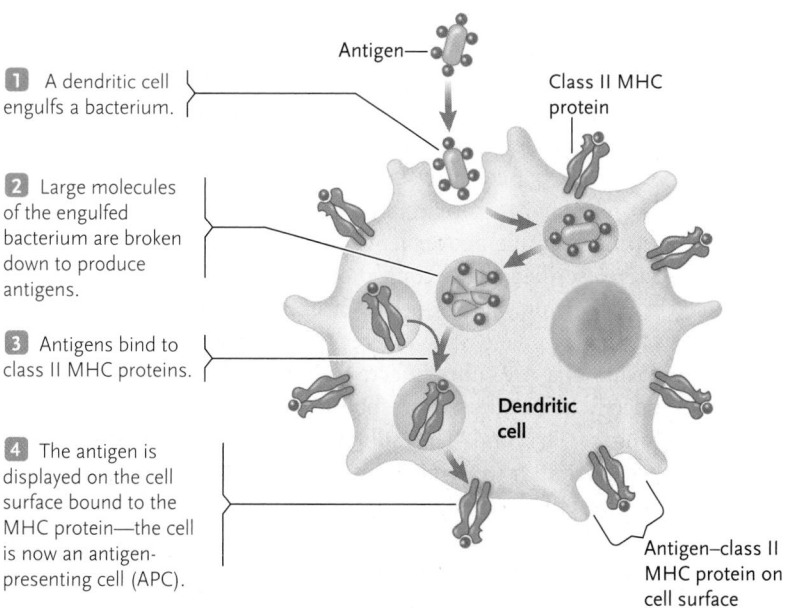

1 A dendritic cell engulfs a bacterium.

2 Large molecules of the engulfed bacterium are broken down to produce antigens.

3 Antigens bind to class II MHC proteins.

4 The antigen is displayed on the cell surface bound to the MHC protein—the cell is now an antigen-presenting cell (APC).

Antigen

Class II MHC protein

Dendritic cell

Antigen–class II MHC protein on cell surface

Figure 51.6

Generation of an antigen-presenting cell after a dendritic cell engulfs a bacterium.

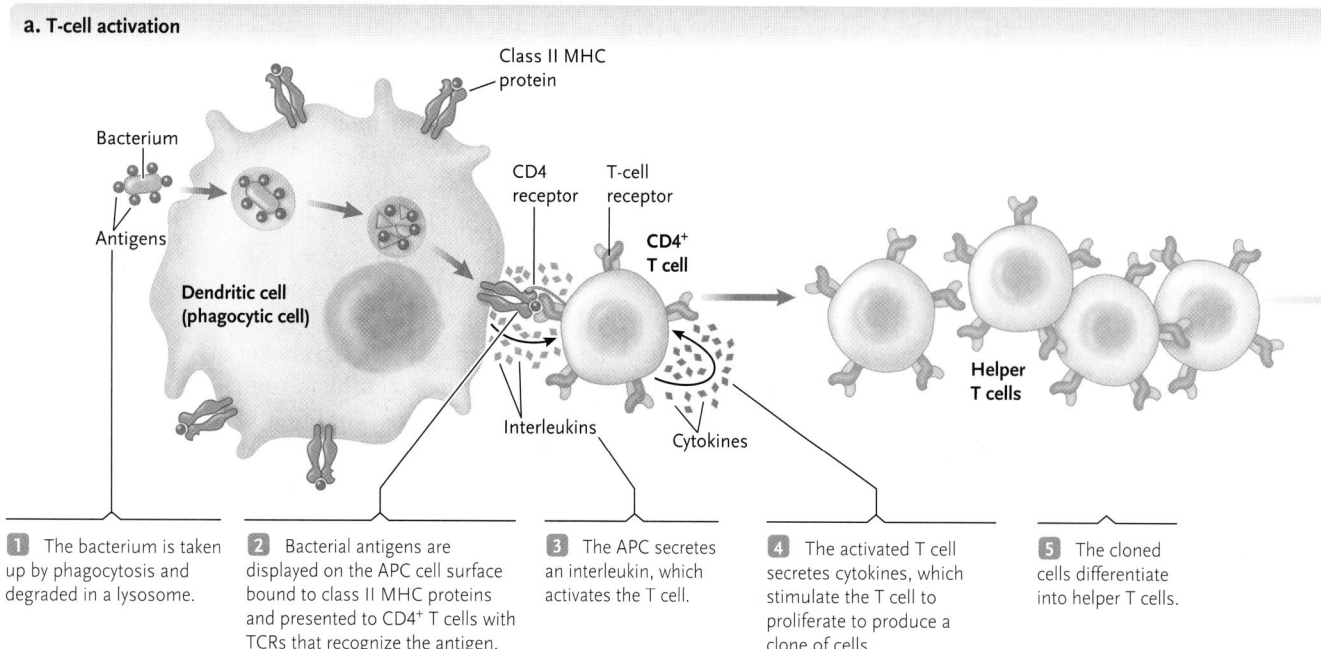

a. T-cell activation

Class II MHC protein

Bacterium

Antigens

CD4 receptor

T-cell receptor

CD4+ T cell

Dendritic cell (phagocytic cell)

Interleukins

Cytokines

Helper T cells

1 The bacterium is taken up by phagocytosis and degraded in a lysosome.

2 Bacterial antigens are displayed on the APC cell surface bound to class II MHC proteins and presented to CD4+ T cells with TCRs that recognize the antigen.

3 The APC secretes an interleukin, which activates the T cell.

4 The activated T cell secretes cytokines, which stimulate the T cell to proliferate to produce a clone of cells.

5 The cloned cells differentiate into helper T cells.

Figure 51.7
Steps involved in T-cell and B-cell activation within the antibody-mediated immune response.

When the APC binds to the CD4+ T cell, the APC secretes an *interleukin* (= between leukocytes), a type of cytokine, which activates the T cell (see Figure 51.7, step 3). The activated T cell then secretes cytokines (step 4), which act in an autocrine manner (see Chapter 43) to stimulate **clonal expansion**, the proliferation of the activated CD4+ T cell by cell division to produce a clone of cells. These clonal cells differentiate into **helper T cells** (step 5), so named because they assist with the activation of B cells. A helper T cell is an example of an **effector T cell**, meaning that it is involved in effecting—bringing about—the specific immune response to the antigen. Effector T cells can differentiate into two types of helper T-cells: Th1 and Th2 subtypes, whose functions have evolved to target different types of pathogens. Differentiation to these subtypes is done in response to signals produced by an APC during T-cell activation. In turn, the APC receives its instructions from innate PRR–PAMP interactions. In the example above, activation of an effector T-cell was described in the context of an extracellular bacterial infection. Some species of bacteria are optimally cleared by phagocytes, whereas other species produce toxins that are more effectively removed by antigen-specific antibodies. To respond to these unique challenges, Th1 and Th2 subtypes skew the adaptive response in a functional direction. Activation of Th1 T-cells enhances the phagocytic activity of macrophages, whereas Th2 cells preferentially activate B-cells and initiate antibody production.

B-Cell Activation. Antibodies are produced and secreted by B cells. The activation of a B cell that makes

the specific antibody against an antigen requires the B cell to present the antigen on its surface and then to link with a helper T cell that has differentiated as a result of encountering and recognizing the same antigen. The process is outlined in Figure 51.7, and **Figure 51.8.**

The process of antigen presentation on a B-cell surface begins when BCRs on the B cell interact directly with soluble bacterial (in our example) antigens in the blood or lymph. Once the antigen binds to a BCR, the complex is taken into the cell and the antigen is processed in the same way as in dendritic cells, being broken down into smaller fragments, culminating with a presentation of each antigen-derived peptide fragment on the B-cell surface in a complex with class II MHC proteins (see Figure 51.7, step 6).

When one of the helper T cells produced above encounters a B cell displaying the same antigen, usually in a lymph node or in the spleen, the T and B cells become tightly linked (see Figure 51.7, step 7). The linkage depends on the TCRs, which recognize and bind the antigen displayed by the class II MHC molecules on the surface of the B cell, and on CD4, which stabilizes the binding as it did for T-cell binding to the dendritic cell. The linkage between the cells first stimulates the helper T cell to secrete interleukins that activate the B cell and then stimulates the B cell to proliferate, producing a clone of these B cells with identical B-cell receptors (step 8). Some of the cloned cells differentiate into relatively short-lived **plasma cells**, which now secrete the same antibody that was displayed on the parental B cell's

b. B-cell activation and antibody production

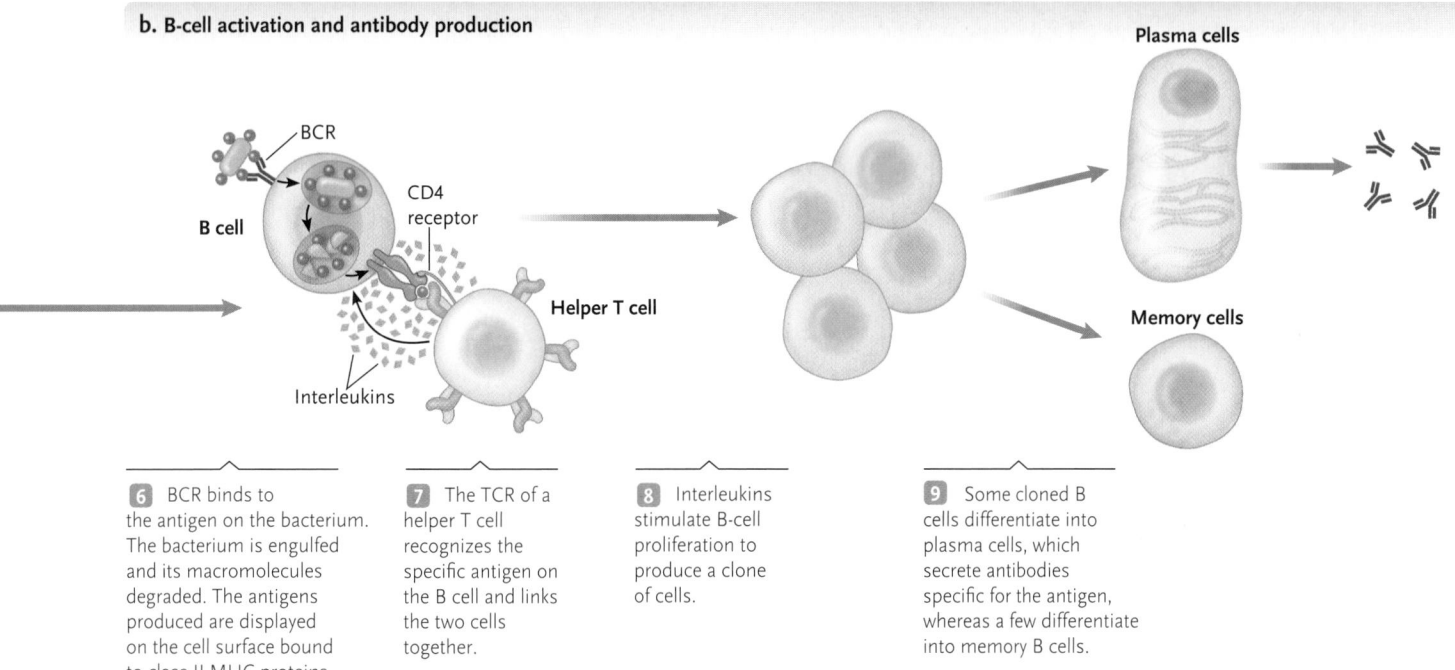

6 BCR binds to the antigen on the bacterium. The bacterium is engulfed and its macromolecules degraded. The antigens produced are displayed on the cell surface bound to class II MHC proteins.

7 The TCR of a helper T cell recognizes the specific antigen on the B cell and links the two cells together.

8 Interleukins stimulate B-cell proliferation to produce a clone of cells.

9 Some cloned B cells differentiate into plasma cells, which secrete antibodies specific for the antigen, whereas a few differentiate into memory B cells.

Antibody-mediated immune response: B-cell activation

A BCR on a B cell recognizes an antigen on the surface of a bacterium, and the bacterium is engulfed.

↓

Pathogen macromolecules are degraded in the B cell, producing antigens.

↓

The B cell displays antigens on its surface bound to class II MHC proteins.

↓

A helper T cell with TCR that recognizes the same antigen links to the B cell.

↓

The helper T cell secretes interleukins that activate the B cell.

↓

The B cell proliferates to produce a clone of cells.

↓

Some B-cell clones differentiate into plasma cells, which secrete antibodies specific to the antigen, and others differentiate into memory B cells.

Figure 51.8
An outline of B-cell activation in antibody-mediated immunity.

surface to circulate in lymph and blood. Others differentiate into **memory B cells**, which are long-lived cells that set the stage for a much more rapid response should the same antigen be encountered later in life (step 9).

Clonal selection is the process by which a lymphocyte is specifically selected for cloning when it encounters a foreign antigen from among a randomly generated, enormous diversity of lymphocytes with receptors that specifically recognize the antigen **(Figure 51.9, p. 1280)**. The process of clonal selection was proposed in the 1950s by several scientists, most notably F. Macfarlane Burnet, Niels Jerne, and David Talmage. Their proposals, made long before the mechanism was understood, described clonal selection as a form of natural selection operating in miniature: antigens select the cells recognizing them, which reproduce and become dominant in the B-cell population. Burnet received the Nobel Prize in 1960 for his research in immunology.

Clearing the Body of Foreign Antigens. How do the antibodies produced in an antibody-mediated immune response clear different types of foreign antigens from the body? Toxins produced by invading bacteria, such as tetanus toxin, can be *neutralized* by antibodies **(Figure 51.10a, p. 1280)**. The antibodies bind to the toxin molecules, inactivating them.

Antibodies bind to antigens on the surfaces of intact bacteria at an infection site or in the circulatory system. Because the two arms of an antibody molecule bind to different copies of the antigen molecule, an antibody molecule may bind to two bacteria with the same antigen. A population of antibodies can link

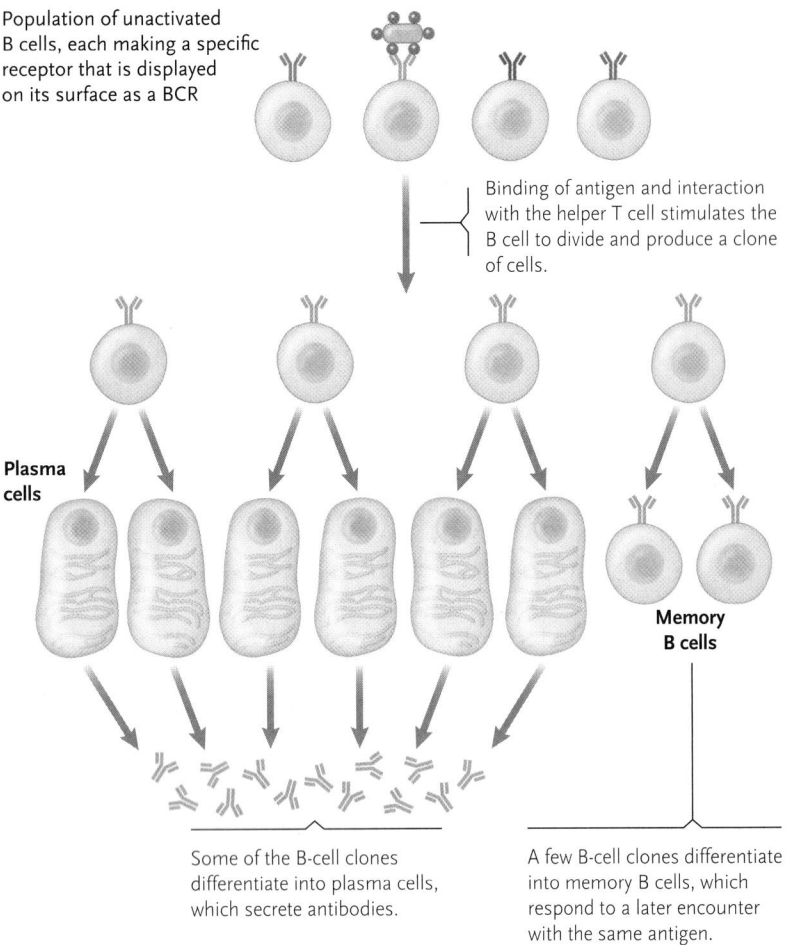

Population of unactivated B cells, each making a specific receptor that is displayed on its surface as a BCR

Binding of antigen and interaction with the helper T cell stimulates the B cell to divide and produce a clone of cells.

Plasma cells

Memory B cells

Some of the B-cell clones differentiate into plasma cells, which secrete antibodies.

A few B-cell clones differentiate into memory B cells, which respond to a later encounter with the same antigen.

Figure 51.9

Clonal selection. The binding of an antigen to a B cell that already displays a specific antibody to that antigen stimulates the B cell to divide and differentiate into plasma cells, which secrete the antibody, and memory cells, which remain in circulation ready to mount a response against the antigen at a later time.

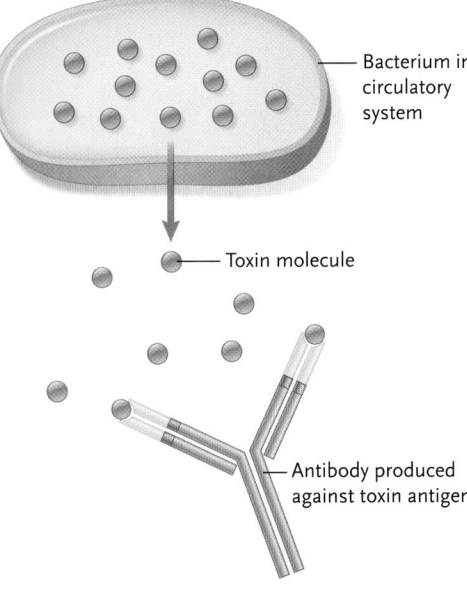

a. **Neutralization**

Bacterium in circulatory system

Toxin molecule

Antibody produced against toxin antigen

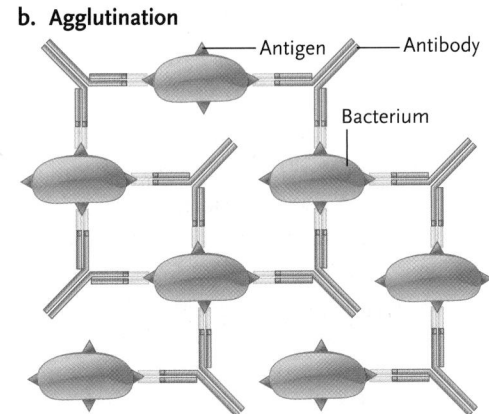

b. **Agglutination**

Antigen   Antibody

Bacterium

Figure 51.10

Examples of clearing antigens from the body.

many bacteria together into a lattice, causing *agglutination,* or clumping of the bacteria (Figure 51.10b). Agglutination immobilizes the bacteria, preventing them from infecting cells. Antibodies can also agglutinate viruses to prevent them from infecting cells.

More important, antibodies bound to antigens aid the innate immune response that was initially set off by the pathogens by stimulating the complement system. Membrane attack complexes are formed and insert themselves into the plasma membranes of the bacteria, leading to their lysis and death. In the case of virus infections, membrane attack complexes can insert themselves into the membranes surrounding enveloped viruses, which disrupts the membrane and prevents the viruses from infecting cells.

Antibodies also enhance phagocytosis of bacteria and viruses. Phagocytic cells have receptors on their surfaces that recognize the heavy-chain end of antibodies (the end of the molecule opposite the antigen-binding sites). Antibodies bound to bacteria or viruses therefore bind to phagocytic cells, which then engulf the pathogens and destroy them.

For simplicity, the adaptive immune response has been described here in terms of a single antigen. Pathogens have many different types of antigens on their surfaces, which means that many different B cells are stimulated to proliferate and many different antibodies are produced. Pathogens are therefore attacked by many different antibodies, each targeted to one antigen on the pathogen's surface.

Immunological Memory. Once an immune reaction has run its course, and the invading pathogen or toxic molecule has been eliminated from the body, division of the plasma cells and T-cell clones stops. Most or all of the clones die and are eliminated from the bloodstream and other body fluids. However, long-lived memory B cells and **memory helper T cells** (which differentiated from helper T cells) remain in an inactive state in the lymphatic system. Their persistence provides an **immunological memory** of the foreign antigen.

Immunological memory is illustrated in **Figure 51.11.** When the body is exposed to a foreign antigen for the first time, a **primary immune response** results, following the steps already described. The first antibodies appear in the blood in 3 to 14 days, and by week 4, the primary response has essentially gone away. IgM is the first antibody type produced and secreted into the bloodstream in a primary immune response. This primary immune response curve is followed whenever a new foreign antigen enters the body.

When a foreign antigen enters the body for a second time, a **secondary immune response** results (see Figure 51.11). The secondary response is more rapid than the primary response because it involves the memory B cells and memory T cells that have been stored. It does not have to initiate the clonal selection of a new B cell and T cell. Moreover, less antigen is needed to elicit a secondary response than a primary response, and many more antibodies are produced. The predominant antibody produced in a secondary immune response is IgG; the switch occurs at the gene level in the memory B cells.

Immunological memory forms the basis of vaccinations, in which antigens in the form of living or dead pathogens or antigenic molecules themselves are introduced into the body. After the immune response, memory B cells and memory T cells remaining in the body can mount an immediate and intense immune reaction against similar antigens. As mentioned in "Why It Matters," Edward Jenner introduced the cowpox virus—a virus closely related to, but less virulent than, the smallpox virus—into healthy individuals, initiating a primary immune response. After the response ran its course, a bank of memory B cells and memory T cells remained in the body, able to quickly recognize the similar antigens of the smallpox virus and initiate a secondary immune response. Similarly, the polio vaccine developed by Jonas Salk uses polioviruses that have been inactivated by exposing them to formaldehyde. Although the viruses are inactive, their surface groups can still act as antigens. The antigens trigger an immune response, leaving memory B and T cells able to mount an intense immune response against active polioviruses.

Active and Passive Immunity. **Active immunity** is the production of antibodies in response to exposure to a foreign antigen, as has just been described. **Passive immunity** is the acquisition of antibodies as a result of direct transfer from another person. This form of immunity provides immediate protection against the antigens that the antibodies recognize without the person receiving the antibodies having developed a primary immune response. Examples of passive immunity include the transfer of IgG antibodies from the mother to the fetus through the placenta and the transfer of IgA antibodies in the first breast milk fed from the mother to the baby. We can use this passive

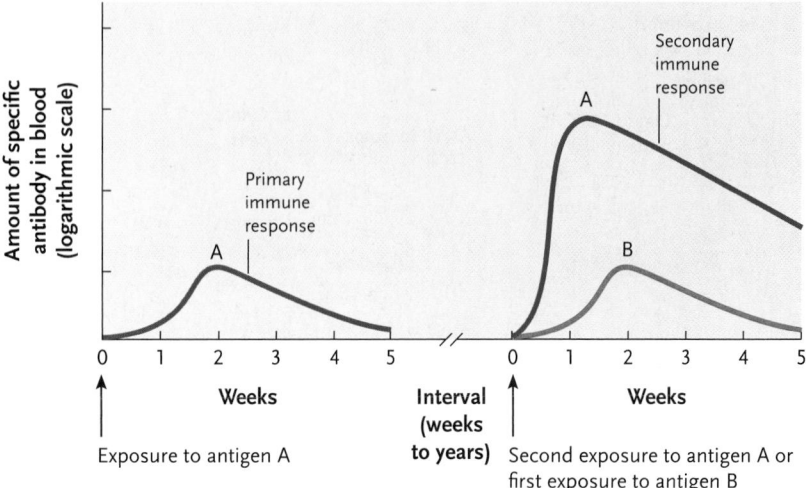

Figure 51.11
Immunological memory: primary and secondary responses to the same antigen.

immunity to our advantage by injecting intravenous immunoglobulin (IVIG) that contains IgG from many donors as treatment for specific conditions. Compared with active immunity, passive immunity is a short-lived phenomenon with no memory, in that the antibodies typically break down within a month. However, in that time, the protection plays an important role. For example, a breast-fed baby is protected until it is able to mount an immune response itself, an ability that is not present until about a month after birth.

Drug Effects on Antibody-Mediated Immunity. Several drugs used to reduce the rejection of transplanted organs target helper T cells. Cyclosporin A, used routinely after organ transplants, blocks the activation of helper T cells and, in turn, the activation of B cells. Although very successful, cyclosporin and other immunosuppressive drugs also leave the treated individual more susceptible to infection by pathogens.

51.3c Immunity Can Be Mediated by Cells

In cell-mediated immunity, cytotoxic T cells directly destroy host cells infected by intracellular pathogens **(Figure 51.12, p. 1282).** The killing process begins when some of the pathogens are broken down by cytoplasmic enzymes inside infected host cells, and the smaller protein fragments (or antigen-derived peptide fragments) themselves act as antigens. These antigens bind to class I MHC proteins, which are delivered to the cell surface by essentially the same mechanisms as in B cells (step 1). At the surface, the antigens are displayed by the class I MHC protein and the cell then functions as an APC.

The APC presents the antigen to a type of T cell in the lymphatic system called a **CD8$^+$ T cell** because it has receptors named CD8 on its surface in addition to the TCRs. The presence of a CD8 receptor distinguishes

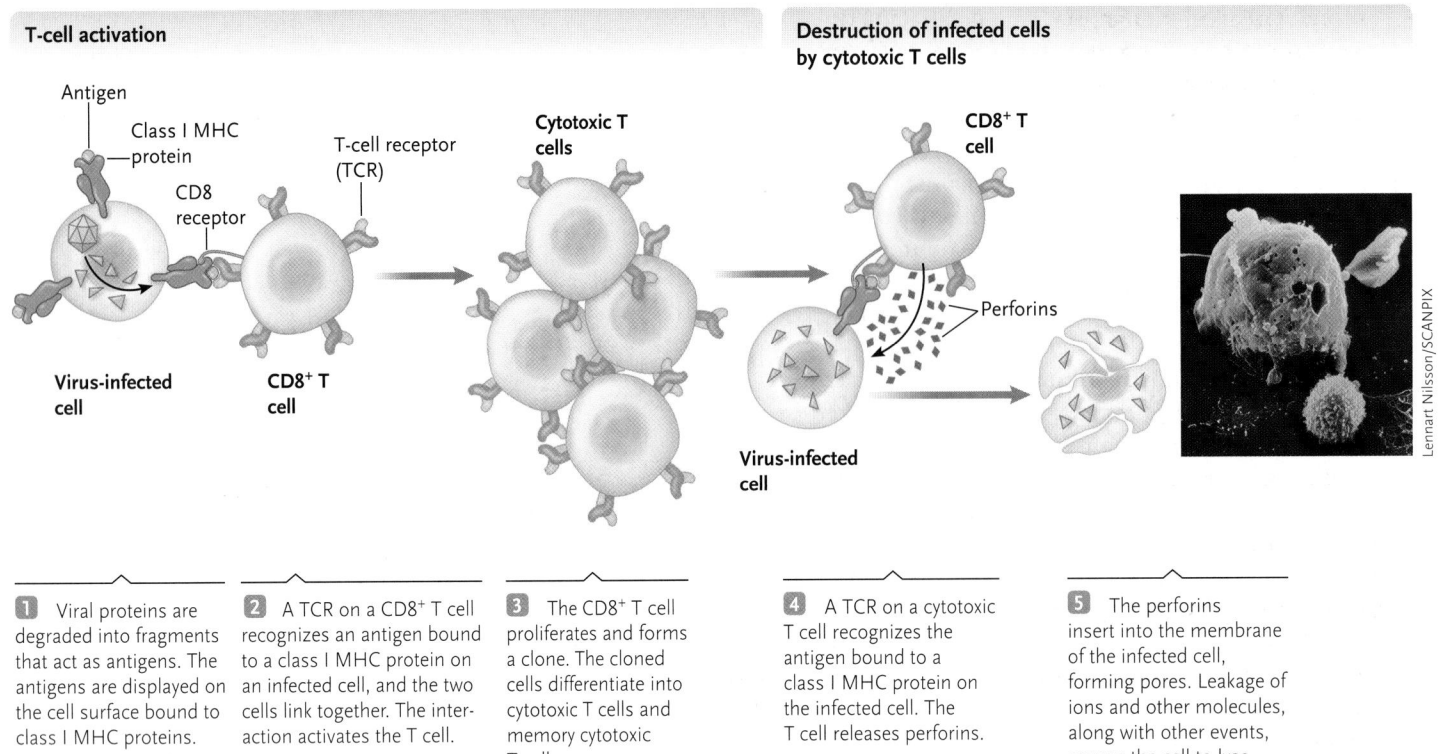

T-cell activation

Antigen

Class I MHC protein

CD8 receptor

T-cell receptor (TCR)

Cytotoxic T cells

Virus-infected cell

CD8⁺ T cell

Destruction of infected cells by cytotoxic T cells

CD8⁺ T cell

Perforins

Virus-infected cell

Lennart Nilsson/SCANPIX

1 Viral proteins are degraded into fragments that act as antigens. The antigens are displayed on the cell surface bound to class I MHC proteins.

2 A TCR on a CD8⁺ T cell recognizes an antigen bound to a class I MHC protein on an infected cell, and the two cells link together. The interaction activates the T cell.

3 The CD8⁺ T cell proliferates and forms a clone. The cloned cells differentiate into cytotoxic T cells and memory cytotoxic T cells.

4 A TCR on a cytotoxic T cell recognizes the antigen bound to a class I MHC protein on the infected cell. The T cell releases perforins.

5 The perforins insert into the membrane of the infected cell, forming pores. Leakage of ions and other molecules, along with other events, causes the cell to lyse.

Figure 51.12
The cell-mediated immune response.

Some Cancer Cells Kill Cytotoxic T Cells to Defeat the Immune System

Among the arsenal of weapons employed by cytotoxic T cells to eliminate abnormal, infected, or cancerous body cells is apoptosis mediated by the *Fas–FasL* system. Fas is a receptor that occurs on the surfaces of many cells; FasL is a ligand displayed on the surfaces of some cell types, including cytotoxic T cells. If a cell carrying the Fas receptor contacts a cytotoxic T cell with the FasL signal displayed on its surface, a cascade of internal reactions initiates apoptosis and kills the cell with the Fas receptor.

Surprisingly, cytotoxic T cells also carry the Fas receptor, so they can kill each other by displaying the FasL signal. This mutual killing plays an important role in reducing the level of an immune reaction after a pathogen has been eliminated. In the case of **immune privilege,** cells in specific regions such as the cornea, nervous tissue, and testes express FasL to

induce the apoptosis of infiltrating cytotoxic T cells and reduce inflammation.

Some cancer cells survive elimination by the immune system by making and displaying FasL and killing any cytotoxic T cells that attack the tumour. This was found first in patients suffering from malignant melanoma, a dangerous skin cancer, who had a breakdown product associated with FasL in their bloodstream.

Proteins extracted from melanoma cells, or sections made from melanoma tissue, tested positive (using antibodies) for the presence of FasL, indicating that FasL was present in the tumour cells. Similarly, the expression of FasL mRNA was also detected in the tumour cells. However, no Fas receptor was found in these samples, suggesting that Fas synthesis was turned off in the tumour cells.

FasL in melanoma cells kills cytotoxic T cells that invade the tumour, whereas the absence of Fas receptors ensures that the tumour cells do not kill each other. The presence of FasL and absence of the Fas receptor may explain why melanomas, and many other types of cancer, are rarely destroyed by the immune system.

Melanoma cells originate from pigment cells in the skin called *melanocytes*. Normal melanocytes do not contain FasL, indicating that synthesis of the protein is turned on as part of the transformation from normal melanocytes into cancer cells.

This research could lead to an effective treatment for cancer using the Fas–FasL system. If melanoma cells could be induced to make Fas as well as FasL, for example, they might eliminate a tumour by killing each other!

this type of T cell from that involved in antibody-mediated immunity. A specific CD8+ T cell that has a TCR that recognizes the antigen binds to that antigen on the APC (see Figure 51.12, step 2). The CD8 receptor on the T cell helps the two cells link together by binding to the MHC I molecule on the infected cell.

The link between the APC and the CD8+ T cell activates the T cell, which then proliferates to form a clone. In some cases, this interaction is sufficient to activate the CD8+ T-cell. Often, helper T-cells assist in the activation of CD8+ T-cells; and enhance the interaction between APCs and CD8+ cells by increasing the production of costimulatory molecules on the surface of the APC. In addition, helper T-cells secrete interleukins that drive the proliferation of the CD8+ clones. This step limits the activation of CD8+ T cells, which could inflict damage in an unregulated environment. Once activated, some of the cells differentiate to become **cytotoxic T cells** (see Figure 51.12, step 3), whereas a few differentiate into *memory cytotoxic T cells*. Cytotoxic T cells are another type of effector T cell. TCRs on the cytotoxic T cells again recognize the antigen bound to class I MHC proteins on the infected cells (the APCs) (step 4). The cytotoxic T cell then destroys the infected cell using mechanisms similar to those used by NK cells. That is, an activated cytotoxic T cell releases perforin, which creates pores in the membrane of the target cell. The leakage of ions and other molecules through the pores causes the infected cell to rupture. The cytotoxic T cell also secretes proteases that enter infected cells through the newly created pores and cause it to self-destruct by apoptosis (see Figure 51.12, step 5 and photo inset). The rupture of dead infected cells releases the pathogens to the interstitial fluid, where they are open to attack by antibodies and phagocytes.

Cytotoxic T cells can also kill cancer cells if their class I MHC molecules display fragments of altered cellular proteins that do not normally occur in the body. Another mechanism used by cytotoxic T cells to kill cells, and a process used by some cancer cells to defeat the mechanism, is described in "Some Cancer Cells Kill Cytotoxic T Cells to Defeat the Immune System."

51.3d How Antibodies Are Used in Research

The ability to generate antibodies against essentially any antigen provides an invaluable research tool for scientists. Most antibodies are obtained by injecting a molecule into a test animal such as a mouse, rabbit, or goat and collecting and purifying the antibodies from the blood. Scientists can then attach a visible marker such as a dye molecule or heavy metal atom to the antibody and determine when and where specific biological molecules are found in cells or tissues. Antibodies can also be used to "grab" a molecule of interest from

a mixture of molecules by attaching antibodies generated against that molecule to plastic beads that are packed into a glass column. When the mixture is poured through the column, the molecule remains bound to the antibody in the column. The molecules can then be obtained from the column in purified form by adding a reagent that breaks the antigen–antibody bonds.

Injecting a molecule of interest into a test animal typically produces a wide spectrum of antibodies that react with different parts of the antigen. Some antibodies may cross-react with other similar antigens, producing false results that can complicate the research. These problems have been solved by producing **monoclonal antibodies**, each of which reacts only against the same segment (epitope) of a single antigen. In addition to their use in scientific research, monoclonal antibodies are also widely used in medical applications such as pregnancy tests, screening for prostate cancer, and testing for HIV and other sexually transmitted diseases.

STUDY BREAK

1. How, in general, do the antibody-mediated and cell-mediated immune responses help clear the body of antigens?
2. Describe the general structure of an antibody molecule.
3. What is clonal selection?
4. How does immunological memory work?

51.4 Malfunctions and Failures of the Immune System

The immune system is highly effective, but it is not foolproof. Some malfunctions of the immune system cause the body to react against its own proteins or cells, producing *autoimmune diseases*. In addition, some viruses and other pathogens have evolved means of avoiding destruction by the immune system. A number of these pathogens, including HIV, even use parts of the immune response to promote infection. Another malfunction causes the *allergic reactions* that many of us experience from time to time.

51.4a The Immune System Normally Protects against Attack

B cells and T cells are involved in the development of **immunological tolerance**, which protects the body's own molecules from attack by the immune system. Although the process is not understood, molecules present in an individual from birth are not recognized as foreign by circulating B and T cells and do not elicit an immune response. During their initial differentiation in the bone marrow and thymus, any B and T cells that react with *self* molecules carried by MHC proteins become suppressed or are induced to kill themselves by apoptosis. The process of excluding self-reactive B and T cells goes on throughout the life of an individual.

Evidence that immunological tolerance is established early in life comes from experiments with mice. For example, if a foreign protein is injected into a mouse at birth, during the period in which tolerance is established, the mouse will not develop antibodies against the protein if it is injected later in life. Similarly, if mutant mice are produced that lack a given complement protein, so that the protein is absent during embryonic development, they will produce antibodies against that protein if it is injected during adult life. Normal mice do not produce antibodies if the protein is injected.

51.4b Immunological Tolerance Sometimes Fails

The mechanisms setting up immunological tolerance sometimes fail, leading to an **autoimmune reaction**— the production of antibodies against molecules of the body. In most cases, the effects of such antiself antibodies are not serious enough to produce recognizable disease. However, in some individuals—about 5 to 10% of the human population—antiself antibodies cause serious problems.

For example, *type 1 diabetes* (see Chapter 43) is an autoimmune reaction against the pancreatic beta cells that produce insulin. The antiself antibodies gradually eliminate the beta cells until the individual is incapable of producing insulin. *Systemic lupus erythematosus (lupus)* is caused by production of a wide variety of antiself antibodies against blood cells, blood platelets, and internal cell structures and molecules such as mitochondria and proteins associated with DNA in the cell nucleus. People with lupus often become anemic and have problems with blood circulation and kidney function because the antibodies, combined with body molecules, accumulate and clog capillaries and the microscopic filtering tubules of the kidneys. Lupus patients may also develop antiself antibodies against the heart and kidneys. *Rheumatoid arthritis* is caused by a self-attack on connective tissues, particularly in the joints, causing pain and inflammation. *Multiple sclerosis* results from an autoimmune attack against a protein of the myelin sheaths that insulate the surfaces of neurons. Multiple sclerosis can seriously disrupt nervous function, producing such symptoms as muscle weakness and paralysis, impaired coordination, and pain.

The causes of most autoimmune diseases are unknown. In some cases, an autoimmune reaction can be traced to injuries that expose body cells or proteins that are normally inaccessible to the immune system, such as the lens protein of the eye, to B and T cells. In other cases, as in type 1 diabetes, an invading virus stimulates the production of antibodies that can also react with self proteins. Antibodies against the Epstein-Barr and hepatitis B viruses can react against myelin basic protein, the protein attacked in multiple sclerosis. Sometimes, environmental chemicals, drugs, or mutations alter body proteins so that they appear foreign to the immune system and come under attack.

Some viruses use parts of the immune system to get a free ride to the cell interior. For example, HIV has a surface molecule that is recognized and bound by the CD4 receptor on the surface of helper T cells. Binding to CD4 locks the virus to the cell surface and stimulates the membrane covering the virus to fuse with the plasma membrane of the helper T cell. (The protein coat of the virus is wrapped in a membrane derived from the plasma membrane of the host cell in which it was produced.) The fusion introduces the virus into the cell, initiating the infection and leading to the destruction and death of the T cell. (Further details on HIV infection and AIDS are presented in "Observational Research: HIV and AIDS," p. 1286.)

51.4c Allergies Result from Overactivity of the Immune System

The substances responsible for allergic reactions form a distinct class of antigens called **allergens**, which induce B cells to secrete an overabundance of IgE antibodies **(Figure 51.13)**. IgE antibodies, in turn, bind to receptors on mast cells in connective tissue and on **basophils**, a type of leukocyte in the blood (see Table 51.1, p. 1270), inducing them to secrete histamine, which produces a severe inflammation. Most of the inflammation occurs in tissues directly exposed to the allergen, such as the surfaces of the eyes, the lining of the nasal passages, and the air passages of the lungs. Signal molecules released by activated mast cells also stimulate mucosal cells to secrete floods of mucus and cause smooth muscle in airways to constrict (histamine also causes airway constriction). The resulting allergic reaction can vary in severity from a mild irritation to serious and even life-threatening debilitation. *Asthma* is a severe response to allergens involving constriction of airways in the lungs. Antihistamines, medications that block histamine receptors, are usually effective in countering the effects of the histamine released by mast cells.

An individual is *sensitized* by a first exposure to an allergen, which may produce only mild allergic symptoms or no reaction at all (see Figure 51.13a). However, the sensitization produces memory B and T cells. At subsequent exposures, the system is poised to produce a greatly intensified allergic response (see Figure 51.13b).

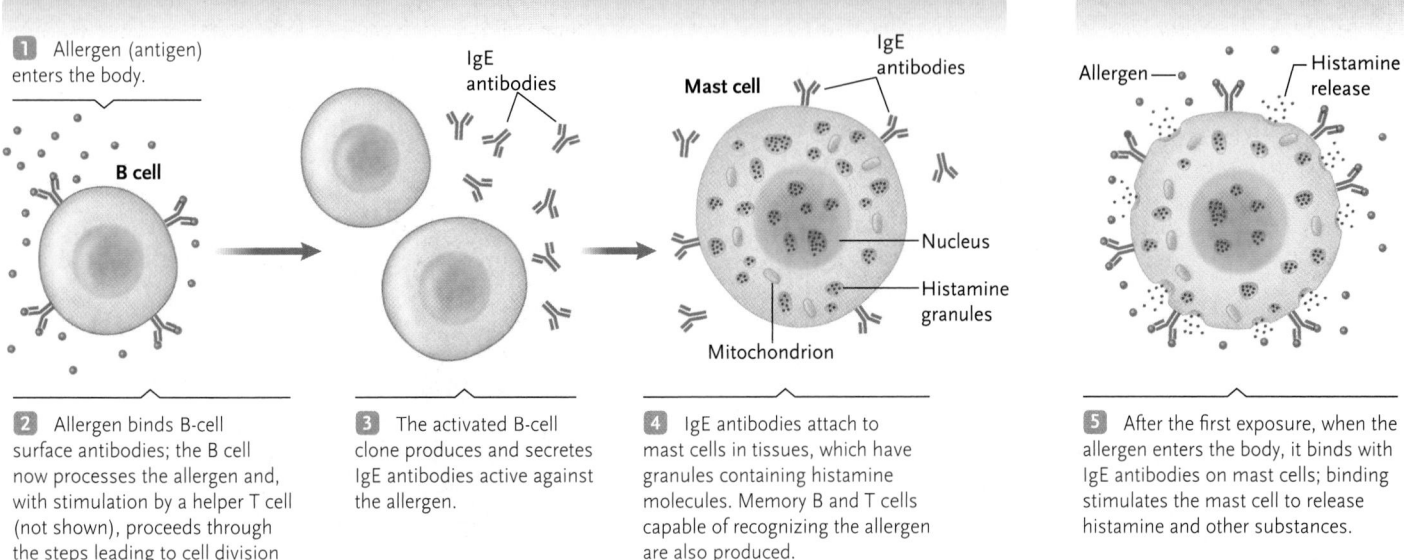

a. Initial exposure to allergen

1 Allergen (antigen) enters the body.

IgE antibodies

B cell

Mast cell

IgE antibodies

Nucleus

Histamine granules

Mitochondrion

2 Allergen binds B-cell surface antibodies; the B cell now processes the allergen and, with stimulation by a helper T cell (not shown), proceeds through the steps leading to cell division and antibody production.

3 The activated B-cell clone produces and secretes IgE antibodies active against the allergen.

4 IgE antibodies attach to mast cells in tissues, which have granules containing histamine molecules. Memory B and T cells capable of recognizing the allergen are also produced.

b. Further exposures to allergen

Allergen

Histamine release

5 After the first exposure, when the allergen enters the body, it binds with IgE antibodies on mast cells; binding stimulates the mast cell to release histamine and other substances.

Figure 51.13

The response of the body to allergens. **(a)** The steps in sensitization after initial exposure to an allergen. **(b)** Production of an allergic response by further exposures to the allergen.

In some persons, inflammation stimulated by an allergen is so severe that the reaction brings on a life-threatening condition called **anaphylactic shock**. Extreme swelling of air passages in the lungs interferes with breathing, and massive leakage of fluid from capillaries causes the blood pressure to drop precipitously. Death may result in minutes if the condition is not treated promptly. In individuals who have become sensitized to the venom of wasps and bees, for example, a single sting may bring on anaphylactic shock within minutes. Allergies developed against drugs such as penicillin and certain foods can have the same drastic effects. Anaphylactic shock can be controlled by immediate injection of epinephrine (adrenaline), which reverses the condition by constricting blood vessels and dilating air passages in the lungs.

CONCEPT FIX From reading this chapter you may believe that all parasites are bad for us. There is newer evidence that some "autoimmune" diseases are a result of our cleanliness and the relatively sterile environments in which we live, no longer exposed to dirt and parasites that used to infect us. Through the millennia we were exposed routinely to and maintained a constant fauna of parasites within our bodies. Research suggests that when parasites are present, components of the immune system, whose purpose is to reduce or suppress activities of other immune cells, are activated. When these parasites are absent, no suppression takes place, and the activated immune system attacks our own bodies. While the subject is still controversial, there are studies in which if you edit it in both places it does not make sense. People suffering from Crohn's disease, irritable bowel syndrome, inflammatory bowel disease, multiple sclerosis, and some allergies are treated with intestinal parasitic worms. The presence of the worms serves to regulate some aspects of our immune systems, and in some people reduce their symptoms, which emphasizes the mutualistic relationships and interactions between gut microorganisms and the immune system. How this works, which molecules are involved, and how the interactions between immune cells are regulated are currently under study. ⬡

STUDY BREAK

1. What is immunological tolerance?
2. Explain how a failure in the immune system can result in an allergy.

51.5 Defences in Other Organisms

All organisms must be able to defend themselves, and we can compare what we know in mammals with what we know about the immune systems of other organisms. Molecular studies in sharks and rays have revealed DNA sequences that are clearly related to the sequences coding for antibodies in mammals, and sharks produce antibodies capable of recognizing and binding specific antigens. Antibody diversity is produced by the same kinds of genetic rearrangements in both sharks and mammals, although the embryonic

Observational Research: HIV and AIDS

Acquired immune deficiency syndrome (AIDS) is a result of infection by the human immunodeficiency virus, HIV **(Figure 1),** and is defined as having a very specific level of immune deficiency. First reported in the late 1970s, HIV now infects more than 40 million people worldwide, 64% of them in Africa. AIDS is a potentially lethal disease, although drug therapy has reduced the death rate for HIV-infected individuals.

HIV is transmitted when an infected person's body fluids, especially blood or semen, enter the blood or tissue fluids of another person. The entry may occur during vaginal, anal, or oral intercourse, or the virus may be transmitted via contaminated needles shared by intravenous drug users or from infected mothers to their infants during pregnancy, birth, and nursing. HIV is rarely transmitted through casual contact; food; or body products such as saliva, tears, urine, or feces.

The primary cellular hosts for HIV are macrophages and helper T cells, which are ultimately destroyed by the virus. Infection makes helper T cells unavailable for the stimulation and proliferation of B cells and cytotoxic T cells. The assault on lymphocytes and macrophages cripples the immune system and makes the body highly vulnerable to otherwise non–life threatening infections.

In 1996, researchers confirmed the process by which HIV initially infects its primary target, the helper T cells. First, a glycoprotein of the viral coat, called *gp120,* attaches the virus to a helper T cell by binding to its CD4 receptor. Another viral protein triggers fusion of the viral surface membrane with the T-cell plasma membrane, releasing the virus into the cell. Once inside, a viral enzyme, *reverse transcriptase,* uses the viral RNA as a template for making a DNA copy. (When it is outside a host cell, the genetic material of HIV is RNA rather than DNA.) Another viral enzyme, *integrase,* then splices the viral DNA into the host cell's DNA **(Figure 2).** Once it is part of the host cell DNA,

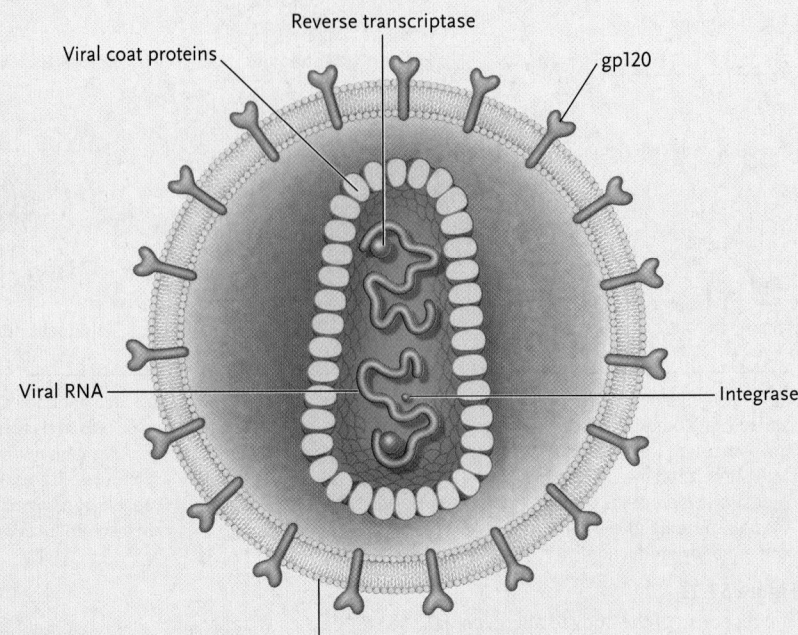

Figure 1
Structure of a free HIV viral particle.

the viral DNA is replicated and passed on as the cell divides. As part of the host cell DNA, the virus is effectively hidden in the helper T cell and protected from attack by the immune system.

When the infected helper T cell is stimulated by an antigen, the viral DNA is copied into new viral RNA molecules and into mRNAs that direct host cell ribosomes to make viral proteins. The viral RNAs are added to the viral proteins to make infective HIV particles, which are released from the host cell by budding **(Figure 3).** The infection also leads uninfected helper T cells to destroy themselves in large numbers by apoptosis, through mechanisms that are still unknown.

Initially, infected people suffer a mild fever and other symptoms that may be mistaken for the flu or the common cold. The symptoms disappear as antibodies against viral proteins appear in the body, and the number of viral particles drops in the bloodstream. An infected person may remain apparently healthy for years yet can infect others. Both the transmitter and the recipient of the virus may be unaware that the virus is present, making it difficult to control the spread of HIV.

Eventually, more and more helper T cells and macrophages are destroyed, wiping out the body's immune response. The infected person becomes susceptible to secondary, opportunistic infections, such as a pneumonia caused by a fungus (*Pneumocystis carinii*); tuberculosis; persistent yeast (*Candida albicans*) infections of the mouth, throat, rectum, or vagina; and infection by many common bacteria and viruses that rarely infect healthy humans. These infections signal the appearance of full-blown AIDS, the characteristic very low lymphocyte count and immune function that allow these opportunistic pathogens to establish and proliferate. If untreated, this results in steady debilitation and death, typically within five years.

Currently, there is no cure for HIV and no vaccine that can prevent infection. The viral envelope proteins (gp120 and gp41) of HIV mutate constantly, making a vaccine developed against one form of the virus useless when the next form appears. Most mutations occur during replication of the virus during reverse transcription of viral RNA.

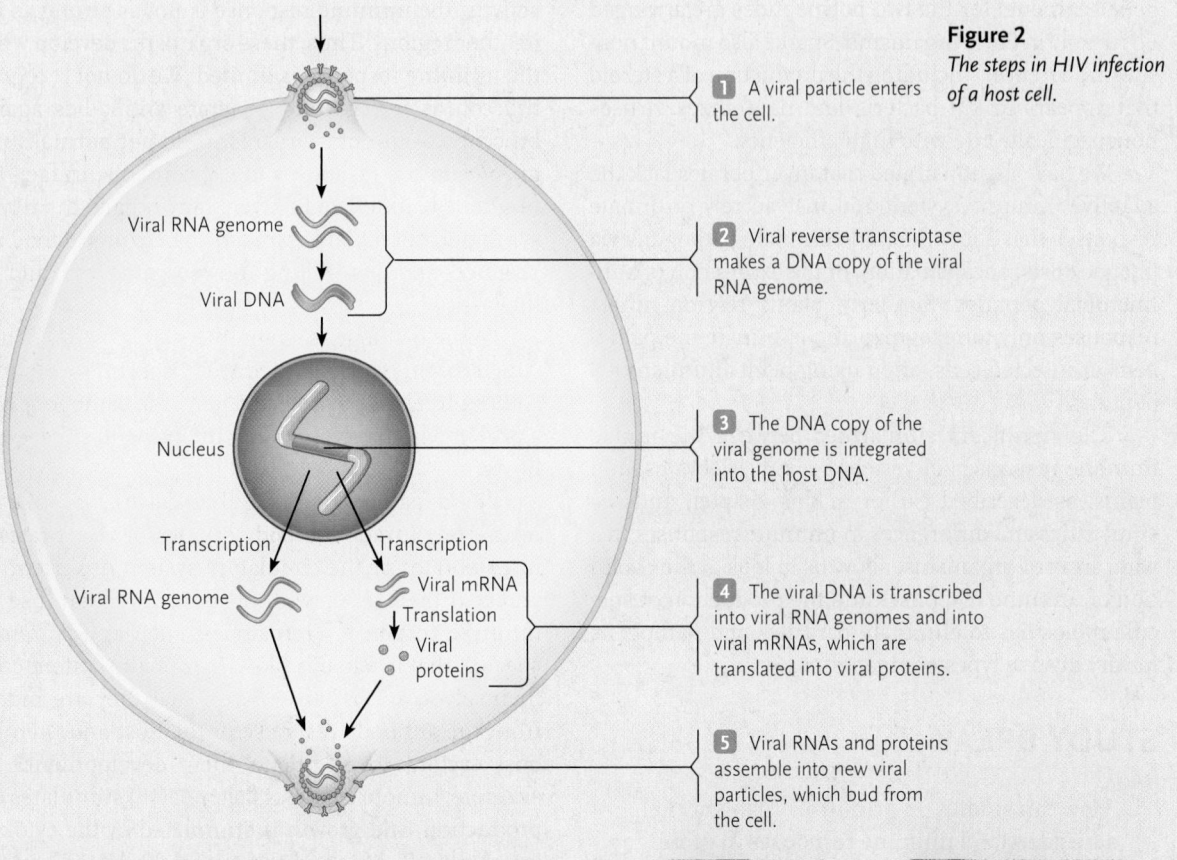

Figure 2
The steps in HIV infection of a host cell.

Viral RNA genome

Viral DNA

Nucleus

Transcription   Transcription

Viral RNA genome   Viral mRNA
Translation
Viral proteins

1 A viral particle enters the cell.

2 Viral reverse transcriptase makes a DNA copy of the viral RNA genome.

3 The DNA copy of the viral genome is integrated into the host DNA.

4 The viral DNA is transcribed into viral RNA genomes and into viral mRNAs, which are translated into viral proteins.

5 Viral RNAs and proteins assemble into new viral particles, which bud from the cell.

The development of AIDS can be greatly slowed by drugs that interfere with reverse transcription of the viral RNA. Infected people may be treated with a "cocktail" of several drugs, including *reverse transcriptase inhibitors*, which inhibit viral reproduction and destruction of helper T cells, or compounds that inhibit fusion of the virus to host cells, thus extending their lives. Drug cocktails are not a cure, however, because the virus is still present. If the therapy is stopped, the virus again replicates and the T-cell population drops.

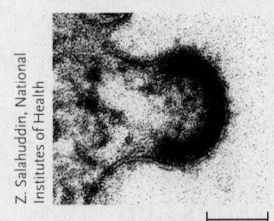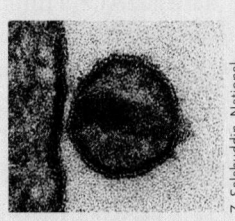

150 nm

Figure 3
An HIV particle budding from a host cell. As it passes from the host cell, it acquires a membrane coat derived from the host cell plasma membrane.

PEOPLE BEHIND BIOLOGY 51.2
Élie Metchnikoff, *Pasteur Institute*

In 1882, a Russian zoologist named Élie Metchnikoff, working in Italy, pierced a larva of the common starfish with a rose thorn. Later he observed cells collecting around the thorn. He recognized that these cells were trying to isolate and eliminate the thorn through what we now term phagocytosis. Although phagocytosis was known to occur in human bacterial infections, Metchnikoff's initial studies with the starfish, and later with animal models injected with human pathogens, demonstrated that this response was used universally by animals to eliminate pathogens. These initial experiments have led to scientific disciplines such as cellular immunology, innate immunity, and comparative immunology and have contributed to our understanding of the evolution of immune responses. For this pioneering work and keen observation, Metchnikoff shared the Nobel Prize in medicine in 1908 with Paul Ehrlich.

gene segments for the two polypeptides are arranged differently in these organisms. Sharks also mount non-specific defences, including the production of a steroid that appears to kill bacteria and neutralizes viruses nonspecifically and with high efficiency.

We have demonstrated that invertebrates lack the adaptive immune system and instead rely on innate responses that allow them to eliminate pathogens via phagocytosis, encapsulation, or the expression of anti-microbial peptides. Similarly, plants rely on innate responses only to recognize and eliminate the pathogens or infected cells, often using plant antimicrobial peptides.

The significant similarities between the innate immune responses of vertebrates, invertebrates, and plants, as described earlier in this chapter, and the similarities and differences in immune responses in a wide array of organisms, allow us to look at the evolution of immune responses and the production of specific molecules to eliminate parasites and pathogens in very diverse types of organisms.

STUDY BREAK

1. How can studies on primitive organisms help us understand immune responses in more advanced organisms such as humans?
2. Why is understanding the evolution of immune responses important?

51.6 How Do Parasites and Pathogens Circumvent Host Responses?

The immune systems of all organisms arose to recognize and eliminate pathogens. Some of the complexities of these systems have been detailed in this chapter. However, the fact that all organisms still suffer from diseases indicates that as strong as our immune systems are, pathogens and parasites are always looking for weak points to exploit.

51.6a Parasites and Pathogens Hide in, Confuse, and Manipulate the Host

Disease-causing organisms can develop in regions of the host's body that do not have a strong immune response, hide in host defence cells, confuse the immune response, or directly manipulate host responses.

Many bacteria, and parasites such as nematodes (roundworms), trematodes (flatworms), and cestodes (tapeworms), enter via the mouth, establish in some region of the alimentary tract, and allow their eggs or offspring to exit with the feces. Although the alimentary tract of most organisms has some level of immune

activity, the immune response is not as strong as it is in other regions. Thus, these organisms develop where the immune response is limited. We do not recognize broccoli as foreign and generate antibodies against broccoli in our guts. Organisms do not normally produce immune responses to gut contents; in fact, this might be counterproductive. Many organisms rely on symbiotic microbial organisms to help digest food and produce vitamins. Killing these organisms could kill the host.

Some pathogens enter the cerebrospinal fluid, which often shows a reduced immune response. Although relatively protected here, the pathogens have a problem in dispersing their offspring to subsequent hosts.

Many pathogens make themselves look like the host. They invade the body cavities of vertebrates or invertebrates or the circulatory system of vertebrates and coat themselves with host factors to confuse the immune response. Trematodes, such as *Schistosoma* species, that live in our blood/lymphatic system cover themselves in host proteins so that they are hidden from the antibody response of the host and then use host factors to stimulate their development. For example, tumour necrosis factor (TNF) stimulates egg production, and growth is stimulated by the cytokine interleukin-7. Much of the pathology caused by this parasite is due to the immune response of the host rather than to any direct damage by the parasite.

Intracellular pathogens such as *Leishmania* species live and reproduce in the macrophages that normally eliminate pathogens, developing in the last place the host would look. They survive by producing anti-oxidant enzymes to detoxify superoxide molecules, they downregulate the signal transduction pathways that produce lethal molecules, and they produce surface glycoproteins that are refractory to host lysosomal enzymes. Therefore, the very cells the host relies on to eliminate parasites serve as incubators for these organisms. Because infected cells are considered self, they are not eliminated by other host immune responses.

Infections with the bacterium *Listeria monocytogenes,* an opportunistic food-borne bacterium that has been found to contaminate deli meats and cold cuts, soft-ripened cheese, and undercooked meats, can cause severe disease in humans. In the past few years in Canada there have been large recalls of processed (ready-to-eat) meats from manufacturers whose products were found contaminated with *L. monocytogenes.* This bacterium is ubiquitous and can continue to grow slowly even at refrigeration temperatures. Listeriosis can be very serious in pregnant women, newborns, the elderly, and individuals who have a compromised immune system due to chronic infection (e.g., HIV, cancer, diabetes). *L. monocytogenes* directly penetrates the intestinal mucosa through enterocytes by binding directly to host cells using listerial internalins and host cell adhesion factors. Alternatively, *L. monocytogenes* can invade or be phagocytosed by human macrophages that normally eliminate human pathogens. By reproducing within these cells, *L. monocytogenes* can avoid host immune molecules and may become inaccessible to antibiotics. This bacterium is also an important animal model of disease to study how bacteria invade intestinal tissues and move from cell to cell, which immune mechanisms are best at eliminating the pathogen, and which aspects of the bacterium we might target with new drugs.

Another potent bacterial disease in Canada is "flesh-eating bacteria" or necrotizing fasciitis. This infection is rare, but in severe cases can destroy the deeper layers of muscles, skin, and underlying soft tissue and can spread within the subcutaneous tissue. This disease is caused by several bacteria, such as *Streptococcus pyogenes, Staphylococcus aureus, Vibrio vulnificus, Clostridium perfringens,* and *Bacteroides fragilis,* although the most common cause is *S. pyogenes.* These bacteria usually enter the body through minor cuts or scrapes and produce toxins that directly kill host tissues, affect blood flow to the afflicted region, and break down and kill host tissues, producing expanding regions of dead tissue. While anyone can be infected and suffer symptoms, people who suffer from chronic health problems (diabetes, cancer, liver and kidney disease) or have a compromised immune system have a greater chance of developing severe illness. Once established in the body, the infection may spread rapidly and can become life-threatening through major organ failure. *S. pyogenes* has evolved numerous molecular mechanisms to modulate both the adaptive and innate immune responses by humans including the secretion of proteolytic enzymes that degrade immunoglobulins. Treatment with broad-spectrum antibiotics and in some cases surgery is required to eliminate the bacteria and remove dead tissue. Intravenous treatment with immunoglobulin (IVIG) to boost the immune system and neutralize the bacterial toxins may be used in addition to surgery and antibiotics to treat necrotizing fasciitis. In some instances amputation of a limb is required to stop the spread of infection. In 1994, Lucien Bouchard, a charismatic politician and lawyer from Quebec, suffered from necrotizing fasciitis. He was treated with antibiotics and IVIG but still had a leg amputated. His illness and treatment brought much publicity to the disease.

Many foodborne microbes also may cause us significant disease. In 2012, a meat-processing plant in Alberta was closed due to contamination of the meat by a pathogenic strain of *Escherichia coli: E. coli* O157:H7. This strain of *E. coli* may cause relatively insignificant symptoms, such as cramps, diarrhea, and dehydration, but in severe cases it can cause kidney failure and death. The very young, the elderly, and individuals with weak or compromised immune systems are most susceptible. Approximately 4000 tonnes of beef and beef products were recalled from Canadian and international markets. We may never know all the reasons and

conditions that contributed to this outbreak. Perhaps all cattle from an area were contaminated, or possibly the *E. coli* from one infected animal contaminated all the meat products during processing and was not discovered during laboratory tests. Regardless, this outbreak required a massive recall and disposal of the meat, financial losses to the companies, restrictions on the movement of Canadian beef to other countries, and the direct impact in terms of illness and death of people who ingested the contaminated meat products.

Some parasites, such as *Plasmodium* sp., which causes malaria in humans, have such complex life cycles within their host that they confuse the host's immune responses and recognition systems, allowing the parasites to evade the immune response of the host as they move from one host cell to another.

Other pathogens regularly change their surface coats to avoid recognition and destruction by the immune system. When the host develops antibodies against one version of the surface proteins, the pathogens produce different surface proteins, and the host produces new antibodies. These changes continue indefinitely, allowing the pathogens to keep one step ahead of the immune system. This strategy, called **antigenic variation**, is used by organisms such as the protozoan parasite that causes African sleeping sickness; the bacterium that causes gonorrhea; and the viruses that cause influenza, the common cold, and AIDS.

In insect vectors of viruses such as dengue, the virus manipulates the insect's immune response. Normally, intracellular pathogens are eliminated via apoptosis (see Section 51.2). The dengue virus, however, induces the expression of *inhibitors of apoptosis* (IAPs) to prevent cell degradation until it has established and replicated, after which it allows the cells to rupture, releasing virus particles that then enter new cells.

Many viruses also produce RNAi suppressors that prevent the host cells from stopping the viruses' multiplication using RNAi strategies, as described in Section 51.2.

51.6b Parasites and Pathogens Engage in Symbiotic Partnerships to Evade Host Immune Responses

Several very different groups of organisms work together symbiotically to overcome the response of a host. Whereas individually they would be eliminated by the host, the combination overwhelms the ability of the host to eliminate either organism.

Nematodes + Bacteria. Soil-living nematodes that enter the body cavity of insects are killed by the combination of encapsulation and melanization described previously, and many soil bacteria cannot enter the insects. Some nematodes have teamed up with specific bacteria in a symbiotic relationship to overcome host defences. Nematodes penetrate the insect and release bacteria that are phagocytosed but not killed. These bacteria multiply within hours and produce compounds that inactivate the insect's immune response. The nematodes feed and reproduce using the bacteria and host tissues as food. When resources for growth and reproduction become limited, millions of infective nematodes store some of these specific bacteria and leave the dead insect en masse in search of new hosts.

Wasps and Viruses. Some insect **parasitoids**, such as wasps, lay their eggs in other insects. Normally, these would be eliminated via encapsulation. However, some parasitoid wasps have formed an allegiance with a polyDNA virus. The virus is injected into the host insect along with the wasp egg. This virus inactivates the host's immune response, allowing the wasp larva to develop, mature, and emerge from the host, carrying some of the virus within its reproductive tract.

In the continuing battle between host and parasite/pathogen, there are constant interactions and feedback mechanisms. Each development by the host to eliminate the pathogen is countered by pathogen factors to ensure the pathogen's survival. And some of the mechanisms used are very ingenious!

51.7 Parasite–Host Interactions and the Successes and Limitations of Vaccines

51.7a Hosts and Parasites/Pathogens Continually Look for New Ways to Survive

The interactions between parasites and pathogens and their hosts are not static. Parasites must use their hosts, plants or animals, to survive, replicate, and continue their life cycles. The host, on the other hand, continues to develop mechanisms to eliminate the parasites and pathogens that cause disease. In general, the presence of parasites can impose selection pressures on the hosts to develop new immune responses to the parasites. In a population that is frequently exposed to a parasite, the number of individuals with parasite resistance or tolerance factors may increase, but this is not so in a population with infrequent exposure. The development of resistance through immune factors, however, may present a physiological cost to the host. Maintaining and expressing these factors may come at the expense of other processes, such as reproduction, life history traits, etc. There are trade-offs in these biological processes: where and how should limited resources and energy be allocated to maximize host fitness? There is often a difference in the life spans of parasites and their hosts. Many parasites have very fast life cycles—bacteria may replicate in hours, fungi in days, other parasites in months—but often the hosts do not reproduce for decades. Therefore, in the constant arms race between parasite and host,

the parasites have the opportunity to evolve much more rapidly than their hosts. There is a constant back and forth between host and parasite that will continue to be modified through changes in both the host and the parasite over time.

51.7b Why Do We Not Have Vaccines to Prevent All Diseases?

Since Jenner's time, vaccines have become an essential tool for controlling infectious disease. Widespread vaccination campaigns have eliminated smallpox and reduced polio to a few isolated locations. Vaccines and implementation programs have reduced the impact of measles, mumps, diphtheria, tetanus, and other diseases. However, there are many diseases for which effective vaccines do not exist. In such cases, vaccine development may be confounded by the ability of a pathogen to mutate and evade immune responses. For example, HIV-1 infects helper T cells, enabling the virus to kill cells that would normally seek out and destroy it. By incorporating its genome into host DNA, the virus is undetected by immune cells. In addition, the high mutation rate of the virus leads to exceptional variability, even within a single person, creating a pool of virus that escapes the immune response. Combined, these characteristics have impaired the development of an HIV-1 vaccine, though some promising vaccine candidates are emerging. Other infections provide their own unique challenges to vaccine development. Parasites such as Plasmodium, which causes malaria, have complex life cycles and change their surface antigens, enabling them to evade the immune system. Rhinoviruses (a.k.a. the common cold) have many different serotypes or variants, so that building a vaccine against all versions is very challenging. Some pathogens only have human hosts, so it is difficult to test vaccine candidates in a lab setting. In many of these cases, new vaccine technologies will have to emerge to make effective vaccines for these challenging pathogens.

51.7c Is Universal Vaccination Required to Eliminate Diseases?

Does every single individual have to be vaccinated for vaccines to eliminate diseases? Ideally, to eliminate specific diseases on a local or global level, everyone would be vaccinated. However, the mathematical modellers of diseases can predict what levels of vaccine coverage are required to give good protection, even if not 100% of the population can be vaccinated. The concept of herd immunity or community immunity is used to describe the measure of protection in a community where some members have not received vaccinations. When large proportions of the population are immune to a disease through previous exposure or through vaccination programs, there are few susceptible individuals remaining to maintain and amplify the infection, and a low probability that a susceptible person will come in contact with an infectious person. Essentially, if there are not enough new people becoming infected, the disease outbreak will die out on its own. This occurs naturally in many disease outbreaks in animals, humans, and plants. If this susceptible proportion of the population increases, outbreaks may return. Therefore, most vaccination programs vaccinate as many people as possible, recognizing that some individuals will remain unvaccinated, including individuals who live far away from vaccination sites, cannot receive vaccines for safety issues, are immunocompromised, have allergies to components of the vaccine, or have religious beliefs that do not allow them to be vaccinated.

STUDY BREAK

1. Compare invertebrate and mammalian immune defences.
2. How do pathogens avoid the immune responses of their hosts?
3. What mechanisms can pathogens use to avoid detection by the host?
4. How can different organisms work together to overcome host defences?
5. How can we compare immune responses between animals, and what benefit would such comparisons provide?

Review

To access course materials such as Aplia and other companion resources, please visit www.NELSONbrain.com.

51.1 Three Lines of Defence against Invasion

- Humans and other vertebrates have three lines of defence against pathogens. The first, which is nonspecific, is the barrier set up by the skin and mucus membranes.

- The second line of defence, also nonspecific, is innate immunity, an innate system that defends the body against pathogens and toxins penetrating the first line. This is the only kind of immune system found in invertebrates.

- The third line of defence, adaptive immunity, is specific: it recognizes and eliminates particular pathogens and retains a memory of that exposure so as to respond rapidly if the pathogen is encountered again. The response is carried out by lymphocytes, a specialized group of leukocytes.

51.2 Nonspecific Defences: Innate Immunity

- In the innate immune system, molecules on the surfaces of pathogens are recognized as foreign by receptors on host cells. This is common to all animals and plants. In invertebrates, this activates systems to remove the pathogen via phagocytosis, encapsulation, or the production of antimicrobial peptides. In vertebrates, the pathogen is combatted by the inflammation and complement systems.

- In both vertebrates and invertebrates, epithelial surfaces and specific tissues secrete antimicrobial peptides in response to attack by microbial pathogens. These disrupt the plasma membranes of pathogens, killing them.

- Inflammation is characterized by heat, pain, redness, and swelling at the infection site. Several interconnecting mechanisms initiate inflammation, including pathogen engulfment; histamine secretion; and cytokine release, which dilates the local blood vessels, increases their permeability, and allows for leakage into body tissues.

- Large arrays of complement proteins are activated when they recognize molecules on the surfaces of pathogens. Some complement proteins form membrane attack complexes, which insert themselves into the plasma membrane of many types of bacteria and cause their lysis. Fragments of other complement proteins coat pathogens, stimulating phagocytes to engulf them.

- Four nonspecific defences are used to combat viral pathogens: RNA interference, interferons, natural killer cells, and apoptosis.

51.3 Specific Defences: Adaptive Immunity

- Adaptive immunity, which is carried out by B and T cells, targets particular pathogens or toxin molecules.

- Antibodies consist of two light and two heavy polypeptide chains, each with variable and constant regions. The variable regions of the chains combine to form the specific antigen-binding site.

- Antibodies occur in five different classes: IgM, IgD, IgG, IgA, and IgE. Each class is determined by its constant region.

- Antibody diversity is produced by genetic rearrangements in developing B cells that combine gene segments into intact genes encoding the light and heavy chains. The rearrangements producing heavy-chain genes and T-cell receptor genes are similar. The light- and heavy-chain genes are transcribed into precursor mRNAs, which are processed into finished mRNAs, which are translated on ribosomes into the antibody polypeptides.

- The antibody-mediated immune response has two general phases: (1) T-cell activation and (2) B-cell activation and antibody production. T-cell activation begins when a dendritic cell engulfs a pathogen and produces antigens, making the cell an antigen-presenting cell (APC). The APC secretes interleukins, which activate the T cell. The T cell then secretes cytokines, which stimulate the T cell to proliferate, producing a clone of cells. The clonal cells differentiate into helper T cells.

- B-cell receptors (BCRs) on B cells recognize antigens on a pathogen and engulf it. The B cells then display the antigens. The TCR on a helper T cell activated by the same antigen binds to the antigen on the B cell. Interleukins from the T cell stimulate the B cell to produce a clone of cells with identical BCRs. The clonal cells differentiate into plasma cells, which secrete antibodies specific for the antigen, and memory B cells, which provide immunological memory of the antigen encounter.

- Clonal expansion is the process of selecting a lymphocyte specifically for cloning when it encounters an antigen from among a randomly generated, large population of lymphocytes with receptors that specifically recognize the antigen.

- Antibodies clear the body of antigens by neutralizing or agglutinating them or by aiding the innate immune response.

- In immunological memory, the first encounter of an antigen elicits a primary immune response. Later exposure to the same antigen elicits a more rapid secondary response with a greater production of antibodies.

- Active immunity is the production of antibodies in response to an antigen. Passive immunity is the acquisition of antibodies by direct transfer from another person.

- In cell-mediated immunity, cytotoxic T cells recognize and bind to antigens displayed on the surfaces of infected body cells or to cancer cells. They then kill the infected cell.

- Antibodies are widely used in research to identify, locate, and determine the functions of molecules in biological systems.

51.4 Malfunctions and Failures of the Immune System

- In immunological tolerance, molecules present in an individual at birth normally do not elicit an immune response.

- In some people, the immune system malfunctions and reacts against the body's own proteins or cells, producing autoimmune disease.

- The first exposure to an allergen sensitizes an individual by leading to the production of memory B and T cells, which cause a greatly intensified response to subsequent exposures.

- Most allergies result when antigens act as allergens by stimulating B cells to produce IgE antibodies, which lead to the release of histamine. Histamines induce the symptoms characteristic of allergies (see Figure 51.13).

51.5 Defences in Other Organisms

- Antibodies, complement proteins, and other molecules with defensive functions have been identified in all vertebrates.

- Invertebrates and plants rely on nonspecific defences, including surface barriers, phagocytes, encapsulation, melanization, and antimicrobial molecules.

51.6 How Do Parasites and Pathogens Circumvent Host Responses?

- Pathogens develop in regions of the host where the immune response is limited.

- Parasites and pathogens cover themselves with host material so that they are not recognized as nonself.

- Parasites and pathogens develop in host immune cells by inactivating the immune response.

- Some pathogens use antigenic variation to keep one step ahead of the host response.

- Some pathogens manipulate the gene expression of host molecules.

- Pathogens may work together to overcome host responses that would kill either of the participants.

Questions

Self-Test Questions

1. Viruses are controlled by
   a. $CD8^+$ T cells that bind class I MHC proteins holding viral antigens
   b. $CD4^+$ T cells that bind free viruses in the blood
   c. B cells secreting perforin
   d. antibodies that bind the viruses with their constant ends
   e. NK cells secreting antiviral antibodies

2. Which of the following is not a component of the inflammatory response?
   a. macrophages
   b. neutrophils
   c. B cells
   d. mast cells
   e. eosinophils

3. When a person resists infection by a pathogen after being vaccinated against it, this is the result of
   a. innate immunity
   b. immunological memory
   c. a response with defensins
   d. an autoimmune reaction
   e. an allergy

4. Which of the following is among the characteristics of a B cell?
   a. It has the same structure in both invertebrates and vertebrates.
   b. It recognizes antigens held on class I MHC proteins.
   c. It binds viral infected cells and directly kills them.
   d. It makes many different BCRs on its surface.
   e. It has a BCR on its surface, which is the IgM molecule.

5. Antibodies
   a. are each composed of four heavy and four light chains
   b. display a variable end, which determines the antibody's location in the body
   c. that belong to the IgE group are the major antibody class in the blood
   d. that are found in large numbers in the mucous membranes belong to class IgG
   e. function primarily to identify and bind antigens free in body fluids

6. The generation of antibody diversity includes the
   a. joining of V to C to J segments to make a functional light-chain gene
   b. choice from several different types of C segments to make a functional light-chain gene
   c. deletion of the J segment to make a functional light-chain gene
   d. joining of V to J to C segments to make a functional light-chain gene
   e. initial generation of IgG followed later by IgM on a given cell

7. An APC
   a. can be a CD8⁺ T cell
   b. derives from a phagocytic cell and is lymphocyte stimulating
   c. secretes antibodies
   d. cannot be a B cell
   e. cannot stimulate helper T cells

8. Which one of the following is a function of antibodies?
   a. deactivate the complement system
   b. neutralize natural killer cells
   c. clump bacteria and viruses for easy phagocytosis by macrophages
   d. eliminate the chance for a secondary response
   e. kill viruses inside of cells

9. Suppose Jen punctured her hand with a muddy nail. In the emergency room, she received both a vaccine and someone else's antibodies against tetanus toxin. Which of the following describes the immunity conferred here?
   a. both active and passive
   b. active only
   c. passive only
   d. first active and later passive
   e. innate

10. Medicine attempts to enhance the immune response when treating
   a. organ transplant recipients
   b. anaphylactic shock
   c. rheumatoid arthritis
   d. HIV infection
   e. type 1 diabetes

Questions for Discussion

1. HIV wreaks havoc with the immune system by attacking helper T cells and macrophages. Would the impact be altered if the virus attacked only macrophages? Explain.

2. Given what you know about how foreign invaders trigger immune responses, explain why mutated forms of viruses, which have altered surface proteins, pose a monitoring problem for memory cells.

3. Cats, dogs, and humans may develop myasthenia gravis, an autoimmune disease in which antibodies develop against acetylcholine receptors in the synapses between neurons and skeletal muscle fibres. Based on what you know of the biochemistry of muscle contraction (see Chapter 46), explain why people with this disease typically experience severe fatigue with even small levels of exertion, drooping of facial muscles, and trouble keeping their eyelids open.

NEL

CHAPTER 51 DEFENCES AGAINST DISEASE | 1293

52

© Plantography/Alamy

CampCrazy Photography/Shutterstock.com

Species at risk in Canada include the white prairie gentian and the beluga whale.

Conservation and Evolutionary Physiology

WHY IT MATTERS

The tundra in northern Canada is a barren land, lacking trees. It has short summers with almost continuous daylight and long winters of almost continuous night. Despite strong winds and winter blizzards, there is little snow, as this region is so dry that it is often referred to as a polar desert. During the short summer, life is plentiful, but during the winter, signs of life are scarce. Most birds and mammals migrate to more hospitable regions for the winter. The plants and many of the animals do remain year round, however **(Figure 52.1).** These plants and animals survive these conditions because they have accrued beneficial adaptations that allow them to do so. Tundra plants must survive freezing in winter and desiccation in summer. They must reproduce during the short window of opportunity the spring and summer provide and must depend on the few pollinators and seed-dispersal agents available in this habitat. While some plants survive over winter and are perennial, many rely on the production of seeds that can withstand freezing to produce new plants annually in the spring. Arctic insects use a similar strategy, producing eggs that can withstand freezing to produce a new generation when more favourable conditions return. Some mammals survive the winter in states of dormancy, such as the Arctic ground squirrel (*Urocitellus parryii*) and the polar bear (*Ursus maritimus*) (Figure 52.1), or on stored food and energy, such as the Arctic lemming (*Dicrostonyx torquatus*). A very few mammalian species remain active through the winter, such as the musk ox (*Ovibos moschatus*), which browses on lichens and mosses that it finds under the shallow snow. For all of these organisms, plant and animal, adaptations will have arisen in different systems in the body and, given the dependence of physiological systems on each another, over time these must have become integrated.

Figure 52.1

(a) Plant life in summer in the tundra. **(b)** Frozen insect in winter. **(c)** Arctic ground squirrels. **(d)** Polar bear mother and cub.

For example, Arctic ground squirrels can nearly double their body weight by depositing large quantities of fat quickly during short summers. Their reproductive season is also extremely short, so pregnancy, parturition, and weaning all occur quickly. The young of the year must fatten up and be ready to hibernate by the time the short summer season ends. The kidneys of these squirrels must be able to excrete metabolic wastes in a concentrated urine while conserving water, and the squirrels' nervous system must be able to regulate circadian rhythms despite the rapidly changing lengths of day and night. Importantly, when the animal enters hibernation, every cell in its body must reduce its metabolic rate in an orchestrated fashion, and during arousal from hibernation, the ground squirrels must be able to rapidly generate heat from tissues that are often only one or two degrees away from freezing. Physiological adaptation to harsh environments requires adaptations at all levels throughout the body.

52.1 Physiological Integration

Adaptations in organisms arise primarily through evolution by natural selection, although they may occur by other means (see Chapter 17). Understanding how basic physiological processes have evolved is a first step in understanding adaptation. The second step is understanding how these basic physiological systems can further evolve to deal with extreme demands. This knowledge is essential for an understanding of habitat selection and range distribution in any species. Organisms can only live in areas where they can fulfill the basic essentials of life: growing, developing, and reproducing.

Our ability to design conservation measures to protect species threatened by environmental change is based on our understanding of these physiological processes and their constraints. This understanding determines our ability to predict the consequences of

environmental change on a species' survival and of the measures that will or will not be beneficial as conservation policy. The adaptations that allow organisms to inhabit unusual environments generally come at the cost of restricting their ability to inhabit others. Understanding biodiversity and species distribution ultimately requires an understanding of organisms' overall physiology.

In the preceding chapters, we have looked at the physiological processes in plants and animals in isolation, system by system. In this chapter, we will first discuss our understanding of how these physiological processes are integrated within organisms and then present a few case histories to illustrate how we can use this information to understand the evolution of physiological processes and to inform conservation policies.

52.1a Physiological Integration within Pathways

Most physiological processes consist of multiple steps. Earlier, you learned about photosynthesis and glycolysis, two critically important biochemical pathways; you traced signalling pathways in cells and hormonal action on cells; and you learned how oxygen was transported from the environment to the mitochondria. All of these processes and pathways require multiple steps (see, e.g., Figure 49.2). Not surprisingly, adaptations that alter the rates of flow along these pathways should occur at every step along the pathway. If this does not occur, then one step will become limiting and act as a bottleneck, preventing the adaptations at other steps from having much effect. For example, there is no sense in an animal breathing harder if the oxygen is not delivered any more rapidly to the tissues. Similarly, a plant's photosynthetic rate won't increase with an increase in light availability if the rate at which a plant fixes CO_2 lags far behind the rate at which light energy is captured.

One would expect that biological processes should be optimized, such that each element in a functional pathway matches the maximal need for flow along that pathway. While this optimization may appear to be the case in principle, it frequently is not, reflecting several sets of constraints. First, integration of these systems is the result of a long series of evolutionary changes that fine-tune these activities. Each of these changes will have occurred randomly. Each potentially advantageous change will only have been selected for if it provided an advantage to the individual that carried it at that time and in that environment, and if the genetic or developmental programming of the species permitted it. This type of constraint is often termed a "historical" or phylogenetic constraint. A classical demonstration of this comes from experiments conducted by Richard E. Lenski and colleagues. In these experiments, 12 populations of the bacterium *Escherichia coli* were allowed to evolve in a nutrient medium containing small amounts of glucose that they could use as a carbon source, and large amounts of citrate, a compound they could not utilize as a carbon source. Any mutation that allowed cells to utilize citrate would provide a large selective advantage to those bacteria since the cells could then utilize the more abundant carbon source. Interestingly, the ability to utilize citrate only evolved in 1 of the 12 populations and only after more than 30 thousand generations! The ability to use citrate as a carbon source may have been the result of a very rare mutation or a series of mutations that needed to occur in a specific order. It turns out that the order in which mutations occurred did matter. The evolution of this unique trait (the ability to utilize citrate as a carbon source) was contingent on the particular history of prior mutations that occurred in each population.

Another set of constraints stems from the fact that most biochemical, morphological, and physiological systems serve multiple functions. As a result, both functions cannot be simultaneously optimized. For example, increased surface areas for gas exchange in both terrestrial plants and animals also lead to water loss, bringing about selection pressure both for and against such an increase unless other solutions exist.

52.1b Physiological Integration between Systems

Just as all steps in a pathway must be integrated for proper function, so must all physiological systems within the body. For example, for tundra or alpine plants to grow and reproduce in a short growing season, they must rapidly increase rates of photosynthesis and carbon fixation in the spring and maintain these rates throughout the summer. For this increased carbon fixation to occur, there must be matching increases in the ability to harvest sunlight and to take up carbon dioxide, water, and minerals. These require biochemical, anatomical, and physiological changes in all cells, tissues, organs, and systems of the plant body. For young salmon migrating from the freshwater streams of their birth into the salty waters of the oceans where they grow and mature, biochemical, anatomical, and physiological changes in the cells of their gills, kidneys, bladders, and digestive systems are needed to accommodate the osmotic challenges associated with the transfer from a freshwater to a marine environment.

52.1c Evolutionary Physiology

Physiology is the study of function, of how plants and animals work. Comparative physiology compares how different groups of plants and animals work. Ecological physiology examines how adaptations in physiological traits match organisms to their ecological circumstances, while evolutionary physiology investigates how organisms have come to be the way they are and how they may change in the future as the environments in which they live change. Some physiological and morphological adaptations appear to have evolved independently many times, while others appear to have evolved only once. Air breathing in fishes has arisen independently multiple times in almost every major group, while the amniotic egg that allowed terrestrial vertebrates to break free of any need to return to aquatic environments to reproduce appears to have evolved just once. Components of all of these subdisciplines (comparative, ecological, and evolutionary physiology) examine plasticity—the extent to which physiological processes are fixed, or can change as circumstances demand. The kidneys of fish, amphibians, birds, and mammals all serve similar functions, but their anatomy and functional characteristics are quite different. How did this come about? Given these anatomical and physiological differences between organisms, what would be the consequence of long-term exposure to high levels of Na^+ ions, to the presence of pollutants such as copper or mercury, to rising annual temperatures, or to increasing levels of atmospheric CO_2? Research into these and similar questions is the realm of comparative, ecological, and evolutionary physiology.

52.1d Conservation Physiology

Conservation physiology is a new subdiscipline that strives to understand the physiological basis of conservation problems and to use that information to evaluate and develop conservation strategies. It is the study of the physiological responses of organisms to environmental changes due to anthropogenic threats (i.e., human activities), threats that

PEOPLE BEHIND BIOLOGY 52.1
Claude Bernard and August Krogh

Claude Bernard was a French scientist considered by many to have initiated the science of experimental medicine in the late 1800s. August Krogh, a Danish scientist, received the Nobel Prize in medicine in 1920 for his pioneering work demonstrating that blood flow is regulated through capillaries that open and close according to the tissue's need for oxygen **(Figure 1)**. Apparently these two scientists—over 50 years apart—came to the same conclusion (now known as the Bernard/Krogh principle), "Among all animals on which physiologists may experiment, some are better suited than others, depending on the problems to be studied, and the solution of a physiological problem often depends on 'the happy choice of an animal' for experiment" (Claude Bernard, 1865).

Examples where the Bernard/Krogh principle, i.e., the choice of the right animal to study, have advanced science abound. The major neurotransmitter in the human body, acetylcholine (ACh), was first discovered in frogs. The hormone calcitonin was discovered in fish. Our fundamental understanding of the ionic basis of the action potential in nerves was developed working with the giant axon of the squid. Our early understanding of the basis of learning and memory came from work on the sea slug *Aplysia*. Both of these last two bodies of research led to Nobel prizes for the researchers involved (Alan Hodgkin and Andrew Huxley for contributions to our understanding of the action potential and Eric Kandel for contributions to our understanding of learning and memory). Major advances in cell and molecular biology currently stem from the use of "model organisms," organisms for which the genome is well understood and can be manipulated easily. These include fruit flies (*Drosophila*); the nematode worm (*Caenorhabditis elegans*); zebrafish (*Danio rerio*); and mice (*Mus musculus*), for which many genes can be eliminated or "knocked out" (see "Knockouts: Genes and Behaviour," Chapter 47). Even the ability to manipulate DNA through the process of polymerase chain reaction (PCR), a process that amplifies gene sequences by rapid exposure to high heat, takes advantage of the Bernard/Krogh principle. It uses polymerase enzymes taken from bacteria from hydrothermal vents that can better tolerate high temperatures. Much of our understanding of basic physiology comes from the use of the Bernard/Krogh principle in the comparative study of evolutionary adaptations in species with unusual adaptations.

a.

Licensed under Public domain via Wikimedia Commons—http://commons.wikimedia.org/wiki/File:Claude_Bernard.jpg#mediaviewer/File:Claude_Bernard.jpg

b.

Bain News Service—This image is available from the United States Library of Congress's Prints and Photographs division under the digital ID ggbain.32006.

Figure 1
(a) Claude Bernard. (b) August Krogh.

could disrupt their population ecology. Threats to biodiversity arising from human activities are a worldwide problem. Knowing how plants and animals work is essential for identifying which species, populations, and communities will be most vulnerable to environmental stressors and for determining how we can rebuild those populations and restore their ecosystems if they are impacted. This information will help determine priorities for conservation measures and could be used to guide policymakers and the public in protecting the environment and its inhabitants.

In the remainder of this chapter we present several case studies of plant and animal adaptations to harsh environments. Where possible, we discuss the insights these examples have provided about the evolution of basic physiological processes as well as insights into the consequences of environmental change for the abundance and distribution of these species. More often we will describe how the basic physiological processes have been adapted to deal with specific problems. The examples we could draw from are enormously abundant. This material is the source of many books, films, and TV documentaries on plants and animals across this planet. We all have our favourite examples; we have chosen a small selection of our own favourites to illustrate basic principles. A humbling thing to keep in mind is that we humans could not survive in any of these harsh environments!

STUDY BREAK

1. Give an example of a physiological pathway consisting of multiple steps other than those mentioned in this chapter.
2. Why must adaptations usually occur at every step in a pathway to alter performance?
3. What are some potential differences between the questions asked by ecological, evolutionary, and conservation physiologists?

52.2 Animal Case History #1: *Artemia*, the Brine Shrimp

Amongst the harshest environments on Earth are the hypersaline (salty) lakes that develop in regions where evaporation is high and the ions in the water slowly become concentrated. Perhaps the most familiar of the hypersaline lakes are the Great Salt Lake in Utah in the United States and the Dead Sea in the Middle East, bordering Israel and Jordan. These lakes, however, can be found on all continents. While they generally occur in tropical and subtropical regions where high evaporation rates are common due to intense solar radiation, they also occur at high altitudes in Tibet, Chile, and even Antarctica. While all hypersaline lakes have a high ionic content (i.e., high salinity), the primary salts differ tremendously. Some of these lakes are alkaline and some are acidic. The primary anion may be chloride, sulfate, or carbonate, or different combinations of these.

Some salty lakes are permanent, and some are seasonal. The size of a permanent lake is a balance between inflow (rain, snow melt, streams) and evaporative water loss. Because evaporation removes only water, leaving behind the ions from the water that flows into the lake, the salinity gradually increases, even if the lake level remains constant. The salinity of seasonal lakes may decrease during the rainy season or snow melt but increase again in the dry season.

The high salt concentrations and ionic compositions in these lakes pose severe osmotic challenges to anything that lives in them, tending to cause dehydration by drawing water out of cells. On top of these challenges, most of these lakes experience low oxygen levels and often extremes of temperature. Finally, the source of evaporation that produces these saline lakes is intense UV radiation, which also damages DNA. Not surprisingly, few organisms can survive such environments, but one that can is *Artemia,* the brine shrimp **(Figure 52.2)**. Brine shrimp are crustaceans, commonly known as sea monkeys. Most of the biodiversity in hypersaline lakes is from prokaryotic organisms, and in many cases *Artemia* is the only eukaryotic organism present. These salt-loving extremophiles are referred to as halophiles (salt-lovers). How do they manage to live in these extremes? The answer lies in multiple levels of adaptation, and in this example we will see adaptations at the molecular, cellular, organ, individual, and population level.

52.2a Adaptation at the Population Level

Artemia are believed to have evolved initially in the Mediterranean region. Over time, multiple species have evolved. Speciation was due to isolation of populations inhabiting different lakes and subsequent adaptation to conditions found in a particular lake. Each new species arose from cysts dispersed by wind or birds, in geographic isolation from each other. Six *Artemia* species reproduce sexually, found mainly in Eurasia close to the Mediterranean home of the ancestral species. Other species reproduce by parthenogenesis (asexual reproduction), two of which are found in the Americas, including Canada.

Not only do we find genetic diversity between species, but there is also a high genetic diversity within each species, heterogeneously distributed over different populations along the distribution range for that species. There is extensive population subdivision, with little gene flow among populations. Each population has acquired and retains adaptations to its local environment. This results in genetic differentiation between populations.

These variations in life history traits and reproductive mode are important adaptations to extreme environments in *Artemia,* as we shall see in the next section.

Figure 52.2
(a) A male and a female (carrying eggs) brine shrimp. **(b)** Commercial brine shrimp breeding ponds. **(c)** Nauplii emerging from encysted eggs. **(d)** A salt lake formed in a region with no outflow and extensive evaporation, the natural habitat of brine shrimp.

a.
b.
c.
d.

© blickwinkel/Alamy
© Aerial Archives/Alamy
© Nature Picture Library/Alamy
Galyna Andrushko/Shutterstock.com

52.2b Adaptations at the Individual Level

One of the most important keys to the success of brine shrimp (Figure 52.2a) is the ability of female *Artemia* to sense the stability of their environment. When the environment is stable (little change in abiotic factors over time), females produce larvae called nauplii via ovoviviparity—live birth from internal eggs (see Chapter 41). These nauplii (singular *nauplius*) are free swimming and grow and moult quickly to become adults. Like the adults, the nauplii must be able to osmoregulate effectively to maintain water balance. However, if the females perceive the environment to be changing to more stressful conditions (lack of food, low oxygen, high or low salinity, high or low temperature, or lack of available mates), instead of producing free-swimming nauplii, they produce cysts (oviparity—in this case, birth from dormant eggs, Figure 52.2c). These cysts each consist of an egg encapsulated in a coat that makes them extremely resistant to environmental stress. As the eggs enter the uterus of the female, three unique peptides are added to the eggshell coat, one to the outer layer and two to the inner layer. This coat may be one of the most environmentally resistant materials in the animal kingdom.

The cysts undergo a process called anhydrobiosis (or cryptobiosis), in which they become metabolically inactive and can survive freezing, near-boiling temperatures, and lack of oxygen in a condition of apparent lifelessness. Depending on the severity of the conditions, they may remain in this state for anywhere from two years to well over two decades. However, when placed in favourable conditions (abundant food, sufficient oxygen, and favourable salinity and temperature), the eggs hatch into nauplii within a few hours. Anhydrobiosis is also found in a wide variety of unicellular organisms (both prokaryotes and eukaryotes) and plants (many plant seeds but also some mature plants; see "Life on the Edge," Box 26.1, Chapter 26) as well as a few other invertebrates.

52.2c Adaptations at the Organ Level

Artemia live in salinities ranging from 1 mM to 5 M and at temperatures ranging from below freezing to 40°C. They can be found with other planktonic animals or alone. Their optimum environment is one in which the salinity is 1 M. In a 5 M environment, where salt begins to precipitate, they can live but not reproduce.

Both nauplii and adult brine shrimp are osmoregulators that maintain their hemolymph (extracellular fluid) at a concentration of 200–400 mOsm—considerably lower and qualitatively different (different ionic composition) from the external hyperosmotic medium. A cuticle covering their integument helps to retard water loss, but even so they continuously lose significant amounts of water by osmosis. To counter this, they drink continuously to restore body volume. This means they must eliminate excess ions across a very large concentration gradient. The salts that are taken up in the water are absorbed across the gut and transported to specialized salt glands, where they are actively pumped out of the animal.

The site of the salt glands is different in the nauplii and the adults. This gland is on the dorsal surface of the cephalothorax in nauplii and on the maxilla of the adults. Salt glands contain Na^+/K^+ ATPase in the cell membranes of the secretory cells, an enzyme that pumps sodium ions out of the shrimp against large concentration and charge gradients. Na^+/K^+ ATPase is a protein composed of two subunits, and in *Artemia*, there is a broader spatial and temporal expression of one of the genes coding for one of these subunits, enhancing the role of Na^+/K^+ ATPase. This is a metabolically demanding process, one that makes the need for oxygen acute. To assist in this, hemoglobin synthesis increases in *Artemia* as dissolved oxygen is reduced and as salinity increases. This increases the oxygen-carrying capacity of the hemolymph.

52.2d Adaptations at the Cellular Level

Studies designed to understand the biochemical and biophysical mechanisms that underlie anhydrobiosis have given rise to two hypotheses. The water replacement hypothesis maintains that the critical step in resisting dehydration is to replace water in membranes and macromolecules with sugars. In *Artemia*, the insertion of trehalose, a disaccharide of glucose, into cell membranes is believed to maintain cell membrane fluidity, protecting the cells from the destructive effects of dehydration. Trehalose is very abundant in cysts, accounting for about 15% of their dry weight. Another hypothesis maintains that hydrophilic (water-repellent) molecules enter a glassy state during desiccation, providing an inert protective matrix that immobilizes macromolecules, preventing denaturation and other structural disruptions. A protein that appears in late embryogenesis (*late embryogenesis abundant (LEA) group 1 protein*) has been proposed to play an important role in glass formation. These two hypotheses are not mutually exclusive.

52.2e Adaptations at the Molecular Level

Environmental stressors such as desiccation, anoxia, and temperature extremes all tend to disrupt protein structure and cause proteins to unfold and to aggregate. In cells that resist such stressors, much focus has been on the roles of special *stress proteins* that stabilize other proteins and help them retain their structure. (These special proteins act as molecular chaperones; see "Molecule behind Biology," Box 52.2.) Three such stress proteins have been suggested to play this role in *Artemia*: Hsp70 (heat shock protein 70), artemin, and p26. While most molecular chaperones are only produced when an organism encounters stressful situations, Hsp70, artemin, and p26 are not induced by stress but are genetically programmed in *Artemia* to appear during early development. They appear in encysted embryos proactively, in anticipation of stress.

When stress arises, half of the p26 is translocated to the nucleus as well as to the mitochondria and the endoplasmic reticulum (ER), where it stabilizes matrix proteins. Artemin appears and disappears at the same times as p26 during development. It is an RNA-binding protein that is very stable at high temperatures. It is possible that artemin is an RNA chaperone; thus, while p26 protects proteins, artemin protects RNA. In the absence of stress, p26 is freely diffusing and not associated with any cytoplasmic elements. Later in

development, nauplii stop synthesizing p26, artemin, and trehalose. Hsp70 continues to act as a chaperone that is induced under stressful situations.

Resistance to Ultraviolet Radiation. It was once felt that molecular chaperones also played a key role in protecting *Artemia* from the high ultraviolet (UV) radiation common to the environments in which they live. Intense UV light leads to altered DNA–protein relationships that impair DNA replication. It was once hypothesized that the ability of molecular chaperones to restrict and/or repair the mutations induced by UV light explained in part the differential survival observed between individuals and populations. It now appears that it is the tough shell that protects against not only mechanical damage but also UV damage.

MOLECULE BEHIND BIOLOGY 52.2

Stress Proteins

As has been covered in earlier chapters, the actions of functional proteins such as enzymes, receptors, transporters, channels, and signalling molecules all depend on their shape and how they are altered by other molecules. Their three-dimensional structures dictate their functional activity. The proper folding of newly synthesized proteins and the maintenance of this proper shape are the role of a complex network of molecular chaperones, a group of molecules that includes the *heat shock proteins* (HSPs; see Section 38.2, Chapter 38). This family of proteins **(Figure 1)** has now been observed in all cells, from yeast to the most highly developed plants and animals. They were first discovered in fruit flies as a response to accidental overheating and subsequently shown to be important for the homeostatic regulation of protein shape under conditions of heat stress, hence their name. These proteins have been shown to be synthesized in response to many other stresses, including physiological stresses encountered during cell division and differentiation, responses to toxic substances, responses to radiation, and responses to inflammation and aging. Because protein molecules are highly dynamic, constant monitoring by the molecular chaperones is required to ensure proper protein function. Recent evidence indicates that reductions in the stress response are involved in aging and aging-related diseases such as Alzheimer's, Parkinson's, and Huntington disease.

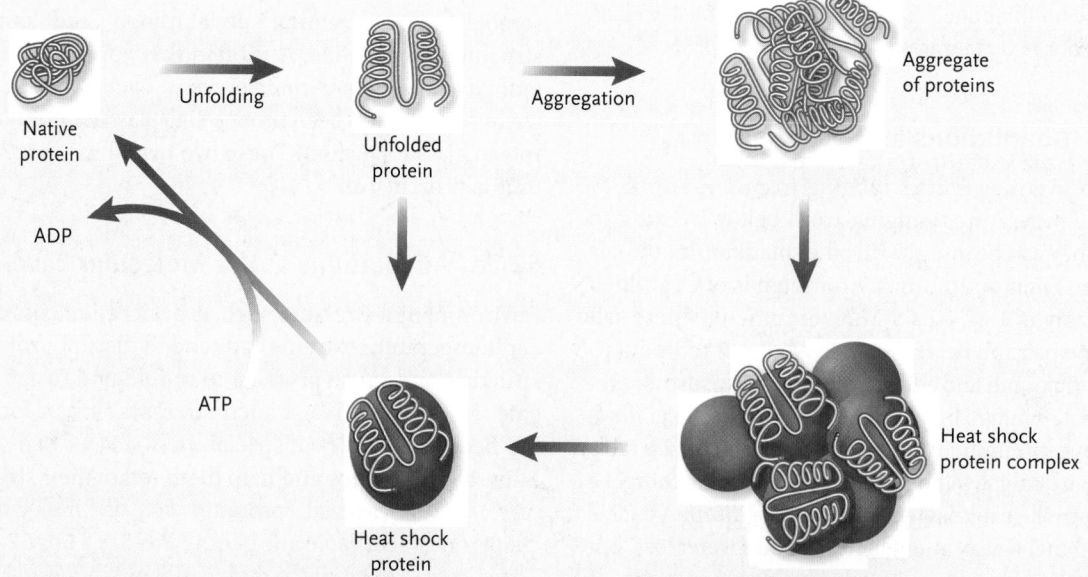

Figure 1
Schematic diagram illustrating how HSPs bind to native proteins to maintain and protect their structure.

52.2f Multiple Levels of Adaptation

In this one example, the brine shrimp *Artemia,* we see adaptations at the population level (populations of the same species exhibiting unique adaptations specific to their local environments), the individual level (females being able to sense stable versus unstable environments), the organ level (evolution of a resistant epithelial coat and the presence of specialized active transport pumps in the salt gland), the cellular level (stabilization of the cell membranes of the cysts), and the molecular level (role of molecular chaperones in protecting proteins).

52.3 Animal Case History #2: The Bar-Headed Goose: Sustained Flight at High Altitude

There are many environments on the planet where available oxygen is limited. Examples of these hypoxic environments are marine tide pools and tropical ponds where during the night, plant respiration exceeds oxygen production by photosynthesis. Frozen lakes in winter and salt marshes are generally low in oxygen. In the terrestrial realm, the most common examples of oxygen-poor environments are animal burrows and high altitudes (see Chapter 41). Many organisms are highly adapted to live in these oxygen-poor environments; that is, they are able not only to tolerate extremely hypoxic conditions but also to perform in them. Within the aquatic world, the champions of hypoxic survival among vertebrate animals include the crucian carp (*Carassius carassius*) and the painted turtle (*Chrysemys picta*), two species that can survive for months with no available oxygen (anoxia). The mole rat (*Heterocephalus glaber*) is a burrower that is one of the most hypoxia tolerant of all terrestrial species. Species of birds and mammals adapted to live at altitude are also very hypoxia tolerant. Among these, a species that has caught the attention and imagination of scientists is the bar-headed goose.

52.3a The Bar-Headed Goose

Bar-Headed Geese, *Anser indicus* (**Figure 52.3**), breed in selected wetlands on the high plateaus of central Asia (Figure 52.3a). Between 25 and 50% of the world population of these geese overwinter in India on the southern side of the Himalayan mountain range. As a result, twice a

a.

Ainars Aunins/Shutterstock.com

b.

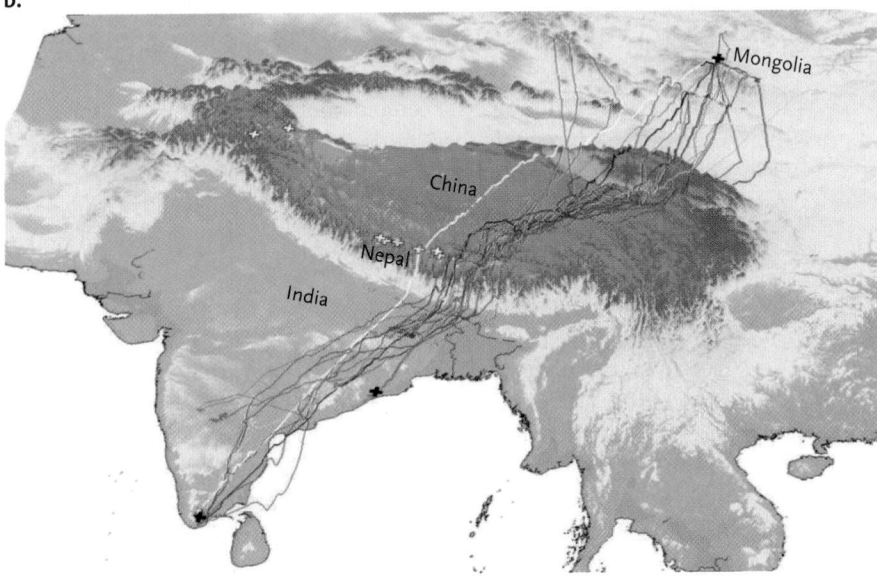

L. A. Hawkes, S. Balachandran, N. Batbayar, P. J. Butler, B. Chua, D. C. Douglas, P. B. Frappell, Y. Hou, W. K. Milsom, S. H. Newman, D. J. Prosser, P. Sathiyaselvam, G. R. Scott, J. Y. Takekawa, T. Natsagdorj, M. Wikelski, M. J. Witt, B. Yan, C. M. Bishop, "The paradox of extreme high-altitude migration in bar-headed geese *Anser indicus,*" *Proceedings A,* Copyright © 2014, The Royal Society, by permission of the Royal Society.

Figure 52.3

(a) Bar-Headed Geese. **(b)** Three-dimensional map showing release locations (black crosses) of Bar-Headed Geese in India (n. 2 sites) and Mongolia (n. 1 site). Coloured lines represent 16 individual geese, and coloured background shading indicates elevation. Solid thick white line shows the great circle route. White crosses show locations of the 14 "eight-thousanders" (the world's highest mountains, each over 8000 m in elevation). **(c)** Cross-section of land elevation followed by Bar-Headed Goose during their northward migration from their Indian wintering grounds (left side of plot) to their Chinese and Mongolian breeding grounds (right side of plot).

year these birds must fly over the world's highest mountains. While there have been unconfirmed reports of them flying over Mount Everest (9000 m above sea level), data from tracked birds indicate that most travel through the mountain passes between the valleys of the Himalayas and the Tibetan Plateau (Figure 52.3b). These passes, however, are still above 5000 m, and there are some confirmed observations of birds flying just above 7000 m. Air at 5000 m contains only 50% of the oxygen at sea level, while that at 7000 m contains only 40%. Most of us are barely able to walk slowly under these conditions. Flight, however, is one of the most energetically demanding forms of locomotion, and the challenges of climbing flight in low-density air are even more metabolically expensive. Most impressive is that on their northward migration to breed in the spring, the geese cross over the Himalayas in one day, starting from near sea level and ascending rapidly to altitude with no acclimatization. This is truly amazing, as climbers that go to elevation must ascend slowly, allowing their bodies to adjust, or they encounter acute mountain sickness or, worse, pulmonary or cerebral edema and death. Even more surprisingly, these birds generally make the crossing during the still of night, using their own aerobic power and avoiding the thermal updrafts created as air warms and rises during the day!

52.3b Anatomical Adaptations

Geese are relatively large, heavy birds. One might expect to see anatomical adaptations in the wing structure of those that fly at high elevations to promote lift in thin air. The data suggest, however, that while for their size, the wingspan (wing length, **Figure 52.4a**) of Bar-Headed Geese is at the top end of the distribution seen for geese in general, and the wing loading (a function of the width and surface area of the wing and weight of the bird) is at the low end, these features are not outside the range of distribution seen in low-altitude species. As a result, their wings do not provide more lift, and thus the wingbeat frequencies of Bar-Headed Geese during steady flight are also roughly the same as those of low-altitude geese. This lack of anatomical adaptation in the wing emphasizes the importance of physiological adaptations for increasing oxygen transport in this species.

52.3c Physiological Adaptations

Lack of Metabolic Suppression. In many hypoxia-adapted animals, oxygen supply and oxygen demand are balanced, at least in part, by reducing oxygen demand. This is achieved by a combination of physiological and biochemical responses that depress metabolism both at the level of physiological control systems (usually by reducing body temperature (T_b) and thus metabolism and the need for oxygen) and at the level of individual cells. This would not be a good strategy

for birds trying to sustain flight at altitude; thus, not surprisingly, data suggest that the falls in T_b and metabolic rate under hypoxic conditions in Bar-Headed Geese are minimal as a result of evolutionary adaptations that enhance O_2 transport.

Enhanced Ventilatory Responses. When exposed to low-oxygen environments, Bar-Headed Geese increase their total ventilation significantly more than low-altitude species of geese. This is entirely due to greater increases in tidal volume (breath size) rather than breathing frequency. This is a more effective way of increasing breathing, so O_2 uptake from the lungs into the blood is much greater than that seen in other birds. To achieve these larger breaths, Bar-Headed Geese must be capable of generating higher inspiratory airflows.

Enhanced Lung Diffusion. The lungs of all birds have several characteristics that provide them with greater hypoxia tolerance. In mammals, pulmonary blood vessels normally constrict in response to low O_2. This strategy diverts blood flow from poorly oxygenated areas of the lungs toward better-oxygenated regions. During hypoxia, when low O_2 occurs throughout the lungs, a massive constriction seen in the lungs (hypoxic pulmonary vasoconstriction) results in increased pulmonary blood pressure that leads to fluid exuding from the capillaries into the alveoli (pulmonary edema). Collection of fluid in the lungs impairs gas exchange. Birds do not exhibit hypoxic pulmonary vasoconstriction. Increased pulmonary blood pressure is also less likely to cause edema in birds than in mammals, because the avian blood–gas barrier is mechanically stronger. Finally, as described in Chapter 49, the unique cross-current arrangement of air and blood capillaries in bird lungs is inherently more efficient for gas exchange than the arrangement in mammalian lungs. Data show that this advantage actually decreases at altitude, however, and this is true for all birds. The only distinct advantage seen in Bar-Headed Geese is that, relative to their low-altitude relatives, the Bar-Headed Geese have larger lungs, giving them greater surface area for gas exchange and an ability to take deeper breaths (Figure 52.4b).

Blood Gas Transport. In general, birds have larger hearts and cardiac stroke volumes than mammals of similar body size. Under hypoxic conditions, they are capable of large increases in heart rate and stroke volume, greatly increasing the delivery of oxygen from the lungs to the tissues. Bird hearts also have increased capillarity compared to mammalian hearts. This increases the capacity for delivering oxygen to the myofibrils of the heart. This adaptation is enhanced in the Bar-Headed Goose. A greater supply of oxygen to the cardiac muscle cells of the heart results in an increased ability to sustain cardiac output during exercise in hypoxia.

The hemoglobin of the Bar-Headed Goose also has a greater oxygen affinity than that of its lower-altitude

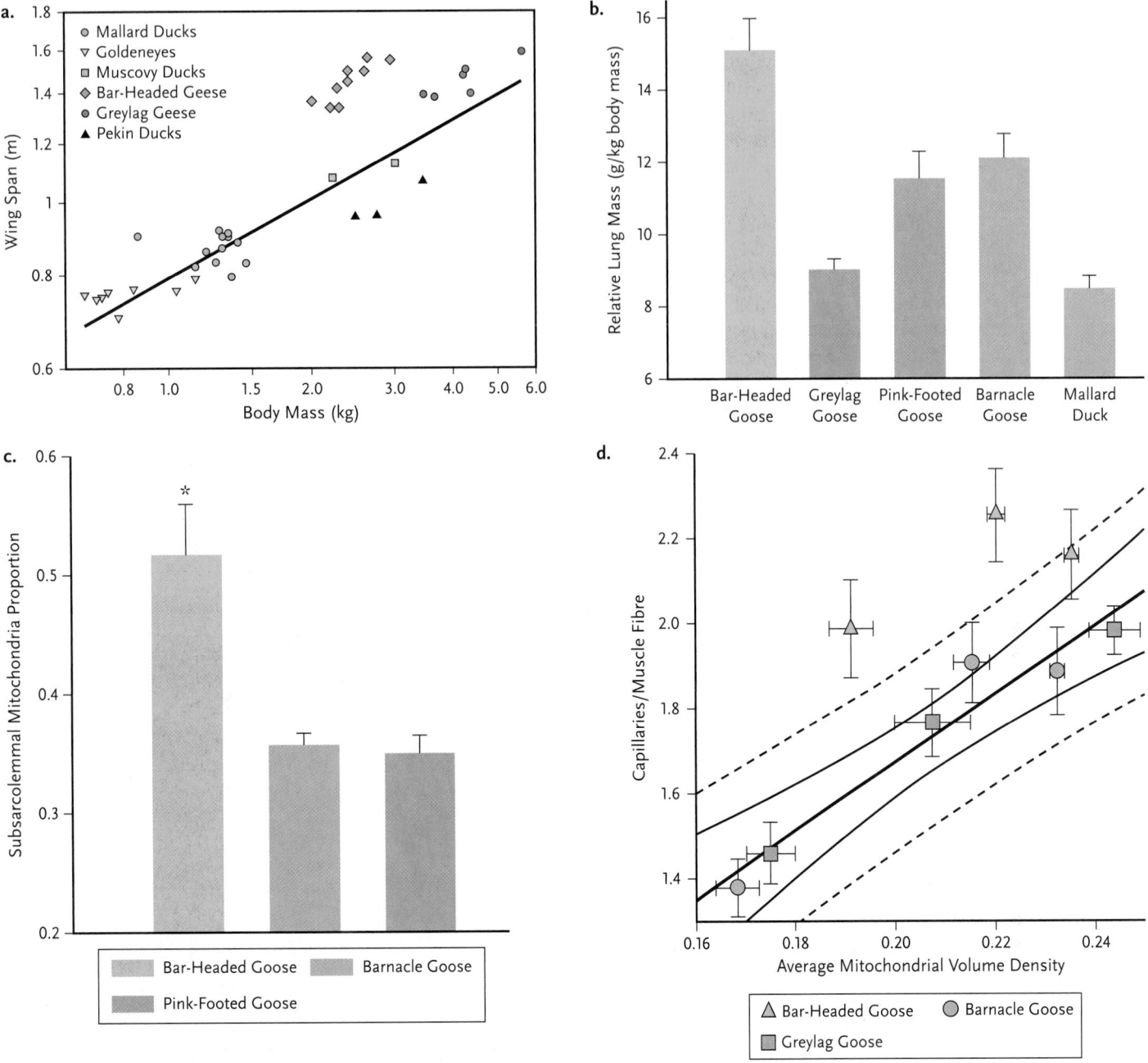

Figure 52.4

(a) Relationship between wingspan and body mass for six species of geese and ducks. **(b)** Relative lung mass (lung mass relative to body mass) for five species of geese and ducks. **(c)** Proportion of mitochondria in a subsarcolemmal position (immediately below the cell membrane) for three species of geese. **(d)** Relationship between the number of capillaries per muscle fibre and mitochondrial volume density for three species of geese.

relatives. This means that it binds oxygen more readily under hypoxic conditions. The same has been found for the Andean Goose (*Chloephaga melanoptera*), a species that is a year-round resident at high altitude. Interestingly, in both groups, the increased affinity of hemoglobin for oxygen is due to a single amino acid substitution in one of the heme molecules. This substitution alters the three-dimensional shape of the protein and its ability to bind and hold oxygen. Equally interesting is that the single amino acid substitution is different in the Bar-Headed Goose and the Andean Goose—a wonderful example of convergent evolution.

While the ability of each millilitre of blood in the Bar-Headed Goose to deliver oxygen is greater than that of its low-altitude relatives, its affinity for oxygen is not enhanced relative to that of the blood of most mammals. A feature that may be of significance for the Bar-Headed Geese when flying at altitude, however, could be the combined effects of changes in temperature and pH on the ability of hemoglobin to transport oxygen (see Chapter 49). Blood in the lung has a high pH due to the loss of CO_2, resulting from the high respiratory rates that may also cool the lungs as a result of the very low temperatures found at altitude. Both of these changes increase the oxygen-binding capacity of the blood. When this blood arrives in the working muscle, it warms up and becomes more acidotic due to the metabolic production of CO_2. This in turn greatly reduces the oxygen-binding affinity of the blood, causing the hemoglobin to release much greater quantities of O_2.

Tissue Diffusion. Diffusion of oxygen from capillary blood into the mitochondria in Bar-Headed Geese is facilitated by two factors, a high degree of capillary branching and the location of most mitochondria at the surface of the muscle fibres just below the cell membrane. Both the greater capillary density and the location of the mitochondria at this subsarcolemmal location reduce the distance the oxygen has to diffuse, increasing the overall ability of these birds to supply O_2 to the active muscle during exercise in hypoxia (Figure 52.4c and d).

Just as pulmonary (lung) edema is a major problem to humans at altitude, so too is cerebral edema. This often-fatal condition arises in humans during high-altitude hypoxia due to increases in cerebral blood flow accompanied by vasoconstriction of cerebral blood vessels. The increase in blood flow is an attempt to maintain cerebral O_2 delivery, while the vasoconstriction is a result of the reduced CO_2 due to the heavy breathing. This combination of events leads to a rise in intracranial pressure and severe edema, causing neurological dysfunction.

This is a perfect example of the need for, and trade-offs associated with, multiple changes in a physiological pathway. In the case of the geese, the reduced effects of CO_2 on vascular tone better enables them to increase cerebral blood flow to retain O_2 while not incurring intracranial hypertension. In the case of low-altitude species, where oxygen is not limiting, enhanced effects of CO_2 on vascular tone better enable them to match cerebral blood flow to metabolic CO_2 production.

Once again, however, we see adaptations at all levels, from the molecular level (hemoglobin adaptations), to the cell level (mitochondrial distribution; neuron and other cellular responses), and the organ system level (capillaries, heart, brain, lung, etc.). As we will see below, adaptations that occur at the level of the individual and the population are also found in Bar-Headed Geese.

52.3d Conservation Issues

Differences in Migratory Patterns between Populations.
The world population of Bar-Headed Geese is approximately 60 thousand.

This population is vulnerable to decline as a result of habitat loss in overwintering areas, climate change–induced habitat alteration in their breeding range, hunting pressure, and susceptibility to emerging infectious diseases such as highly pathogenic avian influenza H5N1. There appear to be distinct subpopulations of these geese, each having their own distinct summer and winter refuges. Because these geese acquire lifetime mates during the winter, the number of distinct subpopulations is probably determined by the number of distinct wintering areas. Distinct differences have been observed in the time and nature of the migration patterns of different populations. Geese from Qinghai Lake in China now winter in the southern Tibetan Qinghai plateau near Lhasa north of the Himalaya and do not cross the mountains **(Figure 52.5)**. Their numbers are increasing in that region possibly due to the effects of climate change and agricultural development. If migratory short-stopping results in larger congregations restricted to a smaller number of wintering areas, the resilience of the geese to environmental change may be lost, as may their ability to make

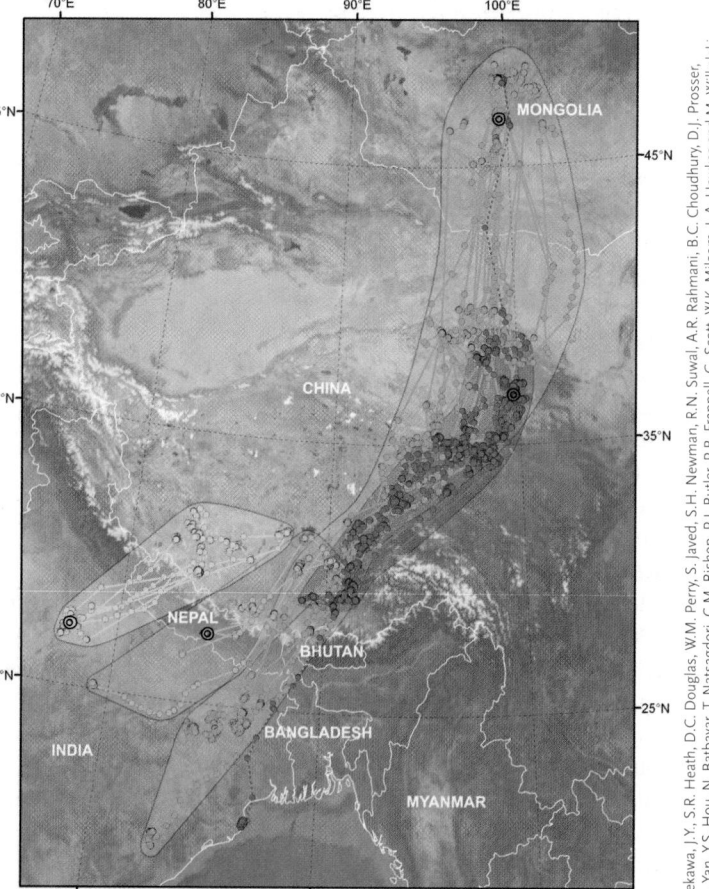

Takekawa, J.Y., S.R. Heath, D.C. Douglas, W.M. Perry, S. Javed, S.H. Newman, R.N. Suwal, A.R. Rahmani, B.C. Choudhury, D.J. Prosser, B.P. Yan, Y.S. Hou, N. Batbayar, T. Natsagdorj, C.M. Bishop, P.J. Butler, P.B. Frappell, G. Scott, W.K. Milsom, L.A. Hawkes and M. Wilkelski. 2009. Geographic variation in Bar-headed geese *Anser indicus*: connectivity of wintering and breeding areas and migration across a broad front. *Wildfowl*, 59: 102–125. This work is licensed under a Creative Commons Attribution 3.0 License: http://creativecommons.org/licenses/by/3.0/

Figure 52.5

Locations (circles), migration pathways (lines), and capture areas (rings) for 60 Bar-Headed Geese marked with satellite transmitters in China (red), India (yellow), Mongolia (green), and Nepal (blue). The Himalayas are highlighted in white. The red-dotted line between China and Mongolia represents a goose marked (April) at Qinghai Lake, China, which migrated to Mongolia for breeding; all other red lines represent birds marked post-breeding at Qinghai Lake.

the trans-mountain flights. Understanding the extent to which exchange occurs between subpopulations as a result of variation in migratory patterns is essential to understanding the population dynamics of the entire range of the species. In turn, understanding the differences in the physiology of the different subpopulations is essential to understanding their migratory patterns and the consequences of climate change for species distribution, and for identifying appropriate management strategies to conserve unique subpopulations.

52.4 Animal Case History #3: Coho Salmon

Human activities are posing threats to biodiversity around the world. Understanding how organisms will respond to environmental change requires a thorough understanding of their physiology and of the ability of their functional systems to accommodate such change. A question that frequently arises is, "How does knowing the effects of low oxygen on an animal's heart rate, or of temperature on an animal's metabolic rate, help us develop better ways to save or rebuild populations and restore ecosystems?" How does an understanding of the material presented in Chapters 39 through 50 help us conserve a species? In this final animal example, we present one case where this has occurred.

Some stocks of Pacific coho salmon (*Oncorhynchus kisutch;* **Figure 52.6**) have been reduced to very low levels in recent years. As a conservation measure, strict regulations were imposed on the harvest of this species. Nonetheless, many coho salmon were still captured accidently as by-catch, both by recreational fishery and by the commercial fishing industry. Live release of "nontarget species" (i.e., by-catch) became a requirement, but even so, many of the fish being thrown back died from the stress of capture and release. Estimates of 0 to 75% of the fish died postcapture, depending on the fishing method used and environmental conditions.

Physiological measurements indicated that captured fish were severely exhausted. Their muscle lactate levels were as high as those found in fish either chased to exhaustion or angled to exhaustion. Their muscle glucose and glycogen levels were low and often below detectable limits. Fish also exhibited an increased hematocrit (red blood cell concentration, a response to low oxygen levels in the blood), increased levels of cortisol (a stress hormone), and increased osmolality and levels of Na^+ and Cl^- (a result of increased ventilation leading to increased salt uptake across the gills).

As a result, in 1998, the Ministry of Fisheries and Oceans imposed the regulation that all British Columbia commercial fishing vessels be required to carry and use fish-recovery boxes (called "Blue Boxes").

Figure 52.6
Migrating coho salmon.

It was believed that if the coho salmon were revived on board vessels prior to release, they would have a better chance of survival when returned to the wild. The challenge was evaluating whether this was a successful strategy since it was also known that many exhausted salmon could survive initially but would succumb to a phenomenon termed "postcapture delayed mortality" up to 24 hours later.

It was also known at the time that fish recovered from exhausting exercise more rapidly if they were allowed to swim slowly rather than remain stationary during recovery. This suggested that salmon caught as by-catch might recover more successfully if held in recovery boxes that were irrigated with flowing water that allowed fish to swim in a stationary position against a current. The water flow also provided sufficient oxygen to meet the metabolic demands of the fish. A door at the head end of the box, in the form of a sluice gate, could be lifted to return the fish to the ocean via a water slide with minimal handling and air exposure. These new boxes were named "Fraser Boxes."

Physiological studies on fish held in both types of boxes showed clear advantages (reduced metabolic, ionic, and osmotic disturbance) to holding fish in flowing water (i.e., in the Fraser Boxes). Over two hours there were significant increases in muscle phosphocreatine and in ATP concentration (indicative of increased energy status) and plasma glucose, and decreases in muscle lactate. There was also a significant decrease in hematocrit, as well as decreases in osmolality and Na^+ and Cl^- concentrations. The rates of recovery varied, with muscle metabolites recovering quickly and levels of osmolytes recovering more slowly. Not only did the procedure promote significant metabolic, ionic, and osmotic recovery, but fish were able to perform well in swim trials following the recovery period. The swim trial results were particularly important because they indicated that once these coho salmon were returned to the wild they would be able to successfully avoid predation and complete their spawning migration. One of the greatest benefits of the Fraser Box was that fish that appeared almost dead at the time of capture due to asphyxiation could also be revived, largely because the flowing water assisted their ventilation and improved their oxygen uptake.

However, there was one downside to the Fraser Box. Many of the fish recovered so quickly that they began to struggle and attempt to escape, restressing themselves. This led to the recommendation that fish be released as soon as they began to exhibit signs of increased activity rather than at a fixed time.

The net result of these studies was that by using a careful analysis of physiological changes, researchers could greatly enhance the survival of the coho salmon and inform optimal conservation policy. This is but one of many such examples of how an understanding of fundamental structure/function relationships in plants and animals is essential to our understanding of the consequences of environmental change (natural or human-made) on a species' survival and of the measures that will or will not be beneficial as conservation policy.

STUDY BREAK

1. What are the fundamental challenges of living in a salt lake?
2. What are the fundamental challenges for a bird flying at high altitudes?
3. What are possible causes of mortality in fish released after capture?

52.5 Plant Case History #1: Plant Species Adapted to Ancient, Low-Phosphorus Soils

The world's richest regions in terms of plant and animal species are also locations where those organisms are under the highest threat of extinction. Many of these organisms are endemic to these global hotspots—that is, they are not found anywhere else on Earth. One of these biodiversity hotspots is southwestern Australia (Figure 52.7a, b). This region is characterized by ancient and infertile soils: because the soils are so old, they have been extensively weathered and have thus lost many of their nutrients. In particular, phosphorus (P) is in very short supply in these soils; since it is an essential element for plants, the low level of phosphorus is a major limiting factor for plant growth. As described in Chapter 37, plants can only take up soil nutrients that are in soluble form (i.e., dissolved in the water fraction of soil). Unlike other soil nutrients, such as nitrogen, phosphorus is not

abundant in the soil solution and thus does not move readily toward plant roots for uptake. Instead, much of the P in the soil is bound to the surface of minerals; over geological time, iron and aluminum also complex with this bound P, making it unavailable to plants. Given this limitation to plant growth, it seems counterintuitive that these low-P soils support a wide diversity of plant species. But, in fact, this pattern of high plant diversity on nutrient-poor soils holds true for other global hotspots, such as the fynbos of South Africa's Cape Floristic Region, which has the highest diversity of plant species outside of the tropics (Figure 52.7c and d). One hypothesis to explain this possible contradiction is that P limitation in soil actually promotes the coexistence of plants; P can take many different forms in soil, thus allowing for a wide range of P-capturing strategies by plants.

One common P-capturing strategy of many plants, described in Chapter 25, is symbiotic association with mycorrhizal fungi. These fungi colonize both the plant roots and the soil around the roots, greatly increasing the surface area for uptake of P from the soil. While mycorrhizal plants do occur in the P-limiting soils of southwest Australia and the fynbos of South Africa, nonmycorrhizal plants are more abundant, particularly those in the Proteaceae family (Figure 52.8a), which use a different strategy than do mycorrhizal plants. Many plants in this family are slow-growing evergreen shrubs and trees that produce dense clusters of short roots (Figure 52.8b) known as cluster roots; there can be up to 1000 short roots (known as rootlets) per centimetre of root axis. Cluster roots release large quantities of carboxylic acids (e.g., citrate and malate) into the soil around the roots, which solubilize the P complexed with iron and aluminum in soil, making it more available. Cluster roots are found in a wide range of plant families, in species that occur in low-nutrient soils. Many plants that form cluster roots have other adaptations allowing them to survive in low-P habitats, such as the ability to reclaim P from older leaves as they die, thus preventing the loss of P from the plant as older leaves are shed.

While these cluster-rooted plants obtain P effectively in P-deficient soils and use it efficiently, they are very sensitive to P and are easily injured or killed by higher levels of P. Evidently, they are unable to reduce P uptake and thus are unable to prevent the accumulation of toxic P concentrations. This P sensitivity threatens the survival of these plants in regions where P enrichment is occurring, due to P-enriched stormwater runoff from urban areas and from fertilizers in agricultural areas. And, ironically, one of the major threats to the biodiversity of southwestern Australia is the one tool available to control another major threat to the plants native to this region. The pathogen *Phytophthora cinnamomi*, an oomycete (see Chapter 24), attacks many of the plant species of this region, causing

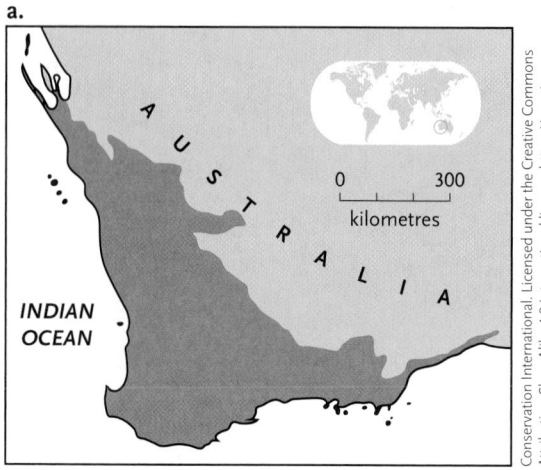

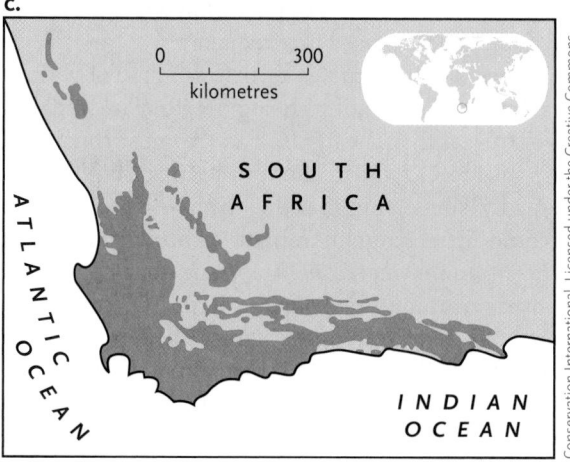

Figure 52.7

Global biodiversity hotspots. **(a)** Southwestern Australia hotspot. **(b)** Sand plain, Southwestern Australia. **(c)** Location of Cape Floristic Region, South Africa. **(d)** Example of fynbos vegetation.

root rot. The only weapon currently available to combat this pathogen is the chemical phosphite, a reduced form of phosphate. Phosphite does not kill the pathogen, but just slows its growth; it is thought to do this by mimicking phosphate in metabolic reactions of the pathogen, thus impairing its normal metabolism and growth. Soil microorganisms oxidize phosphite to phosphate, thus increasing the P availability in regions where phosphite is sprayed, and threatening the ability of cluster-rooted plants to survive in these regions.

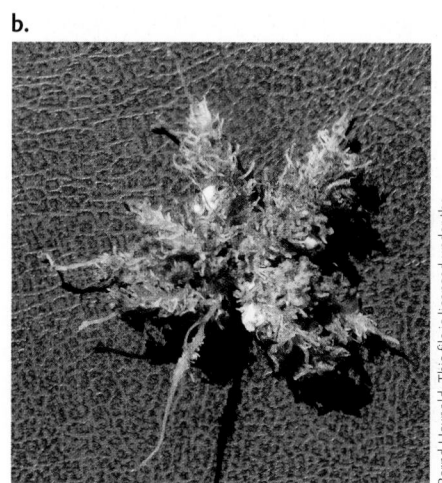

Figure 52.8

Example of plant species in the Proteaceae family. **(a)** Banksia species, southwestern Australia. **(b)** Cluster roots.

Conservation of biodiversity in this global hotspot requires that we find new approaches to combat *P. cinnamomi*, so that we are not just replacing one serious threat to extinction with another.

52.6 Plant Case History #2: Understanding the Threat of Cheatgrass (*Bromus tectorum*)

Conservation physiology can also be used to help predict the behaviour of invasive species and suggest strategies for the control of these organisms. Invasive species are a major threat to plant diversity in many ecosystems, as these invaders effectively compete with native plants for essential resources such as nutrients, water, and light through a combination of reproductive success and stress tolerance. Environmental stresses such as salinity, drought, and extreme hot or cold temperatures all involve osmotic stress. The mechanisms by which plants survive osmotic stress are similar to many of the strategies used by animals, such as those described in Sections 52.2 through 52.4. For example, during exposure to osmotic stress, many plants synthesize osmoprotectants, soluble compounds such as sugars and polyols that accumulate in cells, stabilizing enzymes and protecting membrane integrity during desiccation. Not all plant species are able to synthesize and accumulate osmolytes, and thus an understanding of the relative capacity of native and invasive species to cope with osmotic stress will help us predict invasive success in many habitats. Current models of climate change suggest that these stresses will become more prevalent, and thus climate change may favour species tolerant of warmer, drier, or more saline environments. Another key feature of physiology that will predict relative success in changed climates is the response to CO_2 enrichment. It might be expected that increasing ambient CO_2 concentration will mean enhanced CO_2 fixation by all plants, but will all plants benefit equally? Or are some plants more responsive to increased CO_2 levels? Recent research suggests that some invasive plants in particular will be helped by increasing CO_2 levels, giving them a further advantage over native plants.

One of the most problematic invasive plants in western North America, including southern Alberta and the Great Basin region of the United States, is cheatgrass (*Bromus tectorum*). Cheatgrass germinates quickly to produce seedlings with extensive root systems that use up the available soil moisture before the seedlings of native grasses have a chance to germinate. It produces seeds and then dries out completely in early summer, creating dense mats that burn easily **(Figure 52.9)**. Cheatgrass seeds survive these fires and germinate rapidly postdisturbance. However, the more-frequent fire cycle caused by cheatgrass kills other plants, which are unable to reinvade due to competition

Figure 52.9
Cheatgrass (*Bromus tectorum*) forms dense mats that burn easily when dry.

from dense populations of cheatgrass. It is very difficult to remove cheatgrass once it has initiated this cycle of invasion and increased fire frequency.

Physiological studies of the response of various grass species, including cheatgrass, to increased CO_2 levels reveal that cheatgrass has the most consistent response to CO_2 enrichment in terms of higher productivity, suggesting that this invasive species could become increasingly dominant as atmospheric CO_2 concentrations increase. It appears that cheatgrass also has a unique response to high CO_2 concentrations, different from other plants. Cheatgrass is a C_3 plant, meaning that when it fixes atmospheric CO_2, it first produces a 6-C molecule as CO_2 reacts with ribulose bisphosphate (see Chapter 7); this 6-C molecule quickly splits into two 3-C molecules, giving C_3 plants their name. Unlike other C_3 plants, at high CO_2 concentrations cheatgrass diverts carbon to make a protein that strengthens its cell walls, making them better able to resist desiccation without collapse. This difference, along with the greater productivity of cheatgrass at elevated CO_2 concentrations relative to native species, could enhance the success of this highly invasive species in semiarid regions of western North America. These regions are characterized by variable precipitation that can result in droughts; climate change models predict that these droughts are likely to increase in duration, creating conditions that could give cheatgrass a greater advantage over native species.

Successful strategies for control of cheatgrass require an understanding of the processes that allow cheatgrass to successfully invade native grasslands. One factor that may favour invasion by fast-growing species such as cheatgrass is an increase in nitrogen (N) availability with disturbance. In undisturbed grasslands of these semiarid regions, soil N levels are typically low. Under these low-nutrient conditions, native grasses can successfully outcompete invasive species due to their relatively low growth rates and efficient use of limited nutrients. However, higher N availability following disturbance benefits invasive species such as cheatgrass. For example, after a fire, the burned plant material adds nutrients to the soil either immediately or as it decomposes increasing the pool of available N in the soil. In addition, a recent study reveals that cheatgrass seedlings have a higher rate of N uptake than do some native grasses at the warmer temperatures that promote cheatgrass germination. Both the rate of N uptake per unit root mass and allocation of biomass to roots was also very plastic in seedlings of cheatgrass, meaning that seedlings of this plant could respond quickly to episodes of nutrient availability during warmer periods. In contrast, the native species seemed to have a fixed trade-off between rate of nutrient uptake and allocation of biomass to structural tissues. The greater flexibility of cheatgrass allowed it to be opportunistic in acquiring N that the native species were unable to access.

Understanding of the physiological traits of native and invasive species, such as how they differ in the timing and extent of N uptake, will help us devise strat-egies to control plant invasions and to restore native communities. As with the case studies of animals, the cheatgrass story provides an example of how an understanding of fundamental physiological processes is key to predicting the consequences of environmental change on a species' survival and to developing successful conservation strategies.

STUDY BREAK

1. Why does phosphorus become less available to plants growing in older soils? What adaptations allow some plants to grow successfully in these soils?
2. How do cheatgrass seedlings outcompete native grasses?
3. How might climate change enhance the ability of cheatgrass to invade native grasslands?

Review

 To access course materials such as Aplia and other companion resources, please visit www.NELSONbrain.com.

52.1 Physiological Integration

- Adaptations in organisms arise primarily through evolution by natural selection Understanding how basic physiological processes have evolved and how these basic physiological systems can further evolve to deal with extreme demands increases our ability to design conservation measures to protect species threatened by environmental change.

- Most physiological processes consist of multiple steps. Adaptations that alter the rates of flow along these pathways should occur at every step along the pathway.

- Biological processes are not always optimized. Integration of physiological systems is the result of a long series of evolutionary changes that fine-tune these activities. Each of these changes will have occurred randomly. Each potentially advantageous change only will have been selected for if it provided an advantage to the individual that carried it at that time and in that environment, and if the genetic or developmental programming of the species permitted it. Furthermore, most biochemical, morphological, and physiological systems serve multiple functions. As a result, both functions can't be simultaneously optimized.

- Just as all steps in a pathway must be integrated for proper function, so must all physiological systems within the body.

- Comparative physiology compares how different groups of plants and animals work. Ecological physiology examines how adaptations in physiological traits match organisms to their ecological circumstances, while evolutionary physiology investigates how organisms came to be the way they are and how they may change in the future as the environments in which they live change.

52.2 Animal Case History #1: *Artemia*, the Brine Shrimp

- Few organisms can survive in hypersaline environments, but one that can is *Artemia*, the brine shrimp. These salt-loving extremophiles are referred to as halophiles (salt-lovers). To survive high salt conditions in brine shrimp we see adaptations at the population level (populations of the same species exhibiting unique adaptations specific to their local environments), the individual level (females being able to sense stable versus unstable environments), the organ level (evolution of a resistant epithelial coat and the presence of specialized active transport pumps in the salt gland), the cellular level (stabilization of the cell membranes of the cysts), and the molecular level (role of molecular chaperones in protecting proteins).

52.3 Animal Case History #2: The Bar-Headed Goose: Sustained Flight at High Altitude

- The Bar-Headed Goose, *Anser indicus,* is a species that breeds on the high plateaus of central Asia but over-winters in India on the southern side of the Himalayan mountain range. As a result, twice a year these birds must fly over the world's highest mountains through an environment that is cold and lacking in the oxygen essential to fuel strenuous flight.

- To accomplish these flights under such harsh conditions, we see adaptations at all levels, from the molecular level (hemoglobin adaptations), to the cell level (mitochondrial distribution, adaptations to neuron and other cellular responses), to the organ system level (density of capillaries in tissues, and adaptations to the heart, brain, and lungs) (Figure 52.4).

- There appear to be distinct subpopulations of these geese, each having their own distinct summer and winter refuges. Some populations do not cross the mountains. Their numbers are increasing in that region possibly due to the effects of climate change and agricultural development. If migratory short-stopping results in larger congregations restricted to a smaller number of wintering areas, the resilience of the geese to environmental change may be lost, as may their ability to make the transmountain flights. Understanding the differences in the physiology of the different subpopulations is essential to understanding their migratory patterns and the consequences of climate change for species distribution, and for identifying appropriate management strategies to conserve unique subpopulations.

52.4 Animal Case History #3: Coho Salmon

- Some stocks of Pacific coho salmon (*Oncorhynchus kisutch*) have been reduced to very low levels in recent years. As a conservation measure, strict regulations were imposed on the harvest of this species. Many coho salmon are still captured accidentally as by-catch, however.
- By using a careful analysis of the physiological changes that occur in fish that have been caught and subsequently released, researchers have developed ways to enhance the survival of coho salmon released back into the wild and have been able to develop optimal conservation policy.

52.5 Plant Case History #1: Plant Species Adapted to Ancient, Low-Phosphorus Soils

- Ancient, nutrient-poor soils often support a wide diversity of plant species, perhaps because such

habitats enable a range of nutrient-capturing strategies. Two examples of these nutrient-poor but species-rich habitats, the fynbos of South Africa and the sand plains of southwestern Australia, are home to numerous plants that produce dense clusters of short roots. These cluster roots release organic acids that solubilize phosphorus (P), which is limiting in these environments, making the nutrient more available to the plant (Figures 52.7 and 52.8).

- Cluster-rooted plants cannot survive in soils with higher P levels. P enrichment from agricultural activities, particularly spraying of a chemical used to control a plant pathogen, represents a critical threat to plant biodiversity in southwestern Australia. Understanding of the physiological mechanisms by which these plants obtain and metabolize the form of P present in the herbicide is key to devising a strategy that will control the pathogen while protecting the native cluster-rooted plant species.

52.6 Plant Case History #2: Understanding the Threat of Cheatgrass (*Bromus tectorum*)

- An understanding of ecological plant physiology is necessary to predict the behaviour of invasive plant species and to develop effective strategies for their control. Cheatgrass is a severe problem in western North America; once it invades grasslands in this region, it promotes increased fire frequency, and its rapid germination after fires allows it to further outcompete native grasses. The response of cheatgrass to elevated CO_2 concentrations suggests that it will become an even more serious problem in coming years.

Questions

Self-Test Questions

For multiple choice questions, any number of answers from a to e may be correct.

1. Describe a physiological process that consists of multiple steps. There are many described in earlier chapters in this text. Choose one and describe it fully.

2. Match the following terms (a to d) with the following definitions (i to iv).
   a. comparative physiology
   b. ecological physiology
   c. evolutionary physiology
   d. conservation physiology
      i. study of how physiological traits allow animals to exploit different habitats
      ii. study of how body processes in plants and animals have evolved and how they may change in the future as the environment changes
      iii. study of physiological responses of plants and animals to environmental changes due to anthropogenic threats
      iv. study of how different species of plants and animals function

3. Match the following terms (a to e) with the following definitions (i to v)
   a. cyst
   b. ovoviviparity
   c. anhydrobiosis
   d. trehalose
   e. artemin
      i. describes the manner in which nauplii, the larvae of *Artemia*, are produced
      ii. a molecular chaperone that acts to stabilize proteins under stressful conditions
      iii. eggs encapsulated in a stress-resistant coat
      iv. two glucose molecules bound together
      v. also known as cryptobiosis, a condition of apparent lifelessness

4. Hypoxia is a condition in which the availability of oxygen is reduced. It can be found in which of the following places?
   a. at altitude
   b. in burrows
   c. in tide pools during the day
   d. in the water under the ice of frozen lakes
   e. under waterfalls

5. Which of the following would enhance the delivery of oxygen from the environment to the mitochondria in the tissues?
    a. having larger muscles
    b. having larger lungs
    c. increasing the size of each breath
    d. having more capillaries in each gram of tissue
    e. increasing heart rate

6. Why is phosphorus (P) not very available to plants growing in southwestern Australia?
    a. The soil is very old and has been extensively weathered.
    b. P forms complexes with soil minerals and thus can't be taken up by plant roots.
    c. P is abundant in the soil solution and thus gets leached quickly from the soil.
    d. Both a and b are true.
    e. a, b, and c are all true.

7. Which of the following statements about cluster-rooted plants is true?
    a. They are typically very fast-growing plants.
    b. They increase P availability through the release of basic compounds that increase the solubility of P.
    c. They are quite rare, occurring in only a few plant families.

    d. They have been shown to grow faster when they are supplied with greater amounts of P.
    e. Many of them have adaptations to conserve P, such as the ability to capture P from older leaves before the leaves are shed by the plant.

8. Which of the following is NOT one of the ecophysiological adaptations that make cheatgrass such a successful invasive plant?
    a. Its seedlings grow quickly and develop extensive root systems.
    b. It forms dense mats that burn very slowly, if at all.
    c. Its seeds are not damaged by fire.
    d. At high CO_2 concentrations, it diverts C away from photosynthesis and into production of stronger cell walls.

Questions for Discussion

1. Explain why not all physiological processes are optimized. Describe at least two reasons.

2. Describe a conservation issue that interests you, and explain how an understanding of the physiology of the plants or animals involved could help you develop conservation policy to protect them.

Appendix A
Answers to Self-Test Questions

Chapter 1

1. c 2. a 3. a 4. a 5. c 6. b 7. d 8. d 9. d 10. a

Chapter 2

1. b 2. e 3. d 4. d 5. c 6. b 7. d 8. a 9. b 10. a

Chapter 3

1. c 2. b 3. e 4. d 5. c 6. a 7. d 8. e 9. b 10. a

Chapter 4

1. c 2. d 3. d 4. b 5. d 6. c 7. b 8. a 9. c 10. d

Chapter 5

1. a 2. c 3. b 4. c 5. b 6. b 7. b 8. a 9. d 10. e

Chapter 6

1. a 2. c 3. d 4. d 5. c 6. a 7. b 8. c 9. e 10. d

Chapter 7

1. d 2. c 3. c 4. c 5. a 6. d 7. a 8. c 9. c 10. c

Chapter 8

1. c 2. d 3. c 4. a 5. b 6. b 7. b 8. d 9. b 10. b

Chapter 9

1. d 2. d 3. c 4. d 5. b 6. a 7. a 8. b 9. a 10. c

Chapter 10

1. c

2. (a) The CC parent produces all C gametes, and the Cc parent produces $1/2$ C and $1/2$ c gametes. All offspring would have coloured seeds—half homozygous CC and half heterozygous Cc. (b) Both parents produce $1/2$ C and $1/2$ c gametes. Of the offspring, three-fourths would have coloured seeds ($1/4$ CC + $1/2$ Cc) and one-fourth would have colourless seeds ($1/4$ cc). (c) The Cc parent produces $1/2$ C gametes and $1/2$ c gametes, and the cc parent produces all c gametes. Half of the offspring are coloured ($1/2$ Cc), and half are colourless ($1/2$ cc).

3. The genotypes of the parents are Tt and tt.

4. The taster parents could have a nontaster child, but nontaster parents are not expected to have a child who can taste PTC. The chance that they might have a taster child is $3/4$. The chance that they might have a taster child is $3/4$. The chance of a nontaster child being born to the taster couple is $1/4$. Because each combination of gametes is an independent event, the chance of the couple having a second child, or any child, who cannot taste PTC is expected to be $1/4$.

5. (a) All A B (b) $1/2$ A B + $1/2$ a B (c) $1/2$ A b + $1/2$ a b (d) $1/4$ A B + $1/4$ A b + $1/4$ a B + $1/4$ a b

6. (a) All $AaBB$ (b) $AABB$ ($1/16$) + $AaBB$ ($1/8$) + $aaBB$ ($1/16$) + $AABb$ ($1/8$) + $AaBb$ ($1/4$) + $aaBb$ ($1/8$) + $AAbb$ ($1/16$) + $Aabb$ ($1/8$) + $aabb$ ($1/16$) (c) $AaBb$ ($1/4$) + $Aabb$ ($1/4$) + $aaBb$ ($1/4$) + $aabb$ ($1/4$) (d) $AABB$ ($1/4$) + $AABb$ ($1/4$) + $AaBB$ ($1/4$) + $AaBb$ ($1/4$)

7. (a) All A B C (b) $1/2$ A B c + $1/2$ a B c (c) $1/4$ A B C + $1/4$ A B c + $1/4$ a B C + $1/4$ a B c (d) $1/8$ A B C + $1/8$ A B c + $1/8$ A b C + $1/8$ A b c + $1/8$ a B C + $1/8$ a B c + $1/8$ a b C + $1/8$ a b c

8. This diagram is incorrect because it does not show that each gamete will contain one allele from each of the two genes involved in this cross. The gametes should be Mh, MH, mH, and mh.

9. Because the man can produce only 1 type of allele for each of the 10 genes, he can produce only 1 type of sperm cell with respect to these genes. The woman can produce 2 types of alleles for each of her 2 heterozygous genes, so she can produce $2 \times 2 = 4$ different types of eggs with respect to the 10 genes. In general, as the number of heterozygous genes increases, the number of possible types of gametes increases as $2n$, where $n =$ the number of heterozygous genes.

10. Use a standard testcross; that is, cross the guinea pig with rough, black fur with a double-recessive individual, rr bb (smooth, white fur). If your animal is homozygous RR BB, you would expect all the offspring to have rough, black fur.

11. One gene probably controls pod colour. One allele, for green pods, is dominant; the other allele, for yellow pods, is recessive.

12. The cross $RR \times Rr$ will produce $1/2$ RR and $1/2$ Rr offspring. The cross $Rr \times Rr$ will produce $1/4$ RR, $1/2$ Rr, and $1/4$ rr as combinations of alleles. However, the $1/4$ rr combination is lethal, so it does not appear among the offspring. Therefore, the offspring will be born with only two types, RR and Rr, with twice as many Rr as rr in a 1:2 ratio (or $1/3$ RR + $2/3$ Rr).

13. The parental cross is GG TT $RR \times gg$ tt rr. All offspring of this cross are expected to be tall plants with green pods and round seeds, or Gg Tt Rr. When crossed, this heterozygous F_1 generation is expected to produce eight

different phenotypes among the offspring: green-tall-round, green-dwarf-round, yellow-tall-round, green-tall-wrinkled, yellow-dwarf-round, green-dwarf-wrinkled, yellow-tall-wrinkled, and yellow-dwarf-wrinkled, in a 27:9:9:9:3:3:3:1 ratio.

14. The genotypes are bird 1, *Ff Pp*; bird 2, FF PP; bird 3, *Ff PP*; and bird 4, *Ff Pp*.

15. Yes, it can be determined that the child is not hers, because the father must be AB to have both an A and B child with a type O wife; none of the woman's children could have type O blood with an AB father.

16. The cross is expected to produce white, tabby, and black kittens in a 12:3:1 ratio.

17. The mother is homozygous recessive for both genes, and the father must be heterozygous for both genes. The child is homozygous recessive for both genes. The chance of having a child with normal hands is 1/2, and that of having a child with woolly hair is 1/2. Using the product rule of probability, the probability of having a child with normal hands and woolly hair is $1/2 \times 1/2 = 1/4$.

Chapter 11

1. All sons will be colour-blind, but none of the daughters will be. However, all daughters will be heterozygous carriers of the trait.

2. The chance that her son will be colour-blind is 1/2, regardless of whether she marries a normal or colour-blind male.

3. All of these questions can be answered from the pedigree. Polydactyly is caused by a dominant allele, and the trait is not sex linked. The genotypes of each person are as shown below:

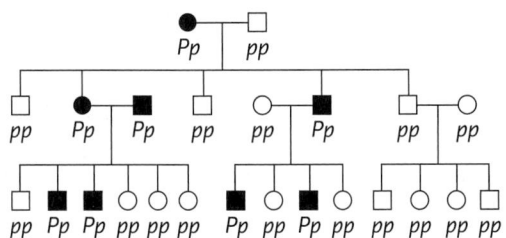

4. The sequence of the genes is *ADBC*.

5. Let the allele for wild-type grey body colour be b^+, and the allele for black body be b. Let the allele for wild-type red eye colour be p^+, and the allele for purple eyes be p. Then the parents are as follows:

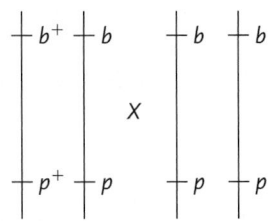

The F₁ flies with black bodies and red eyes are as follows:

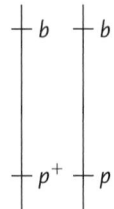

The flies with grey bodies and purple eyes are as shown:

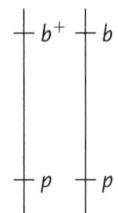

6. The genes are linked by their presence on the same chromosome (an autosome), but they are not sex linked. Because the F₁ females must have produced 600 gametes to give these 600 progeny, and because $42 + 30$ of these were recombinant, the percentage of recombinant gametes is 72/600, or 12%, which implies that 12 map units separate the two genes.

7. Because this trait is probably carried on the Y chromosome, which a man transmits to all his sons, all will have hairy ears. None of the daughters will have hairy ears because they do not have a Y chromosome.

8. You might suspect that a recessive allele is sex linked and is carried on one of the two X chromosomes of the female parent in the cross. When present on the single X of the male (or if present on both X's of a female), the gene is lethal.

Chapter 12

1. c 2. b 3. a 4. a 5. d 6. c 7. b 8. a 9. d 10. d

Chapter 13

1. c 2. a 3. d 4. b 5. d 6. b 7. b 8. c 9. a 10. b

Chapter 14

1. c 2. c 3. b 4. b 5. d 6. c 7. d 8. a 9. a 10. d

Chapter 15

1. c 2. d 3. a 4. a 5. a 6. b 7. a 8. d 9. c 10. b

Chapter 16

1. b 2. a 3. a 4. a 5. d 6. b 7. d 8. a 9. c 10. d

Chapter 17

1. c 2. c 3. d 4. b 5. c 6. b 7. a 8. b 9. d 10. d

Chapter 18
1.c 2.b 3.c 4.d 5.b 6.e 7.a 8.b 9.c 10.d

Chapter 19
1.a 2.e 3.e 4.d 5.d 6.e 7.b 8.c 9.a 10.b

Chapter 20
1.a 2.c 3.e 4.d 5.b 6.c 7.a 8.d 9.c 10.b

Chapter 21
1.d 2.c 3.d 4.d 5.c 6.a 7.c 8.d 9.b 10.c

Chapter 22
1.b 2.c 3.c 4.b 5.d 6.a 7.a 8.c 9.b 10.a

Chapter 23
1.c 2.a 3.a 4.c 5.d 6.b 7.c 8.c 9.d 10.d

Chapter 24
1.b 2.a 3.d 4.b 5.d 6.c 7.b 8.c 9.d 10.c

Chapter 25
1.c 2.b 3.d 4.b 5.b 6.c 7.d 8.c 9.a 10.c

Chapter 26
1.c 2.d 3.c 4.c 5.b 6.b 7.d 8.a 9.d 10.d

Chapter 27
1.c 2.d 3.d 4.d 5.b 6.d 7.c 8.b 9.c 10.d

Chapter 28
1.a 2.b 3.c 4.a 5.c 6.c 7.c 8.b 9.c 10.a

Chapter 29
1.a 2.b 3.b 4.c 5.a 6.a 7.a 8.d 9.d 10.a

Chapter 30
1.c 2.b 3.a 4.a 5.b 6.d 7.b 8.d 9.c 10.b

Chapter 31
1.d 2.c 3.a 4.d 5.d 6.a 7.c 8.d 9.c 10.a

Chapter 32
1.d 2.c 3.d 4.c 5.b 6.b 7.a 8.b 9.b 10.b

Chapter 33
1.c 2.d 3.b 4.b 5.c 6.d 7.b 8.b 9.b 10.b

Chapter 34
1.c 2.d 3.d 4.b 5.a 6.b 7.a 8.b 9.c 10.c

Chapter 35
1.d 2.c 3.a 4.c 5.d 6.c 7.b 8.c 9.c 10.c

Chapter 36
1.b 2.c 3.b 4.a 5.b 6.c 7.d 8.a 9.c 10.c

Chapter 37
1.e 2.c 3.d 4.c 5.b 6.c 7.a 8.d 9.a 10.e

Chapter 38
1.e 2.c 3.d 4.c 5.b 6.c 7.a 8.d 9.a 10.e

Chapter 39
1.a 2.b 3.e 4.a 5.d 6.b 7.d 8.c 9.d 10.c, e

Chapter 40
1.a, b, c, d, e 2.d 3.b 4.d 5.a, d, e 6.d 7.e 8.a, b, c, d 9.b, d 10.d

Chapter 41
1.a, d 2.d 3.a, c 4.d 5.d 6.d 7.b 8.a 9.c 10.b

Chapter 42
1.d 2.d 3.b 4.c 5.d 6.d 7.d 8.b 9.c 10.b

Chapter 43
1.a 2.c 3.e 4.c 5.a 6.d 7.a, b, c, d 8.b, c 9.a 10.c

Chapter 44
1.b, c, d, e 2.a, c, d 3.a, c 4.a, c 5.e 6.b 7.c, d 8.a 9.c 10.b

Chapter 45
1.a 2.d 3.b, d 4.c 5.e 6.a, b, c, d, e 7.d 8.b, d 9.a, b 10.b

Chapter 46
1.d 2.d 3.e 4.e 5.a, b 6.c 7.a, b 8.a 9.e 10.e

Chapter 47
1.b 2.c 3.a 4.b 5.d 6.c 7.d 8.b 9.c 10.b

Chapter 48
1.a 2.b 3.e 4.b 5.d 6.b 7.d 8.c 9.d 10.c

Chapter 49

1. b, c 2. a 3. a, b, c 4. d 5. a, c, e 6. a, b, d 7. a
8. b 9. e 10. a, e

Chapter 50

1. a 2. d 3. c 4. a, c, d, e 5. a 6. c 7. b 8. b 9. b 10. b

Chapter 51

1. a 2. c 3. b 4. e 5. e 6. d 7. b 8. c 9. a 10. d

Chapter 52

1. Given that students can choose any of a number of processes, there is no one answer. A good answer, however, should take a multistep process and go through each step in the process in as much detail as possible. The learning objective of this exercise is to see if students appreciate the following:

Most physiological processes consist of multiple steps. Not surprisingly, adaptations that alter the rates of flow along these pathways should occur at every step along the pathway. If this does not occur, then one step will become limiting and act as a bottleneck, preventing the adaptations at other steps from having much effect. There is no sense of an animal breathing harder if the oxygen is not delivered any more rapidly to the tissues. Similarly, a plant's photosynthetic rate won't increase with an increase in light availability if the rate at which a plant fixes CO_2 lags far behind the rate at which light energy is captured. One would expect that biological design should be optimized, such that each element in a functional pathway matched the maximal need for flow along that pathway. While this optimization may appear to be the case in principle, it frequently is not, reflecting at least two sets of constraints. The first set, often termed "historical" or phylogenetic constraints, are when the genetic or developmental programming of a species will not permit such optimal evolutionary solutions. The second set of constraints stems from the fact that most biochemical, morphological, and physiological systems serve multiple functions, and conflicting functional demands may produce solutions that appear over- or underdesigned for any one specific function. For example, increased surface areas for gas exchange in both terrestrial plants and animals also lead to water loss, bringing about selection pressure both for and against such an increase unless other solutions exist.

2. (a) iv (b) i (c) ii (d) iii 3. (a) iii (b) i (c) v (d) iv (e) ii
4. a, b, d 5. b, c, d, e 6. d 7. e 8. b

Glossary

3′ end The end of a polynucleotide chain at which a hydroxyl group is bonded to the 3 carbon of a deoxyribose sugar. p. 268

5′ cap In eukaryotes, a guanine-containing nucleotide attached in a reverse orientation to the 5 end of pre-mRNA and retained in the mRNA produced from it. The 5′ cap on an mRNA is the site where ribosomes attach to initiate translation. p. 297

5′ end The end of a polynucleotide chain at which a phosphate group is bound to the 5 carbon of a deoxyribose sugar. p. 268

10 nm chromatin fibre The most fundamental level of chromatin packing of a eukaryotic chromosome, in which DNA winds for almost two turns around an eight-protein nucleosome core particle to form a nucleosome, and linker DNA extends between adjacent nucleosomes. The result is a beads-on-a-string type of structure with a 10 nm diameter. p. 283

30 nm chromatin fibre Level of chromatin packing of a eukaryotic chromosome in which histone H1 binds to the 10 nm chromatin fibre, causing it to package into a coiled structure about 30 nm in diameter and with about six nucleosomes per turn. Also referred to as a *solenoid*. p. 284

A site The site where the incoming aminoacyl-tRNA carrying the next amino acid to be added to the polypeptide chain binds to the mRNA. p. 302

abdomen In insects, the region behind the thorax. p. 622

abiotic Nonbiological, often in reference to physical factors in the environment. p. 53

abscisic acid (ABA) A plant hormone involved in the abscission of leaves, flowers, and fruits; dormancy of buds and seeds; and closing of stomata. p. 928

abscission In plants, the dropping of flowers, fruits, and leaves in response to environmental signals. p. 927

absorption spectrum Curve representing the amount of light absorbed at each wavelength. p. 149

absorptive nutrition Mode of nutrition in which an organism secretes digestive enzymes into its environment and then absorbs the small molecules thus produced. p. 541

accommodation A process by which the lens changes to enable the eye to focus on objects at different distances. p. 1108

acid Proton donor that releases H (and anions) when dissolved in water. p. F-18

acid-growth hypothesis A hypothesis to explain how the hormone auxin promotes the growth of plant cells; it suggests that auxin stimulates H pumps in the plasma membrane to move H from the cell interior into the cell wall, which increases wall acidity, making the wall expandable. p. 923

acetabulum Socket of hip joint, receives head of femur. p. 657

acid precipitation Rainfall with low pH, primarily created when gaseous sulfur dioxide (SO_2) dissolves in water vapour in the atmosphere, forming sulfuric acid. p. 909

acidity The concentration of H^+ in a water solution, compared with the concentration of OH^-. p. 93

acoelomate A body plan of bilaterally symmetrical animals that lack a body cavity (coelom) between the gut and the body wall. p. 596

acorn worms Sedentary marine animals living in U-shaped tubes or burrows in coastal sand or mud. p. 639

acquired immune deficiency syndrome (AIDS) A constellation of disorders that follows infection by the HIV virus. pp. 213, 507

acrosome A specialized secretory vesicle on the head of an animal sperm, which helps the sperm penetrate the egg. p. 992

acrosome reaction The process in which enzymes contained in the acrosome are released from an animal sperm and digest a path through the egg coats. p. 1007

action potential The abrupt and transient change in membrane potential that occurs when a neuron conducts an electrical impulse. p. 938

action spectrum Graph produced by plotting the effectiveness of light at each wavelength in driving photosynthesis. p. 150

activation energy The initial input of energy required to start a reaction. p. 85

activator A regulatory protein that controls the expression of one or more genes. pp. 316, 323

active immunity The production of antibodies in the body in response to exposure to a foreign antigen. p. 1281

active parental care Parents' investment of time and energy in caring for offspring after they are born or hatched. p. 687

active site The region of an enzyme that recognizes and combines with a substrate molecule. p. 87

active transport The mechanism by which ions and molecules move against the concentration gradient across a membrane, from the side with the lower concentration to the side with the higher concentration. pp. 109, 865

adaptation, evolutionary Characteristic or suite of characteristics that helps an organism survive longer or reproduce more under a particular set of environmental conditions; the accumulation of adaptive traits over time. p. 423

adaptation, sensory *See* sensory adaptation. p. 1099

adaptive (acquired) immunity A specific line of defence against invasion of the body in which individual pathogens are recognized and attacked to neutralize and eliminate them. p. 1267

adaptive radiation (diversification) A cluster of closely related species that are each adaptively specialized to a specific habitat or food source. p. 451

adaptive trait A genetically based characteristic, preserved by natural selection, that increases an organism's likelihood of survival or its reproductive output. pp. 398, 423

adaptive zone A part of a habitat that may be occupied by a group of species exploiting the same resources in a similar manner. p. 661

adductor muscle A muscle that pulls inward toward the median line of the body; in bivalve molluscs, it pulls the shell closed. p. 615

adenine A purine that base-pairs with either thymine in DNA or uracil in RNA. p. 268

adherens junction Animal cell junction in which intermediate filaments are the anchoring cytoskeletal component. p. 47

adhesion The adherence of molecules to the walls of conducting tubes, as in plants. p. 47

adipose tissue Connective tissue containing large, densely clustered cells called adipocytes that are specialized for fat storage. p. 954

adrenal cortex The outer region of the adrenal glands, which contains endocrine cells that secrete two major types of steroid hormones, the glucocorticoids and the mineralocorticoids. p. 1058

adrenal medulla The central region of the adrenal glands, which contains neurosecretory neurons that secrete the catecholamine hormones epinephrine and norepinephrine. p. 1058

adrenocorticotropic hormone (ACTH) A hormone that triggers hormone secretion by cells in the adrenal cortex. p. 1055

adult stem cells Mammalian stem cells that can differentiate into a limited number of cell types associated with the tissue in which they occur. p. 354

adventitious Formed in an unusual position. p. 853

adventitious root A root that develops from the stem or leaves of a plant. p. 853

aerobe An organism that requires oxygen for cellular respiration. p. 487

afferent arteriole The vessel that delivers blood to the glomerulus of the kidney. p. 1247

afferent neuron A neuron that transmits stimuli collected by a sensory receptor to an interneuron. 1070

African emergence hypothesis A hypothesis proposing that modern humans first evolved in Africa and then dispersed to other continents. p. 477

agar A gelatinous product extracted from certain red algae or seaweed used as a culture medium in the laboratory and as a gelling or stabilizing agent in foods. p. 532

agarose gel electrophoresis Technique by which DNA, RNA, or molecules are separated in a gel subjected to an electric field. p. 346

age structure A statistical description or graph of the relative numbers of individuals in each age class in a population. p. 683

age-specific fecundity The average number of offspring produced by surviving females of a particular age. p. 685

age-specific mortality The proportion of individuals alive at the start of an age interval that died during that age interval. p. 685

age-specific survivorship The proportion of individuals alive at the start of an age interval that survived until the start of the next age interval. p. 685

aggregate fruit A fruit that develops from multiple separate carpels of a single flower, such as a raspberry or strawberry. p. 891

albumin The most abundant protein in blood plasma, important for osmotic balance and pH buffering; also, the portion of an egg that serves as the main source of nutrients and water for the embryo. pp. 654, 970

alcohol A molecule of the form R—OH in which R is a chain of one or more carbon atoms, each of which is linked to hydrogen atoms. p. 100

alcohol fermentation Reaction in which pyruvate is converted into ethyl alcohol and CO_2 in a two-step series that also converts NADH into NAD^+. p. 137

aldosterone A mineralocorticoid hormone released from the adrenal cortex that increases the amount of Na reabsorbed from the urine in the kidneys and absorbed from foods in the intestine, reduces the amount of Na secreted by salivary and sweat glands, and increases the rate of K excretion by the kidneys, keeping Na and K balanced at the levels required for normal cellular function. p. 1059

algin Alginic acid, found in the cell walls of brown algae. p. 526

all-or-nothing principle The principle that an action potential is produced only if the stimulus is strong enough to cause depolarization to reach the threshold. p. 1075

allantoic membrane Forms from mesoderm and endoderm that has bulged outward from the gut and encloses the allantois. p. 1021

allantois In an amniote egg, an extraembryonic membrane sac that fills much of the space between the chorion and the yolk sac and stores the embryo's nitrogenous wastes. p. 1021

allele One of two or more versions of a gene. pp. 201, 220

allele frequency The abundance of one allele relative to others at the same gene locus in individuals of a population. p. 414

allergen A type of antigen responsible for allergic reactions, which induces B cells to secrete an overabundance of IgE antibodies. p. 1284

allopatric speciation The evolution of reproductive isolating mechanisms between two populations that are geographically separated. p. 435

allopolyploidy The genetic condition of having two or more complete sets of chromosomes from different parent species. p. 441

allosteric activator Molecule that converts an enzyme with an allosteric site, a regulatory site outside the active site, from the inactive form to the active form. p. 92

allosteric inhibitor Molecule that converts an enzyme with an allosteric site, a regulatory site outside the active site, from the active form to the inactive form. p. 92

allosteric regulation Specialized control mechanism for enzymes with an allosteric site, a regulatory site outside the active site, that may either slow or accelerate activity depending on the enzyme. p. 92

allosteric site A regulatory site outside the active site. p. 91

alpine tundra A biome that occurs on high mountaintops throughout the world, in which dominant plants form cushions and mats. p. 572

alternation of generations The regular alternation of mode of reproduction in the life cycle of an organism, such as the alternation between diploid (sporophyte) and haploid (gametophyte) phases in plants. pp. 563, 884

alternative hypothesis An explanation of an observed phenomenon that is different from the explanation being tested. p. 58

alternative splicing Mechanism that joins exons in different combinations to produce different mRNAs from a single gene. p. 299

altricial Helpless at birth. p. 668

altruism A behavioural phenomenon in which individuals appear to sacrifice their own reproductive success to help other individuals. p. 1181

alveolus (plural, alveoli) One of the millions of tiny air pockets in mammalian lungs, each surrounded by dense capillary networks. p. 1223

amacrine cell A type of neuron that forms lateral connections in the retina of the eye, connecting bipolar cells and ganglion cells. p. 1112

amino acid A molecule that contains both an amino and a carboxyl group. p. 32

amino group Group that acts as an organic base, consisting of a nitrogen atom bonded on one side to two hydrogen atoms and on the other side to a carbon chain. p. 80

aminoacyl–tRNA A tRNA linked to its "correct" amino acid, which is the finished product of charging. p. 302

aminoacyl–tRNA synthetase An enzyme that catalyzes aminoacylation. p. 302

aminoacylation The process of adding an amino acid to a tRNA. Also referred to as *charging*. p. 302

ammocoetes Larval lamprey eel. p. 645

ammonification A metabolic process in which bacteria and fungi convert organic nitrogen compounds into ammonia and ammonium ions; part of the nitrogen cycle. pp. 766, 911

amniocentesis Technique of prenatal diagnosis in which cells are obtained from the amniotic fluid. p. 258

amnion In an amniote egg, an extraembryonic membrane that encloses the embryo, forming the amniotic cavity and secreting amniotic fluid, which provides an aquatic environment in which the embryo develops. pp. 654, 1020

Amniota The monophyletic group of vertebrates that have an amnion during embryonic development. p. 1021

amniote (amniotic) egg A shelled egg that can survive and develop on land. p. 654

amoeboid Similar to an amoeba, particularly in type of movement. p. 601

amphipathic Containing a region that is hydrophobic and a region that is hydrophilic. p. 100

amplification An increase in the magnitude of each step as a signal transduction pathway proceeds. pp. 116, 1049

amygdala A grey-matter centre of the brain that works as a switchboard, routing information about experiences that have an emotional component through the limbic system. p. 1126

amyloplast Colourless plastid that stores starch in plants. p. 45

anabolic pathway Type of metabolic pathway in which energy is consumed to build complicated molecules from simpler ones; often called a biosynthetic pathway. p. 82

anabolic reaction Metabolic reaction that requires energy to assemble simple substances into more complex molecules. p. 121

anabolic steroid A steroid hormone that stimulates muscle development. p. 1060

anaerobe *See* facultative anaerobe and strict anaerobe.

anaerobic respiration The process by which molecules are oxidized to produce ATP via an electron transport chain and ATP synthase, but unlike aerobic respiration, oxygen is not the final electron acceptor. p. 487

anaphase The phase of mitosis during which the spindle separates sister chromatids and pulls them to opposite spindle poles. p. 175

anaphase-promoting complex (APC) An enzyme complex activated by M phase–promoting factor that controls the separation of sister chromatids and the onset of daughter chromosome separation in anaphase of mitosis. p. 183

anaphylactic shock A severe inflammation stimulated by an allergen, involving extreme swelling of air passages in the lungs, interfering with breathing, and massive leakage of fluid from capillaries, causing blood pressure to drop precipitously. p. 1285

anapsid (lineage Anapsida) A member of the group of amniote vertebrates with no temporal arches and no spaces on the sides of the skull (includes turtles). p. 655

Anapsida An extinct group of fossil fishes. p. 655

anatomy The study of the structures of organisms. p. 946

ancestral character A trait that was present in a distant common ancestor. p. 458

anchoring junction Cell junction that forms belts that run entirely around cells, "welding" adjacent cells together. p. 47

androgen One of a family of hormones that promote the development and maintenance of sex characteristics. pp. 1002, 1059

aneuploid An individual with extra or missing chromosomes. p. 253

angiosperm A flowering plant. Its egg-containing ovules mature into seeds within protected chambers called ovaries. p. 586

angiotensin A peptide hormone that raises blood pressure quickly by constricting arterioles in most parts of the body; it also stimulates release of the steroid hormone aldosterone. p. 1049

animal behaviour The responses of animals to specific internal and external stimuli. p. 1155

animal pole The end of the egg where the egg nucleus is located, which typically gives rise to surface structures and the anterior end of the embryo. p. 1015

Animalia The taxonomic kingdom that includes all living and extinct animals. p. 594

anion A negatively charged ion. pp. 1073, F-11

annual A herbaceous plant that completes its life cycle in one growing season and then dies. p. 840

annulus In ferns, a ring of thick-walled cells that nearly encircles the sporangium and functions in spore release. p. 579

antagonistic pair Two skeletal muscles, one of which flexes as the other extends to move joints. p. 1148

antenna complex (light-harvesting complex) In photosystems, the sites at which light is absorbed and converted into chemical energy during photosynthesis, an aggregate of many chlorophyll pigments and a number of carotenoid pigments that serve as the primary site of absorbing light energy in the form of photons. p. 151

antennal glands Excretory structures at the base of the antennae in some crustaceans. p. 624

anterior Indicating the head end of an animal. p. 596

anterior pituitary The glandular part of the pituitary, composed of endocrine cells that synthesize and secrete several tropic and nontropic hormones. p. 1053

anther The pollen-bearing part of a stamen. p. 885

antheridium (plural, antheridia) In plants, a structure in which sperm are produced. p. 572

Anthocerophyta The phylum comprising hornworts. p. 575

Anthophyta The phylum comprising flowering plants. p. 587

antibiotic A natural or synthetic substance that kills or inhibits the growth of bacteria and other microorganisms. p. 489

antibody A highly specific soluble protein molecule that circulates in the blood and lymph, recognizing and binding to antigens and clearing them from the body. p. 233

antibody-mediated immunity Adaptive immune response in which plasma cells secrete antibodies. p. 1273

anticodon The three-nucleotide segment in tRNAs that pairs with a codon in mRNAs. p. 300

antidiuretic hormone (ADH) A hormone secreted by the posterior pituitary that increases water absorption in the kidneys, thereby increasing the volume of the blood. p. 1056

antigen A foreign molecule that triggers an adaptive immunity response. p. 1273

antigen-presenting cell (APC) A cell that presents an antigen to T cells in antibody-mediated immunity and cell-mediated immunity. p. 1277

antigenic variation The process by which an infectious organism alters its surface proteins to evade a host immune response. Parasites such as the trypanosomes that cause sleeping sickness in humans have 10% of their genes dedicated to generating new surface glycoproteins. p. 1289

antimicrobial peptides Small, potent, broad-spectrum antibiotic peptides that are used by hosts collectively to eliminate bacterial and fungal pathogens. Some antimicrobial peptides may also act as immunomodulators. p. 1268

anti-Müllerian hormone Anti-Müllerian hormone (AMH) is used in testing fertility of women. The AMH gene encodes the protein. p. 1029

antiparallel Refers to strands of DNA that run in opposite directions. p. 269

antiport A secondary active transport mechanism in which a molecule moves through a membrane channel into a cell and powers the active transport of a second molecule out of the cell. Also referred to as *exchange diffusion*. p. 112

aorta A large artery from the heart that branches into arteries leading to all body regions except the lungs. p. 974

aortic body One of several small clusters of chemoreceptors, baroreceptors, and supporting cells located along the aortic arch that measures changes in blood pressure and the composition of arterial blood flowing past it. p. 1228

apical dominance Inhibition of the growth of lateral buds in plants due to auxin diffusing down a shoot tip from the terminal bud. p. 848

apical growth Growth from the tip of a cell or tissue. p. 542

apical meristem A region of unspecialized dividing cells at the shoot tips and root tips of a plant. pp. 568, 839

apical surface The outer surface of epithelial cells. p. 949

apicomplexan A group of parasitic organisms with specific structures in their apical complex to penetrate and enter the cells they parasitize. p. 522

apomixis In plants, the production of offspring without meiosis or formation of gametes. p. 896

apoplast The nonliving component of plant tissues, composed of cell walls and intercellular spaces. p. 868

apoplastic pathway The route followed by water moving through plant cell walls and intercellular spaces (the apoplast). *Compare* symplastic pathway. p. 868

apoptosis Programmed cell death. pp. 185, 1021

aposematic Refers to bright, contrasting patterns that advertise the unpalatability of poisonous or repellent species. p. 715

appendicular skeleton The bones constituting the pectoral (shoulder) and pelvic (hip) girdles and limbs of a vertebrate. p. 1146

appendix A fingerlike sac that extends from the cecum of the large intestine. p. 1206

applied ecology Application of ecological theory and principles to management of natural resources. p. 679

applied research Research conducted with the goal of solving specific practical problems. p. 340

aquaporin A specialized protein channel that facilitates diffusion of water through cell membranes. p. 865

aquatic succession A process in which debris from rivers and runoff accumulates in a body of fresh water, causing it to fill in at the margins. p. 737

aqueous humour A clear fluid that fills the space between the cornea and the lens of the eye. p. 1109

arbuscular mycorrhizas Symbiotic association between a glomeromycete fungus and the roots of a wide range of plants, including nonvascular, nonseed, and seed plants. p. 557

arbuscule Highly branched hypha produced inside root cells by arbuscular mycorrhizal fungi; nutrient exchange site between plant and fungus. p. 548

Archaea One of two domains of prokaryotes; archaeans have some unique molecular and biochemical traits, but they also share some traits with Bacteria and other traits with Eukarya. p. 482

archaeocytes A major group of the domain Archaea, members of which are found in different extreme environments. They include methanogens, extreme halophiles, and some extreme thermophiles. *See* Euryarchaeota. p. 601

archegonium (plural, archegonia) The flask-shaped structure in which bryophyte eggs form. p. 572

archenteron The central endoderm-lined cavity of an embryo at the gastrula stage, which forms the primitive gut. pp. 597, 1016

Archosauromorpha A diverse group of diapsids that comprises crocodilians, pterosaurs, and dinosaurs (including birds). p. 655

arctic tundra A treeless biome that stretches from the boreal forests to the polar ice cap in Europe, Asia, and North America. p. 572

arteries In vertebrates, vessels conducting blood away from the heart at relatively high pressure. pp. 231, 967

arteriole A branch from a small artery at the point where it reaches the organ it supplies. p. 977

artificial selection Selective breeding of animals or plants to ensure that certain desirable traits appear at higher frequency in successive generations. p. 398

ascocarp A reproductive body that bears or contains asci. p. 548

ascospore Spore formed by meiosis in the ascus, a saclike cell produced by ascomycete fungi. p. 549

ascus (plural, asci) A saclike cell in ascomycetes (sac fungi) in which meiosis gives rise to haploid sexual spores (meiospores). p. 548

asexual reproduction Any mode of reproduction in which a single individual gives rise to offspring without fusion of gametes, that is, without genetic input from another individual. *See also* vegetative reproduction. p. 989

assimilation efficiency The ratio of the energy absorbed from consumed food to the total energy content of the food. p. 752

association area One of several areas surrounding the sensory and motor areas of the cerebral cortex that integrate information from the sensory areas, formulate responses, and pass them on to the primary motor area. p. 1128

assumption of parsimony Assumption that the simplest explanation should be the most accurate. p. 459

aster Radiating array produced as microtubules extending from the centrosomes of cells grow in length and extent. p. 179

astrocyte A star-shaped glial cell that provides support to neurons in the vertebrate central nervous system. p. 1070

asymmetrical Characterized by a lack of proportion in the spatial arrangement or placement of parts. p. 596

atmosphere The component of the biosphere that includes the gases and airborne particles enveloping the planet. p. 13

atom The smallest unit that retains the chemical and physical properties of an element. p. F-9

atomic nucleus The nucleus of an atom, containing protons and neutrons. p. 121

atomic number The number of protons in the nucleus of an atom. p. F-9

ATP (adenosine triphosphate) The primary agent that couples exergonic and endergonic reactions. p. 7

ATP cycle Continued breakdown and resynthesis of ATP. p. 85

ATP synthase A membrane-spanning protein complex that couples the energetically favourable transport of protons across a membrane to the synthesis of ATP. p. 132

atrial siphon A tube through which invertebrate chordates expel digestive and metabolic wastes. p. 641

atriopore The hole in the body wall of a cephalochordate through which water is expelled from the body. p. 641

atrioventricular node (AV node) A region of the heart wall that receives signals from the sinoatrial node and conducts them to the ventricle. p. 976

atrioventricular valve (AV valve) A valve composed of endocardium and connective tissue between each atrium and ventricle that prevents backflow of blood from the ventricle to the atrium during emptying of the heart. p. 974

atrium (plural, atria) A body cavity or chamber surrounding the perforated pharynx of invertebrate chordates; also one of the chambers that receive blood returning to the heart. p. 641

autoimmune reaction The production of antibodies against molecules of the body. p. 1284

autonomic nervous system A subdivision of the peripheral nervous system that controls largely involuntary processes, including digestion, secretion by sweat glands, circulation of the blood, many functions of the reproductive and excretory systems, and contraction of smooth muscles in all parts of the body. p. 1092

autopolyploidy The genetic condition of having more than two sets of chromosomes from the same parent species. p. 441

autosomal dominant inheritance Pattern in which the allele that causes a trait is dominant, and only homozygous recessives are unaffected. p. 257

autosomal recessive inheritance Pattern in which individuals with a trait are homozygous for a recessive allele. p. 256

autosome Chromosome other than a sex chromosome. p. 246

autotroph An organism that produces its own food using CO_2 and other simple inorganic compounds from its environment and energy from the Sun or from oxidation of inorganic substances. pp. 63, 145, 487

auxin Any of a family of plant hormones that stimulate growth by promoting cell elongation in stems and coleoptiles, inhibit abscission, govern responses to light and gravity, and have other developmental effects. p. 919

auxotrophs Mutant strains that are unable to synthesize amino acids. p. 191

Avogadro's number The number 6.022×10^{23}, derived by dividing the atomic weight of any element by the weight of an atom of that element. p. F-18

Avr gene A gene in certain plant pathogens that encodes a product triggering a defensive response in the plant. p. 932

axial skeleton The bones constituting the head and trunk of a vertebrate: the cranium, vertebral column, ribs, and sternum (breastbone). pp. 641, 1146

axillary buds Embryonic shoots that develop where a leaf meets the stem. p. 847

axon The single elongated extension of a neuron that conducts signals away from the cell body to another neuron or an effector. p. 1069

axon hillock A junction with the cell body of a neuron from which the axon arises. p. 1069

axon terminal A branch at the tip of an axon that ends as a small, buttonlike swelling. p. 1069

axopods Slender, raylike strands of cytoplasm supported internally by long bundles of microtubules. p. 527

B cell A lymphocyte that recognizes antigens in the body. p. 1273

B-cell receptor (BCR) The receptor on B cells that is specific for a particular antigen. p. 1273

backbone (spine) Vertebral column of vertebrates. p. 641

background extinction rate The average rate of extinction of taxa through time. p. 781

Bacteria One of the two domains of prokaryotes; collectively, bacteria are the most metabolically diverse organisms. p. 482

bacterial chromosome DNA molecule in bacteria in which hereditary information is encoded. p. 285

bacterial flagellum *See* flagellum. p. 33

bacteriophage A virus that infects bacteria. Also referred to as a *phage*. pp. 196, 266, 502

bacteroid A rod-shaped or branched bacterium in the root nodules of nitrogen-fixing plants. p. 912

balanced polymorphism The maintenance of two or more phenotypes in fairly stable proportions over many generations. p. 421

bark The tough outer covering of woody stems and roots, comprising all the living and nonliving tissues between the vascular cambium and the stem surface. p. 856

Barr body The inactive, condensed X chromosome seen in the nucleus of female mammals. p. 250

basal body Structure that anchors cilia and flagella to the surface of a cell. p. 45

basal lamina A membrane secreted at the inner surface of epithelial cells. p. 949

basal nucleus (plural, **basal nuclei**) One of several grey-matter centres that surround the thalamus on both sides of the brain and moderate voluntary movements directed by motor centres in the cerebrum. p. 1126

basal surface The inner surface of epithelial cells. p. 949

base Proton acceptor that reduces the H^+ concentration of a solution. p. F-19

base-pair mismatch An error in the assembly of a new nucleotide chain in which bases other than the correct ones pair together. p. 280

base-pair substitution mutation A particular mutation involving a change from one base pair to another in DNA. p. 310

basement membrane A membrane at the inner surface of epithelia in vertebrates. It consists of the basal lamina and a layer of connective tissue. p. 949

basic research Research conducted to search for explanations about natural phenomena to satisfy curiosity and advance collective knowledge of living systems. p. 340

basidiocarp A fruiting body of a basidiomycete; mushrooms are examples. p. 553

basidiospore A haploid sexual spore produced by basidiomycete fungi. p. 553

basidium (plural, **basidia**) A small, club-shaped structure in which sexual spores of basidiomycetes arise. p. 550

basilar membrane A stiff structural element within the cochlea. p. 1106

basophil A type of leukocyte that is induced to secrete histamine by allergens. p. 1284

Batesian mimicry The form of defence in which a palatable or harmless species resembles an unpalatable or poisonous one. p. 716

behavioural isolation A prezygotic reproductive isolating mechanism in which two species do not mate because of differences in courtship behaviour; also known as ethological isolation. p. 431

biennial A plant that completes its life cycle in two growing seasons and then dies; limited secondary growth occurs in some biennials. p. 840

bilateral symmetry The body plan of animals in which the body can be divided into mirror image right and left halves by a plane passing through the midline of the body. p. 596

bilayer A membrane with two molecular layers. p. 28

bile A mixture of substances including bile salts, cholesterol, and bilirubin that is made in the liver, stored in the gallbladder, and used in the digestion of fats. p. 1197

binary fission Prokaryotic cell division—splitting or dividing into two parts. p. 169

binomial Relating to or consisting of two names or terms. p. 427

binomial nomenclature The naming of species with a two-part scientific name, the first indicating the genus and the second indicating the species. p. 427

biodiversity The richness of living systems as reflected in genetic variability within and among species, the number of species living on Earth, and the variety of communities and ecosystems. p. 780

biofilm A microbial community consisting of a complex aggregation of microorganisms attached to a surface. p. 490

biogeochemical cycle Any of several global processes in which a nutrient circulates between the abiotic environment and living organisms. p. 488

biogeography The study of the geographic distributions of plants and animals. p. 392

bioinformatics Field that fuses biology with mathematics and computer science and is used for the analysis of genome sequences. p. 368

biological clock An internal time-measuring mechanism that adapts an organism to recurring environmental changes. p. 938

biological evolution The process by which some individuals in a population experience changes in their DNA and pass those modified instructions to their offspring. p. 392

biological lineage An evolutionary sequence of ancestral organisms and their descendants. p. 401

biological magnification The increasing concentration of nondegradable poisons in the tissues of animals at higher trophic levels. p. 757

biological research The collective effort of individuals who have worked to understand how living systems function. p. F-2

biological species concept The definition of species based on the ability of populations to interbreed and produce fertile offspring. p. 429

bioluminescent Refers to an organism that glows or releases a flash of light, particularly when disturbed. p. 521

biomass The dry weight of biological material per unit area or volume of habitat. pp. 482, 750

biome A large-scale vegetation type and its associated microorganisms, fungi, and animals. p. F-49

bioremediation Applications of chemical and biological knowledge to decontaminate polluted environments. p. 481

biosphere All regions of Earth's crust, waters, and atmosphere that sustain life. p. 7

biota The total collection of organisms in a geographic region. p. 773

biotechnology The manipulation of living organisms to produce useful products. p. 339

biotic Biological, often in reference to living components of the environment. p. 53

bipedalism The habit in animals of walking upright on two legs. p. 469

bipolar cell A type of neuron in the retina of the eye that connects the rods and cones with the ganglion cells. p. 1112

blade The expanded part of a leaf that provides a large surface area for absorbing sunlight and carbon dioxide. p. 850

blastocoel A fluid-filled cavity in the blastula embryo. p. 1015

blastocyst An embryonic stage in mammals; a single cell–layered hollow ball of about 120 cells with a fluid-filled blastocoel in which a dense mass of cells is localized to one side. p. 1025

blastodisc A disclike layer of cells at the surface of the yolk produced by early cleavage divisions. p. 1019

blastomere A small cell formed during cleavage of the embryo. p. 1015

blastopore The opening at one end of the archenteron in the gastrula that gives rise to the mouth in protostomes and the anus in deuterostomes. p. 597

blastula The hollow ball of cells that is the result of cleavage divisions in an early embryo. p. 1015

blending theory of inheritance Theory suggesting that hereditary traits blend evenly in offspring through mixing of the blood of the two parents. p. 218

blood A fluid connective tissue composed of blood cells suspended in a fluid extracellular matrix, plasma. p. 952

blood–brain barrier A specialized arrangement of capillaries in the brain that prevents most substances dissolved in the blood from entering the cerebrospinal fluid and thus protects the brain and spinal cord from viruses, bacteria, and toxic substances that may circulate in the blood. p. 1124

bolting Rapid formation of a floral shoot in plant species that form rosettes, such as lettuce. p. 925

bolus The food mass after chewing. p. 1201

bone The densest form of connective tissue, in which living cells secrete the mineralized matrix of collagen and calcium salts that surrounds them; forms the skeleton. p. 953

book lungs Pocketlike respiratory organs found in some arachnids consisting of several parallel membrane folds arranged like the pages of a book. p. 622

boreal forest A biome that is a circumpolar expanse of evergreen coniferous trees in Europe, Asia, and North America. p. 557

Bowman's capsule An infolded region at the proximal end of a nephron that cups around the glomerulus and collects the water and solutes filtered out of the blood. p. 1246

brain A single, organized collection of nervous tissue in an organism's head that forms the control centre of the nervous system and major sensory structures. p. 1089

brain hormone A peptide hormone secreted by neurosecretory neurons in the brain of insects. p. 1064

brain stem A stalklike structure formed by the pons and medulla, along with the midbrain, which connects the forebrain with the spinal cord. p. 1124

brassinosteroid Any of a family of plant hormones that stimulate cell division and elongation and differentiation of vascular tissue. p. 927

bronchiole One of the small, branching airways in the lungs that lead into the alveoli. p. 1225

bronchus (plural, **bronchi**) An airway that leads from the trachea to the lungs. p. 1225

brown adipose tissue A specialized tissue in which the most intense heat generation by nonshivering thermogenesis takes place. p. 1258

Bryophyta The phylum of nonvascular plants, including mosses and their relatives. p. 574

bryophyte A general term for plants (such as mosses) that lack internal transport vessels. p. 572

budding A mode of asexual reproduction in which a new individual grows and develops while attached to the parent. pp. 542, 989

buffer Substance that compensates for pH changes by absorbing or releasing H^+. p. 347

bulbourethral gland One of two pea-sized glands on either side of the prostate gland that secrete a mucous fluid that is added to semen. p. 1003

bulk feeder An animal that consumes sizable food items whole or in large chunks. p. 1195

bulk flow The group movement of molecules in response to a difference in pressure between two locations. p. 869

bulk-phase endocytosis (pinocytosis) Mechanism by which extracellular water is taken into a cell together with any molecules that happen to be in solution in the water. p. 112

C-terminal end The end of an amino acid chain with a —COO group. p. 300

Ca^{2+} pump (calcium pump) Pump that pushes Ca^{2+} from the cytoplasm to the cell exterior and from the cytosol into the vesicles of the endoplasmic reticulum. p. 109

cadherin (calcium-dependent adhesion molecule) A cell surface protein responsible for selective cell adhesions that require calcium ions to set up adhesions. p. 1032

calcitonin A nontropic peptide hormone that lowers the level of Ca^{2+} in the blood by inhibiting the ongoing dissolution of calcium from bone. p. 1057

callus An undifferentiated tissue that develops on or around a cut plant surface or in tissue culture. p. 897

Calvin cycle *See* light-independent reaction. p. 7

CAM plant A C_4 plant that runs the Calvin and C_4 cycles at different times to circumvent photorespiration. CAM stands for "crassulacean acid metabolism." p. 161

capillary The smallest-diameter blood vessel, with a wall that is one cell thick, which forms highly branched networks well adapted for diffusion of substances. p. 967

capsid *See* coat. p. 501

capsule An external layer of sticky or slimy polysaccharides coating the cell wall in many prokaryotes. pp. 32, 485

carapace A protective outer covering that extends backward behind the head on the dorsal side of an animal, such as the shell of a turtle or lobster. p. 624

carbon cycle The global circulation of carbon atoms, especially via the processes of photosynthesis and respiration. p. 760

cardiac cycle The systole–diastole sequence of the heart. p. 974

cardiac muscle The contractile tissue of the heart. p. 954

carnivore An animal that primarily eats other animals. p. 670

carotenoid Molecule of yellow-orange pigment by which light is absorbed in photosynthesis. p. 19

carpel The reproductive organ of a flower that houses an ovule and its associated structures. p. 885

carrageenan A chemical extracted from the red alga *Eucheuma* that is used to thicken and stabilize paints, dairy products such as pudding and ice cream, and many other creams and emulsions. p. 532

carrier An individual who carries a mutant allele and could pass it on to offspring but does not display its symptoms. p. 249

carrier protein Transport protein that binds a specific single solute and transports it across the lipid bilayer. p. 106

carrying capacity The maximum size of a population that an environment can support indefinitely. p. 691

Cartagena Protocol on Biosafety An international agreement that promotes biosafety as it relates to genetically modified organisms. p. 359

cartilage A tissue composed of sparsely distributed chondrocytes surrounded by networks of collagen fibres embedded in a tough but elastic matrix of the glycoprotein. p. 953

Casparian strip A thin, waxy, impermeable band that seals abutting cell walls in roots; the strip helps control the type and amount of solutes that enter the stele by blocking the apoplastic pathway at the endodermis and forcing substances to pass through cells (the symplast). p. 868

caspase A protease involved in programmed cell death. p. 185

catabolic pathway Type of metabolic pathway in which energy is released by the breakdown of complex molecules to simpler compounds. p. 82

catabolic reaction Cellular reaction that breaks down complex molecules such as sugar to make their energy available for cellular work. p. 85

catalyst Substance with the ability to accelerate a spontaneous reaction without being changed by the reaction. p. 86

catastrophism The theory that Earth has been affected by sudden, violent events that were sometimes worldwide in scope. p. 393

catecholamine Any of a class of compounds derived from the amino acid tyrosine that circulates in the bloodstream, including epinephrine and norepinephrine. p. 1058

cation A positively charged ion. p. F-11

cation exchange Replacement of one cation with another, as on a soil particle. p. 908

$CD4^+$ T cell A type of T cell in the lymphatic system that has CD4 receptors on its surface. This type of T cell binds to an antigen-presenting cell in antibody-mediated immunity. p. 1277

$CD8^+$ T cell A type of T cell in the lymphatic system that has CD8 receptors on its surface. This type of T cell binds to an antigen-presenting cell in cell-mediated immunity. p. 1281

cDNA library The entire collection of cloned cDNAs made from the mRNAs isolated from a cell. p. 344

cecum A blind pouch formed at the junction of the large and small intestines. p. 1206

cell Smallest unit with the capacity to live and reproduce. p. 1

cell adhesion molecule A cell surface protein responsible for selectively binding cells together. pp. 47, 1032

cell body The portion of the neuron containing genetic material and cellular organelles. p. 1069

cell centre *See* centrosome. p. 42

cell culture A living cell grown in a laboratory vessel. p. 173

cell cycle The sequence of events during which a cell experiences a period of growth followed by nuclear division and cytokinesis. p. 69

cell differentiation A process in which changes in gene expression establish cells with specialized structure and function. p. 883

cell expansion A mechanism that enlarges the cells in specific directions in a developing organ. p. 894

cell junction Junction that seals the spaces between cells and provides direct communication between cells. p. 47

cell lineage Cell derivation from the undifferentiated tissues of the embryo. p. 1035

cell-mediated immunity An adaptive immune response in which a subclass of T cells—cytotoxic T cells—becomes activated and, with other cells of the immune system, attacks host cells infected by pathogens, particularly those infected by a virus. p. 1273

cell plate In cytokinesis in plants, a new cell wall that forms between the daughter nuclei and grows laterally until it divides the cytoplasm. p. 176

cell theory Three generalizations yielded by microscopic observations: all organisms are composed of one or more cells, the cell is the smallest unit that has the properties of life, and cells arise only from the growth and division of preexisting cells. p. 28

cell wall A rigid external layer of material surrounding the plasma membrane of cells in plants, fungi, bacteria, and some protists, providing cell protection and support. p. 32

cellular respiration The process by which energy-rich molecules are broken down to produce energy in the form of ATP. p. 121

cellular senescence Loss of proliferative ability over time. p. 184

cellular slime mould Any of a variety of primitive organisms of the phylum Acrasiomycota, especially of the genus *Dictyostelium*; the life cycle is characterized by a slimelike amoeboid stage and a multicellular reproductive stage. p. 529

cellulose One of the primary constituents of plant cell walls, formed by chains of carbohydrate subunits. p. 837

centimorgan *See* map unit. p. 244

central canal The central portion of the vertebral column in which the spinal cord is found. p. 1090

central nervous system (CNS) One of the two major divisions of the nervous system containing the brain and spinal cord. p. 1068

central vacuole A large, water-filled organelle in plant cells that maintains the turgor of the cell and controls movement of molecules between the cytosol and sap. pp. 46, 866

centriole A cylindrical structure consisting of nine triplets of microtubules in the centrosomes of most animal cells. pp. 42, 178

centromere A specialized chromosomal region that connects sister chromatids and attaches them to the mitotic spindle. p. 174

centrosome (cell centre) The main microtubule organizing centre of a cell, which organizes the microtubule cytoskeleton during interphase and positions many of the cytoplasmic organelles. pp. 42, 178

cephalization The development of an anterior head where sensory organs and nervous system tissue are concentrated. p. 596

cephalothorax The anterior section of an arachnid, consisting of a fused head and thorax. p. 622

cerebellum The portion of the brain that receives sensory input from receptors in muscles and joints, from balance receptors in the inner ear, and from the receptors of touch, vision, and hearing. p. 1125

cerebral cortex A thin outer shell of grey matter covering a thick core of white matter within each hemisphere of the brain; the part of the forebrain responsible for information processing and learning. p. 1124

cerebrospinal fluid Fluid that circulates through the central canal of the spinal cord and the ventricles of the brain, cushioning the brain and spinal cord from jarring movements and impacts, as well as nourishing the CNS and protecting it from toxic substances. p. 1123

cervix The lower end of the uterus. p. 961

channel protein Transport protein that forms a hydrophilic channel in a cell membrane through which water, ions, or other molecules can pass, depending on the protein. p. 106

chaperone protein (chaperonin) "Guide" protein that binds temporarily with newly synthesized proteins, directing their conformation toward the correct tertiary structure and inhibiting incorrect arrangements as the new proteins fold. p. F-36

character A heritable characteristic. p. 218

character displacement The phenomenon in which allopatric populations are morphologically similar and use similar resources, but sympatric populations are morphologically different and use different resources; may also apply to characters influencing mate choice. p. 721

charging *See* aminoacylation. p. 302

charophyte A member of the group of green algae most similar to the algal ancestors of land plants. p. 533

checkpoint Internal control of the cell cycle that prevents a critical phase from beginning until the previous phase is complete. p. 181

chelicerae The first pair of fanglike appendages near the mouth of an arachnid, used for biting prey and often modified for grasping and piercing. p. 622

chemical bond Link formed when atoms of reactive elements combine into molecules. p. F-11

chemical equation A chemical reaction written in balanced form. p. F-14

chemical reaction A reaction that occurs when atoms or molecules interact to form new chemical bonds or break old ones. p. F-14

chemical signal Any secretion from one cell type that can alter the behaviour of a different cell that bears a receptor for it; a means of cell communication. p. 529

chemical synapse A type of communicating connection between two neurons or a neuron and an effector cell in which an electrical impulse arriving at an axon terminal of the presynaptic cell triggers release of a neurotransmitter that crosses the gap and binds to a receptor on the postsynaptic cell, triggering an electrical impulse in that cell. p. 1073

chemiosmosis Ability of cells to use the proton-motive force to do work. p. 132

chemoautotroph An organism that obtains energy by oxidizing inorganic substances such as hydrogen, iron, sulfur, ammonia, nitrites, and nitrates and uses carbon dioxide as a carbon source. p. 487

chemoheterotroph An organism that oxidizes organic molecules as an energy source and obtains carbon in organic form. p. 487

chemokine A protein secreted by activated macrophages that attracts other cells, such as neutrophils. p. 1269

chemoreceptor A sensory receptor that detects specific molecules or chemical conditions such as acidity. p. 1077

chemotroph An organism that obtains energy by oxidizing inorganic or organic substances. p. 487

chiasmata *See* crossover. p. 208

chitin A polysaccharide that contains nitrogen and is present in the cell walls of fungi and the exoskeletons of arthropods. p. 47

chlorophyll Molecule of green pigment that absorbs photons of light in photosynthesis. p. 7

chloroplast The site of photosynthesis in plant cells. p. 45

chlorosis An abnormal yellowing of plant tissues due to lack of chlorophyll; a sign of nutrient deficiency or infection by a pathogen. p. 905

choanocyte One of the inner layer of flagellated cells lining the body cavity of a sponge. p. 601

choanoflagellata A group of minute, single-celled protists found in water; the flask-shaped body has a collar of closely packed microvilli that surrounds the single flagellum by which it moves and takes in food. p. 531

cholesterol The predominant sterol of animal cell membranes. p. 102

chondrocyte A cartilage-producing cell. p. 953

chorion In an amniote egg, an extraembryonic membrane that surrounds the embryo and yolk sac completely and exchanges oxygen and carbon dioxide with the environment; becomes part of the placenta in mammals. p. 1020

chorionic villus (plural, **villi**) One of many treelike extensions from the chorion, which greatly increase the surface area of the chorion. p. 1027

chorionic villus sampling Technique of prenatal diagnosis in which cells are obtained from portions of the placenta that develop from tissues of the embryo. p. 258

chromatids One half of a replicated chromosome. Each chromatid is one double helix of DNA. p. 170

chromatin remodelling Process in which the state of the chromatin is changed so that the proteins that initiate transcription can bind to their promoters. p. 327

chromatin The structural building block of a chromosome, which includes the complex of DNA and its associated proteins. pp. 35, 283

chromoplast Plastid containing red and yellow pigments. p. 45

chromosomal protein The histone and nonhistone protein associated with DNA structure and regulation in the nucleus. p. 283

chromosome The nuclear unit of genetic information, consisting of a DNA molecule and associated proteins. p. 170

chromosome alterations Changes in the structure of chromosomes involving insertion, deletion, inversion, or translocation of significant amounts of DNA sequence. p. 439

chromosome segregation The equal distribution of daughter chromosomes to each of the two cells that result from cell division. p. 171

chromosome theory of inheritance The principle that genes and their alleles are carried on the chromosomes. p. 228

chylomicron A small triglyceride droplet covered by a protein coat. p. 1205

chyme Digested content of the stomach released for further digestion in the small intestine. p. 1203

ciliary body A fine ligament in the eye that anchors the lens to a surrounding layer of connective tissue and muscle. p. 1109

cilium Motile structure, extending from a cell surface, that moves a cell through fluid or fluid over a cell. p. 43

circadian rhythm Any biological activity that is repeated in cycles, each about 24 hours long, independent of any shifts in environmental conditions. pp. 16, 938

circulatory system An organ system consisting of a fluid, a heart, and vessels for moving important molecules, and often cells, from one tissue to another. p. 965

circulatory vessel An element of the circulatory system through which fluid flows and carries nutrients and oxygen to tissues and removes wastes. p. 611

circumcision Removal of the prepuce for religious, cultural, or hygienic reasons. p. 1004

cisternae (singular, cisterna) Membranous channels and vesicles that make up the endoplasmic reticulum. p. 36

citric acid cycle Series of reactions in which acetyl groups are oxidized completely to carbon dioxide and some ATP molecules are synthesized. Also referred to as *Krebs cycle* and *tricarboxylic acid cycle*. p. 126

clade A monophyletic group of organisms that share homologous features derived from a common ancestor. p. 461

cladistics An approach to systematics that uses shared derived characters to infer the phylogenetic relationships and evolutionary history of groups of organisms. p. 461

cladogenesis The evolution of two or more descendent species from a common ancestor. p. 457

cladogram A branching diagram in which the endpoints of the branches represent different species of organisms, used to illustrate phylogenetic relationships. p. 461

claspers A pair of organs on the pelvic fins of male crustaceans and sharks, which help transfer sperm into the reproductive tract of the female. p. 648

class A Linnaean taxonomic category that ranks below a phylum and above an order. p. 428

class II major histocompatibility complex (MHC) A collection of proteins that present antigens on the cell surface of an antigen-presenting cell in an antibody-mediated immune response. p. 1277

classical conditioning A type of learning in which an animal develops a mental association between two phenomena that are usually unrelated. p. 1159

classification An arrangement of organisms into hierarchical groups that reflect their relatedness. p. 428

clathrin The network of proteins that coat and reinforce the cytoplasmic surface of cell membranes. p. 114

cleavage Mitotic cell divisions of the zygote that produce a blastula from a fertilized ovum. pp. 596, 1015

climate The weather conditions prevailing over an extended period of time. p. 160

climax community A relatively stable, late successional stage in which the dominant vegetation replaces itself and persists until an environmental disturbance eliminates it, allowing other species to invade. p. 735

cline A pattern of smooth variation in a characteristic along a geographic gradient. p. 434

clitoris The structure at the junction of the labia minora in front of the vulva, homologous to the penis in the male. p. 999

clonal expansion The proliferation of the activated CD4 T cell by cell division to produce a clone of cells. p. 1278

clonal selection The process by which a lymphocyte is specifically selected for cloning when it encounters a foreign antigen from among a randomly generated, enormous diversity of lymphocytes with receptors that specifically recognize the antigen. p. 1279

clone An individual genetically identical to an original cell from which it descended. pp. 171, 191

closed circulatory system A circulatory system in which the fluid, blood, is confined in blood vessels and is distinct from the interstitial fluid. pp. 616, 966

clumped dispersion A pattern of distribution in which individuals in a population are grouped together. p. 681

cnidocyte A prey-capturing and defensive cell in the epidermis of cnidarians. p. 603

coactivator (mediator) In eukaryotes, a large multiprotein complex that bridges between activators at an enhancer and proteins at the promoter and promoter proximal region to stimulate transcription. p. 324

coat The protective layer of protein that surrounds the nucleic acid core of a virus in free form. *See* capsid. p. 501

coated pit A depression in the plasma membrane that contains receptors for macromolecules to be taken up by endocytosis. p. 114

coccoid Spherical prokaryotic cell. p. 482

cochlea A snail-shaped structure (in vertebrates) in the inner ear containing the organ of hearing. p. 1106

codominance Condition in which alleles have approximately equal effects in individuals, making the alleles equally detectable in heterozygotes. p. 232

codon Each three-letter word (triplet) of the genetic code. p. 292

coelom A fluid-filled body cavity in bilaterally symmetrical animals that is completely lined with derivatives of mesoderm. p. 596

coelomate A body plan of bilaterally symmetrical animals that have a coelom. p. 596

coenzymes Organic cofactors that include complex chemical groups of various kinds. p. 89

coevolution The evolution of genetically based, reciprocal adaptations in two or more species that interact closely in the same ecological setting. pp. 588, 711

cofactor An inorganic or organic nonprotein group that is necessary for catalysis to take place. p. 88

cohesion The high resistance of water molecules to separation. p. F-16

cohesion–tension mechanism of water transport A model of how water is transported from roots to leaves in vascular plants; the evaporation of water from leaves pulls water up in the xylem by creating a continuous negative pressure (tension) that extends to roots. p. 870

cohesion–tension theory of water transport *See* cohesion-tension mechanism of water transport. p. 870

cohort A group of individuals of similar age. p. 685

coleoptile A protective sheath that covers the shoot apical meristem and plumule of the embryo in monocots, such as grasses, as it pushes up through soil. p. 892

coleorhiza A sheath that encloses the radicle of an embryo until it breaks out of the seed coat and enters the soil as the primary root. p. 893

collagen Fibrous glycoprotein—very rich in carbohydrates—embedded in a network of proteoglycans. p. 951

collecting duct A location where urine leaving individual nephrons is processed further. p. 1245

collenchyma One of three simple plant tissues. Flexibly supports rapidly growing plant parts. Its elongated cells are alive at maturity and often collectively form strands or a sheathlike cylinder under the dermal tissue of growing shoot regions and leaf stalks. p. 842

colon The main part of the large intestine. p. 1206

colony Multiple individual organisms of the same species living in a group. p. 516

combinatorial gene regulation The combining of a few regulatory proteins in particular ways so that the transcription of a wide array of genes can be controlled and a large number of cell types can be specified. p. 325

commaless The sequential nature of the words of the nucleic acid code, with no indicators such as commas or spaces to mark the end of one codon and the beginning of the next. p. 294

commensalism A symbiotic interaction in which one species benefits and the other is unaffected. pp. 722, 1267

community Populations of all species that occupy the same area. p. 315

community ecology The ecological discipline that examines groups of populations occurring together in one area. p. 680

companion cell A specialized parenchyma cell that is connected to a mature sieve tube member by plasmodesmata and assists sieve tube members with both the uptake of sugars and the unloading of sugars in tissues. p. 846

comparative genomics A technique for discovering relatedness among organisms by considering the similarity of their respective genome sequences. p. 364

comparative morphology Analysis of the structure of living and extinct organisms. p. 393

compass orientation A wayfinding mechanism that allows animals to move in a particular direction, often over a specific distance or for a prescribed length of time. p. 1171

competitive exclusion principle The ecological principle stating that populations of two or more species cannot coexist indefinitely if they rely on the same limiting resources and exploit them in the same way. p. 717

competitive inhibition Inhibition of an enzyme reaction by an inhibitor molecule that resembles the normal substrate closely enough that it fits into the active site of the enzyme. p. 90

complement system A nonspecific defence mechanism activated by invading pathogens, made up of more than 30 interacting soluble plasma proteins circulating in the blood and interstitial fluid. p. 1271

complementary base-pairing Feature of DNA in which the specific purine–pyrimidine base pairs A–T (adenine–thymine) and G–C (guanine–cytosine) occur to bridge the two sugar–phosphate backbones. p. 269

complementary DNA (cDNA) A DNA molecule that is complementary to an mRNA molecule, synthesized by reverse transcriptase. p. 344

complete digestive system A digestive system with a mouth at one end, through which food enters, and an anus at the other end, through which undigested waste is voided. p. 610

complete metamorphosis The form of metamorphosis in which an insect passes through four separate stages of growth: egg, larva, pupa, and adult. p. 629

compound A molecule whose component atoms are different. p. F-8

compound eye The eye of most insects and some crustaceans, composed of many-faceted, light-sensitive units called ommatidia fitted closely together, each with its own refractive system and each forming a portion of an image. pp. 622, 1108

concentration The number of molecules or ions of a substance in a unit volume of space. p. F-18

concentration gradient The concentration difference that drives diffusion. p. 105

conduction The flow of heat between atoms or molecules in direct contact. pp. 1079, 1252

cone In the vertebrate eye, a photoreceptor in the retina that is specialized for detection of different wavelengths (colours). In cone-bearing plants, a cluster of sporophylls. pp. 577, 1109

conformation The overall three-dimensional shape of a protein. p. 88

conformational change Alteration in the three-dimensional shape of a protein. p. 1137

conformers Animals having internal environments that change as the external environment changes. p. 958

conidium (plural, conidia) An asexually produced fungal spore. p. 549

Coniferophyta The major phylum of cone-bearing gymnosperms, most of which are substantial trees; includes pines, firs, and other conifers. p. 583

conjugation In bacteria, the process by which a copy of part of the DNA of a donor cell moves through the cytoplasmic bridge into the recipient cell where genetic recombination can occur. In ciliate protozoans, a process of sexual reproduction in which individuals of the same species temporarily couple and exchange genetic material. p. 193

connective tissue Tissue with cells scattered through an extracellular matrix; forms layers in and around body structures that support other body tissues, transmit mechanical and other forces, and in some cases act as filters. p. 951

conodont An abundant, bonelike fossil dating from the early Paleozoic era through the early Mesozoic era, now described as a feeding structure of some of the earliest vertebrates. p. 645

consciousness Awareness of oneself, one's identity, and one's surroundings, with understanding of the significance and likely consequences of events. p. 1129

conservation biology An interdisciplinary science that focuses on the maintenance and preservation of biodiversity. p. 419

consumer An organism that consumes other organisms in a community or ecosystem. p. 728

contact inhibition The inhibition of movement or proliferation of normal cells that results from cell–cell contact. p. 183

contractile vacuole A specialized cytoplasmic organelle that pumps fluid in a cyclical manner from within the cell to the outside by alternately filling and then contracting to release its contents at various points on the surface of the cell. p. 517

control Treatment that tells what would be seen in the absence of the experimental manipulation. p. F-3

convection The transfer of heat from a body to a fluid, such as air or water, that passes over its surface. p. 1252

convergent evolution The evolution of similar adaptations in distantly related organisms that occupy similar environments. p. 403

coral reef A structure made from the hard skeletons of coral animals or polyps; found largely in tropical and subtropical marine environments. p. 604

core The nucleic acid centre of a virus in the free form. p. 284

corepressor In the regulation of gene expression in bacteria, a regulatory molecule that combines with a repressor to activate it and shut off an operon. p. 319

cork A nonliving, impermeable secondary tissue that is one element of bark. p. 855

cork cambium A lateral meristem in plants that forms periderm, which in turn produces cork. p. 855

cornea The transparent layer that forms the front wall of the eye, covering the iris. p. 1108

corona The ciliated crownlike organ at the anterior end of rotifers used for feeding or locomotion. p. 610

corpus callosum A structure formed of thick axon bundles that connect the two cerebral hemispheres and coordinate their functions. p. 1127

corpus luteum Cells remaining at the surface of the ovary during the luteal phase; the structure acts as an endocrine gland, secreting several hormones: estrogens, large quantities of progesterone, and inhibin. p. 1000

cortex Generally, an outer, rindlike layer. In mammals, the outer layer of the brain, the kidneys, or the adrenal glands. In plants, the outer region of tissue in a root or stem lying between the epidermis and the vascular tissue, composed mainly of parenchyma. pp. 381, 667, 911

cortical granule A secretory vesicle just under the plasma membrane of an egg cell. p. 995

cortisol The major glucocorticoid steroid hormone secreted by the adrenal cortex, which increases blood glucose by promoting breakdown of proteins and fats. p. 1059

cotranslational import A mechanism in which proteins end up on the inside (lumen) of the endoplasmic reticulum (ER) as they are translated by ribosome associated with the ER. p. 308

cotransport *See* symport. p. 112

cotyledon A leaf of a seed plant embryo; also known as a seed leaf. p. 892

countercurrent exchange A mechanism in which the water flowing over the gills moves in a direction opposite to the flow of blood under the respiratory surface (can also apply to transfer of heat). p. 1222

coupled reaction Reaction that occurs when an exergonic reaction is joined to an endergonic reaction, producing an overall reaction that is exergonic. p. 84

courtship display A behaviour performed by males to attract potential mates or to reinforce the bond between a male and a female. pp. 397, 431

covalent bond Bond formed by electron sharing between atoms. p. F-12

cranial nerve A nerve that connects the brain directly to the head, neck, and body trunk. p. 642

cranium The part of the skull that encloses the brain. p. 641

crassulacean acid metabolism (CAM) A biochemical variation of photosynthesis that was discovered in a member of the plant family Crassulaceae. Carbon dioxide is taken up and stored during the night to allow the stomata to remain closed during the daytime, decreasing water loss. pp. 161, 875

Crenarchaeota A major group of the domain Archaea, separated from the other archaeans based mainly on rRNA sequences. p. 497

crista (plural, **cristae**) Fold that expands the surface area of the inner mitochondrial membrane. p. 41

critical period A restricted stage of development early in life during which an animal has the capacity to respond to specific environmental stimuli. p. 1159

crop Of birds, an enlargement of the digestive tube where the digestive contents are stored and mixed with lubricating mucus. p. 823

cross-pollination Fertilization of one plant by a different plant. p. 219

crossing-over The recombination process in meiosis, in which chromatids exchange segments. p. 208

crossover Site of recombination during meiosis. Also referred to as a *chiasmata*. p. 208

cryptochrome A light-absorbing protein that is sensitive to blue light and that may also be an important early step in various light-based growth responses. p. 934

cupula In certain mechanoreceptors, a gelatinous structure with stereocilia extending into it that moves with pressure changes in the surrounding water; movement of the cupula bends the stereocilia, which triggers release of neurotransmitters. p. 1101

cuticle The outer layer of plants and some animals, which helps prevent desiccation by slowing water loss. pp. 564, 847

Cycadophyta A phylum of palmlike gymnosperms known as cycads; the pollen-bearing and seed-bearing cones (strobili) occur on separate plants. p. 583

cyclic AMP (cAMP) In particular signal transduction pathways, a second messenger that activates protein kinases, which elicit the cellular response by adding phosphate groups to specific target proteins. cAMP functions in one of two major G protein–coupled receptor response pathways. p. 318

cyclic electron transport An electron transport pathway associated with photosystem I in photosynthesis that produces ATP without the synthesis of NADPH. p. 155

cyclin In eukaryotes, protein that regulates the activity of Cdk (cyclin-dependent kinase) and controls progression through the cell cycle. p. 181

cyclin-dependent kinase (Cdk) A protein kinase that controls the cell cycle in eukaryotes. p. 181

cytochrome Protein with a heme prosthetic group that contains an iron atom. p. 125

cytokine A molecule secreted by one cell type that binds to receptors on other cells and, through signal transduction pathways, triggers a response. In innate immunity, cytokines are secreted by activated macrophages. p. 1269

cytokinesis Division of the cytoplasm into two daughter cells following the nuclear division stage of mitosis. p. 176

cytokinin A hormone that promotes and controls growth responses of plants. p. 925

cytoplasm All parts of the cell that surround the central nuclear or nucleoid region. p. 30

cytoplasmic determinants The mRNA and proteins stored in the egg cytoplasm that direct the first stages of animal development in the period before genes of the zygote become active. p. 1014

cytoplasmic inheritance Pattern in which inheritance follows that of genes in the cytoplasmic organelles, mitochondria, or chloroplasts. p. 259

cytoplasmic streaming Intracellular movement of cytoplasm. p. 43

cytosine A pyrimidine that base-pairs with guanine in nucleic acids. p. 268

cytoskeleton The interconnected system of protein fibres and tubes that extends throughout the cytoplasm of a eukaryotic cell. p. 30

cytosol Aqueous solution in the cytoplasm containing ions and various organic molecules. p. 30

cytotoxic T cell A T lymphocyte that functions in cell-mediated immunity to kill body cells infected by viruses or transformed by cancer. p. 1283

daily torpor A period of inactivity and lowered metabolic rate that allows an endotherm to conserve energy when environmental temperatures are low. p. 1260

dalton A standard unit of mass, about 1.66×10^{24} g. p. 333

day-neutral plant A plant that flowers without regard to photoperiod. p. 940

decomposer A small organism, such as a bacterium or fungus, that feeds on the remains of dead organisms, breaking down complex biological molecules or structures into simpler raw materials. p. 145

dehydration synthesis reaction Reaction during which the components of a water molecule are removed, usually as part of the assembly of a larger molecule from smaller subunits. Also referred to as a *condensation reaction*. p. F-21

degeneracy (redundancy) The feature of the genetic code in which, with two exceptions, more than one codon represents each amino acid. p. 294

deletion Chromosomal alteration that occurs if a broken segment is lost from a chromosome. p. 251

demographic transition model A graphic depiction of the historical relationship between a country's economic development and its birth and death rates. p. 701

demography The statistical study of the processes that change a population's size and density through time. p. 684

denaturation A loss of both the structure and function of a protein due to extreme conditions that unfold it from its conformation. p. F-36

dendrite The branched extension of the nerve cell body that receives signals from other nerve cells. p. 1069

dendritic cell A type of phagocyte, so called because it has many surface projections that resemble dendrites of neurons, that engulfs a bacterium in infected tissue by phagocytosis. p. 1277

denitrification A metabolic process in which certain bacteria convert nitrites or nitrates into nitrous oxide and then into molecular nitrogen, which enters the atmosphere. p. 766

density dependent Description of environmental factors for which the strength of their effect on a population varies with the population's density. p. 694

density independent Description of environmental factors for which the strength of their effect on a population does not vary with the population's density. p. 696

deoxyribonucleic acid (DNA) The large, double-stranded, helical molecule that contains the genetic material of all living organisms. p. 264

deoxyribose A five-carbon sugar to which the nitrogenous bases in nucleotides of DNA link covalently. p. F-37

depolarized State of the membrane (which was polarized at rest) as the membrane potential becomes less negative. p. 1074

deposit feeder An animal that consumes particles of organic matter from the solid substrate on which it lives. p. 1195

derived character A new version of a trait found in the most recent common ancestor of a group. p. 458

dermal tissue system The plant tissue system that comprises the outer tissues of the plant body, including the epidermis and periderm; it serves as a protective covering for the plant body. p. 841

dermis The skin layer below the epidermis; it is packed with connective tissue fibres such as collagen, which resist compression, tearing, or puncture of the skin. p. 1256

descent with modification Biological evolution. p. 400

desert A sparsely vegetated biome that forms where precipitation averages less than 25 cm per year. p. 750

desmosome Anchoring junction for which microfilaments anchor the junction in the underlying cytoplasm. p. 47

determinate cleavage A type of cleavage in protostomes in which each cell's developmental path is determined as the cell is produced. p. 597

determinate growth The pattern of growth in most animals in which individuals grow to a certain size and then their growth slows dramatically or stops. p. 839

determination Mechanism in which the developmental fate of a cell is set. p. 1017

detritivore An organism that extracts energy from the organic detritus (refuse) produced at other trophic levels. p. 594

deuterostome A division of the Bilateria in which blastopore forms the anus during development and the mouth appears later (includes Echinodermata and Chordata). p. 596

development A series of programmed changes encoded in DNA, through which a fertilized egg divides into many cells that are ultimately transformed into an adult, which is itself capable of reproduction. p. 2

diabetes mellitus A disease that results from problems with insulin production or action. p. 1060

diapsid (lineage Diapsida) A member of a group within the amniote vertebrates with a skull with two temporal arches. Their living descendants include lizards and snakes, crocodilians, and birds. p. 655

diastole The period of relaxation and filling of the heart between contractions. p. 974

diatom Photosynthetic single-celled organisms with a glassy silica shell; also called bacillariophytes. p. 524

differentiation Follows determination and involves the establishment of a cell-specific developmental program in the cells. Differentiation results in cell types with clearly defined structures and functions. p. 1017

diffusion The net movement of ions or molecules from a region of higher concentration to a region of lower concentration. p. 105

digestion The splitting of carbohydrates, proteins, lipids, and nucleic acids in foods into chemical subunits small enough to be absorbed into the body fluids and cells of an animal. p. 1189

dihybrid A zygote produced from a cross that involves two characters. p. 225

dihybrid cross A cross between two individuals that are heterozygous for two pairs of alleles. p. 225

dikaryon The life stage in certain fungi in which a cell contains two genetically distinct haploid nuclei. p. 553

dikaryotic hyphae Hyphae containing two separate nuclei in one cell. p. 549

dioecious Having male flowers and female flowers on different plants of the same species. p. 885

diphyodont Having two generations of teeth, milk (baby) teeth and adult teeth. p. 667

diploblastic An animal body plan in which adult structures arise from only two cell layers, the ectoderm and the endoderm. p. 595

diploid An organism or cell with two copies of each type of chromosome in its nucleus. p. 170

direct neurotransmitter A neurotransmitter that binds directly to a ligand-gated ion channel in the postsynaptic membrane, opening or closing the channel gate and altering the flow of a specific ion or ions in the postsynaptic cell. p. 1082

directional selection A type of selection in which individuals near one end of the phenotypic spectrum have the highest relative fitness. p. 412

discontinuous replication Replication in which a DNA strand is formed in short lengths that are synthesized in the direction opposite to DNA unwinding. p. 275

dispersal 1. The movement of organisms away from their place of origin, as well as the movement from one breeding site to another; 2. The movement of material that is used by an organism to move to the next stage in their life cycle. p. 601

dispersed duplication Gene copies that are found in different places in the genome, often on two different chromosomes. p. 382

dispersion The spatial distribution of individuals within a population's geographic range. p. 681

disruptive selection A type of natural selection in which extreme phenotypes have higher relative fitness than intermediate phenotypes. p. 414

dissociation The separation of water to produce hydrogen ions and hydroxide ions. p. 1230

distal convoluted tubule The tubule in the human nephron that drains urine into a collecting duct that leads to the renal pelvis. p. 1246

disturbance climax (disclimax) community An ecological community in which regular disturbance inhibits successional change. p. 740

DNA *See* deoxyribonucleic acid. p. 264

DNA chip *See* DNA microarray. p. 376

DNA fingerprinting Technique in which DNA samples are used to distinguish between individuals of the same species. p. 349

DNA helicase An enzyme that catalyzes the unwinding of DNA template strands. p. 274

DNA hybridization Technique in which a gene or sequence of interest is identified in a set of clones when it base-pairs with a single-stranded DNA or RNA molecule called a nucleic acid probe. p. 344

DNA ligase In DNA replication, an enzyme that seals the nicks left after RNA primers are replaced with DNA. p. 276

DNA methylation Process in which a methyl group is added enzymatically to cytosine bases in the DNA. p. 327

DNA microarray A solid surface divided into a microscopic grid of thousands of spaces each containing thousands of copies of a DNA probe. DNA microarrays are used commonly for analysis of gene activity and for detecting differences between cell types. Also referred to as a *DNA chip*. p. 376

DNA polymerase An enzyme that assembles complementary nucleotide chains during DNA replication. p. 271

DNA polymerase I A specialized polymerase responsible for removing RNA primers and replacing them with DNA. p. 276

DNA polymerase III The main, "general-purpose" polymerase for replicating DNA. p. 276

DNA repair mechanism Mechanism to correct base-pair mismatches that escape proofreading. p. 282

DNA technologies Techniques to isolate, purify, analyze, and manipulate DNA sequences. p. 339

domain In protein structure, a distinct, large structural subdivision produced in many proteins by the folding of the amino acid chain. In systematics, the highest taxonomic category; a group of cellular organisms with characteristics that set it apart as a major branch of the evolutionary tree. pp. 104, 428

domestication Selective breeding of other species to increase desirable characteristics in progeny. p. 808

dominance The masking effect of one allele over another. p. 221

dominance hierarchy A social system in which the behaviour of each individual is constrained by that individual's status in a highly structured social ranking. p. 1178

dominant Refers to the allele expressed when more than one allele is present. p. 221

dormancy A period in the life cycle in which biological activity is suspended. pp. 894, 941

dorsal Indicating the back side of an animal. p. 596

dorsal lip of the blastopore A crescent-shaped depression rotated clockwise 90° on the embryo surface that marks the region derived from the grey crescent, to which cells from the animal pole move as gastrulation begins. p. 1018

double fertilization The characteristic feature of sexual reproduction in flowering plants. In the embryo sac, one sperm nucleus unites with the egg to form a diploid zygote from which the embryo develops, and another unites with two polar nuclei to form the primary endosperm nucleus. p. 890

double helix Two nucleotide chains wrapped around each other in a spiral. p. 2

double-helix model Model of DNA consisting of two complementary sugar–phosphate backbones. p. 269

duodenum A short region of the small intestine where secretions from the pancreas and liver enter a common duct. p. 1204

duplication Chromosomal alteration that occurs if a segment is broken from one chromosome and inserted into its homologue. p. 251

E site The site where an exiting tRNA binds prior to its release from the ribosome. p. 302

ecdysis Shedding of the cuticle, exoskeleton, or skin; moulting. p. 600

ecdysone A steroid hormone that controls cuticle formation in insects and crustaceans and possibly nematodes. p. 1049

echolocation A behaviour in which an animal compares echoes of sounds it produced to the original signals. Differences between pulses and echoes allow location of obstacles and prey. p. 1097

ecological community An assemblage of species living in the same place. p. 680

ecological efficiency The ratio of net productivity at one trophic level to net productivity at the trophic level below it. p. 752

ecological isolation A prezygotic reproductive isolating mechanism in which species that live in the same geographic region occupy different habitats. p. 431

ecological niche The resources a population uses and the environmental conditions it requires over its lifetime. p. 717

ecological pyramid A diagram illustrating the effects of energy transfer from one trophic level to the next. p. 754

ecological succession A somewhat predictable series of changes in the species composition of a community over time. p. 561

ecology The study of the interactions between organisms and their environments. p. 679

ecosystem A group of biological communities interacting with their shared physical environment. p. 749

ecosystem ecology An ecological discipline that explores the cycling of nutrients and the flow of energy between the biotic components of an ecological community and the abiotic environment. p. 680

ecotone A wide transition zone between adjacent communities. p. 725

ectoderm The outermost of the three primary germ layers of an embryo, which develops into epidermis and nervous tissue. pp. 595, 1015

ectomycorrhiza A mycorrhiza that grows between and around the young roots of trees and shrubs but does not enter root cells. p. 557

ectoparasite A parasite that lives on the exterior of its host organism. p. 609

ectotherm An animal that obtains its body heat primarily from the external environment. p. 1253

effector In signal transduction, a plasma membrane–associated enzyme, activated by a G protein, that generates one or more second messengers. In homeostatic feedback, the system that returns the condition to the set point if it has strayed away. pp. 960, 1070

effector T cell A cell involved in effecting—bringing about—the specific immune response to an antigen. p. 1278

efferent arteriole The arteriole that receives blood from the glomerulus. p. 1247

efferent neuron A neuron that carries the signals indicating a response away from the interneuron networks to the effectors. p. 1070

egg cell The female reproductive cell. p. 888

eggs Nonmotile gametes. p. 990

Elasmobranchii Cartilaginous fishes, including the skates and rays. p. 647

elastin A rubbery protein in some connective tissues that adds elasticity to the extracellular matrix. It is able to return to its original shape after being stretched, bent, or compressed. p. 952

electrical signalling A means of animal communication in which a signaller emits an electric discharge that can be received by another individual. p. 1167

electrical synapse A mechanical and electrically conductive link between two abutting neurons that is formed at the gap junction. p. 1072

electrocardiogram (ECG) Graphic representation of the electrical activity within the heart, detected by electrodes placed on the body. p. 977

electrochemical gradient A difference in chemical concentration and electric potential across a membrane. pp. 111, 1073

electromagnetic spectrum The range of wavelengths or frequencies of electromagnetic radiation extending from gamma rays to the longest radio waves and including visible light. p. 4

electron Negatively charged particle outside the nucleus of an atom. p. 5

electron microscope Microscope that uses electrons to illuminate the specimen. p. 29

electron shell In chemistry and physics, may be thought of as an orbit followed by electrons around an atom's nucleus. Also known as *principal energy level*. p. F-10

electron transfer system Stage of cellular respiration in which high-energy electrons produced from glycolysis, pyruvate oxidation, and the citric acid cycle are delivered to oxygen by a sequence of electron carriers. p. 162

electronegativity The measure of an atom's attraction for the electrons it shares in a chemical bond with another atom. p. F-13

electroreceptor A specialized sensory receptor that detects electrical fields. pp. 648, 1121

element A pure substance that cannot be broken down into simpler substances by ordinary chemical or physical techniques. p. F-8

elongation factor Proteins that promote various steps in the elongation of peptides during translation. p. 305

embryo An organism in its early stage of reproductive development, beginning in the first moments after fertilization. p. 47

embryo sac The female gametophyte of angiosperms, within which the embryo develops; it usually consists of seven cells: an egg cell, an endosperm mother cell, and five other cells with fleeting reproductive roles. p. 888

embryonic stem cell Stem cells in the mammalian embryo that can differentiate into any cell type. p. 354

emigration The movement of individuals out of a population. p. 684

endangered species A species in immediate danger of extinction throughout all or a significant portion of its range. p. 419

endemic species A species that occurs in only one place on Earth. p. 785

endergonic process Reaction that can proceed only if free energy is supplied. p. 79

endocrine gland Any of several ductless secretory organs that secrete hormones into the blood or extracellular fluid. pp. 950, 1047

endocrine system The system of glands that release their secretions (hormones) directly into the circulatory system. p. 1046

endocytic vesicle Vesicle that carries proteins and other molecules from the plasma membrane to destinations within the cell. p. 38

endocytosis In eukaryotes, the process by which molecules are brought into the cell from the exterior involving a bulging in of the plasma membrane that pinches off to form an endocytic vesicle. p. 38

endoderm The innermost of the three primary germ layers of an embryo, which develops into the gastrointestinal tract and, in some animals, the respiratory organs. pp. 595, 1015

endodermis The innermost layer of the root cortex; a selectively permeable barrier that helps control the movement of water and dissolved minerals into the stele. pp. 854, 868

endomembrane system In eukaryotes, a collection of interrelated internal membranous sacs that divide a cell into functional and structural compartments. pp. 36, 68

endoparasite A parasite that lives in the internal organs of its host organism. p. 723

endoplasmic reticulum (ER) In eukaryotes, an extensive interconnected network of cisternae that is responsible for the synthesis, transport, and initial modification of proteins and lipids. p. 36

endorphin One of a group of small proteins occurring naturally in the brain and around nerve endings that bind to opiate receptors and thus can raise the pain threshold. p. 1055

endoskeleton A supportive internal body structure, such as bones, that provides support. p. 1146

endosperm Nutritive tissue inside the seeds of flowering plants. p. 890

endosporous Pattern of development in some plants (e.g., seed plants) in which the gametophyte develops inside the spore wall. p. 580

endosymbiosis A symbiotic association in which one symbiont or partner lives inside the other. p. 66

endotherm An animal that obtains most of its body heat from internal physiological sources. p. 1253

endothermic Refers to reactions that absorb energy. pp. 78, 667

endotoxin A lipopolysaccharide released from the outer membrane of the cell wall when a bacterium dies and lyses. p. 489

energy The capacity to do work. p. 75

energy budget The total amount of energy that an organism can accumulate and use to fuel its activities. p. 686

energy coupling The process by which ATP is brought in close contact with a reactant molecule involved in an endergonic reaction, and when the ATP is hydrolyzed, the terminal phosphate group is transferred to the reactant molecule. p. 84

energy levels Regions of space within an atom where electrons are found. Also referred to as *energy shells*. p. 121

enhancer In eukaryotes, a region at a significant distance from the beginning of a gene, containing regulatory sequences that determine whether the gene is transcribed at its maximum possible rate. p. 322

enterocoelom In deuterostomes, the body cavity pinched off by outpocketings of the archenteron. p. 598

enthalpy Potential energy in a system. p. 78

entropy Disorder, in thermodynamics. p. 77

envelope Outer glycoprotein layer surrounding the capsid of some viruses, derived in part from host cell plasma membrane. p. 501

enveloped virus A virus that has a surface membrane derived from its host cell. p. 502

enzymatic hydrolysis A process in which chemical bonds are broken by the addition of H^+ and OH^-, the components of a molecule of water. p. 1195

enzyme Protein that accelerates the rate of a cellular reaction. p. 86

eosinophil A type of leukocyte that targets extracellular parasites too large for phagocytosis in the inflammatory response. p. 1269

epiblast The top layer of the blastodisc. p. 1019

epicotyl The upper part of the axis of an early plant embryo, located between the cotyledons and the first true leaves. p. 893

epidermis A complex tissue that covers an organism's body in a single continuous layer or sometimes in multiple layers of tightly packed cells. pp. 603, 846, 1256

epididymis A coiled storage tubule attached to the surface of each testis. p. 1003

epigenetics The study of changes to gene expression that do not arise from changes in the DNA sequence (i.e., mutations). Epigenetic changes may arise from chemical modification of bases (e.g., methylation), chromatin remodelling, protein or RNA binding, etc. p. 1041

epiglottis A flaplike valve at the top of the trachea. p. 1202

epinephrine A nontropic amine hormone secreted by the adrenal medulla. p. 1058

epiphyte A plant that grows independently on other plants and obtains nutrients and water from the air. p. 913

epistasis Interaction of genes, with one or more alleles of a gene at one locus inhibiting or masking the effects of one or more alleles of a gene at a different locus. p. 234

epithelial tissue Tissue formed of sheetlike layers of cells that are usually joined tightly together, with little extracellular matrix material between them. They protect body surfaces from invasion by bacteria and viruses and secrete or absorb substances. p. 949

epitope The small region of an antigen molecule to which B-cell receptors or T-cell receptors bind. p. 1274

equilibrium theory of island biogeography A hypothesis suggesting that the number of species on an island is governed by a give and take between the immigration of new species to the island and the extinction of species already there. p. 741

ER (endoplasmic reticulum) lumen The enclosed space surrounded by a cisterna. p. 36

erythrocyte A red blood cell that contains hemoglobin, a protein that transports O2 in blood. p. 970

erythropoietin (EPO) A hormone that stimulates stem cells in bone marrow to increase erythrocyte production. p. 971

esophagus A connecting passage of the digestive tube. p. 638

essential element Any of a number of elements required by living organisms to ensure normal reproduction, growth, development, and maintenance. p. 903

essential fatty acid Any fatty acid that the body cannot synthesize but needs for normal metabolism. p. 1190

essential nutrient Any of the essential amino acids, fatty acids, vitamins, and minerals required in the diet of an animal. p. 1190

estivation Seasonal torpor in an animal that occurs in summer. p. 1261

estradiol A form of estrogen. p. 1060

estrogen Any of the group of female sex hormones. p. 1059

ethology A discipline that focuses on how animals behave. p. 1159

ethylene A plant hormone that helps regulate seedling growth; stem elongation; the ripening of fruit; and the abscission of fruits, leaves, and flowers. p. 926

euchromatin In eukaryotes, regions of loosely packed chromatin fibres in interphase nuclei. p. 285

eudicot A plant belonging to the Eudicotyledones, one of the two major classes of angiosperms; their embryos generally have two seed leaves (cotyledons), and their pollen grains have three grooves. pp. 587, 840

Eukarya The domain that includes all eukaryotes, organisms that contain a membrane-bound nucleus within each of their cells; all protists, plants, fungi, and animals. p. 482

eukaryote Organism in which the DNA is enclosed in a nucleus. p. 31

eukaryotic chromosome A DNA molecule, with its associated proteins, in the nucleus of a eukaryotic cell. p. 35

euploid An individual with a normal set of chromosomes. p. 253

Euryarchaeota A major group of the domain Archaea, members of which are found in different extreme environments. They include methanogens, extreme halophiles, and some extreme thermophiles. p. 497

eusocial A form of social organization, observed in some insect species, in which numerous related individuals—a large percentage of them sterile female workers—live and work together in a colony for the reproductive benefit of a single queen and her mate(s). p. 1181

eustachian tube A duct leading from the air-filled middle ear to the throat that protects the eardrum from damage caused by changes in environmental atmospheric pressure. p. 1106

evaporation Heat transfer through the energy required to change a liquid to a gas. p. 1252

evolution The main unifying concept in biology, explaining how the diversity of life on Earth arose and how species change over time in response to changes in their abiotic and biotic environment. p. 392

evolutionary developmental biology A field of biology that compares the genes controlling the developmental processes of different animals to determine the evolutionary origin of morphological novelties and developmental processes. p. 456

evolutionary divergence A process whereby natural selection or genetic drift causes populations to become more different over time. p. 399

exchange diffusion *See* antiport. p. 112

excitatory postsynaptic potential (EPSP) The change in membrane potential caused when a neurotransmitter opens a ligand-gated Na^+ channel and Na^+ enters the cell, making it more likely that the postsynaptic neuron will generate an action potential. p. 1085

excretion The process that helps maintain the body's water and ion balance while ridding the body of metabolic wastes. p. 587

exergonic process Reaction that has a negative ΔG because it releases free energy. p. 79

exocrine gland A gland that is connected to the epithelium by a duct and that empties its secretion at the epithelial surface. p. 950

exocytosis In eukaryotes, the process by which a secretory vesicle fuses with the plasma membrane and releases the vesicle contents to the exterior. p. 38

exodermis In the roots of some plants, an outer layer of root cortex that may limit water losses from roots and help regulate the absorption of ions. p. 854

exon An amino acid–coding sequence present in premRNA that is retained in a spliced mRNA that is translated to produce a polypeptide. p. 298

exon shuffling Molecular evolutionary process that combines exons of two or more existing genes to produce a gene that encodes a protein with an unprecedented function. p. 382

exoskeleton A hard external covering of an animal's body that blocks the passage of water and provides support and protection. pp. 620, 1145

exothermic Refers to processes that release energy. p. 78

exotoxin A toxic protein that leaks from or is secreted from a bacterium and interferes with the biochemical processes of body cells in various ways. p. 489

experimental data Information that describes the result of a careful manipulation of the system under study. p. 415

experimental variable The variable to which any difference in observations of experimental treatment subjects and control treatment subjects is attributed. p. F-3

exploitative competition Form of competition in which two or more individuals or populations use the same limiting resources. p. 717

exponential (model of population growth) Model that describes unlimited population growth. p. 689

expression vector A plasmid that can not only carry cloned genes but also drive their expression. p. 352

external fertilization The process in which sperm and eggs are shed into the surrounding water, occurring in most aquatic invertebrates, bony fishes, and amphibians. p. 993

external gill A gill that extends out from the body and lacks a protective covering. p. 1222

extinction The death of the last individual in a species or the last species in a lineage. p. 781

extracellular digestion Digestion that takes place outside body cells, in a pouch or tube enclosed within the body. p. 603

extracellular fluid The fluid occupying the spaces between cells in multicellular animals. p. 958

extracellular matrix (ECM) A molecular system that supports and protects cells and provides mechanical linkages. pp. 47, 948

extra-embryonic membrane A primary tissue layer extended outside the embryo that conducts nutrients from the yolk to the embryo, exchanges gases with the environment outside the egg, or stores metabolic wastes removed from the embryo. p. 1020

eye The organ animals use to sense light. p. 9

F pilus Structure on the cell surface that allows an F^+ donor bacterial cell to attach to an F^- recipient bacterial cell. Also referred to as a *sex pilus*. p. 193

F^+ cell Donor cell in conjugation between bacteria. p. 193

F^- cell Recipient cell in conjugation between bacteria. p. 193

F_1 generation The first generation of offspring from a genetic cross. p. 219

F_2 generation The second generation of offspring from a genetic cross. p. 220

facilitated diffusion Mechanism by which polar and charged molecules diffuse across membranes with the help of transport proteins. p. 106

facilitation hypothesis A hypothesis that explains ecological succession, suggesting that species modify the local environment in ways that make it less suitable for themselves but more suitable for colonization by species typical of the next successional stage. p. 737

facultative anaerobe An organism that can live in the presence or absence of oxygen, using oxygen when it is present and living by fermentation under anaerobic conditions. pp. 139, 487

familial (hereditary) cancer Cancer that runs in a family. p. 333

family A Linnaean taxonomic category that ranks below an order and above a genus. p. 428

family planning program A program that educates people about ways to produce an optimal family size on an economically feasible schedule. p. 702

fast block (to polyspermy) The barrier set up by the wave of depolarization triggered when sperm and egg fuse, making it impossible for other sperm to enter the egg. p. 994

fast muscle fibre A muscle fibre that contracts relatively quickly and powerfully. p. 1142

fat Neutral lipid that is semisolid at biological temperatures. p. 99

fat-soluble vitamin A vitamin that dissolves in liquid fat or fatty oils, in addition to water. p. 1193

fate mapping Mapping of adult or larval structures onto the region of the embryo from which each structure developed. p. 1035

fatty acid One of two components of a neutral lipid, containing a single hydrocarbon chain with a carboxyl group linked at one end. p. 39

feather A sturdy, lightweight structure of birds, derived from scales in the skin of their ancestors. p. 662

feces Condensed and compacted digestive contents in the large intestine. p. 1207

feedback inhibition In enzyme reactions, regulation in which the product of a reaction acts as a regulator of the reaction. Also referred to as *end-product inhibition*. p. 92

fermentation Process in which electrons carried by NADH are transferred to an organic acceptor molecule rather than to the electron transfer system. p. 137

fertilization The fusion of the nuclei of an egg and sperm cell, which initiates development of a new individual. pp. 199, 990

fetus A developing human from the eighth week of gestation onward, at which point, the major organs and organ systems have formed. p. 1025

fibre In sclerenchyma, an elongated, tapered, thick-walled cell that gives plant tissue its flexible strength. p. 843

fibrin A protein necessary for blood clotting; fibrin forms a weblike mesh that traps platelets and red blood cells and holds a clot together. p. 973

fibrinogen A plasma protein that plays a central role in the blood-clotting mechanism. p. 970

fibroblast The type of cell that secretes most of the collagen and other proteins in the loose connective tissue. p. 952

fibronectin A class of glycoproteins that aids in the attachment of cells to the extracellular matrix and helps hold the cells in position. p. 952

fibrous connective tissue Tissue in which fibroblasts are sparsely distributed among dense masses of collagen and elastin fibres that are lined up in highly ordered, parallel bundles, producing maximum tensile strength and elasticity. p. 952

fibrous root system A root system that consists of branching roots rather than a main taproot; roots tend to spread laterally from the base of the stem. p. 853

filament In flowers, the stalk of a stamen, which supports the anther. p. 885

filtration The nonselective movement of some water and a number of solutes—ions and small molecules, but not large molecules such as proteins—into the proximal end of the renal tubules through spaces between cells. p. 1238

first law of thermodynamics The principle that energy can be transferred and transformed but cannot be created or destroyed. p. 76

fission The mode of asexual reproduction in which the parent separates into two or more offspring of approximately equal size. p. 989

fixed action pattern A highly stereotyped instinctive behaviour; when triggered by a specific cue, it is performed over and over in almost exactly the same way. p. 1156

flagellum (plural, flagella) A long, threadlike, cellular appendage responsible for movement; found in both prokaryotes and eukaryotes, but with different structures and modes of locomotion. pp. 43, 486, 601

flame cell The cell that forms the primary filtrate in the excretory system of many bilateria. The urine is propelled through ducts by the synchronous beating of cilia, resembling a flickering flame. p. 608

flower The reproductive structure of angiosperms, consisting of floral parts grouped on a stem; the structure in which seeds develop. p. 586

fluid feeder An animal that obtains nourishment by ingesting liquids that contain organic molecules in solution. p. 1194

fluid mosaic model Model proposing that the membrane consists of a fluid phospholipid bilayer in which proteins are embedded and float freely. p. 98

follicle The ovum and follicle cells. p. 1000

follicle cell A cell that grows from ovarian tissue and nourishes the developing egg. p. 993

follicle-stimulating hormone (FSH) The pituitary hormone that stimulates oocytes in the ovaries to continue meiosis and become follicles. During follicle enlargement, FSH interacts with luteinizing hormone to stimulate follicular cells to secrete estrogens. p. 1055

food chain A depiction of the trophic structure of a community; a portrait of who eats whom. p. 357

food web A set of interconnected food chains with multiple links. p. 572

forebrain The largest division of the brain, which includes the cerebral cortex and basal ganglia. It is credited with the highest intellectual functions. p. 1090

foreskin A loose fold of skin that covers the glans of the penis. *See* prepuce. p. 1004

fossil The remains or traces of an organism of a past geologic age embedded and preserved in Earth's crust. p. 12

founder effect An evolutionary phenomenon in which a population that was established by just a few colonizing individuals has only a fraction of the genetic diversity seen in the population from which it was derived. p. 419

fovea The small region of the retina around which cones are concentrated in mammals and birds with eyes specialized for daytime vision. p. 1109

fragmentation A type of vegetative reproduction in plants in which cells or a piece of the parent break off and then develop into new individuals. pp. 895, 989

frameshift mutation Mutation in a protein-coding gene that causes the reading frame of an mRNA transcribed from the gene to be altered, resulting in the production of a different, and nonfunctional, amino acid sequence in the polypeptide. p. 310

free energy The energy in a system that is available to do work. p. 75

freeze-fracture technique Technique in which experimenters freeze a block of cells rapidly and then fracture the block to split the lipid bilayer and expose the hydrophobic membrane interior. p. 99

fruit A mature ovary, often with accessory parts, from a flower. pp. 586, 891

fruiting body In some fungi, a stalked, spore-producing structure such as a mushroom. p. 529

functional genomics The study of the functions of genes and of other parts of the genome. p. 364

functional groups The atoms in reactive groups. p. F-22

fundamental niche The range of conditions and resources that a population can possibly tolerate and use. p. 721

furculum Wishbone in birds. p. 661

furrow In cytokinesis, a groove that girdles the cell and gradually deepens until it cuts the cytoplasm into two parts. p. 176

G protein–coupled receptor In signal transduction, a surface receptor that responds to a signal by activating a G protein. p. 1049

G0 phase The phase of the cell cycle in eukaryotes in which many cell types stop dividing. p. 172

G1 phase The initial growth stage of the cell cycle in eukaryotes, during which the cell makes proteins and other types of cellular molecules but not nuclear DNA. p. 171

G2 phase The phase of the cell cycle in eukaryotes during which the cell continues to synthesize proteins and grow, completing interphase. p. 171

gallbladder The organ that stores bile between meals, when no digestion is occurring. p. 1205

gametangium (plural, gametangia) A cell or organ in which gametes are produced. pp. 546, 572

gamete A haploid cell; an egg or sperm. Haploid cells fuse during sexual reproduction to form a diploid zygote. pp. 199, 595

gametic isolation A prezygotic reproductive isolating mechanism caused by incompatibility between the sperm of one species and the eggs of another; may prevent fertilization. p. 432

gametogenesis The formation of male and female gametes. p. 990

gametophyte An individual of the haploid generation produced when a spore germinates and grows directly by mitotic divisions in organisms that undergo alternation of generations. pp. 200, 526, 563, 884

ganglion A functional concentration of nervous system tissue composed principally of nerve cell bodies, usually lying outside the central nervous system. pp. 608, 640

ganglion cell A type of neuron in the retina of the eye that receives visual information from photoreceptors via various intermediate cells such as bipolar cells, amacrine cells, and horizontal cells. p. 1112

gap gene In *Drosophila* embryonic development, the first activated set of segmentation genes that progressively subdivide the embryo into regions, determining the segments of the embryo and the adult. p. 1037

gap junction Junction that opens direct channels allowing ions and small molecules to pass directly from one cell to another. p. 48

gastric juice A substance secreted by the stomach that contains the digestive enzyme pepsin. 1202

gastrodermis The derivative of endoderm that lines the gastrovascular cavity of radially symmetrical animals and forms the epithelial lining of the midgut in bilaterally symmetrical anmals. p. 603

gastrovascular cavity A saclike body cavity with a single opening, a mouth, which serves both digestive and circulatory functions. p. 602

gastrula The developmental stage resulting when the cells of the blastula migrate and divide once cleavage is complete. p. 1015

gastrulation The second major process of early development in most animals, which produces an embryo with three distinct primary tissue layers. p. 1015

gated channel Ion transporter in a membrane that switches between open, closed, and intermediate states. p. 106

gemma (plural, gemmae) Small cell mass that forms in cuplike growths on a thallus. p. 574

gemmules Clusters of cells with a resistant covering that allows them to survive unfavourable conditions. p. 602

gene A unit containing the code for a protein molecule or one of its parts, or for functioning RNA molecules such as tRNA and rRNA. p. 30

gene flow The transfer of genes from one population to another through the movement of individuals or their gametes. p. 418

gene-for-gene recognition A mechanism in which plants can detect an attack by a specific pathogen; the product of a specific plant gene interacts with the product of a specific pathogen gene, triggering the plant's defensive response. p. 932

gene pool The sum of all alleles at all gene loci in all individuals in a population. p. 414

gene therapy Correction of genetic disorders using genetic engineering techniques. p. 354

general transcription factor (basal transcription factor) In eukaryotes, a protein that binds to the promoter of a gene in the area of the TATA box and recruits and orients RNA polymerase II to initiate transcription at the correct place. p. 322

generalized compartment model A model used to describe nutrient cycling in which two criteria—organic versus inorganic nutrients and available versus unavailable nutrients—define four compartments where nutrients accumulate. p. 759

generalized transduction Transfer of bacterial genes between bacteria using virulent phages that have incorporated random DNA fragments of the bacterial genome. p. 196

generation time The average time between the birth of an organism and the birth of its offspring. p. 684

generative cell A cell in the pollen grain (male gametophyte) of seed plants that will give rise to sperm. p. 886

genetic code The nucleotide information that specifies the amino acid sequence of a polypeptide. p. 292

genetic counselling Counselling that allows prospective parents to assess the possibility that they might have a child affected by a genetic disorder. p. 258

genetic drift Random fluctuations in allele frequencies as a result of chance events; usually reduces genetic variation in a population. p. 418

genetic engineering The use of DNA technologies to alter genes for practical purposes. p. 339

genetic equilibrium The point at which neither the allele frequencies nor the genotype frequencies in a population change in succeeding generations. p. 415

genetic recombination The process by which the combinations of alleles for different genes in two parental individuals become shuffled into new combinations in offspring individuals. p. 190

genetic screening Biochemical or molecular tests for identifying inherited disorders after a child is born. p. 259

genetically modified organism (GMO) A transgenic organism. p. 359

genome The entire collection of DNA sequence for a given organism. p. 67

genomic imprinting Pattern of inheritance in which the expression of a nuclear gene is based on whether an individual organism inherits the gene from the male or the female parent. p. 259

genomic library A collection of clones that contains a copy of every DNA sequence in a genome. p. 344

genotype The genetic constitution of an organism. p. 221

genotype frequency The percentage of individuals in a population possessing a particular genotype. p. 414

genus A Linnaean taxonomic category ranking below a family and above a species. p. 427

geographic range The overall spatial boundaries within which a population lives. p. 680

germ cell An animal cell that is set aside early in embryonic development and gives rise to the gametes. pp. 991, 1020

germ layer The layers (up to three) of cells produced during the early development of the embryo of most animals. p. 595

germ-line gene therapy Therapy in which a gene is introduced into germ-line cells of an animal to correct a genetic disorder. p. 355

gestation The period of mammalian development in which the embryo develops in the uterus of the mother. p. 668

gibberellin Any of a large family of plant hormones that regulate aspects of growth, including cell elongation. p. 924

gill A respiratory organ formed as an evagination of the body that extends outward into the respiratory medium. p. 611

gill arch One of the series of curved supporting structures between the slits in the pharynx of a chordate. p. 646

gill slit One of the openings in the pharynx of a chordate through which water passes out of the pharynx. p. 640

Ginkgophyta A plant phylum with a single living species, the ginkgo (or maidenhair) tree. p. 583

gizzard The part of the digestive tube that grinds ingested material into fine particles by muscular contractions of the wall. p. 1195

gland A cell or group of cells that produces and releases substances nearby, in another part of the body, or to the outside. p. 950

glans A soft, caplike structure at the end of the penis, containing most of the nerve endings producing erotic sensations. p. 1004

glial cell A nonneuronal cell contained in the nervous tissue that physically supports and provides nutrients to neurons, provides electrical insulation between them, and scavenges cellular debris and foreign matter. pp. 955, 1070

globulin A plasma protein that transports lipids (including cholesterol) and fat-soluble vitamins; a specialized subgroup of globulins, the immunoglobulins, constitute antibodies and other molecules contributing to the immune response. p. 970

glomerulus A ball of blood capillaries surrounded by Bowman's capsule in the human nephron. p. 1246

glucagon A pancreatic hormone with effects opposite to those of insulin: it stimulates glycogen, fat, and protein degradation. p. 1060

glucocorticoid A steroid hormone secreted by the adrenal cortex that helps maintain the blood concentration of glucose and other fuel molecules. p. 1058

glycocalyx A carbohydrate coat covering the cell surface. p. 32

glycogen Energy-providing carbohydrates stored in animal cells. p. 135

glycolysis Stage of cellular respiration in which sugars such as glucose are partially oxidized and broken down into smaller molecules. p. 125

Gnathostomata The group of vertebrates with movable jaws. p. 644

Golgi complex In eukaryotes, the organelle responsible for the final modification, sorting, and distribution of proteins and lipids. p. 37

Golgi tendon organ A proprioceptor of tendons. p. 1103

gonad A specialized gamete-producing organ in which the germ cells collect. Gonads are the primary source of sex hormones in vertebrates: ovaries in the female and testes in the male. p. 1059

gonadotropin A hormone that regulates the activity of the gonads (ovaries and testes). p. 1055

gonadotropin-releasing hormone (GnRH) A tropic hormone secreted by the hypothalamus that causes the pituitary to make luteinizing hormone (LH) and follicle-stimulating hormone (FSH). p. 1060

graded potential A change in membrane potential that does not necessarily trigger an action potential. p. 1085

gradualism The view that Earth and its living systems changed slowly over its history. p. 395

Gram-negative Describing bacteria that do not retain the stain used in the Gram stain procedure. p. 485

Gram-positive Describing bacteria that appear purple when stained using the Gram stain technique. p. 485

Gram stain procedure A procedure of staining bacteria to distinguish between types of bacteria with different cell wall compositions. p. 484

granum (plural, **grana**) Structure in the chloroplasts of higher plants formed by thylakoids stacked one on top of another. p. 46

gravitropism A directional growth response to Earth's gravitational pull that is induced by mechanical and hormonal influences. p. 934

greater vestibular gland One of two glands located slightly below and to the left and right of the opening of the vagina in women. They secrete mucus to provide lubrication, especially when the woman is sexually aroused. p. 999

greenhouse effect A phenomenon in which certain gases foster the accumulation of heat in the lower atmosphere, maintaining warm temperatures on Earth. p. 764

grey crescent A crescent-shaped region of the underlying cytoplasm at the side opposite the point of sperm entry exposed after fertilization when the pigmented layer of cytoplasm rotates toward the site of sperm entry. p. 1018

grey matter Areas of densely packed nerve cell bodies and dendrites in the brain and spinal cord. p. 1123

gross primary productivity The rate at which producers convert solar energy into chemical energy. p. 750

ground meristem The primary meristematic tissue in plants that gives rise to ground tissues, mostly parenchyma. p. 849

ground tissue system One of the three basic tissue systems in plants; includes all tissues other than dermal and vascular tissues. p. 841

growth factor Any of a large group of peptide hormones that regulates the division and differentiation of many cell types in the body. p. 1047

growth hormone (GH) A hormone that stimulates cell division, protein synthesis, and bone growth in children and adolescents, thereby causing body growth. p. 1055

guanine A purine that base-pairs with cytosine in nucleic acids. p. 268

guard cell Either of a pair of specialized crescent-shaped cells that control the opening and closing of stomata in plant tissue. p. 847

guttation The exudation of water from leaves as a result of strong root pressure. p. 872

gymnosperm A seed plant that produces "naked" seeds not enclosed in an ovary. p. 581

H^+ pump *See* proton pump. p. 109

habitat fragmentation A process in which remaining areas of intact habitat are reduced to small, isolated patches. p. 653

habitat The specific environment in which a population lives, as characterized by its biotic and abiotic features. p. 680

habituation The learned loss of responsiveness to stimuli. p. 1159

half-life The time it takes for half of a given amount of a radioisotope to decay. p. F-10

haplodiploidy A pattern of sex determination in insects in which females are diploid and males are haploid. p. 1182

haploid An organism or cell with only one copy of each type of chromosome in its nuclei. p. 170

Hardy–Weinberg principle An evolutionary rule of thumb that specifies the conditions under which a population of diploid organisms achieves genetic equilibrium. p. 415

harvesting efficiency The ratio of the energy content of food consumed to with the energy content of food available. p. 752

head-foot In molluscs, the region of the body that provides the major means of locomotion and contains concentrations of nervous system tissues and sense organs. p. 611

heartwood The inner core of a woody stem; composed of dry tissue and nonliving cells that no longer transport water and solutes and may store resins, tannins, and other defensive compounds. p. 856

heat of vaporization The heat required to give water molecules enough energy of motion to break loose from liquid water and form a gas. p. F-16

heat-shock protein (HSP) Any of a group of chaperone proteins that are present in all cells in all life forms. They are induced when a cell undergoes various types of environmental stresses such as heat, cold, and oxygen deprivation. p. 933

heavy chain The heavier of the two types of polypeptide chains that are found in immunoglobulin and antibody molecules. p. 1274

helical virus A virus in which the protein subunits of the coat assemble in a rodlike spiral around the genome. p. 502

helper T cell A clonal cell that assists with the activation of B cells. p. 1278

hemocoel A cavity in the body of some coelomic invertebrates (arthropods and some molluscs) filled with blood. The hemocoel displaces the coelom, which persists as a small chamber surrounding the gonads or heart. p. 596

hemolymph The circulatory fluid of invertebrates with open circulatory systems, including molluscs and arthropods. pp. 611, 965

hepatic portal vein The blood vessel that leads to capillary networks in the liver. p. 1206

Hepatophyta The phylum that includes liverworts and their bryophyte relatives. p. 573

herbicide A compound that, at proper concentration, kills plants. p. 921

herbivore An animal that obtains energy and nutrients primarily by eating plants. p. 712

herbivory The interaction between herbivorous animals and the plants they eat. p. 712

hermaphroditism The mechanism in which both mature egg-producing and mature sperm-producing tissue are present in the same individual. p. 997

heterochromatin In eukaryotes, regions of densely packed chromatin fibres in interphase nuclei. p. 285

heterodont Having different teeth specialized for different jobs. p. 667

heterosporous Producing two types of spores, "male" microspores and "female" megaspores. p. 570

heterotroph An organism that acquires energy and nutrients by eating other organisms or their remains. pp. 63, 487, 594

heterozygote An individual with two different alleles of a gene. p. 221

heterozygote advantage An evolutionary circumstance in which individuals that are heterozygous at a particular locus have higher relative fitness than either homozygote. p. 421

heterozygous The state of possessing two different alleles of a gene. p. 221

Hfr cell A special donor cell that can transfer genes on a bacterial chromosome to a recipient bacterium. p. 195

hibernation Extended torpor during winter. p. 1260

hindbrain The lower area of the brain that includes the brain stem, medulla oblongata, and pons. p. 1090

hippocampus A grey-matter centre that is involved in sending information. 1126

histone A small, positively charged (basic) protein that is complexed with DNA in the chromosomes of eukaryotes. p. 283

histone code A regulatory mechanism for altering chromatin structure and, therefore, gene activity, based on signals in histone tails represented by chemical modification patterns. p. 327

Holocephali The chimeras, another group of cartilaginous fishes. p. 647

homeobox A region of a homeotic gene that corresponds to an amino acid section of the homeodomain. p. 394

homeodomain An encoded transcription factor of each protein that binds to a region in the promoters of the genes whose transcription it regulates. p. 1039

homeostasis A steady internal condition maintained by responses that compensate for changes in the external environment. pp. 946, 1236

homeostatic mechanism Any process or activity responsible for homeostasis. p. 946

homeotic gene Any of the family of genes that determines the structure of body parts during embryonic development. p. 1039

hominin A member of a monophyletic group of primates, characterized by an erect bipedal stance, that includes modern humans and their recent ancestors. p. 469

hominoid (Hominoidea) The monophyletic group of primates that includes apes and humans. p. 469

homologous Similar. p. 190

homologous genes Genes that are related by descent from a common ancestor. p. 369

homologous traits Characteristics that are similar in two species because they inherited the genetic basis of the trait from their common ancestor. p. 402

homoplasies Characteristics shared by a set of species, often because they live in similar environments, but not present in their common ancestor; often the product of convergent evolution. p. 458

homosporous Producing only one type of spore. p. 570

homozygote An individual with two copies of the same allele. p. 221

homozygous State of possessing two copies of the same allele. p. 221

horizon A noticeable layer of soil, such as topsoil, with a distinct texture and composition that varies with soil type. p. 907

horizontal cell A type of neuron that forms lateral connections among photoreceptor cells in the retina of the eye. p. 1112

hormone A signalling molecule secreted by a cell that can alter the activities of any cell with receptors for it; in animals, typically a molecule produced by one tissue and transported via the bloodstream to another specific tissue to alter its physiological activity. pp. 326, 918, 1046

host A species that is fed upon by a parasite. p. 66

host race A population of insects that may be reproductively isolated from other populations of the same species as a consequence of their adaptation to feed on a specific host plant species. p. 438

human chorionic gonadotropin (hCG) A hormone that keeps the corpus luteum in the ovary from breaking down. p. 1008

human immunodeficiency virus (HIV) A retrovirus that causes acquired immune deficiency syndrome (AIDS). p. 213

humus The organic component of soil remaining after decomposition of plants and animals, animal droppings, and other organic matter. p. 906

hybrid breakdown A postzygotic reproductive isolating mechanism in which hybrids are capable of reproducing, but their offspring have either reduced fertility or reduced viability. p. 433

hybrid inviability A postzygotic reproductive isolating mechanism in which a hybrid individual has a low probability of survival to reproductive age. p. 432

hybrid sterility A postzygotic reproductive isolating mechanism in which hybrid offspring cannot form functional gametes. p. 432

hybrid zone A geographic area where the hybrid offspring of two divergent populations or species are common. p. 436

hybridization When two species interbreed and produce fertile offspring. p. 427

hydration shell A shell of any chemical species that acts as a solvent and surrounds a solute species. When the solvent is water it is often referred to as a *hydration shell* or *hydration sphere*. A classic example is when water molecules form a sphere around a metal ion. F-17

hydrocarbon Molecule consisting of carbon linked only to hydrogen atoms. p. 100

hydrogen bond Noncovalent bond formed by unequal electron sharing between hydrogen atoms and oxygen, nitrogen, or sulfur atoms. p. F-13

hydrogeologic cycle The global cycling of water between the ocean, the atmosphere, land, freshwater ecosystems, and living organisms. p. 760

hydrolysis Reaction in which the components of a water molecule are added to functional groups as molecules are broken into smaller subunits. p. F-21

hydrophilic Refers to polar molecules that associate readily with water. p. F-13

hydrophobic Refers to nonpolar substances that are excluded by water and other polar molecules. p. F-13

hydroponic culture A method of growing plants not in soil but with the roots bathed in a solution that contains water and mineral nutrients. p. 902

hydrostatic skeleton A structure consisting of muscles and fluid that, by themselves, provide support for the animal or part of the animal; no rigid support, such as a bone, is involved. pp. 596, 1143

hydroxyl group Group consisting of an oxygen atom linked to a hydrogen atom on one side and to a carbon chain on the other side. p. 268

hymen A thin flap of tissue that partially covers the opening of the vagina. p. 999

hyomandibular bones Bones that support the hyoid and throat. p. 646

hyperpolarized The condition of a neuron when its membrane potential is more negative than the resting value. p. 1074

hypersensitive response A plant defence that physically cordons off an infection site by surrounding it with dead cells. p. 930

hyperthermia The condition resulting when the heat gain of the body is too great to be counteracted. p. 1259

hypertonic Solution containing dissolved substances at higher concentrations than the cells it surrounds. p. 109

hypha (plural, hyphae) Any of the threadlike filaments that form the mycelium of a fungus. pp. 524, 542

hypoblast The bottom layer of a blastodisc. p. 1019

hypocotyl The region of a plant embryo's vertical axis between the cotyledons and the radicle. p. 893

hypodermis The innermost layer of the skin that contains larger blood vessels and additional reinforcing connective tissue. p. 1257

hypothalamus The portion of the brain that contains centres regulating basic homeostatic functions of the body and contributing to the release of hormones. p. 1126

hypothermia A condition in which the core temperature falls below normal for a prolonged period. p. 1259

hypothesis A "working explanation" of observed facts. p. F-2

hypotonic Solution containing dissolved substances at lower concentrations than the cells it surrounds. p. 108

imbibition The movement of water into a seed as the water molecules are attracted to hydrophilic groups of stored proteins; the first step in germination. p. 894

immigration Movement of organisms into a population. p. 684

immune privilege The situation in which certain sites in the body tolerate the presence of an antigen without mounting an inflammatory immune response. These sites include the brain, eyes, and testicles. p. 1282

immune response The defensive reactions of the immune system. p. 1267

immune system The combined defences, innate and acquired, a body uses to eliminate infections. p. 1266

immunoglobulin A specific protein substance produced by plasma cells to aid in fighting infection. p. 970

immunological memory The capacity of the immune system to respond more rapidly and vigorously to the second contact with a specific antigen than to the primary contact. p. 1280

immunological tolerance The process that protects the body's own molecules from attack by the immune system. p. 1283

imperfect flower A type of incomplete flower that has stamens or carpels, but not both. p. 885

imprinting The process of learning the identity of a caretaker and potential future mate during a critical period. p. 1159

inbreeding A special form of nonrandom mating in which genetically related individuals mate with each other. p. 420

incisors Flattened, chisel-shaped teeth of mammals, located at the front of the mouth, that are used to nip or cut food. p. 464

incomplete dominance Condition in which the effects of recessive alleles can be detected to some extent in heterozygotes. p. 229

incomplete metamorphosis In certain insects, a life cycle characterized by the absence of a pupal stage between the immature and adult stages. p. 629

incurrent siphon A muscular tube that brings water containing oxygen and food into the body of an invertebrate. p. 615

incus The second of the three sound-conducting middle ear bones in vertebrates, located between the malleus and the stapes. p. 1105

independent assortment Mendel's principle that the alleles of the genes that govern two characters segregate independently during formation of gametes. p. 226

indeterminate cleavage A type of cleavage, observed in many deuterostomes, in which the developmental fates of the first few cells produced by mitosis are not determined as soon as cells are produced. p. 597

indeterminate growth Growth that is not limited by an organism's genetic program, so that the organism grows for as long as it lives; typical of many plants. *Compare* determinate growth. p. 839

indirect neurotransmitter A neurotransmitter that acts as a first messenger, binding to a G protein–coupled receptor in the postsynaptic membrane, which activates the receptor and triggers generation of a second messenger such as cyclic AMP or other processes. p. 1082

inducer Concerning regulation of gene expression in bacteria, a molecule that turns on the transcription of the genes in an operon. p. 317

inducible operon Operon whose expression is increased by an inducer molecule. p. 317

induction A mechanism in which one group of cells (the inducer cells) causes or influences another nearby group of cells (the responder cells) to follow a particular developmental pathway. p. 1016

infection thread In the formation of root nodules on nitrogen-fixing plants, the tube formed by the plasma membrane of root hair cells as bacteria enter the cell. p. 912

inflammation The heat, pain, redness, and swelling that occur at the site of an infection. p. 1268

inflorescence The arrangement of flowers on a stem. p. 883

ingestion The feeding methods used to take food into the digestive cavity. p. 1189

inheritance The transmission of DNA (i.e., genetic information) from one generation to the next. p. 187

inhibin A peptide that, in females, is an inhibitor of follicle-stimulating hormone (FSH) secretion from the pituitary, thereby diminishing the signal for follicular growth. In males, inhibin inhibits FSH secretion from the pituitary, thereby decreasing spermatogenesis. p. 1000

inhibiting hormone (IH) A hormone released by the hypothalamus that inhibits the secretion of a particular anterior pituitary hormone. p. 1053

inhibition hypothesis A hypothesis suggesting that new species are prevented from occupying a community by whatever species are already present. p. 737

inhibitory postsynaptic potential (IPSP) A change in membrane potential caused when hyperpolarization occurs, pushing the neuron farther from threshold. p. 1085

initiator codon *See* start codon. p. 294

innate immunity A nonspecific line of defence against pathogens that includes inflammation, which creates internal conditions that inhibit or kill many pathogens, and specialized cells that engulf or kill pathogens or infected body cells. p. 1266

inner boundary membrane Membrane lying just inside the outer boundary membrane of a chloroplast, enclosing the stroma. p. 46

inner cell mass The dense mass of cells within the blastocyst that will become the embryo. p. 1025

inner ear That part of the ear, particularly the cochlea, that converts mechanical vibrations (sound) into neural messages that are sent to the brain. p. 1106

inner mitochondrial membrane Membrane surrounding the mitochondrial matrix. p. 41

inorganic molecule Molecule without carbon atoms in its structure. p. F-20

inositol triphosphate (IP3) In particular, signal transduction pathways, a second messenger that activates transport proteins in the endoplasmic reticulum to release Ca^{2+} into the cytoplasm. IP3 is involved in one of two major G protein–coupled receptor response pathways. p. 919

insertion sequence (IS) A transposable element that contains only genes for its transposition. p. 211

insight learning A phenomenon in which animals can solve problems without apparent trial-and-error attempts at the solution. p. 1159

instar The stage between successive moults in insects and other arthropods. p. 628

instinctive behaviour A genetically "programmed" response that appears in complete and functional form the first time it is used. p. 1155

insulin A hormone secreted by beta cells in the islets of Langerhans, acting mainly on cells of nonworking skeletal muscles, liver cells, and adipose tissue (fat) to lower blood glucose, fatty acid, and amino acid levels and promote the storage of those molecules. pp. 1060, 1207

insulin-like growth factor (IGF) A peptide that directly stimulates growth processes. p. 1055

integral membrane protein Protein embedded in a phospholipid bilayer. p. 103

integration The sorting and interpretation of neural messages and the determination of the appropriate response(s). p. 1069

integrator In homeostatic feedback, the control centre that compares a detected environmental change with a set point. p. 960

integument Skin. In plants, the outer layer of an ovule. pp. 886, 1256

interference competition Form of competition in which individuals fight over resources or otherwise harm each other directly. p. 717

interferon A cytokine produced by infected host cells affected by viral dsRNA, which acts on both the infected cell that produces it, an autocrine effect, and neighbouring uninfected cells, a paracrine effect. p. 1272

interkinesis A brief interphase separating the two meiotic divisions. p. 202

intermediate disturbance hypothesis Hypothesis proposing that species richness is greatest in communities that experience fairly frequent disturbances of moderate intensity. p. 735

intermediate filament A cytoskeletal filament about 10 nm in diameter that provides mechanical strength to cells in tissues. p. 42

intermediate-day plant A plant that flowers only when day length falls between the values for long-day and short-day plants. p. 940

internal fertilization The process in which sperm are released by the male close to or inside the entrance of the reproductive tract of the female. p. 994

internal gill A gill located within the body that has a cover providing physical protection for the gills. Water must be brought to internal gills. p. 1222

interneuron A neuron that integrates information to formulate an appropriate response. p. 1070

internode The region between two nodes on a plant stem. p. 847

interphase The first stage of the mitotic cell cycle, during which the cell grows and replicates its DNA before undergoing mitosis and cytokinesis. p. 171

interspecific competition The competition for resources between species. p. 717

interstitial fluid The fluid occupying the spaces between cells in multicellular animals. p. 958

intertidal zone The shoreline that is alternately submerged and exposed by tides. p. 563

intestine The portion of the digestive system where organic matter is hydrolyzed by enzymes secreted into the digestive tube. As muscular contractions of the intestinal wall move the mixture along, cells lining the intestine absorb the molecular subunits produced by digestion. pp. 32, 93

intracellular digestion The process in which cells take in food particles by endocytosis. p. 603

intraspecific competition The dependence of two or more individuals in a population on the same limiting resource. p. 693

intrinsic rate of increase The maximum possible per capita population growth rate in a population living under ideal conditions. p. 691

intron A non–protein-coding sequence that interrupts the protein-coding sequence in a eukaryotic gene. Introns are removed by splicing in the processing of pre-mRNA to mRNA. p. 298

invagination The process in which cells changing shape and pushing inward from the surface produce an indentation, such as the dorsal lip of the blastopore. p. 33

inversion Chromosomal alteration that occurs if a broken segment reattaches to the same chromosome from which it was lost, but in reversed orientation, so that the order of genes in the segment is reversed with respect to the other genes of the chromosome. p. 252

invertebrate An animal without a vertebral column. p. 9

inverted repeat Enables the transposase enzyme to identify the ends of the transposable element when it catalyzes transposition. p. 211

involution The process by which cells migrate into the blastopore. p. 1018

ion A positively or negatively charged atom. pp. 8, F-11

ionic bond Bond that results from electrical attractions between atoms that have lost or gained electrons. p. 105

iris Of the eye, the coloured muscular membrane that lies behind the cornea and in front of the lens, which by opening or closing determines the size of the pupil and hence the amount of light entering the eye. p. 1108

islets of Langerhans Endocrine cells that secrete the peptide hormones insulin and glucagon into the bloodstream. p. 1060

isotonic Refers to the state of equal concentration of water inside and outside cells. p. 109

isotope A distinct form of the atoms of an element, with the same number of protons but a different number of neutrons. p. F-9

jasmonate Any of a group of plant hormones that help regulate aspects of growth and responses to stress, including attacks by predators and pathogens. p. 929

juvenile hormones A family of fatty acid hormones that govern metamorphosis and reproduction in insects and crustaceans. p. 1063

karyogamy In plants, the fusion of two sexually compatible haploid nuclei after cell fusion (plasmogamy). p. 543

karyotype A characteristic of a species consisting of the shapes and sizes of all of the chromosomes at metaphase. p. 175

keeled sternum The ventrally extended breastbone of a bird to which the flight muscles attach. p. 661

ketone Molecule in which the carbonyl group is linked to a carbon atom in the interior of a carbon chain. p. 1061

keystone species A species that has a greater effect on community structure than its numbers might suggest. p. 732

kin selection Altruistic behaviour to close relatives, allowing them to produce proportionately more surviving copies of the altruist's genes than the altruist might otherwise have produced on its own. p. 1180

kinesis A change in the rate of movement or the frequency of turning movements in response to environmental stimuli. p. 1169

kinetic energy The energy of motion. p. 75

kinetochore A specialized structure consisting of proteins attached to a centromere that mediates the attachment and movement of chromosomes along the mitotic spindle. p. 174

kingdom A Linnaean taxonomic category that ranks below a domain and above a phylum. p. 428

kingdom Animalia The taxonomic kingdom that includes all living and extinct animals. p. 594

kingdom Fungi The taxonomic kingdom that includes all living or extinct fungi. p. 543

kingdom Plantae The taxonomic kingdom encompassing all living or extinct plants. p. 564

Korarchaeota A group of Archaea recognized solely on the basis of rRNA coding sequences in DNA taken from environmental samples. p. 497

Krebs cycle *See* citric acid cycle. p. 129

K-selected species Long-lived, slow-reproducing species that thrive in more stable environments. p. 697

labia majora A pair of fleshy, fat-padded folds that partially cover the labia minora. p. 999

labia minora Two folds of tissue that run from front to rear on either side of the opening to the vagina. p. 999

lactate fermentation Reaction in which pyruvate is converted into lactate. p. 137

lagging strand A DNA strand assembled discontinuously in the direction opposite to DNA unwinding. p. 276

lagging strand template The DNA template strand for the lagging strand. p. 276

larva (larval form) A sexually immature stage in the life cycle of many animals that is morphologically distinct from the adult. p. 595

larynx The voice box. p. 1225

latent phase The time during which a virus remains in the cell in an inactive form. p. 506

lateral bud A bud on the side of a plant stem from which a branch may grow. p. 848

lateral geniculate nuclei Clusters of neurons located in the thalamus that receive visual information from the optic nerves and send it on to the visual cortex. p. 1113

lateral inhibition Visual processing in which lateral movement of signals from a rod or cone proceeds to a horizontal cell and continues to bipolar cells with which the horizontal cell makes inhibitory connections, serving both to sharpen the edges of objects and enhance contrast in an image. p. 1112

lateral line system The complex of mechanoreceptors along the sides of some fishes and aquatic amphibians that detect vibrations in the water. pp. 648, 1101

lateral meristem A plant meristem that gives rise to secondary tissue growth. *Compare* primary meristem. p. 839

lateral root A root that extends away from the main root (or taproot). p. 852

lateralization A phenomenon in which some brain functions are more localized in one of the two hemispheres. p. 1129

leaching The process by which soluble materials in soil are washed into a lower layer of soil or are dissolved and carried away by water. p. 901

leading strand A DNA strand assembled in the direction of DNA unwinding. p. 276

leading strand template The "old" DNA used as a template for synthesis of "new" DNA in the direction of DNA unwinding. p. 276

leaf primordium (plural, **primordia**) A lateral outgrowth from the apical meristem that develops into a young leaf. p. 851

learned behaviour A response of an animal that depends on having a particular kind of experience during development. p. 1155

learning A process in which experiences stored in memory change the behavioural responses of an animal. p. 1129

left aortic arch In mammals, leads blood way from the heart to the aorta. p. 667

leghemoglobin An iron-containing, red-pigmented protein produced in root nodules during the symbiotic association between *Bradyrhizobium* or *Rhizobium* and legumes. p. 912

lek A display ground where males each possess a small territory from which they court attentive females. p. 1176

lens The transparent, biconvex intraocular tissue that helps bring rays of light to a focus on the retina. p. 1108

Lepidosauromorpha A monophyletic lineage of diapsids that includes both marine and terrestrial animals, represented today by sphenodontids, lizards, and snakes. p. 655

leukocyte A white blood cell, which eliminates dead and dying cells from the body, removes cellular debris, and participates in defending the body against invading organisms. p. 971

Leydig cell A cell that produces the male sex hormones. p. 1002

lichen A single vegetative body that is the result of an association between a fungus and a photosynthetic partner, often an alga. p. 555

life cycle The sequential stages through which individuals develop, grow, maintain themselves, and reproduce. p. 517

life history The lifetime pattern of growth, maturation, and reproduction that is characteristic of a population or species. p. 686

life table A chart that summarizes the demographic characteristics of a population. p. 685

ligament A fibrous connective tissue that connects bones to each other at a joint. p. 952

ligand-gated ion channel A channel that opens or closes when a specific chemical, such as a neurotransmitter, binds to the channel. p. 1082

light The portion of the electromagnetic spectrum that humans can detect with their eyes. p. 4

light chain The lighter of the two types of polypeptide chains found in immunoglobulin and antibody molecules. p. 1274

light-independent reaction The second stage of photosynthesis, in which electrons are used as a source of energy to convert inorganic CO_2 to an organic form. Also referred to as the *Calvin cycle*. p. 156

light microscope Microscope that uses light to illuminate the specimen. p. 29

lignin A tough, rather inert polymer that strengthens the secondary walls of various plant cells and thus helps vascular plants grow taller and stay erect on land. p. 837

limbic system A functional network formed by parts of the thalamus, hypothalamus, and basal nuclei, along with other nearby grey-matter centres—the amygdala, hippocampus, and olfactory bulbs—sometimes called the "emotional brain." p. 1126

limiting nutrient An element in short supply within an ecosystem, the shortage of which limits productivity. p. 750

linkage The phenomenon of genes being located on the same chromosome. p. 241

linkage map Map of a chromosome showing the relative locations of genes based on recombination frequencies. p. 243

linked genes Genes on the same chromosome. p. 241

linker A short segment of DNA extending between one nucleosome and the next in a eukaryotic chromosome. p. 283

lipopolysaccharide (LPS) A large molecule that consists of a lipid and a carbohydrate joined by a covalent bond. p. 485

loam Any well-aerated soil composed of a mixture of sand, clay, silt, and organic matter. p. 906

locus The particular site on a chromosome at which a gene is located. p. 229

logistic (model of population growth) Model of population growth that assumes that a population's per capita growth rate decreases as the population gets larger. p. 689

long-day plant A plant that flowers in spring when dark periods become shorter and day length becomes longer. p. 939

long-term memory Memory that stores information from days to years or even for life. p. 1129

long-term potentiation A long-lasting increase in the strength of synaptic connections in activated neural pathways following brief periods of repeated stimulation. p. 1130

loop of Henle In mammals, a U-shaped bend of the proximal convoluted tubule. p. 1246

loose connective tissue A tissue formed of sparsely distributed cells surrounded by a more or less open network of collagen and other glycoprotein fibres. p. 952

lophophore The circular or U-shaped fold with one or two rows of hollow, ciliated tentacles that surrounds the mouth of brachiopods, bryozoans, and phoronids and is used to gather food. p. 607

loss of imprinting A phenomenon in which the imprinting mechanism for a gene does not work, resulting in both alleles of the gene being active. p. 260

lumbar vertebrae In mammals, the vertebrae from the thoracic (bearing ribs) to the sacral (junction with pelvis). p. 670

lumen The inside of the digestive tube. p. 1047

lung One of a pair of invaginated respiratory surfaces, buried in the body interior where they are less susceptible to drying out; the organs of respiration in mammals, birds, reptiles, and most amphibians. p. 1223

luteinizing hormone (LH) A hormone secreted by the pituitary that stimulates the growth and maturation of eggs in females and the secretion of testosterone in males. p. 1055

Lycophyta The plant phylum that includes club mosses and their close relatives. p. 577

lymph The interstitial fluid picked up by the lymphatic system. p. 981

lymph node One of many small, bean-shaped organs spaced along the lymph vessels that contain macrophages and other leukocytes that attack invading disease organisms. p. 983

lymphatic system An accessory system of vessels and organs that helps balance the fluid content of the blood and surrounding tissues and participates in the body's defences against invading disease organisms. p. 981

lymphocyte A leukocyte that carries out most of its activities in the tissues and organs of the lymphatic system. Lymphocytes play major roles in immune responses. p. 983

lysed Refers to a cell that has ruptured or undergone lysis. p. 504

lysogenic cycle Cycle in which the DNA of the bacteriophage is integrated into the DNA of the host bacterial cell and may remain for many generations. pp. 197, 505

lysosome Membrane-bound vesicle containing hydrolytic enzymes for the digestion of many complex molecules. p. 38

lytic cycle The series of events from infection of one bacterial cell by a phage through the release of progeny phages from lysed cells. pp. 196, 504

M phase–promoting factor A complex of M cyclin and cyclin-dependent kinase 1 (Cdk1). The complex initiates mitosis and orchestrates some of its key events. p. 182

macroevolution Large-scale evolutionary patterns in the history of life, producing major changes in species and higher taxonomic groups. p. 401

macromolecule A very large molecule assembled by the covalent linkage of smaller subunit molecules. p. 53

macronucleus In ciliophorans, a single large nucleus that develops from a micronucleus but loses all genes except those required for basic "housekeeping" functions of the cell and for ribosomal RNAs. p. 521

macronutrient In humans, a mineral required in amounts ranging from 50 mg to more than 1 g per day. In plants, a nutrient needed in large amounts for normal growth and development. pp. 903, 1190

macrophage A phagocyte that takes part in nonspecific defences and adaptive immunity. p. 1268

magnetoreceptor A receptor found in some animals that navigate long distances that allows them to detect and use Earth's magnetic field as a source of directional information. p. 1121

magnification The ratio of an object as viewed to its real size. p. 29

major histocompatibility complex A large cluster of genes encoding the MHC proteins. p. 1277

malleus The outermost of the sound-conducting bones of the middle ear in vertebrates. p. 1105

malnutrition A condition resulting from a diet that lacks one or more essential nutrients. p. 1190

Malpighian tubule The main organ of excretion and osmoregulation in insects, helping them maintain water and electrolyte balance. p. 626

mammary glands Specialized organs of female mammals that produce energy-rich milk, a watery mixture of fats, sugars, proteins, vitamins, and minerals. pp. 329, 1013, 1055

mandible In arthropods, one of the paired head appendages posterior to the mouth used for feeding. In vertebrates, the lower jaw. p. 624

mantle One or two folds of the body wall that lines the shell and secretes the substance that forms the shell in molluscs. p. 611

mantle cavity The protective chamber produced by the mantle in many molluscs. p. 611

map unit The unit of a linkage map, equivalent to a recombination frequency of 1%. Also referred to as a *centimorgan*. p. 244

marsupium An external pouch on the abdomen of many female marsupials, containing the mammary glands, and within which the young continue to develop after birth. p. 997

mass number The total number of protons and neutrons in the atomic nucleus. p. F-9

mast cell A type of cell dispersed through connective tissue that releases histamine when activated by the death of cells, caused by a pathogen at an infection site. p. 1269

mastax The toothed grinding organ at the anterior of the digestive tract in rotifers. p. 610

maternal chromosome The chromosome derived from the female parent of an organism. p. 201

maternal-effect gene One of a class of genes that regulate the expression of other genes expressed by the mother during oogenesis and that control the polarity of the egg and, therefore, of the embryo. p. 1037

mating systems The social systems describing how males and females pair up. p. 1175

mating type A genetically defined strain of an organism (such as a fungus) that can only mate with an organism of the opposite mating type; mating types are often designated + and −. p. 540

matter Anything that occupies space and has mass. p. F-8

maxilla (plural, **maxillae**) One of the paired head appendages posterior to the mouth used for feeding in arthropods. pp. 624, 646

mechanical isolation A prezygotic reproductive isolating mechanism caused by differences in the structure of reproductive organs or other body parts. p. 432

mechanoreceptor A sensory receptor that detects mechanical energy, such as changes in pressure, body position, or acceleration. The auditory receptors in the ears are examples of mechanoreceptors. p. 1077

medusa (plural, **medusae**) The tentacled, usually bell-shaped, free-swimming sexual stage in the life cycle of a coelenterate. p. 603

megapascal A unit of pressure used to measure water potential. p. 865

megaspore A plant spore that develops into a female gametophyte; usually larger than a microspore. pp. 570, 886

meiocytes Cells that are destined to divide by meiosis. p. 201

meiosis The division of diploid cells to haploid progeny, consisting of two sequential rounds of nuclear and cellular division. pp. 170, 199, 990

meiosis I The first division of the meiotic cell cycle in which homologous chromosomes pair and undergo an exchange of chromosome segments, and then the homologous chromosomes separate, resulting in two cells, each with the haploid number of chromosomes and with each chromosome still consisting of two chromatids. p. 201

meiosis II The second division of the meiotic cell cycle in which the sister chromatids in each of the two cells produced by meiosis I separate and segregate into different cells, resulting in four cells each with the haploid number of chromosomes. p. 201

melanocyte-stimulating hormone (MSH) A hormone secreted by the anterior pituitary that controls the degree of pigmentation in melanocytes. p. 1055

melanotic encapsulation The mechanism by which hemocytes move toward and form a capsule around pathogens that are too big to phagocytose. The capsule may then be melanized by the deposition of phenolic compounds that further isolate the pathogen. p. 1268

melatonin A peptide hormone secreted by the pineal gland that helps maintain daily biorhythms. p. 1062

membrane attack complexes An abnormal activation of the complement (protein) portion of the blood, forming a cascade reaction that brings blood proteins together, binds them to the cell wall, and then inserts them through the cell membrane. p. 1271

membrane potential An electrical voltage that measures the potential inside a cell membrane relative to the fluid just outside; it is negative under resting conditions and becomes positive during an action potential. pp. 111, 1073

memory The storage and retrieval of a sensory or motor experience or a thought. p. 1124

memory B cell In antibody-mediated immunity, a long-lived cell expressing an antibody on its surface that can bind to a specific antigen. A memory B cell is activated the next time the antigen is encountered, producing a rapid secondary immune response. p. 1279

memory cell An activated lymphocyte that circulates in the blood and lymph, ready to initiate a rapid immune response on subsequent exposure to the same antigen. p. 1273

memory helper T cell In cell-mediated immunity, a long-lived cell differentiated from a helper T cell, which remains in an inactive state in the lymphatic system after an immune reaction has run its course and is ready to be activated on subsequent exposure to the same antigen. p. 1280

meninges Three layers of connective tissue that surround and protect the spinal cord and brain. p. 1123

menstrual cycle A cycle of approximately 1 month in the human female during which an egg is released from an ovary and the uterus is prepared to receive the fertilized egg; if fertilization does not occur, the endometrium breaks down, which releases blood and tissue breakdown products from the uterus to the outside through the vagina. p. 1000

meristem An undifferentiated, permanently embryonic plant tissue that gives rise to new cells forming tissues and organs. p. 839

mesenteries Sheets of loose connective tissue, covered on both surfaces with epithelial cells, which suspend the abdominal organs in the coelom and provide lubricated, smooth surfaces that prevent chafing or abrasion between adjacent structures as the body moves. pp. 596, 952

mesoderm The middle layer of the three primary germ layers of an animal embryo, from which the muscular, skeletal, vascular, and connective tissues develop. pp. 595, 1015

mesoglea A layer of gel-like connective tissue separating the gastrodermis and epidermis in radially symmetrical animals. It contains widely dispersed amoeboid cells. p. 603

mesohyl The gelatinous middle layer of cells lining the body cavity of a sponge. p. 601

mesophyll The ground tissue located between the two outer leaf tissues, composed of loosely packed parenchyma cells that contain chloroplasts. p. 851

messenger RNA (mRNA) An RNA molecule that serves as a template for protein synthesis. p. 292

metabolism The biochemical reactions that allow a cell or organism to extract energy from its surroundings and use that energy to maintain itself, grow, and reproduce. p. 82

metagenomics The study of all DNA sequences, regardless of origin, isolated "in bulk" from ecosystems such as decaying animals, ocean water, termite gut, etc. pp. 380, 482

metamorphosis A reorganization of the form of certain animals during postembryonic development. pp. 602, 1057

metanephridium (plural, **metanephridia**) The excretory tubule of most annelids and molluscs. p. 617

metaphase The phase of mitosis during which the spindle reaches its final form and the spindle microtubules move the chromosomes into alignment at the spindle midpoint. p. 175

micelle A sphere composed of a single layer of lipid molecules. p. 100

microclimate The abiotic conditions immediately surrounding an organism. p. 737

microevolution Small-scale genetic changes within populations, often in response to shifting environmental circumstances or chance events. pp. 401, 410

micronucleus In ciliophorans, one or more diploid nuclei that contain a complete complement of genes, functioning primarily in cellular reproduction. p. 521

micronutrient Any mineral required by an organism only in trace amounts. pp. 903, 1190

micropyle A small opening at one end of an ovule through which the pollen tube passes prior to fertilization. p. 886

microRNAs (miRNAs) Small RNAs that regulate gene expression by binding to specific mRNAs and decreasing their translation. p. 329

microscope Instrument of microscopy with different magnifications and resolutions of specimens. p. 29

microscopy Technique for producing visible images of objects that are too small to be seen by the human eye. p. 28

microspore A plant spore from which a male gametophyte develops; usually smaller than a megaspore. pp. 570, 583, 885

microtubule A cytoskeletal component formed by the polymerization of tubulin into rigid, hollow rods about 25 nm in diameter. p. 41

microtubule organizing centre (MTOC) An anchoring point near the centre of a eukaryotic cell from which most microtubules extend outward. p. 178

microvilli Fingerlike projections forming a brush border in epithelial cells that cover the villi. p. 601

midbrain The uppermost of the three segments of the brain stem, serving primarily as an intermediary between the rest of the brain and the spinal cord. p. 1090

middle ear The air-filled cavity containing three small, interconnected bones: the malleus, incus, and stapes. p. 1105

migration The predictable seasonal movement of animals from the area where they are born to a distant and initially unfamiliar destination, returning to their birth site later. p. 1171

mimic The species in Batesian mimicry that resembles the model. p. 716

mimicry A form of defence in which one species evolves an appearance resembling that of another. p. 716

mineralocorticoid A steroid hormone secreted by the adrenal cortex that regulates the levels of Na and K in the blood and extracellular fluid. p. 1059

minimal medium A growth medium containing the minimal ingredients that enable a nonmutant organism, such as *E. coli,* to grow. p. 191

mismatch repair Repair system that removes mismatched bases from newly synthesized DNA strands. p. 282

missense mutation A base-pair substitution mutation in a protein-coding gene that results in a different amino acid in the encoded polypeptide than the normal one. p. 310

mitochondrial matrix The innermost compartment of the mitochondrion. p. 41

mitochondrion Membrane-bound organelle responsible for synthesis of most of the ATP in eukaryotic cells. p. 40

mitosis Nuclear division that produces daughter nuclei that are exact genetic copies of the parental nucleus. p. 170

mitotic spindle The complex of microtubules that orchestrate the separation of chromosomes during mitosis. p. 174

mobile elements Particular segments of DNA that can move from one place to another; they cut and paste DNA backbones using a type of recombination that does not require homology. p. 210

model The species in Batesian mimicry that is resembled by the mimic. p. 716

model organism An organism with characteristics that make it a particularly useful subject of research because it is likely to produce results widely applicable to other organisms. p. F-52

modern synthesis A unified theory of evolution developed in the middle of the twentieth century. p. 401

molarity (M) The number of moles of a substance dissolved in 1 L of solution. p. F-18

molars Posteriormost teeth of mammals, with a broad chewing surface for grinding food. p. 464

mole (mol) The atomic weight of an element or the molecular weight of a compound. p. F-18

molecular clock A technique for dating the time of divergence of two species or lineages, based on the number of molecular sequence differences between them. p. 471

molecular weight The weight of a molecule in grams, equal to the total mass number of its atoms. p. F-18

molecule A unit composed of atoms combined chemically in fixed numbers and ratios. p. F-12

monoclonal antibody An antibody that reacts only against the same segment (epitope) of a single antigen. p. 1283

monocot A plant belonging to the Monocotyledones, one of the two major classes of angiosperms; monocot embryos have a single seed leaf (cotyledon) and pollen grains with a single groove. pp. 587, 840

monocyte A type of leukocyte that enters damaged tissue from the bloodstream through the endothelial wall of the blood vessel. p. 1268

monoecious Having both "male" flowers (which possess only stamens) and "female" flowers (which possess only carpels). pp. 601, 885

monogamy A mating system in which one male and one female form a long-term association. p. 1175

monohybrid An F_1 heterozygote produced from a genetic cross that involves a single character. p. 221

monohybrid cross A genetic cross between two individuals that are each heterozygous for the same pair of alleles. p. 221

monomers Identical or nearly identical subunits that link together to form polymers during polymerization. p. 59

monophyletic taxon A group of organisms that includes a single ancestral species and all of its descendants. p. 459

monosaccharides The smallest carbohydrates, containing three to seven carbon atoms. p. 85

monotreme A lineage of mammals that lay eggs instead of bearing live young. p. 668

morphological species concept The concept that all individuals of a species share measurable traits that distinguish them from individuals of other species. p. 428

morphology The form or shape of an organism or part of an organism. p. 66

morula The first stage of animal development, a solid ball or layer of blastomeres. p. 1015

mosaic evolution The tendency of characteristics to undergo different rates of evolutionary change within the same lineage. p. 458

motif A highly specialized region in a protein produced by the three-dimensional arrangement of amino acid chains within and between domains. p. 323

motile Capable of self-propelled movement. p. 595

motor neuron An efferent neuron that carries signals to skeletal muscle. p. 1070

motor unit A block of muscle fibres that is controlled by branches of the axon of a single efferent neuron. p. 1141

moult-inhibiting hormone (MIH) A peptide neurohormone secreted by cells in the eyestalks (extensions of the brain leading to the eyes). p. 1063

mRNA splicing Process that removes introns from pre-mRNAs and joins exons together. p. 298

mucosa The lining of the gut that contains epithelial and glandular cells. p. 1198

Müllerian duct The bipotential primitive duct associated with the gonads that leads to a cloaca. p. 1029

Müllerian mimicry A form of defence in which two or more unpalatable species share a similar appearance. p. 716

multicellular organism Individual consisting of interdependent cells. p. 28

multigene family A family of homologous genes in a genome. The members of a multigene family have all evolved from one ancestral gene, and therefore have similar DNA sequences and produce proteins with similar structures and functions. p. 383

multiple alleles More than two different alleles of a gene. p. 232

multiple fruit A fruit that develops from several ovaries in multiple flowers; examples are pineapples and mulberries. p. 891

multiregional hypothesis A hypothesis proposing that after archaic humans migrated from Africa to many regions on Earth, their different populations evolved into modern humans simultaneously. p. 477

muscle fibre A bundle of elongated, cylindrical cells that make up skeletal muscle. pp. 954, 1134

muscle spindle A stretch receptor in muscle; a bundle of small, specialized muscle cells wrapped with the dendrites of afferent neurons and enclosed in connective tissue. p. 1103

muscle tissue Cells that have the ability to contract (shorten) forcibly. p. 954

muscle twitch A single, weak contraction of a muscle fibre. p. 1140

muscularis The muscular coat of a hollow organ or tubular structure. p. 1198

mutation A spontaneous and heritable change in DNA. pp. 282, 310, 417

mutualism A symbiotic interaction between species in which both partners benefit. pp. 541, 722, 1267

mycelium A network of branching hyphae that constitutes the body of a multicellular fungus. pp. 524, 542

mycobiont The fungal component of a lichen. p. 555

mycorrhiza A mutualistic symbiosis in which fungal hyphae associate intimately with plant roots. pp. 548, 910

myoblast An undifferentiated muscle cell. p. 1036

myofibril A cylindrical contractile element about 1 m in diameter that runs lengthwise inside the muscle fibre cell. p. 1135

myogenic heart A heart that maintains its contraction rhythm with no requirement for signals from the nervous system. p. 976

myoglobin An oxygen-storing protein closely related to hemoglobin. p. 1142

N-terminal end The end of a polypeptide chain with an —NH_3 group. p. 300

Na^+/K^+ pump Pump that pushes 3 Na^+ out of the cell and 2 K^+ into the cell in the same pumping cycle. Also referred to as the *sodium–potassium pump*. p. 110

nastic movement In plants, a reversible response to nondirectional stimuli, such as mechanical pressure or humidity. p. 937

natural killer (NK) cell A type of lymphocyte that destroys pathogen-infected cells. p. 1271

natural selection The evolutionary process by which alleles that increase the likelihood of survival and the reproductive output of the individuals that carry them become more common in subsequent generations. p. 398

natural theology A belief that knowledge of God may be acquired through the study of natural phenomena. p. 392

navigation A wayfinding mechanism in which an animal moves toward a specific destination, using both a compass and a "mental map" of where it is in relation to the destination. p. 1171

negative feedback The primary mechanism of homeostasis, in which a stimulus—a change in the external or internal environment—triggers a response that compensates for the environmental change. p. 959

negative pressure breathing Muscular contractions that expand the lungs, lowering the pressure of the air in the lungs and causing air to be pulled inward. p. 1223

nematocyst A coiled thread, encapsulated in a cnidocyte, that cnidarians fire at prey or predators, sometimes releasing a toxin through its tip. p. 603

nephron A specialized excretory tubule that contributes to osmoregulation and carries out excretion, found in all vertebrates. p. 1243

nerve A bundle of axons enclosed in connective tissue and all following the same pathway. p. 1069

nerve cord A bundle of nerves that extends from the central ganglia to the rest of the body, connected to smaller nerves. p. 1089

nerve net A simple nervous system that coordinates responses to stimuli but has no central control organ or brain. pp. 603, 1088

nervous tissue Tissue that contains neurons, which serve as lines of communication and control between body parts. p. 955

net primary productivity The chemical energy remaining in an ecosystem after a producer's cellular respiration is deducted. p. 750

neural crest A band of cells that arises early in the embryonic development of vertebrates near the region where the neural tube pinches off from the ectoderm; later, the cells migrate and develop into unique structures. p. 1021

neural plate Ectoderm thickened and flattened into a longitudinal band, induced by notochord cells. p. 1021

neural signalling The process by which an animal responds appropriately to a stimulus. p. 1069

neural tube A hollow tube in vertebrate embryos that develops into the brain, spinal cord, spinal nerves, and spinal column. p. 1090

neurogenic heart A heart that beats under the control of signals from the nervous system. p. 976

neuromuscular junction The junction between a nerve fibre and the muscle it supplies. p. 1136

neuron An electrically active cell of the nervous system responsible for controlling behaviour and body functions. pp. 955, 1047, 1069

neuronal circuit The connection between axon terminals of one neuron and the dendrites or cell body of a second neuron. p. 1070

neuropile The region of a ganglion in which branching axons and dendrites make interconnections. p. 1089

neurosecretory neuron A neuron that releases a neurohormone into the circulatory system when appropriately stimulated. p. 1047

neurotransmitter A chemical released by an axon terminal at a chemical synapse. p. 1071

neurulation The process in vertebrates by which organogenesis begins with development of the nervous system from ectoderm. p. 1021

neutral mutation hypothesis An evolutionary hypothesis that some variation at gene loci coding for enzymes and other soluble proteins is neither favoured nor eliminated by natural selection. p. 425

neutron Uncharged particle in the nucleus of an atom. p. F-9

neutrophil A type of phagocytic leukocyte that attaches to blood vessel walls in massive numbers when attracted to the infection site by chemokines. p. 1269

nitrification A metabolic process in which certain soil bacteria convert ammonia or ammonium ions into nitrites that are then converted by other bacteria to nitrates, a form usable by plants. pp. 488, 766, 911

nitrogen cycle A biogeochemical cycle that moves nitrogen between the huge atmospheric pool of gaseous molecular nitrogen and several much smaller pools of nitrogen-containing compounds in soils, marine and freshwater ecosystems, and living organisms. p. 762

nitrogen fixation A metabolic process in which certain bacteria and cyanobacteria convert molecular nitrogen into ammonia and ammonium ions, forms usable by plants. pp. 488, 762, 911

nitrogenous base A nitrogen-containing molecule with the properties of a base. p. 83

nociceptor A sensory receptor that detects tissue damage or noxious chemicals; their activity registers as pain. p. 1077

node The point on a stem where one or more leaves are attached. p. 847

node of Ranvier The gap between two Schwann cells, which exposes the axon membrane directly to extracellular fluids. p. 1070

nondisjunction The failure of homologous pairs to separate during the first meiotic division or of chromatids to separate during the second meiotic division. p. 252

nonhistone protein All of the proteins associated with DNA in a eukaryotic chromosome that are not histones. p. 285

nonsense codon *See* stop codon. p. 294

nonsense mutation A base-pair substitution mutation in a gene in which the base-pair change results in a change from a sense codon to a nonsense codon in the mRNA. The polypeptide translated from the mRNA is shorter than the normal polypeptide because of the mutation. p. 310

nonshivering thermogenesis The generation of heat by oxidative mechanisms in nonmuscle tissue throughout the body. p. 1258

nonvascular plant *See* bryophyte. p. 567

norepinephrine A nontropic amine hormone secreted by the adrenal medulla. p. 1058

notochord A flexible rodlike structure constructed of fluid-filled cells surrounded by tough connective tissue, which supports a chordate embryo from head to tail. p. 640

nucellus The inner part of an ovule, containing the embryo sac; equivalent to a megasporangium. p. 886

nuclear envelope In eukaryotes, membranes separating the nucleus from the cytoplasm. p. 33

nuclear localization signal A short amino acid sequence in a protein that directs the protein to the nucleus. p. 309

nuclear pore Opening in the membrane of the nuclear envelope through which large molecules, such as RNA and proteins, move between the nucleus and the cytoplasm. p. 35

nuclear pore complex A large, octagonally symmetrical, cylindrical structure that functions to exchange molecules between the nucleus and cytoplasm and prevents the transport of material not meant to cross the nuclear membrane. A nuclear pore—a channel through the complex—is the path for the exchange of molecules. p. 35

nucleoid The central region of a prokaryotic cell with no boundary membrane separating it from the cytoplasm, where DNA replication and RNA transcription occur. pp. 31, 169, 285, 482

nucleolus The nuclear site of rRNA transcription, processing, and ribosome assembly in eukaryotes. p. 35

nucleoplasm The liquid or semiliquid substance within the nucleus. p. 35

nucleosome The basic structural unit of chromatin in eukaryotes, consisting of DNA wrapped around a histone core. p. 283

nucleosome core particle An eight-protein particle formed by the combination of two molecules each of H2A, H2B, H3, and H4, around which DNA winds for almost two turns. p. 283

nucleosome remodelling complex A multiprotein structure that moves, or modifies, nucleosomes in such a way that exposes promoters to the transcription machinery. p. 321

nucleotide The monomer of nucleic acids consisting of a five-carbon sugar, a nitrogenous base, and a phosphate. p. F-37

nucleus The central region of eukaryotic cells, separated by membranes from the surrounding cytoplasm, where DNA replication and messenger RNA transcription occur. p. 31

null model A conceptual model that predicts what one would see if a particular factor had no effect. p. 415

nutrition The processes by which an organism takes in, digests, absorbs, and converts food into organic compounds. p. 1189

obligate aerobe A microorganism that must use oxygen for cellular respiration and requires oxygen in its surroundings to support growth. p. 487

obligate anaerobe A microorganism that cannot use oxygen and can grow only in the absence of oxygen. p. 487

ocellus (plural, **ocelli**) The simplest eye, which detects light but does not form an image. p. 1108

Okazaki fragments Relatively short segments of DNA synthesized on the lagging strand at a replication fork. p. 275

olfactory bulb A grey-matter centre that relays inputs from odour receptors to both the cerebral cortex and the limbic system. p. 1126

oligodendrocyte A type of glial cell that populates the central nervous system and is responsible for producing myelin. p. 1070

oligosaccharin A complex carbohydrate that in plants serves as a signalling molecule and as a defence against pathogens. p. 929

ommatidium (plural, ommatidia) A faceted visual unit of a compound eye. p. 1108

omnivore An animal that feeds at several trophic levels, consuming plants, animals, and other sources of organic matter. p. 1000

oncogene A gene that, when deregulated, is capable of inducing one or more characteristics of cancer cells. pp. 185, 333

one gene–one enzyme hypothesis Hypothesis showing the direct relationship between genes and enzymes. p. 292

one gene–one polypeptide hypothesis Restatement of the one gene–one enzyme hypothesis, taking into account that some proteins consist of more than one polypeptide and not all proteins are enzymes. p. 292

oocyte A developing gamete that becomes an ootid at the end of meiosis. p. 601

oogenesis The process of producing eggs. p. 992

oogonium (plural, oogonia) A cell that enters meiosis and gives rise to gametes, produced by mitotic divisions of the germ cells in females. p. 991

open circulatory system An arrangement of internal transport in some invertebrates in which the vascular fluid, hemolymph, is released into sinuses, bathing organs directly, and is not always retained within vessels. pp. 611, 965

open reading frames (ORFs) Segments of DNA sequence that contain start and stop codons. Such sequences are candidate genes. p. 369

operant conditioning A form of associative learning in which animals learn to link a voluntary activity, an operant, with its favourable consequences, the reinforcement. p. 1159

operator A DNA regulatory sequence that controls transcription of an operon. p. 315

operculum A lid or flap of the bone serving as the gill cover in some fishes. pp. 615, 649

operon A cluster of prokaryotic genes and the DNA sequences involved in their regulation. p. 315

opsin One of several different proteins that bond covalently with the light-absorbing pigment of rods and cones (retinal). p. 1111

optic chiasm Location just behind the eyes where the optic nerves converge before entering the base of the brain, witth a portion of each optic nerve crossing over to the opposite side. p. 1113

optimal foraging theory A set of mathematical models that predict the diet choices of animals as they encounter a range of potential food items. p. 712

oral hood Soft fleshy structure at the anterior end of a cephalochordate that frames the opening of the mouth. p. 641

orbital The region of space where the electron "lives" most of the time. p. F-10

order A Linnaean taxonomic category of organisms that ranks above a family and below a class. p. 428

organ Two or more different tissues integrated into a structure that carries out a specific function. pp. 836, 947

organ of Corti An organ within the cochlear duct that contains the sensory hair cells detecting sound vibrations transmitted to the inner ear. p. 1106

organ system The coordinated activities of two or more organs to carry out a major body function such as movement, digestion, or reproduction. p. 947

organelles The nucleus and other specialized internal structures and compartments of eukaryotic cells. p. 30

organic molecule Molecule based on carbon. p. 30

organismal ecology An ecological discipline in which researchers study the genetic, biochemical, physiological, morphological, and behavioural adaptations of organisms to their abiotic environments. p. 680

organogenesis The development of the major organ systems, giving rise to a free-living individual with the body organization characteristic of its species. p. 1015

origin of replication (ori) A specific region at which replication of a bacterial chromosome commences. p. 169

osculum (plural, oscula) An opening in a sponge through which water is expelled. p. 601

osmoconformer An animal in which the osmolarity of the cellular and extracellular solutions matches the osmolarity of the environment. p. 1237

osmolality A measure of the osmotic concentration of a solution. It is measured in osmoles (the number of solute molecules and ions) per kilogram of solvent. p. 1237

osmolarity The total solute concentration of a solution, measured in osmoles—the number of solute molecules and ions (in moles)—per litre of solution. p. 1237

osmoregulator An animal that uses control mechanisms to keep the osmolarity of cellular and extracellular fluids the same but at levels that may differ from the osmolarity of the surroundings. p. 1237

osmosis The passive transport of water across a selectively permeable membrane in response to solute concentration gradients, a pressure gradient, or both. pp. 108, 865

osteoblast A cell that produces the collagen and mineral of bone. p. 954

osteoclast A cell that removes bone minerals and recycles them through the bloodstream. p. 954

osteocyte A mature bone cell. p. 954

osteon The structural unit of bone, consisting of a minute central canal surrounded by osteocytes embedded in concentric layers of mineral matter. p. 954

ostracoderm One of an assortment of extinct, jawless fishes that were covered with bony armour. p. 645

otolith One of many small crystals of calcium carbonate embedded in the otolithic membrane of the hair cells. p. 1102

outer boundary membrane A smooth membrane that surrounds a chloroplast, enclosing the stroma. p. 46

outer ear The external structure of the ear, consisting of the pinna and meatus. p. 1105

outer membrane In Gram-negative bacteria, an additional boundary membrane that covers the peptidoglycan layer of the cell wall. p. 485

outer mitochondrial membrane The smooth membrane covering the outside of a mitochondrion. p. 40

outgroup comparison A technique used to identify ancestral and derived characters by comparing the group under study with more distantly related species that are not otherwise included in the analysis. p. 458

oval window An opening in the bony wall that separates the middle ear from the inner ear. p. 1106

ovarian cycle The cyclic events in the ovary leading to ovulation. p. 999

ovary In animals, the female gonad, which produces female gametes and reproductive hormones. In flowering plants, the enlarged base of a carpel in which one or more ovules develop into seeds. pp. 885, 991, 1060

overnutrition The condition caused by excessive intake of specific nutrients. p. 1190

oviduct The tube through which the egg moves from the ovary to the outside of the body. p. 993

oviparous Referring to animals that lay eggs containing the nutrients needed for development of the embryo outside the mother's body. p. 996

ovoviviparous Referring to animals in which fertilized eggs are retained within the body and the embryo develops using nutrients provided by the egg; eggs hatch inside the mother. p. 996

ovulation The process in which oocytes are released into the oviducts as immature eggs. p. 999

ovule In plants, the structure in a carpel in which a female gametophyte develops and fertilization takes place. pp. 582, 885

ovum (plural, **ova**) A female sex cell, or egg. p. 992

oxidation The removal of electrons from a substance. p. 62

oxidative phosphorylation Synthesis of ATP in which ATP synthase uses an H^+ gradient built by the electron transfer system as the energy source to make the ATP. p. 132

oxidized Refers to a substance from which the electrons are removed during oxidation. p. 122

oxytocin A hormone that stimulates the ejection of milk from the mammary glands of a nursing mother. p. 1056

P generation The parental individuals used in an initial cross. p. 220

P site The site in the ribosome where the tRNA carrying the growing polypeptide chain is bound. p. 302

pacemaker cell A specialized cardiac muscle cell in the upper wall of the right atrium that sets the rate of contraction in the heart. p. 976

pairing Process in meiosis in which homologous chromosomes come together and pair. Also referred to as *synapsis*. p. 202

pair-rule genes In *Drosophila* embryonic development, the set of segmentation regulatory genes activated by gap genes that divide the embryo into units of two segments each. p. 1038

pancreas A mixed gland composed of an exocrine portion that secretes digestive enzymes into the small intestine and an endocrine portion, the islets of Langerhans, that secretes insulin and glucagon. p. 1060

parapatric speciation Speciation between populations with adjacent geographic distributions. p. 436

paraphyletic taxon A group of organisms that includes an ancestral species and some, but not all, of its descendants. p. 459

parapodium (plural, **parapodia**) A fleshy lateral extension of the body wall of aquatic annelids, used for locomotion and gas exchange. p. 618

parasite An organism that feeds on the tissues of or otherwise exploits its host. p. 516

parasitism A symbiotic interaction in which one species, the parasite, uses another, the host, in a way that is harmful to the host. pp. 541, 723

parasitoid An insect species in which a female lays eggs in the larva or pupa of another insect species, and her young consume the tissues of the living host. p. 1290

parasympathetic division The division of the autonomic nervous system that predominates during quiet, low-stress situations, such as while relaxing. p. 1093

parathyroid gland One of a pair of glands that produce parathyroid hormone (PTH) (found only in tetrapod vertebrates). p. 1057

parathyroid hormone (PTH) The hormone secreted by the parathyroid glands in response to a fall in blood Ca^{2+} levels. p. 1057

parental Phenotypes identical to the original parental individuals. p. 170

parental investment The time and energy devoted to the production and rearing of offspring. p. 1012

parthenogenesis A mode of asexual reproduction in which animals produce offspring by the growth and development of an egg without fertilization. pp. 610, 989

partial diploid A condition in which part of the genome of a haploid organism is diploid. Recipients in bacterial conjugation between an Hfr and an F cell become partial diploids for part of the Hfr bacterial chromosome. p. 195

partial pressure The individual pressure exerted by each gas within a mixture of gases. p. 1218

parturition The process of giving birth. p. 1028

passive immunity The acquisition of antibodies as a result of direct transfer from another person. p. 1281

passive parental care The amount of energy invested in offspring—in the form of the energy stored in eggs or seeds or energy transferred to developing young through a placenta—before they are born. p. 687

passive transport The transport of substances across cell membranes without expenditure of energy, as in diffusion. pp. 105, 865

paternal chromosome The chromosome derived from the male parent of an organism. p. 201

pathogenesis-related (PR) protein A hydrolytic enzyme that breaks down components of a pathogen's cell wall. p. 930

pattern formation The arrangement of organs and body structures in their proper three-dimensional relationships. p. 1036

pectoral girdle A bony or cartilaginous structure in vertebrates that supports and is attached to the forelimbs. p. 641

pedicellariae Small pincers at the base of short spines in starfishes and sea urchins. p. 638

pedigree Chart that shows all parents and offspring for as many generations as possible, the sex of individuals in the different generations, and the presence or absence of a trait of interest. p. 249

pedipalps The second pair of appendages in the head of chelicerates. p. 622

pellicle A layer of supportive protein fibres located inside the cell, just under the plasma membrane, providing strength and flexibility instead of a cell wall. p. 517

pelvic girdle A bony or cartilaginous structure in vertebrates that supports and is attached to the hindlimbs. p. 641

pepsin An enzyme made in the stomach that breaks down proteins. p. 93

pepsinogen The inactive precursor molecule for pepsin. p. 308

peptide bond A link formed by a dehydration synthesis reaction between the —NH2 group of one amino acid and the —COOH group of a second. p. 302

peptidoglycan A polymeric substance formed from a polysaccharide backbone tied together by short polypeptides, which is the primary structural molecule of bacterial cell walls. p. 484

peptidyl transferase An enzyme that catalyzes the reaction in which an amino acid is cleaved from the tRNA in the P site of the ribosome and forms a peptide bond with the amino acid on the tRNA in the A site of the ribosome. p. 305

peptidyl–tRNA A tRNA linked to a growing polypeptide chain containing two or more amino acids. p. 304

per capita growth rate The difference between the per capita birth rate and the per capita death rate of a population. p. 690

perception The conscious awareness of our external and internal environments derived from the processing of sensory input. p. 1099

perennial A plant in which vegetative growth and reproduction continue year after year. p. 840

perfusion The flow of blood or other body fluids on the internal side of the respiratory surface. p. 1220

pericarp The fruit wall. p. 891

pericycle A tissue of plant roots, located between the endodermis and the phloem, which gives rise to lateral roots. p. 854

periderm The outermost portion of bark; consists of cork, cork cambium, and secondary cortex. p. 856

peripheral membrane protein Protein held to membrane surfaces by noncovalent bonds formed with the polar parts of integral membrane proteins or membrane lipids. p. 105

peripheral nervous system (PNS) All nerve roots and nerves (motor and sensory) that supply the muscles of the body and transmit information about sensation (including pain) to the central nervous system. p. 1055

peristalsis The rippling motion of muscles in the intestine or other tubular organs characterized by the alternate contraction and relaxation of the muscles that propel the contents onward. p. 1198

peritoneum The thin tissue derived from mesoderm that lines the abdominal wall and covers most of the organs in the abdomen. p. 596

peritubular capillary A capillary of the network surrounding the glomerulus. p. 1246

permafrost Perpetually frozen ground below the topsoil. p. 263

peroxisome Microbody that produces hydrogen peroxide as a by-product. p. 308

petal Part of the corolla of a flower, often brightly coloured. p. 885

petiole The stalk by which a leaf is attached to a stem. p. 850

phage *See* bacteriophage. pp. 266, 502

phagocytosis Process in which some types of cells engulf bacteria or other cellular debris to break them down. pp. 39, 114, 1268

pharynx The throat. In some invertebrates, a protrusible tube used to bring food into the mouth for passage to the gastrovascular cavity; in mammals, the common pathway for air entering the larynx and food entering the esophagus. pp. 609, 1225

phenotype The outward appearance of an organism. p. 221

phenotypic variation Differences in appearance or function between individual organisms. p. 410

pheromone A distinctive volatile chemical released in minute amounts to influence the behaviour of members of the same species. p. 405

phloem The food-conducting tissue of a vascular plant. pp. 567, 844

phloem sap The solution of water and organic compounds that flows rapidly through the sieve tubes of flowering plants. p. 877

phosphate group Group consisting of a central phosphorus atom held in four linkages: two that bind —OH groups to the central phosphorus atom, a third that binds an oxygen atom to the central phosphorus atom, and a fourth that links the phosphate group to an oxygen atom. p. 74

phosphodiester bond The linkage of nucleotides in polynucleotide chains by a bridging phosphate group between the 5 carbon of one sugar and the 3 carbon of the next sugar in line. p. 268

phospholipid A phosphate-containing lipid. p. 99

phosphorus cycle A biogeochemcial cycle in which weathering and erosion carry phosphate ions from rocks to soil and into streams and rivers, which eventually transport them to the ocean, where they are slowly incorporated into rocks. p. 768

photoautotroph A photosynthetic organism that uses light as its energy source and carbon dioxide as its carbon source. p. 145

photobiont The photosynthetic component of a lichen. p. 555

photoheterotroph An organism that uses light as the ultimate energy source but obtains carbon in organic form rather than as carbon dioxide. p. 487

photons Discrete particles or packets of energy. p. 5

photoperiodism The response of plants to changes in the relative lengths of light and dark periods in their environment during each 24 hour period. p. 938

photophosphorylation The synthesis of ATP coupled to the transfer of electrons energized by photons of light. p. 154

photopigment Light-absorbing pigment. p. 1108

photopsin One of three photopigments in which retinal is combined with different opsins. p. 1113

photoreceptor A sensory receptor that detects the energy of light. p. 1077

photorespiration A process that metabolizes a by-product of photosynthesis. p. 158

photosynthesis The conversion of light energy to chemical energy in the form of sugar and other organic molecules. p. 145

photosystem A large complex into which the light-absorbing pigments for photosynthesis are organized with proteins and other molecules. p. 147

photosystem I In photosynthesis, a protein complex in the thylakoid membrane that uses energy absorbed from sunlight to synthesize NADPH. p. 151

photosystem II In photosynthesis, a protein complex in the thylakoid membrane that uses energy absorbed from sunlight to synthesize ATP. p. 151

phototroph An organism that obtains energy from light. p. 487

phototropism The tendency of a plant shoot to bend toward a source of light. p. 920

PhyloCode A formal set of rules governing phylogenetic nomenclature. p. 461

phylogenetic species concept A concept that seeks to delineate species as the smallest aggregate population that can be united by shared derived characters. p. 429

phylogenetic tree A branching diagram depicting the evolutionary relationships of groups of organisms. p. 455

phylogeny The evolutionary history of a group of organisms. p. 455

phylum (plural, phyla) A major Linnaean division of a kingdom, ranking above a class. p. 428

physiology The study of the functions of organisms—the physicochemical processes of organisms. pp. 946, 1296

phytoalexin A biochemical that functions as an antibiotic in plants. p. 930

phytochrome A blue-green pigmented plant chromoprotein involved in the regulation of light-dependent growth processes. p. 938

phytoplankton Microscopic, free-flowing aquatic plants and protists. p. 516

pigment A molecule that can absorb photons of light. p. 5

piloting A wayfinding mechanism in which animals use familiar landmarks to guide their journey. p. 1171

pilus (plural, **pili**) A hair or hairlike appendage on the surface of a prokaryote. pp. 33, 486

pinacoderm In sponges, an unstratified outer layer of cells. p. 601

pineal gland A light-sensitive, melatonin-secreting gland that regulates some biological rhythms. p. 1061

pinna The external structure of the outer ear, which concentrates and focuses sound waves. p. 1105

pinocytosis *See* bulk-phase endocytosis. p. 112

pith The soft, spongelike, central cylinder of the stems of most flowering plants, composed mainly of parenchyma. p. 849

pituitary A gland consisting mostly of two fused lobes suspended just below the hypothalamus by a slender stalk of tissue that contains both neurons and blood vessels; it interacts with the hypothalamus to control many physiological functions, including the activity of some other glands. p. 1053

placenta A specialized temporary organ that connects the embryo and fetus with the uterus in mammals, mediating the delivery of oxygen and nutrients. Analagous structures occur in other animals. pp. 668, 996

plasma The clear, yellowish fluid portion of the blood in which cells are suspended. Plasma consists of water, glucose and other sugars, amino acids, plasma proteins, dissolved gases, ions, lipids, vitamins, hormones and other signal molecules, and metabolic wastes. pp. 958, 970

plasma cell A large antibody-producing cell that develops from B cells. p. 1278

plasma membrane The outer limit of the cytoplasm responsible for the regulation of substances moving into and out of cells. pp. 29, 102

plasmid A DNA molecule in the cytoplasm of certain prokaryotes, which often contains genes with functions that supplement those in the nucleoid and can replicate independently of the nucleoid DNA and be passed along during cell division. pp. 286, 484

plasmodesma (plural, **plasmodesmata**) A minute channel that perforates a cell wall and contains extensions of the cytoplasm that directly connect adjacent plant cells. p. 837

plasmodial slime mould A slime mould of the class Myxomycetes. p. 529

plasmodium The composite mass of plasmodial slime moulds consisting of individual nuclei suspended in a common cytoplasm surrounded by a single plasma membrane. p. 529

plasmogamy The sexual stage of fungi during which the cytoplasms of two genetically different partners fuse. p. 543

plasmolysis Condition due to outward osmotic movement of water, in which plant cells shrink so much that they retract from their walls. p. 867

plastids A family of plant organelles. p. 45

plastron The ventral part of the shell of a turtle. p. 659

platelet An oval or rounded cell fragment enclosed in its own plasma membrane, which is found in the blood; they are produced in red bone marrow by the division of stem cells and contain enzymes and other factors that take part in blood clotting. p. 973

pleiotropy Condition in which single genes affect more than one character of an organism. p. 236

pleura The double layer of epithelial tissue covering the lungs. p. 1225

ploidy The number of chromosome sets of a cell or species. p. 171

plumule The rudimentary terminal bud of a plant embryo located at the end of the hypocotyl, consisting of the epicotyl and a cluster of tiny foliage leaves. p. 893

poikilohydric Having little control over internal water content. p. 565

polar body A nonfunctional cell produced in oogenesis. p. 991

polar covalent bond Bond in which electrons are shared unequally. p. F-13

polar nucleus In the embryo sac of a flowering plant, one of two nuclei that migrate into the centre of the sac, become housed in a central cell, and eventually give rise to endosperm. p. 888

polar transport Unidirectional movement of a substance from one end of a cell (or other structure) to the other. p. 922

polarity The unequal distribution of yolk and other components in a mature egg. p. 1015

pollen grain The male gametophyte of a seed plant. pp. 582, 886

pollen sac The microsporangium of a seed plant, in which pollen develops. p. 589

pollen tube A tube that grows from a germinating pollen grain through the tissues of a carpel and carries the sperm cells to the ovary. p. 582

pollination The transfer of pollen to a flower's reproductive parts by air currents or on the bodies of animal pollinators. p. 582

poly(A) tail The string of A nucleotides added posttranscriptionally to the 3 end of a pre-mRNA molecule and retained in the mRNA produced from it that enables the mRNA to be translated efficiently and protects it from attack by RNA-digesting enzymes in the cytoplasm. p. 298

polyandry A polygamous mating system in which one female mates with multiple males. p. 1175

polygamy A mating system in which either males or females may have many mating partners. p. 1175

polygenic inheritance Inheritance in which several to many different genes contribute to the same character. p. 235

polygyny A polygamous mating system in which one male mates with many females. p. 1175

polyhedral virus A virus in which the coat proteins form triangular units that fit together like the parts of a geodesic sphere. p. 502

polymerase chain reaction (pCR) Process that amplifies a specific DNA sequence from a DNA mixture to an extremely large number of copies. p. 345

polymorphic development The production during development of one or more morphologically distinct forms. p. 595

polymorphism The existence of discrete variants of a character among individuals in a population. p. 411

polyp The tentacled, usually sessile stage in the life cycle of a coelenterate. p. 603

polypeptide The chain of amino acids formed by sequential peptide bonds. p. F-30

polyphyletic taxon A group of organisms that belong to different evolutionary lineages and do not share a recent common ancestor. p. 459

polyploid An individual with one or more extra copies of the entire haploid complement of chromosomes. p. 253

polyploidy The condition of having one or more extra copies of the entire haploid complement of chromosomes. p. 439

polysaccharide Chain with more than 10 linked monosaccharide subunits. p. 32

polysome The entire structure of an mRNA molecule and the multiple associated ribosomes that are translating it simultaneously. p. 306

population All individuals of a single species that live together in the same place and time. p. 410

population bottleneck An evolutionary event that occurs when a stressful factor reduces population size greatly and eliminates some alleles from a population. p. 418

population density The number of individuals per unit area or per unit volume of habitat. p. 680

population ecology The ecological discipline that focuses on how a population's size and other characteristics change in space and time. p. 680

population genetics The branch of science that studies the prevalence and variation in genes among populations of individuals. p. 401

population size The number of individuals in a population at a specified time. p. 680

positive feedback A mechanism that intensifies or adds to a change in internal or external environmental condition. p. 961

positive pressure breathing A gulping or swallowing motion that forces air into the lungs. p. 1223

posterior Indicating the tail end of an animal. p. 596

posterior pituitary The neural portion of the pituitary, which stores and releases two hormones made by the hypothalamus, antidiuretic hormone and oxytocin. p. 1053

postsynaptic cell The neuron or the surface of an effector after a synapse that receives the signal from the presynaptic cell. p. 1071

postsynaptic membrane The plasma membrane of the postsynaptic cell. p. 1082

posttranslational import A process for sorting proteins that are translated on cytosolic ribosomes and then moved into organelles. p. 309

postzygotic isolating mechanism A reproductive isolating mechanism that acts after zygote formation. p. 431

potential energy Stored energy. p. 75

precocial Born with fur and quickly mobile. p. 668

precursor mRNA (pre-mRNA) The primary transcript of a eukaryotic protein-coding gene, which is processed to form messenger RNA. p. 297

predation The interaction between predatory animals and the animal prey they consume. p. 712

prediction A statement about what the researcher expects to happen to one variable if another variable changes. p. F-3

pregnancy The period of mammalian development in which the embryo develops in the uterus of the mother. p. 1025

premolars Teeth located in pairs on each side of the upper and lower jaws of mammals, positioned behind the canines and in front of the molars. p. 668

prenatal diagnosis Techniques in which cells derived from a developing embryo or its surrounding tissues or fluids are tested for the presence of mutant alleles or chromosomal alterations. p. 258

prepuce Foreskin; a loose fold of skin that covers the glans of the penis. p. 1004

pressure flow mechanism In vascular plants, pressure that builds up at the source end of a sieve tube system and pushes solutes by bulk flow toward a sink, where they are removed. p. 878

presynaptic cell The neuron with an axon terminal on one side of the synapse that transmits the signal across the synapse to the dendrite or cell body of the postsynaptic cell. p. 1071

presynaptic membrane The plasma membrane of the axon terminal of a presynaptic cell, which releases neurotransmitter molecules into the synapse in response to the arrival of an action potential. p. 1082

prezygotic isolating mechanism A reproductive isolating mechanism that acts prior to the production of a zygote, or fertilized egg. p. 431

primary active transport Transport in which the same protein that transports a substance also hydrolyzes ATP to power the transport directly. p. 109

primary cell layers The ectoderm, mesoderm, and endoderm layers that form the embryonic tissues. p. 1015

primary cell wall The initial cell wall laid down by a plant cell. p. 837

primary consumer A herbivore, a member of the second trophic level. p. 1190

primary endosymbiosis In the model for the origin of plastids in eukaryotes, the first event in which a eukaryotic cell engulfed a photosynthetic cyanobacterium. p. 535

primary growth The growth of plant tissues derived from apical meristems. *Compare* secondary growth. p. 839

primary immune response The response of the immune system to the first challenge by an antigen. p. 1281

primary meristem Root and shoot apical meristems, from which a plant's primary tissues develop. *Compare* lateral meristem. p. 848

primary motor area The area of the cerebral cortex that runs in a band just in front of the primary somatosensory area and is responsible for voluntary movement. p. 1127

primary plant body The portion of a plant that is made up of primary tissues. p. 839

primary producer An autotroph, usually a photosynthetic organism, a member of the first trophic level. p. 145

primary somatosensory area The area of the cerebral cortex that runs in a band across the parietal lobes of the brain and registers information on touch, pain, temperature, and pressure. p. 1127

primary structure The sequence of amino acids in a protein. pp. 104, F-31

primary succession Predictable change in species composition of an ecological community that develops on bare ground. p. 735

primary tissue A plant tissue that develops from an apical meristem. p. 839

primase An enzyme that assembles the primer for a new DNA strand during DNA replication. p. 275

primer A short nucleotide chain made of RNA that is laid down as the first series of nucleotides in a new DNA strand or made of DNA for use in the polymerase chain reaction (pCR). p. 275

primitive groove In the development of birds, the sunken midline of the primitive streak that acts as a conduit for migrating cells to move into the blastocoel. p. 1020

primitive streak In the development of birds, the thickened region of the embryo produced by cells of the epiblast streaming toward the midline of the blastodisc. p. 1019

principle of independent assortment Mendel's principle that the alleles of the genes that govern two characters segregate independently during formation of gametes. p. 226

principle of monophyly A guiding principle of systematic biology that defines monophyletic taxa, each of which contains a single ancestral species and all of its descendants. p. 459

principle of segregation Mendel's principle that the pairs of alleles that control a character segregate as gametes are formed, with half the gametes carrying one allele and the other half carrying the other allele. p. 221

prion An infectious agent that contains only protein and does not include a nucleic acid molecule. p. 509

probability The possibility that an outcome will occur if it is a matter of chance. p. 222

procambium The primary meristem of a plant that develops into primary vascular tissue. p. 849

product An atom or molecule leaving a chemical reaction. p. F-14

product rule Mathematical rule in which the final probability is found by multiplying individual probabilities. p. 223

production efficiency The ratio of the energy content of new tissue produced to the energy assimilated from food. p. 753

progesterone A female sex hormone that stimulates growth of the uterine lining and inhibits contractions of the uterus. p. 1060

progestin A class of sex hormones synthesized by the gonads of vertebrates and active predominantly in females. p. 1059

proglottid One of the segmentlike repeating units that constitute the body of a tapeworm. p. 609

prokaryote Organism in which the DNA is suspended in the cell interior without separation from other cellular components by a discrete membrane. p. 28

prokaryotic chromosome A single, typically circular DNA molecule. p. 32

prolactin (pRL) A peptide hormone secreted by the anterior pituitary that stimulates breast development and milk secretion in mammals. pp. 1029, 1055

prometaphase A transition period between prophase and metaphase during which the microtubules of the mitotic spindle attach to the kinetochores and the chromosomes shuffle until they align in the centre of the cell. p. 174

promiscuity A mating system in which individuals do not form close pair bonds, and both males and females mate with multiple partners. p. 1175

promoter The site to which RNA polymerase binds for initiating transcription of a gene. p. 296

promoter proximal elements Regulatory sequence within the promoter proximal region, a region upstream of a eukaryotic protein-coding gene. Regulatory proteins bind to promoter proximal elements. p. 322

promoter proximal region Upstream of a eukaryotic gene, a region containing regulatory sequences for transcription called promoter proximal elements. p. 322

proofreading mechanism Mechanism of DNA polymerase to back up and remove mispaired nucleotides from a newly synthesized DNA strand. p. 280

propagation (conduction) In animal nervous systems, the concept that the action potential does not need further trigger events to keep going. p. 1079

prophage A viral genome inserted in the host cell DNA. pp. 197, 505

prophase The beginning phase of mitosis during which the duplicated chromosomes within the nucleus condense from a greatly extended state into compact, rodlike structures. p. 173

proprioceptor A mechanoreceptor that detects stimuli used in the central nervous system to maintain body balance and equilibrium and to monitor the position of the head and limbs. p. 1101

prostaglandin One of a group of local regulators derived from fatty acids that are involved in paracrine and autocrine regulation. p. 1002

prostate gland An accessory sex gland in males that adds a thin, milky fluid to the semen and adjusts the pH of the semen to the level of acidity best tolerated by sperm. p. 1003

protein Molecules that carry out most of the activities of life, including the synthesis of all other biological molecules. A protein consists of one or more polypeptides depending on the protein. p. 2

protein kinase Enzyme that transfers a phosphate group from ATP to one or more sites on particular proteins. p. 116

protein phosphatase Enzyme that removes phosphate groups from target proteins. p. 116

proteome The complete set of proteins that can be expressed by the genome of an organism. p. 376

proteomics The study of the proteome. p. 376

protist Organism currently classified in the kingdom Protista. p. 513

protobiont The term given to a group of abiotically produced organic molecules that are surrounded by a membrane or membranelike structure. p. 60

protoderm The primary meristem that will produce stem epidermis. p. 848

proton Positively charged particle in the nucleus of an atom. p. F-9

proton-motive force Stored energy that contributes to ATP synthesis and to the cotransport of substances to and from mitochondria. p. 132

proton pump Pump that moves hydrogen ions across membranes and pushes hydrogen ions across the plasma membrane from the cytoplasm to the cell exterior. Also referred to as H^+ *pump*. p. 7

protonema The structure that arises when a liverwort or moss spore germinates and eventually gives rise to a mature gametophyte. p. 574

protonephridium The simplest form of invertebrate excretory tubule. p. 1241

proto-oncogene A gene that encodes various kinds of proteins that stimulate cell division. Mutated proto-oncogenes contribute to the development of cancer. p. 333

protostome A division of the Bilateria in which the blastopore forms the mouth during development of the embryo and the anus appears later. p. 596

prototrophs Strains that are able to synthesize the necessary amino acids. p. 191

provirus The inserted viral DNA. p. 213

proximal convoluted tubule The tubule between the Bowman's capsule and the loop of Henle in the nephron of the kidney, which carries and processes the filtrate. p. 1246

pseudocoelom A fluid- or organ-filled body cavity between the gut (a derivative of endoderm) and the muscles of the body wall (a derivative of mesoderm). p. 596

pseudocoelomate A body plan of bilaterally symmetrical animals with a body cavity that lacks a complete lining derived from mesoderm. p. 596

pseudogene A gene that is very similar to a functional gene at the DNA sequence level but that has one or more inactivating mutations that prevent it from producing a functional gene product. p. 368

pseudopod (plural, pseudopodia) A temporary cytoplasmic extension of a cell. p. 517

psychrophile An archaean or bacterium that grows optimally at temperatures in the range of -10 to $-20°C$. p. 497

Pterophyta The plant phylum of ferns and their close relatives. p. 577

pulmonary circuit The circuit of the cardiovascular system that supplies the lungs. p. 968

pulvinus (plural, pulvini) A jointlike, thickened pad of tissue at the base of a leaf or petiole; flexes when the leaf makes nastic movements. p. 937

Punnett square Method for determining the genotypes and phenotypes of offspring and their expected proportions. p. 224

pupa The nonfeeding stage between the larva and adult in the complete metamorphosis of some insects, during which the larval tissues are completely reorganized within a protective cocoon or hardened case. p. 629

pupil The dark centre in the middle of the iris through which light passes to the back of the eye. p. 1108

purine A type of nitrogenous base with two carbon–nitrogen rings. p. 58

pyramid of biomass A diagram that illustrates differences in standing crop biomass in a series of trophic levels. p. 754

pyramid of energy A diagram that illustrates the amount of energy that flows through a series of trophic levels. p. 754

pyramid of numbers A diagram that illustrates the number of individual organisms present in a series of trophic levels. p. 755

pyrimidine A type of nitrogenous base with one carbon–nitrogen ring. p. 58

qualitative variation Variation that exists in two or more discrete states, with intermediate forms often being absent. p. 411

quantitative trait A character that displays a continuous distribution of the phenotype involved, typically resulting from several to many contributing genes. p. 235

quantitative variation Variation that is measured on a continuum (such as height in human beings) rather than in discrete units or categories. p. 410

quorum sensing The use of signalling molecules by prokaryotes to communicate and to coordinate their behaviour. p. 491

R gene A resistance gene in a plant; dominant R alleles confer enhanced resistance to plant pathogens. p. 932

r-selected species A short-lived species adapted to function well in a rapidly changing environment. p. 697

radial cleavage A cleavage pattern in deuterostomes in which newly formed cells lie directly above and below other cells of the embryo. p. 596

radial symmetry A body plan of organisms in which structures are arranged regularly around a central axis, like spokes radiating out from the centre of a wheel. p. 602

radiation The transfer of heat energy as electromagnetic radiation. p. 1252

radicle The rudimentary root of a plant embryo. p. 892

radioactivity The giving off of particles of matter and energy by decaying nuclei. p. 266

radioisotope An unstable, radioactive isotope. p. 267

radiometric dating A dating method that uses measurements of certain radioactive isotopes to calculate the absolute ages in years of rocks and minerals. p. 55

radula The tooth-lined "tongue" of molluscs that scrapes food into small particles or drills through the shells of prey. p. 611

random dispersion A pattern of distribution in which the individuals in a population are distributed unpredictably in their habitat. p. 682

rapid eye movement (REM) sleep The period during deep sleep when the delta wave pattern is replaced by rapid, irregular beta waves characteristic of the waking state. The person's heartbeat and breathing rate increase, the limbs twitch, and the eyes move rapidly behind the closed eyelids. p. 1130

reabsorption The process in which some molecules (e.g, glucose and amino acids) and ions are transported by the transport epithelium back into the body fluid (animals with open circulatory systems) or into the blood in capillaries surrounding the tubules (animals with closed circulatory systems) as the filtered solution moves through the excretory tubule. p. 1050

reactants The atoms or molecules entering a chemical reaction. p. F-14

reaction centre Part of photosystems I and II in chloroplasts of plants. In the light-dependent reactions of photosynthesis, the reaction centre receives light energy absorbed by the antenna complex in the same photosystem. p. 151

reading frame A particular grouping of triplet bases read by transfer RNA during translation. pp. 294, 304

realized niche The range of conditions and resources that a population actually uses in nature. p. 721

receptacle The expanded tip of a flower stalk that bears floral organs. p. 885

reception In signal transduction, the binding of a signal molecule with a specific receptor in a target cell. p. 1069

receptor-mediated endocytosis The selective uptake of macromolecules that bind to cell surface receptors concentrated in clathrin-coated pits. p. 112

receptor protein Protein that recognizes and binds molecules from other cells that act as chemical signals. p. 49

recessive An allele that is masked by a dominant allele. p. 221

reciprocal altruism Form of altruistic behaviour in which individuals help nonrelatives if they are likely to return the favour in the future. p. 1181

recognition protein Protein in the plasma membrane that identifies a cell as part of the same individual or as foreign. p. 504

recombinant Phenotype with a different combination of traits from those of the original parents. p. 195

recombinant DNA DNA from two or more different sources joined together. p. 340

recombination frequency In the construction of linkage maps of diploid eukaryotic organisms, the percentage of testcross progeny that are recombinants. p. 203

rectum The final segment of the large intestine. p. 1206

red tide A growth in dinoflagellate populations that causes red, orange, or brown discoloration of coastal ocean waters. p. 521

redox reaction Coupled oxidation–reduction reaction in which electrons are removed from a donor molecule and simultaneously added to an acceptor molecule. p. 122

reduced Refers to a substance that receives electrons during reduction. p. 122

reduction The addition of electrons to a substance. p. 122

reflex A programmed movement that takes place without conscious effort, such as the sudden withdrawal of a hand from a hot surface. p. 1123

refractory period A period that begins at the peak of an action potential and lasts a few milliseconds, during which the threshold required for generation of an action potential is much higher than normal. p. 1075

regulators Animals that maintain factors of the internal environment in a relatively constant state. p. 958

regulatory protein DNA-binding protein that binds to a regulatory sequence and affects the expression of an associated gene or genes. p. 315

reinforcement 1. The enhancement of reproductive isolation that had begun to develop while populations were geographically separated; 2. Encouraging or establishing a pattern of behaviour using a positive or negative stimulus. pp. 436, 1159

relative abundance The relative commonness of populations within a community. p. 414

relative fitness The number of surviving offspring that an individual produces compared with the number left by others in the population. p. 420

release factor A protein that recognizes stop codons in the A site of a ribosome translating an mRNA and terminates translation. Also referred to as the *termination factor*. p. 306

releasing hormone (RH) A peptide neurohormone that controls the secretion of hormones from the anterior pituitary. p. 1053

renal artery An artery that carries bodily fluids into the kidney. p. 1245

renal cortex The outer region of the mammalian kidney that surrounds the renal medulla. p. 1245

renal medulla The inner region of the mammalian kidney. p. 1245

renal pelvis The central cavity in the kidney where urine drains from collecting ducts. p. 1245

renal vein The vein that routes filtered blood away from the kidney. p. 1245

renaturation The reformation of a denatured protein into its folded, functional state. p. F-36

replica plating Technique for identifying and counting genetic recombinants in conjugation, transformation, or transduction experiments in which the colony pattern on a plate containing solid growth medium is pressed onto sterile velveteen and transferred to other plates containing different combinations of nutrients. p. 191

replicates Multiple subjects that receive either the same experimental treatment or the same control treatment. p. F-3

replication bubble A structure resulting from bidirectional DNA replication from a given origin. Two forks, travelling in opposite directions, create a bubble. p. 276

replication fork The region of DNA synthesis where the parental strands separate and two new daughter strands elongate. p. 275

replication origin The site at which DNA replication begins. *See* origin of replication. p. 193

repressible operon Operon whose expression is prevented by a repressor molecule. p. 319

repressor A regulatory protein that prevents the operon genes from being expressed. p. 316

reproduction The process in which a parent or parents produce offspring. p. 989

reproductive isolating mechanism A biological characteristic that prevents the gene pools of two species from mixing. p. 430

reproductive strategy A set of behaviours that lead to reproductive success. p. 989

residual volume The air that remains in lungs after exhalation. p. 1227

resolution The minimum distance two points in a specimen can be separated and still be seen as two points. p. 29

resource partitioning The use of different resources or the use of resources in different ways by species living in the same place. p. 721

respiratory medium The environmental source of O_2 and the sink for released CO_2. For aquatic animals, the respiratory medium is water; for terrestrial animals, it is air. p. 1220

respiratory surface A layer of epithelial cells that provides the interface between the body and the respiratory medium. p. 1219

response In signal transduction, the last stage in which the transduced signal causes the cell to change according to the signal and to the receptors on the cell. In the nervous system, the output resulting from the integration of neural messages. p. 1069

resting potential A steady negative membrane potential exhibited by the membrane of a neuron that is not stimulated—that is, not conducting an impulse. p. 1073

restriction endonuclease (restriction enzyme) An enzyme that cuts DNA at a specific sequence. p. 341

restriction fragment A DNA fragment produced by cutting a long DNA molecule with a restriction enzyme. p. 341

restriction fragment length polymorphisms (RFLPs) When comparing different individuals, restriction enzyme–generated DNA fragments of different lengths from the same region of the genome. p. 349

retina A light-sensitive membrane lining the posterior part of the inside of the eye. p. 1108

retrotransposon A transposable element that transposes via an intermediate RNA copy of the transposable element. p. 212

retrovirus A virus with an RNA genome that replicates via a DNA intermediate. p. 213

reverse transcriptase An enzyme that uses RNA as a template to make a DNA copy of the retrotransposon. Reverse transcriptase is used to make DNA copies of RNA in test tube reactions. p. 212

reversible Refers to a reaction may go from left to right or from right to left, depending on conditions. p. 80

rhizoid A modified hypha that anchors a fungus to its substrate and absorbs moisture. p. 572

rhizome A horizontal, modified stem that can penetrate a substrate and anchor the plant. p. 577

rhodopsin The retinal–opsin photopigment. p. 1111

rhynchocoel A coelomic cavity that contains the proboscis of nemerteans. p. 611

ribonucleic acid (RNA) A polymer assembled from repeating nucleotide monomers in which the five-carbon sugar is ribose. Cellular RNAs are mRNA (which is translated to produce a polypeptide), tRNA (which brings an amino acid to the ribosome for assembly into a polypeptide during translation), and rRNA (which is a structural component of ribosomes). The genetic material of some viruses is RNA. p. 28

ribose A five-carbon sugar to which the nitrogenous bases in nucleotides link covalently. p. F-37

ribosomal RNA (rRNA) The RNA component of ribosomes. p. 32

ribosome A ribonucleoprotein particle that carries out protein synthesis by translating mRNA into chains of amino acids. pp. 32, 292, 302

ribosome binding site In translation initiation in prokaryotes, a sequence just upstream of the start codon that directs the small ribosomal subunit to bind and orient correctly for the complete ribosome to assemble and start translating in the correct spot. p. 304

ribozyme An RNA-based catalyst that is part of the biochemical machinery of all cells. p. 61

ribulose-1,5-bisphosphate (RuBP) carboxylase/oxygenase (rubisco) An enzyme that catalyzes the key reaction of the Calvin cycle, carbon fixation, in which CO_2 combines with RuBP (ribulose 1,5-bisphosphate) to form 3-phosphoglycerate. p. 157

ring species A species with a geographic distribution that forms a ring around uninhabitable terrain. p. 433

RNA interference (RNAi) The phenomenon of silencing a gene posttranscriptionally by a small, single-stranded RNA that is complementary to part of an mRNA. p. 329

RNA polymerase An enzyme that catalyzes the assembly of nucleotides into an RNA strand. p. 292

RNA-seq A technique for whole transcriptome sequencing. One method converts mRNA to cDNA for next-generation sequencing. p. 376

rod In the vertebrate eye, a type of photoreceptor in the retina that is specialized for detection of light at low intensities. pp. 482, 1109

root An anchoring structure in land plants that also absorbs water and nutrients and (in some plant species) stores food. p. 568

root cap A dome-shaped cell mass that forms a protective covering over the apical meristem in the tip of a plant root. p. 853

root hair A tubular outgrowth of the outer wall of a root epidermal cell; root hairs absorb much of a plant's water and minerals from the soil. p. 847

root nodule A localized swelling on a root in which symbiotic nitrogen-fixing bacteria reside. p. 912

root pressure The pressure that develops in plant roots as the result of osmosis, forcing xylem sap upward and out through leaves. *See also* guttation. p. 872

root system An underground (or submerged) network of roots with a large surface area that favours the rapid uptake of soil water and dissolved mineral ions. pp. 568, 839

rough ER Endoplasmic reticulum with many ribosomes studding its outer surface. p. 36

round window A thin membrane that faces the middle ear. p. 1106

ruminant An animal that has a complex, four-chambered stomach. p. 103

S phase The phase of the cell cycle during which DNA replication occurs. p. 171

saccule A fluid-filled chamber in the vestibular apparatus that provides information about the position of the head with respect to gravity (up versus down), as well as changes in the rate of linear movement of the body. p. 1102

salicylic acid (SA) In plants, a chemical synthesized following a wound that has multiple roles in plant defences, including interaction with jasmonates in signalling cascades. p. 929

salivary amylase A substance that hydrolyzes starches to the disaccharide maltose. p. 1201

salivary gland A gland that secretes saliva through a duct on the inside of the cheek or under the tongue; the saliva lubricates food and begins digestion. p. 1201

saltatory conduction A mechanism that allows small-diameter axons to conduct impulses rapidly. p. 1080

saprotroph An organism nourished by dead or decaying organic matter. p. 540

sapwood The newly formed outer wood located between heartwood and the vascular cambium. Compared with heartwood, it is wet, lighter in colour, and not as strong. p. 856

sarcomere The basic unit of contraction in a myofibril. p. 1136

saturated fatty acid Fatty acid with only single bonds linking the carbon atoms. p. 101

savannah A biome comprising grasslands with few trees, which grows in areas adjacent to tropical deciduous forests. p. 727

schizocoelom In protostomes, the body cavity that develops as inner and outer layers of mesoderm separate. p. 598

Schwann cell A type of glial cell in the peripheral nervous system that wraps nerve fibres with myelin and also secretes regulatory factors. p. 1070

scientific method An investigative approach in which scientists make observations about the natural world, develop working explanations about what they observe, and then test those explanations by collecting more information. p. F-2

sclereid A type of sclerenchyma cell; sclereids are typically short and have thick, lignified walls. p. 843

sclerenchyma A ground tissue in which cells develop thick secondary walls, which are commonly lignified and perforated by pits through which water can pass. p. 843

sclerotium Tough mass of hyphae, often serving as a survival or overwintering structure. p. 550

scolex The anterior (head) of a tapeworm, adapted for fastening the worm to the intestinal epithelium of its host. p. 609

scrotum The baglike sac in which the testes are suspended in many mammals. p. 1002

second law of thermodynamics Principle that for any process in which a system changes from an initial to a final state, the total disorder of the system and its surroundings always increases. p. 77

second messenger In particular, in signal transduction pathways, an internal, nonprotein signal molecule that directly or indirectly activates protein kinases, which elicit the cellular response. p. 919

secondary active transport Transport indirectly driven by ATP hydrolysis. p. 109

secondary cell wall A layer added to the cell wall of plants that is more rigid and may become many times thicker than the primary cell wall. p. 837

secondary consumer A carnivore that feeds on herbivores, a member of the third trophic level. p. 1190

secondary endosymbiosis In the model for the origin of plastids in eukaryotes, the second event, in which a nonphotosynthetic eukaryote engulfed a photosynthetic eukaryote. p. 535

secondary growth Plant growth that originates at lateral meristems and increases the diameter of older roots and stems. *Compare* primary growth. p. 839

secondary immune response The rapid immune response that occurs during the second (and subsequent) encounters of the immune system of a mammal with a specific antigen. p. 1281

secondary metabolite Organic compound not required for the growth or survival of an organism; tends to be biologically active. p. 543

secondary productivity Energy stored in new consumer biomass as energy is transferred from producers to consumers. p. 752

secondary structure Regions of alpha helix, beta strand, or random coil in a polypeptide chain. pp. 104, F-31

secondary succession Predictable changes in species composition in an ecological community that develops after existing vegetation is destroyed or disrupted by an environmental disturbance. p. 737

secondary tissue In plants, the tissue that develops from lateral meristems. p. 839

secretory vesicle Vesicle that transports proteins to the plasma membrane. p. 38

seed The structure that forms when an ovule matures after a pollen grain reaches it and a sperm fertilizes the egg. p. 582

seed coat The outer protective covering of a seed. p. 892

segment polarity genes In *Drosophila* embryonic development, the set of segmentation regulatory genes activated by pair-rule genes that set the boundaries and anterior–posterior axis of each segment in the embryo. p. 1039

segmentation The production of body parts and some organ systems in repeating units. p. 599

segmentation genes Genes that work sequentially, progressively subdividing the embryo into regions, determining the segments of the embryo and the adult. p. 1037

segregate *See* principle of segregation. p. 221

segregation The separation of the pairs of alleles that control a character as gametes are formed. p. 169

selective cell adhesion A mechanism in which cells make and break specific connections to other cells or to the extracellular matrix. p. 1016

selectively permeable Membranes that selectively allow, impede, or block the passage of atoms and molecules. p. 60

self-fertilization (self-pollination) Fertilization in which sperm nuclei in pollen produced by anthers fertilize egg cells housed in the carpel of the same flower. Self-fertilization can also occur in hermaphroditic animals. pp. 420, 888

self-incompatibility In plants, the inability of a plant's pollen to fertilize ovules of the same plant. p. 888

semen The secretions of several accessory glands in which sperm are mixed prior to ejaculation. 1003

semicircular canal A part of the vestibular apparatus that detects rotational (spinning) motions. p. 1102

semiconservative replication The process of DNA replication in which the two parental strands separate and each serves as a template for the synthesis of new progeny double-stranded DNA molecules. p. 270

semilunar valve (SL valve) A flap of endocardium and connective tissue reinforced by fibres that prevent the valve from turning inside out. p. 974

seminal vesicle A vesicle that secretes seminal fluid. p. 1003

seminiferous tubule One of the tiny tubes in the testes where sperm cells are produced, grow, and mature. p. 1002

senescence The biologically complex process of aging in mature organisms that leads to the death of cells and eventually the whole organism. p. 926

sense codon A codon that specifies an amino acid. p. 302

sensitization Increased responsiveness to mild stimuli after experiencing a strong stimulus; one of the simplest forms of memory. p. 1130

sensor A tissue or organ that detects a change in an external or internal factor such as pH, temperature, or the concentration of a molecule such as glucose. p. 959

sensory adaptation A condition in which the effect of a stimulus is reduced if it continues at a constant level. p. 1099

sensory hair cell A hair cell that sends impulses along the auditory nerve to the brain when alternating changes of pressure agitate the basilar membrane on which the organ of Corti rests, moving the hair cells. p. 1101

sensory neuron A neuron that transmits stimuli collected by their sensory receptors to interneurons. p. 1070

sensory receptor (transducer) A receptor formed by the dendrites of afferent neurons or by specialized receptor cells making synapses with afferent neurons that pick up information about the external and internal environments of the animal. p. 1075

sensory transduction The conversion of a stimulus into a change in membrane potential. p. 1076

sepal One of the separate, usually green parts forming the calyx of a flower. p. 885

septum (plural, septa) A thin partition or cross wall that separates body segments. pp. 542, 617

sequential hermaphroditism The form of hermaphroditism in which individuals change from one sex to the other. p. 997

serosa The serous membrane: a thin membrane lining the closed cavities of the body; has two layers with a space between that is filled with serous fluid. p. 1199

Sertoli cell One of the supportive cells that completely surrounds developing spermatocytes in the seminiferous tubules. Follicle-stimulating hormone stimulates Sertoli cells to secrete a protein and other molecules that are required for spermatogenesis. p. 1002

sessile Unable to move from one place to another. p. 595

set point The level at which the condition controlled by a homeostatic pathway is to be maintained. p. 960

seta (plural, setae) A chitin-reinforced bristle that protrudes outward from the body wall in some annelid worms. p. 617

sex chromosomes Chromosomes that are different in male and female individuals of the same species. p. 203

sex-linked gene Gene located on a sex chromosome. p. 246

sex pilus *See* F pilus. p. 193

sex ratio The relative proportions of males and females in a population. p. 684

sexual dimorphism Differences in the size or appearance of males and females. p. 1176

sexual reproduction The mode of reproduction in which male and female parents produce offspring through the union of egg and sperm generated by meiosis. pp. 199, 989

sexual selection A form of natural selection established by male competition for access to females and by the females' choice of mates. p. 405

shoot system The stems and leaves of a plant. pp. 568, 839

short-day plant A plant that flowers in late summer or early autumn when dark periods become longer and light periods become shorter. p. 939

short interfering RNAs (siRNAs) Small RNA molecules that regulate expression of certain genes by binding to their mRNA and reducing translation. p. 329

short-term memory Memory that stores information for seconds. p. 1129

sieve tube A series of phloem cells joined end to end, forming a long tube through which nutrients are transported; seen mainly in flowering plants. p. 846

sieve tube member Any of the main conducting cells of phloem that connect end to end, forming a sieve tube. p. 844

sign stimulus A simple cue that triggers a fixed action pattern. p. 1156

signal peptide A short segment of amino acids to which the signal recognition particle binds, temporarily blocking further translation. A signal peptide is found on polypeptides that are sorted to the endoplasmic reticulum. Also referred to as *signal sequence*. p. 309

signal recognition particle (SRP) Protein–RNA complex that binds to signal sequences and targets polypeptide chains to the endoplasmic reticulum. p. 309

signal sequence *See* signal peptide. p. 309

signal transduction The series of events by which a signal molecule released from a controlling cell causes a response (affects the function) of target cells with receptors for the signal. Target cells process the signal in the three sequential steps of reception, transduction, and response. p. 9

silencing Phenomenon in which methylation of cytosines in eukaryotic promoters inhibits transcription and turns the genes off. p. 327

silent mutation A base-pair substitution mutation in a protein-coding gene that does not alter the amino acid specified by the gene. p. 310

simple diffusion Mechanism by which certain small substances diffuse through the lipid part of a biological membrane. p. 105

simple fruit A fruit that develops from a single ovary; in many of them, at least one layer of the pericarp is fleshy and juicy. p. 891

simulation modelling An analytical method in which researchers gather detailed information about a system and then create a series of mathematical equations that predict how the components of the system interact and respond to change. p. 770

simultaneous hermaphroditism A form of hermaphroditism in which individuals develop functional ovaries and testes at the same time. p. 997

single-lens eye An eye type that works by changing the amount of light allowed to enter into the eye and by focusing this incoming light with a lens. p. 1108

single-stranded binding protein (SSB) Protein that coats single-stranded segments of DNA, stabilizing the DNA for the replication process. p. 275

sink Any region of a plant where organic substances are being unloaded from the sieve tube system and used or stored. p. 877

sinoatrial node (SA node) The region of the heart that controls the rate and timing of cardiac muscle cell contraction. p. 976

sinus (plural, **sinuses**) A body space that surrounds an organ. p. 611

siRNA-induced silencing complex (siRISC) A group of proteins, recruited when siRNA binds to mRNA, that degrade the target mRNA. p. 330

sister chromatid One of two exact copies of a chromosome duplicated during replication. p. 171

skeletal muscle A muscle that connects to bones of the skeleton, typically made up of long and cylindrical cells that contain many nuclei. p. 954

sliding DNA clamp A protein that encircles the DNA and binds to the DNA polymerase to tether the enzyme to the template, thereby making replication more efficient. p. 272

slime layer A coat typically composed of polysaccharides that is loosely associated with bacterial cells. p. 32

slow block (to polyspermy) The process in which enzymes released from cortical granules alter the egg coats within minutes after fertilization so that no other sperm can attach and penetrate to the egg. p. 994

slow muscle fibre A muscle fibre that contracts relatively slowly and with low intensity. p. 1142

small interfering RNA (siRNA) A class of single-stranded RNAs that cause RNA interference. p. 323

small ribonucleoprotein particle A complex of RNA and proteins. p. 298

smooth ER Endoplasmic reticulum with no ribosomes attached to its membrane surfaces. Smooth ER has various functions, including synthesis of lipids that become part of cell membranes. p. 37

smooth muscle A relatively small and spindle-shaped muscle cell in which actin and myosin molecules are arranged in a loose network rather than in bundles. p. 955

social behaviour The interactions that animals have with other members of their species. p. 469

sodium–potassium pump *See* Na^+/K^+ pump. p. 110

soil solution A combination of water and dissolved substances that coats soil particles and partially fills pore spaces. p. 907

solenoid *See* 30 nm chromatin fibre. p. 284

solute The molecules of a substance dissolved in water. p. 105

solution Substance formed when molecules and ions separate and are suspended individually, surrounded by water molecules. p. F-17

solvent The water in a solution in which the hydration layer prevents polar molecules or ions from reassociating. p. F-17

somatic cell Any of the cells of an organism's body other than reproductive cells. pp. 199, 991

somatic embryo A plant embryo that is genetically identical to the parent because it arose through asexual means. p. 896

somatic gene therapy Gene therapy in which genes are introduced into somatic cells. p. 355

somatic nervous system A subdivision of the peripheral nervous system controlling body movements that are primarily conscious and voluntary. p. 1091

somites Paired blocks of mesoderm cells along the vertebrate body axis that form during early vertebrate development and differentiate into dermal skin, bone, and muscle. p. 1022

soredium (plural, soredia) A specialized cell cluster produced by lichens, consisting of a mass of algal cells surrounded by fungal hyphae; soredia function like reproductive spores and can give rise to a new lichen. p. 555

sorus (plural, sori) A cluster of sporangia on the underside of a fern frond; reproductive spores arise by meiosis inside each sporangium. p. 578

source In plants, any region (such as a leaf) where organic substances are being loaded into the sieve tube system of phloem. p. 877

source population In metapopulation analyses, a population that is either stable or increasing in size. p. 738

Southern blot analysis Technique in which labelled probes are used to detect specific DNA fragments that have been separated by gel electrophoresis. p. 349

spatial summation The summation of excitatory postsynaptic potentials produced by firing of different presynaptic neurons. p. 1087

specialized transduction Transfer of bacterial genes between bacteria using temperate phages that have incorporated fragments of the bacterial genome as they make the transition from the lysogenic cycle to the lytic cycle. pp. 196, 506

speciation The process of species formation. p. 427

species A group of populations in which the individuals are so closely related in structure, biochemistry, and behaviour that they can successfully interbreed. p. 427

species cluster A group of closely related species recently descended from a common ancestor. p. 436

species composition The particular combination of species that occupy a site. p. 730

species diversity A community characteristic defined by species richness and the relative abundance of species. p. 727

species richness The number of species that live within an ecological community. p. 726

specific epithet The species name in a binomial. p. 427

specific heat The amount of heat required to increase the temperature of a given quantity of water. p. F-16

sperm Motile gamete. p. 990

spermatocyte A developing gamete that becomes a spermatid at the end of meiosis. p. 813

spermatogenesis The process of producing sperm. p. 992

spermatogonium (plural, spermatogonia) A cell that enters meiosis and gives rise to gametes, produced by mitotic divisions of the germ cells in males. p. 991

spermatozoon (plural, spermatozoa) Also called sperm; a haploid cell that develops into a mature sperm cell when meiosis is complete. p. 992

sphincter A powerful ring of smooth muscle that forms a valve between major regions of the digestive tract. p. 1201

spinal cord A column of nervous tissue located within the vertebral column and directly connected to the brain. p. 1090

spinal nerve A nerve that carries signals between the spinal cord and the body trunk and limbs. p. 1123

spindle The structure that separates sister chromatids and moves them to opposite spindle poles. p. 170

spindle pole One of the pair of centrosomes in a cell undergoing mitosis from which bundles of microtubules radiate to form the part of the spindle from that pole. p. 174

spinneret A modified abdominal appendage from which spiders secrete silk threads. p. 623

spiracle An opening in the chitinous exoskeleton of an insect through which air enters and leaves the tracheal system. p. 1221

spiral cleavage The cleavage pattern in many protostomes in which newly produced cells lie in the space between the two cells immediately below them. p. 596

spiral valve A corkscrew-shaped fold of mucous membrane in the digestive system of elasmobranchs, which slows the passage of material and increases the surface area available for digestion and absorption. p. 648

spliceosome A complex formed between the pre-mRNA and small ribonucleoprotein particles, in which mRNA splicing takes place. p. 298

spongocoel The central cavity in a sponge. p. 601

spontaneous reaction Chemical or physical reaction that occurs without outside help. p. 79

sporadic (nonhereditary) cancer Cancer that is not inherited. p. 333

sporangium (plural, sporangia) A single-celled or multicellular structure in fungi and plants in which spores are produced. pp. 545, 570

spore A haploid reproductive structure, usually a single cell, that can develop into a new individual without fusing with another cell; found in plants, fungi, and certain protists. pp. 200, 884

sporophyll A specialized leaf that bears sporangia (spore-producing structures). p. 577

sporophyte An individual of the diploid generation produced through fertilization in organisms that undergo alternation of generations; it produces haploid spores. pp. 200, 526, 563, 883

squalene A liver oil found in sharks that is lighter than water, which increases their buoyancy. p. 648

stability The ability of a community to maintain its species composition and relative abundances when environmental disturbances eliminate some species from the community. p. 725

stabilizing selection A type of natural selection in which individuals expressing intermediate phenotypes have the highest relative fitness. p. 413

stamen A "male" reproductive organ in flowers, consisting of an anther (pollen producer) and a slender filament. p. 885

standing crop biomass The total dry weight of plants present in an ecosystem at a given time. p. 750

stapes The smallest of three sound-conducting bones in the middle ear of tetrapod vertebrates. pp. 652, 1105

starch Energy-providing carbohydrates stored in plant cells. p. 45

start codon The first codon read in an mRNA in translation—AUG. Also referred to as the *initiator codon*. p. 294

statocyst A mechanoreceptor in invertebrates that senses gravity and motion using statoliths. p. 1101

statolith A movable starch- or carbonate-containing stonelike body involved in sensing gravitational pull. pp. 934, 1101

stele The central core of vascular tissue in roots and shoots of vascular plants; it consists of the xylem and phloem together with supporting tissues. p. 854

stem cell Undifferentiated cells in most multicellular organisms that can divide without differentiating and also can divide and differentiate into specialized cell types. p. 353

stereocilia Microvilli covering the surface of hair cells clustered in the base of neuromasts. p. 1101

steroid A type of lipid derived from cholesterol. p. 326

steroid hormone receptor Internal receptor that turns on specific genes when it is activated by binding a signal molecule. p. 326

steroid hormone response element The DNA sequence to which the hormone receptor complex binds. p. 326

sterol Steroid with a single polar —OH group linked to one end of the ring framework and a complex, nonpolar hydrocarbon chain at the other end. p. 102

sticky end End of a DNA fragment, with a single-stranded structure that can form hydrogen bonds with a complementary sticky end on any other DNA molecule cut with the same enzyme. p. 341

stigma The receptive end of a carpel where deposited pollen germinates. p. 885

stimulus A component of a negative feedback control system maintaining homeostasis, specifically an environmental change that triggers a response. p. 959

stoma (plural, stomata) The opening between a pair of guard cells in the epidermis of a plant leaf or stem, through which gases and water vapour pass. pp. 565, 847

stomach The portion of the digestive system in which food is stored and digestion begins. p. 48

stop codon A codon that does not specify amino acids. The three nonsense codons are UAG, UAA, and UGA. Also referred to as the *nonsense codon* and *termination codon*. p. 294

stretch receptor A proprioceptor in the muscles and tendons of vertebrates that detects the position and movement of the limbs. p. 980

strict aerobe Cell with an absolute requirement for oxygen to survive, unable to live solely by fermentations. p. 139

strict anaerobe Organism in which fermentation is the only source of ATP. p. 139

strobilus *See* cone (of a plant). p. 577

stroma An inner compartment of a chloroplast, enclosed by two boundary membranes and containing a third membrane system. p. 46

stromatolite Fossilized remains of ancient cyanobacterial mats that carried out photosynthesis by the water-splitting reaction. p. 63

style The slender stalk of a carpel situated between the ovary and the stigma in plants. p. 885

suberin A waxy, waterproof substance present in cork cells. pp. 856, 868

submucosa A thick layer of elastic connective tissue that contains neuron networks and blood and lymph vessels. p. 1198

subsoil The region of soil beneath the topsoil that contains relatively little organic matter. p. 907

subspecies A taxonomic subdivision of a species. p. 433

substrate The particular reacting molecule or molecular group that an enzyme catalyzes. p. 23

substrate-level phosphorylation An enzyme-catalyzed reaction that transfers a phosphate group from a substrate to ADP. p. 126

succession The change from one community type to another. p. 735

sugar–phosphate backbone Structure in a polynucleotide chain that is formed when deoxyribose sugars are linked by phosphate groups in an alternating sugar–phosphate–sugar–phosphate pattern. p. 268

sum rule Mathematical rule in which the final probability is found by summing individual probabilities. p. 223

surface tension The force that places surface water molecules under tension, making them more resistant to separation than the underlying water molecules. p. F-16

survivorship curve Graphic display of the rate of survival of individuals over a species' life span. p. 685

suspension (filter) feeder An animal that ingests small food items suspended in water. pp. 601, 1195

suspensor In seed plants, a stalklike row of cells that develops from a zygote and helps position the embryo close to the nourishing endosperm. p. 892

swim bladder A gas-filled internal organ that helps fish maintain buoyancy. p. 649

symbiont An organism living in symbiosis with another organism; the symbionts are not usually closely related. p. 540

symbiosis An interspecific interaction in which the ecological relations of two or more species are intimately tied together. p. 66

symmetry (adj., symmetrical) Exact correspondence of form and constituent configuration on opposite sides of a dividing line or plane. p. 595

sympathetic division Division of the autonomic nervous system that predominates in situations involving stress, danger, excitement, or strenuous physical activity. p. 1093

sympatric Occupying the same spaces at the same time. p. 680

sympatric speciation Speciation that occurs without the geographic isolation of populations. p. 438

symplast The living component of plant tissue, composed of protoplasts interconnected by plasmodesmata. p. 868

symplastic pathway The route taken by water that moves through the cytoplasm of plant cells (the symplast). *Compare* apoplastic pathway. p. 868

symport The transport of two molecules in the same direction across a membrane. Also referred to as *cotransport*. p. 107

synapse A site where a neuron makes a communicating connection with another neuron or an effector such as a muscle fibre or gland. p. 1071

synapsid One of a group of amniotes with one temporal arch on each side of the head, which includes living mammals. p. 655

synapsis *See* pairing. p. 202

synaptic cleft A narrow gap that separates the plasma membranes of the presynaptic and postsynaptic cells. p. 1073

systemic acquired resistance A plant defence response to microbial invasion; defensive chemicals including salicylic acid may spread throughout a plant, rendering healthy tissues less vulnerable to infection. p. 932

systemic circuit In amphibians, the branch of a double blood circuit that receives oxygenated blood and provides the blood supply for most of the tissues and cells of a body. p. 968

systemin A plant peptide hormone that functions in defence responses to wounds. p. 929

systems biology An area of biology that studies the organism as a whole to unravel the integrated and interacting network of genes, proteins, and biochemical reactions responsible for life. p. 364

systole The period of contraction and emptying of the heart. p. 974

T cell A lymphocyte produced by the division of stem cells in the bone marrow and then released into the blood and carried to the thymus. T cells participate in adaptive immunity. p. 1273

T-cell receptor (TCR) A receptor that covers the plasma membrane of a T cell, specific for a particular antigen. p. 1273

T-even bacteriophage Virulent bacteriophages, T2, T4, and T6, that have been valuable for genetic studies of bacteriophage structure and function. p. 504

T (transverse) tubule The tubule that passes in a transverse manner from the sarcolemma across a myofibril of striated muscle. p. 1136

tactile signal A means of animal communication in which the signaller uses touch to convey a message to the signal receiver. p. 1167

taiga *See* boreal forest. p. 751

taproot system A root system consisting of a single main root from which lateral roots can extend; often stores starch. p. 852

TATA box A regulatory DNA sequence found in the promoters of many eukaryotic genes transcribed by RNA polymerase II. pp. 295, 321

taxis A behavioural response that is directed either toward or away from a specific stimulus. p. 1169

taxon (plural, taxa) A name designating a group of organisms included within a category in the Linnaean taxonomic hierarchy. p. 428

taxonomic hierarchy A system of classification based on arranging organisms into ever more inclusive categories. p. 428

taxonomy The science of the classification of organisms into an ordered system that indicates natural relationships. p. 427

telomerase An enzyme that adds telomere repeats to chromosome ends. p. 279

telomeres Repeats of simple-sequence DNA that maintain the ends of linear chromosomes. p. 279

telophase The final phase of mitosis, during which the spindle disassembles, the chromosomes decondense, and the nuclei re-form. p. 175

temperate bacteriophage Bacteriophage that may enter an inactive phase (lysogenic cycle) in which the host cell replicates and passes on the bacteriophage DNA for generations before the phage becomes active and kills the host (lytic cycle). pp. 197, 504

temperate deciduous forest A forested biome found at low to middle altitudes at temperate latitudes, with warm summers, cold winters, and annual precipitation between 75 and 250 cm. p. 726

temperate grassland A nonforested biome that stretches across the interiors of most continents, where winters are cold and snowy and summers are warm and fairly dry. p. 751

temperate rain forest A coniferous forest biome supported by heavy rain and fog, which grows where winters are mild and wet and the summers are cool. p. 525

template A nucleotide chain used in DNA replication for the assembly of a complementary chain. p. 212

template strand The DNA strand that is copied into an RNA molecule during gene transcription. p. 292

temporal isolation A prezygotic reproductive isolating mechanism in which species live in the same habitat but breed at different times of day or different times of year. p. 431

temporal summation The summation of several excitatory postsynaptic potentials produced by successive firing of a single presynaptic neuron over a short period of time. p. 1087

tendon A type of fibrous connective tissue that attaches muscles to bones. p. 952

terminal bud A bud that develops at the apex of a shoot. p. 847

termination codon *See* stop codon. p. 294

termination factor *See* release factor. p. 306

terminator Specific DNA sequence for a gene that signals the end of transcription of a gene. Terminators are common for prokaryotic genes. p. 296

territory A plot of habitat, defended by an individual male or a breeding pair of animals, within which the territory holders have exclusive access to food and other necessary resources. p. 1160

tertiary consumer A carnivore that feeds on other carnivores, a member of the fourth trophic level. p. 729

testcross A genetic cross between an individual with the dominant phenotype and a homozygous recessive individual. p. 224

testis (plural, testes) The male gonad. In male vertebrates, secretes androgens and steroid hormones that stimulate and control the development and maintenance of male reproductive systems. pp. 991, 1059

testosterone A hormone produced by the testes, responsible for the development of male secondary sex characteristics and the functioning of the male reproductive organs. pp. 1002, 1059

tetanus A situation in which a muscle fibre cannot relax between stimuli, and twitch summation produces a peak level of continuous contraction. p. 1140

tetrad Homologous pair consisting of four chromatids. p. 202

Tetrapoda A monophyletic lineage of vertebrates that includes animals with four feet, legs, or leglike appendages. pp. 461, 644

thalamus A major switchboard of the brain that receives sensory information and relays it to the regions of the cerebral cortex concerned with motor responses to sensory information of that type. p. 1126

thallus (plural, thalli) A plant body not differentiated into stems, roots, or leaves. pp. 555, 573

theory A broadly applicable idea or hypothesis that has been confirmed by every conceivable test. p. F-4

thermal acclimatization A set of physiological changes in ectotherms in response to seasonal shifts in environmental temperature, allowing the animals to attain good physiological performance at both winter and summer temperatures. p. 1256

thermodynamics The study of the energy flow during chemical and physical reactions. p. 75

thermoreceptor A sensory receptor that detects the flow of heat energy. p. 1077

thermoregulation The control of body temperature. p. 1251

thick filament A type of filament in striated muscle composed of myosin molecules; they interact with thin filaments to shorten muscle fibres during contraction. p. 1135

thigmomorphogenesis A plant response to a mechanical disturbance, such as frequent strong winds; includes inhibition of cellular elongation and production of thick-walled supportive tissue. p. 936

thigmotropism Growth in response to contact with a solid object. p. 936

thin filament A type of filament in striated muscle composed of actin, tropomyosin, and troponin molecules; interacts with thick filaments to shorten muscle fibres during contraction. p. 1135

thorax The central part of an animal's body, between the head and the abdomen. p. 624

threshold potential In signal conduction by neurons, the membrane potential at which the action potential fires. p. 1074

thylakoids Flattened, closed sacs that make up a membrane system within the stroma of a chloroplast. p. 46

thymine A pyrimidine that base-pairs with adenine. p. 268

thymus An organ of the lymphatic system that plays a role in filtering viruses, bacteria, damaged cells, and cellular debris from the lymph and bloodstream and in defending the body against infection and cancer. p. 983

thyroid gland A gland located beneath the voice box (larynx) that secretes hormones regulating growth and metabolism. p. 1056

thyroid-stimulating hormone (TSH) A hormone that stimulates the thyroid gland to grow in size and secrete thyroid hormones. p. 1055

thyroxine (T4) The main hormone of the thyroid gland, responsible for controlling the rate of metabolism in the body. p. 1056

Ti (tumour-inducing) plasmid A plasmid used to make transgenic plants. p. 357

tidal volume The volume of air entering and leaving the lungs during inhalation and exhalation. p. 1226

tight junction Region of tight connection between membranes of adjacent cells. p. 48

time lag The delayed response of organisms to changes in environmental conditions. p. 693

tissue A group of cells and intercellular substances with the same structure that function as a unit to carry out one or more specialized tasks. p. 947

tolerance hypothesis Hypothesis asserting that ecological succession proceeds because competitively superior species replace competitively inferior ones. p. 739

tonoplast The membrane that surrounds the central vacuole in a plant cell. pp. 46, 866

topoisomerase An enzyme that relieves the over-twisting and strain of DNA ahead of the replication fork. p. 275

topsoil The rich upper layer of soil where most plant roots are located; it generally consists of sand, clay particles, and humus. p. 907

torpor A sleeplike state produced when a lowered set point greatly reduces the energy required to maintain body temperature, accompanied by reductions in metabolic, nervous, and physical activity. p. 1260

torsion The realignment of body parts in gastropod molluscs that is independent of shell coiling. p. 615

totipotency The ability to develop into any type of cells. p. 666

totipotent Having the capacity to produce cells that can develop into or generate a new organism or body part. pp. 601, 895

trace element An element that occurs in organisms in very small quantities (0.01%); in nutrition, a mineral required by organisms only in small amounts. p. 1191

trachea In insects, an extensively branched, air-conducting tube formed by invagination of the outer epidermis of the animal and reinforced by rings of chitin. In vertebrates, the windpipe, which branches into the bronchi. p. 1225

tracheal system A branching network of tubes that carries air from small openings in the exoskeleton of an insect to tissues throughout its body. p. 626

tracheid A conducting cell of xylem, usually elongated and tapered. p. 844

traditional evolutionary systematics An approach to systematics that uses phenotypic similarities and differences to infer evolutionary relationships, grouping species that share both ancestral and derived characters. p. 460

trait A particular variation in a genetic or phenotypic character. p. 218

transcription The mechanism by which the information encoded in DNA is made into a complementary RNA copy. p. 292

transcription factors Proteins that recognize and bind to the TATA box and then recruit the polymerase. pp. 296, 321

transcription initiation complex Combination of general transcription factors with RNA polymerase II. p. 322

transcription unit A region of DNA that transcribes a single primary transcript. pp. 296, 316

transcriptome The complete set of RNA transcripts from a given cell under given conditions. p. 376

transcriptomics The study of the transcriptome. p. 376

transduction In cell signalling, the process of changing a signal into the form necessary to cause the cellular response. In prokaryotes, the process in which DNA is transferred from donor to recipient bacterial cells by an infecting bacteriophage. p. 196

transfer cell Any of the specialized cells that form when large amounts of solutes must be loaded or unloaded into the phloem; they facilitate the short-distance transport of organic solutes from the apoplast into the symplast. p. 878

transfer RNA (tRNA) The RNA that brings amino acids to the ribosome for addition to the polypeptide chain. p. 66

transformation The conversion of the hereditary type of a cell by the uptake of DNA released by the breakdown of another cell. pp. 196, 266

transgenic Refers to an organism that has been modified to contain genetic information from an external source. p. 352

transit sequence A part of a gene sequence that targets the protein product to an organelle, endoplasmic lumen, etc. p. 309

transition state An intermediate arrangement of atoms and bonds that both the reactants and the products of a reaction can assume. p. 85

translation The use of the information encoded in the RNA to assemble amino acids into a polypeptide. p. 292

translocation In genetics, a chromosomal alteration that occurs if a broken segment is attached to a different, nonhomologous chromosome. In vascular plants, the long-distance transport of substances by xylem and phloem. pp. 252, 877

transmission In neural signalling, the sending of a message along a neuron and then to another neuron or to a muscle or gland. p. 1069

transpiration The evaporation of water from a plant, principally from the leaves. p. 869

transport The controlled movement of ions and molecules from one side of a membrane to the other. p. 7

transport epithelium A layer of cells with specialized transport proteins in their plasma membranes. p. 1238

transport protein A protein embedded in the cell membrane that forms a channel allowing selected polar molecules and ions to pass across the membrane. p. 30

transposable element (TE) A sequence of DNA that can move from one place to another within the genome of a cell. p. 210

transposase An enzyme that catalyzes some of the reactions inserting or removing the transposable element from the DNA. p. 211

transposition Mechanism of movement of transposable elements involving nonhomologous recombination. p. 210

transposon A bacterial transposable element with an inverted repeat sequence at each end enclosing a central region with one or more genes. p. 211

trichocyst A dartlike protein thread that can be discharged from a surface organelle for defence or to capture prey. p. 521

trichome A single-celled or multicellular outgrowth from the epidermis of a plant that provides protection and shade and often gives the stems or leaves a hairy appearance. p. 847

triglyceride A nonpolar compound produced when a fatty acid binds by a dehydration synthesis reaction at each of glycerol's three —OH-bearing sites. p. 135

triiodothyronine (T3) A hormone secreted by the thyroid gland that regulates metabolism. p. 1056

trimester A division of human gestation, three months in length. p. 1025

triploblastic An animal body plan in which adult structures arise from three primary germ layers: endoderm, mesoderm, and ectoderm. p. 595

trochophore The small, free-swimming, ciliated aquatic larva of various invertebrates, including certain molluscs and annelids. p. 614

trophic cascade The effects of predator–prey interactions that reverberate through other population interactions at two or more trophic levels in an ecosystem. p. 757

trophic level A position in a food chain or web that defines the feeding habits of organisms. p. 728

trophoblast The outer single layer of cells of the blastocyst. p. 1025

trophozoite Motile, feeding stage of *Giardia* and other single-celled protists. p. 513

tropic hormone A hormone that regulates hormone secretion by another endocrine gland. p. 1053

tropical deciduous forest A tropical forest biome that occurs where winter drought reduces photosynthesis and most trees drop their leaves seasonally. p. 751

tropical forest Any forest that grows between the Tropics of Capricorn and Cancer, a region characterized by high temperature and rainfall and thin, nutrient-poor topsoil. p. 582

tropical rain forest A dense tropical forest biome that grows where some rain falls every month, mean annual rainfall exceeds 250 cm, mean annual temperature is at least 25°C, and humidity is above 80%. p. 557

tropism The turning or bending of an organism or one of its parts toward or away from an external stimulus, such as light, heat, or gravity. p. 920

true-breeding Refers to an individual that passes traits without change from one generation to the next. p. 219

tube cell The cell in a pollen grain (male gametophyte) of a seed plant that will give rise to the pollen tube. p. 886

tumour suppressor gene A gene that encodes proteins that inhibit cell division. p. 333

turgid A cell with high internal hydrostatic pressure. p. 867

turgor pressure The internal hydrostatic pressure within plant cells; the normal fullness or tension produced by the fluid content of plant and animal cells. p. 867

turnover rate The rate at which one generation of producers in an ecosystem is replaced by the next. p. 755

tympanum A thin membrane in the auditory canal that vibrates back and forth when struck by sound waves. pp. 652, 1105

umbilical cord A long tissue with blood vessels linking the embryo and the placenta. p. 355

umbilicus Navel; the scar left when the short length of umbilical cord still attached to the infant after birth dries and shrivels within a few days. p. 1027

undernutrition A condition in animals in which intake of organic fuels is inadequate or whose assimilation of such fuels is abnormal. p. 1190

undulating membrane In parabasalid protists, a fin-like structure formed by a flagellum buried in a fold of the cytoplasm that facilitates movement through thick and viscous fluids. An expansion of the plasma membrane in some flagellates that is usually associated with a flagellum. p. 520

uniform dispersion A pattern of distribution in which the individuals in a population are evenly spaced in their habitat. p. 682

uniformitarianism The concept that the geologic processes that sculpted Earth's surface over long periods of time—such as volcanic eruptions, earthquakes, erosion, and the formation and movement of glaciers—are exactly the same as the processes observed today. p. 395

universal A feature of the nucleic acid code, with the same codons specifying the same amino acids in all living organisms. p. 294

unreduced gamete A gamete that contains the same number of chromosomes as a somatic cell. p. 441

unsaturated Fatty acid with one or more double bonds linking the carbons. p. 100

ureter The tube through which urine flows from the renal pelvis to the urinary bladder. p. 1245

urethra The tube through which urine leaves the bladder. In most animals, the urethra opens to the outside. p. 1245

urinary bladder A storage sac located outside the kidneys. p. 1245

uterine cycle The menstrual cycle. p. 1000

uterus A specialized saclike organ in which the embryo develops in viviparous animals. p. 996

utricle A fluid-filled chamber of the vestibular apparatus that provides information about the position of the head with respect to gravity (up versus down), as well as changes in the rate of linear movement of the body. p. 1102

vaccination The process of administering a weakened form of a pathogen to patients as a means of giving them immunity to subsequent infection, and disease, caused by that pathogen. p. 1266

valence electron An electron in the outermost energy level of an atom. p. F-10

van der Waals forces Weak molecular attractions over short distances. p. F-14

variable An environmental factor that may differ among places or an organismal characteristic that may differ among individuals. p. 412

vas deferens (plural, vasa deferentia) The tube through which sperm travel from the epididymis to the urethra in the male reproductive system. p. 1003

vascular bundle A cord of plant vascular tissue; often multistranded with both xylem and phloem. p. 849

vascular cambium A lateral meristem that produces secondary vascular tissues in plants. p. 855

vascular plant A plant with xylem, phloem, and usually well-developed roots, stems, and leaves. p. 567

vascular tissue In plants, tissue that transports water and nutrients or the products of photosynthesis through the plant body. p. 567

vascular tissue system One of the three tissue systems in plants that provide the foundation for plant organs; it consists of transport tubes for water and nutrients. p. 841

vegetal pole The end of the egg opposite the animal pole, which typically gives rise to internal structures such as the gut and the posterior end of the embryo. p. 1015

vegetative reproduction Asexual reproduction in plants by which new individuals arise (or are created) without seeds or spores; examples include fragmentation from the parent plant or the use of cuttings by gardeners. p. 895

vein In a plant, a vascular bundle that forms part of the branching network of conducting and supporting tissues in a leaf or other expanded plant organ. In an animal, a vessel that carries the blood back to the heart. pp. 852, 967

veliger A second larva that occurs after the trochophore in some molluscs. p. 599

ventilation The flow of the respiratory medium (air or water, depending on the animal) over the respiratory surface. p. 1220

ventral Indicating the lower or "belly" side of an animal. p. 596

ventricle In the brain, an irregularly shaped cavity containing cerebrospinal fluid. In the heart, a chamber that pumps blood out of the heart. p. 1090

venule A capillary that merges into the small veins leaving an organ. p. 977

vernalization The stimulation of flowering by a period of low temperature. p. 940

vesicle A small, membrane-bound compartment that transfers substances between parts of the endomembrane system. p. 36

vessel In plants, one of the tubular conducting structures of xylem, typically several centimetres long; most angiosperms and some other vascular plants have xylem vessels. p. 844

vessel member Any of the short cells joined end to end in tubelike columns in xylem. p. 844

vestibular apparatus The specialized sensory structure of the inner ear of most terrestrial vertebrates that is responsible for perceiving the position and motion of the head and, therefore, for maintaining equilibrium and for coordinating head and body movements. p. 1102

vestigial structure An anatomical feature of living organisms that no longer retains its function. p. 393

virion A complete virus particle. p. 503

viroid A plant pathogen that consists of strands or circles of RNA, smaller than any viral DNA or RNA molecule, that have no protein coat. p. 509

virulent bacteriophage Bacteriophage that kills its host bacterial cells during each cycle of infection. pp. 196, 504

virus An infectious agent that contains either DNA or RNA surrounded by a protein coat. p. 266

visceral mass In molluscs, the region of the body containing the internal organs. p. 611

visual signal A means of communication in which animals use facial expressions or body language to send messages to other individuals. p. 1166

vital capacity The maximum tidal volume of air that an individual can inhale and exhale. p. 1227

vitamin An organic molecule required in small quantities that the animal cannot synthesize for itself. p. 1190

vitamin D A steroidlike molecule that increases the absorption of Ca^{2+} and phosphates from ingested food by promoting the synthesis of a calcium-binding protein in the intestine; it also increases the release of Ca^{2+} from bone in response to parathyroid hormone. p. 1057

vitelline coat A gel-like matrix of proteins, glycoproteins, or polysaccharides immediately outside the plasma membrane of an egg cell. p. 993

vitreous humour The jellylike substance that fills the main chamber of the eye, between the lens and the retina. p. 1109

viviparous Referring to animals that retain the embryo within the mother's body and nourish it during at least early embryo development. pp. 668, 996

voltage-gated ion channel A membrane-embedded protein that opens and closes as the membrane potential changes. p. 1075

vulva The external female sex organs. p. 999

water lattice An arrangement formed when a water molecule in liquid water establishes an average of 3.4 hydrogen bonds with its neighbours. p. F-15

water potential The potential energy of water, representing the difference in free energy between pure water and water in cells and solutions; it is the driving force for osmosis. p. 865

watershed An area of land from which precipitation drains into a single stream or river. p. 761

water-soluble vitamin A vitamin with a high proportion of oxygen and nitrogen able to form hydrogen bonds with water. p. 1193

wavelength The distance between two successive peaks of electromagnetic radiation. p. 4

wetland A highly productive ecotone often at the border between a freshwater biome and a terrestrial biome. p. 586

white matter The myelinated axons that surround the grey matter of the central nervous system. p. 1123

wilting The drooping of leaves and stems caused by a loss of turgor. p. 867

wobble hypothesis Hypothesis stating that the complete set of 61 sense codons can be read by fewer than 61 distinct tRNAs because of particular pairing properties of the bases in the anticodons. p. 302

Wolffian duct A bipotential primitive duct associated with the gonads that leads to a cloaca. p. 1029

wood The secondary xylem of trees and shrubs, lying under the bark and consisting largely of cellulose and lignin. p. 855

X chromosome Sex chromosome that occurs paired in female cells and single in male cells. p. 246

X-linked recessive inheritance Pattern in which displayed traits are due to inheritance of recessive alleles carried on the X chromosome. p. 257

X-ray diffraction Method for deducing the position of atoms in a molecule. p. 268

xylem The plant vascular tissue that distributes water and nutrients. p. 843

xylem sap The dilute solution of water and solutes that flows in the xylem. p. 869

Y chromosome Sex chromosome that is paired with an X chromosome in male cells. p. 203

yeast A single-celled fungus that reproduces by budding or fission. p. 28

yolk The portion of an egg that serves as the main energy source for the embryo. p. 654

yolk sac In an amniote egg, an extraembryonic membrane that encloses the yolk. p. 1020

zero population growth A circumstance in which the birth rate of a population equals the death rate. p. 691

zona pellucida A gel-like matrix of proteins, glycoproteins, or polysaccharides immediately outside the plasma membrane of the egg cell. p. 993

zone of cell division The region in a growing root that consists of the root apical meristem and the actively dividing cells behind it. p. 853

zone of elongation The region in a root where newly formed cells grow and elongate. p. 854

zone of maturation The region in a root above the zone of elongation where cells do not increase in length but may differentiate further and take on specialized roles. p. 854

zooplankton Small, usually microscopic, animals that float in aquatic habitats. p. 516

zygospore A multinucleate, thick-walled sexual spore in some fungi that is formed from the union of two gametes. p. 545

zygote A fertilized egg. pp. 199, 595, 992

Index

The letter i designates illustration; t designates table; b designates box; **bold** *designates defined or introduced term.*

Agriculture (*continued*)
    rationale for species selection, 813–821
    research, 829*b*
    technology increasing yields, 807, 821–823, 822*i*, 824*i*, 831
    use of 2,4-D, 828
Agrobacterium, 926
AI-2, 321*b*
Airways, in mammals, **1225**, 1226*i*
Albumin, **654**, 970
Alcohol (ethanol), **1084**–1085*t*
Alcohol fermentation, **137**–138
Aldosterone, **1050**
Algae, 29*i*
    carbon dioxide and, 159–160, 160*i*
    life cycles, 199*i*
Algal ancestors (of plants), **564**, 571*i*
Algin, **526**
ALH84001, 52
Alkaptonuria, **290**
Allantoic membrane, **1021**
Allantois, **1020**–1021, 1021*i*
Allard, Henry, 938
Allele frequencies, **414**, 414*t*, 415, 415*b*–416*b*
Alleles, **412**
Allelopathy, **682**
Allergen, **1284**
Allergies, as a result of overactivity of immune system, 1284–1285, 1285*i*
Allium species, 371, 809*i*
Allopatric speciation, **435**–436, 435*i*
Allopolyploidy, **441**–442, 441*i*
All-or-nothing principle, **1075**
Allosteric activator, **92**
Allosteric inhibitor, **92**
Allosteric regulation, **92**
Allosteric site, **91**
Alpha cells, **1060**
Alpha helix, **104**
Alpha waves, **1130**, 1130*i*
A-amanitin, **552**
Alpine angiosperms, 586*i*
Alpine tundra, **572**
Alprazolam (Xanax), 1084–1085*t*
Alternation of generations life cycle, **200**, 563, 563*i*
Alternative hypothesis, **58**
Alternative splicing, **299**, 299*i*, 328
Altitude, increasing, 1232*b*
Altman, Sydney, 61
Altricial, **668**
Altriciality, **668**
Altruism, **1180**–1181, 1180*i*, 1182*b*, 1187
Alveolus (plural, alveoli), **1225**, 1226*i*
Amacrine cell, **1112**
Amanitin, 307*b*
Ames, Jack, 1154
Amine hormones, **1047**
Amino acids, **1084**–1085*t*
    adding to tRNAs, 302, 303*i*
    as essential elements for animals, 1190
Amino group, **80**
Aminoacyl-tRNA, **302**
Aminoacyl-tRNA synthetase, **302**

Aminocylation, **302**
Ammocoetes, **645**
Ammonia (NH$_3$), **1239**
Ammonification, **911**
Ammonium, **911**, 911*i*
Amniocentesis, 258*i*, **1027**
Amnion, **654**, 1020–1021, 1021*i*
Amniota, **1021**
Amniote (amniotic) egg, **654**
    ancestry, 656*f*
    egg, 655*f*
    origin, 654–659
Amniotic cavity, **1025**–1029, 1026*i*
Amoeba proteus, 1124
Amoeboid cell, **601**
Amphetamines, **1084**–1085*t*
Amphibians, **651**–654, 653*f*, 654*f*
    ectothermy in, 1255
    excretions from, 1243–1244, 1244*i*
    gastrulation in, 1018–1019, 1018*i*, 1019*i*
    positive pressure breathing, 1223–1224, 1224*i*
    reproductive systems of, 993*i*
Amphipathic, **100**
Amplexus, **994**
Amplification, **116**–117, 116*i*, 345, 349
Amp^R gene, **342**, 343*i*
Ampulla, **638**
AMY21 gene, 1212*b*
Amygdala, **1126**
Amylases, **1195**
Amyloplast, **45**
Anabolic pathways, **82**, 83*i*
Anabolic reactions, **136**
Anabolic steroid, **1060**
Anaerobic metabolism, **63**–64
Anaerobic respiration, **138**–139
Anal sphincter, **1207**
Analogous characters, **458**, 458*i*
Anaphase, **175**, 175*i*
Anaphase I, **203**, 205*i*
Anaphase II, **203**, 205*i*
Anaphase-promoting complex (APC), **183**
Anaphylactic shock, **1285**
Anapsid (lineage Anapsida), **655**
Anapsida, **655**
Anatomy, **836**
Ancestral characters, **458**, 459*i*
Ancestral green alga, **564**, 571*i*
Ancestry, DNA fingerprinting in establishment of, **350**
Anchoring junctions, **47**–48, 48*i*
Anderson, W. French, 355
Andometrium, **1025**–1029, 1026*i*
Androgens, **1060**
Aneuploids, **253**–255
Angiosperms, 20*i*, 568*t*, **570**, 571*i*, 576*t*, 586–590, 586*i*
    See also Flowering plants
Angiotensin, **1049**
Angiotensin-converting enzyme (ACE) inhibitors, **1049**
Anglerfish, 1211–1212, 1212*i*
Animal behaviour, **1154**–1155, 1188
    communication, 1165–1169, 1166*i*, 1167*i*, 1168*i*, 1186–1187
    environment and, 1155–1156, 1186

Channel proteins, **106**
Chaperone protein (chaperonin), **F-36**
Character, **218**
Character displacement, **721**
Chargaff's rules, **268**, 269
Charging. *See* Aminocylation
Charophytes, **564**, 564*i*
Chase, Martha, 266, 267*i*
Chase, Ron, 611
Cheatgrass *(Bromus tectorum)* case history, 1308–1309, 1308*i*, 1310
Checkpoints, **181**
Cheetah, African, 419
Chelicerae, **622**
Chelicerata, subphylum, **622**–623, 623*f*
Chemical basis, of cellular respiration, 121–123, 141
Chemical bond, **F-11**
Chemical defences, of plants, **929**–934, 930*t*, 931*i*, 933*i*, 934*i*, 943
Chemical equation, **F-14**
Chemical reaction, **F-14**
Chemical signals, **1166**, 1166*i*
Chemical synapses, **1082**–1083, 1082*i*
    action potentials and, 1094
    conduction of action potentials across, 1079–1083
    summation and, 1085–1087, 1086*i*, 1087*i*
Chemicals
    from plants, 827–828, 827*i*, 831
    from plants and animals, 827*i*, 831
Chemiosmosis, **130**–134, 131*i*, 153–154, 153*i*
    ATP synthesis and, 130, 132, 132*i*
    electron transport and, 141
    uncoupling electron transport and, 133–134, 134*i*
Chemoautotroph, **487**
Chemoheterotroph, **487**
Chemokines, **1269**
Chemoreceptors, **1114**–1117, 1132
Chemotroph, **487**
Chiasmata. *See* Crossover
Chickadees, 1169
Chief cells, **1202**, 1203*i*
Chili peppers, domestication of, 819
Chimpanzees, 1183–1184, 1183*i*, 1184*i*
    See also Pan troglodytes
Chitin, **542**, 543, 1145*b*
Chloride cells, **1243**
Chlorine, 903
    as essential animal nutrient, 1191–1192*b*
    as essential plant nutrient, 904*t*
Chlorocebus pygerythrus, 476
Chlorophyll, **5**–6, 6*i*, 46
    in light absorption, 149–151, 150*i*, 151*i*
    oxidizing, 152–153, 154*i*
Chloroplasts, **45**
    as biochemical factories powered by sunlight, 45–46
    endosymbiosis and, 66
    photosynthesis in, 146, 147*i*
    protein sorting to the, 309
Chlorosis, **905**
Choanocytes, **601**
Choanoflagellates, **384**, 595
Cholecystokinin (CCK), **1208**
Cholesterol, **1205**

Chondrichthyes, class, **647**–648
Chondricthye, 647*f*
Chondrocyte, **953**
Chordata, phylum, **640**–642, 640*f*
Chorion, **1020**–1021, 1021*i*, 1025–1029, 1026*i*
Chorionic villus (plural, villi), 1203–1204, 1204*i*
Chromatids, 209*i*
Chromatin, **35**, 327–328, 328*i*
Chromatin fibers, **284**, 284*i*
Chromatin packaging, **332**
Chromatin remodeling, **327**
Chromium, as essential animal nutrient, 1191–1192*b*
Chromoplasts, **45**
Chromosomal alterations, **251**–255, 252*i*, 261, 442
Chromosomal protein, **283**
Chromosome banding, 472*b*
Chromosome segregation, **171**
Chromosome theory of inheritance, **228**–229
Chromosomes, **35**
    as genetic units divided by mitosis, 170–171
    mapping, 243–245, 245*i*
    mitotic spindles and, 179–180, 179*i*, 180*i*
    number of, 252–255, 256*t*
    replication of, 169, 276–277, 279
    segregation of, 203
    sex, 203, 206
    structure of, 256*t*
    See also Genes, chromosomes, and human genetics
Chylomicron, **1205**
Chyme, **1203**
Chytridiomycota, **544**, 544*i*, 544*t*, 545, 545*i*
Chytrids, 544*t*, **545**, 548
Cicada, 621*f*
Cilia, **43**–45, 44*i*
Ciliary body, **1109**
Cilium, **43**
Circadian rhythm, **16**–17, 16*i*, 938
Circuits, neurons functioning as, **1071**–1073
Circular bacterial chromosome, 285*i*
Circulatory system, 956–957*i*, **964**–965, 984, 985–986
    basic elements of, 965–966
    blood, 970–973, 984
    blood vessels, 977–980, 984
    closed, 616, 967–968
    heart, 973–977, 984
    lymphatic system, 981–983
    maintaining blood flow and pressure, 980–981, 984–985
    open, **611**, 966–967
    single to double blood, 968–970
Circulatory vessel, **611**
Circumcision, **1004**
Cisternae (singular, cisterna), **36**
Citric acid cycle, **126**
    pyruvate oxidation and, 124, 124*i*, 126–129, 128*i*, 129*i*, 141
    reactions to, 128*i*, 129*i*
Clade, **461**
Cladistics, 460*i*, **461**
Cladogenesis, **457**
Cladogram, 462*b*, 462*t*
Claspers, **648**
Class II, **1277**

Cyclic guanosine monophosphate (cGMP), **919**
Cyclin genes, **380**
Cyclin-dependent kinase (Cdk), 181–183
Cyclins, 181–183
CYP2C9 gene, 374
Cystic fibrosis (CF), **97**, 256
Cystic fibrosis transmembrane conductance regulator (CFTR), **97**
Cytochrome, **125**
Cytochrome oxidase, 972*b*
Cytokine, **1269**
Cytokinesis, **176**–177, 177*i*, 1269
Cytokinins, 919*t*, **925**–926, 926*i*
Cytoplasm, **308**
Cytoplasmic determinants, **1014**
Cytoplasmic inheritance, **259**
Cytoplasmic organelles
    of eukaroytic cells, 33
    vesicle traffic in, 40*i*
Cytoplasmic streaming, **43**, 542
Cytosine, **268**
Cytoskeletons, **30**, 41–43, 42*i*
Cytosol, **30**, 1137
Cytotoxic T cells, **1281**, 1282*b*, 1282*i*, 1283

D

D1 protein, 152*b*
Daily torpor, **1260**, 1260*i*
Dakosaurus andiniensis, 661
Dalton, **333**
Dalton's law of partial pressure, **1218**
Daly, Martin, 1184
Dark field microscopy, 30*i*
Darwin, Charles, 390, 392, 455, 456*i*, 586, 920–921, 921*b*, 922*i*
    Beagle voyage, 396, 396*i*
    The Descent of Man, and Selection in Relation to Sex, 400
    descent with modification, 400, 400*i*
    on the eye, 11–13
    Fertilisation of Orchids, 19
    heritable variation, importance of, 410
    On the Origin of Species by Means of Natural Selection, 399, 405
    speciation, 427
    theory of natural selection, 396–399, 398*i*, 400
    tree of life image, 400, 400*i*
Darwin, Francis, 920–921, 922*i*
Dasypeltis scabra, 1213–1214, 1213*i*
Datura stramonium, domestication of, 819
Daucus carota, 827
Daughter cells, in meiosis, **201**–206, 202*i*, 204–205*i*
Davies, Peter, 1261*b*
Day-neutral plant, **940**
De Vries, Hugo, 228
Deadly nightshade, 818
Death, 685–686
Death cap mushroom, 552
Decomposer, **145**
Dedifferentiation, **332**, 333*i*, 895
Deep sea vents, **58**, 58*i*
Deer fly, 11*i*
Deer mouse, 1260, 1260*i*

Defecation reflex, **1207**
Defensins, **1268**, 1269*b*
Degeneracy (redundancy), **294**
Dehydration (of plants), **566**
Dehydration synthesis reaction, **F-21**
Dehydrogenases, **123**
Deleterious mutations, **417**
Deletion, **251**, 252, 252*i*
Delta waves, **1130**
Demographic transition model, **701**
Demography, **684**–686
Denaturation, **F-36**
Dendogram, 478*f*
Dendraster excentricus, 638
Dendrites, **956**
Dendritic, **836**
Dendritic cells (DCs), **1271**, 1277, 1277*i*
Denisovans, **479**
Denitrification, **766**
Density, population. *See* Population density
Density dependent, **694**
Density independent, **969**
Deoxyribonucleic acid (DNA), **263**–264, 287–288
    bacterial conjugation and, 192–196, 193*i*, 194*i*
    connection between RNA, protein and, 290–294, 311
    damage of from ultraviolet light, 15*i*
    in eurkaroytic cells, 33, 35–36
    flow of information to RNA from, 60–61, 60*i*
    as the hereditary molecule, 264–266, 286
    mechanisms to correct replication errors, 280–283, 287
    methylation of, 327
    nucleotide subunits of, 268*i*
    organizing in eukaryotic *versus* prokaryotic cells, 283–286, 287
    as replacement for RNA, 61–62
    replication, 270–280, 287
    structure of, 267–270, 287
    synthesis of, 344*i*
Deoxyribonucleoside triphosphates, **271**
Deoxyribose, **F-37**
Depolarization, **996**
Deposit feeders, **1194**, 1194*i*, 1195, 1196
Derived characters, **458**, 459*i*
Dermal tissue system, 842*i*, 842*t*, 846–847
Dermis, **1256**, 1257*i*
Descartes, René, 392
Descending segment of loop of Henle, 1248*t*, 1249
The Descent of Man, and Selection in Relation to Sex (Darwin), 400
Descent with modification, **400**
Desert, **750**
Desert locust, 1217
Desiccation-tolerant plants, **566**, 566*i*
Desmosome, **47**
Desmostylus, **633**, 671
Determinate cleavage, **597**
Determinate growth, **839**
Determination, **1017**
Detritivore, **594**
Deuteromycetes, **550**
Deuteromycota, **549**
Deuterostome, **596**, 598*f*, 634–635*f*, 635, 636*f*

DNA ligase, 279*t*, **342**, 348
DNA methylation, **327**
DNA microarrays, **363**, 376, 377*i*, 381, 942*b*
DNA molecule, **269**
DNA packing, histones and, **283**, 284*i*
DNA polymerase, **271**–274, 275, 275*i*, 276*i*, 280–282, 282*i*
DNA polymerase I, 279*t*
DNA polymerase III, 279*t*
DNA repair mechanisms, **282**, 283*i*
DNA replication, 272*i*, **287**
  mechanisms to correct errors in, 280–283, 287
  multiple enzmes in, 276, 278*i*
  proteins of, 279*t*
DNA sequencing, **340**, 345*i*, 365–368
  taxonomic uses, 430*b*
  whale studies, 409–410
  wooly mammoth lineage, 404*b*, 404*i*
DNA synthesis, **274**–275
DNA technologies, **339**
  application of, 348–361
  DNA cloning. *See* DNA cloning
  as subject of public concern, 359–360
  why it matters, 339–340
DNA-directed RNA synthesis, **294**–296, 295*i*, 311
Dobbenstein, B., 308–309
Dodders, **913**, 914*i*
Dogbane, 403
Dogs
  domestication of, 809, 820
  selective breeding, 397–398
Do-it-yourself (DIY) biology, 339
Dolly (cloned sheep), 356, 356*i*
Dolphins, 393, 393*i*, 1181
Domains, **65**–66
Domestication, **830**
  of Atlantic salmon, 820–821
  of barley, 816, 816*i*, 817*i*
  of cats, 825, 826*i*
  of cattle, 813, 814*i*, 821
  of corn, 816, 817–818, 818*i*
  of cotton, 814, 814*i*, 821
  cultivation without domestication, 821
  of dogs, 809, 820
  of genus *Brassica*, 823*b*, 823*i*, 824*i*
  of grapes, 818
  of honeybees, 814
  of lentils, 816–817
  for more than one use, 821
  of plants in Solonaceae family, 818–820, 818*i*
  rationale for species selection, 813–821, 830–831
  of rice, 812–813, 812*i*, 815–816, 816*i*
  of squash, 818, 820, 820*i*
  timing and locations of, 808–813, 809*i*, 810*i*, 811*i*, 812*i*, 813, 816*i*, 817*i*, 830
  of wheat, 809–810, 810*i*, 815, 816, 816*i*, 819*b*
  yeast, 814–815
Dominance, **1178**–1179
Dominance hierarchy, **1178**
Dominant, **221**
Dominant alleles, 221
Dopamine, 1084–1085*t*
Dormancy, **894**, 941, 942*i*, 943*i*
Dorsal, **596**

Dorsal aspect, **596**
Dorsal lip of the blastopore, **1018**
D'Ortous de Mairan, Jean-Jacques, 16
Double blood circuits, **968**–969, 968*i*
Double fertilization, **890**
Double helix, **2**
Double-helix model, **269**–270, 270*i*
Double-stranded DNA, six reading frames of, 369*t*
Down syndrome, **253**–254, 254*i*
Downey, Richard Keith, 815*b*
Driscoll, Carlos A., 825
Dromaeosaurus, 401, 402*i*
Drosophila spp. (fruit fly), 241, 242*i*, 244, 245*i*, 247–249, 247*i*, 248*i*, 249–250, 250*i*, 439, 992, 1274*b*
  embryogenesis in, 1036–1037, 1037*i*
  gametic isolation, 432
  hybrid breakdown, 433
  infection of, 1267
  reproductive systems of, 993*i*
Drosopterin, 247*b*
Drugs
  affecting neurotransmission, 1084–1085*t*
  effects on antibody-mediated immunity of, 1281
Dry rot, **554**
Duboule, Denis, 394*b*
Duchenne muscular dystrophy, **349**
Duck-billed platypus, 428*i*
Ducks, sexual selection, 440, 440*i*
Dugesia, 1241, 1241*i*
Dumb rabies, **678**
Dung, fossilized, **476**
Duodenum, **1203**–1204, 1204*i*
Duplication, **251**–252, 252*i*
Dvl gene, 1158*b*

E

E. coli, **32**, 32*i*, 505–506, 505*i*, 690, 1289, 1296
  comparison of K12 and human genomes, 372*t*
  genetic engineering of to make insulin, 353
  genetic recombination in, 191–192, 193*i*, 194*i*
  genome of, 371–372, 371*i*
  Hershey-Chase experiments, 266
  transforming of DNA into, 342, 353*i*
  transforming of plasmids into, 343*i*
  use of DNA fingerprinting in testing for, 350
  use of to make protein, 352
E site, **302**
"Eagle" alarm, **476**
Earth, **55**–56, 56*i*, 395
Earthworms, **618**, 618*f*, 1144*i*, 1195, 1196, 1197*i*
Ecdysis, **600**, 621*f*
Ecdysone, **1049**, 1064*i*
Ecdysozoan protostome, **619**–629
Echinodermata, phylum, **636**–639, 637*f*
Echinoderms, **1089**–1090, 1144
Echinoidea, class, 638–639
Echolocation, **1163**, 1163*i*, 1164*b*
Eclectus parrot, 18–19, 19*i*
Ecological community, **680**
Ecological efficiency, **752**
Ecological isolation, **431**, 431*t*
Ecological light pollution, **20**–21, 21*i*
Ecological niche, **717**

Ecological pyramid, **754**
Ecological succession, **561**
Ecology, **679**
   role of light in, 18–22, 25
   science, 679–680
*Eco*RI, **341**, 342*i*
Ecosystem, **749**
Ecosystem ecology, **680**
Ecosystem engineering, **812**
Ecotone, **725**
Ectoderm, **595**, 1015, 1016*t*, 1021–1022, 1022*i*, 1033*b*
Ectodysplasin gene, 422*b*
Ectomy corrhizas, **557**, 557*i*
Ectoparasite, **609**
Ectoprocta, phylum, 607, 607*f*
Ectotherm/ectothermy, **1253**, 1253*i*, 1254–1256, 1263
Edidin, Michael A., 99
Effector, **960**
Effector T cell, **1278**
Efferent arteriole, **1247**
Efferent neuron, **1070**
Egg cell, **888**
Egg coats, **993**
Egg-eating snakes, **1213**–1214, 1213*i*
Eggplant, domestication of, 818, 818*i*, 819
Eggs, 664*f*
Egyptian vultures, 1213
Einkorn, 442*i*
Ejaculation, **1006**
Elasmobranch, 648*f*
Elasmobranchii, **647**
Elastin, **952**
Electric fishes, 1120*i*, 1121*i*
Electrical signalling, **1167**
Electrical synapses, **1072**, 1072*i*
Electrocardiogram (ECG), **964**–965, 965*i*, 977
Electrochemical gradient, **111**
Electroencephalogram, **1130**
Electromagnetic radiation, 4, 13, 13*i*
Electromagnetic spectrum, 4–5, 5*i*, 148–149, 148*i*
Electron microscope, 29
Electron microscopy, 30*i*
Electron shell, **F-10**
Electron transfer system, **162**
Electron transport/electron transport chain, **67**,
     130–134, 131*i*
   chemiosmosis and, 141
   electrons moving along, 130, 132*i*
   uncoupling chemiosmosis and, 133–134, 134*i*
Electronegativity, **F-13**
Electrons, 148–149, 148*i*, 149*i*
Electroreceptors, **1121**, 1132
Element, **F-8**
Elephant, African, 404*b*, 404*i*
Elephant, Asian, 404*b*, 404*i*
Elephants, 1258
Elimination, as a process from ingestion of food, **1196**
Ellis-van Creveld syndrome, **419**, 420
Elongation, **302**
Elongation factor, **305**
Elongation stage, of translation, 304–306, 305*i*
Elysia chlorotica, **1189**, 1190*i*
Embryo sac, **888**

Embryogenesis, in *Drosophila*, **1036**–1037, 1037*i*
Embryonic development
   of humans/mammals, 1025–1029, 1025*i*, 1026*i*,
     1042–1043
   mechanisms of, 1014–1017, 1015*i*, 1016*i*, 1042
Embryonic sporophyte, development of, **892**–893
Embryonic stem cells, **354**
Embryophyta, 571*i*
Embryos, **890**–891
Emergence, of characteristics of life, 53, 54*i*
Emigration, **684**
Emulsification, 1204*i*, **1205**
Encdocrine system, 956–957*i*
Encephalomyopathy, 259*t*
Endangered species, **419**
Endemic Balkan nephropathy (EN), **825**–827
Endemic species, **785**
Endergonic process, **79**
Endergonic reaction, 79*i*
Endocrine control, **1045**–1046, 1046*i*, 1066–1067
   adrenal cortex, 1058–1059
   adrenal medulla, 1058, 1065–1066
   gonadal sex hormones, 1059–1069, 1066
   hormones, 1046–1049, 1065
   hypothalamus, 1053–1056, 1065
   in invertebrates, 1062–1064, 1066
   mechanisms of hormone action, 1049–1053, 1065
   pancreatic islets of Langerhans, 1060–1061, 1066
   parathyroid gland, 1057–1058, 1065
   pineal gland, 1061–1062, 1066
   pituitary, 1053–1056, 1065
   thyroid gland, 1056–1057, 1065
Endocrine glands, **950**, 951*i*, 1047, 1047*i*, 1051*i*, 1052–1053*t*
Endocrine system, **1046**
Endocytic vesicles, **112**, 114
Endocytosis, **112**–114, 113*i*, 118
Endoderm, **595**, 1015, 1016*t*, 1033*b*
Endodermis, **854**
Endogenous antigens, **1273**
Endoliths, **558**
Endomembrane system, **36**–39, 68, 308–309, 309*i*
Endoparasite, **723**
Endophytes, **558**–559
Endoplasmic reticulum (ER), **36**–37, 37*i*, 40*i*, 69*i*
Endorphins, **1055**, 1084–1085*t*
Endoskeleton, **1146**, 1147*i*, 1148, 1148*i*, 1149*i*
Endosperm, **890**–891
Endosporous, **580**
Endosymbiosis, **66**–67, 67*i*
Endothelial cells, **1269**
Endotherm, **1253**
Endothermia, **667**
Endothermic, **1253**, 1253*i*
Endothermy, **1256**–1261, 1263
Endotoxin, **489**
Energy and enzymes, **74**–75, 96
   energy, defined, 75
   energy as characteristic of life, 54*i*
   energy required by active transport, 109
   enzymes, defined, 86
   factors affecting enzyme activity, 90–94, 95
   free energy, 78–80, 94–95
   laws of thermodynamics, 75–78, 94

Energy and enzymes (*continued*)
  light as a source of energy, 7, 25
  metabolism, 95
  role of enzymes in biological reactions, 85–89, 95
  spontaneous processes, 78–80, 94–95
  thermodynamics and life, 80–85, 95
Energy budget, **686**
Energy coupling, **84**, 84*i*
Energy levels, **121**
Energy return on investment (EROI), **829–830**
Engrailed, **1039**
Enhancer, **322**
Enkephalins, 1084–1085*t*
Enterocoelom, **598**
Enterovirus, 503*t*
Enthalpy, **78**
Entrez, 379
Entropy, **78**–79, 79*i*
Envelope, **501**
Envelope, viral, 502*i*
Enveloped virus, **502**
Environment
  animal behaviour and, 1155–1156, 1186
  genetic variation, 421–423
  phenotypic variation, 411–412
Enzymatic activity, as key function of membrane
    proteins, **103**
Enzymatic hydrolysis, **1196**
Enzymes. *See* Energy and enzymes
Eocene epoch, 447*t*
Eosinophil, 1270*t*
Ephedra, **584**, 585*i*, 890
Epiblast cells, **1025**–1029, 1026*i*
Epicotyl, **893**
Epidermal tissue, 847*i*
Epidermis, **603**, 1256, 1257*i*
Epididymis, **1003**
Epigenetic regulation, cell division and, **331–332**
Epigenetics, **1041**
Epiglottis, 1226*i*
Epinephrine, **981**, 1051, 1058, 1258
Epiphytes, **913**
Epistasis, **233**–234, 234*i*
Epithelial cells, **1195**
  absorption of digestive products by, 1206*i*
  in humans, 1205*i*
Epithelial tissue, **949**–951, 950*i*
Epithelium
  as a barrier to infection, 1266
  in mammals, 1225, 1226*i*
Epitope, **1274**
Epstein-Barr virus, 503*t*
Equilibrium, **80**, 80*i*
Ergot alkaloids, **550**
Ergotism, **549**, 551
Erosion, **395**
Eryops, **652**
Erythrocytes, **970**–971, 971*i*
Erythropoietin (EPO), 1230*b*
Escherich, Theodor, 191
Escovopsis (fungus), 541
Esophagus, **638**, 1198*b*
Essay on the Principles of Population (Malthus), 398

Essential element, **903**
Essential fatty acids, **1190**
Essential nutrient, **1190**
Estivation, **1261**
Estradiol, **1060**
Estrogen, 1003*i*, **1161**
Ethidium bromide, 350*b*
Ethology, **1159**
Ethylene, 919*t*, **926**–927, 927*i*
Eucalyptus trees, 586
Euchromatin, **285**
Eudicot embryos, **892**–893, 893*i*, 895*i*
Eudicot seeds, 893*i*
Eudicots (most fruit trees, roses, beans, potatoes, etc.),
    576*t*, **587**–588, 840, 841*i*, 849*i*, 851–852, 851*i*, 855*i*
Eukaroytic cells, **33**
  mitosis and, 187
  nucleus of, 33
Eukarya domain, **28**, 65–66, 66*i*
Eukaryote genomes, **370**, 374
Eukaryotes, **31**
  cellular respiration in, 124–125, 124*i*
  energy crisis and, 69
  fermentation in, 137–138, 138*i*
  genetic recombination in, 199–210, 201*i*, 202*i*, 204*i*,
    205*i*, 208f, 209*i*, 210*i*, 214–215
  land plants as, 563
  photosynthesis in, 146, 147*i*
  processing mRNAs in, 311
  processing of mRNAs in, 297–300, 297*i*
  regulation of transcription in, 321–328, 336
  transposable elements in, 211–212
Eukaryotic cells, **31**, 50, 72
  cytoplasmic organelles of, 33
  cytoskeletons of, 42*i*
  DNA organization in prokaryotic cells *versus*,
    283–286, 287
  endomembrane system, 36–39
  mitochondria, 40–41
  mitosis in, 170–177, 172*i*, 174*i*, 175*i*, 176*i*, 177*i*
  ribosomes, 36, 36*i*
  rise of multicellularity and, 66–70
Eukaryotic chromosome, **35**
Eukaryotic flagellum, 44*i*
Eukaryotic genes, **344**, 352, 373
Eukaryotic protein, 353*i*
Eukaryotic protein-coding gene, **321**–322, 322*i*
Euplectella aspergillum, **602**
Euploid, **253**
Europe, political opposition to GMOs in, 359
European barn swallow, 19, 19*i*
European fire-bellied toad, 439
European garden snails, 410*i*, 411, 421, 423, 428
Euryarchaeota, **497**
Eusocial animals, **1181**–1183, 1187–1188
Eustachian tube, **1106**
Eutherian, 669*f*
Evans, David, 452
Evaporation, **1252**, 1252*i*
Even-skipped, **1038**
Evolution, **389**–390, 391–392, 406–408
  adaptive, 423–424
  animal, 595–599

biological, 395–400, 401
of birds, 461–464, 463*i*, 466
as characteristic of life, 54*i*
constraints, 423–424
of life, 55*i*
limb development, 394*b*
macroevolution, 401
microevolution, 401
misconception about, 564
modern synthesis, 401
natural selection, 391–392
natural selection, impact of theory, 400
pre-Darwinian views, 392–393
theory, development, 391–408
Evolutionary agents, **417–420**, 417*t*
gene flow, 417*t*, 418
genetic drift, 417*t*, 418–419
genetic variations within populations, 419–420
mutations, 417, 417*t*
natural selection, 417*t*, 420
nonrandom mating, 417*t*, 420
Evolutionary biology, **400–401**
modern synthesis, 401
molecular techniques, 402–403, 404
research, 401–402
Evolutionary convergence, **670–672**
Evolutionary developmental biology, **456**
Evolutionary divergence, **399**
Evolutionary physiology, **1294–1295**, 1296
Ewer, R.F. (Griff), 1185–1186*b*
Exchange diffusion. *See* Antiport
Excitable cells, **1073**
Excitatory postsynaptic potential (EPSP), **1085–1087**,
1087*i*
Excretion, **587**
microscopic tubules and, 1238, 1238*i*
osmoregulation and, 1236–1239, 1262
osmoregulation and excretion in invertebrates,
1240–1242, 1262
osmoregulation and excretion in mammals, 1244–1251,
1262–1263
osmoregulation and excretion in nonmammalian
invertebrates, 1243–1244, 1243*i*, 1244*i*, 1262
Excretory system, 956–957*i*
Exergonic processes, **80**
Exergonic reaction, 79*i*
Exhaled air, **1226–1227**
Exocrine glands, **950**, 951*i*, 1047, 1047*i*
Exocytosis, **112–114**, 113*i*, 118
neurotransmitters and, 1083
Exodermis, **854**
Exogenous antigens, **1273**
Exome Variant Server, 468
Exon shuffling, **299–300**, 384–386, 384*i*
Exons, **373**, 384
Exoskeleton, **621**, 1145–1146, 1146*i*
Exothermic, **78**
Exotoxin, **489**
Expansins, **924**
Experimental data, **415**
Experimental variable, **F-3**
Exploitative competition, **717**
Exponential model of population growth, **689–691**

Expression vectors, **352**, 353, 353*i*
External fertilization, **993**
External genitalia, hormones and, 1030*b*
External gills, **1222**, 1222*i*
Extinction, **781**
Extracellular digestion, **603**, 1195–1196
Extracellular fluid, **958**
Extracellular matrix (ECM), **49**, 49*i*, 380, 951–952
Extracellular structure, **33**
Extra-embryonic membrane, **1020–1021**, 1021*i*
Extraterrestrial life, search for, **70–71**, 70*i*, 72
Extraterrestrial origins, **58**
Extremophile, **64–65**
Eyes, **9–11**
compound, **622**
Darwin on, 11–13
development of, 1022–1024, 1022*i*, 1023*i*, 1024*i*
evolution of, 12*i*
of invertebrates, 1108, 1109*i*
sensing light without, 8–9
of vertebrates, 1109, 1110*i*

F
F^+ cell, **193**
F^- cell, **193**
F factor, **193–195**, 194*i*, 286*i*
F pilus, **193**
F1 generation, **219**
F2 generation, **220**
Facial expression, 477*f*
Facilitated diffusion, **106**, 107*i*, 108*i*
Facilitated passive transport, **105–106**
Facilitation hypothesis, **737**
Facultative anaerobe, **139**
$FADH_2$ converting potential energy in, **130**, 131*i*
Faloona, F., 345
Familial (hereditary) cancer, **333**
Family, **428**
Family planning program, **702**
Fanworms, 1080
Fas-FasL system, 1282*b*
Fast anaerobic fibres, **1142**
Fast block (to polyspermy), **994**
Fast muscle fibres, 1141*t*, **1142**
Fat, **99**
enzymatic digestion of, 1205*i*
oxidizing, 135, 136*i*
Fate mapping, **1035**
Fat-soluble vitamins, **1193**, 1193*b*
Fatty acids, **101–102**, 101*i*, 102*i*, 1048, 1192
Feather, **662**
Feather follicles, **1257**, 1257*i*
Feather star, **639**
Feces, **1207**
Fecundity, **687**
Feedback inhibition, 93*i*
Feedback loops, **332**
Feedback pathways, regulating hormones by, **1048**, 1048*i*
Feeding structure, **607**
Felis silvestris bieti, 826*i*
Felis silvestris cafra, 825, 826*i*
Felis silvestris catus, 825, 826*i*
Felis silvestris lybica, 825

Jensen, Keith, 1183
Jet lag, **17**–18, 18*i*
Jimson weed, 819
John, Dick, 186
Johnson, Robert T., 180–181
Jumping genes, **190**, 210
    See also Mobile elements
Junipers, 557*i*
Junk DNA, **373**
Jurassic period, 448*t*
Juvenile hormone (JH), 629*f*, 1064*i*, **1161**
Juxtamedullary nephrons, **1246**, 1246*i*

K

K⁺ channels, **1083**
K⁺ ions, **1073**
Kandel, Eric, 1297*b*
Kangaroo rat, 1250–1251, 1250*i*
Karnik, Rohit, 845–846*b*
Karyogamy, **543**
Karyotype, **175**
Kearns-Sayre syndrome, 259*t*
Keeled sternum, **661**
Kepler, Johannes, 71
Keratin, **1256**
Ketone, **1061**
Keystone species, **732**
Kidneys, 1247*i*
    in mammals, 1245–1246, 1245*i*, 1246*i*
    in marine fish, 1243
Kin selection, **1180**–1181, 1180*i*, 1187
Kinesin molecule, 43*i*
Kinesis, **1169**
Kinetic energy, **75**
Kinetochore, **174**
King, Thomas, 1034–1035
King penguin, 19, 19*i*
Kingdom Animalia, **594**
Kingdom Fungi, **543**
Kingdom Plantae, **564**
Kirshenbaum, Lorrie, 975*b*
Kirtland's Warbler, 1169
Kitlg gene, 422*b*
Knockout mouse, **354**, 1158*b*
Korarchaeota, **497**
Krebs cycle. *See* Citric acid cycle
Krogh, August, 1297*b*
K-selected species, **697**

L

Labia majora, **999**
Labia minora, **999**
Labium, **627**
Lac operon, **316**–317, 316*i*
Lac repressor protein, **378**
Lactate fermentation, **137**–138
Lactic acid, **1142**
Lactose intolerance, **477**
LacZ+ gene, **342**, 343*i*
Lagging strand, **276**
Lagging strand template, **276**
Lamarck, Jean Baptiste de, 395–396
Lambda, **505**–506, 505*i*

Lamellae, **1209**
Lamprey, **644**–646
Lancelet, 641*f*
Land, movement of organisms from, 453
Land plants
    defining characteristics of, 563
    phylogenetic relationships between major groups of, 571*i*
Language, **476**
Large ground-finch, 397*i*, 399
Large intestine, in humans, 1198*b*, **1206**–1207
Larrea tridentata, **917**–918, 918*i*
Larson, Doug, 858*b*
Larus, **678**
Larva (larval form), **595**
Larval molluscs, 1241, 1241*i*
Larynx, **1225**
    in mammalian respiratory system, 1226*i*
    in mammals, 1225, 1226*i*
Last Universal Common Ancestor (LUCA), **2**, 65*i*
Latent phase, **506**
Lateral bud, **848**
Lateral geniculate nuclei, **1113**
Lateral inhibition, **1112**
Lateral line system, 1102*i*
Lateral meristem, **839**
Lateral root, **852**
Lateralization, **1129**
Laws of thermodynamics, **75**–78, 77*i*, 94
Layne, John, 1163–1164
Leaching, **901**
Leading strand, **276**
Leading strand template, **276**
Leaf endophytes, 559*i*
Leaf frog, 1244, 1244*i*, 1255
Leaf primordium (plural, primordia), **851**
Leaf-cutter ants, 540, 540*i*, 541*i*
Leakey, Mary, 473*f*
Learned behaviour, **1155**
Learning, **1130**, 1133, 1159, 1186
Leaves, **569**, 569*i*, 850–852, 850*i*, 851*i*, 852*i*, 871, 871*i*
Leber hereditary optic neuropathy, 259*t*
Leder, Philip, 293
Lederberg, Joshua, 191
Left aortic arch, **667**
Leg, **473**–474
Leghemoglobin, **913**
Leishmania species, **1288**
Lekking behaviour, **1176**, 1176*i*
Lens, **1108**
Lens culinaris, 816–817, 816–817*i*
Lens orientalis, 816–817, 816–817*i*
Lenski, Richard E., 1296
Lentils, domestication of, 816–817, 817*i*
Leopard, 1179, 1180*i*
"Leopard" alarm, **476**
Lepeophtheirus salmonis, 821
Lepidodendron (lycophyte tree), 578*i*
Lepidosaura, infraclass, **660**–661
Lepidosauromorpha, **655**
Leptin, **1208**–1209, 1210*b*
Lethal mutations, **417**
Leukocytes, **954**, 971, 971*t*, 973, 1268, 1270*t*

oxidizing, 135, 136*i*
  in plasma, 970
  Rubisco, 157–158, 158*i*
Proteoglycans, **49**, 49f
Proteomes, **363**–364, 376
Proteomics, **339**, 376, 377
Proterozoic eon, 448*t*
Prothoracicotropic hormone (PTTH), **1063**, 1064*i*
Protist cells, 29*i*
Protobiont, **60**
Protoderm, **851**–852
Proton, **F-9**
Proton gradient, **130**, 132, 132*i*
Proton pump, **7**
Protonema, **574**, 574*i*, 575
Protonephridia, **1241**, 1241*i*
Protonephridium, **1241**
Proton-motive force, **132**
Proto-oncogenes, **333**
Protosome
  developmental polymorphism, 612–613f
  ecdysozoan, 619–629
  lophotrochozoan, 606–619
Protostome, **596**, 598f
Prototrophs, **191**
Provoirus, **213**
Proximal convoluted tubule, **1246**, 1246*i*, 1248*t*, 1249
Pruetz, Jill, 1154–1155
Prusiner, Stanley, 511
Pseudocoelom, **596**
Pseudocoelomate, **596**
Pseudocoelum, **596**
Pseudogenes, **373**
Pseudopod (plural, pseudopodia), **517**
Psychrophile, **497**
Pterophyta, 571*i*, **576**, 576*t*, 577–580
Pulmonary arteries, **973**
Pulmonary circuit, **968**
Pulmonary veins, **973**–974
Pulvinus (plural, pulvini), **937**
Pumpkin, 820
Punnett square, 258*i*
Pupa, **629**
Pupil, **1108**
Purines, **268**, 1084–1085*t*
Purkinje fibres, **976**, 976*i*
Purple monkey-flower, 432, 432*i*, 440*b*
Pyloric sphincter, **1203**
Pyramid of biomass, **754**
Pyramid of energy, **754**
Pyramid of numbers, **755**
Pyrimidines, **268**
Pyruvate oxidation, **126**
  citric acid cycle and, 124, 124*i*, 126–129, 128*i*, 129*i*, 141
  metabolic pathway of, 138*i*
  reactions to, 128*i*
Pythons, 402*i*, 403, 1194*i*
PYY gene, 1210*b*

Q

Qualitative variation, **411**, 411*i*
Quantitative trait, **235**
Quantitative variation, **410**, 411*i*

Quaternary period, 447*t*
Quebecol, 873*b*
Queen Anne's lace, 827
Quinghao, 829*b*
Quorum sensing, **23**

R

R allele, **932**
R gene, **932**
Rabies, **678**–679
Rabies baiting, **679**, 679f
Radial canal, **638**
Radial cleavage, **596**
Radial symmetry, **602**, 602–606
Radiant energy, detection and perception of, **1107**–1108
Radiation, **1252**
Radicle, **892**
Radioactivity, **266**
Radioisotope, **267**
Radiometric dating, **55**
Radula, **611**
Ramaria (coral fungus), 552*i*
Random dispersion, **682**
Random fertilization, **209**–210
Random segregation, **208**–209
Ranunculus ficaria, 20*i*
Rao, Potu N., 180–181
Raphanobrassica, **442**
Rapid eye movement (REM) sleep, **1130**
Rat snake, 433*i*
Ratites, 393*i*
Rays, **1239**, 1243
Reabsorption, **1050**
  in nephrons and collecting ducts, 1248*t*, 1249–1250
  as a tubule function in excretion, 1238
Reactants, **87**–89
Reaction centre, **151**
Reading frame, **294**
Realized niche, **721**
Receptacle, **885**
Reception, **1069**
  in nervous systems, 1069–1070, 1070*i*
  in signal pathway, 114
Receptor controls, 1207*i*
Receptor proteins, **1047**
Receptor-mediated endocytosis, **112**
Receptors
  adaptable nature of, 1099
  control of digestion by, 1207*i*
  for touch and pressure, 1100, 1100*i*
  types of, 1077, 1078*i*
Recessive, **221**
Reciprocal altruism, **1181**
Recognition of nonself, **1267**
Recognition protein, **504**
Recombinant DNA, **340**, 342*i*, 348, 359
Recombination, reproduction with, **990**
Recombination frequency, **243**–245, 244*i*
Rectum, 1198*b*, **1206**, 1207
Red blood cells (RBCs), 1230*b*
Red fox, 678
Red tide, **521**
Redox reaction, **122**, 132*i*

Reduced, **122**
Reducing atmosphere, **57**
Reduction reactions, **122**
Red-winged Blackbird, 19, 19*i*, 1175
Reeve, Kern, 1185*b*
Reflexes, **1123**–1124, 1123*i*
Refractory period, **1006**, 1080
Regenerating ATP, **84**–85, 85*i*
Regulators, **958**
Regulatory protein, **315**
Regulatory sequences, **373**
Reinforcement, **436**,
Relatedness, 1180*i*, **1182**–1183
Relative abundance, **414**
Relative fitness, **420**
Relaxation, contraction and, **1139**
Relaxin, 1012*b*
Release, as a tubule function in excretion, **1238**
Release factor, **306**
Releasing hormone (RH), **1053**
Renal artery, **1245**, 1245*i*
Renal cortex, **1245**, 1245*i*
Renal medulla, **1245**, 1245*i*
Renal pelvis, **1245**, 1245*i*
Renal vein, **1245**
Renaturation, **F-36**
Replica plating, **191**
Replicates, **F-3**
Replication
    in cell cycle, 169
    DNA, 270–280, 287
Replication bubble, 280*i*
Replication fork, **275**
Replication origin, **193**
Repressible operon, **319**
Repressor, **316**
Reproduction, **989**
    age, 687–688
    biological species concept, 429
    breeding frequency, 687
    as characteristic of life, 54*i*
    controlling, 1010
    courtship displays, 431
    in flowering plants. *See* Flowering plants
    geographical speciation, 433–439
    inbreeding, 420
    isolating mechanisms, 430–433
    nonrandom mating, 417*t*, 420
    polyploidy speciation, 441*i*
    proportion, 684
    sexual selection, 405–406
    supporting theory of endosymbiosis, 66
    See also Speciation
Reproduction in animals, **987**, 989, 1010
    asexual, 989–990, 990*i*, 1009
    controlling, 1008–1009, 1010
    drive to reproduce, 989, 1009
    mechanisms of sexual reproduction, 990–998, 1009
    sexual, 989–990, 1009
    sexual reproduction in mammals, 998–1008, 1009–1010
Reproductive isolating mechanism, **430**–433
Reproductive shoot, **839**

Reproductive strategy, **989**
Reproductive system, 956–957*i*
Reptiles, 657*f*, 659*f*
    ectothermy in, 1255
    uric acid from, 1244
Residual volume, **1227**
Resilin, 952*b*
Resolution, **29**
Resource partitioning, **721**
Respiratory intermediates, utilizing for anabolic
        reactions, **136**
Respiratory medium, **1220**
Respiratory surface, **1219**
Respiratory system, 956–957*i*
    in mammals, 1233
    See also Gas exchange
Response, **1069**
    in nervous systems, 1069–1070, 1070*i*
    in signal pathway, 114
Resting membrane potential, **1073**–1074, 1074*i*
Resting potential, **1073**
Restriction endonucleases, **341**, 348
Restriction endonucleases (restriction enzyme), **341**
Restriction enzymes, **341**, 342, 348
Restriction fragment length polymorphisms (RFLPs), **349**
Restriction fragments, **341**
Restriction sites, **341**, 342*i*, 344, 349
Retina, **1108**
    of mammals and birds, 1109–1112, 1110*i*, 1111*i*, 1113*i*
    structure of, 1113*i*
    visual processing in, 1112, 1113*i*
Retrotransposons, **212**, 212*i*, 213, 213*i*
Retrovirus, 503*t*
Reverse transcriptase, **340**, 344, 344*i*, 370, 507–508*i*
Reverse-transcribed RNA viruses, **373**
Reversible, **80**
Rhabdovirus, 503*t*
Rhamphorynchus meunsteri, 657*f*
Rhea darwinii, **428**
Rheumatoid arthritis, **1284**
Rhinovirus, 503*t*
Rhizobium, **912**–913
Rhizoids, **572**, 573, 574*i*, 577*i*, 578
Rhizomes, **577**, 577*i*, 579*i*, 850
Rhizopus nigricans, 547*i*
Rhodopsin, **8**, 10*i*
Rhynchocoel, **611**
Rhynia, 577
Rhynia gwynne-vaughnii, 577*i*
Ribbon worm, 611, 611*f*
Ribonucleic acid (RNA), **28**
Ribose, **271**
Ribosomal RNA (rRNA), **32**
Ribosomal subunits, **302**, 303*i*
Ribosome binding site, **304**
Ribosomes, **36**, 36*i*, 302, 306, 307*i*
Ribozyme, 61*i*
Ribulose-1,5-bisphosphate (RuBP) carboxylase/oxygenase
        (rubisco), **157**
Rice
    domestication of, 812–813, 812*i*, 815–816
    genetic modification of, 829*b*
    genetically engineered, 358*i*

Secondary structure, **104**

Secondary succession, **737**

Secondary tissue, **839**

Secretion of enzymes
    in nephrons and collecting ducts, 1248*t*, 1249–1250
    as a process from ingestion of food, 1196
    as a tubule function in excretion, 1238

Secretory vesicles, **112**

Security of food sources, **808**

Sedative drugs, 1084–1085*t*

Sedimentation, 447*i*

Seed coat, 582i, **583**, 584*i*

Seed ferns, **577**, 587

Seedless vascular plants, **575–580**, 576t

Seeds, **581**, 582–583, 582i
    development of embryonic sporophytes in, 892–893, 893i
    germination of, 894
    suppressing growth in, 928

Segment polarity genes, **1039**

Segmentation, **599**, 617*f*

Segmentation genes, **1037–1039**, 1038*i*

Segregate, **221**

Segregation, principle of, **220–221**, 222*i*, 227*i*

Selaginella densa, **577**, 578*i*

Selaginella strobilus, 570*i*

Selection, **403–406**

Selective adhesion properties, of cells, 1033*b*

Selective breeding, **808–811**, 809*i*, 810*i*, 811*i*, 812*i*

Selective cell adhesions, **1016**

Selective pressures, **391**, 397–398, 401

Selectively permeable, **60**

Selenium, as essential animal nutrient, 1191–1192*b*

"Self-feeders," **836**

Self-fertilization (self-pollination), **420**

Self-incompatibility, **888–890**, 891*i*

Selfing, **890**

Self-pollination, **888–890**

Semen, **1003**

Semiconservative replication, **270**

Semilunar valve (SL valve), **974**

Seminal vesicle, **1003**

Seminiferous tubules, **1003–1004**, 1005*i*

Senescence, **926–927**, 927*i*

Sense codon, **392**

Sensitization, **1130**

Sensory adaptation, **1099**

Sensory cell membrane proteins, 1078*i*

Sensory hair cell, **1101**

Sensory integration, **959**, 1131

Sensory neuron, **1070**

Sensory receptor (transducer), **1076**, 1078*i*

Sensory regions, of cerebral cortex, **1127**, 1128*i*

Sensory transduction, **1076**, 1111–1112, 1111*i*, 1112*i*

Sepal, **885**

Septum (plural, septa), **542**, 542*i*, 617

Sequence analysis, supporting theory of endosymbiosis, **67**

Sequence similarity searches, **374**, 375

Sequential hermaphroditism, **997**

Serosa, **1199**

Serotonin, 1084–1085*t*

Sertoli cells, **1002**, 1004, 1005*i*

Sertraline (Zoloft), 1084–1085*t*

Sessile (stationary), **563**, 588

Sessile animal, **595**

Set point, **1251**, 1257–1259, 1258–1259

Seta (plural, setae), **617**

Sex chromosomes, **203**, 206, 254*t*

Sex pilus. *See* F pilus

Sex ratio, **684**

Sex-linked genes, **246–251**, 246*i*, 247*i*, 248*i*, 250*i*, 251*i*, 261

Sexual dimorphism, **405**, 406*i*

Sexual reproduction, **206–207**, 1009, 1016*i*
    in animals, 989–990
    in mammals, 998–1008, 1009–1010
    mechanisms of, 990–998, 1009
    speed of domestication and, 810

Sexual selection, **405–406**, 1176–1177, 1177*i*, 1187

Sharks, 1239, 1243, 1254

Sheep, domestication of, 821

Sheep cloning, 1035

Shivering, **1258**

Shoot systems, **568**, 836, 839

Short interfering RNAs (siRNAs), **329–330**, 329*i*

Short tandem repeat (STR) sequence, **349**

Short-day plant, **939**

Short-term memory, **1129–1130**

Shoulder, **475**

Shoulder blade, **475**, 475*f*

Siamese fighting fish, 405

Sickle cell *(HbS)*, **421**, 421*i*

Sickle cell disease, **217–218**, 218*i*, 256, 349, 349*i*

Sieve tube, **846**

Sieve tube members, **844**, 844*i*, 846

Sieve tubes, **879**

Sign stimulus, **1156**

Signal initiation, by neurons, **1072–1079**, 1074*i*, 1075*i*, 1076*i*, 1077*i*, 1078*i*, 1094

Signal peptide, **309**

Signal recognition particle (SRP), **309**

Signal sequence. *See* Signal peptide

Signal transduction, **103**, 114, 115*i*

Signal transduction pathways, plant hormones and, **918–919**, 920*i*

Signalling. *See* Cell membranes

Signalling cascade, **114**

Signals and responses (plants), **917–918**, 944–945
    biological clocks, 938–943, 944
    chemical defences, 929–934, 943
    hormones, 918–929, 943
    movements, 934–938, 943–944

Sildenafil citrate (Viagra), 1084–1085*t*

Silencing, **327**

Silent mutation, **310**

Silkworm moths, 1166–1167

Silurian period, 448*t*, **575**

Simple diffusion, **105**

Simple epithelia, **950**, 950*i*

Simple fruits, **891**, 892*i*

Simple passive transport, **105–106**

Simple tissues, **841**

Simulation modeling, **770**

Simultaneous hermaphroditism, **997**, 997*i*